Algebra A Combined Approach

SECOND EDITION

K. Elayn Martin-Gay
University of New Orleans

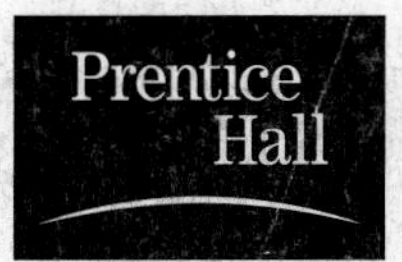

PRENTICE HALL
Upper Saddle River, New Jersey 07458

Library of Congress Cataloging-in-Publication Data

Martin-Gay, K. Elayn
Algebra a combined approach/K. Elayn Martin-Gay.—2nd ed.
p. cm.
Includes index.
ISBN 0-13-067450-8—ISBN 0-13-06745-6—ISBN 0-13-100267-8
1. Algebra. I. Title.

QA152.3. M35 2003
512—dc21 2002074301

Executive Acquisition Editor: Karin E. Wagner
Editor in Chief: Christine Hoag
Project Manager: Mary Beckwith
Vice President/Director of Production and Manufacturing: David W. Riccardi
Executive Managing Editor: Kathleen Schiaparelli
Senior Managing Editor: Linda Mihatov Behrens
Production Management: Elm Street Publishing Services, Inc.
Manufacturing Buyer: Alan Fischer
Manufacturing Manager: Trudy Pisciotti
Executive Marketing Manager: Eilish Collins Main
Marketing Assistant: Annett Uebel
Development Editor: Kathy Sessa-Federico
Editor in Chief, Development: Carol Trueheart
Media Project Manager, Developmental Math: Audra J. Walsh
Editorial Assistant: Heather Balderson
Art Director: Maureen Eide
Assistant to the Art Director: John Christiana
Interior Designer: Circa 86
Cover Designer: Jack Robol
Art Editor: Thomas Benfatti
Creative Director: Carole Anson
Director of Creative Services: Paul Belfanti
Photo Researcher: Melinda Alexander
Photo Editor: Beth Boyd
Cover Art: Seaform details: Handblown glass, by Dale Chihuly. Photo by Terry Rishel.
Art Studio: Scientific Illustrators
Compositor: Preparé/Emilcomp

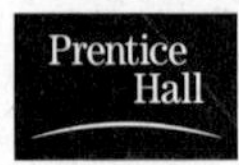

Photo Credits appear on page I14, which constitutes a continuation of the copyright page.

Printed in the United States of America
10 9 8 7 6 5 4 3 2

ISBN: 0-13-067451-6 (paperback) 0-13-100267-8 (case bound)

Pearson Education Ltd.
Pearson Education Australia Pty., Limited
Pearson Education Singapore, Pte. Ltd.
Pearson Education North Asia Ltd.
Pearson Education Canada, Ltd.
Pearson Educacíon de Mexico, S.A. de C.V.
Pearson Education—Japan
Pearson Education Malaysia, Pte. Ltd.

CONTENTS

CHAPTER 8 Systems of Equations and Inequalities 559

CHAPTER 9 Rational Exponents, Radicals, and Complex Numbers 651

CHAPTER 10 Quadratic Equations and Functions 735

PREFACE

About This Book

Algebra A Combined Approach, Second Edition is intended for a two semester course in introductory and intermediate algebra. It was written to provide a **solid foundation** in algebra for students who might have had no previous experience in algebra. Specific care was taken to ensure that students have the most **up-to-date relevant** text preparation for their next mathematics course or for nonmathematical courses that require an understanding of algebraic fundamentals. I have tried to achieve this by writing a user-friendly text that is keyed to objectives and contains many worked-out examples. As suggested by the AMATYC Crossroads Document and the NCTM Standards (plus Addenda), real-life and real-data applications, data interpretation, conceptual understanding, problem solving, writing, cooperative learning, appropriate use of technology, mental mathematics, number sense, critical thinking, and geometric concepts are emphasized and integrated throughout the book.

The many factors that contributed to the success of the first edition have been retained. In preparing the Second Edition, I considered comments and suggestions of colleagues, students, and many users of the prior edition throughout the country.

Algebra A Combined Approach, Second Edition is part of a series of texts that can include *Basic College Mathematics,* Second Edition; *Prealgebra,* Third Edition; *Introductory Algebra,* Second Edition; and *Intermediate Algebra*, Second Edition. Throughout the series pedagogical features are designed to develop student proficiency in algebra and problem solving, and to prepare students for future courses.

Key Pedagogical Features and Changes in the Second Edition

Readability and Connections I have tried to make the writing style as clear as possible while still retaining the mathematical integrity of the content. When a new topic is presented, an effort has been made to relate the new ideas to those that students may already know. Constant reinforcement and connections within problem-solving strategies, data interpretation, geometry, patterns, graphs, and situations from everyday life can help students gradually master both new and old information. In addition, each section begins with a list of objectives covered in the section. Clear organization of section material based on objectives further enhances readability.

Problem-Solving Process This is formally introduced in Chapter 2 with a four-step process that is integrated throughout the text. The four steps are **Understand**, **Translate**, **Solve**, and **Interpret**. The repeated use of these steps in a variety of examples shows their wide applicability. Reinforcing the steps can increase students' comfort level and confidence in tackling problems.

Applications and Connections Every effort was made to include as many interesting and relevant real-life applications as possible throughout the text in both worked-out examples and exercise sets. In the Second Edition, the applications have been thoroughly revised and updated, and the number of applications has increased. The applications help to motivate students and strengthen their understanding of mathematics in the real world. They show connections to a wide range of fields including agriculture, allied health, anthropology, art, astronomy, biology, business, chemistry, construction, consumer affairs, earth science, education, entertainment, environmental issues, finance, geography, government, history, medicine, music, nutrition, physics, sports, travel, and weather. Many of the applications are based on recent real data. Sources for data include newspapers, magazines, publicly held companies, government agencies, special-interest groups, research organizations, and reference books. Opportunities for obtaining your own real data are also included. See the Applications Index on page xx.

Practice Problems Throughout the text, each worked-out example has a parallel Practice Problem placed next to the example in the margin. These invite students to be actively involved in the learning process before beginning the end-of-section exercise set. Practice Problems immediately reinforce a skill after it is developed. Answers appear at the bottom of the page for quick reference.

Concept Checks These margin exercises are appropriately placed throughout the text. They allow students to gauge their grasp of an idea as it is being explained in the text. Concept Checks stress conceptual understanding at the point of use and help suppress misconceived notions before they start. Answers appear at the bottom of the page.

Increased Integration of Geometry Concepts In addition to the traditional topics in introductory and intermediate algebra courses, this text contains a strong emphasis on problem solving and geometric concepts, which are integrated throughout. The geometry concepts presented are those most important to a student's understanding of algebra, and I have included many applications and exercises devoted to this topic. These are marked with the the geometry icon. Also, geometric figures, and a review of angles, lines, and special triangles are covered in the appendices.

Helpful Hints Helpful Hints contain practical advice on applying mathematical concepts. These are found throughout the text and strategically placed where students are most likely to need immediate reinforcement. Helpful Hints are highlighted for quick reference.

Visual Reinforcement of Concepts The Second Edition contains a wealth of graphics, models, photographs, and illustrations to visually clarify and reinforce concepts. These include new and updated bar graphs, line graphs, calculator screens, application illustrations, and geometric figures.

Calculator and Graphing Calculator Explorations These optional explorations offer point-of-use intruction, through examples and exercises, on the proper use of scientific and graphing calculators as tools in the mathematical problem-solving process. Placed appropriately throughout the text, Calculator and Graphing Calculator Explorations also reinforce concepts learned in the corresponding section and motivate discovery-based learning.

Additional exercises building on the skill developed in the Explorations may be found in exercise sets throughout the text. Exercises requiring a calculator are marked with the icon. Exercises requiring a graphing calculator are marked with the icon. An Introduction to Using a Graphing Utility is included in the appendix.

Study Skills Reminders New Study Skills Reminder boxes are integrated throughout the text. They are strategically placed to constantly remind and encourage students as they hone their study skills. A new **Section 1.1**, Tips on Success in Mathematics, provides an overview of the Study Skills needed to succeed in math. These are reinforced by the Study Skills Reminder boxes throughout the text.

Focus On Appropriately placed throughout each chapter, these are divided into Focus on Mathematical Connections, Focus on Business and Career, Focus on the Real World, and Focus on History. They are written to help students develop effective habits for engaging in investigations of other branches of mathematics, understanding the importance of mathematics in various careers and in the world of business, and seeing the relevance of mathematics in both the present and past through critical thinking exercises and group activities.

Chapter Highlights Found at the end of each chapter, these contain key definitions, concepts, and examples to help students understand and retain what they have learned and help them organize their notes and study for tests.

Chapter Activity These features occur once per chapter at the end of the chapter, often serving as a chapter wrap-up. For individual or group completion, the Chapter Activity, usually hands-on or data-based, complements and extends to concepts of the chapter, allowing students to make decisions and interpretations and to think and write about algebra.

Integrated Reviews These "mid-chapter reviews" are appropriately placed once per chapter. Integrated Reviews allow students to review and assimilate the many different skills learned separately over several sections before moving on to related material in the chapter.

Pretests Each chapter begins with a pretest that is designed to help students identify areas where they need to pay special attention in the upcoming chapter.

Chapter Review and Test The end of each chapter contains a review of topics introduced in the chapter. The Chapter Review offers exercises that are keyed to sections of the chapter. The Chapter Test is a practice test and is not keyed to sections of the chapter.

Cumulative Review These features are found at the end of each chapter (except Chapters R and 1). Each problem contained in the Cumulative Review is an earlier worked example in the text that is referenced in the back of the book along with the answer. Students who need to see a complete worked-out solution, with explanation, can do so by turning to the appropriate example in the text.

Student Resource Icons At the beginning of each section, videotape and CD, tutorial software, Prentice Hall Tutor Center, and solutions manual icons are displayed. These icons help reinforce that these learning aids are available should students wish to use them to help them review concepts and skills at their own pace. These items have direct correlation to the text and emphasize the text's methods of solution.

Functional Use of Color and New Design Elements of this text are highlighted with color or design to make it easier for students to read and study. Special care has been taken to use color within solutions to examples or in the art to **help clarify, distinguish, or connect concepts**.

Exercise Sets Each text section ends with an Exercise Set. Each exercise in the set, except those found in parts labeled Review and Preview or Combining Concepts, is keyed to one of the objectives of the section. Wherever possible, a specific example is also referenced. In addition to the approximately 5500 exercises in end-of-section exercise sets, exercises may also be found in the Pretests, Integrated Reviews, Chapter Reviews, Chapter Tests, and Cumulative Reviews.

Exercises and examples marked with a video icon have been worked out step-by-step by the author in the videos that accompany this text.

Throughout the exercises in the text there is an emphasis on data and graphical interpretation via tables, charts, and graphs. The ability to interpret data and read and create a variety of types of graphs is developed gradually so students become comfortable with it. Similarly, geometric concepts—such as perimeter and area—are integrated throughout the text. Exercises and examples marked with a geometry icon △ have been identified for convenience.

Each exercise set contains one or more of the following features.

Mental Math Found at the beginning of an exercise set, these mental warmups reinforce concepts found in the accompanying section and increase students' confidence before they tackle an exercise set. By relying on their own mental skills, students increase not only their confidence in themselves but also their number sense and estimation ability.

Review and Preview These exercises occur in each exercise set (except for those in Chapters R and 1) after the exercises keyed to the objectives of the section. Review and Preview problems are keyed to previous sections and review concepts learned earlier in the text that are needed in the next section or in the next chapter. These exercises show the links between earlier topics and later material.

Combining Concepts These exercises are found at the end of each exercise set after the Review and Preview exercises. Combining Concepts exercises require students to combine several concepts from that section or to take the concepts of the section a step further by combining them with concepts learned in previous sections. For instance, sometimes students are required to combine the concepts of the section with the problem-solving process they learned in Chapter 2 to try their hand at solving an application problem.

Writing Exercises These exercises occur in almost every exercise set and are marked with an icon. They require students to assimilate information and provide a written response to explain concepts or justify their thinking. Guidelines recommended by the American Mathematical Association of Two Year Colleges (AMATYC) and other professional groups recommend incorporating writing in mathematics courses to reinforce concepts.

Vocabulary Checks Vocabulary Checks, **new to this edition**, provide an opportunity for students to become more familiar with the use of mathematical terms as they strengthen their verbal skills. These appear at the end of the chapter before the Chapter Highlights.

Data and Graphical Interpretation There is an emphasis on data interpretation in exercises via tables and graphs. The ability to interpret data and read and create a variety of types of graphs is developed gradually so students become comfortable with it.

Internet Excursions These exercises occur once per chapter. Internet Excursions require students to use the Internet as a data-collection tool to complete the exercises, allowing students first-hand experience with manipulating and working with real data.

Key Content Features in the Second Edition

Overview This new edition retains many of the factors that have contributed to its success. Even so, **every section of the text was carefully re-examined**. Throughout the new edition you will find numerous new applications, examples, and many real-life applications and exercises. For example, look at the exercise sets of Sections 2.5, 2.6, 3.1, 3.4, 7.4, 8.4, and 9.6. Some sections have internal re-organization to better clarify and enhance the presentation.

Chapter 1 now begins with Tips for Success in Mathematics (Section 1.1). **New applications** and real data enhance the chapter.

New Study Skills Reminder boxes have been inserted throughout the text. These boxes reinforce the tips from Section 1.1. They are placed strategically to encourage students to hone their study skills.

Increased Integration of Geometry Concepts The geometry concepts presented are those most important to a student's understanding of algebra, and I have included many **applications and exercises** devoted to this topic. These are marked with a geometry icon. Also, geometric figures and a review of angles, lines, and special triangles are covered in the appendices.

New Examples Detailed step-by-step examples were added, deleted, replaced, or updated as needed. Many of these reflect real life.

Exercise Sets Revised and Updated The exercise sets have been carefully examined and extensively revised. The **real-world and real-data applications** have been thoroughly updated and many new applications are included. In addition, an **increased number of challenging problems** have been included in the new edition. **Writing exercises** are now included in most exercise sets and new **Vocabulary Checks** have been added to the end of the chapter to help students become proficient in the language of mathematics.

Additional changes in content include:

- Determinants and Cramer's Rule is now covered in Appendix F.
- To facilitate the transition from introductory to intermediate algebra, two new review appendices have been added. These are: Appendix A, Transition Review: Exponents, Polynomials, and Factoring Strategies and Appendix B, Transition Review: Solving Linear and Quadratic Equations.
- Reading Graphs, formerly Section 1.8, has been combined into Chapter 3 to streamline the presentation.

Enhanced Supplements Package The Second Edition is supported by a wealth of supplements designed for **added effectiveness and efficiency**. New items include the MathPro 5 on-line tutorial with diagnostic and unique video clip feature, a new computerized testing system (TestGen-EQ with Quiz-Master), Prentice Hall Tutor Center, digitized videos on CD, and Instructor to Instructor, Teaching Mathematics CD Series. Please see the list of supplements for descriptions.

Options for On-line and Distance Learning

For maximum convenience, Prentice Hall offers on-line interactivity and delivery options for a variety of distance learning needs. Instructors may access or adopt these in conjunction with this text.

http://www.prenhall.com/martin-gay_algebra 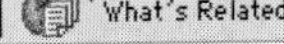

The **Companion Web** site includes basic distance learning access to provide links to the text's Internet Excursions and a selection of on-line self quizzes. Email is available.

WebCT WebCT includes distance learning access to content found in the Martin-Gay companion Web site plus more. WebCT provides tools to create, manage, and use on-line course materials. Save time and take advantage of items such as on-line help, communication tools, and access to instructor and student manuals. Your college may already have WebCT's software installed on their server or you may choose to download it. Contact your local Prentice Hall sales representative for details.

BlackBoard Visit http://www.prenhall.com/demo. For distance learning access to content and features from the Martin-Gay companion Web site plus more, Blackboard provides simple templates and tools.

Course Compass™ Powered by BlackBoard. Visit http://www.prenhall.com/demo.

Supplements for the Instructor

Printed Supplements

Annotated Instructor's Edition (0-13-067450-8)

- Answers to all exercises printed on the same text page.
- Teaching Tips throughout the text placed at key points in the margin.

Instructor's Solution Manual (0-13-067461-3)

- Solutions to even-numbered section exercises.
- Solutions to every (even and odd) Mental Math exercise.
- Solutions to every (even and odd) Practice Problem (margin exercise).
- Solutions to every (even and odd) exercise found in the Pretests, Integrated Reviews (mid-chapter reviews), Chapter Reviews, Chapter Tests, and Cumulative Reviews.

Instructor's Resource Manual with Tests (0-13-067462-1)

- Notes to the Instructor that include an introduction to Interactive Learning, Interpreting Graphs and Data, Alternative Assessment, Using Technology, and Helping Students Succeed.
- Two free-response Pretests per chapter.
- Eight Chapter Tests per chapter (3 multiple-choice, 5 free-response).
- Two Cumulative Review Tests (one multiple-choice, one free-response) every two chapters.
- Eight Final Exams (3 multiple-choice, 5 free-response).
- Twenty additional exercises per section for added test exercises if needed.
- Group Activities (an average of two per chapter; providing short group activities in a convenient, ready-to-use format).
- Answers to all items.

Media Supplements

TestGen-EQ with QuizMaster CD-ROM (Windows/Macintosh) (0-13-067454-0)

- Algorithmically driven, text-specific testing program.
- Networkable for administering tests and capturing grades on-line.
- Edit and add your own questions to create a nearly unlimited number of tests and worksheets.
- Use the new "Function Plotter" to create graphs.
- Tests can be easily exported to HTML so they can be posted to the Web for student practice.

- Includes an email function for network users, enabling instructors to send a message to a specific student or an entire group.
- Network-based reports and summaries for a class or student and for cumulative or selected scores are available.

MathPro 5 Instructor Version

- On-line, customizable tutorial, diagnostic, and assessment program for anytime, anywhere tutorial support.
- Text specific at the learning objective level.
- Diagnostic option identifies student skills, provides individual learning plan, and tutorial reinforcement.
- Integration of TestGen-EQ allows for testing to operate within the tutorial environment.
- Course management tracking of tutorial and testing activity.

MathPro Explorer 4.0

- Network Version IBM/Mac **0-13-067457-5**.
- Enables instructors to create either customized or algorithmically generated practice quizzes from any section of a chapter.
- Includes email function for network users, enabling instructors to send a message to a specific student or to an entire group.
- Network-based reports and summaries for a class or student and for cumulative or selected scores.

Instructor to Instructor Teaching Mathematics CD Series

- Written and presented by Elayn Martin-Gay.
- Contains suggestions for presenting course material, utilizing the integrated resource package, time-saving tips, and much more.

Companion Web Site http://www.prenhall.com/martin-gay_algebra

- Create a customized on-line syllabus with Syllabus Manager.
- Links related to the Internet Excursions in each chapter allow students to find and retrieve real data for use in guided problem solving.
- Assign quizzes or monitor student self quizzes by having students email results, such as true/false reading quizzes or vocabulary check quizzes.
- Destination links provide additional opportunities to explore related sites.

Supplements for the Student

Printed Supplements

Student's Solution Manual (0-13-067459-1)

- Solutions to odd-numbered section exercises.
- Solutions to every (even and odd) Mental Math exercise.
- Solutions to every (even and odd) Practice Problem (margin exercise).
- Solutions to every (even and odd) exercise found in the Pretests, Integrated Reviews (mid-chapter reviews), Chapter Reviews, Chapter Tests, and Cumulative Reviews.

Media Supplements

MathPro 5 (Student Version)

- Online, customizable tutorial, diagnostic, and assessment software.
- Text specific to the learning objective level, providing anytime, anywhere tutorial support.
- Algorithmically driven for virtually unlimited practice problems with immediate feedback.

- "Watch" screen videoclips by K. Elayn Martin-Gay.
- Step-by Step solutions.
- Summary of Progress.

MathPro 4.0 Student Version (0-13-067456-7)

- Available on CD-ROM for stand alone use or can be networked in the school laboratory.
- Text specific tutorial exercises and instructions at the objective level.
- Algorithmically generated Practice Problems.
- "Watch" screen videoclips by K. Elayn Martin-Gay.

Videotape Series (0-13-067453-2)

- Written and presented by Elayn Martin-Gay.
- Keyed to each section of the text.
- Step-by-step solutions to exercises from each section of the text. Exercises that are worked in the videos are marked with a video icon.

New Digitized Lecture Videos on CD-ROM (0-13-047345-6)

- The entire set of *Algebra A Combined Approach* , Second Edition lecture videotapes in digital form.
- Convenient access anytime to video tutorial support from a computer at home or on campus.
- Available shrink-wrapped with the text or stand-alone.

New Prentice Hall Tutor Center

- Staffed with developmental math instructors and open 5 days a week, 7 hours per day.
- Obtain help for examples and exercises in Martin-Gay, *Algebra A Combined Approach,* Second Edition via toll-free telephone, fax, or email.
- The Prentice Hall Tutor Center is accessed through a registration number that may be bundled with a new text or purchased separately with a used book. Visit http://www.prenhall.com/tutorcenter to learn more.

Companion Web Site www.prenhall.com/martin-gay_algebra

- Links related to the Internet Excursions in each chapter allow you to collect data to solve specific internet exercises.

Acknowledgments

First, as usual, I would like to thank my husband, Clayton, for his constant encouragement. I would also like to thank my children, Eric and Bryan, for their sense of humor and for learning how to cook.

I would also like to thank my extended family for their invaluable help and also their sense of humor. Their contributions are too numerous to list. They are Rod, Karen, and Adam Pasch; Michael, Christopher, Matthew, and Jessica Callac; Stuart, Earline, Melissa, Mandy, Bailey, and Ethan Martin; Mark, Sabrina, and Madison Martin; Leo and Barbara Miller; and Jewett Gay.

I would like to thank the following reviewers for their input and suggestions on the previous edition and the new edition.

Peter Arvanites, *Rockland Community College*
David Boni, *Monroe Community College*

Paulette Botley, *Edmonds Community College*
James W. Brewer, *Florida Atlantic University*
Joan Burke, *Montclair State University*
Janet Cater, *Bakersfield College*
Patrick Cross, *University of Oklahoma*
Marjorie Darrah, *Alderson-Broaddus College*
Karen Driskell, *South Plains College*
Dorothy French, *Community College of Philadelphia*
Mary Ellen Gallegos, *Santa Fe Community College*
James W. Harris, *John A. Logan College*
Brian Hayes, *Triton College*
Kayana Hoagland, *South Puget Sound Community College*
Rosa Kavanaugh, *Ozarks Technical Community College*
Deanna Li, *North Seattle Community College*
Marcel Maupin, *Oklahoma State University, Oklahoma City*
Wendy McGuire, *Santa Fe Community College*
Christofer McNally, *Tallahassee Community College*
Julie Miller, *Daytona Beach Community College*
Michael Montaño, *Riverside Community College*
Cameron Neal, Jr., *Temple College*
Ellen O'Connell, *Triton College*
Ted Panitz, *Cape Cod Community College*
Marilyn Garrett Platt, *Gaston College*
Antony Ponder, *Sinclair Community College*
R. B. Pruitt, *South Plains College*
Flauren Ricketts, *Normandale Community College*
Len Ruth, *Sinclair Community College*
Lynne Sage, *Bellevue Community College*
Susan Santolucito, *Delgado Community College*
Rebecca M. Schantz, *Prairie State College*
Sue Sharkey, *Waukeska County Technical College*
Linda Shoesmith, *Scott Community College*
Susan Shulman, *Middlesex County College*

There were many people who helped me develop this text and I will attempt to thank some of them here. Laurie Semarne was invaluable for contributing to the overall accuracy of this text. Emily Keaton and Kathy Sessa-Federico were also invaluable for their many suggestions and contributions during the development and writing of this first edition. Ingrid Mount at Elm Street Publishing Services provided guidance throughout the production process. I thank Jenny Crawford and Richard Semmler for all their work on the solutions, text, and accuracy. Lastly, a special thank you to my project manager Mary Beckwith and executive editor Karin Wagner, for their support and assistance throughout the development and production of this text and to all the staff at Prentice Hall: Linda Behrens, Alan Fischer, Maureen Eide, Grace Hazeldine, Tom Benfatti, Eilish Main, John Tweeddale, Chris Hoag, Paul Corey, and Tim Bozik.

K. Elayn Martin-Gay

About the Author

K. Elayn Martin-Gay has taught mathematics at the University of New Orleans for more than 20 years and has received numerous teaching awards including the local University Alumni Association's Award for Excellence in Teaching.

Over the years, Elayn has developed a videotaped lecture series to help her students understand algebra better. This highly successful video material is the basis for her books: *Basic College Mathematics*, Second Edition; *Prealgebra*, Third Edition; *Introductory Algebra*, Second Edition; *Intermediate Algebra*, Second Edition; *Algebra A Combined Approach*, Second Edition; and her hardback series: *Beginning Algebra*, Third Edition; *Intermediate Algebra*, Third Edition; *Beginning and Intermediate Algebra*, Second Edition; and *Intermediate Algebra: A Graphing Approach*, Second Edition.

To my mother, Barbara M. Miller,
and her husband, Leo Miller,
and to the memory of my father,
Robert J. Martin

APPLICATIONS INDEX

HIGHLIGHTS OF ALGEBRA A COMBINED APPROACH, SECOND EDITION

Algebra A Combined Approach, Second Edition is the primary learning tool in a fully integrated learning package to help you succeed in this course. Author K. Elayn Martin-Gay focuses on enhancing the traditional emphasis of mastering the basics with innovative pedagogy and a meaningful learning program. There are three goals that drive her authorship:

- ▲ **Master and apply skills and concepts**
- ▲ **Build confidence**
- ▲ **Increase motivation**

Take a few moments now to examine some of the features that have been incorporated into *Algebra A Combined Approach, Second Edition* to help students excel.

Exponents and Polynomials

C H A P T E R

Recall from Chapter 1 that an exponent is a shorthand notation for repeated factors. This chapter explores additional concepts about exponents and exponential expressions. An especially useful type of exponential expression is a polynomial. Polynomials model many real-world phenomena. Our goal in this chapter is to become proficient with operations on polynomials.

Niagara Falls is visited by millions of tourists each year. Located between Niagara Falls, New York, and Niagara Falls, Ontario, Canada, on the Niagara River linking Lake Erie and Lake Ontario, the Falls consist of the American Falls, Bridal Veil Falls, and the Canadian, or Horseshoe, Falls. Together, they are known simply as Niagara Falls. The Falls were formed about 12,000 years ago and have since receded upstream 7 miles to their present location. Together, the Falls are roughly 3660 feet wide along their brinks. Water flowing over Niagara Falls drops about 167 feet to the river below and has the potential of generating 4.4 million kilowatts in hydroelectric power. In Exercise 99 on page 209 in Section 3.2, we will use exponents, through scientific notation, to compute the phenomenal volume of water flowing over Niagara Falls in an hour.

267

Page 267

◀ Real World Applications

Chapter-opening real-world applications introduce you to everyday situations that are applicable to the mathematics you will learn in the upcoming chapter, showing the relevance of mathematics in daily life.

Become a Confident Problem Solver

A goal of this text is to help you develop problem-solving abilities.

EXAMPLE 4 Calculating Cellular Phone Usage

A local cellular phone company charges Elaine Chapoton \$50 per month and \$0.36 per minute of phone use in her usage category. If Elaine was charged \$99.68 for a month's cellular phone use, determine the number of whole minutes of phone use.

Solution:

1. UNDERSTAND. Read and reread the problem. Let's propose that Elaine uses the phone for 70 minutes. Pay careful attention as to how we calculate her bill. For 70 minutes of use, Elaine's phone bill will be \$50 plus \$0.36 per minute of use. This is $\$50 + 0.36(70) = \$75.20$, less than \$99.68. We now understand the problem and know that the number of minutes is greater than 70.

   If we let

   x = number of minutes, then

   $0.36x$ = charge per minute of phone use

2. TRANSLATE.

\$50	added to	minute charge	is equal to	\$99.68
↓	↓	↓	↓	↓
50	+	$0.36x$	=	99.68

3. SOLVE.

$$50 + 0.36x = 99.68$$

$$50 + 0.36x - 50 = 99.68 - 50 \quad \text{Subtract 50 from both sides.}$$

$$0.36x = 49.68 \quad \text{Simplify.}$$

$$\frac{0.36x}{0.36} = \frac{49.68}{0.36} \quad \text{Divide both sides by 0.36.}$$

$$x = 138 \quad \text{Simplify.}$$

4. INTERPRET.

Check: If Elaine spends 138 minutes on her cellular phone, her bill is $\$50 + \$0.36(138) = \$99.68$.

State: Elaine spent 138 minutes on her cellular phone this month. ●

Page 124

◀ GENERAL STRATEGY FOR PROBLEM-SOLVING

Save time by having a plan. This text's organization can help you. Note the outlined problem-solving steps, *Understand, Translate, Solve,* and *Interpret.*

Problem solving is introduced early, emphasized and integrated throughout the book. The author provides patient explanations and illustrates how to apply the problem-solving procedure to the in-text examples.

GEOMETRY ▶

Geometric concepts are integrated throughout the text. Examples and exercises involving geometric concepts are now identified with a triangle icon. △ The text includes appendices on geometry as well.

△ **29.** The CART Fed Ex Championship Series is an open-wheeled race car competition based in the United States. A CART car has a maximum length of 199 inches, a maximum width of 78.5 inches, and a maximum height of 33 inches. When the CART series travels to another country for a grand prix, teams must ship their cars. Find the volume of the smallest shipping crate needed to ship a CART car of maximum dimensions. (*Source:* Championship Auto Racing Teams, Inc.)

CART Racing Car

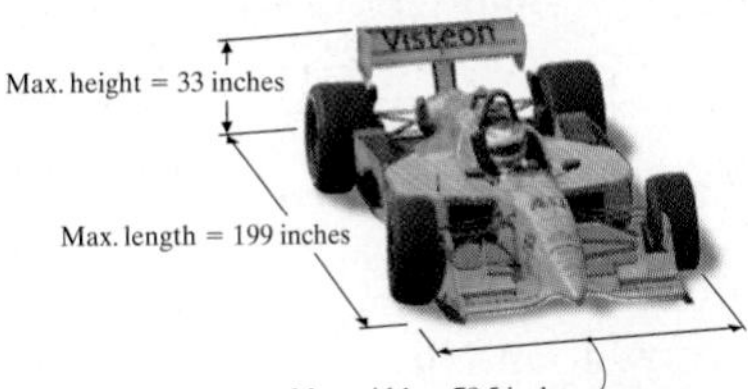

Page 139

Master and Apply Basic Skills and Concepts

K. Elayn Martin-Gay provides thorough explanations of key concepts and enlivens the content by integrating successful and innovative pedagogy. *Algebra A Combined Approach, Second Edition* integrates skill building throughout the text and provides problem-solving strategies and hints along the way. These features have been included to enhance your understanding of algebraic concepts.

Concept Check

The points $(-2, -5)$, $(0, -2)$, $(4, 4)$, and $(10, 13)$ all lie on the same line. Work with a partner and verify that the slope is the same no matter which points are used to find slope.

Page 226

◀ Concept Checks

Concept Checks are special margin exercises found in most sections. Work these to help gauge your grasp of the concept being developed in the text.

Page 200

Combining Concepts ▶

Combining Concepts exercises are found at the end of each exercise set. Solving these exercises will expose you to the way mathematical ideas build upon each other.

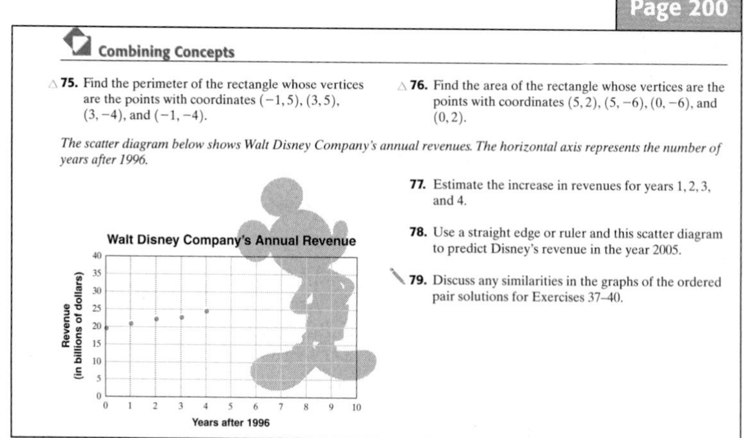

Combining Concepts

75. Find the perimeter of the rectangle whose vertices are the points with coordinates $(-1, 5)$, $(3, 5)$, $(3, -4)$, and $(-1, -4)$.

76. Find the area of the rectangle whose vertices are the points with coordinates $(5, 2)$, $(5, -6)$, $(0, -6)$, and $(0, 2)$.

The scatter diagram below shows Walt Disney Company's annual revenues. The horizontal axis represents the number of years after 1996.

77. Estimate the increase in revenues for years 1, 2, 3, and 4.

78. Use a straight edge or ruler and this scatter diagram to predict Disney's revenue in the year 2005.

79. Discuss any similarities in the graphs of the ordered pair solutions for Exercises 37–40.

Practice Problems ▶

Practice Problems occur in the margins next to every Example. Work these problems after an example to immediately reinforce your understanding.

EXAMPLE 8 Finding the Slope of a Line

The following graph shows the cost y (in cents) of an in-state long-distance telephone call in Massachusetts where x is the length of the call in minutes. Find the slope of the line and attach the proper units for the rate of change. Then write a sentence explaining the meaning of slope in this application.

Solution: Use $(2, 48)$ and $(5, 81)$ to calculate slope.

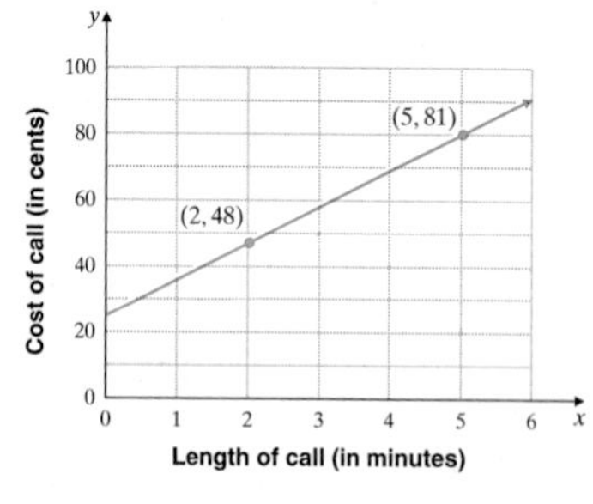

$$m = \frac{81 - 48}{5 - 2} = \frac{33}{3} = \frac{11 \text{ cents}}{1 \text{ minute}}$$

This means that the rate of change of a phone call is 11 cents per 1 minute or the cost of the phone call increases 11 cents per minute. ●

Practice Problem 8

Find the slope of the line and write the slope as a rate of change. This graph represents annual food and drink sales y (in billions of dollars) for year x.

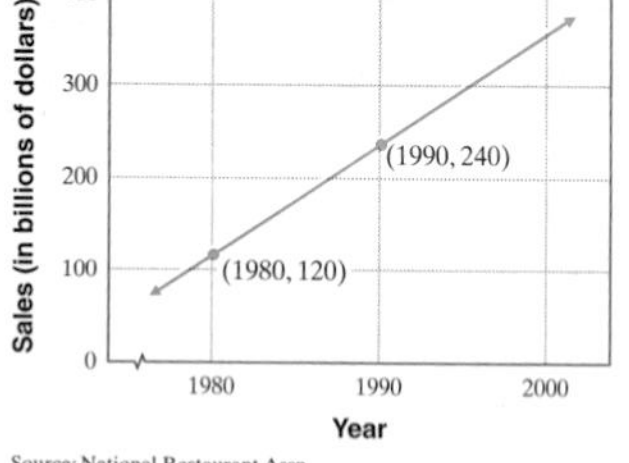

Answers

7. 15%, 8. $m = 12$; Each year the sales of food and drink from restaurants increases by $12 billion dollars.

Page 233

Writing Exercises ▶

New Writing Exercises, marked by an icon, are now found in most practice sets.

41. Explain why a point on the boundary line should not be chosen as the test point.

Page 251

Test Yourself and Check Your Understanding

Good exercise sets and an abundance of worked-out examples are essential for building student confidence. The exercises you will find in this worktext are intended to help you build skills and understand concepts as well as motivate and challenge you. In addition, features like Chapter Highlights, Chapter Reviews, Chapter Tests, and Cumulative Reviews are found at the end of each chapter to help you study and organize your notes.

Chapter 2 Pretest

Simplify.

1. $3c - 4 + 6c - 9$
2. $-5(2y - 3) - 7y + 1$

Solve.

3. $3 - x = -12$
4. $12 - (5 - 4b) = 9 + 3b$

Page 82

◀ PRETESTS

Pretests open each chapter. Take a Pretest to evaluate where you need the most help before beginning a new chapter.

INTEGRATED REVIEWS ▶

Integrated Reviews serve as mid-chapter reviews and help you to assimilate the new skills you have learned separately over several sections.

Page 119

Name ______________ Section ________ Date ________

Integrated Review—Linear Equations and Inequalities

Solve each equation or inequality.

1. $-4x = 20$
2. $-4x < 20$
3. $\frac{3x}{4} \geq 2$
4. $5x + 3 \geq 2 + 4x$

Review and Preview

Write each algebraic expression described. See Section 2.1.

△ **57.** A plot of land is in the shape of a triangle. If one side is x meters, a second side is $(2x - 3)$ meters, and a third side is $(3x - 5)$ meters, express the perimeter of the lot as a simplified expression in x.

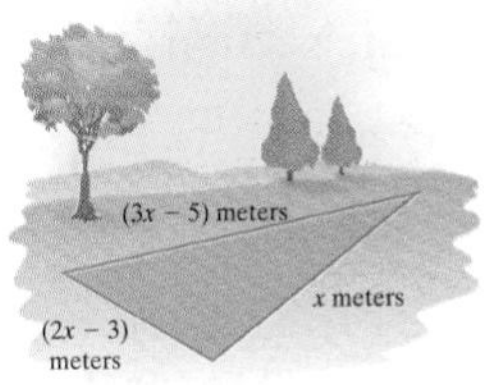

58. A portion of a board has length x feet. The other part has length $(7x - 9)$ feet. Express the total length of the board as a simplified expression in x.

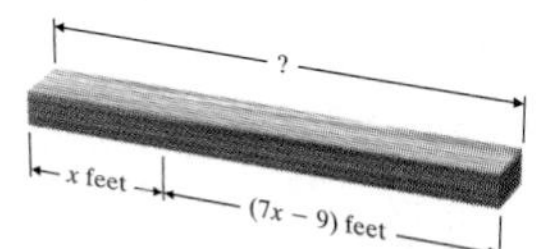

Page 117

◀ REVIEW AND PREVIEW

Review and Preview exercises review concepts learned earlier in the text that are needed in the next section or chapters.

Page 169

CHAPTER 2 Highlights

DEFINITIONS AND CONCEPTS	EXAMPLES
Section 2.1 Simplifying Expressions	
The **numerical coefficient** of a **term** is its numerical factor.	TERM / NUMERICAL COEFFICIENT $-7y$ / -7 x / 1 $\frac{1}{5}a^2b$ / $\frac{1}{5}$
Terms with the same variables raised to exactly the same powers are **like terms**.	LIKE TERMS / UNLIKE TERMS $12x, -x$ / $3y, 3y^2$ $-2xy, 5yx$ / $7a^2b, -2ab^2$
To combine like terms, add the numerical coefficients and multiply the result by the common variable factor.	$9y + 3y = 12y$ $-4z^2 + 5z^2 - 6z^2 = -5z^2$
To remove parentheses, apply the distributive property.	$-4(x + 7) + 10(3x - 1)$ $= -4x - 28 + 30x - 10$ $= 26x - 38$

CHAPTER HIGHLIGHTS ▶

Found at the end of every chapter, the Chapter Highlights contain key definitions, concepts, and examples to help students understand and retain what they have learned.

Increase Motivation

Throughout *Algebra A Combined Approach, Second Edition*, K. Elayn Martin-Gay provides interesting real-world applications to strengthen your understanding of the relevance of math in everyday life. When a new topic is presented, an effort has been made to relate the new ideas to those that students may already know. The Second Edition increases emphasis on visualization to clarify and reinforce key concepts.

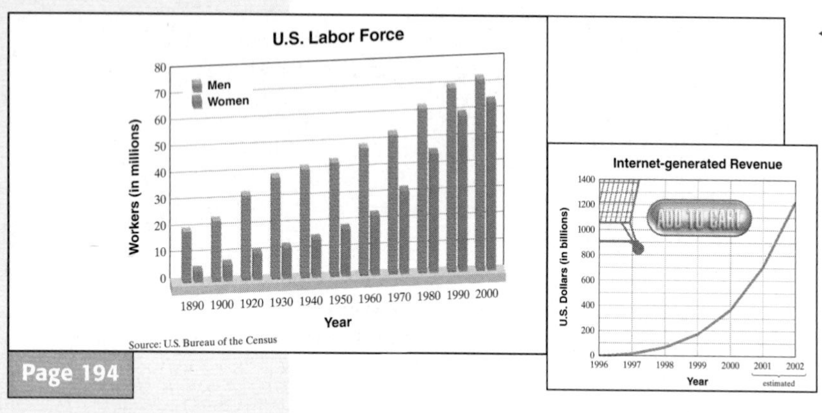

Page 194

◀ Real data is integrated throughout the worktext, drawn from current and familiar sources.

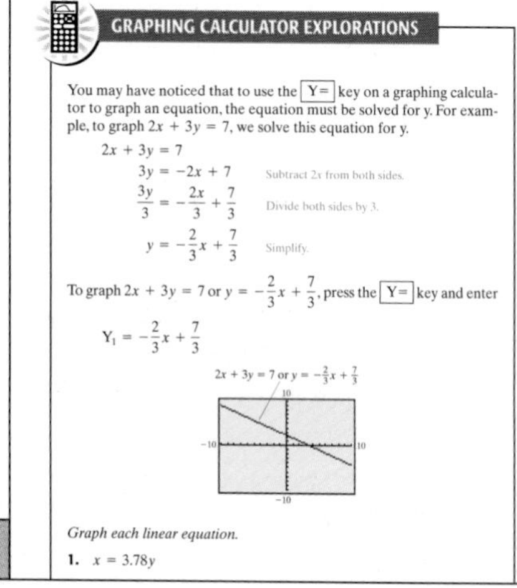

GRAPHING CALCULATOR EXPLORATIONS

You may have noticed that to use the $\boxed{Y=}$ key on a graphing calculator to graph an equation, the equation must be solved for y. For example, to graph $2x + 3y = 7$, we solve this equation for y.

$$2x + 3y = 7$$
$$3y = -2x + 7 \quad \text{Subtract } 2x \text{ from both sides.}$$
$$\frac{3y}{3} = -\frac{2x}{3} + \frac{7}{3} \quad \text{Divide both sides by 3.}$$
$$y = -\frac{2}{3}x + \frac{7}{3} \quad \text{Simplify.}$$

To graph $2x + 3y = 7$ or $y = -\frac{2}{3}x + \frac{7}{3}$, press the $\boxed{Y=}$ key and enter

$$Y_1 = -\frac{2}{3}x + \frac{7}{3}$$

Graph each linear equation.

1. $x = 3.78y$

Page 220

Calculator Explorations

▲ Optional Calculator Explorations and exercises appear in appropriate sections.

CHAPTER 3 ACTIVITY **Finding a Linear Model**

This activity may be completed by working in groups or individually.

The following table shows the estimated number of foreign visitors (in millions) to the United States for the years 2000 through 2003.

Year	Foreign Visitors to the United States (in millions)
2000	51.5
2001	54.2
2002	56.9
2003	59.6

(*Source:* Tourism Industries/International Trade Administration, U.S. Department of Commerce)

1. Make a scatter diagram of the paired data in the table.
2. Use what you have learned in this chapter to write an equation of the line representing the paired data in the table. Explain how you found the equation, and what each variable represents.
3. What is the slope of your line? What does the slope mean in this context?
4. Use your linear equation to predict the number of foreign visitors to the United States in 2010.
5. Compare your linear equation to that found by other students or groups. Is it the same, similar, or different? How?
6. Compare your prediction from question 3 to that of other students or groups. Describe what you find.

Page 252

▲ Graphics, models, and illustrations provide visual reinforcement.

Focus On Boxes ▶

Focus On boxes found throughout each chapter help you see the relevance of math through critical-thinking exercises and group activities. Try these on your own or with a classmate. Focus On covers the areas of: History, Mathematical Connections, Real World, and Business and Career.

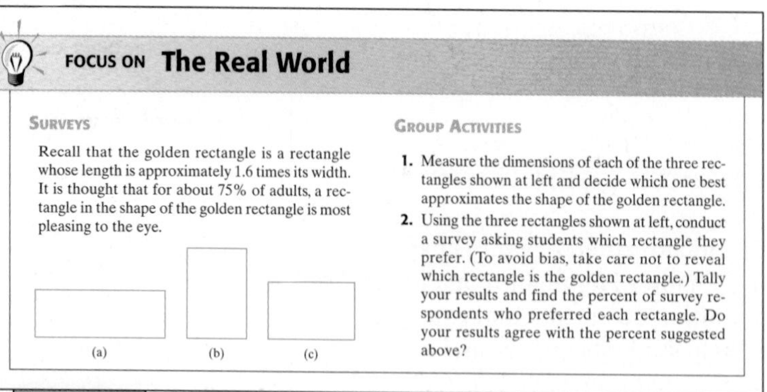

FOCUS ON **The Real World**

SURVEYS

Recall that the golden rectangle is a rectangle whose length is approximately 1.6 times its width. It is thought that for about 75% of adults, a rectangle in the shape of the golden rectangle is most pleasing to the eye.

GROUP ACTIVITIES

1. Measure the dimensions of each of the three rectangles shown at left and decide which one best approximates the shape of the golden rectangle.
2. Using the three rectangles shown at left, conduct a survey asking students which rectangle they prefer. (To avoid bias, take care not to reveal which rectangle is the golden rectangle.) Tally your results and find the percent of survey respondents who preferred each rectangle. Do your results agree with the percent suggested above?

Page 126

Build Confidence

Several features of this text can be helpful in building your confidence and mathematical competence. As you study, also notice the connections the author makes to relate new material to ideas that you may already know.

1.1 Tips for Success in Mathematics

Before reading this section, remember that your instructor is your best source for information. Please see your instructor for any additional help or information.

A Getting Ready for This Course

Now that you have decided to take this course, remember that a **positive attitude** will make all the difference in the world. Your belief that you can succeed is just as important as your commitment to this course. Make sure that you are ready for this course by having the time and positive attitude that it takes to succeed.

Next make sure that you have scheduled your math course at a time that will give you the best chance for success. For example, if you are also working, you may want to check with your employer to make sure that your work hours will not conflict with your course schedule.

OBJECTIVES

- A Getting ready for this course.
- B General tips for success.
- C How to use this text.
- D Get help as soon as you need it.
- E How to prepare for and take an exam.
- F Tips on time management.

Page 3

◀ TIPS FOR SUCCESS

New coverage of study skills in Section 1.1 reinforces this important component to success in this course.

Page 126

STUDY SKILLS REMINDER

How Are Your Homework Assignments Going?

It is so important in mathematics to keep up with homework. Why? Many concepts build on each other. Oftentimes, your understanding of a day's lecture in mathematics depends on an understanding of the previous day's material.

Remember that completing your homework assignment involves a lot more than attempting a few of the problems assigned.

To complete a homework assignment, remember these four things:

1. Attempt all of it.
2. Check it.
3. Correct it.
4. If needed, ask questions about it.

STUDY SKILLS REMINDERS ▶

New Study Skills Reminders are integrated throughout the book to reinforce section 1.1 and encourage the development of strong study skills.

Mental Math

Decide whether a line with the given slope is upward-sloping, downward-sloping, horizontal, or vertical.

1. $m = \frac{7}{6}$ **2.** $m = -3$ **3.** $m = 0$ **4.** m is undefined

Page 235

◀ MENTAL MATH

Mental Math warm-up exercises reinforce concepts found in the accompanying section and can increase your confidence before beginning an exercise set.

HELPFUL HINTS ▶

Found throughout the text, these contain practical advice on applying mathematical concepts. They are strategically placed where you are most likely to need immediate reinforcement.

Page 246

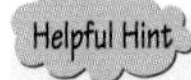

When graphing an inequality, make sure the test point is substituted into the **original inequality**. For Example 3, we substituted the test point $(0, 0)$ into the **original inequality** $2x - y \geq 3$, *not* $2x - y = 3$.

Chapter 2 VOCABULARY CHECK

Fill in each blank with one of the words or phrases listed below.

like terms	numerical coefficient	linear equation in one variable
equivalent equations	formula	proportion
linear inequality in one variable	ratio	

1. Terms with the same variables raised to exactly the same powers are called ______
2. A ______ can be written in the form $ax + b = c$.
3. Equations that have the same solution are called ______
4. An equation that describes a known relationship among quantities is called a ______
5. A ______ can be written in the form $ax + b < c$, (or $>, \leq, \geq$).
6. The ______ of a term is its numerical factor.
7. A ______ is the quotient of two numbers or two quantities.
8. A ______ is a mathematical statement that two ratios are equal.

Page 169

◀ VOCABULARY CHECKS

New Vocabulary Checks allow you to write your answers to questions about chapter content and strengthen verbal skills.

Enrich Your Learning

Seek out these additional Student Resources to match your personal learning style.

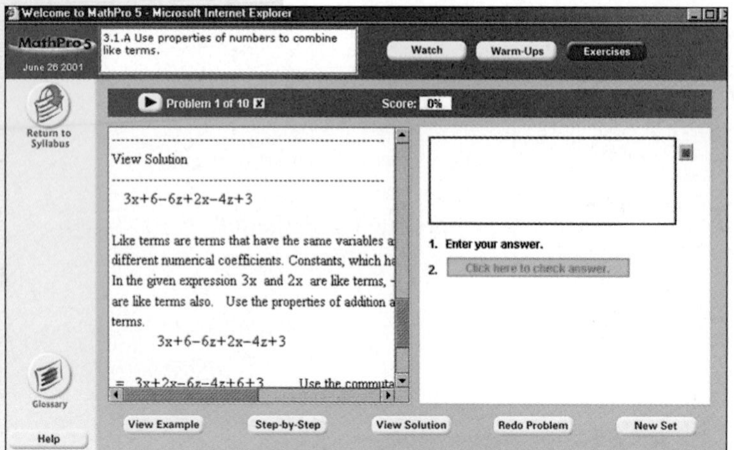

MathPro 5 is the online customizable tutorial, diagnostic and assessment software. It is text-specific to the learning objective level and provides anytime, anywhere tutorial support. It provides:

- Diagnostic review of student skills
- Virtually unlimited practice problems with immediate feedback
- Video clips by K. Elayn Martin-Gay
- Step-by-step solutions
- Summary of progress

MathPro 4 is available on CD-ROM for standalone use or can be networked in the school laboratory.

Text-specific videos, available on CD or VHS, are hosted by the award-winning teacher and author of *Algebra A Combined Approach,* Second Edition. They cover each objective in every chapter section as a supplementary review.

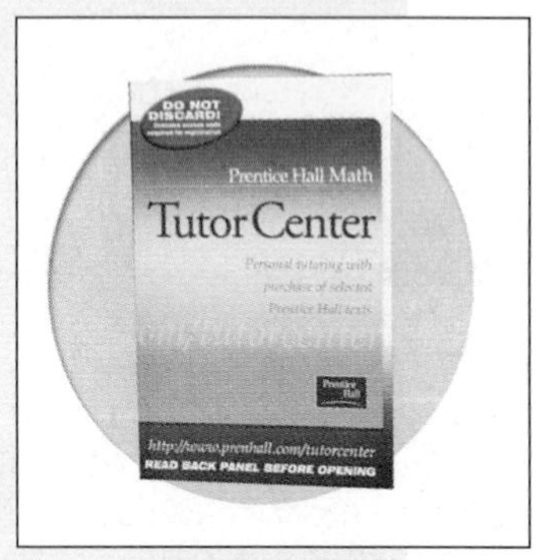

Prentice Hall Tutor Center provides text-specific tutoring via phone, fax, and e-mail. Visit http://prenhall.com/tutorcenter for details.

Also Available:

- ▲ Student Solutions Manual
- ▲ How to Study Math
- ▲ Math on the Internet
- ▲ *The New York Times/ Themes of the Times*

Ask your instructor or bookstore about these additional study aids.

Prealgebra Review

Mathematics is an important tool for everyday life. Knowing basic mathematic skills can simplify many tasks. For example, we use fractions to represent parts of a whole, such as "half an hour" or "third of a cup." Understanding decimals helps us work efficiently in our money system. Percent is a concept used virtually every day in ordinary and business life.

This optional review chapter covers basic topics and skills from prealgebra. Knowledge of these topics is needed for success in algebra.

Donald and Doris Fisher opened the first Gap store (named after the "generation gap") in 1969 near the campus of what is now the San Francisco State University. This single store, which sold mostly Levi's blue jeans, has evolved into an international clothing giant. Gap Inc. now sells a variety of clothing through its store chains: Banana Republic, Gap (including babyGap and GapKids), and Old Navy. In 2000, Gap Inc. had approximately $13.7 billion in sales. In Exercise 86 on page R-17, we will look at the breakdown of Gap stores by brand and find the fraction of Old Navy stores.

Answers

1. ______
2. ______
3. ______
4. ______
5. ______
6. ______
7. ______
8. ______
9. ______
10. ______
11. ______
12. ______
13. ______
14. ______
15. ______
16. ______
17. ______
18. ______
19. ______
20. ______
21. ______

Name ______ Section ______ Date ______

Chapter R Pretest

1. List the factors of 12.

2. Write the prime factorization of 150.

3. Find the LCM of 8, 14, and 20.

4. Write $\frac{7}{8}$ as an equivalent fraction with a denominator of 40.

Simplify each fraction.

5. $\frac{24}{40}$

6. $\frac{120}{250}$

Perform each indicated operation and simplify.

7. $\frac{2}{9} \cdot \frac{3}{8}$

8. $\frac{1}{4} + \frac{5}{6}$

9. $\frac{3}{7} \div \frac{7}{10}$

10. $\frac{2}{3} - \frac{5}{9}$

Perform each indicated operation.

11. $76 + 0.5 + 2.03$

12. $18 - 12.67$

13. $\begin{array}{r} 12.8 \\ \times\ 0.19 \\ \hline \end{array}$

14. $7.5\overline{)261.75}$

15. Write 7.16 as a fraction. Do not simplify.

16. Write $\frac{3}{16}$ as a decimal.

17. Write $\frac{5}{6}$ as a decimal.

18. Round 78.6159 to the nearest tenth.

19. Round 78.6159 to the nearest hundredth.

20. Write 80.6% as a decimal.

21. Write 0.3 as a percent.

R.1 Factors and the Least Common Multiple

OBJECTIVES

- (A) Write the factors of a number.
- (B) Write the prime factorization of a number.
- (C) Find the LCM of a list of numbers.

(A) Factoring Numbers

To **factor** means to write as a product.

In arithmetic we factor numbers, and in algebra we factor expressions containing variables. Throughout this text, you will encounter the word *factor* often. Always remember that factoring means writing as a product.

Since $2 \cdot 3 = 6$, we say that 2 and 3 are **factors** of 6. Also, $2 \cdot 3$ is a **factorization** of 6.

EXAMPLE 1 List the factors of 6.

Solution: First we write the different factorizations of 6.

$$6 = 1 \cdot 6, \quad 6 = 2 \cdot 3$$

The factors of 6 are 1, 2, 3, and 6. ●

Practice Problem 1

List the factors of 4.

EXAMPLE 2 List the factors of 20.

Solution: $20 = 1 \cdot 20, \quad 20 = 2 \cdot 10, \quad 20 = 4 \cdot 5, \quad 20 = 2 \cdot 2 \cdot 5$

The factors of 20 are 1, 2, 4, 5, 10, and 20. ●

Practice Problem 2

List the factors of 18.

In this section, we will concentrate on **natural numbers** only. The natural numbers (also called counting numbers) are

Natural Numbers: 1, 2, 3, 4, 5, 6, 7, and so on

Every natural number except 1 is either a prime number or a composite number.

Prime and Composite Numbers

A **prime number** is a natural number greater than 1 whose only factors are 1 and itself. The first few prime numbers are 2, 3, 5, 7, 11, 13, 17, 19, 23, 29, ...
A **composite number** is a natural number greater than 1 that is not prime.

EXAMPLE 3 Identify each number as prime or composite: 3, 20, 7, 4

Solution:

3 is a prime number. Its factors are 1 and 3 only.
20 is a composite number. Its factors are 1, 2, 4, 5, 10, and 20.
7 is a prime number. Its factors are 1 and 7 only.
4 is a composite number. Its factors are 1, 2, and 4. ●

Practice Problem 3

Identify each number as prime or composite: 5, 18, 11, 6

(B) Writing Prime Factorizations

When a number is written as a product of primes, this product is called the **prime factorization** of the number. For example, the prime factorization of 12 is $2 \cdot 2 \cdot 3$ since

$$12 = 2 \cdot 2 \cdot 3$$

and all the factors are prime.

Answers

1. 1, 2, 4, **2.** 1, 2, 3, 6, 9, 18, **3.** 5, 11 prime; 6, 18 composite

Practice Problem 4

Write the prime factorization of 28.

EXAMPLE 4 Write the prime factorization of 45.

Solution: We can begin by writing 45 as the product of two numbers, say 9 and 5.

$$45 = 9 \cdot 5$$

The number 5 is prime, but 9 is not. So we write 9 as $3 \cdot 3$.

$$45 = 9 \cdot 5$$
$$= 3 \cdot 3 \cdot 5$$

Each factor is now a prime number, so the prime factorization of 45 is $3 \cdot 3 \cdot 5$.

Helpful Hint

Recall that order is not important when multiplying numbers. For example,

$$3 \cdot 3 \cdot 5 = 3 \cdot 5 \cdot 3 = 5 \cdot 3 \cdot 3 = 45$$

For this reason, any of the products shown can be called *the* prime factorization of 45.

Practice Problem 5

Write the prime factorization of 60.

EXAMPLE 5 Write the prime factorization of 80.

Solution: We first write 80 as a product of two numbers. We continue this process until all factors are prime.

$$80 = 8 \cdot 10$$
$$4 \cdot 2 \cdot 2 \cdot 5$$
$$= 2 \cdot 2 \cdot 2 \cdot 2 \cdot 5$$

All factors are now prime, so the prime factorization of 80 is

$$2 \cdot 2 \cdot 2 \cdot 2 \cdot 5.$$

Try the Concept Check in the margin.

Concept Check

Suppose that you choose $80 = 4 \cdot 20$ as your first step in Example 5 and another student chooses $80 = 5 \cdot 16$. Will you both end up with the same prime factorization as in Example 5? Explain.

Helpful Hint

There are a few quick **divisibility tests** to determine if a number is divisible by the primes 2, 3, or 5.

A whole number is divisible by

- **2** if the ones digit is 0, 2, 4, 6, or 8.

  ↓

  132 is divisible by 2
- **3** if the sum of the digits is divisible by 3.

  144 is divisible by 3 since $1 + 4 + 4 = 9$ is divisible by 3
- **5** if the ones digit is 0 or 5.

  ↓

  1115 is divisible by 5

Answers

4. $28 = 2 \cdot 2 \cdot 7$, **5.** $60 = 2 \cdot 2 \cdot 3 \cdot 5$

Concept Check: yes; answers may vary

When finding the prime factorization of larger numbers, you may want to use the procedure shown in Example 6.

EXAMPLE 6 Write the prime factorization of 252.

Solution: Since the ones digit of 252 is 2, we know that 252 is divisible by 2.

$$2\overline{)252}^{\;126}$$

126 is divisible by 2 also.

$$\begin{array}{r} 63 \\ 2\overline{)126} \\ 2\overline{)252} \end{array}$$

63 is not divisible by 2 but is divisible by 3. We divide 63 by 3 and continue in this same manner until the quotient is a prime number.

$$\begin{array}{r} 7 \\ 3\overline{)\ 21} \\ 3\overline{)\ 63} \\ 2\overline{)126} \\ 2\overline{)252} \end{array}$$

The prime factorization of 252 is $2 \cdot 2 \cdot 3 \cdot 3 \cdot 7$.

Practice Problem 6

Write the prime factorization of 297.

C Finding the Least Common Multiple

A **multiple** of a number is the product of that number and any natural number. For example, the multiples of 3 are

$$\underbrace{3 \cdot 1}_{3,} \quad \underbrace{3 \cdot 2}_{6,} \quad \underbrace{3 \cdot 3}_{9,} \quad \underbrace{3 \cdot 4}_{12,} \quad \underbrace{3 \cdot 5}_{15,} \quad \underbrace{3 \cdot 6}_{18,} \quad \underbrace{3 \cdot 7}_{21,} \quad \text{and so on.}$$

The multiples of 2 are

$$\underbrace{2 \cdot 1}_{2,} \quad \underbrace{2 \cdot 2}_{4,} \quad \underbrace{2 \cdot 3}_{6,} \quad \underbrace{2 \cdot 4}_{8,} \quad \underbrace{2 \cdot 5}_{10,} \quad \underbrace{2 \cdot 6}_{12,} \quad \underbrace{2 \cdot 7}_{14,} \quad \text{and so on.}$$

Notice that 2 and 3 have multiples that are common to both.

Multiples of 2: 2, 4, 6, 8, 10, 12, 14, 16, 18, and so on

Multiples of 3: 3, 6, 9, 12, 15, 18, 21, and so on

The least or smallest common multiple of 2 and 3 is 6. The number 6 is called the **least common multiple** or **LCM** of 2 and 3. It is the smallest number that is a multiple of both 2 and 3.

Finding the LCM by the method above can sometimes be time-consuming. Let's look at another method that uses prime factorization.

To find the LCM of 4 and 10, for example, we write the prime factorization of each.

$$4 = 2 \cdot 2$$
$$10 = 2 \cdot 5$$

If the LCM is to be a multiple of 4, it must contain the factors $2 \cdot 2$. If the LCM is to be a multiple of 10, it must contain the factors $2 \cdot 5$. Since we decide whether the LCM is a multiple of 4 and 10 separately, the LCM does not need to contain three factors of 2. The LCM only needs to contain a factor

Answer

6. $3 \cdot 3 \cdot 3 \cdot 11$

the greatest number of times that the factor appears in any **one** prime factorization.

The LCM is a multiple of 4.

$$\text{LCM} = \overbrace{2 \cdot \underbrace{2 \cdot 5}} = 20$$

The LCM is a multiple of 10.

The number 2 is a factor twice since that is the greatest number of times that 2 is a factor in either of the prime factorizations.

To Find the LCM of a List of Numbers

Step 1. Write the prime factorization of each number.

Step 2. Write the product containing each different prime factor (from Step 1) the greatest number of times that it appears in any one factorization. This product is the LCM.

Practice Problem 7

Find the LCM of 14 and 35.

EXAMPLE 7 Find the LCM of 18 and 24.

Solution: First we write the prime factorization of each number.

$$18 = 2 \cdot 3 \cdot 3$$
$$24 = 2 \cdot 2 \cdot 2 \cdot 3$$

Now we write each factor the greatest number of times that it appears in any **one** prime factorization.

The greatest number of times that 2 appears is **3** times.
The greatest number of times that 3 appears is **2** times.

$$\text{LCM} = \underbrace{2 \cdot 2 \cdot 2}_{\text{2 is a factor 3 times.}} \cdot \underbrace{3 \cdot 3}_{\text{3 is a factor 2 times.}} = 72$$

Practice Problem 8

Find the LCM of 5 and 9.

EXAMPLE 8 Find the LCM of 11 and 6.

Solution: 11 is a prime number, so we simply rewrite it. Then we write the prime factorization of 6.

$$11 = 11$$
$$6 = 2 \cdot 3$$
$$\text{LCM} = 2 \cdot 3 \cdot 11 = 66.$$

Practice Problem 9

Find the LCM of 4, 15, and 10.

EXAMPLE 9 Find the LCM of 5, 6, and 12.

Solution:

$$5 = 5$$
$$6 = 2 \cdot 3$$
$$12 = 2 \cdot 2 \cdot 3$$
$$\text{LCM} = 2 \cdot 2 \cdot 3 \cdot 5 = 60.$$

Answers

7. 70, **8.** 45, **9.** 60

Name ______________________ Section __________ Date ____________

EXERCISE SET R.1

A *List the factors of each number. See Examples 1 and 2.*

1. 9 **2.** 8 **3.** 24 **4.** 36 **5.** 42

6. 50 **7.** 80 **8.** 63 **9.** 19 **10.** 31

Identify each number as prime or composite. See Example 3.

11. 13 **12.** 21 **13.** 39 **14.** 17 **15.** 37

16. 41 **17.** 51 **18.** 53 **19.** 2065 **20.** 1798

B *Write each prime factorization. See Examples 4 through 6.*

21. 18 **22.** 12 **23.** 20 **24.** 30

25. 56 **26.** 48 **27.** 300 **28.** 500

29. 81 **30.** 64 **31.** 588 **32.** 315

Multiple choice. Select the best choice to complete each statement.

33. The prime factorization of 24 is
a. $2 \cdot 2 \cdot 6$ **b.** $2 \cdot 2 \cdot 3$
c. $2 \cdot 2 \cdot 2 \cdot 3$ **d.** 1, 2, 3, 4, 6, 8, 12, 24

34. The factors of 63 are
a. 1, 3, 7, 9, 63 **b.** 1, 3, 7, 9, 21, 63
c. $3 \cdot 3 \cdot 7$ **d.** 1, 3, 21, 63

C *Find the LCM of each list of numbers. See Examples 7 through 9.*

35. 6, 14

36. 9, 15

37. 3, 4

38. 4, 5

39. 20, 30

40. 30, 40

41. 5, 7

42. 2, 11

43. 6, 12

44. 6, 18

45. 12, 20

46. 18, 30

47. 50, 70

48. 20, 90

49. 24, 36

50. 18, 21

51. 5, 10, 12

52. 3, 9, 20

53. 2, 3, 5

54. 3, 5, 7

55. 8, 18, 30

56. 4, 14, 35

57. 4, 8, 24

58. 5, 15, 45

Combining Concepts

Find the LCM of each pair of numbers.

59. 315, 504

60. 1000, 1125

61. The LCM of 6 and 7 is 42. In general, describe when the LCM of two numbers is equal to their product.

62. Is the following statement true or false? The number 45 is a prime number.

63. Craig Campanella and Edie Hall both have night jobs. Craig has every fifth night off and Edie has every seventh night off. How often will they have the same night off?

64. Elizabeth Kaster and Lori Sypher are both publishing company representatives in Louisiana. Elizabeth spends a day in New Orleans every 35 days, and Lori spends a day in New Orleans every 20 days. How often are they in New Orleans on the same day?

R.2 Fractions

OBJECTIVES

- (A) Write equivalent fractions.
- (B) Write fractions in simplest form.
- (C) Multiply and divide fractions.
- (D) Add and subtract fractions.
- (E) Perform operations on mixed numbers.

SSM TUTOR CENTER SG CD & VIDEO MATH PRO WEB

A quotient of two numbers such as $\frac{2}{9}$ is called a **fraction**. The parts of a fraction are:

Fraction bar → $\frac{2}{9}$ ← Numerator ← Denominator

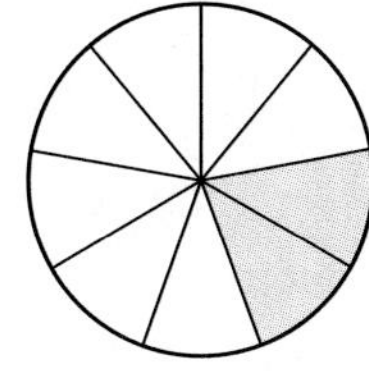

$\frac{2}{9}$ of the circle is shaded.

A fraction may be used to refer to part of a whole. For example, $\frac{2}{9}$ of the circle in the figure is shaded. The denominator 9 tells us how many equal parts the whole circle is divided into and the numerator 2 tells us how many equal parts are shaded.

In this section, we will use numerators that are **whole numbers** and denominators that are nonzero whole numbers. The whole numbers consist of 0 and the natural numbers.

Whole Numbers: 0, 1, 2, 3, 4, 5, and so on

(A) Writing Equivalent Fractions

More than one fraction can be used to name the same part of a whole. Such fractions are called **equivalent fractions**.

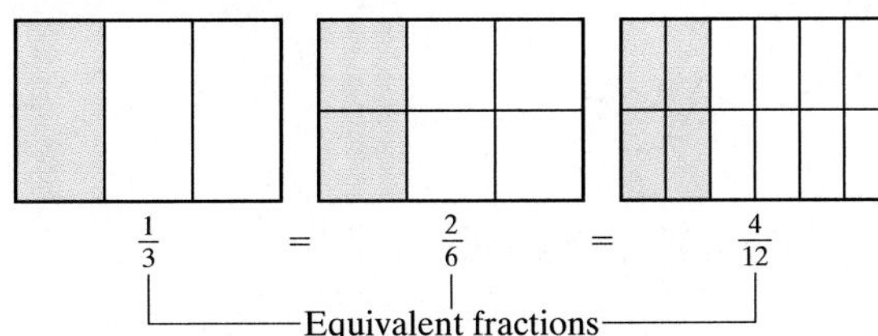

Equivalent Fractions

Fractions that represent the same portion of a whole are called **equivalent fractions.**

To write equivalent fractions, we use the **fundamental principle of fractions.** This principle guarantees that, if we multiply both the numerator and the denominator by the same nonzero number, the result is an equivalent fraction. For example, if we multiply the numerator and denominator of $\frac{1}{3}$ by the same number, 2, the result is the equivalent fraction $\frac{2}{6}$.

$$\frac{1 \cdot 2}{3 \cdot 2} = \frac{2}{6}$$

Fundamental Principle of Fractions

If a, b, and c are numbers, then

$$\frac{a}{b} = \frac{a \cdot c}{b \cdot c} \quad \text{or} \quad \frac{a \cdot c}{b \cdot c} = \frac{a}{b}$$

as long as b and c are not 0.

Practice Problem 1

Write $\frac{1}{4}$ as an equivalent fraction with a denominator of 20.

EXAMPLE 1 Write $\frac{2}{5}$ as an equivalent fraction with a denominator of 15.

Solution: Since $5 \cdot 3 = 15$, we use the fundamental principle of fractions and multiply the numerator and denominator of $\frac{2}{5}$ by 3.

$$\frac{2}{5} = \frac{2 \cdot 3}{5 \cdot 3} = \frac{6}{15}$$

Then $\frac{2}{5}$ is equivalent to $\frac{6}{15}$. They both represent the same part of a whole. ●

B Simplifying Fractions

A fraction is said to be **simplified** or in **lowest terms** when the numerator and the denominator have no factors in common other than 1. For example, the fraction $\frac{5}{11}$ is in lowest terms since 5 and 11 have no common factors other than 1.

One way to simplify fractions is to write both the numerator and the denominator as a product of primes and then apply the fundamental principle of fractions.

Practice Problem 2

Simplify: $\frac{20}{35}$

EXAMPLE 2 Simplify: $\frac{42}{49}$

Solution: We write the numerator and the denominator as products of primes. Then we apply the fundamental principle of fractions to the common factor 7.

$$\frac{42}{49} = \frac{2 \cdot 3 \cdot 7}{7 \cdot 7} = \frac{2 \cdot 3}{7} = \frac{6}{7}$$ ●

Concept Check

Explain the error in the following steps.

a. $\frac{15}{55} = \frac{15}{55} = \frac{1}{5}$

b. $\frac{6}{7} = \frac{5+1}{5+2} = \frac{1}{2}$

Try the Concept Check in the margin.

EXAMPLES Simplify each fraction.

3. $\frac{11}{27} = \frac{11}{3 \cdot 3 \cdot 3}$ There are no common factors other than 1, so $\frac{11}{27}$ is already simplified.

4. $\frac{88}{20} = \frac{2 \cdot 2 \cdot 2 \cdot 11}{2 \cdot 2 \cdot 5} = \frac{22}{5}$ ●

Practice Problems 3–4

Simplify each fraction.

3. $\frac{7}{20}$

4. $\frac{12}{40}$

A **proper fraction** is a fraction whose numerator is less than its denominator. The fraction $\frac{22}{5}$ from Example 4 is called an improper fraction. An **improper fraction** is a fraction whose numerator is greater than or equal to its denominator.

The improper fraction $\frac{22}{5}$ may be written as the mixed number $4\frac{2}{5}$. Notice that a **mixed number** has a whole number part and a fraction part. We review operations on mixed numbers in objective E in this section.

We may simplify some fractions by recalling that the fraction bar means division.

$$\frac{6}{6} = 6 \div 6 = 1 \quad \text{and} \quad \frac{3}{1} = 3 \div 1 = 3$$

Answers

1. $\frac{5}{20}$, **2.** $\frac{4}{7}$, **3.** $\frac{7}{20}$, **4.** $\frac{3}{10}$

Concept Check: answers may vary

EXAMPLES Simplify by dividing the numerator by the denominator.

5. $\frac{3}{3} = 1$ Since $3 \div 3 = 1$.

6. $\frac{4}{2} = 2$ Since $4 \div 2 = 2$.

7. $\frac{7}{7} = 1$ Since $7 \div 7 = 1$.

8. $\frac{8}{1} = 8$ Since $8 \div 1 = 8$.

In general, if the numerator and the denominator are the same nonzero number, the fraction is equivalent to 1. Also, if the denominator of a fraction is 1, the fraction is equivalent to the numerator.

If a is any number other than 0, then $\frac{a}{a} = 1$.

Also, if a is any number, $\frac{a}{1} = a$.

C Multiplying and Dividing Fractions

To multiply two fractions, we multiply numerator times numerator to obtain the numerator of the product. Then we multiply denominator times denominator to obtain the denominator of the product.

Multiplying Fractions

$$\frac{a}{b} \cdot \frac{c}{d} = \frac{a \cdot c}{b \cdot d} \quad \text{if } b \neq 0 \text{ and } d \neq 0$$

EXAMPLE 9 Multiply: $\frac{2}{15} \cdot \frac{5}{13}$. Simplify the product if possible.

Solution: $\frac{2}{15} \cdot \frac{5}{13} = \frac{2 \cdot 5}{15 \cdot 13}$ Multiply numerators. Multiply denominators.

To simplify the product, we divide the numerator and the denominator by any common factors.

$$\frac{2}{15} \cdot \frac{5}{13} = \frac{2 \cdot 5}{3 \cdot 5 \cdot 13} = \frac{2}{39}$$

Before we divide fractions, we first define **reciprocals**. Two numbers are reciprocals of each other if their product is 1.

The reciprocal of $\frac{2}{3}$ is $\frac{3}{2}$ because $\frac{2}{3} \cdot \frac{3}{2} = \frac{6}{6} = 1$.

The reciprocal of 5 is $\frac{1}{5}$ because $5 \cdot \frac{1}{5} = \frac{5}{1} \cdot \frac{1}{5} = \frac{5}{5} = 1$.

Practice Problems 5–8

Simplify by dividing the numerator by the denominator.

5. $\frac{4}{4}$ 6. $\frac{9}{3}$

7. $\frac{10}{10}$ 8. $\frac{5}{1}$

Practice Problem 9

Multiply: $\frac{3}{7} \cdot \frac{3}{5}$. Simplify the product if possible.

Answers

5. 1, **6.** 3, **7.** 1, **8.** 5, **9.** $\frac{9}{35}$

To divide fractions, we multiply the first fraction by the reciprocal of the second fraction. For example,

$$\frac{1}{2} \div \frac{5}{7} = \frac{1}{2} \cdot \frac{7}{5} = \frac{1 \cdot 7}{2 \cdot 5} = \frac{7}{10}$$

Helpful Hint

To divide, multiply by the reciprocal.

Dividing Fractions

$$\frac{a}{b} \div \frac{c}{d} = \frac{a}{b} \cdot \frac{d}{c}, \quad \text{if } b \neq 0, d \neq 0, \text{and } c \neq 0$$

Practice Problems 10–12

Divide and simplify.

10. $\frac{2}{9} \div \frac{3}{4}$

11. $\frac{8}{11} \div 24$

12. $\frac{5}{4} \div \frac{5}{8}$

EXAMPLES Divide and simplify.

10. $\frac{4}{5} \div \frac{5}{16} = \frac{4}{5} \cdot \frac{16}{5} = \frac{4 \cdot 16}{5 \cdot 5} = \frac{64}{25}$

11. $\frac{7}{10} \div 14 = \frac{7}{10} \div \frac{14}{1} = \frac{7}{10} \cdot \frac{1}{14} = \frac{7 \cdot 1}{2 \cdot 5 \cdot 2 \cdot 7} = \frac{1}{20}$

12. $\frac{3}{8} \div \frac{3}{10} = \frac{3}{8} \cdot \frac{10}{3} = \frac{3 \cdot 2 \cdot 5}{2 \cdot 2 \cdot 2 \cdot 3} = \frac{5}{4}$

D Adding and Subtracting Fractions

To add or subtract fractions with the same denominator, we combine numerators and place the sum or difference over the common denominator.

Adding and Subtracting Fractions with the Same Denominator

$$\frac{a}{b} + \frac{c}{b} = \frac{a+c}{b}, \quad \text{if } b \neq 0$$

$$\frac{a}{b} - \frac{c}{b} = \frac{a-c}{b}, \quad \text{if } b \neq 0$$

Practice Problems 13–16

Add or subtract as indicated. Then simplify if possible.

13. $\frac{2}{11} + \frac{5}{11}$

14. $\frac{1}{8} + \frac{3}{8}$

15. $\frac{13}{10} - \frac{3}{10}$

16. $\frac{7}{6} - \frac{2}{6}$

EXAMPLES Add or subtract as indicated. Then simplify if possible.

13. $\frac{2}{7} + \frac{4}{7} = \frac{2+4}{7} = \frac{6}{7}$

14. $\frac{3}{10} + \frac{2}{10} = \frac{3+2}{10} = \frac{5}{10} = \frac{5}{2 \cdot 5} = \frac{1}{2}$

15. $\frac{9}{7} - \frac{2}{7} = \frac{9-2}{7} = \frac{7}{7} = 1$

16. $\frac{5}{3} - \frac{1}{3} = \frac{5-1}{3} = \frac{4}{3}$

To add or subtract with different denominators, we first write the fractions as **equivalent fractions** with the same denominator. We use the smallest or **least common denominator**, or **LCD**. The LCD is the same as the least common multiple we reviewed in Section R.1.

Answers

10. $\frac{8}{27}$, **11.** $\frac{1}{33}$, **12.** 2, **13.** $\frac{7}{11}$, **14.** $\frac{1}{2}$, **15.** 1, **16.** $\frac{5}{6}$

EXAMPLE 17 Add: $\frac{2}{5} + \frac{1}{4}$

Solution: We first must find the least common denominator before the fractions can be added. The least common multiple for the denominators 5 and 4 is 20. This is the LCD we will use.

We write both fractions as equivalent fractions with denominators of 20. Since

$$\frac{2}{5} = \frac{2 \cdot 4}{5 \cdot 4} = \frac{8}{20} \quad \text{and} \quad \frac{1}{4} = \frac{1 \cdot 5}{4 \cdot 5} = \frac{5}{20}$$

then

$$\frac{2}{5} + \frac{1}{4} = \frac{8}{20} + \frac{5}{20} = \frac{13}{20}$$

Practice Problem 17

Add: $\frac{3}{8} + \frac{1}{20}$

EXAMPLE 18 Subtract and simplify: $\frac{19}{6} - \frac{23}{12}$

Solution: The LCD is 12. We write both fractions as equivalent fractions with denominators of 12.

$$\frac{19}{6} - \frac{23}{12} = \frac{19 \cdot 2}{6 \cdot 2} - \frac{23}{12}$$
$$= \frac{38}{12} - \frac{23}{12}$$
$$= \frac{15}{12} = \frac{3 \cdot 5}{2 \cdot 2 \cdot 3} = \frac{5}{4}$$

Practice Problem 18

Subtract and simplify: $\frac{8}{15} - \frac{1}{3}$

E Performing Operations on Mixed Numbers

To perform operations on mixed numbers, first write each mixed number as an improper fraction. To recall how this is done, let's write $3\frac{1}{5}$ as an improper fraction.

$$3\frac{1}{5} = 3 + \frac{1}{5} = \frac{15}{5} + \frac{1}{5} = \frac{16}{5}$$

Because of the steps above, notice we can use a shortcut process for writing a mixed number as an improper fraction.

$$3\frac{1}{5} = \frac{5 \cdot 3 + 1}{5} = \frac{16}{5}$$

EXAMPLE 19 Divide: $2\frac{1}{8} \div 1\frac{2}{3}$

Solution: First write each mixed number as an improper fraction.

$$2\frac{1}{8} = \frac{8 \cdot 2 + 1}{8} = \frac{17}{8}; \qquad 1\frac{2}{3} = \frac{3 \cdot 1 + 2}{3} = \frac{5}{3}$$

Now divide as usual.

$$2\frac{1}{8} \div 1\frac{2}{3} = \frac{17}{8} \div \frac{5}{3} = \frac{17}{8} \cdot \frac{3}{5} = \frac{51}{40} \quad \text{or} \quad 1\frac{11}{40}$$

Practice Problem 19

Multiply: $5\frac{1}{6} \cdot 4\frac{2}{5}$

Answers

17. $\frac{17}{40}$, **18.** $\frac{1}{5}$, **19.** $22\frac{11}{15}$

As a general rule, if the original exercise contains mixed numbers, write the result as a mixed number, if possible.

Practice Problem 20

Subtract: $7\frac{3}{8} - 6\frac{1}{4}$

EXAMPLE 20 Add: $2\frac{1}{8} + 1\frac{2}{3}$.

Solution: $2\frac{1}{8} + 1\frac{2}{3} = \frac{17}{8} + \frac{5}{3} = \frac{51}{24} + \frac{40}{24} = \frac{91}{24}$ or $3\frac{19}{24}$

When adding or subtracting larger mixed numbers, you might want to use the following method.

Practice Problem 21

Add: $76\frac{1}{9} + 35\frac{3}{4}$

EXAMPLE 21 Subtract: $50\frac{1}{6} - 38\frac{1}{3}$

Solution:

$$\begin{array}{rcrcr} 50\frac{1}{6} & = & 50\frac{1}{6} & = & 49\frac{7}{6} \\ -38\frac{1}{3} & = & -38\frac{2}{6} & = & -38\frac{2}{6} \\ \hline & & & & 11\frac{5}{6} \end{array}$$

$50\frac{1}{6} = 49 + 1 + \frac{1}{6} = 49\frac{7}{6}$

Answers

20. $1\frac{1}{8}$, **21.** $111\frac{31}{36}$

FOCUS ON History

FACTORING MACHINE

Small numbers can be broken down into their prime factors relatively easily. However, factoring larger numbers can be difficult and time-consuming. The first known successful attempt to automate the process of factoring whole numbers is credited to a French infantry officer and mathematics enthusiast, Eugène Olivier Carissan. In 1919, he designed and built a machine that uses gears and a hand crank to factor numbers.

Carissan's factoring machine had been all but forgotten after his death in 1925. In 1989, a Canadian researcher came across a description of the machine in an article printed in an obscure French journal in 1920. This led to a five-year search for traces of the machine. Eventually, the factoring machine was found in a French astronomical observatory which had received the invention from Carissan's family after his death.

Mathematical historians agree that the factoring machine was a remarkable achievement in its pre-computer era. Up to 40 numbers per second could be processed by the machine while its operator turned the crank at two revolutions per minute. Carissan was able to use his machine to prove that the number 708,158,977 was prime in under 10 minutes. He could also find the prime factorizations of up to 13-digit numbers with the machine.

Name ______________________ Section __________ Date __________

EXERCISE SET R.2

A *Write each fraction as an equivalent fraction with the given denominator. See Example 1.*

1. $\frac{7}{10}$ with a denominator of 30

2. $\frac{2}{3}$ with a denominator of 9

3. $\frac{2}{9}$ with a denominator of 18

4. $\frac{8}{7}$ with a denominator of 56

5. $\frac{4}{5}$ with a denominator of 20

6. $\frac{4}{5}$ with a denominator of 25

B *Simplify each fraction. See Examples 2 through 8.*

7. $\frac{2}{4}$ **8.** $\frac{3}{6}$ **9.** $\frac{10}{15}$ **10.** $\frac{15}{20}$ **11.** $\frac{3}{7}$

12. $\frac{5}{9}$ **13.** $\frac{20}{20}$ **14.** $\frac{24}{24}$ **15.** $\frac{35}{7}$ **16.** $\frac{42}{6}$

17. $\frac{18}{30}$ **18.** $\frac{42}{45}$ **19.** $\frac{16}{20}$ **20.** $\frac{8}{40}$ **21.** $\frac{66}{48}$

22. $\frac{64}{24}$ **23.** $\frac{120}{244}$ **24.** $\frac{360}{700}$ **25.** $\frac{192}{264}$ **26.** $\frac{455}{525}$

C **E** *Multiply or divide as indicated. See Examples 9 through 12 and 19.*

27. $\frac{1}{2}\cdot\frac{3}{4}$ **28.** $\frac{10}{6}\cdot\frac{3}{5}$ **29.** $\frac{2}{3}\cdot\frac{3}{4}$ **30.** $\frac{7}{8}\cdot\frac{3}{21}$ **31.** $5\frac{1}{9}\cdot 3\frac{2}{3}$

32. $2\frac{3}{4}\cdot 1\frac{7}{8}$ **33.** $7\frac{2}{5} \div \frac{1}{5}$ **34.** $9\frac{5}{6} \div \frac{1}{6}$ **35.** $\frac{1}{2} \div \frac{7}{12}$ **36.** $\frac{7}{12} \div \frac{1}{2}$

37. $\frac{3}{4} \div \frac{1}{20}$ **38.** $\frac{3}{5} \div \frac{9}{10}$ **39.** $\frac{7}{10}\cdot\frac{5}{21}$ **40.** $\frac{3}{35}\cdot\frac{10}{63}$ **41.** $\frac{9}{20} \div 12$

42. $\frac{25}{36} \div 10$ **43.** $4\frac{2}{11}\cdot 2\frac{1}{2}$ **44.** $6\frac{6}{7}\cdot 3\frac{1}{2}$ **45.** $8\frac{3}{5} \div 2\frac{9}{10}$ **46.** $1\frac{7}{8} \div 3\frac{8}{9}$

D **E** *Add or subtract as indicated. See Examples 13 through 18, 20 and 21.*

47. $\frac{4}{5}+\frac{1}{5}$

48. $\frac{6}{7}+\frac{1}{7}$

49. $\frac{4}{5}-\frac{1}{5}$

50. $\frac{6}{7}-\frac{1}{7}$

51. $\frac{23}{105}+\frac{4}{105}$

52. $\frac{13}{132}+\frac{35}{132}$

53. $\frac{17}{21}-\frac{10}{21}$

54. $\frac{18}{35}-\frac{11}{35}$

55. $9\frac{7}{8}+2\frac{3}{8}$

56. $8\frac{1}{8}-6\frac{3}{8}$

57. $5\frac{2}{5}-3\frac{4}{5}$

58. $7\frac{3}{4}+2\frac{1}{4}$

59. $\frac{2}{3}+\frac{3}{7}$

60. $\frac{3}{4}+\frac{1}{6}$

61. $\frac{10}{3}-\frac{5}{21}$

62. $\frac{11}{7}-\frac{3}{35}$

63. $\frac{10}{21}+\frac{5}{21}$

64. $\frac{11}{35}+\frac{3}{35}$

65. $\frac{5}{22}-\frac{5}{33}$

66. $\frac{7}{10}-\frac{8}{15}$

67. $8\frac{11}{12}-1\frac{5}{6}$

68. $4\frac{7}{8}-2\frac{3}{16}$

69. $17\frac{2}{5}+30\frac{2}{3}$

70. $26\frac{11}{20}+40\frac{7}{10}$

71. $\frac{12}{5}-1$

72. $2-\frac{3}{8}$

73. $\frac{2}{3}-\frac{5}{9}+\frac{5}{6}$

74. $\frac{8}{11}-\frac{1}{4}+\frac{1}{2}$

75. $28\frac{1}{5}-19\frac{2}{3}$

76. $40\frac{2}{7}-20\frac{1}{2}$

77. In your own words, describe how to add or subtract fractions.

78. In your own words, describe how to divide fractions.

Combining Concepts

Each circle below represents a whole, or 1. Determine the unknown part of the circle.

79.

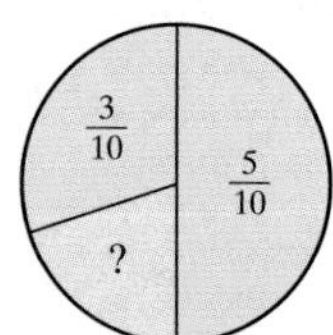

80.

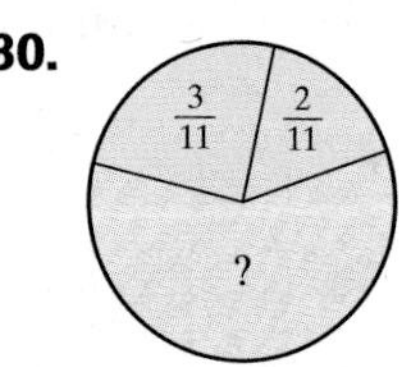

81.

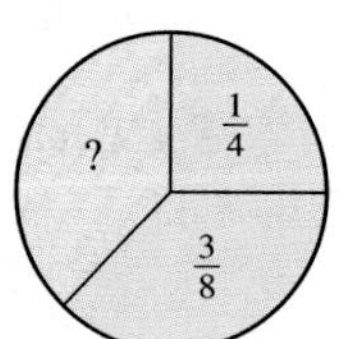

82.

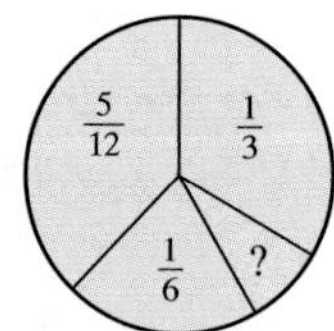

83. During the 2000 Summer Olympic Games, Ellina Zvereva of Belarus took the gold medal in the women's discus throw with a distance of $224\frac{5}{12}$ feet. However, the Olympic record for the women's discus throw was set in 1988 by Martina Hellmann of East Germany with a distance of $237\frac{1}{6}$ feet. How much longer was the Olympic record discus throw than the gold medal throw in 2000? (*Source: World Almanac and Book of Facts, 2001*)

84. Approximately $\frac{41}{50}$ of all American adults agree that the U.S. federal government should support basic scientific research. What fraction of American adults do *not* agree that the U.S. federal government should support such research? (*Source:* National Science Foundation)

85. The breakdown of science and engineering doctorate degrees awarded in the United States is summarized in the graph shown, called a circle graph or a pie chart. Use the graph to answer the questions. (*Source:* National Science Foundation)

Science and Engineering Doctorates Awarded, by Field of Study

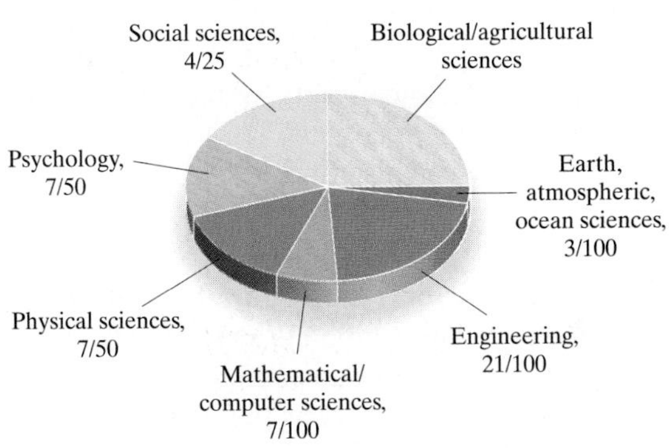

a. What fraction of science and engineering doctorates are awarded in the physical sciences?

b. Engineering doctorates make up what fraction of all science and engineering doctorates awarded in the United States?

c. What fraction of all science and engineering doctorates are awarded in the biological and agricultural sciences?

d. Social sciences and psychology doctorates together make up what fraction of all science and engineering doctorates awarded in the United States?

86. As of February 2001, Gap Inc. operated a total of 3676 stores worldwide. The following chart shows the store breakdown by brand. (*Source:* Gap Inc.)

Brand	Number of Stores
Gap (Domestic)	2079
Gap (International)	529
Banana Republic	402
Old Navy	666
Total	3676

a. What fraction of Gap-brand stores were Old Navy stores? Simplify this fraction.

b. What fraction of Gap-brand stores were either domestic or international Gap stores? Simplify this fraction.

The area of a plane figure is a measure of the amount of surface of the figure. Find the area of each figure. (The area of a rectangle is the product of its length and width. The area of a triangle is $\frac{1}{2}$ the product of its base and height. Recall that area is measured in square units.)

 **87.**

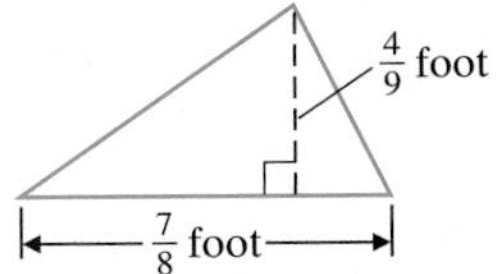

△ **88.**

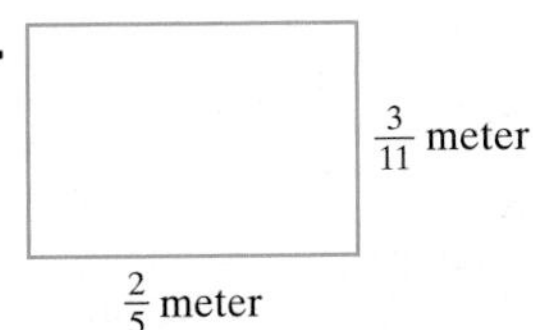

Internet Excursions

Go To: http://www.prenhall.com/martin-gay_algebra

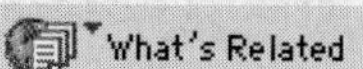

Publicly held corporations sell shares of their company's stock on a stock exchange such as the New York Stock Exchange. Stock prices are reported as decimals. Many sites on the World Wide Web allow you to look up stock prices if you know their ticker symbols. For instance, by visiting the given World Wide Web address you will be directed to the CNN Financial Network Web site, or a related site, where you can find today's opening, high, and low prices and the previous day's closing price for any stock by entering its ticker symbol. You can also find the stock's highest price and lowest price in the past 52 weeks.

89. Look up current stock prices for Coca-Cola Co. (ticker symbol: KO). Use the information on the Web site to complete the table below. What is the difference between today's high price and today's low price?

Date	
Time	
Today's high	
Today's low	

90. Look up the current stock prices for Wal-Mart Stores, Inc. (ticker symbol: WMT). Use the information on the Web site to complete the table below. How much more was the 52-week high price than the 52-week low price?

Date	
Time	
52-week high	
52-week low	

R.3 Decimals and Percents

OBJECTIVES

- **A** Write decimals as fractions.
- **B** Add, subtract, multiply, and divide decimals.
- **C** Round decimals to a given decimal place.
- **D** Write fractions as decimals.
- **E** Write percents as decimals and decimals as percents.

A Writing Decimals as Fractions

Like fractional notation, **decimal notation** is used to denote a part of a whole. Below is a **place value chart** that shows the value of each place.

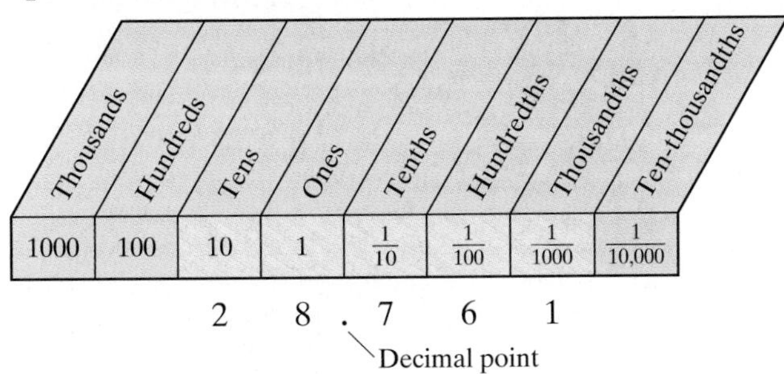

Try the Concept Check in the margin.

Concept Check

Fill in the blank: In the number 52.634, the 3 is in the ______ place.

a. Tens
b. Ones
c. Tenths
d. Hundredths
e. Thousandths

The next chart shows decimals written as fractions.

Decimal Form	Fractional Form	
0.1	$\frac{1}{10}$	tenths
0.07	$\frac{7}{100}$	hundredths
2.31	$\frac{231}{100}$	hundredths
0.9862	$\frac{9862}{10,000}$	ten-thousandths

EXAMPLES Write each decimal as a fraction. Do not simplify.

1. $0.37 = \frac{37}{100}$

2 decimal places — 2 zeros

2. $1.3 = \frac{13}{10}$

1 decimal place — 1 zero

3. $2.649 = \frac{2649}{1000}$

3 decimal places — 3 zeros

Practice Problems 1–3

Write each decimal as a fraction. Do not simplify.

1. 0.27
2. 5.1
3. 7.685

Answers

1. $\frac{27}{100}$, **2.** $\frac{51}{10}$, **3.** $\frac{7685}{1000}$

Concept Check: d

B Adding, Subtracting, Multiplying, and Dividing Decimals

To **add** or **subtract** decimals, we write the numbers vertically with decimal points lined up. We then add the digits in the like place values from right to left. We place the decimal point in the answer directly below the decimal points in the problem.

Practice Problem 4

Add.

a. $7.19 + 19.782 + 1.006$

b. $12 + 0.79 + 0.03$

EXAMPLE 4 Add.

a. $5.87 + 23.279 + 0.003$ **b.** $7 + 0.23 + 0.6$

Solution:

a.
$$\begin{array}{r} 5.87 \\ 23.279 \\ +\ 0.003 \\ \hline 29.152 \end{array}$$

b.
$$\begin{array}{r} 7. \\ 0.23 \\ +\ 0.6 \\ \hline 7.83 \end{array}$$

Practice Problem 5

Subtract.

a. $84.23 - 26.982$

b. $90 - 0.19$

EXAMPLE 5 Subtract.

a. $32.15 - 11.237$ **b.** $70 - 0.48$

Solution:

a.
$$\begin{array}{r} \overset{1}{3}\overset{11}{\cancel{2}}.\overset{4}{\cancel{1}}\overset{10}{\cancel{5}}\cancel{0} \\ -\ 1\ 1.\ 2\ 3\ 7 \\ \hline 2\ 0.\ 9\ 1\ 3 \end{array}$$

b.
$$\begin{array}{r} \overset{6}{\cancel{7}}\overset{9}{\cancel{0}}.\overset{9}{\cancel{0}}\overset{10}{\cancel{0}} \\ -\ 0.\ 4\ 8 \\ \hline 6\ 9.\ 5\ 2 \end{array}$$

Now let's study the following product of decimals. Notice the pattern in the decimal points.

$$0.03 \times 0.6 = \frac{3}{100} \times \frac{6}{10} = \frac{18}{1000} \quad \text{or} \quad 0.018$$

0.03: 2 decimal places; 0.6: 1 decimal place; 0.018: 3 decimal places

In general, to **multiply** decimals we multiply the numbers as if they were whole numbers. The decimal point in the product is placed so that the number of decimal places in the product is the same as the *sum* of the number of decimal places in the factors.

Practice Problem 6

Multiply.

a. 0.31×4.6

b. 1.26×0.03

EXAMPLE 6 Multiply.

a. 0.072×3.5 **b.** 0.17×0.02

Solution:

a.
$$\begin{array}{rl} 0.072 & \text{3 decimal places} \\ \times\ \ 3.5 & \text{1 decimal place} \\ \hline 360 & \\ 216\ & \\ \hline 0.2520 & \text{4 decimal places} \end{array}$$

b.
$$\begin{array}{rl} 0.17 & \text{2 decimal places} \\ \times\ 0.02 & \text{2 decimal places} \\ \hline 0.0034 & \text{4 decimal places} \end{array}$$

Answers

4. a. 27.978, **b.** 12.82, **5. a.** 57.248, **b.** 89.81, **6. a.** 1.426, **b.** 0.0378

To divide a decimal by a whole number using long division, we place the decimal point in the quotient directly above the decimal point in the dividend. For example,

```
   2.47
3)7.41
 -6          To check, see that 2.47 × 3 = 7.41
  14
 -12
   21
  -21
    0
```

In general, to **divide** decimals we move the decimal point in the divisor to the right until the divisor is a whole number. Then we move the decimal point in the dividend the same number of places that the decimal point in the divisor was moved. The decimal point in the quotient lies directly above the moved decimal point in the dividend.

EXAMPLE 7 Divide.

a. 9.46 ÷ 0.04 **b.** 31.5 ÷ 0.007

Solution:

a.

```
      236.5
04.)946.0
   -8
    14
   -12
     26
    -24
      20
     -20
       0
```

b.

```
          4500.
0007.)31500.
      -28
        35
       -35
         0
```

Practice Problem 7

Divide.

a. 21.75 ÷ 0.5

b. 15.6 ÷ 0.006

C Rounding Decimals

We **round** the decimal part of a decimal number in nearly the same way as we round the whole numbers. The only difference is that we drop digits to the right of the rounding place, instead of replacing these digits by 0 s. For example,

24.954 rounded to the nearest hundredth is 24.95

To Round Decimals to a Place Value to the Right of the Decimal Point

Step 1. Locate the digit to the right of the given place value.

Step 2.
- If this digit is 5 or greater, add 1 to the digit in the given place value and drop all digits to its right.
- If this digit is less than 5, drop all digits to the right of the given place.

Answers

7. a. 43.5, **b.** 2600

EXAMPLE 8 Round 7.8265 to the nearest hundredth.

hundredths place

Solution: 7.8265

Step 1. Locate the digit to the right of the hundredths place.
Step 2. This digit is 5 or greater, so we add 1 to the hundredths place digit and drop all digits to its right.

Thus, 7.8265 rounded to the nearest hundredth is 7.83.

Practice Problem 8

Round 12.9187 to the nearest hundredth.

EXAMPLE 9 Round 19.329 to the nearest tenth.

tenths place

Solution: 19.329

Step 1. Locate the digit to the right of the tenths place.
Step 2. This digit is less than 5, so we drop this digit and all digits to its right.

Thus, 19.329 rounded to the nearest tenth is 19.3.

Practice Problem 9

Round 245.348 to the nearest tenth.

D Writing Fractions as Decimals

To write fractions as decimals, interpret the fraction bar as division and find the quotient.

Writing Fractions as Decimals

To write fractions as decimals, divide the numerator by the denominator.

EXAMPLE 10 Write $\frac{1}{4}$ as a decimal.

Solution:

$$\begin{array}{r} 0.25 \\ 4\overline{)1.00} \\ -8 \\ \hline 20 \\ -20 \\ \hline 0 \end{array}$$

$$\frac{1}{4} = 0.25$$

Practice Problem 10

Write $\frac{2}{5}$ as a decimal.

EXAMPLE 11 Write $\frac{2}{3}$ as a decimal.

Solution:

$$\begin{array}{r} 0.666 \\ 3\overline{)2.000} \\ -1\,8 \\ \hline 20 \\ -18 \\ \hline 20 \\ -18 \\ \hline 2 \end{array}$$

This pattern will continue so that $\frac{2}{3} = 0.6666\ldots$.

Practice Problem 11

Write $\frac{5}{6}$ as a decimal.

Answers

8. 12.92, **9.** 245.3, **10.** 0.4, **11.** $0.8\overline{3}$

A bar can be placed over the digit 6 to indicate that it repeats. We call this a **repeating decimal**.

$$\frac{2}{3} = 0.666\ldots = 0.\overline{6}$$

We can also write a decimal approximation for $\frac{2}{3}$. For example, $\frac{2}{3}$ rounded to the nearest hundredth is 0.67. This can be written as $\frac{2}{3} \approx 0.67$. The $\approx$ sign means "is approximately equal to."

Try the Concept Check in the margin.

Concept Check

The notation $0.5\overline{2}$ is the same as

a. $\frac{52}{100}$

b. $\frac{52\ldots}{100}$

c. $0.52222222\ldots$

EXAMPLE 12

Write $\frac{22}{7}$ as a decimal. Round to the nearest hundredth. (The fraction $\frac{22}{7}$ is an approximation for π.)

Solution:

$$\begin{array}{r} 3.142 \approx 3.14 \\ 7\overline{)22.000} \\ -21 \\ \hline 1\,0 \\ -7 \\ \hline 30 \\ -28 \\ \hline 20 \\ -14 \\ \hline 6 \end{array}$$

If rounding to the nearest hundredth, carry the division process out to one more decimal place, the thousandths place.

The fraction $\frac{22}{7}$ in decimal form is approximately 3.14.

Practice Problem 12

Write $\frac{1}{9}$ as a decimal. Round to the nearest thousandth.

E Writing Percents as Decimals and Decimals as Percents

The word **percent** comes from the Latin phrase *per centum*, which means **"per 100."** Thus, 53% means 53 per 100, or

$$53\% = \frac{53}{100}$$

When solving problems containing percents, it is often necessary to write a percent as a decimal. To see how this is done, study the chart below.

Percent	Fraction	Decimal
7%	$\frac{7}{100}$	0.07
63%	$\frac{63}{100}$	0.63
109%	$\frac{109}{100}$	1.09

To convert directly from a percent to a decimal, notice that

$$7\% = 0.07$$

To Write a Percent as a Decimal

Drop the percent symbol and move the decimal point two places to the left.

Answers

12. 0.111

Concept Check: c

Practice Problem 13

Write each percent as a decimal.

a. 20%

b. 1.2%

c. 465%

EXAMPLE 13 Write each percent as a decimal.

a. 25% **b.** 2.6% **c.** 195%

Solution: We drop the % and move the decimal point two places to the left. Recall that the decimal point of a whole number is to the right of the ones place digit.

a. 25% = 25.% = 0.25

b. 2.6% = 02.6% = 0.026

c. 195% = 195.% = 1.95

To write a decimal as a percent, we simply reverse the preceding steps. That is, we move the decimal point two places to the right and attach the percent symbol, %.

To Write a Decimal as a Percent

Move the decimal point two places to the right and attach the percent symbol,%.

Practice Problem 14

Write each decimal as a percent.

a. 0.42

b. 0.003

c. 2.36

d. 0.7

EXAMPLE 14 Write each decimal as a percent.

a. 0.85 **b.** 1.25 **c.** 0.012 **d.** 0.6

Solution: We move the decimal point two places to the right and attach the percent symbol, %.

a. 0.85 = 0.85 = 85%

b. 1.25 = 1.25 = 125%

c. 0.012 = 0.012 = 1.2%

d. 0.6 = 0.60 = 60%

Answers

13. a. 0.20, **b.** 0.012, **c.** 4.65,
14. a. 42%, **b.** 0.3%, **c.** 236%, **d.** 70%

Name ______________ Section ________ Date ________

EXERCISE SET R.3

 *Write each decimal as a fraction. Do not simplify. See Examples 1 through 3.*

1. 0.6 **2.** 0.9 **3.** 1.86 **4.** 7.23

5. 0.114 **6.** 0.239 **7.** 123.1 **8.** 892.7

 *Add or subtract as indicated. See Examples 4 and 5.*

9. $5.7 + 1.13$ **10.** $2.31 + 6.4$ **11.** $24.6 + 2.39 + 0.0678$ **12.** $32.4 + 1.58 + 0.0934$

13. $8.8 - 2.3$ **14.** $7.6 - 2.1$ **15.** $18 - 2.78$ **16.** $28 - 3.31$

17. $\begin{array}{r} 45.02 \\ 3.006 \\ +\ 8.405 \\ \hline \end{array}$ **18.** $\begin{array}{r} 65.0028 \\ 5.0903 \\ +\ 6.9 \\ \hline \end{array}$ **19.** $\begin{array}{r} 654.9 \\ -\ 56.67 \\ \hline \end{array}$ **20.** $\begin{array}{r} 863.2 \\ -\ 39.45 \\ \hline \end{array}$

Multiply or divide as indicated. See Examples 6 and 7.

21. $\begin{array}{r} 0.2 \\ \times\ 0.6 \\ \hline \end{array}$ **22.** $\begin{array}{r} 0.7 \\ \times\ 0.9 \\ \hline \end{array}$ **23.** $\begin{array}{r} 6.75 \\ \times\ 10 \\ \hline \end{array}$ **24.** $\begin{array}{r} 8.91 \\ \times\ 100 \\ \hline \end{array}$

25. $\begin{array}{r} 5.62 \\ \times\ 7.7 \\ \hline \end{array}$ **26.** $\begin{array}{r} 8.03 \\ \times\ 5.5 \\ \hline \end{array}$ **27.** $\begin{array}{r} 16.003 \\ \times\ 5.31 \\ \hline \end{array}$ **28.** $\begin{array}{r} 31.006 \\ \times\ 3.71 \\ \hline \end{array}$

29. $5\overline{)0.47}$ **30.** $2\overline{)11.7}$ **31.** $0.6\overline{)42}$ **32.** $0.9\overline{)36}$

33. $0.82\overline{)4.756}$ **34.** $0.92\overline{)3.312}$ **35.** $0.063\overline{)52.92}$ **36.** $0.054\overline{)51.84}$

37. In your own words, describe how to add or subtract decimal numbers.

38. In your own words, describe how to multiply decimal numbers.

C *Round each decimal to the given place value. See Examples 8 and 9.*

39. 0.57, nearest tenth

40. 0.58, nearest tenth

41. 0.234, nearest hundredth

42. 0.452, nearest hundredth

43. 0.5942, nearest thousandth

44. 63.4523, nearest thousandth

45. 98,207.23, nearest tenth

46. 68,936.543, nearest tenth

47. 12.347, nearest hundredth

48. 42.9878, nearest thousandth

D *Write each fraction as a decimal. If the decimal is a repeating decimal, write using the bar notation and then round to the nearest hundredth. See Examples 10 through 12.*

49. $\frac{3}{4}$

50. $\frac{9}{25}$

51. $\frac{1}{3}$

52. $\frac{7}{9}$

53. $\frac{7}{16}$

54. $\frac{5}{8}$

55. $\frac{6}{11}$

56. $\frac{1}{6}$

E *Write each percent as a decimal. See Example 13.*

57. 28%

58. 36%

59. 3.1%

60. 2.2%

61. 135%

62. 417%

63. 96.55%

64. 81.49%

65. During the 2000–2001 season, grapefruit accounted for 52% of all Florida fresh citrus shipments. Write this percent as a decimal. (*Source:* Citrus Administrative Committee)

66. The average one-year survival rate for a heart transplant recipient is 82.3%. The average one-year survival rate for a liver transplant patient is 81.6%. Write each percent as a decimal. (*Source:* Bureau of Health Resources Development)

Write each decimal as a percent. See Example 14.

67. 0.68

68. 0.32

69. 0.876

70. 0.521

71. 1

72. 3

73. 0.5

74. 0.1

Combining Concepts

75. The passenger volume in a 2001 Dodge Intrepid Sedan is 104.40 cubic feet. The passenger volume in a 2001 Dodge Caravan SE is 140.30 cubic feet. How much more passenger space is in the Caravan than in the Intrepid? (*Source:* DaimlerChrysler)

76. The chart shows the average number of pounds of various flour and cereal products consumed by each United States citizen annually. (*Source:* National Agricultural Statistics Service)

Flour/Cereal product	Pounds
Wheat flour	147.8
Milled rice products	20.1
Corn products	23.1
Oat products	6.5
Barley products	1.3

a. How much more corn products than oat products does the average U.S. citizen consume?
b. What is the total amount of flour/cereal products consumed by the average U.S. citizen annually?

77. An estimated $\frac{7}{20}$ of all candy sold in the United States each year is used to give, share, or enjoy during major holidays. What percent of candy is purchased in conjunction with holidays? (*Source:* National Confectioners Association)

CHAPTER R ACTIVITY

Interpreting Survey Results

This activity may be completed by working in groups or individually.

Conduct the following survey with 12 students in one of your classes and record the results.

a. What is your age?
Under 20 20s 30s 40s 50s 60 and older

b. What is your gender?
Female Male

c. How did you arrive on campus today?
Walked Drove Bicycled
Took public transportation Other

1. For each survey question, tally the results for each category.

Age

Category	Tally
Under 20	
20s	
30s	
40s	
50s	
60+	
Total	

Gender

Category	Tally
Female	
Male	
Total	

Mode of Transportation

Category	Tally
Walk	
Drive	
Bicycle	
Public Transit	
Other	
Total	

2. For each survey question, find the fraction of the total number of responses that fall in each answer category. Use the tallies from Question 1 to complete the Fraction columns of the tables at the right.

3. For each survey question, convert the fraction of the total number of responses that fall in each answer category to a decimal number. Use the fractions from Question 2 to complete the Decimal columns of the tables below.

4. For each survey question, find the percent of the total number of responses that fall in each answer category. Complete the Percent columns of the tables below.

5. Study the tables. What may you conclude from them? What do they tell you about your survey respondents? Write a paragraph summarizing your findings.

Age

Category	Fraction	Decimal	Percent
Under 20			
20s			
30s			
40s			
50s			
60+			

Gender

Category	Fraction	Decimal	Percent
Female			
Male			

Mode of Transportation

Category	Fraction	Decimal	Percent
Walk			
Drive			
Bicycle			
Public Transit			
Other			

Chapter R VOCABULARY CHECK

Fill in each blank with one of the words or phrases listed below.

mixed number	factor	improper fraction	percent
multiple	composite number	proper fraction	simplified
prime number	equivalent		

1. To ________ means to write as a product.
2. A __________ of a number is the product of that number and any natural number.
3. A __________________ is a natural number greater than 1 that is not prime.
4. The word __________ mean per 100.
5. Fractions that represent the same portion of a whole are called ____________ fractions.
6. An __________________ is a fraction whose numerator is greater than or equal to its denominator.
7. A _______________ is a natural number greater than 1 whose only factors are 1 and itself.
8. A fraction is ____________ when the numerator and the denominator have no factors in common other than 1.
9. A ________________ is one whose numerator is less than its denominator.
10. A _______________ contains a whole number part and a fraction part.

FOCUS ON Study Skills

CRITICAL THINKING

What Is Critical Thinking?

Although exact definitions often vary, thinking critically usually refers to evaluating, analyzing, and interpreting information in order to make a decision, draw a conclusion, reach a goal, make a prediction, or form an opinion. Critical thinking often involves problem solving, communication, and reasoning skills. Critical thinking is more than a technique that helps students pass their courses. Critical thinking skills are life skills. Developing these skills can help you solve problems in your workplace and in everyday life. For instance, well-developed critical thinking skills would be useful in the following situation:

> Suppose you work as a medical lab technician. Your lab supervisor has decided that some lab equipment should be replaced. She asks you to collect information on several different models from equipment manufacturers. Your assignment is to study the data and then make a recommendation on which model the lab should buy.

How Can Critical Thinking Be Developed?

Just as physical exercise can help to develop and strengthen certain muscles of the body, mental exercise can help to develop critical thinking skills. Mathematics is ideal for helping to develop critical thinking skills because it requires using logic and reasoning, recognizing patterns, making conjectures and educated guesses, and drawing conclusions. You will find many opportunities to build your critical thinking skills throughout *Algebra A Combined Approach*:

- In real-life application problems (see Exercise 32 in Section 2.5)
- In writing exercises marked with the ✎ icon (see Exercise 43 in Section 1.3)
- In the Combining Concepts subsection of exercise sets (see Exercise 59 in Section 3.4)
- In the Chapter Activities (see the Chapter 7 Activity)
- In the Critical Thinking and Group Activities found in Focus On features like this one throughout the book.

CHAPTER R Highlights

Definitions and Concepts	**Examples**
Section R.1 Factors and the Least Common Multiple	
To **factor** means to write as a product.	Since $1 \cdot 12 = 12, 2 \cdot 6 = 12$, and $3 \cdot 4 = 12$, the factors of 12 are $1, 2, 3, 4, 6, 12$
When a number is written as a product of primes, this product is called the **prime factorization** of a number.	Write the prime factorization of 60. $60 = 6 \cdot 10$ $= 2 \cdot 3 \cdot 2 \cdot 5$ The prime factorization of 60 is $2 \cdot 2 \cdot 3 \cdot 5$.
The **least common multiple (LCM)** of a list of numbers is the smallest number that is a multiple of all the numbers in the list.	
To Find the LCM of a List of Numbers Step 1. Write the prime factorization of each number. Step 2. Write the product containing each different prime factor (from Step 1) the greatest number of times that it appears in any one factorization. This product is the LCM.	Find the LCM of 12 and 40. $12 = 2 \cdot 2 \cdot 3$ $40 = 2 \cdot 2 \cdot 2 \cdot 5$ $\text{LCM} = 2 \cdot 2 \cdot 2 \cdot 3 \cdot 5 = 120$
Section R.2 Fractions	
Fractions that represent the same quantity are called **equivalent fractions**.	$\frac{1}{5} = \frac{1 \cdot 4}{5 \cdot 4} = \frac{4}{20}$ $\frac{1}{5}$ and $\frac{4}{20}$ are equivalent fractions.
Fundamental Principle of Fractions If a, b, and c are numbers, then $\frac{a}{b} = \frac{a \cdot c}{b \cdot c}$ or $\frac{a \cdot c}{b \cdot c} = \frac{a}{b}$ as long as b and c are not 0.	
A fraction is **simplified** when the numerator and the denominator have no factors in common other than 1.	$\frac{13}{17}$ is simplified.
To simplify a fraction, factor the numerator and the denominator; then apply the fundamental principle of fractions to divide out common factors.	Simplify. $\frac{6}{14} = \frac{2 \cdot 3}{2 \cdot 7} = \frac{3}{7}$

DEFINITIONS AND CONCEPTS	**EXAMPLES**
Section R.2 Fractions (*continued*)	
Two fractions are **reciprocals** if their product is 1. The reciprocal of $\frac{a}{b}$ is $\frac{b}{a}$, as long as a and b are not 0.	The reciprocal of $\frac{6}{25}$ is $\frac{25}{6}$.
To multiply fractions, multiply numerator times numerator to find the numerator of the product and denominator times denominator to find the denominator of the product.	$\frac{2}{5}\cdot\frac{3}{7} = \frac{6}{35}$
To divide fractions, multiply the first fraction by the reciprocal of the second fraction.	$\frac{5}{9} \div \frac{2}{7} = \frac{5}{9}\cdot\frac{7}{2} = \frac{35}{18}$
To add fractions with the same denominator, add the numerators and place the sum over the common denominator.	$\frac{5}{11} + \frac{3}{11} = \frac{8}{11}$
To subtract fractions with the same denominator, subtract the numerators and place the difference over the common denominator.	$\frac{13}{15} - \frac{3}{15} = \frac{10}{15} = \frac{2}{3}$
To add or subtract fractions with different denominators, first write each fraction as an equivalent fraction with the LCD as denominator.	$\frac{2}{9} + \frac{3}{6} = \frac{2\cdot 2}{9\cdot 2} + \frac{3\cdot 3}{6\cdot 3} = \frac{4+9}{18} = \frac{13}{18}$
Section R.3 Decimals and Percents	
To write decimals as fractions, use place values.	$0.12 = \frac{12}{100} = \frac{3}{25}$ simplified
TO ADD OR SUBTRACT DECIMALS Step 1. Write the decimals so that the decimal points line up vertically. Step 2. Add or subtract as for whole numbers. Step 3. Place the decimal point in the sum or difference so that it lines up vertically with the decimal points in the problem.	Subtract: $2.8 - 1.04$ Add: $25 + 0.02$ $\begin{array}{r} \overset{7\ 10}{2.\cancel{8}\cancel{0}} \\ -1.04 \\ \hline 1.76 \end{array}$ $\begin{array}{r} 25. \\ +\ 0.02 \\ \hline 25.02 \end{array}$
TO MULTIPLY DECIMALS Step 1. Multiply the decimals as though they are whole numbers. Step 2. The decimal point in the product is placed so that the number of decimal places in the product is equal to the **sum** of the number of decimal places in the factors.	Multiply: 1.48×5.9 $\begin{array}{r} 1.48 \\ \times\ \ 5.9 \\ \hline 1332 \\ 740\ \ \\ \hline 8.732 \end{array}$ ← 2 decimal places; ← 1 decimal place; ← 3 decimal places

DEFINITIONS AND CONCEPTS	**EXAMPLES**
Section R.3 Decimals and Percents (*continued*)	
TO DIVIDE DECIMALS Step 1. Move the decimal point in the divisor to the right until the divisor is a whole number. Step 2. Move the decimal point in the dividend to the right the **same number of places** as the decimal point was moved in Step 1. Step 3. Divide. The decimal point in the quotient is directly over the moved decimal point in the dividend.	Divide: $1.118 \div 2.6$ $$\begin{array}{r} 0.43 \\ 2.6\overline{)1.118} \\ -104 \\ \hline 78 \\ -78 \\ \hline 0 \end{array}$$
To write fractions as decimals, divide the numerator by the denominator.	Write $\frac{3}{8}$ as a decimal. $$\begin{array}{r} 0.375 \\ 8\overline{)3.000} \\ -2\,4 \\ \hline 60 \\ -56 \\ \hline 40 \\ -40 \\ \hline 0 \end{array}$$
To write a percent as a decimal, drop the % symbol and move the decimal point two places to the left. **To write a decimal as a percent**, move the decimal point two places to the right and attach the % symbol.	$25\% = 25.\% = 0.25$ $0.7 = 0.70 = 70\%$

Name ______________________ Section __________ Date ______________

Chapter R Review

(R.1) *Write the prime factorization of each number.*

1. 42

2. 800

Find the least common multiple (LCM) of each list of numbers.

3. 12, 30

4. 7, 42

5. 4, 6, 10

6. 2, 5, 7

(R.2) *Write each fraction as an equivalent fraction with the given denominator.*

7. $\frac{5}{8}$ with a denominator of 24

8. $\frac{2}{3}$ with a denominator of 60

Simplify each fraction.

9. $\frac{8}{20}$

10. $\frac{15}{100}$

11. $\frac{12}{6}$

12. $\frac{8}{8}$

Perform each indicated operation and simplify.

13. $\frac{1}{7} \cdot \frac{8}{11}$

14. $\frac{5}{12} + \frac{2}{15}$

15. $\frac{3}{10} \div 6$

16. $\frac{7}{9} - \frac{1}{6}$

17. $3\frac{3}{8} \cdot 4\frac{1}{4}$

18. $2\frac{1}{3} - 1\frac{5}{6}$

19. $16\frac{9}{10} + 3\frac{2}{3}$

20. $6\frac{2}{7} \div 2\frac{1}{5}$

The area of a plane figure is a measure of the amount of surface of the figure. Find the area of each figure below. (The area of a rectangle is the product of its length and width. The area of a triangle is $\frac{1}{2}$ the product of its base and height.)

△ **21.**

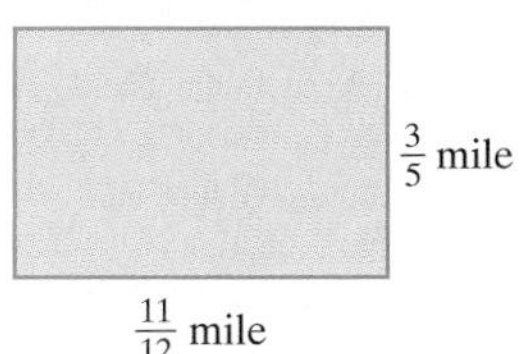

△ **22.**

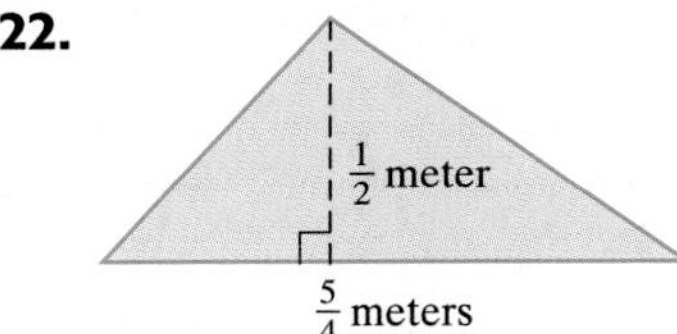

(R.3) *Write each decimal as a fraction. Do not simplify.*

23. 1.81

24. 0.035

Perform each indicated operation.

25. $\begin{array}{r} 76.358 \\ +18.76 \\ \hline \end{array}$

26. 35 + 0.02 + 1.765

27. 18 − 4.62

28. $\begin{array}{r} 804.062 \\ -112.489 \\ \hline \end{array}$

29. $\begin{array}{r} 7.6 \\ \times\ 12 \\ \hline \end{array}$

30. $\begin{array}{r} 14.63 \\ \times\quad 3.2 \\ \hline \end{array}$

31. $27\overline{)772.2}$

32. $0.06\overline{)13.8}$

Round each decimal to the given place value.

33. 0.7652, nearest hundredth

34. 25.6293, nearest tenth

Write each fraction as a decimal. If the decimal is a repeating decimal, write it using the bar notation and then round to the nearest thousandth.

35. $\frac{1}{2}$

36. $\frac{3}{8}$

37. $\frac{4}{11}$

38. $\frac{5}{6}$

Write each percent as a decimal.

39. 29%

40. 1.4%

Write each decimal as a percent.

41. 0.39

42. 1.2

43. In 2000, the home ownership rate in the United States was 67.4%. Write this percent as a decimal.

44. Choose the true statement.
a. 2.3% = 0.23
b. 5 = 500%
c. 40% = 4

Name ______________________ Section __________ Date __________

Chapter R Test

1. Write the prime factorization of 72.

2. Find the LCM of 5, 18, 20.

3. Write $\frac{5}{12}$ as an equivalent fraction with a denominator of 60.

Simplify each fraction.

4. $\frac{15}{20}$

5. $\frac{48}{100}$

6. Write 1.3 as a fraction.

Perform each indicated operation and simplify.

7. $\frac{5}{8} + \frac{7}{10}$

8. $\frac{2}{3} \cdot \frac{27}{49}$

9. $\frac{9}{10} \div 18$

10. $\frac{8}{9} - \frac{1}{12}$

11. $1\frac{2}{9} + 3\frac{2}{3}$

12. $5\frac{6}{11} - 3\frac{7}{22}$

13. $6\frac{7}{8} \div \frac{1}{8}$

14. $2\frac{1}{10} \cdot 6\frac{1}{2}$

Perform each indicated operation.

15. $43 + 0.21 + 1.9$

16. $123.6 - 57.72$

17. $\begin{array}{r} 7.93 \\ \times\ 1.6 \\ \hline \end{array}$

18. $0.25\overline{)80}$

19. Round 23.7272 to the nearest hundredth.

20. Write $\frac{7}{8}$ as a decimal.

21. Write $\frac{1}{6}$ as a repeating decimal. Then approximate the result to the nearest thousandth.

22. Write 63.2% as a decimal.

23. Write 0.09 as a percent.

24. Write $\frac{3}{4}$ as a percent. (*Hint:* Write $\frac{3}{4}$ as a decimal, and then write the decimal as a percent.)

Answers

1. __________
2. __________
3. __________
4. __________
5. __________
6. __________
7. __________
8. __________
9. __________
10. __________
11. __________
12. __________
13. __________
14. __________
15. __________
16. __________
17. __________
18. __________
19. __________
20. __________
21. __________
22. __________
23. __________
24. __________

25. ______

26. ______

27. ______

28. ______

29. ______

30. ______

Most of the water on Earth is in the form of oceans. Only a small part is fresh water. The graph below is called a circle graph or pie chart. This particular circle graph shows the distribution of fresh water on Earth. Use this graph to answer Questions 25 through 28. (Source: Philip's World Atlas)

Fresh Water Distribution

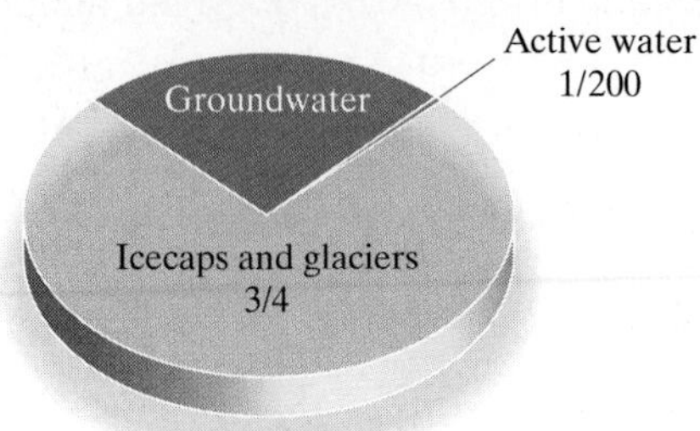

25. What fractional part of fresh water is icecaps and glaciers?

26. What fractional part of fresh water is active water?

27. What fractional part of fresh water is groundwater?

28. What fractional part of fresh water is groundwater or icecaps and glaciers?

Find the area of each figure.

△ **29.**

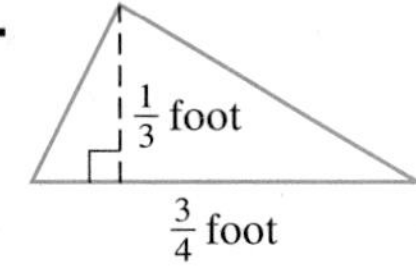

△ **30.**

Rectangle — $\frac{7}{8}$ centimeters by $\frac{9}{8}$ centimeters

Real Numbers and Introduction to Algebra

CHAPTER 1

The power of mathematics is its flexibility. We apply numbers to almost every aspect of our lives. The power of algebra is its generality. In algebra, we use letters to represent numbers.

In this chapter, we begin with a review of the basic symbols—the language—of arithmetic. We then introduce the use of a variable in place of a number. From there, we translate phrases to algebraic expressions and sentences to equations. This is the beginning of problem solving, which we formally study in Chapter 2.

The stars have been a source of interest to different cultures for centuries. Polaris, the North Star, guided ancient sailors. The Egyptians honored Sirius, the brightest star in the sky, in temples. Around 150 B.C., a Greek astronomer, Hipparchus, devised a system of classifying the brightness of stars. He called the brightest stars "first magnitude" and the faintest stars "sixth magnitude." Hipparchus's system is the basis of the apparent magnitude scale used by modern astronomers. This modern scale has been modified to include negative numbers. In Exercises 81 through 86 on page 20, we shall see how this scale is used to describe the brightness of objects such as the sun, the moon, and some planets.

Answers

1. ______
2. ______
3. ______
4. ______
5. ______
6. ______
7. ______
8. ______
9. ______
10. ______
11. ______
12. ______
13. ______
14. ______
15. ______
16. ______
17. ______
18. ______
19. ______
20. ______
21. ______
22. ______
23. ______
24. ______
25. ______

Name ______________________ Section __________ Date __________

Chapter 1 Pretest

Insert <, >, or = to form a true statement.

1. 0 −3 **2.** −10 −8 **3.** 1.7 1.07

Find the absolute value.

4. $|5|$ **5.** $|-1.2|$ **6.** $|0|$

7. Evaluate $xy - x^2$ when $x = 2$ and $y = 5$.

8. Write the phrase below as an algebraic expression. Let x represent the unknown number. Twice a number decreased by 10.

Evaluate the following.

9. 4^3

10. -3^2

11. Find the opposite of $-\frac{3}{5}$.

12. Find the reciprocal of $\frac{1}{8}$.

Perform the indicated operations and simplify.

13. $3 + 2 \cdot 5^2$ **14.** $-10 + 13$ **15.** $-6 - 21$ **16.** $(-7)(-8)$

17. $-2.8 \div 0.04$

18. $\dfrac{-4 - 6^2}{5(-2)}$

19. Evaluate $x^2 - 2xy$ when $x = -4$ and $y = -7$.

Decide whether the given number is a solution to the given equation.

20. $x - 12 = 5$; 7

21. $\frac{x}{8} + 2 = 7$; 40

22. At the beginning of the week, the balance in Trina Zimmerman's checkbook was −$21. By the end of the week, her checkbook balance was $34. How much had her checkbook balance increased by the end of the week?

23. Use the commutative property of addition to complete the statement: $2y + 5 =$ ____.

24. Use the distributive property to write the expression without parentheses: $4(3 + 2t) =$ ____.

25. Identify the property illustrated by the expression $6 + (-6) = 0$.

1.1 Tips for Success in Mathematics

OBJECTIVES

- Ⓐ Getting ready for this course.
- Ⓑ General tips for success.
- Ⓒ How to use this text.
- Ⓓ Get help as soon as you need it.
- Ⓔ How to prepare for and take an exam.
- Ⓕ Tips on time management.

SSM TUTOR CENTER SG CD & VIDEO MATH PRO WEB

Before reading this section, remember that your instructor is your best source for information. Please see your instructor for any additional help or information.

Ⓐ Getting Ready for This Course

Now that you have decided to take this course, remember that a **positive attitude** will make all the difference in the world. Your belief that you can succeed is just as important as your commitment to this course. Make sure that you are ready for this course by having the time and positive attitude that it takes to succeed.

Next make sure that you have scheduled your math course at a time that will give you the best chance for success. For example, if you are also working, you may want to check with your employer to make sure that your work hours will not conflict with your course schedule.

Now you are ready for your first class period. Double-check your schedule and allow yourself extra time to arrive in case of traffic or in case you have trouble locating your classroom. Make sure that you bring at least your textbook, paper, and a writing instrument with you. Are you required to have a lab manual, graph paper, calculator, or some other supply besides this text? If so, bring this material with you also.

Ⓑ General Tips for Success

Below are some general tips that will increase your chance for success in a mathematics class. Many of these tips will also help you in other courses you may be registered for.

Exchange names and phone numbers with at least one other person in class. This contact person can be a great help in case you miss the class assignment or want to discuss math concepts or exercises that you find difficult.

Choose to attend all class periods. If possible, sit near the front of the classroom. This way, you will see and hear the presentation better. It may also be easier for you to participate in classroom activities.

Do your homework. You've probably heard the phrase "practice makes perfect" in relation to music and sports. It also applies to mathematics. You will find that the more time you spend solving mathematics problems, the easier the process becomes. Be sure to schedule enough time to complete your assignments before the next class period.

Check your work. Review the steps you made while working a problem. Learn to check your answers in the original problems. You may also compare your answers to the answers to selected exercises listed in the back of the book. If you have made a mistake, try to figure out what went wrong. Then correct your mistake. If you can't find your mistake, don't erase your work or throw it away. Bring your work to your instructor, a tutor in a math lab, or a classmate. It is easier for someone to find where you had trouble if they look at your original work.

Learn from your mistakes. Everyone, even your instructor, makes mistakes. (That definitely includes me—Elayn Martin-Gay. You usually don't see my mistakes because many other people double-check my work in this text. If I make a mistake on a videotape, it is edited out so that you are not confused by it.) Use your errors to learn and to become a better math student. The key is finding and understanding your errors. Was your mistake a careless one or did you make it because you can't read your own "math" writing? If so, try to work more slowly or write more neatly and make a conscious effort to carefully check your work. Did you make a mistake because you don't understand a concept? Take the time to review the concept or ask questions to better understand the concept.

Know how to get help if you need it. It's OK to ask for help. In fact, it's a good idea to ask for help whenever there is something that you don't understand. Make sure you know when your instructor has office hours and how to find his or her office. Find out if math tutoring services are available on your campus. Check out the hours, location, and requirements of the tutoring service. Know whether videotapes or software are available and how to access these resources.

Organize your class materials, including homework assignments, graded quizzes and tests, and notes from your class or lab. All of these items will make valuable references throughout your course and as you study for upcoming tests and your final exam. Make sure that you can locate any of these materials when you need them.

Read your textbook before class. Reading a mathematics textbook is unlike entertainment reading such as reading a newspaper. Your pace will be much slower. It is helpful to have a pencil and paper with you when you read. Try to work out examples on your own as you encounter them in your text. You may also write down any questions that you want to ask in class. I know that when you read a mathematics textbook, sometimes some of the information in a section will still be unclear. But once you hear a lecture or watch a video on that section, you will understand it much more easily than if you had not read your text.

Don't be afraid to ask questions. From experience, I can tell you that you are not the only person in class with questions. Other students are normally grateful that someone has spoken up.

Hand in assignments on time. This way you can be sure that you will not lose points needlessly for being late. Show every step of a problem and be neat and organized. Also be sure that you understand which problems are assigned for homework. You can always double-check this assignment with another student in your class.

C Using This Text

There are many helpful resources that are available to you in this text. It is important that you become familiar with and use these resources. This should increase your chances for success in this course. For example:

- Each example in the section has a parallel Practice Problem. As you read a section, try each Practice Problem after you've finished the corresponding example. This "learn-by-doing" approach will help you grasp ideas before you move on to other concepts.
- The main section of exercises in an exercise set are referenced by an objective, such as A or B and also an example(s). Use this referencing in case you have trouble completing an assignment from the exercise set.
- If you need extra help in a particular section, check at the beginning of the section to see what videotapes and software are available.
- Make sure that you understand the meaning of the icons that are beside many exercises. The video icon tells you that the corresponding exercise may be viewed on the videotape that corresponds to that section. The pencil icon tells you that this exercise is a writing exercise in which you should answer in complete sentences. The △ icon simply tells you that this exercise involves geometric concepts.

- Integrated Reviews in each chapter offer you a chance to practice—in one place—the many concepts that you have learned separately over several sections.
- There are many opportunities at the end of each chapter to help you understand the concepts of the chapter.

  Vocabulary Checks provide a vocabulary self-check to make sure that you know the vocabulary in that chapter.

  Chapter Highlights contain chapter summaries with examples.

  Chapter Reviews contain review problems organized by section.

  Chapter Tests are sample tests to help you prepare for an exam.

  Cumulative Reviews are reviews consisting of material from the beginning of the book to the end of that particular chapter.

See the preface at the beginning of this text for a more thorough explanation of the features of this text.

D Getting Help

If you have trouble completing assignments or understanding the mathematics, get help as soon as you need it! This tip is presented as an objective on its own because it is *so* important. In mathematics, usually the material presented in one section builds on your understanding of the previous section. What does this mean? It means that if you don't understand the concepts covered during a class period, there is a good chance that you will not understand the concepts covered during the next class period. If this happens to you, get help as soon as you can.

Where can you get help? Many suggestions have been made in this section on where to get help, and now it is up to you to do it. Try your instructor, a tutoring center, or a math lab, or you may want to form a study group with fellow classmates. If you do decide to see your instructor or go to a tutoring center, make sure that you have a neat notebook and be ready with your questions.

E Preparing for and Taking an Exam

Make sure that you allow yourself plenty of time to prepare for a test. If you think that you are a little "math anxious," it may be that you are not preparing for a test in a way that will ensure success. The way that you prepare for a test in mathematics is important. To prepare for a test,

1. Review your previous homework assignments.
2. Review any notes from class and section level quizzes you may have taken. (If this is a final exam, also review chapter tests you have taken.)
3. Review concepts and definitions by reading the Highlights at the end of each chapter.
4. Practice working exercises by completing the Chapter Review found at the end of each chapter. (If this is a final exam, work a few Cumulative Reviews. There is one found at the end of each chapter (except Chapter 1). Choose the review found at the end of the latest chapter that you have covered in your course.) **Don't stop here!**
5. It is important that you place yourself in conditions similar to test conditions to see how you will perform. In other words, once you feel that you know the material, get out a few blank sheets of paper and take a sample test. There is a Chapter Test available at the end of each chapter,

or you can work selected problems from the Chapter Review, or your instructor may provide you with a review sheet. During this sample test, do not use your notes or your textbook. Then check your sample test. If you are not satisfied with the results, study the areas that you are weak in and try again.

6. On the day of the test, allow yourself plenty of time to arrive to where you will be taking your exam.

When taking your test,

1. Read the directions on the test carefully.
2. Read each problem carefully as you take your test. Make sure that you answer the question asked.
3. Watch your time and pace yourself so that you may attempt each problem on your test.
4. If you have time, check your work and answers.
5. Do not turn your test in early. If you have extra time, spend it double-checking your work.

F Managing Your Time

As a college student, you know the demands that classes, homework, work, and family place on your time. Some days you probably wonder how you'll ever get everything done. One key to managing your time is developing a schedule. Here are some hints for making a schedule:

1. Make a list of all of your weekly commitments for the term. Include classes, work, regular meetings, extracurricular activities, etc. You may also find it helpful to list such things as doing laundry, regular workouts, grocery shopping, etc.
2. Next, estimate the time needed for each item on the list. Also make a note of how often you will need to do each item. Don't forget to include time estimates for reading, studying, and homework you do outside of your classes. You may want to ask your instructor for help estimating the time needed for this item.
3. In the exercise set on the next page, you are asked to block out a typical week on the schedule grid given. Start with items with fixed time slots, like classes and work.
4. Next, include the items on your list with flexible time slots. Think carefully about how best to schedule some items such as study time.
5. Don't fill up every time slot on the schedule. Remember that you need to allow time for eating, sleeping, and relaxing! You should also allow a little extra time in case things take longer than planned.
6. If you find that your weekly schedule is too full for you to handle, you may need to make some changes in your workload, class load, or in other areas of your life. You may want to talk to your advisor, manager or supervisor at work, or someone in your college's academic counseling center for help with such decisions.

Note: In this chapter, we begin a feature called Study Skills Reminder. The purpose of this feature is to remind you of some of the information given in this section and to further expand on some topics in this section.

Name ______________________ Section __________ Date __________

EXERCISE SET 1.1

1. What is your instructor's name?

2. What are your instructor's office location and office hours?

3. What is the best way to contact your instructor?

4. What does the ✎ icon mean?

5. What does the 📼 icon mean?

6. What does the △ icon mean?

7. Do you have the name and contact information of at least one other student in class?

8. Will your instructor allow you to use a calculator in this class?

9. Are videotapes and/or tutorial software available to you? If so, where?

10. Is there a tutoring service available? If so, what are its hours?

11. Have you attempted this course before? If so, write down ways that you may improve your chances of success during this attempt.

12. List some steps that you may take in case you begin having trouble understanding the material or completing an assignment.

13. Read or reread objective F and fill out the schedule grid below.

	Monday	Tuesday	Wednesday	Thursday	Friday	Saturday	Sunday
7:00 a.m.							
8:00 a.m.							
9:00 a.m.							
10:00 a.m.							
11:00 a.m.							
12:00 p.m.							
1:00 p.m.							
2:00 p.m.							
3:00 p.m.							
4:00 p.m.							
5:00 p.m.							
6:00 p.m.							
7:00 p.m.							
8:00 p.m.							
9:00 p.m.							

1.2 Symbols and Sets of Numbers

OBJECTIVES

A. Define the meaning of the symbols $=$, $\neq$, $<$, $>$, $\leq$, and $\geq$.

B. Translate sentences into mathematical statements.

C. Identify integers, rational numbers, irrational numbers, and real numbers.

D. Find the absolute value of a real number.

We begin with a review of the set of natural numbers and the set of whole numbers and how we use symbols to compare these numbers. A **set** is a collection of objects, each of which is called a **member** or **element** of the set. A pair of brace symbols { } encloses the list of elements and is translated as "the set of" or "the set containing."

Natural Numbers

$\{1, 2, 3, 4, 5, 6, \ldots\}$

Whole Numbers

$\{0, 1, 2, 3, 4, 5, 6, \ldots\}$

Helpful Hint

The three dots (an ellipsis) at the end of the list of elements of a set means that the list continues in the same manner indefinitely.

These numbers can be pictured on a **number line**. To draw a number line, first draw a line. Choose a point on the line and label it 0. To the right of 0, label any other point 1. Being careful to use the same distance as from 0 to 1, mark off equally spaced distances to the right of 1. Label these points 2, 3, 4, 5, and so on. Since the whole numbers continue indefinitely, it is not possible to show every whole number on the number line. The arrow at the right end of the line indicates that the pattern continues indefinitely.

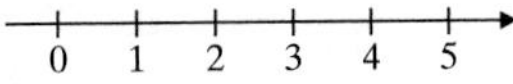

A Equality and Inequality Symbols

Picturing natural numbers and whole numbers on a number line helps us to see the order of the numbers. Symbols can be used to describe in writing the order of two quantities. We will use equality symbols and inequality symbols to compare quantities.

Below is a review of these symbols. The letters a and b are used to represent quantities. Letters such as a and b that are used to represent numbers or quantities are called **variables**.

		Meaning
Equality symbol:	$a = b$	a is equal to b.
Inequality symbols:	$a \neq b$	a is not equal to b.
	$a < b$	a is less than b.
	$a > b$	a is greater than b.
	$a \leq b$	a is less than or equal to b.
	$a \geq b$	a is greater than or equal to b.

These symbols may be used to form **mathematical statements** such as

$2 = 2$ and $2 \neq 6$

On the number line, we see that a number **to the right of** another number is **larger**. Similarly, a number **to the left of** another number is **smaller**. For example, 3 is to the left of 5 on the number line, which means that 3 is less than 5, or $3 < 5$. Similarly, 2 is to the right of 0 on the number line, which means 2 is greater than 0, or $2 > 0$. Since 0 is to the left of 2, we can also say that 0 is less than 2, or $0 < 2$.

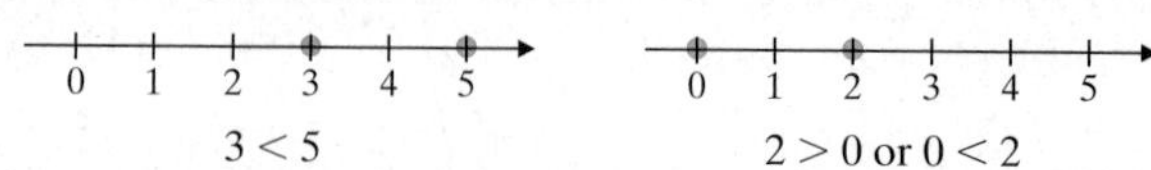

Helpful Hint

Notice that $2 > 0$ has exactly the same meaning as $0 < 2$. Switching the order of the numbers and reversing the "direction of the inequality symbol" does not change the meaning of the statement.

$5 > 3$ has the same meaning as $3 < 5$.

Also notice that when the statement is true, the inequality arrow points to the smaller number.

Practice Problems 1–6

Determine whether each statement is true or false.

1. $8 < 6$
2. $100 > 10$
3. $21 \leq 21$
4. $21 \geq 21$
5. $0 \leq 5$
6. $25 \geq 22$

EXAMPLES Determine whether each statement is true or false.

1.	$2 < 3$	True.	Since 2 is to the left of 3 on the number line
2.	$72 > 27$	True.	Since 72 is to the right of 27 on the number line
3.	$8 \geq 8$	True.	Since $8 = 8$ is true
4.	$8 \leq 8$	True.	Since $8 = 8$ is true
5.	$23 \leq 0$	False.	Since neither $23 < 0$ nor $23 = 0$ is true
6.	$0 \leq 23$	True.	Since $0 < 23$ is true

Helpful Hint

If either $3 < 3$ or $3 = 3$ is true, then $3 \leq 3$ is true.

B Translating Sentences into Mathematical Statements

Now, let's use the symbols discussed above to translate sentences into mathematical statements.

Practice Problem 7

Translate each sentence into a mathematical statement.

a. Fourteen is greater than or equal to fourteen.

b. Zero is less than five.

c. Nine is not equal to ten.

EXAMPLE 7 Translate each sentence into a mathematical statement.

a. Nine is less than or equal to eleven.
b. Eight is greater than one.
c. Three is not equal to four.

Solution:

a.

nine	is less than or equal to	eleven
↓	↓	↓
9	$\leq$	11

b.

eight	is greater than	one
↓	↓	↓
8	$>$	1

Answers

1. false, **2.** true, **3.** true, **4.** true, **5.** true, **6.** true, **7. a.** $14 \geq 14$, **b.** $0 < 5$, **c.** $9 \neq 10$

c. three → 3; is not equal to → $\neq$; four → 4

Ⓒ Identifying Common Sets of Numbers

Whole numbers are not sufficient to describe many situations in the real world. For example, quantities smaller than zero must sometimes be represented, such as temperatures less than 0 degrees.

We can place numbers less than zero on the number line as follows: Numbers less than 0 are to the left of 0 and are labeled -1, -2, -3, and so on. The numbers we have labeled on the number line below are called the set of **integers**.

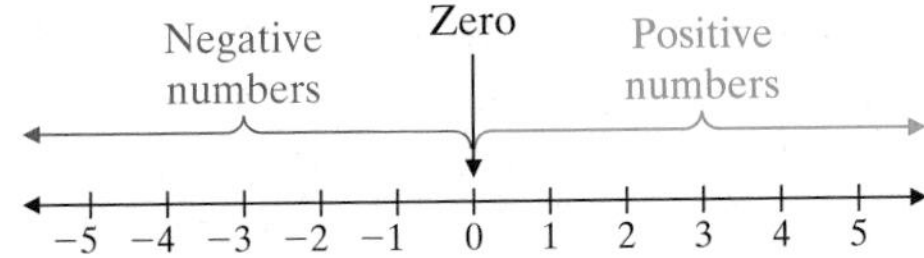

Integers to the left of 0 are called **negative integers**; integers to the right of 0 are called **positive integers**. The integer 0 is neither positive nor negative.

Integers

$\{\ldots, -3, -2, -1, 0, 1, 2, 3, \ldots\}$

Helpful Hint

A $-$ sign, such as the one in -1, tells us that the number is to the left of 0 on the number line.

-1 is read "negative one."

A $+$ sign or no sign tells us that a number lies to the right of 0 on the number line. For example, 3 and $+3$ both mean positive three.

EXAMPLE 8

Use an integer to express the number in the following. "The lowest temperature ever recorded at South Pole Station, Antarctica, occurred during the month of June. The record-low temperature was 117 degrees below zero." (*Source*: The National Oceanic and Atmospheric Administration)

Solution: The integer -117 represents 117 degrees below zero.

Practice Problem 8

Use an integer to express the number in the following. "The lowest altitude in North America is found in Death Valley, California. Its altitude is 282 feet below sea level." (*Source: The World Almanac, 2001*)

Answer

8. -282

A problem with integers in real-life settings arises when quantities are smaller than some integer but greater than the next smallest integer. On the number line, these quantities may be visualized by points between integers. Some of these quantities between integers can be represented as a quotient of integers. For example,

The point on the number line halfway between 0 and 1 can be represented by $\frac{1}{2}$, a quotient of integers.

The point on the number line halfway between 0 and -1 can be represented by $-\frac{1}{2}$. Other quotients of integers and their graphs are shown in the margin.

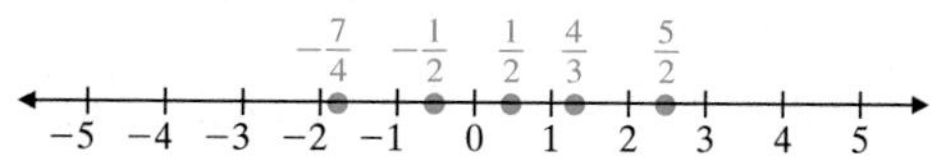

These numbers, each of which can be represented as a quotient of integers, are examples of rational numbers. It's not possible to list the set of rational numbers using the notation that we have been using. For this reason, we will use a different notation.

Rational Numbers

$$\left\{\frac{a}{b}\,\middle|\, a \text{ and } b \text{ are integers and } b \neq 0\right\}$$

We read this set as "the set of numbers $\frac{a}{b}$ such that a and b are integers and **b is not 0**."

Helpful Hint

We commonly refer to rational numbers as fractions.

Notice that every integer is also a rational number since each integer can be written as a quotient of integers. For example, the integer 5 is also a rational number since $5 = \frac{5}{1}$. For the rational number $\frac{5}{1}$, recall that the top number, 5, is called the numerator and the bottom number, 1, is called the denominator.

Let's practice **graphing** numbers on a number line.

EXAMPLE 9 Graph the numbers on the number line.

$$-\frac{4}{3},\ \frac{1}{4},\ \frac{3}{2},\ 2\frac{1}{8},\ 3.5$$

Solution: To help graph the improper fractions in the list, we first write them as mixed numbers.

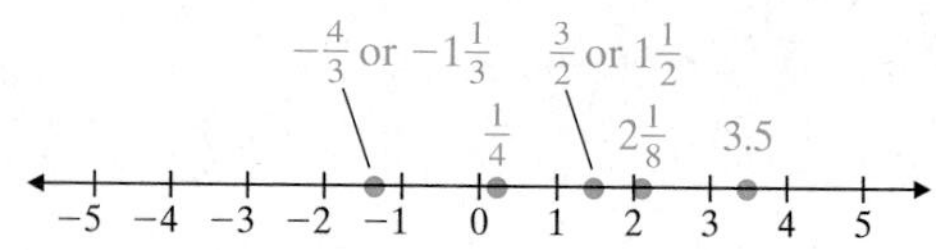

Practice Problem 9

Graph the numbers on the number line.

$$-2.5,\ -\frac{2}{3},\ \frac{1}{5},\ \frac{5}{4},\ 2.25$$

−5 −4 −3 −2 −1 0 1 2 3 4 5

Answer

9. −2.5, $-\frac{2}{3}$, $\frac{1}{5}$, $\frac{5}{4}$, 2.25

−5 −4 −3 −2 −1 0 1 2 3 4 5

Every rational number has a point on the number line that corresponds to it. But not every point on the number line corresponds to a rational number. Those points that do not correspond to rational numbers correspond instead to **irrational numbers**.

Irrational Numbers

{nonrational numbers that correspond to points on the number line}

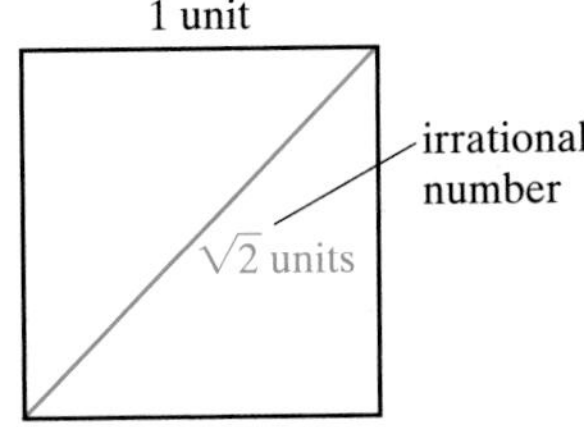

An irrational number that you have probably seen is π. Also, $\sqrt{2}$, the length of the diagonal of the square shown in the margin, is an irrational number.

Both rational and irrational numbers can be written as decimal numbers. The decimal equivalent of a rational number will either terminate or repeat in a pattern. For example, upon dividing we find that

$$\frac{3}{4} = 0.75 \qquad \text{(Decimal number terminates or ends.)}$$

$$\frac{2}{3} = 0.66666\ldots \qquad \text{(Decimal number repeats in a pattern.)}$$

The decimal representation of an irrational number will neither terminate nor repeat. (For further review of decimals, see Section R.3.)

The set of numbers, each of which corresponds to a point on the number line, is called the set of **real numbers**. One and only one point on the number line corresponds to each real number.

Real Numbers

{Numbers that correspond to points on the number line}

Several different sets of numbers have been discussed in this section. The following diagram shows the relationships among these sets of real numbers. Notice that, together, the rational numbers and the irrational numbers make up the real numbers.

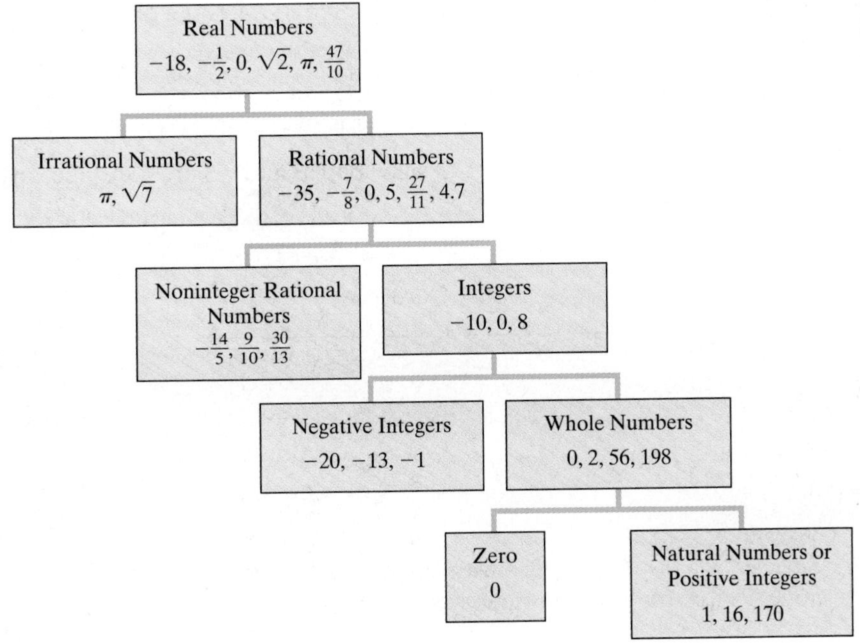

Practice Problem 10

Given the set $\{-100, -\frac{2}{5}, 0, \pi, 6, 10.6, 913\}$, list the numbers in this set that belong to the set of:

a. Natural numbers
b. Whole numbers
c. Integers
d. Rational numbers
e. Irrational numbers
f. Real numbers

Answers

10. a. 6, 913, **b.** 0, 6, 913, **c.** −100, 0, 6, 913, **d.** $-100, -\frac{2}{5}, 0, 6, 10.6, 913$, **e.** π, **f.** all numbers in the given set

EXAMPLE 10

Given the set $\{-2, 0, \frac{1}{4}, 112, -3, 11, \sqrt{2}, 5.1\}$, list the numbers in this set that belong to the set of:

a. Natural numbers
b. Whole numbers
c. Integers
d. Rational numbers
e. Irrational numbers
f. Real numbers

Solution:

a. The natural numbers are 11 and 112.
b. The whole numbers are 0, 11, and 112.
c. The integers are −3, −2, 0, 11, and 112.
d. Recall that integers are rational numbers also. The rational numbers are $-3, -2, 0, \frac{1}{4}, 5.1, 11$, and 112.
e. The irrational number is $\sqrt{2}$.
f. The real numbers are all numbers in the given set. ●

D Finding the Absolute Value of a Number

The number line not only gives us a picture of the real numbers, it also helps us visualize the distance between numbers. The distance between a real number a and 0 is given a special name called the **absolute value** of a. "The absolute value of a" is written in symbols as $|a|$.

Absolute Value

The **absolute value** of a real number a, denoted by $|a|$, is the distance between a and 0 on a number line.

For example, $|3| = 3$ and $|-3| = 3$ since both 3 and −3 are a distance of 3 units from 0 on the number line.

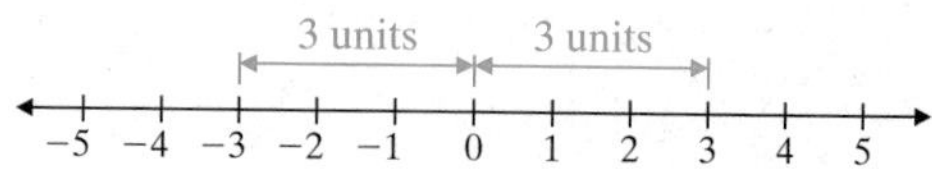

Helpful Hint

Since $|a|$ is a distance, $|a|$ is always either positive or 0. It is never negative. That is, **for any real number a, $|a| \geq 0$.**

EXAMPLE 11 Find the absolute value of each number.

a. $|4|$
b. $|-5|$
c. $|0|$

Solution:

a. $|4| = 4$ since 4 is 4 units from 0 on the number line.
b. $|-5| = 5$ since -5 is 5 units from 0 on the number line.
c. $|0| = 0$ since 0 is 0 units from 0 on the number line.

Practice Problem 11

Find the absolute value of each number.

a. $|7|$, b. $|-8|$, c. $\left|-\frac{2}{3}\right|$

EXAMPLE 12

Insert $<$, $>$, or $=$ in the appropriate space to make each statement true.

a. $|0| \quad 2$
b. $|-5| \quad 5$
c. $|-3| \quad |-2|$
d. $|5.9| \quad |6|$
e. $|-7| \quad |6|$

Solution:

a. $|0| < 2$ since $|0| = 0$ and $0 < 2$.
b. $|-5| = 5$.
c. $|-3| > |-2|$ since $3 > 2$.
d. $|5.9| < |6|$ since $5.9 < 6$.
e. $|-7| > |6|$ since $7 > 6$.

Practice Problem 12

Insert $<$, $>$, or $=$ in the appropriate space to make each statement true.

a. $|-4| \quad 4$, b. $-3 \quad |0|$,
c. $|-2.7| \quad |-2|$, d. $|6| \quad |16|$,
e. $|-6| \quad |-16|$

Answers

11. a. 7, **b.** 8, **c.** $\frac{2}{3}$, **12. a.** $=$, **b.** $<$, **c.** $>$, **d.** $<$, **e.** $<$

STUDY SKILLS REMINDER

Are you getting all the mathematics help that you need?

Remember that in addition to your instructor, there are many places to get help with your mathematics course. For example, see the list below.

- There is an accompanying video lesson for every section in this text.
- The back of this book contains answers to odd-numbered exercises as well as answers to every exercise in the Chapter Pretests, Integrated Reviews, Chapter Reviews, Chapter Tests, and Cumulative Reviews. The back of the book also contains selected solutions.
- MathPro is available with this text. It is a tutorial software program with lessons corresponding to each section in the text.
- There is a student solutions manual available that contains worked-out solutions to odd-numbered exercises as well as solutions to every exercise in the Chapter Pretests, Integrated Reviews, Chapter Reviews, Chapter Tests, and Cumulative Reviews.
- Check with your instructor for other local resources available to you, such as a tutor center.

Name ______________________ Section __________ Date __________

EXERCISE SET 1.2

A *Insert <, >, or = in the space between the paired numbers to make each statement true. See Examples 1 through 6.*

1. 4 10

2. 8 5

3. 7 3

4. 9 15

5. 6.26 6.26

6. 2.13 1.13

7. 0 7

8. 20 0

9. The freezing point of water is 32° Fahrenheit. The boiling point of water is 212° Fahrenheit. Write an inequality statement using < or > comparing the numbers 32 and 212.

10. The freezing point of water is 0° Celsius. The boiling point of water is 100° Celsius. Write an inequality statement using < or > comparing the numbers 0 and 100.

Determine whether each statement is true or false. See Examples 1 through 6.

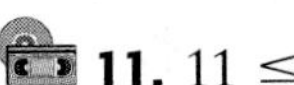

11. $11 \leq 11$

12. $8 \geq 9$

13. $10 > 11$

14. $17 > 16$

15. $3 + 8 \geq 3(8)$

16. $8 \cdot 8 \leq 8 \cdot 7$

17. $7 > 0$

18. $4 < 7$

△ **19.** An angle measuring 30° and an angle measuring 45° are shown. Use the inequality symbols ≤ or ≥ to write a statement comparing the numbers 30 and 45.

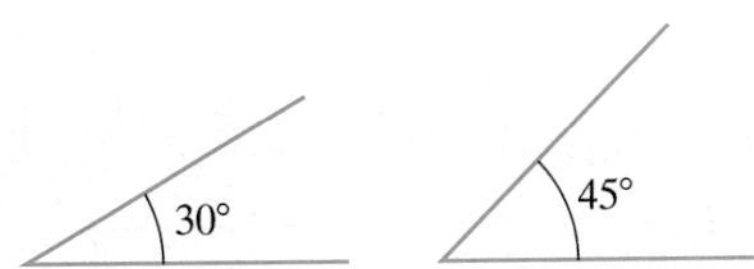

△ **20.** The sum of the measures of the angles of a triangle is 180°. The sum of the measures of the angles of a parallelogram is 360°. Use the inequality symbols ≤ or ≥ to write a statement comparing the numbers 360 and 180.

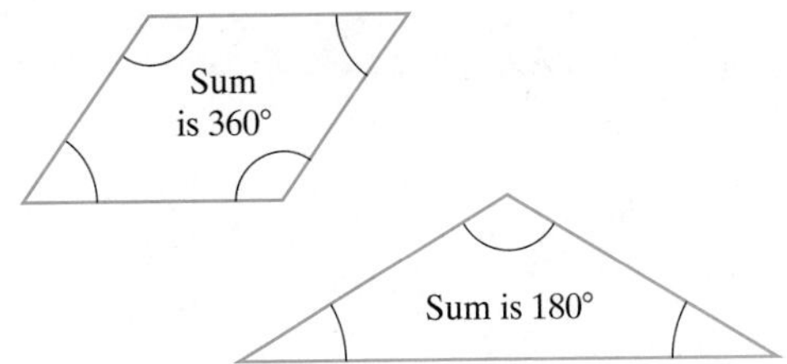

Rewrite each inequality so that the inequality symbol points in the opposite direction and the resulting statement has the same meaning as the given one.

21. $25 \geq 20$ **22.** $-13 \leq 13$ **23.** $0 < 6$ **24.** $5 > 3$ **25.** $-10 > -12$ **26.** $-4 < -2$

B *Write each sentence as a mathematical statement. See Example 7.*

27. Seven is less than eleven.

28. Twenty is greater than two.

29. Five is greater than or equal to four.

30. Negative ten is less than or equal to thirty-seven.

31. Fifteen is not equal to negative two.

32. Negative seven is not equal to seven.

 *Use integers to represent the values in each statement. See Example 8.*

33. The highest elevation in California is Mt. Whitney with an altitude of 14,494 feet. The lowest elevation in California is Death Valley with an altitude of 282 feet below sea level. (*Source:* U.S. Geological Survey)

34. Driskill Mountain, in Louisiana, has an altitude of 535 feet. New Orleans, Louisiana, lies 8 feet below sea level. (*Source:* U.S. Geological Survey)

35. From 1990 to 2000, the population of Washington, D.C., decreased by 34,841. (*Source:* U.S. Census Bureau)

36. From 1999 to 2000, the enrollment in public and private U.S. high schools increased by 143,000 students. (*Source:* National Center for Education Statistics)

37. Gretchen Bertani deposited \$475 in her savings account. She later withdrew \$195.

38. David Lopez was deep-sea diving. During his dive, he ascended 17 feet and later descended 15 feet.

Graph each set of numbers on the number line. See Example 9.

39. $-4, 0, 2, 5$

40. $-3, 0, 1, 5$

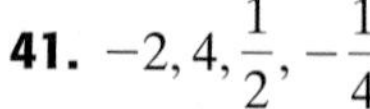
41. $-2, 4, \frac{1}{2}, -\frac{1}{4}$

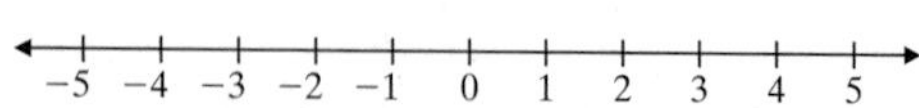

42. $-5, 3, -\frac{1}{2}, \frac{1}{4}$

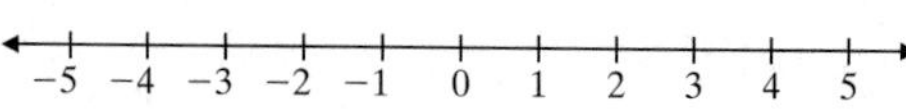

43. $-2.5, \frac{7}{4}, 3.25, -\frac{3}{2}$

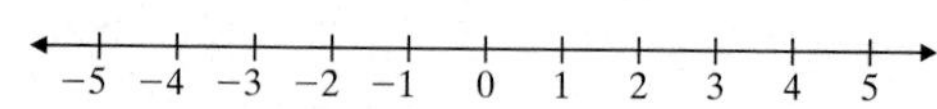

44. $4.5, -\frac{9}{4}, 1.75, \frac{5}{2}$

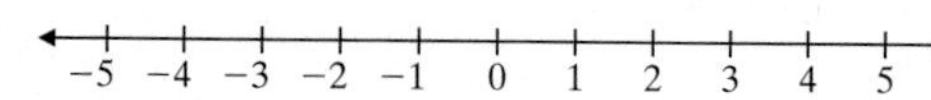

Tell which set or sets each number belongs to: natural numbers, whole numbers, integers, rational numbers, irrational numbers, and real numbers. See Example 10.

45. 0 **46.** $\frac{1}{4}$ **47.** -7 **48.** $-\frac{1}{7}$

49. 265 **50.** 7941 **51.** $\frac{2}{3}$ **52.** $\sqrt{3}$

Determine whether each statement is true or false.

53. Every rational number is also an integer.

54. Every natural number is positive.

55. 0 is a real number.

56. Every whole number is an integer.

57. Every negative number is also a rational number.

58. Every rational number is also a real number.

59. Every real number is also a rational number.

60. $\frac{1}{2}$ is an integer.

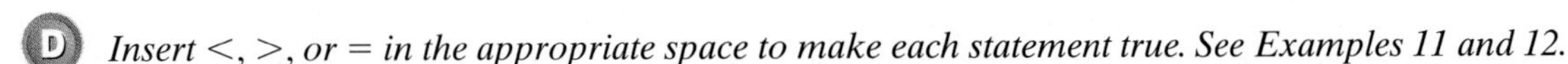

Insert $<$, $>$, or $=$ in the appropriate space to make each statement true. See Examples 11 and 12.

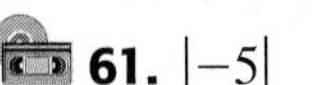

61. $|-5| \quad -4$ **62.** $0 \quad |0|$ **63.** $|-1| \quad |1|$ **64.** $\left|\frac{2}{5}\right| \quad \left|-\frac{2}{5}\right|$

65. $|-2| \quad |-3|$ **66.** $-5.00 \quad |-5.0|$ **67.** $|0| \quad |-8|$ **68.** $|-12| \quad \frac{24}{2}$

Combining Concepts

Tell whether each statement is true or false.

69. $\frac{1}{2} < \frac{1}{3}$ **70.** $\frac{3}{6} \geq \frac{1}{2}$ **71.** $|-5.3| \geq |5.3|$ **72.** $-1\frac{1}{2} > -\frac{1}{2}$

73. $-9.6 > -9.1$ **74.** $-7.3 < -7.1$ **75.** $-\frac{2}{3} \leq -\frac{1}{5}$ **76.** $-\frac{5}{6} > -\frac{1}{6}$

This graph shows the number of visitors, in millions, at the three most popular U.S. national parks. Each bar represents a different park in the year 2000 and the height of each bar represents the number of visitors (in millions) for that particular park in the year 2000. Use this graph to answer Exercises 77 through 80.

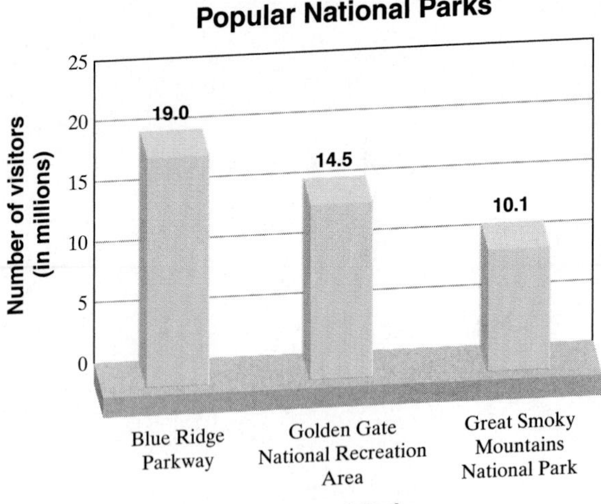

77. What is the park with the most visitors?

78. What is the number of visitors for the Great Smoky Mountains National Park?

79. Write an inequality statement using $\leq$ or $\geq$ comparing the number of visitors for Blue Ridge Parkway and Golden Gate National Recreation Area.

80. Write an inequality statement using $<$ or $>$ comparing the number of visitors for Great Smoky Mountains National Park and Blue Ridge Parkway.

The apparent magnitude of a star is the measure of its brightness as seen by someone on Earth. The smaller the apparent magnitude, the brighter the star. Below, the apparent magnitudes of some stars are listed. Use this table to answer Exercises 81 through 86.

Star	Apparent Magnitude	Star	Apparent Magnitude
Arcturus	−0.04	Spica	0.98
Sirius	−1.46	Rigel	0.12
Vega	0.03	Regulus	1.35
Antares	0.96	Canopus	−0.72
Sun	−26.7	Hadar	0.61

(*Source: Norton's 2000.0: Star Atlas and Reference Handbook*, 18th ed., Longman Group, UK, 1989)

81. The apparent magnitude of the sun is -26.7. The apparent magnitude of the star Arcturus is -0.04. Write an inequality statement comparing the numbers -0.04 and -26.7.

82. The apparent magnitude of Antares is 0.96. The apparent magnitude of Spica is 0.98. Write an inequality statement comparing the numbers 0.96 and 0.98.

83. Which is brighter, the sun or Arcturus?

84. Which is dimmer, Antares or Spica?

85. Which star listed is the brightest?

86. Which star listed is the dimmest?

87. In your own words, explain how to find the absolute value of a number.

88. Give an example of a real-life situation that can be described with integers but not with whole numbers.

1.3 Exponents, Order of Operations, and Variable Expressions

OBJECTIVES

- (A) Define and use exponents and the order of operations.
- (B) Evaluate algebraic expressions, given replacement values for variables.
- (C) Determine whether a number is a solution of a given equation.
- (D) Translate phrases into expressions and sentences into equations.

(A) Exponents and the Order of Operations

Frequently in algebra, products occur that contain repeated multiplication of the same factor. For example, the volume of a cube whose sides each measure 2 centimeters is $(2 \cdot 2 \cdot 2)$ cubic centimeters. We may use **exponential notation** to write such products in a more compact form. For example,

$$2 \cdot 2 \cdot 2 \text{ may be written as } 2^3.$$

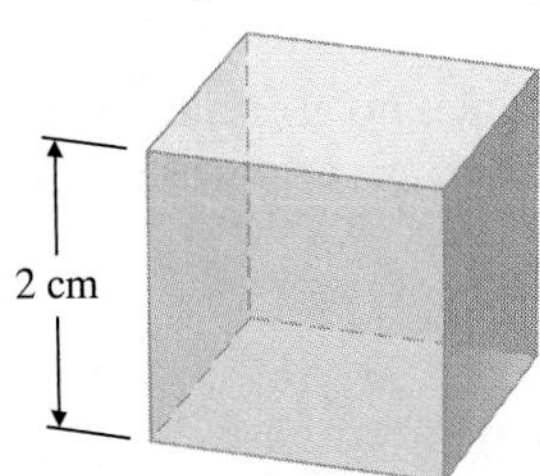

Volume is $(2 \cdot 2 \cdot 2)$ cubic centimeters.

The 2 in 2^3 is called the **base**; it is the repeated factor. The 3 in 2^3 is called the **exponent** and is the number of times the base is used as a factor. The expression 2^3 is called an **exponential expression**.

$$2^3 = 2 \cdot 2 \cdot 2 = 8$$

(exponent: 3; base: 2; 2 is a factor 3 times.)

EXAMPLE 1 Evaluate (find the value of) each expression.

a. 3^2 [read as "3 squared" or as "3 to the second power"]
b. 5^3 [read as "5 cubed" or as "5 to the third power"]
c. 2^4 [read as "2 to the fourth power"]
d. 7^1 **e.** $\left(\frac{3}{7}\right)^2$

Solution:

a. $3^2 = 3 \cdot 3 = 9$
b. $5^3 = 5 \cdot 5 \cdot 5 = 125$
c. $2^4 = 2 \cdot 2 \cdot 2 \cdot 2 = 16$
d. $7^1 = 7$
e. $\left(\frac{3}{7}\right)^2 = \left(\frac{3}{7}\right)\left(\frac{3}{7}\right) = \frac{3 \cdot 3}{7 \cdot 7} = \frac{9}{49}$

Practice Problem 1

Evaluate each expression.

a. 4^2
b. 2^2
c. 3^4
d. 9^1
e. $\left(\frac{2}{5}\right)^2$

Helpful Hint

$2^3 \neq 2 \cdot 3$ since 2^3 indicates **repeated multiplication of the same factor.**

$2^3 = 2 \cdot 2 \cdot 2 = 8$, whereas $2 \cdot 3 = 6$

Answers

1. a. 16, **b.** 4, **c.** 81, **d.** 9, **e.** $\frac{4}{25}$

Using symbols for mathematical operations is a great convenience. The more operation symbols presented in an expression, the more careful we must be when performing the indicated operation. For example, in the expression $2 + 3 \cdot 7$, do we add first or multiply first? To eliminate confusion, **grouping symbols** may be used. Examples of grouping symbols are parentheses ( ), brackets [ ], braces { }, absolute value bars | |, and the fraction bar. If we wish $2 + 3 \cdot 7$ to be simplified by adding first, we enclose $2 + 3$ in parentheses.

$$(2 + 3) \cdot 7 = 5 \cdot 7 = 35$$

If we wish to multiply first, $3 \cdot 7$ may be enclosed in parentheses.

$$2 + (3 \cdot 7) = 2 + 21 = 23$$

To eliminate confusion when no grouping symbols are present, we use the following agreed-upon order of operations.

Order of Operations

If grouping symbols such as parentheses are present, simplify expressions within those first, starting with the innermost set. If fraction bars are present, simplify the numerator and the denominator separately.

1. Evaluate exponential expressions.
2. Perform multiplications or divisions in order from left to right.
3. Perform additions or subtractions in order from left to right.

Using this order of operations, we now simplify $2 + 3 \cdot 7$. There are no grouping symbols and no exponents, so we multiply and then add.

$$2 + 3 \cdot 7 = 2 + 21 \quad \text{Multiply.}$$
$$= 23 \quad \text{Add.}$$

Practice Problems 2–5

Simplify each expression.

2. $3 + 2 \cdot 4^2$
3. $17 - 3 + 4$
4. $\frac{9}{5} \cdot \frac{1}{3} - \frac{1}{3}$
5. $8[2(6 + 3) - 9]$

EXAMPLES Simplify each expression

2. $6 \div 3 + 5^2 = 6 \div 3 + 25$ Evaluate 5^2

$= 2 + 25$ Divide.

$= 27$ Add.

3. $20 \div 5 \cdot 4 = 4 \cdot 4$

$= 16$

Helpful Hint

Remember to multiply or divide in order from left to right.

4. $\frac{3}{2} \cdot \frac{1}{2} - \frac{1}{2} = \frac{3}{4} - \frac{1}{2}$ Multiply.

$= \frac{3}{4} - \frac{2}{4}$ The least common denominator is 4.

$= \frac{1}{4}$ Subtract.

5. $3[4(5 + 2) - 10] = 3[4(7) - 10]$ Simplify the expression in parentheses. They are the innermost grouping symbols.

$= 3[28 - 10]$ Multiply 4 and 7.

$= 3[18]$ Subtract inside the brackets.

$= 54$ Multiply.

In the next example, the fraction bar serves as a grouping symbol and separates the numerator and denominator. Simplify each separately.

Answers

2. 35, **3.** 18, **4.** $\frac{4}{15}$, **5.** 72

EXAMPLE 6 Simplify: $\dfrac{3 + |4 - 3| + 2^2}{6 - 3}$

Solution:

$$\frac{3 + |4 - 3| + 2^2}{6 - 3} = \frac{3 + |1| + 2^2}{6 - 3}$$ Simplify the expression inside the absolute value bars.

$$= \frac{3 + 1 + 2^2}{3}$$ Find the absolute value and simplify the denominator.

$$= \frac{3 + 1 + 4}{3}$$ Evaluate the exponential expression.

$$= \frac{8}{3}$$ Simplify the numerator.

Practice Problem 6

Simplify: $\dfrac{1 + |7 - 4| + 3^2}{8 - 5}$

Helpful Hint

Be careful when evaluating an exponential expression.

$3 \cdot 4^2 = 3 \cdot 16 = 48$ (Base is 4.) $\qquad (3 \cdot 4)^2 = (12)^2 = 144$ (Base is $3 \cdot 4$.)

B Evaluating Algebraic Expressions

Recall that letters used to represent quantities are called **variables**. An **algebraic expression** is a collection of numbers, variables, operation symbols, and grouping symbols. For example,

$$2x, \quad -3, \quad 2x - 10, \quad 5(p^2 + 1), \qquad \text{and} \qquad \frac{3y^2 - 6y + 1}{5}$$

are algebraic expressions. The expression $2x$ means $2 \cdot x$. Also, $5(p^2 + 1)$ means $5 \cdot (p^2 + 1)$ and $3y^2$ means $3 \cdot y^2$. If we give a specific value to a variable, we can **evaluate an algebraic expression**. To evaluate an algebraic expression means to find its numerical value once we know the values of the variables.

Algebraic expressions are often used in problem solving. For example, the expression

$$16t^2$$

gives the distance in feet (neglecting air resistance) that an object will fall in t seconds.

Answer

6. $\dfrac{13}{3}$

Practice Problem 7

Evaluate each expression when $x = 1$ and $y = 4$.

a. $2y - x$

b. $\frac{8x}{3y}$

c. $\frac{x}{y} + \frac{5}{y}$

d. $y^2 - x^2$

EXAMPLE 7 Evaluate each expression when $x = 3$ and $y = 2$.

a. $2x - y$ **b.** $\frac{3x}{2y}$ **c.** $\frac{x}{y} + \frac{y}{2}$ **d.** $x^2 - y^2$

Solution:

a. Replace x with 3 and y with 2.

$$\begin{aligned} 2x - y &= 2(3) - 2 && \text{Let } x = 3 \text{ and } y = 2. \\ &= 6 - 2 && \text{Multiply.} \\ &= 4 && \text{Subtract.} \end{aligned}$$

b. Replace x with 3 and y with 2.

$$\frac{3x}{2y} = \frac{3 \cdot 3}{2 \cdot 2} = \frac{9}{4} \quad \text{Let } x = 3 \text{ and } y = 2.$$

c. Replace x with 3 and y with 2. Then simplify.

$$\frac{x}{y} + \frac{y}{2} = \frac{3}{2} + \frac{2}{2} = \frac{5}{2}$$

d. Replace x with 3 and y with 2.

$$x^2 - y^2 = 3^2 - 2^2 = 9 - 4 = 5$$

C Solutions of Equations

Many times a problem-solving situation is modeled by an equation. An **equation** is a mathematical statement that two expressions have equal value. The equal symbol "=" is used to equate the two expressions. For example, $3 + 2 = 5$, $7x = 35$, $\frac{2(x - 1)}{3} = 0$, and $I = PRT$ are all equations.

Helpful Hint

An equation contains the equal symbol "=". An algebraic expression does not.

Concept Check

Which of the following are equations? Which are expressions?

a. $5x = 8$

b. $5x - 8$

c. $12y + 3x$

d. $12y = 3x$

Try the Concept Check in the margin.

When an equation contains a variable, deciding which values of the variable make the equation a true statement is called **solving** the equation for the variable. A **solution** of an equation is a value for the variable that makes the equation true. For example, 3 is a solution of the equation $x + 4 = 7$, because if x is replaced with 3 the statement is true.

$$\begin{aligned} x + 4 &= 7 \\ \downarrow & \\ 3 + 4 &\stackrel{?}{=} 7 && \text{Replace } x \text{ with 3.} \\ 7 &= 7 && \text{True.} \end{aligned}$$

Similarly, 1 is not a solution of the equation $x + 4 = 7$, because $1 + 4 = 7$ is **not** a true statement.

Practice Problem 8

Decide whether 3 is a solution of $5x - 10 = x + 2$.

EXAMPLE 8 Decide whether 2 is a solution of $3x + 10 = 8x$.

Solution: Replace x with 2 and see if a true statement results.

$$\begin{aligned} 3x + 10 &= 8x && \text{Original equation} \\ 3(2) + 10 &\stackrel{?}{=} 8(2) && \text{Replace } x \text{ with 2.} \\ 6 + 10 &\stackrel{?}{=} 16 && \text{Simplify each side.} \\ 16 &= 16 && \text{True.} \end{aligned}$$

Since we arrived at a true statement after replacing x with 2 and simplifying both sides of the equation, 2 is a solution of the equation.

Answers

7. a. 7, **b.** $\frac{2}{3}$, **c.** $\frac{3}{2}$, **d.** 15, **8.** It is a solution.

Concept Check: equations: a, d; expressions: b, c

D Translating Words to Symbols

Now that we know how to represent an unknown number by a variable, let's practice translating phrases into algebraic expressions and sentences into equations. Oftentimes solving problems involves the ability to translate word phrases and sentences into symbols. Below is a list of key words and phrases to help us translate.

Addition (+)	Subtraction (−)	Multiplication (·)	Division (÷)	Equality (=)
Sum	Difference of	Product	Quotient	Equals
Plus	Minus	Times	Divide	Gives
Added to	Subtracted from	Multiply	Into	Is/was/ should be
More than	Less than	Twice	Ratio	Yields
Increased by	Decreased by	Of	Divided by	Amounts to
Total	Less			Represents Is the same as

EXAMPLE 9

Write an algebraic expression that represents each phrase. Let the variable x represent the unknown number.

a. The sum of a number and 3
b. The product of 3 and a number
c. Twice a number
d. 10 decreased by a number
e. 5 times a number increased by 7

Solution:

a. $x + 3$ since "sum" means to add
b. $3 \cdot x$ and $3x$ are both ways to denote the product of 3 and x
c. $2 \cdot x$ or $2x$
d. $10 - x$ because "decreased by" means to subtract
e. $\underbrace{5x}_{\text{5 times a number}} + 7$

Practice Problem 9

Write an algebraic expression that represents each phrase. Let the variable x represent the unknown number.

a. The product of a number and 5
b. A number added to 7
c. Three times a number
d. A number subtracted from 8
e. Twice a number plus 1

Helpful Hint

Make sure you understand the difference when translating phrases containing "decreased by," "subtracted from," and "less than."

Phrase	Translation
A number decreased by 10	$x - 10$
A number subtracted from 10	$10 - x$
10 less than a number	$x - 10$
A number less 10	$x - 10$

Notice the order.

Now let's practice translating sentences into equations.

Answers

9. a. $5x$, **b.** $7 + x$, **c.** $3x$, **d.** $8 - x$, **e.** $2x + 1$

Practice Problem 10

Write each sentence as an equation. Let x represent the unknown number.

a. The product of a number and 6 is 24.
b. The difference of 10 and a number is 18.
c. Twice a number decreased by 1 is 99.

Answers

10. a. $6x = 24$, **b.** $10 - x = 18$, **c.** $2x - 1 = 99$

EXAMPLE 10

Write each sentence as an equation. Let x represent the unknown number.

a. The quotient of 15 and a number is 4.
b. Three subtracted from 12 is a number.
c. Four times a number added to 17 is 21.

Solution:

a. In words: the quotient of 15 and a number | is | 4
↓ ↓ ↓
Translate: $\frac{15}{x} = 4$

b. In words: three subtracted **from** 12 | is | a number
↓ ↓ ↓
Translate: $12 - 3 = x$

Care must be taken when the operation is subtraction. The expression $3 - 12$ would be incorrect. Notice that $3 - 12 \neq 12 - 3$.

c. In words: four times a number | added to | 17 | is | 21
↓ ↓ ↓ ↓ ↓
Translate: $4x + 17 = 21$

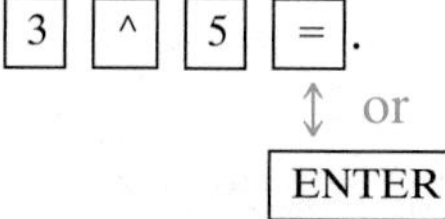

CALCULATOR EXPLORATIONS

Exponents

To evaluate exponential expressions on a calculator, find the key marked [y^x] or [^]. To evaluate, for example, 3^5, press the following keys: [3] [y^x] [5] [=] or [3] [^] [5] [=].

↕ or

[ENTER]

The display should read [243].

Order of Operations

Some calculators follow the order of operations, and others do not. To see whether or not your calculator has the order of operations built in, use your calculator to find $2 + 3 \cdot 4$. To do this, press the following sequence of keys:

[2] [+] [3] [×] [4] [=].

↕ or

[ENTER]

The correct answer is 14 because the order of operations is to multiply before we add. If the calculator displays [14], then it has the order of operations built in.

Even if the order of operations is built in, parentheses must sometimes be inserted. For example, to simplify $\frac{5}{12 - 7}$, press the keys

[5] [÷] [(] [1] [2] [−] [7] [)] [=].

↕ or

[ENTER]

The display should read [1].

Use a calculator to evaluate each expression.

1. 5^3
2. 7^4
3. 9^5
4. 8^6
5. $2(20 - 5)$
6. $3(14 - 7) + 21$
7. $24(862 - 455) + 89$
8. $99 + (401 + 962)$
9. $\frac{4623 + 129}{36 - 34}$
10. $\frac{956 - 452}{89 - 86}$

Name ____________________ Section __________ Date __________

EXERCISE SET 1.3

A *Evaluate. See Example 1.*

1. 3^5

2. 5^3

3. 3^3

4. 4^4

5. 1^5

6. 1^8

7. 5^1

8. 8^1

9. $\left(\frac{1}{5}\right)^3$

10. $\left(\frac{6}{11}\right)^2$

11. $\left(\frac{2}{3}\right)^4$

12. $\left(\frac{1}{2}\right)^5$

13. 7^2

14. 9^2

15. $(1.2)^2$

16. $(0.07)^2$

△ **17.** The area of a square whose sides each measure 5 meters is $(5 \cdot 5)$ square meters. Write this area using exponential notation.

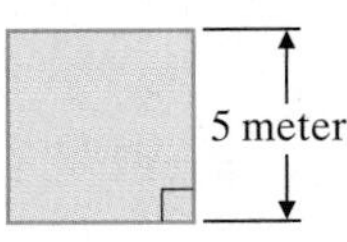

△ **18.** The volume of a solid is a measure of the space it encloses. The volume of a sphere whose radius is 5 meters is $\left(\frac{4}{3}\pi \cdot 5 \cdot 5 \cdot 5\right)$ cubic meters. Write this volume using exponential notation.

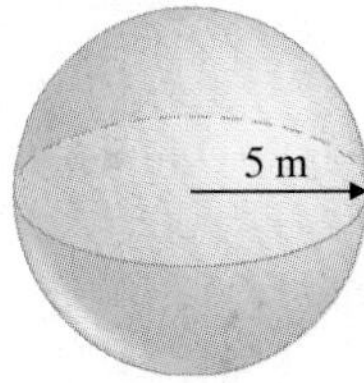

Simplify each expression. See Examples 2 through 6.

19. $5 + 6 \cdot 2$

20. $8 + 5 \cdot 3$

21. $4 \cdot 8 - 6 \cdot 2$

22. $12 \cdot 5 - 3 \cdot 6$

23. $2(8 - 3)$

24. $5(6 - 2)$

25. $2 + (5 - 2) + 4^2$

26. $6 - 2 \cdot 2 + 2^5$

27. $5 \cdot 3^2$

28. $2 \cdot 5^2$

29. $\frac{1}{4} \cdot \frac{2}{3} - \frac{1}{6}$

30. $\frac{3}{4} \cdot \frac{1}{2} + \frac{2}{3}$

31. $\frac{6 - 4}{9 - 2}$

32. $\frac{8 - 5}{24 - 20}$

33. $2[5 + 2(8 - 3)]$

34. $3[4 + 3(6 - 4)]$

35. $\frac{19 - 3 \cdot 5}{6 - 4}$

36. $\frac{4 \cdot 3 + 2}{4 + 3 \cdot 2}$

37. $\frac{|6 - 2| + 3}{8 + 2 \cdot 5}$

38. $\frac{15 - |3 - 1|}{12 - 3 \cdot 2}$

39. $\frac{3 + 3(5 + 3)}{3^2 + 1}$

40. $\frac{3 + 6(8 - 5)}{4^2 + 2}$

41. $\frac{6 + |8 - 2| + 3^2}{18 - 3}$

42. $\frac{16 + |13 - 5| + 4^2}{17 - 5}$

43. Are parentheses necessary in the expression $2 + (3 \cdot 5)$? Explain your answer.

44. Are parentheses necessary in the expression $(2 + 3) \cdot 5$? Explain your answer.

For Exercises 45 and 46, match each expression in the first column with its value in the second column.

45.	
a. $(6 + 2) \cdot (5 + 3)$	19
b. $(6 + 2) \cdot 5 + 3$	22
c. $6 + 2 \cdot 5 + 3$	64
d. $6 + 2 \cdot (5 + 3)$	43

46.	
a $(1 + 4) \cdot 6 - 3$	15
b $1 + 4 \cdot (6 - 3)$	13
c $1 + 4 \cdot 6 - 3$	27
d $(1 + 4) \cdot (6 - 3)$	22

B *Evaluate each expression when $x = 1$, $y = 3$, and $z = 5$. See Example 7.*

47. $3y$

48. $4x$

49. $\frac{z}{5x}$

50. $\frac{y}{2z}$

51. $3x - 2$

52. $6y - 8$

53. $|2x + 3y|$

54. $|5z - 2y|$

55. $xy + z$

56. $yz - x$

57. $5y^2$

58. $2z^2$

Evaluate each expression when $x = 2$, $y = 6$, and $z = 3$. See Example 8.

59. $5z$

60. $7x$

61. $\frac{y}{x}$

62. $\frac{y}{x \cdot z}$

63. $\frac{y}{x} + \frac{y}{x}$

64. $\frac{9}{z} + \frac{4z}{y}$

Recall that perimeter measures the distance around a plane figure and area measures the amount of surface of a plane figure. The expression $2l + 2w$ gives the perimeter of the rectangle below, and the expression lw gives its area (measured in square units).

△ **65.** Complete the chart below for the given lengths and widths. Be sure to include units.

Length: l	Width: w	Perimeter of Rectangle: $2l + 2w$	Area of Rectangle: lw
3 in.	4 in.		
1 in.	6 in.		
2 in.	5 in.		

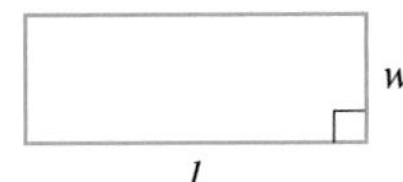

66. Study the perimeters and areas found in the chart to the left. Do you notice any trends?

C *Decide whether the given number is a solution of the given equation. See Example 8.*

67. $3x - 6 = 9; 5$

68. $2x + 7 = 3x; 6$

69. $2x + 6 = 5x - 1; 0$

70. $4x + 2 = x + 8; 2$

71. $2x - 5 = 5; 8$

72. $3x - 10 = 8; 6$

73. $x + 6 = x + 6; 2$

74. $x + 6 = x + 6; 10$

75. $x = 5x + 15; 0$

76. $4 = 1 - x; 1$

77. $\frac{1}{3}x = 9; 27$

78. $\frac{2}{7}x = \frac{3}{14}; 6$

D *Write each phrase as an algebraic expression. Let x represent the unknown number. See Example 9.*

79. Fifteen more than a number

80. One-half times a number

81. Five subtracted from a number

82. The quotient of a number and 9

83. Five decreased by a number

84. A number less twenty

85. Three times a number, increased by 22

86. The product of 8 and a number

Write each sentence as an equation. Use x to represent any unknown number. See Example 10.

87. One increased by two equals the quotient of nine and three.

88. Four subtracted from eight is equal to two squared.

89. Three is not equal to four divided by two.

90. The difference of sixteen and four is greater than ten.

91. The sum of 5 and a number is 20.

92. Twice a number is 17.

93. Thirteen minus three times a number is 13.

94. Seven subtracted from a number is 0.

95. The quotient of 12 and a number is $\frac{1}{2}$.

96. The sum of 8 and twice a number is 42.

Combining Concepts

97. Insert parentheses so that the following expression simplifies to 32.

$$20 - 4 \cdot 4 \div 2$$

98. Insert parentheses so that the following expression simplifies to 28.

$$2 \cdot 5 + 3^2$$

99. In your own words, explain the difference between an expression and an equation.

100. Determine whether each is an expression or an equation.
a. $3x^2 - 26$
b. $3x^2 - 26 = 1$
c. $2x - 5 = 7x - 5$
d. $9y + x - 8$

101. Why is 8^2 usually read as "eight squared"? (*Hint*: What is the area of the **square** below?)

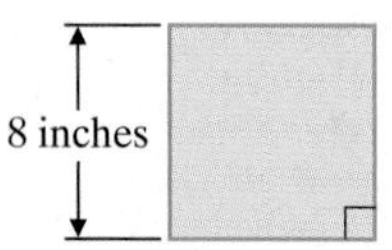

102. Why is 4^3 usually read as "four cubed"? (*Hint*: What is the volume of the **cube** below?)

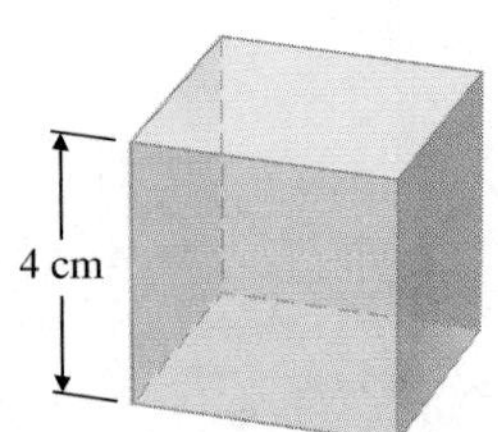

1.4 Adding Real Numbers

Real numbers can be added, subtracted, multiplied, divided, and raised to powers, just as whole numbers can.

OBJECTIVES

A Add real numbers.

B Solve problems that involve addition of real numbers.

C Find the opposite of a number.

A Adding Real Numbers

To begin, we will use the number line to help picture the addition of real numbers.

EXAMPLE 1 Add: $3 + 2$

Solution: We start at 0 on a number line, and draw an arrow representing 3. This arrow is three units long and points to the right since 3 is positive. From the tip of this arrow, we draw another arrow representing 2. The number below the tip of this arrow is the sum, 5.

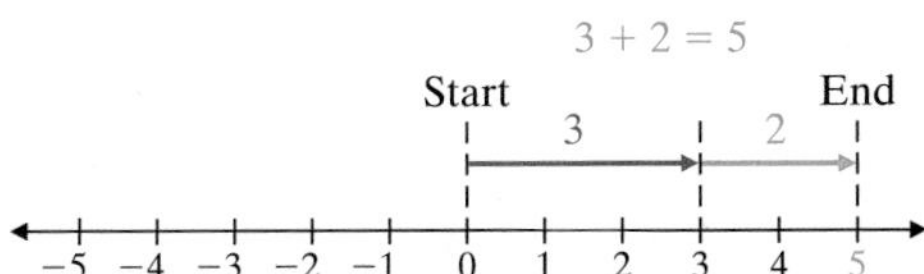

Practice Problem 1

Add using a number line: $1 + 5$

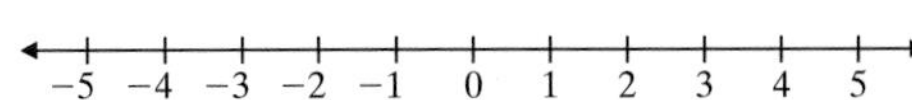

EXAMPLE 2 Add: $-1 + (-2)$

Solution: We start at 0 on a number line, and draw an arrow representing -1. This arrow is one unit long and points to the left since -1 is negative. From the tip of this arrow, we draw another arrow representing -2. The number below the tip of this arrow is the sum, -3.

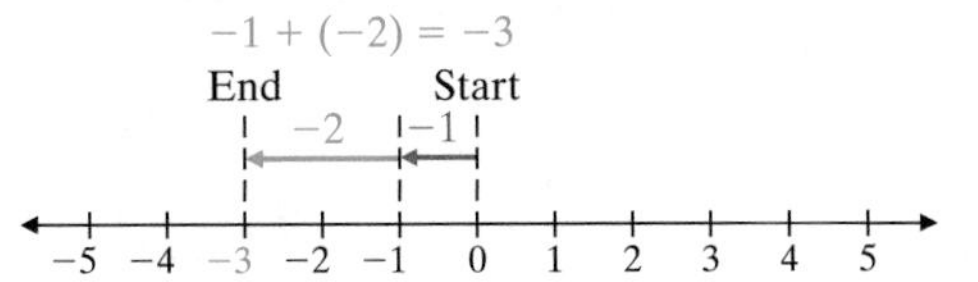

Practice Problem 2

Add using a number line: $-2 + (-4)$

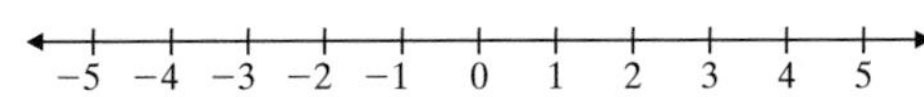

Thinking of integers as money earned or lost might help make addition more meaningful. Earnings can be thought of as positive numbers. If \$1 is earned and later another \$3 is earned, the total amount earned is \$4. In other words, $1 + 3 = 4$.

On the other hand, losses can be thought of as negative numbers. If \$1 is lost and later another \$3 is lost, a total of \$4 is lost. In other words, $(-1) + (-3) = -4$.

In Examples 1 and 2, we added numbers with the same sign. Adding numbers whose signs are not the same can be pictured on a number line also.

EXAMPLE 3 Add: $-4 + 6$

Solution:

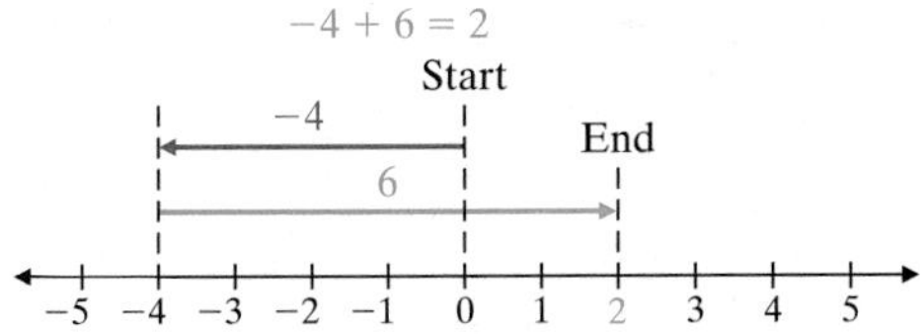

Practice Problem 3

Add using a number line: $-5 + 8$

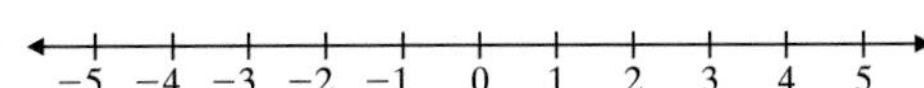

Answers

1. 6, **2.** −6, **3.** 3

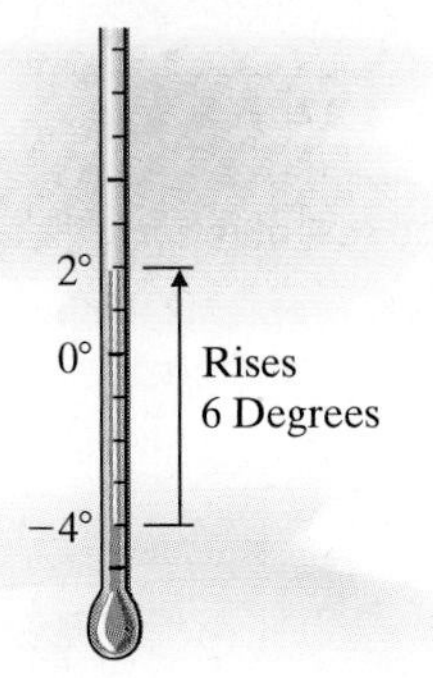

Let's use temperature as an example. If the thermometer registers 4 degrees below 0 degrees and then rises 6 degrees, the new temperature is 2 degrees above 0 degrees. Thus, it is reasonable that $-4 + 6 = 2$. (See the diagram in the margin.)

Practice Problem 4

Add using a number line: $5 + (-4)$

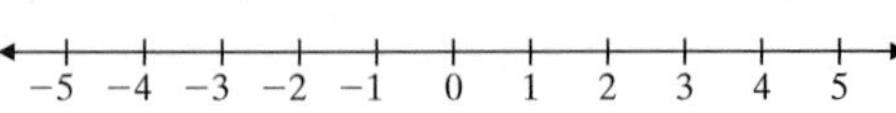

EXAMPLE 4 Add: $4 + (-6)$

Solution:

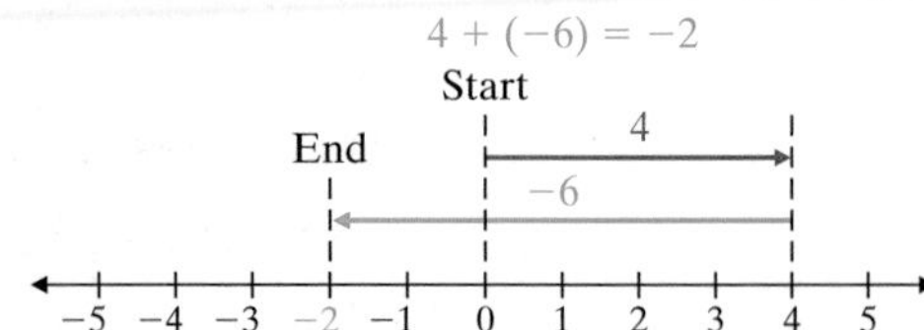

Using a number line each time we add two numbers can be time consuming. Instead, we can notice patterns in the previous examples and write rules for adding real numbers.

Adding Real Numbers

To add two real numbers

1. with the *same sign*, add their absolute values. Use their common sign as the sign of the answer.
2. with *different signs*, subtract their absolute values. Give the answer the same sign as the number with the larger absolute value.

Practice Problem 5

Add without using a number line: $(-8) + (-5)$

EXAMPLE 5 Add without using a number line: $(-7) + (-6)$

Solution: Here, we are adding two numbers with the same sign.

$$(-7) + (-6) = -13$$

↖ sum of absolute values

same sign

Practice Problem 6

Add without using a number line: $(-14) + 6$

EXAMPLE 6 Add without using a number line: $(-10) + 4$

Solution: Here, we are adding two numbers with different signs.

$$(-10) + 4 = -6$$

↖ difference of absolute values

sign of number with larger absolute value, −10

Practice Problems 7–12

Add without using a number line.

7. $(-17) + (-10)$
8. $(-4) + 12$
9. $1.5 + (-3.2)$
10. $-\frac{6}{11} + \left(-\frac{3}{11}\right)$
11. $12.8 + (-3.6)$
12. $-\frac{4}{5} + \frac{2}{3}$

EXAMPLES Add without using a number line.

7. $(-8) + (-11) = -19$

8. $(-2) + 10 = 8$

9. $0.2 + (-0.5) = -0.3$

10. $-\frac{7}{10} + \left(-\frac{1}{10}\right) = -\frac{8}{10} = -\frac{4}{5}$

11. $11.4 + (-4.7) = 6.7$

12. $-\frac{3}{8} + \frac{2}{5} = -\frac{15}{40} + \frac{16}{40} = \frac{1}{40}$

Answers

4. 1, **5.** −13, **6.** −8, **7.** −27, **8.** 8, **9.** −1.7, **10.** $-\frac{9}{11}$, **11.** 9.2, **12.** $-\frac{2}{15}$

EXAMPLE 13 Find each sum.

a. $3 + (-7) + (-8)$

b. $[7 + (-10)] + [-2 + (-4)]$

Solution:

a. Perform the additions from left to right.

$$3 + (-7) + (-8) = -4 + (-8) \quad \text{Adding numbers with different signs}$$
$$= -12 \quad \text{Adding numbers with like signs}$$

b. Simplify inside the brackets first.

$$[7 + (-10)] + [-2 + (-4)] = [-3] + [-6]$$
$$= -9 \quad \text{Add.}$$

Practice Problem 13

Find each sum.

a. $16 + (-9) + (-9)$

b. $[3 + (-13)] + [-4 + (-7)]$

Helpful Hint

Don't forget that brackets are grouping symbols. We simplify within them first.

B Solving Problems That Involve Addition

Positive and negative numbers are used in everyday life. Stock market returns show gains and losses as positive and negative numbers. Temperatures in cold climates often dip into the negative range, commonly referred to as "below zero" temperatures. Bank statements report deposits and withdrawals as positive and negative numbers.

EXAMPLE 14 Calculating Gain or Loss

During a three-day period, a share of Lamplighter's International stock recorded the following gains and losses:

Monday	**Tuesday**	**Wednesday**
a gain of \$2	a loss of \$1	a loss of \$3

Find the overall gain or loss for the stock for the three days.

Solution: Gains can be represented by positive numbers. Losses can be represented by negative numbers. The overall gain or loss is the sum of the gains and losses.

In words:	gain	plus	loss	plus	loss	
	↓	↓	↓	↓	↓	
Translate	2	+	(−1)	+	(−3)	= −2

The overall loss is \$2.

Practice Problem 14

During a four-day period, a share of Walco stock recorded the following gains and losses:

Tuesday	**Wednesday**
a loss of \$2	a loss of \$1
Thursday	**Friday**
a gain of \$3	a gain of \$3

Find the overall gain or loss for the stock for the four days.

C Finding Opposites

To help us subtract real numbers in the next section, we first review what we mean by opposites. The graphs of 4 and −4 are shown on the number line below.

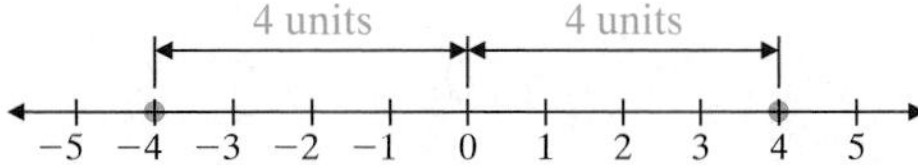

Notice that the graph of 4 and −4 lie on opposite sides of 0, and each is 4 units away from 0. Such numbers are known as **opposites** or **additive inverses** of each other.

Answers

13. a. −2, **b.** −21, **14.** a gain of \$3

Opposite or Additive Inverse

Two numbers that are the same distance from 0 but lie on opposite sides of 0 are called **opposites** or **additive inverses** of each other.

Practice Problems 15–18

Find the opposite of each number.

15. -35
16. 12
17. $-\frac{3}{11}$
18. 1.9

EXAMPLES Find the opposite of each number.

15. 10 The opposite of 10 is -10.

16. -3 The opposite of -3 is 3.

17. $\frac{1}{2}$ The opposite of $\frac{1}{2}$ is $-\frac{1}{2}$.

18. -4.5 The opposite of -4.5 is 4.5.

We use the symbol "–" to represent the phrase "the opposite of" or "the additive inverse of." In general, if a is a number, we write the opposite or additive inverse of a as $-a$. We know that the opposite of -3 is 3. Notice that this translates as

the opposite of	-3	is	3
↓	↓	↓	↓
$-$	(-3)	$=$	**3**

This is true in general.

If a is a number, then $-(-a) = a$.

Practice Problem 19

Simplify each expression.

a. $-(-22)$
b. $-\left(-\frac{2}{7}\right)$
c. $-(-x)$
d. $-|-14|$

EXAMPLE 19 Simplify each expression.

a. $-(-10)$ **b.** $-\left(-\frac{1}{2}\right)$ **c.** $-(-2x)$ **d.** $-|-6|$

Solution:

a. $-(-10) = 10$ **b.** $-\left(-\frac{1}{2}\right) = \frac{1}{2}$

c. $-(-2x) = 2x$

d. Since $|-6| = 6$, then $-|-6| = -6$.

Let's discover another characteristic about opposites. Notice that the sum of a number and its opposite is 0.

$$10 + (-10) = 0$$
$$-3 + 3 = 0$$
$$\frac{1}{2} + \left(-\frac{1}{2}\right) = 0$$

In general, we can write the following:

The sum of a number a and its opposite $-a$ is 0.

$$a + (-a) = 0$$

Notice that this means that the opposite of 0 is then 0 since $0 + 0 = 0$.

Answers

15. 35, **16.** -12, **17.** $\frac{3}{11}$, **18.** -1.9, **19. a.** 22, **b.** $\frac{2}{7}$, **c.** x, **d.** -14

Name ______________________ Section __________ Date __________

EXERCISE SET 1.4

A *Add. See Examples 1 through 13.*

1. $6 + 3$

2. $9 + (-12)$

3. $-6 + (-8)$

4. $-6 + (-14)$

5. $8 + (-7)$

6. $6 + (-4)$

7. $-14 + 2$

8. $-10 + 5$

9. $-2 + (-3)$

10. $-7 + (-4)$

11. $-9 + (-3)$

12. $7 + (-5)$

13. $-7 + 3$

14. $-5 + 9$

15. $10 + (-3)$

16. $8 + (-6)$

17. $5 + (-7)$

18. $3 + (-6)$

19. $-16 + 16$

20. $23 + (-23)$

21. $27 + (-46)$

22. $53 + (-37)$

23. $-18 + 49$

24. $-26 + 14$

25. $-33 + (-14)$

26. $-18 + (-26)$

27. $6.3 + (-8.4)$

28. $9.2 + (-11.4)$

29. $|-8| + (-16)$

30. $|-6| + (-61)$

31. $117 + (-79)$

32. $144 + (-88)$

33. $-9.6 + (-3.5)$

34. $-6.7 + (-7.6)$

35. $-\frac{3}{8} + \frac{5}{8}$

36. $-\frac{5}{12} + \frac{7}{12}$

37. $-\frac{7}{16} + \frac{1}{4}$

38. $-\frac{5}{9} + \frac{1}{3}$

39. $-\frac{7}{10} + \left(-\frac{3}{5}\right)$

40. $-\frac{5}{6} + \left(-\frac{2}{3}\right)$

41. $-15 + 9 + (-2)$

42. $-9 + 15 + (-5)$

43. $-21 + (-16) + (-22)$

44. $-18 + (-6) + (-40)$ **45.** $-23 + 16 + (-2)$ **46.** $-14 + (-3) + 11$ **47.** $|5 + (-10)|$

48. $|7 + (-17)|$ **49.** $6 + (-4) + 9$ **50.** $8 + (-2) + 7$

51. $[-17 + (-4)] + [-12 + 15]$ **52.** $[-2 + (-7)] + [-11 + 22]$ **53.** $|9 + (-12)| + |-16|$

54. $|43 + (-73)| + |-20|$ **55.** $-13 + [5 + (-3) + 4]$ **56.** $-30 + [1 + (-6) + 8]$

57. Explain why adding a negative number to another negative number always gives a negative sum.

58. When a positive and a negative number are added, sometimes the sum is positive, sometimes it is zero, and sometimes it is negative. Explain why this happens.

B *Solve each of the following. See Example 14.*

59. The lowest temperature ever recorded in Massachusetts was −35°F. The highest recorded temperature in Massachusetts was 142° higher than the record low temperature. Find Massachusetts' highest recorded temperature. (*Source*: National Climatic Data Center)

60. On January 2, 1943, the temperature was −4° at 7:30 a.m. in Spearfish, South Dakota. Incredibly, it got 49° warmer in the next 2 minutes. To what temperature did it rise by 7:32?

61. The lowest elevation on Earth is −411 meters (that is, 411 meters below sea level) at the Dead Sea. If you are standing 316 meters above the Dead Sea, what is your elevation? (*Source*: National Geographic Society)

62. The lowest elevation in Australia is −52 feet at Lake Eyre. If you are standing at a point 439 feet above Lake Eyre, what is your elevation? (*Source*: National Geographic Society)

63. When checking the stock listings in the newspaper, LaTonda finds that one of her stocks posted net changes of $-2\frac{1}{2}$ points and $-\frac{7}{16}$ point over the last two days. What is the combined change?

64. Yesterday your stock posted a net change of $+\frac{11}{16}$ point, but today it showed a loss of $-1\frac{1}{8}$ points. Find the overall change for the two days.

65. In golf, scores that are under par for the entire round are shown as negative scores; scores that are over par are positive, and par is 0. Tiger Woods won the 2001 Players Championship with round scores of 0, −3, −6, and −5. What was his total overall score? (*Source*: PGA of America)

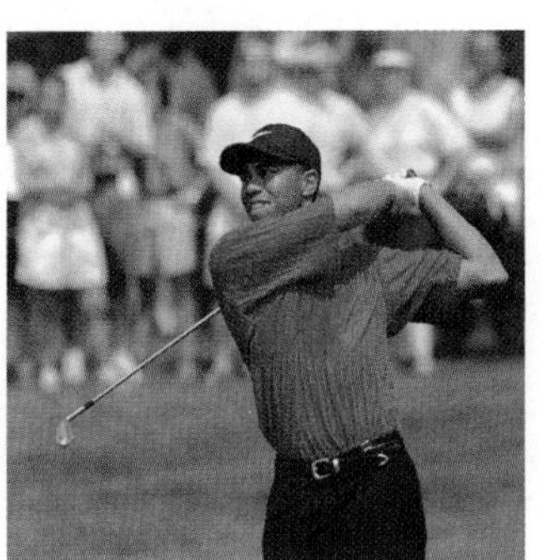

66. During the LPGA 2001 Nabisco Championship, Annika Sorenstam won with the following scores: 0, −2, −2, and −3. What was her total overall score? (*Source*: PGA of America)

67. A negative net income results when a company spends more money than it brings in. JCPenney Company, Inc. had the following quarterly net incomes during its 2000 fiscal year.

Quarter of Fiscal 2000	Net Income (in millions)
First	−\$118
Second	\$23
Third	−\$30
Fourth	−\$284

(*Source*: JCPenney Company, Inc.)

What was the total net income for 2000?

68. Amazon.com had the following quarterly net incomes during its 2000 fiscal year.

Quarter of Fiscal 2000	Net Income (in millions)
First	−\$308
Second	−\$317
Third	−\$241
Fourth	−\$545

(*Source*: Amazon.com, Inc.)

What was the total net income for 2000?

C *Find each additive inverse or opposite. See Examples 15 through 18.*

69. 6 **70.** 4 **71.** -2 **72.** -8

73. 0 **74.** $-\frac{1}{4}$ **75.** $|-6|$ **76.** $|-11|$

77. In your own words, explain how to find the opposite of a number.

78. In your own words, explain why 0 is the only number that is its own opposite.

Simplify each of the following. See Example 19.

79. $-|-2|$ **80.** $-(-3)$ **81.** $-|0|$ **82.** $\left|-\frac{2}{3}\right|$ **83.** $-\left|-\frac{2}{3}\right|$ **84.** $-(-7)$

Combining Concepts

The following bar graph shows each month's average daily low temperature in degrees Fahrenheit for Barrow, Alaska. Use this graph to answer Exercises 85 through 90.

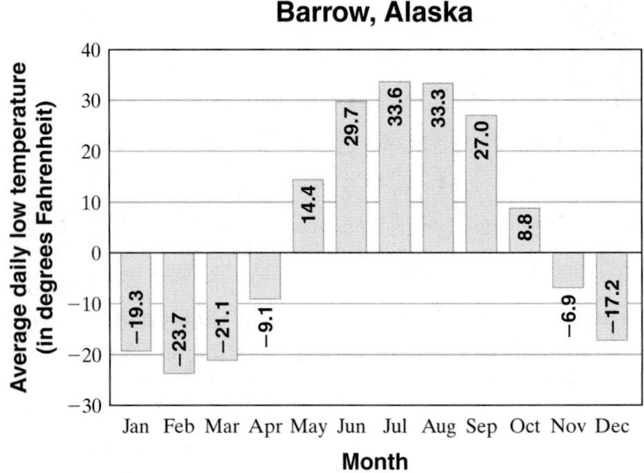

Source: National Climatic Data Center

85. For what month is the graphed temperature the highest?

86. For what month is the graphed temperature the lowest?

87. For what month is the graphed temperature positive *and* closest to 0°?

88. For what month is the graphed temperature negative *and* closest to 0°?

89. Find the average of the temperatures shown for the months of April, May, and October. (To find the average of three temperatures, find their sum and divide by 3.)

90. Find the average of the temperatures shown for the months of January, September, and October.

If a is a positive number and b is a negative number, fill in the blanks with the words positive or negative.

91. $-a$ is a __________ number.

92. $-b$ is a __________ number.

93. $a + a$ is a __________ number.

94. $b + b$ is a __________ number.

1.5 Subtracting Real Numbers

OBJECTIVES

- **A** Subtract real numbers.
- **B** Evaluate algebraic expressions using real numbers.
- **C** Determine whether a number is a solution of a given equation.
- **D** Solve problems that involve subtraction of real numbers.
- **E** Find complementary and supplementary angles.

SSM TUTOR CENTER SG CD & VIDEO MATH PRO WEB

A Subtracting Real Numbers

Now that addition of real numbers has been discussed, we can explore subtraction. We know that $9 - 7 = 2$. Also, $9 + (-7) = 2$. This means that

$$9 - 7 = 9 + (-7)$$

Notice that the *difference* of 9 and 7 is the same as the *sum* of 9 and the opposite of 7. This is how we can subtract real numbers.

Subtracting Real Numbers

If a and b are real numbers, then $a - b = a + (-b)$.

In other words, to find the difference of two numbers, we add the opposite of the number being subtracted.

EXAMPLE 1 Subtract.

a. $-13 - 4$ **b.** $5 - (-6)$ **c.** $3 - 6$ **d.** $-1 - (-7)$

Solution:

a. $-13 - 4 = -13 + (-4)$ (add; opposite) Add −13 to the opposite of 4, which is −4.

$= -17$

b. $5 - (-6) = 5 + (6)$ (add; opposite) Add 5 to the opposite of −6, which is 6.

$= 11$

c. $3 - 6 = 3 + (-6)$ Add 3 to the opposite of 6, which is −6.

$= -3$

d. $-1 - (-7) = -1 + (7) = 6$

Helpful Hint

Study the patterns indicated.

No change; Change to addition; Change to opposite.

$5 - 11 = 5 + (-11) = -6$

$-3 - 4 = -3 + (-4) = -7$

$7 - (-1) = 7 + (1) = 8$

EXAMPLES Subtract.

2. $5.3 - (-4.6) = 5.3 + (4.6) = 9.9$

3. $-\frac{3}{10} - \frac{5}{10} = -\frac{3}{10} + \left(-\frac{5}{10}\right) = -\frac{8}{10} = -\frac{4}{5}$

4. $-\frac{2}{3} - \left(-\frac{4}{5}\right) = -\frac{2}{3} + \left(\frac{4}{5}\right) = -\frac{10}{15} + \frac{12}{15} = \frac{2}{15}$

Practice Problem 1

Subtract.

a. $-20 - 6$
b. $3 - (-5)$
c. $7 - 17$
d. $-4 - (-9)$

Practice Problems 2–4

Subtract.

2. $9.6 - (-5.7)$
3. $-\frac{4}{9} - \frac{2}{9}$
4. $-\frac{1}{4} - \left(-\frac{2}{5}\right)$

Answers

1. a. −26, **b.** 8, **c.** −10, **d.** 5, **2.** 15.3, **3.** $-\frac{2}{3}$, **4.** $\frac{3}{20}$

Practice Problem 5

Subtract 7 from -11.

EXAMPLE 5 Subtract 8 from -4.

Solution: Be careful when interpreting this. The order of numbers in subtraction is important. 8 is to be subtracted **from** -4.

$$-4 - 8 = -4 + (-8) = -12$$

If an expression contains additions and subtractions, just write the subtractions as equivalent additions. Then simplify from left to right.

Practice Problem 6

Simplify each expression.

a. $-20 - 5 + 12 - (-3)$
b. $5.2 - (-4.4) + (-8.8)$

EXAMPLE 6 Simplify each expression.

a. $-14 - 8 + 10 - (-6)$
b. $1.6 - (-10.3) + (-5.6)$

Solution:

a. $-14 - 8 + 10 - (-6) = -14 + (-8) + 10 + 6 = -6$
b. $1.6 - (-10.3) + (-5.6) = 1.6 + 10.3 + (-5.6) = 6.3$

When an expression contains parentheses and brackets, remember the order of operations. Start with the innermost set of parentheses or brackets and work your way outward.

Practice Problem 7

Simplify each expression.

a. $-9 + [(-4 - 1) - 10]$
b. $5^2 - 20 + [-11 - (-3)]$

EXAMPLE 7 Simplify each expression.

a. $-3 + [(-2 - 5) - 2]$
b. $2^3 - 10 + [-6 - (-5)]$

Solution:

a. Start with the innermost set of parentheses. Rewrite $-2 - 5$ as an addition.

$-3 + [(-2 - 5) - 2] = -3 + [(-2 + (-5)) - 2]$
$= -3 + [(-7) - 2]$ Add: $-2 + (-5)$.
$= -3 + [-7 + (-2)]$ Write $-7 - 2$ as an addition.
$= -3 + [-9]$ Add.
$= -12$ Add.

b. Start simplifying the expression inside the brackets by writing $-6 - (-5)$ as an addition.

$2^3 - 10 + [-6 - (-5)] = 2^3 - 10 + [-6 + 5]$
$= 2^3 - 10 + [-1]$ Add.
$= 8 - 10 + (-1)$ Evaluate 2^3.
$= 8 + (-10) + (-1)$ Write $8 - 10$ as an addition.
$= -2 + (-1)$ Add.
$= -3$ Add.

B Evaluating Algebraic Expressions

It is important to be able to evaluate expressions for given replacement values. This helps, for example, when checking solutions of equations.

Answers

5. -18, **6. a.** -10, **b.** 0.8, **7. a.** -24, **b.** -3

EXAMPLE 8 Find the value of each expression when $x = 2$ and $y = -5$.

a. $\dfrac{x - y}{12 + x}$ **b.** $x^2 - y$

Solution:

a. Replace x with 2 and y with -5. Be sure to put parentheses around -5 to separate signs. Then simplify the resulting expression.

$$\frac{x - y}{12 + x} = \frac{2 - (-5)}{12 + 2} = \frac{2 + 5}{14} = \frac{7}{14} = \frac{1}{2}$$

b. Replace the x with 2 and y with -5 and simplify.

$$x^2 - y = 2^2 - (-5) = 4 - (-5) = 4 + 5 = 9$$

Practice Problem 8

Find the value of each expression when $x = 1$ and $y = -4$.

a. $\dfrac{x - y}{14 + x}$

b. $x^2 - y$

C Solutions of Equations

Recall from Section 1.3 that a solution of an equation is a value for the variable that makes the equation true.

EXAMPLE 9 Determine whether -4 is a solution of $x - 5 = -9$.

Solution: Replace x with -4 and see if a true statement results.

$$\begin{aligned} x - 5 &= -9 && \text{Original equation} \\ -4 - 5 &\stackrel{?}{=} -9 && \text{Replace } x \text{ with } -4. \\ -4 + (-5) &\stackrel{?}{=} -9 \\ -9 &= -9 && \text{True} \end{aligned}$$

Thus -4 is a solution of $x - 5 = -9$.

Practice Problem 9

Determine whether -2 is a solution of $-1 + x = 1$.

D Solving Problems That Involve Subtraction

Another use of real numbers is in recording altitudes above and below sea level, as shown in the next example.

EXAMPLE 10 Finding the Difference in Elevations

The lowest point on the surface of the Earth is the Dead Sea, at an elevation of 1349 feet below sea level. The highest point is Mt. Everest, at an elevation of 29,035 feet. How much of a variation in elevation is there between these two world extremes? (*Source:* National Geographic Society)

Solution: To find the variation in elevation between the two heights, find the difference of the high point and the low point.

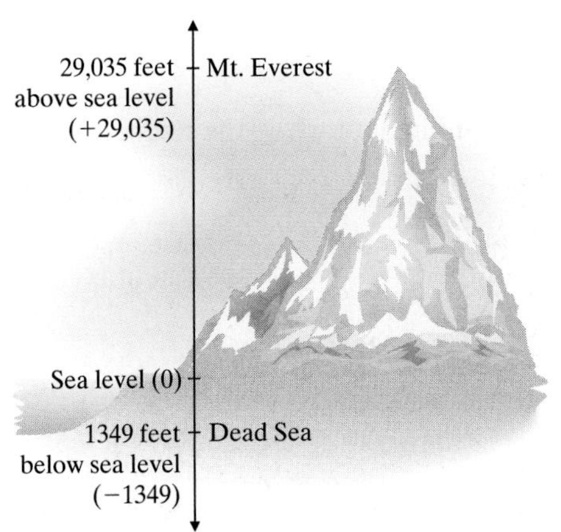

Practice Problem 10

At 6.00 p.m., the temperature at the Winter Olympics was 14°; by morning the temperature dropped to −23°. Find the overall change in temperature.

Answers

8. a. $\frac{1}{3}$, **b.** 5, **9.** −2 is not a solution, **10.** −37°

In words:	high point	minus	low point	
	↓	↓	↓	
Translate:	29,035	$-$	(-1349)	$= 29{,}035 + 1349$ feet
				$= 30{,}384$ feet

Thus, the variation in elevation is 30,384 feet. ●

E Finding Complementary and Supplementary Angles

A knowledge of geometric concepts is needed by many professionals, such as doctors, carpenters, electronic technicians, gardeners, machinists, and pilots, just to name a few. With this in mind, we review the geometric concepts of **complementary** and **supplementary angles**.

Complementary and Supplementary Angles

Two angles are **complementary** if their sum is 90°.

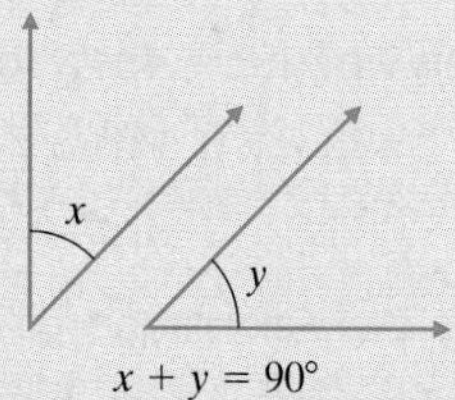

Two angles are **supplementary** if their sum is 180°.

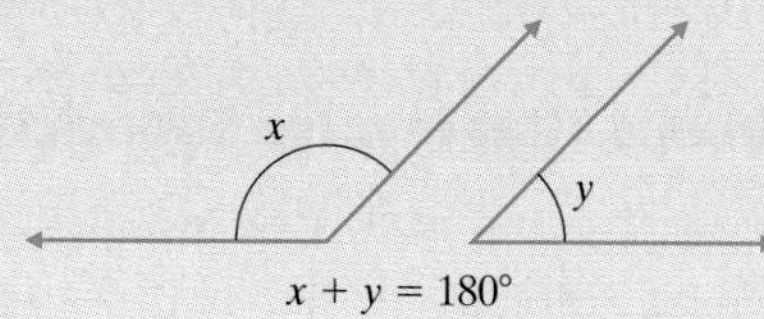

EXAMPLE 11 Find each unknown complementary or supplementary angle.

a.

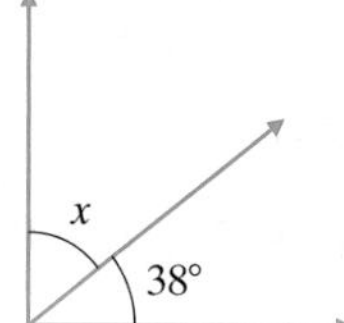

b.

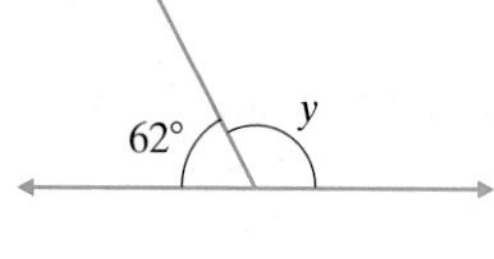

Solution:

a. These angles are complementary, so their sum is 90°. This means that x is $90° - 38°$.

$$x = 90° - 38° = 52°$$

b. These angles are supplementary, so their sum is 180°. This means that y is $180° - 62°$.

$$y = 180° - 62° = 118°$$

●

Practice Problem 11

Find each unknown complementary or supplementary angle.

a.

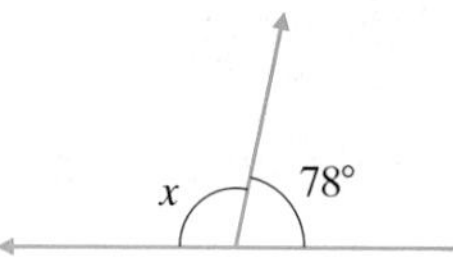

b.

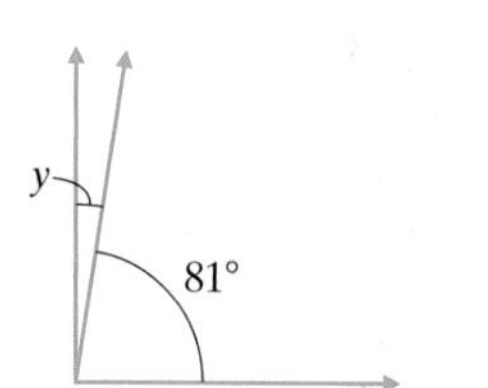

Answers

11. a. 102°, **b.** 9°

Name ______________________ Section __________ Date __________

EXERCISE SET 1.5

A *Subtract. See Examples 1 through 4.*

1. $-6 - 4$

2. $-12 - 8$

3. $4 - 9$

4. $8 - 11$

5. $16 - (-3)$

6. $12 - (-5)$

7. $\frac{1}{2} - \frac{1}{3}$

8. $\frac{3}{4} - \frac{7}{8}$

9. $-16 - (-18)$

10. $-20 - (-48)$

11. $-6 - 5$

12. $-8 - 4$

13. $7 - (-4)$

14. $3 - (-6)$

15. $-6 - (-11)$

16. $-4 - (-16)$

17. $16 - (-21)$

18. $15 - (-33)$

19. $9.7 - 16.1$

20. $8.3 - 11.2$

21. $-44 - 27$

22. $-36 - 51$

23. $-21 - (-21)$

24. $-17 - (-17)$

25. $-2.6 - (-6.7)$

26. $-6.1 - (-5.3)$

27. $-\frac{3}{11} - \left(-\frac{5}{11}\right)$

28. $-\frac{4}{7} - \left(-\frac{1}{7}\right)$

29. $-\frac{1}{6} - \frac{3}{4}$

30. $-\frac{1}{10} - \frac{7}{8}$

31. $8.3 - (-0.62)$

32. $4.3 - (-0.87)$

33. If a and b are positive numbers, then is $a - b$ always positive, always negative, or sometimes positive and sometimes negative?

34. If a and b are negative numbers, then is $a - b$ always positive, always negative, or sometimes positive and sometimes negative?

Write each phrase as an expression and simplify. See Example 5.

35. Subtract -5 from 8.

36. Subtract 3 from -2.

37. Subtract -1 from -6.

38. Subtract 17 from 1.

39. Subtract 8 from 7.

40. Subtract 9 from -4.

41. Decrease -8 by 15.

42. Decrease 11 by -14.

Simplify each expression. (Remember the order of operations.) See Examples 6 and 7.

43. $-10 - (-8) + (-4) - 20$

44. $-16 - (-3) + (-11) - 14$

45. $5 - 9 + (-4) - 8 - 8$

46. $7 - 12 + (-5) - 2 + (-2)$

47. $-6 - (2 - 11)$

48. $-9 - (3 - 8)$

49. $3^3 - 8 \cdot 9$

50. $2^3 - 6 \cdot 3$

51. $2 - 3(8 - 6)$

52. $4 - 6(7 - 3)$

53. $(3 - 6) + 4^2$

54. $(2 - 3) + 5^2$

55. $-2 + [(8 - 11) - (-2 - 9)]$

56. $-5 + [(4 - 15) - (-6) - 8]$

57. $|-3| + 2^2 + [-4 - (-6)]$

58. $|-2| + 6^2 + (-3 - 8)$

B *Evaluate each expression when $x = -5$, $y = 4$, and $t = 10$. See Example 8.*

59. $x - y$

60. $y - x$

61. $|x| + 2t - 8y$

62. $|x + t - 7y|$

63. $\dfrac{9 - x}{y + 6}$

64. $\dfrac{15 - x}{y + 2}$

65. $y^2 - x$

66. $t^2 - x$

67. $\dfrac{|x - (-10)|}{2t}$

68. $\dfrac{|5y - x|}{6t}$

C *Decide whether the given number is a solution of the given equation. See Example 9.*

69. $x - 9 = 5;\quad -4$

70. $x - 10 = -7;\quad 3$

71. $-x + 6 = -x - 1;\quad -2$

72. $-x - 6 = -x - 1;\quad -10$

73. $-x - 13 = -15;\quad 2$

74. $4 = 1 - x;\quad 5$

D *Solve. See Example 10.*

75. Within 24 hours in 1916, the temperature in Browning, Montana, fell from 44° to −56°. How large a drop in temperature was this?

76. The coldest temperature ever recorded in Louisiana was −16°F. The hottest temperature ever recorded in Louisiana was 114°F. How much of a variation in temperature is there between these two extremes? (*Source:* National Climatic Data Center)

77. In a series of plays, the San Francisco 49ers gain 2 yards, lose 5 yards, and then lose another 20 yards. What is their total gain or loss of yardage?

78. In some card games, it is possible to have a negative score. Lavonne Schultz currently has a score of 15 points. She then loses 24 points. What is her new score?

79. Pythagoras died in the year −475 (or 475 B.C.). When was he born, if he was 94 years old when he died?

80. The Greek astronomer and mathematician Geminus died in 60 A.D. at the age of 70. When was he born?

81. A commercial jet liner hits an air pocket and drops 250 feet. After climbing 120 feet, it drops another 178 feet. What is its overall vertical change?

82. Tyson Industries stock posted a loss of 1.625 points yesterday. If it drops another 0.75 point today, find its overall change for the two days.

83. The highest point in Africa is Mt. Kilimanjaro, Tanzania, at an elevation of 19,340 feet. The lowest point is Lake Assal, Djibouti, at 512 feet below sea level. How much higher is Mt. Kilimanjaro than Lake Assal? (*Source:* National Geographic Society)

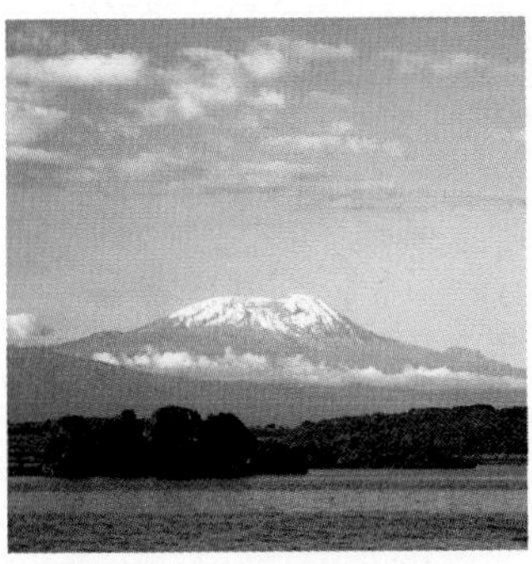

84. The airport in Bishop, California, is at an elevation of 4101 feet above sea level. The nearby Furnace Creek Airport in Death Valley, California, is at an elevation of 226 feet below sea level. How much higher in elevation is the Bishop Airport than the Furnace Creek Airport? (*Source:* National Climatic Data Center)

E *Find each unknown complementary or supplementary angle. See Example 11.*

△ **85.**

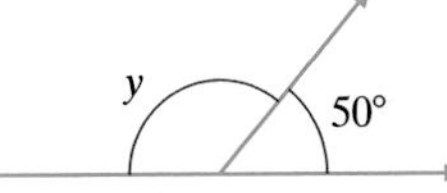

△ **86.**

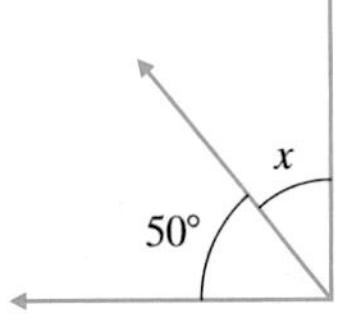

△ **87.** Complementary angles:

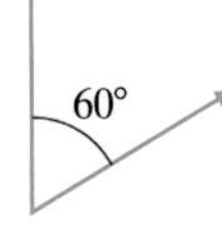

△ **88.** Supplementary angles:

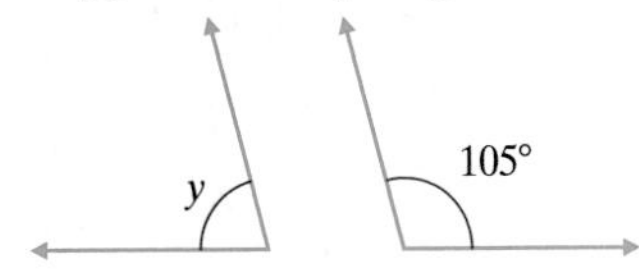

The following bar graph shows each month's average daily low temperature in degrees Fahrenheit for Barrow, Alaska. Use this graph to answer Exercises 89 through 91.

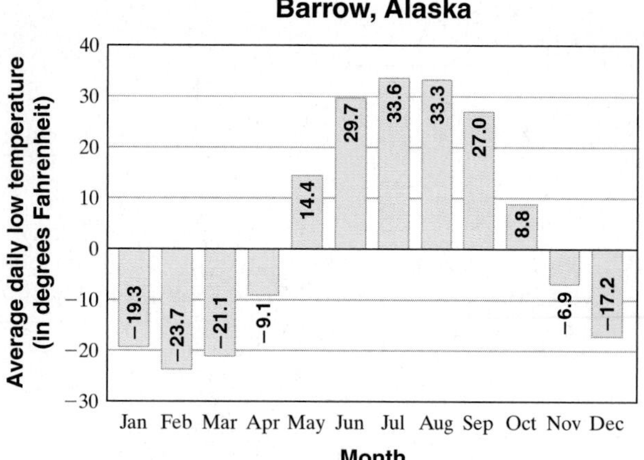

Source: National Climatic Data Center

89. Record the monthly increases and decreases in the low temperature from the previous month.

Month	Monthly Increase or Decrease
February	
March	
April	
May	
June	
July	
August	
September	
October	
November	
December	

90. Which month had the greatest increase in temperature?

91. Which month had the greatest decrease in temperature?

Combining Concepts

If a is a positive number and b is a negative number, determine whether each statement is true or false. Explain your answer.

92. $a - b$ is always a positive number.

93. $b - a$ is always a negative number.

94. $|b| - |a|$ is always a positive number.

95. $|b - a|$ is always a positive number.

Without calculating, determine whether each answer is positive or negative. Then use a calculator to find the exact difference.

96. 56,875 − 87,262

97. 4.362 − 7.0086

Name ______________________________

Answers

1. ______

2. ______

3. ______

4. ______

5. ______

6. ______

7. ______

8. ______

9. ______

10. ______

11. ______

12. ______

13. ______

14. ______

15. ______

16. ______

17. ______

18. ______

Integrated Review—Operations on Real Numbers

Answer the following with "a positive number," "a negative number," or "sometimes a positive number, sometimes a negative number."

1. The opposite of a positive number is ______________.

2. The sum of two negative numbers is ______________.

3. The absolute value of a negative number is ______________.

4. The difference of two positive numbers is ______________.

5. The sum of a positive number and a negative number is ______________.

6. The absolute value of a positive number is ______________.

Perform each indicated operation and simplify.

7. $7 - (-3)$ **8.** $-8 - 10$ **9.** $-14 - (-12)$ **10.** $-3 - (-1)$

11. $-19 + (-23)$ **12.** $18 + (-25)$ **13.** $-15 + 17$ **14.** $-2 + (-37)$

15. $4.5 - 7.9$ **16.** $-8.6 - 1.2$ **17.** $-\frac{3}{4} - \frac{1}{7}$ **18.** $\frac{2}{3} - \frac{7}{8}$

19. ______
20. ______
21. ______
22. ______
23. ______
24. ______
25. ______
26. ______
27. ______
28. ______
29. ______
30. ______
31. ______
32. ______
33. ______
34. ______
35. ______
36. ______
37. ______
38. ______
39. ______
40. ______

19. $-9 - (-7) + 4 - 6$

20. $11 - 20 + (-3) - 12$

21. $24 - 6(14 - 11)$

22. $30 - 5(10 - 8)$

23. $(7 - 17) + 4^2$

24. $9^2 + (10 - 30)$

25. $|-9| + 3^2 + (-4 - 20)$

26. $|-4 - 5| + 5^2 + (-50)$

27. $-7 + [(1 - 2) + (-2 - 9)]$

28. $-6 + [(-3 + 7) + (4 - 15)]$

29. Subtract 5 from 1.

30. Subtract -2 from -3.

31. Subtract $-\frac{2}{5}$ from $\frac{1}{4}$.

32. Subtract $\frac{1}{10}$ from $-\frac{5}{8}$.

33. $2(19 - 17)^3 - 3(-7 + 9)^2$

34. $3(10 - 9)^2 + 6(20 - 19)^3$

Evaluate each expression when $x = -2$, $y = -1$, and $z = 9$.

35. $x - y$

36. $x + y$

37. $y + z$

38. $z - y$

39. $\frac{|5z - x|}{y - x}$

40. $\frac{|-x - y + z|}{2z}$

B Finding Reciprocals

Addition and subtraction are related. Every difference of two numbers $a - b$ can be written as the sum $a + (-b)$. Multiplication and division are related also. For example, the quotient $6 \div 3$ can be written as the product $6 \cdot \frac{1}{3}$. Recall that the pair of numbers 3 and $\frac{1}{3}$ has a special relationship. Their product is 1 and they are called **reciprocals** or **multiplicative inverses** of each other.

Reciprocal or Multiplicative Inverse

Two numbers whose product is 1 are called **reciprocals** or **multiplicative inverses** of each other.

EXAMPLE 9 Find the reciprocal of each number.

a. 22 Reciprocal is $\frac{1}{22}$ since $22 \cdot \frac{1}{22} = 1$.

b. $\frac{3}{16}$ Reciprocal is $\frac{16}{3}$ since $\frac{3}{16} \cdot \frac{16}{3} = 1$.

c. -10 Reciprocal is $-\frac{1}{10}$ since $-10 \cdot -\frac{1}{10} = 1$.

d. $-\frac{9}{13}$ Reciprocal is $-\frac{13}{9}$ since $-\frac{9}{13} \cdot -\frac{13}{9} = 1$.

e. 1.7 Reciprocal is $\frac{1}{1.7}$ since $1.7 \cdot \frac{1}{1.7} = 1$. ●

Does the number 0 have a reciprocal? If it does, it is a number n such that $0 \cdot n = 1$. Notice that this can never be true since $0 \cdot n = 0$. This means that 0 has no reciprocal.

Quotients Involving Zero

The number 0 does not have a reciprocal.

C Dividing Real Numbers

We may now write a quotient as an equivalent product.

Quotient of Two Real Numbers

If a and b are real numbers and b is not 0, then

$$a \div b = \frac{a}{b} = a \cdot \frac{1}{b}$$

In other words, the quotient of two real numbers is the product of the first number and the multiplicative inverse or reciprocal of the second number.

EXAMPLE 10

Use the definition of the quotient of two numbers to find each quotient.

a. $-18 \div 3$ **b.** $\frac{-14}{-2}$ **c.** $\frac{20}{-4}$

Practice Problem 9

Find the reciprocal of each number.

a. 13 b. $\frac{7}{15}$ c. -5

d. $-\frac{8}{11}$ e. 7.9

Practice Problem 10

Use the definition of the quotient of two numbers to find each quotient.

a. $-12 \div 4$ b. $\frac{-20}{-10}$

c. $\frac{36}{-4}$

Answers

9. a. $\frac{1}{13}$, **b.** $\frac{15}{7}$, **c.** $-\frac{1}{5}$, **d.** $-\frac{11}{8}$, **e.** $\frac{1}{7.9}$,
10. a. -3, **b.** 2, **c.** -9

Solution:

a. $-18 \div 3 = -18 \cdot \frac{1}{3} = -6$

b. $\frac{-14}{-2} = -14 \cdot -\frac{1}{2} = 7$

c. $\frac{20}{-4} = 20 \cdot -\frac{1}{4} = -5$

Since the quotient $a \div b$ can be written as the product $a \cdot \frac{1}{b}$, it follows that sign patterns for dividing two real numbers are the same as sign patterns for multiplying two real numbers.

Dividing Real Numbers

1. The quotient of two numbers with the same sign is a positive number.
2. The quotient of two numbers with different signs is a negative number.

Practice Problem 11

Divide.

a. $\frac{-25}{5}$ b. $\frac{-48}{-6}$

c. $\frac{50}{-2}$ d. $\frac{-72}{0.2}$

EXAMPLE 11 Divide.

a. $\frac{-30}{-10} = 3$ Same sign, so the quotient is positive.

b. $\frac{-100}{5} = -20$

c. $\frac{20}{-2} = -10$

d. $\frac{42}{-0.6} = -70$

Unlike signs, so the quotient is negative.

$0.6\overline{)42.0}$ = 70.

In the examples above, we divided mentally or by long division. When we divide by a fraction, it is usually easier to multiply by its reciprocal.

Practice Problems 12–13

Divide.

12. $-\frac{5}{9} \div \frac{2}{3}$

13. $-\frac{2}{7} \div \left(-\frac{1}{5}\right)$

EXAMPLES Divide.

12. $\frac{2}{3} \div \left(-\frac{5}{4}\right) = \frac{2}{3} \cdot \left(-\frac{4}{5}\right) = -\frac{8}{15}$

13. $-\frac{1}{6} \div \left(-\frac{2}{3}\right) = -\frac{1}{6} \cdot \left(-\frac{3}{2}\right) = \frac{3}{12} = \frac{1}{4}$

Our definition of the quotient of two real numbers does not allow for division by 0 because 0 does not have a reciprocal. How then do we interpret $\frac{3}{0}$? We say that an expression such as this one is **undefined**. Can we divide 0 by a number other than 0? Yes; for example,

$$\frac{0}{3} = 0 \cdot \frac{1}{3} = 0$$

Division Involving Zero

Division by 0 is undefined. For example, $\frac{-5}{0}$ is undefined.

0 divided by a nonzero number is 0. For example, $\frac{0}{-5} = 0$.

Answers

11. a. -5, **b.** 8, **c.** -25, **d.** -360, **12.** $-\frac{5}{6}$, **13.** $\frac{10}{7}$

EXAMPLE 14 Perform each indicated operation.

a. $\frac{1}{0}$ is undefined. **b.** $\frac{0}{-3} = 0$

c. $\frac{0(-8)}{2} = \frac{0}{2} = 0$

Notice that $\frac{12}{-2} = -6, -\frac{12}{2} = -6$, and $\frac{-12}{2} = -6$. This means that $\frac{12}{-2} = -\frac{12}{2} = \frac{-12}{2}$

In other words, a single negative sign in a fraction can be written in the denominator, in the numerator, or in front of the fraction without changing the value of the fraction.

If a and b are real numbers, then $b \neq 0, \frac{a}{-b} = \frac{-a}{b} = -\frac{a}{b}$.

Examples combining basic arithmetic operations along with the principles of the order of operations help us to review these concepts.

EXAMPLE 15 Simplify each expression.

a. $\frac{(-12)(-3) + 4}{-7 - (-2)}$

b. $\frac{2(-3)^2 - 20}{-5 + 4}$

Solution:

a. First, simplify the numerator and denominator separately; then divide.

$$\frac{(-12)(-3) + 4}{-7 - (-2)} = \frac{36 + 4}{-7 + 2}$$
$$= \frac{40}{-5}$$
$$= -8 \quad \text{Divide.}$$

b. Simplify the numerator and denominator separately; then divide.

$$\frac{2(-3)^2 - 20}{-5 + 4} = \frac{2 \cdot 9 - 20}{-5 + 4} = \frac{18 - 20}{-5 + 4} = \frac{-2}{-1} = 2$$

D Evaluating Algebraic Expressions

Using what we have learned about dividing real numbers, we continue to practice evaluating algebraic expressions.

EXAMPLE 16 Evaluate $\frac{3x^2}{4y}$ when $x = -2$ and $y = -4$.

Solution: Replace x with -2 and y with -4 and simplify.

$$\frac{3x^2}{4y} = \frac{3(-2)^2}{4(-4)} = \frac{3 \cdot 4}{-16} = \frac{12}{-16} = -\frac{3}{4}$$

Practice Problem 14

Perform each indicated operation.

a. $\frac{-7}{0}$

b. $\frac{0}{-2}$

c. $\frac{0(-5)}{3}$

Practice Problem 15

Simplify each expression.

a. $\frac{-7(-4) + 2}{-10 - (-5)}$

b. $\frac{5(-2)^3 + 52}{-4 + 1}$

Practice Problem 16

Evaluate $\frac{x + y}{3x}$ when $x = -1$ and $y = -5$.

Answers

14. a. undefined, **b.** 0, **c.** 0,
15. a. -6, **b.** -4, **16.** 2

E Solutions of Equations

We use our skills in multiplying and dividing real numbers to check possible solutions for an equation.

Practice Problem 17

Determine whether -8 is a solution of $\frac{x}{4} - 3 = x + 3$.

Answer

17. -8 is a solution.

EXAMPLE 17 Determine whether -10 is a solution of $\frac{-20}{x} + 15 = 2x$.

Solution:

$$\frac{-20}{x} + 15 = 2x \quad \text{Original equation}$$
$$\frac{-20}{-10} + 15 \stackrel{?}{=} 2(-10) \quad \text{Replace } x \text{ with } -10.$$
$$2 + 15 \stackrel{?}{=} -20 \quad \text{Divide.}$$
$$17 = -20 \quad \textbf{False}$$

Since we have a false statement, -10 is *not* a solution of the equation.

CALCULATOR EXPLORATIONS

Entering Negative Numbers on a Scientific Calculator

To enter a negative number on a scientific calculator, find a key marked [+/−]. (On some calculators, this key is marked [CHS] for "change sign.") To enter -8, for example, press the keys [8] [+/−]. The display will read [−8].

Entering Negative Numbers on a Graphing Calculator

To enter a negative number on a graphing calculator, find a key marked [(−)]. Do not confuse this key with the key [−], which is used for subtraction. To enter -8, for example, press the keys [(−)] [8]. The display will read [−8].

Operations with Real Numbers

To evaluate $-2(7 - 9) - 20$ on a calculator, press the keys

[2] [+/−] [×] [(] [7] [−] [9] [)] [−] [2] [0] [=], or

[(−)] [2] [(] [7] [−] [9] [)] [−] [2] [0] [ENTER].

The display will read [−16] or [−2(7 − 9) − 20 −16]

Use a calculator to simplify each expression.

1. $-38(26 - 27)$

2. $-59(-8) + 1726$

3. $134 + 25(68 - 91)$

4. $45(32) - 8(218)$

5. $\frac{-50(294)}{175 - 265}$

6. $\frac{-444 - 444.8}{-181 - 324}$

7. $9^5 - 4550$

8. $5^8 - 6259$

9. $(-125)^2$ (Be careful.)

10. -125^2 (Be careful.)

Name ______________________ Section __________ Date ____________

EXERCISE SET 1.6

A *Multiply. See Examples 1 through 6.*

1. $-6(4)$ **2.** $-8(5)$ **3.** $2(-1)$ **4.** $7(-4)$

5. $-5(-10)$ **6.** $-6(-11)$ **7.** $-3 \cdot 4$ **8.** $-2 \cdot 8$

9. $-\frac{1}{2}\left(-\frac{3}{5}\right)$ **10.** $-\frac{1}{8}\left(-\frac{1}{3}\right)$ **11.** $-\frac{3}{4}\left(-\frac{8}{9}\right)$ **12.** $-\frac{5}{6}\left(-\frac{3}{10}\right)$

13. $5(-1.4)$ **14.** $6(-2.5)$ **15.** $-0.2(-0.7)$ **16.** $-0.5(-0.3)$

17. $(-5)(-5)$ **18.** $(-7)(-7)$ **19.** $\frac{2}{3}\left(-\frac{4}{9}\right)$ **20.** $\frac{2}{7}\left(-\frac{2}{11}\right)$

21. $-\frac{20}{25}\left(\frac{5}{16}\right)$ **22.** $-\frac{25}{36}\left(\frac{6}{15}\right)$ **23.** $-2.1(-0.4)$ **24.** $-1.3(-0.6)$

Simplify. See Examples 7 and 8.

25. $(-1)(2)(-3)(-5)$ **26.** $(-2)(-3)(-4)(-2)$ **27.** $(2)(-1)(-3)(5)(3)$ **28.** $(3)(-5)(-2)(-1)(-2)$

29. $(-4)^2$ **30.** $(-3)^3$ **31.** -4^2 **32.** -6^2

33. $-3(2-8)$ **34.** $-4(3-9)$ **35.** $-3[(2-8)-(-6-8)]$

36. $-2[(3-5)-(2-9)]$ **37.** $\left(-\frac{3}{4}\right)^2$ **38.** $\left(-\frac{2}{7}\right)^2$

39. $(-2)^4$ **40.** -2^4 **41.** -1^5 **42.** $(-1)^5$

State whether each statement is true or false.

43. The product of three negative integers is negative.

44. The product of three positive integers is positive.

45. The product of four negative integers is negative.

46. The product of four positive integers is positive.

B *Find each reciprocal. See Example 9.*

47. 9

48. 100

49. $\frac{2}{3}$

50. $\frac{1}{7}$

51. -14

52. -8

53. $-\frac{3}{11}$

54. $-\frac{6}{13}$

55. 0.2

56. 1.5

57. $\frac{1}{-6.3}$

58. $\frac{1}{-8.9}$

59. Find any real numbers that are their own reciprocals.

60. Explain why 0 has no reciprocal.

C *Divide. See Examples 10 through 14.*

61. $\frac{18}{-2}$

62. $\frac{20}{-10}$

63. $\frac{-48}{12}$

64. $\frac{-60}{5}$

65. $\frac{0}{-4}$

66. $\frac{0}{-9}$

67. $\frac{5}{0}$

68. $\frac{3}{0}$

69. $\frac{-12}{-4}$

70. $\frac{-45}{-9}$

71. $\frac{30}{-2}$

72. $\frac{14}{-2}$

73. $\frac{6}{7} \div \left(-\frac{1}{3}\right)$

74. $\frac{4}{5} \div \left(-\frac{1}{2}\right)$

75. $-\frac{5}{9} \div \left(-\frac{3}{4}\right)$

76. $-\frac{1}{10} \div \left(-\frac{8}{11}\right)$

77. $-\frac{4}{9} \div \frac{4}{9}$

78. $-\frac{5}{12} \div \frac{5}{12}$

79. $-\frac{5}{8} \div \frac{3}{4}$

80. $-\frac{5}{6} \div \frac{2}{3}$

81. $-48 \div 1.2$

82. $-86 \div 2.5$

83. $-3.2 \div -0.02$

84. $-4.9 \div -0.07$

Simplify. See Example 15.

85. $\dfrac{-9(-3)}{-6}$

86. $\dfrac{-6(-3)}{-4}$

87. $\dfrac{12}{9-12}$

88. $\dfrac{-15}{1-4}$

89. $\dfrac{-6^2+4}{-2}$

90. $\dfrac{3^2+4}{5}$

91. $\dfrac{8+(-4)^2}{4-12}$

92. $\dfrac{6+(-2)^2}{4-9}$

93. $\dfrac{22+(3)(-2)}{-5-2}$

94. $\dfrac{-20+(-4)(3)}{1-5}$

95. $\dfrac{-3-5^2}{2(-7)}$

96. $\dfrac{-2-4^2}{3(-6)}$

97. $\dfrac{6-2(-3)}{4-3(-2)}$

98. $\dfrac{8-3(-2)}{2-5(-4)}$

99. $\dfrac{-3-2(-9)}{-15-3(-4)}$

100. $\dfrac{-4-8(-2)}{-9-2(-3)}$

101. $\dfrac{|5-9|+|10-15|}{|2(-3)|}$

102. $\dfrac{|-3+6|+|-2+7|}{|-2\cdot 2|}$

103. $\dfrac{-7(-1)+(-3)4}{(-2)(5)+(-6)(-8)}$

104. $\dfrac{8(-7)+(-2)(-6)}{(-9)(3)+(-10)(-11)}$

D *Evaluate each expression when $x = -5$ and $y = -3$. See Example 16.*

105. $\dfrac{2x-5}{y-2}$

106. $\dfrac{2y-12}{x-4}$

107. $\dfrac{6-y}{x-4}$

108. $\dfrac{4 - 2x}{y + 3}$

109. $\dfrac{x^2 + y}{3y}$

110. $\dfrac{y^2 - x}{2x}$

Decide whether the given number is a solution of the given equation. See Example 17.

111. $-3x - 5 = -20;\quad 5$

112. $17 - 4x = x + 27;\quad -2$

113. $\dfrac{x}{5} + 2 = -1;\quad 15$

114. $\dfrac{x}{6} - 3 = 5;\quad 48$

115. $\dfrac{x + 4}{5} = -6;\quad -30$

116. $\dfrac{x - 3}{7} = -2;\quad -11$

Combining Concepts

Write each as an algebraic expression. Then simplify the expression.

117. 7 subtracted from the quotient of 0 and 5

118. Twice the sum of −3 and −4

119. −1 added to the product of −8 and −5

120. The difference of −9 and the product of −4 and −6

121. The quotient of −8 and −20

122. The quotient of −9 and −30

If a is a positive number and b is a negative number, determine whether each expression simplifies to a positive number or a negative number.

123. $\dfrac{a}{b}$

124. $\dfrac{b}{a}$

125. $\dfrac{b + b}{a + a}$

126. $\dfrac{-a}{-b}$

127. Phar-Mor is a retail drugstore chain. For fiscal year 2000, Phar-Mor posted a net income of −$11 million. Phar-Mor reports its earnings to shareholders at the end of each of four quarters throughout its fiscal year. What was Phar-Mor's average net income per quarter during fiscal year 2000? (*Source*: Phar-Mor Inc.)

If q is a negative number, r is a negative number, and t is a positive number, determine whether each expression simplifies to a positive or negative number. If it is not possible to determine, so state.

128. $q \cdot r \cdot t$

129. $q^2 \cdot r \cdot t$

130. $q + t$

131. $t + r$

132. $t(q + r)$

133. $r(q - t)$

134. During the fourth quarter of fiscal year 2000, Ames Department Stores, Inc., posted a net income of −$152 million. If this result continued for the next four quarters, what would Ames' net income be the following year? (*Source:* Ames Department Stores, Inc.)

135. If a and b are any real numbers, is the statement $a \cdot b = b \cdot a$ always true? Why or why not?

136. If a and b are any real numbers, is the statement $a - b = b - a$ always true? Why or why not?

Internet Excursions

Go To: http://www.prenhall.com/martin-gay_algebra

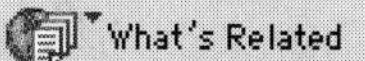

A major stock market in the United States is the National Association of Securities Dealers Automated Quotations (NASDAQ). The given World Wide Web address will provide you with access to the NASDAQ site, or a related site, where you can get an InfoQuote for any stock traded on the NASDAQ exchange. You can then choose to make a stock chart of the closing share price for the past 6 months. By clicking on this graph, you can view a table of the closing prices by date.

137. Get an InfoQuote for Microsoft Corporation (ticker symbol: MSFT).

a. Make a stock chart of the closing share price for the past 6 months. Describe any trends you see.

b. Complete the following table. (You will need to click on the graph to find the closing price one month ago.)

Date of previous day	
Previous day's close	
Date one month ago	
Close one month ago	

c. What is the difference between the previous day's closing price and the closing price one month ago?

d. If you had bought 100 shares of Microsoft stock one month ago and sold all your shares at the previous day's closing price, how much money would you have gained or lost?

138. Get an InfoQuote for Applebee's International, Inc. (ticker symbol: APPB).

a. Make a stock chart of the closing share price for the past 6 months. Describe any trends you see.

b. Complete the following table. (You will need to click on the graph to find the closing price one month ago.)

Date of previous day	
Previous day's close	
Date one month ago	
Close one month ago	

c. What is the difference between the previous day's closing price and the closing price one month ago?

d. If you had bought 100 shares of Applebee's stock one month ago and sold all your shares at the previous day's closing price, how much money would you have gained or lost?

FOCUS ON History

Magic Squares

A magic square is a set of numbers arranged in a square table so that the sum of the numbers in each column, row, and diagonal is the same. For instance, in the magic square to the right, the sum of each column, row, and diagonal is 15. Notice that no number is used more than once in the magic square.

2	9	4
7	5	3
6	1	8

The properties of magic squares have been known for a very long time and once were thought to be good luck charms. The ancient Egyptians and Greeks understood their patterns. A magic square even made it into a famous work of art. The engraving titled *Melencolia I*, created by German artist Albrecht Dürer in 1514, features the following four-by-four magic square on the building behind the central figure.

16	3	2	13
5	10	11	8
9	6	7	12
4	15	14	1

Critical Thinking

1. Verify that what is shown in the Dürer engraving is, in fact, a magic square. What is the common sum of the columns, rows, and diagonals?
2. Negative numbers can also be used in magic squares. Complete the following magic square:

		−2
	−1	
0		−4

3. Use the numbers −12, −9, −6, −3, 0, 3, 6, 9, and 12 to form a magic square.

1.7 Properties of Real Numbers

OBJECTIVES

- (A) Use the commutative and associative properties.
- (B) Use the distributive property.
- (C) Use the identity and inverse properties.

(A) Using the Commutative and Associative Properties

In this section we give names to properties of real numbers with which we are already familiar. Throughout this section, the variables *a, b,* and *c* represent real numbers.

We know that order does not matter when adding numbers. For example, we know that $7 + 5$ is the same as $5 + 7$. This property is given a special name—the **commutative property of addition**. We also know that order does not matter when multiplying numbers. For example, we know that $-5(6) = 6(-5)$. This property means that multiplication is commutative also and is called the **commutative property of multiplication**.

Commutative Properties

Addition: $a + b = b + a$

Multiplication: $a \cdot b = b \cdot a$

These properties state that the *order* in which any two real numbers are added or multiplied does not change their sum or product. For example, if we let $a = 3$ and $b = 5$, then the commutative properties guarantee that

$$3 + 5 = 5 + 3 \quad \text{and} \quad 3 \cdot 5 = 5 \cdot 3$$

Helpful Hint

Is subtraction also commutative? Try an example. Is $3 - 2 = 2 - 3$? **No!** The left side of this statement equals 1; the right side equals -1. There is no commutative property of subtraction. Similarly, there is no commutative property for division. For example, $10 \div 2$ does not equal $2 \div 10$.

EXAMPLE 1 Use a commutative property to complete each statement.

a. $x + 5 =$ ______ **b.** $3 \cdot x =$ ______

Solution:

a. $x + 5 = 5 + x$ By the commutative property of addition

b. $3 \cdot x = x \cdot 3$ By the commutative property of multiplication

Practice Problem 1

Use a commutative property to complete each statement.

a. $7 \cdot y =$ ________

b. $4 + x =$ ________

Concept Check

Which of the following pairs of actions are commutative?

a. "raking the leaves" and "bagging the leaves"

b. "putting on your left glove" and "putting on your right glove"

c. "putting on your coat" and "putting on your shirt"

d. "reading a novel" and "reading a newspaper"

Try the Concept Check in the margin.

Let's now discuss grouping numbers. We know that when we add three numbers, the way in which they are grouped or associated does not change their sum. For example, we know that $2 + (3 + 4) = 2 + 7 = 9$. This result is the same if we group the numbers differently. In other words, $(2 + 3) + 4 = 5 + 4 = 9$, also. Thus, $2 + (3 + 4) = (2 + 3) + 4$. This property is called the **associative property of addition**.

We also know that changing the grouping of numbers when multiplying does not change their product. For example, $2 \cdot (3 \cdot 4) = (2 \cdot 3) \cdot 4$ (check it). This is the **associative property of multiplication**.

Answers

1. a. $y \cdot 7$, **b.** $x + 4$

Concept Check: b, d

Associative Properties

Addition: $(a + b) + c = a + (b + c)$
Multiplication: $(a \cdot b) \cdot c = a \cdot (b \cdot c)$

These properties state that the way in which three numbers are *grouped* does not change their sum or their product.

EXAMPLE 2 Use an associative property to complete each statement.

a. $5 + (4 + 6) =$ _____
b. $(-1 \cdot 2) \cdot 5 =$ _____

Solution:

a. $5 + (4 + 6) = (5 + 4) + 6$ By the associative property of addition
b. $(-1 \cdot 2) \cdot 5 = -1 \cdot (2 \cdot 5)$ By the associative property of multiplication

Practice Problem 2

Use an associative property to complete each statement.

a. $5 \cdot (-3 \cdot 6) =$ _______
b. $(-2 + 7) + 3 =$ _______

Helpful Hint

Remember the difference between the commutative properties and the associative properties. The commutative properties have to do with the *order* of numbers and the associative properties have to do with the *grouping* of numbers.

EXAMPLES

Determine whether each statement is true by an associative property or a commutative property.

3. $(7 + 10) + 4 = (10 + 7) + 4$ Since the order of two numbers was changed and their grouping was not, this is true by the commutative property of addition.

4. $2 \cdot (3 \cdot 1) = (2 \cdot 3) \cdot 1$ Since the grouping of the numbers was changed and their order was not, this is true by the associative property of multiplication.

Practice Problems 3–4

Determine whether each statement is true by an associative property or a commutative property.

3. $5 \cdot (4 \cdot 7) = 5 \cdot (7 \cdot 4)$
4. $-2 + (4 + 9) = (-2 + 4) + 9$

Let's now illustrate how these properties can help us simplify expressions.

EXAMPLES Simplify each expression.

5. $10 + (x + 12) = 10 + (12 + x)$ By the commutative property of addition
$= (10 + 12) + x$ By the associative property of addition
$= 22 + x$ Add.

6. $-3(7x) = (-3 \cdot 7)x$ By the associative property of multiplication
$= -21x$ Multiply.

Practice Problems 5–6

Simplify each expression.

5. $(-3 + x) + 17$
6. $4(5x)$

B Using the Distributive Property

The **distributive property of multiplication over addition** is used repeatedly throughout algebra. It is useful because it allows us to write a product as a sum or a sum as a product.

We know that $7(2 + 4) = 7(6) = 42$. Compare that with $7(2) + 7(4) = 14 + 28 = 42$. Since both original expressions equal 42, they must equal each other, or

$$7(2 + 4) = 7(2) + 7(4)$$

This is an example of the distributive property. The product on the left side of the equal sign is equal to the sum on the right side. We can think of the 7 as being distributed to each number inside the parentheses.

Answers

2. a. $(5 \cdot -3) \cdot 6$, **b.** $-2 + (7 + 3)$, **3.** commutative, **4.** associative, **5.** $14 + x$, **6.** $20x$

Distributive Property of Multiplication Over Addition

$a(b + c) = ab + ac$

Since multiplication is commutative, this property can also be written as

$$(b + c)a = ba + ca$$

The distributive property can also be extended to more than two numbers inside the parentheses. For example,

$$\begin{aligned} 3(x + y + z) &= 3(x) + 3(y) + 3(z) \\ &= 3x + 3y + 3z \end{aligned}$$

Since we define subtraction in terms of addition, the distributive property is also true for subtraction. For example,

$$\begin{aligned} 2(x - y) &= 2(x) - 2(y) \\ &= 2x - 2y \end{aligned}$$

EXAMPLES

Use the distributive property to write each expression without parentheses. Then simplify the result.

7. $2(x + y) = 2(x) + 2(y)$
$= 2x + 2y$

8. $-5(-3 + 2z) = -5(-3) + (-5)(2z)$
$= 15 - 10z$

9. $5(x + 3y - z) = 5(x) + 5(3y) - 5(z)$
$= 5x + 15y - 5z$

10. $-1(2 - y) = (-1)(2) - (-1)(y)$
$= -2 + y$

11. $-(3 + x - w) = -1(3 + x - w)$
$= (-1)(3) + (-1)(x) - (-1)(w)$
$= -3 - x + w$

12. $4(3x + 7) + 10 = 4(3x) + 4(7) + 10$ Apply the distributive property.
$= 12x + 28 + 10$ Multiply.
$= 12x + 38$ Add.

Helpful Hint

Notice in Example 11 that $-(3 + x - w)$ is first rewritten as $-1(3 + x - w)$.

Practice Problems 7–12

Use the distributive property to write each expression without parentheses. Then simplify the result.

7. $5(x + y)$
8. $-3(2 + 7x)$
9. $4(x + 6y - 2z)$
10. $-1(3 - a)$
11. $-(8 + a - b)$
12. $9(2x + 4) + 9$

The distributive property can also be used to write a sum as a product.

EXAMPLES

Use the distributive property to write each sum as a product.

13. $8 \cdot 2 + 8 \cdot x = 8(2 + x)$

14. $7s + 7t = 7(s + t)$

Practice Problems 13–14

Use the distributive property to write each sum as a product.

13. $9 \cdot 3 + 9 \cdot y$
14. $4x + 4y$

C Using the Identity and Inverse Properties

Next, we look at the **identity properties**.

The number 0 is called the identity for addition because when 0 is added to any real number, the result is the same real number. In other words, the *identity* of the real number is not changed.

Answers

7. $5x + 5y$, **8.** $-6 - 21x$, **9.** $4x + 24y - 8z$, **10.** $-3 + a$, **11.** $-8 - a + b$, **12.** $18x + 45$, **13.** $9(3 + y)$, **14.** $4(x + y)$

The number 1 is called the identity for multiplication because when a real number is multiplied by 1, the result is the same real number. In other words, the *identity* of the real number is not changed.

Identities for Addition and Multiplication

0 is the identity element for addition.

$$a + 0 = a \quad \text{and} \quad 0 + a = a$$

1 is the identity element for multiplication.

$$a \cdot 1 = a \quad \text{and} \quad 1 \cdot a = a$$

Notice that 0 is the *only* number that can be added to any real number with the result that the sum is the same real number. Also, 1 is the *only* number that can be multiplied by any real number with the result that the product is the same real number.

Additive inverses or **opposites** were introduced in Section 1.4. Two numbers are called additive inverses or opposites if their sum is 0. The additive inverse or opposite of 6 is -6 because $6 + (-6) = 0$. The additive inverse or opposite of -5 is 5 because $-5 + 5 = 0$.

Reciprocals or **multiplicative inverses** were introduced in Section R.2. Two nonzero numbers are called reciprocals or multiplicative inverses if their product is 1. The reciprocal or multiplicative inverse of $\frac{2}{3}$ is $\frac{3}{2}$ because $\frac{2}{3} \cdot \frac{3}{2} = 1$. Likewise, the reciprocal of -5 is $-\frac{1}{5}$ because $-5\left(-\frac{1}{5}\right) = 1$.

Additive or Multiplicative Inverses

The numbers a and $-a$ are additive inverses or opposites of each other because their sum is 0; that is,

$$a + (-a) = 0$$

The numbers b and $\frac{1}{b}$ (for $b \neq 0$) are reciprocals or multiplicative inverses of each other because their product is 1; that is,

$$b \cdot \frac{1}{b} = 1$$

Try the Concept Check in the margin.

EXAMPLES Name the property illustrated by each true statement.

15. $3 \cdot y = y \cdot 3$ — Commutative property of multiplication (order changed)

16. $(x + 7) + 9 = x + (7 + 9)$ — Associative property of addition (grouping changed)

17. $(b + 0) + 3 = b + 3$ — Identity element for addition

18. $2 \cdot (z \cdot 5) = 2 \cdot (5 \cdot z)$ — Commutative property of multiplication (order changed)

19. $-2 \cdot \left(-\frac{1}{2}\right) = 1$ — Multiplicative inverse property

20. $-2 + 2 = 0$ — Additive inverse property

21. $-6 \cdot (y \cdot 2) = (-6 \cdot 2) \cdot y$ — Commutative and associative properties of multiplication (order and grouping changed)

Concept Check

Which of the following is the

a. opposite of $-\frac{3}{10}$, and which is the

b. reciprocal of $-\frac{3}{10}$?

$$1, -\frac{10}{3}, \frac{3}{10}, 0, \frac{10}{3}, -\frac{3}{10}$$

Practice Problems 15–21

Name the property illustrated by each true statement.

15. $5 + (-5) = 0$
16. $12 + y = y + 12$
17. $-4 \cdot (6 \cdot x) = (-4 \cdot 6) \cdot x$
18. $6 + (z + 2) = 6 + (2 + z)$
19. $3\left(\frac{1}{3}\right) = 1$
20. $(x + 0) + 23 = x + 23$
21. $(7 \cdot y) \cdot 10 = y \cdot (7 \cdot 10)$

Answers

15. additive inverse property, **16.** commutative property of addition, **17.** associative property of multiplication, **18.** commutative property of addition, **19.** multiplicative inverse property, **20.** identity element for addition, **21.** commutative and associative properties of multiplication

Concept Check:

a. $\frac{3}{10}$, **b.** $-\frac{10}{3}$

Name ____________________ Section __________ Date __________

EXERCISE SET 1.7

Use a commutative property to complete each statement. See Examples 1 and 3.

1. $x + 16 =$ ________ **2.** $4 + y =$ ________ **3.** $-4 \cdot y =$ ________ **4.** $-2 \cdot x =$ ________

5. $xy =$ ________ **6.** $ab =$ ________ **7.** $2x + 13 =$ ________ **8.** $19 + 3y =$ ________

Use an associative property to complete each statement. See Examples 2 and 4.

9. $(xy) \cdot z =$ ________ **10.** $3 \cdot (x \cdot y) =$ ________ **11.** $2 + (a + b) =$ ________

12. $(y + 4) + z =$ ________ **13.** $4 \cdot (ab) =$ ________ **14.** $(-3y) \cdot z =$ ________

15. $(a + b) + c =$ ________ **16.** $6 + (r + s) =$ ________

Use the commutative and associative properties to simplify each expression. See Examples 5 and 6.

17. $8 + (9 + b)$ **18.** $(r + 3) + 11$ **19.** $4(6y)$ **20.** $2(42x)$

21. $\frac{1}{5}(5y)$ **22.** $\frac{1}{8}(8z)$ **23.** $(13 + a) + 13$ **24.** $7 + (x + 4)$

25. $-9(8x)$ **26.** $-3(12y)$ **27.** $\frac{3}{4}\left(\frac{4}{3}s\right)$ **28.** $\frac{2}{7}\left(\frac{7}{2}r\right)$

29. Write an example that shows that division is not commutative.

30. Write an example that shows that subtraction is not commutative.

Use the distributive property to write each expression without parentheses. Then simplify the result. See Examples 7 through 12.

31. $4(x + y)$ **32.** $7(a + b)$ **33.** $9(x - 6)$ **34.** $11(y - 4)$

35. $2(3x + 5)$ **36.** $5(7 + 8y)$ **37.** $7(4x - 3)$ **38.** $3(8x - 1)$

39. $3(6 + x)$ **40.** $2(x + 5)$ **41.** $-2(y - z)$ **42.** $-3(z - y)$

43. $-7(3y + 5)$ **44.** $-5(2r + 11)$ **45.** $5(x + 4m + 2)$ **46.** $8(3y + z - 6)$

47. $-4(1 - 2m + n)$ **48.** $-4(4 + 2p + 5)$ **49.** $-(5x + 2)$ **50.** $-(9r + 5)$

51. $-(r - 3 - 7p)$ **52.** $-(q - 2 + 6r)$ **53.** $\frac{1}{2}(6x + 8)$ **54.** $\frac{1}{4}(4x - 2)$

55. $-\frac{1}{3}(3x - 9y)$ **56.** $-\frac{1}{5}(10a - 25b)$ **57.** $3(2r + 5) - 7$ **58.** $10(4s + 6) - 40$

59. $-9(4x + 8) + 2$ **60.** $-11(5x + 3) + 10$ **61.** $-4(4x + 5) - 5$ **62.** $-6(2x + 1) - 1$

Use the distributive property to write each sum as a product. See Examples 13 and 14.

63. $4 \cdot 1 + 4 \cdot y$ **64.** $14 \cdot z + 14 \cdot 5$ **65.** $11x + 11y$ **66.** $9a + 9b$

67. $(-1) \cdot 5 + (-1) \cdot x$ **68.** $(-3)a + (-3)y$ **69.** $30a + 30b$ **70.** $25x + 25y$

C *Find the additive inverse or opposite of each of the following numbers. See Example 20.*

71. 16 **72.** 14 **73.** -8 **74.** -3

75. $-(-1.2)$ **76.** $-(7.9)$ **77.** $-|-2|$ **78.** $-|-9|$

Find the multiplicative inverse or reciprocal of each number. See Example 19.

79. $\frac{2}{3}$ **80.** $\frac{3}{4}$ **81.** $-\frac{5}{6}$ **82.** $-\frac{7}{8}$

83. $3\frac{5}{6}$ **84.** $2\frac{3}{5}$ **85.** -2 **86.** -5

Name the properties illustrated by each true statement. See Examples 15 through 21.

87. $3 \cdot 5 = 5 \cdot 3$

88. $4(3 + 8) = 4 \cdot 3 + 4 \cdot 8$

89. $2 + (x + 5) = (2 + x) + 5$

90. $(x + 9) + 3 = (9 + x) + 3$

91. $9(3 + 7) = 9 \cdot 3 + 9 \cdot 7$

92. $1 \cdot 9 = 9$

93. $(4 \cdot y) \cdot 9 = 4 \cdot (y \cdot 9)$

94. $6 \cdot \frac{1}{6} = 1$

95. $0 + 6 = 6$

96. $(a + 9) + 6 = a + (9 + 6)$

97. $-4(y + 7) = -4 \cdot y + (-4) \cdot 7$

98. $(11 + r) + 8 = (r + 11) + 8$

99. $-4 \cdot (8 \cdot 3) = (8 \cdot -4) \cdot 3$

100. $r + 0 = r$

Combining Concepts

Fill in the table with the opposite (additive inverse), the reciprocal (multiplicative inverse), or the expression. Assume that the value of each expression is not 0.

	101.	**102.**	**103.**	**104.**	**105.**	**106.**
Expression	8	$-\frac{2}{3}$	x	$4y$		
Opposite						$7x$
Reciprocal					$\frac{1}{2x}$	

Name the property illustrated by each step.

107. **a.** △ + (□ + ○) = (□ + ○) + △

b. = (○ + □) + △

c. = ○ + (□ + △)

108. **a.** $(x + y) + z = x + (y + z)$

b. $= (y + z) + x$

c. $= (z + y) + x$

109. Explain why 0 is called the identity element for addition.

110. Explain why 1 is called the identity element for multiplication.

Determine which pairs of actions are commutative.

111. "taking a test" and "studying for the test"

112. "putting on your shoes" and "putting on your socks"

113. "putting on your left shoe" and "putting on your right shoe"

114. "reading the sports section" and "reading the comics section"

CHAPTER 1 ACTIVITY Analyzing Wealth

This activity may be completed by working in groups or individually.

Tom Stanley and Bill Danko are experts on millionaires. While doing research for their book, *The Millionaire Next Door*, they found that over 3 million households in the United States have net worths* of over $1,000,000. Most of these millionaires are regular people who have made their fortunes not overnight in risky business deals, but over the years through steady saving and wise investing. Stanley and Danko say that almost anyone can become a millionaire—the key is to live below your means. They give the following rule of thumb to test your progress in becoming a millionaire.

> Your net worth should be 10% of your annual income, multiplied by your age, and then doubled. (*Source: Parade Magazine*, June 22, 1997)

Stanley and Danko caution people to not worry if their net worths don't meet this goal. With some discipline, they can still catch up.

* Net worth is the total value of all your assets (such as property, bank accounts, and investments) minus what you owe (such as a car loan, mortgage, or education loan).

1. Suppose you are a financial planner. A 30-year-old client would like to eventually become a millionaire. He makes $36,000 each year. What should his net worth be?

2. Let N represent net worth. Choose variables to represent annual income and age. Write an algebraic equation that represents Stanley and Danko's rule of thumb for becoming a millionaire.

3. Simplify the equation you wrote in Question 2.

4. Suppose that a 42-year-old client hopes to retire as a millionaire. She earns $64,000 each year and has a net worth of $500,000. Using the equation you wrote in Questions 2 and 3, is she on track? What is the difference between her net worth and the goal?

5. Suppose that a 35-year-old client earns $50,000 each year. To become a millionaire, what should his net worth be at this age? According to Stanley and Danko's rule of thumb, if he continues to earn $50,000 during the next year, how much additional net worth will he need to still be on track at age 36?

6. (Optional) Using your own age and annual income, compute the net worth goal for becoming a millionaire in your own situation.

FOCUS ON History

MULTICULTURALISM

Numbers have a long history. The numbers we are accustomed to using probably originated in India in the 3rd century and were later adapted by Arabic cultures. Many other ancient civilizations developed their own unique number systems.

Roman Numerals

I	V	X	L	C	D	M
1	5	10	50	100	500	1000

If numerals decrease in value from left to right, the values are added. If a smaller numeral appears to the left of a larger numeral, the smaller value is subtracted. For example,

$$\text{XVII} = 10 + 5 + 1 + 1 = 17, \quad \text{but} \quad \text{XLIV} = 50 - 10 + 5 - 1 = 44.$$

Chinese Numerals

一	二	三	四	五	六	七	八	九	十	百	千	萬	億
1	2	3	4	5	6	7	8	9	10	100	1000	10,000	100,000

Numerals are written vertically. If a digit representing 2–9 appears before a digit representing 10, 100, 1000, 10,000, or 100,000, multiplication is indicated.

七
千 } (7 × 1000)
三
百 } (3 × 100)
八
十 } (8 × 10)
五 } 5

This number is $7000 + 300 + 80 + 5 = 7385$

Egyptian Hieroglyphic Numerals

𓏺	𓎆	𓍢	𓆼	𓂭	𓆐
1	10	100	1000	10,000	100,000

The Egyptian system is also multiplicative. For example, 3 𓎆 hieroglyphs represents 3×10.

𓏺𓏺𓏺𓏺𓎆𓎆𓍢𓍢𓍢𓍢𓍢𓆼𓆼𓂭 $= 4 \times 1 + 2 \times 10 + 5 \times 100 + 2 \times 1000 + 1 \times 10{,}000$
$= 12{,}524$

GROUP ACTIVITIES

- Write several numbers using each of the Roman, Chinese, and Egyptian hieroglyphic systems. Trade your numbers with another student in your group to translate into our numerals. Check one another's work.
- Research the number system of another ancient culture (such as the Babylonian, Mayan Indian, or Ionic Greek culture). Write the numbers 712, 4690, 5113, and 208 using that system. Demonstrate the system to the rest of your group.

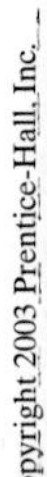

Chapter 1 VOCABULARY CHECK

Fill in each blank with one of the words or phrases listed below.

inequality symbols	opposites	absolute value	numerator	denominator
grouping symbols	exponent	base	reciprocals	variable
equation	solution			

1. The symbols $\neq$, $<$, and $>$ are called __________________.
2. A mathematical statement that two expressions are equal is called an __________.
3. The ________________ of a number is the distance between that number and 0 on the number line.
4. A symbol used to represent a number is called a __________.
5. Two numbers that are the same distance from 0 but lie on opposite sides of 0 are called ___________.
6. The number in a fraction above the fraction bar is called the ____________.
7. A __________ of an equation is a value for the variable that makes the equation a true statement.
8. Two numbers whose product is 1 are called ____________.
9. In 2^3, the 2 is called the _______ and the 3 is called the ___________.
10. The number in a fraction below the fraction bar is called the ______________.
11. Parentheses and brackets are examples of _________________.

CHAPTER 1 Highlights

DEFINITIONS AND CONCEPTS	**EXAMPLES**
Section 1.2 Symbols and Sets of Numbers	
A **set** is a collection of objects, called **elements**, enclosed in braces.	$\{a, c, e\}$
Natural numbers: $\{1, 2, 3, 4, \ldots\}$ **Whole numbers**: $\{0, 1, 2, 3, 4, \ldots\}$ **Integers**: $\{\ldots, -3, -2, -1, 0, 1, 2, 3, \ldots\}$ **Rational numbers**: {real numbers that can be expressed as a quotient of integers} **Irrational numbers**: {real numbers that cannot be expressed as a quotient of integers} **Real numbers**: {all numbers that correspond to a point on the number line}	Given the set $\left\{-3.4, \sqrt{3}, 0, \frac{2}{3}, 5, -4\right\}$ list the numbers that belong to the set of Natural numbers 5 Whole numbers $0, 5$ Integers $-4, 0, 5$ Rational numbers $-3.4, 0, \frac{2}{3}, 5, -4$ Irrational numbers $\sqrt{3}$ Real numbers $-3.4, \sqrt{3}, 0, \frac{2}{3}, 5, -4$
A line used to picture numbers is called a **number line**.	−5 −4 −3 −2 −1 0 1 2 3 4 5
The **absolute value** of a real number a denoted by $\lvert a \rvert$ is the distance between a and 0 on the number line.	$\lvert 5 \rvert = 5 \quad \lvert 0 \rvert = 0 \quad \lvert -2 \rvert = 2$
SYMBOLS: $=$ is equal to $\neq$ is not equal to $>$ is greater than $<$ is less than $\leq$ is less than or equal to $\geq$ is greater than or equal to	$-7 = -7$ $3 \neq -3$ $4 > 1$ $1 < 4$ $6 \leq 6$ $18 \geq -\frac{1}{3}$

Definitions and Concepts	**Examples**
Section 1.2 Symbols and Sets of Numbers (*continued*)	
Order Property for Real Numbers For any two real numbers a and b, a is less than b if a is to the left of b on the number line.	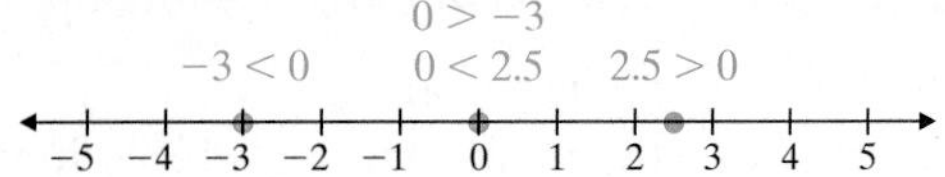
Section 1.3 Exponents, Order of Operations, and Variable Expressions	
The expression a^n is an **exponential expression**. The number a is called the **base**; it is the repeated factor. The number n is called the **exponent**; it is the number of times that the base is a factor.	$4^3 = 4 \cdot 4 \cdot 4 = 64$ $7^2 = 7 \cdot 7 = 49$
Order of Operations Simplify expressions in the following order. If grouping symbols are present, simplify expressions within those first, starting with the innermost set. Also, simplify the numerator and the denominator of a fraction separately. **1.** Simplify exponential expressions. **2.** Multiply or divide in order from left to right. **3.** Add or subtract in order from left to right.	$\frac{8^2 + 5(7-3)}{3 \cdot 7} = \frac{8^2 + 5(4)}{21}$ $= \frac{64 + 5(4)}{21}$ $= \frac{64 + 20}{21}$ $= \frac{84}{21}$ $= 4$
A symbol used to represent a number is called a **variable**.	Examples of variables are q, x, z
An **algebraic expression** is a collection of numbers, variables, operation symbols, and grouping symbols.	Examples of algebraic expressions are $5x,\quad 2(y-6),\quad \frac{q^2 - 3q + 1}{6}$
To **evaluate an algebraic expression** containing a variable, substitute a given number for the variable and simplify.	Evaluate $x^2 - y^2$ when $x = 5$ and $y = 3$. $x^2 - y^2 = (5)^2 - 3^2$ $= 25 - 9$ $= 16$
A mathematical statement that two expressions are equal is called an **equation**.	Equations: $3x - 9 = 20$ $A = \pi r^2$
A **solution** of an equation is a value for the variable that makes the equation a true statement.	Determine whether 4 is a solution of $5x + 7 = 27$. $5x + 7 = 27$ $5(4) + 7 \stackrel{?}{=} 27$ $20 + 7 \stackrel{?}{=} 27$ $27 = 27$ True 4 is a solution.
Section 1.4 Adding Real Numbers	
To Add Two Numbers with the Same Sign **1.** Add their absolute values. **2.** Use their common sign as the sign of the sum.	Add. $10 + 7 = 17$ $-3 + (-8) = -11$

DEFINITIONS AND CONCEPTS	EXAMPLES
Section 1.4 Adding Real Numbers (*continued*)	
TO ADD TWO NUMBERS WITH DIFFERENT SIGNS 1. Subtract their absolute values. 2. Use the sign of the number whose absolute value is larger as the sign of the sum.	$-25 + 5 = -20$ $14 + (-9) = 5$
Two numbers that are the same distance from 0 but lie on opposite sides of 0 are called **opposites** or **additive inverses**. The opposite of a number a is denoted by $-a$.	The opposite of -7 is 7. The opposite of 123 is -123.
Section 1.5 Subtracting Real Numbers	
To subtract two numbers a and b, add the first number a to the opposite of the second number, b. $a - b = a + (-b)$	Subtract. $3 - (-44) = 3 + 44 = 47$ $-5 - 22 = -5 + (-22) = -27$ $-30 - (-30) = -30 + 30 = 0$
Section 1.6 Multiplying and Dividing Real Numbers	
MULTIPLYING REAL NUMBERS The product of two numbers with the same sign is a positive number. The product of two numbers with different signs is a negative number.	Multiply. $7 \cdot 8 = 56$ $-7 \cdot (-8) = 56$ $-2 \cdot 4 = -8$ $2 \cdot (-4) = -8$
PRODUCTS INVOLVING ZERO The product of 0 and any number is 0. $b \cdot 0 = 0$ and $0 \cdot b = 0$	$-4 \cdot 0 = 0$ $0 \cdot \left(-\frac{3}{4}\right) = 0$
QUOTIENT OF TWO REAL NUMBERS $\frac{a}{b} = a \cdot \frac{1}{b}$	Divide. $\frac{42}{2} = 42 \cdot \frac{1}{2} = 21$
DIVIDING REAL NUMBERS The quotient of two numbers with the same sign is a positive number. The quotient of two numbers with different signs is a negative number.	$\frac{90}{10} = 9$ $\frac{-90}{-10} = 9$ $\frac{42}{-6} = -7$ $\frac{-42}{6} = -7$
QUOTIENTS INVOLVING ZERO The quotient of a nonzero number and 0 is undefined. $\frac{b}{0}$ is undefined. The quotient of 0 and any nonzero number is 0. $\frac{0}{b} = 0$	$\frac{-85}{0}$ is undefined. $\frac{0}{18} = 0$ $\frac{0}{-47} = 0$

DEFINITIONS AND CONCEPTS	**EXAMPLES**
Section 1.7 Properties of Real Numbers	
COMMUTATIVE PROPERTIES Addition: $a + b = b + a$ Multiplication: $a \cdot b = b \cdot a$	$3 + (-7) = -7 + 3$ $-8 \cdot 5 = 5 \cdot (-8)$
ASSOCIATIVE PROPERTIES Addition: $(a + b) + c = a + (b + c)$ Multiplication: $(a \cdot b) \cdot c = a \cdot (b \cdot c)$	$(5 + 10) + 20 = 5 + (10 + 20)$ $(-3 \cdot 2) \cdot 11 = -3 \cdot (2 \cdot 11)$
Two numbers whose product is 1 are called **multiplicative inverses** or **reciprocals**. The reciprocal of a nonzero number a is $\frac{1}{a}$ because $a \cdot \frac{1}{a} = 1$.	The reciprocal of 3 is $\frac{1}{3}$. The reciprocal of $-\frac{2}{5}$ is $-\frac{5}{2}$.
DISTRIBUTIVE PROPERTY $a(b + c) = a \cdot b + a \cdot c$	$5(6 + 10) = 5 \cdot 6 + 5 \cdot 10$ $-2(3 + x) = -2 \cdot 3 + (-2)(x)$
IDENTITIES $a + 0 = a$ $\quad 0 + a = a$ $a \cdot 1 = a$ $\quad 1 \cdot a = a$ **INVERSES** Additive or opposite: $a + (-a) = 0$ Multiplicative or reciprocal: $b \cdot \frac{1}{b} = 1, \quad b \neq 0$	$5 + 0 = 5$ $\quad 0 + (-2) = -2$ $-14 \cdot 1 = -14$ $\quad 1 \cdot 27 = 27$ $7 + (-7) = 0$ $3 \cdot \frac{1}{3} = 1$

STUDY SKILLS REMINDER

Are you preparing for a test on Chapter 1?

Below I have listed some *common trouble areas* for students in Chapter 1. After studying for your test—but before taking your test—read these.

- Do you know the difference between $|-3|$, $-|-3|$, and $-(-3)$?

  $|-3| = 3;\quad -|-3| = -3;\quad$ and $\quad -(-3) = 3$ (Section 1.2)

- Evaluate $x - y$ if $x = 7$ and $y = -3$.

  $x - y = 7 - (-3) = 10$ (Section 1.3)

- Make sure you are familiar with order of operations. Sometimes the simplest-looking expressions can give you the most trouble.

  $1 + 2(3 + 6) = 1 + 2(9) = 1 + 18 = 19$ (Section 1.3)

- Do you know the difference between $(-3)^2$ and -3^2?

  $(-3)^2 = 9$ and $-3^2 = -9$ (Section 1.6)

- Do you know that these fractions are equivalent?

  $-\frac{1}{3} = \frac{-1}{3} = \frac{1}{-3}$ (Section 1.6)

Remember: This is simply a checklist of selected topics given to check your understanding. For a review of Chapter 1 in the text, see the material at the end of Chapter 1.

Name ______________ Section ________ Date __________

Chapter 1 Review

(1.2) *Insert $<$, $>$, or $=$ in the appropriate space to make each statement true.*

1. $8 \quad 10$ **2.** $7 \quad 2$ **3.** $-4 \quad -5$ **4.** $\frac{12}{2} \quad -8$

5. $|-7| \quad |-8|$ **6.** $|-9| \quad -9$ **7.** $-|-1| \quad -1$ **8.** $|-14| \quad -(-14)$

9. $1.2 \quad 1.02$ **10.** $-\frac{3}{2} \quad -\frac{3}{4}$

Translate each statement into symbols.

11. Four is greater than or equal to negative three.

12. Six is not equal to five.

13. 0.03 is less than 0.3.

14. New York City has 155 museums and 400 art galleries. Write an inequality statement comparing the numbers 155 and 400. (*Source*: Absolute Trivia.com)

Given the sets of numbers below, list the numbers in each set that also belong to the set of:

a. Natural numbers **b.** Whole numbers
c. Integers **d.** Rational numbers
e. Irrational numbers **f.** Real numbers

15. $\left\{-6, 0, 1, 1\frac{1}{2}, 3, \pi, 9.62\right\}$

16. $\left\{-3, -1.6, 2, 5, \frac{11}{2}, 15.1, \sqrt{5}, 2\pi\right\}$

The following chart shows the gains and losses in dollars of Density Oil and Gas stock for a particular week. Use this chart to answer Exercises 17 and 18.

Day	Gain or Loss (in dollars)
Monday	+1
Tuesday	−2
Wednesday	+5
Thursday	+1
Friday	−4

17. Which day showed the greatest loss?

18. Which day showed the greatest gain?

(1.3) *Choose the correct answer for each statement.*

19. The expression $6 \cdot 3^2 + 2 \cdot 8$ simplifies to
a. -52 **b.** 440 **c.** 70 **d.** 64

20. The expression $68 - 5 \cdot 2^3$ simplifies to
a. -232 **b.** 28 **c.** 38 **d.** 504

Simplify each expression.

21. $3(1 + 2 \cdot 5) + 4$

22. $8 + 3(2 \cdot 6 - 1)$

23. $\dfrac{4 + |6 - 2| + 8^2}{4 + 6 \cdot 4}$

24. $5[3(2 + 5) - 5]$

Translate each word statement to symbols.

25. The difference of twenty and twelve is equal to the product of two and four.

26. The quotient of nine and two is greater than negative five.

Evaluate each expression when $x = 6$, $y = 2$, and $z = 8$.

27. $2x + 3y$

28. $x(y + 2z)$

29. $\dfrac{x}{y} + \dfrac{z}{2y}$

30. $x^2 - 3y^2$

△ **31.** The expression $180 - a - b$ represents the measure of the unknown angle of the given triangle. Replace a with 37 and b with 80 to find the measure of the unknown angle.

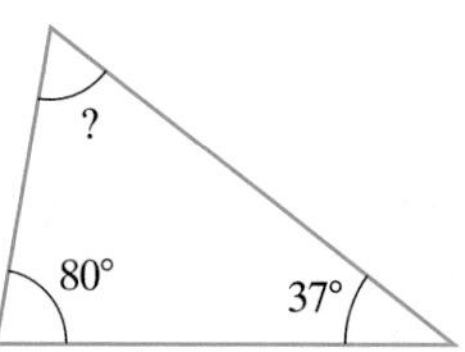

Decide whether the given number is a solution to the given equation.

32. $7x - 3 = 18; 3$

33. $3x^2 + 4 = x - 1; 1$

(1.4) *Find the additive inverse or opposite of each number.*

34. -9

35. $\dfrac{2}{3}$

36. $|-2|$

37. $-|-7|$

Add.

38. $-15 + 4$

39. $-6 + (-11)$

40. $\dfrac{1}{16} + \left(-\dfrac{1}{4}\right)$

41. $-8 + |-3|$

42. $-4.6 + (-9.3)$

43. $-2.8 + 6.7$

(1.5) *Perform each indicated operation.*

44. $6 - 20$

45. $-3.1 - 8.4$

46. $-6 - (-11)$

47. $4 - 15$

48. $-21 - 16 + 3(8 - 2)$

49. $\dfrac{11 - (-9) + 6(8 - 2)}{2 + 3 \cdot 4}$

Evaluate each expression for $x = 3$, $y = -6$, and $z = -9$. Then choose the correct evaluation.

50. $2x^2 - y + z$

a. 15 **b.** 3 **c.** 27 **d.** -3

51. $\dfrac{y - 4x}{2x}$

a. 3 **b.** 1 **c.** -1 **d.** -3

52. At the beginning of the week the price of Density Oil and Gas stock from Exercises 17 and 18 is $50 per share. Find the price of a share of stock at the end of the week.

Find each multiplicative inverse or reciprocal.

53. -6

54. $\dfrac{3}{5}$

(1.6) *Simplify each expression.*

55. $6(-8)$

56. $(-2)(-14)$

57. $\dfrac{-18}{-6}$

58. $\dfrac{42}{-3}$

59. $-3(-6)(-2)$

60. $(-4)(-3)(0)(-6)$

61. $\dfrac{4 \cdot (-3) + (-8)}{2 + (-2)}$

62. $\dfrac{3(-2)^2 - 5}{-14}$

(1.7) *Name the property illustrated in each equation.*

63. $-6 + 5 = 5 + (-6)$

64. $6 \cdot 1 = 6$

65. $3(8 - 5) = 3 \cdot 8 + 3 \cdot (-5)$

66. $4 + (-4) = 0$

67. $2 + (3 + 9) = (2 + 3) + 9$

68. $2 \cdot 8 = 8 \cdot 2$

69. $6(8 + 5) = 6 \cdot 8 + 6 \cdot 5$

70. $(3 \cdot 8) \cdot 4 = 3 \cdot (8 \cdot 4)$

71. $4 \cdot \frac{1}{4} = 1$

72. $8 + 0 = 8$

73. $4(8 + 3) = 4(3 + 8)$

Name ______________________ Section __________ Date __________

Chapter 1 Test

Translate each statement into symbols.

1. The absolute value of negative seven is greater than five.

2. The sum of nine and five is greater than or equal to four.

Simplify each expression.

3. $-13 + 8$

4. $-13 - (-2)$

5. $6 \cdot 3 - 8 \cdot 4$

6. $(13)(-3)$

7. $(-6)(-2)$

8. $\dfrac{|-16|}{-8}$

9. $\dfrac{-8}{0}$

10. $\dfrac{|-6| + 2}{5 - 6}$

11. $\dfrac{1}{2} - \dfrac{5}{6}$

12. $-1\dfrac{1}{8} + 5\dfrac{3}{4}$

13. $-\dfrac{3}{5} + \dfrac{15}{8}$

14. $3(-4)^2 - 80$

15. $6[5 + 2(3 - 8) - 3]$

16. $\dfrac{-12 + 3 \cdot 8}{4}$

17. $\dfrac{(-2)(0)(-3)}{-6}$

Insert <, >, or = in the appropriate space to make each statement true.

18. $-3 \quad -7$

19. $4 \quad -8$

20. $|-3| \quad 2$

21. $|-2| \quad -1 - (-3)$

22. Given $\left\{-5, -1, \frac{1}{4}, 0, 1, 7, 11.6, \sqrt{7}, 3\pi\right\}$, list the numbers in this set that also belong to the set of:

a. Natural numbers

b. Whole numbers

c. Integers

d. Rational numbers

e. Irrational numbers

f. Real numbers

Evaluate each expression when $x = 6$, $y = -2$, and $z = -3$.

23. $x^2 + y^2$

24. $x + yz$

25. $2 + 3x - y$

26. $\dfrac{y + z - 1}{x}$

Answers

1. ______
2. ______
3. ______
4. ______
5. ______
6. ______
7. ______
8. ______
9. ______
10. ______
11. ______
12. ______
13. ______
14. ______
15. ______
16. ______
17. ______
18. ______
19. ______
20. ______
21. ______
22. a. ______
b. ______
c. ______
d. ______
e. ______
f. ______
23. ______
24. ______
25. ______
26. ______

27. ____________

28. ____________

29. ____________

30. ____________

31. ____________

32. ____________

33. ____________

34. ____________

35. ____________

36. ____________

Identify the property illustrated by each expression.

27. $8 + (9 + 3) = (8 + 9) + 3$

28. $6 \cdot 8 = 8 \cdot 6$

29. $-6(2 + 4) = -6 \cdot 2 + (-6) \cdot 4$

30. $\frac{1}{6}(6) = 1$

31. Find the opposite of -9.

32. Find the reciprocal of $-\frac{1}{3}$.

The New Orleans Saints were 22 yards from the goal when the series of gains and losses shown in the chart occurred. Use this chart to answer Questions 33 and 34.

	Gains and Losses (in yards)
First down	5
Second down	−10
Third down	−2
Fourth down	29

33. During which down did the greatest loss of yardage occur?

34. Was a touchdown scored?

35. The temperature at the Winter Olympics was a frigid 14° below zero in the morning, but by noon it had risen 31°. What was the temperature at noon?

36. Jean Avarez decided to sell 280 shares of stock, which decreased in value by \$1.50 per share yesterday. How much money did she lose?

Equations, Inequalities, and Problem Solving

In this chapter, we solve equations and inequalities. Once we know how to solve equations and inequalities, we may solve word problems. Of course, problem solving is an integral topic in algebra and its discussion is continued throughout this text.

Crayola crayons have a long history starting in 1903. In that year, cousins Edwin Binney and C. Harold Smith produced their first Crayola product: a box of eight petroleum-based paraffin crayons, in the colors red, orange, yellow, green, blue, violet, brown, and black, that sold for 5¢. The "Crayola" brand name was coined by Edwin's wife, Alice, by combining the French word *craie* (chalk) with *ola* (short for oleaginous). Today, the Crayola brand name is recognized by 99% of Americans, and Crayola crayons are available in 120 colors. In Exercise 17 on page 128 (Section 2.5), you will find the number of votes cast for America's two favorite Crayola crayon colors.

Answers

1. ______
2. ______
3. ______
4. ______
5. ______
6. ______
7. ______
8. ______
9. ______
10. ______
11. ______
12. ______
13. ______
14. ______
15. ______
16. ______
17. ______
18. ______
19. ______
20. ______

Name ______ Section ______ Date ______

Chapter 2 Pretest

Simplify.

1. $3c - 4 + 6c - 9$

2. $-5(2y - 3) - 7y + 1$

Solve.

3. $3 - x = -12$

4. $12 - (5 - 4b) = 9 + 3b$

5. $\frac{2}{3}m = -8$

6. $-7 - 3y = 17 + 5y$

7. $3(1 - 4x) + 2(5x) = 9$

8. $0.20x + 0.15(60) = 0.75(18)$

9. $2(x - 1) = 2x + 5$

10. Three times the sum of a number and -2 is the same as 2 more than the number. Find the number.

11. Find two consecutive even integers such that three times the smaller is 16 more than twice the larger.

12. Substitute the given values into the given formula and solve for the unknown variable.

$V = \frac{1}{3}Ah; V = 60, h = 4$

△ **13.** If the area of a right-triangularly shaped sign is 18 square feet and its height is 4 feet, find the base of the sign.

14. Solve the given formula for the specified variable.

$2x + y = 8$ for y

15. What number is 22% of 90?

16. Write the ratio "4 quarts to 5 gallons" in fractional notation in lowest terms.

17. Solve the following proportion.

$\frac{3x}{8} = \frac{9}{7}$

Solve each inequality. Graph the solutions.

18. $-4 + x \leq 2$

−10 −8 −6 −4 −2 0 2 4 6 8 10

19. $-\frac{3}{2}y > 6$

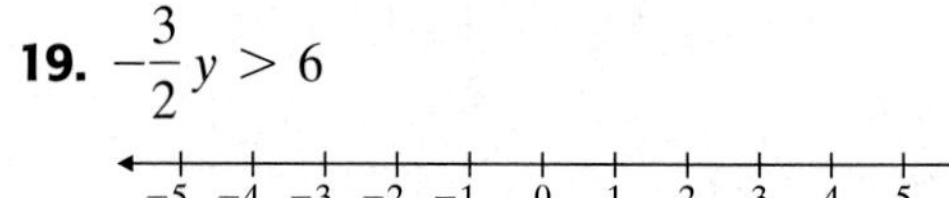

20. $-5x + 3 \leq 4(x - 6)$

−5 −4 −3 −2 −1 0 1 2 3 4 5

2.1 Simplifying Expressions

As we explore in this section, we will see that an expression such as $3x + 2x$ is not written as simply as possible. This is because—even without replacing x by a value—we can perform the indicated addition.

OBJECTIVES

A Identify terms, like terms, and unlike terms.

B Combine like terms.

C Simplify expressions containing parentheses.

D Write word phrases as algebraic expressions.

A Identifying Terms, Like Terms, and Unlike Terms

Before we practice simplifying expressions, we must learn some new language. A **term** is a number or the product of a number and variables raised to powers.

Terms

$$-y,\quad 2x^3,\quad -5,\quad 3xz^2,\quad \frac{2}{y},\quad 0.8z$$

The **numerical coefficient** of a term is the numerical factor. The numerical coefficient of $3x$ is 3. Recall that $3x$ means $3 \cdot x$.

Term	Numerical Coefficient
$3x$	3
$\frac{y^3}{5}$	$\frac{1}{5}$ since $\frac{y^3}{5}$ means $\frac{1}{5} \cdot y^3$
$-0.7ab^3c^5$	-0.7
z	1
$-y$	-1
-5	-5

Helpful Hint

The term $-y$ means $-1y$ and thus has a numerical coefficient of -1.
The term z means $1z$ and thus has a numerical coefficient of 1.

EXAMPLE 1 Identify the numerical coefficient in each term.

a. $-3y$ **b.** $22z^4$ **c.** y **d.** $-x$ **e.** $\frac{x}{7}$

Solution:

a. The numerical coefficient of $-3y$ is -3.
b. The numerical coefficient of $22z^4$ is 22.
c. The numerical coefficient of y is 1, since y is $1y$.
d. The numerical coefficient of $-x$ is -1, since $-x$ is $-1x$.
e. The numerical coefficient of $\frac{x}{7}$ is $\frac{1}{7}$, since $\frac{x}{7}$ is $\frac{1}{7} \cdot x$.

Practice Problem 1

Identify the numerical coefficient in each term.

a. $-4x$ b. $15y^3$ c. x

d. $-y$ e. $\frac{z}{4}$

Terms with the same variables raised to exactly the same powers are called **like terms**. Terms that aren't like terms are called **unlike terms**.

Like Terms	Unlike Terms	
$3x, 2x$	$5x, 5x^2$	Why? Same variable x, but different powers of x and x^2
$-6x^2y, 2x^2y, 4x^2y$	$7y, 3z, 8x^2$	Why? Different variables
$2ab^2c^3, ac^3b^2$	$6abc^3, 6ab^2$	Why? Different variables and different powers

Answers

1. a. -4, **b.** 15, **c.** 1, **d.** -1, **e.** $\frac{1}{4}$

Helpful Hint

In like terms, each variable and its exponent must match exactly, but these factors don't need to be in the same order.

$2x^2y$ and $3yx^2$ are like terms.

Practice Problem 2

Determine whether the terms are like or unlike.

a. $7x, -6x$ b. $3x^2y^2, -x^2y^2, 4x^2y^2$
c. $-5ab, 3ba$

EXAMPLE 2 Determine whether the terms are like or unlike.

a. $2x, 3x^2$ **b.** $4x^2y, x^2y, -2x^2y$ **c.** $-2yz, -3zy$ **d.** $-x^4, x^4$

Solution:

a. Unlike terms, since the exponents on x are not the same.
b. Like terms, since each variable and its exponent match.
c. Like terms, since $zy = yz$ by the commutative property.
d. Like terms.

B Combining Like Terms

An algebraic expression containing the sum or difference of like terms can be simplified by applying the distributive property. For example, by the distributive property, we rewrite the sum of the like terms $3x + 2x$ as

$$3x + 2x = (3 + 2)x = 5x$$

Also,

$$-y^2 + 5y^2 = (-1 + 5)y^2 = 4y^2$$

Simplifying the sum or difference of like terms is called **combining like terms.**

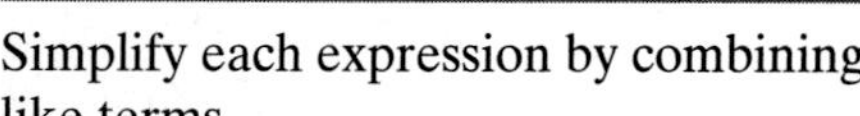

Practice Problem 3

Simplify each expression by combining like terms.

a. $9y - 4y$ b. $11x^2 + x^2$
c. $5y - 3x + 6x$

EXAMPLE 3 Simplify each expression by combining like terms.

a. $7x - 3x$ **b.** $10y^2 + y^2$ **c.** $8x^2 + 2x - 3x$

Solution:

a. $7x - 3x = (7 - 3)x = 4x$
b. $10y^2 + y^2 = (10 + 1)y^2 = 11y^2$
c. $8x^2 + 2x - 3x = 8x^2 + (2 - 3)x = 8x^2 - x$

Practice Problems 4–7

Simplify each expression by combining like terms.

4. $7y + 2y + 6 + 10$
5. $-2x + 4 + x - 11$
6. $3z - 3z^2$
7. $8.9y + 4.2y - 3$

EXAMPLES Simplify each expression by combining like terms.

4. $2x + 3x + 5 + 2 = (2 + 3)x + (5 + 2)$
$= 5x + 7$

5. $-5a - 3 + a + 2 = -5a + 1a + (-3 + 2)$
$= (-5 + 1)a + (-3 + 2)$
$= -4a - 1$

6. $4y - 3y^2$ These two terms cannot be combined because they are unlike terms.

7. $2.3x + 5x - 6 = (2.3 + 5)x - 6$
$= 7.3x - 6$

The examples above suggest the following.

Answers

2. a. like, **b.** like, **c.** like,
3. a. $5y$, **b.** $12x^2$, **c.** $5y + 3x$,
4. $9y + 16$, **5.** $-x - 7$, **6.** $3z - 3z^2$,
7. $13.1y - 3$

Combining Like Terms

To **combine like terms**, combine the numerical coefficients and multiply the result by the common variable factors.

C Simplifying Expressions Containing Parentheses

In simplifying expressions we make frequent use of the distributive property to remove parentheses.

EXAMPLES

Find each product by using the distributive property to remove parentheses.

8. $5(x + 2) = 5(x) + 5(2)$ Apply the distributive property.
$= 5x + 10$ Multiply.

9. $-2(y + 0.3z - 1) = -2(y) + (-2)(0.3z) + (-2)(-1)$ Apply the distributive property.
$= -2y - 0.6z + 2$ Multiply.

10. $-(x + y - 2z + 6) = -1(x + y - 2z + 6)$ Distribute -1 over each term.
$= -1(x) - 1(y) - 1(-2z) - 1(6)$
$= -x - y + 2z - 6$

Helpful Hint

If a "$-$" sign precedes parentheses, the sign of each term inside the parentheses is changed when the distributive property is applied to remove the parentheses.

Examples:

$-(2x + 1) = -2x - 1$

$-(x - 2y) = -x + 2y$

$-(-5x + y - z) = 5x - y + z$

$-(-3x - 4y - 1) = 3x + 4y + 1$

When simplifying an expression containing parentheses, we often use the distributive property first to remove parentheses and then again to combine any like terms.

EXAMPLES Simplify each expression.

11. $3(2x - 5) + 1 = 6x - 15 + 1$ Apply the distributive property.
$= 6x - 14$ Combine like terms.

12. $8 - (7x + 2) + 3x = 8 - 7x - 2 + 3x$ Apply the distributive property.
$= -7x + 3x + 8 - 2$
$= -4x + 6$ Combine like terms.

13. $-2(4x + 7) - (3x - 1) = -8x - 14 - 3x + 1$ Apply the distributive property.
$= -11x - 13$ Combine like terms.

Practice Problems 8–10

Find each product by using the distributive property to remove parentheses.

8. $3(y + 6)$
9. $-4(x + 0.2y - 3)$
10. $-(3x + 2y + z - 1)$

Practice Problems 11–14

Simplify each expression.

11. $4(x - 6) + 20$
12. $5 - (3x + 9)$
13. $-3(7x + 1) - (4x - 2)$
14. $8 + 11(2y - 9)$

Answers

8. $3y + 18$, **9.** $-4x - 0.8y + 12$, **10.** $-3x - 2y - z + 1$, **11.** $4x - 4$, **12.** $-3x - 4$, **13.** $-25x - 1$, **14.** $-91 + 22y$

Helpful Hint

Don't forget to use the distributive property and multiply before adding or subtracting like terms.

14. $9 + 3(4x - 10) = 9 + 12x - 30$ Apply the distributive property.
$= -21 + 12x$ Combine like terms.

Practice Problem 15

Subtract $9x - 10$ from $4x - 3$.

EXAMPLE 15 Subtract $4x - 2$ from $2x - 3$.

Solution: We first note that "subtract $4x - 2$ **from** $2x - 3$" translates to $(2x - 3) - (4x - 2)$. Next, we simplify the algebraic expression.

$(2x - 3) - (4x - 2) = 2x - 3 - 4x + 2$ Apply the distributive property.
$= -2x - 1$ Combine like terms.

D Writing Algebraic Expressions

To prepare for problem solving, we next practice writing word phrases as algebraic expressions.

Practice Problems 16–18

Write each phrase as an algebraic expression and simplify if possible. Let x represent the unknown number.

16. Three times a number, *subtracted from* 10
17. The sum of a number and 2, divided by 5
18. Three times a number, added to the sum of twice a number and 6

EXAMPLES

Write each phrase as an algebraic expression and simplify if possible. Let x represent the unknown number.

16. Twice a number, plus 6

$2x \quad + \quad 6$

This expression cannot be simplified.

17. The difference of a number and 4, divided by 7

$(x - 4) \quad \div \quad 7$

This expression cannot be simplified.

18. Five plus the sum of a number and 1

$5 \quad + \quad (x + 1)$

Next, we simplify this expression.

$5 + (x + 1) = 5 + x + 1$
$= 6 + x$

Answers

15. $-5x + 7$, **16.** $10 - 3x$, **17.** $\frac{(x + 2)}{5}$, **18.** $5x + 6$

Name ______________________ Section __________ Date __________

Mental Math

 *Identify the numerical coefficient of each term. See Example 1.*

1. $-7y$ **2.** $3x$ **3.** x **4.** $-y$ **5.** $17x^2y$ **6.** $1.2xyz$

Indicate whether the terms in each list are like or unlike. See Example 2.

7. $5y, -y$ **8.** $-2x^2y, 6xy$ **9.** $2z, 3z^2$

10. $ab^2, -7ab^2$ **11.** $8wz, \frac{1}{7}zw$ **12.** $7.4p^3q^2, 6.2p^3q^2r$

EXERCISE SET 2.1

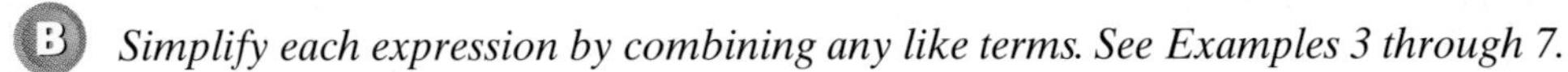 *Simplify each expression by combining any like terms. See Examples 3 through 7.*

1. $7y + 8y$ **2.** $3x + 2x$ **3.** $8w - w + 6w$

4. $c - 7c + 2c$ **5.** $3b - 5 - 10b - 4$ **6.** $6g + 5 - 3g - 7$

7. $m - 4m + 2m - 6$ **8.** $a + 3a - 2 - 7a$ **9.** $5g - 3 - 5 - 5g$

10. $8p + 4 - 8p - 15$ **11.** $6.2x - 4 + x - 1.2$ **12.** $7.9y - 0.7 - y + 0.2$

13. $2k - k - 6$ **14.** $7c - 8 - c$ **15.** $-9x + 4x + 18 - 10x$

16. $5y - 14 + 7y - 20y$ **17.** $6x - 5x + x - 3 + 2x$ **18.** $8h + 13h - 6 + 7h - h$

19. $7x^2 + 8x^2 - 10x^2$ **20.** $8x^3 + x^3 - 11x^3$ **21.** $3.4m - 4 - 3.4m - 7$

22. $2.8w - 0.9 - 0.5 - 2.8w$

23. $6x + 0.5 - 4.3x - 0.4x + 3$

24. $0.4y - 6.7 + y - 0.3 - 2.6y$

C *Simplify each expression. Use the distributive property to remove any parentheses. See Examples 8 through 10.*

25. $5(y + 4)$

26. $7(r + 3)$

27. $-2(x + 2)$

28. $-4(y + 6)$

29. $-5(2x - 3y + 6)$

30. $-2(4x - 3z - 1)$

31. $-(3x - 2y + 1)$

32. $-(y + 5z - 7)$

Remove parentheses and simplify each expression. See Examples 11 through 14.

33. $7(d - 3) + 10$

34. $9(z + 7) - 15$

35. $-4(3y - 4) + 12y$

36. $-3(2x + 5) - 6x$

37. $3(2x - 5) - 5(x - 4)$

38. $2(6x - 1) - (x - 7)$

39. $-2(3x - 4) + 7x - 6$

40. $8y - 2 - 3(y + 4)$

41. $5k - (3k - 10)$

42. $-11c - (4 - 2c)$

43. $(3x + 4) - (6x - 1)$

44. $(8 - 5y) - (4 + 3y)$

45. $5(x + 2) - (3x - 4)$

46. $4(2x - 3) - (x + 1)$

47. $-3(7y - 1) + 4(4y + 7)$

48. $-5(6y + 2) + 5(2y - 1)$

49. $2 + 4(6x - 6)$

50. $8 + 4(3x - 4)$

51. $0.5(m + 2) + 0.4m$

52. $0.2(k + 8) - 0.1k$

53. $10 - 3(2x + 3y)$

54. $14 - 11(5m + 3n)$

55. $6(3x - 6) - 2(x + 1) - 17x$

56. $7(2x + 5) - 4(x + 2) - 20x$

57. $\frac{1}{2}(12x - 4) - (x + 5)$

58. $\frac{1}{3}(9x - 6) - (x - 2)$

59. In your own words, explain how to combine like terms.

60. Do like terms contain the same numerical coefficients? Explain your answer.

Perform each indicated operation. Don't forget to simplify if possible. See Example 15.

61. Add $6x + 7$ to $4x - 10$

62. Add $3y - 5$ to $y + 16$

63. Subtract $7x + 1$ from $3x - 8$

64. Subtract $4x - 7$ from $12 + x$

65. Subtract $5m - 6$ from $m - 9$

66. Subtract $m - 3$ from $2m - 6$

D *Write each phrase as an algebraic expression and simplify if possible. Let x represent the unknown number. See Examples 16 through 18.*

67. Twice a number, decreased by four

68. The difference of a number and two, divided by five

69. Three-fourths of a number, increased by twelve

70. Eight more than triple a number

71. The sum of 5 times a number and -2, added to 7 times the number

72. The sum of 3 times a number and 10, **subtracted from** 9 times the number

73. Eight times the sum of a number and six

74. Five, subtracted from four times a number

75. Double a number minus the sum of the number and ten

76. Half a number minus the product of the number and eight

Review and Preview

Evaluate each expression for the given values. See Section 1.6.

77. If $x = -1$ and $y = 3$, find $y - x^2$

78. If $g = 0$ and $h = -4$, find $gh - h^2$

79. If $a = 2$ and $b = -5$, find $a - b^2$

80. If $x = -3$, find $x^3 - x^2 + 4$

81. If $y = -5$ and $z = 0$, find $yz - y^2$

82. If $x = -2$, find $x^3 - x^2 - x$

Combining Concepts

Given the following information, determine whether each scale is balanced or not.

1 cone balances 1 cube

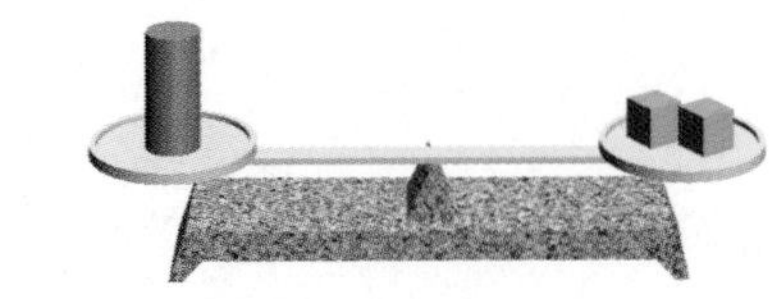

1 cylinder balances 2 cubes

83.

84.

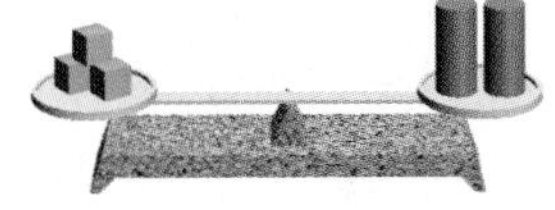

85.

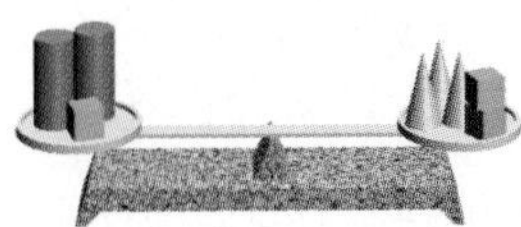

86.

Write each algebraic expression described.

△ **87.** Recall that the perimeter of a figure is the total distance around the figure. Given the following rectangle, express the perimeter as an algebraic expression containing the variable x.

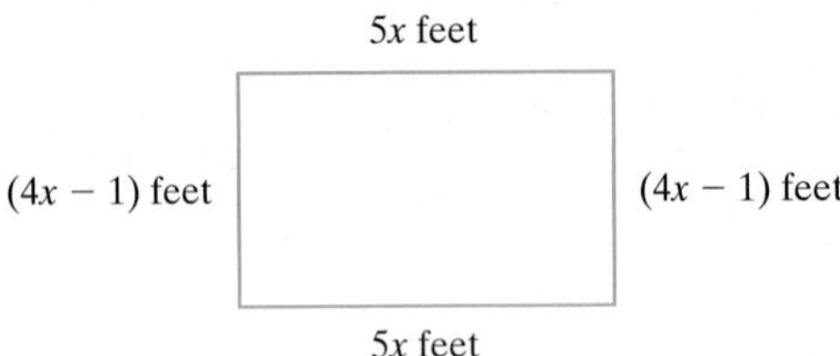

△ **88.** Given the following triangle, express its perimeter as an algebraic expression containing the variable x.

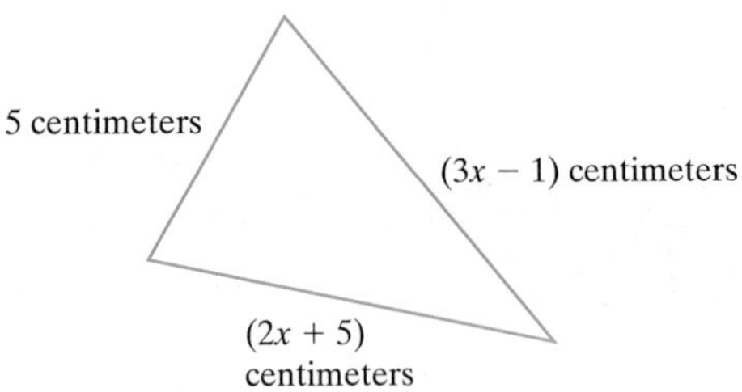

△ **89.** To convert from feet to inches, we multiply by 12. For example, the number of inches in 2 feet is $12 \cdot 2$ inches. If one board has a length of $(x + 2)$ *feet* and a second board has a length of $(3x - 1)$ *inches*, express their total length in inches as an algebraic expression.

90. The value of 7 nickels is $5 \cdot 7$ cents. Likewise, the value of x nickels is $5x$ cents. If the money box in a drink machine contains x *nickels*, $3x$ *dimes*, and $(30x - 1)$ *quarters*, express their total value in cents as an algebraic expression.

2.2 The Addition Property of Equality

OBJECTIVES

- A Use the addition property of equality to solve linear equations.
- B Simplify an equation and then use the addition property of equality.
- C Write word phrases as algebraic expressions.

A Using the Addition Property

Recall from Section 1.3 that an equation is a statement in which two expressions have the same value. Also, a value of the variable that makes an equation a true statement is called a solution or root of the equation. The process of finding the solution of an equation is called **solving** the equation for the variable. In this section, we concentrate on solving *linear equations* in one variable.

Linear Equation in One Variable

A **linear equation in one variable** can be written in the form

$$Ax + B = C$$

where A, B, and C are real numbers and $A \neq 0$.

Evaluating a linear equation for a given value of the variable, as we did in Section 1.3, can tell us whether that value is a solution. But we can't rely on evaluating an equation as our method of solving it—with what value would we start?

Instead, to solve a linear equation in x, we write a series of simpler equations, all *equivalent* to the original equation, so that the final equation has the form

$$x = \text{number} \qquad \text{or} \qquad \text{number} = x$$

Equivalent equations are equations that have the same solution. This means that the "number" above is the solution to the original equation.

The first property of equality that helps us write simpler equivalent equations is the **addition property of equality**.

Addition Property of Equality

If a, b, and c are real numbers, then

$$a = b \qquad \text{and} \qquad a + c = b + c$$

are equivalent equations.

This property guarantees that adding the same number to both sides of an equation does not change the solution of the equation. Since subtraction is defined in terms of addition, we may also **subtract the same number from both sides** without changing the solution.

A good way to picture a true equation is as a balanced scale. Since it is balanced, each side of the scale weighs the same amount.

If the same weight is added to or subtracted from each side, the scale remains balanced.

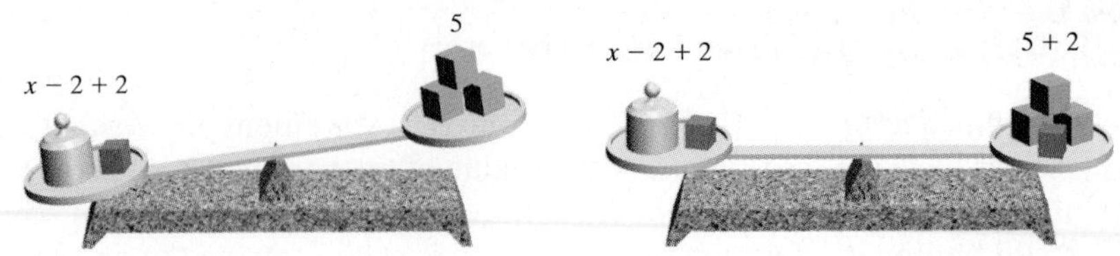

We use the addition property of equality to write equivalent equations until the variable is alone (by itself on one side of the equation) and the equation looks like "x = number" or "number = x."

Try the Concept Check in the margin.

Concept Check

Use the addition property to fill in the blank so that the middle equation simplifies to the last equation.

$$\begin{aligned} x - 5 &= 3 \\ x - 5 + \_\_ &= 3 + \_\_ \\ x &= 8 \end{aligned}$$

EXAMPLE 1 Solve $x - 7 = 10$ for x.

Solution: To solve for x, we first get x alone on one side of the equation. To do this, we add 7 to both sides of the equation.

$$\begin{aligned} x - 7 &= 10 \\ x - 7 + 7 &= 10 + 7 && \text{Add 7 to both sides.} \\ x &= 17 && \text{Simplify.} \end{aligned}$$

The solution of the equation $x = 17$ is obviously 17.
Since we are writing equivalent equations, the solution of the equation $x - 7 = 10$ is also 17.

Check: To check, replace x with 17 in the original equation.

$$\begin{aligned} x - 7 &= 10 && \text{Original equation.} \\ 17 - 7 &\stackrel{?}{=} 10 && \text{Replace } x \text{ with 17.} \\ 10 &= 10 && \text{True} \end{aligned}$$

Since the statement is true, 17 is the solution.

Practice Problem 1

Solve $x - 5 = 8$ for x.

EXAMPLE 2 Solve: $y + 0.6 = -1.0$

Solution: To solve for y, we subtract 0.6 from both sides of the equation.

$$\begin{aligned} y + 0.6 &= -1.0 \\ y + 0.6 - 0.6 &= -1.0 - 0.6 && \text{Subtract 0.6 from both sides.} \\ y &= -1.6 && \text{Combine like terms.} \end{aligned}$$

Check:

$$\begin{aligned} y + 0.6 &= -1.0 && \text{Original equation.} \\ -1.6 + 0.6 &\stackrel{?}{=} -1.0 && \text{Replace } y \text{ with } -1.6. \\ -1.0 &= -1.0 && \text{True} \end{aligned}$$

The solution is -1.6.

Practice Problem 2

Solve: $y + 1.7 = 0.3$

Answers

1. $x = 13$, **2.** $y = -1.4$

Concept Check: 5

EXAMPLE 3 Solve: $\frac{1}{2} = x - \frac{3}{4}$

Solution: To get x alone, we add $\frac{3}{4}$ to both sides.

$$\frac{1}{2} = x - \frac{3}{4}$$

$$\frac{1}{2} + \frac{3}{4} = x - \frac{3}{4} + \frac{3}{4} \quad \text{Add } \frac{3}{4} \text{ to both sides.}$$

$$\frac{1}{2} \cdot \frac{2}{2} + \frac{3}{4} = x \quad \text{The LCD is 4.}$$

$$\frac{2}{4} + \frac{3}{4} = x \quad \text{Add the fractions.}$$

$$\frac{5}{4} = x$$

Check:

$$\frac{1}{2} = x - \frac{3}{4} \quad \text{Original equation.}$$

$$\frac{1}{2} \stackrel{?}{=} \frac{5}{4} - \frac{3}{4} \quad \text{Replace } x \text{ with } \frac{5}{4}.$$

$$\frac{1}{2} \stackrel{?}{=} \frac{2}{4} \quad \text{Subtract.}$$

$$\frac{1}{2} = \frac{1}{2} \quad \text{True}$$

The solution is $\frac{5}{4}$.

Practice Problem 3

Solve: $\frac{7}{8} = y - \frac{1}{3}$

Helpful Hint

We may solve an equation so that the variable is alone on *either* side of the equation. For example, $\frac{5}{4} = x$ is equivalent to $x = \frac{5}{4}$.

EXAMPLE 4 Solve: $5t - 5 = 6t$

Solution: To solve for t, we first want all terms containing t on one side of the equation. To do this, we subtract $5t$ from both sides of the equation.

$$5t - 5 = 6t$$

$$5t - 5 - 5t = 6t - 5t \quad \text{Subtract } 5t \text{ from both sides.}$$

$$-5 = t \quad \text{Combine like terms.}$$

Check:

$$5t - 5 = 6t \quad \text{Original equation.}$$

$$5(-5) - 5 \stackrel{?}{=} 6(-5) \quad \text{Replace } t \text{ with } -5.$$

$$-25 - 5 \stackrel{?}{=} -30$$

$$-30 = -30 \quad \text{True}$$

The solution is -5.

Practice Problem 4

Solve: $3x + 10 = 4x$

Answers

3. $y = \frac{29}{24}$, **4.** $x = 10$

B Simplifying Equations

Many times, it is best to simplify one or both sides of an equation before applying the addition property of equality.

Practice Problem 5

Solve:
$10w + 3 - 4w + 4 = -2w + 3 + 7w$

EXAMPLE 5 Solve: $2x + 3x - 5 + 7 = 10x + 3 - 6x - 4$

Solution: First we simplify both sides of the equation.

$$2x + 3x - 5 + 7 = 10x + 3 - 6x - 4$$

$$5x + 2 = 4x - 1 \quad \text{Combine like terms on each side of the equation.}$$

Next, we want all terms with a variable on one side of the equation and all numbers on the other side.

$$5x + 2 - 4x = 4x - 1 - 4x \quad \text{Subtract } 4x \text{ from both sides.}$$

$$x + 2 = -1 \quad \text{Combine like terms.}$$

$$x + 2 - 2 = -1 - 2 \quad \text{Subtract 2 from both sides to get } x \text{ alone.}$$

$$x = -3 \quad \text{Combine like terms.}$$

Check:

$$2x + 3x - 5 + 7 = 10x + 3 - 6x - 4 \quad \text{Original equation.}$$

$$2(-3) + 3(-3) - 5 + 7 \stackrel{?}{=} 10(-3) + 3 - 6(-3) - 4 \quad \text{Replace } x \text{ with } -3.$$

$$-6 - 9 - 5 + 7 \stackrel{?}{=} -30 + 3 + 18 - 4 \quad \text{Multiply.}$$

$$-13 = -13 \quad \text{True}$$

The solution is -3. ●

If an equation contains parentheses, we use the distributive property to remove them, as before. Then we combine any like terms.

Practice Problem 6

Solve: $3(2w - 5) - (5w + 1) = -3$

EXAMPLE 6 Solve: $6(2a - 1) - (11a + 6) = 7$

Solution:

$$6(2a - 1) - 1(11a + 6) = 7$$

$$6(2a) + 6(-1) - 1(11a) - 1(6) = 7 \quad \text{Apply the distributive property.}$$

$$12a - 6 - 11a - 6 = 7 \quad \text{Multiply.}$$

$$a - 12 = 7 \quad \text{Combine like terms.}$$

$$a - 12 + 12 = 7 + 12 \quad \text{Add 12 to both sides.}$$

$$a = 19 \quad \text{Simplify.}$$

Check: Check by replacing a with 19 in the original equation. ●

Answers

5. $w = -4$, **6.** $w = 13$

EXAMPLE 7 Solve: $3 - x = 7$

Solution: First we subtract 3 from both sides.

$$3 - x = 7$$
$$3 - x - 3 = 7 - 3 \quad \text{Subtract 3 from both sides.}$$
$$-x = 4 \quad \text{Simplify.}$$

We have not yet solved for x since x is not alone. However, this equation does say that the opposite of x is 4. If the opposite of x is 4, then x is the opposite of 4, or $x = -4$.

If $-x = 4$,
then $x = -4$.

Check:

$$3 - x = 7 \quad \text{Original equation.}$$
$$3 - (-4) \stackrel{?}{=} 7 \quad \text{Replace } x \text{ with } -4.$$
$$3 + 4 \stackrel{?}{=} 7 \quad \text{Add.}$$
$$7 = 7 \quad \text{True}$$

The solution is -4.

Practice Problem 7

Solve: $12 - y = 9$

C Writing Algebraic Expressions

In this section, we continue to practice writing algebraic expressions.

EXAMPLE 8

a. The sum of two numbers is 8. If one number is 3, find the other number.
b. The sum of two numbers is 8. If one number is x, write an expression representing the other number.

Solution:

a. If the sum of two numbers is 8 and one number is 3, we find the other number by subtracting 3 from 8. The other number is $8 - 3$, or 5.

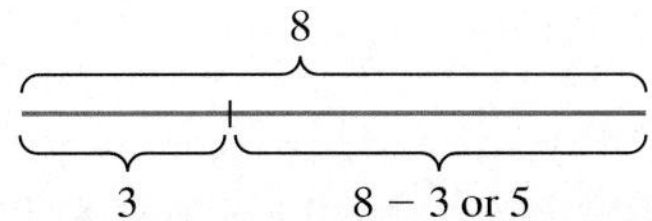

b. If the sum of two numbers is 8 and one number is x, we find the other number by subtracting x from 8. The other number is represented by $8 - x$.

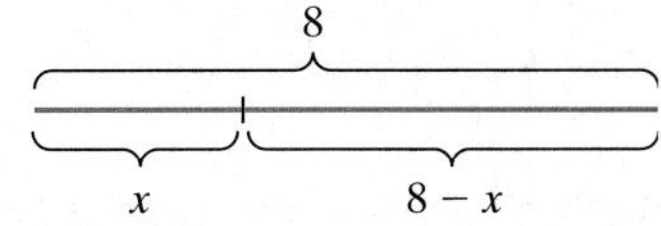

Practice Problem 8

The sum of two numbers is 11. If one number is x, write an expression representing the other number.

Answers

7. $y = 3$, **8.** $11 - x$

Practice Problem 9

In a recent House of Representatives race in California, Lucille Roybal-Allard received 49,489 more votes than Wayne Miller. If Wayne received n votes, how many did Lucille receive? (*Source:* Voter News Service)

Answer

9. $(n + 49{,}489)$ votes

EXAMPLE 9 The Verrazano-Narrows Bridge in New York City is the longest suspension bridge in North America. The Golden Gate Bridge in San Francisco is 60 feet shorter than the Verrazano-Narrows Bridge. If the length of the Verrazano-Narrows Bridge is m feet, express the length of the Golden Gate Bridge as an algebraic expression in m. (*Source:* Survey of State Highway Engineers)

Solution: Since the Golden Gate is 60 feet shorter than the Verrazano-Narrows Bridge, we have that its length is

In words:	Length of Verrazano-Narrows Bridge	minus	60
	↓	↓	↓
Translate:	m	$-$	60

The Golden Gate Bridge is $(m - 60)$ feet long. ●

STUDY SKILLS REMINDER

Have you decided to successfully complete this course?

Ask yourself if one of your current goals is to successfully complete this course.

If it is not a goal of yours, ask yourself why not. One common reason is fear of failure. Amazingly enough, fear of failure alone can be strong enough to keep many of us from doing our best in any endeavor. Another common reason is that you simply haven't taken the time to make successfully completing this course one of your goals.

If you are taking this mathematics course, then successfully completing this course probably should be one of your goals. To make it a goal, start by writing this goal in your mathematics notebook. Then read or reread Section 1.1 and make a commitment to try the suggestions in this section.

If successfully completing this course is already a goal of yours, also read or reread Section 1.1 and try some of the suggestions in this section so that you are actively working toward your goal.

Good luck and don't forget that a positive attitude will make a big difference, also.

Name ______________________ Section __________ Date __________

Mental Math

Solve each equation mentally. See Examples 1 and 2.

1. $x + 4 = 6$ **2.** $x + 7 = 10$ **3.** $n + 18 = 30$

4. $z + 22 = 40$ **5.** $b - 11 = 6$ **6.** $d - 16 = 5$

EXERCISE SET 2.2

 *Solve each equation. Check each solution. See Examples 1 through 4.*

1. $x + 7 = 10$ **2.** $x + 14 = 25$ **3.** $x - 2 = -4$ **4.** $y - 9 = 1$

5. $3 + x = -11$ **6.** $8 + z = -8$ **7.** $r - 8.6 = -8.1$ **8.** $t - 9.2 = -6.8$

9. $\frac{1}{3} + f = \frac{3}{4}$ **10.** $c + \frac{1}{6} = \frac{3}{8}$ **11.** $x - \frac{2}{5} = -\frac{3}{20}$ **12.** $y - \frac{4}{7} = -\frac{3}{14}$

13. $5b - 0.7 = 6b$ **14.** $9x + 5.5 = 10x$ **15.** $7x - 3 = 6x$ **16.** $18x - 9 = 19x$

17. In your own words, explain what is meant by the solution of an equation.

18. In your own words, explain how to check a solution of an equation.

B *Solve each equation. Don't forget to first simplify each side of the equation, if possible. Check each solution. See Examples 5 through 7.*

19. $7x + 2x = 8x - 3$ **20.** $3n + 2n = 7 + 4n$ **21.** $\frac{5}{6}x + \frac{1}{6}x = -9$

22. $\frac{13}{11}y - \frac{2}{11}y = -3$ **23.** $2y + 10 = 5y - 4y$ **24.** $4x - 4 = 10x - 7x$

25. $3x - 6 = 2x + 5$ **26.** $7y + 2 = 6y + 2$ **27.** $\frac{3}{7}x + 2 = -\frac{4}{7}x - 5$

28. $\frac{1}{5}x - 1 = -\frac{4}{5}x - 13$ **29.** $5x - 6 = 6x - 5$ **30.** $2x + 7 = x - 10$

31. $8y + 2 - 6y = 3 + y - 10$

32. $4p - 11 - p = 2 + 2p - 20$

33. $13x - 9 + 2x - 5 = 12x - 1 + 2x$

34. $15x + 20 - 10x - 9 = 25x + 8 - 21x - 7$

35. $-6.5 - 4x - 1.6 - 3x = -6x + 9.8$

36. $-1.4 - 7x - 3.6 - 2x = -8x + 4.4$

37. $\frac{3}{8}x - \frac{1}{6} = -\frac{5}{8}x - \frac{2}{3}$

38. $\frac{2}{5}x - \frac{1}{12} = -\frac{3}{5}x - \frac{3}{4}$

39. $2(x - 4) = x + 3$

40. $3(y + 7) = 2y - 5$

41. $7(6 + w) = 6(2 + w)$

42. $6(5 + c) = 5(c - 4)$

43. $10 - (2x - 4) = 7 - 3x$

44. $15 - (6 - 7k) = 2 + 6k$

45. $-5(n - 2) = 8 - 4n$

46. $-4(z - 3) = 2 - 3z$

47. $-3(x - 4) = -4x$

48. $-2(x - 1) = -3x$

49. $3(n - 5) - (6 - 2n) = 4n$

50. $5(3 + z) - (8z + 9) = -4z$

51. $-2(x + 6) + 3(2x - 5) = 3(x - 4) + 10$

52. $-5(x + 1) + 4(2x - 3) = 2(x + 2) - 8$

C *Write each algebraic expression described. See Examples 8 and 9.*

53. Two numbers have a sum of 20. If one number is p, express the other number in terms of p.

54. Two numbers have a sum of 13. If one number is y, express the other number in terms of y.

55. A 10-foot board is cut into two pieces. If one piece is x feet long, express the other length in terms of x.

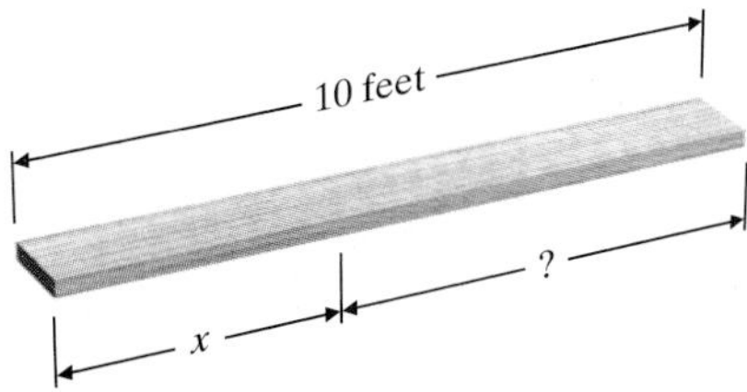

56. A 5-foot piece of string is cut into two pieces. If one piece is x feet long, express the other length in terms of x.

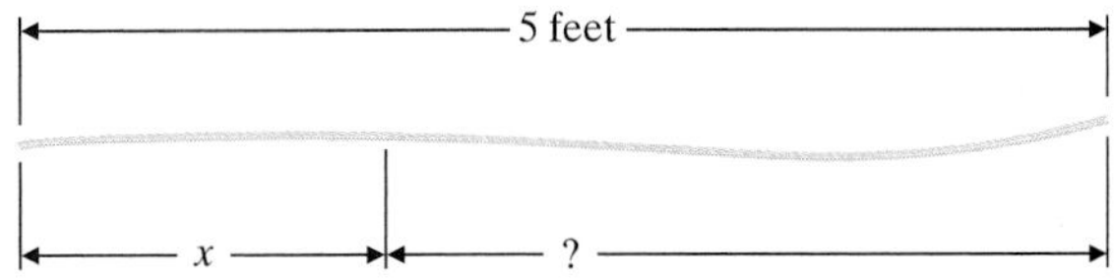

△ **57.** Recall that two angles are *supplementary* if their sum is 180°. If one angle measures $x°$, express the measure of its supplement in terms of x.

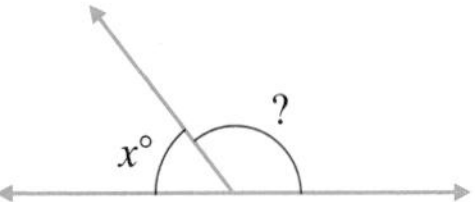

△ **58.** Recall that two angles are *complementary* if their sum is 90°. If one angle measures $x°$, express the measure of its complement in terms of x.

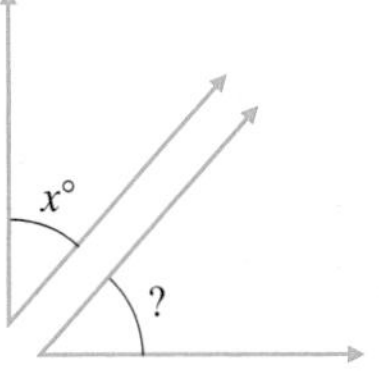

59. In Election 2000, Pat Ahumada ran against Solomon P. Ortiz for one of Texas's seats in the U.S. House of Representatives. Ahumada received 47,628 fewer votes than Ortiz. If Ahumada received n votes, how many did Ortiz receive? (*Source:* Voter News Service)

60. The longest interstate highway in the U.S. is I-90, which connects Seattle, Washington, and Boston, Massachusetts. The second longest interstate highway, I-80 (connecting San Francisco, California, and Teaneck, New Jersey), is 178.5 miles shorter than I-90. If the length of I-80 is m miles, express the length of I-90 as an algebraic expression in m. (*Source:* U.S. Department of Transportation–Federal Highway Administration)

61. The area of the Sahara Desert in Africa is 7 times the area of the Gobi Desert in Asia. If the area of the Gobi Desert is x square miles, express the area of the Sahara Desert as an algebraic expression in x.

62. The largest meteorite in the world is the Hoba West located in Namibia. Its weight is 3 times the weight of the Armanty meteorite located in Outer Mongolia. If the weight of the Armanty meteorite is y kilograms, express the weight of the Hoba West meteorite as an algebraic expression in y.

Review and Preview

Find each multiplicative inverse or reciprocal. See Section 1.7.

63. $\frac{5}{8}$

64. $\frac{7}{6}$

65. 2

66. 5

67. $-\frac{1}{9}$

68. $-\frac{3}{5}$

Perform each indicated operation and simplify. See Section 1.6.

69. $\frac{3x}{3}$

70. $\frac{-2y}{-2}$

71. $-5\left(-\frac{1}{5}y\right)$

72. $7\left(\frac{1}{7}r\right)$

73. $\frac{3}{5}\left(\frac{5}{3}x\right)$

74. $\frac{9}{2}\left(\frac{2}{9}x\right)$

Combining Concepts

Use a calculator to determine the solution of each equation.

75. $36.766 + x = -108.712$

76. $-85.325 = x - 97.985$

Solve

△ **77.** The sum of the angles of a triangle is 180°. If one angle of a triangle measures $x°$ and a second angle measures $(2x + 7)°$, express the measure of the third angle in terms of x. Simplify the expression.

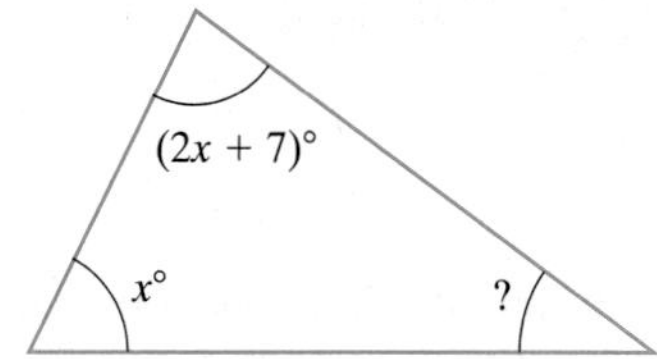

△ **78.** A quadrilateral is a four-sided figure (like the one shown in the figure) whose angle sum is 360°. If one angle measures $x°$, a second angle measures $3x°$, and a third angle measures $5x°$, express the measure of the fourth angle in terms of x. Simplify the expression.

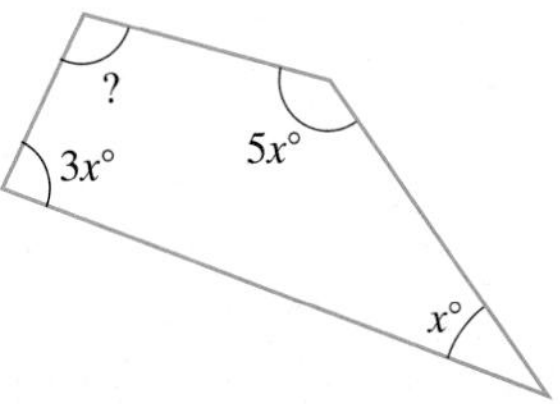

2.3 The Multiplication Property of Equality

OBJECTIVES

- **A** Use the multiplication property of equality to solve linear equations.
- **B** Use both the addition and multiplication properties of equality to solve linear equations.
- **C** Write word phrases as algebraic expressions.

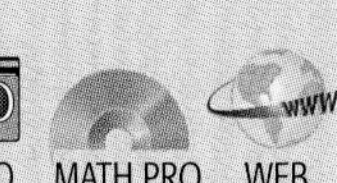

SSM TUTOR CENTER SG CD & VIDEO MATH PRO WEB

A Using the Multiplication Property

As useful as the addition property of equality is, it cannot help us solve every type of linear equation in one variable. For example, adding or subtracting a value on both sides of the equation does not help solve

$$\frac{5}{2}x = 15$$

because the variable x is being multiplied by a number (other than 1). Instead, we apply another important property of equality, the **multiplication property of equality**.

Multiplication Property of Equality

If a, b, and c are real numbers and $c \neq 0$, then

$$a = b \quad \text{and} \quad ac = bc$$

are equivalent equations.

This property guarantees that multiplying both sides of an equation by the same nonzero number does not change the solution of the equation. Since division is defined in terms of multiplication, we may also **divide both sides of the equation by the same nonzero number** without changing the solution.

EXAMPLE 1 Solve: $\frac{5}{2}x = 15$

Solution: To get x alone, we multiply both sides of the equation by the reciprocal (or multiplicative inverse) of $\frac{5}{2}$, which is $\frac{2}{5}$.

$$\frac{5}{2}x = 15$$

$$\frac{2}{5} \cdot \left(\frac{5}{2}x\right) = \frac{2}{5} \cdot 15 \qquad \text{Multiply both sides by } \frac{2}{5}.$$

$$\left(\frac{2}{5} \cdot \frac{5}{2}\right)x = \frac{2}{5} \cdot 15 \qquad \text{Apply the associative property.}$$

$$1x = 6 \qquad \text{Simplify.}$$

or

$$x = 6$$

Check: Replace x with 6 in the original equation.

$$\frac{5}{2}x = 15 \qquad \text{Original equation.}$$

$$\frac{5}{2}(6) \stackrel{?}{=} 15 \qquad \text{Replace } x \text{ with 6.}$$

$$15 = 15 \qquad \text{True}$$

The solution is 6.

Practice Problem 1

Solve: $\frac{3}{7}x = 9$

Answer

1. $x = 21$

In the equation $\frac{5}{2}x = 15$, $\frac{5}{2}$ is the coefficient of x. When the coefficient of x is a *fraction*, we will get x alone by multiplying by the reciprocal. When the coefficient of x is an integer or a decimal, it is usually more convenient to divide both sides by the coefficient. (Dividing by a number is, of course, the same as multiplying by the reciprocal of the number.)

Practice Problem 2

Solve: $7x = 42$

EXAMPLE 2 Solve: $5x = 30$

Solution: To get x alone, we divide both sides of the equation by 5, the coefficient of x.

$$5x = 30$$

$$\frac{5x}{5} = \frac{30}{5} \quad \text{Divide both sides by 5.}$$

$$1 \cdot x = 6 \quad \text{Simplify.}$$

$$x = 6$$

Check:

$$5x = 30 \quad \text{Original equation.}$$

$$5 \cdot 6 \stackrel{?}{=} 30 \quad \text{Replace } x \text{ with 6.}$$

$$30 = 30 \quad \text{True}$$

The solution is 6. ●

Practice Problem 3

Solve: $-4x = 52$

EXAMPLE 3 Solve: $-3x = 33$

Solution:

Recall that $-3x$ means $-3 \cdot x$. To get x alone, we divide both sides by the coefficient of x, that is, -3.

$$-3x = 33$$

$$\frac{-3x}{-3} = \frac{33}{-3} \quad \text{Divide both sides by } -3.$$

$$1x = -11 \quad \text{Simplify.}$$

$$x = -11$$

Check:

$$-3x = 33 \quad \text{Original equation.}$$

$$-3(-11) \stackrel{?}{=} 33 \quad \text{Replace } x \text{ with } -11.$$

$$33 = 33 \quad \text{True}$$

The solution is -11. ●

Practice Problem 4

Solve: $\frac{y}{5} = 13$

EXAMPLE 4 Solve: $\frac{y}{7} = 20$

Solution:

Recall that $\frac{y}{7} = \frac{1}{7}y$. To get y alone, we multiply both sides of the equation by 7, the reciprocal of $\frac{1}{7}$.

$$\frac{y}{7} = 20$$

$$\frac{1}{7}y = 20$$

$$7 \cdot \frac{1}{7}y = 7 \cdot 20 \quad \text{Multiply both sides by 7.}$$

$$1y = 140 \quad \text{Simplify.}$$

$$y = 140$$

Answers

2. $x = 6$, **3.** $x = -13$, **4.** $y = 65$

Check:
$$\frac{y}{7} = 20 \quad \text{Original equation.}$$
$$\frac{140}{7} \stackrel{?}{=} 20 \quad \text{Replace } y \text{ with 140.}$$
$$20 = 20 \quad \text{True}$$

The solution is 140.

EXAMPLE 5 Solve: $3.1x = 4.96$

Solution:
$$3.1x = 4.96$$
$$\frac{3.1x}{3.1} = \frac{4.96}{3.1} \quad \text{Divide both sides by 3.1.}$$
$$1x = 1.6 \quad \text{Simplify.}$$
$$x = 1.6$$

Check: Check by replacing x with 1.6 in the original equation. The solution is 1.6.

EXAMPLE 6 Solve: $-\frac{2}{3}x = -\frac{5}{2}$

Solution: To get x alone, we multiply both sides of the equation by $-\frac{3}{2}$, the reciprocal of the coefficient of x.
$$-\frac{2}{3}x = -\frac{5}{2}$$
$$-\frac{3}{2}\cdot -\frac{2}{3}x = -\frac{3}{2}\cdot -\frac{5}{2} \quad \text{Multiply both sides by } -\frac{3}{2}\text{, the reciprocal of } -\frac{2}{3}.$$
$$x = \frac{15}{4} \quad \text{Simplify.}$$

Check: Check by replacing x with $\frac{15}{4}$ in the original equation. The solution is $\frac{15}{4}$.

B Using Both the Addition and Multiplication Properties

We are now ready to combine the skills learned in the last section with the skills learned from this section to solve equations by applying more than one property.

EXAMPLE 7 Solve: $-z - 4 = 6$

Solution: First, to get $-z$, the term containing the variable alone, we add 4 to both sides of the equation.
$$-z - 4 + 4 = 6 + 4 \quad \text{Add 4 to both sides.}$$
$$-z = 10 \quad \text{Simplify.}$$

Next, recall that $-z$ means $-1 \cdot z$. Thus to get z alone, we either multiply or divide both sides of the equation by -1. In this example, we divide.
$$-z = 10$$
$$\frac{-z}{-1} = \frac{10}{-1} \quad \text{Divide both sides by the coefficient } -1.$$
$$1z = -10 \quad \text{Simplify.}$$
$$z = -10$$

Check:
$$-z - 4 = 6 \quad \text{Original equation.}$$
$$-(-10) - 4 \stackrel{?}{=} 6 \quad \text{Replace } z \text{ with } -10.$$
$$10 - 4 \stackrel{?}{=} 6$$
$$6 = 6 \quad \text{True}$$

The solution is -10.

Practice Problem 5

Solve: $2.6x = 13.52$

Practice Problem 6

Solve: $-\frac{5}{6}y = -\frac{3}{5}$

Practice Problem 7

Solve: $-x + 7 = -12$

Answers

5. $x = 5.2$, **6.** $y = \frac{18}{25}$, **7.** $x = 19$

Practice Problem 8

Solve: $-7x + 2x + 3 - 20 = -2$

EXAMPLE 8 Solve: $a + a - 10 + 7 = -13$

Solution: First, we simplify both sides of the equation by combining like terms.

$$a + a - 10 + 7 = -13$$

$$2a - 3 = -13 \quad \text{Combine like terms.}$$

$$2a - 3 + 3 = -13 + 3 \quad \text{Add 3 to both sides.}$$

$$2a = -10 \quad \text{Simplify.}$$

$$\frac{2a}{2} = \frac{-10}{2} \quad \text{Divide both sides by 2.}$$

$$a = -5 \quad \text{Simplify.}$$

Check: To check, replace a with -5 in the original equation. The solution is -5.

C Writing Algebraic Expressions

We continue to sharpen our problem-solving skills by writing algebraic expressions.

Practice Problem 9

If x is the first of two consecutive integers, express the sum of the first and the second integer in terms of x. Simplify if possible.

EXAMPLE 9 Writing an Expression for Consecutive Integers

If x is the first of three consecutive integers, express the sum of the three integers in terms of x. Simplify if possible.

Solution: An example of three consecutive integers is 7, 8, and 9.

+1 +2
7 8 9

The second consecutive integer is always 1 more than the first, and the third consecutive integer is 2 more than the first. If x is the first of three consecutive integers, the three consecutive integers are x, $x + 1$, and $x + 2$.

+1 +2
x $x + 1$ $x + 2$

Their sum is shown below.

In words:	first integer	+	second integer	+	third integer
Translate:	x	+	$(x + 1)$	+	$(x + 2)$

This simplifies to $3x + 3$.

Study these examples of consecutive even and consecutive odd integers.

Consecutive even integers:

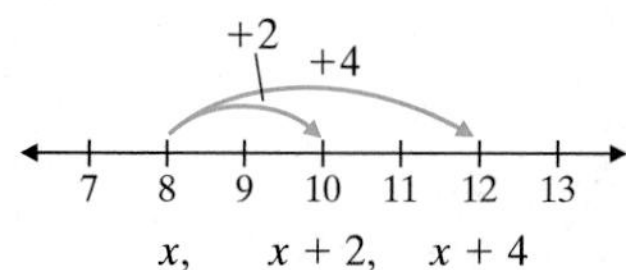

Answers

8. $x = -3$, **9.** $2x + 1$

Consecutive odd integers:

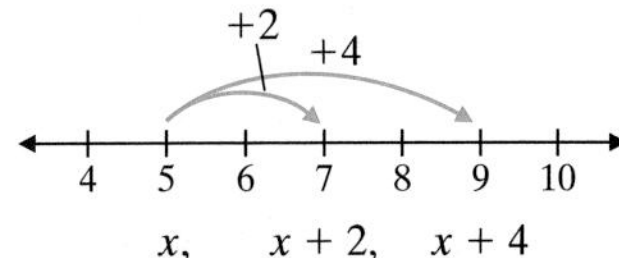

If x is an odd integer, then $x + 2$ is the next odd integer. This 2 simply means that odd integers are always 2 units from each other.

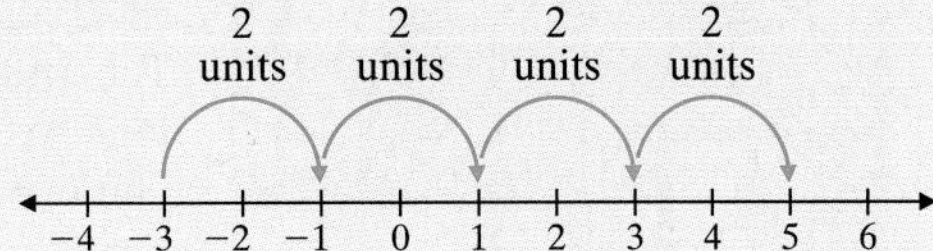

STUDY SKILLS REMINDER

Are you organized?

Have you ever had trouble finding a completed assignment? When it's time to study for a test, are your notes neat and organized? Have you ever had trouble reading your own mathematics handwriting? (Be honest—I have had trouble reading my own handwriting before.)

When any of these things happen, it's time to get organized. Here are a few suggestions:

Write your notes and complete your homework assignment in a notebook with pockets (spiral or ring binder). Take class notes in this notebook, and then follow the notes with your completed homework assignment. When you receive graded papers or handouts, place them in the notebook pocket so that you will not lose them.

Place a mark (possibly an exclamation point) beside any note(s) that seem especially important to you. Also place a mark (possibly a question mark) beside any note(s) or homework that you are having trouble with. Don't forget to see your intructor, a tutor, or your fellow classmates to help you understand the concepts or exercises you have marked.

Also, if you are having trouble reading your own handwriting, *slow down* and write your mathematics work clearly!

FOCUS ON History

THE GOLDEN RECTANGLE IN ART

Piet Mondrian, *Composition with Grid #1*, 1918. Oil on canvas, $31\frac{9}{16} \times 19\frac{5}{8}$ in. (80.2×49.9 cm). The Museum of Fine Arts, Houston, Gift of Mr. and Mrs. Pierre Schlumberger. © 2003 Mondrian/ Holtzman Trust, c/o Beeldrecht/ Artists Rights Society (ARS), New York.

The golden rectangle is a rectangle whose length is approximately 1.6 times its width. The early Greeks thought that a rectangle with these dimensions was the most pleasing to the eye. Examples of the golden rectangle are found in many ancient, as well as modern, works of art. For example, the Parthenon in Athens, Greece, shows the golden rectangle in many aspects of its design. Modern-era artists, including Piet Mondrian (1872–1944) and Georges Seurat (1859–1891), also frequently used the proportions of a golden rectangle in their paintings.

To test whether a rectangle is a golden rectangle, divide the rectangle's length by its width. If the result is approximately 1.6, we can consider the rectangle to be a golden rectangle. For instance, consider Mondrian's *Composition with Grid #1*, which was painted on an 80.2×49.9 cm canvas. Because $\frac{80.2}{49.9} \approx 1.6$, the dimensions of the canvas form a golden rectangle. In what other ways are golden rectangles connected with this painting?

Examples of golden rectangles can be found in the designs of many everyday objects. Visual artists, from architects to product and package designers, use the golden rectangle shape in such things as the face of a building, the floor of a room, the front of a food package, the front cover of a book, and even the shape of a credit card.

GROUP ACTIVITY

Find an example of a golden rectangle in a building or an everyday object. Use a ruler to measure its dimensions and verify that the length is approximately 1.6 times the width.

Name ______________________ Section __________ Date __________

Mental Math

Solve each equation mentally. See Examples 2 and 3.

1. $3a = 27$ **2.** $9c = 54$ **3.** $5b = 10$ **4.** $7t = 14$ **5.** $6x = -30$ **6.** $8r = -64$

EXERCISE SET 2.3

A *Solve each equation. Check each solution. See Examples 1 through 6.*

1. $-5x = 20$

2. $7x = 49$

3. $3x = 0$

4. $2x = 0$

5. $-x = -12$

6. $-y = 8$

7. $\frac{2}{3}x = -8$

8. $\frac{3}{4}n = -15$

9. $\frac{1}{6}d = \frac{1}{2}$

10. $\frac{1}{8}v = \frac{1}{4}$

11. $\frac{a}{2} = 1$

12. $\frac{d}{15} = 2$

13. $\frac{k}{-7} = 0$

14. $\frac{f}{-5} = 0$

15. $1.7x = 10.71$

16. $8.5y = 18.7$

17. $42 = 7x$

18. $81 = 3x$

19. $4.4 = -0.8x$

20. $6.3 = -0.6x$

21. $-\frac{3}{7}p = -2$

22. $-\frac{4}{5}r = -5$

23. $-\frac{4}{3}x = 12$

24. $-\frac{10}{3}x = 30$

B *Solve each equation. Check each solution. See Examples 7 and 8.*

25. $2x - 4 = 16$

26. $3x - 1 = 26$

27. $-x + 2 = 22$

28. $-x + 4 = -24$

29. $6a + 3 = 3$

30. $8t + 5 = 5$

31. $6x + 10 = -20$

32. $-10y + 15 = 5$

33. $5 - 0.3k = 5$

34. $2 + 0.4p = 2$

35. $-2x + \frac{1}{2} = \frac{7}{2}$

36. $-3n - \frac{1}{3} = \frac{8}{3}$

37. $\frac{x}{3} + 2 = -5$

38. $\frac{b}{4} - 1 = -7$

39. $10 = 2x - 1$

40. $12 = 3j - 4$

41. $6z - 8 - z + 3 = 0$

42. $4a + 1 + a - 11 = 0$

43. $10 - 3x - 6 - 9x = 7$

44. $12x + 30 + 8x - 6 = 10$

45. $1 = 0.4x - 0.6x - 5$

46. $19 = 0.4x - 0.9x - 6$

47. $z - 5z = 7z - 9 - z$

48. $t - 6t = -13 + t - 3t$

Write each algebraic expression described. Simplify if possible. See Example 9.

49. If x represents the first of two consecutive odd integers, express the sum of the two integers in terms of x.

50. If x is the first of four consecutive even integers, write their sum as an algebraic expression in x.

51. If x is the first of three consecutive integers, express the sum of the first integer and the third integer as an algebraic expression containing the variable x.

52. If x is the first of two consecutive integers, express the sum of 20 and the second consecutive integer as an algebraic expression containing the variable x.

Review and Preview

Simplify each expression. See Section 2.1.

53. $5x + 2(x - 6)$

54. $-7y + 2y - 3(y + 1)$

55. $6(2z + 4) + 20$

56. $-(3a - 3) + 2a - 6$

57. $-(x - 1) + x$

58. $8(z - 6) + 7z - 1$

Combining Concepts

Solve.

59. $0.07x - 5.06 = -4.92$

60. $0.06y + 2.63 = 2.5562$

61. The equation $3x + 6 = 2x + 10 + x - 4$ is true for all real numbers. Substitute a few real numbers for x to see that this is so and then try solving the equation. Describe what happens.

62. The equation $6x + 2 - 2x = 4x + 1$ has no solution. Try solving this equation for x and describe what happens.

63. From the results of Exercises 61 and 62, when do you think an equation has all real numbers as its solutions?

64. From the results of Exercises 61 and 62, when do you think an equation has no solution?

2.4 Further Solving Linear Equations

A Solving Linear Equations

We now combine our knowledge from the previous sections into a general strategy for solving linear equations. One new piece in this strategy is a suggestion to "clear an equation of fractions" as a first step. Doing so makes the equation more manageable, since working with integers is more convenient than working with fractions. We will discuss this further in Example 3.

OBJECTIVES

A Apply the general strategy for solving a linear equation.

B Solve equations containing fractions or decimals.

C Recognize identities and equations with no solution.

To Solve Linear Equations in One Variable

Step 1. If an equation contains fractions, multiply both sides by the LCD to clear the equation of fractions.

Step 2. Use the distributive property to remove parentheses if they occur.

Step 3. Simplify each side of the equation by combining like terms.

Step 4. Get all variable terms on one side and all numbers on the other side by using the addition property of equality.

Step 5. Get the variable alone by using the multiplication property of equality.

Step 6. Check the solution by substituting it into the original equation.

EXAMPLE 1 Solve: $4(2x - 3) + 7 = 3x + 5$

Solution: There are no fractions, so we begin with Step 2.

$$4(2x - 3) + 7 = 3x + 5$$

Step 2. $8x - 12 + 7 = 3x + 5$ Apply the distributive property.

Step 3. $8x - 5 = 3x + 5$ Combine like terms.

Step 4. Get all variable terms on the same side of the equation by subtracting $3x$ from both sides and then adding 5 to both sides.

$8x - 5 - 3x = 3x + 5 - 3x$ Subtract $3x$ from both sides.

$5x - 5 = 5$ Simplify.

$5x - 5 + 5 = 5 + 5$ Add 5 to both sides.

$5x = 10$ Simplify.

Step 5. Use the multiplication property of equality to get x alone.

$\frac{5x}{5} = \frac{10}{5}$ Divide both sides by 5.

$x = 2$ Simplify.

Step 6. Check.

$4(2x - 3) + 7 = 3x + 5$ Original equation

$4[2(2) - 3] + 7 \stackrel{?}{=} 3(2) + 5$ Replace x with 2.

$4(4 - 3) + 7 \stackrel{?}{=} 6 + 5$

$4(1) + 7 \stackrel{?}{=} 11$

$4 + 7 \stackrel{?}{=} 11$

$11 = 11$ True

The solution is 2.

Practice Problem 1

Solve: $5(3x - 1) + 2 = 12x + 6$

Answer

1. $x = 3$

Helpful Hint

When checking solutions, use the original written equation.

Practice Problem 2

Solve: $9(5 - x) = -3x$

EXAMPLE 2 Solve: $8(2 - t) = -5t$

Solution: First, we apply the distributive property.

$$8(2 - t) = -5t$$

Step 2. $16 - 8t = -5t$ Use the distributive property.

Step 4. $16 - 8t + 8t = -5t + 8t$ Add $8t$ to both sides.

$16 = 3t$ Combine like terms.

Step 5. $\frac{16}{3} = \frac{3t}{3}$ Divide both sides by 3.

$\frac{16}{3} = t$ Simplify.

Step 6. Check.

$8(2 - t) = -5t$ Original equation

$8\left(2 - \frac{16}{3}\right) \stackrel{?}{=} -5\left(\frac{16}{3}\right)$ Replace t with $\frac{16}{3}$.

$8\left(\frac{6}{3} - \frac{16}{3}\right) \stackrel{?}{=} -\frac{80}{3}$ The LCD is 3.

$8\left(-\frac{10}{3}\right) \stackrel{?}{=} -\frac{80}{3}$ Subtract fractions.

$-\frac{80}{3} = -\frac{80}{3}$ True

The solution is $\frac{16}{3}$.

B Solving Equations Containing Fractions or Decimals

If an equation contains fractions, we can clear the equation of fractions by multiplying both sides by the LCD of all denominators. By doing this, we avoid working with time-consuming fractions.

Practice Problem 3

Solve: $\frac{5}{2}x - 1 = \frac{3}{2}x - 4$

EXAMPLE 3 Solve: $\frac{x}{2} - 1 = \frac{2}{3}x - 3$

Solution: We begin by clearing fractions. To do this, we multiply both sides of the equation by the LCD of 2 and 3, which is 6.

$$\frac{x}{2} - 1 = \frac{2}{3}x - 3$$

Step 1. $6\left(\frac{x}{2} - 1\right) = 6\left(\frac{2}{3}x - 3\right)$ Multiply both sides by the LCD, 6.

Step 2. $6\left(\frac{x}{2}\right) - 6(1) = 6\left(\frac{2}{3}x\right) - 6(3)$ Apply the distributive property.

$3x - 6 = 4x - 18$ Simplify.

Helpful Hint

Don't forget to multiply *each* term by the LCD.

Answers

2. $x = \frac{15}{2}$, **3.** $x = -3$

There are no longer grouping symbols and no like terms on either side of the equation, so we continue with Step 4.

$$3x - 6 = 4x - 18$$

Step 4. $3x - 6 - 3x = 4x - 18 - 3x$ Subtract $3x$ from both sides.

$-6 = x - 18$ Simplify.

$-6 + 18 = x - 18 + 18$ Add 18 to both sides.

$12 = x$ Simplify.

Step 5. The variable is now alone, so there is no need to apply the multiplication property of equality.

Step 6. Check.

$\frac{x}{2} - 1 = \frac{2}{3}x - 3$ Original equation

$\frac{12}{2} - 1 \stackrel{?}{=} \frac{2}{3} \cdot 12 - 3$ Replace x with 12.

$6 - 1 \stackrel{?}{=} 8 - 3$ Simplify.

$5 = 5$ True

The solution is 12.

EXAMPLE 4 Solve: $\frac{2(a + 3)}{3} = 6a + 2$

Solution: We clear the equation of fractions first.

$$\frac{2(a + 3)}{3} = 6a + 2$$

Step 1. $3 \cdot \frac{2(a + 3)}{3} = 3(6a + 2)$ Clear the fraction by multiplying both sides by the LCD, 3.

Step 2. Next, we use the distributive property and remove parentheses.

$2a + 6 = 18a + 6$ Apply the distributive property.

Step 4. $2a + 6 - 6 = 18a + 6 - 6$ Subtract 6 from both sides.

$2a = 18a$

$2a - 18a = 18a - 18a$ Subtract $18a$ from both sides.

$-16a = 0$

Step 5. $\frac{-16a}{-16} = \frac{0}{-16}$ Divide both sides by -16.

$a = 0$ Write the fraction in simplest form.

Step 6. To check, replace a with 0 in the original equation. The solution is 0.

When solving a problem about money, you may need to solve an equation containing decimals. If you choose, you may multiply to clear the equation of decimals.

Practice Problem 4

Solve: $\frac{3(x - 2)}{5} = 3x + 6$

Answer

4. $x = -3$

Practice Problem 5

Solve:
$0.06x - 0.10(x - 2) = -0.02(8)$

EXAMPLE 5 Solve: $0.25x + 0.10(x - 3) = 0.05(22)$

Solution: First we clear this equation of decimals by multiplying both sides of the equation by 100. Recall that multiplying a decimal number by 100 has the effect of moving the decimal point 2 places to the right.

$$0.25x + 0.10(x - 3) = 0.05(22)$$

Step 1. $0.25x + 0.10(x - 3) = 0.05(22)$ Multiply both sides by 100.

$$25x + 10(x - 3) = 5(22)$$

Step 2. $25x + 10x - 30 = 110$ Apply the distributive property.

Step 3. $35x - 30 = 110$ Combine like terms.

Step 4. $35x - 30 + 30 = 110 + 30$ Add 30 to both sides.

$35x = 140$ Combine like terms.

Step 5. $\frac{35x}{35} = \frac{140}{35}$ Divide both sides by 35.

$$x = 4$$

Step 6. To check, replace x with 4 in the original equation. The solution is 4.

C Recognizing Identities and Equations with No Solution

So far, each equation that we have solved has had a single solution. However, not every equation in one variable has a single solution. Some equations have no solution, while others have an infinite number of solutions. For example,

$$x + 5 = x + 7$$

has no solution since no matter which **real number** we replace x with, the equation is false.

real number + 5 = same real number + 7 FALSE

On the other hand,

$$x + 6 = x + 6$$

has infinitely many solutions since x can be replaced by any real number and the equation is always true.

real number + 6 = same real number + 6 TRUE

The equation $x + 6 = x + 6$ is called an **identity**. The next few examples illustrate special equations like these.

Answer

5. $x = 9$

EXAMPLE 6 Solve: $-2(x - 5) + 10 = -3(x + 2) + x$

Solution: $-2(x - 5) + 10 = -3(x + 2) + x$

$-2x + 10 + 10 = -3x - 6 + x$ Apply the distributive property on both sides.

$-2x + 20 = -2x - 6$ Combine like terms.

$-2x + 20 + 2x = -2x - 6 + 2x$ Add $2x$ to both sides.

$20 = -6$ Combine like terms.

The final equation contains no variable terms, and there is no value for x that makes $20 = -6$ a true equation. We conclude that there is **no solution** to this equation.

EXAMPLE 7 Solve: $3(x - 4) = 3x - 12$

Solution: $3(x - 4) = 3x - 12$

$3x - 12 = 3x - 12$ Apply the distributive property.

The left side of the equation is now identical to the right side. Every real number may be substituted for x and a true statement will result. We arrive at the same conclusion if we continue.

$3x - 12 = 3x - 12$

$3x - 12 + 12 = 3x - 12 + 12$ Add 12 to both sides.

$3x = 3x$ Combine like terms.

$3x - 3x = 3x - 3x$ Subtract $3x$ from both sides.

$0 = 0$

Again, one side of the equation is identical to the other side. Thus, $3(x - 4) = 3x - 12$ is an **identity** and **every real number** is a solution.

Try the Concept Check in the margin.

Practice Problem 6

Solve: $5(2 - x) + 8x = 3(x - 6)$

Practice Problem 7

Solve: $-6(2x + 1) - 14 = -10(x + 2) - 2x$

Concept Check

Suppose you have simplified several equations and obtain the following results. What can you conclude about the solutions to the original equation?

a. $7 = 7$ b. $x = 0$ c. $7 = -4$

Answers

6. no solution, **7.** Every real number is a solution.

Concept Check: **a.** Every real number is a solution. **b.** The solution is 0. **c.** There is no solution.

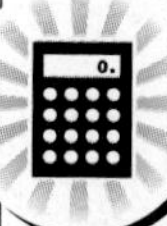

CALCULATOR EXPLORATIONS

Checking Equations

We can use a calculator to check possible solutions of equations. To do this, replace the variable by the possible solution and evaluate both sides of the equation separately.

Equation: $3x - 4 = 2(x + 6)$ Solution: $x = 16$

$$3x - 4 = 2(x + 6) \quad \text{Original equation}$$

$$3(16) - 4 \stackrel{?}{=} 2(16 + 6) \quad \text{Replace } x \text{ with 16.}$$

Now evaluate each side with your calculator.

Evaluate left side: [3] [×] [16] [−] [4] [=] or [ENTER] Display: [44]

Evaluate right side: [2] [(] [16] [+] [6] [)] [=] or [ENTER] Display: [44]

Since the left side equals the right side, the equation checks.

Use a calculator to check the possible solutions to each equation.

1. $2x = 48 + 6x;\quad x = -12$

2. $-3x - 7 = 3x - 1;\quad x = -1$

3. $5x - 2.6 = 2(x + 0.8);\quad x = 4.4$

4. $-1.6x - 3.9 = -6.9x - 25.6;\quad x = 5$

5. $\dfrac{564x}{4} = 200x - 11(649);\quad x = 121$

6. $20(x - 39) = 5x - 432;\quad x = 23.2$

Name ______________________ Section __________ Date __________

EXERCISE SET 2.4

 *Solve each equation. See Examples 1 and 2.*

1. $-4y + 10 = -2(3y + 1)$

2. $-3x + 1 = -2(x - 2)$

3. $9x - 8 = 10 + 15x$

4. $15x - 5 = 7 + 12x$

5. $-2(3x - 4) = 2x$

6. $-(5x - 10) = 5x$

7. $4(2n - 1) = (6n + 4) + 1$

8. $4(4y + 2) = 2(1 + 6y) + 8$

9. $5(2x - 1) - 2(3x) = 1$

10. $3(2 - 5x) + 4(6x) = 12$

11. $6(x - 3) + 10 = -8$

12. $-4(2 + n) + 9 = 1$

13. $8 - 2(a - 1) = 7 + a$

14. $5 - 6(2 + b) = b - 14$

15. $4x + 3 = 2x + 11$

16. $6y - 8 = 3y + 7$

17. $-2y - 10 = 5y + 18$

18. $7n + 5 = 10n - 10$

19. $-3(t - 5) + 2t = 5t - 4$

20. $-(4a - 7) - 5a = 10 + a$

21. $5y + 2(y - 6) = 4(y + 1) - 2$

22. $9x + 3(x - 4) = 10(x - 5) + 7$

B *Solve each equation. See Examples 3 through 5.*

23. $\frac{3}{4}x - \frac{1}{2} = 1$

24. $\frac{2}{3}x + \frac{5}{3} = \frac{5}{3}$

25. $x + \frac{5}{4} = \frac{3}{4}x$

26. $\frac{7}{8}x + \frac{1}{4} = \frac{3}{4}x$

27. $\frac{x}{2} - 1 = \frac{x}{5} + 2$

28. $\frac{x}{5} - 2 = \frac{x}{3}$

29. $\frac{6(3 - z)}{5} = -z$

30. $\frac{4(5 - w)}{3} = -w$

31. $0.06 - 0.01(x + 1) = -0.02(2 - x)$

32. $-0.01(5x + 4) = 0.04 - 0.01(x + 4)$

33. $\frac{3(x - 5)}{2} = \frac{2(x + 5)}{3}$

34. $\frac{5(x - 1)}{4} = \frac{3(x + 1)}{2}$

35. $0.50x + 0.15(70) = 0.25(142)$

36. $0.40x + 0.06(30) = 0.20(49)$

37. $0.12(y - 6) + 0.06y = 0.08y - 0.07(10)$

38. $0.60(z - 300) + 0.05z = 0.70z - 0.41(500)$

39. $\frac{2(x + 1)}{4} = 3x - 2$

40. $\frac{3(y + 3)}{5} = 2y + 6$

41. $x + \frac{7}{6} = 2x - \frac{7}{6}$

42. $\frac{5}{2}x - 1 = x + \frac{1}{4}$

43. $\frac{9}{2} + \frac{5}{2}y = 2y - 4$

44. $3 - \frac{1}{2}x = 5x - 8$

45. Explain the difference between simplifying an expression and solving an equation.

46. When solving an equation, if an equivalent equation is $0 = 5$, what can we conclude? If an equivalent equation is $-2 = -2$, what can we conclude?

C *Solve each equation. See Examples 6 and 7.*

47. $5x - 5 = 2(x + 1) + 3x - 7$

48. $3(2x - 1) + 5 = 6x + 2$

49. $\frac{x}{4} + 1 = \frac{x}{4}$

50. $\frac{x}{3} - 2 = \frac{x}{3}$

51. $3x - 7 = 3(x + 1)$

52. $2(x - 5) = 2x + 10$

53. $2(x + 3) - 5 = 5x - 3(1 + x)$

54. $4(2 + x) + 1 = 7x - 3(x - 2)$

55. On your own, construct an equation for which every real number is a solution.

56. On your own, construct an equation that has no solution.

Review and Preview

Write each algebraic expression described. See Section 2.1.

△ **57.** A plot of land is in the shape of a triangle. If one side is x meters, a second side is $(2x - 3)$ meters, and a third side is $(3x - 5)$ meters, express the perimeter of the lot as a simplified expression in x.

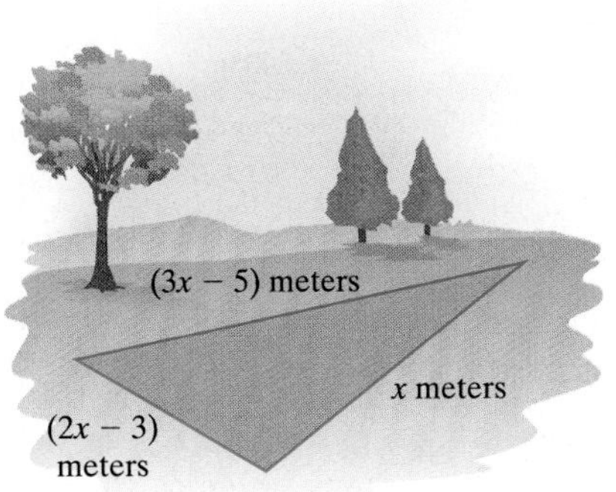

58. A portion of a board has length x feet. The other part has length $(7x - 9)$ feet. Express the total length of the board as a simplified expression in x.

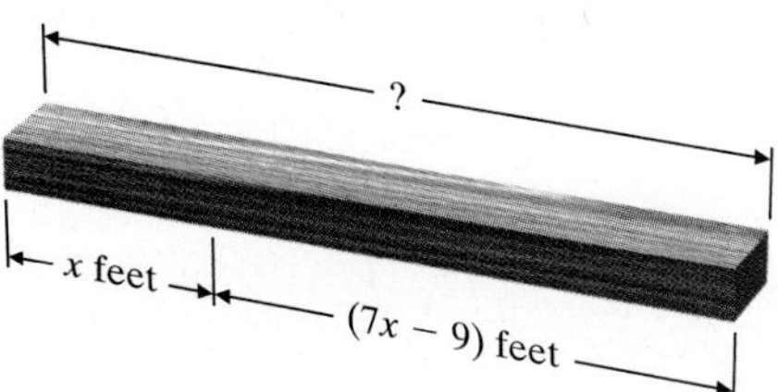

Write each phrase as an algebraic expression. Use x for the unknown number. See Section 2.1.

59. A number subtracted from -8

60. Three times a number

61. The sum of -3 and twice a number

62. The difference of 8 and twice a number

63. The product of 9 and the sum of a number and 20

64. The quotient of -12 and the difference of a number and 3

Combining Concepts

Solve.

65. $1000(7x - 10) = 50(412 + 100x)$

66. $1000(x + 40) = 100(16 + 7x)$

67. $0.035x + 5.112 = 0.010x + 5.107$

68. $0.127x - 2.685 = 0.027x - 2.38$

△ **69.** The perimeter of a geometric figure is the sum of the lengths of its sides. If the perimeter of the following pentagon (five-sided figure) is 28 centimeters, find the length of each side.

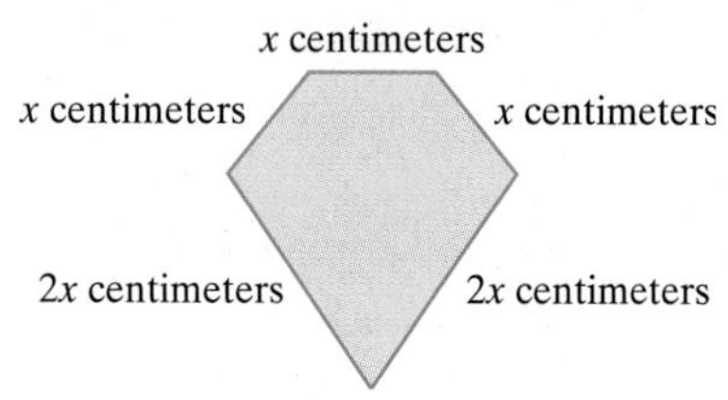

△ **70.** The perimeter of the following triangle is 35 meters. Find the length of each side.

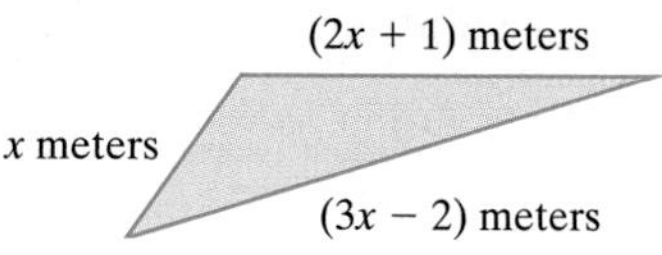

Name ______________________ Section __________ Date __________

Integrated Review—Solving Linear Equations

Solve. Feel free to use the steps given in Section 2.4.

1. $x - 10 = -4$

2. $y + 14 = -3$

3. $9y = 108$

4. $-3x = 78$

5. $-6x + 7 = 25$

6. $5y - 42 = -47$

7. $\frac{2}{3}x = 9$

8. $\frac{4}{5}z = 10$

9. $\frac{r}{-4} = -2$

10. $\frac{y}{-8} = 8$

11. $6 - 2x + 8 = 10$

12. $-5 - 6y + 6 = 19$

13. $2x - 7 = 6x - 27$

14. $3 + 8y = 3y - 2$

15. $-3a + 6 + 5a = 7a - 8a$

16. $4b - 8 - b = 10b - 3b$

17. $-\frac{2}{3}x = \frac{5}{9}$

18. $-\frac{3}{8}y = -\frac{1}{16}$

19. $10 = -6n + 16$

20. $-5 = -2m + 7$

Answers

1. ______
2. ______
3. ______
4. ______
5. ______
6. ______
7. ______
8. ______
9. ______
10. ______
11. ______
12. ______
13. ______
14. ______
15. ______
16. ______
17. ______
18. ______
19. ______
20. ______

21. ______

22. ______

23. ______

24. ______

25. ______

26. ______

27. ______

28. ______

29. ______

30. ______

31. ______

32. ______

21. $3(5c - 1) - 2 = 13c + 3$

22. $4(3t + 4) - 20 = 3 + 5t$

23. $\frac{2(z + 3)}{3} = 5 - z$

24. $\frac{3(w + 2)}{4} = 2w + 3$

25. $-2(2x - 5) = -3x + 7 - x + 3$

26. $-4(5x - 2) = -12x + 4 - 8x + 4$

27. $0.02(6t - 3) = 0.04(t - 2) + 0.02$

28. $0.03(m + 7) = 0.02(5 - m) + 0.03$

29. $-3y = \frac{4(y - 1)}{5}$

30. $-4x = \frac{5(1 - x)}{6}$

31. $\frac{5}{3}x - \frac{7}{3} = x$

32. $\frac{7}{5}n + \frac{3}{5} = -n$

2.5 An Introduction to Problem Solving

OBJECTIVE

A. Translate a problem to an equation, then use the equation to solve the problem.

SSM TUTOR CENTER SG CD & VIDEO MATH PRO WEB

In the preceding sections, we practiced translating phrases into expressions and sentences into equations as well as solving linear equations. We are now ready to put our skills to practical use. To begin, we present a general strategy for problem solving.

General Strategy for Problem Solving

1. UNDERSTAND the problem. During this step, become comfortable with the problem. Some ways of doing this are:
   Read and reread the problem.
   Choose a variable to represent the unknown.
   Construct a drawing.
   Propose a solution and check. Pay careful attention to how you check your proposed solution. This will help when writing an equation to model the problem.
2. TRANSLATE the problem into an equation.
3. SOLVE the equation.
4. INTERPRET the results: *Check* the proposed solution in the stated problem and *state* your conclusion.

A Translating and Solving Problems

Much of problem solving involves a direct translation from a sentence to an equation.

EXAMPLE 1 Finding an Unknown Number

Twice the sum of a number and 4 is the same as four times the number, decreased by 12. Find the number.

Solution:

1. UNDERSTAND. Read and reread the problem. If we let
   $x =$ the unknown number, then
   "the sum of a number and 4" translates to "$x + 4$" and
   "four times the number" translates to "$4x$"
2. TRANSLATE.

twice	sum of a number and 4	is the same as	four times the number	decreased by	12
↓	↓	↓	↓	↓	↓
2	$(x + 4)$	$=$	$4x$	$-$	12

3. SOLVE

$$2(x + 4) = 4x - 12$$
$$2x + 8 = 4x - 12 \quad \text{Apply the distributive property.}$$
$$2x + 8 - 4x = 4x - 12 - 4x \quad \text{Subtract } 4x \text{ from both sides.}$$
$$-2x + 8 = -12$$
$$-2x + 8 - 8 = -12 - 8 \quad \text{Subtract 8 from both sides.}$$
$$-2x = -20$$
$$\frac{-2x}{-2} = \frac{-20}{-2} \quad \text{Divide both sides by } -2.$$
$$x = 10$$

Practice Problem 1

Three times the difference of a number and 5 is the same as twice the number decreased by 3. Find the number.

Answer

1. The number is 12.

4. INTERPRET.

Check: Check this solution in the problem as it was originally stated. To do so, replace "number" with 10. Twice the sum of "10" and 4 is 28, which is the same as 4 times "10" decreased by 12.

State: The number is 10.

Practice Problem 2

An 18-foot wire is to be cut so that the longer piece is 5 times longer than the shorter piece. Find the length of each piece.

EXAMPLE 2 Finding the Length of a Board

A 10-foot board is to be cut into two pieces so that the longer piece is 4 times the shorter. Find the length of each piece.

Solution:

1. UNDERSTAND the problem. To do so, read and reread the problem. You may also want to propose a solution. For example, if 3 feet represents the length of the shorter piece, then $4(3) = 12$ feet is the length of the longer piece, since it is 4 times the length of the shorter piece. This guess gives a total board length of 3 feet + 12 feet = 15 feet, which is too long. However, the purpose of proposing a solution is not to guess correctly, but to help better understand the problem and how to model it.

In general, if we let

x = length of shorter piece, then
$4x$ = length of longer piece

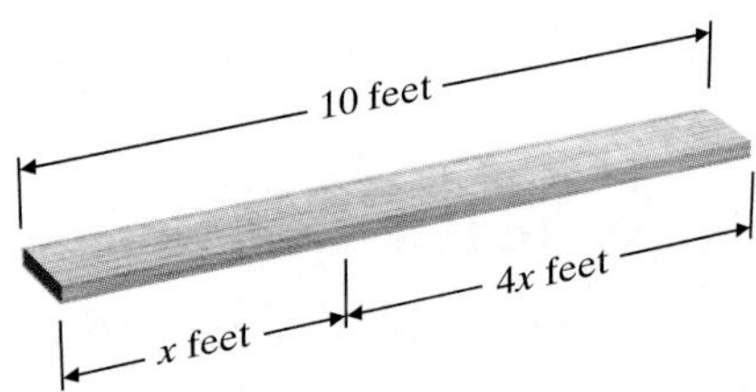

2. TRANSLATE the problem. First, we write the equation in words.

length of shorter piece	added to	length of longer piece	equals	total length of board
↓	↓	↓	↓	↓
x	$+$	$4x$	$=$	10

3. SOLVE.

$$x + 4x = 10$$
$$5x = 10 \quad \text{Combine like terms.}$$
$$\frac{5x}{5} = \frac{10}{5} \quad \text{Divide both sides by 5.}$$
$$x = 2$$

4. INTERPRET.

Check: Check the solution in the stated problem. If the shorter piece of board is 2 feet, the longer piece is $4 \cdot (2 \text{ feet}) = 8$ feet and the sum of the two pieces is 2 feet + 8 feet = 10 feet.

Answer

2. shorter piece = 3 ft; longer piece = 15 ft

State: The shorter piece of board is 2 feet and the longer piece of board is 8 feet.

Helpful Hint

Make sure that units are included in your answer, if appropriate.

EXAMPLE 3 Finding the Number of Republican and Democratic Senators

In the 107th Congress, the U.S. House of Representatives had a total of 430 Democrats and Republicans. There were 10 more Republican representatives than Democratic. Find the number of representatives from each party. (*Source:* Office of the Clerk of the U.S. House of Representatives)

Solution:

1. UNDERSTAND the problem. Read and reread the problem. Let's suppose that there are 200 Democratic representatives. Since there are 10 more Republicans than Democrats, there must be $200 + 10 = 210$ Republicans. The total number of Democrats and Republicans is then $200 + 210 = 410$. This is incorrect since the total should be 430, but we now have a better understanding of the problem.

In general, if we let

$$x = \text{number of Democrats, then}$$
$$x + 10 = \text{number of Republicans}$$

2. TRANSLATE the problem. First, we write the equation in words.

number of Democrats	added to	number of Republicans	equals	430
↓	↓	↓	↓	↓
x	$+$	$(x + 10)$	$=$	430

3. SOLVE.

$$x + (x + 10) = 430$$
$$2x + 10 = 430 \quad \text{Combine like terms.}$$
$$2x + 10 - 10 = 430 - 10 \quad \text{Subtract 10 from both sides.}$$
$$2x = 420$$
$$\frac{2x}{2} = \frac{420}{2} \quad \text{Divide both sides by 2.}$$
$$x = 210$$

4. INTERPRET.

Check: If there are 210 Democratic representatives, then there are $210 + 10 = 220$ Republican representatives. The total number of representatives is then $210 + 220 = 430$. The results check.

State: There are 210 Democratic and 220 Republican representatives in the 107th Congress.

Practice Problem 3

Starting in the year 2004, the state of California will have 21 more electoral votes for president than the state of Texas. If the total electoral votes for these two states will be 89, find the number of electoral votes for each state.

Answer

3. Texas = 34 electoral votes; California = 55 electoral votes

Practice Problem 4

Enterprise Car Rental charges a daily rate of \$34 plus \$0.20 per mile. Suppose that you rent a car for a day and your bill (before taxes) is \$104. How many miles did you drive?

Practice Problem 5

The measure of the second angle of a triangle is twice the measure of the smallest angle. The measure of the third angle of the triangle is three times the measure of the smallest angle. Find the measures of the angles.

Answers

4. 350 miles, **5.** smallest = 30°; second = 60°; third = 90°

EXAMPLE 4 Calculating Cellular Phone Usage

A local cellular phone company charges Elaine Chapoton \$50 per month and \$0.36 per minute of phone use in her usage category. If Elaine was charged \$99.68 for a month's cellular phone use, determine the number of whole minutes of phone use.

Solution:

1. UNDERSTAND. Read and reread the problem. Let's propose that Elaine uses the phone for 70 minutes. Pay careful attention as to how we calculate her bill. For 70 minutes of use, Elaine's phone bill will be \$50 plus \$0.36 per minute of use. This is $\$50 + 0.36(70) = \$75.20$, less than \$99.68. We now understand the problem and know that the number of minutes is greater than 70.

If we let

x = number of minutes, then

$0.36x$ = charge per minute of phone use

2. TRANSLATE.

\$50	added to	minute charge	is equal to	\$99.68
↓	↓	↓	↓	↓
50	+	$0.36x$	=	99.68

3. SOLVE.

$$50 + 0.36x = 99.68$$

$$50 + 0.36x - 50 = 99.68 - 50 \quad \text{Subtract 50 from both sides.}$$

$$0.36x = 49.68 \quad \text{Simplify.}$$

$$\frac{0.36x}{0.36} = \frac{49.68}{0.36} \quad \text{Divide both sides by 0.36.}$$

$$x = 138 \quad \text{Simplify.}$$

4. INTERPRET.

Check: If Elaine spends 138 minutes on her cellular phone, her bill is $\$50 + \$0.36(138) = \$99.68$.

State: Elaine spent 138 minutes on her cellular phone this month.

EXAMPLE 5 Finding Angle Measures

If the two walls of the Vietnam Veterans Memorial in Washington, D.C., were connected, an isosceles triangle would be formed. The measure of the third angle is 97.5° more than the measure of either of the other two equal angles. Find the measure of the third angle. (*Source:* National Park Service)

Solution:

1. UNDERSTAND. Read and reread the problem. We then draw a diagram (recall that an isosceles triangle has two angles with the same measure) and let

x = degree measure of one angle
x = degree measure of the second equal angle
$x + 97.5$ = degree measure of the third angle

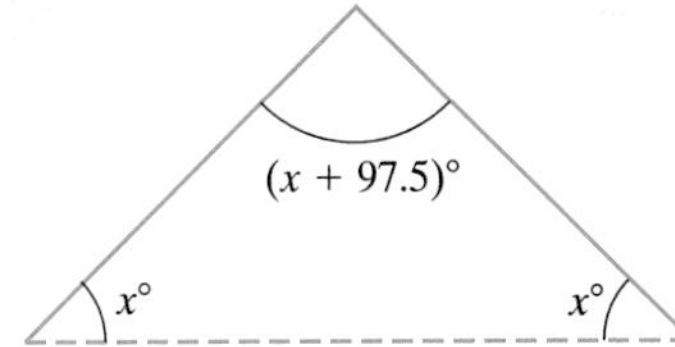

2. TRANSLATE. Recall that the sum of the measures of the angles of a triangle equals 180.

measure of first angle		measure of second angle		measure of third angle	equals	180
↓		↓		↓	↓	↓
x	$+$	x	$+$	$(x + 97.5)$	$=$	180

3. SOLVE.

$$x + x + (x + 97.5) = 180$$
$$3x + 97.5 = 180 \quad \text{Combine like terms.}$$
$$3x + 97.5 - 97.5 = 180 - 97.5 \quad \text{Subtract 97.5 from both sides.}$$
$$3x = 82.5$$
$$\frac{3x}{3} = \frac{82.5}{3} \quad \text{Divide both sides by 3.}$$
$$x = 27.5$$

4. INTERPRET.

Check: If $x = 27.5$, then the measure of the third angle is $x + 97.5 = 125$. The sum of the angles is then $27.5 + 27.5 + 125 = 180$, the correct sum.

State: The third angle measures 125°.*

*(The two walls actually meet at an angle of 125 degrees 12 minutes. The measurement of 97.5° given in the problem is an approximation.)

STUDY SKILLS REMINDER

How Are Your Homework Assignments Going?

It is so important in mathematics to keep up with homework. Why? Many concepts build on each other. Oftentimes, your understanding of a day's lecture in mathematics depends on an understanding of the previous day's material.

Remember that completing your homework assignment involves a lot more than attempting a few of the problems assigned.

To complete a homework assignment, remember these four things:

1. Attempt all of it.
2. Check it.
3. Correct it.
4. If needed, ask questions about it.

FOCUS ON The Real World

SURVEYS

Recall that the golden rectangle is a rectangle whose length is approximately 1.6 times its width. It is thought that for about 75% of adults, a rectangle in the shape of the golden rectangle is most pleasing to the eye.

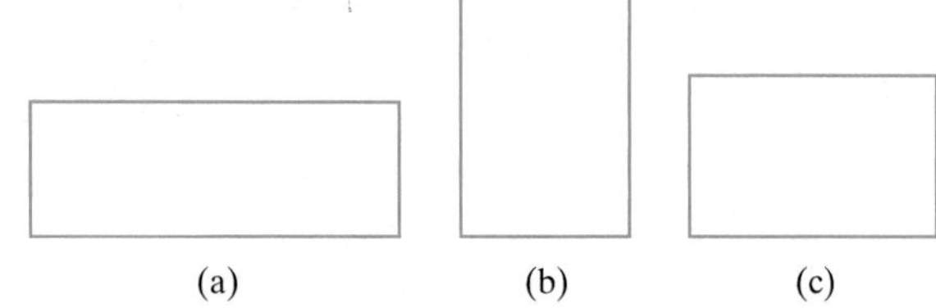

GROUP ACTIVITIES

1. Measure the dimensions of each of the three rectangles shown at left and decide which one best approximates the shape of the golden rectangle.
2. Using the three rectangles shown at left, conduct a survey asking students which rectangle they prefer. (To avoid bias, take care not to reveal which rectangle is the golden rectangle.) Tally your results and find the percent of survey respondents who preferred each rectangle. Do your results agree with the percent suggested above?

Name ______________________ Section __________ Date ____________

EXERCISE SET 2.5

 *Solve. See Example 1.*

1. The sum of twice a number and $\frac{1}{5}$ is equal to the difference between three times the number and $\frac{4}{5}$. Find the number.

2. The sum of four times a number and $\frac{2}{3}$ is equal to the difference of five times the number and $\frac{5}{6}$. Find the number.

 **3.** Twice the difference of a number and 8 is equal to three times the sum of the number and 3. Find the number.

4. Five times the sum of a number and -1 is the same as 6 times the number. Find the number.

5. The product of twice a number and three is the same as the difference of five times the number and $\frac{3}{4}$. Find the number.

6. If the difference of a number and four is doubled, the result is $\frac{1}{4}$ less than the number. Find the number.

7. If the sum of a number and five is tripled, the result is one less than twice the number. Find the number.

8. Twice the sum of a number and six equals three times the sum of the number and four. Find the number.

Solve. See Examples 2 through 5.

9. The governor of Michigan makes \$39,000 more per year than the governor of Oregon. If the total of their salaries is \$215,600, find the salary of each. (*Source: The World Almanac, 2001*)

10. In the 2000 Summer Olympics, the United States Team won 12 more gold medals than the China Team. If the total number of gold medals for both is 68, find the number of gold medals that each team won. (*Source: The World Almanac, 2001*)

 **11.** A 40-inch board is to be cut into three pieces so that the second piece is twice as long as the first piece and the third piece is 5 times as long as the first piece. If x represents the length of the first piece, find the lengths of all three pieces.

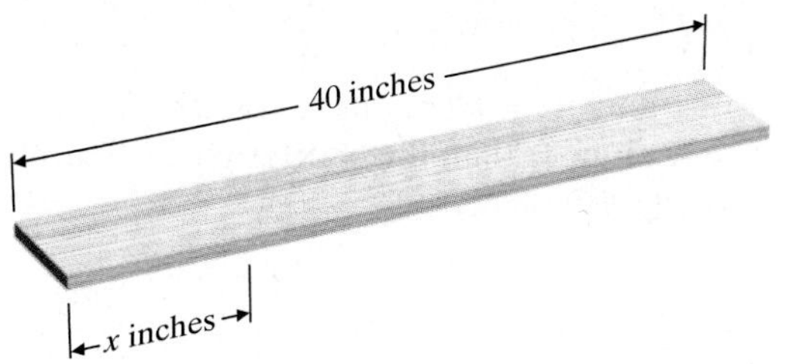

12. A 21-foot beam is to be divided so that the longer piece is 1 foot more than 3 times the shorter piece. If x represents the length of the shorter piece, find the lengths of both pieces.

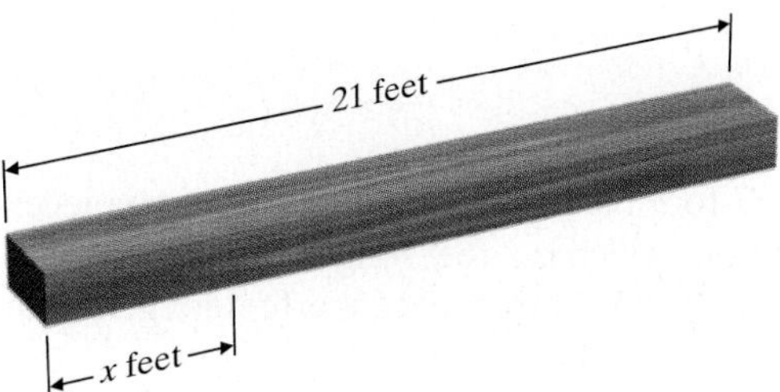

13. A car rental agency advertised renting a Buick Century for \$24.95 per day and \$0.29 per mile. If you rent this car for 2 days, how many whole miles can you drive on a \$100 budget?

14. A plumber gave an estimate for the renovation of a kitchen. Her hourly pay is \$27 per hour and the plumber's parts will cost \$80. If her total estimate is \$404, how many hours does she expect this job to take?

△ **15.** The flag of Equatorial Guinea contains an isosceles triangle. (Recall that an isosceles triangle contains two angles with the same measure.) If the measure of the third angle of the triangle is 30° more than twice the measure of either of the other two angles, find the measure of each angle of the triangle. (*Hint*: Recall that the sum of the measures of the angles of a triangle is 180°.)

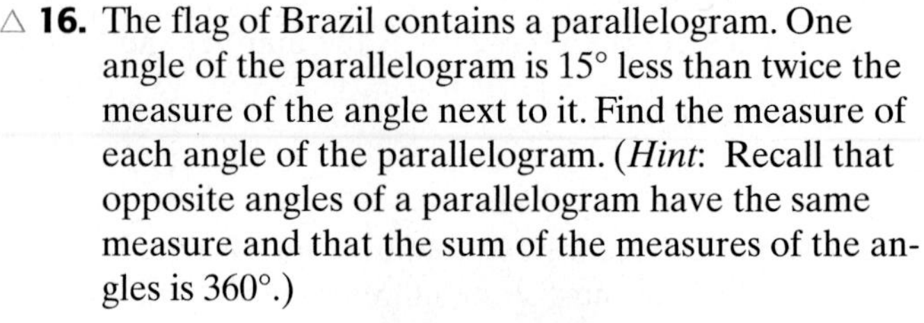

△ **16.** The flag of Brazil contains a parallelogram. One angle of the parallelogram is 15° less than twice the measure of the angle next to it. Find the measure of each angle of the parallelogram. (*Hint*: Recall that opposite angles of a parallelogram have the same measure and that the sum of the measures of the angles is 360°.)

17. In the Crayola Color Census 2000, Crayola crayon users were asked to vote for their favorite Crayola color. The color blue was ranked first, followed by cerulean in second place. Blue received 3366 more votes than cerulean. Together, both colors received a total of 19,278 votes. Find the number of votes each color received. (*Source:* Binney & Smith, Inc.)

18. In 2000 the U.S. poverty threshold for a family of four with two children was \$568 more than the poverty threshold for the same family in 1999. the sum of the poverty thresholds for 1999 and 2000 was \$34,358. What was the poverty threshold in each year? (*Source:* U.S. Census Bureau)

△ **19.** Two angles are supplementary if their sum is 180°. One angle measures three times the measure of a smaller angle. If x represents the measure of the smaller angle and these two angles are supplementary, find the measure of each angle.

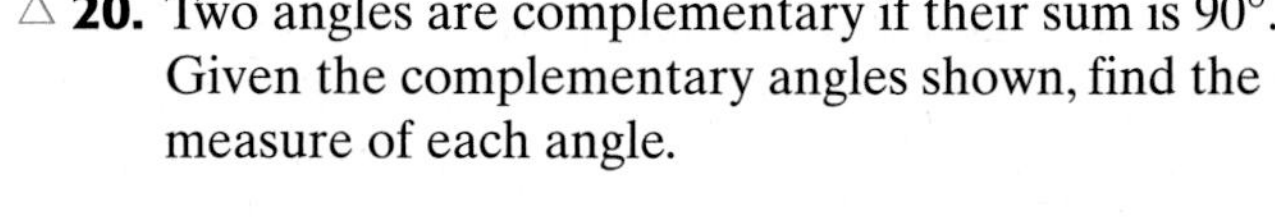

△ **20.** Two angles are complementary if their sum is 90°. Given the complementary angles shown, find the measure of each angle.

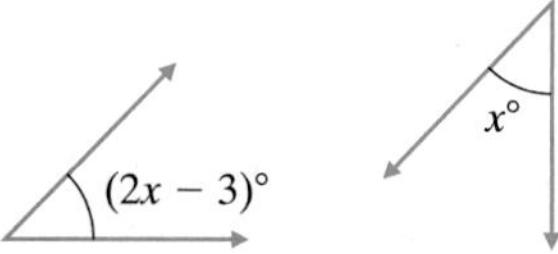

21. A 17-foot piece of string is cut into two pieces so that one piece is 2 feet longer than twice the shorter piece. If the shorter piece is x feet long, find the lengths of both pieces.

22. An 18-foot wire is to be cut so that the longer piece is 5 times longer than the shorter piece. Find the length of each piece.

23. On April 7, 2001, the Mars Odyssey spacecraft was launched, beginning a multi-year mission to observe and map the planet Mars. Mars Odyssey was launched on Boeing's Delta II 7925 launch vehicle using nine strap-on solid rocket motors. Each solid rocket motor has a height that is 8 meters more than 5 times its diameter. If the sum of the height and the diameter for a single solid rocket motor is 14 meters, find each dimension. (*Source:* NASA)

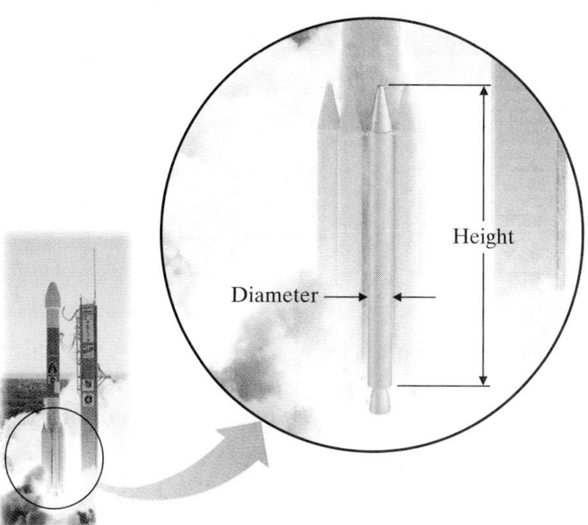

24. Over the past few years the satellite Voyager II has passed by the planets Saturn, Uranus, and Neptune, continually updating information about these planets, including the number of moons for each. Uranus is now believed to have 13 more moons than Neptune. Also, Saturn is now believed to have 2 more than twice the number of moons of Neptune. If the total number of moons for these planets is 47, find the number of moons for each planet. (*Source:* National Space Science Data Center)

25. The area of the Sahara Desert is 7 times the area of the Gobi Desert. If the sum of their areas is 4,000,000 square miles, find the area of each desert.

26. The largest meteorite in the world is the Hoba West located in Namibia. Its weight is 3 times the weight of the Armanty meteorite located in Outer Mongolia. If the sum of their weights is 88 tons, find the weight of each.

The graph below shows the states with the highest tourism budgets for a recent year. The height of the bar for each state corresponds to the amount of money spent on tourism. Use the graph for Exercises 27 through 32.

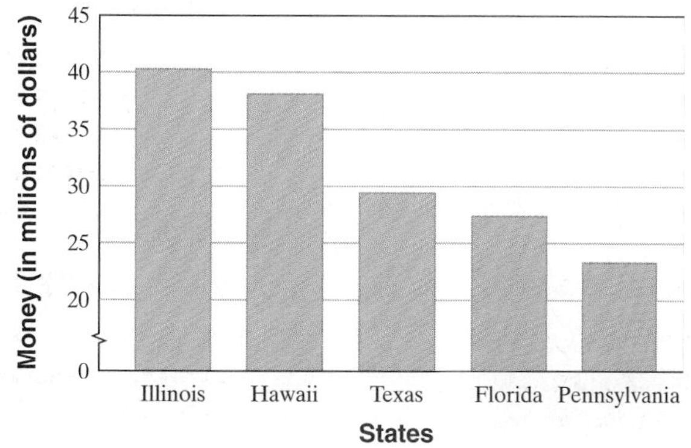

27. In your own words, describe what the word *tourism* means.

28. Which state spends the most money on tourism?

29. Which states spend between \$25 and \$30 million on tourism?

30. The states of Texas and Florida spend a total of \$56.6 million for tourism. The state of Texas spends \$2.2 million more than the state of Florida. Find the amount that each state spends on tourism.

31. The states of Hawaii and Pennsylvania spend a total of \$60.9 million for tourism. The state of Hawaii spends \$8.1 million less than twice the amount of money that the state of Pennsylvania spends. Find the amount that each state spends on tourism.

32. Compare the heights of the bars in the graph with your results of Exercises 30 and 31. Are your answers reasonable?

33. The Pentagon in Washington, D.C., is the headquarters for the U.S. Department of Defense. The Pentagon is also the world's largest office building in terms of ground space with a floor area of over 6.5 million square feet. This is three times the floor area of the Empire State Building. About how much floor space does the Empire State Building have? Round to the nearest tenth of a million.

34. Hertz Car Rental charges a daily rate of \$39 plus \$0.20 per mile for a certain car. Suppose that you rent that car for a day and your bill (before taxes) is \$95. How many miles did you drive?

Solve. These applications have to do with consecutive integers. For a review of consecutive integers, see Section 2.3.

35. On April 1, 2001, Notre Dame defeated Purdue in the 2001 NCAA Division I Women's Basketball Championship. The two teams' final scores for the game were two consecutive even integers whose sum was 134. Find each final score. (*Source:* National Collegiate Athletic Association)

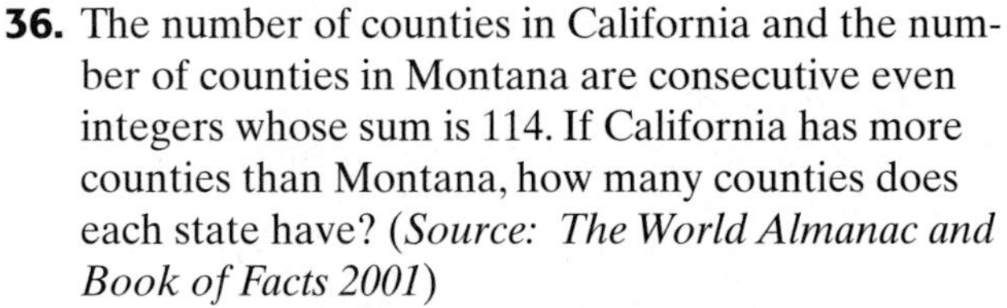

36. The number of counties in California and the number of counties in Montana are consecutive even integers whose sum is 114. If California has more counties than Montana, how many counties does each state have? (*Source: The World Almanac and Book of Facts 2001*)

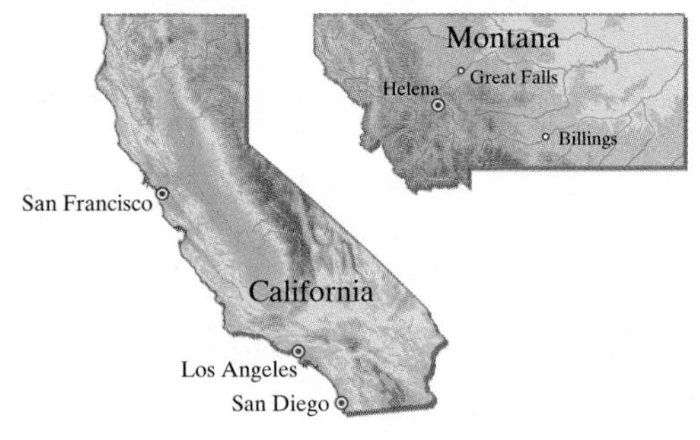

37. In the 2000 Summer Olympics, China won more medals than Australia, who won more medals than Germany. If the number of medals won by each country is three consecutive integers whose sum is 174, how many medals did each country win? (*Source: The World Almanac and Book of Facts 2001*)

38. To make an international telephone call, you need the code for the country you are calling. The codes for Mali Republic, Côte d'Ivoire, and Niger are three consecutive odd integers whose sum is 675. Find the code for each country.

△ **39.** The measures of the angles of a triangle are 3 consecutive even integers. Find the measure of each angle.

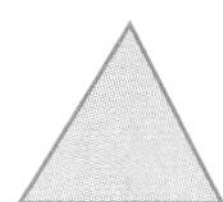

△ **40.** A quadrilateral is a polygon with 4 sides. The sum of the measures of the 4 angles in a quadrilateral is 360°. If the measures of the angles of a quadrilateral are consecutive odd integers, find the measures.

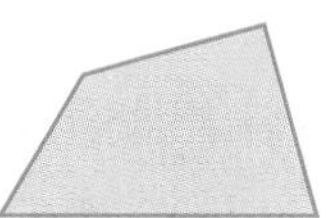

Review and Preview

Translate each sentence into an equation. See Sections 1.2 and 2.5.

41. Half of the difference of a number and one is thirty-seven.

42. Five times the opposite of a number is the number plus sixty.

43. If three times the sum of a number and 2 is divided by 5, the quotient is 0.

44. If the sum of a number and 9 is subtracted from 50, the result is 0.

Evaluate each expression for the given values. See Section 1.2.

45. $2W + 2L$; $W = 7$ and $L = 10$

46. $\frac{1}{2}Bh$; $B = 14$ and $h = 22$

47. πr^2; $r = 15$

48. $r \cdot t$; $r = 15$ and $t = 2$

Combining Concepts

49. Give an example of how you recently solved a problem using mathematics.

50. In your own words, explain why a solution of a word problem should be checked using the original wording of the problem and not the equation written from the wording.

△ **51.** The golden rectangle is a rectangle whose length is approximately 1.6 times its width. The early Greeks thought that a rectangle with these dimensions was the most pleasing to the eye and examples of the golden rectangle are found in many early works of art. For example, the Parthenon in Athens contains many examples of golden rectangles. Mike Hallahan would like to plant a rectangular garden in the shape of a golden rectangle. If he has 78 feet of fencing available, find the dimensions of the garden.

The length-width rectangle approximates the golden rectangle as well as the width-height rectangle.

△ **53.** Examples of golden rectangles can be found today in architecture and manufacturing packaging. Find an example of a golden rectangle in your home. A few suggestions: the front face of a book, the floor of a room, the front of a box of food.

△ **52.** It is thought that for about 75% of adults, a rectangle in the shape of the golden rectangle is the most pleasing to the eye. Draw 3 rectangles, one in the shape of the golden rectangle, and poll your class. Do the results agree with the percentage given above?

2.6 Formulas and Problem Solving

A Using Formulas to Solve Problems

A **formula** describes a known relationship among quantities. Many formulas are given as equations. For example, the formula

$$d = r \cdot t$$

stands for the relationship

$$\text{distance} = \text{rate} \cdot \text{time}$$

Let's look at one way that we can use this formula.

If we know we traveled a distance of 100 miles at a rate of 40 miles per hour, we can replace the variables d and r in the formula $d = rt$ and find our travel time, t.

$d = rt$ Formula

$100 = 40t$ Replace d with 100 and r with 40.

To solve for t, we divide both sides of the equation by 40.

$\frac{100}{40} = \frac{40t}{40}$ Divide both sides by 40.

$\frac{5}{2} = t$ Simplify.

The travel times was $\frac{5}{2}$ hours, or $2\frac{1}{2}$ hours.

In this section, we solve problems that can be modeled by known formulas. We use the same problem-solving strategy that was introduced in the previous section.

OBJECTIVES

A Use formulas to solve problems.

B Solve a formula or equation for one of its variables.

EXAMPLE 1 Finding Time Given Rate and Distance

A glacier is a giant mass of rocks and ice that flows downhill like a river. Portage Glacier in Alaska is about 6 miles, or 31,680 *feet*, long and moves 400 *feet* per year. Icebergs are created when the front end of the glacier flows into Portage Lake. How long does it take for ice at the head (beginning) of the glacier to reach the lake?

Practice Problem 1

A family is planning their vacation to visit relatives. They will drive from Cincinnati, Ohio to Rapid City, South Dakota, a distance of 1180 miles. They plan to average a rate of 50 miles per hour. How much time will they spend driving?

Answer

1. 23.6 hours

Solution:

1. UNDERSTAND. Read and reread the problem. The appropriate formula needed to solve this problem is the distance formula, $d = rt$. To become familiar with this formula, let's find the distance that ice traveling at a rate of 400 feet per year travels in 100 years. To do so, we let time t be 100 years and rate r be the given 400 feet per year, and substitute these values into the formula $d = rt$. We then have that distance $d = 400(100) = 40{,}000$ feet. Since we are interested in finding how long it takes ice to travel 31,680 feet, we now know that it is less than 100 years.

   Since we are using the formula $d = rt$, we let

   $t =$ the time in years for ice to reach the lake

   $r =$ rate or speed of ice

   $d =$ distance from beginning of glacier to lake

2. TRANSLATE. To translate to an equation, we use the formula $d = rt$ and let distance $d = 31{,}680$ feet and rate $r = 400$ feet per year.

   $$d = r \cdot t$$

   $$31{,}680 = 400 \cdot t \qquad \text{Let } d = 31{,}680 \text{ and } r = 400.$$

3. SOLVE. Solve the equation for t. To solve for t, divide both sides by 400.

   $$\frac{31{,}680}{400} = \frac{400 \cdot t}{400} \qquad \text{Divide both sides by 400.}$$

   $$79.2 = t \qquad \text{Simplify.}$$

4. INTERPRET.

Check: To check, substitute 79.2 for t and 400 for r in the distance formula and check to see that the distance is 31,680 feet.

State: It takes 79.2 years for the ice at the head of Portage Glacier to reach the lake.

Helpful Hint

Don't forget to include units, if appropriate.

Practice Problem 2

A wood deck is being built behind a house. The width of the deck must be 18 feet because of the shape of the house. If there is 450 square feet of decking material, find the length of the deck.

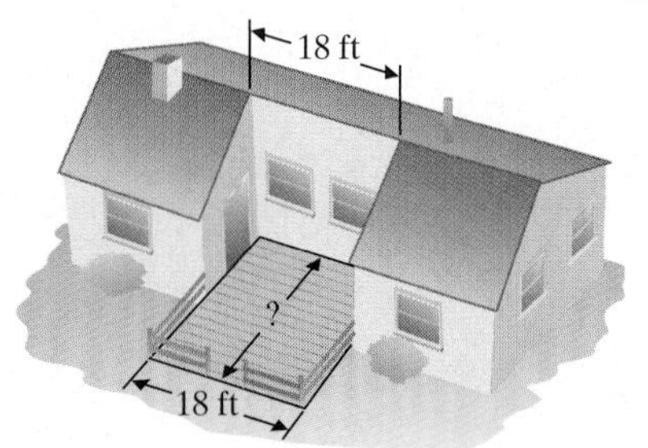

Answer

2. 25 feet

EXAMPLE 2 Calculating the Length of a Garden

Charles Pecot can afford enough fencing to enclose a rectangular garden with a perimeter of 140 feet. If the width of his garden is to be 30 feet, find the length.

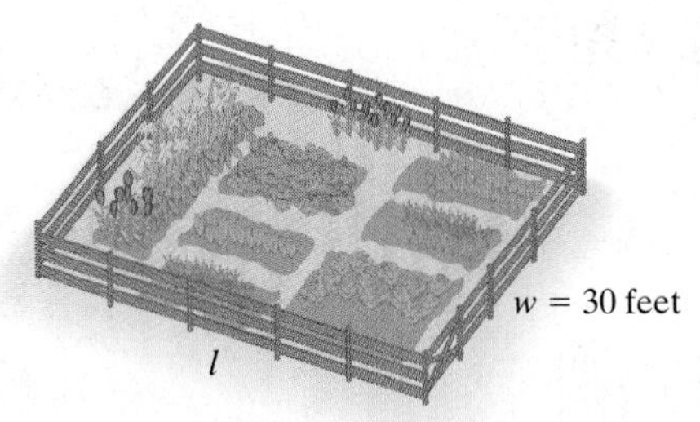

Solution:

1. UNDERSTAND. Read and reread the problem. The formula needed to solve this problem is the formula for the perimeter of a rectangle, $P = 2l + 2w$. Before continuing, let's become familar with this formula.

$l =$ the length of the rectangular garden

$w =$ the width of the rectangular garden

$P =$ perimeter of the garden

2. TRANSLATE. To translate to an equation, we use the formula $P = 2l + 2w$ and let perimeter $P = 140$ feet and width $w = 30$ feet.

$P = 2l + 2w$ Let $P = 140$ and $w = 30$.

$140 = 2l + 2(30)$

3. SOLVE.

$$140 = 2l + 2(30)$$
$$140 = 2l + 60 \quad \text{Multiply } 2(30).$$
$$140 - 60 = 2l + 60 - 60 \quad \text{Subtract 60 from both sides.}$$
$$80 = 2l \quad \text{Combine like terms.}$$
$$40 = l \quad \text{Divide both sides by 2.}$$

4. INTERPRET.

Check: Substitute 40 for l and 30 for w in the perimeter formula and check to see that the perimeter is 140 feet.

State: The length of the rectangular garden is 40 feet.

Solving a Formula for a Variable

We say that the formula

$$d = rt$$

is solved for d because d is alone on one side of the equation and the other side contains no d's. Suppose that we have a large number of problems to solve where we are given distance d and rate r and asked to find time t. In this case, it may be easier to first solve the formula $d = rt$ for t. To solve for t, we divide both sides of the equation by r.

$$d = rt$$
$$\frac{d}{r} = \frac{rt}{r} \quad \text{Divide both sides by } r.$$
$$\frac{d}{r} = t \quad \text{Simplify.}$$

To solve a formula or an equation for a specified variable, we use the same steps as for solving a linear equation. These steps are listed next.

To Solve Equations for a Specified Variable

Step 1. If an equation contains fractions, multiply both sides by the LCD to clear the equation of fractions.

Step 2. Use the distributive property to remove parentheses if they occur.

Step 3. Simplify each side of the equation by combining like terms.

Step 4. Get all terms containing the specified variable on one side and all other terms on the other side by using the addition property of equality.

Step 5. Get the specified variable alone by using the multiplication property of equality.

Practice Problem 3

Solve $C = 2\pi r$ for r. (This formula is used to find the circumference C of a circle given its radius r.)

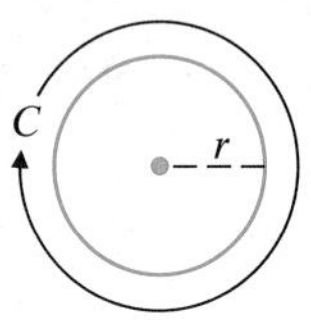

EXAMPLE 3 Solve $V = lwh$ for l.

Solution: This formula is used to find the volume of a box. To solve for l, we divide both sides by wh.

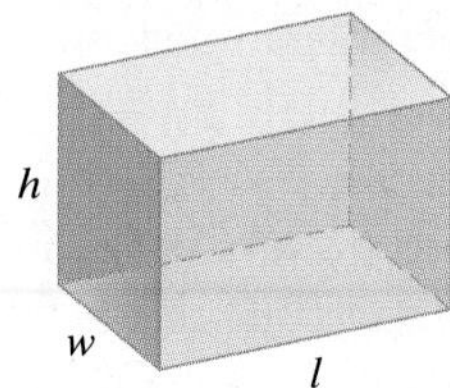

$$V = lwh$$

$$\frac{V}{wh} = \frac{lwh}{wh} \quad \text{Divide both sides by } wh.$$

$$\frac{V}{wh} = l \quad \text{Simplify.}$$

Since we have l alone on one side of the equation, we have solved for l in terms of V, w, and h. Remember that it does not matter on which side of the equation we get the variable alone.

Practice Problem 4 

Solve $P = 2l + 2w$ for w.

EXAMPLE 4 Solve $y = mx + b$ for x.

Solution: First we get mx alone by subtracting b from both sides.

$$y = mx + b$$

$$y - b = mx + b - b \quad \text{Subtract } b \text{ from both sides.}$$

$$y - b = mx \quad \text{Combine like terms.}$$

Next we solve for x by dividing both sides by m.

$$\frac{y-b}{m} = \frac{mx}{m}$$

$$\frac{y-b}{m} = x \quad \text{Simplify.}$$

Practice Problem 5 

Solve $A = \dfrac{a+b}{2}$ for b.

EXAMPLE 5 Solve $A = \dfrac{bh}{2}$ for h.

Solution: First let's clear the equation of fractions by multiplying both sides by 2.

$$A = \frac{bh}{2}$$

$$2 \cdot A = 2\left(\frac{bh}{2}\right) \quad \text{Multiply both sides by 2 to clear fractions.}$$

$$2A = bh$$

$$\frac{2A}{b} = \frac{bh}{b} \quad \text{Divide both sides by } b \text{ to get } h \text{ alone.}$$

$$\frac{2A}{b} = h \quad \text{Simplify.}$$

Answers

3. $r = \dfrac{C}{2\pi}$, **4.** $w = \dfrac{P - 2l}{2}$, **5.** $b = 2A - a$

Name ______________________ Section ____________ Date ____________

EXERCISE SET 2.6

Substitute the given values into each given formula and solve for the unknown variable. See Examples 1 and 2.

△ **1.** $A = bh$; $A = 45$, $b = 15$
(Area of a parallelogram)

2. $d = rt$; $d = 195$, $t = 3$
(Distance formula)

△ **3.** $S = 4lw + 2wh$; $S = 102$, $l = 7$, $w = 3$
(Surface area of a special rectangular box)

△ **4.** $V = lwh$; $l = 14$, $w = 8$, $h = 3$
(Volume of a rectangular box)

△ **5.** $A = \frac{1}{2}(B + b)h$; $A = 180$, $B = 11$, $b = 7$
(Area of a trapezoid)

△ **6.** $A = \frac{1}{2}(B + b)h$; $A = 60$, $B = 7$, $b = 3$
(Area of a trapezoid)

△ **7.** $P = a + b + c$; $P = 30$, $a = 8$, $b = 10$
(Perimeter of a triangle)

△ **8.** $V = \frac{1}{3}Ah$; $V = 45$, $h = 5$
(Volume of a pyramid)

△ **9.** $C = 2\pi r$; $C = 15.7$ (use the approximation 3.14 for π)
(Circumference of a circle)

△ **10.** $A = \pi r^2$; $r = 4.5$ (use the approximation 3.14 for π)
(Area of a circle)

11. $I = PRT$; $I = 3750$, $P = 25{,}000$, $R = 0.05$
(Simple interest formula)

12. $I = PRT$; $I = 1{,}056{,}000$, $R = 0.055$, $T = 6$
(Simple interest formula)

△ **13.** $V = \frac{1}{3}\pi r^2 h$; $V = 565.2$, $r = 6$ (use the approximation 3.14 for π)
(Volume of a cone)

△ **14.** $V = \frac{4}{3}\pi r^3$; $r = 3$ (use the approximation 3.14 for π)
(Volume of a sphere)

Solve. See Examples 1 and 2.

△ **15.** The world's largest sign for Coca-Cola is located in Arica, Chile. The rectangular sign has a length of 400 feet and has an area of 52,400 square feet. Find the width of the sign. (*Source:* Fabulous Facts about Coca-Cola, Atlanta, GA)

△ **16.** The length of a rectangular garden is 6 meters. If 21 meters of fencing are required to fence the garden, find its width.

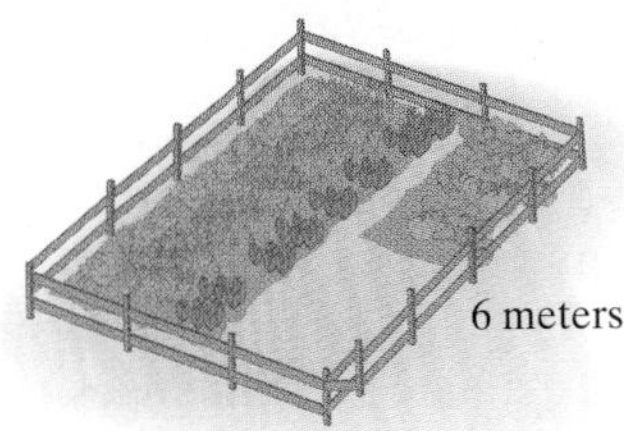

17. The Cat is a high-speed catamaran auto ferry that operates between Bar Harbor, Maine, and Yarmouth, Nova Scotia. The Cat can make the trip in about $2\frac{1}{2}$ hours at a speed of 55 mph. About how far apart are Bar Harbor and Yarmouth? (*Source:* Bay Ferries)

18. A family is planning their vacation to Disney World. They will drive from a small town outside New Orleans, Louisiana, to Orlando, Florida, a distance of 700 miles. They plan to average a rate of 55 miles per hour. How long will this trip take?

19. The highest temperature ever recorded in Europe was 122°F in Seville, Spain, in August 1881. Convert this record high temperature to Celsius. (*Source:* National Climatic Data Center)

20. The lowest temperature ever recorded in Oceania was −10°C at the Haleakala Summit in Maui, Hawaii, in January 1961. Convert this record low temperature to Fahrenheit. (*Source:* National Climatic Data Center)

△ **21.** Piranha fish require 1.5 cubic feet of water per fish to maintain a healthy environment. Find the maximum number of piranhas you could put in a tank measuring 8 feet by 3 feet by 6 feet.

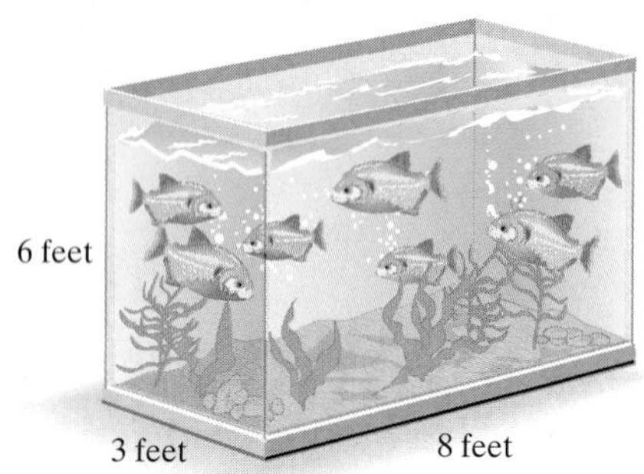

△ **22.** Find how many goldfish you can put in a cylindrical tank whose diameter is 8 meters and whose height is 3 meters if each goldfish needs 2 cubic meters of water.

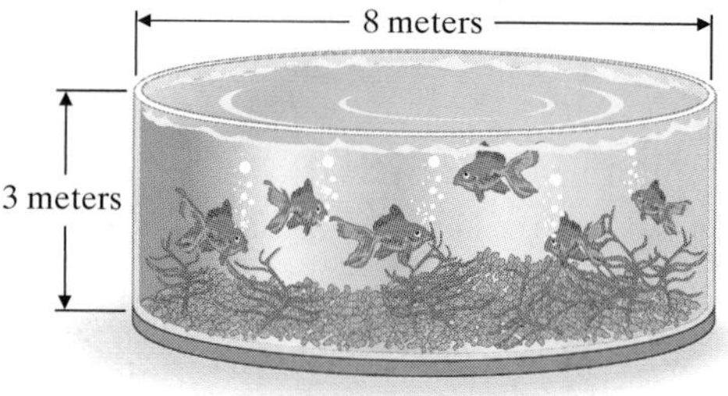

△ **23.** The longest runway at Los Angeles International Airport has the shape of a rectangle and an area of 1,813,500 square feet. This runway is 150 feet wide. How long is the runway? (*Source:* Los Angeles World Airports)

24. Beaumont, Texas, is about 150 miles from Toledo Bend. If Leo Miller leaves Beaumont at 4 a.m. and averages 45 mph, when should he arrive at Toledo Bend?

25. The X-30 is a new "space plane" being developed that will skim the edge of space at 4000 miles per hour. Neglecting altitude, if the circumference of the Earth is approximately 25,000 miles, how long will it take for the X-30 to travel around the Earth?

26. In the United States, the longest hang glider flight was a 303-mile, $8\frac{1}{2}$-hour flight from New Mexico to Kansas. What was the average rate during this flight?

△ **27.** A lawn is in the shape of a trapezoid with a height of 60 feet and bases of 70 feet and 130 feet. How many bags of fertilizer must be purchased to cover the lawn if each bag covers 4000 square feet?

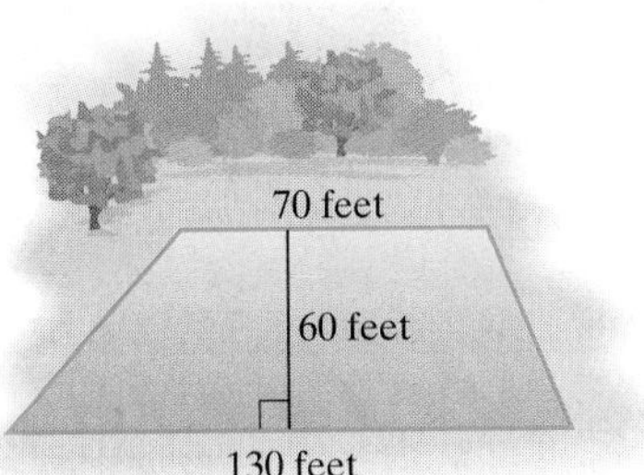

△ **28.** If the area of a right-triangularly shaped sail is 20 square feet and its base is 5 feet, find the height of the sail.

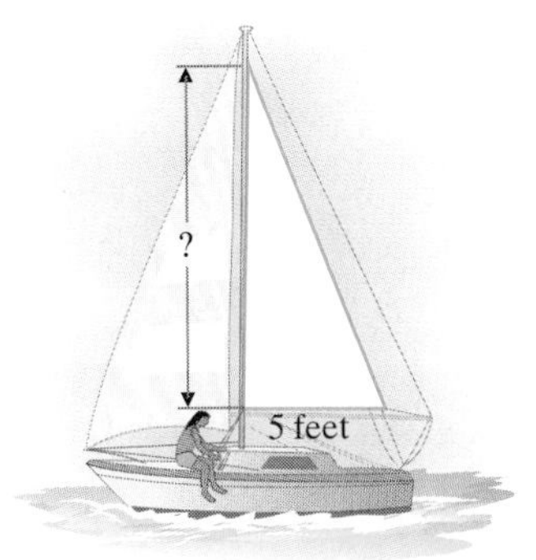

△ **29.** The CART Fed Ex Championship Series is an open-wheeled race car competition based in the United States. A CART car has a maximum length of 199 inches, a maximum width of 78.5 inches, and a maximum height of 33 inches. When the CART series travels to another country for a grand prix, teams must ship their cars. Find the volume of the smallest shipping crate needed to ship a CART car of maximum dimensions. (*Source:* Championship Auto Racing Teams, Inc.)

CART Racing Car

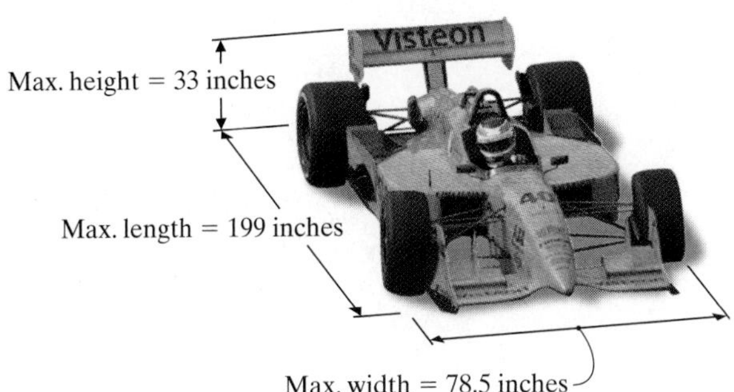

30. On a road course, a CART car's speed can average up to around 105 mph. Based on this speed, how long would it take a CART driver to travel from Los Angeles to New York City, a distance of about 2810 miles by road, without stopping? Round to the nearest tenth of an hour.

△ **31.** Maria's Pizza sells one 16-inch cheese pizza or two 10-inch cheese pizzas for $9.99. Determine which size gives more pizza.

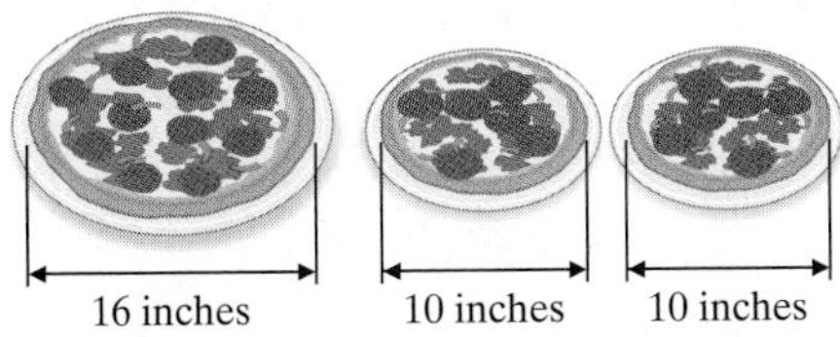

△ **32.** Find how much rope is needed to wrap the Earth at the equator, if the radius of the Earth is 4000 miles. (*Hint*: Use 3.14 for π and the formula for circumference.)

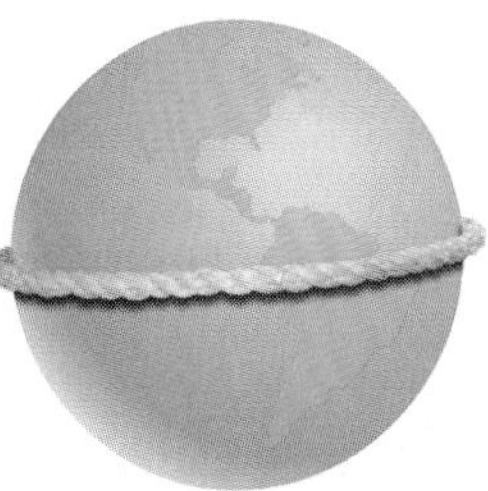

33. Dry ice is a name given to solidified carbon dioxide. At $-78.5°C$ it changes directly from a solid to a gas. Convert this temperature to degrees Fahrenheit.

34. Lightning bolts can reach a temperature of 50,000°F. Convert this temperature to degrees Celsius.

35. The distance from the sun to the Earth is approximately 93,000,000 miles. If light travels at a rate of 186,000 miles per second, how long does it take light from the sun to reach us?

36. Light travels at a rate of 186,000 miles per second. If our moon is 238,860 miles from the Earth, how long does it take light reflected off the moon to reach us? (Round to the nearest tenth of a second.)

△ **37.** The Hoberman Sphere is a toy ball that expands and contracts. When it is completely closed, it has a diameter of 9.5 inches. Find the volume of the Hoberman Sphere when it is completely closed. Use 3.14 for π. Round to the nearest whole cubic inch. (*Source:* Hoberman Designs, Inc.)

△ **38.** When the Hoberman Sphere (see Exercise 37) is completely expanded, its diameter is 30 inches. Find the volume of the Hoberman Sphere when it is completely expanded. Use 3.14 for π. Round to the nearest whole cubic inch. (*Source:* Hoberman Designs, Inc.)

39. The average temperature on the planet Mercury is 167°C. Convert this temperature to degrees Fahrenheit. (*Source:* National Space Science Data Center)

40. The average temperature on the planet Jupiter is −227°F. Convert this temperature to degrees Celsius. Round to the nearest degree. (*Source:* National Space Science Data Center)

41. Bolts of lightning can travel at 270,000 miles per second. How many times can a lightning bolt travel around the world at the equator, in one second? (See Exercise 32. Round to the nearest tenth.)

42. A glacier is a giant mass of rocks and ice that flows downhill like a river. Exit Glacier, near Seward, Alaska, moves at a rate of 20 inches a day. Find the distance in feet the glacier moves in a year. (Assume 365 days a year. Round to 2 decimal places.)

43. Flying fish do not *actually* fly, but glide. They have been known to travel a distance of 1300 feet at a rate of 20 miles per hour. How many seconds did it take to travel this distance? (*Hint:* First convert miles per hour to feet per second. Recall that 1 mile = 5280 feet. Round to the nearest tenth of a second.)

△ **44.** Stalactites join stalagmites to form columns. A column found at Natural Bridge Caverns near San Antonio, Texas, rises 15 feet and has a *diameter* of only 2 inches. Find the volume of this column in cubic inches. (*Hint*: Use the formula for volume of a cylinder and use 3.14 for π.)

45. A Japanese "bullet" train set a new world record for train speed at 552 kilometers per hour during a manned test run on the Yamanashi Maglev Test Line in April 1999. The Yamanashi Maglev Test Line is 42.8 kilometers long. How many *minutes* would a test run on the Yamanashi Line last at this record-setting speed? Round to the nearest hundredth of a minute. (*Source:* Japan Railways Central Co.)

46. In 1983, the Hawaiian volcano Kilauea began erupting in a series of episodes still occurring at the time of this writing. At times, the lava flows advanced at speeds of up to 0.5 kilometer per hour. In 1983 and 1984 lava flows destroyed 16 homes in the Royal Gardens subdivision, about 6 km away from the eruption site. Roughly how long did it take the lava to reach Royal Gardens? Round to the nearest hour. (*Source:* U.S. Geological Survey Hawaiian Volcano Observatory)

B *Solve each formula for the specified variable. See Examples 3 through 5.*

47. $f = 5gh$ for h

△ **48.** $C = 2\pi r$ for r

△ **49.** $V = LWH$ for W

50. $T = mnr$ for n

51. $3x + y = 7$ for y

52. $-x + y = 13$ for y

53. $A = P + PRT$ for R

54. $A = P + PRT$ for T

△ **55.** $V = \frac{1}{3}Ah$ for A

56. $D = \frac{1}{4}fk$ for k

△ 57. $P = a + b + c$ for a

58. $PR = s_1 + s_2 + s_3 + s_4$ for s_3

59. $S = 2\pi rh + 2\pi r^2$ for h

△ 60. $S = 4lw + 2wh$ for h

Review and Preview

Write each percent as a decimal. See Section R.3.

61. 32%

62. 8%

63. 200%

64. 0.5%

Write each decimal as a percent. See Section R.3.

65. 0.17

66. 0.03

67. 7.2

68. 5

Combining Concepts

Solve.

69. $N = R + \frac{V}{G}$ for V
(Urban forestry: tree plantings per year)

70. $B = \frac{F}{P - V}$ for V
(Business: break-even point)

71. The formula $V = LWH$ is used to find the volume of a box. If the length of a box is doubled, the width is doubled, and the height is doubled, how does this affect the volume? Explain your answer.

72. The formula $A = bh$ is used to find the area of a parallelogram. If the base of a parallelogram is doubled and its height is doubled, how does this affect the area? Explain your answer.

73. Find the temperature at which the Celsius measurement and Fahrenheit measurement are the same number.

2.7 Percent, Ratio, and Proportion

OBJECTIVES

- A Solve percent equations.
- B Solve problems involving percents.
- C Write ratios as fractions.
- D Solve proportions.
- E Solve problems modeled by proportions.

SSM TUTOR CENTER SG CD & VIDEO MATH PRO WEB

A Solving Percent Equations

Many of today's statistics are given in terms of percent: a basketball player's free throw percent, current interest rates, stock market trends, and nutrition labeling, just to name a few. In this section, we first explore percent, percent equations, and applications involving percents. See Section R.3 if a further review of percents is needed.

EXAMPLE 1 The number 63 is what percent of 72?

Solution:

1. UNDERSTAND. Read and reread the problem. Next, let's suppose that the percent is 80%. To check, we find 80% of 72.

   $80\% \text{ of } 72 = 0.80(72) = 57.6$

   This is close, but not 63. At this point, though, we have a better understanding of the problem, we know the correct answer is close to and greater than 80%, and we know how to check our proposed solution later.

   Let $x =$ the unknown percent.

2. TRANSLATE. Recall that "is" means "equals" and "of" signifies multiplying. Let's translate the sentence directly.

the number 63	is	what percent	of	72
↓	↓	↓	↓	↓
63	=	x	·	72

3. SOLVE.

$$63 = 72x$$
$$0.875 = x \quad \text{Divide both sides by 72.}$$
$$87.5\% = x \quad \text{Write as a percent.}$$

4. INTERPRET.

Check: Verify that 87.5% of 72 is 63.

State: The number 63 is 87.5% of 72.

Practice Problem 1

The number 22 is what percent of 40?

EXAMPLE 2 The number 120 is 15% of what number?

Solution:

1. UNDERSTAND. Read and reread the problem.

   Let $x =$ the unknown number.

2. TRANSLATE.

the number 120	is	15%	of	what number
↓	↓	↓	↓	↓
120	=	15%	·	x

3. SOLVE.

$$120 = 0.15x \quad \text{Write 15\% as 0.15.}$$
$$800 = x \quad \text{Divide both sides by 0.15.}$$

4. INTERPRET.

Practice Problem 2

The number 150 is 40% of what number?

Answers

1. 55%, **2.** 375

Check: Check the proposed solution by finding 15% of 800 and verifying that the result is 120.

State: Thus, 120 is 15% of 800.

B Solving Problems Involving Percent

As mentioned earlier, percents are often used in statistics. Recall that the graph below is called a circle graph or a pie chart. The circle or pie represents a whole, or 100%. Each circle is divided into sectors (shaped like pieces of a pie) that represent various parts of the whole 100%.

EXAMPLE 3

The circle graph below shows the purpose of trips made by American travelers. Use this graph to answer the questions below.

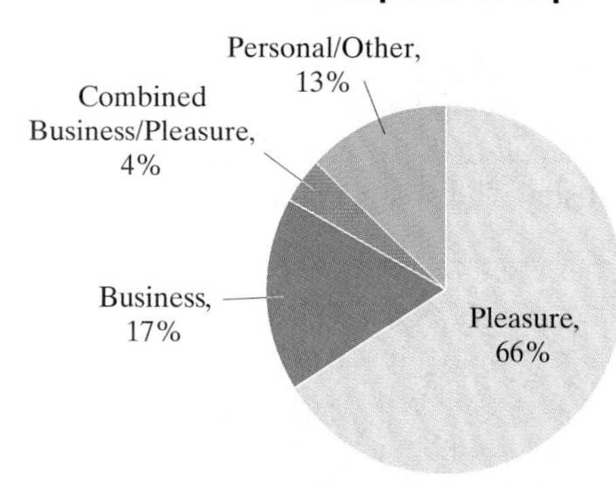

Source: Travel Industry Association of America

a. What percent of trips made by American travelers are solely for the purpose of business?

b. What percent of trips made by American travelers are for the purpose of business or combined business/pleasure?

c. On an airplane flight of 253 Americans, how many of these people might we expect to be traveling solely for business?

Solution:

a. From the circle graph, we see that 17% of trips made by American travelers are solely for the purpose of business.

b. From the circle graph, we know that 17% of trips are solely for business and 4% of trips are for combined business/pleasure. The sum 17% + 4% or 21% of trips made by American travelers are for the purpose of business or combined business/pleasure.

c. Since 17% of trips made by American travelers are for business, we find 17% of 253.

$$\begin{aligned} 17\% \text{ of } 253 &= 0.17(253) \\ &= 43.01 \end{aligned}$$

We might then expect that about 43 American travelers on the flight are traveling solely for business.

Practice Problem 3

Use the circle graph from Example 3 to answer each question.

a. What percent of trips made by American travelers are solely for pleasure?

b. What percent of trips made by American travelers are for the purpose of pleasure or combined business/pleasure?

c. On an airplane flight of 250 Americans, how many of these might we expect to be traveling solely for pleasure?

Answers

3. a. 66%, **b.** 70%, **c.** 165

 Writing Ratios as Fractions

A **ratio** is the quotient of two numbers or two quantities. For example, a percent can be thought of as a ratio, since it is the quotient of a number and 100.

$$53\% = \frac{53}{100} \quad \text{or} \quad \text{the ratio of 53 to 100}$$

Ratio

The ratio of a number a to a number b is their quotient. Ways of writing ratios are

$$a \text{ to } b, \qquad a:b, \quad \text{and} \quad \frac{a}{b}$$

Whenever possible, we will convert quantities in a ratio to the same unit of measurement.

EXAMPLE 4 Write a ratio for each phrase. Use fractional notation.

a. The ratio of 2 parts salt to 5 parts water

b. The ratio of 18 inches to 2 feet

Solution:

a. The ratio of 2 parts salt to 5 parts water is $\frac{2}{5}$.

b. First we convert to the same unit of measurement. For example,

$$2 \text{ feet} = 2 \cdot 12 \text{ inches} = 24 \text{ inches}$$

The ratio of 18 inches to 2 feet is then $\frac{18}{24}$, or $\frac{3}{4}$ in lowest terms. ●

Practice Problem 4

Write a ratio for each phrase. Use fractional notation.

a. The ratio of 3 parts oil to 7 parts gasoline

b. The ratio of 40 minutes to 3 hours

 Solving Proportions

Ratios can be used to form proportions. A **proportion** is a mathematical statement that two ratios are equal.

For example, the equation

$$\frac{1}{2} = \frac{4}{8}$$

is a proportion that says that the ratios $\frac{1}{2}$ and $\frac{4}{8}$ are equal.

Notice that a proportion contains four numbers. If any three numbers are known, we can solve and find the fourth number. One way to do so is to use cross products. To understand cross products, let's start with the proportion

$$\frac{a}{b} = \frac{c}{d}$$

and multiply both sides by the LCD, bd.

$$\frac{a}{b} = \frac{c}{d}$$

$$bd\left(\frac{a}{b}\right) = bd\left(\frac{c}{d}\right) \quad \text{Multiply both sides by the LCD, } bd.$$

$$\underbrace{ad}_{\text{cross product}} = \underbrace{bc}_{\text{cross product}} \quad \text{Simplify.}$$

Answers

4. a. $\frac{3}{7}$, **b.** $\frac{2}{9}$

Notice why ad and bc are called cross products.

$$\frac{a}{b} = \frac{c}{d}$$

bc

ad

Cross Products

If $\frac{a}{b} = \frac{c}{d}$, then $ad = bc$.

EXAMPLE 5 Solve for x: $\frac{45}{x} = \frac{5}{7}$

Solution: To solve, we set cross products equal.

$$\frac{45}{x} = \frac{5}{7}$$

$$\begin{aligned} 45 \cdot 7 &= x \cdot 5 && \text{Set cross products equal.} \\ 315 &= 5x && \text{Multiply.} \\ \frac{315}{5} &= \frac{5x}{5} && \text{Divide both sides by 5.} \\ 63 &= x && \text{Simplify.} \end{aligned}$$

Check: To check, substitute 63 for x in the original proportion. The solution is 63.

EXAMPLE 6 Solve for x: $\frac{x - 5}{3} = \frac{x + 2}{5}$

Solution:

$$\frac{x - 5}{3} = \frac{x + 2}{5}$$

$$\begin{aligned} 5(x - 5) &= 3(x + 2) && \text{Set cross products equal.} \\ 5x - 25 &= 3x + 6 && \text{Multiply.} \\ 5x &= 3x + 31 && \text{Add 25 to both sides.} \\ 2x &= 31 && \text{Subtract } 3x \text{ from both sides.} \\ \frac{2x}{2} &= \frac{31}{2} && \text{Divide both sides by 2.} \\ x &= \frac{31}{2} \end{aligned}$$

Check: Verify that $\frac{31}{2}$ is the solution.

Practice Problem 5

Solve for x: $\frac{3}{8} = \frac{63}{x}$

Practice Problem 6

Solve for x: $\frac{2x + 1}{7} = \frac{x - 3}{5}$

Answers

5. $x = 168$, **6.** $x = -\frac{26}{3}$

Try the Concept Check in the margin.

E Solving Problems Modeled by Proportions

Proportions can be used to model and solve many real-life problems. When using proportions in this way, it is important to judge whether the solution is reasonable. Doing so helps us to decide if the proportion has been formed correctly. We use the same problem-solving strategy that was introduced in Section 2.5.

EXAMPLE 7 Calculating Cost with a Proportion

Three Zip disks cost \$37.47. How much should 5 disks cost?

Solution:

1. UNDERSTAND. Read and reread the problem. We know that the cost of 5 Zip disks is more than the cost of 3 disks, or \$37.47, and less than the cost of 6 disks, which is double the cost of 3 disks, or 2(\$37.47) = \$74.94. Let's suppose that 5 disks cost \$60.00. To check, we see if 3 disks is to 5 disks as the *price* of 3 disks is to the *price* of 5 disks. In other words, we see if

$$\frac{3 \text{ disks}}{5 \text{ disks}} = \frac{\text{price of 3 disks}}{\text{price of 5 disks}}$$

or

$$\frac{3}{5} = \frac{37.47}{60.00}$$

$3(60.00) = 5(37.47)$ Set cross products equal.

$180.00 = 187.35$ Not a true statement.

Thus, \$60 is not correct but we now have a better understanding of the problem.

Let x = price of 5 disks.

2. TRANSLATE.

$$\frac{3 \text{ disks}}{5 \text{ disks}} = \frac{\textit{price of } 3 \textit{ disks}}{\textit{price of } 5 \textit{ disks}}$$

$$\frac{3}{5} = \frac{37.47}{x}$$

3. SOLVE.

$$\frac{3}{5} = \frac{37.47}{x}$$

$3x = 5(37.47)$ Set cross products equal.

$3x = 187.35$

$x = 62.45$ Divide both sides by 3.

4. INTERPRET.

Concept Check

For which of the following equations can we immediately use cross products to solve for x?

a. $\frac{2 - x}{5} = \frac{1 + x}{3}$

b. $\frac{2}{5} - x = \frac{1 + x}{3}$

Practice Problem 7

To estimate the number of people in Jackson, population 50,000, who have no health insurance, 250 people were polled. Of those polled, 39 had no insurance. How many people in the city might we expect to be uninsured?

Answers

7. 7800 people

Concept Check: a

Check: Verify that 3 disks is to 5 disks as \$37.47 is to \$62.45. Also, notice that our solution is a reasonable one as discussed in Step 1.

State: Five Zip disks should cost about \$62.45.

Helpful Hint

The proportion $\frac{5 \text{ disks}}{3 \text{ disks}} = \frac{\text{price of 5 disks}}{\text{price of 3 disks}}$ could also have been used to solve the problem above. Notice that the cross products are the same.

When shopping for an item offered in many different sizes, it is important to be able to determine the best buy, or the best price per unit. To find the **unit price** of an item, divide the total price of the item by the total number of units.

$$\text{unit price} = \frac{\text{total price}}{\text{number of units}}$$

For example, if a 16-ounce can of green beans is priced at \$0.88, its unit price is

$$\text{unit price} = \frac{\$0.88}{16} = \$0.055$$

Practice Problem 8

Which is the better buy for the same brand of toothpaste?
8 ounces for \$2.59
10 ounces for \$3.11

EXAMPLE 8 Finding the Better Buy

A supermarket offers a 14-ounce box of cereal for \$3.79 and an 18-ounce box of the same brand of cereal for \$4.99. Which is the better buy?

Solution: To find the better buy, we compare unit prices. The following unit prices were rounded to three decimal places.

Size	Price	Unit Price
14 ounce	\$3.79	$\frac{\$3.79}{14} \approx \$0.271$
18 ounce	\$4.99	$\frac{\$4.99}{18} \approx \$0.277$

The 14-ounce box of cereal has the lower unit price so it is the better buy.

Answer

8. 10 ounces

Name ____________________ Section __________ Date __________

Mental Math

Tell whether the percent labels in the circle graphs are correct.

1.
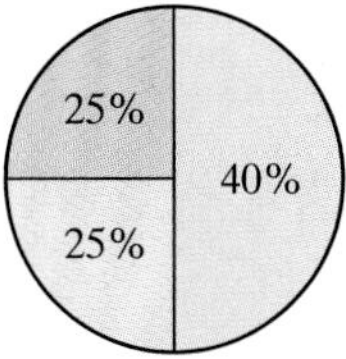

2.
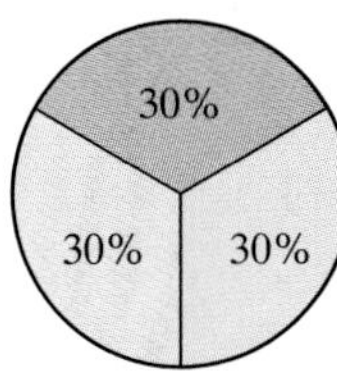

3.
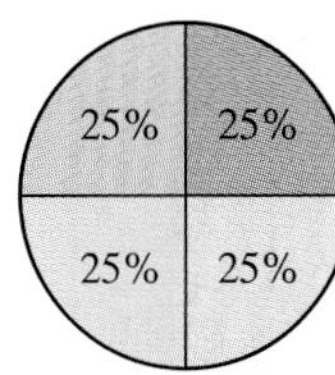

4.
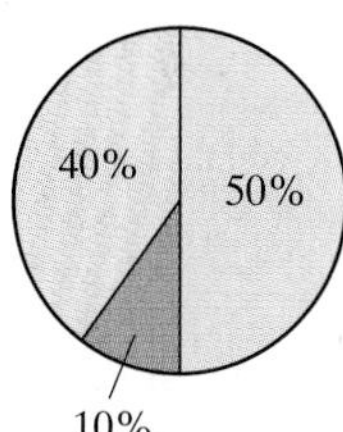

EXERCISE SET 2.7

 *Find each number described. See Examples 1 and 2.*

1. What number is 16% of 70?

2. What number is 88% of 1000?

3. The number 28.6 is what percent of 52?

4. The number 87.2 is what percent of 436?

 **5.** The number 45 is 25% of what number?

6. The number 126 is 35% of what number?

7. Find 23% of 20.

8. Find 140% of 86.

9. The number 40 is 80% of what number?

10. The number 56.25 is 45% of what number?

11. The number 144 is what percent of 480?

12. The number 42 is what percent of 35?

 *Solve. See Examples 1 through 3. Many applications in this exercise group may be solved more efficiently with the use of a calculator.*

13. Nordstrom's advertised a 25% off sale. If a London Fog coat originally sold for $256, find the decrease in price and the sale price.

14. Time Saver increased the price of a $0.95 cola by 15%. Find the increase in price and the new price.

15. Scoville units are used to measure the hotness of a pepper. Measuring 577 thousand Scoville units, the "Red Savina" habanero pepper was known as the hottest chili pepper. That has recently changed with the discovery of Naga Jolokia pepper from India. It measures 48% hotter than the habanero. Find the measure of the Naga Jolokia pepper. Approximate to the nearest thousand units.

16. At this writing, the women's world record for throwing a disc (like a heavy Frisbee) was set by Jennifer Griffin of the United States in 2000. Her throw was 138.56 meters. The men's world record was set by Christian Voigt of Germany in 2001. His throw was 56.6% farther than Jennifer's. Find the length of his throw. Round to the nearest meter. (*Source:* World Flying Disc Federation)

The graph shows the communities in the United States that have the highest percentages of citizens that shop by catalog. The height of the bar for each city corresponds to the percent of citizens that shop by catalog. Use the graph to answer Exercises 17 through 22.

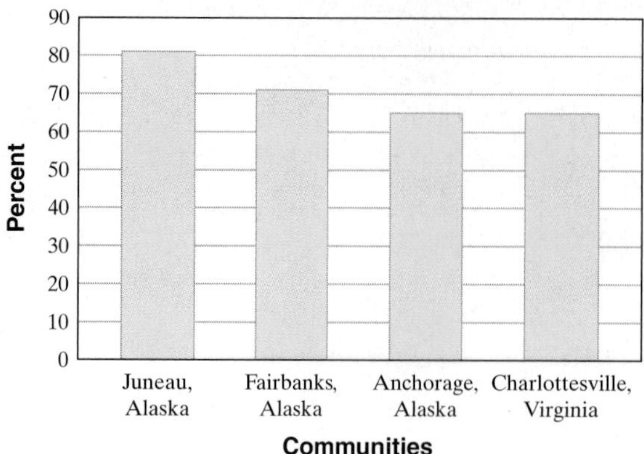

17. What percent of the population in Juneau, Alaska, shops by catalog?

18. What percent of the population in Charlottesville, Virginia, shops by catalog?

19. According to the 2000 Census, Anchorage has a population of 260,283. How many catalog shoppers live in Anchorage? Round to the nearest whole number. (*Source:* U.S. Census Bureau)

20. According to the 2000 Census, Juneau has a population of 30,711. How many catalog shoppers live in Juneau? Round to the nearest whole number. (*Source:* U.S. Census Bureau)

21. Do the percents shown in the graph have a sum of 100%? Why or why not?

22. Survey your algebra class and find what percent of the class has shopped by catalog.

The circle graph below shows the number of minutes that adults spend on their home phone each day. Use this graph for Exercises 23 through 26. See Example 3.

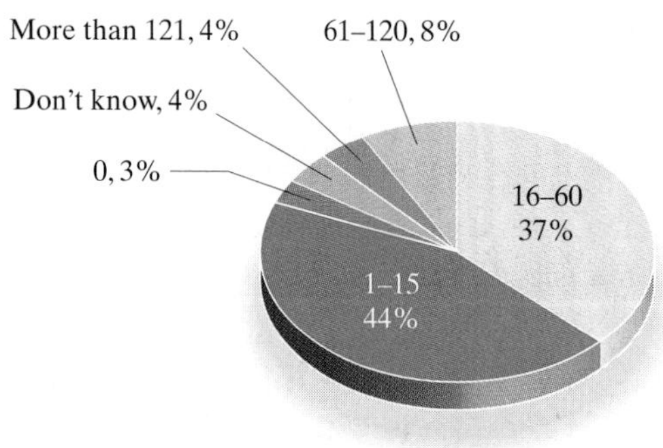

Source: Bruskin/Goldring Research for Sony Electronics

23. What percent of adults spend more than 121 minutes on the phone each day?

24. What percent of adults spend no time on the phone each day?

25. Liberty is a town whose adult population is approximately 135,000. How many of these adults might you expect to talk 16–60 minutes on the phone each day?

26. Poll the students in your algebra class. Find what percent of students spend 1–15 minutes on the phone each day. Is this percent close to 44%? Why or why not?

27. A recent survey showed that 42% of recent college graduates named flexible hours as their most desired employment benefit. In a graduating class of 860 college students, how many would you expect to rank flexible hours as their top priority in job benefits? (Round to the nearest whole.) (*Source:* JobTrak.com)

28. A recent survey showed that 64% of U.S. colleges have Internet access in their classrooms. There are approximately 9800 post-secondary institutions in the United States. How many of these would you expect to have Internet access in their classrooms? (*Source:* Market Data Retrieval, National Center for Education Statistics)

Fill in the percent column in each table. See Examples 1 through 3.

29.

Ford Motor Company
Year 2000 Vehicle Sales in North America

	Thousands of Vehicles	Percent of Total (Round to nearest whole percent)
United States	4486	
Canada	300	
Mexico	147	$147/4933 \approx 3\%$
TOTAL	4933	

(*Source: Ford Motor Company 2000 Annual Report*)

30.

Kellogg Company
Year 2000 Net Sales

	Millions of Dollars	Percent of Total (Round to nearest whole percent)
United States	4067	
Europe	1463	$1463/6954 \approx 21\%$
Latin America	627	
Other	797	
TOTAL	6954	

(*Source: Kellogg Company 2000 Annual Report*)

 *Write each ratio in fractional notation in lowest terms. See Example 4.*

31. 2 megabytes to 15 megabytes

32. 18 disks to 41 disks

33. 10 inches to 12 inches

34. 15 miles to 40 miles

35. 5 quarts to 3 gallons

36. 8 inches to 3 feet

37. 4 nickels to 2 dollars

38. 12 quarters to 2 dollars

39. 175 centimeters to 5 meters

40. 90 centimeters to 4 meters

41. 190 minutes to 3 hours

42. 60 hours to 2 days

43. Suppose someone tells you that the ratio of 11 inches to 2 feet is $\frac{11}{2}$. How would you correct that person and explain the error?

44. Write a ratio that can be written in fractional notation as $\frac{3}{2}$.

D *Solve each proportion. See Examples 5 and 6.*

45. $\frac{2}{3} = \frac{x}{6}$

46. $\frac{x}{2} = \frac{16}{6}$

47. $\frac{x}{10} = \frac{5}{9}$

48. $\frac{9}{4x} = \frac{6}{2}$

49. $\frac{4x}{6} = \frac{7}{2}$

50. $\frac{a}{5} = \frac{3}{2}$

51. $\frac{x-3}{x} = \frac{4}{7}$

52. $\frac{y}{y-16} = \frac{5}{3}$

53. $\frac{x+1}{2x+3} = \frac{2}{3}$

54. $\frac{x+1}{x+2} = \frac{5}{3}$

55. $\frac{9}{5} = \frac{12}{3x+2}$

56. $\frac{6}{11} = \frac{27}{3x-2}$

57. $\frac{3}{x+1} = \frac{5}{2x}$

58. $\frac{7}{x-3} = \frac{8}{2x}$

59. $\frac{15}{3x-4} = \frac{5}{x}$

60. $\frac{x}{3} = \frac{2x+5}{6}$

E *Solve. See Example 7.*

61. The ratio of the weight of an object on Earth to the weight of the same object on Pluto is 100 to 3. If an elephant weighs 4100 pounds on Earth, find the elephant's weight on Pluto.

62. If a 170-pound person weighs approximately 65 pounds on Mars, how much does a 9000-pound satellite weigh on Mars? Round to the nearest pound.

63. There are 110 calories per 28.4 grams of Crispy Rice cereal. Find how many calories are in 42.6 grams of this cereal.

64. On an architect's blueprint, 1 inch corresponds to 4 feet. Find the length of a wall represented by a line that is $3\frac{7}{8}$ inches long on the blueprint.

65. A recent headline read, "Women Earn Bigger Check in 1 of Every 6 Couples." If there are 23,000 couples in a nearby metropolitan area, how many women would you expect to earn bigger paychecks?

66. A human factors expert recommends that there be at least 9 square feet of floor space in a college classroom for every student in the class. Find the minimum floor space that 40 students need.

67. To mix weed killer with water correctly, it is necessary to mix 8 teaspoons of weed killer with 2 gallons of water. Find how many gallons of water are needed to mix with the entire box if it contains 36 teaspoons of weed killer.

68. Ken Hall, a tailback, holds the high school sports record for total yards rushed in a season. In 1953, he rushed for 4045 total yards in 12 games. Find his average rushing yards per game.

Given the following prices charged for various sizes of an item, find the best buy. See Example 8.

69. Laundry detergent
110 ounces for \$5.79
240 ounces for \$13.99

70. Jelly
10 ounces for \$1.14
15 ounces for \$1.69

71. Tuna (in cans)
6 ounces for \$0.69
8 ounces for \$0.90
16 ounces for \$1.89

72. Picante sauce
10 ounces for \$0.99
16 ounces for \$1.69
30 ounces for \$3.29

Review and Preview

Place $<$, $>$, or $=$ in the appropriate space to make each a true statement. See Sections 1.1 and 1.6.

73. $-5 \quad -7$

74. $\frac{12}{3} \quad 2^2$

75. $|-5| \quad -(-5)$

76. $-3^3 \quad (-3)^3$

77. $(-3)^2 \quad -3^2$

78. $|-2| \quad -|-2|$

Combining Concepts

Standardized nutrition labels like the one below have been displayed on food items since 1994. The percent column on the right shows the percent of daily values based on a 2000-calorie diet shown at the bottom of the label. For example, a serving of this food contains 4 grams of total fat, where the recommended daily fat based on a 2000-calorie diet is 65 grams of fat. This means that $\frac{4}{65}$ *or approximately 6% (as shown) of your daily recommended fat is taken in by eating a serving of this food.*

Use this nutrition label to answer Exercises 79 through 81.

Nutrition Facts

Serving Size	**18 Crackers (31g)**
Servings Per Container	**About 9**
Amount Per Serving	
Calories 130	Calories from Fat 35
	% Daily Value*
Total Fat 4g	**6%**
Saturated Fat 0.5g	**3%**
Polyunsaturated Fat 0g	
Monounsaturated Fat 1.5g	
Cholesterol 0mg	**0%**
Sodium 230mg	*x*
Total Carbohydrate 23g	*y*
Dietary Fiber 2g	**8%**
Sugars 3g	
Protein 2g	
Vitamin A 0% •	Vitamin C 0%
Calcium 2% •	Iron 6%

* Percent Daily Values are based on a 2,000 calorie diet. Your daily values may be higher or lower depending on your calorie needs.

	Calories	2,000	2,500
Total Fat	Less than	65g	80g
Sat. Fat	Less than	20g	25g
Cholesterol	Less than	300mg	300mg
Sodium	Less than	2400mg	2400mg
Total Carbohydrate		300g	375g
Dietary Fiber		25g	30g

79. Based on a 2000-calorie diet, what percent of daily value of sodium is contained in a serving of this food? In other words, find x in the label. (Round to the nearest tenth of a percent.)

80. Based on a 2000-calorie diet, what percent of daily value of total carbohydrate is contained in a serving of this food? In other words, find y in the label. (Round to the nearest tenth of a percent.)

81. Notice on the nutrition label that one serving of this food contains 130 calories and 35 of these calories are from fat. Find the percent of calories from fat. (Round to the nearest tenth of a percent.) It is recommended that no more than 30% of calorie intake come from fat. Does this food satisfy this recommendation?

Use the nutrition label below to answer Exercises 82 through 84.

NUTRITIONAL INFORMATION PER SERVING

Serving Size: 9.8 oz.		**Servings Per Container: 1**	
Calories	280	Polyunsaturated Fat	1g
Protein	12g	Saturated Fat	3g
Carbohydrate	45g	Cholesterol	20mg
Fat	6g	Sodium	520mg
Percent of Calories from Fat	?	Potassium	220mg

82. If fat contains approximately 9 calories per gram, find the percent of calories from fat in one serving of this food. (Round to the nearest tenth of a percent.)

83. If protein contains approximately 4 calories per gram, find the percent of calories from protein from one serving of this food. (Round to the nearest tenth of a percent.)

84. Find a food that contains more than 30% of its calories per serving from fat. Analyze the nutrition label and verify that the percents shown are correct.

2.8 Linear Inequalities and Problem Solving

OBJECTIVES

A. Use interval notation.
B. Solve linear inequalities using the addition property of inequality.
C. Solve linear inequalities using the multiplication property of inequality.
D. Solve linear inequalities using both properties of inequality.
E. Solve problems that can be modeled by linear inequalities.

Relationships among measurable quantities are not always described by equations. For example, suppose that a salesperson earns a base of \$600 per month plus a commission of 20% of sales. Suppose we want to find the minimum amount of sales needed to receive a total income of *at least* \$1500 per month. Here, the phrase "at least" implies that an income of \$1500 *or more* is acceptable. In symbols, we can write

$$\text{income} \geq 1500$$

This is an example of an inequality, which we will solve in Example 11.

A *linear inequality* is similar to a linear equation except that the equality symbol is replaced with an inequality symbol, such as $<, >, \leq$, or $\geq$.

Linear Inequality in One Variable

A **linear inequality in one variable** is an inequality that can be written in the form

$$ax + b < c$$

where a, b, and c are real numbers and $a \neq 0$.

For example,

$$3x + 5 \geq 4 \qquad 2y < 0 \qquad 4n \geq n - 3$$

$$3(x - 4) < 5x \qquad \frac{x}{3} \leq 5$$

In this section, when we make definitions, state properties, or list steps about an inequality containing the symbol $<$, we mean that the definition, property, or steps apply to an inequality containing the symbols $>, \leq$, and $\geq$ also.

A Using Interval Notation

A **solution** of an inequality is a value of the variable that makes the inequality a true statement. The **solution set** of an inequality is the set of all solutions. Notice that the solution set of the inequality $x > 2$, for example, contains all numbers greater than 2. Its graph is an interval on the number line since an infinite number of values satisfy the variable. If we use open/closed-circle notation, the graph of $\{x|x > 2\}$ looks like:

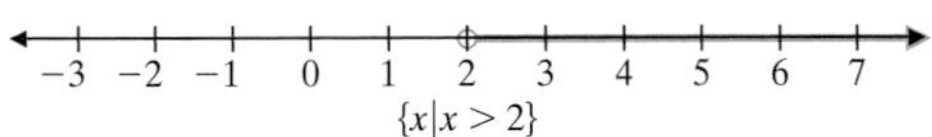

In this text, a different graphing notation will be used to help us understand **interval notation**. Instead of an open circle, we use a parenthesis; instead of a closed circle, we use a bracket. With this new notation, the graph of $\{x|x > 2\}$ now looks like:

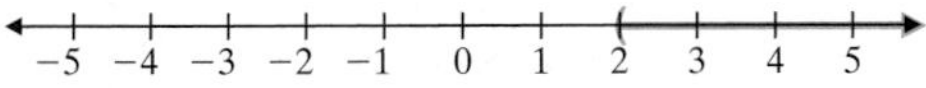

and can be represented in interval notation as $(2, \infty)$. The symbol ∞ is read "infinity" and indicates that the interval includes *all* numbers greater than 2. The left parenthesis indicates that 2 *is not* included in the interval. Using a left bracket, [, would indicate that 2 *is* included in the interval. The following table shows three equivalent ways to describe an interval: in set notation, as a graph, and in interval notation.

Set Notation	Graph	Interval Notation
$\{x \mid x < a\}$	a	$(-\infty, a)$
$\{x \mid x > a\}$	a	(a, ∞)
$\{x \mid x \le a\}$	a	$(-\infty, a]$
$\{x \mid x \ge a\}$	a	$[a, \infty)$
$\{x \mid a < x < b\}$	a b	(a, b)
$\{x \mid a \le x \le b\}$	a b	$[a, b]$
$\{x \mid a < x \le b\}$	a b	$(a, b]$
$\{x \mid a \le x < b\}$	a b	$[a, b)$

Helpful Hint

Notice that a parenthesis is always used to enclose ∞ and $-\infty$.

EXAMPLES

Graph each set on a number line and then write it in interval notation.

1. $\{x \mid x \ge 2\}$ 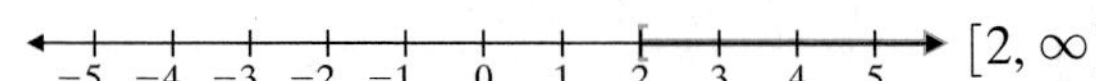 $[2, \infty)$

2. $\{x \mid x < -1\}$ 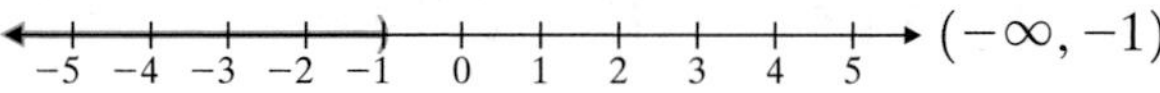 $(-\infty, -1)$

3. $\{x \mid 0.5 < x \le 3\}$ 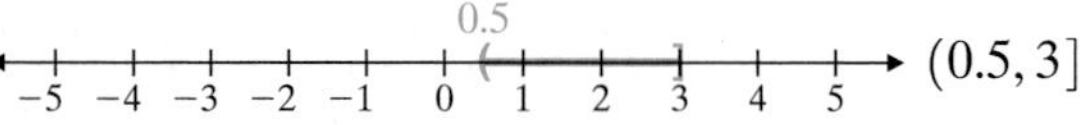 $(0.5, 3]$

Try the Concept Check in the margin.

B Using the Addition Property of Inequality

Interval notation can be used to write solutions of linear inequalities. To solve a linear inequality, we use a process similar to the one used to solve a linear equation. We use properties of inequalities to write equivalent inequalities until the variable is alone on one side of the inequality.

Practice Problems 1–3

Graph each set on a number line and then write it in interval notation.

1. $\{x \mid x > -3\}$

−5 −4 −3 −2 −1 0 1 2 3 4 5

2. $\{x \mid x \le 0\}$

−5 −4 −3 −2 −1 0 1 2 3 4 5

3. $\{x \mid -0.5 \le x < 2\}$

−5 −4 −3 −2 −1 0 1 2 3 4 5

Concept Check

Explain what is wrong with writing the interval $(5, \infty]$.

Answers

1. $(-3, \infty)$,

−5 −4 −3 −2 −1 0 1 2 3 4 5

2. $(-\infty, 0]$,

−5 −4 −3 −2 −1 0 1 2 3 4 5

3. $[-0.5, 2)$,

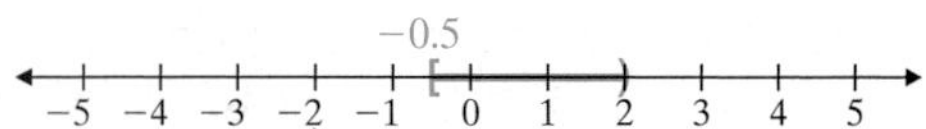

Concept Check: should be $(5, \infty)$ since a parenthesis is always used to enclose ∞

Addition Property of Inequality

If a, b, and c are real numbers, then

$$a < b \quad \text{and} \quad a + c < b + c$$

are equivalent inequalities.

In other words, we may add the same real number to both sides of an inequality, and the resulting inequality will have the same solution set. This property also allows us to subtract the same real number from both sides.

EXAMPLE 4 Solve: $x - 2 < 5$. Graph the solution set.

Solution:

$x - 2 < 5$

$x - 2 + 2 < 5 + 2$ Add 2 to both sides.

$x < 7$ Simplify.

The solution set is $\{x|x < 7\}$, which in interval notation is $(-\infty, 7)$. The graph of the solution set is

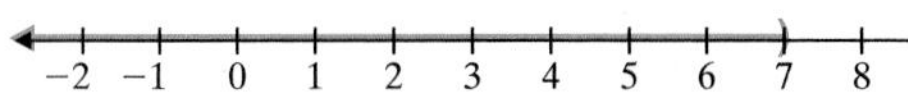

Helpful Hint

In Example 4, the solution set is $\{x|x < 7\}$. This means that *all* numbers less than 7 are solutions. For example, 6.9, 0, $-\pi$, 1, and -56.7 are solutions, just to name a few. To see this, replace x in $x - 2 < 5$ with each of these numbers and see that the result is a true inequality.

EXAMPLE 5 Solve: $4x - 2 < 5x$. Graph the solution set.

Solution: To get x alone on one side of the inequality, we subtract $4x$ from both sides.

$4x - 2 < 5x$

$4x - 2 - 4x < 5x - 4x$ Subtract $4x$ from both sides.

$-2 < x$ or $x > -2$ Simplify.

Helpful Hint

Don't forget that $-2 < x$ means the same as $x > -2$.

The solution set is $\{x|x > -2\}$, which in interval notation is $(-2, \infty)$. The graph is

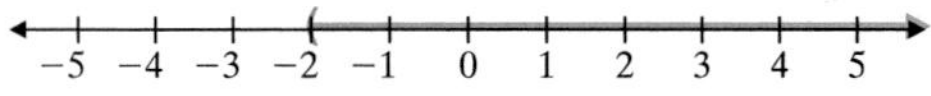

Practice Problem 4

Solve: $x + 3 < 1$. Graph the solution set.

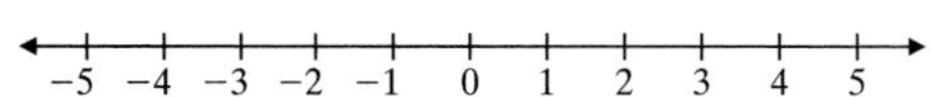

Practice Problem 5

Solve: $3x - 4 < 4x$. Graph the solution set.

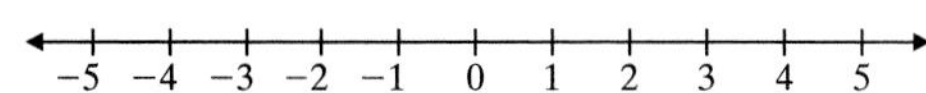

Answers

4. $\{x|x < -2\}, (-\infty, -2)$,

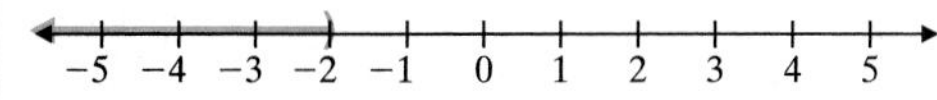

5. $\{x|x > -4\}, (-4, \infty)$,

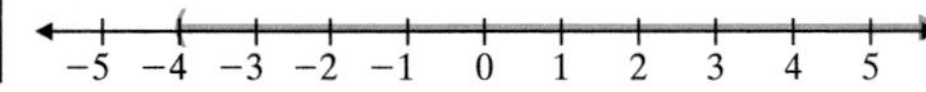

Practice Problem 6

Solve: $5x - 1 \geq 4x + 4$. Graph the solution set.

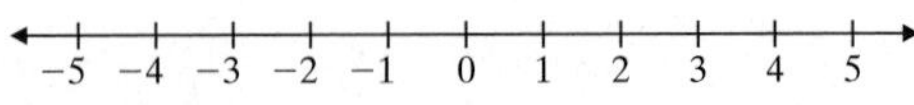

EXAMPLE 6 Solve: $3x + 4 \geq 2x - 6$. Graph the solution set.

Solution:

$$
\begin{aligned}
3x + 4 &\geq 2x - 6 \\
3x + 4 - 2x &\geq 2x - 6 - 2x && \text{Subtract } 2x \text{ from both sides.} \\
x + 4 &\geq -6 && \text{Combine like terms.} \\
x + 4 - 4 &\geq -6 - 4 && \text{Subtract 4 from both sides.} \\
x &\geq -10 && \text{Simplify.}
\end{aligned}
$$

The solution set is $\{x|x \geq -10\}$, which in interval notation is $[-10, \infty)$. The graph of the solution set is

C Using the Multiplication Property of Inequality

Next, we introduce and use the multiplication property of inequality to solve linear inequalities. To understand this property, let's start with the true statement $-3 < 7$ and multiply both sides by 2.

$$
\begin{aligned}
-3 &< 7 \\
-3(2) &< 7(2) && \text{Multiply both sides by 2.} \\
-6 &< 14 && \text{True.}
\end{aligned}
$$

The statement remains true.

Notice what happens if both sides of $-3 < 7$ are multiplied by -2.

$$
\begin{aligned}
-3 &< 7 \\
-3(-2) &< 7(-2) \\
6 &< -14 && \text{False.}
\end{aligned}
$$

The inequality $6 < -14$ is a false statement. However, *if the direction of the inequality sign is reversed*, the result is

$$6 > -14 \quad \text{True.}$$

These examples suggest the following property.

Multiplication Property of Inequality

If a, b, and c are real numbers and c is **positive**, then $a < b$ and $ac < bc$ are equivalent inequalities.

If a, b, and c are real numbers and c is **negative**, then $a < b$ and $ac > bc$ are equivalent inequalities.

In other words, we may multiply both sides of an inequality by the same positive real number, and the result is an equivalent inequality. We may also multiply both sides of an inequality by the same *negative number* and *reverse the direction of the inequality symbol*, and the result is an equivalent inequality. The multiplication property holds for division also since division is defined in terms of multiplication.

Helpful Hint

Whenever both sides of an inequality are multiplied or divided by a negative number, the direction of the inequality symbol *must be* reversed to form an equivalent inequality.

Answer

6. $\{x|x \geq 5\}$, $[5, \infty)$,

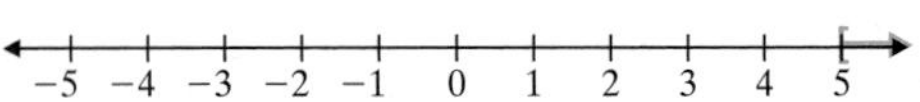

EXAMPLE 7 Solve: $\frac{1}{4}x \leq \frac{3}{2}$. Graph the solution set.

Solution:

$$\frac{1}{4}x \leq \frac{3}{2}$$

$$4 \cdot \frac{1}{4}x \leq 4 \cdot \frac{3}{2} \quad \text{Multiply both sides by 4.}$$

$$x \leq 6 \quad \text{Simplify.}$$

Helpful Hint

The inequality symbol is the same since we are multiplying by a *positive* number.

The solution set is $\{x|x \leq 6\}$, which in interval notation is $(-\infty, 6]$. The graph of this solution set is

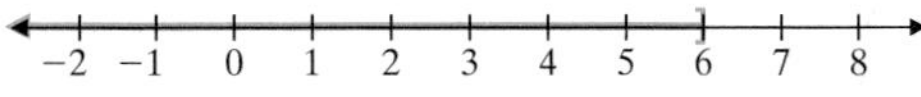

Practice Problem 7

Solve: $\frac{1}{6}x \leq \frac{2}{3}$. Graph the solution set.

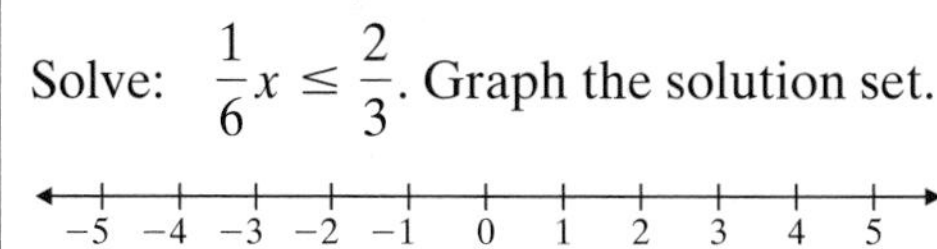

EXAMPLE 8 Solve: $-2.3x < 6.9$. Graph the solution set.

Solution:

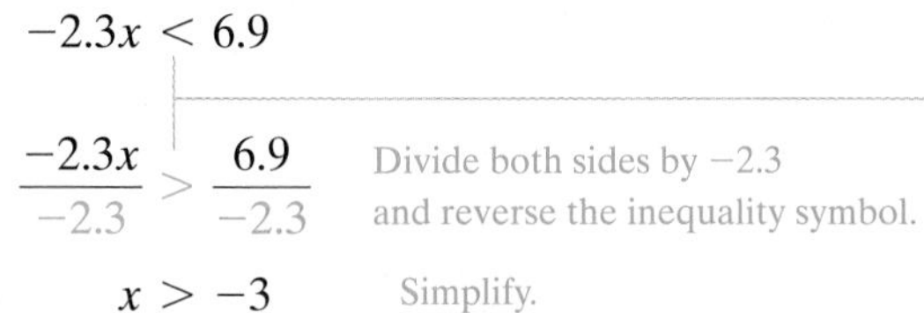

$$-2.3x < 6.9$$

$$\frac{-2.3x}{-2.3} > \frac{6.9}{-2.3} \quad \text{Divide both sides by } -2.3 \text{ and reverse the inequality symbol.}$$

$$x > -3 \quad \text{Simplify.}$$

Helpful Hint

The inequality symbol is *reversed* since we divided by a *negative* number.

The solution set is $\{x|x > -3\}$, which is $(-3, \infty)$ in interval notation. The graph of the solution set is

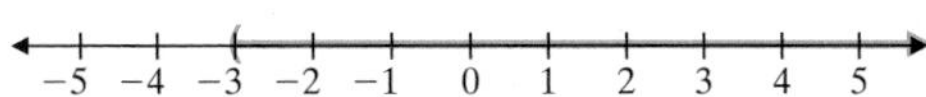

Practice Problem 8

Solve: $-1.1x < 5.5$. Graph the solution set.

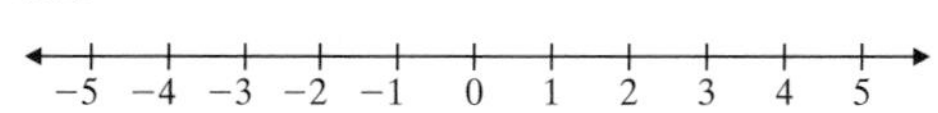

Try the Concept Check in the margin.

To solve linear inequalities in general, we follow steps similar to those for solving linear equations.

Concept Check

In which of the following inequalities must the inequality symbol be reversed during the solution process?

a. $-2x > 7$
b. $2x - 3 > 10$
c. $-x + 4 + 3x < 5$
d. $-x + 4 < 5$

Solving a Linear Inequality in One Variable

Step 1. Clear the equation of fractions or decimals by multiplying both sides of the inequality by an appropriate number.

Step 2. Use the distributive property to remove grouping symbols such as parentheses.

Step 3. Combine like terms on each side of the inequality.

Step 4. Use the addition property of inequality to write the inequality as an equivalent inequality with variable terms on one side and numbers on the other side.

Step 5. Use the multiplication property of inequality to get the variable alone on one side of the inequality.

D Using Both Properties of Inequality

Many problems require us to use both properties of inequality.

Answers

7. $\{x|x \leq 4\}, (-\infty, 4,]$

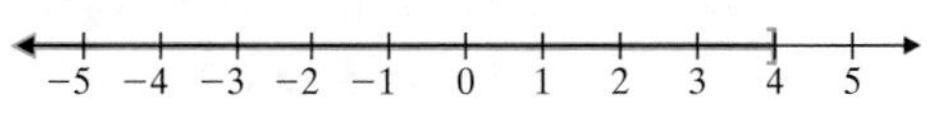

8. $\{x|x > -5\}, (-5, \infty),$

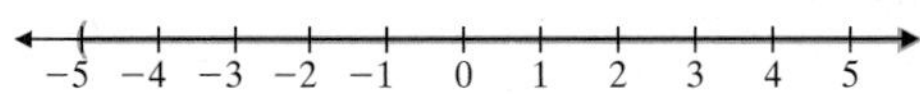

Concept Check: a, d

Practice Problem 9

Solve: $6 - 2x \leq 8x - 14$. Write the solution set in interval notation.

Practice Problem 10

Solve: $\frac{3}{4}(x + 2) \geq x - 6$. Write the solution set in interval notation.

Practice Problem 11

A salesperson earns \$1000 a month plus a commission of 15% of sales. Find the minimum amount of sales needed to receive a total income of at least \$4000 per month.

Answers

9. $[2, \infty)$, **10.** $(-\infty, 30]$, **11.** \$20,000

EXAMPLE 9

Solve: $5 - x \leq 4x - 15$. Write the solution set in interval notation.

Solution:

$$5 - x \leq 4x - 15$$

$$5 - x + x \leq 4x - 15 + x \quad \text{Add } x \text{ to both sides.}$$

$$5 \leq 5x - 15 \quad \text{Combine like terms.}$$

$$5 + 15 \leq 5x - 15 + 15 \quad \text{Add 15 to both sides.}$$

$$20 \leq 5x \quad \text{Combine like terms.}$$

$$\frac{20}{5} \leq \frac{5x}{5} \quad \text{Divide both sides by 5.}$$

$$4 \leq x \quad \text{or} \quad x \geq 4 \quad \text{Simplify.}$$

The solution set is $[4, \infty)$.

EXAMPLE 10

Solve: $\frac{2}{5}(x - 6) \geq x - 1$. Write the solution set in interval notation.

Solution:

$$\frac{2}{5}(x - 6) \geq x - 1$$

$$5\left[\frac{2}{5}(x - 6)\right] \geq 5(x - 1) \quad \text{Multiply both sides by 5 to eliminate fractions.}$$

$$2x - 12 \geq 5x - 5 \quad \text{Use the distributive property.}$$

$$-3x - 12 \geq -5 \quad \text{Subtract } 5x \text{ from both sides.}$$

$$-3x \geq 7 \quad \text{Add 12 to both sides.}$$

$$\frac{-3x}{-3} \leq \frac{7}{-3} \quad \text{Divide both sides by } -3 \text{ and reverse the inequality symbol.}$$

$$x \leq -\frac{7}{3} \quad \text{Simplify.}$$

The solution set is $\left(-\infty, -\frac{7}{3}\right]$.

E Linear Inequalities and Problem Solving

Problems containing words such as "at least," "at most," "between," "no more than," and "no less than" usually indicate that an inequality is to be solved instead of an equation. In solving applications involving linear inequalities, we use the same four-step strategy as when we solved applications involving linear equations.

EXAMPLE 11 Calculating Income with Commission

A salesperson earns \$600 per month plus a commission of 20% of sales. Find the minimum amount of sales needed to receive a total income of at least \$1500 per month.

Solution:

1. UNDERSTAND. Read and reread the problem. Let
   $x =$ amount of sales
2. TRANSLATE. As stated in the beginning of this section, we want the income to be greater than or equal to \$1500. To write an inequality, notice that the salesperson's income consists of \$600 plus a commission (20% of sales).

In words:	600	+	commission (20% of sales	$\geq$	1500
	↓		↓		↓
Translate:	600	+	$0.20x$	$\geq$	1500

3. SOLVE the inequality for x.

$$600 + 0.20x \geq 1500$$
$$600 + 0.20x - 600 \geq 1500 - 600$$
$$0.20x \geq 900$$
$$x \geq 4500$$

4. INTERPRET.

Check: The income for sales of \$4500 is

$600 + 0.20(4500)$, or 1500

Thus, if sales are greater than or equal to \$4500, income is greater than or equal to \$1500.

State: The minimum amount of sales needed for the salesperson to earn at least \$1500 per month is \$4500. ●

EXAMPLE 12 Finding the Annual Consumption

In the United States, the annual consumption of cigarettes is declining. The consumption c in billions of cigarettes per year since the year 1985 can be approximated by the formula

$$c = -14.25t + 598.69$$

where t is the number of years after 1985. Use this formula to predict the first year that the consumption of cigarettes will be less than 200 billion per year.

Solution:

1. UNDERSTAND. Read and reread the problem. To become familiar with the given formula, let's find the cigarette consumption after 20 years, which would be the year 1985 + 20, or 2005. To do so, we substitute 20 for t in the given formula.

$$c = -14.25(20) + 598.69 = 313.69$$

Thus, in 2005, we predict cigarette consumption to be about 313.69 billion.

Variables have already been assigned in the given formula. For review, they are

$c =$ the annual consumption of cigarettes in the United States in billions of cigarettes

$t =$ the number of years after 1985

2. TRANSLATE. We are looking for the first year that the consumption of cigarettes c is less than 200. Since we are finding years t, we substitute the expression in the formula given for c, or

$$-14.25t + 598.69 < 200$$

3. SOLVE the inequality.

$$-14.25t + 598.69 < 200$$ Subtract 598.69 from both sides.

$$-14.25t < -398.69$$ Divide both sides by −14.25 and round the result.

$$t > 27.98$$

4. INTERPRET.

Check: We substitute a number greater than 27.98 and see that c is less than 200.

State: The annual consumption of cigarettes will be less than 200 billion for more than 27.98 years after 1985, or in approximately 28 + 1985 = 2013. ●

Practice Problem 12

Use the formula given in Example 12 to predict when the consumption of cigarettes will be less than 100 billion per year.

Answer

12. after the year 2020

FOCUS ON The Real World

ENERGY AUDIT

Have you ever been surprised by high electric bills? Has it made you wonder where all of your electricity expenditure is going or how to lower your bill? If so, one approach to learning more about your electricity consumption is performing an energy audit of your home or apartment. Once you understand your patterns of electricity usage, you can make informed decisions on where to cut back or whether or not to replace an older appliance with a newer, more energy efficient one.

To perform your own energy audit, fill out the table below. First, make a list of all the electrical appliances in your home. Be sure to include components of your heating, hot water, and/or air-conditioning systems, major appliances, indoor and outdoor lights, computer and audio-visual components, and small kitchen or personal care appliances. Don't forget to include easily overlooked items such as room space heaters, ceiling fans, and water bed heaters.

Next, estimate how many hours each item is run per 30-day month. For items used nearly every day, estimate daily usage in hours and multiply by 30. For items used less often, estimate how many hours they are used per week. Then divide by 7 and multiply by 30 to get an estimate for a 30-day month.

For each item on the list, record its wattage. This information can usually be found on its serial number plate. Wattage is abbreviated W, so look for a number like 13W or 200W. If wattage is not listed on the plate, look for information on voltage (abbreviated V for volts) and amperage (abbreviated A for amps). Wattage can be estimated by multiplying volts times amps. (*Note*: Sometimes a range of values is listed for voltage. If the range includes 120 V, use 120 in the wattage calculation. Otherwise, use the maximum value of the voltage range for the wattage calculation.)

Fill in the fourth column of the table by multiplying the number of hours each appliance is run during a month by its wattage to find watt-hours. Then fill in the fifth column of the table by dividing watt-hours by 1000 to find kilowatt-hours. For the last column, consult your electricity bill to find the price charged by your electric company per kilowatt-hour (often abbreviated KWH). Alternatively, contact the local electric company to ask its standard charge per kilowatt-hour for residential customers. Fill in the last column of the table by figuring the cost to run an appliance for one month: Multiply the number of kilowatt-hours per month by what the electric company charges per kilowatt-hour.

Appliance	Hours Run per 30-Day Month	Wattage (or use volts × amps)	Watt-Hours (hours × wattage)	Kilowatt-Hours (watt-hours ÷ 1000)	Cost to Run Appliance for One Month (kilowatt-hours × cost per KWH)

CRITICAL THINKING

Do an energy audit of your home, apartment, or dormitory room.

1. Which item is the most expensive to operate over the course of the month? Does this surprise you? Why or why not?
2. Which item is the least expensive to operate? Does this surprise you? Why or why not?
3. Are there any items on the list whose usage could be cut back to save energy costs? Which ones would be the most viable choices? How much could usage be cut and how much money would that save? Explain.
4. Are there any appliances that could be replaced with more energy-efficient models? Conduct research to find a more recent model that would be a better choice. If the usage of the new model is the same as the old model, how much money could be saved each month by switching? How long will it take for the monthly energy cost savings to "pay off" the price of buying the new appliance? Explain.
5. In what other ways could you lower your electric bill?

Name ______________________ Section __________ Date ____________

Mental Math

Solve each inequality.

1. $x - 2 < 4$

2. $x - 1 > 6$

3. $x + 5 \geq 15$

4. $x + 1 \leq 8$

5. $3x > 12$

6. $5x < 20$

7. $\frac{x}{2} \leq 1$

8. $\frac{x}{4} \geq 2$

EXERCISE SET 2.8

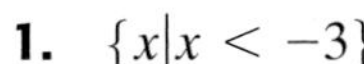

Graph the solution set of each inequality on a number line and then write it in interval notation. See Examples 1 through 3.

1. $\{x|x < -3\}$

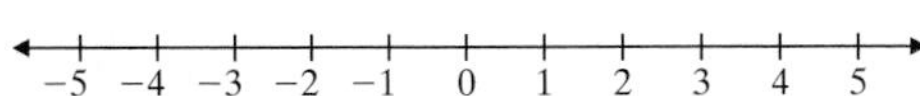

2. $\{x|x \geq -7\}$

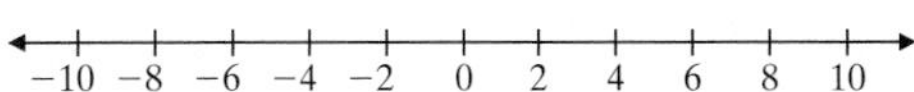

3. $\{x|x \geq 0.3\}$

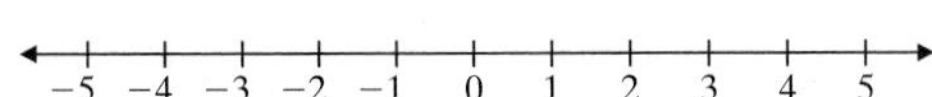

4. $\{x|x < -0.2\}$

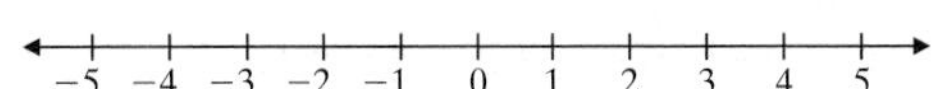

5. $\{x|5 < x\}$

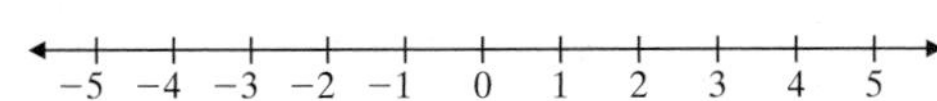

6. $\{x|-7 \geq x\}$

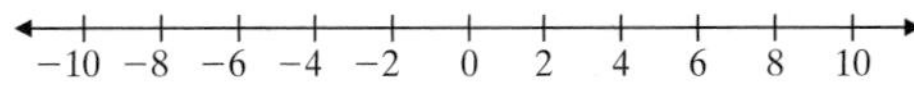

7. $\{x|-2 < x < 5\}$

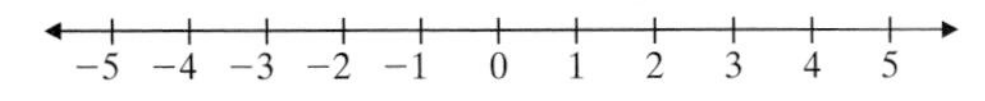

8. $\{x|-5 \leq x \leq -1\}$

−5 −4 −3 −2 −1 0 1 2 3 4 5

9. $\{x|5 > x > -1\}$

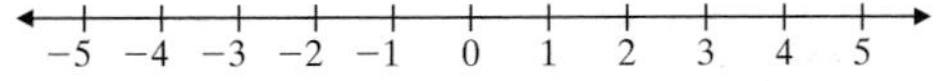

10. $\{x|3 \geq x \geq -7\}$

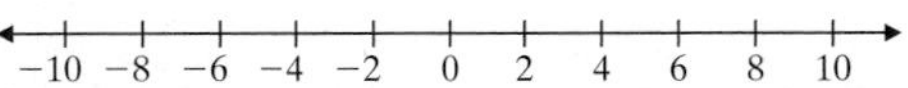

11. When graphing the solution set of an inequality, explain how you know whether to use a parenthesis or a bracket.

12. Explain what is wrong with the interval notation $(-6, -\infty)$.

B *Solve. Graph the solution set and write it in interval notation. See Examples 4 through 6.*

13. $x - 7 \geq -9$

14. $x + 2 \leq -1$

15. $7x < 6x + 1$

16. $11x < 10x + 5$

17. $8x - 7 \leq 7x - 5$

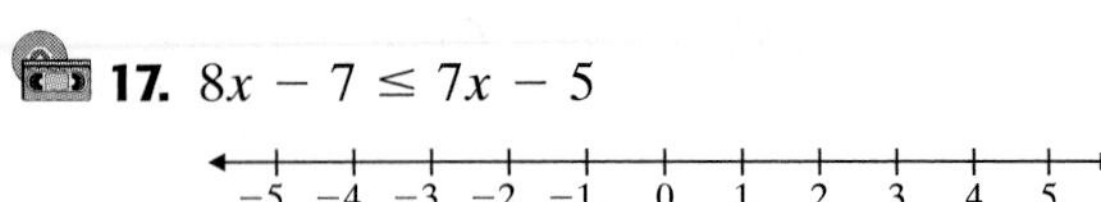

18. $7x - 1 \geq 6x - 1$

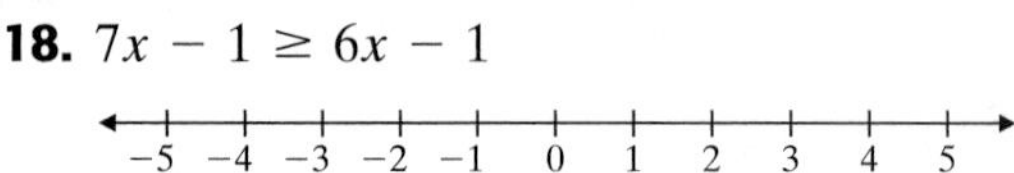

C *Solve. Graph the solution set and then write it in interval notation. See Examples 7 and 8.*

19. $\frac{3}{4}x \geq 2$

20. $\frac{5}{6}x \geq -8$

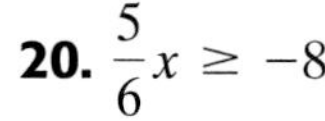

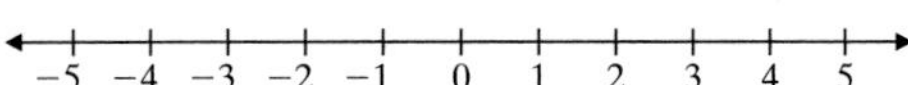

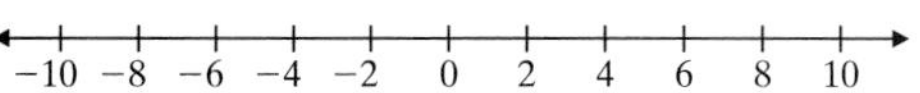

21. $5x < -23.5$

22. $4x > -11.2$

23. $-3x \geq 9$

24. $-4x \geq 15$

25. $-x < -4$

26. $-x > -2$

D *Solve. Write the solution set using interval notation. See Examples 9 and 10.*

27. $-2x + 7 \geq 9$

28. $8 - 5x \leq 23$

29. $15 + 2x \geq 4x - 7$

30. $20 + x < 6x$

31. $3(x - 5) < 2(2x - 1)$

32. $5(x + 4) \leq 4(2x + 3)$

33. $\frac{1}{2} + \frac{2}{3} \geq \frac{x}{6}$

34. $\frac{3}{4} - \frac{2}{3} > \frac{x}{6}$

35. $-5x + 4 \leq -4(x - 1)$

36. $-6x + 2 < -3(x + 4)$

37. $\frac{1}{4}(x - 7) \geq x + 2$

38. $\frac{3}{5}(x + 1) \leq x + 1$

39. $0.8x + 0.6x \geq 4.2$

40. $0.7x - x > 0.45$

41. $4(2x + 1) > 4$

42. $6(2 - x) \geq 12$

43. $\frac{5x + 1}{7} - \frac{2x - 6}{4} \geq -4$

44. $\frac{1 - 2x}{3} + \frac{3x + 7}{7} > 1$

45. $4(x - 6) + 2x - 4 \geq 3(x - 7) + 10x$

46. $7(2x + 3) + 4x \leq 7 + 5(3x - 4)$

47. $-3(2x - 1) < -4[2 + 3(x + 2)]$

48. $-2(4x + 2) > -5[1 + 2(x - 1)]$

49. $14 - (5x - 6) \geq -6(x + 1) - 5$

50. $13y - (9y + 2) \leq 5(y - 6) + 10$

51. $\frac{1}{2}(3x - 4) \leq \frac{3}{4}(x - 6) + 1$

52. $\frac{2}{3}(x + 3) < \frac{1}{6}(2x - 8) + 2$

E *Solve. See Examples 11 and 12.*

53. Shureka Washburn has scores of 72, 67, 82, and 79 on her algebra tests. Use an inequality to find the minimum score she must make on the final exam to pass the course with an average of 60 or higher, given that the final exam counts as two tests.

54. In a Winter Olympics 5000-meter speed-skating event, Hans Holden scored times of 6.85, 7.04, and 6.92 minutes on his first three trials. Use an inequality to find the maximum time he can score on his last trial so that his average time is under 7.0 minutes.

55. A small plane's maximum takeoff weight is 2000 pounds. Six passengers weigh an average of 160 pounds each. Use an inequality to find the maximum weight of luggage and cargo the plane can carry.

56. A clerk must use the elevator to move boxes of paper. The elevator's weight limit is 1500 pounds. If each box of paper weighs 66 pounds and the clerk weighs 147 pounds, use an inequality to find the maximum number of boxes she can move on the elevator at one time.

57. To mail an envelope first class, the U.S. Post Office charges 34 cents for the first ounce and 23 cents per ounce for each additional ounce. Use an inequality to find the maximum weight that can be mailed for $3.10.

58. A shopping mall parking garage charges $1 for the first half-hour and 60 cents for each additional half-hour or a portion of a half-hour. Use an inequality to find how long you can park if you have only $4.00 in cash.

59. Northeast Telephone Company offers two billing plans for local calls. Plan 1 charges $25 per month for unlimited calls, and plan 2 charges $13 per month plus 6 cents per call. Use an inequality to find the number of monthly calls for which plan 1 is more economical than plan 2.

60. A car rental company offers two subcompact rental plans. Plan A charges $32 per day for unlimited mileage, and plan B charges $24 per day plus 15 cents per mile. Use an inequality to find the number of daily miles for which plan A is more economical than plan B.

61. At room temperature, glass used in windows actually has some properties of a liquid. It has a very slow, viscous flow. (Viscosity is the property of a fluid that resists internal flow. For example, lemonade flows more easily than fudge syrup. Fudge syrup has a higher viscosity than lemonade.) Glass does not become a true liquid until temperatures are greater than or equal to 500°C. Find the Fahrenheit temperatures for which glass is a liquid. (Use the formula $F = \frac{9}{5}C + 32$.)

62. Stibnite is a silvery white mineral with a metallic luster. It is one of the few minerals that melts easily in a match flame or at temperatures of approximately 977°F or greater. Find the Celsius temperatures for which stibnite melts. [Use the formula $C = \frac{5}{9}(F - 32)$.]

63. Although beginning salaries vary greatly according to your field of study, the equation

$$s = 651.2t + 27{,}821$$

can be used to approximate and to predict average beginning salaries for candidates with bachelor's degrees. The variable s is the starting salary and t is the number of years after 1989.

a. Approximate when beginning salaries for candidates will be greater than $35,000.

b. Determine the year you plan to graduate from college. Use this year to find the corresponding value of t and approximate your beginning salary.

64. Use the formula in Example 12 to estimate the years that the consumption of cigarettes will be less than 50 billion per year.

This graph shows the growth in the number of Starbucks coffee bar locations from 1991 through 2000. For each year shown, there is a corresponding point on the graph. Each point is labelled with a pair of numbers in parentheses. The meaning of these numbers is below. Notice that the height of the graph for each year shown corresponds to the number of Starbucks locations. Use the points shown on this graph to answer Exercises 57 through 62.

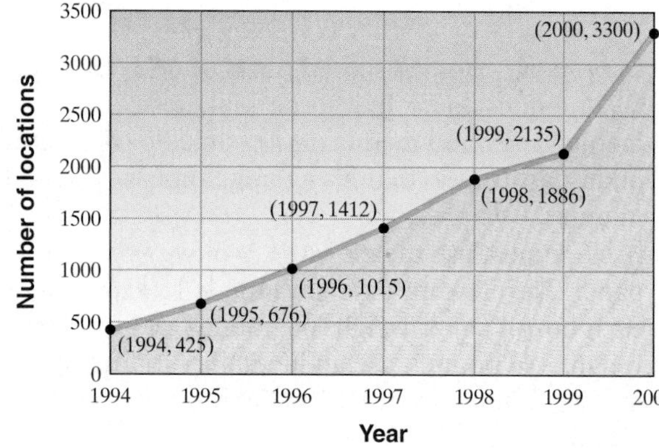

Source: Starbucks Corporation

65. How many Starbucks locations were there in 1996?

66. How many Starbucks locations were there in 1999?

67. Between which two years did the greatest increase in the number of Starbucks locations occur?

68. In what year were there approximately 1900 Starbucks locations?

69. During what year(s) was the number of Starbucks locations greater than 1400?

70. During what year(s) was the number of Starbucks locations greater than 2000?

71. Is the number of Starbucks locations increasing or decreasing over time? Explain how you arrived at your answer.

72. Between what years was the increase in the number of Starbucks locations the least? Explain how you arrived at your answer.

73. During what year(s) was the number of Starbucks locations less than 500?

Review and Preview

List or describe the integers that make both inequalities true.

74. $x < 5$ and $x > 1$

75. $x \geq 0$ and $x \leq 7$

76. $x \geq -2$ and $x \geq 2$

77. $x < 6$ and $x < -5$

Graph each set on a number line and write it in interval notation. See Section 2.8.

78. $\{x|0 \leq x \leq 5\}$

−5 −4 −3 −2 −1 0 1 2 3 4 5

79. $\{x|-7 < x \leq 1\}$

−10 −8 −6 −4 −2 0 2 4 6 8 10

80. $\left\{x \middle| -\frac{1}{2} < x < \frac{3}{2}\right\}$

−5 −4 −3 −2 −1 0 1 2 3 4 5

81. $\{x|-2.5 \leq x < 5.3\}$

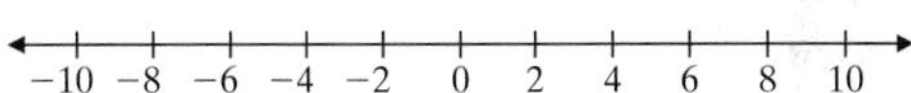

Combining Concepts

Solve each inequality.

82. $4(x - 1) \geq 4x - 8$

83. $3x + 1 < 3(x - 2)$

84. $7x < 7(x - 2)$

85. $8(x+3) \leq 7(x+5)+x$

86. Explain how solving a linear inequality is similar to solving a linear equation.

87. Explain how solving a linear inequality is different from solving a linear equation.

Solve.

88. Eric Daly has scores of 75, 83, and 85 on his history tests. Use an inequality to find the scores he can make on his final exam to receive a B in the class. The final exam counts as two tests, and a B is received if the final course average is greater than or equal to 80.

89. Maria Lipco has scores of 85, 95, and 92 on her algebra tests. Use an inequality to find the scores she can make on her final exam to receive an A in the course. The final exam counts as three tests, and an A is received if the final course average is greater than or equal to 90. Round to one decimal place.

Internet Excursions

WWW Go To: http://www.prenhall.com/martin-gay_algebra What's Related

The National Climatic Data Center (NCDC) is the world's largest active archive of weather data. The given World Wide Web address will provide you with access to NCDC's Climate Visualization site, or a related site, where you can find data on precipitation, temperature, and drought conditions for any state. Most states are subdivided into geographical divisions. Data from 1895 through the present is available for many states.

Choose the state in which your college or university is located. Select temperature as the parameter, and then choose a year. The graph that is then generated will show the average monthly temperatures for that year for all divisions of the state your specified. At the bottom of the Web page, note the link to the data file listing all the values shown in the graphs. Use the information in these Web pages to complete the following exercises.

90. Fill in the following blanks. In the year ________, the lowest average monthly temperature of ________ °F occurred in Division ________ of the state of ________. Let F represent the average monthly temperature in Fahrenheit degrees. Write an inequality that describes the minimum average monthly temperature in Fahrenheit degrees in your state for this year. Then use the relationship $F = \frac{9}{5}C + 32$ to convert this inequality to one that describes the minimum average monthly temperature in degrees Celsius.

91. Fill in the following blanks. In the year ________, the highest average monthly temperature of ________ °F occurred in Division ________ of the state of ________. Let F represent the average monthly temperature in Fahrenheit degrees. Write an inequality that describes the maximum average monthly temperature in Fahrenheit degrees in your state for this year. Then use the relationship $F = \frac{9}{5}C + 32$ to convert this inequality to one that describes the maximum average monthly temperature in degrees Celsius.

CHAPTER 2 ACTIVITY Investigating Averages

MATERIALS:

- small rubber ball or crumpled paper ball
- bucket or waste can

This activity may be completed by working in groups or individually.

1. Try shooting the ball into the bucket or waste can 5 times. Record your results below.

Shots Made **Shots Missed**

2. Find your shooting percent for the 5 shots (that is, the percent of the shots you actually made out of the number you tried).

3. Suppose you are going to try an additional 5 shots. How many of the next 5 shots will you have to make to have a 50% shooting percent for all 10 shots? An 80% shooting percent?

4. Did you solve an equation in Question 3? If so, explain what you did. If not, explain how you could use an equation to find the answers.

5. Now suppose you are going to try an additional 22 shots. How many of the next 22 shots will you have to make to have at least a 50% shooting percent for all 27 shots? At least a 70% shooting percent?

6. Choose one of the sports played at your college that is currently in season. How many regular-season games are scheduled? What is the team's current percentage of games won?

7. Suppose the team has a goal of finishing the season with a winning percentage better than 110% of their current wins. At least how many of the remaining games must they win to achieve their goal?

Chapter 2 VOCABULARY CHECK

Fill in each blank with one of the words or phrases listed below.

like terms	numerical coefficient	linear equation in one variable
equivalent equations	formula	proportion
linear inequality in one variable	ratio	

1. Terms with the same variables raised to exactly the same powers are called ____________
2. A ____________________________ can be written in the form $ax + b = c$.
3. Equations that have the same solution are called ____________________
4. An equation that describes a known relationship among quantities is called a __________
5. A ____________________________ can be written in the form $ax + b < c$, (or $>, \leq, \geq$).
6. The ____________________ of a term is its numerical factor.
7. A ________ is the quotient of two numbers or two quantities.
8. A ____________ is a mathematical statement that two ratios are equal.

CHAPTER 2 Highlights

DEFINITIONS AND CONCEPTS	EXAMPLES
Section 2.1 Simplifying Expressions	
The **numerical coefficient** of a **term** is its numerical factor.	**TERM** / **NUMERICAL COEFFICIENT** $-7y$ — -7 x — 1 $\frac{1}{5}a^2b$ — $\frac{1}{5}$
Terms with the same variables raised to exactly the same powers are **like terms**.	**LIKE TERMS** / **UNLIKE TERMS** $12x, -x$ — $3y, 3y^2$ $-2xy, 5yx$ — $7a^2b, -2ab^2$
To combine like terms, add the numerical coefficients and multiply the result by the common variable factor.	$9y + 3y = 12y$ $-4z^2 + 5z^2 - 6z^2 = -5z^2$
To remove parentheses, apply the distributive property.	$-4(x + 7) + 10(3x - 1)$ $= -4x - 28 + 30x - 10$ $= 26x - 38$
Section 2.2 The Addition Property of Equality	
A **linear equation in one variable** can be written in the form $Ax + B = C$ where $A, B,$ and C are real numbers and $A \neq 0$.	$-3x + 7 = 2$ $3(x - 1) = -8(x + 5) + 4$
Equivalent equations are equations that have the same solution.	$x - 7 = 10$ and $x = 17$ are equivalent equations.
ADDITION PROPERTY OF EQUALITY Adding the same number to or subtracting the same number from both sides of an equation does not change its solution.	$y + 9 = 3$ $y + 9 - 9 = 3 - 9$ $y = -6$

Definitions and Concepts | Examples

Section 2.3 The Multiplication Property of Equality

Multiplication Property of Equality

Multiplying both sides or dividing both sides of an equation by the same nonzero number does not change its solution.

$$\frac{2}{3}a = 18$$

$$\frac{3}{2}\left(\frac{2}{3}a\right) = \frac{3}{2}(18)$$

$$a = 27$$

Section 2.4 Further Solving Linear Equations

To Solve Linear Equations

1. Clear the equation of fractions.
2. Remove any grouping symbols such as parentheses.
3. Simplify each side by combining like terms.
4. Get all variable terms on one side and all numbers on the other side by using the addition property of equality.
5. Get the variable alone by using the multiplication property of equality.
6. Check the solution by substituting it into the original equation.

Solve: $\frac{5(-2x + 9)}{6} + 3 = \frac{1}{2}$

1. $6 \cdot \frac{5(-2x + 9)}{6} + 6 \cdot 3 = 6 \cdot \frac{1}{2}$
2. $5(-2x + 9) + 18 = 3$ Apply the distributive property.
   $-10x + 45 + 18 = 3$
3. $-10x + 63 = 3$ Combine like terms.
4. $-10x + 63 - 63 = 3 - 63$ Subtract 63.
   $-10x = -60$
5. $\frac{-10x}{-10} = \frac{-60}{-10}$ Divide by -10.
   $x = 6$

Section 2.5 An Introduction to Problem Solving

Problem-Solving Steps

The height of the Hudson volcano in Chile is twice the height of the Kiska volcano in the Aleutian Islands. If the sum of their heights is 12,870 feet, find the height of each.

1. UNDERSTAND the problem.

1. Read and reread the problem. Guess a solution and check your guess.
   Let x be the height of the Kiska volcano. Then $2x$ is the height of the Hudson volcano.

x  $2x$ 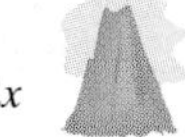

2. TRANSLATE the problem.

2.
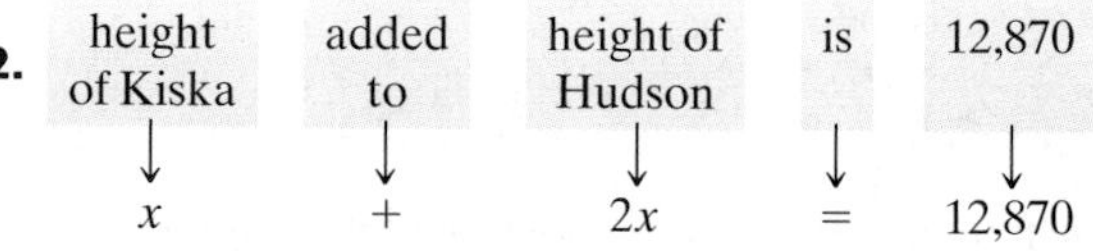

height of Kiska	added to	height of Hudson	is	12,870
↓	↓	↓	↓	↓
x	$+$	$2x$	$=$	12,870

3. SOLVE the equation.

3. $x + 2x = 12{,}870$
   $3x = 12{,}870$
   $x = 4290$

4. INTERPRET the results.

4. *Check:* If x is 4290, then $2x$ is 2(4290) or 8580. Their sum is 4290 + 8580 or 12,870, the required amount.
   State: The Kiska volcano is 4290 feet high, and the Hudson volcano is 8580 feet high.

Definitions and Concepts	**Examples**
Section 2.6 Formulas and Problem Solving	
An equation that describes a known relationship among quantities is called a **formula**.	$A = lw$ (area of a rectangle) $I = PRT$ (simple interest)
To solve a formula for a specified variable, use the same steps as for solving a linear equation. Treat the specified variable as the only variable of the equation.	Solve $P = 2l + 2w$ for l. $P = 2l + 2w$ $P - 2w = 2l + 2w - 2w$ Subtract $2w$. $P - 2w = 2l$ $\frac{P - 2w}{2} = \frac{2l}{2}$ Divide by 2. $\frac{P - 2w}{2} = l$
Section 2.7 Percent, Ratio, and Proportion	
Use the same problem-solving steps to solve a problem containing percents.	32% of what number is 36.8?
1. UNDERSTAND.	**1.** Read and reread. Propose a solution and check. Let $x =$ the unknown number.
2. TRANSLATE.	**2.** 32% of what number is 36.8 32% · x = 36.8
3. SOLVE.	**3.** Solve $32\% \cdot x = 36.8$ $0.32x = 36.8$ $\frac{0.32x}{0.32} = \frac{36.8}{0.32}$ Divide by 0.32. $x = 115$ Simplify.
4. INTERPRET.	**4.** 32% of 115 is 36.8.
A **ratio** is the quotient of two numbers or two quantities. **The ratio of *a* to *b*** can also be written as $\frac{a}{b}$ or $a:b$	Write the ratio of 5 hours to 1 day using fractional notation. $\frac{5 \text{ hours}}{1 \text{ day}} = \frac{5 \text{ hours}}{24 \text{ hours}} = \frac{5}{24}$
A **proportion** is a mathematical statement that two ratios are equal.	$\frac{2}{3} = \frac{8}{12}$ $\frac{x}{7} = \frac{15}{35}$
In the proportion $\frac{a}{b} = \frac{c}{d}$, the products ad and bc are called **cross products**.	$\frac{2}{3} = \frac{8}{12}$ → $3 \cdot 8$ or 24; $2 \cdot 12$ or 24
If $\frac{a}{b} = \frac{c}{d}$ then $ad = bc$.	Solve: $\frac{3}{4} = \frac{x}{x - 1}$ $\frac{3}{4} = \frac{x}{x - 1}$ $3(x - 1) = 4x$ Set cross products equal. $3x - 3 = 4x$ $-3 = x$

DEFINITIONS AND CONCEPTS	**EXAMPLES**
Section 2.8 Linear Inequalities and Problem Solving	
Properties of inequalities are similar to properties of equations. Don't forget that if you multiply or divide both sides of an inequality by the same *negative* number, you must reverse the direction of the inequality symbol.	$-2x \le 4$ $\frac{-2x}{-2} \ge \frac{4}{-2}$ Divide by -2; reverse the inequality symbol. $x \ge -2$
TO SOLVE LINEAR INEQUALITIES **1.** Clear the inequality of fractions. **2.** Remove grouping symbols. **3.** Simplify each side by combining like terms. **4.** Write all variable terms on one side and all numbers on the other side using the addition property of inequality. **5.** Get the variable alone by using the multiplication property of inequality.	Solve: $3(x + 2) \le -2 + 8$ **1.** $3(x + 2) \le -2 + 8$ No fractions to clear. **2.** $3x + 6 \le -2 + 8$ Apply the distributive property. **3.** $3x + 6 \le 6$ Combine like terms. **4.** $3x + 6 - 6 \le 6 - 6$ Subtract 6. $3x \le 0$ **5.** $\frac{3x}{3} \le \frac{0}{3}$ Divide by 3. $x \le 0$ The solution set is $\{x \mid x \le 0\}$ or $(-\infty, 0]$.

Name ______________________ Section __________ Date __________

Chapter 2 Review

(2.1) *Simplify each expression.*

1. $5x - x + 2x$

2. $0.2z - 4.6z - 7.4z$

3. $\frac{1}{2}x + 3 + \frac{7}{2}x - 5$

4. $\frac{4}{5}y + 1 + \frac{6}{5}y + 2$

5. $2(n - 4) + n - 10$

6. $3(w + 2) - (12 - w)$

7. Subtract $7x - 2$ from $x + 5$

8. Subtract $1.4y - 3$ from $y - 0.7$

Write each phrase as an algebraic expression. Simplify if possible.

9. Three times a number decreased by 7

10. Twice the sum of a number and 2.8, added to 3 times the number

(2.2) *Solve each equation.*

11. $8x + 4 = 9x$

12. $5y - 3 = 6y$

13. $\frac{2}{7}x + \frac{5}{7}x = 6$

14. $3x - 5 = 4x + 1$

15. $2x - 6 = x - 6$

16. $4(x + 3) = 3(1 + x)$

17. $6(3 + n) = 5(n - 1)$

18. $5(2 + x) - 3(3x + 2) = -5(x - 6) + 2$

Choose the correct algebraic expression.

19. The sum of two numbers is 10. If one number is x, express the other number in terms of x.

a. $x - 10$
b. $10 - x$
c. $10 + x$
d. $10x$

20. Mandy is 5 inches taller than Melissa. If x inches represents the height of Mandy, express Melissa's height in terms of x.

a. $x - 5$
b. $5 - x$
c. $5 + x$
d. $5x$

△**21.** If one angle measures $(x + 5)°$, express the measure of its supplement in terms of x.

a. $185 + x$
b. $95 + x$
c. $175 - x$
d. $x - 170$

$(x + 5)°$?

(2.3) *Solve each equation.*

22. $\frac{3}{4}x = -9$

23. $\frac{x}{6} = \frac{2}{3}$

24. $-5x = 0$

25. $-y = 7$

26. $0.2x = 0.15$

27. $\frac{-x}{3} = 1$

28. $-3x + 1 = 19$

29. $5x + 25 = 20$

30. $5x - 6 + x = 4x$

31. $-y + 4y = -y$

32. $-5x + \frac{3}{7} = \frac{10}{7}$

33. Write the sum of three consecutive integers as an expression in x. Let x be the first even integer.

(2.4) *Solve each equation.*

34. $\frac{5}{3}x + 4 = \frac{2}{3}x$

35. $-(5x + 1) = -7x + 3$

36. $-4(2x + 1) = -5x + 5$

37. $-6(2x - 5) = -3(9 + 4x)$

38. $3(8y - 1) = 6(5 + 4y)$

39. $\frac{3(2 - z)}{5} = z$

40. $\frac{4(n + 2)}{5} = -n$

41. $0.5(2n - 3) - 0.1 = 0.4(6 + 2n)$

42. $-9 - 5a = 3(6a - 1)$

43. $\frac{5(c + 1)}{6} = 2c - 3$

44. $\frac{2(8 - a)}{3} = 4 - 4a$

45. $200(70x - 3560) = -179(150x - 19{,}300)$

46. $1.72y - 0.04y = 0.42$

(2.5) *Solve each of the following.*

47. The height of the Washington Monument is 50.5 inches more than 10 times the length of a side of its square base. If the sum of these two dimensions is 7327 inches, find the height of the Washington Monument. (*Source:* National Park Service)

48. A 12-foot board is to be divided into two pieces so that one piece is twice as long as the other. If x represents the length of the shorter piece, find the length of each piece.

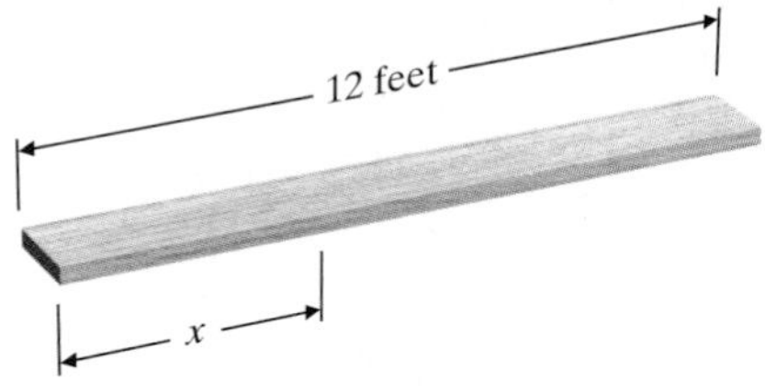

49. In March 2001, Kellogg Company acquired Keebler Foods Company. After the merger, the total number of Kellogg and Keebler manufacturing plants was 53. The number of Kellogg plants was one less than twice the number of Keebler plants. How many of each type of plant were there? (*Source: Kellogg Company 2000 Annual Report*)

50. Find three consecutive integers whose sum is -114.

51. The quotient of a number and 3 is the same as the difference of the number and two. Find the number.

52. Double the sum of a number and 6 is the opposite of the number. Find the number.

(2.6) *Substitute the given values into the given formulas and solve for the unknown variable.*

53. $P = 2l + 2w$; $P = 46, l = 14$

54. $V = lwh$; $V = 192, l = 8, w = 6$

Solve each equation for the indicated variable.

55. $y = mx + b$ for m

56. $r = vst - 5$ for s

57. $2y - 5x = 7$ for x

58. $3x - 6y = -2$ for y

△ **59.** $C = \pi D$ for π

△ **60.** $C = 2\pi r$ for π

△ **61.** A swimming pool holds 900 cubic meters of water. If its length is 20 meters and its height is 3 meters, find its width.

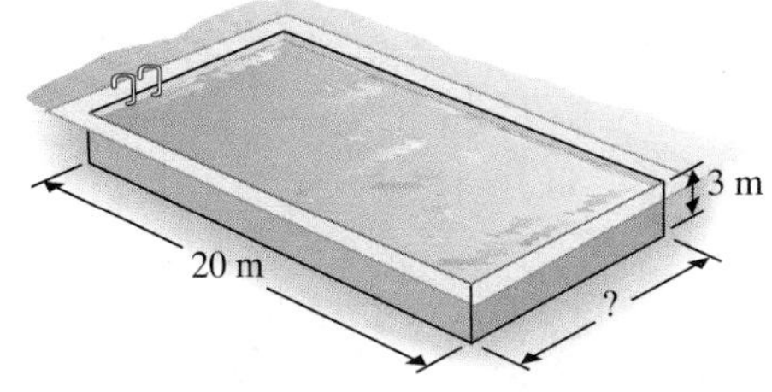

62. On April 28, 2001, the highest temperature recorded in the United States was 104°F, which occurred in Death Valley. California. Convert this temperature to degrees Celsius. (*Source:* National Weather Service)

63. A charity 10K race is given annually to benefit a local hospice organization. How long will it take to run/walk a 10K race (10 kilometers or 10,000 meters) if your average pace is 125 **meters** per minute?

(2.7) *Find each of the following.*

64. The number 9 is what percent of 45?

65. The number 59.5 is what percent of 85?

66. The number 137.5 is 125% of what number?

67. The number 768 is 60% of what number?

68. A recent survey found that 66.9% of Americans use the Internet. If a city has a population of 76,000 how many people in that city would you expect to use the Internet? (*Source:* UCLA Center for Communication Policy)

The graph below shows the percent(s) of cell phone users who have engaged in various behaviors while driving and talking on their cell phones. Use this graph to answer Exercises 69 through 72.

Effects of Cell phone Use on Driving

Percent of motorists who use a cell phone while driving

	Swerved into another lane	Sped up	Cut off someone	Almost hit a car
Percent	46%	41%	21%	18%

Source: Progressive Insurance

69. What percent of motorists who use a cell phone while driving have almost hit another car?

70. What is the most common effect of cell phone use on driving?

71. If a cell-phone service has an estimated 4600 customers who use their cell phones while driving, how many of these customers would you expect to have cut someone off while driving and talking on their cell phones?

72. Do the percents in the graph to the left have a sum of 100%? Why or why not?

Write each phrase as a ratio in fractional notation.

73. 20 cents to 1 dollar

74. four parts red to six parts white

Solve each proportion.

75. $\frac{x}{2} = \frac{12}{4}$

76. $\frac{20}{1} = \frac{x}{25}$

77. $\frac{32}{100} = \frac{100}{x}$

78. $\frac{20}{2} = \frac{c}{5}$

79. $\frac{2}{x - 1} = \frac{3}{x + 3}$

80. $\frac{4}{y - 3} = \frac{2}{y - 3}$

81. $\frac{y + 2}{y} = \frac{5}{3}$

82. $\frac{x - 3}{3x + 2} = \frac{2}{6}$

Given the following prices charged for various sizes of an item, find the best buy.

83. Shampoo
10 ounces for \$1.29
16 ounces for \$2.15

84. Frozen green beans
8 ounces for \$0.89
15 ounces for \$1.63
20 ounces for \$2.36

Solve.

85. A machine can process 300 parts in 20 minutes. Find how many parts can be processed in 45 minutes.

86. As his consulting fee, a student charges \$90.00 per day. Find how much he charges for 3 hours of consulting. Assume an 8-hour work day.

87. One fund raiser can address 100 letters in 35 minutes. Find how many he can address in 55 minutes.

(2.8) *Graph on a number line and then write it in interval notation.*

88. $\{x | x \leq -2\}$

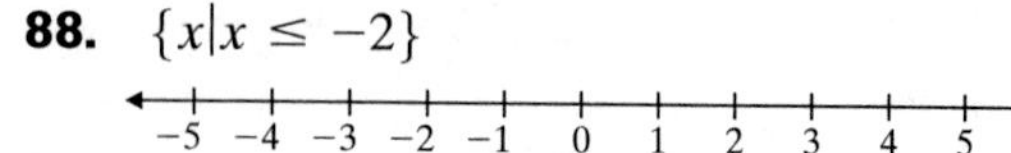

89. $\{x | x > 0\}$

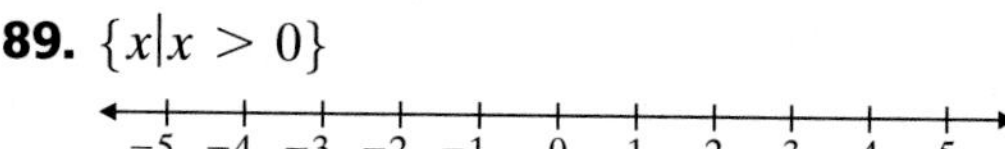

Solve. Write your answers in interval notation.

90. $3(x - 5) > -(x + 3)$

91. $-2(x + 7) \geq 3(x + 2)$

92. $4x - (5 + 2x) < 3x - 1$

93. $3(x - 8) < 7x + 2(5 - x)$

94. $24 \geq 6x - 2(3x - 5) + 2x$

95. $48 + x \geq 5(2x + 4) - 2x$

96. $\frac{x}{3} + \frac{1}{2} > \frac{2}{3}$

97. $x + \frac{3}{4} < \frac{-x}{2} + \frac{9}{4}$

98. $\frac{x - 5}{2} \leq \frac{3}{8}(2x + 6)$

99. $\frac{3(x - 2)}{5} > \frac{-5(x - 2)}{3}$

Solve.

100. George Boros can pay his housekeeper \$15 per week to do his laundry, or he can have the laundromat do it at a cost of 50 cents per pound for the first 10 pounds and 40 cents for each additional pound. Use an inequality to find the weight at which it is more economical to use the housekeeper than the laundromat.

101. In the Olympic gymnastics competition, Nana must average a score of 9.65 to win the silver medal. Seven of the eight judges have reported scores of 9.5, 9.7, 9.9, 9.7, 9.7, 9.6, and 9.5. Use an inequality to find the minimum score that Nana must receive from the last judge to win the silver medal.

Name ____________ Section ______ Date ______

Chapter 2 Test

Simplify each expression.

1. $2y - 6 - y - 4$

2. 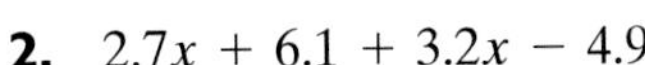 $2.7x + 6.1 + 3.2x - 4.9$

3. $4(x - 2) - 3(2x - 6)$

4. $-5(y + 1) + 2(3 - 5y)$

Solve each equation.

5. $-\frac{4}{5}x = 4$

6. $4(n - 5) = -(4 - 2n)$

7. $5y - 7 + y = -(y + 3y)$

8. $4z + 1 - z = 1 + z$

9. $\frac{2(x + 6)}{3} = x - 5$

10. $\frac{4(y - 1)}{5} = 2y + 3$

11. $\frac{1}{2} - x + \frac{3}{2} = x - 4$

12. $\frac{5}{y + 1} = \frac{4}{y + 2}$

13. $\frac{1}{3}(y + 3) = 4y$

14. $-0.3(x - 4) + x = 0.5(3 - x)$

15. $-4(a + 1) - 3a = -7(2a - 3)$

Solve each application.

16. A number increased by two-thirds of the number is 35. Find the number.

△ **17.** A gallon of water seal covers 200 square feet. How many gallons are needed to paint two coats of water seal on a deck that measures 20 feet by 35 feet?

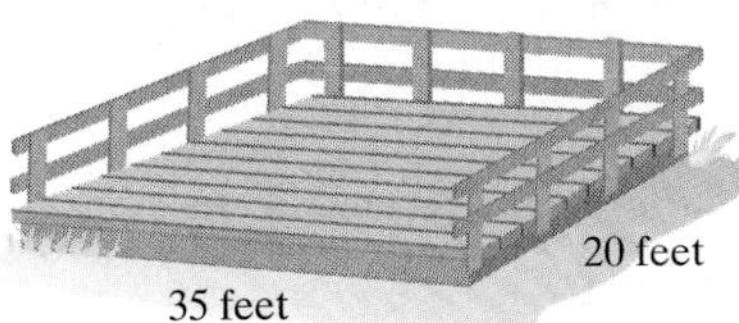

18. Decide which is the best buy in crackers.
6 ounces for \$1.19
10 ounces for \$2.15
16 ounces for \$3.25

19. In a sample of 85 fluorescent bulbs, 3 were found to be defective. At this rate, how many defective bulbs should be found in 510 bulbs?

20. Find the value of x if $y = -14$, $m = -2$, and $b = -2$ in the formula $y = mx + b$.

Answers

1. ____
2. ____
3. ____
4. ____
5. ____
6. ____
7. ____
8. ____
9. ____
10. ____
11. ____
12. ____
13. ____
14. ____
15. ____
16. ____
17. ____
18. ____
19. ____
20. ____

21. ____
22. ____
23. ____
24. ____
25. ____
26. ____
27. ____
28. ____
29. ____
30. ____
31. ____
32. ____

Solve each equation for the indicated variable.

21. $V = \pi r^2 h$ for h

22. $3x - 4y = 10$ for y

Solve each inequality. Graph the solution set and write it in interval notation.

23. $3x - 5 > 7x + 3$

−5 −4 −3 −2 −1 0 1 2 3 4 5

24. $x + 6 > 4x - 6$

−5 −4 −3 −2 −1 0 1 2 3 4 5

Solve each inequality. Write your answers in interval notation.

25. $-0.3x \geq 2.4$

26. $-5(x - 1) + 6 \leq -3(x + 4) + 1$

27. $\dfrac{2(5x + 1)}{3} > 2$

The following graph shows the breakdown of tornadoes occurring in the United States by strength. The corresponding Fujita Tornado Scale categories are shown in parentheses. Use this graph to answer Exercises 28 and 29.

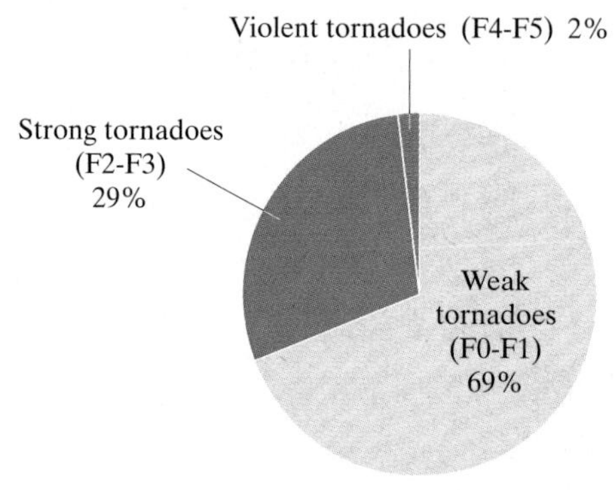

Source: National Climatic Data Center

28. What percent of tornadoes occurring in the United States are classified as "strong," that is, F2 or F3 on the Fujita Scale?

29. According to the National Climatic Data Center, in an average year, about 800 tornadoes are reported in the United States. How many of these would you expect to be classified as "weak" tornadoes?

30. The number 72 is what percent of 180?

31. New York State has more public libraries than any other state. It has 650 more public libraries than Indiana does. If the total number of public libraries for these states is 1504, find the number of public libraries in New York and the number in Indiana. (*Source: The World Almanac and Book of Facts*, 2001)

32. The number of NFL football teams (as of September 2002) is 3 more than the number of NBA basketball teams. If the total number of NFL and NBA teams is 61, find the number of each type of team. (*Sources:* NFL and NBA)

Name ______________________ Section __________ Date __________

Cumulative Review

Determine whether each statement is true or false.

1. $8 \geq 8$ **2.** $8 \leq 8$ **3.** $23 \leq 0$ **4.** $0 \leq 23$

5. Insert $<$, $>$, or $=$ in the appropriate space to make each statement true.

a. $|0| \quad 2$ **d.** $|5.9| \quad |6|$

b. $|-5| \quad 5$ **e.** $|-7| \quad |6|$

c. $|-3| \quad |-2|$

6. Simplify the expression $\dfrac{3 + |4 - 3| + 2^2}{6 - 3}$.

Add without using number lines.

7. $(-8) + (-11)$ **8.** $(-2) + 10$ **9.** $0.2 + (-0.5)$

10. Simplify each expression.

a. $-3 + [(-2 - 5) - 2]$ **b.** $2^3 - 10 + [-6 - (-5)]$

11. Use order of operations and simplify each expression.

a. $7(0)(-6)$ **c.** $(-1)(5)(-9)$

b. $(-2)(-3)(-4)$ **d.** $-4(-11) - 5(-2)$

12. Use the definition of the quotient of two numbers to find each quotient.

a. $-18 \div 3$ **b.** $\dfrac{-14}{-2}$ **c.** $\dfrac{20}{-4}$

Use the distributive property to write each expression without parentheses. Then simplify the result.

13. $-5(-3 + 2z)$ **14.** $4(3x + 7) + 10$

Answers

1. ______
2. ______
3. ______
4. ______
5. a. ______
   b. ______
   c. ______
   d. ______
   e. ______
6. ______
7. ______
8. ______
9. ______
10. a. ______
    b. ______
11. a. ______
    b. ______
    c. ______
    d. ______
12. a. ______
    b. ______
    c. ______
13. ______
14. ______

Name the property illustrated by each true statement.

15. $3 \cdot y = y \cdot 3$

16. $(x + 7) + 9 = x + (7 + 9)$

17. $(b + 0) + 3 = b + 3$

18. $2 \cdot (z \cdot 5) = 2 \cdot (5 \cdot z)$

19. $-2 \cdot \left(-\frac{1}{2}\right) = 1$

20. $-2 + 2 = 0$

21. $-6 \cdot (y \cdot 2) = (-6 \cdot 2) \cdot y$

22. Tell whether the terms are like or unlike.

a. $2x, 3x^2$ **b.** $4x^2y, x^2y, -2x^2y$ **c.** $-2yz, -3zy$ **d.** $-x^4, x^4$

23. Subtract $4x - 2$ from $2x - 3$.

24. Solve $x - 7 = 10$ for x.

25. Solve: $-z - 4 = 6$

26. Solve: $\frac{2(a + 3)}{3} = 6a + 2$

27. In the 107th Congress, the U.S. House of Representatives had a total of 430 Democrats and Republicans. There were 10 more Republican representatives than Democratic. Find the number of representatives from each party. (*Source:* Office of the Clerk of the U.S. House of Representatives)

28. A glacier is a giant mass of rocks and ice that flows downhill like a river. Portage Glacier in Alaska is about 6 miles, or 31,680 feet, long and moves 400 feet per year. Icebergs are created when the front end of the glacier flows into Portage Lake. How long does it take for ice at the head (beginning) of the glacier to reach the lake?

29. The number 63 is what percent of 72?

30. Solve for x: $\frac{45}{x} = \frac{5}{7}$

31. Graph $\{x | x < -1\}$ and write it in interval notation.

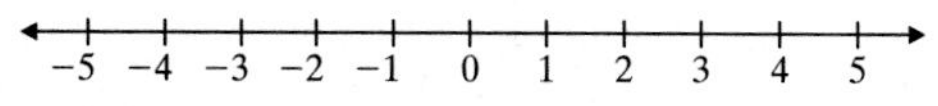

32. Solve $5 - x \le 4x - 15$. Write the solution set in interval notation.

15. ____

16. ____

17. ____

18. ____

19. ____

20. ____

21. ____

22. a. ____

b. ____

c. ____

d. ____

23. ____

24. ____

25. ____

26. ____

27. ____

28. ____

29. ____

30. ____

31. ____

32. ____

Graphing Equations and Inequalities

CHAPTER 3

In Chapter 2 we learned to solve and graph the solutions of linear equations and inequalities in one variable on number lines. Now we define and present techniques for solving and graphing linear equations and inequalities in two variables on grids.

Ninety-three percent of American children go trick-or-treating at Halloween. Each year, Americans spend roughly $2 billion on Halloween candy, including 20 million pounds of the ever-popular candy corn. While that means that Americans purchase approximately 8.3 billion kernels of candy corn to help celebrate Halloween, adults prefer to hand out chocolate to trick-or-treaters. According to a recent survey, nearly 80% of adults said they planned to give out chocolate treats at Halloween. The survey also revealed that SNICKERS® Bars are the top chocolate pick for distributing to little beggars on All Saints' Eve. In Exercise 46 on page 184, we will examine the trends in Halloween candy sales.

Answers

1. see graph

2.

3. see graph

4. see graph

5. see graph

6. see graph

7.

8.

9.

10.

11.

Name ______________ Section ________ Date ________

Chapter 3 Pretest

1. Plot each ordered pair: $(-4, 3)$, $(0, -2)$, and $(5, 0)$

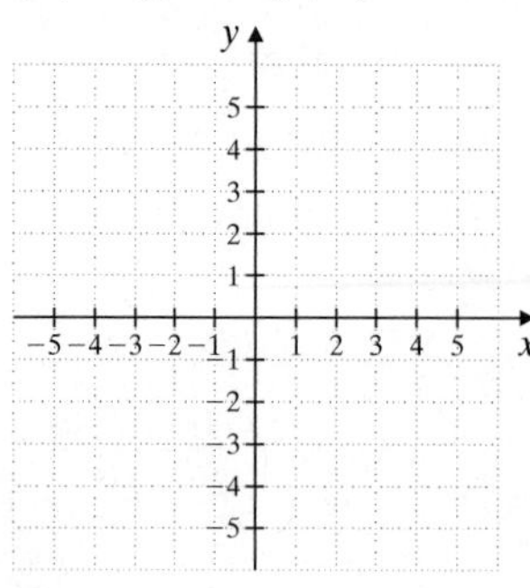

2. Complete the ordered pair solution for the given linear equation.
$8x - 3y = 2; (-2, \ )$

Graph.

3. $3x - y = 6$

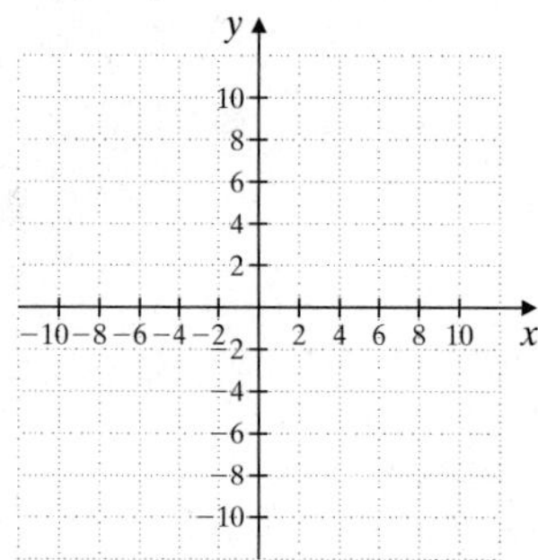

4. The line with x-intercept: $(-1, 0)$, y-intercept: $(0, 4)$

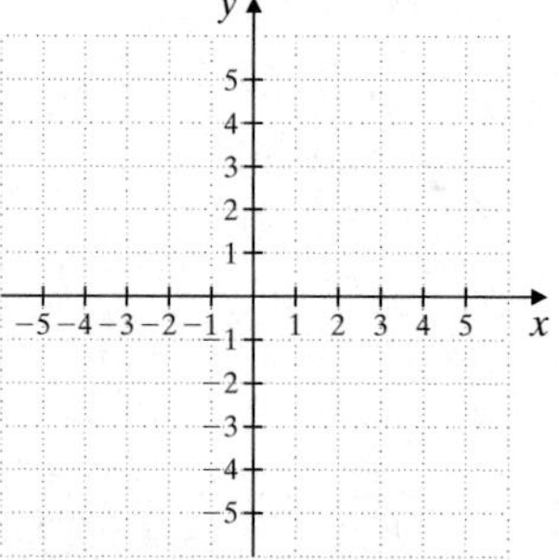

5. $3x + 2y \leq 6$

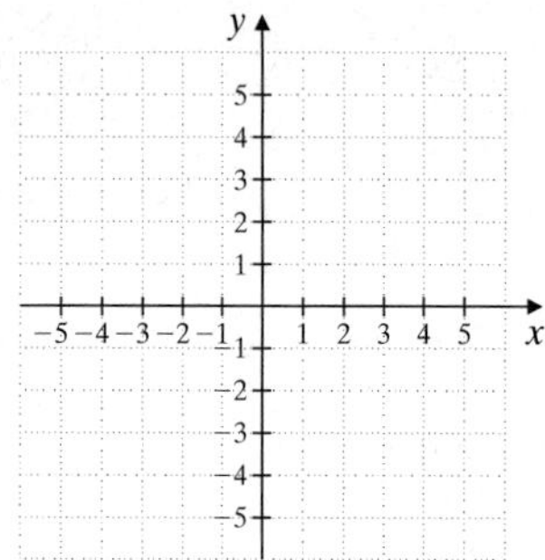

6. $y + 2 = 0$

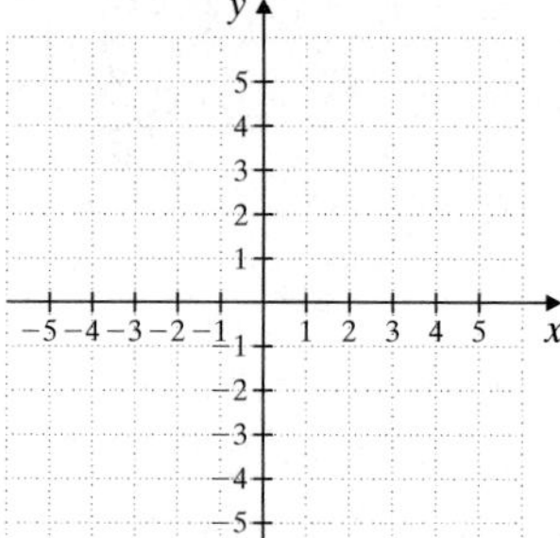

Find the slope of each line.

7. Passes through $(-7, 8)$ and $(3, 5)$

8. $4x - 5y = 20$

9. $x = 10$

Determine whether each pair of lines is parallel, perpendicular, or neither.

10. $-4x + 2y = 5$
$2x + y = 7$

11. $-2x + 3y = 1$
$3x + 2y = 12$

3.1 Reading Graphs and the Rectangular Coordinate System

In today's world, where the exchange of information must be fast and entertaining, graphs are becoming increasingly popular. They provide a quick way of making comparisons, drawing conclusions, and approximating quantities.

OBJECTIVES

- **A** Read bar and line graphs.
- **B** Plot ordered pairs of numbers on the rectangular coordinate system.
- **C** Graph paired data to create a scatter diagram.
- **D** Find the missing coordinate of an ordered pair solution, given one coordinate of the pair.

A Reading Bar and Line Graphs

A **bar graph** consists of a series of bars arranged vertically or horizontally. The bar graph in Example 1 shows a comparison of worldwide Internet users by region. The names of the regions are listed vertically and a bar is shown for each region. Corresponding to the length of the bar for each region is a number along a horizontal axis. These horizontal numbers are number of Internet users in millions.

EXAMPLE 1

The following bar graph shows the estimated worldwide number of Internet users by region as of November 2000.

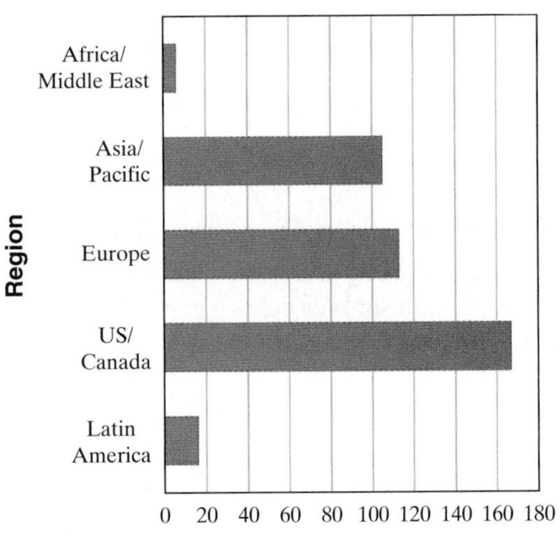

a. Find the region that has the most Internet users and approximate the number of users.

b. How many more users are in the US/Canada region than the Europe region?

Solution:

a. Since these bars are arranged horizontally, we look for the longest bar, which is the bar representing the US/Canada region. To approximate the number associated with this region, we move from the right edge of this

Practice Problem 1

Use the graph from Example 1 to answer the following.

a. Find the region with the fewest Internet users and approximate the number of users.

b. How many more users are in the Asia/Pacific region than in the Latin America region?

Answers

1. a. Africa/Middle East region, 7 million Internet users, **b.** about 89 million Internet users

bar vertically downward to the Internet user axis. This region has approximately 167 million Internet users.

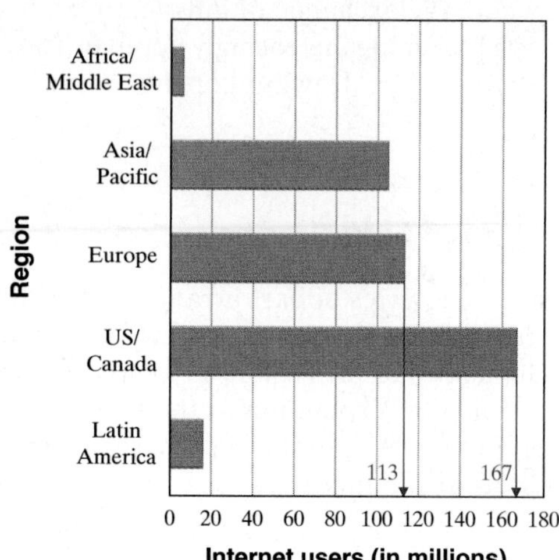

Source: Nua Internet Surveys

b. The US/Canada region has approximately 167 million Internet users. The Europe region has approximately 113 million Internet users. To find how many more users are in the US/Canada region, we subtract $167 - 113 = 54$ or 54 million more Internet users.

A **line graph** consists of a series of points connected by a line. The next graph is an example of a line graph. It is also sometimes called a **broken line graph**.

Practice Problem 2 

Use the graph from Example 2 to answer the following.

a. What is the pulse rate 40 minutes after lighting a cigarette?

b. What is the pulse rate when the cigarette is being lit?

c. When is the pulse rate the highest?

EXAMPLE 2

The line graph shows the relationship between time spent smoking a cigarette and pulse rate. Time is recorded along the horizontal axis in minutes, with 0 minutes being the moment a smoker lights a cigarette. Pulse is recorded along the vertical axis in heartbeats per minute.

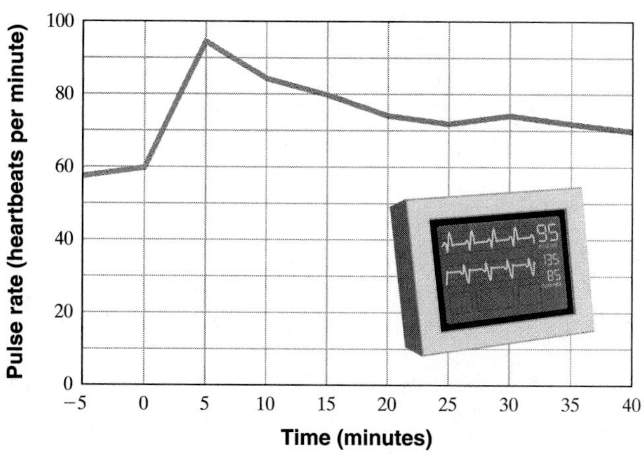

a. What is the pulse rate 15 minutes after a cigarette is lit?

b. When is the pulse rate the lowest?

c. When does the pulse rate show the greatest change?

Answers

2. a. 70, **b.** 60, **c.** 5 min. after lighting

Solution:

a. We locate the number 15 along the time axis and move vertically upward until the line is reached. From this point on the line, we move horizontally to the left until the pulse rate axis is reached. Reading the number of beats per minute, we find that the pulse rate is 80 beats per minute 15 minutes after a cigarette is lit.

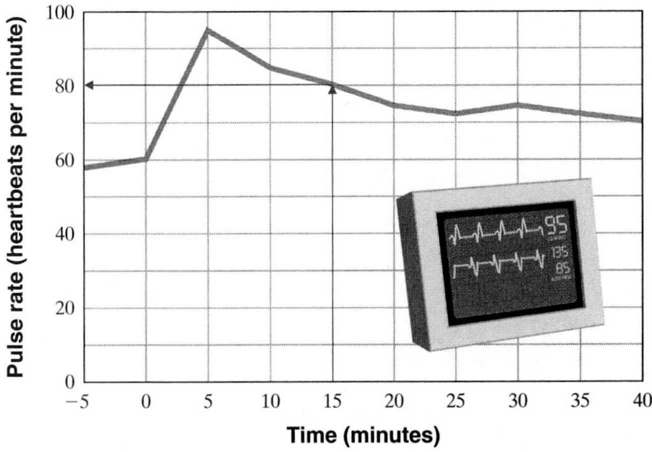

b. We find the lowest point of the line graph, which represents the lowest pulse rate. From this point, we move vertically downward to the time axis. We find that the pulse rate is the lowest at −5 minutes, which means 5 minutes *before* lighting a cigarette.

c. The pulse rate shows the greatest change during the 5 minutes between 0 and 5. Notice that the line graph is *steepest* between 0 and 5 minutes. ●

Notice in the graph above that there are two numbers associated with each point of the graph. For example, we discussed earlier that 15 minutes after "lighting up," the pulse rate is 80 beats per minute. If we agree to write the time first and the pulse rate second, we can say there is a point on the graph corresponding to the **ordered pair** of numbers (15, 80). A few more ordered pairs are shown alongside their corresponding points.

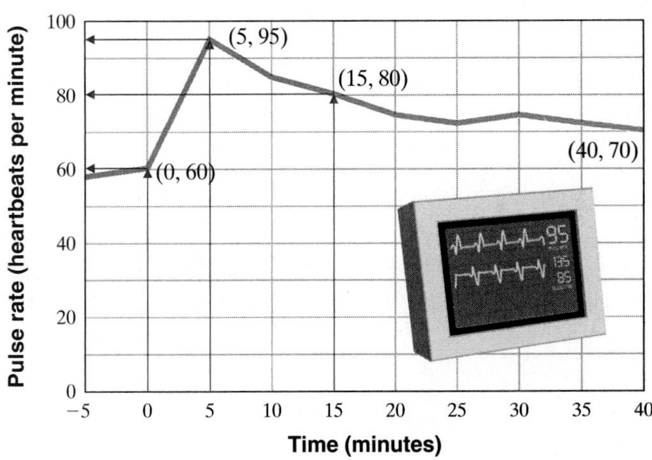

B Plotting Ordered Pairs of Numbers

In general, we use the idea of ordered pairs to describe the location of a point in a plane (such as a piece of paper). We start with a horizontal and a vertical axis. Each axis is a number line, and for the sake of consistency we construct our axes to intersect at the 0 coordinate of both. This point of intersection is called the **origin**. Notice that these two number lines or axes divide the plane into four regions called **quadrants**. The quadrants are usually numbered with Roman numerals as shown. The axes are not considered to be in any quadrant.

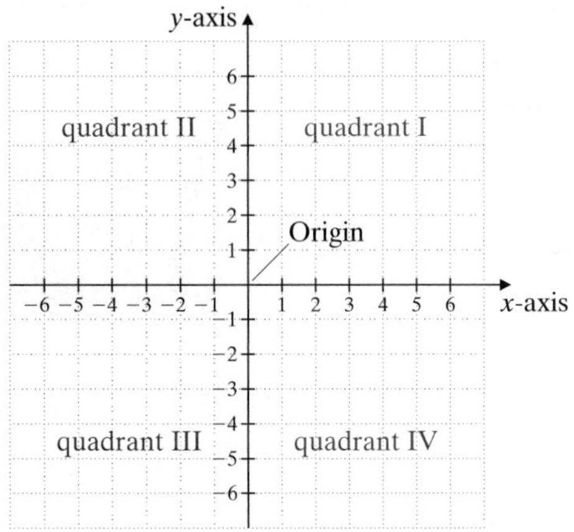

It is helpful to label axes, so we label the horizontal axis the ***x*-axis** and the vertical axis the ***y*-axis**. We call the system described above the **rectangular coordinate system**, or the **coordinate plane**. Just as with other graphs shown, we can then describe the locations of points by ordered pairs of numbers. We list the horizontal ***x*-axis** measurement first and the vertical ***y*-axis** measurement second.

To plot or graph the point corresponding to the ordered pair

$$(a, b),$$

we start at the origin. We then move a units left or right (right if a is positive, left if a is negative). From there, we move b units up or down (up if b is positive, down if b is negative). For example, to plot the point corresponding to the ordered pair $(3, 2)$, we start at the origin, move 3 units right, and from there move 2 units up. (See the figure below.) The x-value, 3, is also called the ***x*-coordinate** and the y-value, 2, is also called the ***y*-coordinate**. From now on, we will call the point with coordinates $(3, 2)$ simply the point $(3, 2)$. The point $(-2, 5)$ is graphed below also.

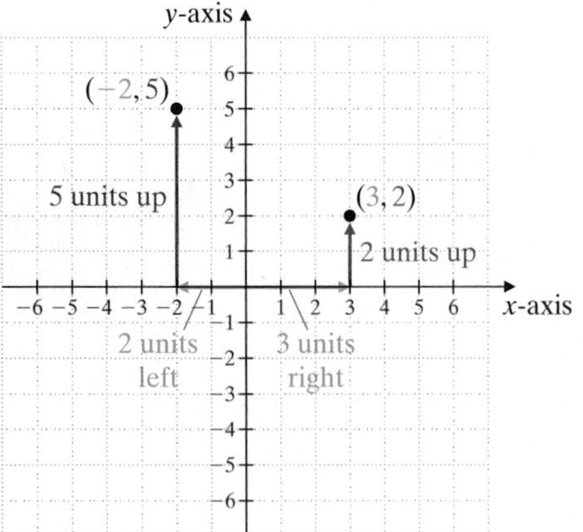

Don't forget that **each ordered pair corresponds to exactly one point in the plane and that each point in the plane corresponds to exactly one ordered pair**.

Try the Concept Check in the margin.

EXAMPLE 3 On a single coordinate system, plot each ordered pair. State in which quadrant, if any, each point lies.

a. $(5, 3)$ **b.** $(-2, -4)$ **c.** $(1, -2)$ **d.** $(-5, 3)$

e. $(0, 0)$ **f.** $(0, 2)$ **g.** $(-5, 0)$ **h.** $\left(0, -5\frac{1}{2}\right)$

Solution:

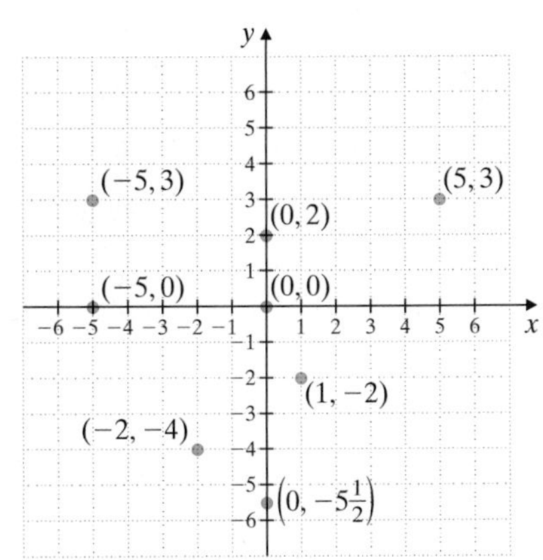

a. Point $(5, 3)$ lies in quadrant I.

b. Point $(-2, -4)$ lies in quadrant III.

c. Point $(1, -2)$ lies in quadrant IV.

d. Point $(-5, 3)$ lies in quadrant II.

e.–h. Points $(0, 0)$, $(0, 2)$, $(-5, 0)$, and $\left(0, -5\frac{1}{2}\right)$ lie on axes, so they are not in any quadrant.

Concept Check

Is the graph of the point $(-5, 1)$ in the same location as the graph of the point $(1, -5)$? Explain.

Practice Problem 3

On a single coordinate system, plot each ordered pair. State in which quadrant, if any, each point lies.

a. $(4, 2)$ b. $(-1, -3)$
c. $(2, -2)$ d. $(-5, 1)$
e. $(0, 3)$ f. $(3, 0)$
g. $(0, -4)$ h. $\left(-2\frac{1}{2}, 0\right)$

Answers

3.

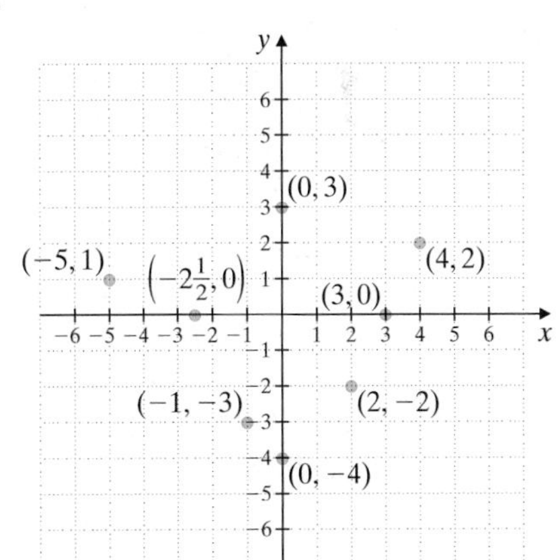

a. Point $(4, 2)$ lies in quadrant I.
b. Point $(-1, -3)$ lies in quadrant III.
c. Point $(2, -2)$ lies in quadrant IV.
d. Point $(-5, 1)$ lies in quadrant II.
e.–h. Points $(0, 3)$, $(3, 0)$, $(0, -4)$, and $\left(-2\frac{1}{2}, 0\right)$ lie on axes, so they are not in any quadrant.

Concept Check: The graph of point $(-5, 1)$ lies in quadrant II and the graph of point $(1, -5)$ lies in quadrant IV. They are *not* in the same location.

Concept Check

For each description of a point in the rectangular coordinate system, write an ordered pair that represents it.

a. Point A is located three units to the left of the y-axis and five units above the x-axis.
b. Point B is located six units below the origin.

Practice Problem 4

The table gives the number of tornadoes that have occurred in the United States for the years shown. (*Source:* Storm Prediction Center, National Weather Service)

Year	Tornadoes
1995	1234
1996	1173
1997	1148
1998	1424
1999	1343
2000	997

a. Write this paired data as a set of ordered pairs of the form (year, tornadoes).
b. Create a scatter diagram of the paired data.

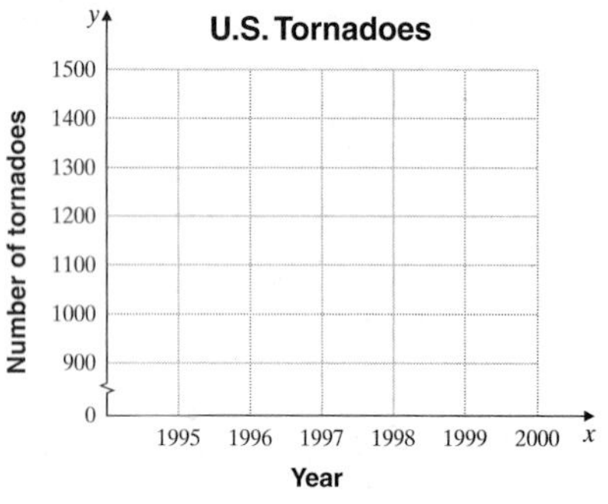

c. What trend in the paired data, if any, does the scatter diagram show?

Answers

4. a. $(1995, 1234)$, $(1996, 1173)$, $(1997, 1148)$, $(1998, 1424)$, $(1999, 1343)$, $(2000, 997)$

b.

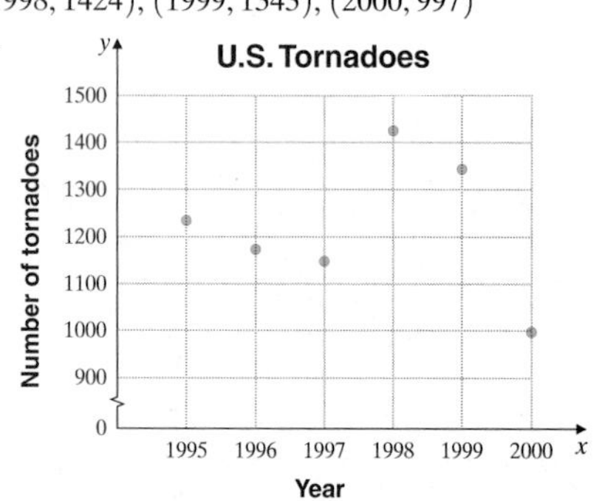

c. No trend; the number of tornadoes varies greatly from year to year.

Concept Check: **a.** $(-3, 5)$, **b.** $(0, -6)$

Try the Concept Check in the margin.

From Example 1, notice that the y-coordinate of any point on the x-axis is 0. For example, the point $(-5, 0)$ lies on the x-axis. Also, the x-coordinate of any point on the y-axis is 0. For example, the point $(0, 2)$ lies on the y-axis.

C Creating Scatter Diagrams

Data that can be represented as ordered pairs are called **paired data**. Many types of data collected from the real world are paired data. For instance, the annual measurements of a child's height can be written as ordered pairs of the form (year, height in inches) and are paired data. The graph of paired data as points in the rectangular coordinate system is called a **scatter diagram**. Scatter diagrams can be used to look for patterns and trends in paired data.

EXAMPLE 4 The table gives the annual net sales for Wal-Mart Stores for the years shown. (*Source:* Wal-Mart Stores, Inc.)

Year	Wal-Mart Net Sales (in billions of dollars)
1997	105
1998	118
1999	138
2000	165
2001	191

a. Write this paired data as a set of ordered pairs of the form (year, revenue in billions of dollars).

b. Create a scatter diagram of the paired data.

c. What trend in the paired data does the scatter diagram show?

Solution:

a. The ordered pairs are $(1997, 105)$, $(1998, 118)$, $(1999, 138)$, $(2000, 165)$, and $(2001, 191)$.

b. We begin by plotting the ordered pairs. Because the x-coordinate in each ordered pair is a year, we label the x-axis "Year" and mark the horizontal axis with the years given. Then we label the y-axis or vertical axis "Wal-Mart Net Sales (in billions of dollars)." In this case, it is convenient to mark the vertical axis in multiples of 20, starting with 0. Since no net sale is less than 100, we use the notation ⚡ to skip to 100, then proceed by multiples of 20.

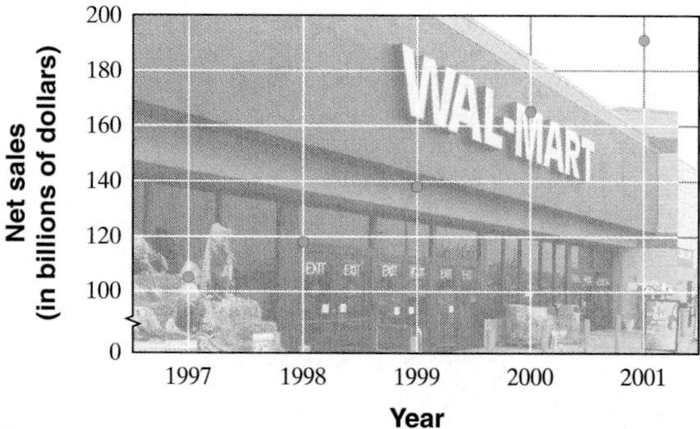

c. The scatter diagram shows that Wal-Mart net sales steadily increased over the years 1997–2001.

D Completing Ordered Pair Solutions

Let's see how we can use ordered pairs to record solutions of equations containing two variables. An equation in one variable such as $x + 1 = 5$ has one solution, which is 4: the number 4 is the value of the variable x that makes the equation true.

An equation in two variables, such as $2x + y = 8$, has solutions consisting of two values, one for x and one for y. For example, $x = 3$ and $y = 2$ is a solution of $2x + y = 8$ because, if x is replaced with 3 and y with 2, we get a true statement.

$$\begin{aligned} 2x + y &= 8 \\ 2(3) + 2 &\stackrel{?}{=} 8 \\ 8 &= 8 \quad \text{True} \end{aligned}$$

The solution $x = 3$ and $y = 2$ can be written as $(3, 2)$, an ordered pair of numbers.

In general, an ordered pair is a **solution** of an equation in two variables if replacing the variables by the values of the ordered pair results in a *true statement*. For example, another ordered pair solution of $2x + y = 8$ is $(5, -2)$. Replacing x with 5 and y with -2 results in a true statement.

$$\begin{aligned} 2x + y &= 8 \\ 2(5) + (-2) &\stackrel{?}{=} 8 \quad \text{Replace } x \text{ with 5 and } y \text{ with } -2. \\ 10 - 2 &\stackrel{?}{=} 8 \\ 8 &= 8 \quad \text{True} \end{aligned}$$

EXAMPLE 5 Complete each ordered pair so that it is a solution to the equation $3x + y = 12$.

a. $(0, \ )$ **b.** $(\ , 6)$ **c.** $(-1, \ )$

Solution:

a. In the ordered pair $(0, \ )$, the x-value is 0. We let $x = 0$ in the equation and solve for y.

$$\begin{aligned} 3x + y &= 12 \\ 3(0) + y &= 12 \quad \text{Replace } x \text{ with 0.} \\ 0 + y &= 12 \\ y &= 12 \end{aligned}$$

The completed ordered pair is $(0, 12)$.

b. In the ordered pair $(\ , 6)$, the y-value is 6. We let $y = 6$ in the equation and solve for x.

$$\begin{aligned} 3x + y &= 12 \\ 3x + 6 &= 12 \quad \text{Replace } y \text{ with 6.} \\ 3x &= 6 \quad \text{Subtract 6 from both sides.} \\ x &= 2 \quad \text{Divide both sides by 3.} \end{aligned}$$

The ordered pair is $(2, 6)$.

c. In the ordered pair $(-1, \ )$, the x-value is -1. We let $x = -1$ in the equation and solve for y.

$$\begin{aligned} 3x + y &= 12 \\ 3(-1) + y &= 12 \quad \text{Replace } x \text{ with } -1. \\ -3 + y &= 12 \\ y &= 15 \quad \text{Add 3 to both sides.} \end{aligned}$$

The ordered pair is $(-1, 15)$.

Practice Problem 5

Complete each ordered pair so that it is a solution to the equation $x + 2y = 8$.

a. $(0, \ )$
b. $(\ , 3)$
c. $(-4, \ )$

Answers

5. a. $(0, 4)$, **b.** $(2, 3)$, **c.** $(-4, 6)$

Solutions of equations in two variables can also be recorded in a **table of paired values**, as shown in the next example.

Practice Problem 6

Complete the table for the equation $y = -2x$.

	x	y
a.	-3	
b.		0
c.		10

EXAMPLE 6 Complete the table for the equation $y = 3x$.

	x	y
a.	-1	
b.		0
c.		-9

Solution:

a. We replace x with -1 in the equation and solve for y.

$y = 3x$

$y = 3(-1)$ Let $x = -1$.

$y = -3$

The ordered pair is $(-1, -3)$.

b. We replace y with 0 in the equation and solve for x.

$y = 3x$

$0 = 3x$ Let $y = 0$.

$0 = x$ Divide both sides by 3.

The ordered pair is $(0, 0)$.

c. We replace y with -9 in the equation and solve for x.

$y = 3x$

$-9 = 3x$ Let $y = -9$.

$-3 = x$ Divide both sides by 3.

The ordered pair is $(-3, -9)$. The completed table is shown to the right.

x	y
-1	-3
0	0
-3	-9

Practice Problem 7

Complete the table for the equation $x = 5$.

x	y
	-2
	0
	4

EXAMPLE 7 Complete the table for the equation $y = 3$.

x	y
-2	
0	
-5	

Solution: The equation $y = 3$ is the same as $0x + y = 3$. No matter what value we replace x by, y always equals 3. The completed table is shown to the right.

x	y
-2	3
0	3
-5	3

Answers

6.

	x	y
a.	-3	6
b.	0	0
c.	-5	10

7.

x	y
5	-2
5	0
5	4

By now, you have noticed that equations in two variables often have more than one solution. We discuss this more in the next section.

A table showing ordered pair solutions may be written vertically, or horizontally as shown in the next example.

EXAMPLE 8 A small business purchased a computer for \$2000. The business predicts that the computer will be used for 5 years and the value in dollars y of the computer in x years is $y = -300x + 2000$. Complete the table.

x	0	1	2	3	4	5
y						

Solution: To find the value of y when x is 0, we replace x with 0 in the equation. We use this same procedure to find y when x is 1 and when x is 2.

When $x = 0$,

$y = -300x + 2000$

$y = -300 \cdot 0 + 2000$

$y = 0 + 2000$

$y = 2000$

When $x = 1$,

$y = -300x + 2000$

$y = -300 \cdot 1 + 2000$

$y = -300 + 2000$

$y = 1700$

When $x = 2$,

$y = -300x + 2000$

$y = -300 \cdot 2 + 2000$

$y = -600 + 2000$

$y = 1400$

We have the ordered pairs $(0, 2000)$, $(1, 1700)$, and $(2, 1400)$. This means that in 0 years the value of the computer is \$2000, in 1 year the value of the computer is \$1700, and in 2 years the value is \$1400. To complete the table of values, we continue the procedure for $x = 3$, $x = 4$, and $x = 5$.

When $x = 3$,

$y = -300x + 2000$

$y = -300 \cdot 3 + 2000$

$y = -900 + 2000$

$y = 1100$

When $x = 4$,

$y = -300x + 2000$

$y = -300 \cdot 4 + 2000$

$y = -1200 + 2000$

$y = 800$

When $x = 5$,

$y = -300x + 2000$

$y = -300 \cdot 5 + 2000$

$y = -1500 + 2000$

$y = 500$

The completed table is shown below.

x	0	1	2	3	4	5
y	2000	1700	1400	1100	800	500

Practice Problem 8

A company purchased a fax machine for \$400. The business manager of the company predicts that the fax machine will be used for 7 years and the value in dollars y of the machine in x years is $y = -50x + 400$. Complete the table.

x	1	2	3	4	5	6	7
y							

Answer

8.

x	1	2	3	4	5	6	7
y	350	300	250	200	150	100	50

The ordered pair solutions recorded in the completed table for Example 6 are another set of paired data. They are graphed next. Notice that this scatter diagram gives a visual picture of the decrease in value of the computer.

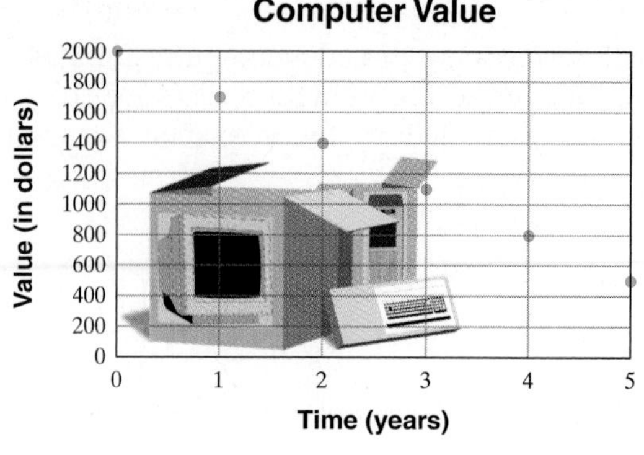

x	y
0	2000
1	1700
2	1400
3	1100
4	800
5	500

Name ______________________ Section __________ Date __________

Mental Math

Give two ordered pair solutions for each linear equation.

1. $x + y = 10$

2. $x + y = 6$

EXERCISE SET 3.1

The following bar graph shows the top 10 tourist destinations and the number of tourists that visit each country per year. Use this graph to answer Exercises 1 through 6. See Example 1.

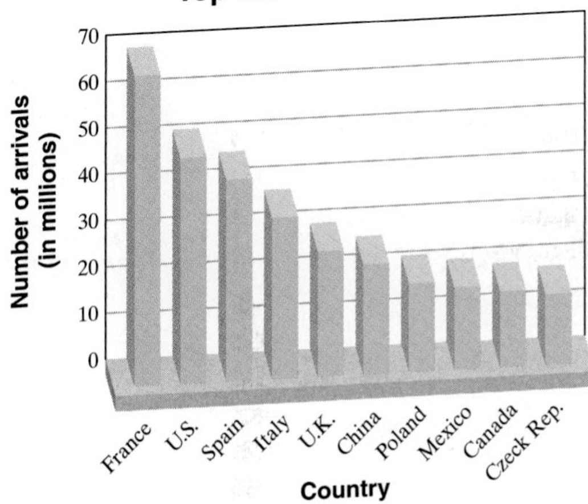

Source: World Tourism Organization

1. Which country is the most popular tourist destination?

2. Which country shown is the least popular tourist destination?

3. Which countries have more than 30 million tourists per year?

4. Which countries shown have fewer than 20 million tourists per year?

5. Estimate the number of tourists per year whose destination is Italy.

6. Estimate the number of tourists per year whose destination is France.

Use the line graph in Example 2 to answer Exercises 7 through 10.

7. Approximate the pulse rate 5 minutes before lighting a cigarette.

8. Approximate the pulse rate 10 minutes after lighting a cigarette.

9. Find the difference in pulse rate between 5 minutes before and 10 minutes after lighting a cigarette.

10. When is the pulse rate fewer than 60 heartbeats per minute?

The line graph below shows the number of students per computer in U.S. public schools. Use this graph for Exercises 11 through 16. See Example 2.

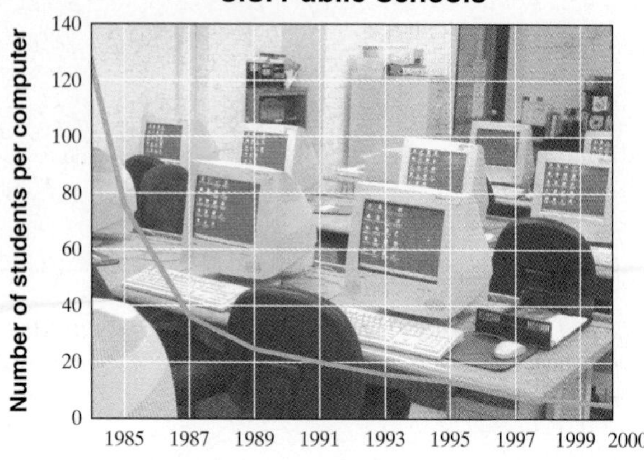

Source: QED's Technology in Public Schools, 10th Edition

11. Approximate the number of students per computer in 1991.

12. Approximate the number of students per computer in 2000.

13. During what year was the greatest decrease in number of students per computer?

14. What was the first year that the number of students per computer fell below 20?

15. What was the first year shown that the number of students per computer fell below 10?

16. Discuss any trends shown by this line graph.

The special bar graph shown is called a double bar graph. This double bar graph is used to compare men and women in the U.S. labor force per year. Use this graph for Exercises 17 through 26.

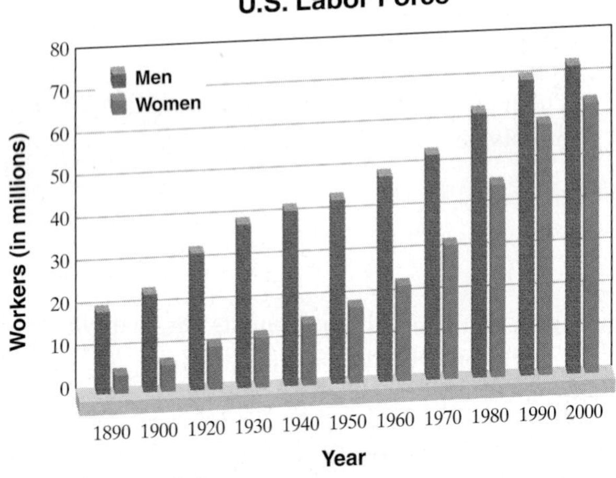

Source: U.S. Bureau of the Census

17. Estimate the number of men in the workforce in 1890.

18. Estimate the number of women in the workforce in 1890.

19. Estimate the number of women in the workforce in 2000.

20. Estimate the number of men in the workforce in 2000.

21. Give the first year that the number of men in the workforce rose above 20 million.

22. Give the first year that the number of women in the workforce rose above 20 million.

23. Estimate the difference in the number of men and women in the workforce in 1940.

24. Estimate the difference in the number of men and women in the workforce in 2000.

25. Discuss any trends shown by this graph.

26. List an advantage of using a double bar graph.

B *Plot each ordered pair. State in which quadrant, if any, each point lies. See Example 3.*

27. $(1, 5)$ $(-5, -2)$ $(-3, 0)$
$(0, -1)$ $(2, -4)$ $\left(-1, 4\frac{1}{2}\right)$

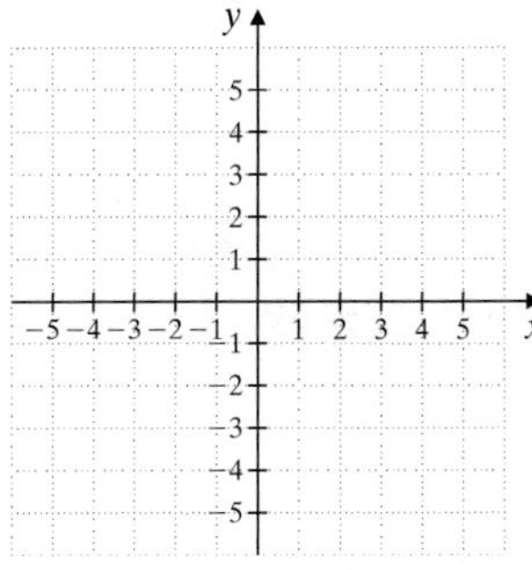

28. $(2, 4)$ $(0, 2)$ $(-2, 1)$
$(-3, -3)$ $\left(3\frac{3}{4}, 0\right)$ $(5, -4)$

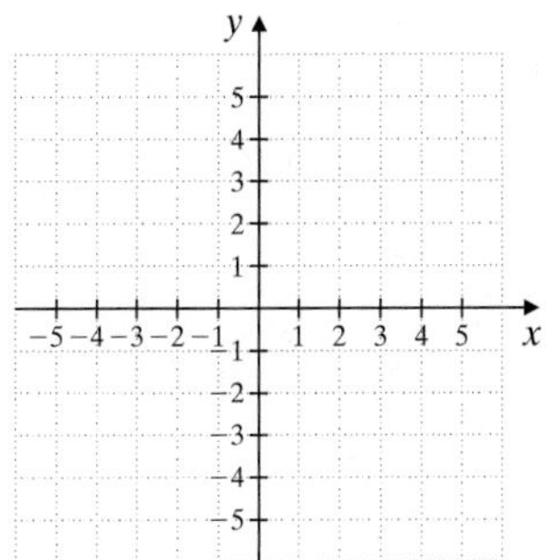

29. When is the graph of the ordered pair (a, b) the same as the graph of the ordered pair (b, a)?

30. In your own words, describe how to plot an ordered pair.

Find the x- and y-coordinates of each labeled point. See Example 3.

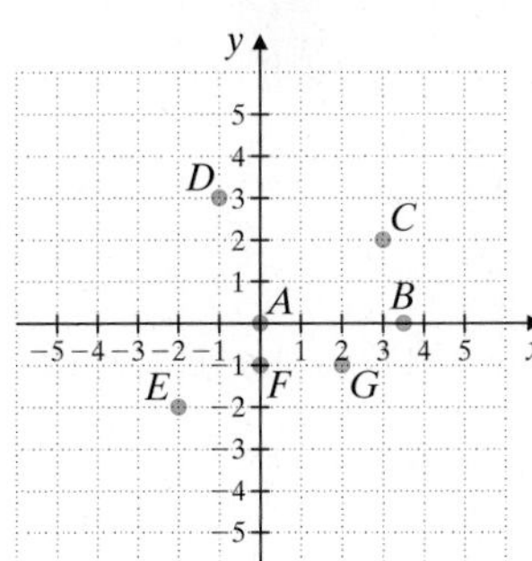

31. *A* **32.** *B* **33.** *C*

34. *D* **35.** *E* **36.** *F*

37. *G*

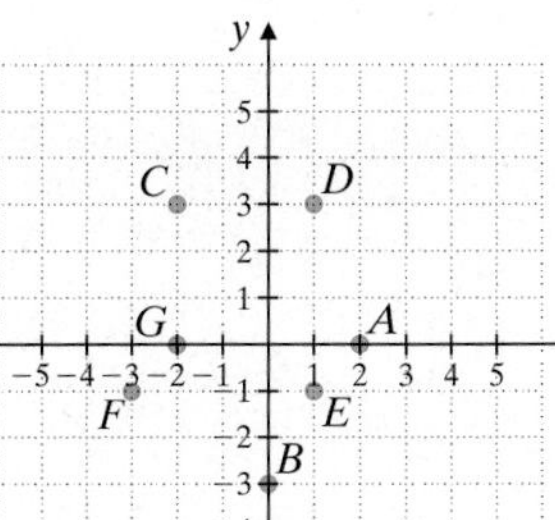

38. *A* **39.** *B* **40.** *C*

41. *D* **42.** *E* **43.** *F*

44. *G*

C *Solve. See Example 4.*

45. The table shows the average farm size (in acres) in the United States during the years shown. (*Source:* National Agricultural Statistics Service)

Year	Average Farm Size (in acres)
1995	438
1996	438
1997	436
1998	435
1999	432
2000	434

a. Write this paired data as a set of ordered pairs of the form (year, average farm size).

b. Create a scatter diagram of the paired data. Be sure to label the axes appropriately.

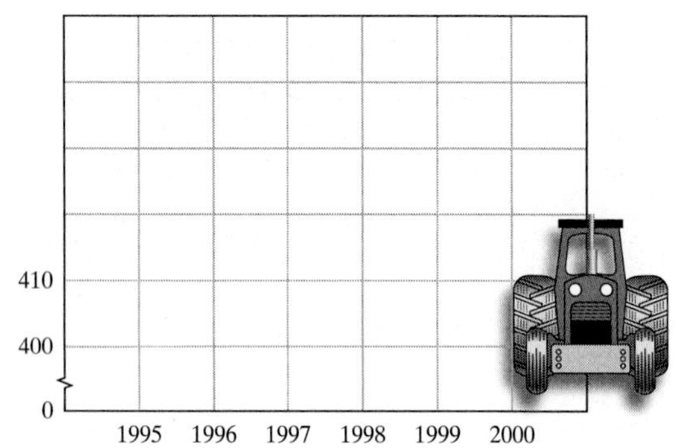

46. In the United States, Halloween is the top holiday for candy sales. The table shows the sales of chocolate and candy at Halloween (in billions of dollars) for the years shown. (*Source:* National Confectioners Association)

Year	Halloween Candy Sales (in billions of dollars)
1995	1.47
1996	1.66
1997	1.71
1998	1.79
1999	1.90
2000	1.99
2001	2.04

a. Write this paired data as a set of ordered pairs of the form (year, Halloween candy sales).

b. Create a scatter diagram of the paired data. Be sure to label the axes appropriately.

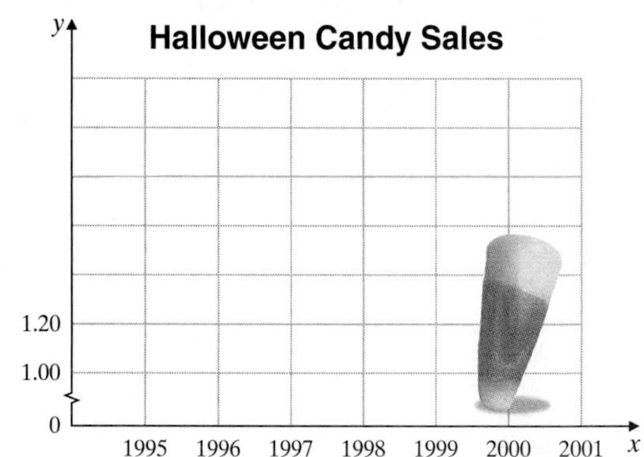

c. What trend in the paired data does the scatter diagram show?

47. The table shows the distance from the equator (in miles) and the average annual snowfall (in inches) for each of eight selected U.S. cities. (*Sources*: National Climatic Data Center, Wake Forest University Albatross Project)

City	Distance from Equator (in miles)	Average Annual Snowfall (in inches)
Atlanta, GA	2313	2
Austin, TX	2085	1
Baltimore, MD	2711	21
Chicago, IL	2869	39
Detroit, MI	2920	42
Juneau, AK	4038	99
Miami, FL	1783	0
Winston-Salem, NC	2493	9

a. Write this paired data as a set of ordered pairs of the form (distance from equator, average annual snowfall).

b. Create a scatter diagram of the paired data. Be sure to label the axes appropriately.

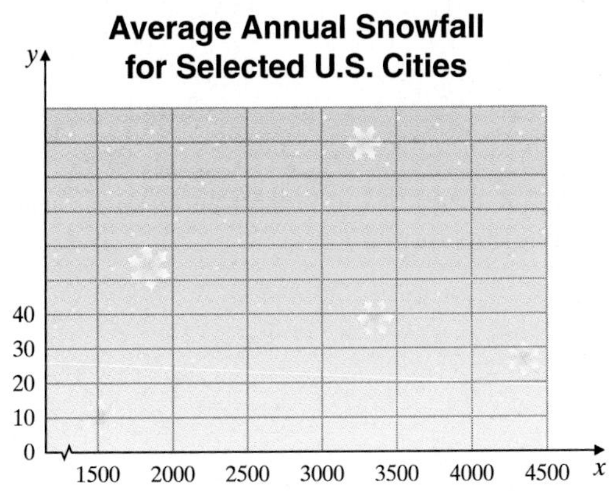

c. What trend in the paired data does the scatter diagram show?

48. The table shows the opening day payroll (in millions of dollars) and the number of games won by the ten Major League Baseball teams with the highest payrolls during the 2000 baseball season. (*Sources*: The Associated Press, Major League Baseball)

Team	2000 Opening Day Payroll (in millions of dollars)	2000 Games Won
NY Yankees	92.5	87
Los Angeles Dodgers	88.1	86
Atlanta Braves	84.5	95
Baltimore Orioles	81.4	74
Arizona Diamondbacks	81.0	85
NY Mets	79.5	94
Boston Red Sox	77.9	85
Cleveland Indians	75.9	90
Texas Rangers	70.8	71
Tampa Bay Devil Rays	62.8	69

a. Write this paired data as a set of ordered pairs of the form (opening day payroll, games won).

b. Create a scatter diagram of the paired data. Be sure to label the axes appropriately.

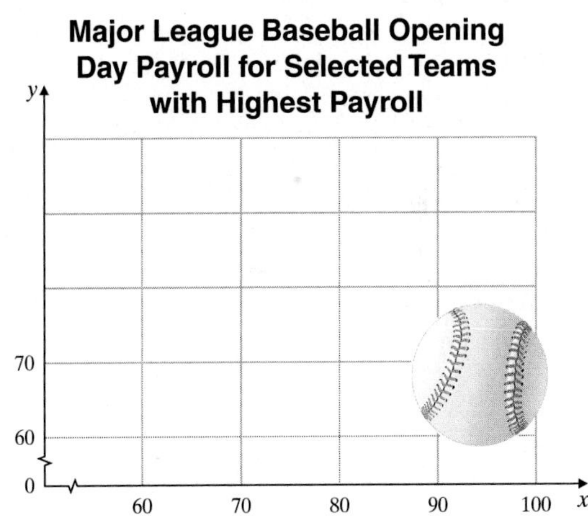

49. Minh, a psychology student, kept a record of how much time she spent studying for each of her 20-point psychology quizzes and her score on each quiz.

Hours Spent Studying	Quiz Score
0.50	10
0.75	12
1.00	15
1.25	16
1.50	18
1.50	19
1.75	19
2.00	20

a. Write each paired data as an ordered pair of the form (hours spent studying, quiz score).
b. Create a scatter diagram of the paired data. Be sure to label the axes appropriately.

Minh's Chart for Psychology

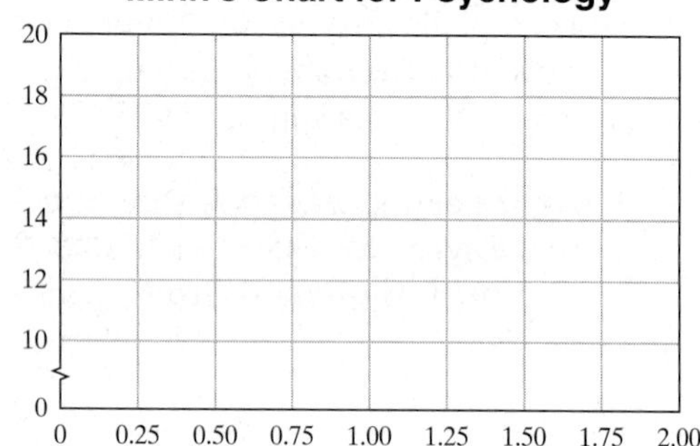

c. What might Minh conclude from the scatter diagram?

50. A local lumberyard uses quantity pricing. The table shows the price per board for different amounts of lumber purchased.

Price per Board (in dollars)	Number of Boards Purchased
8.00	1
7.50	10
6.50	25
5.00	50
2.00	100

a. Write each paired data as an ordered pair of the form (price per board, number of boards purchased).
b. Create a scatter diagram of the paired data. Be sure to label the axes appropriately.

Lumberyard Board Pricing

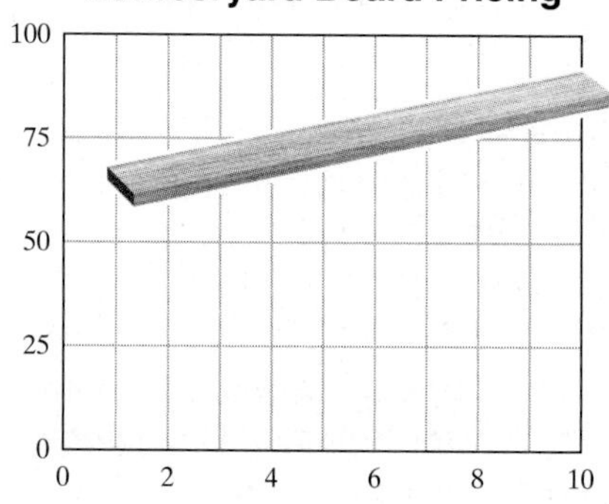

c. What trend in the paired data does the scatter diagram show?

D *Complete each ordered pair so that it is a solution of the given linear equation. See Example 5.*

51. $x - 4y = 4; (\ , -2), (4, \ )$

52. $x - 5y = -1; (\ , -2), (4, \ )$

53. $3x + y = 9; (0, \ ), (\ , 0)$

54. $x + 5y = 15; (0, \ ), (\ , 0)$

55. $y = -7; (11, \ ), (\ , -7)$

56. $x = \frac{1}{2}; (\ , 0), \left(\frac{1}{2}, \ \right)$

Complete the table of ordered pairs for each linear equation. See Examples 6 through 8.

57. $x + 3y = 6$

x	y
0	
	0
	1

58. $2x + y = 4$

x	y
0	
	0
	2

59. $2x - y = 12$

x	y
0	
	-2
3	

60. $-5x + y = 10$

x	y
	0
	5
0	

61. $2x + 7y = 5$

x	y
0	
	0
	1

62. $x - 6y = 3$

x	y
0	
1	
	-1

Complete the table of ordered pairs for each equation. Then plot the ordered pair solutions. See Examples 6 through 8.

63. $x = 3$

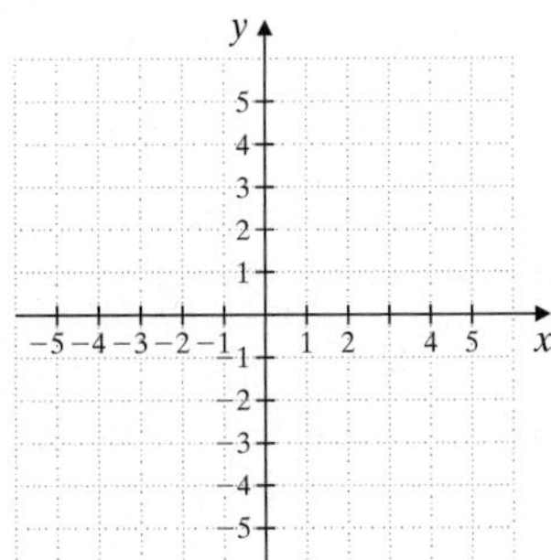

x	y
	0
	-0.5
	$\frac{1}{4}$

64. $y = -1$

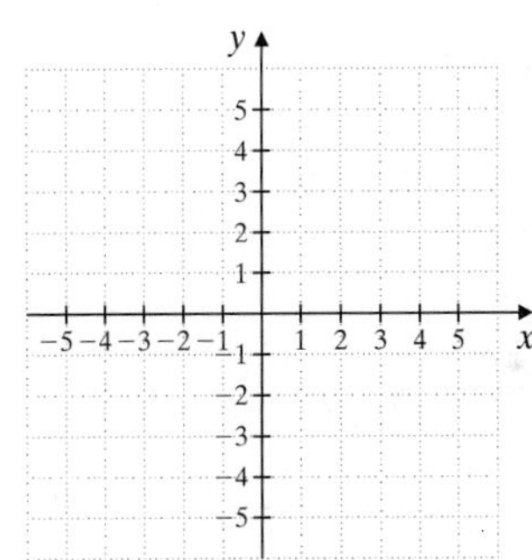

x	y
-2	
0	
-1	

65. $x = -5y$

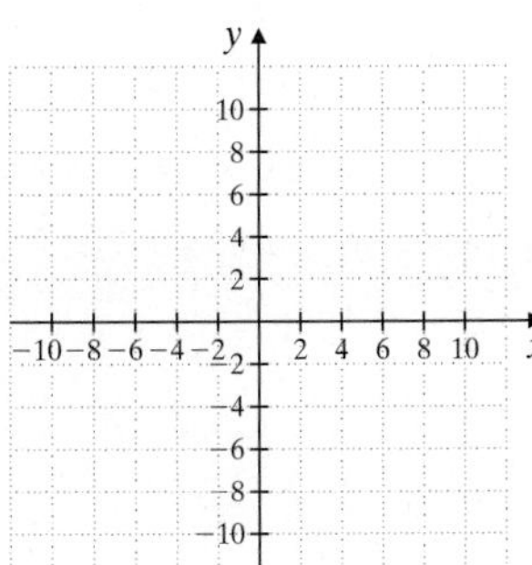

x	y
	0
	1
10	

66. $y = -3x$

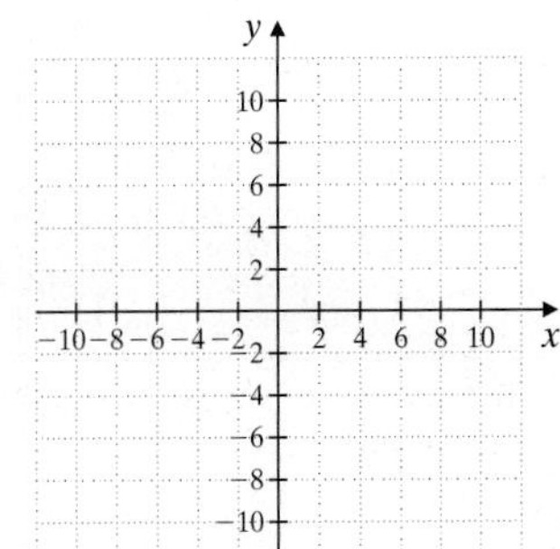

x	y
0	
-2	
	9

Solve. See Example 8.

67. The cost in dollars y of producing x computer desks is given by $y = 80x + 5000$.

a. Complete the table.

x	100	200	300
y			

b. Find the number of computer desks that can be produced for \$8600. (*Hint*: Find x when $y = 8600$.)

68. The hourly wage y of an employee at a certain production company is given by $y = 0.25x + 9$ where x is the number of units produced by the employee in an hour.

a. Complete the table.

x	0	1	5	10
y				

b. Find the number of units that an employee must produce each hour to earn an hourly wage of \$12.25. (*Hint*: Find x when $y = 12.25$.)

Review and Preview

Solve each equation for y. See Section 2.4.

69. $x + y = 5$

70. $x - y = 3$

71. $2x + 4y = 5$

72. $5x + 2y = 7$

73. $10x = -5y$

74. $4y = -8x$

Combining Concepts

△ **75.** Find the perimeter of the rectangle whose vertices are the points with coordinates $(-1, 5)$, $(3, 5)$, $(3, -4)$, and $(-1, -4)$.

△ **76.** Find the area of the rectangle whose vertices are the points with coordinates $(5, 2)$, $(5, -6)$, $(0, -6)$, and $(0, 2)$.

The scatter diagram below shows Walt Disney Company's annual revenues. The horizontal axis represents the number of years after 1996.

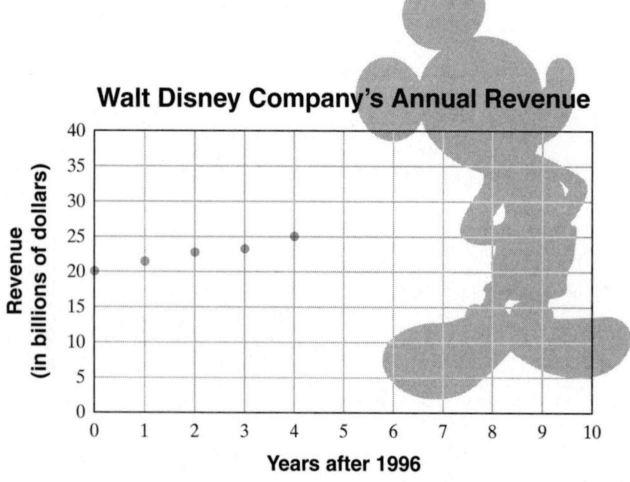

77. Estimate the *increase* in revenues for years 1, 2, 3, and 4.

78. Use a straight edge or ruler and this scatter diagram to predict Disney's revenue in the year 2005.

79. Discuss any similarities in the graphs of the ordered pair solutions for Exercises 37–40.

80. The percent y of recorded music sales that were in cassette format from 1995 through 2000 is given by $y = -3.95x + 24.93$. In the equation, x represents the number of years after 1995. (*Source:* Recording Industry Association of America)

a. Complete the table.

x	1	3	5
y			

b. Find the year in which approximately 17% of recorded music sales were cassettes. (*Hint*: Find x when $y = 17$ and round to the nearest whole number.)

81. The amount y of land operated by farms in the United States (in million acres) from 1990 through 2000 is given by $y = -4.22x + 985.02$. In the equation, x represents the number of years after 1990. (*Source:* National Agricultural Statistics Service)

a. Complete the table.

x	4	7	10
y			

b. Find the year in which there were approximately 947 million acres of land operated by farms. (*Hint*: Find x when $y = 947$ and round to the nearest whole number.)

FOCUS ON The Real World

READING A MAP

How do you find a location on a map? Most maps we use today have a grid that is based on the rectangular coordinate system we use in algebra. After finding the coordinates of cities and other landmarks from the map index, the grid can help us find places on the map. To eliminate confusion, many maps use letters to label the grid along one edge and numbers along the other. However, the coordinates are still pairs of numbers and letters. For instance, the coordinates for Toledo on the map are A-2.

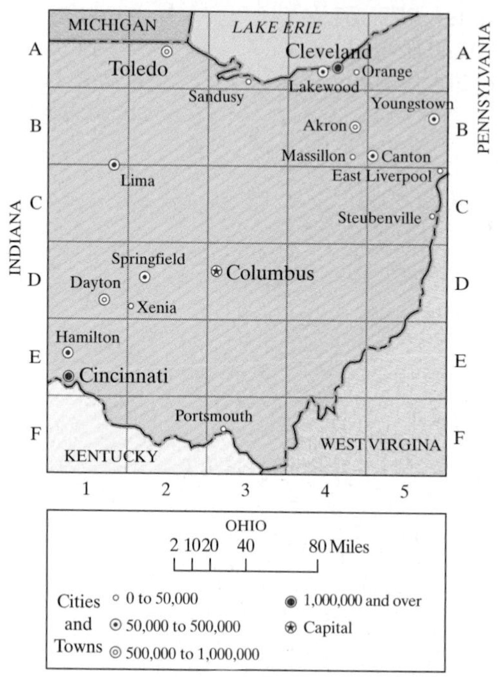

CRITICAL THINKING

1. Find the coordinates of the following cities: Hamilton, Columbus, Youngstown, and Cincinnati.
2. What cities correspond to the following coordinates: F-3, A-3, B-4, and D-2?
3. How are the map's coordinate system and the rectangular coordinate system we use in algebra the same? How are they different? What are the advantages of each?

3.2 Graphing Linear Equations

In the previous section, we found that equations in two variables may have more than one solution. For example, both $(2, 2)$ and $(0, 4)$ are solutions of the equation $x + y = 4$. In fact, this equation has an infinite number of solutions. Other solutions include $(-2, 6)$, $(4, 0)$, and $(6, -2)$. Notice the pattern that appears in the graph of these solutions.

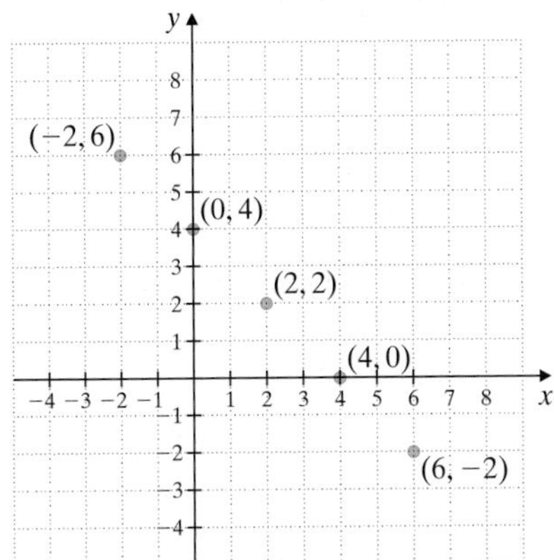

OBJECTIVE

A Graph a linear equation by finding and plotting ordered pair solutions.

These solutions all appear to lie on the same line, as seen in the second graph. It can be shown that every ordered pair solution of the equation corresponds to a point on this line, and every point on this line corresponds to an ordered pair solution. Thus, we say that this line is the **graph of the equation** $x + y = 4$. Notice that we can only show a part of a line on a graph. The arrowheads on each end of the line below remind us that the line actually extends indefinitely in both directions.

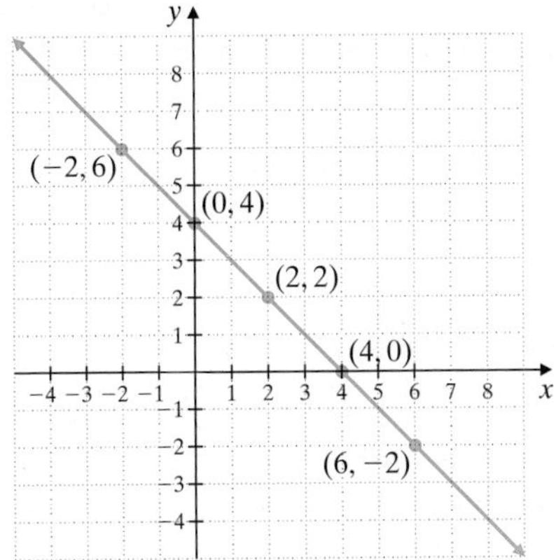

The equation $x + y = 4$ is called a *linear equation in two variables* and *the graph of every linear equation in two variables is a straight line.*

Linear Equation in Two Variables

A **linear equation in two variables** is an equation that can be written in the form

$$Ax + By = C$$

where A, B, and C are real numbers and A and B are not both 0. **The graph of a linear equation in two variables is a straight line.**

The form $Ax + By = C$ is called **standard form**. Following are examples of linear equations in two variables.

$$2x + y = 8 \qquad -2x = 7y \qquad y = \frac{1}{3}x + 2 \qquad y = 7$$

(A) Graphing Linear Equations

From geometry, we know that a straight line is determined by just two points. Thus, to graph a linear equation in two variables we need to find just two of its infinitely many solutions. Once we do so, we plot the solution points and draw the line connecting the points. Usually, we find a third solution as well, as a check.

Practice Problem 1

Graph the linear equation $x + 3y = 6$.

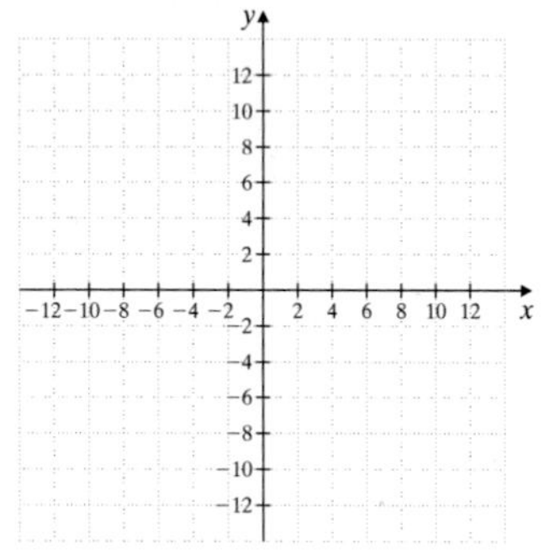

Answer

1.

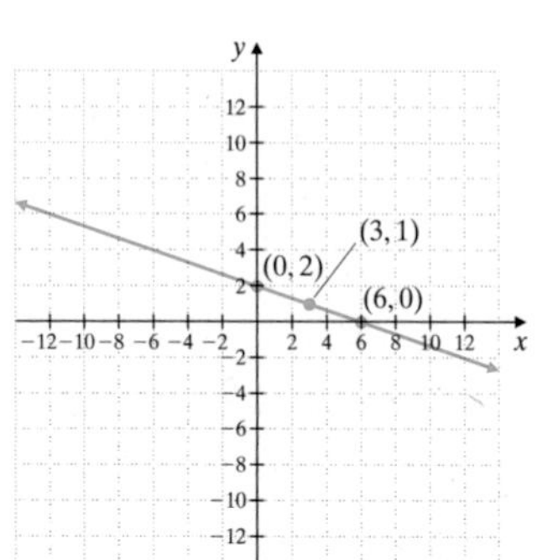

EXAMPLE 1 Graph the linear equation $2x + y = 5$.

Solution: To graph this equation, we find three ordered pair solutions of $2x + y = 5$. To do this, we choose a value for one variable, x or y, and solve for the other variable. For example, if we let $x = 1$, then $2x + y = 5$ becomes

$$
\begin{aligned}
2x + y &= 5 \\
2(1) + y &= 5 && \text{Replace } x \text{ with 1.} \\
2 + y &= 5 && \text{Multiply.} \\
y &= 3 && \text{Subtract 2 from both sides.}
\end{aligned}
$$

Since $y = 3$ when $x = 1$, the ordered pair $(1, 3)$ is a solution of $2x + y = 5$. Next, we let $x = 0$.

$$
\begin{aligned}
2x + y &= 5 \\
2(0) + y &= 5 && \text{Replace } x \text{ with 0.} \\
0 + y &= 5 \\
y &= 5
\end{aligned}
$$

The ordered pair $(0, 5)$ is a second solution.

The two solutions found so far allow us to draw the straight line that is the graph of all solutions of $2x + y = 5$. However, we will find a third ordered pair as a check. Let $y = -1$.

$$
\begin{aligned}
2x + y &= 5 \\
2x + (-1) &= 5 && \text{Replace } y \text{ with } -1. \\
2x - 1 &= 5 \\
2x &= 6 && \text{Add 1 to both sides.} \\
x &= 3 && \text{Divide both sides by 2.}
\end{aligned}
$$

The third solution is $(3, -1)$. These three ordered pair solutions are listed in the table and plotted on the coordinate plane. The graph of $2x + y = 5$ is the line through the three points.

x	y
1	3
0	5
3	−1

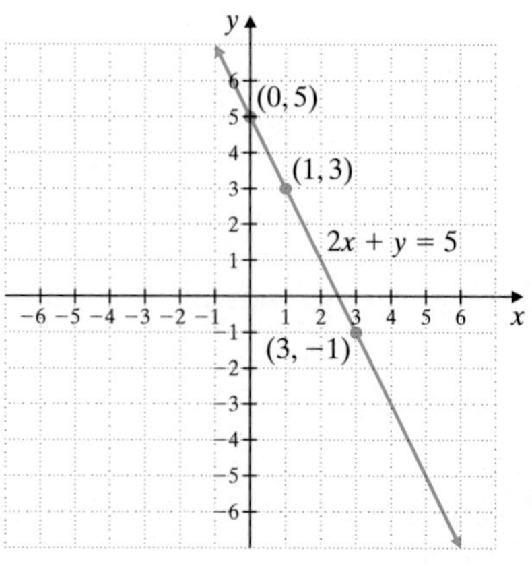

Helpful Hint

All three points should fall on the same straight line. If not, check your ordered pair solutions for a mistake.

EXAMPLE 2 Graph the linear equation $-5x + 3y = 15$.

Solution: We find three ordered pair solutions of $-5x + 3y = 15$.

Let $x = 0$.

$$-5x + 3y = 15$$
$$-5 \cdot 0 + 3y = 15$$
$$0 + 3y = 15$$
$$3y = 15$$
$$y = 5$$

Let $y = 0$.

$$-5x + 3y = 15$$
$$-5x + 3 \cdot 0 = 15$$
$$-5x + 0 = 15$$
$$-5x = 15$$
$$x = -3$$

Let $x = -2$.

$$-5x + 3y = 15$$
$$-5 \cdot -2 + 3y = 15$$
$$10 + 3y = 15$$
$$3y = 5$$
$$y = \frac{5}{3}$$

The ordered pairs are $(0, 5)$, $(-3, 0)$, and $\left(-2, \frac{5}{3}\right)$. The graph of $-5x + 3y = 15$ is the line through the three points.

x	y
0	5
−3	0
−2	$\frac{5}{3} = 1\frac{2}{3}$

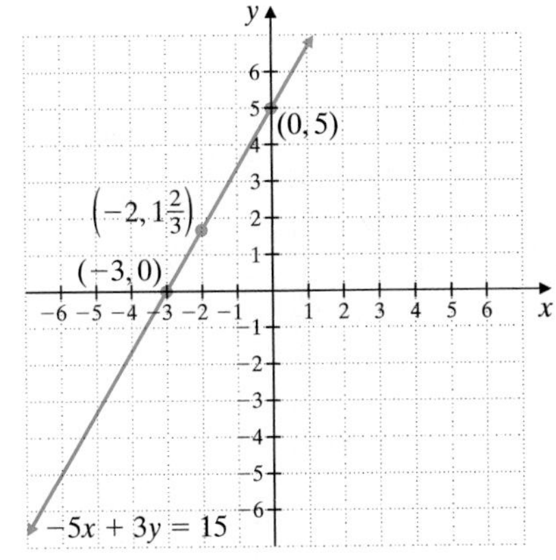

EXAMPLE 3 Graph the linear equation $y = 3x$.

Solution: We find three ordered pair solutions. Since this equation is solved for y, we'll choose three x values.

If $x = 2$, $y = 3 \cdot 2 = 6$.
If $x = 0$, $y = 3 \cdot 0 = 0$.
If $x = -1$, $y = 3 \cdot -1 = -3$.

Next, we plot the ordered pair solutions and draw a line through the plotted points. The line is the graph of $y = 3x$. Every point on the graph represents an ordered pair solution of the equation and every ordered pair solution is a point on this line.

Practice Problem 2

Graph the linear equation $-2x + 4y = 8$.

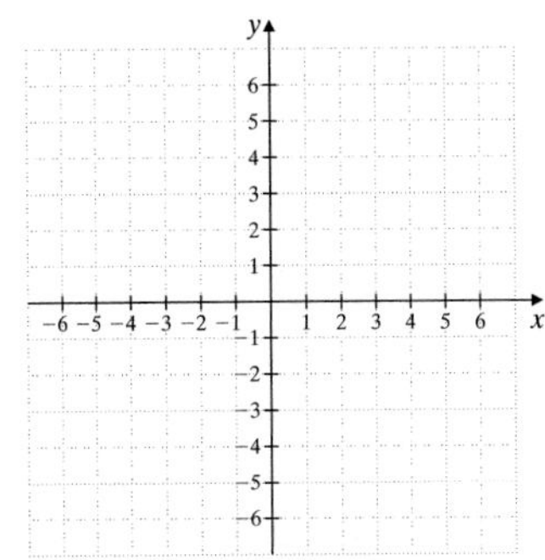

Practice Problem 3

Graph the linear equation $y = 2x$.

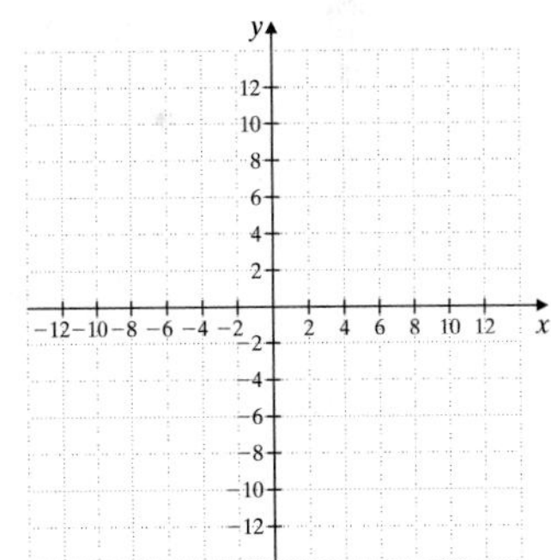

Answers

2.

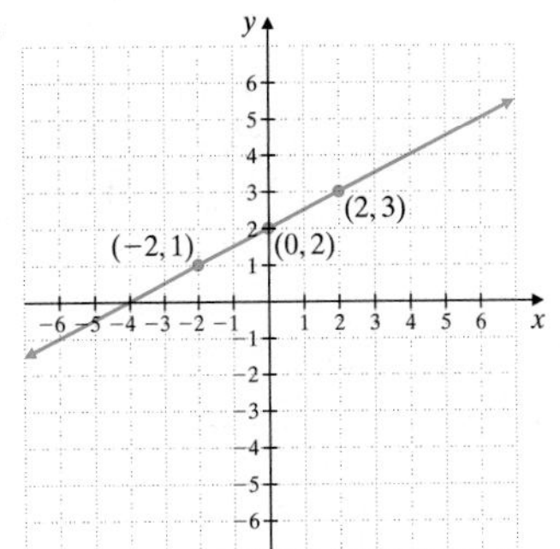

3. See page 206.

x	y
2	6
0	0
−1	−3

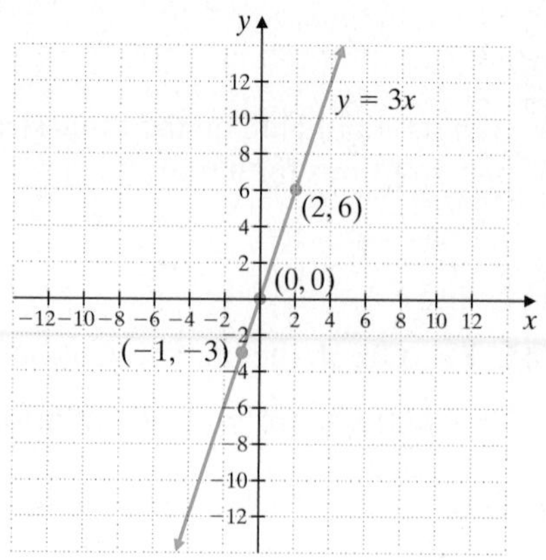

Practice Problem 4

Graph the linear equation $y = -\frac{1}{2}x$.

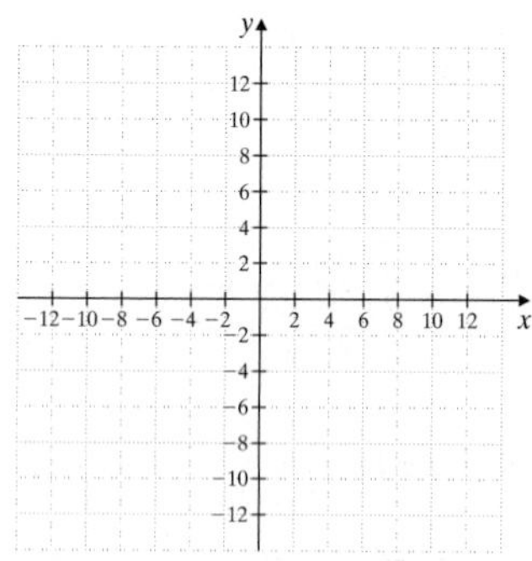

EXAMPLE 4 Graph the linear equation $y = -\frac{1}{3}x$.

Solution: We find three ordered pair solutions, plot the solutions, and draw a line through the plotted solutions. To avoid fractions, we'll choose x values that are multiples of 3 to substitute into the equation.

If $x = 6$, then $y = -\frac{1}{3} \cdot 6 = -2$.

If $x = 0$, then $y = -\frac{1}{3} \cdot 0 = 0$.

If $x = -3$, then $y = -\frac{1}{3} \cdot -3 = 1$.

x	y
6	−2
0	0
−3	1

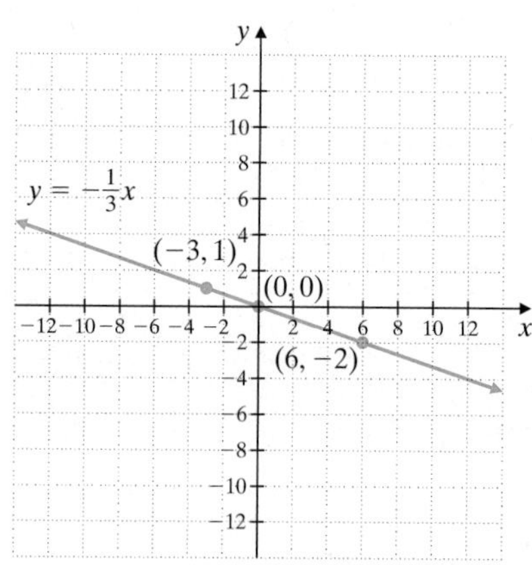

Let's compare the graphs in Examples 3 and 4. The graph of $y = 3x$ tilts upward (as we follow the line from left to right) and the graph of $y = -\frac{1}{3}x$ tilts downward (as we follow the line from left to right). Also notice that both lines go through the origin or that $(0, 0)$ is an ordered pair solution of both equations. We will learn more about the tilt, or slope, of a line in Section 3.4.

EXAMPLE 5 Graph the linear equation $y = 3x + 6$ and compare this graph with the graph of $y = 3x$ in Example 3.

Answers

3.

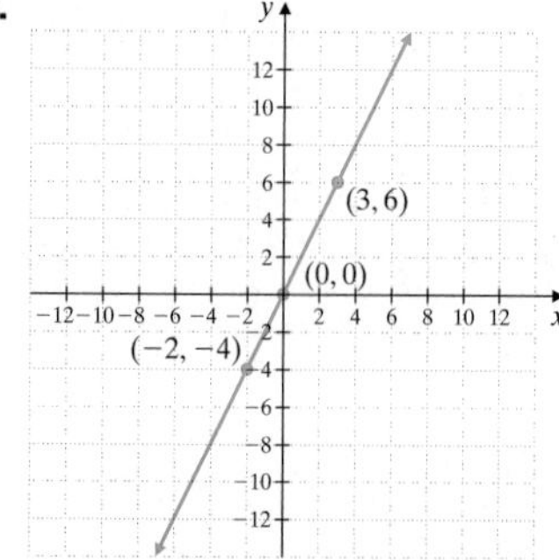

4.

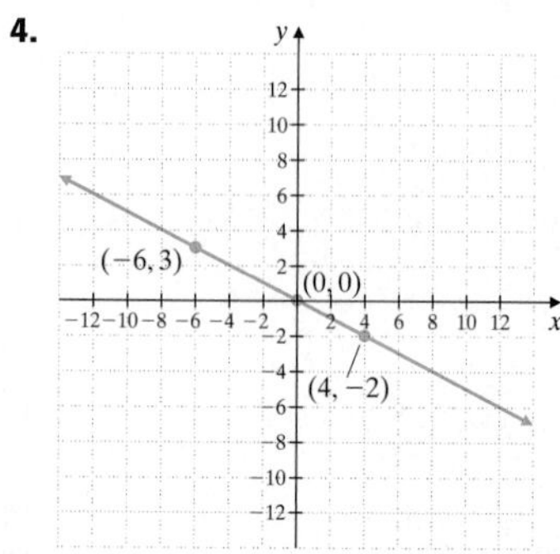

Solution: We find three ordered pair solutions, plot the solutions, and draw a line through the plotted solutions. We choose x-values and substitute them into the equation $y = 3x + 6$.

If $x = -3$, then $y = 3(-3) + 6 = -3$.
If $x = 0$, then $y = 3(0) + 6 = 6$.
If $x = 1$, then $y = 3(1) + 6 = 9$.

x	y
−3	−3
0	6
1	9

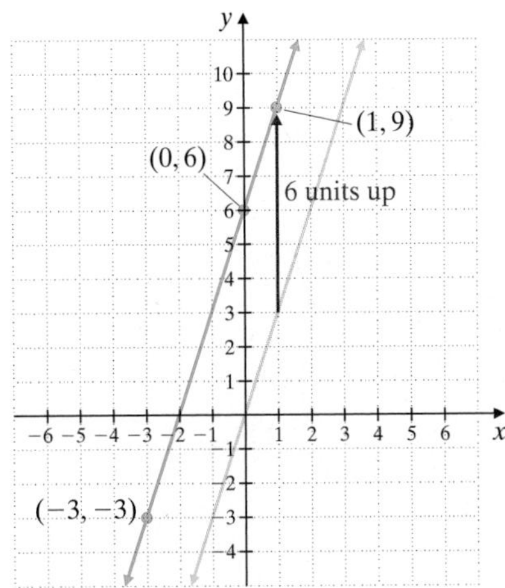

The most startling similarity is that both graphs appear to have the same upward tilt as we move from left to right. Also, the graph of $y = 3x$ crosses the y-axis at the origin, while the graph of $y = 3x + 6$ crosses the y-axis at 6. It appears that the graph of $y = 3x + 6$ is the same as the graph of $y = 3x$ except that the graph of $y = 3x + 6$ is moved 6 units upward.

EXAMPLE 6 Graph the linear equation $y = -2$.

Solution: Recall from Section 3.1 that the equation $y = -2$ is the same as $0x + y = -2$. No matter what value we replace x with, y is -2.

x	y
0	−2
3	−2
−2	−2

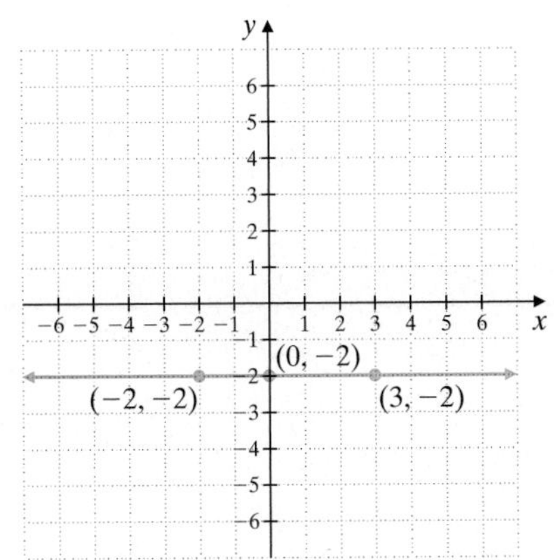

Notice that the graph of $y = -2$ is a horizontal line.

Practice Problem 5

Graph the linear equation $y = 2x + 3$ and compare this graph with the graph of $y = 2x$ in Practice Problem 3.

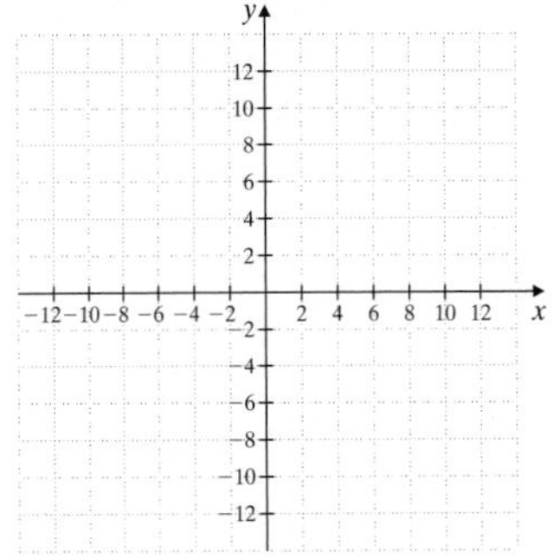

Practice Problem 6

Graph the linear equation $x = 3$.

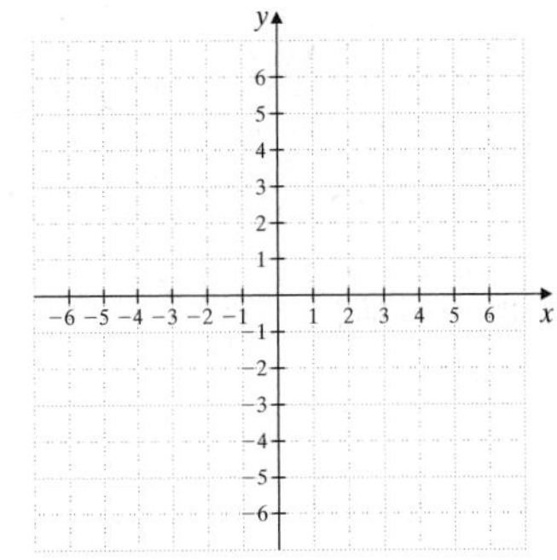

Answers

5.

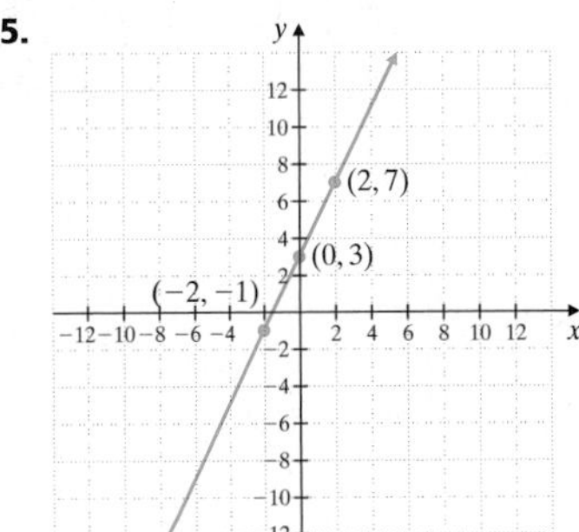

Same as the graph of $y = 2x$ except that the graph of $y = 2x + 3$ is moved 3 units upward.

6.

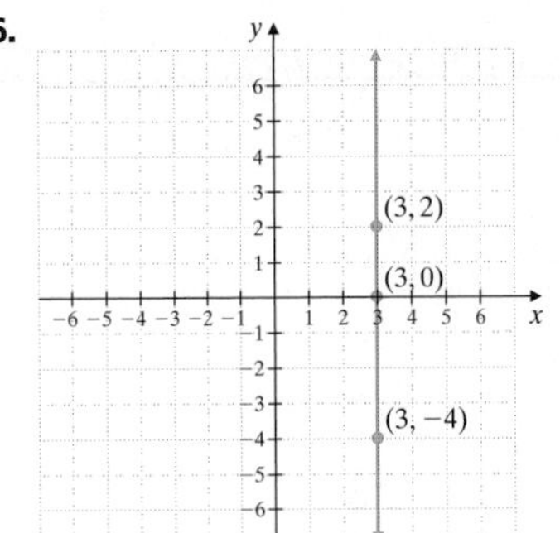

GRAPHING CALCULATOR EXPLORATIONS

In this section, we begin an optional study of graphing calculators and graphing software packages for computers. These graphers use the same point plotting technique that was introduced in this section. The advantage of this graphing technology is, of course, that graphing calculators and computers can find and plot ordered pair solutions much faster than we can. Note, however, that the features described in these boxes may not be available on all graphing calculators.

The rectangular screen where a portion of the rectangular coordinate system is displayed is called a **window**. We call it a **standard window** for graphing when both the x- and y-axes show coordinates between -10 and 10. This information is often displayed in the window menu on a graphing calculator as follows.

Xmin = −10

Xmax = 10

Xscl = 1 — The scale on the x-axis is one unit per tick mark.

Ymin = −10

Ymax = 10

Yscl = 1 — The scale on the y-axis is one unit per tick mark.

To use a graphing calculator to graph the equation $y = 2x + 3$, press the [Y=] key and enter the keystrokes [2] [x] [+] [3]. The top row should now read $Y_1 = 2x + 3$. Next press the [GRAPH] key, and the display should look like this:

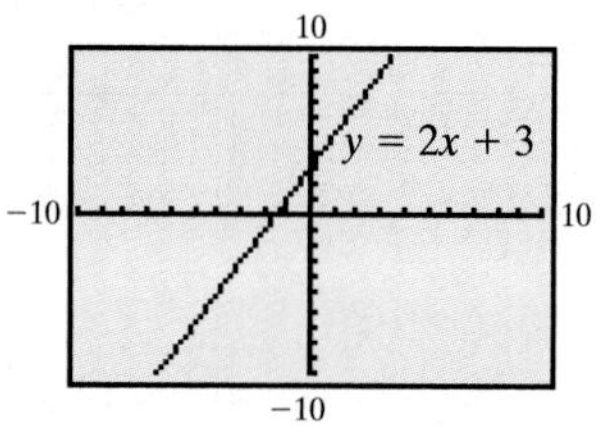

Use a standard window and graph the following linear equations. (Unless otherwise stated, use a standard window when graphing.)

1. $y = -3x + 7$

2. $y = -x + 5$

3. $y = 2.5x - 7.9$

4. $y = -1.3x + 5.2$

5. $y = -\frac{3}{10}x + \frac{32}{5}$

6. $y = \frac{2}{9}x - \frac{22}{3}$

Name ______________________ Section __________ Date __________

EXERCISE SET 3.2

A *For each equation, find three ordered pair solutions by completing the table. Then use the ordered pairs to graph the equation. See Examples 1 through 6.*

1. $x - y = 6$

x	y
	0
4	
	-1

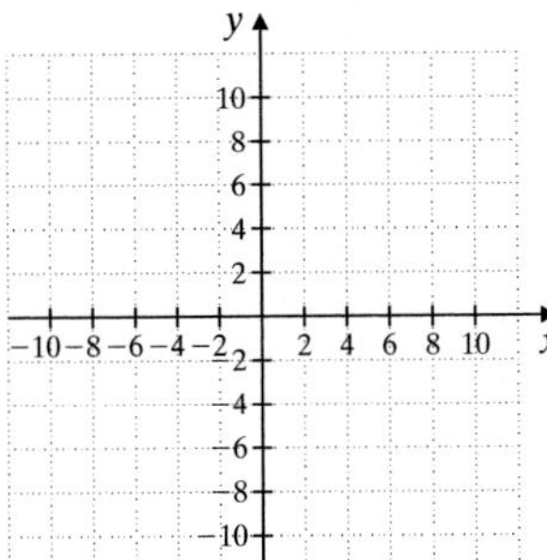

2. $x - y = 4$

x	y
0	
	2
-1	

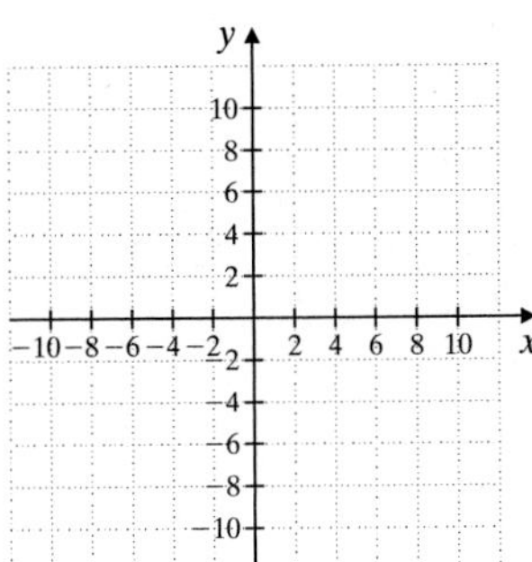

3. $y = -4x$

x	y
1	
0	
-1	

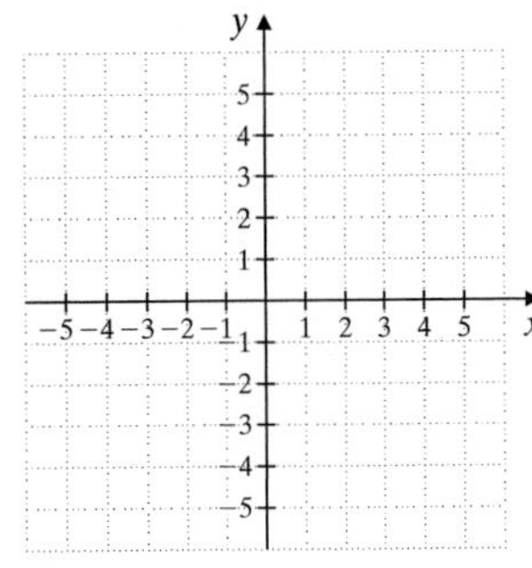

4. $y = -5x$

x	y
1	
0	
-1	

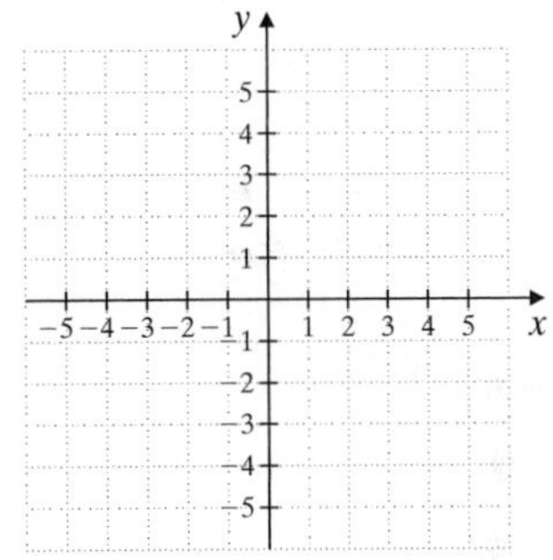

5. $y = \frac{1}{3}x$

x	y
0	
6	
-3	

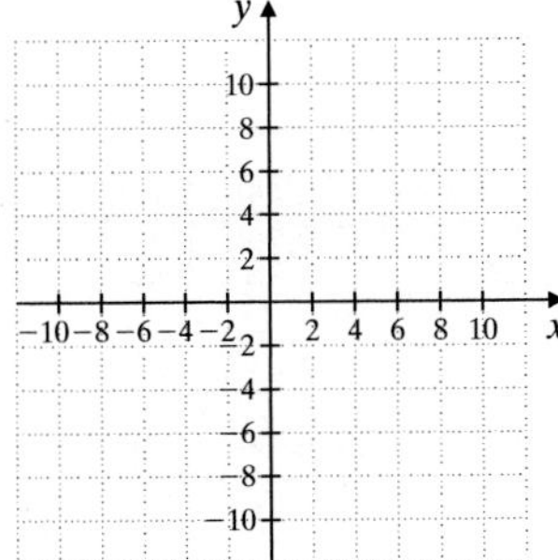

6. $y = \frac{1}{2}x$

x	y
0	
-4	
2	

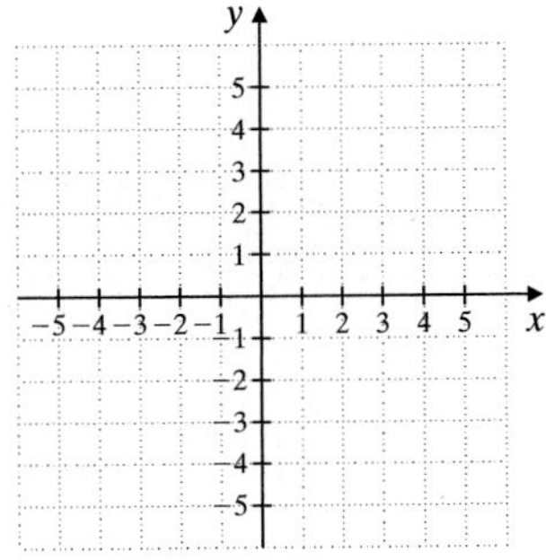

7. $y = -4x + 3$

x	y
0	
1	
2	

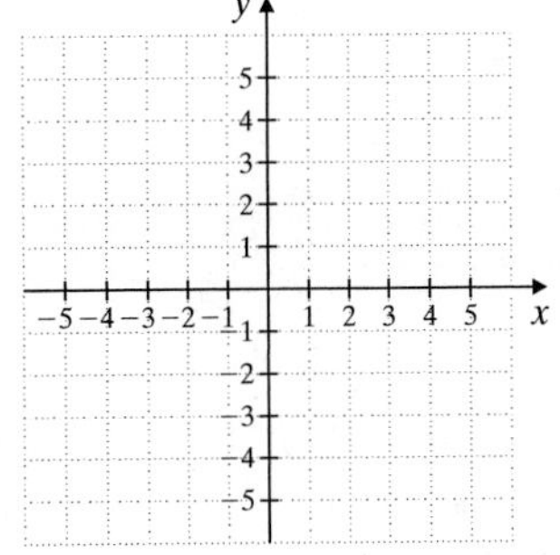

8. $y = -5x + 2$

x	y
0	
1	
2	

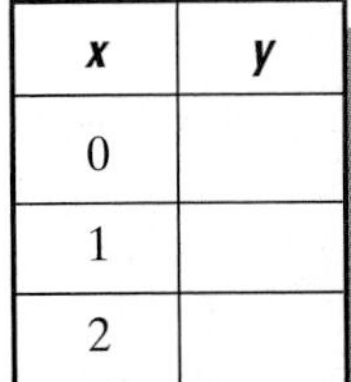

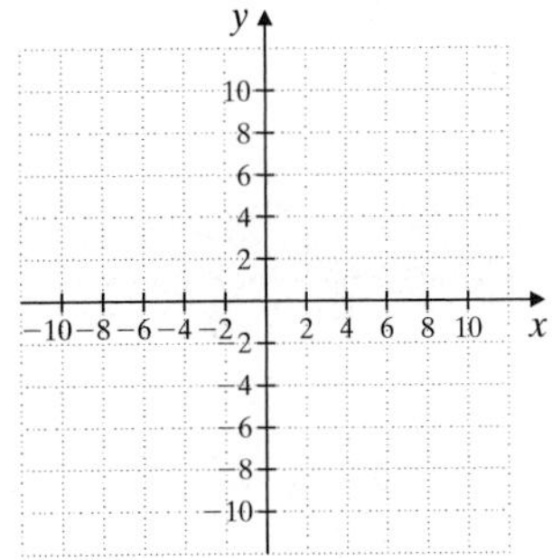

Graph each linear equation. See Examples 1 through 6.

9. $x + y = 1$

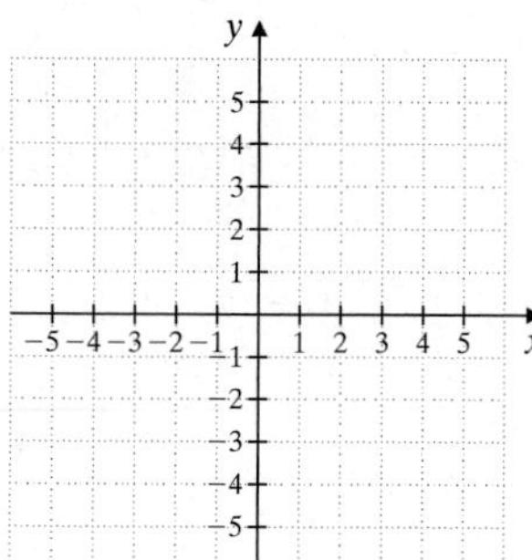

10. $x + y = 7$

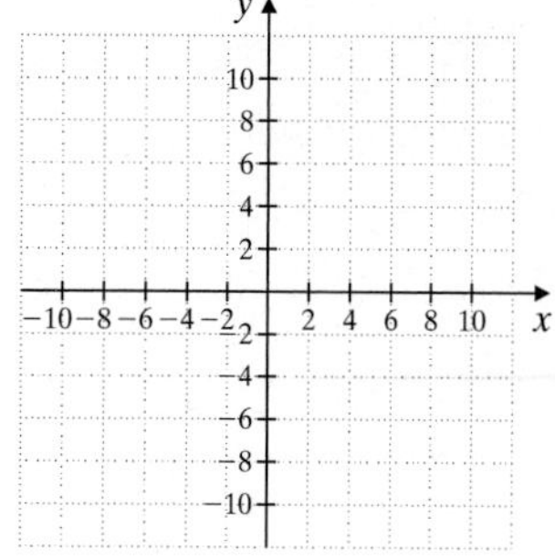

 **11.** $-x + y = 6$

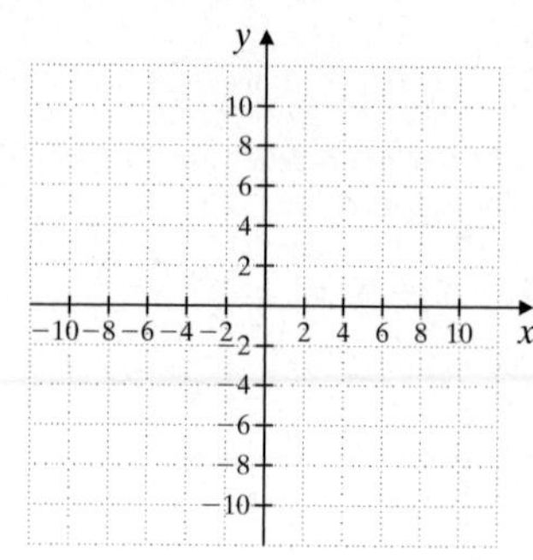

12. $x - y = -2$

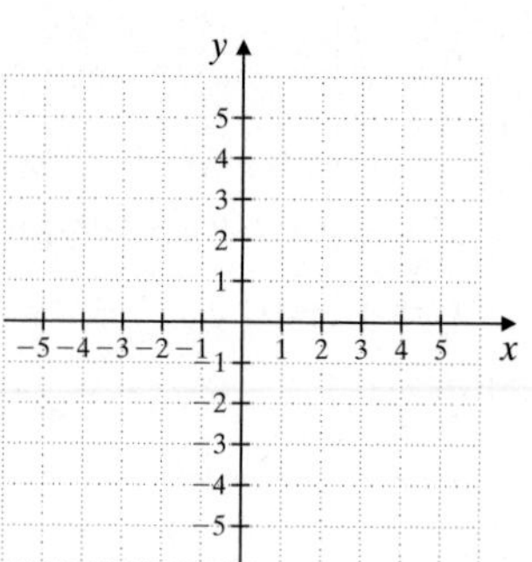

13. $x - 2y = 6$

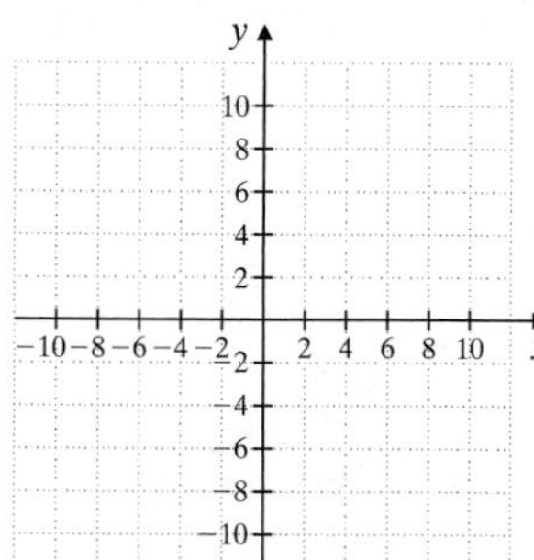

14. $-x + 5y = 5$

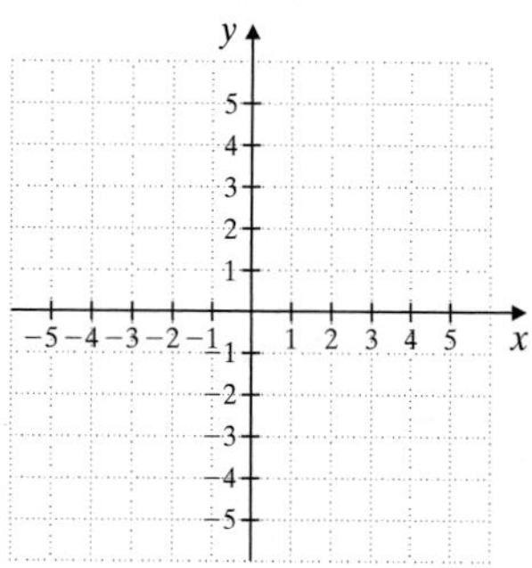

15. $y = 6x + 3$

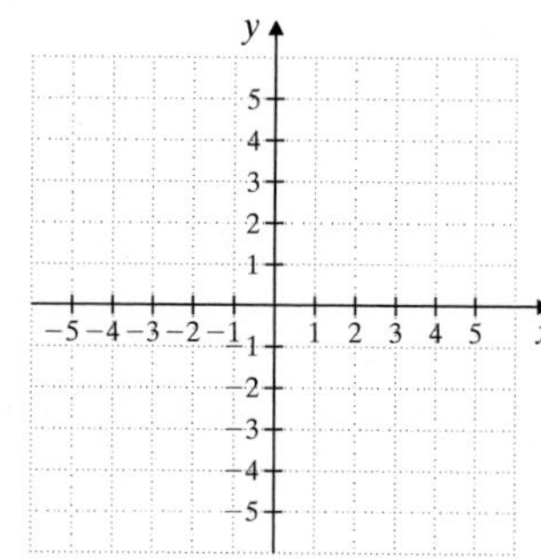

16. $y = -2x + 7$

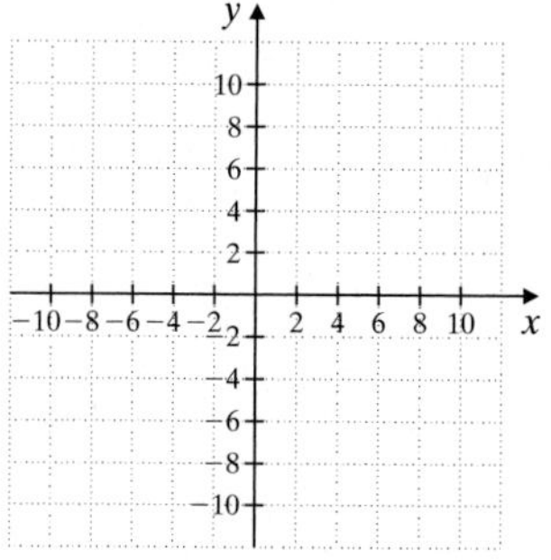

17. $x = -4$

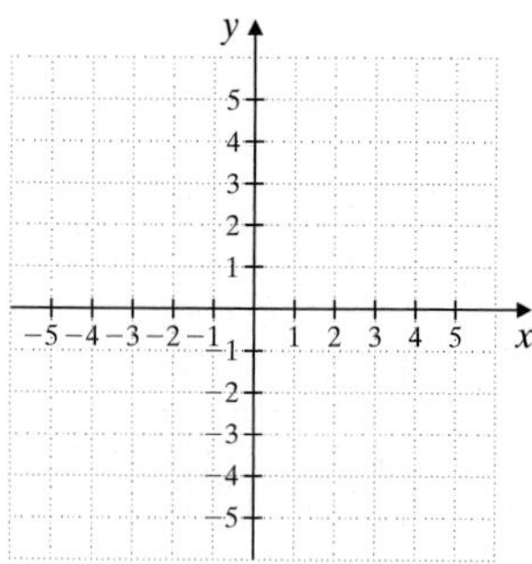

18. $y = 5$

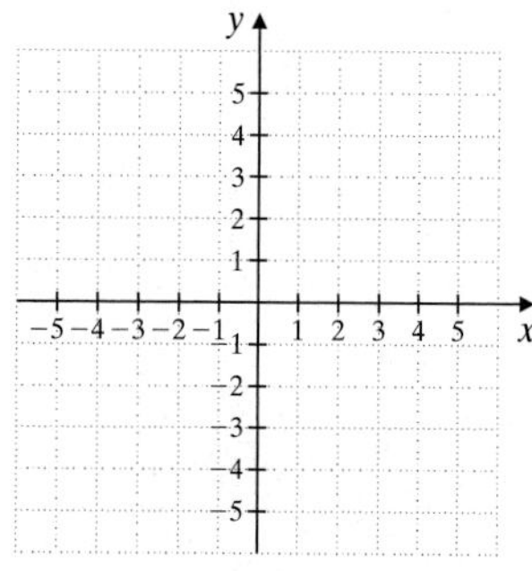

19. $y = 3$

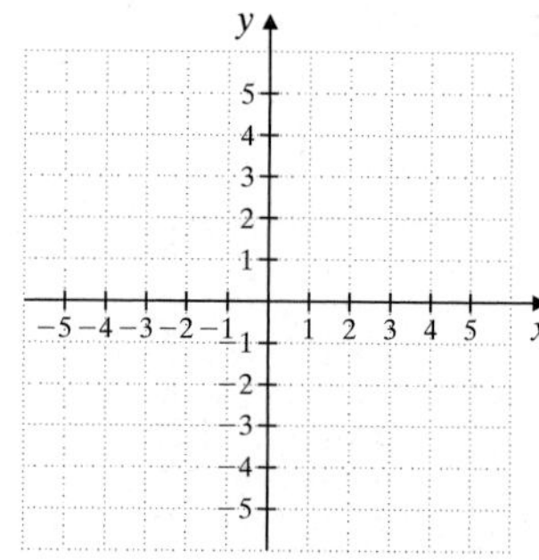

20. $x = -1$

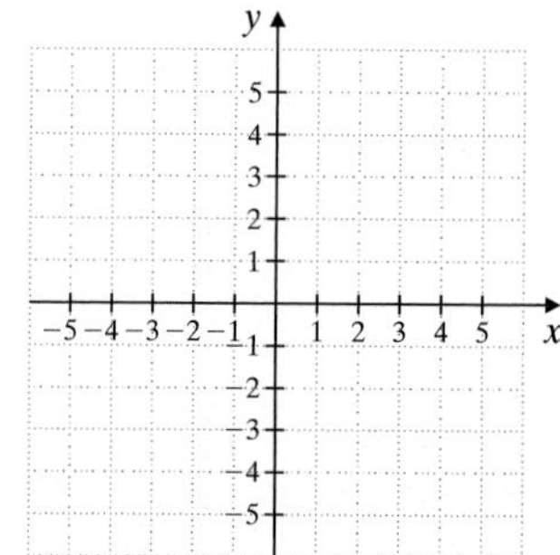

21. $y = x$

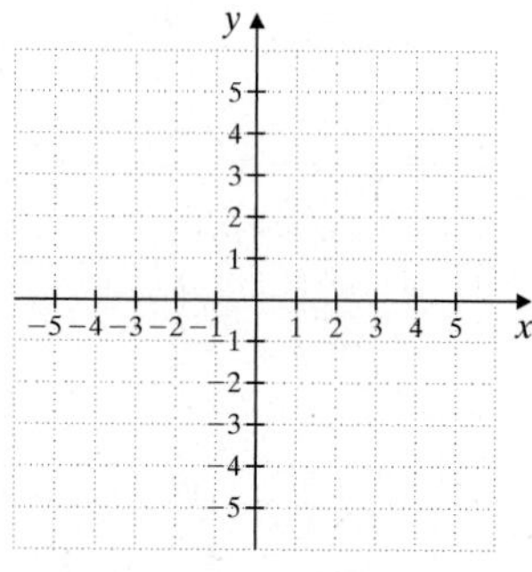

22. $y = -x$

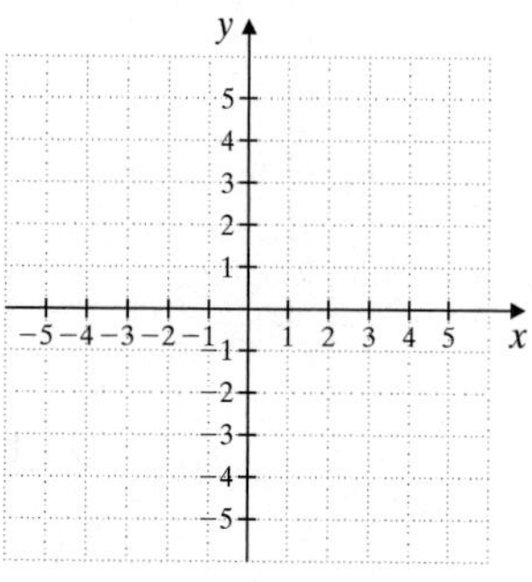

23. $y = 5x$

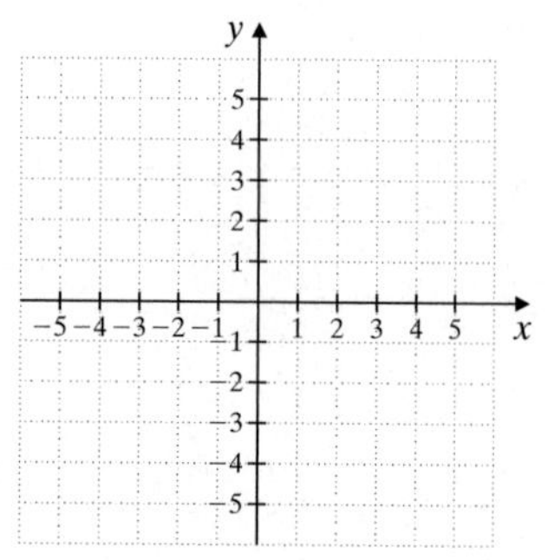

24. $y = 4x$

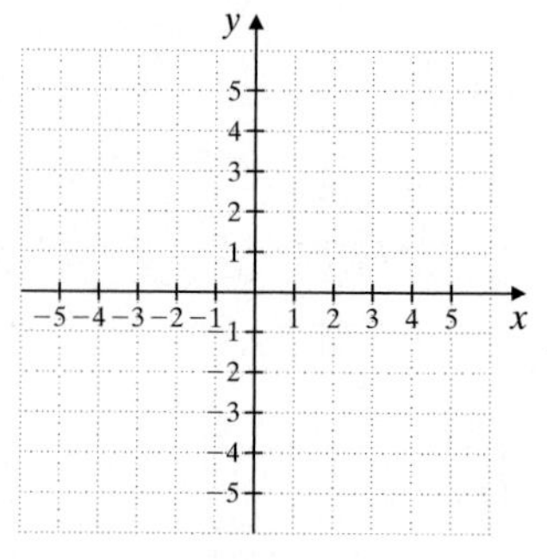

25. $x + 3y = 9$

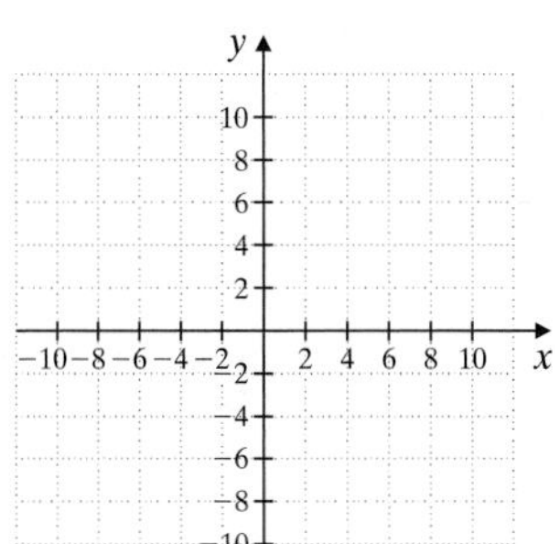

26. $2x + y = 2$

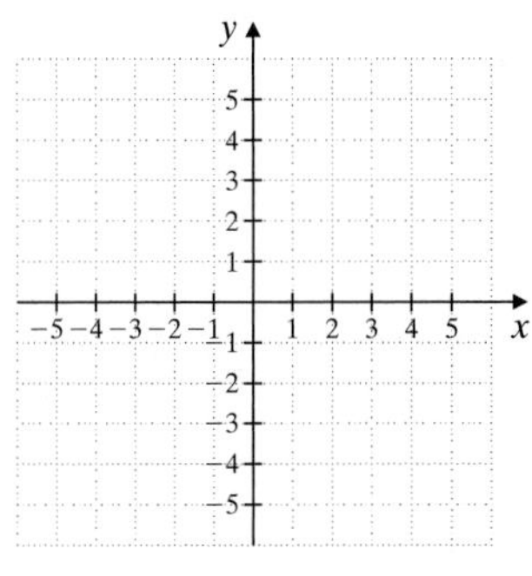

27. $y = \frac{1}{2}x - 1$

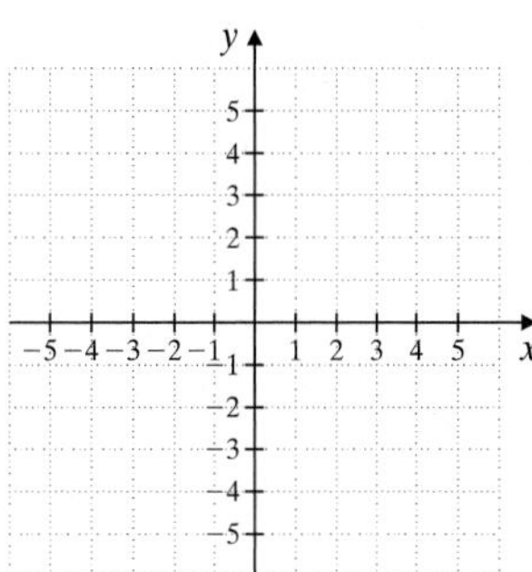

28. $y = \frac{1}{4}x + 3$

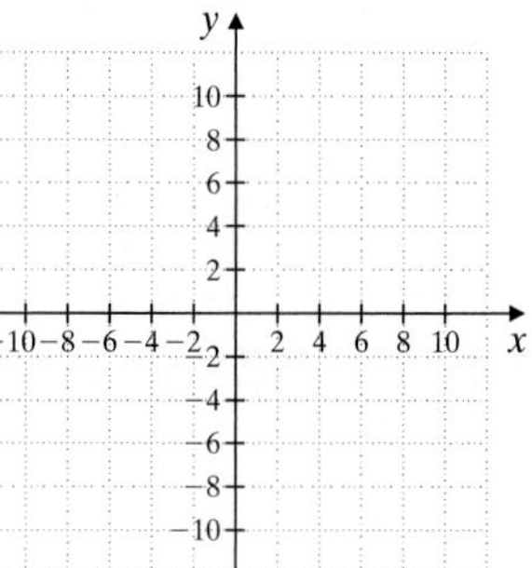

29. $3x - 2y = 12$

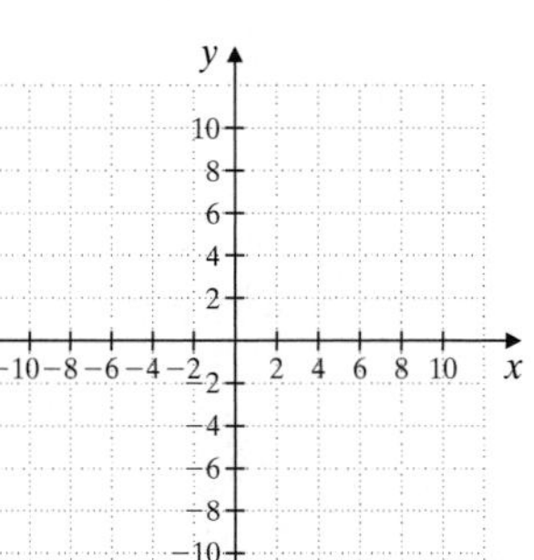

30. $2x - 7y = 14$

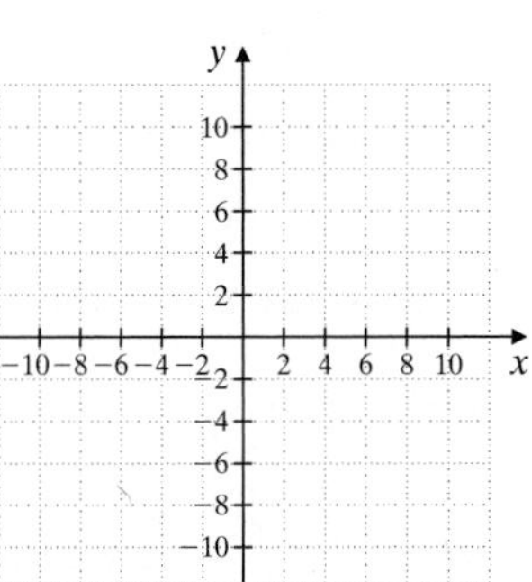

Graph each pair of linear equations on the same set of axes. Discuss how the graphs are similar and how they are different. See Example 5.

31. $y = 5x$
$y = 5x + 4$

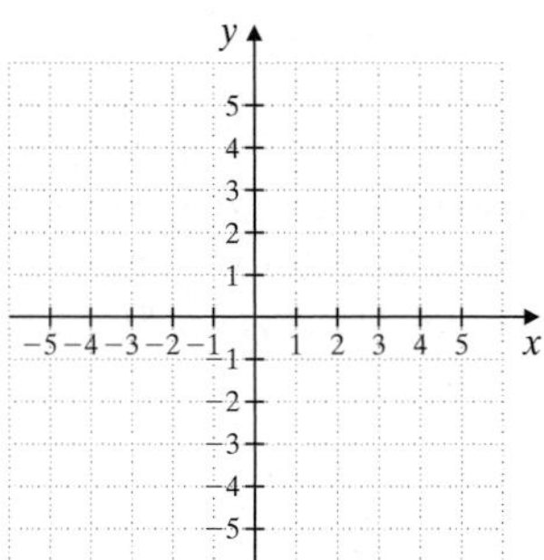

32. $y = 2x$
$y = 2x + 5$

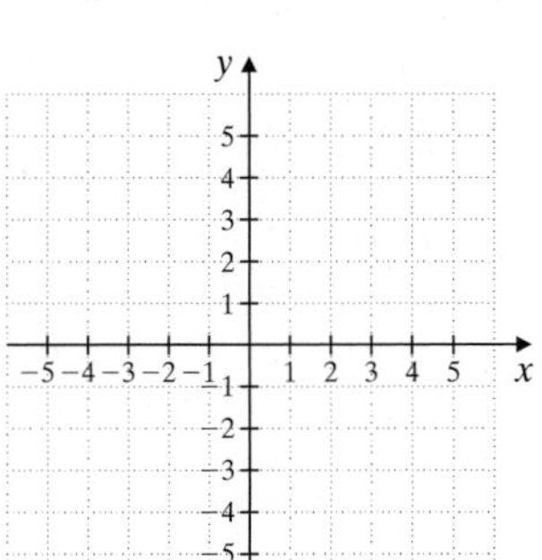

33. $y = -2x$
$y = -2x - 3$

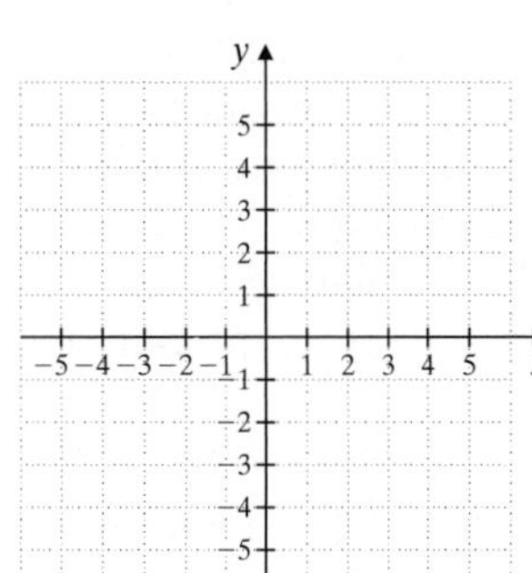

34. $y = x$
$y = x - 7$

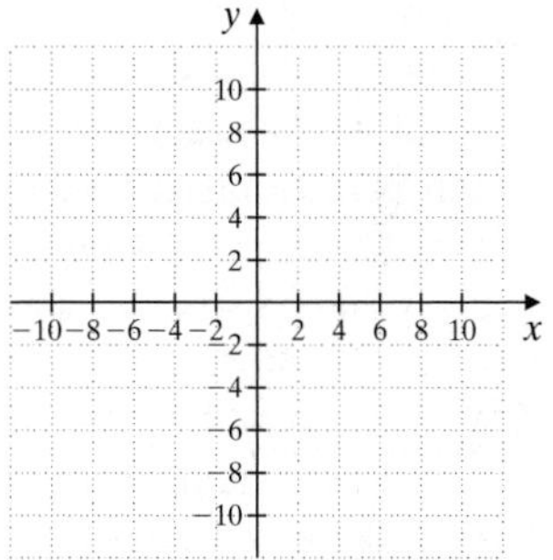

Review and Preview

△ **35.** The coordinates of three vertices of a rectangle are $(-2, 5)$, $(4, 5)$, and $(-2, -1)$. Find the coordinates of the fourth vertex. See Section 3.1.

△ **36.** The coordinates of two vertices of a square are $(-3, -1)$ and $(2, -1)$. Find the coordinates of two pairs of points possible for the third and fourth vertices. See Section 3.1.

Complete each table. See Section 3.1.

37. $x - y = -3$

x	y
0	
	0

38. $y - x = 5$

x	y
0	
	0

39. $y = 2x$

x	y
0	
	0

40. $x = -3y$

x	y
0	
	0

Combining Concepts

41. Graph the nonlinear equation $y = x^2$ by completing the table shown. Plot the ordered pairs and connect them with a smooth curve.

x	y
0	
1	
−1	
2	
−2	

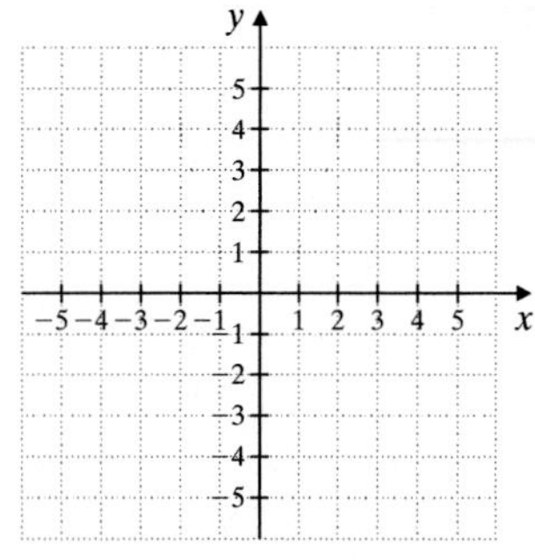

42. Graph the nonlinear equation $y = |x|$ by completing the table shown. Plot the ordered pairs and connect them. This curve is "V" shaped.

x	y
0	
1	
−1	
2	
−2	

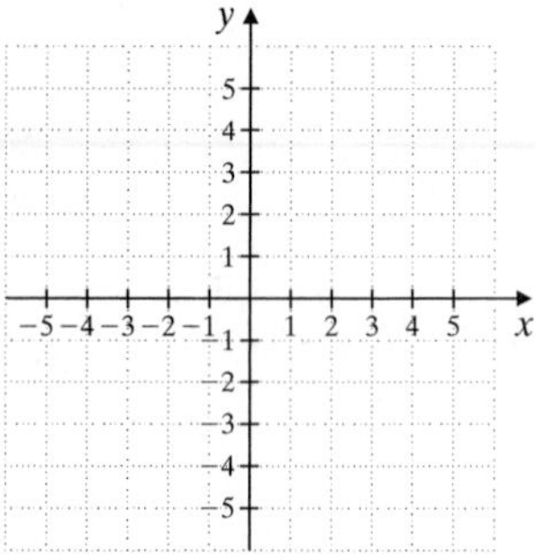

△ **43.** The perimeter of the trapezoid is 22 centimeters. Write a linear equation in two variables for the perimeter. Find y if x is 3 centimeters.

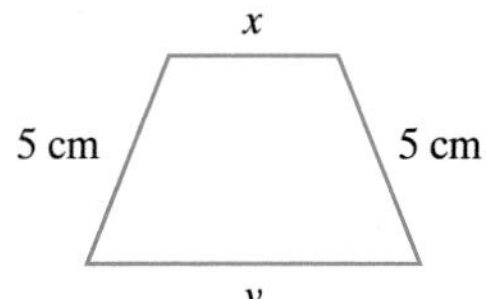

44. If (a, b) is an ordered pair solution of $x + y = 5$, is (b, a) also a solution? Explain why or why not.

45. One of the top five occupations in terms of growth in the next few years is expected to be registered nursing. The number of people y in thousands employed as registered nurses in the United States can be estimated by the linear equation $y = 45x + 2214$ where x is the number of years after 2001. (*Source:* Based on data from the Bureau of Labor Statistics)

a. Graph the linear equation. The break in the vertical axis means that the numbers between 0 and 2200 have been skipped.

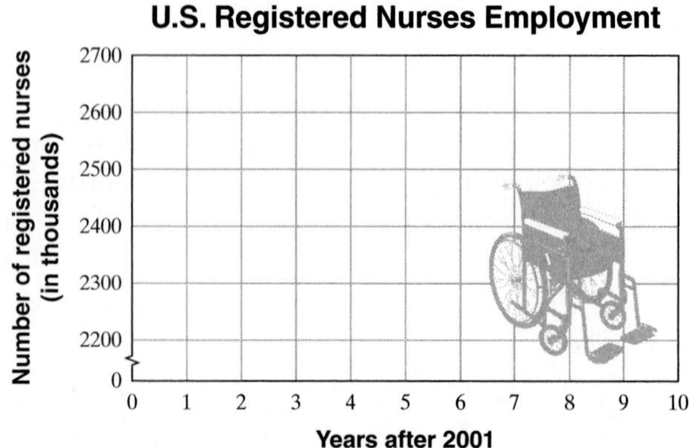

b. Does the point $(6, 2484)$ lie on the line? If so, what does this ordered pair mean?

46. Head Start is a comprehensive child development program serving young children in low-income families. The number of children y (in thousands) enrolled in Head Start from 1993–2000 can be approximated by the linear equation $y = 20x + 712$, where x is the number of years after 1993. (*Source:* Head Start Bureau, the Administration on Children, Youth and Families)

a. Graph the linear equation.

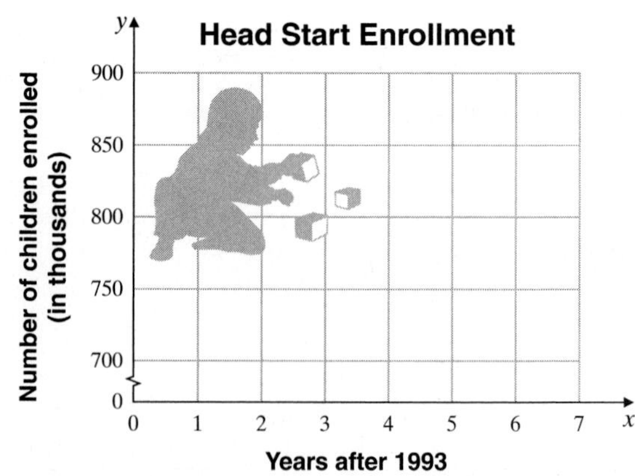

b. Does the point $(4, 792)$ lie on the line? If so, what does this ordered pair mean?

47. The number of U.S. households y in millions that have at least one television set can be estimated by the linear equation $y = 1.43x + 95$ where x is the number of years after 1995. (*Source:* Nielsen Media Research)

a. Graph the linear equation.

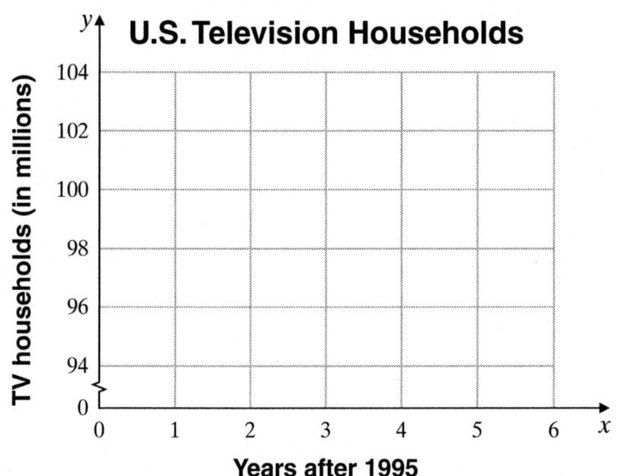

b. Complete the ordered pair (5,).

c. Write a sentence explaining the meaning of the ordered pair found in part b.

48. The restaurant industry is busier than ever. The yearly revenue for restaurants in the U.S. can be estimated by $y = 11.58x + 348$ where x is the number of years after 1998 and y is the revenue in billions of dollars. (*Source:* National Restaurant Assn.)

a. Graph the linear equation.

b. Complete the ordered pair (3,).

c. Write a sentence explaining the meaning of the ordered pair found in part b.

STUDY SKILLS REMINDER

Tips for studying for an exam

To prepare for an exam, try the following study techniques.

- Start the study process days before your exam.
- Make sure that you are current and up-to-date on your assignments.
- If there is a topic that you are unsure of, use one of the many resources that are available to you. For example,

  See your instructor.
  Visit a learning resource center on campus where math tutors are available.
  Read the textbook material and examples on the topic.
  View a videotape on the topic.
- Reread your notes and carefully review the Chapter Highlights at the end of the chapter.
- Work the review exercises at the end of the chapter and check your answers. Make sure that you correct any missed exercises. If you have trouble on a topic, use a resource listed above.
- Find a quiet place to take the Chapter Test found at the end of the chapter. Do not use any resources when taking this sample test. This way you will have a clear indication of how prepared you are for your exam. Check your answers and make sure that you correct any missed exercises.
- Get lots of rest the night before the exam. It's hard to show how well you know the material if your brain is foggy from lack of sleep.

Good luck and keep a positive attitude.

3.3 Intercepts

A Identifying Intercepts

The graph of $y = 4x - 8$ is shown below. Notice that this graph crosses the y-axis at the point $(0, -8)$. This point is called the **y-intercept**. Likewise the graph crosses the x-axis at $(2, 0)$. This point is called the **x-intercept**.

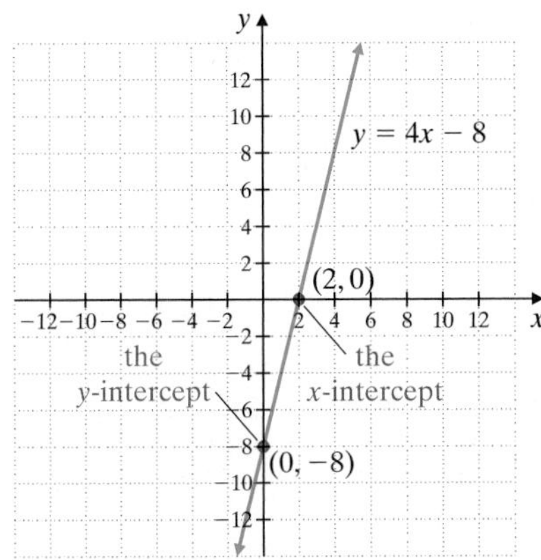

The intercepts are $(2, 0)$ and $(0, -8)$.

Helpful Hint

If a graph crosses the x-axis at $(-3, 0)$ and the y-axis at $(0, 7)$, then

$\underbrace{(-3, 0)}_{\uparrow \text{ x-intercept}}$ $\quad$ $\underbrace{(0, 7)}_{\uparrow \text{ y-intercept}}$

Notice that for the y-intercept, the x-value is 0 and for the x-intercept, the y-value is 0.

EXAMPLES Identify the x- and y-intercepts.

1.

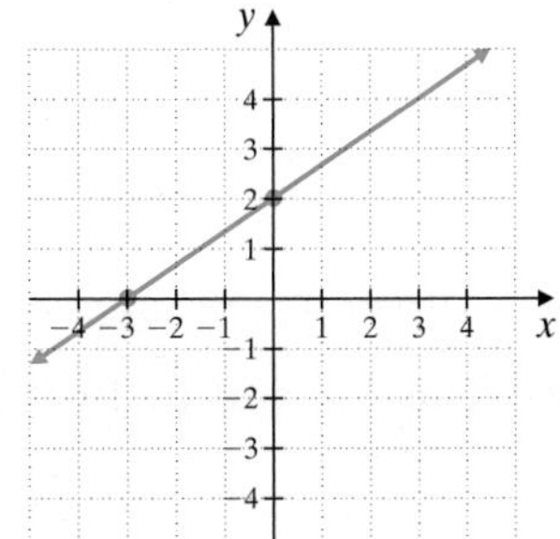

Solution:

x-intercept: $(-3, 0)$

y-intercept: $(0, 2)$

OBJECTIVES

A Identify intercepts of a graph.

B Graph a linear equation by finding and plotting intercept points.

C Identify and graph vertical and horizontal lines.

Practice Problem 1

Identify the x- and y-intercepts.

1.

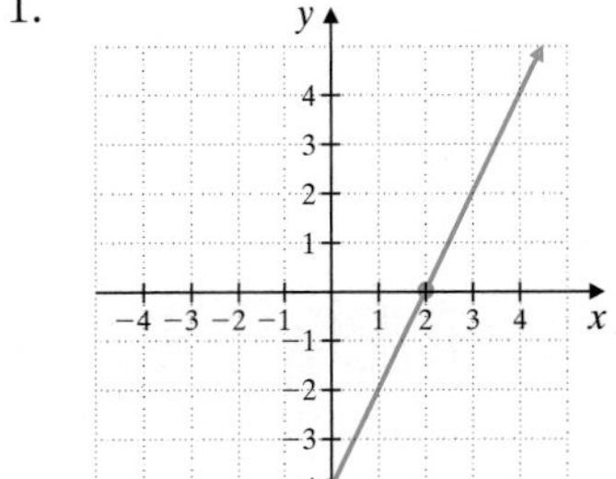

Answer

1. x-intercept: $(2, 0)$; y-intercept: $(0, -4)$

Practice Problems 2-3

Identify the x- and y-intercepts.

2.

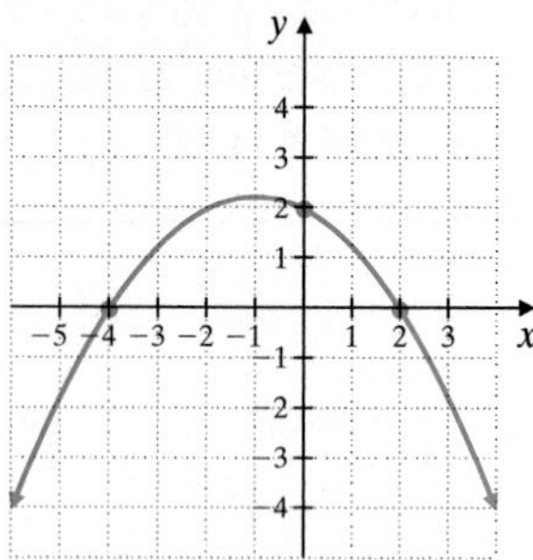

3.

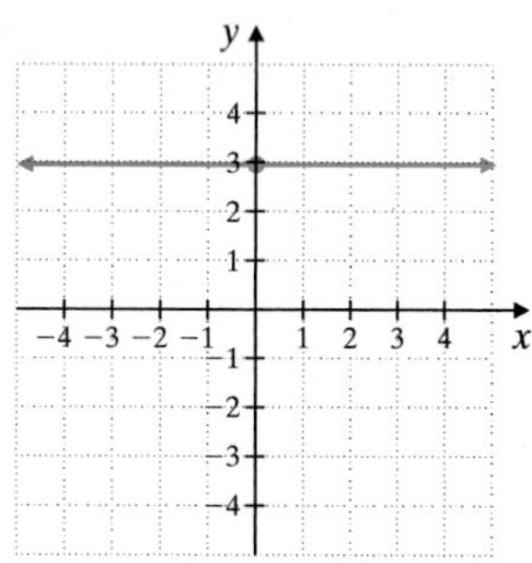

Practice Problem 4

Graph $2x - y = 4$ by finding and plotting its intercepts.

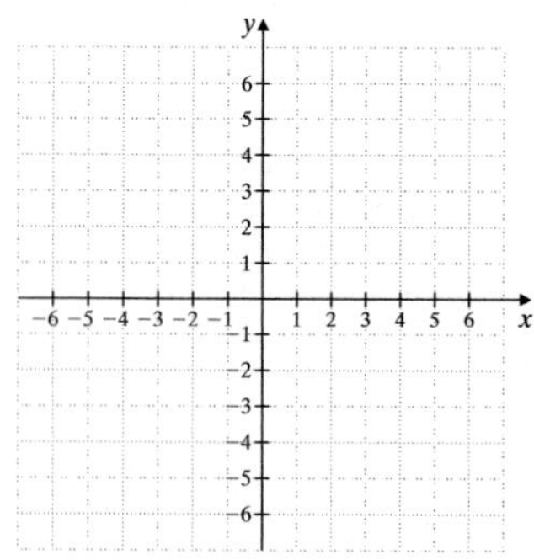

Answers

2. x-intercepts: $(-4, 0)$, $(2, 0)$; y-intercept: $(0, 2)$, **3.** x-intercept: none; y-intercept: $(0, 3)$, **4.** See page 217.

2.

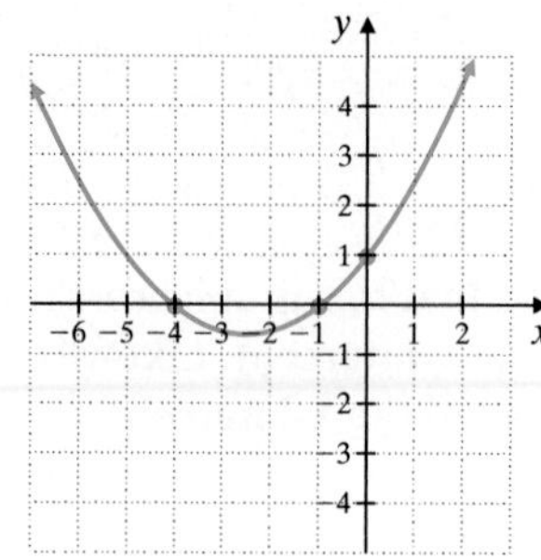

Solution:

x-intercepts: $(-4, 0)(-1, 0)$
y-intercept: $(0, 1)$

3.

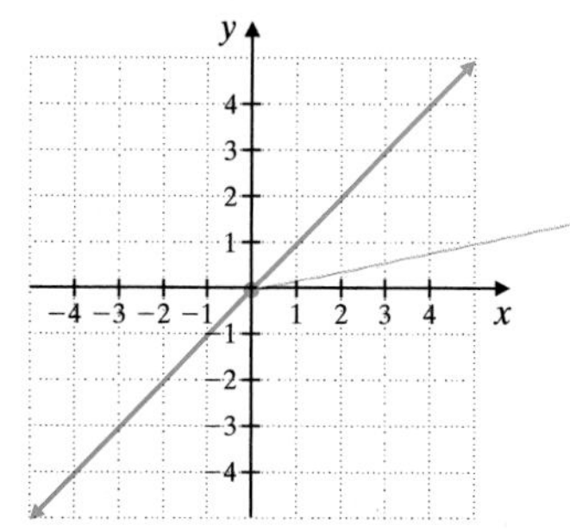

Helpful Hint

Notice that any time $(0, 0)$ is a point of a graph, then it is an x-intercept and a y-intercept.

Solution:

x-intercept: $(0, 0)$
y-intercept: $(0, 0)$

Here, the x- and y-intercept happen to be the same point.

B Finding and Plotting Intercepts

Given an equation of a line, we can usually find intercepts easily since one coordinate is 0.

One way to find the y-intercept of a line, from its equation is to let $x = 0$, since a point on the y-axis has an x-coordinate of 0. To find the x-intercept of a line, let $y = 0$, since a point on the x-axis has a y-coordinate of 0.

Finding x- and y-Intercepts

To find the x-intercept, let $y = 0$ and solve for x.
To find the y-intercept, let $x = 0$ and solve for y.

EXAMPLE 4 Graph $x - 3y = 6$ by finding and plotting its intercepts.

Solution: We let $y = 0$ to find the x-intercept and $x = 0$ to find the y-intercept.

Let $y = 0$.	Let $x = 0$.
$x - 3y = 6$	$x - 3y = 6$
$x - 3(0) = 6$	$0 - 3y = 6$
$x - 0 = 6$	$-3y = 6$
$x = 6$	$y = -2$

The x-intercept is $(6, 0)$ and the y-intercept is $(0, -2)$. We find a third ordered pair solution to check our work. If we let $y = -1$, then $x = 3$. We plot the points $(6, 0)$, $(0, -2)$, and $(3, -1)$. The graph of $x - 3y = 6$ is the line drawn through these points as shown.

x	y
6	0
0	−2
3	−1

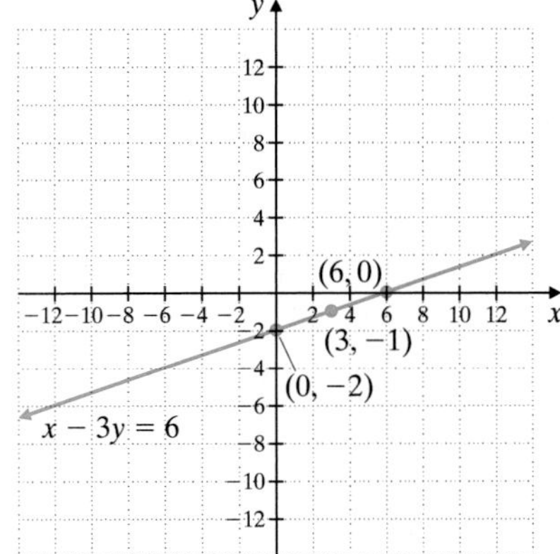

EXAMPLE 5 Graph $x = -2y$ by finding and plotting its intercepts.

Solution: We let $y = 0$ to find the x-intercept and $x = 0$ to find the y-intercept.

Let $y = 0$. Let $x = 0$.

$$x = -2y \qquad x = -2y$$
$$x = -2(0) \qquad 0 = -2y$$
$$x = 0 \qquad 0 = y$$

Both the x-intercept and y-intercept are $(0, 0)$. In other words, when $x = 0$, then $y = 0$, which gives the ordered pair $(0, 0)$. Also, when $y = 0$, then $x = 0$, which gives the same ordered pair $(0, 0)$. This happens when the graph passes through the origin. Since two points are needed to determine a line, we must find at least one more ordered pair that satisfies $x = -2y$. We let $y = -1$ to find a second ordered pair solution and let $y = 1$ as a checkpoint.

Let $y = -1$. Let $y = 1$.

$$x = -2(-1) \qquad x = -2(1)$$
$$x = 2 \qquad x = -2$$

The ordered pairs are $(0, 0)$, $(2, -1)$, and $(-2, 1)$. We plot these points to graph $x = -2y$.

x	y
0	0
2	−1
−2	1

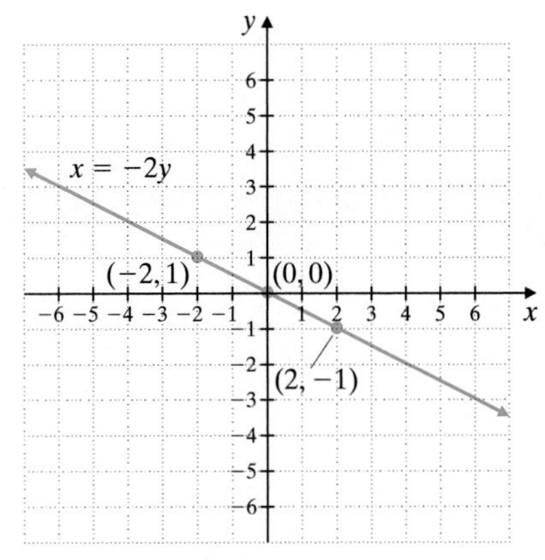

Practice Problem 5

Graph $y = 3x$ by finding and plotting its intercepts.

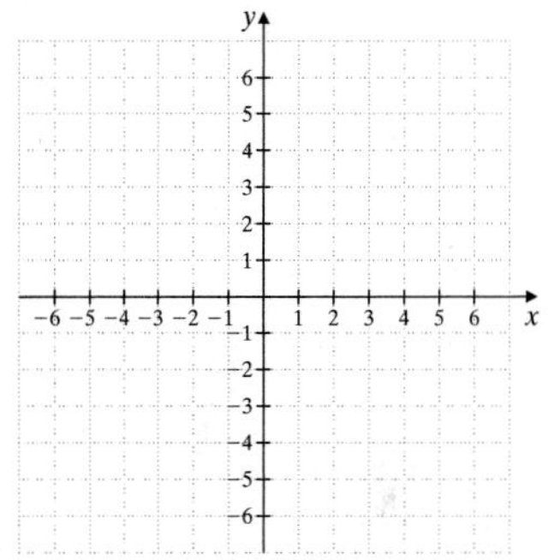

Answers

4.

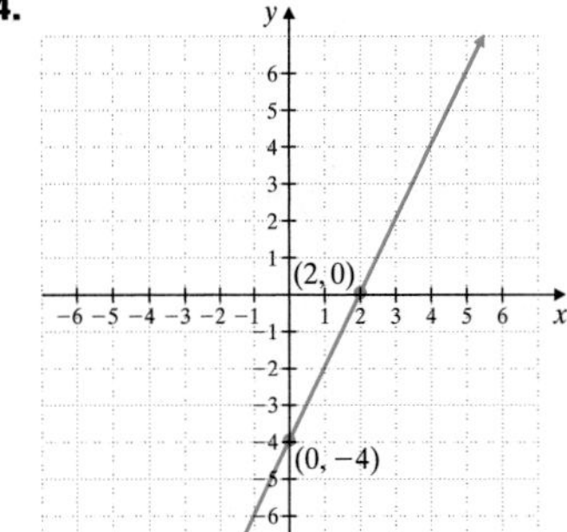

5.

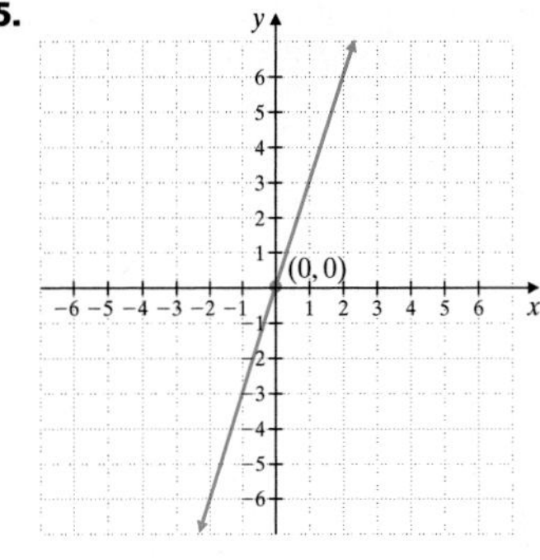

C Graphing Vertical and Horizontal Lines

The equation $x = 2$, for example, is a linear equation in two variables because it can be written in the form $x + 0y = 2$. The graph of this equation is a vertical line, as shown in the next example.

Practice Problem 6

Graph: $x = -3$

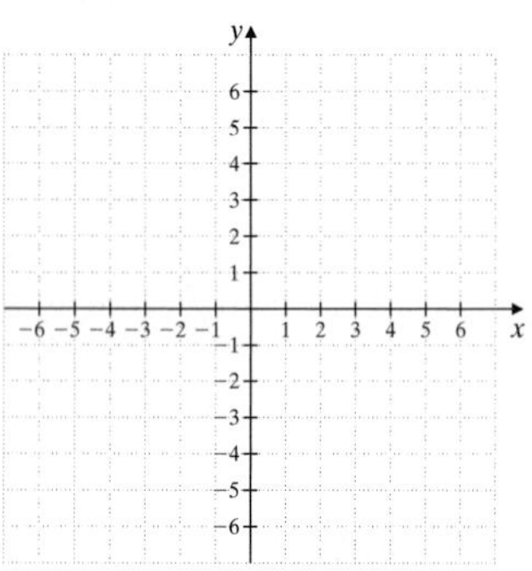

EXAMPLE 6 Graph: $x = 2$

Solution: The equation $x = 2$ can be written as $x + 0y = 2$. For any y-value chosen, notice that x is 2. No other value for x satisfies $x + 0y = 2$. Any ordered pair whose x-coordinate is 2 is a solution of $x + 0y = 2$. We will use the ordered pair solutions $(2, 3)$, $(2, 0)$, and $(2, -3)$ to graph $x = 2$.

x	y
2	3
2	0
2	−3

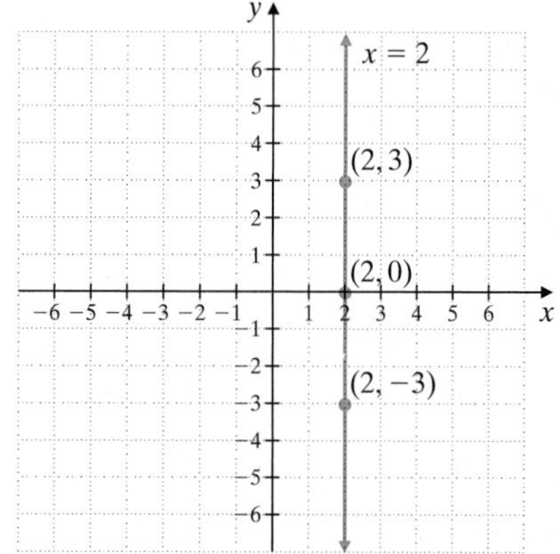

The graph is a vertical line with x-intercept $(2, 0)$. Note that this graph has no y-intercept because x is never 0.

In general, we have the following.

Vertical Lines

The graph of $x = c$, where c is a real number, is a vertical line with x-intercept $(c, 0)$.

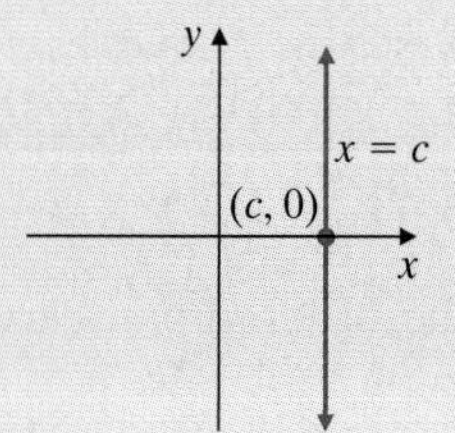

Answer

6.

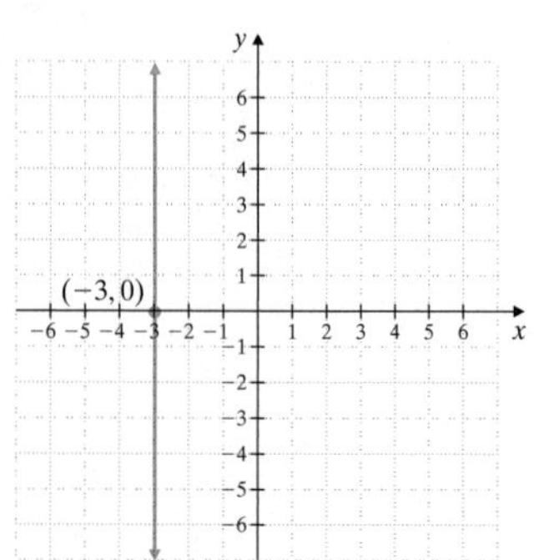

EXAMPLE 7 Graph: $y = -3$

Solution: The equation $y = -3$ can be written as $0x + y = -3$. For any x-value chosen, y is -3. If we choose 4, 1, and -2 as x-values, the ordered pair solutions are $(4, -3)$, $(1, -3)$, and $(-2, -3)$. We use these ordered pairs to graph $y = -3$. The graph is a horizontal line with y-intercept $(0, -3)$ and no x-intercept.

x	y
4	-3
1	-3
-2	-3

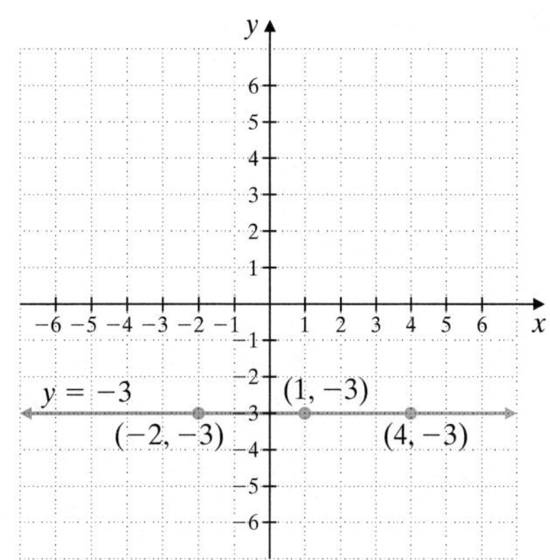

In general, we have the following.

Horizontal Lines

The graph of $y = c$, where c is a real number, is a horizontal line with y-intercept $(0, c)$.

Practice Problem 7

Graph: $y = 4$

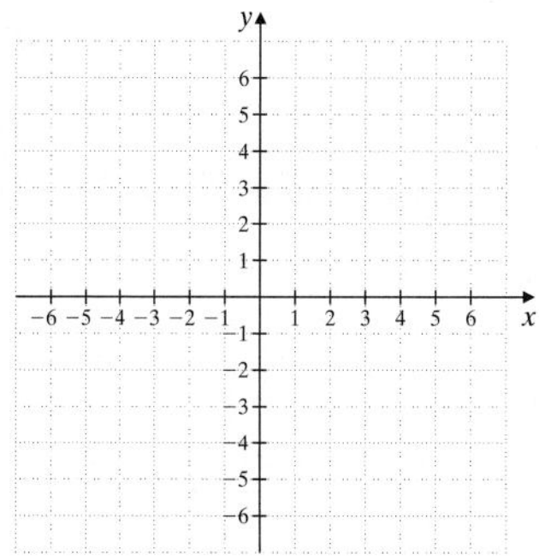

Answer

7.

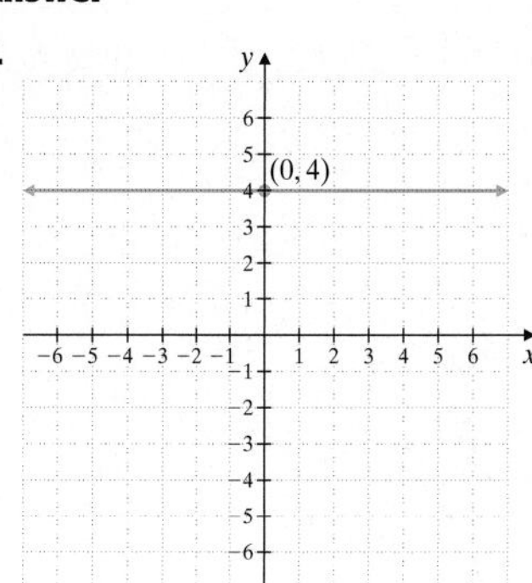

GRAPHING CALCULATOR EXPLORATIONS

You may have noticed that to use the $\boxed{Y=}$ key on a graphing calculator to graph an equation, the equation must be solved for y. For example, to graph $2x + 3y = 7$, we solve this equation for y.

$$2x + 3y = 7$$

$$3y = -2x + 7 \quad \text{Subtract } 2x \text{ from both sides.}$$

$$\frac{3y}{3} = -\frac{2x}{3} + \frac{7}{3} \quad \text{Divide both sides by 3.}$$

$$y = -\frac{2}{3}x + \frac{7}{3} \quad \text{Simplify.}$$

To graph $2x + 3y = 7$ or $y = -\frac{2}{3}x + \frac{7}{3}$, press the $\boxed{Y=}$ key and enter

$$Y_1 = -\frac{2}{3}x + \frac{7}{3}$$

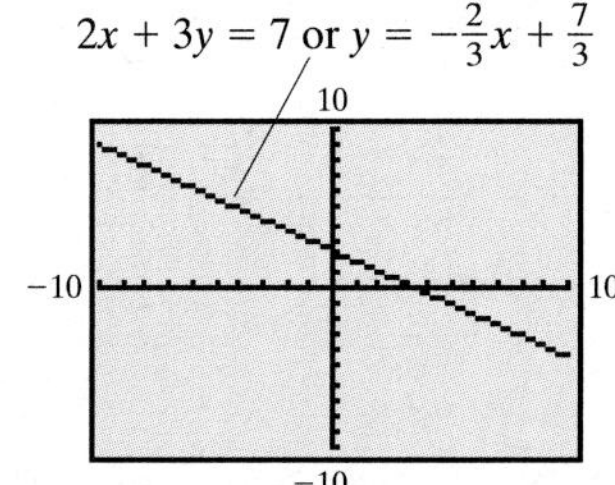

Graph each linear equation.

1. $x = 3.78y$

2. $-2.61y = x$

3. $-2.2x + 6.8y = 15.5$

4. $5.9x - 0.8y = -10.4$

Name ______________________ Section __________ Date ____________

Mental Math

Answer the following true or false.

1. The graph of $x = 2$ is a horizontal line.

2. All lines have an x-intercept *and* a y-intercept.

3. The graph of $y = 4x$ contains the point $(0, 0)$.

4. The graph of $x + y = 5$ has an x-intercept of $(5, 0)$ and a y-intercept of $(0, 5)$.

EXERCISE SET 3.3

A *Identify the intercepts. See Examples 1 through 3.*

1.

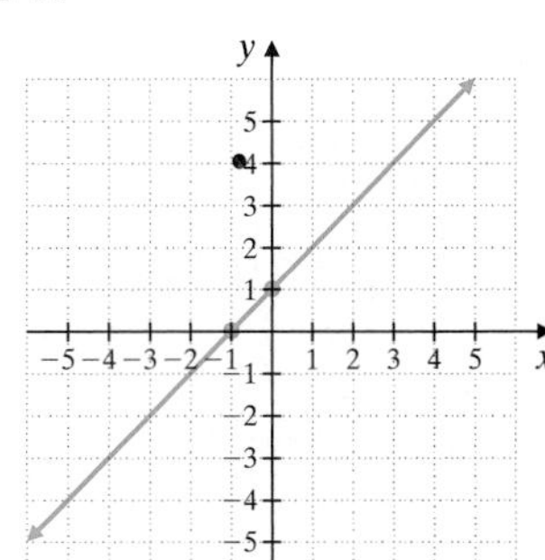

2.

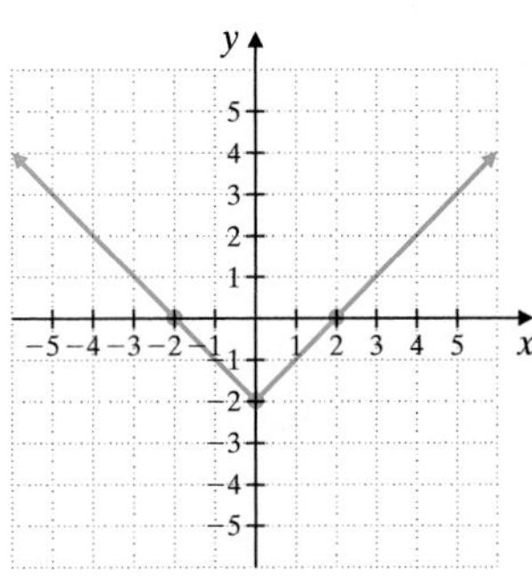

3.

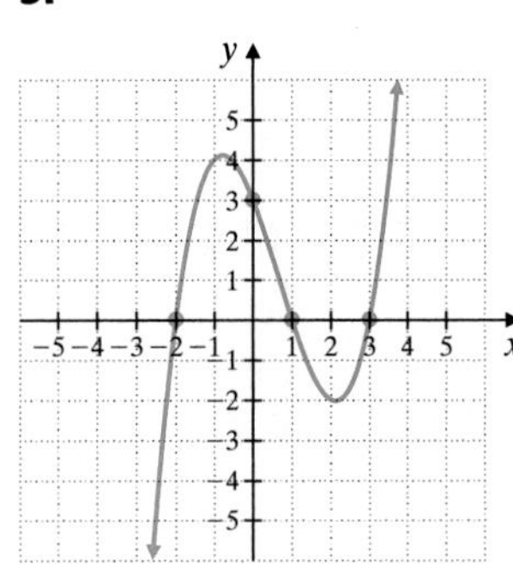

4.

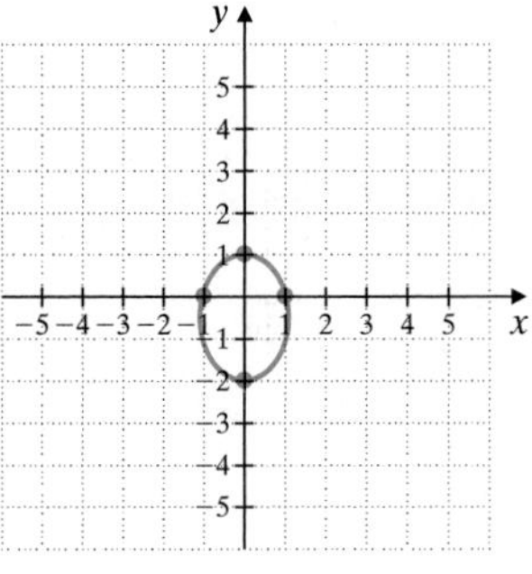

5. What is the greatest number of x- and y-intercepts that a line can have?

6. What is the smallest number of x- and y-intercepts that a line can have?

7. What is the smallest number of x- and y-intercepts that a circle can have?

8. What is the greatest number of x- and y-intercepts that a circle can have?

B *Graph each linear equation by finding and plotting its intercepts. See Examples 4 and 5.*

9. $x - y = 3$

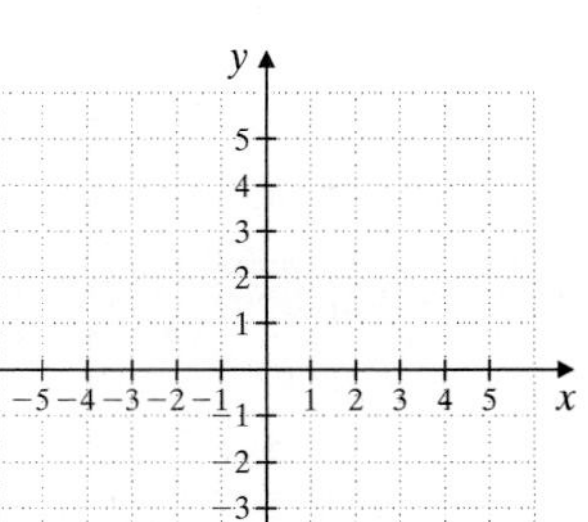

10. $x - y = -4$

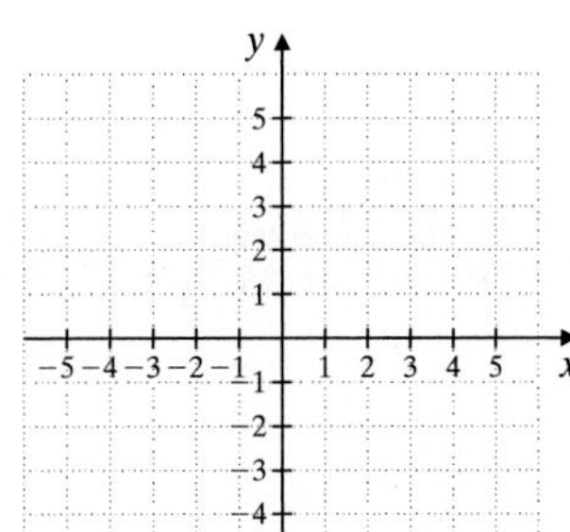

11. $x = 5y$

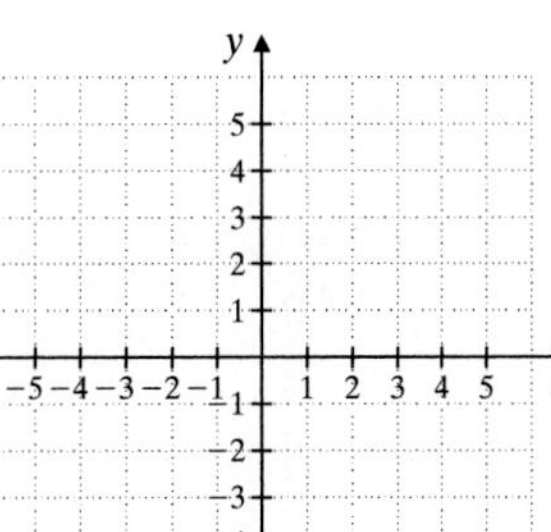

12. $2x = y$

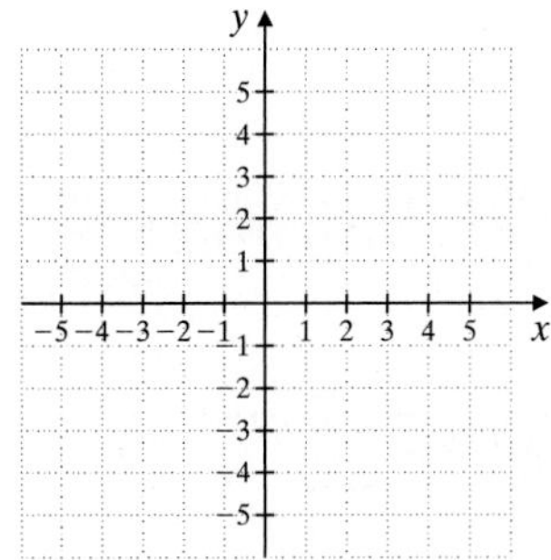

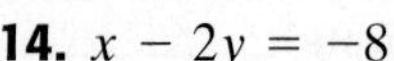

13. $-x + 2y = 6$

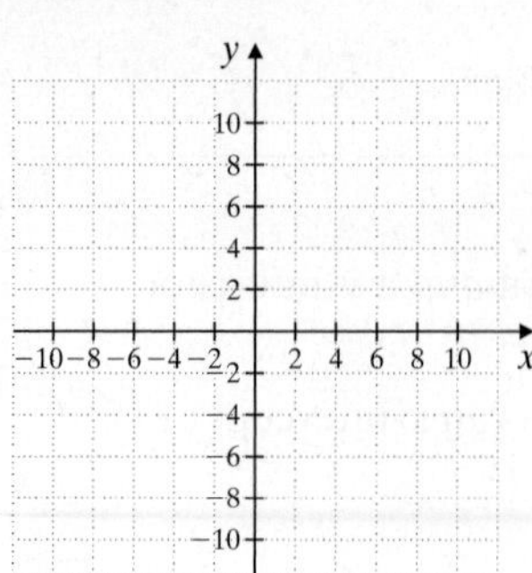

14. $x - 2y = -8$

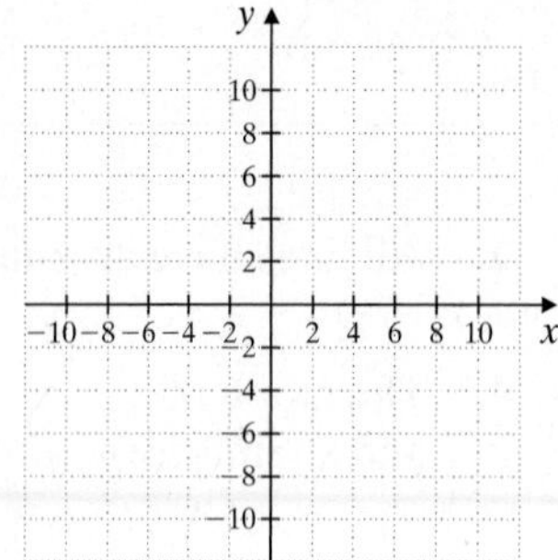

15. $2x - 4y = 8$

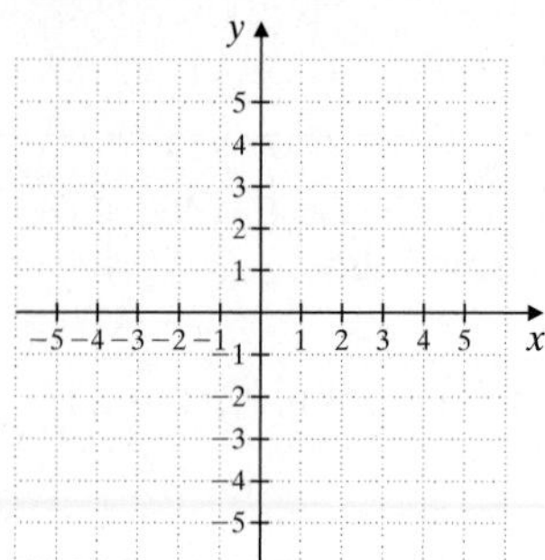

16. $2x + 3y = 6$

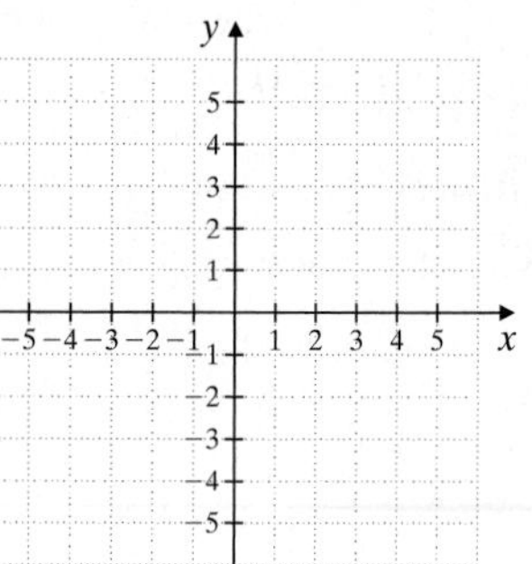

17. $x = 2y$

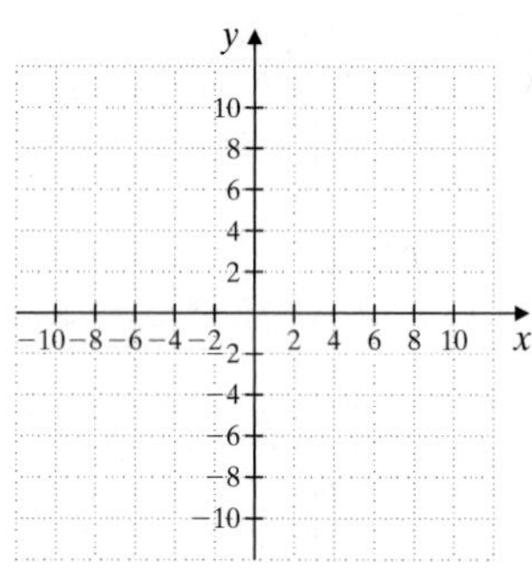

18. $y = -2x$

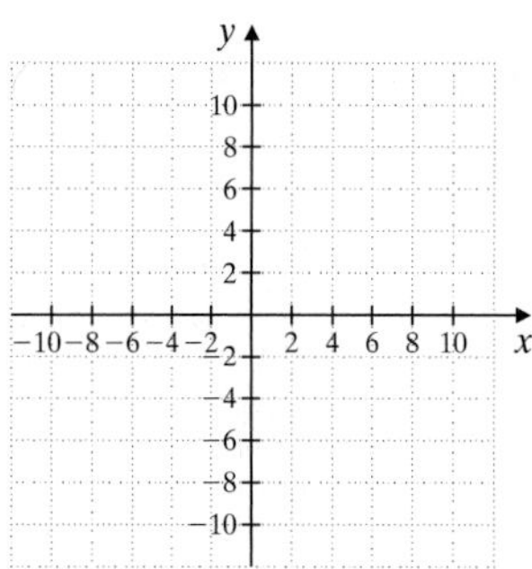

19. $y = 3x + 6$

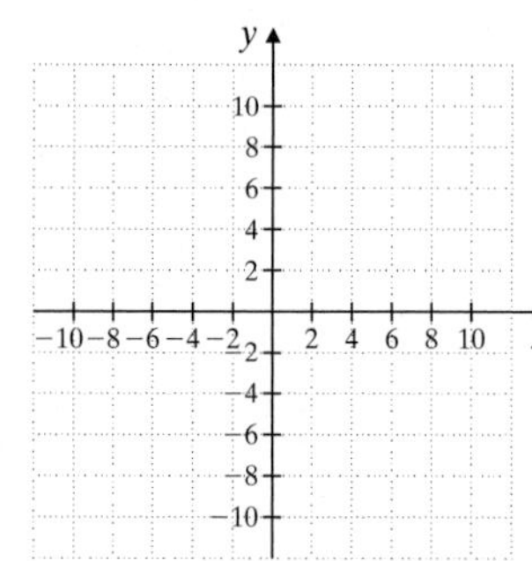

20. $y = 2x + 10$

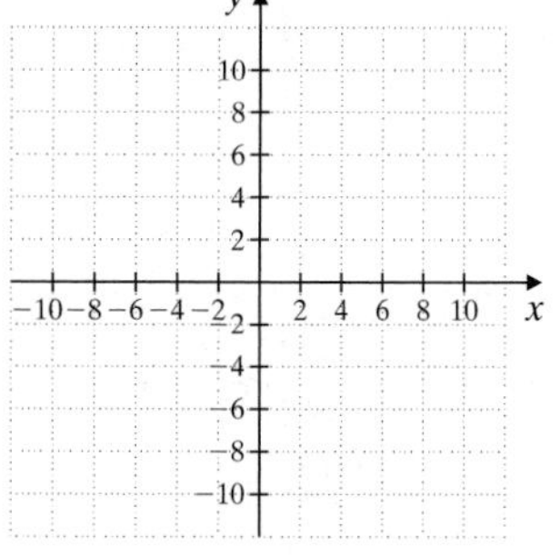

21. $x = y$

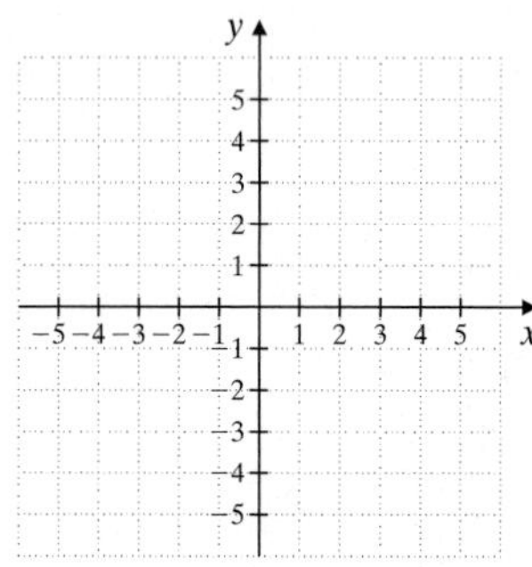

22. $x = -y$

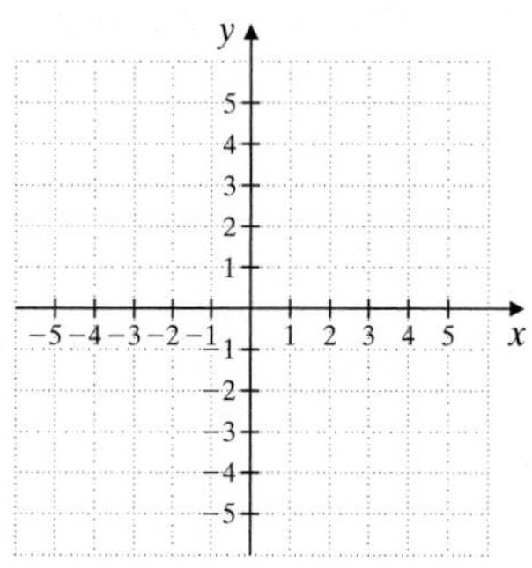

23. $x + 8y = 8$

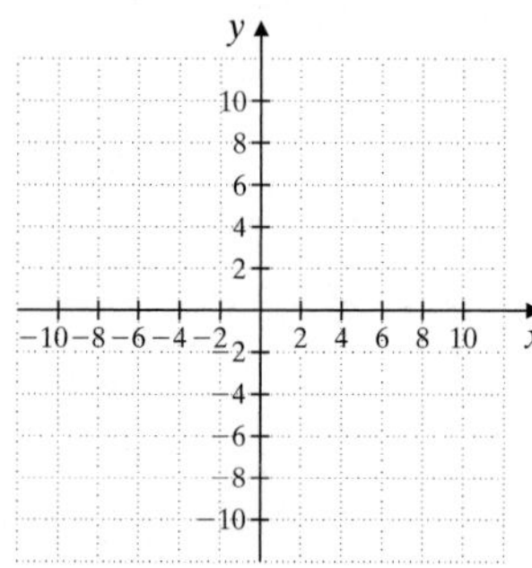

24. $x + 3y = 9$

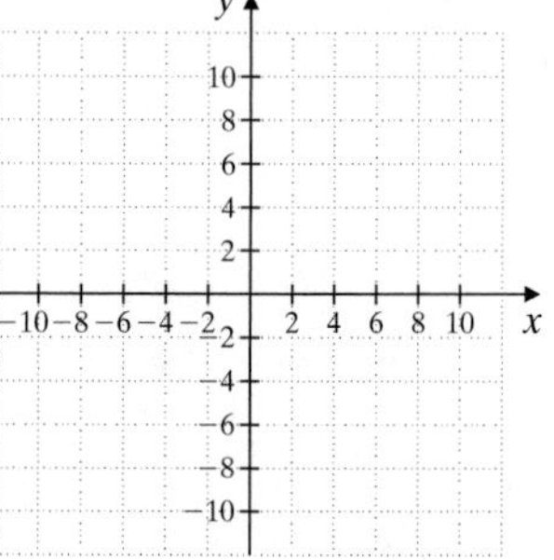

25. $5 = 6x - y$

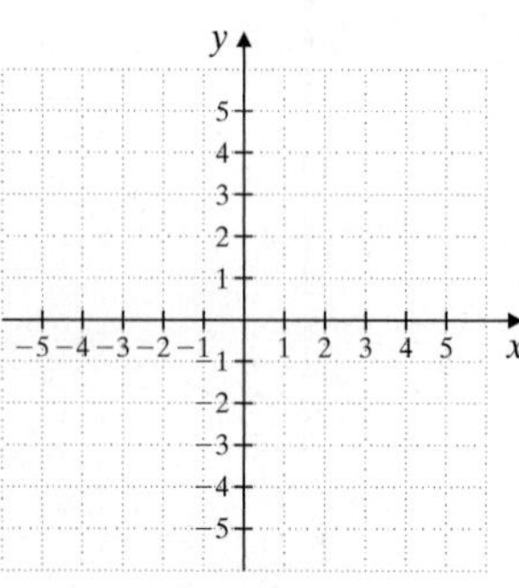

26. $4 = x - 3y$

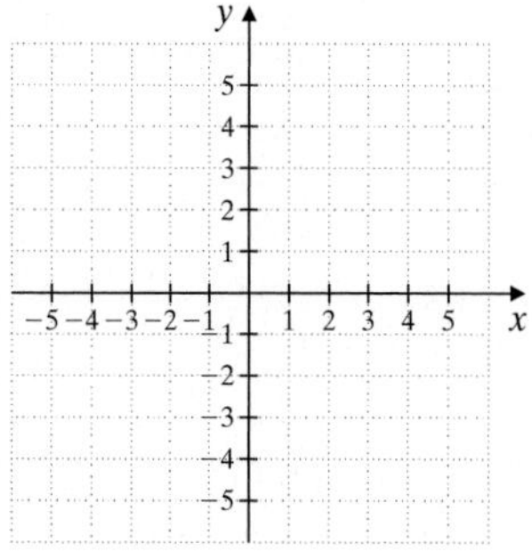

27. $-x + 10y = 11$

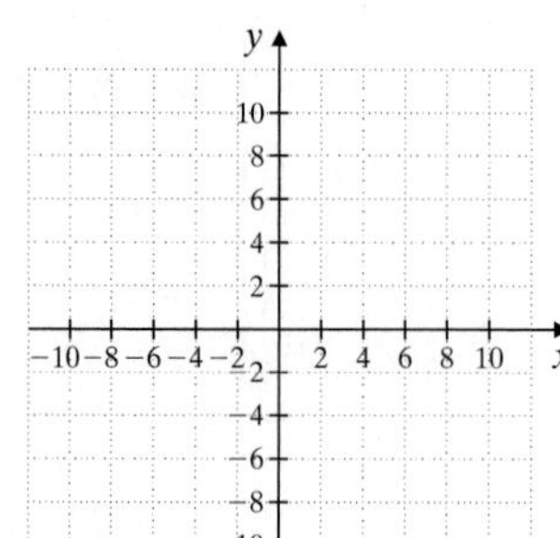

28. $-x + 9 = -y$

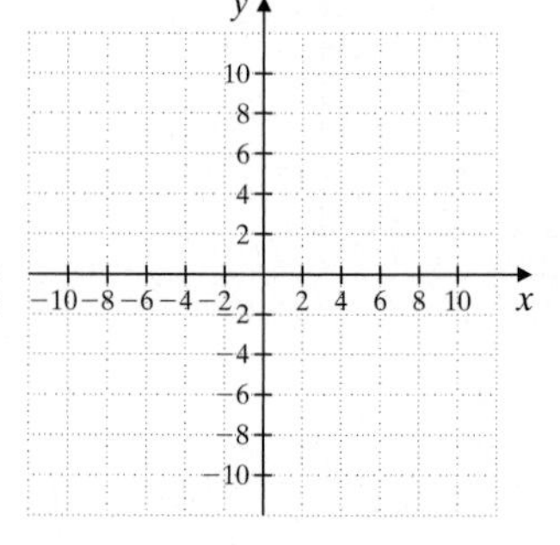

C *Graph each linear equation. See Examples 6 and 7.*

29. $x = -1$

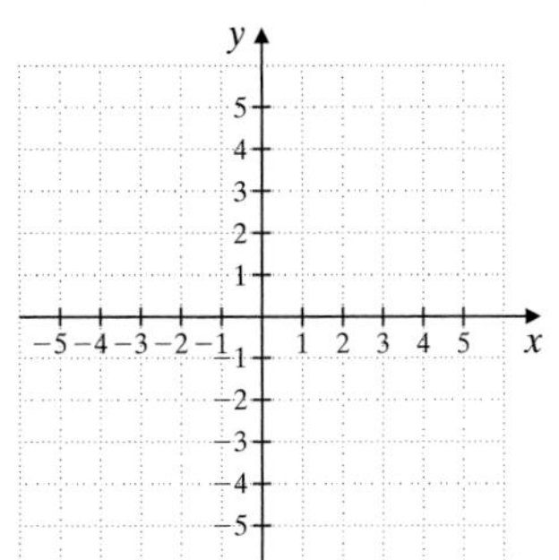

30. $y = 5$

31. $y = 0$

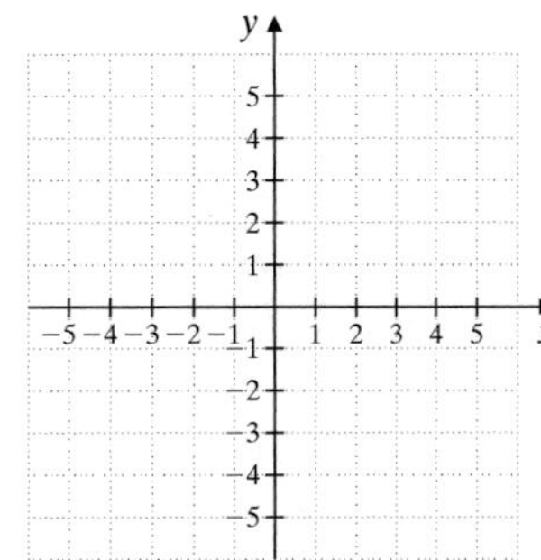

32. $x = 0$

33. $y + 7 = 0$

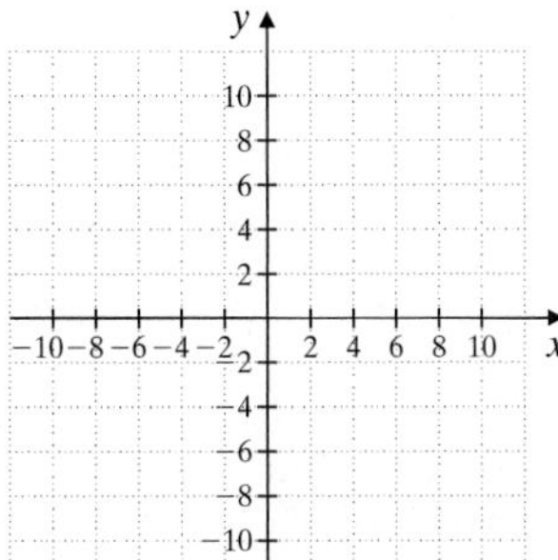

34. $x - 2 = 0$

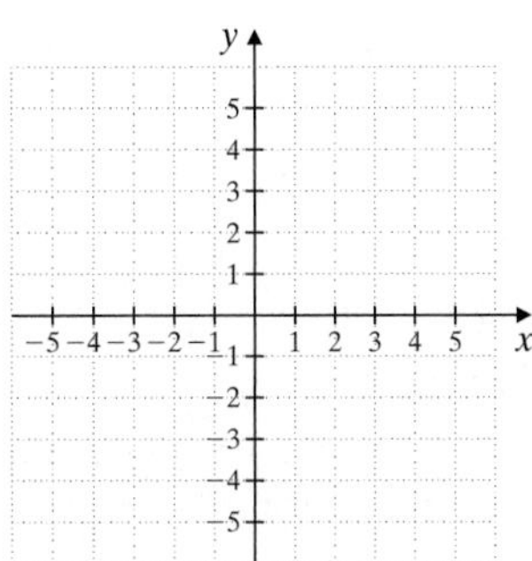

35. $x + 3 = 0$

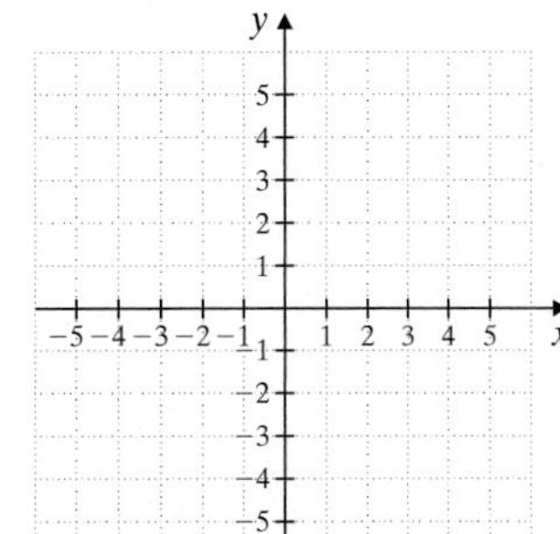

36. $y - 6 = 0$

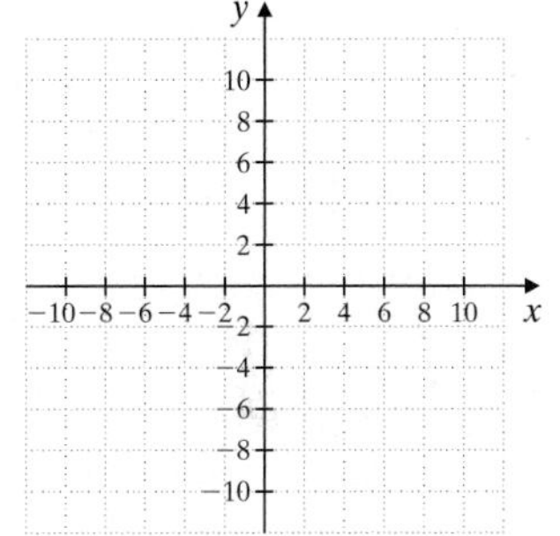

Review and Preview

Simplify. See Sections 1.4 through 1.6.

37. $\dfrac{-6 - 3}{2 - 8}$

38. $\dfrac{4 - 5}{-1 - 0}$

39. $\dfrac{-8 - (-2)}{-3 - (-2)}$

40. $\dfrac{12 - 3}{10 - 9}$

41. $\dfrac{0 - 6}{5 - 0}$

42. $\dfrac{2 - 2}{3 - 5}$

Combining Concepts

Match each equation with its graph.

43. $y = 3$

44. $y = 2x + 2$

45. $x = 3$

46. $y = 2x + 3$

A.

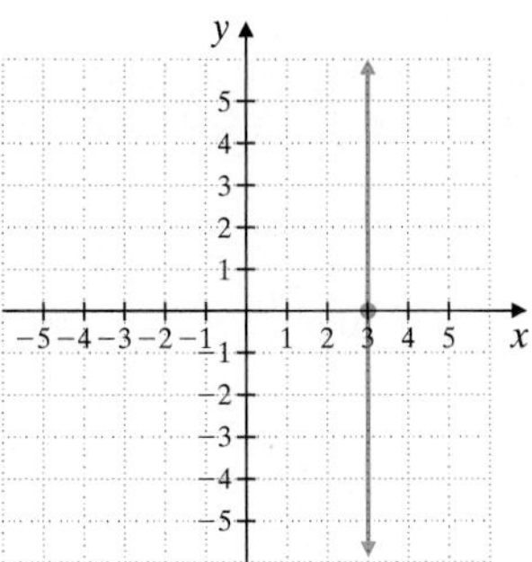

B.

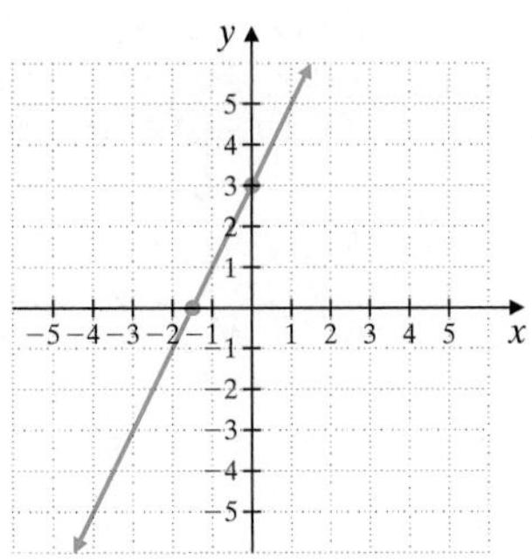

C.

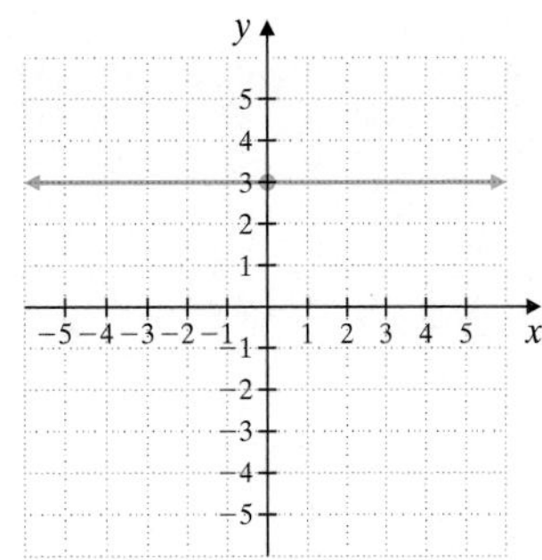

D.

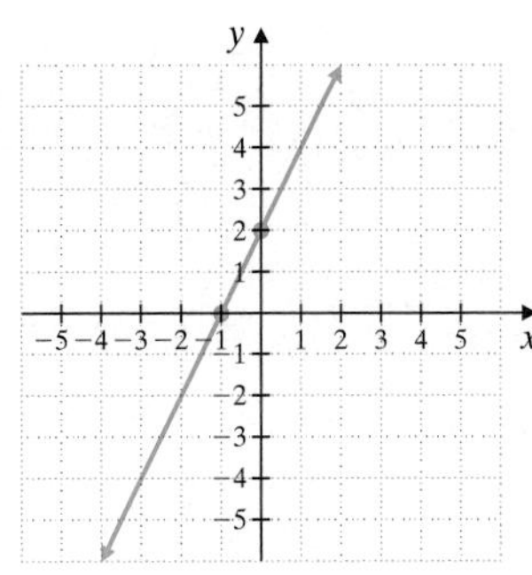

47. Discuss whether a vertical line ever has a y-intercept.

48. Discuss whether a horizontal line ever has an x-intercept.

49. The production supervisor at Alexandra's Office Products finds that it takes 3 hours to manufacture a particular office chair and 6 hours to manufacture an office desk. A total of 1200 hours is available to produce office chairs and desks of this style. The linear equation that models this situation is $3x + 6y = 1200$, where x represents the number of chairs produced and y the number of desks manufactured.
   a. Complete the ordered pair solution $(0, \ )$ of this equation. Describe the manufacturing situation that corresponds to this solution.
   b. Complete the ordered pair solution $(\ , 0)$ of this equation. Describe the manufacturing situation that corresponds to this solution.
   c. If 50 desks are manufactured, find the greatest number of chairs that can be made.

Two lines in the same plane that do not intersect are called ***parallel lines.***

50. Use your own graph paper to draw a line parallel to the line $x = 5$ that intersects the x-axis at 1. What is the equation of this line?

51. Use your own graph paper to draw a line parallel to the line $y = -1$ that intersects the y-axis at -4. What is the equation of this line?

52. The number of music cassettes y (in millions) shipped to retailers in the United States from 1995 through 2000 can be modeled by the equation $y = -37.2x + 264.4$, where x represents the number of years after 1995. (*Source:* Recording Industry Association of America)
   a. Find the x-intercept of this equation. (Round to the nearest tenth.)
   b. What does this x-intercept mean?

53. The number of Disney Stores y for the years 1996–2000 can by modeled by the equation $y = 51.6x + 560.2$, where x represents the number of years after 1996. (*Source: The Walt Disney Company Fact Book 2000*)
   a. Find the y-intercept of this equation.
   b. What does this y-intercept mean?

3.4 Slope

OBJECTIVES

- (A) Find the slope of a line given two points of the line.
- (B) Find the slope of a line given its equation.
- (C) Find the slopes of horizontal and vertical lines.
- (D) Compare the slopes of parallel and perpendicular lines.
- (E) Solve problems of slope.

(A) Finding the Slope of a Line Given Two Points

Thus far, much of this chapter has been devoted to graphing lines. You have probably noticed by now that a key feature of a line is its slant or steepness. In mathematics, the slant or steepness of a line is formally known as its **slope**. We measure the slope of a line by the ratio of vertical change (rise) to the corresponding horizontal change (run) as we move along the line.

On the line below, for example, suppose that we begin at the point $(1, 2)$ and move to the point $(4, 6)$. The vertical change is the change in y-coordinates: $6 - 2$ or 4 units. The corresponding horizontal change is the change in x-coordinates: $4 - 1 = 3$ units. The ratio of these changes is

$$\text{slope} = \frac{\text{change in } y \text{ (vertical change or rise)}}{\text{change in } x \text{ (horizontal change or run)}} = \frac{4}{3}$$

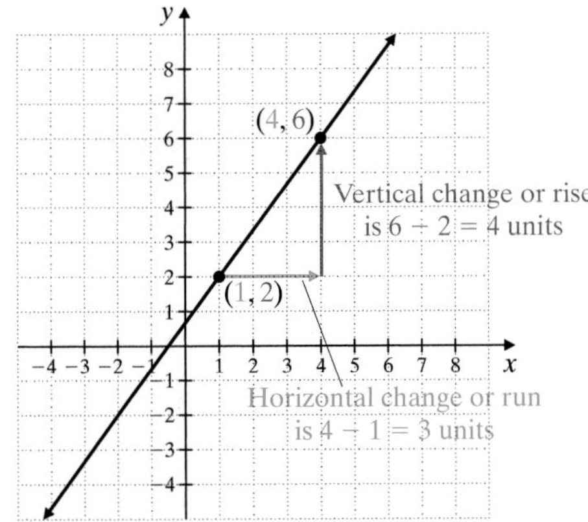

The slope of this line, then, is $\frac{4}{3}$. This means that for every 4 units of change in y-coordinates, there is a corresponding change of 3 units in x-coordinates.

Helpful Hint

It makes no difference what two points of a line are chosen to find its slope. The slope of a line is the same everywhere on the line.

Slope of a Line

The slope m of the line containing the points (x_1, y_1) and (x_2, y_2) is given by

$$m = \frac{\text{rise}}{\text{run}} = \frac{\text{change in } y}{\text{change in } x} = \frac{y_2 - y_1}{x_2 - x_1}, \quad \text{as long as } x_2 \neq x_1$$

Practice Problem 1

Find the slope of the line through $(-2, 3)$ and $(4, -1)$. Graph the line.

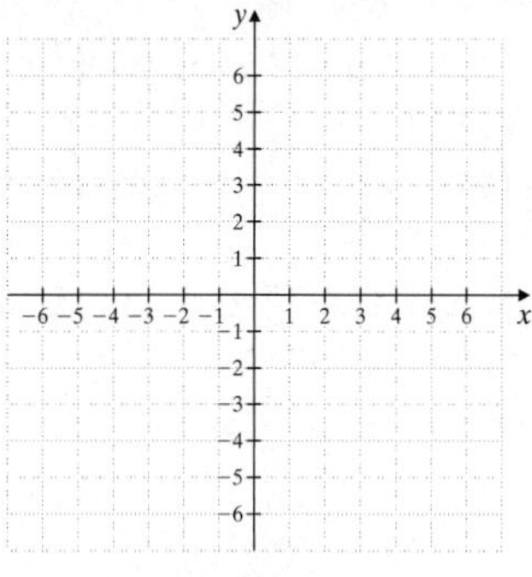

EXAMPLE 1

Find the slope of the line through $(-1, 5)$ and $(2, -3)$. Graph the line.

Solution: Let (x_1, y_1) be $(-1, 5)$ and (x_2, y_2) be $(2, -3)$. Then, by the definition of slope, we have the following.

$$m = \frac{y_2 - y_1}{x_2 - x_1}$$

$$= \frac{-3 - 5}{2 - (-1)}$$

$$= \frac{-8}{3} = -\frac{8}{3}$$

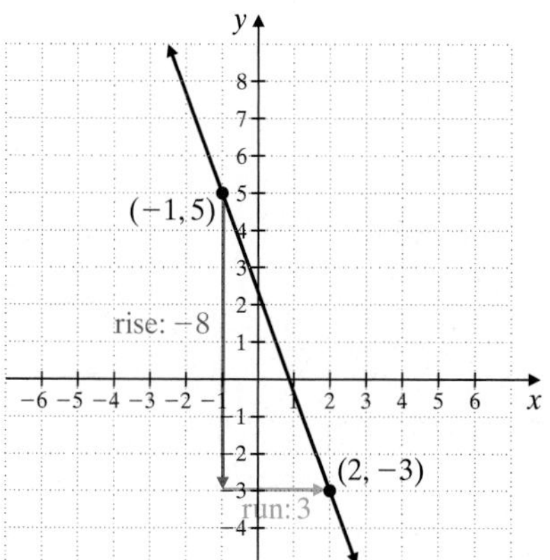

The slope of the line is $-\frac{8}{3}$.

In Example 1, we could just as well have identified (x_1, y_1) with $(2, -3)$ and (x_2, y_2) with $(-1, 5)$. It makes no difference which point is called (x_1, y_1) or (x_2, y_2).

Try the Concept Check in the margin.

Helpful Hint

When finding the slope of a line through two given points, it makes no difference which given point is called (x_1, y_1) and which is called (x_2, y_2). However, once an x-coordinate is called x_1, make sure its corresponding y-coordinate is called y_1.

Concept Check

The points $(-2, -5)$, $(0, -2)$, $(4, 4)$, and $(10, 13)$ all lie on the same line. Work with a partner and verify that the slope is the same no matter which points are used to find slope.

Answers

1. $-\frac{2}{3}$

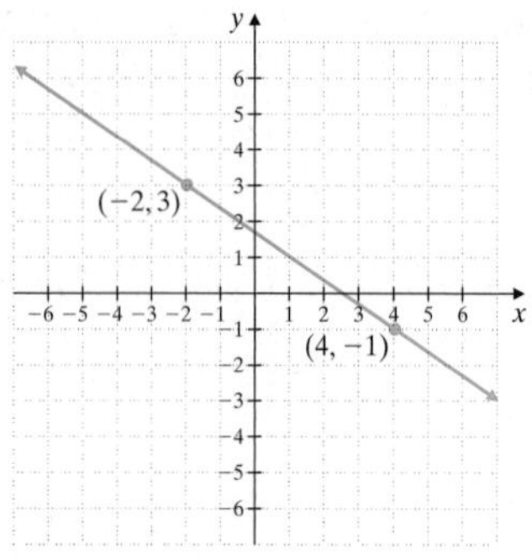

Concept Check: $m = \frac{3}{2}$

EXAMPLE 2

Find the slope of the line through $(-1, -2)$ and $(2, 4)$. Graph the line.

Solution: Let (x_1, y_1) be $(2, 4)$ and (x_2, y_2) be $(-1, -2)$.

$$m = \frac{y_2 - y_1}{x_2 - x_1} = \frac{-2 - 4}{-1 - 2} = \frac{-6}{-3} = 2$$

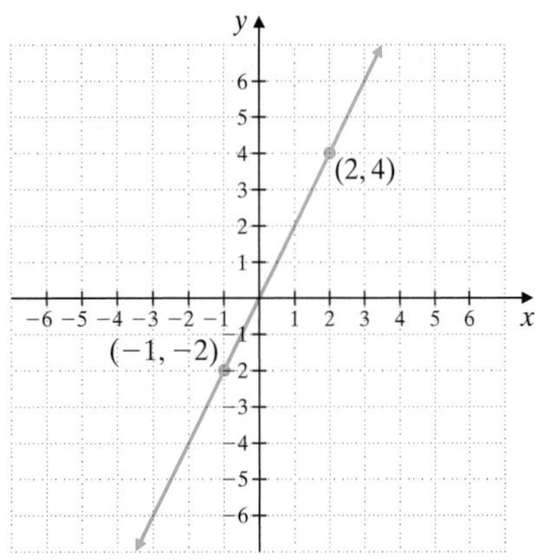

The slope is 2.

Practice Problem 2

Find the slope of the line through $(-2, 1)$ and $(3, 5)$. Graph the line.

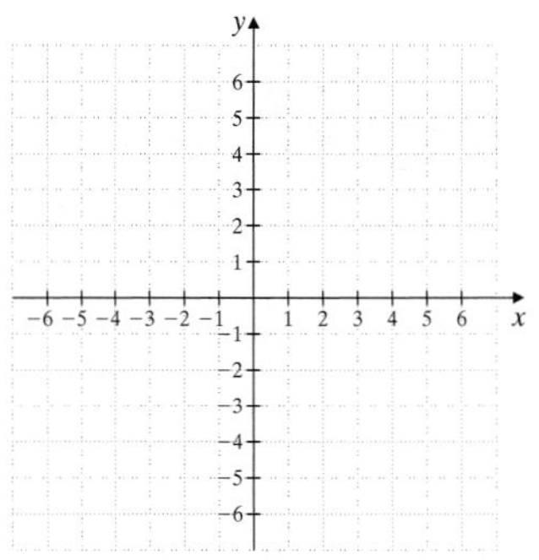

Try the Concept Check in the margin.

Notice that the slope of the line in Example 1 is negative, and the slope of the line in Example 2 is positive. Let your eye follow the line with negative slope from left to right and notice that the line "goes down." If you follow the line with positive slope from left to right, you will notice that the line "goes up." This is true in general.

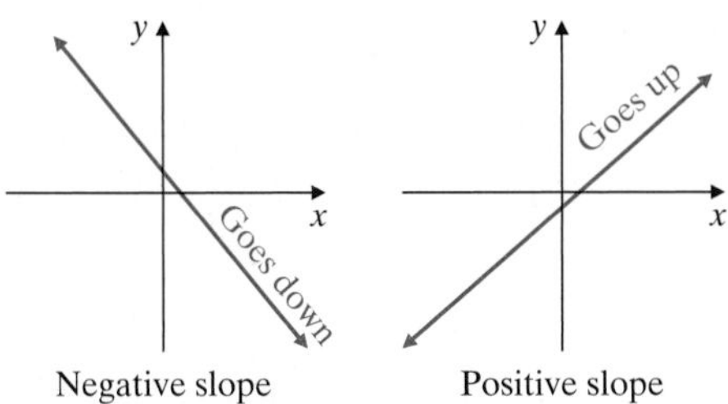

Negative slope Positive slope

Concept Check

What is wrong with the following slope calculation for the points $(3, 5)$ and $(-2, 6)$?

$$m = \frac{5 - 6}{-2 - 3} = \frac{-1}{-5} = \frac{1}{5}$$

B Finding the Slope of a Line Given Its Equation

As we have seen, the slope of a line is defined by two points on the line. Thus, if we know the equation of a line, we can find its slope by finding two of its points. For example, let's find the slope of the line

$$y = 3x - 2$$

To find two points, we can choose two values for x and substitute to find corresponding y-values. If $x = 0$, for example, $y = 3 \cdot 0 - 2$ or $y = -2$. If $x = 1$, $y = 3 \cdot 1 - 2$ or $y = 1$. This gives the ordered pairs $(0, -2)$ and $(1, 1)$. Using the definition for slope, we have

$$m = \frac{1 - (-2)}{1 - 0} = \frac{3}{1} = 3 \quad \text{The slope is 3.}$$

Notice that the slope, 3, is the same as the coefficient of x in the equation $y = 3x - 2$. This is true in general.

Answers

2. $\frac{4}{5}$

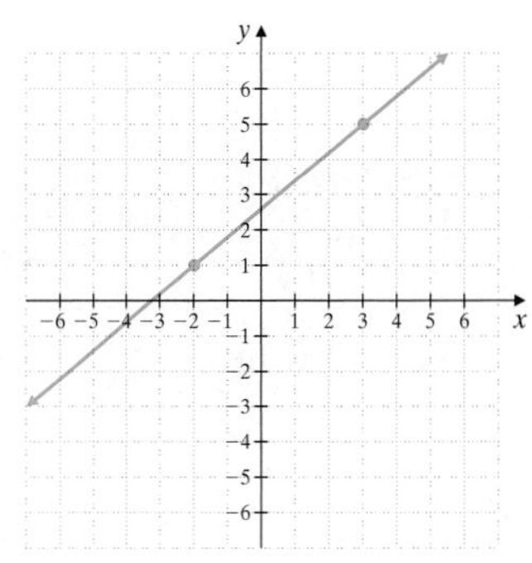

Concept Check: $m = \frac{5 - 6}{3 - (-2)} = \frac{-1}{5} = -\frac{1}{5}$

If a linear equation is solved for y, the coefficient of x is the line's slope. In other words, the slope of the line given by $y = mx + b$ is m, the coefficient of x. [In Section 7.1, we will learn that the y-intercept of this equation is $(0, b)$.]

Practice Problem 3

Find the slope of the line $5x + 4y = 10$.

EXAMPLE 3 Find the slope of the line $-2x + 3y = 11$.

Solution: When we solve for y, the coefficient of x is the slope.

$$-2x + 3y = 11$$

$$3y = 2x + 11 \quad \text{Add } 2x \text{ to both sides.}$$

$$y = \frac{2}{3}x + \frac{11}{3} \quad \text{Divide both sides by 3.}$$

The slope is $\frac{2}{3}$.

C Finding Slopes of Horizontal and Vertical Lines

Practice Problem 4

Find the slope of $y = 3$.

EXAMPLE 4 Find the slope of the line $y = -1$.

Solution: Recall that $y = -1$ is a horizontal line with y-intercept -1. To find the slope, we find two ordered pair solutions of $y = -1$, knowing that solutions of $y = -1$ must have a y-value of -1. We will use $(2, -1)$ and $(-3, -1)$. We let (x_1, y_1) be $(2, -1)$ and (x_2, y_2) be $(-3, -1)$.

$$m = \frac{y_2 - y_1}{x_2 - x_1} = \frac{-1 - (-1)}{-3 - 2} = \frac{0}{-5} = 0$$

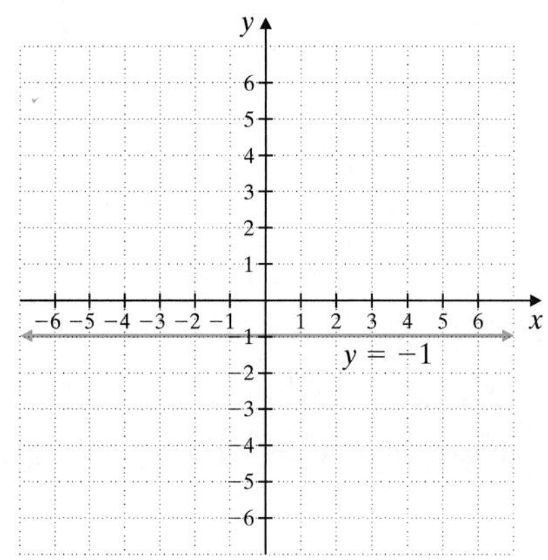

The slope of the line $y = -1$ is 0. Since the y-values will have a difference of 0 for every horizontal line, we can say that all **horizontal lines have a slope of 0.**

Practice Problem 5

Find the slope of the line $x = -2$.

EXAMPLE 5 Find the slope of the line $x = 5$.

Solution: Recall that the graph of $x = 5$ is a vertical line with x-intercept 5. To find the slope, we find two ordered pair solutions of $x = 5$. Ordered pair solutions of $x = 5$ must have an x-value of 5. We will use $(5, 0)$ and $(5, 4)$. We let $(x_1, y_1) = (5, 0)$ and $(x_2, y_2) = (5, 4)$.

$$m = \frac{y_2 - y_1}{x_2 - x_1} = \frac{4 - 0}{5 - 5} = \frac{4}{0}$$

Answers

3. $-\frac{5}{4}$, **4.** 0, **5.** undefined slope

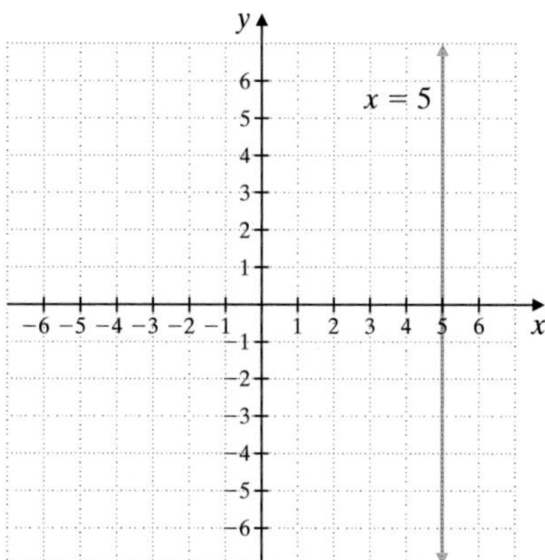

Since $\frac{4}{0}$ is undefined, we say the slope of the vertical line $x = 5$ is undefined.

Since the x-values will have a difference of 0 for every vertical line, we can say that all **vertical lines have undefined slope.**

Helpful Hint

Slope of 0 and undefined slope are not the same. Vertical lines have undefined slope, while horizontal lines have a slope of 0.

Here is a general review of slope.
Slope m of the line through (x_1, y_1) and (x_2, y_2) is given by the equation

$$m = \frac{y_2 - y_1}{x_2 - x_1}.$$

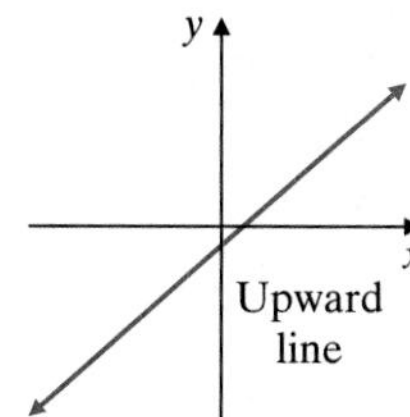

Positive slope: $m > 0$

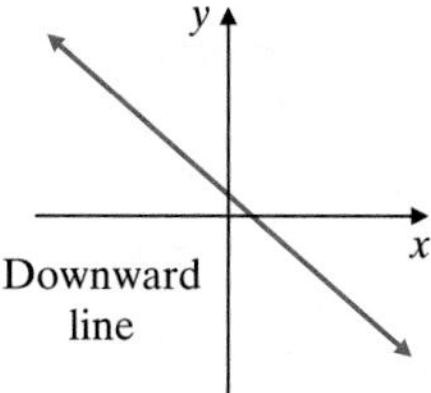

Negative slope: $m < 0$

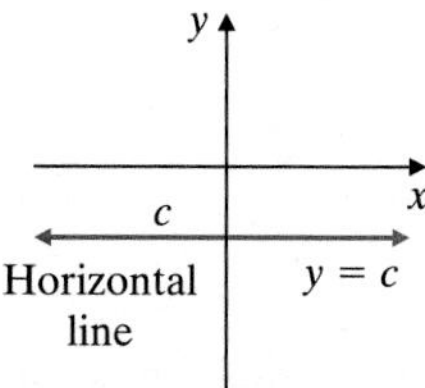

Zero slope: $m = 0$

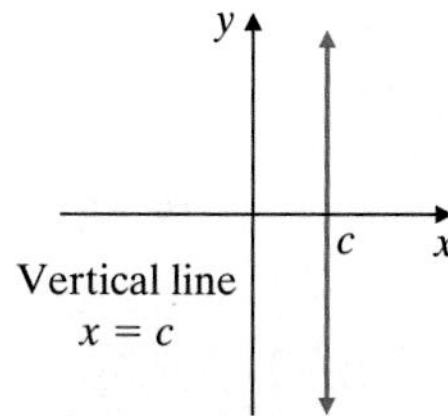

No slope or undefined slope

D Slopes of Parallel and Perpendicular Lines

Two lines in the same plane are **parallel** if they do not intersect. Slopes of lines can help us determine whether lines are parallel. Since parallel lines have the same steepness, it follows that they have the same slope.

For example, the graphs of

$$y = -2x + 4$$

and

$$y = -2x - 3$$

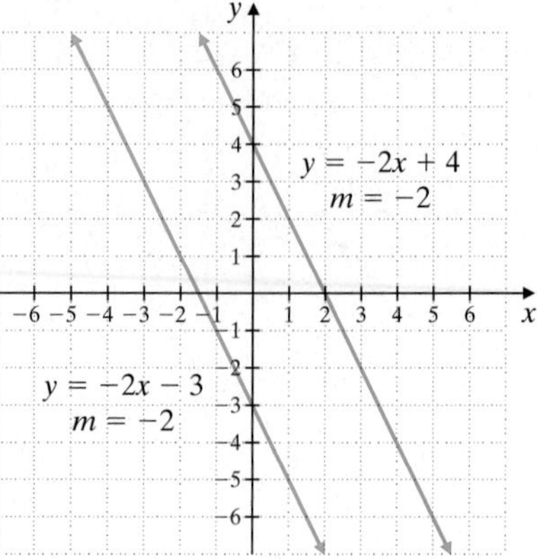

are shown. These lines have the same slope, −2. They also have different y-intercepts, so the lines are parallel. (If the y-intercepts were the same also, the lines would be the same.)

Parallel Lines

Nonvertical parallel lines have the same slope and different y-intercepts.

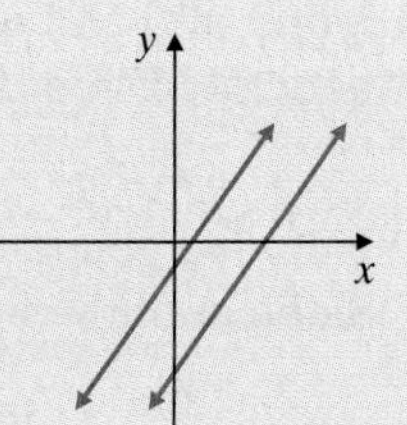

Two lines are **perpendicular** if they lie in the same plane and meet at a 90° (right) angle. How do the slopes of perpendicular lines compare? The product of the slopes of two perpendicular lines is −1.

For example, the graphs of

$$y = 4x + 1$$

and

$$y = -\frac{1}{4}x - 3$$

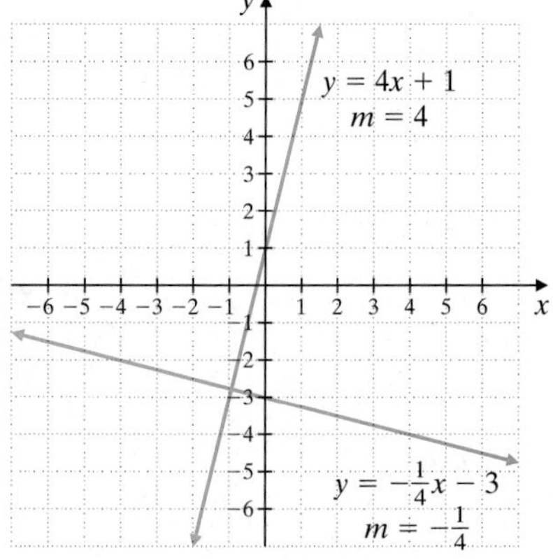

are shown. The slopes of the lines are 4 and $-\frac{1}{4}$. Their product is $4\left(-\frac{1}{4}\right) = -1$, so the lines are perpendicular.

Perpendicular Lines

If the product of the slopes of two lines is −1, then the lines are perpendicular. (Two nonvertical lines are perpendicular if the slope of one is the negative reciprocal of the slope of the other.)

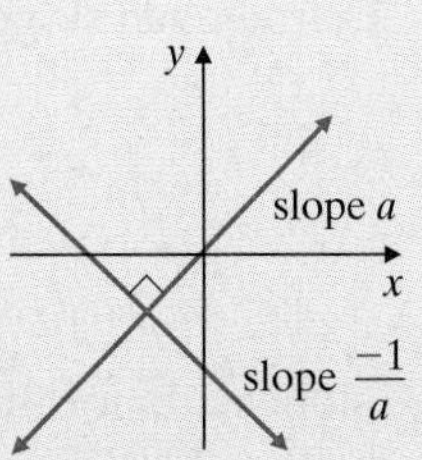

Helpful Hint

Here are examples of numbers that are negative (opposite) reciprocals.

Number	Negative Reciprocal	Their product is −1.
$\frac{2}{3}$	$-\frac{3}{2}$	$\frac{2}{3} \cdot -\frac{3}{2} = -\frac{6}{6} = -1$
-5 or $-\frac{5}{1}$	$\frac{1}{5}$	$-5 \cdot \frac{1}{5} = -\frac{5}{5} = -1$

Helpful Hint

Here are a few important points about vertical and horizontal lines.

- Two distinct vertical lines are parallel.
- Two distinct horizontal lines are parallel.
- A horizontal line and a vertical line are always perpendicular.

EXAMPLE 6

Determine whether each pair of lines is parallel, perpendicular, or neither.

a. $y = -\frac{1}{5}x + 1$
$2x + 10y = 3$

b. $x + y = 3$
$-x + y = 4$

c. $3x + y = 5$
$2x + 3y = 6$

Solution:

a. The slope of the line $y = -\frac{1}{5}x + 1$ is $-\frac{1}{5}$. We find the slope of the second line by solving its equation for y.

$$2x + 10y = 3$$

$$10y = -2x + 3 \quad \text{Subtract } 2x \text{ from both sides.}$$

$$y = \frac{-2}{10}x + \frac{3}{10} \quad \text{Divide both sides by 10.}$$

$$y = -\frac{1}{5}x + \frac{3}{10} \quad \text{Simplify.}$$

The slope of this line is $-\frac{1}{5}$ also. Since the lines have the same slope and different y-intercepts, they are parallel, as shown in the figure on the next page.

Practice Problem 6

Determine whether each pair of lines is parallel, perpendicular, or neither.

a. $x + y = 5$
$2x + y = 5$

b. $5y = 2x - 3$
$5x + 2y = 1$

c. $y = 2x + 1$
$4x - 2y = 8$

Answers

6. a. neither, **b.** perpendicular, **c.** parallel

b. To find each slope, we solve each equation for y.

$$x + y = 3 \qquad\qquad -x + y = 4$$
$$y = -x + 3 \qquad\qquad y = x + 4$$

The slope is -1. The slope is 1.

The slopes are not the same, so the lines are not parallel. Next we check the product of the slopes: $(-1)(1) = -1$. Since the product is -1, the lines are perpendicular, as shown in the figure below to the right.

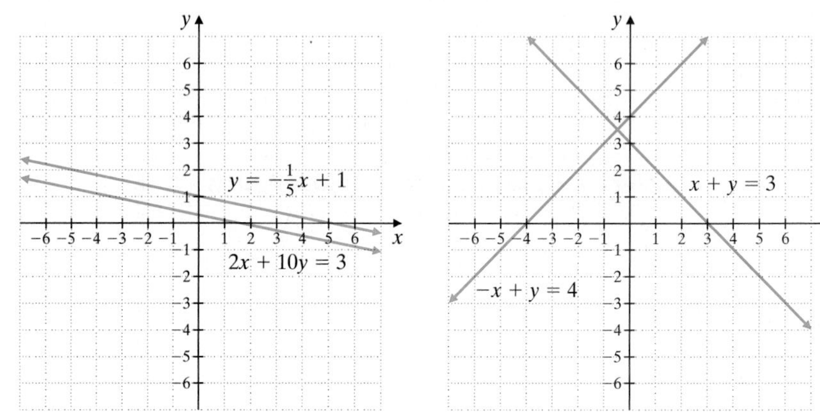

c. We solve each equation for y to find each slope. The slopes are -3 and $-\frac{2}{3}$. The slopes are not the same and their product is not -1. Thus, the lines are neither parallel nor perpendicular.

Concept Check

Write the equations of three parallel lines.

Try the Concept Check in the margin.

E Solving Problems of Slope

There are many real-world applications of slope. For example, the pitch of a roof used by builders and architects, is its slope. The pitch of the roof is $\frac{7}{10}\left(\frac{\text{rise}}{\text{run}}\right)$. This means that the roof rises vertically 7 feet for every horizontal 10 feet.

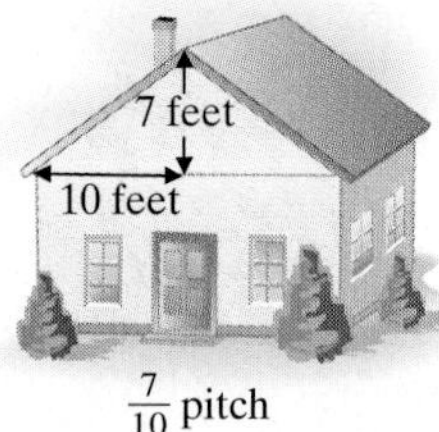

$\frac{7}{10}$ pitch

The grade of a road is its slope written as a percent. A 7% grade, as shown below, means that the road rises (or falls) 7 feet for every horizontal 100 feet. $\left(\text{Recall that } 7\% = \frac{7}{100}.\right)$

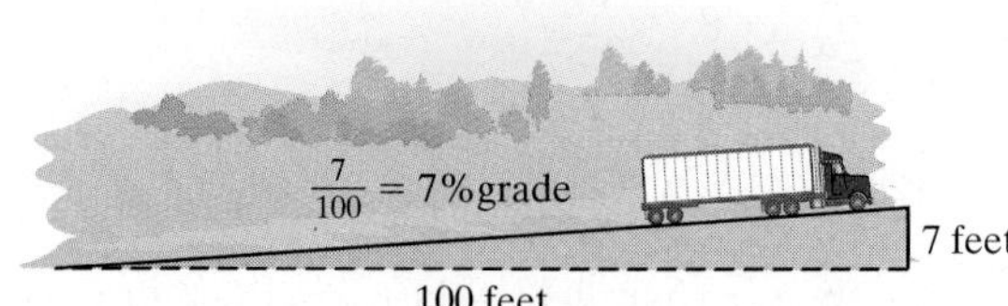

Answer

Concept Check: for example, $y = 2x - 3, y = 2x - 1, y = 2x$

EXAMPLE 7 Finding the Grade of a Road

At one part of the road to the summit of Pike's Peak, the road rises 15 feet for a horizontal distance of 250 feet. Find the grade of the road.

Solution: Recall that the grade of a road is its slope written as a percent.

$$\text{grade} = \frac{\text{rise}}{\text{run}} = \frac{15}{250} = 0.06 = 6\%$$

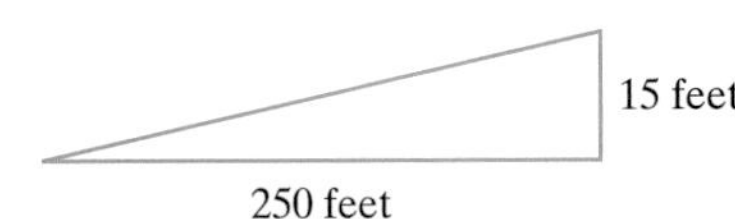

The grade is 6%.

Slope can also be interpreted as a rate of change. In other words, slope tells us how fast y is changing with respect to x.

EXAMPLE 8 Finding the Slope of a Line

The following graph shows the cost y (in cents) of an in-state long-distance telephone call in Massachusetts where x is the length of the call in minutes. Find the slope of the line and attach the proper units for the rate of change. Then write a sentence explaining the meaning of slope in this application.

Solution: Use $(2, 48)$ and $(5, 81)$ to calculate slope.

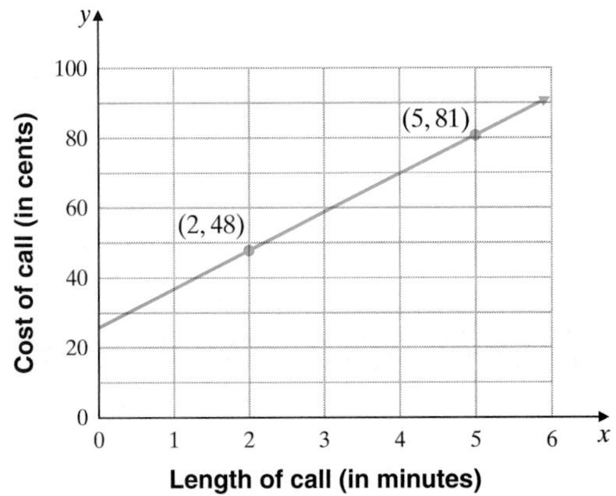

$$m = \frac{81 - 48}{5 - 2} = \frac{33}{3} = \frac{11 \text{ cents}}{1 \text{ minute}}$$

This means that the rate of change of a phone call is 11 cents per 1 minute or the cost of the phone call increases 11 cents per minute.

Practice Problem 7

Find the grade of the road shown.

Practice Problem 8

Find the slope of the line and write the slope as a rate of change. This graph represents annual food and drink sales y (in billions of dollars) for year x.

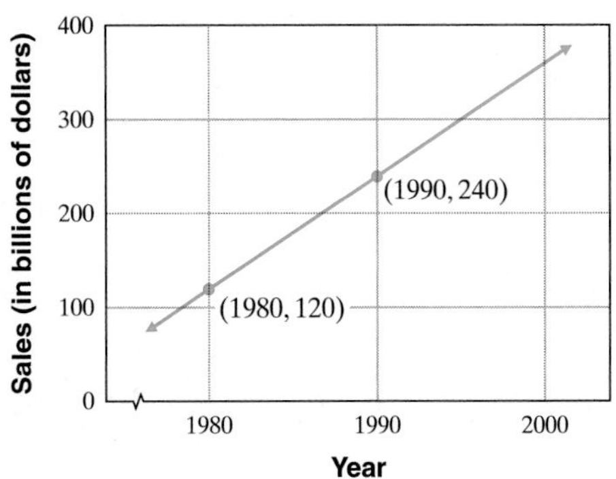

Answers

7. 15%, **8.** $m = 12$; Each year the sales of food and drink from restaurants increases by $12 billion dollars.

GRAPHING CALCULATOR EXPLORATIONS

It is possible to use a graphing calculator and sketch the graph of more than one equation on the same set of axes. This feature can be used to see parallel lines with the same slope. For example, graph the equations $y = \frac{2}{5}x$, $y = \frac{2}{5}x + 7$, and $y = \frac{2}{5}x - 4$ on the same set of axes. To do so, press the [Y=] key and enter the equations on the first three lines.

$$Y_1 = \left(\frac{2}{5}\right)x$$

$$Y_2 = \left(\frac{2}{5}\right)x + 7$$

$$Y_3 = \left(\frac{2}{5}\right)x - 4$$

The displayed equations should look like:

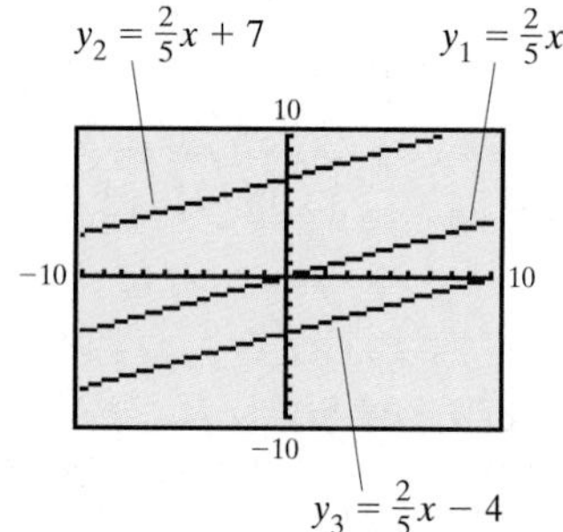

These lines are parallel as expected since they all have a slope of $\frac{2}{5}$. The graph of $y = \frac{2}{5}x + 7$ is the graph of $y = \frac{2}{5}x$ moved 7 units upward with a y-intercept of 7. Also, the graph of $y = \frac{2}{5}x - 4$ is the graph of $y = \frac{2}{5}x$ moved 4 units downward with a y-intercept of −4.

Graph the parallel lines on the same set of axes. Describe the similarities and differences in their graphs.

1. $y = 3.8x$, $y = 3.8x - 3$, $y = 3.8x + 9$

2. $y = -4.9x$, $y = -4.9x + 1$, $y = -4.9x + 8$

3. $y = \frac{1}{4}x$, $y = \frac{1}{4}x + 5$, $y = \frac{1}{4}x - 8$

4. $y = -\frac{3}{4}x$, $y = -\frac{3}{4}x - 5$, $y = -\frac{3}{4}x + 6$

Name ______________________ Section __________ Date __________

Mental Math

Decide whether a line with the given slope is upward-sloping, downward-sloping, horizontal, or vertical.

1. $m = \frac{7}{6}$ **2.** $m = -3$ **3.** $m = 0$ **4.** m is undefined

EXERCISE SET 3.4

Use the points shown on each graph to find the slope of each line. See Examples 1 and 2.

1.

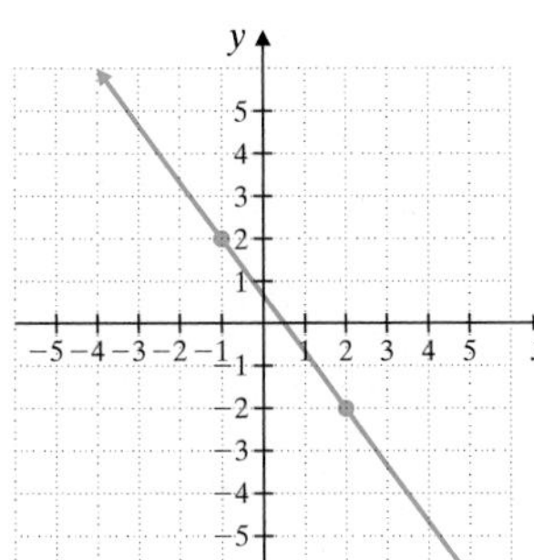

2.

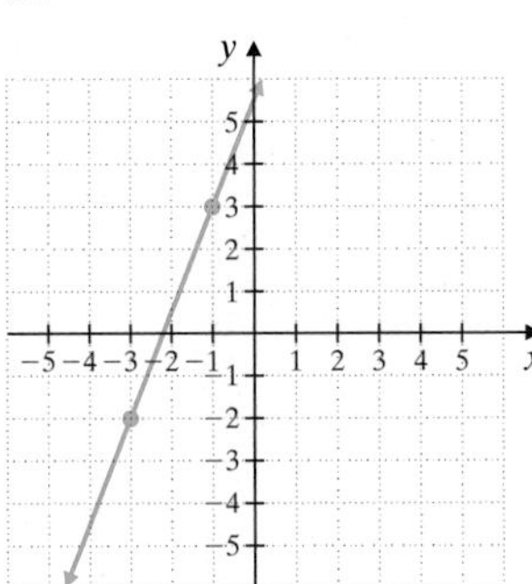

3.

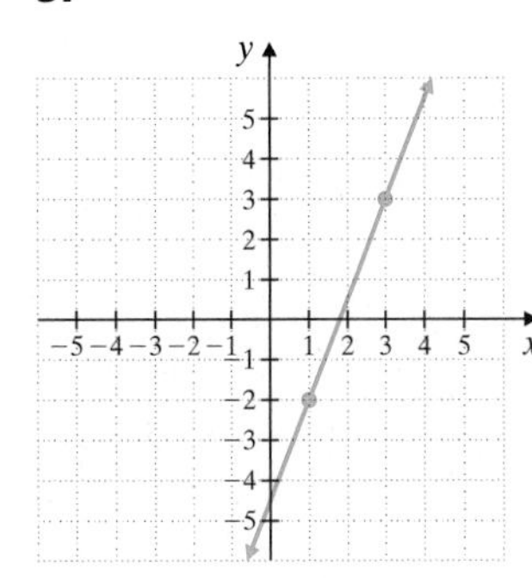

4.

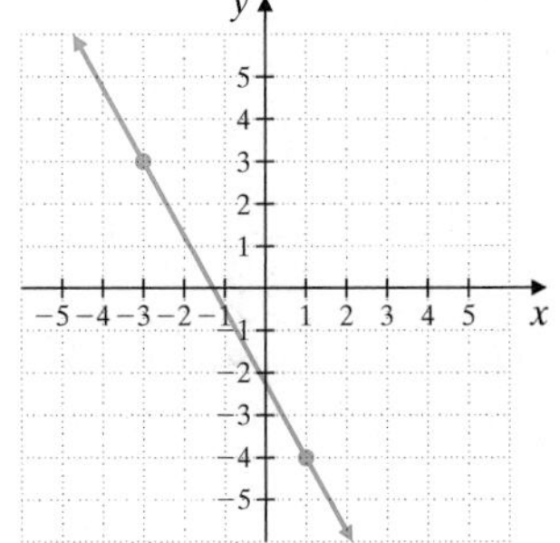

Find the slope of the line that passes through the given points. See Examples 1 and 2.

5. $(0, 0)$ and $(7, 8)$ **6.** $(-1, 5)$ and $(0, 0)$ **7.** $(-1, 5)$ and $(6, -2)$ **8.** $(-1, 9)$ and $(-3, 4)$

9. $(1, 4)$ and $(5, 3)$ **10.** $(3, 1)$ and $(2, 6)$ **11.** $(-2, 8)$ and $(1, 6)$ **12.** $(4, -3)$ and $(2, 2)$

13. $(5, 1)$ and $(-2, 1)$ **14.** $(5, 4)$ and $(5, 0)$

For each graph, determine which line has the larger slope.

15.

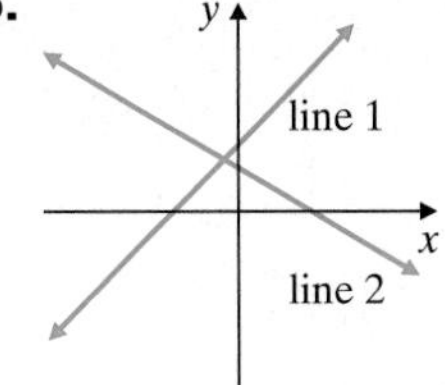

16.

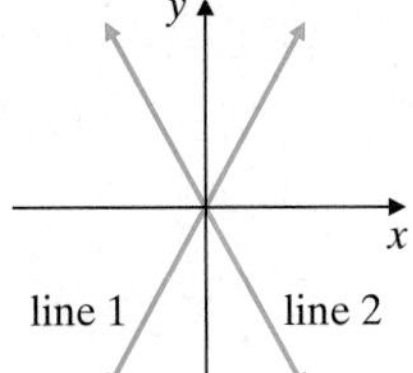

17.

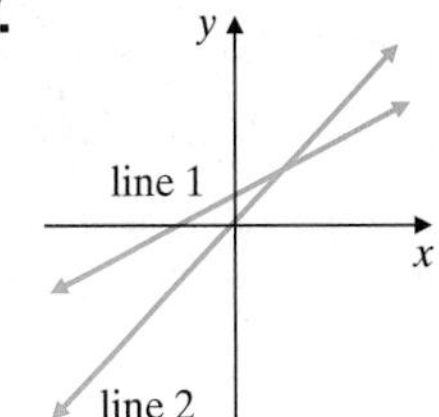

18.

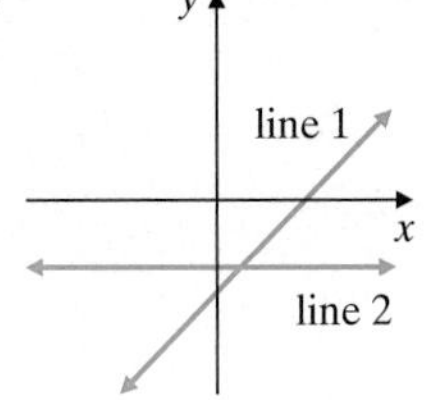

B *Find the slope of each line. See Example 3.*

19. $y = 5x - 2$

20. $y = -2x + 6$

21. $2x + y = 7$

22. $-5x + y = 10$

23. $2x - 3y = 10$

24. $-3x - 4y = 6$

25. $x = 2y$

26. $x = -4y$

C *Find the slope of each line. See Examples 4 and 5.*

27.

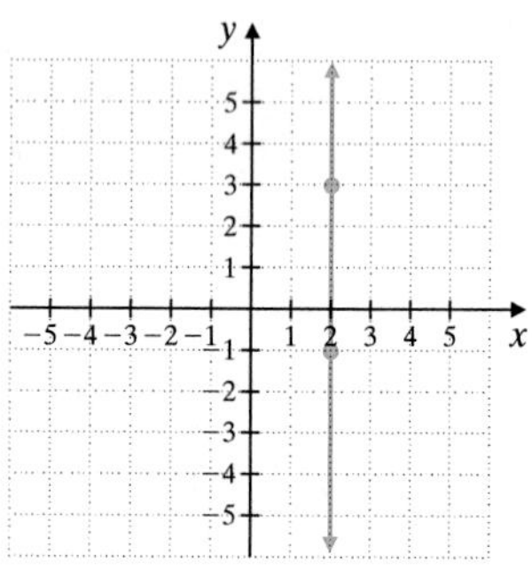

28.

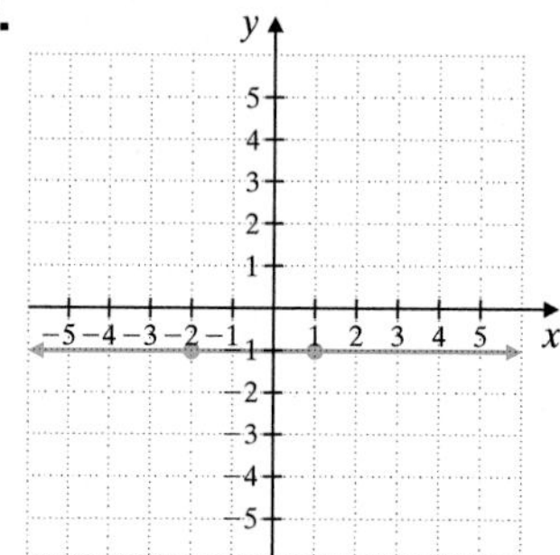

29. $x = 1$

30. $y = -2$

31. $y = -3$

32. $x = 5$

△ **D** *Determine whether each pair of lines is parallel, perpendicular, or neither. See Example 6.*

33. $x - 3y = -6$
$3x - y = 0$

34. $-5x + y = -6$
$x + 5y = 5$

35. $10 + 3x = 5y$
$5x + 3y = 1$

36. $y = 4x - 2$
$4x + y = 5$

37. $6x = 5y + 1$
$-12x + 10y = 1$

38. $-x + 2y = -2$
$2x = 4y + 3$

△ *Find the slope of the line that is (**a**) parallel and (**b**) perpendicular to the line through each pair of points. See Example 6.*

39. $(-3, -3)$ and $(0, 0)$

40. $(6, -2)$ and $(1, 4)$

41. $(-8, -4)$ and $(3, 5)$

42. $(6, -1)$ and $(-4, -10)$

E *The pitch of a roof is its slope. Find the pitch of each roof shown. See Example 7.*

43.

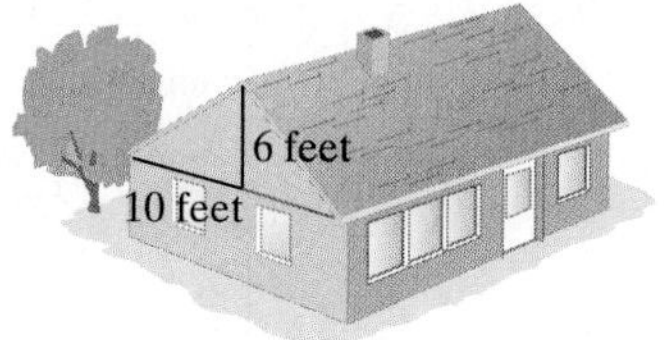

44.

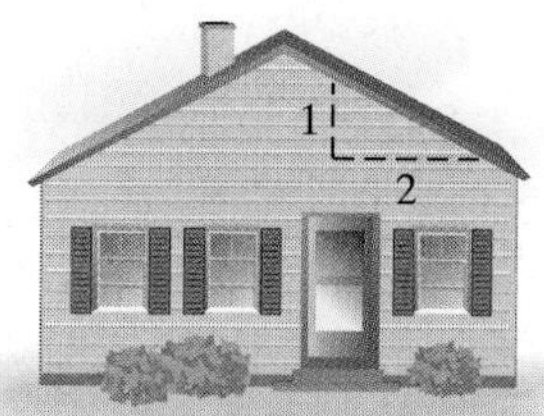

The grade of a road is its slope written as a percent. Find the grade of each road shown. See Example 7.

45.

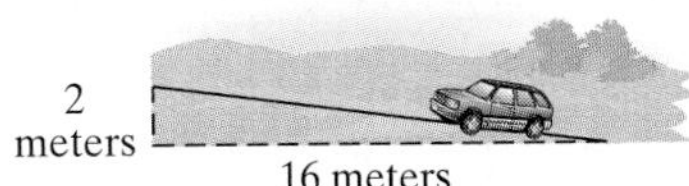

46.

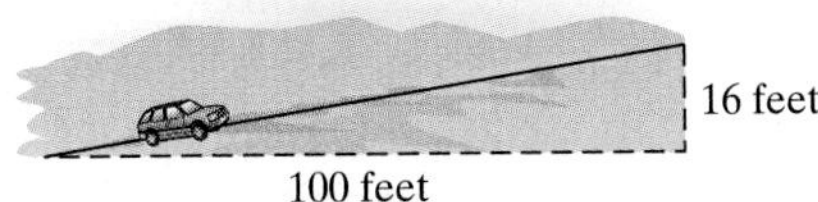

47. One of Japan's superconducting "bullet" trains is researched and tested at the Yamanashi Maglev Test Line near Otsuki City. The steepest section of the track has a rise of 2580 meters for a horizontal distance of 6450 meters. What is the grade for this section of track? (*Source:* Japan Railways Central Co.)

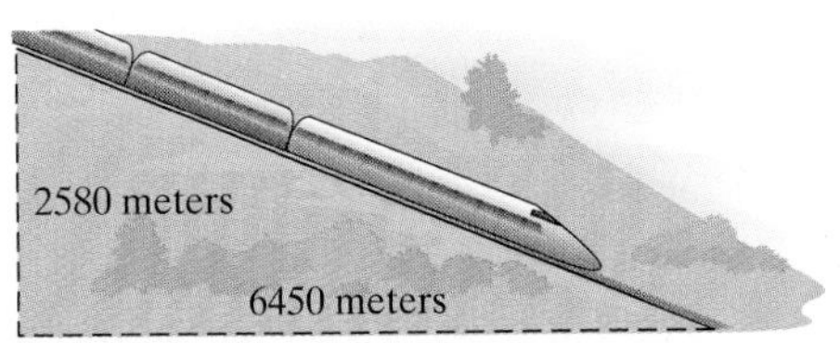

48. The steepest street is Baldwin Street in Dunedin, New Zealand. It has a maximum rise of 10 meters for a horizontal distance of 12.66 meters. Find the grade for this section of road. Round to the nearest whole percent. (*Source: The Guinness Book of Records*)

49. Professional plumbers suggest that a sewer pipe should rise 0.25 inch for every horizontal foot. Find the recommended slope for a sewer pipe. Round to the nearest hundredth.

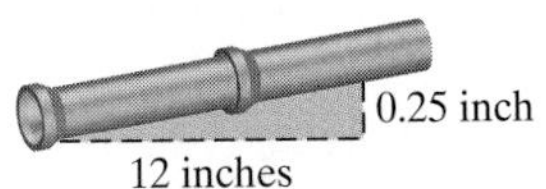

50. According to federal regulations, a wheelchair ramp should rise no more than 1 foot for a horizontal distance of 12 feet. Write the slope as a grade. Round to the nearest tenth of a percent.

Find the slope of each line and write the slope as a rate of change. Don't forget to attach the proper units. See Example 8.

51. This graph approximates the number of U.S. Internet users y (in millions) for year x.

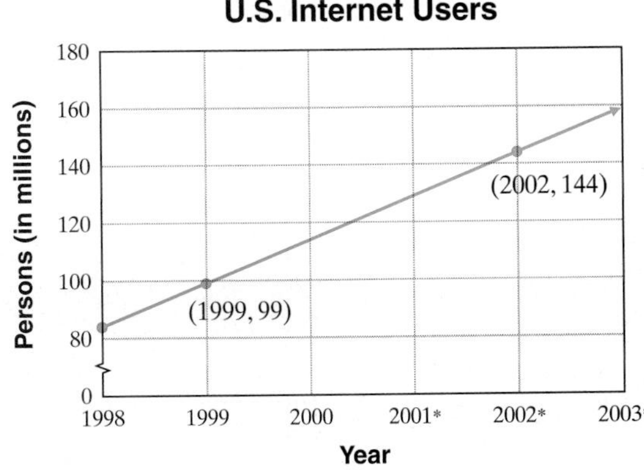

52. This graph approximates the total number of cosmetic surgeons for year x.

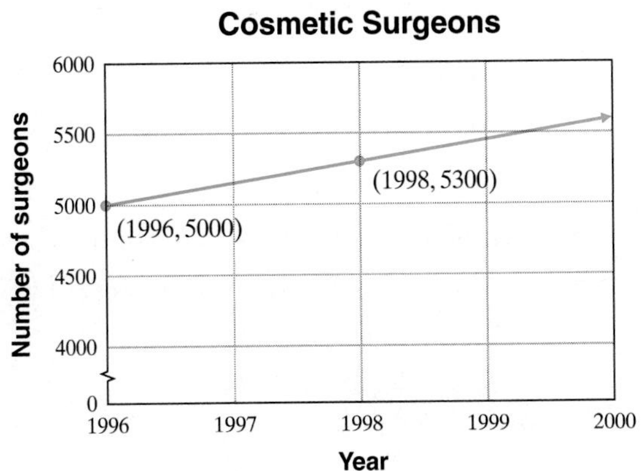

53. The graph below shows the total cost y (in dollars) of owning and operating a compact car where x is the number of miles driven.

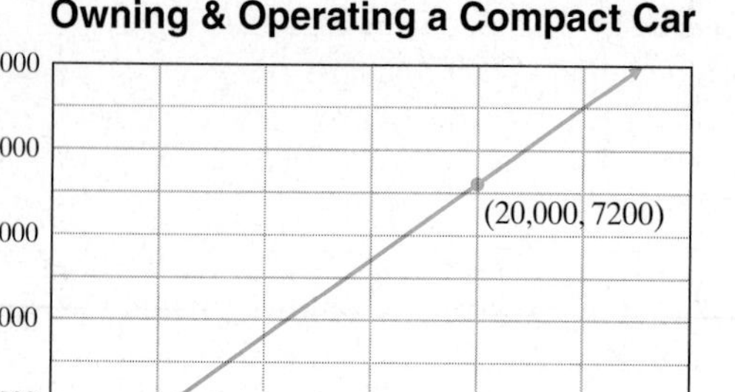

Source: Federal Highway Administration

54. The graph below shows the total cost y (in dollars) of owning and operating a full-size pickup truck, where x is the number of miles driven.

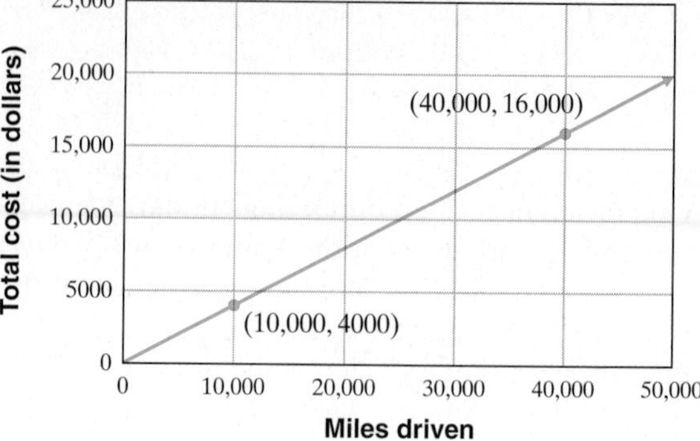

Source: Federal Highway Administration

Review and Preview

Solve each equation for y. See Section 2.6.

55. $y - (-6) = 2(x - 4)$ **56.** $y - 7 = -9(x - 6)$ **57.** $y - 1 = -6(x - (-2))$ **58.** $y - (-3) = 4(x - (-5))$

Combining Concepts

Match each line with its slope.

A. $m = 0$ **B.** undefined slope **C.** $m = 3$

D. $m = 1$ **E.** $m = -\frac{1}{2}$ **F.** $m = -\frac{3}{4}$

59.

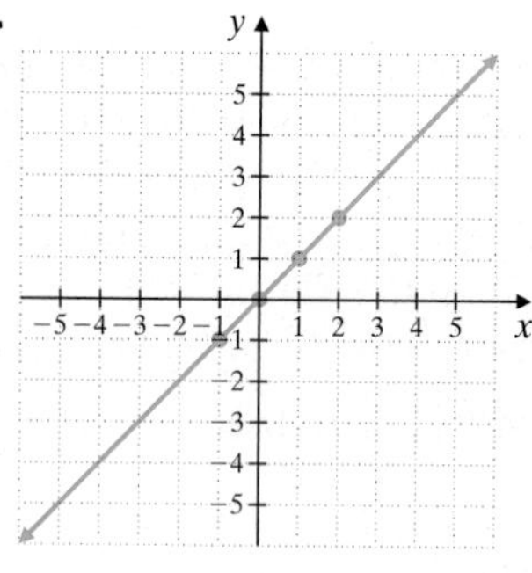

60.

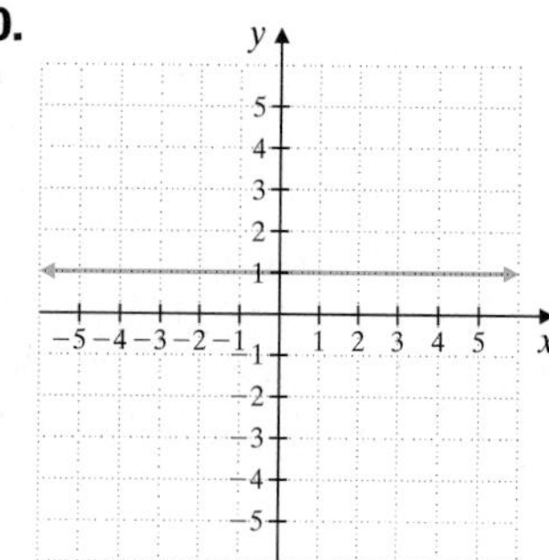

61.

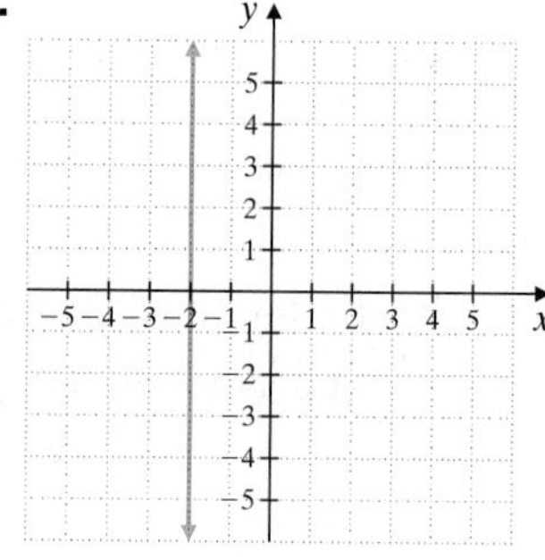

62.

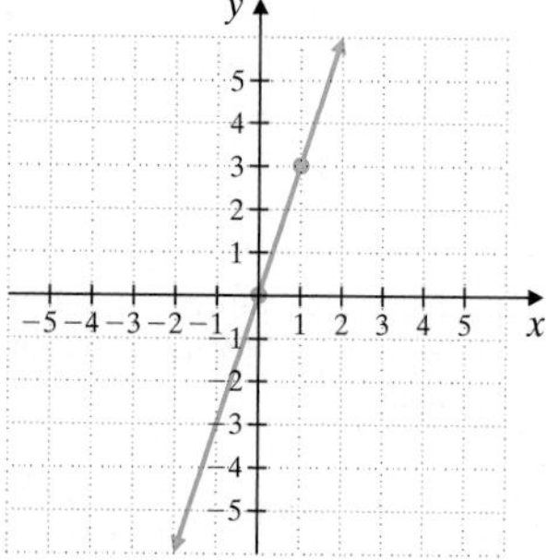

63.

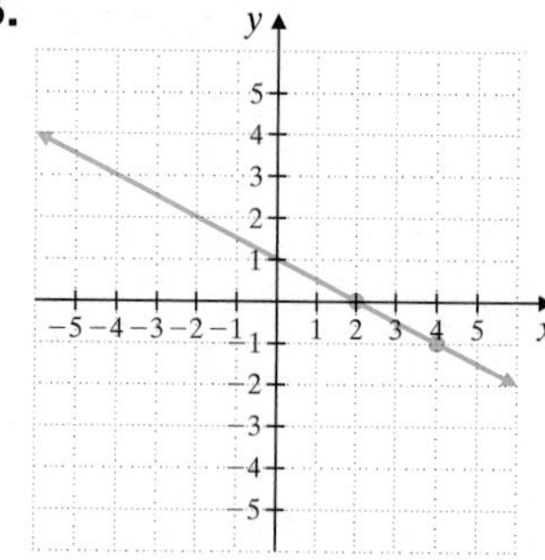

64.

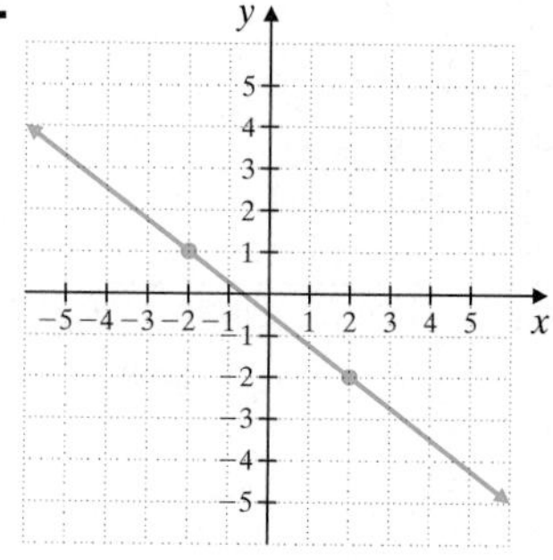

The following line graph shows the average fuel economy (in miles per gallon) by passenger automobiles produced during each of the model years shown. Use this graph to answer Exercises 65 through 70.

65. What was the average fuel economy (in miles per gallon) for automobiles produced during 1995?

66. Find the decrease in average fuel economy for automobiles for the years 1998 to 2000.

67. During which of the model years shown was average fuel economy the lowest?
What was the average fuel economy for that year?

68. During which of the model years shown was average fuel economy the highest?
What was the average fuel economy for that year?

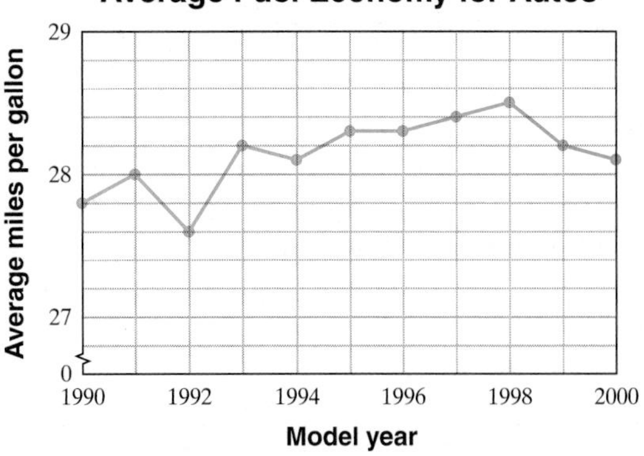

Source: U.S. Environmental Protection Agency, Office of Transportation and Air Quality

69. What line segment has the greatest slope?

70. What line segment has the smallest slope?

71. Find x so that the pitch of the roof is $\frac{1}{3}$.

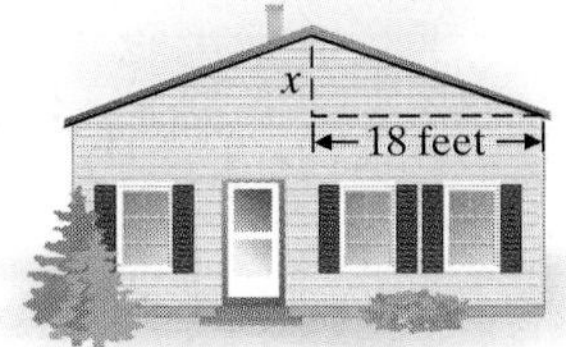

72. Find x so that the pitch of the roof is $\frac{2}{5}$.

73. The average price of an acre of U.S. farmland was \$782 in 1994. In 2001, the price of an acre rose to approximately \$1,132. (*Source:* National Agricultural Statistics Service)

a. Write two ordered pairs of the form (year, price of acre).

b. Find the slope of the line through the two points.

c. Write a sentence explaining the meaning of the slope as a rate of change.

74. There were approximately 9420 kidney transplants performed in the United States in 1990. In 2000, the number of kidney transplants performed in the United States rose to 13,290. (*Source:* Organ Procurement and Transplantation Network)

a. Write two ordered pairs of the form (year, number of kidney transplants).

b. Find the slope of the line between the two points.

c. Write a sentence explaining the meaning of the slope as a rate of change.

75. In the years 1998 through 2000, (the number of admissions to the movie theater in the U.S. and Canada can be modeled by the linear equation $y = -30x + 1485$ where x is years after 1998 and y is admissions in millions. (*Source:* Motion Picture Assn. of America)

a. Find the y-intercept of this line.

b. Write a sentence explaining the meaning of this intercept.

c. Find the slope of this line.

d. Write a sentence explaining the meaning of the slope as a rate of change.

76. The table below shows weight and fuel economy information for selected Dodge and Ford 2001 model year passenger vehicles. The linear equation $y = -0.003x + 34.7$ models the relationship between weight and fuel economy for these vehicles, where x = weight in pounds and y = combined city/highway fuel economy in miles per gallon. (*Sources*: Automotive Information Center, U.S. Environmental Protection Agency)

2001 Model	Weight (in pounds)	Combined Fuel Economy (in miles per gallon)
DODGE		
Caravan SE	3908	22
Grand Caravan ES	4252	20
Intrepid Sedan	3480	23
Dodge Neon	2635	27
Stratus Sedan SE	3227	23
FORD		
Focus SE Sedan	2564	31
Mustang Coupe	3114	23
Taurus SE Sedan	3354	22
Crown Victoria Sedan	3946	20
Escape XLS	2991	25
Expedition XLT	4891	18
Excursion XLT	6650	15
Windstar	4058	20

a. Find the y-intercept of this line.

b. Find the slope of this line.

c. Write a sentence explaining the meaning of the slope as a rate of change.

△ **77.** Show that a triangle with vertices at the points $(1, 1)$, $(-4, 4)$, and $(-3, 0)$ is a right triangle.

△ **78.** Show that the quadrilateral with vertices $(1, 3)$, $(2, 1)$, $(-4, 0)$, and $(-3, -2)$ is a parallelogram.

Find the slope of the line through the given points.

79. $(2.1, 6.7)$ and $(-8.3, 9.3)$

80. $(-3.8, 1.2)$ and $(-2.2, 4.5)$

81. $(2.3, 0.2)$ and $(7.9, 5.1)$

82. $(14.3, -10.1)$ and $(9.8, -2.9)$

83. The graph of $y = -\frac{1}{3}x + 2$ has a slope of $-\frac{1}{3}$. The graph of $y = -2x + 2$ has a slope of -2. The graph of $y = -4x + 2$ has a slope of -4. Graph all three equations on a single coordinate system. As the absolute value of the slope becomes larger, how does the steepness of the line change?

84. The graph of $y = \frac{1}{2}x$ has a slope of $\frac{1}{2}$. The graph of $y = 3x$ has a slope of 3. The graph of $y = 5x$ has a slope of 5. Graph all three equations on a single coordinate system. As slope becomes larger, how does the steepness of the line change?

Name ____________________ Section __________ Date __________

Integrated Review—Summary on Linear Equations

Find the slope of each line.

1.

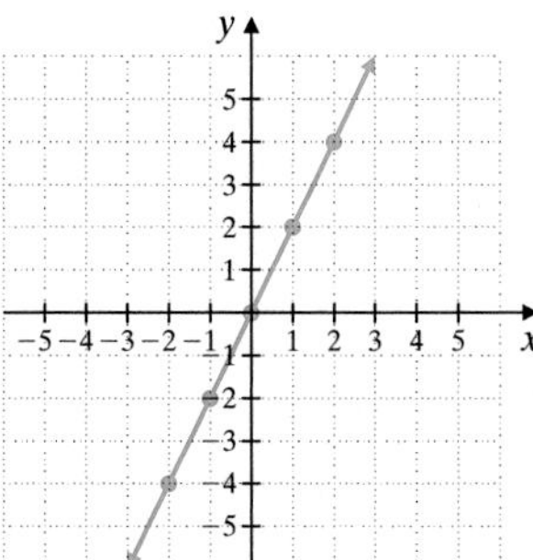

2.

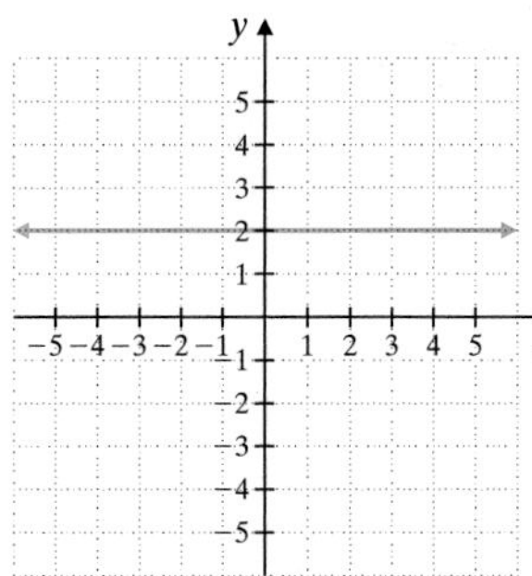

3.

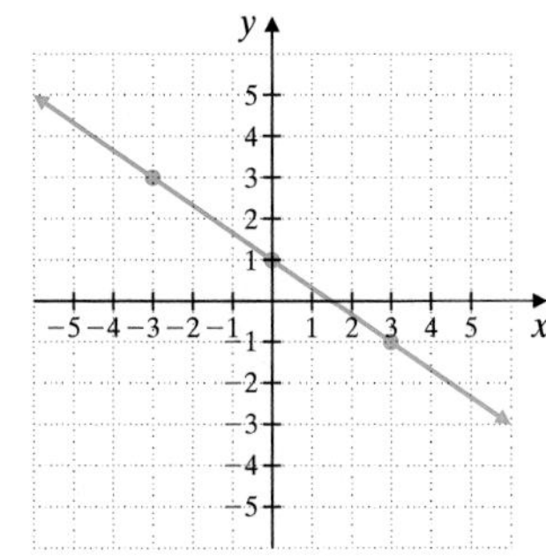

4.

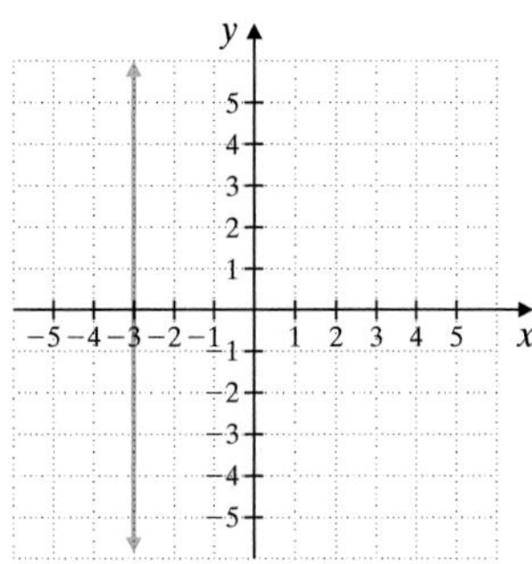

Graph each linear equation.

5. $y = -2x$

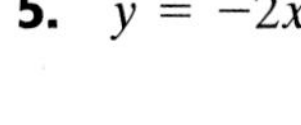

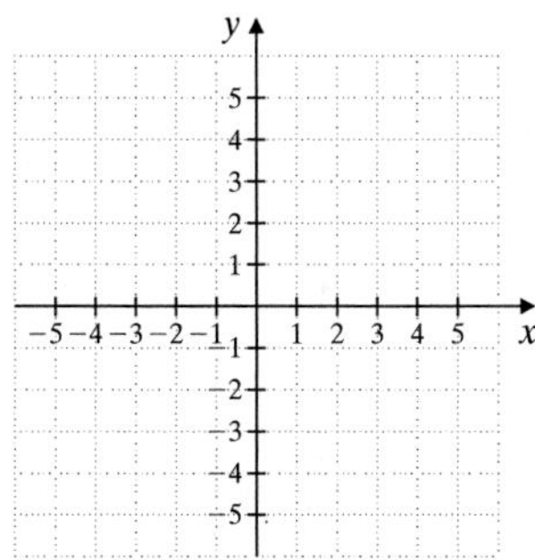

6. $x + y = 3$

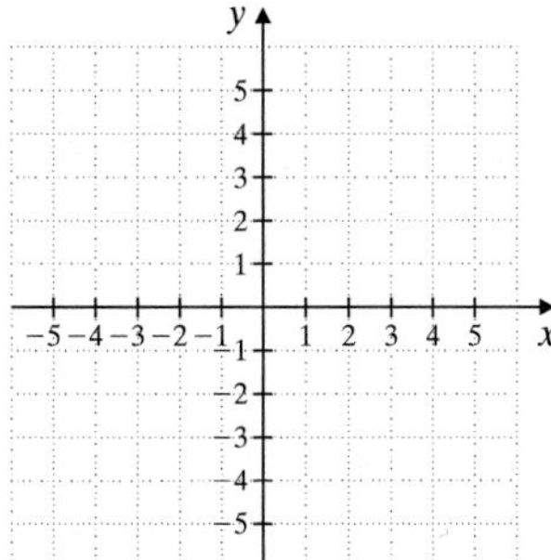

7. $x = -1$

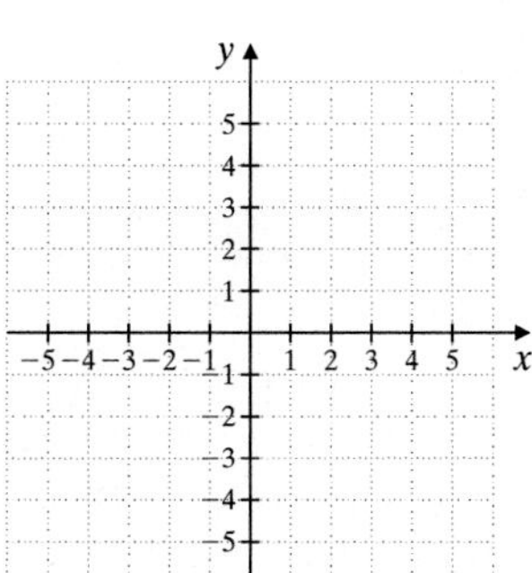

8. $y = 4$

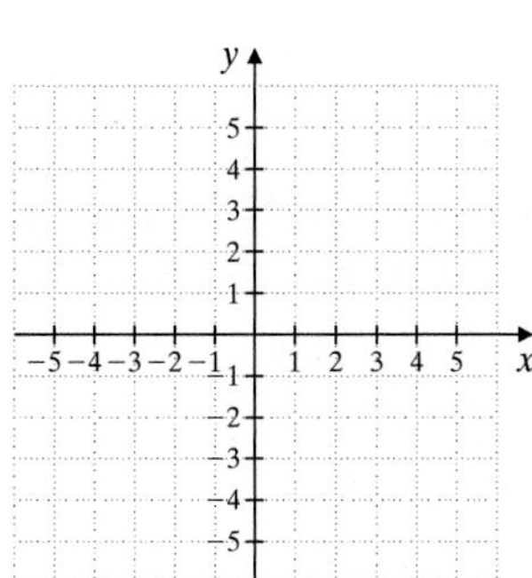

Answers

1. ______
2. ______
3. ______
4. ______
5. see graph
6. see graph
7. see graph
8. see graph

9. see graph
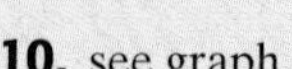

10. see graph

11. ____

12. ____

13. ____

14. ____

15. ____

16. ____

17. ____

18. ____

19. a. ____

b. ____

c. ____

9. $x - 2y = 6$

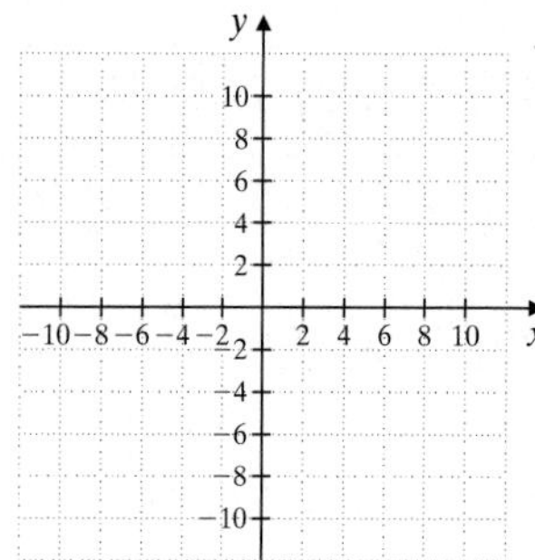

10. $y = 3x + 2$

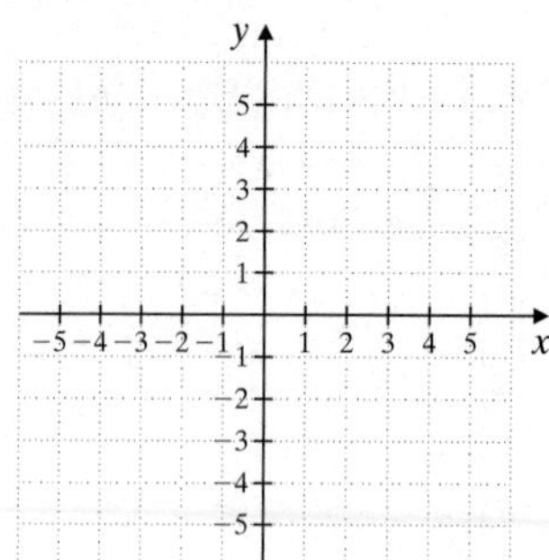

Find the slope of each line.

11. $y = 3x - 1$ **12.** $y = -6x + 2$ **13.** $7x + 2y = 11$ **14.** $2x - y = 0$

Find the slope of each line.

15. $x = 2$ **16.** $y = -4$

Determine whether each pair of lines is parallel, perpendicular, or neither.

17. $6x - y = 7$
$2x + 3y = 4$

18. $3x - 6y = 4$
$y = -2x$

19. Seventy-five percent of U.S. households own an outdoor barbecue grill. In 1997, the number of grill units shipped was 11.6 million. In 2001, the number of grill units shipped was 15.4 million. (*Source:* Barbecue Industry Assn.)

a. Write two ordered pairs of the form (year, millions of grill units shipped).

b. Find the slope of the line between the two points.

c. Write a sentence explaining the meaning of the slope as a rate of change.

3.5 Graphing Linear Inequalities in Two Variables

OBJECTIVES

A. Determine whether an ordered pair is a solution of a linear inequality in two variables.

B. Graph a linear inequality in two variables.

Recall that a linear equation in two variables is an equation that can be written in the form $Ax + By = C$ where A, B, and C are real numbers and A and B are not both 0. A **linear inequality in two variables** is an inequality that can be written in one of the forms

$$Ax + By < C \qquad Ax + By \le C$$
$$Ax + By > C \qquad Ax + By \ge C$$

where A, B, and C are real numbers and A and B are not both 0.

A Determining Solutions of Linear Inequalities in Two Variables

Just as for linear equations in x and y, an ordered pair is a **solution** of an inequality in x and y if replacing the variables with the coordinates of the ordered pair results in a true statement.

EXAMPLE 1

Determine whether each ordered pair is a solution of the equation $2x - y < 6$.

a. $(5, -1)$ **b.** $(2, 7)$

Solution:

a. We replace x with 5 and y with -1 and see if a true statement results.

$$2x - y < 6$$
$$2(5) - (-1) < 6 \quad \text{Replace } x \text{ with 5 and } y \text{ with } -1.$$
$$10 + 1 < 6$$
$$11 < 6 \quad \text{False}$$

The ordered pair $(5, -1)$ is not a solution since $11 < 6$ is a false statement.

b. We replace x with 2 and y with 7 and see if a true statement results.

$$2x - y < 6$$
$$2(2) - (7) < 6 \quad \text{Replace } x \text{ with 2 and } y \text{ with 7.}$$
$$4 - 7 < 6$$
$$-3 < 6 \quad \text{True}$$

The ordered pair $(2, 7)$ is a solution since $-3 < 6$ is a true statement. ●

Practice Problem 1

Determine whether each ordered pair is a solution of $x - 4y > 8$.

a. $(-3, 2)$ b. $(9, 0)$

B Graphing Linear Inequalities in Two Variables

The linear equation $x - y = 1$ is graphed next. Recall that all points on the line correspond to ordered pairs that satisfy the equation $x - y = 1$.

Notice the line defined by $x - y = 1$ divides the rectangular coordinate system plane into 2 sides. All points on one side of the line satisfy the inequality $x - y < 1$ and all points on the other side satisfy the inequality $x - y > 1$. The graph on the next page shows a few examples of this.

Answers

1. a. no, **b.** yes

$x - y$	< 1
$1 - 3$	< 1 True
$-2 - 1$	< 1 True
$-4 - (-4)$	< 1 True

$x - y$	> 1
$4 - 1$	> 1 True
$2 - (-2)$	> 1 True
$0 - (-4)$	> 1 True

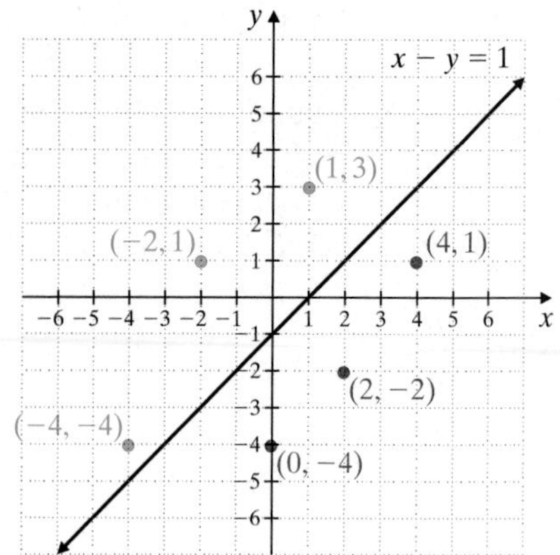

The graph of $x - y < 1$ is the region shaded blue and the graph of $x - y > 1$ is the region shaded red below.

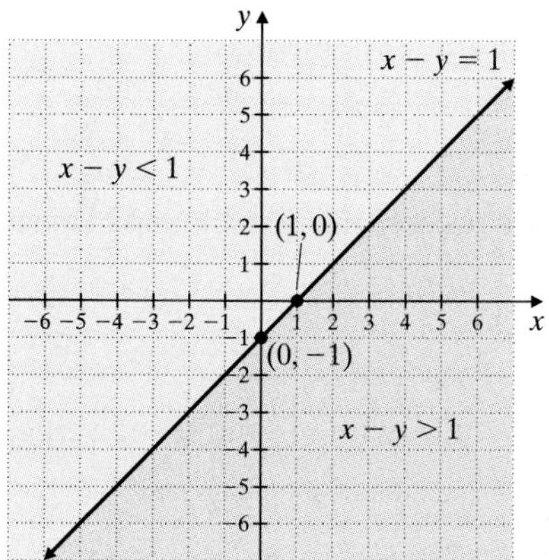

The region to the left of the line and the region to the right of the line are called **half-planes**. Every line divides the plane (similar to a sheet of paper extending indefinitely in all directions) into two half-planes; the line is called the **boundary**.

Recall that the inequality $x - y \leq 1$ means

$$x - y = 1 \quad \text{or} \quad x - y < 1$$

Thus, the graph of $x - y \leq 1$ is the half-plane $x - y < 1$ along with the boundary line $x - y = 1$.

To Graph a Linear Inequality in Two Variables

Step 1. Graph the boundary line found by replacing the inequality sign with an equal sign. If the inequality sign is $>$ or $<$, graph a dashed boundary line (indicating that the points on the line are not solutions of the inequality). If the inequality sign is $\geq$ or $\leq$, graph a solid boundary line (indicating that the points on the line are solutions of the inequality).

Step 2. Choose a point, *not* on the boundary line, as a test point. Substitute the coordinates of this test point into the *original* inequality.

Step 3. If a true statement is obtained in Step 2, shade the half-plane that contains the test point. If a false statement is obtained, shade the half-plane that does not contain the test point.

EXAMPLE 2 Graph: $x + y < 7$

Solution:

Step 1. First we graph the boundary line by graphing the equation $x + y = 7$. We graph this boundary as a *dashed line* because the inequality sign is $<$, and thus the points on the line are not solutions of the inequality $x + y < 7$.

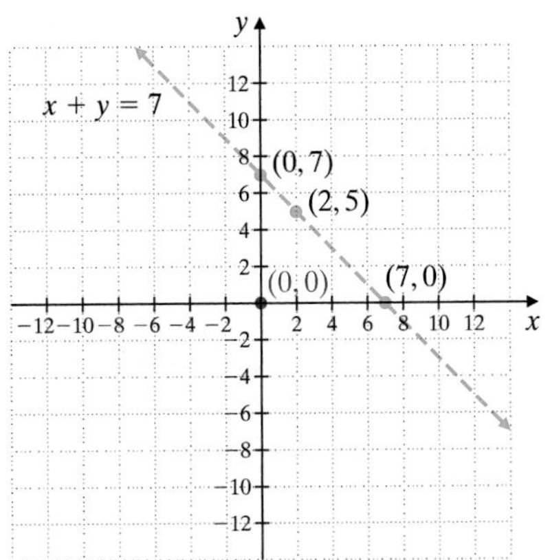

Step 2. Next we choose a test point, being careful *not* to choose a point on the boundary line. We choose $(0, 0)$, and substitute the coordinates of $(0, 0)$ into $x + y < 7$.

$x + y < 7$ Original inequality

$0 + 0 < 7$ Replace x with 0 and y with 0.

$0 < 7$ True

Step 3. Since the result is a true statement, $(0, 0)$ is a solution of $x + y < 7$, and every point in the same half-plane as $(0, 0)$ is also a solution. To indicate this, we shade the entire half-plane containing $(0, 0)$, as shown.

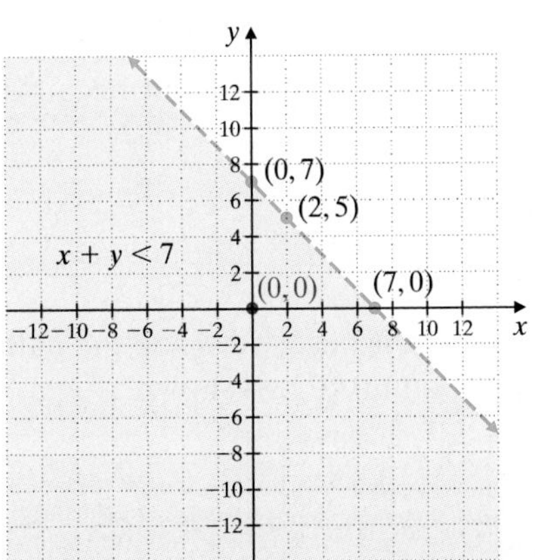

EXAMPLE 3 Graph: $2x - y \geq 3$

Solution:

Step 1. We graph the boundary line by graphing $2x - y = 3$. We draw this line as a solid line because the inequality sign is $\geq$, and thus the points on the line are solutions of $2x - y \geq 3$.

Step 2. Once again, $(0, 0)$ is a convenient test point since it is not on the boundary line.

Practice Problem 2

Graph: $x - y > 3$

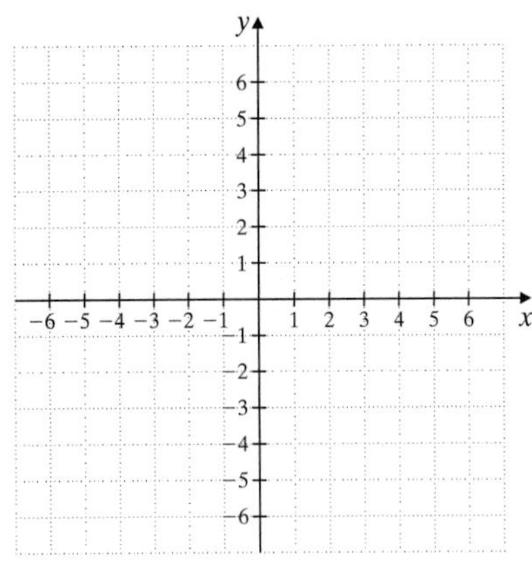

Practice Problem 3

Graph: $x - 4y \leq 4$

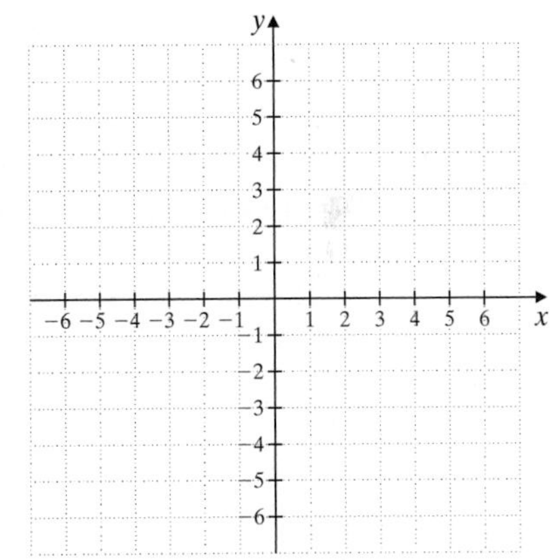

Answers

2.

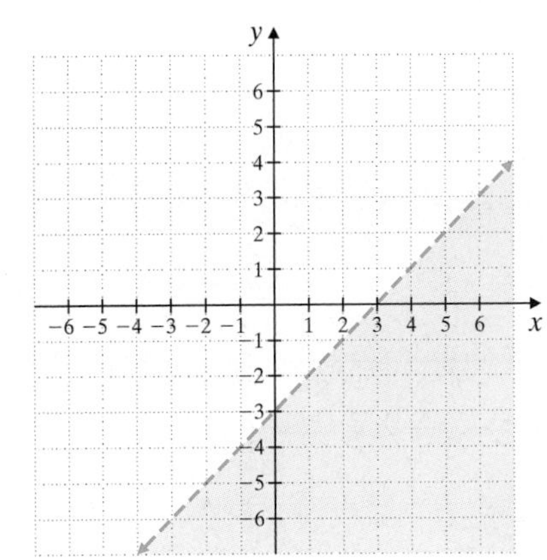

3. See page 246.

Practice Problem 4

Graph: $y < 3x$

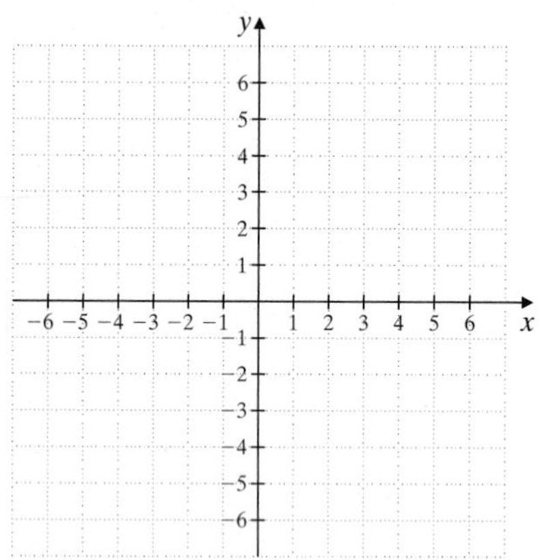

We substitute 0 for x and 0 for y into the original inequality.

$$2x - y \geq 3$$
$$2(0) - 0 \geq 3 \quad \text{Let } x = 0 \text{ and } y = 0.$$
$$0 \geq 3 \quad \text{False}$$

Step 3. Since the statement is false, no point in the half-plane containing $(0, 0)$ is a solution. Therefore, we shade the half-plane that does not contain $(0, 0)$. Every point in the shaded half-plane and every point on the boundary line is a solution of $2x - y \geq 3$.

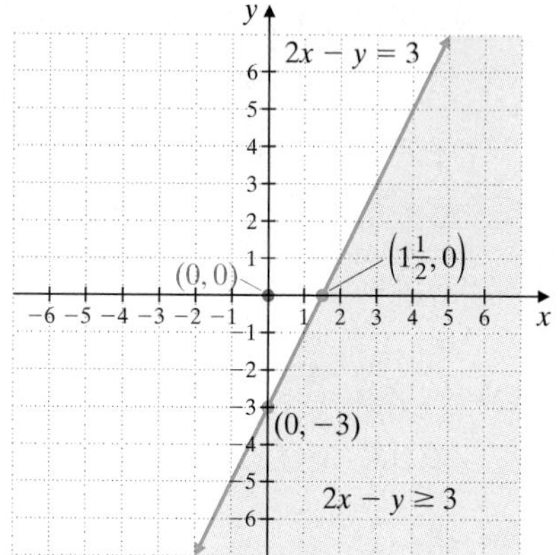

Helpful Hint

When graphing an inequality, make sure the test point is substituted into the **original inequality**. For Example 3, we substituted the test point $(0, 0)$ into the **original inequality** $2x - y \geq 3$, *not* $2x - y = 3$.

EXAMPLE 4 Graph: $x > 2y$

Solution:

Step 1. We find the boundary line by graphing $x = 2y$. The boundary line is a dashed line since the inequality symbol is $>$.

Step 2. We cannot use $(0, 0)$ as a test point because it is a point on the boundary line. We choose instead $(0, 2)$.

$$x > 2y$$
$$0 > 2(2) \quad \text{Let } x = 0 \text{ and } y = 2.$$
$$0 > 4 \quad \text{False}$$

Step 3. Since the statement is false, we shade the half-plane that does not contain the test point $(0, 2)$, as shown.

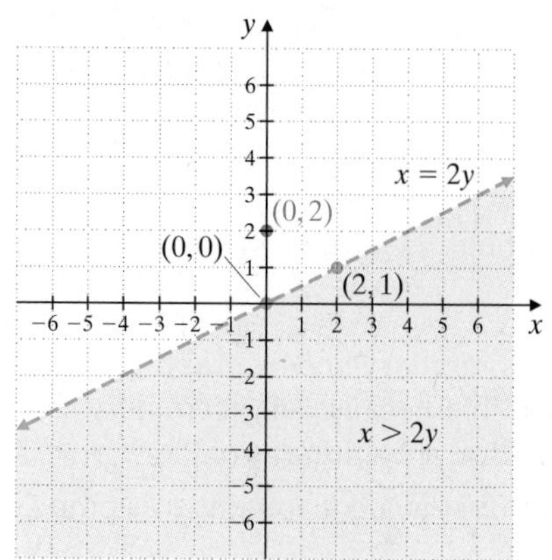

Answers

3.

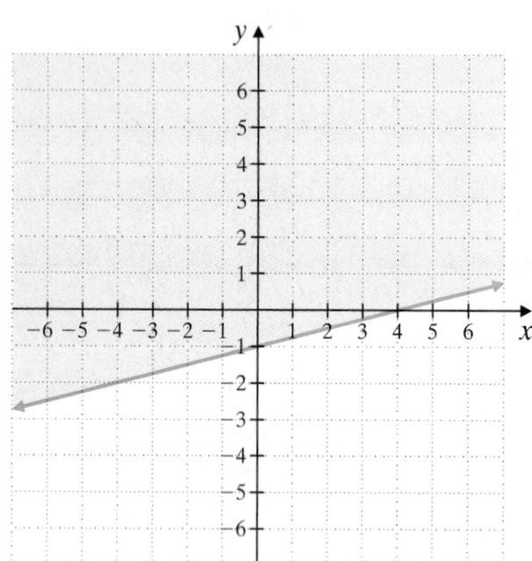

4.

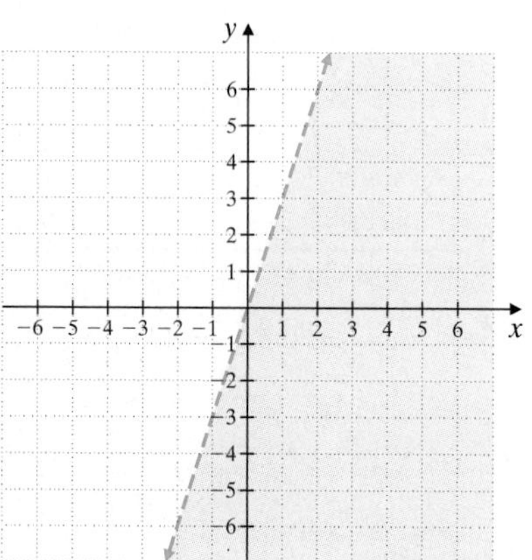

EXAMPLE 5 Graph: $5x + 4y \le 20$

Solution: We graph the solid boundary line $5x + 4y = 20$ and choose $(0, 0)$ as the test point.

$$5x + 4y \le 20$$
$$5(0) + 4(0) \le 20 \quad \text{Let } x = 0 \text{ and } y = 0.$$
$$0 \le 20 \quad \text{True}$$

We shade the half-plane that contains $(0, 0)$, as shown.

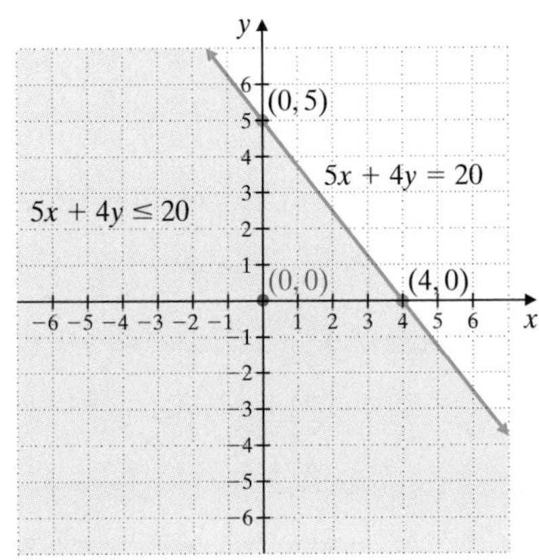

EXAMPLE 6 Graph: $y > 3$

Solution: We graph the dashed boundary line $y = 3$ and choose $(0, 0)$ as the test point. (Recall that the graph of $y = 3$ is a horizontal line with y-intercept 3.)

$$y > 3$$
$$0 > 3 \quad \text{Let } y = 0.$$
$$0 > 3 \quad \text{False}$$

We shade the half-plane that does not contain $(0, 0)$, as shown.

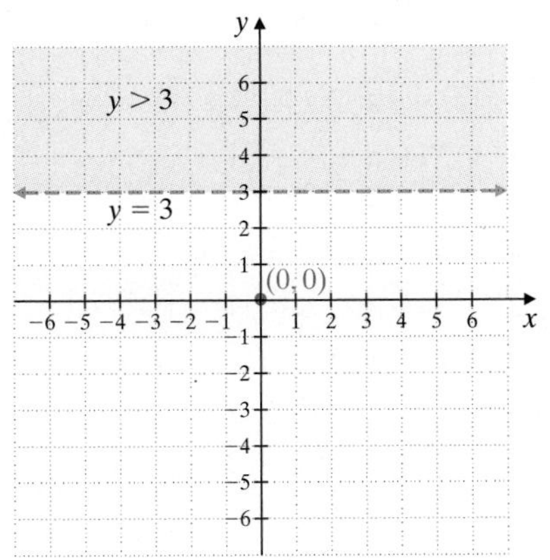

Practice Problem 5

Graph: $3x + 2y \ge 12$

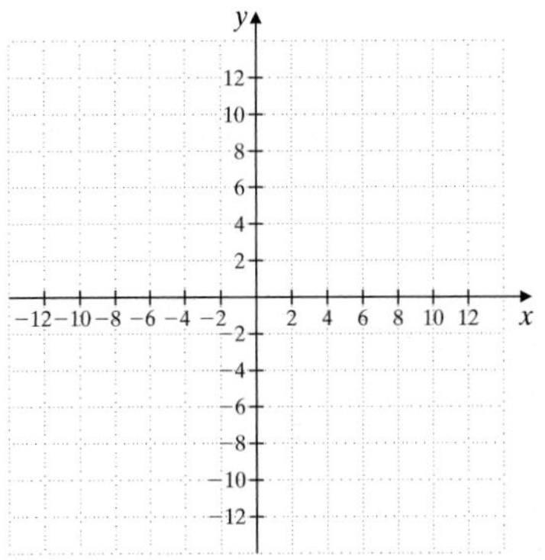

Practice Problem 6

Graph: $x < 2$

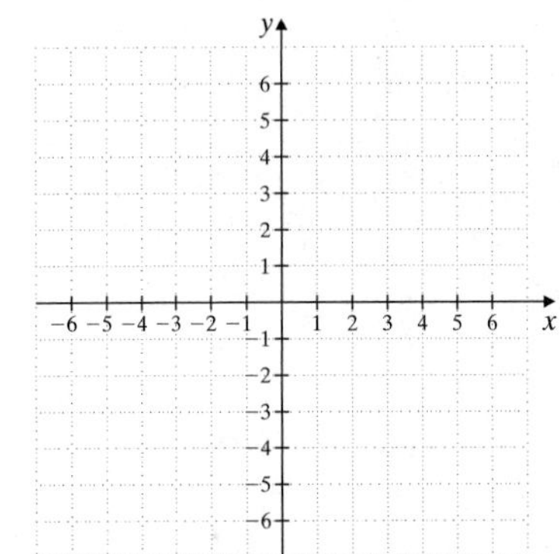

Answers

5.

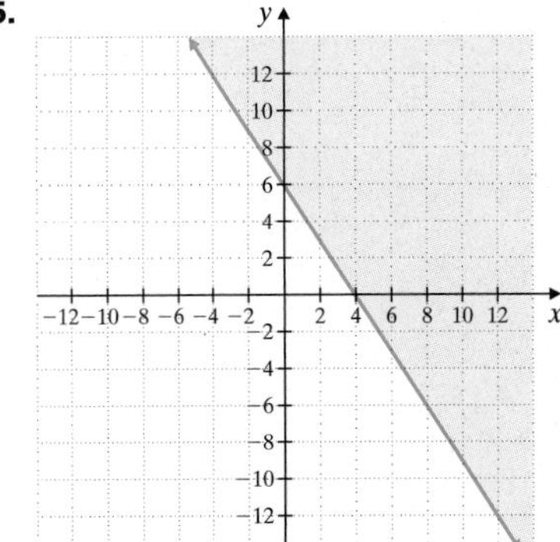

6.

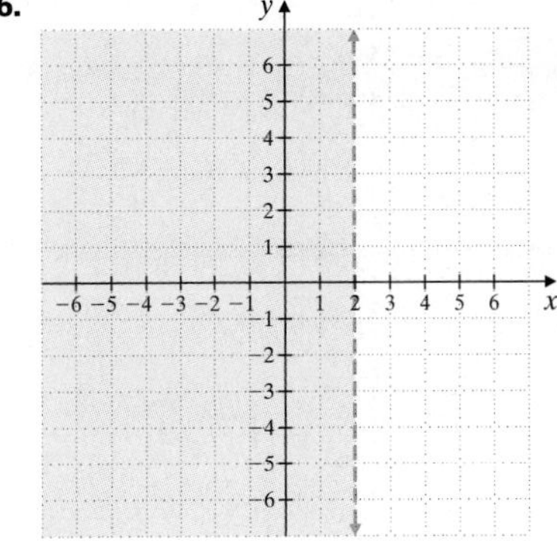

Practice Problem 7

Graph: $y \geq \frac{1}{4}x + 3$

EXAMPLE 7 Graph: $y \leq \frac{2}{3}x - 4$

Solution: Graph the solid boundary line $y = \frac{2}{3}x - 4$. This equation is in slope-intercept form with slope $\frac{2}{3}$ and y-intercept -4.

We use this information to graph the line. Then we choose $(0, 0)$ as our test point.

$$y < \frac{2}{3}x - 4$$

$$0 < \frac{2}{3} \cdot 0 - 4$$

$$0 < -4 \quad \text{False}$$

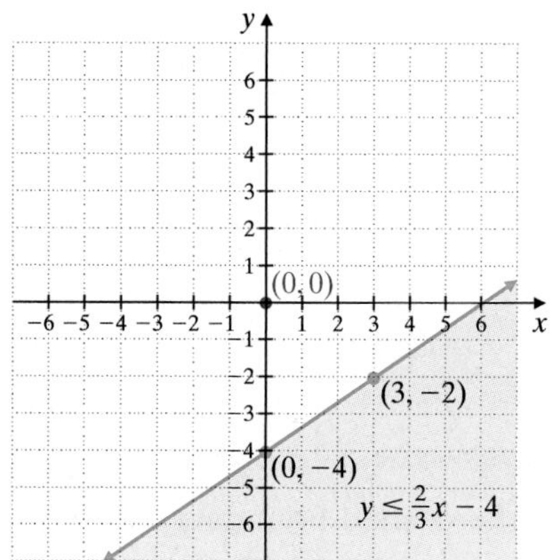

We shade the half-plane that does not contain $(0, 0)$, as shown.

Answer

7.

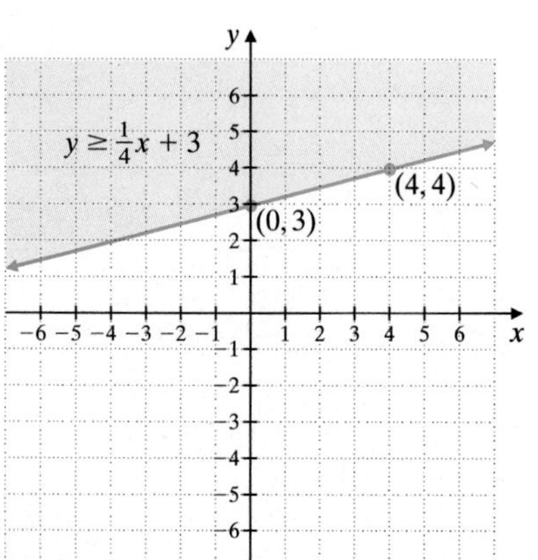

Name ______________________ Section __________ Date __________

Mental Math

State whether the graph of each inequality includes its corresponding boundary line.

1. $y \geq x + 4$ **2.** $x - y > -7$ **3.** $y \geq x$ **4.** $x > 0$

Decide whether $(0, 0)$ is a solution of each given inequality.

5. $x + y > -5$ **6.** $2x + 3y < 10$ **7.** $x - y \leq -1$ **8.** $\frac{2}{3}x + \frac{5}{6}y > 4$

EXERCISE SET 3.5

A *Determine whether the ordered pairs given are solutions of the linear inequality in two variables. See Example 1.*

1. $x - y > 3$; $(0, 3), (2, -1)$ **2.** $y - x < -2$; $(2, 1), (5, -1)$ **3.** $3x - 5y \leq -4$; $(2, 3), (-1, -1)$

4. $2x + y \geq 10$; $(0, 11), (5, 0)$ **5.** $x < -y$; $(0, 2), (-5, 1)$ **6.** $y > 3x$; $(0, 0), (1, 4)$

B *Graph each inequality. See Examples 2 through 7.*

7. $x + y \leq 1$

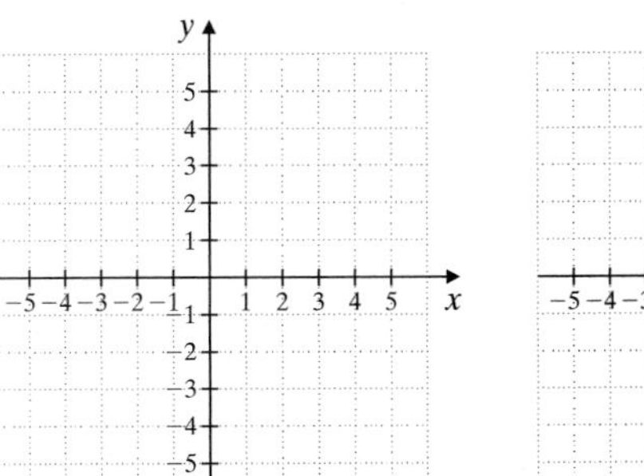

8. $x + y \geq -2$

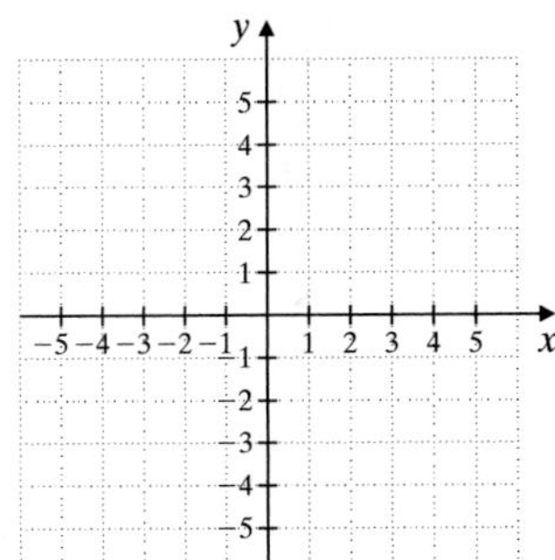

9. $2x - y > -4$

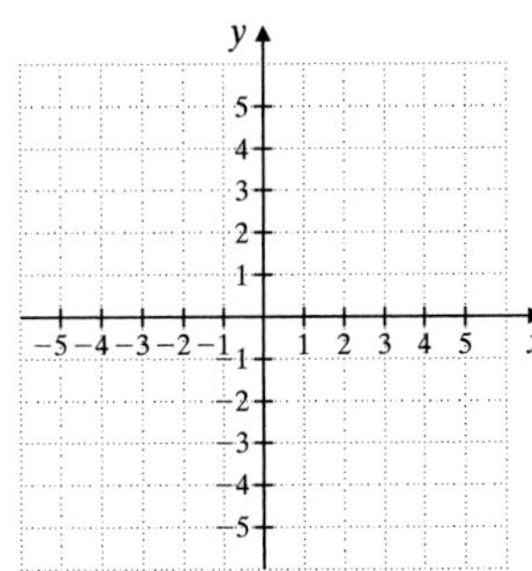

10. $x - 3y < 3$

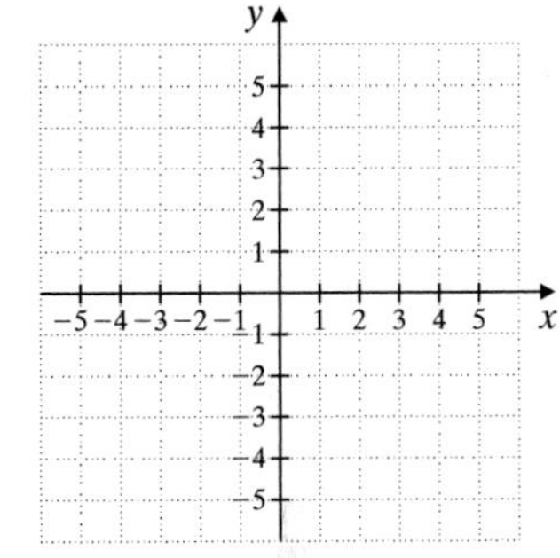

11. $y > 2x$

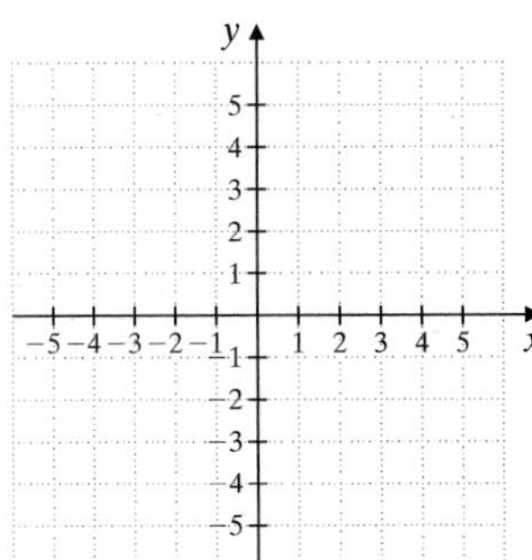

12. $y < 3x$

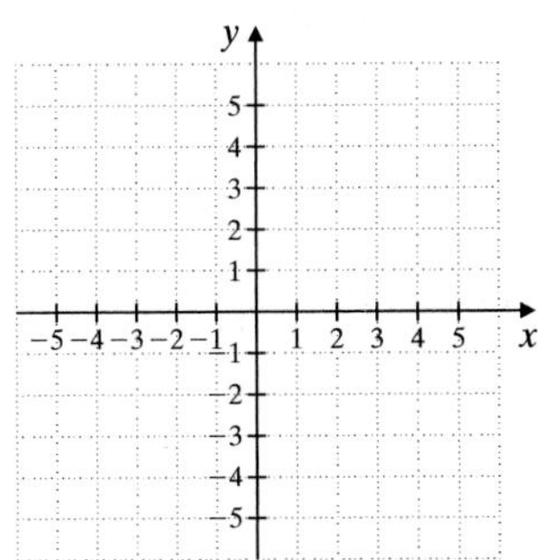

13. $x \leq -3y$

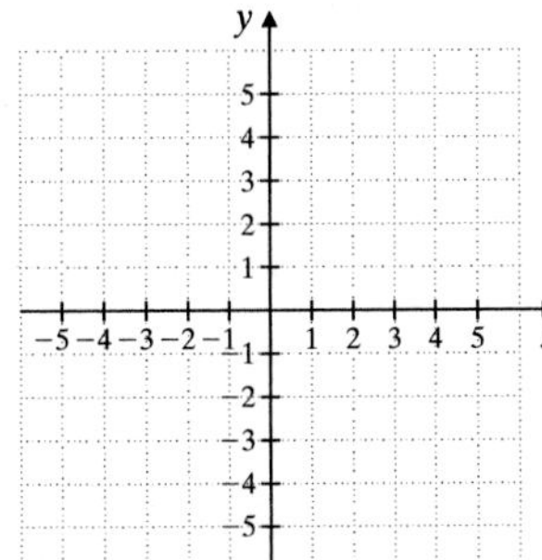

14. $x \geq -2y$

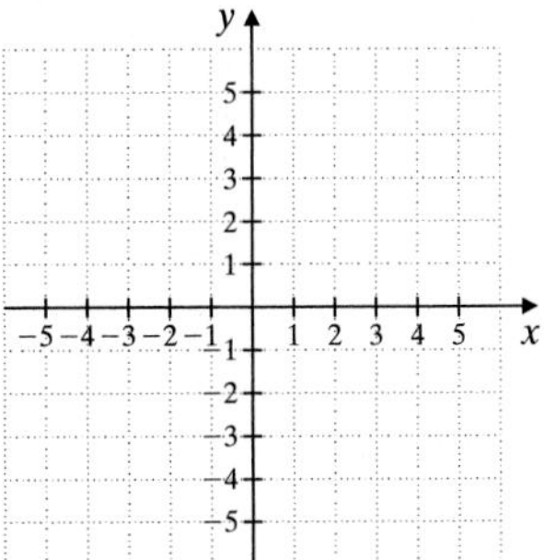

15. $y \geq x + 5$

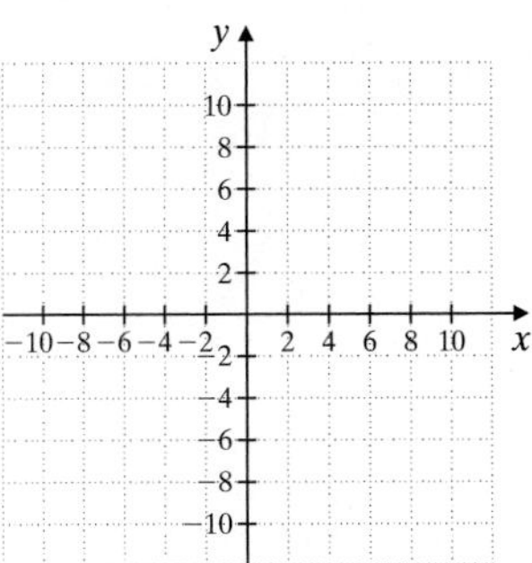

16. $y \leq x + 1$

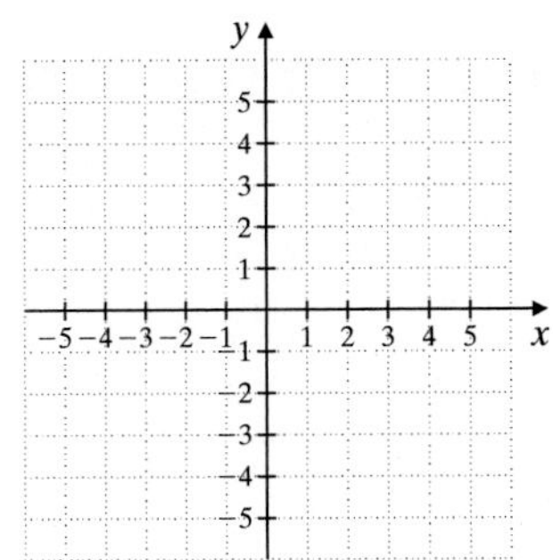

17. $y < 4$

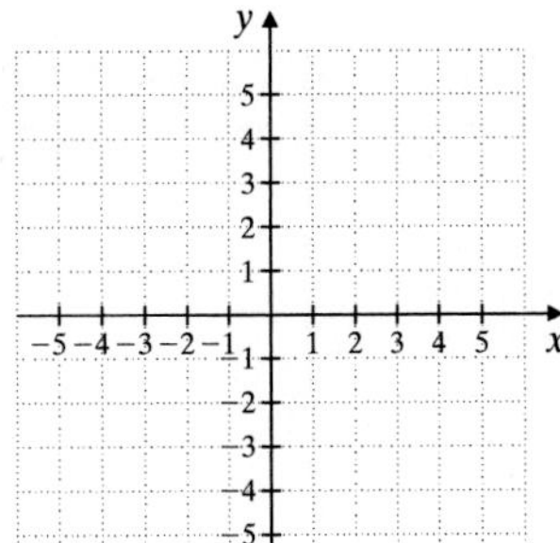

18. $y > 2$

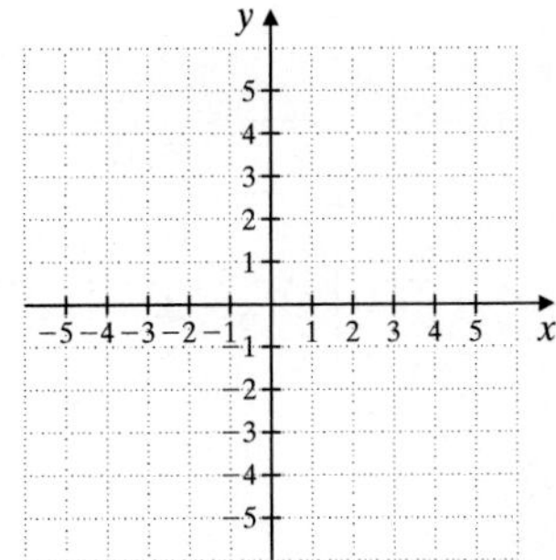

19. $x \geq -3$

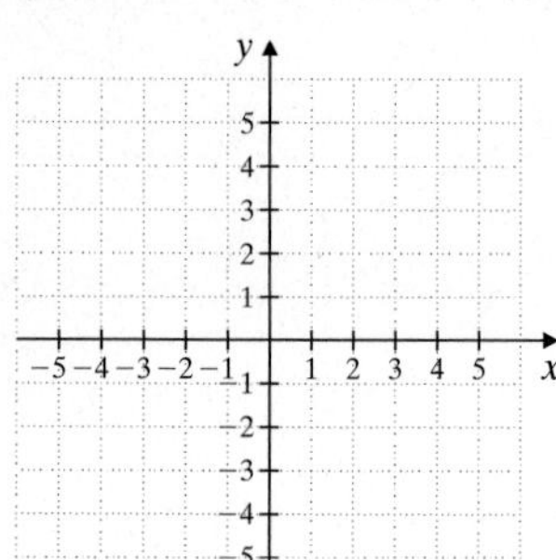

20. $x \leq -1$

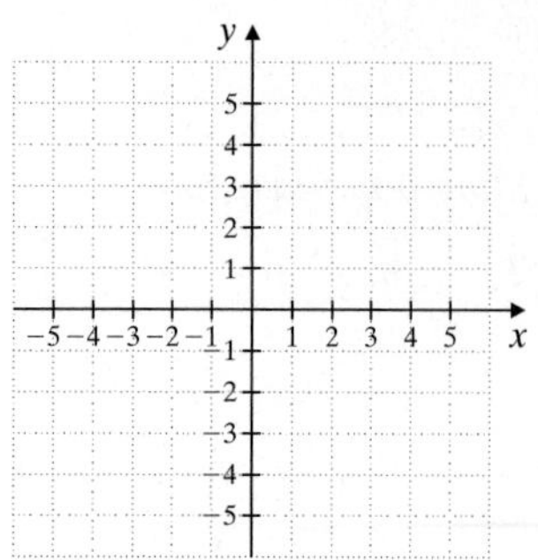

21. $5x + 2y \leq 10$

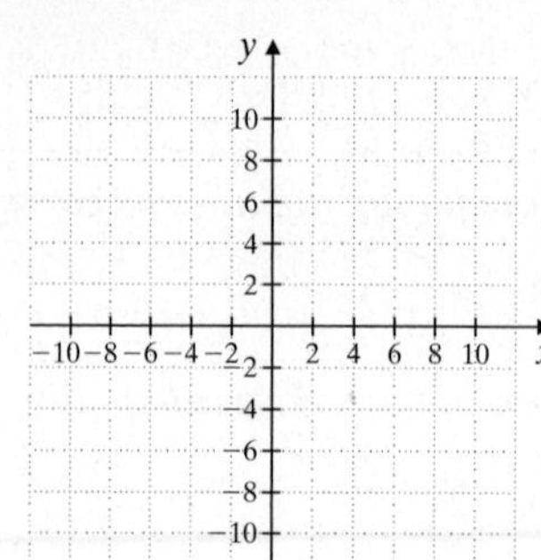

22. $4x + 3y \geq 12$

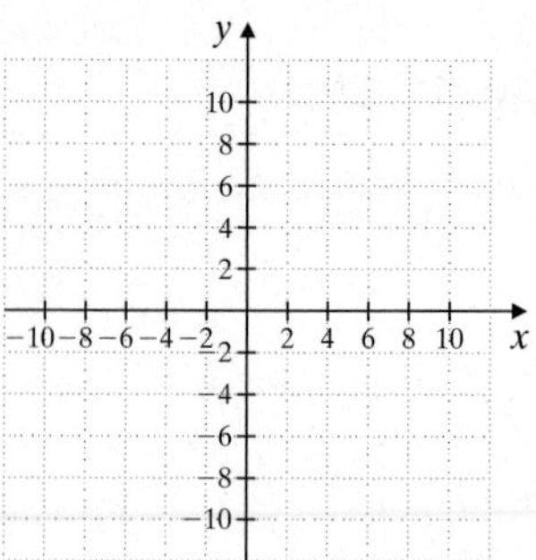

23. $x > y$

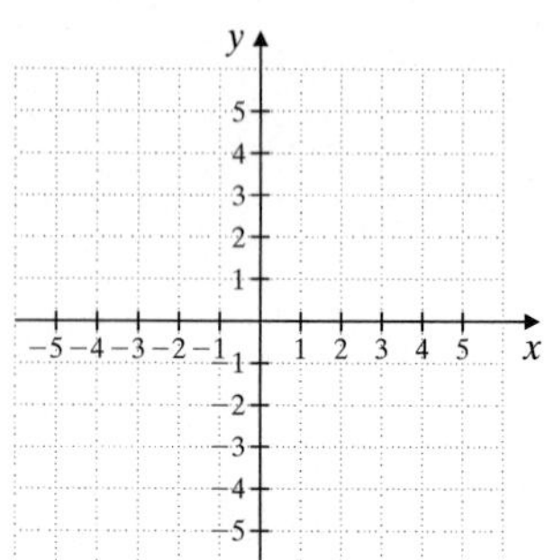

24. $x \leq -y$

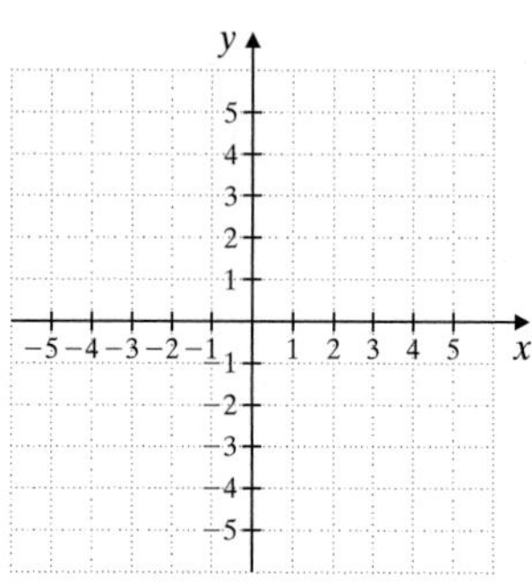

25. $x - y \leq 6$

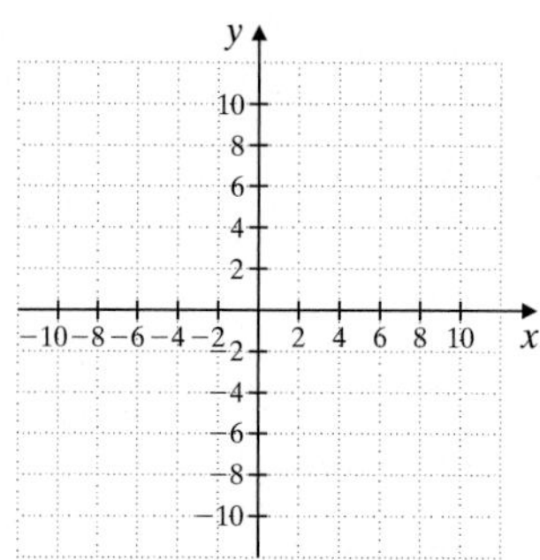

26. $x - y > 10$

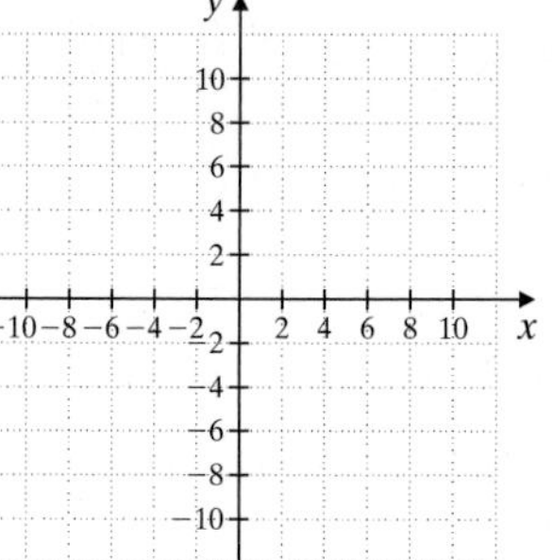

27. $x \geq 0$

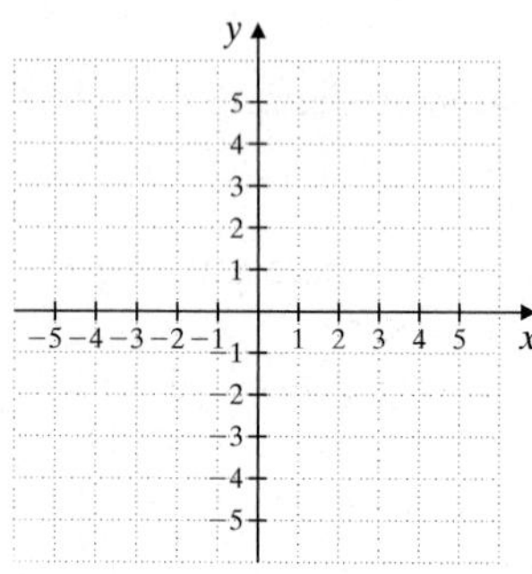

28. $y \leq 0$

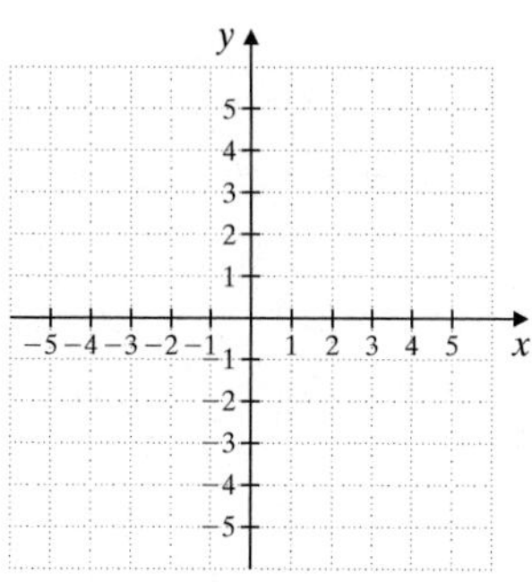

29. $2x + 7y > 5$

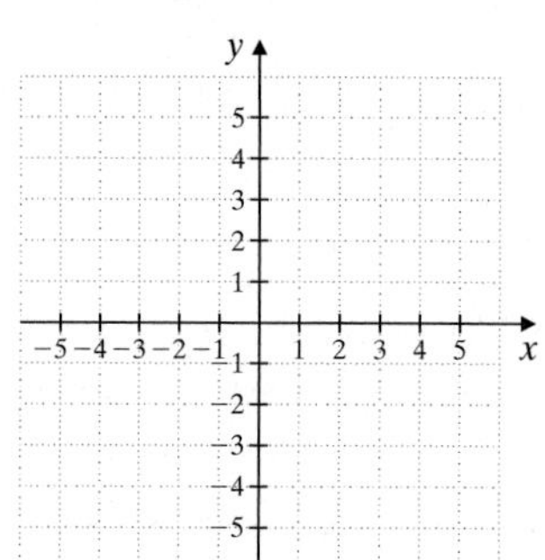

30. $3x + 5y \leq -2$

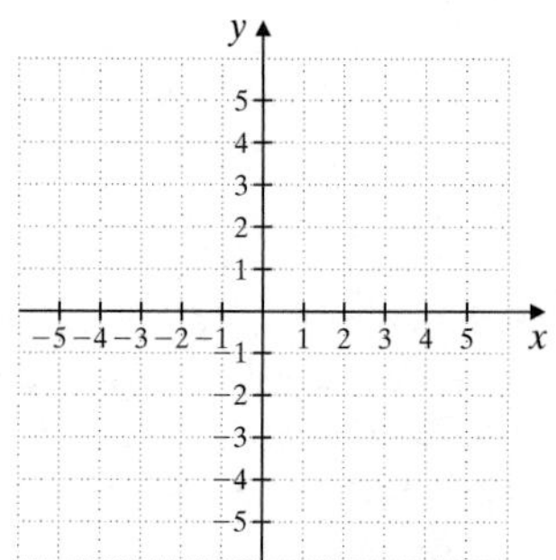

31. $y \geq \frac{1}{2}x - 4$

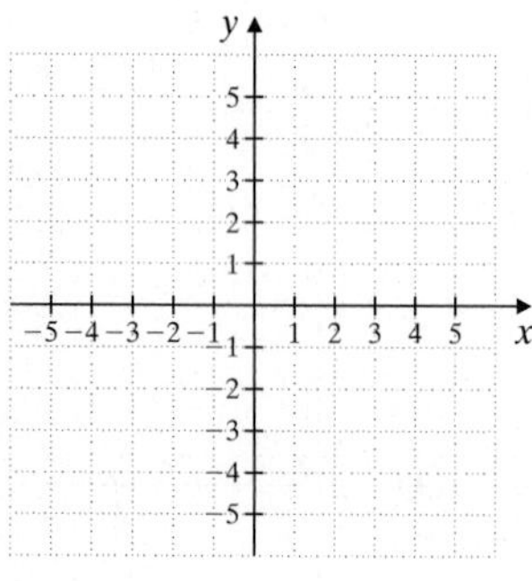

32. $y < \frac{2}{5}x - 3$

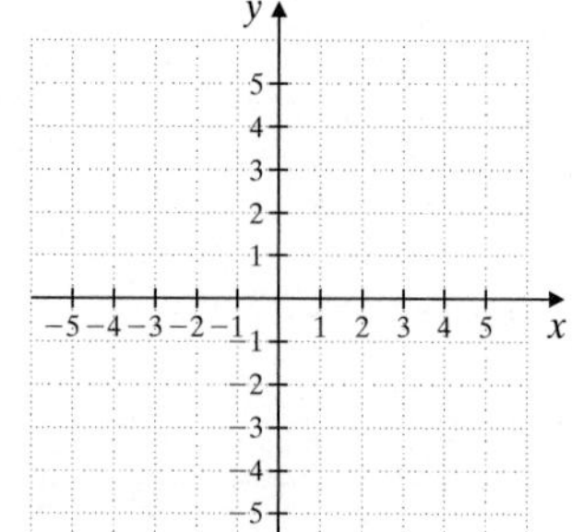

Review and Preview

Approximate the coordinates of each point of intersection. See Section 3.1.

33.

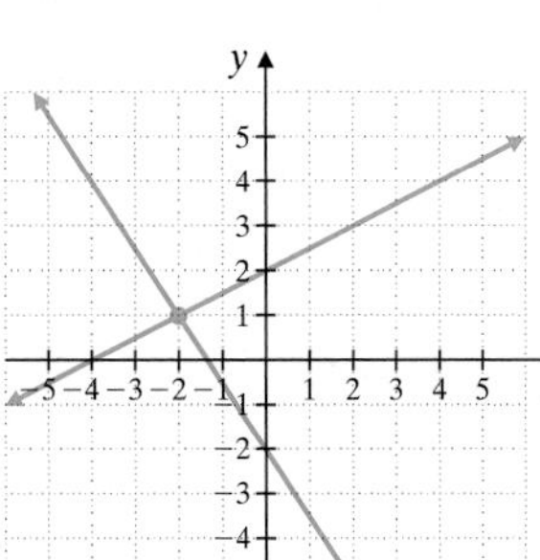

34.

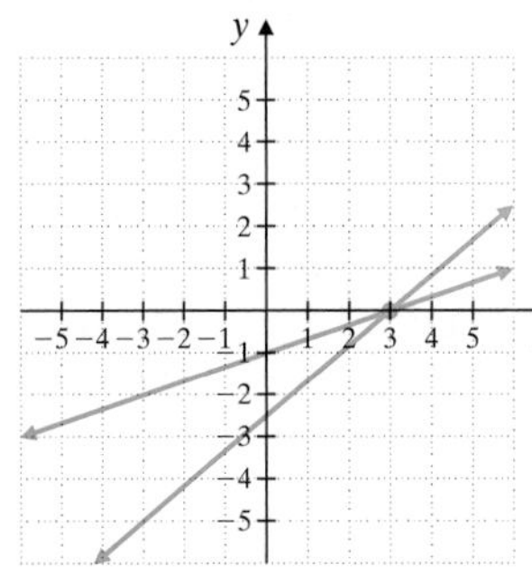

35.

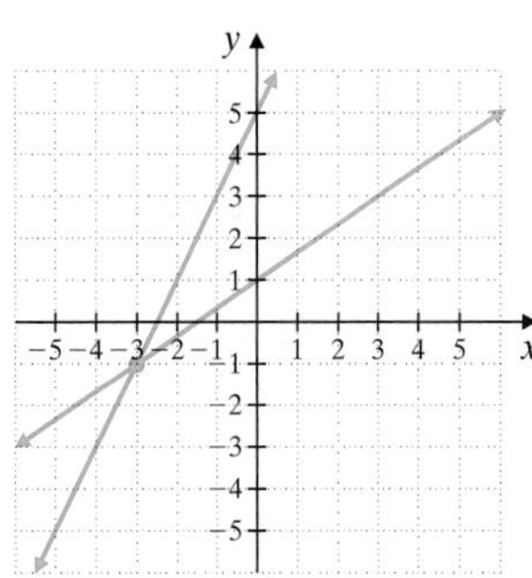

36.

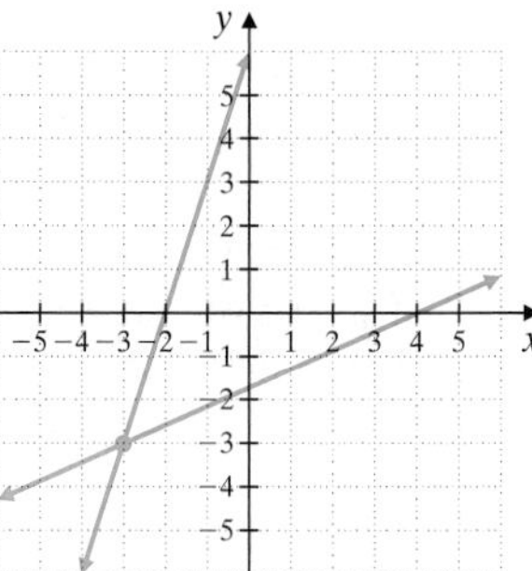

Combining Concepts

Match each inequality with its graph.

A. $x > 2$ **B.** $y < 2$ **C.** $y \leq 2x$ **D.** $y \leq -3x$

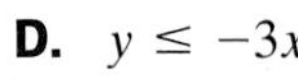

37.

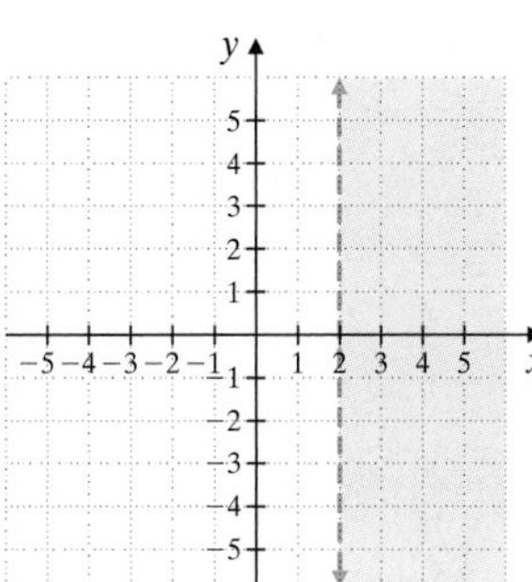

38.

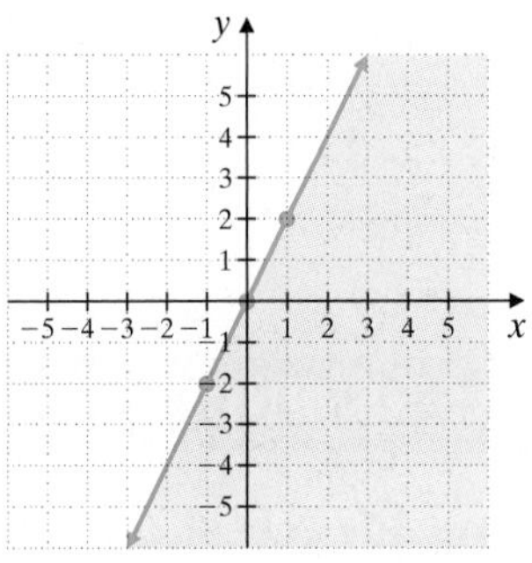

39.

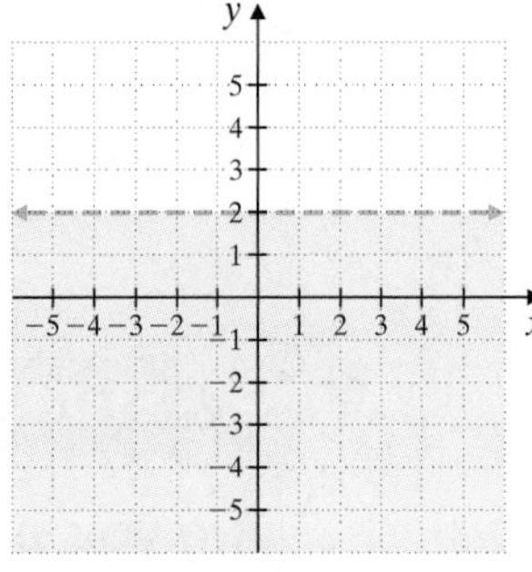

40.

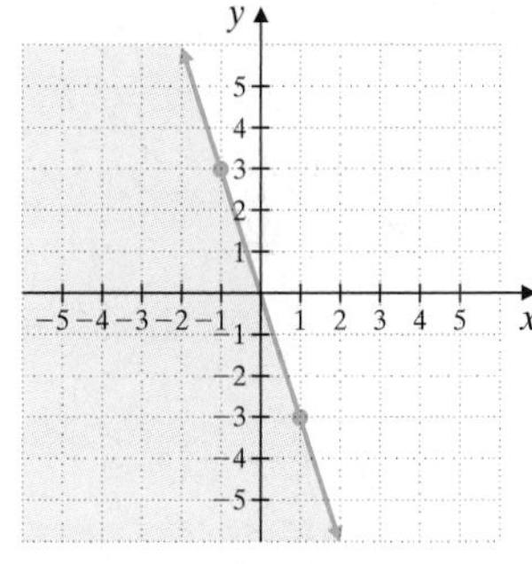

41. Explain why a point on the boundary line should not be chosen as the test point.

43. It's the end of the budgeting period for Dennis Fernandes and he has \$500 left in his budget for car rental expenses. He plans to spend this budget on a sales trip throughout southern Texas. He will rent a car that costs \$30 per day and \$0.15 per mile and he can spend no more than \$500.

a. Write an inequality describing this situation. Let x = number of days and let y = number of miles.

b. Graph this inequality below.

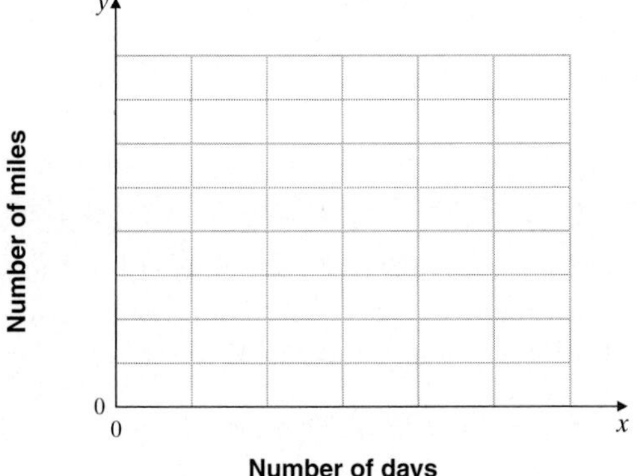

c. Why is the grid showing quadrant I only?

42. Write an inequality whose solutions are all points of numbers whose sum is at least 13.

44. Scott Sambracci and Sara Thygeson are planning their wedding. They have calculated that they want the cost of their wedding ceremony x plus the cost of their reception y to be no more than \$5000.

a. Write an inequality describing this relationship.

b. Graph this inequality below.

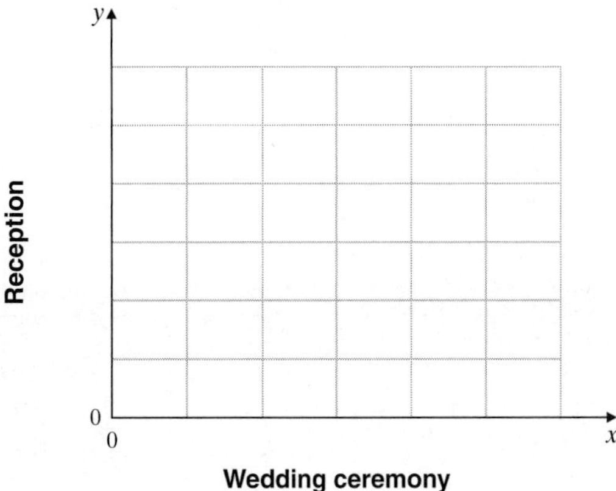

c. Why is the grid showing quadrant I only?

CHAPTER 3 ACTIVITY Finding a Linear Model

This activity may be completed by working in groups or individually.

The following table shows the estimated number of foreign visitors (in millions) to the United States for the years 2000 through 2003.

Year	Foreign Visitors to the United States (in millions)
2000	51.5
2001	54.2
2002	56.9
2003	59.6

(*Source:* Tourism Industries/International Trade Administration, U.S. Department of Commerce)

1. Make a scatter diagram of the paired data in the table.
2. Use what you have learned in this chapter to write an equation of the line representing the paired data in the table. Explain how you found the equation, and what each variable represents.
3. What is the slope of your line? What does the slope mean in this context?
4. Use your linear equation to predict the number of foreign visitors to the United States in 2010.
5. Compare your linear equation to that found by other students or groups. Is it the same, similar, or different? How?
6. Compare your prediction from question 3 to that of other students or groups. Describe what you find.

STUDY SKILLS REMINDER

Are you preparing for a test on Chapter 3?

Below I have listed some common trouble areas for students in Chapter 3. After studying for your test—but before taking your test—read these.

- If you are having trouble with graphing, you might want to ask your instructor if you can use graph paper on your test. This will save you time and keep your graphs neat.
- Don't forget that the graph of an ordered pair is a *single* point in the rectangular coordinate system.
- Make sure you remember that to find the slope of a linear equation using its equation, *first* solve the equation for y. *Then* the coefficient of x is its slope.

$$2x + 3y = 7$$

$$3y = -2x + 7 \quad \text{Subtract } 2x \text{ from both sides.}$$

$$\frac{3y}{3} = -\frac{2}{3}x + \frac{7}{3} \quad \text{Divide both sides by 3.}$$

$$y = -\frac{2}{3}x + \frac{7}{3} \quad \text{slope}$$

- Remember that a point that is an x-intercept will have a y-value of 0 and a point that is a y-intercept will have an x-value of 0. Also—the point $(0, 0)$ will be both an x- and y-intercept.

Chapter 3 VOCABULARY CHECK

Fill in each blank with one of the words listed below.

y-axis	x-axis	solution	linear	standard
x-intercept	y-intercept	y	x	slope

1. An ordered pair is a __________ of an equation in two variables if replacing the variables by the coordinates of the ordered pair results in a true statement.
2. The vertical number line in the rectangular coordinate system is called the ________.
3. A ________ equation can be written in the form $Ax + By = C$.
4. A(n) ____________ is a point of the graph where the graph crosses the x-axis.
5. The form $Ax + By = C$ is called __________ form.
6. A(n) ____________ is a point of the graph where the graph crosses the y-axis.
7. To find an x-intercept of a graph, let _____ = 0.
8. The horizontal number line in the rectangular coordinate system is called the ________.
9. To find a y-intercept of a graph, let _____ = 0.
10. The ________ of a line measures the steepness or tilt of a line.

CHAPTER 3 Highlights

DEFINITIONS AND CONCEPTS | **EXAMPLES**

Section 3.1 Reading Graphs and the Rectangular Coordinate System

The **rectangular coordinate system** consists of a plane and a vertical and a horizontal number line intersecting at their 0 coordinate. The vertical number line is called the **y-axis** and the horizontal number line is called the **x-axis**. The point of intersection of the axes is called the **origin**.

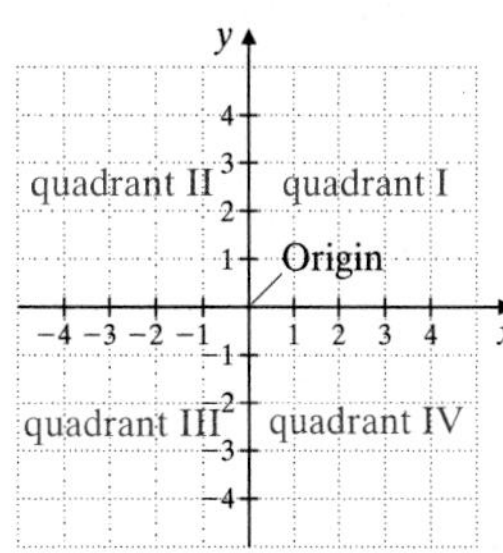

To **plot** or **graph** an ordered pair means to find its corresponding point on a rectangular coordinate system.

To plot or graph an ordered pair such as $(3, -2)$, start at the origin. Move 3 units to the right and from there, 2 units down.

To plot or graph $(-3, 4)$; start at the origin. Move 3 units to the left and from there, 4 units up.

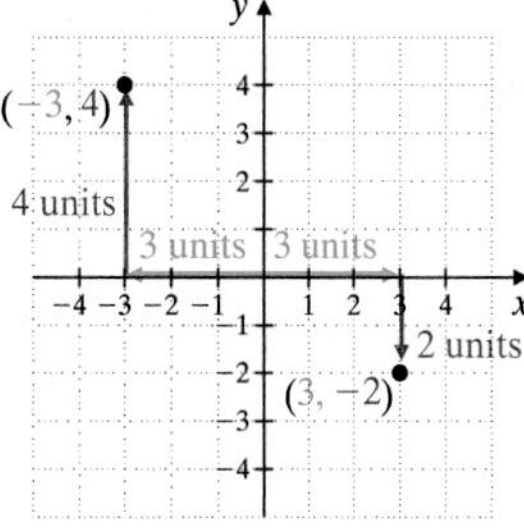

An ordered pair is a **solution** of an equation in two variables if replacing the variables with the coordinates of the ordered pair results in a true statement.

DEFINITIONS AND CONCEPTS | EXAMPLES

Section 3.1 Reading Graphs and the Rectangular Coordinate System *(continued)*

If one coordinate of an ordered pair solution is known, the other value can be determined by substitution.

Complete the ordered pair (0,) for the equation $x - 6y = 12$.

$$x - 6y = 12$$

$$0 - 6y = 12 \quad \text{Let } x = 0.$$

$$\frac{-6y}{-6} = \frac{12}{-6} \quad \text{Divide by } -6.$$

$$y = -2$$

The ordered pair solution is $(0, -2)$.

Section 3.2 Graphing Linear Equations

A **linear equation in two variables** is an equation that can be written in the form $Ax + By = C$, where A and B are not both 0. The form $Ax + By = C$ is called **standard form**.

$3x + 2y = -6 \qquad x = -5$

$y = 3 \qquad y = -x + 10$

$x + y = 10$ is in standard form.

To graph a linear equation in two variables, find three ordered pair solutions. Plot the solution points and draw the line connecting the points.

Graph: $x - 2y = 5$

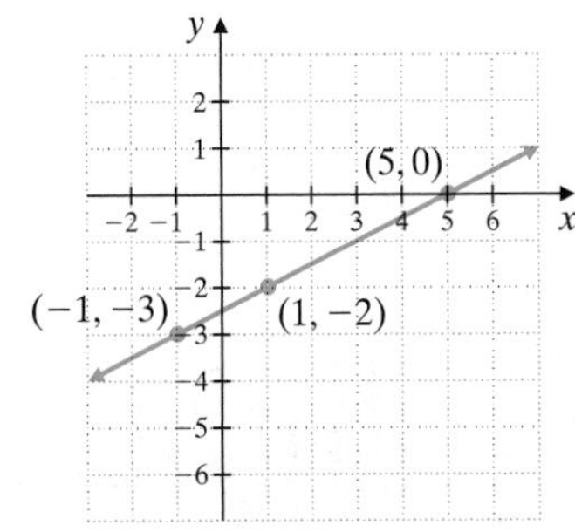

x	y
5	0
1	−2
−1	−3

Section 3.3 Intercepts

An **intercept** of a graph is a point where the graph intersects an axis. If a graph intersects the x-axis at a, then $(a, 0)$ is the ***x*-intercept**. If a graph intersects the y-axis at b, then $(0, b)$ is the ***y*-intercept**.

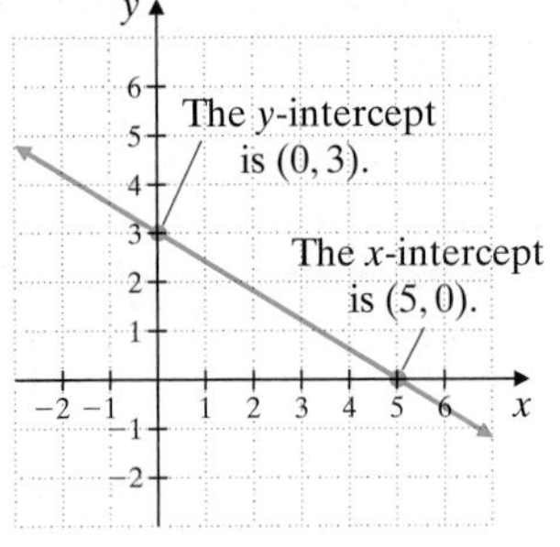

To find the *x*-intercept, let $y = 0$ and solve for x.
To find the *y*-intercept, let $x = 0$ and solve for y.

Find the intercepts for $2x - 5y = -10$.

If $y = 0$, then

$$2x - 5 \cdot 0 = -10$$

$$2x = -10$$

$$\frac{2x}{2} = \frac{-10}{2}$$

$$x = -5$$

If $x = 0$, then

$$2 \cdot 0 - 5y = -10$$

$$-5y = -10$$

$$\frac{-5y}{-5} = \frac{-10}{-5}$$

$$y = 2$$

DEFINITIONS AND CONCEPTS	EXAMPLES

Section 3.3 Intercepts *(continued)*

The x-intercept is $(-5, 0)$. The y-intercept is $(0, 2)$.

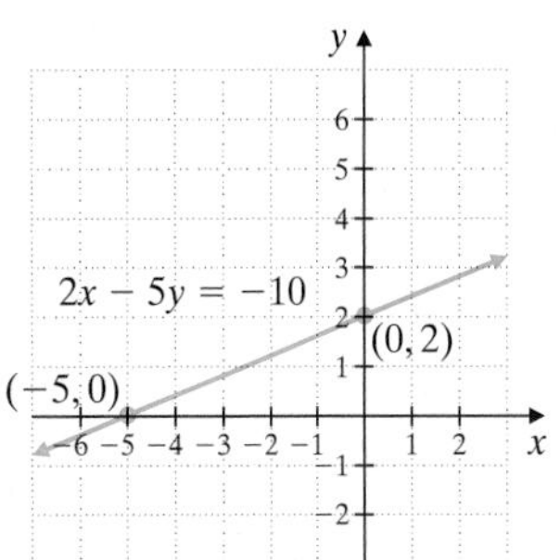

The graph of $x = c$ is a vertical line with x-intercept $(c, 0)$.
The graph of $y = c$ is a horizontal line with y-intercept $(0, c)$.

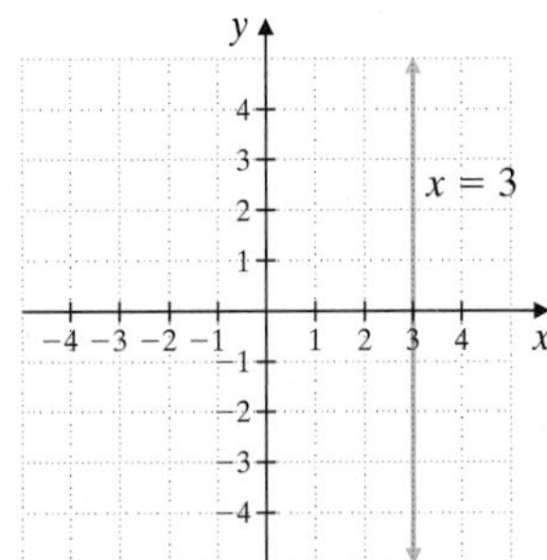

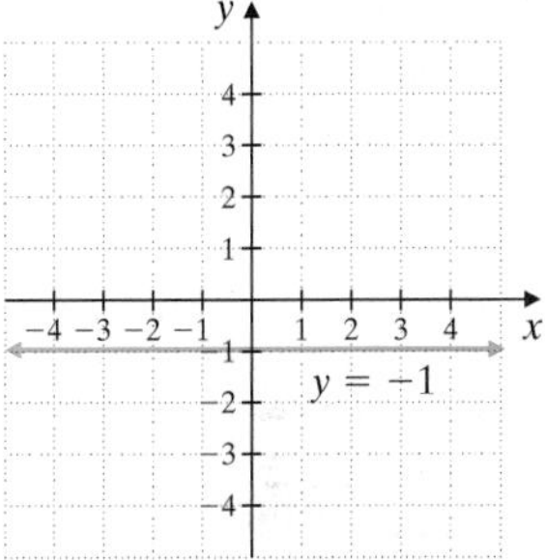

Section 3.4 Slope

The **slope *m*** of the line through points (x_1, y_1) and (x_2, y_2) is given by

$$m = \frac{y_2 - y_1}{x_2 - x_1} \quad \text{as long as } x_2 \neq x_1$$

The slope of the line through points $(-1, 6)$ and $(-5, 8)$ is

$$m = \frac{y_2 - y_1}{x_2 - x_1} = \frac{8 - 6}{-5 - (-1)} = \frac{2}{-4} = -\frac{1}{2}$$

A horizontal line has slope 0.
The slope of a vertical line is undefined.
Nonvertical parallel lines have the same slope.
Two nonvertical lines are perpendicular if the slope of one is the negative reciprocal of the slope of the other.

The slope of the line $y = -5$ is 0.
The line $x = 3$ has undefined slope.

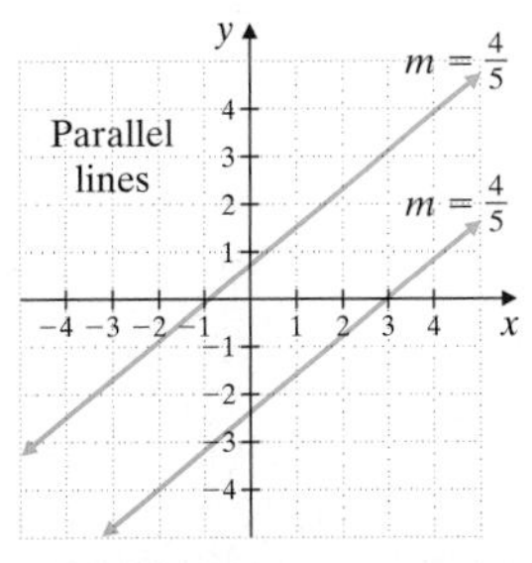

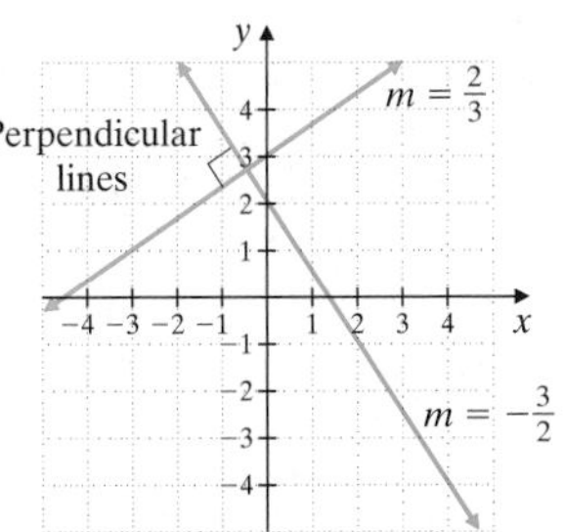

Section 3.5 Graphing Linear Inequalities in Two Variables

A **linear inequality in two variables** is an inequality that can be written in one of these forms:

$$Ax + By < C \qquad Ax + By \leq C$$

$$Ax + By > C \qquad Ax + By \geq C$$

where A and B are not both 0.

$$2x - 5y < 6 \qquad x \geq -5$$

$$y > -8x \qquad y \leq 2$$

DEFINITIONS AND CONCEPTS | EXAMPLES

Section 3.5 Graphing Linear Inequalities in Two Variables *(continued)*

TO GRAPH A LINEAR INEQUALITY

1. Graph the boundary line by graphing the related equation. Draw the line solid if the inequality symbol is $\le$ or $\ge$. Draw the line dashed if the inequality symbol is $<$ or $>$.
2. Choose a test point not on the line. Substitute its coordinates into the original inequality.
3. If the resulting inequality is true, shade the half-plane that contains the test point. If the inequality is not true, shade the half-plane that does not contain the test point.

Graph: $2x - y \le 4$

1. Graph $2x - y = 4$. Draw a solid line because the inequality symbol is $\le$.
2. Check the test point $(0, 0)$ in the original inequality, $2x - y \le 4$.

$2 \cdot 0 - 0 \le 4$ Let $x = 0$ and $y = 0$.

$0 \le 4$ True

3. The inequality is true, so shade the half-plane containing $(0, 0)$ as shown.

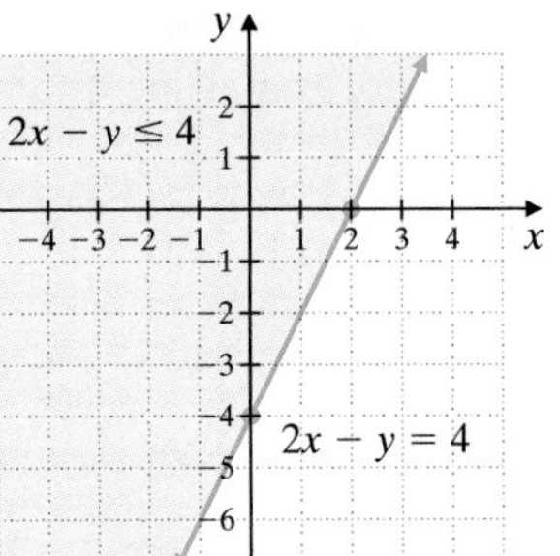

STUDY SKILLS REMINDER

How are your homework assignments going?

By now, you should have good homework habits. If not, it's never too late to begin. Why is it so important in mathematics to keep up with homework? You probably now know the answer to that question. You may have realized by now that many concepts in mathematics build on each other. Your understanding of one chapter in mathematics usually depends on your understanding of the previous chapter's material.

Don't forget that completing your homework assignment involves a lot more than attempting a few of the problems assigned.

To complete a homework assignment, remember these four things:

1. Attempt all of it.
2. Check it.
3. Correct it.
4. If needed, ask questions about it.

Name ______________________ Section __________ Date ____________

Chapter 3 Review

(3.1) *Use the graph below showing Disney's consumer products revenues to answer Exercises 1 through 4.*

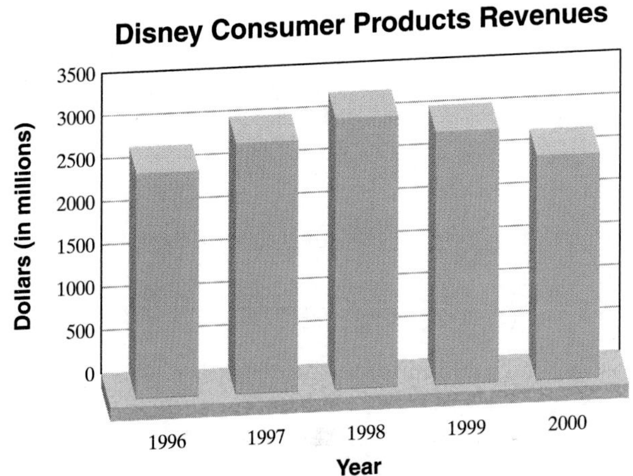

Source: The Walt Disney Company

1. Approximate Disney's consumer products revenue in 2000.

2. Approximate the increase in consumer products revenue in 1998.

3. What year shows the greatest revenue?

4. What trend is shown by this graph?

Use the following graph to answer Exercises 5 through 8.

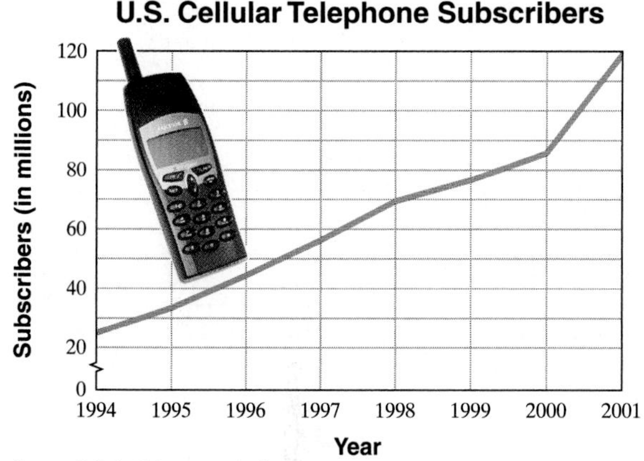

Source: Cellular Telecommunications Industry Association

5. Approximate the number of cellular phone subscribers in 2000.

6. Approximate the increase in cellular phone subscribers in 2001.

7. What year shows the greatest number of subscribers?

8. What trend is shown by this graph?

Plot each pair on the same rectangular coordinate system.

9. $(-7, 0)$

10. $\left(0, 4\frac{4}{5}\right)$

11. $(-2, -5)$

12. $(1, -3)$

13. $(0.7, 0.7)$

14. $(-6, 4)$

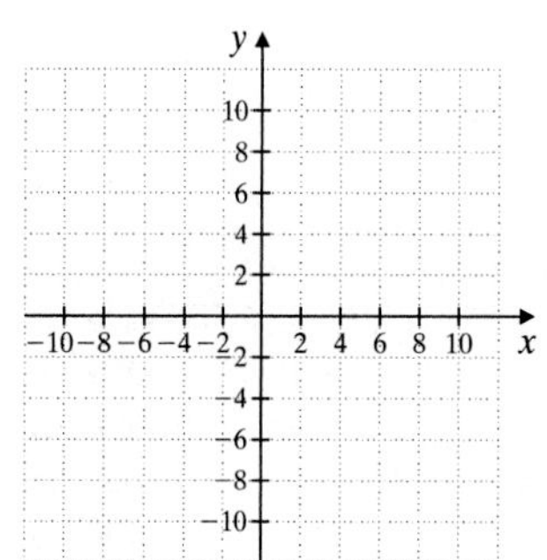

Complete each ordered pair so that it is a solution of the given equation.

15. $-2 + y = 6x$; (7,)

16. $y = 3x + 5$; (, −8)

Complete the table of values for each given equation.

17. $9 = -3x + 4y$

x	y
	0
	3
9	

18. $y = 5$

x	y
7	
−7	
0	

19. $x = 2y$

x	y
	0
	5
	−5

20. The cost in dollars of producing x compact disc holders is given by $y = 5x + 2000$.

a. Complete the table.

x	1	100	1000
y			

b. Find the number of compact disc holders that can be produced for $6430.

(3.2) *Graph each linear equation.*

21. $x - y = 1$

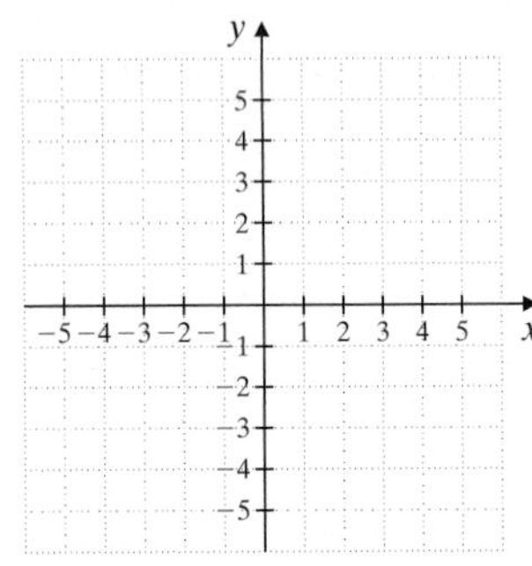

22. $x + y = 6$

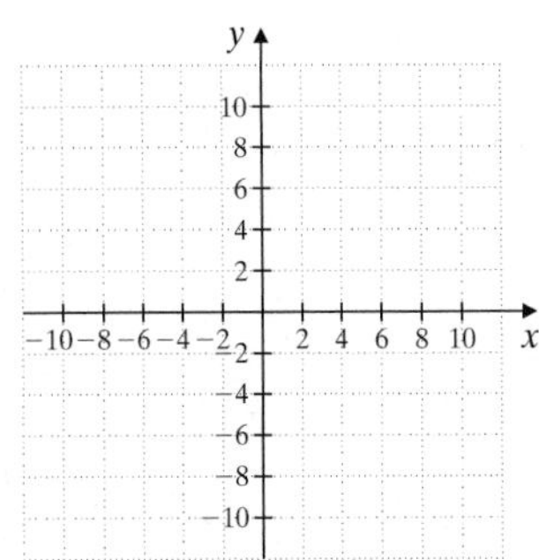

23. $x - 3y = 12$

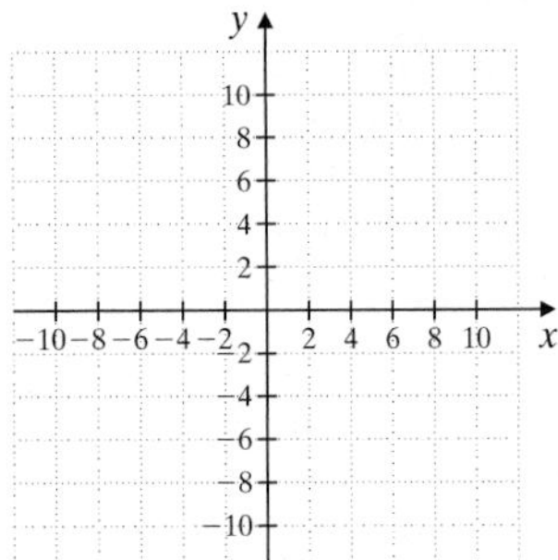

24. $5x - y = -8$

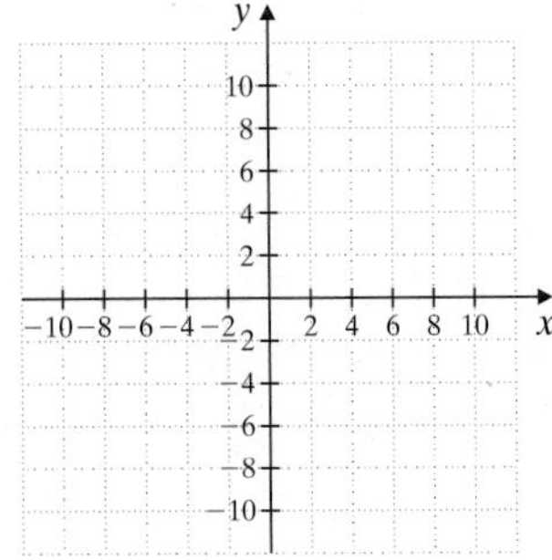

25. $x = 3y$

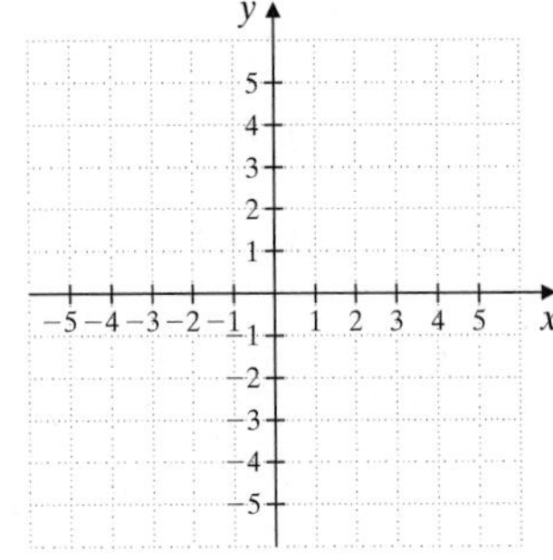

26. $y = -2x$

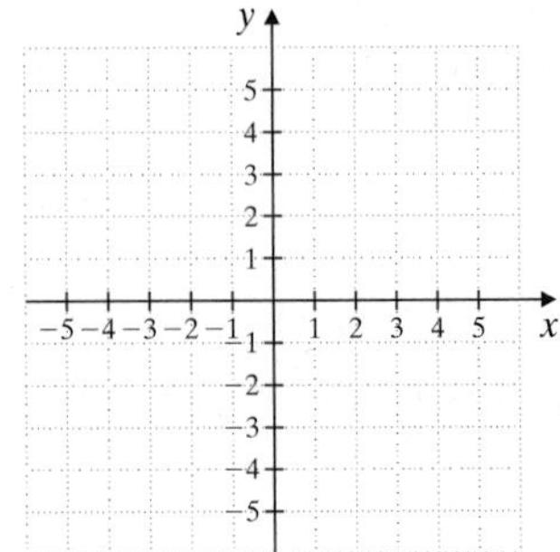

(3.3) *Identify the intercepts in each graph.*

27.

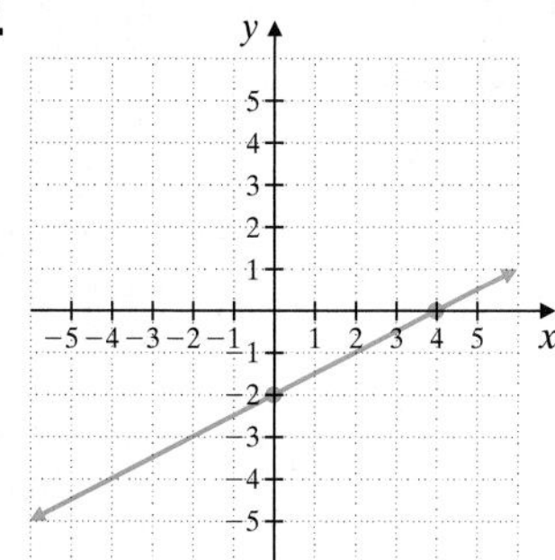

28.

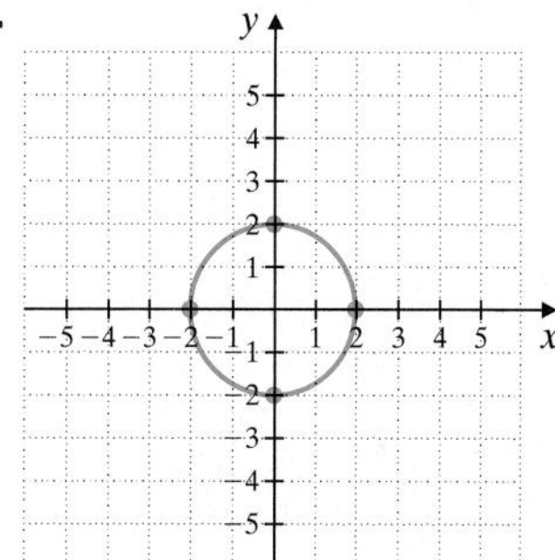

Graph each linear equation.

29. $y = -3$

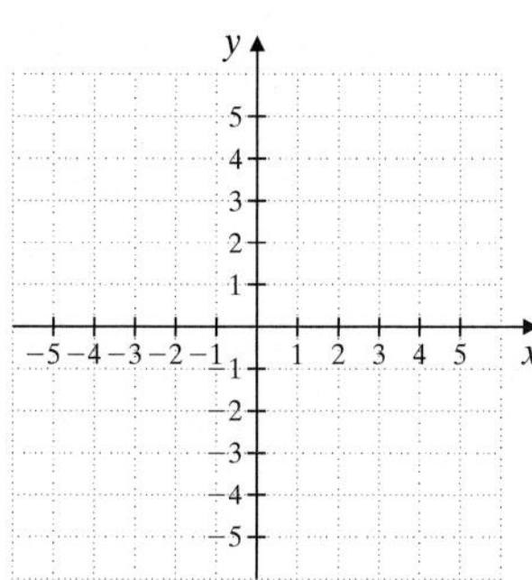

30. $x = 5$

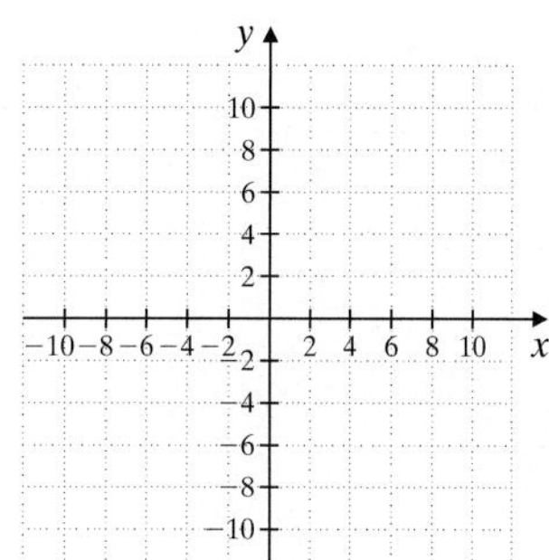

Find the intercepts for each equation.

31. $x - 3y = 12$

32. $-4x + y = 8$

(3.4) *Find the slope of each line.*

33.

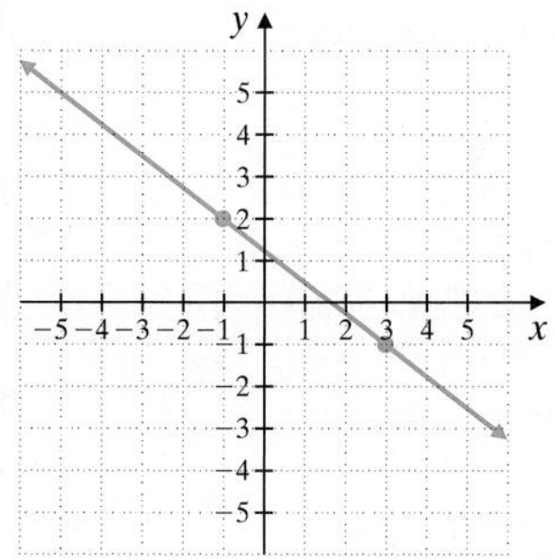

34.

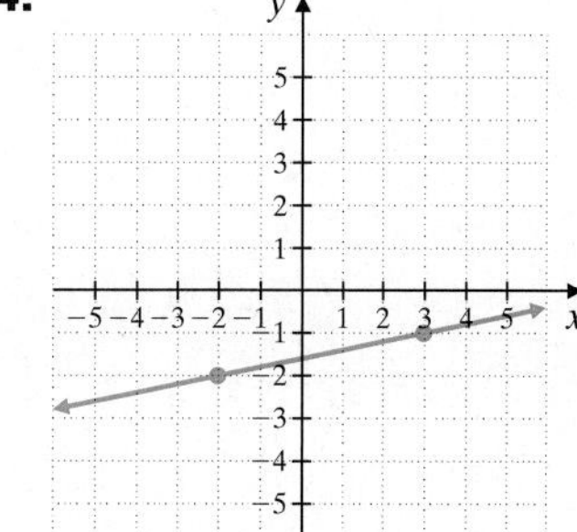

Match each line with its slope.

A.

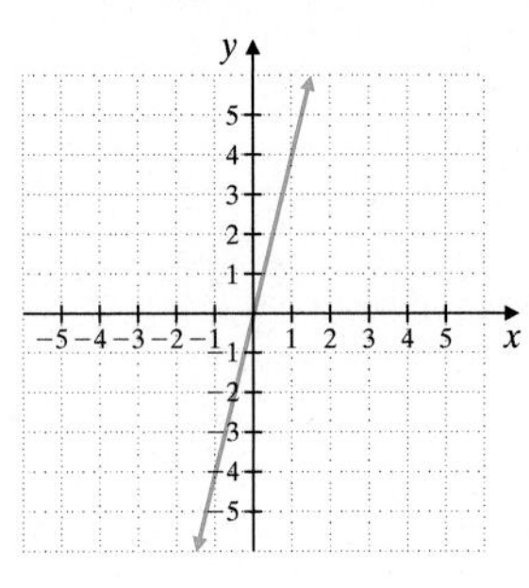

B.

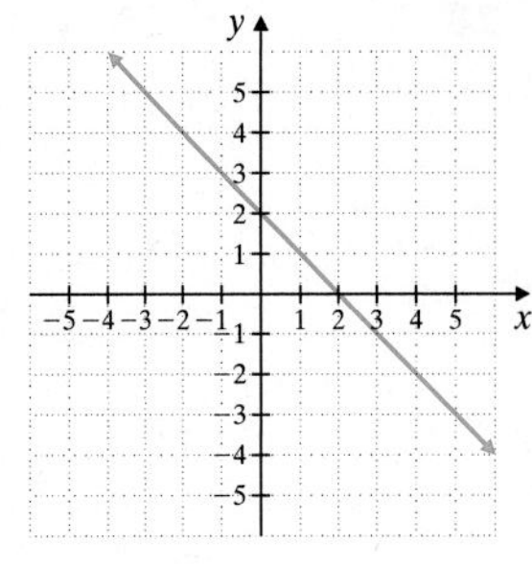

C.

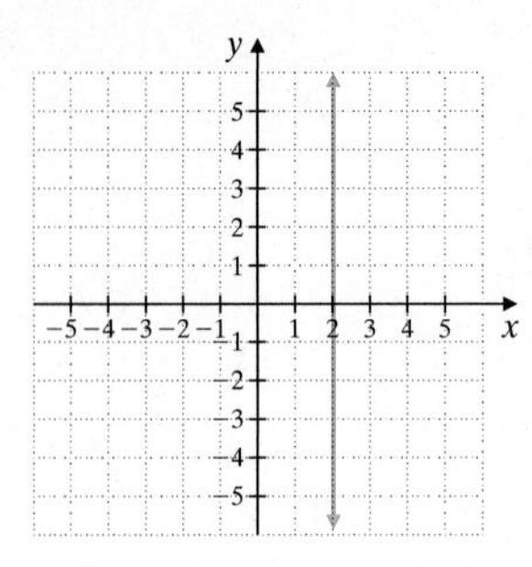

D.

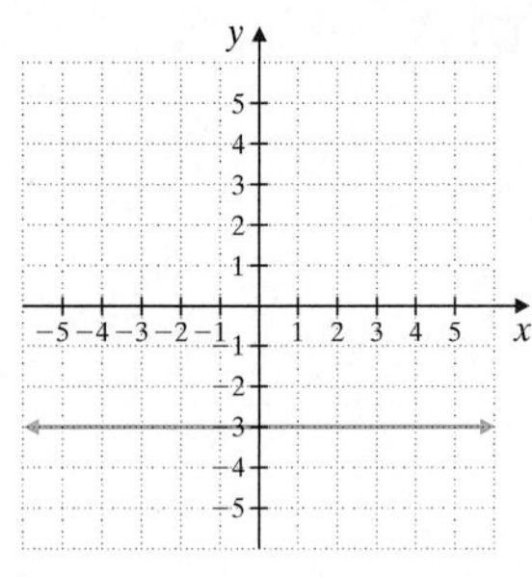

35. $m = 0$

36. $m = -1$

37. undefined slope

38. $m = 4$

Find the slope of the line that passes through each pair of points.

39. $(2, 5)$ and $(6, 8)$

40. $(4, 7)$ and $(1, 2)$

41. $(1, 3)$ and $(-2, -9)$

42. $(-4, 1)$ and $(3, -6)$

Find the slope of each line.

43. $y = 3x + 7$

44. $x - 2y = 4$

45. $y = -2$

46. $x = 0$

△ *Determine whether each pair of lines is parallel, perpendicular, or neither.*

47. $x - y = -6$
$x + y = 3$

48. $3x + y = 7$
$-3x - y = 10$

49. $y = 4x + \frac{1}{2}$
$4x + 2y = 1$

Find the slope of each line and write the slope as a rate of change. Don't forget to attach the proper units.

50. The graph below approximates the number of U.S. persons y (in millions) who have a bachelor's degree or higher per year x.

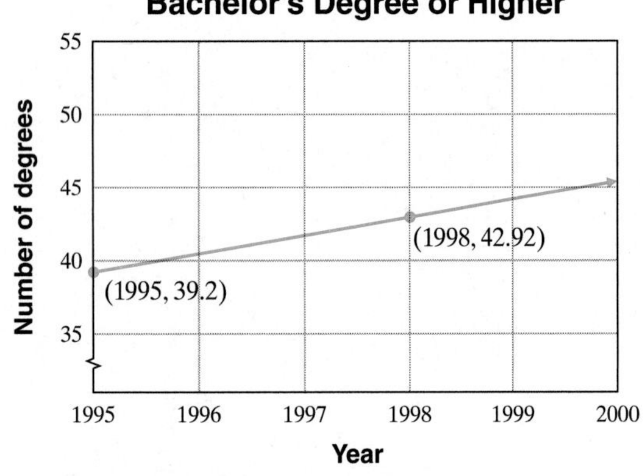

51. The graph below approximates the number of U.S. travelers y (in millions) that are vacationing per year x.

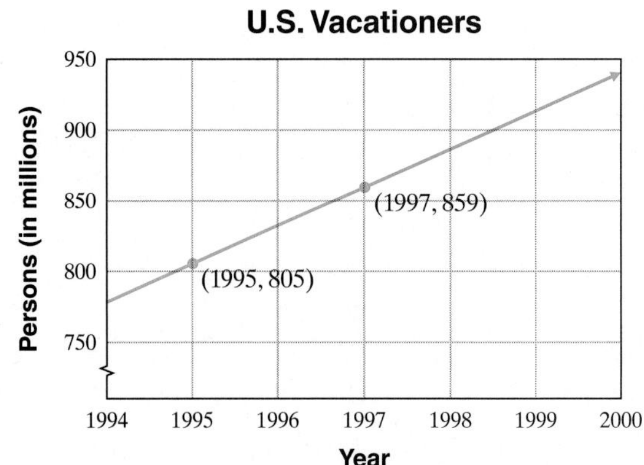

(3.5) *Graph each inequality.*

52. $x + 6y < 6$

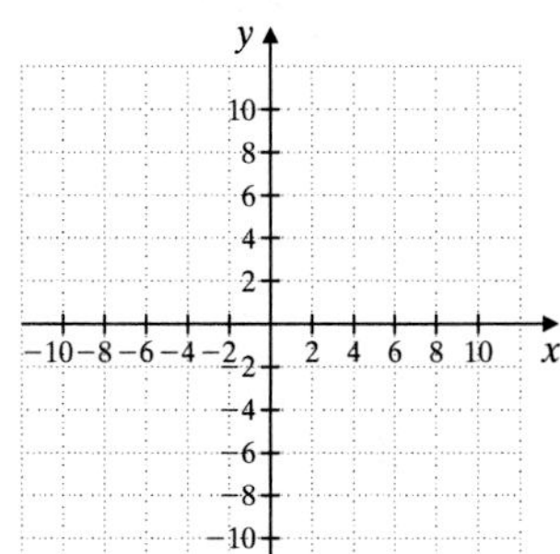

53. $x + y > -2$

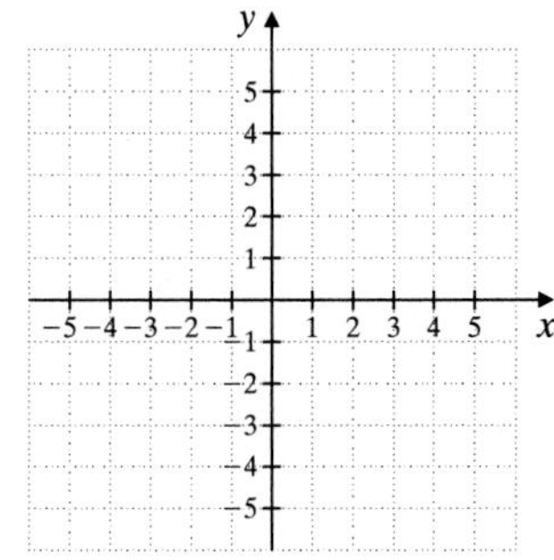

54. $y \geq -7$

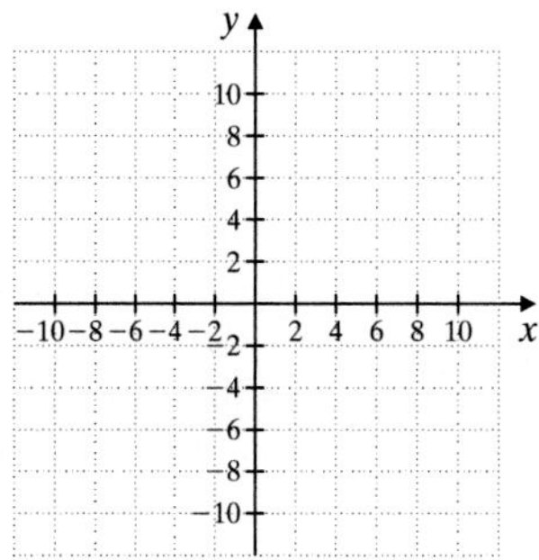

55. $y \leq -4$

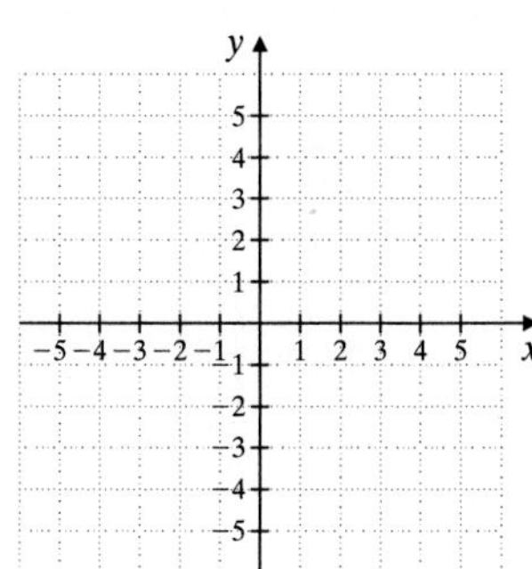

56. $-x \leq y$

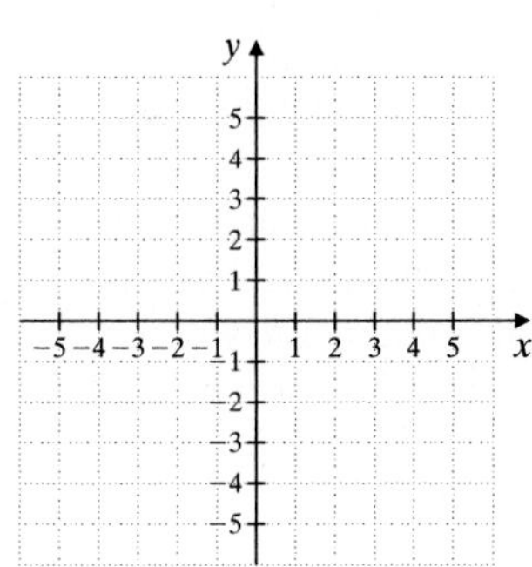

57. $x \geq -y$

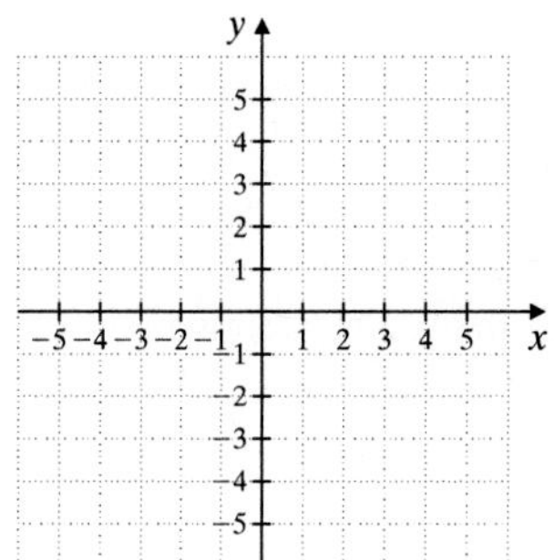

FOCUS ON The Real World

MISLEADING GRAPHS

Graphs are very common in magazines and in newspapers such as *USA Today*. Graphs can be a convenient way to get an idea across because, as the old saying goes, "a picture is worth a thousand words." However, some graphs can be deceptive, which may or may not be intentional. It is important to know some of the ways that graphs can be misleading.

Beware of graphs like the one at the right. Notice that the graph shows a company's profit for various months. It appears that profit is growing quite rapidly. However, this impressive picture tells us little without knowing what units of profit are being graphed. Does the graph show profit in dollars or millions of dollars? An unethical company with profit increases of only a few pennies could use a graph like this one to make the profit increase seem much more substantial than it really is. A truthful graph describes the size of the units used along the vertical axis.

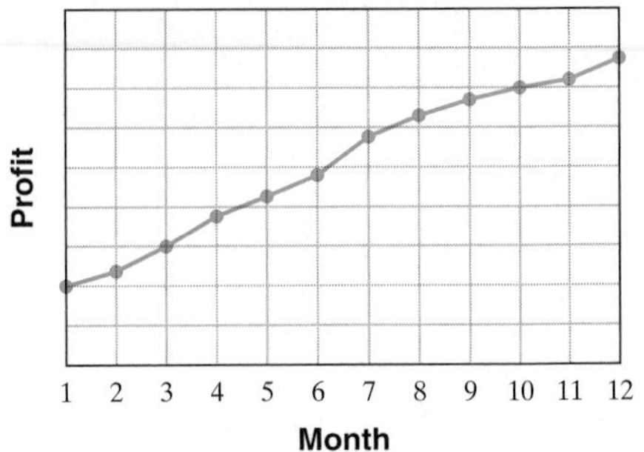

Another type of graph to watch for is one that misrepresents relationships. For example, the bar graph at the right shows the number of men and women employees in the accounting and shipping departments of a certain company. In the accounting department, the bar representing the number of women is shown twice as tall as the bar representing the number of men. However, the number of women (13) is not twice the number of men (10). This set of bars misrepresents the relationship between the number of men and women. Do you see how the relationship between the number of men and women in the shipping department is distorted by the heights of the bars used? A truthful graph will use bar heights that are proportional to the numbers they represent.

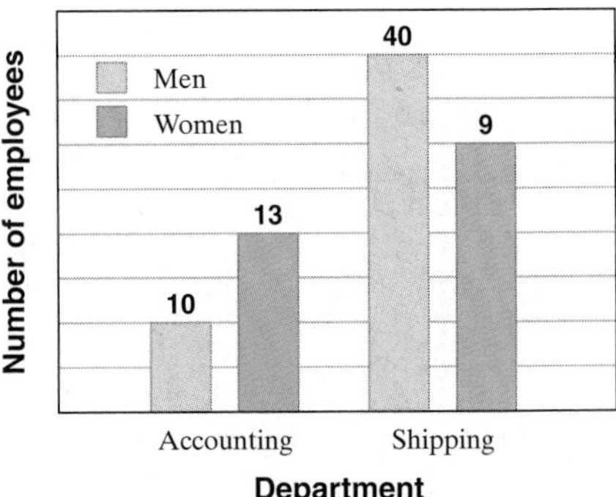

The impression a graph can give also depends on its vertical scale. The two graphs below represent exactly the same data. The only difference between the two graphs is the vertical scale—one shows enrollments from 246 to 260 students and the other shows enrollments between 0 and 300 students. If you were trying to convince readers that algebra enrollment at UPH had changed drastically over the period 1996–2000, which graph would you use? Which graph do you think gives the more honest representation?

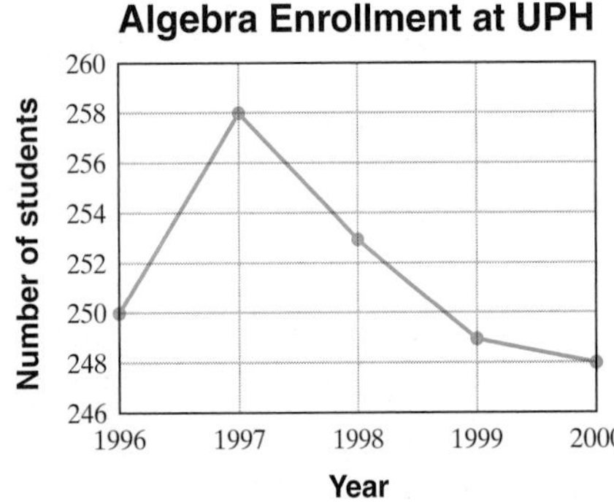

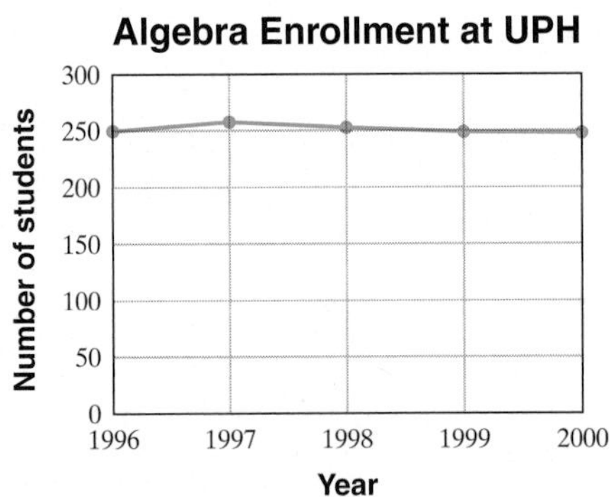

Name ______________________ Section __________ Date __________

Chapter 3 Test

Complete each ordered pair so that it is a solution of the given equation.

1. $12y - 7x = 5; (1, \quad)$

2. $y = 17; (-4, \quad)$

Find the slope of each line.

3.

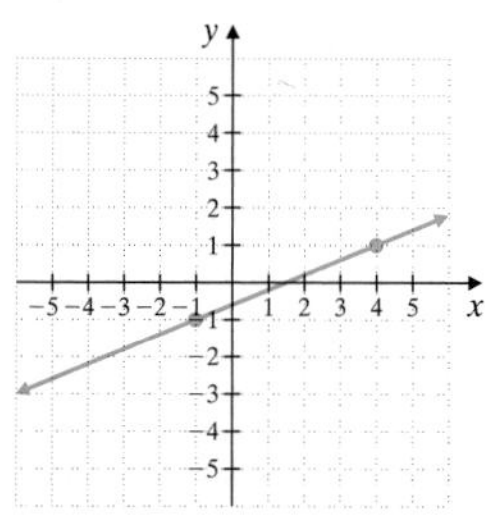

4.

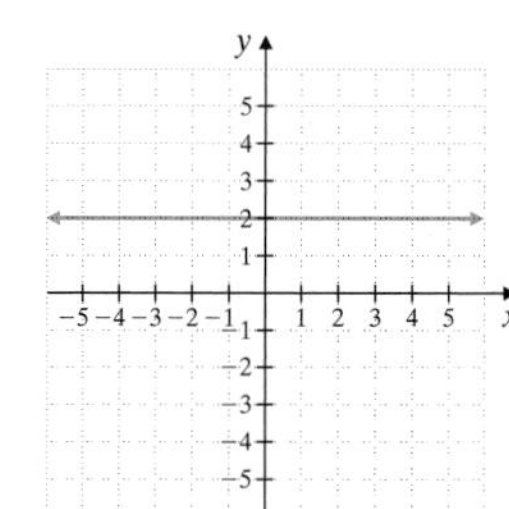

5. Passes through $(6, -5)$ and $(-1, 2)$

6. Passes through $(0, -8)$ and $(-1, -1)$

7. $-3x + y = 5$

8. $x = 6$

Graph.

9. $2x + y = 8$

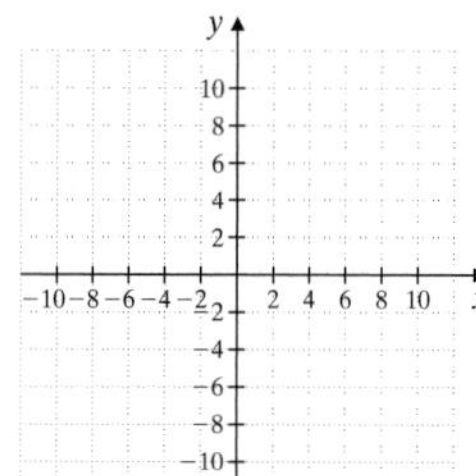

10. $-x + 4y = 5$

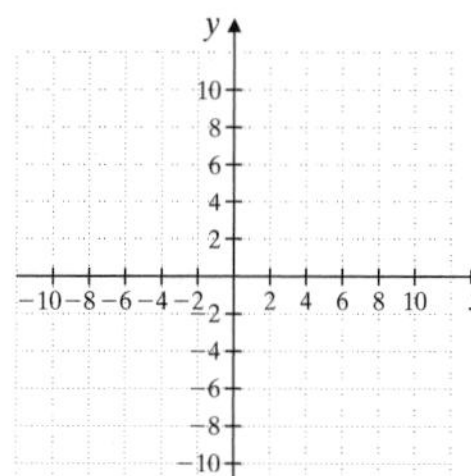

11. $x - y \geq -2$

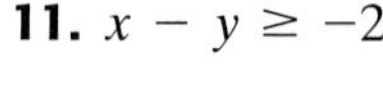

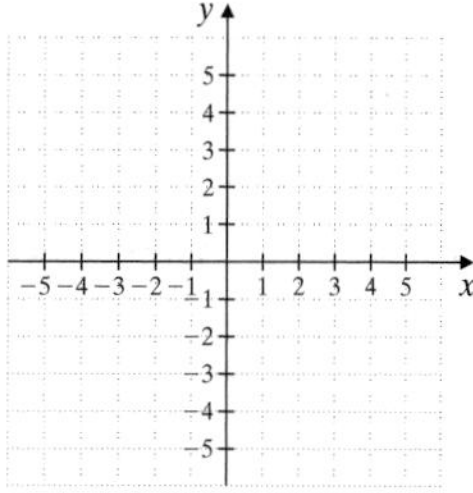

12. $y \geq -4x$

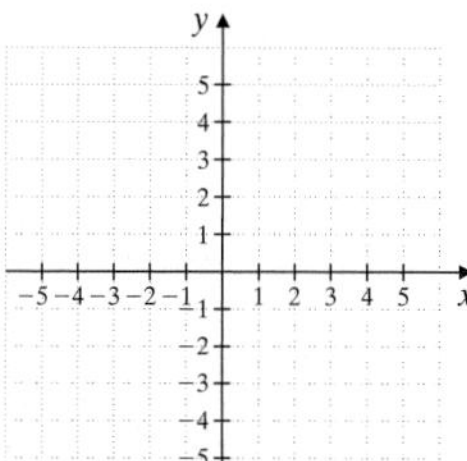

13. $5x - 7y = 10$

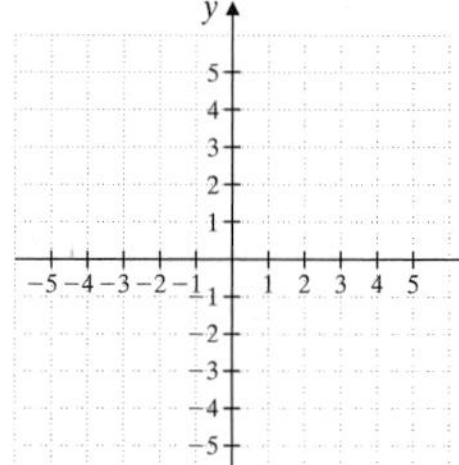

14. $2x - 3y > -6$

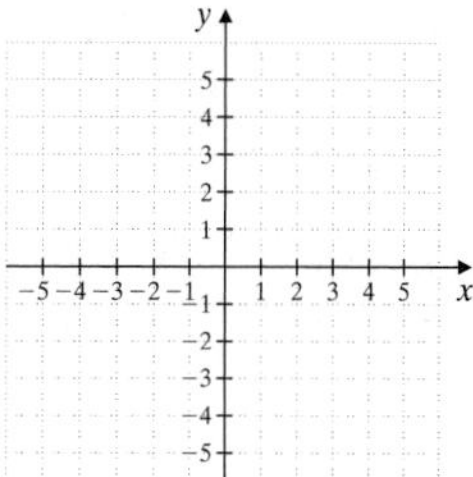

15. $6x + y > -1$

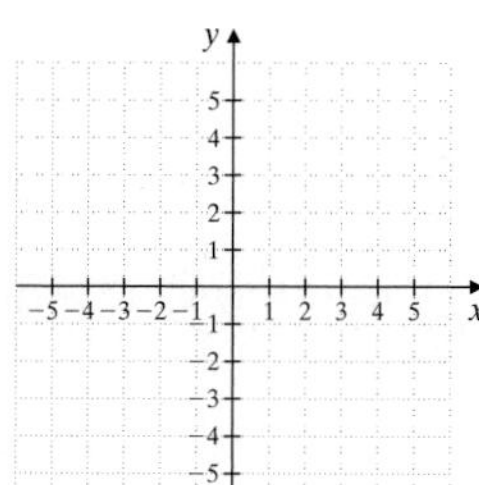

16. $y = -1$

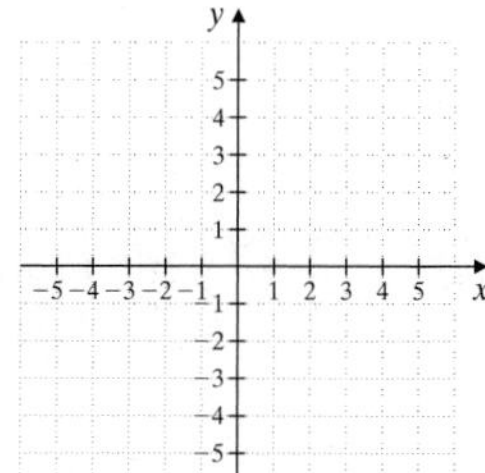

Determine whether each pair of lines is parallel, perpendicular, or neither.

17. $y = 2x - 6$
$-4x = 2y$

18. $-x + 3y = 2$
$6x - 18y = 3$

Answers

1. __________

2. __________

3. __________

4. __________

5. __________

6. __________

7. __________

8. __________

9. see graph

10. see graph

11. see graph

12. see graph

13. see graph

14. see graph

15. see graph

16. see graph

17. __________

18. __________

19. ______

20. a. ______

21. ______

22. ______

23. ______

24. ______

25. ______

△ **19.** The perimeter of the parallelogram below is 42 meters. Write a linear equation in two variables for the perimeter. Use this equation to find x when y is 8.

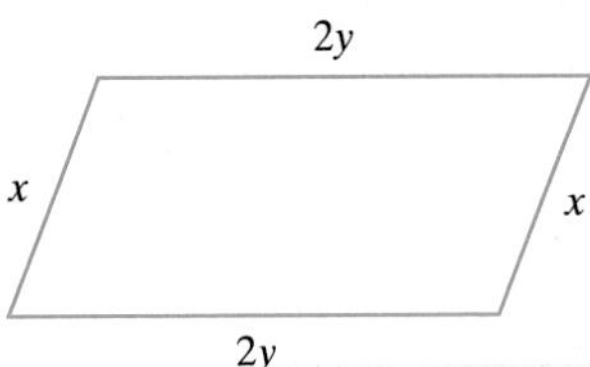

20. The table gives the number of cable TV subscribers (in millions) for the years shown. (*Source:* Nielsen Media Research)

Year	Cable TV Subscribers (in millions)
1986	38
1988	44
1990	50
1992	53
1994	57
1996	62
1998	67
2000	69

a. Write this data as a set of ordered pairs of the form (year, number of cable TV subscribers in millions).

b. Create a scatter diagram of the data. Be sure to label the axes properly.

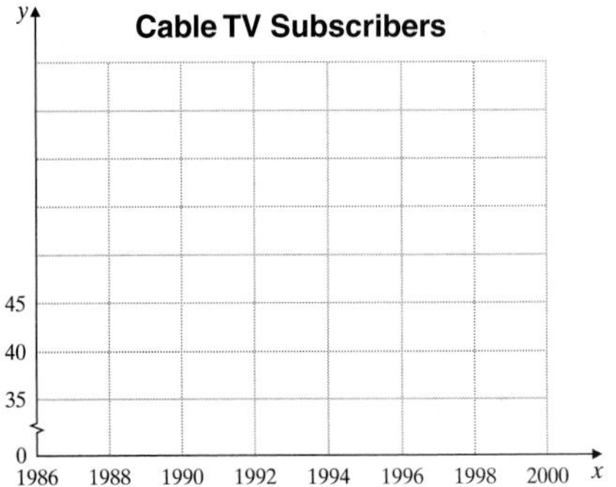

21. This graph approximates the movie ticket sales y (in millions) for the year x. Find the slope of the line and write the slope as a rate of change. Don't forget to attach the proper units.

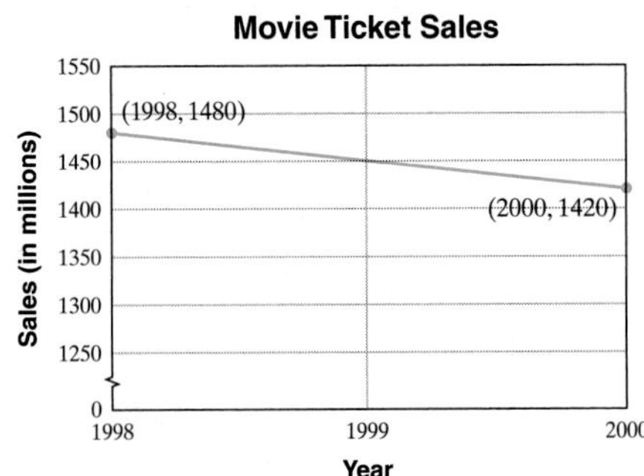

The line graph shows the total amount of revenue generated by the Internet for the years 1996–2002. Use this graph to answer Questions 22 through 25.

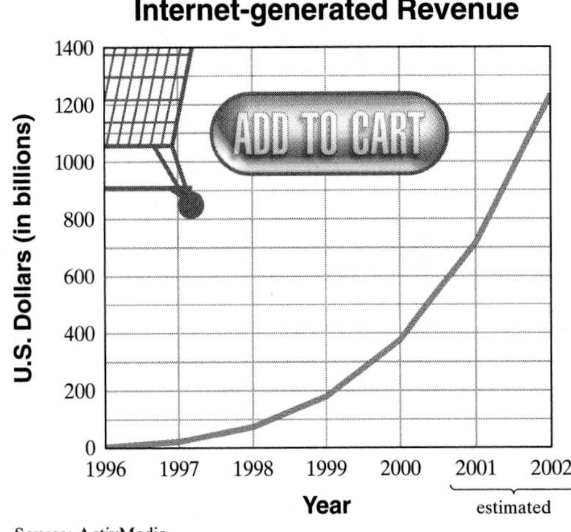

22. Estimate Internet revenue in 2000.

23. Estimate predicted Internet revenue in 2002.

24. Find the increase in Internet revenue from 1996 to 2000.

25. What year shows the greatest increase in revenue?

Name ______________________ Section __________ Date __________

Cumulative Review

1. Given the set $\left\{-2, 0, \frac{1}{4}, 112, -3, 11, \sqrt{2}, 5.1\right\}$, list the numbers in this set that belong to the set of:

   a. Natural numbers

   b. Whole numbers

   c. Integers

   d. Rational numbers

   e. Irrational numbers

   f. Real numbers

2. Evaluate (find the value of) the following:

   a. 3^2

   b. 5^3

   c. 2^4

   d. 7^1

   e. $\left(\frac{3}{7}\right)^2$

3. Simplify: $\frac{3}{2} \cdot \frac{1}{2} - \frac{1}{2}$

4. Write an algebraic expression that represents each phrase. Let the variable x represent the unknown number.

   a. The sum of a number and 3

   b. The product of 3 and a number

   c. Twice a number

   d. 10 decreased by a number

   e. 5 times a number increased by 7

Simplify each expression.

5. $6 \div 3 + 5^2$

6. $3[4(5 + 2) - 10]$

7. Add: $11.4 + (-4.7)$

8. If $x = 2$ and $y = -5$, find the value of each expression.

   a. $\frac{x - y}{12 + x}$

   b. $x^2 - y$

Divide.

9. $\frac{-30}{-10}$

10. $\frac{42}{-0.6}$

Find each product by using the distributive property to remove parentheses.

11. $5(x + 2)$

12. $-2(y + 0.3z - 1)$

13. $-(x + y - 2z + 6)$

Answers

1. a. ______
   b. ______
   c. ______
   d. ______
   e. ______
   f. ______

2. a. ______
   b. ______
   c. ______
   d. ______
   e. ______

3. ______

4. a. ______
   b. ______
   c. ______
   d. ______
   e. ______

5. ______

6. ______

7. ______

8. a. ______
   b. ______

9. ______

10. ______

11. ______

12. ______

13. ______

14.

15.

16.

17.

18.

19.

20.

21.

22.

23.

24. a.

b.

25. a.

b.

c.

26.

27.

Write each phrase as an algebraic expression and simplify if possible. Let x represent the unknown number.

14. Twice a number, plus 6.

15. The difference of a number and 4, divided by 7.

16. Five plus the sum of a number and 1.

17. Solve for x: $\frac{5}{2}x = 15$

18. Solve: $6(2a - 1) - (11a + 6) = 7$

19. Solve: $\frac{y}{7} = 20$

20. Solve: $0.25x + 0.10(x - 3) = 0.05(22)$

21. Twice the sum of a number and 4 is the same as four times the number decreased by 12. Find the number.

22. Charles Pecot can afford enough fencing to enclose a rectangular garden with a perimeter of 140 feet. If the width of his garden is to be 30 feet, find the length.

23. The number 120 is 15% of what number?

24. The following bar graph shows the estimated worldwide number of Internet users by region as of November 2000.

a. Find the region that has the most Internet users and approximate the number of users.

b. How many more users are in the US/Canada region than the Europe region?

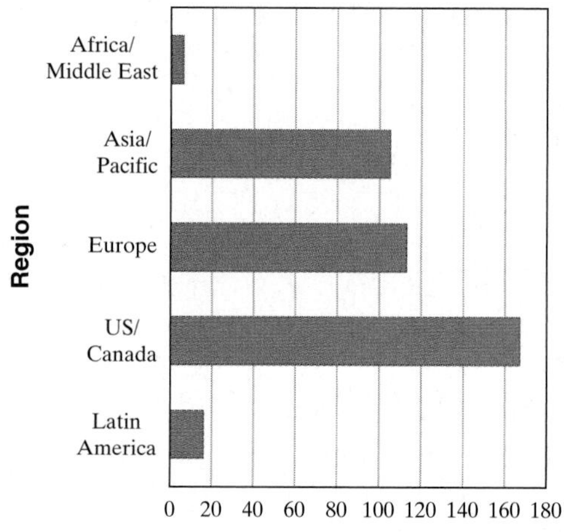

Source: Nua Internet Surveys

25. Complete each ordered pair solution so that it is a solution of the equation $3x + y = 12$.

a. $(0, \ )$

b. $(\ , 6)$

c. $(-1, \ )$

26. Graph: $2x + y = 5$

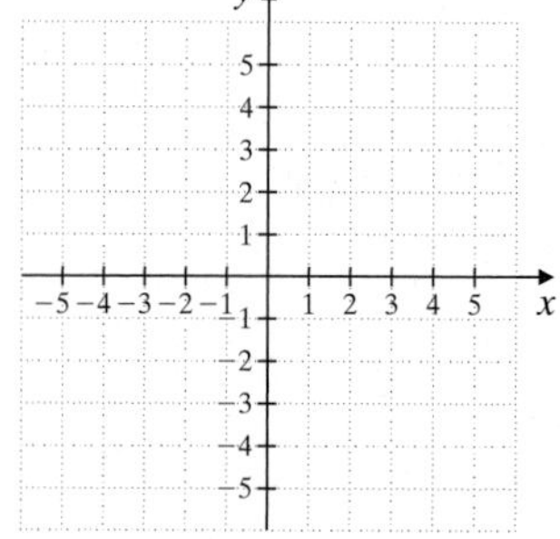

27. Find the slope of the line: $-2x + 3y = 11$

Exponents and Polynomials

CHAPTER 4

Recall from Chapter 1 that an exponent is a shorthand notation for repeated factors. This chapter explores additional concepts about exponents and exponential expressions. An especially useful type of exponential expression is a polynomial. Polynomials model many real-world phenomena. Our goal in this chapter is to become proficient with operations on polynomials.

Niagara Falls is visited by millions of tourists each year. Located between Niagara Falls, New York, and Niagara Falls, Ontario, Canada, on the Niagara River linking Lake Erie and Lake Ontario, the Falls consist of the American Falls, Bridal Veil Falls, and the Canadian, or Horseshoe, Falls. Together, they are known simply as Niagara Falls. The Falls were formed about 12,000 years ago and have since receded upstream 7 miles to their present location. Together, the Falls are roughly 3660 feet wide along their brinks. Water flowing over Niagara Falls drops about 167 feet to the river below and has the potential of generating 4.4 million kilowatts in hydroelectric power. In Exercise 99 on page 277 in Section 4.2, we will use exponents, through scientific notation, to compute the phenomenal volume of water flowing over Niagara Falls in an hour.

Answers

1. ____
2. ____
3. ____
4. ____
5. ____
6. ____
7. ____
8. ____
9. ____
10. ____
11. ____
12. ____
13. ____
14. ____
15. ____
16. ____
17. ____
18. ____
19. ____
20. ____

Name ____ Section ____ Date ____

Chapter 4 Pretest

1. Evaluate: $\left(-\frac{3}{4}\right)^2$

Simplify.

2. $(4y^6)(2y^7)$

3. $\frac{a^9b^{16}}{a^{12}b^5}$

4. $4^0 + 2x^0$

5. $\left(-\frac{1}{6}\right)^{-3}$

6. $\left(\frac{m^{-2}n}{m^6n^{-8}}\right)^{-2}$

7. $12x^2 + 3x - 5 - 8x^2 + 7x$

8. Express in scientific notation:
0.000000814

9. Find the degree of the following polynomial.
$8x - 4x^5 + 6x^3 + 10$

10. Find the value of the given polynomial when $x = -1$.
$-3x^3 + 2x^2 - 4$

Perform the indicated operations.

11. $(4x^2 - 3x + 9) + (6x^2 + 3x - 8)$

12. $(6y^2 - 4) - (-3y^2 + 5y - 1)$

13. $(2a^2 + 3ab - 7b^2) - (3a^2 + 3ab + 9b^2)$

14. $\left(-\frac{2}{7}n^6\right)\left(\frac{21}{16}n^3\right)$

15. $-2t^2(3t^5 + 4t^3 - 8)$

16. $(2y - 1)(5y + 6)$

17. $(7a - 5)^2$

18. $(4b + 9)(4b - 9)$

19. $\frac{16p^4 - 8p^3 + 20p^2}{4p}$

20. $\frac{5x^2 - 28x - 12}{x - 6}$

4.1 Exponents

OBJECTIVES

- **A** Evaluate exponential expressions.
- **B** Use the product rule for exponents.
- **C** Use the power rule for exponents.
- **D** Use the power rules for products and quotients.
- **E** Use the quotient rule for exponents, and define a number raised to the 0 power.
- **F** Decide which rule(s) to use to simplify an expression.

A Evaluating Exponential Expressions

In this section, we continue our work with integer exponents. As we reviewed in Section 1.3, for example,

$$2 \cdot 2 \cdot 2 \cdot 2 \cdot 2 = 2^5$$

The exponent 5 tells us how many times that 2 is a factor. The expression 2^5 is called an **exponential expression**. It is also called the fifth **power** of 2, or we can say that 2 is **raised** to the fifth power.

$$5^6 = \underbrace{5 \cdot 5 \cdot 5 \cdot 5 \cdot 5 \cdot 5}_{\text{6 factors; each factor is 5}} \quad \text{and} \quad (-3)^4 = \underbrace{(-3) \cdot (-3) \cdot (-3) \cdot (-3)}_{\text{4 factors; each factor is } -3}$$

The base of an exponential expression is the repeated factor. The exponent is the number of times that the base is used as a factor.

$$\overset{\text{exponent or power}}{a^n} = \underbrace{a \cdot a \cdot a \ldots a}_{n \text{ factors of } a}$$

(base: a)

Practice Problems 1–6

Evaluate (find the value of) each expression.

1. 3^4
2. 7^1
3. $(-2)^3$
4. -2^3
5. $\left(\frac{2}{3}\right)^2$
6. $5 \cdot 6^2$

EXAMPLES Evaluate (find the value of) each expression.

1. $2^3 = 2 \cdot 2 \cdot 2 = 8$

2. $3^1 = 3$. To raise 3 to the first power means to use 3 as a factor only once. When no exponent is shown, the exponent is assumed to be 1.

3. $(-4)^2 = (-4)(-4) = 16$

4. $-4^2 = -(4 \cdot 4) = -16$

5. $\left(\frac{1}{2}\right)^4 = \frac{1}{2} \cdot \frac{1}{2} \cdot \frac{1}{2} \cdot \frac{1}{2} = \frac{1}{16}$

6. $4 \cdot 3^2 = 4 \cdot 9 = 36$

Notice how similar -4^2 is to $(-4)^2$ in the examples above. The difference between the two is the parentheses. In $(-4)^2$, the parentheses tell us that the base, or the repeated factor, is -4. In -4^2, only 4 is the base.

Helpful Hint

Be careful when identifying the base of an exponential expression. Pay close attention to the use of parentheses.

$(-3)^2$	-3^2	$2 \cdot 3^2$
The base is -3.	The base is 3.	The base is 3.
$(-3)^2 = (-3)(-3) = 9$	$-3^2 = -(3 \cdot 3) = -9$	$2 \cdot 3^2 = 2 \cdot 3 \cdot 3 = 18$

An exponent has the same meaning whether the base is a number or a variable. If x is a real number and n is a positive integer, then x^n is the product of n factors, each of which is x.

$$x^n = \underbrace{x \cdot x \cdot x \cdot x \cdot x \ldots x}_{n \text{ factors; each factor is } x}$$

Answers

1. 81, **2.** 7, **3.** -8, **4.** -8, **5.** $\frac{4}{9}$, **6.** 180

Practice Problem 7

Evaluate each expression for the given value of x.

a. $3x^2$ when x is 4

b. $\frac{x^4}{-8}$ when x is -2

EXAMPLE 7 Evaluate each expression for the given value of x.

a. $2x^3$ when x is 5 **b.** $\frac{9}{x^2}$ when x is -3

Solution:

a. When x is 5, $2x^3 = 2 \cdot 5^3$

$= 2 \cdot (5 \cdot 5 \cdot 5)$

$= 2 \cdot 125$

$= 250$

b. When x is -3, $\frac{9}{x^2} = \frac{9}{(-3)^2}$

$= \frac{9}{(-3)(-3)}$

$= \frac{9}{9} = 1$

B Using the Product Rule

Exponential expressions can be multiplied, divided, added, subtracted, and themselves raised to powers. By our definition of an exponent,

$$5^4 \cdot 5^3 = \underbrace{(5 \cdot 5 \cdot 5 \cdot 5)}_{\text{4 factors of 5}} \cdot \underbrace{(5 \cdot 5 \cdot 5)}_{\text{3 factors of 5}}$$

$$= \underbrace{5 \cdot 5 \cdot 5 \cdot 5 \cdot 5 \cdot 5 \cdot 5}_{\text{7 factors of 5}}$$

$$= 5^7$$

Also,

$$x^2 \cdot x^3 = (x \cdot x) \cdot (x \cdot x \cdot x)$$

$$= x \cdot x \cdot x \cdot x \cdot x$$

$$= x^5$$

In both cases, notice that the result is exactly the same if the exponents are added.

$$5^4 \cdot 5^3 = 5^{4+3} = 5^7 \quad \text{and} \quad x^2 \cdot x^3 = x^{2+3} = x^5$$

This suggests the following rule.

Product Rule for Exponents

If m and n are positive integers and a is a real number, then

$$a^m \cdot a^n = a^{m+n}$$

For example,

$$3^5 \cdot 3^7 = 3^{5+7} = 3^{12}$$

In other words, to multiply two exponential expressions with the **same base**, we keep the base and add the exponents. We call this **simplifying** the exponential expression.

Practice Problems 8–12

Use the product rule to simplify each expression.

8. $7^3 \cdot 7^2$
9. $x^4 \cdot x^9$
10. $r^5 \cdot r$
11. $s^6 \cdot s^2 \cdot s^3$
12. $(-3)^9 \cdot (-3)$

Answers

7. a. 48, **b.** -2, **8.** 7^5, **9.** x^{13}, **10.** r^6, **11.** s^{11}, **12.** $(-3)^{10}$

EXAMPLES Use the product rule to simplify each expression.

8. $4^2 \cdot 4^5 = 4^{2+5} = 4^7$

9. $x^2 \cdot x^5 = x^{2+5} = x^7$

10. $y^3 \cdot y = y^3 \cdot y^1$

$= y^{3+1}$

$= y^4$

11. $y^3 \cdot y^2 \cdot y^7 = y^{3+2+7} = y^{12}$

12. $(-5)^7 \cdot (-5)^8 = (-5)^{7+8} = (-5)^{15}$

Helpful Hint

Don't forget that if no exponent is written, it is assumed to be 1.

EXAMPLE 13 Use the product rule to simplify $(2x^2)(-3x^5)$.

Solution: Recall that $2x^2$ means $2 \cdot x^2$ and $-3x^5$ means $-3 \cdot x^5$.

$$\begin{aligned}(2x^2)(-3x^5) &= 2 \cdot x^2 \cdot -3 \cdot x^5 && \text{Remove parentheses.}\\ &= 2 \cdot -3 \cdot x^2 \cdot x^5 && \text{Group factors with common bases.}\\ &= -6x^7 && \text{Simplify.}\end{aligned}$$

Practice Problem 13

Use the product rule to simplify $(6x^3)(-2x^9)$.

Helpful Hint

These examples will remind you of the difference between adding and multiplying terms.

Addition

$5x^3 + 3x^3 = (5 + 3)x^3 = 8x^3$ By the distributive property.

$7x + 4x^2 = 7x + 4x^2$ Cannot be combined.

Multiplication

$(5x^3)(3x^3) = 5 \cdot 3 \cdot x^3 \cdot x^3 = 15x^{3+3} = 15x^6$ By the product rule.

$(7x)(4x^2) = 7 \cdot 4 \cdot x \cdot x^2 = 28x^{1+2} = 28x^3$ By the product rule.

C Using the Power Rule

Exponential expressions can themselves be raised to powers. Let's try to discover a rule that simplifies an expression like $(x^2)^3$. By the definition of a^n,

$$(x^2)^3 = (x^2)(x^2)(x^2) \quad (x^2)^3 \text{ means 3 factors of } (x^2).$$

which can be simplified by the product rule for exponents.

$$(x^2)^3 = (x^2)(x^2)(x^2) = x^{2+2+2} = x^6$$

Notice that the result is exactly the same if we multiply the exponents.

$$(x^2)^3 = x^{2\cdot3} = x^6$$

The following rule states this result.

Power Rule for Exponents

If m and n are positive integers and a is a real number, then

$$(a^m)^n = a^{mn}$$

For example,

$$(7^2)^5 = 7^{2\cdot5} = 7^{10}$$

In other words, to raise an exponential expression to a power, we keep the base and multiply the exponents.

EXAMPLES Use the power rule to simplify each expression.

14. $(5^3)^6 = 5^{3\cdot6} = 5^{18}$

15. $(y^8)^2 = y^{8\cdot2} = y^{16}$

Practice Problems 14–15

Use the power rule to simplify each expression.

14. $(9^4)^{10}$ 15. $(z^6)^3$

Answers

13. $-12x^{12}$, **14.** 9^{40}, **15.** z^{18}

Take a moment to make sure that you understand when to apply the product rule and when to apply the power rule.

Product Rule → Add Exponents
$x^5 \cdot x^7 = x^{5+7} = x^{12}$
$y^6 \cdot y^2 = y^{6+2} = y^8$

Power Rule → Multiply Exponents
$(x^5)^7 = x^{5 \cdot 7} = x^{35}$
$(y^6)^2 = y^{6 \cdot 2} = y^{12}$

D Using the Power Rules for Products and Quotients

When the base of an exponential expression is a product, the definition of a^n still applies. For example, simplify $(xy)^3$ as follows.

$$(xy)^3 = (xy)(xy)(xy) \quad (xy)^3 \text{ means 3 factors of } (xy).$$
$$= x \cdot x \cdot x \cdot y \cdot y \cdot y \quad \text{Group factors with common bases.}$$
$$= x^3y^3 \quad \text{Simplify.}$$

Notice that to simplify the expression $(xy)^3$, we raise each factor within the parentheses to a power of 3.

$$(xy)^3 = x^3y^3$$

In general, we have the following rule.

Power of a Product Rule

If n is a positive integer and a and b are real numbers, then

$$(ab)^n = a^nb^n$$

For example:

$$(3x)^5 = 3^5x^5$$

In other words, to raise a product to a power, we raise each factor to the power.

EXAMPLES Simplify each expression.

16. $(st)^4 = s^4 \cdot t^4 = s^4t^4$ — Use the power of a product rule.

17. $(2a)^3 = 2^3 \cdot a^3 = 8a^3$ — Use the power of a product rule.

18. $(-5x^2y^3z)^2 = (-5)^2 \cdot (x^2)^2 \cdot (y^3)^2 \cdot (z^1)^2$ — Use the power of a product rule.

$= 25x^4y^6z^2$ — Use the power rule for exponents.

Practice Problems 16–18

Simplify each expression.

16. $(xy)^7$ 17. $(3y)^4$

18. $(-2p^4q^2r)^3$

Answers

16. x^7y^7, **17.** $81y^4$, **18.** $-8p^{12}q^6r^3$

Let's see what happens when we raise a quotient to a power. For example, we simplify $\left(\frac{x}{y}\right)^3$ as follows.

$$\left(\frac{x}{y}\right)^3 = \left(\frac{x}{y}\right)\left(\frac{x}{y}\right)\left(\frac{x}{y}\right) \quad \left(\frac{x}{y}\right)^3 \text{ means 3 factors of } \left(\frac{x}{y}\right).$$

$$= \frac{x \cdot x \cdot x}{y \cdot y \cdot y} \quad \text{Multiply fractions.}$$

$$= \frac{x^3}{y^3} \quad \text{Simplify.}$$

Notice that to simplify the expression, $\left(\frac{x}{y}\right)^3$, we raise both the numerator and the denominator to a power of 3.

$$\left(\frac{x}{y}\right)^3 = \frac{x^3}{y^3}$$

In general, we have the following rule.

Power of a Quotient Rule

If n is a positive integer and a and c are real numbers, then

$$\left(\frac{a}{c}\right)^n = \frac{a^n}{c^n}, \quad c \neq 0$$

For example:

$$\left(\frac{y}{7}\right)^3 = \frac{y^3}{7^3}$$

In other words, to raise a quotient to a power, we raise both the numerator and the denominator to the power.

EXAMPLES Simplify each expression.

19. $\left(\frac{m}{n}\right)^7 = \frac{m^7}{n^7}, \quad n \neq 0$ Use the power of a quotient rule.

20. $\left(\frac{2x^4}{3y^5}\right)^4 = \frac{2^4 \cdot (x^4)^4}{3^4 \cdot (y^5)^4}$ Use the power of a quotient rule.

$= \frac{16x^{16}}{81y^{20}}, \quad y \neq 0$ Use the power rule for exponents.

Practice Problems 19–20

Simplify each expression.

19. $\left(\frac{r}{s}\right)^6$

20. $\left(\frac{5x^6}{9y^3}\right)^2$

E Using the Quotient Rule and Defining the Zero Exponent

Another pattern for simplifying exponential expressions involves quotients.

$$\frac{x^5}{x^3} = \frac{x \cdot x \cdot x \cdot x \cdot x}{x \cdot x \cdot x}$$

$$= \frac{x \cdot x \cdot x \cdot x \cdot x}{x \cdot x \cdot x}$$

$$= 1 \cdot 1 \cdot 1 \cdot x \cdot x$$

$$= x \cdot x$$

$$= x^2$$

Answers

19. $\frac{r^6}{s^6}, \quad s \neq 0,$ **20.** $\frac{25x^{12}}{81y^6}, \quad y \neq 0$

Notice that the result is exactly the same if we subtract exponents of the common bases.

$$\frac{x^5}{x^3} = x^{5-3} = x^2$$

The following rule states this result in a general way.

Quotient Rule for Exponents

If m and n are positive integers and a is a real number, then

$$\frac{a^m}{a^n} = a^{m-n}, \quad a \neq 0$$

For example,

$$\frac{x^6}{x^2} = x^{6-2} = x^4, \quad x \neq 0$$

In other words, to divide one exponential expression by another with a common base, we keep the base and subtract the exponents.

Practice Problems 21–24

Simplify each quotient.

21. $\frac{y^7}{y^3}$ 22. $\frac{5^9}{5^6}$

23. $\frac{(-2)^{14}}{(-2)^{10}}$ 24. $\frac{7a^4b^{11}}{ab}$

EXAMPLES Simplify each quotient.

21. $\frac{x^5}{x^2} = x^{5-2} = x^3$ Use the quotient rule.

22. $\frac{4^7}{4^3} = 4^{7-3} = 4^4 = 256$ Use the quotient rule.

23. $\frac{(-3)^5}{(-3)^2} = (-3)^3 = -27$ Use the quotient rule.

24. $\frac{2x^5y^2}{xy} = 2 \cdot \frac{x^5}{x^1} \cdot \frac{y^2}{y^1}$

$= 2 \cdot (x^{5-1}) \cdot (y^{2-1})$ Use the quotient rule.

$= 2x^4y^1$ or $2x^4y$

Let's now give meaning to an expression such as x^0. To do so, we will simplify $\frac{x^3}{x^3}$ in two ways and compare the results.

$\frac{x^3}{x^3} = x^{3-3} = x^0$ Apply the quotient rule.

$\frac{x^3}{x^3} = \frac{x \cdot x \cdot x}{x \cdot x \cdot x} = 1$ Apply the fundamental principle for fractions.

Since $\frac{x^3}{x^3} = x^0$ and $\frac{x^3}{x^3} = 1$, we define that $x^0 = 1$ as long as x is not 0.

Answers

21. y^4, **22.** 125, **23.** 16, **24.** $7a^3b^{10}$

Zero Exponent

$a^0 = 1$, as long as a is not 0.

For example, $5^0 = 1$.

In other words, a base raised to the 0 power is 1, as long as the base is not 0.

EXAMPLES Simplify each expression.

25. $3^0 = 1$

26. $(5x^3y^2)^0 = 1$

27. $(-4)^0 = 1$

28. $-4^0 = -1 \cdot 4^0 = -1 \cdot 1 = -1$

Try the Concept Check in the margin.

F Deciding Which Rule to Use

Let's practice deciding which rule to use to simplify.

EXAMPLE 29 Simplify each expression.

a. $x^7 \cdot x^4$

b. $\left(\frac{1}{2}\right)^4$

c. $(9y^5)^2$

Solution:

a. Here, we have a product, so we use the product rule to simplify.

$$x^7 \cdot x^4 = x^{7+4} = x^{11}$$

b. This is a quotient raised to a power, so we use the power of a quotient rule.

$$\left(\frac{1}{2}\right)^4 = \frac{1^4}{2^4} = \frac{1}{16}$$

c. This is a product raised to a power, so we use the power of a product rule.

$$(9y^5)^2 = 9^2(y^5)^2 = 81y^{10}$$

Practice Problems 25–28

Simplify each expression.

25. 8^0

26. $(2r^2s)^0$

27. $(-5)^0$

28. -5^0

Concept Check

Suppose you are simplifying each expression. Tell whether you would *add* the exponents, *subtract* the exponents, *multiply* the exponents, *divide* the exponents, or *none of these.*

a. $(x^{63})^{21}$

b. $\frac{y^{15}}{y^3}$

c. $z^{16} + z^8$

d. $w^{45} \cdot w^9$

Practice Problem 29

Simplify each expression.

a. $\frac{x^7}{x^4}$

b. $(3y^4)^4$

c. $\left(\frac{x}{4}\right)^3$

Answers

25. 1, **26.** 1, **27.** 1, **28.** -1, **29. a.** x^3, **b.** $81y^{16}$, **c.** $\frac{x^3}{64}$

Concept Check: **a.** multiply, **b.** subtract, **c.** none of these, **d.** add

STUDY SKILLS REMINDER

Are you satisfied with your performance on a particular quiz or exam?

If not, analyze your quiz or exam like you would a good mystery novel. Look for common themes in your errors.

Were most of your errors a result of

- *Carelessness?* If your errors were careless, did you turn in your work before the allotted time expired? If so, resolve next time to use the entire time allotted. Any extra time can be spent checking your work.
- *Running out of time?* If so, make a point to better manage your time on your next exam. A few suggestions are to work any questions that you are unsure of last and to check your work after all questions have been answered.
- *Not understanding a concept?* If so, review that concept and correct your work. Remember next time to make sure that all concepts on a quiz or exam are understood before the exam.

Name ______________________ Section ________ Date ________

Mental Math

State the bases and the exponents for each expression.

1. 3^2 **2.** 5^4 **3.** $(-3)^6$ **4.** -3^7 **5.** -4^2

6. $(-4)^3$ **7.** $5 \cdot 3^4$ **8.** $9 \cdot 7^6$ **9.** $5x^2$ **10.** $(5x)^2$

EXERCISE SET 4.1

A *Evaluate each expression. See Examples 1 through 6.*

1. 7^2 **2.** -3^2 **3.** $(-5)^1$ **4.** $(-3)^2$ **5.** -2^4 **6.** -4^3

7. $(-2)^4$ **8.** $(-4)^3$ **9.** $\left(\frac{1}{3}\right)^3$ **10.** $\left(-\frac{1}{9}\right)^2$ **11.** $7 \cdot 2^4$ **12.** $9 \cdot 1^2$

13. Explain why $(-5)^4 = 625$, while $-5^4 = -625$.

14. Explain why $5 \cdot 4^2 = 80$, while $(5 \cdot 4)^2 = 400$.

Evaluate each expression with the given replacement values. See Example 7.

15. x^2 when $x = -2$ **16.** x^3 when $x = -2$ **17.** $5x^3$ when $x = 3$

18. $4x^2$ when $x = -1$ **19.** $2xy^2$ when $x = 3$ and $y = 5$ **20.** $-4x^2y^3$ when $x = 2$ and $y = -1$

21. $\frac{2z^4}{5}$ when $z = -2$ **22.** $\frac{10}{3y^3}$ when $y = 5$

B *Use the product rule to simplify each expression. Write the results using exponents. See Examples 8 through 13.*

23. $x^2 \cdot x^5$ **24.** $y^2 \cdot y$ **25.** $(-3)^3 \cdot (-3)^9$ **26.** $(-5)^7 \cdot (-5)^6$

27. $(5y^4)(3y)$ **28.** $(-2z^3)(-2z^2)$ **29.** $(4z^{10})(-6z^7)(z^3)$ **30.** $(12x^5)(-x^6)(x^4)$

△ **31.** The rectangle below has width $4x^2$ feet and length $5x^3$ feet. Find its area.

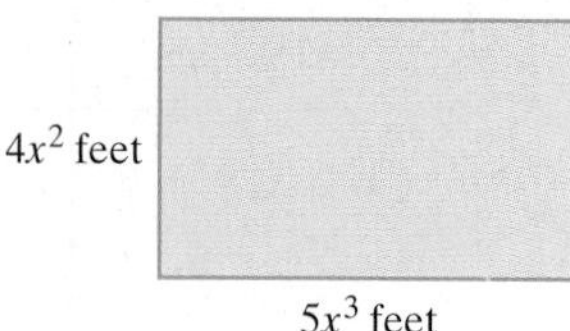

△ **32.** The parallelogram below has base length $9y^7$ meters and height $2y^{10}$ meters. Find its area.

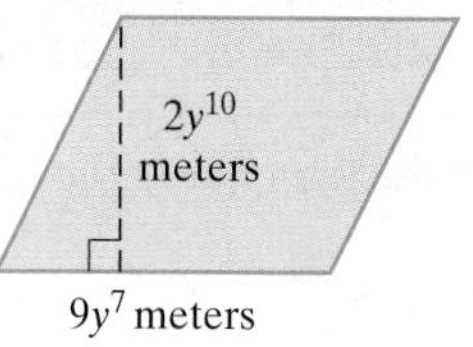

C D *Use the power rule and the power of a product or quotient rule to simplify each expression. See Examples 14 through 20.*

33. $(x^9)^4$ **34.** $(y^7)^5$ **35.** $(pq)^7$ **36.** $(ab)^6$ **37.** $(2a^5)^3$ **38.** $(4x^6)^2$

39. $\left(\frac{m}{n}\right)^9$ **40.** $\left(\frac{xy}{7}\right)^2$ **41.** $(x^2y^3)^5$ **42.** $(a^4b)^7$ **43.** $\left(\frac{-2xz}{y^5}\right)^2$ **44.** $\left(\frac{y^4}{-3z^3}\right)^3$

45. The square shown has sides of length $8z^5$ decimeters. Find its area.

46. Given the circle below with radius $5y$ centimeters, find its area. Do not approximate π.

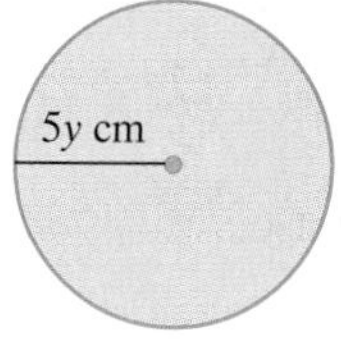

47. The vault below is in the shape of a cube. If each side is $3y^4$ feet, find its volume.

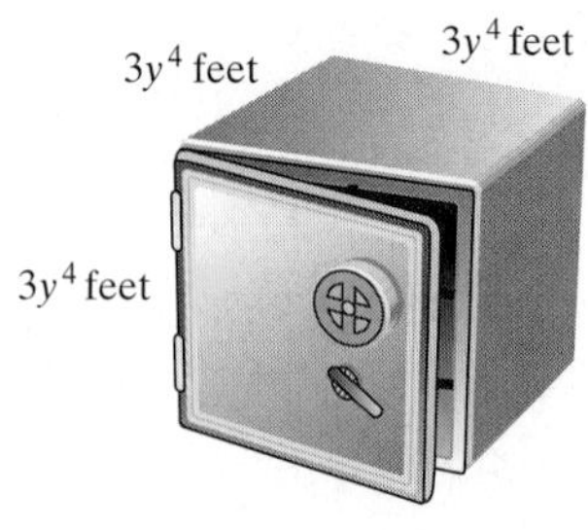

48. The silo shown is in the shape of a cylinder. If its radius is $4x$ meters and its height is $5x^3$ meters, find its volume. Do not approximate π.

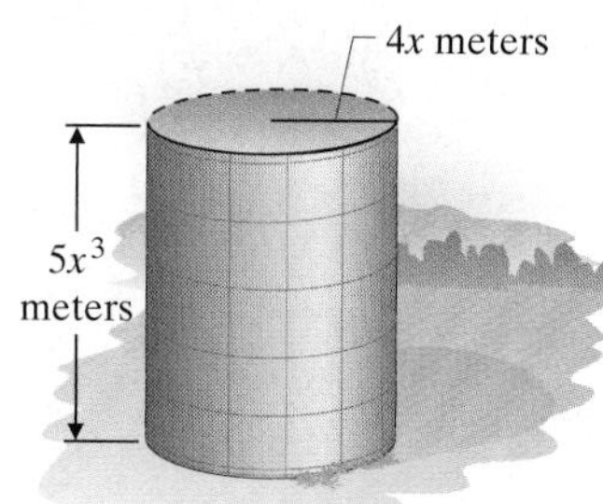

E *Use the quotient rule and simplify each expression. See Examples 21 through 24.*

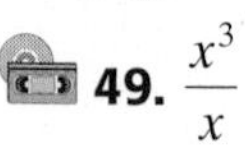
49. $\frac{x^3}{x}$ **50.** $\frac{y^{10}}{y^9}$ **51.** $\frac{(-2)^5}{(-2)^3}$ **52.** $\frac{(-5)^{14}}{(-5)^{11}}$

53. $\dfrac{p^7q^{20}}{pq^{15}}$

54. $\dfrac{x^8y^6}{xy^5}$

55. $\dfrac{7x^2y^6}{14x^2y^3}$

56. $\dfrac{9a^4b^7}{3ab^2}$

Simplify each expression. See Examples 25 through 28.

57. $(2x)^0$

58. $-4x^0$

59. $-2x^0$

60. $(4y)^0$

61. $5^0 + y^0$

62. $-3^0 + 4^0$

63. In your own words, explain why $5^0 = 1$.

64. In your own words, explain when $(-3)^n$ is positive and when it is negative.

F *Simplify each expression. See Example 29.*

65. -5^2

66. $(-5)^2$

67. $\left(\dfrac{1}{4}\right)^3$

68. $\left(\dfrac{2}{3}\right)^3$

69. $\dfrac{z^{12}}{z^4}$

70. $\dfrac{b^4}{b}$

71. $(9xy)^2$

72. $(2ab)^5$

73. $(6b)^0$

74. $(5ab)^0$

75. $2^3 + 2^5$

76. $7^2 - 7^0$

77. b^4b^2

78. y^4y^1

79. $a^2a^3a^4$

80. $x^2x^{15}x^9$

81. $(2x^3)(-8x^4)$

82. $(3y^4)(-5y)$

83. $(4a)^3$

84. $(2ab)^4$

85. $(-6xyz^3)^2$

86. $(-3xy^2a^3b)^3$

87. $\left(\dfrac{3y^5}{6x^4}\right)^3$

88. $\left(\dfrac{2ab}{6yz}\right)^4$

89. $\dfrac{3x^5}{x^4}$

90. $\dfrac{5x^9}{x^3}$

91. $\dfrac{2x^3y^2z}{xyz}$

92. $\dfrac{x^{12}y^{13}}{x^5y^7}$

Review and Preview

Subtract. See Section 1.5.

93. $5 - 7$

94. $9 - 12$

95. $3 - (-2)$

96. $5 - (-10)$

97. $-11 - (-4)$

98. $-15 - (-21)$

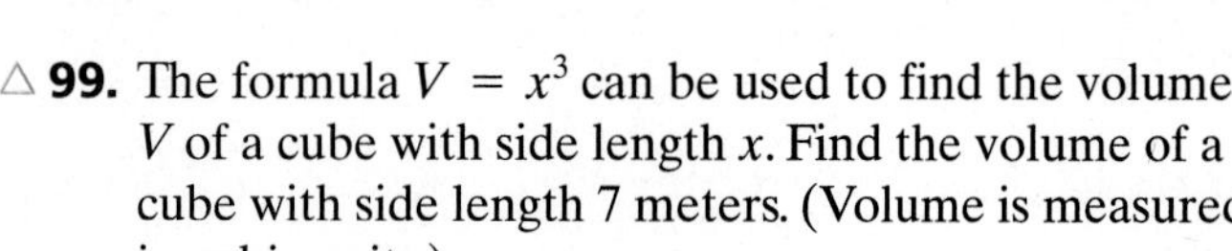

Combining Concepts

△ **99.** The formula $V = x^3$ can be used to find the volume V of a cube with side length x. Find the volume of a cube with side length 7 meters. (Volume is measured in cubic units.)

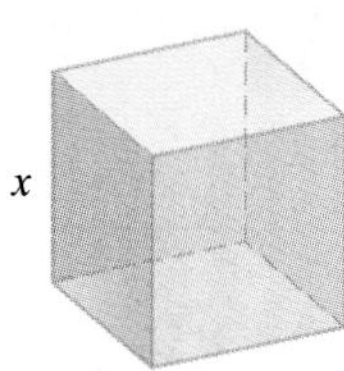

△ **100.** The formula $S = 6x^2$ can be used to find the surface area S of a cube with side length x. Find the surface area of a cube with side length 5 meters. (Surface area is measured in square units.)

△ **101.** To find the amount of water that a swimming pool in the shape of a cube can hold, do we use the formula for volume of the cube or surface area of the cube? (See Exercises 99 and 100.)

△ **102.** To find the amount of material needed to cover an ottoman in the shape of a cube, do we use the formula for volume of the cube or surface area of the cube? (See Exercises 99 and 100.)

Simplify each expression. Assume that variables represent positive integers.

103. $x^{5a}x^{4a}$

104. $b^{9a}b^{4a}$

105. $(a^b)^5$

106. $(2a^{4b})^4$

107. $\dfrac{x^{9a}}{x^{4a}}$

108. $\dfrac{y^{15b}}{y^{6b}}$

109. Suppose you borrow money for 6 months. If the interest rate is compounded monthly, the formula $A = P\left(1 + \frac{r}{12}\right)^6$ gives the total amount A to be repaid at the end of 6 months. For a loan of $P = \$1000$ and interest rate of 9% ($r = 0.09$), how much money will you need to pay off the loan?

110. On April 18, 2001, the Federal Reserve discount rate was set at 4%. (*Source:* Federal Reserve Board) The discount rate is the interest rate at which banks can borrow money from the Federal Reserve System. Suppose a bank needs to borrow money from the Federal Reserve System for 3 months. If the interest is compounded monthly, the formula $A = P\left(1 + \frac{r}{12}\right)^3$ gives the total amount A to be repaid at the end of 3 months. For a loan of $P = \$500{,}000$ and interest rate of $r = 0.04$, how much money will the bank repay to the Federal Reserve at the end of 3 months? Round to the nearest dollar.

4.2 Negative Exponents and Scientific Notation

OBJECTIVES

- **A** Simplify expressions containing negative exponents.
- **B** Use the rules and definitions for exponents to simplify exponential expressions.
- **C** Write numbers in scientific notation.
- **D** Convert numbers in scientific notation to standard form.

SSM TUTOR CENTER SG CD & VIDEO MATH PRO WEB

A Simplifying Expressions Containing Negative Exponents

Our work with exponential expressions so far has been limited to exponents that are positive integers or 0. Here we will also give meaning to an expression like x^{-3}.

Suppose that we wish to simplify the expression $\frac{x^2}{x^5}$. If we use the quotient rule for exponents, we subtract exponents:

$$\frac{x^2}{x^5} = x^{2-5} = x^{-3}, \quad x \neq 0$$

But what does x^{-3} mean? Let's simplify $\frac{x^2}{x^5}$ using the definition of a^n.

$$\frac{x^2}{x^5} = \frac{x \cdot x}{x \cdot x \cdot x \cdot x \cdot x}$$

$$= \frac{x \cdot x}{x \cdot x \cdot x \cdot x \cdot x}$$ Divide numerator and denominator by common factors by applying the fundamental principle for fractions.

$$= \frac{1}{x^3}$$

If the quotient rule is to hold true for negative exponents, then x^{-3} must equal $\frac{1}{x^3}$.

From this example, we state the definition for negative exponents.

Negative Exponents

If a is a real number other than 0 and n is an integer, then

$$a^{-n} = \frac{1}{a^n}$$

For example,

$$x^{-3} = \frac{1}{x^3}$$

In other words, another way to write a^{-n} is to take its reciprocal and change the sign of its exponent.

EXAMPLES Simplify by writing each expression with positive exponents only.

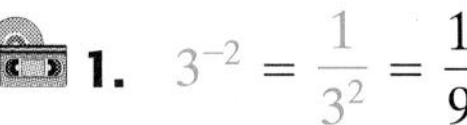

1. $3^{-2} = \frac{1}{3^2} = \frac{1}{9}$ Use the definition of negative exponent.

2. $2x^{-3} = 2 \cdot \frac{1}{x^3} = \frac{2}{x^3}$ Use the definition of negative exponent.

Helpful Hint

Don't forget that since there are no parentheses, only x is the base for the exponent -3.

3. $2^{-1} + 4^{-1} = \frac{1}{2} + \frac{1}{4} = \frac{2}{4} + \frac{1}{4} = \frac{3}{4}$

4. $(-2)^{-4} = \frac{1}{(-2)^4} = \frac{1}{(-2)(-2)(-2)(-2)} = \frac{1}{16}$

Practice Problems 1–4

Simplify by writing each expression with positive exponents only.

1. 5^{-3}
2. $7x^{-4}$
3. $5^{-1} + 3^{-1}$
4. $(-3)^{-4}$

Answers

1. $\frac{1}{125}$, **2.** $\frac{7}{x^4}$, **3.** $\frac{8}{15}$, **4.** $\frac{1}{81}$

Helpful Hint

A negative exponent *does not affect* the sign of its base.

Remember: Another way to write a^{-n} is to take its reciprocal and change the sign of its exponent: $a^{-n} = \frac{1}{a^n}$. For example,

$$x^{-2} = \frac{1}{x^2}, \qquad 2^{-3} = \frac{1}{2^3} \text{ or } \frac{1}{8}$$

$$\frac{1}{y^{-4}} = \frac{1}{\frac{1}{y^4}} = y^4, \qquad \frac{1}{5^{-2}} = 5^2 \text{ or } 25$$

Practice Problems 5–8

Simplify each expression. Write each result using positive exponents only.

5. $\left(\frac{6}{7}\right)^{-2}$ 6. $\frac{x}{x^{-4}}$

7. $\frac{y^{-9}}{z^{-5}}$ 8. $\frac{y^{-4}}{y^6}$

EXAMPLES Simplify each expression. Write each result using positive exponents only.

5. $\left(\frac{2}{3}\right)^{-3} = \frac{2^{-3}}{3^{-3}} = \frac{3^3}{2^3} = \frac{27}{8}$ Use the negative exponent rule.

6. $\frac{y}{y^{-2}} = \frac{y^1}{y^{-2}} = y^{1-(-2)} = y^3$ Use the quotient rule.

7. $\frac{p^{-4}}{q^{-9}} = \frac{q^9}{p^4}$ Use the negative exponent rule.

8. $\frac{x^{-5}}{x^7} = x^{-5-7} = x^{-12} = \frac{1}{x^{12}}$

B Simplifying Exponential Expressions

All the previously stated rules for exponents apply for negative exponents also. Here is a summary of the rules and definitions for exponents.

Summary of Exponent Rules

If m and n are integers and a, b, and c are real numbers, then:

Product rule for exponents:	$a^m \cdot a^n = a^{m+n}$
Power rule for exponents:	$(a^m)^n = a^{m \cdot n}$
Power of a product:	$(ab)^n = a^n b^n$
Power of a quotient:	$\left(\frac{a}{c}\right)^n = \frac{a^n}{c^n}$, $c \neq 0$
Quotient rule for exponents:	$\frac{a^m}{a^n} = a^{m-n}$, $a \neq 0$
Zero exponent:	$a^0 = 1$, $a \neq 0$
Negative exponent:	$a^{-n} = \frac{1}{a^n}$, $a \neq 0$

Answers

5. $\frac{49}{36}$, **6.** x^5, **7.** $\frac{z^5}{y^9}$, **8.** $\frac{1}{y^{10}}$

EXAMPLES

Simplify each expression. Write each result using positive exponents only.

9. $\dfrac{(x^3)^4x}{x^7} = \dfrac{x^{12}\cdot x}{x^7} = \dfrac{x^{12+1}}{x^7} = \dfrac{x^{13}}{x^7} = x^{13-7} = x^6$ Use the power rule.

10. $\left(\dfrac{3a^2}{b}\right)^{-3} = \dfrac{3^{-3}(a^2)^{-3}}{b^{-3}}$ Raise each factor in the numerator and the denominator to the −3 power.

$= \dfrac{3^{-3}a^{-6}}{b^{-3}}$ Use the power rule.

$= \dfrac{b^3}{3^3a^6}$ Use the negative exponent rule.

$= \dfrac{b^3}{27a^6}$ Write 3^3 as 27.

11. $(y^{-3}z^6)^{-6} = (y^{-3})^{-6}(z^6)^{-6}$ Raise each factor to the −6 power.

$= y^{18}z^{-36} = \dfrac{y^{18}}{z^{36}}$

12. $\dfrac{(2x)^5}{x^3} = \dfrac{2^5\cdot x^5}{x^3} = 2^5\cdot x^{5-3} = 32x^2$ Raise each factor in the numerator to the fifth power.

13. $\dfrac{x^{-7}}{(x^4)^3} = \dfrac{x^{-7}}{x^{12}} = x^{-7-12} = x^{-19} = \dfrac{1}{x^{19}}$

14. $(5y^3)^{-2} = 5^{-2}(y^3)^{-2}$ Raise each factor to the −2 power.

$= 5^{-2}y^{-6} = \dfrac{1}{5^2y^6} = \dfrac{1}{25y^6}$

15. $\dfrac{(2xy)^{-3}}{(x^2y^3)^2} = \dfrac{2^{-3}x^{-3}y^{-3}}{(x^2)^2\cdot(y^3)^2} = \dfrac{2^{-3}x^{-3}y^{-3}}{x^4y^6}$

$= 2^{-3}x^{-3-4}y^{-3-6} = 2^{-3}x^{-7}y^{-9}$

$= \dfrac{1}{2^3x^7y^9}$ or $\dfrac{1}{8x^7y^9}$

Practice Problems 9–15

Simplify each expression. Write each result using positive exponents only.

9. $\dfrac{(x^5)^3x}{x^4}$
10. $\left(\dfrac{9x^3}{y}\right)^{-2}$
11. $(a^{-4}b^7)^{-5}$
12. $\dfrac{(2x)^4}{x^8}$
13. $\dfrac{y^{-10}}{(y^5)^4}$
14. $(4a^2)^{-3}$
15. $\dfrac{(3x^{-2}y)^{-2}}{4x^7y}$

C Writing Numbers in Scientific Notation

Both very large and very small numbers frequently occur in many fields of science. For example, the distance between the sun and the planet Pluto is approximately 5,906,000,000 kilometers, and the mass of a proton is approximately 0.00000000000000000000000165 gram. It can be tedious to write these numbers in this standard decimal notation, so **scientific notation** is used as a convenient shorthand for expressing very large and very small numbers.

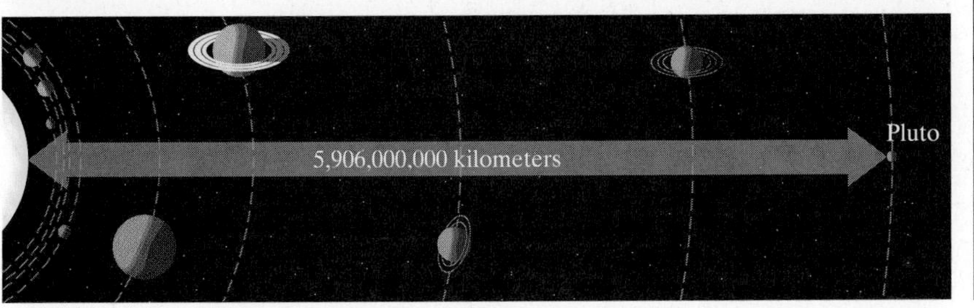

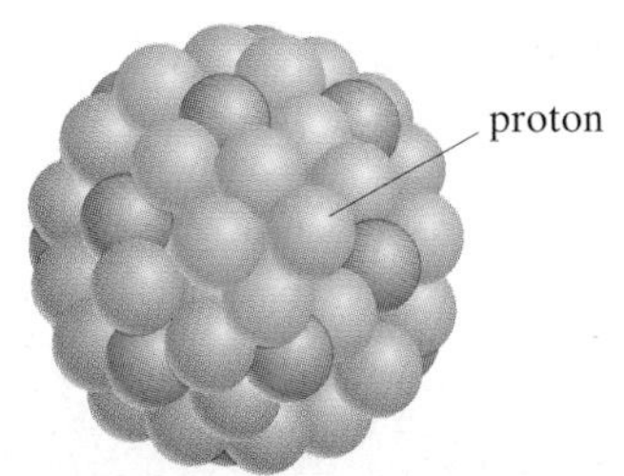

Mass of proton is approximately 0.000 000 000 000 000 000 000 001 65 gram

Answers

9. x^{12}, **10.** $\dfrac{y^2}{81x^6}$, **11.** $\dfrac{a^{20}}{b^{35}}$, **12.** $\dfrac{16}{x^4}$, **13.** $\dfrac{1}{y^{30}}$, **14.** $\dfrac{1}{64a^6}$, **15.** $\dfrac{1}{36x^3y^3}$

Scientific Notation

A positive number is written in scientific notation if it is written as the product of a number a, where $1 \leq a < 10$, and an integer power r of 10: $a \times 10^r$

The following numbers are written in scientific notation. The $\times$ sign for multiplication is used as part of the notation.

2.03×10^2	7.362×10^7	5.906×10^9	(Distance between the sun and Pluto)
1×10^{-3}	8.1×10^{-5}	1.65×10^{-24}	(Mass of a proton)

The following steps are useful when writing numbers in scientific notation.

To Write a Number in Scientific Notation

Step 1. Move the decimal point in the original number to the right so that the new number has a value between 1 and 10.

Step 2. Count the number of decimal places the decimal point is moved in Step 1. If the original number is 10 or greater, the count is positive. If the original number is less than 1, the count is negative.

Step 3. Multiply the new number in Step 1 by 10 raised to an exponent equal to the count found in Step 2.

Practice Problem 16

Write each number in scientific notation.

a. 420,000 b. 0.00017
c. 9,060,000,000 d. 0.000007

EXAMPLE 16 Write each number in scientific notation.

a. 367,000,000 **b.** 0.000003
c. 20,520,000,000 **d.** 0.00085

Solution:

a. Step 1. Move the decimal point until the number is between 1 and 10.

367,000,000
8 places

Step 2. The decimal point is moved 8 places and the original number is 10 or greater, so the count is positive 8.

Step 3. $367{,}000{,}000 = 3.67 \times 10^8$.

b. Step 1. Move the decimal point until the number is between 1 and 10.

0.000003
6 places

Step 2. The decimal point is moved 6 places and the original number is less than 1, so the count is -6.

Step 3. $0.000003 = 3.0 \times 10^{-6}$

c. $20{,}520{,}000{,}000 = 2.052 \times 10^{10}$

d. $0.00085 = 8.5 \times 10^{-4}$

D Converting Numbers to Standard Form

A number written in scientific notation can be rewritten in standard form. For example, to write 8.63×10^3 in standard form, recall that $10^3 = 1000$.

$$8.63 \times 10^3 = 8.63(1000) = 8630$$

Notice that the exponent on the 10 is positive 3, and we moved the decimal point 3 places to the right.

Answers

16. a. 4.2×10^5, **b.** 1.7×10^{-4}, **c.** 9.06×10^9, **d.** 7×10^{-6}

To write 7.29×10^{-3} in standard form, recall that $10^{-3} = \frac{1}{10^3} = \frac{1}{1000}$.

$$7.29 \times 10^{-3} = 7.29\left(\frac{1}{1000}\right) = \frac{7.29}{1000} = 0.00729$$

The exponent on the 10 is negative 3, and we moved the decimal to the left 3 places.

In general, **to write a scientific notation number in standard form**, move the decimal point the same number of places as the exponent on 10. If the exponent is positive, move the decimal point to the right; if the exponent is negative, move the decimal point to the left.

Try the Concept Check in the margin.

Concept Check

Which number in each pair is larger?

a. 7.8×10^3 or 2.1×10^5

b. 9.2×10^{-2} or 2.7×10^4

c. 5.6×10^{-4} or 6.3×10^{-5}

EXAMPLE 17

Write each number in standard notation, without exponents.

a. 1.02×10^5 **b.** 7.358×10^{-3}

c. 8.4×10^7 **d.** 3.007×10^{-5}

Solution:

a. Move the decimal point 5 places to the right.

$$1.02 \times 10^5 = 102{,}000.$$

b. Move the decimal point 3 places to the left.

$$7.358 \times 10^{-3} = 0.007358$$

c. $8.4 \times 10^7 = 84{,}000{,}000.$ 7 places to the right

d. $3.007 \times 10^{-5} = 0.00003007$ 5 places to the left

Performing operations on numbers written in scientific notation makes use of the rules and definitions for exponents.

Practice Problem 17

Write the numbers in standard notation, without exponents.

a. 3.062×10^{-4} b. 5.21×10^4

c. 9.6×10^{-5} d. 6.002×10^6

EXAMPLE 18

Perform each indicated operation. Write each result in standard decimal notation.

a. $(8 \times 10^{-6})(7 \times 10^3)$

b. $\frac{12 \times 10^2}{6 \times 10^{-3}}$

Solution:

a. $(8 \times 10^{-6})(7 \times 10^3) = 8 \cdot 7 \cdot 10^{-6} \cdot 10^3$

$= 56 \times 10^{-3}$

$= 0.056$

b. $\frac{12 \times 10^2}{6 \times 10^{-3}} = \frac{12}{6} \times 10^{2-(-3)} = 2 \times 10^5 = 200{,}000$

Practice Problem 18

Perform each indicated operation. Write each result in standard decimal notation.

a. $(9 \times 10^7)(4 \times 10^{-9})$

b. $\frac{8 \times 10^4}{2 \times 10^{-3}}$

Answers

17. a. 0.0003062, **b.** 52,100, **c.** 0.000096, **d.** 6,002,000, **18. a.** 0.36, **b.** 40,000,000

Concept Check: **a.** 2.1×10^5, **b.** 2.7×10^4, **c.** 5.6×10^{-4}

CALCULATOR EXPLORATIONS

Scientific Notation

To enter a number written in scientific notation on a scientific calculator, locate the scientific notation key, which may be marked [EE] or [EXP]. To enter 3.1×10^7, press [3.1] [EE] [7]. The display should read [3.1 07].

Enter each number written in scientific notation on your calculator.

1. 5.31×10^3
2. -4.8×10^{14}
3. 6.6×10^{-9}
4. -9.9811×10^{-2}

Multiply each of the following on your calculator. Notice the form of the result.

5. $3{,}000{,}000 \times 5{,}000{,}000$
6. $230{,}000 \times 1{,}000$

Multiply each of the following on your calculator. Write the product in scientific notation.

7. $(3.26 \times 10^6)(2.5 \times 10^{13})$
8. $(8.76 \times 10^{-4})(1.237 \times 10^9)$

Name ______________________ Section __________ Date ____________

Mental Math

State each expression using positive exponents only.

1. $5x^{-2}$
2. $3x^{-3}$
3. $\frac{1}{y^{-6}}$
4. $\frac{1}{x^{-3}}$
5. $\frac{4}{y^{-3}}$
6. $\frac{16}{y^{-7}}$

EXERCISE SET 4.2

A *Simplify each expression. Write each result using positive exponents only. See Examples 1 through 8.*

1. 4^{-3}
2. 6^{-2}
3. $7x^{-3}$
4. $(7x)^{-3}$
5. $\left(-\frac{1}{4}\right)^{-3}$
6. $\left(-\frac{1}{8}\right)^{-2}$
7. $3^{-1} + 2^{-1}$
8. $4^{-1} + 4^{-2}$
9. $\frac{1}{p^{-3}}$
10. $\frac{1}{q^{-5}}$
11. $\frac{p^{-5}}{q^{-4}}$
12. $\frac{r^{-5}}{s^{-2}}$
13. $\frac{x^{-2}}{x}$
14. $\frac{y}{y^{-3}}$
15. $\frac{z^{-4}}{z^{-7}}$
16. $\frac{x^{-4}}{x^{-1}}$
17. $2^0 + 3^{-1}$
18. $4^{-2} - 4^{-3}$
19. $(-3)^{-2}$
20. $(-2)^{-6}$
21. $\frac{-1}{p^{-4}}$
22. $\frac{-1}{y^{-6}}$
23. $-2^0 - 3^0$
24. $5^0 + (-5)^0$

B *Simplify each expression. Write each result using positive exponents only. See Examples 9 through 15.*

25. $\frac{x^2x^5}{x^3}$
26. $\frac{y^4y^5}{y^6}$
27. $\frac{p^2p}{p^{-1}}$
28. $\frac{y^3y}{y^{-2}}$
29. $\frac{(m^5)^4m}{m^{10}}$
30. $\frac{(x^2)^8x}{x^9}$
31. $\frac{r}{r^{-3}r^{-2}}$
32. $\frac{p}{p^{-3}q^{-5}}$
33. $(x^5y^3)^{-3}$
34. $(z^5x^5)^{-3}$
35. $\frac{(x^2)^3}{x^{10}}$
36. $\frac{(y^4)^2}{y^{12}}$
37. $\frac{(a^5)^2}{(a^3)^4}$
38. $\frac{(x^2)^5}{(x^4)^3}$
39. $\frac{8k^4}{2k}$
40. $\frac{27r^4}{3r^6}$
41. $\frac{-6m^4}{-2m^3}$
42. $\frac{15a^4}{-15a^5}$
43. $\frac{-24a^6b}{6ab^2}$
44. $\frac{-5x^4y^5}{15x^4y^2}$
45. $\frac{6x^2y^3}{-7xy^5}$
46. $\frac{-8xa^2b}{-5xa^5b}$
47. $(a^{-5}b^2)^{-6}$
48. $(4^{-1}x^5)^{-2}$
49. $\left(\frac{x^{-2}y^4}{x^3y^7}\right)^2$
50. $\left(\frac{a^5b}{a^7b^{-2}}\right)^{-3}$
51. $\frac{4^2z^{-3}}{4^3z^{-5}}$
52. $\frac{3^{-1}x^4}{3^3x^{-7}}$
53. $\frac{2^{-3}x^{-4}}{2^2x}$
54. $\frac{5^{-1}z^7}{5^{-2}z^9}$
55. $\frac{7ab^{-4}}{7^{-1}a^{-3}b^2}$
56. $\frac{6^{-5}x^{-1}y^2}{6^{-2}x^{-4}y^4}$
57. $\left(\frac{a^{-5}b}{ab^3}\right)^{-4}$
58. $\left(\frac{r^{-2}s^{-3}}{r^{-4}s^{-3}}\right)^{-3}$
59. $\frac{(xy^3)^5}{(xy)^{-4}}$
60. $\frac{(rs)^{-3}}{(r^2s^3)^2}$

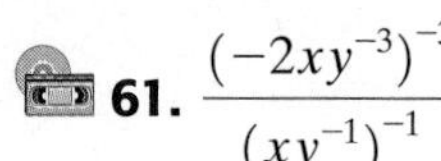

61. $\dfrac{(-2xy^{-3})^{-3}}{(xy^{-1})^{-1}}$

62. $\dfrac{(-3x^2y^2)^{-2}}{(xyz)^{-2}}$

63. $\dfrac{(a^4b^{-7})^{-5}}{(5a^2b^{-1})^{-2}}$

64. $\dfrac{(a^6b^{-2})^4}{(4a^{-3}b^{-3})^3}$

△ 65. Find the volume of the cube.

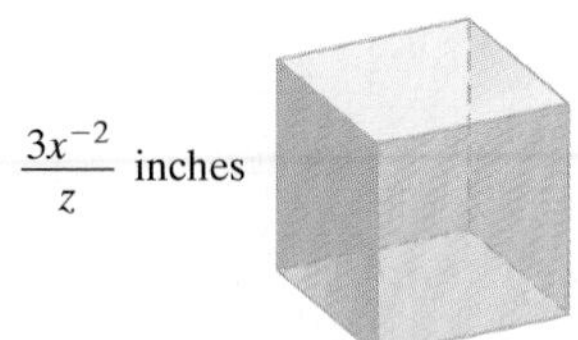

△ 66. Find the area of the triangle.

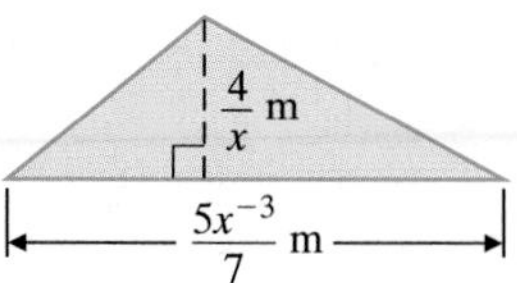

C *Write each number in scientific notation. See Example 16.*

67. 78,000

68. 9,300,000,000

69. 0.00000167

70. 0.00000017

71. 0.00635

72. 0.00194

73. 1,160,000

74. 700,000

75. The temperature of the Sun at its core is 15,600,000 degrees Kelvin. Write 15,600,000 in scientific notation. (*Source:* Students for the Exploration and Development of Space)

76. Google.com is an Internet search engine that allows users to search over 1,346,966,000 Web pages. Write 1,346,966,000 in scientific notation. (*Source:* Google, Inc.)

77. At this writing, the world's largest optical telescopes are the twin Keck Telescopes located near the summit of Mauna Kea in Hawaii. The elevation of the Keck Telescopes is about 13,600 feet above sea level. Write 13,600 in scientific notation. (*Source:* W.M. Keck Observatory)

78. More than 2,000,000,000 pencils are manufactured in the U.S. annually. Write this number in scientific notation. (*Source:* AbsoluteTrivia.com)

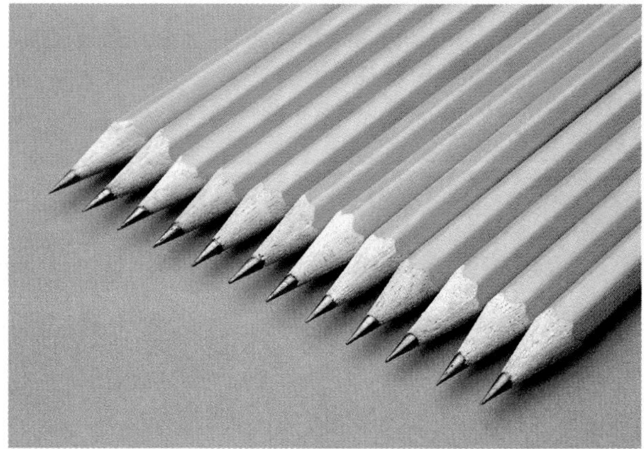

79. In May 2001, the population of the United States was roughly 284,000,000. Write 284,000,000 in scientific notation. (*Source:* U.S. Census Bureau)

80. Pioneer 10 became the first spacecraft to leave the solar system eleven years after it was launched in 1972. When it was contacted in April 2001 to verify that it could still transmit a radio signal, Pioneer 10 was 7,290,000,000 miles from Earth. Write 7,290,000,000 in scientific notation. (*Source:* NASA Ames Research Center)

D *Write each number in standard notation. See Example 17.*

81. 8.673×10^{-10}

82. 9.056×10^{-4}

83. 3.3×10^{-2}

84. 4.8×10^{-6}

85. 2.032×10^{4}

86. 9.07×10^{10}

87. The mass of an atom of the uranium isotope U-238 is 3.97×10^{-22} grams. Write this number in standard notation.

88. The mass of a hydrogen atom is 1.7×10^{-24} grams. Write this number in standard notation.

89. Each second, the Sun converts 7.0×10^{8} tons of hydrogen into helium and energy in the form of gamma rays. Write this number in standard notation. (*Source:* Students for the Exploration and Development of Space)

90. In chemistry, Avogadro's number is the number of atoms in one mole of an element. Avogadro's number is $6.02214199 \times 10^{23}$. Write this number in standard notation. (*Source:* National Institute of Standards and Technology)

Evaluate each expression using exponential rules. Write each result in standard notation. See Example 18.

91. $(1.2 \times 10^{-3})(3 \times 10^{-2})$

92. $(2.5 \times 10^{6})(2 \times 10^{-6})$

93. $(4 \times 10^{-10})(7 \times 10^{-9})$

94. $(5 \times 10^{6})(4 \times 10^{-8})$

95. $\dfrac{8 \times 10^{-1}}{16 \times 10^{5}}$

96. $\dfrac{25 \times 10^{-4}}{5 \times 10^{-9}}$

97. $\dfrac{1.4 \times 10^{-2}}{7 \times 10^{-8}}$

98. $\dfrac{0.4 \times 10^{5}}{0.2 \times 10^{11}}$

99. Although the actual amount varies by season and time of day, the average volume of water that flows over Niagara Falls (the American and Canadian falls combined) each second is 7.5×10^{5} gallons. How much water flows over Niagara Falls in an hour? Write the result in scientific notation. (*Hint*: 1 hour equals 3600 seconds) (*Source:* niagarafallslive.com)

100. A beam of light travels 9.460×10^{12} kilometers per year. How far does light travel in 10,000 years? Write the result in scientific notation.

Review and Preview

Simplify each expression by combining any like terms. See Section 2.1.

101. $3x - 5x + 7$

102. $7w + w - 2w$

103. $y - 10 + y$

104. $-6z + 20 - 3z$

105. $7x + 2 - 8x - 6$

106. $10y - 14 - y - 14$

Combining Concepts

Simplify each expression. Write each result in standard notation.

107. $(2.63 \times 10^{12})(-1.5 \times 10^{-10})$

108. $(6.785 \times 10^{-4})(4.68 \times 10^{10})$

Light travels at a rate of 1.86×10^5 *miles per second. Use this information and the distance formula* $d = r \cdot t$ *to answer Exercises 109 and 110.*

109. If the distance from the moon to the Earth is 238,857 miles, find how long it takes the reflected light of the moon to reach the Earth. (Round to the nearest tenth of a second.)

110. If the distance from the sun to the Earth is 93,000,000 miles, find how long it takes the light of the sun to reach the Earth. (Round to the nearest tenth of a second.)

Simplify each expression. Assume that variables represent positive integers.

111. $a^{-4m} \cdot a^{5m}$

112. $(x^{-3s})^3$

113. $(3y^{2z})^3$

114. $a^{4m+1} \cdot a^4$

115. It was stated earlier that for an integer n,

$$x^{-n} = \frac{1}{x^n}, \quad x \neq 0$$

Explain why x may not equal 0.

116. Determine whether each statement is true or false.

a. $5^{-1} < 5^{-2}$

b. $\left(\frac{1}{5}\right)^{-1} < \left(\frac{1}{5}\right)^{-2}$

c. $a^{-1} < a^{-2}$ for all nonzero numbers.

Internet Excursions

Go To: http://www.prenhall.com/martin-gay_algebra What's Related

The Bureau of the Public Debt is part of the U.S. Department of the Treasury. The Bureau of the Public Debt borrows the money needed to run the federal government and keeps track of the debt. The given World Wide Web address will provide you with access to the Bureau of the Public Debt's Web Site, or a related site, where you can find the current size of the U.S. public debt to the penny. This site is updated daily. It also lists the amount of the public debt for the past month, as well as for selected dates in prior months and years.

117. Find the size of the public debt that is listed most recently on the Bureau of the Public Debt's Web site. Use the information on the Web site to record the debt amount and its date. Then write the debt amount in scientific notation, rounded to the nearest hundredth.
Debt amount: ______________
Date: ______________
Scientific notation (rounded to nearest hundredth): ______________

118. Look up the size of the public debt at the end of the month six months ago. Use the information on the Web site to record the debt amount and its date. Then write the debt amount in scientific notation, rounded to the nearest hundredth.
Debt amount: ______________
Date: ______________
Scientific notation (rounded to nearest hundredth): ______________

4.3 Introduction to Polynomials

A Defining Term and Coefficient

In this section, we introduce a special algebraic expression called a polynomial. Let's first review some definitions presented in Section 2.1.

Recall that a term is a number or the product of a number and variables raised to powers. The terms of an expression are separated by plus signs. The terms of the expression $4x^2 + 3x$ are $4x^2$ and $3x$. The terms of the expression $9x^4 - 7x - 1$, or $9x^4 + (-7x) + (-1)$, are $9x^4$, $-7x$, and -1.

Expression	Terms
$4x^2 + 3x$	$4x^2, 3x$
$9x^4 - 7x - 1$	$9x^4, -7x, -1$
$7y^3$	$7y^3$

The **numerical coefficient** of a term, or simply the **coefficient**, is the numerical factor of each term. If no numerical factor appears in the term, then the coefficient is understood to be 1. If the term is a number only, it is called a **constant term** or simply a **constant**.

Term	Coefficient
x^5	1
$3x^2$	3
$-4x$	-4
$-x^2y$	-1
3 (constant)	3

OBJECTIVES

- A Define term and coefficient of a term.
- B Define polynomial, monomial, binomial, trinomial, and degree.
- C Evaluate polynomials for given replacement values.
- D Simplify a polynomial by combining like terms.
- E Simplify a polynomial in several variables.

EXAMPLE 1

Complete the table for the expression $7x^5 - 8x^4 + x^2 - 3x + 5$.

Term	Coefficient
x^2	
	-8
$-3x$	
	7
5	

Solution: The completed table is shown below.

Term	Coefficient
x^2	1
$-8x^4$	-8
$-3x$	-3
$7x^5$	7
5	5

Practice Problem 1

Complete the table for the expression $-6x^6 + 4x^5 + 7x^3 - 9x^2 - 1$.

Term	Coefficient
$7x^3$	
	-9
$-6x^6$	
	4
-1	

Answer

1. term: $-9x^2$; $4x^5$, coefficient: 7; -6; -1

B Defining Polynomial, Monomial, Binomial, Trinomial, and Degree

Now we are ready to define what we mean by a polynomial.

Polynomial

A **polynomial in x** is a finite sum of terms of the form ax^n, where a is a real number and n is a whole number.

For example,

$$x^5 - 3x^3 + 2x^2 - 5x + 1$$

is a polynomial in x. Notice that this polynomial is written in **descending powers** of x because the powers of x decrease from left to right. (Recall that the term 1 can be thought of as $1x^0$.)

On the other hand,

$$x^{-5} + 2x - 3$$

is **not** a polynomial because one of its terms contains a variable with an exponent, -5, that is not a whole number.

Types of Polynomials

A **monomial** is a polynomial with exactly one term.
A **binomial** is a polynomial with exactly two terms.
A **trinomial** is a polynomial with exactly three terms.

The following are examples of monomials, binomials, and trinomials. Each of these examples is also a polynomial.

Polynomials			
Monomials	Binomials	Trinomials	More Than Three Terms
ax^2	$x + y$	$x^2 + 4xy + y^2$	$5x^3 - 6x^2 + 3x - 6$
$-3z$	$3p + 2$	$x^5 + 7x^2 - x$	$-y^5 + y^4 - 3y^3 - y^2 + y$
4	$4x^2 - 7$	$-q^4 + q^3 - 2q$	$x^6 + x^4 - x^3 + 1$

Each term of a polynomial has a degree. The **degree of a term in one variable** is the exponent on the variable.

Practice Problem 2

Identify the degree of each term of the trinomial $-15x^3 + 2x^2 - 5$.

Answer

2. 3; 2; 0

EXAMPLE 2

Identify the degree of each term of the trinomial $12x^4 - 7x + 3$.

Solution: The term $12x^4$ has degree 4.

The term $-7x$ has degree 1 since $-7x$ is $-7x^1$.

The term 3 has degree 0 since 3 is $3x^0$.

Each polynomial also has a degree.

Degree of a Polynomial

The **degree of a polynomial** is the greatest degree of any term of the polynomial.

EXAMPLE 3

Find the degree of each polynomial and tell whether the polynomial is a monomial, binomial, trinomial, or none of these.

a. $-2t^2 + 3t + 6$ **b.** $15x - 10$ **c.** $7x + 3x^3 + 2x^2 - 1$

Solution:

a. The degree of the trinomial $-2t^2 + 3t + 6$ is 2, the greatest degree of any of its terms.

b. The degree of the binomial $15x - 10$ or $15x^1 - 10$ is 1.

c. The degree of the polynomial $7x + 3x^3 + 2x^2 - 1$ is 3. ●

Practice Problem 3

Find the degree of each polynomial and tell whether the polynomial is a monomial, binomial, trinomial, or none of these.

a. $-6x + 14$

b. $9x - 3x^6 + 5x^4 + 2$

c. $10x^2 - 6x - 6$

C Evaluating Polynomials

Polynomials have different values depending on the replacement values for the variables. When we find the value of a polynomial for a given replacement value, we are evaluating the polynomial for that value.

EXAMPLE 4 Evaluate each polynomial when $x = -2$.

a. $-5x + 6$

b. $3x^2 - 2x + 1$

Solution:

a. $-5x + 6 = -5(-2) + 6$ Replace x with -2.
$= 10 + 6$
$= 16$

b. $3x^2 - 2x + 1 = 3(-2)^2 - 2(-2) + 1$ Replace x with -2.
$= 3(4) + 4 + 1$
$= 12 + 4 + 1$
$= 17$ ●

Many physical phenomena can be modeled by polynomials.

Practice Problem 4

Evaluate each polynomial when $x = -1$.

a. $-2x + 10$

b. $6x^2 + 11x - 20$

EXAMPLE 5 Finding Free-Fall Time

The CN Tower in Toronto, Ontario, is 1821 feet tall and is the world's tallest self-supporting structure. An object is dropped from the top of this building. Neglecting air resistance, the height in feet of the object at time t seconds is given by the polynomial $-16t^2 + 1821$. Find the height of the object when $t = 1$ second and when $t = 10$ seconds. (*Source:* World Almanac)

Solution: To find each height, we evaluate the polynomial when $t = 1$ and when $t = 10$.

$-16t^2 + 1821 = -16(1)^2 + 1821$ Replace t with 1.
$= -16(1) + 1821$
$= -16 + 1821$
$= 1805$

Practice Problem 5

Find the height of the object in Example 5 when $t = 3$ seconds and when $t = 7$ seconds.

Answers

3. a. binomial, 1, **b.** none of these, 6, **c.** trinomial, 2,
4. a. 12, **b.** −25, **5.** 1677 ft; 1037 ft

The height of the object at 1 second is 1805 feet.

$$\begin{aligned} -16t^2 + 1821 &= -16(10)^2 + 1821 \quad \text{Replace } t \text{ with 10.} \\ &= -16(100) + 1821 \\ &= -1600 + 1821 \\ &= 221 \end{aligned}$$

The height of the object at 10 seconds is 221 feet.

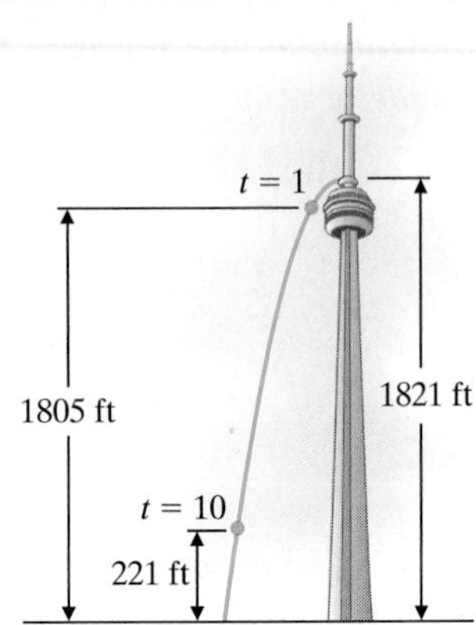

D Simplifying Polynomials by Combining Like Terms

We can simplify polynomials with like terms by combining the like terms. Recall that like terms are terms that contain exactly the same variables raised to exactly the same powers.

Like Terms	Unlike Terms
$5x^2, -7x^2$	$3x, 3y$
$y, 2y$	$-2x^2, -5x$
$\frac{1}{2}a^2b, -a^2b$	$6st^2, 4s^2t$

Only like terms can be combined. We combine like terms by applying the distributive property.

Practice Problems 6–10

Simplify each polynomial by combining any like terms.

6. $-6y + 8y$
7. $14y^2 + 3 - 10y^2 - 9$
8. $7x^3 + x^3$
9. $23x^2 - 6x - x - 15$
10. $\frac{2}{7}x^3 - \frac{1}{4}x + 2 - \frac{1}{2}x^3 + \frac{3}{8}x$

EXAMPLES Simplify each polynomial by combining any like terms.

6. $-3x + 7x = (-3 + 7)x = 4x$

7. $11x^2 + 5 + 2x^2 - 7 = 11x^2 + 2x^2 + 5 - 7$
$= 13x^2 - 2$

8. $9x^3 + x^3 = 9x^3 + 1x^3$ Write x^3 as $1x^3$.
$= 10x^3$

9. $5x^2 + 6x - 9x - 3 = 5x^2 - 3x - 3$ Combine like terms $6x$ and $-9x$.

10. $\frac{2}{5}x^4 + \frac{2}{3}x^3 - x^2 + \frac{1}{10}x^4 - \frac{1}{6}x^3$

$$= \left(\frac{2}{5} + \frac{1}{10}\right)x^4 + \left(\frac{2}{3} - \frac{1}{6}\right)x^3 - x^2$$

$$= \left(\frac{4}{10} + \frac{1}{10}\right)x^4 + \left(\frac{4}{6} - \frac{1}{6}\right)x^3 - x^2$$

$$= \frac{5}{10}x^4 + \frac{3}{6}x^3 - x^2$$

$$= \frac{1}{2}x^4 + \frac{1}{2}x^3 - x^2$$

Answers

6. $2y$, **7.** $4y^2 - 6$, **8.** $8x^3$, **9.** $23x^2 - 7x - 15$, **10.** $-\frac{3}{14}x^3 + \frac{1}{8}x + 2$

△ **EXAMPLE 11** Write a polynomial that describes the total area of the squares and rectangles shown below. Then simplify the polynomial.

Solution:

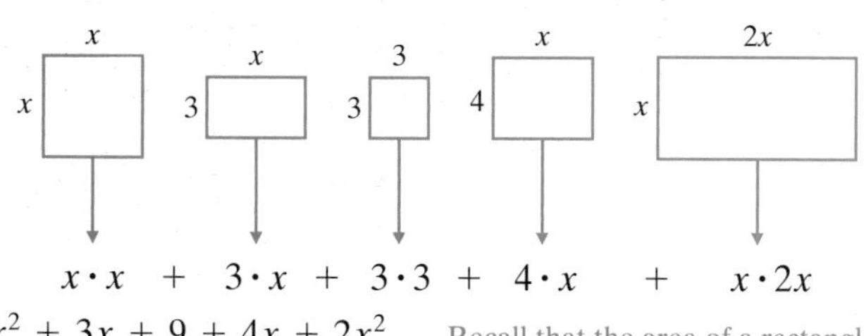

Area:

$x \cdot x + 3 \cdot x + 3 \cdot 3 + 4 \cdot x + x \cdot 2x$

$= x^2 + 3x + 9 + 4x + 2x^2$ Recall that the area of a rectangle is length times width.

$= 3x^2 + 7x + 9$ Combine like terms.

E Simplifying Polynomials Containing Several Variables

A polynomial may contain more than one variable. One example is

$$5x + 3xy^2 - 6x^2y^2 + x^2y - 2y + 1$$

We call this expression a polynomial in several variables.

The **degree of a term** with more than one variable is the sum of the exponents on the variables. The **degree of the polynomial** in several variables is still the greatest degree of the terms of the polynomial.

EXAMPLE 12

Identify the degrees of the terms and the degree of the polynomial $5x + 3xy^2 - 6x^2y^2 + x^2y - 2y + 1$.

Solution: To organize our work, we use a table.

Terms of Polynomial	Degree of Term	Degree of Polynomial
$5x$	1	
$3xy^2$	1 + 2 or 3	
$-6x^2y^2$	2 + 2 or 4	4 (highest degree)
x^2y	2 + 1 or 3	
$-2y$	1	
1	0	

To simplify a polynomial containing several variables, we combine any like terms.

EXAMPLES Simplify each polynomial by combining any like terms.

13. $3xy - 5y^2 + 7xy - 9x^2 = (3 + 7)xy - 5y^2 - 9x^2$
$= 10xy - 5y^2 - 9x^2$

This term can be written as $10xy$ or $10yx$.

14. $9a^2b - 6a^2 + 5b^2 + a^2b - 11a^2 + 2b^2$
$= 10a^2b - 17a^2 + 7b^2$

Practice Problem 11 △

Write a polynomial that describes the total area of the squares and rectangles shown below. Then simplify the polynomial.

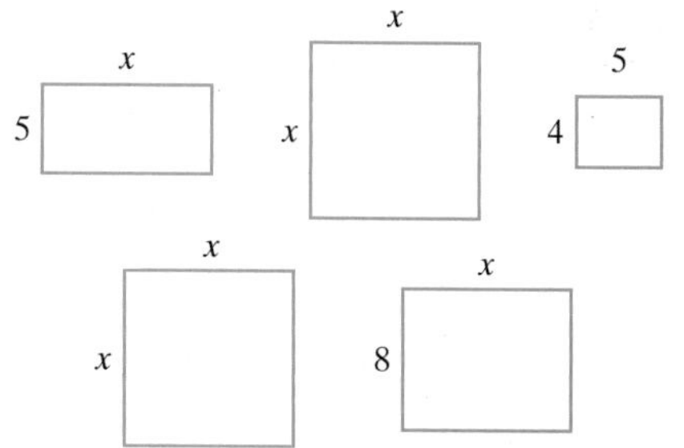

Practice Problem 12

Identify the degrees of the terms and the degree of the polynomial $-2x^3y^2 + 4 - 8xy + 3x^3y + 5xy^2$.

Practice Problems 13–14

Simplify each polynomial by combining any like terms.

13. $11ab - 6a^2 - ba + 8b^2$

14. $7x^2y^2 + 2y^2 - 4y^2x^2 + x^2 - y^2 + 5x^2$

Answers

11. $2x^2 + 13x + 20$, **12.** 5, 0, 2, 4, 3; 5, **13.** $10ab - 6a^2 + 8b^2$, **14.** $3x^2y^2 + y^2 + 6x^2$

FOCUS ON The Real World

SPACE EXPLORATION

From scientific observations on Earth, we know that Saturn, the second largest planet in our solar system, is a giant ball of gas surrounded by rings and orbited by 19 moons. We also know that Saturn has a diameter of 120,000 kilometers and a mass of 569,000,000,000,000,000,000,000,000 kilograms. But what is Saturn like below its outer layer of clouds? What are Saturn's rings made of? What is the surface of Saturn's largest moon, Titan, like? Could life ever be supported on Titan?

NASA is hoping to answer these questions and more about the sixth planet from the sun in our solar system with its $3,400,000,000 Cassini mission. The Cassini spacecraft is scheduled to arrive in orbit around Saturn in June 2004. The goal of this mission is to study Saturn, its rings, and its moons. The Cassini spacecraft will also launch the Huygens probe to study Titan.

The Cassini mission began on October 15, 1997, with the launch of a mighty Titan IV booster rocket. The entire launch vehicle including rocket fuel weighed more than 2,000,000 pounds before launch. The Titan IV flung Cassini into space at a speed of 14,400 kilometers per hour. To take advantage of something called *gravity assist*, Cassini is taking a roundabout path to Saturn past Venus (twice), Earth, and Jupiter. Altogether, Cassini will travel 3,540,000,000 kilometers before reaching Saturn, a planet that is only 1,430,000,000 kilometers from the sun.

Once Cassini reaches Saturn, it will begin collecting all kinds of data about the planet and its moons. Over the course of Cassini's mission, it will collect over 2,000,000,000,000 bits of scientific information, or about the same amount of data in 800 sets of the *Encyclopedia Britannica*. About once per day, Cassini will use its 4-meter antenna to transmit the latest data that it has collected back to Earth at a frequency of 8,400,000,000 cycles per second. For comparison, the FM band on a radio is centered around 100,000,000 cycles per second. It will take from 70 to 90 minutes for Cassini's transmissions to reach Earth and by then the signals will be very weak. The power of the signal transmitted by the spacecraft is 20 watts but, even with the huge antennas used on Earth, only 0.0000000000000001 watt can be received. (*Source:* based on data from National Aeronautics and Space Administration)

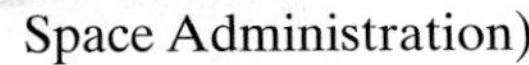

CRITICAL THINKING

1. Make a list of the numbers (other than those in dates) used in the article. Rewrite each number in scientific notation.
2. What are the advantages of scientific notation?
3. What are the disadvantages of scientific notation?
4. In your opinion, how large or small should a number be to make using scientific notation worthwhile?

Name ____________________ Section __________ Date __________

EXERCISE SET 4.3

A *Complete each table for each polynomial. See Example 1.*

1. $x^2 - 3x + 5$

Term	Coefficient
x^2	
	-3
5	

2. $2x^3 - x + 4$

Term	Coefficient
	2
$-x$	
4	

3. $-5x^4 + 3.2x^2 + x - 5$

Term	Coefficient
$-5x^4$	
$3.2x^2$	
x	
-5	

4. $9.7x^7 - 3x^5 + x^3 - \frac{1}{4}x^2$

Term	Coefficient
$9.7x^7$	
$-3x^5$	
x^3	
$-\frac{1}{4}x^2$	

B *Find the degree of each polynomial and determine whether it is a monomial, binomial, trinomial, or none of these. See Examples 2 and 3.*

5. $x + 2$

6. $-6y + y^2 + 4$

 **7.** $9m^3 - 5m^2 + 4m - 8$

8. $5a^2 + 3a^3 - 4a^4$

9. $12x^4 - x^2 - 12x^2$

10. $7r^2 + 2r - 3r^5$

11. $3z - 5$

12. $5y + 2$

13. Describe how to find the degree of a term.

14. Describe how to find the degree of a polynomial.

15. Explain why xyz is a monomial while $x + y + z$ is a trinomial.

16. Explain why the degree of the term $5y^3$ is 3 and the degree of the polynomial $2y + y + 2y$ is 1.

C *Evaluate each polynomial when* **(a)** $x = 0$ *and* **(b)** $x = -1$. *See Examples 4 and 5.*

17. $x + 6$

18. $2x - 10$

19. $x^2 - 5x - 2$

20. $x^2 - 4$

21. $x^3 - 15$

22. $-2x^3 + 3x^2 - 6$

A rocket is fired upward from the ground with an initial velocity of 200 feet per second. Neglecting air resistance, the height of the rocket at any time t can be described in feet by the polynomial $-16t^2 + 200t$. *Find the height of the rocket at the time given in Exercises 23 through 26. See Example 5.*

23. $t = 1$ second

24. $t = 5$ seconds

25. $t = 7.6$ seconds

26. $t = 10.3$ seconds

27. The number of wireless telephone subscribers (in millions) x years after 1990 is given by the polynomial $0.97x^2 - 0.91x + 7.46$ for 1993 through 2000. Use this model to predict the number of wireless telephone subscribers in 2005 ($x = 15$). (*Source:* Based on data from Cellular Telecommunications & Internet Association)

28. The annual per capita consumption of chicken in pounds in the United States x years after 1990 is given by the polynomial $0.08x^3 - 1.19x^2 + 6.45x + 69.93$ for 1991 through 2000. Use this model to predict the per capita consumption of chicken in 2003 ($x = 13$). (*Source:* Based on data from U.S. Department of Agriculture, Economic Research Service)

D *Simplify each expression by combining like terms. See Examples 6 through 10.*

29. $14x^2 + 9x^2$

30. $18x^3 - 4x^3$

31. $15x^2 - 3x^2 - y$

32. $12k^3 - 9k^3 + 11$

33. $8s - 5s + 4s$

34. $5y + 7y - 6y$

35. $0.1y^2 - 1.2y^2 + 6.7 - 1.9$

36. $7.6y + 3.2y^2 - 8y - 2.5y^2$

Recall that the perimeter of a figure such as the ones shown in Exercises 37 and 38 is the sum of the lengths of its sides. Write each perimeter as a polynomial. Then simplify the polynomial.

37.

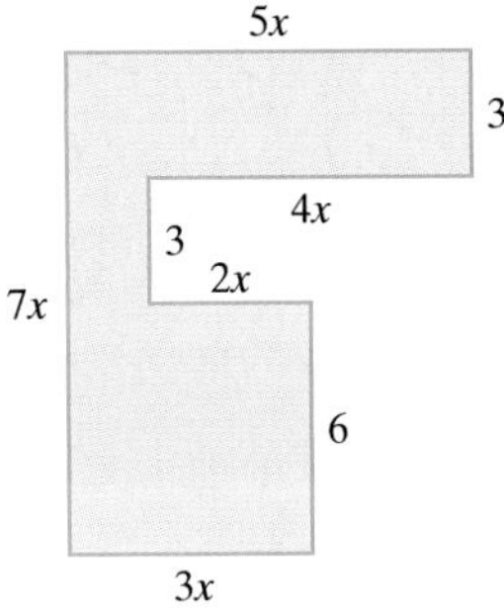

38.

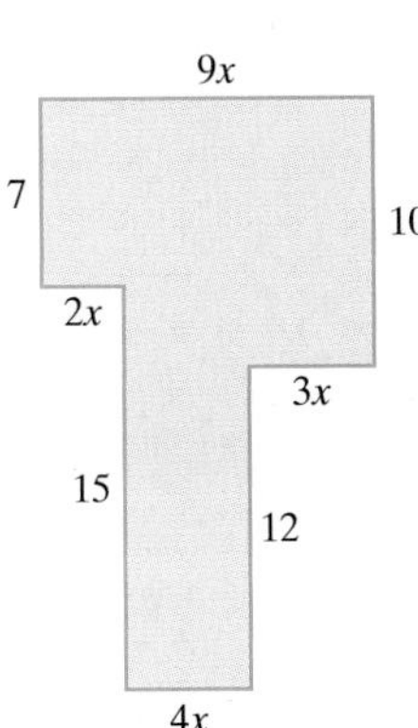

Write a polynomial that describes the total area of each set of rectangles and squares shown in Exercises 39 and 40. Then simplify the polynomial. See Example 11.

39.

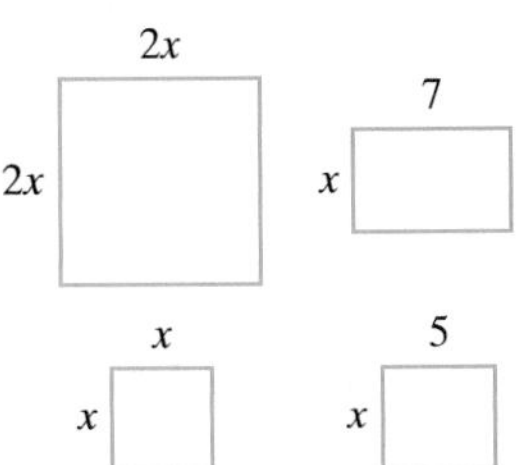

40.

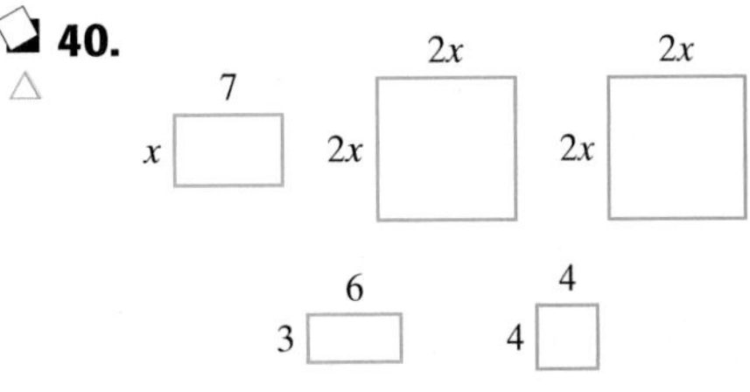

E *Identify the degrees of the terms and the degree of the polynomial. See Example 12.*

41. $9ab - 6a + 5b - 3$

42. $y^4 - 6y^3x + 2x^2y^2 - 5y^2 + 3$

43. $x^3y - 6 + 2x^2y^2 + 5y^3$

44. $2a^2b + 10a^4b - 9ab + 6$

Simplify each polynomial by combining any like terms. See Examples 13 and 14.

45. $3ab - 4a + 6ab - 7a$

46. $-9xy + 7y - xy - 6y$

47. $4x^2 - 6xy + 3y^2 - xy$

48. $3a^2 - 9ab + 4b^2 - 7ab$

49. $5x^2y + 6xy^2 - 5yx^2 + 4 - 9y^2x$

50. $17a^2b - 16ab^2 + 3a^3 + 4ba^3 - b^2a$

51. $14y^3 - 9 + 3a^2b^2 - 10 - 19b^2a^2$

52. $18x^4 + 2x^3y^3 - 1 - 2y^3x^3 - 17x^4$

Review and Preview

Simplify each expression. See Section 2.1.

53. $4 + 5(2x + 3)$

54. $9 - 6(5x + 1)$

55. $2(x - 5) + 3(5 - x)$

56. $-3(w + 7) + 5(w + 1)$

Combining Concepts

57. Explain why the height of the rocket in Exercises 23 through 26 increases and then decreases as time passes.

58. Approximate (to the nearest tenth of a second) how long before the rocket in Exercises 23 through 26 hits the ground.

Simplify each polynomial by combining like terms.

59. $1.85x^2 - 3.76x + 9.25x^2 + 10.76 - 4.21x$

60. $7.75x + 9.16x^2 - 1.27 - 14.58x^2 - 18.34$

4.4 Adding and Subtracting Polynomials

A Adding Polynomials

To add polynomials, we use commutative and associative properties and then combine like terms. To see if you are ready to add polynomials,

Try the Concept Check in the margin.

To Add Polynomials

To add polynomials, combine all like terms.

EXAMPLES Add.

1. $(4x^3 - 6x^2 + 2x + 7) + (5x^2 - 2x)$
   $= 4x^3 - 6x^2 + 2x + 7 + 5x^2 - 2x$ Remove parentheses.
   $= 4x^3 + (-6x^2 + 5x^2) + (2x - 2x) + 7$ Combine like terms.
   $= 4x^3 - x^2 + 7$ Simplify.
2. $(-2x^2 + 5x - 1)$ and $(-2x^2 + x + 3)$ translates to
   $(-2x^2 + 5x - 1) + (-2x^2 + x + 3)$
   $= -2x^2 + 5x - 1 - 2x^2 + x + 3$ Remove parentheses.
   $= (-2x^2 - 2x^2) + (5x + 1x) + (-1 + 3)$ Combine like terms.
   $= -4x^2 + 6x + 2$ Simplify.

Polynomials can be added vertically if we line up like terms underneath one another.

EXAMPLE 3 Add $(7y^3 - 2y^2 + 7)$ and $(6y^2 + 1)$ using a vertical format.

Solution: Vertically line up like terms and add.

$$\begin{array}{r} 7y^3 - 2y^2 + 7 \\ 6y^2 + 1 \\ \hline 7y^3 + 4y^2 + 8 \end{array}$$

B Subtracting Polynomials

To subtract one polynomial from another, recall the definition of subtraction. To subtract a number, we add its opposite: $a - b = a + (-b)$. To subtract a polynomial, we also add its opposite. Just as $-b$ is the opposite of b, $-(x^2 + 5)$ is the opposite of $(x^2 + 5)$.

EXAMPLE 4 Subtract: $(5x - 3) - (2x - 11)$

Solution: From the definition of subtraction, we have

$(5x - 3) - (2x - 11) = (5x - 3) + [-(2x - 11)]$ Add the opposite.
$= (5x - 3) + (-2x + 11)$ Apply the distributive property.
$= 5x - 3 - 2x + 11$ Remove parentheses.
$= 3x + 8$ Combine like terms.

OBJECTIVES

A Add polynomials.
B Subtract polynomials.
C Add or subtract polynomials in one variable.
D Add or subtract polynomials in several variables.

Concept Check

When combining like terms in the expression $5x - 8x^2 - 8x$, which of the following is the proper result?

a. $-11x^2$ b. $-3x - 8x^2$
c. $-11x$ d. $-11x^4$

Practice Problems 1–2

Add.

1. $(3x^5 - 7x^3 + 2x - 1) + (3x^3 - 2x)$
2. $(5x^2 - 2x + 1)$ and $(-6x^2 + x - 1)$

Practice Problem 3

Add $(9y^2 - 6y + 5)$ and $(4y + 3)$ using a vertical format.

To Subtract Polynomials

To subtract two polynomials, change the signs of the terms of the polynomial being subtracted and then add.

Practice Problem 4

Subtract: $(9x + 5) - (4x - 3)$

Answers

1. $3x^5 - 4x^3 - 1$, **2.** $-x^2 - x$, **3.** $9y^2 - 2y + 8$, **4.** $5x + 8$

Concept Check: **b**

Practice Problem 5

Subtract:
$(4x^3 - 10x^2 + 1) - (-4x^3 + x^2 - 11)$

EXAMPLE 5 Subtract: $(2x^3 + 8x^2 - 6x) - (2x^3 - x^2 + 1)$

Solution: First, we change the sign of each term of the second polynomial; then we add.

$$\begin{aligned}&(2x^3 + 8x^2 - 6x) - (2x^3 - x^2 + 1)\\ &= (2x^3 + 8x^2 - 6x) + (-2x^3 + x^2 - 1)\\ &= 2x^3 + 8x^2 - 6x - 2x^3 + x^2 - 1\\ &= 2x^3 - 2x^3 + 8x^2 + x^2 - 6x - 1\\ &= 9x^2 - 6x - 1 \quad \text{Combine like terms.}\end{aligned}$$

Just as polynomials can be added vertically, so can they be subtracted vertically.

Practice Problem 6

Subtract $(6y^2 - 3y + 2)$ from $(2y^2 - 2y + 7)$ using a vertical format.

EXAMPLE 6

Subtract $(5y^2 + 2y - 6)$ from $(-3y^2 - 2y + 11)$ using a vertical format.

Solution: Arrange the polynomials in a vertical format, lining up like terms.

$$\begin{array}{r} -3y^2 - 2y + 11\\ -(5y^2 + 2y - 6)\\ \hline \end{array} \qquad \begin{array}{r} -3y^2 - 2y + 11\\ -5y^2 - 2y + 6\\ \hline -8y^2 - 4y + 17 \end{array}$$

Helpful Hint

Don't forget to change the sign of each term in the polynomial being subtracted.

C Adding and Subtracting Polynomials in One Variable

Let's practice adding and subtracting polynomials in one variable.

Practice Problem 7

Subtract $(3x + 1)$ from the sum of $(4x - 3)$ and $(12x - 5)$.

EXAMPLE 7 Subtract $(5z - 7)$ from the sum of $(8z + 11)$ and $(9z - 2)$.

Solution: Notice that $(5z - 7)$ is to be subtracted **from** a sum. The translation is

$$\begin{aligned}&[(8z + 11) + (9z - 2)] - (5z - 7)\\ &= 8z + 11 + 9z - 2 - 5z + 7 \quad \text{Remove grouping symbols.}\\ &= 8z + 9z - 5z + 11 - 2 + 7 \quad \text{Group like terms.}\\ &= 12z + 16 \quad \text{Combine like terms.}\end{aligned}$$

D Adding and Subtracting Polynomials in Several Variables

Now that we know how to add or subtract polynomials in one variable, we can also add and subtract polynomials in several variables.

Practice Problems 8–9

Add or subtract as indicated.

8. $(2a^2 - ab + 6b^2) - (-3a^2 + ab - 7b^2)$
9. $(5x^2y^2 + 3 - 9x^2y + y^2) - (-x^2y^2 + 7 - 8xy^2 + 2y^2)$

Answers

5. $8x^3 - 11x^2 + 12$, **6.** $-4y^2 + y + 5$, **7.** $13x - 9$, **8.** $5a^2 - 2ab + 13b^2$, **9.** $6x^2y^2 - 4 - 9x^2y + 8xy^2 - y^2$

EXAMPLES Add or subtract as indicated.

8. $(3x^2 - 6xy + 5y^2) + (-2x^2 + 8xy - y^2)$
$= 3x^2 - 6xy + 5y^2 - 2x^2 + 8xy - y^2$
$= x^2 + 2xy + 4y^2$ Combine like terms.

9. $(9a^2b^2 + 6ab - 3ab^2) - (5b^2a + 2ab - 3 - 9b^2)$ Change the sign of each term of the polynomial being subtracted.
$= 9a^2b^2 + 6ab - 3ab^2 - 5b^2a - 2ab + 3 + 9b^2$
$= 9a^2b^2 + 4ab - 8ab^2 + 9b^2 + 3$ Combine like terms.

Name ______________________ Section __________ Date __________

EXERCISE SET 4.4

 *Add. See Examples 1 through 3.*

1. $(3x + 7) + (9x + 5)$

2. $(3x^2 + 7) + (3x^2 + 9)$

3. $(-7x + 5) + (-3x^2 + 7x + 5)$

4. $(3x - 8) + (4x^2 - 3x + 3)$

5. $(-5x^2 + 3) + (2x^2 + 1)$

6. $(-y - 2) + (3y + 5)$

 7. $(-3y^2 - 4y) + (2y^2 + y - 1)$

8. $(7x^2 + 2x - 9) + (-3x^2 + 5)$

Add using a vertical format. See Example 3.

9. $\begin{array}{r} 3t^2 + 4 \\ \underline{5t^2 - 8} \end{array}$

10. $\begin{array}{r} 7x^3 + 3 \\ \underline{2x^3 + 1} \end{array}$

11. $\begin{array}{r} 10a^3 - 8a^2 + 9 \\ \underline{5a^3 + 9a^2 + 7} \end{array}$

12. $\begin{array}{r} 2x^3 - 3x^2 + x - 4 \\ \underline{5x^3 + 2x^2 - 3x + 2} \end{array}$

B *Subtract. See Examples 4 and 5.*

13. $(2x + 5) - (3x - 9)$

14. $(5x^2 + 4) - (-2y^2 + 4)$

15. $3x - (5x - 9)$

16. $4 - (-y - 4)$

17. $(2x^2 + 3x - 9) - (-4x + 7)$

18. $(-7x^2 + 4x + 7) - (-8x + 2)$

19. $(-7y^2 + 5) - (-8y^2 + 12)$

20. $(4 + 5a) - (-a - 5)$

21. $(5x + 8) - (-2x^2 - 6x + 8)$

22. $(-6y^2 + 3y - 4) - (9y^2 - 3y)$

Subtract using a vertical format. See Example 6.

23. $\begin{array}{r} 4z^2 - 8z + 3 \\ \underline{-\ (6z^2 + 8z - 3)} \end{array}$

24. $\begin{array}{r} 7a^2 - 9a + 6 \\ \underline{-(11a^2 - 4a + 2)} \end{array}$

25. $\begin{array}{r} 5u^5 - 4u^2 + 3u - 7 \\ \underline{-(3u^5 + 6u^2 - 8u + 2)} \end{array}$

26. $\begin{array}{r} 5x^3 - 4x^2 + 6x - 2 \\ \underline{-(3x^3 - 2x^2 - x - 4)} \end{array}$

C *Add or subtract as indicated. See Example 7.*

27. $(3x + 5) + (2x - 14)$

28. $(9x - 1) - (5x + 2)$

29. $(7y + 7) - (y - 6)$

30. $(14y + 12) + (-3y - 5)$

31. $(x^2 + 2x + 1) - (3x^2 - 6x + 2)$

32. $(5y^2 - 3y - 1) - (2y^2 + y + 1)$

33. $(3x^2 + 5x - 8) + (5x^2 + 9x + 12) - (x^2 - 14)$

34. $(-a^2 + 1) - (a^2 - 3) + (5a^2 - 6a + 7)$

Perform each indicated operation. See Examples 2, 6, and 7.

35. Subtract $4x$ from $7x - 3$.

36. Subtract y from $y^2 - 4y + 1$.

37. Add $(4x^2 - 6x + 1)$ and $(3x^2 + 2x + 1)$.

38. Add $(-3x^2 - 5x + 2)$ and $(x^2 - 6x + 9)$.

39. Subtract $(5x + 7)$ from $(7x^2 + 3x + 9)$.

40. Subtract $(5y^2 + 8y + 2)$ from $(7y^2 + 9y - 8)$.

41. Subtract $(4y^2 - 6y - 3)$ from the sum of $(8y^2 + 7)$ and $(6y + 9)$.

42. Subtract $(4x^2 - 2x + 2)$ from the sum of $(x^2 + 7x + 1)$ and $(7x + 5)$.

43. Subtract $(3x^2 - 4)$ from the sum of $(x^2 - 9x + 2)$ and $(2x^2 - 6x + 1)$.

44. Subtract $(y^2 - 9)$ from the sum of $(3y^2 + y + 4)$ and $(2y^2 - 6y - 10)$.

 *Add or subtract as indicated. See Examples 8 and 9.*

45. $(9a + 6b - 5) + (-11a - 7b + 6)$

46. $(3x - 2 + 6y) + (7x - 2 - y)$

47. $(4x^2 + y^2 + 3) - (x^2 + y^2 - 2)$

48. $(7a^2 - 3b^2 + 10) - (-2a^2 + b^2 - 12)$

49. $(x^2 + 2xy - y^2) + (5x^2 - 4xy + 20y^2)$

50. $(a^2 - ab + 4b^2) + (6a^2 + 8ab - b^2)$

51. $(11r^2s + 16rs - 3 - 2r^2s^2) - (3sr^2 + 5 - 9r^2s^2)$

52. $(3x^2y - 6xy + x^2y^2 - 5) - (11x^2y^2 - 1 + 5yx^2)$

Review and Preview

Multiply. See Section 4.1.

53. $3x(2x)$

54. $-7x(x)$

55. $(12x^3)(-x^5)$

56. $6r^3(7r^{10})$

57. $10x^2(20xy^2)$

58. $-z^2y(11zy)$

Combining Concepts

59. Given the following triangle, find its perimeter.

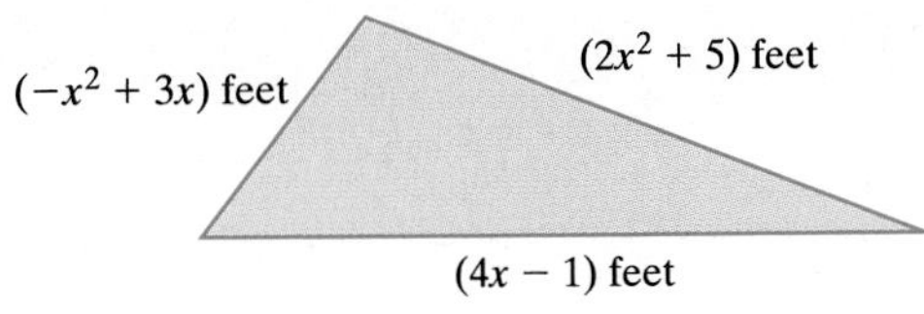

△ **60.** Given the following quadrilateral, find its perimeter.

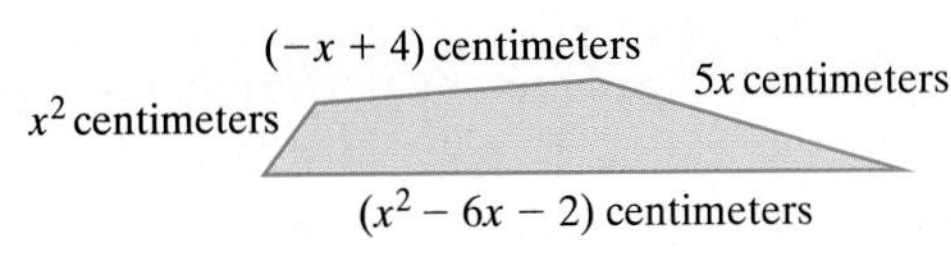

61. A wooden beam is $(4y^2 + 4y + 1)$ meters long. If a piece $(y^2 - 10)$ meters is removed, express the length of the remaining piece of beam as a polynomial in y.

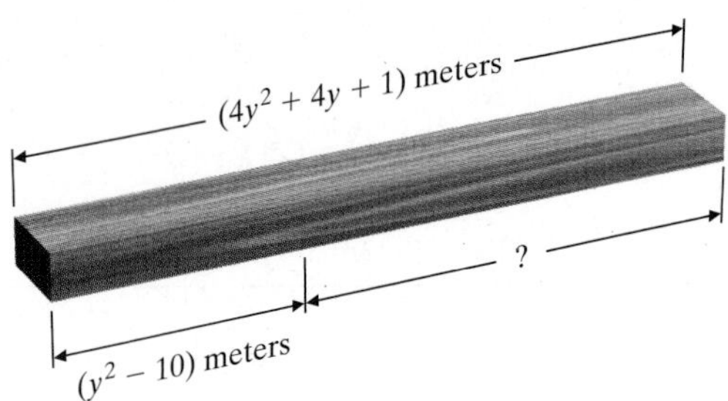

62. A piece of quarter-round molding is $(13x - 7)$ inches long. If a piece $(2x + 2)$ inches is removed, express the length of the remaining piece of molding as a polynomial in x.

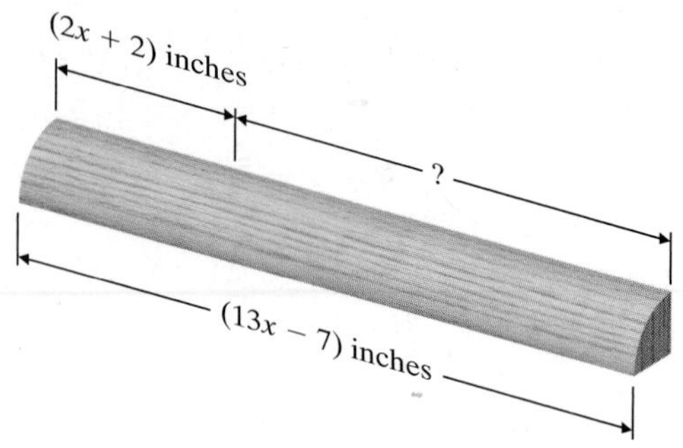

Perform each indicated operation.

63. $[(1.2x^2 - 3x + 9.1) - (7.8x^2 - 3.1 + 8)] + (1.2x - 6)$

64. $[(7.9y^4 - 6.8y^3 + 3.3y) + (6.1y^3 - 5)] - (4.2y^4 + 1.1y - 1)$

65. The polynomial $-2.85x^2 + 8.75x + 26.7$ represents the number of Americans (in millions) enrolled in individual-practice-association HMOs during 1997–1999. The polynomial $0.35x^2 + 3.55x + 40$ represents the number of Americans (in millions) enrolled in all other types of HMOs during 1997–1999. In both polynomials, x represents the number of years after 1997. Find a polynomial for the total enrollment (in millions) in HMOs of all kinds during this period. (*Source:* Based on data from the Public Health Service)

66. The polynomial $0.015x^2 + 0.002x + 1.128$ represents the sales of electricity (in trillion kilowatt hours) in the U.S. residential sector during 1998–2000. The polynomial $-0.0075x^2 + 0.0615x + 2.113$ represents the sales of electricity (in trillion kilowatt hours) in all other U.S. sectors during 1998–2000. In both polynomials, x represents the number of years after 1998. Find a polynomial for the total sales of electricity (in trillion kilowatt hours) to all sectors in the United States during this period. (Source: Based on data from the Energy Information Administration)

67. Simplify each expression by performing the indicated operation. Explain how you arrived at each answer.

a. $x + x$

b. $x \cdot x$

c. $-x - x$

d. $(-x)(-x)$

4.5 Multiplying Polynomials

A Multiplying Monomials

Recall from Section 4.1 that to multiply two monomials such as $(-5x^3)$ and $(-2x^4)$, we use the associative and commutative properties and regroup. Remember also that to multiply exponential expressions with a common base, we add exponents.

$(-5x^3)(-2x^4) = (-5)(-2)(x^3 \cdot x^4)$ Use the commutative and associative properties.

$= 10x^7$ Multiply.

EXAMPLES Multiply.

1. $6x \cdot 4x = (6 \cdot 4)(x \cdot x)$ Use the commutative and associative properties.

$= 24x^2$ Multiply.

2. $-7x^2 \cdot 2x^5 = (-7 \cdot 2)(x^2 \cdot x^5)$

$= -14x^7$

3. $(-12x^5)(-x) = (-12x^5)(-1x)$

$= (-12)(-1)(x^5 \cdot x)$

$= 12x^6$

B Multiplying Monomials by Polynomials

To multiply a monomial such as $7x$ by a trinomial such as $x^2 + 2x + 5$, we use the distributive property.

EXAMPLES Multiply.

4. $7x(x^2 + 2x + 5) = 7x(x^2) + 7x(2x) + 7x(5)$ Apply the distributive property.

$= 7x^3 + 14x^2 + 35x$ Multiply.

5. $5x(2x^3 + 6) = 5x(2x^3) + 5x(6)$ Apply the distributive property.

$= 10x^4 + 30x$ Multiply.

6. $-3x^2(5x^2 + 6x - 1)$

$= (-3x^2)(5x^2) + (-3x^2)(6x) + (-3x^2)(-1)$ Apply the distributive property.

$= -15x^4 - 18x^3 + 3x^2$ Multiply.

C Multiplying Two Polynomials

We also use the distributive property to multiply two binomials.

EXAMPLE 7 Multiply: $(3x + 2)(2x - 5)$

Solution:

$(3x + 2)(2x - 5) = 3x(2x - 5) + 2(2x - 5).$ Use the distributive property.

$= 3x(2x) + 3x(-5) + 2(2x) + 2(-5)$

$= 6x^2 - 15x + 4x - 10$ Multiply.

$= 6x^2 - 11x - 10$ Combine like terms.

This idea can be expanded so that we can multiply any two polynomials.

OBJECTIVES

A Multiply monomials.

B Multiply a monomial by a polynomial.

C Multiply two polynomials.

D Multiply polynomials vertically.

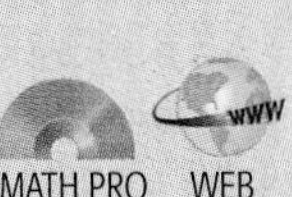

SSM TUTOR CENTER SG CD & VIDEO MATH PRO WEB

Practice Problems 1–3

Multiply.

1. $10x \cdot 9x$
2. $8x^3(-11x^7)$
3. $(-5x^4)(-x)$

Practice Problems 4–6

Multiply.

4. $4x(x^2 + 4x + 3)$
5. $8x(7x^4 + 1)$
6. $-2x^3(3x^2 - x + 2)$

Practice Problem 7

Multiply: $(4x + 5)(3x - 4)$

Answers

1. $90x^2$, **2.** $-88x^{10}$, **3.** $5x^5$, **4.** $4x^3 + 16x^2 + 12x$, **5.** $56x^5 + 8x$, **6.** $-6x^5 + 2x^4 - 4x^3$, **7.** $12x^2 - x - 20$

To Multiply Two Polynomials

Multiply each term of the first polynomial by each term of the second polynomial, and then combine like terms.

Practice Problems 8–9

Multiply.

8. $(3x - 2y)^2$
9. $(x + 3)(2x^2 - 5x + 4)$

EXAMPLES Multiply.

8. $(2x - y)^2$

$= (2x - y)(2x - y)$

$= 2x(2x) + 2x(-y) + (-y)(2x) + (-y)(-y)$

$= 4x^2 - 2xy - 2xy + y^2$ Multiply.

$= 4x^2 - 4xy + y^2$ Combine like terms.

9. $(t + 2)(3t^2 - 4t + 2)$

$= t(3t^2) + t(-4t) + t(2) + 2(3t^2) + 2(-4t) + 2(2)$

$= 3t^3 - 4t^2 + 2t + 6t^2 - 8t + 4$

$= 3t^3 + 2t^2 - 6t + 4$ Combine like terms.

D Multiplying Polynomials Vertically

Another convenient method for multiplying polynomials is to multiply vertically, similar to the way we multiply real numbers. This method is shown in the next examples.

Practice Problem 10

Multiply vertically:

$$(3y^2 + 1)(y^2 - 4y + 5)$$

EXAMPLE 10 Multiply vertically: $(2y^2 + 5)(y^2 - 3y + 4)$

Solution:

$$\begin{array}{rrrrrl}
 & & y^2 & - 3y & + 4 & \\
 & & & 2y^2 & + 5 & \\
\hline
 & & 5y^2 & - 15y & + 20 & \text{Multiply } y^2 - 3y + 4 \text{ by } 5 \\
2y^4 & - 6y^3 & + 8y^2 & & & \text{Multiply } y^2 - 3y + 4 \text{ by } 2y^2 \\
\hline
2y^4 & - 6y^3 & + 13y^2 & - 15y & + 20 & \text{Combine like terms.}
\end{array}$$

Practice Problem 11

Find the product of $(4x^2 - x - 1)$ and $(3x^2 + 6x - 2)$ using a vertical format.

EXAMPLE 11

Find the product of $(2x^2 - 3x + 4)$ and $(x^2 + 5x - 2)$ using a vertical format.

Solution: First, we arrange the polynomials in a vertical format. Then we multiply each term of the second polynomial by each term of the first polynomial.

$$\begin{array}{rrrrrl}
 & & 2x^2 & - 3x & + 4 & \\
 & & x^2 & + 5x & - 2 & \\
\hline
 & & -4x^2 & + 6x & - 8 & \text{Multiply } 2x^2 - 3x + 4 \text{ by } -2. \\
 & 10x^3 & - 15x^2 & + 20x & & \text{Multiply } 2x^2 - 3x + 4 \text{ by } 5x. \\
2x^4 & - 3x^3 & + 4x^2 & & & \text{Multiply } 2x^2 - 3x + 4 \text{ by } x^2. \\
\hline
2x^4 & + 7x^3 & - 15x^2 & + 26x & - 8 & \text{Combine like terms.}
\end{array}$$

Answers

8. $9x^2 - 12xy + 4y^2$,
9. $2x^3 + x^2 - 11x + 12$,
10. $3y^4 - 12y^3 + 16y^2 - 4y + 5$,
11. $12x^4 + 21x^3 - 17x^2 - 4x + 2$

Name ______________________ Section __________ Date __________

Mental Math

Find each product.

1. $x^3 \cdot x^5$ **2.** $x^2 \cdot x^6$ **3.** $y^4 \cdot y$ **4.** $y^9 \cdot y$ **5.** $x^7 \cdot x^7$ **6.** $x^{11} \cdot x^{11}$

EXERCISE SET 4.5

 *Multiply. See Examples 1 through 3.*

1. $8x^2 \cdot 3x$

2. $6x \cdot 3x^2$

3. $(-3.1x^3)(4x^9)$

4. $(-5.2x^4)(3x^4)$

5. $(-x^3)(-x)$

6. $(-x^6)(-x)$

7. $\left(-\frac{1}{3}y^2\right)\left(\frac{2}{5}y\right)$

8. $\left(-\frac{3}{4}y^7\right)\left(\frac{1}{7}y^4\right)$

9. $(2x)(-3x^2)(4x^5)$

10. $(x)(5x^4)(-6x^7)$

B *Multiply. See Examples 4 through 6.*

11. $3x(2x + 5)$

12. $2x(6x + 3)$

13. $7x(x^2 + 2x - 1)$

14. $5y(y^2 + y - 10)$

15. $-2a(a + 4)$

16. $-3a(2a + 7)$

17. $3x(2x^2 - 3x + 4)$

18. $4x(5x^2 - 6x - 10)$

19. $3a(a^2 + 2)$

20. $x^3(x + 12)$

21. $-2a^2(3a^2 - 2a + 3)$

22. $-4b^2(3b^3 - 12b^2 - 6)$

23. $3x^2y(2x^3 - x^2y^2 + 8y^3)$

24. $4xy^2(7x^3 + 3x^2y^2 - 9y^3)$

25. The area of the largest rectangle below is $x(x + 3)$. Find another expression for this area by finding the sum of the areas of the smaller rectangles.

26. Write an expression for the area of the largest rectangle below in two different ways.

x
1 2x

C *Multiply. See Examples 7 through 9.*

27. $(x + 4)(x + 3)$

28. $(x + 2)(x + 9)$

29. $(a + 7)(a - 2)$

30. $(y - 10)(y + 11)$

31. $\left(x + \frac{2}{3}\right)\left(x - \frac{1}{3}\right)$

32. $\left(x + \frac{3}{5}\right)\left(x - \frac{2}{5}\right)$

33. $(3x^2 + 1)(4x^2 + 7)$

34. $(5x^2 + 2)(6x^2 + 2)$

35. $(4x - 3)(3x - 5)$

36. $(8x - 3)(2x - 4)$

37. $(1 - 3a)(1 - 4a)$

38. $(3 - 2a)(2 - a)$

39. $(2y - 4)^2$

40. $(6x - 7)^2$

41. $(x - 2)(x^2 - 3x + 7)$

42. $(x + 3)(x^2 + 5x - 8)$

43. $(x + 5)(x^3 - 3x + 4)$

44. $(a + 2)(a^3 - 3a^2 + 7)$

45. $(2a - 3)(5a^2 - 6a + 4)$

46. $(3 + b)(2 - 5b - 3b^2)$

47. $(7xy - y)^2$

48. $(x^2 - 4)^2$

49. The area of the figure below is $(x + 2)(x + 3)$. Find another expression for this area by finding the sum of the areas of the smaller rectangles.

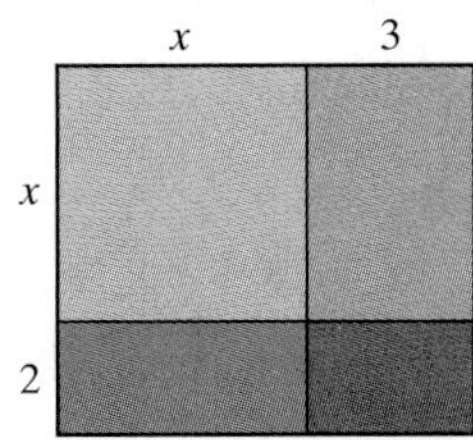

50. Write an expression for the area of the figure below in two different ways

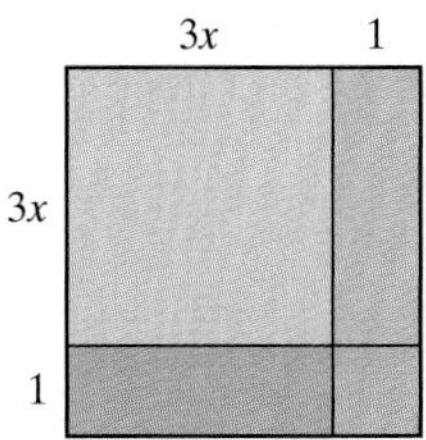

Multiply vertically. See Examples 10 and 11.

51. $(2x - 11)(6x + 1)$

52. $(4x - 7)(5x + 1)$

53. $(x + 3)(2x^2 + 4x - 1)$

54. $(4x - 5)(8x^2 + 2x - 4)$

55. $(x^2 + 5x - 7)(x^2 - 7x - 9)$

56. $(3x^2 - x + 2)(x^2 + 2x + 1)$

Review and Preview

Perform each indicated operation. See Section 4.1.

57. $(5x)^2$

58. $(4p)^2$

59. $(-3y^3)^2$

60. $(-7m^2)^2$

For income tax purposes, Rob Calcutta, the owner of Copy Services, uses a method called ***straight-line depreciation*** *to show the depreciated (or decreased) value of a copy machine he recently purchased. Rob assumes that he can use the machine for 7 years. The graph below shows the depreciated (or decreased) value of the machine over the years. Use this graph to answer Exercises 61 through 66. See Section 3.1.*

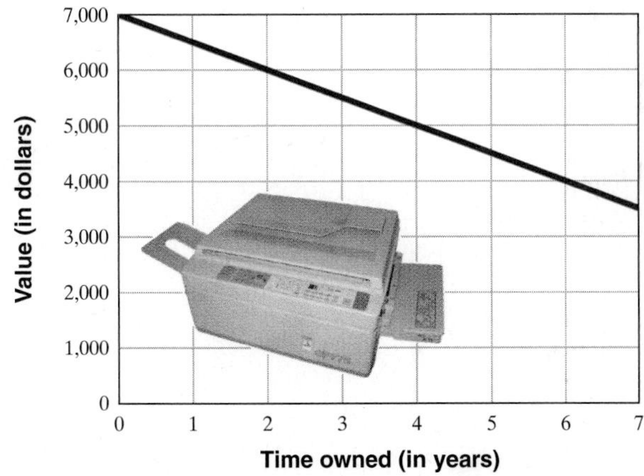

61. What was the purchase price of the copy machine? (*Hint*: This is when time owned is 0 years.)

62. What is the depreciated value of the machine in 7 years?

63. What loss in value occurred during the first year?

64. What loss in value occurred during the second year?

65. Why do you think this method of depreciating is called straight-line depreciation?

66. Why is the line tilted downward?

Combining Concepts

Express as the product of polynomials. Then multiply.

△ **67.** Find the area of the rectangle.

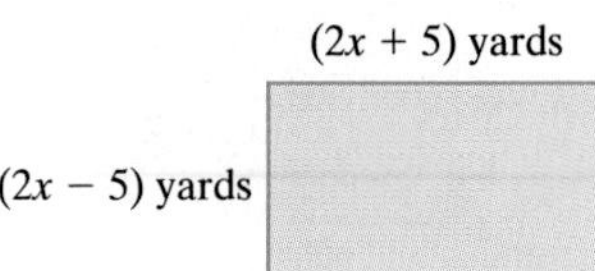

△ **68.** Find the area of the square field.

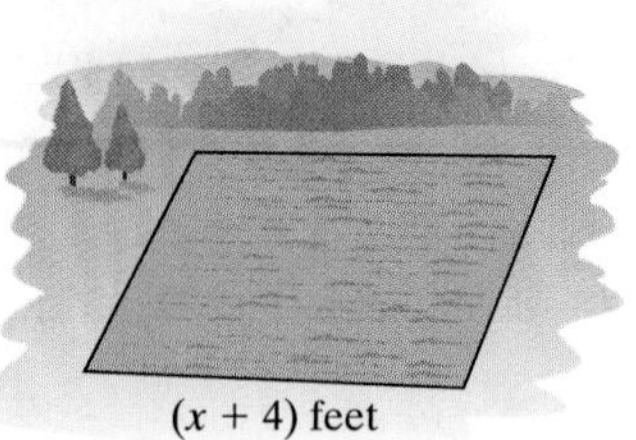

△ **69.** Find the area of the triangle.

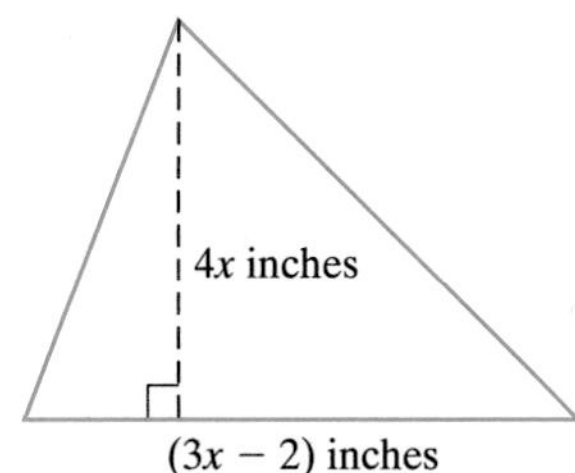

△ **70.** Find the volume of the cube-shaped glass block.

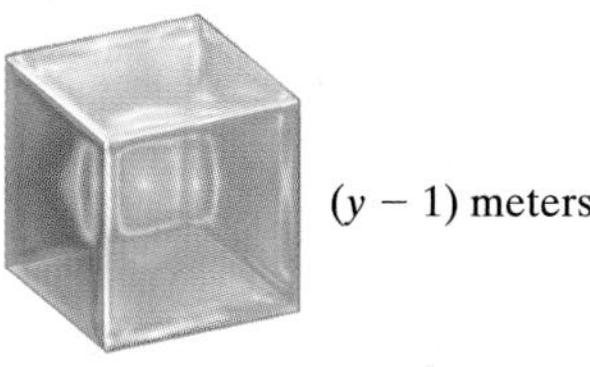

71. Perform each indicated operation. Explain the difference between the two expressions.
a. $(3x + 5) + (3x + 7)$
b. $(3x + 5)(3x + 7)$

72. Evaluate each of the following.
a. $(2 + 3)^2; 2^2 + 3^2$
b. $(8 + 10)^2; 8^2 + 10^2$
Does $(a + b)^2 = a^2 + b^2$ no matter what the values of a and b are? Why or why not?

73. Multiply each of the following polynomials.
a. $(a + b)(a - b)$
b. $(2x + 3y)(2x - 3y)$
c. $(4x + 7)(4x - 7)$
d. Can you make a general statement about all products of the form $(x + y)(x - y)$?

4.6 Special Products

OBJECTIVES

- A Multiply two binomials using the FOIL method.
- B Square a binomial.
- C Multiply the sum and difference of two terms.
- D Use special products to multiply binomials.

SSM TUTOR CENTER SG CD & VIDEO MATH PRO WEB

A Using the FOIL Method

In this section, we multiply binomials using special products. First, we introduce a special order for multiplying binomials called the FOIL order or method. We demonstrate by multiplying $(3x + 1)$ by $(2x + 5)$.

The FOIL Method

F stands for the product of the **First** terms.	$(3x + 1)(2x + 5)$ $(3x)(2x) = 6x^2$	F
O stands for the product of the **Outer** terms.	$(3x + 1)(2x + 5)$ $(3x)(5) = 15x$	O
I stands for the product of the **Inner** terms.	$(3x + 1)(2x + 5)$ $(1)(2x) = 2x$	I
L stands for the product of the **Last** terms.	$(3x + 1)(2x + 5)$ $(1)(5) = 5$	L

$$\begin{aligned}(3x + 1)(2x + 5) &= \overset{F}{6x^2} + \overset{O}{15x} + \overset{I}{2x} + \overset{L}{5}\\ &= 6x^2 + 17x + 5 \quad \text{Combine like terms.}\end{aligned}$$

Let's practice multiplying binomials using the FOIL method.

EXAMPLE 1 Multiply: $(x - 3)(x + 4)$

Solution:

$$\begin{aligned}(x - 3)(x + 4) &= \overset{F}{(x)(x)} + \overset{O}{(x)(4)} + \overset{I}{(-3)(x)} + \overset{L}{(-3)(4)}\\ &= x^2 + 4x - 3x - 12\\ &= x^2 + x - 12 \quad \text{Combine like terms.}\end{aligned}$$

Practice Problem 1

Multiply: $(x + 7)(x - 5)$

EXAMPLE 2 Multiply: $(5x - 7)(x - 2)$

Solution:

$$\begin{aligned}(5x - 7)(x - 2) &= \overset{F}{5x(x)} + \overset{O}{5x(-2)} + \overset{I}{(-7)(x)} + \overset{L}{(-7)(-2)}\\ &= 5x^2 - 10x - 7x + 14\\ &= 5x^2 - 17x + 14 \quad \text{Combine like terms.}\end{aligned}$$

Practice Problem 2

Multiply: $(6x - 1)(x - 4)$

EXAMPLE 3 Multiply: $(y^2 + 6)(2y - 1)$

Solution: $(y^2 + 6)(2y - 1) = \overset{F}{2y^3} - \overset{O}{1y^2} + \overset{I}{12y} - \overset{L}{6}$

Notice in this example that there are no like terms that can be combined, so the product is $2y^3 - y^2 + 12y - 6$.

Practice Problem 3

Multiply: $(2y^2 + 3)(y - 4)$

Answers

1. $x^2 + 2x - 35$, **2.** $6x^2 - 25x + 4$, **3.** $2y^3 - 8y^2 + 3y - 12$

B Squaring Binomials

An expression such as $(3y + 1)^2$ is called the square of a binomial. Since $(3y + 1)^2 = (3y + 1)(3y + 1)$, we can use the FOIL method to find this product.

Practice Problem 4

Multiply: $(2x + 9)^2$

EXAMPLE 4 Multiply: $(3y + 1)^2$

Solution:

$$
\begin{aligned}
(3y + 1)^2 &= (3y + 1)(3y + 1) \\
&\quad\;\, \text{F} \qquad\quad \text{O} \qquad\quad \text{I} \qquad\quad \text{L} \\
&= (3y)(3y) + (3y)(1) + 1(3y) + 1(1) \\
&= 9y^2 + 3y + 3y + 1 \\
&= 9y^2 + 6y + 1
\end{aligned}
$$

Notice the pattern that appears in Example 4.

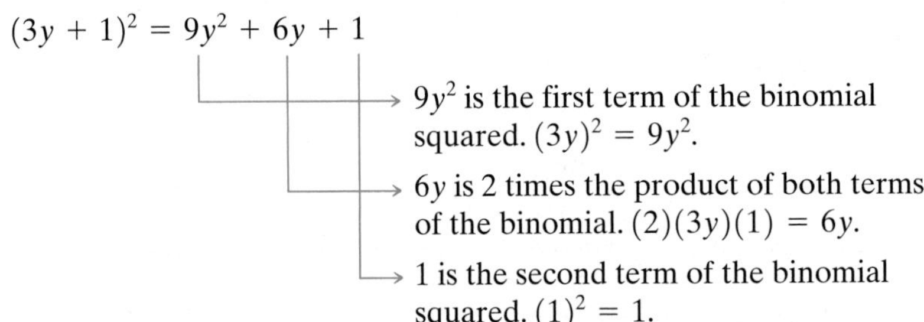

This pattern leads to the formulas below, which can be used when squaring a binomial. We call these **special products**.

Squaring a Binomial

A binomial squared is equal to the square of the first term plus or minus twice the product of both terms plus the square of the second term.

$$(a + b)^2 = a^2 + 2ab + b^2$$

$$(a - b)^2 = a^2 - 2ab + b^2$$

This product can be visualized geometrically.

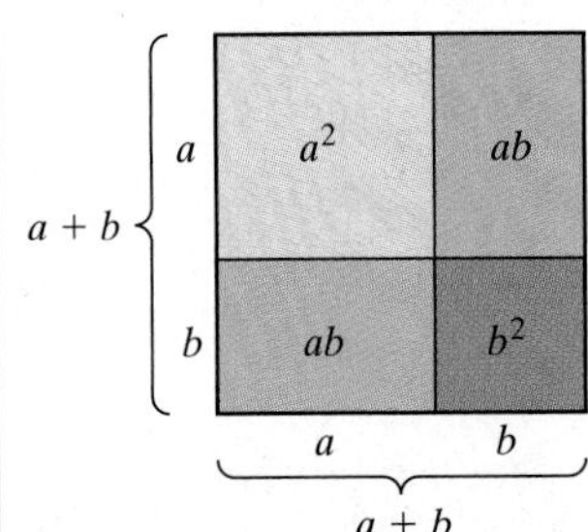

The area of the large square is side · side.

Area $= (a + b)(a + b) = (a + b)^2$

The area of the large square is also the sum of the areas of the smaller rectangles.

Area $= a^2 + ab + ab + b^2 = a^2 + 2ab + b^2$

Thus, $(a + b)^2 = a^2 + 2ab + b^2$.

Answer

4. $4x^2 + 36x + 81$

EXAMPLES Use a special product to square each binomial.

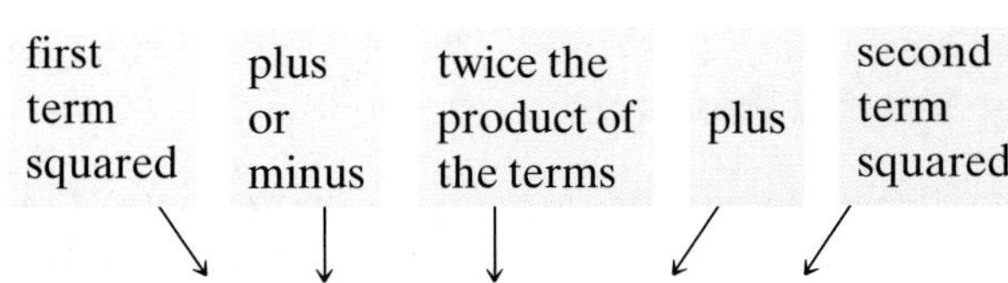

5. $(t + 2)^2 = t^2 + 2(t)(2) + 2^2 = t^2 + 4t + 4$

6. $(p - q)^2 = p^2 - 2(p)(q) + q^2 = p^2 - 2pq + q^2$

7. $(2x + 5)^2 = (2x)^2 + 2(2x)(5) + 5^2 = 4x^2 + 20x + 25$

8. $(x^2 - 7y)^2 = (x^2)^2 - 2(x^2)(7y) + (7y)^2 = x^4 - 14x^2y + 49y^2$

Helpful Hint

Notice that

$$(a + b)^2 \neq a^2 + b^2 \quad \text{The middle term } 2ab \text{ is missing.}$$

$$(a + b)^2 = (a + b)(a + b) = a^2 + 2ab + b^2$$

Likewise,

$$(a - b)^2 \neq a^2 - b^2$$

$$(a - b)^2 = (a - b)(a - b) = a^2 - 2ab + b^2$$

C Multiplying the Sum and Difference of Two Terms

Another special product is the product of the sum and difference of the same two terms, such as $(x + y)(x - y)$. Finding this product by the FOIL method, we see a pattern emerge.

$$\begin{aligned}(x + y)(x - y) &= \overset{F}{x^2} - \overset{O}{xy} + \overset{I}{xy} - \overset{L}{y^2} \\ &= x^2 - y^2\end{aligned}$$

Notice that the two middle terms subtract out. This is because the **O**uter product is the opposite of the **I**nner product. Only the **difference of squares** remains.

Multiplying the Sum and Difference of Two Terms

The product of the sum and difference of two terms is the square of the first term minus the square of the second term.

$$(a + b)(a - b) = a^2 - b^2$$

EXAMPLES Use a special product to multiply.

first term squared — minus — second term squared

9. $(x + 4)(x - 4) = x^2 - 4^2 = x^2 - 16$

10. $(6t + 7)(6t - 7) = (6t)^2 - 7^2 = 36t^2 - 49$

11. $\left(x - \frac{1}{4}\right)\left(x + \frac{1}{4}\right) = x^2 - \left(\frac{1}{4}\right)^2 = x^2 - \frac{1}{16}$

12. $(2p - q)(2p + q) = (2p)^2 - q^2 = 4p^2 - q^2$

13. $(3x^2 - 5y)(3x^2 + 5y) = (3x^2)^2 - (5y)^2 = 9x^4 - 25y^2$

Practice Problems 5–8

Use a special product to square each binomial.

5. $(y + 3)^2$
6. $(r - s)^2$
7. $(6x + 5)^2$
8. $(x^2 - 3y)^2$

Practice Problems 9–13

Use a special product to multiply.

9. $(x + 7)(x - 7)$
10. $(4y + 5)(4y - 5)$
11. $\left(x - \frac{1}{3}\right)\left(x + \frac{1}{3}\right)$
12. $(3a - b)(3a + b)$
13. $(2x^2 - 6y)(2x^2 + 6y)$

Answers

5. $y^2 + 6y + 9$, **6.** $r^2 - 2rs + s^2$, **7.** $36x^2 + 60x + 25$, **8.** $x^4 - 6x^2y + 9y^2$, **9.** $x^2 - 49$, **10.** $16y^2 - 25$, **11.** $x^2 - \frac{1}{9}$, **12.** $9a^2 - b^2$, **13.** $4x^4 - 36y^2$

Concept Check

Match each expression on the left to the equivalent expression or expressions in the list on the right.

$(a + b)^2$

$(a + b)(a - b)$

a. $(a + b)(a + b)$
b. $a^2 - b^2$
c. $a^2 + b^2$
d. $a^2 - 2ab + b^2$
e. $a^2 + 2ab + b^2$

Try the Concept Check in the margin.

D Using Special Products

Let's now practice using our special products on a variety of multiplication problems. This practice will help us recognize when to apply what special product formula.

Practice Problems 14–16

Use a special product to multiply.

14. $(7x - 1)^2$
15. $(5y + 3)(2y - 5)$
16. $(2a - 1)(2a + 1)$

EXAMPLES Use a special product to multiply.

14. $(x - 9)(x + 9)$ This is the sum and difference of the same two terms.

$= x^2 - 9^2 = x^2 - 81$

15. $(3y + 2)^2$ This is a binomial squared.

$= (3y)^2 + 2(3y)(2) + 2^2$

$= 9y^2 + 12y + 4$

16. $(6a + 1)(a - 7)$ No special product applies. Use the FOIL method.

F O I L

$= 6a \cdot a + 6a(-7) + 1 \cdot a + 1(-7)$

$= 6a^2 - 42a + a - 7$

$= 6a^2 - 41a - 7$

Answers

14. $49x^2 - 14x + 1$, **15.** $10y^2 - 19y - 15$, **16.** $4a^2 - 1$

Concept Check: a or e, b

Name ______________________ Section __________ Date ____________

EXERCISE SET 4.6

A *Multiply using the FOIL method. See Examples 1 through 3.*

1. $(x + 3)(x + 4)$ **2.** $(x + 5)(x + 1)$ **3.** $(x - 5)(x + 10)$ **4.** $(y - 12)(y + 4)$

5. $(5x - 6)(x + 2)$ **6.** $(3y - 5)(2y - 7)$ **7.** $(y - 6)(4y - 1)$ **8.** $(2x - 9)(x - 11)$

9. $(2x + 5)(3x - 1)$ **10.** $(6x + 2)(x - 2)$ **11.** $(y^2 + 7)(6y + 4)$ **12.** $(y^2 + 3)(5y + 6)$

13. $\left(x - \frac{1}{3}\right)\left(x + \frac{2}{3}\right)$ **14.** $\left(x - \frac{2}{5}\right)\left(x + \frac{1}{5}\right)$ **15.** $(4 - 3a)(2 - 5a)$ **16.** $(3 - 2a)(6 - 5a)$

17. $(x + 5y)(2x - y)$ **18.** $(x + 4y)(3x - y)$

B *Multiply. See Examples 4 through 8.*

19. $(x + 2)^2$ **20.** $(x + 7)^2$ **21.** $(2x - 1)^2$ **22.** $(7x - 3)^2$ **23.** $(3a - 5)^2$

24. $(5a + 2)^2$ **25.** $(x^2 + 5)^2$ **26.** $(x^2 + 3)^2$ **27.** $\left(y - \frac{2}{7}\right)^2$ **28.** $\left(y - \frac{3}{4}\right)^2$

29. $(2a - 3)^2$ **30.** $(5b - 4)^2$ **31.** $(5x + 9)^2$ **32.** $(6s + 2)^2$ **33.** $(3x - 7y)^2$

34. $(4s - 2y)^2$ **35.** $(4m + 5n)^2$ **36.** $(3n + 5m)^2$

37. Using your own words, explain how to square a binomial such as $(a + b)^2$.

38. Explain how to find the product of two binomials using the FOIL method.

C *Multiply. See Examples 9 through 13.*

39. $(a - 7)(a + 7)$ **40.** $(b + 3)(b - 3)$ **41.** $(x + 6)(x - 6)$ **42.** $(x - 8)(x + 8)$

43. $(3x - 1)(3x + 1)$ **44.** $(4x - 5)(4x + 5)$ **45.** $(x^2 + 5)(x^2 - 5)$ **46.** $(a^2 + 6)(a^2 - 6)$

47. $(2y^2 - 1)(2y^2 + 1)$ **48.** $(3x^2 + 1)(3x^2 - 1)$ **49.** $(4 - 7x)(4 + 7x)$ **50.** $(8 - 7x)(8 + 7x)$

51. $\left(3x - \frac{1}{2}\right)\left(3x + \frac{1}{2}\right)$ **52.** $\left(10x + \frac{2}{7}\right)\left(10x - \frac{2}{7}\right)$ **53.** $(9x + y)(9x - y)$ **54.** $(2x - y)(2x + y)$

55. $(2m + 5n)(2m - 5n)$ **56.** $(5m + 4n)(5m - 4n)$

D *Multiply. See Examples 14 through 16.*

57. $(a + 5)(a + 4)$

58. $(a + 5)(a + 7)$

59. $(a - 7)^2$

60. $(b - 2)^2$

61. $(4a + 1)(3a - 1)$

62. $(6a + 7)(6a + 5)$

63. $(x + 2)(x - 2)$

64. $(x - 10)(x + 10)$

65. $(3a + 1)^2$

66. $(4a + 2)^2$

67. $(x + y)(4x - y)$

68. $(3x + 2)(4x - 2)$

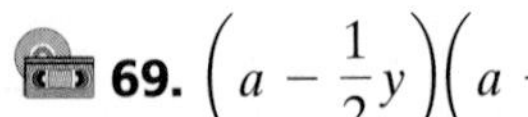

69. $\left(a - \frac{1}{2}y\right)\left(a + \frac{1}{2}y\right)$

70. $\left(\frac{a}{2} + 4y\right)\left(\frac{a}{2} - 4y\right)$

71. $(3b + 7)(2b - 5)$

72. $(3y - 13)(y - 3)$

73. $(x^2 + 10)(x^2 - 10)$

74. $(x^2 + 8)(x^2 - 8)$

75. $(4x + 5)(4x - 5)$

76. $(3x + 5)(3x - 5)$

77. $(5x - 6y)^2$

78. $(4x - 9y)^2$

79. $(2r - 3s)(2r + 3s)$

80. $(6r - 2x)(6r + 2x)$

Review and Preview

Simplify each expression. See Sections 4.1 and 4.2.

81. $\frac{50b^{10}}{70b^5}$

82. $\frac{60y^6}{80y^2}$

83. $\frac{8a^{17}b^5}{-4a^7b^{10}}$

84. $\frac{-6a^8y}{3a^4y}$

85. $\frac{2x^4y^{12}}{3x^4y^4}$

86. $\frac{-48ab^6}{32ab^3}$

Combining Concepts

Express each as a product of polynomials in x. Then multiply and simplify.

△ **87.** Find the area of the square rug if its side is $(2x + 1)$ feet.

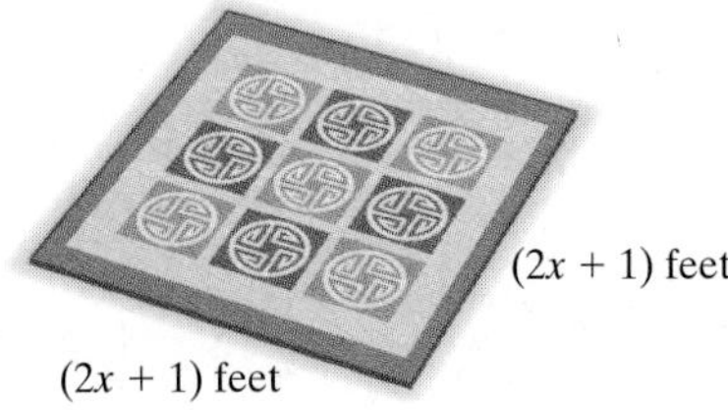

△ **88.** Find the area of the rectangular canvas if its length is $(3x - 2)$ inches and its width is $(x - 4)$ inches.

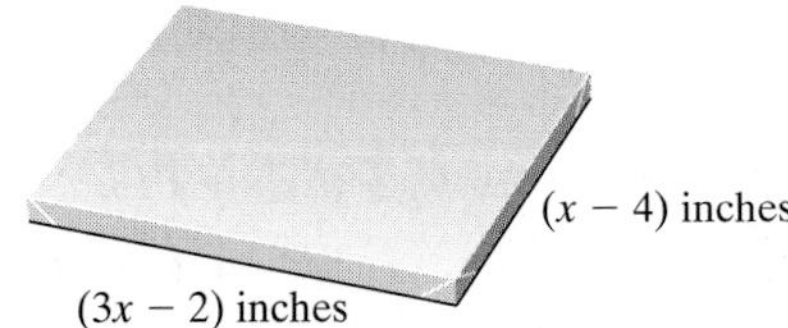

△ **89.** Find the area of the shaded region.

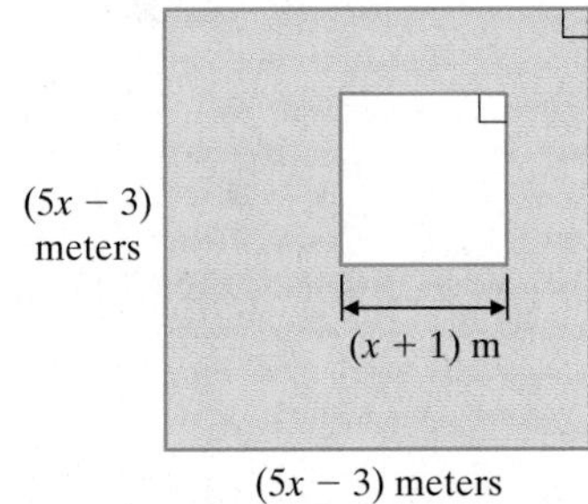

△ **90.** Find the area of the shaded region.

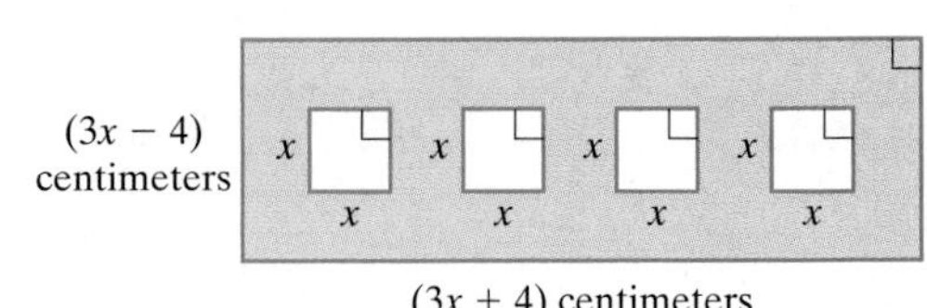

Name ______________________ Section __________ Date __________

Integrated Review—Exponents and Operations on Polynomials

Perform operations and simplify.

1. $(5x^2)(7x^3)$

2. $(4y^2)(8y^7)$

3. -4^2

4. $(-4)^2$

5. $(x - 5)(2x + 1)$

6. $(3x - 2)(x + 5)$

7. $(x - 5) + (2x + 1)$

8. $(3x - 2) + (x + 5)$

9. $\dfrac{7x^9y^{12}}{x^3y^{10}}$

10. $\dfrac{20a^2b^8}{14a^2b^2}$

11. $(12m^7n^6)^2$

12. $(4y^9z^{10})^3$

13. $(4y - 3)(4y + 3)$

14. $(7x - 1)(7x + 1)$

15. $(x^{-7}y^5)^9$

16. $(3^{-1}x^9)^3$

Answers

1. ______
2. ______
3. ______
4. ______
5. ______
6. ______
7. ______
8. ______
9. ______
10. ______
11. ______
12. ______
13. ______
14. ______
15. ______
16. ______

17. ______

18. ______

19. ______

20. ______

21. ______

22. ______

23. ______

24. ______

25. ______

26. ______

27. ______

28. ______

29. ______

30. ______

31. ______

32. ______

17. $(7x^2 - 2x + 3) - (5x^2 + 9)$

18. $(10x^2 + 7x - 9) - (4x^2 - 6x + 2)$

19. $0.7y^2 - 1.2 + 1.8y^2 - 6y + 1$

20. $7.8x^2 - 6.8x - 3.3 + 0.6x^2 - 0.9$

21. $(x + 4)^2$

22. $(y - 9)^2$

23. $(x + 4) + (x + 4)$

24. $(y - 9) + (y - 9)$

25. $7x^2 - 6xy + 4(y^2 - xy)$

26. $5a^2 - 3ab + 6(b^2 - a^2)$

27. $(x - 3)(x^2 + 5x - 1)$

28. $(x + 1)(x^2 - 3x - 2)$

29. $(2x - 7)(3x + 10)$

30. $(5x - 1)(4x + 5)$

31. $(2x - 7)(x^2 - 6x + 1)$

32. $(5x - 1)(x^2 + 2x - 3)$

4.7 Dividing Polynomials

OBJECTIVES

A Divide a polynomial by a monomial.

B Use long division to divide a polynomial by a polynomial other than a monomial.

C Use synthetic division.

A Dividing by a Monomial

To divide a polynomial by a monomial, recall addition of fractions. Fractions that have a common denominator are added by adding the numerators:

$$\frac{a}{c} + \frac{b}{c} = \frac{a + b}{c}$$

If we read this equation from right to left and let a, b, and c be monomials, $c \neq 0$, we have the following.

To Divide a Polynomial by a Monomial

Divide each term of the polynomial by the monomial.

$$\frac{a + b}{c} = \frac{a}{c} + \frac{b}{c}, \quad c \neq 0$$

Throughout this section, we assume that denominators are not 0.

EXAMPLE 1 Divide: $6m^2 + 2m$ by $2m$

Solution: We begin by writing the quotient in fraction form. Then we divide each term of the polynomial $6m^2 + 2m$ by the monomial $2m$.

$$\frac{6m^2 + 2m}{2m} = \frac{6m^2}{2m} + \frac{2m}{2m}$$

$$= 3m + 1 \qquad \text{Simplify.}$$

Check: To check, we multiply.

$$2m(3m + 1) = 2m(3m) + 2m(1) = 6m^2 + 2m$$

The quotient $3m + 1$ checks.

Practice Problem 1

Divide: $25x^3 + 5x^2$ by $5x^2$

Concept Check

In which of the following is $\frac{x + 5}{5}$ simplified correctly?

a. $\frac{x}{5} + 1$ b. x c. $x + 1$

Try the Concept Check in the margin.

EXAMPLE 2 Divide: $\frac{9x^5 - 12x^2 + 3x}{3x^2}$

Solution:

$$\frac{9x^5 - 12x^2 + 3x}{3x^2} = \frac{9x^5}{3x^2} - \frac{12x^2}{3x^2} + \frac{3x}{3x^2} \qquad \text{Divide each term by } 3x^2.$$

$$= 3x^3 - 4 + \frac{1}{x} \qquad \text{Simplify.}$$

Notice that the quotient is not a polynomial because of the term $\frac{1}{x}$. This expression is called a rational expression—we will study rational expressions in Chapter 6. Although the quotient of two polynomials is not always a polynomial, we may still check by multiplying.

Check:

$$3x^2\left(3x^3 - 4 + \frac{1}{x}\right) = 3x^2(3x^3) - 3x^2(4) + 3x^2\left(\frac{1}{x}\right)$$

$$= 9x^5 - 12x^2 + 3x$$

Practice Problem 2

Divide: $\frac{30x^7 + 10x^2 - 5x}{5x^2}$

Answers

1. $5x + 1$, **2.** $6x^5 + 2 - \frac{1}{x}$

Concept Check: a

Practice Problem 3

Divide: $\dfrac{12x^3y^3 - 18xy + 6y}{3xy}$

EXAMPLE 3 Divide: $\dfrac{8x^2y^2 - 16xy + 2x}{4xy}$

Solution:

$$\frac{8x^2y^2 - 16xy + 2x}{4xy} = \frac{8x^2y^2}{4xy} - \frac{16xy}{4xy} + \frac{2x}{4xy}$$ Divide each term by $4xy$.

$$= 2xy - 4 + \frac{1}{2y}$$ Simplify.

Check: $$4xy\left(2xy - 4 + \frac{1}{2y}\right) = 4xy(2xy) - 4xy(4) + 4xy\left(\frac{1}{2y}\right)$$
$$= 8x^2y^2 - 16xy + 2x$$

B Dividing by a Polynomial Other Than a Monomial

To divide a polynomial by a polynomial other than a monomial, we use a process known as long division. Polynomial long division is similar to number long division, so we review long division by dividing 13 into 3660.

Helpful Hint

Recall that 3660 is called the dividend.

```
     281
13)3660
   26↓        2·13 = 26
   106        Subtract and bring down the next digit in the dividend.
   104↓       8·13 = 104
     20       Subtract and bring down the next digit in the dividend.
     13       1·13 = 13
      7       Subtract. There are no more digits to bring down, so the remainder is 7.
```

The quotient is 281 R 7, which can be written as $281\frac{7}{13}$ (7 ← remainder, 13 ← divisor).

Recall that division can be checked by multiplication. To check this division problem, we see that

$13 \cdot 281 + 7 = 3660$, the dividend.

Now we demonstrate long division of polynomials.

Practice Problem 4

Divide: $x^2 + 12x + 35$ by $x + 5$

EXAMPLE 4 Divide $x^2 + 7x + 12$ by $x + 3$ using long division.

Solution:

```
              x
      x + 3)x² + 7x + 12     How many times does x divide x²? x²/x = x.
           -x² - 3x    ↓     Multiply: x(x + 3).
                 4x + 12     Subtract and bring down the next term.
```

To subtract, change the signs of these terms and add.

Now we repeat this process.

```
              x + 4
      x + 3)x² + 7x + 12     How many times does x divide 4x? 4x/x = 4.
            x² + 3x
                 4x + 12
                -4x - 12     Multiply: 4(x + 3).
                       0     Subtract. The remainder is 0.
```

To subtract, change the signs of these terms and add.

The quotient is $x + 4$.

Answers

3. $4x^2y^2 - 6 + \frac{2}{x}$, **4.** $x + 7$

Check: We check by multiplying.

	divisor	·	quotient	+	remainder	=	dividend
	↓		↓		↓		↓
or	$(x+3)$	·	$(x+4)$	+	0	=	$x^2+7x+12$

The quotient checks.

EXAMPLE 5 Divide $6x^2 + 10x - 5$ by $3x - 1$ using long division.

Solution:

$$\begin{array}{r l}
2x + 4 & \\
3x - 1\overline{)6x^2 + 10x - 5} & \frac{6x^2}{3x} = 2x\text{, so } 2x \text{ is a term of the quotient.} \\
\underline{6x^2 - 2x} & \text{Multiply: } 2x(3x-1). \\
12x - 5 & \text{Subtract and bring down the next term.} \\
\underline{12x - 4} & \frac{12x}{3x} = 4. \text{ Multiply: } 4(3x-1). \\
-1 & \text{Subtract. The remainder is } -1.
\end{array}$$

Thus $(6x^2 + 10x - 5)$ divided by $(3x - 1)$ is $(2x + 4)$ with a remainder of -1. This can be written as follows.

$$\frac{6x^2 + 10x - 5}{3x - 1} = 2x + 4 + \frac{-1}{3x - 1} \quad \begin{array}{l}\leftarrow \text{remainder} \\ \leftarrow \text{divisor}\end{array}$$

Check: To check, we multiply $(3x - 1)(2x + 4)$. Then we add the remainder, -1, to this product.

$$\begin{aligned}(3x - 1)(2x + 4) + (-1) &= (6x^2 + 12x - 2x - 4) - 1 \\ &= 6x^2 + 10x - 5\end{aligned}$$

The quotient checks.

Notice that the division process is continued until the degree of the remainder polynomial is less than the degree of the divisor polynomial.

EXAMPLE 6 Divide: $\dfrac{4x^2 + 7 + 8x^3}{2x + 3}$

Solution: Before we begin the division process, we rewrite $4x^2 + 7 + 8x^3$ as $8x^3 + 4x^2 + 0x + 7$. Notice that we have written the polynomial in descending order and have represented the missing x term by $0x$.

$$\begin{array}{r l}
4x^2 - 4x + 6 & \\
2x + 3\overline{)8x^3 + 4x^2 + 0x + 7} & \\
\underline{8x^3 + 12x^2} & \\
-8x^2 + 0x & \\
\underline{-8x^2 - 12x} & \\
12x + 7 & \\
\underline{12x + 18} & \\
-11 & \text{Remainder.}
\end{array}$$

Thus, $\dfrac{4x^2 + 7 + 8x^3}{2x + 3} = 4x^2 - 4x + 6 + \dfrac{-11}{2x + 3}$.

Practice Problem 5

Divide: $6x^2 + 7x - 5$ by $2x - 1$

Practice Problem 6

Divide: $\dfrac{5 - x + 9x^3}{3x + 2}$

Answers

5. $3x + 5$, **6.** $3x^2 - 2x + 1 + \dfrac{3}{3x + 2}$

Practice Problem 7

Divide: $x^3 - 1$ by $x - 1$

EXAMPLE 7 Divide $x^3 - 8$ by $x - 2$.

Solution: Notice that the polynomial $x^3 - 8$ is missing an x^2 term and an x term. We'll represent these terms by inserting $0x^2$ and $0x$.

$$\begin{array}{r} x^2 + 2x + 4 \\ x - 2\overline{)x^3 + 0x^2 + 0x - 8} \\ \underline{-x^3 + 2x^2} \qquad\qquad \\ 2x^2 + 0x - 8 \\ \underline{-2x^2 + 4x} \qquad \\ 4x - 8 \\ \underline{-4x + 8} \\ 0 \end{array}$$

Thus, $\dfrac{x^3 - 8}{x - 2} = x^2 + 2x + 4$.

Check: To check, see that $(x^2 + 2x + 4)(x - 2) = x^3 - 8$.

C Using Synthetic Division

When a polynomial is to be divided by a binomial of the form $x - c$, a short-cut process called **synthetic division** may be used. On the left is an example of long division, and on the right is the same example showing the coefficients of the variables only.

$$\begin{array}{r} 2x^2 + 5x + 2 \\ x - 3\overline{)2x^3 - x^2 - 13x + 1} \\ \underline{2x^3 - 6x^2} \qquad\qquad\quad \\ 5x^2 - 13x \qquad \\ \underline{5x^2 - 15x} \qquad \\ 2x + 1 \\ \underline{2x - 6} \\ 7 \end{array} \qquad \begin{array}{r} 2 \quad 5 \quad 2 \\ 1 - 3\overline{)2 - 1 - 13 + 1} \\ \underline{2 - 6} \qquad\qquad \\ 5 - 13 \qquad \\ \underline{5 - 15} \qquad \\ 2 + 1 \\ \underline{2 - 6} \\ 7 \end{array}$$

Notice that as long as we keep coefficients of powers of x in the same column, we can perform division of polynomials by performing algebraic operations on the coefficients only. This shorter process of dividing with coefficients only in a special format is called synthetic division. To find $(2x^3 - x^2 - 13x + 1) \div (x - 3)$ by synthetic division, follow the next example.

Practice Problem 8

Use synthetic division to divide $3x^3 - 2x^2 + 5x + 4$ by $x - 2$.

Answers

7. $x^2 + x + 1$ **8.** $3x^2 + 4x + 13 + \dfrac{30}{x - 2}$

EXAMPLE 8

Use synthetic division to divide $2x^3 - x^2 - 13x + 1$ by $x - 3$.

Solution: To use synthetic division, the divisor must be in the form $x - c$. Since we are dividing by $x - 3$, c is 3. We write down 3 and the coefficients of the dividend.

```
c
 3| 2  -1  -13  1
    ↓
   ---------------
    2
```

Next, draw a line and bring down the first coefficient of the dividend.

```
 3| 2  -1  -13  1
        6
   ---------------
    2
```

Multiply $3 \cdot 2$ and write down the product, 6.

```
 3| 2  -1  -13  1
        6
   ---------------
    2   5
```

Add $-1 + 6$. Write down the sum, 5.

```
 3| 2  -1  -13  1
        6   15
   ---------------
    2   5    2
```

$3 \cdot 5 = 15$

$-13 + 15 = 2$

```
 3| 2  -1  -13  1
        6   15  6
   ---------------
    2   5    2  7
```

$3 \cdot 2 = 6$

$1 + 6 = 7$

The quotient is found in the bottom row. The numbers 2, 5, and 2 are the coefficients of the quotient polynomial, and the number 7 is the remainder. The degree of the quotient polynomial is one less than the degree of the dividend. In our example, the degree of the dividend is 3, so the degree of the quotient polynomial is 2. As we found when we performed the long division, the quotient is

$$2x^2 + 5x + 2, \quad \text{remainder } 7$$

or

$$2x^2 + 5x + 2 + \frac{7}{x - 3}$$

When using synthetic division, if there are missing powers of the variable, insert 0s as coefficients.

EXAMPLE 9

Use synthetic division to divide $x^4 - 2x^3 - 11x^2 + 34$ by $x + 2$.

Solution: The divisor is $x + 2$, which in the form $x - c$ is $x - (-2)$. Thus, c is -2. There is no x-term in the dividend, so we insert a coefficient of 0. The dividend coefficients are 1, -2, -11, 0, and 34.

```
c
-2| 1  -2  -11  0   34
       -2    8  6  -12
   --------------------
    1  -4   -3  6   22
```

The dividend is a fourth-degree polynomial, so the quotient polynomial is a third-degree polynomial. The quotient is $x^3 - 4x^2 - 3x + 6$ with a remainder of 22. Thus,

$$\frac{x^4 - 2x^3 - 11x^2 + 34}{x + 2} = x^3 - 4x^2 - 3x + 6 + \frac{22}{x + 2}$$

Practice Problem 9

Use synthetic division to divide $x^4 + 3x^3 - 5x + 4$ by $x + 1$.

Answer

9. $x^3 + 2x^2 - 2x - 3 + \frac{7}{x + 1}$

Helpful Hint

Before dividing by synthetic division, write the dividend in descending order of variable exponents. Any "missing powers" of the variable must be represented by 0 times the variable raised to the missing power.

Try the Concept Check in the margin.

Concept Check

Which division problems are candidates for the synthetic division process?

a. $(3x^2 + 5) \div (x + 4)$
b. $(x^3 - x^2 + 2) \div (3x^3 - 2)$
c. $(y^4 + y - 3) \div (x^2 + 1)$
d. $x^5 \div (x - 5)$

Answer
Concept Check: a and d

FOCUS ON **History**

EXPONENTIAL NOTATION

The French mathematician and philosopher René Descartes (1596–1650) is generally credited with devising the system of exponents that we use in math today. His book *La Géométrie* was the first to show successive powers of an unknown quantity x as x, xx, x^3, x^4, x^5, and so on. No one knows why Descartes preferred to write xx instead of x^2. However, the use of xx for the square of the quantity x continued to be popular. Those who used the notation defended it by saying that xx takes up no more space when written than x^2 does.

Before Descartes popularized the use of exponents to indicate powers, other less convenient methods were used. Some mathematicians preferred to write out the Latin words *quadratus* and *cubus* whenever they wanted to indicate that a quantity was to be raised to the second power or the third power. Other mathematicians used the abbreviations of *quadratus* and *cubus*, Q and C, to indicate second and third powers of a quantity.

Name ______________________ Section __________ Date __________

Mental Math

Simplify each expression.

1. $\frac{a^6}{a^4}$ 2. $\frac{y^2}{y}$ 3. $\frac{a^3}{a}$ 4. $\frac{p^8}{p^3}$ 5. $\frac{k^5}{k^2}$ 6. $\frac{k^7}{k^5}$

EXERCISE SET 4.7

A *Perform each division. See Examples 1 through 3.*

1. $\frac{20x^2 + 5x + 9}{5}$
2. $\frac{8x^3 - 4x^2 + 6x + 2}{2}$
3. $\frac{12x^4 + 3x^2}{x}$
4. $\frac{15x^2 - 9x^5}{x}$
5. $\frac{15p^3 + 18p^2}{3p}$
6. $\frac{14m^2 - 27m^3}{7m}$
7. $\frac{-9x^4 + 18x^5}{6x^5}$
8. $\frac{6x^5 + 3x^4}{3x^4}$
9. $\frac{-9x^5 + 3x^4 - 12}{3x^3}$
10. $\frac{6a^2 - 4a + 12}{-2a^2}$
11. $\frac{4x^4 - 6x^3 + 7}{-4x^4}$
12. $\frac{-12a^3 + 36a - 15}{3a}$
13. $\frac{a^2b^2 - ab^3}{ab}$
14. $\frac{m^3n^2 - mn^4}{mn}$
15. $\frac{2x^2y + 8x^2y^2 - xy^2}{2xy}$
16. $\frac{11x^3y^3 - 33xy + x^2y^2}{11xy}$

B *Find each quotient using long division. See Examples 4 through 6.*

17. $\frac{x^2 + 4x + 3}{x + 3}$
18. $\frac{x^2 + 7x + 10}{x + 5}$
19. $\frac{2x^2 + 13x + 15}{x + 5}$
20. $\frac{3x^2 + 8x + 4}{x + 2}$
21. $\frac{2x^2 - 7x + 3}{x - 4}$
22. $\frac{3x^2 - x - 4}{x - 1}$
23. $\frac{8x^2 + 6x - 27}{2x - 3}$
24. $\frac{18w^2 + 18w - 8}{3w + 4}$
25. $\frac{9a^3 - 3a^2 - 3a + 4}{3a + 2}$
26. $\frac{4x^3 + 12x^2 + x - 12}{2x + 3}$
27. $\frac{2b^3 + 9b^2 + 6b - 4}{b + 4}$
28. $\frac{2x^3 + 3x^2 - 3x + 4}{x + 2}$
29. $\frac{8x^2 + 10x + 1}{2x + 1}$
30. $\frac{3x^2 + 17x + 7}{3x + 2}$
31. $\frac{2x^3 + 2x^2 - 17x + 8}{x - 2}$
32. $\frac{4x^3 + 11x^2 - 8x - 10}{x + 3}$

Find each quotient using long division. Don't forget to write the polynomials in descending order and fill in any missing terms. See Examples 6 and 7.

33. $\frac{x^3 - 27}{x - 3}$
34. $\frac{x^3 + 64}{x + 4}$
35. $\frac{1 - 3x^2}{x + 2}$
36. $\frac{7 - 5x^2}{x + 3}$

37. $\dfrac{-4b + 4b^2 - 5}{2b - 1}$

38. $\dfrac{-3y + 2y^2 - 15}{2y + 5}$

C *Use synthetic division to divide. See Examples 8 and 9.*

39. $\dfrac{x^2 + 3x - 40}{x - 5}$

40. $\dfrac{x^2 - 14x + 24}{x - 2}$

41. $\dfrac{x^2 + 5x - 6}{x + 6}$

42. $\dfrac{x^2 + 12x + 32}{x + 4}$

43. $\dfrac{x^3 - 7x^2 - 13x + 5}{x - 2}$

44. $\dfrac{x^3 + 6x^2 + 4x - 7}{x + 5}$

45. $\dfrac{4x^2 - 9}{x - 2}$

46. $\dfrac{3x^2 - 4}{x - 1}$

47. $\dfrac{2x^4 - 13x^3 + 16x^2 - 9x + 20}{x - 5}$

48. $\dfrac{3x^4 + 5x^3 - x^2 + x - 2}{x + 2}$

49. $\dfrac{7x^2 - 4x + 12 + 3x^3}{x + 1}$

50. $\dfrac{x^4 + 4x^3 - x^2 - 16x - 4}{x - 2}$

Review and Preview

Fill in each blank. See Sections 4.1 and 4.2.

51. $12 = 4 \cdot$ ______

52. $12 = 2 \cdot$ ______

53. $20 = -5 \cdot$ ______

54. $20 = -4 \cdot$ ______

55. $9x^2 = 3x \cdot$ ______

56. $9x^2 = 9x \cdot$ ______

57. $36x^2 = 4x \cdot$ ______

58. $36x^2 = 2x \cdot$ ______

Combining Concepts

Divide.

59. $\dfrac{x^5 + x^2}{x^2 + x}$

60. $\dfrac{x^6 - x^4}{x^3 + 1}$

Solve.

△ **61.** The perimeter of a square is $(12x^3 + 4x - 16)$ feet. Find the length of its side.

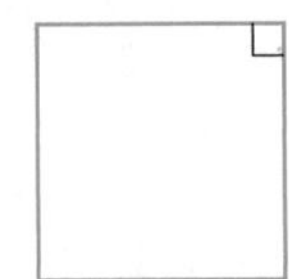

Perimeter is $(12x^3 + 4x - 16)$ feet

△ **62.** The volume of the swimming pool shown is $(36x^5 - 12x^3 + 6x^2)$ cubic feet. If its height is $2x$ feet and its width is $3x$ feet, find its length.

△ **63.** The area of the following parallelogram is $(10x^2 + 31x + 15)$ square meters. If its base is $(5x + 3)$ meters, find its height.

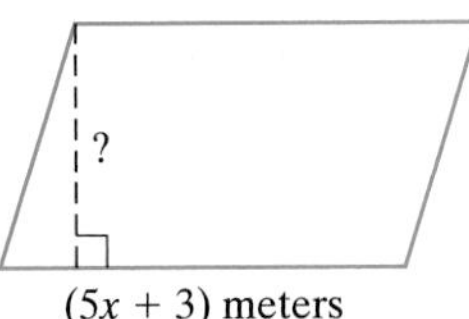

△ **64.** The area of the top of the Ping-Pong table shown is $(49x^2 + 70x - 200)$ square inches. If its length is $(7x + 20)$ inches, find its width.

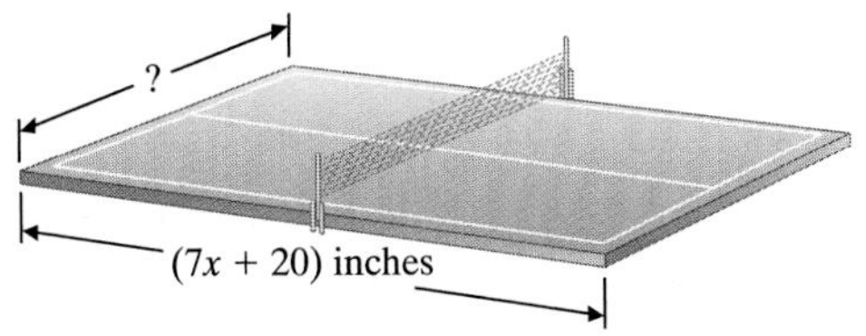

65. Explain how to check a polynomial long division result when the remainder is 0.

66. Explain how to check a polynomial long division result when the remainder is not 0.

CHAPTER 4 ACTIVITY Modeling with Polynomials

MATERIALS:

- Calculator

This activity may be completed by working in groups or individually.

The polynomial model $-40x^2 + 140x + 8393$ gives the average daily total supply of motor gasoline (in thousand barrels per day) in the United States for the period 1998–2000. The polynomial model $7x^2 + 22x + 8082$ gives the average daily supply of domestically produced motor gasoline (in thousand barrels per day) in the United States for the same period. In both models, x is the number of years after 1998. The other source of motor gasoline in the United States, contributing to the total supply, is imported motor gasoline. (*Source:* Based on data from the Energy Information Administration)

1. Use the given polynomials to complete the following table showing the average daily supply (both total and domestic) over the period 1998–2000 by evaluating each polynomial at the given values of x. Then subtract each value in the fourth column from the corresponding value in the third column. Record the result in the last column, titled "Difference." What do you think these values represent?

Year	x	Average Daily Total Supply (thousand barrels per day)	Average Daily Domestic Supply (thousand barrels per day)	Difference
1998	0			
1999	1			
2000	2			

2. Use the polynomial models to find a new polynomial model representing the average daily supply of imported motor gasoline. Then evaluate your new polynomial model to complete the accompanying table.

Year	x	Average Daily Imported Supply (thousand barrels per day)
1998	0	
1999	1	
2000	2	

3. Compare the values in the last column of the table in question 1 to the values in the last column of the table in question 2. What do you notice? What can you conclude?

4. Make a bar graph of the data in the table in question 2. Describe what you see

Average Daily Imported Supply (thousand barrels a day)

0

STUDY SKILLS REMINDER

What should you do on the day of an exam?

On the day of an exam, try the following:

- Allow yourself plenty of time to arrive at your classroom.
- Read the directions on the test carefully.
- Read each problem carefully as you take your test. Make sure that you answer the question asked.
- Watch your time and pace yourself so that you may attempt each problem on your test.
- If you have time, check your work and answers.
- Do not turn your test in early. If you have extra time, spend it double-checking your work.

Good luck!

FOCUS ON **History**

NEGATIVE EXPONENTS

Negative exponents were the invention of the English mathematician John Wallis (1616–1703). His book *Arithmetica Infinitorum*, published in 1656, begins with proofs of the laws of exponents. He extended these to cover negative exponents as well and showed that $x^0 = 1$, $x^{-1} = \frac{1}{x}$, $x^{-2} = \frac{1}{x^2}$, and so on. He also showed that, in general, x^{-n} represents the reciprocal of x^n.

Not long after Wallis published his *Arithmetica Infinitorum*, Sir Issac Newton (1642–1727) adopted Wallis's definition and use of negative exponents. Newton's widely circulated mathematical and scientific writings helped the use of negative exponents become universally accepted.

Chapter 4 VOCABULARY CHECK

Fill in each blank with one of the words or phrases listed below.

term	coefficient	monomial	binomial	trinomial
polynomials	degree of a term	degree of a polynomial	FOIL	

1. A ________ is a number or the product of numbers and variables raised to powers.
2. The ________ method may be used when multiplying two binomials.
3. A polynomial with exactly 3 terms is called a ___________.
4. The ______________________ is the greatest degree of any term of the polynomial.
5. A polynomial with exactly 2 terms is called a ___________.
6. The ____________ of a term is its numerical factor.
7. The ________________ is the sum of the exponents on the variables in the term.
8. A polynomial with exactly 1 term is called a ____________.
9. Monomials, binomials, and trinomials are all examples of ______________.

CHAPTER 4 Highlights

DEFINITIONS AND CONCEPTS	EXAMPLES
Section 4.1 Exponents	
a^n means the product of n factors, each of which is a.	$3^2 = 3 \cdot 3 = 9$ $(-5)^3 = (-5)(-5)(-5) = -125$ $\left(\frac{1}{2}\right)^4 = \frac{1}{2}\cdot\frac{1}{2}\cdot\frac{1}{2}\cdot\frac{1}{2} = \frac{1}{16}$
Let m and n be integers and no denominators be 0.	
Product Rule: $a^m \cdot a^n = a^{m+n}$	$x^2 \cdot x^7 = x^{2+7} = x^9$
Power Rule: $(a^m)^n = a^{mn}$	$(5^3)^8 = 5^{3\cdot 8} = 5^{24}$
Power of a Product Rule: $(ab)^n = a^n b^n$	$(7y)^4 = 7^4 y^4$
Power of a Quotient Rule: $\left(\frac{a}{b}\right)^n = \frac{a^n}{b^n}$	$\left(\frac{x}{8}\right)^3 = \frac{x^3}{8^3}$
Quotient Rule: $\frac{a^m}{a^n} = a^{m-n}$	$\frac{x^9}{x^4} = x^{9-4} = x^5$
Zero Exponent: $a^0 = 1, a \neq 0$	$5^0 = 1; x^0 = 1, x \neq 0$
Section 4.2 Negative Exponents and Scientific Notation	
If $a \neq 0$ and n is an integer, $a^{-n} = \frac{1}{a^n}$	$3^{-2} = \frac{1}{3^2} = \frac{1}{9}; 5x^{-2} = \frac{5}{x^2}$ Simplify: $\left(\frac{x^{-2}y}{x^5}\right)^{-2} = \frac{x^4y^{-2}}{x^{-10}} = x^{4-(-10)}y^{-2} = \frac{x^{14}}{y^2}$
A positive number is written in scientific notation if it is written as the product of a number a, where $1 \leq a < 10$, and an integer power r of 10. $a \times 10^r$	$1200 = 1.2 \times 10^3$ $0.000000568 = 5.68 \times 10^{-7}$

DEFINITIONS AND CONCEPTS	EXAMPLES

Section 4.3 Introduction to Polynomials

A **term** is a number or the product of a number and variables raised to powers.

$-5x, 7a^2b, \frac{1}{4}y^4, 0.2$

The **numerical coefficient** or **coefficient** of a term is its numerical factor.

TERM	COEFFICIENT
$7x^2$	7
y	1
$-a^2b$	-1

A **polynomial** is a finite sum of terms of the form ax^n where a is a real number and n is a whole number.

$5x^3 - 6x^2 + 3x - 6$ (Polynomial)

A **monomial** is a polynomial with exactly 1 term.

$\frac{5}{6}y^3$ (Monomial)

A **binomial** is a polynomial with exactly 2 terms.

$-0.2a^2b - 5b^2$ (Binomial)

A **trinomial** is a polynomial with exactly 3 terms.

$3x^2 - 2x + 1$ (Trinomial)

The **degree of a polynomial** is the greatest degree of any term of the polynomial.

POLYNOMIAL	DEGREE
$5x^2 - 3x + 2$	2
$7y + 8y^2z^3 - 12$	$2 + 3 = 5$

Section 4.4 Adding and Subtracting Polynomials

To add polynomials, combine like terms.

Add.

$$(7x^2 - 3x + 2) + (-5x - 6)$$
$$= 7x^2 - 3x + 2 - 5x - 6$$
$$= 7x^2 - 8x - 4$$

To subtract two polynomials, change the signs of the terms of the second polynomial, and then add.

Subtract.

$$(17y^2 - 2y + 1) - (-3y^3 + 5y - 6)$$
$$= (17y^2 - 2y + 1) + (3y^3 - 5y + 6)$$
$$= 17y^2 - 2y + 1 + 3y^3 - 5y + 6$$
$$= 3y^3 + 17y^2 - 7y + 7$$

Section 4.5 Multiplying Polynomials

To multiply two polynomials, multiply each term of one polynomial by each term of the other polynomial, and then combine like terms.

Multiply.

$$(2x + 1)(5x^2 - 6x + 2)$$
$$= 2x(5x^2 - 6x + 2) + 1(5x^2 - 6x + 2)$$
$$= 10x^3 - 12x^2 + 4x + 5x^2 - 6x + 2$$
$$= 10x^3 - 7x^2 - 2x + 2$$

Definitions and Concepts	**Examples**
Section 4.6 Special Products	
The **FOIL method** may be used when multiplying two binomials.	Multiply: $(5x - 3)(2x + 3)$ First, Outer, Inner, Last $= (5x)(2x) + (5x)(3) + (-3)(2x) + (-3)(3)$ (F, O, I, L) $= 10x^2 + 15x - 6x - 9$ $= 10x^2 + 9x - 9$
Squaring a Binomial $(a + b)^2 = a^2 + 2ab + b^2$	Square each binomial. $(x + 5)^2 = x^2 + 2(x)(5) + 5^2$ $= x^2 + 10x + 25$
$(a - b)^2 = a^2 - 2ab + b^2$	$(3x - 2y)^2 = (3x)^2 - 2(3x)(2y) + (2y)^2$ $= 9x^2 - 12xy + 4y^2$
Multiplying the Sum and Difference of Two Terms $(a + b)(a - b) = a^2 - b^2$	Multiply. $(6y + 5)(6y - 5) = (6y)^2 - 5^2$ $= 36y^2 - 25$
Section 4.7 Dividing Polynomials	
To divide a polynomial by a monomial, $\frac{a + b}{c} = \frac{a}{c} + \frac{b}{c}, c \neq 0$	Divide. $\frac{15x^5 - 10x^3 + 5x^2 - 2x}{5x^2}$ $= \frac{15x^5}{5x^2} - \frac{10x^3}{5x^2} + \frac{5x^2}{5x^2} - \frac{2x}{5x^2}$ $= 3x^3 - 2x + 1 - \frac{2}{5x}$
To divide a polynomial by a polynomial other than a monomial, use long division.	$2x + 3 \overline{)10x^2 + 13x - 7}$ gives quotient $5x - 1 + \frac{-4}{2x + 3}$ or $5x - 1 - \frac{4}{2x + 3}$ $10x^2 + 15x$ $-2x - 7$ $-2x - 3$ -4
A shortcut method called **synthetic division** may be used to divide a polynomial by a binomial of the form $x - c$.	Use synthetic division to divide $2x^3 - x^2 - 8x - 1$ by $x - 2$. `2	2 -1 -8 -1` `	4 6 -4` ` 2 3 -2 -5` The quotient is $2x^2 + 3x - 2 - \frac{5}{x - 2}$.

STUDY SKILLS REMINDER

Are you preparing for a test on Chapter 4?

Below is a list of some *common trouble areas* for students in Chapter 4. After studying for your test—but before taking your test—read these.

- Do you know that a negative exponent does not make the base a negative number? For example,

  $$3^{-2} = \frac{1}{3^2} = \frac{1}{9}$$

- Make sure you remember that x has an understood coefficient of 1 and an understood exponent of 1. For example,

  $$2x + x = 2x + 1x = 3x; \quad x^5 \cdot x = x^5 \cdot x^1 = x^6$$

- Do you know the difference between $5x^2$ and $(5x)^2$?

  $5x^2$ is $5 \cdot x^2$; $(5x)^2 = 5^2 \cdot x^2$ or $25 \cdot x^2$

- Can you evaluate $x^2 - x$ when $x = -2$?

  $$x^2 - x = (-2)^2 - (-2) = 4 - (-2) = 4 + 2 = 6$$

- Can you subtract $5x^2 + 1$ from $3x^2 - 6$?

  $$(3x^2 - 6) - (5x^2 + 1) = 3x^2 - 6 - 5x^2 - 1 = -2x^2 - 7$$

- Make sure you are familiar with squaring a binomial.

  $$(3x - 4)^2 = (3x)^2 - 2(3x)(4) + 4^2 = 9x^2 - 24x + 16$$

  or

  $$(3x - 4)^2 = (3x - 4)(3x - 4) = 9x^2 - 24x + 16$$

Remember: This is simply a checklist of common trouble areas. For a review of Chapter 4, see the Highlights and Chapter Review.

Name ______________________ Section __________ Date __________

Chapter 4 Review

(4.1) *State the base and the exponent for each expression.*

1. 3^2 **2.** $(-5)^4$ **3.** -5^4 **4.** x^6

Evaluate each expression.

5. 8^3 **6.** $(-6)^2$ **7.** -6^2 **8.** $-4^3 - 4^0$ **9.** $(3b)^0$ **10.** $\frac{8b}{8b}$

Simplify each expression.

11. $y^2 \cdot y^7$ **12.** $x^9 \cdot x^5$ **13.** $(2x^5)(-3x^6)$ **14.** $(-5y^3)(4y^4)$ **15.** $(x^4)^2$

16. $(y^3)^5$ **17.** $(3y^6)^4$ **18.** $(2x^3)^3$ **19.** $\frac{x^9}{x^4}$ **20.** $\frac{z^{12}}{z^5}$

21. $\frac{a^5b^4}{ab}$ **22.** $\frac{x^4y^6}{xy}$ **23.** $\frac{12xy^6}{3x^4y^{10}}$ **24.** $\frac{2x^7y^8}{8xy^2}$ **25.** $5a^7(2a^4)^3$

26. $(2x)^2(9x)$ **27.** $(-5a)^0 + 7^0 + 8^0$ **28.** $8x^0 + 9^0$

Simplify the given expression and choose the correct result.

29. $\left(\frac{3x^4}{4y}\right)^3$

a. $\frac{27x^{64}}{64y^3}$

b. $\frac{27x^{12}}{64y^3}$

c. $\frac{9x^{12}}{12y^3}$

d. $\frac{3x^{12}}{4y^3}$

30. $\left(\frac{5a^6}{b^3}\right)^2$

a. $\frac{10a^{12}}{b^6}$

b. $\frac{25a^{36}}{b^9}$

c. $\frac{25a^{12}}{b^6}$

d. $25a^{12}b^6$

(4.2) *Simplify each expression.*

31. 7^{-2} **32.** -7^{-2} **33.** $2x^{-4}$ **34.** $(2x)^{-4}$

35. $\left(\frac{1}{5}\right)^{-3}$ **36.** $\left(\frac{-2}{3}\right)^{-2}$ **37.** $2^0 + 2^{-4}$ **38.** $6^{-1} - 7^{-1}$

Simplify each expression. Assume that variables in an exponent represent positive integers only. Write each answer using positive exponents only.

39. $\dfrac{x^5}{x^{-3}}$

40. $\dfrac{z^4}{z^{-4}}$

41. $\dfrac{r^{-3}}{r^{-4}}$

42. $\dfrac{y^{-2}}{y^{-5}}$

43. $\left(\dfrac{bc^{-2}}{bc^{-3}}\right)^4$

44. $\left(\dfrac{x^{-3}y^{-4}}{x^{-2}y^{-5}}\right)^{-3}$

45. $\dfrac{x^{-4}y^{-6}}{x^2y^7}$

46. $\dfrac{a^5b^{-5}}{a^{-5}b^5}$

Write each number in scientific notation.

47. 0.00027

48. 0.8868

49. 80,800,000

50. −868,000

51. In January 2001, the United States imported approximately 109,379,000 kilograms of coffee. Write this number in scientific notation. (*Source:* International Coffee Organization)

52. The approximate diameter of the Milky Way galaxy is 150,000 light years. Write this number in scientific notation. (*Source:* NASA IMAGE/POETRY Education and Public Outreach Program)

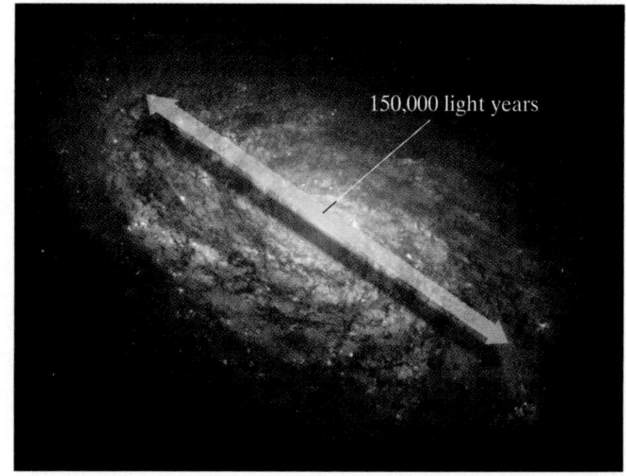

Write each number in standard form.

53. 8.67×10^5

54. 3.86×10^{-3}

55. 8.6×10^{-4}

56. 8.936×10^5

57. The volume of the planet Jupiter is 1.43128×10^{15} cubic kilometers. Write this number in standard notation. (*Source:* National Space Science Data Center)

58. An angstrom is a unit of measure, equal to 1×10^{-10} meter, used for measuring wavelengths or the diameters of atoms. Write this number in standard notation. (*Source:* National Institute of Standards and Technology)

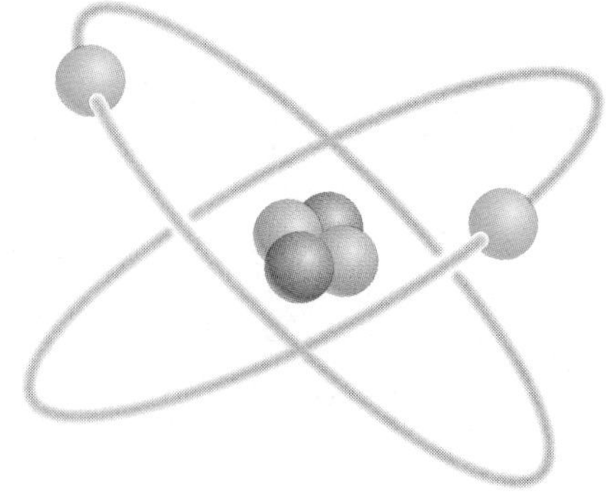

Simplify. Express each result in standard form.

59. $(8 \times 10^4)(2 \times 10^{-7})$

60. $\dfrac{8 \times 10^4}{2 \times 10^{-7}}$

(4.3) *Find the degree of each polynomial.*

61. $y^5 + 7x - 8x^4$

62. $9y^2 + 30y + 25$

63. $-14x^2y - 28x^2y^3 - 42x^2y^2$

64. $6x^2y^2z^2 + 5x^2y^3 - 12xyz$

△ **65.** The surface area of a box with a square base and a height of 5 units is given by the polynomial $2x^2 + 20x$. Fill in the table below by evaluating $2x^2 + 20x$ for the given values of x.

x	1	3	5.1	10
$2x^2 + 20x$				

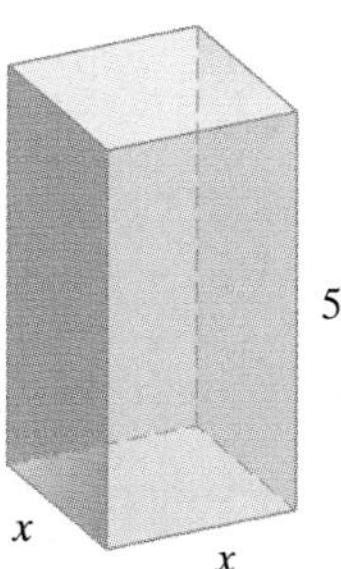

Combine like terms in each expression.

66. $7a^2 - 4a^2 - a^2$

67. $9y + y - 14y$

68. $6a^2 + 4a + 9a^2$

69. $21x^2 + 3x + x^2 + 6$

70. $4a^2b - 3b^2 - 8q^2 - 10a^2b + 7q^2$

71. $2s^{14} + 3s^{13} + 12s^{12} - s^{10}$

(4.4) *Add or subtract as indicated.*

72. $(3x^2 + 2x + 6) + (5x^2 + x)$

73. $(2x^5 + 3x^4 + 4x^3 + 5x^2) + (4x^2 + 7x + 6)$

74. $(-5y^2 + 3) - (2y^2 + 4)$

75. $(2m^7 + 3x^4 + 7m^6) - (8m^7 + 4m^2 + 6x^4)$

76. $(3x^2 - 7xy + 7y^2) - (4x^2 - xy + 9y^2)$

77. Add $(-9x^2 + 6x + 2)$ and $(4x^2 - x - 1)$.

78. Subtract $(4x^2 + 8x - 7)$ from the sum of $(x^2 + 7x + 9)$ and $(x^2 + 4)$.

(4.5) *Multiply each expression.*

79. $6(x + 5)$

80. $9(x - 7)$

81. $4(2a + 7)$

82. $9(6a - 3)$

83. $-7x(x^2 + 5)$

84. $-8y(4y^2 - 6)$

85. $-2(x^3 - 9x^2 + x)$

86. $-3a(a^2b + ab + b^2)$

87. $(3a^3 - 4a + 1)(-2a)$

88. $(6b^3 - 4b + 2)(7b)$

89. $(2x + 2)(x - 7)$

90. $(2x - 5)(3x + 2)$

91. $(4a - 1)(a + 7)$

92. $(6a - 1)(7a + 3)$

93. $(x + 7)(x^3 + 4x - 5)$

94. $(x + 2)(x^5 + x + 1)$

95. $(x^2 + 2x + 4)(x^2 + 2x - 4)$

96. $(x^3 + 4x + 4)(x^3 + 4x - 4)$

97. $(x + 7)^3$

98. $(2x - 5)^3$

(4.6) *Use special products to multiply each of the following.*

99. $(x + 7)^2$

100. $(x - 5)^2$

101. $(3x - 7)^2$

102. $(4x + 2)^2$

103. $(5x - 9)^2$

104. $(5x + 1)(5x - 1)$

105. $(7x + 4)(7x - 4)$

106. $(a + 2b)(a - 2b)$

107. $(2x - 6)(2x + 6)$

108. $(4a^2 - 2b)(4a^2 + 2b)$

Express each as a product of polynomials in x. Then multiply and simplify.

△ **109.** Find the area of the square if its side is $(3x - 1)$ meters.

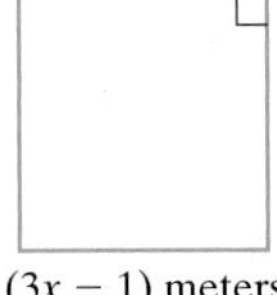

△ **110.** Find the area of the rectangle.

$(x - 1)$ miles

$(5x + 2)$ miles

(4.7) *Divide.*

111. $\dfrac{x^2 + 21x + 49}{7x^2}$

112. $\dfrac{5a^3b - 15ab^2 + 20ab}{-5ab}$

113. $(a^2 - a + 4) \div (a - 2)$

114. $(4x^2 + 20x + 7) \div (x + 5)$

115. $\dfrac{a^3 + a^2 + 2a + 6}{a - 2}$

116. $\dfrac{9b^3 - 18b^2 + 8b - 1}{3b - 2}$

117. $\dfrac{4x^4 - 4x^3 + x^2 + 4x - 3}{2x - 1}$

118. $\dfrac{-10x^2 - x^3 - 21x + 18}{x - 6}$

△ **119.** The area of the rectangle below is $(15x^3 - 3x^2 + 60)$ square feet. If its length is $3x^2$ feet, find its width.

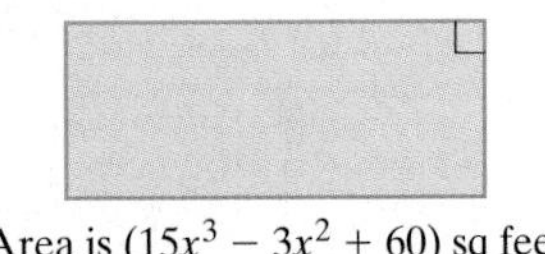

Area is $(15x^3 - 3x^2 + 60)$ sq feet

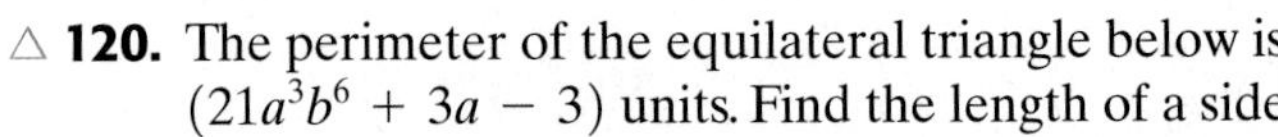

△ **120.** The perimeter of the equilateral triangle below is $(21a^3b^6 + 3a - 3)$ units. Find the length of a side.

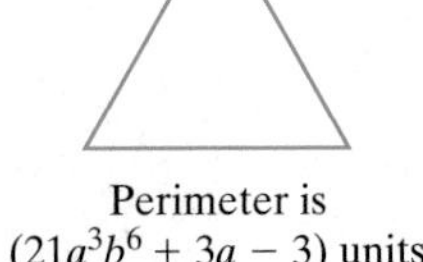

Perimeter is $(21a^3b^6 + 3a - 3)$ units

Name ______________________ Section __________ Date __________

Chapter 4 Test

Evaluate each expression.

1. 2^5 **2.** $(-3)^4$ **3.** -3^4 **4.** 4^{-3}

Simplify each exponential expression.

5. $(3x^2)(-5x^9)$ **6.** $\frac{y^7}{y^2}$ **7.** $\frac{r^{-8}}{r^{-3}}$

Simplify each expression. Write the result using only positive exponents.

8. $\left(\frac{x^2y^3}{x^3y^{-4}}\right)^2$ **9.** $\frac{6^2x^{-4}y^{-1}}{6^3x^{-3}y^7}$

Express each number in scientific notation.

10. 563,000 **11.** 0.0000863

Write each number in standard form.

12. 1.5×10^{-3} **13.** 6.23×10^4

14. Simplify. Write the answer in standard form.
$(1.2 \times 10^5)(3 \times 10^{-7})$

15. Find the degree of the following polynomial.
$4xy^2 + 7xyz + 9x^3yz$

16. Simplify by combining like terms.
$5x^2 + 4x - 7x^2 + 11 + 8x$

Perform each indicated operation.

17. $(8x^3 + 7x^2 + 4x - 7) + (8x^3 - 7x - 6)$

18.
$$\begin{array}{r} 5x^3 + x^2 + 5x - 2 \\ -\ (8x^3 - 4x^2 + x - 7) \\ \hline \end{array}$$

19. Subtract $(4x + 2)$ from the sum of $(8x^2 + 7x + 5)$ and $(x^3 - 8)$.

20. Multiply: $(3x + 7)(x^2 + 5x + 2)$

Answers

1. ______
2. ______
3. ______
4. ______
5. ______
6. ______
7. ______
8. ______
9. ______
10. ______
11. ______
12. ______
13. ______
14. ______
15. ______
16. ______
17. ______
18. ______
19. ______
20. ______

21. ______

22. ______

23. ______

24. ______

25. ______

26. ______

27. see table

28. ______

29. ______

30. ______

31. ______

21. Multiply $x^3 - x^2 + x + 1$ by $2x^2 - 3x + 7$ using a vertical format.

22. Use the FOIL method to multiply $(x + 7)(3x - 5)$.

Use special products to multiply each of the following.

23. $(3x - 7)(3x + 7)$

24. $(4x - 2)^2$

25. $(8x + 3)^2$

26. $(x^2 - 9b)(x^2 + 9b)$

27. When it is completed, the Suyong Bay Tower in Pusan, Korea, will be the world's tallest building at 1516 feet tall. Neglecting air resistance, the height of an object dropped from this building at time t seconds is given by the polynomial $-16t^2 + 1516$. Find the height of the object at the given times below. (*Source:* Council on Tall Buildings and Urban Habitat)

t	0 Seconds	3 Seconds	6 Seconds	9 Seconds
$-16t^2 + 1516$				

△ **28.** Find the area of the top of the table. Express the area as a product, then multiply and simplify.

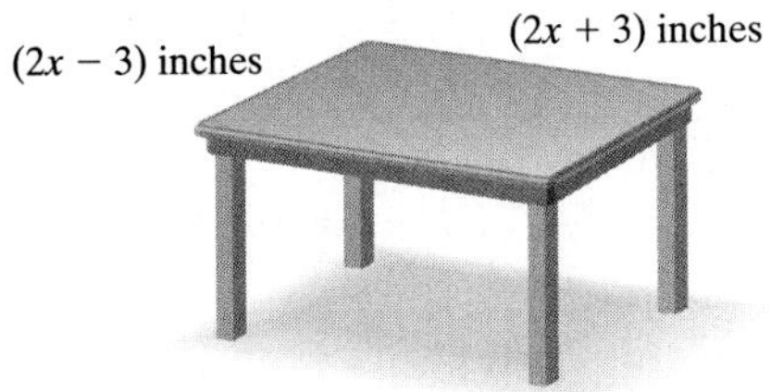

Divide.

29. $\dfrac{4x^2 + 2xy - 7x}{8xy}$

30. $(x^2 + 7x + 10) \div (x + 5)$

31. $\dfrac{27x^3 - 8}{3x + 2}$

Name ______________________ Section __________ Date __________

Answers
1. a.
b.
c.
2.
3.
4.
5.
6.
7.
8.
9.
10.
11.
12. see graph

Cumulative Review

1. Translate each sentence into a mathematical statement.

a. Nine is less than or equal to eleven.

b. Eight is greater than one.

c. Three is not equal to four.

2. Decide whether 2 is a solution of $3x + 10 = 8x$.

3. Subtract 8 from -4.

4. If $x = -2$ and $y = -4$, evaluate $\frac{3x^2}{4y}$.

Simplify each expression by combining like terms.

5. $2x + 3x + 5 + 2$

6. $-5a - 3 + a + 2$

7. $2.3x + 5x - 6$

8. Solve: $-3x = 33$

9. Solve: $3(x - 4) = 3x - 12$

10. Solve for l: $V = lwh$

11. Solve for x: $\frac{x-5}{3} = \frac{x+2}{5}$

12. Graph the linear equation $-5x + 3y = 15$.

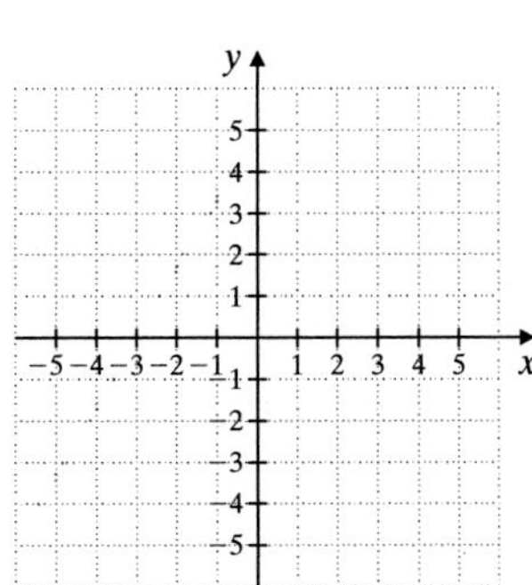

13. see graph

14. a.

b.

c.

15.

16.

17.

18.

19.

20.

21.

22.

23.

24.

25.

13. Graph $x = -2y$ by finding and plotting its intercepts.

14. Determine whether each pair of lines is parallel, perpendicular, or neither.

a. $y = -\frac{1}{5}x + 1$
$2x + 10y = 3$

b. $x + y = 3$
$-x + 5 = 4$

c. $3x + y = 5$
$2x + 3y = 6$

Simplify each expression.

15. $(5^3)^6$

16. $(y^8)^2$

Simplify the following expression. Write each result using positive exponents only.

17. $\frac{(x^3)^4 x}{x^7}$

18. $(y^{-3}z^6)^{-6}$

19. $\frac{x^{-7}}{(x^4)^3}$

Simplify each polynomial by combining any like terms.

20. $-3x + 7x$

21. $11x^2 + 5 + 2x^2 - 7$

22. Multiply: $(2x - y)^2$

Use a special product to square each binomial.

23. $(t + 2)^2$

24. $(x^2 - 7y)^2$

25. Divide: $\frac{8x^2y^2 - 16xy + 2x}{4xy}$

Factoring Polynomials

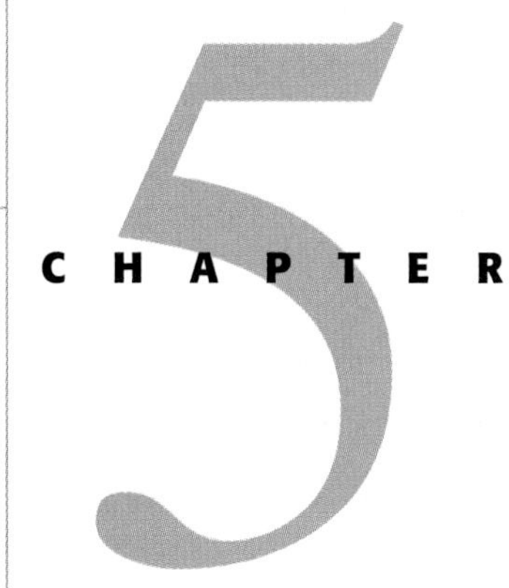

In Chapter 4, we learned how to multiply polynomials. Now we will deal with an operation that is the reverse process of multiplying—factoring. Factoring is an important algebraic skill because it allows us to write a sum as a product. As we will see in Sections 5.6 and 5.7, factoring can be used to solve equations other than linear equations. In Chapter 6, we will also use factoring to simplify and perform arithmetic operations on rational expressions.

When recently completed, the Petronas Twin Towers in Kuala Lumpur, Malaysia, became the world's tallest building. At a height of 1483 feet, these Islamic-influenced towers beat out the previous record holder, the Sears Tower in Chicago, by 33 feet. Each of the twin towers has 88 stories, and together they have over 32,000 windows. This colossal building complex was designed by American architect Cesar Pelli and cost over $1.2 billion to build. In Exercise 89 on page 377, a polynomial expression for the height of an object dropped from the Petronas Twin Towers is factored.

Answers

1. ______
2. ______
3. ______
4. ______
5. ______
6. ______
7. ______
8. ______
9. ______
10. ______
11. ______
12. ______
13. ______
14. ______
15. ______
16. ______
17. ______
18. ______
19. ______
20. ______

Name ______ Section ______ Date ______

Chapter 5 Pretest

Factor each polynomial completely. If a polynomial cannot be factored, write "prime."

1. $2x^3y - 6x^2y^2$

2. $xy + 6x - 4y - 24$

3. $a^2 + 8a + 12$

4. $m^2 + 4m - 3$

5. $3x^3 - 18x^2 + 15x$

6. $2x^2 + 5x - 12$

7. $14x^2 + 63x + 70$

8. $24b^2 - 25b + 6$

9. $15y^2 + 38y + 7$

10. $x^2 + 24x + 144$

11. $4x^2 - 12xy + 9y^2$

12. $a^2 - 49b^2$

13. $1 - 64t^2$

14. $x^3 + 125$

15. $160a^3 - 20b^3$

Solve each equation.

16. $(x - 12)(x + 5) = 0$

17. $y^2 - 13y = 0$

18. $2m^3 - 2m^2 - 24m = 0$

△ **19.** The length of a rectangle is 7 inches more than its width. Its area is 120 square inches. Find the dimensions of the rectangle.

20. The sum of a number and its square is 240. Find the number.

5.1 The Greatest Common Factor

In the product $2 \cdot 3 = 6$, the numbers 2 and 3 are called **factors** of 6 and $2 \cdot 3$ is a **factored form** of 6. This is true of polynomials also. Since $(x + 2)(x + 3) = x^2 + 5x + 6$, then $(x + 2)$ and $(x + 3)$ are factors of $x^2 + 5x + 6$, and $(x + 2)(x + 3)$ is a factored form of the polynomial.

The process of writing a polynomial as a product is called **factoring** the polynomial.

Study the examples below and look for a pattern.

Try the Concept Check in the margin.

Multiplying: $5(x^2 + 3) = 5x^2 + 15$ $\quad 2x(x - 7) = 2x^2 - 14x$

Factoring: $5x^2 + 15 = 5(x^2 + 3)$ $\quad 2x^2 - 14x = 2x(x - 7)$

Do you see that factoring is the reverse process of multiplying?

$$x^2 + 5x + 6 \underset{\text{multiplying}}{\overset{\text{factoring}}{=}} (x + 2)(x + 3)$$

A Finding the Greatest Common Factor

The first step in factoring a polynomial is to see whether the terms of the polynomial have a common factor. If there is one, we can write the polynomial as a product by **factoring out** the common factor. We will usually factor out the *greatest* common factor (GCF).

The **greatest common factor (GCF) of a list of terms** is the product of the GCF of the numerical coefficients and the GCF of the variable factors.

$$20x^2y^2 = 2 \cdot 2 \cdot 5 \cdot x \cdot x \cdot y \cdot y$$
$$6xy^3 = 2 \cdot 3 \cdot x \cdot y \cdot y \cdot y$$
$$\text{GCF} = 2 \cdot x \cdot y \cdot y = 2xy^2$$

Helpful Hint

Notice below that the GCF of a list of terms contains the smallest exponent on each common variable.

The GCF of x^5y^6, x^2y^7 and x^3y^4 is x^2y^4. (Smallest exponent on x. Smallest exponent on y.)

EXAMPLE 1 Find the greatest common factor of each list of terms.

a. $6x^2$, $10x^3$, and $-8x$
b. $-18y^2$, $-63y^3$, and $27y^4$
c. a^3b^2, a^5b, and a^6b^2

Solution:

a.
$$6x^2 = 2 \cdot 3 \cdot x^2$$
$$10x^3 = 2 \cdot 5 \cdot x^3$$
$$-8x = -1 \cdot 2 \cdot 2 \cdot 2 \cdot x^1$$
$$\text{GCF} = 2 \cdot x^1 \text{ or } 2x$$

The GCF of x^2, x^3, and x is x.

OBJECTIVES

- **A** Find the greatest common factor of a list of terms.
- **B** Factor out the greatest common factor from the terms of a polynomial.
- **C** Factor by grouping.

Concept Check

Multiply: $2(x - 4)$
What do you think the result of factoring $2x - 8$ would be? Why?

Practice Problem 1

Find the greatest common factor of each list of terms.

a. $6x^2$, $9x^4$, and $-12x^5$
b. $-16y$, $-20y^6$, and $40y^4$
c. a^5b^4, ab^3, and a^3b^2

Answers

1. a. $3x^2$, **b.** $4y$, **c.** ab^2

Concept Check: $2x - 8$; The result would be $2(x - 4)$ because factoring is the reverse process of multiplying.

b. $-18y^2 = -1 \cdot 2 \cdot 3 \cdot 3 \cdot y^2$
$-63y^3 = -1 \cdot 3 \cdot 3 \cdot 7 \cdot y^3$
$27y^4 = 3 \cdot 3 \cdot 3 \cdot y^4$
$\text{GCF} = 3 \cdot 3 \cdot y^2$ or $9y^2$

→ The GCF of y^2, y^3, and y^4 is y^2.

c. The GCF of a^3, a^5, and a^6 is a^3.
The GCF of b^2, b, and b^2 is b. Thus,
the GCF of a^3b^2, a^5b, and a^6b^2 is a^3b.

B Factoring Out the Greatest Common Factor

To factor a polynomial such as $8x + 14$, we first see whether the terms have a greatest common factor other than 1. In this case, they do: The GCF of $8x$ and 14 is 2.

We factor out 2 from each term by writing each term as a product of 2 and the term's remaining factors.

$$8x + 14 = 2 \cdot 4x + 2 \cdot 7$$

Using the distributive property, we can write

$$\begin{aligned} 8x + 14 &= 2 \cdot 4x + 2 \cdot 7 \\ &= 2(4x + 7) \end{aligned}$$

Thus, a factored form of $8x + 14$ is $2(4x + 7)$. We can check by multiplying:

$$2(4x + 7) = 2 \cdot 4x + 2 \cdot 7 = 8x + 14.$$

Try the Concept Check in the margin.

Concept Check

Which of the following is/are factored form(s) of $6t + 18$?

a. 6
b. $6 \cdot t + 6 \cdot 3$
c. $6(t + 3)$
d. $3(t + 6)$

Helpful Hint

A factored form of $8x + 14$ is *not*

$$2 \cdot 4x + 2 \cdot 7$$

Although the *terms* have been factored (written as a product), the *polynomial* $8x + 14$ has not been factored. A factored form of $8x + 14$ is the *product* $2(4x + 7)$.

Practice Problem 2

Factor each polynomial by factoring out the greatest common factor (GCF).

a. $10y + 25$
b. $x^4 - x^9$

EXAMPLE 2

Factor each polynomial by factoring out the greatest common factor (GCF).

a. $6t + 18$ **b.** $y^5 - y^7$

Solution:

a. The GCF of terms $6t$ and 18 is 6. Thus,

$$\begin{aligned} 6t + 18 &= 6 \cdot t + 6 \cdot 3 \\ &= 6(t + 3) \end{aligned}$$ Apply the distributive property.

We can check our work by multiplying 6 and $(t + 3)$.
$6(t + 3) = 6 \cdot t + 6 \cdot 3 = 6t + 18$, the original polynomial.

b. The GCF of y^5 and y^7 is y^5. Thus,

$$\begin{aligned} y^5 - y^7 &= (y^5)1 - (y^5)y^2 \\ &= y^5(1 - y^2) \end{aligned}$$

Helpful Hint
Don't forget the 1.

Answers

2. a. $5(2y + 5)$, **b.** $x^4(1 - x^5)$

Concept Check: c

EXAMPLE 3 Factor: $-9a^5 + 18a^2 - 3a$

Solution:

$$-9a^5 + 18a^2 - 3a = (3a)(-3a^4) + (3a)(6a) + (3a)(-1)$$
$$= 3a(-3a^4 + 6a - 1)$$

Helpful Hint

Don't forget the -1.

In Example 3, we could have chosen to factor out $-3a$ instead of $3a$. If we factor out $-3a$, we have

$$-9a^5 + 18a^2 - 3a = (-3a)(3a^4) + (-3a)(-6a) + (-3a)(1)$$
$$= -3a(3a^4 - 6a + 1)$$

Helpful Hint

Notice the changes in signs when factoring out $-3a$.

EXAMPLES Factor.

4. $6a^4 - 12a = 6a(a^3 - 2)$

5. $\frac{3}{7}x^4 + \frac{1}{7}x^3 - \frac{5}{7}x^2 = \frac{1}{7}x^2(3x^2 + x - 5)$

6. $15p^2q^4 + 20p^3q^5 + 5p^3q^3 = 5p^2q^3(3q + 4pq^2 + p)$

EXAMPLE 7 Factor: $5(x + 3) + y(x + 3)$

Solution: The binomial $(x + 3)$ is present in both terms and is the greatest common factor. We use the distributive property to factor out $(x + 3)$.

$$5(x + 3) + y(x + 3) = (x + 3)(5 + y)$$

C Factoring by Grouping

Once the GCF is factored out, we can often continue to factor the polynomial, using a variety of techniques. We discuss here a technique called **factoring by grouping.** This technique can be used to factor some polynomials with four terms.

EXAMPLE 8 Factor $xy + 2x + 3y + 6$ by grouping.

Solution: The GCF of the first two terms is x, and the GCF of the last two terms is 3.

$$xy + 2x + 3y + 6 = (xy + 2x) + (3y + 6)$$
$$= x(y + 2) + 3(y + 2)$$

Helpful Hint

Notice that this *not* a factored form of the original polynomial. It is a sum, not a product.

Next we factor out the common binomial factor, $(y + 2)$.

$$x(y + 2) + 3(y + 2) = (y + 2)(x + 3)$$

Practice Problem 3

Factor: $-10x^3 + 8x^2 - 2x$

Practice Problems 4–6

Factor.

4. $4x^3 + 12x$
5. $\frac{2}{5}a^5 - \frac{4}{5}a^3 + \frac{1}{5}a^2$
6. $6a^3b + 3a^3b^2 + 9a^2b^4$

Practice Problem 7

Factor: $7(p + 2) + q(p + 2)$

Practice Problem 8

Factor $ab + 7a + 2b + 14$ by grouping.

Answers

3. $-2x(5x^2 - 4x + 1)$, **4.** $4x(x^2 + 3)$, **5.** $\frac{1}{5}a^2(2a^3 - 4a + 1)$, **6.** $3a^2b(2a + ab + 3b^3)$, **7.** $(p + 2)(7 + q)$, **8.** $(b + 7)(a + 2)$

Check: Multiply $(y + 2)$ by $(x + 3)$.

$$(y + 2)(x + 3) = xy + 2x + 3y + 6,$$

the original polynomial.

Thus, the factored form of $xy + 2x + 3y + 6$ is the product $(y + 2)(x + 3)$. ●

Practice Problems 9–10

Factor by grouping.

9. $28x^3 - 7x^2 + 12x - 3$
10. $2xy + 5y^2 - 4x - 10y$

EXAMPLES Factor by grouping.

9. $15x^3 - 10x^2 + 6x - 4$

$= (15x^3 - 10x^2) + (6x - 4)$

$= 5x^2(3x - 2) + 2(3x - 2)$ Factor each group.

$= (3x - 2)(5x^2 + 2)$ Factor out the common factor, $(3x - 2)$.

10. $3x^2 + 4xy - 3x - 4y$

$= (3x^2 + 4xy) + (-3x - 4y)$ Factor each group. A -1 is factored from the second pair of terms so that there is a common factor, $(3x + 4y)$.

$= x(3x + 4y) - 1(3x + 4y)$

$= (3x + 4y)(x - 1)$ Factor out the common factor, $(3x + 4y)$. ●

Practice Problems 11–13

Factor by grouping.

11. $4x^3 + x - 20x^2 - 5$
12. $2x - 2 + x^3 - 3x^2$
13. $3xy - 4 + x - 12y$

EXAMPLES Factor by grouping.

11. $3x^3 - 2x - 9x^2 + 6$ Factor each group. A -3 is factored from the second pair of terms so that there is a common factor, $(3x^2 - 2)$.

$= x(3x^2 - 2) - 3(3x^2 - 2)$

$= (3x^2 - 2)(x - 3)$ Factor out the common factor, $(3x^2 - 2)$.

12. $5x - 10 + x^3 - x^2 = 5(x - 2) + x^2(x - 1)$

There is no common binomial factor that can now be factored out. No matter how we rearrange the terms, no grouping will lead to a common factor. Thus, this polynomial is not factorable by grouping.

13. $3xy + 2 - 3x - 2y$

Notice that the first two terms have no common factor other than 1. However, if we rearrange these terms, a grouping emerges that does lead to a common factor.

$3xy + 2 - 3x - 2y$

$= (3xy - 3x) + (-2y + 2)$

$= 3x(y - 1) - 2(y - 1)$ Factor -2 from the second group.

$= (y - 1)(3x - 2)$ Factor out the common factor, $(y - 1)$. ●

Helpful Hint

Throughout this chapter, we will be factoring polynomials. Even when the instructions do not so state, it is always a good idea to check your answers by multiplying.

Answers

9. $(4x - 1)(7x^2 + 3)$, **10.** $(2x + 5y)(y - 2)$, **11.** $(4x^2 + 1)(x - 5)$, **12.** can't be factored, **13.** $(3y + 1)(x - 4)$

Name ______________________ Section __________ Date __________

Mental Math

Find the GCF of each pair of integers.

1. 2, 16 **2.** 3, 18 **3.** 6, 15 **4.** 20, 15 **5.** 14, 35 **6.** 27, 36

EXERCISE SET 5.1

 *Find the GCF for each list. See Example 1.*

1. y^2, y^4, y^7

2. x^3, x^2, x^3

3. $x^{10}y^2, xy^2, x^3y^3$

4. p^7q, p^8q^2, p^9q^3

5. $8x, 4$

6. $9y, y$

7. $12y^4, 20y^3$

8. $32x, 18x^2$

9. $-10x^2, 15x^3$

10. $-21x^3, 14x$

11. $12x^3, -6x^4, 3x^5$

12. $15y^2, 5y^7, -20y^3$

13. $-18x^2y, 9x^3y^3, 36x^3y$

14. $7x, -21x^2y^2, 14xy$

B *Factor out the GCF from each polynomial. See Examples 2 through 7.*

15. $3a + 6$

16. $18a + 12$

17. $30x - 15$

18. $42x - 7$

19. $x^3 + 5x^2$

20. $y^5 - 6y^4$

21. $6y^4 - 2y$

22. $5x^2 + 10x^6$

23. $32xy - 18x^2$

24. $10xy - 15x^2$

25. $4x - 8y + 4$

26. $7x + 21y - 7$

27. $6x^3 - 9x^2 + 12x$

28. $12x^3 + 16x^2 - 8x$

29. $a^7b^6 - a^3b^2 + a^2b^5 - a^2b^2$

30. $x^9y^6 + x^3y^5 - x^4y^3 + x^3y^3$

31. $5x^3y - 15x^2y + 10xy$

32. $14x^3y + 7x^2y - 7xy$

33. $8x^5 + 16x^4 - 20x^3 + 12$

34. $9y^6 - 27y^4 + 18y^2 + 6$

35. $\frac{1}{3}x^4 + \frac{2}{3}x^3 - \frac{4}{3}x^5 + \frac{1}{3}x$

36. $\frac{2}{5}y^7 - \frac{4}{5}y^5 + \frac{3}{5}y^2 - \frac{2}{5}y$

37. $y(x + 2) + 3(x + 2)$

38. $z(y + 4) + 3(y + 4)$

39. $8(x + 2) - y(x + 2)$

40. $x(y^2 + 1) - 3(y^2 + 1)$

41. Construct a binomial whose greatest common factor is $5a^3$. (*Hint:* Multiply $5a^3$ by a binomial whose terms contain no common factor other than 1. $5a^3(\square + \square)$.)

42. Construct a trinomial whose greatest common factor is $2x^2$. See the hint for Exercise 41.

C *Factor each four-term polynomial by grouping. See Examples 8 through 13.*

43. $x^3 + 2x^2 + 5x + 10$

44. $x^3 + 4x^2 + 3x + 12$

45. $5x + 15 + xy + 3y$

46. $xy + y + 2x + 2$

47. $6x^3 - 4x^2 + 15x - 10$

48. $16x^3 - 28x^2 + 12x - 21$

49. $2y - 8 + xy - 4x$

50. $6x - 42 + xy - 7y$

51. $2x^3 + x^2 + 8x + 4$

52. $2x^3 - x^2 - 10x + 5$

53. $4x^2 - 8xy - 3x + 6y$

54. $5xy - 15x - 6y + 18$

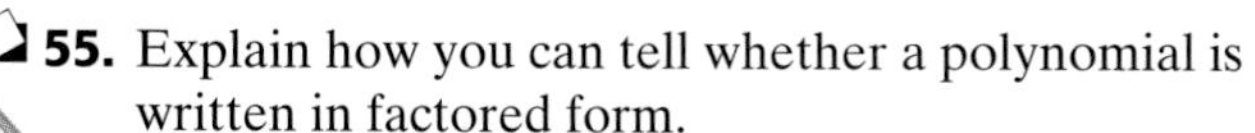

55. Explain how you can tell whether a polynomial is written in factored form.

56. Construct a four-term polynomial that can be factored by grouping.

Review and Preview

Multiply. See Section 4.5.

57. $(x + 2)(x + 5)$

58. $(y + 3)(y + 6)$

59. $(b + 1)(b - 4)$

60. $(x - 5)(x + 10)$

Fill in the chart by finding two numbers that have the given product and sum. The first column is filled in for you.

		61.	**62.**	**63.**	**64.**	**65.**	**66.**	**67.**	**68.**
Two Numbers	4, 7								
Their Product	28	12	20	8	16	−10	−9	−24	−36
Their Sum	11	8	9	−9	−10	3	0	−5	−5

Combining Concepts

Factor out the GCF from each polynomial. Then factor by grouping.

69. $12x^2y - 42x^2 - 4y + 14$

70. $90 + 15y^2 - 18x - 3xy^2$

Write an expression for the area of each shaded region. Then write the expression as a factored polynomial.

△ **71.**

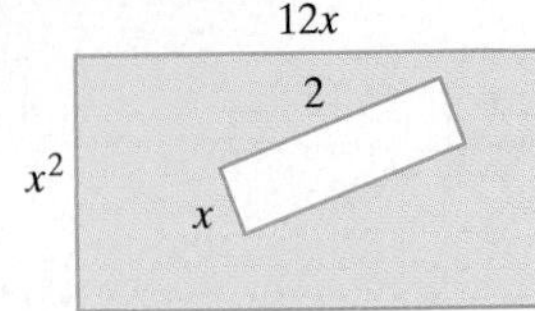

△ **72.**

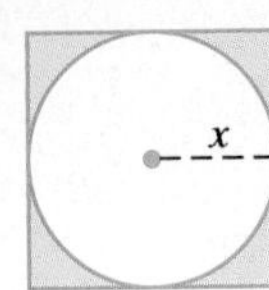

Write an expression for the length of each rectangle. (Hint: Factor the area binomial and recall that Area = width · length.)

△ **73.**

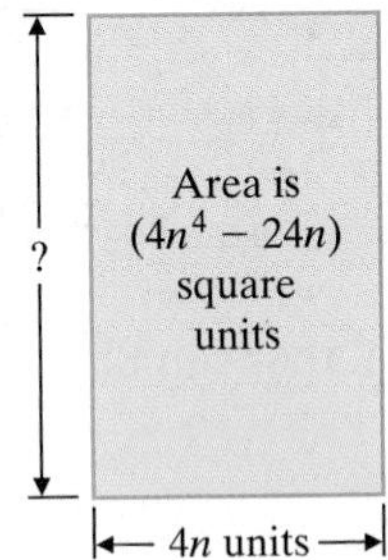

△ **74.**

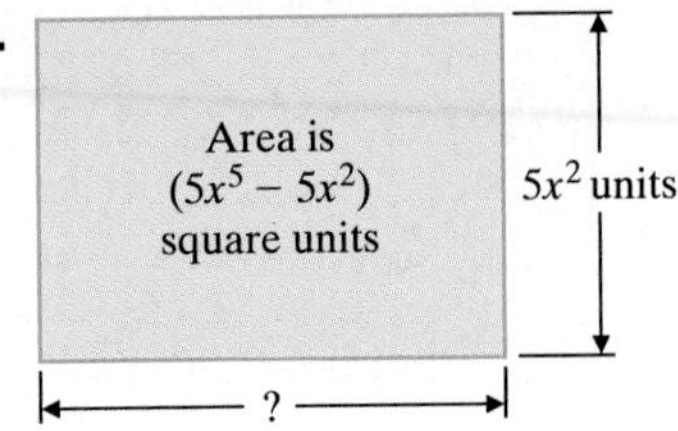

75. The nonresidential sales of electricity (in billion kilowatt hours) in the United States during 1998–2000 can be approximated by the polynomial $-8x^2 + 60x + 2000$, where x is the number of years after 1998. (*Source:* Energy Information Administration)

a. Find the amount of electricity sold in 1998. To do so, let $x = 0$ and evaluate $-8x^2 + 60x + 2000$.

b. Find the amount of electricity sold in 2000.

c. Factor the polynomial $-8x^2 + 60x + 2000$.

76. The average daily total supply of motor gasoline (in thousand barrels per day) in the United States for the period 1998–2000 can be approximated by the polynomial $-40x^2 + 140x + 8000$, where x is the number of years after 1998. (*Source:* Energy Information Administration)

a. Find the average daily total supply of motor gasoline in 1999. To do so, let $x = 1$ and evaluate $-40x^2 + 140x + 8000$.

b. Find the average daily total supply of motor gasoline in 2000.

c. Factor the polynomial $-40x^2 + 140x + 8000$.

5.2 Factoring Trinomials of the Form $x^2 + bx + c$

OBJECTIVES

Ⓐ Factor trinomials of the form $x^2 + bx + c$.

Ⓑ Factor out the greatest common factor and then factor a trinomial of the form $x^2 + bx + c$.

Ⓐ Factoring Trinomials of the Form $x^2 + bx + c$

In this section, we factor trinomials of the form $x^2 + bx + c$, such as

$$x^2 + 4x + 3, \quad x^2 - 8x + 15, \quad x^2 + 4x - 12, \quad \text{and} \quad r^2 - r - 42$$

Notice that for these trinomials, the coefficient of the squared variable is 1.

Recall that factoring means to write as a product and that factoring and multiplying are reverse processes. Using the FOIL method of multiplying binomials, we have the following.

$$\begin{aligned}(x + 3)(x + 1) &= \overset{F}{x^2} + \overset{O}{1x} + \overset{I}{3x} + \overset{L}{3} \\ &= x^2 + 4x + 3\end{aligned}$$

Thus, a factored form of $x^2 + 4x + 3$ is $(x + 3)(x + 1)$.

Notice that the product of the first terms of the binomials is $x \cdot x = x^2$, the first term of the trinomial. Also, the product of the last two terms of the binomials is $3 \cdot 1 = 3$, the third term of the trinomial. The sum of these same terms is $3 + 1 = 4$, the coefficient of the middle, x, term of the trinomial.

The product of these numbers is 3.

$$x^2 + 4x + 3 = (x + 3)(x + 1)$$

The sum of these numbers is 4.

Many trinomials, such as the one above, factor into two binomials. To factor $x^2 + 7x + 10$, let's assume that it factors into two binomials and begin by writing two pairs of parentheses. The first term of the trinomial is x^2, so we use x and x as the first terms of the binomial factors.

$$x^2 + 7x + 10 = (x + \square)(x + \square)$$

To determine the last term of each binomial factor, we look for two integers whose product is 10 and whose sum is 7. The integers are 2 and 5. Thus,

$$x^2 + 7x + 10 = (x + 2)(x + 5)$$

To see if we have factored correctly, we multiply.

$$\begin{aligned}(x + 2)(x + 5) &= x^2 + 5x + 2x + 10 \\ &= x^2 + 7x + 10 \qquad \text{Combine like terms.}\end{aligned}$$

Helpful Hint

Since multiplication is commutative, the factored form of $x^2 + 7x + 10$ can be written as either $(x + 2)(x + 5)$ or $(x + 5)(x + 2)$.

To Factor a Trinomial of the Form $x^2 + bx + c$

The product of these numbers is c.

$$x^2 + bx + c = (x + \square)(x + \square)$$

The sum of these numbers is b.

EXAMPLE 1 Factor: $x^2 + 7x + 12$

Solution: We begin by writing the first terms of the binomial factors.

$$(x + \square)(x + \square)$$

Next we look for two numbers whose product is 12 and whose sum is 7. Since our numbers must have a positive product and a positive sum, we look at pairs of positive factors of 12 only.

Factors of 12	Sum of Factors
1, 12	13
2, 6	8
3, 4	7

Correct sum, so the numbers are 3 and 4.

$$x^2 + 7x + 12 = (x + 3)(x + 4)$$

Check: Multiply $(x + 3)$ by $(x + 4)$.

EXAMPLE 2 Factor: $x^2 - 8x + 15$

Solution: Again, we begin by writing the first terms of the binomials.

$$(x + \square)(x + \square)$$

Now we look for two numbers whose product is 15 and whose sum is -8. Since our numbers must have a positive product and a negative sum, we look at pairs of negative factors of 15 only.

Factors of 15	Sum of Factors
−1, −15	−16
−3, −5	−8

Correct sum, so the numbers are -3 and -5.

$$x^2 - 8x + 15 = (x - 3)(x - 5)$$

EXAMPLE 3 Factor: $x^2 + 4x - 12$

Solution: $x^2 + 4x - 12 = (x + \square)(x + \square)$

Practice Problem 1

Factor: $x^2 + 9x + 20$

Practice Problem 2

Factor each trinomial.

a. $x^2 - 13x + 22$
b. $x^2 - 27x + 50$

Practice Problem 3

Factor: $x^2 + 5x - 36$

Answers

1. $(x + 4)(x + 5)$, **2. a.** $(x - 2)(x - 11)$, **b.** $(x - 2)(x - 25)$, **3.** $(x + 9)(x - 4)$

We look for two numbers whose product is -12 and whose sum is 4. Since our numbers must have a negative product, we look at pairs of factors with opposite signs.

Factors of -12	Sum of Factors
$-1, 12$	11
$1, -12$	-11
$-2, 6$	4
$2, -6$	-4
$-3, 4$	1
$3, -4$	-1

Correct sum, so the numbers are -2 and 6.

$$x^2 + 4x - 12 = (x - 2)(x + 6)$$

EXAMPLE 4 Factor: $r^2 - r - 42$

Solution: Because the variable in this trinomial is r, the first term of each binomial factor is r.

$$r^2 - r - 42 = (r + \square)(r + \square)$$

Now we look for two numbers whose product is -42 and whose sum is -1, the numerical coefficient of r. The numbers are 6 and -7. Therefore,

$$r^2 - r - 42 = (r + 6)(r - 7)$$

Practice Problem 4

Factor each trinomial.

a. $q^2 - 3q - 40$
b. $y^2 + 2y - 48$

EXAMPLE 5 Factor: $a^2 + 2a + 10$

Solution: Look for two numbers whose product is 10 and whose sum is 2. Neither 1 and 10 nor 2 and 5 give the required sum, 2. We conclude that $a^2 + 2a + 10$ is not factorable with integers. A polynomial such as $a^2 + 2a + 10$ is called a **prime polynomial.**

Practice Problem 5

Factor: $x^2 + 6x + 15$

EXAMPLE 6 Factor: $x^2 + 5xy + 6y^2$

Solution: $x^2 + 5xy + 6y^2 = (x + \square)(x + \square)$

Recall that the middle term $5xy$ is the same as $5yx$. Notice that $5y$ is the "coefficient" of x. We then look for two terms whose product is $6y^2$ and whose sum is $5y$. The terms are $2y$ and $3y$ because $2y \cdot 3y = 6y^2$ and $2y + 3y = 5y$. Therefore,

$$x^2 + 5xy + 6y^2 = (x + 2y)(x + 3y)$$

Practice Problem 6

Factor each trinomial.

a. $x^2 + 6xy + 8y^2$
b. $a^2 - 13ab + 30b^2$

EXAMPLE 7 Factor: $x^4 + 5x^2 + 6$

Solution: As usual, we begin by writing the first terms of the binomials. Since the greatest power of x in this polynomial is x^4, we write

$$(x^2 + \square)(x^2 + \square) \quad \text{since } x^2 \cdot x^2 = x^4$$

Now we look for two factors of 6 whose sum is 5. The numbers are 2 and 3. Thus,

$$x^4 + 5x^2 + 6 = (x^2 + 2)(x^2 + 3)$$

Practice Problem 7

Factor: $x^4 + 8x^2 + 12$

Answers

4. a. $(q - 8)(q + 5)$, **b.** $(y + 8)(y - 6)$, **5.** prime polynomial, **6. a.** $(x + 2y)(x + 4y)$, **b.** $(a - 3b)(a - 10b)$, **7.** $(x^2 + 6)(x^2 + 2)$

The following sign patterns may be useful when factoring trinomials.

Helpful Hint

A positive constant in a trinomial tells us to look for two numbers with the same sign. The sign of the coefficient of the middle term tells us whether the signs are both positive or both negative.

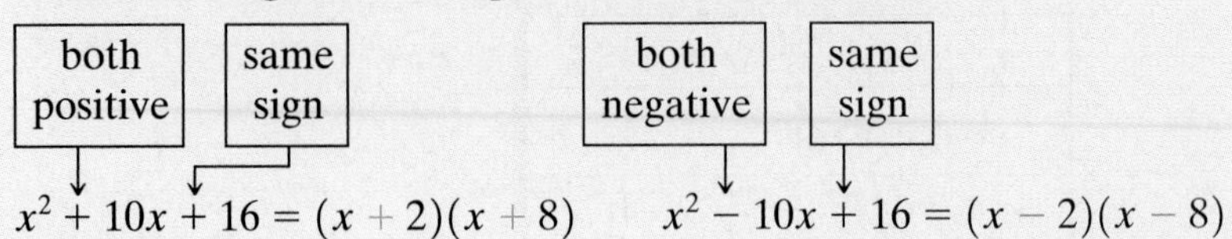

$x^2 + 10x + 16 = (x + 2)(x + 8)$ $\quad$ $x^2 - 10x + 16 = (x - 2)(x - 8)$

A negative constant in a trinomial tells us to look for two numbers with opposite signs.

opposite signs $\quad$ opposite signs

$x^2 + 6x - 16 = (x + 8)(x - 2)$ $\quad$ $x^2 - 6x - 16 = (x - 8)(x + 2)$

B Factoring Out the Greatest Common Factor

Remember that the first step in factoring any polynomial is to factor out the greatest common factor (if there is one other than 1 or -1).

Practice Problem 8

Factor each trinomial.

a. $x^3 + 3x^2 - 4x$

b. $4x^2 - 24x + 36$

EXAMPLE 8 Factor: $3m^2 - 24m - 60$

Solution: First we factor out the greatest common factor, 3, from each term.

$$3m^2 - 24m - 60 = 3(m^2 - 8m - 20)$$

Now we factor $m^2 - 8m - 20$ by looking for two factors of -20 whose sum is -8. The factors are -10 and 2. Therefore, the complete factored form is

$$3m^2 - 24m - 60 = 3(m + 2)(m - 10)$$

Helpful Hint

Remember to write the common factor 3 as part of the factored form.

Practice Problem 9

Factor: $5x^5 - 25x^4 - 30x^3$

EXAMPLE 9 Factor: $2x^4 - 26x^3 + 84x^2$

Solution:

$$2x^4 - 26x^3 + 84x^2 = 2x^2(x^2 - 13x + 42) \quad \text{Factor out common factor of } 2x^2.$$
$$= 2x^2(x - 6)(x - 7) \quad \text{Factor } x^2 - 13x + 42.$$

Answers

8. a. $x(x + 4)(x - 1)$, **b.** $4(x - 3)(x - 3)$, **9.** $5x^3(x + 1)(x - 6)$

Name ______________________ Section __________ Date __________

Mental Math

Complete each factored form.

1. $x^2 + 9x + 20 = (x + 4)(x \quad)$ **2.** $x^2 + 12x + 35 = (x + 5)(x \quad)$ **3.** $x^2 - 7x + 12 = (x - 4)(x \quad)$

4. $x^2 - 13x + 22 = (x - 2)(x \quad)$ **5.** $x^2 + 4x + 4 = (x + 2)(x \quad)$ **6.** $x^2 + 10x + 24 = (x + 6)(x \quad)$

EXERCISE SET 5.2

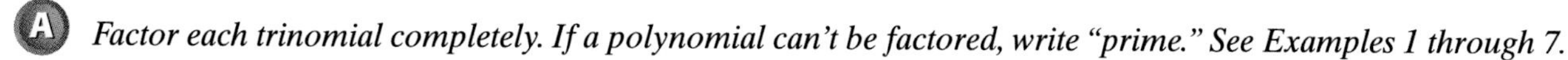

A *Factor each trinomial completely. If a polynomial can't be factored, write "prime." See Examples 1 through 7.*

1. $x^2 + 7x + 6$ **2.** $x^2 + 6x + 8$ **3.** $x^2 - 10x + 9$ **4.** $x^2 - 6x + 9$ **5.** $x^2 - 3x - 18$

6. $x^2 - x - 30$ **7.** $x^2 + 3x - 70$ **8.** $x^2 + 4x - 32$ **9.** $x^2 + 5x + 2$ **10.** $x^2 - 7x + 5$

11. $x^2 + 8xy + 15y^2$ **12.** $x^2 + 6xy + 8y^2$ **13.** $a^4 - 2a^2 - 15$ **14.** $y^4 - 3y^2 - 70$

15. Write a polynomial that factors as $(x - 3)(x + 8)$.

16. To factor $x^2 + 13x + 42$, think of two numbers whose ________ is 42 and whose ________ is 13.

Complete each sentence in your own words.

17. If $x^2 + bx + c$ is factorable and c is negative, then the signs of the last-term factors of the binomials are opposite because....

18. If $x^2 + bx + c$ is factorable and c is positive, then the signs of the last-term factors of the binomials are the same because....

B *Factor each trinomial completely. See Examples 1 through 9.*

19. $2z^2 + 20z + 32$ **20.** $3x^2 + 30x + 63$ **21.** $2x^3 - 18x^2 + 40x$ **22.** $x^3 - x^2 - 56x$

23. $x^2 - 3xy - 4y^2$ **24.** $x^2 - 4xy - 77y^2$ **25.** $x^2 + 15x + 36$ **26.** $x^2 + 19x + 60$

27. $x^2 - x - 2$ **28.** $x^2 - 5x - 14$ **29.** $r^2 - 16r + 48$ **30.** $r^2 - 10r + 21$

31. $x^2 + xy - 2y^2$ **32.** $x^2 - xy - 6y^2$ **33.** $3x^2 + 9x - 30$ **34.** $4x^2 - 4x - 48$

35. $3x^2 - 60x + 108$ **36.** $2x^2 - 24x + 70$ **37.** $x^2 - 18x - 144$ **38.** $x^2 + x - 42$

39. $r^2 - 3r + 6$

40. $x^2 + 4x - 10$

41. $x^2 - 8x + 15$

42. $x^2 - 9x + 14$

43. $6x^3 + 54x^2 + 120x$

44. $3x^3 + 3x^2 - 126x$

45. $4x^2y + 4xy - 12y$

46. $3x^2y - 9xy + 45y$

47. $x^2 - 4x - 21$

48. $x^2 - 4x - 32$

49. $x^2 + 7xy + 10y^2$

50. $x^2 - 3xy - 4y^2$

51. $64 + 24t + 2t^2$

52. $50 + 20t + 2t^2$

53. $x^3 - 2x^2 - 24x$

54. $x^3 - 3x^2 - 28x$

55. $2t^5 - 14t^4 + 24t^3$

56. $3x^6 + 30x^5 + 72x^4$

57. $5x^3y - 25x^2y^2 - 120xy^3$

58. $3x^2 - 6xy - 72y^2$

Review and Preview

Multiply. See Section 4.5.

59. $(2x + 1)(x + 5)$

60. $(3x + 2)(x + 4)$

61. $(5y - 4)(3y - 1)$

62. $(4z - 7)(7z - 1)$

63. $(a + 3)(9a - 4)$

64. $(y - 5)(6y + 5)$

Combining Concepts

Write the perimeter of each rectangle as a simplified polynomial. Then factor the polynomial.

△ **65.**

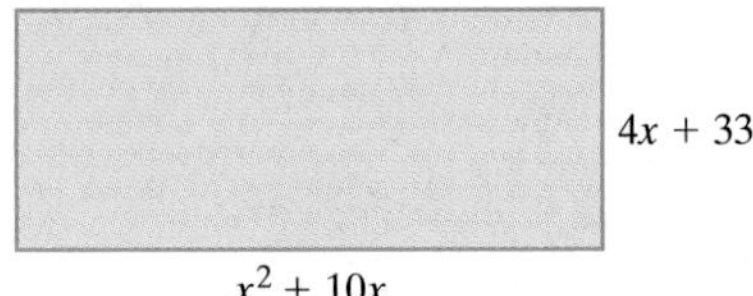

△ **66.**

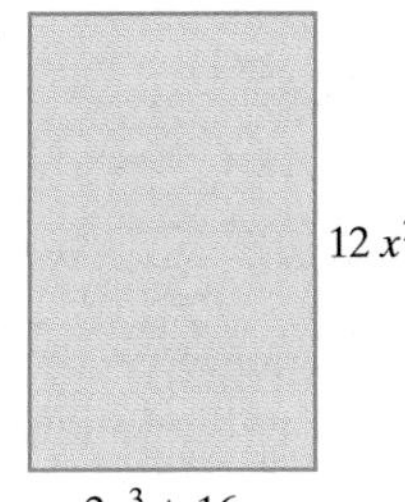

Factor each trinomial completely.

67. $y^2(x + 1) - 2y(x + 1) - 15(x + 1)$

68. $z^2(x + 1) - 3z(x + 1) - 70(x + 1)$

Find a positive value of c so that each trinomial is factorable.

69. $y^2 - 4y + c$

70. $n^2 - 16n + c$

Find a positive value of b so that each trinomial is factorable.

71. $x^2 + bx + 15$

72. $y^2 + by + 20$

Factor each trinomial. (Hint: Notice that $x^{2n} + 4x^n + 3$ factors as $(x^n + 1)(x^n + 3)$.)

73. $x^{2n} + 5x^n + 6$

74. $x^{2n} + 8x^n - 20$

5.3 Factoring Trinomials of the Form $ax^2 + bx + c$

OBJECTIVES

Ⓐ Factor trinomials of the form $ax^2 + bx + c$, where $a \neq 1$.

Ⓑ Factor out the GCF before factoring a trinomial of the form $ax^2 + bx + c$.

Ⓐ Factoring Trinomials of the Form $ax^2 + bx + c$

In this section, we factor trinomials of the form $ax^2 + bx + c$, such as

$$3x^2 + 11x + 6, \quad 8x^2 - 22x + 5, \quad \text{and} \quad 2x^2 + 13x - 7$$

Notice that the coefficient of the squared variable in these trinomials is a number other than 1. We will factor these trinomials using a trial-and-check method based on our work in the last section.

To begin, let's review the relationship between the numerical coefficients of the trinomial and the numerical coefficients of its factored form. For example, since $(2x + 1)(x + 6) = 2x^2 + 13x + 6$, the factored form of $2x^2 + 13x + 6$ is

$$2x^2 + 13x + 6 = (2x + 1)(x + 6)$$

Notice that $2x$ and x are factors of $2x^2$, the first term of the trinomial. Also, 6 and 1 are factors of 6, the last term of the trinomial, as shown:

$$2x^2 + 13x + 6 = (2x + 1)(x + 6)$$

$2x \cdot x$

$1 \cdot 6$

Also notice that $13x$, the middle term, is the sum of the following products:

$$2x^2 + 13x + 6 = (2x + 1)(x + 6)$$

$$\begin{array}{r} 1x \\ +\,12x \\ \hline 13x \end{array}$$ Middle term

Let's use this pattern to factor $5x^2 + 7x + 2$. First, we find factors of $5x^2$. Since all numerical coefficients in this trinomial are positive, we will use factors with positive numerical coefficients only. Thus, the factors of $5x^2$ are $5x$ and x. Let's try these factors as first terms of the binomials. Thus far, we have

$$5x^2 + 7x + 2 = (5x + \square)(x + \square)$$

Next, we need to find positive factors of 2. Positive factors of 2 are 1 and 2. Now we try possible combinations of these factors as second terms of the binomials until we obtain a middle term of $7x$.

$$(5x + 1)(x + 2) = 5x^2 + 11x + 2$$

$$\begin{array}{r} 1x \\ +\,10x \\ \hline 11x \end{array}$$ → **Incorrect** middle term

Let's try switching factors 2 and 1.

$$(5x + 2)(x + 1) = 5x^2 + 7x + 2$$

$$\begin{array}{r} 2x \\ +\,5x \\ \hline 7x \end{array}$$ → **Correct** middle term

Thus the factored form of $5x^2 + 7x + 2$ is $(5x + 2)(x + 1)$. To check, we multiply $(5x + 2)$ and $(x + 1)$. The product is $5x^2 + 7x + 2$.

Practice Problem 1

Factor each trinomial.

a. $4x^2 + 12x + 5$
b. $5x^2 + 27x + 10$

EXAMPLE 1 Factor: $3x^2 + 11x + 6$

Solution: Since all numerical coefficients are positive, we use factors with positive numerical coefficients. We first find factors of $3x^2$.

Factors of $3x^2$: $3x^2 = 3x \cdot x$

If factorable, the trinomial will be of the form

$$3x^2 + 11x + 6 = (3x + \square)(x + \square)$$

Next we factor 6.

Factors of 6: $6 = 1 \cdot 6, \quad 6 = 2 \cdot 3$

Now we try combinations of factors of 6 until a middle term of $11x$ is obtained. Let's try 1 and 6 first.

$$(3x + 1)(x + 6) = 3x^2 + 19x + 6$$

$1x$
$+ 18x$
$19x$ → **Incorrect** middle term

Now let's next try 6 and 1.

$$(3x + 6)(x + 1)$$

Before multiplying, notice that the terms of the factor $3x + 6$ have a common factor of 3. The terms of the original trinomial $3x^2 + 11x + 6$ have no common factor other than 1, so the terms of its factors will also contain no common factor other than 1. This means that $(3x + 6)(x + 1)$ is not a factored form.

Next let's try 2 and 3 as last terms.

$$(3x + 2)(x + 3) = 3x^2 + 11x + 6$$

$2x$
$+ 9x$
$11x$ → **Correct** middle term

Thus the factored form of $3x^2 + 11x + 6$ is $(3x + 2)(x + 3)$. ●

Helpful Hint

If the terms of a trinomial have no common factor (other than 1), then the terms of each of its binomial factors will contain no common factor (other than 1).

Try the Concept Check in the margin.

Concept Check

Do the terms of $3x^2 + 29x + 18$ have a common factor? Without multiplying, decide which of the following factored forms could not be a factored form of $3x^2 + 29x + 18$.

a. $(3x + 18)(x + 1)$
b. $(3x + 2)(x + 9)$
c. $(3x + 6)(x + 3)$
d. $(3x + 9)(x + 2)$

EXAMPLE 2 Factor: $8x^2 - 22x + 5$

Solution: Factors of $8x^2$: $8x^2 = 8x \cdot x, \quad 8x^2 = 4x \cdot 2x$

We'll try $8x$ and x.

$$8x^2 - 22x + 5 = (8x + \square)(x + \square)$$

Practice Problem 2

Factor each trinomial.

a. $6x^2 - 5x + 1$
b. $2x^2 - 11x + 12$

Answers

1. a. $(2x + 5)(2x + 1)$, **b.** $(5x + 2)(x + 5)$, **2. a.** $(3x - 1)(2x - 1)$, **b.** $(2x - 3)(x - 4)$

Concept Check: no; a, c, d

Since the middle term, $-22x$, has a negative numerical coefficient, we factor 5 into negative factors.

Factors of 5: $5 = -1 \cdot -5$

Let's try -1 and -5.

$$(8x - 1)(x - 5) = 8x^2 + 41x + 5$$

$$\begin{array}{r} -1x \\ +(-40x) \\ \hline -41x \end{array} \longrightarrow \text{Incorrect middle term}$$

Now let's try -5 and -1.

$$(8x - 5)(x - 1) = 8x^2 - 13x + 5$$

$$\begin{array}{r} -5x \\ +(-8x) \\ \hline -13x \end{array} \longrightarrow \text{Incorrect middle term}$$

Don't give up yet! We can still try other factors of $8x^2$. Let's try $4x$ and $2x$ with -1 and -5.

$$(4x - 1)(2x - 5) = 8x^2 - 22x + 5$$

$$\begin{array}{r} -2x \\ +(-20x) \\ \hline -22x \end{array} \longrightarrow \text{Correct middle term}$$

The factored form of $8x^2 - 22x + 5$ is $(4x - 1)(2x - 5)$.

EXAMPLE 3 Factor: $2x^2 + 13x - 7$

Solution: Factors of $2x^2$: $2x^2 = 2x \cdot x$

Factors of -7: $-7 = -1 \cdot 7$, $-7 = 1 \cdot -7$

We try possible combinations of these factors:

$(2x + 1)(x - 7) = 2x^2 - 13x - 7$ **Incorrect** middle term

$(2x - 1)(x + 7) = 2x^2 + 13x - 7$ **Correct** middle term

The factored form of $2x^2 + 13x - 7$ is $(2x - 1)(x + 7)$.

EXAMPLE 4 Factor: $10x^2 - 13xy - 3y^2$

Solution: Factors of $10x^2$: $10x^2 = 10x \cdot x$, $10x^2 = 2x \cdot 5x$

Factors of $-3y^2$: $-3y^2 = -3y \cdot y$, $-3y^2 = 3y \cdot -y$

We try some combinations of these factors:

$(10x - 3y)(x + y) = 10x^2 + 7xy - 3y^2$

$(x + 3y)(10x - y) = 10x^2 + 29xy - 3y^2$

$(5x + 3y)(2x - y) = 10x^2 + xy - 3y^2$

$(2x - 3y)(5x + y) = 10x^2 - 13xy - 3y^2$ **Correct** middle term

The factored form of $10x^2 - 13xy - 3y^2$ is $(2x - 3y)(5x + y)$.

Practice Problem 3

Factor each trinomial.

a. $35x^2 + 4x - 4$

b. $4x^2 + 3x - 7$

Practice Problem 4

Factor each trinomial.

a. $14x^2 - 3xy - 2y^2$

b. $12a^2 - 16ab - 3b^2$

Answers

3. a. $(5x + 2)(7x - 2)$, **b.** $(4x + 7)(x - 1)$,
4. a. $(7x + 2y)(2x - y)$, **b.** $(6a + b)(2a - 3b)$

B Factoring Out the Greatest Common Factor

Don't forget that the first step in factoring any polynomial is to look for a common factor to factor out.

Practice Problem 5

Factor each trinomial.

a. $3x^3 + 17x^2 + 10x$

b. $6xy^2 + 33xy - 18x$

EXAMPLE 5 Factor: $24x^4 + 40x^3 + 6x^2$

Solution: Notice that all three terms have a common factor of $2x^2$. Thus we factor out $2x^2$ first.

$$24x^4 + 40x^3 + 6x^2 = 2x^2(12x^2 + 20x + 3)$$

Next we factor $12x^2 + 20x + 3$.

Factors of $12x^2$: $12x^2 = 4x \cdot 3x$, $12x^2 = 12x \cdot x$, $12x^2 = 6x \cdot 2x$

Since all terms in the trinomial have positive numerical coefficients, we factor 3 using positive factors only.

Factors of 3: $3 = 1 \cdot 3$

We try some combinations of the factors.

$$2x^2(4x + 3)(3x + 1) = 2x^2(12x^2 + 13x + 3)$$
$$2x^2(12x + 1)(x + 3) = 2x^2(12x^2 + 37x + 3)$$
$$2x^2(2x + 3)(6x + 1) = 2x^2(12x^2 + 20x + 3)$$ **Correct** middle term

The factored form of $24x^4 + 40x^3 + 6x^2$ is $2x^2(2x + 3)(6x + 1)$. ●

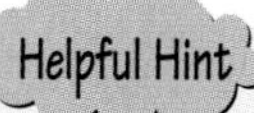

Don't forget to include the common factor in the factored form.

When the term containing the squared variable has a negative coefficient, you may want to first factor out a common factor of -1.

Practice Problem 6

Factor: $-5x^2 - 19x + 4$

EXAMPLE 6 Factor: $-6x^2 - 13x + 5$

Solution: We begin by factoring out a common factor of -1.

$$-6x^2 - 13x + 5 = -1(6x^2 + 13x - 5)$$ Factor out -1.
$$= -1(3x - 1)(2x + 5)$$ Factor $6x^2 + 13x - 5$. ●

Answers

5. a. $x(3x + 2)(x + 5)$, **b.** $3x(2y - 1)(y + 6)$, **6.** $-1(x + 4)(5x - 1)$

Name ______________________ Section __________ Date ____________

EXERCISE SET 5.3

 *Complete each factored form.*

1. $5x^2 + 22x + 8 = (5x + 2)(\quad)$

2. $2y^2 + 15y + 25 = (2y + 5)(\quad)$

3. $50x^2 + 15x - 2 = (5x + 2)(\quad)$

4. $6y^2 + 11y - 10 = (2y + 5)(\quad)$

5. $20x^2 - 7x - 6 = (5x + 2)(\quad)$

6. $8y^2 - 2y - 55 = (2y + 5)(\quad)$

Factor each trinomial completely. See Examples 1 through 4.

7. $2x^2 + 13x + 15$

8. $3x^2 + 8x + 4$

9. $8y^2 - 17y + 9$

10. $21x^2 - 41x + 10$

 **11.** $2x^2 - 9x - 5$

12. $36r^2 - 5r - 24$

13. $20r^2 + 27r - 8$

14. $3x^2 + 20x - 63$

15. $10x^2 + 17x + 3$

16. $2x^2 + 7x + 5$

17. $x + 3x^2 - 2$

18. $y + 8y^2 - 9$

19. $6x^2 - 13xy + 5y^2$

20. $8x^2 - 14xy + 3y^2$

21. $15x^2 - 16x - 15$

22. $25x^2 - 5x - 6$

23. $-9x + 20 + x^2$

24. $-7x + 12 + x^2$

25. $2x^2 - 7x - 99$

26. $2x^2 + 7x - 72$

27. $-27t + 7t^2 - 4$

28. $4t^2 - 7 - 3t$

29. $3a^2 + 10ab + 3b^2$

30. $2a^2 + 11ab + 5b^2$

31. $49x^2 - 7x - 2$

32. $3x^2 + 10x - 8$

33. $18x^2 - 9x - 14$

34. $42a^2 - 43a + 6$

 *Factor each trinomial completely. See Examples 1 through 6.*

35. $12x^3 + 11x^2 + 2x$

36. $8a^3 + 14a^2 + 3a$

37. $21x^2 - 48x - 45$

38. $12x^2 - 14x - 10$

39. $7x + 12x^2 - 12$

40. $16x + 15x^2 - 15$

41. $6x^2y^2 - 2xy^2 - 60y^2$

42. $8x^2y + 34xy - 84y$

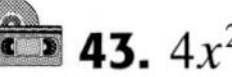 **43.** $4x^2 - 8x - 21$

44. $6x^2 - 11x - 10$

45. $3x^2 - 42x + 63$

46. $5x^2 - 75x + 60$

47. $8x^2 + 6x - 27$

48. $-x^2 + 4x + 21$

49. $-x^2 + 2x + 24$

50. $54a^2 + 39ab - 8b^2$

51. $4x^3 - 9x^2 - 9x$

52. $6x^3 - 31x^2 + 5x$

53. $24x^2 - 58x + 9$

54. $36x^2 + 55x - 14$

55. $40a^2b + 9ab - 9b$

56. $24y^2x + 7yx - 5x$

57. $15x^4 + 19x^2 + 6$

58. $6x^3 - 28x^2 + 16x$

59. $6y^3 - 8y^2 - 30y$

60. $12x^3 - 34x^2 + 24x$

61. $10x^3 + 25x^2y - 15xy^2$

62. $42x^4 - 99x^3y - 15x^2y^2$

63. $-14x^2 + 39x - 10$

64. $-15x^2 + 26x - 8$

Review and Preview

The following graph shows the national unemployment rate for the United States for the months November 2000 through April 2001. Use the graph to answer Exercises 65 through 68. See Section 3.1.

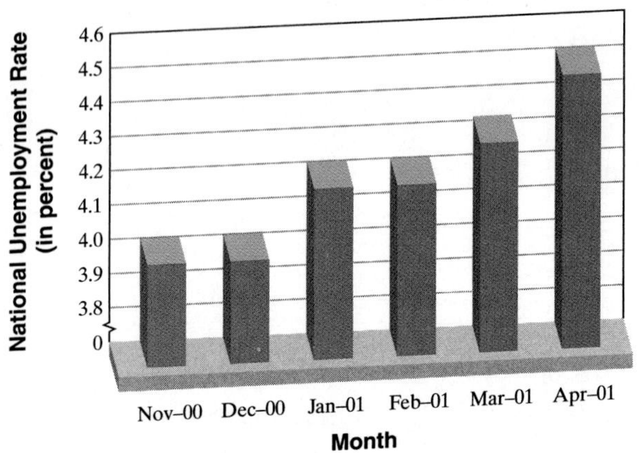

Source: U.S. Bureau of Labor Statistics

65. During which month(s) was the unemployment rate 4.2%?

66. During which month was the unemployment rate the highest?

67. By how much did the unemployment rate change from November 2000 to April 2001?

68. Describe any trend you notice from this graph.

Combining Concepts

Factor each trinomial completely.

69. $4x^2(y - 1)^2 + 10x(y - 1)^2 + 25(y - 1)^2$

70. $3x^2(a + 3)^3 - 28x(a + 3)^3 + 25(a + 3)^3$

71. $-12x^3y^2 + 3x^2y^2 + 15xy^2$
(*Hint:* Begin by factoring out $-3xy^2$.)

Find a positive value of b so that each trinomial is factorable.

72. $3x^2 + bx - 5$

73. $2z^2 + bz - 7$

Find a positive value of c so that each trinomial is factorable.

74. $5x^2 + 7x + c$

75. $3x^2 - 8x + c$

76. In your own words, describe the steps you will use to factor a trinomial.

5.4 Factoring Trinomials of the Form $ax^2 + bx + c$, by Grouping

OBJECTIVE

Ⓐ Use the grouping method to factor trinomials of the form $ax^2 + bx + c$.

SSM TUTOR CENTER SG CD & VIDEO MATH PRO WEB

Ⓐ Using the Grouping Method

There is an alternative method that can be used to factor trinomials of the form $ax^2 + bx + c, a \neq 1$. This method is called the **grouping method** because it uses factoring by grouping as we learned in Section 5.1.

To see how this method works, let's multiply the following:

$$(2x + 1)(3x + 5) = 6x^2 + 10x + 3x + 5$$
$$10 \cdot 3 = 30$$
$$6 \cdot 5 = 30$$
$$= 6x^2 + 13x + 5$$

Notice that the product of the coefficients of the first and last terms is $6 \cdot 5 = 30$. This is the same as the product of the coefficients of the two middle terms, $10 \cdot 3 = 30$.

Let's use this pattern to write $2x^2 + 11x + 12$ as a four-term polynomial. We will then factor the polynomial by grouping.

$$2x^2 + 11x + 12 \quad \text{Find two numbers whose product is } 2 \cdot 12 = 24$$
$$= 2x^2 + \square x + \square x + 12 \quad \text{and whose sum is 11.}$$

Since we want a positive product and a positive sum, we consider pairs of positive factors of 24 only.

Factors of 24	Sum of Factors	
1, 24	25	
2, 12	14	
3, 8	11	Correct sum

The factors are 3 and 8. Now we use these factors to write the middle term $11x$ as $3x + 8x$ (or $8x + 3x$). We replace $11x$ with $3x + 8x$ in the original trinomial and then we can factor by grouping.

$$\begin{aligned} 2x^2 + 11x + 12 &= 2x^2 + 3x + 8x + 12 \\ &= (2x^2 + 3x) + (8x + 12) && \text{Group the terms.} \\ &= x(2x + 3) + 4(2x + 3) && \text{Factor each group.} \\ &= (2x + 3)(x + 4) && \text{Factor out } (2x + 3). \end{aligned}$$

In general, we have the following procedure.

To Factor Trinomials by Grouping

Step 1. Factor out a greatest common factor, if there is one other than 1.
Step 2. For the resulting trinomial $ax^2 + bx + c$, find two numbers whose product is $a \cdot c$ and whose sum is b.
Step 3. Write the middle term, bx, using the factors found in Step 2.
Step 4. Factor by grouping.

Practice Problem 1

Factor each trinomial by grouping.

a. $3x^2 + 14x + 8$

b. $12x^2 + 19x + 5$

EXAMPLE 1 Factor $8x^2 - 14x + 5$ by grouping.

Solution:

Step 1. The terms of this trinomial contain no greatest common factor other than 1.

Step 2. This trinomial is of the form $ax^2 + bx + c$ with $a = 8$, $b = -14$, and $c = 5$. Find two numbers whose product is $a \cdot c$ or $8 \cdot 5 = 40$, and whose sum is b or -14. The numbers are -4 and -10.

Step 3. Write $-14x$ as $-4x - 10x$ so that

$$8x^2 - 14x + 5 = 8x^2 - 4x - 10x + 5$$

Step 4. Factor by grouping.

$$\begin{aligned} 8x^2 - 4x - 10x + 5 &= 4x(2x - 1) - 5(2x - 1) \\ &= (2x - 1)(4x - 5) \end{aligned}$$

Practice Problem 2

Factor each trinomial by grouping.

a. $6x^2y - 7xy - 5y$

b. $30x^2 - 26x + 4$

EXAMPLE 2 Factor $6x^2 - 2x - 20$ by grouping.

Solution:

Step 1. First factor out the greatest common factor, 2.

$$6x^2 - 2x - 20 = 2(3x^2 - x - 10)$$

Step 2. Next notice that $a = 3$, $b = -1$, and $c = -10$ in the resulting trinomial. Find two numbers whose product is $a \cdot c$ or $3(-10) = -30$ and whose sum is b, -1. The numbers are -6 and 5.

Step 3. $3x^2 - x - 10 = 3x^2 - 6x + 5x - 10$

Step 4. $= 3x(x - 2) + 5(x - 2)$

$= (x - 2)(3x + 5)$

The factored form of $6x^2 - 2x - 20 = 2(x - 2)(3x + 5)$.

Don't forget to include the common factor of 2.

Answers

1. a. $(x + 4)(3x + 2)$, **b.** $(4x + 5)(3x + 1)$,
2. a. $y(2x + 1)(3x - 5)$, **b.** $2(5x - 1)(3x - 2)$

Name ______________________ Section ________ Date ________

EXERCISE SET 5.4

A *Factor each polynomial by grouping. Notice that Step 3 has already been done in these exercises. See Examples 1 and 2.*

1. $x^2 + 3x + 2x + 6$

2. $x^2 + 5x + 3x + 15$

3. $x^2 - 4x + 7x - 28$

4. $x^2 - 6x + 2x - 12$

5. $y^2 + 8y - 2y - 16$

6. $z^2 + 10z - 7z - 70$

7. $3x^2 + 4x + 12x + 16$

8. $2x^2 + 5x + 14x + 35$

9. $8x^2 - 5x - 24x + 15$

10. $4x^2 - 9x - 32x + 72$

11. $5x^4 - 3x^2 + 25x^2 - 15$

12. $2y^4 - 10y^2 + 7y^2 - 35$

Factor each trinomial by grouping. Exercises 13–16 are broken into parts to help you get started. See Examples 1 and 2.

13. $6x^2 + 11x + 3$
- **a.** Find two numbers whose product is $6 \cdot 3 = 18$ and whose sum is 11.
- **b.** Write $11x$ using the factors from part (a).
- **c.** Factor by grouping.

14. $8x^2 + 14x + 3$
- **a.** Find two numbers whose product is $8 \cdot 3 = 24$ and whose sum is 14.
- **b.** Write $14x$ using the factors from part (a).
- **c.** Factor by grouping.

15. $15x^2 - 23x + 4$
- **a.** Find two numbers whose product is $15 \cdot 4 = 60$ and whose sum is -23.
- **b.** Write $-23x$ using the factors from part (a).
- **c.** Factor by grouping.

16. $6x^2 - 13x + 5$
- **a.** Find two numbers whose product is $6 \cdot 5 = 30$ and whose sum is -13.
- **b.** Write $-13x$ using the factors from part (a).
- **c.** Factor by grouping.

17. $21y^2 + 17y + 2$

18. $15x^2 + 11x + 2$

19. $7x^2 - 4x - 11$

20. $8x^2 - x - 9$

21. $10x^2 - 9x + 2$

22. $30x^2 - 23x + 3$

23. $2x^2 - 7x + 5$

24. $2x^2 - 7x + 3$

25. $12x + 4x^2 + 9$

26. $20x + 25x^2 + 4$

27. $4x^2 - 8x - 21$

28. $6x^2 - 11x - 10$

29. $10x^2 - 23x + 12$

30. $21x^2 - 13x + 2$

31. $2x^3 + 13x^2 + 15x$

32. $3x^3 + 8x^2 + 4x$

33. $16y^2 - 34y + 18$

34. $4y^2 - 2y - 12$

35. $-13x + 6 + 6x^2$

36. $-25x + 12 + 12x^2$

37. $54a^2 - 9a - 30$

38. $30a^2 + 38a - 20$

39. $20a^3 + 37a^2 + 8a$

40. $10a^3 + 17a^2 + 3a$

41. $12x^3 - 27x^2 - 27x$

42. $30x^3 - 155x^2 + 25x$

Review and Preview

Multiply. See Section 4.6.

43. $(x - 2)(x + 2)$

44. $(y - 5)(y + 5)$

45. $(y + 4)(y + 4)$

46. $(x + 7)(x + 7)$

47. $(9z + 5)(9z - 5)$

48. $(8y + 9)(8y - 9)$

49. $(4x - 3)^2$

50. $(2z - 1)^2$

Combining Concepts

Factor each polynomial by grouping.

51. $x^{2n} + 2x^n + 3x^n + 6$
(*Hint:* Don't forget that $x^{2n} = x^n \cdot x^n$.)

52. $x^{2n} + 6x^n + 10x^n + 60$

53. $3x^{2n} + 16x^n - 35$

54. $12x^{2n} - 40x^n + 25$

55. In your own words, explain how to factor a trinomial by grouping.

5.5 Factoring by Special Products

OBJECTIVES

- Ⓐ Factor perfect square trinomials.
- Ⓑ Factor the difference of two squares.
- Ⓒ Factor the sum or difference of two cubes.

SG

MATH PRO WEB

Ⓐ Factoring Perfect Square Trinomials

A trinomial that is the square of a binomial is called a **perfect square trinomial**. For example,

$$(x + 3)^2 = (x + 3)(x + 3)$$
$$= x^2 + 6x + 9$$

Thus $x^2 + 6x + 9$ is a perfect square trinomial.

In Chapter 4, we discovered special product formulas for squaring binomials.

$$(a + b)^2 = a^2 + 2ab + b^2 \quad \text{and} \quad (a - b)^2 = a^2 - 2ab + b^2$$

Because multiplication and factoring are reverse processes, we can now use these special products to help us factor perfect square trinomials. If we reverse these equations, we have the following.

Factoring Perfect Square Trinomials

$$a^2 + 2ab + b^2 = (a + b)^2$$
$$a^2 - 2ab + b^2 = (a - b)^2$$

To use these equations to help us factor, we must first be able to recognize a perfect square trinomial. A trinomial is a perfect square trinomial if it can be written so that its first term is the square of some quantity a, its last term is the square of some quantity b, and its middle term is twice the product of the quantities a and b.

EXAMPLE 1 Factor: $m^2 + 10m + 25$

Solution: Notice that the first term is a square: $m^2 = (m)^2$, the last term is a square: $25 = 5^2$, and the middle term $10m = 2 \cdot 5 \cdot m$.

This is a perfect square trinomial. Thus,

$$m^2 + 10m + 25 = m^2 + 2(m)(5) + 5^2 = (m + 5)^2$$

Practice Problem 1

Factor: $x^2 + 8x + 16$

EXAMPLES Factor each trinomial.

2. $4x^2 + 4x + 1 = (2x)^2 + 2 \cdot 2x \cdot 1 + 1^2$ See whether it is a perfect square trinomial.

$= (2x + 1)^2$ Factor.

3. $9x^2 - 12x + 4 = (3x)^2 - 2(3x)(2) + 2^2$ See whether it is a perfect square trinomial.

$= (3x - 2)^2$ Factor.

Practice Problems 2–3

Factor.

2. $9x^2 + 6x + 1$
3. $25x^2 - 20x + 4$

Answers

1. $(x + 4)^2$, **2.** $(3x + 1)^2$, **3.** $(5x - 2)^2$

Practice Problem 4

Factor: $4x^3 - 32x^2y + 64xy^2$

EXAMPLE 4 Factor: $3a^2x - 12abx + 12b^2x$

Solution: The terms of this trinomial have a greatest common factor of $3x$, which we factor out first.

$$3a^2x - 12abx + 12b^2x = 3x(a^2 - 4ab + 4b^2)$$

The polynomial $a^2 - 4ab + 4b^2$ is a perfect square trinomial. Notice that the first term is a square: $a^2 = (a)^2$, the last term is a square: $4b^2 = (2b)^2$, and $4ab = 2(a)(2b)$. The factoring can now be completed as

$$3x(a^2 - 4ab + 4b^2) = 3x(a - 2b)^2$$

Helpful Hint

If you recognize a trinomial as a perfect square trinomial, use the special formulas to factor. However, methods for factoring trinomials in general from Sections 5.3 and 5.4 will also result in the correct factored form.

B Factoring the Difference of Two Squares

We now factor special types of binomials, beginning with the **difference of two squares**. The special product pattern presented in Section 4.6 for the product of a sum and a difference of two terms is used again here. However, the emphasis is now on factoring rather than on multiplying.

Difference of Two Squares

$$a^2 - b^2 = (a + b)(a - b)$$

Notice that a binomial is a difference of two squares when it is the difference of the square of some quantity *a* and the square of some quantity *b*.

Practice Problems 5–8

Factor.

5. $x^2 - 49$
6. $4y^2 - 81$
7. $12 - 3a^2$
8. $y^2 - \frac{1}{25}$

EXAMPLES Factor.

5. $x^2 - 9 = x^2 - 3^2$

$= (x + 3)(x - 3)$

6. $16y^2 - 9 = (4y)^2 - 3^2$

$= (4y + 3)(4y - 3)$

7. $50 - 8y^2 = 2(25 - 4y^2)$ Factor out the common factor of 2.

$= 2[5^2 - (2y)^2]$

$= 2(5 + 2y)(5 - 2y)$

Answers

4. $4x(x - 4y)^2$, **5.** $(x + 7)(x - 7)$, **6.** $(2y + 9)(2y - 9)$, **7.** $3(2 + a)(2 - a)$, **8.** $\left(y + \frac{1}{5}\right)\left(y - \frac{1}{5}\right)$

8. $x^2 - \frac{1}{4} = x^2 - \left(\frac{1}{2}\right)^2$

$= \left(x + \frac{1}{2}\right)\left(x - \frac{1}{2}\right)$

The binomial $x^2 + 9$ is a **sum of two squares** and cannot be factored by using real numbers. *In general, except for factoring out a greatest common factor, the sum of two squares usually cannot be factored by using real numbers.*

Helpful Hint

The sum of two squares whose greatest common factor is 1 usually cannot be factored by using real numbers.

EXAMPLE 9 Factor: $p^4 - 16$

Solution: $p^4 - 16 = (p^2)^2 - 4^2$

$= (p^2 + 4)(p^2 - 4)$

The binomial factor $p^2 + 4$ cannot be factored by using real numbers, but the binomial factor $p^2 - 4$ is a difference of squares.

$$(p^2 + 4)(p^2 - 4) = (p^2 + 4)(p + 2)(p - 2)$$

Helpful Hint

1. Don't forget to first see whether there's a greatest common factor (other than 1) that can be factored out.
2. Factor completely. In other words, check to see whether any factors can be factored further (as in Example 9).

EXAMPLES Factor each difference of two squares.

10. $4x^3 - 49x = x(4x^2 - 49)$ Factor out the common factor, x.

$= x[(2x)^2 - 7^2]$

$= x(2x + 7)(2x - 7)$ Factor the difference of two squares.

11. $162x^4 - 2 = 2(81x^4 - 1)$ Factor out the common factor, 2.

$= 2(9x^2 + 1)(9x^2 - 1)$ Factor the difference of two squares.

$= 2(9x^2 + 1)(3x + 1)(3x - 1)$ Factor the difference of two squares.

Practice Problem 9

Factor: $a^4 - 81$

Practice Problems 10–12

Factor each difference of two squares.

10. $9x^3 - 25x$
11. $48x^4 - 3$
12. $-9x^2 + 100$

Answers

9. $(a^2 + 9)(a + 3)(a - 3)$, **10.** $x(3x - 5)(3x + 5)$, **11.** $3(4x^2 + 1)(2x + 1)(2x - 1)$, **12.** $-1(3x - 10)(3x + 10)$

12. $-49x^2 + 16 = -1(49x^2 - 16)$ Factor out -1.

$= -1(7x + 4)(7x - 4)$ Factor the difference of two squares.

Try the Concept Check in the margin.

Concept Check

Is $(x - 4)(y^2 - 9)$ completely factored? Why or why not?

EXAMPLE 13 Factor: $(x + 3)^2 - 36$

Solution:

$$(x + 3)^2 - 36 = (x + 3)^2 - 6^2$$

$= [(x + 3) + 6][(x + 3) - 6]$ Factor as the difference of two squares.

$= [x + 3 + 6][x + 3 - 6]$ Remove parentheses.

$= (x + 9)(x - 3)$ Simplify.

Practice Problem 13

Factor: $(x + 1)^2 - 9$

EXAMPLE 14 Factor: $x^2 + 4x + 4 - y^2$

Solution: Factoring by grouping comes to mind since the sum of the first three terms of this polynomial is a perfect square trinomial.

$x^2 + 4x + 4 - y^2 = (x^2 + 4x + 4) - y^2$ Group the first three terms.

$= (x + 2)^2 - y^2$ Factor the perfect square trinomial.

This is not completely factored yet since we have a *difference*, not a *product*. Since $(x + 2)^2 - y^2$ is a difference of squares, we have

$$(x + 2)^2 - y^2 = [(x + 2) + y][(x + 2) - y]$$
$$= (x + 2 + y)(x + 2 - y)$$

Practice Problem 14

Factor: $a^2 + 2a + 1 - b^2$

C Factoring the Sum or Difference of Two Cubes

Although the sum of two squares usually cannot be factored, the sum of two cubes, as well as the difference of two cubes, can be factored as follows.

Sum and Difference of Two Cubes

$$a^3 + b^3 = (a + b)(a^2 - ab + b^2)$$
$$a^3 - b^3 = (a - b)(a^2 + ab + b^2)$$

Answers

13. $(x - 2)(x + 4)$,
14. $(a + 1 + b)(a + 1 - b)$

Concept Check: no; $(y^2 - 9)$ can be factored

To check the first pattern, let's find the product of $(a + b)$ and $(a^2 - ab + b^2)$.

$$(a + b)(a^2 - ab + b^2) = a(a^2 - ab + b^2) + b(a^2 - ab + b^2)$$
$$= a^3 - a^2b + ab^2 + a^2b - ab^2 + b^3$$
$$= a^3 + b^3$$

EXAMPLE 15 Factor: $x^3 + 8$

Solution: First we write the binomial in the form $a^3 + b^3$. Then we use the formula

$$a^3 + b^3 = (a + b)(a^2 - a \cdot b + b^2), \quad \text{where } a \text{ is } x \text{ and } b \text{ is } 2.$$
$$x^3 + 8 = x^3 + 2^3 = (x + 2)(x^2 - x \cdot 2 + 2^2)$$

Thus, $x^3 + 8 = (x + 2)(x^2 - 2x + 4)$

Practice Problem 15

Factor: $x^3 + 27$

EXAMPLE 16 Factor: $p^3 + 27q^3$

Solution: $p^3 + 27q^3 = p^3 + (3q)^3$

$$= (p + 3q)[p^2 - (p)(3q) + (3q)^2]$$
$$= (p + 3q)(p^2 - 3pq + 9q^2)$$

Practice Problem 16

Factor: $x^3 + 64y^3$

EXAMPLE 17 Factor: $y^3 - 64$

Solution: This is a difference of cubes since $y^3 - 64 = y^3 - 4^3$.

From $a^3 - b^3 = (a - b)(a^2 + a \cdot b + b^2)$ we have that

$$y^3 - 4^3 = (y - 4)(y^2 + y \cdot 4 + 4^2)$$
$$= (y - 4)(y^2 + 4y + 16)$$

Practice Problem 17

Factor: $y^3 - 8$

Helpful Hint

When factoring sums or differences of cubes, be sure to notice the sign patterns.

Same sign

$$x^3 + y^3 = (x + y)(x^2 - xy + y^2)$$

Opposite sign Always positive

Same sign

$$x^3 - y^3 = (x - y)(x^2 + xy + y^2)$$

Opposite sign Always positive

Answers

15. $(x + 3)(x^2 - 3x + 9)$,
16. $(x + 4y)(x^2 - 4xy + 16y^2)$,
17. $(y - 2)(y^2 + 2y + 4)$

Practice Problem 18

Factor: $27a^2 - b^3a^2$

Answer

18. $a^2(3 - b)(9 + 3b + b^2)$

EXAMPLE 18 Factor: $125q^2 - n^3q^2$

Solution: First we factor out a common factor of q^2.

$$125q^2 - n^3q^2 = q^2(125 - n^3)$$
$$= q^2(5^3 - n^3)$$

Opposite sign — Positive

$$= q^2(5 - n)[5^2 + (5)(n) + (n^2)]$$
$$= q^2(5 - n)(25 + 5n + n^2)$$

Thus, $125q^2 - n^3q^2 = q^2(5 - n)(25 + 5n + n^2)$. The trinomial $25 + 5n + n^2$ cannot be factored further.

GRAPHING CALCULATOR EXPLORATIONS

Graphing

A graphing calculator is a convenient tool for evaluating an expression at a given replacement value. For example, let's evaluate $x^2 - 6x$ when $x = 2$. To do so, store the value 2 in the variable x and then enter and evaluate the algebraic expression.

```
2→X
                2
X²-6X
               -8
```

The value of $x^2 - 6x$ when $x = 2$ is -8. You may want to use this method for evaluating expressions as you explore the following.

We can use a graphing calculator to explore factoring patterns numerically. Use your calculator to evaluate $x^2 - 2x + 1$, $x^2 - 2x - 1$, and $(x - 1)^2$ for each value of x given in the table. What do you observe?

	$x^2 - 2x + 1$	$x^2 - 2x - 1$	$(x - 1)^2$
$x = 5$			
$x = -3$			
$x = 2.7$			
$x = -12.1$			
$x = 0$			

Notice in each case that $x^2 - 2x - 1 \neq (x - 1)^2$. Because for each x in the table the value of $x^2 - 2x + 1$ and the value of $(x - 1)^2$ are the same, we might guess that $x^2 - 2x + 1 = (x - 1)^2$. We can verify our guess algebraically with multiplication:

$$(x - 1)(x - 1) = x^2 - x - x + 1 = x^2 - 2x + 1$$

Name ____________________ Section __________ Date __________

Mental Math

State each number as a square.

1. 1 **2.** 25 **3.** 81 **4.** 64 **5.** 9 **6.** 100

State each term as a square.

7. $9x^2$ **8.** $16y^2$ **9.** $25a^2$ **10.** $81b^2$ **11.** $36p^4$ **12.** $4q^4$

EXERCISE SET 5.5

A *Determine whether each trinomial is a perfect square trinomial. See Examples 1 through 4.*

1. $x^2 + 16x + 64$ **2.** $x^2 + 22x + 121$ **3.** $y^2 + 5y + 25$ **4.** $y^2 + 4y + 16$

5. $4x^2 + 12xy + 8y^2$ **6.** $25x^2 + 20xy + 2y^2$ **7.** $25a^2 - 40ab + 16b^2$ **8.** $36a^2 - 12ab + b^2$

9. Fill in the blank so that $x^2 +$ ______ $x + 16$ is a perfect square trinomial.

10. Fill in the blank so that $9x^2 +$ ______ $x + 25$ is a perfect square trinomial.

B *Factor each trinomial completely. See Examples 1 through 4.*

11. $x^2 + 22x + 121$ **12.** $x^2 + 18x + 81$ **13.** $x^2 - 16x + 64$ **14.** $x^2 - 12x + 36$

15. $16a^2 - 24a + 9$ **16.** $25x^2 + 20x + 4$ **17.** $3x^2 - 24x + 48$ **18.** $2n^2 - 28n + 98$

19. $x^2y^2 - 10xy + 25$ **20.** $4x^2y^2 - 28xy + 49$ **21.** $m^3 + 18m^2 + 81m$ **22.** $y^3 + 12y^2 + 36y$

23. $1 + 6x^2 + x^4$ **24.** $1 + 16x^2 + x^4$ **25.** $9x^2 - 24xy + 16y^2$ **26.** $25x^2 - 60xy + 36y^2$

27. Describe a perfect square trinomial.

28. Write a perfect square trinomial that factors as $(x + 3y)^2$.

C *Factor each binomial completely. See Examples 5 through 14.*

29. $x^2 - 25$ **30.** $y^2 - 100$ **31.** $9 - 4z^2$ **32.** $16x^2 - y^2$

33. $16r^2 + 1$ **34.** $49y^2 + 9$ **35.** $x^3y - 121xy^3$ **36.** $25xy^2 - 4x$

37. $(y + 2)^2 - 49$

38. $(x - 1)^2 - z^2$

39. $64x^2 - 100$

40. $4x^2 - 36$

41. $18x^2y - 2y$

42. $12xy^2 - 108x$

43. $9x^2 - 49$

44. $25x^2 - 4$

45. $x^4 - 81$

46. $x^4 - 256$

47. $(x + 2y)^2 - 9$

48. $(3x + y)^2 - 25$

49. $x^2 + 16x + 64 - x^4$

50. $x^2 + 20x + 100 - x^4$

51. $x^2 - 10x + 25 - y^2$

52. $x^2 - 18x + 81 - y^2$

53. $4x^2 + 4x + 1 - z^2$

54. $9y^2 + 12y + 4 - x^2$

55. $m^4 - 1$

56. $n^4 - 16$

C *Factor. See Examples 15 through 18.*

57. $x^3 + 27$

58. $y^3 + 1$

59. $z^3 - 1$

60. $x^3 - 8$

61. $m^3 + n^3$

62. $r^3 + 125$

63. $x^3y^2 - 27y^2$

64. $64 - p^3$

65. $a^3b + 8b^4$

66. $8ab^3 + 27a^4$

67. $125y^3 - 8x^3$

68. $54y^3 - 128$

69. $x^6 - y^3$

70. $x^3 - y^6$

71. $8x^3 + 27y^3$

72. $125x^3 + 8y^3$

73. $x^3 - 1$

74. $x^3 - 8$

75. $x^3 + 125$

76. $x^3 + 216$

77. $3x^6y^2 + 81y^2$

78. $x^2y^9 + x^2y^3$

Review and Preview

Solve each equation. See Section 2.4.

79. $x - 5 = 0$

80. $x + 7 = 0$

81. $3x + 1 = 0$

82. $5x - 15 = 0$

83. $-2x = 0$

84. $3x = 0$

85. $-5x + 25 = 0$

86. $-4x - 16 = 0$

Combining Concepts

△ **87.** A manufacturer of metal washers needs to determine the cross-sectional area of each washer. If the outer radius of the washer is R and the radius of the hole is r, express the area of the washer as a polynomial. Factor this polynomial completely.

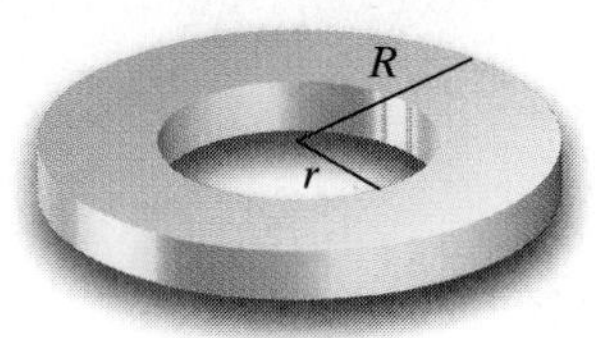

△ **88.** Express the area of the shaded region as a polynomial. Factor the polynomial completely.

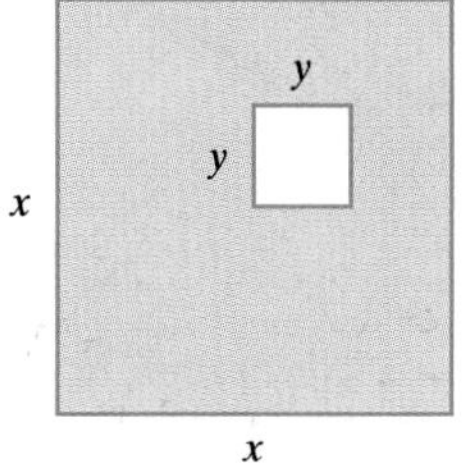

89. At this writing, the world's tallest building is the Petronas Twin Towers in Kuala Lumpur, Malaysia, at a height of 1483 feet. (*Source:* Council on Tall Buildings and Urban Habitat) Suppose a worker is suspended 39 feet below the tip of the pinnacle atop one of the towers, at a height of 1444 feet above the ground. If the worker accidentally drops a bolt, the height of the bolt after t seconds is given by the expression $1444 - 16t^2$.

a. Find the height of the bolt after 3 seconds.

b. Find the height of the bolt after 7 seconds.

c. To the nearest whole second, estimate when the bolt hits the ground.

d. Factor $1444 - 16t^2$.

90. A performer with the Moscow Circus is planning a stunt involving a free fall from the top of the Moscow State University building, which is 784 feet tall. (*Source:* Council on Tall Buildings and Urban Habitat) Neglecting air resistance, the performer's height above gigantic cushions positioned at ground level after t seconds is given by the expression $784 - 16t^2$.

a. Find the performer's height after 2 seconds.

b. Find the performer's height after 5 seconds.

c. To the nearest whole second, estimate when the performer reaches the cushions positioned at ground level.

d. Factor $784 - 16t^2$.

Find the value of c that makes each trinomial a perfect square trinomial.

91. $x^2 + 6x + c$

92. $y^2 + 10y + c$

93. $m^2 - 14m + c$

94. $n^2 - 2n + c$

95. Factor $x^6 - 1$ completely, using the following methods from this chapter.

a. Factor the expression by treating it as the difference of two squares, $(x^3)^2 - 1^2$.

b. Factor the expression treating it as the difference of two cubes, $(x^2)^3 - 1^3$.

c. Are the answers to parts (a) and (b) the same? Why or why not?

Factor. Assume that variables used as exponents represent positive integers.

96. $x^{2n} - 25$

97. $x^{2n} - 36$

98. $36x^{2n} - 49$

99. $25x^{2n} - 81$

100. $x^{4n} - 16$

101. $x^{4n} - 625$

Name ______________ Section ________ Date ________

Integrated Review—Choosing a Factoring Strategy

The key to proficiency in factoring polynomials is to practice until you are comfortable with each technique. A strategy for factoring polynomials completely is given next.

Factoring a Polynomial

Step 1. Are there any common factors? If so, factor out the greatest common factor.

Step 2. How many terms are in the polynomial?

a. If there are *two* terms, decide if one of the following formulas may be applied:

i. Difference of two squares: $a^2 - b^2 = (a - b)(a + b)$

ii. Difference of two cubes: $a^3 - b^3 = (a - b)(a^2 + ab + b^2)$

iii. Sum of two cubes: $a^3 + b^3 = (a + b)(a^2 - ab + b^2)$

b. If there are *three* terms, try one of the following:

i. Perfect square trinomial: $a^2 + 2ab + b^2 = (a + b)^2$

$a^2 - 2ab + b^2 = (a - b)^2$

ii. If not a perfect square trinomial, factor by using the methods presented in Sections 5.2 through 5.4.

c. If there are *four* or more terms, try factoring by grouping.

Step 3. See whether any factors in the factored polynomial can be factored further.

Factor each polynomial completely.

1. $x^2 + x - 12$ **2.** $x^2 - 10x + 16$ **3.** $x^2 - x - 6$

4. $x^2 + 2x + 1$ **5.** $x^2 - 6x + 9$ **6.** $x^2 + x - 2$

7. $x^2 + x - 6$ **8.** $x^2 + 7x + 12$ **9.** $x^2 - 7x + 10$

10. $x^2 - x - 30$ **11.** $2x^2 - 98$ **12.** $3x^2 - 75$

13. $x^2 + 3x + 5x + 15$ **14.** $3y - 21 + xy - 7x$ **15.** $x^2 + 6x - 16$

16. $x^2 - 3x - 28$ **17.** $4x^3 + 20x^2 - 56x$ **18.** $6x^3 - 6x^2 - 120x$

19. $12x^2 + 34x + 24$ **20.** $8a^2 + 6ab - 5b^2$ **21.** $4a^2 - b^2$

22. $x^2 - 25y^2$ **23.** $28 - 13x - 6x^2$ **24.** $20 - 3x - 2x^2$

25. $x^2 - 2x + 4$ **26.** $a^2 + a - 3$ **27.** $6y^2 + y - 15$

28. $4x^2 - x - 5$ **29.** $18x^3 - 63x^2 + 9x$ **30.** $12a^3 - 24a^2 + 4a$

31. $16a^2 - 56a + 49$ **32.** $25p^2 - 70p + 49$ **33.** $14 + 5x - x^2$

34. $3 - 2x - x^2$ **35.** $3x^4y + 6x^3y - 72x^2y$ **36.** $2x^3y + 8x^2y^2 - 10xy^3$

37. $12x^3y + 243xy$ **38.** $6x^3y^2 + 8xy^2$ **39.** $2xy - 72x^3y$

Answers

1. ________
2. ________
3. ________
4. ________
5. ________
6. ________
7. ________
8. ________
9. ________
10. ________
11. ________
12. ________
13. ________
14. ________
15. ________
16. ________
17. ________
18. ________
19. ________
20. ________
21. ________
22. ________
23. ________
24. ________
25. ________
26. ________
27. ________
28. ________
29. ________
30. ________
31. ________
32. ________
33. ________
34. ________
35. ________
36. ________
37. ________
38. ________
39. ________

40.
41.
42.
43.
44.
45.
46.
47.
48.
49.
50.
51.
52.
53.
54.
55.
56.
57.
58.
59.
60.
61.
62.
63.
64.
65.
66.
67.
68.
69.
70.
71.
72.
73.
74.
75.
76.
77.
78.
79.
80.

40. $2x^3 - 18x$

41. $x^3 + 6x^2 - 4x - 24$

42. $x^3 - 2x^2 - 36x + 72$

43. $6a^3 + 10a^2$

44. $4n^2 - 6n$

45. $3x^3 - x^2 + 12x - 4$

46. $x^3 - 2x^2 + 3x - 6$

47. $6x^2 + 18xy + 12y^2$

48. $12x^2 + 46xy - 8y^2$

49. $5(x + y) + x(x + y)$

50. $7(x - y) + y(x - y)$

51. $14t^2 - 9t + 1$

52. $3t^2 - 5t + 1$

53. $3x^2 + 2x - 5$

54. $7x^2 + 19x - 6$

55. $1 - 8a - 20a^2$

56. $1 - 7a - 60a^2$

57. $x^4 - 10x^2 + 9$

58. $x^4 - 13x^2 + 36$

59. $x^2 - 23x + 120$

60. $y^2 + 22y + 96$

61. $x^2 - 14x - 48$

62. $16a^2 - 56ab + 49b^2$

63. $25p^2 - 70pq + 49q^2$

64. $7x^2 + 24xy + 9y^2$

65. $-x^2 - x + 30$

66. $-x^2 + 6x - 8$

67. $3rs - s + 12r - 4$

68. $x^3 - 2x^2 + 3x - 6$

69. $4x^2 - 8xy - 3x + 6y$

70. $4x^2 - 2xy - 7yz + 14xz$

71. $x^2 + 9xy - 36y^2$

72. $3x^2 + 10xy - 8y^2$

73. $x^4 - 14x^2 - 32$

74. $x^4 - 22x^2 - 75$

75. $x^4 - x$

76. $x^6 + x^3$

77. $8x^3 + 125y^3$

78. $27x^3 - 64y^3$

79. Explain why it makes good sense to factor out the GCF first, before using other methods of factoring.

80. The sum of two squares usually does not factor. Is the sum of two squares $9x^2 + 81y^2$ factorable?

5.6 Solving Quadratic Equations by Factoring

In this section, we introduce a new type of equation—the **quadratic equation**.

Quadratic Equation

A quadratic equation is one that can be written in the form

$$ax^2 + bx + c = 0$$

where a, b, and c are real numbers and $a \neq 0$.

OBJECTIVES

A. Solve quadratic equations by factoring.

B. Solve equations with degree greater than 2 by factoring.

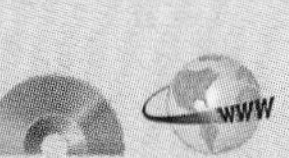

SSM TUTOR CENTER SG CD & VIDEO MATH PRO WEB

Some examples of quadratic equations are shown below.

$$3x^2 + 5x + 6 = 0 \qquad x^2 = 9 \qquad y^2 + y = 1$$

The form $ax^2 + bx + c = 0$ is called the **standard form** of a quadratic equation. The quadratic equation $3x^2 + 5x + 6 = 0$ is the only equation above that is in standard form.

Quadratic equations model many real-life situations. For example, let's suppose an object is dropped from the top of a 256-foot cliff and we want to know how long before the object strikes the ground. The answer to this question is found by solving the quadratic equation $-16t^2 + 256 = 0$. (See Example 1 in Section 5.7.)

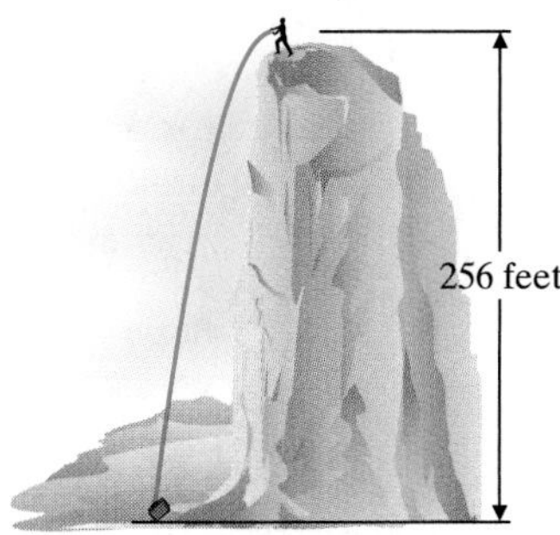

A Solving Quadratic Equations by Factoring

Some quadratic equations can be solved by making use of factoring and the **zero factor property**.

Zero Factor Property

If a and b are real numbers and if $ab = 0$, then $a = 0$ or $b = 0$.

In other words, if the product of two numbers is 0, then at least one of the numbers must be 0.

EXAMPLE 1 Solve: $(x - 3)(x + 1) = 0$

Solution: If this equation is to be a true statement, then either the factor $x - 3$ must be 0 or the factor $x + 1$ must be 0. In other words, either

$$x - 3 = 0 \quad \text{or} \quad x + 1 = 0$$

If we solve these two linear equations, we have

$$x = 3 \quad \text{or} \quad x = -1$$

Practice Problem 1

Solve: $(x - 7)(x + 2) = 0$

Answer

1. 7 and -2

Thus, 3 and -1 are both solutions of the equation $(x - 3)(x + 1) = 0$. To check, we replace x with 3 in the original equation. Then we replace x with -1 in the original equation.

Check:

$(x - 3)(x + 1) = 0$		$(x - 3)(x + 1) = 0$	
$(3 - 3)(3 + 1) \stackrel{?}{=} 0$	Replace x with 3.	$(-1 - 3)(-1 + 1) \stackrel{?}{=} 0$	Replace x with -1.
$0(4) = 0$	True	$(-4)(0) = 0$	True

The solutions are 3 and -1.

Helpful Hint

The zero factor property says that *if a product is 0, then a factor is 0.*

If $a \cdot b = 0$, then $a = 0$ or $b = 0$.
If $x(x + 5) = 0$, then $x = 0$ or $x + 5 = 0$.
If $(x + 7)(2x - 3) = 0$, then $x + 7 = 0$ or $2x - 3 = 0$.

Use this property only when the product is 0. For example, if $a \cdot b = 8$, we do not know the value of a or b. The values may be $a = 2, b = 4$ or $a = 8, b = 1$, or any other two numbers whose product is 8.

Practice Problem 2

Solve: $(x - 10)(3x + 1) = 0$

EXAMPLE 2 Solve: $(x - 5)(2x + 7) = 0$

Solution: The product is 0. By the zero factor property, this is true only when a factor is 0. To solve, we set each factor equal to 0 and solve the resulting linear equations.

$$(x - 5)(2x + 7) = 0$$

$$x - 5 = 0 \quad \text{or} \quad 2x + 7 = 0$$

$$x = 5 \qquad\qquad 2x = -7$$

$$x = -\frac{7}{2}$$

Check: Let $x = 5$.

$$(x - 5)(2x + 7) = 0$$

$$(5 - 5)(2 \cdot 5 + 7) \stackrel{?}{=} 0 \quad \text{Replace } x \text{ with } 5.$$

$$0 \cdot 17 \stackrel{?}{=} 0$$

$$0 = 0 \quad \text{True}$$

Let $x = -\frac{7}{2}$.

$$(x - 5)(2x + 7) = 0$$

$$\left(-\frac{7}{2} - 5\right)\left(2\left(-\frac{7}{2}\right) + 7\right) \stackrel{?}{=} 0 \quad \text{Replace } x \text{ with } -\frac{7}{2}.$$

$$\left(-\frac{17}{2}\right)(-7 + 7) \stackrel{?}{=} 0$$

$$\left(-\frac{17}{2}\right) \cdot 0 \stackrel{?}{=} 0$$

$$0 = 0 \quad \text{True}$$

The solutions are 5 and $-\frac{7}{2}$.

Answer

2. 10 and $-\frac{1}{3}$

EXAMPLE 3 Solve: $x(5x - 2) = 0$

Solution: $x(5x - 2) = 0$

$$x = 0 \quad \text{or} \quad 5x - 2 = 0 \qquad \text{Use the zero factor property.}$$
$$5x = 2$$
$$x = \frac{2}{5}$$

Check these solutions in the original equation. The solutions are 0 and $\frac{2}{5}$.

Practice Problem 3

Solve each equation.

a. $y(y + 3) = 0$
b. $x(4x - 3) = 0$

EXAMPLE 4 Solve: $x^2 - 9x - 22 = 0$

Solution: One side of the equation is 0. However, to use the zero factor property, one side of the equation must be 0 *and* the other side must be written as a product (must be factored). Thus, we must first factor this polynomial.

$$x^2 - 9x - 22 = 0$$
$$(x - 11)(x + 2) = 0 \qquad \text{Factor.}$$

Now we can apply the zero factor property.

$$x - 11 = 0 \quad \text{or} \quad x + 2 = 0$$
$$x = 11 \qquad\qquad x = -2$$

Check:

Let $x = 11$.

$$x^2 - 9x - 22 = 0$$
$$11^2 - 9 \cdot 11 - 22 \stackrel{?}{=} 0$$
$$121 - 99 - 22 \stackrel{?}{=} 0$$
$$22 - 22 \stackrel{?}{=} 0$$
$$0 = 0 \qquad \text{True}$$

Let $x = -2$.

$$x^2 - 9x - 22 = 0$$
$$(-2)^2 - 9(-2) - 22 \stackrel{?}{=} 0$$
$$4 + 18 - 22 \stackrel{?}{=} 0$$
$$22 - 22 \stackrel{?}{=} 0$$
$$0 = 0 \qquad \text{True.}$$

The solutions are 11 and -2.

Practice Problem 4

Solve: $x^2 - 3x - 18 = 0$

EXAMPLE 5 Solve: $x^2 - 9x = -20$

Solution: First we rewrite the equation in standard form so that one side is 0. Then we factor the polynomial.

$$x^2 - 9x = -20$$
$$x^2 - 9x + 20 = 0 \qquad \text{Write in standard form by adding 20 to both sides.}$$
$$(x - 4)(x - 5) = 0 \qquad \text{Factor.}$$

Next we use the zero factor property and set each factor equal to 0.

$$x - 4 = 0 \quad \text{or} \quad x - 5 = 0 \qquad \text{Set each factor equal to 0.}$$
$$x = 4 \qquad\qquad x = 5 \qquad \text{Solve.}$$

Check: Check these solutions in the original equation. The solutions are 4 and 5.

Practice Problem 5

Solve: $x^2 - 14x = -24$

Answers

3. a. 0 and -3, **b.** 0 and $\frac{3}{4}$, **4.** 6 and -3, **5.** 12 and 2

The following steps may be used to solve a quadratic equation by factoring.

To Solve Quadratic Equations by Factoring

Step 1. Write the equation in standard form so that one side of the equation is 0.
Step 2. Factor the quadratic equation completely.
Step 3. Set each factor containing a variable equal to 0.
Step 4. Solve the resulting equations.
Step 5. Check each solution in the original equation.

Since it is not always possible to factor a quadratic polynomial, not all quadratic equations can be solved by factoring. Other methods of solving quadratic equations are presented in Chapter 10.

Practice Problem 6

Solve each equation.

a. $x(x - 4) = 5$
b. $x(3x + 7) = 6$

EXAMPLE 6 Solve: $x(2x - 7) = 4$

Solution: First we write the equation in standard form; then we factor.

$$x(2x - 7) = 4$$
$$2x^2 - 7x = 4 \quad \text{Multiply.}$$
$$2x^2 - 7x - 4 = 0 \quad \text{Write in standard form.}$$
$$(2x + 1)(x - 4) = 0 \quad \text{Factor.}$$
$$2x + 1 = 0 \quad \text{or} \quad x - 4 = 0 \quad \text{Set each factor equal to zero.}$$
$$2x = -1 \qquad x = 4 \quad \text{Solve.}$$
$$x = -\frac{1}{2}$$

Check the solutions in the original equation. The solutions are $-\frac{1}{2}$ and 4. ●

To solve the equation $x(2x - 7) = 4$, do **not** set each factor equal to 4. Remember that to apply the zero factor property, one side of the equation must be 0 and the other side of the equation must be in factored form.

B Solving Equations with Degree Greater than Two by Factoring

Some equations with degree greater than 2 can be solved by factoring and then using the zero factor property.

Practice Problem 7

Solve: $2x^3 - 18x = 0$

EXAMPLE 7 Solve: $3x^3 - 12x = 0$

Solution: To factor the left side of the equation, we begin by factoring out the greatest common factor, $3x$.

$$3x^3 - 12x = 0$$
$$3x(x^2 - 4) = 0 \quad \text{Factor out the GCF, } 3x.$$
$$3x(x + 2)(x - 2) = 0 \quad \text{Factor } x^2 - 4 \text{, a difference of two squares.}$$
$$3x = 0 \quad \text{or} \quad x + 2 = 0 \quad \text{or} \quad x - 2 = 0 \quad \text{Set each factor equal to 0.}$$
$$x = 0 \qquad x = -2 \qquad x = 2 \quad \text{Solve.}$$

Answers

6. a. 5 and −1, **b.** $\frac{2}{3}$ and −3, **7.** 0, 3, and −3

Thus, the equation $3x^3 - 12x = 0$ has three solutions: 0, −2, and 2.

Check: Replace x with each solution in the original equation.

Let $x = 0$.	Let $x = -2$.	Let $x = 2$.
$3(0)^3 - 12(0) \stackrel{?}{=} 0$	$3(-2)^3 - 12(-2) \stackrel{?}{=} 0$	$3(2)^3 - 12(2) \stackrel{?}{=} 0$
$0 = 0$	$3(-8) + 24 \stackrel{?}{=} 0$	$3(8) - 24 \stackrel{?}{=} 0$
True	$0 = 0$ True	$0 = 0$ True

The solutions are 0, −2, and 2.

EXAMPLE 8 Solve: $(5x - 1)(2x^2 + 15x + 18) = 0$

Solution:

$$(5x - 1)(2x^2 + 15x + 18) = 0$$

$$(5x - 1)(2x + 3)(x + 6) = 0 \quad \text{Factor the trinomial.}$$

$$5x - 1 = 0 \quad \text{or} \quad 2x + 3 = 0 \quad \text{or} \quad x + 6 = 0 \quad \text{Set each factor equal to 0.}$$

$$5x = 1 \qquad 2x = -3 \qquad x = -6 \quad \text{Solve.}$$

$$x = \frac{1}{5} \qquad x = -\frac{3}{2}$$

Check each solution in the original equation. The solutions are $\frac{1}{5}$, $-\frac{3}{2}$, and −6.

Practice Problem 8

Solve: $(x + 3)(3x^2 - 20x - 7) = 0$

Answer

8. $-3, -\frac{1}{3}$, and 7

STUDY SKILLS REMINDER

How well do you know your textbook?

See if you can answer the questions below.

1. What does the [icon] icon mean?
2. What does the [icon] icon mean?
3. What does the △ icon mean?
4. Where can you find a review for each chapter? What answers to this review can be found in the back of your text?
5. Each chapter contains an overview of the chapter along with examples. What is this feature called?
6. Does this text contain any solutions to exercises? If so, where?

FOCUS ON The Real World

FINDING PRIME NUMBERS

Now that you have discovered a way to identify relatively small prime numbers with the Sieve of Eratosthenes, perhaps you are wondering whether there are very large primes. The answer is yes. Another ancient Greek mathematician, Euclid, proved that there are an infinite number of primes. Thus, there do exist huge numbers that are prime. Researchers call prime numbers with more than 1000 digits *titanic primes*.

Knowing that very large prime numbers exist and finding them, and proving that they are in fact prime are two very different things. The Great Internet Mersenne Prime Search, or GIMPS, is a distributed computing project whose aim is finding large prime numbers. Founded in 1996, GIMPS now utilizes a network of over 21,500 personal computers that perform prime-locating arithmetic computations during each computer's "spare" computing time. These computers are owned by over twelve thousand different individuals, businesses, and schools around the world who donate computing time to the project. Together, this vast and varied array of individual personal computers form a virtual supercomputer capable of making 720 billion calculations per second.

The GIMPS project has been highly successful. Since its founding, it has been responsible for finding the four most recent largest-known prime numbers. At the time that this book was written, the largest-known prime number is $2^{6,972,593} - 1$. It was discovered by GIMPS participant Nayan Hajratwala of Plymouth, Michigan, on June 1, 1999. This record prime has 2,098,960 digits! It is in a special class of prime numbers called the Mersenne primes, named after a 17th-century French monk and mathematician, Father Marin Mersenne. A Mersenne prime is a prime number of the form $2^p - 1$, where p is a positive integer. For instance, $2^2 - 1 = 3, 2^3 - 1 = 7$, and $2^5 - 1 = 31$ are all Mersenne primes. Hajratwala's record prime is only the 38th known Mersenne prime.

GROUP ACTIVITIES

1. The following World Wide Web site contains information on the current status of the largest-known prime numbers: http://www.utm.edu/research/primes/largest.html. Visit this site and report on the five largest-known prime numbers. Be sure to include information about the numbers themselves (their form, whether they are Mersenne primes, how many digits, and so on), who found them, when they were found, and how they were found (if possible). Have any primes larger than Hajratwala's record prime in 1999 been found? If so, how many?
2. Research and report on the uses of prime numbers. Explain why there is an ongoing search for the largest-known prime number.

STUDY SKILLS REMINDER

Are you satisfied with your performance in this course thus far?

If not, ask yourself the following questions:

- Am I attending all class periods and arriving on time?
- Am I working and checking my homework assignments?
- Am I getting help when I need it?
- In addition to my instructor, am I using the supplements to this text that could help me? For example, the tutorial video lessons? MathPro, the tutorial software?
- Am I satisfied with my performance on quizzes and tests?

If you answered no to *any* of these questions, read or reread Section 1.1 for suggestions in these areas. Also, you may want to contact your instructor for additional feedback.

Name ______________________ Section __________ Date ______________

Mental Math

Solve each equation by inspection.

1. $(a - 3)(a - 7) = 0$
2. $(a - 5)(a - 2) = 0$
3. $(x + 8)(x + 6) = 0$
4. $(x + 2)(x + 3) = 0$
5. $(x + 1)(x - 3) = 0$
6. $(x - 1)(x + 2) = 0$

EXERCISE SET 5.6

A *Solve each equation. See Examples 1 through 3.*

1. $(x - 2)(x + 1) = 0$
2. $(x + 3)(x + 2) = 0$
3. $(x - 6)(x - 7) = 0$
4. $(x + 4)(x - 10) = 0$
5. $(x + 9)(x + 17) = 0$
6. $(x - 11)(x - 1) = 0$
7. $x(x + 6) = 0$
8. $x(x - 7) = 0$
9. $3x(x - 8) = 0$
10. $2x(x + 12) = 0$
11. $(2x + 3)(4x - 5) = 0$
12. $(3x - 2)(5x + 1) = 0$
13. $(2x - 7)(7x + 2) = 0$
14. $(9x + 1)(4x - 3) = 0$
15. $\left(x - \frac{1}{2}\right)\left(x + \frac{1}{3}\right) = 0$
16. $\left(x + \frac{2}{9}\right)\left(x - \frac{1}{4}\right) = 0$
17. $(x + 0.2)(x + 1.5) = 0$
18. $(x + 1.7)(x + 2.3) = 0$
19. Write a quadratic equation that has two solutions, 6 and −1. Leave the polynomial in the equation in factored form.
20. Write a quadratic equation that has two solutions, 0 and −2. Leave the polynomial in the equation in factored form.

A **B** *Solve each equation. See Examples 4 through 8.*

21. $x^2 - 13x + 36 = 0$
22. $x^2 + 2x - 63 = 0$
23. $x^2 + 2x - 8 = 0$
24. $x^2 - 5x + 6 = 0$
25. $x^2 - 7x = 0$
26. $x^2 - 3x = 0$
27. $x^2 + 20x = 0$
28. $x^2 + 15x = 0$
29. $x^2 = 16$
30. $x^2 = 9$
31. $x^2 - 4x = 32$
32. $x^2 - 5x = 24$

33. $x(3x - 1) = 14$

34. $x(4x - 11) = 3$

35. $3x^2 + 19x - 72 = 0$

36. $36x^2 + x - 21 = 0$

37. $4x^3 - x = 0$

38. $4y^3 - 36y = 0$

39. $4(x - 7) = 6$

40. $5(3 - 4x) = 9$

41. $(4x - 3)(16x^2 - 24x + 9) = 0$

42. $(2x + 5)(4x^2 - 10x + 25) = 0$

43. $4y^2 - 1 = 0$

44. $4y^2 - 81 = 0$

45. $(2x + 3)(2x^2 - 5x - 3) = 0$

46. $(2x - 9)(x^2 + 5x - 36) = 0$

47. $x^2 - 15 = -2x$

48. $x^2 - 26 = -11x$

49. $5x^3 - 6x - 8 = 0$

50. $12x^2 + 7x - 12 = 0$

51. $30x^2 - 11x = 30$

52. $9x^2 + 6x = -2$

53. $6y^2 - 22y - 40 = 0$

54. $3x^2 - 6x - 9 = 0$

55. $(y - 2)(y + 3) = 6$

56. $(y - 5)(y - 2) = 28$

57. $x^3 - 12x^2 + 32x = 0$

58. $x^3 - 14x^2 + 49x = 0$

59. Write a quadratic equation in standard form that has two solutions, 5 and 7.

60. Write an equation that has three solutions, 0, 1, and 2.

Review and Preview

Perform each indicated operation. Write all results in lowest terms. See Section R.2.

61. $\frac{3}{5} + \frac{4}{9}$

62. $\frac{2}{3} + \frac{3}{7}$

63. $\frac{7}{10} - \frac{5}{12}$

64. $\frac{5}{9} - \frac{5}{12}$

65. $\frac{4}{5} \cdot \frac{7}{8}$

66. $\frac{3}{7} \cdot \frac{12}{17}$

Combining Concepts

67. Explain the error and solve correctly:

$$x(x - 2) = 8$$
$$x = 8 \quad \text{or} \quad x - 2 = 8$$
$$x = 10$$

68. Explain the error and solve correctly:

$$(x - 4)(x + 2) = 0$$
$$x = -4 \quad \text{or} \quad x = 2$$

69. A compass is accidentally thrown upward and out of an air balloon at a height of 300 feet. The height, y, of the compass at time x is given by the equation $y = -16x^2 + 20x + 300$.

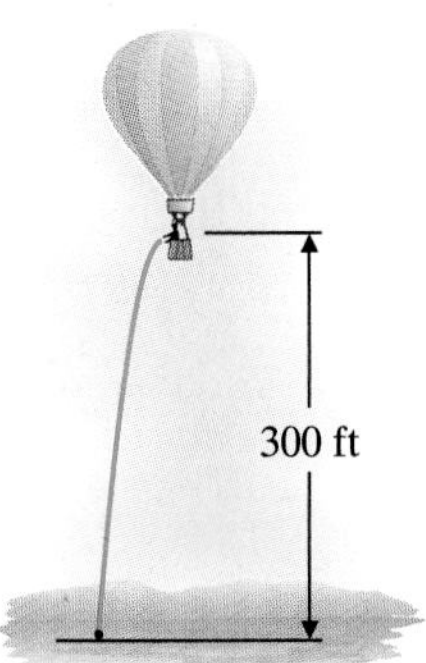

a. Find the height of the compass at the given times by filling in the table below.

Time, x (in seconds)	0	1	2	3	4	5	6
Height, y (in feet)							

b. Use the table to determine when the compass strikes the ground.

c. Use the table to approximate the maximum height of the compass.

70. A rocket is fired upward from the ground with an initial velocity of 100 feet per second. The height, y, of the rocket at any time x is given by the equation $y = -16x^2 + 100x$.

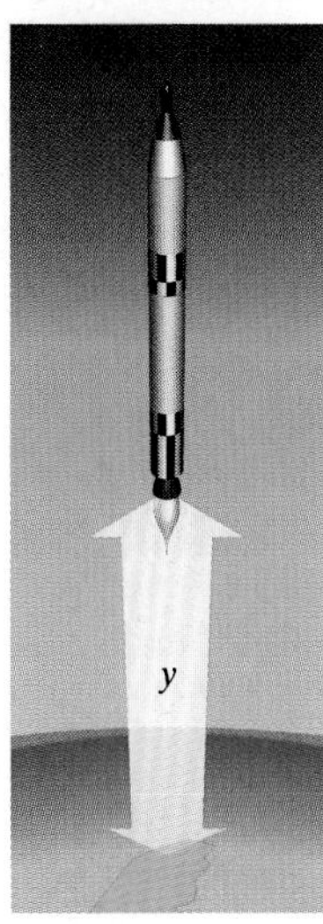

a. Find the height of the rocket at the given times by filling in the table below.

Time, x (in seconds)	0	1	2	3	4	5	6	7
Height, y (in feet)								

b. Use the equation to approximate when the rocket strikes the ground to the nearest tenth of a second.

c. Use the table to approximate the maximum height of the rocket.

Solve each equation.

71. $(x - 3)(3x + 4) = (x + 2)(x - 6)$

72. $(2x - 3)(x + 6) = (x - 9)(x + 2)$

73. $(2x - 3)(x + 8) = (x - 6)(x + 4)$

74. $(x + 6)(x - 6) = (2x - 9)(x + 4)$

FOCUS ON Mathematical Connections

Geometry

Factoring polynomials can be visualized using areas of rectangles. To see this, let's first find the areas of the following squares and rectangles. (Recall that Area = Length · Width.)

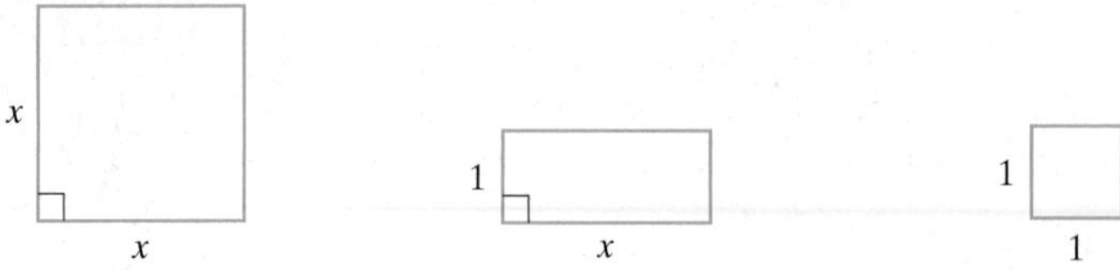

To use these areas to visualize factoring the polynomial $x^2 + 3x + 2$, for example, use the shapes below to form a rectangle. The factored form is found by reading the length and the width of the rectangle as shown below.

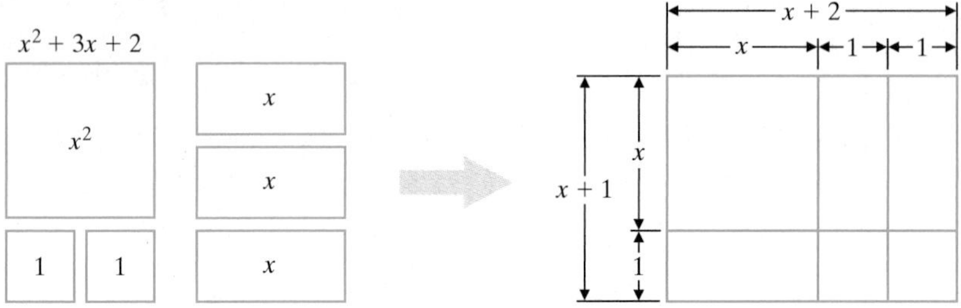

Thus, $x^2 + 3x + 2 = (x + 2)(x + 1)$.
Try using this method to visualize the factored form of each polynomial below.

Group Activity

Work in a group and use tiles to find the factored form of the polynomials below. (Tiles can be hand made from index cards.)

1. $x^2 + 6x + 5$
2. $x^2 + 5x + 6$
3. $x^2 + 5x + 4$
4. $x^2 + 4x + 3$
5. $x^2 + 6x + 9$
6. $x^2 + 4x + 4$

5.7 Quadratic Equations and Problem Solving

OBJECTIVE

Ⓐ Solve problems that can be modeled by quadratic equations.

Ⓐ Solving Problems Modeled by Quadratic Equations

Some problems may be modeled by quadratic equations. To solve these problems, we use the same problem-solving steps that were introduced in Section 2.5. When solving these problems, keep in mind that a solution of an equation that models a problem may not be a solution to the problem. For example, a person's age or the length of a rectangle is always a positive number. Thus we discard solutions that do not make sense as solutions of the problem.

EXAMPLE 1 Finding Free-Fall Time

For a TV commercial, a piece of luggage is dropped from a cliff 256 feet above the ground to show the durability of the luggage. Neglecting air resistance, the height h in feet of the luggage above the ground after t seconds is given by the quadratic equation

$$h = -16t^2 + 256$$

Find how long it takes for the luggage to hit the ground.

Solution:

1. UNDERSTAND. Read and reread the problem. Then draw a picture of the problem.

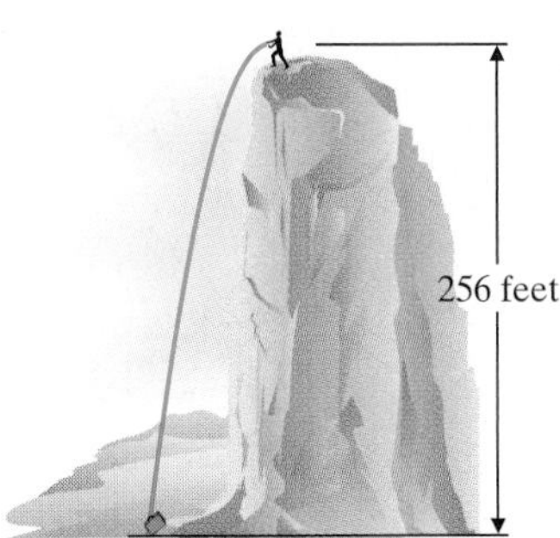

The equation $h = -16t^2 + 256$ models the height of the falling luggage at time t. Familiarize yourself with this equation by finding the height of the luggage at $t = 1$ second and $t = 2$ seconds.

When $t = 1$ second, the height of the suitcase is

$h = -16(1)^2 + 256 = 240$ feet.

When $t = 2$ seconds, the height of the suitcase is

$h = -16(2)^2 + 256 = 192$ feet.

2. TRANSLATE. To find how long it takes the luggage to hit the ground, we want to know the value of t for which the height $h = 0$.

$$0 = -16t^2 + 256$$

3. SOLVE. We solve the quadratic equation by factoring.

$$0 = -16t^2 + 256$$
$$0 = -16(t^2 - 16)$$
$$0 = -16(t - 4)(t + 4)$$
$$t - 4 = 0 \quad \text{or} \quad t + 4 = 0$$
$$t = 4 \qquad\qquad t = -4$$

4. INTERPRET. Since the time t cannot be negative, the proposed solution is 4 seconds.

Check: Verify that the height of the luggage when t is 4 seconds is 0.

When $t = 4$ seconds, $h = -16(4)^2 + 256 = -256 + 256 = 0$ feet.

Practice Problem 1

An object is dropped from the roof of a 144-foot-tall building. Neglecting air resistance, the height h in feet of the object above ground after t seconds is given by the quadratic equation

$$h = -16t^2 + 144$$

Find how long it takes the object to hit the ground.

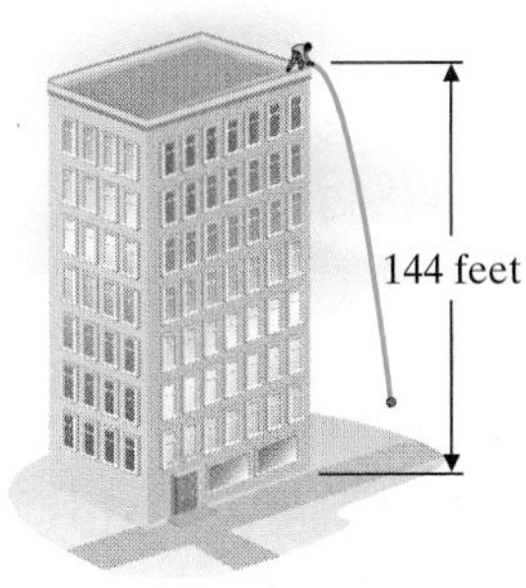

Answer

1. 3 seconds

State: The solution checks and the luggage hits the ground 4 seconds after it is dropped.

Practice Problem 2

The square of a number minus twice the number is 63. Find the number.

EXAMPLE 2 Finding a Number

The square of a number plus three times the number is 70. Find the number.

Solution:

1. UNDERSTAND. Read and reread the problem. Suppose that the number is 5. The square of 5 is 5^2 or 25. Three times 5 is 15. Then $25 + 15 = 40$, not 70, so the number must be greater than 5. Remember, the purpose of proposing a number, such as 5, is to better understand the problem. Now that we do, we will let $x =$ the number.
2. TRANSLATE.

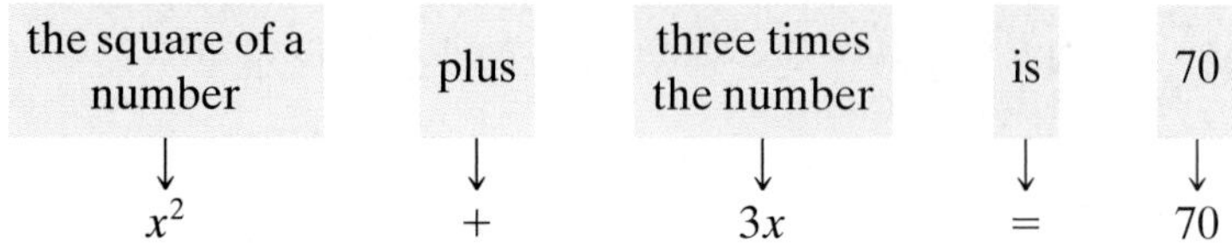

$$x^2 \quad + \quad 3x \quad = \quad 70$$

3. SOLVE.

$$\begin{aligned} x^2 + 3x &= 70 \\ x^2 + 3x - 70 &= 0 && \text{Subtract 70 from both sides.} \\ (x + 10)(x - 7) &= 0 && \text{Factor.} \\ x + 10 = 0 \quad &\text{or} \quad x - 7 = 0 && \text{Set each factor equal to 0.} \\ x = -10 \quad & \qquad\quad x = 7 && \text{Solve.} \end{aligned}$$

4. INTERPRET.

Check: The square of -10 is $(-10)^2$, or 100. Three times -10 is $3(-10)$ or -30. Then $100 + (-30) = 70$, the correct sum, so -10 checks.

The square of 7 is 7^2 or 49. Three times 7 is $3(7)$, or 21. Then $49 + 21 = 70$, the correct sum, so 7 checks.

State: There are two numbers. They are -10 and 7.

Practice Problem 3

The length of a rectangle is 5 feet more than its width. The area of the rectangle is 176 square feet. Find the length and the width of the rectangle.

EXAMPLE 3 Finding the Dimensions of a Sail

The height of a triangular sail is 2 meters less than twice the length of the base. If the sail has an area of 30 square meters, find the length of its base and the height.

Solution:

1. UNDERSTAND. Read and reread the problem. Since we are finding the length of the base and the height, we let

   $x =$ the length of the base

   Since the height is 2 meters less than twice the base,

   $2x - 2 =$ the height

   An illustration is shown in the margin on the next page.
2. TRANSLATE. We are given that the area of the triangle is 30 square meters, so we use the formula for area of a triangle.

area of triangle	=	$\frac{1}{2}$	·	base	·	height
↓		↓		↓		↓
30	=	$\frac{1}{2}$	·	x	·	$(2x - 2)$

Answers

2. 9 or -7, **3.** length $= 16$ ft; width $= 11$ ft

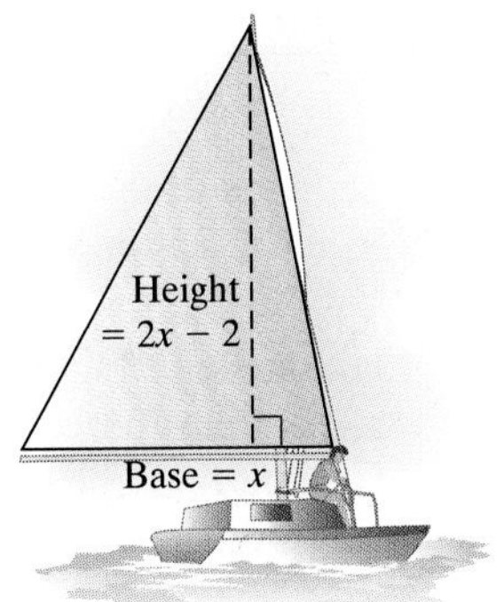

3. SOLVE. Now we solve the quadratic equation.

$$30 = \frac{1}{2}x(2x - 2)$$

$$30 = x^2 - x \qquad \text{Multiply.}$$

$$x^2 - x - 30 = 0 \qquad \text{Write in standard form.}$$

$$(x - 6)(x + 5) = 0 \qquad \text{Factor.}$$

$$x - 6 = 0 \quad \text{or} \quad x + 5 = 0 \qquad \text{Set each factor equal to 0.}$$

$$x = 6 \qquad\qquad x = -5$$

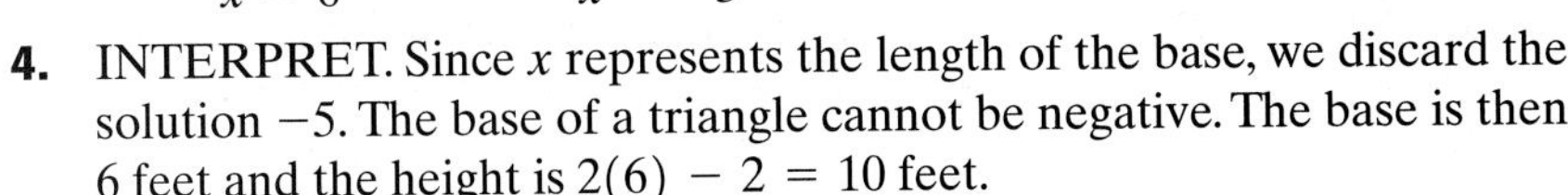

4. INTERPRET. Since x represents the length of the base, we discard the solution -5. The base of a triangle cannot be negative. The base is then 6 feet and the height is $2(6) - 2 = 10$ feet.

Check: To check this problem, we recall that $\frac{1}{2}\text{base} \cdot \text{height} = \text{area}$, or

$$\frac{1}{2}(6)(10) = 30 \qquad \text{The required area}$$

State: The base of the triangular sail is 6 meters and the height is 10 meters. ●

The next example makes use of the **Pythagorean theorem** and consecutive integers. Before we review this theorem, recall that a **right triangle** is a triangle that contains a 90° or right angle. The **hypotenuse** of a right triangle is the side opposite the right angle and is the longest side of the triangle. The **legs** of a right triangle are the other sides of the triangle.

Pythagorean Theorem

In a right triangle, the sum of the squares of the lengths of the two legs is equal to the square of the length of the hypotenuse.

$$(\text{leg})^2 + (\text{leg})^2 = (\text{hypotenuse})^2 \quad \text{or} \quad a^2 + b^2 = c^2$$

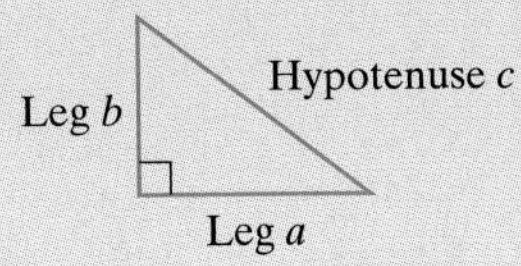

Study the following diagrams for a review of consecutive integers.

Examples

If x is the first integer, then consecutive integers are $x, x + 1, x + 2, \ldots$

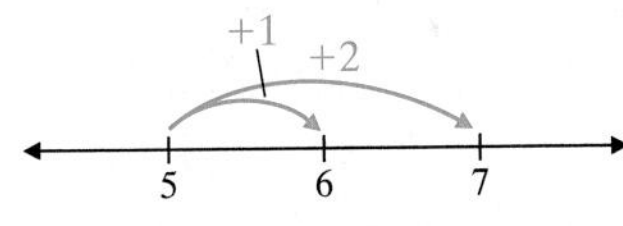

If x is the first even integer, then consecutive even integers are $x, x + 2, x + 4, \ldots$

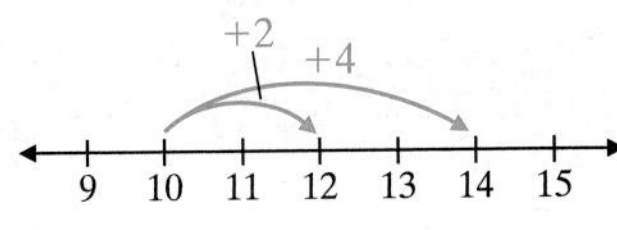

If x is the first odd integer, then consecutive odd integers are $x, x + 2, x + 4, \ldots$

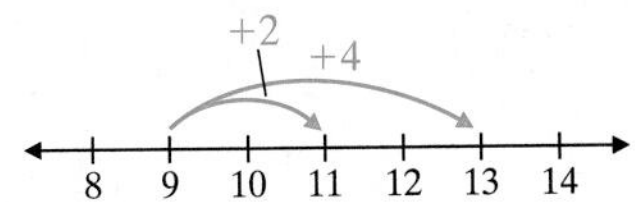

Practice Problem 4

Solve.

a. Find two consecutive odd integers whose product is 23 more than their sum.

b. The length of one leg of a right triangle is 7 meters less than the length of the other leg. The length of the hypotenuse is 13 meters. Find the lengths of the legs.

Answers

4. a. 5 and 7 or -5 and -3, **b.** 5 meters, 12 meters

EXAMPLE 4 Finding the Dimensions of a Triangle

Find the lengths of the sides of a right triangle if the lengths can be expressed as three consecutive even integers.

Solution:

1. UNDERSTAND. Read and reread the problem. Let's suppose that the length of one leg of the right triangle is 4 units. Then the other leg is the next even integer, or 6 units, and the hypotenuse of the triangle is the next even integer, or 8 units. Remember that the hypotenuse is the longest side. Let's see if a triangle with sides of these lengths forms a right triangle. To do this, we check to see whether the Pythagorean theorem holds true.

$$4^2 + 6^2 \stackrel{?}{=} 8^2$$
$$16 + 36 \stackrel{?}{=} 64$$
$$52 = 64 \quad \text{False.}$$

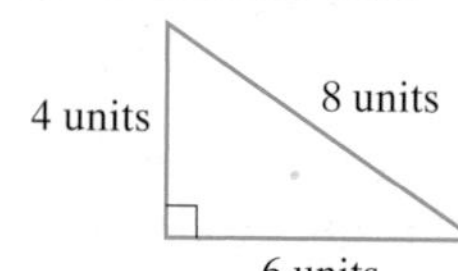

Our proposed numbers do not check, but we now have a better understanding of the problem.

We let $x, x + 2$, and $x + 4$ be three consecutive even integers. Since these integers represent lengths of the sides of a right triangle, we have the following.

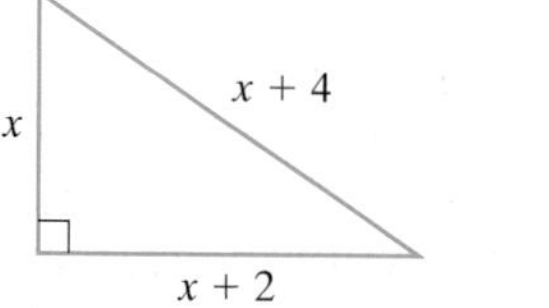

x = one leg
$x + 2$ = other leg
$x + 4$ = hypotenuse (longest side)

2. TRANSLATE. By the Pythagorean theorem, we have that

$$(\text{hypotenuse})^2 = (\text{leg})^2 + (\text{leg})^2$$
$$(x + 4)^2 = (x)^2 + (x + 2)^2$$

3. SOLVE. Now we solve the equation.

$$(x + 4)^2 = x^2 + (x + 2)^2$$
$$x^2 + 8x + 16 = x^2 + x^2 + 4x + 4 \quad \text{Multiply.}$$
$$x^2 + 8x + 16 = 2x^2 + 4x + 4 \quad \text{Combine like terms.}$$
$$x^2 - 4x - 12 = 0 \quad \text{Write in standard form.}$$
$$(x - 6)(x + 2) = 0 \quad \text{Factor.}$$
$$x - 6 = 0 \quad \text{or} \quad x + 2 = 0 \quad \text{Set each factor equal to 0.}$$
$$x = 6 \qquad x = -2$$

4. INTERPRET. We discard $x = -2$ since length cannot be negative. If $x = 6$, then $x + 2 = 8$ and $x + 4 = 10$.

Check: Verify that

$$(\text{hypotenuse})^2 = (\text{leg})^2 + (\text{leg})^2$$
$$10^2 \stackrel{?}{=} 6^2 + 8^2$$
$$100 \stackrel{?}{=} 36 + 64$$
$$100 = 100 \quad \text{True}$$

State: The sides of the right triangle have lengths 6 units, 8 units, and 10 units.

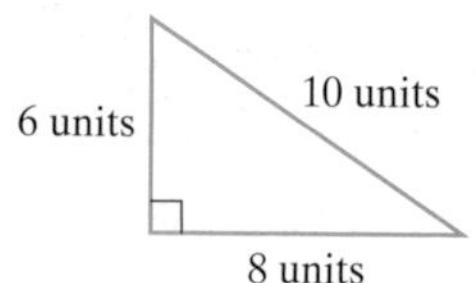

Name ______________________ Section __________ Date ____________

EXERCISE SET 5.7

A *See Examples 1 through 4 for all exercises. For Exercises 1 through 6, represent each given condition using a single variable, x.*

△ **1.** The length and width of a rectangle whose length is 4 centimeters more than its width

△ **2.** The length and width of a rectangle whose length is twice its width

3. Two consecutive odd integers

4. Two consecutive even integers

△ **5.** The base and height of a triangle whose height is one more than four times its base

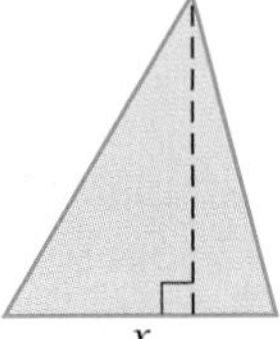

△ **6.** The base and height of a trapezoid whose base is three less than five times its height

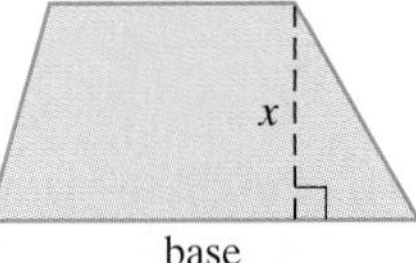

Use the information given to find the dimensions of each figure.

△ **7.**

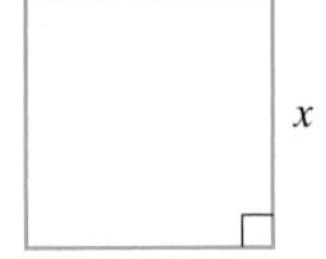

The *area* of the square is 121 square units. Find the length of its sides.

△ **8.**

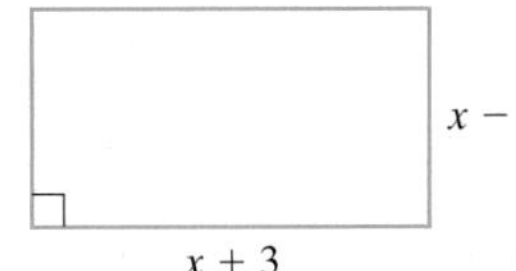

The *area* of the rectangle is 84 square inches. Find its length and width.

△ **9.**

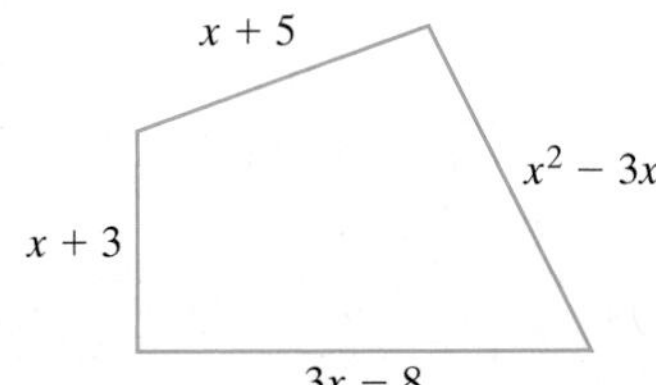

The *perimeter* of the quadrilateral is 120 centimeters. Find the lengths of the sides.

△ **10.**

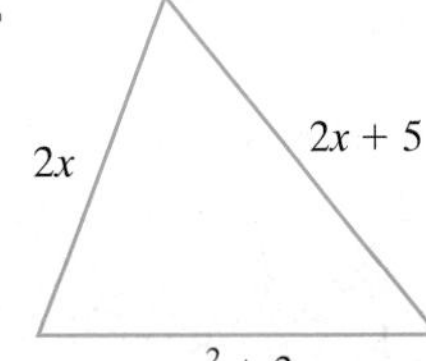

The *perimeter* of the triangle is 85 feet. Find the lengths of its sides.

△ **11.**

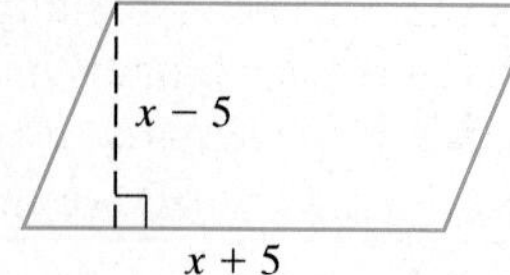

The *area* of the parallelogram is 96 square miles. Find its base and height.

△ **12.**

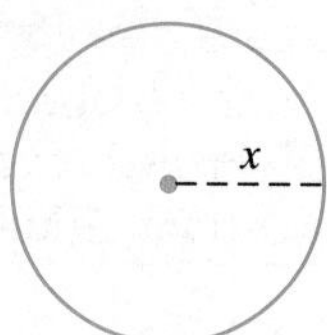

The *area* of the circle is 25π square kilometers. Find its radius.

Solve.

13. An object is thrown upward from the top of an 80-foot building with an initial velocity of 64 feet per second. The height h of the object after t seconds is given by the quadratic equation $h = -16t^2 + 64t + 80$. When will the object hit the ground?

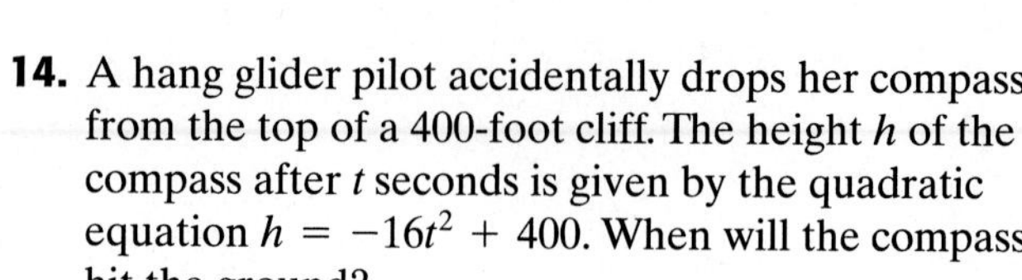

14. A hang glider pilot accidentally drops her compass from the top of a 400-foot cliff. The height h of the compass after t seconds is given by the quadratic equation $h = -16t^2 + 400$. When will the compass hit the ground?

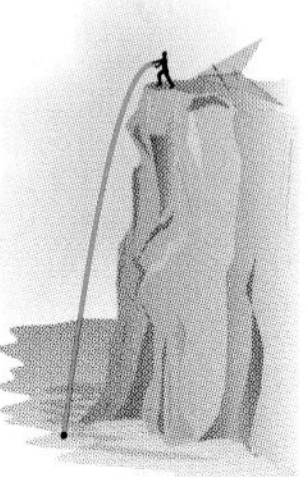

15. The length of a rectangle is 7 centimeters less than twice its width. Its area is 30 square centimeters. Find the dimensions of the rectangle.

16. The length of a rectangle is 9 inches more than its width. Its area is 112 square inches. Find the dimensions of the rectangle.

△ *The equation* $D = \frac{1}{2}n(n - 3)$ *gives the number of diagonals D for a polygon with n sides. For example, a polygon with 6 sides has* $D = \frac{1}{2} \cdot 6(6 - 3)$ *or* $D = 9$ *diagonals. (See if you can count all 9 diagonals. Some are shown in the figure.) Use this equation,* $D = \frac{1}{2}n(n - 3)$, *for Exercises 17 through 20.*

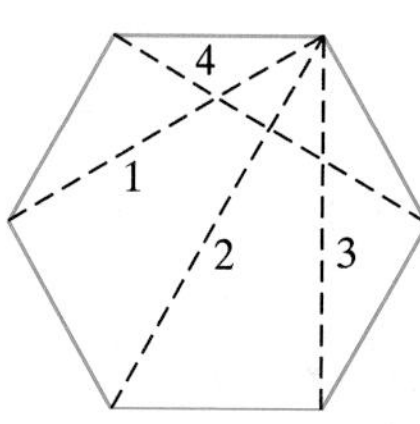

17. Find the number of diagonals for a polygon that has 12 sides.

18. Find the number of diagonals for a polygon that has 15 sides.

19. Find the number of sides n for a polygon that has 35 diagonals.

20. Find the number of sides n for a polygon that has 14 diagonals.

Solve.

21. The sum of a number and its square is 132. Find the number.

22. The sum of a number and its square is 182. Find the number.

△ **23.** Two boats travel at a right angle to each other after leaving the same dock at the same time. One hour later the boats are 17 miles apart. If one boat travels 7 miles per hour faster than the other boat, find the rate of each boat.

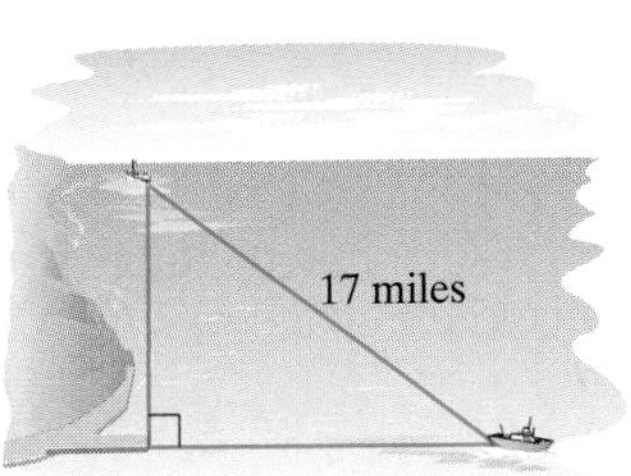

△ **24.** The side of a square equals the width of a rectangle. The length of the rectangle is 6 meters longer than its width. The sum of the areas of the square and the rectangle is 176 square meters. Find the side of the square.

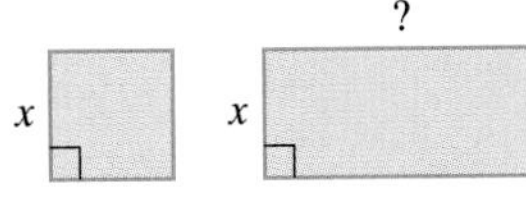

25. The sum of two numbers is 20, and the sum of their squares is 218. Find the numbers.

26. The sum of two numbers is 25, and the sum of their squares is 325. Find the numbers.

△ **27.** If the sides of a square are increased by 3 inches, the area becomes 64 square inches. Find the length of the sides of the original square.

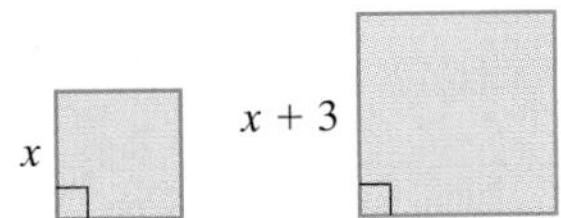

△ **28.** If the sides of a square are increased by 5 meters, the area becomes 100 square meters. Find the length of the sides of the original square.

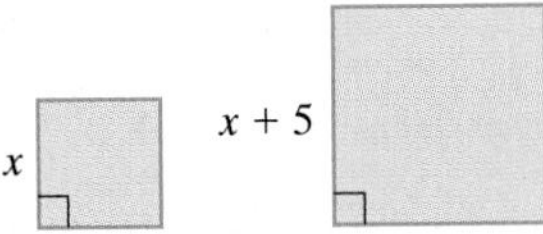

△ **29.** One leg of a right triangle is 4 millimeters longer than the smaller leg and the hypotenuse is 8 millimeters longer than the smaller leg. Find the lengths of the sides of the triangle.

△ **30.** One leg of a right triangle is 9 centimeters longer than the other leg and the hypotenuse is 45 centimeters. Find the lengths of the legs of the triangle.

△ **31.** The length of the base of a triangle is twice its height. If the area of the triangle is 100 square kilometers, find the height.

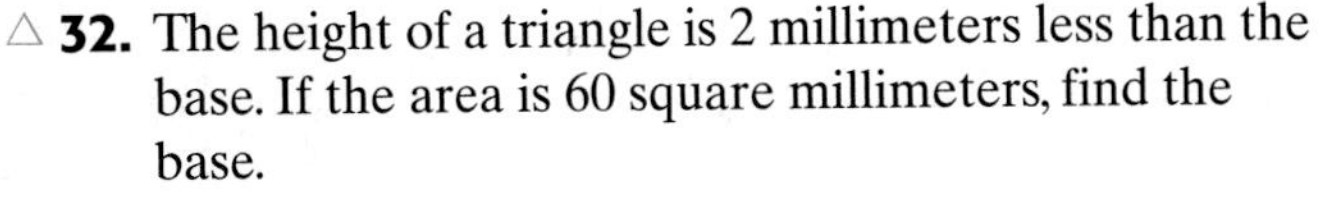

△ **32.** The height of a triangle is 2 millimeters less than the base. If the area is 60 square millimeters, find the base.

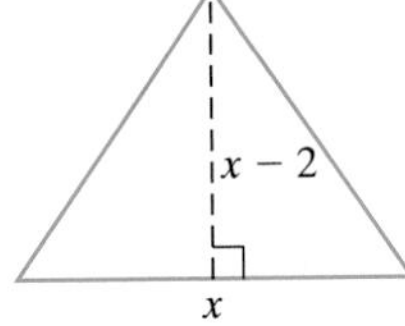

△ **33.** Find the length of the shorter leg of a right triangle if the longer leg is 12 feet more than the shorter leg and the hypotenuse is 12 feet less than twice the shorter leg.

△ **34.** Find the length of the shorter leg of a right triangle if the longer leg is 10 miles more than the shorter leg and the hypotenuse is 10 miles less than twice the shorter leg.

35. An object is dropped from 39 feet below the tip of the pinnacle atop one of the 1483-foot-tall Petronas Twin Towers in Kuala Lumpor, Malaysia. (*Source:* Council on Tall Buildings and Urban Habitat) The height h of the object after t seconds is given by the equation $h = -16t^2 + 1444$. Find how many seconds pass before the object reaches the ground.

36. An object is dropped from the top of 311 South Wacker Drive, a 961-foot-tall office building in Chicago. (*Source:* Council on Tall Buildings and Urban Habitat) The height h of the object after t seconds is given by the equation $h = -16t^2 + 961$. Find how many seconds pass before the object reaches the ground.

37. At the end of 2 years, P dollars invested at an interest rate r compounded annually increases to an amount, A dollars, given by

$$A = P(1 + r)^2$$

Find the interest rate if \$100 increased to \$144 in 2 years.

38. At the end of 2 years, P dollars invested at an interest rate r compounded annually increases to an amount, A dollars, given by

$$A = P(1 + r)^2$$

Find the interest rate if \$2000 increased to \$2420 in 2 years.

△ **39.** Find the dimensions of a rectangle whose width is 7 miles less than its length and whose area is 120 square miles.

△ **40.** Find the dimensions of a rectangle whose width is 2 inches less than half its length and whose area is 160 square inches.

41. If the cost, C, for manufacturing x units of a certain product is given by $C = x^2 - 15x + 50$, find the number of units manufactured at a cost of \$9500.

42. If a switchboard handles n telephones, the number C of telephone connections it can make simultaneously is given by the equation $C = \frac{n(n-1)}{2}$. Find how many telephones are handled by a switchboard making 120 telephone connections simultaneously.

Review and Preview

The following double line graph shows a comparison of the amount of land (in thousand acres) operated by farms in Alabama during the years shown with the amount of land operated by farms in North Carolina. Use this graph to answer Exercises 43 through 49. See Section 3.1.

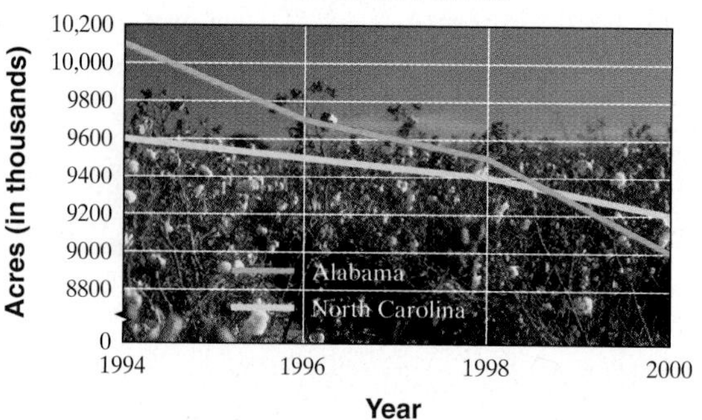

Source: U.S. Department of Agriculture, National Agricultural Statistics Service

43. Approximate the amount of land operated by farms in North Carolina in 1994.

44. Approximate the amount of land operated by farms in North Carolina in 1996.

45. Approximate the amount of land operated by farms in Alabama in 1998.

46. Approximate the amount of land operated by farms in Alabama in 2000.

47. Approximate the year that the colored lines in this graph intersect.

48. In your own words, explain the meaning of the point of intersection in the graph.

49. Describe the trends shown in this graph and speculate as to why these trends have occurred.

Combining Concepts

△ **50.** According to the International America's Cup Class (IACC) rule, a sailboat competing in the America's Cup match must have a 110-foot-tall mast and a combined mainsail and jib sail area of 3000 square feet. (*Source:* America's Cup Organizing Committee) A design for an IACC-class sailboat calls for the mainsail to be 60% of the combined sail area. If the height of the triangular mainsail is 28 feet more than twice the length of the boom, find the length of the boom and the height of the mainsail.

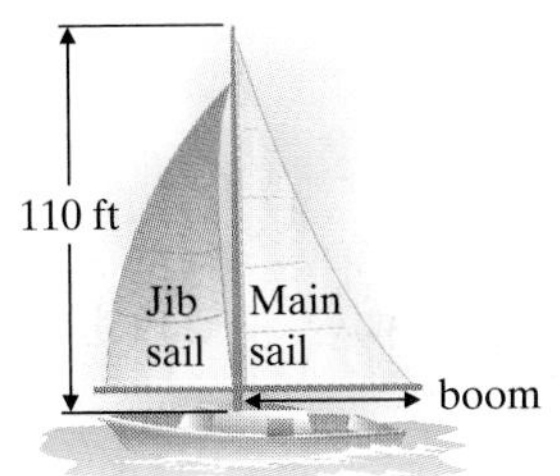

△ **51.** A rectangular pool is surrounded by a walk 4 meters wide. The pool is 6 meters longer than its width. If the total area is 576 square meters more than the area of the pool, find the dimensions of the pool.

△ **52.** A rectangular garden is surrounded by a walk of uniform width. The area of the garden is 180 square yards. If the dimensions of the garden plus the walk are 16 yards by 24 yards, find the width of the walk.

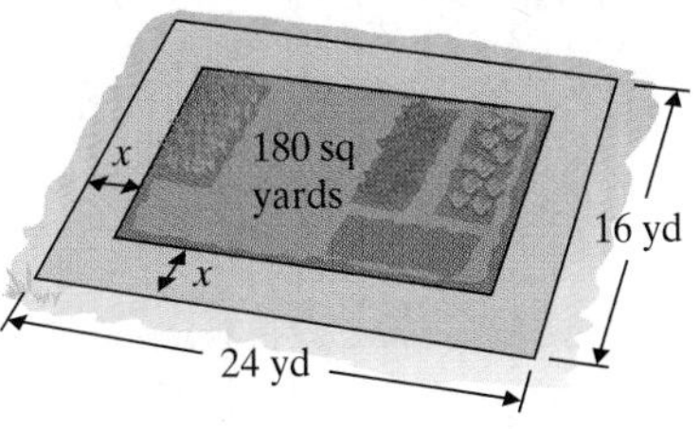

Internet Excursions

Go To: http://www.prenhall.com/martin-gay_algebra

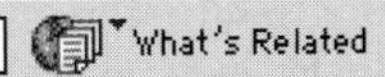

The United States Postal Service uses five-digit ZIP codes to simplify mail distribution. Each ZIP code corresponds to a unique post office location. If you know a ZIP code, you can look up the city and state associated with it by visiting the above World Wide Web address where you will gain access to the United States Postal Service's City/State/ZIP Code Associations Web site, or a related site.

53. The first three digits in the ZIP codes of two cities are 704. The remaining two-digit numbers in each ZIP code have a sum of 54, and the sum of their squares is 1556. What are the two ZIP codes? Use the given link to look up the cities and states associated with these ZIP codes.

54. The first three digits in the ZIP codes of two cities are 347. The remaining two-digit numbers in each ZIP code have a sum of 40, and the sum of their squares is 962. What are the two ZIP codes? Use the given link to look up the cities and states associated with these ZIP codes.

CHAPTER 5 ACTIVITY Factoring

Choosing Among Building Options: Shaunesa has just had a 10-foot-by-15-foot, in-ground swimming pool installed in her backyard. She has $3000 left from the building project that she would like to spend on surrounding the pool with a patio of constant width (see the figure). She has talked to several local suppliers about options for building this patio and must choose among the following.

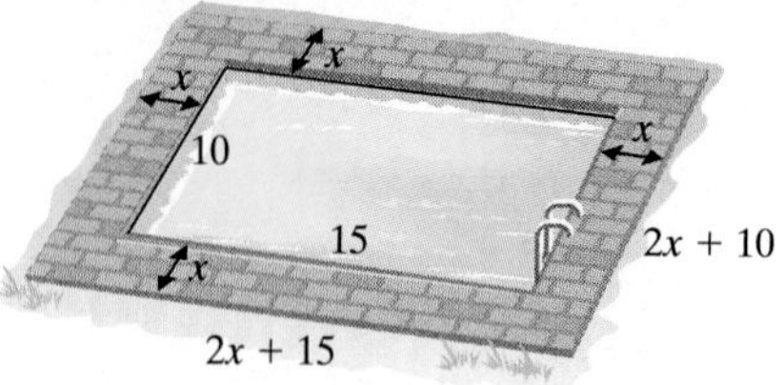

Option	Material	Price
A	Poured cement	$5 per square foot
B	Brick	$7.50 per square foot plus a $30 flat fee for delivering the bricks
C	Outdoor carpeting	$4.50 per square foot plus $10.86 per foot of the pool's perimeter to install an edging

1. Find the area of the swimming pool.
2. Write an algebraic expression for the total area of the region containing both the pool and patio.
3. Use subtraction to find an algebraic expression for the area of just the patio (not including the pool).
4. Find the perimeter of the swimming pool alone.
5. For each patio material option, write an algebraic expression for the total cost of installing the patio based on its area and the given price information.
6. If Shaunesa plans to spend the entire $3000 she has saved for the patio, how wide can the patio in option A be?
7. If Shaunesa plans to spend the entire $3000 she has saved for the patio, how wide can the patio in option B be?
8. If Shaunesa plans to spend the entire $3000 she has saved for the patio, how wide can the patio in option C be?
9. Which option should Shaunesa choose? Why? Discuss the pros and cons of each option.

Chapter 5 VOCABULARY CHECK

Fill in each blank with one of the words or phrases listed below.

factoring   quadratic equation   perfect square trinomial
greatest common factor

1. An equation that can be written in the form $ax^2 + bx + c = 0$ (with a not 0) is called a ________________.
2. ____________ is the process of writing an expression as a product.
3. The ______________________ of a list of terms is the product of all common factors.
4. A trinomial that is the square of some binomial is called a ____________________.

CHAPTER 5 Highlights

DEFINITIONS AND CONCEPTS	EXAMPLES
Section 5.1 The Greatest Common Factor	
Factoring is the process of writing an expression as a product.	Factor: $6 = 2 \cdot 3$ Factor: $x^2 + 5x + 6 = (x + 2)(x + 3)$
The GCF of a list of variable terms contains the smallest exponent on each common variable.	The GCF of z^5, z^3, and z^{10} is z^3.
The GCF of a list of terms is the product of all common factors.	Find the GCF of $8x^2y$, $10x^3y^2$, and $50x^2y^3$. $8x^2y = 2 \cdot 2 \cdot 2 \cdot x^2 \cdot y$ $10x^3y^2 = 2 \cdot 5 \cdot x^3 \cdot y^2$ $50x^2y^3 = 2 \cdot 5 \cdot 5 \cdot x^2 \cdot y^3$ $\text{GCF} = 2 \cdot x^2 \cdot y$ or $2x^2y$
TO FACTOR BY GROUPING	Factor: $10ax + 15a - 6xy - 9y$
Step 1. Group the terms into two groups of two terms.	**Step 1.** $(10ax + 15a) + (-6xy - 9y)$
Step 2. Factor out the GCF from each group.	**Step 2.** $5a(2x + 3) - 3y(2x + 3)$
Step 3. If there is a common binomial factor, factor it out.	**Step 3.** $(2x + 3)(5a - 3y)$
Step 4. If not, rearrange the terms and try Steps 1–3 again.	
Section 5.2 Factoring Trinomials of the Form $x^2 + bx + c$	
The sum of these numbers is b. $x^2 + bx + c = (x + \square)(x + \square)$ The product of these numbers is c.	Factor: $x^2 + 7x + 12$ $3 + 4 = 7$ $3 \cdot 4 = 12$ $x^2 + 7x + 12 = (x + 3)(x + 4)$
Section 5.3 Factoring Trinomials of the Form $ax^2 + bx + c$	
To factor $ax^2 + bx + c$, try various combinations of factors of ax^2 and c until a middle term of bx is obtained when checking.	Factor: $3x^2 + 14x - 5$ Factors of $3x^2$: $3x, x$ Factors of -5: $-1, 5$ and $1, -5$. $(3x - 1)(x + 5)$ $-1x$ $+ 15x$ $14x$ Correct middle term

Definitions and Concepts | Examples

Section 5.4 Factoring Trinomials of the Form $ax^2 + bx + c$ by Grouping

To Factor $ax^2 + bx + c$ by Grouping

Step 1. Find two numbers whose product is $a \cdot c$ and whose sum is b.

Step 2. Rewrite bx, using the factors found in Step 1.

Step 3. Factor by grouping.

Factor: $3x^2 + 14x - 5$

Step 1. Find two numbers whose product is $3 \cdot (-5)$ or -15 and whose sum is 14. They are 15 and -1.

Step 2. $3x^2 + 14x - 5$

$= 3x^2 + 15x - 1x - 5$

Step 3. $= 3x(x + 5) - 1(x + 5)$

$= (x + 5)(3x - 1)$

Section 5.5 Factoring by Special Products

Perfect Square Trinomial

$$a^2 + 2ab + b^2 = (a + b)^2$$
$$a^2 - 2ab + b^2 = (a - b)^2$$

Factor.

$$25x^2 + 30x + 9 = (5x + 3)^2$$
$$49z^2 - 28z + 4 = (7z - 2)^2$$

Difference of Two Squares

$$a^2 - b^2 = (a + b)(a - b)$$

Factor.

$$36x^2 - y^2 = (6x + y)(6x - y)$$

Sum and Difference of Two Cubes

$$a^3 + b^3 = (a + b)(a^2 - ab + b^2)$$
$$a^3 - b^3 = (a - b)(a^2 + ab + b^2)$$

Factor.

$$8y^3 + 1 = (2y + 1)(4y^2 - 2y + 1)$$
$$27p^3 - 64q^3 = (3p - 4q)(9p^2 + 12pq + 16q^2)$$

Section 5.6 Solving Quadratic Equations by Factoring

A **quadratic equation** is an equation that can be written in the form $ax^2 + bx + c = 0$ with a not 0.

The form $ax^2 + bx + c = 0$ is called the **standard form** of a quadratic equation.

Quadratic Equation	**Standard Form**
$x^2 = 16$	$x^2 - 16 = 0$
$y = -2y^2 + 5$	$2y^2 + y - 5 = 0$

Zero Factor Property

If a and b are real numbers and if $ab = 0$, then $a = 0$ or $b = 0$.

If $(x + 3)(x - 1) = 0$, then $x + 3 = 0$ or $x - 1 = 0$.

To Solve Quadratic Equations by Factoring

Step 1. Write the equation in standard form so that one side of the equation is 0.

Step 2. Factor completely.

Step 3. Set each factor containing a variable equal to 0.

Step 4. Solve the resulting equations.

Step 5. Check solutions in the original equation.

Solve: $3x^2 = 13x - 4$

Step 1. $3x^2 - 13x + 4 = 0$

Step 2. $(3x - 1)(x - 4) = 0$

Step 3. $3x - 1 = 0$ or $x - 4 = 0$

Step 4. $3x = 1$ $\quad$ $x = 4$

$x = \frac{1}{3}$

Step 5. Check both $\frac{1}{3}$ and 4 in the original equation.

DEFINITIONS AND CONCEPTS	**EXAMPLES**
Section 5.7 Quadratic Equations and Problem Solving	
PROBLEM-SOLVING STEPS	A garden is in the shape of a rectangle whose length is two feet more than its width. If the area of the garden is 35 square feet, find its dimensions.
1. UNDERSTAND the problem.	**1.** Read and reread the problem. Guess a solution and check your guess. Draw a diagram. Let x be the width of the rectangular garden. Then $x + 2$ is the length. 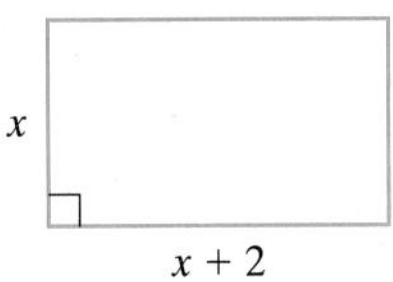
2. TRANSLATE.	**2.** length · width = area $(x + 2) \cdot x = 35$
3. SOLVE.	**3.** $(x + 2)x = 35$ $x^2 + 2x - 35 = 0$ $(x - 5)(x + 7) = 0$ $x - 5 = 0$ or $x + 7 = 0$ $x = 5$ $x = -7$
4. INTERPRET.	**4.** Discard the solution of -7 since x represents width. *Check:* If x is 5 feet, then $x + 2 = 5 + 2 = 7$ feet. The area of a rectangle whose width is 5 feet and whose length is 7 feet is (5 feet)(7 feet) or 35 square feet. *State:* The garden is 5 feet by 7 feet.

STUDY SKILLS REMINDER

Are you prepared for a test on Chapter 5?

Below is a list of some *common trouble areas* for students in Chapter 5. After studying for your test—but before taking your test—read these.

- Don't forget that the first step to factor any polynomial is to first factor out any common factors.

  $9x^2 - 36 = 9(x^2 - 4) = 9(x + 2)(x - 2)$

- Can you completely factor $x^4 - 24x^2 - 25$?

  $$x^4 - 24x^2 - 25 = (x^2 - 25)(x^2 + 1) = (x + 5)(x - 5)(x^2 + 1)$$

- Remember that to use the zero factor property to solve a quadratic equation, one side of the equation must be 0 and the other side must be a factored polynomial.

  $x(x - 2) = 3$ Cannot use zero factor property.

  $x^2 - 2x - 3 = 0$

  $(x - 3)(x + 1) = 0$ Now we can use zero factor property.

  $x - 3 = 0$ or $x + 1 = 0$

  $x = 3$ or $x = -1$

Remember: This is simply a sampling of selected topics given to check your understanding. For a review of Chapter 5 in your text, see the material at the end of this chapter.

FOCUS ON Mathematical Connections

Number Theory

By now, you have realized that being able to write a number as the product of prime numbers is very useful in the process of factoring polynomials. You probably also know at least a few numbers that are prime (such as 2, 3, and 5). But what about the other prime numbers? When we come across a number, how will we know if it is a prime number? Apparently, the ancient Greek mathematician Eratosthenes had a similar question because in the third century B.C. he devised a simple method for identifying primes. The method is called the Sieve of Eratosthenes because it "sifts out" the primes in a list of numbers. The Sieve of Eratosthenes is generally considered to be the most useful for identifying primes less than 1,000,000.

Here's how the sieve works: suppose you want to find the prime numbers in the first n natural numbers. Write the numbers, in order, from 2 to n. We know that 2 is prime, so circle it. Now, cross out each

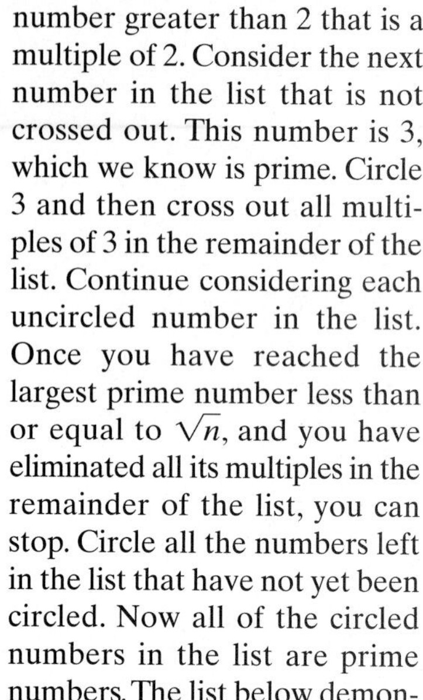

number greater than 2 that is a multiple of 2. Consider the next number in the list that is not crossed out. This number is 3, which we know is prime. Circle 3 and then cross out all multiples of 3 in the remainder of the list. Continue considering each uncircled number in the list. Once you have reached the largest prime number less than or equal to $\sqrt{n}$, and you have eliminated all its multiples in the remainder of the list, you can stop. Circle all the numbers left in the list that have not yet been circled. Now all of the circled numbers in the list are prime numbers. The list below demonstrates the Sieve of Eratosthenes on the numbers 2 through 30. Because $\sqrt{30} \approx 5.477$, we need only eck and eliminate the multiples of primes up to and including 5, which is the largest prime less than or equal to the square root of 30.

	(2)	(3)	~~4~~	(5)	~~6~~	(7)	~~8~~	~~9~~	~~10~~
(11)	~~12~~	(13)	~~14~~	~~15~~	~~16~~	(17)	~~18~~	(19)	~~20~~
~~21~~	~~22~~	(23)	~~24~~	~~25~~	~~26~~	~~27~~	~~28~~	(29)	~~30~~

We can see that the prime numbers less than 30 are 2, 3, 5, 7, 11, 13, 17, 19, 23, and 29.

Group Activity

Work with your group to identify the prime numbers less than 300 using the Sieve of Eratosthenes. What is the largest prime number that you will need to check in this process?

Name ______________________ Section __________ Date ____________

Chapter 5 Review

(5.1) *Complete each factoring.*

1. $6x^2 - 15x = 3x(\qquad)$

2. $4x^5 + 2x - 10x^4 = 2x(\qquad\qquad)$

Factor out the GCF from each polynomial.

3. $5m + 30$

4. $20x^3 + 12x^2 + 24x$

5. $3x(2x + 3) - 5(2x + 3)$

6. $5x(x + 1) - (x + 1)$

Factor each polynomial by grouping.

7. $3x^2 - 3x + 2x - 2$

8. $6x^2 + 10x - 3x - 5$

9. $3a^2 + 9ab + 3b^2 + ab$

(5.2) *Factor each trinomial.*

10. $x^2 + 6x + 8$

11. $x^2 - 11x + 24$

12. $x^2 + x + 2$

13. $x^2 - 5x - 6$

14. $x^2 + 2x - 8$

15. $x^2 + 4xy - 12y^2$

16. $x^2 + 8xy + 15y^2$

17. $72 - 18x - 2x^2$

18. $32 + 12x - 4x^2$

19. $5y^3 - 50y^2 + 120y$

20. To factor $x^2 + 2x - 48$, think of two numbers whose product is _______ and whose sum is _______.

21. What is the first step to factoring $3x^2 + 15x + 30$?

(5.3) or (5.4) *Factor each trinomial.*

22. $2x^2 + 13x + 6$

23. $4x^2 + 4x - 3$

24. $6x^2 + 5xy - 4y^2$

25. $x^2 - x + 2$

26. $2x^2 - 23x - 39$

27. $18x^2 - 9xy - 20y^2$

28. $10y^3 + 25y^2 - 60y$

Write the perimeter of each figure as a simplified polynomial. Then factor each polynomial.

△ **29.**

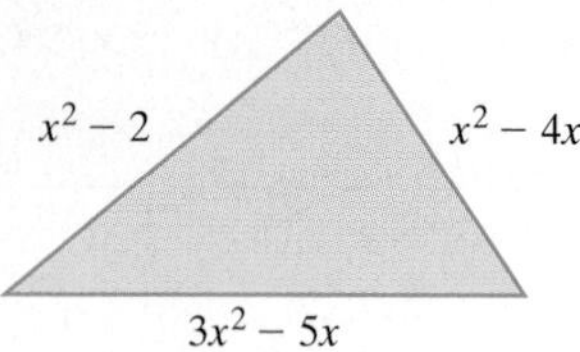

△ **30.**

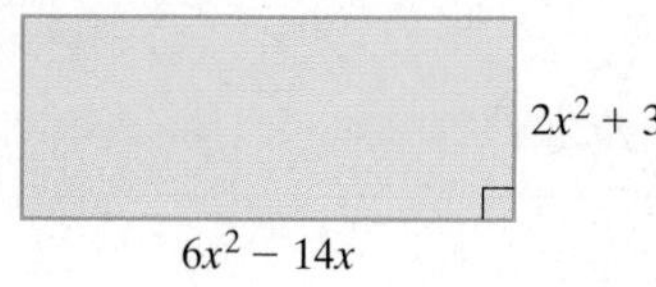

(5.5) *Factor each polynomial completely.*

31. $x^2 - 81$

32. $x^2 + 12x + 36$

33. $4x^2 - 9$

34. $9t^2 - 25s^2$

35. $16x^2 + y^2$

36. $n^2 - 18n + 81$

37. $3r^2 + 36r + 108$

38. $9y^2 - 42y + 49$

39. $5m^8 - 5m^6$

40. $4x^2 - 28xy + 49y^2$

41. $3x^2y + 6xy^2 + 3y^3$

42. $16x^4 - 1$

43. $(y + 2)^2 - 25$

44. $(x - 3)^2 - 16$

45. $8 - 27y^3$

46. $1 - 64y^3$

47. $6x^4y + 48xy$

48. $2x^5 + 16x^2y^3$

49. $x^2 - 2x + 1 - y^2$

△ **50.** The volume of the cylindrical shell is $\pi R^2h - \pi r^2h$ cubic units. Write this volume as a factored expression.

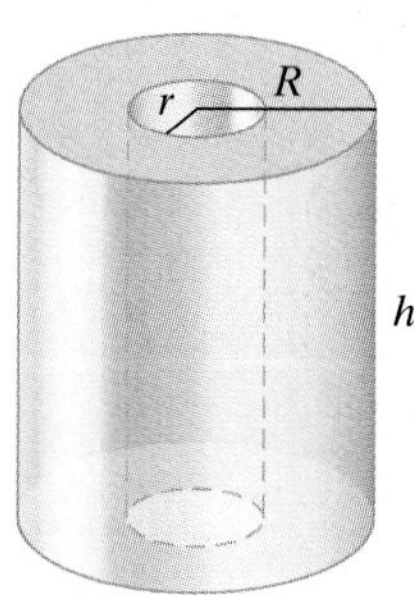

(5.6) *Solve each equation.*

51. $(x + 6)(x - 2) = 0$

52. $3x(x + 1)(7x - 2) = 0$

53. $4(5x + 1)(x + 3) = 0$

54. $x^2 + 8x + 7 = 0$

55. $x^2 - 2x - 24 = 0$

56. $x^2 + 10x = -25$

57. $x(x - 10) = -16$

58. $(3x - 1)(9x^2 + 3x + 1) = 0$

59. $56x^2 - 5x - 6 = 0$

60. $m^2 = 6m$

61. $r^2 = 25$

62. Write a quadratic equation that has the two solutions 4 and 5.

(5.7) *Use the given information to choose the correct dimensions.*

△ **63.** The perimeter of a rectangle is 24 inches. The length is twice the width. Find the dimensions of the rectangle.

a. 5 inches by 7 inches
b. 5 inches by 10 inches
c. 4 inches by 8 inches
d. 2 inches by 10 inches

△ **64.** The area of a rectangle is 80 meters. The length is one more than three times the width. Find the dimensions of the rectangle.

a. 8 meters by 10 meters
b. 4 meters by 13 meters
c. 4 meters by 20 meters
d. 5 meters by 16 meters

Use the given information to find the dimensions of each figure.

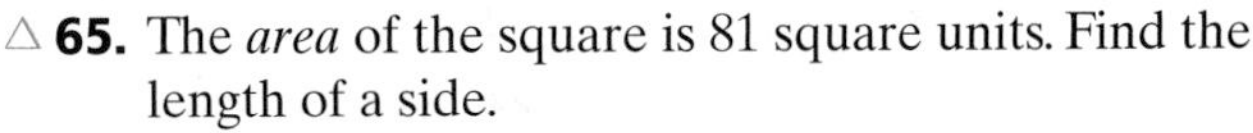

△ **65.** The *area* of the square is 81 square units. Find the length of a side.

△ **66.** The *perimeter* of the quadrilateral is 47 units. Find the lengths of the sides.

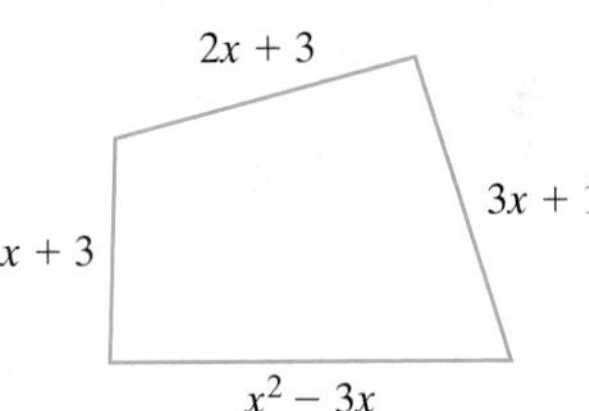

Solve.

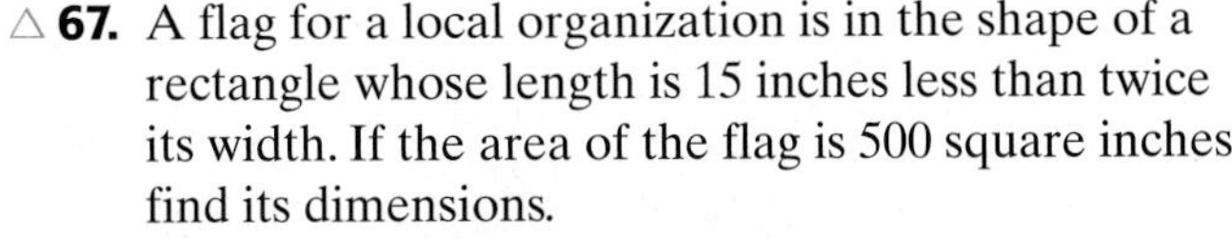

△ **67.** A flag for a local organization is in the shape of a rectangle whose length is 15 inches less than twice its width. If the area of the flag is 500 square inches, find its dimensions.

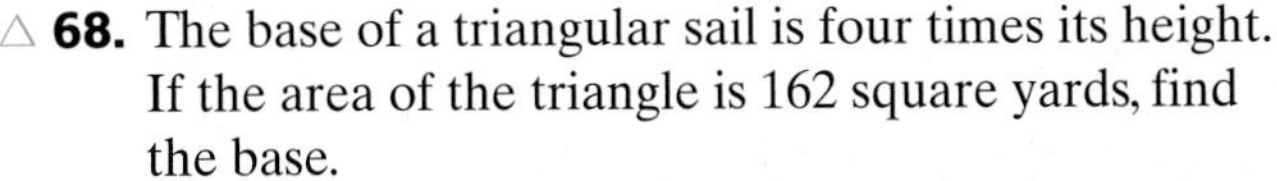

△ **68.** The base of a triangular sail is four times its height. If the area of the triangle is 162 square yards, find the base.

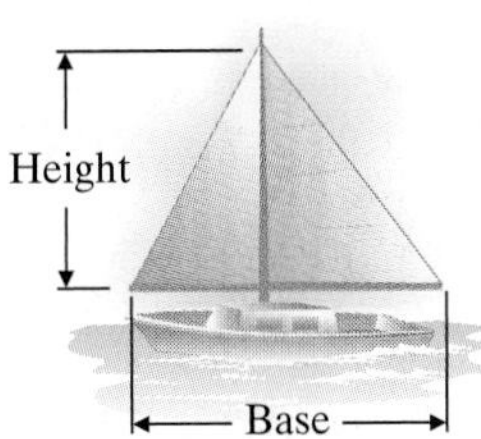

69. Find two consecutive positive integers whose product is 380.

70. A rocket is fired from the ground with an initial velocity of 440 feet per second. Its height h after t seconds is given by the equation $h = -16t^2 + 440t$.

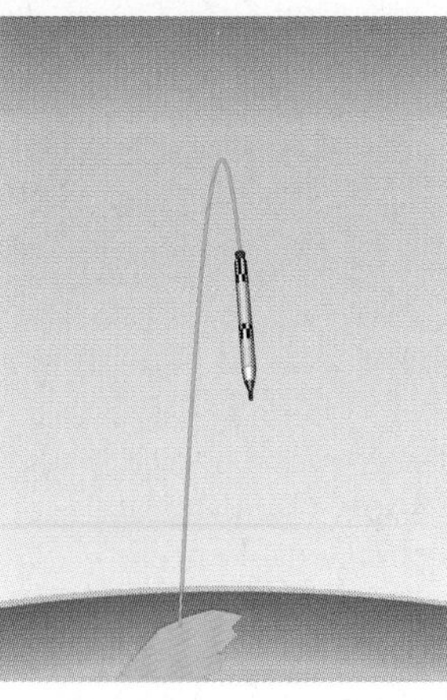

a. Find how many seconds pass before the rocket reaches a height of 2800 feet. Explain why two answers are obtained.

b. Find how many seconds pass before the rocket reaches the ground again.

△ **71.** An architect's squaring instrument is in the shape of a right triangle. Find the length of the longer leg of the right triangle if the hypotenuse is 8 centimeters longer than the longer leg and the shorter leg is 8 centimeters shorter than the longer leg.

Name ______________________ Section ________ Date ________

Chapter 5 Test

Factor each polynomial completely. If a polynomial cannot be factored, write "prime."

1. $9x^2 - 3x$

2. $x^2 + 11x + 28$

3. $49 - m^2$

4. $y^2 + 22y + 121$

5. $x^4 - 16$

6. $4(a + 3) - y(a + 3)$

7. $x^2 + 4$

8. $y^2 - 8y - 48$

9. $3a^2 + 3ab - 7a - 7b$

10. $3x^2 - 5x + 2$

11. $180 - 5x^2$

12. $3x^3 - 21x^2 + 30x$

13. $(x + 5)^2 - y^2$

14. $xy^2 - 7y^2 - 4x + 28$

15. $x - x^5$

16. $x^2 + 14xy + 24y^2$

17. $x^3 + 64$

18. $81xy^3 - 3xz^3$

Solve each equation.

19. $(x - 3)(x + 9) = 0$

20. $x^2 + 10x + 24 = 0$

21. $x^2 + 5x = 14$

22. $3x(2x - 3)(3x + 4) = 0$

23. $5t^3 - 45t = 0$

24. $3x^2 = -12x$

25. $t^2 - 2t - 15 = 0$

26. $(x - 1)(3x^2 - x - 2) = 0$

Answers

1. ________
2. ________
3. ________
4. ________
5. ________
6. ________
7. ________
8. ________
9. ________
10. ________
11. ________
12. ________
13. ________
14. ________
15. ________
16. ________
17. ________
18. ________
19. ________
20. ________
21. ________
22. ________
23. ________
24. ________
25. ________
26. ________

27. ______

△ **27.** The *area* of the rectangle is 54 square units. Find the dimensions of the rectangle.

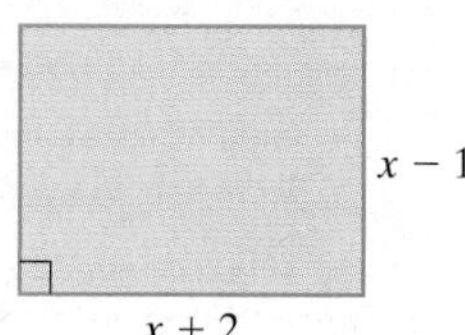

Solve.

28. ______

△ **28.** A deck for a home is in the shape of a triangle. The length of the base of the triangle is 9 feet longer than its height. If the area of the triangle is 68 square feet find the length of the base.

Base
Altitude

29. The sum of two numbers is 17, and the sum of their squares is 145. Find the numbers.

29. ______

30. ______

30. A window washer is suspended 38 feet below the roof of the 1127-foot-tall John Hancock Center in Chicago. (*Source:* Council on Tall Buildings and Urban Habitat) If the window washer drops an object from this height, the object's height h after t seconds is given by the equation $h = -16t^2 + 1089$. Find how many seconds pass before the object reaches the ground.

△ **31.** Find the lengths of the sides of a right triangle if the hypotenuse is 10 centimeters longer than the shorter leg and 5 centimeters longer than the longer leg.

31. ______

Name ______________________ Section __________ Date __________

Cumulative Review

1. Write each sentence as an equation. Let x represent the unknown number.
   a. The quotient of 15 and a number is 4.
   b. Three subtracted from 12 is a number.
   c. Four times a number added to 17 is 21.

2. Find the sums.
   a. $3 + (-7) + (-8)$
   b. $[7 + (-10)] + [-2 + (-4)]$

Simplify each expression.

3. $-3 + [(-2 - 5)-2]$

4. $2^3 - 10 + [-6 - (-5)]$

5. Solve: $3 - x = 7$

6. A 10-foot board is to be cut into two pieces so that the longer piece is 4 times longer than the shorter. Find the length of each piece.

7. Solve $y = mx + b$ for x.

8. Solve: $-2.3x < 6.9$. Graph the solution set.

   −5 −4 −3 −2 −1 0 1 2 3 4 5

9. Complete the table for the equation $y = 3x$.

x	y
−1	
	0
	−9

10. Graph the linear equation $-5x + 3y = 15$.

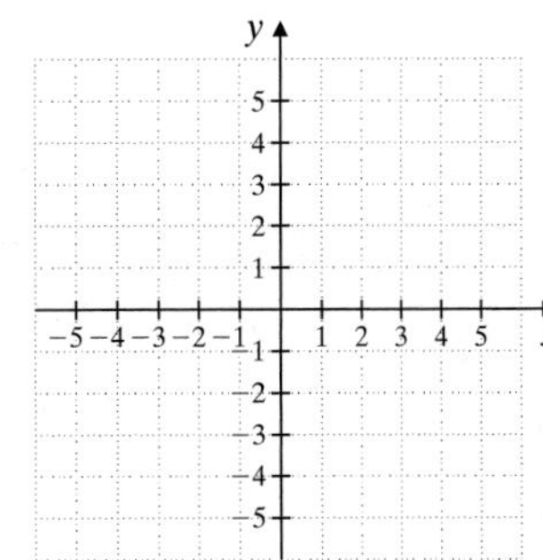

Answers

1. a. ______
   b. ______
   c. ______
2. a. ______
   b. ______
3. ______
4. ______
5. ______
6. ______
7. ______
8. ______
9. see table
10. see graph

11. __________
12. __________
13. __________
14. __________
15. __________
16. __________
17. __________
18. __________
19. __________
20. __________
21. __________
22. __________
23. __________
24. __________

Simplify each quotient.

11. $\frac{x^5}{x^2}$ **12.** $\frac{4^7}{4^3}$ **13.** $\frac{(-3)^5}{(-3)^2}$ **14.** $\frac{2x^5y^2}{xy}$

Simplify by writing each expression with positive exponents only.

15. $2x^{-3}$ **16.** $(-2)^{-4}$

Multiply.

17. $5x(2x^3 + 6)$ **18.** $-3x^2(5x^2 + 6x - 1)$

19. Divide: $\frac{4x^2 + 7 + 8x^3}{2x + 3}$ **20.** Factor $x^2 + 7x + 12$

Factor completely.

21. $9x^2 - 12x + 4$ **22.** $x^2 - \frac{1}{4}$

23. $p^3 + 27q^3$ **24.** Solve: $x^2 - 9x - 22 = 0$

Rational Expressions

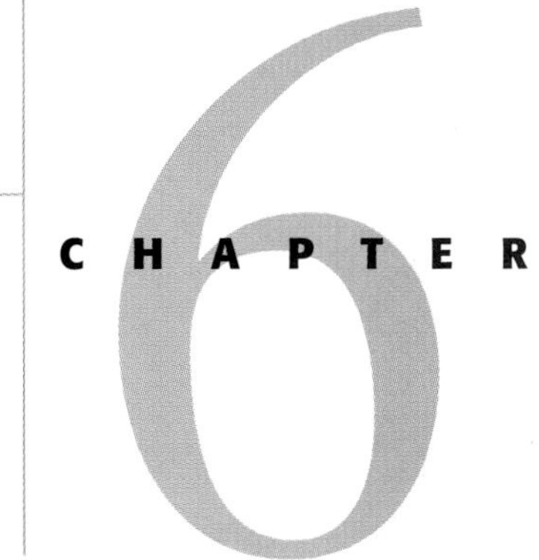

In this chapter, we expand our knowledge of algebraic expressions to include algebraic fractions, called **rational expressions**. We explore the operations of addition, subtraction, multiplication, and division using principles similar to the principles for numerical fractions.

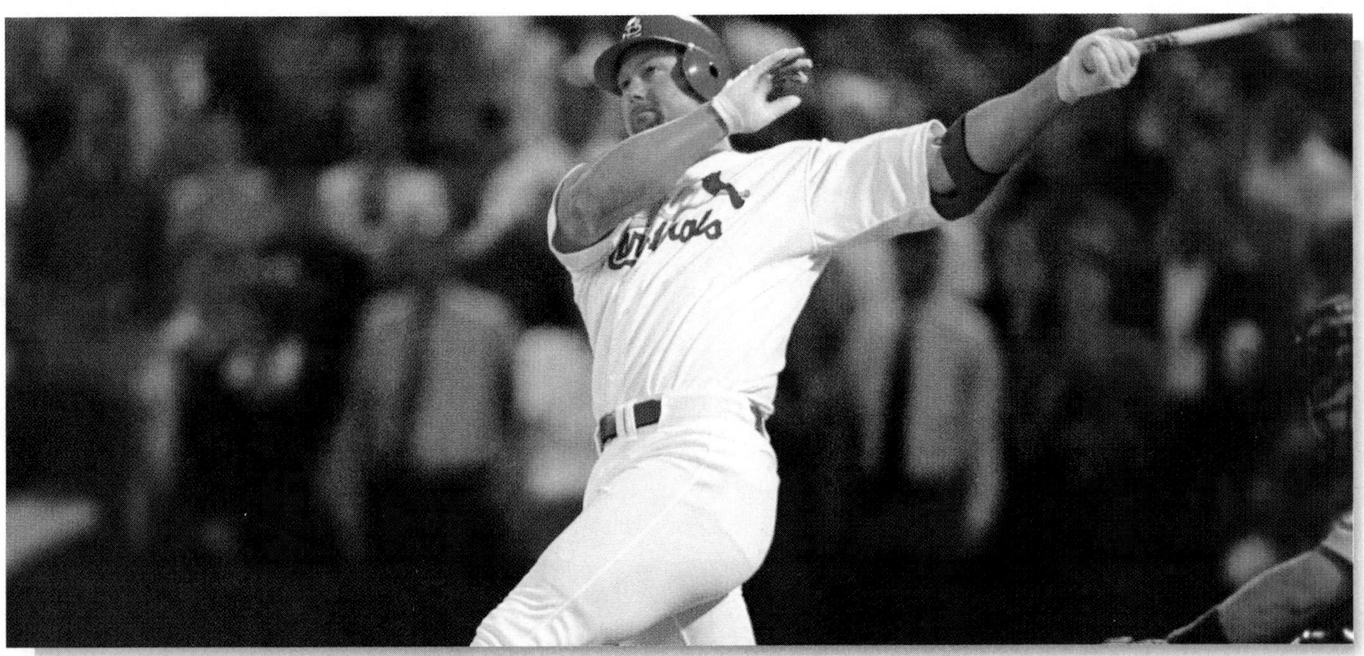

Rational expressions can be found in formulas for many real-world situations. In Exercise 71 on page 422, you will find a rational expression used in a formula to calculate a baseball player's slugging percentage. Slugging percentage is a statistic that gauges a hitter's power. In essence, slugging percentage gives the average number of bases a hitter takes per batting attempt. A strong batter should be able to hit the ball farther, allowing him or her to get farther around the bases on a hit, which will result in a higher slugging percentage. Babe Ruth holds the record for slugging percentage in a single major league season with 0.8472 in 1920, as well as the record for career slugging percentage with 0.6897 spanning his major league career from 1914 to 1935. More recently, single-season slugging percentage leaders in the major leagues have included Todd Helton (0.6983) in 2000, Larry Walker (0.7100) in 1999, and Mark McGwire (0.7525) in 1998.

Answers

1. ______

2. ______

3. ______

4. ______

5. ______

6. ______

7. ______

8. ______

9. ______

10. ______

11. ______

12. ______

13. ______

14. ______

15. ______

16. ______

17. ______

18. ______

19. ______

Name ______ Section ______ Date ______

Chapter 6 Pretest

1. Find any real numbers for which the following expression is undefined.

$$\frac{x+2}{x^2-9x-10}$$

2. Simplify: $\dfrac{4x+32}{x^2+10x+16}$

3. Find the LCD for the following rational expressions.

$$\frac{1}{5x+10}, \frac{3}{2x^2+10x+12}$$

Perform the indicated operations and simplify if possible.

4. $\dfrac{y^2-8y+7}{2y-14} \cdot \dfrac{6y+18}{y^2+2y-3}$

5. $\dfrac{5x^3}{x^2-25} \div \dfrac{x^6}{(x+5)^2}$

6. $\dfrac{b}{b^2-9b-22} + \dfrac{2}{b^2-9b-22}$

7. $\dfrac{3}{x-1} - 4$

8. $\dfrac{2}{x-5} - \dfrac{7}{5-x}$

9. $\dfrac{x}{x^2-16} + \dfrac{3}{x^2-7x+12}$

Solve each equation.

10. $\dfrac{5}{b} + \dfrac{3}{5} = \dfrac{4}{5b}$

11. $9 + \dfrac{7}{d-7} = \dfrac{d}{d-7}$

12. $\dfrac{4y+5}{y^2+5y+6} + \dfrac{3}{y+3} = \dfrac{2}{y+2}$

13. Solve the equation for the indicated variable. $\dfrac{2A}{b} = h$; for b.

Simplify each complex fraction.

14. $\dfrac{\dfrac{12m^3}{5n^2}}{\dfrac{4m^6}{25n^8}}$

15. $\dfrac{16 - \dfrac{1}{a^2}}{\dfrac{4}{a} + \dfrac{1}{a^2}}$

△ **16.** Given that the two triangles are similar, find x.

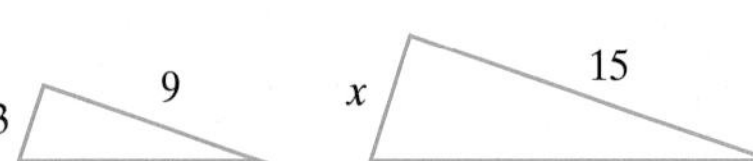

17. A number added to the product of 10 and the reciprocal of the number equals 7. Find the number.

18. Sonya can wash the windows in her house in 5 hours. Her daughter completes the same job in 8 hours. Find how long it takes if they work together.

19. Tom flies his airplane 495 miles with a tail wind of 25 miles per hour. Against the wind, he flies only 405 miles in the same amount of time. Find the rate of the plane in still air.

6.1 Simplifying Rational Expressions

OBJECTIVES

A. Find the value of a rational expression given a replacement number.

B. Identify values for which a rational expression is undefined.

C. Simplify, or write rational expressions in lowest terms.

A Evaluating Rational Expressions

A rational number is a number that can be written as a quotient of integers. A *rational expression* is also a quotient; it is a quotient of polynomials. Examples are

$$\frac{2}{3}, \quad \frac{3y^3}{8}, \quad \frac{-4p}{p^3 + 2p + 1}, \quad \text{and} \quad \frac{5x^2 - 3x + 2}{3x + 7}$$

Rational Expression

A **rational expression** is an expression that can be written in the form

$$\frac{P}{Q}$$

where P and Q are polynomials and $Q \neq 0$.

Rational expressions have different numerical values depending on what values replace the variables.

EXAMPLE 1

Find the numerical value of $\dfrac{x + 4}{2x - 3}$ for each replacement value.

a. $x = 5$ **b.** $x = -2$

Solution:

a. We replace each x in the expression with 5 and then simplify.

$$\frac{x + 4}{2x - 3} = \frac{5 + 4}{2(5) - 3} = \frac{9}{10 - 3} = \frac{9}{7}$$

b. We replace each x in the expression with -2 and then simplify.

$$\frac{x + 4}{2x - 3} = \frac{-2 + 4}{2(-2) - 3} = \frac{2}{-7} \quad \text{or} \quad -\frac{2}{7}$$

In the example above, we wrote $\dfrac{2}{-7}$ as $-\dfrac{2}{7}$. For a negative fraction such as $\dfrac{2}{-7}$, recall from Section 1.6 that

$$\frac{2}{-7} = \frac{-2}{7} = -\frac{2}{7}$$

In general, for any fraction,

$$\frac{-a}{b} = \frac{a}{-b} = -\frac{a}{b}, \quad b \neq 0$$

This is also true for rational expressions. For example,

$$\frac{-(x + 2)}{x} = \frac{x + 2}{-x} = -\frac{x + 2}{x}$$

↑ Notice the parentheses.

Practice Problem 1

Find the value of $\dfrac{x - 3}{5x + 1}$ for each replacement value.

a. $x = 4$

b. $x = -3$

Answers

1. a. $\dfrac{1}{21}$, **b.** $\dfrac{-6}{-14} = \dfrac{3}{7}$

B Identifying When a Rational Expression Is Undefined

In the preceding box, notice that we wrote $b \neq 0$ for the denominator b. The denominator of a rational expression must not equal 0 since division by 0 is not defined. This means we must be careful when replacing the variable in a rational expression by a number. For example, suppose we replace x with 5 in the rational expression $\frac{2+x}{x-5}$. The expression becomes

$$\frac{2+x}{x-5} = \frac{2+5}{5-5} = \frac{7}{0}$$

But division by 0 is undefined. Therefore, in this expression we can allow x to be any real number *except* 5. **A rational expression is undefined for values that make the denominator 0.**

EXAMPLE 2

Are there any values for x for which each expression is undefined?

a. $\frac{x}{x-3}$ **b.** $\frac{x^2+2}{x^2-3x+2}$ **c.** $\frac{x^3-6x^2-10x}{3}$

Solution: To find values for which a rational expression is undefined, we find values that make the denominator 0.

a. The denominator of $\frac{x}{x-3}$ is 0 when $x - 3 = 0$ or when $x = 3$. Thus, when $x = 3$, the expression $\frac{x}{x-3}$ is undefined.

b. We set the denominator equal to 0.

$$x^2 - 3x + 2 = 0$$

$$(x-2)(x-1) = 0 \quad \text{Factor.}$$

$$x - 2 = 0 \quad \text{or} \quad x - 1 = 0 \quad \text{Set each factor equal to 0.}$$

$$x = 2 \qquad\qquad x = 1 \quad \text{Solve.}$$

Thus, when $x = 2$ or $x = 1$, the denominator $x^2 - 3x + 2$ is 0. So the rational expression $\frac{x^2+2}{x^2-3x+2}$ is undefined when $x = 2$ or when $x = 1$.

c. The denominator of $\frac{x^3-6x^2-10x}{3}$ is never 0, so there are no values of x for which this expression is undefined.

Practice Problem 2

Are there any values for x for which each rational expression is undefined?

a. $\frac{x}{x+2}$ b. $\frac{x-3}{x^2+5x+4}$ c. $\frac{x^2-3x+2}{5}$

C Simplifying Rational Expressions

A fraction is said to be written in lowest terms or simplest form when the numerator and denominator have no common factors other than 1 (or -1). For example, the fraction $\frac{7}{10}$ is written in lowest terms since the numerator and denominator have no common factors other than 1 (or -1).

The process of writing a rational expression in lowest terms or simplest form is called **simplifying** a rational expression. The following **fundamental principle of rational expressions** is used to simplify a rational expression.

> **Fundamental Principle of Rational Expressions**
>
> If $\frac{P}{Q}$ is a rational expression and R is a nonzero polynomial, then
>
> $$\frac{PR}{QR} = \frac{P}{Q}$$

Answers

2. a. $x = -2$, **b.** $x = -4, x = -1$, **c.** no

Simplifying a rational expression is similar to simplifying a fraction.

Simplify: $\frac{15}{20}$

$$\frac{15}{20} = \frac{3 \cdot 5}{2 \cdot 2 \cdot 5}$$ Factor the numerator and the denominator.

$$= \frac{3 \cdot 5}{2 \cdot 2 \cdot 5}$$ Look for common factors.

$$= \frac{3}{2 \cdot 2} = \frac{3}{4}$$ Apply the fundamental principle.

Simplify: $\frac{x^2 - 9}{x^2 + x - 6}$

$$\frac{x^2 - 9}{x^2 + x - 6} = \frac{(x - 3)(x + 3)}{(x - 2)(x + 3)}$$ Factor the numerator and the denominator.

$$= \frac{(x - 3)(x + 3)}{(x - 2)(x + 3)}$$ Look for common factors.

$$= \frac{x - 3}{x - 2}$$ Apply the fundamental principle.

Thus, the rational expression $\frac{x^2 - 9}{x^2 + x - 6}$ has the same value as the rational expression $\frac{x - 3}{x - 2}$ for all values of x except 2 and -3. (Remember that when x is 2, the denominator of both rational expressions is 0 and when x is -3, the original rational expression has a denominator of 0.)

As we simplify rational expressions, we will assume that the simplified rational expression is equal to the original rational expression for all real numbers except those for which either denominator is 0. The following steps may be used to simplify rational expressions.

To Simplify a Rational Expression

Step 1. Completely factor the numerator and denominator.

Step 2. Apply the fundamental principle of rational expressions to divide out common factors.

EXAMPLE 3 Simplify: $\frac{5x - 5}{x^3 - x^2}$

Solution: To begin, we factor the numerator and denominator if possible. Then we apply the fundamental principle.

$$\frac{5x - 5}{x^3 - x^2} = \frac{5(x - 1)}{x^2(x - 1)} = \frac{5}{x^2}$$

EXAMPLE 4 Simplify: $\frac{x^2 + 8x + 7}{x^2 - 4x - 5}$

Solution: We factor the numerator and denominator and then apply the fundamental principle.

$$\frac{x^2 + 8x + 7}{x^2 - 4x - 5} = \frac{(x + 7)(x + 1)}{(x - 5)(x + 1)} = \frac{x + 7}{x - 5}$$

Practice Problem 3

Simplify: $\frac{x^4 + x^3}{5x + 5}$

Practice Problem 4

Simplify: $\frac{x^2 + 11x + 18}{x^2 + x - 2}$

Answers

3. $\frac{x^3}{5}$, **4.** $\frac{x + 9}{x - 1}$

Practice Problem 5

Simplify: $\dfrac{x^2 + 10x + 25}{x^2 + 5x}$

EXAMPLE 5 Simplify: $\dfrac{x^2 + 4x + 4}{x^2 + 2x}$

Solution: We factor the numerator and denominator and then apply the fundamental principle.

$$\frac{x^2 + 4x + 4}{x^2 + 2x} = \frac{(x + 2)(x + 2)}{x(x + 2)} = \frac{x + 2}{x}$$

Helpful Hint

When simplifying a rational expression, the fundamental principle applies to **common *factors*, not common *terms*.**

$$\frac{x \cdot (x + 2)}{x \cdot x} = \frac{x + 2}{x}$$

Common factors. These can be divided out.

$$\frac{x + 2}{x}$$

Common terms. Fundamental principle does not apply. This is in simplest form.

Try the Concept Check in the margin.

Practice Problem 6

Simplify: $\dfrac{x + 5}{x^2 - 25}$

EXAMPLE 6 Simplify: $\dfrac{x + 9}{x^2 - 81}$

Solution: We factor and then apply the fundamental principle.

$$\frac{x + 9}{x^2 - 81} = \frac{x + 9}{(x + 9)(x - 9)} = \frac{1}{x - 9}$$

Practice Problem 7

Simplify each rational expression.

a. $\dfrac{x + 4}{4 + x}$ b. $\dfrac{x - 4}{4 - x}$

Concept Check

Recall that the fundamental principle applies to common factors only. Which of the following are *not* true? Explain why.

a. $\dfrac{3 - 1}{3 + 5} = -\dfrac{1}{5}$

b. $\dfrac{2x + 10}{2} = x + 5$

c. $\dfrac{37}{72} = \dfrac{3}{2}$

d. $\dfrac{2x + 3}{2} = x + 3$

EXAMPLE 7 Simplify each rational expression.

a. $\dfrac{x + y}{y + x}$ **b.** $\dfrac{x - y}{y - x}$

Solution:

a. The expression $\dfrac{x + y}{y + x}$ can be simplified by using the commutative property of addition to rewrite the denominator $y + x$ as $x + y$.

$$\frac{x + y}{y + x} = \frac{x + y}{x + y} = 1$$

b. The expression $\dfrac{x - y}{y - x}$ can be simplified by recognizing that $y - x$ and $x - y$ are opposites. In other words, $y - x = -1(x - y)$. We proceed as follows:

$$\frac{x - y}{y - x} = \frac{1 \cdot (x - y)}{(-1)(x - y)} = \frac{1}{-1} = -1$$

Answers

5. $\dfrac{x + 5}{x}$, **6.** $\dfrac{1}{x - 5}$, **7. a.** 1, **b.** -1

Concept Check: a, c, d

Name ______________ Section ________ Date __________

Mental Math

Find any real numbers for which each rational expression is undefined. See Example 2.

1. $\dfrac{x+5}{x}$

2. $\dfrac{x^2-5x}{x-3}$

3. $\dfrac{x^2+4x-2}{x(x-1)}$

4. $\dfrac{x+2}{(x-5)(x-6)}$

EXERCISE SET 6.1

A *Find the value of the following expressions when $x = 2$, $y = -2$, and $z = -5$. See Example 1.*

1. $\dfrac{x+5}{x+2}$

2. $\dfrac{x+8}{2x+5}$

3. $\dfrac{y^3}{y^2-1}$

4. $\dfrac{z}{z^2-5}$

5. $\dfrac{x^2+8x+2}{x^2-x-6}$

6. $\dfrac{x+5}{x^2+4x-8}$

7. The total revenue R from the sale of a popular music compact disc is approximately given by the equation

$$R = \frac{150x^2}{x^2+3}$$

where x is the number of years since the CD has been released and revenue R is in millions of dollars.
   a. Find the total revenue generated by the end of the first year.
   b. Find the total revenue generated by the end of the second year. Round to the nearest tenth.
   c. Find the total revenue generated in the second year only.

8. For a certain model fax machine, the manufacturing cost C per machine is given by the equation

$$C = \frac{250x + 10{,}000}{x}$$

where x is the number of fax machines manufactured and cost C is in dollars per machine.
   a. Find the cost per fax machine when manufacturing 100 fax machines.
   b. Find the cost per fax machine when manufacturing 1000 fax machines.
   c. Does the cost per machine decrease or increase when more machines are manufactured? Explain why this is so.

B *Find any real numbers for which each rational expression is undefined. See Example 2.*

9. $\dfrac{7}{2x}$

10. $\dfrac{3}{5x}$

11. $\dfrac{x+3}{x+2}$

12. $\dfrac{5x+1}{x-3}$

13. $\dfrac{4x^2+9}{2x-8}$

14. $\dfrac{9x^3+4x}{15x+45}$

15. $\dfrac{9x^3+4}{15x+30}$

16. $\dfrac{19x^3+2}{x^3-x}$

17. $\dfrac{x^2-5x-2}{4}$

18. $\dfrac{9y^5+y^3}{9}$

19. Explain why the denominator of a fraction or a rational expression must not equal 0.

20. Does $\dfrac{(x-3)(x+3)}{x-3}$ have the same value as $x+3$ for all real numbers? Explain why or why not.

C *Simplify each expression. See Examples 3 through 7.*

21. $\frac{2}{8x + 16}$

22. $\frac{3}{9x + 6}$

23. $\frac{x - 2}{x^2 - 4}$

24. $\frac{x + 5}{x^2 - 25}$

25. $\frac{2x - 10}{3x - 30}$

26. $\frac{3x - 12}{4x - 16}$

27. $\frac{x + 7}{7 + x}$

28. $\frac{y + 9}{9 + y}$

29. $\frac{x - 7}{7 - x}$

30. $\frac{y - 9}{9 - y}$

31. $\frac{-5a - 5b}{a + b}$

32. $\frac{7x + 35}{x^2 + 5x}$

33. $\frac{x + 5}{x^2 - 4x - 45}$

34. $\frac{x - 3}{x^2 - 6x + 9}$

35. $\frac{5x^2 + 11x + 2}{x + 2}$

36. $\frac{12x^2 + 4x - 1}{2x + 1}$

37. $\frac{x + 7}{x^2 + 5x - 14}$

38. $\frac{x - 10}{x^2 - 17x + 70}$

39. $\frac{2x^2 + 3x - 2}{2x - 1}$

40. $\frac{4x^2 + 24x}{x + 6}$

41. $\frac{x^2 + 7x + 10}{x^2 - 3x - 10}$

42. $\frac{2x^2 + 7x - 4}{x^2 + 3x - 4}$

43. $\frac{3x^2 + 7x + 2}{3x^2 + 13x + 4}$

44. $\frac{4x^2 - 4x + 1}{2x^2 + 9x - 5}$

45. $\frac{2x^2 - 8}{4x - 8}$

46. $\frac{5x^2 - 500}{35x + 350}$

47. $\frac{11x^2 - 22x^3}{6x - 12x^2}$

48. $\frac{16r^2 - 4s^2}{4r - 2s}$

49. $\frac{2 - x}{x - 2}$

50. $\frac{7 - y}{y - 7}$

51. $\frac{x^2 - 1}{x^2 - 2x + 1}$

52. $\frac{x^2 - 16}{x^2 - 8x + 16}$

53. $\frac{m^2 - 6m + 9}{m^2 - 9}$

54. $\frac{m^2 - 4m + 4}{m^2 + m - 6}$

Review and Preview

Perform each indicated operation. See Section R.2.

55. $\frac{1}{3} \cdot \frac{9}{11}$

56. $\frac{5}{27} \cdot \frac{2}{5}$

57. $\frac{5}{6} \cdot \frac{10}{11} \cdot \frac{2}{3}$

58. $\frac{4}{3} \cdot \frac{1}{7} \cdot \frac{10}{13}$

59. $\frac{1}{3} \div \frac{1}{4}$

60. $\frac{7}{8} \div \frac{1}{2}$

61. $\frac{13}{20} \div \frac{2}{9}$

62. $\frac{8}{15} \div \frac{5}{8}$

Combining Concepts

Simplify each expression. Each exercise contains a four-term polynomial that should be factored by grouping.

63. $\frac{x^2 + xy + 2x + 2y}{x + 2}$

64. $\frac{ab + ac + b^2 + bc}{b + c}$

65. $\frac{5x + 15 - xy - 3y}{2x + 6}$

66. $\frac{xy - 6x + 2y - 12}{y^2 - 6y}$

67. Explain how to write a fraction in lowest terms.

68. Explain how to write a rational expression in lowest terms.

69. A company's gross profit margin P can be computed with the formula $P = \frac{R - C}{R}$, where $R =$ the company's revenue and $C =$ the cost of goods sold. For fiscal year 2001, consumer electronics retailer Best Buy had revenues of \$15.3 billion and cost of goods sold of \$12.3 billion. (*Source:* Best Buy Company, Inc.) What was Best Buy's gross profit margin in 2001? Express the answer as a percent rounded to the nearest tenth of a percent.

70. During a storm, water treatment engineers monitor how quickly rain is falling. If too much rain comes too fast, there is a danger of sewers backing up. A formula that gives the rainfall intensity i in millimeters per hour for a certain strength storm in eastern Virginia is

$$i = \frac{5840}{t + 29}$$

where t is the duration of the storm in minutes. What rainfall intensity should engineers expect for a storm of this strength in eastern Virginia that lasts for 80 minutes? Round answer to one decimal place.

71. A baseball player's slugging percentage S can be calculated with the following formula: $S = \dfrac{h + d + 2t + 3r}{b}$, where h = number of hits, d = number of doubles, t = number of triples, r = number of home runs, and b = number of at bats. In 2000, Manny Ramirez of the Boston Red Sox led the American League in slugging percentage. During the 2000 season, Ramirez had 439 at bats, 154 hits, 34 doubles, 2 triples, and 38 home runs. (*Source:* Major League Baseball) Calculate Ramirez's 2000 slugging percentage. Round to the nearest tenth of a percent.

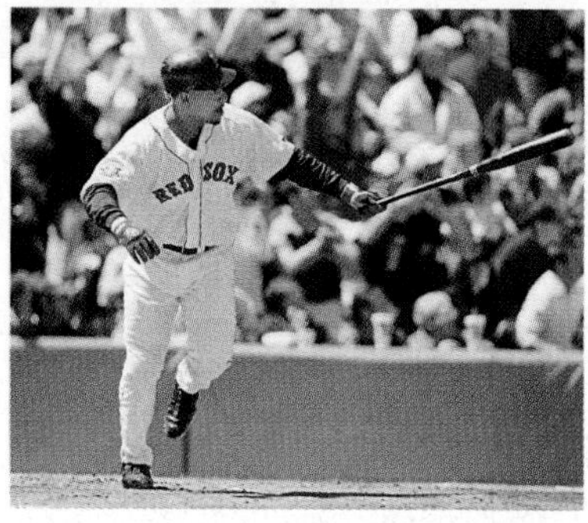

72. Anthropologists and forensic scientists use a measure called the cephalic index to help classify skulls. The cephalic index of a skull with width W and length L from front to back is given by the formula

$$C = \frac{100W}{L}$$

A long skull has an index value less than 75, a medium skull has an index value between 75 and 85, and a broad skull has an index value over 85. Find the cephalic index of a skull that is 5 inches wide and 6.4 inches long. Classify the skull.

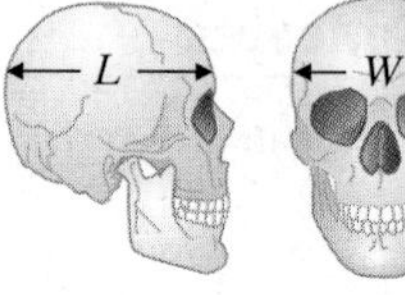

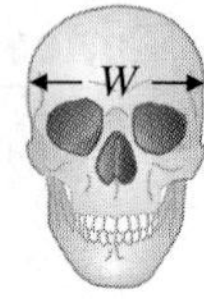

6.2 Multiplying and Dividing Rational Expressions

OBJECTIVES

- **A** Multiply rational expressions.
- **B** Divide rational expressions.
- **C** Multiply and divide rational expressions.
- **D** Convert between units of measure.

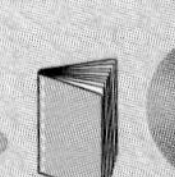

A Multiplying Rational Expressions

Just as simplifying rational expressions is similar to simplifying number fractions, multiplying and dividing rational expressions is similar to multiplying and dividing number fractions.

Multiply: $\frac{3}{5} \cdot \frac{10}{11}$	Multiply: $\frac{x-3}{x+5} \cdot \frac{2x+10}{x^2-9}$

Multiply numerators and then multiply denominators.

$\frac{3}{5} \cdot \frac{10}{11} = \frac{3 \cdot 10}{5 \cdot 11}$	$\frac{x-3}{x+5} \cdot \frac{2x+10}{x^2-9} = \frac{(x-3) \cdot (2x+10)}{(x+5) \cdot (x^2-9)}$

Simplify by factoring numerators and denominators.

$= \frac{3 \cdot 2 \cdot 5}{5 \cdot 11}$	$= \frac{(x-3) \cdot 2(x+5)}{(x+5)(x+3)(x-3)}$

Apply the fundamental principle.

$= \frac{3 \cdot 2}{11}$ or $\frac{6}{11}$	$= \frac{2}{x+3}$

Multiplying Rational Expressions

If $\frac{P}{Q}$ and $\frac{R}{S}$ are rational expressions, then

$$\frac{P}{Q} \cdot \frac{R}{S} = \frac{PR}{PR}$$

To multiply rational expressions, multiply the numerators and then multiply the denominators.

EXAMPLE 1 Multiply.

a. $\frac{25x}{2} \cdot \frac{1}{y^3}$ **b.** $\frac{-7x^2}{5y} \cdot \frac{3y^5}{14x^2}$

Solution: To multiply rational expressions, we first multiply the numerators and then multiply the denominators of both expressions. Then we write the product in lowest terms.

a. $\frac{25x}{2} \cdot \frac{1}{y^3} = \frac{25x \cdot 1}{2 \cdot y^3} = \frac{25x}{2y^3}$

The expression $\frac{25x}{2y^3}$ is in lowest terms.

b. $\frac{-7x^2}{5y} \cdot \frac{3y^5}{14x^2} = \frac{-7x^2 \cdot 3y^5}{5y \cdot 14x^2}$ Multiply.

Practice Problem 1

Multiply.

a. $\frac{16y}{3} \cdot \frac{1}{x^2}$ b. $\frac{-5a^3}{3b^3} \cdot \frac{2b^2}{15a}$

Answers

1. a. $\frac{16y}{3x^2}$, **b.** $-\frac{2a^2}{9b}$

The expression $\frac{-7x^2 \cdot 3y^5}{5y \cdot 14x^2}$ is not in lowest terms, so we factor the numerator and the denominator and apply the fundamental principle.

$$= \frac{-1 \cdot 7 \cdot 3 \cdot x^2 \cdot y \cdot y^4}{5 \cdot 2 \cdot 7 \cdot x^2 \cdot y}$$

$$= -\frac{3y^4}{10}$$

When multiplying rational expressions, it is usually best to factor each numerator and denominator first. This will help us when we apply the fundamental principle to write the product in lowest terms.

Practice Problem 2

Multiply: $\frac{6x + 6}{7} \cdot \frac{14}{x^2 - 1}$

EXAMPLE 2 Multiply: $\frac{x^2 + x}{3x} \cdot \frac{6}{5x + 5}$

Solution:

$$\frac{x^2 + x}{3x} \cdot \frac{6}{5x + 5} = \frac{x(x + 1)}{3x} \cdot \frac{2 \cdot 3}{5(x + 1)}$$ Factor numerators and denominators.

$$= \frac{x(x + 1) \cdot 2 \cdot 3}{3x \cdot 5(x + 1)}$$ Multiply.

$$= \frac{2}{5}$$ Apply the fundamental principle.

The following steps may be used to multiply rational expressions.

To Multiply Rational Expressions

Step 1. Completely factor numerators and denominators.

Step 2. Multiply numerators and multiply denominators.

Step 3. Simplify or write the product in lowest terms by applying the fundamental principle to all common factors.

Concept Check

Which of the following is a true statement?

a. $\frac{1}{3} \cdot \frac{1}{2} = \frac{1}{5}$ b. $\frac{2}{x} \cdot \frac{5}{x} = \frac{10}{x}$

c. $\frac{3}{x} \cdot \frac{1}{2} = \frac{3}{2x}$

d. $\frac{x}{7} \cdot \frac{x + 5}{4} = \frac{2x + 5}{28}$

Try the Concept Check in the margin.

Practice Problem 3

Multiply: $\frac{4x + 8}{7x^2 - 14x} \cdot \frac{3x^2 - 5x - 2}{9x^2 - 1}$

EXAMPLE 3 Multiply: $\frac{3x + 3}{5x^2 - 5x} \cdot \frac{2x^2 + x - 3}{4x^2 - 9}$

Solution:

$$\frac{3x + 3}{5x^2 - 5x} \cdot \frac{2x^2 + x - 3}{4x^2 - 9} = \frac{3(x + 1)}{5x(x - 1)} \cdot \frac{(2x + 3)(x - 1)}{(2x - 3)(2x + 3)}$$ Factor.

$$= \frac{3(x + 1)(2x + 3)(x - 1)}{5x(x - 1)(2x - 3)(2x + 3)}$$ Multiply.

$$= \frac{3(x + 1)}{5x(2x - 3)}$$ Simplify.

Answers

2. $\frac{12}{x - 1}$, **3.** $\frac{4(x + 2)}{7x(3x - 1)}$

Concept Check: c

B Dividing Rational Expressions

We can divide by a rational expression in the same way we divide by a number fraction. Recall that to divide by a fraction, we multiply by its reciprocal.

For example, to divide $\frac{3}{2}$ by $\frac{7}{8}$, we multiply $\frac{3}{2}$ by $\frac{8}{7}$.

$$\frac{3}{2} \div \frac{7}{8} = \frac{3}{2} \cdot \frac{8}{7} = \frac{3 \cdot 4 \cdot 2}{2 \cdot 7} = \frac{12}{7}$$

Helpful Hint

Don't forget how to find reciprocals. The reciprocal of $\frac{a}{b}$ is $\frac{b}{a}$, $a \neq 0$, $b \neq 0$.

Dividing Rational Expressions

If $\frac{P}{Q}$ and $\frac{R}{S}$ are rational expressions and $\frac{R}{S}$ is not 0, then

$$\frac{P}{Q} \div \frac{R}{S} = \frac{P}{Q} \cdot \frac{S}{R} = \frac{PS}{QR}$$

To divide two rational expressions, multiply the first rational expression by the reciprocal of the second rational expression.

EXAMPLE 4 Divide: $\frac{3x^3}{40} \div \frac{4x^3}{y^2}$

Solution:

$$\frac{3x^3}{40} \div \frac{4x^3}{y^2} = \frac{3x^3}{40} \cdot \frac{y^2}{4x^3} \quad \text{Multiply by the reciprocal of } \frac{4x^3}{y^2}.$$

$$= \frac{3x^3y^2}{160x^3}$$

$$= \frac{3y^2}{160} \quad \text{Simplify.}$$

Practice Problem 4

Divide: $\frac{7x^2}{6} \div \frac{x}{2y}$

EXAMPLE 5 Divide: $\frac{(x-1)(x+2)}{10} \div \frac{2x+4}{5}$

Solution:

$$\frac{(x-1)(x+2)}{10} \div \frac{2x+4}{5} = \frac{(x-1)(x+2)}{10} \cdot \frac{5}{2x+4} \quad \text{Multiply by the reciprocal of } \frac{2x+4}{5}.$$

$$= \frac{(x-1)(x+2) \cdot 5}{5 \cdot 2 \cdot 2 \cdot (x+2)} \quad \text{Factor and multiply.}$$

$$= \frac{x-1}{4} \quad \text{Simplify.}$$

Practice Problem 5

Divide: $\frac{(2x+3)(x-4)}{6} \div \frac{3x-12}{2}$

The following may be used to divide by a rational expression.

To Divide by a Rational Expression

Multiply by its reciprocal.

Answers

4. $\frac{7xy}{3}$, **5.** $\frac{2x+3}{9}$

EXAMPLE 6 Divide: $\dfrac{6x+2}{x^2-1} \div \dfrac{3x^2+x}{x-1}$

Solution:

$$\frac{6x+2}{x^2-1} \div \frac{3x^2+x}{x-1} = \frac{6x+2}{x^2-1} \cdot \frac{x-1}{3x^2+x} \quad \text{Multiply by the reciprocal.}$$

$$= \frac{2(3x+1)(x-1)}{(x+1)(x-1)\cdot x(3x+1)} \quad \text{Factor and multiply.}$$

$$= \frac{2}{x(x+1)} \quad \text{Simplify.}$$

Practice Problem 6

Divide: $\dfrac{10x+4}{x^2-4} \div \dfrac{5x^3+2x^2}{x+2}$

EXAMPLE 7 Divide: $\dfrac{2x^2-11x+5}{5x-25} \div \dfrac{4x-2}{10}$

Solution:

$$\frac{2x^2-11x+5}{5x-25} \div \frac{4x-2}{10} = \frac{2x^2-11x+5}{5x-25} \cdot \frac{10}{4x-2} \quad \text{Multiply by the reciprocal.}$$

$$= \frac{(2x-1)(x-5)\cdot 2\cdot 5}{5(x-5)\cdot 2(2x-1)} \quad \text{Factor and multiply.}$$

$$= \frac{1}{1} \text{ or } 1 \quad \text{Simplify.}$$

Practice Problem 7

Divide: $\dfrac{3x^2-10x+8}{7x-14} \div \dfrac{9x-12}{21}$

C Multiplying and Dividing Rational Expressions

Let's make sure that we understand the difference between multiplying and dividing rational expressions.

Rational Expressions	
Multiplication	Multiply the numerators and multiply the denominators.
Division	Multiply by the reciprocal of the divisor.

EXAMPLE 8 Multiply or divide as indicated.

a. $\dfrac{x-4}{5} \cdot \dfrac{x}{x-4}$ **b.** $\dfrac{x-4}{5} \div \dfrac{x}{x-4}$

c. $\dfrac{x^2-4}{2x+6} \cdot \dfrac{x^2+4x+3}{2-x}$

Solution:

a. $\dfrac{x-4}{5} \cdot \dfrac{x}{x-4} = \dfrac{(x-4)\cdot x}{5\cdot(x-4)} = \dfrac{x}{5}$

b. $\dfrac{x-4}{5} \div \dfrac{x}{x-4} = \dfrac{x-4}{5} \cdot \dfrac{x-4}{x} = \dfrac{(x-4)^2}{5x}$

c. $\dfrac{x^2-4}{2x+6} \cdot \dfrac{x^2+4x+3}{2-x} = \dfrac{(x-2)(x+2)\cdot(x+1)(x+3)}{2(x+3)\cdot(2-x)}$ Factor and multiply.

Practice Problem 8

Multiply or divide as indicated.

a. $\dfrac{x+3}{x} \cdot \dfrac{7}{x+3}$

b. $\dfrac{x+3}{x} \div \dfrac{7}{x+3}$

c. $\dfrac{3-x}{x^2+6x+5} \cdot \dfrac{2x+10}{x^2-7x+12}$

Answers

6. $\dfrac{2}{x^2(x-2)}$, **7.** 1, **8. a.** $\dfrac{7}{x}$, **b.** $\dfrac{(x+3)^2}{7x}$, **c.** $-\dfrac{2}{(x+1)(x-4)}$

Recall from Section 6.1 that $x - 2$ and $2 - x$ are opposites. This means that $\frac{x-2}{2-x} = -1$. Thus,

$$\frac{(x-2)(x+2)\cdot(x+1)(x+3)}{2(x+3)\cdot(2-x)} = \frac{-1(x+2)(x+1)}{2}$$

$$= -\frac{(x+2)(x+1)}{2}$$

D Converting between Units of Measure

Now that we know how to multiply fractions and rational expressions, we can use this knowledge to help us convert between units of measure. To do so, we will use **unit fractions**. A unit fraction is a fraction that equals 1. For example, since 12 in. = 1 ft, we have the unit fractions

$$\frac{12\text{ in.}}{1\text{ ft}} = 1 \quad \text{and} \quad \frac{1\text{ ft}}{12\text{ in.}} = 1$$

EXAMPLE 9 Converting from Cubic Feet to Cubic Yards

The largest building in the world by volume is The Boeing Company's Everett, Washington, factory complex where Boeing's wide-body jetliners, the 747, 767, and 777, are built. The volume of this factory complex is 472,370,319 cubic feet. Find the volume of this Boeing facility in cubic yards. (*Source:* The Boeing Company)

Solution: There are 27 cubic feet in 1 cubic yard. (See the diagram.)

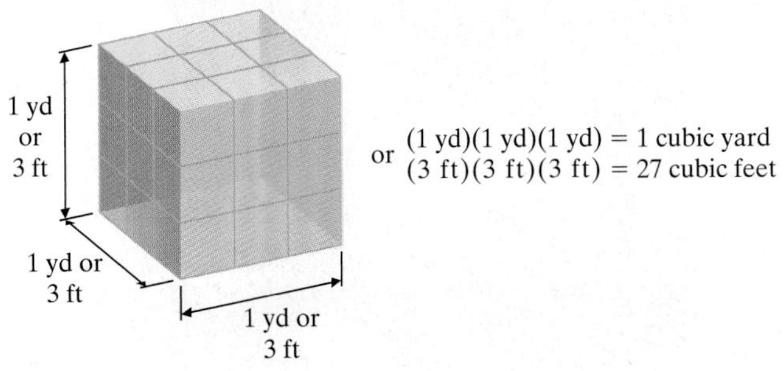

$$472{,}370{,}319\text{ cu ft} = 472{,}370{,}319\,\cancel{\text{cu ft}}\cdot\frac{1\text{ cu yd}}{27\,\cancel{\text{cu ft}}}$$

$$= \frac{472{,}370{,}319}{27}\text{ cu yd}$$

$$= 17{,}495{,}197\text{ cu yd}$$

Practice Problem 9

The largest casino in the world is the Foxwoods Resort Casino in Ledyard, CT. The gaming area for this casino is 21,444 *square yards*. Find the size of the gaming area in *square feet*. (*Source: The Guinness Book of Records*)

Answer

9. 192,996 sq ft

Helpful Hint

When converting among units of measurement, if possible, write the unit fraction so that **the numerator contains the units you are converting to** and **the denominator contains the original units.**

$$48 \text{ in.} = \frac{48 \cancel{\text{in.}}}{1} \cdot \underbrace{\frac{1 \text{ ft}}{12 \cancel{\text{in.}}}}_{\text{Unit fraction}} \begin{array}{l} \leftarrow \text{Units converting to} \\ \leftarrow \text{Original units} \end{array}$$

$$= \frac{48}{12} \text{ ft} = 4 \text{ ft}$$

Practice Problem 10

Man has been timed at a speed of approximately 40.9 feet per second. Convert this to miles per hour. Round to the nearest tenth. (*Source: World Almanac and Book of Facts, 2001*)

EXAMPLE 10 Converting from Feet per Second to Miles per Hour

At the 2000 Summer Olympics, U.S. athlete Marian Jones won the gold medal in the women's 100-meter track event. She ran the distance at an average speed of 30.5 feet per second. Convert this speed to miles per hour. (*Source: World Almanac and Book of Facts, 2001*)

Solution: Recall that 1 mile = 5280 feet and 1 hour = 3600 seconds$(60 \cdot 60)$.

Unit fractions

$$30.5 \text{ feet/second} = \frac{30.5 \cancel{\text{feet}}}{1 \cancel{\text{second}}} \cdot \frac{3600 \cancel{\text{seconds}}}{1 \text{ hour}} \cdot \frac{1 \text{ mile}}{5280 \cancel{\text{feet}}}$$

$$= \frac{30.5 \cdot 3600}{5280} \text{ miles/hour}$$

$$\approx 20.8 \text{ miles/hour (rounded to the nearest tenth)}$$

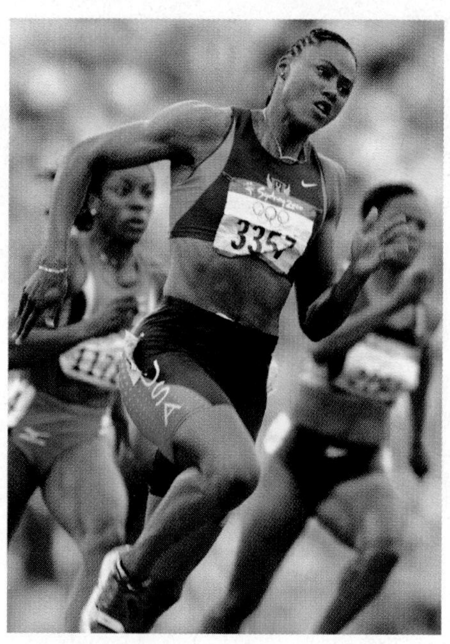

Answer

10. 27.9 mph

Name ______________________ Section __________ Date __________

Mental Math

Find each product. See Example 1.

1. $\frac{2}{y} \cdot \frac{x}{3}$

2. $\frac{3x}{4} \cdot \frac{1}{y}$

3. $\frac{5}{7} \cdot \frac{y^2}{x^2}$

4. $\frac{x^5}{11} \cdot \frac{4}{z^3}$

5. $\frac{9}{x} \cdot \frac{x}{5}$

6. $\frac{y}{7} \cdot \frac{3}{y}$

EXERCISE SET 6.2

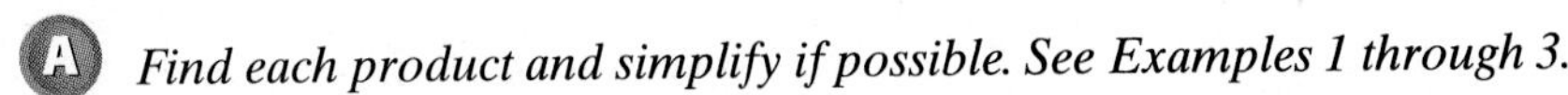

A *Find each product and simplify if possible. See Examples 1 through 3.*

1. $\frac{3x}{y^2} \cdot \frac{7y}{4x}$

2. $\frac{9x^2}{y} \cdot \frac{4y}{3x^2}$

3. $\frac{8x}{2} \cdot \frac{x^5}{4x^2}$

4. $\frac{6x^2}{10x^3} \cdot \frac{5x}{12}$

5. $-\frac{5a^2b}{30a^2b^2} \cdot b^3$

6. $-\frac{9x^3y^2}{18xy^5} \cdot y^3$

7. $\frac{x}{2x - 14} \cdot \frac{x^2 - 7x}{5}$

8. $\frac{4x - 24}{20x} \cdot \frac{5}{x - 6}$

9. $\frac{6x + 6}{5} \cdot \frac{10}{36x + 36}$

10. $\frac{x^2 + x}{8} \cdot \frac{16}{x + 1}$

11. $\frac{m^2 - n^2}{m + n} \cdot \frac{m}{m^2 - mn}$

12. $\frac{(m - n)^2}{m + n} \cdot \frac{m}{m^2 - mn}$

13. $\frac{x^2 - 25}{x^2 - 3x - 10} \cdot \frac{x + 2}{x}$

14. $\frac{a^2 + 6a + 9}{a^2 - 4} \cdot \frac{a + 3}{a - 2}$

△ **15.** Find the area of the rectangle.

$\frac{2x}{x^2 - 25}$ feet

$\frac{x + 5}{9x}$ feet

△ **16.** Find the area of the square.

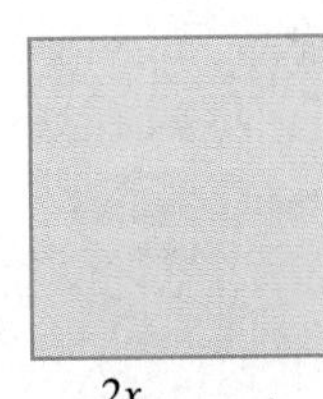

$\frac{2x}{5x^2 + 3x}$ meters

B *Find each quotient and simplify. See Examples 4 through 7.*

17. $\frac{5x^7}{2x^5} \div \frac{10x}{4x^3}$

18. $\frac{9y^4}{6y} \div \frac{y^2}{3}$

19. $\frac{8x^2}{y^3} \div \frac{4x^2y^3}{6}$

20. $\frac{7a^2b}{3ab^2} \div \frac{21a^2b^2}{14ab}$

21. $\frac{(x-6)(x+4)}{4x} \div \frac{2x-12}{8x^2}$

22. $\frac{(x+3)^2}{5} \div \frac{5x+15}{25}$

23. $\frac{3x^2}{x^2-1} \div \frac{x^5}{(x+1)^2}$

24. $\frac{x+1}{(x+1)(2x+3)} \div \frac{20}{2x+3}$

25. $\frac{m^2-n^2}{m+n} \div \frac{m}{m^2+nm}$

26. $\frac{(m-n)^2}{m+n} \div \frac{m^2-mn}{m}$

27. $\frac{x+2}{7-x} \div \frac{x^2-5x+6}{x^2-9x+14}$

28. $(x-3) \div \frac{x^2+3x-18}{x}$

29. $\frac{x^2+7x+10}{x-1} \div \frac{x^2+2x-15}{x-1}$

30. $\frac{a^2-b^2}{9} \div \frac{3b-3a}{27x^2}$

C *Multiply or divide as indicated. See Example 8.*

31. $\frac{5x-10}{12} \div \frac{4x-8}{8}$

32. $\frac{6x+6}{5} \div \frac{3x+3}{10}$

33. $\frac{x^2+5x}{8} \cdot \frac{9}{3x+15}$

34. $\frac{3x^2+12x}{6} \cdot \frac{9}{2x+8}$

35. $\frac{7}{6p^2+q} \div \frac{14}{18p^2+3q}$

36. $\frac{3x+6}{20} \div \frac{4x+8}{8}$

37. $\frac{3x+4y}{x^2+4xy+4y^2} \cdot \frac{x+2y}{2}$

38. $\frac{x^2-y^2}{3x^2+3xy} \cdot \frac{3x^2+6x}{3x^2-2xy-y^2}$

39. $\frac{(x+2)^2}{x-2} \div \frac{x^2-4}{2x-4}$

40. $\frac{x^2-4}{2y} \div \frac{2-x}{6xy}$

41. $\frac{3y}{3-x} \div \frac{12xy}{x^2-9}$

42. $\frac{x+3}{x^2-9} \div \frac{5x+15}{(x-3)^2}$

43. $\dfrac{a^2 + 7a + 12}{a^2 + 5a + 6} \cdot \dfrac{a^2 + 8a + 15}{a^2 + 5a + 4}$

44. $\dfrac{b^2 + 2b - 3}{b^2 + b - 2} \cdot \dfrac{b^2 - 4}{b^2 + 6b + 8}$

D *Convert as indicated. See Examples 9 and 10.*

45. 10 square feet = ________ square inches.

46. 1008 square inches = _____ square feet.

47. 90,000 cubic inches = ________ cubic yards (round to the nearest hundredth).

48. 2 cubic yards = __________ cubic inches.

49. 50 miles per hour = ______ feet per second (round to the nearest whole).

50. 10 feet per second = _______ miles per hour (round to the nearest tenth).

51. The Pentagon, headquarters for the Department of Defense, contains 3,705,793 square feet of office and storage space. Convert this to square yards. Round to the nearest square yard. (*Source:* U.S. Department of Defense)

52. The world's tallest building, the Petronas Twin Towers in Kuala Lumpur, Malaysia, has 408,742 square yards of floor space. Convert this to square feet. (*Source:* KLCC Group)

53. In October 1997, driver Andy Green set the current world land speed record of 763 miles per hour in a specially designed car, Thrust SSC. Find this speed in feet per second. Round to the nearest whole. (*Source:* Federation International de L'Automobile)

54. Maurice Greene of the United States holds the current world record for the men's 100-meter track event. In 1999, he covered the distance at an average speed of 33.5 feet per second. Convert this speed to miles per hour. Round to the nearest tenth. (*Source:* International Amateur Athletic Federation)

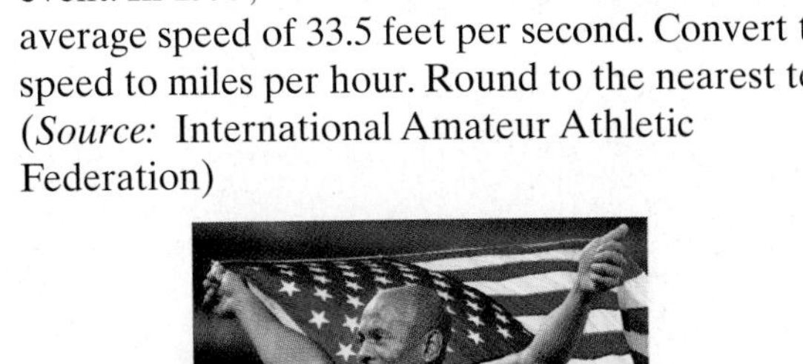

Review and Preview

Perform each indicated operation. See Section R.2.

55. $\dfrac{1}{5} + \dfrac{4}{5}$

56. $\dfrac{3}{15} + \dfrac{6}{15}$

57. $\dfrac{9}{9} - \dfrac{19}{9}$

58. $\dfrac{4}{3} - \dfrac{8}{3}$

59. $\dfrac{6}{5} + \left(\dfrac{1}{5} - \dfrac{8}{5}\right)$

60. $-\dfrac{3}{2} + \left(\dfrac{1}{2} - \dfrac{3}{2}\right)$

Combining Concepts

Multiply or divide as indicated.

61. $\left(\dfrac{x^2 - y^2}{x^2 + y^2} \div \dfrac{x^2 - y^2}{3x}\right) \cdot \dfrac{x^2 + y^2}{6}$

62. $\left(\dfrac{x^2 - 9}{x^2 - 1} \cdot \dfrac{x^2 + 2x + 1}{2x^2 + 9x + 9}\right) \div \dfrac{2x + 3}{1 - x}$

63. $\left(\dfrac{2a + b}{b^2} \cdot \dfrac{3a^2 - 2ab}{ab + 2b^2}\right) \div \dfrac{a^2 - 3ab + 2b^2}{5ab - 10b^2}$

64. $\left(\dfrac{x^2y^2 - xy}{4x - 4y} \div \dfrac{3y - 3x}{8x - 8y}\right) \cdot \dfrac{y - x}{8}$

65. In your own words, explain how you multiply rational expressions.

66. Explain how dividing rational expressions is similar to dividing rational numbers.

67. On June 1, 2001, 1 euro was equivalent to 0.85 U.S. dollar. If you had wanted to exchange \$2000 U.S. to euros on that day for a European vacation, how much would you have received? Round to the nearest hundredth. (*Source:* International Monetary Fund)

68. An environmental technician finds that warm water from an industrial process is being discharged into a nearby pond at a rate of 30 gallons per minute. Plant regulations state that the flow rate should be no more than 0.1 cubic feet per second. Is the flow rate of 30 gallons per minute in violation of the plant regulations? (*Hint:* 1 cubic foot is equivalent to 7.48 gallons.)

6.3 Adding and Subtracting Rational Expressions with the Same Denominator and Least Common Denominator

OBJECTIVES

- (A) Add and subtract rational expressions with common denominators.
- (B) Find the least common denominator of a list of rational expressions.
- (C) Write a rational expression as an equivalent expression whose denominator is given.

SSM TUTOR CENTER SG CD & VIDEO MATH PRO WEB

(A) Adding and Subtracting Rational Expressions with the Same Denominator

Like multiplication and division, addition and subtraction of rational expressions is similar to addition and subtraction of rational numbers. In this section, we add and subtract rational expressions with a common denominator.

Add: $\frac{6}{5} + \frac{2}{5}$ | Add: $\frac{9}{x+2} + \frac{3}{x+2}$

Add the numerators and place the sum over the common denominator.

$$\frac{6}{5} + \frac{2}{5} = \frac{6+2}{5}$$

$$= \frac{8}{5} \quad \text{Simplify.}$$

$$\frac{9}{x+2} + \frac{3}{x+2} = \frac{9+3}{x+2}$$

$$= \frac{12}{x+2} \quad \text{Simplify.}$$

Adding and Subtracting Rational Expressions with Common Denominators

If $\frac{P}{R}$ and $\frac{Q}{R}$ are rational expressions, then

$$\frac{P}{R} + \frac{Q}{R} = \frac{P+Q}{R} \quad \text{and} \quad \frac{P}{R} - \frac{Q}{R} = \frac{P-Q}{R}$$

To add or subtract rational expressions, add or subtract numerators and place the sum or difference over the common denominator.

EXAMPLE 1 Add: $\frac{5m}{2n} + \frac{m}{2n}$

Solution: $\frac{5m}{2n} + \frac{m}{2n} = \frac{5m+m}{2n}$ Add the numerators.

$= \frac{6m}{2n}$ Simplify the numerator by combining like terms.

$= \frac{3m}{n}$ Simplify by applying the fundamental principle.

Practice Problem 1

Add: $\frac{8x}{3y} + \frac{x}{3y}$

EXAMPLE 2 Subtract: $\frac{2y}{2y-7} - \frac{7}{2y-7}$

Solution: $\frac{2y}{2y-7} - \frac{7}{2y-7} = \frac{2y-7}{2y-7}$ Subtract the numerators.

$= \frac{1}{1}$ or 1 Simplify.

Practice Problem 2

Subtract: $\frac{3x}{3x-7} - \frac{7}{3x-7}$

Answers

1. $\frac{3x}{y}$, **2.** 1

Practice Problem 3

Subtract: $\dfrac{2x^2 + 5x}{x + 2} - \dfrac{4x + 6}{x + 2}$

EXAMPLE 3 Subtract: $\dfrac{3x^2 + 2x}{x - 1} - \dfrac{10x - 5}{x - 1}$

Solution:

$$\frac{3x^2 + 2x}{x - 1} - \frac{10x - 5}{x - 1} = \frac{3x^2 + 2x - (10x - 5)}{x - 1}$$ Subtract the numerators. Notice the parentheses.

$$= \frac{3x^2 + 2x - 10x + 5}{x - 1}$$ Use the distributive property.

$$= \frac{3x^2 - 8x + 5}{x - 1}$$ Combine like terms.

$$= \frac{(x - 1)(3x - 5)}{x - 1}$$ Factor.

$$= 3x - 5$$ Simplify.

Helpful Hint

Notice how the numerator $10x - 5$ was subtracted in Example 3.

This − sign applies to the entire numerator of $10x - 5$.

So parentheses are inserted here to indicate this.

$$\frac{3x^2 + 2x}{x - 1} - \frac{10x - 5}{x - 1} = \frac{3x^2 + 2x - (10x - 5)}{x - 1}$$

B Finding the Least Common Denominator

Recall from Chapter R that to add and subtract fractions with different denominators, we first find a least common denominator (LCD). Then we write all fractions as equivalent fractions with the LCD.

For example, suppose we add $\frac{8}{3}$ and $\frac{2}{5}$. The LCD of denominators 3 and 5 is 15, since 15 is the smallest number that both 3 and 5 divide into evenly. So we rewrite each fraction so that its denominator is 15. (Notice how we apply the fundamental principle.)

$$\frac{8}{3} + \frac{2}{5} = \frac{8(5)}{3(5)} + \frac{2(3)}{5(3)} = \frac{40}{15} + \frac{6}{15} = \frac{40 + 6}{15} = \frac{46}{15}$$

To add or subtract rational expressions with different denominators, we also first find an LCD and then write all rational expressions as equivalent expressions with the LCD. The **least common denominator (LCD) of a list of rational expressions** is a polynomial of least degree whose factors include all the factors of the denominators in the list.

To Find the Least Common Denominator (LCD)

Step 1. Factor each denominator completely.

Step 2. The least common denominator (LCD) is the product of all unique factors found in Step 1, each raised to a power equal to the greatest number of times that the factor appears in any one factored denominator.

Answer

3. $2x - 3$

EXAMPLE 4 Find the LCD for each pair.

a. $\frac{1}{8}, \frac{3}{22}$ **b.** $\frac{7}{5x}, \frac{6}{15x^2}$

Solution:

a. We start by finding the prime factorization of each denominator.

$$8 = 2^3 \quad \text{and}$$
$$22 = 2 \cdot 11$$

Next we write the product of all the unique factors, each raised to a power equal to the greatest number of times that the factor appears.

The greatest number of times that the factor 2 appears is 3.

The greatest number of times that the factor 11 appears is 1.

$$\text{LCD} = 2^3 \cdot 11^1 = 8 \cdot 11 = 88$$

b. We factor each denominator.

$$5x = 5 \cdot x \quad \text{and}$$
$$15x^2 = 3 \cdot 5 \cdot x^2$$

The greatest number of times that the factor 5 appears is 1.

The greatest number of times that the factor 3 appears is 1.

The greatest number of times that the factor x appears is 2.

$$\text{LCD} = 3^1 \cdot 5^1 \cdot x^2 = 15x^2$$

EXAMPLE 5 Find the LCD of $\frac{7x}{x+2}$ and $\frac{5x^2}{x-2}$.

Solution: The denominators $x + 2$ and $x - 2$ are completely factored already. The factor $x + 2$ appears once and the factor $x - 2$ appears once.

$$\text{LCD} = (x + 2)(x - 2)$$

EXAMPLE 6 Find the LCD of $\frac{6m^2}{3m + 15}$ and $\frac{2}{(m + 5)^2}$.

Solution: We factor each denominator.

$$3m + 15 = 3(m + 5)$$
$$(m + 5)^2 = (m + 5)^2 \quad \text{This denominator is already factored.}$$

The greatest number of times that the factor 3 appears is 1.

The greatest number of times that the factor $m + 5$ appears *in any one denominator* is 2.

$$\text{LCD} = 3(m + 5)^2$$

Try the Concept Check in the margin.

EXAMPLE 7 Find the LCD of $\frac{t - 10}{t^2 - t - 6}$ and $\frac{t + 5}{t^2 + 3t + 2}$.

Solution:

$$t^2 - t - 6 = (t - 3)(t + 2)$$
$$t^2 + 3t + 2 = (t + 1)(t + 2)$$
$$\text{LCD} = (t - 3)(t + 2)(t + 1)$$

Practice Problem 4

Find the LCD for each pair.

a. $\frac{2}{9}, \frac{7}{15}$ b. $\frac{5}{6x^3}, \frac{11}{8x^5}$

Practice Problem 5

Find the LCD of $\frac{3a}{a + 5}$ and $\frac{7a}{a - 5}$.

Practice Problem 6

Find the LCD of $\frac{7x^2}{(x - 4)^2}$ and $\frac{5x}{3x - 12}$.

Concept Check

Choose the correct LCD of $\frac{x}{(x + 1)^2}$ and $\frac{5}{x + 1}$.

a. $x + 1$ b. $(x + 1)^2$
c. $(x + 1)^3$ d. $5x(x + 1)^2$

Practice Problem 7

Find the LCD of $\frac{y + 5}{y^2 + 2y - 3}$ and $\frac{y + 4}{y^2 - 3y + 2}$.

Answers

4. a. 45, **b.** $24x^5$ **5.** $(a + 5)(a - 5)$, **6.** $3(x - 4)^2$, **7.** $(y + 3)(y - 2)(y - 1)$

Concept Check: b

EXAMPLE 8 Find the LCD of $\dfrac{2}{x-2}$ and $\dfrac{10}{2-x}$.

Solution: The denominators $x - 2$ and $2 - x$ are opposites. That is, $2 - x = -1(x - 2)$. We can use either $x - 2$ or $2 - x$ as the LCD.

$$\text{LCD} = x - 2 \quad \text{or} \quad \text{LCD} = 2 - x$$

Practice Problem 8

Find the LCD of $\dfrac{6}{x-4}$ and $\dfrac{9}{4-x}$.

C Writing Equivalent Rational Expressions

Next we practice writing a rational expression as an equivalent rational expression with a given denominator. To do this, we apply the fundamental principle, which says that $\dfrac{PR}{QR} = \dfrac{P}{Q}$, or equivalently that $\dfrac{P}{Q} = \dfrac{PR}{QR}$. This can be seen by recalling that multiplying an expression by 1 produces an equivalent expression. In other words,

$$\frac{P}{Q} = \frac{P}{Q}\cdot 1 = \frac{P}{Q}\cdot\frac{R}{R} = \frac{PR}{QR}$$

EXAMPLE 9 Write the rational expression as an equivalent rational expression with the given denominator.

$$\frac{4b}{9a} = \frac{\quad}{27a^2b}$$

Solution: We can ask ourselves: "What do we multiply $9a$ by to get $27a^2b$?" The answer is $3ab$, since $9a(3ab) = 27a^2b$. So we multiply the numerator and denominator by $3ab$.

$$\frac{4b}{9a} = \frac{4b(3ab)}{9a(3ab)} = \frac{12ab^2}{27a^2b}$$

Practice Problem 9

Write the rational expression as an equivalent rational expression with the given denominator.

$$\frac{2x}{5y} = \frac{\quad}{20x^2y^2}$$

EXAMPLE 10 Write the rational expression as an equivalent rational expression with the given denominator.

$$\frac{5}{x^2-4} = \frac{\quad}{(x-2)(x+2)(x-4)}$$

Solution: First we factor the denominator $x^2 - 4$ as $(x - 2)(x + 2)$. If we multiply the original denominator $(x - 2)(x + 2)$ by $x - 4$, the result is the new denominator $(x - 2)(x + 2)(x - 4)$. Thus, we multiply the numerator and the denominator by $x - 4$.

$$\frac{5}{x^2-4} = \frac{5}{(x-2)(x+2)} = \frac{5(x-4)}{(x-2)(x+2)(x-4)} = \frac{5x-20}{(x-2)(x+2)(x-4)}$$

Practice Problem 10

Write the rational expression as an equivalent rational expression with the given denominator.

$$\frac{3}{x^2-25} = \frac{\quad}{(x+5)(x-5)(x-3)}$$

Answers

8. $(x-4)$ or $(4-x)$, **9.** $\dfrac{8x^3y}{20x^2y^2}$, **10.** $\dfrac{3x-9}{(x+5)(x-5)(x-3)}$

Name ______________________ Section __________ Date __________

Mental Math

Perform each indicated operation.

1. $\frac{2}{3} + \frac{1}{3}$ **2.** $\frac{5}{11} + \frac{1}{11}$ **3.** $\frac{3x}{9} + \frac{4x}{9}$ **4.** $\frac{3y}{8} + \frac{2y}{8}$

5. $\frac{8}{9} - \frac{7}{9}$ **6.** $\frac{14}{12} - \frac{3}{12}$ **7.** $\frac{7y}{5} + \frac{10y}{5}$ **8.** $\frac{12x}{7} - \frac{4x}{7}$

EXERCISE SET 6.3

A *Add or subtract as indicated. Simplify the result if possible. See Examples 1 through 3.*

1. $\frac{a}{13} + \frac{9}{13}$ **2.** $\frac{x+1}{7} + \frac{6}{7}$ **3.** $\frac{4m}{3n} + \frac{5m}{3n}$ **4.** $\frac{3p}{2} + \frac{11p}{2}$ **5.** $\frac{4m}{m-6} - \frac{24}{m-6}$

6. $\frac{8y}{y-2} - \frac{16}{y-2}$ **7.** $\frac{9}{3+y} + \frac{y+1}{3+y}$ **8.** $\frac{9}{y+9} + \frac{y}{y+9}$ **9.** $\frac{5x+4}{x-1} - \frac{2x+7}{x-1}$

10. $\frac{x^2+9x}{x+7} - \frac{4x+14}{x+7}$ **11.** $\frac{a}{a^2+2a-15} - \frac{3}{a^2+2a-15}$ **12.** $\frac{3y}{y^2+3y-10} - \frac{6}{y^2+3y-10}$

13. $\frac{2x+3}{x^2-x-30} - \frac{x-2}{x^2-x-30}$ **14.** $\frac{3x-1}{x^2+5x-6} - \frac{2x-7}{x^2+5x-6}$

△ **15.** A square has a side of length $\frac{5}{x-2}$ meters. Express its perimeter as a rational expression.

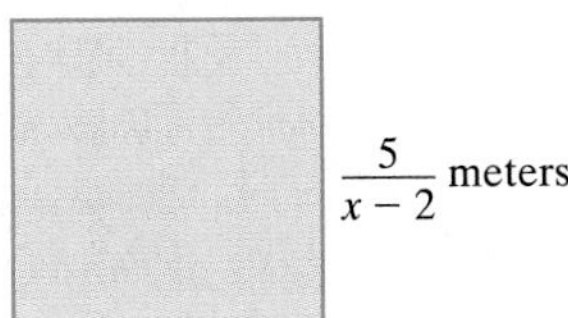

△ **16.** A trapezoid has sides of the indicated lengths. Find its perimeter.

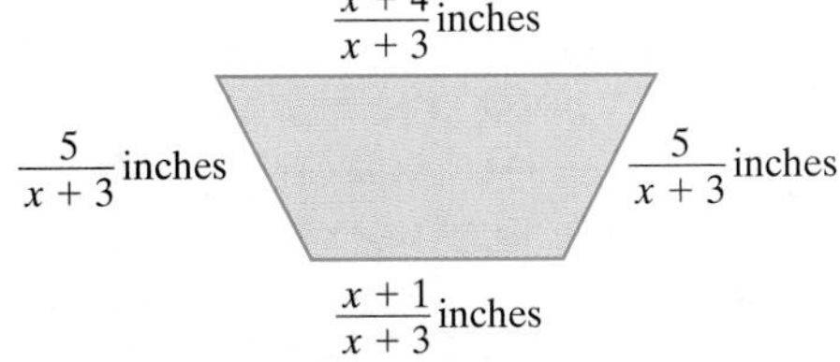

17. In your own words, describe how to add or subtract two rational expressions with the same denominators.

18. Explain the similarities between subtracting $\frac{3}{8}$ from $\frac{7}{8}$ and subtracting $\frac{6}{x+3}$ from $\frac{9}{x+3}$.

B *Find the LCD for each list of rational expressions. See Examples 4 through 8.*

19. $\frac{19}{2x}, \frac{5}{4x^3}$

20. $\frac{17x}{4y^5}, \frac{2}{8y}$

21. $\frac{9}{8x}, \frac{3}{2x + 4}$

22. $\frac{1}{6y}, \frac{3x}{4y + 12}$

23. $\frac{2}{x + 3}, \frac{5}{x - 2}$

24. $\frac{-6}{x - 1}, \frac{4}{x + 5}$

25. $\frac{x}{x + 6}, \frac{10}{3x + 18}$

26. $\frac{12}{x + 5}, \frac{x}{4x + 20}$

27. $\frac{1}{3x + 3}, \frac{8}{2x^2 + 4x + 2}$

28. $\frac{19x + 5}{4x - 12}, \frac{3}{2x^2 - 12x + 18}$

29. $\frac{5}{x - 8}, \frac{3}{8 - x}$

30. $\frac{2x + 5}{3x - 7}, \frac{5}{7 - 3x}$

31. $\frac{5x + 1}{x^2 + 3x - 4}, \frac{3x}{x^2 + 2x - 3}$

32. $\frac{4}{x^2 + 4x + 3}, \frac{4x - 2}{x^2 + 10x + 21}$

33. Write some instructions to help a friend who is having difficulty finding the LCD of two rational expressions.

34. Explain why the LCD of the rational expressions $\frac{7}{x + 1}$ and $\frac{9x}{(x + 1)^2}$ is $(x + 1)^2$ and not $(x + 1)^3$.

C *Rewrite each rational expression as an equivalent rational expression with the given denominator. See Examples 9 and 10.*

35. $\frac{3}{2x} = \frac{\quad}{4x^2}$

36. $\frac{3}{9y^5} = \frac{\quad}{72y^9}$

37. $\frac{6}{3a} = \frac{\quad}{12ab^2}$

38. $\frac{17a}{4y^2x} = \frac{\quad}{32y^3x^2z}$

39. $\frac{9}{x + 3} = \frac{\quad}{2(x + 3)}$

40. $\frac{4x + 1}{3x + 6} = \frac{\quad}{3y(x + 2)}$

41. $\frac{9a + 2}{5a + 10} = \frac{\quad}{5b(a + 2)}$

42. $\frac{5 + y}{2x^2 + 10} = \frac{\quad}{4(x^2 + 5)}$

43. $\frac{x}{x^3 + 6x^2 + 8x} = \frac{\quad}{x(x + 4)(x + 2)(x + 1)}$

44. $\frac{5x}{x^2 + 2x - 3} = \frac{\quad}{(x - 1)(x - 5)(x + 3)}$

45. $\frac{9y - 1}{15x^2 - 30} = \frac{\quad}{30x^2 - 60}$

46. $\frac{6}{x^2 - 9} = \frac{\quad}{(x + 3)(x - 3)(x + 2)}$

Review and Preview

Perform each indicated operation. See Section R.2.

47. $\frac{2}{3} + \frac{5}{7}$

48. $\frac{9}{10} - \frac{3}{5}$

49. $\frac{2}{6} - \frac{3}{4}$

50. $\frac{11}{15} + \frac{5}{9}$

51. $\frac{1}{12} + \frac{3}{20}$

52. $\frac{7}{30} + \frac{3}{18}$

Combining Concepts

53. You are throwing a barbecue and you want to make sure that you purchase the same number of hot dogs as hot dog buns. Hot dogs come 8 to a package and hot dog buns come 12 to a package. What is the least number of each type of package you should buy?

54. The planet Mercury revolves around the sun in 88 Earth days. It takes Jupiter 4332 Earth days to make one revolution around the sun. (*Source:* National Space Science Data Center) If the two planets are aligned as shown in the figure, how long will it take for them to align again?

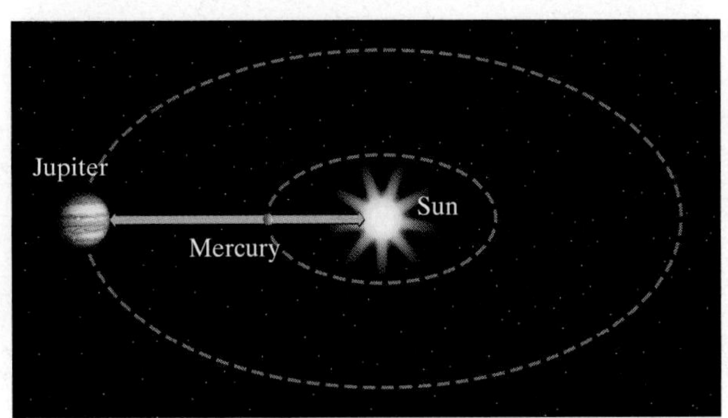

55. An algebra student approaches you with a problem. He's tried to subtract two rational expressions, but his result does not match the book's. Check to see if the student has made an error. If so, correct his work shown below.

$$\frac{2x-6}{x-5} - \frac{x+4}{x-5}$$
$$= \frac{2x-6-x+4}{x-5}$$
$$= \frac{x-2}{x-5}$$

FOCUS ON Business and Career

Fast-Growing Careers

According to U.S. Bureau of Labor Statistics projections, the careers listed below will have the largest job growth in the next decade.

	Employment (Numbers in thousands)		
Occupation	**1998**	**2008**	**Change**
1. Systems analysts	617	1,194	+577
2. Retail salespersons	4,056	4,620	+563
3. Cashiers	3,198	3,754	+556
4. General managers and top executives	3,362	3,913	+551
5. Truck drivers, light and heavy	2,970	3,463	+493
6. Office clerks, general	3,021	3,484	+463
7. Registered nurses	2,079	2,530	+451
8. Computer support specialists	429	869	+439
9. Personal care and home health aides	746	1,179	+433
10. Teacher assistants	1,192	1,567	+375

(*Source:* Bureau of Labor Statistics, U.S. Department of Labor)

What do all of these in-demand occupations have in common? They all require a knowledge of math! For some careers like systems analysts, salespersons, cashiers, and nurses, the ways math is used on the job may be obvious. For other occupations, the use of math may not be quite as obvious. However, tasks common to many jobs like filling in a time sheet or a mileage log, writing up an expense report, planning a budget, figuring a bill, ordering supplies, completing a packing list, and even making a work schedule all require math.

Critical Thinking

Suppose that your college placement office is planning to publish an occupational handbook on math in popular occupations. Choose one of the occupations from the list above that interests you. Research the occupation. Then write a brief entry for the occupational handbook that describes how a person in that career would use math in his or her job. Include an example if possible.

6.4 Adding and Subtracting Rational Expressions with Different Denominators

OBJECTIVE

Ⓐ Add and subtract rational expressions with different denominators.

Ⓐ Adding and Subtracting Rational Expressions with Different Denominators

In the previous section, we practiced all the skills we need to add and subtract rational expressions with different denominators. The steps are as follows.

To Add or Subtract Rational Expressions with Different Denominators

Step 1. Find the LCD of the rational expressions.

Step 2. Rewrite each rational expression as an equivalent expression whose denominator is the LCD found in Step 1.

Step 3. Add or subtract numerators and write the sum or difference over the common denominator.

Step 4. Simplify or write the rational expression in lowest terms.

EXAMPLE 1 Perform each indicated operation.

a. $\frac{a}{4} - \frac{2a}{8}$ **b.** $\frac{3}{10x^2} + \frac{7}{25x}$

Solution:

a. First, we must find the LCD. Since $4 = 2^2$ and $8 = 2^3$, the LCD $= 2^3 = 8$. Next we write each fraction as an equivalent fraction with the denominator 8, and then we subtract.

$$\frac{a}{4} - \frac{2a}{8} = \frac{a(2)}{4(2)} - \frac{2a}{8} = \frac{2a}{8} - \frac{2a}{8} = \frac{2a - 2a}{8} = \frac{0}{8} = 0$$

b. Since $10x^2 = 2 \cdot 5 \cdot x \cdot x$ and $25x = 5 \cdot 5 \cdot x$, the LCD $= 2 \cdot 5^2 \cdot x^2 = 50x^2$. We write each fraction as an equivalent fraction with a denominator of $50x^2$.

$$\frac{3}{10x^2} + \frac{7}{25x} = \frac{3(5)}{10x^2(5)} + \frac{7(2x)}{25x(2x)}$$

$$= \frac{15}{50x^2} + \frac{14x}{50x^2}$$

$$= \frac{15 + 14x}{50x^2}$$ Add numerators. Write the sum over the common denominator.

Practice Problem 1

Perform each indicated operation.

a. $\frac{y}{5} - \frac{3y}{15}$ b. $\frac{5}{8x} + \frac{11}{10x^2}$

EXAMPLE 2 Subtract: $\frac{6x}{x^2 - 4} - \frac{3}{x + 2}$

Solution: Since $x^2 - 4 = (x + 2)(x - 2)$, the LCD $= (x - 2)(x + 2)$. We write equivalent expressions with the LCD as denominators.

Practice Problem 2

Subtract: $\frac{10x}{x^2 - 9} - \frac{5}{x + 3}$

Answers

1. a. 0, **b.** $\frac{25x + 44}{40x^2}$, **2.** $\frac{5}{x - 3}$

$$\frac{6x}{x^2-4}-\frac{3}{x+2}=\frac{6x}{(x-2)(x+2)}-\frac{3(x-2)}{(x+2)(x-2)}$$

$$=\frac{6x-3(x-2)}{(x+2)(x-2)}$$ Subtract numerators. Write the difference over the common denominator.

$$=\frac{6x-3x+6}{(x+2)(x-2)}$$ Apply the distributive property in the numerator.

$$=\frac{3x+6}{(x+2)(x-2)}$$ Combine like terms in the numerator.

Next we factor the numerator to see if this rational expression can be simplified.

$$\frac{3x+6}{(x+2)(x-2)}=\frac{3(x+2)}{(x+2)(x-2)}$$ Factor.

$$=\frac{3}{x-2}$$ Apply the fundamental principle to simplify.

Practice Problem 3

Add: $\frac{5}{7x}+\frac{2}{x+1}$

EXAMPLE 3 Add: $\frac{2}{3t}+\frac{5}{t+1}$

Solution: The LCD is $3t(t+1)$. We write each rational expression as an equivalent rational expression with a denominator of $3t(t+1)$.

$$\frac{2}{3t}+\frac{5}{t+1}=\frac{2(t+1)}{3t(t+1)}+\frac{5(3t)}{(t+1)(3t)}$$

$$=\frac{2(t+1)+5(3t)}{3t(t+1)}$$ Add numerators. Write the sum over the common denominator.

$$=\frac{2t+2+15t}{3t(t+1)}$$ Apply the distributive property in the numerator.

$$=\frac{17t+2}{3t(t+1)}$$ Combine like terms in the numerator.

Practice Problem 4

Subtract: $\frac{10}{x-6}-\frac{15}{6-x}$

EXAMPLE 4 Subtract: $\frac{7}{x-3}-\frac{9}{3-x}$

Solution: To find a common denominator, we notice that $x-3$ and $3-x$ are opposites. That is, $3-x=-(x-3)$. We write the denominator $3-x$ as $-(x-3)$ and simplify.

$$\frac{7}{x-3}-\frac{9}{3-x}=\frac{7}{x-3}-\frac{9}{-(x-3)}$$

$$=\frac{7}{x-3}-\frac{-9}{x-3}$$ Apply $\frac{a}{-b}=\frac{-a}{b}$.

$$=\frac{7-(-9)}{x-3}$$ Subtract numerators. Write the difference over the common denominator.

$$=\frac{16}{x-3}$$

Practice Problem 5

Add: $2+\frac{x}{x+5}$

EXAMPLE 5 Add: $1+\frac{m}{m+1}$

Solution: Recall that 1 is the same as $\frac{1}{1}$. The LCD of $\frac{1}{1}$ and $\frac{m}{m+1}$ is $m+1$.

Answers

3. $\frac{19x+5}{7x(x+1)}$, **4.** $\frac{25}{x-6}$, **5.** $\frac{3x+10}{x+5}$

$$1 + \frac{m}{m+1} = \frac{1}{1} + \frac{m}{m+1}$$ Write 1 as $\frac{1}{1}$.

$$= \frac{1(m+1)}{1(m+1)} + \frac{m}{m+1}$$ Multiply both the numerator and the denominator of $\frac{1}{1}$ by $m + 1$.

$$= \frac{m+1+m}{m+1}$$ Add numerators. Write the sum over the common denominator.

$$= \frac{2m+1}{m+1}$$ Combine like terms in the numerator.

EXAMPLE 6 Subtract: $\frac{3}{2x^2+x} - \frac{2x}{6x+3}$

Solution: First, we factor the denominators.

$$\frac{3}{2x^2+x} - \frac{2x}{6x+3} = \frac{3}{x(2x+1)} - \frac{2x}{3(2x+1)}$$

The LCD is $3x(2x + 1)$. We write equivalent expressions with denominators of $3x(2x + 1)$.

$$\frac{3}{x(2x+1)} - \frac{2x}{3(2x+1)} = \frac{3(3)}{x(2x+1)(3)} - \frac{2x(x)}{3(2x+1)(x)}$$

$$= \frac{9-2x^2}{3x(2x+1)}$$ Subtract numerators. Write the difference over the common denominator.

Practice Problem 6

Subtract: $\frac{4}{3x^2+2x} - \frac{3x}{12x+8}$

EXAMPLE 7 Add: $\frac{2x}{x^2+2x+1} + \frac{x}{x^2-1}$

Solution: First we factor the denominators.

$$\frac{2x}{x^2+2x+1} + \frac{x}{x^2-1}$$

$$= \frac{2x}{(x+1)(x+1)} + \frac{x}{(x+1)(x-1)}$$

Rewrite each expression with LCD $(x + 1)(x + 1)(x - 1)$.

$$= \frac{2x(x-1)}{(x+1)(x+1)(x-1)} + \frac{x(x+1)}{(x+1)(x-1)(x+1)}$$

$$= \frac{2x(x-1)+x(x+1)}{(x+1)^2(x-1)}$$ Add numerators. Write the sum over the common denominator.

$$= \frac{2x^2-2x+x^2+x}{(x+1)^2(x-1)}$$ Apply the distributive property in the numerator.

$$= \frac{3x^2-x}{(x+1)^2(x-1)} \text{ or } \frac{x(3x-1)}{(x+1)^2(x-1)}$$

The numerator was factored as a last step to see if the rational expression could be simplified further. Since there are no factors common to the numerator and the denominator, we can't simplify further.

Practice Problem 7

Add: $\frac{6x}{x^2+4x+4} + \frac{x}{x^2-4}$

Answers

6. $\frac{16-3x^2}{4x(3x+2)}$, **7.** $\frac{x(7x-10)}{(x+2)^2(x-2)}$

STUDY SKILLS REMINDER

How are you doing?

If you haven't done so yet, take a few moments and think about how you are doing in this course. Are you working toward your goal of successfully completing this course? Is your performance on homework, quizzes, and tests satisfactory? If not, you might want to see your instructor to see if he/she has any suggestions on how you can improve your performance. Let me once again remind you that, in addition to your instructor, there are many places to get help with your mathematics course. A few suggestions are below.

- This text has an accompanying video lesson for every section in this text.
- The back of this book contains answers to odd-numbered exercises and selected solutions.
- MathPro is available with this text. It is a tutorial software program with lessons corresponding to each section in the text.
- There is a student solutions manual available that contains worked-out solutions to odd-numbered exercises as well as solutions to every exercise in the Chapter Pretests, Integrated Reviews, Chapter Reviews, Chapter Tests, and Cumulative Reviews.
- Don't forget to check with your instructor for other local resources available to you, such as a tutor center.

Name ______________________ Section ____________ Date ____________

EXERCISE SET 6.4

A *Perform each indicated operation. Simplify if possible. See Examples 1 through 7.*

1. $\frac{4}{2x} + \frac{9}{3x}$

2. $\frac{15}{7a} + \frac{8}{6a}$

3. $\frac{15a}{b} + \frac{6b}{5}$

4. $\frac{4c}{d} - \frac{8x}{5}$

5. $\frac{3}{x} + \frac{5}{2x^2}$

6. $\frac{14}{3x^2} + \frac{6}{x}$

7. $\frac{6}{x+1} + \frac{10}{2x+2}$

8. $\frac{8}{x+4} - \frac{3}{3x+12}$

9. $\frac{3}{x+2} - \frac{1}{x^2-4}$

10. $\frac{15}{2x-4} + \frac{x}{x^2-4}$

11. $\frac{3}{4x} + \frac{8}{x-2}$

12. $\frac{5}{y^2} - \frac{y}{2y+1}$

13. $\frac{6}{x-3} + \frac{8}{3-x}$

14. $\frac{9}{x-3} + \frac{9}{3-x}$

15. $\frac{-8}{x^2-1} - \frac{7}{1-x^2}$

16. $\frac{-9}{25x^2-1} + \frac{7}{1-25x^2}$

17. $\frac{5}{x} + 2$

18. $\frac{7}{x^2} - 5x$

19. $\frac{5}{x-2} + 6$

20. $\frac{6y}{y+5} + 1$

21. $\frac{y+2}{y+3} - 2$

22. $\frac{7}{2x-3} - 3$

23. $\frac{-x+2}{x} - \frac{x-6}{4x}$

24. $\frac{-y+1}{y} - \frac{2y-5}{3y}$

25. $\frac{5x}{x+2} - \frac{3x-4}{x+2}$

26. $\frac{7x}{x-3} - \frac{4x+9}{x-3}$

27. $\frac{3x^4}{x} - \frac{4x^2}{x^2}$

28. $\frac{5x}{6} + \frac{15x^2}{2}$

29. $\frac{1}{x+3} - \frac{1}{(x+3)^2}$

30. $\frac{5x}{(x-2)^2} - \frac{3}{x-2}$

31. $\frac{4}{5b} + \frac{1}{b-1}$

32. $\frac{1}{y+5} + \frac{2}{3y}$

33. $\frac{2}{m} + 1$

34. $\frac{6}{x} - 1$

35. $\frac{6}{1-2x} - \frac{4}{2x-1}$

36. $\dfrac{10}{3n-4} - \dfrac{5}{4-3n}$

37. $\dfrac{7}{(x+1)(x-1)} + \dfrac{8}{(x+1)^2}$

38. $\dfrac{5x+2}{(x+1)(x+5)} - \dfrac{2}{x+5}$

39. $\dfrac{x}{x^2-1} - \dfrac{2}{x^2-2x+1}$

40. $\dfrac{x}{x^2-4} - \dfrac{5}{x^2-4x+4}$

41. $\dfrac{3a}{2a+6} - \dfrac{a-1}{a+3}$

42. $\dfrac{1}{x+y} - \dfrac{y}{x^2-y^2}$

43. $\dfrac{y-1}{2y+3} + \dfrac{3}{(2y+3)^2}$

44. $\dfrac{x-6}{5x+1} + \dfrac{6}{(5x+1)^2}$

45. $\dfrac{5}{2-x} + \dfrac{x}{2x-4}$

46. $\dfrac{-1}{a-2} + \dfrac{4}{4-2a}$

47. $\dfrac{-7}{y^2-3y+2} - \dfrac{2}{y-1}$

48. $\dfrac{2}{x^2+4x+4} + \dfrac{1}{x+2}$

49. $\dfrac{13}{x^2-5x+6} - \dfrac{5}{x-3}$

50. $\dfrac{27}{y^2-81} + \dfrac{3}{2(y+9)}$

51. $\dfrac{x+8}{x^2-5x-6} + \dfrac{x+1}{x^2-4x-5}$

52. $\dfrac{x}{x^2+12x+20} - \dfrac{1}{x^2+8x-20}$

53. In your own words, explain how to add two rational expressions with different denominators.

54. In your own words, explain how to subtract two rational expressions with different denominators.

Review and Preview

Solve each linear or quadratic equation. See Sections 2.4 and 5.6.

55. $3x + 5 = 7$

56. $5x - 1 = 8$

57. $2x^2 - x - 1 = 0$

58. $4x^2 - 9 = 0$

59. $4(x + 6) + 3 = -3$

60. $2(3x + 1) + 15 = -7$

Combining Concepts

Perform each indicated operation.

61. $\dfrac{3}{x} - \dfrac{2x}{x^2 - 1} + \dfrac{5}{x + 1}$

62. $\dfrac{5}{x - 2} + \dfrac{7x}{x^2 - 4} - \dfrac{11}{x}$

63. $\dfrac{5}{x^2 - 4} + \dfrac{2}{x^2 - 4x + 4} - \dfrac{3}{x^2 - x - 6}$

64. $\dfrac{8}{x^2 + 6x + 5} - \dfrac{3x}{x^2 + 4x - 5} + \dfrac{2}{x^2 - 1}$

65. $\dfrac{9}{x^2 + 9x + 14} - \dfrac{3x}{x^2 + 10x + 21} + \dfrac{x + 4}{x^2 + 5x + 6}$

66. $\dfrac{x + 10}{x^2 - 3x - 4} - \dfrac{8}{x^2 + 6x + 5} - \dfrac{9}{x^2 + x - 20}$

67. A board of length $\dfrac{3}{x + 4}$ inches was cut into two pieces. If one piece is $\dfrac{1}{x - 4}$ inches, express the length of the other board as a rational expression.

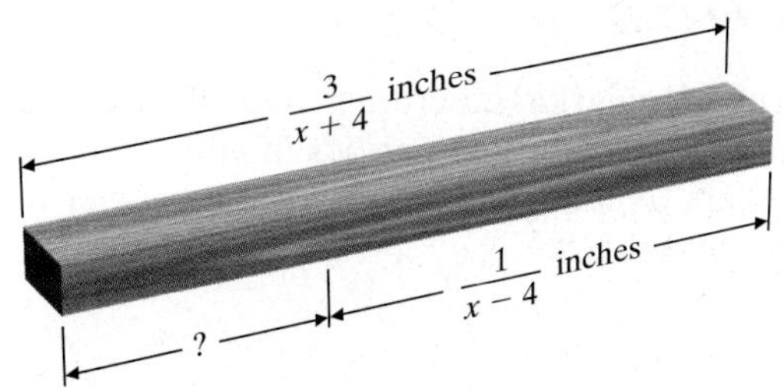

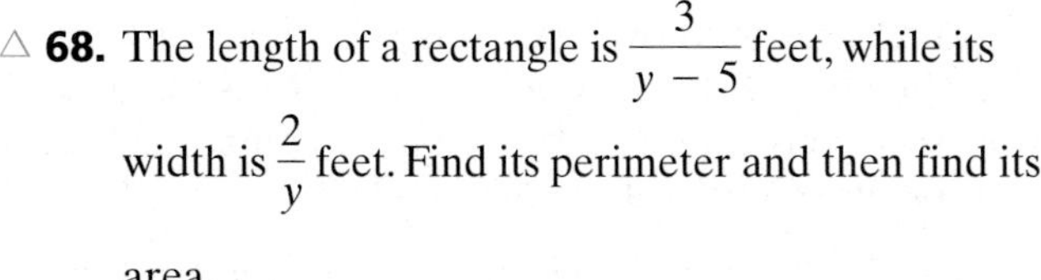

△ **68.** The length of a rectangle is $\dfrac{3}{y - 5}$ feet, while its width is $\dfrac{2}{y}$ feet. Find its perimeter and then find its area.

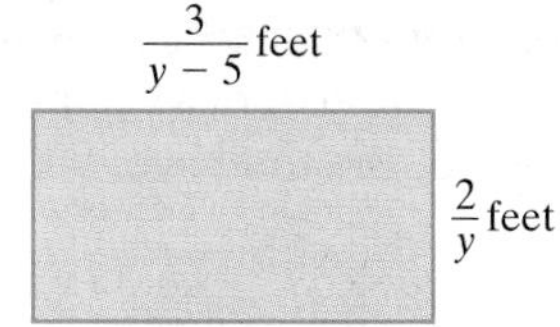

69. In ice hockey, penalty killing percentage is a statistic calculated as $1 - \dfrac{G}{P}$, where G = opponent's power play goals and P = opponent's power play opportunities. Simplify this expression.

70. The dose of medicine prescribed for a child depends on the child's age A in years and the adult dose D for the medication. Two expressions that give a child's dose are Young's Rule, $\dfrac{DA}{A + 12}$, and Cowling's Rule, $\dfrac{D(A + 1)}{24}$. Find an expression for the difference in the doses given by these expressions.

71. Explain when the LCD of the rational expressions in a sum is the product of the denominators.

72. Explain when the LCD is the same as one of the denominators of a rational expression to be added or subtracted.

FOCUS ON History

EPIGRAM OF DIOPHANTUS

One of the great algebraists of ancient times was a man named Diophantus. Little is known of his life other than that he lived and worked in Alexandria. Some historians believe he lived during the first century of the Christian era, about the time of Nero. The only clue to his personal life is the following epigram found in a collection called the Palatine Anthology.

> God granted him youth for a sixth of his life and added a twelfth part to this. He clothed his cheeks in down. He lit him the light of wedlock after a seventh part, and five years after his marriage, He granted him a son. Alas, lateborn wretched child. After attaining the measure of half his father's life, cruel fate overtook him, thus leaving Diophantus during the last four years of his life only such consolation as the science of numbers. How old was Diophantus at his death?*

*From *The Nature and Growth of Modern Mathematics*, Edna Kramer, 1970, Fawcett Premier Books, Vol. 1, pages 107–108.

We are looking for Diophantus' age when he died, so let x represent that age. If we sum the parts of his life, we should get the total age.

Parts of his life:

- $\frac{1}{6} \cdot x + \frac{1}{12} \cdot x$ is the time of his youth.
- $\frac{1}{7} \cdot x$ is the time between his youth and when he married.
- 5 years is the time between his marriage and the birth of his son.
- $\frac{1}{2} \cdot x$ is the time Diophantus had with his son.
- 4 years is the time between his son's death and his own.

The sum of these parts should equal Diophantus' age when he died.

$$\frac{1}{6} \cdot x + \frac{1}{12} \cdot x + \frac{1}{7} \cdot x + 5 + \frac{1}{2} \cdot x + 4 = x$$

CRITICAL THINKING

1. Solve the epigram by solving the equation.

2. How old was Diophantus when his son was born? How old was the son when he died?

3. Solve the following epigram:

   I was four when my mother packed my lunch and sent me off to school. Half my life was spent in school and another sixth was spent on a farm. Alas, hard times befell me. My crops and cattle fared poorly and my land was sold. I returned to school for 3 years and have spent one tenth of my life teaching. How old am I?

GROUP ACTIVITY

4. Write an epigram describing your life. Be sure that none of the time periods in your epigram overlap. Exchange epigrams with a partner to solve and check.

6.5 Solving Equations Containing Rational Expressions

OBJECTIVES

A. Solve equations containing rational expressions.

B. Solve equations containing rational expressions for a specified variable.

A Solving Equations Containing Rational Expressions

In Chapter 2, we solved equations containing fractions. In this section, we continue the work we began in Chapter 2 by solving equations containing rational expressions. For example,

$$\frac{x}{5} + \frac{x+2}{9} = 8 \quad \text{and} \quad \frac{x+1}{9x-5} = \frac{2}{3x}$$

are equations containing rational expressions. To solve equations such as these, we use the multiplication property of equality to clear the equation of fractions by multiplying both sides of the equation by the LCD.

EXAMPLE 1 Solve: $\frac{x}{2} + \frac{8}{3} = \frac{1}{6}$

Solution: The LCD of denominators 2, 3, and 6 is 6, so we multiply both sides of the equation by 6.

$$6\left(\frac{x}{2} + \frac{8}{3}\right) = 6\left(\frac{1}{6}\right)$$

$$6\left(\frac{x}{2}\right) + 6\left(\frac{8}{3}\right) = 6\left(\frac{1}{6}\right) \quad \text{Use the distributive property.}$$

$$3 \cdot x + 16 = 1 \quad \text{Multiply and simplify.}$$

$$3x = -15 \quad \text{Subtract 16 from both sides.}$$

$$x = -5 \quad \text{Divide both sides by 3.}$$

Check: To check, we replace x with -5 in the original equation.

$$\frac{-5}{2} + \frac{8}{3} \stackrel{?}{=} \frac{1}{6} \quad \text{Replace } x \text{ with } -5.$$

$$\frac{1}{6} = \frac{1}{6} \quad \text{True}$$

This number checks, so the solution is -5. ●

Practice Problem 1

Solve: $\frac{x}{4} + \frac{4}{5} = \frac{1}{20}$

Helpful Hint

Make sure that *each* term is multiplied by the LCD.

EXAMPLE 2 Solve: $\frac{t-4}{2} - \frac{t-3}{9} = \frac{5}{18}$

Solution: The LCD of denominators 2, 9, and 18 is 18, so we multiply both sides of the equation by 18.

$$18\left(\frac{t-4}{2} - \frac{t-3}{9}\right) = 18\left(\frac{5}{18}\right)$$

$$18\left(\frac{t-4}{2}\right) - 18\left(\frac{t-3}{9}\right) = 18\left(\frac{5}{18}\right) \quad \text{Use the distributive property.}$$

$$9(t-4) - 2(t-3) = 5 \quad \text{Simplify.}$$

$$9t - 36 - 2t + 6 = 5 \quad \text{Use the distributive property.}$$

$$7t - 30 = 5 \quad \text{Combine like terms.}$$

$$7t = 35$$

$$t = 5 \quad \text{Solve for } t.$$

Practice Problem 2

Solve: $\frac{x+2}{3} - \frac{x-1}{5} = \frac{1}{15}$

Helpful Hint

Multiply *each* term by 18.

Answers

1. $x = -3$, 2. $x = -6$

Check: $\frac{t-4}{2} - \frac{t-3}{9} = \frac{5}{18}$

$\frac{5-4}{2} - \frac{5-3}{9} \stackrel{?}{=} \frac{5}{18}$ Replace t with 5.

$\frac{1}{2} - \frac{2}{9} \stackrel{?}{=} \frac{5}{18}$ Simplify.

$\frac{5}{18} = \frac{5}{18}$ True

The solution is 5.

Recall from Section 6.1 that a rational expression is defined for all real numbers except those that make the denominator of the expression 0. This means that if an equation contains *rational expressions with variables in the denominator*, we must be certain that the proposed solution does not make the denominator 0. If replacing the variable with the proposed solution makes the denominator 0, the rational expression is undefined and this proposed solution must be rejected.

Practice Problem 3

Solve: $2 + \frac{6}{x} = x + 7$

EXAMPLE 3 Solve: $3 - \frac{6}{x} = x + 8$

Solution: In this equation, 0 cannot be a solution because if x is 0, the rational expression $\frac{6}{x}$ is undefined. The LCD is x, so we multiply both sides of the equation by x.

$$x\left(3 - \frac{6}{x}\right) = x(x + 8)$$

$$x(3) - x\left(\frac{6}{x}\right) = x \cdot x + x \cdot 8 \quad \text{Use the distributive property.}$$

Helpful Hint

Multiply *each* term by x.

$$3x - 6 = x^2 + 8x \quad \text{Simplify.}$$

Now we write the quadratic equation in standard form and solve for x.

$$0 = x^2 + 5x + 6$$

$$0 = (x + 3)(x + 2) \quad \text{Factor.}$$

$$x + 3 = 0 \quad \text{or} \quad x + 2 = 0 \quad \text{Set each factor equal to 0 and solve.}$$

$$x = -3 \qquad x = -2$$

Notice that neither -3 nor -2 makes the denominator in the original equation equal to 0.

Check: To check these solutions, we replace x in the original equation by -3, and then by -2.

If $x = -3$:

$3 - \frac{6}{x} = x + 8$

$3 - \frac{6}{-3} \stackrel{?}{=} -3 + 8$

$3 - (-2) \stackrel{?}{=} 5$

$5 = 5$ True

If $x = -2$:

$3 - \frac{6}{x} = x + 8$

$3 - \frac{6}{-2} \stackrel{?}{=} -2 + 8$

$3 - (-3) \stackrel{?}{=} 6$

$6 = 6$ True

Both -3 and -2 are solutions.

Answer

3. $x = -6, x = 1$

The following steps may be used to solve an equation containing rational expressions.

To Solve an Equation Containing Rational Expressions

Step 1. Multiply both sides of the equation by the LCD of all rational expressions in the equation.

Step 2. Remove any grouping symbols and solve the resulting equation.

Step 3. Check the solution in the original equation.

EXAMPLE 4 Solve: $\frac{4x}{x^2 + x - 30} + \frac{2}{x - 5} = \frac{1}{x + 6}$

Solution: The denominator $x^2 + x - 30$ factors as $(x + 6)(x - 5)$. The LCD is then $(x + 6)(x - 5)$, so we multiply both sides of the equation by this LCD.

$$(x + 6)(x - 5)\left(\frac{4x}{x^2 + x - 30} + \frac{2}{x - 5}\right) = (x + 6)(x - 5)\left(\frac{1}{x + 6}\right) \quad \text{Multiply by the LCD.}$$

$$(x + 6)(x - 5) \cdot \frac{4x}{x^2 + x - 30} + (x + 6)(x - 5) \cdot \frac{2}{x - 5} \quad \text{Use the distributive property.}$$

$$= (x + 6)(x - 5) \cdot \frac{1}{x + 6}$$

$$4x + 2(x + 6) = x - 5 \quad \text{Simplify.}$$

$$4x + 2x + 12 = x - 5 \quad \text{Use the distributive property.}$$

$$6x + 12 = x - 5 \quad \text{Combine like terms.}$$

$$5x = -17$$

$$x = -\frac{17}{5} \quad \text{Divide both sides by 5.}$$

Check: Check by replacing x with $-\frac{17}{5}$ in the original equation. The solution is $-\frac{17}{5}$.

Practice Problem 4

Solve: $\frac{2}{x + 3} + \frac{3}{x - 3} = \frac{-2}{x^2 - 9}$

EXAMPLE 5 Solve: $\frac{2x}{x - 4} = \frac{8}{x - 4} + 1$

Solution: Multiply both sides by the LCD, $x - 4$.

$$(x - 4)\left(\frac{2x}{x - 4}\right) = (x - 4)\left(\frac{8}{x - 4} + 1\right) \quad \text{Multiply by the LCD.}$$

$$(x - 4) \cdot \frac{2x}{x - 4} = (x - 4) \cdot \frac{8}{x - 4} + (x - 4) \cdot 1 \quad \text{Use the distributive property.}$$

$$2x = 8 + (x - 4) \quad \text{Simplify.}$$

$$2x = 4 + x$$

$$x = 4$$

Notice that 4 makes the denominator 0 in the original equation. Therefore, 4 is *not* a solution and this equation has *no solution*.

Practice Problem 5

Solve: $\frac{5x}{x - 1} = \frac{5}{x - 1} + 3$

Answers

4. $x = -1$, **5.** No solution

Concept Check

When can we clear fractions by multiplying through by the LCD?

a. When adding or subtracting rational expressions
b. When solving an equation containing rational expressions
c. Both of these
d. Neither of these

Practice Problem 6

Solve: $x - \dfrac{6}{x+3} = \dfrac{2x}{x+3} + 2$

Practice Problem 7

Solve $\dfrac{1}{a} + \dfrac{1}{b} = \dfrac{1}{x}$ for a.

Answers

6. $x = 4$, **7.** $a = \dfrac{bx}{b-x}$

Concept Check: b

Try the Concept Check in the margin.

As we can see from Example 5, it is important to check the proposed solution(s) in the original equation.

EXAMPLE 6 Solve: $x + \dfrac{14}{x-2} = \dfrac{7x}{x-2} + 1$

Solution: Notice the denominators in this equation. We can see that 2 can't be a solution. The LCD is $x - 2$, so we multiply both sides of the equation by $x - 2$.

$$(x-2)\left(x + \frac{14}{x-2}\right) = (x-2)\left(\frac{7x}{x-2} + 1\right)$$

$$(x-2)(x) + (x-2)\left(\frac{14}{x-2}\right) = (x-2)\left(\frac{7x}{x-2}\right) + (x-2)(1)$$

$$x^2 - 2x + 14 = 7x + x - 2 \quad \text{Simplify.}$$

$$x^2 - 2x + 14 = 8x - 2 \quad \text{Combine like terms.}$$

$$x^2 - 10x + 16 = 0 \quad \text{Write the quadratic equation in standard form.}$$

$$(x-8)(x-2) = 0 \quad \text{Factor.}$$

$$x - 8 = 0 \quad \text{or} \quad x - 2 = 0 \quad \text{Set each factor equal to 0.}$$

$$x = 8 \qquad x = 2 \quad \text{Solve.}$$

As we have already noted, 2 can't be a solution of the original equation. So we need only replace x with 8 in the original equation. We find that 8 is a solution; the only solution is 8.

B Solving Equations for a Specified Variable

The last example in this section is an equation containing several variables, and we are directed to solve for one of the variables. The steps used in the preceding examples can be applied to solve equations for a specified variable as well.

EXAMPLE 7 Solve $\dfrac{1}{a} + \dfrac{1}{b} = \dfrac{1}{x}$ for x.

Solution: (This type of equation often models a work problem, as we shall see in the next section.) The LCD is abx, so we multiply both sides by abx.

$$abx\left(\frac{1}{a} + \frac{1}{b}\right) = abx\left(\frac{1}{x}\right)$$

$$abx\left(\frac{1}{a}\right) + abx\left(\frac{1}{b}\right) = abx \cdot \frac{1}{x}$$

$$bx + ax = ab \quad \text{Simplify.}$$

$$x(b + a) = ab \quad \text{Factor out } x \text{ from each term on the left side.}$$

$$\frac{x(b+a)}{b+a} = \frac{ab}{b+a} \quad \text{Divide both sides by } b + a.$$

$$x = \frac{ab}{b+a} \quad \text{Simplify.}$$

This equation is now solved for x.

Name ______________________ Section __________ Date ____________

Mental Math

Solve each equation for the variable.

1. $\frac{x}{5} = 2$
2. $\frac{x}{8} = 4$
3. $\frac{z}{6} = 6$
4. $\frac{y}{7} = 8$

EXERCISE SET 6.5

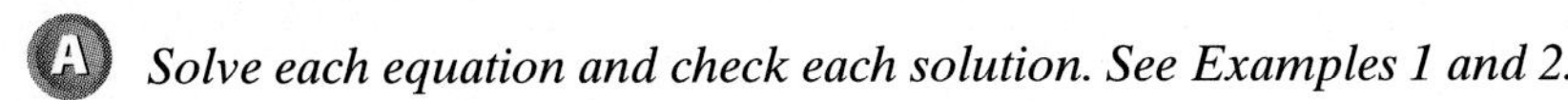

A *Solve each equation and check each solution. See Examples 1 and 2.*

1. $\frac{x}{5} + 3 = 9$
2. $\frac{x}{5} - 2 = 9$
3. $\frac{x}{2} + \frac{5x}{4} = \frac{x}{12}$
4. $\frac{x}{6} + \frac{4x}{3} = \frac{x}{18}$
5. $2 - \frac{8}{x} = 6$
6. $5 + \frac{4}{x} = 1$
7. $2 + \frac{10}{x} = x + 5$
8. $6 + \frac{5}{y} = y - \frac{2}{y}$
9. $\frac{a}{5} = \frac{a - 3}{2}$
10. $\frac{b}{5} = \frac{b + 2}{6}$
11. $\frac{x - 3}{5} + \frac{x - 2}{2} = \frac{1}{2}$
12. $\frac{a + 5}{4} + \frac{a + 5}{2} = \frac{a}{8}$

Solve each equation and check each answer. See Examples 3 through 6.

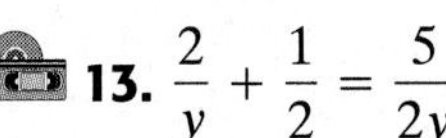

13. $\frac{2}{y} + \frac{1}{2} = \frac{5}{2y}$
14. $\frac{6}{3y} + \frac{3}{y} = 1$
15. $\frac{11}{2x} + \frac{2}{3} = \frac{7}{2x}$
16. $\frac{5}{3} - \frac{3}{2x} = \frac{3}{2}$
17. $2 + \frac{3}{a - 3} = \frac{a}{a - 3}$
18. $\frac{2y}{y - 2} - \frac{4}{y - 2} = 4$
19. $\frac{3}{2a - 5} = -1$
20. $\frac{6}{4 - 3x} = -3$
21. $\frac{y}{y + 4} + \frac{4}{y + 4} = 3$
22. $\frac{5y}{y + 1} - \frac{3}{y + 1} = 4$
23. $\frac{a}{a - 6} = \frac{-2}{a - 1}$
24. $\frac{5}{x - 6} = \frac{x}{x - 2}$
25. $\frac{2x}{x + 2} - 2 = \frac{x - 8}{x - 2}$
26. $\frac{4y}{y - 3} - 3 = \frac{3y - 1}{y + 3}$
27. $\frac{4y}{y - 4} + 5 = \frac{5y}{y - 4}$

28. $\frac{2a}{a+2} - 5 = \frac{7a}{a+2}$

29. $\frac{2}{x-2} + 1 = \frac{x}{x+2}$

30. $1 + \frac{3}{x+1} = \frac{x}{x-1}$

31. $\frac{t}{t-4} = \frac{t+4}{6}$

32. $\frac{15}{x+4} = \frac{x-4}{x}$

33. $\frac{x+1}{3} - \frac{x-1}{6} = \frac{1}{6}$

34. $\frac{3x}{5} - \frac{x-6}{3} = -\frac{2}{5}$

35. $\frac{y}{2y+2} + \frac{2y-16}{4y+4} = \frac{2y-3}{y+1}$

36. $\frac{1}{x+2} = \frac{4}{x^2-4} - \frac{1}{x-2}$

37. $\frac{4r-4}{r^2+5r-14} + \frac{2}{r+7} = \frac{1}{r-2}$

38. $\frac{3}{x+3} = \frac{12x+19}{x^2+7x+12} - \frac{5}{x+4}$

39. $\frac{x+1}{x+3} = \frac{x^2-11x}{x^2+x-6} - \frac{x-3}{x-2}$

40. $\frac{2t+3}{t-1} - \frac{2}{t+3} = \frac{5-6t}{t^2+2t-3}$

B *Solve each equation for the indicated variable. See Example 7.*

41. $R = \frac{E}{I}$ for I (Electronics: resistance of a circuit)

△ **42.** $\frac{A}{W} = L$ for W (Geometry: area of a rectangle)

43. $T = \frac{V}{Q}$ for Q (Water purification: settling time)

44. $T = \frac{2U}{B+E}$ for B (Merchandising: stock turnover rate)

45. $i = \frac{A}{t+B}$ for t (Hydrology: rainfall intensity)

46. $C = \frac{D(A+1)}{24}$ for A (Medicine: Cowling's Rule for child's dose)

47. $N = R + \frac{V}{G}$ for G (Urban forestry: tree plantings per year)

48. $B = \frac{705w}{h^2}$ for w (Health: body-mass index)

△ **49.** $\frac{C}{\pi r} = 2$ for r (Geometry: circumference of a circle)

50. $W = \frac{CE^2}{2}$ for C (Electronics: energy stored in a capacitor)

51. $\frac{1}{y} + \frac{1}{3} = \frac{1}{x}$ for x

52. $\frac{1}{5} + \frac{2}{y} = \frac{1}{x}$ for x

Review and Preview

Write each phrase as an expression. See Section 1.6.

53. The reciprocal of x

54. The reciprocal of the expression $x + 1$

55. The reciprocal of x, added to the reciprocal of 2

56. The reciprocal of x, subtracted from the reciprocal of 5

Answer each question.

57. If a tank is filled in 3 hours, what part of the tank is filled in 1 hour?

58. If a strip of beach is cleaned in 4 hours, what part of the beach is cleaned in 1 hour?

Combining Concepts

Recall that two angles are supplementary if the sum of their measures is 180°. Find the measures of the supplementary angles.

△ **59.**

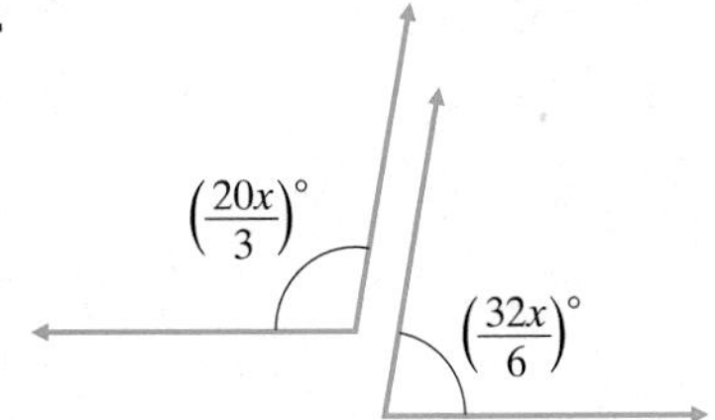

△ **60.**

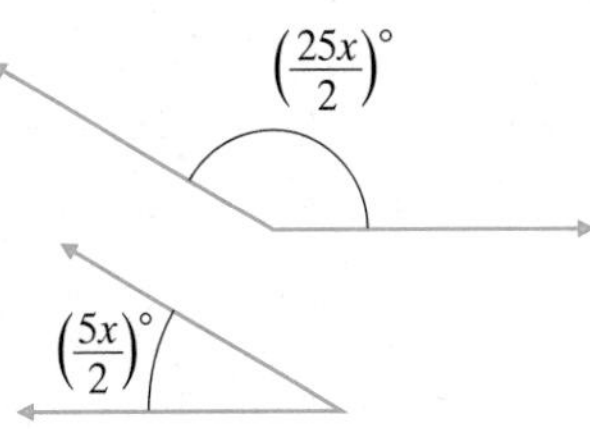

Recall that two angles are complementary if the sum of their measures is 90°. Find the measures of the complementary angles.

△ **61.**

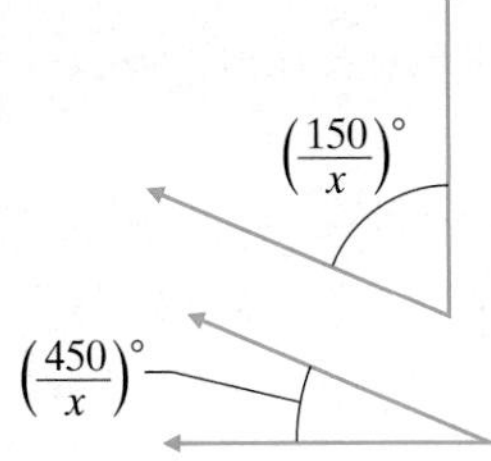

△ **62.**

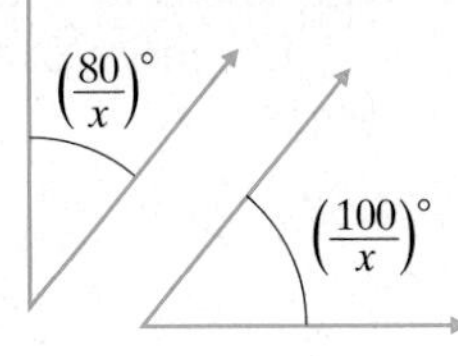

Solve each equation.

63. $\dfrac{4}{a^2 + 4a + 3} + \dfrac{2}{a^2 + a - 6} - \dfrac{3}{a^2 - a - 2} = 0$

64. $\dfrac{-4}{a^2 + 2a - 8} + \dfrac{1}{a^2 + 9a + 20} = \dfrac{-4}{a^2 + 3a - 10}$

65. When adding the expressions in $\dfrac{3x}{2} + \dfrac{x}{4}$, can you multiply each term by 4? Why or why not?

66. When solving the equation $\dfrac{3x}{2} + \dfrac{x}{4} = 1$, can you multiply both sides of the equation by 4? Why or why not?

STUDY SKILLS REMINDER

Are you preparing for a test on Chapter 6?

Below I have listed *a common trouble* area for students in Chapter 6. After studying for your test—but before taking your test—read this.

Do you know the differences between how to perform operations such as $\dfrac{4}{x} + \dfrac{2}{3}$ or $\dfrac{4}{x} \div \dfrac{2}{x}$ and how to solve an equation such as $\dfrac{4}{x} + \dfrac{2}{3} = 1$?

$$\mathbf{\frac{4}{x}} + \mathbf{\frac{2}{3}} = \frac{4 \cdot 3}{x \cdot 3} + \frac{2 \cdot x}{3 \cdot x} = \frac{12}{3x} + \frac{2x}{3x} = \frac{12 + 2x}{3x} \quad \text{or} \quad \frac{2(6 + x)}{3x}, \text{ the sum.}$$

Addition—write each expression as an equivalent expression with the same LCD denominator.

$$\mathbf{\frac{4}{x}} \div \mathbf{\frac{2}{x}} = \frac{4}{x} \cdot \frac{x}{2} = \frac{4 \cdot x}{x \cdot 2} = \frac{4}{2} = 2, \text{ the quotient.}$$

Division—multiply the first rational expression by the reciprocal of the second.

$\mathbf{\frac{4}{x}} + \mathbf{\frac{2}{3}} = 1$	Equation to be solved.
$3x\left(\frac{4}{x} + \frac{2}{3}\right) = 3x \cdot 1$	Multiply both sides of the equation by the LCD $3x$.
$3x\left(\frac{4}{x}\right) + 3x\left(\frac{2}{3}\right) = 3x \cdot 1$	Use the distributive property.
$12 + 2x = 3x$	Multiply and simplify.
$12 = x$	Subtract $2x$ from both sides.

The solution is 12.

For more examples and exercises see the Integrated Review on the next page.

Name ______________________ Section __________ Date __________

Integrated Review—Summary on Rational Expressions

It is important to know the difference between performing operations with rational expressions and solving an equation containing rational expressions. Study the examples below.

Performing Operations with Rational Expressions

Adding: $\frac{1}{x}+\frac{1}{x+5}=\frac{1\cdot(x+5)}{x(x+5)}+\frac{1\cdot x}{x(x+5)}=\frac{x+5+x}{x(x+5)}=\frac{2x+5}{x(x+5)}$

Subtracting: $\frac{3}{x}-\frac{5}{x^2y}=\frac{3\cdot xy}{x\cdot xy}-\frac{5}{x^2y}=\frac{3xy-5}{x^2y}$

Multiplying: $\frac{2}{x}\cdot\frac{5}{x-1}=\frac{2\cdot 5}{x(x-1)}=\frac{10}{x(x-1)}$

Dividing: $\frac{4}{2x+1}\div\frac{x-3}{x}=\frac{4}{2x+1}\cdot\frac{x}{x-3}=\frac{4x}{(2x+1)(x-3)}$

Solving an Equation Containing Rational Expressions

To solve an equation containing rational expressions, we clear the equation of fractions by multiplying both sides by the LCD.

$$\frac{3}{x}-\frac{5}{x-1}=\frac{1}{x(x-1)}$$ Note that x can't be 0 or 1.

$$x(x-1)\left(\frac{3}{x}\right)-x(x-1)\left(\frac{5}{x-1}\right)=x(x-1)\cdot\frac{1}{x(x-1)}$$ Multiply both sides by the LCD.

$$3(x-1)-5x=1$$ Simplify.

$$3x-3-5x=1$$ Use the distributive property.

$$-2x-3=1$$ Combine like terms.

$$-2x=4$$ Add 3 to both sides.

$$x=-2$$ Divide both sides by -2.

Determine whether each of the following is an equation or an expression. If it is an equation, solve it for its variable. If it is an expression, perform the indicated operation.

1. $\frac{1}{x}+\frac{2}{3}$

2. $\frac{3}{a}+\frac{5}{6}$

3. $\frac{1}{x}+\frac{2}{3}=\frac{3}{x}$

4. $\frac{3}{a}+\frac{5}{6}=1$

5. $\frac{2}{x+1}-\frac{1}{x}$

6. $\frac{4}{x-3}-\frac{1}{x}$

7. $\frac{2}{x+1}-\frac{1}{x}=1$

8. $\frac{4}{x-3}-\frac{1}{x}=\frac{6}{x(x-3)}$

Answers

1. ______
2. ______
3. ______
4. ______
5. ______
6. ______
7. ______
8. ______

9. ____________

10. ____________

11. ____________

12. ____________

13. ____________

14. ____________

15. ____________

16. ____________

17. ____________

18. ____________

19. ____________

20. ____________

21. ____________

22. ____________

23. ____________

24. ____________

9. $\dfrac{15x}{x+8} \cdot \dfrac{2x+16}{3x}$

10. $\dfrac{9z+5}{15} \cdot \dfrac{5z}{81z^2-25}$

11. $\dfrac{2x+1}{x-3} + \dfrac{3x+6}{x-3}$

12. $\dfrac{4p-3}{2p+7} + \dfrac{3p+8}{2p+7}$

13. $\dfrac{x+5}{7} = \dfrac{8}{2}$

14. $\dfrac{1}{2} = \dfrac{x+1}{8}$

15. $\dfrac{5a+10}{18} \div \dfrac{a^2-4}{10a}$

16. $\dfrac{9}{x^2-1} \div \dfrac{12}{3x+3}$

17. $\dfrac{x+2}{3x-1} + \dfrac{5}{(3x-1)^2}$

18. $\dfrac{4}{(2x-5)^2} + \dfrac{x+1}{2x-5}$

19. $\dfrac{x-7}{x} - \dfrac{x+2}{5x}$

20. $\dfrac{9}{x^2-4} + \dfrac{2}{x+2} = \dfrac{-1}{x-2}$

21. $\dfrac{3}{x+3} = \dfrac{5}{x^2-9} - \dfrac{2}{x-3}$

22. $\dfrac{10x-9}{x} - \dfrac{x-4}{3x}$

23. Explain the difference between solving an equation such as $\dfrac{x}{2} + \dfrac{3}{4} = \dfrac{x}{4}$ for x and performing an operation such as adding $\dfrac{x}{2} + \dfrac{3}{4}$.

24. When solving an equation such as $\dfrac{y}{4} = \dfrac{y}{2} - \dfrac{1}{4}$, we may multiply all terms by 4. When subtracting two rational expressions such as $\dfrac{y}{2} - \dfrac{1}{4}$, we may not. Explain why.

6.6 Rational Equations and Problem Solving

OBJECTIVES

- A Solve problems about numbers.
- B Solve problems about work.
- C Solve problems about distance, rate, and time.
- D Solve problems about similar triangles.

A Solving Problems about Numbers

In this section, we solve problems that can be modeled by equations containing rational expressions. To solve these problems, we use the same problem-solving steps that were first introduced in Section 2.5. In our first example, our goal is to find an unknown number.

EXAMPLE 1 Finding an Unknown Number

The quotient of a number and 6, minus $\frac{5}{3}$ is the quotient of the number and 2. Find the number.

Solution:

1. UNDERSTAND. Read and reread the problem. Suppose that the unknown number is 2, then we see if the quotient of 2 and 6, or $\frac{2}{6}$, minus $\frac{5}{3}$ is equal to the quotient of 2 and 2, or $\frac{2}{2}$.

   $$\frac{2}{6} - \frac{5}{3} = \frac{1}{3} - \frac{5}{3} = -\frac{4}{3}, \text{ not } \frac{2}{2}$$

   Don't forget that the purpose of a proposed solution is to better understand the problem.
   Let $x =$ the unknown number.

2. TRANSLATE.

In words:	the quotient of x and 6	minus	$\frac{5}{3}$	is	the quotient of x and 2
	↓	↓	↓	↓	↓
Translate:	$\frac{x}{6}$	$-$	$\frac{5}{3}$	$=$	$\frac{x}{2}$

3. SOLVE. Here, we solve the equation $\frac{x}{6} - \frac{5}{3} = \frac{x}{2}$. We begin by multiplying both sides of the equation by the LCD 6.

$$6\left(\frac{x}{6} - \frac{5}{3}\right) = 6\left(\frac{x}{2}\right)$$

$$6\left(\frac{x}{6}\right) - 6\left(\frac{5}{3}\right) = 6\left(\frac{x}{2}\right) \quad \text{Apply the distributive property.}$$

$$x - 10 = 3x \quad \text{Simplify.}$$

$$-10 = 2x \quad \text{Subtract } x \text{ from both sides.}$$

$$-\frac{10}{2} = \frac{2x}{2} \quad \text{Divide both sides by 2.}$$

$$-5 = x \quad \text{Simplify.}$$

4. INTERPRET.

Check: To check, we verify that "the quotient of -5 and 6 minus $\frac{5}{3}$ is the quotient of -5 and 2, or $-\frac{5}{6} - \frac{5}{3} = -\frac{5}{2}$. The statement is true.

State: The unknown number is -5.

Practice Problem 1

The quotient of a number and 2 minus $\frac{1}{3}$ is the quotient of the number and 6.

Answer

1. $x = 1$

B Solving Problems about Work

The next example is often called a work problem. Work problems usually involve people or machines doing a certain task.

EXAMPLE 2 Finding Work Rates

Sam Waterton and Frank Schaffer work in a plant that manufactures automobiles. Sam can complete a quality control tour of the plant in 3 hours while his assistant, Frank, needs 7 hours to complete the same job. The regional manager is coming to inspect the plant facilities, so both Sam and Frank are directed to complete a quality control tour together. How long will this take?

Solution:

1. UNDERSTAND. Read and reread the problem. The key idea here is the relationship between the **time** (hours) it takes to complete the job and the **part of the job** completed in 1 unit of time (hour). For example, if the **time** it takes Sam to complete the job is 3 hours, the **part of the job** he can complete in 1 hour is $\frac{1}{3}$.

Similarly, Frank can complete $\frac{1}{7}$ of the job in 1 hour.

Let x = the **time** in hours it takes Sam and Frank to complete the job together.

Then $\frac{1}{x}$ = the **part of the job** they complete in 1 hour.

	Hours to Complete Total Job	Part of Job Completed in 1 Hour
Sam	3	$\frac{1}{3}$
Frank	7	$\frac{1}{7}$
Together	x	$\frac{1}{x}$

2. TRANSLATE.

In words:	part of job Sam completed in 1 hour	added to	part of job Frank completed in 1 hour	is equal to	part of job they completed together in 1 hour
	↓	↓	↓	↓	↓
Translate:	$\frac{1}{3}$	$+$	$\frac{1}{7}$	$=$	$\frac{1}{x}$

Practice Problem 2

Andrew and Timothy Larson volunteer at a local recycling plant. Andrew can sort a batch of recyclables in 2 hours alone while his brother Timothy needs 3 hours to complete the same job. If they work together, how long will it take them to sort one batch?

Answer

2. $1\frac{1}{5}$ hours

3. SOLVE. Here, we solve the equation $\frac{1}{3} + \frac{1}{7} = \frac{1}{x}$. We begin by multiplying both sides of the equation by the LCD, $21x$.

$$21x\left(\frac{1}{3}\right) + 21x\left(\frac{1}{7}\right) = 21x\left(\frac{1}{x}\right)$$

$$7x + 3x = 21 \qquad \text{Simplify.}$$

$$10x = 21$$

$$x = \frac{21}{10} \quad \text{or} \quad 2\frac{1}{10} \text{ hours}$$

4. INTERPRET.

Check: Our proposed solution is $2\frac{1}{10}$ hours. This proposed solution is reasonable since $2\frac{1}{10}$ hours is more than half of Sam's time and less than half of Frank's time. Check this solution in the originally *stated* problem.

State: Sam and Frank can complete the quality control tour in $2\frac{1}{10}$ hours.

C Solving Problems about Distance, Rate, and Time

Next we look at a problem solved by the distance/rate/time formula.

EXAMPLE 3 Finding Speeds of Vehicles

A car travels 180 miles in the same time that a truck travels 120 miles. If the car's speed is 20 miles per hour faster than the truck's, find the car's speed and the truck's speed.

Solution:

1. UNDERSTAND. Read and reread the problem. Suppose that the truck's speed is 45 miles per hour. Then the car's speed is 20 miles per hour more, or 65 miles per hour.

We are given that the car travels 180 miles in the same time that the truck travels 120 miles. To find the time it takes the car to travel 180 miles, we use the formula $d = rt$ solved for t: $\frac{d}{r} = t$.

Car's Time

$$t = \frac{d}{r} = \frac{180}{65} = 2\frac{50}{65} = 2\frac{10}{13} \text{ hours}$$

Truck's Time

$$t = \frac{d}{r} = \frac{120}{45} = 2\frac{30}{45} = 2\frac{2}{3} \text{ hours}$$

Since the times are not the same, our proposed solution is not correct. But we have a better understanding of the problem.

Let $x =$ the speed of the truck.

Since the car's speed is 20 miles per hour faster than the truck's, then

$x + 20 =$ the speed of the car

Practice Problem 3

A car travels 280 miles in the same time that a motorcycle travels 240 miles. If the car's speed is 10 miles per hour more than the motorcycle's, find the speed of the car and the speed of the motorcycle.

Answer

3. car: 70 mph; motorcycle: 60 mph

Use the formula $d = r \cdot t$ or **d**istance = **r**ate · **t**ime. Prepare a chart to organize the information in the problem.

	Distance	= Rate	· Time
Truck	120	x	$\frac{120}{x}$ ← distance / ← rate
Car	180	$x + 20$	$\frac{180}{x+20}$ ← distance / ← rate

Helpful Hint

If $d = r \cdot t$, then $t = \frac{d}{r}$ or $time = \frac{distance}{rate}$.

2. TRANSLATE. Since the car and the truck traveled the same amount of time, we have the following.

In words: car's time = truck's time

Translate: $\frac{180}{x + 20} = \frac{120}{x}$

3. SOLVE. We begin by multiplying both sides of the equation by the LCD, $x(x + 20)$, or cross multiplying.

$$\frac{180}{x + 20} = \frac{120}{x}$$

$$180x = 120(x + 20)$$

$$180x = 120x + 2400 \quad \text{Use the distributive property.}$$

$$60x = 2400 \quad \text{Subtract } 120x \text{ from both sides.}$$

$$x = 40 \quad \text{Divide both sides by 60.}$$

4. INTERPRET. The speed of the truck is 40 miles per hour. The speed of the car must then be $x + 20$ or 60 miles per hour.

Check: Find the time it takes the car to travel 180 miles and the time it takes the truck to travel 120 miles.

Car's Time

$$t = \frac{d}{r} = \frac{180}{60} = 3 \text{ hours}$$

Truck's Time

$$t = \frac{d}{r} = \frac{120}{40} = 3 \text{ hours}$$

Since both travel the same amount of time, the proposed solution is correct.

State: The car's speed is 60 miles per hour and the truck's speed is 40 miles per hour.

D Solving Problems about Similar Triangles

Similar triangles have the same shape but not necessarily the same size. In similar triangles, the measures of corresponding angles are equal, and corresponding sides are in proportion.

If triangle ABC and triangle XYZ shown on the next page are similar, then we know that the measure of angle A = the measure of angle X, the measure of angle B = the measure of angle Y, and the measure of angle C = the measure of angle Z. We also know that corresponding sides are in proportion: $\frac{a}{x} = \frac{b}{y} = \frac{c}{z}$.

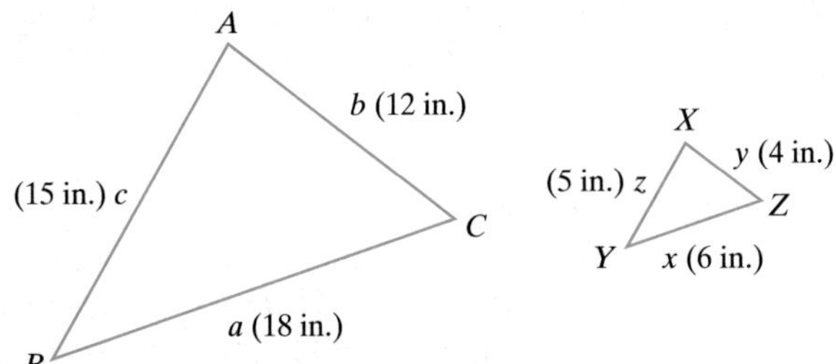

In this section, we will position similar triangles so that they have the same orientation.

To show that corresponding sides are in proportion for the triangles above, we write the ratios of the corresponding sides.

$$\frac{a}{x} = \frac{18}{6} = 3 \qquad \frac{b}{y} = \frac{12}{4} = 3 \qquad \frac{c}{z} = \frac{15}{5} = 3$$

 EXAMPLE 4 Finding the Length of a Side of a Triangle

If the following two triangles are similar, find the missing length x.

2 yards 10 yards 3 yards x yards

Solution: Since the triangles are similar, their corresponding sides are in proportion and we have

$$\frac{2}{3} = \frac{10}{x}$$

To solve, we multiply both sides by the LCD, $3x$, or cross multiply.

$$2x = 30$$

$$x = 15 \quad \text{Divide both sides by 2.}$$

The missing length x is 15 yards.

Practice Problem 4 

If the following two triangles are similar, find the missing length x.

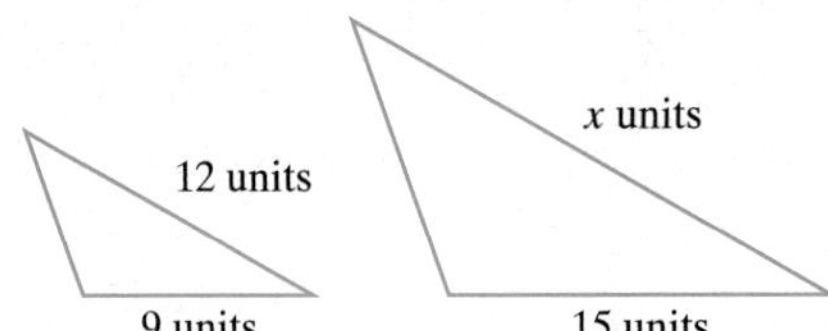

Answer

4. $x = 20$ units

FOCUS ON Business and Career

MORTGAGES

A loan to buy a house or other property is called a **mortgage**. When you are thinking of getting a mortgage to buy a house, it is helpful to know how much your monthly mortgage payment will be. One way to calculate the monthly payment P is to use the formula

$$P = \frac{\dfrac{Ar}{12}}{1 - \dfrac{1}{\left(1 + \dfrac{r}{12}\right)^{12t}}}$$

where A = the amount of the mortgage, r = the annual interest rate (written as a decimal), and t = the loan term in years. Try the exercises to the right and below.

CRITICAL THINKING

1. The average interest rate for a 30-year fixed mortgage in the United States in January 2001 was 7.30%. (*Source:* HSH Associates) Suppose you had borrowed \$90,000 to buy a house in January 2001. If your loan term was 30 years, calculate your monthly mortgage payment.
2. The average interest rate for a 15-year fixed mortgage in the United States in May 2000 was 8.40%. (*Source:* HSH Associates) Suppose you had borrowed \$80,000 to buy a house in May 2000. If your loan term was 15 years, calculate your monthly mortgage payment.

Another way to calculate a monthly mortgage payment is to use one of the many sites on the World Wide Web that offers an interactive mortgage calculator. For instance, by visiting the given World Wide Web address, you will be able to access the Interest.com Web site, or a related site, where you can calculate a monthly mortgage payment by entering the annual interest rate as a percent, the term of the loan in years, and the total home loan amount. Use the site below to solve the given exercises.

Internet Excursions

Go To: http://www.prenhall.com/martin-gay_algebra

3. Suppose you would like to borrow \$55,000 to buy a vacation cottage. If the annual interest rate is 6.85% and you plan to take out a 15-year loan, what will be your monthly mortgage payment?
4. Suppose you would like to borrow \$120,000 to buy a house. If the annual interest rate is 7.12% and you plan to take out a 20-year loan, what will be your monthly mortgage payment?

Name ______________________ Section __________ Date __________

Mental Math

Without solving algebraically, select the best choice for each exercise.

1. One person can complete a job in 7 hours. A second person can complete the same job in 5 hours. How long will it take them to complete the job if they work together?
   a. more than 7 hours
   b. between 5 and 7 hours
   c. less than 5 hours

2. One inlet pipe can fill a pond in 30 hours. A second inlet pipe can fill the same pond in 25 hours. How long before the pond is filled if both inlet pipes are on?
   a. less than 25 hours
   b. between 25 and 30 hours
   c. more than 30 hours

EXERCISE SET 6.6

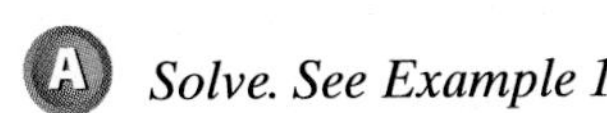

Solve. See Example 1.

1. Three times the reciprocal of a number equals 9 times the reciprocal of 6. Find the number.

2. Twelve divided by the sum of a number and 2 equals the quotient of 4 and the difference of the number and 2. Find the number.

3. If twice a number added to 3 is divided by the number plus 1, the result is three halves. Find the number.

4. A number added to the product of 6 and the reciprocal of the number equals -5. Find the number.

5. Two divided by the difference of a number and 3, minus 4 divided by the number plus 3, equals 8 times the reciprocal of the difference of the number squared and 9. What is the number?

6. If 15 times the reciprocal of a number is added to the ratio of 9 times the number minus 7 and the number plus 2, the result is 9. What is the number?

7. One-fourth equals the quotient of a number and 8. Find the number.

8. Four times a number added to 5 is divided by 6. The result is $\frac{7}{2}$. Find the number.

B *Solve. See Example 2.*

9. Smith Engineering found that an experienced surveyor can survey a roadbed in 4 hours. An apprentice surveyor needs 5 hours to survey the same stretch of road. If the two work together, find how long it takes them to complete the job.

10. An experienced bricklayer can construct a small wall in 3 hours. The apprentice can complete the job in 6 hours. Find how long it takes if they work together.

11. In 2 minutes, a conveyor belt can move 300 pounds of recyclable aluminum from the delivery truck to a storage area. A smaller belt can move the same quantity of cans the same distance in 6 minutes. If both belts are used, find how long it takes to move the cans to the storage area.

12. Find how long it takes the conveyor belts described in Exercise 11 to move 1200 pounds of cans. (*Hint:* Think of 1200 pounds as four 300-pound jobs.)

13. Marcus and Tony work for Lombardo's Pipe and Concrete. Mr. Lombardo is preparing an estimate for a customer. He knows that Marcus can lay a slab of concrete in 6 hours. Tony can lay the same size slab in 4 hours. If both work on the job and the cost of labor is \$45.00 per hour, decide what the labor estimate should be.

14. Mr. Dodson can paint his house by himself in 4 days. His son will need an additional day to complete the job if he works by himself. If they work together, find how long it takes to paint the house.

15. One custodian can clean a suite of offices in 3 hours. When a second worker is asked to join the regular custodian, the job takes only $1\frac{1}{2}$ hours. How long would it take the second worker to do the same job alone?

16. One person can proofread copy for a small newspaper in 4 hours. If a second proofreader is also employed, the job can be done in $2\frac{1}{2}$ hours. How long would it take the second proofreader to do the same job alone?

17. One pipe fills a storage pond in 20 hours. A second pipe fills the same pond in 15 hours. When a third pipe is added and all three are used to fill the pond, it takes only 6 hours. Find how long it would take the third pipe alone to do the job.

18. One pump fills a tank 3 times as fast as another pump. If the pumps work together, they fill the tank in 21 minutes. How long would it take each pump alone to fill the tank?

C *Solve. See Example 3.*

19. A runner begins her workout by jogging to the park, a distance of 3 miles. She then jogs home at the same speed but along a different route. This return trip is 9 miles and her time is one hour longer. Complete the accompanying chart and use it to find the runner's jogging speed.

	Distance	= Rate	· Time
Trip to park	3		x
Return trip	9		$x + 1$

20. A marketing manager travels 1080 miles in a corporate jet and then an additional 240 miles by car. If the car ride takes one hour longer than the jet ride, and if the rate of the jet is 6 times the rate of the car, find the time the manager travels by jet and find the time the manager travels by car.

21. A cyclist rode the first 20-mile portion of his workout at a constant speed. For the 16-mile cooldown portion of his workout, he reduced his speed by 2 miles per hour. Each portion of the workout took the same time. Find the cyclist's speed during the first portion and find his speed during the cooldown portion.

22. A tractor-trailer travels 300 miles through the flatland in the same amount of time that it travels 180 miles through mountains. The rate of the tractor-trailer is 20 miles per hour slower in the mountains than in the flatland. Find both the flatland rate and mountain rate.

23. A boat can travel 9 miles upstream in the same amount of time it takes to travel 11 miles downstream. If the current of the river is 3 miles per hour, complete the chart below and use it to find the speed r of the boat in still water.

	Distance	= Rate	· Time
Upstream	9	$r - 3$	
Downstream	11	$r + 3$	

24. A pilot flies 630 miles with a tail wind of 35 miles per hour. Against the wind, she flies only 455 miles in the same amount of time. Find the rate of the plane in still air.

25. A cyclist rides 16 miles per hour on level ground on a still day. He finds that he rides 48 miles with the wind behind him in the same amount of time that he rides 16 miles into the wind. Find the rate of the wind.

26. The current on a portion of the Mississippi River is 3 miles per hour. A barge can go 6 miles upstream in the same amount of time it takes to go 10 miles downstream. Find the speed of the boat in still water.

27. While road testing a new make of car, the editor of a consumer magazine finds that she can go 10 miles into a 3-mile-per-hour wind in the same amount of time she can go 11 miles with a 3-mile-per-hour wind behind her. Find the speed of the car in still air.

28. A fisherman on Pearl River rows 9 miles downstream in the same amount of time he rows 3 miles upstream. If the current is 6 miles per hour, find how long it takes him to cover the 12 miles.

 **D** *Given that the following pairs of triangles are similar, find each missing length. See Example 4.*

△ **29.**

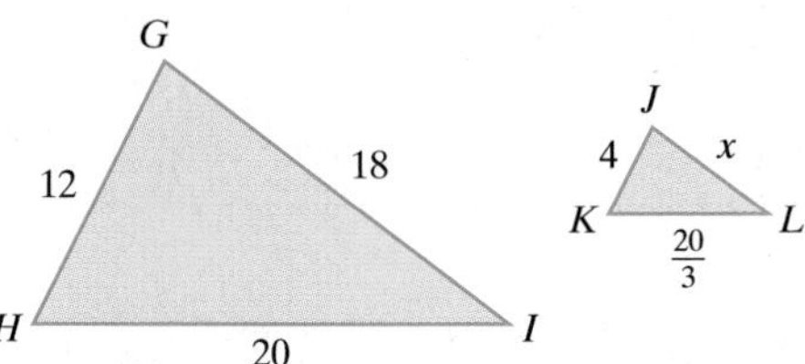

△ **30.**

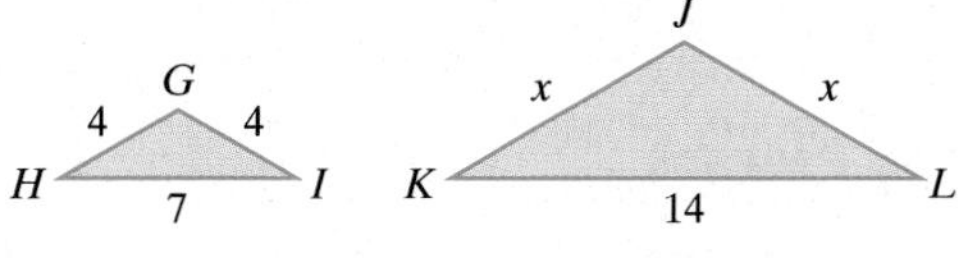

△ **31.**

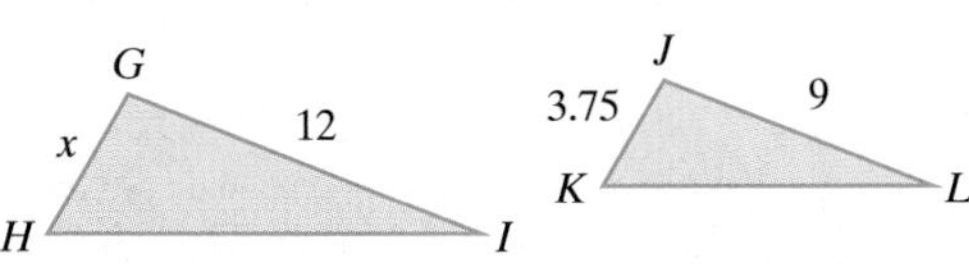

△ **32.**

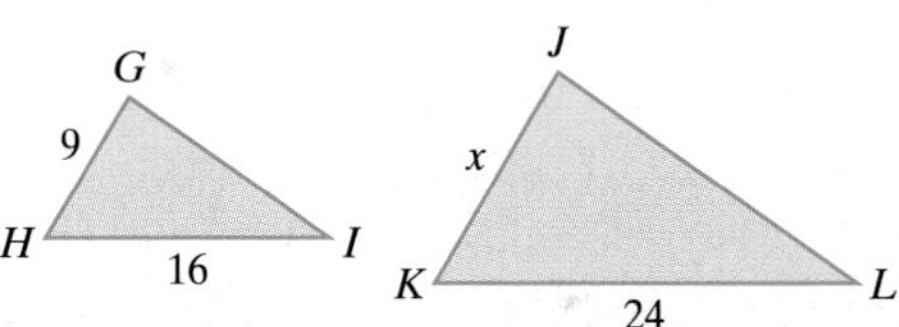

△ **33.**

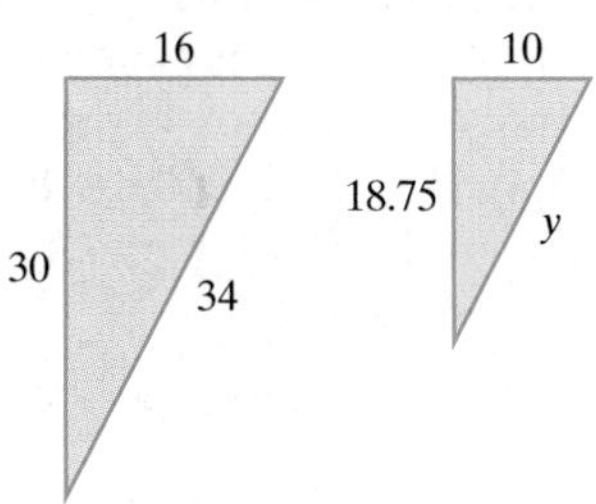

△ **34.**

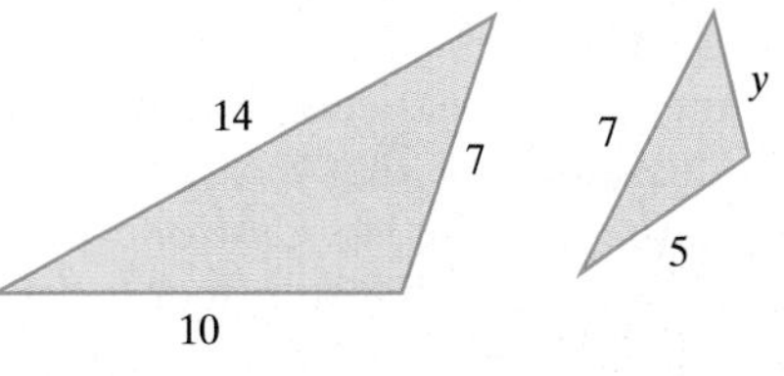

△ **35.**

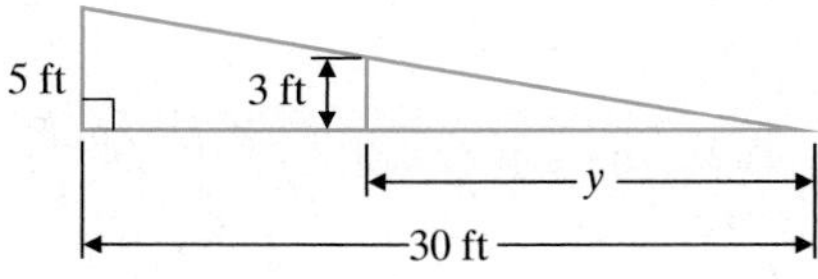

△ **36.**

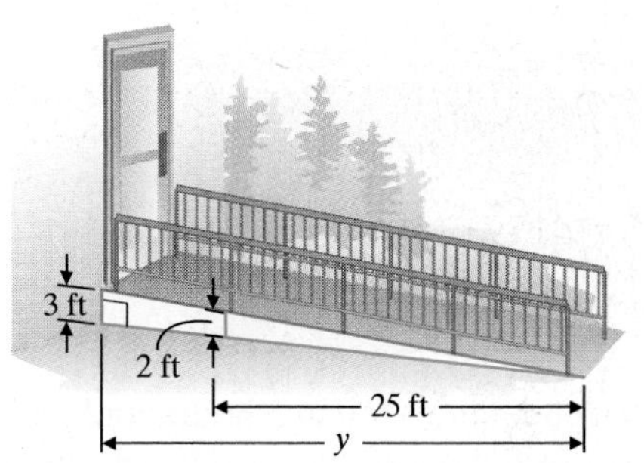

△ **37.** An architect is completing the plans for a triangular deck. Use the diagram below to find the missing dimension.

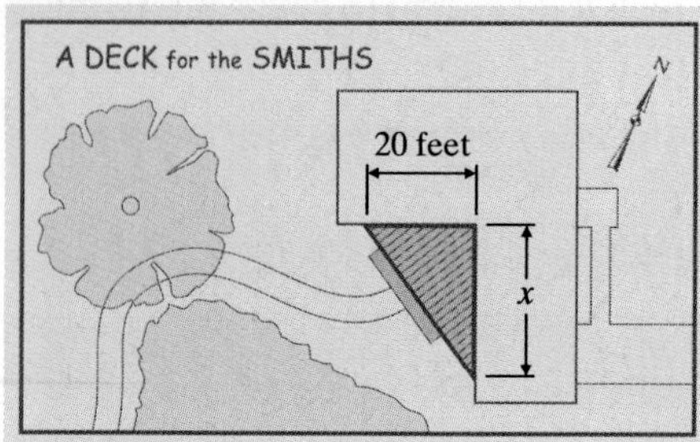

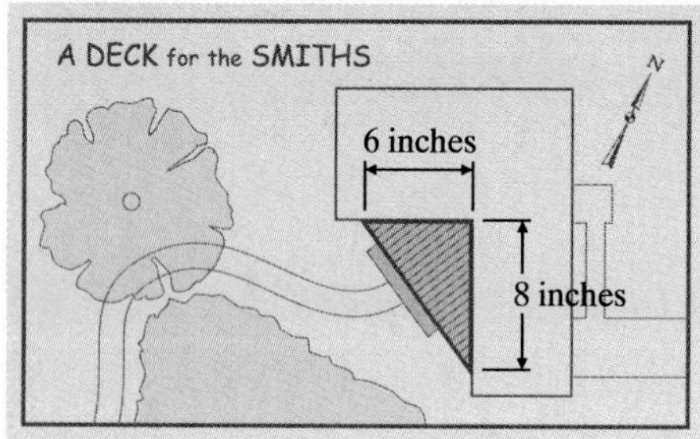

△ **38.** A student wishes to make a small model of a triangular mainsail in order to study the effects of wind on the sail. The smaller model will be the same shape as a regular size sailboat's mainsail. Use the following diagram to find the missing dimensions.

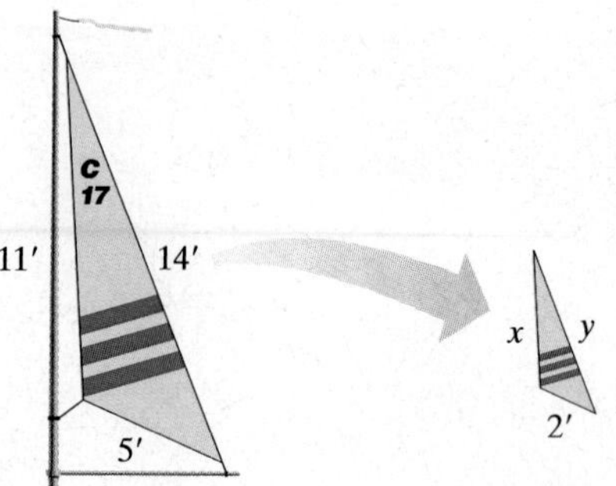

Review and Preview

Simplify. Follow the circled steps in the order shown. See Section R.2.

39. $\dfrac{\dfrac{3}{4}+\dfrac{1}{4}}{\dfrac{3}{8}+\dfrac{13}{8}}$ ① Add. ←③ Divide. ② Add.

40. $\dfrac{\dfrac{9}{5}+\dfrac{6}{5}}{\dfrac{17}{6}+\dfrac{7}{6}}$ ① Add. ←③ Divide. ② Add.

41. $\dfrac{\dfrac{2}{5}+\dfrac{1}{5}}{\dfrac{7}{10}+\dfrac{7}{10}}$ ① Add. ←③ Divide. ② Add.

42. $\dfrac{\dfrac{1}{4}+\dfrac{5}{4}}{\dfrac{3}{8}+\dfrac{7}{8}}$ ① Add. ←③ Divide. ② Add.

Combining Concepts

43. Brazilians Helio Castroneves and Bruno Junqueira placed first and fifth, respectively, in the 2001 Indianapolis 500. The track is 2.5 miles long. When traveling at their fastest lap speeds, Junqueira drove 2.459 miles in the same time that Castroneves completed an entire 2.5-mile lap. Castroneves' fastest lap speed was 3.6 mph faster than Junqueira's fastest lap speed. Find each driver's fastest lap speed. Round each speed to the nearest tenth. (*Source:* Indy Racing League)

44. A hyena spots a giraffe 0.5 mile away and begins running toward it. The giraffe starts running away from the hyena just as the hyena begins running toward it. A hyena can run at a speed of 40 mph and a giraffe can run at 32 mph. How long will it take for the hyena to overtake the giraffe? (*Source: The World Almanac and Book of Facts, 2001*)

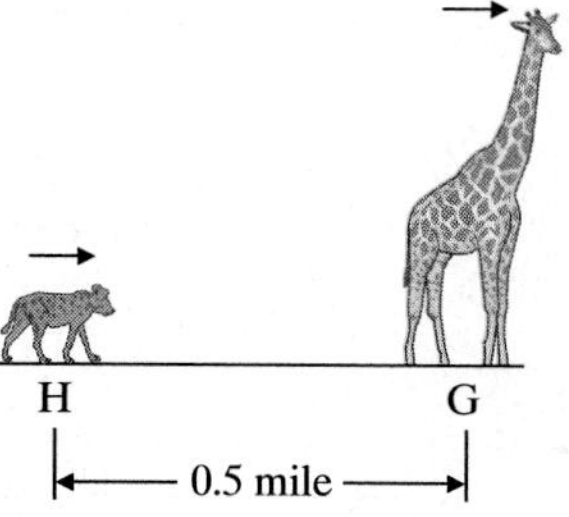

45. Person A can complete a job in 5 hours, and person B can complete the same job in 3 hours. Without solving algebraically, discuss reasonable and unreasonable answers for how long it would take them to complete the job together.

6.7 Simplifying Complex Fractions

A rational expression whose numerator or denominator or both numerator and denominator contain fractions is called a **complex rational expression** or a **complex fraction**. Some examples are

$$\frac{4}{2-\frac{1}{2}} \qquad \frac{\frac{3}{2}}{\frac{4}{7}-x} \qquad \frac{\frac{1}{x+2}}{x+2-\frac{1}{x}}$$

← Numerator of complex fraction
← Main fraction bar
← Denominator of complex fraction

Our goal in this section is to write complex fractions in simplest form. A complex fraction is in simplest form when it is in the form $\frac{P}{Q}$, where P and Q are polynomials that have no common factors.

OBJECTIVES

- **A** Simplify complex fractions using method 1.
- **B** Simplify complex fractions using method 2.

A Simplifying Complex Fractions—Method 1

In this section, two methods of simplifying complex fractions are represented. The first method presented uses the fact that the main fraction bar indicates division.

Method 1: To Simplify a Complex Fraction

Step 1. Add or subtract fractions in the numerator or denominator so that the numerator is a single fraction and the denominator is a single fraction.

Step 2. Perform the indicated division by multiplying the numerator of the complex fraction by the reciprocal of the denominator of the complex fraction.

Step 3. Write the rational expression in lowest terms.

EXAMPLE 1 Simplify the complex fraction $\frac{\frac{5}{8}}{\frac{2}{3}}$.

Solution: Since the numerator and denominator of the complex fraction are already single fractions, we proceed to step 2: perform the indicated division by multiplying the numerator $\frac{5}{8}$ by the reciprocal of the denominator $\frac{2}{3}$.

$$\frac{\frac{5}{8}}{\frac{2}{3}} = \frac{5}{8}\cdot\frac{3}{2} = \frac{15}{16}$$

The reciprocal of $\frac{2}{3}$ is $\frac{3}{2}$.

EXAMPLE 2 Simplify: $\dfrac{\frac{2}{3}+\frac{1}{5}}{\frac{2}{3}-\frac{2}{9}}$

Solution: We simplify the expressions above and below the main fraction bar separately. First we add $\frac{2}{3}$ and $\frac{1}{5}$ to obtain a single fraction in the numerator. Then we subtract $\frac{2}{9}$ from $\frac{2}{3}$ to obtain a single fraction in the denominator.

Practice Problem 1

Simplify the complex fraction $\frac{\frac{3}{7}}{\frac{5}{9}}$.

Practice Problem 2

Simplify: $\dfrac{\frac{3}{4}-\frac{2}{3}}{\frac{1}{2}+\frac{3}{8}}$

Answers

1. $\frac{27}{35}$, **2.** $\frac{2}{21}$

$$\frac{\frac{2}{3}+\frac{1}{5}}{\frac{2}{3}-\frac{2}{9}} = \frac{\frac{2(5)}{3(5)}+\frac{1(3)}{5(3)}}{\frac{2(3)}{3(3)}-\frac{2}{9}}$$ The LCD of the numerator's fractions is 15. The LCD of the denominator's fractions is 9

$$= \frac{\frac{10}{15}+\frac{3}{15}}{\frac{6}{9}-\frac{2}{9}}$$ Simplify.

$$= \frac{\frac{13}{15}}{\frac{4}{9}}$$ Add the numerator's fractions. Subtract the denominator's fractions.

Next we perform the indicated division by multiplying the numerator of the complex fraction by the reciprocal of the denominator of the complex fraction.

$$\frac{\frac{13}{15}}{\frac{4}{9}} = \frac{13}{15}\cdot\frac{9}{4}$$ The reciprocal of $\frac{4}{9}$ is $\frac{9}{4}$.

$$= \frac{13\cdot 3\cdot 3}{3\cdot 5\cdot 4} = \frac{39}{20}$$ ●

Practice Problem 3

Simplify: $\dfrac{\frac{2}{5}-\frac{1}{x}}{\frac{x}{10}-\frac{1}{3}}$

EXAMPLE 3 Simplify: $\dfrac{\frac{1}{z}-\frac{1}{2}}{\frac{1}{3}-\frac{z}{6}}$

Solution: Subtract to get a single fraction in the numerator and a single fraction in the denominator of the complex fraction.

$$\frac{\frac{1}{z}-\frac{1}{2}}{\frac{1}{3}-\frac{z}{6}} = \frac{\frac{2}{2z}-\frac{z}{2z}}{\frac{2}{6}-\frac{z}{6}}$$ The LCD of the numerator's fractions is $2z$. The LCD of the denominator's fractions is 6.

$$= \frac{\frac{2-z}{2z}}{\frac{2-z}{6}}$$

$$= \frac{2-z}{2z}\cdot\frac{6}{2-z}$$ Multiply by the reciprocal of $\frac{2-z}{6}$.

$$= \frac{2\cdot 3\cdot(2-z)}{2\cdot z\cdot(2-z)}$$ Factor.

$$= \frac{3}{z}$$ Write in lowest terms. ●

B Simplifying Complex Fractions—Method 2

Next we study a second method for simplifying complex fractions. In this method, we multiply the numerator and the denominator of the complex fraction by the LCD of all fractions in the complex fraction.

Answer

3. $\dfrac{6(2x-5)}{x(3x-10)}$

Method 2: To Simplify a Complex Fraction

Step 1. Find the LCD of all the fractions in the complex fraction.
Step 2. Multiply both the numerator and the denominator of the complex fraction by the LCD from Step 1.
Step 3. Perform the indicated operations and write the result in lowest terms.

We use method 2 to rework Example 2.

EXAMPLE 4 Simplify: $\dfrac{\dfrac{2}{3}+\dfrac{1}{5}}{\dfrac{2}{3}-\dfrac{2}{9}}$

Solution: The LCD of $\frac{2}{3}, \frac{1}{5}, \frac{2}{3}$, and $\frac{2}{9}$ is 45, so we multiply the numerator and the denominator of the complex fraction by 45. Then we perform the indicated operations, and write in lowest terms.

$$\frac{\dfrac{2}{3}+\dfrac{1}{5}}{\dfrac{2}{3}-\dfrac{2}{9}} = \frac{45\left(\dfrac{2}{3}+\dfrac{1}{5}\right)}{45\left(\dfrac{2}{3}-\dfrac{2}{9}\right)}$$

$$= \frac{45\left(\dfrac{2}{3}\right)+45\left(\dfrac{1}{5}\right)}{45\left(\dfrac{2}{3}\right)-45\left(\dfrac{2}{9}\right)} \quad \text{Apply the distributive property.}$$

$$= \frac{30+9}{30-10} = \frac{39}{20} \quad \text{Simplify.}$$

Helpful Hint

The same complex fraction was simplified using two different methods in Examples 2 and 4. Notice that the simplified results are the same.

EXAMPLE 5 Simplify: $\dfrac{\dfrac{x+1}{y}}{\dfrac{x}{y}+2}$

Solution: The LCD of $\frac{x+1}{y}$ and $\frac{x}{y}$ is y, so we multiply the numerator and the denominator of the complex fraction by y.

$$\frac{\dfrac{x+1}{y}}{\dfrac{x}{y}+2} = \frac{y\left(\dfrac{x+1}{y}\right)}{y\left(\dfrac{x}{y}+2\right)}$$

$$= \frac{y\left(\dfrac{x+1}{y}\right)}{y\left(\dfrac{x}{y}\right)+y\cdot 2} \quad \text{Apply the distributive property in the denominator.}$$

$$= \frac{x+1}{x+2y} \quad \text{Simplify.}$$

Practice Problem 4

Use method 2 to simplify the complex fraction in Practice Problem 2:

$$\frac{\dfrac{3}{4}-\dfrac{2}{3}}{\dfrac{1}{2}+\dfrac{3}{8}}$$

Practice Problem 5

Simplify: $\dfrac{1+\dfrac{x}{y}}{\dfrac{2x+1}{y}}$

Answers

4. $\frac{2}{21}$, **5.** $\frac{y+x}{2x+1}$

Practice Problem 6

Simplify: $\dfrac{\dfrac{5}{6y} + \dfrac{y}{x}}{\dfrac{y}{3} - x}$

Answer

6. $\dfrac{5x + 6y^2}{2yx(y - 3x)}$

EXAMPLE 6 Simplify: $\dfrac{\dfrac{x}{y} + \dfrac{3}{2x}}{\dfrac{x}{2} + y}$

Solution: The LCD of $\dfrac{x}{y}, \dfrac{3}{2x}, \dfrac{x}{2}$, and $\dfrac{y}{1}$ is $2xy$, so we multiply both the numerator and the denominator of the complex fraction by $2xy$.

$$\frac{\dfrac{x}{y} + \dfrac{3}{2x}}{\dfrac{x}{2} + y} = \frac{2xy\left(\dfrac{x}{y} + \dfrac{3}{2x}\right)}{2xy\left(\dfrac{x}{2} + y\right)}$$

$$= \frac{2xy\left(\dfrac{x}{y}\right) + 2xy\left(\dfrac{3}{2x}\right)}{2xy\left(\dfrac{x}{2}\right) + 2xy(y)}$$ Apply the distributive property.

$$= \frac{2x^2 + 3y}{x^2y + 2xy^2}$$

$$\text{or } \frac{2x^2 + 3y}{xy(x + 2y)}$$

STUDY SKILLS REMINDER

Is your notebook still organized?

Is your notebook still organized? If it's not, it's not too late to start organizing it. Start writing your notes and completing your homework assignment in a notebook with pockets (spiral or ring binder). Take class notes in this notebook, and then follow the notes with your completed homework assignment. When you receive graded papers or handouts, place them in the notebook pocket so that you will not lose them.

Remember to mark (possibly with an exclamation point) any note(s) that seems extra important to you. Also remember to mark (possibly with a question mark) any notes or homework that you are having trouble with. Don't forget to see your instructor or a math tutor to help you with the concepts or exercises that you are having trouble understanding.

Also—don't forget to write neatly.

Name ________________________ Section __________ Date __________

EXERCISE SET 6.7

 B *Simplify each complex fraction. See Examples 1 through 6.*

1. $\dfrac{\frac{1}{2}}{\frac{3}{4}}$

2. $\dfrac{\frac{1}{8}}{-\frac{5}{12}}$

3. $\dfrac{-\frac{4x}{9}}{-\frac{2x}{3}}$

4. $\dfrac{-\frac{6y}{11}}{\frac{4y}{9}}$

5. $\dfrac{-\frac{5}{12x^2}}{\frac{25}{16x^3}}$

6. $\dfrac{-\frac{7}{8y}}{\frac{21}{4y}}$

7. $\dfrac{\frac{1}{3}}{\frac{1}{2}-\frac{1}{4}}$

8. $\dfrac{\frac{7}{10}-\frac{3}{5}}{\frac{1}{2}}$

9. $\dfrac{2+\frac{7}{10}}{1+\frac{3}{5}}$

10. $\dfrac{4-\frac{11}{12}}{5+\frac{1}{4}}$

11. $\dfrac{\frac{m}{n}-1}{\frac{m}{n}+1}$

12. $\dfrac{\frac{x}{2}+2}{\frac{x}{2}-2}$

13. $\dfrac{\frac{1}{5}-\frac{1}{x}}{\frac{7}{10}+\frac{1}{x^2}}$

14. $\dfrac{\frac{1}{y^2}+\frac{2}{3}}{\frac{1}{y}-\frac{5}{6}}$

15. $\dfrac{1+\frac{1}{y-2}}{y+\frac{1}{y-2}}$

16. $\dfrac{x-\frac{1}{2x+1}}{1-\frac{x}{2x+1}}$

17. $\dfrac{\frac{4y-8}{16}}{\frac{6y-12}{4}}$

18. $\dfrac{\frac{7y+21}{3}}{\frac{3y+9}{8}}$

19. $\dfrac{\frac{x}{y}+1}{\frac{x}{y}-1}$

20. $\dfrac{\frac{3}{5y}+8}{\frac{3}{5y}-8}$

21. $\dfrac{1}{2 + \dfrac{1}{3}}$

22. $\dfrac{3}{1 - \dfrac{4}{3}}$

23. $\dfrac{\dfrac{ax + ab}{x^2 - b^2}}{\dfrac{x + b}{x - b}}$

24. $\dfrac{\dfrac{m + 2}{m - 2}}{\dfrac{2m + 4}{m^2 - 4}}$

25. $\dfrac{\dfrac{8}{x + 4} + 2}{\dfrac{12}{x + 4} - 2}$

26. $\dfrac{\dfrac{25}{x + 5} + 5}{\dfrac{3}{x + 5} - 5}$

27. $\dfrac{\dfrac{s}{r} + \dfrac{r}{s}}{\dfrac{s}{r} - \dfrac{r}{s}}$

28. $\dfrac{\dfrac{2}{x} + \dfrac{x}{2}}{\dfrac{2}{x} - \dfrac{x}{2}}$

29. Explain how to simplify a complex fraction using method 1.

30. Explain how to simplify a complex fraction using method 2.

Review and Preview

Use the bar graph below to answer Exercises 31 through 34. See Section 3.1.

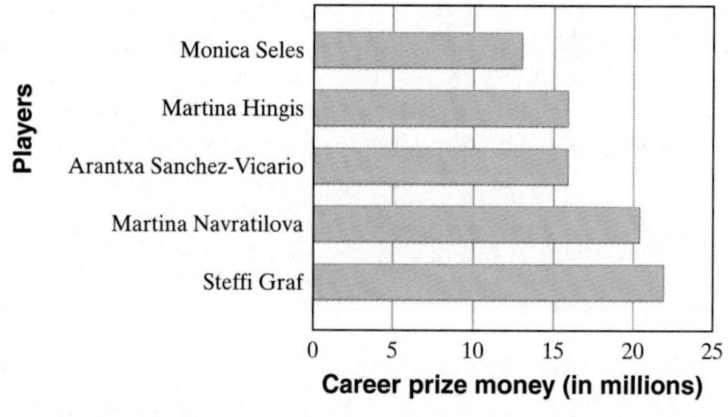

Source: Sanex WTA Tour Media Information System

31. Which women's tennis player has earned the most prize money in her career?

32. Estimate how much more prize money Arantxa Sanchez-Vicario has earned over her career than Monica Seles.

33. Which of the players shown have earned less than \$18 million in prize money over their careers?

 **34.** During her career, Martina Navratilova won over 300 singles and doubles tournament titles. Assuming her prize money was earned only for tournament titles, how much prize money did she earn per tournament title, on average?

Combining Concepts

To find the average of two numbers, we find their sum and divide by 2. For example, the average of 65 and 81 is found by simplifying $\frac{65 + 81}{2}$. *This simplifies to* $\frac{146}{2} = 73$.

35. Find the average of $\frac{1}{3}$ and $\frac{3}{4}$.

36. Write the average of $\frac{3}{n}$ and $\frac{5}{n^2}$ as a simplified rational expression.

37. In electronics, when two resistors R_1 (read R sub 1) and R_2 (read R sub 2) are connected in parallel, the total resistance is given by the complex fraction

$$\frac{1}{\frac{1}{R_1} + \frac{1}{R_2}}.$$

Simplify this expression.

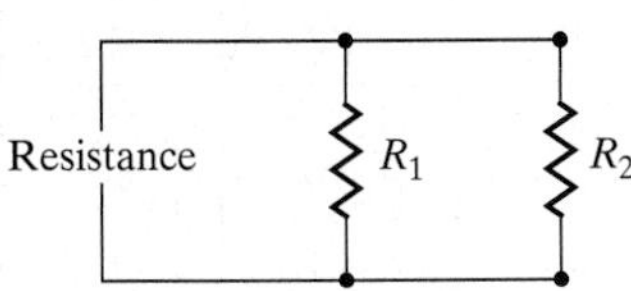

38. Astronomers occasionally need to know the day of the week a particular date fell on. The complex fraction

$$\frac{J + \frac{3}{2}}{7}$$

where J is the *Julian day number*, is used to make this calculation. Simplify this expression.

Simplify each of the following. First, write each expression without exponents. Then simplify the complex fraction. The first step has been completed for Exercise 39.

39. $\dfrac{x^{-1} + 2^{-1}}{x^{-2} - 4^{-1}} = \dfrac{\frac{1}{x} + \frac{1}{2}}{\frac{1}{x^2} - \frac{1}{4}}$

40. $\dfrac{3^{-1} - x^{-1}}{9^{-1} - x^{-2}}$

41. $\dfrac{y^{-2}}{1 - y^{-2}}$

42. $\dfrac{4 + x^{-1}}{3 + x^{-1}}$

CHAPTER 6 ACTIVITY Analyzing a Table

MATERIALS:

- Calculator

This activity may be completed by working in groups or individually.

A person's body-mass index (BMI) can be used as an indicator of whether the person should lose weight. BMI can be calculated with the formula $B = \frac{705w}{h^2}$ where w is weight in pounds and h is height in inches. Doctors recommend that body-mass index values fall between 19 and 25. BMI values of 25 to 29.9 are considered overweight, and BMI values of 30 and over are considered obese.

1. Use the BMI formula to complete the table for the given combinations of height and weight. (*Hint:* You may want to use spreadsheet software or a calculator that has a table feature to help fill in the table.)
2. Use the table to find the BMI for (a) a person who is 66 inches tall and weighs 160 pounds, (b) a person who is 62 inches tall and weighs 120 pounds, (c) a person who is 70 inches tall and weighs 170 pounds. Do these BMI values indicate that any of these people should try to lose weight?
3. Examine the table. What pattern do you notice as you look across the rows? What pattern do you notice as you look down the columns?
4. Mark the table to show weight and height combinations with corresponding BMI values that fall within the recommended range.
5. Why would having a table like this on hand be beneficial to a doctor? Explain.

Height

Weight	60	62	64	66	68	70	72	74
100								
110								
120								
130								
140								
150								
160								
170								
180								
190								
200								

Chapter 6 VOCABULARY CHECK

Fill in each blank with one of the words or phrases listed below.

rational expression complex fraction

1. A ________________ is an expression that can be written in the form $\frac{P}{Q}$, where P and Q are polynomials and Q is not 0.

2. In a ________________ the numerator or denominator or both may contain fractions.

Highlights

DEFINITIONS AND CONCEPTS	EXAMPLES
Section 6.1 Simplifying Rational Expressions	
A **rational expression** is an expression that can be written in the form $\frac{P}{Q}$, where P and Q are polynomials and Q does not equal 0.	$\frac{7y^3}{4}, \frac{x^2+6x+1}{x-3}, \frac{-5}{s^3+8}$
To find values for which a rational expression is undefined, find values for which the denominator is 0.	Find any values for which the expression $\frac{5y}{y^2-4y+3}$ is undefined. $y^2 - 4y + 3 = 0$ Set the denominator equal to 0. $(y-3)(y-1) = 0$ Factor. $y - 3 = 0$ or $y - 1 = 0$ Set each factor equal to 0. $y = 3$ $y = 1$ Solve. The expression is undefined when y is 3 and when y is 1.
FUNDAMENTAL PRINCIPLE OF RATIONAL EXPRESSIONS If P, Q, and R are polynomials, and Q and R are not 0, then $\frac{PR}{QR} = \frac{P}{Q}$	By the fundamental principle, $\frac{(x-3)(x+1)}{x(x+1)} = \frac{x-3}{x}$ as long as $x \neq 0$ and $x \neq -1$.
TO SIMPLIFY A RATIONAL EXPRESSION **Step 1.** Factor the numerator and denominator. **Step 2.** Apply the fundamental principle to divide out common factors.	Simplify: $\frac{4x+20}{x^2-25}$ $\frac{4x+20}{x^2-25} = \frac{4(x+5)}{(x+5)(x-5)} = \frac{4}{x-5}$

Definitions and Concepts	**Examples**

Section 6.2 Multiplying and Dividing Rational Expressions

To Multiply Rational Expressions

Step 1. Factor numerators and denominators.

Step 2. Multiply numerators and multiply denominators.

Step 3. Write the product in lowest terms.

$$\frac{P}{Q}\cdot\frac{R}{S}=\frac{PR}{QS}$$

Multiply: $\frac{4x+4}{2x-3}\cdot\frac{2x^2+x-6}{x^2-1}$

$$\frac{4x+4}{2x-3}\cdot\frac{2x^2+x-6}{x^2-1}$$
$$=\frac{4(x+1)}{2x-3}\cdot\frac{(2x-3)(x+2)}{(x+1)(x-1)}$$
$$=\frac{4(x+1)(2x-3)(x+2)}{(2x-3)(x+1)(x-1)}$$
$$=\frac{4(x+2)}{x-1}$$

To divide by a rational expression, multiply by the reciprocal.

$$\frac{P}{Q}\div\frac{R}{S}=\frac{P}{Q}\cdot\frac{S}{R}=\frac{PS}{QR}$$

Divide: $\frac{15x+5}{3x^2-14x-5}\div\frac{15}{3x-12}$

$$\frac{15x+5}{3x^2-14x-5}\div\frac{15}{3x-12}$$
$$=\frac{5(3x+1)}{(3x+1)(x-5)}\cdot\frac{3(x-4)}{(3\cdot 5)}$$
$$=\frac{x-4}{x-5}$$

Section 6.3 Adding and Subtracting Rational Expressions with the Same Denominator and Least Common Denominator

To add or subtract rational expressions with the same denominator, add or subtract numerators, and place the sum or difference over the common denominator.

$$\frac{P}{R}+\frac{Q}{R}=\frac{P+Q}{R}$$
$$\frac{P}{R}-\frac{Q}{R}=\frac{P-Q}{R}$$

Perform each indicated operation.

$$\frac{5}{x+1}+\frac{x}{x+1}=\frac{5+x}{x+1}$$
$$\frac{2y+7}{y^2-9}-\frac{y+4}{y^2-9}$$
$$=\frac{(2y+7)-(y+4)}{y^2-9}$$
$$=\frac{2y+7-y-4}{y^2-9}$$
$$=\frac{y+3}{(y+3)(y-3)}$$
$$=\frac{1}{y-3}$$

To Find the Least Common Denominator (LCD)

Step 1. Factor the denominators.

Step 2. The LCD is the product of all unique factors, each raised to a power equal to the greatest number of times that it appears in any one factored denominator.

Find the LCD for

$$\frac{7x}{x^2+10x+25} \text{ and } \frac{11}{3x^2+15x}$$
$$x^2+10x+25=(x+5)(x+5)$$
$$3x^2+15x=3x(x+5)$$
$$\text{LCD}=3x(x+5)(x+5) \text{ or } 3x(x+5)^2$$

Definitions and Concepts	**Examples**

Section 6.4 Adding and Subtracting Rational Expressions With Different Denominators

To Add or Subtract Rational Expressions With Different Denominators

Step 1. Find the LCD.

Step 2. Rewrite each rational expression as an equivalent expression whose denominator is the LCD.

Step 3. Add or subtract numerators and place the sum or difference over the common denominator.

Step 4. Write the result in lowest terms.

Perform the indicated operation.

$$\frac{9x+3}{x^2-9} - \frac{5}{x-3}$$
$$= \frac{9x+3}{(x+3)(x-3)} - \frac{5}{x-3}$$

LCD is $(x+3)(x-3)$.

$$= \frac{9x+3}{(x+3)(x-3)} - \frac{5(x+3)}{(x-3)(x+3)}$$
$$= \frac{9x+3-5(x+3)}{(x+3)(x-3)}$$
$$= \frac{9x+3-5x-15}{(x+3)(x-3)}$$
$$= \frac{4x-12}{(x+3)(x-3)}$$
$$= \frac{4(x-3)}{(x+3)(x-3)} = \frac{4}{x+3}$$

Section 6.5 Solving Equations Containing Rational Expressions

To Solve an Equation Containing Rational Expressions

Step 1. Multiply both sides of the equation by the LCD of all rational expressions in the equation.

Step 2. Remove any grouping symbols and solve the resulting equation.

Step 3. Check the solution in the original equation.

Solve: $\frac{5x}{x+2} + 3 = \frac{4x-6}{x+2}$ The LCD is $x+2$.

$$(x+2)\left(\frac{5x}{x+2} + 3\right) = (x+2)\left(\frac{4x-6}{x+2}\right)$$
$$(x+2)\left(\frac{5x}{x+2}\right) + (x+2)(3) = (x+2)\left(\frac{4x-6}{x+2}\right)$$
$$5x + 3x + 6 = 4x - 6$$
$$4x = -12$$
$$x = -3$$

The solution checks; the solution is -3.

Section 6.6 Rational Equations and Problem Solving

Problem-Solving Steps

1. UNDERSTAND. Read and reread the problem.

A small plane and a car leave Kansas City, Missouri, and head for Minneapolis, Minnesota, a distance of 450 miles. The speed of the plane is 3 times the speed of the car, and the plane arrives 6 hours ahead of the car. Find the speed of the car.

Let x = the speed of the car.

Then $3x$ = the speed of the plane.

	Distance =	Rate ·	Time
Car	450	x	$\frac{450}{x}\left(\frac{\text{distance}}{\text{rate}}\right)$
Plane	450	$3x$	$\frac{450}{3x}\left(\frac{\text{distance}}{\text{rate}}\right)$

DEFINITIONS AND CONCEPTS	**EXAMPLES**

Section 6.6 Rational Equations and Problem Solving *(continued)*

2. TRANSLATE.

In words: plane's time + 6 hours = car's time

Translate: $\frac{450}{3x} + 6 = \frac{450}{x}$

3. SOLVE.

$$\frac{450}{3x} + 6 = \frac{450}{x}$$

$$3x\left(\frac{450}{3x}\right) + 3x(6) = 3x\left(\frac{450}{x}\right)$$

$$450 + 18x = 1350$$

$$18x = 900$$

$$x = 50$$

4. INTERPRET.

Check the solution by replacing x with 50 in the original equation. **State** the conclusion: The speed of the car is 50 miles per hour.

Section 6.7 Simplifying Complex Fractions

METHOD 1: TO SIMPLIFY A COMPLEX FRACTION

Step 1. Add or subtract fractions in the numerator and the denominator of the complex fraction.

Step 2. Perform the indicated division.

Step 3. Write the result in lowest terms.

Simplify:

$$\frac{\frac{1}{x} + 2}{\frac{1}{x} - \frac{1}{y}} = \frac{\frac{1}{x} + \frac{2x}{x}}{\frac{y}{xy} - \frac{x}{xy}}$$

$$= \frac{\frac{1 + 2x}{x}}{\frac{y - x}{xy}}$$

$$= \frac{1 + 2x}{x} \cdot \frac{xy}{y - x}$$

$$= \frac{y(1 + 2x)}{y - x}$$

METHOD 2. TO SIMPLIFY A COMPLEX FRACTION

Step 1. Find the LCD of all fractions in the complex fraction.

Step 2. Multiply the numerator and the denominator of the complex fraction by the LCD.

Step 3. Perform the indicated operations and write the result in lowest terms.

$$\frac{\frac{1}{x} + 2}{\frac{1}{x} - \frac{1}{y}} = \frac{xy\left(\frac{1}{x} + 2\right)}{xy\left(\frac{1}{x} - \frac{1}{y}\right)}$$

$$= \frac{xy\left(\frac{1}{x}\right) + xy(2)}{xy\left(\frac{1}{x}\right) - xy\left(\frac{1}{y}\right)}$$

$$= \frac{y + 2xy}{y - x} \quad \text{or} \quad \frac{y(1 + 2x)}{y - x}$$

Name ______________________ Section __________ Date ____________

Chapter 6 Review

(6.1) *Find any real number for which each rational expression is undefined.*

1. $\dfrac{x + 5}{x^2 - 4}$

2. $\dfrac{5x + 9}{4x^2 - 4x - 15}$

Find the value of each rational expression when $x = 5$, $y = 7$, and $z = -2$.

3. $\dfrac{2 - z}{z + 5}$

4. $\dfrac{x^2 + xy - y^2}{x + y}$

Simplify each rational expression.

5. $\dfrac{2x + 6}{x^2 + 3x}$

6. $\dfrac{3x - 12}{x^2 - 4x}$

7. $\dfrac{x + 2}{x^2 - 3x - 10}$

8. $\dfrac{x + 4}{x^2 + 5x + 4}$

9. $\dfrac{x^3 - 4x}{x^2 + 3x + 2}$

10. $\dfrac{5x^2 - 125}{x^2 + 2x - 15}$

11. $\dfrac{x^2 - x - 6}{x^2 - 3x - 10}$

12. $\dfrac{x^2 - 2x}{x^2 + 2x - 8}$

Simplify each expression. First, factor the four-term polynomials by grouping.

13. $\dfrac{x^2 + xa + xb + ab}{x^2 - xc + bx - bc}$

14. $\dfrac{x^2 + 5x - 2x - 10}{x^2 - 3x - 2x + 6}$

(6.2) *Perform each indicated operation and simplify.*

15. $\dfrac{15x^3y^2}{z} \cdot \dfrac{z}{5xy^3}$

16. $\dfrac{-y^3}{8} \cdot \dfrac{9x^2}{y^3}$

17. $\dfrac{x^2 - 9}{x^2 - 4} \cdot \dfrac{x - 2}{x + 3}$

18. $\dfrac{2x + 5}{x - 6} \cdot \dfrac{2x}{-x + 6}$

19. $\dfrac{x^2 - 5x - 24}{x^2 - x - 12} \div \dfrac{x^2 - 10x + 16}{x^2 + x - 6}$

20. $\dfrac{4x + 4y}{xy^2} \div \dfrac{3x + 3y}{x^2y}$

21. $\dfrac{x^2 + x - 42}{x - 3} \cdot \dfrac{(x - 3)^2}{x + 7}$

22. $\dfrac{2a + 2b}{3} \cdot \dfrac{a - b}{a^2 - b^2}$

23. $\dfrac{x^2 - 9x + 14}{x^2 - 5x + 6} \cdot \dfrac{x + 2}{x^2 - 5x - 14}$

24. $(x - 3) \cdot \dfrac{x}{x^2 + 3x - 18}$

25. $\dfrac{2x^2 - 9x + 9}{8x - 12} \div \dfrac{x^2 - 3x}{2x}$

26. $\dfrac{x^2 - y^2}{x^2 + xy} \div \dfrac{3x^2 - 2xy - y^2}{3x^2 + 6x}$

(6.3) *Perform each indicated operation and simplify.*

27. $\dfrac{x}{x^2 + 9x + 14} + \dfrac{7}{x^2 + 9x + 14}$

28. $\dfrac{x}{x^2 + 2x - 15} + \dfrac{5}{x^2 + 2x - 15}$

29. $\dfrac{4x - 5}{3x^2} - \dfrac{2x + 5}{3x^2}$

30. $\dfrac{9x + 7}{6x^2} - \dfrac{3x + 4}{6x^2}$

Find the LCD of each pair of rational expressions.

31. $\dfrac{x + 4}{2x}, \dfrac{3}{7x}$

32. $\dfrac{x - 2}{x^2 - 5x - 24}, \dfrac{3}{x^2 + 11x + 24}$

Rewrite each rational expression as an equivalent expression whose denominator is the given polynomial.

33. $\dfrac{5}{7x} = \dfrac{}{14x^3y}$

34. $\dfrac{9}{4y} = \dfrac{}{16y^3x}$

35. $\dfrac{x + 2}{x^2 + 11x + 18} = \dfrac{}{(x + 2)(x - 5)(x + 9)}$

36. $\dfrac{3x - 5}{x^2 + 4x + 4} = \dfrac{}{(x + 2)^2(x + 3)}$

(6.4) *Perform each indicated operation and simplify.*

37. $\dfrac{4}{5x^2} - \dfrac{6}{y}$

38. $\dfrac{2}{x - 3} - \dfrac{4}{x - 1}$

39. $\dfrac{x + 7}{x + 3} - \dfrac{x - 3}{x + 7}$

40. $\dfrac{4}{x + 3} - 2$

41. $\dfrac{3}{x^2 + 2x - 8} + \dfrac{2}{x^2 - 3x + 2}$

42. $\dfrac{2x - 5}{6x + 9} - \dfrac{4}{2x^2 + 3x}$

43. $\dfrac{x - 1}{x^2 - 2x + 1} - \dfrac{x + 1}{x - 1}$

44. $\dfrac{x - 1}{x^2 + 4x + 4} + \dfrac{x - 1}{x + 2}$

Find the perimeter and the area of each figure.

45.

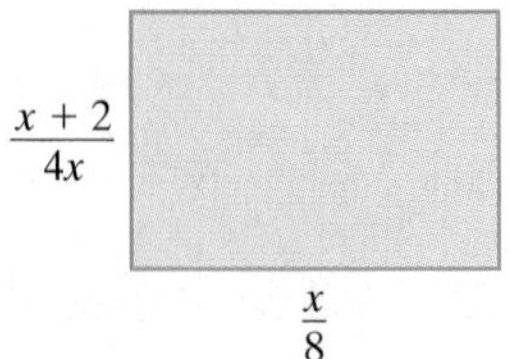

△ **46.**

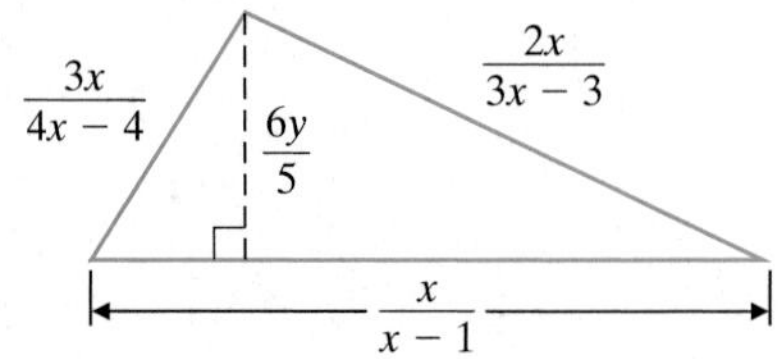

(6.5) *Solve each equation.*

47. $\frac{x+4}{9} = \frac{5}{9}$

48. $\frac{n}{10} = 9 - \frac{n}{5}$

49. $\frac{5y-3}{7} = \frac{15y-2}{28}$

50. $\frac{2}{x+1} - \frac{1}{x-2} = -\frac{1}{2}$

51. $\frac{1}{a+3} + \frac{1}{a-3} = -\frac{5}{a^2-9}$

52. $\frac{y}{2y+2} + \frac{2y-16}{4y+4} = \frac{y-3}{y+1}$

53. $\frac{4}{x+3} + \frac{8}{x^2-9} = 0$

54. $\frac{2}{x-3} - \frac{4}{x+3} = \frac{8}{x^2-9}$

55. $\frac{x-3}{x+1} - \frac{x-6}{x+5} = 0$

56. $x + 5 = \frac{6}{x}$

Solve each equation for the indicated variable.

57. $\frac{4A}{5b} = x^2$ for b

58. $\frac{x}{7} + \frac{y}{8} = 10$ for y

(6.6) *Solve each problem.*

59. Five times the reciprocal of a number equals the sum of $\frac{3}{2}$ the reciprocal of the number and $\frac{7}{6}$. What is the number?

60. The reciprocal of a number equals the reciprocal of the difference of 4 and the number. Find the number.

61. A car travels 90 miles in the same time that a car traveling 10 miles per hour slower travels 60 miles. Find the speed of each car.

62. The current in a bayou near Lafayette, Louisiana, is 4 miles per hour. A paddle boat travels 48 miles upstream in the same amount of time it takes to travel 72 miles downstream. Find the speed of the boat in still water.

63. When Mark and Maria manicure Mr. Stergeon's lawn, it takes them 5 hours. If Mark works alone, it takes 7 hours. Find how long it takes Maria alone.

64. It takes pipe A 20 days to fill a fish pond. Pipe B takes 15 days. Find how long it takes both pipes together to fill the pond.

Given that the pairs of triangles are similar, find each missing length x.

△ **65.**

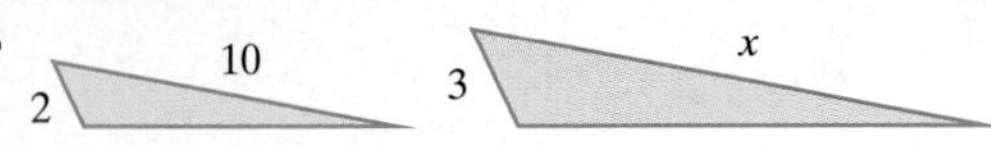

△ **66.**

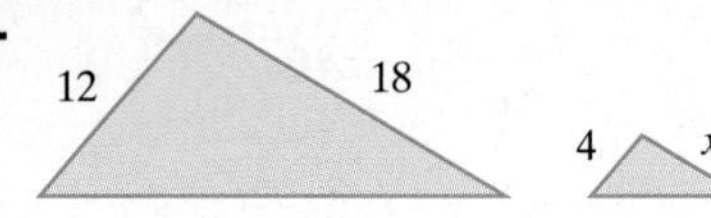

△ **67.**

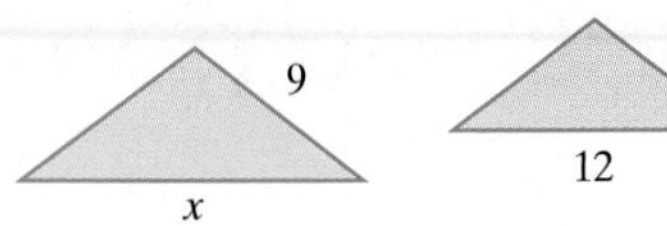

△ **68.**

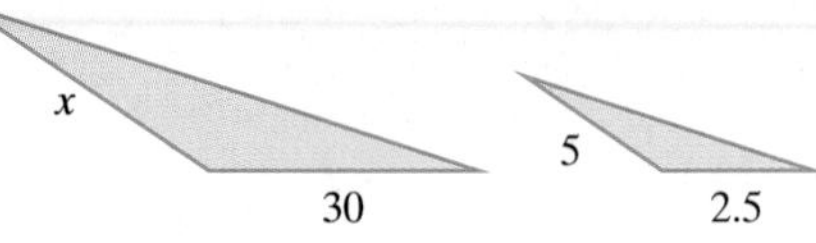

(6.7) *Simplify each complex fraction.*

69. $\dfrac{\dfrac{5x}{27}}{-\dfrac{10xy}{21}}$

70. $\dfrac{\dfrac{8x}{x^2 - 9}}{\dfrac{4}{x + 3}}$

71. $\dfrac{\dfrac{3}{5} + \dfrac{2}{7}}{\dfrac{1}{5} + \dfrac{5}{6}}$

72. $\dfrac{2 + \dfrac{1}{x^2}}{\dfrac{1}{x} + \dfrac{2}{x^2}}$

73. $\dfrac{3 - \dfrac{1}{y}}{2 - \dfrac{1}{y}}$

74. $\dfrac{\dfrac{6}{x + 2} + 4}{\dfrac{8}{x + 2} - 4}$

Name ______________________ Section __________ Date __________

Chapter 6 Test

1. Find any real numbers for which the following expression is undefined.
$$\frac{x+5}{x^2+4x+3}$$

2. For a certain computer desk, the manufacturing cost C per desk (in dollars) is
$$C=\frac{100x+3000}{x}$$
where x is the number of desks manufactured.
   a. Find the average cost per desk when manufacturing 200 computer desks.
   b. Find the average cost per desk when manufacturing 1000 computer desks.

Simplify each rational expression.

3. $\dfrac{3x-6}{5x-10}$

4. $\dfrac{x+10}{x^2-100}$

5. $\dfrac{x+6}{x^2+12x+36}$

6. $\dfrac{7-x}{x-7}$

7. $\dfrac{2m^3-2m^2-12m}{m^2-5m+6}$

8. $\dfrac{y-x}{x^2-y^2}$

Perform each indicated operation and simplify if possible.

9. $\dfrac{x^2-13x+42}{x^2+10x+21}\div\dfrac{x^2-4}{x^2+x-6}$

10. $\dfrac{3}{x-1}\cdot(5x-5)$

11. $\dfrac{y^2-5y+6}{2y+4}\cdot\dfrac{y+2}{2y-6}$

12. $\dfrac{5}{2x+5}-\dfrac{6}{2x+5}$

13. $\dfrac{5a}{a^2-a-6}-\dfrac{2}{a-3}$

14. $\dfrac{6}{x^2-1}+\dfrac{3}{x+1}$

15. $\dfrac{x^2-9}{x^2-3x}\div\dfrac{x^2+4x+1}{2x+10}$

16. $\dfrac{x+2}{x^2+11x+18}+\dfrac{5}{x^2-3x-10}$

17. $\dfrac{4y}{y^2+6y+5}-\dfrac{3}{y^2+5y+4}$

Answers

1. ________
2. a. ________
   b. ________
3. ________
4. ________
5. ________
6. ________
7. ________
8. ________
9. ________
10. ________
11. ________
12. ________
13. ________
14. ________
15. ________
16. ________
17. ________

18.

19.

20.

21.

22.

23.

24.

25.

26.

27.

28.

Solve each equation.

18. $\dfrac{4}{y} - \dfrac{5}{3} = \dfrac{-1}{5}$

19. $\dfrac{5}{y+1} = \dfrac{4}{y+2}$

20. $\dfrac{a}{a-3} = \dfrac{3}{a-3} - \dfrac{3}{2}$

21. $\dfrac{10}{x^2-25} = \dfrac{3}{x+5} + \dfrac{1}{x-5}$

Simplify each complex fraction.

22. $\dfrac{\dfrac{5x^2}{yz^2}}{\dfrac{10x}{z^3}}$

23. $\dfrac{\dfrac{b}{a} - \dfrac{a}{b}}{\dfrac{1}{b} + \dfrac{1}{a}}$

24. $\dfrac{5 - \dfrac{1}{y^2}}{\dfrac{1}{y} + \dfrac{2}{y^2}}$

△ **25.** Given that the two triangles are similar, find x.

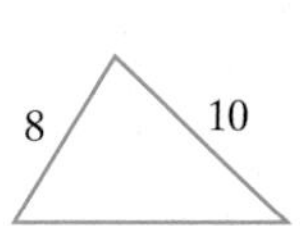

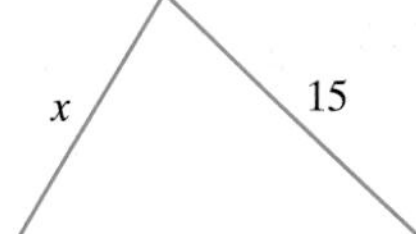

26. A number plus five times its reciprocal is equal to six. Find the number.

27. A pleasure boat traveling down the Red River takes the same time to go 14 miles upstream as it takes to go 16 miles downstream. If the current of the river is 2 miles per hour, find the speed of the boat in still water.

28. An inlet pipe can fill a tank in 12 hours. A second pipe can fill the tank in 15 hours. If both pipes are used, find how long it takes to fill the tank.

Name ______________ Section ________ Date ____________

Cumulative Review

Simplify each expression.

1. $6 \div 3 + 5^2$

2. $3[4(5 + 2) - 10]$

Write each phrase as an algebraic expression and simplify if possible. Let x represent the unknown number.

3. Twice a number, plus 6.

4. The difference of a number and 4, divided by 7.

5. Five plus the sum of a number and 1.

6. Solve for x: $\frac{5}{2}x = 15$

7. Solve: $\frac{2}{5}(x - 6) \geq x - 1$. Write the solution set in interval notation.

8. The following bar graph shows the estimated worldwide number of Internet users by region as of November 2000.

Worldwide Internet Users

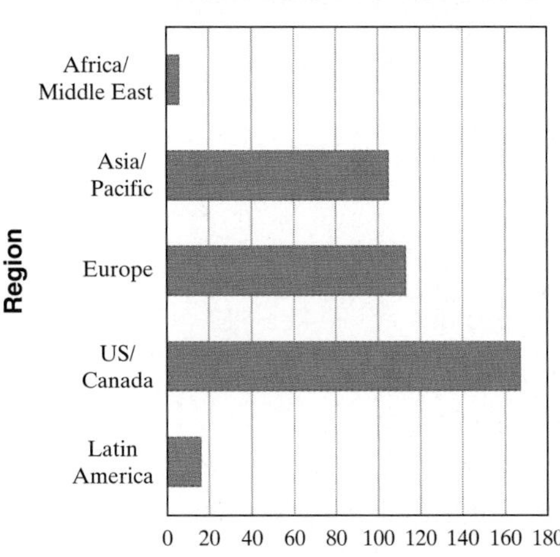

Internet users (in millions)

Source: Nua Internet Surveys

a. Find the region that has the most Internet users and approximate the number of users.

b. How many more users are in the US/Canada region than the Europe region?

9. A small business purchased a computer for \$2000. The business predicts that the computer will be used for 5 years and the value in dollars y of the computer in x years is $y = -300x + 2000$. Complete the table.

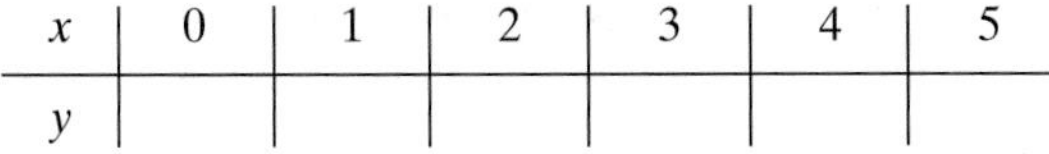

x	0	1	2	3	4	5
y						

Answers

1. ____________

2. ____________

3. ____________

4. ____________

5. ____________

6. ____________

7 ____________

8. a. ____________

b. ____________

9. see table

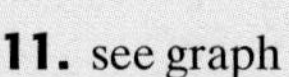

10. see graph

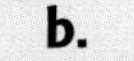

11. see graph

12. a ________

b. ________

c. ________

13. ________

14. ________

15. ________

16. ________

17. ________

18. ________

19. ________

20. ________

21. ________

22. ________

23. ________

10. Graph: $y = -3$

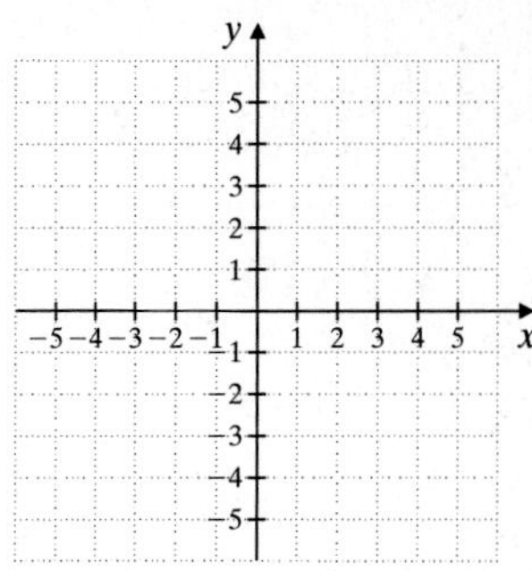

11. Find the slope of the line through $(-1, -2)$ and $(2, 4)$. Graph the line.

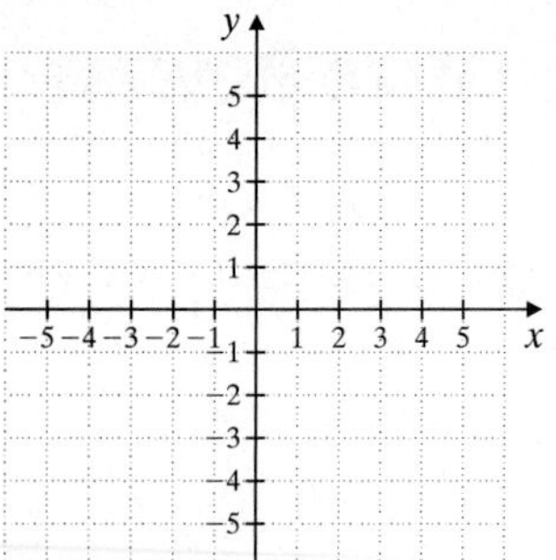

12. Find the degree of each polynomial and tell whether the polynomial is a monomial, binomial, trinomial, or none of these.

a. $-2t^2 + 3t + 6$

b. $15x - 10$

c. $7x + 3x^3 + 2x^2 - 1$

13. Add: $(-2x^2 + 5x - 1)$ and $(-2x^2 + x + 3)$

14. Multiply: $(3y + 1)^2$

15. Factor: $-9a^5 + 18a^2 - 3a$

16. Factor: $x^2 + 4x - 12$

17. Factor: $8x^2 - 22x + 5$

18. Solve: $x^2 - 9x = -20$

19. Divide: $\dfrac{2x^2 - 11x + 5}{5x - 25} \div \dfrac{4x - 2}{10}$

20. Write the rational expression as an equivalent rational expression with the given denominator.

$$\frac{4b}{9a} = \frac{\quad}{27a^2b}$$

21. Add: $1 + \dfrac{m}{m + 1}$

22. Solve: $3 - \dfrac{6}{x} = x + 8$

23. Simplify: $\dfrac{\dfrac{x + 1}{y}}{\dfrac{x}{y} + 2}$

Graphs and Functions

In this chapter, we continue our investigation of graphs of equations in two variables, which began in Chapter 3. These equations and their graphs lead to the notion of relation and to the notion of function, perhaps the single most important and useful concept in all of mathematics.

At the beginning of the 20th century, there were approximately 237,600 students enrolled in the 977 institutions of higher education in the United States. At that time, only 19% of bachelor's degree recipients were women. By the year 2010, the projected 4700 colleges and universities in the United States will have an estimated 17,490,000 students. Roughly 59% of bachelor's degree recipients are expected to be women. The phenomenal growth of colleges and universities can also be seen in the average tuition costs at these institutions of higher learning. For instance, the average annual tuition at a private four-year college or university has increased from $1809 in 1970 to $15,380 in 2000, an increase of about 750%! In Exercises 33 and 34 on page 498 we will use linear equations to predict the future cost of annual tuition at both two-year and four-year public colleges and universities.

Answers

1. ________
2. ________
3. ________
4. ________
5. ________
6. ________
7. ________
8. ________
9. ________
10. ________
11. ________
12. ________
13. ________
14. ________

Name ____________ Section ________ Date ________

Chapter 7 Pretest

Use the slope-intercept form of the linear equation to write the equation of each line with the given slope and y-intercept.

1. Slope $\frac{1}{3}$; y-intercept 6

2. Slope -7; y-intercept 0

Find an equation of each line satisfying the conditions given. Write the equations in standard form.

3. Slope 2; through $(-3, -7)$

4. Through $(5, 4)$ and $(-1, 6)$

5. Horizontal; through $(9, 10)$

6. Perpendicular to $3y - x = 6$; through $(8, 0)$

7. Find the domain and the range of the given relation. Also determine whether the relation is a function. $\{(-2, 5), (3, -7), (2, 5)\}$

If $f(x) = 7x - 1$ and $g(x) = 2x^2 + x - 5$, find the following.

8. $f(-3)$

9. $g(0)$

10. If $P(x) = -x^2 + 2x + 6$, find $P(-1)$.

11. If $f(x) = \frac{x - 5}{x + 2}$, find $f(8)$.

12. If y varies directly as x, find the constant of variation and the direct variation equation for $y = 3$ when $x = 12$.

13. If y varies inversely as x, find the constant of variation and the inverse variation equation for $y = 7$ when $x = 2$.

14. Write the statement below as an equation. Use k as the constant of variation.

y, varies jointly as x, and the square of z.

7.1 The Slope-Intercept Form

A Graphing a Line Using Its Slope and y-Intercept

From Section 3.4, we know that when a linear equation is solved for y, the coefficient of x is the slope of the line. For example, the slope of the line whose equation is $y = 3x + 1$ is 3. In this equation, $y = 3x + 1$, what does 1 represent? To find out, let $x = 0$ and watch what happens.

$$y = 3x + 1$$
$$y = 3 \cdot 0 + 1 \quad \text{Let } x = 0.$$
$$y = 1$$

We now have the ordered pair $(0, 1)$, which is the y-intercept.

This is true in general. To see this, let $x = 0$ and solve for y in $y = mx + b$.

$$y = m \cdot 0 + b \quad \text{Let } x = 0$$
$$y = b$$

We obtain the ordered pair $(0, b)$, which is the y-intercept. (You may also see b alone called the y-intercept.)

The form $y = mx + b$ is appropriately called the **slope-intercept** form of a linear equation. (slope: m; y-intercept: b)

Slope-Intercept Form

When a linear equation in two variables is written in slope-intercept form,

$$y = mx + b$$

then m is the slope of the line and $(0, b)$ is the y-intercept of the line.

We can use the slope-intercept form to graph a linear equation.

EXAMPLE 1 Graph: $y = \frac{1}{4}x - 3$

Solution: Recall that the slope of the graph of $y = \frac{1}{4}x - 3$ is $\frac{1}{4}$ and the y-intercept is $(0, -3)$. To graph the line, we first plot the y-intercept $(0, -3)$. To find another point on the line, we recall that slope is $\frac{\text{rise}}{\text{run}} = \frac{1}{4}$. We may then plot another point by starting at $(0, -3)$, rising 1 unit up, and then running 4 units to the right. We are now at the point $(4, -2)$. The graph of $y = \frac{1}{4}x - 3$ is the line through points $(0, -3)$ and $(4, -2)$, as shown.

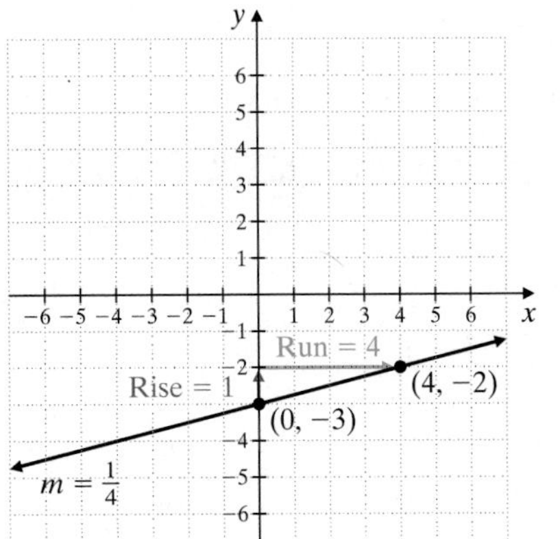

OBJECTIVES

A Graph a line using its slope and y-intercept.

B Use the slope-intercept form to write an equation of the line.

C Interpret the slope-intercept form in an application.

Practice Problem 1

Graph: $y = \frac{2}{3}x + 1$

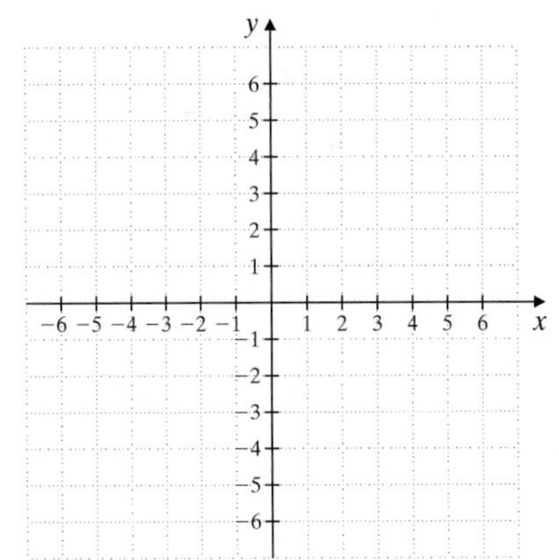

Answer

1.

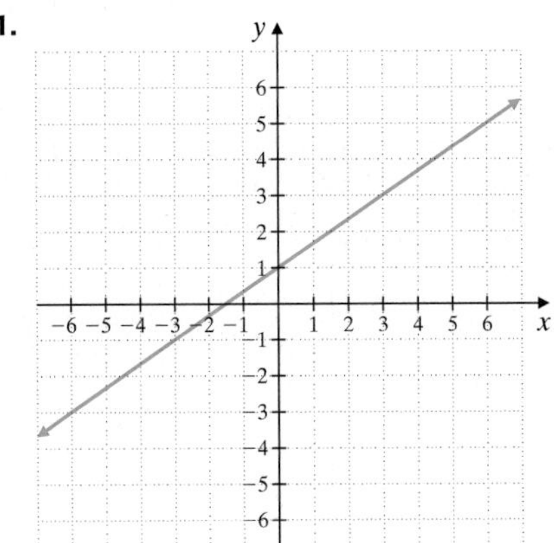

Practice Problem 2

Graph: $3x + y = -2$

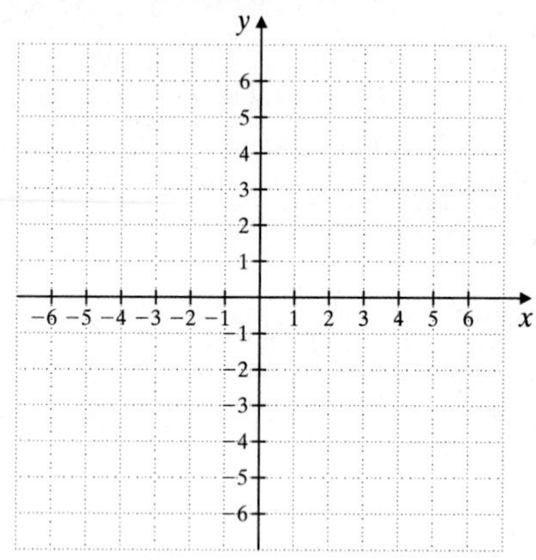

EXAMPLE 2 Graph: $2x + y = 3$

Solution: First, we solve the equation for y to write it in slope-intercept form. In slope-intercept form, the equation is $y = -2x + 3$. Next we plot the y-intercept $(0, 3)$. To find another point on the line, we use the slope -2, which can be written as $\frac{\text{rise}}{\text{run}} = \frac{-2}{1}$. We start at $(0, 3)$ and move vertically 2 units down, since the numerator of the slope is -2; then we move horizontally 1 unit to the right since the denominator of the slope is 1. We arrive at the point $(1, 1)$. The line through $(1, 1)$ and $(0, 3)$ will have the required slope of -2.

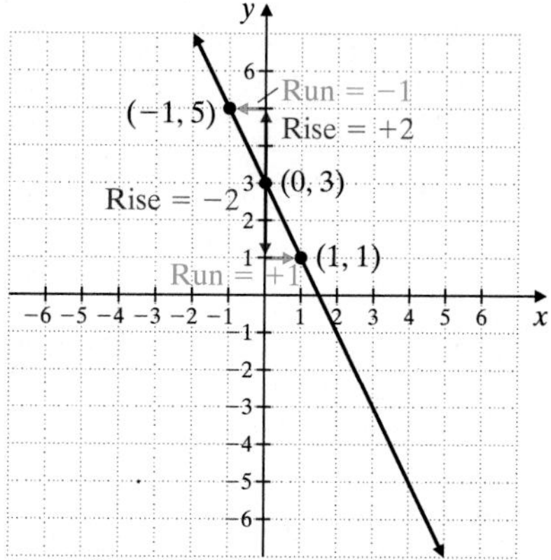

The slope -2 can also be written as $\frac{2}{-1}$, so to find another point for Example 2 we could start at $(0, 3)$ and move 2 units up and then 1 unit left. We would stop at the point $(-1, 5)$. The line through $(-1, 5)$ and $(0, 3)$ will have the required slope and will be the same line as shown previously through $(1, 1)$ and $(0, 3)$.

B Using the Slope-Intercept Form to Write an Equation

Given the slope and y-intercept of a line, we may write its equation as well as graph the line.

Practice Problem 3

Write an equation of the line with slope $\frac{2}{3}$ and y-intercept $(0, 1)$.

EXAMPLE 3

Write an equation of the line with y-intercept $(0, -3)$ and slope of $\frac{1}{4}$.

Solution: We are given the slope and the y-intercept. We let $m = \frac{1}{4}$ and $b = -3$, and write the equation in slope-intercept form, $y = mx + b$.

$$y = mx + b$$

$$y = \frac{1}{4}x + (-3) \qquad \text{Let } m = \frac{1}{4} \text{ and } b = -3.$$

$$y = \frac{1}{4}x - 3 \qquad \text{Simplify.}$$

Notice that the graph of this equation has slope $\frac{1}{4}$ and y-intercept $(0, -3)$ as desired.

Answers

2.

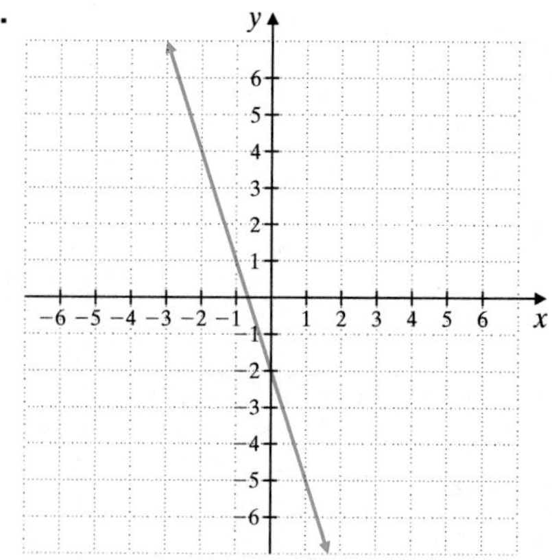

3. $y = \frac{2}{3}x + 1$

Try the Concept Check in the margin.

C Interpreting the Slope-Intercept Form

The graph of an adult one-day pass price for Disney World is shown below. Notice that the graph resembles the graph of a line. Often, businesses depend on equations that "closely fit" lines like this one to model the data and predict future trends. For example, by a method called least squares regression, the linear equation $y = 1.568x + 29.00$ approximates the data shown, where x is the number of years since 1988 and y is the ticket price for that year.

Ticket Price at Disney World

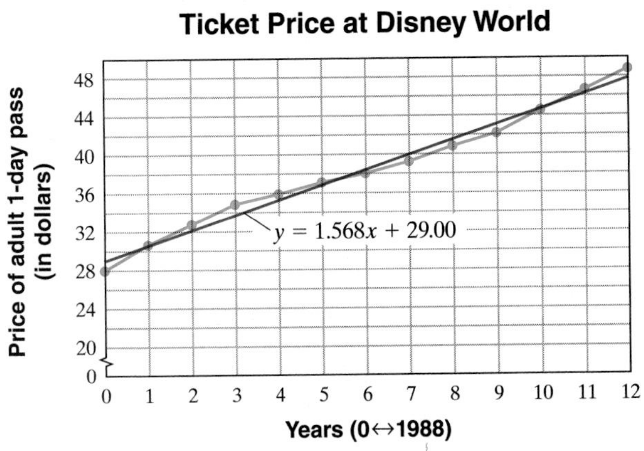

Source: The Walt Disney Company

Concept Check

What is wrong with the following equation of a line with y-intercept $(0, 4)$ and slope 2?

$$y = 4x + 2$$

The notation 0 ↔ 1988 means that the number 0 corresponds to the year 1988, 1 corresponds to the year 1989, and so on.

EXAMPLE 4 Predicting Future Prices

The adult one-day pass price y for Disney World is given by

$$y = 1.568x + 29.00$$

where x is the number of years since 1988.

a. Use this equation to predict the ticket price for 2004.
b. What does the slope of this equation mean?
c. What does the y-intercept of this equation mean?

Solution:

a. To predict the price of a pass in 2004, we need to find y when x is 16. (Since year 1988 corresponds to $x = 0$, year 2004 corresponds to $x = 2004 - 1988 = 16$.)

$$y = 1.568x + 29.00$$
$$= 1.568(16) + 29.00 \quad \text{Let } x = 16.$$
$$= 54.088$$

We predict that in 2004 the price of an adult one-day pass to Disney World will be about $54.09.

b. The slope of $y = 1.568x + 29.00$ is 1.568. We can think of this number as $\frac{\text{rise}}{\text{run}}$ or $\frac{1.568}{1}$. This means that the ticket price increases on the average by $1.568 every 1 year.

c. The y-intercept of $y = 1.568x + 29.00$ is $(0, 29.00)$ ← year, price

This means that at year $x = 0$, or 1988, the ticket price was about $29.00. ●

Practice Problem 4

For the period 1980 through 2020, the number of people y age 85 or older living in the United States is given by the equation $y = 110{,}520x + 2{,}127{,}400$, where x is the number of years since 1980. (*Source*: Based on data and estimates from the U.S. Bureau of the Census)

a. Estimate the number of people age 85 or older living in the United States in 2010.

b. What does the slope of this equation mean?

c. What does the y-intercept of this equation mean?

Answers

a. 5,443,000 **b.** The number of people age 85 or older in the United States increases at a rate of 110,520 per year. **c.** At year $x = 0$, or 1980, there were 2,127,400 people age 85 or older in the United States.

Concept Check: y-intercept and slope were switched, should be $y = 2x + 4$

GRAPHING CALCULATOR EXPLORATIONS

You may have noticed by now that to use the [Y=] key on a grapher to graph an equation, the equation must be solved for y.

Graph each equation by first solving the equation for y.

1. $x = 3.5y$

2. $-2.7y = x$

3. $5.78x + 2.31y = 10.98$

4. $-7.22x + 3.89y = 12.57$

5. $y - x = 3.78$

6. $3y - 5x = 6x - 4$

7. $y - 5.6x = 7.7x + 1.5$

8. $y + 2.6x = -3.2$

Name ______________________ Section __________ Date __________

Mental Math

Find the slope and the y-intercept of each line.

1. $y = -4x + 12$

2. $y = \frac{2}{3}x - \frac{7}{2}$

3. $y = 5x$

4. $y = -x$

5. $y = \frac{1}{2}x + 6$

6. $y = -\frac{2}{3}x + 5$

EXERCISE SET 7.1

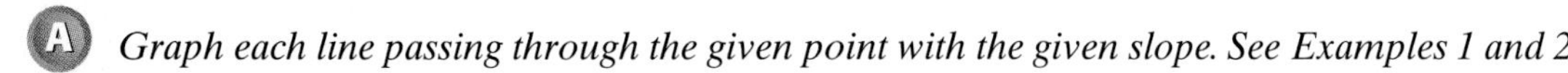

A *Graph each line passing through the given point with the given slope. See Examples 1 and 2.*

1. Through $(1, 3)$ with slope $\frac{3}{2}$

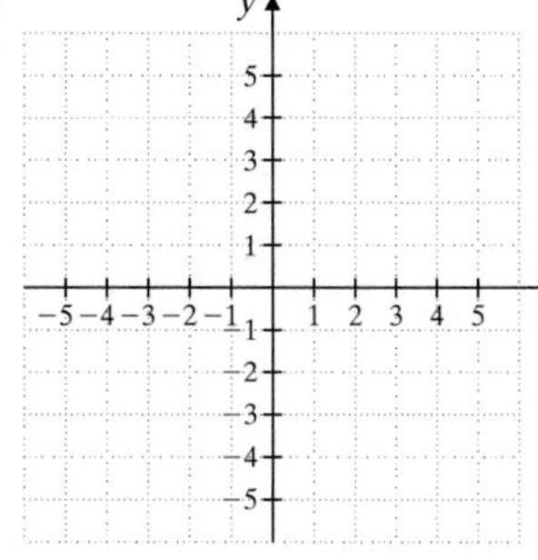

2. Through $(-2, -4)$ with slope $\frac{2}{5}$

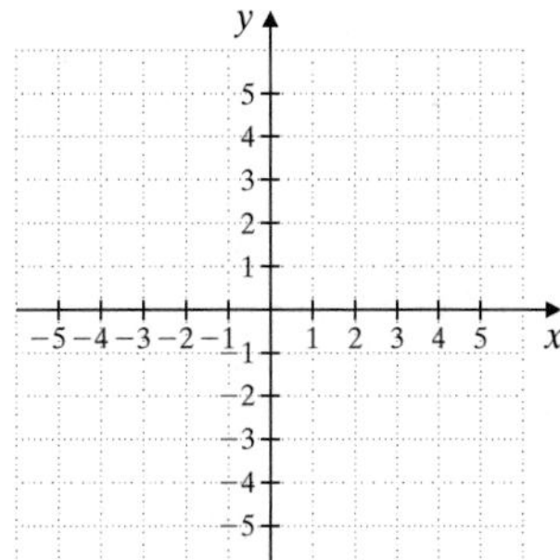

3. Through $(0, 0)$ with slope 5

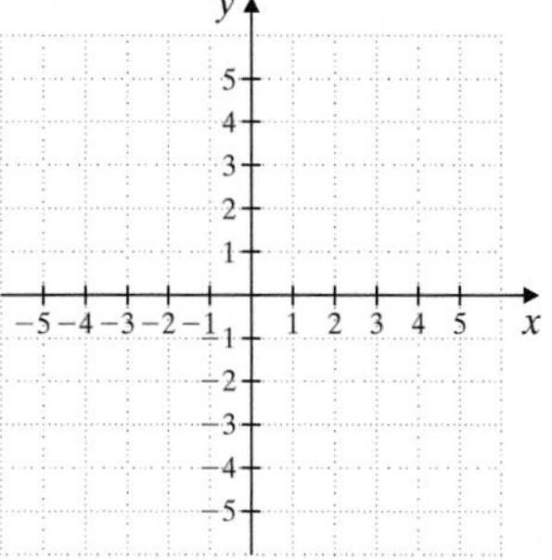

4. Through $(-5, 2)$ with slope 2

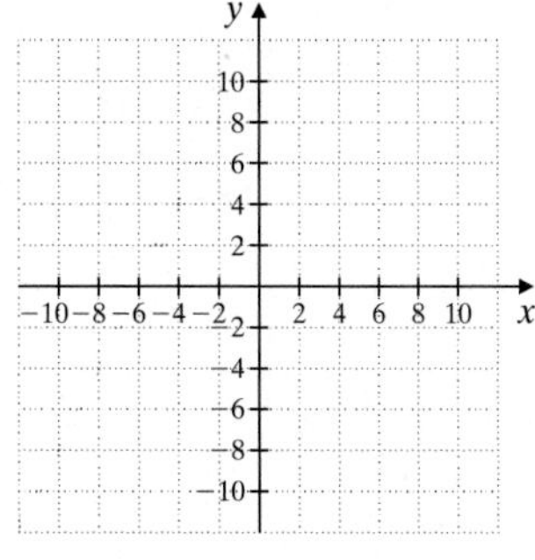

5. Through $(0, 7)$ with slope -1

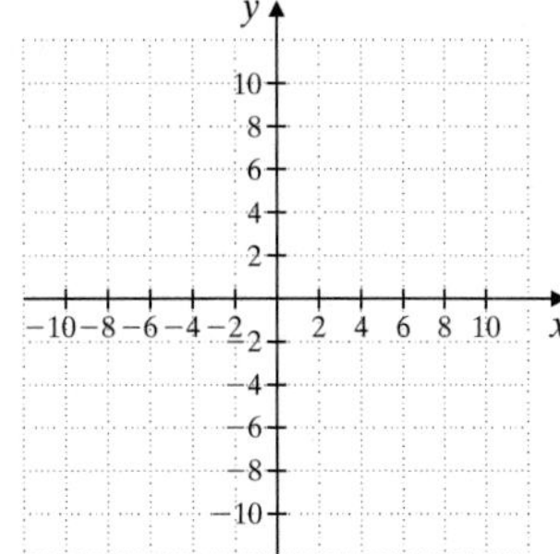

6. Through $(3, 0)$ with slope -3

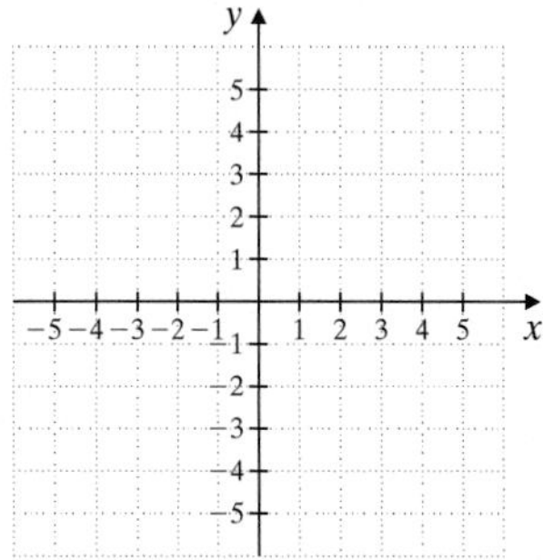

Graph each linear equation using the slope and y-intercept. See Examples 1 and 2.

7. $y = -2x$

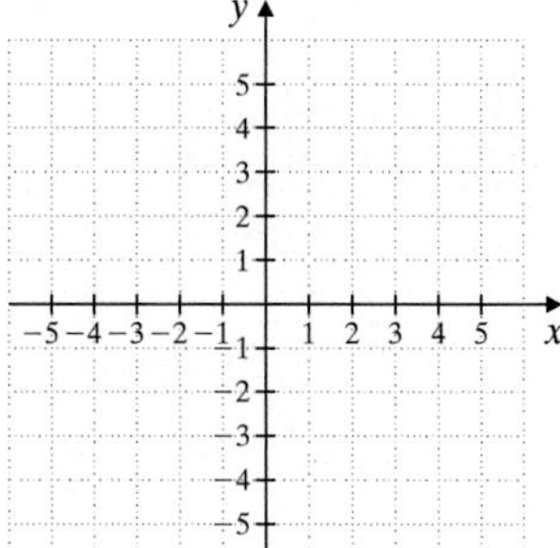

8. $y = 2x$

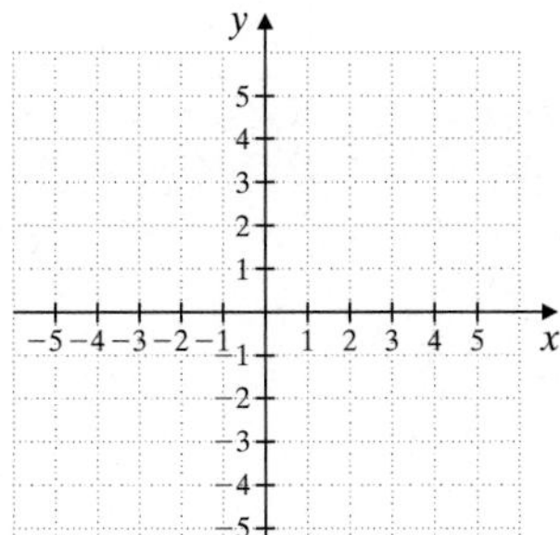

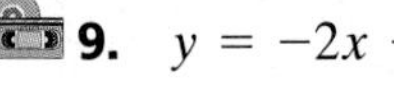

9. $y = -2x + 3$

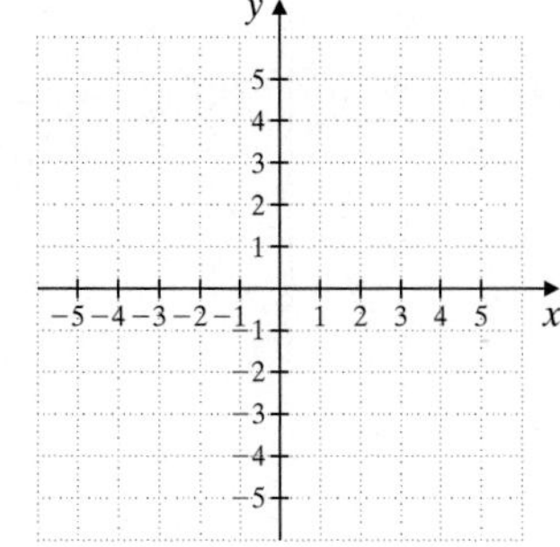

10. $y = 2x + 6$

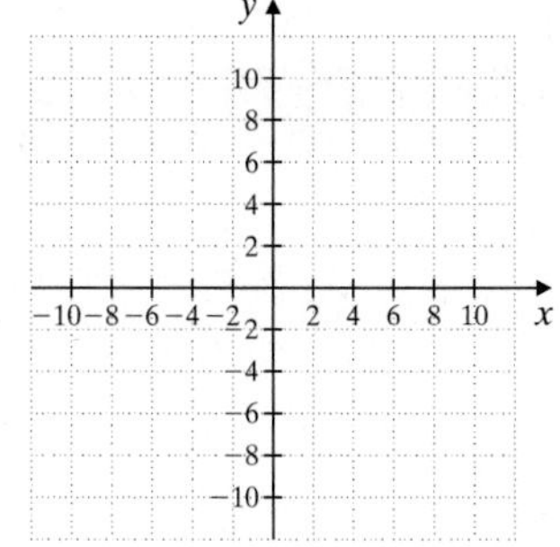

11. $y = \frac{1}{2}x$

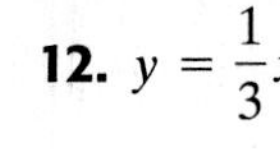

12. $y = \frac{1}{3}x$

13. $y = \frac{1}{2}x - 4$

14. $y = \frac{1}{3}x - 2$

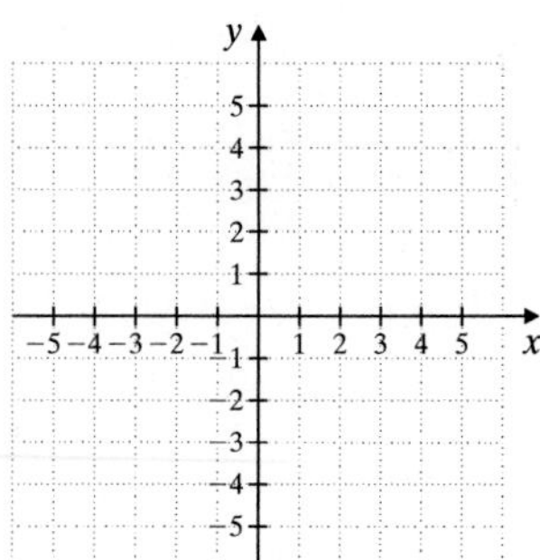

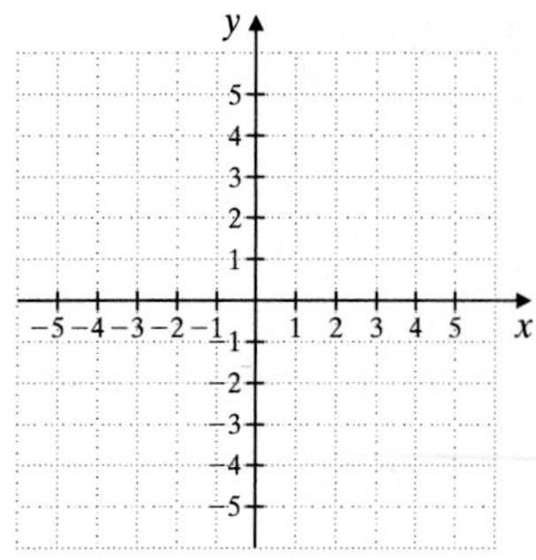

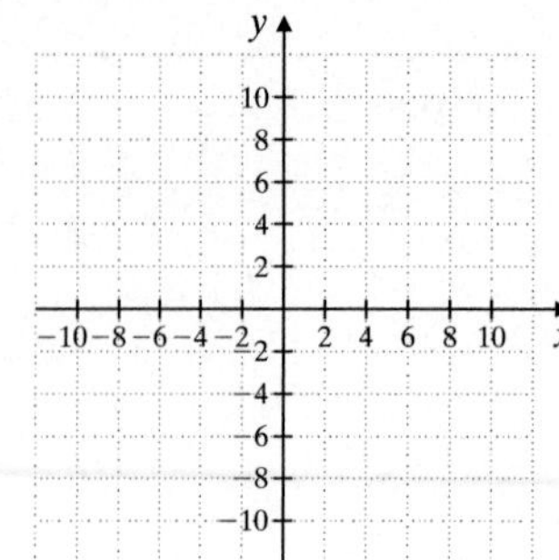

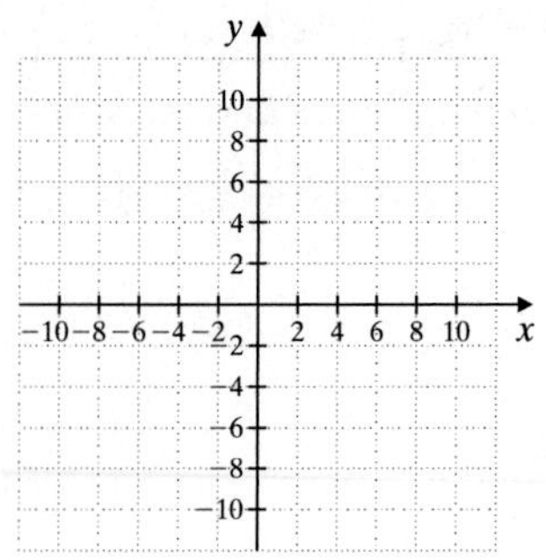

15. $x - y = 3$

16. $x - y = -4$

17. $x + 2y = 8$

18. $x - 3y = 3$

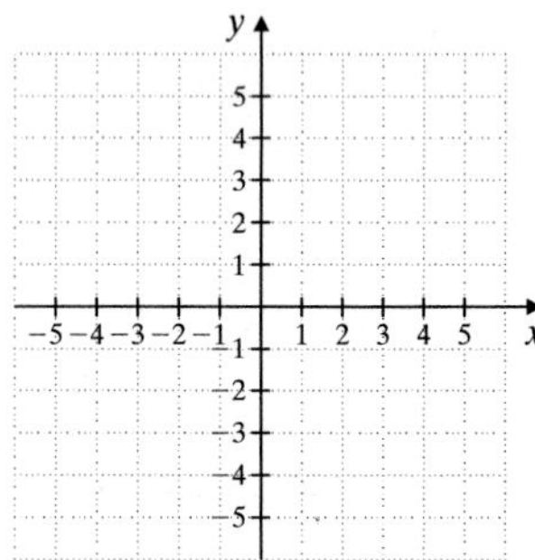

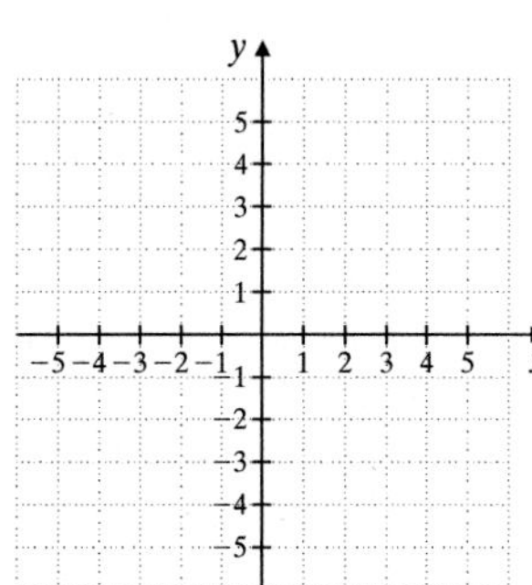

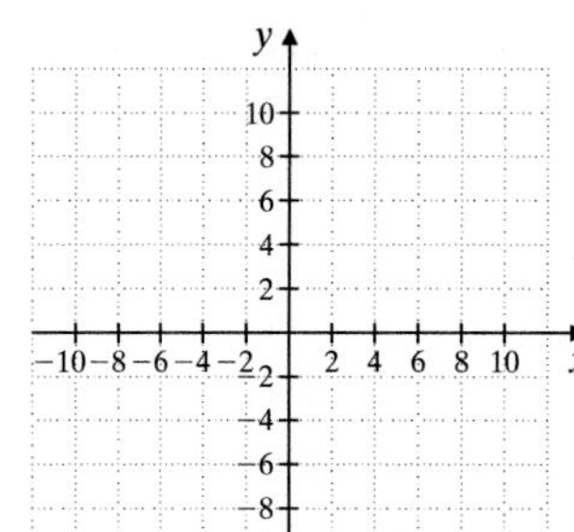

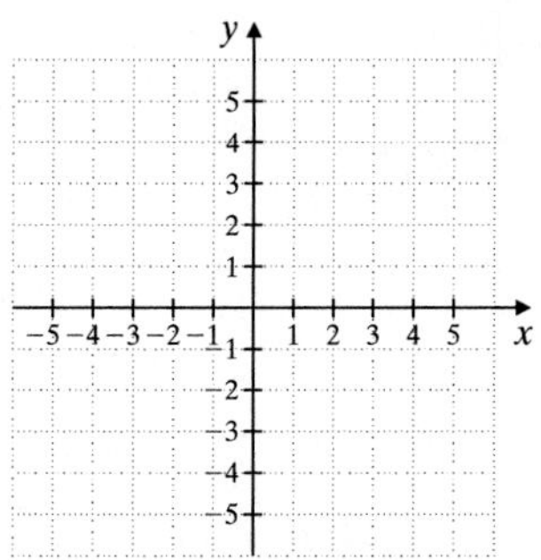

The graph of $y = 5x$ follows. Use this graph to match each linear equation with its graph. See Examples 1 and 2.

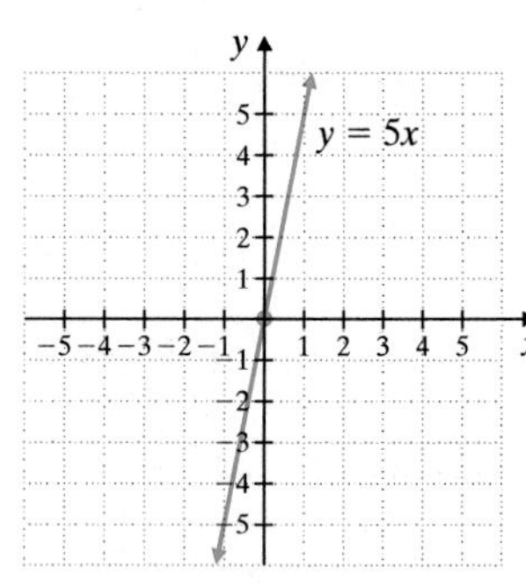

19. $y = 5x - 3$

20. $y = 5x - 2$

21. $y = 5x + 1$

22. $y = 5x + 3$

A

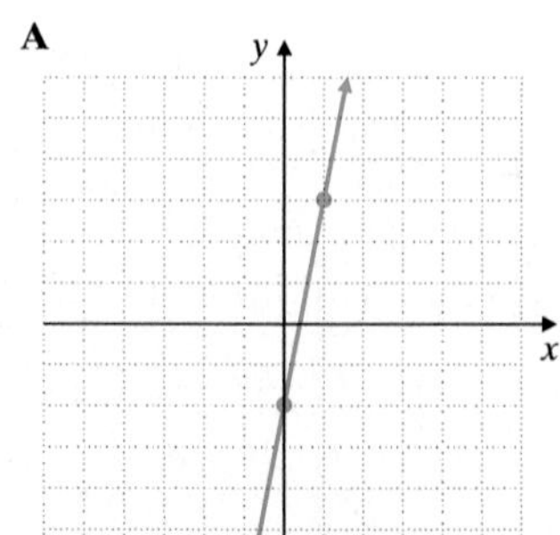

B

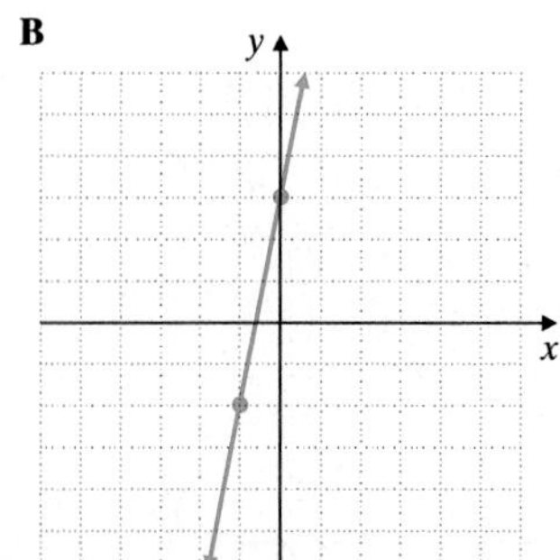

C

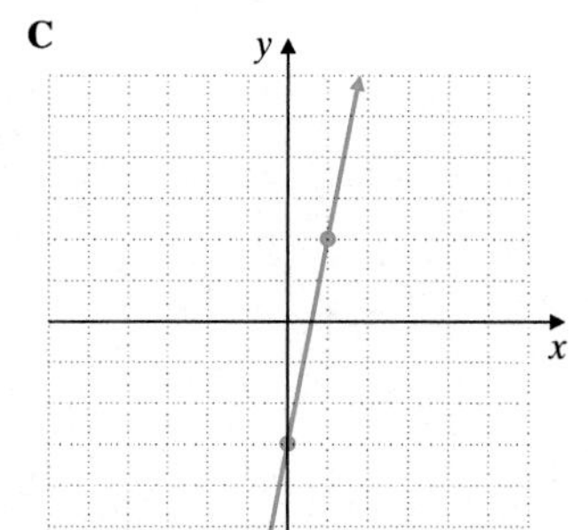

D

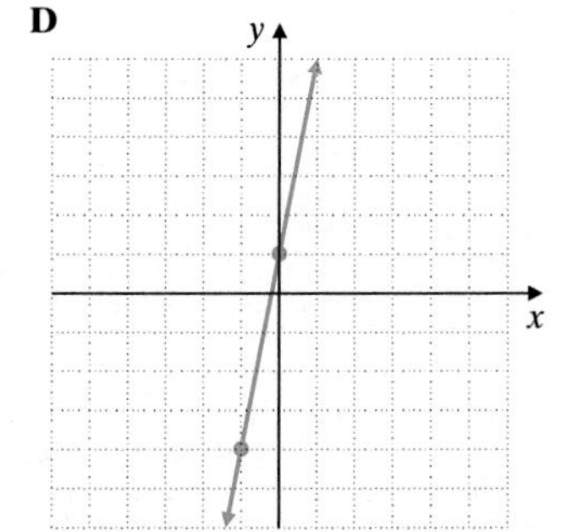

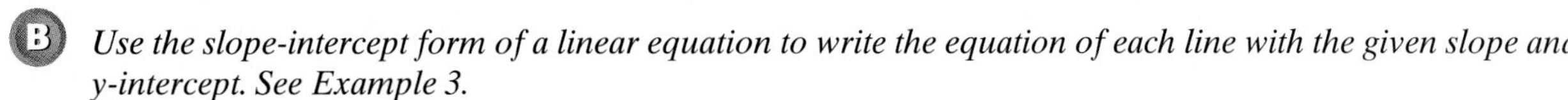

B *Use the slope-intercept form of a linear equation to write the equation of each line with the given slope and y-intercept. See Example 3.*

23. Slope -1; y-intercept $(0, 1)$

24. Slope $\frac{1}{2}$; y-intercept $(0, -6)$

25. Slope 2; y-intercept $\left(0, \frac{3}{4}\right)$

26. Slope -3; y-intercept $\left(0, -\frac{1}{5}\right)$

27. Slope $\frac{2}{7}$; y-intercept $(0, 0)$

28. Slope $-\frac{4}{5}$; y-intercept $(0, 0)$

C *Solve. See Example 4.*

29. The annual average income y of an American man with an associate's degree is given by the linear equation $y = 1765.1x + 35{,}815.0$, where x is the number of years after 1995. (*Source*: Based on data from the U.S. Bureau of the Census, 1995–2000)

a. Find the average income of an American man with an associate's degree in 2000.
b. Find and interpret the slope of the equation.
c. Find and interpret the y-intercept of the equation.

30. The annual average income y of an American woman with a bachelor's degree is given by the linear equation $y = 1733.8x + 26{,}914.7$, where x is the number of years after 1995. (*Source*: Based on data from the U.S. Bureau of the Census, 1995–2000)

a. Find the average income of an American woman with a bachelor's degree in 2000.
b. Find and interpret the slope of the equation.
c. Find and interpret the y-intercept of the equation.

31. One of the top ten occupations in terms of job growth in the next few years is expected to be computer support specialist. The number of people y, in thousands, employed as computer support specialists in the United States can be estimated by the linear equation $220x - 5y = -2145$, where x is the number of years after 1998. (*Source*: Based on projections from the U.S. Bureau of Labor Statistics, 1998–2008)

a. Find the slope and y-intercept of the linear equation.
b. What does the slope mean in this context?
c. What does the y-intercept mean in this context?

32. One of the faster growing occupations over the next few years is expected to be medical assistant. The number of people y, in thousands, employed as medical assistants in the United States can be estimated by the linear equation $-146x + 10y = 2520$, where x is the number of years after 1998. (*Source*: Based on projections from the U.S. Bureau of Labor Statistics, 1998–2008)

a. Find the slope and y-intercept of the linear equation.
b. What does the slope mean in this context?
c. What does the y-intercept mean in this context?

33. The yearly cost of undergraduate tuition and required fees for attending a public four-year college full-time can be estimated by the linear equation $y = 136.2x + 2827.6$, where x is the number of years after 1996 and y is the total cost in dollars. (*Source*: Based on data from The College Board, 1996–2000)

a. Use this equation to approximate the yearly cost of attending a four-year public college in 2010.

b. Use this equation to predict in what year the yearly cost of tuition and required fees will exceed \$5000. (*Hint:* Let $y = 5000$ and solve for x.)

c. Use this equation to approximate the yearly cost of attending a four-year college in the present year. If you attend a four-year college, is this amount greater or less than the amount currently charged by the college you attend?

34. The yearly cost of tuition and required fees for attending a public two-year college full-time can be estimated by the linear equation $y = 68.3x + 1372$, where x is the number of years after 1996 and y is the total cost in dollars. (*Source:* Based on data from The College Board, 1996–2000)

a. Use this equation to approximate the yearly cost of attending a two-year public college in 2010.

b. Use this equation to predict in what year the yearly cost of tuition and required fees will exceed \$3000. (*Hint:* Let $y = 3000$ and solve for x.)

c. Use this equation to approximate the yearly cost of attending a two-year college in the present year. If you attend a two-year college, is this amount greater or less than the amount currently charged by the college you attend?

Review and Preview

Simplify and solve for y. See Section 2.6.

35. $y - 2 = 5(x + 6)$

36. $y - 0 = -3[x - (-10)]$

37. $y - (-1) = 2(x - 0)$

38. $y - 9 = -8[x - (-4)]$

Combining Concepts

39. In your own words, explain how to graph an equation using its slope and y-intercept.

40. Suppose that the revenue of a company has increased at a steady rate of \$42,000 per year since 1995. Also the company's revenue in 1995 was \$2,900,000. Write an equation that describes the company's revenue since 1995.

41. Suppose that a bird dives off a 500-foot cliff and descends at a rate of 7 feet per second. Write an equation that describes the bird's height at any time x.

7.2 More Equations of Lines

OBJECTIVES

- **A** Use the point-slope form to write the equation of a line.
- **B** Write equations of vertical and horizontal lines.
- **C** Write equations of parallel and perpendicular lines.
- **D** Use the point-slope form in real-world applications.

A Using the Point-Slope Form to Write an Equation

When the slope of a line and a point on the line are known, the equation of the line can also be found. To do this, we use the slope formula to write the slope of a line that passes through points (x, y), and (x_1, y_1). We have

$$m = \frac{y - y_1}{x - x_1}$$

We multiply both sides of this equation by $x - x_1$ to obtain

$$y - y_1 = m(x - x_1)$$

This form is called the *point-slope form* of the equation of a line.

Point-Slope Form of the Equation of a Line

The **point-slope form** of the equation of a line is

$$y - y_1 = \overset{\text{slope}}{\overset{\downarrow}{m}}(x - x_1)$$

(arrows labeled "point" indicate y_1 and x_1)

where m is the slope of the line and (x_1, y_1) is a point on the line.

EXAMPLE 1

Write an equation of the line with slope -3 and containing the point $(1, -5)$.

Solution: Because we know the slope and a point on the line, we use the point-slope form with $m = -3$ and $(x_1, y_1) = (1, -5)$.

$$\begin{aligned} y - y_1 &= m(x - x_1) && \text{Point-slope form} \\ y - (-5) &= -3(x - 1) && \text{Let } m = -3 \text{ and } (x_1, y_1) = (1, -5). \\ y + 5 &= -3x + 3 && \text{Use the distributive property.} \\ y &= -3x - 2 \end{aligned}$$

The equation is $y = -3x - 2$.

Practice Problem 1

Write an equation of the line with slope -2 and containing the point $(2, -4)$. Write the equation in slope-intercept form, $y = mx + b$.

EXAMPLE 2

Write an equation of the line through points $(4, 0)$ and $(-4, -5)$.

Solution: First we find the slope of the line.

$$m = \frac{-5 - 0}{-4 - 4} = \frac{-5}{-8} = \frac{5}{8}$$

Next we make use of the point-slope form. We replace (x_1, y_1) by either $(4, 0)$ or $(-4, -5)$ in the point-slope equation. We will choose the point $(4, 0)$. The line through $(4, 0)$ with slope $\frac{5}{8}$ is as follows.

Practice Problem 2

Write an equation of the line through points $(3, 0)$ and $(-2, 4)$. Write the equation in slope-intercept form, $y = mx + b$.

Answers

1. $y = -2x$, **2.** $y = -\frac{4}{5}x + \frac{12}{5}$

$$y - y_1 = m(x - x_1) \quad \text{Point-slope form}$$

$$y - 0 = \frac{5}{8}(x - 4) \quad \text{Let } m = \frac{5}{8} \text{ and } (x_1, y_1) = (4, 0).$$

$$y = \frac{5}{8}x - \frac{5}{8}\cdot 4 \quad \text{Use the distributive property.}$$

$$y = \frac{5}{8}x - \frac{5}{2} \quad \text{Simplify.}$$

The equation is $y = \frac{5}{8}x - \frac{5}{2}$. If we had chosen to use the point $(-4, -5)$, we would have obtained $y - (-5) = \frac{5}{8}[x - (-4)]$, which also simplifies to $y = \frac{5}{8}x - \frac{5}{2}$.

Helpful Hint

If two points of a line are given, either one may be used with the point-slope form to write an equation of the line.

B Writing Equations of Vertical and Horizontal Lines

A few special types of linear equations are those whose graphs are vertical and horizontal lines.

Practice Problem 3

Write an equation of the horizontal line containing the point $(-1, 6)$.

Practice Problem 4

Write an equation of the line containing the point $(4, 7)$ with undefined slope.

EXAMPLE 3

Write an equation of the horizontal line containing the point $(2, 3)$.

Solution: Recall, from Section 3.3, that a horizontal line has an equation of the form $y = b$. Since the line contains the point $(2, 3)$, the equation is $y = 3$.

EXAMPLE 4

Write an equation of the line containing the point $(2, 3)$ with undefined slope.

Solution: Since the line has undefined slope, the line must be vertical. A vertical line has an equation of the form $x = c$, and since the line contains the point $(2, 3)$, the equation is $x = 2$.

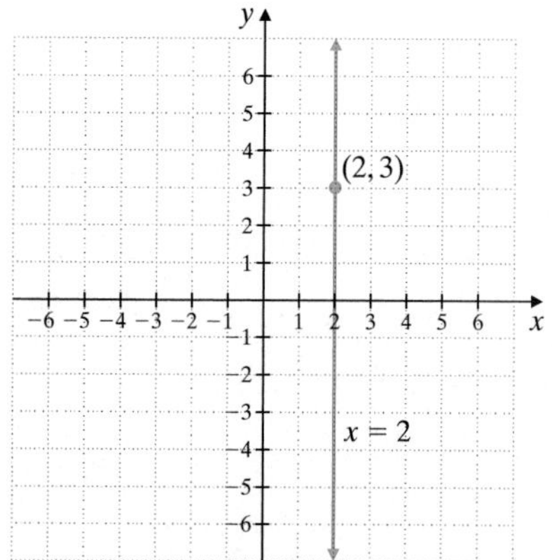

C Writing Equations of Parallel and Perpendicular Lines

Next, we write equations of parallel and perpendicular lines.

Answers

3. $y = 6$, **4.** $x = 4$

EXAMPLE 5

Write an equation of the line containing the point $(4, 4)$ and parallel to the line $2x + y = -6$.

Solution: Because the line we want to find is *parallel* to the line $2x + y = -6$, the two lines must have equal slopes. So we first find the slope of $2x + y = -6$ by solving it for y to write it in the form $y = mx + b$. Here $y = -2x - 6$, so the slope is -2.

Now we use the point-slope form to write the equation of a line through $(4, 4)$ with slope -2.

$$y - y_1 = m(x - x_1)$$

$$y - 4 = -2(x - 4) \qquad \text{Let } m = -2, x_1 = 4, \text{ and } y_1 = 4.$$

$$y - 4 = -2x + 8 \qquad \text{Use the distributive property.}$$

$$y = -2x + 12$$

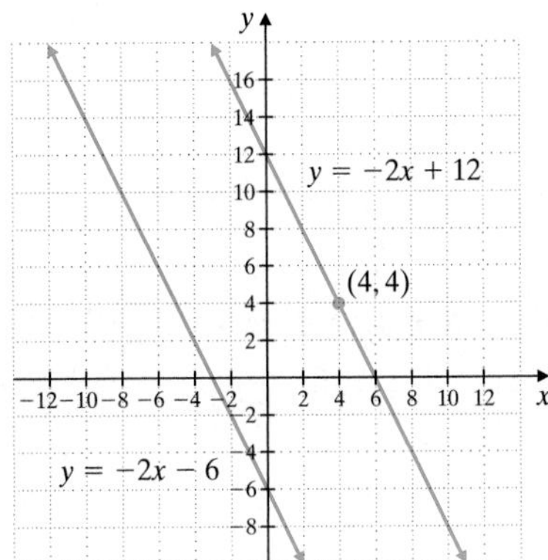

The equation, $y = -2x - 6$, and the new equation, $y = -2x + 12$, have the same slope but different y-intercepts so their graphs are parallel. Also, the graph of $y = -2x + 12$ contains the point $(4, 4)$, as desired. ●

Practice Problem 5

Write an equation of the line containing the point $(-1, 2)$ and parallel to the line $3x + y = 5$. Write the equation in the form $y = mx + b$.

EXAMPLE 6

Write an equation of the line containing the point $(-2, 1)$ and perpendicular to the line $3x + 5y = 4$.

Solution: First we find the slope of $3x + 5y = 4$ by solving it for y.

$$5y = -3x + 4$$

$$y = -\frac{3}{5}x + \frac{4}{5}$$

The slope of the given line is $-\frac{3}{5}$. A line perpendicular to this line will have a slope that is the negative reciprocal of $-\frac{3}{5}$, or $\frac{5}{3}$. We use the point-slope form to write an equation of a new line through $(-2, 1)$ with slope $\frac{5}{3}$.

$$y - 1 = \frac{5}{3}[x - (-2)]$$

$$y - 1 = \frac{5}{3}(x + 2) \qquad \text{Simplify.}$$

$$y - 1 = \frac{5}{3}x + \frac{10}{3} \qquad \text{Use the distributive property.}$$

$$y = \frac{5}{3}x + \frac{13}{3} \qquad \text{Add 1 to both sides.}$$

Practice Problem 6

Write an equation of the line containing the point $(3, 4)$ and perpendicular to the line $2x + 4y = 5$. Write the equation in standard form.

Answers

5. $y = -3x - 1$, **6.** $2x - y = 2$

The equation $y = -\frac{3}{5}x + \frac{4}{5}$ and the new equation $y = \frac{5}{3}x + \frac{13}{3}$ have negative reciprocal slopes, so their graphs are perpendicular. Also, the graph of $y = \frac{5}{3}x + \frac{13}{3}$ contains the point $(-2, 1)$, as desired.

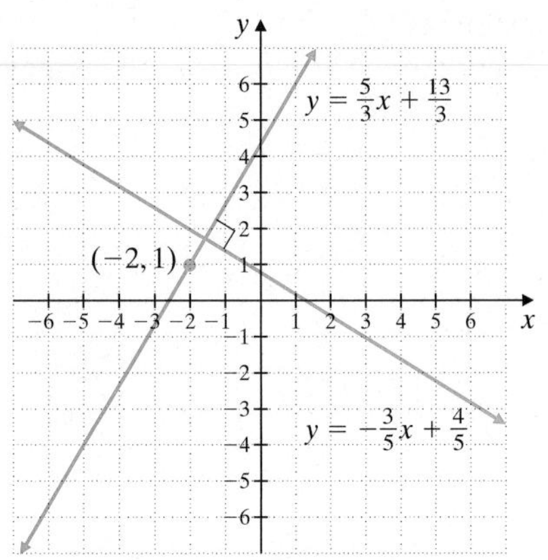

D Using the Point-Slope Form in Applications

The point-slope form of an equation is very useful for solving real-world problems.

Practice Problem 7

Southwest Regional is an established office product maintenance company that has enjoyed constant growth in new maintenance contracts since 1985. In 1993, the company obtained 15 new contracts and in 2000, the company obtained 36 new contracts. Use these figures to predict the number of new contracts this company can expect in 2007.

Answer

7. 57 new contracts

EXAMPLE 7 Predicting Sales

Southern Star Realty is an established real estate company that has enjoyed constant growth in sales since 1993. In 1995 the company sold 200 houses, and in 2000 the company sold 275 houses. Use these figures to predict the number of houses this company may sell in 2005.

Solution:

1. UNDERSTAND. Read and reread the problem. Then let

$x =$ the number of years after 1993 and

$y =$ the number of houses sold in the year corresponding to x.

The information provided then gives the ordered pairs $(2, 200)$ and $(7, 275)$. To better visualize the sales of Southern Star Realty, we graph the line that passes through the points $(2, 200)$ and $(7, 275)$.

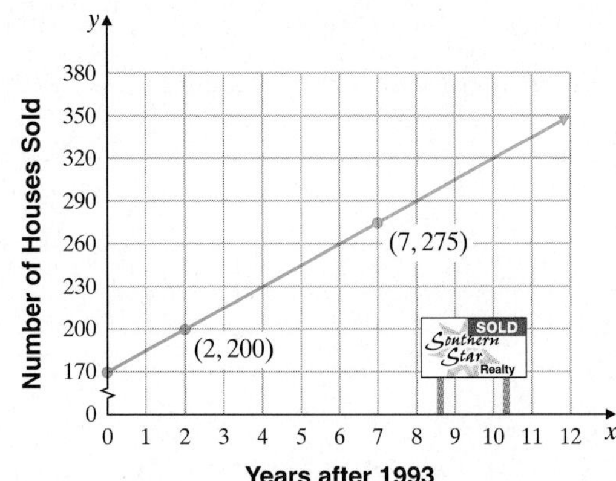

2. TRANSLATE. We write an equation of the line that passes through the points $(2, 200)$ and $(7, 275)$. To do so, we first find the slope of the line.

$$m = \frac{275 - 200}{7 - 2} = \frac{75}{5} = 15$$

Then, using the point-slope form to write the equation, we have

$$y - y_1 = m(x - x_1)$$

$$y - 200 = 15(x - 2) \quad \text{Let } m = 15 \text{ and } (x_1, y_1) = (2, 200).$$

$$y - 200 = 15x - 30 \quad \text{Multiply.}$$

$$y = 15x + 170 \quad \text{Add 200 to both sides.}$$

3. SOLVE. To predict the number of houses sold in 2005, we use $y = 15x + 170$ and complete the ordered pair (12,) since $2005 - 1993 = 12$.

$$y = 15(12) + 170 \quad \text{Let } x = 12.$$

$$y = 350$$

4. INTERPRET.

Check: Verify that the point $(12, 350)$ is a point on the line graphed in Step 1.
State: Southern Star Realty should expect to sell 350 houses in 2005.

GRAPHING CALCULATOR EXPLORATIONS

Many graphing calculators have a TRACE feature. This feature allows you to trace along a graph and see the corresponding x- and y-coordinates appear on the screen. Use this feature for the following exercises.

Graph each equation and then use the TRACE feature to complete each ordered pair solution. (Many times the tracer will not show the exact x- or y-value asked for. In each case, trace as closely as you can to the given x- or y-coordinate and approximate the other, unknown coordinate to one decimal place.)

1. $y = 2.3x + 6.7$;
$x = 5.1, y = ?$

2. $y = -4.8x + 2.9$;
$x = -1.8, y = ?$

3. $y = -5.9x - 1.6$;
$x = ?, y = 7.2$

4. $y = 0.4x - 8.6$;
$x = ?, y = -4.4$

5. $y = 5.2x - 3.3$;
$x = 2.3, y = ?$
$x = ?, y = 36$

6. $y = -6.2x - 8.3$;
$x = 3.2, y = ?$
$x = ?, y = 12$

FOCUS ON The Real World

CENTERS OF MASS

The **center of mass**, also known as **center of gravity**, of an object is the point at which the mass of the object may be considered to be concentrated. For a two-dimensional object or surface, such as a flat board, the center of mass can be described as the point on which the surface would balance.

The idea of center of mass is an important one in many disciplines, especially physics and its applications. The following list describes situations in which an object's center of mass is important.

- Geographers are sometimes concerned with pinpointing the *geographic center* of a county, state, country, or continent. A geographic center is actually the center of mass of a geographic region if it is considered as a two-dimensional surface. The geographic center of the 48 contiguous United States is near Lebanon, Kansas. The geographic center of the North American continent is 6 miles west of Balta, North Dakota.
- A top-loading washing machine is designed so its center of mass is located within its agitator post. During the spin cycle, the washer tub spins around its center of mass. If clothes aren't carefully distributed within the tub, they can bunch up and throw off the center of mass. This causes the machine to vibrate and sometimes jump or shake wildly. Some washing machine models will stop operating when the loads become unbalanced in this way.
- Single-hulled boats are normally designed so that their centers of mass are below the water line. This provides a boat with stability. Otherwise, with a center of mass above the water line, the boat would have a tendency to tip over in the water.
- Most small airplanes must be carefully loaded so as not to affect the location of the airplane's center of mass. The center of mass of an airplane must be near the center of the wings. Pilots of small aircraft usually try to balance the weight of their cargo around the airplane's center of mass.

GROUP ACTIVITY

Attach a piece of graph paper to a piece of cardboard. Cut out a triangle and label the vertices with their coordinates. Lay the triangle on a horizontal table top with the graph paper face down. Slide the triangle toward the edge of the table until it is balanced on the edge, just about to tip over the side of the table. Firmly hold the triangle in place while another group member uses the straight edge of the table to draw a line on the graph paper side of the triangle marking the position of the table edge. Rotate the triangle a quarter turn and rebalance the triangle on the edge of the table. Draw a second line on the graph paper side of the triangle marking the position of the table edge.

1. The point where the two lines drawn on the triangle intersect is the center of mass of the triangle. Find the coordinates of this point.
2. Verify that the point you have located is roughly the center of mass of the triangle by balancing it at this point on the tip of a pencil or pen.
3. List the coordinates of the vertices of your triangle. What is the relationship between the coordinates of the center of mass and the coordinates of the triangle's vertices? (*Hint*: You may find it helpful to examine the sum of the x-coordinates and the sum of the y-coordinates of the vertices of the triangle.)
4. Test your observation in Question 3. Cut out another triangle. Label its vertices and, using your observation from Question 3, predict the location of the center of mass of the triangle. Use the balancing procedure to find the center of mass. How close was your prediction?

Name ______________________ Section __________ Date __________

Mental Math

Find the slope of and a point on the line described by each equation.

1. $y - 4 = -2(x - 1)$

2. $y - 6 = -3(x - 4)$

3. $y - 0 = \frac{1}{4}(x - 2)$

4. $y - 1 = -\frac{2}{3}(x - 0)$

5. $y + 2 = 5(x - 3)$

6. $y - 7 = 4(x + 6)$

EXERCISE SET 7.2

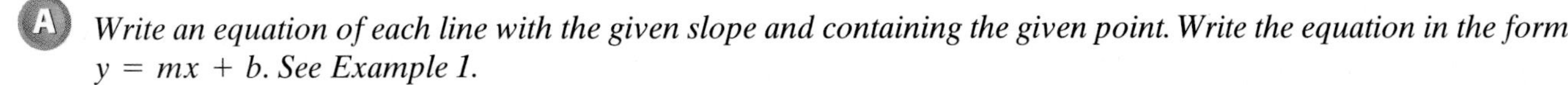

A *Write an equation of each line with the given slope and containing the given point. Write the equation in the form $y = mx + b$. See Example 1.*

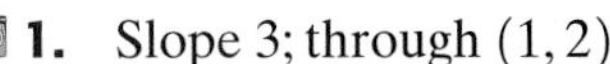

1. Slope 3; through $(1, 2)$

2. Slope 4; through $(5, 1)$

3. Slope -2; through $(1, -3)$

4. Slope -4; through $(2, -4)$

5. Slope $\frac{1}{2}$; through $(-6, 2)$

6. Slope $\frac{2}{3}$; through $(-9, 4)$

7. Slope $-\frac{9}{10}$; through $(-3, 0)$

8. Slope $-\frac{1}{5}$; through $(4, -6)$

9. Slope 2; through $(-2, 3)$

10. Slope 3; through $(-4, 2)$

11. Slope $-\frac{4}{3}$; through $(-5, 0)$

12. Slope $-\frac{3}{5}$; through $(4, -1)$

Write an equation of the line passing through the given points. Write the equation in the form $y = mx + b$. See Example 2.

13. $(2, 0)$ and $(4, 6)$

14. $(3, 0)$ and $(7, 8)$

15. $(-2, 5)$ and $(-6, 13)$

16. $(7, -4)$ and $(2, 6)$

17. $(-2, -4)$ and $(-4, -3)$

18. $(-9, -2)$ and $(-3, 10)$

19. $(-3, -8)$ and $(-6, -9)$

20. $(8, -3)$ and $(4, -8)$

21. $(-7, -4)$ and $(0, -6)$

22. $(2, -8)$ and $(-4, -3)$

23. $\left(\frac{3}{5}, \frac{4}{10}\right)$ and $\left(-\frac{1}{5}, \frac{7}{10}\right)$

24. $\left(\frac{1}{2}, -\frac{1}{4}\right)$ and $\left(\frac{3}{2}, \frac{3}{4}\right)$

B *Write an equation of each line. See Examples 3 and 4.*

25. Vertical; through $(2, 6)$

26. Slope 0; through $(-2, -4)$

27. Horizontal; through $(-3, 1)$

28. Vertical; through $(4, 7)$

29. Undefined slope; through $(0, 5)$

30. Horizontal; through $(0, 5)$

C *Write an equation of each line. Write the equation in the form $x = a$, $y = b$, or $y = mx + b$. See Examples 5 and 6.*

31. Through $(3, 8)$; parallel to $y = 4x - 2$

32. Through $(1, 5)$; parallel to $y = 3x - 4$

33. Through $(2, -5)$; perpendicular to $y = -2x - 6$

34. Through $(-4, 8)$; perpendicular to $y = -4x - 1$

35. Through $(1, 4)$; parallel to $y = 7$

36. Through $(-2, 6)$; perpendicular to $y = 7$

37. Through $(-2, -3)$; parallel to $3x + 2y = 5$

38. Through $(-2, -3)$; perpendicular to $3x + 2y = 5$

39. Through $(3, 5)$; perpendicular to $2x - y = 8$

40. Through $(6, 1)$; parallel to $8x - y = 9$

41. Through $(-1, -5)$; perpendicular to $x = 3$

42. Through $(4, -6)$; parallel to $x = -2$

43. Through $(6, -2)$; parallel to $2x + 4y = 9$

44. Through $(8, -3)$; parallel to $6x + 2y = 5$

45. Through $(-1, 5)$; perpendicular to $x - 4y = 4$

46. Through $(2, -3)$; perpendicular to $x - 5y = 10$

D *Solve. See Example 7.*

47. A rock is dropped from the top of a 400-foot building. After 1 second, the rock is traveling 32 feet per second. After 3 seconds, the rock is traveling 96 feet per second. Let y be the rate of descent and x be the number of seconds since the rock was dropped.

a. Write a linear equation that relates time x to rate y. [*Hint:* Use the ordered pairs $(1, 32)$ and $(3, 96)$.]

b. Use this equation to determine the rate of travel of the rock 4 seconds after it was dropped.

48. The Whammo Company has learned that by pricing a newly released Frisbee at \$6, sales will reach 2000 per day. Raising the price to \$8 will cause the sales to fall to 1500 per day. Assume that the ratio of change in price to change in daily sales is constant, and let x be the price of the Frisbee and y be number of sales.

a. Find the linear equation that models the price–sales relationship for this Frisbee. [*Hint:* The line must pass through $(6, 2000)$ and $(8, 1500)$.]

b. Use this equation to predict the daily sales of Frisbees if the price is set at \$7.50.

49. A fruit company recently released a new applesauce. By the end of its first year, profits on this product amounted to \$30,000. The anticipated profit for the end of the fourth year is \$66,000. The ratio of change in time to change in profit is constant. Let x be years and y be profit.

a. Write a linear equation that relates profit and time.

b. Use this equation to predict the company's profit at the end of the seventh year.

c. Predict when the profit should reach \$126,000.

50. The Pool Fun Company has learned that, by pricing a newly released Fun Noodle at \$3, sales will reach 10,000 Fun Noodles per day during the summer. Raising the price to \$5 will cause the sales to fall to 8000 Fun Noodles per day. Let x be price and y be the number sold.

a. Assume that the relationship between sales price and number of Fun Noodles sold is linear and write an equation describing this relationship.

b. Use this equation to predict the daily sales of Fun Noodles if the price is \$3.50.

51. In 1996, the median price of an existing home in the United States was \$115,800. In 2000, the median price of an existing home was \$142,200. Let y be the median price of an existing home in the year x, where $x = 0$ represents 1996. (*Source:* National Association of REALTORS®)

a. Write a linear equation that models the median existing home price in terms of the year x. [*Hint:* The line must pass through $(0, 115{,}800)$ and $(4, 142{,}200)$.]

b. Use this equation to predict the median existing home price for 2005.

52. The number of commercial airplanes delivered to customers by Boeing in 1997 was 374. In 2000, Boeing delivered a total of 489 commercial airplanes to customers. Let y be the number of Boeing commercial aircraft delivered to customers in the year x, where $x = 0$ represents 1997. (*Source:* The Boeing Company)

a. Write a linear equation that models the number of Boeing commercial aircraft delivered to customers in terms of the year x. [*Hint:* The line must pass through $(0, 374)$ and $(3, 489)$.]

b. Use this equation to predict the number of commercial aircraft Boeing delivers to customers in 2006.

53. The number of DVD players sold in the United States in 1998 was 1,089,261. In 2000, there were 8,498,545 DVD players sold in the United States. Let y be the number of DVD players sold in the year x, where $x = 0$ represents 1998. (*Source:* Consumer Electronics Association)

a. Write a linear equation that models the number of DVD players sold in the United States in the year x. [*Hint:* The line must pass through $(0, 1{,}089{,}261)$ and $(2, 8{,}498{,}545)$.]

b. Use this equation to estimate the number of DVD players that will be sold in 2005. [*Hint:* The line must pass through $(0, 1{,}089{,}261)$ and $(2, 8{,}498{,}545)$.]

54. The number of people employed in the United States as teacher assistants was 1192 thousand in 1998. By 2008, this number is expected to rise to 1567 thousand. Let y be the number of teacher assistants (in thousands) employed in the United States in the year x, where $x = 0$ represents 1998. (*Source:* U.S. Bureau of Labor Statistics)

a. Write a linear equation that models the number of people (in thousands) employed as teacher assistants in the year x. [*Hint:* The line must pass through $(0, 1192)$ and $(10, 1567)$.]

b. Use this equation to estimate the number of people who will be employed as teacher assistants in 2007.

Review and Preview

Complete each ordered pair for the given equation. See Section 3.1.

55. $y = 7x + 3; (4, \quad)$

56. $y = 2x - 6; (2, \quad)$

57. $y = 4.2x; (-2, \quad)$

58. $y = -1.3x; (6, \quad)$

59. $y = x^2 + 2x + 1; (1, \quad)$

60. $y = x^2 - 6x + 4; (0, \quad)$

Combining Concepts

Answer true or false.

61. A vertical line is always perpendicular to a horizontal line.

62. A vertical line is always parallel to a vertical line.

Write an equation of each line.

63. Through $(5, -6)$; perpendicular to $y = 9$

64. Through $(-3, -5)$; parallel to $y = 9$

Use a grapher with a TRACE feature to see the results of each exercise.

65. Exercise 31: Graph the equation and verify that it passes through $(3, 8)$ and is parallel to $y = 4x - 2$.

66. Exercise 32: Graph the equation and verify that it passes through $(1, 5)$ and is parallel to $y = 3x - 4$.

Internet Excursions

Go To: http://www.prenhall.com/martin-gay_algebra

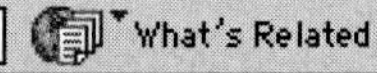

The U.S. Bureau of Labor Statistics (BLS) is the principal fact-finding agency for the federal government in the broad field of labor economics and statistics. The BLS regularly makes data such as unemployment figures, average earnings, and job growth numbers available to the American public, government, and businesses. The World Wide Web address listed above will provide you with access to the BLS page of Labor Force Statistics from the Current Population Survey, or a related site. You will be able to research information needed to answer the following questions.

67. From the listing of links, choose a table or data set that interests you. Write down two ordered pairs from that set of data. Describe what each ordered pair represents. Then use the ordered pairs to find an equation for the line between these two points.

68. Now choose a different table or data set of interest. Write down two ordered pairs and describe what each ordered pair represents. Use the ordered pairs to find an equation for the line between these two points. Then make a prediction using your equation and explain its significance.

Name ______________________ Section __________ Date __________

Answers

1. see graph

2. see graph

3. see graph

4. see graph

5. ______

6. ______

Integrated Review—Linear Equations in Two Variables

Below is a review of equations of lines.

Forms of Linear Equations

$Ax + By = C$	**Standard form** of a linear equation A and B are not both 0.
$y = mx + b$	**Slope-intercept form** of a linear equation The slope is m, and the y-intercept is $(0, b)$.
$y - y_1 = m(x - x_1)$	**Point-slope form** of a linear equation The slope is m, and (x_1, y_1) is a point on the line.
$y = c$	**Horizontal line** The slope is 0, and the y-intercept is $(0, c)$.
$x = c$	**Vertical line** The slope is undefined and the x-intercept is $(c, 0)$.

Parallel and Perpendicular Lines

Nonvertical parallel lines have the same slope.
The product of the slopes of two nonvertical perpendicular lines is -1.

Graph each linear equation.

1. $y = -2x$

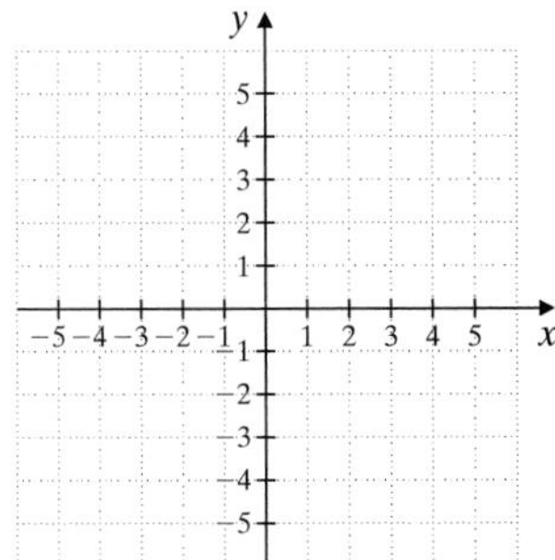

2. $3x - 2y = 6$

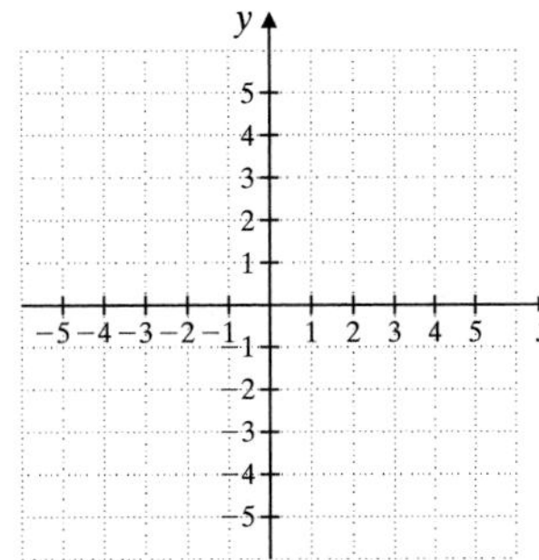

3. $x = -3$

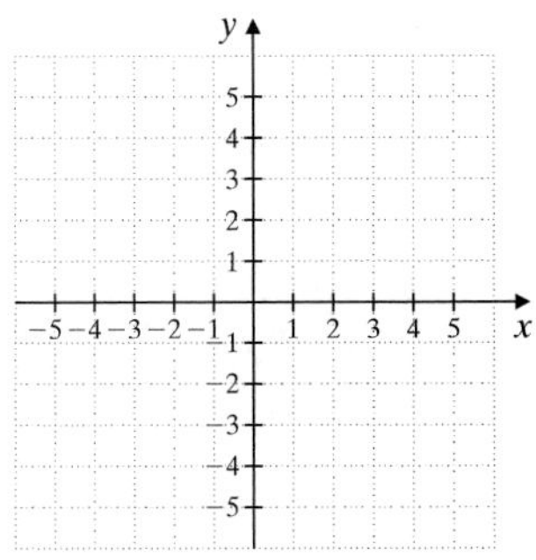

4. $y = 1.5$

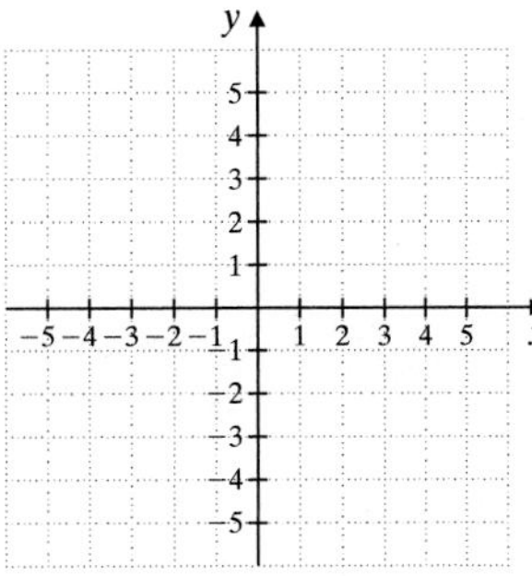

Find the slope of the line containing each pair of points.

5. $(-2, -5), (3, -5)$

6. $(5, 2), (0, 5)$

7. ______

8. ______

9. ______

10. ______

11. ______

12. ______

13. ______

14. ______

15. ______

16. ______

17. ______

18. ______

19. ______

20. ______

21. ______

22. ______

23. ______

24. ______

Find the slope and y-intercept of each line.

7. $y = 3x - 5$

8. $5x - 2y = 7$

Determine whether each pair of lines is parallel, perpendicular, or neither.

9. $y = 8x - 6$
$y = 8x + 6$

10. $y = \frac{2}{3}x + 1$
$2y + 3x = 1$

Find the equation of each line. Write the equation in the form $x = a$, $y = b$, or $y = mx + b$.

11. Through $(1, 6)$ and $(5, 2)$

12. Vertical line; through $(-2, -10)$

13. Horizontal line; through $(1, 0)$

14. Through $(2, -8)$ and $(-6, -5)$

15. Through $(-2, 4)$ with slope -5

16. Slope -4; y-intercept $\left(0, \frac{1}{3}\right)$

17. Slope $\frac{1}{2}$; y-intercept $(0, -1)$

18. Through $\left(\frac{1}{2}, 0\right)$ with slope 3

19. Through $(-1, -5)$; parallel to $3x - y = 5$

20. Through $(0, 4)$; perpendicular to $4x - 5y = 10$

21. Through $(2, -3)$; perpendicular to $4x + y = \frac{2}{3}$

22. Through $(-1, 0)$; parallel to $5x + 2y = 2$

23. Undefined slope; through $(-1, 3)$

24. $m = 0$; through $(-1, 3)$

7.3 Introduction to Functions

OBJECTIVES

- (A) Define relation, domain, and range.
- (B) Identify functions.
- (C) Use the vertical line test for functions.
- (D) Find the domain and range of a relation from its graph.
- (E) Use function notation.
- (F) Graph a linear function.

SSM TUTOR CENTER SG CD & VIDEO MATH PRO WEB

(A) Defining Relation, Domain, and Range

Equations in two variables, such as $y = 2x + 1$, describe **relations** between x-values and y-values. For example, if $x = 1$, then this equation describes how to find the y-value related to $x = 1$. In words, the equation $y = 2x + 1$ says that twice the x-value increased by 1 gives the corresponding y-value. The x-value of 1 corresponds to the y-value of $2(1) + 1 = 3$ for this equation, and we have the ordered pair $(1, 3)$.

There are other ways of describing relations or correspondences between two numbers or, in general, a first set (sometimes called the set of *inputs*) and a second set (sometimes called the set of *outputs*). For example,

First Set: Input	Correspondence	Second Set: Output
People in a certain city	Each person's age	The set of nonnegative integers

A few examples of ordered pairs from this relation might be (Ana, 4); (Bob, 36); (Trey, 21); and so on.

Below are just a few other ways of describing relations between two sets and the ordered pairs that they generate.

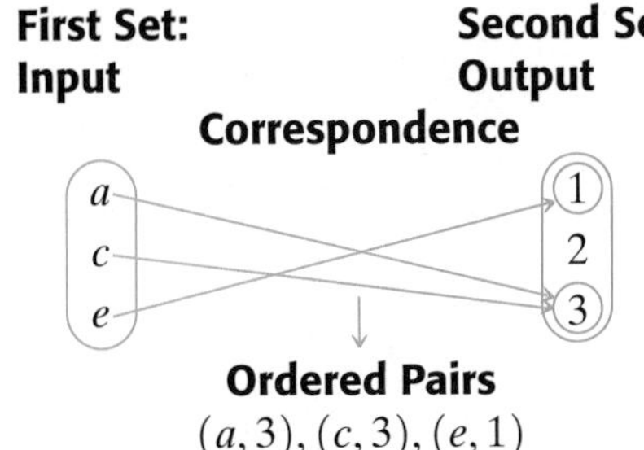

Ordered Pairs

$(a, 3), (c, 3), (e, 1)$

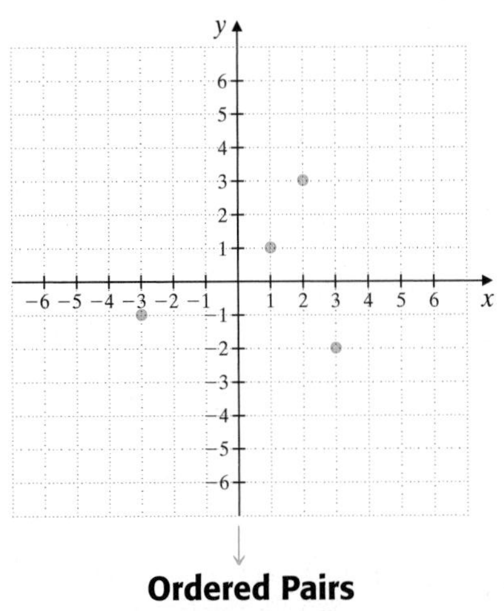

Ordered Pairs

$(-3, -1), (1, 1), (2, 3), (3, -2)$

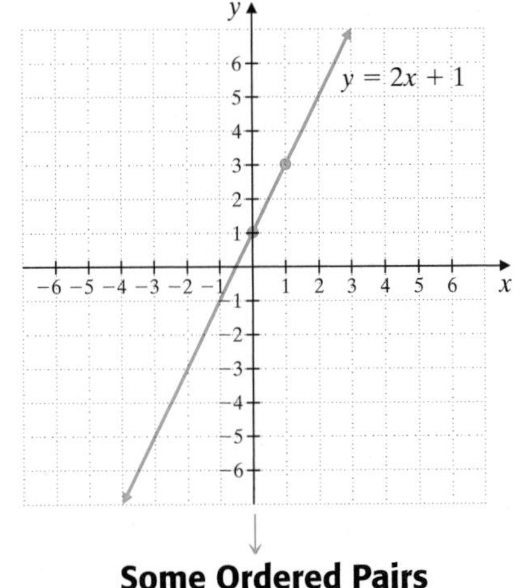

Some Ordered Pairs

$(1, 3), (0, 1)$, and so on

Relation, Domain, and Range

A **relation** is a set of ordered pairs.
The **domain** of the relation is the set of all first components of the ordered pairs.
The range of the relation is the set of all second components of the ordered pairs.

For example, the domain for our relation in the middle of the previous page is $\{a, c, e\}$ and the range is $\{1, 3\}$. Notice that the range does not include the element 2 of the second set. This is because no element of the first set is assigned to this element. If a relation is defined in terms of x- and y-values, we will agree that the domain corresponds to x-values and that the range corresponds to y-values.

EXAMPLES Determine the domain and range of each relation.

1. $\{(2,3), (2,4), (0,-1), (3,-1)\}$

The domain is the set of all first coordinates of the ordered pairs, $\{2, 0, 3\}$.

The range is the set of all second coordinates, $\{3, 4, -1\}$.

2.

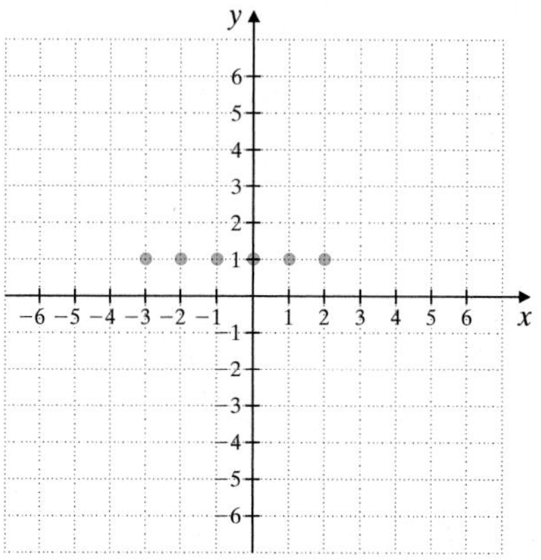

The relation is $\{(-3,1), (-2,1), (-1,1), (0,1), (1,1), (2,1)\}$.

The domain is $\{-3, -2, -1, 0, 1, 2\}$.

The range is $\{1\}$.

3.

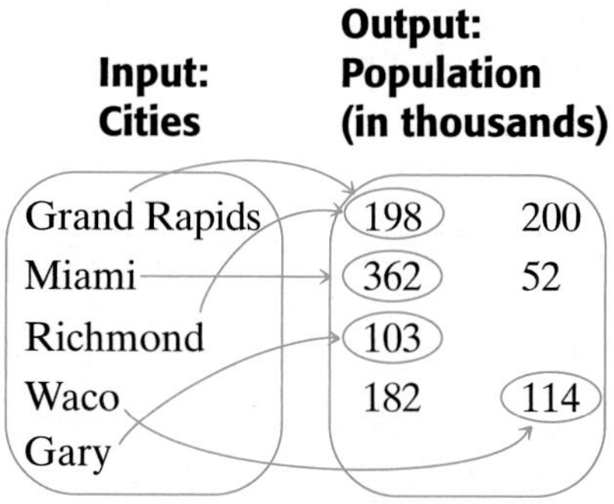

The domain is the first set, {Gary, Grand Rapids, Miami, Richmond, Waco}. The range is the numbers in the second set that correspond to elements in the first set, $\{103, 114, 198, 362\}$.

B Identifying Functions

Now we consider a special kind of relation called a *function*.

Function

A **function** is a relation in which each first component in the ordered pairs corresponds to *exactly one* second component.

Practice Problems 1–3

Determine the domain and range of each relation.

1. $\{(1,6), (2,8), (0,3), (0,-2)\}$
2.

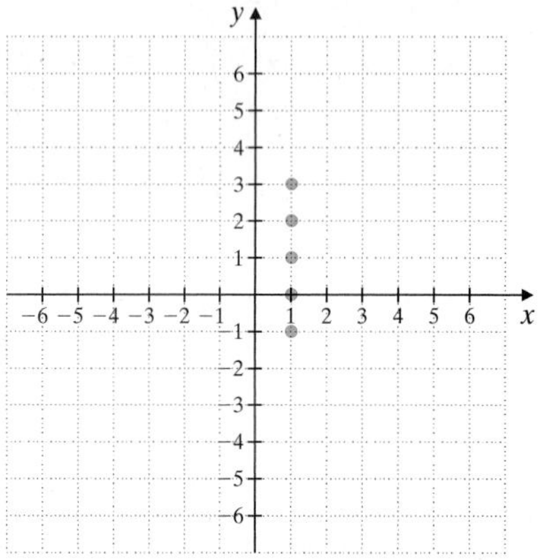

3.

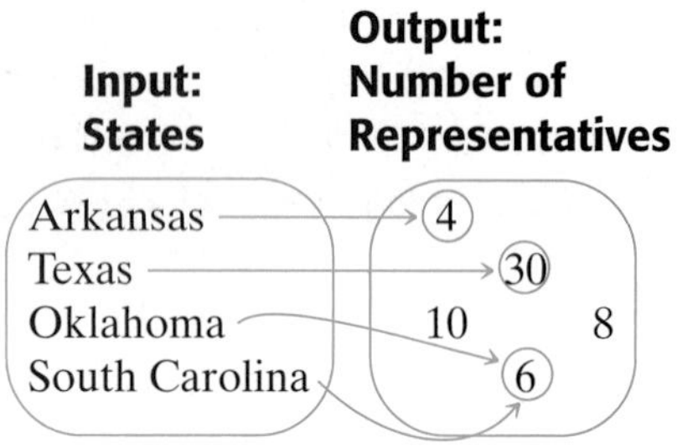

Answers

1. domain: $\{1, 2, 0\}$; range: $\{6, 8, 3, -2\}$, **2.** domain: $\{1\}$; range: $\{-1, 0, 1, 2, 3\}$, **3.** domain: {Arkansas, Texas, Oklahoma, South Carolina} range: $\{4, 30, 6\}$

A function is a special type of relation, so all functions are relations. But not all relations are functions.

EXAMPLES Determine whether each relation is also a function.

4. $\{(-2,5),(2,7),(-3,5),(9,9)\}$
Although the ordered pairs $(-2,5)$ and $(-3,5)$ have the same y-value, each x-value is assigned to only one y-value, so this set of ordered pairs is a function.

5.

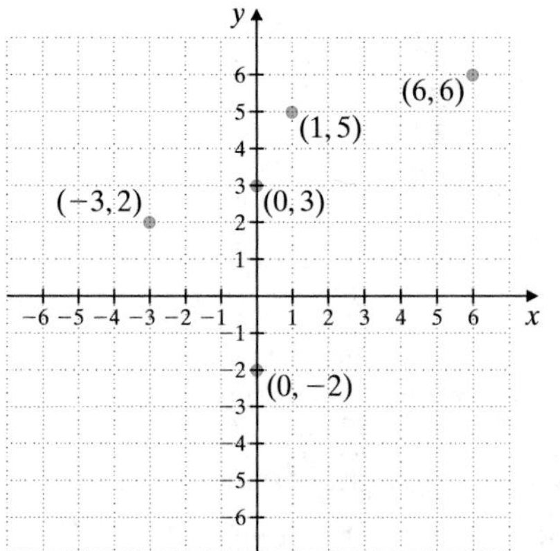

The x-value 0 is assigned to two y-values, -2 and 3, in this graph, so this relation is not a function.

6.

Input	Correspondence	Output
People in a certain city	Each person's age	The set of nonnegative integers

This relation is a function because although two different people may have the same age, each person has only one age. This means that each element in the first set is assigned to only one element in the second set.

Practice Problems 4–6

Determine whether each relation is also a function.

4. $\{(-3,7),(1,7),(2,2)\}$
5.

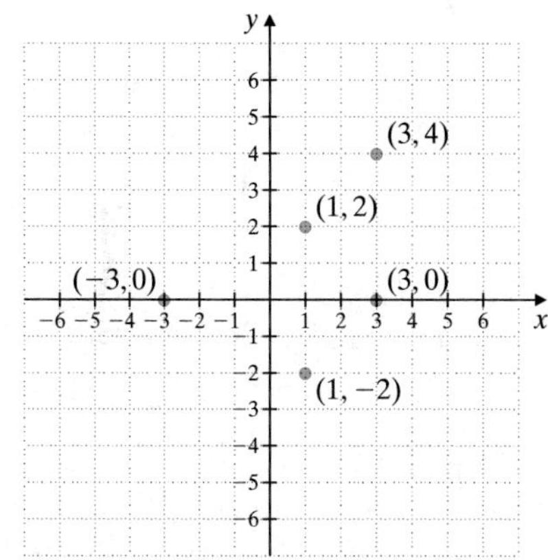

6.

Input	Correspondence	Output
People in a certain state	County/parish that a person lives in	Counties of that state

Try the Concept Check in the margin.

We will call an equation such as $y = 2x + 1$ a relation since this equation defines a set of ordered pair solutions.

Concept Check

Explain why a function can contain both the ordered pairs $(1,3)$ and $(2,3)$ but not both $(3,1)$ and $(3,2)$.

EXAMPLE 7

Determine whether the relation $y = 2x + 1$ is also a function.

Solution: The relation $y = 2x + 1$ is a function if each x-value corresponds to just one y-value. For each x-value substituted in the equation $y = 2x + 1$, the multiplication and addition performed gives a single result, so only one y-value will be associated with each x-value. Thus, $y = 2x + 1$ is a function.

Practice Problem 7

Determine whether the relation $y = 3x + 2$ is also a function.

EXAMPLE 8 Determine whether the relation $x = y^2$ is also a function.

Solution: In $x = y^2$, if $y = 3$, then $x = 9$. Also, if $y = -3$, then $x = 9$. In other words, we have the ordered pairs $(9,3)$ and $(9,-3)$. Since the x-value 9 corresponds to two y-values, 3 and -3, $x = y^2$ is not a function.

Practice Problem 8

Determine whether the relation $x = y^2 + 1$ is also a function.

Answers

4. function, **5.** not a function, **6.** function, **7.** yes, **8.** no

Concept Check: Two different ordered pairs can have the same y-value, but not the same x-value in a function.

Practice Problems 9–13

Use the vertical line test to determine which are graphs of functions.

9.

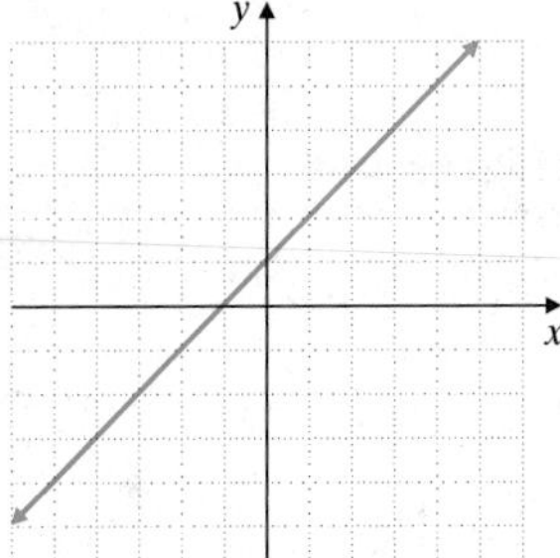

10.

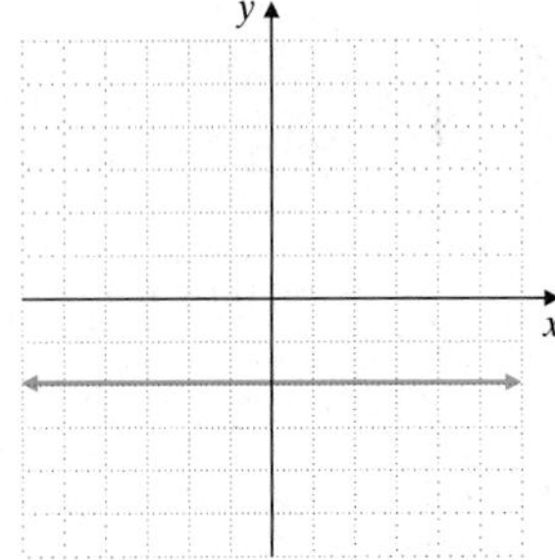

11.

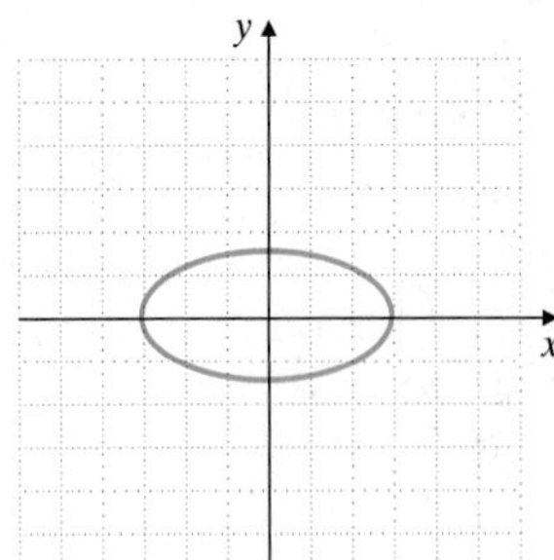

12.

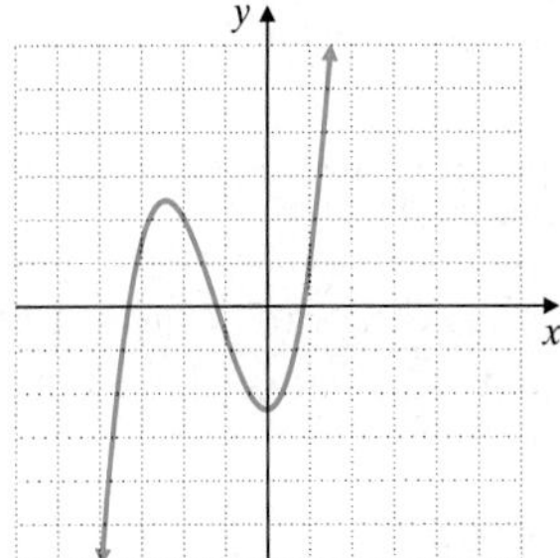

13.

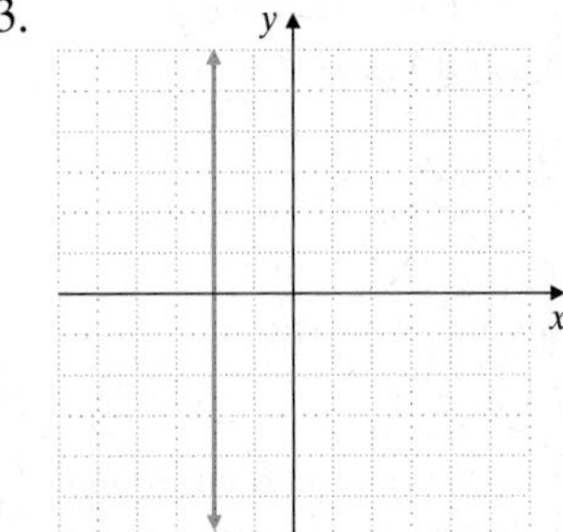

Answers

9. function, **10.** function, **11.** not a function, **12.** function, **13.** not a function

C Using the Vertical Line Test

As we have seen, not all relations are functions. Consider the graphs of $y = 2x + 1$ and $x = y^2$ shown next. On the graph of $y = 2x + 1$, notice that each x-value corresponds to only one y-value. Recall from Example 7 that $y = 2x + 1$ is a function.

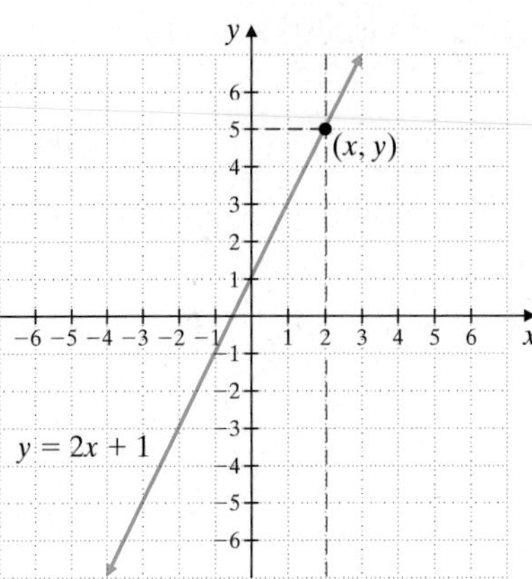

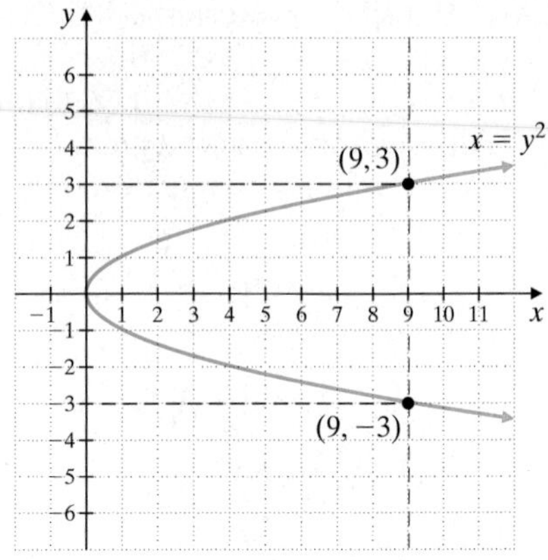

On the graph of $x = y^2$, the x-value 9, for example, corresponds to two y-values, 3 and -3, as shown by the vertical line. Recall from Example 8 that $x = y^2$ is not a function.

Graphs can be used to help determine whether a relation is also a function by the following **vertical line test**.

Vertical Line Test

If no vertical line can be drawn so that it intersects a graph more than once, the graph is the graph of a function. If such a line can be drawn, the graph is not that of a function.

EXAMPLES

Use the vertical line test to determine which are graphs of functions.

9.

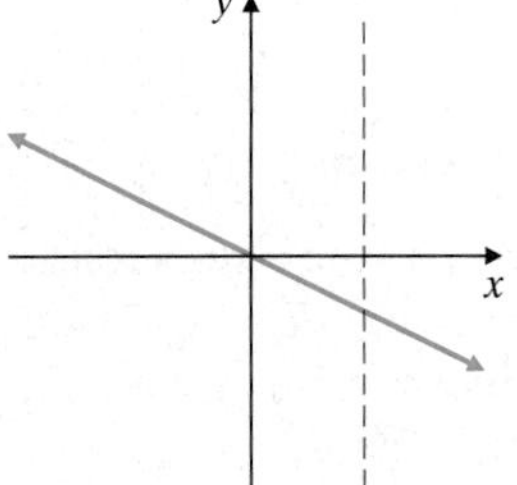

This is the graph of a function since no vertical line will intersect this graph more than once.

10.

This is the graph of a function.

11.

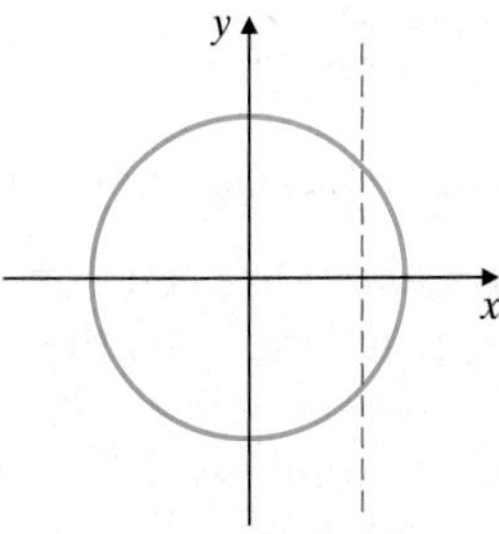

This is not the graph of a function. Note that vertical lines can be drawn that intersect the graph in two points.

12.

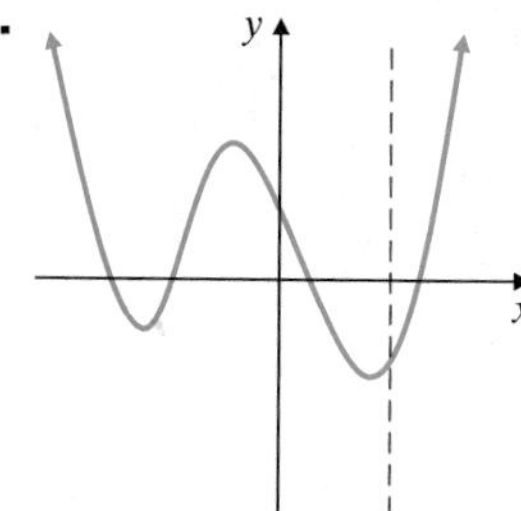

This is the graph of a function.

13.

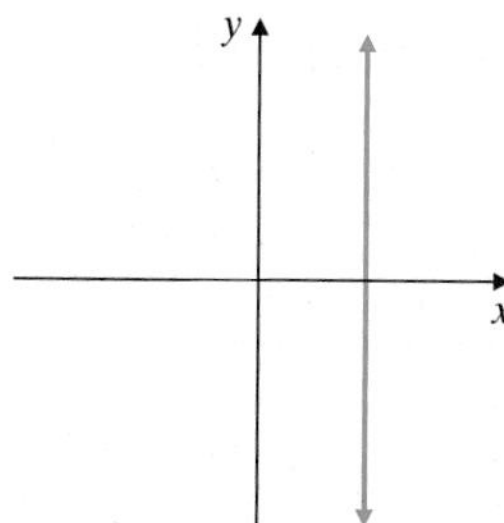

This is not the graph of a function. A vertical line can be drawn that intersects this line at every point.

Try the Concept Check in the margin.

D Finding Domain and Range from a Graph

Next we practice finding the domain and range of a relation from its graph.

EXAMPLES Find the domain and range of each relation.

14.

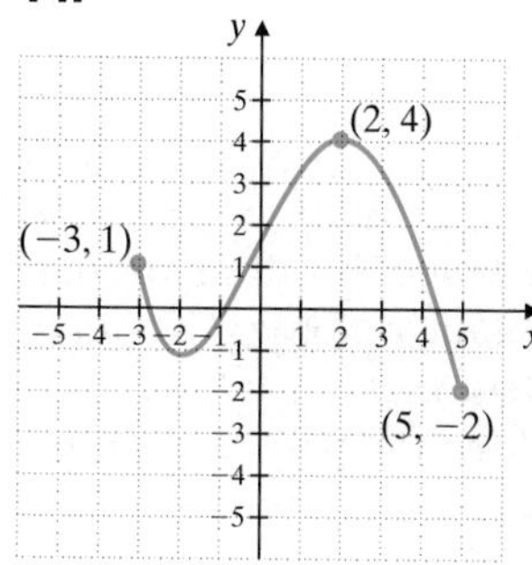

(2, 4)
(−3, 1)
(5, −2)

Range: The y-values graphed are from −2 to 4, or [−2, 4].

Domain: The x-values graphed are from −3 to 5, or [−3, 5].

15.

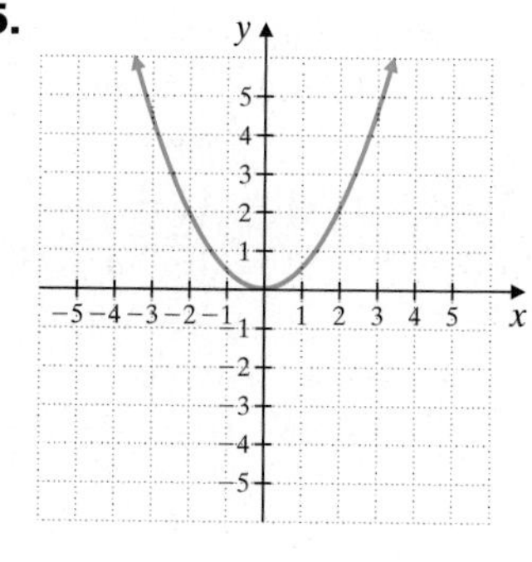

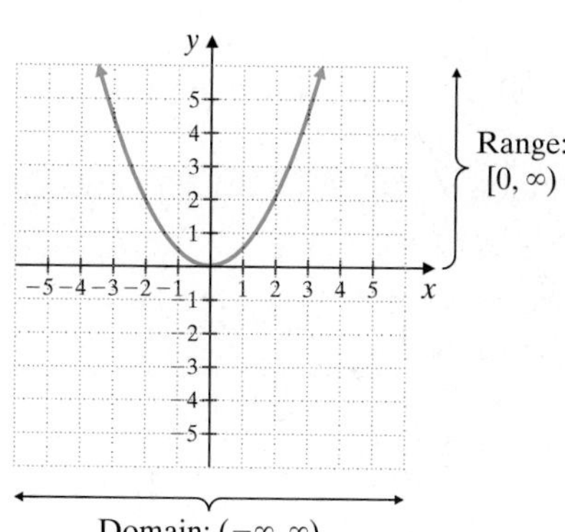

Concept Check

Determine which equations represent functions. Explain your answer.

a. $y = 14$

b. $x = -5$

c. $x + y = 6$

Practice Problems 14–17

Find the domain and range of each relation.

14.

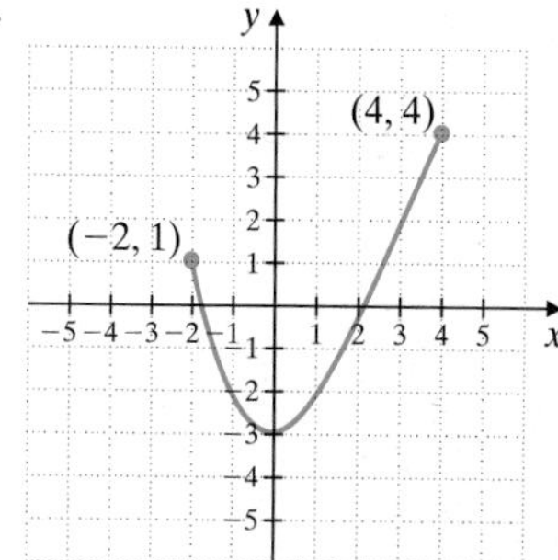

15.

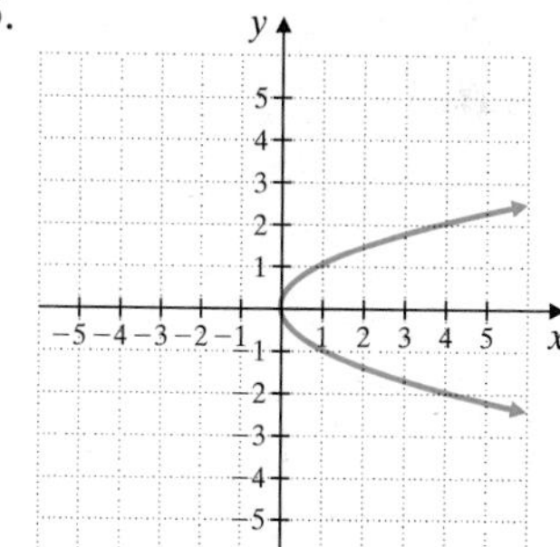

16.

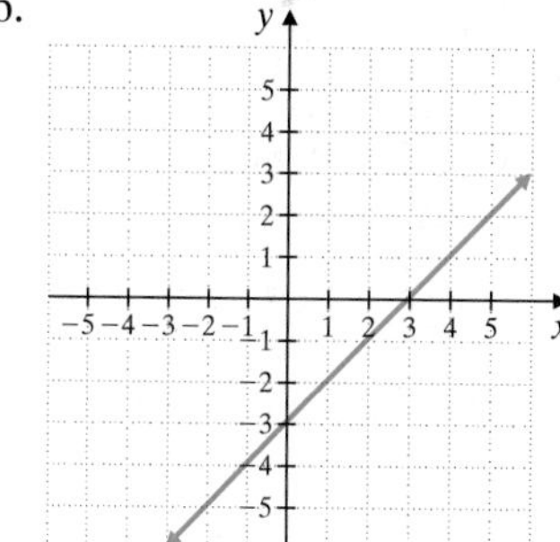

17.

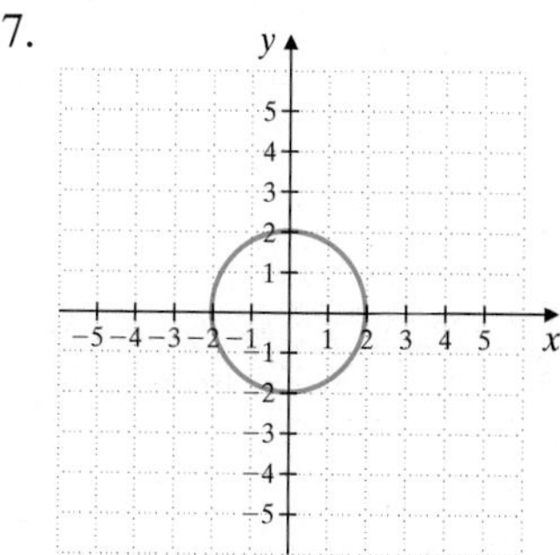

Answers

14. domain: $[-2, 4]$; range: $[-3, 4]$, **15.** domain: $[0, \infty)$; range: $(-\infty, \infty)$, **16.** domain: $(-\infty, \infty)$; range: $(-\infty, \infty)$, **17.** domain: $[-2, 2]$; range: $[-2, 2]$

Concept Check: a, c

16.

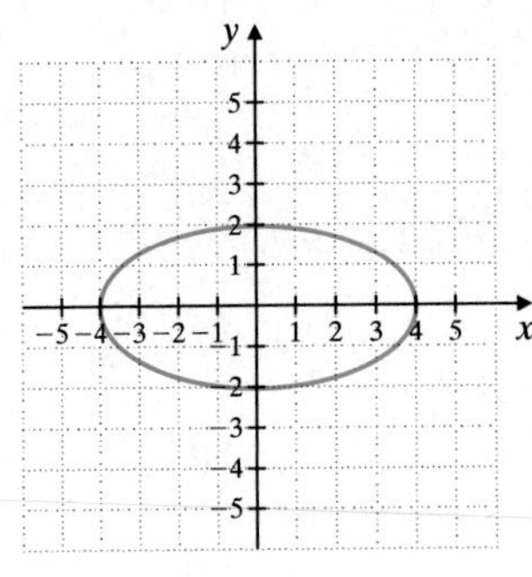

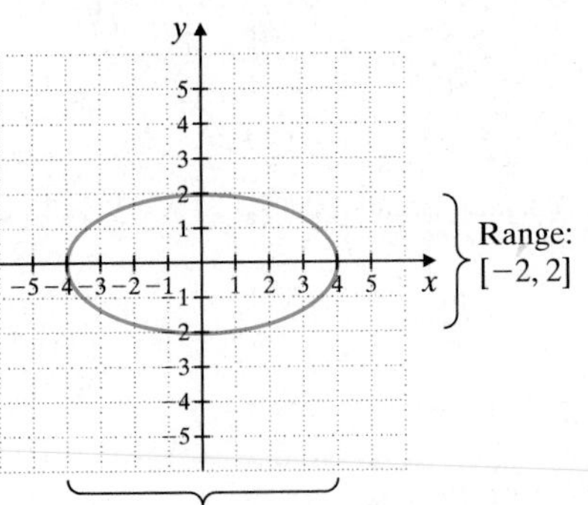

Domain: [−4, 4]

17.

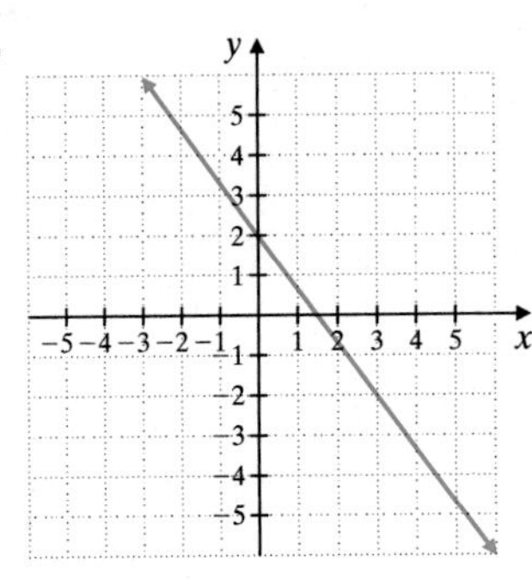

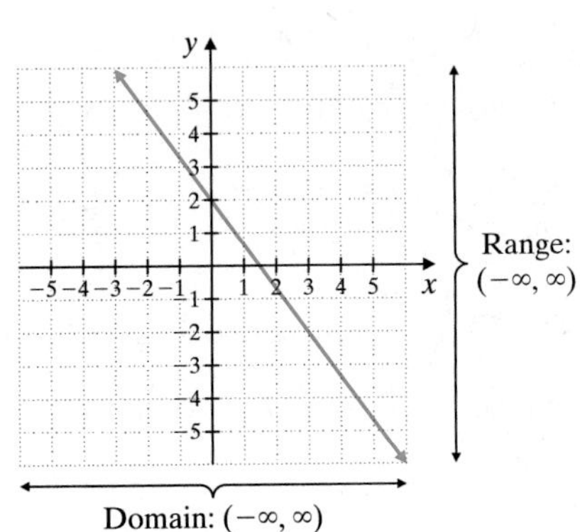

Domain: (−∞, ∞)

E Using Function Notation

Many times letters such as f, g, and h are used to name functions. To denote that y is a function of x, we can write

$$y = f(x)$$

This means that **y is a function of x** or that y *depends on* x. For this reason, y is called the **dependent variable** and x the **independent variable**. The notation $f(x)$ is read "f of x" and is called **function notation.**

For example, to use function notation with the function $y = 4x + 3$, we write $f(x) = 4x + 3$. The notation $f(1)$ means to replace x with 1 and find the resulting y- or function value. Since

$$f(x) = 4x + 3$$

then

$$f(1) = 4(1) + 3 = 7$$

This means that when $x = 1$, y or $f(x) = 7$. The corresponding ordered pair is $(1, 7)$. Here, the input is 1 and the output is $f(1)$ or 7. Now let's find $f(2)$, $f(0)$, and $f(-1)$.

$$\begin{aligned} f(x) &= 4x + 3 \\ f(2) &= 4(2) + 3 \\ &= 8 + 3 \\ &= 11 \end{aligned} \qquad \begin{aligned} f(x) &= 4x + 3 \\ f(0) &= 4(0) + 3 \\ &= 0 + 3 \\ &= 3 \end{aligned} \qquad \begin{aligned} f(x) &= 4x + 3 \\ f(-1) &= 4(-1) + 3 \\ &= -4 + 3 \\ &= -1 \end{aligned}$$

Ordered Pairs:

$(2, 11)$ $\qquad$ $(0, 3)$ $\qquad$ $(-1, -1)$

Note that $f(x)$ is a special symbol in mathematics used to denote a function. The symbol $f(x)$ is read "f of x." It does *not* mean $f \cdot x$ (f times x).

EXAMPLES Find each function value.

18. If $g(x) = 3x - 2$, find $g(1)$.

$g(1) = 3(1) - 2 = 1$

19. If $g(x) = 3x - 2$, find $g(0)$.

$g(0) = 3(0) - 2 = -2$

20. If $f(x) = 7x^2 - 3x + 1$, find $f(1)$.

$f(1) = 7(1)^2 - 3(1) + 1 = 5$

21. If $f(x) = 7x^2 - 3x + 1$, find $f(-2)$.

$f(-2) = 7(-2)^2 - 3(-2) + 1 = 35$

Try the Concept Check in the margin.

F Graphing Linear Functions

Recall that the graph of a linear equation in two variables is a line, and a line that is not vertical will always pass the vertical line test. Thus, *all linear equations are functions except those whose graphs are vertical lines.* We call such functions *linear functions.*

Linear Function

A **linear function** is a function that can be written in the form

$$f(x) = mx + b$$

EXAMPLE 22 Graph the function $f(x) = 2x + 1$.

Solution: Since $y = f(x)$, we can replace $f(x)$ with y and graph as usual. The graph of $y = 2x + 1$ has slope 2 and y-intercept $(0, 1)$ Its graph is shown.

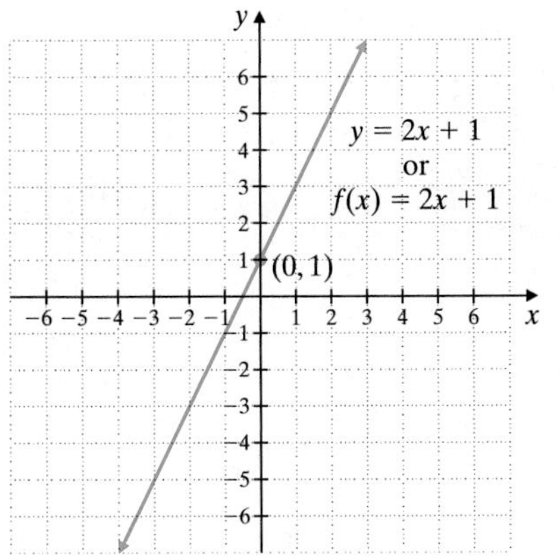

Practice Problems 18–21

Find each function value.

18. If $g(x) = 4x + 5$, find $g(0)$.
19. If $g(x) = 4x + 5$, find $g(-5)$.
20. If $f(x) = 3x^2 - x + 2$, find $f(2)$.
21. If $f(x) = 3x^2 - x + 2$, find $f(-1)$.

Concept Check

Suppose $y = f(x)$ and we are told that $f(3) = 9$. Which is not true?

a. When $x = 3$, $y = 9$.
b. A possible function is $f(x) = x^2$.
c. A point on the graph of the function is $(3, 9)$.
d. A possible function is $f(x) = 2x + 4$.

Practice Problem 22

Graph the function $f(x) = 3x - 2$.

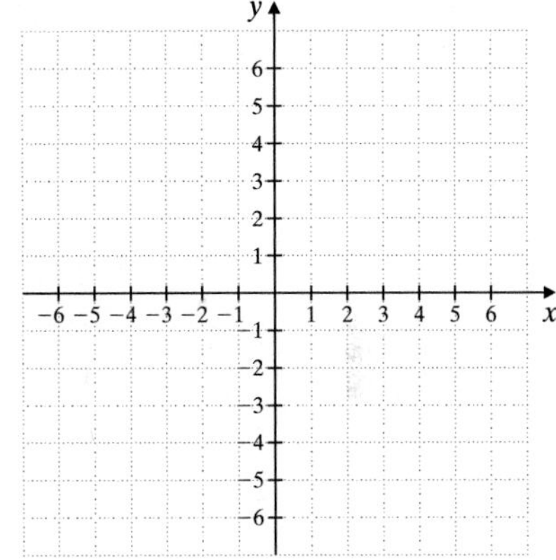

Answers

18. $g(0) = 5$, **19.** $g(-5) = -15$, **20.** $f(2) = 12$, **21.** $f(-1) = 6$, **22.**

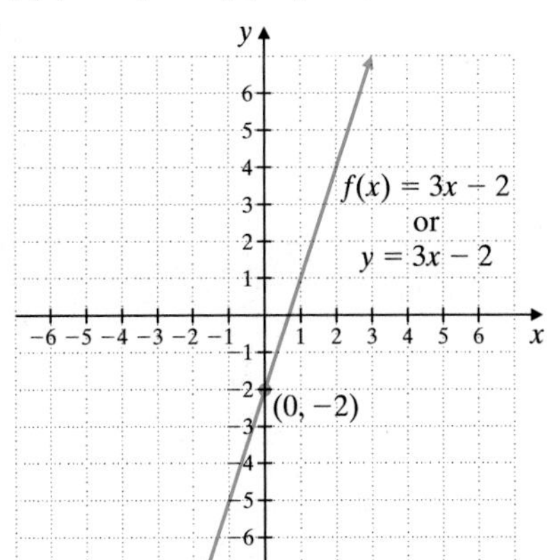

Concept Check: d

FOCUS ON Mathematical Connections

PERPENDICULAR BISECTORS

A **perpendicular bisector** is a line that is perpendicular to a given line segment and divides the segment into two equal lengths. A perpendicular bisector crosses the line segment at the point that is located exactly halfway between the two endpoints of the line segment. That point is called the **midpoint** of the line segment. If a line segment has the endpoints (x_1, y_1) and (x_2, y_2), then the midpoint of this line segment is the point with coordinates $\left(\dfrac{x_1 + x_2}{2}, \dfrac{y_1 + y_2}{2}\right)$.

An example of a line segment and its perpendicular bisector is shown in the figure.

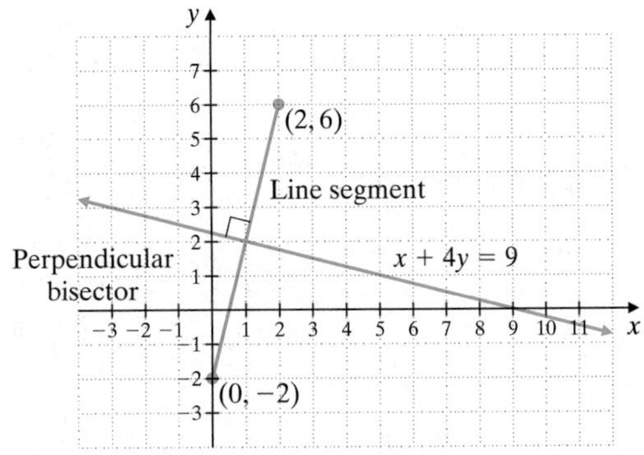

To find the equation of a line segment's perpendicular bisector, follow these steps:

Step 1. Find the midpoint of the line segment. (See the formula in Section 9.1.)

Step 2. Find the slope of the line segment.

Step 3. Find the slope of a line that is perpendicular to the line segment.

Step 4. Use the midpoint and the slope of the perpendicular line to find the equation of the perpendicular bisector.

CRITICAL THINKING

Use the steps given above and what you have learned in this chapter to find the equation of the perpendicular bisector of each line segment whose endpoints are given.

1. $(3, -1), (-5, 1)$
2. $(-6, -3), (-8, -1)$
3. $(-2, 6), (-22, -4)$
4. $(5, 8), (7, 2)$
5. $(2, 3), (-4, 7)$
6. $(-6, 8), (-4, -2)$

Name ______________________ Section __________ Date __________

EXERCISE SET 7.3

 *Find the domain and the range of each relation. Also determine whether the relation is a function. See Examples 1 through 6.*

1. $\{(-1,7),(0,6),(-2,2),(5,6)\}$

2. $\{(4,9),(-4,9),(2,3),(10,-5)\}$

3. $\{(-2,4),(6,4),(-2,-3),(-7,-8)\}$

4. $\{(6,6),(5,6),(5,-2),(7,6)\}$

5. $\{(1,1),(1,2),(1,3),(1,4)\}$

6. $\{(1,1),(2,1),(3,1),(4,1)\}$

7. $\left\{\left(\frac{3}{2},\frac{1}{2}\right),\left(1\frac{1}{2},-7\right),\left(0,\frac{4}{5}\right)\right\}$

8. $\{(\pi,0),(0,\pi),(-2,4),(4,-2)\}$

9. $\{(-3,-3),(0,0),(3,3)\}$

10. $\left\{\left(\frac{1}{2},\frac{1}{4}\right),\left(0,\frac{7}{8}\right),(0.5,\pi)\right\}$

 **11.**

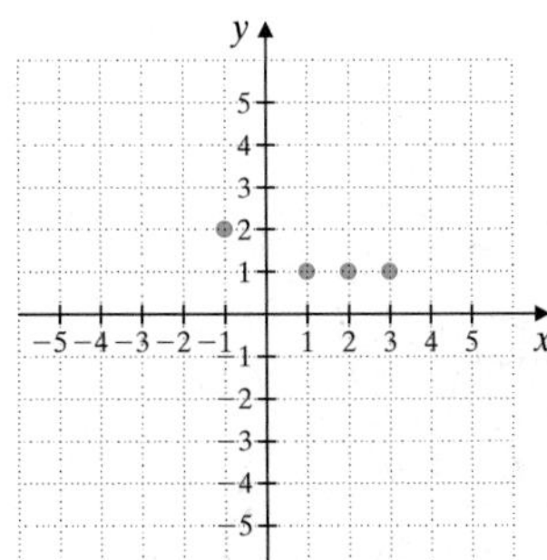

12.

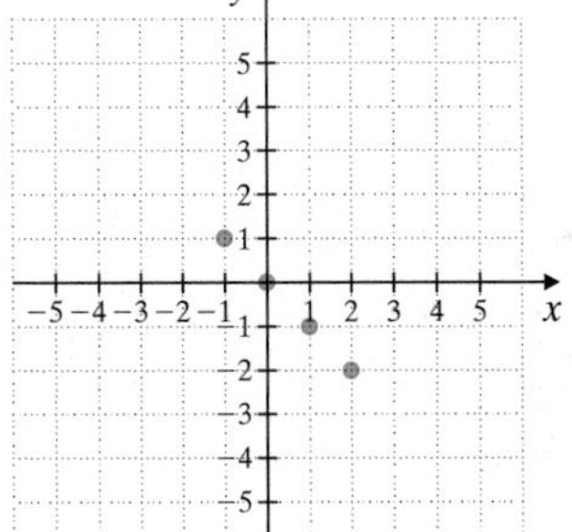

13.

Input: State

Output: Number of Congressional Representatives

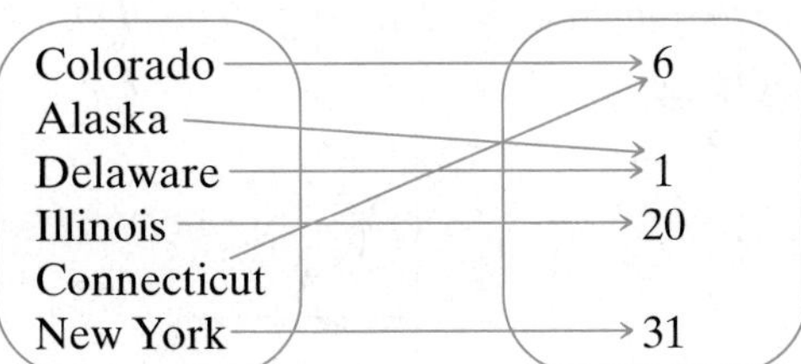

14.

Input: Animal

Output: Average Life Span (in years)

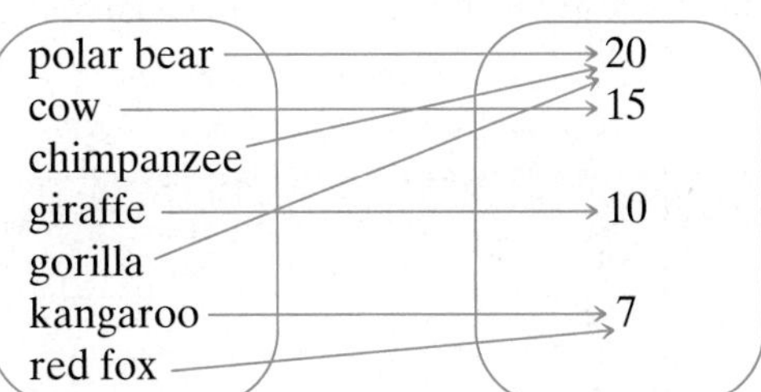

15. **Input: Degrees Fahrenheit** **Output: Degrees Celsius**

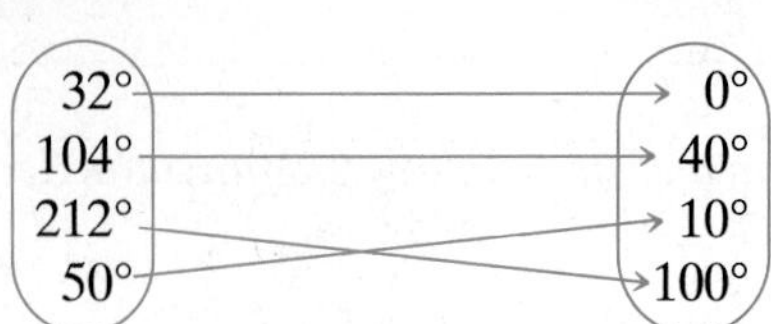

16.

Input: Words

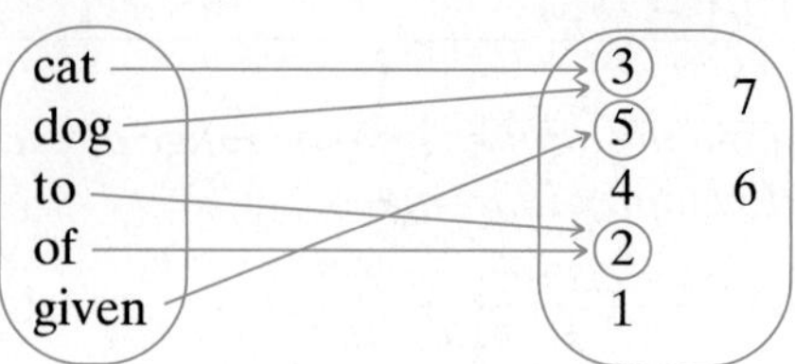

17. **Input:** **Output:**

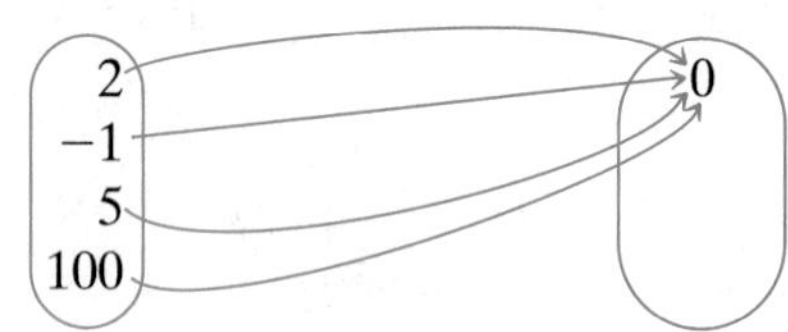

18. **Input:** **Output:**

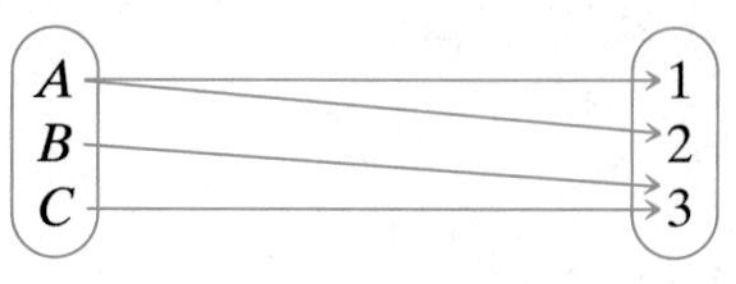

Determine whether each relation is a function. See Examples 4 through 6.

19.

First Set: Input	Correspondence	Second Set: Output
Class of algebra students	Grade average	Set of nonnegative numbers

20.

First Set: Input	Correspondence	Second Set: Output
People in New Orleans (population 500,000)	Birthdate	Days of the year

Determine whether each relation is also a function. See Examples 7 and 8.

21. $y = x + 1$

22. $y = x - 1$

23. $x = 2y^2$

24. $y = x^2$

25. $y - x = 7$

26. $2x - 3y = 9$

C *Use the vertical line test to determine whether each graph is the graph of a function. See Examples 9 through 13.*

27.

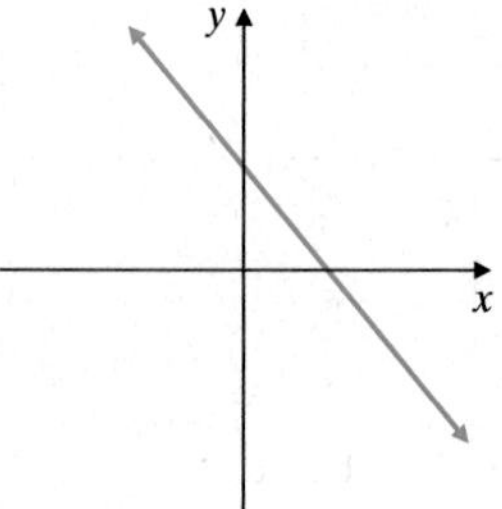

28.

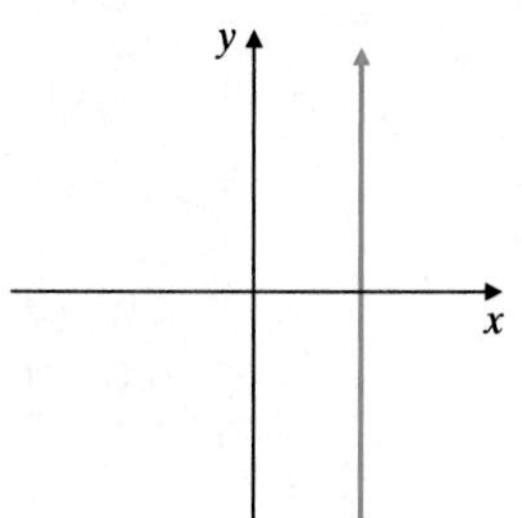

29.

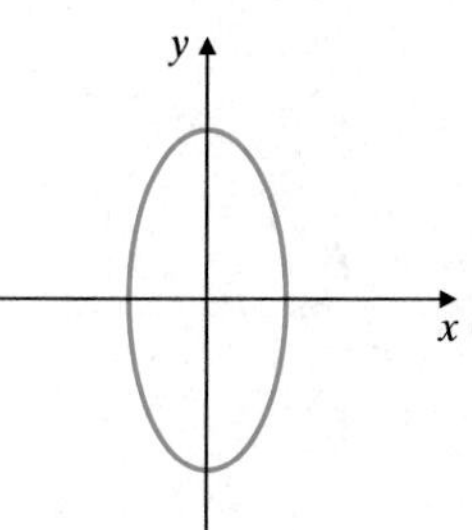

30.

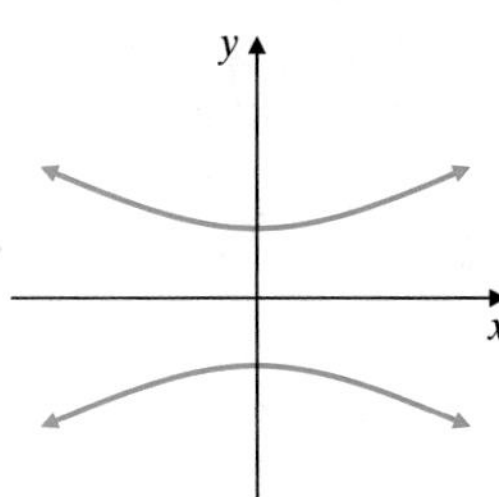

31.

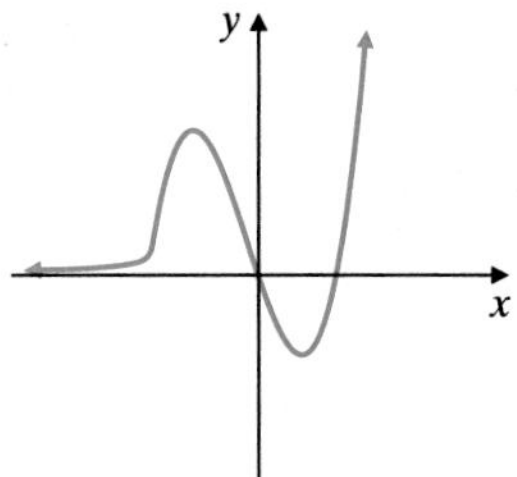

32.

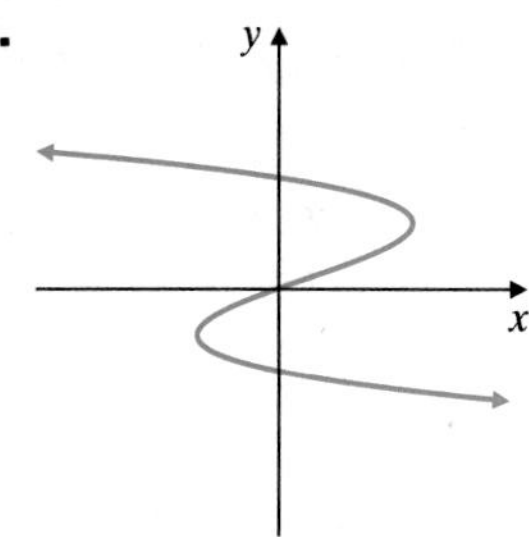

D *Find the domain and the range of each relation. Use the vertical line test to determine whether each graph is the graph of a function. See Examples 14 through 17.*

33.

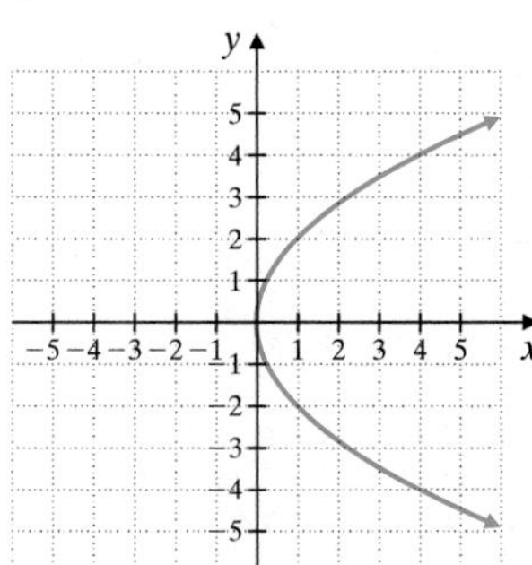

34.

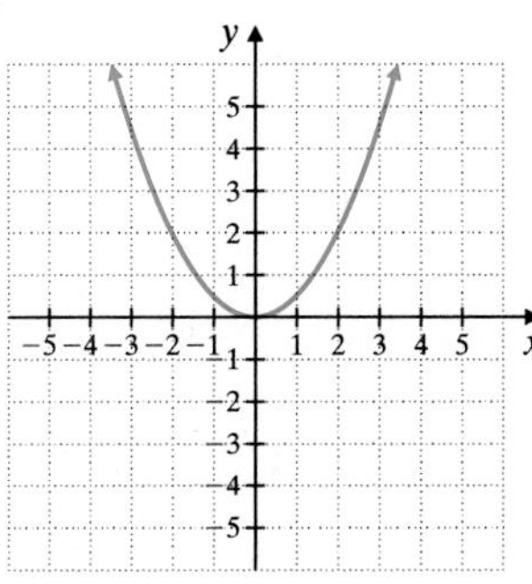

35.

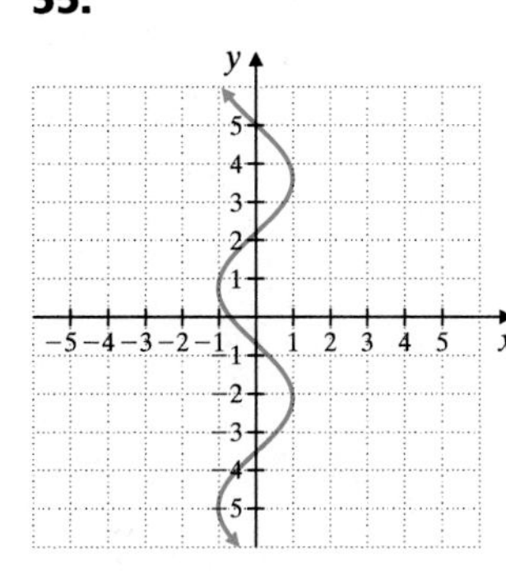

36.

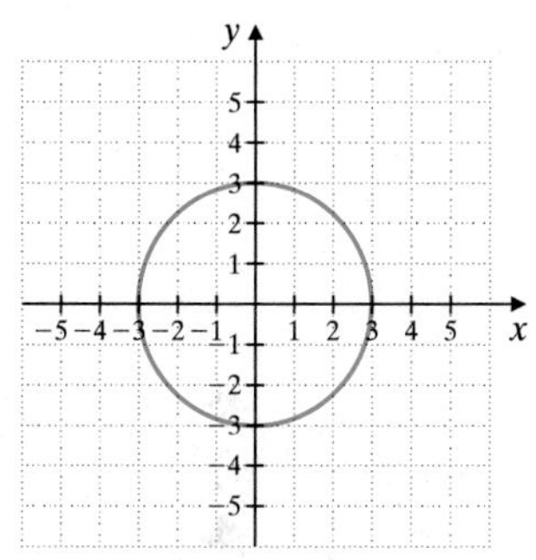

37.

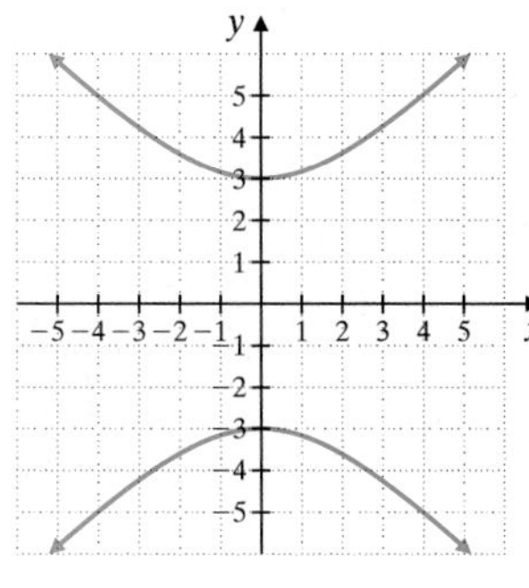

38.

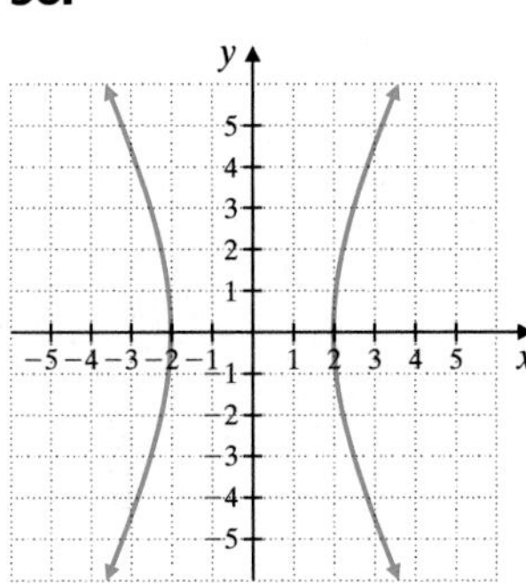

39.

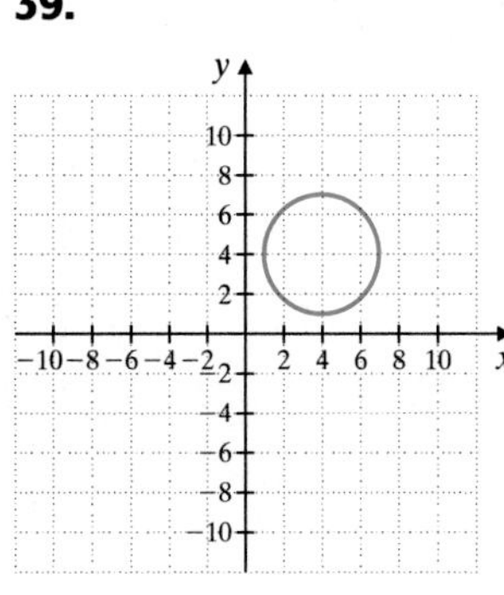

40.

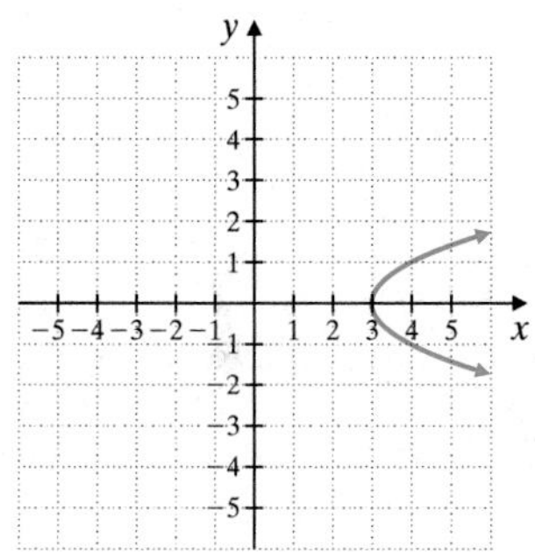

41.

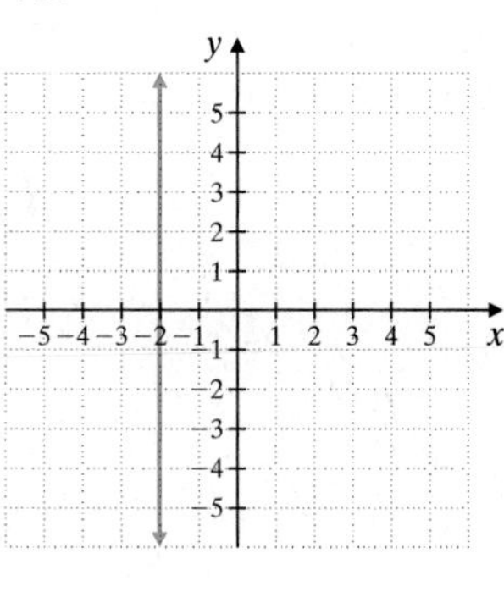

42.

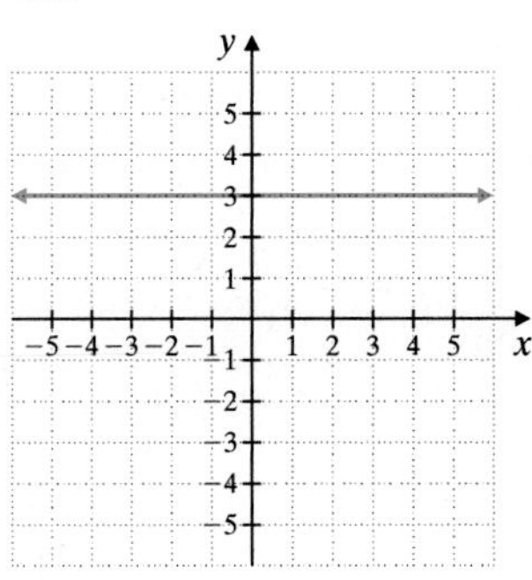

43.

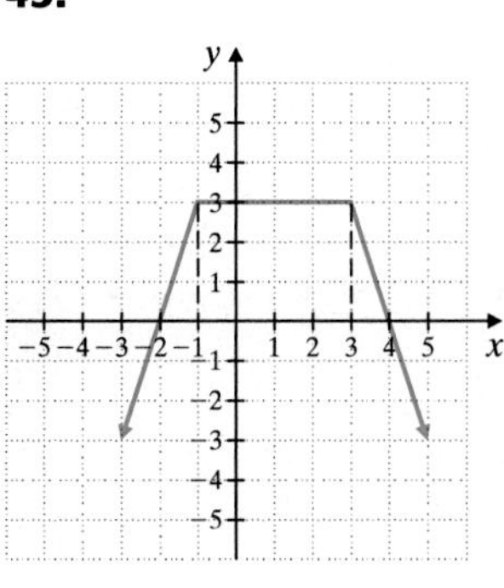

44.

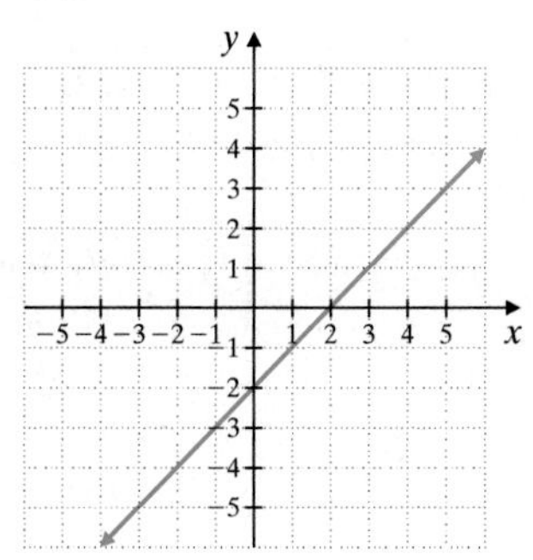

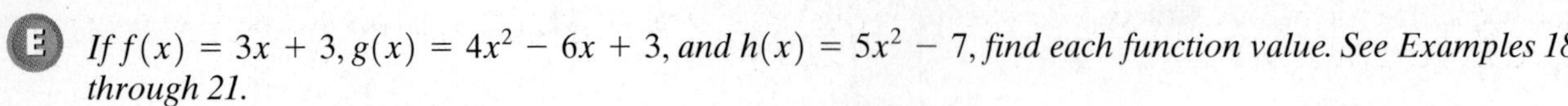
If $f(x) = 3x + 3$, $g(x) = 4x^2 - 6x + 3$, *and* $h(x) = 5x^2 - 7$, *find each function value. See Examples 18 through 21.*

45. $f(4)$

46. $f(-1)$

47. $h(-3)$

48. $h(0)$

49. $g(2)$

50. $g(1)$

51. $g(0)$

52. $h(-2)$

For each function, find the indicated values. See Examples 18 through 21.

53. $f(x) = \frac{1}{2}x$;

a. $f(0)$

b. $f(2)$

c. $f(-2)$

54. $g(x) = -\frac{1}{3}x$;

a. $g(0)$

b. $g(-1)$

c. $g(3)$

55. $f(x) = -5$;

a. $f(2)$

b. $f(0)$

c. $f(606)$

56. $h(x) = 7$;

a. $h(7)$

b. $h(542)$

c. $h\left(-\frac{3}{4}\right)$

The function $A(r) = \pi r^2$ *may be used to find the area of a circle if we are given its radius. Use this function to answer Exercises 57 and 58.*

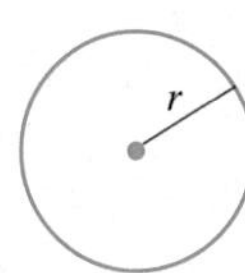

△ **57.** Find the area of a circle whose radius is 5 centimeters. (Do not approximate π.)

△ **58.** Find the area of a circular garden whose radius is 8 feet. (Do not approximate π.)

The function $V(x) = x^3$ *may be used to find the volume of a cube if we are given the length x of a side. Use this function to answer Exercises 59 and 60.*

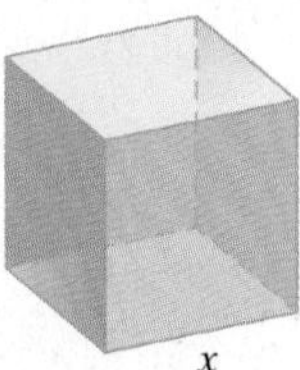

△ **59.** Find the volume of a cube whose side is 14 inches.

△ **60.** Find the volume of a die whose side is 1.7 centimeters.

Forensic scientists use the following functions to find the height of a woman if they are given the height of her femur bone (f) or her tibia bone (t) in centimeters.

$$H(f) = 2.59f + 47.24$$
$$H(t) = 2.72t + 61.28$$

Use these functions to answer Exercises 61 and 62.

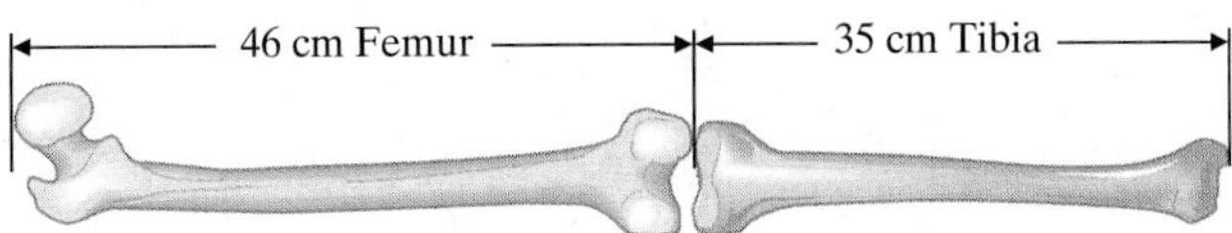

61. Find the height of a woman whose femur measures 46 centimeters.

62. Find the height of a woman whose tibia measures 35 centimeters.

The dosage in milligrams D of Ivermectin, a heartworm preventive, for a dog who weighs x pounds is given by

$$D(x) = \frac{136}{25}x$$

Use this function to answer Exercises 63 and 64.

63. Find the proper dosage for a dog that weighs 30 pounds.

64. Find the proper dosage for a dog that weighs 50 pounds.

65. The per capita consumption (in pounds) of all poultry in the United States is given by the function $C(x) = 1.69x + 87.54$, where x is the number of years since 1995. (*Source*: Based on actual and estimated data from the Economic Research Service, U.S. Department of Agriculture, 1995–2002)

a. Find and interpret $C(5)$.

b. Estimate the per capita consumption of all poultry in the United States in 2002.

66. The amount of money (in billions of dollars) spent by U.S. biotechnology companies on research and development annually is given by the function $R(x) = 0.8x + 5.0$, where x is the number of years since 1993. (*Source*: Based on data from Ernst & Young LLP, Annual Biotechnology Industry Reports, 1993–2000)

a. Find and interpret $R(7)$.

b. Estimate the amount of money spent on biotechnology research and development in 1998.

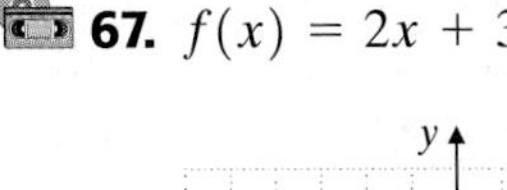

F *Graph each linear function. See Example 22.*

67. $f(x) = 2x + 3$

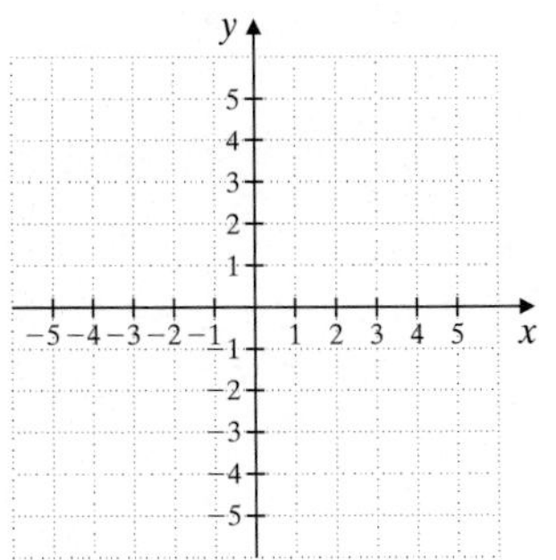

68. $f(x) = 5x - 1$

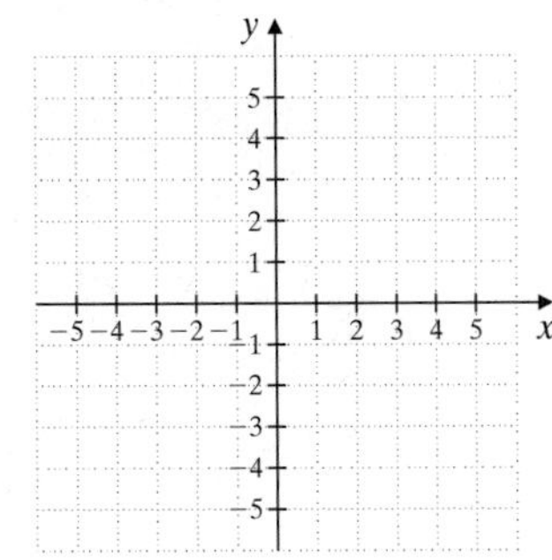

69. $f(x) = -3x$

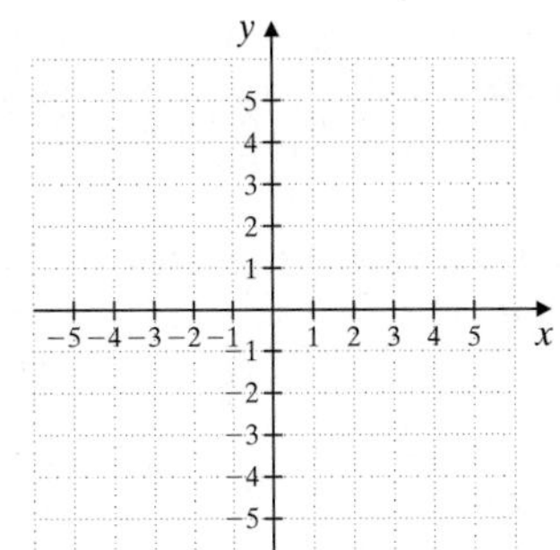

70. $f(x) = -4x$

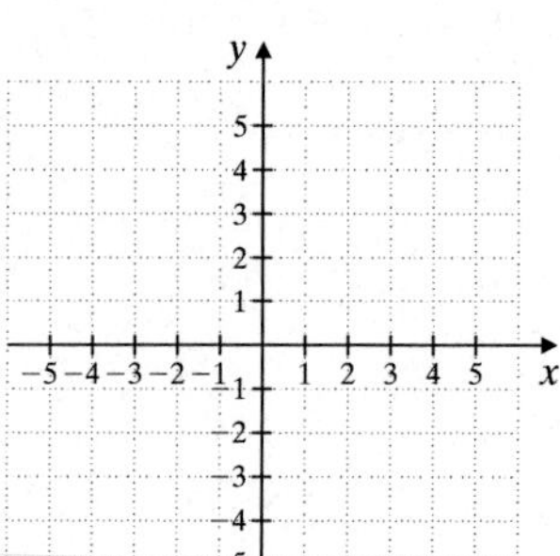

71. $f(x) = -x + 2$

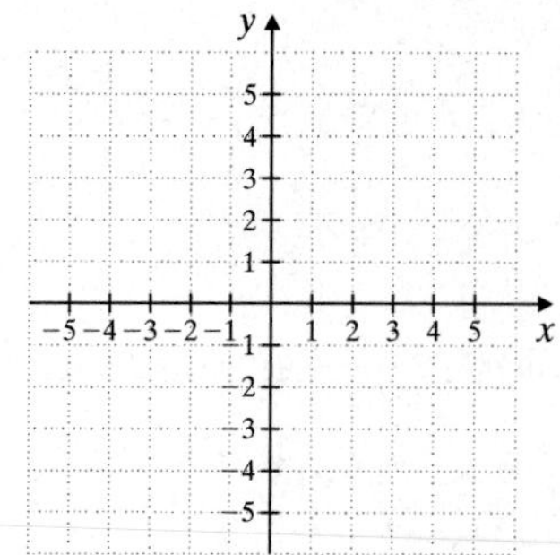

72. $f(x) = -x + 1$

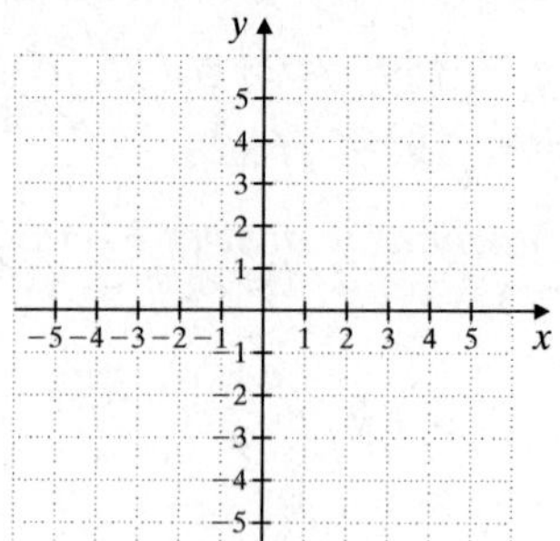

Review and Preview

Solve. See Section 2.8.

73. $2x - 7 \leq 21$

74. $-3x + 1 > 0$

75. $5(x - 2) \geq 3(x - 1)$

76. $-2(x + 1) \leq -x + 10$

77. $\frac{x}{2} + \frac{1}{4} < \frac{1}{8}$

78. $\frac{x}{5} - \frac{3}{10} \geq \frac{x}{2} - 1$

Combining Concepts

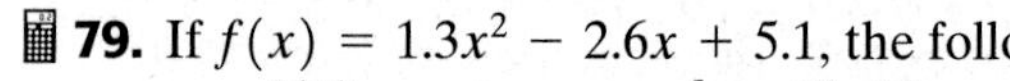

79. If $f(x) = 1.3x^2 - 2.6x + 5.1$, the following.
a. $f(2)$ **b.** $f(-2)$ **c.** $f(3.1)$

For each function, find the indicated values.

80. $f(x) = 2x + 7$;
a. $f(2)$ **b.** $f(a)$

81. $f(x) = x^2 - 12$;
a. $f(12)$ **b.** $f(a)$

82. Describe a function whose domain is the set of people in your home town.

83. Describe a function whose domain is the set of people in your algebra class.

84. Since $y = x + 7$ describes a function, rewrite the equation using function notation.

85. In your own words, explain how to find the domain of a function given its graph.

86. Explain the vertical line test and how it is used.

7.4 Polynomial and Rational Functions

OBJECTIVES

- A Evaluate polynomial functions.
- B Evaluate rational functions.

SSM TUTOR CENTER SG CD & VIDEO MATH PRO WEB

A Evaluating Polynomial Functions

At times it is convenient to use function notation to represent polynomials. For example, we may write $P(x)$ to represent the polynomial $3x^2 - 2x - 5$. In symbols, we would write

$$P(x) = 3x^2 - 2x - 5$$

This function is called a **polynomial function** because the expression $3x^2 - 2x - 5$ is a polynomial.

Helpful Hint

Recall that the symbol $P(x)$ **does not mean** P times x. It is a special symbol used to denote a function.

EXAMPLES If $P(x) = 3x^2 - 2x - 5$, find each function value.

1. $P(1) = 3(1)^2 - 2(1) - 5 = -4$ Let $x = 1$ in the function $P(x)$.
2. $P(-2) = 3(-2)^2 - 2(-2) - 5 = 11$ Let $x = -2$ in the function $P(x)$.

Practice Problems 1–2

If $P(x) = 5x^2 - 3x + 7$, find each function value.

1. $P(2)$
2. $P(-1)$

Many real-world phenomena are modeled by polynomial functions. If the polynomial function model is given, we can often find the solution of a problem by evaluating the function at a certain value.

EXAMPLE 3 Finding the Height of an Object

The world's highest bridge, Royal Gorge suspension bridge in Colorado, is 1053 feet above the Arkansas River. An object is dropped from the top of this bridge. Neglecting air resistance, the height of the object at time t seconds is given by the polynomial function $P(t) = -16t^2 + 1053$. Find the height of the object when $t = 1$ second and when $t = 8$ seconds.

Solution: To find the height of the object at 1 second, we find $P(1)$.

$$P(t) = -16t^2 + 1053$$
$$P(1) = -16(1)^2 + 1053$$
$$P(1) = 1037$$

When $t = 1$ second, the height of the object is 1037 feet.

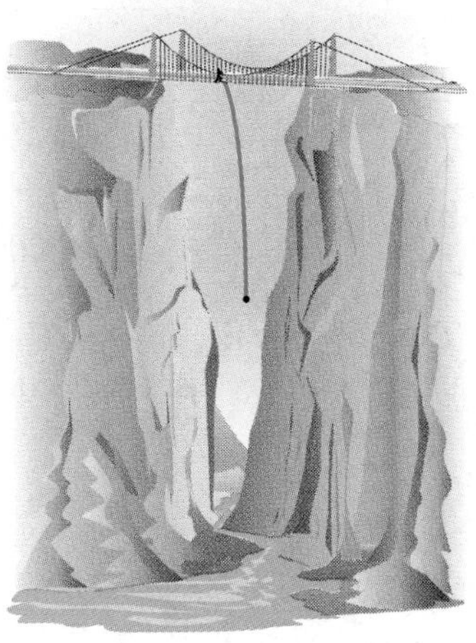

Practice Problem 3

Use the polynomial function in Example 3 to find the height of the object when $t = 3$ seconds and $t = 7$ seconds.

Answers

1. $P(2) = 21$, **2.** $P(-1) = 15$, **3.** At 3 seconds, height is 909 feet; at 7 seconds, height is 269 feet

To find the height of the object at 8 seconds, we find $P(8)$.

$$P(t) = -16t^2 + 1037$$
$$P(8) = -16(8)2 + 1037$$
$$P(8) = -1024 + 1037$$
$$P(8) = 13$$

When $t = 8$ seconds, the height of the object is 13 feet. Notice that as time t increases, the height of the object decreases. ●

B Evaluating Rational Functions

Functions can also represent rational expressions. For example, we call the function $f(x) = \frac{x^2 + 2}{x - 3}$ a **rational function** since $\frac{x^2 + 2}{x - 3}$ is a rational expression in one variable.

The domain of a rational function such as $f(x) = \frac{x^2 + 2}{x - 3}$ is the set of all possible replacement values for x. In other words, since the rational expression $\frac{x^2 + 2}{x - 3}$ is not defined when $x = 3$, we say that the domain of $f(x) = \frac{x^2 + 2}{x - 3}$ is all real numbers except 3. We can write the domain as:

$$\{x | x \text{ is a real number and } x \neq 3\}$$

Practice Problem 4

A company's cost per book for printing x particular books is given by the rational function $C(x) = \frac{0.8x + 5000}{x}$.

Find the cost per book for printing:

a. 100 books
b. 1000 books

EXAMPLE 4 Finding Unit Cost

For the ICL Production Company, the rational function $C(x) = \frac{2.6x + 10{,}000}{x}$ describes the company's cost per disc of pressing x compact discs. Find the cost per disc for pressing:

a. 100 compact discs
b. 1000 compact discs

Solution:

a. $C(100) = \frac{2.6(100) + 10{,}000}{100} = \frac{10{,}260}{100} = 102.6$

The cost per disc for pressing 100 compact discs is $102.60.

b. $C(1000) = \frac{2.6(1000) + 10{,}000}{1000} = \frac{12{,}600}{1000} = 12.6$

The cost per disc for pressing 1000 compact discs is $12.60. Notice that as more compact discs are produced, the cost per disc decreases. ●

Answers

4. a. $50.80, **b.** $5.80

GRAPHING CALCULATOR EXPLORATIONS

Recall that since the rational expression $\frac{7x - 2}{(x - 2)(x + 5)}$ is not defined when $x = 2$ or when $x = -5$, we say that the domain of the rational function $f(x) = \frac{7x - 2}{(x - 2)(x + 5)}$ is all real numbers except 2 and -5. This domain can be written as $\{x | x \text{ is a real number and } x \neq 2, x \neq -5\}$. This means that the graph of $f(x)$ should not cross the vertical lines $x = 2$ and $x = -5$. The graph of $f(x)$ in *connected* mode follows. In connected mode the grapher tries to connect all dots of the graph so that the result is a smooth curve. This is what has happened in the graph. Notice that the graph appears to contain vertical lines at $x = 2$ and at $x = -5$. We know that this cannot happen because the function is not defined at $x = 2$ and at $x = -5$. We also know that this cannot happen because the graph of this function would not pass the vertical line test.

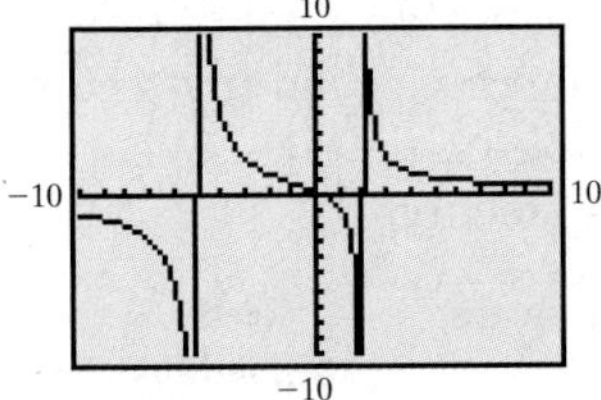

If we graph $f(x)$ in *dot* mode, the graph appears as follows. In dot mode the grapher will not connect dots with a smooth curve. Notice that the vertical lines have disappeared, and we have a better picture of the graph. The graph, however, actually appears more like the hand-drawn graph to its right. By using a Table feature, a Calculate Value feature, or by tracing, we can see that the function is not defined at $x = 2$ and at $x = -5$.

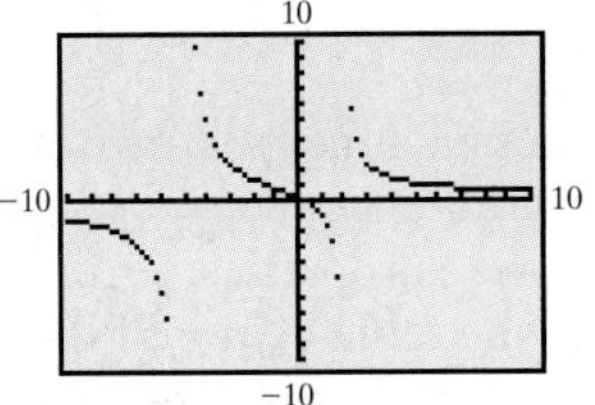

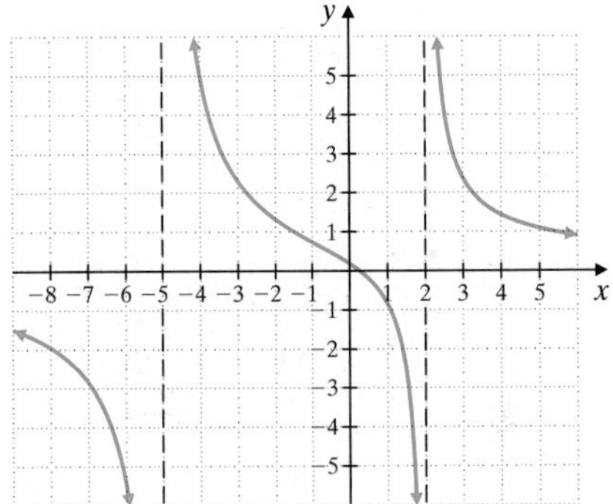

Find the domain of each rational function. Then graph each rational function and use the graph to confirm the domain.

1. $f(x) = \frac{x + 1}{x^2 - 4}$

2. $g(x) = \frac{5x}{x^2 - 9}$

3. $h(x) = \frac{x^2}{2x^2 + 7x - 4}$

4. $f(x) = \frac{3x + 2}{4x^2 - 19x - 5}$

FOCUS ON Mathematical Connections

MULTICULTURAFINITE DIFFERENCES

When polynomial functions are evaluated at successive integer values, a list of values called a **sequence** is generated. The differences between successive pairs of numbers in such a sequence have special properties. Let's investigate these properties, beginning with a first-degree polynomial function, the linear function.

Notice in the table below on the left that *first differences* are the differences between the successive pairs of numbers in the original sequence. Find the first differences for any other linear function and fill in the table on the right. What do you notice? (*Note:* You may wish to try several different linear functions.)

x	Original Sequence $f(x) = 3x + 4$	First Differences
8	28	
		3
7	25	
		3
6	22	
		3
5	19	
		3
4	16	
		3
3	13	
		3
2	10	
		3
1	7	

x	Original Sequence $f(x) =$	First Differences

Now let's look at differences for a second-degree polynomial. Notice in the table below on the left that *second differences* are the differences between successive pairs of first differences. Find first and second differences for any other second-degree polynomial function and fill in the table on the right. What do you notice? (*Note:* You may wish to try several different second-degree polynomial functions.)

x	Original Sequence $f(x) = 2x^2 - 3x + 4$	First Differences	Second Differences
8	108		
		27	
7	81		4
		23	
6	58		4
		19	
5	39		4
		15	
4	24		4
		11	
3	13		4
		7	
2	6		4
		3	
1	3		

x	Original Sequence $f(x) =$	First Differences	Second Differences

CRITICAL THINKING

1. As you might guess, third differences are the differences between successive pairs of second differences. Find the first, second, and third differences for any two third-degree polynomial functions. What do you notice?
2. What would you expect to be true about the differences for a fourth-degree polynomial function?
3. What would you expect to be true about the differences for an nth-degree polynomial function?

Name ______________________ Section __________ Date __________

EXERCISE SET 7.4

A *If* $P(x) = x^2 + x + 1$ *and* $Q(x) = 5x^2 - 1$, *find each function value. See Examples 1 and 2.*

1. $P(7)$ **2.** $Q(4)$ **3.** $Q(-10)$ **4.** $P(-4)$ **5.** $P(0)$ **6.** $Q(0)$

If $P(x) = x^3 + 2x - 3$ *and* $Q(x) = 7x + 5$, *find each function value. See Examples 1 and 2.*

7. $P(2)$ **8.** $Q(6)$ **9.** $Q(-3)$ **10.** $P(-1)$ **11.** $P(-2)$ **12.** $Q(0)$

Solve. See Example 3.

13. A projectile is fired upward from the ground with an initial velocity of 300 feet per second. Neglecting air resistance, the height of the projectile at any time t can be described by the polynomial function $P(t) = -16t^2 + 300t$. Find the height of the projectile at each given time.

a. $t = 1$ second
b. $t = 2$ seconds
c. $t = 3$ seconds
d. $t = 4$ seconds
e. Explain why the height increases and then decreases as time passes.
f. Approximate (to the nearest second) how long before the object hits the ground.

14. An object is thrown upward with an initial velocity of 25 feet per second from the top of the 984-foot-high Eiffel Tower in Paris, France. The height of the object at any time t can be described by the polynomial function $P(t) = -16t^2 + 25t + 984$. Find the height of the projectile at each given time. (*Source:* Council on Tall Buildings and Urban Habitat, Lehigh University)

a. $t = 1$ second
b. $t = 3$ seconds
c. $t = 5$ seconds
d. Approximate (to the nearest second) how long before the object hits the ground.

15. The polynomial function $P(x) = 45x - 100{,}000$ models the relationship between the number of computer briefcases x that a company sells and the profit the company makes, $P(x)$. Find $P(4000)$, the profit from selling 4000 computer briefcases.

16. The total cost (in dollars) for MCD, Inc., Manufacturing Company to produce x blank audiocassette tapes per week is given by the polynomial function $C(x) = 0.8x + 10{,}000$. Find the total cost of producing 20,000 tapes per week.

17. The total revenues (in dollars) for MCD, Inc., Manufacturing Company to sell x blank audiocasette tapes per week is given by the polynomial function $R(x) = 2x$. Find the total revenue from selling 20,000 tapes per week.

18. In business, profit equals revenue minus cost, or $P(x) = R(x) - C(x)$. Find the profit function for MCD, Inc. by subtracting the given functions in Exercises 16 and 17.

19. Sport utility vehicle (SUV) sales in the U.S. have increased since 1992. The function $f(x) = 0.014x^2 + 0.12x + 0.85$ can be used to approximate the number of SUV sales during the years 1990–2000 where x is the number of years after 1990 and $f(x)$ is the SUV sales (in millions). Round answers to the nearest tenth of a million. (*Source:* Wards Communications)

a. Approximate the number of SUV's sold in 1999.

b. Use the function to predict the number of SUV's sold in 2005.

20. Digital camera sales have increased since 1996. The function $f(x) = 34.7x^2 + 68.2x + 377.3$ can be used to approximate the revenue from selling digital cameras for the years 1996–2001 where x is the number of years after 1996 and $f(x)$ is the sales revenue (in million of dollars). Round answers to the nearest whole million. (*Source:* International Data Corporation)

a. Approximate the revenue from digital camera sales in 2000.

b. Use the function to predict the revenue from digital camera sales in 2004.

21. The number of computers for instructional use in public schools is increasing. The polynomial function $f(x) = 98x^2 + 514x + 4746$ models the number of these computers for the years 1994–2000, where x represents the number of years after 1994 and $f(x)$ is the number of computers used for instruction. Write an equivalent expression for $f(x)$ by factoring the greatest common factor from the terms of $98x^2 + 514x + 4746$.

22. The polynomial function $f(x) = 2900x^2 - 3500x + 120{,}000$ models the number of applications for trademark registrations for the years 1990–2000 where x represents the number of years after 1990 and $f(x)$ is the number of trademark registrations. Write an equivalent expression for $f(x)$ by factoring the greatest common factor from the terms of $2900x^2 - 3500x + 120{,}000$. (*Source:* International Trademark Association)

23. Suppose that a movie is being filmed in New York City. An action shot requires an object to be thrown upward with an initial velocity of 80 feet per second off the top of 1 Madison Square Plaza, a height of 576 feet. The height $h(t)$ in feet of the object after t seconds is given by the function $h(t) = -16t^2 + 80t + 576$. (*Source:* The World Almanac, 2001)

a. Find the height of the object at $t = 0$ seconds, $t = 2$ seconds, $t = 4$ seconds, and $t = 6$ seconds.

b. Explain why the height of the object increases and then decreases as time passes.

c. Factor the polynomial $-16t^2 + 80t + 576$.

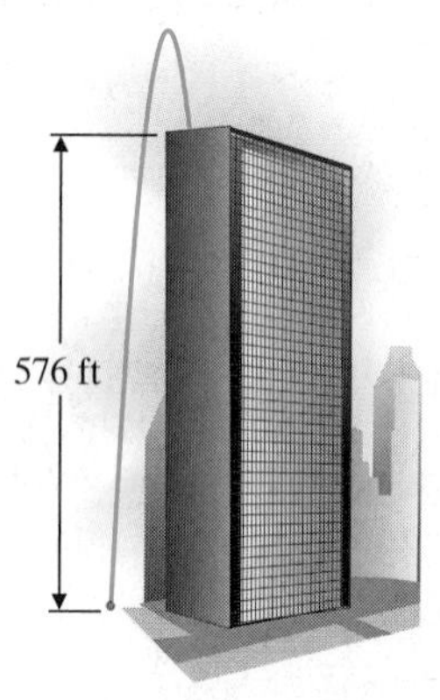

24. Suppose that an object is thrown upward with an initial velocity of 64 feet per second off the edge of a 960-foot-cliff. The height $h(t)$ in feet of the object after t seconds is given by the function

$$h(t) = -16t^2 + 64t + 960$$

a. Find the height of the object at $t = 0$ seconds, $t = 3$ seconds, $t = 6$ seconds, and $t = 9$ seconds.

b. Explain why the height of the object increases and then decreases as time passes.

c. Factor the polynomial $-16t^2 + 64t + 960$.

25. The function $f(x) = 136.7x^2 + 327.6x + 21.6$ can be used to approximate the increasing number of radio stations on the Internet during the years 1996–2001 where x is the number of years after 1996 and $f(x)$ is the number of stations. Round answers to the nearest whole. (*Source:* BRS Media, Inc.)

a. Approximate the number of radio stations on the Internet in 1996.

b. Approximate the number of radio stations on the Internet in 2000.

c. Use the function to predict the number of radio stations on the Internet in 2004.

d. From parts (a), (b), and (c), determine whether the number of radio stations on the Internet is increasing at a steady rate. Explain why or why not.

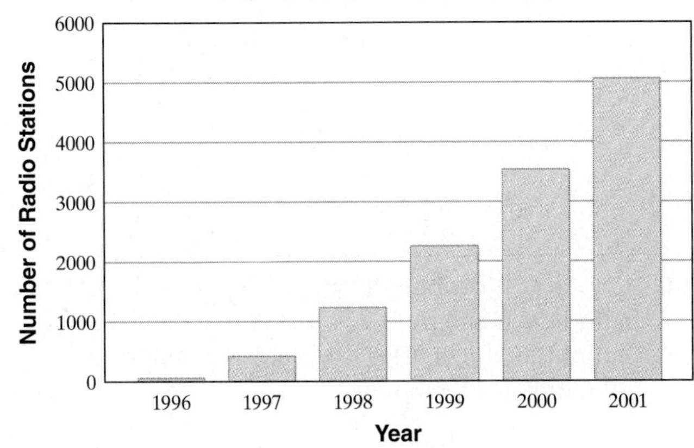

26. The function $f(x) = -x^2 + 11.4x + 49.8$ can be used to approximate the number of Americans enrolled in health maintenance organizations (HMOs) during the period 1995–2000, where x is the number of years after 1995 and $f(x)$ is the number of millions of Americans. Round answers to the nearest tenth of a million. (*Source:* Based on data from *Health, United States, 2001*, National Center for Health Statistics)

a. Approximate the number of Americans enrolled in HMOs in 1995.

b. Approximate the number of Americans enrolled in HMOs in 2000.

c. Use the function to predict the number of Americans enrolled in HMOs in 2005.

d. From parts (a), (b), and (c), determine whether the number of Americans enrolled in HMOs is changing at a steady rate. Explain why or why not.

B *If* $f(x) = \dfrac{x + 8}{2x - 1}$ *and* $g(x) = \dfrac{x - 2}{x - 5}$, *find each function value. See Example 4.*

27. $f(2)$ **28.** $g(10)$ **29.** $g(0)$ **30.** $f(0)$ **31.** $f(-1)$

32. $g(-5)$ **33.** Find the domain of $g(x)$. **34.** Find the domain of $f(x)$.

If $f(x) = \dfrac{x^2 + 5}{x}$ *and* $g(x) = \dfrac{x^2 + 2x}{x + 3}$, *find each function value. See Example 4.*

35. $f(1)$ **36.** $g(3)$ **37.** $g(-6)$ **38.** $f(-2)$ **39.** $g(0)$

40. $f(2)$ **41.** Find the domain of $f(x)$. **42.** Find the domain of $g(x)$.

Solve. See Example 4.

43. The total revenue from the sale of a popular book is approximated by the rational function $R(x) = \frac{1000x^2}{x^2 + 4}$, where x is the number of years since publication and $R(x)$ is the total revenue in millions of dollars.
a. Find the total revenue at the end of the first year.
b. Find the total revenue at the end of the second year.
c. Find the revenue during the second year only.

44. The function $f(x) = \frac{100{,}000x}{100 - x}$ models the cost in dollars for removing x percent of the pollutants from a bayou in which a nearby company dumped creosol.
a. Find the cost of removing 20% of the pollutants from the bayou. (*Hint:* Find $f(20)$.)
b. Find the cost of removing 60% of the pollutants and then 80% of the pollutants.
c. Find $f(90)$, then $f(95)$, and then $f(99)$. What happens to the cost as x approaches 100%?

Review and Preview

Solve each equation for x. See Section 6.5.

45. $\frac{x}{5} = \frac{x + 2}{3}$

46. $\frac{x}{4} = \frac{x + 3}{6}$

47. $\frac{x - 3}{2} = \frac{x - 5}{6}$

48. $\frac{x - 6}{4} = \frac{x - 2}{5}$

Combining Concepts

49. An object is thrown upward from the ground with an initial velocity of 64 feet per second. The height $h(t)$ in feet of the object after t seconds is given by the polynomial function

$$h(t) = -16t^2 + 64t$$

a. Write an equivalent factored expression for the function $h(t)$ by factoring $-16t^2 + 64t$.
b. Find $h(1)$ by using

$$h(t) = -16t^2 + 64t$$

and then by using the factored form of $h(t)$.
c. Explain why the values found in part (b) are the same.

50. An object is dropped from the gondola of a hot-air balloon at a height of 224 feet. The height $h(t)$ in feet of the object after t seconds is given by the polynomial function

$$h(t) = -16t^2 + 224$$

a. Write an equivalent factored expression for the function $h(t)$ by factoring $-16t^2 + 224$.
b. Find $h(2)$ by using

$$h(t) = -16t^2 + 224$$

and then by using the factored form of the function.
c. Explain why the values found in part (b) are the same.

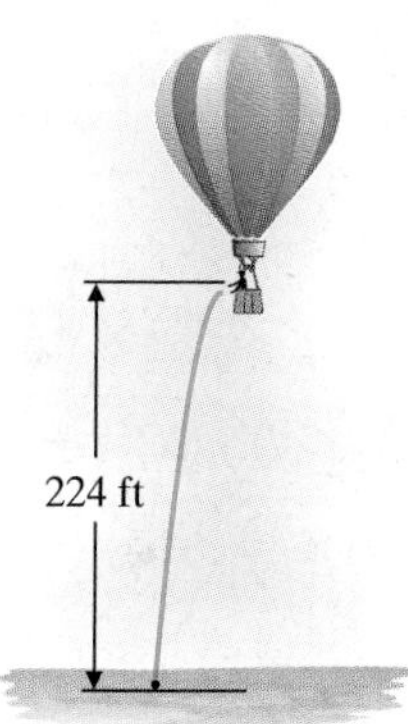

If $P(x) = 3x + 3$, $Q(x) = 4x^2 - 6x + 3$, and $R(x) = 5x^2 - 7$, find each function.

51. $P(x) + Q(x)$

52. $Q(x) - R(x)$

53. If $P(x) = 2x - 3$, find $P(a)$, $P(-x)$, and $P(x + h)$.

54. If $P(x) = 5x + 1$, find $P(a)$, $P(-x)$, and $P(x + h)$.

If $R(x) = x + 5$, $Q(x) = x^2 - 2$, and $P(x) = 5x$, find each function.

55. $P(x) \cdot R(x)$

56. $P(x) \cdot Q(x)$

If $f(x) = x^2 - 3x$, find each function value.

57. $f(a)$

58. $f(a + h)$

7.5 Variation and Problem Solving

In the previous section, we saw many real-life examples of polynomial and rational functions. Other examples of these functions include types of variation, which we introduce in this section.

OBJECTIVES

- A Solve problems involving direct variation.
- B Solve problems involving inverse variation.
- C Solve problems involving joint variation.
- D Solve problems involving combined variation.

A Solving Problems Involving Direct Variation

A very familiar example of **direct variation** is the relationship of the circumference C of a circle to its radius r. The formula $C = 2\pi r$ expresses that the circumference is always 2π times the radius. In other words, C is always a constant multiple (2π) of r. Because it is, we say that C *varies directly as* r, that C *varies directly with* r, or that C *is directly proportional to* r.

Direct Variation

y* varies directly as *x, or ***y* is directly proportional to *x***, if there is a nonzero constant k such that

$$y = kx$$

The number k is called the **constant of variation** or the **constant of proportionality**.

In the above definition, the relationship described between x and y is a linear one. In other words, the graph of $y = kx$ is a line. The slope of the line is k, and the line passes through the origin.

For example, the graph of the direct variation equation $C = 2\pi r$ is shown. The horizontal axis represents the radius r, and the vertical axis is the circumference C. From the graph we can read that when the radius is 6 units, the circumference is approximately 38 units. Also, when the circumference is 45 units, the radius is between 7 and 8 units. Notice that as the radius increases, the circumference increases.

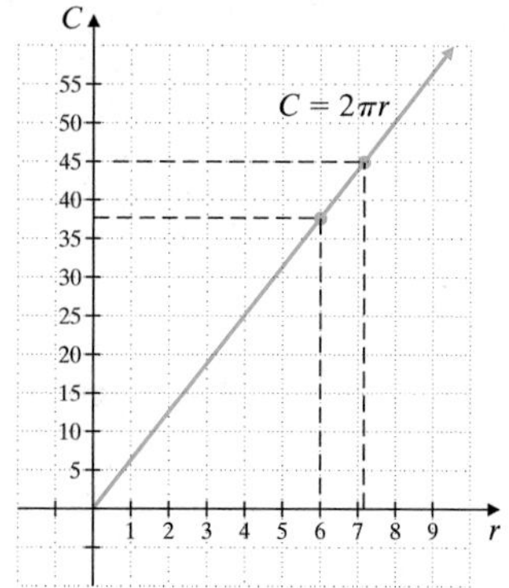

EXAMPLE 1

Suppose that y varies directly as x. If y is 5 when x is 30, find the constant of variation and the direct variation equation.

Solution: Since y varies directly as x, we write $y = kx$. If $y = 5$ when $x = 30$, we have that

$$y = kx$$
$$5 = k(30) \quad \text{Replace } y \text{ with 5 and } x \text{ with 30.}$$
$$\frac{1}{6} = k \quad \text{Solve for } k.$$

The constant of variation is $\frac{1}{6}$. After finding the constant of variation k, the direct variation equation can be written as $y = \frac{1}{6}x$.

Practice Problem 1

Suppose that y varies directly as x. If y is 24 when x is 8, find the constant of variation and the direct variation equation.

Answer

1. $k = 3$; $y = 3x$

Practice Problem 2

Use Hooke's law as stated in Example 2. If a 56-pound weight attached to a spring stretches the spring 8 inches, find the distance that an 85-pound weight attached to the spring stretches the spring.

EXAMPLE 2 Using Direct Variation and Hooke's Law

Hooke's law states that the distance a spring stretches is directly proportional to the weight attached to the spring. If a 40-pound weight attached to the spring stretches the spring 5 inches, find the distance that a 65-pound weight attached to the spring stretches the spring.

Solution:

1. UNDERSTAND. Read and reread the problem. Notice that we are given that the distance a spring stretches is *directly proportional* to the weight attached. We let

   d = the distance stretched

   w = the weight attached

   The constant of variation is represented by k.

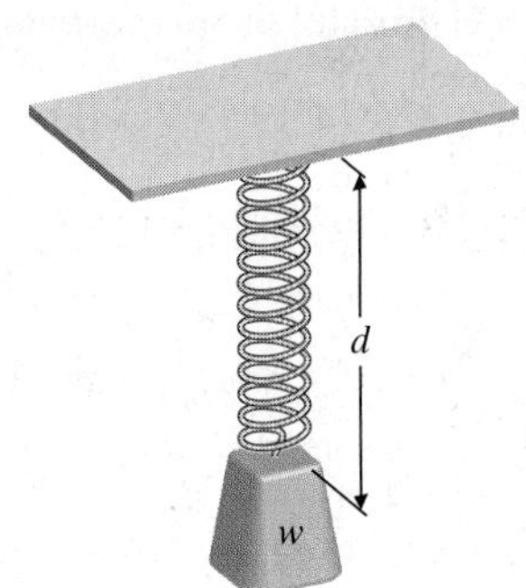

2. TRANSLATE. Because d is directly proportional to w, we write

   $d = kw$

3. SOLVE. When a weight of 40 pounds is attached, the spring stretches 5 inches. That is, when $w = 40$, $d = 5$.

   $5 = k(40)$ Replace d with 5 and w with 40.

   $\frac{1}{8} = k$ Solve for k.

   Now when we replace k with $\frac{1}{8}$ in the equation $d = kw$, we have

   $d = \frac{1}{8}w$

   To find the stretch when a weight of 65 pounds is attached, we replace w with 65 to find d.

   $$d = \frac{1}{8}(65)$$

   $$= \frac{65}{8} = 8\frac{1}{8} \text{ or } 8.125$$

4. INTERPRET.

Check: Check the proposed solution of 8.125 inches in the original problem.

State: The spring stretches 8.125 inches when a 65-pound weight is attached.

Answer

2. $12\frac{1}{7}$ in.

B Solving Problems Involving Inverse Variation

When y is proportional to the *reciprocal* of another variable x, we say that *y varies inversely as x*, or that *y is inversely proportional to x*. An example of the **inverse variation** relationship is the relationship between the pressure that a gas exerts and the volume of its container. As the volume of a container decreases, the pressure of the gas it contains increases.

Inverse Variation

y varies inversely as x, or **y is inversely proportional to x**, if there is a nonzero constant k such that

$$y = \frac{k}{x}$$

The number k is called the **constant of variation** or the **constant of proportionality**.

Notice that $y = \frac{k}{x}$ is an equation containing a rational expression. Its graph for $k > 0$ and $x > 0$ is shown. From the graph, we can see that as x increases, y decreases.

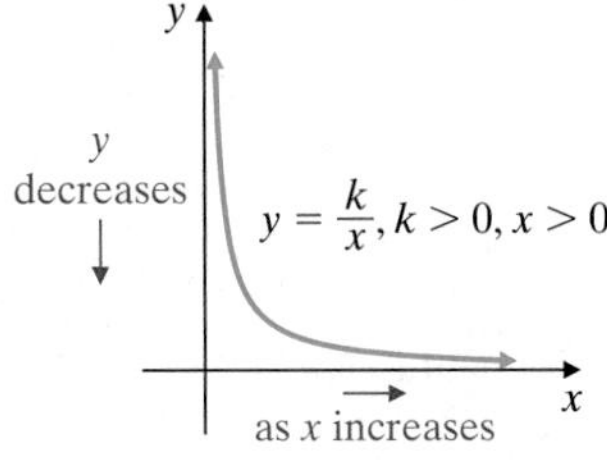

EXAMPLE 3

Suppose that u varies inversely as w. If u is 3 when w is 5, find the constant of variation and the inverse variation equation.

Solution: Since u varies inversely as w, we have $u = \frac{k}{w}$. We let $u = 3$ and $w = 5$, and we solve for k.

$$u = \frac{k}{w}$$

$$3 = \frac{k}{5} \quad \text{Let } u = 3 \text{ and } w = 5.$$

$$15 = k \quad \text{Multiply both sides by 5.}$$

The constant of variation k is 15. This gives the inverse variation equation

$$u = \frac{15}{w}$$

Practice Problem 3

Suppose that y varies inversely as x. If y is 6 when x is 3, find the constant of variation and the inverse variation equation.

Answer

3. $k = 18$; $y = \frac{18}{x}$

Practice Problem 4

The speed r at which one needs to drive in order to travel a constant distance is inversely proportional to the time t. A fixed distance can be driven in 5 hours at a rate of 24 mph. Find the rate needed to drive the same distance in 4 hours.

Answer

4. 30 mph

EXAMPLE 4 Using Inverse Variation and Boyle's Law

Boyle's law says that if the temperature stays the same, the pressure P of a gas is inversely proportional to the volume V. If a cylinder in a steam engine has a pressure of 960 kilopascals when the volume is 1.4 cubic meters, find the pressure when the volume increases to 2.5 cubic meters.

Solution:

1. UNDERSTAND. Read and reread the problem. Notice that we are given that the pressure of a gas is *inversely proportional* to the volume. We will let P = the pressure and V = the volume. The constant of variation is represented by k.
2. TRANSLATE. Because P is inversely proportional to V, we write

$$P = \frac{k}{V}$$

When $P = 960$ kilopascals, the volume $V = 1.4$ cubic meters. We use this information to find k.

$$960 = \frac{k}{1.4} \quad \text{Let } P = 960 \text{ and } V = 1.4.$$

$$1344 = k \quad \text{Multiply both sides by 1.4.}$$

Thus, the value of k is 1344. Replacing k with 1344 in the variation equation, we have

$$P = \frac{1344}{V}$$

Next we find P when V is 2.5 cubic meters.

3. SOLVE.

$$P = \frac{1344}{2.5} \quad \text{Let } V = 2.5$$

$$= 537.6$$

4. INTERPRET. *Check* the proposed solution in the original problem.

State: When the volume is 2.5 cubic meters, the pressure is 537.6 kilopascals.

C Solving Problems Involving Joint Variation

Sometimes the ratio of a variable to the product of many other variables is constant. For example, the ratio of distance traveled to the product of speed and time traveled is constantly 1:

$$\frac{d}{rt} = 1 \quad \text{or} \quad d = rt$$

Such a relationship is called **joint variation**.

Joint Variation

If the ratio of a variable y to the product of two or more variables is constant, then **y varies jointly as**, or **is jointly proportional to**, the other variables. If

$$y = kxz$$

then the number k is the **constant of variation** or the **constant of proportionality**.

Try the Concept Check in the margin.

Concept Check

Which type of variation is represented by the equation $xy = 8$? Explain.

a. Direct variation
b. Inverse variation
c. Joint variation

EXAMPLE 5

The lateral surface area of a cylinder varies jointly as its radius and height. Express surface area S in terms of radius r and height h.

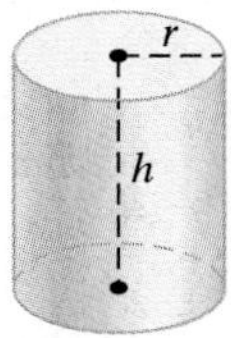

Solution: Because the surface area varies jointly as the radius r and the height h, we equate S to a constant multiple of r and h:

$$S = krh$$

Note: From actual values of S, r, and h, it can be determined that the constant k is 2π, and we then have the formula $S = 2\pi rh$.

Practice Problem 5

The area of a triangle varies jointly as its base and height. Express the area in terms of base b and height h.

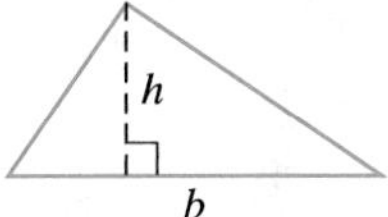

 Solving Problems Involving Combined Variation

Some examples of variation involve combinations of direct, inverse, and joint variation. We will call these variations **combined variation**.

EXAMPLE 6

Suppose that y varies directly as the square of x. If y is 24 when x is 2, find the constant of variation and the variation equation.

Solution: Since y varies directly as the square of x, we have

$$y = kx^2$$

Now let $y = 24$ and $x = 2$ and solve for k.

$$y = kx^2$$
$$24 = k \cdot 2^2$$
$$24 = 4k$$
$$6 = k$$

The constant of variation is 6, so the variation equation is

$$y = 6x^2$$

Practice Problem 6

Suppose that y varies inversely as the square of x. If y is 24 when x is 2, find the constant of variation and the variation equation.

Answers

5. $A = kbh$, **6.** $k = 96$; $y = \frac{96}{x^2}$

Concept Check: b; answers may vary

Practice Problem 7 △

The maximum weight that a rectangular beam can support varies jointly as its width and the square of its height and inversely as its length. If a beam $\frac{1}{3}$ foot wide, 1 foot high, and 10 feet long can support 3 tons, find how much weight a similar beam can support if it is 1 foot wide, $\frac{1}{3}$ foot high, and 9 feet long.

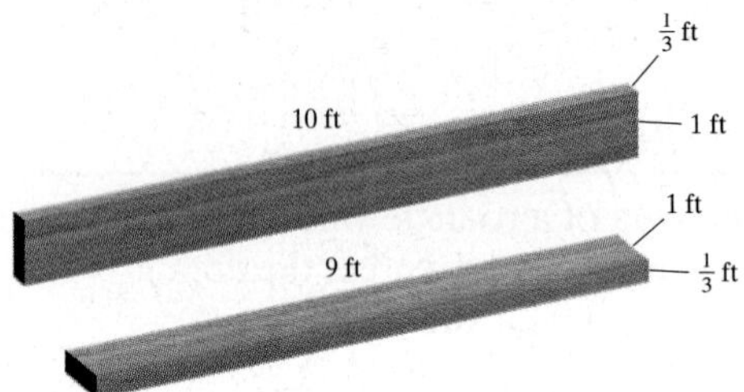

EXAMPLE 7 Using Combined Variation

The maximum weight that a circular column can support is directly proportional to the fourth power of its diameter and inversely proportional to the square of its height. A 2-meter-wide column that is 8 meters in height can support 1 ton. Find the weight that a 1-meter-wide column that is 4 meters in height can support.

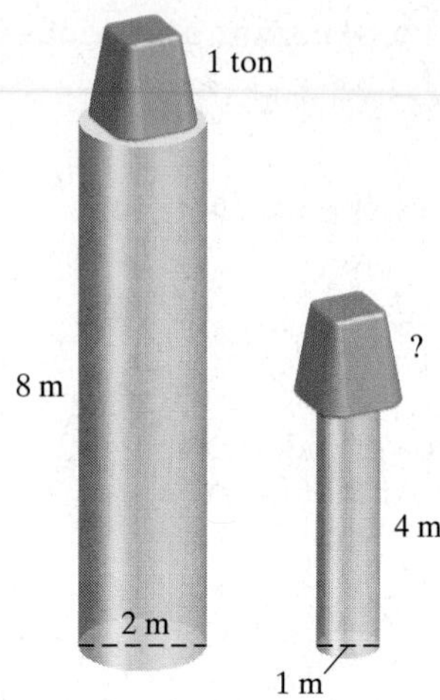

Solution:

1. UNDERSTAND. Read and reread the problem. Let w = weight, d = diameter, h = height, and k = the constant of variation.
2. TRANSLATE. Since w is directly proportional to d^4 and inversely proportional to h^2, we have

$$w = \frac{kd^4}{h^2}$$

3. SOLVE. To find k, we are given that a 2-meter-wide column that is 8 meters in height can support 1 ton. That is, $w = 1$ when $d = 2$ and $h = 8$, or

$$1 = \frac{k \cdot 2^4}{8^2} \quad \text{Let } w = 1, d = 2, \text{ and } h = 8.$$

$$1 = \frac{k \cdot 16}{64}$$

$$4 = k \quad \text{Solve for } k.$$

Now we replace k with 4 in the equation $w = \frac{kd^4}{h^2}$:

$$w = \frac{4d^4}{h^2}$$

To find weight, w, for a 1-meter-wide column that is 4 meters in height, we let $d = 1$ and $h = 4$.

$$w = \frac{4 \cdot 1^4}{4^2}$$

$$w = \frac{4}{16} = \frac{1}{4}$$

4. INTERPRET. *Check* the proposed solution in the original problem.

State: The 1-meter-wide column that is 4 meters in height can hold $\frac{1}{4}$ ton of weight. ●

Answer

7. $1\frac{1}{9}$ tons

Name ______________________ Section __________ Date __________

EXERCISE SET 7.5

A *If y varies directly as x, find the constant of variation and the direct variation equation for each situation. See Example 1.*

1. $y = 4$ when $x = 20$

2. $y = 5$ when $x = 30$

3. $y = 6$ when $x = 4$

4. $y = 12$ when $x = 8$

5. $y = 7$ when $x = \frac{1}{2}$

6. $y = 11$ when $x = \frac{1}{3}$

7. $y = 0.2$ when $x = 0.8$

8. $y = 0.4$ when $x = 2.5$

Solve. See Example 2.

9. The weight of a synthetic ball varies directly with the cube of its radius. A ball with a radius of 2 inches weighs 1.20 pounds. Find the weight of a ball of the same material with a 3-inch radius.

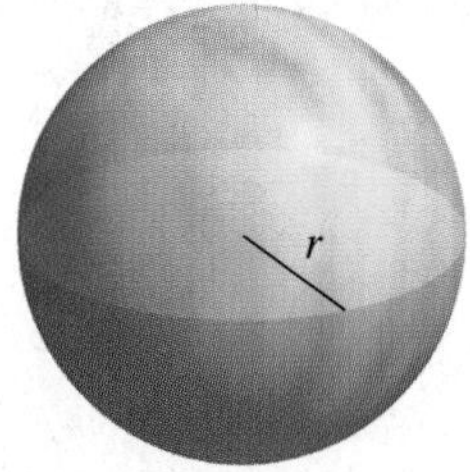

10. At sea, the distance to the horizon is directly proportional to the square root of the elevation of the observer. If a person who is 36 feet above the water can see 7.4 miles, find how far a person 64 feet above the water can see. Round to the nearest tenth of a mile.

11. The amount P of pollution varies directly with the population N of people. Kansas City has a population of 442,000 and produces 260,000 tons of pollutants. Find how many tons of pollution we should expect St. Louis to produce, if we know that its population is 348,000. Round to the nearest whole ton. (*Population Source: The World Almanac*, 2002)

12. Charles's law states that if the pressure P stays the same, the volume V of a gas is directly proportional to its temperature T. If a balloon is filled with 20 cubic meters of a gas at a temperature of 300 K, find the new volume if the temperature rises to 360 K while the pressure stays the same.

B *If y varies inversely as x, find the constant of variation and the inverse variation equation for each situation. See Example 3.*

13. $y = 6$ when $x = 5$ **14.** $y = 20$ when $x = 9$ **15.** $y = 100$ when $x = 7$ **16.** $y = 63$ when $x = 3$

17. $y = \frac{1}{8}$ when $x = 16$ **18.** $y = \frac{1}{10}$ when $x = 40$ **19.** $y = 0.2$ when $x = 0.7$ **20.** $y = 0.6$ when $x = 0.3$

Solve. See Example 4.

21. Pairs of markings a set distance apart are made on highways so that police can detect drivers exceeding the speed limit. Over a fixed distance, the speed R varies inversely with the time T. In one particular pair of markings, R is 45 mph when T is 6 seconds. Find the speed of a car that travels the given distance in 5 seconds.

22. The weight of an object on or above the surface of Earth varies inversely as the square of the distance between the object and Earth's center. If a person weighs 160 pounds on Earth's surface, find the individual's weight if he moves 200 miles above Earth. Round to the nearest whole pound. (Assume that Earth's radius is 4000 miles.)

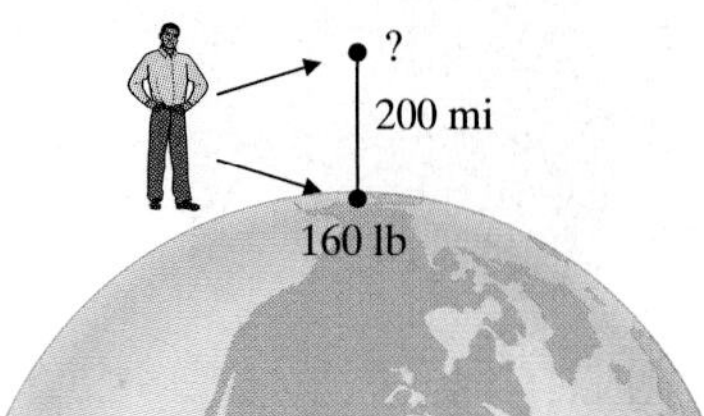

23. If the voltage V in an electric circuit is held constant, the current I is inversely proportional to the resistance R. If the current is 40 amperes when the resistance is 270 ohms, find the current when the resistance is 150 ohms.

24. Because it is more efficient to produce larger numbers of items, the cost of producing Dysan computer disks is inversely proportional to the number produced. If 4000 can be produced at a cost of $1.20 each, find the cost per disk when 6000 are produced.

25. The intensity I of light varies inversely as the square of the distance d from the light source. If the distance from the light source is doubled (see the figure), determine what happens to the intensity of light at the new location.

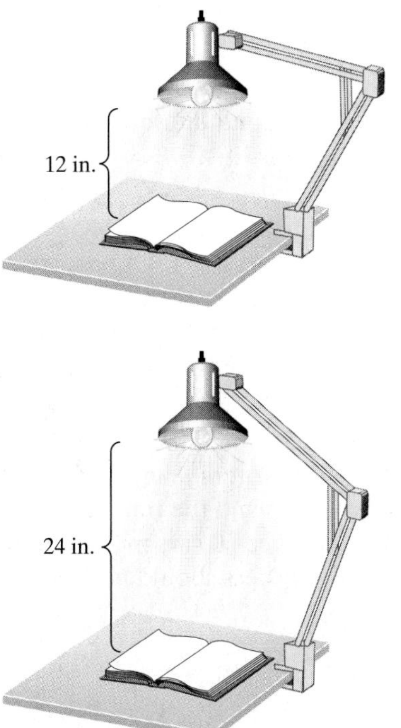

△ **26.** The maximum weight that a circular column can hold is inversely proportional to the square of its height. If an 8-foot column can hold 2 tons, find how much weight a 10-foot column can hold.

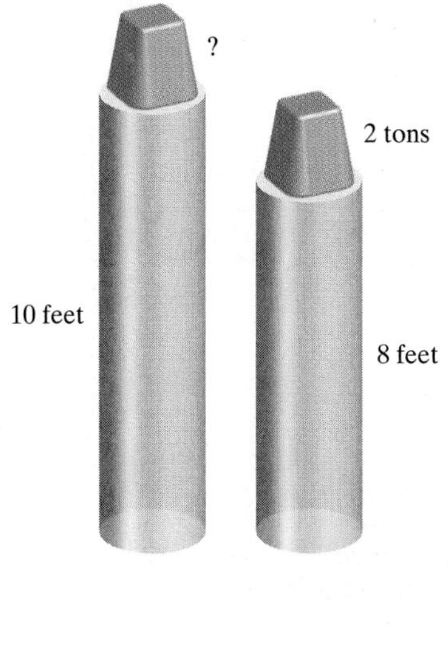

Write each statement as an equation. Use k as the constant of variation. See Example 5.

27. x varies jointly as y and z.

28. P varies jointly as R and the square of S.

29. r varies jointly as s and the cube of t.

30. a varies jointly as b and c.

For each statement, find the constant of variation and the variation equation. See Examples 5 and 6.

31. y varies directly as the cube of x; $y = 9$ when $x = 3$

32. y varies directly as the cube of x; $y = 32$ when $x = 4$

33. y varies directly as the square root of x; $y = 0.4$ when $x = 4$

34. y varies directly as the square root of x; $y = 2.1$ when $x = 9$

35. y varies inversely as the square of x; $y = 0.052$ when $x = 5$

36. y varies inversely as the square of x; $y = 0.011$ when $x = 10$

37. y varies jointly as x and the cube of z; $y = 120$ when $x = 5$ and $z = 2$

38. y varies jointly as x and the square of z; $y = 360$ when $x = 4$ and $z = 3$

Solve. See Example 7.

△ **39.** The maximum weight that a rectangular beam can support varies jointly as its width and the square of its height and inversely as its length. If a beam $\frac{1}{2}$ foot wide, $\frac{1}{3}$ foot high, and 10 feet long can support 12 tons find how much a similar beam can support if the beam is $\frac{2}{3}$ foot wide, $\frac{1}{2}$ foot high, and 16 feet long.

40. The number of cars manufactured on an assembly line at a General Motors plant varies jointly as the number of workers and the time they work. If 200 workers can produce 60 cars in 2 hours, find how many cars 240 workers should be able to make in 3 hours.

△ **41.** The volume of a cone varies jointly as its height and the square of its radius. If the volume of a cone is 32π cubic inches when the radius is 4 inches and the height is 6 inches, find the volume of a cone when the radius is 3 inches and the height is 5 inches.

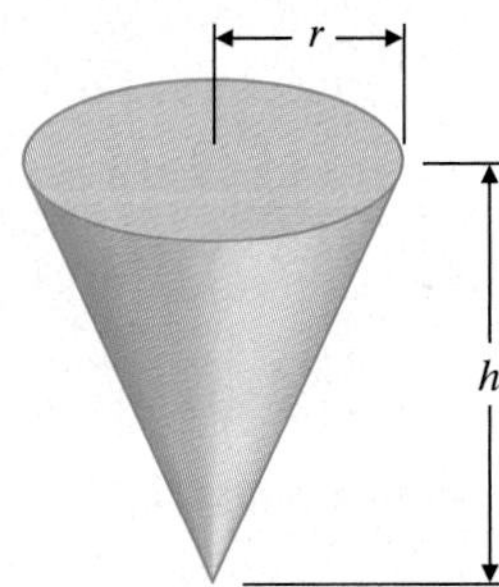

△ **42.** When a wind blows perpendicularly against a flat surface, its force is jointly proportional to the surface area and the speed of the wind. A sail whose surface area is 12 square feet experiences a 20-pound force when the wind speed is 10 miles per hour. Find the force on an 8-square-foot sail if the wind speed is 12 miles per hour.

43. The intensity of light (in foot-candles) varies inversely as the square of x, the distance in feet from the light source. The intensity of light 2 feet from the source is 80 foot-candles. How far away is the source if the intensity of light is 5 foot-candles?

44. The horsepower that can be safely transmitted to a shaft varies jointly as the shaft's angular speed of rotation (in revolutions per minute) and the cube of its diameter. A 2-inch shaft making 120 revolutions per minute safely transmits 40 horsepower. Find how much horsepower can be safely transmitted by a 3-inch shaft making 80 revolutions per minute.

Review and Preview

Determine whether the given ordered pair is a solution of both equations. See Section 3.1.

45. $(3, -1)$; $x - y = 4$
$x + 2y = 1$

46. $(0, 2)$; $x + 3y = 6$
$4x - y = -2$

47. $(-4, 0)$; $3x + 2y = -12$
$x = 4y$

48. $(-5, 2)$; $x + y = -3$
$2x - y = -8$

Combining Concepts

△ **49.** The volume of a cylinder varies jointly as the height and the square of the radius. If the height is halved and the radius is doubled, determine what happens to the volume.

50. The horsepower to drive a boat varies directly as the cube of the speed of the boat. If the speed of the boat is to double, determine the corresponding increase in horsepower required.

51. Suppose that y varies directly as x^2. If x is doubled, what is the effect on y?

52. Suppose that y varies directly as x. If x is doubled, what is the effect on y?

STUDY SKILLS REMINDER

Are you satisfied with your performance on a particular quiz or exam?

If not, analyze your quiz or exam like you would a good mystery novel. Look for common themes in your errors.

Were most of your errors a result of

- *Carelessness*? If your errors were careless, did you turn in your work before the allotted time expired? If so, resolve next time to use the entire time allotted. Any extra time can be spent checking your work.
- *Running out of time*? If so, make a point to better manage your time on your next exam. A few suggestions are to work any questions that you are unsure of last and to check your work after all of the questions have been answered.
- *Not understanding a concept*? If so, review that concept and correct your work. Remember next time to make sure that all concepts on a quiz or exam are understood before the exam.

CHAPTER 7 ACTIVITY

Measuring Slope

Materials:

- cardboard
- tape or glue
- ruler
- metal washer
- scissors
- string
- rug needle

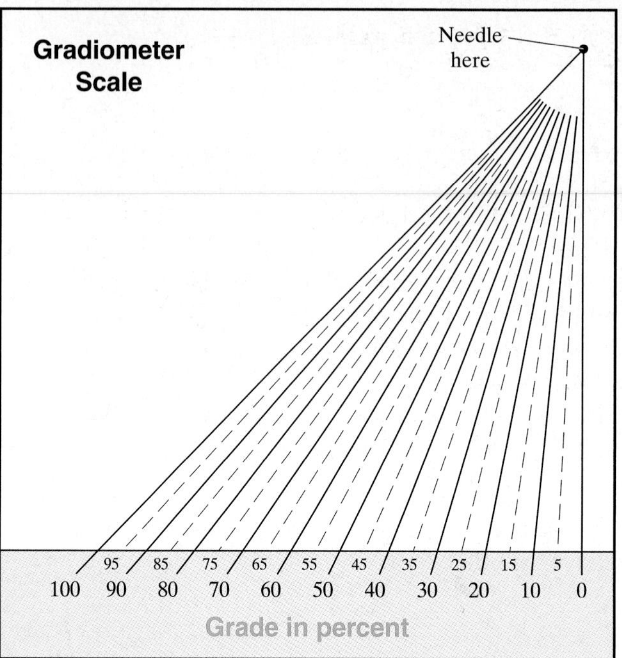

This activity may be completed by working in groups or individually.

The grade of an incline is the same as its slope given as a percent. A 6% grade means that for every horizontal run of 100 units, there is a vertical rise of 6 units. This can also be written as a slope of 0.06.

A gradiometer is a device that measures the grade of an incline. You can build your own gradiometer by following these steps

- Cut out the gradiometer scale given at the right.
- Attach the scale to a piece of rigid cardboard. Trim the cardboard so it is even with the bottom edge of the scale.
- Thread the rug needle with string. Poke the needle through the gradiometer scale and cardboard at the large dot in the upper-right corner.
- Tie a large knot in the portion of the string hanging out the back. Pull the string from the front until the knot blocks the hole.
- Tie the washer to the portion of the string hanging across the front of the scale so that, when the gradiometer is held upright, the washer hangs in the portion marked "Grade (percent)" at the bottom.
- Attach the gradiometer to a ruler (roughly in the middle) so the bottom edge of the gradiometer is aligned with the bottom edge of the ruler.
- To use your gradiometer to measure the grade of an incline, place the bottom edge of the ruler along the incline. The point at which the string attached to the washer crosses the scale at the bottom of the gradiometer corresponds to the grade of the incline. (See figure.)

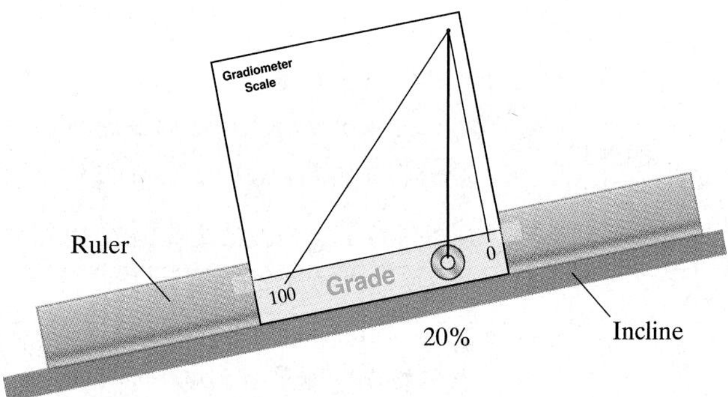

1. Refer to the graph on page 197. Express the slope of the line as a percent (grade). Use your gradiometer to measure the grade of the line. How close is your gradiometer reading to the actual grade? (*Hint:* You will need to hold the textbook upright on a flat surface to measure the grade of the line.)
2. Use your gradiometer to measure the grade of three inclines in your classroom. Interpret each measurement. (*Hint:* Consider measuring the inclines of desks, lecterns, ramps, or steps).
3. Notice that your gradiometer directly measures only positive slopes—inclines that rise from left to right. Explain how you could use your gradiometer to measure a negative slope (an incline that falls from left to right).
4. (Optional) According to the Americans with Disabilities Act (1990), handicapped-accessible ramps should have a grade of no more than 8.3%. Ramps with vertical rises greater than 6 inches and grades greater than 5% must provide handrails. Use your gradiometer to measure the grades of several wheelchair ramps on campus. Do they comply with the Americans with Disabilities Act guidelines?

Chapter 7 VOCABULARY CHECK

Fill in each blank with one of the words or phrases listed below.

relation	line	function	standard	slope	domain
slope-intercept	perpendicular	point-slope	range	parallel	linear function

1. A __________ is a set of ordered pairs.
2. The graph of every linear equation in two variables is a _______.
3. ___________ form of linear equation in two variables is $Ax + By = C$.
4. The ________ of a relation is the set of all second components of the ordered pairs of the relation.
5. __________ lines have the same slope and different y-intercepts.
6. ________________ form of a linear equation in two variables is $y = mx + b$.
7. A __________ is a relation in which each first component in the ordered pairs corresponds to exactly one second component.
8. In the equation $y = 4x - 2$, the coefficient of x is the ________ of its corresponding graph.
9. Two lines are _______________ if the product of their slopes is -1.
10. The __________ of a relation is the set of all first components of the ordered pairs of the relation.
11. A ________________ is a function that can be written in the form $f(x) = mx + b$.
12. The equation $y - 8 = -5(x + 1)$ is written in _____________ form.

CHAPTER 7 Highlights

DEFINITIONS AND CONCEPTS	EXAMPLES
Section 7.1 The Slope-Intercept Form	
We can use the slope-intercept form to write an equation of a line given its slope and y-intercept.	Write an equation of the line with y-intercept $(0, -1)$ and slope $\frac{2}{3}$. $y = mx + b$ $y = \frac{2}{3}x - 1$
Section 7.2 More Equations of Lines	
The **point-slope form** of the equation of a line is $y - y_1 = m(x - x_1)$ where m is the slope of the line and (x_1, y_1) is a point on the line.	Find an equation of the line with slope 2 containing the point $(1, -4)$. Write the equation in standard form: $Ax + By = C$. $y - y_1 = m(x - x_1)$ $y - (-4) = 2(x - 1)$ $y + 4 = 2x - 2$ $-2x + y = -6$ Standard form

DEFINITIONS AND CONCEPTS — EXAMPLES

Section 7.3 Introduction to Functions

A **relation** is a set of ordered pairs. The **domain** of the relation is the set of all first components of the ordered pairs. The **range** of the relation is the set of all second components of the ordered pairs.

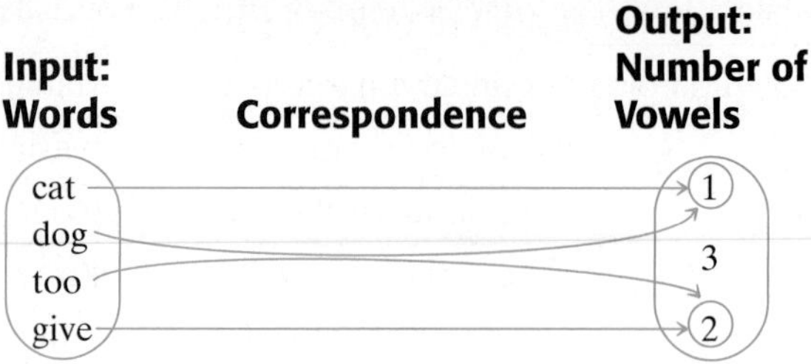

Domain: {cat, dog, too, give}
Range: {1, 2}

A **function** is a relation in which each element of the first set corresponds to exactly one element of the second set.

The previous relation is a function. Each word contains one exact number of vowels.

VERTICAL LINE TEST

If no vertical line can be drawn so that it intersects a graph more than once, the graph is the graph of a function. If such a line can be drawn, the graph is not that of a function.

Find the domain and the range of the relation. Also determine whether the relation is a function.

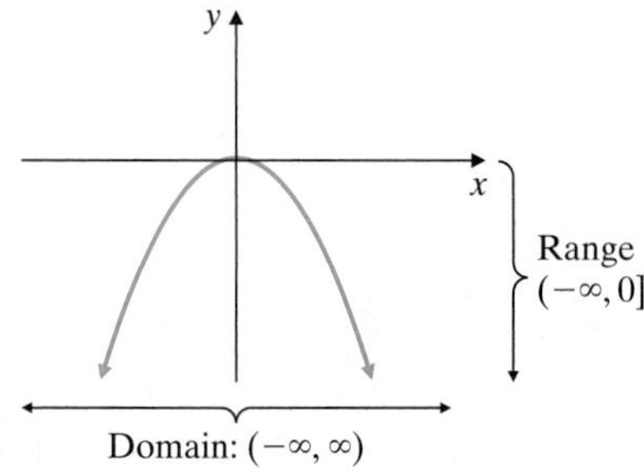

By the vertical line test, this is the graph of a function.

The symbol $f(x)$ means **function of *x*** and is called **function notation**.

If $f(x) = 2x^2 - 5$, find $f(-3)$.

$$f(-3) = 2(-3)^2 - 5 = 2(9) - 5 = 13$$

A **linear function** is a function that can be written in the form

$$f(x) = mx + b$$

$$f(x) = -3, g(x) = 5x, h(x) = -\frac{1}{3}x - 7$$

To graph a linear function, use the slope and y-intercept.

Graph: $f(x) = -2x$
(or $y = -2x + 0$)

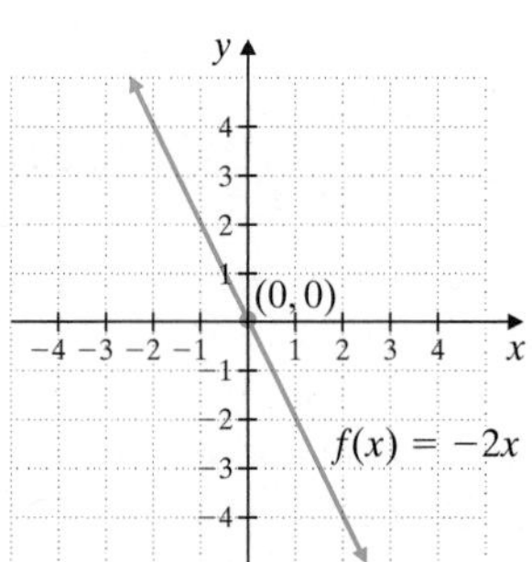

The slope is $\frac{2}{-1}$.

The y-intercept is $(0, 0)$.

DEFINITIONS AND CONCEPTS	EXAMPLES
Section 7.4 Polynomial and Rational Functions	
A function P is a **polynomial function** if $P(x)$ is a polynomial.	For the polynomial function $P(x) = -x^2 + 6x - 12$ find $P(-2)$. $P(-2) = -(-2)^2 + 6(-2) - 12 = -28$
A **rational function** is a function described by a rational expression.	$f(x) = \frac{2x - 6}{7}$, $h(t) = \frac{t^2 - 3t + 5}{t - 1}$
Section 7.5 Variation and Problem Solving	
y **varies directly** as x, or y is **directly proportional** to x, if there is a nonzero constant k such that $y = kx$	The circumference of a circle C varies directly as its radius r. $C = \underbrace{2\pi}_{k} r$
y **varies inversely** as x, or y is **inversely proportional** to x, if there is a nonzero constant k such that $y = \frac{k}{x}$	Pressure P varies inversely with volume V. $P = \frac{k}{V}$
y **varies jointly** as x and z, or y is **jointly proportional** to x and z, if there is a nonzero constant k such that $y = kxz$	The lateral surface area S of a cylinder varies jointly as its radius r and height h. $S = \underbrace{2\pi}_{k} rh$

FOCUS ON History

CARTESIAN COORDINATE SYSTEM

The French mathematician and philosopher Rene Descartes (1596–1650) is generally credited with devising the rectangular coordinate system that we use in mathematics today. It is said that Descartes thought of describing the location of a point in a plane using a fixed frame of reference while watching a fly crawl on his ceiling as he laid in bed one morning meditating. He incorporated this idea of defining a point's position in the plane by giving its distances, x and y, to two fixed axes in his text *La Géométrie.*

Although Descartes is credited with the concept of the rectangular coordinate system, nowhere in his written works does an example of the modern grid-like coordinate system appear. He also never referred to a point's location as we do today with (x, y)-notation giving an ordered pair of coordinates. In fact, Descartes never even used the term *coordinate*! Instead, his basic ideas were expanded upon by later mathematicians. The Dutch mathematician Frans van Schooten (1615–1660) is credited with making Descartes' concept of a coordinate system widely accepted with his text, *Geometria a Renato Des Cartes* (*Geometry by René Descartes*). The German mathematician Gottfried Wilhelm Leibniz (1646–1716) later contributed the terms *abscissa* (the x-axis in the modern rectangular coordinate system), ordinate (the y-axis), and coordinate to the development of the rectangular coordinate system.

FOCUS ON Business and Career

Linear Modeling

As we saw in Section 7.1, businesses often depend on equations that "closely fit" data. To *model* the data means to find an equation that describes the relationship between the paired data of two variables, such as time in years and profit. A model that accurately summarizes the relationship between two variables can be used to replace a potentially lengthy listing of the raw data. An accurate model might also be used to predict future trends by answering questions such as "If the trend seen in our company's performance in the last several years continues, what level of profit can we reasonably expect in 3 years?"

There are several ways to find a linear equation that models a set of data. If only two ordered pair data points are involved, an exact equation that contains both points can be found using the methods of Section 3.4. When more than two ordered pair data points are involved, it may be impossible to find a linear equation that contains all of the data points. In this case, the graph of the **best fit equation** should have a majority of the plotted ordered pair data points on the graph or close to it. In statistics, a technique called least squares regression is used to determine an equation that best fits a set of data. Various graphing utilities have built-in capabilities for finding an equation (called a regression equation) that best fits a set of ordered pair data points. Regression capabilities are often found with a graphing utility's statistics features.* A best fit equation can also be estimated using an algebraic method, which is outlined in the Group Activity below. In either case, a useful first step when finding a linear equation that models a set of data is creating a scatter diagram of the ordered pair data points to verify that a linear equation is an appropriate model.

Group Activity

Phoenix Sky Harbor International Airport, located in Phoenix, Arizona, is the fifth busiest airport in the United States in terms of aircraft operations. In terms of passenger traffic, Phoenix Sky Harbor is the ninth busiest airport in the United States. The table shows the total passenger traffic (in millions) for the years 1994 through 2000. Use the table along with your answers to the questions below to find a linear equation $y = mx + b$ that represents total passenger traffic y (in millions) at Phoenix Sky Harbor International Airport, where x represents the number of years after 1994.

Year	1994	1995	1996	1997	1998	1999	2000
Total Passenger Traffic (in millions)	26	28	30	31	32	34	36

(*Source:* The City of Phoenix Aviation Department)

1. Create a scatter diagram of the paired data given in the table. Does a linear model seem appropriate for the data?
2. Use a straightedge to draw on your graph what appears to be the line that "best fits" the data you plotted.
3. Estimate the coordinates of two points that fall on your best fit line. Use these points to find the equation of the line that passes through both points.
4. Use this equation to find the value of y for $x = 11$. Interpret the meaning of this pair of data.
5. How could this equation be useful to those who operate Phoenix Sky Harbor International Airport?
6. Compare your group's linear equation with other groups' equations. Are they the same or different? Explain why.
7. (Optional) Enter the data from the table into a graphing utility and use the linear regression feature to find a linear equation that models the data. Compare this equation with the one you found in Question 3. How are they alike or different?
8. (Optional) Using corporation annual reports or articles from magazines or newspapers, search for a set of business-related data that could be modeled with a linear equation. Explain how modeling this data could be useful to a business. Then find the best fit equation for the data.

*To find out more about using a graphing utility to find a regression equation, consult the user's manual for your graphing utility.

Name ______________________ Section __________ Date __________

Chapter 7 Review

(7.1) *Graph each line passing through the given point with the given slope.*

1. Through $(2, -3)$ with slope $\frac{2}{3}$

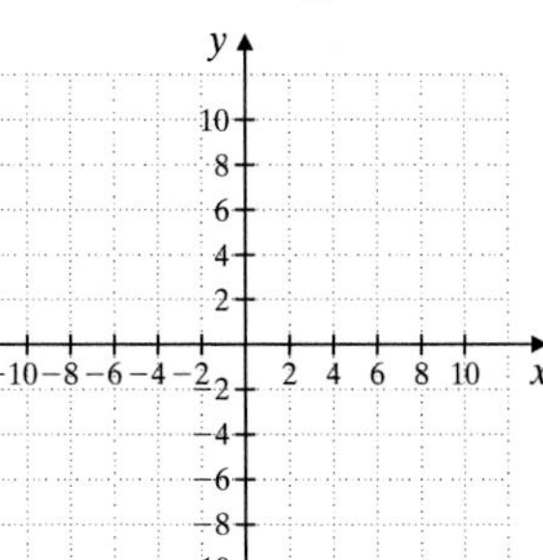

2. Through $(1, -4)$ with slope $\frac{1}{2}$

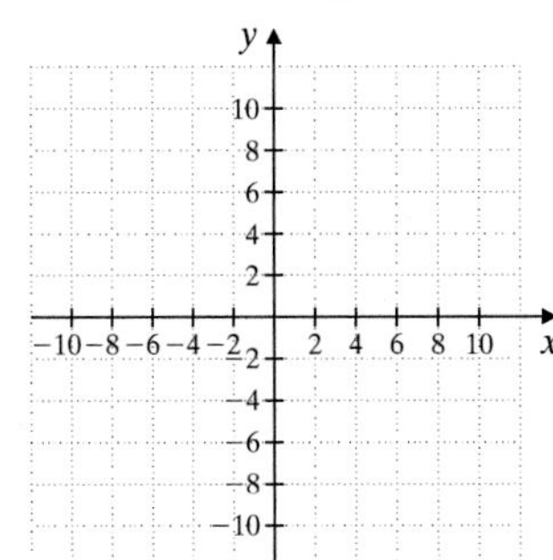

3. Through $(0, 1)$ with slope 2

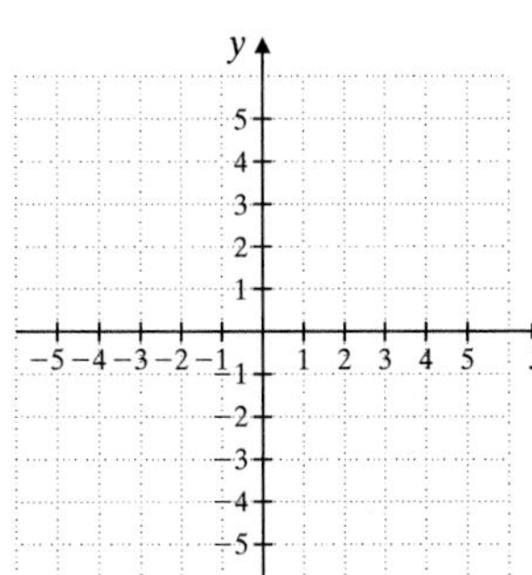

4. Through $(-2, 0)$ with slope -3

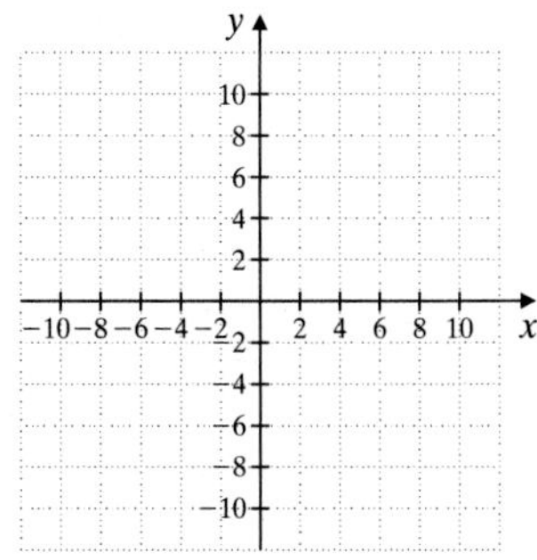

Graph each linear equation using the slope and y-intercept.

5. $y = -x + 1$

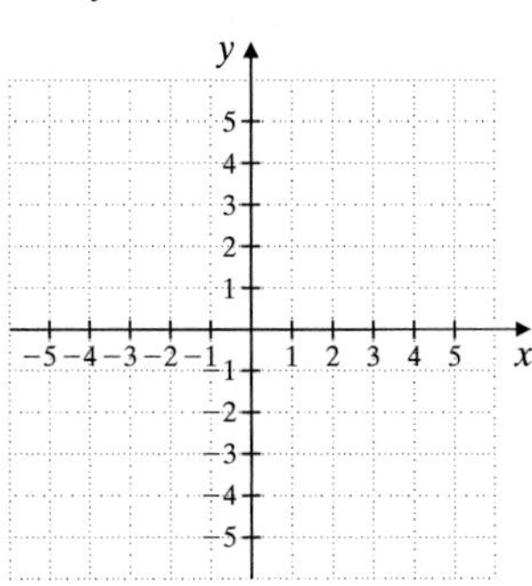

6. $y = 4x - 3$

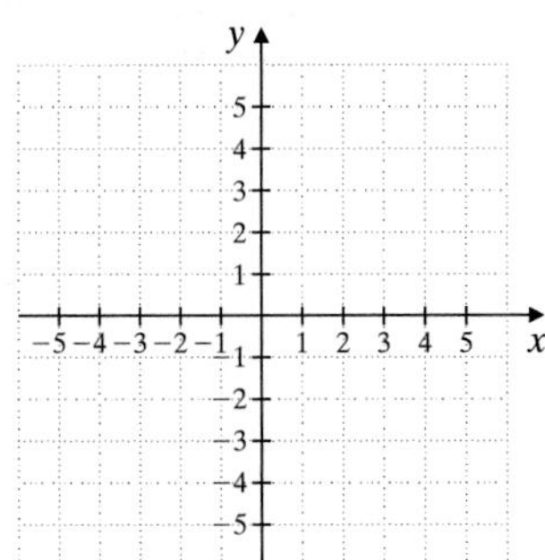

7. $3x - y = 6$

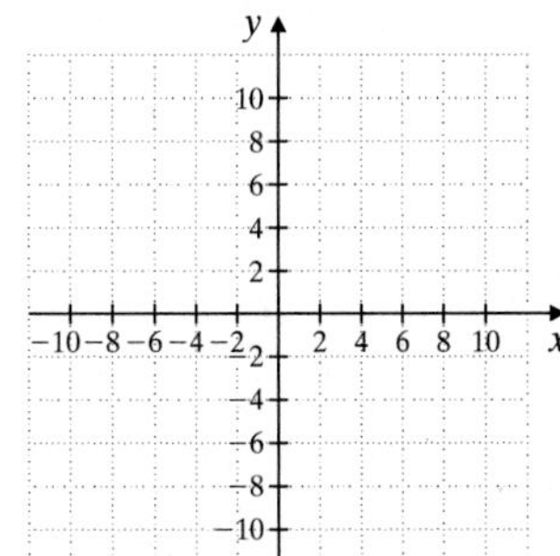

8. $y = -5x$

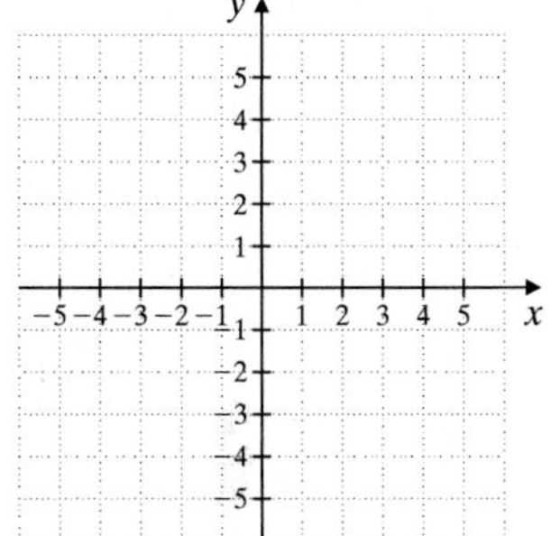

9. The cost C, in dollars, of renting a minivan for a day is given by the linear equation $C = 0.3x + 42$, where x is number of miles driven.

a. Find the cost of renting the minivan for a day and driving it 150 miles.

b. Find and interpret the slope of this equation.

c. Find and interpret the y-intercept of this equation.

(7.2) *Write an equation of the line satisfying each set of conditions.*

10. Horizontal; through $(3, -1)$

11. Vertical; through $(-2, -4)$

12. Slope undefined; through $(-4, -3)$

13. Slope 0; through $(2, 5)$

Write the equation of the line satisfying each set of conditions. Write the equation in the form $y = mx + b$.

14. Through $(-3, 5)$; slope 3

15. Slope 2; through $(5, -2)$

16. Through $(-6, -1)$ and $(-4, -2)$

17. Through $(-5, 3)$ and $(-4, -8)$

18. Through $(2, -6)$; parallel to $y = -2x + 3$

19. Through $(-4, -2)$; parallel to $y = -\frac{3}{2}x + 1$

20. Through $(-6, -1)$; perpendicular to $4x + 3y = 5$

21. Through $(-4, 5)$; perpendicular to $2x - 3y = 6$

22. The value of a computer bought in 1996 continues to depreciate, or decrease, as time passes. Two years after the computer was bought, it was worth \$2600; four years after it was bought, it was worth \$1000.

a. Assuming that this relationship between the number of years past 1996 and value of computer is linear, write an equation describing this relationship. [*Hint*: Use ordered pairs of the form (years past 1996, value of computer).]

b. Use this equation to estimate the value of the computer in 2001.

23. The value of a building bought in 1980 continues to appreciate, or increase, as time passes. Seven years after the building was bought, it was worth \$165,000; 12 years after it was bought, it was worth \$180,000.

a. Assuming that this relationship between the number of years past 1980 and value of the building is linear, write an equation describing this relationship. [*Hint:* Use ordered pairs of the form (years past 1980, value of building).]

b. Use this equation to estimate the value of the building in 2005.

(7.3) *Find the domain and range for each relation. Then determine whether the relation is also a function.*

24. $\left\{\left(-\frac{1}{2}, \frac{3}{4}\right), (6, 0.65), (0, -12), (25, 25)\right\}$

25. $\left\{\left(\frac{3}{4}, -\frac{1}{2}\right), (0.65, 6), (-12, 0), (25, 25)\right\}$

26.

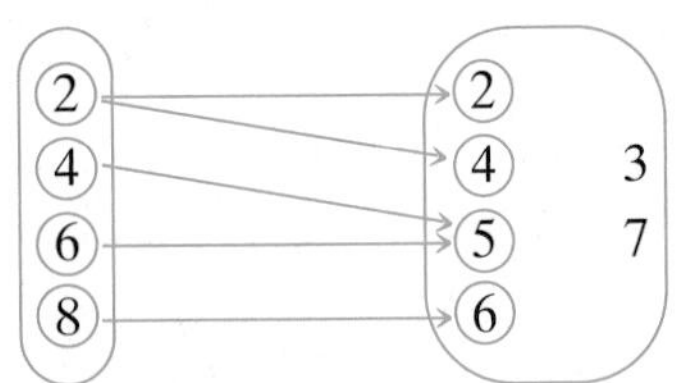

27.

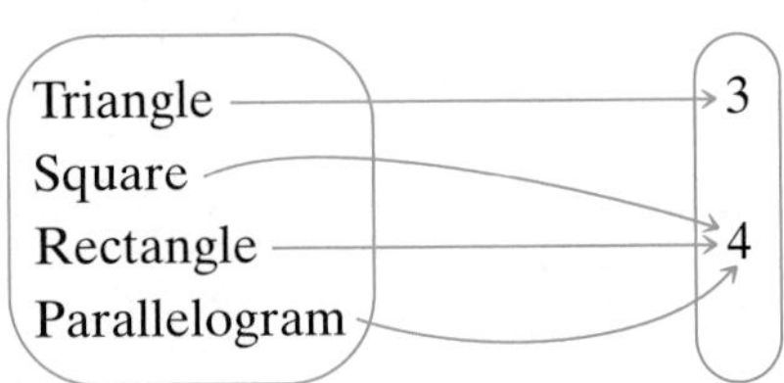

28.  **29.**  **30.** **31.**

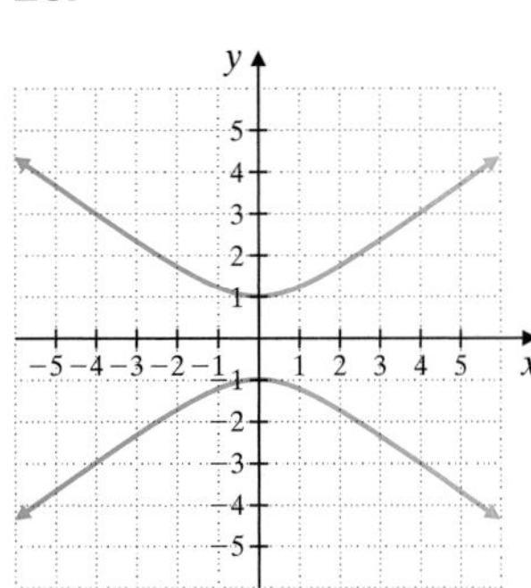

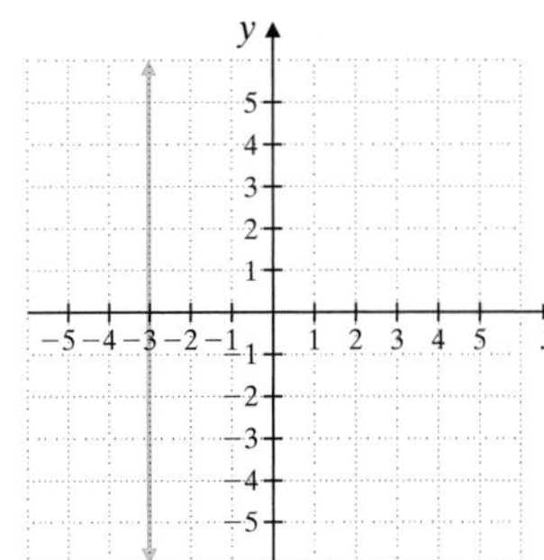

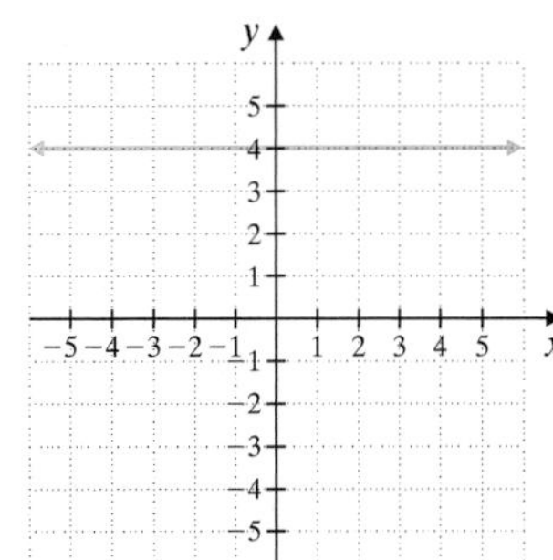

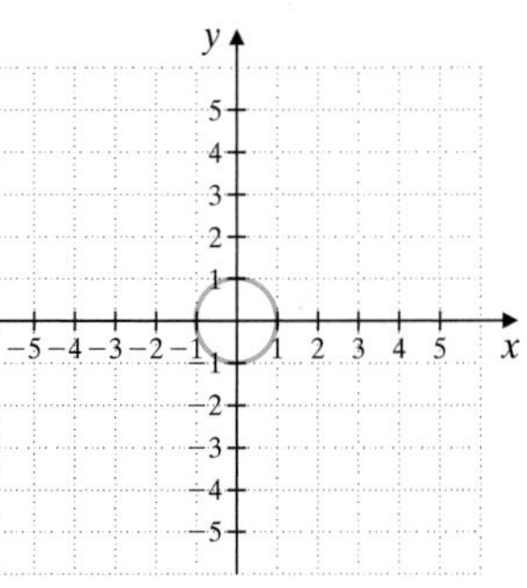

If $f(x) = x - 5$, $g(x) = -3x$, *and* $h(x) = 2x^2 - 6x + 1$, *find each function value.*

32. $f(2)$

33. $g(0)$

34. $g(-6)$

35. $h(-1)$

36. $h(1)$

37. $f(5)$

The function $J(x) = 2.54x$ *may be used to approximate the weight of an object on Jupiter* (J) *given its weight on Earth* (x).

38. If a person weighs 150 pounds on Earth, find the equivalent weight on Jupiter.

39. A 2000-pound probe on Earth weighs how many pounds on Jupiter?

Graph each linear function.

40. $f(x) = x + 2$

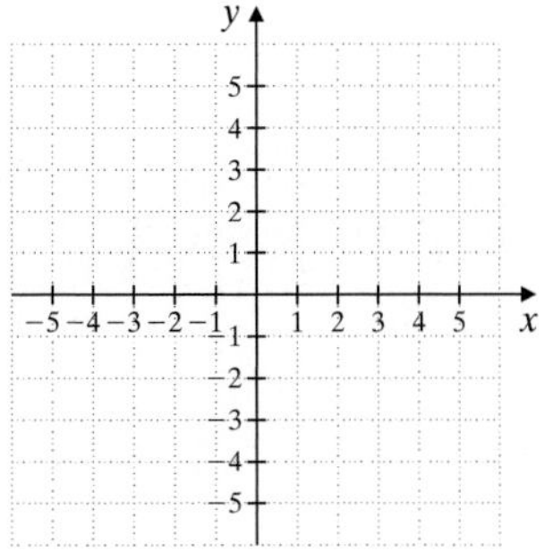

41. $f(x) = -\frac{1}{2}x + 3$

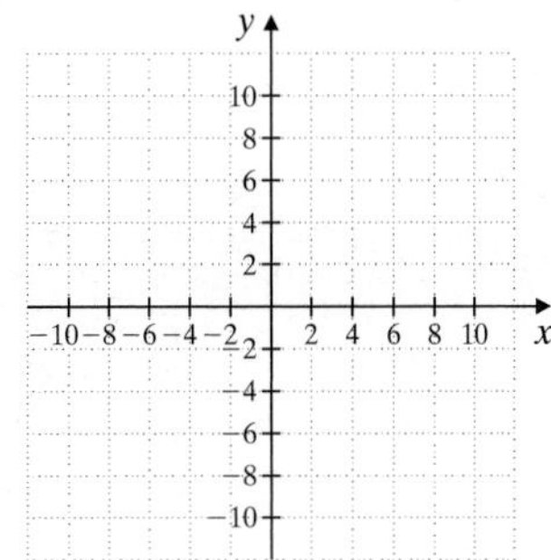

(7.4) *If* $p(x) = 9x^2 - 7x + 8$, *find each function value.*

42. $P(6)$

43. $P(-2)$

44. $P(-3)$

If $P(x) = 2x - 1$ *and* $Q(x) = x^2 + 2x - 5$, *find each function.*

45. $P(x) + Q(x)$

46. $2P(x) - Q(x)$

47. The average cost of manufacturing x bookcases is given by the rotional function.

$$C(x) = \frac{35x + 4200}{x}$$

a. Find the average cost per bookcase of manufacturing 50 bookcases.

b. Find the average cost per bookcase of manufacturing 100 bookcases.

c. As the number of bookcases increases, does the average cost per bookcase increase or decrease? (See parts (a) and (b).)

(7.5) *Solve each variation problem.*

48. A is directly proportional to B. If $A = 6$ when $B = 14$, find A when $B = 21$.

49. C is inversely proportional to D. If $C = 12$ when $D = 8$, find C when $D = 24$.

50. According to Boyle's law, the pressure exerted by a gas is inversely proportional to the volume, as long as the temperature stays the same. If a gas exerts a pressure of 1250 pounds per square inch when the volume is 2 cubic feet, find the volume when the pressure is 800 pounds per square inch.

△ **51.** The surface area of a sphere varies directly as the square of its radius. If the surface area is 36π square inches when the radius is 3 inches, find the surface area when the radius is 4 inches.

Name ______________________ Section __________ Date __________

Chapter 7 Test

1. Find the slope and the y-intercept of the line $3x + 12y = 8$.

Match each equation with its graph.

2. $f(x) = 3x + 1$

3. $f(x) = 3x - 2$

4. $f(x) = 3x + 2$

5. $f(x) = 3x - 5$

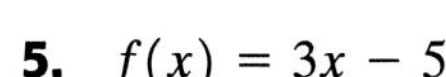

A

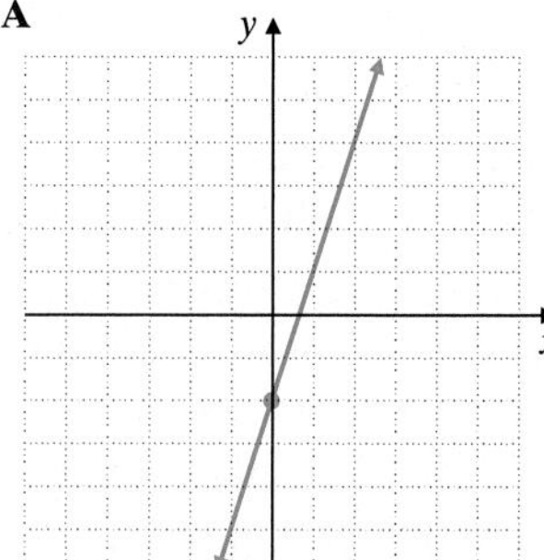

B

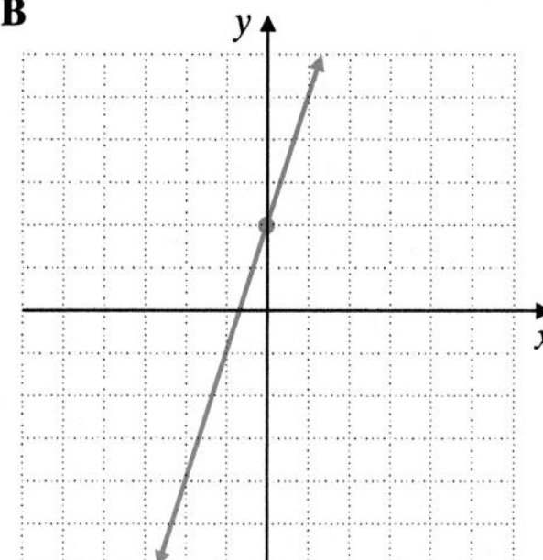

C

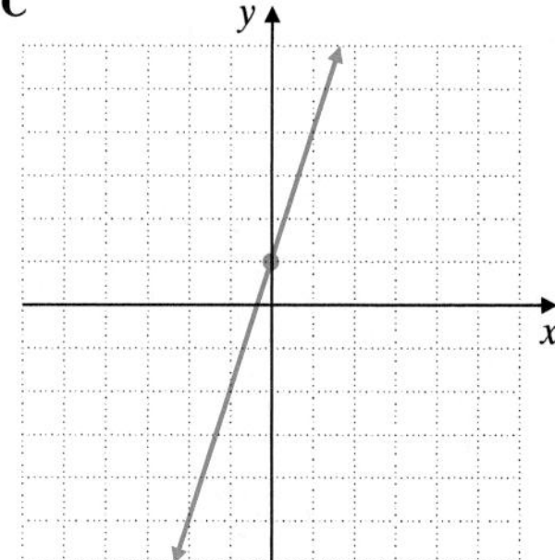

D

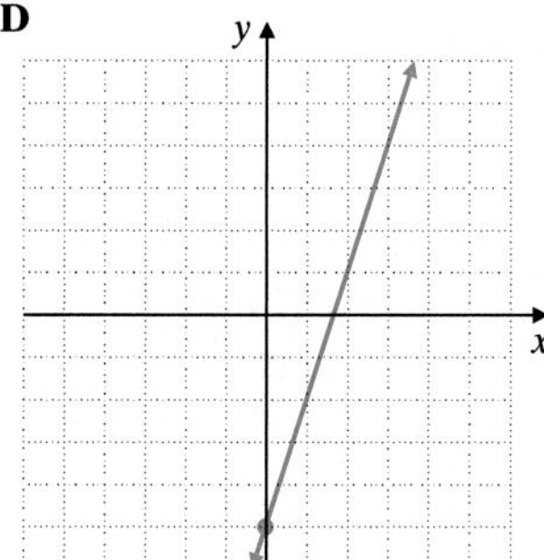

Find an equation of the line satisfying each set of conditions. Write the equations in the form $x = a$, $y = b$, or $y = mx + b$.

6. Horizontal; through $(2, -8)$

7. Vertical; through $(-4, -3)$

8. Perpendicular to $x = 5$; through $(3, -2)$

9. Through $(4, -1)$; slope -3

10. Through $(0, -2)$; slope 5

11. Through $(4, -2)$ and $(6, -3)$

12. Through $(-1, 2)$; perpendicular to $3x - y = 4$

13. Parallel to $2y + x = 3$; through $(3, -2)$

Answers

1. ______
2. ______
3. ______
4. ______
5. ______
6. ______
7. ______
8. ______
9. ______
10. ______
11. ______
12. ______
13. ______

14. ______

15. a. ______

b. ______

c. ______

d. ______

16. ______

17. ______

18. ______

19. ______

14. Line L_1 has the equation $2x - 5y = 8$. Line L_2 passes through the points $(1, 4)$ and $(-1, -1)$. Determine whether these lines are parallel, perpendicular, or neither.

15. For the 2000 Major League Baseball season, the following linear equation describes the relationship between a team's opening-day player payroll x (in millions of dollars) and the number of games y that team won during the regular season: $y = 0.154x + 72.379$. Round answers to the nearest wholes. (*Sources*: Based on data from The Associated Press and Major League Baseball)

a. According to this equation, how many games would have been won during the 2000 season by a team with an opening-day payroll of \$60 million?

b. The Arizona Diamondbacks had an opening-day payroll of \$81 million in 2000. According to this equation, how many games would they have won during the season?

c. According to this equation, what opening-day payroll would have been necessary in 2000 to have won 90 games during the season?

d. Find and interpret the slope of the equation.

Find the domain and range of each relation. Also determine whether the relation is a function.

16.

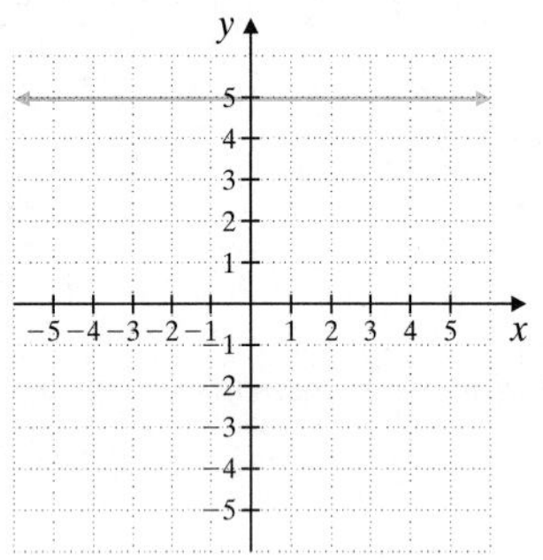

17.

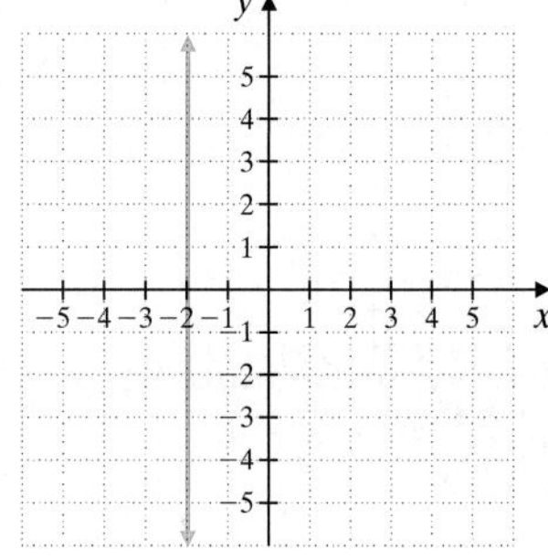

18.

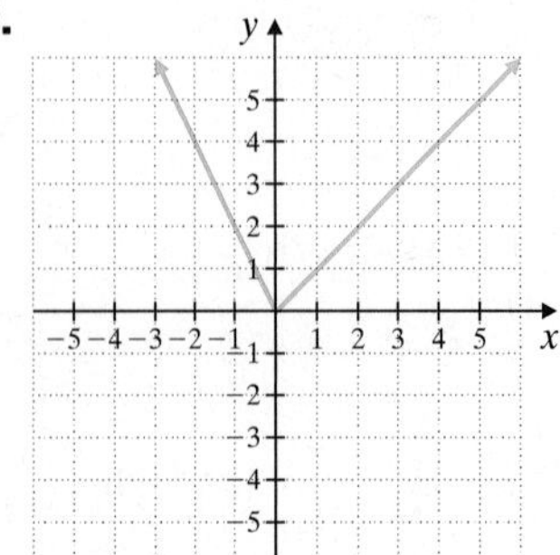

19.

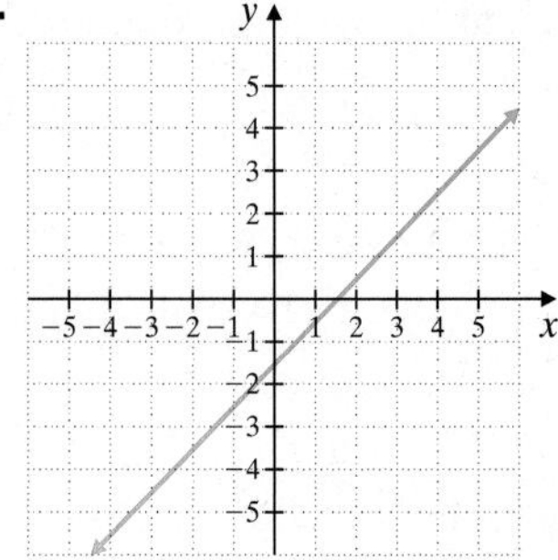

20. A pebble is hurled upward from the top of the 880-foot-tall Canada Trust Tower with an initial velocity of 96 feet per second. Neglecting air resistance, the height of the pebble after t seconds is given by the polynomial function.

$$h(t) = -16t^2 + 96t + 880$$

a. Find $h(1)$.

b. Find $h(5.1)$.

c. Find $h(11)$.

21. If $f(x) = \dfrac{5x^2}{1 - x}$, find $f(2)$.

22. Suppose that W is inversely proportional to V. If $W = 20$ when $V = 12$, find W when $V = 15$.

23. Suppose that Q is jointly proportional to R and the square of S. If $Q = 24$ when $R = 3$ and $S = 4$, find Q when $R = 2$ and $S = 3$.

24. When an anvil is dropped into a gorge, the speed at which it strikes the ground is directly proportional to the square root of the distance it falls. An anvil that falls 400 feet hits the ground at a speed of 160 feet per second. Find the height of a cliff over the gorge if a dropped anvil hits the ground at a speed of 128 feet per second.

20. a. ____________

b. ____________

c. ____________

21. ____________

22. ____________

23. ____________

24. ____________

STUDY SKILLS REMINDER

How are your homework assignments going?

It is so important in mathematics to keep up with homework. Why? Many concepts build on each other. Often, your understanding of a day's lecture in mathematics depends on an understanding of the previous day's material.

Remember that completing your homework assignment involves a lot more than attempting a few of the problems assigned.

To complete a homework assignment, remember these four things:

1. Attempt all of it.
2. Check it.
3. Correct it.
4. If needed, ask questions about it.

FOCUS ON The Real World

Road Grades

Have you ever driven on a hilly highway and seen a sign like the one below? The 7% on the sign refers to the grade of the road. The grade of a road is the same as its slope given as a percent. A 7 percent grade means that for every rise of 7 units there is a run of 100 units. The type of units doesn't matter as long as they are the same. For instance, we could say that a 7 percent grade represents a rise of 7 feet for every run of 100 feet or a rise of 7 meters for every run of 100 meters, and so on.

$$7\% = 0.07 = \frac{7}{100} \qquad 7\% \text{ grade} = \frac{\text{rise of } 7}{\text{run of } 100}$$

Most highways are designed to have grades of 6 percent or less. If a portion of a highway has a grade that is steeper than 6 percent, a sign is usually posted giving the grade and the number of miles for the grade. Truck drivers need to know when the road is particularly steep. They may need to take precautions such as using a different gear, reducing their speed, or testing their brakes.

Here is a sampling of road grades:

- A portion of the John Scott Highway in Steubenville, Ohio, has a 10% grade. (*Source:* Ohio Department of Transportation)
- Joaquin Road in Portola Valley, California, has an average grade of 15%. (*Source:* Western Wheelers Bicycle Club)
- The steepest grade in Seattle, Washington, is 26% on East Roy Street between 25th Avenue and 26th Avenue. (*Source: Seattle Post-Intelligencer*, Nov. 21, 1994)
- The steepest street in Pittsburgh, Pennsylvania, is Canton Avenue with a 37% grade. (*Source:* Pittsburgh Department of Public Works)
- The steepest street in the world is Baldwin Street in Dunedin, New Zealand. Its maximum grade is 79%. (*Source: Guinness Book of Records*, 1996)

Cooperative Learning Activity

Try to find a road sign with a percent grade warning or the name and grade of a steep road in your area. Describe its slope and make a scale drawing to represent its grade.

Name ____________ Section ________ Date ________

Cumulative Review

1. Simplify each expression.

   a. $-14 - 8 + 10 - (-6)$

   b. $1.6 - (-10.3) + (-5.6)$

Find the reciprocal of each number.

2. 22

3. $\frac{3}{16}$

4. -10

5. $-\frac{9}{13}$

6. 1.7

7. a. The sum of two numbers is 8. If one number is 3, find the other number.

   b. The sum of two numbers is 8. If one number is x, write an expression representing the other number.

8. Solve:

   $-2(x - 5) + 10 = -3(x + 2) + x$

9. Write a ratio for each phrase. Use fractional notation.

   a. The ratio of 2 parts salt to 5 parts water

   b. The ratio of 18 inches to 2 feet

Graph each set on a number line and then write it in interval notation.

10. $\{x|x \geq 2\}$

−5 −4 −3 −2 −1 0 1 2 3 4 5

11. $\{x|0.5 < x \leq 3\}$

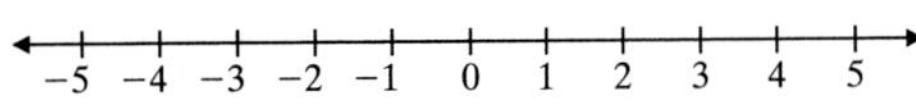

12. Graph the linear equation $y = 3x$.

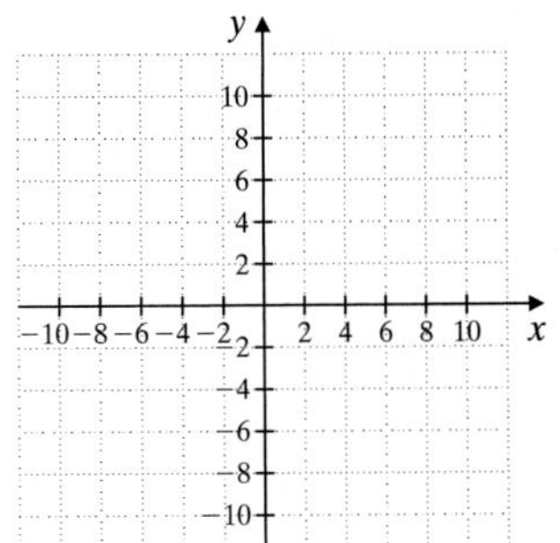

13. Graph the linear equation $y = -2$.

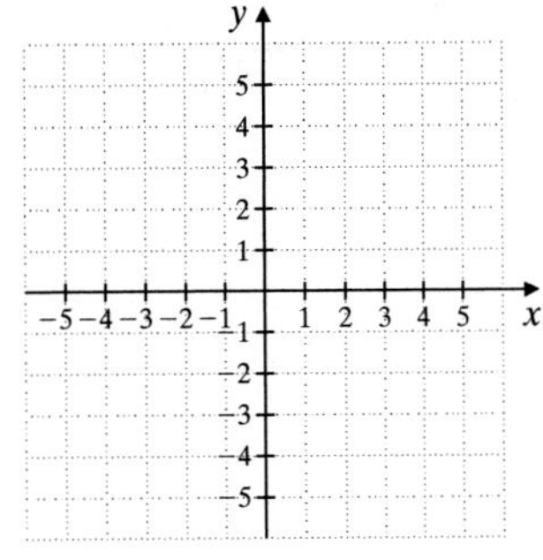

Answers

1. a. ____________

   b. ____________

2. ____________

3. ____________

4. ____________

5. ____________

6. ____________

7. a. ____________

   b. ____________

8. ____________

9. a. ____________

   b. ____________

10. ____________

11. ____________

12. ____________

13. ____________

14. ______________

15. ______________

16. ______________

17. ______________

18. ______________

19. ______________

20. ______________

21. ______________

22. ______________

23. ______________

24. ______________

25. ______________

Simplify each expression.

14. $\left(\dfrac{m}{n}\right)^7$

15. $\left(\dfrac{2x^4}{3y^5}\right)^4$

16. Subtract: $(2x^3 + 8x^2 - 6x) - (2x^3 - x^2 + 1)$

17. Divide $6x^2 + 10x - 5$ by $3x - 1$.

18. Solve: $x(2x - 7) = 4$

19. Find the lengths of the sides of a right triangle if the lengths can be expressed by three consecutive even integers.

20. Subtract: $\dfrac{2y}{2y - 7} - \dfrac{7}{2y - 7}$

21. Simplify: $\dfrac{\dfrac{x}{y} + \dfrac{3}{2x}}{\dfrac{x}{2} + y}$

22. Write an equation of the line with y-intercept $(0, -3)$ and slope of $\dfrac{1}{4}$.

23. Find an equation of the horizontal line containing the point $(2, 3)$.

Find the following.

24. If $f(x) = 7x^2 - 3x + 1$, find $f(1)$.

25. If $g(x) = 3x - 2$, find $g(0)$.

Systems of Equations and Inequalities

CHAPTER 8

In this chapter, two or more equations in two or more variables are solved simultaneously. Such a collection of equations is called a **system of equations**. Systems of equations are good mathematical models for many real-world problems because these problems may involve several related patterns. We will study various methods for solving systems of equations and will conclude with a look at systems of inequalities.

Today, television is a fact of life. In 1946, a year after World War II ended, the television industry came to life. Broadcasting was initially dominated by two radio companies, Columbia Broadcasting System (CBS) and National Broadcasting Company (NBC). Originally, the Federal Communications Commission (FCC) restricted television broadcasts in the U.S. to the very high frequency (VHF) channels 2–13. The use of channels 14–83, the ultra-high frequency (UHF) channels, was banned. Competition for the VHF channels was fierce, and together three major networks. ABC (American Broadcasting Company), CBS, and NBC, controlled 95% of viewership well into the 1970s. The FCC gradually began making UHF channels available for broadcasting and the number of television stations (particularly UHF stations) exploded. Exercise 53 on page 587 uses a system of equations to find the year in which the number of UHF stations caught up to and surpassed the number of VHF stations in the U.S.

Answers
1. see graph
2. see graph
3.
4.
5.
6.
7.
8.
9.
10.
11.
12.
13.
14.
15. see graph
16. see graph
17.
18.

Name ______________________ Section __________ Date __________

Chapter 8 Pretest

Solve each system by graphing.

1. $\begin{cases} x + y = 5 \\ x - y = 7 \end{cases}$

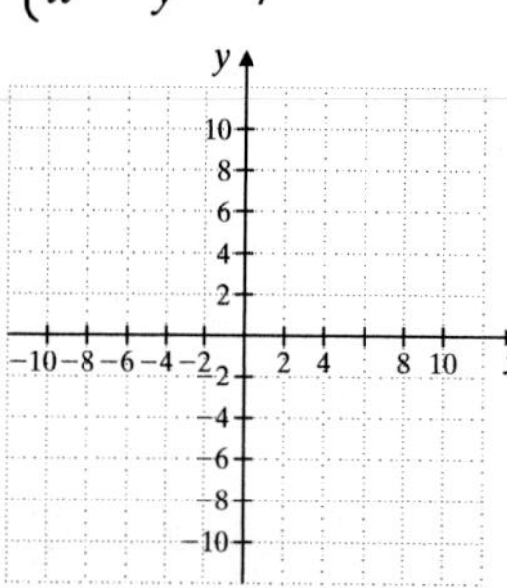

2. $\begin{cases} y = 4x \\ x = 1 \end{cases}$

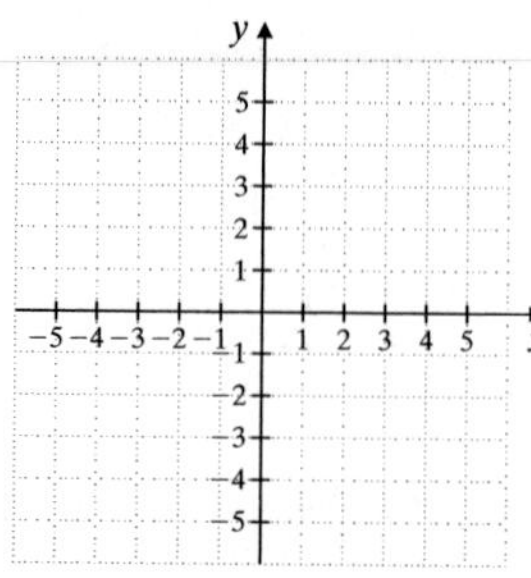

Solve each system using the substitution method.

3. $\begin{cases} x + y = 6 \\ x = 3y - 2 \end{cases}$

4. $\begin{cases} 5x + y = 13 \\ 4x - 5y = 22 \end{cases}$

5. $\begin{cases} 4x = y + 6 \\ 8x - 2y = 12 \end{cases}$

Solve each system by the addition method.

6. $\begin{cases} 4x - 3y = 13 \\ 5x - 9y = 53 \end{cases}$

7. $\begin{cases} \dfrac{x}{2} + \dfrac{y}{3} = 2 \\ \dfrac{x}{6} - \dfrac{y}{4} = 5 \end{cases}$

8. $\begin{cases} 6x + 10y = -4 \\ -x + y = -1 \end{cases}$

9. $\begin{cases} x - y = 3 \\ 7y = -7 \\ 2x + y + 3z = -6 \end{cases}$

10. $\begin{cases} 3x - y + 2z = -1 \\ 2x + 5y - z = 4 \\ 4x + 6y + z = 2 \end{cases}$

Use matrices to solve each system.

11. $\begin{cases} 3x + 4y = -10 \\ x - y = 20 \end{cases}$

12. $\begin{cases} -2x + y = 6 \\ 4x - 2y = 12 \end{cases}$

13. $\begin{cases} 3x + 2z = -5 \\ x + y + z = -2 \\ -x + 2y - 3z = -5 \end{cases}$

14. $\begin{cases} x - 6y + z = -48 \\ 5x + y + 3z = 8 \\ 2x - y - z = -8 \end{cases}$

Graph the solution of each system of linear inequalities.

15. $\begin{cases} y \leq x + 1 \\ y > 3x - 2 \end{cases}$

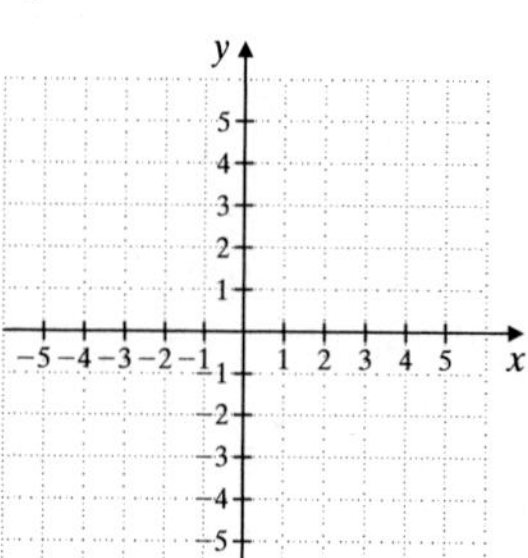

16. $\begin{cases} -6x + 3y \geq 0 \\ y \leq 3 \end{cases}$

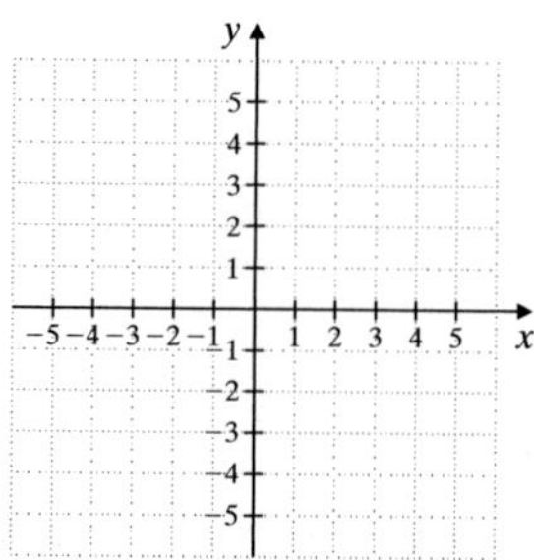

Solve.

17. Six times one number minus a second is 12, and the sum of the numbers is 16. Find the numbers.

18. The measure of the largest angle of a triangle is five times the measure of the smallest angle, and the measure of the remaining angle is 40° more than the measure of the smallest angle. Find the measure of each angle.

8.1 Solving Systems of Linear Equations by Graphing

OBJECTIVES

- **A** Decide whether an ordered pair is a solution of a system of linear equations.
- **B** Solve a system of linear equations by graphing.
- **C** Identify special systems: those with no solution and those with an infinite number of solutions.

A **system of linear equations** consists of two or more linear equations. In this section, we focus on solving systems of linear equations containing two equations in two variables. Examples of such linear systems are

$$\begin{cases} 3x - 3y = 0 \\ x = 2y \end{cases} \qquad \begin{cases} x - y = 0 \\ 2x + y = 10 \end{cases} \qquad \begin{cases} y = 7x - 1 \\ y = 4 \end{cases}$$

A Deciding Whether an Ordered Pair Is a Solution

A **solution** of a system of two equations in two variables is an ordered pair of numbers that is a solution of both equations in the system.

EXAMPLE 1 Determine whether $(12, 6)$ is a solution of the system

$$\begin{cases} 2x - 3y = 6 \\ x = 2y \end{cases}$$

Solution: To determine whether $(12, 6)$ is a solution of the system, we replace x with 12 and y with 6 in both equations.

$2x - 3y = 6$	First equation	$x = 2y$	Second equation
$2(12) - 3(6) \stackrel{?}{=} 6$	Let $x = 12$ and $y = 6$.	$12 \stackrel{?}{=} 2(6)$	Let $x = 12$ and $y = 6$.
$24 - 18 \stackrel{?}{=} 6$	Simplify.	$12 = 12$	True
$6 = 6$	True		

Since $(12, 6)$ is a solution of both equations, it is a solution of the system. ●

Practice Problem 1

Determine whether $(3, 9)$ is a solution of the system

$$\begin{cases} 5x - 2y = -3 \\ y = 3x \end{cases}$$

EXAMPLE 2 Determine whether $(-1, 2)$ is a solution of the system

$$\begin{cases} x + 2y = 3 \\ 4x - y = 6 \end{cases}$$

Solution: We replace x with -1 and y with 2 in both equations.

$x + 2y = 3$	First equation	$4x - y = 6$	Second equation
$-1 + 2(2) \stackrel{?}{=} 3$	Let $x = -1$ and $y = 2$.	$4(-1) - 2 \stackrel{?}{=} 6$	Let $x = -1$ and $y = 2$.
$-1 + 4 \stackrel{?}{=} 3$	Simplify.	$-4 - 2 \stackrel{?}{=} 6$	Simplify.
$3 = 3$	True	$-6 = 6$	False

$(-1, 2)$ is not a solution of the second equation, $4x - y = 6$, so it is not a solution of the system. ●

Practice Problem 2

Determine whether $(3, -2)$ is a solution of the system

$$\begin{cases} 2x - y = 8 \\ x + 3y = 4 \end{cases}$$

B Solving Systems of Equations by Graphing

Since a solution of a system of two equations in two variables is a solution common to both equations, it is also a point common to the graphs of both equations. Let's practice finding solutions of both equations in a system—that is, solutions of the system—by graphing and identifying points of intersection.

Answers

1. $(3, 9)$ is a solution of the system, 2. $(3, -2)$ is not a solution of the system

Practice Problem 3

Solve the system of equations by graphing.

$$\begin{cases} -3x + y = -10 \\ x - y = 6 \end{cases}$$

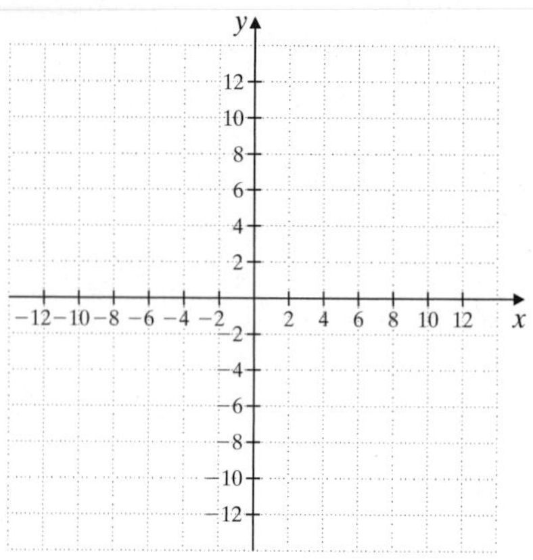

EXAMPLE 3 Solve the system of equations by graphing.

$$\begin{cases} -x + 3y = 10 \\ x + y = 2 \end{cases}$$

Solution: On a single set of axes, graph each linear equation.

$-x + 3y = 10$

x	y
0	$\frac{10}{3}$
−4	2
2	4

$x + y = 2$

x	y
0	2
2	0
1	1

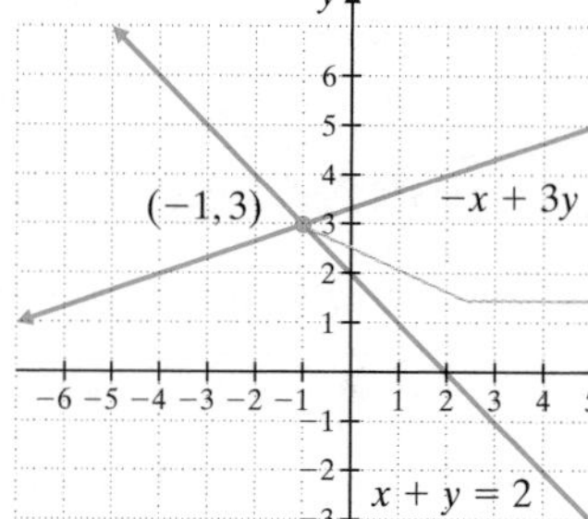

Helpful Hint

The point of intersection gives the solution of the system.

The two lines appear to intersect at the point $(-1, 3)$. To check, we replace x with -1 and y with 3 in both equations.

$-x + 3y = 10$	First equation	$x + y = 2$	Second equation
$-(-1) + 3(3) \stackrel{?}{=} 10$	Let $x = -1$ and $y = 3$.	$-1 + 3 \stackrel{?}{=} 2$	Let $x = -1$ and $y = 3$.
$1 + 9 \stackrel{?}{=} 10$	Simplify.	$2 = 2$	True
$10 = 10$	True		

$(-1, 3)$ checks, so it is the solution of the system. ●

Helpful Hint

Neatly drawn graphs can help when "guessing" the solution of a system of linear equations by graphing.

Practice Problem 4

Solve the system of equations by graphing.

$$\begin{cases} x + 3y = -1 \\ y = 1 \end{cases}$$

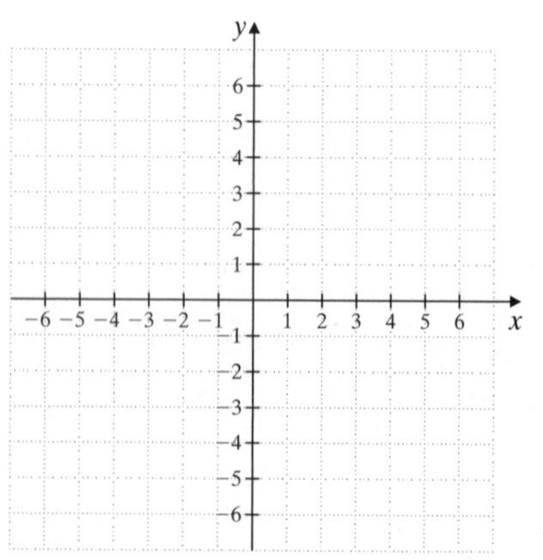

EXAMPLE 4 Solve the system of equations by graphing.

$$\begin{cases} 2x + 3y = -2 \\ x = 2 \end{cases}$$

Solution: We graph each linear equation on a single set of axes.

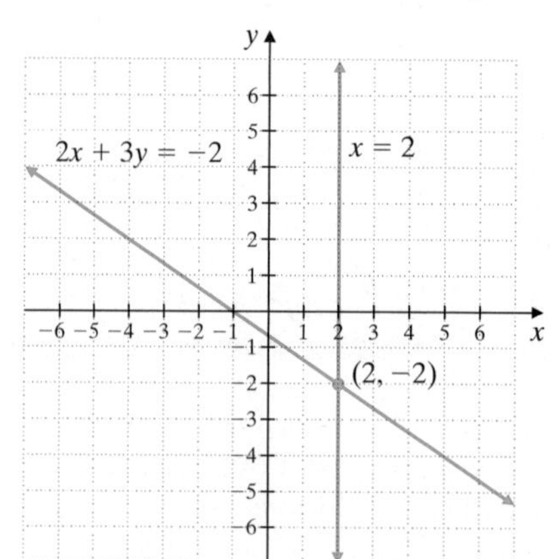

Answers

3. please see page 564, **4.** please see page 564

The two lines appear to intersect at the point $(2, -2)$. To determine whether $(2, -2)$ is the solution, we replace x with 2 and y with -2 in both equations.

$2x + 3y = -2$	First equation	$x = 2$	Second equation
$2(2) + 3(-2) \stackrel{?}{=} -2$	Let $x = 2$ and $y = -2$.	$2 = 2$	Let $x = 2$.
$4 + (-6) \stackrel{?}{=} -2$	Simplify.	$2 = 2$	True
$-2 = -2$	True		

Since a true statement results in both equations, $(2, -2)$ is the solution of the system. ●

C Identifying Special Systems of Linear Equations

Not all systems of linear equations have a single solution. Some systems have no solution and some have an infinite number of solutions.

EXAMPLE 5 Solve the system of equations by graphing.

$$\begin{cases} 2x + y = 7 \\ 2y = -4x \end{cases}$$

Solution: We graph the two equations in the system. The equations in slope-intercept form are $y = -2x + 7$ and $y = -2x$. Notice from the equations that the lines have the same slope, -2, and different y-intercepts. This means that the lines are parallel.

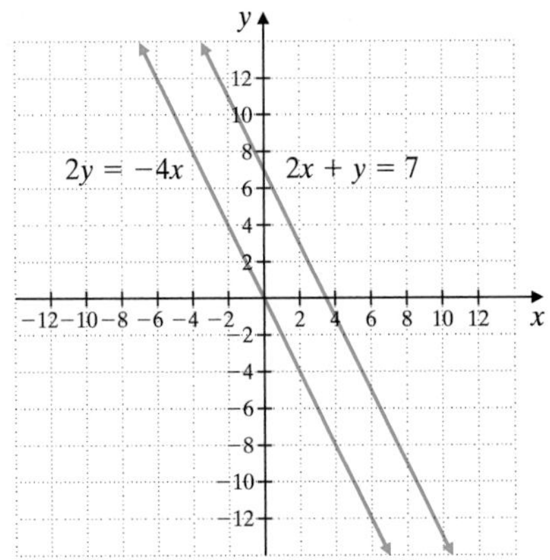

Since the lines are parallel, they do not intersect. This means that the system has *no solution*. ●

EXAMPLE 6 Solve the system of equations by graphing.

$$\begin{cases} x - y = 3 \\ -x + y = -3 \end{cases}$$

Solution: We graph each equation.

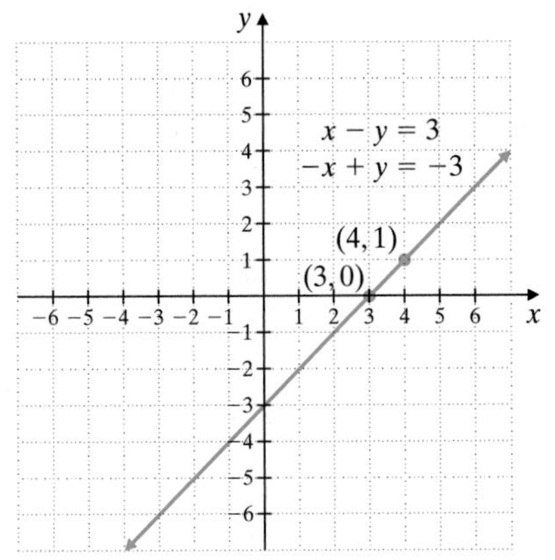

Practice Problem 5

Solve the system of equations by graphing.

$$\begin{cases} 3x - y = 6 \\ 6x = 2y \end{cases}$$

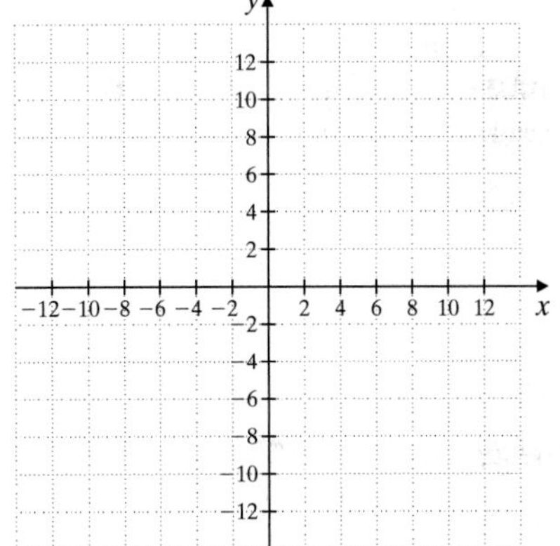

Practice Problem 6

Solve the system of equations by graphing.

$$\begin{cases} 3x + 4y = 12 \\ 9x + 12y = 36 \end{cases}$$

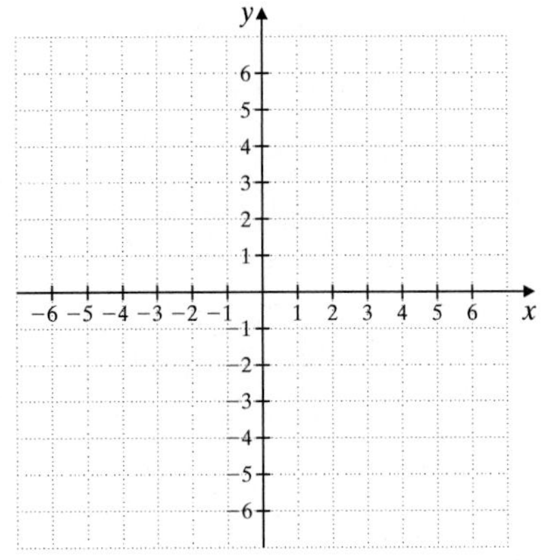

Answers

5. please see page 564, **6.** please see page 564

Answers

3. $(2, -4)$

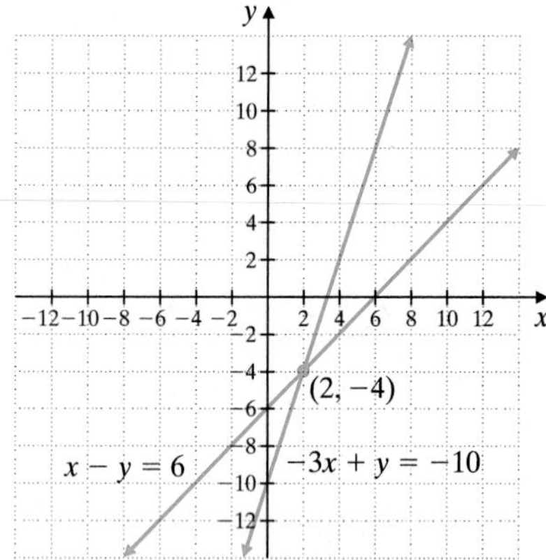

4. $(-4, 1)$

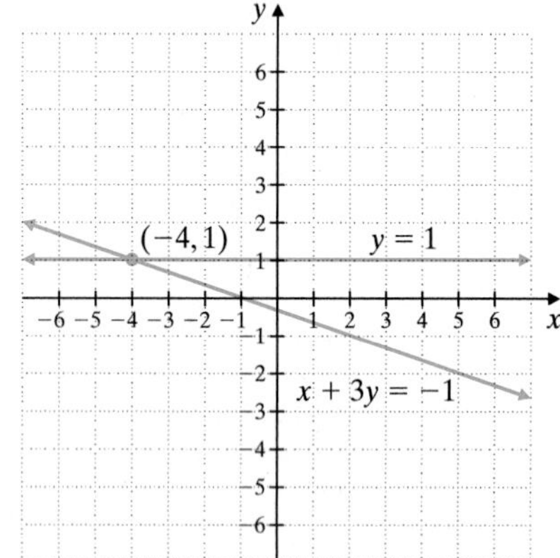

5. no solution

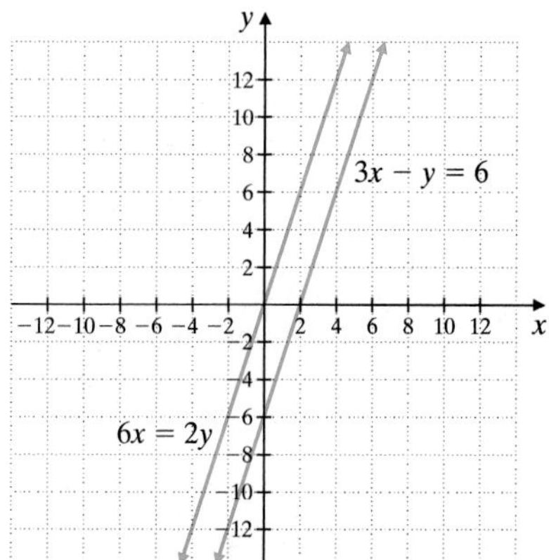

6. infinite number of solutions

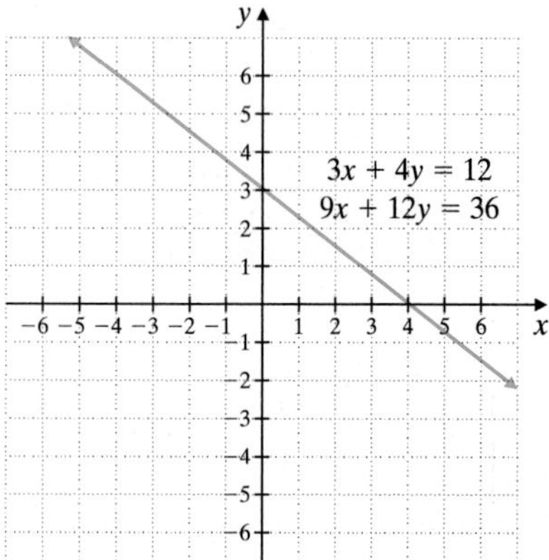

The graphs of the equations are the same line. To see this, notice that if both sides of the first equation in the system are multiplied by -1, the result is the second equation.

$$x - y = 3 \quad \text{First equation}$$
$$-1(x - y) = -1(3) \quad \text{Multiply both sides by } -1.$$
$$-x + y = -3 \quad \text{Simplify. This is the second equation.}$$

This means that the system has an infinite number of solutions. Any ordered pair that is a solution of one equation is a solution of the other and is then a solution of the system.

Examples 5 and 6 are special cases of systems of linear equations. A system that has no solution is said to be an **inconsistent system**. If the graphs of the two equations of a system are identical, we call the equations **dependent equations**.

As we have seen, three different situations can occur when graphing the two lines associated with the equations in a linear system. These situations are shown in the figures.

One point of intersection: one solution

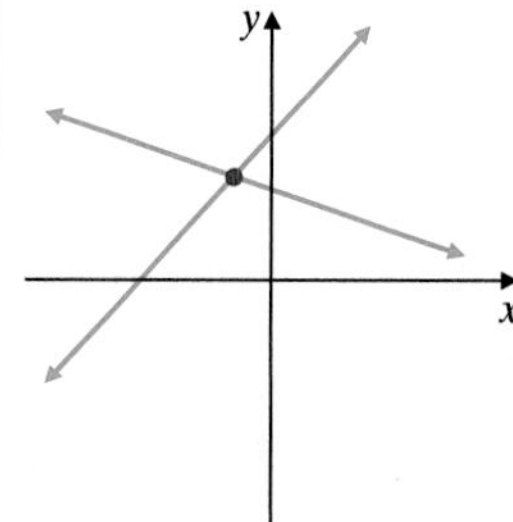

Consistent system
(at least one solution)
Independent equations
(graphs of equations differ)

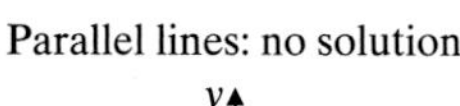

Parallel lines: no solution

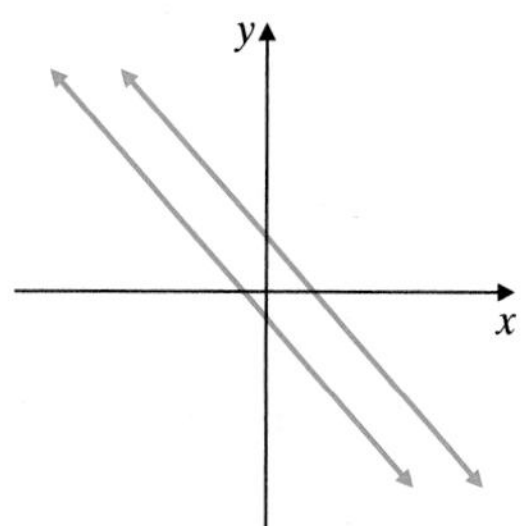

Inconsistent system
(no solution)
Independent equations
(graphs of equations differ)

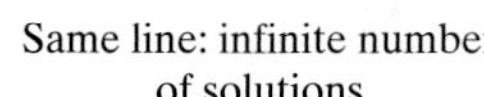

Same line: infinite number of solutions

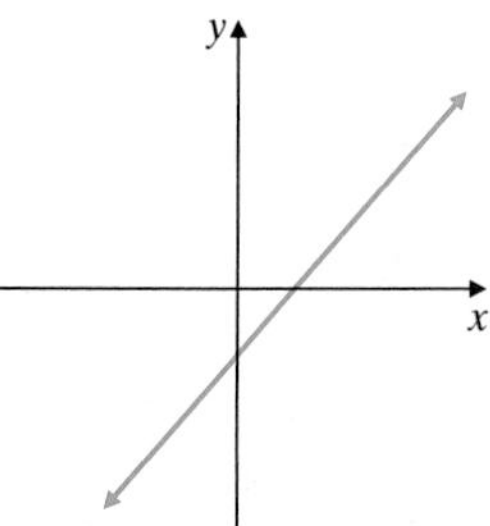

Consistent system
(at least one solution)
Dependent equations
(graphs of equations identical)

GRAPHING CALCULATOR EXPLORATIONS

We may use a grapher to approximate solutions of systems of equations by graphing both equations on the same set of axes and approximating any points of intersection. For example, let's approximate the solution of the system

$$\begin{cases} y = -2.6x + 5.6 \\ y = 4.3x - 4.9 \end{cases}$$

We use a standard window and graph the equations on a single screen.

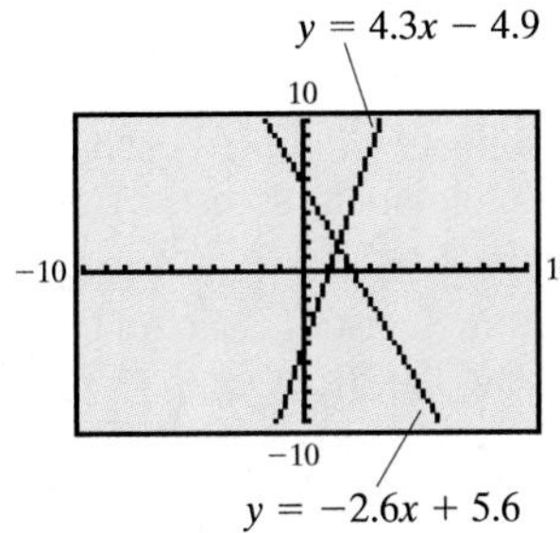

The two lines intersect. To approximate the point of intersection, we trace to the point of intersection and use an INTERSECT feature of the grapher, a ZOOM IN feature of the grapher, or redefine the window to $[0, 3]$ by $[0, 3]$. If we redefine the window to $[0, 3]$ by $[0, 3]$, the screen should look like the following:

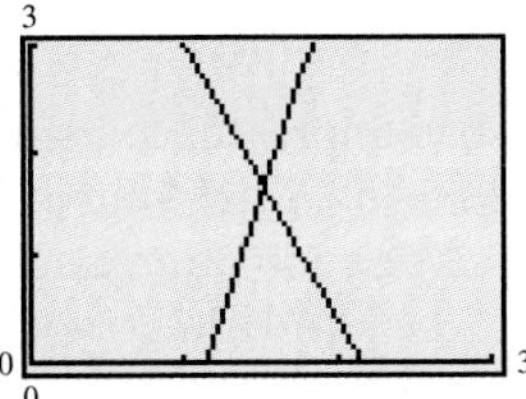

By tracing along the curves, we can see that the point of intersection has an x-value between 1.5 and 1.532. We can continue to zoom and trace or redefine the window until the coordinates of the point of intersection can be determined to the nearest hundredth. The approximate point of intersection is $(1.52, 1.64)$.

Solve each system of equations. Approximate each solution to two decimal places.

1. $y = -1.65x + 3.65$
$y = 4.56x - 9.44$

2. $y = 7.61x + 3.48$
$y = -1.26x - 6.43$

3. $2.33x - 4.72y = 10.61$
$5.86x - 6.22y = -8.89$

4. $-7.89x - 5.68y = 3.26$
$-3.65x + 4.98y = 11.77$

FOCUS ON Business and Career

ASSESSING JOB OFFERS

When you finish your present course of study, you will probably look for a job. When your job search has paid off and you receive a job offer, how will you decide whether to take the job? How do you decide between two or more job offers? These decisions are an important part of the job search and may not be easy to make. To evaluate the job offer, you should consider the nature of the work involved in the job, the type of company or organization that has offered the job, and the salary and benefits offered by the employer. You may also need to compare the compensation packages of two or more job offers. The following hints on assessing a job's compensation package were included in the Bureau of Labor Statistics' *Occupational Outlook Handbook*, 2000–01 edition.

> ***Salaries and benefits***. Wait for the employer to introduce these subjects. Some companies will not talk about pay until they have decided to hire you. In order to know if their offer is reasonable, you need a rough estimate of what the job should pay. You may have to go to several sources for this information. Try to find family, friends, or acquaintances who recently were hired in similar jobs. Ask your teachers and the staff in placement offices about starting pay for graduates with your qualifications. Help-wanted ads in newspapers sometimes give salary ranges for similar positions. Check the library or your school's career center for salary surveys such as those conducted by the National Association of Colleges and Employers or various professional associations.
>
> If you are considering the salary and benefits for a job in another geographic area, make allowances for differences in the cost of living, which may be significantly higher in a large metropolitan area than in a smaller city, town, or rural area.
>
> You also should learn the organization's policy regarding overtime. Depending on the job, you may or may not be exempt from laws requiring the employer to compensate you for overtime. Find out how many hours you will be expected to work each week and whether you receive overtime pay or compensatory time off for working more than the specified number of hours in a week.
>
> Also take into account that the starting salary is just that—the start. Your salary should be reviewed on a regular basis; many organizations do it every year. How much can you expect to earn after 1, 2, or 3 or more years? An employer cannot be specific about the amount of pay if it includes commissions and bonuses.
>
> Benefits can also add a lot to your base pay, but they vary widely. Find out exactly what the benefit package includes and how much of the costs you must bear.

National, state, and metropolitan area data from the National Compensation Survey, which integrates data from three existing Bureau of Labor Statistics programs—the Employment Cost Index, the Occupational Compensation Survey, and the Employee Benefits Survey—are available from:

> Bureau of Labor Statistics, Office of Compensation and Working Conditions, 2 Massachusetts Ave. NE, Room 4130, Washington, DC 20212-0001. Telephone: (202) 691-6199.
> Internet: http://stats.bls.gov/comhome.htm

Data on earnings by detailed occupation from the Occupational Employment Statistics (OES) Survey are available from:

> Bureau of Labor Statistics, Office of Employment and Unemployment Statistics, Occupational Employment Statistics, 2 Massachusetts Ave. NE, Room 4840, Washington, DC 20212-0001. Telephone: (202) 691-6569.
> Internet: http://stats.bls.gov/oeshome.htm

CRITICAL THINKING

1. Suppose you have been searching for a position as an electronics sales associate. You have received two job offers. The first job pays a monthly salary of \$2000 per month plus a commission of 4% on all sales made. The second pays a monthly salary of \$2300 per month plus a commission of 2% on all sales made. At what level of monthly sales would the jobs pay the same amount? Based only on the given information about the jobs, which job would you choose? Why? What other information would you want to have about the jobs before making a decision?
2. Suppose you have been searching for an entry-level bookkeeping position. You have received two job offers. The first company offers you a starting hourly wage of \$7.50 per hour and says that each year entry-level workers receive a raise of \$0.75 per hour. The second company offers you a starting hourly wage of \$8.50 per hour and says that you can expect a \$0.50 per hour raise each year. After how many years will the two jobs pay the same hourly wage? Based only on the given information about the jobs, which job would you choose? Why? What other information would you want to have about the jobs before making a decision?

Name ______________________ Section __________ Date __________

Mental Math

Each rectangular coordinate system shows the graph of the equations in a system of equations. Use each graph to determine the number of solutions for each associated system. If the system has only one solution, give its coordinates.

1.

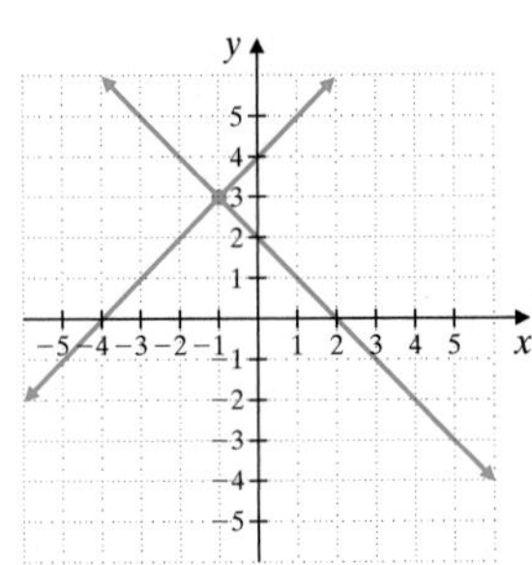

2.

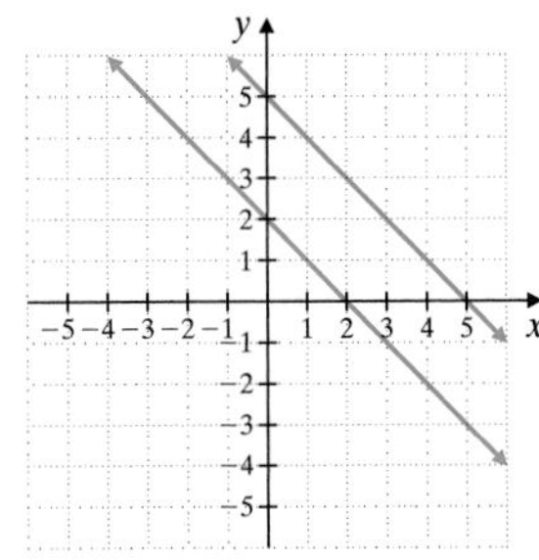

3.

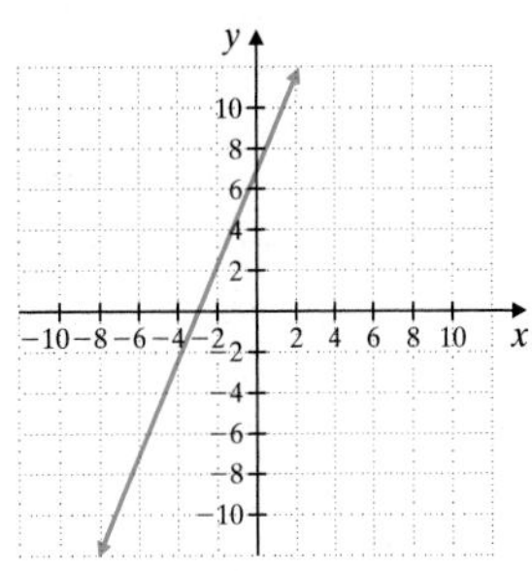

4.

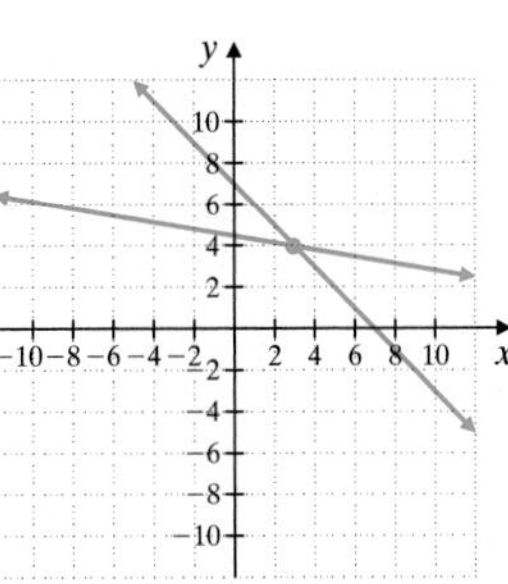

5.

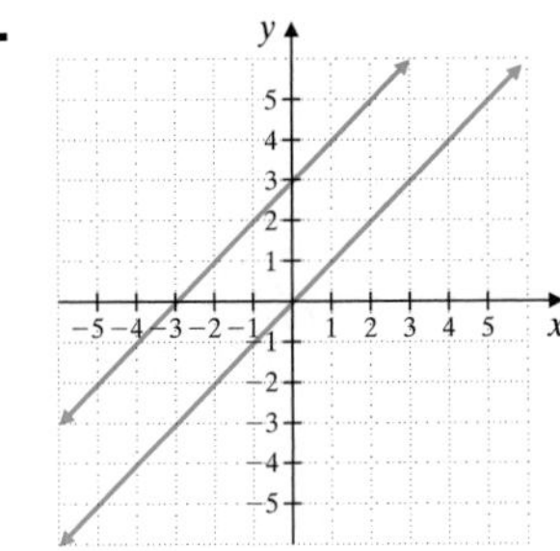

6.

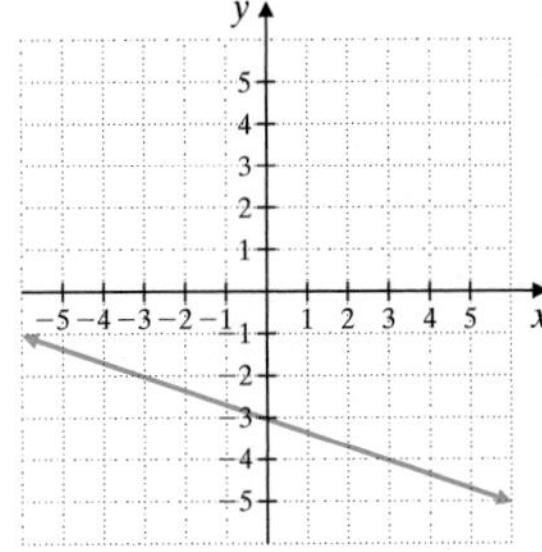

7.

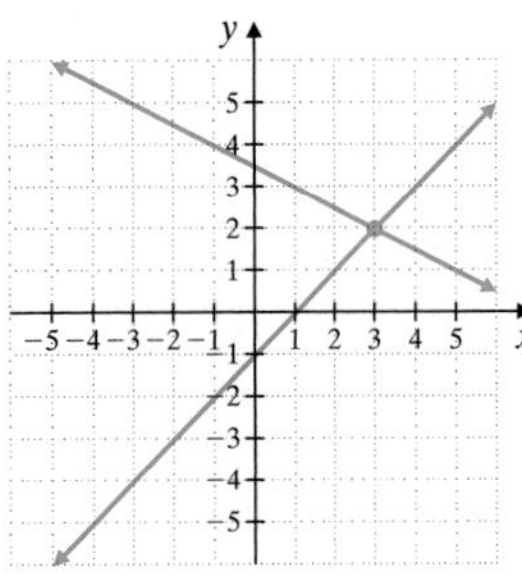

8.

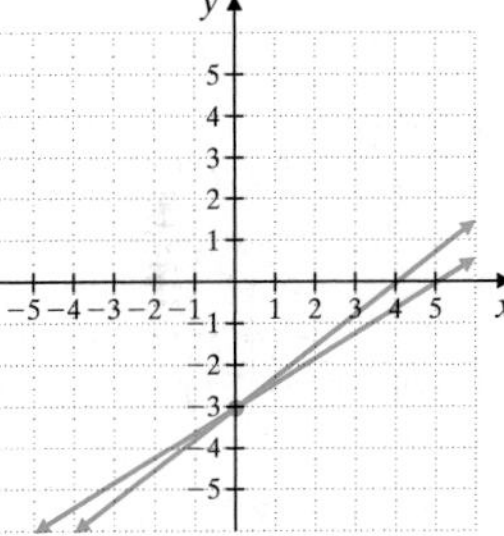

EXERCISE SET 8.1

A *Determine whether each ordered pair is a solution of the system of linear equations. See Examples 1 and 2.*

1. $\begin{cases} x + y = 8 \\ 3x + 2y = 21 \end{cases}$

a. $(2, 4)$

b. $(5, 3)$

2. $\begin{cases} 2x + y = 5 \\ x + 3y = 5 \end{cases}$

a. $(5, 0)$

b. $(2, 1)$

3. $\begin{cases} 3x - y = 5 \\ x + 2y = 11 \end{cases}$

a. $(3, 4)$

b. $(0, -5)$

4. $\begin{cases} 2x - 3y = 8 \\ x - 2y = 6 \end{cases}$

a. $(-2, -4)$

b. $(7, 2)$

5. $\begin{cases} 2y = 4x \\ 2x - y = 0 \end{cases}$

a. $(-3, -6)$

b. $(0, 0)$

6. $\begin{cases} 4x = 1 - y \\ x - 3y = -8 \end{cases}$

a. $(0, 1)$

b. $(1, -3)$

B **C** *Solve each system of linear equations by graphing. See Examples 3 through 6.*

7. $\begin{cases} x + y = 4 \\ x - y = 2 \end{cases}$

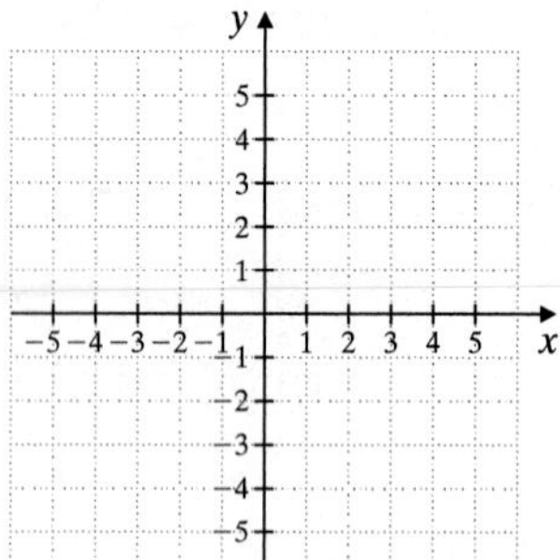

8. $\begin{cases} x + y = 3 \\ x - y = 5 \end{cases}$

9. $\begin{cases} x + y = 6 \\ -x + y = -6 \end{cases}$

10. $\begin{cases} x + y = 1 \\ -x + y = -3 \end{cases}$

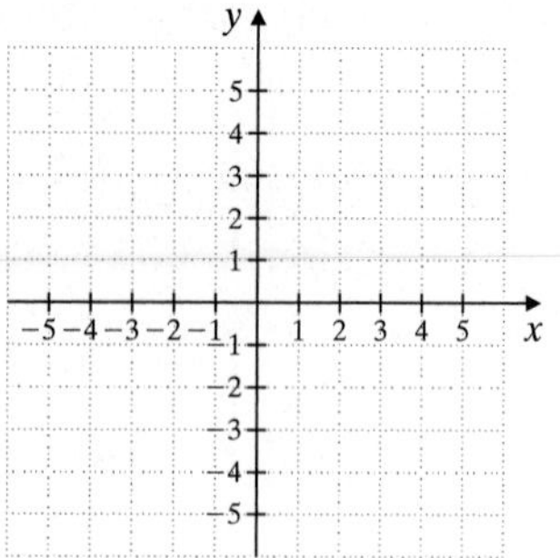

11. $\begin{cases} y = 2x \\ 3x - y = -2 \end{cases}$

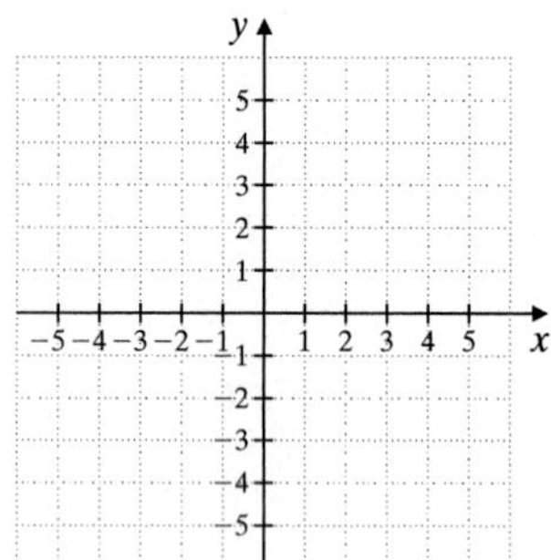

12. $\begin{cases} y = -3x \\ 2x - y = -5 \end{cases}$

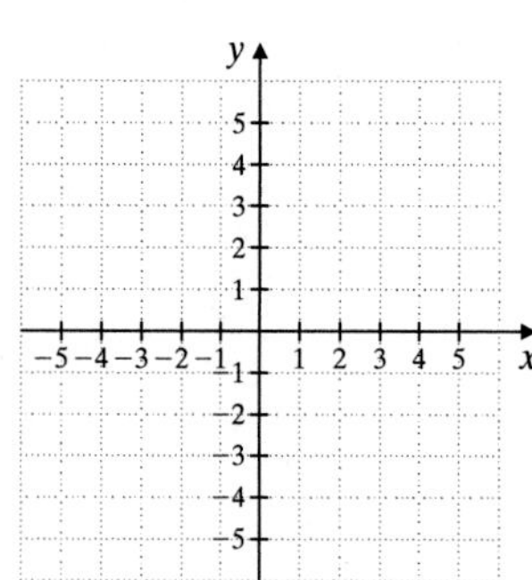

13. $\begin{cases} y = x + 1 \\ y = 2x - 1 \end{cases}$

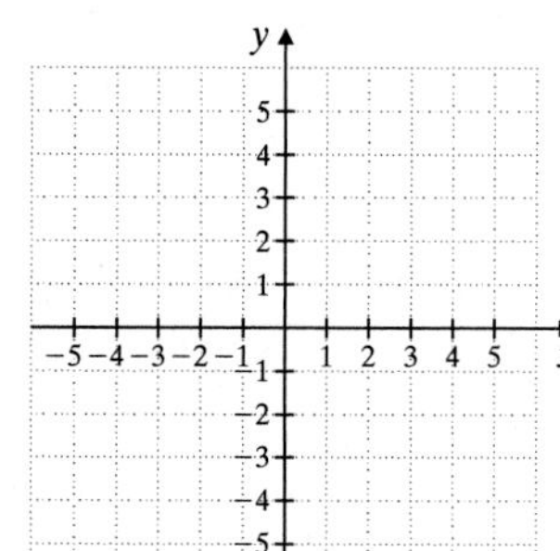

14. $\begin{cases} y = 3x - 4 \\ y = x + 2 \end{cases}$

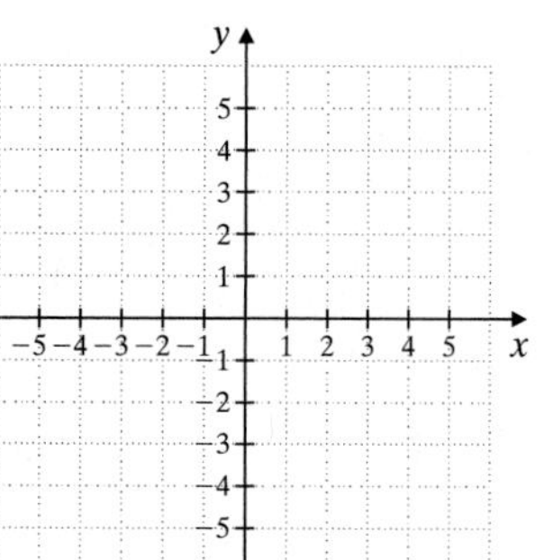

15. $\begin{cases} 2x + y = 0 \\ 3x + y = 1 \end{cases}$

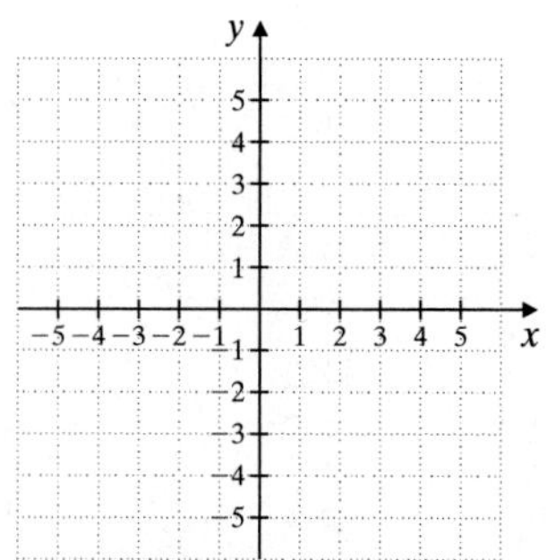

16. $\begin{cases} 2x + y = 1 \\ 3x + y = 0 \end{cases}$

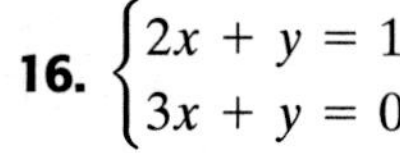

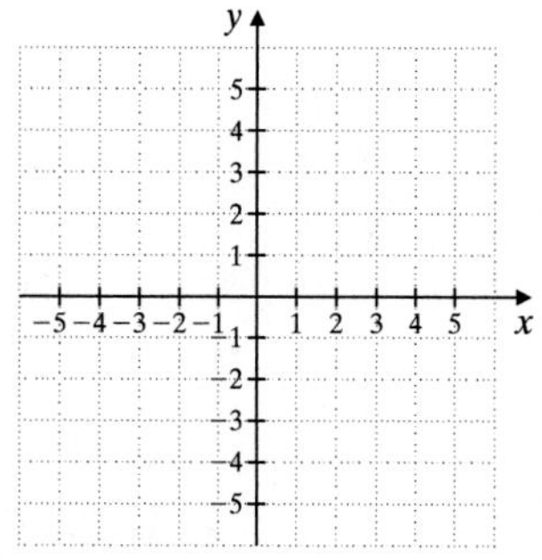

17. $\begin{cases} y = -x - 1 \\ y = 2x + 5 \end{cases}$

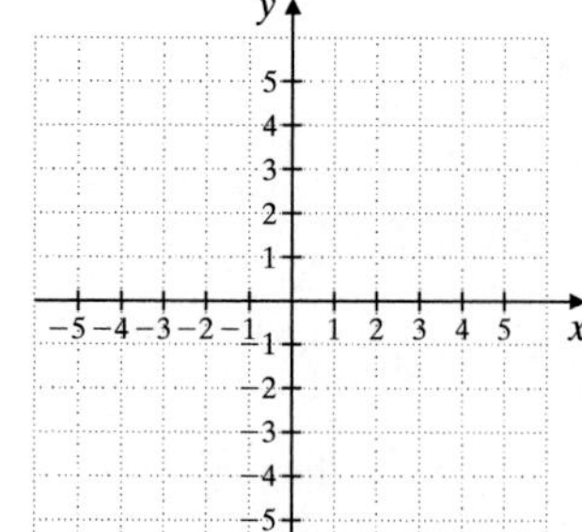

18. $\begin{cases} y = x - 1 \\ y = -3x - 5 \end{cases}$

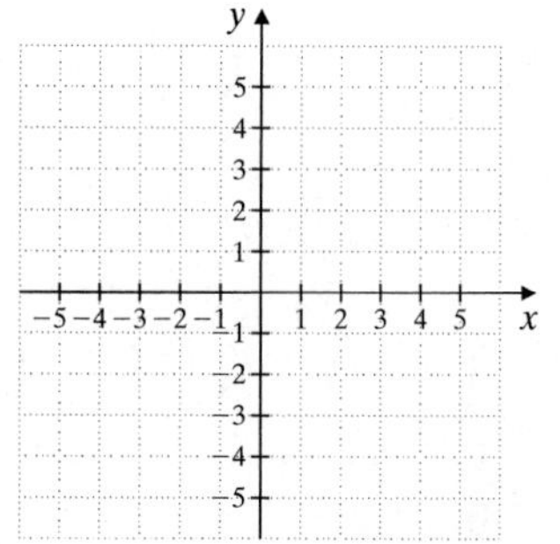

19. $\begin{cases} 2x - y = 6 \\ y = 2 \end{cases}$

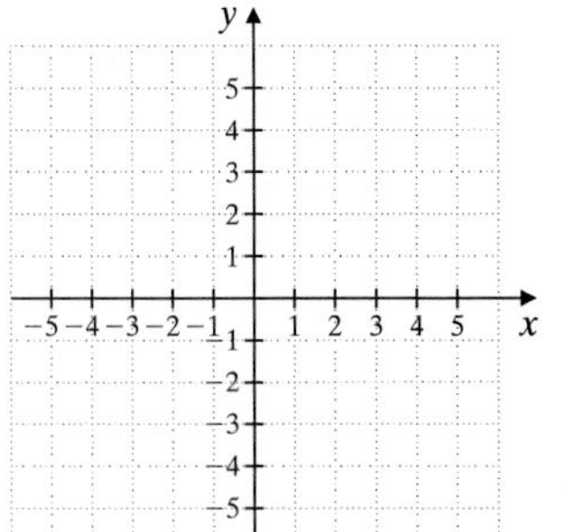

20. $\begin{cases} x + y = 5 \\ x = 4 \end{cases}$

21. $\begin{cases} x + y = 5 \\ x + y = 6 \end{cases}$

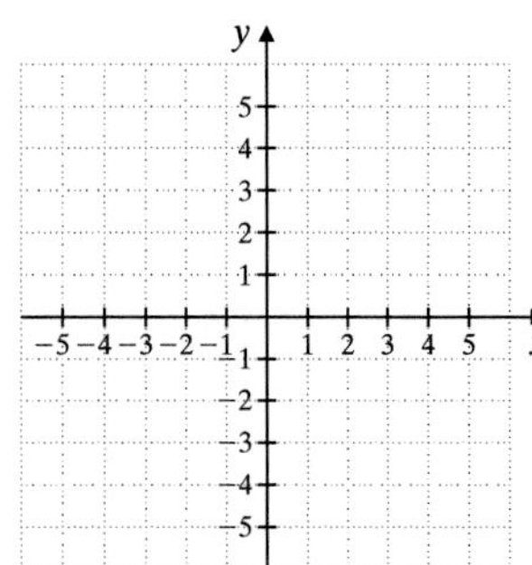

22. $\begin{cases} x - y = 4 \\ x - y = 1 \end{cases}$

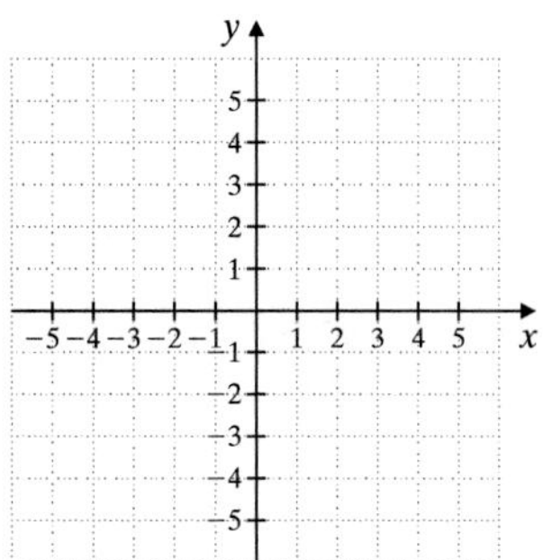

23. $\begin{cases} 2x + y = 4 \\ x + y = 2 \end{cases}$

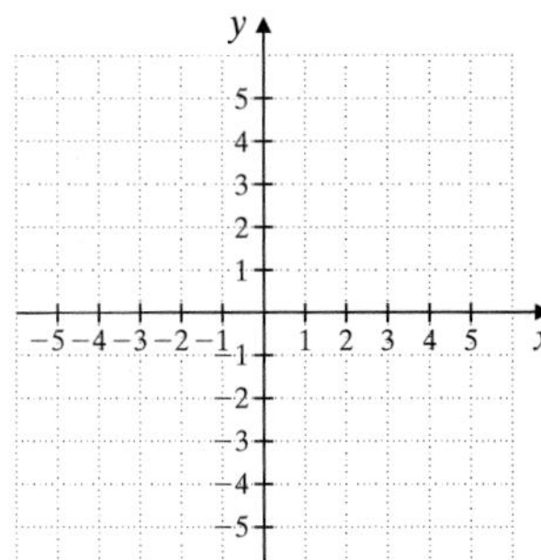

24. $\begin{cases} y + 2x = 3 \\ 4x = 2 - 2y \end{cases}$

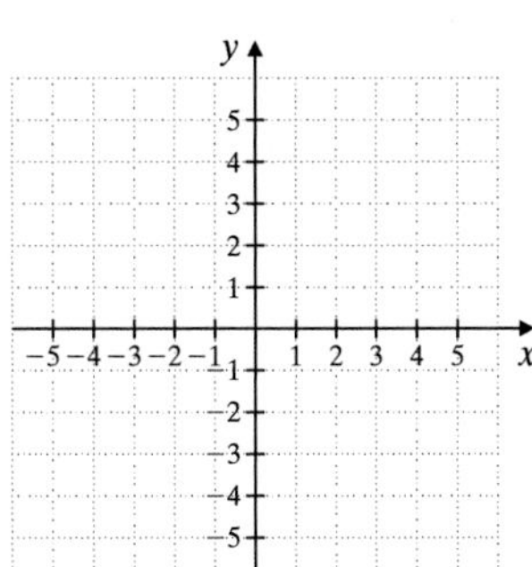

25. $\begin{cases} x - 2y = 2 \\ 3x + 2y = -2 \end{cases}$

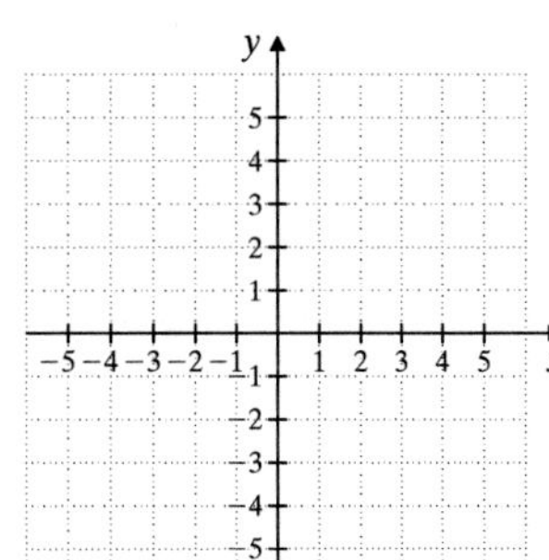

26. $\begin{cases} x + 3y = 7 \\ 2x - 3y = -4 \end{cases}$

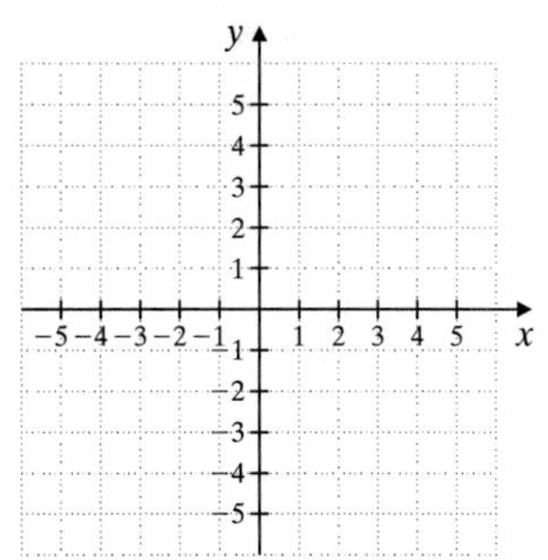

27. $\begin{cases} y - 3x = -2 \\ 6x - 2y = 4 \end{cases}$

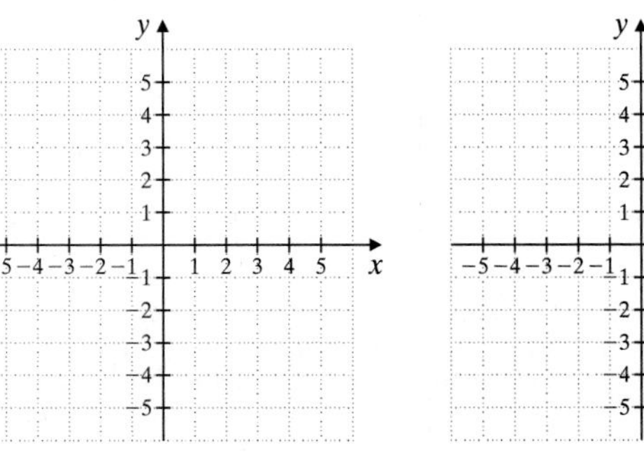

28. $\begin{cases} x - 2y = -6 \\ -2x + 4y = 12 \end{cases}$

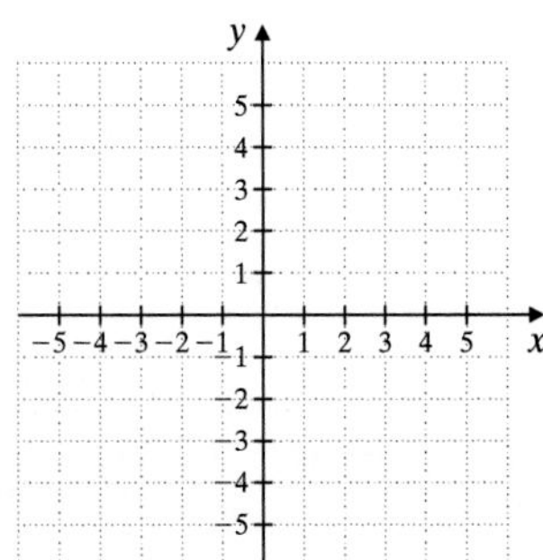

29. $\begin{cases} x = 3 \\ y = -1 \end{cases}$

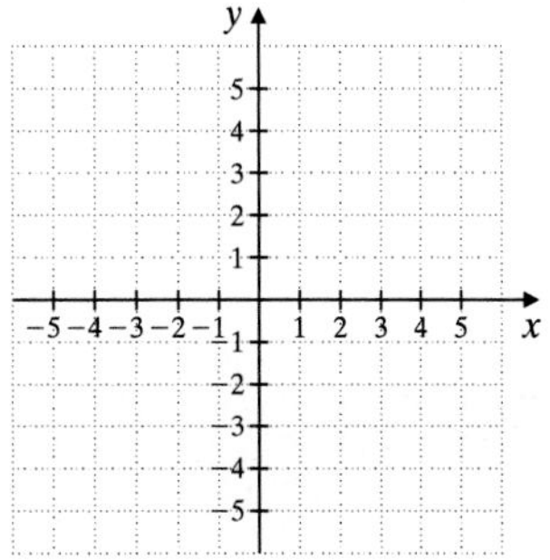

30. $\begin{cases} x = -5 \\ y = 3 \end{cases}$

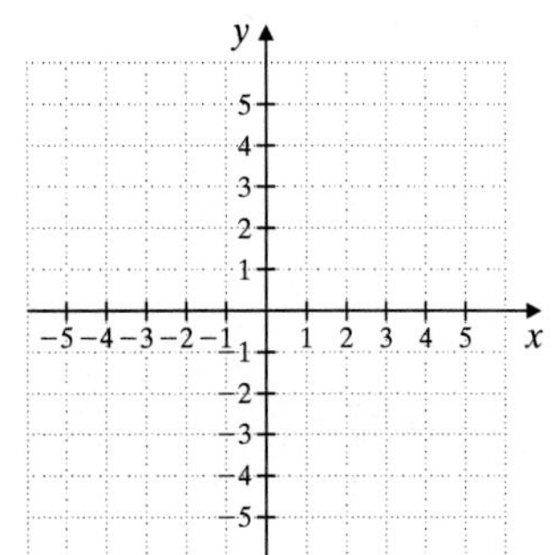

31. $\begin{cases} y = x - 2 \\ y = 2x + 3 \end{cases}$

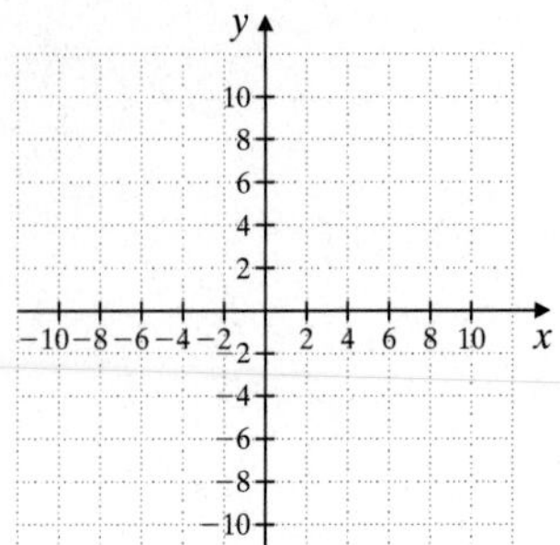

32. $\begin{cases} y = x + 5 \\ y = -2x - 4 \end{cases}$

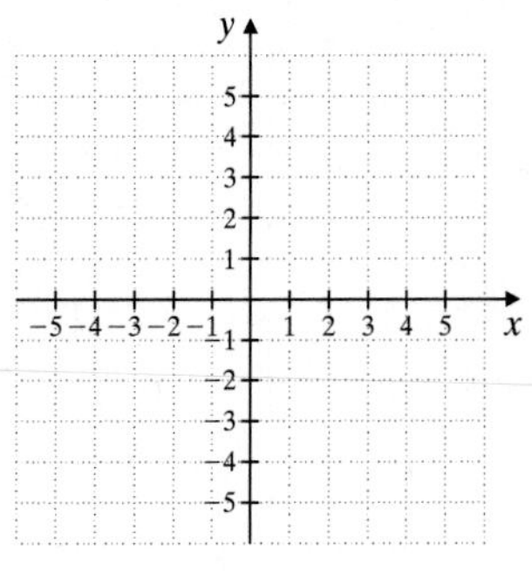

33. $\begin{cases} 2x - 3y = -2 \\ -3x + 5y = 5 \end{cases}$

34. $\begin{cases} 4x - y = 7 \\ 2x - 3y = -9 \end{cases}$

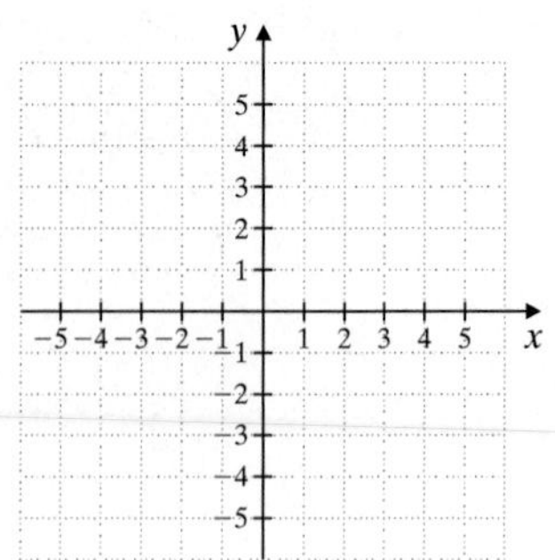

35. Draw a graph of two linear equations whose associated system has the solution $(-1, 4)$.

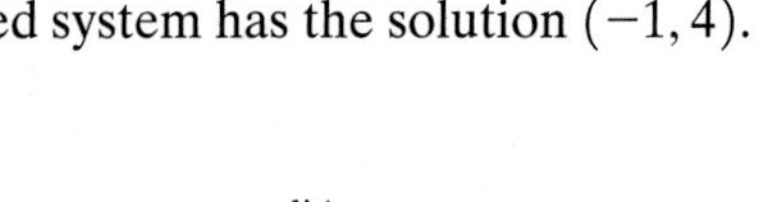

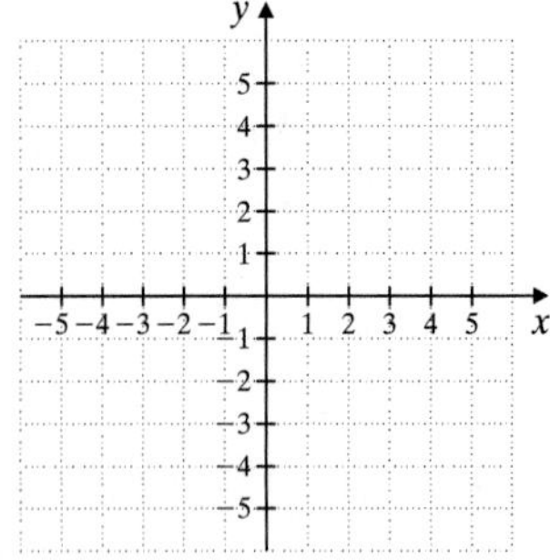

36. Draw a graph of two linear equations whose associated system has the solution $(3, -2)$.

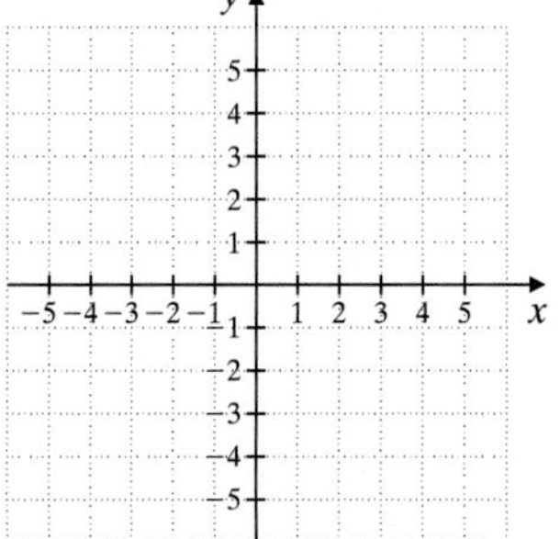

37. Draw a graph of two linear equations whose associated system has no solution.

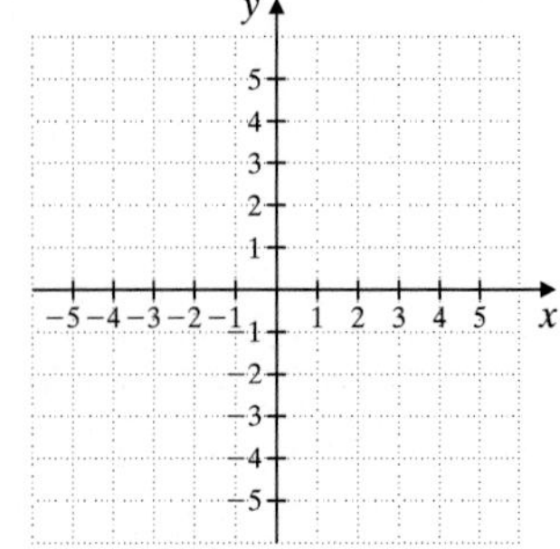

38. Draw a graph of two linear equations whose associated system has an infinite number of solutions.

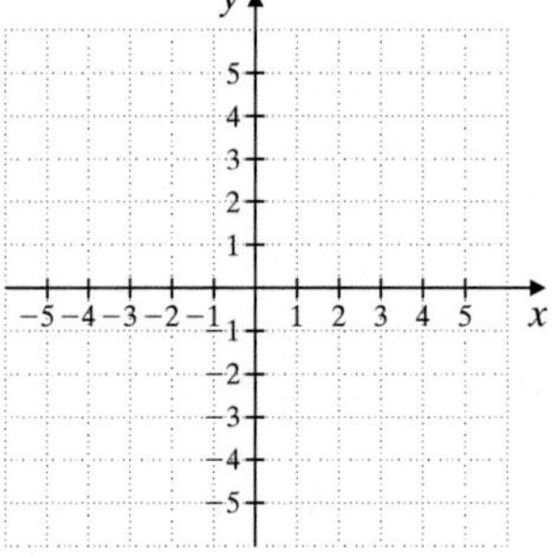

The double line graph below shows the number of pounds of fishery products from U.S. domestic catch and from imports. Use this graph to answer Exercises 39 and 40.

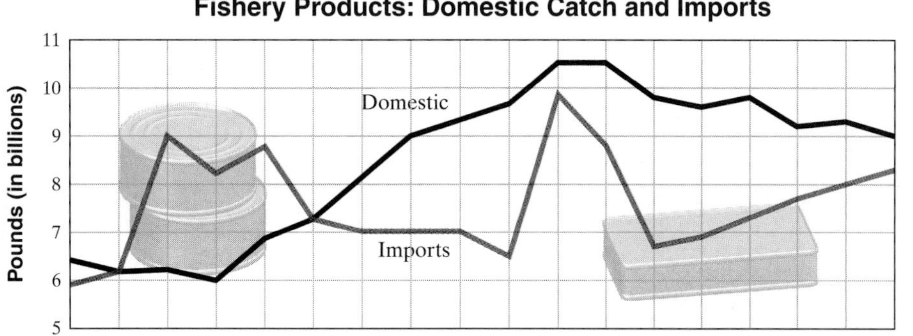

Source: U.S. Bureau of the Census, *Statistical Abstract of the United States*: 1995, 115th ed., Washington, DC, 1995.

39. In what year(s) was the number of pounds of imported fishery products equal to the number of pounds of domestic catch?

40. In what year(s) was the number of pounds of imported fishery products less than or equal to the number of pounds of domestic catch?

The double line graph below shows the number of Kmart stores vs. the number of Wal-Mart and Wal-Mart Supercenter stores. Use this graph to answer Exercises 41 and 42.

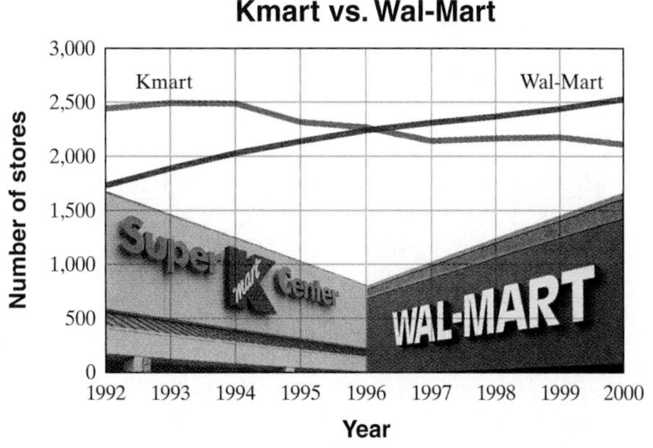

Sources: Kmart Corporation, Wal-Mart Stores, Inc.

41. In what year was the number of Kmart stores approximately equal to the number of Wal-Mart stores?

42. In what years was the number of Wal-Mart stores greater than the number of Kmart stores?

43. In the next column are tables of values for two linear equations.

a. Find a solution of the corresponding system.

b. Graph several ordered pairs from each table and sketch the two lines.

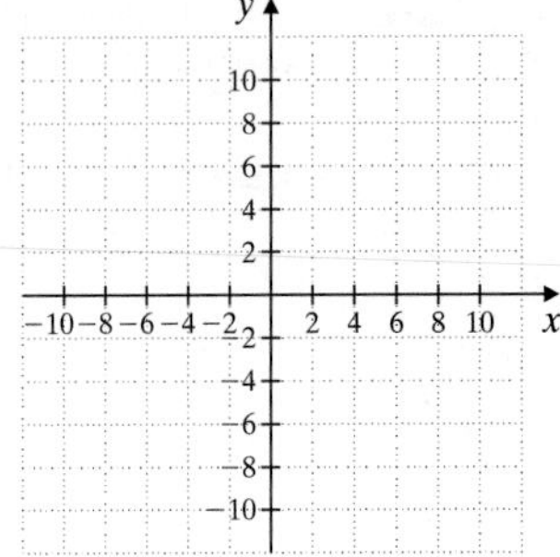

x	y
1	3
2	5
3	7
4	9
5	11

x	y
1	6
2	7
3	8
4	9
5	10

c. Does your graph confirm the solution from part (**a**)?

Review and Preview

Solve each equation. See Section 2.4.

44. $5(x - 3) + 3x = 1$

45. $-2x + 3(x + 6) = 17$

46. $4\left(\frac{y + 1}{2}\right) + 3y = 0$

47. $-y + 12\left(\frac{y - 1}{4}\right) = 3$

48. $8a - 2(3a - 1) = 6$

49. $3z - (4z - 2) = 9$

Combining Concepts

50. Construct a system of two linear equations that has $(2, 5)$ as a solution.

51. Construct a system of two linear equations that has $(0, 1)$ as a solution.

52. The ordered pair $(-2, 3)$ is a solution of the three linear equations below:

$$x + y = 1$$
$$2x - y = -7$$
$$x + 3y = 7$$

If each equation has a distinct graph, describe the graph of all three equations on the same axes.

53. Explain how to use a graph to determine the number of solutions of a system.

8.2 Solving Systems of Linear Equations by Substitution

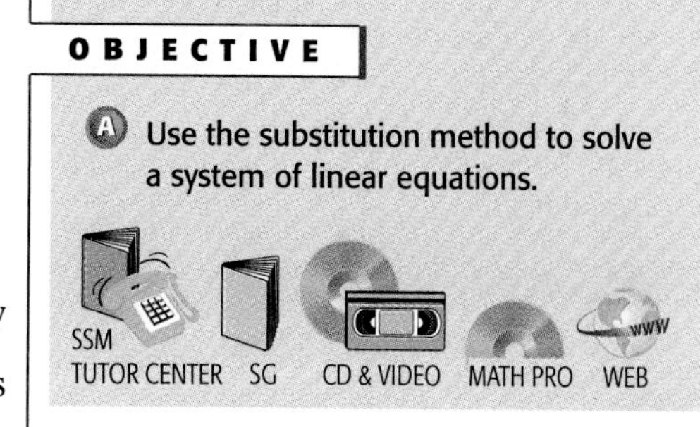

A Using the Substitution Method

You may have suspected by now that graphing alone is not an accurate way to solve a system of linear equations. For example, a solution of $\left(\frac{1}{2}, \frac{2}{9}\right)$ is unlikely to be read correctly from a graph. In this section, we discuss a second, more accurate method for solving systems of equations. This method is called the **substitution method** and is introduced in the next example.

EXAMPLE 1 Solve the system:

$$\begin{cases} 2x + y = 10 & \text{First equation} \\ x = y + 2 & \text{Second equation} \end{cases}$$

Solution: The second equation in this system is $x = y + 2$. This tells us that x and $y + 2$ have the same value. This means that we may substitute $y + 2$ for x in the first equation.

$$2x + y = 10 \quad \text{First equation}$$

$$2(y + 2) + y = 10 \quad \text{Substitute } y + 2 \text{ for } x \text{ since } x = y + 2.$$

Notice that this equation now has one variable, y. Let's now solve this equation for y.

Helpful Hint

Don't forget the distributive property.

$$\begin{aligned} 2(y + 2) + y &= 10 \\ 2y + 4 + y &= 10 && \text{Use the distributive property.} \\ 3y + 4 &= 10 && \text{Combine like terms.} \\ 3y &= 6 && \text{Subtract 4 from both sides.} \\ y &= 2 && \text{Divide both sides by 3.} \end{aligned}$$

Now we know that the y-value of the ordered pair solution of the system is 2. To find the corresponding x-value, we replace y with 2 in the equation $x = y + 2$ and solve for x.

$$\begin{aligned} x &= y + 2 \\ x &= 2 + 2 && \text{Let } y = 2. \\ x &= 4 \end{aligned}$$

The solution of the system is the ordered pair $(4, 2)$. Since an ordered pair solution must satisfy both linear equations in the system, we could have chosen the equation $2x + y = 10$ to find the corresponding x-value. The resulting x-value is the same.

Check: We check to see that $(4, 2)$ satisfies both equations of the original system.

First Equation

$$\begin{aligned} 2x + y &= 10 \\ 2(4) + 2 &\stackrel{?}{=} 10 \\ 10 &= 10 && \text{True} \end{aligned}$$

Second Equation

$$\begin{aligned} x &= y + 2 \\ 4 &\stackrel{?}{=} 2 + 2 && \text{Let } x = 4 \text{ and } y = 2. \\ 4 &= 4 && \text{True} \end{aligned}$$

Practice Problem 1

Use the substitution method to solve the system:

$$\begin{cases} 2x + 3y = 13 \\ x = y + 4 \end{cases}$$

Answer

1. $(5, 1)$

The solution of the system is $(4, 2)$.
A graph of the two equations shows the two lines intersecting at the point $(4, 2)$.

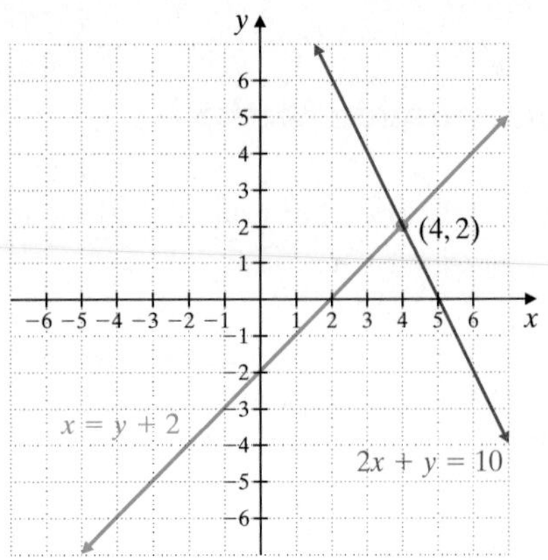

Practice Problem 2

Use the substitution method to solve the system:

$$\begin{cases} 4x - y = 2 \\ y = 5x \end{cases}$$

EXAMPLE 2 Solve the system:

$$\begin{cases} 5x - y = -2 \\ y = 3x \end{cases}$$

Solution: The second equation is solved for y in terms of x. We substitute $3x$ for y in the first equation.

$5x - y = -2$ First equation

$5x - (3x) = -2$

Now we solve for x.

$5x - 3x = -2$

$2x = -2$ Combine like terms.

$x = -1$ Divide both sides by 2.

The x-value of the ordered pair solution is -1. To find the corresponding y-value, we replace x with -1 in the equation $y = 3x$.

$y = 3x$

$y = 3(-1)$ Let $x = -1$.

$y = -3$

Check to see that the solution of the system is $(-1, -3)$.

To solve a system of equations by substitution, we first need an equation solved for one of its variables.

Practice Problem 3

Solve the system:

$$\begin{cases} 3x + y = 5 \\ 3x - 2y = -7 \end{cases}$$

Answers

2. $(-2, -10)$, **3.** $\left(\frac{1}{3}, 4\right)$

EXAMPLE 3 Solve the system:

$$\begin{cases} x + 2y = 7 \\ 2x + 2y = 13 \end{cases}$$

Solution: We choose one of the equations and solve for x or y. We will solve the first equation for x by subtracting $2y$ from both sides.

$x + 2y = 7$ First equation

$x = 7 - 2y$ Subtract $2y$ from both sides.

Since $x = 7 - 2y$, we now substitute $7 - 2y$ for x in the second equation and solve for y.

Don't forget to insert parentheses when substituting $7 - 2y$ for x.

$$\begin{aligned} 2x + 2y &= 13 && \text{Second equation} \\ 2(7 - 2y) + 2y &= 13 && \text{Let } x = 7 - 2y. \\ 14 - 4y + 2y &= 13 && \text{Apply the distributive property.} \\ 14 - 2y &= 13 && \text{Simplify.} \\ -2y &= -1 && \text{Subtract 14 from both sides.} \\ y &= \frac{1}{2} && \text{Divide both sides by } -2. \end{aligned}$$

To find x, we let $y = \frac{1}{2}$ in the equation $x = 7 - 2y$.

$$\begin{aligned} x &= 7 - 2y \\ x &= 7 - 2\left(\frac{1}{2}\right) && \text{Let } y = \frac{1}{2}. \\ x &= 7 - 1 \\ x &= 6 \end{aligned}$$

Check the solution in both equations of the original system. The solution is $\left(6, \frac{1}{2}\right)$.

The following steps summarize how to solve a system of equations by the substitution method.

To Solve a System of Two Linear Equations by the Substitution Method

Step 1. Solve one of the equations for one of its variables.
Step 2. Substitute the expression for the variable found in Step 1 into the other equation.
Step 3. Solve the equation from Step 2 to find the value of one variable.
Step 4. Substitute the value found in Step 3 in any equation containing both variables to find the value of the other variable.
Step 5. Check the proposed solution in the original system.

Try the Concept Check in the margin.

EXAMPLE 4 Solve the system:

$$\begin{cases} 7x - 3y = -14 \\ -3x + y = 6 \end{cases}$$

Solution: To avoid introducing fractions, we will solve the second equation for y.

$$\begin{aligned} -3x + y &= 6 && \text{Second equation} \\ y &= 3x + 6 \end{aligned}$$

Concept Check

As you solve the system

$$\begin{cases} 2x + y = -5 \\ x - y = 5 \end{cases}$$

you find that $y = -5$. Is this the solution of the system?

Practice Problem 4

Solve the system:

$$\begin{cases} 5x - 2y = 6 \\ -3x + y = -3 \end{cases}$$

Answers

4. $(0, -3)$

Concept Check: no, the solution will be an ordered pair

Next, we substitute $3x + 6$ for y in the first equation.

$$\begin{aligned} 7x - 3y &= -14 && \text{First equation} \\ 7x - 3(3x + 6) &= -14 && \text{Let } y = 3x + 6. \\ 7x - 9x - 18 &= -14 && \text{Use the distributive property.} \\ -2x - 18 &= -14 && \text{Simplify.} \\ -2x &= 4 && \text{Add 18 to both sides.} \\ \frac{-2x}{-2} &= \frac{4}{-2} && \text{Divide both sides by } -2. \\ x &= -2 \end{aligned}$$

To find the corresponding y-value, we substitute -2 for x in the equation $y = 3x + 6$. Then $y = 3(-2) + 6$ or $y = 0$. The solution of the system is $(-2, 0)$. Check this solution in both equations of the system.

Helpful Hint

When solving a system of equations by the substitution method, begin by solving an equation for one of its variables. If possible, solve for a variable that has a coefficient of 1 or -1 to avoid working with time-consuming fractions.

Practice Problem 5

Solve the system:

$$\begin{cases} -x + 3y = 6 \\ y = \frac{1}{3}x + 2 \end{cases}$$

EXAMPLE 5 Solve the system: $\begin{cases} \frac{1}{2}x - y = 3 \\ x = 6 + 2y \end{cases}$

Solution: The second equation is already solved for x in terms of y. Thus we substitute $6 + 2y$ for x in the first equation and solve for y.

$$\begin{aligned} \frac{1}{2}x - y &= 3 && \text{First equation} \\ \frac{1}{2}(6 + 2y) - y &= 3 && \text{Let } x = 6 + 2y. \\ 3 + y - y &= 3 && \text{Apply the distributive property.} \\ 3 &= 3 && \text{Simplify.} \end{aligned}$$

Arriving at a true statement such as $3 = 3$ indicates that the two linear equations in the original system are equivalent. This means that their graphs are identical, as shown in the figure. There is an infinite number of solutions to the system, and any solution of one equation is also a solution of the other.

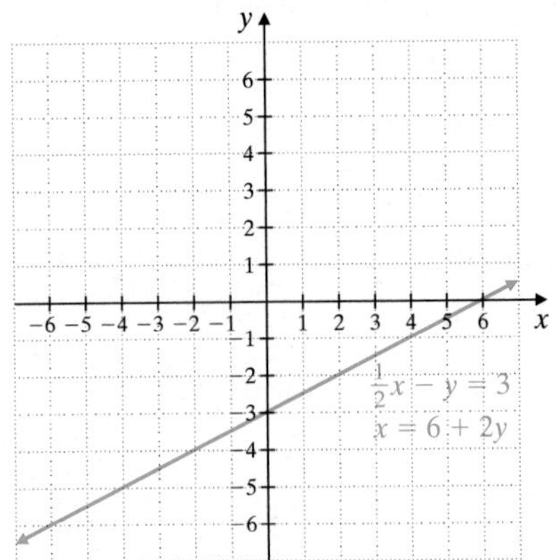

Answer

5. infinite number of solutions

EXAMPLE 6 Solve the system:

$$\begin{cases} 6x + 12y = 5 \\ -4x - 8y = 0 \end{cases}$$

Solution: We choose the second equation and solve for y.

$$-4x - 8y = 0 \quad \text{Second equation}$$
$$-8y = 4x \quad \text{Add } 4x \text{ to both sides.}$$
$$\frac{-8y}{-8} = \frac{4x}{-8} \quad \text{Divide both sides by } -8.$$
$$y = -\frac{1}{2}x \quad \text{Simplify.}$$

Now we replace y with $-\frac{1}{2}x$ in the first equation.

$$6x + 12y = 5 \quad \text{First equation}$$
$$6x + 12\left(-\frac{1}{2}x\right) = 5 \quad \text{Let } y = -\frac{1}{2}x.$$
$$6x + (-6x) = 5 \quad \text{Simplify.}$$
$$0 = 5 \quad \text{Combine like terms.}$$

The false statement $0 = 5$ indicates that this system has no solution. The graph of the linear equations in the system is a pair of parallel lines, as shown in the figure.

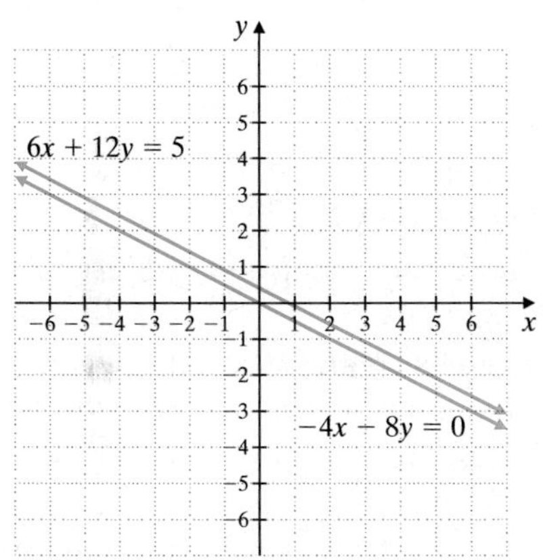

Try the Concept Check in the margin.

Practice Problem 6

Solve the system:

$$\begin{cases} 2x - 3y = 6 \\ -4x + 6y = -12 \end{cases}$$

Concept Check

Describe how the graphs of the equations in a system appear if the system has

a. no solution
b. one solution
c. an infinite number of solutions

Answers

6. infinite number of solutions

Concept Check: **a.** parallel lines, **b.** intersect at one point, **c.** identical graphs

FOCUS ON Business and Career

BREAK-EVEN POINT

When a business sells a new product, it generally does not start making a profit right away. There are usually many expenses associated with creating a new product. These expenses might include an advertising blitz to introduce the product to the public. These start-up expenses might also include the cost of market research and product development or any brand-new equipment needed to manufacture the product. Start-up costs like these are generally called *fixed costs* because they don't depend on the number of items manufactured. Expenses that depend on the number of items manufactured, such as the cost of materials and shipping, are called *variable costs*. The total cost of manufacturing the new product is given by the cost equation: Total cost = Fixed costs + Variable costs.

For instance, suppose a greeting card company is launching a new line of greeting cards. The company spent \$7000 doing product research and development for the new line and spent \$15,000 on advertising the new line. The company does not need to buy any new equipment to manufacture the cards, but the paper and ink needed to make each card will cost \$0.20 per card. The total cost y in dollars for manufacturing x cards is $y = 22{,}000 + 0.20x$.

Once a business sets a price for the new product, the company can find the product's expected *revenue*. Revenue is the amount of money the company takes in from the sales of its product. The revenue from selling a product is given by the revenue equation: Revenue = Price per item × Number of items sold.

For instance, suppose that the card company plans to sell its new cards for \$1.50 each. The revenue y, in dollars, that the company can expect to receive from the sales of x cards is $y = 1.50x$.

If the total cost and revenue equations are graphed on the same coordinate system, the graphs should intersect. The point of intersection is where total cost equals revenue and is called the *break-even point*. The break-even point gives the number of items x that must be manufactured and sold for the company to recover its expenses. If fewer than this number of items are produced and sold, the company loses money. If more than this number of items are produced and sold, the company makes a profit. In the case of the greeting card company, approximately 16,923 cards must be manufactured and sold for the company to break even on this new card line. The total cost and revenue of producing and selling 16,923 cards is the same. It is approximately \$25,385.

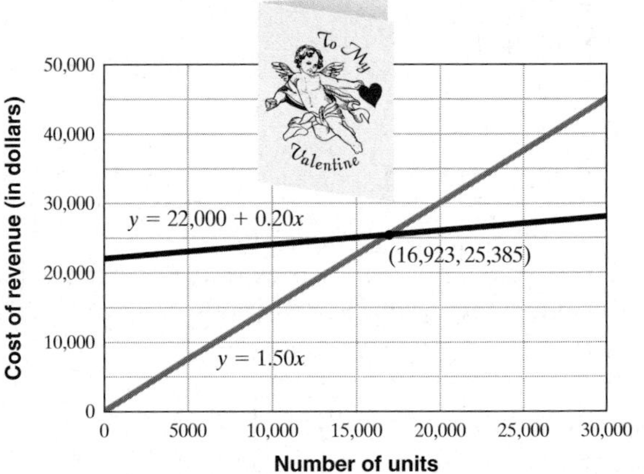

GROUP ACTIVITY

Suppose your group is starting a small business near your campus.

a. Choose a business and decide what campus-related product or service you will provide.

b. Research the fixed costs of starting up such a business.

c. Research the variable costs of producing such a product or providing such a service.

d. Decide how much you would charge per unit of your product or service.

e. Find a system of equations for the total cost and revenue of your product or service.

f. How many units of your product or service must be sold before your business will break even?

Name ______________________ Section __________ Date ____________

EXERCISE SET 8.2

A *Solve each system of equations by the substitution method. See Examples 1 and 2.*

1. $\begin{cases} x + y = 3 \\ x = 2y \end{cases}$ **2.** $\begin{cases} x + y = 20 \\ x = 3y \end{cases}$ **3.** $\begin{cases} x + y = 6 \\ y = -3x \end{cases}$ **4.** $\begin{cases} x + y = 6 \\ y = -4x \end{cases}$

5. $\begin{cases} 3x + 2y = 16 \\ x = 3y - 2 \end{cases}$ **6.** $\begin{cases} 2x + 3y = 18 \\ x = 2y - 5 \end{cases}$ **7.** $\begin{cases} 3x - 4y = 10 \\ x = 2y \end{cases}$ **8.** $\begin{cases} 3x - 4y = 10 \\ y = 2x \end{cases}$

9. $\begin{cases} y = 3x + 1 \\ 4y - 8x = 12 \end{cases}$ **10.** $\begin{cases} y = 2x + 3 \\ 5y - 7x = 18 \end{cases}$ **11.** $\begin{cases} y = 2x + 9 \\ y = 7x + 10 \end{cases}$ **12.** $\begin{cases} y = 5x - 3 \\ y = 8x + 4 \end{cases}$

Solve each system of equations by the substitution method. See Examples 3 through 6.

13. $\begin{cases} x + 2y = 6 \\ 2x + 3y = 8 \end{cases}$ **14.** $\begin{cases} x + 3y = -5 \\ 2x + 2y = 6 \end{cases}$ **15.** $\begin{cases} 2x - 5y = 1 \\ 3x + y = -7 \end{cases}$ **16.** $\begin{cases} 4x + 2y = 5 \\ 2x + y = -4 \end{cases}$

17. $\begin{cases} 2y = x + 2 \\ 6x - 12y = 0 \end{cases}$ **18.** $\begin{cases} 3y = x + 6 \\ 4x + 12y = 0 \end{cases}$ **19.** $\begin{cases} 4x + y = 11 \\ 2x + 5y = 1 \end{cases}$ **20.** $\begin{cases} 3x + y = -14 \\ 4x + 3y = -22 \end{cases}$

21. $\begin{cases} 2x - 3y = -9 \\ 3x = y + 4 \end{cases}$ **22.** $\begin{cases} 8x - 3y = -4 \\ 7x = y + 3 \end{cases}$ **23.** $\begin{cases} 6x - 3y = 5 \\ x + 2y = 0 \end{cases}$ **24.** $\begin{cases} 10x - 5y = -21 \\ x + 3y = 0 \end{cases}$

25. $\begin{cases} 3x - y = 1 \\ 2x - 3y = 10 \end{cases}$ **26.** $\begin{cases} 2x - y = -7 \\ 4x - 3y = -11 \end{cases}$ **27.** $\begin{cases} -x + 2y = 10 \\ -2x + 3y = 18 \end{cases}$ **28.** $\begin{cases} -x + 3y = 18 \\ -3x + 2y = 19 \end{cases}$

29. $\begin{cases} 5x + 10y = 20 \\ 2x + 6y = 10 \end{cases}$ **30.** $\begin{cases} 2x + 4y = 6 \\ 5x + 10y = 15 \end{cases}$ **31.** $\begin{cases} 3x + 6y = 9 \\ 4x + 8y = 16 \end{cases}$

32. $\begin{cases} 6x + 3y = 12 \\ 9x + 6y = 15 \end{cases}$ **33.** $\begin{cases} \frac{1}{3}x - y = 2 \\ x - 3y = 6 \end{cases}$ **34.** $\begin{cases} \frac{1}{4}x - 2y = 1 \\ x - 8y = 4 \end{cases}$

35. Explain how to identify a system with no solution when using the substitution method.

36. Occasionally, when using the substitution method, we obtain the equation $0 = 0$. Explain how this result indicates that the graphs of the equations in the system are identical.

Review and Preview

Write equivalent equations by multiplying both sides of each given equation by the given nonzero number. See Section 2.3.

37. $3x + 2y = 6$ by -2 **38.** $-x + y = 10$ by 5 **39.** $-4x + y = 3$ by 3 **40.** $5a - 7b = -4$ by -4

Add the binomials. See Section 4.4.

41. $\begin{array}{r} 3n + 6m \\ \underline{2n - 6m} \end{array}$

42. $\begin{array}{r} -2x + 5y \\ \underline{2x + 11y} \end{array}$

43. $\begin{array}{r} -5a - 7b \\ \underline{5a - 8b} \end{array}$

44. $\begin{array}{r} 9q + p \\ \underline{-9q - p} \end{array}$

Combining Concepts

Solve each system by the substitution method. First simplify each equation by combining like terms.

45. $\begin{cases} -5y + 6y = 3x + 2(x - 5) - 3x + 5 \\ 4(x + y) - x + y = -12 \end{cases}$

46. $\begin{cases} 5x + 2y - 4x - 2y = 2(2y + 6) - 7 \\ 3(2x - y) - 4x = 1 + 9 \end{cases}$

Use a graphing calculator to solve each system.

47. $\begin{cases} y = 5.1x + 14.56 \\ y = -2x - 3.9 \end{cases}$

48. $\begin{cases} y = 3.1x - 16.35 \\ y = -9.7x + 28.45 \end{cases}$

49. $\begin{cases} 3x + 2y = 14.04 \\ 5x + y = 18.5 \end{cases}$

50. $\begin{cases} x + y = -15.2 \\ -2x + 5y = -19.3 \end{cases}$

51. For the years 1970 through 1999, the annual percentage y of U.S. households that used fuel oil to heat their homes is given by the equation $y = -0.52x + 24.89$, where x is the number of years after 1970. For the same period the annual percentage y of U.S. households that used electricity to heat their homes is given by the equation $y = 0.76x + 8.97$, where x is the number of years after 1970. (*Source:* U.S. Census Bureau, American Housing Survey Branch)

a. Use the substitution method to solve this system of equations. (Round your final results to the nearest whole numbers.)

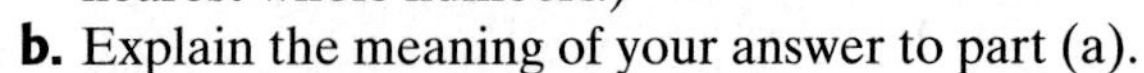

b. Explain the meaning of your answer to part (a).

c. Sketch a graph of the system of equations. Write a sentence describing the use of fuel oil and electricity for heating homes between 1970 and 1999.

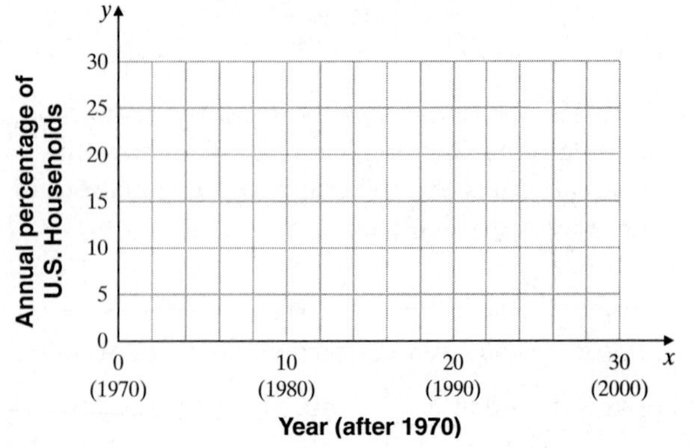

52. The number y of music CDs (in millions) shipped to retailers in the United States from 1990 through 2000 is given by the equation $y = 69.6x + 303.8$, where x is the number of years since 1990. The number y of music cassettes (in millions) shipped to retailers in the United States from 1990 through 2000 is given by the equation $y = -35.0x + 437.2$, where x is the number of years since 1990. (*Source:* Recording Industry Association of America)

a. Use the substitution method to solve this system of equations. (Round x to the nearest tenth and y to the nearest whole.)

b. Explain the meaning of your answer to part (a).

c. Sketch a graph of the system of equations. Write a sentence describing the trends in the popularity of these two types of music formats.

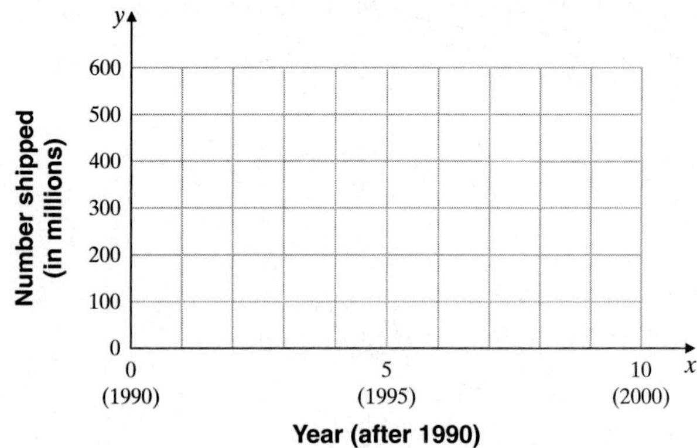

8.3 Solving Systems of Linear Equations by Addition

OBJECTIVE

A Use the addition method to solve a system of linear equations.

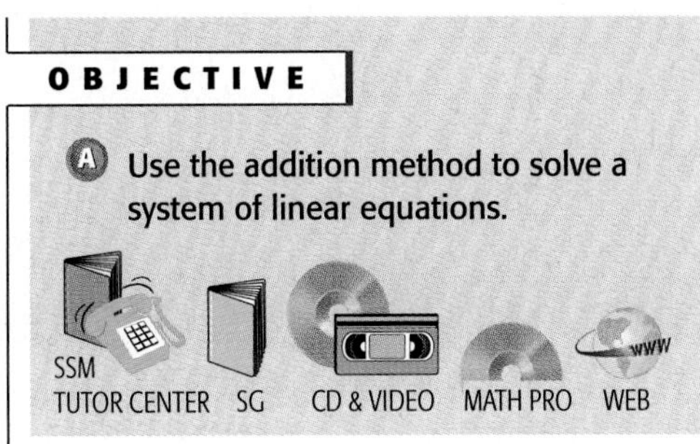

A Using the Addition Method

We have seen that substitution is an accurate method for solving a system of linear equations. Another accurate method is the **addition** or **elimination method**. The addition method is based on the addition property of equality: Adding equal quantities to both sides of an equation does not change the solution of the equation. In symbols,

if $A = B$ and $C = D$, then $A + C = B + D$

EXAMPLE 1 Solve the system: $\begin{cases} x + y = 7 \\ x - y = 5 \end{cases}$

Solution: Since the left side of each equation is equal to its right side, we are adding equal quantities when we add the left sides of the equations together and the right sides of the equations together. This adding eliminates the variable y and gives us an equation in one variable, x. We can then solve for x.

$x + y = 7$ First equation

$x - y = 5$ Second equation

$2x = 12$ Add the equations to eliminate y.

$x = 6$ Divide both sides by 2.

The x-value of the solution is 6. To find the corresponding y-value, we let $x = 6$ in either equation of the system. We will use the first equation.

$x + y = 7$ First equation

$6 + y = 7$ Let $x = 6$.

$y = 1$ Solve for y.

The solution is $(6, 1)$.

Check: Check the solution in both equations.

First Equation

$x + y = 7$

$6 + 1 \stackrel{?}{=} 7$ Let $x = 6$ and $y = 1$.

$7 = 7$ True

Second Equation

$x - y = 5$

$6 - 1 \stackrel{?}{=} 5$ Let $x = 6$ and $y = 1$.

$5 = 5$ True

Thus, the solution of the system is $(6, 1)$ and the graphs of the two equations intersect at the point $(6, 1)$ as shown.

Practice Problem 1

Use the addition method to solve the system: $\begin{cases} x + y = 13 \\ x - y = 5 \end{cases}$

Answer

1. $(9, 4)$

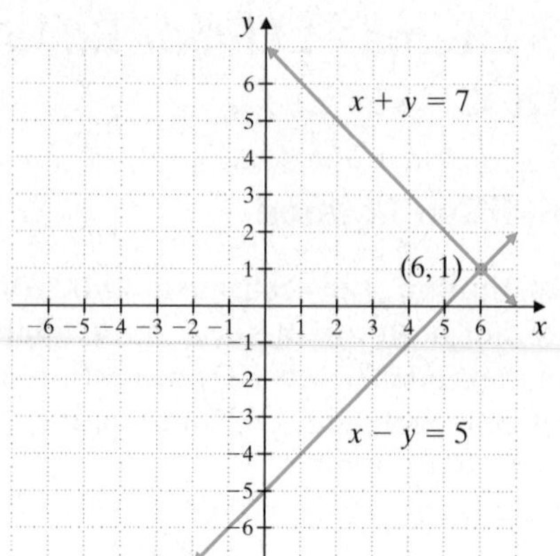

Practice Problem 2

Solve the system: $\begin{cases} 2x - y = -6 \\ -x + 4y = 17 \end{cases}$

EXAMPLE 2 Solve the system: $\begin{cases} -2x + y = 2 \\ -x + 3y = -4 \end{cases}$

Solution: If we simply add these two equations, the result is still an equation in two variables. However, our goal is to eliminate one of the variables so that we have an equation in the other variable. To do this, notice what happens if we multiply *both sides* of the first equation by −3. We are allowed to do this by the multiplication property of equality. Then the system

$$\begin{cases} -3(-2x + y) = -3(2) \\ -x + 3y = -4 \end{cases} \quad \text{simplifies to} \quad \begin{cases} 6x - 3y = -6 \\ -x + 3y = -4 \end{cases}$$

When we add the resulting equations, the y variable is eliminated.

$$\begin{aligned} 6x - 3y &= -6 \\ -x + 3y &= -4 \\ \hline 5x \qquad &= -10 \quad \text{Add.} \\ x &= -2 \quad \text{Divide both sides by 5.} \end{aligned}$$

To find the corresponding y-value, we let $x = -2$ in either of the original equations. We use the first equation of the original system.

$$\begin{aligned} -2x + y &= 2 \quad \text{First equation} \\ -2(-2) + y &= 2 \quad \text{Let } x = -2. \\ 4 + y &= 2 \\ y &= -2 \end{aligned}$$

Check the ordered pair $(-2, -2)$ in both equations of the *original* system. The solution is $(-2, -2)$.

Helpful Hint

When finding the second value of an ordered pair solution, any equation equivalent to one of the original equations in the system may be used.

In Example 2, the decision to multiply the first equation by −3 was no accident. **To eliminate a variable** when adding two equations, **the coefficient**

Answer

2. $(-1, 4)$

of the variable in one equation must be the opposite of its coefficient in the other equation.

Be sure to multiply *both sides* of an equation by a chosen number when solving by the addition method. A common mistake is to multiply only the side containing the variables.

EXAMPLE 3 Solve the system: $\begin{cases} 2x - y = 7 \\ 8x - 4y = 1 \end{cases}$

Solution: When we multiply both sides of the first equation by -4, the resulting coefficient of x is -8. This is the opposite of 8, the coefficient of x in the second equation. Then the system

$\begin{cases} -4(2x - y) = -4(7) \\ 8x - 4y = 1 \end{cases}$ simplifies to

$$\begin{cases} -8x + 4y = -28 \\ 8x - 4y = 1 \end{cases}$$
$$0 = -27 \quad \text{Add the equations.}$$

Helpful Hint

Don't forget to multiply both sides by -4.

When we add the equations, both variables are eliminated and we have $0 = -27$, a false statement. This means that the system has no solution. The equations, if graphed, represent parallel lines.

EXAMPLE 4 Solve the system: $\begin{cases} 3x - 2y = 2 \\ -9x + 6y = -6 \end{cases}$

Solution: First we multiply both sides of the first equation by 3 and then we add the resulting equations.

$\begin{cases} 3(3x - 2y) = 3 \cdot 2 \\ -9x + 6y = -6 \end{cases}$ simplifies to $\begin{cases} 9x - 6y = 6 \\ -9x + 6y = -6 \end{cases}$ Add the equations.

$$0 = 0$$

Both variables are eliminated and we have $0 = 0$, a true statement. This means that the system has an infinite number of solutions.

Try the Concept Check in the margin.

EXAMPLE 5 Solve the system: $\begin{cases} 3x + 4y = 13 \\ 5x - 9y = 6 \end{cases}$

Solution: We can eliminate the variable y by multiplying the first equation by 9 and the second equation by 4. Then we add the resulting equations.

$\begin{cases} 9(3x + 4y) = 9(13) \\ 4(5x - 9y) = 4(6) \end{cases}$ simplifies to $\begin{cases} 27x + 36y = 117 \\ 20x - 36y = 24 \end{cases}$

$$47x = 141 \quad \text{Add the equations.}$$
$$x = 3 \quad \text{Solve for } x.$$

To find the corresponding y-value, we let $x = 3$ in one of the original equations of the system. Doing so in any of these equations will give $y = 1$. Check to see that $(3, 1)$ satisfies each equation in the original system. The solution is $(3, 1)$.

Practice Problem 3

Solve the system: $\begin{cases} x - 3y = -2 \\ -3x + 9y = 5 \end{cases}$

Practice Problem 4

Solve the system: $\begin{cases} 2x + 5y = 1 \\ -4x - 10y = -2 \end{cases}$

Concept Check

Suppose you are solving the system

$$\begin{cases} 3x + 8y = -5 \\ 2x - 4y = 3 \end{cases}$$

You decide to use the addition method by multiplying both sides of the second equation by 2. In which of the following was the multiplication performed correctly? Explain.

a. $4x - 8y = 3$ b. $4x - 8y = 6$

Practice Problem 5

Solve the system: $\begin{cases} 4x + 5y = 14 \\ 3x - 2y = -1 \end{cases}$

Answers

3. no solution, **4.** infinite number of solutions, **5.** $(1, 2)$

Concept Check: b

If we had decided to eliminate x instead of y in Example 5, the first equation could have been multiplied by 5 and the second by -3. Try solving the original system this way to check that the solution is $(3, 1)$.

The following steps summarize how to solve a system of linear equations by the addition method.

To Solve a System of Two Linear Equations by the Addition Method

Step 1. Rewrite each equation in standard form $Ax + By = C$.

Step 2. If necessary, multiply one or both equations by a nonzero number so that the coefficients of a chosen variable in the system are opposites.

Step 3. Add the equations.

Step 4. Find the value of one variable by solving the resulting equation from Step 3.

Step 5. Find the value of the second variable by substituting the value found in Step 4 into either of the original equations.

Step 6. Check the proposed solution in the original system.

Try the Concept Check in the margin.

Concept Check

Suppose you are solving the system

$$\begin{cases} -4x + 7y = 6 \\ x + 2y = 5 \end{cases}$$

by the addition method.

a. What step(s) should you take if you wish to eliminate x when adding the equations?

b. What step(s) should you take if you wish to eliminate y when adding the equations?

EXAMPLE 6 Solve the system: $\begin{cases} -x - \dfrac{y}{2} = \dfrac{5}{2} \\ \dfrac{x}{6} - \dfrac{y}{2} = 0 \end{cases}$

Solution: We begin by clearing each equation of fractions. To do so, we multiply both sides of the first equation by the LCD 2 and both sides of the second equation by the LCD 6. Then the system

$$\begin{cases} 2\left(-x - \dfrac{y}{2}\right) = 2\left(\dfrac{5}{2}\right) \\ 6\left(\dfrac{x}{6} - \dfrac{y}{2}\right) = 6(0) \end{cases} \quad \text{simplifies to} \quad \begin{cases} -2x - y = 5 \\ x - 3y = 0 \end{cases}$$

We can now eliminate the variable x by multiplying the second equation by 2.

$$\begin{cases} -2x - y = 5 \\ 2(x - 3y) = 2 \cdot 0 \end{cases} \quad \text{simplifies to} \quad \begin{cases} -2x - y = 5 \\ 2x - 6y = 0 \end{cases}$$

$$-7y = 5 \quad \text{Add the equations.}$$

$$y = -\frac{5}{7} \quad \text{Solve for } y.$$

To find x, we could replace y with $-\frac{5}{7}$ in one of the equations with two variables. Instead, let's go back to the simplified system and multiply by appropriate factors to eliminate the variable y and solve for x. To do this, we multiply the first equation by -3. Then the system

$$\begin{cases} -3(-2x - y) = -3(5) \\ x - 3y = 0 \end{cases} \quad \text{simplifies to} \quad \begin{cases} 6x + 3y = -15 \\ x - 3y = 0 \end{cases}$$

$$7x = -15 \quad \text{Add the equations.}$$

$$x = -\frac{15}{7} \quad \text{Solve for } x.$$

Check the ordered pair $\left(-\frac{15}{7}, -\frac{5}{7}\right)$ in both equations of the original system.

The solution is $\left(-\frac{15}{7}, -\frac{5}{7}\right)$.

Practice Problem 6

Solve the system: $\begin{cases} -\dfrac{x}{3} + y = \dfrac{4}{3} \\ \dfrac{x}{2} - \dfrac{5}{2}y = -\dfrac{1}{2} \end{cases}$

Answers

6. $\left(-\frac{17}{2}, -\frac{3}{2}\right)$

Concept Check: **a.** multiply the second equation by 4, **b.** possible answer: multiply the first equation by -2 and the second equation by 7

Name ______________________ Section ____________ Date ______________

EXERCISE SET 8.3

 *Solve each system of equations by the addition method. See Example 1.*

1. $\begin{cases} 3x + y = 5 \\ 6x - y = 4 \end{cases}$

2. $\begin{cases} 4x + y = 13 \\ 2x - y = 5 \end{cases}$

3. $\begin{cases} x - 2y = 8 \\ -x + 5y = -17 \end{cases}$

4. $\begin{cases} x - 2y = -11 \\ -x + 5y = 23 \end{cases}$

5. $\begin{cases} 3x + 2y = 11 \\ 5x - 2y = 29 \end{cases}$

6. $\begin{cases} 4x + 2y = 2 \\ 3x - 2y = 12 \end{cases}$

7. $\begin{cases} x + y = 6 \\ x - y = 6 \end{cases}$

8. $\begin{cases} x - y = 1 \\ -x + 2y = 0 \end{cases}$

Solve each system of equations by the addition method. See Examples 2 through 5.

9. $\begin{cases} 3x + y = -11 \\ 6x - 2y = -2 \end{cases}$

10. $\begin{cases} 4x + y = -13 \\ 6x - 3y = -15 \end{cases}$

11. $\begin{cases} x + 5y = 18 \\ 3x + 2y = -11 \end{cases}$

12. $\begin{cases} x + 4y = 14 \\ 5x + 3y = 2 \end{cases}$

13. $\begin{cases} 2x - 5y = 4 \\ 3x - 2y = 4 \end{cases}$

14. $\begin{cases} 6x - 5y = 7 \\ 4x - 6y = 7 \end{cases}$

15. $\begin{cases} 2x + 3y = 0 \\ 4x + 6y = 3 \end{cases}$

16. $\begin{cases} -x + 5y = -1 \\ 3x - 15y = 3 \end{cases}$

17. $\begin{cases} 3x + y = 4 \\ 9x + 3y = 6 \end{cases}$

18. $\begin{cases} 2x + y = 6 \\ 4x + 2y = 12 \end{cases}$

19. $\begin{cases} 3x - 2y = 7 \\ 5x + 4y = 8 \end{cases}$

20. $\begin{cases} 6x - 5y = 25 \\ 4x + 15y = 13 \end{cases}$

21. $\begin{cases} \frac{2}{3}x + 4y = -4 \\ 5x + 6y = 18 \end{cases}$

22. $\begin{cases} \frac{3}{2}x + 4y = 1 \\ 9x + 24y = 5 \end{cases}$

23. $\begin{cases} 4x - 6y = 8 \\ 6x - 9y = 12 \end{cases}$

24. $\begin{cases} 9x - 3y = 12 \\ 12x - 4y = 18 \end{cases}$

25. $\begin{cases} 8x = -11y - 16 \\ 2x + 3y = -4 \end{cases}$

26. $\begin{cases} 10x + 3y = -12 \\ 5x = -4y - 16 \end{cases}$

27. When solving a system of equations by the addition method, how do we know when the system has no solution?

28. Explain why the addition method might be preferred over the substitution method for solving the system $\begin{cases} 2x - 3y = 5 \\ 5x + 2y = 6. \end{cases}$

Solve each system of equations by the addition method. See Example 6.

29. $\begin{cases} \frac{x}{3} + \frac{y}{6} = 1 \\ \frac{x}{2} - \frac{y}{4} = 0 \end{cases}$

30. $\begin{cases} \frac{x}{2} + \frac{y}{8} = 3 \\ x - \frac{y}{4} = 0 \end{cases}$

31. $\begin{cases} x - \frac{y}{3} = -1 \\ -\frac{x}{2} + \frac{y}{8} = \frac{1}{4} \end{cases}$

32. $\begin{cases} 2x - \frac{3y}{4} = -3 \\ x + \frac{y}{9} = \frac{13}{3} \end{cases}$

33. $\begin{cases} \dfrac{x}{3} - y = 2 \\ -\dfrac{x}{2} + \dfrac{3y}{2} = -3 \end{cases}$

34. $\begin{cases} \dfrac{x}{2} + \dfrac{y}{4} = 1 \\ -\dfrac{x}{4} - \dfrac{y}{8} = 1 \end{cases}$

35. $\begin{cases} \dfrac{3}{5}x - y = -\dfrac{4}{5} \\ 3x + \dfrac{y}{2} = -\dfrac{9}{5} \end{cases}$

36. $\begin{cases} 3x + \dfrac{7}{2}y = \dfrac{3}{4} \\ -\dfrac{x}{2} + \dfrac{5}{3}y = -\dfrac{5}{4} \end{cases}$

37. $\begin{cases} 3.5x + 2.5y = 17 \\ -1.5x - 7.5y = -33 \end{cases}$

38. $\begin{cases} -2.5x - 6.5y = 47 \\ 0.5x - 4.5y = 37 \end{cases}$

39. $\begin{cases} 0.02x + 0.04y = 0.09 \\ -0.1x + 0.3y = 0.8 \end{cases}$

40. $\begin{cases} 0.04x - 0.05y = 0.105 \\ 0.2x - 0.6y = 1.05 \end{cases}$

Review and Preview

Rewrite each sentence using mathematical symbols. Do not solve the equations. See Sections 2.3 and 2.5.

41. Twice a number, added to 6, is 3 less than the number.

42. The sum of three consecutive integers is 66.

43. Three times a number, subtracted from 20, is 2.

44. Twice the sum of 8 and a number is the difference of the number and 20.

45. The product of 4 and the sum of a number and 6 is twice the number.

46. If the quotient of twice a number and 7 is subtracted from the reciprocal of the number, the result is 2.

Combining Concepts

47. Use the system of linear equations below to answer the questions.

$$\begin{cases} x + y = 5 \\ 3x + 3y = b \end{cases}$$

a. Find the value of b so that the system has an infinite number of solutions.

b. Find a value of b so that there are no solutions to the system.

48. Use the system of linear equations below to answer the questions.

$$\begin{cases} x + y = 4 \\ 2x + by = 8 \end{cases}$$

a. Find the value of b so that the system has an infinite number of solutions.

b. Find a value of b so that the system has a single solution.

Solve each system by the addition method.

49. $\begin{cases} 2x + 3y = 14 \\ 3x - 4y = -69.1 \end{cases}$

50. $\begin{cases} 5x - 2y = -19.8 \\ -3x + 5y = -3.7 \end{cases}$

3. SOLVE. Now we solve the system.

$$\begin{cases} x + y = 37 \\ x - y = 21 \end{cases}$$

Notice that the coefficients of the variable y are opposites. Let's then solve by the addition method and begin by adding the equations.

$$\begin{aligned} x + y &= 37 \\ x - y &= 21 \\ \hline 2x \quad &= 58 \qquad \text{Add the equations.} \\ x &= \frac{58}{2} = 29 \qquad \text{Divide both sides by 2.} \end{aligned}$$

Now we let $x = 29$ in the first equation to find y.

$$\begin{aligned} x + y &= 37 \qquad \text{First equation} \\ 29 + y &= 37 \\ y &= 37 - 29 = 8 \end{aligned}$$

4. INTERPRET. The solution of the system is $(29, 8)$.

Check: Notice that the sum of 29 and 8 is $29 + 8 = 37$, the required sum. Their difference is $29 - 8 = 21$, the required difference.
State: The numbers are 29 and 8.

EXAMPLE 3 Solving a Problem about Prices

The Cirque du Soleil show Alegria is performing locally. Matinee admission for 4 adults and 2 children is $374, while admission for 2 adults and 3 children is $285.

a. What is the price of an adult's ticket?

b. What is the price of a child's ticket?

c. Suppose that a special rate of $1000 is offered for groups of 20 persons. Should a group of 4 adults and 16 children use the group rate? Why or why not?

Solution:

1. UNDERSTAND. Read and reread the problem and guess a solution. Let's suppose that the price of an adult's ticket is $50 and the price of a child's ticket is $40. To check our proposed solution, let's see if admission for 4 adults and 2 children is $374. Admission for 4 adults is 4($50) or $200 and admission for 2 children is 2($40) or $80. This gives a total admission of $200 + $80 = $280, not the required $374. Again though, we have

Practice Problem 3

Admission prices at a local weekend fair were $5 for children and $7 for adults. The total money collected was $3379, and 587 people attended the fair. How many children and how many adults attended the fair?

Answer

3. 365 children and 222 adults

accomplished the purpose of this process: We have a better understanding of the problem. To continue, we let

$A =$ the price of an adult's ticket and

$C =$ the price of a child's ticket

2. TRANSLATE. We translate the problem into two equations using both variables.

In words:	admission for 4 adults	and	admission for 2 children	is	\$374
	↓	↓	↓	↓	↓
Translate:	$4A$	$+$	$2C$	$=$	374

In words:	admission for 2 adults	and	admission for 3 children	is	\$285
	↓	↓	↓	↓	↓
Translate:	$2A$	$+$	$3C$	$=$	285

3. SOLVE. We solve the system.

$$\begin{cases} 4A + 2C = 374 \\ 2A + 3C = 285 \end{cases}$$

Since both equations are written in standard form, we solve by the addition method. First we multiply the second equation by -2 so that when we add the equations we eliminate the variable A. Then the system

$$\begin{cases} 4A + 2C = 374 \\ -2(2A + 3C) = -2(285) \end{cases} \quad \text{simplifies to} \quad \begin{cases} 4A + 2C = 374 \\ -4A - 6C = -570 \end{cases}$$

$$\text{Add the equations.} \qquad -4C = -196$$

$$-4C = -196$$

$$C = \frac{-196}{-4} = 49 \text{ or \$49, the children's ticket price.}$$

To find A, we replace C with 49 in the first equation.

$$4A + 2C = 374 \qquad \text{First equation}$$

$$4A + 2(49) = 374 \qquad \text{Let } C = 49.$$

$$4A + 98 = 374$$

$$4A = 276$$

$$A = \frac{276}{4} = 69 \text{ or \$69, the adult's ticket price.}$$

4. INTERPRET.

Check: Notice that 4 adults and 2 children will pay

$4(\$69) + 2(\$49) = \$276 + \$98 = \$374$, the required amount. Also, the price for 2 adults and 3 children is $2(\$69) + 3(\$49) = \$138 + \$147 = \$285$, the required amount.

State: Answer the three original questions.

a. Since $A = 69$, the price of an adult's ticket is \$69.

b. Since $C = 49$, the price of a child's ticket is \$49.

c. The regular admission price for 4 adults and 16 children is

$$4(\$69) + 16(\$49) = \$276 + \$784$$
$$= \$1060$$

This is \$60 more than the special group rate of \$1000, so they should request the group rate.

EXAMPLE 4 Finding Rates

Betsy Beasley and Alfredo Drizarry live 15 miles away from each other. They decide to meet one day by walking toward one another. After 2 hours they meet. If Betsy walks one mile per hour faster than Alfredo, find both walking speeds.

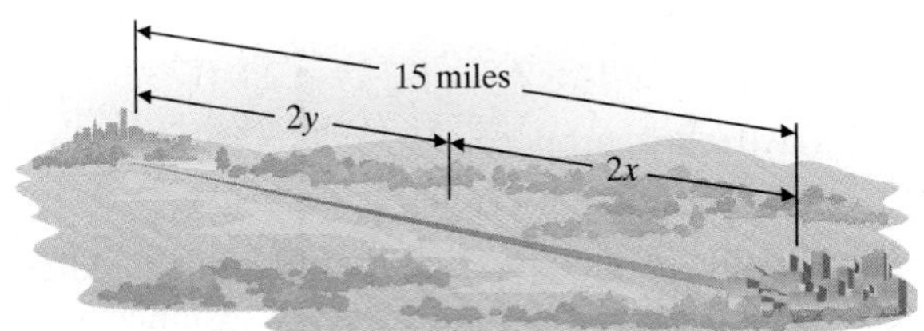

Solution:

1. UNDERSTAND. Read and reread the problem. Let's propose a solution and use the formula $d = r \cdot t$ to check. Suppose that Betsy's rate is 4 miles per hour. Since Betsy's rate is 1 mile per hour faster, Alfredo's rate is 3 miles per hour. To check, see if they can walk a total of 15 miles in 2 hours. Betsy's distance is rate · time $= 4(2) = 8$ miles and Alfredo's distance is rate · time $= 3(2) = 6$ miles. Their total distance is 8 miles + 6 miles = 14 miles, not the required 15 miles. Now that we have a better understanding of the problem, let's model it with a system of equations.

   First, we let

   x = Alfredo's rate in miles per hour and

   y = Betsy's rate in miles per hour

   Now we use the facts stated in the problem and the formula $d = rt$ to fill in the following chart.

	r	· t	= d
Alfredo	x	2	$2x$
Betsy	y	2	$2y$

2. TRANSLATE. We translate the problem into two equations using both variables.

In words:	Alfredo's distance	+	Betsy's distance	=	15
	↓		↓		↓
Translate:	$2x$	+	$2y$	=	15

In words:	Betsy's rate	is	1 mile per hour faster than Alfredo's
	↓	↓	↓
Translate:	y	=	$x + 1$

Practice Problem 4

Two cars are 440 miles apart and traveling toward each other. They meet in 3 hours. If one car's speed is 10 miles per hour faster than the other car's speed, find the speed of each car.

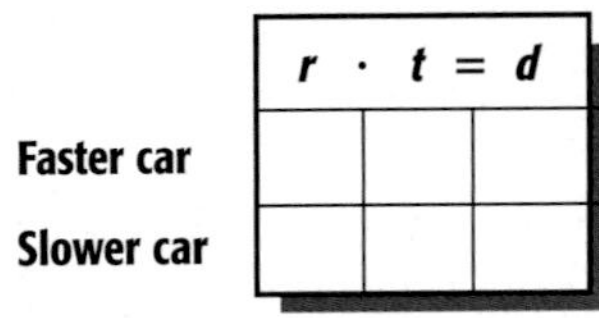

	r	· t	= d
Faster car			
Slower car			

Answer

4. One car's speed is $68\frac{1}{3}$ mph and the other car's speed is $78\frac{1}{3}$ mph.

3. SOLVE. The system of equations we are solving is

$$\begin{cases} 2x + 2y = 15 \\ y = x + 1 \end{cases}$$

Let's use substitution to solve the system since the second equation is solved for y.

$$2x + 2y = 15 \qquad \text{First equation}$$

$$2x + 2(x + 1) = 15 \qquad \text{Replace } y \text{ with } x + 1.$$

$$2x + 2x + 2 = 15$$

$$4x = 13$$

$$x = \frac{13}{4} = 3.25$$

$$y = x + 1 = 3.25 + 1 = 4.25$$

4. INTERPRET. Alfredo's proposed rate is 3.25 miles per hour and Betsy's proposed rate is 4.25 miles per hour.

Check: Use the formula $d = rt$ and find that in 2 hours, Alfredo's distance is (3.25)(2) miles or 6.5 miles. In 2 hours, Betsy's distance is (4.25)(2) miles or 8.5 miles. The total distance walked is 6.5 miles + 8.5 miles or 15 miles, the given distance.

State: Alfredo walks at a rate of 3.25 miles per hour and Betsy walks at a rate of 4.25 miles per hour. ●

Practice Problem 5

Barb Hayes, a pharmacist, needs 50 liters of a 60% alcohol solution. She currently has available a 20% solution and a 70% solution. How many liters of each must she use to make the needed 50 liters of 60% alcohol solution?

EXAMPLE 5 Finding Amounts of Solutions

Eric Daly, a chemistry teaching assistant, needs 10 liters of a 20% saline solution (salt water) for his 2 p.m. laboratory class. Unfortunately, the only mixtures on hand are a 5% saline solution and a 25% saline solution. How much of each solution should he mix to produce the 20% solution?

Solution:

1. UNDERSTAND. Read and reread the problem. Suppose that we need 4 liters of the 5% solution. Then we need 10 − 4 = 6 liters of the 25% solution. To see if this gives us 10 liters of a 20% saline solution, let's find the amount of pure salt in each solution.

	concentration rate	×	amount of solution	=	amount of pure salt
	↓		↓		↓
5% solution:	0.05	×	4 liters	=	0.2 liters
25% solution:	0.25	×	6 liters	=	1.5 liters
20% solution:	0.20	×	10 liters	=	2 liters

Since 0.2 liters + 1.5 liters = 1.7 liters, not 2 liters, our proposed solution is incorrect. But we have gained some insight into how to model and check this problem.

We let

x = number of liters of 5% solution and

y = number of liters of 25% solution

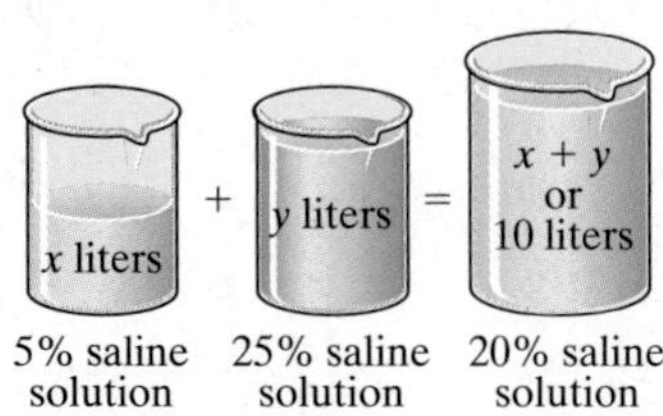

Answer

5. 10 liters of the 20% alcohol solution and 40 liters of the 70% alcohol solution

Now we use a table to organize the given data.

	Concentration Rate	Liters of Solution	Liters of Pure Salt
First solution	5%	x	$0.05x$
Second solution	25%	y	$0.25y$
Mixture needed	20%	10	$(0.20)(10)$

2. TRANSLATE. We translate into two equations using both variables.

In words:	liters of 5% solution	+	liters of 25% solution	=	10
	↓		↓		↓
Translate:	x	+	y	=	10

In words:	salt in 5% solution	+	salt in 25% solution	=	salt in mixture
	↓		↓		↓
Translate:	$0.05x$	+	$0.25y$	=	$(0.20)(10)$

3. SOLVE. Here we solve the system

$$\begin{cases} x + y = 10 \\ 0.05x + 0.25y = 2 \end{cases}$$

To solve by the addition method, we first multiply the first equation by -25 and the second equation by 100. Then the system

$$\begin{cases} -25(x + y) = -25(10) \\ 100(0.05x + 0.25y) = 100(2) \end{cases} \quad \text{simplifies to} \quad \begin{cases} -25x - 25y = -250 \\ 5x + 25y = 200 \\ \hline -20x \qquad = -50 \quad \text{Add.} \\ x = 2.5 \end{cases}$$

To find y, we let $x = 2.5$ in the first equation of the original system.

$$\begin{aligned} x + y &= 10 \\ 2.5 + y &= 10 \qquad \text{Let } x = 2.5. \\ y &= 7.5 \end{aligned}$$

4. INTERPRET. Thus, we propose that Eric needs to mix 2.5 liters of 5% saline solution with 7.5 liters of 25% saline solution.

Check: Notice that $2.5 + 7.5 = 10$, the required number of liters. Also, the sum of the liters of salt in the two solutions equals the liters of salt in the required mixture:

$$\begin{aligned} 0.05(2.5) + 0.25(7.5) &= 0.20(10) \\ 0.125 + 1.875 &= 2 \end{aligned}$$

State: Eric needs 2.5 liters of the 5% saline solution and 7.5 liters of the 25% solution. ●

Try the Concept Check in the margin.

Concept Check

Suppose you mix an amount of a 30% acid solution with an amount of a 50% acid solution. Which of the following acid strengths would be possible for the resulting acid mixture?

a. 22% b. 44% c. 63%

Answer

Concept Check: b

FOCUS ON Mathematical Connections

SOLVING NONLINEAR SYSTEMS

Recall that a linear equation in two variables is an equation that can be written in the form $Ax + By = C$. By this definition, we can see that an equation of the form $\frac{A}{x} + \frac{B}{y} = C$ is clearly not linear. However, with a slight adjustment, we can solve a nonlinear system such as

$$\begin{cases} \dfrac{A}{x} + \dfrac{B}{y} = C \\ \dfrac{D}{x} + \dfrac{E}{y} = F \end{cases}$$

using the methods we already know for solving linear systems.

To solve such a system, first make the following substitutions. Let $w = \frac{1}{x}$ and $z = \frac{1}{y}$ in both equations. Then

$$\begin{cases} \dfrac{A}{x} + \dfrac{B}{y} = C \\ \dfrac{D}{x} + \dfrac{E}{y} = F \end{cases} \quad \text{becomes} \quad \begin{cases} Aw + Bz = C \\ Dw + Ez = F \end{cases}$$

This new system of equations is linear and can be solved with any of the techniques we already know. Once the values of w and z have been found, simply substitute them into the equations $w = \frac{1}{x}$ and $z = \frac{1}{y}$. Then solve each equation to find the value of x and the value of y.

CRITICAL THINKING

Apply the method described above to solve each nonlinear system.

1. $\begin{cases} \dfrac{2}{x} + \dfrac{3}{y} = 5 \\ \dfrac{5}{x} - \dfrac{3}{y} = 2 \end{cases}$

2. $\begin{cases} x + \dfrac{2}{y} = 7 \\ 3x + \dfrac{3}{y} = 6 \end{cases}$

3. $\begin{cases} \dfrac{3}{x} - \dfrac{2}{y} = -18 \\ \dfrac{2}{x} + \dfrac{3}{y} = 1 \end{cases}$

4. $\begin{cases} \dfrac{2}{x} - \dfrac{4}{y} = 5 \\ \dfrac{1}{x} - \dfrac{2}{y} = \dfrac{3}{2} \end{cases}$

Name ____________________ Section __________ Date __________

Mental Math

Without actually solving each problem, choose the correct solution by deciding which choice satisfies the given conditions.

△ **1.** The length of a rectangle is 3 feet longer than the width. The perimeter is 30 feet. Find the dimensions of the rectangle.
a. length = 8 feet; width = 5 feet
b. length = 8 feet; width = 7 feet
c. length = 9 feet; width = 6 feet

△ **2.** An isosceles triangle, a triangle with two sides of equal length, has a perimeter of 20 inches. Each of the equal sides is one inch longer than the third side. Find the lengths of the three sides.
a. 6 inches, 6 inches, and 7 inches
b. 7 inches, 7 inches, and 6 inches
c. 6 inches, 7 inches, and 8 inches

3. Two computer disks and three notebooks cost $17. However, five computer disks and four notebooks cost $32. Find the price of each.
a. notebook = $4;
computer disk = $3
b. notebook = $3;
computer disk = $4
c. notebook = $5;
computer disk = $2

4. Two music CDs and four music cassette tapes cost a total of $40. However, three music CDs and five cassette tapes cost $55. Find the price of each.
a. CD = $12; cassette = $4
b. CD = $15; cassette = $2
c. CD = $10; cassette = $5

5. Kesha has a total of 100 coins, all of which are either dimes or quarters. The total value of the coins is $13.00. Find the number of each type of coin.
a. 80 dimes; 20 quarters
b. 20 dimes; 44 quarters
c. 60 dimes; 40 quarters

6. Yolanda has 28 gallons of saline solution available in two large containers at her pharmacy. One container holds three times as much as the other container. Find the capacity of each container.
a. 15 gallons; 5 gallons
b. 20 gallons; 8 gallons
c. 21 gallons; 7 gallons

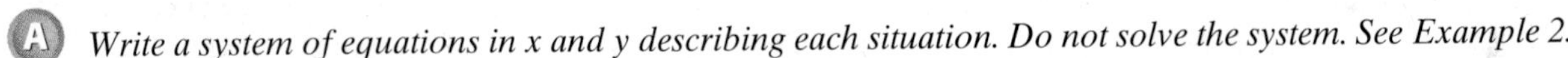

EXERCISE SET 8.4

A *Write a system of equations in x and y describing each situation. Do not solve the system. See Example 2.*

1. Two numbers add up to 15 and have a difference of 7. Let x represent the larger number.

2. The total of two numbers is 16. The first number plus 2 more than 3 times the second equals 18. Let x represent the first number.

3. Keiko has a total of $6500, which she has invested in two accounts. The larger account is $800 greater than the smaller account. Let x represent the larger account.

4. Dominique has four times as much money in his savings account as in his checking account. The total amount is $2300. Let x represent the amount of money in his checking account.

Solve. See Example 2.

5. Two numbers total 83 and have a difference of 17. Find the two numbers.

6. The sum of two numbers is 76 and their difference is 52. Find the two numbers.

7. A first number plus twice a second number is 8. Twice the first number plus the second totals 25. Find the numbers.

8. One number is 4 more than twice the second number. Their total is 25. Find the numbers.

9. The highest scorer during the WNBA 2000 regular season was Katie Smith of the Minnesota Lynx. Over the season, Smith scored 3 more points than the second-highest scorer, Sheryl Swoopes of the Houston Comets. Together, Smith and Swoopes scored 1289 points during the 2000 regular season. How many points did each player score over the course of the season? (*Source:* Women's National Basketball Association)

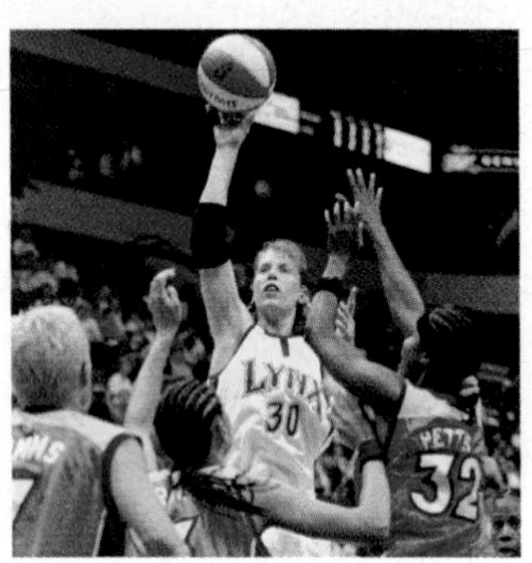

10. Pavel Bure of the Florida Panthers was the NHL's leading goal scorer during the 2000–2001 regular season. Alexei Kovalev of the Pittsburgh Penguins, who was ranked fifth for goals, scored 15 fewer goals than Bure. Together, these two players made a total of 103 goals during the 2000–2001 regular season. How many goals each did Bure and Kovalev make? (*Source:* National Hockey League)

Solve. See Example 3.

11. Ann Marie Jones has been pricing Amtrak train fares for a group trip to New York. Three adults and four children must pay \$159. Two adults and three children must pay \$112. Find the price of an adult's ticket, and find the price of a child's ticket.

12. Last month, Jerry Papa purchased five cassettes and two compact discs at Wall-to-Wall Sound for \$65. This month he bought three cassettes and four compact discs for \$81. Find the price of each cassette, and find the price of each compact disc.

13. Johnston and Betsy Waring have a jar containing 80 coins, all of which are either quarters or nickels. The total value of the coins is \$14.60. How many of each type of coin do they have?

14. Art and Bette Meish purchased 40 stamps, a mixture of 34¢ and 21¢ stamps. Find the number of each type of stamp if they spent \$12.95.

15. David and Jacquelyn Bick own 30 shares of General Electric Co. stock and 55 shares of The Ohio Art Company (makers of Etch A Sketch and other toys). At the close of the markets on a particular day in 2001, their stock portfolio consisting of these two stocks was worth \$2348.10. The closing price of General Electric stock was \$35.77 more per share than the closing price of The Ohio Art Company stock on that day. What was the closing price of each stock on that day? (*Source:* Bridge Information Services)

16. Caitlin Jackson has an investment in Polaroid and AOL Time Warner stock. On a particular day in 2001, Polaroid stock closed at \$3 per share and AOL Time Warner stock closed at \$52.80 per share. Caitlin's portfolio made up of these two stocks was worth \$3255 at the end of the day. If Caitlin owns 31 more shares of AOL Time Warner stock than Polaroid stock, how many shares of each type of stock does she own? (*Source:* Bridge Information Services)

17. Cyril and Anoa Nantambu operate a small construction and supply company. In July they charged the Shaffers \$1702.50 for 65 hours of labor and 3 tons of material. In August the Shaffers paid \$1349 for 49 hours of labor and $2\frac{1}{2}$ tons of material. Find the cost per hour of labor and the cost per ton of material.

18. Joan Gundersen rented a car from Hertz, which rents its cars for a daily fee plus an additional charge per mile driven. Joan recalls that a car rented for 5 days and driven for 300 miles cost her \$178, while a car rented for 4 days and driven for 500 miles cost \$197. Find the daily fee, and find the mileage charge.

Solve. See Example 4.

19. Pratap Puri rowed 18 miles down the Delaware River in 2 hours, but the return trip took him $4\frac{1}{2}$ hours. Find the rate Pratap can row in still water, and find the rate of the current.
Let x = rate Pratap can row in still water and
y = rate of the current

	d =	r	· t
Downstream	18	$x + y$	2
Upstream	18	$x - y$	$4\frac{1}{2}$

20. The Jonathan Schultz family took a canoe 10 miles down the Allegheny River in 1 hour and 15 minutes. After lunch it took them 4 hours to return. Find the rate of the current.
Let x = rate the family can row in still water and
y = rate of the current

	d =	r	· t
Downstream	10	$x + y$	$1\frac{1}{4}$
Upstream	10	$x - y$	4

21. Dave and Sandy Hartranft are frequent flyers with Delta Airlines. They often fly from Philadelphia to Chicago, a distance of 780 miles. On one particular trip they fly into the wind, and the flight takes 2 hours. The return trip, with the wind behind them, only takes $1\frac{1}{2}$ hours. Find the speed of the wind and find the speed of the plane in still air.

22. With a strong wind behind it, a United Airlines jet flies 2400 miles from Los Angeles to Orlando in 4 hours and 45 minutes. The return trip takes 6 hours, as the plane flies into the wind. Find the speed of the plane in still air, and find the wind speed to the nearest tenth of a mile per hour.

23. Jim Williamson began a 186-mile bicycle trip to build up stamina for a triathlete competition. Unfortunately, his bicycle chain broke, so he finished the trip walking. The whole trip took 6 hours. If Jim walks at a rate of 4 miles per hour and rides at 40 mph, find the amount of time he spent on the bicycle.

24. In Canada, eastbound and westbound trains travel along the same track, with sidings to pull onto to avoid accidents. Two trains are now 150 miles apart, with the westbound train traveling twice as fast as the eastbound train. A warning must be issued to pull one train onto a siding or else the trains will crash in $1\frac{1}{4}$ hours. Find the speed of the eastbound train and the speed of the westbound train.

Solve. See Example 5.

25. Dorren Schmidt is a chemist with Gemco Pharmaceutical. She needs to prepare 12 ounces of a 9% hydrochloric acid solution. Find the amount of a 4% solution and the amount of a 12% solution she should mix to get this solution.

Concentration Rate	Liters of Solution	Liters of Pure Acid
0.04	x	$0.04x$
0.12	y	?
0.09	12	?

26. Elise Everly is preparing 15 liters of a 25% saline solution. Elise has two other saline solutions with strengths of 40% and 10%. Find the amount of 40% solution and the amount of 10% solution she should mix to get 15 liters of a 25% solution.

Concentration Rate	Liters of Solution	Liters of Pure Salt
0.40	x	$0.40x$
0.10	y	?
0.25	15	?

27. Wayne Osby blends coffee for a local coffee café. He needs to prepare 200 pounds of blended coffee beans selling for \$3.95 per pound. He intends to do this by blending together a high-quality bean costing \$4.95 per pound and a cheaper bean costing \$2.65 per pound. To the nearest pound, find how much high-quality coffee bean and how much cheaper coffee bean he should blend.

28. Macadamia nuts cost an astounding \$16.50 per pound, but research by an independent firm says that mixed nuts sell better if macadamias are included. The standard mix costs \$9.25 per pound. Find how many pounds of macadamias and how many pounds of the standard mix should be combined to produce 40 pounds that will cost \$10 per pound. Find the amounts to the nearest tenth of a pound.

Solve. See Examples 1 through 5.

△ **29.** Recall that two angles are complementary if their sum is 90°. Find the measures of two complementary angles if one angle is twice the other.

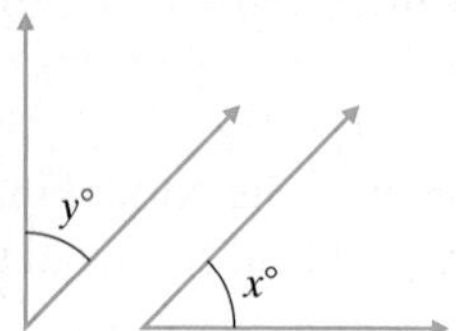

△ **30.** Recall that two angles are supplementary if their sum is 180°. Find the measures of two supplementary angles if one angle is 20° more than four times the other.

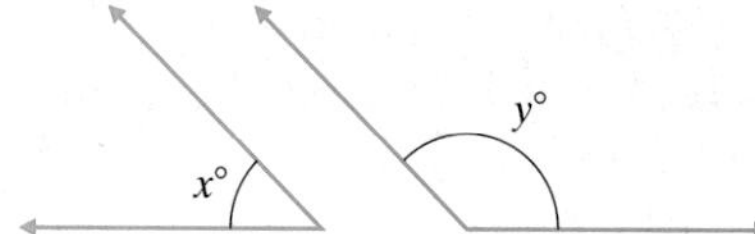

△ **31.** Find the measures of two complementary angles if one angle is 10° more than three times the other.

△ **32.** Find the measures of two supplementary angles if one angle is 18° more than twice the other.

33. In the United States, the percent of women using the Internet is increasing faster than the percent of men. For the years 1996–2001, the function $y = 7x + 18.7$ can be used to estimate the percent of females using the Internet while the function $y = 6x + 27.7$ can be used to estimate the percent of males. For both functions, x is the number of years since 1996. If this trend continues, predict the year in which the percent of females using the Internet is equal to the percent of males. (*Source:* Pew Internet & American Life Project)

34. The percent of car-vehicle sales is decreasing while the percent of light-truck (pickups, sport-utility vans and minivans) vehicle sales is increasing. For the years 1997–2000, the function $y = -x + 54.5$ can be used to estimate the percent of vehicle sales being cars while the function $y = x + 45.5$ can be used to estimate the percent of vehicle sales being light trucks. For both functions, x is the number of years since 1997.

a. If this trend continues, predict the year in which the percent of car sales equals the percent of light-truck sales.

b. Before the actual 2001 vehicle sales data was published, USA today predicted that light-truck sales would likely be greater than car sales in the year 2001. Does your prediction from part a agree with this statement? (*Source: USA Today* and Autodata)

35. Carrie and Raymond McCormick had a pottery stand at the annual Skippack Craft Fair. They sold some of their pottery at the original price of \$9.50 each, but later decreased the price of each by \$2. If they sold all 90 pieces and took in \$721, find how many they sold at the original price and how many they sold at the reduced price.

36. Trinity Church held its annual spaghetti supper and fed a total of 387 people. They charged \$6.80 for adults and half-price for children. If they took in \$2444.60, find how many adults and how many children attended the supper.

37. The Santa Fe National Historic Trail is approximately 1200 miles between Old Franklin, Missouri, and Santa Fe, New Mexico. Suppose that a group of hikers start from each town and walk the trail toward each other. They meet after a total hiking time of 240 hours. If one group travels $\frac{1}{2}$ mile per hour slower than the other group, find the rate of each group. (*Source:* National Park Service)

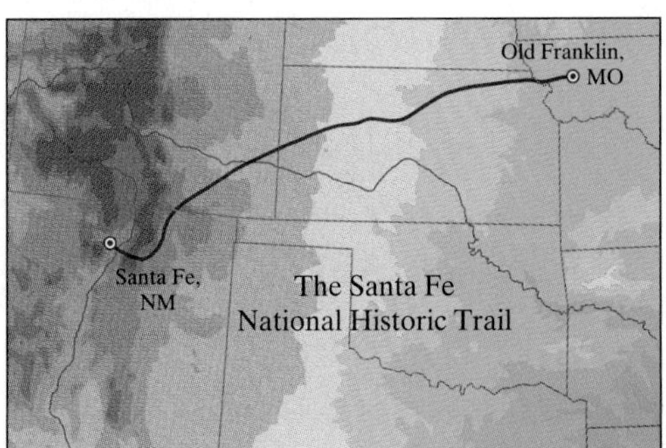

38. California 1 South is a historic highway that stretches 123 miles along the coast from Monterey to Morro Bay. Suppose that two cars start driving this highway, one from each town. They meet after 3 hours. Find the rate of each car if one car travels 1 mile per hour faster than the other car. (*Source: National Geographic*)

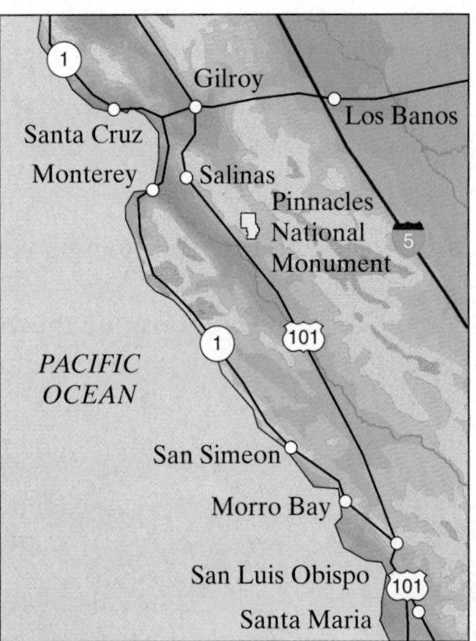

39. A 30% solution of fertilizer is to be mixed with a 60% solution of fertilizer in order to get 150 gallons of a 50% solution. How many gallons of the 30% solution and 60% solution should be mixed?

40. A 10% acid solution is to be mixed with a 50% acid solution in order to get 120 ounces of a 20% acid solution. How many ounces of the 10% solution and 50% solution should be mixed?

41. Traffic signs are regulated by the *Manual on Uniform Traffic Control Devices* (MUTCD). According to this manual, if the sign below is placed on a freeway, its perimeter must be 144 inches. Also, its length is 12 inches longer than its width. Find the dimensions of this sign.

42. According to the MUTCD (see Exercise 41), this sign must have a perimeter of 60 inches. Also, its length must be 6 inches longer than its width. Find the perimeter of this sign.

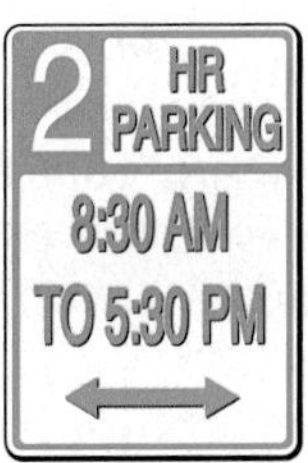

43. The percent of men 65 years of age or older in the labor force has decreased most years since 1900 while the percent of women 65 years of age or older has slightly increased. For the years 1900–2000, the function $y = -0.52x + 64.5$ approximates the percent of men 65 years of age or older in the labor force while the function $y = 0.02x + 7.2$ approximates the percent of women 65 years of age or older in the labor force. For both functions, x is the number of years after 1900. (*Source: The World Almanac*, 2002)

a. Explain how the decrease of men described can be verified by their given function while the slight increase of women described can be verified by their given function.

b. If this trend continues, determine the year when the percent of men and woman 65 years of age or older in the labor force are the same.

44. The annual U.S. per capita consumption of butter has remained about the same since 1980 while consumption of margarine has decreased. For the years 1995–1999, the function $y = 0.05x + 4.3$ approximates the annual U.S. per capita consumption of butter in pounds and the function $y = -0.31x + 9.3$ approximates the annual U.S. per capita consumption of margarine in pounds. For both functions, x is the number of years after 1995. If this trend continues, determine the year when the pounds of butter consumed equals the pounds of margarine consumed.

△ **45.** In the figure, line l and line m are parallel lines cut by transversal t. Find the values of x and y.

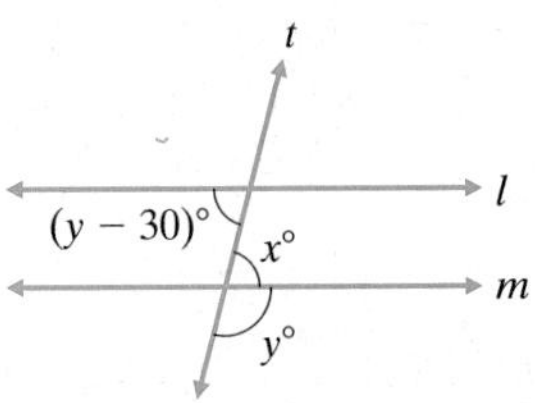

△ **46.** Find the values of x and y in the following isosceles triangle.

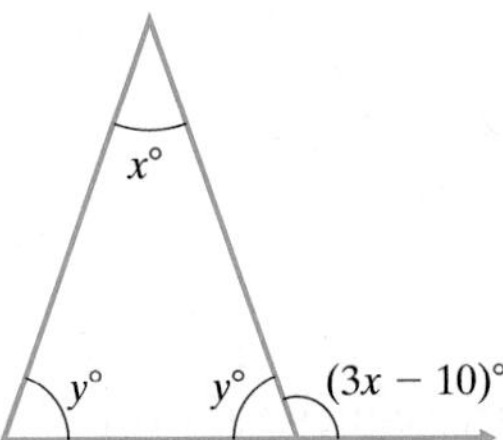

Review and Preview

Find the square of each expression. For example, the square of 7 is 7^2 or 49. The square of 5x is $(5x)^2$ or $25x^2$. See Section 4.1.

47. 4 **48.** 3 **49.** $6x$ **50.** $11y$ **51.** $10y^3$ **52.** $8x^5$

Combining Concepts

△**53.** Dale and Sharon Mahnke have decided to fence off a garden plot behind their house, using their house as the "fence" along one side of the garden. The length (which runs parallel to the house) is 3 feet less than twice the width. Find the dimensions if 33 feet of fencing is used along the three sides requiring it.

△**54.** Judy McElroy plans to erect 152 feet of fencing around her rectangular horse pasture. A river bank serves as one side length of the rectangle. If each width is 4 feet longer than half the length, find the dimensions.

Internet Excursions

Go To: http://www.prenhall.com/martin-gay_algebra

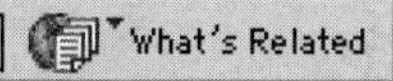

Major League Soccer (MLS) had its inaugural season in 1996, and by 2001 had expanded to 12 teams. The given World Wide Web address will provide you with access to the Major League Soccer site, or a related site, for a listing of statistics for the current regular season and links to statistics for past seasons.

55. Using actual data for the current (or most recent) season, write a problem similar to Exercises 9 and 10 involving statistics for MLS leading scorers. When you have finished writing your problem, trade with another student in your class and solve each other's problem. Then check each other's work.

56. Using actual data for the current (or most recent) season, write a problem similar to Exercises 9 and 10 involving statistics for MLS game attendance (found under Regular Season League Stats). When you have finished writing your problem, trade with another student in your class and solve each other's problem. Then check each other's work.

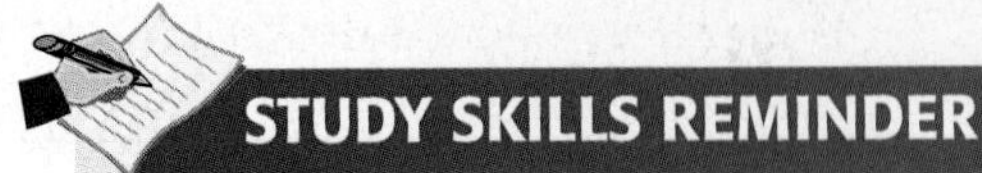

Are you satisfied with your performance on a particular quiz or exam?

If not, don't forget to analyze your quiz or exam and look for common errors.

Were most of your errors a result of

- *Carelessness?* If your errors were careless, did you turn in your work before the allotted time expired? If so, resolve next time to use the entire time allotted. Any extra time can be spent checking your work.
- *Running out of time?* If so, make a point to better manage your time on your next exam. A few suggestions are to work any questions that you are unsure of last and to check your work after all questions have been answered.
- *Not understanding a concept?* If so, review that concept and correct your work. Remember next time to make sure that all concepts on a quiz or exam are understood before the exam.

8.5 Solving Systems of Linear Equations in Three Variables

OBJECTIVES

A. Solve a system of three linear equations in three variables.

B. Solve problems that can be modeled by a system of three linear equations.

In this section, we solve systems of linear equations in three variables. We call the equation $3x - y + z = -15$, for example, a **linear equation in three variables** since there are three variables and each variable is raised only to the power 1. A solution of this equation is an **ordered triple (x, y, z)** that makes the equation a true statement.

For example, the ordered triple $(2, 0, -21)$ is a solution of $3x - y + z = -15$ since replacing x with 2, y with 0, and z with -21 yields the true statement

$$3(2) - 0 + (-21) = -15$$

The graph of this equation is a plane in three-dimensional space, just as the graph of a linear equation in two variables is a line in two-dimensional space.

Although we will not discuss the techniques for graphing equations in three variables, visualizing the possible patterns of intersecting planes gives us insight into the possible patterns of solutions of a system of three three-variable linear equations. There are four possible patterns.

1. Three planes have a single point in common. This point represents the single solution of the system. This system is **consistent**.

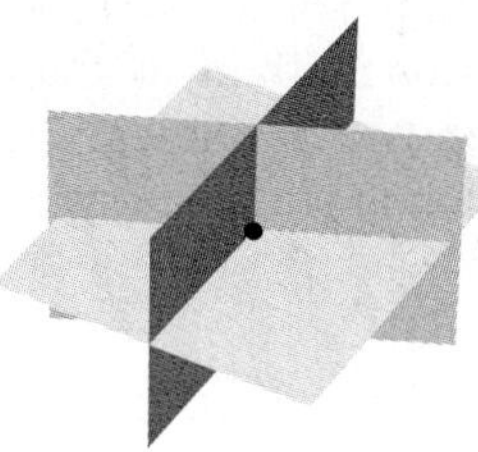

2. Three planes intersect at no point common to all three. This system has no solution. A few ways that this can occur are shown. This system is **inconsistent**.

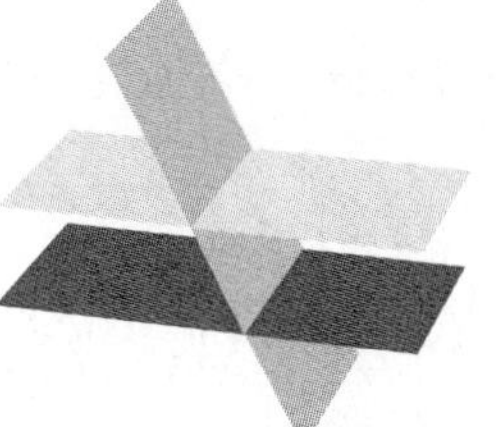

3. Three planes intersect at all the points of a single line. The system has infinitely many solutions. This system is **consistent**.

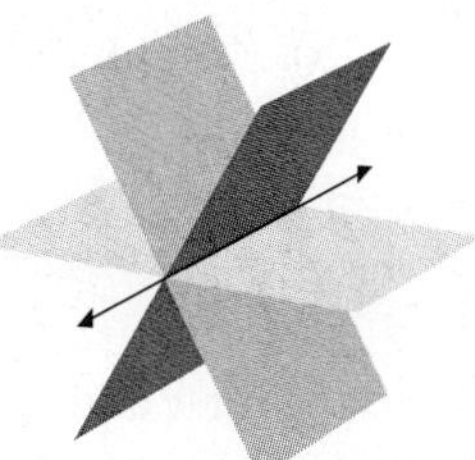

4. Three planes coincide at all points on the plane. The system is consistent, and the equations are **dependent**.

Solving a System of Three Linear Equations by the Elimination Method

Step 1. Write each equation in standard form, $Ax + By + Cz = D$.

Step 2. Choose a pair of equations and use them to eliminate a variable.

Step 3. Choose any other pair of equations and eliminate the *same variable* as in Step 2.

Step 4. Two equations in two variables should be obtained from Step 2 and Step 3. Use methods from Section 8.3 to solve this system for both variables.

Step 5. To solve for the third variable, substitute the values of the variables found in Step 4 into any of the original equations containing the third variable.

Step 6. Check the ordered triple solution in *all three* original equations.

A Solving a System of Three Linear Equations in Three Variables

Just as with systems of two equations in two variables, we can use the elimination or substitution method to solve a system of three equations in three variables. To use the elimination method, we eliminate a variable and obtain a system of two equations in two variables. Then we use the methods we learned in the previous section to solve the system of two equations.

Practice Problem 1

Solve the system:

$$\begin{cases} 2x - y + 3z = 13 \\ x + y - z = -2 \\ 3x + 2y + 2z = 13 \end{cases}$$

EXAMPLE 1 Solve the system:

$$\begin{cases} 3x - y + z = -15 & \text{Equation (1)} \\ x + 2y - z = 1 & \text{Equation (2)} \\ 2x + 3y - 2z = 0 & \text{Equation (3)} \end{cases}$$

Solution: We add equations (1) and (2) to eliminate z.

$$\begin{array}{r} 3x - y + z = -15 \\ x + 2y - z = 1 \\ \hline 4x + y = -14 \end{array} \quad \text{Equation (4)}$$

Next we add two *other* equations and *eliminate z again*. To do so, we multiply both sides of equation (1) by 2 and add this resulting equation to equation (3). Then

$$\begin{cases} 2(3x - y + z) = 2(-15) \\ 2x + 3y - 2z = 0 \end{cases} \text{ simplifies to } \begin{cases} 6x - 2y + 2z = -30 \\ 2x + 3y - 2z = 0 \end{cases}$$
$$\overline{8x + y = -30} \quad \text{Equation (5)}$$

Now we solve equations (4) and (5) for x and y. To solve by elimination, we multiply both sides of equation (4) by -1 and add this resulting equation to equation (5). Then

$$\begin{cases} -1(4x + y) = -1(-14) \\ 8x + y = -30 \end{cases} \text{ simplifies to } \begin{cases} -4x - y = 14 \\ 8x + y = -30 \end{cases}$$
$$\begin{array}{rl} 4x = -16 & \text{Add the equations.} \\ x = -4 & \text{Solve for } x. \end{array}$$

Answer

1. $(1, 1, 4)$

We now replace x with -4 in equation (4) or (5).

$$4x + y = -14 \quad \text{Equation (4)}$$
$$4(-4) + y = -14 \quad \text{Let } x = -4.$$
$$y = 2 \quad \text{Solve for } y.$$

Finally, we replace x with -4 and y with 2 in equation (1), (2), or (3).

$$x + 2y - z = 1 \quad \text{Equation (2)}$$
$$-4 + 2(2) - z = 1 \quad \text{Let } x = -4 \text{ and } y = 2.$$
$$-4 + 4 - z = 1$$
$$-z = 1$$
$$z = -1$$

The ordered triple solution is $(-4, 2, -1)$. To check, we let $x = -4$, $y = 2$, and $z = -1$ in *all three* original equations of the system.

Equation (1)

$$3x - y + z = -15$$
$$3(-4) - 2 + (-1) = -15$$
$$-12 - 2 - 1 = -15$$
$$-15 = -15$$

True.

Equation (2)

$$x + 2y - z = 1$$
$$-4 + 2(2) - (-1) = 1$$
$$-4 + 4 + 1 = 1$$
$$1 = 1$$

True.

Equation (3)

$$2x + 3y - 2z = 0$$
$$2(-4) + 3(2) - 2(-1) = 0$$
$$-8 + 6 + 2 = 0$$
$$0 = 0$$

True.

All three statements are true, so the ordered triple solution is $(-4, 2, -1)$. ●

EXAMPLE 2 Solve the system:

$$\begin{cases} 2x - 4y + 8z = 2 & (1) \\ -x - 3y + z = 11 & (2) \\ x - 2y + 4z = 0 & (3) \end{cases}$$

Solution: When we add equations (2) and (3) to eliminate x, the new equation is

$$-5y + 5z = 11 \quad (4)$$

To eliminate x again, we multiply both sides of equation (2) by 2 and add the resulting equation to equation (1). Then

$$\begin{cases} 2x - 4y + 8z = 2 \\ 2(-x - 3y + z) = 2(11) \end{cases} \text{ simplifies to } \begin{cases} 2x - 4y + 8z = 2 \\ -2x - 6y + 2z = 22 \end{cases}$$
$$-10y + 10z = 24 \quad (5)$$

Next we solve for y and z using equations (4) and (5). To do so, we multiply both sides of equation (4) by -2 and add the resulting equation to equation (5).

$$\begin{cases} -2(-5y + 5z) = -2(11) \\ -10y + 10z = 24 \end{cases} \text{ simplifies to } \begin{cases} 10y - 10z = -22 \\ -10y + 10z = 24 \end{cases}$$
$$0 = 2 \quad \text{False.}$$

Since the statement is false, this system is inconsistent and has no solution. The solution set is the empty set { } or $\varnothing$. ●

Practice Problem 2

Solve the system:

$$\begin{cases} 2x + 4y - 2z = 3 \\ -x + y - z = 6 \\ x + 2y - z = 1 \end{cases}$$

Answer

2. $\varnothing$

Concept Check

In the system

$$\begin{cases} x + y + z = 6 & \text{Equation (1)} \\ 2x - y + z = 3 & \text{Equation (2)} \\ x + 2y + 3z = 14 & \text{Equation (3)} \end{cases}$$

equations (1) and (2) are used to eliminate y. Which action could be used to finish solving? Why?
(a) Use (1) and (2) to eliminate z.
(b) Use (2) and (3) to eliminate y.
(c) Use (1) and (3) to eliminate x.

Practice Problem 3

Solve the system:

$$\begin{cases} 3x + 2y = -1 \\ 6x - 2z = 4 \\ y - 3z = 2 \end{cases}$$

Answers

3. $\left(\frac{1}{3}, -1, -1\right)$

Concept Check: b

Try the Concept Check in the margin.

EXAMPLE 3 Solve the system:

$$\begin{cases} 2x + 4y = 1 & (1) \\ 4x - 4z = -1 & (2) \\ y - 4z = -3 & (3) \end{cases}$$

Solution: Notice that equation (2) has no term containing the variable y. Let's eliminate y using equations (1) and (3). We multiply both sides of equation (3) by -4 and add the resulting equation to equation (1). Then

$$\begin{cases} 2x + 4y = 1 \\ -4(y - 4z) = -4(-3) \end{cases} \text{ simplifies to } \begin{cases} 2x + 4y = 1 \\ -4y + 16z = 12 \\ \hline 2x + 16z = 13 \quad (4) \end{cases}$$

Next we solve for z using equations (4) and (2). We multiply both sides of equation (4) by -2 and add the resulting equation to equation (2).

$$\begin{cases} -2(2x + 16z) = -2(13) \\ 4x - 4z = -1 \end{cases} \text{ simplifies to } \begin{cases} -4x - 32z = -26 \\ 4x - 4z = -1 \\ \hline -36z = -27 \end{cases}$$

$$z = \frac{3}{4}$$

Now we replace z with $\frac{3}{4}$ in equation (3) and solve for y.

$$y - 4\left(\frac{3}{4}\right) = -3 \qquad \text{Let } z = \frac{3}{4} \text{ in equation (3).}$$

$$y - 3 = -3$$

$$y = 0$$

Finally, we replace y with 0 in equation (1) and solve for x.

$$2x + 4(0) = 1 \qquad \text{Let } y = 0 \text{ in equation (1).}$$

$$2x = 1$$

$$x = \frac{1}{2}$$

The ordered triple solution is $\left(\frac{1}{2}, 0, \frac{3}{4}\right)$. Check to see that this solution satisfies *all three* equations of the system.

EXAMPLE 4 Solve the system:

$$\begin{cases} x - 5y - 2z = 6 & (1) \\ -2x + 10y + 4z = -12 & (2) \\ \frac{1}{2}x - \frac{5}{2}y - z = 3 & (3) \end{cases}$$

Solution: We multiply both sides of equation (3) by 2 to eliminate fractions, and we multiply both sides of equation (2) by $-\frac{1}{2}$ so that the coefficient of x is 1. The resulting system is then

$$\begin{cases} x - 5y - 2z = 6 & (1) \\ x - 5y - 2z = 6 & \text{Multiply (2) by } -\frac{1}{2}. \\ x - 5y - 2z = 6 & \text{Multiply (3) by 2.} \end{cases}$$

All three resulting equations are identical, and therefore equations (1), (2), and (3) are all equivalent. There are infinitely many solutions of this system. The equations are dependent. The solution set can be written as $\{(x, y, z)|x - 5y - 2z = 6\}$. ●

As mentioned earlier, we can also use the substitution method to solve a system of linear equations in three variables.

EXAMPLE 5 Solve the system:

$$\begin{cases} x - 4y - 5z = 35 & (1) \\ x - 3y \qquad = 0 & (2) \\ \quad -y + z = -25 & (3) \end{cases}$$

Solution: Notice in equations (2) and (3) that a variable is missing. Also notice that both equations contain the variable y. Let's use the substitution method by solving equation (2) for x and equation (3) for z and substituting the results in equation (1).

$$\begin{aligned} x - 3y &= 0 && (2) \\ x &= 3y && \text{Solve equation (2) for } x. \\ -y + z &= -55 && (3) \\ z &= y - 55 && \text{Solve equation (3) for } z. \end{aligned}$$

Now substitute $3y$ for x and $y - 55$ for z in equation (1).

$$\begin{aligned} x - 4y - 5z &= 35 && (1) \\ 3y - 4y - 5(y - 55) &= 35 && \text{Let } x = 3y \text{ and } z = y - 55. \\ 3y - 4y - 5y + 275 &= 35 && \text{Use the distributive law and multiply.} \\ -6y + 275 &= 35 && \text{Combine like terms.} \\ -6y &= -240 && \text{Subtract 275 from both sides.} \\ y &= 40 && \text{Solve.} \end{aligned}$$

To find x, recall that $x = 3y$ and substitute 40 for y. Then $x = 3y$ becomes $x = 3 \cdot 40 = 120$. To find z, recall that $z = y - 55$ and substitute 40 for y, also. Then $z = y - 55$ becomes $z = 40 \cdot 55 = -15$. The solution is $(120, 40, -15)$. ●

Practice Problem 4

Solve the system:

$$\begin{cases} x - 3y + 4z = 2 \\ -2x + 6y - 8z = -4 \\ \frac{1}{2}x - \frac{3}{2}y + 2z = 1 \end{cases}$$

Practice Problem 5

Solve the system:

$$\begin{cases} 2x + 5y - 3z = 30 & (1) \\ x + y \qquad = -3 & (2) \\ 2x \qquad - z = 0 & (3) \end{cases}$$

(Hint: Equations (2) and (3) each contain the variable x and have a variable missing.)

Do not forget to distribute.

Answers

4. $\{(x, y, z)|x - 3y + 4z = 2\}$, **5.** $(-5, 2, -10)$

B Solving Problems Modeled by Systems of Three Equations

To introduce problem solving with systems of three linear equations in three variables, we solve a problem about triangles.

Practice Problem 6 

The measure of the largest angle of a triangle is 90° more than the measure of the smallest angle, and the measure of the remaining angle is 30° more than the measure of the smallest angle. Find the measure of each angle.

EXAMPLE 6 Finding Angle Measures

The measure of the largest angle of a triangle is 80° more than the measure of the smallest angle, and the measure of the remaining angle is 10° more than the measure of the smallest angle. Find the measure of each angle.

Solution:

1. UNDERSTAND. Read and reread the problem. Recall that the sum of the measures of the angles of a triangle is 180°. Then guess a solution. If the smallest angle measures 20°, the measure of the largest angle is 80° more, or $20° + 80° = 100°$. The measure of the remaining angle is 10° more than the measure of the smallest angle, or $20° + 10° = 30°$. The sum of these three angles is $20° + 100° + 30° = 150°$, not the required 180°. We now know that the measure of the smallest angle is greater than 20°.

   To model this problem we will let

   x = degree measure of the smallest angle

   y = degree measure of the largest angle

   z = degree measure of the remaining angle

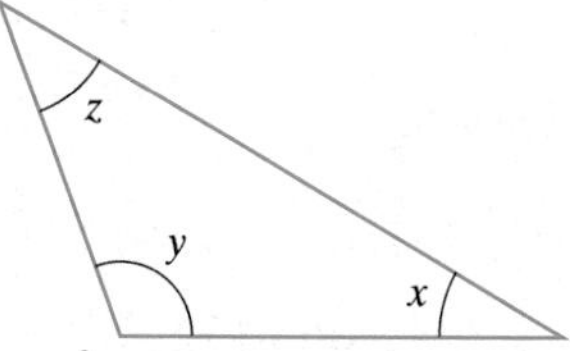

2. TRANSLATE. We translate the given information into three equations.

In words:	the sum of the measures	=	180
	↓		↓
Translate:	$x + y + z$	=	180

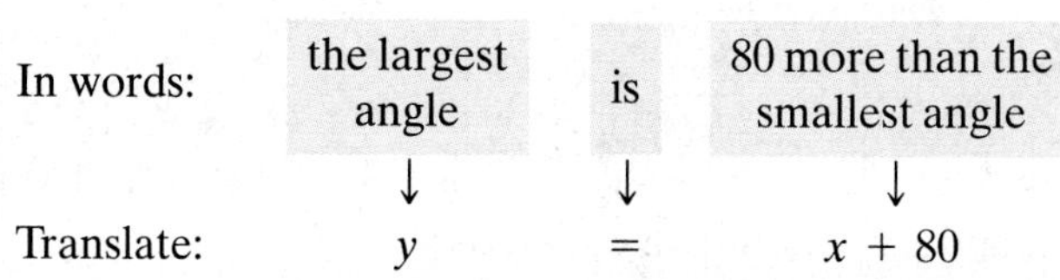

In words:	the largest angle	is	80 more than the smallest angle
	↓	↓	↓
Translate:	y	=	$x + 80$

In words:	the remaining angle	is	10 more than the smallest angle
	↓	↓	↓
Translate:	z	=	$x + 10$

3. SOLVE. We solve the system

$$\begin{cases} x + y + z = 180 \\ y = x + 80 \\ z = x + 10 \end{cases}$$

Answer

6. 20°; 50°; 110°

Since y and z are both expressed in terms of x, we will solve using the substitution method. We substitute $y = x + 80$ and $z = x + 10$ in the first equation.

Then

$$x + y + z = 180 \quad \text{First equation}$$

$$x + (x + 80) + (x + 10) = 180 \quad \text{Let } y = x + 80 \text{ and } z = x + 10.$$

$$3x + 90 = 180$$

$$3x = 90$$

$$x = 30$$

Then $y = x + 80 = 30 + 80 = 110$, and $z = x + 10 = 30 + 10 = 40$. The ordered triple solution is $(30, 110, 40)$.

4. INTERPRET.

Check: Notice that $30° + 40° + 110° = 180°$. Also, the measure of the largest angle, 110°, is 80° more than the measure of the smallest angle, 30°. The measure of the remaining angle, 40°, is 10° more than the measure of the smallest angle, 30°.

State: The angles measure 30°, 110°, and 40°.

FOCUS ON The Real World

ANOTHER MATHEMATICAL MODEL

Sometimes mathematical models other than linear models are appropriate for data. Suppose that an equation of the form $y = ax^2 + bx + c$ is an appropriate model for the ordered pairs (x_1, y_1), (x_2, y_2), and (x_3, y_3). Then it is necessary to find the values of a, b, and c such that the given ordered pairs are solutions of the equation $y = ax^2 + bx + c$. To do so, substitute each ordered pair into the equation. Each time, the result is an equation in three unknowns: a, b, and c. Solving the resulting system of three linear equation in three unknowns will give the required values of a, b, and c.

GROUP ACTIVITY

1. The table gives the total beef supply (in billions of pounds) in the United States in each of the years listed.
   a. Write the data as ordered pairs of the form (x, y), where y is the beef supply (in billions of pounds) in the year x ($x = 0$ represents 1998).
   b. Find the values of a, b, and c such that the equation $y = ax^2 + bx + c$ models this data.
   c. Verify that the model you found in part (b) gives each of the ordered pair solutions from part (a).
   d. According to the model, what was the U.S. beef supply in 2001?

Total U.S. Beef Supply

Year	Beef Supply (billions of pounds)
1998	28.9
2000	30.3
2002	28.8

(*Source:* Economic Research Service, U.S. Department of Agriculture)

2. The table gives Saab production figures for each of the years listed.
   a. Write the data as ordered pairs of the form (x, y), where y is Saab production in the year x ($x = 0$ represents 1996).
   b. Find the values of a, b, and c such that the equation $y = ax^2 + bx + c$ models this data.
   c. According to the model, what was the total Saab production in 2001?

Total Saab Production

Year	Number of Saab Vehicles Produced
1996	28,439
1998	30,756
2000	39,479

(*Source:* Automotive Intelligence, www.autointell.com)

3. a. Make up an equation of the form $y = ax^2 + bx + c$.
   b. Find three ordered pair solutions of the equation.
   c. Without revealing your equation from part (a), exchange lists of ordered pair solutions with another group.
   d. Use the method described above to find the values of a, b, and c such that the equation $y = ax^2 + bx + c$ has the ordered pair solutions you received from the other group.
   e. Check with the other group to see if your equation from part (d) is the correct one.

Name ______________________ Section __________ Date ____________

EXERCISE SET 8.5

A *Solve each system. See Examples 1 through 5.*

1. $\begin{cases} x + y = 3 \\ 2y = 10 \\ 3x + 2y - 3z = 1 \end{cases}$

2. $\begin{cases} 5x = 5 \\ 2x + y = 4 \\ 3x + y - 4z = -15 \end{cases}$

3. $\begin{cases} 2x + 2y + z = 1 \\ -x + y + 2z = 3 \\ x + 2y + 4z = 0 \end{cases}$

4. $\begin{cases} 2x - 3y + z = 5 \\ x + y + z = 0 \\ 4x + 2y + 4z = 4 \end{cases}$

5. $\begin{cases} x - 2y + z = -5 \\ -3x + 6y - 3z = 15 \\ 2x - 4y + 2z = -10 \end{cases}$

6. $\begin{cases} 3x + y - 2z = 2 \\ -6x - 2y + 4z = -2 \\ 9x + 3y - 6z = 6 \end{cases}$

7. $\begin{cases} 4x - y + 2z = 5 \\ 2y + z = 4 \\ 4x + y + 3z = 10 \end{cases}$

8. $\begin{cases} 5y - 7z = 14 \\ 2x + y + 4z = 10 \\ 2x + 6y - 3z = 30 \end{cases}$

9. $\begin{cases} x + 5z = 0 \\ 5x + y + 10z = -24 \\ y - 3z = 0 \end{cases}$

10. $\begin{cases} 4x - y = 0 \\ 3x - y - z = 6 \\ -x + 5z = 0 \end{cases}$

11. $\begin{cases} 6x - 5z = 17 \\ 5x - y + 3z = -1 \\ 2x + y = -41 \end{cases}$

12. $\begin{cases} x + 2y = 6 \\ 7x + 3y + z = -33 \\ x - z = 16 \end{cases}$

13. $\begin{cases} x + y + z = 8 \\ 2x - y - z = 10 \\ x - 2y - 3z = 22 \end{cases}$

14. $\begin{cases} 5x + y + 3z = 1 \\ x - y + 3z = -7 \\ -x + y = 1 \end{cases}$

15. $\begin{cases} x + 2y - z = 5 \\ 6x + y + z = 7 \\ 2x + 4y - 2z = 5 \end{cases}$

16. $\begin{cases} 4x - y + 3z = 10 \\ x + y - z = 5 \\ 8x - 2y + 6z = 10 \end{cases}$

17. $\begin{cases} 2x - 3y + z = 2 \\ x - 5y + 5z = 3 \\ 3x + y - 3z = 5 \end{cases}$

18. $\begin{cases} 4x + y - z = 8 \\ x - y + 2z = 3 \\ 3x - y + z = 6 \end{cases}$

19. $\begin{cases} -2x - 4y + 6z = -8 \\ x + 2y - 3z = 4 \\ 4x + 8y - 12z = 16 \end{cases}$

20. $\begin{cases} -6x + 12y + 3z = -6 \\ 2x - 4y - z = 2 \\ -x + 2y + \dfrac{z}{2} = -1 \end{cases}$

21. $\begin{cases} 2x + 2y - 3z = 1 \\ y + 2z = -14 \\ 3x - 2y = -1 \end{cases}$

22. $\begin{cases} 7x + 4y = 10 \\ x - 4y + 2z = 6 \\ y - 2z = -1 \end{cases}$

23. $\begin{cases} \dfrac{3}{4}x - \dfrac{1}{3}y + \dfrac{1}{2}z = 9 \\ \dfrac{1}{6}x + \dfrac{1}{3}y - \dfrac{1}{2}z = 2 \\ \dfrac{1}{2}x - y + \dfrac{1}{2}z = 2 \end{cases}$

24. $\begin{cases} \dfrac{1}{3}x - \dfrac{1}{4}y + z = -9 \\ \dfrac{1}{2}x - \dfrac{1}{3}y - \dfrac{1}{4}z = -6 \\ x - \dfrac{1}{2}y - z = -8 \end{cases}$

25. The fraction $\frac{1}{24}$ can be written as the following sum:

$$\frac{1}{24} = \frac{x}{8} + \frac{y}{4} + \frac{z}{3}$$

where the numbers x, y, and z are solutions of

$$\begin{cases} x + y + z = 1 \\ 2x - y + z = 0 \\ -x + 2y + 2z = -1 \end{cases}$$

Solve the system and see that the sum of the fractions is $\frac{1}{24}$.

26. The fraction $\frac{1}{18}$ can be written as the following sum:

$$\frac{1}{18} = \frac{x}{2} + \frac{y}{3} + \frac{z}{9}$$

where the numbers x, y, and z are solutions of

$$\begin{cases} x + 3y + z = -3 \\ -x + y + 2z = -14 \\ 3x + 2y - z = 12 \end{cases}$$

Solve the system and see that the sum of the fractions is $\frac{1}{18}$.

B *Solve. See Example 6.*

27. Rabbits in a lab are to be kept on a strict daily diet that includes 30 grams of protein, 16 grams of fat, and 24 grams of carbohydrates. The scientist has only three food mixes available with the following grams of nutrients per unit.

	Protein	Fat	Carbohydrate
Mix A	4	6	3
Mix B	6	1	2
Mix C	4	1	12

Find how many units of each mix are needed daily to meet each rabbit's dietary need.

28. Gerry Gundersen mixes different solutions with concentrations of 25%, 40%, and 50% to get 200 liters of a 32% solution. If he uses twice as much of the 25% solution as of the 40% solution, find how many liters of each kind he uses.

△ **29.** The perimeter of a quadrilateral (four-sided polygon) is 29 inches. The longest side is twice as long as the shortest side. The other two sides are equally long and are 2 inches longer than the shortest side. Find the length of all four sides.

△ **30.** The measure of the largest angle of a triangle is 90° more than the measure of the smallest angle, and the measure of the remaining angle is 30° more than the measure of the smallest angle. Find the measure of each angle.

31. The sum of three numbers is 40. One number is five more than a second number. It is also twice the third. Find the numbers.

32. The sum of the digits of a three-digit number is 15. The tens-place digit is twice the hundreds-place digit, and the ones-place digit is 1 less than the hundreds-place digit. Find the three-digit number.

33. In 2001, the WNBA's top scorer was Katie Smith of the Minnesota Lynx. She scored a total of 739 points during the regular season. The number of two-point field goals Smith made was 51 less than twice the number of three-point field goals she made. The number of free throws (each worth one point) she made was 8 more than twice the number of two-point field goals she made. Find how many free throws, two-point field goals, and three-point field goals Katie Smith made during the 2001 regular season. (*Source:* Women's National Basketball Association)

34. During the 2000–2001 regular NBA season, the top-scoring player was Jerry Stackhouse of the Detroit Pistons. Stackhouse scored a total of 2380 points during the regular season. The number of free throws (each worth one point) he made was 2 more than four times the number of three-point field goals he made. Stackhouse also made 58 more free throws than two-point field goals. How many free throws, two-point field goals, and three-point field goals did Jerry Stackhouse make during the 2000–2001 season? (*Source:* National Basketball Association)

△ **35.** Find the values of x, y, and z in the following triangle.

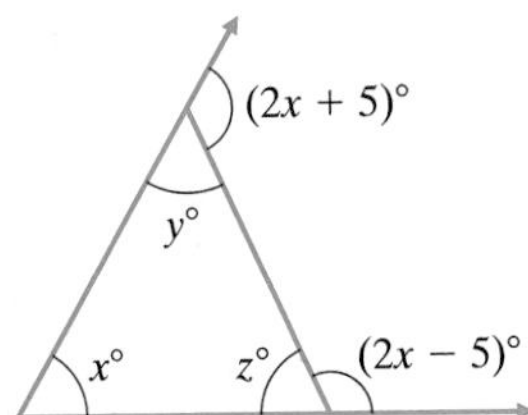

△ **36.** The sum of the measures of the angles of a quadrilateral is 360°. Find the value of x, y, and z in the following quadrilateral.

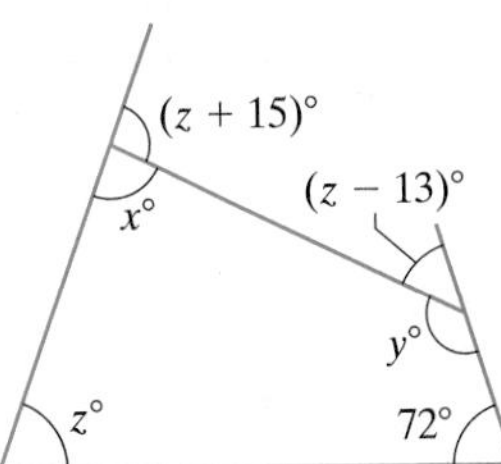

Review and Preview

Multiply both sides of equation (1) by 2, and add the resulting equation to equation (2). See Sections 4.4 and 4.5.

37. $3x - y + z = 2$ (1)
$-x + 2y + 3z = 6$ (2)

38. $2x + y + 3z = 7$ (1)
$-4x + y + 2z = 4$ (2)

Multiply both sides of equation (1) by −3, and add the resulting equation to equation (2). See Sections 4.4 and 4.5.

39. $x + 2y - z = 0$ (1)
$3x + y - z = 2$ (2)

40. $2x - 3y + 2z = 5$ (1)
$x - 9y + z = -1$ (2)

Combining Concepts

41. Write a linear equation in three variables that has $(-1, 2, -4)$ as a solution. (There are many possibilities.) Explain the process you used to write an equation.

42. When solving a system of three equation in three unknowns, explain how to determine that a system has no solution.

Solving systems involving more than three variables can be accomplished with methods similar to those encountered in this section. Apply what you already know to solve each system of equations in four variables.

43. $\begin{cases} x + y \qquad\quad - w = 0 \\ \quad y + 2z + w = 3 \\ x \qquad - z \qquad = 1 \\ 2x - y \qquad - w = -1 \end{cases}$

44. $\begin{cases} 5x + 4y \qquad\qquad = 29 \\ \quad y + z - w = -2 \\ 5x \qquad + z \qquad = 23 \\ \quad y - z + w = 4 \end{cases}$

45. Find the values of a, b, and c such that the equation $y = ax^2 + bx + c$ has ordered pair solutions $(1, 6)$, $(-1, -2)$, and $(0, -1)$. To do so, substitute each ordered pair solution into the equation. Each time, the result is an equation in three unknowns: a, b, and c. Then solve the resulting system of three linear equations in three unknowns, a, b, and c.

46. Find the values of a, b, and c such that the equation $y = ax^2 + bx + c$ has ordered pair solutions $(1, 2)$, $(2, 3)$, and $(-1, 6)$. (*Hint:* See Exercise 53.)

47. Data (x, y) for the total number y (in thousands) of college-bound students who took the ACT assessment in the year x are $(5, 945)$, $(6, 925)$, and $(11, 1070)$, where $x = 5$ represents 1995 and $x = 11$ represents 2001. Find the values of a, b, and c such that the equation $y = ax^2 + bx + c$ models this data. According to your model, how many students will take the ACT in 2007? (*Source:* ACT, Inc.)

48. Monthly normal rainfall data (x, y) for Portland, Oregon, are $(4, 2.47)$, $(7, 0.6)$, $(8, 1.1)$, where x represents time in months (with $x = 1$ representing January) and y represents rainfall in inches. Find the values of a, b, and c rounded to 2 decimal places such that the equation $y = ax^2 + bx + c$ models this data. According to your model, how much rain should Portland expect during September? (*Source:* National Climatic Data Center)

8.6 Solving Systems of Equations Using Matrices

OBJECTIVES

- **A** Use matrices to solve a system of two equations.
- **B** Use matrices to solve a system of three equations.

By now, you may have noticed that the solution of a system of equations depends on the coefficients of the equations in the system and not on the variables. In this section, we introduce how to solve a system of equations using a **matrix**.

A Using Matrices to Solve a System of Two Equations

A **matrix** (plural: **matrices**) is a rectangular array of numbers. The following are examples of matrices.

$$\begin{bmatrix} 1 & 0 \\ 0 & 1 \end{bmatrix} \qquad \begin{bmatrix} 2 & 1 & 3 & -1 \\ 0 & -1 & 4 & 5 \\ -6 & 2 & 1 & 0 \end{bmatrix} \qquad \begin{bmatrix} a & b & c \\ d & e & f \end{bmatrix}$$

The numbers aligned horizontally in a matrix are in the same **row**. The numbers aligned vertically are in the same **column**.

$$\begin{matrix} \text{Row 1} \rightarrow \\ \text{Row 2} \rightarrow \end{matrix} \begin{bmatrix} 2 & 1 & 0 \\ -1 & 6 & 2 \end{bmatrix}$$

This matrix has 2 rows and 3 columns. It is called a 2×3 (read "two by three") matrix.

Column 1 ↑ Column 2 ↑ Column 3 ↑

To see the relationship between systems of equations and matrices, study the example below.

System of Equations

$$\begin{cases} 2x - 3y = 6 & \text{Equation 1} \\ x + y = 0 & \text{Equation 2} \end{cases}$$

Corresponding Matrix

$$\left[\begin{array}{cc:c} 2 & -3 & 6 \\ 1 & 1 & 0 \end{array}\right] \begin{matrix} \text{Row 1} \\ \text{Row 2} \end{matrix}$$

Notice that the rows of the matrix correspond to the equations in the system. The coefficients of the variables are placed to the left of a vertical dashed line. The constants are placed to the right. Each of these numbers in the matrix is called an **element**.

Helpful Hint

Before writing the corresponding matrix associated with a system of equations, make sure that the equations are written in standard form.

The method of solving systems by matrices is to write this matrix as an equivalent matrix from which we can easily identify the solution. Two matrices are equivalent if they represent systems that have the same solution set. The following **row operations** can be performed on matrices, and the result is an equivalent matrix.

Elementary Row Operations

1. Any two rows in a matrix may be interchanged.
2. The elements of any row may be multiplied (or divided) by the same nonzero number.
3. The elements of any row may be multiplied (or divided) by a nonzero number and added to their corresponding elements in any other row.

Helpful Hint

Notice that these *row* operations are the same operations that we can perform on *equations* in a system.

Practice Problem 1

Use matrices to solve the system:

$$\begin{cases} x + 2y = -4 \\ 2x - 3y = 13 \end{cases}$$

EXAMPLE 1 Use matrices to solve the system:

$$\begin{cases} x + 3y = 5 \\ 2x - y = -4 \end{cases}$$

Solution: The corresponding matrix is $\left[\begin{array}{cc|c} 1 & 3 & 5 \\ 2 & -1 & -4 \end{array}\right]$. We use elementary row operations to write an equivalent matrix that looks like $\left[\begin{array}{cc|c} 1 & a & b \\ 0 & 1 & c \end{array}\right]$.

For the matrix given, the element in the first row, first column is already 1, as desired. Next we write an equivalent matrix with a 0 below the 1. To do this, we multiply row 1 by -2 and add to row 2. *We will change only row 2.*

$$\left[\begin{array}{cc|c} 1 & 3 & 5 \\ -2(1) + 2 & -2(3) + (-1) & -2(5) + (-4) \end{array}\right] \text{ simplifies to } \left[\begin{array}{cc|c} 1 & 3 & 5 \\ 0 & -7 & -14 \end{array}\right]$$

(Arrows label the terms: Row 1 element, Row 2 element, Row 1 element, Row 2 element, Row 1 element, Row 2 element)

Now we change the -7 to a 1 by use of an elementary row operation. We divide row 2 by -7, then

$$\left[\begin{array}{cc|c} 1 & 3 & 5 \\ \frac{0}{-7} & \frac{-7}{-7} & \frac{-14}{-7} \end{array}\right] \text{ simplifies to } \left[\begin{array}{cc|c} 1 & 3 & 5 \\ 0 & 1 & 2 \end{array}\right]$$

This last matrix corresponds to the system

$$\begin{cases} x + 3y = 5 \\ y = 2 \end{cases}$$

Thus we know that y is 2. To find x, we let $y = 2$ in the first equation, $x + 3y = 5$.

$$x + 3y = 5 \quad \text{First equation}$$

$$x + 3(2) = 5 \quad \text{Let } y = 2.$$

$$x = -1$$

The ordered pair solution is $(-1, 2)$. Check to see that this ordered pair satisfies both original equations.

Answer

1. $(2, -3)$

EXAMPLE 2 Use matrices to solve the system:

$$\begin{cases} 2x - y = 3 \\ 4x - 2y = 5 \end{cases}$$

Solution: The corresponding matrix is $\left[\begin{array}{cc|c} 2 & -1 & 3 \\ 4 & -2 & 5 \end{array}\right]$. To get 1 in the row 1, column 1 position, we divide the elements of row 1 by 2.

$$\left[\begin{array}{cc|c} \frac{2}{2} & -\frac{1}{2} & \frac{3}{2} \\ 4 & -2 & 5 \end{array}\right] \text{ simplifies to } \left[\begin{array}{cc|c} 1 & -\frac{1}{2} & \frac{3}{2} \\ 4 & -2 & 5 \end{array}\right]$$

To get 0 under the 1, we multiply the elements of row 1 by −4 and add the new elements to the elements of row 2.

$$\left[\begin{array}{cc|c} 1 & -\frac{1}{2} & \frac{3}{2} \\ -4(1) + 4 & -4\left(-\frac{1}{2}\right) - 2 & -4\left(\frac{3}{2}\right) + 5 \end{array}\right] \text{ simplifies to}$$

$$\left[\begin{array}{cc|c} 1 & -\frac{1}{2} & \frac{3}{2} \\ 0 & 0 & -1 \end{array}\right]$$

The corresponding system is $\begin{cases} x - \frac{1}{2}y = \frac{3}{2} \\ 0 = -1 \end{cases}$. The equation $0 = -1$ is false for all y or x values; hence the system is inconsistent and has no solution. ●

Practice Problem 2

Use matrices to solve the system:

$$\begin{cases} -3x + y = 0 \\ -6x + 2y = 2 \end{cases}$$

Try the Concept Check in the margin.

Concept Check

Consider the system

$$\begin{cases} 2x - 3y = 8 \\ x + 5y = -3 \end{cases}$$

What is wrong with its corresponding matrix shown below?

$$\left[\begin{array}{cc|c} 2 & -3 & 8 \\ 0 & 5 & -3 \end{array}\right]$$

B Using Matrices to Solve a System of Three Equations

To solve a system of three equations in three variables using matrices, we will write the corresponding matrix in the form

$$\left[\begin{array}{ccc|c} 1 & a & b & d \\ 0 & 1 & c & e \\ 0 & 0 & 1 & f \end{array}\right]$$

EXAMPLE 3 Use matrices to solve the system:

$$\begin{cases} x + 2y + z = 2 \\ -2x - y + 2z = 5 \\ x + 3y - 2z = -8 \end{cases}$$

Solution: The corresponding matrix is $\left[\begin{array}{ccc|c} 1 & 2 & 1 & 2 \\ -2 & -1 & 2 & 5 \\ 1 & 3 & -2 & -8 \end{array}\right]$.

Our goal is to write an equivalent matrix with 1s along the diagonal (see the numbers in red) and 0s below the 1s. The element in row 1, column 1 is already 1. Next we get 0s for each element in the rest of column 1. To do this, first we multiply the elements of row 1 by 2 and add the new elements to row 2. Also,

Practice Problem 3

Use matrices to solve the system:

$$\begin{cases} x + 3y + z = 5 \\ -3x + y - 3z = 5 \\ x + 2y - 2z = 9 \end{cases}$$

Answers

2. no solution, **3.** $(1, 2, -2)$

Concept Check: matrix should be $\left[\begin{array}{cc|c} 2 & -3 & 8 \\ 1 & 5 & -3 \end{array}\right]$

we multiply the elements of row 1 by -1 and add the new elements to the elements of row 3. *We do not change row 1.* Then

$$\left[\begin{array}{ccc|c} 1 & 2 & 1 & 2 \\ 2(1)-2 & 2(2)-1 & 2(1)+2 & 2(2)+5 \\ -1(1)+1 & -1(2)+3 & -1(1)-2 & -1(2)-8 \end{array}\right] \text{ simplifies to } \left[\begin{array}{ccc|c} 1 & 2 & 1 & 2 \\ 0 & 3 & 4 & 9 \\ 0 & 1 & -3 & -10 \end{array}\right]$$

We continue down the diagonal and use elementary row operations to get 1 where the element 3 is now. To do this, we interchange rows 2 and 3.

$$\left[\begin{array}{ccc|c} 1 & 2 & 1 & 2 \\ 0 & 3 & 4 & 9 \\ 0 & 1 & -3 & -10 \end{array}\right] \text{ is equivalent to } \left[\begin{array}{ccc|c} 1 & 2 & 1 & 2 \\ 0 & 1 & -3 & -10 \\ 0 & 3 & 4 & 9 \end{array}\right]$$

Next we want the new row 3, column 2 element to be 0. We multiply the elements of row 2 by -3 and add the result to the elements of row 3.

$$\left[\begin{array}{ccc|c} 1 & 2 & 1 & 2 \\ 0 & 1 & -3 & -10 \\ -3(0)+0 & -3(1)+3 & -3(-3)+4 & -3(-10)+9 \end{array}\right] \text{ simplifies to } \left[\begin{array}{ccc|c} 1 & 2 & 1 & 2 \\ 0 & 1 & -3 & -10 \\ 0 & 0 & 13 & 39 \end{array}\right]$$

Finally, we divide the elements of row 3 by 13 so that the final diagonal element is 1.

$$\left[\begin{array}{ccc|c} 1 & 2 & 1 & 2 \\ 0 & 1 & -3 & -10 \\ \frac{0}{13} & \frac{0}{13} & \frac{13}{13} & \frac{39}{13} \end{array}\right] \text{ simplifies to } \left[\begin{array}{ccc|c} 1 & 2 & 1 & 2 \\ 0 & 1 & -3 & -10 \\ 0 & 0 & 1 & 3 \end{array}\right]$$

This matrix corresponds to the system

$$\begin{cases} x + 2y + z = 2 \\ \quad\;\; y - 3z = -10 \\ \qquad\quad\;\; z = 3 \end{cases}$$

We identify the z-coordinate of the solution as 3. Next we replace z with 3 in the second equation and solve for y.

$$\begin{aligned} y - 3z &= -10 && \text{Second equation} \\ y - 3(3) &= -10 && \text{Let } z = 3. \\ y &= -1 \end{aligned}$$

To find x, we let $z = 3$ and $y = -1$ in the first equation.

$$\begin{aligned} x + 2y + z &= 2 && \text{First equation} \\ x + 2(-1) + 3 &= 2 && \text{Let } z = 3 \text{ and } y = -1. \\ x &= 1 \end{aligned}$$

The ordered triple solution is $(1, -1, 3)$. Check to see that it satisfies all three equations in the original system. ●

Name ______________________ Section __________ Date __________

Ans p A104

EXERCISE SET 8.6

A *Use matrices to solve each system of linear equations. See Examples 1 and 2.*

1. $\begin{cases} x + y = 1 \\ x - 2y = 4 \end{cases}$

2. $\begin{cases} 2x - y = 8 \\ x + 3y = 11 \end{cases}$

3. $\begin{cases} x + 3y = 2 \\ x + 2y = 0 \end{cases}$

4. $\begin{cases} 4x - y = 5 \\ 3x - 3 = 0 \end{cases}$

5. $\begin{cases} x - 2y = 4 \\ 2x - 4y = 4 \end{cases}$

6. $\begin{cases} -x + 3y = 6 \\ 3x - 9y = 9 \end{cases}$

7. $\begin{cases} 3x - 3y = 9 \\ 2x - 2y = 6 \end{cases}$

8. $\begin{cases} 9x - 3y = 6 \\ -18x + 6y = -12 \end{cases}$

9. $\begin{cases} x - 4 = 0 \\ x + y = 1 \end{cases}$

10. $\begin{cases} 3y = 6 \\ x + y = 7 \end{cases}$

B *Use matrices to solve each system of linear equations. See Example 3.*

11. $\begin{cases} x + y = 3 \\ 2y = 10 \\ 3x + 2y - 4z = 12 \end{cases}$

12. $\begin{cases} 5x = 5 \\ 2x + y = 4 \\ 3x + y - 5z = -15 \end{cases}$

13. $\begin{cases} 2y - z = -7 \\ x + 4y + z = -4 \\ 5x - y + 2z = 13 \end{cases}$

14. $\begin{cases} 4y + 3z = -2 \\ 5x - 4y = 1 \\ -5x + 4y + z = -3 \end{cases}$

15. $\begin{cases} x + y + z = 2 \\ 2x - z = 5 \\ 3y + z = 2 \end{cases}$

16. $\begin{cases} x + 2y + z = 5 \\ x - y - z = 3 \\ y + z = 2 \end{cases}$

17. $\begin{cases} 4x + y + z = 3 \\ -x + y - 2z = -11 \\ x + 2y + 2z = -1 \end{cases}$

18. $\begin{cases} x + y + z = 9 \\ 3x - y + z = -1 \\ -2x + 2y - 3z = -2 \end{cases}$

Review and Preview

Evaluate. See Section 1.6.

19. $(-1)(-5) - (6)(3)$

20. $(2)(-8) - (-4)(1)$

21. $(4)(-10) - (2)(-2)$

22. $(-7)(3) - (-2)(-6)$

23. $(-3)(-3) - (-1)(-9)$

24. $(5)(6) - (10)(10)$

Combining Concepts

25. The percent y of U.S. households that owned a black-and-white television set between the years 1980 and 1993 can be modeled by the linear equation $2.3x + y = 52$, where x represents the number of years after 1980. Similarly, the percent y of U.S. households that owned a microwave oven during this same period can be modeled by the linear equation $-5.4x + y = 14$. (*Source:* Based on data from the Energy Information Administration, U.S. Department of Energy)

a. Determine the year in which the percent of households owning black-and-white television sets was the same as the percent of households owning microwave ovens. Use matrix methods to estimate the year in which this occurred.

b. Did more households own black-and-white television sets or microwave ovens in 1980? In 1993? What trends do these models show? Does this seem to make sense? Why or why not?

c. According to the models, when will the percent of households owning black-and-white television sets reach 0%?

8.7 Systems of Linear Inequalities

OBJECTIVE

A Graph a system of linear inequalities

SSM TUTOR CENTER SG CD & VIDEO MATH PRO WEB

A Graphing Systems of Linear Inequalities

In Section 3.5 we solved linear inequalities in two variables. Just as two linear equations make a system of linear equations, two linear inequalities make a **system of linear inequalities**. Systems of inequalities are very important in a process called linear programming. Many businesses use linear programming to find the most profitable way to use limited resources such as employees, machines, or buildings.

A **solution of a system of linear inequalities** is an ordered pair that satisfies each inequality in the system. The set of all such ordered pairs is the solution set of the system. Graphing this set gives us a picture of the solution set. We can graph a system of inequalities by graphing each inequality in the system and identifying the region of overlap.

Graphing the Solutions of a System of Linear Inequalities

Step 1. Graph each inequality in the system on the same set of axes.

Step 2. The solutions of the system are the points common to the graphs of all the inequalities in the system.

EXAMPLE 1 Graph the solutions of the system: $\begin{cases} 3x \geq y \\ x + 2y \leq 8 \end{cases}$

Solution: We begin by graphing each inequality on the *same* set of axes. The graph of the solutions of the system is the region contained in the graphs of both inequalities. In other words, it is their intersection.

First let's graph $3x \geq y$. The boundary line is the graph of $3x = y$. We sketch a solid boundary line since the inequality $3x \geq y$ means $3x > y$ or $3x = y$. The test point $(1, 0)$ satisfies the inequality, so we shade the half-plane that includes $(1, 0)$.

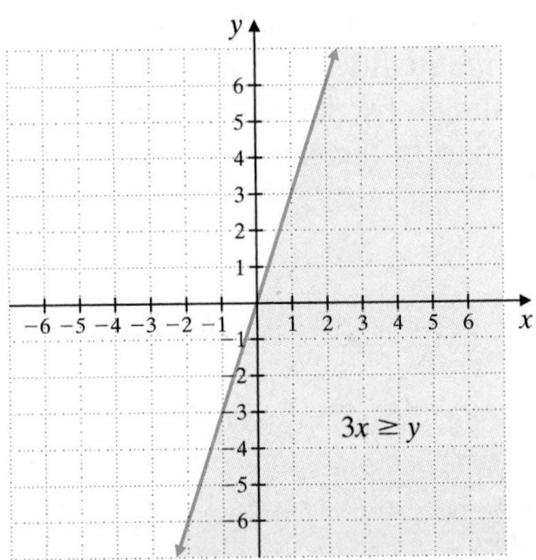

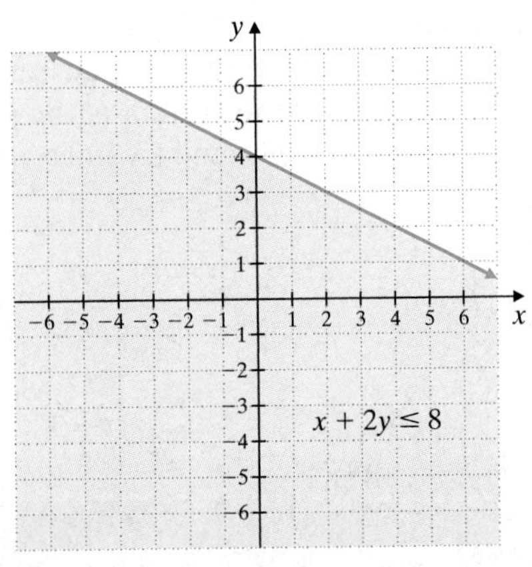

Next we sketch a solid boundary line $x + 2y = 8$ on the same set of axes. The test point $(0, 0)$ satisfies the inequality $x + 2y \leq 8$, so we shade the half-plane that includes $(0, 0)$. (For clarity, the graph of $x + 2y \leq 8$ is shown here on a separate set of axes.)

An ordered pair solution of the system must satisfy both inequalities. These solutions are points that lie in both shaded regions. The solution of the

Practice Problem 1

Graph the solutions of the system:

$$\begin{cases} 2x \leq y \\ x + 4y \geq 4 \end{cases}$$

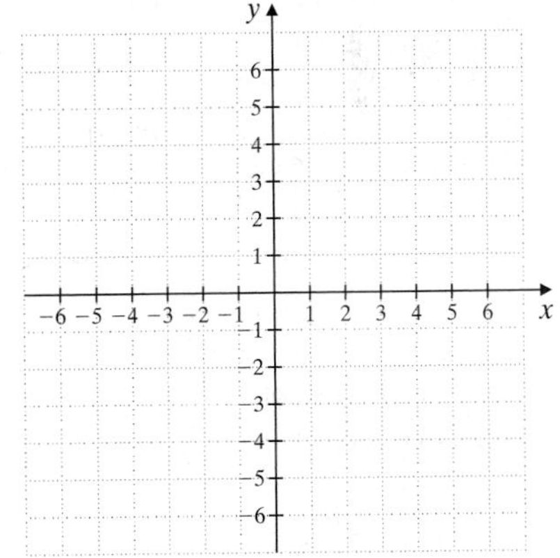

Answer

1.

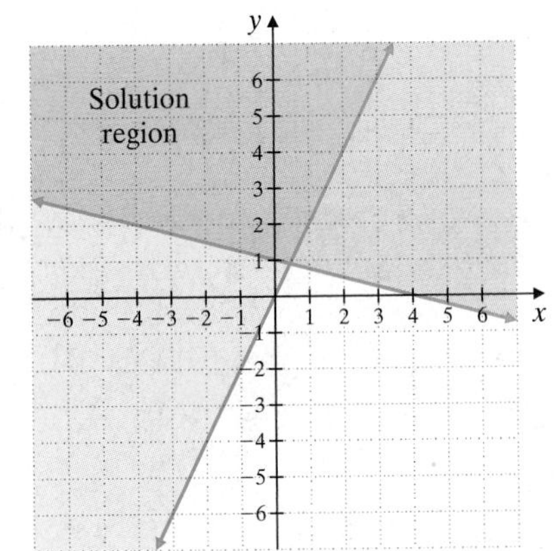

system is the darkest shaded region. This solution includes parts of both boundary lines.

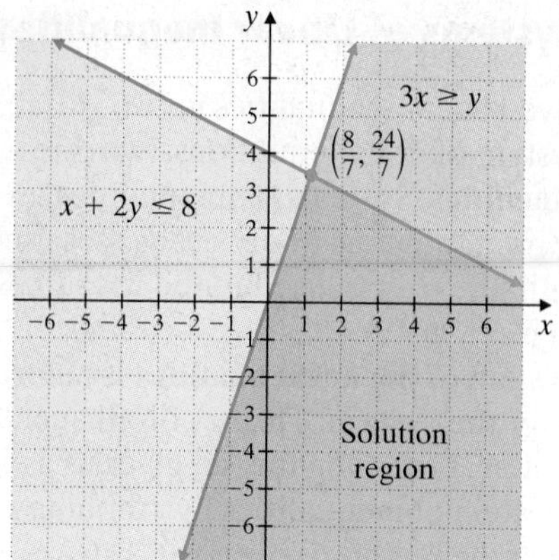

Practice Problem 2

Graph the solutions of the system:

$$\begin{cases} -x + y < 3 \\ y < 1 \\ 2x + y > -2 \end{cases}$$

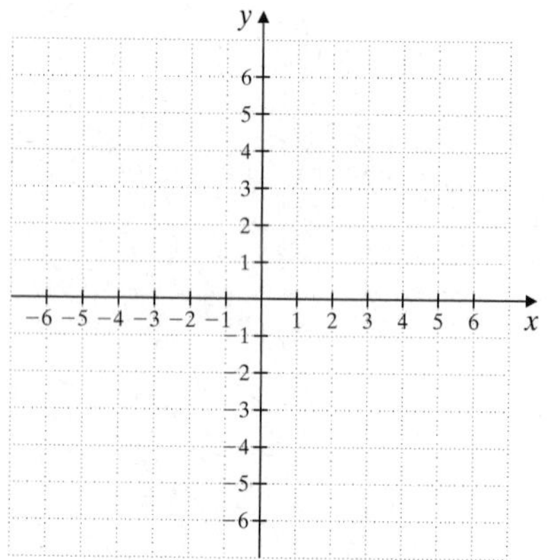

In linear programming, it is sometimes necessary to find the coordinates of the **corner point**: the point at which the two boundary lines intersect. To find the corner point for the system of Example 1, we solve the related linear system

$$\begin{cases} 3x = y \\ x + 2y = 8 \end{cases}$$

using either the substitution or the elimination method. The lines intersect at $\left(\frac{8}{7}, \frac{24}{7}\right)$, the corner point of the graph.

EXAMPLE 2 Graph the solutions of the system: $\begin{cases} x - y < 2 \\ x + 2y > -1 \\ y < 2 \end{cases}$

Solution: First we graph all three inequalities on the same set of axes. All boundary lines are dashed lines since the inequality symbols are < and >. The solution of the system is the region shown by the darkest shading. In this example, the boundary lines are *not* a part of the solution.

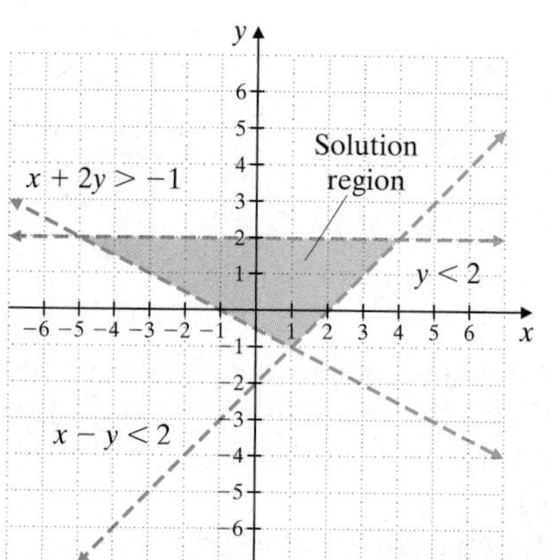

Answer

2.

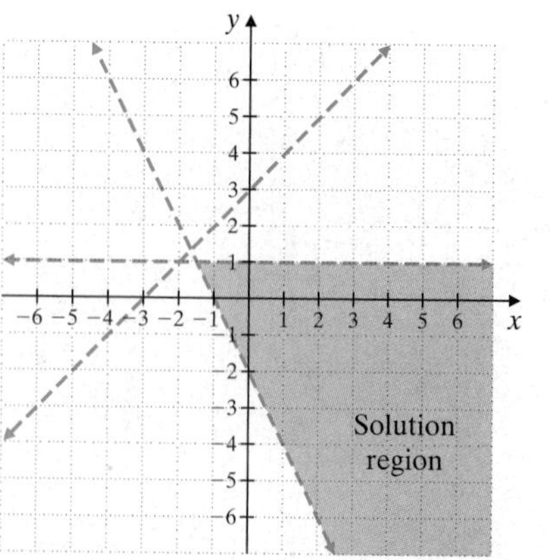

Try the Concept Check in the margin.

EXAMPLE 3 Graph the solutions of the system:
$$\begin{cases} -3x + 4y \le 12 \\ x \le 3 \\ x \ge 0 \\ y \ge 0 \end{cases}$$

Solution: We graph the inequalities on the same set of axes. The intersection of the inequalities is the solution region. It is the only region shaded in this graph and includes the portions of all four boundary lines that border the shaded region.

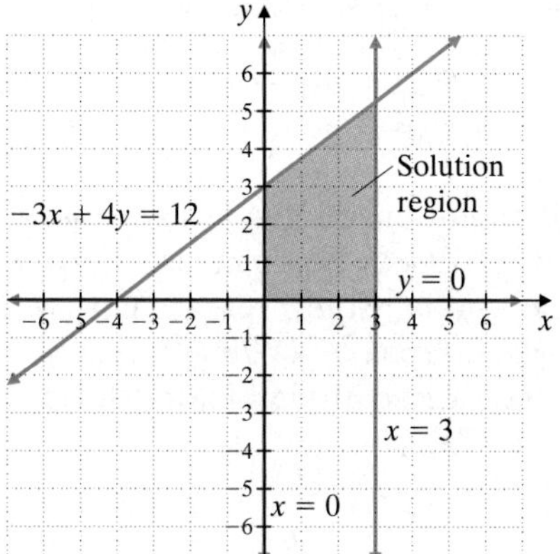

Concept Check

Describe the solution of the system of inequalities:
$$\begin{cases} x \le 2 \\ x \ge 2 \end{cases}$$

Practice Problem 3

Graph the solutions of the system:
$$\begin{cases} 2x - 3y \le 6 \\ y \ge 0 \\ y \le 4 \\ x \ge 0 \end{cases}$$

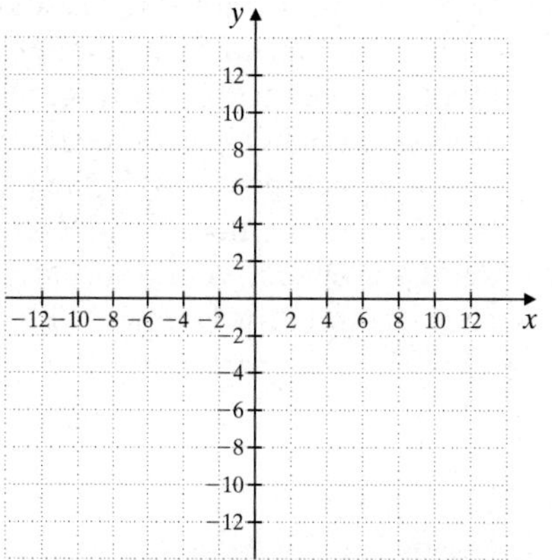

Answers

3.

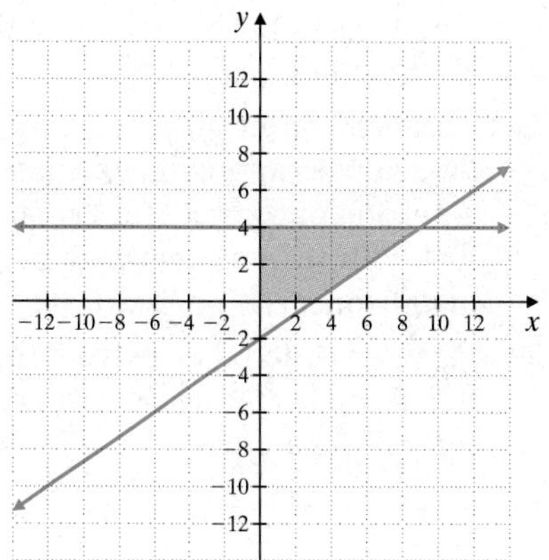

Concept Check: the line $x = 2$

FOCUS ON The Real World

LINEAR MODELING

In Chapter 3, we learned several ways to find a linear model when given either two ordered pairs or an ordered pair and slope. Another way to find a linear model of the form $y = mx + b$ for two ordered pairs (x_1, y_1) and (x_2, y_2) is to solve the following system of linear equations for m and b:

$$\begin{cases} y_1 = mx_1 + b \\ y_2 = mx_2 + b \end{cases}$$

For example, suppose a researcher wishes to find a linear model for the number of traffic accidents involving teenagers. The researcher locates statistics stating that there were 9920 teenage deaths from motor vehicle accidents in the United States in 1979. By 1999, this number had decreased to 5749 teenage deaths in motor vehicle accidents. (*Source:* Insurance Institute for Highway Safety)

This data gives two ordered pairs: (1979, 9920) and (1999, 5749). Alternatively, the ordered pairs could be written as (9, 9920) and (29, 5749), where the x-coordinate represents the number of years after 1970. (Adjusting data given as years in this way often simplifies calculations.) By substituting the coordinates of the second set of ordered pairs into the general linear system, we obtain the system

$$\begin{cases} 9920 = 9m + b \\ 5749 = 29m + b \end{cases}$$

The solution of this system is $m = -208.55$ and $b = 11{,}796.95$. We can use these values to write the model the researcher wished to find: $y = -208.55x + 11{,}796.95$, where y is the number of teenage deaths in motor vehicle accidents x years after 1990.

Internet Excursions

Go To: http://www.prenhall.com/martin-gay_algebra

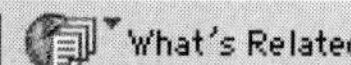

The Insurance Institute for Highway Safety (IIHS) is a nonprofit research organization supported by automobile insurers that collects, studies, and distributes motor vehicle safety statistics. This page offers access to a wide variety of traffic fatality statistics organized by categories such as alcohol-related fatalities, fatalities of children, and fatalities involving bicycles. (Alternatively, you can visit the IIHS homepage at http://www.hwysafety.org and look for the Fatality Facts option.)

1. Browse the list to find a set of data that interests you. Make a list of the ordered pairs that make up the set of data.
2. Create a scatter diagram of the data. Does the data appear approximately linear? If not, is there a portion of the data that appears approximately linear? If so, indicate which portion is approximately linear. If not, start over with Question 1.
3. Pick two ordered pairs from the linear portion of the data. Use these ordered pairs to form a system of linear equations.
4. Solve the system from Question 3. Find the linear equation that models your data.
5. Add the graph of your linear model to the scatter diagram from Question 2. How well does your model "fit" the data?
6. What trend does your model describe over the linear portion of your data?

Name ______________________ Section __________ Date __________

EXERCISE SET 8.7

A *Graph the solutions of each system of linear inequalities. See Examples 1 through 3.*

1. $\begin{cases} y \geq x + 1 \\ y \geq 3 - x \end{cases}$

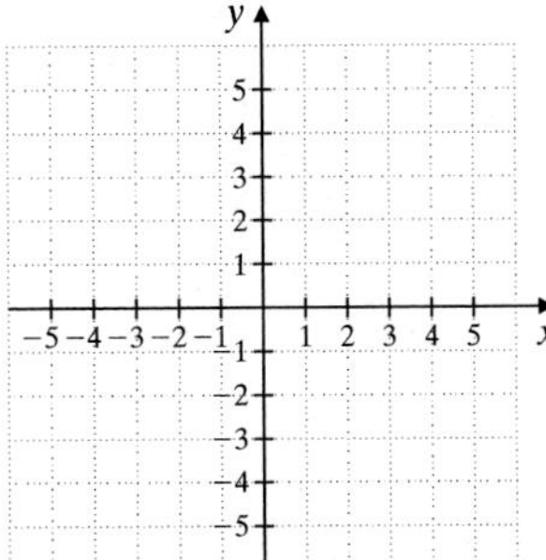

2. $\begin{cases} y \geq x - 3 \\ y \geq -1 - x \end{cases}$

3. $\begin{cases} y < 3x - 4 \\ y \leq x + 2 \end{cases}$

4. $\begin{cases} y \leq 2x + 1 \\ y > x + 2 \end{cases}$

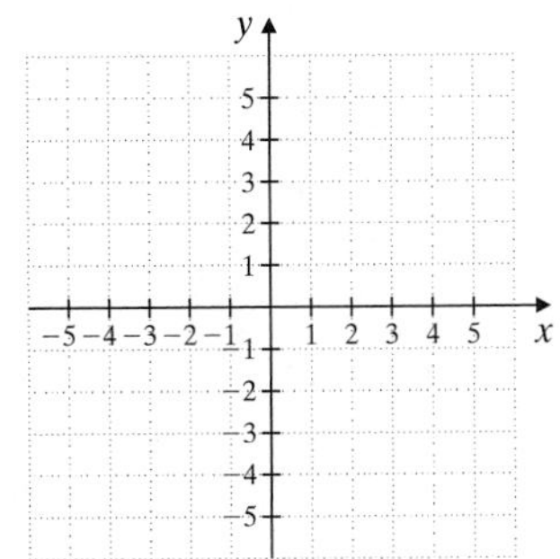

5. $\begin{cases} y \leq -2x - 2 \\ y \geq x + 4 \end{cases}$

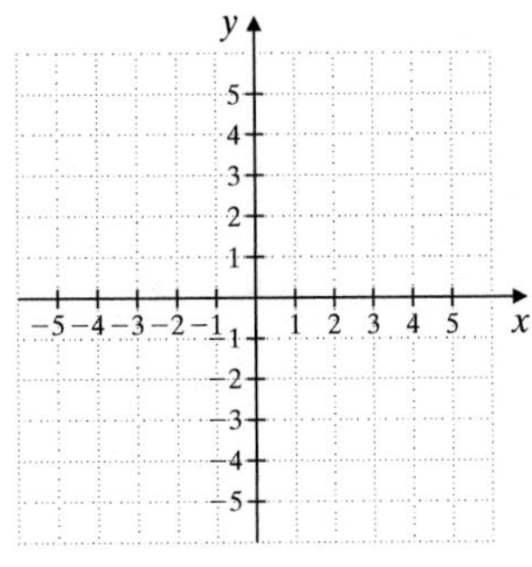

6. $\begin{cases} y \leq 2x + 4 \\ y \geq -x - 5 \end{cases}$

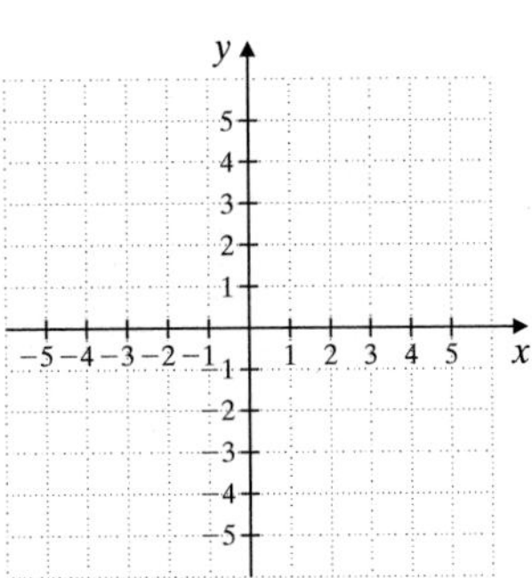

7. $\begin{cases} y \geq -x + 2 \\ y \leq 2x + 5 \end{cases}$

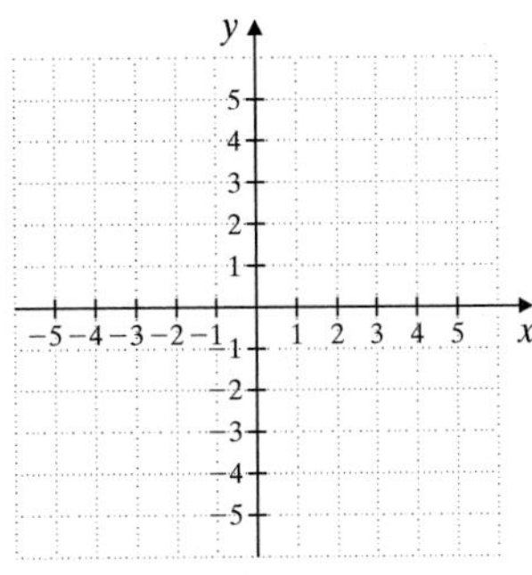

8. $\begin{cases} y \geq x - 5 \\ y \leq -3x + 3 \end{cases}$

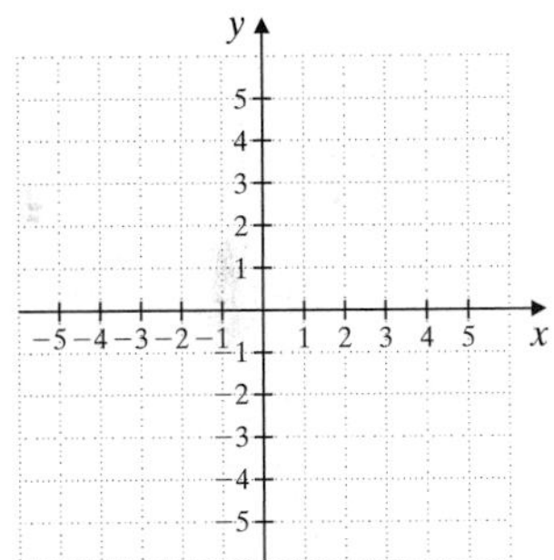

9. $\begin{cases} x \geq 3y \\ x + 3y \leq 6 \end{cases}$

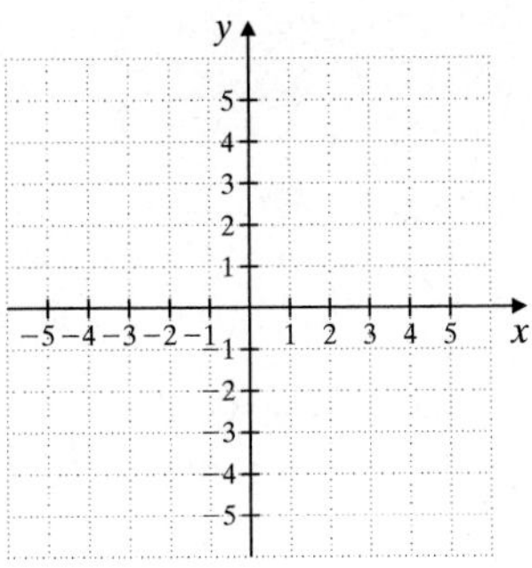

10. $\begin{cases} -2x < y \\ x + 2y < 3 \end{cases}$

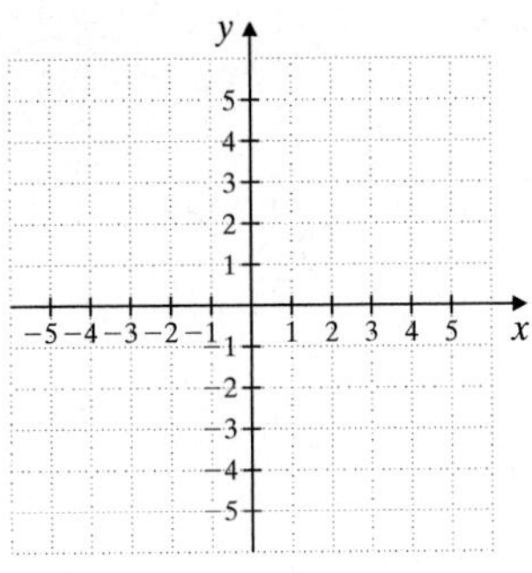

11. $\begin{cases} x \leq 2 \\ y \geq -3 \end{cases}$

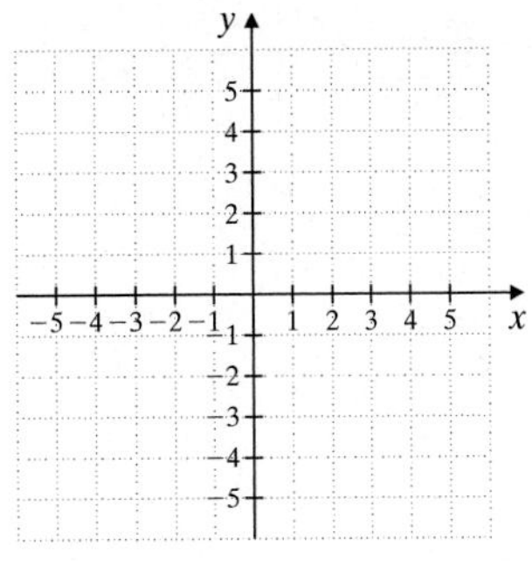

12. $\begin{cases} x \geq -3 \\ y \geq -2 \end{cases}$

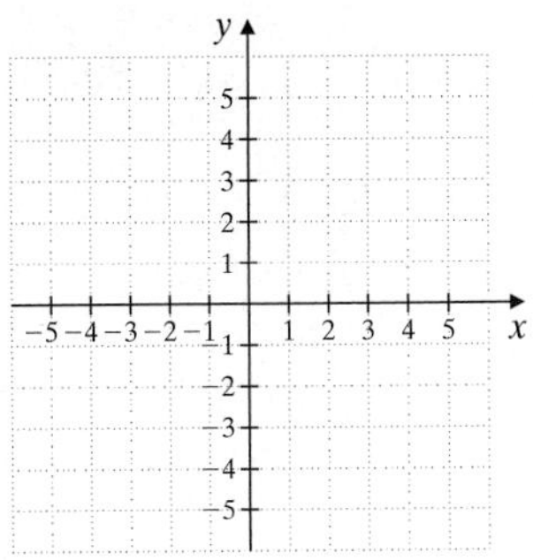

13. $\begin{cases} y \geq 1 \\ x < -3 \end{cases}$

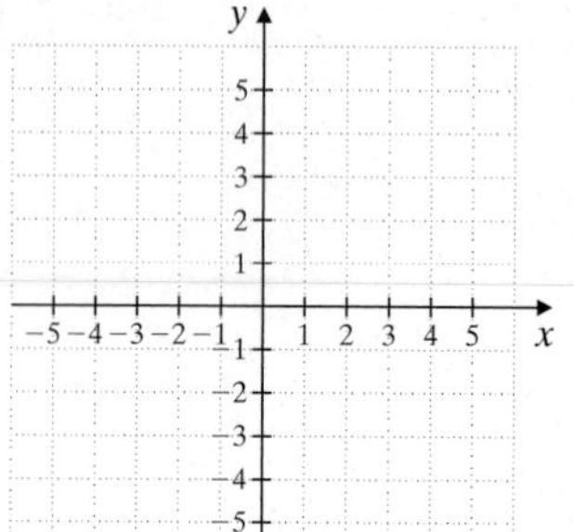

14. $\begin{cases} y > 2 \\ x \geq -1 \end{cases}$

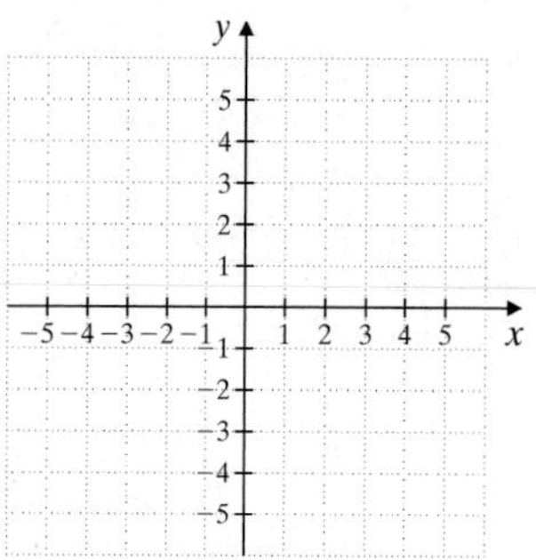

15. $\begin{cases} y + 2x \geq 0 \\ 5x - 3y \leq 12 \\ y \leq 2 \end{cases}$

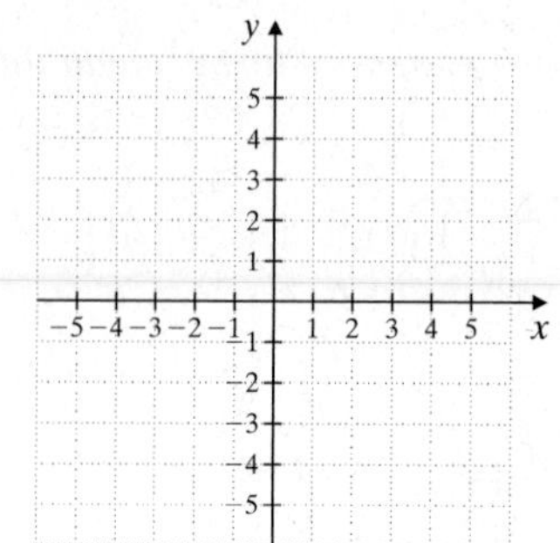

16. $\begin{cases} y + 2x \leq 0 \\ 5x + 3y \geq -2 \\ y \leq 4 \end{cases}$

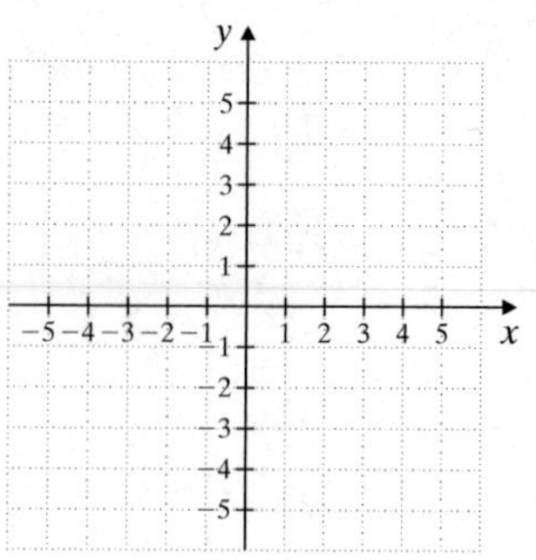

17. $\begin{cases} 3x - 4y \geq -6 \\ 2x + y \leq 7 \\ y \geq -3 \end{cases}$

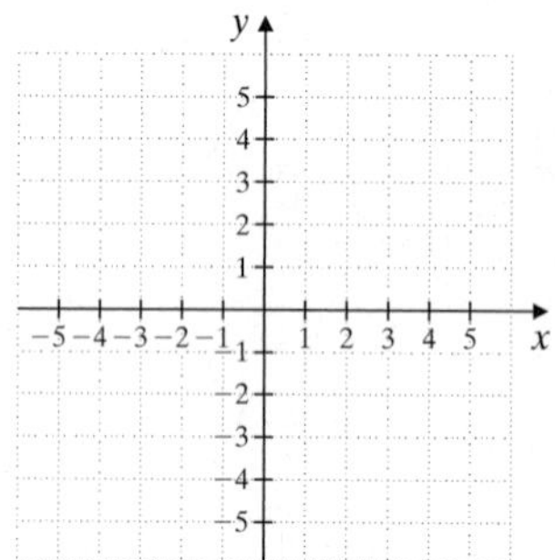

18. $\begin{cases} 4x - y \geq -2 \\ 2x + 3y \leq -8 \\ y \geq -5 \end{cases}$

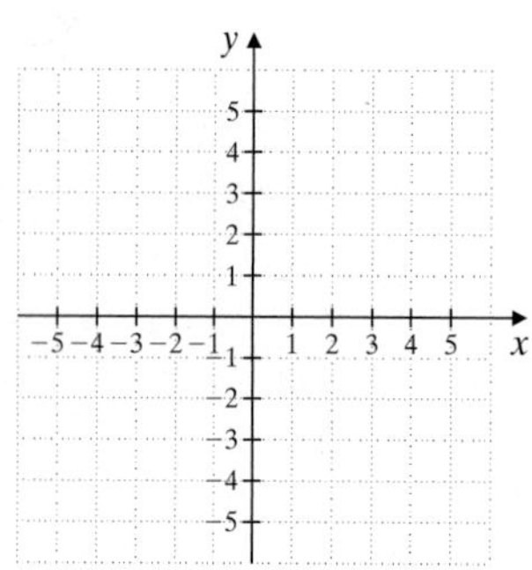

19. $\begin{cases} 2x + y \leq 5 \\ x \leq 3 \\ x \geq 0 \\ y \geq 0 \end{cases}$

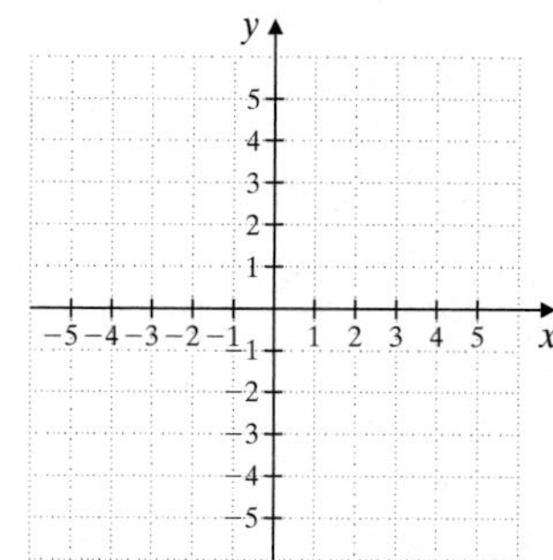

20. $\begin{cases} 3x + y \leq 4 \\ x \leq 4 \\ x \geq 0 \\ y \geq 0 \end{cases}$

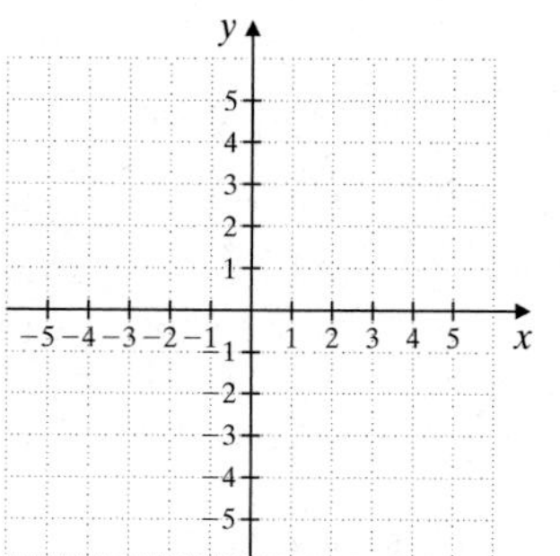

Match each system of inequalities to the corresponding graph.

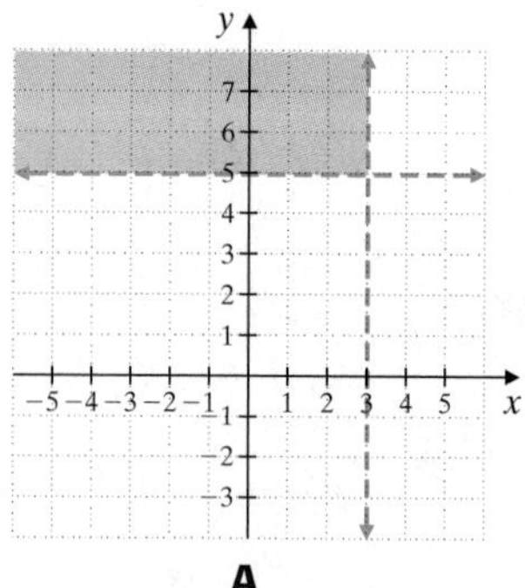

A

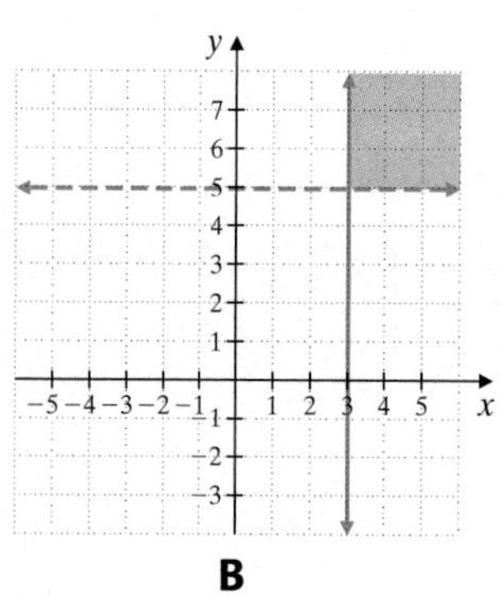

B

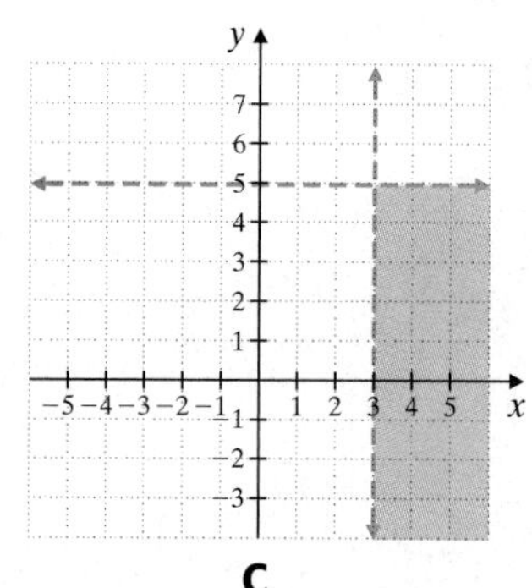

C

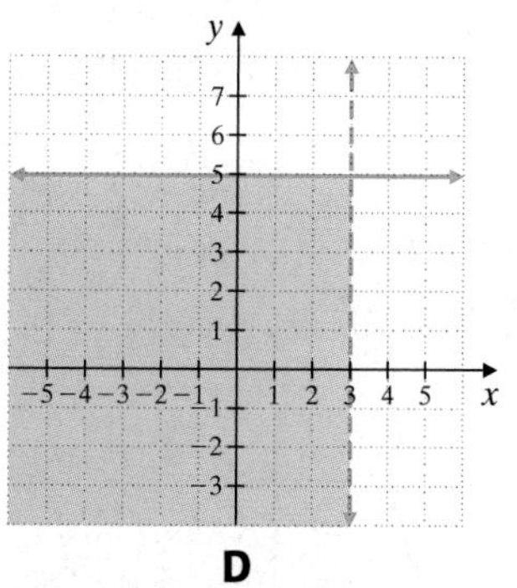

D

21. $\begin{cases} y < 5 \\ x > 3 \end{cases}$

22. $\begin{cases} y > 5 \\ x < 3 \end{cases}$

23. $\begin{cases} y \leq 5 \\ x < 3 \end{cases}$

24. $\begin{cases} y > 5 \\ x \geq 3 \end{cases}$

Review and Preview

Evaluate each expression. See Section 1.6.

25. $(-3)^2$

26. $(-5)^3$

27. $\left(\frac{2}{3}\right)^2$

28. $\left(\frac{3}{4}\right)^3$

Perform each indicated operation. See Section 1.6.

29. $(-2)^2 - (-3) + 2(-1)$

30. $5^2 - 11 + 3(-5)$

31. $8^2 + (-13) - 4(-2)$

32. $(-12)^2 + (-1)(2) - 6$

Combining Concepts

33. Tony Noellert budgets his time at work today. Part of the day he can write bills; the rest of the day he can use to write purchase orders. The total time available is at most 8 hours. Less than 3 hours is to be spent writing bills.

a. Write a system of inequalities to describe the situation. (Let x = hours available for writing bills and y = hours available for writing purchase orders.)

b. Graph the solutions of the system.

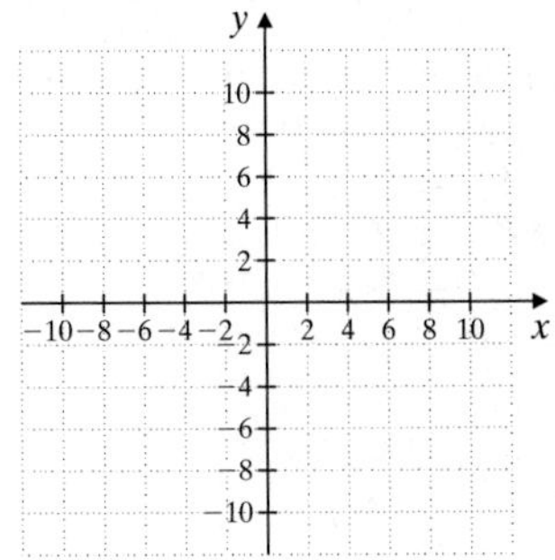

34. Explain how to decide which region to shade to show the solution region of the following system.

$$\begin{cases} x \geq 3 \\ y \geq -2 \end{cases}$$

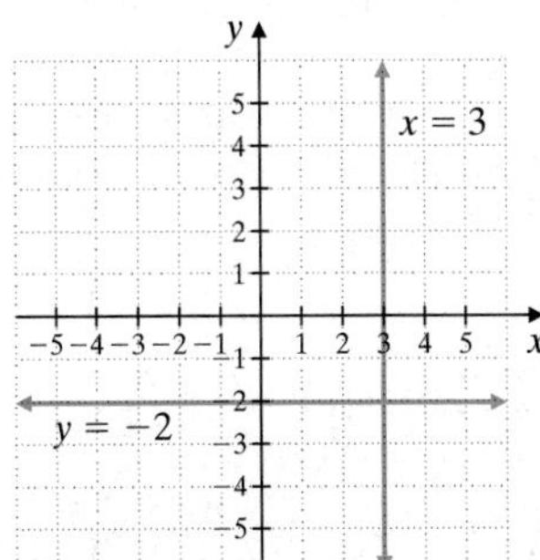

CHAPTER 8 ACTIVITY Analyzing the Courses of Ships

MATERIALS:

- Ruler, graphing calculator (optional)

This activity may be completed by working in groups or individually.

From overhead photographs or satellite imagery of ships on the ocean, defense analysts can tell a lot about a ship's immediate course by looking at its wake. Assuming that two ships will maintain their present courses, it is possible to extend their paths, based on the wakes visible in the photograph, and find possible points of collision.

Investigate the courses and possibility of collision of the two ships shown in the figure. Assume that the ships will maintain their present courses.

1. Using each ship's wake as a guide, extend the paths of the ships on the figure. Estimate the coordinates of the point of intersection of the ships' courses from the grid. If the ships continue in these courses, they could possibly collide at the point of intersection of their paths.

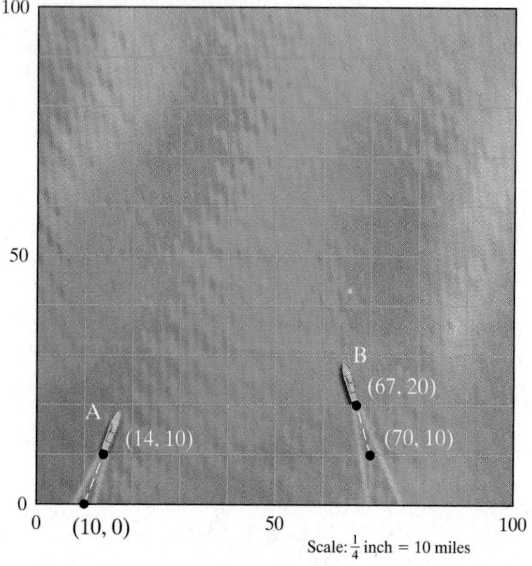

Scale: $\frac{1}{4}$ inch = 10 miles

2. Using the coordinates labeled on each ship's wake, find a linear equation that describes each path.

3. (Optional) Use a graphing calculator to graph both equations in the same window. Use the Intersect or Trace feature to estimate the point of intersection of the two paths. Compare this estimate to your estimate in Question 1.

4. Solve the system of two linear equations using one of the methods in this chapter. The solution is the point of intersection of the two paths. Compare your answer to your estimates from Questions 1 and 3.

5. Plot the point of intersection you found in Question 4 on the figure. Use the figure's scale to find each ship's distance from this point of collision by measuring from the bow (tip) of each ship with a ruler. Suppose that the speed of ship A is r_1 and the speed of ship B is r_2. Given the present positions and courses of the two ships, find a relationship between their speeds that would ensure their collision.

Chapter 8 VOCABULARY CHECK

Fill in each blank with one of the words or phrases listed below.

system of linear equations	solution	consistent	independent	matrix
dependent	inconsistent	substitution	addition	square

1. In a system of linear equations in two variables, if the graphs of the equations are the same, the equations are ____________ equations.
2. Two or more linear equations are called a ________________________.
3. A system of equations that has at least one solution is called a(n) ____________ system.
4. A __________ of a system of two equations in two variables is an ordered pair of numbers that is a solution of both equations in the system.
5. Two algebraic methods for solving systems of equations are __________ and ____________.
6. A system of equations that has no solution is called a(n) ____________ system.
7. In a system of linear equations in two variables, if the graphs of the equations are different, the equations are ____________ equations.
8. If a matrix has the same number of rows and columns, it is called a _________ matrix.
9. A _________ is a rectangular array of numbers.

CHAPTER 8 Highlights

DEFINITIONS AND CONCEPTS | **EXAMPLES**

Section 8.1 Solving Systems of Linear Equations by Graphing

A **system of linear equations** consists of two or more linear equations.

$$\begin{cases} 2x + y = 6 \\ x = -3y \end{cases} \quad \begin{cases} -3x + 5y = 10 \\ x - 4y = -2 \end{cases}$$

A **solution** of a system of two equations in two variables is an ordered pair of numbers that is a solution of both equations in the system.

Determine whether $(-1, 3)$ is a solution of the system.

$$\begin{cases} 2x - y = -5 \\ x = 3y - 10 \end{cases}$$

Replace x with -1 and y with 3 in both equations.

$$2x - y = -5$$
$$2(-1) - 3 \stackrel{?}{=} -5$$
$$-5 = -5 \quad \text{True}$$
$$x = 3y - 10$$
$$-1 \stackrel{?}{=} 3(3) - 10$$
$$-1 = -1 \quad \text{True}$$

$(-1, 3)$ is a solution of the system.

Graphically, a solution of a system is a point common to the graphs of both equations.

Solve by graphing: $\begin{cases} 3x - 2y = -3 \\ x + y = 4 \end{cases}$

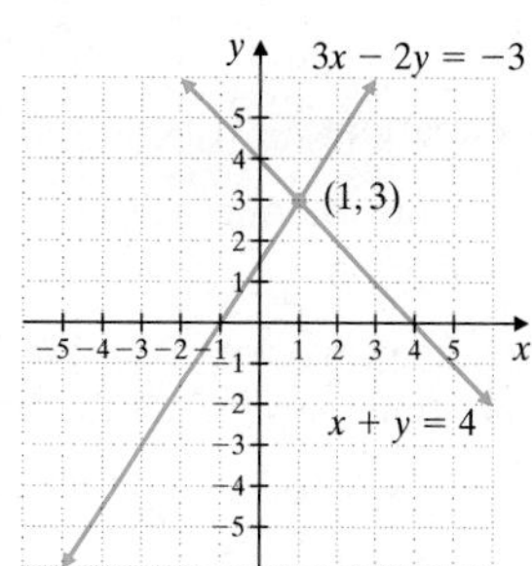

Definitions and Concepts — Examples

Section 8.1 Solving Systems of Linear Equations by Graphing *(continued)*

Three different situations can occur when graphing the two lines associated with the equations in a linear system.

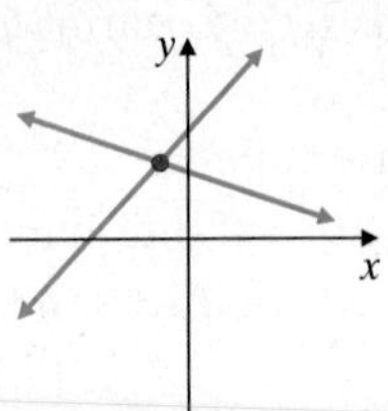

One point of intersection; one solution

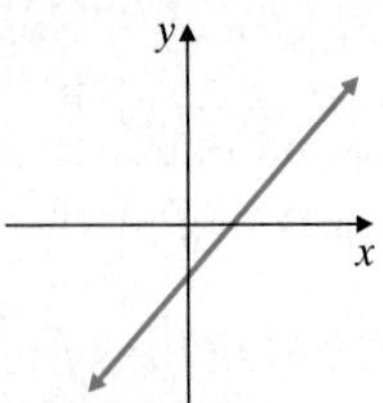

Same line; infinite number of solutions

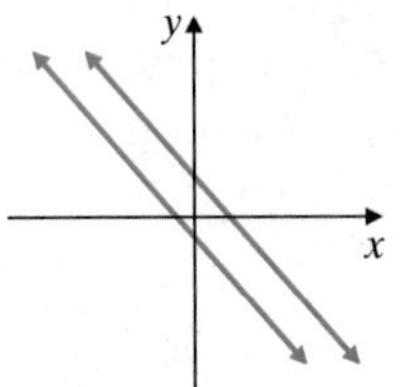

Parallel lines: no solution

Section 8.2 Solving Systems of Linear Equations by Substitution

To Solve a System of Linear Equations by the Substitution Method

Step 1. Solve one equation for a variable.

Step 2. Substitute the expression for the variable into the other equation.

Step 3. Solve the equation from Step 2 to find the value of one variable.

Step 4. Substitute the value from Step 3 in either original equation to find the value of the other variable.

Step 5. Check the solution in both original equations.

Solve by substitution.

$$\begin{cases} 3x + 2y = 1 \\ x = y - 3 \end{cases}$$

Substitute $y - 3$ for x in the first equation.

$$\begin{aligned} 3x + 2y &= 1 \\ 3(y - 3) + 2y &= 1 \\ 3y - 9 + 2y &= 1 \\ 5y &= 10 \\ y &= 2 \qquad \text{Divide by 5.} \end{aligned}$$

To find x, substitute 2 for y in $x = y - 3$ so that $x = 2 - 3$ or -1. The solution $(-1, 2)$ checks.

Section 8.3 Solving Systems of Linear Equations by Addition

To Solve a System of Linear Equations by the Addition Method

Step 1. Rewrite each equation in standard form $Ax + By = C$.

Step 2. Multiply one or both equations by a nonzero number so that the coefficients of a variable are opposites.

Step 3. Add the equations.

Step 4. Find the value of one variable by solving the resulting equation.

Step 5. Substitute the value from Step 4 into either original equation to find the value of the other variable.

Solve by addition.

$$\begin{cases} x - 2y = 8 \\ 3x + y = -4 \end{cases}$$

Multiply both sides of the first equation by -3.

$$\begin{cases} -3x + 6y = -24 \\ 3x + y = -4 \end{cases}$$

$$\begin{aligned} 7y &= -28 \qquad \text{Add.} \\ y &= -4 \qquad \text{Divide by 7.} \end{aligned}$$

To find x, let $y = -4$ in an original equation.

$$\begin{aligned} x - 2(-4) &= 8 \qquad \text{First equation} \\ x + 8 &= 8 \\ x &= 0 \end{aligned}$$

DEFINITIONS AND CONCEPTS / EXAMPLES

Section 8.3 Solving Systems of Linear Equations by Addition *(continued)*

Step 6. Check the solution in both original equations.

The solution $(0, -4)$ checks.

If solving a system of linear equations by substitution or addition yields a true statement such as $-2 = -2$, then the graphs of the equations in the system are identical and there is an infinite number of solutions of the system.

Solve: $\begin{cases} 2x - 6y = -2 \\ x = 3y - 1 \end{cases}$

Substitute $3y - 1$ for x in the first equation.

$$\begin{aligned} 2(3y - 1) - 6y &= -2 \\ 6y - 2 - 6y &= -2 \\ -2 &= -2 \quad \text{True} \end{aligned}$$

The system has an infinite number of solutions.

If solving a system of linear equations yields a false statement such as $0 = 3$, the graphs of the equations in the system are parallel lines and the system has no solution.

Solve: $\begin{cases} 5x - 2y = 6 \\ -5x + 2y = -3 \end{cases}$

$$0 = 3 \quad \text{False}$$

The system has no solution.

Section 8.4 Systems of Linear Equations and Problem Solving

PROBLEM-SOLVING STEPS

1. UNDERSTAND. Read and reread the problem.

Two angles are supplementary if their sum is 180°. The larger of two supplementary angles is three times the smaller, decreased by twelve. Find the measure of each angle. Let

x = measure of smaller angle and

y = measure of larger angle

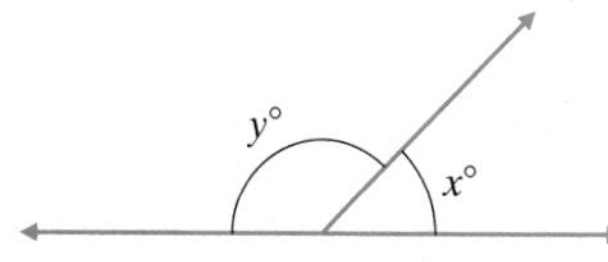

2. TRANSLATE.

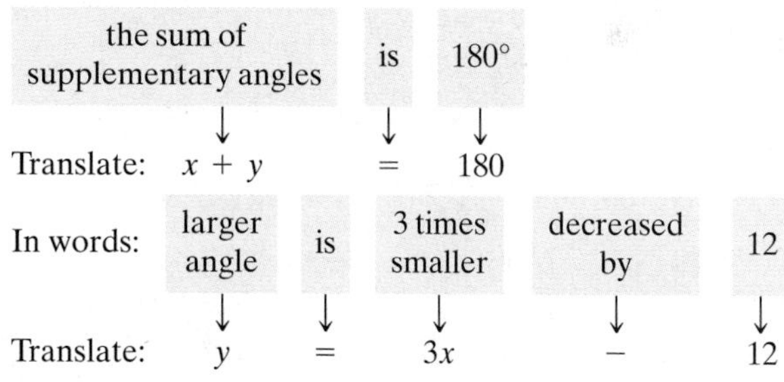

In words: the sum of supplementary angles | is | 180°

Translate: $x + y = 180$

In words: larger angle | is | 3 times smaller | decreased by | 12

Translate: $y = 3x - 12$

3. SOLVE.

Solve the system.

$$\begin{cases} x + y = 180 \\ y = 3x - 12 \end{cases}$$

Use the substitution method and replace y with $3x - 12$ in the first equation.

$$\begin{aligned} x + y &= 180 \\ x + (3x - 12) &= 180 \\ 4x &= 192 \\ x &= 48 \end{aligned}$$

4. INTERPRET.

Since $y = 3x - 12$, then $y = 3 \cdot 48 - 12$ or 132.

The solution checks. The smaller angle measures 48° and the larger angle measures 132°.

DEFINITIONS AND CONCEPTS | **EXAMPLES**

Section 8.5 Solving Systems of Linear Equations in Three Variables

A **solution** of an equation in three variables x, y, and z is an **ordered triple** (x, y, z) that makes the equation a true statement.

Verify that $(-2, 1, 3)$ is a solution of $2x + 3y - 2z = -7$. Replace x with -2, y with 1, and z with 3.

$$\begin{aligned} 2(-2) + 3(1) - 2(3) &= -7 \\ -4 + 3 - 6 &= -7 \\ -7 &= -7 \quad \text{True.} \end{aligned}$$

$(-2, 1, 3)$ is a solution.

SOLVING A SYSTEM OF THREE LINEAR EQUATIONS BY THE ELIMINATION METHOD

Step 1. Write each equation in standard form, $Ax + By + Cz = D$.

Step 2. Choose a pair of equations and use them to eliminate a variable.

Step 3. Choose any other pair of equations and eliminate the same variable.

Step 4. Solve the system of two equations in two variables from Steps 2 and 3.

Step 5. Solve for the third variable by substituting the values of the variables from Step 4 into any of the original equations.

Step 6. Check the solution in all three original equations.

Solve:

$$\begin{cases} 2x + y - z = 0 & (1) \\ x - y - 2z = -6 & (2) \\ -3x - 2y + 3z = -22 & (3) \end{cases}$$

1. Each equation is written in standard form.

2.
$$\begin{array}{ll} 2x + y - z = 0 & (1) \\ \underline{x - y - 2z = -6} & (2) \\ 3x \qquad - 3z = -6 & (4) \text{ Add.} \end{array}$$

3. Eliminate y from equations (1) and (3) also.

$$\begin{array}{lll} 4x + 2y - 2z = 0 & & \text{Multiply equation (1) by 2.} \\ \underline{-3x - 2y + 3z = -22} & (3) & \\ x \qquad + z = -22 & (5) & \text{Add.} \end{array}$$

4. Solve.

$$\begin{cases} 3x - 3z = -6 & (4) \\ x + z = -22 & (5) \end{cases}$$

$$\begin{array}{ll} x - z = -2 & \text{Divide equation (4) by 3.} \\ \underline{x + z = -22} & (5) \\ 2x \quad = -24 & \\ x \quad = -12 & \end{array}$$

To find z, use equation (5).

$$\begin{aligned} x + z &= -22 \\ -12 + z &= -22 \\ z &= -10 \end{aligned}$$

5. To find y, use equation (1).

$$\begin{aligned} 2x + y - z &= 0 \\ 2(-12) + y - (-10) &= 0 \\ -24 + y + 10 &= 0 \\ y &= 14 \end{aligned}$$

6. The solution $(-12, 14, -10)$ checks.

DEFINITIONS AND CONCEPTS | **EXAMPLES**

Section 8.6 Solving Systems of Equations Using Matrices

A **matrix** is a rectangular array of numbers.

$$\begin{bmatrix} -7 & 0 & 3 \\ 1 & 2 & 4 \end{bmatrix} \quad \begin{bmatrix} a & b & c \\ d & e & f \\ g & h & i \end{bmatrix}$$

The **matrix** corresponding to a system is composed of the coefficients of the variables and the constants of the system.

The matrix corresponding to the system

$$\begin{cases} x - y = 1 \\ 2x + y = 11 \end{cases} \text{ is } \left[\begin{array}{cc|c} 1 & -1 & 1 \\ 2 & 1 & 11 \end{array}\right]$$

The following **row operations** can be performed on matrices, and the result is an equivalent matrix.

Elementary row operations:

1. Interchange any two rows.
2. Multiply (or divide) the elements of one row by the same nonzero number.
3. Multiply (or divide) the elements of one row by the same nonzero number and add them to their corresponding elements in any other row.

Use matrices to solve: $\begin{cases} x - y = 1 \\ 2x + y = 11 \end{cases}$

The corresponding matrix is

$$\left[\begin{array}{cc|c} 1 & -1 & 1 \\ 2 & 1 & 11 \end{array}\right]$$

Use row operations to write an equivalent matrix with 1s along the diagonal and 0s below each 1 in the diagonal. Multiply row 1 by -2 and add to row 2. Change row 2 only.

$$\left[\begin{array}{cc|c} 1 & -1 & 1 \\ -2(1) + 2 & -2(-1) + 1 & -2(1) + 11 \end{array}\right]$$

simplifies to $\left[\begin{array}{cc|c} 1 & -1 & 1 \\ 0 & 3 & 9 \end{array}\right]$

Divide row 2 by 3.

$$\left[\begin{array}{cc|c} 1 & -1 & 1 \\ \frac{0}{3} & \frac{3}{3} & \frac{9}{3} \end{array}\right] \text{ simplifies to } \left[\begin{array}{cc|c} 1 & -1 & 1 \\ 0 & 1 & 3 \end{array}\right]$$

This matrix corresponds to the system

$$\begin{cases} x - y = 1 \\ y = 3 \end{cases}$$

Let $y = 3$ in the first equation.

$$x - 3 = 1$$
$$x = 4$$

The ordered pair solution is $(4, 3)$.

Section 8.7 Systems of Linear Inequalities

A **system of linear inequalities** consists of two or more linear inequalities.

$$\begin{cases} x - y \geq 3 \\ y \leq -2x \end{cases}$$

To graph a system of inequalities, graph each inequality in the system. The overlapping region is the solution of the system.

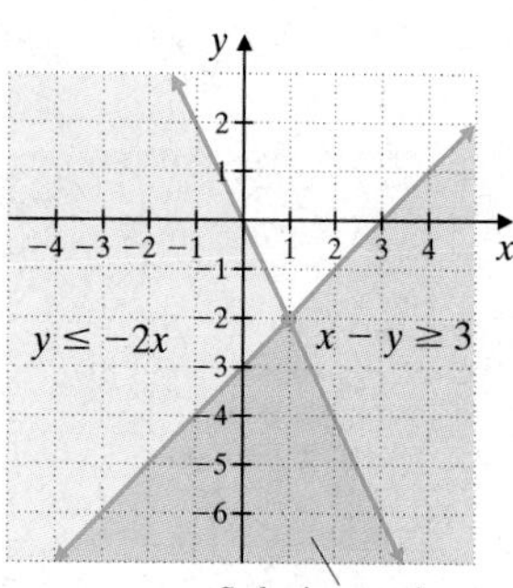

Solution region

STUDY SKILLS REMINDER

How well do you know your textbook?

See if you can answer the questions below.

1. What does the icon mean?
2. What does the icon mean?
3. What does the △ icon mean?
4. Where can you find a review for each chapter? What answers to this review can be found in the back of your text?
5. Each chapter contains an overview of the chapter along with examples. What is this feature called?
6. Does this text contain any solutions to exercises? If so, where?

Name ______________________ Section __________ Date __________

Chapter 8 Review

(8.1) *Determine whether each ordered pair is a solution of the system of linear equations.*

1. $\begin{cases} 2x - 3y = 12 \\ 3x + 4y = 1 \end{cases}$

a. $(12, 4)$

b. $(3, -2)$

2. $\begin{cases} 4x + y = 0 \\ -8x - 5y = 9 \end{cases}$

a. $\left(\frac{3}{4}, -3\right)$

b. $(-2, 8)$

3. $\begin{cases} 5x - 6y = 18 \\ 2y - x = -4 \end{cases}$

a. $(-6, -8)$

b. $3, \frac{5}{2}$

4. $\begin{cases} 2x + 3y = 1 \\ 3y - x = 4 \end{cases}$

a. $(2, 2)$

b. $(-1, 1)$

Solve each system of equations by graphing.

5. $\begin{cases} x + y = 5 \\ x - y = 1 \end{cases}$

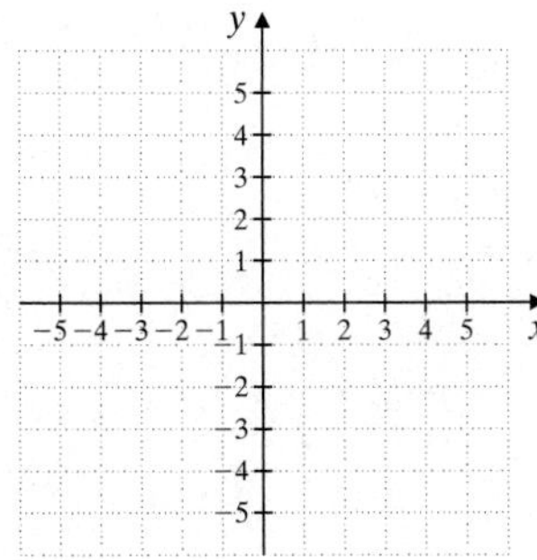

6. $\begin{cases} x + y = 3 \\ x - y = -1 \end{cases}$

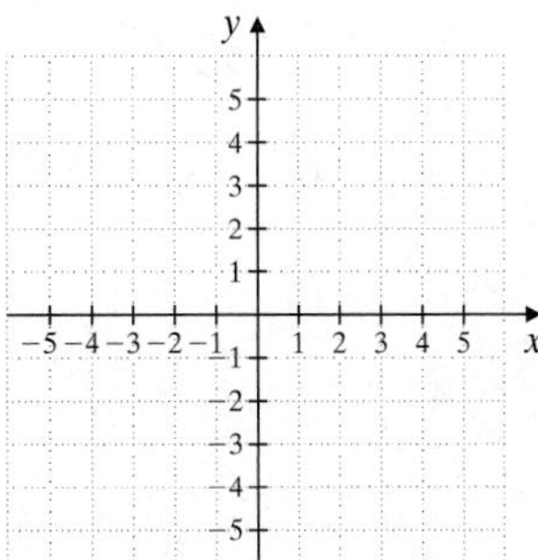

7. $\begin{cases} x = 5 \\ y = -1 \end{cases}$

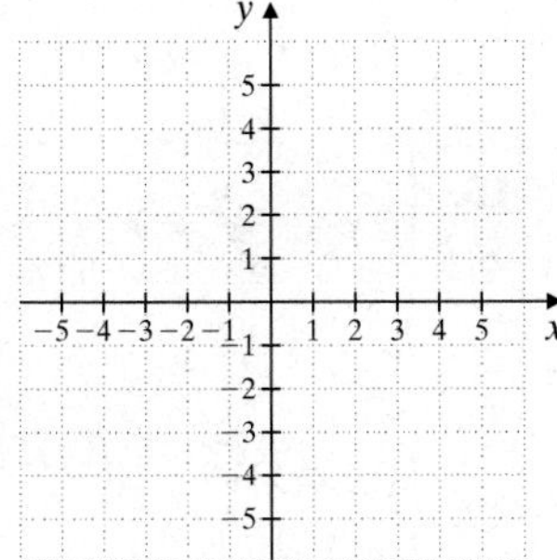

8. $\begin{cases} x = -3 \\ y = 2 \end{cases}$

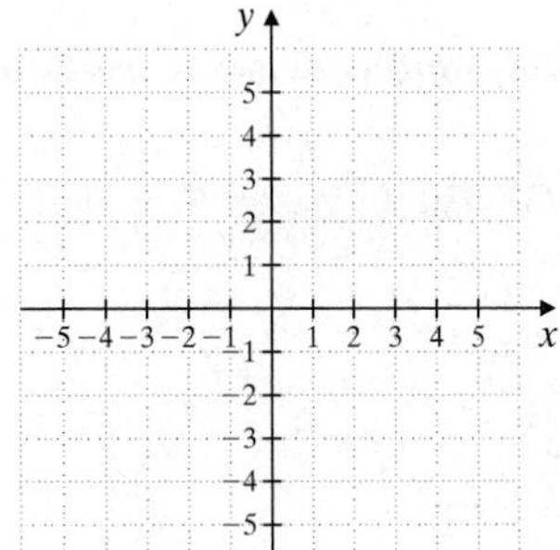

9. $\begin{cases} 2x + y = 5 \\ x = -3y \end{cases}$

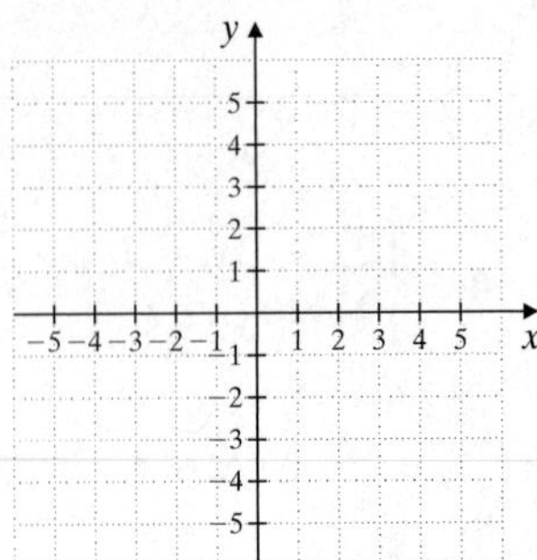

10. $\begin{cases} 3x + y = -2 \\ y = -5x \end{cases}$

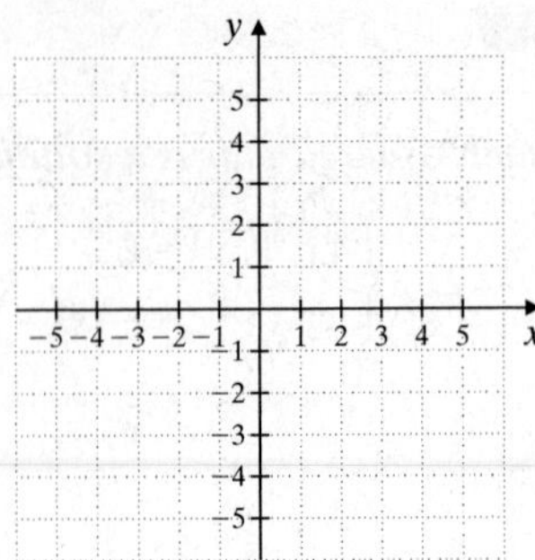

11. $\begin{cases} y = 2x + 4 \\ y = -x - 5 \end{cases}$

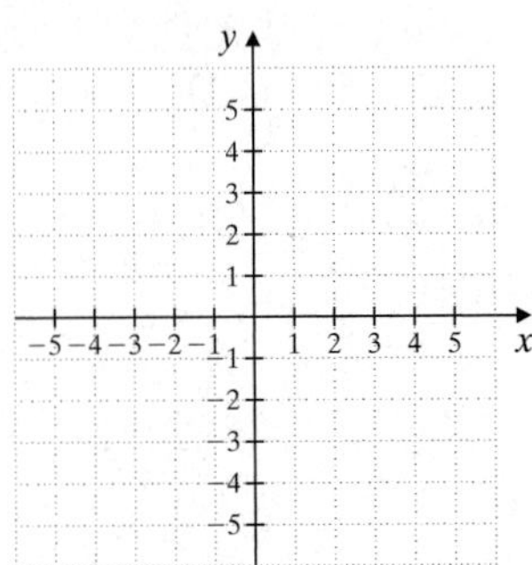

12. $\begin{cases} y = x - 5 \\ y = -2x + 2 \end{cases}$

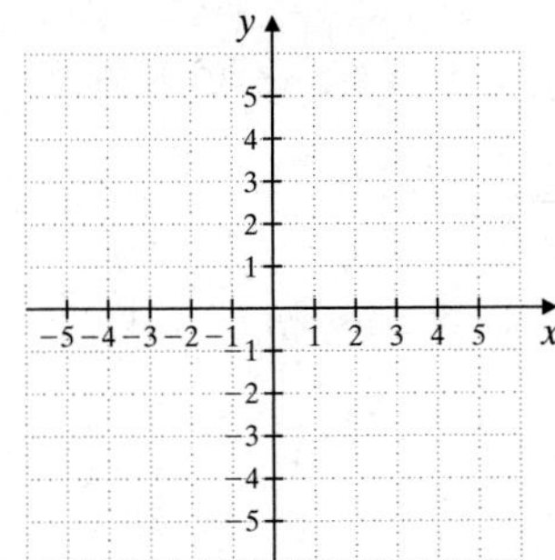

13. $\begin{cases} y = 3x \\ -6x + 2y = 6 \end{cases}$

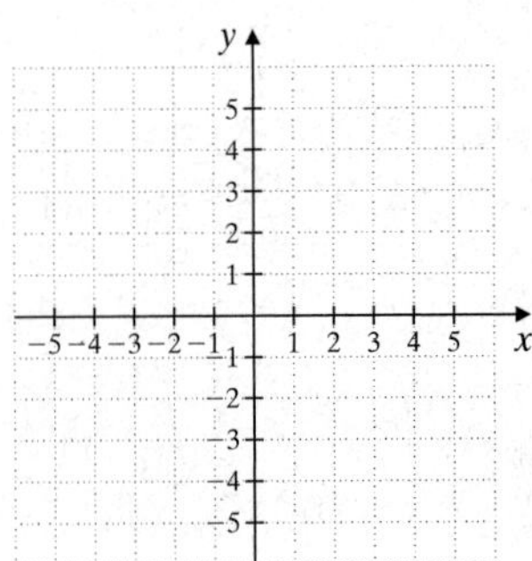

14. $\begin{cases} x - 2y = 2 \\ -2x + 4y = -4 \end{cases}$

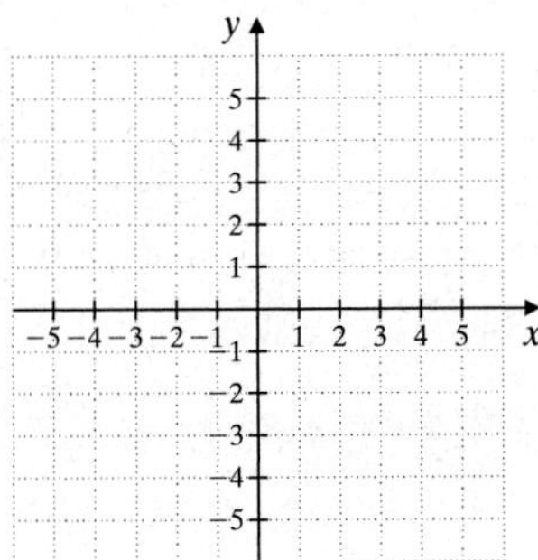

(8.2) *Solve each system of equations by the substitution method.*

15. $\begin{cases} x = 2y \\ 2x - 3y = 2 \end{cases}$

16. $\begin{cases} x = 5y \\ x - 4y = 1 \end{cases}$

17. $\begin{cases} y = 2x + 6 \\ 3x - 2y = -11 \end{cases}$

18. $\begin{cases} y = 3x - 7 \\ 2x - 3y = 7 \end{cases}$

19. $\begin{cases} x + 3y = -3 \\ 2x + y = 4 \end{cases}$

20. $\begin{cases} 3x + y = 11 \\ x + 2y = 12 \end{cases}$

21. $\begin{cases} 4y = 2x - 3 \\ x - 2y = 4 \end{cases}$

22. $\begin{cases} 2x = 3y - 18 \\ x + 4y = 2 \end{cases}$

23. $\begin{cases} x + y = 6 \\ y = -x - 4 \end{cases}$

24. $\begin{cases} -3x + y = 6 \\ y = 3x + 2 \end{cases}$

(8.3) *Solve each system of equations by the addition method.*

25. $\begin{cases} x + y = 14 \\ x - y = 18 \end{cases}$

26. $\begin{cases} x + y = 9 \\ x - y = 13 \end{cases}$

27. $\begin{cases} 2x + 3y = -6 \\ x - 3y = -12 \end{cases}$

28. $\begin{cases} 4x + y = 15 \\ -4x + 3y = -19 \end{cases}$

29. $\begin{cases} 2x - 3y = -15 \\ x + 4y = 31 \end{cases}$

30. $\begin{cases} x - 5y = -22 \\ 4x + 3y = 4 \end{cases}$

31. $\begin{cases} 2x - 6y = -1 \\ -x + 3y = \dfrac{1}{2} \end{cases}$

32. $\begin{cases} -4x - 6y = 8 \\ 2x + 3y = -3 \end{cases}$

33. $\begin{cases} \dfrac{3}{4}x + \dfrac{2}{3}y = 2 \\ x + \dfrac{y}{3} = 6 \end{cases}$

34. $\begin{cases} \dfrac{2}{5}x + \dfrac{3}{4}y = 1 \\ x + 3y = -2 \end{cases}$

35. $\begin{cases} 10x + 2y = 0 \\ 3x + 5y = 33 \end{cases}$

36. $\begin{cases} 0.6x - 0.3y = -1.5 \\ 0.04x - 0.02y = -0.1 \end{cases}$

(8.4) *Solve each problem by writing and solving a system of linear equations.*

37. The sum of two numbers is 16. Three times the larger number decreased by the smaller number is 72. Find the two numbers.

38. The Forrest Theater can seat a total of 360 people. They take in $15,150 when every seat is sold. If orchestra section tickets cost $45 and balcony tickets cost $35, find the number of seats in the orchestra section and the number of seats in the balcony.

39. A riverboat can head 340 miles upriver in 19 hours, but the return trip takes only 14 hours. Find the current of the river and find the speed of the ship in still water to the nearest tenth of a mile.

	d =	***r*** ·	***t***
Upriver	340	$x - y$	19
Downriver	340	$x + y$	14

40. Sam and Cynthia Abney invested $9000 one year ago. Part of the money was invested at 6% and the rest at 10%. If the total interest earned in one year was $652.80, find how much was invested at each rate.

△ **41.** Ancient Greeks thought that a picture had the most pleasing dimensions if the length was approximately 1.6 times as long as the width. This ratio is known as the Golden Ratio. If Sandreka Walker has 6 feet of framing material, find the dimensions of the largest frame she can make that satisfies the Golden Ratio. Find the dimensions to the nearest hundredth of a foot.

42. Find the amount of a 6% acid solution and the amount of a 14% acid solution Pat Mayfield should combine to prepare 50 cc (cubic centimeters) of a 12% solution.

43. A deli charges \$3.80 for a breakfast of three eggs and four strips of bacon. The charge is \$2.75 for two eggs and three strips of bacon. Find the cost of each egg and the cost of each strip of bacon.

44. An exercise enthusiast alternates between jogging and walking. He traveled 15 miles during the past 3 hours. He jogs at a rate of 7.5 miles per hour and walks at a rate of 4 miles per hour. Find how much time, to the nearest hundredth of an hour, he actually spent jogging and how much time he spent walking.

(8.5) *Solve each system of equations in three variables.*

45. $\begin{cases} x + z = 4 \\ 2x - y = 4 \\ x + y - z = 0 \end{cases}$

46. $\begin{cases} 2x + 5y = 4 \\ x - 5y + z = -1 \\ 4x - z = 11 \end{cases}$

47. $\begin{cases} 4y + 2z = 5 \\ 2x + 8y = 5 \\ 6x + 4z = 1 \end{cases}$

48. $\begin{cases} 5x + 7y = 9 \\ 14y - z = 28 \\ 4x + 2z = -4 \end{cases}$

49. $\begin{cases} 3x - 2y + 2z = 5 \\ -x + 6y + z = 4 \\ 3x + 14y + 7z = 20 \end{cases}$

50. $\begin{cases} x + 2y + 3z = 11 \\ y + 2z = 3 \\ 2x + 2z = 10 \end{cases}$

51. $\begin{cases} 7x - 3y + 2z = 0 \\ 4x - 4y - z = 2 \\ 5x + 2y + 3z = 1 \end{cases}$

52. $\begin{cases} x - 3y - 5z = -5 \\ 4x - 2y + 3z = 13 \\ 5x + 3y + 4z = 22 \end{cases}$

Use systems of equations to solve.

53. The sum of three numbers is 98. The sum of the first and second is two more than the third number, and the second is four times the first. Find the numbers.

54. An employee at See's Candy Store needs a special mixture of candy. She has creme-filled chocolates that sell for \$3.00 per pound, chocolate-covered nuts that sell for \$2.70 per pound, and chocolate-covered raisins that sell for \$2.25 per pound. She wants to have twice as many raisins as nuts in the mixture. Find how many pounds of each she should use to make 45 pounds worth \$2.80 per pound.

55. Chris Kringler has \$2.77 in her coin jar—all in pennies, nickels, and dimes. If she has 53 coins in all and four more nickels than dimes, find how many of each type of coin she has.

△ **56.** The perimeter of an isosceles (two sides equal) triangle is 73 centimeters. If the unequal side is 7 centimeters longer than the two equal sides, find the lengths of the three sides.

(8.6) *Use matrices to solve each system.*

57. $\begin{cases} 3x + 10y = 1 \\ x + 2y = -1 \end{cases}$

58. $\begin{cases} 3x - 6y = 12 \\ 2y = x - 4 \end{cases}$

59. $\begin{cases} 3x - 2y = -8 \\ 6x + 5y = 11 \end{cases}$

60. $\begin{cases} 6x - 6y = -5 \\ 10x - 2y = 1 \end{cases}$

61. $\begin{cases} 3x - 6y = 0 \\ 2x + 4y = 5 \end{cases}$

62. $\begin{cases} 5x - 3y = 10 \\ -2x + y = -1 \end{cases}$

63. $\begin{cases} 0.2x - 0.3y = -0.7 \\ 0.5x + 0.3y = 1.4 \end{cases}$

64. $\begin{cases} 3x + 2y = 8 \\ 3x - y = 5 \end{cases}$

65. $\begin{cases} x + z = 4 \\ 2x - y = 0 \\ x + y - z = 0 \end{cases}$

66. $\begin{cases} 2x + 5y = 4 \\ x - 5y + z = -1 \\ 4x - z = 11 \end{cases}$

67. $\begin{cases} 3x - y = 11 \\ x + 2z = 13 \\ y - z = -7 \end{cases}$

68. $\begin{cases} 5x + 7y + 3z = 9 \\ 14y - z = 28 \\ 4x + 2z = -4 \end{cases}$

69. $\begin{cases} 7x - 3y + 2z = 0 \\ 4x - 4y - z = 2 \\ 5x + 2y + 3z = 1 \end{cases}$

70. $\begin{cases} x + 2y + 3z = 14 \\ y + 2z = 3 \\ 2x - 2z = 10 \end{cases}$

(8.7) *Graph the solution of each system of linear inequalities.*

71. $\begin{cases} y \geq 2x - 3 \\ y \leq -2x + 1 \end{cases}$

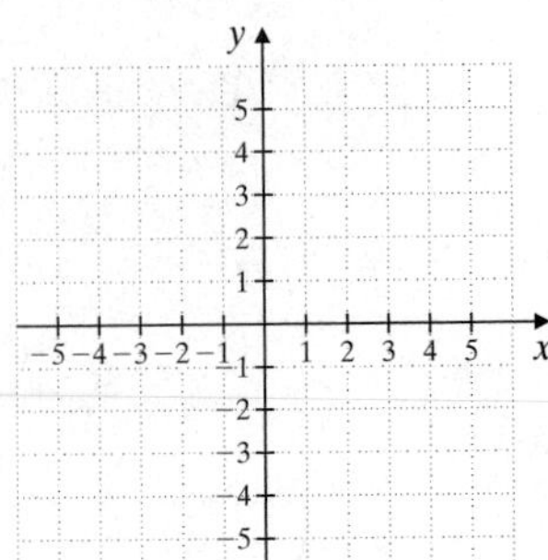

72. $\begin{cases} y \leq -3x - 3 \\ y \leq 2x + 7 \end{cases}$

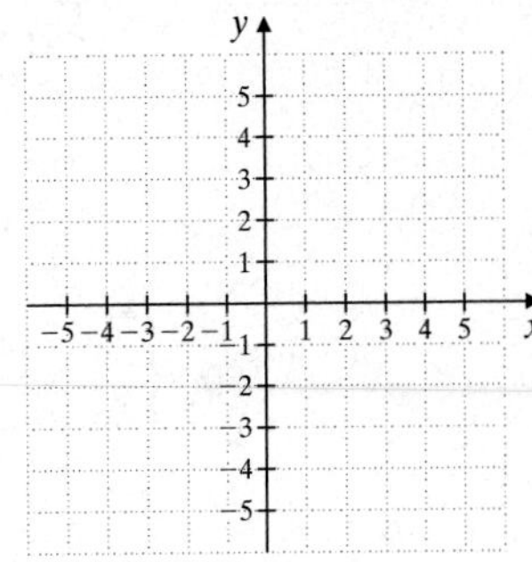

73. $\begin{cases} x + 2y > 0 \\ x - \quad y \leq 6 \end{cases}$

74. $\begin{cases} x - 2y \geq \quad 7 \\ x + \quad y \leq -5 \end{cases}$

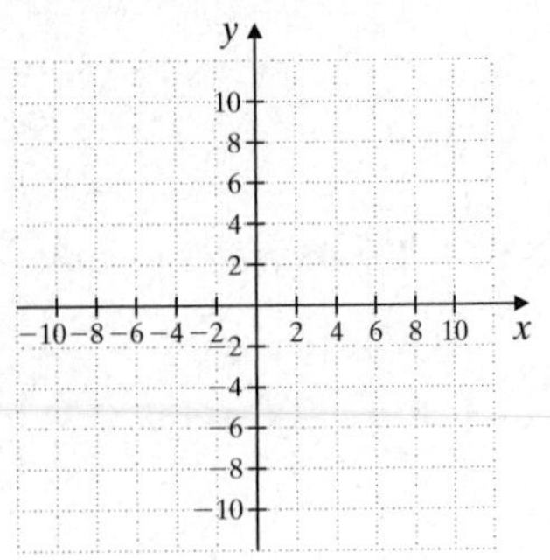

75. $\begin{cases} 3x - 2y \leq 4 \\ 2x + \quad y \geq 5 \\ \qquad\quad y \leq 4 \end{cases}$

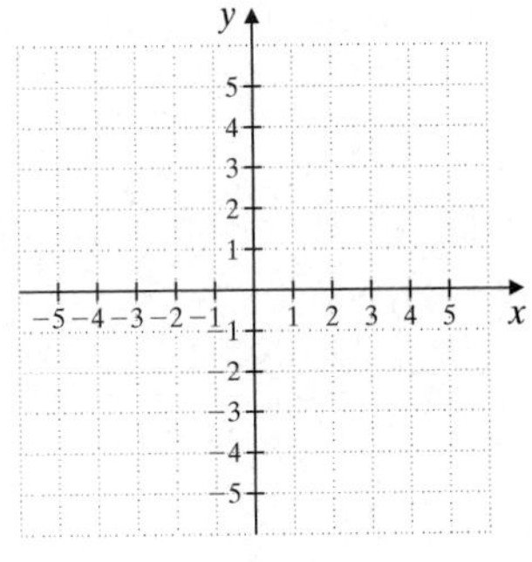

76. $\begin{cases} 4x - \quad y \leq \quad 0 \\ 3x - 2y \geq -5 \\ \qquad\quad y \geq -4 \end{cases}$

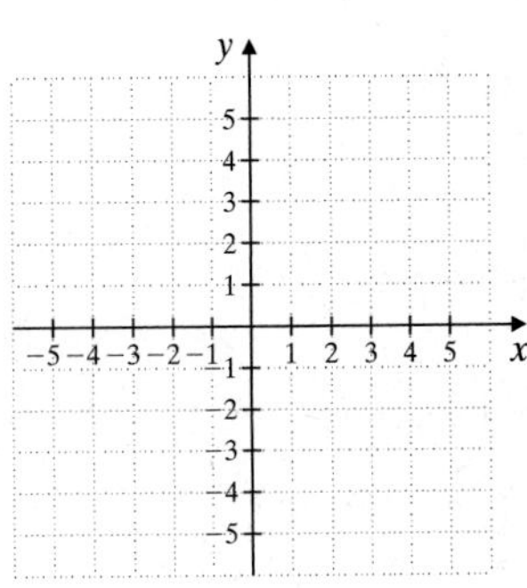

77. $\begin{cases} x + 2y \leq 5 \\ x \leq 2 \\ x \geq 0 \\ y \geq 0 \end{cases}$

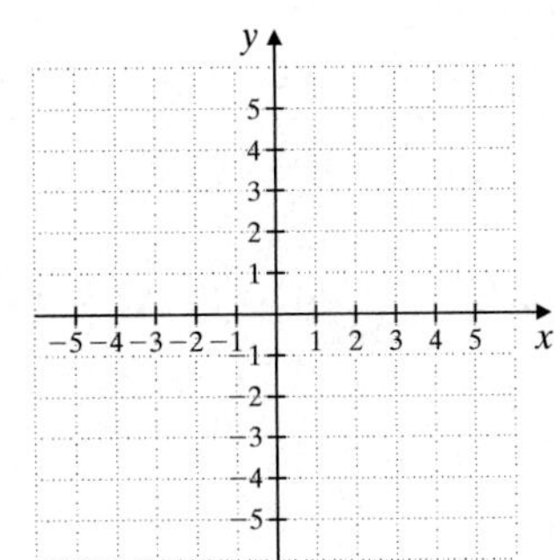

78. $\begin{cases} x + 3y \leq 7 \\ y \leq 5 \\ x \geq 0 \\ y \geq 0 \end{cases}$

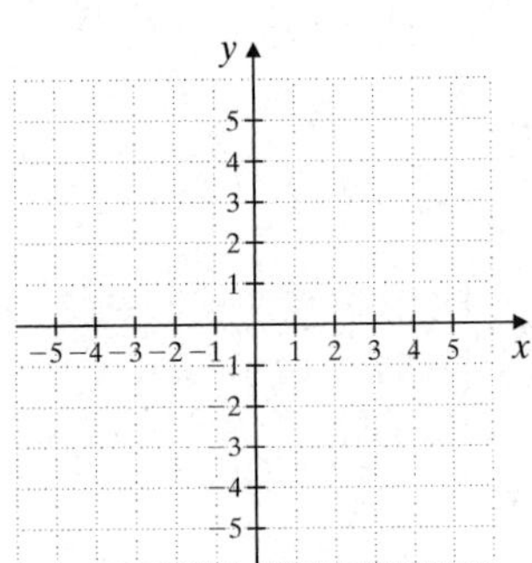

Name ______________________ Section __________ Date __________

Chapter 8 Test

Solve each system by graphing.

1. $\begin{cases} x - y = 2 \\ 3x - y = -2 \end{cases}$

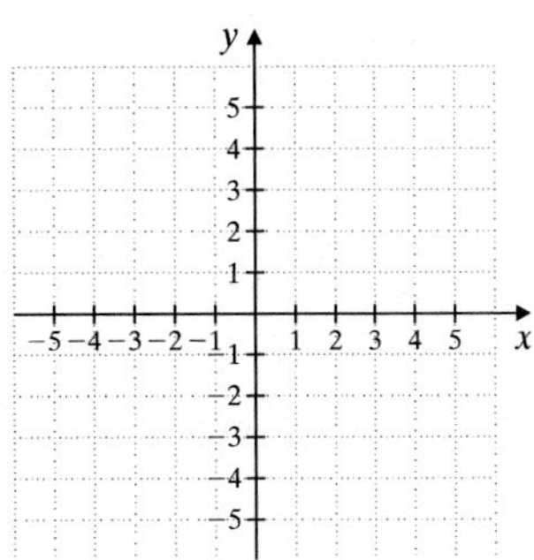

2. $\begin{cases} y = -3x \\ 3x + y = 6 \end{cases}$

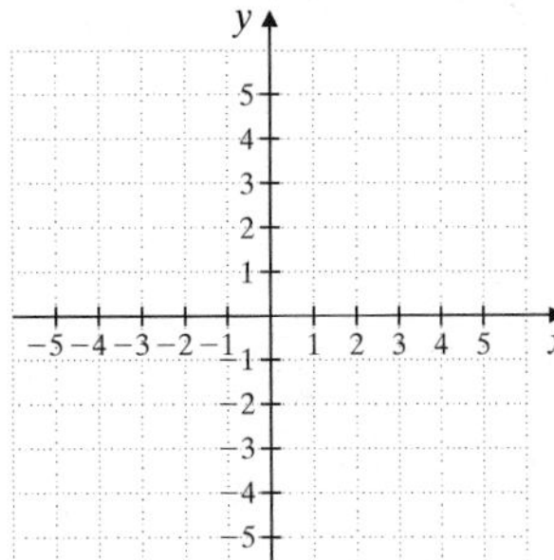

Solve each system.

3. $\begin{cases} 3x - 2y = -14 \\ y = x + 5 \end{cases}$

4. $\begin{cases} 3x + y = 7 \\ 4x + 3y = 1 \end{cases}$

5. $\begin{cases} x - y = 4 \\ x - 2y = 11 \end{cases}$

6. $\begin{cases} 8x - 4y = 12 \\ y = 2x - 3 \end{cases}$

7. $\begin{cases} x + y = 28 \\ x - y = 12 \end{cases}$

8. $\begin{cases} y - x = 6 \\ y + 2x = -6 \end{cases}$

9. $\begin{cases} 5x - 6y = 7 \\ 7x - 4y = 12 \end{cases}$

10. $\begin{cases} x - \frac{2}{3}y = 3 \\ -2x + 3y = 10 \end{cases}$

11. $\begin{cases} 0.01x - 0.06y = -0.23 \\ 0.2x + 0.4y = 0.2 \end{cases}$

12. $\begin{cases} 6x - y = 0 \\ \frac{3}{2}x - \frac{y}{4} = 1 \end{cases}$

Answers

1. see graph
2. see graph
3. ______
4. ______
5. ______
6. ______
7. ______
8. ______
9. ______
10. ______
11. ______
12. ______

13. ______________________

14. ______________________

15. ______________________

16. ______________________

17. ______________________

18. ______________________

19. ______________________

20. ______________________

21. ______________________

22. ______________________

23. ______________________

24. ______________________

13. $\begin{cases} x + y + z = 4 \\ 2x + 5y = 1 \\ x - y - 2z = 0 \end{cases}$

14. $\begin{cases} 3x + 2y + 3z = 3 \\ x - z = 9 \\ 4y + z = -4 \end{cases}$

Use matrices to solve each system.

15. $\begin{cases} x - y = -2 \\ 3x - 3y = -6 \end{cases}$

16. $\begin{cases} x + 2y = -1 \\ 2x + 5y = -5 \end{cases}$

17. $\begin{cases} x - y - z = 0 \\ 3x - y - 5z = -2 \\ 2x + 3y = -5 \end{cases}$

18. $\begin{cases} 2x - y + 3z = 4 \\ 3x - 3z = -2 \\ -5x + y = 0 \end{cases}$

Solve each problem by writing and using a system of linear equations.

19. Two numbers have a sum of 124 and a difference of 32. Find the numbers.

20. Lisa has a bundle of money consisting of \$1 bills and \$5 bills. There are 62 bills in the bundle. The total value of the bundle is \$230. Find the number of \$1 bills and the number of \$5 bills.

21. A 30% alcohol solution is to be mixed with a 70% alcohol solution. How much of each is needed to make 10 liters of a 40% solution?

22. Two hikers start at opposite ends of the St. Tammany Trails and walk toward each other. The trail is 36 miles long and they meet in 4 hours. If one hiker is twice as fast as the other, find both hiking speeds.

23. A motel in New Orleans charges \$90 per day for double occupancy and \$80 per day for single occupancy. If 80 rooms are occupied for a total of \$6930, how many rooms of each kind are occupied?

24. The research department of a company that manufactures children's fruit drinks is experimenting with a new flavor. A 17.5% fructose solution is needed, but only 10% and 20% solutions are available. How many gallons of a 10% fructose solution should be mixed with a 20% fructose solution to obtain 20 gallons of a 17.5% fructose solution?

The graph below shows the percent of recorded music purchases that fell within the rap/hip-hop or country music genres for the years shown. Use this graph to answer Questions 17 and 18.

25. ______

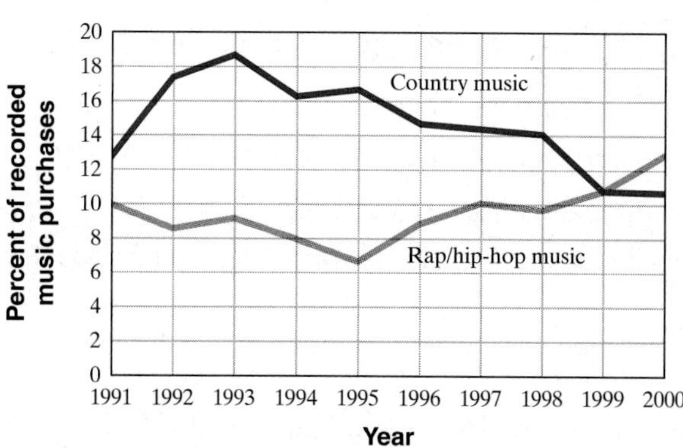

Source: Recording Industry Association of America

26. ______

25. In what year were purchases of country music equal to purchases of rap/hip-hop music?

26. In what year(s) were there more purchases of country music than rap/hip-hop music?

Graph the solutions of each system of linear inequalities.

27. $\begin{cases} 2y - x \geq 1 \\ x + y \geq -4 \\ y \leq 2 \end{cases}$

28. $\begin{cases} y + 2x \leq 4 \\ y \leq 2 \\ y \geq 0 \\ x \geq 0 \end{cases}$

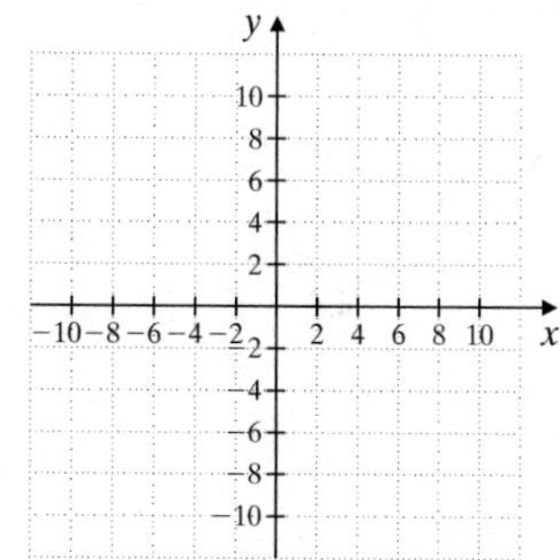

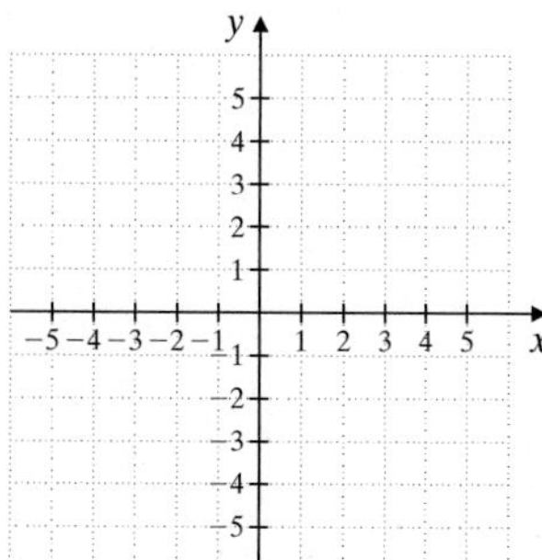

27. see graph

28. see graph

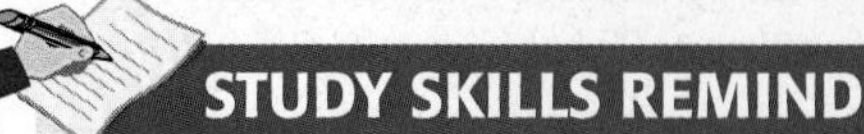

What should you do the day of an exam?

On the day of an exam, try the following:

- Allow yourself plenty of time to arrive.
- Read the directions on the test carefully.
- Read each problem carefully as you take your test. Make sure that you answer the question asked.
- Watch your time and pace yourself so that you may attempt each problem on your test.
- If you have time, check your work and answers.
- Do not turn your test in early. If you have extra time, spend it double-checking your work.

Good luck!

Name ______________ Section ________ Date ________

Cumulative Review

Multiply.

1. $-5(-10)$

2. $-\frac{2}{3}\cdot\frac{4}{7}$

3. Solve: $4(2x - 3) + 7 = 3x + 5$

4. The circle graph below shows the purpose of trips made by American travelers. Use this graph to answer the questions below.

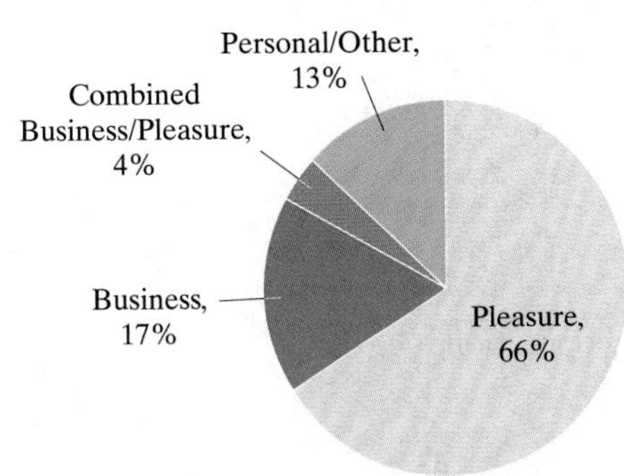

Source: Travel Industry Association of America

   a. What percent of trips made by American travelers are solely for the purpose of business?

   b. What percent of trips made by American travelers are for the purpose of business or combined business/pleasure?

   c. On an airplane flight of 253 Americans, how many of these people might we expect to be traveling solely for business?

5. Graph $y = -3$.

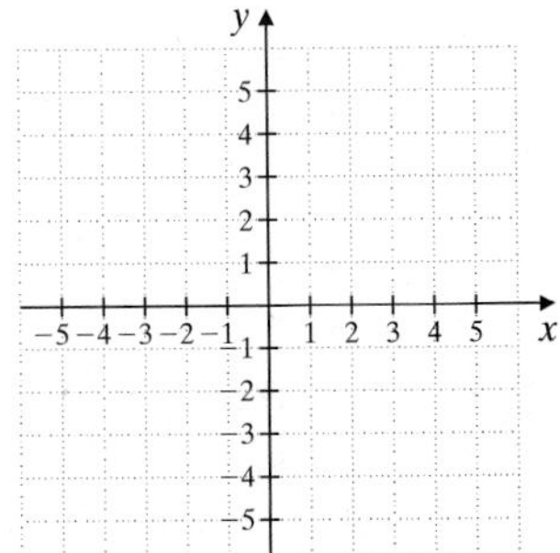

6. Graph: $5x + 4y \leq 20$

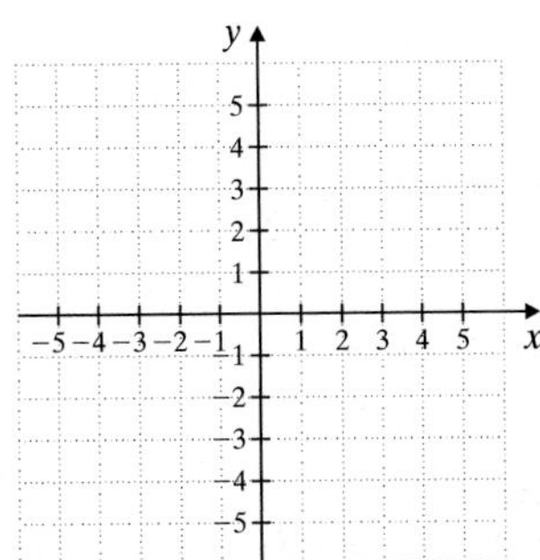

7. Write the following numbers in standard notation, without exponents.

   a. 1.02×10^5

   b. 7.358×10^{-3}

   c. 8.4×10^7

   d. 3.007×10^{-5}

8. Find the product: $(3x + 2)(2x - 5)$

Answers

1. ________
2. ________
3. ________
4. a ________
   b. ________
   c. ________
5. see graph
6. see graph
7. a ________
   b. ________
   c. ________
   d. ________
8. ________

9. ______

10. ______

11. a ______

b. ______

c. ______

12. ______

13. a ______

b. ______

14. ______

15. ______

16. ______

17. ______

18. ______

19. ______

20. ______

9. Factor $xy + 2x + 3y + 6$ by grouping.

10. Factor: $3x^2 + 11x + 6$

11. Are there any values for x for which each expression is undefined?

a. $\dfrac{x}{x-3}$

b. $\dfrac{x^2+2}{x^2-3x+2}$

c. $\dfrac{x^3-6x^2-10x}{3}$

12. Simplify: $\dfrac{x^2+4x+4}{x^2+2x}$

13. Perform each indicated operation.

a. $\dfrac{a}{4} - \dfrac{2a}{8}$

b. $\dfrac{3}{10x^2} + \dfrac{7}{25x}$

14. Solve:

$$\frac{4x}{x^2+x-30} + \frac{2}{x-5} = \frac{1}{x+6}$$

15. Write an equation of the line with slope -3 containing the point $(1, -5)$.

16. Determine wheher the relation $y = 2x + 1$ is also a function.

17. Solve the system: $\begin{cases} 3x + 4y = 13 \\ 5x - 9y = 6 \end{cases}$

18. Solve the system:

$$\begin{cases} 2x - 4y + 8z = 2 \\ -x - 3y + z = 11 \\ x - 2y + 4z = 0 \end{cases}$$

Use matrices to solve.

19. $\begin{cases} 2x - y = 3 \\ 4x - 2y = 5 \end{cases}$

20. $\begin{cases} x + 2y + z = 2 \\ -2x - y + 2z = 5 \\ x + 3y - 2z = -8 \end{cases}$

Rational Exponents, Radicals, and Complex Numbers

In this chapter, radical notation is reviewed, and then rational exponents are introduced. As the name implies, rational exponents are exponents that are rational numbers. We present an interpretation of rational exponents that is consistent with the meaning and rules already established for integer exponents, and we present two forms of notation for roots: radical and exponent. We conclude this chapter with complex numbers, a natural extension of the real number system.

Mount Vesuvius is the only active volcano on the European continent. Although the Romans thought the volcano was extinct, Vesuvius erupted violently in 79 A.D. The eruption buried the cities of Pompeii, Herculaneum, and Stabiae under up to 60 feet of ash and mud, killing approximately 16,000 people. Although Pompeii was completely engulfed, the city was far from destroyed. The blanket of mud and ash perfectly preserved much of the city in a snapshot of daily life of the ancient Romans. Pompeii lay undisturbed for over 1500 years until the first excavations were made and its archaeological significance was proven. The 79 A.D. eruption of Vesuvius seemed to be the volcano's renewal. It has erupted with varying degrees of violence more than 30 times since Pompeii's burial, most recently in 1944. In Exercise 84 on page 678, we will use a radical expression to find the surface area of Mount Vesuvius.

Answers
1.
2.
3.
4.
5.
6.
7.
8.
9.
10.
11.
12.
13.
14.
15.
16.
17.
18.
19.
20.

Name ______________________ Section __________ Date __________

Chapter 9 Pretest

Find each root. Assume that all variables represent positive real numbers.

1. $\sqrt{81x^2}$ **2.** $\sqrt[3]{-27y^9}$

3. $\sqrt[5]{x^{30}}$ **4.** If $f(x) = \sqrt{7x + 2}$, find $f(2)$.

Simplify. Write with positive exponents.

5. $144^{1/2}$ **6.** $81^{3/4}$ **7.** $\dfrac{3}{2m^{-5/6}}$ **8.** $(9x^4y^{-6})^{3/2}$

Perform each indicated operation. Assume that all variables represent positive real numbers.

9. $\sqrt{13} \cdot \sqrt{3x}$ **10.** $\sqrt{\dfrac{6}{49}}$ **11.** $3\sqrt{18} - 6\sqrt{50}$

12. $(4\sqrt{x} + 1)(2\sqrt{x} - 3)$ **13.** $(\sqrt{5} - y)^2$

Simplify.

14. $\sqrt[3]{135a^2b^{12}}$

Rationalize the denominator. Assume that all variables represent positive real numbers.

15. $\dfrac{2}{\sqrt{12x}}$ **16.** $\dfrac{14}{2 + \sqrt{11}}$ **17.** Solve: $\sqrt{2x + 1} - 5 = 0$

Perform each indicated operation and simplify. Write the result in the form $a + bi$.

18. $(10 - 3i) - (6 + 5i)$ **19.** $(2 - 5i)^2$ **20.** $\dfrac{3 + i}{5 + i}$

9.1 Radical Expressions and Functions

OBJECTIVES

A. Find square roots.
B. Approximate roots using a calculator.
C. Find cube roots.
D. Find nth roots.
E. Find $\sqrt[n]{a^n}$ when a is any real number.
F. Find function values of radical functions.

A Finding Square Roots

The opposite of squaring a number is taking the *square root* of a number. For example, since the square of 4, or 4^2, is 16, we say that a square root of 16 is 4.

Square Root

The number b is a **square root** of a if $b^2 = a$.

EXAMPLES Find the real square roots of each number.

1. 25 Since $5^2 = 25$ and $(-5)^2 = 25$, the square roots of 25 are 5 and -5.
2. 49 Since $7^2 = 49$ and $(-7)^2 = 49$, the square roots of 49 are 7 and -7.
3. -4 There is no real number whose square is -4. The number -4 has no real number square root.

Practice Problems 1–3

Find the square roots of each number.

1. 36
2. 81
3. -16

We denote the *nonnegative*, or *principal, square root* with the **radical sign**:

$$\sqrt{25} = 5$$

We denote the *negative square root* with the **negative radical sign**:

$$-\sqrt{25} = -5$$

An expression containing a radical sign is called a **radical expression**. An expression within, or "under," a radical sign is called a **radicand**.

radical expression: $\sqrt{a}$ (radical sign; radicand)

Principal and Negative Square Roots

The **principal square root** of a nonnegative number a is its nonnegative square root. The principal square root is written as $\sqrt{a}$. The **negative square root** of a is written as $-\sqrt{a}$.

EXAMPLES

Find each square root. Assume that all variables represent nonnegative real numbers.

4. $\sqrt{36} = 6$ because $6^2 = 36$.
5. $\sqrt{0} = 0$ because $0^2 = 0$.
6. $\sqrt{\frac{4}{49}} = \frac{2}{7}$ because $\left(\frac{2}{7}\right)^2 = \frac{4}{49}$.
7. $\sqrt{0.25} = 0.5$ because $(0.5)^2 = 0.25$.
8. $\sqrt{x^6} = x^3$ because $(x^3)^2 = x^6$.
9. $\sqrt{9x^{10}} = 3x^5$ because $(3x^5)^2 = 9x^{10}$.
10. $-\sqrt{81} = -9$. The negative in front of the radical indicates the negative square root of 81.
11. $\sqrt{-81}$ is not a real number.

Practice Problems 4–11

Find each square root. Assume that all variables represent nonnegative real numbers.

4. $\sqrt{25}$
5. $\sqrt{0}$
6. $\sqrt{\frac{9}{25}}$
7. $\sqrt{0.36}$
8. $\sqrt{x^{10}}$
9. $\sqrt{36x^6}$
10. $-\sqrt{25}$
11. $\sqrt{-25}$

Answers

1. 6, −6, **2.** 9, −9, **3.** no real number square root, **4.** 5, **5.** 0, **6.** $\frac{3}{5}$, **7.** 0.6, **8.** x^5, **9.** $6x^3$, **10.** −5, **11.** not a real number

Helpful Hints

- Remember: $\sqrt{0} = 0$.
- Don't forget that the square root of a negative number is not a real number. For example,

  $\sqrt{-9}$ is not a real number

  because there is no real number that when multiplied by itself would give a product of -9. In Section 7.7, we will see what kind of a number $\sqrt{-9}$ is.

B Approximating Roots

Recall that numbers such as 1, 4, 9, and 25 are called **perfect squares**, since $1 = 1^2, 4 = 2^2, 9 = 3^2$, and $25 = 5^2$. Square roots of perfect square radicands simplify to rational numbers. What happens when we try to simplify a root such as $\sqrt{3}$? Since 3 is not a perfect square, $\sqrt{3}$ is not a rational number. It is called an **irrational number**, and we can find a decimal **approximation** of it. To find decimal approximations, we can use the table in the appendix or a calculator. For example, an approximation for $\sqrt{3}$ is

$$\sqrt{3} \approx 1.732$$

↑ approximation symbol

To see if the approximation is reasonable, notice that since

$$1 < 3 < 4, \quad \text{then}$$

$$\sqrt{1} < \sqrt{3} < \sqrt{4}, \quad \text{or}$$

$$1 < \sqrt{3} < 2.$$

We found $\sqrt{3} \approx 1.732$, a number between 1 and 2, so our result is reasonable.

Practice Problem 12

Use a calculator or the appendix to approximate $\sqrt{30}$. Round the approximation to three decimal places and check to see that your approximation is reasonable.

EXAMPLE 12

Use a calculator or the appendix to approximate $\sqrt{20}$. Round the approximation to three decimal places and check to see that your approximation is reasonable.

Solution:

$$\sqrt{20} \approx 4.472$$

Is this reasonable? Since $16 < 20 < 25$, then $\sqrt{16} < \sqrt{20} < \sqrt{25}$, or $4 < \sqrt{20} < 5$. The approximation is between 4 and 5 and is thus reasonable.

C Finding Cube Roots

Finding roots can be extended to other roots such as cube roots. For example, since $2^3 = 8$, we call 2 the *cube root* of 8. In symbols, we write

$$\sqrt[3]{8} = 2$$

Answer

12. 5.477

Cube Root

The **cube root** of a real number a is written as $\sqrt[3]{a}$, and

$$\sqrt[3]{a} = b \quad \text{only if} \quad b^3 = a$$

From this definition, we have

$$\sqrt[3]{64} = 4 \text{ since } 4^3 = 64$$

$$\sqrt[3]{-27} = -3 \text{ since } (-3)^3 = -27$$

$$\sqrt[3]{x^3} = x \text{ since } x^3 = x^3$$

Notice that, unlike with square roots, *it is possible to have a negative radicand when finding a cube root.* This is so because the *cube* of a negative number is a negative number. Therefore, the *cube root* of a negative number is a negative number.

EXAMPLES Find each cube root.

13. $\sqrt[3]{1} = 1$ because $1^3 = 1$.

14. $\sqrt[3]{-64} = -4$ because $(-4)^3 = -64$.

15. $\sqrt[3]{\frac{8}{125}} = \frac{2}{5}$ because $\left(\frac{2}{5}\right)^3 = \frac{8}{125}$.

16. $\sqrt[3]{x^6} = x^2$ because $(x^2)^3 = x^6$.

17. $\sqrt[3]{-8x^9} = -2x^3$ because $(-2x^3)^3 = -8x^9$.

Practice Problems 13–17

Find each cube root.

13. $\sqrt[3]{0}$
14. $\sqrt[3]{-8}$
15. $\sqrt[3]{\frac{1}{64}}$
16. $\sqrt[3]{x^9}$
17. $\sqrt[3]{-64x^6}$

D Finding *n*th Roots

Just as we can raise a real number to powers other than 2 or 3, we can find roots other than square roots and cube roots. In fact, we can find the ***n*th root** of a number, where n is any natural number. In symbols, the nth root of a is written as $\sqrt[n]{a}$, where n is called the **index**. The index 2 is usually omitted for square roots.

Helpful Hint

If the index is even, such as in $\sqrt{\ }$, $\sqrt[4]{\ }$, $\sqrt[6]{\ }$, and so on, the radicand must be nonnegative for the root to be a real number. For example,

$\sqrt[4]{16} = 2$, but $\sqrt[4]{-16}$ is not a real number,

$\sqrt[6]{64} = 2$, but $\sqrt[6]{-64}$ is not a real number.

If the index is odd, such as in $\sqrt[3]{\ }$, $\sqrt[5]{\ }$, and so on, the radicand may be any real number. For example,

$\sqrt[3]{64} = 4$ and $\sqrt[3]{-64} = -4$,

$\sqrt[5]{32} = 2$ and $\sqrt[5]{-32} = -2$.

Try the Concept Check in the margin.

Concept Check

Which one is not a real number?

a. $\sqrt[3]{-15}$
b. $\sqrt[4]{-15}$
c. $\sqrt[5]{-15}$
d. $\sqrt{(-15)^2}$

Answers

13. 0, **14.** -2, **15.** $\frac{1}{4}$, **16.** x^3, **17.** $-4x^2$

Concept Check: b

Practice Problems 18–22

Find each root.

18. $\sqrt[4]{16}$
19. $\sqrt[5]{-32}$
20. $-\sqrt{36}$
21. $\sqrt[4]{-16}$
22. $\sqrt[3]{8x^6}$

EXAMPLES Find each root.

18. $\sqrt[4]{81} = 3$ because $3^4 = 81$ and 3 is positive.

19. $\sqrt[5]{-243} = -3$ because $(-3)^5 = -243$.

20. $-\sqrt{25} = -5$ because -5 is the opposite of $\sqrt{25}$.

21. $\sqrt[4]{-81}$ is not a real number. There is no real number that, when raised to the fourth power, is -81.

22. $\sqrt[3]{64x^3} = 4x$ because $(4x)^3 = 64x^3$.

E Finding $\sqrt[n]{a^n}$ When a Is Any Real Number

Recall that the notation $\sqrt{a^2}$ indicates the positive square root of a^2 only. For example,

$$\sqrt{(-5)^2} = \sqrt{25} = 5$$

When variables are present in the radicand and it is *unclear whether the variable represents a positive number or a negative number*, absolute value bars are sometimes needed to ensure that the result is a positive number. For example,

$$\sqrt{x^2} = |x|$$

This ensures that the result is positive. This same situation may occur when the index is any *even* positive integer. When the index is any *odd* positive integer, absolute value bars are not necessary.

Finding $\sqrt[n]{a^n}$

If n is an *even* positive integer, then $\sqrt[n]{a^n} = |a|$.

If n is an *odd* positive integer, then $\sqrt[n]{a^n} = a$.

Practice Problems 23–29

Simplify. Assume that the variables represent any real number.

23. $\sqrt{(-5)^2}$
24. $\sqrt{x^6}$
25. $\sqrt[4]{(x+6)^4}$
26. $\sqrt[3]{(-3)^3}$
27. $\sqrt[5]{(7x-1)^5}$
28. $\sqrt{36x^2}$
29. $\sqrt{x^2+6x+9}$

Answers

18. 2, **19.** -2, **20.** -6, **21.** not a real number, **22.** $2x^2$, **23.** 5, **24.** $|x^3|$, **25.** $|x+6|$, **26.** -3, **27.** $7x-1$, **28.** $6|x|$, **29.** $|x+3|$

EXAMPLES

Simplify. Assume that the variables represent any real number.

23. $\sqrt{(-3)^2} = |-3| = 3$ When the index is even, the absolute value bars ensure that the result is not negative.

24. $\sqrt{x^2} = |x|$

25. $\sqrt[4]{(x-2)^4} = |x-2|$

26. $\sqrt[3]{(-5)^3} = -5$ Absolute value bars are not needed when the index is odd.

27. $\sqrt[5]{(2x-7)^5} = 2x-7$

28. $\sqrt{25x^2} = 5|x|$

29. $\sqrt{x^2+2x+1} = \sqrt{(x+1)^2} = |x+1|$

 Finding Function Values

Functions of the form

$$f(x) = \sqrt[n]{x}$$

are called **radical functions.** Recall that the domain of a function in x is the set of all possible replacement values of x. This means that if n is even, the domain is the set of all nonnegative numbers, or $\{x|x \geq 0\}$. If n is odd, the domain is the set of all real numbers. Keep this in mind as we find function values. In Chapter 12, we will graph these functions and discuss their domains further.

EXAMPLES

If $f(x) = \sqrt{x - 4}$ and $g(x) = \sqrt[3]{x + 2}$, find each function value.

30. $f(8) = \sqrt{8 - 4} = \sqrt{4} = 2$

31. $f(6) = \sqrt{6 - 4} = \sqrt{2}$

32. $g(-1) = \sqrt[3]{-1 + 2} = \sqrt[3]{1} = 1$

33. $g(1) = \sqrt[3]{1 + 2} = \sqrt[3]{3}$

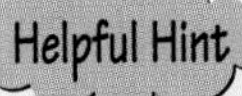

Notice that for the function $f(x) = \sqrt{x - 4}$, the domain includes all real numbers that make the radicand ≥ 0. To see what numbers these are, solve $x - 4 \geq 0$ and find that $x \geq 4$. The domain is $\{x|x \geq 4\}$.

The domain of the cube root function $g(x) = \sqrt[3]{x + 2}$ is the set of real numbers.

See Chapter 12 for further discussions of domains.

Practice Problems 30–33

If $f(x) = \sqrt{x + 2}$ and $g(x) = \sqrt[3]{x - 1}$, find each function value.

30. $f(7)$
31. $g(9)$
32. $f(0)$
33. $g(10)$

Answers

30. 3, **31.** 2, **32.** $\sqrt{2}$, **33.** $\sqrt[3]{9}$

FOCUS ON History

HERON OF ALEXANDRIA

Heron (also Hero) was a Greek mathematician and engineer. He lived and worked in Alexandria, Egypt, around 75 A.D. During his prolific work life, Heron developed a rotary steam engine called an aeolipile, a surveying tool called a dioptra, as well as a wind organ and a fire engine. As an engineer, he must have had the need to approximate square roots because he described an iterative method for doing so in his work *Metrica*. Heron's method for approximating a square root can be summarized as follows:

> Suppose that x is not a perfect square and a^2 is the nearest perfect square to x. For a rough estimate of the value of $\sqrt{x}$, find the value of $y_1 = \frac{1}{2}\left(a + \frac{x}{a}\right)$. This estimate can be improved by calculating a second estimate using the first estimate y_1 in place of a: $y_2 = \frac{1}{2}\left(y_1 + \frac{x}{y_1}\right)$. Repeating this process several times will give more and more accurate estimates of $\sqrt{x}$.

CRITICAL THINKING

1. **a.** Which perfect square is closest to 80?
   b. Use Heron's method for approximating square roots to calculate the first estimate of the square root of 80.
   c. Use the first estimate of the square root of 80 to find a more refined second estimate.
   d. Use a calculator to find the actual value of the square root of 80. List all digits shown on your calculator's display.
   e. Compare the actual value from part (d) to the values of the first and second estimates. What do you notice?
   f. How many iterations of this process are necessary to get an estimate that differs no more than one digit from the actual value recorded in part (d)?
2. Repeat Question 1 for finding an estimate of the square root of 30.
3. Repeat Question 1 for finding an estimate of the square root of 4572.
4. Why would this iterative method have been important to people of Heron's era? Would you say that this method is as important today? Why or why not?

Name ______________________ Section __________ Date __________

EXERCISE SET 9.1

A *Find the real square roots of each number. See Examples 1 through 3.*

1. 4 **2.** 9 **3.** -25 **4.** -49 **5.** 100 **6.** 64

Find each square root. Assume that all variables represent nonnegative real numbers. See Examples 4 through 11.

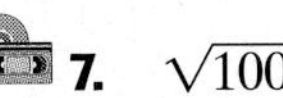

7. $\sqrt{100}$ **8.** $\sqrt{400}$ **9.** $\sqrt{\frac{1}{4}}$ **10.** $\sqrt{\frac{9}{25}}$ **11.** $\sqrt{0.0001}$ **12.** $\sqrt{0.04}$

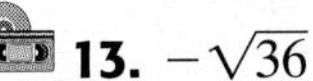

13. $-\sqrt{36}$ **14.** $-\sqrt{9}$ **15.** $\sqrt{x^{10}}$ **16.** $\sqrt{x^{16}}$ **17.** $\sqrt{16y^6}$ **18.** $\sqrt{64y^{20}}$

B *Use a calculator or the appendix to approximate each square root to three decimal places. Check to see that each approximation is reasonable. See Example 12.*

19. $\sqrt{7}$ **20.** $\sqrt{11}$ **21.** $\sqrt{38}$ **22.** $\sqrt{56}$ **23.** $\sqrt{200}$ **24.** $\sqrt{300}$

C *Find each cube root. See Examples 13 through 17.*

25. $\sqrt[3]{64}$ **26.** $\sqrt[3]{27}$ **27.** $\sqrt[3]{\frac{1}{8}}$ **28.** $\sqrt[3]{\frac{27}{64}}$ **29.** $\sqrt[3]{-1}$

30. $\sqrt[3]{-125}$ **31.** $\sqrt[3]{x^{12}}$ **32.** $\sqrt[3]{x^{15}}$ **33.** $\sqrt[3]{-27x^9}$ **34.** $\sqrt[3]{-64x^6}$

D *Find each root. Assume that all variables represent nonnegative real numbers. See Examples 18 through 22.*

35. $-\sqrt[4]{16}$ **36.** $\sqrt[5]{-243}$ **37.** $\sqrt[4]{-16}$ **38.** $\sqrt{-16}$ **39.** $\sqrt[5]{-32}$

40. $\sqrt[5]{-1}$ 41. $\sqrt[5]{x^{20}}$ 42. $\sqrt[4]{x^{20}}$ 43. $\sqrt[6]{64x^{12}}$ 44. $\sqrt[5]{-32x^{15}}$

45. $\sqrt{81x^4}$ 46. $\sqrt[4]{81x^4}$ 47. $\sqrt[4]{256x^8}$ 48. $\sqrt{256x^8}$

E *Simplify. Assume that the variables represent any real number. See Examples 23 through 29.*

49. $\sqrt{(-8)^2}$ 50. $\sqrt{(-7)^2}$ 51. $\sqrt[3]{(-8)^3}$ 52. $\sqrt[5]{(-7)^5}$ 53. $\sqrt{4x^2}$

54. $\sqrt[4]{16x^4}$ 55. $\sqrt[3]{x^3}$ 56. $\sqrt[5]{x^5}$ 57. $\sqrt[4]{(x-2)^4}$ 58. $\sqrt[6]{(2x-1)^6}$

59. $\sqrt{x^2 + 4x + 4}$
(*Hint:* Factor the polynomial first.)

60. $\sqrt{x^2 - 8x + 16}$
(*Hint:* Factor the polynomial first.)

F *If $f(x) = \sqrt{2x+3}$ and $g(x) = \sqrt[3]{x-8}$, find each function value. See Examples 30 through 33.*

61. $f(0)$ 62. $g(0)$ 63. $g(7)$ 64. $f(-1)$

65. $g(-19)$ 66. $f(3)$ 67. $f(2)$ 68. $g(1)$

Review and Preview

Simplify each exponential expression. See Sections 4.1 and 4.2.

69. $(-2x^3y^2)^5$ 70. $(4y^6z^7)^3$ 71. $(-3x^2y^3z^5)(20x^5y^7)$ 72. $(-14a^5bc^2)(2abc^4)$ 73. $\dfrac{7x^{-1}y}{14(x^5y^2)^{-2}}$ 74. $\dfrac{(2a^{-1}b^2)^3}{(8a^2b)^{-2}}$

Combining Concepts

75. Explain why $\sqrt{-64}$ is not a real number.

76. Explain why $\sqrt[3]{-64}$ is a real number.

For Exercises 77 through 80, do not use a calculator.

77. $\sqrt{160}$ is closest to
a. 10 **b.** 13 **c.** 20 **d.** 40

78. $\sqrt{1000}$ is closest to
a. 10 **b.** 30 **c.** 100 **d.** 500

△ 79. The perimeter of the triangle is closest to
a. 12 **b.** 18 **c.** 66 **d.** 132

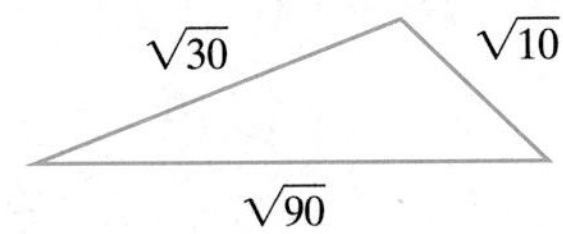

80. The length of the bent wire is closest to
a. 5 **b.** $\sqrt{28}$ **c.** 7 **d.** 14

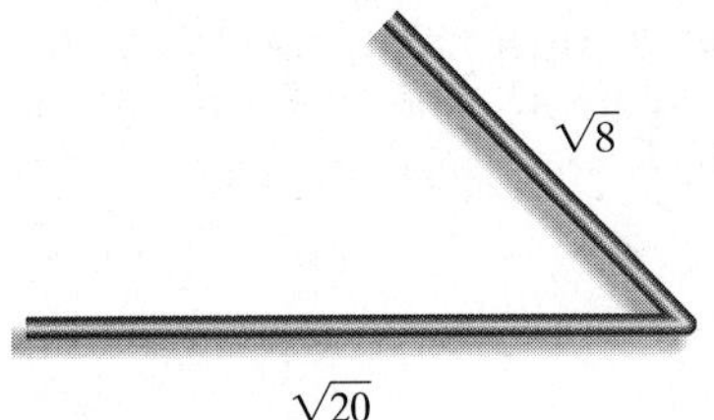

The Mosteller formula for calculating adult body surface area is $B = \sqrt{\frac{hw}{3131}}$*, where B is an individual's body surface area in square meters, h is the individual's height in inches, and w is the individual's weight in pounds. Use this information to answer Exercises 81 and 82. Round answers to 2 decimal places.*

△ 81. Find the body surface area of an individual who is 66 inches tall and who weighs 135 pounds.

△ 82. Find the body surface area of an individual who is 74 inches tall and who weighs 225 pounds.

83. Suppose that a friend tells you that $\sqrt{13} \approx 5.7$. Without a calculator, how can you convince your friend that he or she must have made an error?

84. Escape velocity is the minimum speed that an object must reach to escape a planet's pull of gravity. Escape velocity v is given by the equation $v = \sqrt{\frac{2Gm}{r}}$, where m is the mass of the planet, r is its radius, and G is the universal gravitational constant, which has a value of $G = 6.67 \times 10^{-11}\ \text{m}^3/\text{kg}\cdot\text{s}^2$. The mass of Earth is 5.97×10^{24} kg and its radius is 6.37×10^{6} m. Use this information to find the escape velocity for Earth. Round to the nearest whole number. (*Source:* National Space Science Data Center)

Internet Excursions

Go To: http://www.prenhall.com/martin-gay_algebra

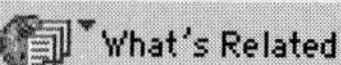

The National Space Science Data Center (NSSDC) is a division of NASA that provides access to information collected from NASA experiments and space flight missions. One of the many areas in which the NSSDC maintains a library of information is the planetary sciences. The given World Wide Web site gives a listing of planetary fact sheets from which the user may choose. (Alternatively, you can visit the NSSDC homepage at http://nssdc.gsfc.nasa.gov and navigate to Planetary Sciences. Then look for the Planetary Fact Sheets option.)

85. Choose any one of the fact sheets for a planet, asteroid, or the sun. Use the information given in the fact sheet and the escape velocity formula given in Exercise 84 to compute the escape velocity for that body. Then compare your calculation to the escape velocity given in the fact sheet. How close is your calculation? (*Note:* You should use the "volumetric mean radius" for the planet's radius in your calculation.)

86. Repeat Exercise 85 for one of the other listed bodies of the solar system.

9.2 Rational Exponents

OBJECTIVES

- **A** Understand the meaning of $a^{1/n}$.
- **B** Understand the meaning of $a^{m/n}$.
- **C** Understand the meaning of $a^{-m/n}$.
- **D** Use rules for exponents to simplify expressions that contain rational exponents.
- **E** Use rational exponents to simplify radical expressions.

A Understanding $a^{1/n}$

So far in this text, we have not defined expressions with rational exponents such as $3^{1/2}$, $x^{2/3}$, and $-9^{-1/4}$. We will define these expressions so that the rules for exponents shall apply to these rational exponents as well.

Suppose that $x = 5^{1/3}$. Then

$$x^3 = (5^{1/3})^3 = 5^{1/3 \cdot 3} = 5^1 \text{ or } 5$$

(using rules for exponents)

Since $x^3 = 5$, then x is the number whose cube is 5, or $x = \sqrt[3]{5}$. Notice that we also know that $x = 5^{1/3}$. This means that

$$5^{1/3} = \sqrt[3]{5}$$

Definition of $a^{1/n}$

If n is a positive integer greater than 1 and $\sqrt[n]{a}$ is a real number, then

$$a^{1/n} = \sqrt[n]{a}$$

Notice that the denominator of the rational exponent corresponds to the index of the radical.

EXAMPLES

Use radical notation to rewrite each expression. Simplify if possible.

1. $4^{1/2} = \sqrt{4} = 2$
2. $64^{1/3} = \sqrt[3]{64} = 4$
3. $x^{1/4} = \sqrt[4]{x}$
4. $-9^{1/2} = -\sqrt{9} = -3$
5. $(81x^8)^{1/4} = \sqrt[4]{81x^8} = 3x^2$
6. $5y^{1/3} = 5\sqrt[3]{y}$

Practice Problems 1–6

Use radical notation to rewrite each expression. Simplify if possible.

1. $25^{1/2}$
2. $27^{1/3}$
3. $x^{1/5}$
4. $-25^{1/2}$
5. $(-27y^6)^{1/3}$
6. $7x^{1/5}$

B Understanding $a^{m/n}$

As we expand our use of exponents to include $\frac{m}{n}$, we define their meaning so that rules for exponents still hold true. For example, by properties of exponents,

$$8^{2/3} = (8^{1/3})^2 = (\sqrt[3]{8})^2 \quad \text{or} \quad 8^{2/3} = (8^2)^{1/3} = \sqrt[3]{8^2}$$

Definition of $a^{m/n}$

If m and n are positive integers greater than 1 with $\frac{m}{n}$ in simplest form, then

$$a^{m/n} = \sqrt[n]{a^m} = (\sqrt[n]{a})^m$$

as long as $\sqrt[n]{a}$ is a real number.

Answers

1. 5, **2.** 3, **3.** $\sqrt[5]{x}$, **4.** -5, **5.** $-3y^2$, **6.** $7\sqrt[5]{x}$

Notice that the denominator n of the rational exponent corresponds to the index of the radical. The numerator m of the rational exponent indicates that the base is to be raised to the mth power. This means that

$$8^{2/3} = \sqrt[3]{8^2} = \sqrt[3]{64} = 4 \quad \text{or} \quad 8^{2/3} = (\sqrt[3]{8})^2 = 2^2 = 4$$

Helpful Hint

Most of the time, $(\sqrt[n]{a})^m$ will be easier to calculate than $\sqrt[n]{a^m}$.

Practice Problems 7–11

Use radical notation to rewrite each expression. Simplify if possible.

7. $9^{3/2}$
8. $-256^{3/4}$
9. $(-32)^{2/5}$
10. $\left(\frac{1}{4}\right)^{3/2}$
11. $(2x + 1)^{2/7}$

EXAMPLES

Use radical notation to rewrite each expression. Simplify if possible.

7. $4^{3/2} = (\sqrt{4})^3 = 2^3 = 8$

8. $-16^{3/4} = -(\sqrt[4]{16})^3 = -(2)^3 = -8$

9. $(-27)^{2/3} = (\sqrt[3]{-27})^2 = (-3)^2 = 9$

10. $\left(\frac{1}{9}\right)^{3/2} = \left(\sqrt{\frac{1}{9}}\right)^3 = \left(\frac{1}{3}\right)^3 = \frac{1}{27}$

11. $(4x - 1)^{3/5} = \sqrt[5]{(4x - 1)^3}$

Helpful Hint

The *denominator* of a rational exponent is the index of the corresponding radical. For example, $x^{1/5} = \sqrt[5]{x}$, and $z^{2/3} = \sqrt[3]{z^2}$ or $z^{2/3} = (\sqrt[3]{z})^2$.

C Understanding $a^{-m/n}$

The rational exponents we have given meaning to exclude negative rational numbers. To complete the set of definitions, we define $a^{-m/n}$.

Definition of $a^{-m/n}$

$$a^{-m/n} = \frac{1}{a^{m/n}}$$

as long as $a^{m/n}$ is a nonzero real number.

Practice Problems 12–13

Write each expression with a positive exponent. Then simplify.

12. $27^{-2/3}$
13. $-256^{-3/4}$

Answers

7. 27, **8.** −64, **9.** 4, **10.** $\frac{1}{8}$, **11.** $\sqrt[7]{(2x + 1)^2}$, **12.** $\frac{1}{9}$, **13.** $-\frac{1}{64}$

EXAMPLES

Write each expression with a positive exponent. Then simplify.

12. $16^{-3/4} = \frac{1}{16^{3/4}} = \frac{1}{(\sqrt[4]{16})^3} = \frac{1}{2^3} = \frac{1}{8}$

13. $(-27)^{-2/3} = \frac{1}{(-27)^{2/3}} = \frac{1}{(\sqrt[3]{-27})^2} = \frac{1}{(-3)^2} = \frac{1}{9}$

Helpful Hint

If an expression contains a negative rational exponent, you may want to first write the expression with a positive exponent, then interpret the rational exponent. Notice that the sign of the base is not affected by the sign of its exponent. For example,

$$9^{-3/2} = \frac{1}{9^{3/2}} = \frac{1}{(\sqrt{9})^3} = \frac{1}{27}$$

Also,

$$(-27)^{-1/3} = \frac{1}{(-27)^{1/3}} = -\frac{1}{3}$$

Try the Concept Check in the margin.

D Using Rules for Exponents

It can be shown that the properties of integer exponents hold for rational exponents. By using these properties and definitions, we can now simplify expressions that contain rational exponents. These rules are repeated here for review.

Summary of Exponent Rules

If m and n are rational numbers, and a, b, and c are numbers for which the expressions below exist, then

Product rule for exponents:	$a^m \cdot a^n = a^{m+n}$
Power rule for exponents:	$(a^m)^n = a^{m \cdot n}$
Power rules for products and quotients:	$(ab)^n = a^n b^n$ and
	$\left(\frac{a}{c}\right)^n = \frac{a^n}{c^n}, \quad c \neq 0$
Quotient rule for exponents:	$\frac{a^m}{a^n} = a^{m-n}, \quad a \neq 0$
Zero exponent:	$a^0 = 1, \quad a \neq 0$
Negative exponent:	$a^{-n} = \frac{1}{a^n}, \quad a \neq 0$

EXAMPLES Use the properties of exponents to simplify.

14. $x^{1/2}x^{1/3} = x^{1/2+1/3} = x^{3/6+2/6} = x^{5/6}$ Use the product rule.

15. $\frac{7^{1/3}}{7^{4/3}} = 7^{1/3-4/3} = 7^{-3/3} = 7^{-1} = \frac{1}{7}$ Use the quotient rule.

16. $y^{-4/7} \cdot y^{6/7} = y^{-4/7+6/7} = y^{2/7}$ Use the product rule.

17. $(5^{3/8})^4 = 5^{3/8 \cdot 4} = 5^{12/8} = 5^{3/2}$ Use the power rule.

18. $\frac{(2x^{2/5})^5}{x^2} = \frac{2^5(x^{2/5})^5}{x^2}$ Use the power rule.

$= \frac{32x^2}{x^2}$ Simplify.

$= 32x^{2-2}$ Use the quotient rule.

$= 32x^0$ Simplify.

$= 32 \cdot 1$ or 32 Substitute 1 for x^0.

Concept Check

Which one is correct?

a. $-8^{2/3} = \frac{1}{4}$

b. $8^{-2/3} = -\frac{1}{4}$

c. $8^{-2/3} = -4$

d. $-8^{-2/3} = -\frac{1}{4}$

Practice Problems 14–18

Use the properties of exponents to simplify.

14. $x^{1/3}x^{1/4}$

15. $\frac{9^{2/5}}{9^{12/5}}$

16. $y^{-3/10} \cdot y^{6/10}$

17. $(11^{2/9})^3$

18. $\frac{(3x^{2/3})^3}{x^2}$

Answers

14. $x^{7/12}$, **15.** $\frac{1}{81}$, **16.** $y^{3/10}$, **17.** $11^{2/3}$, **18.** 27

Concept Check: d

E Using Rational Exponents to Simplify Radical Expressions

We can simplify some radical expressions by first writing the expression with rational exponents. Use the properties of exponents to simplify, and then convert back to radical notation.

Practice Problems 19–21

Use rational exponents to simplify. Assume that all variables represent positive real numbers.

19. $\sqrt[10]{y^5}$

20. $\sqrt[4]{9}$

21. $\sqrt[9]{a^6b^3}$

EXAMPLES

Use rational exponents to simplify. Assume that all variables represent positive real numbers.

19. $\sqrt[8]{x^4} = x^{4/8}$ Write with rational exponents.
$= x^{1/2}$ Simplify the exponent.
$= \sqrt{x}$ Write with radical notation.

20. $\sqrt[6]{25} = 25^{1/6}$ Write with rational exponents.
$= (5^2)^{1/6}$ Write 25 as 5^2.
$= 5^{2/6}$ Use the power rule.
$= 5^{1/3}$ Simplify the exponent.
$= \sqrt[3]{5}$ Write with radical notation.

21. $\sqrt[6]{r^2s^4} = (r^2s^4)^{1/6}$ Write with rational exponents.
$= r^{2/6}s^{4/6}$ Use the power rule.
$= r^{1/3}s^{2/3}$ Simplify the exponents.
$= (rs^2)^{1/3}$ Use $a^nb^n = (ab)^n$.
$= \sqrt[3]{rs^2}$ Write with radical notation. ●

Practice Problems 22–24

Use rational exponents to write as a single radical.

22. $\sqrt{y} \cdot \sqrt[3]{y}$

23. $\dfrac{\sqrt[3]{x}}{\sqrt[4]{x}}$

24. $\sqrt{5} \cdot \sqrt[3]{2}$

EXAMPLES Use rational exponents to write as a single radical.

22. $\sqrt{x} \cdot \sqrt[4]{x} = x^{1/2} \cdot x^{1/4} = x^{1/2+1/4}$
$= x^{3/4} = \sqrt[4]{x^3}$

23. $\dfrac{\sqrt{x}}{\sqrt[3]{x}} = \dfrac{x^{1/2}}{x^{1/3}} = x^{1/2-1/3} = x^{3/6-2/6}$
$= x^{1/6} = \sqrt[6]{x}$

24. $\sqrt[3]{3} \cdot \sqrt{2} = 3^{1/3} \cdot 2^{1/2}$ Write with rational exponents.
$= 3^{2/6} \cdot 2^{3/6}$ Write the exponents so that they have the same denominator.
$= (3^2 \cdot 2^3)^{1/6}$ Use $a^nb^n = (ab)^n$.
$= \sqrt[6]{3^2 \cdot 2^3}$ Write with radical notation.
$= \sqrt[6]{72}$ Multiply $3^2 \cdot 2^3$. ●

Answers

19. $\sqrt{y}$, **20.** $\sqrt{3}$, **21.** $\sqrt[3]{a^2b}$, **22.** $\sqrt[6]{y^5}$, **23.** $\sqrt[12]{x}$, **24.** $\sqrt[6]{500}$

Name ______________ Section ________ Date ________

EXERCISE SET 9.2

A *Use radical notation to rewrite each expression. Simplify if possible. See Examples 1 through 6.*

1. $49^{1/2}$ **2.** $64^{1/3}$ **3.** $27^{1/3}$ **4.** $8^{1/3}$ **5.** $\left(\frac{1}{16}\right)^{1/4}$ **6.** $\left(\frac{1}{64}\right)^{1/2}$

7. $169^{1/2}$ **8.** $81^{1/4}$ **9.** $2m^{1/3}$ **10.** $(2m)^{1/3}$ **11.** $(9x^4)^{1/2}$

12. $(16x^8)^{1/2}$ **13.** $(-27)^{1/3}$ **14.** $-64^{1/2}$ **15.** $-16^{1/4}$ **16.** $(-32)^{1/5}$

B *Use radical notation to rewrite each expression. Simplify if possible. See Examples 7 through 11.*

17. $16^{3/4}$ **18.** $4^{5/2}$ **19.** $(-64)^{2/3}$ **20.** $(-8)^{4/3}$ **21.** $(-16)^{3/4}$ **22.** $(-9)^{3/2}$

23. $(2x)^{3/5}$ **24.** $2x^{3/5}$ **25.** $(7x + 2)^{2/3}$ **26.** $(x - 4)^{3/4}$ **27.** $\left(\frac{16}{9}\right)^{3/2}$ **28.** $\left(\frac{49}{25}\right)^{3/2}$

C *Write with positive exponents. Simplify if possible. See Examples 12 and 13.*

29. $8^{-4/3}$ **30.** $64^{-2/3}$ **31.** $(-64)^{-2/3}$ **32.** $(-8)^{-4/3}$ **33.** $(-4)^{-3/2}$ **34.** $(-16)^{-5/4}$

35. $x^{-1/4}$ **36.** $y^{-1/6}$ **37.** $\frac{1}{a^{-2/3}}$ **38.** $\frac{1}{n^{-8/9}}$ **39.** $\frac{5}{7x^{-3/4}}$ **40.** $\frac{2}{3y^{-5/7}}$

41. Explain how writing x^{-7} with positive exponents is similar to writing $x^{-1/4}$ with positive exponents.

42. Explain how writing $2x^{-5}$ with positive exponents is similar to writing $2x^{-3/4}$ with positive exponents.

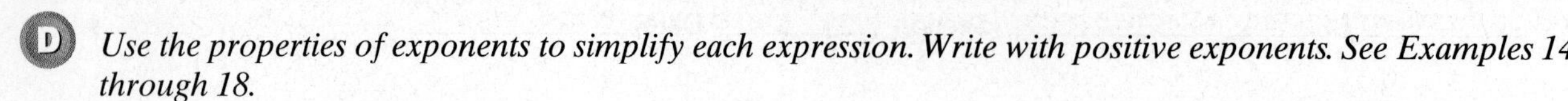

D *Use the properties of exponents to simplify each expression. Write with positive exponents. See Examples 14 through 18.*

43. $a^{2/3}a^{5/3}$ **44.** $b^{9/5}b^{8/5}$ **45.** $x^{-2/5} \cdot x^{7/5}$ **46.** $y^{4/3} \cdot y^{-1/3}$ **47.** $3^{1/4} \cdot 3^{3/8}$

48. $5^{1/2} \cdot 5^{1/6}$ **49.** $\dfrac{y^{1/3}}{y^{1/6}}$ **50.** $\dfrac{x^{3/4}}{x^{1/8}}$ **51.** $(4u^2)^{3/2}$ **52.** $(32^{1/5}x^{2/3})^3$

53. $\dfrac{b^{1/2}b^{3/4}}{-b^{1/4}}$ **54.** $\dfrac{a^{1/4}a^{-1/2}}{a^{2/3}}$ **55.** $\dfrac{(3x^{1/4})^3}{x^{1/12}}$ **56.** $\dfrac{(2x^{1/5})^4}{x^{3/10}}$ **57.** $\dfrac{(y^3z)^{1/6}}{y^{-1/2}z^{1/3}}$

58. $\dfrac{(m^2n)^{1/4}}{m^{-1/2}n^{5/8}}$ **59.** $\dfrac{(x^3y^2)^{1/4}}{(x^{-5}y^{-1})^{-1/2}}$ **60.** $\dfrac{(a^{-2}b^3)^{1/8}}{(a^{-3}b)^{-1/4}}$

E *Use rational exponents to simplify each radical. Assume that all variables represent positive real numbers. See Examples 19 through 21.*

61. $\sqrt[6]{x^3}$ **62.** $\sqrt[9]{a^3}$ **63.** $\sqrt[6]{4}$ **64.** $\sqrt[4]{36}$ **65.** $\sqrt[4]{16x^2}$ **66.** $\sqrt[8]{4y^2}$

67. $\sqrt[4]{(x+3)^2}$ **68.** $\sqrt[8]{(y+1)^4}$ **69.** $\sqrt[8]{x^4y^4}$ **70.** $\sqrt[9]{y^6z^3}$ **71.** $\sqrt[12]{a^8b^4}$ **72.** $\sqrt[10]{a^5b^5}$

Use rational expressions to write as a single radical expression. See Examples 22 through 24.

73. $\sqrt[3]{y} \cdot \sqrt[5]{y^2}$

74. $\sqrt[3]{y^2} \cdot \sqrt[6]{y}$

75. $\dfrac{\sqrt[3]{b^2}}{\sqrt[4]{b}}$

76. $\dfrac{\sqrt[4]{a}}{\sqrt[5]{a}}$

77. $\sqrt[3]{x} \cdot \sqrt[4]{x} \cdot \sqrt[8]{x^3}$

78. $\sqrt[6]{y} \cdot \sqrt[3]{y} \cdot \sqrt[5]{y^2}$

79. $\dfrac{\sqrt[3]{a^2}}{\sqrt[6]{a}}$

80. $\dfrac{\sqrt[5]{b^2}}{\sqrt[10]{b^3}}$

81. $\sqrt{3} \cdot \sqrt[3]{4}$

82. $\sqrt[3]{5} \cdot \sqrt{2}$

83. $\sqrt[5]{7} \cdot \sqrt[3]{y}$

84. $\sqrt[4]{5} \cdot \sqrt[3]{x}$

85. $\sqrt{5r} \cdot \sqrt[3]{s}$

86. $\sqrt[3]{b} \cdot \sqrt[5]{4a}$

Review and Preview

Write each integer as a product of two integers such that one of the factors is a perfect square. For example, write 18 as $9 \cdot 2$, because 9 is a perfect square.

87. 75

88. 20

89. 48

90. 45

Write each integer as a product of two integers such that one of the factors is a perfect cube. For example, write 24 as $8 \cdot 3$, because 8 is a perfect cube.

91. 16

92. 56

93. 54

94. 80

Combining Concepts

Basal metabolic rate (BMR) is the number of calories per day a person needs to maintain life. A person's basal metabolic rate $B(w)$ in calories per day can be estimated with the function $B(w) = 70w^{3/4}$, where w is the person's weight in kilograms. Use this information to answer Exercises 95 and 96.

95. Estimate the BMR for a person who weighs 60 kilograms. Round to the nearest calorie. (*Note:* 60 kilograms is approximately 132 pounds.)

96. Estimate the BMR for a person who weighs 90 kilograms. Round to the nearest calorie. (*Note:* 90 kilograms is approximately 198 pounds.)

The number of cellular telephone subscriptions in the United States from 1994 through 2000 can be modeled by the function $f(x) = 2.5x^{8/5}$, where y is the number of cellular telephone subscriptions in millions, x years after 1990. (Source: Based on data from the Cellular Telecommunications & Internet Association, 1994–2000) Use this information to answer Exercises 97 and 98.

97. Use this model to estimate the number of cellular telephone subscriptions in the United States in 2000. Round to the nearest tenth of a million.

98. Predict the number of cellular telephone subscriptions in the United States in 2007. Round to the nearest tenth of a million.

Fill in each box with the correct expression.

99. $\square \cdot a^{2/3} = a^{3/3}$, or a

100. $\square \cdot x^{1/8} = x^{4/8}$, or $x^{1/2}$

101. $\dfrac{\square}{x^{-2/5}} = x^{3/5}$

102. $\dfrac{\square}{y^{-3/4}} = y^{4/4}$, or y

Use a calculator to write a four-decimal-place approximation of each number.

103. $8^{1/4}$

104. $18^{3/5}$

105. In physics, the speed of a wave traveling over a stretched string with tension t and density u is given by the expression $\dfrac{\sqrt{t}}{\sqrt{u}}$. Write this expression with rational exponents.

106. In electronics, the angular frequency of oscillations in a certain type of circuit is given by the expression $(LC)^{-1/2}$. Use radical notation to write this expression.

9.3 Simplifying Radical Expressions

OBJECTIVES

A Use the product rule for radicals.

B Use the quotient rule for radicals.

B Simplify radicals.

SSM TUTOR CENTER SG CD & VIDEO MATH PRO WEB

A Using the Product Rule

It is possible to simplify some radicals that do not evaluate to rational numbers. To do so, we use a product rule and a quotient rule for radicals. To discover the product rule, notice the following pattern:

$$\sqrt{9}\cdot\sqrt{4} = 3\cdot 2 = 6$$
$$\sqrt{9\cdot 4} = \sqrt{36} = 6$$

Since both expressions simplify to 6, it is true that

$$\sqrt{9}\cdot\sqrt{4} = \sqrt{9\cdot 4}$$

This pattern suggests the following product rule for radicals.

Product Rule for Radicals

If $\sqrt[n]{a}$ and $\sqrt[n]{b}$ are real numbers, then

$$\sqrt[n]{a}\cdot\sqrt[n]{b} = \sqrt[n]{ab}$$

Notice that the product rule is the relationship $a^{1/n}\cdot b^{1/n} = (ab)^{1/n}$ stated in radical notation.

EXAMPLES Use the product rule to multiply.

1. $\sqrt{3}\cdot\sqrt{5} = \sqrt{3\cdot 5} = \sqrt{15}$
2. $\sqrt{21}\cdot\sqrt{x} = \sqrt{21x}$
3. $\sqrt[3]{4}\cdot\sqrt[3]{2} = \sqrt[3]{4\cdot 2} = \sqrt[3]{8} = 2$
4. $\sqrt[4]{5}\cdot\sqrt[4]{2x^3} = \sqrt[4]{5\cdot 2x^3} = \sqrt[4]{10x^3}$
5. $\sqrt{\frac{2}{a}}\cdot\sqrt{\frac{b}{3}} = \sqrt{\frac{2}{a}\cdot\frac{b}{3}} = \sqrt{\frac{2b}{3a}}$

Practice Problems 1–5

Use the product rule to multiply.

1. $\sqrt{2}\cdot\sqrt{7}$
2. $\sqrt{17}\cdot\sqrt{y}$
3. $\sqrt[3]{2}\cdot\sqrt[3]{32}$
4. $\sqrt[4]{6}\cdot\sqrt[4]{3x^2}$
5. $\sqrt{\frac{3}{x}}\cdot\sqrt{\frac{y}{2}}$

B Using the Quotient Rule

To discover the quotient rule for radicals, notice the following pattern:

$$\sqrt{\frac{4}{9}} = \frac{2}{3}$$
$$\frac{\sqrt{4}}{\sqrt{9}} = \frac{2}{3}$$

Since both expressions simplify to $\frac{2}{3}$, it is true that

$$\sqrt{\frac{4}{9}} = \frac{\sqrt{4}}{\sqrt{9}}$$

This pattern suggests the following quotient rule for radicals.

Quotient Rule for Radicals

If $\sqrt[n]{a}$ and $\sqrt[n]{b}$ are real numbers and $\sqrt[n]{b}$ is not zero, then

$$\sqrt[n]{\frac{a}{b}} = \frac{\sqrt[n]{a}}{\sqrt[n]{b}}$$

Answers

1. $\sqrt{14}$, **2.** $\sqrt{17y}$, **3.** 4, **4.** $\sqrt[4]{18x^2}$, **5.** $\sqrt{\frac{3y}{2x}}$

Notice that the quotient rule is the relationship $\left(\frac{a}{b}\right)^{1/n} = \frac{a^{1/n}}{b^{1/n}}$ stated in radical notation. We can use the quotient rule to simplify radical expressions by reading the rule from left to right or to divide radicals by reading the rule from right to left.

For example,

$$\sqrt{\frac{x}{16}} = \frac{\sqrt{x}}{\sqrt{16}} = \frac{\sqrt{x}}{4} \quad \text{Using } \sqrt[n]{\frac{a}{b}} = \frac{\sqrt[n]{a}}{\sqrt[n]{b}}$$

$$\frac{\sqrt{75}}{\sqrt{3}} = \sqrt{\frac{75}{3}} = \sqrt{25} = 5 \quad \text{Using } \frac{\sqrt[n]{a}}{\sqrt[n]{b}} = \sqrt[n]{\frac{a}{b}}$$

Note: *For the remainder of this chapter, we will assume that variables represent positive real numbers. If this is so, we need not insert absolute value bars when we simplify even roots.*

Practice Problems 6–9

Use the quotient rule to simplify. Assume that all variables represent positive real numbers.

6. $\sqrt{\frac{9}{25}}$

7. $\sqrt{\frac{y}{36}}$

8. $\sqrt[3]{\frac{27}{64}}$

9. $\sqrt[5]{\frac{7}{32x^5}}$

EXAMPLES Use the quotient rule to simplify.

6. $\sqrt{\frac{25}{49}} = \frac{\sqrt{25}}{\sqrt{49}} = \frac{5}{7}$

7. $\sqrt{\frac{x}{9}} = \frac{\sqrt{x}}{\sqrt{9}} = \frac{\sqrt{x}}{3}$

8. $\sqrt[3]{\frac{8}{27}} = \frac{\sqrt[3]{8}}{\sqrt[3]{27}} = \frac{2}{3}$

9. $\sqrt[4]{\frac{3}{16y^4}} = \frac{\sqrt[4]{3}}{\sqrt[4]{16y^4}} = \frac{\sqrt[4]{3}}{2y}$

C Simplifying Radicals

Both the product and quotient rules can be used to simplify a radical. If the product rule is read from right to left, we have that $\sqrt[n]{ab} = \sqrt[n]{a} \cdot \sqrt[n]{b}$. We use this to simplify the following radicals.

Practice Problem 10

Simplify: $\sqrt{18}$

EXAMPLE 10 Simplify: $\sqrt{50}$

Solution: We factor 50 such that one factor is the largest perfect square that divides 50. The largest perfect square factor of 50 is 25, so we write 50 as $25 \cdot 2$ and use the product rule for radicals to simplify.

$$\sqrt{50} = \sqrt{25 \cdot 2} = \sqrt{25} \cdot \sqrt{2} = 5\sqrt{2}$$

↑ the largest perfect square factor of 50

Helpful Hint

Don't forget that, for example, $5\sqrt{2}$ means $5 \cdot \sqrt{2}$.

Practice Problems 11–13

Simplify.

11. $\sqrt[3]{40}$

12. $\sqrt{14}$

13. $\sqrt[4]{162}$

EXAMPLES Simplify.

11. $\sqrt[3]{24} = \sqrt[3]{8 \cdot 3} = \sqrt[3]{8} \cdot \sqrt[3]{3} = 2\sqrt[3]{3}$

↑ the largest perfect cube factor of 24

12. $\sqrt{26}$ The largest perfect square factor of 26 is 1, so $\sqrt{26}$ cannot be simplified further.

13. $\sqrt[4]{32} = \sqrt[4]{16 \cdot 2} = \sqrt[4]{16} \cdot \sqrt[4]{2} = 2\sqrt[4]{2}$

↑ the largest 4th power factor of 32

Answers

6. $\frac{3}{5}$, **7.** $\frac{\sqrt{y}}{6}$, **8.** $\frac{3}{4}$, **9.** $\frac{\sqrt[5]{7}}{2x}$, **10.** $3\sqrt{2}$, **11.** $2\sqrt[3]{5}$, **12.** $\sqrt{14}$, **13.** $3\sqrt[4]{2}$

After simplifying a radical such as a square root, always check the radicand to see that it contains no other perfect square factors. It may, if the largest perfect square factor of the radicand was not originally recognized. For example,

$$\sqrt{200} = \sqrt{4 \cdot 50} = \sqrt{4} \cdot \sqrt{50} = 2\sqrt{50}$$

Notice that the radicand 50 still contains the perfect square factor 25. This is because 4 is not the largest perfect square factor of 200. We continue as follows:

$$2\sqrt{50} = 2\sqrt{25 \cdot 2} = 2 \cdot \sqrt{25} \cdot \sqrt{2} = 2 \cdot 5 \cdot \sqrt{2} = 10\sqrt{2}$$

The radical is now simplified since 2 contains no perfect square factors (other than 1).

Helpful Hint

To recognize the largest perfect power factors of a radicand, it will help if you are familiar with some perfect powers. A few are listed below.

Perfect Squares	1, 4, 9, 16, 25, 36, 49, 64, 81, 100, 121, 144 1^2 2^2 3^2 4^2 5^2 6^2 7^2 8^2 9^2 10^2 11^2 12^2
Perfect Cubes	1, 8, 27, 64, 125 1^3 2^3 3^3 4^3 5^3
Perfect 4th powers	1, 16, 81, 256 1^4 2^4 3^4 4^4

Helpful Hint

We say that a radical of the form $\sqrt[n]{a}$ is simplified when the radicand a contains no factors that are perfect nth powers (other than 1 or -1).

EXAMPLES

Simplify. Assume that all variables represent positive real numbers.

14. $\sqrt{25x^3} = \sqrt{25 \cdot x^2 \cdot x}$ Find the largest perfect square factor.

$= \sqrt{25 \cdot x^2} \cdot \sqrt{x}$ Use the product rule.

$= 5x\sqrt{x}$ Simplify.

15. $\sqrt[3]{54x^6y^8} = \sqrt[3]{27 \cdot 2 \cdot x^6 \cdot y^6 \cdot y^2}$ Factor the radicand and identify perfect cube factors.

$= \sqrt[3]{27 \cdot x^6 \cdot y^6 \cdot 2y^2}$

$= \sqrt[3]{27 \cdot x^6 \cdot y^6} \cdot \sqrt[3]{2y^2}$ Use the product rule.

$= 3x^2y^2\sqrt[3]{2y^2}$ Simplify.

16. $\sqrt[4]{81z^{11}} = \sqrt[4]{81 \cdot z^8 \cdot z^3}$ Factor the radicand and identify perfect 4th power factors.

$= \sqrt[4]{81 \cdot z^8} \cdot \sqrt[4]{z^3}$ Use the product rule.

$= 3z^2\sqrt[4]{z^3}$ Simplify.

Practice Problems 14–16

Simplify. Assume that all variables represent positive real numbers.

14. $\sqrt{49a^5}$

15. $\sqrt[3]{24x^9y^7}$

16. $\sqrt[4]{16z^9}$

Answers

14. $7a^2\sqrt{a}$, **15.** $2x^3y^2\sqrt[3]{3y}$, **16.** $2z^2\sqrt[4]{z}$

Practice Problems 17–20

Use the quotient rule to divide. Then simplify if possible. Assume that all variables represent positive real numbers.

17. $\dfrac{\sqrt{75}}{\sqrt{3}}$ 18. $\dfrac{\sqrt{80y}}{3\sqrt{5}}$

19. $\dfrac{5\sqrt[3]{162x^8}}{\sqrt[3]{3x^2}}$ 20. $\dfrac{3\sqrt[4]{243x^9y^6}}{\sqrt[4]{x^{-3}y}}$

Concept Check

Find and correct the error:

$$\frac{\sqrt[3]{27}}{\sqrt{9}} = \sqrt[3]{\frac{27}{9}} = \sqrt[3]{3}$$

Answers

17. 5, **18.** $\frac{4}{3}\sqrt{y}$, **19.** $15x^2\sqrt[3]{2}$, **20.** $9x^3y\sqrt[4]{3y}$

Concept Check: $\dfrac{\sqrt[3]{27}}{\sqrt{9}} = \dfrac{3}{3} = 1$

EXAMPLES

Use the quotient rule to divide. Then simplify if possible. Assume that all variables represent positive real numbers.

17. $\dfrac{\sqrt{20}}{\sqrt{5}} = \sqrt{\dfrac{20}{5}}$ Use the quotient rule.

$= \sqrt{4}$ Simplify.

$= 2$ Simplify.

18. $\dfrac{\sqrt{50x}}{2\sqrt{2}} = \dfrac{1}{2}\cdot\sqrt{\dfrac{50x}{2}}$ Use the quotient rule.

$= \dfrac{1}{2}\cdot\sqrt{25x}$ Simplify.

$= \dfrac{1}{2}\cdot\sqrt{25}\cdot\sqrt{x}$ Factor $25x$.

$= \dfrac{1}{2}\cdot 5\cdot\sqrt{x}$ Simplify.

$= \dfrac{5}{2}\sqrt{x}$

19. $\dfrac{7\sqrt[3]{48y^4}}{\sqrt[3]{2y}} = 7\sqrt[3]{\dfrac{48y^4}{2y}} = 7\sqrt[3]{24y^3} = 7\sqrt[3]{8\cdot y^3\cdot 3}$

$= 7\sqrt[3]{8\cdot y^3}\cdot\sqrt[3]{3} = 7\cdot 2y\sqrt[3]{3} = 14y\sqrt[3]{3}$

20. $\dfrac{2\sqrt[4]{32a^8b^6}}{\sqrt[4]{a^{-1}b^2}} = 2\sqrt[4]{\dfrac{32a^8b^6}{a^{-1}b^2}} = 2\sqrt[4]{32a^9b^4} = 2\sqrt[4]{16\cdot a^8\cdot b^4\cdot 2\cdot a}$

$= 2\sqrt[4]{16\cdot a^8\cdot b^4}\cdot\sqrt[4]{2\cdot a} = 2\cdot 2a^2b\cdot\sqrt[4]{2a} = 4a^2b\sqrt[4]{2a}$ ●

Try the Concept Check in the margin.

Name ______________________ Section __________ Date ____________

EXERCISE SET 9.3

 *Use the product rule to multiply. Assume that all variables represent positive real numbers. See Examples 1 through 5.*

1. $\sqrt{7} \cdot \sqrt{2}$

2. $\sqrt{11} \cdot \sqrt{10}$

3. $\sqrt[4]{8} \cdot \sqrt[4]{2}$

4. $\sqrt[4]{27} \cdot \sqrt[4]{3}$

5. $\sqrt[3]{4} \cdot \sqrt[3]{9}$

6. $\sqrt[3]{10} \cdot \sqrt[3]{5}$

7. $\sqrt{2} \cdot \sqrt{3x}$

8. $\sqrt{3y} \cdot \sqrt{5x}$

9. $\sqrt{\frac{7}{x}} \cdot \sqrt{\frac{2}{y}}$

10. $\sqrt{\frac{6}{m}} \cdot \sqrt{\frac{n}{5}}$

11. $\sqrt[4]{4x^3} \cdot \sqrt[4]{5}$

12. $\sqrt[4]{ab^2} \cdot \sqrt[4]{27ab}$

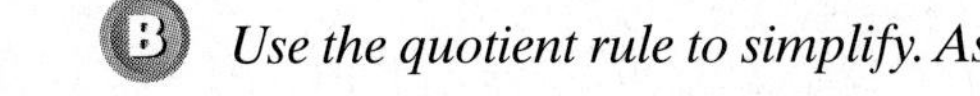 *Use the quotient rule to simplify. Assume that all variables represent positive real numbers. See Examples 6 through 9.*

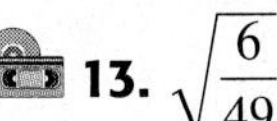 13. $\sqrt{\frac{6}{49}}$

14. $\sqrt{\frac{8}{81}}$

15. $\sqrt{\frac{2}{49}}$

16. $\sqrt{\frac{5}{121}}$

17. $\sqrt[4]{\frac{x^3}{16}}$

18. $\sqrt[4]{\frac{y}{81x^4}}$

19. $\sqrt[3]{\frac{4}{27}}$ 20. $\sqrt[3]{\frac{3}{64}}$ 21. $\sqrt[4]{\frac{8}{x^8}}$ 22. $\sqrt[4]{\frac{a^3}{81}}$ 23. $\sqrt[3]{\frac{2x}{81y^{12}}}$ 24. $\sqrt[3]{\frac{3}{8x^6}}$

25. $\sqrt{\frac{x^2y}{100}}$ 26. $\sqrt{\frac{y^2z}{400}}$ 27. $\sqrt{\frac{5x^2}{169y^2}}$ 28. $\sqrt{\frac{y^{10}}{225x^6}}$ 29. $-\sqrt[3]{\frac{z^7}{125x^3}}$ 30. $-\sqrt[3]{\frac{1000a}{b^9}}$

C *Simplify. Assume that all variables represent positive real numbers. See Examples 10 through 16.*

31. $\sqrt{32}$ 32. $\sqrt{27}$ 33. $\sqrt[3]{192}$ 34. $\sqrt[3]{108}$ 35. $5\sqrt{75}$ 36. $3\sqrt{8}$

37. $\sqrt{24}$ 38. $\sqrt{20}$ 39. $\sqrt{100x^5}$ 40. $\sqrt{64y^9}$ 41. $\sqrt[3]{16y^7}$ 42. $\sqrt[3]{64y^9}$

43. $\sqrt[4]{a^8b^7}$ 44. $\sqrt[5]{32z^{12}}$ 45. $\sqrt{y^5}$ 46. $\sqrt[3]{y^5}$ 47. $\sqrt{25a^2b^3}$ 48. $\sqrt{9x^5y^7}$

49. $\sqrt[5]{-32x^{10}y}$ **50.** $\sqrt[5]{-243z^9}$ **51.** $\sqrt[3]{50x^{14}}$ **52.** $\sqrt[3]{40y^{10}}$ **53.** $-\sqrt{32a^8b^7}$

54. $-\sqrt{20ab^6}$ **55.** $\sqrt{9x^7y^9}$ **56.** $\sqrt{12r^9s^{12}}$ **57.** $\sqrt[3]{125r^9s^{12}}$ **58.** $\sqrt[3]{8a^6b^9}$

Use the quotient rule to divide. Then simplify if possible. Assume that all variables represent positive real numbers. See Examples 17 through 20.

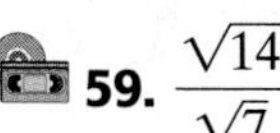

59. $\dfrac{\sqrt{14}}{\sqrt{7}}$ **60.** $\dfrac{\sqrt{45}}{\sqrt{9}}$ **61.** $\dfrac{\sqrt[3]{24}}{\sqrt[3]{3}}$ **62.** $\dfrac{\sqrt[3]{10}}{\sqrt[3]{2}}$ **63.** $\dfrac{5\sqrt[4]{48}}{\sqrt[4]{3}}$

64. $\dfrac{7\sqrt[4]{162}}{\sqrt[4]{2}}$ **65.** $\dfrac{\sqrt{x^5y^3}}{\sqrt{xy}}$ **66.** $\dfrac{\sqrt{a^7b^6}}{\sqrt{a^3b^2}}$ **67.** $\dfrac{8\sqrt[3]{54m^7}}{\sqrt[3]{2m}}$ **68.** $\dfrac{\sqrt[3]{128x^3}}{-3\sqrt[3]{2x}}$

69. $\dfrac{3\sqrt{100x^2}}{2\sqrt{2x^{-1}}}$ **70.** $\dfrac{\sqrt{270y^2}}{5\sqrt{3y^{-4}}}$ **71.** $\dfrac{\sqrt[4]{96a^{10}b^3}}{\sqrt[4]{3a^2b^3}}$ **72.** $\dfrac{\sqrt[5]{64x^{10}y^3}}{\sqrt[5]{2x^3y^{-7}}}$

Review and Preview

Perform each indicated operation. See Sections 4.4 and 4.5.

73. $6x + 8x$

74. $(6x)(8x)$

75. $(2x + 3)(x - 5)$

76. $(2x + 3) + (x - 5)$

77. $9y^2 - 8y^2$

78. $(9y^2)(-8y^2)$

79. $-3(x + 5)$

80. $-3 + x + 5$

81. $(x - 4)^2$

82. $(2x + 1)^2$

Combining Concepts

83. The formula for the surface area A of a cone with height h and radius r is given by

$$A = \pi r\sqrt{r^2 + h^2}$$

a. Find the surface area of a cone whose height is 3 centimeters and whose radius is 4 centimeters.

b. Approximate to two decimal places the surface area of a cone whose height is 7.2 feet and whose radius is 6.8 feet.

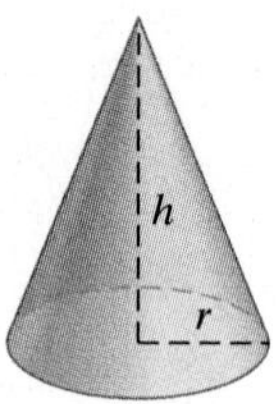

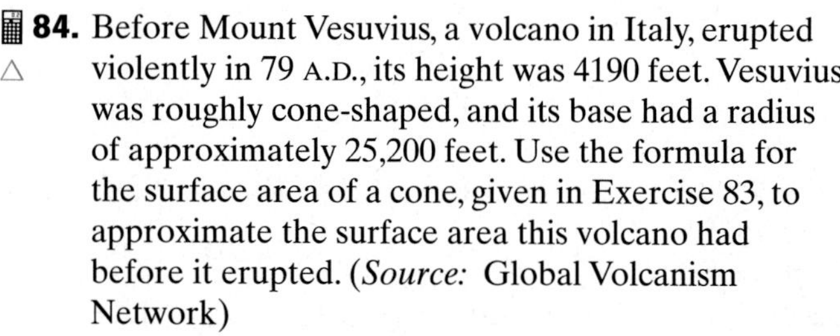

84. Before Mount Vesuvius, a volcano in Italy, erupted violently in 79 A.D., its height was 4190 feet. Vesuvius was roughly cone-shaped, and its base had a radius of approximately 25,200 feet. Use the formula for the surface area of a cone, given in Exercise 83, to approximate the surface area this volcano had before it erupted. (*Source:* Global Volcanism Network)

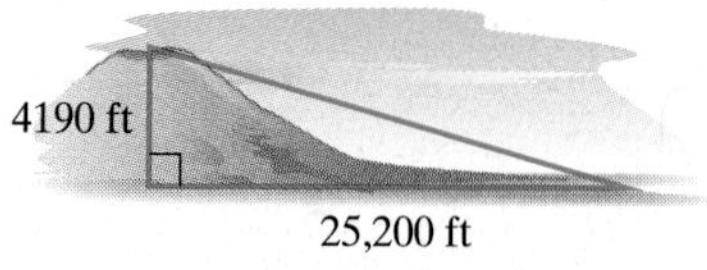

85. The owner of Knightime Video has determined that the demand equation for renting older released tapes is $F(x) = 0.6\sqrt{49 - x^2}$, where x is the price in dollars per two-day rental and $F(x)$ is the number of times the video is demanded per week.

a. Approximate to one decimal place the demand per week of an older released video if the rental price is \$3 per two-day rental.

b. Approximate to one decimal place the demand per week of an older released video if the rental price is \$5 per two-day rental.

c. Explain how the owner of the video store can use this equation to predict the number of copies of each tape that should be in stock.

9.4 Adding, Subtracting, and Multiplying Radical Expressions

OBJECTIVES

A Add or subtract radical expressions.

B Multiply radical expressions.

A Adding or Subtracting Radical Expressions

We have learned that the sum or difference of like terms can be simplified. To simplify these sums or differences, we use the distributive property. For example,

$$2x + 3x = (2 + 3)x = 5x$$

The distributive property can also be used to add *like radicals*.

Like Radicals

Radicals with the same index and the same radicand are **like radicals**. The example below shows how to use the distributive property to simplify an expression containing like radicals.

$$2\sqrt{7} + 3\sqrt{7} = (2 + 3)\sqrt{7} = 5\sqrt{7}$$

Like radicals

Helpful Hint

The expression

$$5\sqrt{7} - 3\sqrt{6}$$

does not contain like radicals and cannot be simplified further.

EXAMPLES Add or subtract as indicated.

1. $4\sqrt{11} + 8\sqrt{11} = (4 + 8)\sqrt{11} = 12\sqrt{11}$

2. $5\sqrt[3]{3x} - 7\sqrt[3]{3x} = (5 - 7)\sqrt[3]{3x} = -2\sqrt[3]{3x}$

3. $2\sqrt{7} + 2\sqrt[3]{7}$ This expression cannot be simplified since $2\sqrt{7}$ and $2\sqrt[3]{7}$ do not contain like radicals.

Try the Concept Check in the margin.

When adding or subtracting radicals, always check first to see whether any radicals can be simplified.

EXAMPLES

Add or subtract as indicated. Assume that all variables represent positive real numbers.

4. $\sqrt{20} + 2\sqrt{45} = \sqrt{4 \cdot 5} + 2\sqrt{9 \cdot 5}$ Factor 20 and 45.

$= \sqrt{4} \cdot \sqrt{5} + 2 \cdot \sqrt{9} \cdot \sqrt{5}$ Use the product rule.

$= 2 \cdot \sqrt{5} + 2 \cdot 3 \cdot \sqrt{5}$ Simplify $\sqrt{4}$ and $\sqrt{9}$.

$= 2\sqrt{5} + 6\sqrt{5}$ Add like radicals.

$= 8\sqrt{5}$

5. $\sqrt[3]{54} - 5\sqrt[3]{16} + \sqrt[3]{2}$

$= \sqrt[3]{27} \cdot \sqrt[3]{2} - 5 \cdot \sqrt[3]{8} \cdot \sqrt[3]{2} + \sqrt[3]{2}$ Factor and use the product rule.

$= 3 \cdot \sqrt[3]{2} - 5 \cdot 2 \cdot \sqrt[3]{2} + \sqrt[3]{2}$ Simplify $\sqrt[3]{27}$ and $\sqrt[3]{8}$.

$= 3\sqrt[3]{2} - 10\sqrt[3]{2} + \sqrt[3]{2}$ Write $5 \cdot 2$ as 10.

$= -6\sqrt[3]{2}$ Combine like radicals.

Practice Problems 1–3

Add or subtract as indicated.

1. $5\sqrt{15} + 2\sqrt{15}$
2. $9\sqrt[3]{2y} - 15\sqrt[3]{2y}$
3. $6\sqrt{10} - 3\sqrt[3]{10}$

Concept Check

True or false:

$\sqrt{a} + \sqrt{b} = \sqrt{a + b}$?

Explain.

Practice Problems 4–8

Add or subtract as indicated. Assume that all variables represent positive real numbers.

4. $\sqrt{50} + 5\sqrt{18}$
5. $\sqrt[3]{24} - 4\sqrt[3]{192} + \sqrt[3]{3}$
6. $\sqrt{20x} - 6\sqrt{16x} + \sqrt{45x}$
7. $\sqrt[4]{32} + \sqrt{32}$
8. $\sqrt[3]{8y^5} + \sqrt[3]{27y^5}$

Answers

1. $7\sqrt{15}$, **2.** $-6\sqrt[3]{2y}$, **3.** $6\sqrt{10} - 3\sqrt[3]{10}$, **4.** $20\sqrt{2}$, **5.** $-13\sqrt[3]{3}$, **6.** $5\sqrt{5x} - 24\sqrt{x}$, **7.** $2\sqrt[4]{2} + 4\sqrt{2}$, **8.** $5y\sqrt[3]{y^2}$

Concept Check: false; answers may vary

6. $\sqrt{27x} - 2\sqrt{9x} + \sqrt{72x}$

$= \sqrt{9} \cdot \sqrt{3x} - 2 \cdot \sqrt{9} \cdot \sqrt{x} + \sqrt{36} \cdot \sqrt{2x}$ Factor and use the product rule.

$= 3 \cdot \sqrt{3x} - 2 \cdot 3 \cdot \sqrt{x} + 6 \cdot \sqrt{2x}$ Simplify $\sqrt{9}$ and $\sqrt{36}$.

$= 3\sqrt{3x} - 6\sqrt{x} + 6\sqrt{2x}$ Write $2 \cdot 3$ as 6.

Helpful Hint

None of these terms contain like radicals. We can simplify no further.

7. $\sqrt[3]{98} + \sqrt{98} = \sqrt[3]{98} + \sqrt{49} \cdot \sqrt{2}$ Factor and use the product rule.

$= \sqrt[3]{98} + 7\sqrt{2}$ No further simplification is possible.

8. $\sqrt[3]{48y^4} + \sqrt[3]{6y^4} = \sqrt[3]{8y^3} \cdot \sqrt[3]{6y} + \sqrt[3]{y^3} \cdot \sqrt[3]{6y}$ Factor and use the product rule.

$= 2y\sqrt[3]{6y} + y\sqrt[3]{6y}$ Simplify $\sqrt[3]{8y^3}$ and $\sqrt[3]{y^3}$.

$= 3y\sqrt[3]{6y}$ Combine like radicals.

Practice Problems 9–10

Add or subtract as indicated. Assume that all variables represent positive real numbers.

9. $\dfrac{\sqrt{75}}{9} - \dfrac{\sqrt{3}}{2}$

10. $\sqrt[3]{\dfrac{5x}{27}} + 4\sqrt[3]{5x}$

Answers

9. $\dfrac{\sqrt{3}}{18}$, **10.** $\dfrac{13\sqrt[3]{5x}}{3}$

EXAMPLES

Add or subtract as indicated. Assume that all variables represent positive real numbers.

9. $\dfrac{\sqrt{45}}{4} - \dfrac{\sqrt{5}}{3} = \dfrac{3\sqrt{5}}{4} - \dfrac{\sqrt{5}}{3}$ To subtract, notice that the LCD is 12.

$= \dfrac{3\sqrt{5} \cdot 3}{4 \cdot 3} - \dfrac{\sqrt{5} \cdot 4}{3 \cdot 4}$ Write each expression as an equivalent expression with a denominator of 12.

$= \dfrac{9\sqrt{5}}{12} - \dfrac{4\sqrt{5}}{12}$ Multiply factors in the numerators and the denominators.

$= \dfrac{5\sqrt{5}}{12}$ Subtract.

10. $\sqrt[3]{\dfrac{7x}{8}} + 2\sqrt[3]{7x} = \dfrac{\sqrt[3]{7x}}{\sqrt[3]{8}} + 2\sqrt[3]{7x}$ Use the quotient rule for radicals.

$= \dfrac{\sqrt[3]{7x}}{2} + 2\sqrt[3]{7x}$ Simplify.

$= \dfrac{\sqrt[3]{7x}}{2} + \dfrac{2\sqrt[3]{7x} \cdot 2}{2}$ Write each expression as an equivalent expression with a denominator of 2.

$= \dfrac{\sqrt[3]{7x}}{2} + \dfrac{4\sqrt[3]{7x}}{2}$

$= \dfrac{5\sqrt[3]{7x}}{2}$ Add.

B Multiplying Radical Expressions

We can multiply radical expressions by using many of the same properties used to multiply polynomial expressions. For instance, to multiply $\sqrt{2}(\sqrt{6} - 3\sqrt{2})$, we use the distributive property and multiply $\sqrt{2}$ by each term inside the parentheses.

$\sqrt{2}(\sqrt{6} - 3\sqrt{2}) = \sqrt{2}(\sqrt{6}) - \sqrt{2}(3\sqrt{2})$ Use the distributive property.

$= \sqrt{2 \cdot 6} - 3\sqrt{2 \cdot 2}$

$= \sqrt{2 \cdot 2 \cdot 3} - 3 \cdot 2$ Use the product rule for radicals.

$= 2\sqrt{3} - 6$

EXAMPLE 11 Multiply: $\sqrt{3}(5 + \sqrt{30})$

Solution: $\sqrt{3}(5 + \sqrt{30}) = \sqrt{3}(5) + \sqrt{3}(\sqrt{30})$

$= 5\sqrt{3} + \sqrt{3 \cdot 30}$

$= 5\sqrt{3} + \sqrt{3 \cdot 3 \cdot 10}$

$= 5\sqrt{3} + 3\sqrt{10}$

Practice Problem 11

Multiply: $\sqrt{2}(6 + \sqrt{10})$

EXAMPLES

Multiply. Assume that all variables represent positive real numbers.

12. $(\sqrt{5} - \sqrt{6})(\sqrt{7} + 1) = \underbrace{\sqrt{5} \cdot \sqrt{7}}_{\text{First}} + \underbrace{\sqrt{5} \cdot 1}_{\text{Outer}} - \underbrace{\sqrt{6} \cdot \sqrt{7}}_{\text{Inner}} - \underbrace{\sqrt{6} \cdot 1}_{\text{Last}}$ Using the FOIL order.

$= \sqrt{35} + \sqrt{5} - \sqrt{42} - \sqrt{6}$ Simplify.

13. $(\sqrt{2x} + 5)(\sqrt{2x} - 5) = (\sqrt{2x})^2 - 5^2$ Multiply the sum and difference of two terms:

$= 2x - 25$ $(a + b)(a - b) = a^2 - b^2$

14. $(\sqrt{3} - 1)^2 = (\sqrt{3})^2 - 2 \cdot \sqrt{3} \cdot 1 + 1^2$ Square the binomial:

$= 3 - 2\sqrt{3} + 1$ $(a - b)^2 = a^2 - 2ab + b^2$

$= 4 - 2\sqrt{3}$

Square the binomial: $(a + b)^2 = a^2 + 2ab + b^2$

15. $(\underbrace{\sqrt{x - 3}}_{a} + \underbrace{5}_{b})^2 = \underbrace{(\sqrt{x - 3})^2}_{a^2} + \underbrace{2 \cdot \sqrt{x - 3}}_{+2 \cdot a} \underbrace{\cdot 5}_{\cdot b} + \underbrace{5^2}_{+b^2}$

$= x - 3 + 10\sqrt{x - 3} + 25$ Simplify.

$= x + 22 + 10\sqrt{x - 3}$ Combine like terms.

Practice Problems 12–15

Multiply. Assume that all variables represent positive real numbers.

12. $(\sqrt{3} - \sqrt{5})(\sqrt{2} + 7)$
13. $(\sqrt{5y} + 2)(\sqrt{5y} - 2)$
14. $(\sqrt{3} - 7)^2$
15. $(\sqrt{x + 1} + 2)^2$

Answers

11. $6\sqrt{2} + 2\sqrt{5}$, **12.** $\sqrt{6} + 7\sqrt{3} - \sqrt{10} - 7\sqrt{5}$, **13.** $5y - 4$, **14.** $52 - 14\sqrt{3}$, **15.** $x + 5 + 4\sqrt{x + 1}$

FOCUS ON History

DEVELOPMENT OF THE RADICAL SYMBOL

The first mathematician to use the symbol we use today to denote a square root was Christoff Rudolff (1499–1545). In 1525, Rudolff wrote and published the first German algebra text, *Die Coss*. In it, he used √ to represent a square root, the symbol w√ to represent a cube root, and the symbol ww√ to represent a fourth root. It was another 100 years before the square root symbol was extended with an overbar called a *vinculum*, $\sqrt{}$, to indicate the inclusion of several terms under the radical symbol. This innovation was introduced by René Descartes (1596–1650) in 1637 in his text *La Géométrie*. The modern use of a numeral as part of a radical sign to indicate the index of the radical for higher roots did not appear until 1690, when this notation was used by French mathematician Michel Rolle (1652–1719) in his text *Traité d'Algébre*.

Name ______________________ Section __________ Date __________

Mental Math

Simplify. Assume that all variables represent positive real numbers.

1. $2\sqrt{3} + 4\sqrt{3}$
2. $5\sqrt{7} + 3\sqrt{7}$
3. $8\sqrt{x} - 5\sqrt{x}$
4. $3\sqrt{y} + 10\sqrt{y}$
5. $7\sqrt[3]{x} + 5\sqrt[3]{x}$
6. $8\sqrt[3]{z} - 2\sqrt[3]{z}$
7. $(\sqrt{3})^2$
8. $(\sqrt{4x+1})^2$

EXERCISE SET 9.4

A *Add or subtract as indicated. Assume that all variables represent positive real numbers. See Examples 1 through 10.*

1. $\sqrt{8} - \sqrt{32}$
2. $\sqrt{27} - \sqrt{75}$
3. $2\sqrt{2x^3} + 4x\sqrt{8x}$
4. $3\sqrt{45x^3} + x\sqrt{5x}$

5. $2\sqrt{50} - 3\sqrt{125} + \sqrt{98}$
6. $4\sqrt{32} - \sqrt{18} + 2\sqrt{128}$
7. $\sqrt[3]{16x} - \sqrt[3]{54x}$
8. $2\sqrt[3]{3a^4} - 3a\sqrt[3]{81a}$
9. $\sqrt{9b^3} - \sqrt{25b^3} + \sqrt{49b^3}$
10. $\sqrt{4x^7} + 9x^2\sqrt{x^3} - 5x\sqrt{x^5}$
11. $\dfrac{5\sqrt{2}}{3} + \dfrac{2\sqrt{2}}{5}$
12. $\dfrac{\sqrt{3}}{2} + \dfrac{4\sqrt{3}}{3}$
13. $\sqrt[3]{\dfrac{11}{8}} - \dfrac{\sqrt[3]{11}}{6}$
14. $\dfrac{2\sqrt[3]{4}}{7} - \dfrac{\sqrt[3]{4}}{14}$
15. $\dfrac{\sqrt{20x}}{9} + \sqrt{\dfrac{5x}{9}}$
16. $\dfrac{3x\sqrt{7}}{5} + \sqrt{\dfrac{7x^2}{100}}$
17. $7\sqrt{9} - 7 + \sqrt{3}$
18. $\sqrt{16} - 5\sqrt{10} + 7$
19. $2 + 3\sqrt{y^2} - 6\sqrt{y^2} + 5$
20. $3\sqrt{7} - \sqrt[3]{x} + 4\sqrt{7} - 3\sqrt[3]{x}$
21. $3\sqrt{108} - 2\sqrt{18} - 3\sqrt{48}$

22. $-\sqrt{75} + \sqrt{12} - 3\sqrt{3}$

23. $-5\sqrt[3]{625} + \sqrt[3]{40}$

24. $-2\sqrt[3]{108} - \sqrt[3]{32}$

25. $\sqrt{9b^3} - \sqrt{25b^3} + \sqrt{16b^3}$

26. $\sqrt{4x^7y^5} + 9x^2\sqrt{x^3y^5} - 5xy\sqrt{x^5y^3}$

27. $5y\sqrt{8y} + 2\sqrt{50y^3}$

28. $3\sqrt{8x^2y^3} - 2x\sqrt{32y^3}$

29. $\sqrt[3]{54xy^3} - 5\sqrt[3]{2xy^3} + y\sqrt[3]{128x}$

30. $2\sqrt[3]{24x^3y^4} + 4x\sqrt[3]{81y^4}$

31. $6\sqrt[3]{11} + 8\sqrt{11} - 12\sqrt{11}$

32. $3\sqrt[3]{5} + 4\sqrt{5}$

33. $-2\sqrt[4]{x^7} + 3\sqrt[4]{16x^7}$

34. $6\sqrt[3]{24x^3} - 2\sqrt[3]{81x^3} - x\sqrt[3]{3}$

35. $\dfrac{4\sqrt{3}}{3} - \dfrac{\sqrt{12}}{3}$

36. $\dfrac{\sqrt{45}}{10} + \dfrac{7\sqrt{5}}{10}$

37. $\dfrac{\sqrt[3]{8x^4}}{7} + \dfrac{3x\sqrt[3]{x}}{7}$

38. $\dfrac{\sqrt[4]{48}}{5x} - \dfrac{2\sqrt[4]{3}}{10x}$

39. $\sqrt{\dfrac{28}{x^2}} + \sqrt{\dfrac{7}{4x^2}}$

40. $\dfrac{\sqrt{99}}{5x} - \sqrt{\dfrac{44}{x^2}}$

41. $\sqrt[3]{\dfrac{16}{27}} - \dfrac{\sqrt[3]{54}}{6}$

42. $\dfrac{\sqrt[3]{3}}{10} + \sqrt[3]{\dfrac{24}{125}}$

43. $-\dfrac{\sqrt[3]{2x^4}}{9} + \sqrt[3]{\dfrac{250x^4}{27}}$

44. $\dfrac{\sqrt[3]{y^5}}{8} + \dfrac{5y\sqrt[3]{y^2}}{4}$

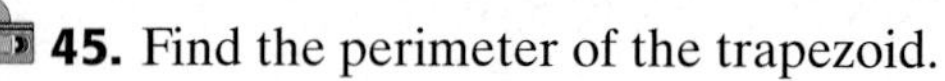

45. Find the perimeter of the trapezoid.

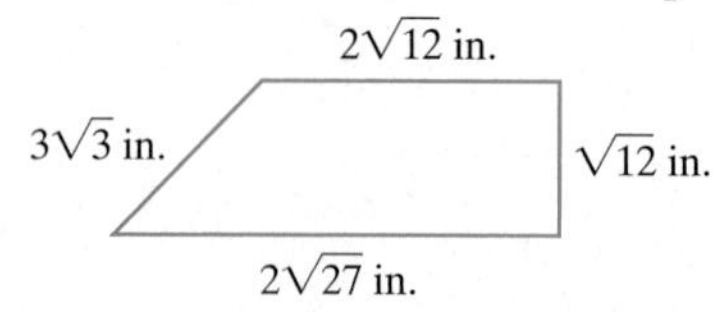

△ **46.** Find the perimeter of the triangle.

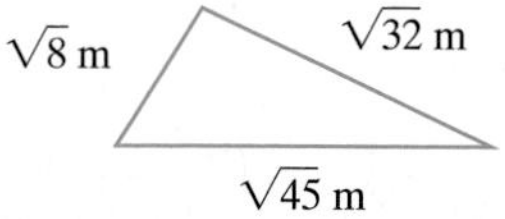

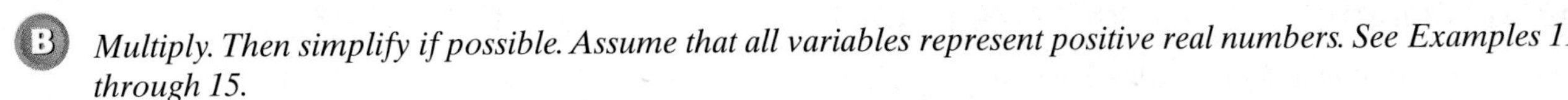

B *Multiply. Then simplify if possible. Assume that all variables represent positive real numbers. See Examples 11 through 15.*

47. $\sqrt{7}(\sqrt{5} + \sqrt{3})$

48. $\sqrt{5}(\sqrt{15} - \sqrt{35})$

49. $(\sqrt{5} - \sqrt{2})^2$

50. $(3x - \sqrt{2})(3x - \sqrt{2})$

51. $\sqrt{3x}(\sqrt{3} - \sqrt{x})$

52. $\sqrt{5y}(\sqrt{y} + \sqrt{5})$

53. $(2\sqrt{x} - 5)(3\sqrt{x} + 1)$

54. $(8\sqrt{y} + z)(4\sqrt{y} - 1)$

55. $(\sqrt[3]{a} - 4)(\sqrt[3]{a} + 5)$

56. $(\sqrt[3]{a} + 2)(\sqrt[3]{a} + 7)$

57. $6(\sqrt{2} - 2)$

58. $\sqrt{5}(6 - \sqrt{5})$

59. $\sqrt{2}(\sqrt{2} + x\sqrt{6})$

60. $\sqrt{3}(\sqrt{3} - 2\sqrt{5x})$

61. $(2\sqrt{7} + 3\sqrt{5})(\sqrt{7} - 2\sqrt{5})$

62. $(\sqrt{6} - 4\sqrt{2})(3\sqrt{6} + 1)$

63. $(\sqrt{x} - y)(\sqrt{x} + y)$

64. $(3\sqrt{x} + 2)(\sqrt{3x} - 2)$

65. $(\sqrt{3} + x)^2$

66. $(\sqrt{y} - 3x)^2$

67. $(\sqrt{5x} - 3\sqrt{2})(\sqrt{5x} - 3\sqrt{3})$

68. $(5\sqrt{3x} - \sqrt{y})(4\sqrt{x} + 1)$

69. $(\sqrt[3]{4} + 2)(\sqrt[3]{2} - 1)$

70. $(\sqrt[3]{3} + \sqrt[3]{2})(\sqrt[3]{9} - \sqrt[3]{4})$

71. $(\sqrt[3]{x} + 1)(\sqrt[3]{x} - 4\sqrt{x} + 7)$

72. $(\sqrt[3]{3x} + 3)(\sqrt[3]{2x} - 3x - 1)$

73. $(\sqrt{x - 1} + 5)^2$

74. $(\sqrt{3x + 1} + 2)^2$

75. $(\sqrt{2x + 5} - 1)^2$

76. $(\sqrt{x - 6} - 7)^2$

Review and Preview

Factor each numerator and denominator. Then simplify if possible. See Section 6.1.

77. $\dfrac{2x - 14}{2}$

78. $\dfrac{8x - 24y}{4}$

79. $\dfrac{7x - 7y}{x^2 - y^2}$

80. $\dfrac{x^3 - 8}{4x - 8}$

81. $\dfrac{6a^2b - 9ab}{3ab}$

82. $\dfrac{14r - 28r^2s^2}{7rs}$

83. $\dfrac{-4 + 2\sqrt{3}}{6}$

84. $\dfrac{-5 + 10\sqrt{7}}{5}$

Combining Concepts

△ **85.** Find the perimeter and area of the rectangle.

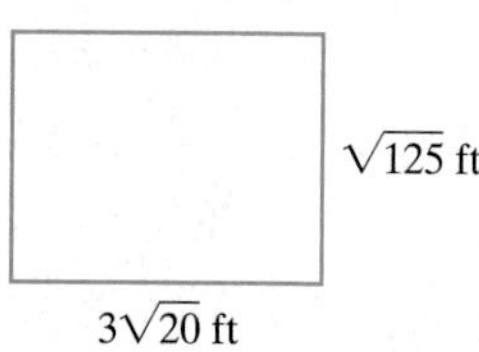

△ **86.** Find the area and perimeter of the trapezoid. (*Hint:* The area of a trapezoid is the product of half the height $6\sqrt{3}$ meters and the sum of the bases $2\sqrt{63}$ and $7\sqrt{7}$ meters.)

$2\sqrt{63}$ m

$2\sqrt{27}$ m $\quad$ $6\sqrt{3}$ m

$7\sqrt{7}$ m

87. **a.** Add: $\sqrt{3} + \sqrt{3}$

b. Multiply: $\sqrt{3} \cdot \sqrt{3}$

c. Describe the differences in parts a and b.

88. Multiply: $(\sqrt{2} + \sqrt{3} - 1)^2$

9.5 Rationalizing Numerators and Denominators of Radical Expressions

OBJECTIVES

A Rationalize denominators.

B Rationalize denominators having two terms.

C Rationalize numerators.

A Rationalizing Denominators

Often in mathematics it is helpful to write a radical expression such as $\frac{\sqrt{3}}{\sqrt{2}}$ either without a radical in the denominator or without a radical in the numerator. The process of writing this expression as an equivalent expression but without a radical in the denominator is called **rationalizing the denominator**. To rationalize the denominator of $\frac{\sqrt{3}}{\sqrt{2}}$, we use the fundamental principle of fractions and multiply the numerator and the denominator by $\sqrt{2}$. Recall that this is the same as multiplying by $\frac{\sqrt{2}}{\sqrt{2}}$, which simplifies to 1.

$$\frac{\sqrt{3}}{\sqrt{2}} = \frac{\sqrt{3}\cdot\sqrt{2}}{\sqrt{2}\cdot\sqrt{2}} = \frac{\sqrt{6}}{\sqrt{4}} = \frac{\sqrt{6}}{2}$$

EXAMPLE 1 Rationalize the denominator of $\frac{2}{\sqrt{5}}$.

Solution: To rationalize the denominator, we multiply the numerator and denominator by a factor that makes the radicand in the denominator a perfect square.

$$\frac{2}{\sqrt{5}} = \frac{2\cdot\sqrt{5}}{\sqrt{5}\cdot\sqrt{5}} = \frac{2\sqrt{5}}{5}$$ The denominator is now rationalized.

Practice Problem 1

Rationalize the denominator of $\frac{7}{\sqrt{2}}$.

EXAMPLE 2 Rationalize the denominator of $\frac{2\sqrt{16}}{\sqrt{9x}}$. Assume that x represents a positive real number.

Solution: First we simplify the radicals; then we rationalize the denominator.

$$\frac{2\sqrt{16}}{\sqrt{9x}} = \frac{2(4)}{\sqrt{9}\cdot\sqrt{x}} = \frac{8}{3\sqrt{x}}$$

To rationalize the denominator, we multiply the numerator and the denominator by $\sqrt{x}$.

$$\frac{8}{3\sqrt{x}} = \frac{8\cdot\sqrt{x}}{3\sqrt{x}\cdot\sqrt{x}} = \frac{8\sqrt{x}}{3x}$$

Practice Problem 2

Rationalize the denominator of $\frac{2\sqrt{9}}{\sqrt{16y}}$.

Practice Problem 3

Rationalize the denominator of $\sqrt[3]{\frac{2}{25}}$.

EXAMPLE 3 Rationalize the denominator of $\sqrt[3]{\frac{1}{2}}$.

Solution: $\sqrt[3]{\frac{1}{2}} = \frac{\sqrt[3]{1}}{\sqrt[3]{2}} = \frac{1}{\sqrt[3]{2}}$

Now we rationalize the denominator. Since $\sqrt[3]{2}$ is a cube root, we want to multiply by a value that will make the radicand 2 a perfect cube. If we multiply by $\sqrt[3]{2^2}$, we get $\sqrt[3]{2^3} = 2$. Thus,

$$\frac{1\cdot\sqrt[3]{2^2}}{\sqrt[3]{2}\cdot\sqrt[3]{2^2}} = \frac{\sqrt[3]{4}}{\sqrt[3]{2^3}} = \frac{\sqrt[3]{4}}{2}$$ Multiply numerator and denominator by $\sqrt[3]{2^2}$ and then simplify.

Try the Concept Check in the margin.

Concept Check

Determine by which number both the numerator and denominator should be multiplied to rationalize the denominator of the radical expression.

a. $\frac{1}{\sqrt[3]{7}}$ b. $\frac{1}{\sqrt[4]{8}}$

Answers

1. $\frac{7\sqrt{2}}{2}$, **2.** $\frac{3\sqrt{y}}{2y}$, **3.** $\frac{\sqrt[3]{10}}{5}$

Concept Check: **a.** $\sqrt[3]{7^2}$ or $\sqrt[3]{49}$, **b.** $\sqrt[4]{2}$

Practice Problem 4

Rationalize the denominator of $\sqrt{\frac{5m}{11n}}$.

Assume that all variables represent positive real numbers.

EXAMPLE 4

Rationalize the denominator of $\sqrt{\frac{7x}{3y}}$. Assume that all variables represent positive real numbers.

Solution:

$$\sqrt{\frac{7x}{3y}} = \frac{\sqrt{7x}}{\sqrt{3y}}$$ Use the quotient rule. No radical may be simplified further.

$$= \frac{\sqrt{7x} \cdot \sqrt{3y}}{\sqrt{3y} \cdot \sqrt{3y}}$$ Multiply numerator and denominator by $\sqrt{3y}$ so that the radicand in the denominator is a perfect square.

$$= \frac{\sqrt{21xy}}{3y}$$ Use the product rule in the numerator and denominator. Remember that $\sqrt{3y} \cdot \sqrt{3y} = 3y$.

Practice Problem 5

Rationalize the denominator of $\frac{\sqrt[5]{a^2}}{\sqrt[5]{32b^{12}}}$.

Assume that all variables represent positive real numbers.

EXAMPLE 5

Rationalize the denominator of $\frac{\sqrt[4]{x}}{\sqrt[4]{81y^5}}$. Assume that all variables represent positive real numbers.

Solution: First we simplify each radical if possible.

$$\frac{\sqrt[4]{x}}{\sqrt[4]{81y^5}} = \frac{\sqrt[4]{x}}{\sqrt[4]{81y^4} \cdot \sqrt[4]{y}}$$ Use the product rule in the denominator.

$$= \frac{\sqrt[4]{x}}{3y\sqrt[4]{y}}$$ Write $\sqrt[4]{81y^4}$ as $3y$.

$$= \frac{\sqrt[4]{x} \cdot \sqrt[4]{y^3}}{3y\sqrt[4]{y} \cdot \sqrt[4]{y^3}}$$ Multiply numerator and denominator by $\sqrt[4]{y^3}$ so that the radicand in the denominator is a perfect 4th power.

$$= \frac{\sqrt[4]{xy^3}}{3y\sqrt[4]{y^4}}$$ Use the product rule in the numerator and denominator.

$$= \frac{\sqrt[4]{xy^3}}{3y^2}$$ In the denominator, $\sqrt[4]{y^4} = y$ and $3y \cdot y = 3y^2$.

B Rationalizing Denominators Having Two Terms

Remember the product of the sum and difference of two terms?

$$(a + b)(a - b) = a^2 - b^2$$

These two expressions are called **conjugates** of each other.

To rationalize a denominator that is a sum or difference of two terms, we use conjugates. To see how and why this works, let's rationalize the denominator of the expression $\frac{5}{\sqrt{3} - 2}$. To do so, we multiply both the numerator and the denominator by $\sqrt{3} + 2$, the *conjugate*, of the denominator $\sqrt{3} - 2$ and see what happens.

$$\frac{5}{\sqrt{3} - 2} = \frac{5(\sqrt{3} + 2)}{(\sqrt{3} - 2)(\sqrt{3} + 2)}$$

$$= \frac{5(\sqrt{3} + 2)}{(\sqrt{3})^2 - 2^2}$$ Multiply the sum and difference of two terms: $(a + b)(a - b) = a^2 - b^2$.

Answers

4. $\frac{\sqrt{55mn}}{11n}$, **5.** $\frac{\sqrt[5]{a^2b^3}}{2b^3}$

$$= \frac{5(\sqrt{3} + 2)}{3 - 4}$$

$$= \frac{5(\sqrt{3} + 2)}{-1}$$

$$= -5(\sqrt{3} + 2) \quad \text{or} \quad -5\sqrt{3} - 10$$

Notice in the denominator that the product of $(\sqrt{3} - 2)$ and its conjugate, $(\sqrt{3} + 2)$, is -1. In general, the product of an expression and its conjugate will contain no radical terms. This is why, when rationalizing a denominator or a numerator containing two terms, we multiply by its conjugate. Examples of conjugates are

$$\sqrt{a} - \sqrt{b} \quad \text{and} \quad \sqrt{a} + \sqrt{b}$$

$$x + \sqrt{y} \quad \text{and} \quad x - \sqrt{y}$$

EXAMPLE 6 Rationalize the denominator of $\frac{2}{3\sqrt{2} + 4}$.

Solution: We multiply the numerator and the denominator by the conjugate of $3\sqrt{2} + 4$.

$$\frac{2}{3\sqrt{2} + 4} = \frac{2(3\sqrt{2} - 4)}{(3\sqrt{2} + 4)(3\sqrt{2} - 4)}$$

$$= \frac{2(3\sqrt{2} - 4)}{(3\sqrt{2})^2 - 4^2}$$ Multiply the sum and difference of two terms: $(a + b)(a - b) = a^2 - b^2$.

$$= \frac{2(3\sqrt{2} - 4)}{18 - 16}$$ Write $(3\sqrt{2})^2$ as $9 \cdot 2$ or 18 and 4^2 as 16.

$$= \frac{2(3\sqrt{2} - 4)}{2} \quad \text{or} \quad 3\sqrt{2} - 4$$

Practice Problem 6

Rationalize the denominator of $\frac{3}{2\sqrt{5} + 1}$.

As we saw in Example 6, it is often helpful to leave a numerator in factored form to help determine whether the expression can be simplified.

EXAMPLE 7 Rationalize the denominator of $\frac{\sqrt{6} + 2}{\sqrt{5} - \sqrt{3}}$.

Solution: We multiply the numerator and the denominator by the conjugate of $\sqrt{5} - \sqrt{3}$.

$$\frac{\sqrt{6} + 2}{\sqrt{5} - \sqrt{3}} = \frac{(\sqrt{6} + 2)(\sqrt{5} + \sqrt{3})}{(\sqrt{5} - \sqrt{3})(\sqrt{5} + \sqrt{3})}$$

$$= \frac{\sqrt{6}\sqrt{5} + \sqrt{6}\sqrt{3} + 2\sqrt{5} + 2\sqrt{3}}{(\sqrt{5})^2 - (\sqrt{3})^2}$$

$$= \frac{\sqrt{30} + \sqrt{18} + 2\sqrt{5} + 2\sqrt{3}}{5 - 3}$$

$$= \frac{\sqrt{30} + 3\sqrt{2} + 2\sqrt{5} + 2\sqrt{3}}{2}$$

Practice Problem 7

Rationalize the denominator of $\frac{\sqrt{5} + 3}{\sqrt{3} - \sqrt{2}}$.

Answers

6. $\frac{3(2\sqrt{5} - 1)}{19}$, **7.** $\sqrt{15} + \sqrt{10} + 3\sqrt{3} + 3\sqrt{2}$

Practice Problem 8

Rationalize the denominator of $\frac{3}{2 - \sqrt{x}}$. Assume that all variables represent positive real numbers.

EXAMPLE 8

Rationalize the denominator of $\frac{2\sqrt{m}}{3\sqrt{x} + \sqrt{m}}$. Assume that all variables represent positive real numbers.

Solution: We multiply by the conjugate of $3\sqrt{x} + \sqrt{m}$ to eliminate the radicals from the denominator.

$$\frac{2\sqrt{m}}{3\sqrt{x} + \sqrt{m}} = \frac{2\sqrt{m}(3\sqrt{x} - \sqrt{m})}{(3\sqrt{x} + \sqrt{m})(3\sqrt{x} - \sqrt{m})} = \frac{6\sqrt{mx} - 2m}{(3\sqrt{x})^2 - (\sqrt{m})^2}$$

$$= \frac{6\sqrt{mx} - 2m}{9x - m}$$

C Rationalizing Numerators

As mentioned earlier, it is also often helpful to write an expression such as $\frac{\sqrt{3}}{\sqrt{2}}$ as an equivalent expression without a radical in the numerator. This process is called **rationalizing the numerator.** To rationalize the numerator of $\frac{\sqrt{3}}{\sqrt{2}}$, we multiply the numerator and the denominator by $\sqrt{3}$.

$$\frac{\sqrt{3}}{\sqrt{2}} = \frac{\sqrt{3} \cdot \sqrt{3}}{\sqrt{2} \cdot \sqrt{3}} = \frac{\sqrt{9}}{\sqrt{6}} = \frac{3}{\sqrt{6}}$$

Practice Problem 9

Rationalize the numerator of $\frac{\sqrt{18}}{\sqrt{75}}$.

EXAMPLE 9 Rationalize the numerator of $\frac{\sqrt{7}}{\sqrt{45}}$.

Solution: First we simplify $\sqrt{45}$.

$$\frac{\sqrt{7}}{\sqrt{45}} = \frac{\sqrt{7}}{\sqrt{9 \cdot 5}} = \frac{\sqrt{7}}{3\sqrt{5}}$$

Next we rationalize the numerator by multiplying the numerator and the denominator by $\sqrt{7}$.

$$\frac{\sqrt{7}}{3\sqrt{5}} = \frac{\sqrt{7} \cdot \sqrt{7}}{3\sqrt{5} \cdot \sqrt{7}} = \frac{7}{3\sqrt{5 \cdot 7}} = \frac{7}{3\sqrt{35}}$$

Practice Problem 10

Rationalize the numerator of $\frac{\sqrt[3]{3a}}{\sqrt[3]{7b}}$.

EXAMPLE 10 Rationalize the numerator of $\frac{\sqrt[3]{2x^2}}{\sqrt[3]{5y}}$.

Solution:

$$\frac{\sqrt[3]{2x^2}}{\sqrt[3]{5y}} = \frac{\sqrt[3]{2x^2} \cdot \sqrt[3]{2^2x}}{\sqrt[3]{5y} \cdot \sqrt[3]{2^2x}}$$ Multiply the numerator and denominator by $\sqrt[3]{2^2x}$ so that the radicand in the numerator is a perfect cube.

$$= \frac{\sqrt[3]{2^3x^3}}{\sqrt[3]{5y \cdot 2^2x}}$$ Use the product rule in the numerator and denominator.

$$= \frac{2x}{\sqrt[3]{20xy}}$$ Simplify.

Just as for denominators, to rationalize a numerator that is a sum or difference of two terms, we use conjugates.

Answers

8. $\frac{6 + 3\sqrt{x}}{4 - x}$, **9.** $\frac{6}{5\sqrt{6}}$, **10.** $\frac{3a}{\sqrt[3]{63a^2b}}$

EXAMPLE 11

Rationalize the numerator of $\frac{\sqrt{x}+2}{5}$. Assume that all variables represent positive real numbers.

Solution: We multiply the numerator and the denominator by the conjugate of $\sqrt{x}+2$, the numerator.

$$\frac{\sqrt{x}+2}{5} = \frac{(\sqrt{x}+2)(\sqrt{x}-2)}{5(\sqrt{x}-2)}$$ Multiply by $\sqrt{x}-2$, the conjugate of $\sqrt{x}+2$.

$$= \frac{(\sqrt{x})^2 - 2^2}{5(\sqrt{x}-2)}$$ $(a+b)(a-b) = a^2 - b^2$.

$$= \frac{x-4}{5(\sqrt{x}-2)}$$

Practice Problem 11

Rationalize the numerator of $\frac{\sqrt{x}+5}{3}$.

Assume that all variables represent positive real numbers.

Answer

11. $\frac{x-25}{3(\sqrt{x}-5)}$

STUDY SKILLS REMINDER

How are your homework assignments going?

By now, you should have good homework habits. If not, it's never too late to begin. Why is it so important in mathematics to keep up with homework? You probably now know the answer to that question. You have probably realized by now that many concepts in mathematics build on each other. Your understanding of one chapter in mathematics usually depends on your understanding of the previous chapter's material.

Don't forget that completing your homework assignment involves a lot more than attempting a few of the problems assigned.

To complete a homework assignment, remember these four things:

1. Attempt all of it.
2. Check it.
3. Correct it.
4. If needed, ask questions about it.

Name ______________________ Section __________ Date __________

Mental Math

Find the conjugate of each expression.

1. $\sqrt{2} + x$
2. $\sqrt{3} + y$
3. $5 - \sqrt{a}$
4. $6 - \sqrt{b}$
5. $7\sqrt{4} + 8\sqrt{x}$
6. $9\sqrt{2} - 6\sqrt{y}$

EXERCISE SET 9.5

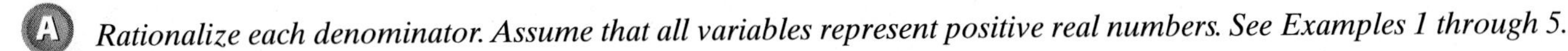

A *Rationalize each denominator. Assume that all variables represent positive real numbers. See Examples 1 through 5.*

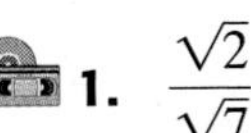

1. $\frac{\sqrt{2}}{\sqrt{7}}$
2. $\frac{\sqrt{3}}{\sqrt{2}}$
3. $\sqrt{\frac{1}{5}}$
4. $\sqrt{\frac{1}{2}}$
5. $\frac{4}{\sqrt[3]{3}}$
6. $\frac{6}{\sqrt[3]{9}}$
7. $\frac{3}{\sqrt{8x}}$
8. $\frac{5}{\sqrt{27a}}$
9. $\frac{3}{\sqrt[3]{4x^2}}$
10. $\frac{5}{\sqrt[3]{3y}}$
11. $\frac{9}{\sqrt{3a}}$
12. $\frac{x}{\sqrt{5}}$
13. $\frac{3}{\sqrt[3]{2}}$
14. $\frac{5}{\sqrt[3]{9}}$
15. $\frac{2\sqrt{3}}{\sqrt{7}}$
16. $\frac{-5\sqrt{2}}{\sqrt{11}}$
17. $\sqrt{\frac{2x}{5y}}$
18. $\sqrt{\frac{13a}{2b}}$
19. $\sqrt[3]{\frac{3}{5}}$
20. $\sqrt[3]{\frac{7}{10}}$
21. $\sqrt{\frac{3x}{50}}$
22. $\sqrt{\frac{11y}{45}}$
23. $\frac{1}{\sqrt{12z}}$
24. $\frac{1}{\sqrt{32x}}$

25. $\dfrac{\sqrt[3]{2y^2}}{\sqrt[3]{9x^2}}$ 26. $\dfrac{\sqrt[3]{3x}}{\sqrt[3]{4y^4}}$ 27. $\sqrt[4]{\dfrac{16}{9x^7}}$ 28. $\sqrt[5]{\dfrac{32}{m^6n^{13}}}$ 29. $\dfrac{5a}{\sqrt[5]{8a^9b^{11}}}$ 30. $\dfrac{9y}{\sqrt[4]{4y^9}}$

B *Rationalize each denominator. Assume that all variables represent positive real numbers. See Examples 6 through 8.*

31. $\dfrac{6}{2 - \sqrt{7}}$ 32. $\dfrac{3}{\sqrt{7} - 4}$ 33. $\dfrac{-7}{\sqrt{x} - 3}$ 34. $\dfrac{-8}{\sqrt{y} + 4}$

35. $\dfrac{\sqrt{2} - \sqrt{3}}{\sqrt{2} + \sqrt{3}}$ 36. $\dfrac{\sqrt{3} + \sqrt{4}}{\sqrt{2} + \sqrt{3}}$ 37. $\dfrac{\sqrt{a} + 1}{2\sqrt{a} - \sqrt{b}}$

38. $\dfrac{2\sqrt{a} - 3}{2\sqrt{a} - \sqrt{b}}$ 39. $\dfrac{8}{1 + \sqrt{10}}$ 40. $\dfrac{-3}{\sqrt{6} - 2}$

41. $\dfrac{\sqrt{x}}{\sqrt{x} + \sqrt{y}}$ 42. $\dfrac{2\sqrt{a}}{2\sqrt{x} - \sqrt{y}}$ 43. $\dfrac{2\sqrt{3} + \sqrt{6}}{4\sqrt{3} - \sqrt{6}}$ 44. $\dfrac{4\sqrt{5} + \sqrt{2}}{2\sqrt{5} - \sqrt{2}}$

C *Rationalize each numerator. Assume that all variables represent positive real numbers. See Examples 9 and 10.*

45. $\sqrt{\dfrac{5}{3}}$ 46. $\sqrt{\dfrac{3}{2}}$ 47. $\sqrt{\dfrac{18}{5}}$ 48. $\sqrt{\dfrac{12}{7}}$ 49. $\dfrac{\sqrt{4x}}{7}$ 50. $\dfrac{\sqrt{3x^5}}{6}$

51. $\frac{\sqrt[3]{5y^2}}{\sqrt[3]{4x}}$

52. $\frac{\sqrt[3]{4x}}{\sqrt[3]{z^4}}$

53. $\sqrt{\frac{2}{5}}$

54. $\sqrt{\frac{3}{7}}$

55. $\frac{\sqrt{2x}}{11}$

56. $\frac{\sqrt{y}}{7}$

57. $\sqrt[3]{\frac{7}{8}}$

58. $\sqrt[3]{\frac{25}{2}}$

59. $\frac{\sqrt[3]{3x^5}}{10}$

60. $\sqrt[3]{\frac{9y}{7}}$

61. $\sqrt{\frac{18x^4y^6}{3z}}$

62. $\sqrt{\frac{8x^5y}{2z}}$

63. When rationalizing the denominator of $\frac{\sqrt{5}}{\sqrt{7}}$, explain why both the numerator and the denominator must be multiplied by $\sqrt{7}$.

64. When rationalizing the numerator of $\frac{\sqrt{5}}{\sqrt{7}}$, explain why both the numerator and the denominator must be multiplied by $\sqrt{5}$.

Rationalize each numerator. Assume that all variables represent positive real numbers. See Example 11.

65. $\frac{2 - \sqrt{11}}{6}$

66. $\frac{\sqrt{15} + 1}{2}$

67. $\frac{2 - \sqrt{7}}{-5}$

68. $\frac{\sqrt{5} + 2}{\sqrt{2}}$

69. $\frac{\sqrt{x} + 3}{\sqrt{x}}$

70. $\frac{5 + \sqrt{2}}{\sqrt{2x}}$

71. $\frac{\sqrt{x} + 1}{\sqrt{x} - 1}$

72. $\frac{\sqrt{x} + \sqrt{y}}{\sqrt{x} - \sqrt{y}}$

Review and Preview

Solve each equation. See Sections 2.4 and 5.6.

73. $2x - 7 = 3(x - 4)$

74. $9x - 4 = 7(x - 2)$

75. $(x - 6)(2x + 1) = 0$

76. $(y + 2)(5y + 4) = 0$

77. $x^2 - 8x = -12$

78. $x^2 = x$

Combining Concepts

△ **79.** The formula of the radius r of a sphere with surface area A is

$$r = \sqrt{\frac{A}{4\pi}}$$

Rationalize the denominator of the radical expression in this formula.

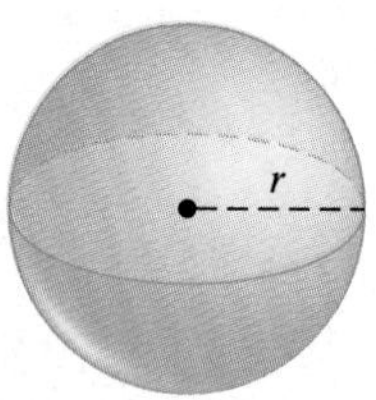

△ **80.** The formula for the radius r of a cone with height 7 centimeters and volume V is

$$r = \sqrt{\frac{3V}{7\pi}}$$

Rationalize the numerator of the radical expression in this formula.

81. Explain why rationalizing the denominator does not change the value of the original expression.

82. Explain why rationalizing the numerator does not change the value of the original expression.

Name ______________________ Section __________ Date __________

Integrated Review—Radicals and Rational Exponents

Throughout this review, assume that all variables represent positive real numbers. Find each root.

1. $\sqrt{81}$ **2.** $\sqrt[3]{-8}$ **3.** $\sqrt[4]{\frac{1}{16}}$ **4.** $\sqrt{x^6}$

5. $\sqrt[3]{y^9}$ **6.** $\sqrt{4y^{10}}$ **7.** $\sqrt[5]{-32y^5}$ **8.** $\sqrt[4]{81b^{12}}$

Use radical notation to rewrite each expression. Simplify if possible.

9. $36^{1/2}$ **10.** $(3y)^{1/4}$ **11.** $64^{-2/3}$ **12.** $(x + 1)^{3/5}$

Use the properties of exponents to simplify each expression. Write with positive exponents.

13. $y^{-1/6} \cdot y^{7/6}$ **14.** $\frac{(2x^{1/3})^4}{x^{5/6}}$ **15.** $\frac{x^{1/4}x^{3/4}}{x^{-1/4}}$ **16.** $4^{1/3} \cdot 4^{2/5}$

Use rational exponents to simplify each radical.

17. $\sqrt[3]{8x^6}$ **18.** $\sqrt[12]{a^9b^6}$

Use rational exponents to write each as a single radical expression.

19. $\sqrt[4]{x} \cdot \sqrt{x}$ **20.** $\sqrt{5} \cdot \sqrt[3]{2}$

Simplify.

21. $\sqrt{40}$ **22.** $\sqrt[4]{16x^7y^{10}}$ **23.** $\sqrt[3]{54x^4}$ **24.** $\sqrt[5]{-64b^{10}}$

Answers

1. ______
2. ______
3. ______
4. ______
5. ______
6. ______
7. ______
8. ______
9. ______
10. ______
11. ______
12. ______
13. ______
14. ______
15. ______
16. ______
17. ______
18. ______
19. ______
20. ______
21. ______
22. ______
23. ______
24. ______

25. ______

26. ______

27. ______

28. ______

29. ______

30. ______

31. ______

32. ______

33. ______

34. ______

35. ______

36. ______

37. ______

38. ______

39. ______

40. ______

Multiply or divide. Then simplify if possible.

25. $\sqrt{5} \cdot \sqrt{x}$

26. $\sqrt[3]{8x} \cdot \sqrt[3]{8x^2}$

27. $\dfrac{\sqrt{98y^6}}{\sqrt{2y}}$

28. $\dfrac{\sqrt[4]{48a^9b^3}}{\sqrt[4]{ab^3}}$

Perform each indicated operation.

29. $\sqrt{20} - \sqrt{75} + 5\sqrt{7}$

30. $\sqrt[3]{54y^4} - y\sqrt[3]{16y}$

31. $\sqrt{3}(\sqrt{5} - \sqrt{2})$

32. $(\sqrt{7} + \sqrt{3})^2$

33. $(2x - \sqrt{5})(2x + \sqrt{5})$

34. $(\sqrt{x+1} - 1)^2$

Rationalize each denominator.

35. $\sqrt{\dfrac{7}{3}}$

36. $\dfrac{5}{\sqrt[3]{2x^2}}$

37. $\dfrac{\sqrt{3} - \sqrt{7}}{2\sqrt{3} + \sqrt{7}}$

Rationalize each numerator.

38. $\sqrt{\dfrac{7}{3}}$

39. $\sqrt[3]{\dfrac{9y}{11}}$

40. $\dfrac{\sqrt{x} - 2}{\sqrt{x}}$

9.6 Radical Equations and Problem Solving

OBJECTIVES

A Solve equations that contain radical expressions.

B Use the Pythagorean theorem to model problems.

A Solving Equations That Contain Radical Expressions

In this section, we present techniques to solve equations containing radical expressions such as

$$\sqrt{2x-3} = 9$$

We use the power rule to help us solve these radical equations.

Power Rule

If both sides of an equation are raised to the same power, *all* solutions of the original equation are *among* the solutions of the new equation.

This property *does not* say that raising both sides of an equation to a power yields an equivalent equation. A solution of the new equation *may* or *may* not be a solution of the original equation. Thus, *each solution of the new equation must be checked* to make sure it is a solution of the original equation. Recall that a proposed solution that is not a solution of the original equation is called an extraneous solution.

Also recall that the *solution set* of an equation is the set of all its solutions. In this section we will write solutions using solution set notation.

EXAMPLE 1 Solve: $\sqrt{2x-3} = 9$

Solution: We use the power rule to square both sides of the equation to eliminate the radical.

$$\begin{aligned} \sqrt{2x-3} &= 9 \\ (\sqrt{2x-3})^2 &= 9 \\ 2x-3 &= 81 \\ 2x &= 84 \\ x &= 42 \end{aligned}$$

Now we check the solution in the original equation.

Check:

$$\begin{aligned} \sqrt{2x-3} &= 9 \\ \sqrt{2(42)-3} &\stackrel{?}{=} 9 \quad \text{Let } x = 42. \\ \sqrt{84-3} &\stackrel{?}{=} 9 \\ \sqrt{81} &\stackrel{?}{=} 9 \\ 9 &= 9 \quad \text{True.} \end{aligned}$$

The solution checks, so we conclude that the solution set is $\{42\}$. ●

Practice Problem 1

Solve: $\sqrt{3x-2} = 5$

To solve a radical equation, first isolate a radical on one side of the equation.

EXAMPLE 2 Solve: $\sqrt{-10x-1} + 3x = 0$

Solution: First we isolate the radical on one side of the equation. To do this, we subtract $3x$ from both sides.

$$\begin{aligned} \sqrt{-10x-1} + 3x &= 0 \\ \sqrt{-10x-1} + 3x - 3x &= 0 - 3x \\ \sqrt{-10x-1} &= -3x \end{aligned}$$

Next we use the power rule to eliminate the radical.

$$\begin{aligned} (\sqrt{-10x-1})^2 &= (-3x)^2 \\ -10x-1 &= 9x^2 \end{aligned}$$

Practice Problem 2

Solve: $\sqrt{9x-2} - 2x = 0$

Answers

1. $\{9\}$, **2.** $\left\{\frac{1}{4}, 2\right\}$

Since this is a quadratic equation, we can set the equation equal to 0 and try to solve by factoring.

$$9x^2 + 10x + 1 = 0$$
$$(9x + 1)(x + 1) = 0 \quad \text{Factor.}$$
$$9x + 1 = 0 \quad \text{or} \quad x + 1 = 0 \quad \text{Set each factor equal to 0.}$$
$$x = -\frac{1}{9} \qquad x = -1$$

Check: Let $x = -\frac{1}{9}$.

$$\sqrt{-10x - 1} + 3x = 0$$
$$\sqrt{-10\left(-\frac{1}{9}\right) - 1} + 3\left(-\frac{1}{9}\right) \stackrel{?}{=} 0$$
$$\sqrt{\frac{10}{9} - \frac{9}{9}} - \frac{3}{9} \stackrel{?}{=} 0$$
$$\sqrt{\frac{1}{9}} - \frac{1}{3} \stackrel{?}{=} 0$$
$$\frac{1}{3} - \frac{1}{3} = 0 \quad \text{True.}$$

Let $x = -1$.

$$\sqrt{-10x - 1} + 3x = 0$$
$$\sqrt{-10(-1) - 1} + 3(-1) \stackrel{?}{=} 0$$
$$\sqrt{10 - 1} - 3 \stackrel{?}{=} 0$$
$$\sqrt{9} - 3 \stackrel{?}{=} 0$$
$$3 - 3 = 0$$

True.

Both solutions check. The solution set is $\left\{-\frac{1}{9}, -1\right\}$. ●

The following steps may be used to solve a radical equation.

Solving a Radical Equation

Step 1. Isolate one radical on one side of the equation.

Step 2. Raise each side of the equation to a power equal to the index of the radical and simplify.

Step 3. If the equation still contains a radical term, repeat Steps 1 and 2. If not, solve the equation.

Step 4. Check all proposed solutions in the original equation.

Practice Problem 3

Solve: $\sqrt[3]{x - 5} + 2 = 1$

EXAMPLE 3 Solve: $\sqrt[3]{x + 1} + 5 = 3$

Solution: First we isolate the radical by subtracting 5 from both sides of the equation.

$$\sqrt[3]{x + 1} + 5 = 3$$
$$\sqrt[3]{x + 1} = -2$$

Next we raise both sides of the equation to the third power to eliminate the radical.

$$\left(\sqrt[3]{x + 1}\right)^3 = (-2)^3$$
$$x + 1 = -8$$
$$x = -9$$

The solution checks in the original equation, so the solution set is $\{-9\}$. ●

Answer

3. $\{4\}$

EXAMPLE 4 Solve: $\sqrt{4 - x} = x - 2$

Solution:

$$\sqrt{4 - x} = x - 2$$
$$(\sqrt{4 - x})^2 = (x - 2)^2$$
$$4 - x = x^2 - 4x + 4$$
$$x^2 - 3x = 0 \quad \text{Write the quadratic equation in standard form.}$$
$$x(x - 3) = 0 \quad \text{Factor.}$$
$$x = 0 \text{ or } x - 3 = 0 \quad \text{Set each factor equal to 0.}$$
$$x = 3$$

Check:

$$\sqrt{4 - x} = x - 2$$
$$\sqrt{4 - 0} \stackrel{?}{=} 0 - 2 \quad \text{Let } x = 0.$$
$$2 = -2 \quad \text{False.}$$

$$\sqrt{4 - x} = x - 2$$
$$\sqrt{4 - 3} \stackrel{?}{=} 3 - 2 \quad \text{Let } x = 3.$$
$$1 = 1 \quad \text{True.}$$

The proposed solution 3 checks, but 0 does not. Since 0 is an extraneous solution, the solution set is $\{3\}$.

Helpful Hint

In Example 4, notice that $(x - 2)^2 = x^2 - 4x + 4$. Make sure binomials are squared correctly.

Try the Concept Check in the margin.

EXAMPLE 5 Solve: $\sqrt{2x + 5} + \sqrt{2x} = 3$

Solution: We get one radical alone by subtracting $\sqrt{2x}$ from both sides.

$$\sqrt{2x + 5} + \sqrt{2x} = 3$$
$$\sqrt{2x + 5} = 3 - \sqrt{2x}$$

Now we use the power rule to begin eliminating the radicals. First we square both sides.

$$(\sqrt{2x + 5})^2 = (3 - \sqrt{2x})^2$$
$$2x + 5 = 9 - 6\sqrt{2x} + 2x \quad \text{Multiply: } (3 - \sqrt{2x})(3 - \sqrt{2x})$$

There is still a radical in the equation, so we get the radical alone again. Then we square both sides.

$$2x + 5 = 9 - 6\sqrt{2x} + 2x$$
$$6\sqrt{2x} = 4 \quad \text{Get the radical alone.}$$
$$(6\sqrt{2x})^2 = 4^2 \quad \text{Square both sides of the equation to eliminate the radical.}$$
$$36(2x) = 16$$
$$72x = 16 \quad \text{Multiply.}$$
$$x = \frac{16}{72} \quad \text{Solve.}$$
$$x = \frac{2}{9} \quad \text{Simplify.}$$

Practice Problem 4

Solve: $\sqrt{9 + x} = x + 3$

Concept Check

How can you immediately tell that the equation $\sqrt{2y + 3} = -4$ has no real solution?

Practice Problem 5

Solve: $\sqrt{3x + 1} + \sqrt{3x} = 2$

Answers

4. $\{0\}$, **5.** $\left\{\frac{3}{16}\right\}$

Concept Check: answers may vary

The proposed solution $\frac{2}{9}$ checks in the original equation. The solution set is $\left\{\frac{2}{9}\right\}$. ●

Helpful Hint

Make sure expressions are squared correctly. In Example 5, we squared $(3 - \sqrt{2x})$ as

$$\begin{aligned}(3 - \sqrt{2x})^2 &= (3 - \sqrt{2x})(3 - \sqrt{2x}) \\ &= 3\cdot 3 - 3\sqrt{2x} - 3\sqrt{2x} + \sqrt{2x}\cdot\sqrt{2x} \\ &= 9 - 6\sqrt{2x} + 2x\end{aligned}$$

Try the Concept Check in the margin.

Concept Check

What is wrong the following solution?

$$\begin{aligned}\sqrt{2x+5} + \sqrt{4-x} &= 8 \\ (\sqrt{2x+5} + \sqrt{4-x})^2 &= 8^2 \\ (2x+5) + (4-x) &= 64 \\ x + 9 &= 64 \\ x &= 55\end{aligned}$$

B Using the Pythagorean Theorem

Recall that the Pythagorean theorem states that in a right triangle, the length of the hypotenuse squared equals the sum of the lengths of each of the legs squared.

Pythagorean Theorem

If a and b are the lengths of the legs of a right triangle and c is the length of the hypotenuse, then $a^2 + b^2 = c^2$.

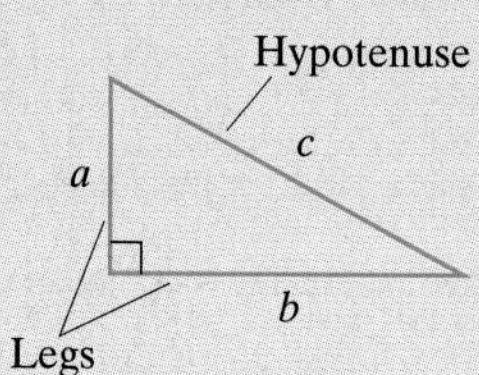

Practice Problem 6

Find the length of the unknown leg of the right triangle.

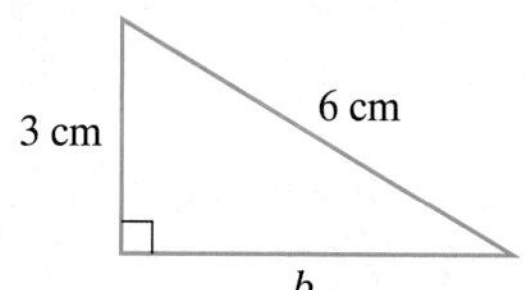

EXAMPLE 6 Find the length of the unknown leg of the right triangle.

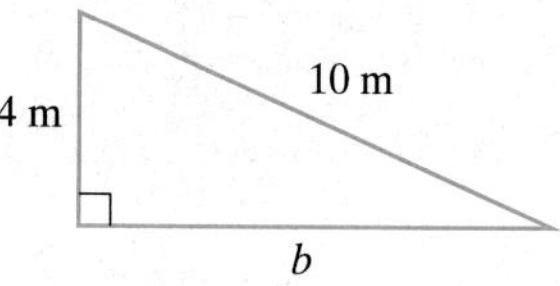

Solution: In the formula $a^2 + b^2 = c^2$, c is the hypotenuse. Here, $c = 10$, the length of the hypotenuse, and $a = 4$. We solve for b. Then $a^2 + b^2 = c^2$ becomes

$$\begin{aligned}4^2 + b^2 &= 10^2 \\ 16 + b^2 &= 100 \\ b^2 &= 84 \quad \text{Subtract 16 from both sides.}\end{aligned}$$

Recall from Section 9.1 our definition of square root that if $b^2 = a$, then b is a square root of a. Since b is a length and thus is positive, we have that

$$b = \sqrt{84} = \sqrt{4\cdot 21} = 2\sqrt{21}$$

The unknown leg of the triangle is exactly $2\sqrt{21}$ meters long. Using a calculator, this is approximately 9.2 meters. ●

Answers

6. $3\sqrt{3}$ cm

Concept Check: answers may vary

△ EXAMPLE 7 Calculating Placement of a Wire

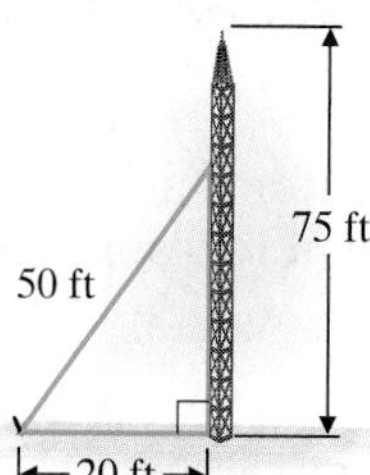

A 50-foot supporting wire is to be attached to a 75-foot antenna. Because of surrounding buildings, sidewalks, and roadways, the wire must be anchored exactly 20 feet from the base of the antenna.

a. How high from the base of the antenna must the wire be attached?

b. Local regulations require that a supporting wire be attached at a height no less than $\frac{3}{5}$ of the total height the antenna. From part (a), have local regulations been met?

Solution:

1. UNDERSTAND. Read and reread the problem. From the diagram we notice that a right triangle is formed with hypotenuse 50 feet and one leg 20 feet. We let $x =$ the height from the base of the antenna to the attached wire.

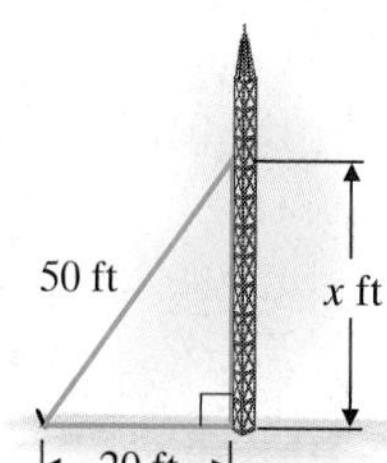

2. TRANSLATE. We'll use the Pythagorean theorem.

$$(a)^2 + (b)^2 = (c)^2$$

$$(20)^2 + x^2 = (50)^2 \quad a = 20, c = 50$$

3. SOLVE.

$$(20)^2 + x^2 = (50)^2$$

$$400 + x^2 = 2500$$

$$x^2 = 2100 \quad \text{Subtract 400 from both sides.}$$

$$x = \sqrt{2100}$$

$$= 10\sqrt{21}$$

4. INTERPRET. *Check* the work and *state* the solution.

a. The wire is attached exactly $10\sqrt{21}$ feet from the base of the pole, or approximately 45.8 feet.

b. The supporting wire must be attached at a height no less than $\frac{3}{5}$ of the total height of the antenna. This height is $\frac{3}{5}$(75 feet), or 45 feet.

Since we know from part (a) that the wire is to be attached at a height of approximately 45.8 feet, local regulations have been met. ●

Practice Problem 7 △

A furniture upholsterer wishes to cut a strip from a piece of fabric that is 45 inches by 45 inches. The strip must be cut on the bias of the fabric. What is the longest strip that can be cut? Give an exact answer and a two-decimal-place approximation.

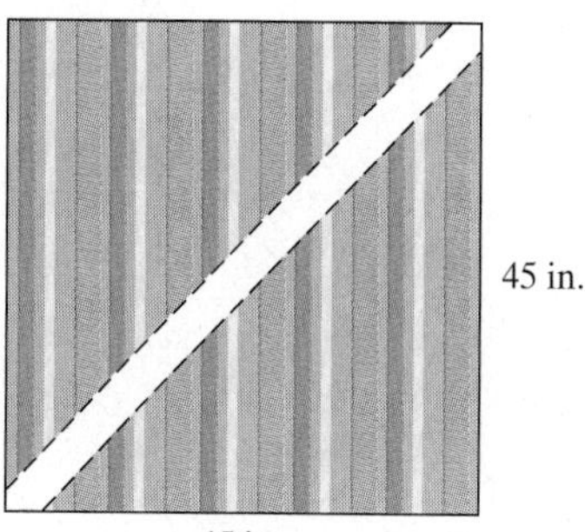

Answer

7. $45\sqrt{2}$ in. $\approx$ 63.64 in.

GRAPHING CALCULATOR EXPLORATIONS

We can use a graphing calculator to solve radical equations. For example, to use a graphing calculator to approximate the solutions of the equation solved in Example 4, we graph the following:

$$Y_1 = \sqrt{4 - x} \quad \text{and} \quad Y_2 = x - 2$$

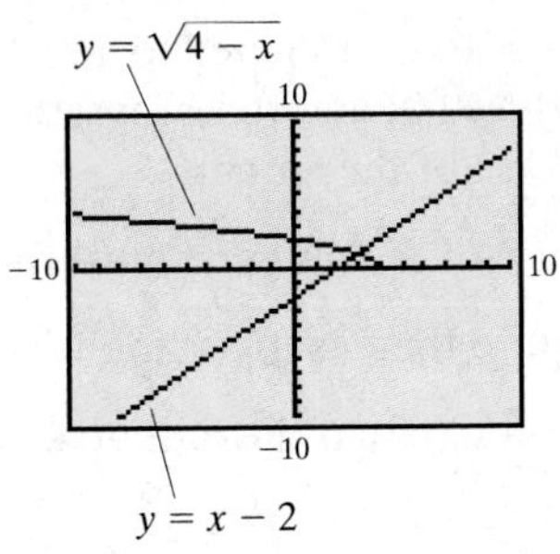

The x-value of the point of intersection is the solution. Use the INTERSECT feature or the ZOOM and TRACE features of your graphing calculator to see that the solution is 3.

Use a graphing calculator to solve each radical equation. Round all solutions to the nearest hundredth.

1. $\sqrt{x + 7} = x$

2. $\sqrt{3x + 5} = 2x$

3. $\sqrt{2x + 1} = \sqrt{2x + 2}$

4. $\sqrt{10x - 1} = \sqrt{-10x + 10} - 1$

5. $1.2x = \sqrt{3.1x + 5}$

6. $\sqrt{1.9x^2 - 2.2} = -0.8x + 3$

Name ______________________ Section __________ Date __________

EXERCISE SET 9.6

 *Solve. See Examples 1 and 2.*

1. $\sqrt{2x} = 4$ **2.** $\sqrt{3x} = 3$ **3.** $\sqrt{x - 3} = 2$ **4.** $\sqrt{x + 1} = 5$

5. $\sqrt{2x} = -4$ **6.** $\sqrt{5x} = -5$ **7.** $\sqrt{4x - 3} - 5 = 0$ **8.** $\sqrt{x - 3} - 1 = 0$

9. $\sqrt{2x - 3} - 2 = 1$ **10.** $\sqrt{3x + 3} - 4 = 8$

Solve. See Example 3.

11. $\sqrt[3]{6x} = -3$ **12.** $\sqrt[3]{4x} = -2$ **13.** $\sqrt[3]{x - 2} - 3 = 0$ **14.** $\sqrt[3]{2x - 6} - 4 = 0$

Solve. See Examples 4 and 5.

15. $\sqrt{13 - x} = x - 1$ **16.** $\sqrt{2x - 3} = 3 - x$ **17.** $x - \sqrt{4 - 3x} = -8$

18. $2x + \sqrt{x + 1} = 8$ **19.** $\sqrt{y + 5} = 2 - \sqrt{y - 4}$ **20.** $\sqrt{x + 3} + \sqrt{x - 5} = 3$

21. $\sqrt{x - 3} + \sqrt{x + 2} = 5$ **22.** $\sqrt{2x - 4} - \sqrt{3x + 4} = -2$

Solve. See Examples 1 through 5.

23. $\sqrt{3x - 2} = 5$ **24.** $\sqrt{5x - 4} = 9$ **25.** $-\sqrt{2x} + 4 = -6$ **26.** $-\sqrt{3x + 9} = -12$

27. $\sqrt{3x + 1} + 2 = 0$ **28.** $\sqrt{3x + 1} - 2 = 0$ **29.** $\sqrt[4]{4x + 1} - 2 = 0$ **30.** $\sqrt[4]{2x - 9} - 3 = 0$

31. $\sqrt{4x - 3} = 5$ **32.** $\sqrt{3x + 9} = 12$ **33.** $\sqrt[3]{6x - 3} - 3 = 0$ **34.** $\sqrt[3]{3x} + 4 = 7$

35. $\sqrt[3]{2x - 3} - 2 = -5$ **36.** $\sqrt[3]{x - 4} - 5 = -7$ **37.** $\sqrt{x + 4} = \sqrt{2x - 5}$

38. $\sqrt{3y + 6} = \sqrt{7y - 6}$ **39.** $x - \sqrt{1 - x} = -5$ **40.** $x - \sqrt{x - 2} = 4$

41. $\sqrt[3]{-6x - 1} = \sqrt[3]{-2x - 5}$ **42.** $x + \sqrt{x + 5} = 7$ **43.** $\sqrt{5x - 1} - \sqrt{x} + 2 = 3$

44. $\sqrt{2x - 1} - 4 = -\sqrt{x - 4}$ **45.** $\sqrt{2x - 1} = \sqrt{1 - 2x}$ **46.** $\sqrt{7x - 4} = \sqrt{4 - 7x}$

47. $\sqrt{3x + 4} - 1 = \sqrt{2x + 1}$ **48.** $\sqrt{x - 2} + 3 = \sqrt{4x + 1}$

49. $\sqrt{y + 3} - \sqrt{y - 3} = 1$ **50.** $\sqrt{x + 1} - \sqrt{x - 1} = 2$

B *Find the length of the unknown side of each triangle. See Example 6.*

△ **51.**

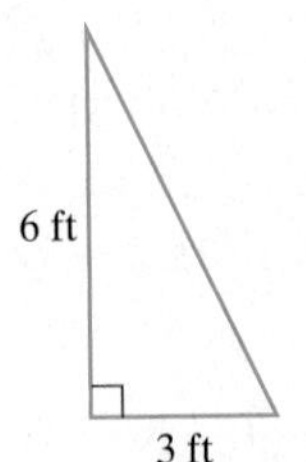

△ **52**

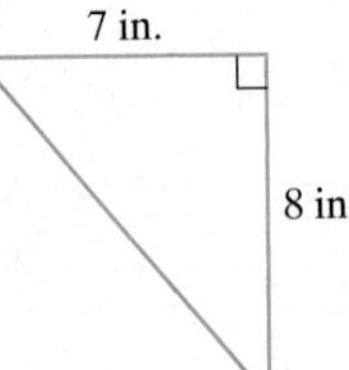

△ **53.**

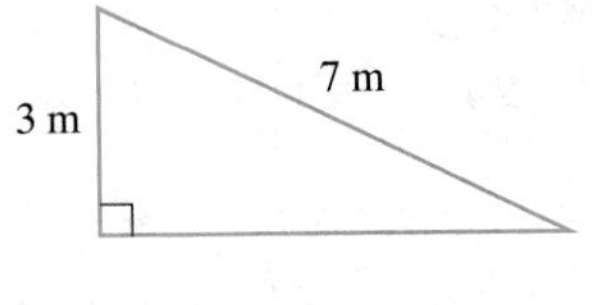

△ **54.**

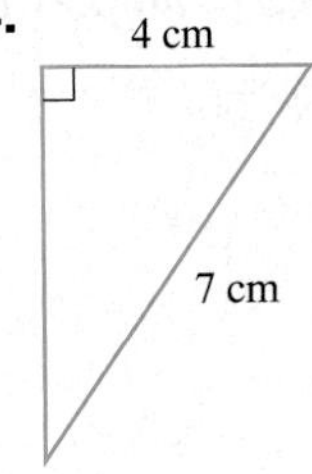

Find the length of the unknown side of each triangle. Give the exact length and a one-decimal-place approximation. See Example 6.

55. △

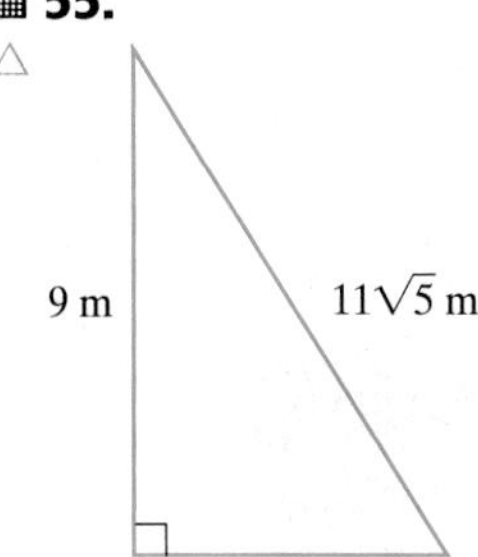

56. △

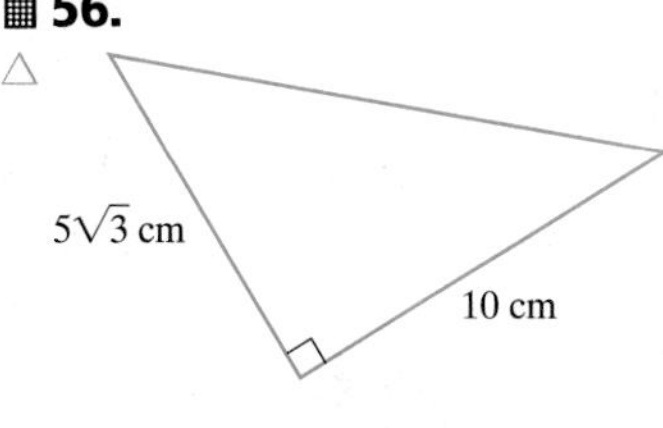

57. △

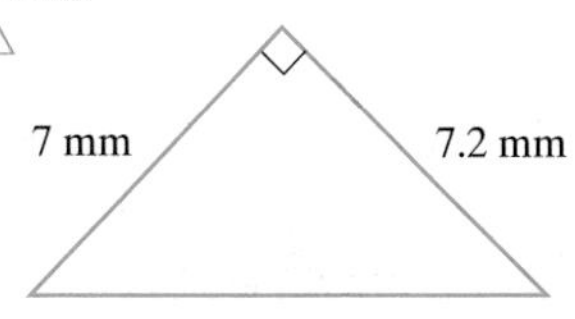

58. △

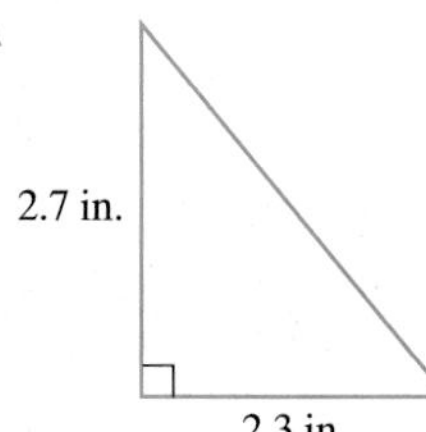

Solve. Give exact answers and two-decimal-place approximations where appropriate. See Example 7.

59. △ A wire is needed to support a vertical pole 15 feet high. The cable will be anchored to a stake 8 feet from the base of the pole. How much cable is needed?

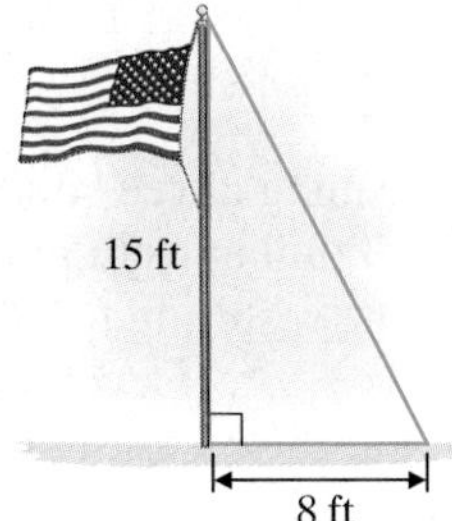

60. △ The tallest structure in the United States is a TV tower in Blanchard, North Dakota. Its height is 2063 feet. A 2382-foot length of wire is to be used as a guy wire attached to the top of the tower. Approximate to the nearest foot how far from the base of the tower the guy wire must be anchored. (*Source:* U.S. Geological Survey)

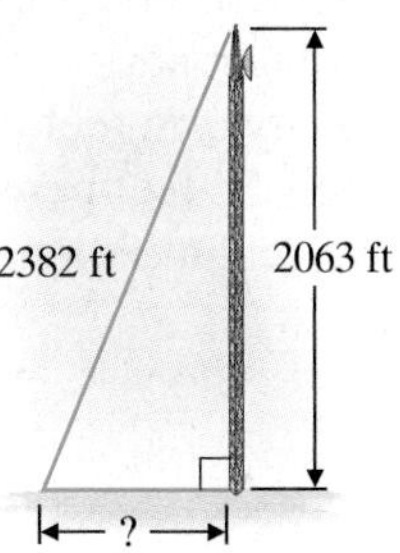

△ **61.** A spotlight is mounted on the eaves of a house 12 feet above the ground. A flower bed runs between the house and the sidewalk, so the closest the ladder can be placed to the house is 5 feet. How long a ladder is needed so that an electrician can reach the place where the light is mounted?

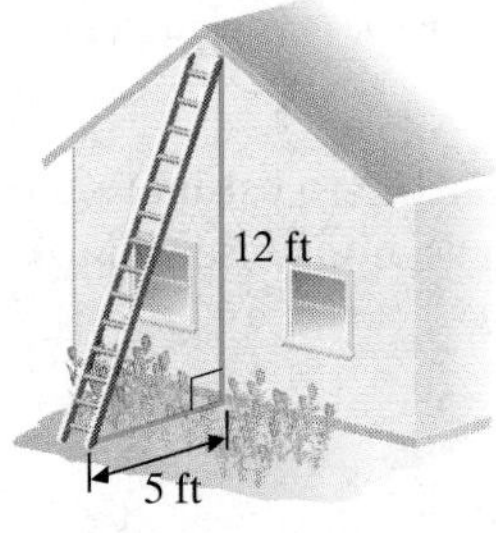

62. △ A wire is to be attached to support a telephone pole. Because of surrounding buildings, sidewalks, and roadway, the wire must be anchored exactly 15 feet from the base of the pole. Telephone company workers have only 30 feet of cable, and 2 feet of that must be used to attach the cable to the pole and to the stake on the ground. How high from the base of the pole can the wire be attached?

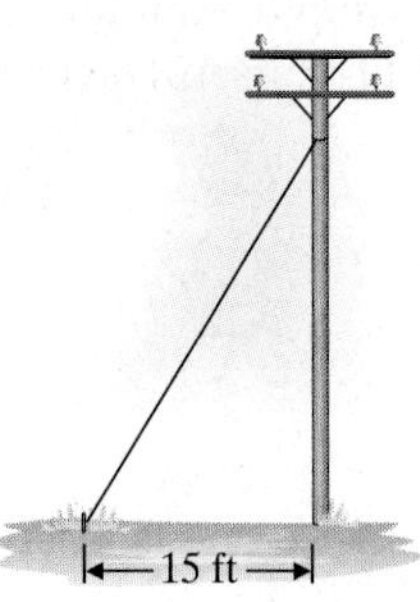

△ **63.** The radius of the moon is 1080 miles. Use the formula for the radius r of a sphere given its surface area A,

$$r = \sqrt{\frac{A}{4\pi}}$$

to find the surface area of the moon. Round to the nearest square mile. (*Source*: National Space Science Data Center)

64. Police departments find it very useful to be able to approximate driving speeds in skidding accidents. If the road surface is wet concrete, the function $S(x) = \sqrt{10.5x}$ is used, where $S(x)$ is the speed of the car in miles per hour and x is the distance skidded in feet. Find how fast a car was moving if it skidded 280 feet on wet concrete.

65. The formula $v = \sqrt{2gh}$ relates the velocity v, in feet per second, of an object after it falls h feet accelerated by gravity g, in feet per second squared. If g is approximately 32 feet per second squared, find how far an object has fallen if its velocity is 80 feet per second.

66. Two tractors are pulling a tree stump from a field. If two forces A and B pull at right angles (90°) to each other, the size of the resulting force R is given by the formula $R = \sqrt{A^2 + B^2}$. If tractor A is exerting 600 pounds of force and the resulting force is 850 pounds, find how much force tractor B is exerting.

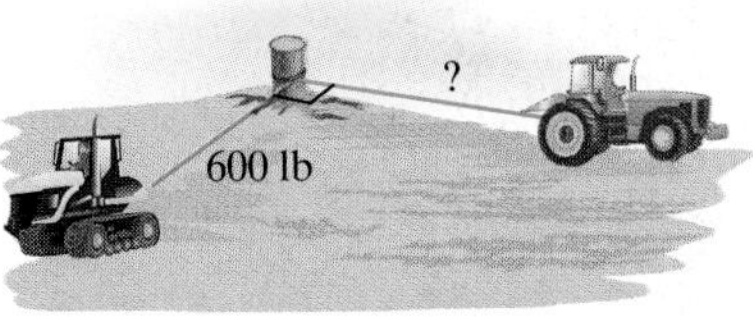

In psychology, it has been suggested that the number S of nonsense syllables that a person can repeat consecutively depends on his or her IQ score I according to the equation $S = 2\sqrt{I} - 9$.

67. Use this relationship to estimate the IQ of a person who can repeat 11 nonsense syllables consecutively.

68. Use this relationship to estimate the IQ of a person who can repeat 15 nonsense syllables consecutively.

The **period** *of a pendulum is the time it takes for the pendulum to make one full back-and-forth swing. The period of a pendulum depends on the length of the pendulum. The formula for the period P, in seconds, is* $P = 2\pi\sqrt{\dfrac{l}{32}}$ *where l is the length of the pendulum in feet. Use this formula for Exercises 69 through 74.*

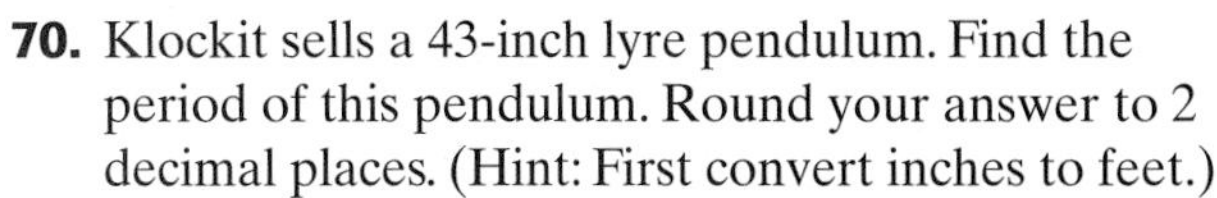

69. Find the period of a pendulum whose length is 2 feet. Give an exact answer and a two-decimal-place approximation.

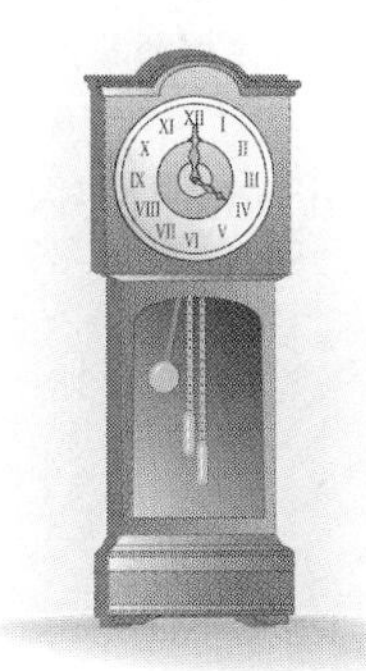

70. Klockit sells a 43-inch lyre pendulum. Find the period of this pendulum. Round your answer to 2 decimal places. (Hint: First convert inches to feet.)

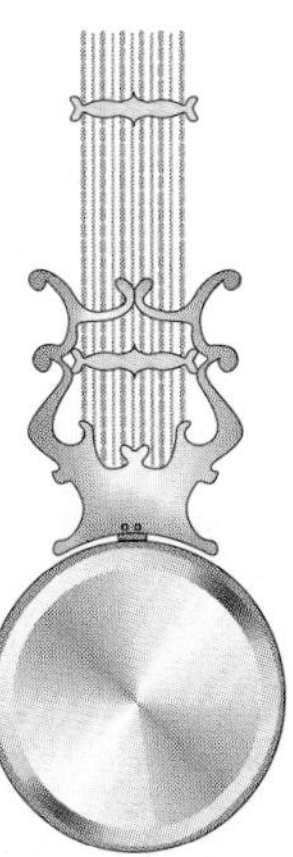

71. Find the length of a pendulum whose period is 4 seconds. Round your answer to 2 decimal places.

72. Find the length of a pendulum whose period is 3 seconds. Round your answer to 2 decimal places.

73. Study the relationship between period and pendulum length in Exercises 69 through 72 and make a conjecture about this relationship.

74. Galileo experimented with pendulums. He supposedly made conjectures about pendulums of equal length with different bob weights. Try this experiment. Make two pendulums 3 feet long. Attach a heavy weight (lead) to one and a light weight (a cork) to the other. Pull both pendulums back the same angle measure and release. Make a conjecture from your observations. (There is more about pendulums in the Chapter 9 Activity.)

If the three lengths of the sides of a triangle are known, Heron's formula can be used to find its area. If a, b, and c are the three lengths of the sides, Heron's formula *for area is:*

$$A = \sqrt{s(s-a)(s-b)(s-c)}$$

where s is half the perimeter of the triangle, or $s = \dfrac{1}{2}(a+b+c)$. *Use this formula to find the area of each triangle. Give an exact answer and then a 2-decimal place approximation.*

△ **75.**

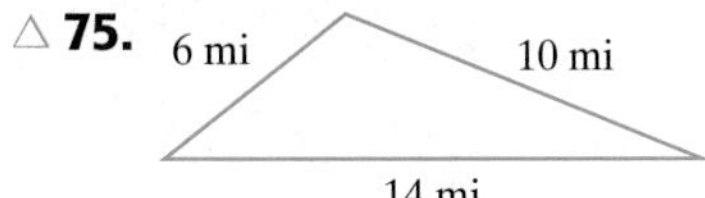

△ **76.**

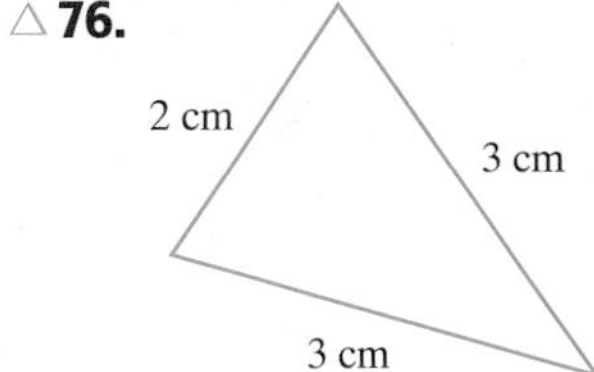

77. Describe when Heron's formula might be useful.

78. In your own words, explain why you think S in *Heron's formula* is called the *semiperimeter.*

The maximum distance $D(h)$ in kilometers that a person can see from a height h kilometers above the ground is given by the function $D(h) = 111.7\sqrt{h}$. Use this function for Exercises 79 and 80. Round your answers to two decimal places.

79. Find the height that would allow a person to see 80 kilometers.

80. Find the height that would allow a person to see 40 kilometers.

Review and Preview

Simplify. See Section 6.7.

81. $\dfrac{\frac{x}{6}}{\frac{2x}{3}+\frac{1}{2}}$

82. $\dfrac{\frac{1}{y}+\frac{4}{5}}{\frac{-3}{20}}$

83. $\dfrac{\frac{z}{5}+\frac{1}{10}}{\frac{z}{20}-\frac{z}{5}}$

84. $\dfrac{\frac{1}{y}+\frac{1}{x}}{\frac{1}{y}-\frac{1}{x}}$

Combining Concepts

85. Solve: $\sqrt{\sqrt{x+3}+\sqrt{x}} = \sqrt{3}$

86. Explain why proposed solutions of radical equations must be checked.

87. The cost $C(x)$ in dollars per day to operate a small delivery service is given by $C(x) = 80\sqrt[3]{x} + 500$, where x is the number of deliveries per day. In July, the manager decides that it is necessary to keep delivery costs below \$1620.00. Find the greatest number of deliveries this company can make per day and still keep overhead below \$1620.00.

88. Consider the equations $\sqrt{2x} = 4$ and $\sqrt[3]{2x} = 4$.
a. Explain the difference in solving these equations.
b. Explain the similarity in solving these equations.

9.7 Complex Numbers

OBJECTIVES

- A Write square roots of negative numbers in the form *bi*.
- B Add or subtract complex numbers.
- C Multiply complex numbers.
- D Divide complex numbers.
- E Raise *i* to powers.

A Writing Numbers in the Form *bi*

Our work with radical expressions has excluded expressions such as $\sqrt{-16}$ because $\sqrt{-16}$ is not a real number; there is no real number whose square is -16. In this section, we discuss a number system that includes roots of negative numbers. This number system is the **complex number system**, and it includes the set of real numbers as a subset. The complex number system allows us to solve equations such as $x^2 + 1 = 0$ that have no real number solutions. The set of complex numbers includes the *imaginary unit*.

Imaginary Unit

The **imaginary unit**, written i, is the number whose square is -1. That is,

$$i^2 = -1 \quad \text{and} \quad i = \sqrt{-1}$$

To write the square root of a negative number in terms of i, we use the property that if a is a positive number, then

$$\begin{aligned}\sqrt{-a} &= \sqrt{-1}\cdot\sqrt{a}\\ &= i\cdot\sqrt{a}\end{aligned}$$

Using i, we can write $\sqrt{-16}$ as

$$\sqrt{-16} = \sqrt{-1\cdot 16} = \sqrt{-1}\cdot\sqrt{16} = i\cdot 4 \text{ or } 4i$$

EXAMPLES Write using i notation.

1. $\sqrt{-36} = \sqrt{-1\cdot 36} = \sqrt{-1}\cdot\sqrt{36} = i\cdot 6 \text{ or } 6i$
2. $\sqrt{-5} = \sqrt{-1(5)} = \sqrt{-1}\cdot\sqrt{5} = i\sqrt{5}.$
3. $-\sqrt{-20} = -\sqrt{-1\cdot 20} = -\sqrt{-1}\cdot\sqrt{4\cdot 5} = -i\cdot 2\sqrt{5} = -2i\sqrt{5}$

Helpful Hint

Since $\sqrt{5}i$ can easily be confused with $\sqrt{5i}$, we write $\sqrt{5}i$ as $i\sqrt{5}$.

Practice Problems 1–3

Write using i notation.

1. $\sqrt{-25}$
2. $\sqrt{-3}$
3. $-\sqrt{-50}$

The product rule for radicals does not necessarily hold true for imaginary numbers. *To multiply square roots of negative numbers, first we write each number in terms of the imaginary unit i.* For example, to multiply $\sqrt{-4}$ and $\sqrt{-9}$, we first write each number in the form bi:

$$\sqrt{-4}\cdot\sqrt{-9} = 2i(3i) = 6i^2 = 6(-1) = -6 \quad \text{Correct.}$$

Make sure you notice that the product rule does not work for this example. In other words, $\sqrt{-4}\cdot\sqrt{-9} = \sqrt{(-4)(-9)} = \sqrt{36} = 6$ is Incorrect!

EXAMPLES Multiply or divide as indicated.

4. $\sqrt{-3}\cdot\sqrt{-5} = i\sqrt{3}(i\sqrt{5}) = i^2\sqrt{15} = -1\sqrt{15} = -\sqrt{15}$
5. $\sqrt{-36}\cdot\sqrt{-1} = 6i(i) = 6i^2 = 6(-1) = -6$
6. $\sqrt{8}\cdot\sqrt{-2} = 2\sqrt{2}(i\sqrt{2}) = 2i(\sqrt{2}\,\sqrt{2}) = 2i(2) = 4i$
7. $\dfrac{\sqrt{-125}}{\sqrt{5}} = \dfrac{i\sqrt{125}}{\sqrt{5}} = i\sqrt{25} = 5i$

Practice Problems 4–7

Multiply or divide as indicated.

4. $\sqrt{-2}\cdot\sqrt{-7}$
5. $\sqrt{-25}\cdot\sqrt{-1}$
6. $\sqrt{27}\cdot\sqrt{-3}$
7. $\dfrac{\sqrt{-8}}{\sqrt{2}}$

Answers

1. $5i$, **2.** $i\sqrt{3}$, **3.** $-5i\sqrt{2}$, **4.** $-\sqrt{14}$, **5.** -5, **6.** $9i$, **7.** $2i$

Now that we have practiced working with the imaginary unit, we define *complex numbers.*

Complex Numbers

A **complex number** is a number that can be written in the form $a + bi$, where a and b are real numbers.

Notice that the set of real numbers is a subset of the complex numbers since any real number can be written in the form of a complex number. For example,

$$16 = 16 + 0i$$

In general, a complex number $a + bi$ is a real number if $b = 0$. Also, a complex number is called an **imaginary number** if $a = 0$. For example,

$$3i = 0 + 3i \qquad \text{and} \qquad i\sqrt{7} = 0 + i\sqrt{7}$$

are imaginary numbers.

The following diagram shows the relationship between complex numbers and their subsets.

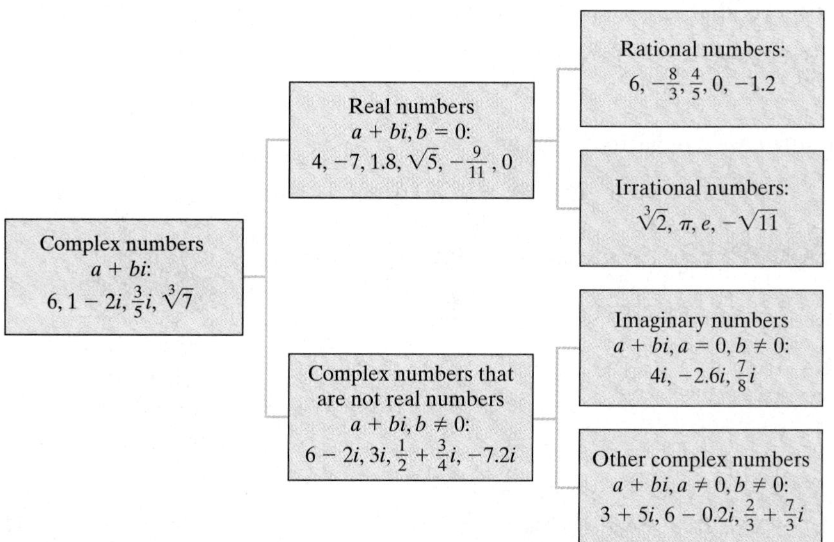

Try the Concept Check in the margin.

Concept Check

True or false? Every complex number is also a real number.

B Adding or Subtracting Complex Numbers

Two complex numbers $a + bi$ and $c + di$ are equal if and only if $a = c$ and $b = d$. Complex numbers can be added or subtracted by adding or subtracting their real parts and then adding or subtracting their imaginary parts.

Sum or Difference of Complex Numbers

If $a + bi$ and $c + di$ are complex numbers, then their sum is

$$(a + bi) + (c + di) = (a + c) + (b + d)i$$

Their difference is

$$(a + bi) - (c + di) = a + bi - c - di = (a - c) + (b - d)i$$

Answer

Concept Check: false

EXAMPLES Add or subtract as indicated.

8. $(2 + 3i) + (-3 + 2i) = (2 - 3) + (3 + 2)i = -1 + 5i$

9. $5i - (1 - i) = 5i - 1 + i$
$= -1 + (5 + 1)i$
$= -1 + 6i$

10. $(-3 - 7i) - (-6) = -3 - 7i + 6$
$= (-3 + 6) - 7i$
$= 3 - 7i$

Practice Problems 8–10

Add or subtract as indicated.

8. $(5 + 2i) + (4 - 3i)$
9. $6i - (2 - i)$
10. $(-2 - 4i) - (-3)$

C Multiplying Complex Numbers

To multiply two complex numbers of the form $a + bi$, we multiply as though they were binomials. Then we use the relationship $i^2 = -1$ to simplify.

EXAMPLES Multiply.

11. $-7i \cdot 3i = -21i^2$
$= -21(-1)$ Replace i^2 with -1.
$= 21$

12. $3i(2 - i) = 3i \cdot 2 - 3i \cdot i$ Use the distributive property.
$= 6i - 3i^2$ Multiply.
$= 6i - 3(-1)$ Replace i^2 with -1.
$= 6i + 3$
$= 3 + 6i$

Use the FOIL order. (First, Outer, Inner, Last)

13. $(2 - 5i)(4 + i) = \underset{F}{2(4)} + \underset{O}{2(i)} - \underset{I}{5i(4)} - \underset{L}{5i(i)}$
$= 8 + 2i - 20i - 5i^2$
$= 8 - 18i - 5(-1)$ $i^2 = -1$
$= 8 - 18i + 5$
$= 13 - 18i$

14. $(2 - i)^2 = (2 - i)(2 - i)$
$= 2(2) - 2(i) - 2(i) + i^2$
$= 4 - 4i + (-1)$ $i^2 = -1$
$= 3 - 4i$

15. $(7 + 3i)(7 - 3i) = 7(7) - 7(3i) + 3i(7) - 3i(3i)$
$= 49 - 21i + 21i - 9i^2$
$= 49 - 9(-1)$ $i^2 = -1$
$= 49 + 9$
$= 58$

Practice Problems 11–15

Multiply.

11. $-5i \cdot 3i$
12. $-2i(6 - 2i)$
13. $(3 - 4i)(6 + i)$
14. $(1 - 2i)^2$
15. $(6 + 5i)(6 - 5i)$

Notice that if you add, subtract, or multiply two complex numbers, the result is a complex number.

D Dividing Complex Numbers

From Example 15, notice that the product of $7 + 3i$ and $7 - 3i$ is a real number. These two complex numbers are called *complex conjugates* of one another. In general, we have the following definition.

Answers

8. $9 - i$, **9.** $-2 + 7i$, **10.** $1 - 4i$, **11.** 15, **12.** $-4 - 12i$, **13.** $22 - 21i$, **14.** $-3 - 4i$, **15.** 61

Complex Conjugates

The complex numbers $(a + bi)$ and $(a - bi)$ are called **complex conjugates** of each other, and

$$(a + bi)(a - bi) = a^2 + b^2$$

To see that the product of a complex number $a + bi$ and its conjugate $a - bi$ is the real number $a^2 + b^2$, we multiply:

$$\begin{aligned}(a + bi)(a - bi) &= a^2 - abi + abi - b^2i^2 \\ &= a^2 - b^2(-1) \\ &= a^2 + b^2\end{aligned}$$

We will use complex conjugates to divide by a complex number.

Practice Problem 16

Divide and write in the form $a + bi$: $\dfrac{3 + i}{2 - 3i}$

EXAMPLE 16 Divide and write in the form $a + bi$: $\dfrac{2 + i}{1 - i}$

Solution: We multiply the numerator and the denominator by the complex conjugate of $1 - i$ to eliminate the imaginary number in the denominator.

$$\begin{aligned}\frac{2 + i}{1 - i} &= \frac{(2 + i)(1 + i)}{(1 - i)(1 + i)} \\ &= \frac{2(1) + 2(i) + 1(i) + i^2}{1^2 - i^2} \\ &= \frac{2 + 3i - 1}{1 + 1} \\ &= \frac{1 + 3i}{2} = \frac{1}{2} + \frac{3}{2}i\end{aligned}$$

●

Practice Problem 17

Divide and write in the form $a + bi$: $\dfrac{6}{5i}$

EXAMPLE 17 Divide and write in the form $a + bi$: $\dfrac{7}{3i}$

Solution: We multiply the numerator and the denominator by the conjugate of $3i$. Note that $3i = 0 + 3i$, so its conjugate is $0 - 3i$ or $-3i$.

$$\frac{7}{3i} = \frac{7(-3i)}{(3i)(-3i)} = \frac{-21i}{-9i^2} = \frac{-21i}{-9(-1)} = \frac{-21i}{9} = \frac{-7i}{3} = -\frac{7}{3}i$$

●

E Finding Powers of *i*

We can use the fact that $i^2 = -1$ to simplify i^3 and i^4.

$$i^3 = i^2 \cdot i = (-1)i = -i$$

$$i^4 = i^2 \cdot i^2 = (-1) \cdot (-1) = 1$$

We continue this process and use the fact that $i^4 = 1$ and $i^2 = -1$ to simplify i^5 and i^6.

$$i^5 = i^4 \cdot i = 1 \cdot i = i$$

$$i^6 = i^4 \cdot i^2 = 1 \cdot (-1) = -1$$

Answers

16. $\frac{3}{13} + \frac{11}{13}i$, **17.** $-\frac{6}{5}i$

If we continue finding powers of i, we generate the following pattern. Notice that the values $i, -1, -i$, and 1 repeat as i is raised to higher and higher powers.

$$\begin{array}{lll} i^1 = i & i^5 = i & i^9 = i \\ i^2 = -1 & i^6 = -1 & i^{10} = -1 \\ i^3 = -i & i^7 = -i & i^{11} = -i \\ i^4 = 1 & i^8 = 1 & i^{12} = 1 \end{array}$$

This pattern allows us to find other powers of i. To do so, we will use the fact that $i^4 = 1$ and rewrite a power of i in terms of i^4. For example,

$$i^{22} = i^{20} \cdot i^2 = (i^4)^5 \cdot i^2 = 1^5 \cdot (-1) = 1 \cdot (-1) = -1$$

EXAMPLES Find each power of i.

18. $i^7 = i^4 \cdot i^3 = 1(-i) = -i$

19. $i^{20} = (i^4)^5 = 1^5 = 1$

20. $i^{46} = i^{44} \cdot i^2 = (i^4)^{11} \cdot i^2 = 1^{11}(-1) = -1$

21. $i^{-12} = \dfrac{1}{i^{12}} = \dfrac{1}{(i^4)^3} = \dfrac{1}{(1)^3} = \dfrac{1}{1} = 1$

Practice Problems 18–21

Find the powers of i.

18. i^{11}
19. i^{40}
20. i^{50}
21. i^{-10}

Answers

18. $-i$, **19.** 1, **20.** -1, **21.** -1

STUDY SKILLS REMINDER

Are you preparing for a test on Chapter 9?

Below I have listed some common trouble areas for students in Chapter 9. After studying for your test, but before taking your test, read these.

- Remember how to convert an expression with rational expressions to one with radicals and one with radicals to one with rational expressions.

$$7^{2/3} = \sqrt[3]{7^2} \text{ or } (\sqrt[3]{7})^2$$
$$\sqrt[5]{4^3} = 4^{3/5}$$

- Remember the difference between $\sqrt{x} + \sqrt{x}$ and $\sqrt{x} \cdot \sqrt{x}$, $x > 0$.

$$\sqrt{x} + \sqrt{x} = 2\sqrt{x}$$
$$\sqrt{x} \cdot \sqrt{x} = x$$

- Don't forget the difference between rationalizing the denominator of $\sqrt{\frac{2}{x}}$ and rationalizing the denominator of $\frac{\sqrt{2}}{\sqrt{x} + 1}$, $x > 0$.

$$\sqrt{\frac{2}{x}} = \frac{\sqrt{2}}{\sqrt{x}} = \frac{\sqrt{2} \cdot \sqrt{x}}{\sqrt{x} \cdot \sqrt{x}} = \frac{\sqrt{2x}}{x}$$
$$\frac{\sqrt{2}}{\sqrt{x} + 1} = \frac{\sqrt{2}(\sqrt{x} - 1)}{(\sqrt{x} + 1)(\sqrt{x} - 1)} = \frac{\sqrt{2}(\sqrt{x} - 1)}{x - 1}$$

Remember: This is simply a checklist of common trouble areas. For a review of Chapter 9, see the Highlights and Chapter Review at the end of this chapter.

Name ____________________ Section __________ Date __________

Mental Math

Simplify. See Example 1.

1. $\sqrt{-81}$
2. $\sqrt{-49}$
3. $\sqrt{-7}$
4. $\sqrt{-3}$
5. $-\sqrt{16}$
6. $-\sqrt{4}$
7. $\sqrt{-64}$
8. $\sqrt{-100}$

EXERCISE SET 9.7

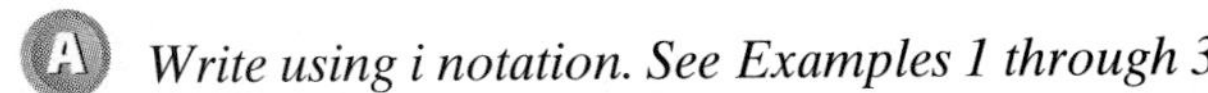

A *Write using i notation. See Examples 1 through 3.*

1. $\sqrt{-24}$
2. $\sqrt{-32}$
3. $-\sqrt{-36}$
4. $-\sqrt{-121}$
5. $8\sqrt{-63}$
6. $4\sqrt{-20}$
7. $-\sqrt{54}$
8. $\sqrt{-63}$

Multiply or divide as indicated. See Examples 4 through 7.

9. $\sqrt{-2} \cdot \sqrt{-7}$
10. $\sqrt{-11} \cdot \sqrt{-3}$
11. $\sqrt{-5} \cdot \sqrt{-10}$
12. $\sqrt{-2} \cdot \sqrt{-6}$
13. $\sqrt{16} \cdot \sqrt{-1}$
14. $\sqrt{3} \cdot \sqrt{-27}$
15. $\dfrac{\sqrt{-9}}{\sqrt{3}}$
16. $\dfrac{\sqrt{49}}{\sqrt{-10}}$
17. $\dfrac{\sqrt{-80}}{\sqrt{-10}}$
18. $\dfrac{\sqrt{-40}}{\sqrt{-8}}$

B *Add or subtract as indicated. Write your answers in the form a + bi. See Examples 8 through 10.*

19. $(4 - 7i) + (2 + 3i)$
20. $(2 - 4i) - (2 - i)$
21. $(6 + 5i) - (8 - i)$
22. $(8 - 3i) + (-8 + 3i)$
23. $6 - (8 + 4i)$
24. $(9 - 4i) - 9$
25. $(6 - 3i) - (4 - 2i)$
26. $(-2 - 4i) - (6 - 8i)$
27. $(5 - 6i) - 4i$
28. $(6 - 2i) + 7i$
29. $(2 + 4i) + (6 - 5i)$
30. $(5 - 3i) + (7 - 8i)$

C *Multiply. Write your answers in the form $a + bi$. See Examples 11 through 15.*

31. $6i \cdot 2i$

32. $5i \cdot 7i$

33. $-9i \cdot 7i$

34. $-6i \cdot 4i$

35. $-10i \cdot -4i$

36. $-2i \cdot -11i$

37. $6i(2 - 3i)$

38. $5i(4 - 7i)$

39. $-3i(-1 + 9i)$

40. $-5i(-2 + i)$

41. $(4 + i)(5 + 2i)$

42. $(3 + i)(2 + 4i)$

43. $(\sqrt{3} + 2i)(\sqrt{3} - 2i)$

44. $(\sqrt{5} - 5i)(\sqrt{5} + 5i)$

45. $(4 - 2i)^2$

46. $(6 - 3i)^2$

47. $(6 - 2i)(3 + i)$

48. $(2 - 4i)(2 - i)$

49. $(1 - i)(1 + i)$

50. $(6 + 2i)(6 - 2i)$

51. $(9 + 8i)^2$

52. $(4 + 7i)^2$

53. $(1 - i)^2$

54. $(2 - 2i)^2$

D *Divide. Write your answers in the form $a + bi$. See Examples 16 and 17.*

55. $\frac{4}{i}$

56. $\frac{5}{6i}$

57. $\frac{7}{4 + 3i}$

58. $\frac{9}{1 - 2i}$

59. $\frac{6i}{1 - 2i}$

60. $\frac{3i}{5 + i}$

61. $\frac{3 + 5i}{1 + i}$

62. $\frac{6 + 2i}{4 - 3i}$

63. $\frac{4 - 5i}{2i}$

64. $\frac{6 + 8i}{3i}$

65. $\frac{16 + 15i}{-3i}$

66. $\frac{2 - 3i}{-7i}$

67. $\frac{2}{3 + i}$

68. $\frac{5}{3 - 2i}$

69. $\frac{2 - 3i}{2 + i}$

70. $\frac{6 + 5i}{6 - 5i}$

E *Find each power of i. See Examples 18 through 21.*

71. i^8

72. i^{10}

73. i^{21}

74. i^{15}

75. i^{11}

76. i^{40}

77. i^{-6}

78. i^{-9}

79. $(2i)^6$

80. $(5i)^4$

81. $(-3i)^5$

82. $(-2i)^7$

Review and Preview

Thirty people were recently polled about the average monthly balance in their checking account. The results of this poll are shown in the bar graph. Use this graph to answer Exercises 83 through 88. See Section 3.1.

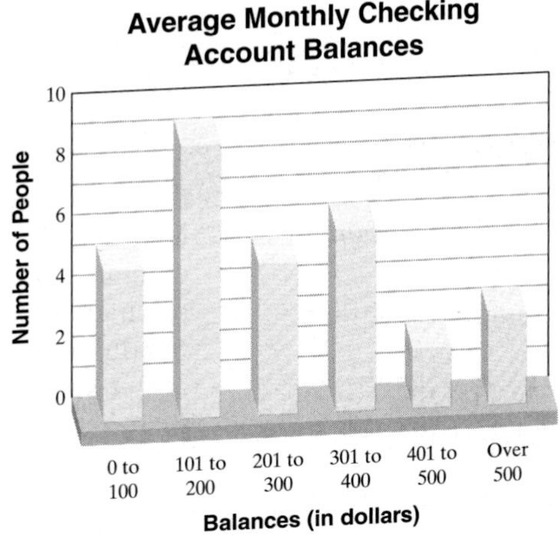

83. How many people polled reported an average checking balance of \$201 to \$300?

84. How many people polled reported an average checking balance of \$0 to \$100?

85. How many people polled reported an average checking balance of \$200 or less?

86. How many people polled reported an average checking balance of \$301 or more?

87. What percent of people polled reported an average checking balance of \$201 to \$300?

88. What percent of people polled reported an average checking balance of 0 to \$100?

Combining Concepts

Write each expression in the form $a + bi$.

89. $i^3 + i^4$

90. $i^8 - i^7$

91. $i^6 + i^8$

92. $i^4 + i^{12}$

93. $2 + \sqrt{-9}$

94. $5 - \sqrt{-16}$

95. $\dfrac{6 + \sqrt{-18}}{3}$

96. $\dfrac{4 - \sqrt{-8}}{2}$

97. $\dfrac{5 - \sqrt{-75}}{10}$

98. Describe how to find the conjugate of a complex number.

99. Explain why the product of a complex number and its complex conjugate is a real number.

Simplify.

100. $(8 - \sqrt{-3}) - (2 + \sqrt{-12})$

101. $(8 - \sqrt{-4}) - (2 + \sqrt{-16})$

102. Determine whether $2i$ is a solution of $x^2 + 4 = 0$.

103. Determine whether $-1 + i$ is a solution of $x^2 + 2x = -2$.

CHAPTER 9 ACTIVITY

Calculating the Length and Period of a Pendulum

MATERIALS:

- string (at least 1 meter long)
- weight
- meter stick
- stopwatch
- calculator

This activity may be completed by working in groups or individually.

Make a simple pendulum by securely tying the string to a weight.

The formula relating a pendulum's period T (in seconds) to its length l (in centimeters) is

$$T = 2\pi\sqrt{\frac{l}{980}}$$

The **period** of a pendulum is defined as the time it takes the pendulum to complete one full back-and-forth swing. In this activity, you will be measuring your simple pendulum's period with a stopwatch. Because the periods will be only a few seconds long, it will be more accurate for you to time a total of five complete swings and then find the average time of one complete swing.

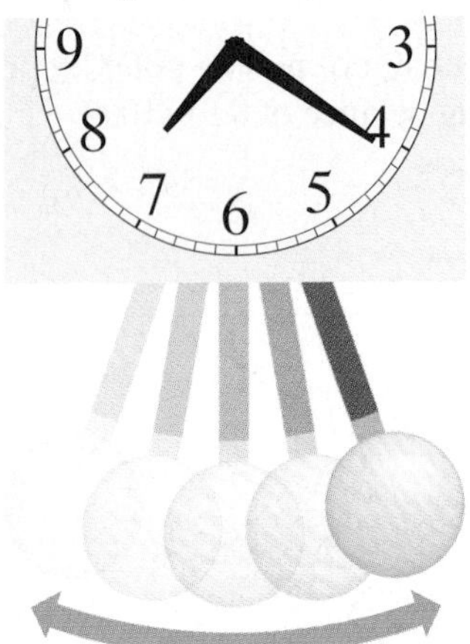

1. For each of the pendulum (string) lengths given in Table 1, measure the time required for 5 complete swings and record it in the appropriate column. Next, divide this value by 5 to find the measured period of the pendulum for the given length and record it in the Measured Period T_m column in the table. Use the given formula to calculate the theoretical period T for the same pendulum length and record it in the appropriate column. (Round to two decimal places.) Find and record in the last column the difference between the measured period and the theoretical period.
2. For each of the periods T given in Table 2, use the given formula and calculate the theoretical pendulum length l required to yield the given period. Record l in the appropriate column; round to one decimal place. Next, use this length l and measure and record the time for 5 complete swings. Divide this value by 5 to find the measured period T_m, and record it. Then find and record in the last column the difference between the theoretical period and the measured period.
3. Use the general trends you find in the tables to describe the relationship between a pendulum's period and its length.
4. Discuss the differences you found between the values of the theoretical period and the measured period. What factors contributed to these differences?

Table 1

Length l (centimeters)	Time for 5 Swings (seconds)	Measured Priod T_m (seconds)	Theoretical Period T (seconds)	Difference $\lvert T - T_m \rvert$
30				
55				
70				

Table 2

Period T (seconds)	Theoretical Length l (centimeters)	Time for 5 Swings (seconds)	Measured Period T_m (seconds)	Difference $\lvert T - T_m \rvert$
1				
1.25				
2				

Chapter 9 VOCABULARY CHECK

Fill in each blank with one of the words or phrases listed below.

index	rationalizing	conjugate	principal square root	cube root
complex number	like radicals	radicand	imaginary unit	

1. The ____________ of $\sqrt{3} + 2$ is $\sqrt{3} - 2$.
2. The ____________ of a nonnegative number a is written as $\sqrt{a}$.
3. The process of writing a radical expression as an equivalent expression but without a radical in the denominator is called ____________ the denominator.
4. The ____________, written i, is the number whose square is -1.
5. The ____________ of a number is written as $\sqrt[3]{a}$.
6. In the notation $\sqrt[n]{a}$, n is called the ________ and a is called the ________.
7. Radicals with the same index and the same radicand are called ____________.
8. A ____________ is a number that can be written in the form $a + bi$, where a and b are real numbers.

CHAPTER 9 Highlights

DEFINITIONS AND CONCEPTS	EXAMPLES
Section 9.1 Radical Expressions and Functions	
The **positive**, or **principal**, **square root** of a nonnegative number a is written as $\sqrt{a}$. $\sqrt{a} = b$ only if $b^2 = a$ and $b \geq 0$	$\sqrt{36} = 6$ $\quad \sqrt{\frac{9}{100}} = \frac{3}{10}$
The **negative square root** of a is written as $-\sqrt{a}$.	$-\sqrt{36} = -6$ $\quad -\sqrt{0.04} = -0.2$
The **cube root** of a real number a is written as $\sqrt[3]{a}$. $\sqrt[3]{a} = b$ only if $b^3 = a$	$\sqrt[3]{27} = 3$ $\quad \sqrt[3]{-\frac{1}{8}} = -\frac{1}{2}$ $\sqrt[3]{y^6} = y^2$ $\quad \sqrt[3]{64x^9} = 4x^3$
If n is an even positive integer, then $\sqrt[n]{a^n} = \lvert a \rvert$. If n is an odd positive integer, then $\sqrt[n]{a^n} = a$.	$\sqrt{(-3)^2} = \lvert -3 \rvert = 3$ $\sqrt[3]{(-7)^3} = -7$
A **radical function** in x is a function defined by an expression containing a root of x.	If $f(x) = \sqrt{x} + 2$, $f(1) = \sqrt{1} + 2 = 1 + 2 = 3$ $f(3) = \sqrt{3} + 2 \approx 3.73$
Section 9.2 Rational Exponents	
$a^{1/n} = \sqrt[n]{a}$ if $\sqrt[n]{a}$ is a real number.	$81^{1/2} = \sqrt{81} = 9$ $(-8x^3)^{1/3} = \sqrt[3]{-8x^3} = -2x$
If m and n are positive integers greater than 1 with $\frac{m}{n}$ in lowest terms and $\sqrt[n]{a}$ is a real number, then $a^{m/n} = (a^{1/n})^m = (\sqrt[n]{a})^m$	$4^{5/2} = (\sqrt{4})^5 = 2^5 = 32$ $27^{2/3} = (\sqrt[3]{27})^2 = 3^2 = 9$
$a^{-m/n} = \frac{1}{a^{m/n}}$ as long as $a^{m/n}$ is a nonzero number.	$16^{-3/4} = \frac{1}{16^{3/4}} = \frac{1}{(\sqrt[4]{16})^3} = \frac{1}{2^3} = \frac{1}{8}$
Exponent rules are true for rational exponents.	$x^{2/3} \cdot x^{-5/6} = x^{2/3-5/6} = x^{-1/6} = \frac{1}{x^{1/6}}$ $(8^{14})^{1/7} = 8^2 = 64$ $\frac{a^{4/5}}{a^{-2/5}} = a^{4/5-(-2/5)} = a^{6/5}$

DEFINITIONS AND CONCEPTS	**EXAMPLES**
Section 9.3 Simplifying Radical Expressions	
PRODUCT AND QUOTIENT RULES If $\sqrt[n]{a}$ and $\sqrt[n]{b}$ are real numbers, $\sqrt[n]{a}\cdot\sqrt[n]{b} = \sqrt[n]{a\cdot b}$ $\dfrac{\sqrt[n]{a}}{\sqrt[n]{b}} = \sqrt[n]{\dfrac{a}{b}}$, provided $\sqrt[n]{b} \neq 0$	Multiply or divide as indicated: $\sqrt{11}\cdot\sqrt{3} = \sqrt{33}$ $\dfrac{\sqrt[3]{40x}}{\sqrt[3]{5x}} = \sqrt[3]{8} = 2$
A radical of the form $\sqrt[n]{a}$ is **simplified** when a contains no factors that are perfect nth powers.	$\sqrt{40} = \sqrt{4\cdot 10} = 2\sqrt{10}$ $\sqrt{36x^5} = \sqrt{36x^4\cdot x} = 6x^2\sqrt{x}$ $\sqrt[3]{24x^7y^3} = \sqrt[3]{8x^6y^3\cdot 3x} = 2x^2y\sqrt[3]{3x}$
Section 9.4 Adding, Subtracting, and Multiplying Radical Expressions	
Radicals with the same index and the same radicand are **like radicals**. The distributive property can be used to add like radicals. Radical expressions are multiplied by using many of the same properties used to multiply polynomials.	$5\sqrt{6} + 2\sqrt{6} = (5 + 2)\sqrt{6} = 7\sqrt{6}$ $-\sqrt[3]{3x} - 10\sqrt[3]{3x} + 3\sqrt[3]{10x}$ $= (-1 - 10)\sqrt[3]{3x} + 3\sqrt[3]{10x}$ $= -11\sqrt[3]{3x} + 3\sqrt[3]{10x}$ Multiply: $(\sqrt{5} - \sqrt{2x})(\sqrt{2} + \sqrt{2x})$ $= \sqrt{10} + \sqrt{10x} - \sqrt{4x} - 2x$ $= \sqrt{10} + \sqrt{10x} - 2\sqrt{x} - 2x$ $(2\sqrt{3} - \sqrt{8x})(2\sqrt{3} + \sqrt{8x})$ $= 4(3) - 8x = 12 - 8x$
Section 9.5 Rationalizing Numerators and Denominators of Radical Expressions	
The **conjugate** of $a + b$ is $a - b$. The process of writing the denominator of a radical expression without a radical is called **rationalizing the denominator**.	The conjugate of $\sqrt{7} + \sqrt{3}$ is $\sqrt{7} - \sqrt{3}$. Rationalize each denominator: $\dfrac{\sqrt{5}}{\sqrt{3}} = \dfrac{\sqrt{5}\cdot\sqrt{3}}{\sqrt{3}\cdot\sqrt{3}} = \dfrac{\sqrt{15}}{3}$ $\dfrac{6}{\sqrt{7} + \sqrt{3}} = \dfrac{6(\sqrt{7} - \sqrt{3})}{(\sqrt{7} + \sqrt{3})(\sqrt{7} - \sqrt{3})}$ $= \dfrac{6(\sqrt{7} - \sqrt{3})}{7 - 3}$ $= \dfrac{6(\sqrt{7} - \sqrt{3})}{4} = \dfrac{3(\sqrt{7} - \sqrt{3})}{2}$
The process of writing the numerator of a radical expression without a radical is called **rationalizing the numerator**.	Rationalize each numerator: $\dfrac{\sqrt[3]{9}}{\sqrt[3]{5}} = \dfrac{\sqrt[3]{9}\cdot\sqrt[3]{3}}{\sqrt[3]{5}\cdot\sqrt[3]{3}} = \dfrac{\sqrt[3]{27}}{\sqrt[3]{15}} = \dfrac{3}{\sqrt[3]{15}}$ $\dfrac{\sqrt{9} + \sqrt{3x}}{12} = \dfrac{(\sqrt{9} + \sqrt{3x})(\sqrt{9} - \sqrt{3x})}{12(\sqrt{9} - \sqrt{3x})}$ $= \dfrac{9 - 3x}{12(\sqrt{9} - \sqrt{3x})}$ $= \dfrac{3(3 - x)}{3\cdot 4(3 - \sqrt{3x})} = \dfrac{3 - x}{4(3 - \sqrt{3x})}$

DEFINITIONS AND CONCEPTS — EXAMPLES

Section 9.6 Radical Equations and Problem Solving

SOLVING A RADICAL EQUATION

Step 1. Write the equation so that one radical is by itself on one side of the equation.

Step 2. Raise each side of the equation to a power equal to the index of the radical and simplify.

Step 3. If the equation still contains a radical, repeat Steps 1 and 2. If not, solve the equation.

Step 4. Check all proposed solutions in the original equation.

Solve: $x = \sqrt{4x + 9} + 3$

1. $x - 3 = \sqrt{4x + 9}$

2. $(x - 3)^2 = (\sqrt{4x + 9})^2$

$x^2 - 6x + 9 = 4x + 9$

3. $x^2 - 10x = 0$

$x(x - 10) = 0$

$x = 0$ or $x = 10$

4. The proposed solution 10 checks, but 0 does not. The solution is $\{10\}$.

Section 9.7 Complex Numbers

A **complex number** is a number that can be written in the form $a + bi$, where a and b are real numbers.

$i^2 = -1$ and $i = \sqrt{-1}$

Simplify: $\sqrt{-9}$

$\sqrt{-9} = \sqrt{-1 \cdot 9} = \sqrt{-1} \cdot \sqrt{9} = i \cdot 3$, or $3i$

Complex Numbers	**Written in Form $a + bi$**
12	$12 + 0i$
$-5i$	$0 + (-5)i$
$-2 - 3i$	$-2 + (-3)i$

Multiply.

$$\sqrt{-3} \cdot \sqrt{-7} = i\sqrt{3} \cdot i\sqrt{7} = i^2\sqrt{21} = -\sqrt{21}$$

To add or subtract complex numbers, add or subtract their real parts and then add or subtract their imaginary parts.

Perform each indicated operation.

$$(-3 + 2i) - (7 - 4i) = -3 + 2i - 7 + 4i = -10 + 6i$$

To multiply complex numbers, multiply as though they were binomials.

$$(-7 - 2i)(6 + i) = -42 - 7i - 12i - 2i^2 = -42 - 19i - 2(-1) = -42 - 19i + 2 = -40 - 19i$$

The complex numbers $(a + bi)$ and $(a - bi)$ are called **complex conjugates**.

The complex conjugate of $(3 + 6i)$ is $(3 - 6i)$.

Their product is a real number:

$$(3 - 6i)(3 + 6i) = 9 - 36i^2 = 9 - 36(-1) = 9 + 36 = 45$$

To divide complex numbers, multiply the numerator and the denominator by the conjugate of the denominator.

Divide: $$\frac{4}{2 - i} = \frac{4(2 + i)}{(2 - i)(2 + i)} = \frac{4(2 + i)}{4 - i^2} = \frac{4(2 + i)}{5} = \frac{8 + 4i}{5} = \frac{8}{5} + \frac{4}{5}i$$

FOCUS ON The Real World

DIFFUSION

Diffusion is the spontaneous movement of the molecules of a substance from a region of higher concentration to a region of lower concentration until a uniform concentration throughout the region is reached. For example, if a drop of food coloring is added to a glass of water, the molecules of the coloring are diffused so that the entire glass of water is colored evenly without any kind of stirring. Diffusion is also mostly responsible for the spread of the smell of baking brownies throughout a house.

Diffusion is used or seen in important aspects of many disciplines. The following list describes situations in which diffusion plays a role.

- In the commercial production of sugar, sugar can be extracted from sugar cane through a diffusion process.
- Solid-state diffusion plays a role in the manufacturing process of silicon computer chips.
- In biology, the diffusion phenomenon allows water molecules, nutrient molecules, and dissolved gas molecules (such as oxygen and carbon dioxide) to pass through the semipermeable membranes of cell walls.
- During a human pregnancy, the fetus is nourished from the mother's blood supply via diffusion through the placenta. Waste materials from the fetus are also diffused through the placenta to be carried away by the mother's circulatory system.
- The medical treatment known as kidney dialysis, in which waste materials are removed from the blood of a patient without kidney function, is made possible by diffusion.
- A diffusion process is widely used to separate the uranium isotope U-235, which can be used as a fuel in nuclear power plants, from the uranium isotope U-238, which cannot be used to create nuclear energy.

In chemistry, Graham's law states that the diffusion rate of a substance in its gaseous state is inversely proportional to the square root of its molecular weight. Another useful property of diffusion is that the distance a material diffuses over time is directly proportional to the square root of the time.

CRITICAL THINKING

1. Write an equation for the relationship described by Graham's law. Be sure to define the variables and constants that you use.
2. According to Graham's law, which molecule will diffuse more rapidly: a molecule with a molecular weight of 58.4 or a molecule with a molecular weight of 180.2? Explain your reasoning.
3. Write an equation for the relationship between the distance that a material diffuses and time. Again, be sure to define the variables and constants that you use.
4. Suppose it takes sugar 1 week to diffuse a distance of 1 cm from its starting point in a particular liquid. How long will it take the sugar to diffuse a total of 3 cm from its starting point in the liquid?

Name ______________________ Section __________ Date __________

Chapter 9 Review

(9.1) *Find each root. Assume that all variables represent positive real numbers.*

1. $\sqrt{81}$

2. $\sqrt[4]{81}$

3. $\sqrt[3]{-8}$

4. $\sqrt[4]{-16}$

5. $-\sqrt{\frac{1}{49}}$

6. $\sqrt{x^{64}}$

7. $-\sqrt{36}$

8. $\sqrt[3]{64}$

9. $\sqrt[3]{-a^6b^9}$

10. $\sqrt{16a^4b^{12}}$

11. $\sqrt[5]{32a^5b^{10}}$

12. $\sqrt[5]{-32x^{15}y^{20}}$

13. $\sqrt{\frac{x^{12}}{36y^2}}$

14. $\sqrt[3]{\frac{27y^3}{z^{12}}}$

Simplify. Use absolute value bars when necessary.

15. $\sqrt{x^2}$

16. $\sqrt[4]{(x^2-4)^4}$

17. $\sqrt[3]{(-27)^3}$

18. $\sqrt[5]{(-5)^5}$

19. $-\sqrt[5]{x^5}$

20. $\sqrt[4]{16(2y+z)^4}$

21. $\sqrt{25(x-y)^2}$

22. $\sqrt[5]{y^5}$

23. $\sqrt[6]{x^6}$

24. If $f(x) = \sqrt{x} + 3$, find $f(0)$ and $f(9)$.

25. If $g(x) = \sqrt[3]{x-3}$, find $g(11)$ and $g(20)$.

(9.2) *Evaluate.*

26. $\left(\frac{1}{81}\right)^{1/4}$ **27.** $\left(-\frac{1}{27}\right)^{1/3}$ **28.** $(-27)^{-1/3}$ **29.** $(-64)^{-1/3}$ **30.** $-9^{3/2}$

31. $64^{-1/3}$ **32.** $(-25)^{5/2}$ **33.** $\left(\frac{25}{49}\right)^{-3/2}$ **34.** $\left(\frac{8}{27}\right)^{-2/3}$ **35.** $\left(-\frac{1}{36}\right)^{-1/4}$

Write with rational exponents.

36. $\sqrt[3]{x^2}$ **37.** $\sqrt[5]{5x^2y^3}$

Write using radical notation.

38. $y^{4/5}$ **39.** $5(xy^2z^5)^{1/3}$ **40.** $(x + 2y)^{-1/2}$

Simplify each expression. Assume that all variables represent positive real numbers. Write with only positive exponents.

41. $a^{1/3}a^{4/3}a^{1/2}$ **42.** $\frac{b^{1/3}}{b^{4/3}}$ **43.** $(a^{1/2}a^{-2})^3$ **44.** $(x^{-3}y^6)^{1/3}$

45. $\left(\frac{b^{3/4}}{a^{-1/2}}\right)^8$ **46.** $\frac{x^{1/4}x^{-1/2}}{x^{2/3}}$ **47.** $\left(\frac{49c^{5/3}}{a^{-1/4}b^{5/6}}\right)^{-1}$ **48.** $a^{-1/4}(a^{5/4} - a^{9/4})$

Use a calculator and write a three-decimal-place approximation of each number.

49. $\sqrt{20}$ **50.** $\sqrt[3]{-39}$ **51.** $\sqrt[4]{726}$ **52.** $56^{1/3}$ **53.** $-78^{3/4}$ **54.** $105^{-2/3}$

Use rational exponents to write each as a single radical.

55. $\sqrt[3]{2} \cdot \sqrt{7}$ **56.** $\sqrt[3]{3} \cdot \sqrt[4]{x}$

(9.3) *Perform each indicated operation and then simplify if possible. Assume that all variables represent positive real numbers.*

57. $\sqrt{3} \cdot \sqrt{8}$ **58.** $\sqrt[3]{7y} \cdot \sqrt[3]{x^2z}$ **59.** $\dfrac{\sqrt{44x^3}}{\sqrt{11x}}$ **60.** $\dfrac{\sqrt[4]{a^6b^{13}}}{\sqrt[4]{a^2b}}$

Simplify.

61. $\sqrt{60}$ **62.** $-\sqrt{75}$ **63.** $\sqrt[3]{162}$ **64.** $\sqrt[3]{-32}$ **65.** $\sqrt{36x^7}$

66. $\sqrt[3]{24a^5b^7}$ **67.** $\sqrt{\dfrac{p^{17}}{121}}$ **68.** $\sqrt[3]{\dfrac{y^5}{27x^6}}$ **69.** $\sqrt[4]{\dfrac{xy^6}{81}}$ **70.** $\sqrt{\dfrac{2x^3}{49y^4}}$

△ **71.** The formula for the radius r of a circle of area A is

$$r = \sqrt{\frac{A}{\pi}}$$

a. Find the exact radius of a circle whose area is 25 square meters.

b. Approximate to two decimal places the radius of a circle whose area is 104 square inches.

(9.4) *Perform each indicated operation. Assume that all variables represent positive real numbers.*

72. $\sqrt{20} + \sqrt{45} - 7\sqrt{5}$

73. $x\sqrt{75x} - \sqrt{27x^3}$

74. $\sqrt[3]{128} + \sqrt[3]{250}$

75. $3\sqrt[4]{32a^5} - a\sqrt[4]{162a}$

76. $\dfrac{5}{\sqrt{4}} + \dfrac{\sqrt{3}}{3}$

77. $\sqrt{\dfrac{8}{x^2}} - \sqrt{\dfrac{50}{16x^2}}$

78. $2\sqrt{50} - 3\sqrt{125} + \sqrt{98}$

79. $2a\sqrt[4]{32b^5} - 3b\sqrt[4]{162a^4b} + \sqrt[4]{2a^4b^5}$

Multiply and then simplify if possible. Assume that all variables represent positive real numbers.

80. $\sqrt{3}(\sqrt{27} - \sqrt{3})$

81. $(\sqrt{x} - 3)^2$

82. $(\sqrt{5} - 5)(2\sqrt{5} + 2)$

83. $(2\sqrt{x} - 3\sqrt{y})(2\sqrt{x} + 3\sqrt{y})$

84. $(\sqrt{a} + 3)(\sqrt{a} - 3)$

85. $(\sqrt[3]{a} + 2)^2$

86. $(\sqrt[3]{5x} + 9)(\sqrt[3]{5x} - 9)$

87. $(\sqrt[3]{a} + 4)\left(\sqrt[3]{a^2} - 4\sqrt[3]{a} + 16\right)$

(9.5) *Rationalize each denominator. Assume that all variables represent positive real numbers.*

88. $\dfrac{3}{\sqrt{7}}$

89. $\sqrt{\dfrac{x}{12}}$

90. $\dfrac{5}{\sqrt[3]{4}}$

91. $\sqrt{\dfrac{24x^5}{3y^2}}$

92. $\sqrt[3]{\dfrac{15x^6y^7}{z^2}}$

93. $\sqrt[4]{\dfrac{81}{8x^{10}}}$

94. $\dfrac{3}{\sqrt{y} - 2}$ **95.** $\dfrac{\sqrt{2} - \sqrt{3}}{\sqrt{2} + \sqrt{3}}$

Rationalize each numerator. Assume that all variables represent positive real numbers.

96. $\dfrac{\sqrt{11}}{3}$ **97.** $\sqrt{\dfrac{18}{y}}$ **98.** $\dfrac{\sqrt[3]{9}}{7}$ **99.** $\sqrt{\dfrac{24x^5}{3y^2}}$ **100.** $\sqrt[3]{\dfrac{xy^2}{10z}}$ **101.** $\dfrac{\sqrt{x} + 5}{-3}$

(9.6) *Solve each equation.*

102. $\sqrt{y - 7} = 5$ **103.** $\sqrt{2x} + 10 = 4$ **104.** $\sqrt[3]{2x - 6} = 4$

105. $\sqrt{x + 6} = \sqrt{x + 2}$ **106.** $2x - 5\sqrt{x} = 3$ **107.** $\sqrt{x + 9} = 2 + \sqrt{x - 7}$

Find each unknown length.

△ **108.**

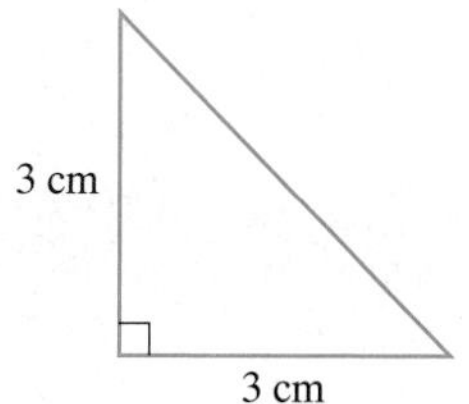

△ **109.**

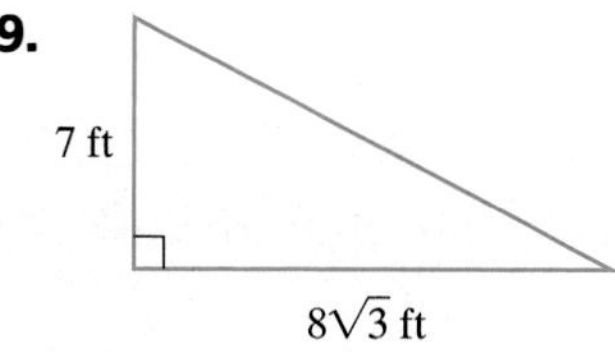

△ **110.** Craig and Daniel Cantwell want to determine the distance x across a pond on their property. They are able to measure the distances shown on the following diagram. Find how wide the lake is at the crossing point indicated by the triangle to the nearest tenth of a foot.

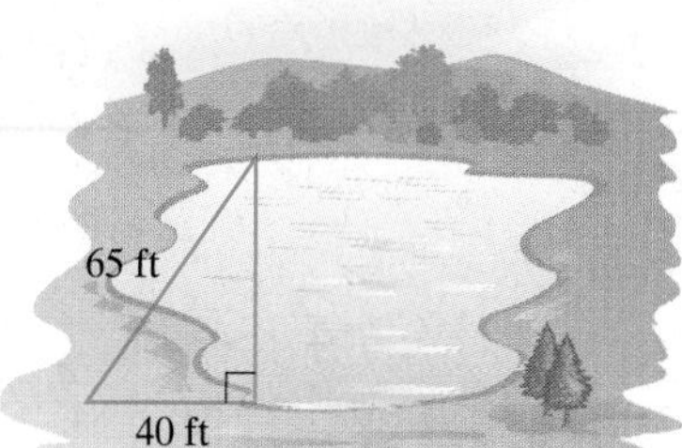

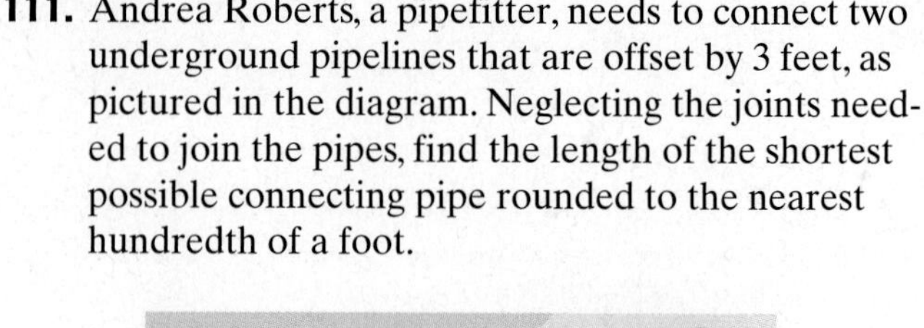

111. Andrea Roberts, a pipefitter, needs to connect two underground pipelines that are offset by 3 feet, as pictured in the diagram. Neglecting the joints needed to join the pipes, find the length of the shortest possible connecting pipe rounded to the nearest hundredth of a foot.

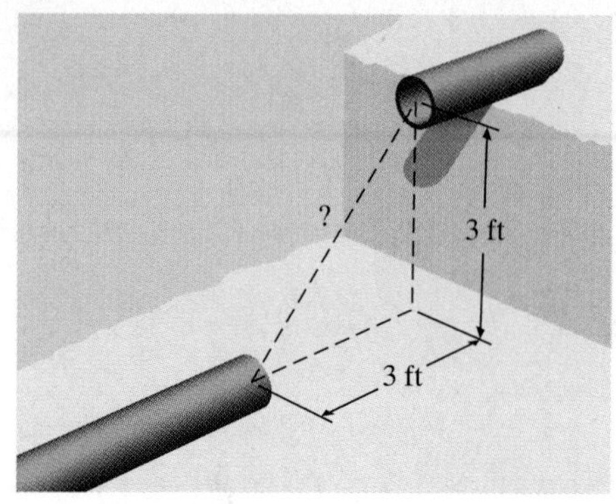

(9.7) *Perform each indicated operation and simplify. Write the results in the form $a + bi$.*

112. $\sqrt{-8}$

113. $-\sqrt{-6}$

114. $\sqrt{-4} + \sqrt{-16}$

115. $\sqrt{-2} \cdot \sqrt{-5}$

116. $(12 - 6i) + (3 + 2i)$

117. $(-8 - 7i) - (5 - 4i)$

118. $(2i)^6$

119. $2i(2 - 5i)$

120. $-3i(6 - 4i)$

121. $(3 + 2i)(1 + i)$

122. $(2 - 3i)^2$

123. $(\sqrt{6} - 9i)(\sqrt{6} + 9i)$

124. $\dfrac{2 + 3i}{2i}$

125. $\dfrac{1 + i}{-3i}$

Name ______________ Section ________ Date ________

Answers

1. ______
2. ______
3. ______
4. ______
5. ______
6. ______
7. ______
8. ______
9. ______
10. ______
11. ______
12. ______
13. ______
14. ______
15. ______
16. ______
17. ______
18. ______
19. ______

Chapter 9 Test

Raise to the power or find the root. Assume that all variables represent positive real numbers. Write with only positive exponents.

1. $\sqrt{216}$ **2.** $-\sqrt[4]{x^{64}}$ **3.** $\left(\frac{1}{125}\right)^{1/3}$ **4.** $\left(\frac{1}{125}\right)^{-1/3}$

5. $\left(\frac{8x^3}{27}\right)^{2/3}$ **6.** $\sqrt[3]{-a^{18}b^9}$ **7.** $\left(\frac{64c^{4/3}}{a^{-2/3}b^{5/6}}\right)^{1/2}$ **8.** $a^{-2/3}(a^{5/4} - a^3)$

Find each root. Use absolute value bars when necessary.

9. $\sqrt[4]{(4xy)^4}$ **10.** $\sqrt[3]{(-27)^3}$

Rationalize each denominator. Assume that all variables represent positive real numbers.

11. $\sqrt{\frac{9}{y}}$ **12.** $\frac{4 - \sqrt{x}}{4 + 2\sqrt{x}}$ **13.** $\sqrt[3]{\frac{8}{9x}}$

14. Rationalize the numerator of $\frac{\sqrt{6} + x}{8}$ and simplify.

Perform each indicated operation. Assume that all variables represent positive real numbers.

15. $\sqrt{125x^3} - 3\sqrt{20x^3}$ **16.** $\sqrt{3}(\sqrt{16} - \sqrt{2})$ **17.** $(\sqrt{x} + 1)^2$

18. $(\sqrt{2} - 4)(\sqrt{3} + 1)$ **19.** $(\sqrt{5} + 5)(\sqrt{5} - 5)$

20.
21.
22.
23.
24.
25.
26.
27.
28.
29.
30.
31.
32.
33.
34.

Use a calculator to approximate each number to three decimal places.

20. $\sqrt{561}$

21. $386^{-2/3}$

Solve.

22. $x = \sqrt{x - 2} + 2$

23. $\sqrt{x^2 - 7} + 3 = 0$

24. $\sqrt{x + 5} = \sqrt{2x - 1}$

Perform each indicated operation and simplify. Write the results in the form $a + bi$.

25. $\sqrt{-2}$

26. $-\sqrt{-8}$

27. $(12 - 6i) - (12 - 3i)$

28. $(6 - 2i)(6 + 2i)$

29. $(4 + 3i)^2$

30. $\dfrac{1 + 4i}{1 - i}$

△ **31.** Find x.

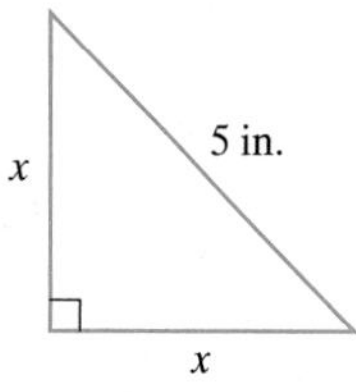

32. If $g(x) = \sqrt{x + 2}$, find $g(0)$ and $g(23)$.

Solve.

33. The function $V = \sqrt{2.5r}$ can be used to estimate the maximum safe velocity, V, in miles per hour, at which a car can travel if it is driven along a curved road with a *radius of curvature*, r, in feet. To the nearest whole number, find the maximum safe speed if a cloverleaf exit on an expressway has a radius of curvature of 300 feet.

34. Use the formula from Exercise 33 to find the radius of curvature if the safe velocity is 30 miles per hour.

Name ______ Section ______ Date ______

Cumulative Review

1. Solve: $y + 0.6 = -1.0$

2. Solve: $8(2 - t) = -5t$

3. A local cellular phone company charges Elaine Chapoton \$50 per month and \$0.36 per minute of phone use in her usage category. If Elaine was charged \$99.68 for month's cellular phone use, determine the number of whole minutes of phone use.

4. Complete the table for the equation $y = 3$.

x	y
-2	
0	
-5	

5. Find the slope of the line through $(-1, 5)$ and $2, -3)$. Graph the line.

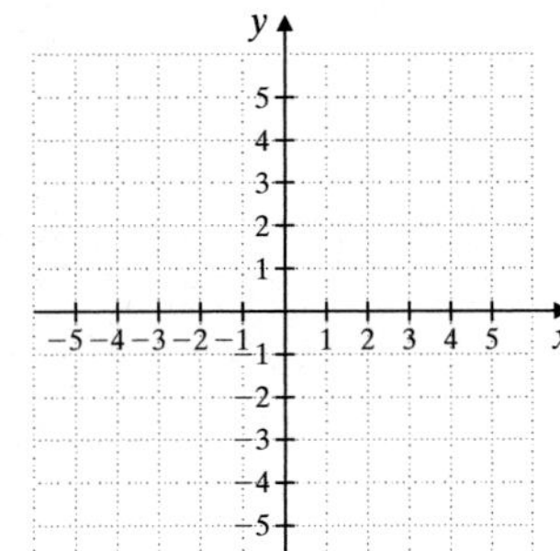

Simplify the following expressions.

6. 3^0

7. $(5x^3y^2)^0$

8. -4^0

9. Multiply: $(3y + 2)^2$

10. Divide $x^2 + 7x + 12$ by $x + 3$ using long division.

11. Factor: $r^2 - r - 42$

12. Factor: $10x^2 - 13xy - 3y^2$

13. Factor $8x^2 - 14x + 5$ by grouping.

14. Factor the difference of squares.

   a. $4x^3 - 49x$

   b. $162x^4 - 2$

15. Solve: $(5x - 1)(2x^2 + 15x + 18) = 0$

16. Simplify: $\dfrac{x^2 + 8x + 7}{x^2 - 4x - 5}$

Answers

1. ______

2. ______

3. ______

4. see table

5. ______

6. ______

7. ______

8. ______

9. ______

10. ______

11. ______

12. ______

13. ______

14. a. ______

b. ______

15. ______

16. ______

17. ______

17. The quotient of a number and 6, minus $\frac{5}{3}$ is the quotient of the number and 2. Find the number.

18. Write an equation of the line containing the point $(4, 4)$ and parallel to the line $2x + y = -6$.

18. ______

If $P(x) = 3x^2 - 2x - 5$, *find the following:*

19. $P(1)$

20. $P(-2)$

19. ______

20. ______

21. Hooke's law states that the distance a spring stretches is directly proportional to the weight attached to the spring. If a 40-pound weight attached to the spring stretches the spring 5 inches, find the distance that a 65-pound weigh attached to the spring stretches the spring.

21. ______

22. ______

23. ______

22. Solve the system

$$\begin{cases} 2x + y = 10 \\ x = y + 2 \end{cases}$$

23. Solve the system

$$\begin{cases} 2x - y = 7 \\ 8x - 4y = 1 \end{cases}$$

24. ______

24. Solve the system:

$$\begin{cases} x - 5y - 2z = 6 \\ -2x + 10y + 4z = -12 \\ \frac{1}{2}x - \frac{5}{2}y - z = 3 \end{cases}$$

25. Use matrices to solve the system:

$$\begin{cases} 2x - y = 3 \\ 4x - 2y = 5 \end{cases}$$

25. ______

26. ______

Write each expression with a positive exponent, and then simplify.

26. $16^{-3/4}$

27. $(-27)^{-2/3}$

27. ______

Quadratic Equations and Functions

An important part of algebra is learning to model and solve problems. Often, the model of a problem is a quadratic equation or a function containing a second-degree polynomial. In this chapter, we continue the work from Chapter 5, solving polynomial equations in one variable by factoring. Two other methods of solving quadratic equations are analyzed in this chapter, with methods of solving nonlinear inequalities in one variable and the graphs of quadratic functions.

The surface of Earth is heated by the sun and then that heat is slowly radiated into outer space. Sometimes, certain gases in the atmosphere reflect the heat radiation back to Earth, preventing it from escaping. The gradual warming of the atmosphere is known as the greenhouse effect. This effect is compounded by the increase of certain gases (greenhouse gases) such as carbon dioxide, methane, and nitrous oxide. Although these gases occur naturally and are needed to keep the surface of Earth at a temperature that is hospitable to life, the recent buildup of these gases is due primarily to human activities. According to the Natural Resources Defense Council, carbon dioxide concentrations have increased by 30% globally over the past century. In Exercise 54 on page 805, we will use a quadratic function to analyze the U.S. level of emissions of methane gas.

Answers

1. ________
2. ________
3. ________
4. ________
5. ________
6. ________
7. ________
8. ________
9. ________
10. ________
11. ________
12. see graph
13. see graph
14. see graph
15. see graph
16. ________

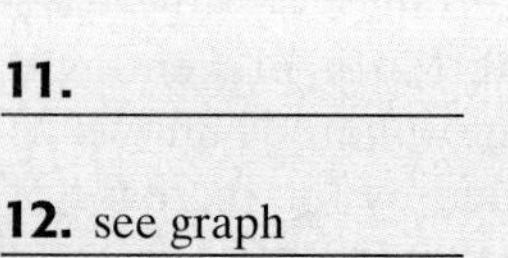

Name ________________ Section ________ Date ________

Chapter 10 Pretest

Solve each equation for the variable.

1. $x^2 = 54$

2. $(3y + 2)^2 = 12$

Solve each equation for the variable by completing the square.

3. $x^2 + 4x = 10$

4. $3y^2 + 18y - 1 = 0$

Use the quadratic formula to solve each equation.

5. $2x^2 - x + 4 = 0$

6. $3y^2 + \frac{1}{2}y - \frac{1}{3} = 0$

Solve.

7. $x^3 = 64$

8. $x^{2/3} - 5x^{1/3} + 4 = 0$

Solve each inequality for x. Write the solution set in interval notation.

9. $(x - 6)(x + 5) \geq 0$

10. $x^2 < -x$

11. $\frac{x + 2}{x + 3} > 1$

Graph each quadratic function. Label the vertex, y-intercept, and x-intercepts (if any).

12. $f(x) = x^2 - 5$

13. $g(x) = (x - 2)^2$

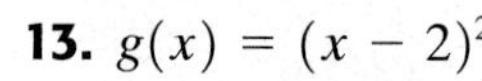

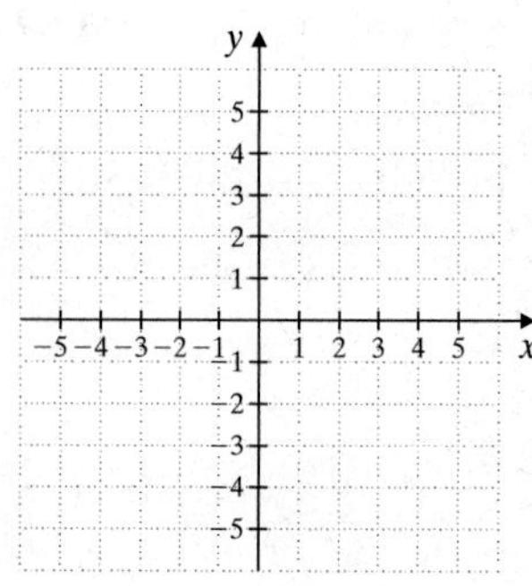

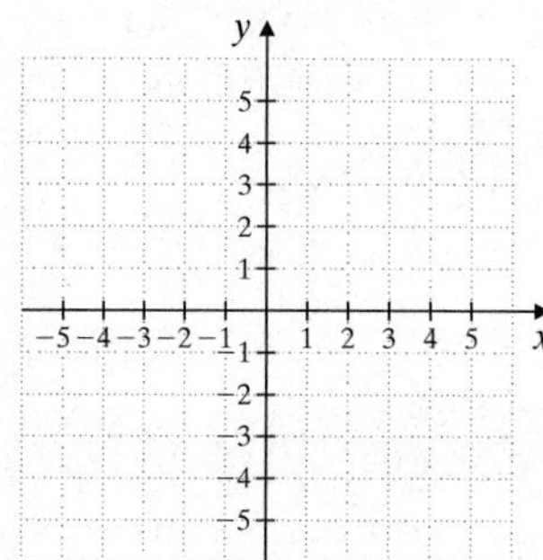

14. $h(x) = (x + 3)^2 - 1$

15. $f(x) = -x^2 - 4x + 2$

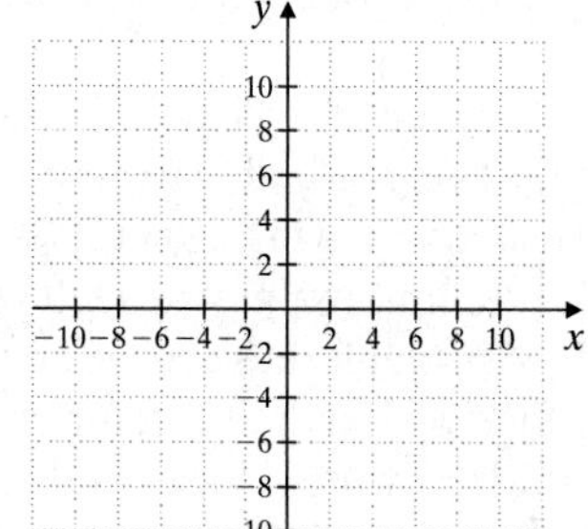

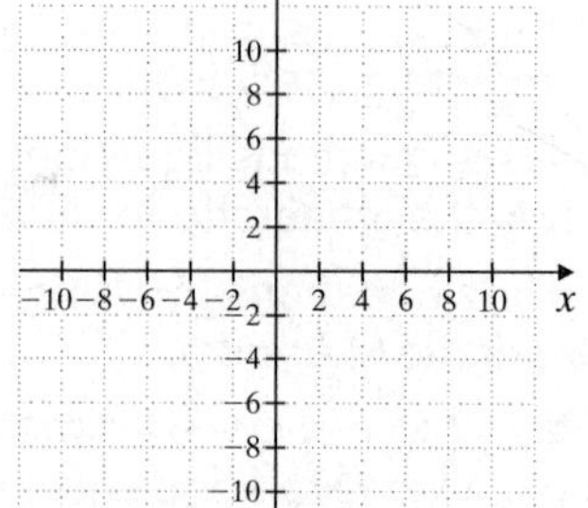

Solve.

△ **16.** The width of a rectangle is $\frac{1}{4}$ its length. If its area is 100 square centimeters, find its length and width.

10.1 Solving Quadratic Equations by Completing the Square

OBJECTIVES

- **A** Use the square root property to solve quadratic equations.
- **B** Write perfect square trinomials.
- **C** Solve quadratic equations by completing the square.
- **D** Use quadratic equations to solve problems.

A Using the Square Root Property

In Chapter 5, we solved quadratic equations by factoring. Recall that a **quadratic**, or **second-degree**, **equation** is an equation that can be written in the form $ax^2 + bx + c = 0$, where a, b, and c are real numbers and a is not 0. To solve a quadratic equation such as $x^2 = 9$ by factoring, we use the zero-factor property. To use the zero-factor property, the equation must first be written in the standard form $ax^2 + bx + c = 0$.

$x^2 = 9$

$x^2 - 9 = 0$ Subtract 9 from both sides to write in standard form.

$(x + 3)(x - 3) = 0$ Factor.

$x + 3 = 0$ or $x - 3 = 0$ Set each factor equal to 0.

$x = -3$ $\quad x = 3$ Solve.

The solution set is $\{-3, 3\}$, the positive and negative square roots of 9.

Not all quadratic equations can be solved by factoring, so we need to explore other methods. Notice that the solutions of the equation $x^2 = 9$ are two numbers whose square is 9:

$3^2 = 9$ and $(-3)^2 = 9$

Thus, we can solve the equation $x^2 = 9$ by taking the square root of both sides. Be sure to include both $\sqrt{9}$ and $-\sqrt{9}$ as solutions since both $\sqrt{9}$ and $-\sqrt{9}$ are numbers whose square is 9.

$x^2 = 9$

$x = \pm\sqrt{9}$ The notation $\pm\sqrt{9}$ (read as plus or minus $\sqrt{9}$) indicates the pair of numbers $+\sqrt{9}$ and $-\sqrt{9}$.

$x = \pm 3$

This illustrates the square root property.

Helpful Hint

The notation ± 3, for example, is read as "plus or minus 3." It is a shorthand notation for the pair of numbers +3 and −3.

Square Root Property

If b is a real number and if $a^2 = b$, then $a = \pm\sqrt{b}$.

EXAMPLE 1 Use the square root property to solve $x^2 = 50$.

Solution:

$x^2 = 50$

$x = \pm\sqrt{50}$ Use the square root property.

$x = \pm 5\sqrt{2}$ Simplify the radical.

Check: Let $x = 5\sqrt{2}$.

$x^2 = 50$

$(5\sqrt{2})^2 \stackrel{?}{=} 50$

$25 \cdot 2 \stackrel{?}{=} 50$

$50 = 50$ True.

Let $x = -5\sqrt{2}$.

$x^2 = 50.$

$(-5\sqrt{2})^2 \stackrel{?}{=} 50$

$25 \cdot 2 \stackrel{?}{=} 50$

$50 = 50$ True.

The solution set is $\{5\sqrt{2}, -5\sqrt{2}\}$.

Practice Problem 1

Use the square root property to solve $x^2 = 20$.

Answer

1. $\{2\sqrt{5}, -2\sqrt{5}\}$

Practice Problem 2

Use the square root property to solve $5x^2 = 55$.

EXAMPLE 2 Use the square root property to solve $2x^2 = 14$.

Solution: First we get the squared variable alone on one side of the equation.

$$2x^2 = 14$$
$$x^2 = 7 \quad \text{Divide both sides by 2.}$$
$$x = \pm\sqrt{7} \quad \text{Use the square root property.}$$

Check: Let $x = \sqrt{7}$.

$$2x^2 = 14$$
$$2(\sqrt{7})^2 \stackrel{?}{=} 14$$
$$2 \cdot 7 \stackrel{?}{=} 14$$
$$14 = 14 \quad \text{True.}$$

Let $x = -\sqrt{7}$.

$$2x^2 = 14$$
$$2(-\sqrt{7})^2 \stackrel{?}{=} 14$$
$$2 \cdot 7 \stackrel{?}{=} 14$$
$$14 = 14 \quad \text{True.}$$

The solution set is $\{\sqrt{7}, -\sqrt{7}\}$.

Practice Problem 3

Use the square root property to solve $(x + 2)^2 = 18$.

EXAMPLE 3 Use the square root property to solve $(x + 1)^2 = 12$.

Solution:

$$(x + 1)^2 = 12$$
$$x + 1 = \pm\sqrt{12} \quad \text{Use the square root property.}$$
$$x + 1 = \pm 2\sqrt{3} \quad \text{Simplify the radical.}$$
$$x = -1 \pm 2\sqrt{3} \quad \text{Subtract 1 from both sides.}$$

Check: Below is a check for $-1 + 2\sqrt{3}$. The check for $-1 - 2\sqrt{3}$ is almost the same and is left for you to do on your own.

$$(x + 1)^2 = 12$$
$$(-1 + 2\sqrt{3} + 1)^2 \stackrel{?}{=} 12$$
$$(2\sqrt{3})^2 \stackrel{?}{=} 12$$
$$4 \cdot 3 \stackrel{?}{=} 12$$
$$12 = 12 \quad \text{True.}$$

The solution set is $\{-1 + 2\sqrt{3}, -1 - 2\sqrt{3}\}$.

Practice Problem 4

Use the square root property to solve $(3x - 1)^2 = -4$.

EXAMPLE 4 Use the square root property to solve $(2x - 5)^2 = -16$.

Solution:

$$(2x - 5)^2 = -16$$
$$2x - 5 = \pm\sqrt{-16} \quad \text{Use the square root property.}$$
$$2x - 5 = \pm 4i \quad \text{Simplify the radical.}$$
$$2x = 5 \pm 4i \quad \text{Add 5 to both sides.}$$
$$x = \frac{5 \pm 4i}{2} \quad \text{Divide both sides by 2.}$$

Check each proposed solution in the original equation to see that the solution set is $\left\{\frac{5 + 4i}{2}, \frac{5 - 4i}{2}\right\}$.

Try the Concept Check in the margin.

Concept Check

How do you know just by looking that $(x - 2)^2 = -4$ has complex solutions?

B Writing Perfect Square Trinomials

Notice from Examples 3 and 4 that, if we write a quadratic equation so that one side is the square of a binomial, we can solve by using the square root property. To write the square of a binomial, we must have a perfect square trinomial. Recall that a perfect square trinomial is a trinomial that can be factored into two identical binomial factors, that is, as a binomial squared.

Answers

2. $\{\sqrt{11}, -\sqrt{11}\}$, **3.** $\{-2 + 3\sqrt{2}, -2 - 3\sqrt{2}\}$, **4.** $\left\{\frac{1 - 2i}{3}, \frac{1 + 2i}{3}\right\}$

Concept Check: answers may vary

Perfect Square Trinomials	**Factored Form**
$x^2 + 8x + 16$	$(x + 4)^2$
$x^2 - 6x + 9$	$(x - 3)^2$
$x^2 + 3x + \frac{9}{4}$	$\left(x + \frac{3}{2}\right)^2$

Notice that for each perfect square trinomial, *the constant term of the trinomial is the square of half the coefficient of the x-term.* For example,

$$x^2 + 8x + 16 \qquad\qquad x^2 - 6x + 9$$

$$\frac{1}{2}(8) = 4 \text{ and } 4^2 = 16 \qquad \frac{1}{2}(-6) = -3 \text{ and } (-3)^2 = 9$$

EXAMPLE 5 Add the proper constant to $x^2 + 6x$ so that the result is a perfect square trinomial. Then factor.

Solution: We add the square of half the coefficient of x.

$$x^2 + 6x + 9 = (x + 3)^2 \quad \text{In factored form}$$

$$\frac{1}{2}(6) = 3 \text{ and } 3^2 = 9$$

Practice Problem 5

Add the proper constant to $x^2 + 12x$ so that the result is a perfect square trinomial. Then factor.

EXAMPLE 6 Add the proper constant to $x^2 - 3x$ so that the result is a perfect square trinomial. Then factor.

Solution: We add the square of half the coefficient of x.

$$x^2 - 3x + \frac{9}{4} = \left(x - \frac{3}{2}\right)^2 \quad \text{In factored form}$$

$$\frac{1}{2}(-3) = -\frac{3}{2} \text{ and } \left(-\frac{3}{2}\right)^2 = \frac{9}{4}$$

Practice Problem 6

Add the proper constant to $y^2 - 5y$ so that the result is a perfect square trinomial. Then factor.

C Solving by Completing the Square

The process of writing a quadratic equation so that one side is a perfect square trinomial is called **completing the square**. We will use this process in the next examples.

Practice Problem 7

Solve $x^2 + 8x = 1$ by completing the square.

EXAMPLE 7 Solve $p^2 + 2p = 4$ by completing the square.

Solution: First we add the square of half the coefficient of p to both sides so that the resulting trinomial will be a perfect square trinomial. The coefficient of p is 2.

$$\frac{1}{2}(2) = 1 \quad \text{and} \quad 1^2 = 1$$

Answers

5. $x^2 + 12x + 36 = (x + 6)^2$,
6. $y^2 - 5y + \frac{25}{4} = \left(x - \frac{5}{2}\right)^2$,
7. $\{-4 - \sqrt{17}, -4 + \sqrt{17}\}$

Now we add 1 to both sides of the original equation.

$$p^2 + 2p = 4$$

$$p^2 + 2p + 1 = 4 + 1 \quad \text{Add 1 to both sides.}$$

$$(p + 1)^2 = 5 \quad \text{Factor the trinomial; simplify the right side.}$$

We may now use the square root property and solve for p.

$$p + 1 = \pm\sqrt{5} \quad \text{Use the square root property.}$$

$$p = -1 \pm \sqrt{5} \quad \text{Subtract 1 from both sides.}$$

Notice that there are two solutions: $-1 + \sqrt{5}$ and $-1 - \sqrt{5}$. The solution set is $\{-1 + \sqrt{5}, -1 - \sqrt{5}\}$.

Practice Problem 8

Solve $y^2 - 5y + 2 = 0$ by completing the square.

EXAMPLE 8 Solve $m^2 - 7m - 1 = 0$ by completing the square.

Solution: First we add 1 to both sides of the equation so that the left side has no constant term. We can then add the constant term on both sides that will make the left side a perfect square trinomial.

$$m^2 = 7m - 1 = 0$$

$$m^2 - 7m = 1$$

Now we find the constant term that makes the left side a perfect square trinomial by squaring half the coefficient of m. We add this constant to both sides of the equation.

$$\frac{1}{2}(-7) = -\frac{7}{2} \quad \text{and} \quad \left(-\frac{7}{2}\right)^2 = \frac{49}{4}$$

$$m^2 - 7m + \frac{49}{4} = 1 + \frac{49}{4} \quad \text{Add } \frac{49}{4} \text{ to both sides of the equation.}$$

$$\left(m - \frac{7}{2}\right)^2 = \frac{53}{4} \quad \text{Factor the perfect square trinomial and simplify the right side.}$$

$$m - \frac{7}{2} = \pm\sqrt{\frac{53}{4}} \quad \text{Use the square root property.}$$

$$m = \frac{7}{2} \pm \frac{\sqrt{53}}{2} \quad \text{Add } \frac{7}{2} \text{ to both sides and simplify } \sqrt{\frac{53}{4}}.$$

$$m = \frac{7 \pm \sqrt{53}}{2} \quad \text{Simplify.}$$

The solution set is $\left\{\frac{7 + \sqrt{53}}{2}, \frac{7 - \sqrt{53}}{2}\right\}$.

The following steps may be used to solve a quadratic equation such as $ax^2 + bx + c = 0$ by completing the square. This method may be used whether or not the polynomial $ax^2 + bx + c$ is factorable.

Answer

8. $\left\{\frac{5 - \sqrt{17}}{2}, \frac{5 + \sqrt{17}}{2}\right\}$

Solving a Quadratic Equation in x by Completing the Square

Step 1. If the coefficient of x^2 is 1, go to Step 2. Otherwise, divide both sides of the equation by the coefficient of x^2.

Step 2. Get all variable terms alone on one side of the equation.

Step 3. Complete the square for the resulting binomial by adding the square of half of the coefficient of x to both sides of the equation.

Step 4. Factor the resulting perfect square trinomial and write it as the square of a binomial.

Step 5. Use the square root property to solve for x.

EXAMPLE 9 Solve $4x^2 - 24x + 41 = 0$ by completing the square.

Solution: First we divide both sides of the equation by 4 so that the coefficient of x^2 is 1.

$$4x^2 - 24x + 41 = 0$$

Step 1. $x^2 - 6x + \frac{41}{4} = 0$ Divide both sides of the equation by 4.

Step 2. $x^2 - 6x = -\frac{41}{4}$ Subtract $\frac{41}{4}$ from both sides.

Since $\frac{1}{2}(-6) = -3$ and $(-3)^2 = 9$, we add 9 to both sides of the equation.

Step 3. $x^2 - 6x + 9 = -\frac{41}{4} + 9$ Add 9 to both sides.

Step 4. $(x - 3)^2 = -\frac{41}{4} + \frac{36}{4}$ Factor the perfect square trinomial.

$(x - 3)^2 = -\frac{5}{4}$

Step 5. $x - 3 = \pm\sqrt{-\frac{5}{4}}$ Use the square root property.

$x - 3 = \pm\frac{i\sqrt{5}}{2}$ Simplify the radical.

$x = 3 \pm \frac{i\sqrt{5}}{2}$ Add 3 to both sides.

$= \frac{6}{2} \pm \frac{i\sqrt{5}}{2}$ Find a common denominator.

$= \frac{6 \pm i\sqrt{5}}{2}$ Simplify.

The solution set is $\left\{\frac{6 + i\sqrt{5}}{2}, \frac{6 - i\sqrt{5}}{2}\right\}$.

Practice Problem 9

Solve $2x^2 - 2x + 7 = 0$ by completing the square.

Answer

9. $\left\{\frac{1 + i\sqrt{13}}{2}, \frac{1 - i\sqrt{13}}{2}\right\}$

D Solving Problems Modeled by Quadratic Equations

Recall the **simple interest** formula $I = Prt$, where I is the interest earned, P is the principal, r is the rate of interest, and t is time. If \$100 is invested at a simple interest rate of 5% annually, at the end of 3 years the total interest I earned is

$$I = P \cdot r \cdot t$$

or

$$I = 100 \cdot 0.05 \cdot 3 = \$15$$

and the new principal is

$$\$100 + \$15 = \$115$$

Most of the time, the interest computed on money borrowed or money deposited is **compound interest**. Unlike simple interest, compound interest is computed on original principal *and* on interest already earned. To see the difference between simple interest and compound interest, suppose that \$100 is invested at a rate of 5% compounded annually. To find the total amount of money at the end of 3 years, we calculate as follows:

$$I = P \cdot r \cdot t$$

First year:	Interest $= \$100 \cdot 0.05 \cdot 1 = \$5.00$
	New principal $= \$100.00 + \$5.00 = \$105.00$
Second year:	Interest $= \$105.00 \cdot 0.05 \cdot 1 = \$5.25$
	New principal $= 105.00 + \$5.25 = \$110.25$
Third year:	Interest $= \$110.25 \cdot 0.05 \cdot 1 \approx \$5.51$
	New principal $= \$110.25 + \$5.51 = \$115.76$

At the end of the third year, the total compound interest earned is \$15.76, whereas the total simple interest earned is \$15.

It is tedious to calculate compound interest as we did above, so we use a compound interest formula. The formula for calculating the total amount of money when interest is compounded annually is

$$A = P(1 + r)^t$$

where P is the original investment, r is the interest rate per compounding period, and t is the number of periods. For example, the amount of money A at the end of 3 years if \$100 is invested at 5% compounded annually is

$$A = \$100(1 + 0.05)^3 \approx 100(1.1576) = \$115.76$$

as we previously calculated.

Practice Problem 10

Use the formula from Example 10 to find the interest rate r if \$1600 compounded annually grows to \$1764 in 2 years.

Answer

10. 5%

EXAMPLE 10 Finding an Interest Rate

Find the interest rate r if \$2000 compounded annually grows to \$2420 in 2 years.

Solution:

1. UNDERSTAND the problem. For this example, make sure that you understand the formula for compounding interest annually.
2. TRANSLATE. We substitute the given values into the formula.

$$A = P(1 + r)^t$$
$$2420 = 2000(1 + r)^2 \quad \text{Let } A = 2420, P = 2000, \text{ and } t = 2.$$

3. SOLVE. We now solve the equation for r.

$$2420 = 2000(1 + r)^2$$

$$\frac{2420}{2000} = (1 + r)^2 \qquad \text{Divide both sides by 2000.}$$

$$\frac{121}{100} = (1 + r)^2 \qquad \text{Simplify the fraction.}$$

$$\pm\sqrt{\frac{121}{100}} = 1 + r \qquad \text{Use the square root property.}$$

$$\pm\frac{11}{10} = 1 + r \qquad \text{Simplify.}$$

$$-1 \pm \frac{11}{10} = r$$

$$-\frac{10}{10} \pm \frac{11}{10} = r$$

$$\frac{1}{10} = r \quad \text{or} \quad -\frac{21}{10} = r$$

4. INTERPRET. The rate cannot be negative, so we reject $-\frac{21}{10}$.

Check: $\frac{1}{10} = 0.10 = 10\%$ per year. If we invest \$2000 at 10% compounded annually, in 2 years the amount in the account would be $2000(1 + 0.10)^2 = 2420$ dollars, the desired amount.

State: The interest rate is 10% compounded annually.

GRAPHING CALCULATOR EXPLORATIONS

In Section 5.8, we showed how we can use a grapher to approximate real number solutions of a quadratic equation written in standard form. We can also use a grapher to solve a quadratic equation when it is not written in standard form. For example, to solve $(x + 1)^2 = 12$, the quadratic equation in Example 3, we graph the following on the same set of axes. We use Xmin $= -10$, Xmax $= 10$, Ymin $= -13$, and Ymax $= 13$.

$$Y_1 = (x + 1)^2 \quad \text{and} \quad Y_2 = 12$$

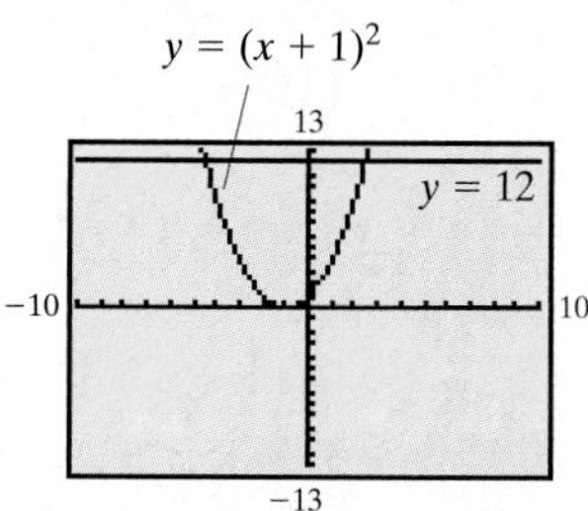

We use the INTERSECT feature or the ZOOM and TRACE features to locate the points of intersection of the graphs. The x-values of these points are the solutions of $(x + 1)^2 = 12$. The solutions, rounded to two decimal points, are 2.46 and −4.46.

Check to see that these numbers are approximations of the exact solutions, $-1 \pm 2\sqrt{3}$.

Use a grapher to solve each quadratic equation. Round all solutions to the nearest hundredth.

1. $x(x - 5) = 8$

2. $x(x + 2) = 5$

3. $x^2 + 0.5x = 0.3x + 1$

4. $x^2 - 2.6x = -2.2x + 3$

5. Use a grapher to solve $(2x - 5)^2 = -16$, (Example 4) using the window

Xmin $= -20$
Xmax $= 20$
Xscl $= 1$
Ymin $= -20$
Ymax $= 20$
Yscl $= 1$

Explain the results. Compare your results with the solution found in Example 4.

6. What are the advantages and disadvantages of using a grapher to solve quadratic equations?

Name ______________________ Section __________ Date __________

EXERCISE SET 10.1

A *Use the square root property to solve each equation. See Examples 1 through 4.*

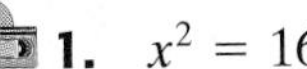

1. $x^2 = 16$

2. $x^2 = 49$

3. $x^2 - 7 = 0$

4. $x^2 - 11 = 0$

5. $x^2 = 18$

6. $y^2 = 20$

7. $3z^2 - 30 = 0$

8. $2x^2 = 4$

9. $(x + 5)^2 = 9$

10. $(y - 3)^2 = 4$

11. $(z - 6)^2 = 18$

12. $(y + 4)^2 = 27$

13. $(2x - 3)^2 = 8$

14. $(4x + 9)^2 = 6$

15. $x^2 + 9 = 0$

16. $x^2 + 4 = 0$

17. $x^2 - 6 = 0$

18. $y^2 - 10 = 0$

19. $2z^2 + 16 = 0$

20. $3p^2 + 36 = 0$

21. $(x - 1)^2 = -16$

22. $(y + 2)^2 = -25$

23. $(z + 7)^2 = 5$

24. $(x + 10)^2 = 11$

25. $(x + 3)^2 = -8$

26. $(y - 4)^2 = -18$

B *Add the proper constant to each binomial so that the resulting trinomial is a perfect square trinomial. Then factor the trinomial. See Examples 5 and 6.*

27. $x^2 + 16x$

28. $y^2 + 2y$

29. $z^2 - 12z$

30. $x^2 - 8x$

31. $p^2 + 9p$

32. $n^2 + 5n$

33. $r^2 - 3r$

34. $p^2 - 7p$

C *Solve each equation by completing the square. See Examples 7 through 9.*

35. $x^2 + 8x = -15$

36. $y^2 + 6y = -8$

37. $x^2 + 6x + 2 = 0$

38. $x^2 - 2x - 2 = 0$

39. $x^2 + x - 1 = 0$

40. $x^2 + 3x - 2 = 0$

41. $x^2 + 2x - 5 = 0$

42. $y^2 + y - 7 = 0$

43. $3p^2 - 12p + 2 = 0$

44. $2x^2 + 14x - 1 = 0$

45. $2x^2 + 7x = 4$

46. $3x^2 - 4x = 4$

47. $x^2 + 8x + 1 = 0$

48. $x^2 - 10x + 2 = 0$

49. $3y^2 + 6y - 4 = 0$

50. $2y^2 + 12y + 3 = 0$

51. $y^2 + 2y + 2 = 0$

52. $x^2 + 4x + 6 = 0$

53. $x^2 - 6x + 3 = 0$

54. $x^2 - 7x - 1 = 0$

55. $2a^2 + 8a = -12$

56. $3x^2 + 12x = -14$

57. $2x^2 - x + 6 = 0$

58. $4x^2 - 2x + 5 = 0$

59. $x^2 + 10x + 28 = 0$

60. $y^2 + 8y + 18 = 0$

61. $z^2 + 3z - 4 = 0$

62. $y^2 + y - 2 = 0$

63. $2x^2 - 4x + 3 = 0$

64. $9x^2 - 36x = -40$

65. $3x^2 + 3x = 5$

66. $5y^2 - 15y = 1$

D *Use the formula $A = P(1 + r)^t$ to solve Exercises 67 through 70. See Example 10.*

67. Find the rate r at which \$3000 grows to \$4320 in 2 years.

68. Find the rate r at which \$800 grows to \$882 in 2 years.

69. Find the rate r at which \$810 grows to approximately \$1000 in 2 years.

70. Find the rate r at which \$2000 grows to \$2880 in 2 years.

71. In your own words, what is the difference between simple interest and compound interest?

72. If you are depositing money in an account that pays 4%, would you prefer the interest to be simple or compound? Explain your answer.

73. If you are borrowing money at a rate of 10%, would you prefer the interest to be simple or compound? Explain your answer.

Review and Preview

Simplify each expression. See Section 9.5.

74. $\frac{6 + 4\sqrt{5}}{2}$

75. $\frac{10 - 20\sqrt{3}}{2}$

76. $\frac{3 - 9\sqrt{2}}{6}$

77. $\frac{12 - 8\sqrt{7}}{16}$

Evaluate $\sqrt{b^2 - 4ac}$ for each set of values. See Section 9.3.

78. $a = 2, b = 4, c = -1$

79. $a = 1, b = 6, c = 2$

80. $a = 3, b = -1, c = -2$

81. $a = 1, b = -3, c = -1$

Combining Concepts

Find two possible missing terms so that each is a perfect square trinomial.

82. $x^2 + \square + 16$

83. $y^2 + \square + 9$

Neglecting air resistance, the distance $s(t)$ in feet traveled by a freely falling object is given by the function $s(t) = 16t^2$, where t is time in seconds. Use this formula to solve Exercises 84 through 87. Round answers to two decimal places.

84. The Petronas Towers in Kuala Lumpur, built in 1997, are the tallest buildings in Malaysia. Each tower is 1483 feet tall. How long would it take an object to fall to the ground from the top of one of the towers? (*Source*: Council on Tall Buildings and Urban Habitat, Lehigh University)

85. The height of the Chicago Beach Tower Hotel, built in 1998 in Dubai, United Arab Emirates, is 1053 feet. How long would it take an object to fall to the ground from the top of the building? (*Source*: Council on Tall Buildings and Urban Habitat, Lehigh University)

86. The height of the Nurek Dam in Tajikistan (part of the former USSR that borders Afghanistan) is 984 feet. How long would it take an object to fall from the top to the base of the dam? (*Source*: U.S. Committee on Large Dams of the International Commission on Large Dams)

87. The Hoover Dam, located on the Colorado River on the border of Nevada and Arizona near Las Vegas, is 725 feet tall. How long would it take an object to fall from the top to the base of the dam? (*Source*: U.S. Committee on Large Dams of the International Commission on Large Dams)

Solve.

△ **88.** The area of a square room is 225 square feet. Find the dimensions of the room.

△ **89.** The area of a circle is 36π square inches. Find the radius of the circle.

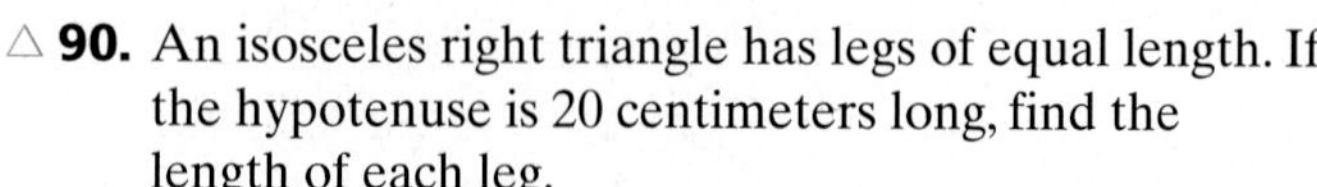

△ **90.** An isosceles right triangle has legs of equal length. If the hypotenuse is 20 centimeters long, find the length of each leg.

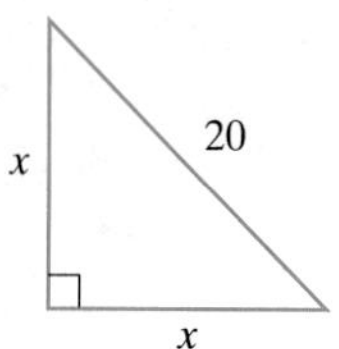

△ **91.** A 27-inch TV is advertised in the *Daily Sentry* newspaper. If 27 inches is the measure of the diagonal of the picture tube, find the measure of the side of the picture tube.

A common equation used in business is a demand equation. It expresses the relationship between the unit price of some commodity and the quantity demanded. For Exercises 92 and 93, p represents the unit price and x represents the quantity demanded in thousands.

92. A manufacturing company has found that the demand equation for a certain type of scissors is given by the equation $p = -x^2 + 47$. Find the demand for the scissors if the price is \$11 per pair.

93. Acme, Inc., sells desk lamps and has found that the demand equation for a certain style of desk lamp is given by the equation $p = -x^2 + 15$. Find the demand for the desk lamp if the price is \$7 per lamp.

Internet Excursions

Go To: http://www.prenhall.com/martin-gay_algebra

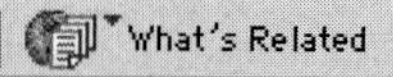

This World Wide Web address will direct you to a Web site that contains current interest rates—both composite rates and those offered by individual savings institutions—on a wide variety of financial products such as savings deposits and auto loans.

94. Choose a financial product. Using actual data for a current interest rate on that type of product, write a problem similar to Exercises 67 through 70. When you have finished writing your problem, trade with another student in your class to solve. Then check each other's work.

95. Choose a different financial product from this Web site. Write another interest rate problem using current interest rates. Then exchange problems with another student in your class to solve. Check each other's work.

10.2 Solving Quadratic Equations by Using the Quadratic Formula

OBJECTIVES

- **A** Solve quadratic equations by using the quadratic formula.
- **B** Determine the number and type of solutions of a quadratic equation by using the discriminant.
- **C** Solve geometric problems modeled by quadratic equations.

SSM TUTOR CENTER SG CD & VIDEO MATH PRO WEB

A Solving Equations by Using the Quadratic Formula

Any quadratic equation can be solved by completing the square. Since the same sequence of steps is repeated each time we complete the square, let's complete the square for a general quadratic equation, $ax^2 + bx + c = 0$. By doing so, we will find a pattern for the solutions of a quadratic equation known as the **quadratic formula**.

Recall that to complete the square for an equation such as $ax^2 + bx + c = 0, a \neq 0$, we first divide both sides by the coefficient of x^2.

$$ax^2 + bx + c = 0$$

$$x^2 + \frac{b}{a}x + \frac{c}{a} = 0 \quad \text{Divide both sides by } a \text{, the coefficient of } x^2.$$

$$x^2 + \frac{b}{a}x = -\frac{c}{a} \quad \text{Subtract the constant } \frac{c}{a} \text{ from both sides.}$$

Next we find the square of half $\frac{b}{a}$, the coefficient of x.

$$\frac{1}{2}\left(\frac{b}{a}\right) = \frac{b}{2a} \quad \text{and} \quad \left(\frac{b}{2a}\right)^2 = \frac{b^2}{4a^2}$$

Now we add this result to both sides of the equation.

$$x^2 + \frac{b}{a}x + \frac{b^2}{4a^2} = -\frac{c}{a} + \frac{b^2}{4a^2} \quad \text{Add } \frac{b^2}{4a^2} \text{ to both sides.}$$

$$x^2 + \frac{b}{a}x + \frac{b^2}{4a^2} = \frac{-c \cdot 4a}{a \cdot 4a} + \frac{b^2}{4a^2} \quad \text{Find a common denominator on the right side.}$$

$$x^2 + \frac{b}{a}x + \frac{b^2}{4a^2} = \frac{b^2 - 4ac}{4a^2} \quad \text{Simplify the right side.}$$

$$\left(x + \frac{b}{2a}\right)^2 = \frac{b^2 - 4ac}{4a^2} \quad \text{Factor the perfect square trinomial on the left side.}$$

$$x + \frac{b}{2a} = \pm\sqrt{\frac{b^2 - 4ac}{4a^2}} \quad \text{Use the square root property.}$$

$$x + \frac{b}{2a} = \pm\frac{\sqrt{b^2 - 4ac}}{2a} \quad \text{Simplify the radical.}$$

$$x = -\frac{b}{2a} \pm \frac{\sqrt{b^2 - 4ac}}{2a} \quad \text{Subtract } \frac{b}{2a} \text{ from both sides.}$$

$$x = \frac{-b \pm \sqrt{b^2 - 4ac}}{2a} \quad \text{Simplify.}$$

The resulting equation identifies the solutions of the general quadratic equation in standard form and is called the quadratic formula. It can be used to solve any equation written in standard form $ax^2 + bx + c = 0$ as long as a is not 0.

Quadratic Formula

A quadratic equation written in the form $ax^2 + bx + c = 0, a \neq 0$, has the solutions

$$x = \frac{-b \pm \sqrt{b^2 - 4ac}}{2a}$$

Practice Problem 1

Solve: $2x^2 + 9x + 10 = 0$

EXAMPLE 1 Solve: $3x^2 + 16x + 5 = 0$

Solution: This equation is in standard form with $a = 3$, $b = 16$, and $c = 5$. We substitute these values into the quadratic formula.

$$x = \frac{-b \pm \sqrt{b^2 - 4ac}}{2a} \quad \text{Quadratic formula}$$

$$= \frac{-16 \pm \sqrt{16^2 - 4(3)(5)}}{2(3)} \quad \text{Let } a = 3, b = 16, \text{ and } c = 5.$$

$$= \frac{-16 \pm \sqrt{256 - 60}}{6}$$

$$= \frac{-16 \pm \sqrt{196}}{6} = \frac{-16 \pm 14}{6}$$

$$x = \frac{-16 + 14}{6} = -\frac{1}{3} \quad \text{or} \quad x = \frac{-16 - 14}{6} = -\frac{30}{6} = -5$$

The solution set is $\left\{-\frac{1}{3}, -5\right\}$.

Helpful Hint

To replace a, b, and c correctly in the quadratic formula, write the quadratic equation in standard form $ax^2 + bx + c = 0$.

Practice Problem 2

Solve: $2x^2 - 6x - 1 = 0$

EXAMPLE 2 Solve: $2x^2 - 4x = 3$

Solution: First we write the equation in standard form by subtracting 3 from both sides.

$$2x^2 - 4x - 3 = 0$$

Now $a = 2$, $b = -4$, and $c = -3$. We substitute these values into the quadratic formula.

$$x = \frac{-b \pm \sqrt{b^2 - 4ac}}{2a}$$

$$= \frac{-(-4) \pm \sqrt{(-4)^2 - 4(2)(-3)}}{2(2)}$$

$$= \frac{4 \pm \sqrt{16 + 24}}{4}$$

$$= \frac{4 \pm \sqrt{40}}{4} = \frac{4 \pm 2\sqrt{10}}{4}$$

$$= \frac{2(2 \pm \sqrt{10})}{2 \cdot 2} = \frac{2 \pm \sqrt{10}}{2}$$

The solution set is $\left\{\frac{2 + \sqrt{10}}{2}, \frac{2 - \sqrt{10}}{2}\right\}$.

Try the Concept Check in the margin.

Concept Check

For the quadratic equation $x^2 = 7$, which substitution is correct?

a. $a = 1, b = 0$, and $c = -7$
b. $a = 1, b = 0$, and $c = 7$
c. $a = 0, b = 0$, and $c = 7$
d. $a = 1, b = 1$, and $c = -7$

Answers

1. $\left\{-\frac{5}{2}, -2\right\}$ **2.** $\left\{\frac{3 + \sqrt{11}}{2}, \frac{3 - \sqrt{11}}{2}\right\}$

Concept Check: a

To simplify the expression $\frac{4 \pm 2\sqrt{10}}{4}$ in Example 2, note that we factored 2 out of both terms of the numerator *before* simplifying.

$$\frac{4 \pm 2\sqrt{10}}{4} = \frac{\boxed{2}(2 \pm \sqrt{10})}{\boxed{2} \cdot 2} = \frac{2 \pm \sqrt{10}}{2}$$

EXAMPLE 3 Solve: $\frac{1}{4}m^2 - m + \frac{1}{2} = 0$

Solution: We could use the quadratic formula with $a = \frac{1}{4}$, $b = -1$, and $c = \frac{1}{2}$. Instead, let's find a simpler, equivalent, standard-form equation whose coefficients are not fractions.

First we multiply both sides of the equation by 4 to clear the fractions.

$$4\left(\frac{1}{4}m^2 - m + \frac{1}{2}\right) = 4 \cdot 0$$

$$m^2 - 4m + 2 = 0 \quad \text{Simplify.}$$

Now we can substitute $a = 1$, $b = -4$, and $c = 2$ into the quadratic formula and simplify.

$$m = \frac{-(-4) \pm \sqrt{(-4)^2 - 4(1)(2)}}{2(1)}$$

$$= \frac{4 \pm \sqrt{16 - 8}}{2}$$

$$= \frac{4 \pm \sqrt{8}}{2} = \frac{4 \pm 2\sqrt{2}}{2} = \frac{2(2 \pm \sqrt{2})}{2} = 2 \pm \sqrt{2}$$

The solution set is $\{2 + \sqrt{2}, 2 - \sqrt{2}\}$.

Practice Problem 3

Solve: $\frac{1}{6}x^2 - \frac{1}{2}x - 1 = 0$

EXAMPLE 4 Solve: $p = -3p^2 - 3$

Solution: The equation in standard form is $3p^2 + p + 3 = 0$. Thus, $a = 3$, $b = 1$, and $c = 3$ in the quadratic formula.

$$p = \frac{-1 \pm \sqrt{1^2 - 4(3)(3)}}{2(3)} = \frac{-1 \pm \sqrt{1 - 36}}{6}$$

$$= \frac{-1 \pm \sqrt{-35}}{6} = \frac{-1 \pm i\sqrt{35}}{6}$$

The solution set is $\left\{\frac{-1 + i\sqrt{35}}{6}, \frac{-1 - i\sqrt{35}}{6}\right\}$

Try the Concept Check in the margin.

Practice Problem 4

Solve: $x = -4x^2 - 4$

Concept Check

What is the first step in solving $-3x^2 = 5x - 4$ using the quadratic formula?

Answers

3. $\left\{\frac{3 + \sqrt{33}}{2}, \frac{3 - \sqrt{33}}{2}\right\}$,

4. $\left\{\frac{-1 - 3i\sqrt{7}}{8}, \frac{-1 + 3i\sqrt{7}}{8}\right\}$

Concept Check: Write the equation in standard form.

B Using the Discriminant

In the quadratic formula $x = \dfrac{-b \pm \sqrt{b^2 - 4ac}}{2a}$, the radicand $b^2 - 4ac$ is called the **discriminant** because when we know its value, we can **discriminate** among the possible number and type of solutions of a quadratic equation. Possible values of the discriminant and their meanings are summarized next.

Discriminant

The following table relates the discriminant $b^2 - 4ac$ of a quadratic equation of the form $ax^2 + bx + c = 0$ with the number and type of solutions of the equation.

$b^2 - 4ac$	Number and Type of Solutions
Positive	Two real solutions
Zero	One real solution
Negative	Two complex but not real solutions

Practice Problem 5

Use the discriminant to determine the number and type of solutions of $x^2 + 4x + 4 = 0$.

EXAMPLE 5

Use the discriminant to determine the number and type of solutions of $x^2 + 2x + 1$.

Solution: In $x^2 + 2x + 1 = 0$, $a = 1$, $b = 2$, and $c = 1$. Thus,

$$b^2 - 4ac = 2^2 - 4(1)(1) = 0$$

Since $b^2 - 4ac = 0$, this quadratic equation has one real solution.

Practice Problem 6

Use the discriminant to determine the number and type of solutions of $5x^2 + 7 = 0$.

EXAMPLE 6

Use the discriminant to determine the number and type of solutions of $3x^2 + 2 = 0$.

Solution: In this equation, $a = 3$, $b = 0$, and $c = 2$. Then

$$b^2 - 4ac = 0^2 - 4(3)(2) = -24$$

Since $b^2 - 4ac$ is negative, this quadratic equation has two complex but not real solutions.

Practice Problem 7

Use the discriminant to determine the number and type of solutions of $3x^2 - 2x - 2 = 0$.

EXAMPLE 7

Use the discriminant to determine the number and type of solutions of $2x^2 - 7x - 4 = 0$.

Solution: In this equation, $a = 2$, $b = -7$, and $c = -4$. Then

$$b^2 - 4ac = (-7)^2 - 4(2)(-4) = 81$$

Since $b^2 - 4ac$ is positive, this quadratic equation has two real solutions.

C Solving Problems Modeled by Quadratic Equations

The quadratic formula is useful in solving problems that are modeled by quadratic equations.

Answers

5. one real solution, **6.** two complex but not real solutions, **7.** two real solutions

△ EXAMPLE 8 Calculating Distance Saved

At a local university, students often leave the sidewalk and cut across the lawn to save walking distance. Given the diagram below of a favorite place to cut across the lawn, approximate to the nearest foot how many feet of walking distance a student saves by cutting across the lawn instead of walking on the sidewalk.

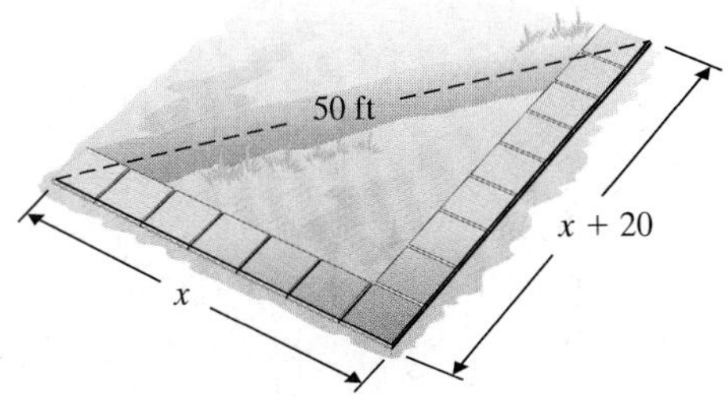

Solution:

1. UNDERSTAND. Read and reread the problem. You may want to review the Pythagorean theorem.
2. TRANSLATE. By the Pythagorean theorem, we have

   In words: $(\text{leg})^2 + (\text{leg})^2 = (\text{hypotenuse})^2$

   Translate: $x^2 + (x + 20)^2 = 50^2$

3. SOLVE. Use the quadratic formula to solve.

$$x^2 + x^2 + 40x + 400 = 2500 \quad \text{Square } (x + 20) \text{ and } 50.$$
$$2x^2 + 40x - 2100 = 0 \quad \text{Set the equation to 0.}$$
$$x^2 + 20x - 1050 = 0 \quad \text{Divide by 2.}$$

Here, $a = 1$, $b = 20$, and $c = -1050$. By the quadratic formula,

$$x = \frac{-20 \pm \sqrt{20^2 - 4(1)(-1050)}}{2 \cdot 1}$$
$$= \frac{-20 \pm \sqrt{400 + 4200}}{2} = \frac{-20 \pm \sqrt{4600}}{2}$$
$$= \frac{-20 \pm \sqrt{100 \cdot 46}}{2} = \frac{-20 \pm 10\sqrt{46}}{2}$$
$$= -10 \pm 5\sqrt{46} \quad \text{Simplify.}$$

Check:

4. INTERPRET. We check our calculations from the quadratic formula. The length of a side of a triangle can't be negative, so we reject $-10 - 5\sqrt{46}$. Since $-10 + 5\sqrt{46} \approx 24$ feet, the walking distance along the sidewalk is

$$x + (x + 20) \approx 24 + (24 + 20) = 68 \text{ feet.}$$

State: A person saves about $68 - 50$ or 18 feet of walking distance by cutting across the lawn.

Practice Problem 8 △

Given the diagram below, approximate to the nearest foot how many feet of walking distance a person saves by cutting across the lawn instead of walking on the sidewalk.

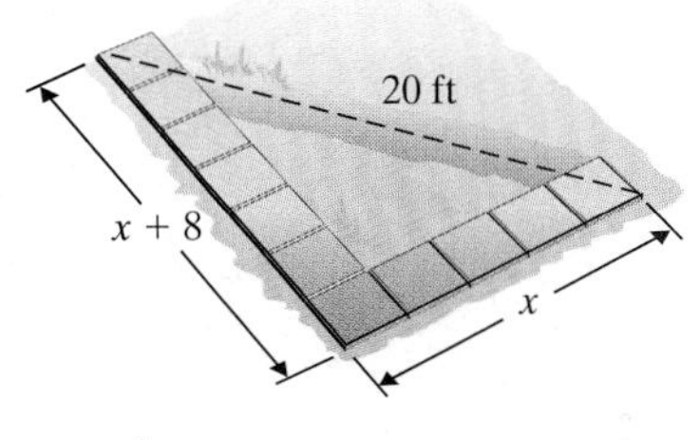

Answer

8. 7 or 8 ft depending upon when you round

Practice Problem 9

How long after the object in Example 9 is thrown will it be 100 feet from the ground? Round to the nearest tenth of a second.

EXAMPLE 9 Calculating Landing Time

An object is thrown upward from the top of a 200-foot cliff with a velocity of 12 feet per second. The height h in feet of the object after t seconds is

$$h = -16t^2 + 12t + 200$$

How long after the object is thrown will it strike the ground? Round to the nearest tenth of a second.

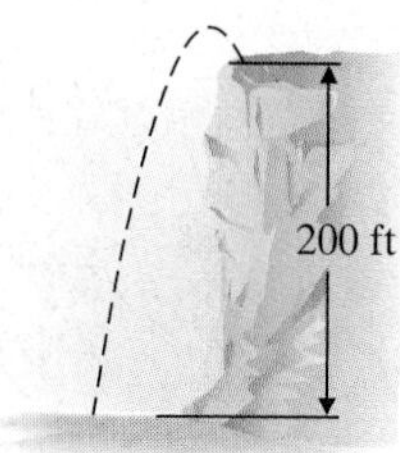

Solution:

1. UNDERSTAND. Read and reread the problem.
2. TRANSLATE. Since we want to know when the object strikes the ground, we want to know when the height $h = 0$, or

$$0 = -16t^2 + 12t + 200$$

3. SOLVE. First we divide both sides of the equation by -4.

$$0 = 4t^2 - 3t - 50 \quad \text{Divide both sides by } -4.$$

Here, $a = 4$, $b = -3$, and $c = -50$. By the quadratic formula,

$$t = \frac{-(-3) \pm \sqrt{(-3)^2 - 4(4)(-50)}}{2 \cdot 4}$$

$$= \frac{3 \pm \sqrt{9 + 800}}{8}$$

$$= \frac{3 \pm \sqrt{809}}{8}$$

Check:

4. INTERPRET. We check our calculations from the quadratic formula. Since the time won't be negative, we reject the proposed solution

$$\frac{3 - \sqrt{809}}{8}.$$

State: The time it takes for the object to strike the ground is exactly $\frac{3 + \sqrt{809}}{8}$ seconds $\approx$ 3.9 seconds.

Answer

9. 2.9 sec

Name ______________________ Section __________ Date __________

Mental Math

Identify the values of a, b, and c in each quadratic equation.

1. $x^2 + 3x + 1 = 0$ **2.** $2x^2 - 5x - 7 = 0$ **3.** $7x^2 - 4 = 0$

4. $x^2 + 9 = 0$ **5.** $6x^2 - x = 0$ **6.** $5x^2 + 3x = 0$

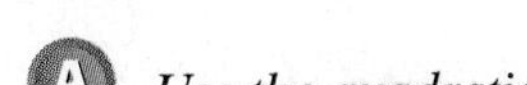

EXERCISE SET 10.2

A *Use the quadratic formula to solve each equation. See Examples 1 through 4.*

1. $m^2 + 5m - 6 = 0$ **2.** $p^2 + 11p - 12 = 0$ **3.** $2y = 5y^2 - 3$ **4.** $5x^2 - 3 = 14x$

5. $x^2 - 6x + 9 = 0$ **6.** $y^2 + 10y + 25 = 0$ **7.** $x^2 + 7x + 4 = 0$ **8.** $y^2 + 5y + 3 = 0$

9. $8m^2 - 2m = 7$ **10.** $11n^2 - 9n = 1$ **11.** $3m^2 - 7m = 3$ **12.** $x^2 - 13 = 5x$

13. $\frac{1}{2}x^2 - x - 1 = 0$ **14.** $\frac{1}{6}x^2 + x + \frac{1}{3} = 0$ **15.** $\frac{2}{5}y^2 + \frac{1}{5}y = \frac{3}{5}$ **16.** $\frac{1}{8}x^2 + x = \frac{5}{2}$

17. $\frac{1}{3}y^2 - y - \frac{1}{6} = 0$ **18.** $\frac{1}{2}y^2 = y + \frac{1}{2}$ **19.** $10y^2 + 10y + 3 = 0$ **20.** $3y^2 + 6y + 5 = 0$

21. $x^2 + 5x = -2$ **22.** $y^2 - 8 = 4y$ **23.** $(m + 2)(2m - 6) = 5(m - 1) - 12$

24. $7p(p - 2) + 2(p + 4) = 3$

25. $\frac{x^2}{3} - x = \frac{5}{3}$

26. $\frac{x^2}{2} - 3 = -\frac{9}{2}x$

27. $x(6x + 2) - 3 = 0$

28. $x(7x + 1) = 2$

29. Solve Exercise 1 by factoring. Explain the result.

30. Solve Exercise 2 by factoring. Explain the result.

Use the quadratic formula to solve each equation. See Examples 1 through 4.

31. $6 = -4x^2 + 3x$

32. $9x^2 + x + 2 = 0$

33. $(x + 5)(x - 1) = 2$

34. $x(x + 6) = 2$

35. $x^2 + 6x + 13 = 0$

36. $x^2 + 2x + 2 = 0$

37. $\frac{2}{5}y^2 + \frac{1}{5}y + \frac{3}{5} = 0$

38. $\frac{1}{8}x^2 + x + \frac{5}{2} = 0$

39. $\frac{1}{2}y^2 = y - \frac{1}{2}$

40. $\frac{2}{3}x^2 - \frac{20}{3}x = -\frac{100}{6}$

41. $(n - 2)^2 = 2n$

42. $\left(p - \frac{1}{2}\right)^2 = \frac{p}{2}$

B *Use the discriminant to determine the number and types of solutions of each equation. See Examples 5 through 7.*

43. $9x - 2x^2 + 5 = 0$

44. $5 - 4x + 12x^2 = 0$

45. $4x^2 + 12x = -9$

46. $9x^2 + 1 = 6x$

47. $3x = -2x^2 + 7$

48. $3x^2 = 5 - 7x$

49. $6 = 4x - 5x^2$

50. $8x = 3 - 9x^2$

C *Solve. See Examples 8 and 9.*

△ **51.** Nancy, Thelma, and John Varner live on a corner lot. Often, neighborhood children cut across their lot to save walking distance. Given the diagram below, approximate to the nearest foot how many feet of walking distance children save by cutting across their property instead of walking around the lot.

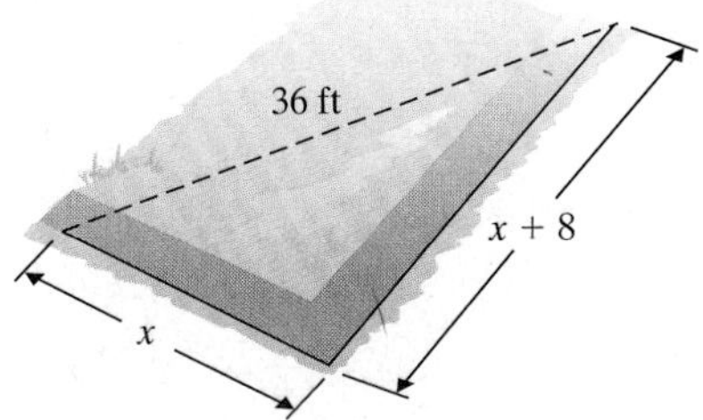

△ **52.** Given the diagram below, approximate to the nearest foot how many feet of walking distance a person saves by cutting across the lawn instead of walking on the sidewalk.

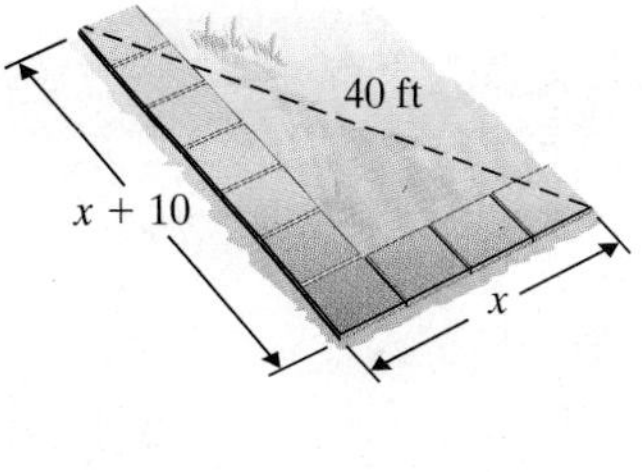

△ **53.** The hypotenuse of an isosceles right triangle is 2 centimeters longer than either of its legs. Find the exact length of each side. (*Hint:* An isosceles right triangle is a right triangle whose legs are the same length.)

△ **54.** The hypotenuse of an isosceles right triangle is one meter longer than either of its legs. Find the length of each side.

△ **55.** Uri Chechov's rectangular dog pen for his Irish setter must have an area of 400 square feet. Also, the length must be 10 feet longer than the width. Find the dimensions of the pen.

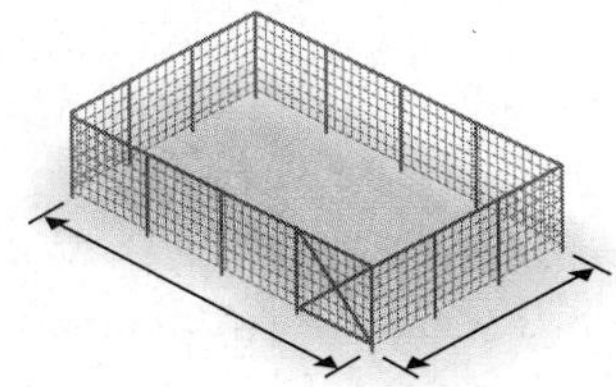

△ **56.** An entry in the Peach Festival Poster Contest must be rectangular and have an area of 1200 square inches. Furthermore, its length must be 20 inches longer than its width. Find the dimensions each entry must have.

△ **57.** A holding pen for cattle must be square and have a diagonal length of 100 meters.

a. Find the length of a side of the pen.

b. Find the area of the pen.

△ **58.** A rectangle is three times longer than it is wide. It has a diagonal of length 50 centimeters.

a. Find the dimensions of the rectangle.

b. Find the perimeter of the rectangle.

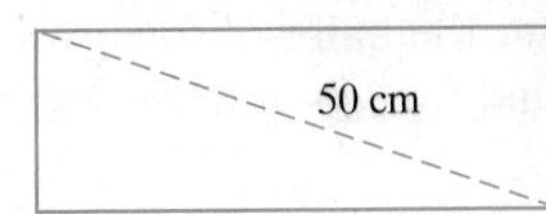

59. The heaviest reported door in the world is the 708.6 ton radiation shield door in the National Institute for Fusion Science at Toki, Japan. If the height of the door is 1.1 feet longer than its width, and its front area (neglecting depth) is 1439.9 square feet, find its width and height [Interesting note: the door is 6.6 feet thick.] (*Source*: *Guiness World Records*, 2001)

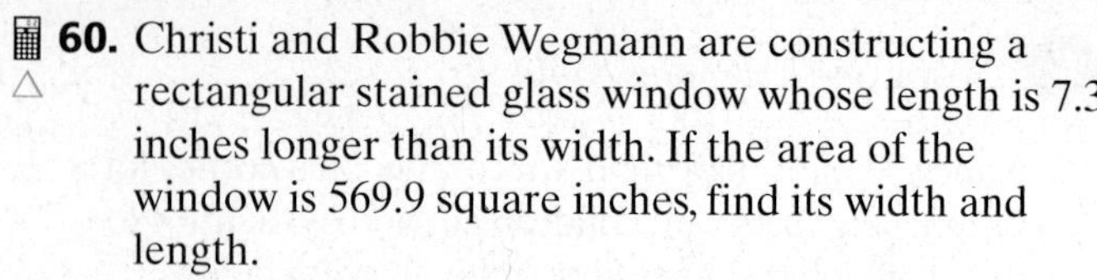
60. Christi and Robbie Wegmann are constructing a rectangular stained glass window whose length is 7.3 inches longer than its width. If the area of the window is 569.9 square inches, find its width and length.

61. If a point B divides a line segment such that the smaller portion is to the larger portion as the larger is to the whole, the whole is the length of the *golden ratio*.

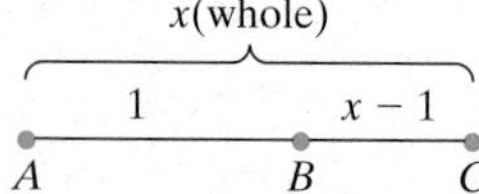

The golden ratio was thought by the Greeks to be the most pleasing to the eye, and many of their buildings contained numerous examples of the golden ratio. The value of the golden ratio is the positive solution of the following equation.

$$\begin{matrix}\text{(smaller)} \\ \text{(larger)}\end{matrix} \quad \frac{x-1}{1} = \frac{1}{x} \quad \begin{matrix}\text{(larger)} \\ \text{(whole)}\end{matrix}$$

Find this value.

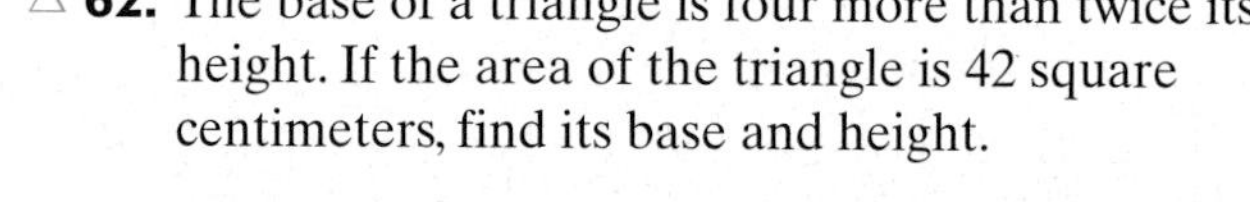
62. The base of a triangle is four more than twice its height. If the area of the triangle is 42 square centimeters, find its base and height.

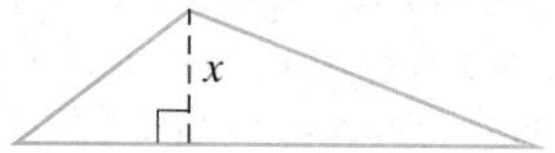

The Wollomombi Falls in Australia have a height of 1100 feet. A pebble is thrown upward from the top of the falls with an initial velocity of 20 feet per second. The height of the pebble h in feet after t seconds is given by the equation $h = -16t^2 + 20t + 1100$. *Use this equation for Exercises 63 and 64.*

63. How long after the pebble is thrown will it hit the ground? Round to the nearest tenth of a second.

64. How long after the pebble is thrown will it be 550 feet from the ground? Round to the nearest tenth of a second.

A ball is thrown downward from the top of a 180-foot building with an initial velocity of 20 feet per second. The height of the ball h in feet after t seconds is given by the equation $h = -16t^2 - 20t + 180$. *Use this equation to answer Exercises 65 and 66.*

65. How long after the ball is thrown will it strike the ground? Round the result to the nearest tenth of a second.

66. How long after the ball is thrown will it be 50 feet from the ground? Round the result to the nearest tenth of a second.

Review and Preview

Solve each equation. See Sections 6.5 and 9.6.

67. $\sqrt{5x - 2} = 3$

68. $\sqrt{y + 2} + 7 = 12$

69. $\frac{1}{x} + \frac{2}{5} = \frac{7}{x}$

70. $\frac{10}{z} = \frac{5}{z} - \frac{1}{3}$

Factor. See Sections 5.2 through 5.5.

71. $x^4 + x^2 - 20$

72. $2y^4 + 11y^2 - 6$

73. $z^4 - 13z^2 + 36$

74. $x^4 - 1$

Combining Concepts

Use the quadratic formula and a calculator to approximate each solution to the nearest tenth.

75. $2x^2 - 6x + 3 = 0$

76. $3.6x^2 + 1.8x - 4.3 = 0$

The graph shows the daily low temperatures for one week in New Orleans, Louisiana. Use this graph to answer Exercises 77 through 80.

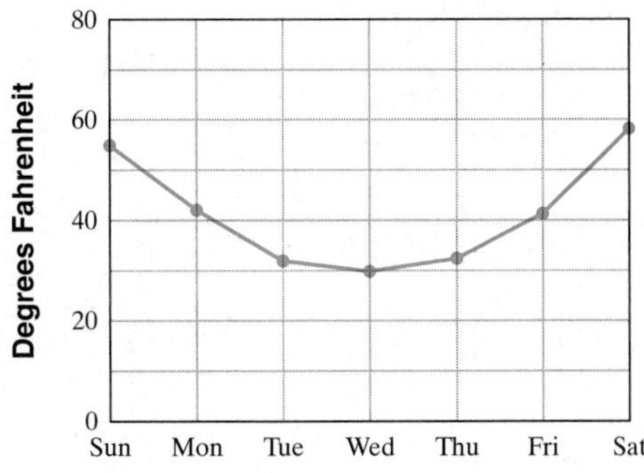

77. Which day of the week shows the greatest decrease in the low temperature?

78. Which day of the week shows the greatest increase in the low temperature?

79. Which day of the week had the lowest temperature?

80. Use the graph to estimate the low temperature on Thursday.

Notice that the shape of the temperature graph is similar to a parabola (see Section 5.9). In fact, this graph can be approximated by the quadratic function $f(x) = 3x^2 - 18x + 57$, where $f(x)$ is the temperature in degrees Fahrenheit and x is the number of days from Sunday. Use this function to answer Exercises 81 and 82.

81. Use the given quadratic function to approximate the low temperature on Thursday. Does your answer agree with the graph above?

82. Use the given function and the quadratic formula to find when the low temperature was 35°F. [*Hint:* Let $f(x) = 35$ and solve for x.] Round your answer to one decimal place and interpret your result. Does your answer agree with the graph above?

83. Use a grapher to solve Exercise 75.

84. Use a grapher to solve Exercise 76.

85. Procter & Gamble's net earnings can be modeled by the quadratic function $f(x) = -199.5x^2 - 21.5x + 3763$, where $f(x)$ is net earnings in millions of dollars and x is the number of years after 1999. (*Source:* Based on data from The Procter & Gamble Co., 1999–2001)

a. Find Procter & Gamble's net earnings in 2001.

b. If the trend described by the model continues, predict the year after 1999 in which Procter & Gamble's net earnings will be \$485 million.

86. The number of inmates in custody in U.S. prisons and jails can be modeled by the quadratic function $p(x) = -x^2 + 93x + 1128$, where $p(x)$ is the number of inmates in thousands and x is the number of years after 1990. (*Source:* Based on data from the Bureau of Justice Statistics, U.S. Department of Justice, 1990–2000)

a. Find the number of prison and jail inmates in the United States in 1995.

b. Find the number of prison and jail inmates in the United States in 2000.

c. If the trend described by the model continues, predict the year in which the number of prisoners will be 3,200,000 (that is, 3200 thousand).

87. The relationship between body weight and the Recommended Dietary Allowance (RDA) for vitamin A in children up to age 10 is modeled by the quadratic equation $y = 0.149x^2 - 4.475x + 406.478$, where y is the RDA for vitamin A in micrograms for a child whose weight is x pounds. (*Source:* Based on data from the Food and Nutrition Board, National Academy of Sciences—Institute of Medicine, 1989)

a. Determine the vitamin A requirements of a child who weighs 35 pounds.

b. What is the weight of a child whose RDA of vitamin A is 600 micrograms? Round your answer to the nearest pound.

88. The total amount of passenger traffic at Phoenix Sky Harbor International Airport in Phoenix, Arizona, during the period 1960 through 2000 can be modeled by the equation $y = 26x^2 - 136x + 917$, where y is the number of passengers enplaned and deplaned in thousands and x is the number of years after 1960. (*Source:* Based on data from The City of Phoenix Aviation Department, 1960–2000)

a. Estimate the passenger traffic at Phoenix Sky Harbor International Airport in 2000.

b. According to this model, in what year will passenger traffic at Phoenix Sky Harbor International Airport reach 60,000,000 passengers?

10.3 Solving Equations by Using Quadratic Methods

OBJECTIVES

A Solve various equations that are quadratic in form.

B Solve problems that lead to quadratic equations.

A Solving Equations That Are Quadratic in Form

In this section, we discuss various types of equations that can be solved in part by using the methods for solving quadratic equations.

Once each equation is simplified, you may want to use these steps when deciding what method to use to solve the quadratic equation.

Solving a Quadratic Equation

Step 1. If the equation is in the form $(ax + b)^2 = c$, use the square root property and solve. If not, go to Step 2.

Step 2. Write the equation in standard form: $ax^2 + bx + c = 0$.

Step 3. Try to solve the equation by the factoring method. If not possible, go to Step 4.

Step 4. Solve the equation by the quadratic formula.

The first example is a radical equation that becomes a quadratic equation once we square both sides.

EXAMPLE 1 Solve: $x - \sqrt{x} - 6 = 0$

Solution: Recall that to solve a radical equation, we first get the radical alone on one side of the equation. Then we square both sides.

$$x - 6 = \sqrt{x} \quad \text{Add } \sqrt{x} \text{ to both sides.}$$
$$x^2 - 12x + 36 = x \quad \text{Square both sides.}$$
$$x^2 - 13x + 36 = 0 \quad \text{Set the equation equal to 0.}$$
$$(x - 9)(x - 4) = 0$$
$$x - 9 = 0 \quad \text{or} \quad x - 4 = 0 \quad \text{Set each factor equal to 0.}$$
$$x = 9 \qquad x = 4 \quad \text{Solve.}$$

Check: Let $x = 9$.

$$x - \sqrt{x} - 6 = 0$$
$$9 - \sqrt{9} - 6 \stackrel{?}{=} 0$$
$$9 - 3 - 6 \stackrel{?}{=} 0$$
$$0 = 0 \quad \text{True.}$$

Let $x = 4$.

$$x - \sqrt{x} - 6 = 0$$
$$4 - \sqrt{4} - 6 \stackrel{?}{=} 0$$
$$4 - 2 - 6 \stackrel{?}{=} 0$$
$$-4 = 0 \quad \text{False.}$$

The solution set is $\{9\}$.

Practice Problem 1

Solve: $x - \sqrt{x - 1} - 3 = 0$

EXAMPLE 2 Solve: $\dfrac{3x}{x - 2} - \dfrac{x + 1}{x} = \dfrac{6}{x(x - 2)}$

Solution: In this equation, x cannot be either 2 or 0 because these values cause denominators to equal zero. To solve for x, we first multiply both sides of the equation by $x(x - 2)$ to clear the fractions. By the distributive property, this means that we multiply each term by $x(x - 2)$.

$$x(x - 2)\left(\frac{3x}{x - 2}\right) - x(x - 2)\left(\frac{x + 1}{x}\right) = x(x - 2)\left[\frac{6}{x(x - 2)}\right]$$
$$3x^2 - (x - 2)(x + 1) = 6 \quad \text{Simplify.}$$
$$3x^2 - (x^2 - x - 2) = 6 \quad \text{Multiply.}$$
$$3x^2 - x^2 + x + 2 = 6$$
$$2x^2 + x - 4 = 0 \quad \text{Simplify.}$$

Practice Problem 2

Solve: $\dfrac{2x}{x - 1} - \dfrac{x + 2}{x} = \dfrac{5}{x(x - 1)}$

Answers

1. $\{5\}$, **2.** $\left\{\dfrac{1 + \sqrt{13}}{2}, \dfrac{1 - \sqrt{13}}{2}\right\}$

This equation cannot be factored using integers, so we solve by the quadratic formula.

$$x = \frac{-1 \pm \sqrt{1^2 - 4(2)(-4)}}{2 \cdot 2}$$ Let $a = 2$, $b = 1$, and $c = -4$, in the quadratic formula.

$$= \frac{-1 \pm \sqrt{1 + 32}}{4}$$ Simplify.

$$= \frac{-1 \pm \sqrt{33}}{4}$$

Neither proposed solution will make the denominators 0.

The solution set is $\left\{\frac{-1 + \sqrt{33}}{4}, \frac{-1 - \sqrt{33}}{4}\right\}$.

Practice Problem 3

Solve: $x^4 - 5x^2 - 36 = 0$

Concept Check

a. True or false? The maximum number of solutions that a quadratic equation can have is 2.
b. True or false? The maximum number of solutions that an equation in quadratic form can have is 2.

EXAMPLE 3 Solve: $p^4 - 3p^2 - 4 = 0$

Solution: First we factor the trinomial.

$$p^4 - 3p^2 - 4 = 0$$

$$(p^2 - 4)(p^2 + 1) = 0$$ Factor.

$$(p - 2)(p + 2)(p^2 + 1) = 0$$ Factor further.

$p - 2 = 0$ or $p + 2 = 0$ or $p^2 + 1 = 0$ Set each factor equal to 0 and solve.

$p = 2$ $\quad p = -2$ $\quad p^2 = -1$

$p = \pm\sqrt{-1} = \pm i$

The solution set is $\{2, -2, i, -i\}$.

Try the Concept Check in the margin.

Practice Problem 4

Solve: $(x + 4)^2 - (x + 4) - 6 = 0$

EXAMPLE 4 Solve: $(x - 3)^2 - 3(x - 3) - 4 = 0$

Solution: Notice that the quantity $(x - 3)$ is repeated in this equation. Sometimes it is helpful to substitute a variable (in this case other than x) for the repeated quantity. We will let $y = x - 3$. Then

$$(x - 3)^2 - 3(x - 3) - 4 = 0$$

becomes

$$y^2 - 3y - 4 = 0$$ Let $x - 3 = y$.

$$(y - 4)(y + 1) = 0$$ Factor.

To solve, we use the zero-factor property.

$y - 4 = 0$ or $y + 1 = 0$ Set each factor equal to 0.

$y = 4$ $\quad y = -1$ Solve.

To find values of x, we substitute back. That is, we substitute $x - 3$ for y.

$x - 3 = 4$ or $x - 3 = -1$

$x = 7$ $\quad x = 2$

Helpful Hint

When using substitution, don't forget to substitute back to the original variable.

Both 2 and 7 check. The solution is $\{2, 7\}$.

Answers

3. $\{3, -3, 2i, -2i\}$, **4.** $\{-1, -6\}$

Concept Check: **a.** true, **b.** false

EXAMPLE 5 Solve: $x^{2/3} - 5x^{1/3} + 6 = 0$

Solution: The key to solving this equation is recognizing that $x^{2/3} = (x^{1/3})^2$. We replace $x^{1/3}$ with m so that

$$(x^{1/3})^2 - 5x^{1/3} + 6 = 0$$

becomes

$$m^2 - 5m + 6 = 0$$

Now we solve by factoring.

$$m^2 - 5m + 6 = 0$$
$$(m - 3)(m - 2) = 0 \quad \text{Factor.}$$
$$m - 3 = 0 \quad \text{or} \quad m - 2 = 0 \quad \text{Set each factor equal to 0.}$$
$$m = 3 \qquad m = 2$$

Since $m = x^{1/3}$, we have

$$x^{1/3} = 3 \quad \text{or} \quad x^{1/3} = 2$$
$$x = 3^3 = 27 \quad \text{or} \quad x = 2^3 = 8$$

Both 8 and 27 check. The solution set is $\{8, 27\}$.

Helpful Hint

Example 3 can be solved using substitution also. Think of $p^4 - 3p^2 - 4 = 0$ as

$$(p^2)^2 - 3p^2 - 4 = 0$$

Then let $x = p^2$, and solve and substitute back. The solution set will be the same.

$$x^2 - 3x - 4 = 0$$

Practice Problem 5

Solve: $x^{2/3} - 7x^{1/3} + 10 = 0$

B Solving Problems That Lead to Quadratic Equations

The next example is a work problem. This problem is modeled by a rational equation that simplifies to a quadratic equation.

EXAMPLE 6 Finding Work Time

Together, an experienced word processor and an apprentice word processor can process a document in 6 hours. Alone, the experienced word processor can process the document 2 hours faster than the apprentice word processor can. Find the time in which each person can process the document alone.

Solution:

1. UNDERSTAND. Read and reread the problem. The key idea here is the relationship between the *time* (hours) it takes to complete the job and the *part of the job* completed in one unit of time (hour). For example, because they can complete the job together in 6 hours, the *part of the job* they can complete in 1 hour is $\frac{1}{6}$. We let

   x = the time in hours it takes the apprentice word processor to complete the job alone

   $x - 2$ = the time in hours it takes the experienced word processor to complete the job alone

Practice Problem 6

Together, Karen and Doug Lewis can clean a strip of beach in 5 hours. Alone, Karen can clean the strip of beach one hour faster than Doug. Find the time that each person can clean the strip of beach alone. Give an exact answer and a one-decimal-place approximation.

Answers

5. $\{8, 125\}$, **6.** Doug: $\frac{11 + \sqrt{101}}{2} \approx 10.5$ hr; Karen: $\frac{9 + \sqrt{101}}{2} \approx 9.5$ hr

We can summarize in a chart the information discussed.

	Total Hours to Complete Job	Part of Job Completed in 1 Hour
Apprentice word processor	x	$\frac{1}{x}$
Experienced word processor	$x - 2$	$\frac{1}{x-2}$
Together	6	$\frac{1}{6}$

2. TRANSLATE.

In words:	part of job completed by apprentice word processor in 1 hour	added to	part of job completed by experienced word processor in 1 hour	added to	part of job completed together in 1 hour
	↓	↓	↓	↓	↓
Translate:	$\frac{1}{x}$	$+$	$\frac{1}{x-2}$	$=$	$\frac{1}{6}$

3. SOLVE.

$$\frac{1}{x} + \frac{1}{x-2} = \frac{1}{6}$$

$$6x(x-2)\left(\frac{1}{x} + \frac{1}{x-2}\right) = 6x(x-2)\cdot\frac{1}{6}$$ Multiply both sides by the LCD $6x(x-2)$.

$$6x(x-2)\cdot\frac{1}{x} + 6x(x-2)\cdot\frac{1}{x-2} = 6x(x-2)\cdot\frac{1}{6}$$ Use the distributive property.

$$6(x-2) + 6x = x(x-2)$$
$$6x - 12 + 6x = x^2 - 2x$$
$$0 = x^2 - 14x + 12$$

Now we can substitute $a = 1$, $b = -14$, and $c = 12$ into the quadratic formula and simplify.

$$x = \frac{-(-14) \pm \sqrt{(-14)^2 - 4(1)(12)}}{2(1)} = \frac{14 \pm \sqrt{148}}{2}$$

Using a calculator or a square root table, we see that $\sqrt{148} \approx 12.2$ rounded to one decimal place. Thus,

$$x \approx \frac{14 \pm 12.2}{2}$$

$$x \approx \frac{14 + 12.2}{2} = 13.1 \quad \text{or} \quad x \approx \frac{14 - 12.2}{2} = 0.9$$

4. INTERPRET.

Check: If the apprentice word processor completes the job alone in 0.9 hours, the experienced word processor completes the job alone in $x - 2 = 0.9 - 2 = -1.1$ hours. Since this is not possible, we reject the solution 0.9. The approximate solution is thus 13.1 hours.

State: The apprentice word processor can complete the job alone in approximately 13.1 hours, and the experienced word processor can complete the job alone in approximately $x - 2 = 13.1 - 2 = 11.1$ hours.

EXAMPLE 7 Calculating Driving Speeds

Beach and Fargo are about 400 miles apart. A salesperson travels from Fargo to Beach one day at a certain speed. She returns to Fargo the next day and drives 10 miles per hour faster. Her total travel time was $14\frac{2}{3}$ hours. Find her speed to Beach and the return speed to Fargo.

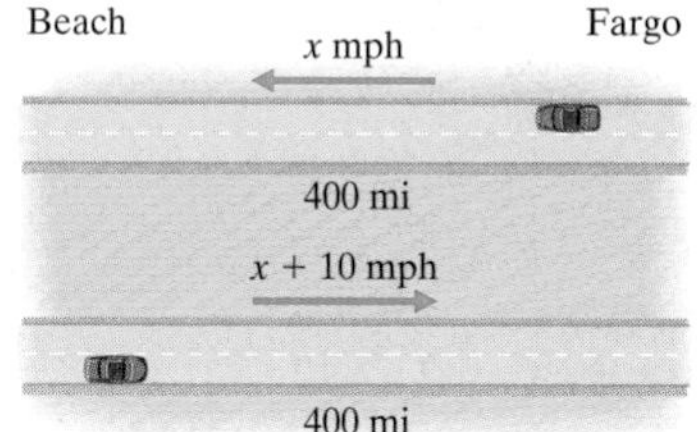

Solution:

1. UNDERSTAND. Read and reread the problem. Let

$x =$ the speed to Beach, so

$x + 10 =$ the return speed to Fargo

Then organize the given information in a table.

	Distance	= Rate	· Time
To Beach	400	x	$\frac{400}{x}$
Return to Fargo	400	$x + 10$	$\frac{400}{x+10}$

2. TRANSLATE.

In words: time to Beach + return time to Fargo = $14\frac{2}{3}$ hours

Translate: $\frac{400}{x} + \frac{400}{x+10} = \frac{44}{3}$

3. SOLVE.

$$\frac{400}{x} + \frac{400}{x+10} = \frac{44}{3}$$

$$\frac{100}{x} + \frac{100}{x+10} = \frac{11}{3}$$ Divide both sides by 4.

$$3x(x+10)\left(\frac{100}{x} + \frac{100}{x+10}\right) = 3x(x+10)\cdot\frac{11}{3}$$ Multiply both sides by the LCD, $3x(x + 10)$.

$$3x(x+10)\frac{100}{x} + 3x(x+10)\frac{100}{x+10} = 3x(x+10)\cdot\frac{11}{3}$$ Use the distributive property.

$$3(x+10)100 + 3x(100) = x(x+10)11$$

$$300x + 3000 + 300x = 11x^2 + 110x$$

$$0 = 11x^2 - 490x - 3000$$ Set equation equal to 0.

$$0 = (11x + 60)(x - 50)$$ Factor.

$11x + 60 = 0$ or $x - 50 = 0$ Set each factor equal to 0.

$x = -\frac{60}{11}$ or $-5\frac{5}{11}$ $x = 50$

Practice Problem 7

A family drives 500 miles to the beach for a vacation. The return trip was made at a speed that was 10 miles per hour faster. The total traveling time was $18\frac{1}{3}$ hours. Find the speed to the beach and the return speed.

Answer

7. 50 mph to the beach; 60 mph returning

4. INTERPRET.

Check: The speed is not negative, so it's not $-5\frac{5}{11}$. The number 50 does check.

State: The speed to Beach was 50 miles per hour and the return speed to Fargo was 60 miles per hour.

FOCUS ON Business and Career

FLOWCHARTS

We saw in the Focus On Business and Career feature in Chapter 6 that four of the top 10 fastest-growing jobs through 2008 are computer related. A useful skill in computer-related careers is *flowcharting*. A **flowchart** is a diagram showing a sequence of procedures used to complete a task. Flowcharts are commonly used in computer programming to help a programmer plan the steps and commands needed to write a program. Flowcharts are also used in other types of careers, such as manufacturing or finance, to describe the sequence of events needed in a certain process.

A flowchart usually uses the following symbols to represent certain types of actions.

Start or end of process | Individual operation or task description | Condition or decision | Connector •

For instance, suppose we want to write a flowchart for the process of computing a household's monthly electric bill. The flowchart might look something like the one here.

CRITICAL THINKING

1. Using the given flowchart as a guide, describe in words this utility company's pricing structure for household electricity usage.

2. Make a flowchart for the process of computing the human-equivalent age for the age of a dog if 1 dog year is equivalent to 7 human years.

3. Make a flowchart for the process of determining the number and type of solutions of a quadratic equation of the form $ax^2 + bx + c = 0$ using the discriminant.

4. (Optional) Write a programmable calculator program for your graphing calculator that determines the number and type of solutions of a quadratic equation of the form $ax^2 + bx + c = 0$.

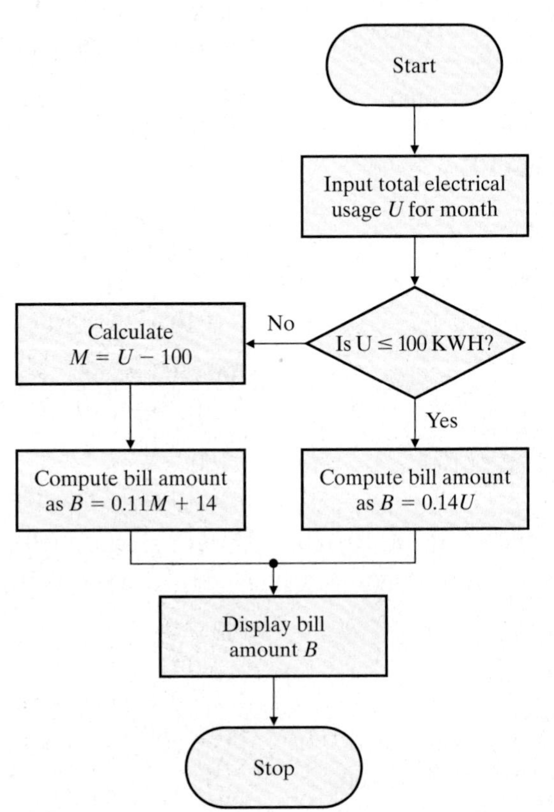

Name ______________________ Section __________ Date __________

EXERCISE SET 10.3

 *Solve. See Example 1.*

1. $2x = \sqrt{10 + 3x}$

2. $3x = \sqrt{8x + 1}$

3. $x - 2\sqrt{x} = 8$

4. $x - \sqrt{2x} = 4$

5. $\sqrt{9x} = x + 2$

6. $\sqrt{16x} = x + 3$

Solve. See Example 2.

7. $\dfrac{2}{x} + \dfrac{3}{x - 1} = 1$

8. $\dfrac{6}{x^2} = \dfrac{3}{x + 1}$

9. $\dfrac{3}{x} + \dfrac{4}{x + 2} = 2$

10. $\dfrac{5}{x - 2} + \dfrac{4}{x + 2} = 1$

11. $\dfrac{7}{x^2 - 5x + 6} = \dfrac{2x}{x - 3} - \dfrac{x}{x - 2}$

12. $\dfrac{11}{2x^2 + x - 15} = \dfrac{5}{2x - 5} - \dfrac{x}{x + 3}$

Solve. See Example 3.

13. $p^4 - 16 = 0$

14. $x^4 + 2x^2 - 3 = 0$

15. $4x^4 + 11x^2 = 3$

16. $z^4 = 81$

17. $z^4 - 13z^2 + 36 = 0$

18. $9x^4 + 5x^2 - 4 = 0$

Solve. See Examples 4 and 5.

19. $x^{2/3} - 3x^{1/3} - 10 = 0$

20. $x^{2/3} + 2x^{1/3} + 1 = 0$

21. $(5n + 1)^2 + 2(5n + 1) - 3 = 0$

22. $(m - 6)^2 + 5(m - 6) + 4 = 0$

23. $2x^{2/3} - 5x^{1/3} = 3$

24. $3x^{2/3} + 11x^{1/3} = 4$

25. $1 + \frac{2}{3t - 2} = \frac{8}{(3t - 2)^2}$

26. $2 - \frac{7}{x + 6} = \frac{15}{(x + 6)^2}$

27. $20x^{2/3} - 6x^{1/3} - 2 = 0$

28. $4x^{2/3} + 16x^{1/3} = -15$

Solve. See Examples 1 through 5.

29. $a^4 - 5a^2 + 6 = 0$

30. $x^4 - 12x^2 + 11 = 0$

31. $\frac{2x}{x - 2} + \frac{x}{x + 3} = \frac{-5}{x + 3}$

32. $\frac{5}{x - 3} + \frac{x}{x + 3} = \frac{19}{x^2 - 9}$

33. $(p + 2)^2 = 9(p + 2) - 20$

34. $2(4m - 3)^2 - 9(4m - 3) = 5$

35. $2x = \sqrt{11x + 3}$

36. $4x = \sqrt{2x + 3}$

37. $x^{2/3} - 8x^{1/3} + 15 = 0$

38. $x^{2/3} - 2x^{1/3} - 8 = 0$

39. $y^3 + 9y - y^2 - 9 = 0$

40. $x^3 + x - 3x^2 - 3 = 0$

41. $2x^{2/3} + 3x^{1/3} - 2 = 0$

42. $6x^{2/3} - 25x^{1/3} - 25 = 0$

43. $x^{-2} - x^{-1} - 6 = 0$

44. $y^{-2} - 8y^{-1} + 7 = 0$

45. $x - \sqrt{x} = 2$

46. $x - \sqrt{3x} = 6$

47. $\frac{x}{x - 1} + \frac{1}{x + 1} = \frac{2}{x^2 - 1}$

48. $\frac{x}{x - 5} + \frac{5}{x + 5} = \frac{-1}{x^2 - 25}$

49. $p^4 - p^2 - 20 = 0$

50. $x^4 - 10x^2 + 9 = 0$

51. $2x^3 = -54$

52. $y^3 - 216 = 0$

53. $1 = \dfrac{4}{x - 7} + \dfrac{5}{(x - 7)^2}$

54. $3 + \dfrac{1}{(2p + 4)} = \dfrac{10}{(2p + 4)^2}$

55. $27y^4 + 15y^2 = 2$

56. $8z^4 + 14z^2 = -5$

B *Solve. See Examples 6 and 7.*

57. A jogger ran 3 miles, decreased her speed by 1 mile per hour and then ran another 4 miles. If her total time jogging was $1\frac{3}{5}$ hours, find her speed for each part of her run.

58. Mark Keaton's workout consists of jogging for 3 miles, and then riding his bike for 5 miles at a speed 4 miles per hour faster than he jogs. If his total workout time is 1 hour, find his jogging speed and his biking speed.

59. A Chinese restaurant in Mandeville, Louisiana, has a large goldfish pond around the restaurant. Suppose that an inlet pipe and a hose together can fill the pond in 8 hours. The inlet pipe alone can complete the job in one hour less time than the hose alone. Find the time that the hose can complete the job alone and the time that the inlet pipe can complete the job alone. Round each to the nearest tenth of an hour.

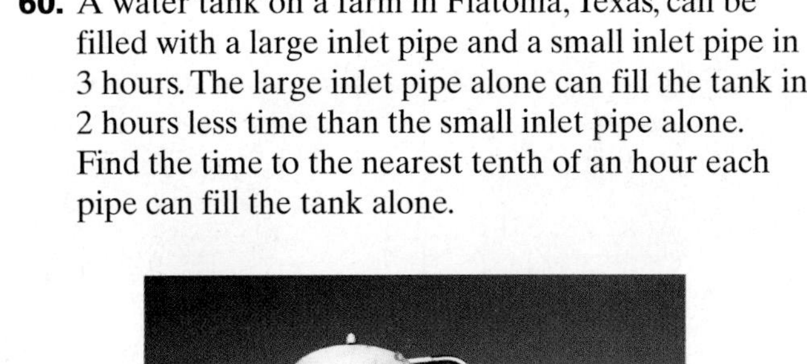

60. A water tank on a farm in Flatonia, Texas, can be filled with a large inlet pipe and a small inlet pipe in 3 hours. The large inlet pipe alone can fill the tank in 2 hours less time than the small inlet pipe alone. Find the time to the nearest tenth of an hour each pipe can fill the tank alone.

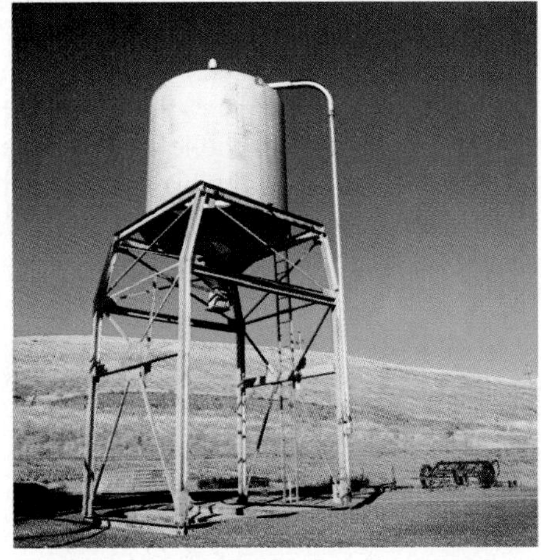

61. Roma Sherry drove 330 miles from her home town to Tucson. During her return trip, she was able to increase her speed by 11 miles per hour. If her return trip took 1 hour less time, find her original speed and her speed returning home.

62. A salesperson drove to Portland, a distance of 300 miles. During the last 80 miles of his trip, heavy rainfall forced him to decrease his speed by 15 miles per hour. If his total driving time was 6 hours, find his original speed and his speed during the rainfall.

63. Bill Shaughnessy and his son Billy can clean the house together in 4 hours. When the son works alone, it takes him an hour longer to clean than it takes his dad alone. Find how long to the nearest tenth of an hour it takes the son to clean alone.

64. Together, Noodles and Freckles eat a 50-pound bag of dog food in 30 days. Noodles by herself eats a 50-pound bag in 2 weeks less time than Freckles does by himself. How many days to the nearest whole day would a 50-pound bag of dog food last Freckles?

65. The product of a number and 4 less than the number is 96. Find the number.

66. A whole number increased by its square is two more than twice itself. Find the number.

△ **67.** Suppose that we want to make an open box from a square sheet of cardboard by cutting out squares from each corner as shown and then folding along the dotted lines. If the box is to have a volume of 300 cubic centimeters, find the original dimensions of the sheet of cardboard.

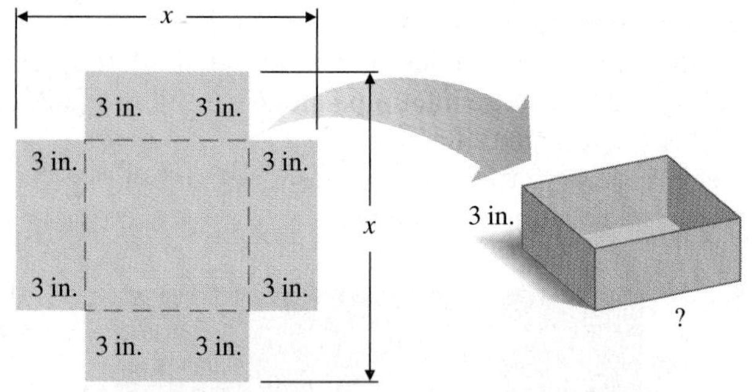

a. The ? in the drawing to the left will be the length (and also the width) of the box as shown in the drawing to the right. Represent this length in terms of x.

b. Use the formula for volume of a box, $V = l \cdot w \cdot h$, to write an equation in x.

c. Solve the equation for x and give the dimensions of the sheet of cardboard. Check your solution.

△ **68.** Suppose that we want to make an open box from a square sheet of cardboard by cutting out squares from each corner as shown and then folding along the dotted lines. If the box is to have a volume of 128 cubic inches, find the original dimensions of the sheet of cardboard. (*Hint:* Use Exercise 67 parts (a), (b), and (c) to help you.)

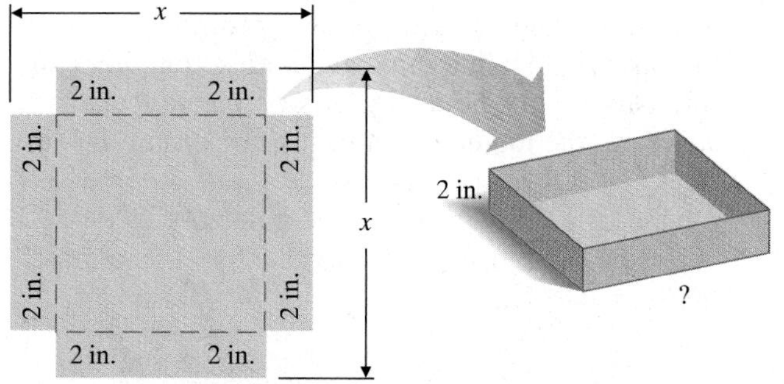

△ **69.** A sprinkler that sprays water in a circular motion is to be used to water a square garden. If the area of the garden is 920 square feet, find the smallest whole number *radius* that the sprinkler can be adjusted to so that the entire garden is watered.

△ **70.** Suppose that a square field has an area of 6270 square feet. See Exercise 69 and find a new sprinkler radius.

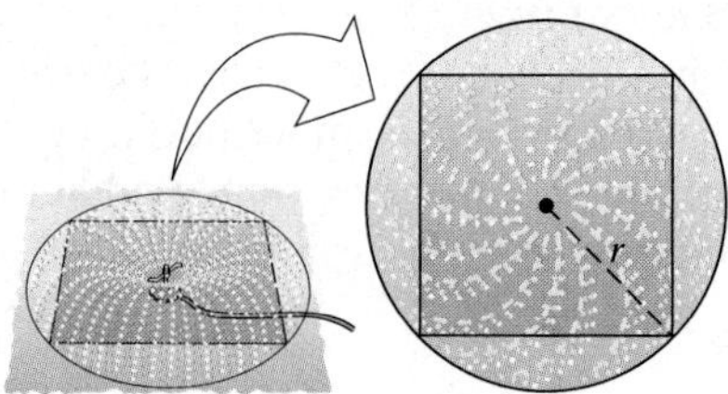

Review and Preview

Solve each inequality. See Section 2.8.

71. $\frac{5x}{3} + 2 \leq 7$

72. $\frac{2x}{3} + \frac{1}{6} \geq 2$

73. $\frac{y-1}{15} > \frac{-2}{5}$

74. $\frac{z-2}{12} < \frac{1}{4}$

Combining Concepts

75. Write a polynomial equation that has three solutions: 2, 5, and -7.

76. Write a polynomial equation that has three solutions: 0, $2i$, and $-2i$.

77. During the 2001 American Memorial auto race held at the Eurospeedway in Lausitz, Germany, Tony Kanaan posted the fastest speed but Kenny Brack won the race. The track is 10,682 feet long. Kanaan's fastest speed was 6.326 feet per second faster than Brack's fastest speed. Traveling at these fastest speeds, Brack would have taken 0.73 seconds longer than Kanaan to complete a lap. (*Source:* Based on data from Championship Auto Racing Teams, Inc.)

a. Find Kenny Brack's fastest speed during the race. Round to three decimal places.

b. Find Tony Kanaan's fastest speed during the race. Round to three decimal places.

c. Convert Kanaan's speed to miles per hour. Round to three decimal places.

78. Use a grapher to solve Exercise 29. Compare the solution with the solution from Exercise 29. Explain any differences.

FOCUS ON Business and Career

FINANCIAL RATIOS

A financial ratio is a number found with a rational expression that tells something about a company's activities. Such ratios allow a comparison between the financial positions of two companies, even if the values of the companies' financial data are very different. Here are some common financial ratios:

- The **current ratio** gauges a company's ability to pay its short-term debts. It is given by the formula

$$\text{current ratio} = \frac{\text{total current assets}}{\text{total current liabilities}}$$

The higher the value of this ratio, the better able the company is to pay off its short-term debts.

- The **total asset turnover ratio** gauges how effectively a company is using all of its resources to generate sales of its products and services. It is given by the formula

$$\text{total asset turnover ratio} = \frac{\text{sales}}{\text{total assets}}$$

The higher the value of this ratio, the more effective the company is at utilizing its resources for sales generation.

- The **gross profit margin ratio** gauges how effectively the company is making pricing decisions as well as controlling production costs. It is given by the formula

$$\text{gross profit margin ratio} = \frac{\text{sales} - \text{cost of sales}}{\text{sales}}$$

The higher the value of this ratio, the better the company is controlling costs and pricing products.

- The **price-to-earnings (P/E) ratio** gauges the stock market's view of a company with respect to risk. It is given by the formula

$$\text{P/E ratio} = \frac{\text{stock market price per share}}{\text{earnings per share}}$$

A high P/E ratio generally means a company with low risk. A high P/E ratio also translates into better growth potential for the company's earnings.

For all of these ratios, a higher-than-industry-average ratio is generally considered to be a sign of good financial health.

ADDITIONAL DEFINITIONS

- **Assets**—things of value that are owned by a company. *Current assets* include cash and assets that can be converted into cash quickly. *Total assets* are all things of value, including property and equipment, owned by the company. Current assets and total assets may be found on a company's consolidated balance sheet or statement of financial position in an annual report.
- **Liabilities**—what a company owes to creditors. *Current liabilities* include any debts expected to come due within the next year. Current liabilities may be found on a company's consolidated balance sheet or statement of financial position in an annual report.
- **Sales**—the total amount of money collected by a company from the sales of its goods or services. Sales (or sometimes noted as "net sales") may be found on a company's consolidated statement of income/earnings/operations in an annual report.
- **Cost of sales**—a company's cost of inventory actually sold to customers. This is also sometimes referred to as "cost of goods/merchandise sold." Cost of sales may be found on a company's consolidated statement of income/earnings/operations in an annual report.
- **Earnings per share**—the value of a company's earnings available for each share of common stock held by stockholders. Earnings per share (or sometimes noted as net income per share) may be found on a company's consolidated statement of income/earnings/operations in an annual report.
- **Stock market price per share**—the current price of a company's share of stock as given on one of the major stock markets. Current share prices may be found in newspapers or on the World Wide Web.

GROUP ACTIVITY

Locate annual reports for two companies involved in similar industries. Using the information and definitions given above, compute these four financial ratios for each company. Then compare the companies' ratios and discuss what the ratios indicate about the two companies. Which company do you think is in better overall financial health? Why?

Name ______________ Section ________ Date ________

Integrated Review—Summary on Solving Quadratic Equations

Use the square root property to solve each equation.

1. $x^2 - 10 = 0$

2. $x^2 - 14 = 0$

3. $(x - 1)^2 = 8$

4. $(x + 5)^2 = 12$

Solve each equation by completing the square.

5. $x^2 + 2x - 12 = 0$

6. $x^2 - 12x + 11 = 0$

7. $3x^2 + 3x = 5$

8. $16y^2 + 16y = 1$

Use the quadratic formula to solve each equation.

9. $2x^2 - 4x + 1 = 0$

10. $\frac{1}{2}x^2 + 3x + 2 = 0$

11. $x^2 + 4x = -7$

12. $x^2 + x = -3$

Solve each equation. Use a method of your choice.

13. $x^2 + 3x + 6 = 0$

14. $2x^2 + 18 = 0$

15. $x^2 + 17x = 0$

Answers

1. ________
2. ________
3. ________
4. ________
5. ________
6. ________
7. ________
8. ________
9. ________
10. ________
11. ________
12. ________
13. ________
14. ________
15. ________

16. $4x^2 - 2x - 3 = 0$

17. $(x - 2)^2 = 27$

18. $\frac{1}{2}x^2 - 2x + \frac{1}{2} = 0$

19. $3x^2 + 2x = 8$

20. $2x^2 = -5x - 1$

21. $x(x - 2) = 5$

22. $x^2 - 31 = 0$

23. $5x^2 - 55 = 0$

24. $5x^2 + 55 = 0$

25. $x(x + 5) = 66$

26. $5x^2 + 6x - 2 = 0$

27. $2x^2 + 3x = 1$

△**28.** The diagonal of a square room measures 20 feet. Find the exact length of a side of the room. Then approximate the length to the nearest tenth of a foot.

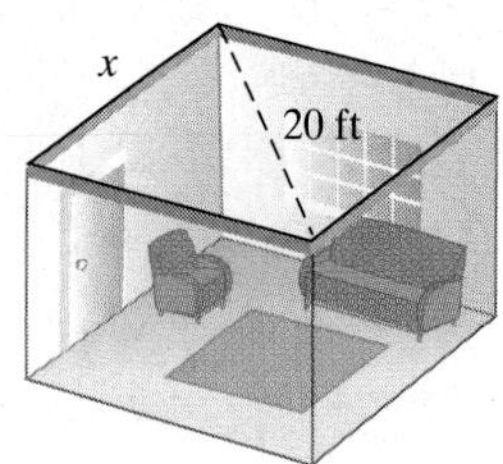

29. Diane Gray and Lucy Hoag together can prepare a crawfish boil for a large party in 4 hours. Lucy alone can complete the job in 2 hours less time than Diane alone. Find the time that each person can prepare the crawfish boil alone. Round each time to the nearest tenth of an hour.

30. Kraig Blackwelder exercises at Total Body Gym. On the treadmill, he runs 5 miles, then increases his speed by 1 mile per hour and runs an additional 2 miles. If his total time on the treadmill is $1\frac{1}{3}$ hours, find his speed during each part of his run.

16. ______

17. ______

18. ______

19. ______

20. ______

21. ______

22. ______

23. ______

24. ______

25. ______

26. ______

27. ______

28. ______

29. ______

30. ______

10.4 Nonlinear Inequalities in One Variable

OBJECTIVES

A Solve polynomial inequalities of degree 2 or greater.

B Solve inequalities that contain rational expressions with variables in the denominator.

A Solving Polynomial Inequalities

Just as we can solve linear inequalities in one variable, we can also solve quadratic inequalities in one variable. A **quadratic inequality** is an inequality that can be written so that one side is a quadratic expression and the other side is 0. Here are examples of quadratic inequalities in one variable. Each is written in **standard form**.

$$x^2 - 10x + 7 \le 0 \qquad 3x^2 + 2x - 6 > 0$$
$$2x^2 + 9x - 2 < 0 \qquad x^2 - 3x + 11 \ge 0$$

A solution of a quadratic inequality in one variable is a value of the variable that makes the inequality a true statement.

The value of an expression such as $x^2 - 3x - 10$ will sometimes be positive, sometimes negative, and sometimes 0, depending on the value substituted for x. To solve the inequality $x^2 - 3x - 10 < 0$, we look for all values of x that make the expression $x^2 - 3x - 10$ **less than 0**, or **negative**. To understand how we find these values, we'll study the graph of the quadratic function $y = x^2 - 3x - 10$.

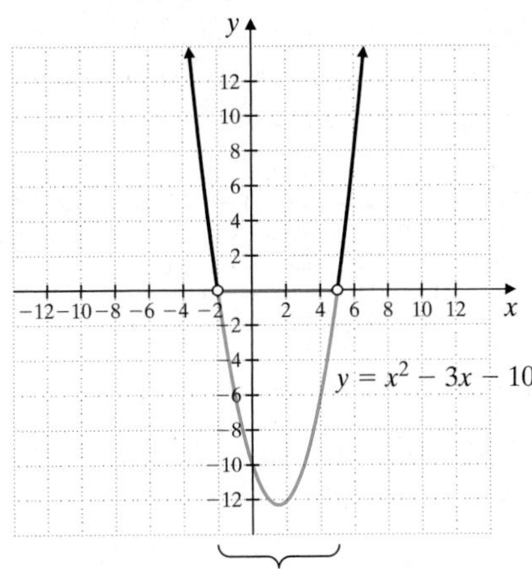

Notice that the x-values for which y or $x^2 - 3x - 10$ is positive are separated from the x values for which y or $x^2 - 3x - 10$ is negative by the values for which y or $x^2 - 3x - 10$ is 0, the x-intercepts. Thus, the solution set of $x^2 - 3x - 10 < 0$ consists of all real numbers from -2 to 5 or, in interval notation, $(-2, 5)$.

It is not necessary to graph $y = x^2 - 3x - 10$ to solve the related inequality $x^2 - 3x - 10 < 0$. Instead, we can draw a number line representing the x-axis and keep the following in mind: *A region on the number line for which the value of $x^2 - 3x - 10$ is positive is separated from a region on the number line for which the value of $x^2 - 3x - 10$ is negative by a value for which the expression is 0.*

Let's find these values for which the expression is 0 by solving the related equation, $x^2 - 3x - 10 = 0$.

$$x^2 - 3x - 10 = 0$$
$$(x - 5)(x + 2) = 0 \qquad \text{Factor.}$$
$$x - 5 = 0 \quad \text{or} \quad x + 2 = 0 \qquad \text{Set each factor equal to 0.}$$
$$x = 5 \qquad\qquad x = -2 \qquad \text{Solve.}$$

These two numbers −2 and 5 divide the number line into three regions. We will call the regions A, B, and C. These regions are important because if the value of $x^2 - 3x - 10$ is negative when a number from a region is substituted for x, then $x^2 - 3x - 10$ is negative when any number in that region is substituted for x. Similarly, if the value of $x^2 - 3x - 10$ is positive when a number from a region is substituted for x, then $x^2 - 3x - 10$ is positive when any number in that region is substituted for x.

To see whether the inequality $x^2 - 3x - 10 < 0$ is true or false in each region, we choose a test point from each region and substitute its value for x in the inequality $x^2 - 3x - 10 < 0$. If the resulting inequality is true, the region containing the test point is a solution region.

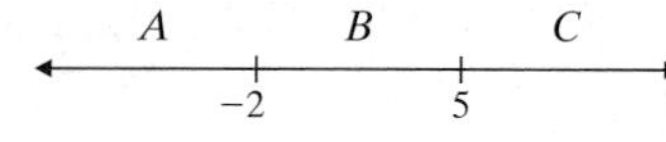

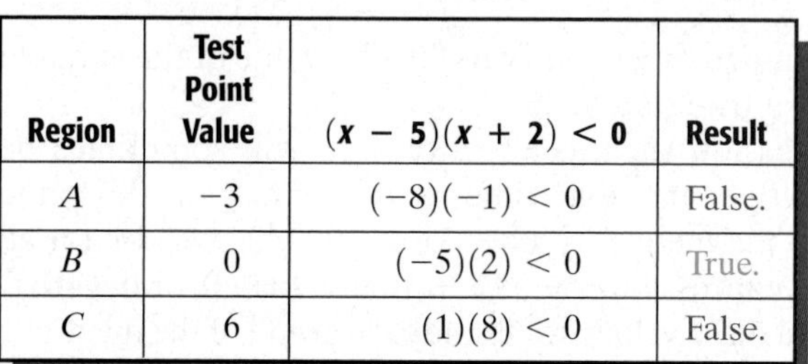

Region	Test Point Value	$(x - 5)(x + 2) < 0$	Result
A	−3	$(-8)(-1) < 0$	False.
B	0	$(-5)(2) < 0$	True.
C	6	$(1)(8) < 0$	False.

The values in region B satisfy the inequality. The numbers −2 and 5 are not included in the solution set since the inequality symbol is <. The solution set is $(-2, 5)$, and its graph is shown.

Practice Problem 1

Solve: $(x - 2)(x + 4) > 0$

EXAMPLE 1 Solve: $(x + 3)(x - 3) > 0$

Solution: First we solve the related equation, $(x + 3)(x - 3) = 0$.

$$(x + 3)(x - 3) = 0$$

$$x + 3 = 0 \quad \text{or} \quad x - 3 = 0$$

$$x = -3 \qquad\qquad x = 3$$

The two numbers −3 and 3 separate the number line into three regions, A, B, and C.

Now we substitute the value of a test point from each region. If the test value satisfies the inequality, every value in the region containing the test value is a solution.

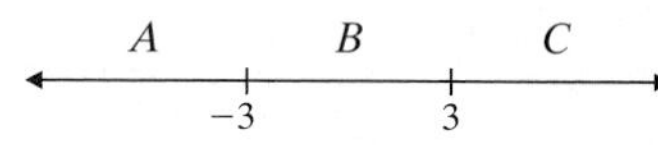

Region	Test Point Value	$(x + 3)(x - 3) > 0$	Result
A	−4	$(-1)(-7) > 0$	True.
B	0	$(3)(-3) > 0$	False.
C	4	$(7)(1) > 0$	True.

The points in regions A and C satisfy the inequality. The numbers −3 and 3 are not included in the solution since the inequality symbol is >. The solution set is $(-\infty, -3) \cup (3, \infty)$, and its graph is shown.

Answer

1. $(-\infty, -4) \cup (2, \infty)$

The steps below may be used to solve a polynomial inequality.

Solving a Polynomial Inequality

Step 1. Write the inequality in standard form and then solve the related equation.
Step 2. Separate the number line into regions with the solutions from Step 1.
Step 3. For each region, choose a test point and determine whether its value satisfies the *original inequality*.
Step 4. The solution set includes the regions whose test point value is a solution. If the inequality symbol is $\leq$ or $\geq$, the values from Step 1 are solutions; if $<$ or $>$, they are not.

Try the Concept Check in the margin.

EXAMPLE 2 Solve: $x^2 - 4x \leq 0$

Solution: First we solve the related equation, $x^2 - 4x = 0$.

$$x^2 - 4x = 0$$
$$x(x - 4) = 0$$
$$x = 0 \quad \text{or} \quad x = 4$$

The numbers 0 and 4 separate the number line into three regions, A, B and C.

A B C
0 4

We check a test value in each region in the original inequality. Values in region B satisfy the inequality. The numbers 0 and 4 are included in the solution since the inequality symbol is $\leq$. The solution set is $[0, 4]$, and its graph is shown.

A B C
F 0 T 4 F

EXAMPLE 3 Solve: $(x + 2)(x - 1)(x - 5) \leq 0$

Solution: First we solve $(x + 2)(x - 1)(x - 5) = 0$. By inspection, we see that the solutions are -2, 1, and 5. They separate the number line into four regions, A, B, C, and D. Next we check test points from each region.

Region	Test Point Value	$(x + 2)(x - 1)(x - 5) \leq 0$	Result
A	-3	$(-1)(-4)(-8) \leq 0$	True.
B	0	$(2)(-1)(-5) \leq 0$	False.
C	2	$(4)(1)(-3) \leq 0$	True.
D	6	$(8)(5)(1) \leq 0$	False.

The solution set is $(-\infty, -2] \cup [1, 5]$, and its graph is shown. We include the numbers -2, 1, and 5 because the inequality symbol is $\leq$.

A B C D
T −2 F 1 T 5 F

Concept Check

When choosing a test point in Step 4, why would the solutions from Step 2 not make good choices for test points?

Practice Problem 2

Solve: $x^2 - 6x \leq 0$

Practice Problem 3

Solve: $(x - 2)(x + 1)(x + 5) \leq 0$

A B C D
−2 1 5

Answers

2. $[0, 6]$, **3.** $(-\infty, -5] \cup [-1, 2]$

Concept Check: The solutions found in Step 2 have a value of 0 in the original inequality.

B Solving Rational Inequalities

Inequalities containing rational expressions with variables in the denominator are solved by using a similar procedure.

Practice Problem 4

Solve: $\dfrac{x-3}{x+5} \le 0$

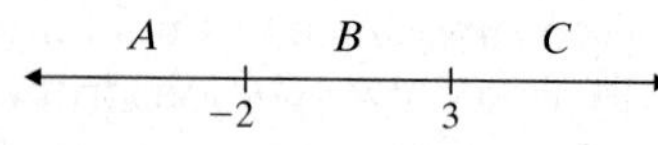

EXAMPLE 4 Solve: $\dfrac{x+2}{x-3} \le 0$

Solution: First we find all values that make the denominator equal to 0. To do this, we solve $x - 3 = 0$, or $x = 3$.

Next, we solve the related equation, $\dfrac{x+2}{x-3} = 0$.

$$\frac{x+2}{x-3} = 0 \qquad \text{Multiply both sides by the LCD, } x - 3.$$
$$x + 2 = 0$$
$$x = -2$$

Now we place these numbers on a number line and proceed as before, checking test point values in the original inequality.

Choose −3 from region *A*.

$$\frac{x+2}{x-3} \le 0$$
$$\frac{-3+2}{-3-3} \le 0$$
$$\frac{-1}{-6} \le 0$$
$$\frac{1}{6} \le 0 \quad \text{False.}$$

Choose 0 from region *B*.

$$\frac{x+2}{x-3} \le 0$$
$$\frac{0+2}{0-3} \le 0$$
$$-\frac{2}{3} \le 0 \quad \text{True.}$$

Choose 4 from region *C*.

$$\frac{x+2}{x-3} \le 0$$
$$\frac{4+2}{4-3} \le 0$$
$$6 \le 0 \quad \text{False.}$$

The solution set is $[-2, 3)$. This interval includes -2 because -2 satisfies the original inequality. This interval does not include 3 because 3 would make the denominator 0.

A B C
F −2 T 3 F

The steps below may be used to solve a rational inequality with variables in the denominator.

Solving a Rational Inequality

Step 1. Solve for values that make all denominators 0.

Step 2. Solve the related equation.

Step 3. Separate the number line into regions with the solutions from Steps 1 and 2.

Step 4. For each region, choose a test point and determine whether its value satisfies the *original inequality*.

Step 5. The solution set includes the regions whose test point value is a solution. Check whether to include values from Step 2. Be sure *not* to include values that make any denominator 0.

Answer

4. $(-5, 3]$

EXAMPLE 5 Solve: $\frac{5}{x+1} < -2$

Solution: First we find values for x that make the denominator equal to 0.

$$x + 1 = 0$$
$$x = -1$$

Next we solve $\frac{5}{x+1} = -2$.

$$(x+1) \cdot \frac{5}{x+1} = (x+1) \cdot -2 \quad \text{Multiply both sides by the LCD, } x + 1.$$
$$5 = -2x - 2 \quad \text{Simplify.}$$
$$7 = -2x$$
$$-\frac{7}{2} = x$$

A B C

$-\frac{7}{2}$ -1

We use these two solutions to divide a number line into three regions and choose test points. Only a test point value from region B satisfies the *original inequality*. The solution set is $\left(-\frac{7}{2}, -1\right)$, and its graph is shown.

A B C

F $-\frac{7}{2}$ T -1 F

Practice Problem 5

Solve: $\frac{3}{x-2} < 2$

Answer

5. $(-\infty, 2) \cup \left(\frac{7}{2}, \infty\right)$

FOCUS ON History

THE EVOLUTION OF SOLVING QUADRATIC EQUATIONS

The ancient Babylonians (circa 2000 B.C.) are sometimes credited with being the first to solve quadratic equations. This is only partially true because the Babylonians had no concept of an equation. However, what they did develop was a method for completing the square to apply to problems that today would be solved with a quadratic equation. The Babylonians only recognized positive solutions to such problems and did not acknowledge the existence of negative solutions at all.

Babylonian mathematical knowledge influenced much of the ancient world, most notably Hindu Indians. The Hindus were the first culture to denote debts in everyday business affairs with negative numbers. With this level of comfort with negative numbers, the Indian mathematician Brahmagupta (598–665 A.D.) extended the Babylonian methods and was the first to recognize negative solutions to quadratic equations. Later Hindu mathematicians noted that every positive number has two square roots: a positive square root and a negative square root. Hindus allowed irrational solutions (quite an innovation in the ancient world!) to quadratic equations and were the first to realize that quadratic equations could have 0, 1, or 2 real number solutions. They did not, however, acknowledge complex numbers and, therefore, could not solve equations with solutions requiring the square root of a negative number.

Complex numbers were finally developed by European mathematicians during the 17th and 18th centuries. Up until that time, what we would today consider complex solutions to quadratic equations were routinely ignored by mathematicians.

Name ______________________ Section __________ Date __________

EXERCISE SET 10.4

A *Solve. See Examples 1 through 3.*

1. $(x + 1)(x + 5) > 0$

2. $(x + 1)(x + 5) \leq 0$

3. $(x - 3)(x + 4) \leq 0$

4. $(x + 4)(x - 1) > 0$

5. $x^2 - 7x + 10 \leq 0$

6. $x^2 + 8x + 15 \geq 0$

7. $3x^2 + 16x < -5$

8. $2x^2 - 5x < 7$

9. $(x - 6)(x - 4)(x - 2) > 0$

10. $(x - 6)(x - 4)(x - 2) \leq 0$

11. $x(x - 1)(x + 4) \leq 0$

12. $x(x - 6)(x + 2) > 0$

13. $(x^2 - 9)(x^2 - 4) > 0$

14. $(x^2 - 16)(x^2 - 1) \leq 0$

15. $x^2 - x - 56 > 0$

16. $x^2 - 4x - 5 < 0$

17. $6x^2 + 5x \leq 4$

18. $12x^2 - 5x \geq 3$

19. $x^2 > x$

20. $x^2 < 25$

21. $(2x - 8)(x + 4)(x - 6) \leq 0$

22. $(3x - 12)(x + 5)(2x - 3) \geq 0$

B *Solve. See Examples 4 and 5.*

23. $\dfrac{x + 7}{x - 2} < 0$

24. $\dfrac{x - 5}{x - 6} > 0$

25. $\dfrac{5}{x + 1} > 0$

26. $\dfrac{3}{y - 5} < 0$

27. $\dfrac{x + 1}{x - 4} \geq 0$

28. $\dfrac{x + 1}{x - 4} \leq 0$

29. $\dfrac{x + 2}{x - 3} < 1$

30. $\dfrac{x - 1}{x + 4} > 2$

31. $\dfrac{x}{x - 10} < 0$

32. $\dfrac{x + 10}{x - 10} > 0$

33. $\dfrac{x - 5}{x + 4} \geq 0$

34. $\dfrac{x - 3}{x + 2} \leq 0$

35. $\dfrac{x(x + 6)}{(x - 7)(x + 1)} \geq 0$

36. $\dfrac{(x - 2)(x + 2)}{(x + 1)(x - 4)} \leq 0$

37. $\dfrac{-1}{x - 1} > -1$

38. $\dfrac{4}{y + 2} < -2$

39. $\dfrac{x}{x + 4} \leq 2$

40. $\dfrac{4x}{x - 3} \geq 5$

Review and Preview

Fill in each table so that each ordered pair is a solution of the given function. See Section 7.3.

41. $f(x) = x^2$

x	y
0	
1	
−1	
2	
−2	

42. $f(x) = 2x^2$

x	y
0	
1	
−1	
2	
−2	

43. $f(x) = -x^2$

x	y
0	
1	
−1	
2	
−2	

44. $f(x) = -3x^2$

x	y
0	
1	
−1	
2	
−2	

Combining Concepts

45. Explain why $\frac{x+2}{x-3} > 0$ and $(x+2)(x-3) > 0$ have the same solutions.

46. Explain why $\frac{x+2}{x-3} \geq 0$ and $(x+2)(x-3) \geq 0$ do not have the same solutions.

Find all numbers that satisfy each statement.

47. A number minus its reciprocal is less than zero. Find the numbers.

48. Twice a number added to its reciprocal is nonnegative. Find the numbers.

49. The total profit $P(x)$ for a company producing x thousand units is given by the function $P(x) = -2x^2 + 26x - 44$. Find the values of x for which the company makes a profit. [*Hint:* The company makes a profit when $P(x) > 0$.]

50. A projectile is fired straight up from the ground with an initial velocity of 80 feet per second. Its height $s(t)$ in feet at any time t in seconds is given by the function $s(t) = -16t^2 + 80t$. Find the interval of time for which the height of the projectile is greater than 96 feet.

Use a graphing calculator to check each exercise.

51. Exercise 15

52. Exercise 16

10.5 Quadratic Functions and Their Graphs

OBJECTIVES

- **A** Graph quadratic functions of the form $f(x) = x^2 + k$.
- **B** Graph quadratic functions of the form $f(x) = (x - h)^2$.
- **C** Graph quadratic functions of the form $f(x) = (x - h)^2 + k$.
- **D** Graph quadratic functions of the form $f(x) = ax^2$.
- **E** Graph quadratic functions of the form $f(x) = a(x - h)^2 + k$.

A Graphing $f(x) = x^2 + k$

In Section 7.4, we defined polynomial functions. A special type of polynomial function is a *quadratic function*.

Quadratic Function

A **quadratic function** is a function that can be written in the form $f(x) = ax^2 + bx + c$, where $a, b,$ and c are real numbers and $a \neq 0$.

Notice that equations of the form $y = ax^2 + bx + c$, where $a \neq 0$, also define quadratic functions since y is a function of x or $y = f(x)$.

As we shall see in Example 1, the graph of a quadratic function such as $f(x) = ax^2 + bx + c$ is a curve called a parabola. In general, if $a > 0$, the parabola opens upward and if $a < 0$, the parabola opens downward. Also, the vertex of a parabola is the lowest point if the parabola opens upward and the highest point if the parabola opens downward. The axis of symmetry is the vertical line that passes through the vertex. Notice that a parabola is the mirror image of itself across its axis of symmetry.

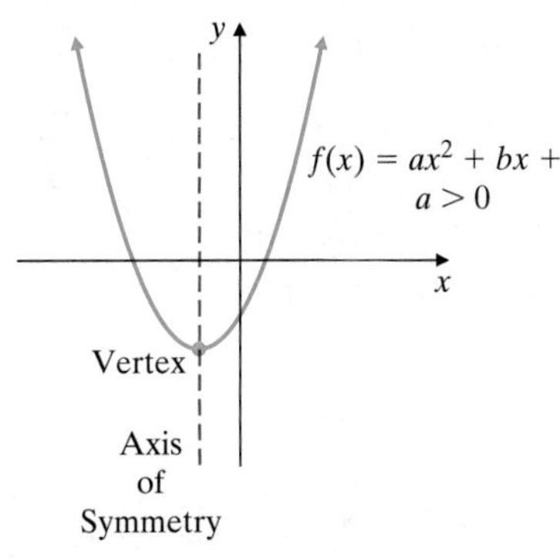

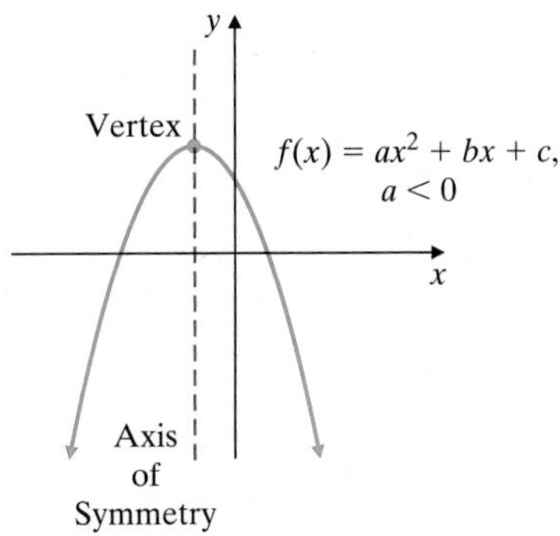

EXAMPLE 1

Graph $f(x) = x^2$ and $g(x) = x^2 + 3$ on the same set of axes.

Solution: First we construct a table of values for f and plot the points. Notice that for each x-value, the corresponding value of $g(x)$ must be 3 more than the corresponding value of $f(x)$ since $f(x) = x^2$ and $g(x) = x^2 + 3$. In other words, the graph of $g(x) = x^2 + 3$ is the same as the graph of $f(x) = x^2$ shifted upward 3 units. The axis of symmetry for both graphs is the y-axis.

x	$f(x) = x^2$	$g(x) = x^2 + 3$
−2	4	7
−1	1	4
0	0	3
1	1	4
2	4	7

Each y-value is increased by 3.

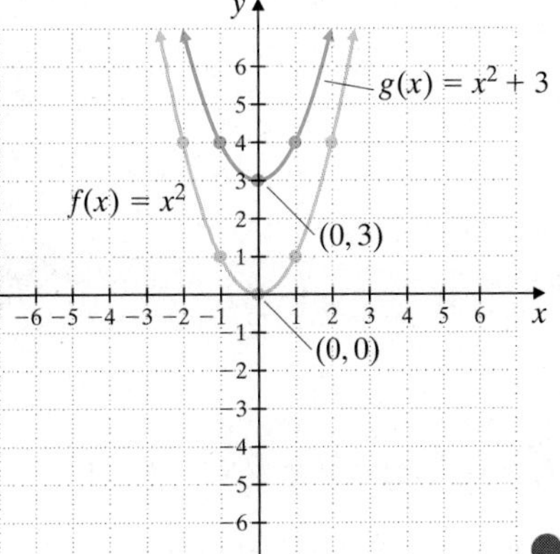

Practice Problem 1

Graph $f(x) = x^2$ and $g(x) = x^2 + 4$ on the same set of axes.

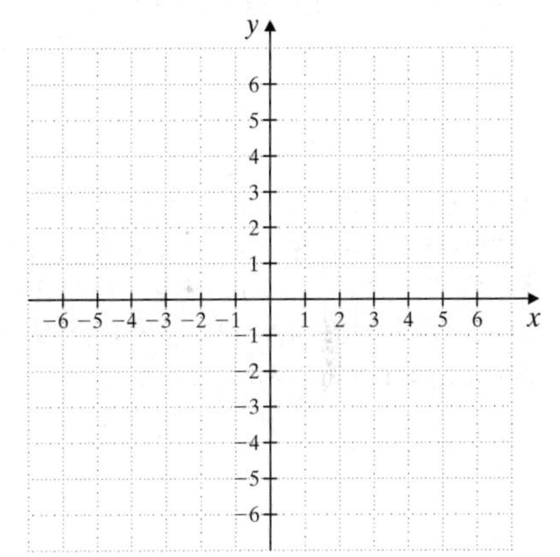

Answer

1.

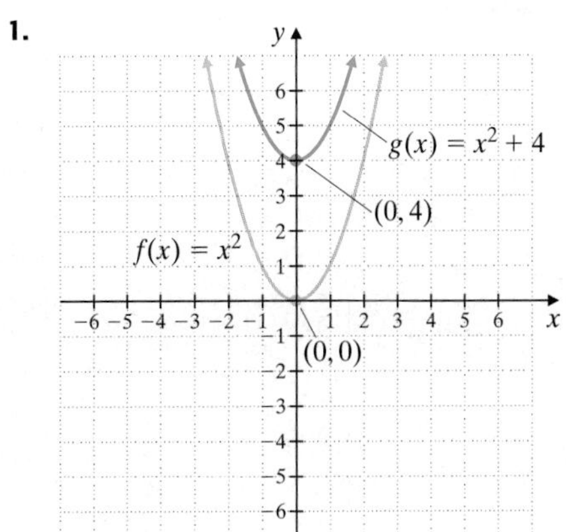

Practice Problems 2–3

Graph each function.

2. $F(x) = x^2 + 1$

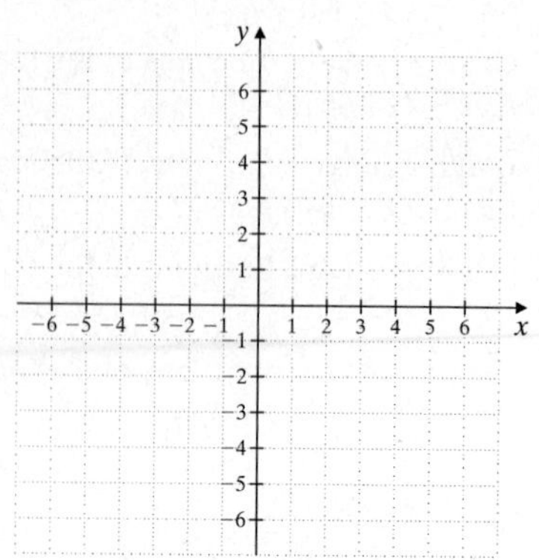

3. $g(x) = x^2 - 2$

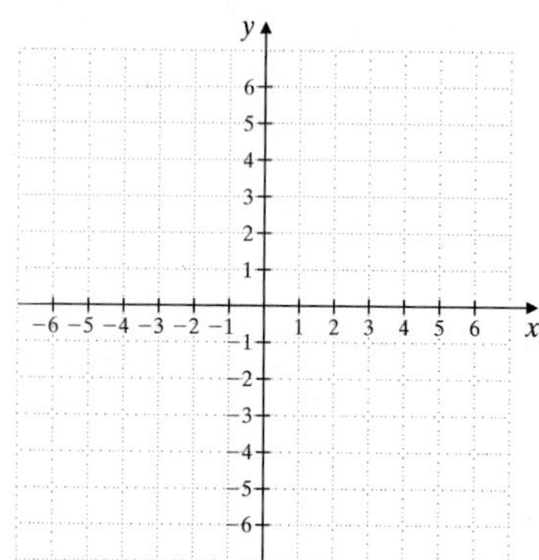

Answers

2.

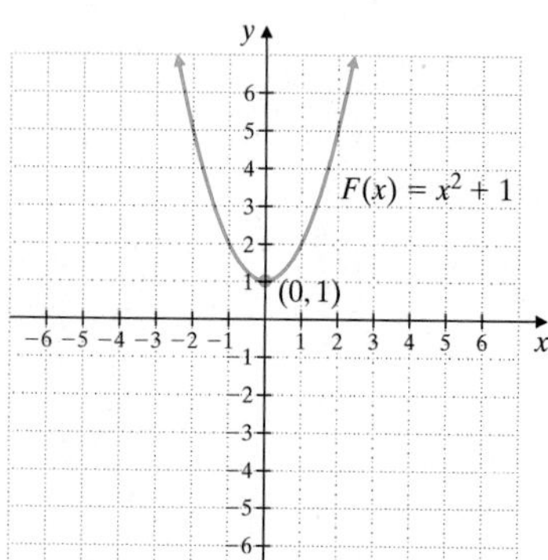

3.

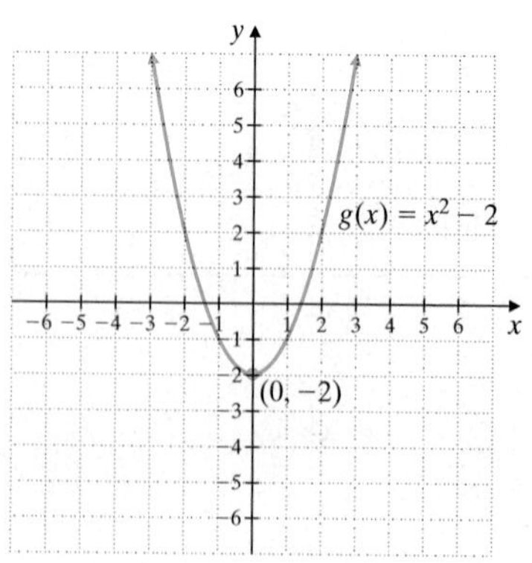

In general, we have the following properties.

Graphing the Parabola Defined by $f(x) = x^2 + k$

If k is positive, the graph of $f(x) = x^2 + k$ is the graph of $y = x^2$ shifted upward k units.

If k is negative, the graph of $f(x) = x^2 + k$ is the graph of $y = x^2$ shifted downward $|k|$ units.

The vertex is $(0, k)$, and the axis of symmetry is the y-axis.

EXAMPLES Graph each function.

2. $F(x) = x^2 + 2$

The graph of $F(x) = x^2 + 2$ is obtained by shifting the graph of $y = x^2$ upward 2 units.

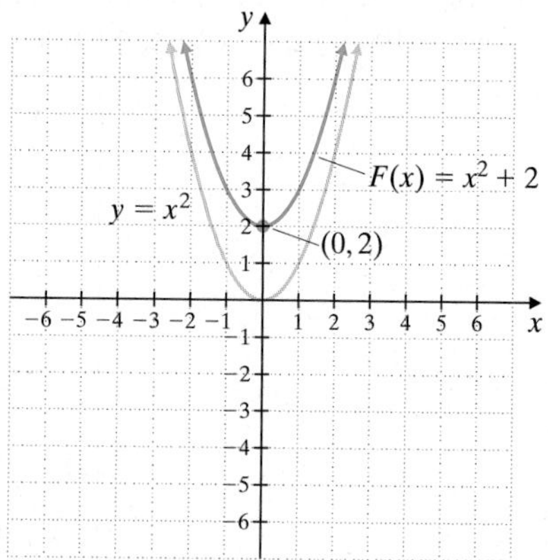

3. $g(x) = x^2 - 3$

The graph of $g(x) = x^2 - 3$ is obtained by shifting the graph of $y = x^2$ downward 3 units.

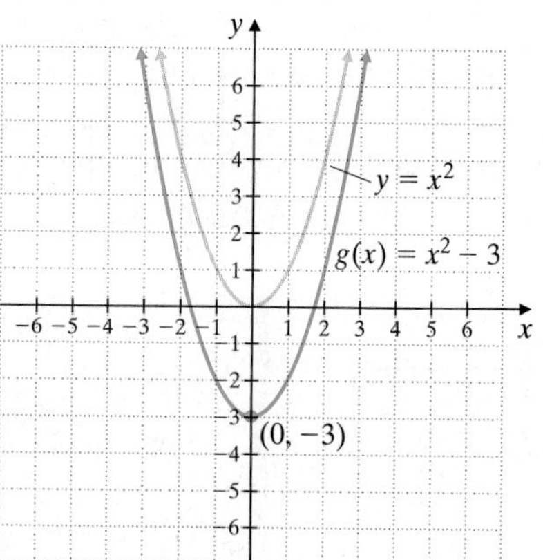

B Graphing $f(x) = (x - h)^2$

Now we will graph functions of the form $f(x) = (x - h)^2$.

EXAMPLE 4

Graph $f(x) = x^2$ and $g(x) = (x - 2)^2$ on the same set of axes.

Solution: By plotting points, we see that for each x-value, the corresponding value of $g(x)$ is the same as the value of $f(x)$ when the x-value is increased by 2. Thus, the graph of $g(x) = (x - 2)^2$ is the graph of $f(x) = x^2$ shifted to the right 2 units. The axis of symmetry for the graph of $g(x) = (x - 2)^2$ is also shifted 2 units to the right and is the line $x = 2$.

x	$f(x) = x^2$	x	$g(x) = (x - 2)^2$
−2	4	0	4
−1	1	1	1
0	0	2	0
1	1	3	1
2	4	4	4

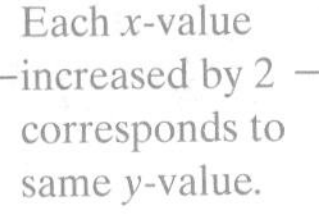

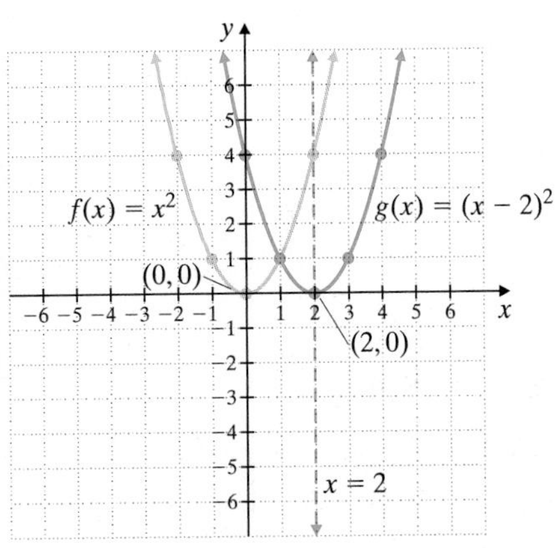

In general, we have the following properties.

Graphing the Parabola Defined by $f(x) = (x - h)^2$

If h is positive, the graph of $f(x) = (x - h)^2$ is the graph of $y = x^2$ shifted to the right h units.

If h is negative, the graph of $f(x) = (x - h)^2$ is the graph of $y = x^2$ shifted to the left $|h|$ units.

The vertex is $(h, 0)$, and the axis of symmetry is the vertical line $x = h$.

EXAMPLES Graph each function.

5. $G(x) = (x - 3)^2$

The graph of $G(x) = (x - 3)^2$ is obtained by shifting the graph of $y = x^2$ to the right 3 units.

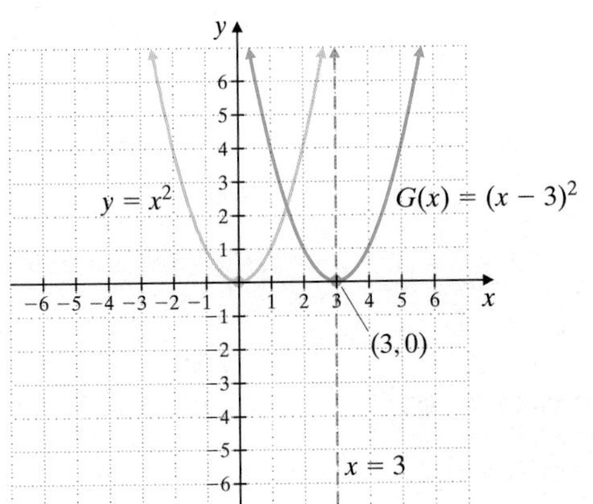

Practice Problem 4

Graph $f(x) = x^2$ and $g(x) = (x - 1)^2$ on the same set of axes.

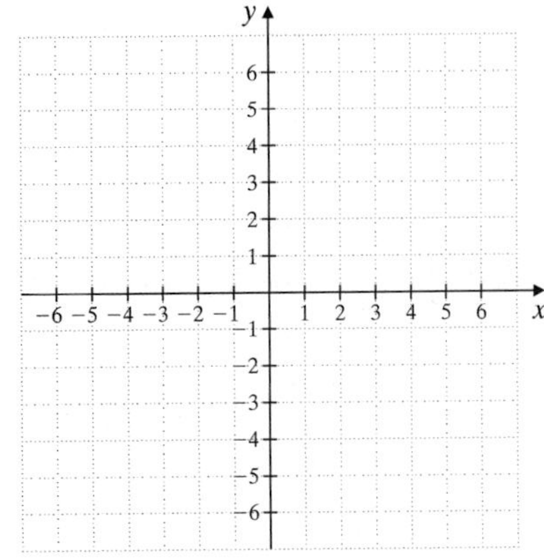

Answer

4.

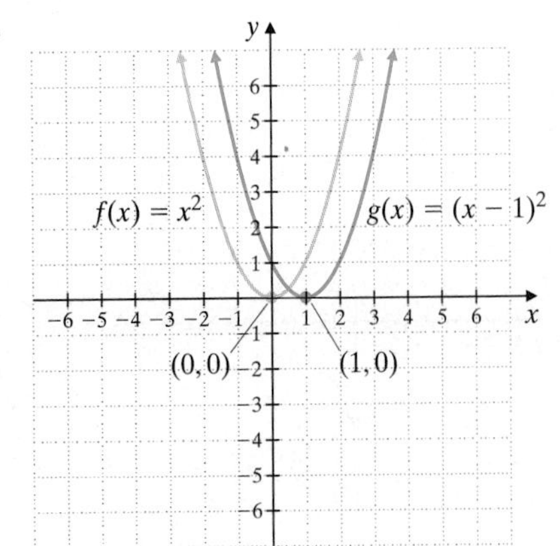

Practice Problems 5–6

Graph each function.

5. $G(x) = (x - 4)^2$

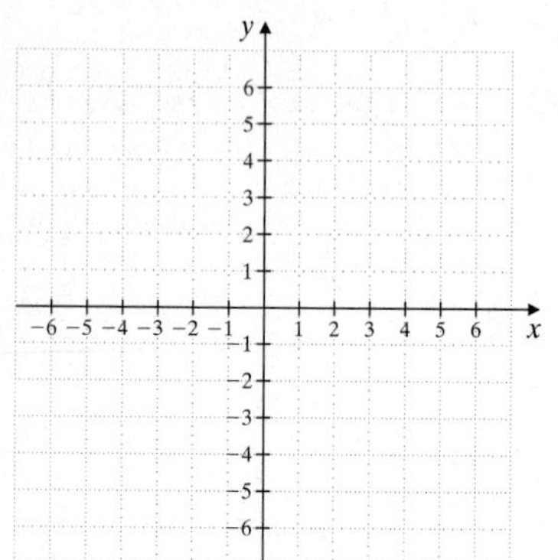

6. $F(x) = (x + 2)^2$

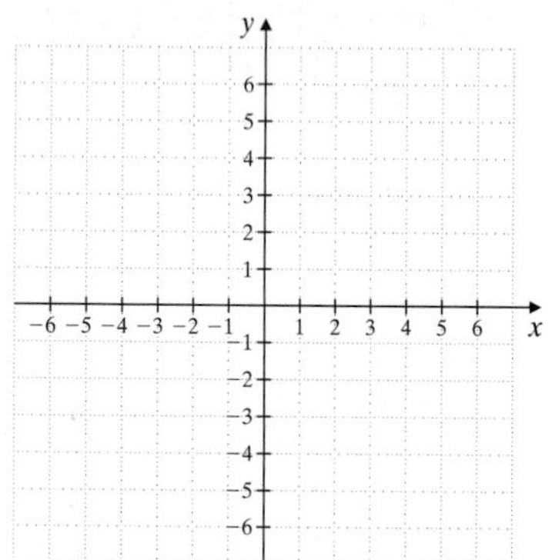

Answers

5.

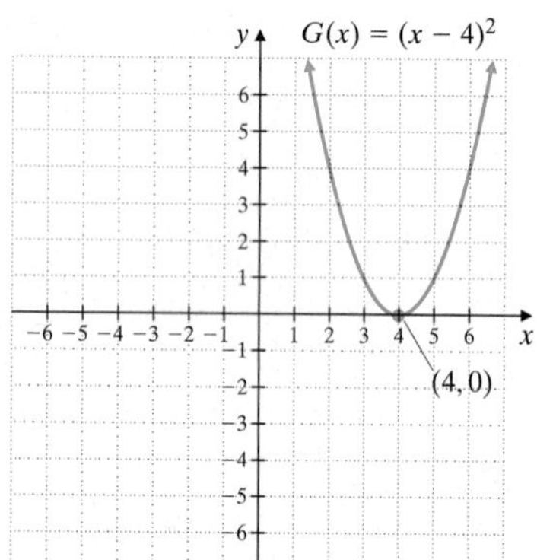

6.

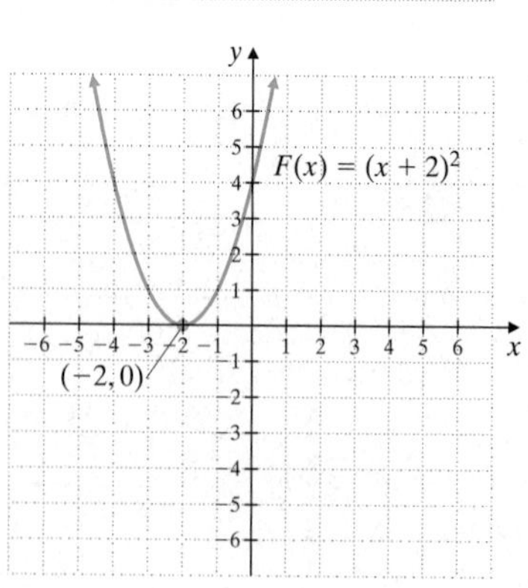

6. $F(x) = (x + 1)^2$

The equation $F(x) = (x + 1)^2$ can be written as $F(x) = [x - (-1)]^2$. The graph of $F(x) = [x - (-1)]^2$ is obtained by shifting the graph of $y = x^2$ to the left 1 unit.

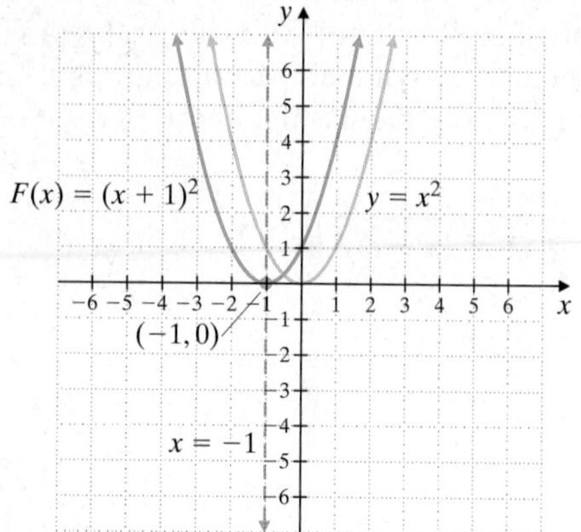

C Graphing $f(x) = (x - h)^2 + k$

As we will see in graphing functions of the form $f(x) = (x - h)^2 + k$, it is possible to combine vertical and horizontal shifts.

Graphing the Parabola Defined by $f(x) = (x - h)^2 + k$

The parabola has the same shape as $y = x^2$.
The vertex is (h, k), and the axis of symmetry is the vertical line $x = h$.

EXAMPLE 7 Graph: $F(x) = (x - 3)^2 + 1$

Solution: The graph of $F(x) = (x - 3)^2 + 1$ is the graph of $y = x^2$ shifted 3 units to the right and 1 unit up. The vertex is then $(3, 1)$, and the axis of symmetry is $x = 3$. A few ordered pair solutions are plotted to aid in graphing.

x	$F(x) = (x - 3)^2 + 1$
1	5
2	2
4	2
5	5

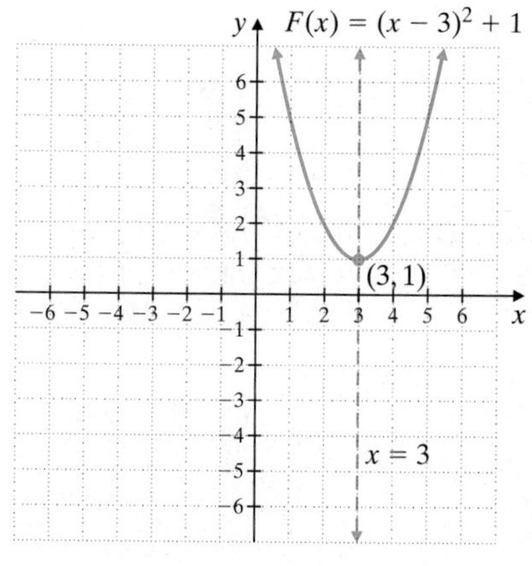

D Graphing $f(x) = ax^2$

Next, we discover the change in the shape of the graph when the coefficient of x^2 is not 1.

EXAMPLE 8

Graph $f(x) = x^2$, $g(x) = 3x^2$, and $h(x) = \frac{1}{2}x^2$ on the same set of axes.

Solution: Comparing the table of values, we see that for each x-value, the corresponding value of $g(x)$ is triple the corresponding value of $f(x)$. Similarly, the value of $h(x)$ is half the value of $f(x)$.

x	$f(x) = x^2$
−2	4
−1	1
0	0
1	1
2	4

x	$g(x) = 3x^2$
−2	12
−1	3
0	0
1	3
2	12

x	$h(x) = \frac{1}{2}x^2$
−2	2
−1	$\frac{1}{2}$
0	0
1	$\frac{1}{2}$
2	2

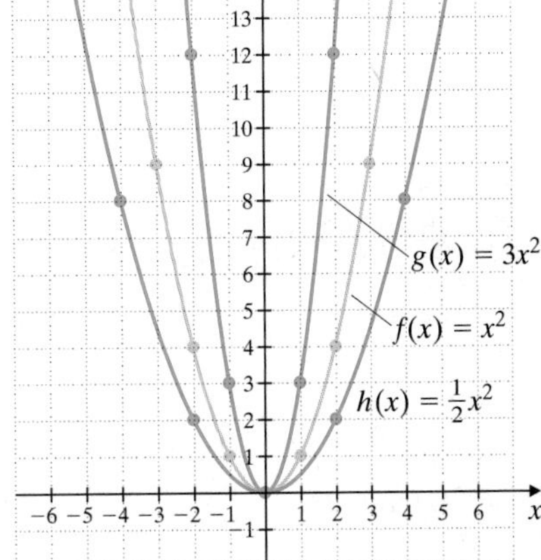

The result is that the graph of $g(x) = 3x^2$ is narrower than the graph of $f(x) = x^2$ and the graph of $h(x) = \frac{1}{2}x^2$ is wider. The vertex for each graph is $(0, 0)$, and the axis of symmetry is the y-axis.

Graphing the Parabola Defined by $f(x) = ax^2$

If a is positive, the parabola opens upward, and if a is negative, the parabola opens downward.
If $|a| > 1$, the graph of the parabola is narrower than the graph of $y = x^2$.
If $|a| < 1$, the graph of the parabola is wider than the graph of $y = x^2$.

EXAMPLE 9 Graph: $f(x) = -2x^2$

Solution: Because $a = -2$, a negative value, this parabola opens downward. Since $|-2| = 2$ and $2 > 1$, the parabola is narrower than the graph of $y = x^2$. The vertex is $(0, 0)$, and the axis of symmetry is the y-axis. We verify this by plotting a few points.

Practice Problem 7

Graph: $F(x) = (x - 2)^2 + 3$

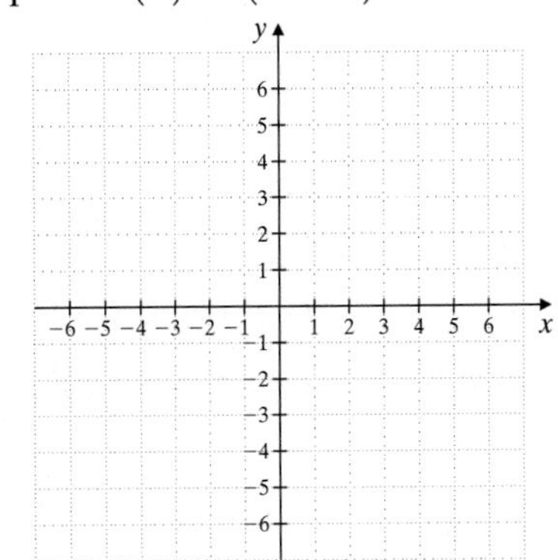

Practice Problem 8

Graph $f(x) = x^2$, $g(x) = 2x^2$, and $h(x) = \frac{1}{3}x^2$ on the same set of axes.

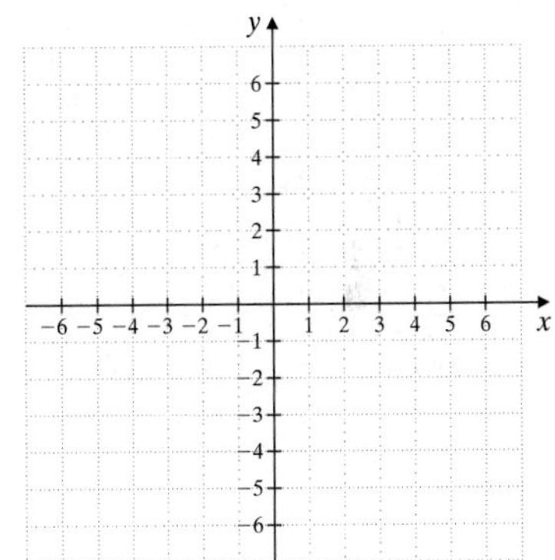

Answers

7.

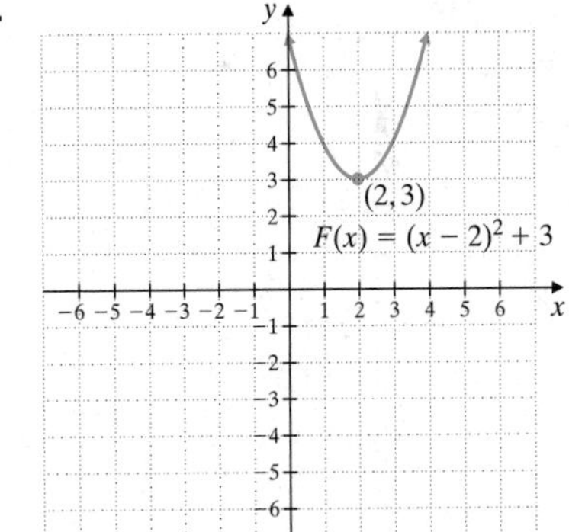

8.

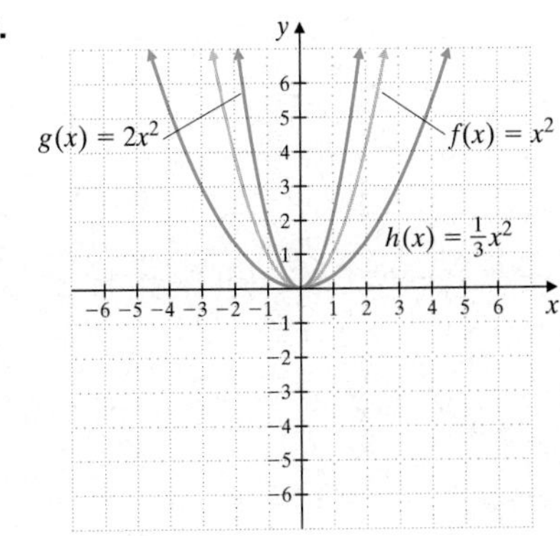

Practice Problem 9

Graph: $f(x) = -3x^2$

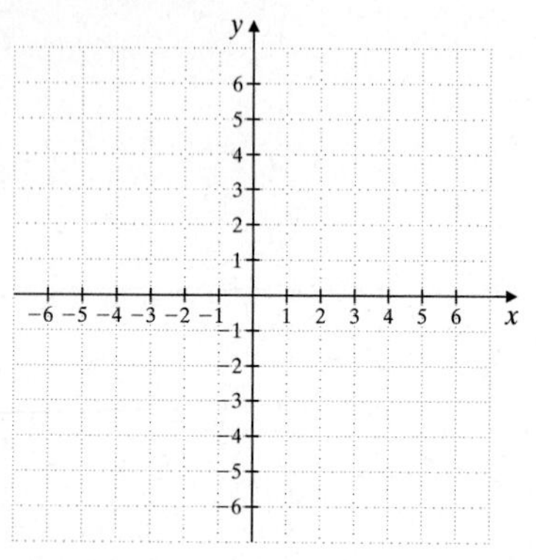

Practice Problem 10

Graph: $f(x) = 2(x + 3)^2 - 4$. Find the vertex and axis of symmetry.

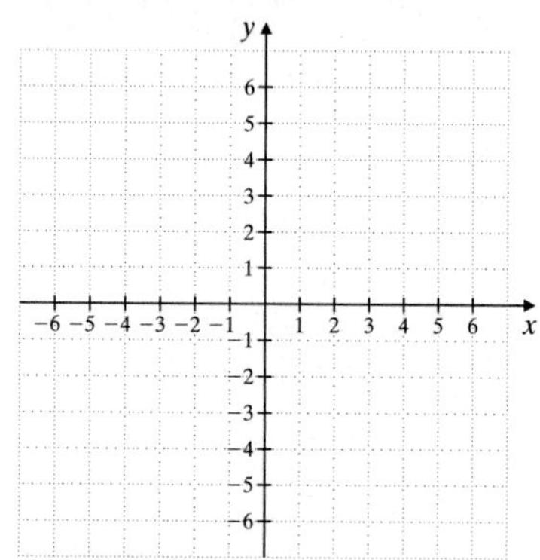

Answers

9.

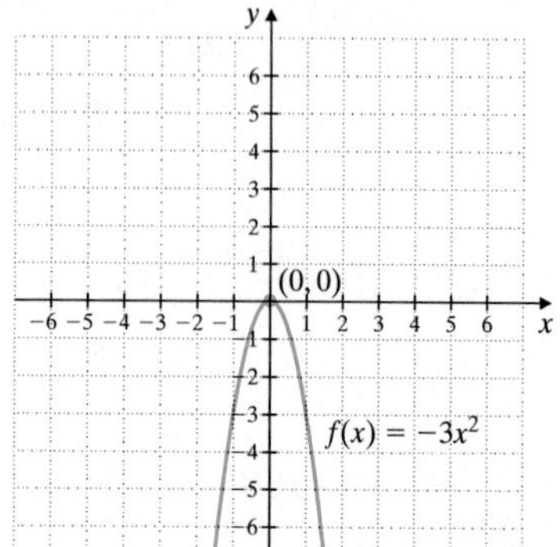

10.

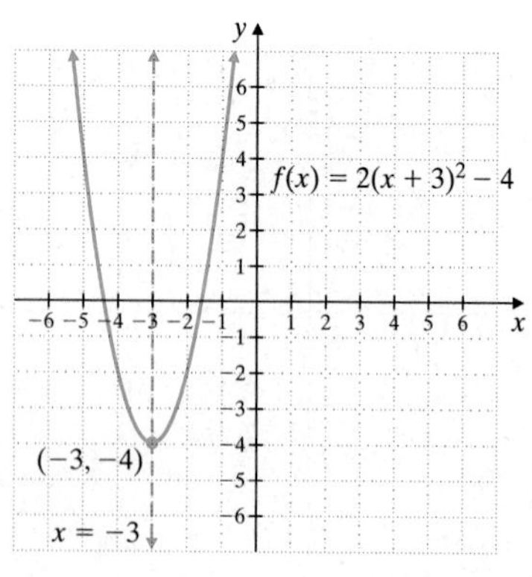

x	$f(x) = -2x^2$
−2	−8
−1	−2
0	0
1	−2
2	−8

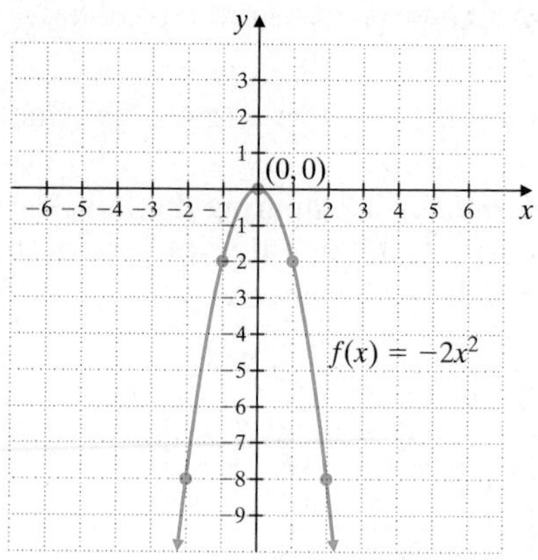

E Graphing $f(x) = a(x - h)^2 + k$

Now we will see the shape of the graph of a quadratic function of the form $f(x) = a(x - h)^2 + k$.

EXAMPLE 10

Graph: $g(x) = \frac{1}{2}(x + 2)^2 + 5$. Find the vertex and the axis of symmetry.

Solution: The function $g(x) = \frac{1}{2}(x + 2)^2 + 5$ may be written as $g(x) = \frac{1}{2}[x - (-2)]^2 + 5$. Thus, this graph is the same as the graph of $y = x^2$ shifted 2 units to the left and 5 units upward and widened because a is $\frac{1}{2}$. The vertex is $(-2, 5)$, and the axis of symmetry is $x = -2$. We plot a few points to verify.

x	$g(x) = \frac{1}{2}(x + 2)^2 + 5$
−4	7
−3	$5\frac{1}{2}$
−2	5
−1	$5\frac{1}{2}$
0	7

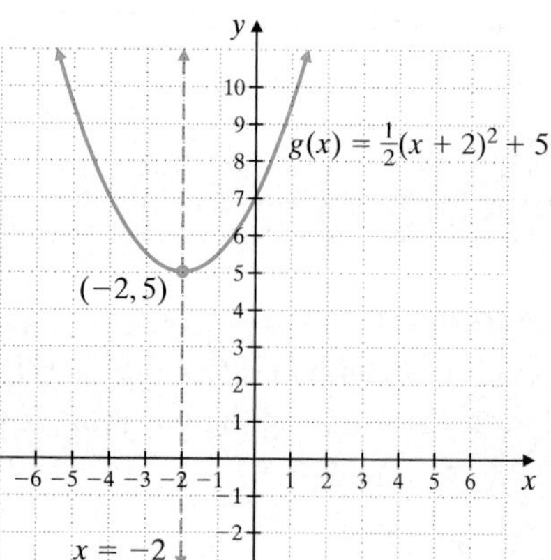

In general, the following holds.

Graphing a Quadratic Function

The graph of a quadratic function written in the form $f(x) = a(x - h)^2 + k$ is a parabola with vertex (h, k).
If $a > 0$, the parabola opens upward.
If $a < 0$, the parabola opens downward. The axis of symmetry is the line whose equation is $x = h$.

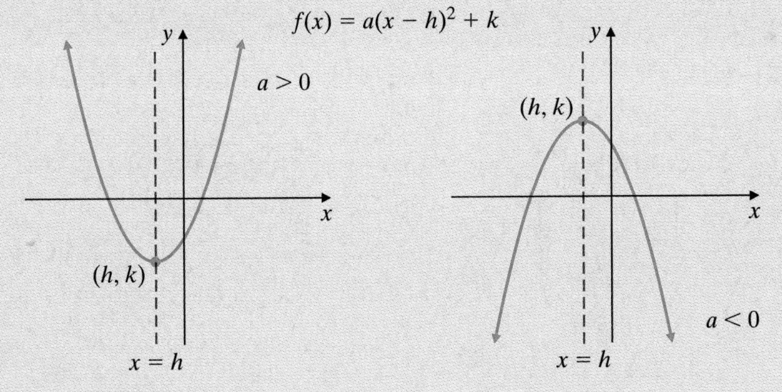

Try the Concept Check in the margin.

Concept Check

Which description of the graph of $f(x) = -0.35(x + 3)^2 - 4$ is correct?

a. The graph opens downward and has its vertex at $(-3, 4)$.

b. The graph opens upward and has its vertex at $(-3, 4)$.

c. The graph opens downward and has its vertex at $(-3, -4)$.

d. The graph is narrower than the graph of $y = x^2$.

Answer

Concept Check: c

GRAPHING CALCULATOR EXPLORATIONS

Use a graphing calculator to graph the first function of each pair. Then use its graph to predict the graph of the second function. Check your prediction by graphing both on the same set of axes.

1. $F(x) = \sqrt{x}; G(x) = \sqrt{x} + 1$

2. $g(x) = x^3; H(x) = x^3 - 2$

3. $H(x) = |x|; f(x) = |x - 5|$

4. $h(x) = x^3 + 2; g(x) = (x - 3)^3 + 2$

5. $f(x) = |x + 4|; F(x) = |x + 4| + 3$

6. $G(x) = \sqrt{x} - 2; g(x) = \sqrt{x - 4} - 2$

Name ______________________ Section __________ Date __________

Mental Math

State the vertex of the graph of each quadratic function.

1. $f(x) = x^2$ **2.** $f(x) = -5x^2$ **3.** $g(x) = (x - 2)^2$ **4.** $g(x) = (x + 5)^2$

5. $f(x) = 2x^2 + 3$ **6.** $h(x) = x^2 - 1$ **7.** $g(x) = (x + 1)^2 + 5$ **8.** $h(x) = (x - 10)^2 - 7$

EXERCISE SET 10.5

  *Graph each quadratic function. Label the vertex and sketch and label the axis of symmetry. See Examples 1 through 6.*

1. $f(x) = x^2 - 1$ **2.** $g(x) = x^2 + 3$ **3.** $f(x) = (x - 5)^2$ **4.** $g(x) = (x + 5)^2$

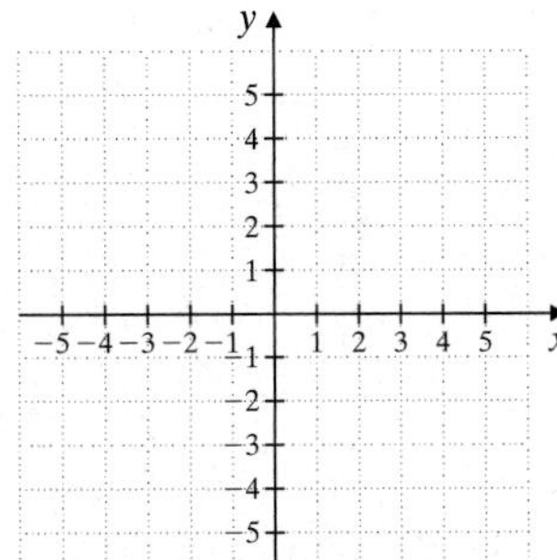

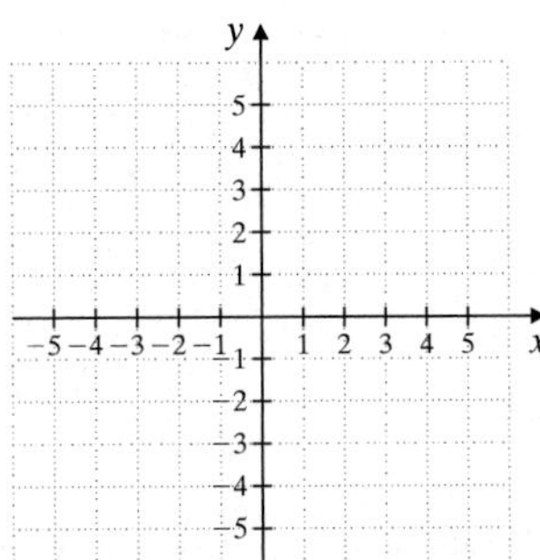

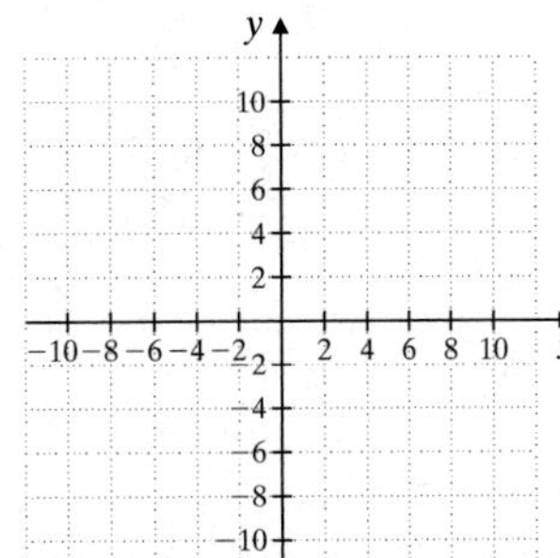

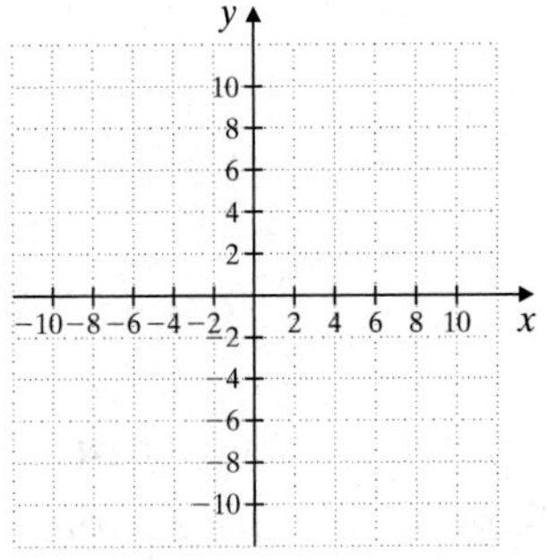

5. $h(x) = x^2 + 5$

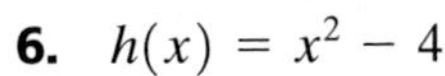

6. $h(x) = x^2 - 4$

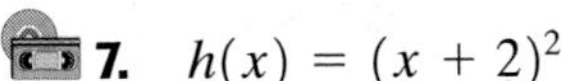

7. $h(x) = (x + 2)^2$

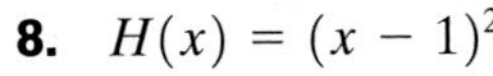

8. $H(x) = (x - 1)^2$

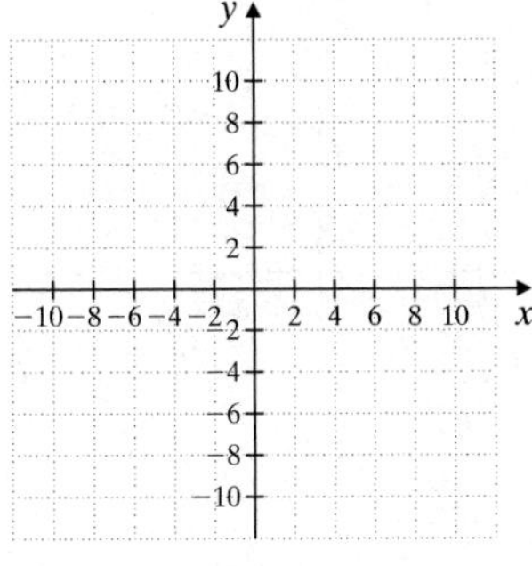

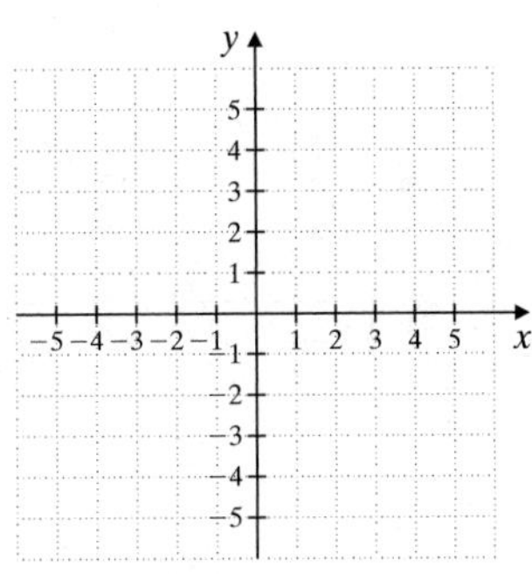

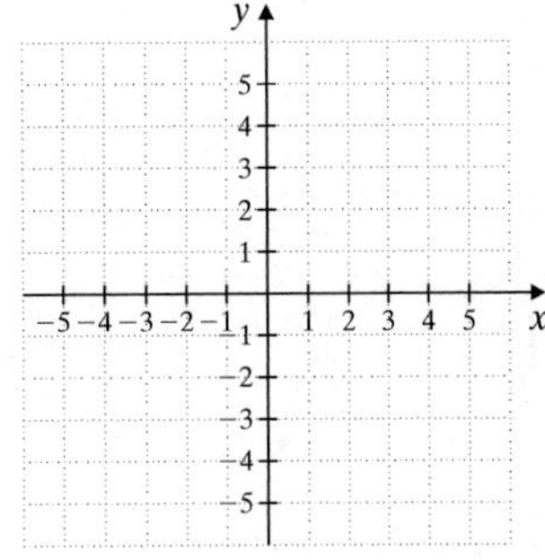

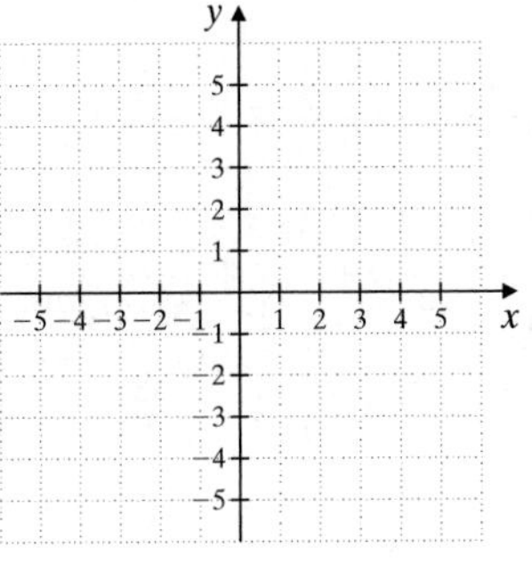

9. $g(x) = x^2 + 7$ **10.** $f(x) = x^2 - 2$ **11.** $G(x) = (x + 3)^2$ **12.** $f(x) = (x - 6)^2$

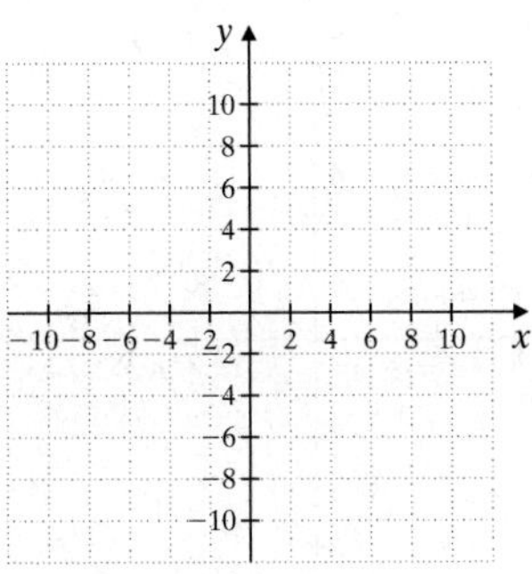

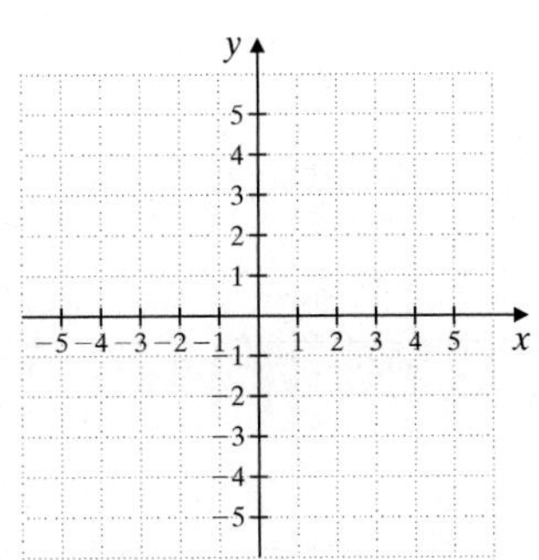

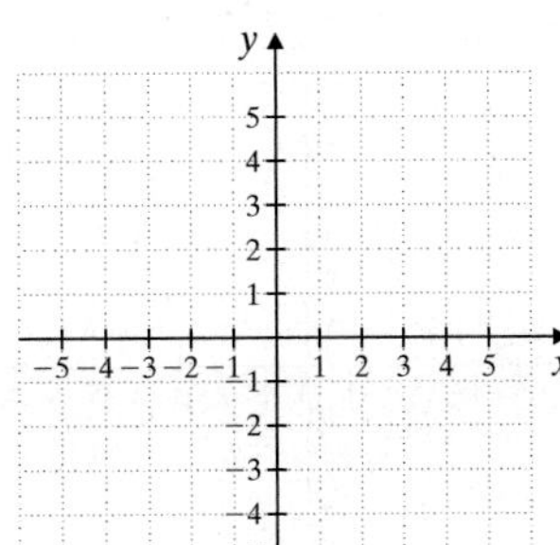

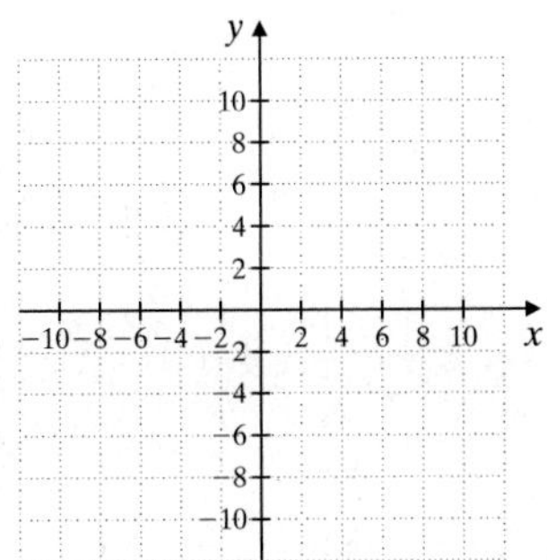

C *Graph each quadratic function. Label the vertex and sketch and label the axis of symmetry. See Example 7.*

13. $f(x) = (x - 2)^2 + 5$

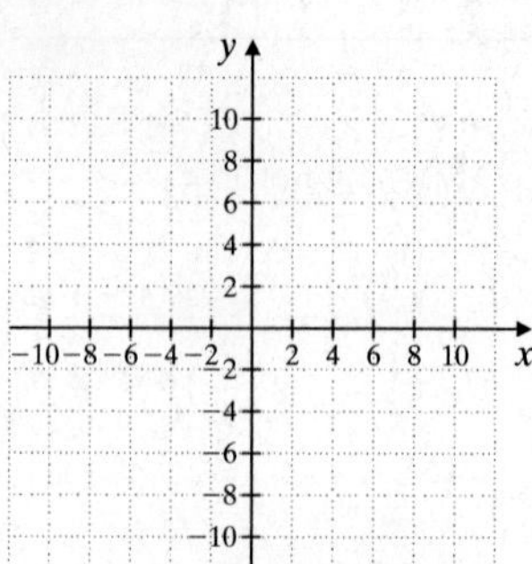

14. $g(x) = (x - 6)^2 + 1$

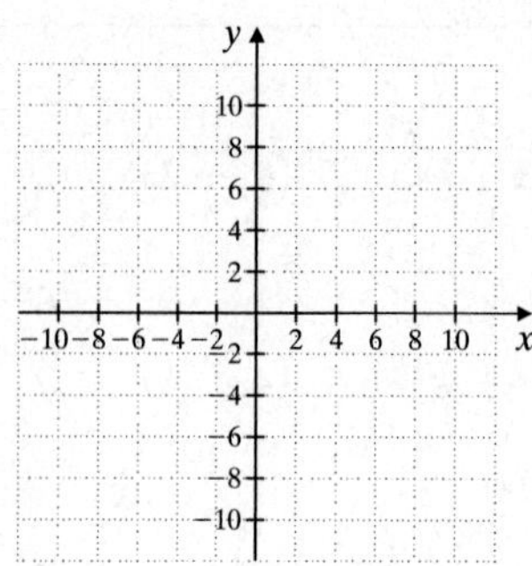

15. $h(x) = (x + 1)^2 + 4$

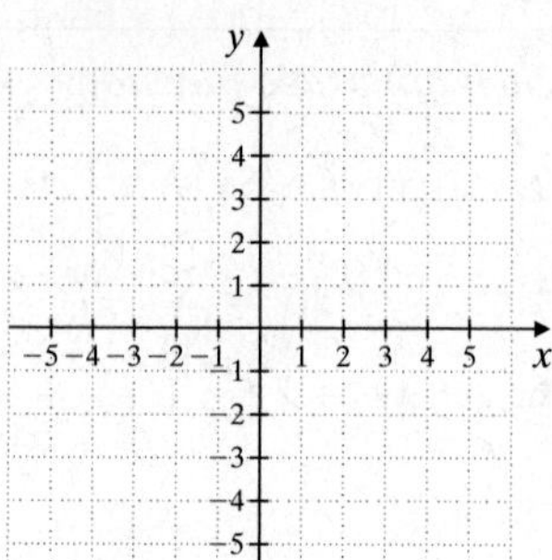

16. $G(x) = (x + 3)^2 + 3$

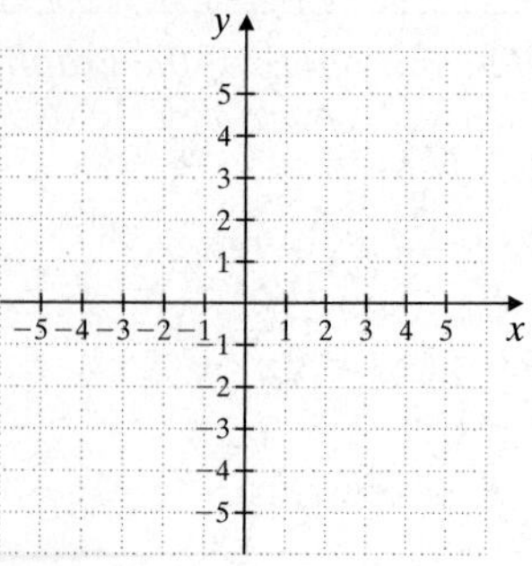

17. $g(x) = (x + 2)^2 - 5$

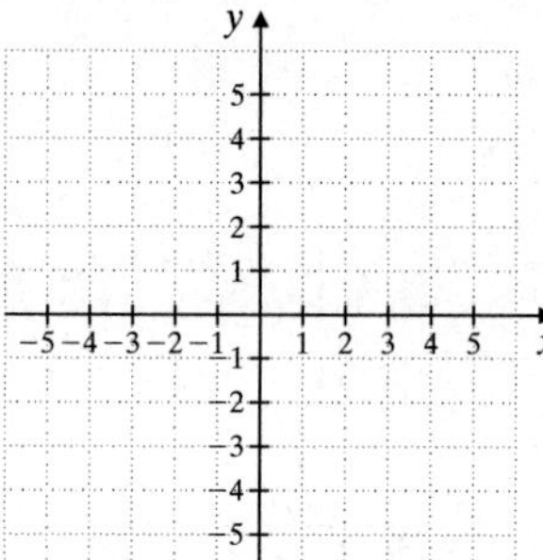

18. $h(x) = (x + 4)^2 - 6$

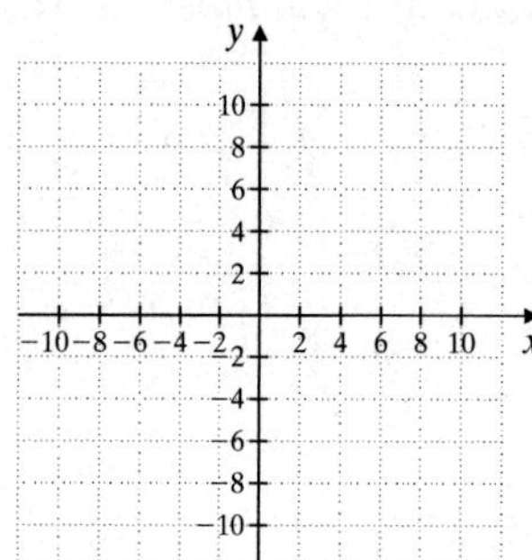

19. $h(x) = (x - 3)^2 + 2$

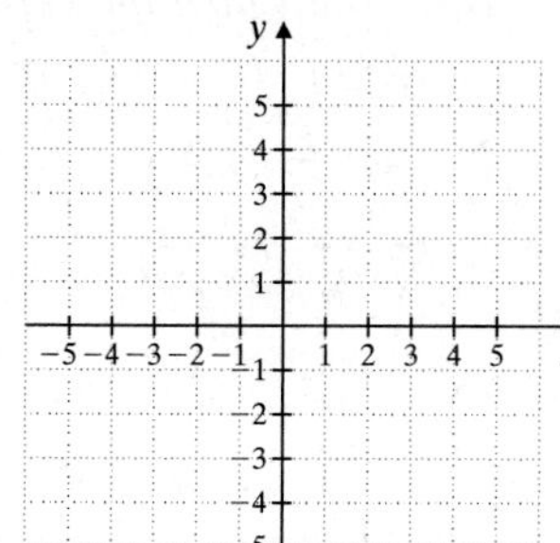

20. $F(x) = (x - 2)^2 - 3$

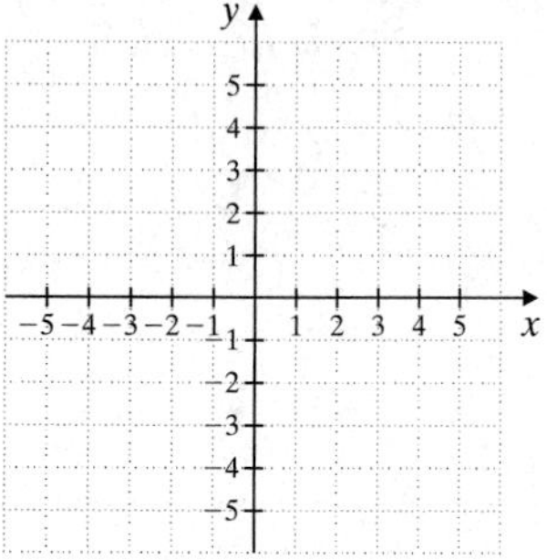

D *Graph each quadratic function. Label the vertex and sketch and label the axis of symmetry. See Examples 8 and 9.*

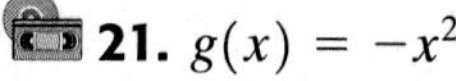

21. $g(x) = -x^2$

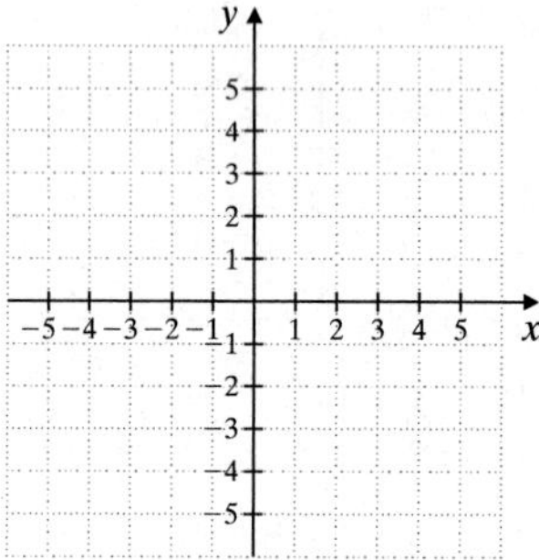

22. $f(x) = 5x^2$

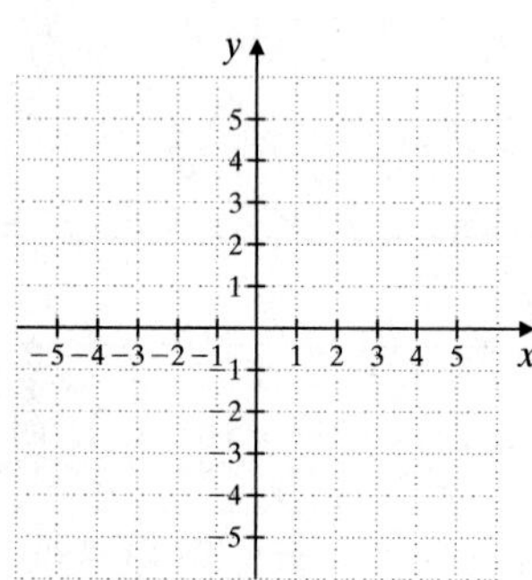

23. $h(x) = \frac{1}{3}x^2$

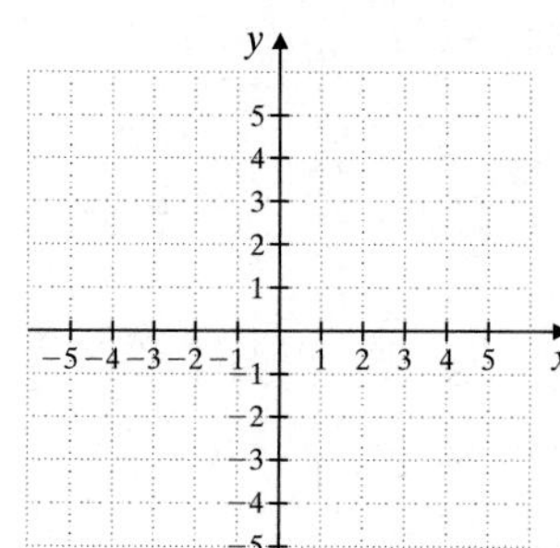

24. $g(x) = -3x^2$

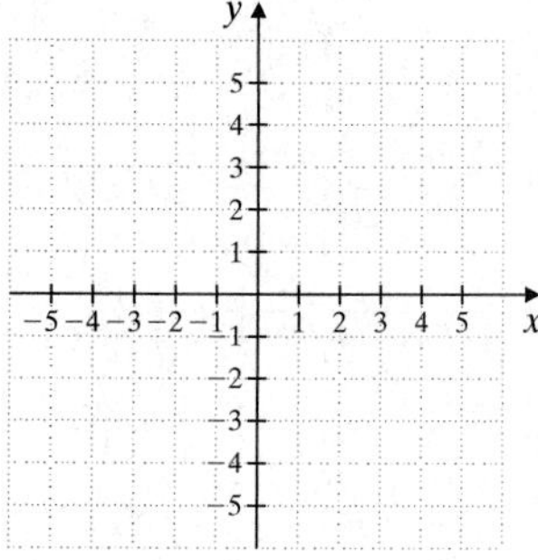

25. $H(x) = 2x^2$

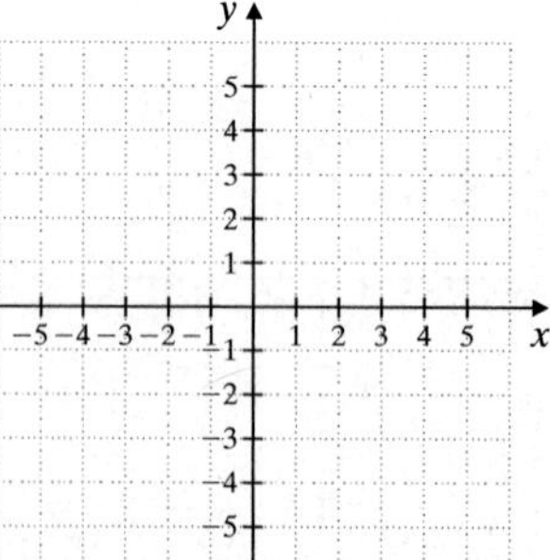

26. $f(x) = -\frac{1}{4}x^2$

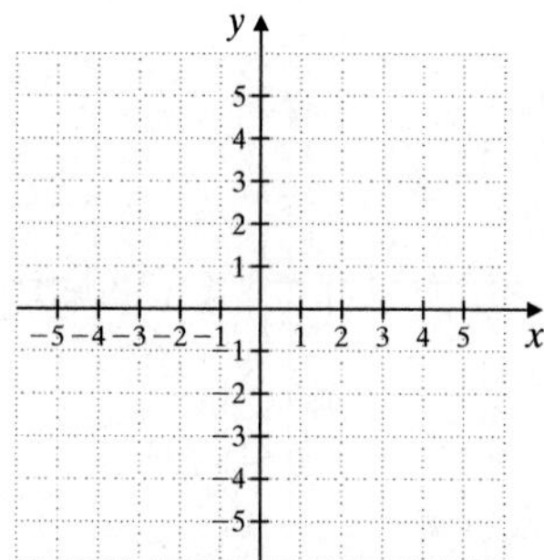

27. $F(x) = -4x^2$

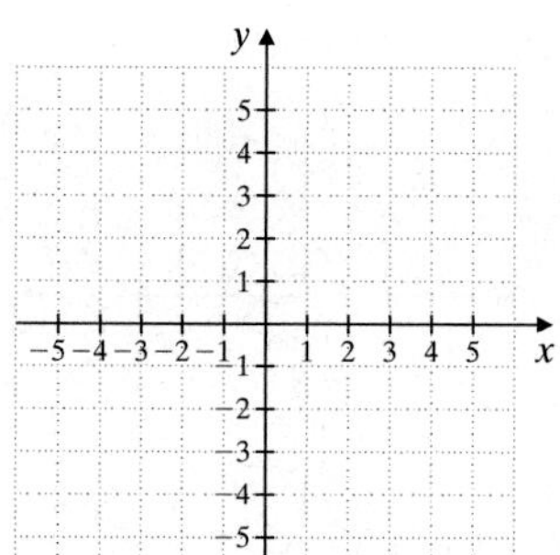

28. $G(x) = \frac{1}{5}x^2$

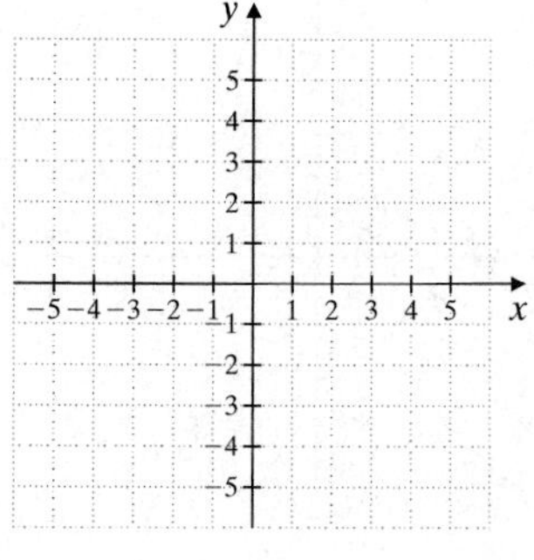

E *Graph each quadratic function. Label the vertex and sketch and label the axis of symmetry. See Example 10.*

29. $f(x) = 10(x + 4)^2 - 6$

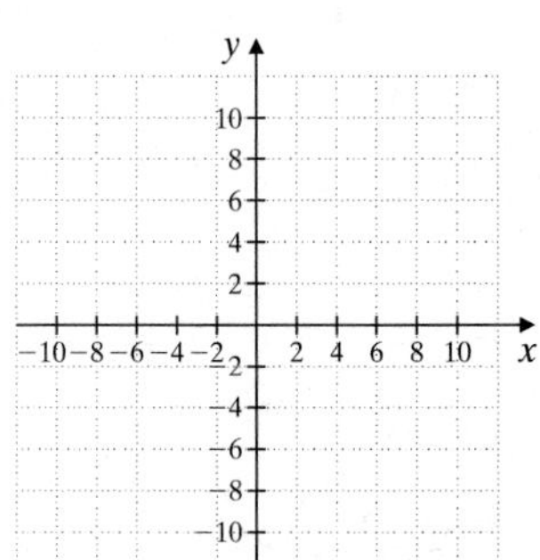

30. $g(x) = 4(x - 4)^2 + 2$

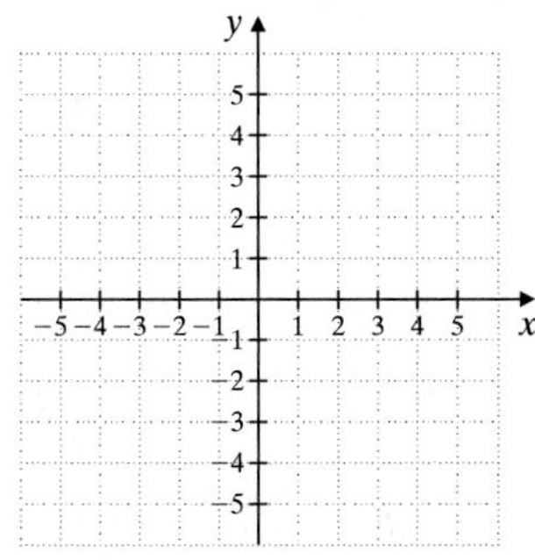

31. $h(x) = -3(x + 3)^2 + 1$

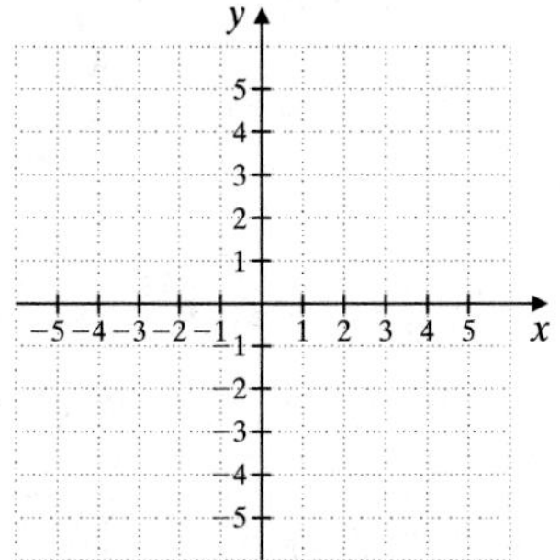

32. $f(x) = -(x - 2)^2 - 6$

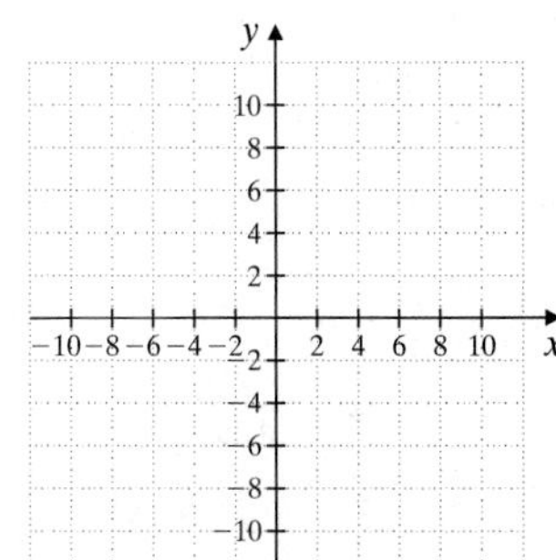

33. $H(x) = \frac{1}{2}(x - 6)^2 - 3$

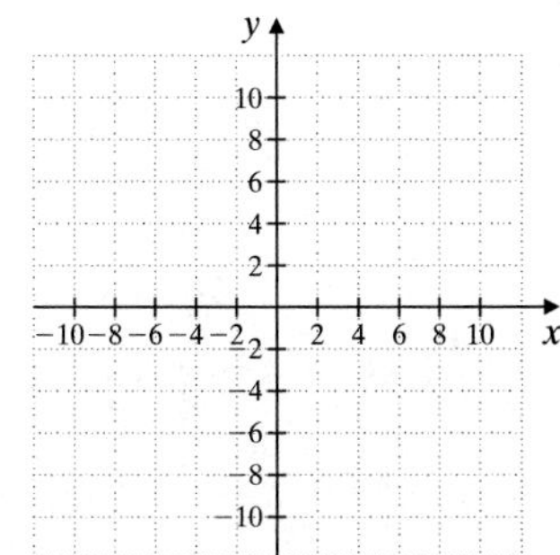

34. $G(x) = \frac{1}{5}(x + 4)^2 + 3$

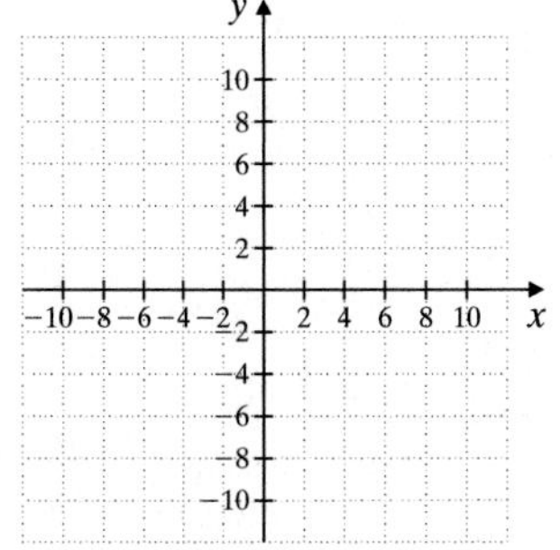

35. $f(x) = -(x - 1)^2$

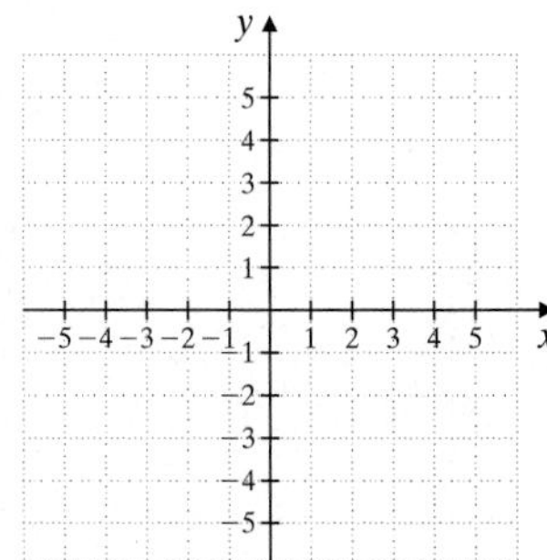

36. $f(x) = 2(x + 3)^2$

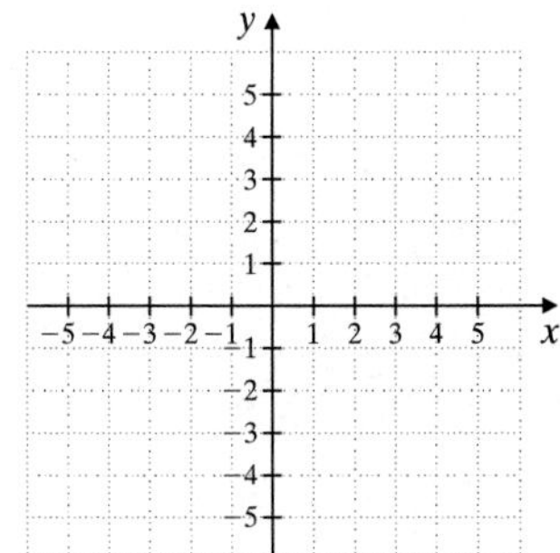

37. $F(x) = \left(x + \frac{1}{2}\right)^2 - 2$

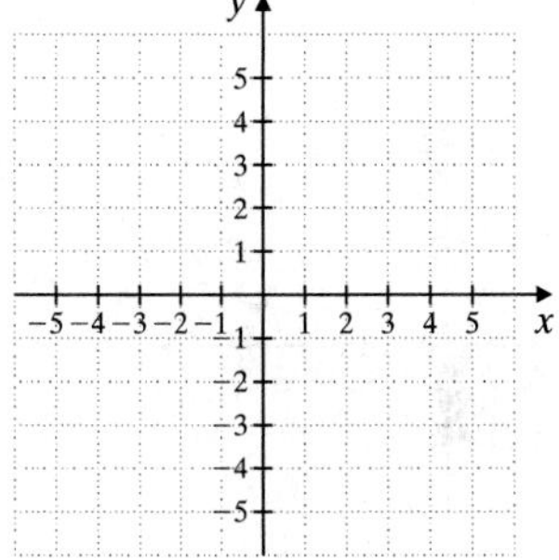

38. $H(x) = \left(x + \frac{1}{4}\right)^2 - 3$

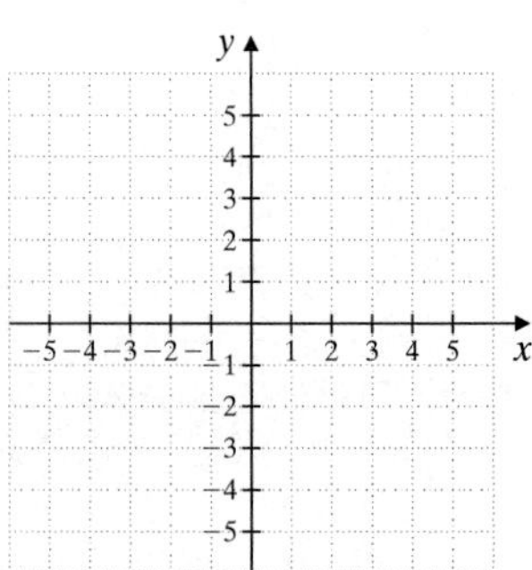

39. $F(x) = -x^2 + 2$

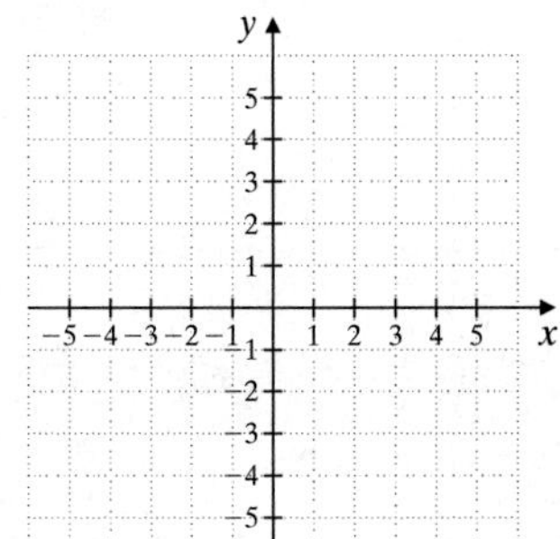

40. $G(x) = 3x^2 + 1$

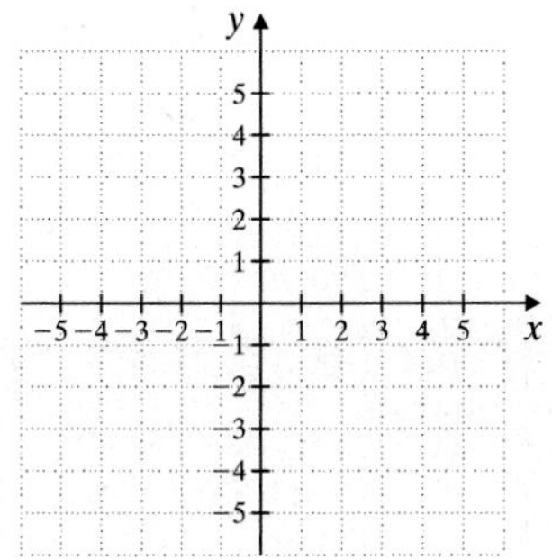

Review and Preview

Add the proper constant to each binomial so that the resulting trinomial is a perfect square trinomial. See Section 10.1.

41. $x^2 + 8x$ **42.** $y^2 + 4y$ **43.** $z^2 - 16z$ **44.** $x^2 - 10x$ **45.** $y^2 + y$ **46.** $z^2 - 3z$

Combining Concepts

Write the equation of the parabola that has the same shape as $f(x) = 5x^2$ but with each given vertex.

47. $(2, 3)$ **48.** $(1, 6)$ **49.** $(-3, 6)$ **50.** $(4, -1)$

The shifting properties covered in this section apply to the graphs of all functions. Given the accompanying graph of $y = f(x)$, graph each function.

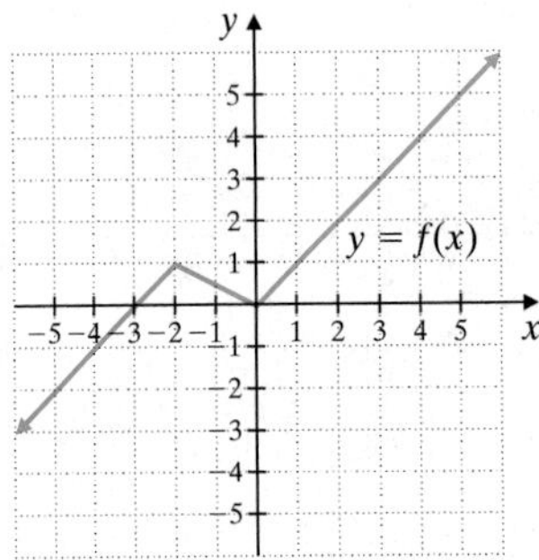

51. $y = f(x) + 1$

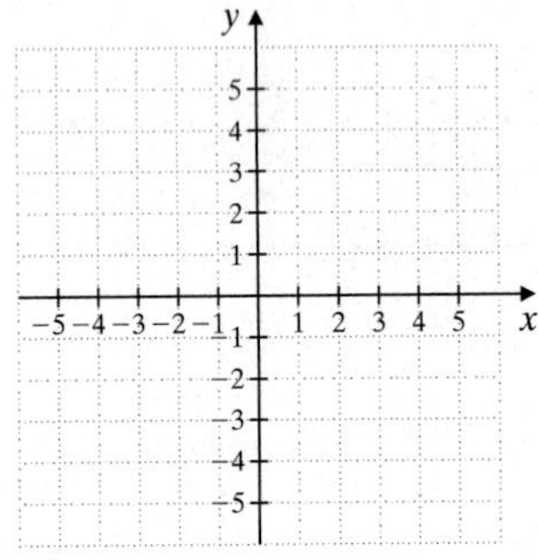

52. $y = f(x) - 2$

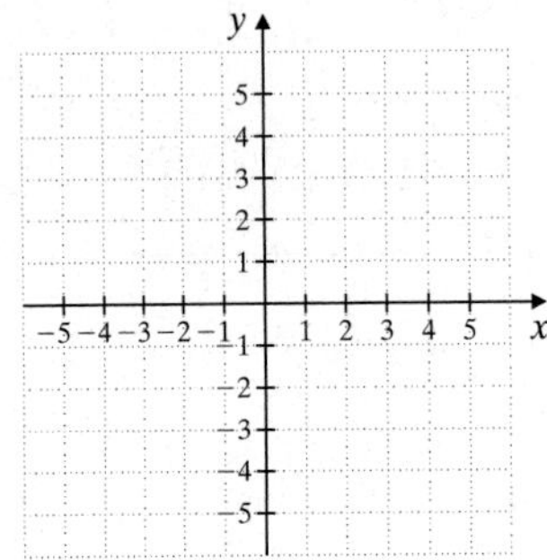

53. $y = f(x - 3)$

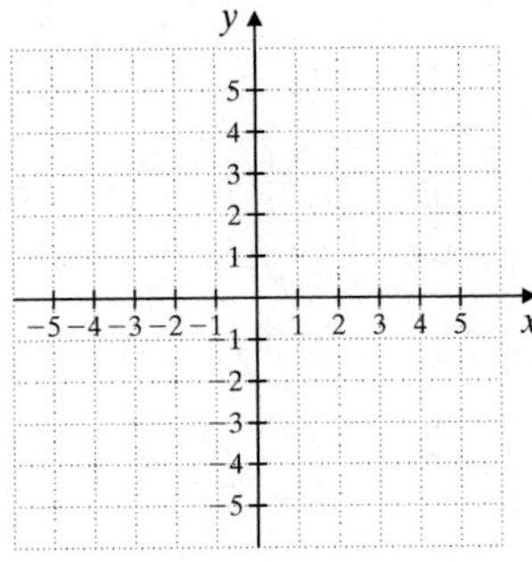

54. $y = f(x + 3)$

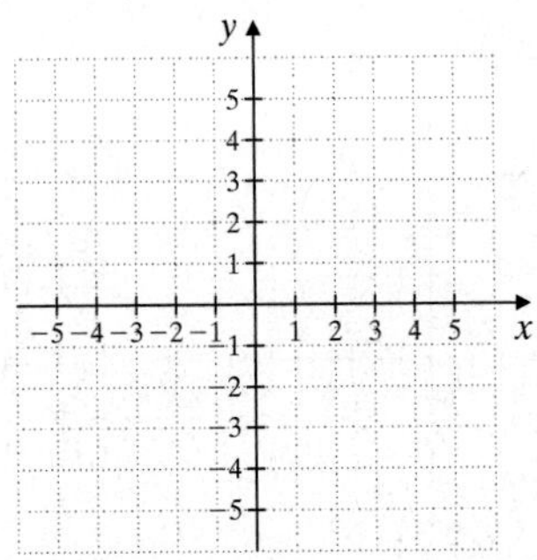

55. $y = f(x + 2) + 2$

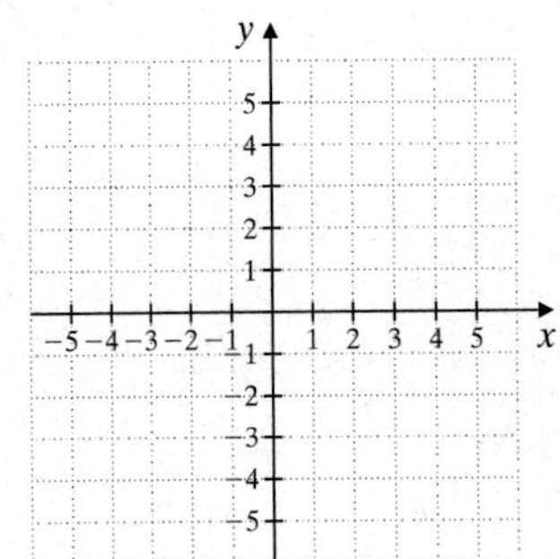

56. $y = f(x - 1) + 1$

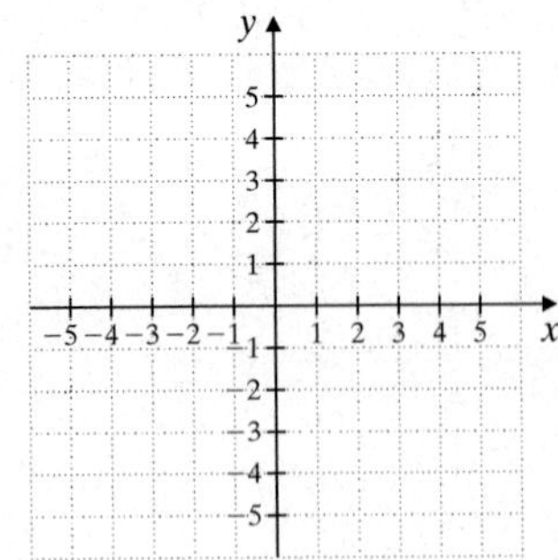

10.6 Further Graphing of Quadratic Functions

A Writing Quadratic Functions in the Form $y = a(x - h)^2 + k$

We know that the graph of a quadratic function is a parabola. If a quadratic function is written in the form

$$f(x) = a(x - h)^2 + k$$

we can easily find the vertex (h, k) and graph the parabola. To write a quadratic function in this form, we need to complete the square. (See Section 10.1 for a review of completing the square.)

EXAMPLE 1

Graph: $f(x) = x^2 - 4x - 12$. Find the vertex and any intercepts.

Solution: The graph of this quadratic function is a parabola. To find the vertex of the parabola, we complete the square on the binomial $x^2 - 4x$. To simplify our work, we let $f(x) = y$.

$$y = x^2 - 4x - 12 \quad \text{Let } f(x) = y.$$
$$y + 12 = x^2 - 4x \quad \text{Add 12 to both sides to get the } x\text{-variable terms alone.}$$

Now we add the square of half of -4 to both sides.

$$\frac{1}{2}(-4) = -2 \quad \text{and} \quad (-2)^2 = 4$$

$$y + 12 + 4 = x^2 - 4x + 4 \quad \text{Add 4 to both sides.}$$
$$y + 16 = (x - 2)^2 \quad \text{Factor the trinomial.}$$
$$y = (x - 2)^2 - 16 \quad \text{Subtract 16 from both sides.}$$
$$f(x) = (x - 2)^2 - 16 \quad \text{Replace y with } f(x).$$

From this equation, we can see that the vertex of the parabola is $(2, -16)$, a point in quadrant IV, and the axis of symmetry is the line $x = 2$.

Notice that $a = 1$. Since $a > 0$, the parabola opens upward. This parabola opening upward with vertex $(2, -16)$ will have two x-intercepts.

To find the x-intercepts, we let $f(x)$ or $y = 0$.

$$0 = x^2 - 4x - 12$$
$$0 = (x - 6)(x + 2)$$
$$0 = x - 6 \quad \text{or} \quad 0 = x + 2$$
$$6 = x \qquad -2 = x$$

The two x-intercepts are $(6, 0)$ and $(-2, 0)$. To find the y-intercept, we let $x = 0$.

$$f(0) = 0^2 - 4 \cdot 0 - 12 = -12$$

The y-intercept is $(0, -12)$. The sketch of $f(x) = x^2 - 4x - 12$ is shown.

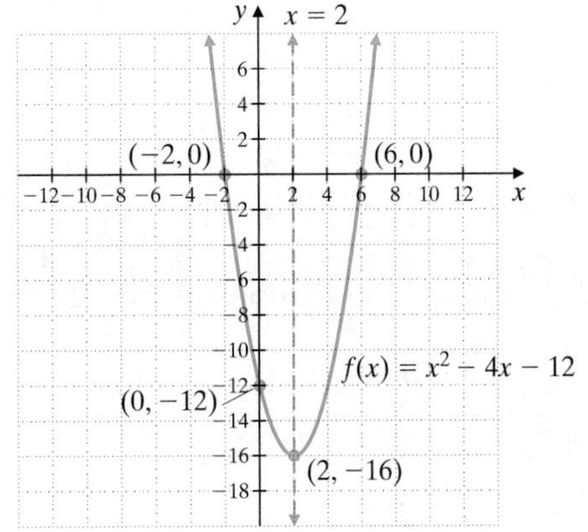

OBJECTIVES

- **A** Write quadratic functions in the form $y = a(x - h)^2 + k$.
- **B** Derive a formula for finding the vertex of a parabola.
- **C** Find the minimum or maximum value of a quadratic function.

Practice Problem 1

Graph: $f(x) = x^2 - 4x - 5$. Find the vertex and any intercepts.

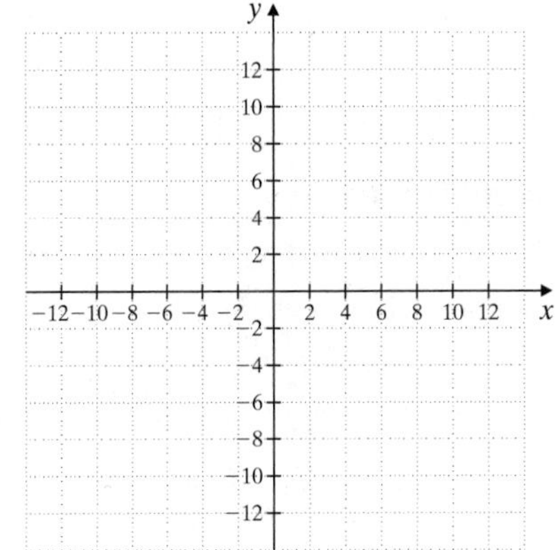

Answer

1. vertex: $(2, -9)$; x-intercepts: $(-1, 0)$, $(5, 0)$; y-intercept: $(0, -5)$

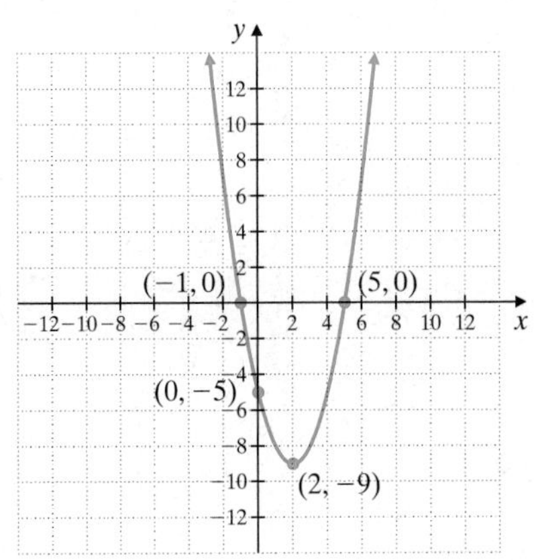

Helpful Hint

Parabola Opens Upward
Vertex in I or II: no x-intercepts
Vertex in III or IV: 2 x-intercepts

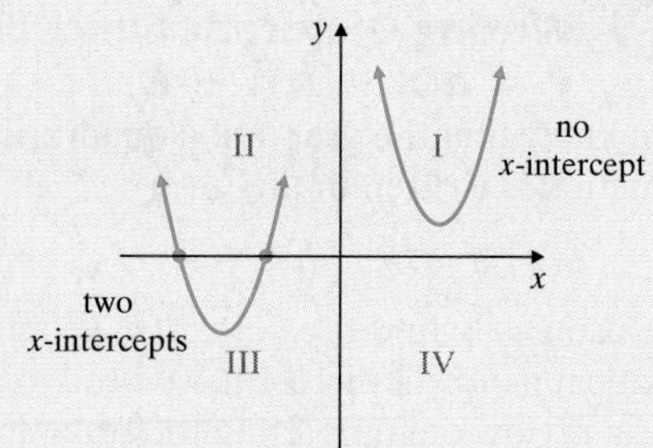

Parabola Opens Downward
Vertex in I or II: 2 x-intercepts
Vertex in III or IV: no x-intercepts

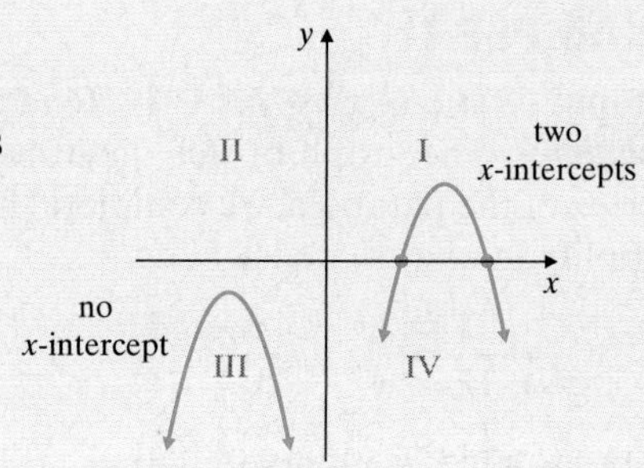

Practice Problem 2

Graph: $f(x) = 2x^2 + 2x + 5$. Find the vertex and any intercepts.

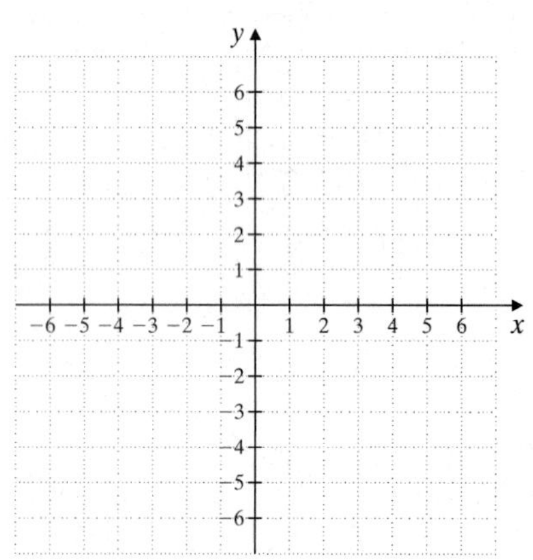

Answer

2. vertex: $\left(-\frac{1}{2}, \frac{9}{2}\right)$; y-intercept: $(0, 5)$

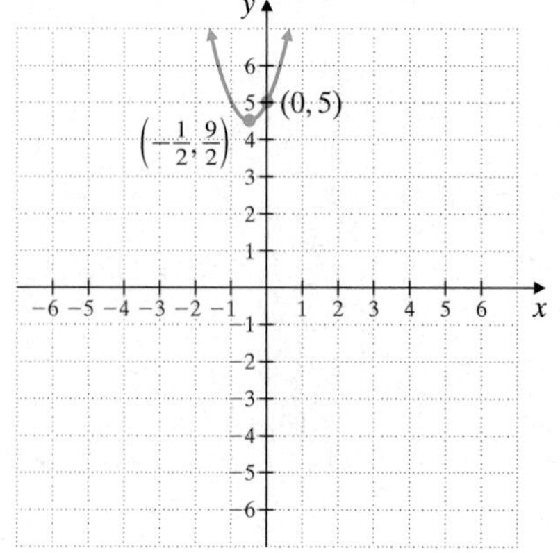

EXAMPLE 2

Graph: $f(x) = 3x^2 + 3x + 1$. Find the vertex and any intercepts.

Solution: We replace $f(x)$ with y and complete the square on x to write the equation in the form $y = a(x - h)^2 + k$.

$$y = 3x^2 + 3x + 1 \qquad \text{Replace } f(x) \text{ with } y.$$

$$y - 1 = 3x^2 + 3x \qquad \text{Get the } x\text{-variable terms alone.}$$

Next we factor 3 from the terms $3x^2 + 3x$ so that the coefficient of x^2 is 1.

$$y - 1 = 3(x^2 + x) \qquad \text{Factor out 3.}$$

The coefficient of x is 1. Then $\frac{1}{2}(1) = \frac{1}{2}$ and $\left(\frac{1}{2}\right)^2 = \frac{1}{4}$. Since we are adding $\frac{1}{4}$ inside the parentheses, we are really adding $3\left(\frac{1}{4}\right)$, so we *must* add $3\left(\frac{1}{4}\right)$ to the left side.

$$y - 1 + 3\left(\frac{1}{4}\right) = 3\left(x^2 + x + \frac{1}{4}\right)$$

$$y - \frac{1}{4} = 3\left(x + \frac{1}{2}\right)^2 \qquad \text{Simplify the left side and factor the right side.}$$

$$y = 3\left(x + \frac{1}{2}\right)^2 + \frac{1}{4} \qquad \text{Add } \frac{1}{4} \text{ to both sides.}$$

$$f(x) = 3\left(x + \frac{1}{2}\right)^2 + \frac{1}{4} \qquad \text{Replace } y \text{ with } f(x).$$

Then $a = 3$, $h = -\frac{1}{2}$, and $k = \frac{1}{4}$. This means that the parabola opens upward with vertex $\left(-\frac{1}{2}, \frac{1}{4}\right)$ and that the axis of symmetry is the line $x = -\frac{1}{2}$.

To find the y-intercept, we let $x = 0$. Then

$$f(0) = 3(0)^2 + 3(0) + 1 = 1$$

This parabola has no x-intercepts since the vertex is in the second quadrant and it opens upward. We use the vertex, axis of symmetry, and y-intercept to graph the parabola.

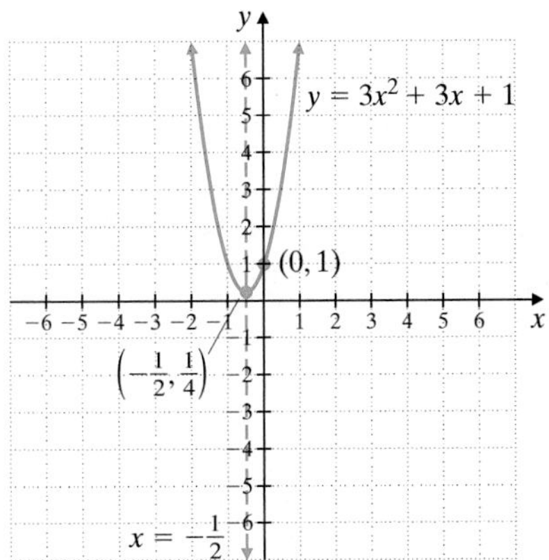

EXAMPLE 3

Graph: $f(x) = -x^2 - 2x + 3$. Find the vertex and any intercepts.

Solution: We write $f(x)$ in the form $a(x - h)^2 + k$ by completing the square. First we replace $f(x)$ with y.

$$f(x) = -x^2 - 2x + 3$$

$$y = -x^2 - 2x + 3$$

$$y - 3 = -x^2 - 2x \quad \text{Subtract 3 from both sides to get the } x\text{-variable terms alone.}$$

$$y - 3 = -1(x^2 + 2x) \quad \text{Factor } -1 \text{ from the terms } -x^2 - 2x.$$

The coefficient of x is 2. Then $\frac{1}{2}(2) = 1$ and $1^2 = 1$. We add 1 to the right side inside the parentheses and add $-1(1)$ to the left side.

$$y - 3 - 1(1) = -1(x^2 + 2x + 1)$$

$$y - 4 = -1(x + 1)^2 \quad \text{Simplify the left side and factor the right side.}$$

$$y = -1(x + 1)^2 + 4 \quad \text{Add 4 to both sides.}$$

$$f(x) = -1(x + 1)^2 + 4 \quad \text{Replace } y \text{ with } f(x).$$

Since $a = -1$, the parabola opens downward with vertex $(-1, 4)$ and axis of symmetry $x = -1$.

To find the y-intercept, we let $x = 0$ and solve for y. Then

$$f(0) = -0^2 - 2(0) + 3 = 3$$

Thus, $(0, 3)$ is the y-intercept.

To find the x-intercepts, we let y or $f(x) = 0$ and solve for x.

$$f(x) = -x^2 - 2x + 3$$

$$0 = -x^2 - 2x + 3 \quad \text{Let } f(x) = 0.$$

Practice Problem 3

Graph: $f(x) = -x^2 - 2x + 8$. Find the vertex and any intercepts.

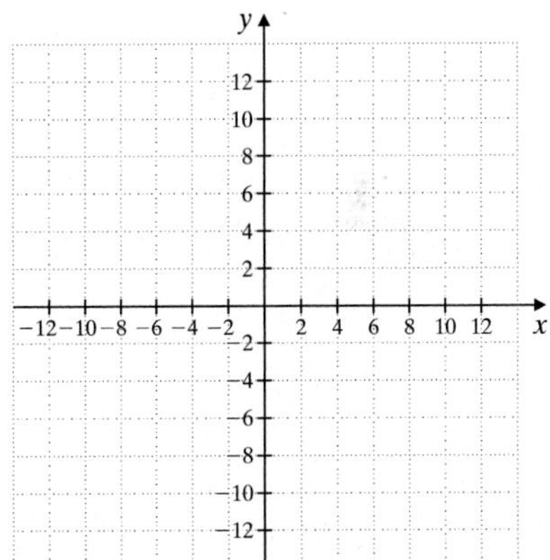

Helpful Hint

This can be written as $f(x) = -1[x - (-1)]^2 + 4$. Notice that the vertex is $(-1, 4)$.

Answer

3. vertex: $(-1, 9)$; x-intercepts: $(-4, 0)$, $(2, 0)$; y-intercept: $(0, 8)$

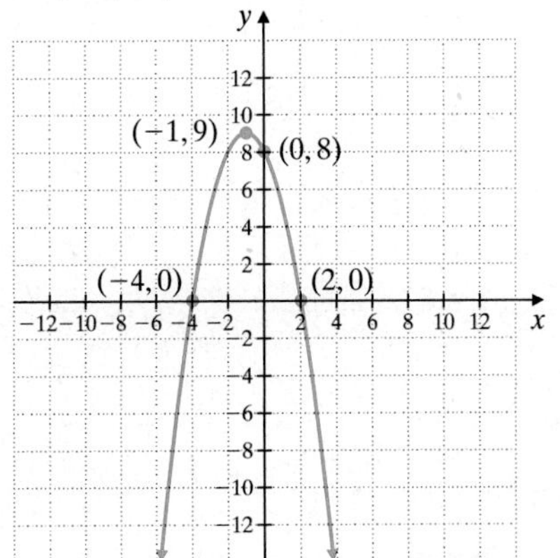

Now we divide both sides by -1 so that the coefficient of x^2 is 1.

$$\frac{0}{-1} = \frac{-x^2}{-1} - \frac{2x}{-1} + \frac{3}{-1} \quad \text{Divide both sides by } -1.$$

$$0 = x^2 + 2x - 3 \quad \text{Simplify.}$$

$$0 = (x + 3)(x - 1) \quad \text{Factor.}$$

$$x + 3 = 0 \quad \text{or} \quad x - 1 = 0 \quad \text{Set each factor equal to 0.}$$

$$x = -3 \qquad\qquad x = 1 \quad \text{Solve.}$$

The x-intercepts are $(-3, 0)$ and $(1, 0)$. We use these points to graph the parabola.

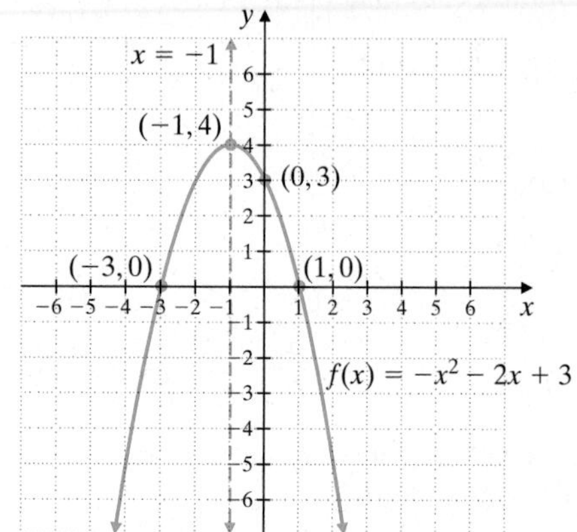

B Deriving a Formula for Finding the Vertex

There is a formula for finding the vertex of a parabola. Now that we have practiced completing the square, we will show that the x-coordinate of the vertex of the graph of $f(x)$ or $y = ax^2 + bx + c$ can be found by the formula $x = \frac{-b}{2a}$. To do so, we complete the square on x and write the equation in the form $y = (x - h)^2 + k$.

First we get the x-variable terms alone by subtracting c from both sides.

$$y = ax^2 + bx + c$$

$$y - c = ax^2 + bx$$

$$y - c = a\left(x^2 + \frac{b}{a}x\right) \quad \text{Factor } a \text{ from the terms } ax^2 + bx.$$

Now we add the square of half of $\frac{b}{a}$, or $\left(\frac{b}{2a}\right)^2 = \frac{b^2}{4a^2}$, to the right side inside the parentheses. Because of the factor a, what we really added is $a\left(\frac{b^2}{4a^2}\right)$ and this must be added to the left side as well.

$$y - c + a\left(\frac{b^2}{4a^2}\right) = a\left(x^2 + \frac{b}{a}x + \frac{b^2}{4a^2}\right)$$

$$y - c + \frac{b^2}{4a} = a\left(x + \frac{b}{2a}\right)^2$$

$$y = a\left(x + \frac{b}{2a}\right)^2 + c - \frac{b^2}{4a}$$

Simplify the left side and factor the right side. Add c to both sides and subtract $\frac{b^2}{4a}$ from both sides.

Compare this form with $f(x)$ or $y = a(x - h)^2 + k$ and see that h is $\frac{-b}{2a}$, which means that the x-coordinate of the vertex of the graph of $f(x) = ax^2 + bx + c$ is $\frac{-b}{2a}$.

Let's use the vertex formula below to find the vertex of the parabola we graphed in Example 1.

Vertex Formula

The graph of $f(x) = ax^2 + bx + c$, when $a \neq 0$, is a parabola with vertex

$$\left(\frac{-b}{2a}, f\left(\frac{-b}{2a}\right)\right)$$

EXAMPLE 4 Find the vertex of the graph of $f(x) = x^2 - 4x - 12$.

Solution: In the quadratic function $f(x) = x^2 - 4x - 12$, notice that $a = 1, b = -4$, and $c = -12$.

$$\frac{-b}{2a} = \frac{-(-4)}{2(1)} = 2$$

The x-value of the vertex is 2. To find the corresponding $f(x)$ or y-value, find $f(2)$. Then

$$f(2) = 2^2 - 4(2) - 12 = 4 - 8 - 12 = -16$$

The vertex is $(2, -16)$. These results agree with our findings in Example 1. ●

Practice Problem 4

Find the vertex of the graph of $f(x) = x^2 - 4x - 5$. Compare your result with the result of Practice Problem 1.

C Finding Minimum and Maximum Values

The quadratic function whose graph is a parabola that opens upward has a minimum value, and the quadratic function whose graph is a parabola that opens downward has a maximum value. The $f(x)$- or y-value of the vertex is the minimum or maximum value of the function.

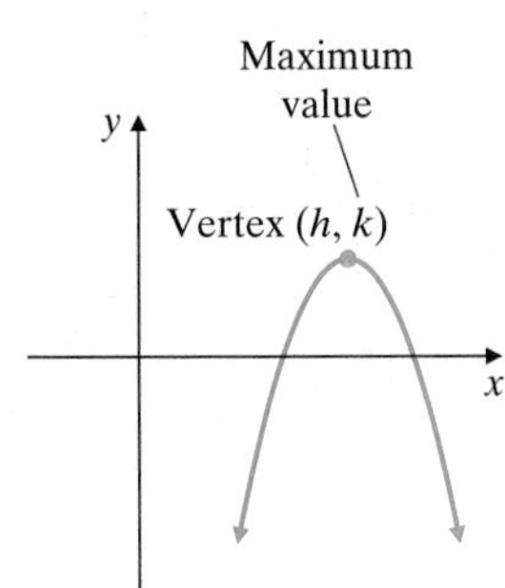

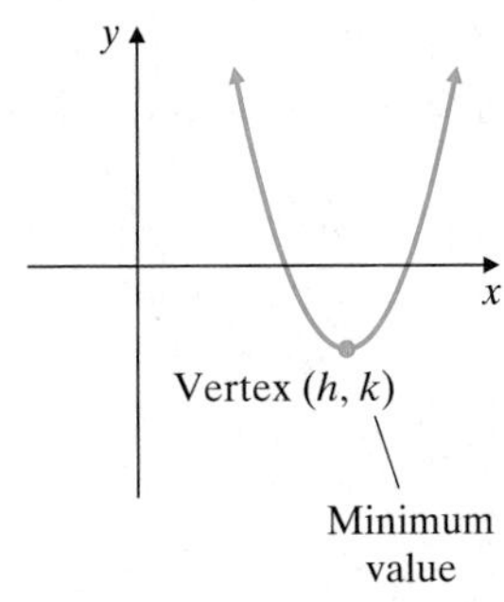

Recall from Section 10.2 that the discriminant, $b^2 - 4ac$, tells us how many solutions the quadratic equation $0 = ax^2 + bx + c$ has. It also tells us how many x-intercepts the graph of a quadratic equation $y = ax^2 + bx + c$ has.

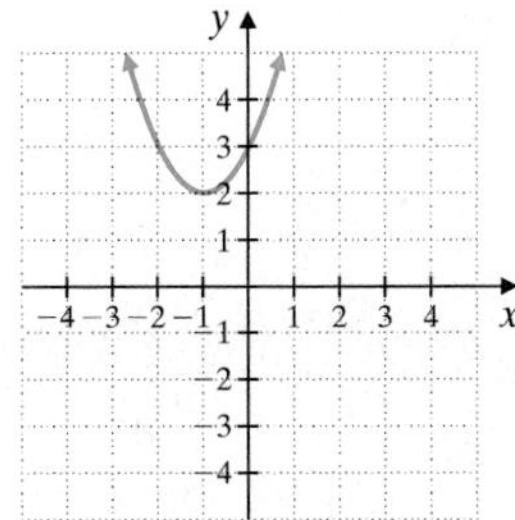

$y = x^2 + 2x + 3$
$b^2 - 4ac < 0$
No x-intercepts

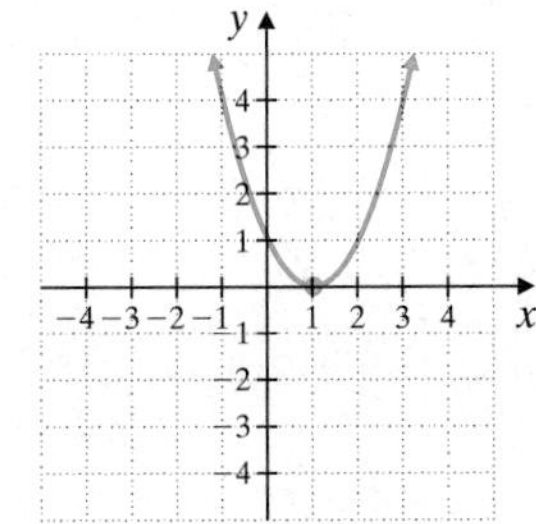

$y = x^2 - 2x + 1$
$b^2 - 4ac = 0$
One x-intercept

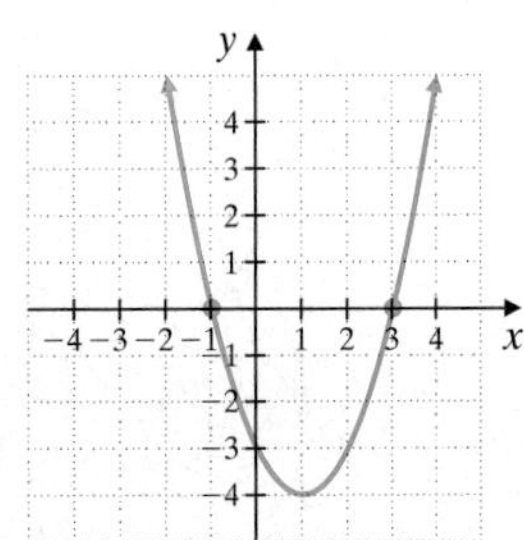

$y = x^2 - 2x - 3$
$b^2 - 4ac > 0$
Two x-intercepts

Answer

4. $(2, -9)$

Concept Check

Without making any calculations, tell whether the graph of $f(x) = 7 - x - 0.3x^2$ has a maximum value or a minimum value. Explain your reasoning.

Practice Problem 5

An object is thrown upward from the top of a 100-foot cliff. Its height in feet above ground after t seconds is given by the function $f(t) = -16t^2 + 10t + 100$. Find the maximum height of the object and the number of seconds it took for the object to reach its maximum height.

Try the Concept Check in the margin.

EXAMPLE 5 Finding Maximum Height

A rock is thrown upward from the ground. Its height in feet above ground after t seconds is given by the function $f(t) = -16t^2 + 20t$. Find the maximum height of the rock and the number of seconds it took for the rock to reach its maximum height.

Solution:

1. UNDERSTAND. The maximum height of the rock is the largest value of $f(t)$. Since the function $f(t) = -16t^2 + 20t$ is a quadratic function, its graph is a parabola. It opens downward since $-16 < 0$. Thus, the maximum value of $f(t)$ is the $f(t)$- or y-value of the vertex of its graph.
2. TRANSLATE. To find the vertex (h, k), we notice that for $f(t) = -16t^2 + 20t$, $a = -16$, $b = 20$, and $c = 0$. We will use these values and the vertex formula

$$\left(\frac{-b}{2a}, f\left(\frac{-b}{2a}\right)\right)$$

3. SOLVE. $h = \frac{-b}{2a} = \frac{-20}{-32} = \frac{5}{8}$

$$f\left(\frac{5}{8}\right) = -16\left(\frac{5}{8}\right)^2 + 20\left(\frac{5}{8}\right) = -16\left(\frac{25}{64}\right) + \frac{25}{2} = -\frac{25}{4} + \frac{50}{4} = \frac{25}{4}$$

4. INTERPRET. The graph of $f(t)$ is a parabola opening downward with vertex $\left(\frac{5}{8}, \frac{25}{4}\right)$. This means that the rock's maximum height is $\frac{25}{4}$ feet, or $6\frac{1}{4}$ feet, which was reached in $\frac{5}{8}$ second.

Answers

5. maximum height: $101\frac{9}{16}$ ft in $\frac{5}{16}$ sec

Concept Check: $f(x)$ has a maximum value since it opens downward.

Name ______________________ Section __________ Date __________

EXERCISE SET 10.6

Find the vertex of the graph of each quadratic function by completing the square or using the vertex formula. See Examples 1 through 4.

1. $f(x) = x^2 + 8x + 7$ **2.** $f(x) = x^2 + 6x + 5$ **3.** $f(x) = -x^2 + 10x + 5$ **4.** $f(x) = -x^2 - 8x + 2$

5. $f(x) = 5x^2 - 10x + 3$ **6.** $f(x) = -3x^2 + 6x + 4$ **7.** $f(x) = -x^2 + x + 1$ **8.** $f(x) = x^2 - 9x + 8$

Match each function with its graph. See Examples 1 through 4.

A.

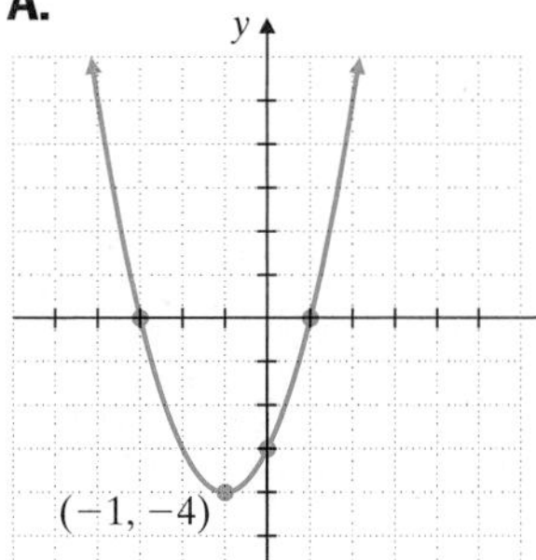

B.

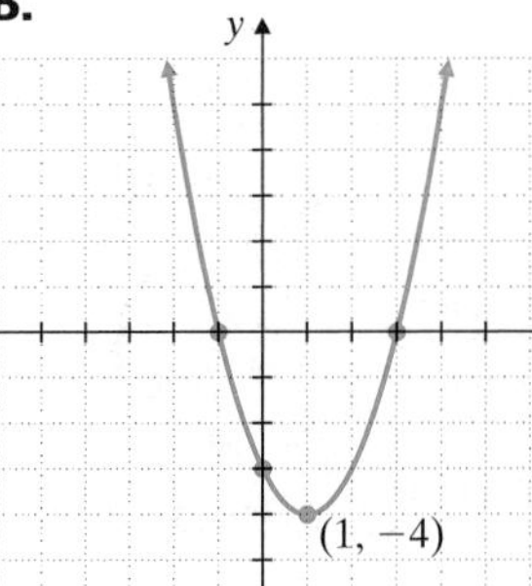

C.

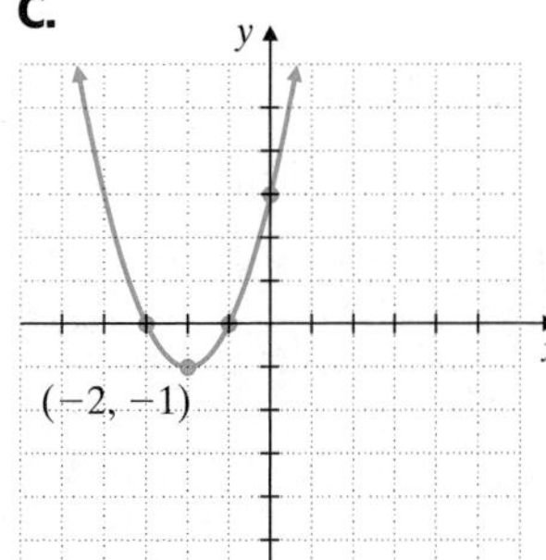

D.

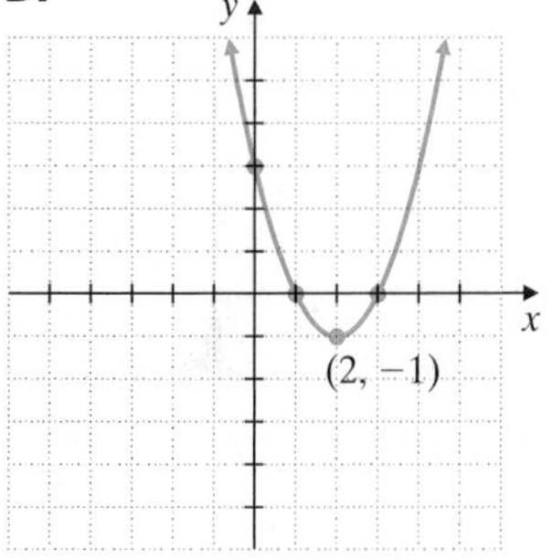

9. $f(x) = x^2 - 4x + 3$ **10.** $f(x) = x^2 + 2x - 3$ **11.** $f(x) = x^2 - 2x - 3$ **12.** $f(x) = x^2 + 4x + 3$

Find the vertex of the graph of each quadratic function. Determine whether the graph opens upward or downward, find any intercepts, and graph the function. See Examples 1 through 4.

13. $f(x) = x^2 + 4x - 5$

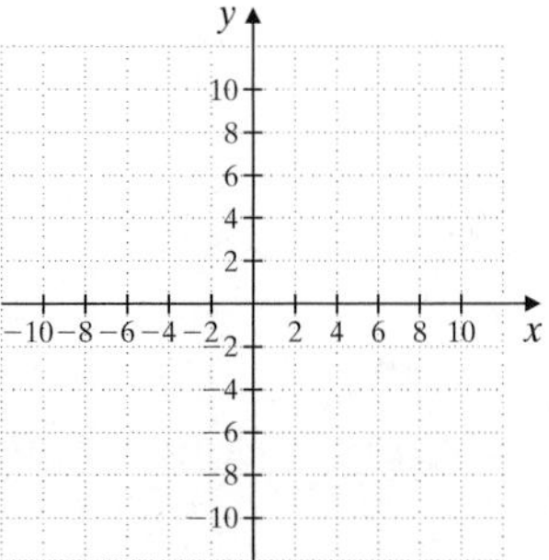

14. $f(x) = x^2 + 2x - 3$

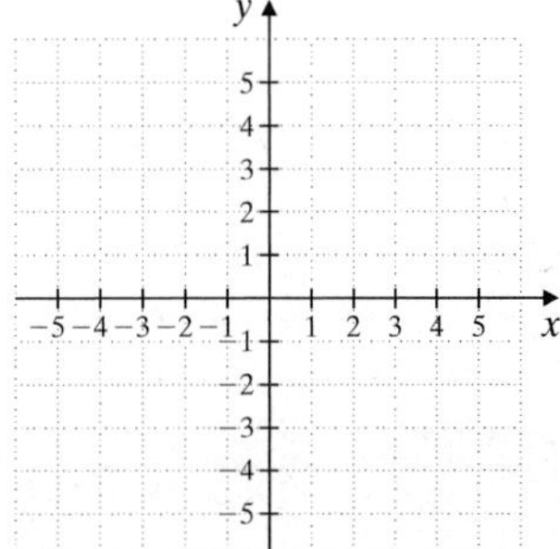

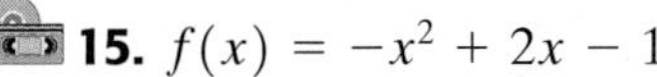

15. $f(x) = -x^2 + 2x - 1$

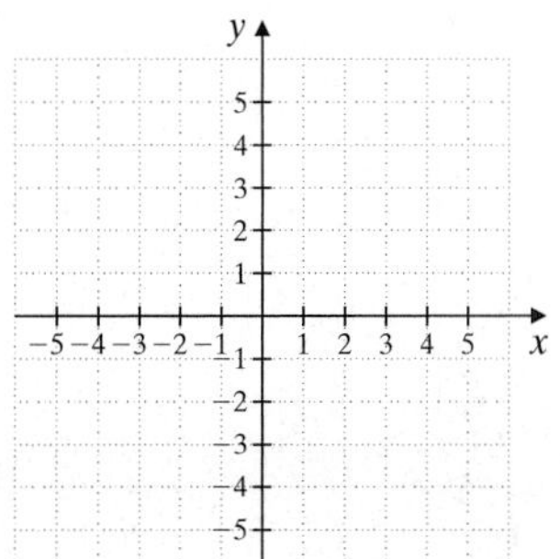

16. $f(x) = -x^2 + 4x - 4$

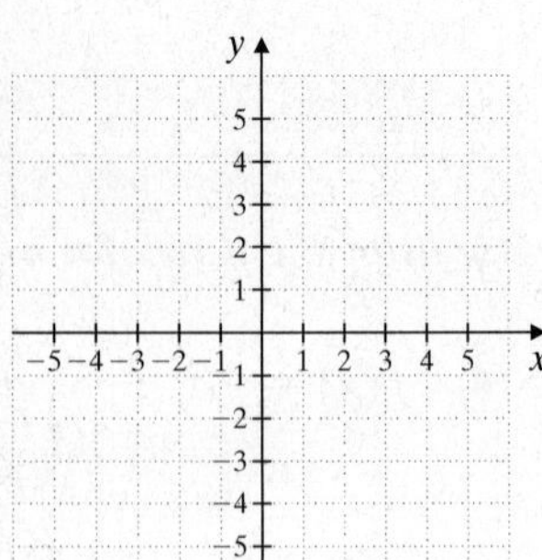

17. $f(x) = x^2 - 4$

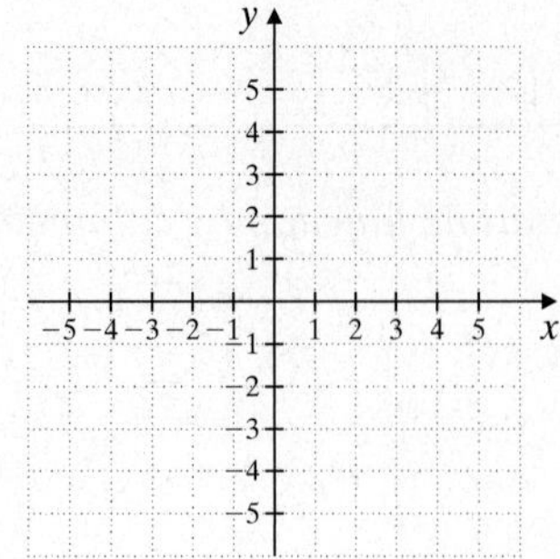

18. $f(x) = x^2 - 1$

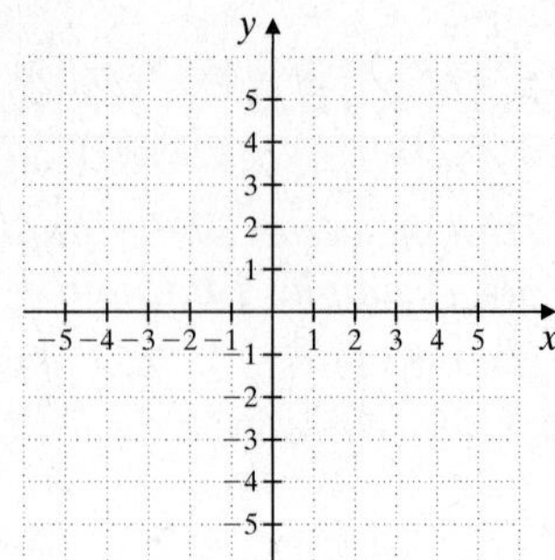

19. $f(x) = 4x^2 + 4x - 3$

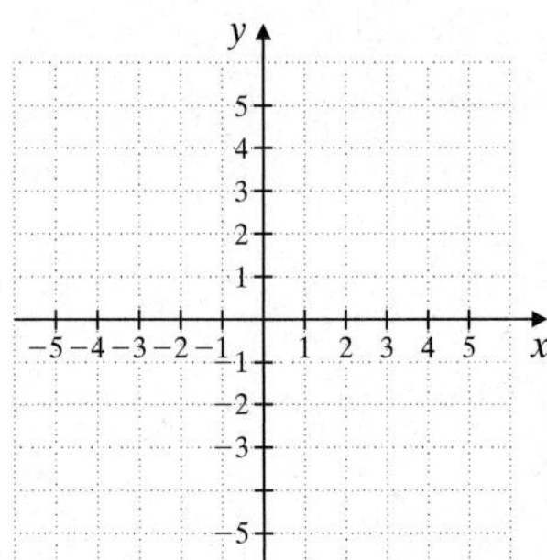

20. $f(x) = 2x^2 - x - 3$

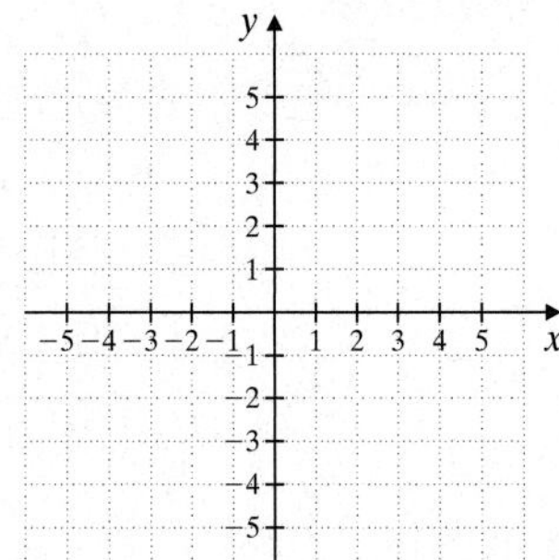

21. $f(x) = \frac{1}{2}x^2 + 4x + \frac{15}{2}$

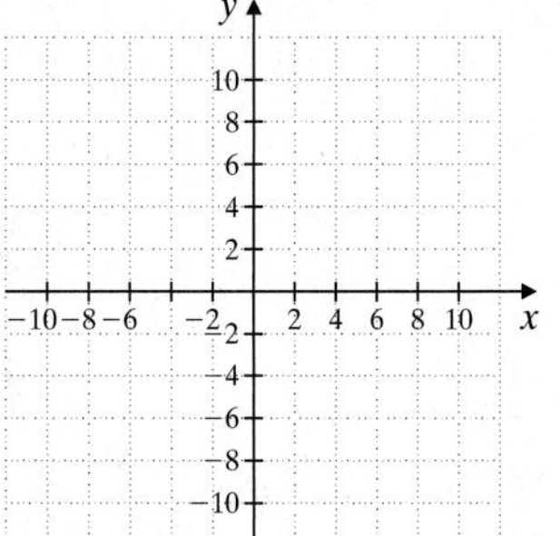

22. $f(x) = \frac{1}{5}x^2 + 2x + \frac{9}{5}$

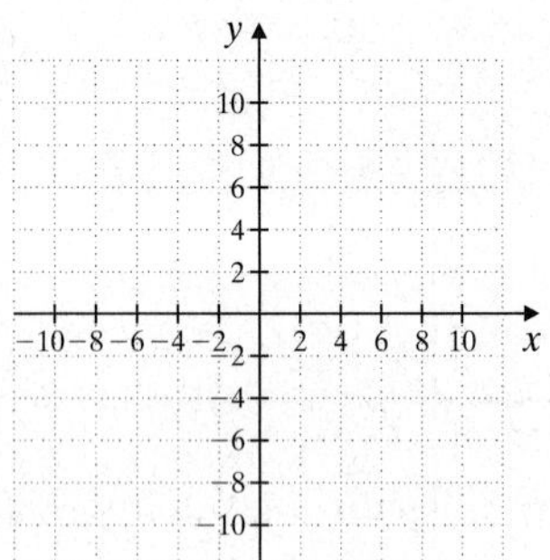

23. $f(x) = x^2 - 4x + 5$

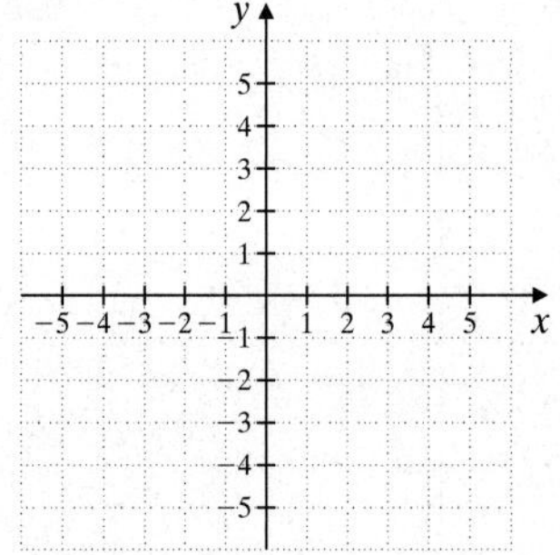

24. $f(x) = x^2 - 6x + 11$

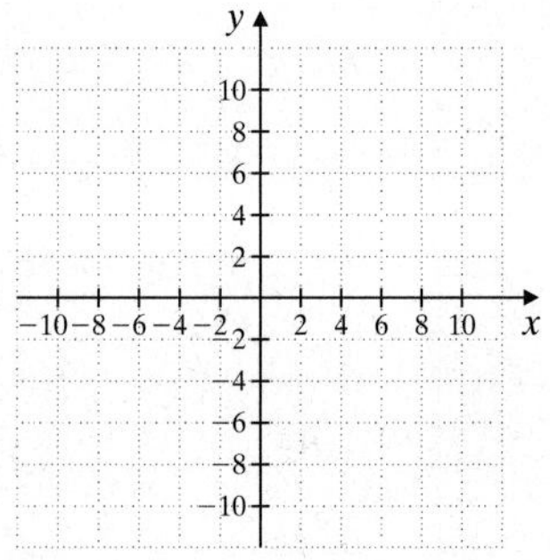

25. $f(x) = 2x^2 + 4x + 5$

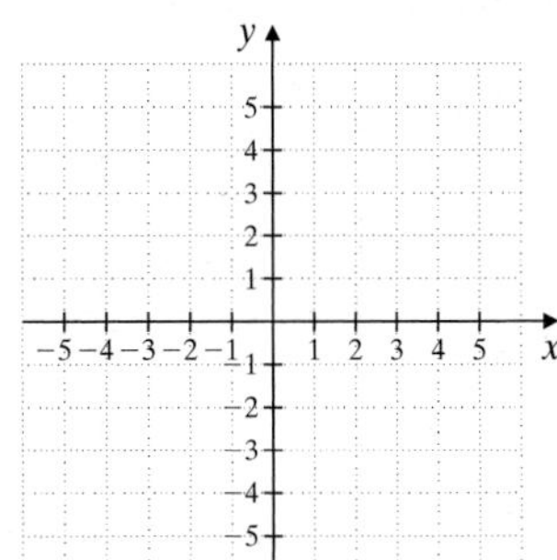

26. $f(x) = 3x^2 + 12x + 16$

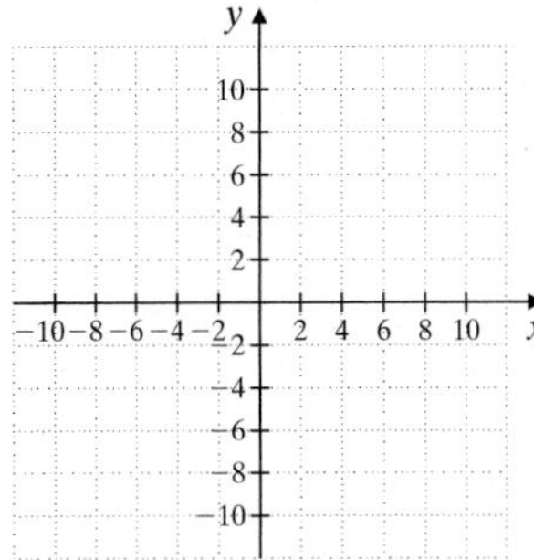

27. $f(x) = -2x^2 + 12x$

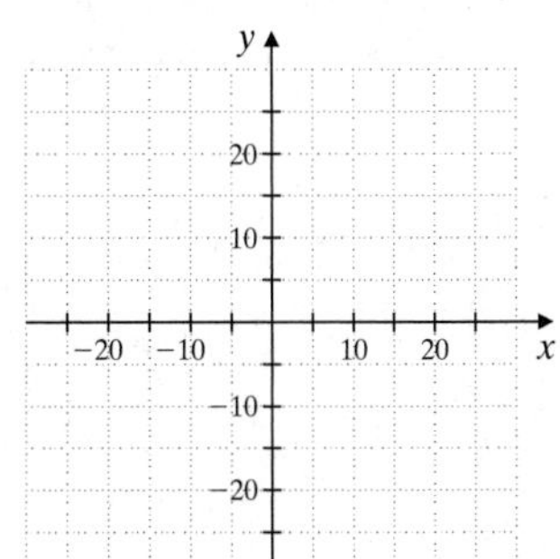

28. $f(x) = -4x^2 + 8x$

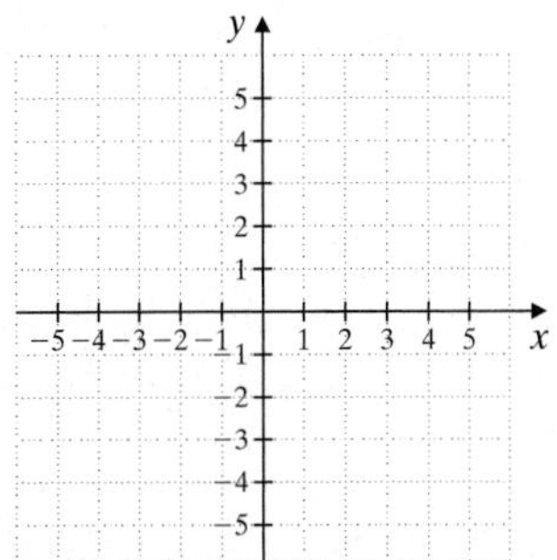

C *Solve. See Example 5.*

29. If a projectile is fired straight upward from the ground with an initial speed of 96 feet per second, then its height h in feet after t seconds is given by the function $h(t) = -16t^2 + 96t$. Find the maximum height of the projectile.

30. The cost C in dollars of manufacturing x bicycles at Holladay's Production Plant is given by the function $C(x) = 2x^2 - 800x + 92{,}000$.

a. Find the number of bicycles that must be manufactured to minimize the cost.

b. Find the minimum cost.

31. If Rheam Gaspar throws a ball upward with an initial speed of 32 feet per second, then its height h in feet after t seconds is given by the function $h(t) = -16t^2 + 32t$. Find the maximum height of the ball.

32. The Utah Ski Club sells calendars to raise money. The profit P, in cents, from selling x calendars is given by the function $P(x) = 360x - x^2$.

a. Find how many calendars must be sold to maximize profit.

b. Find the maximum profit.

33. Find two numbers whose sum is 60 and whose product is as large as possible. [*Hint:* Let x and $60 - x$ be the two positive numbers. Their product can be described by the function $f(x) = x(60 - x)$.]

34. Find two numbers whose sum is 11 and whose product is as large as possible. (Use the hint for Exercise 33.)

35. Find two numbers whose difference is 10 and whose product is as small as possible. (Use the hint for Exercise 33.)

36. Find two numbers whose difference is 8 and whose product is as small as possible.

△ **37.** The length and width of a rectangle must have a sum of 40. Find the dimensions of the rectangle that will have the maximum area. (Use the hint for Exercise 33.)

△ **38.** The length and width of a rectangle must have a sum of 50. Find the dimensions of the rectangle that will have maximum area.

Review and Preview

Find the vertex of the graph of each function. See Section 10.5.

39. $f(x) = x^2 + 2$

40. $f(x) = (x - 3)^2$

41. $g(x) = x + 2$

42. $h(x) = x - 3$

43. $f(x) = (x + 5)^2 + 2$

44. $f(x) = 2(x - 3)^2 + 2$

45. $f(x) = 3(x - 4)^2 + 1$

46. $f(x) = (x + 1)^2 + 4$

Combining Concepts

Find the vertex of the graph of each quadratic function. Determine whether the graph opens upward or downward, find the y-intercept, approximate the x-intercepts to one decimal place, and graph the function.

47. $f(x) = x^2 + 10x + 15$

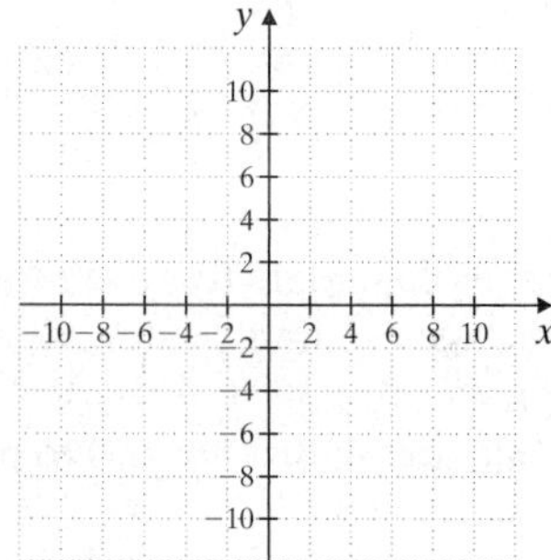

48. $f(x) = 2x^2 + 4x - 1$

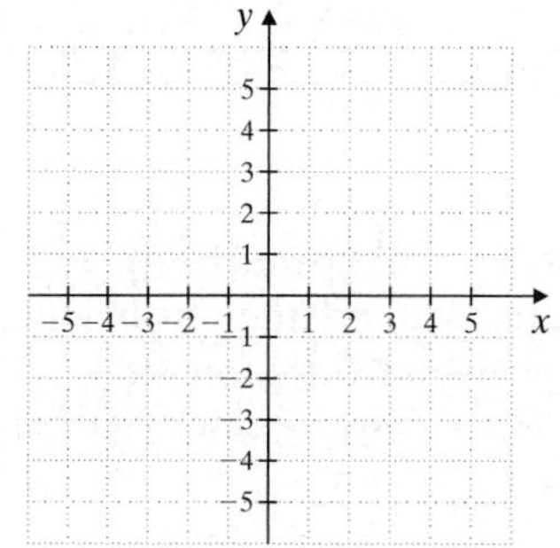

Use a graphing calculator to verify the graph of each exercise.

49. Exercise 21.

50. Exercise 22.

Find the maximum or minimum value of each function. Approximate to two decimal places.

51. $f(x) = 2.3x^2 - 6.1x + 3.2$

52. $f(x) = 7.6x^2 + 9.8x - 2.1$

53. The number of inmates in custody in U.S. prisons and jails can be modeled by the quadratic function

$$p(x) = -x^2 + 93x + 1128$$

where $p(x)$ is the number of inmates in thousands and x is the number of years after 1990. (*Source*: Based on data from the Bureau of Justice Statistics, U.S. Department of Justice, 1990–2000)

a. Will this function have a maximum or a minimum? How can you tell?

b. According to this model, in what year will the number of prison and jail inmates in custody in the United States be at its maximum/minimum?

c. What is the maximum/minimum number of inmates predicted?

54. Methane is a gas produced by landfills, natural gas systems, and coal mining that contributes to the greenhouse effect and global warming. Projected methane emissions in the United States can be modeled by the quadratic function

$$f(x) = -0.072x^2 + 1.93x + 173.9$$

where $f(x)$ is the amount of methane produced in million metric tons and x is the number of years after 2000. (*Source*: Based on data from the U.S. Environmental Protection Agency, 2000–2020)

a. According to this model, what will U.S. emissions of methane be in 2009?

b. Will this function have a maximum or a minimum? How can you tell?

c. In what year will methane emissions in the United States be at their maximum/minimum? Round to the nearest whole year.

d. What is the level of methane emissions for that year? (Use your rounded answer from part c.)

STUDY SKILLS REMINDER

Are you prepared for your final exam?

To start preparing for your final exam, try the following study techniques.

- Review the material that you will be responsible for on your exam. Also check your notebook for any lecture notes that you highlighted.
- Review any formulas that you may need to memorize.
- Check to see if your instructor or math department will be conducting a final exam review.
- Check with your instructor to see whether there are final exams from previous semesters/quarters that are available to students for study.
- Use your previously taken tests as a practice final exam. To do so, rewrite the test questions in mixed order on blank sheets of paper. This will help you prepare for exam conditions.
- If you are unsure of a few topics, see your instructor or visit a learning lab for further assistance. Also, viewing the video segment of a troublesome section will help.
- If you need further exercises to work, try the chapter tests at the end of appropriate chapters.

Good luck! I hope you are enjoying this textbook and your intermediate algebra course.

CHAPTER 10 ACTIVITY Recognizing Linear and Quadratic Models

This activity may be completed by working in groups or individually.

We have seen in this and previous chapters that data can be modeled by both linear models and quadratic models. However, when we are given a set of data to model, how can we tell if a linear or quadratic model is appropriate? The best answer depends on looking at a scatter diagram of the data. If the plotted data points fall roughly on a line, a linear model is usually the better choice. If the plotted data points seem to fall on a definite curve or if a maximum or minimum point is apparent, a quadratic model is usually the better choice.

One of the sets of data shown in the tables is best modeled by a linear function and one is best modeled by a quadratic function. In each case, the variable x represents the number of years after 1994.

Saturn Vehicle Sales (in thousands)

Year	1996	1997	1998	1999	2000
x	2	3	4	5	6
Number of Saturns sold, y (in thousands)	279	251	232	233	272

(*Source:* General Motors)

Number of Domestic Wal-Mart Stores and Supercenters

Year	1994	1995	1996	1997	1998	1999	2000	2001
x	0	1	2	3	4	5	6	7
Number of stores, y	2022	2132	2234	2304	2362	2433	2522	2624

(*Source:* Wal-Mart Stores, Inc.)

1. Make a scatter diagram for each set of data. Which type of model should be used for each set of data?
2. For the set of data that you have determined to be linear, find a linear function that fits the data points. Explain the method that you used. (*Hint*: See the Focus on Business and Career: Linear Modeling in Section 3.4 or the Focus on the Real World: Linear Modeling in Section 4.5 for more information.)
3. For the set of data that you have determined to be quadratic, identify the point on your scatter diagram that appears to be the vertex of the parabola. Use the coordinates of this vertex in the quadratic model $f(x) = a(x - h)^2 + k$.
4. Solve for the remaining unknown constant in the quadratic model by substituting the coordinates for another data point into the function. Write the final form of the quadratic model for this data set.
5. Use your models to estimate the number of Saturns sold and the number of domestic Wal-Mart stores and supercenters in 2002.
6. (Optional) For each set of data, enter the data from the table into a graphing calculator and use either the linear regression feature or the quadratic regression feature to find an appropriate function that models the data.* Compare these functions with the ones you found by hand. How are they alike or different?

*To find out more about using your graphing calculator to find a regression equation, consult your user's manual.

Chapter 10 VOCABULARY CHECK

Fill in each blank with one of the words or phrases listed below.

quadratic formula	quadratic	discriminant	$\pm\sqrt{b}$
completing the square	quadratic inequality		
(h, k) $(0, k)$ $(h, 0)$	$\dfrac{-b}{2a}$		

1. The _____________ helps us know find the number and type of solutions of a quadratic equation.
2. If $a^2 = b$, then $a =$ _________.
3. The graph of $f(x) = ax^2 + bx + c$, where a is not 0, is a parabola whose vertex has an x-value of _____.
4. A(n) _________________ is an inequality that can be written so that one side is a quadratic expression and the other side is 0.
5. The process of writing a quadratic equation so that one side is a perfect square trinomial is called _________________.
6. The graph of $f(x) = x^2 + k$ has vertex _________.
7. The graph of $f(x) = (x - h)^2$ has vertex _________.
8. The graph of $f(x) = (x - h)^2 + k$ has vertex _________.
9. The formula $x = \dfrac{-b \pm \sqrt{b^2 - 4ac}}{2a}$ is called the _________________.
10. A ___________ equation is one that can be written in the form $ax^2 + bx + c = 0$, where $a, b,$ and c are real numbers and a is not 0.

CHAPTER 10 Highlights

DEFINITIONS AND CONCEPTS | **EXAMPLES**

Section 10.1 Solving Quadratic Equations by Completing the Square

SQUARE ROOT PROPERTY

If b is a real number and if $a^2 = b$, then $a = \pm\sqrt{b}$.

Solve: $(x + 3)^2 = 14$

$x + 3 = \pm\sqrt{14}$

$x = -3 \pm \sqrt{14}$

SOLVING A QUADRATIC EQUATION IN x BY COMPLETING THE SQUARE

Step 1. If the coefficient of x^2 is not 1, divide both sides of the equation by the coefficient of x^2.

Step 2. Get the variable terms alone.

Step 3. Complete the square by adding the square of half of the coefficient of x to both sides.

Step 4. Write the resulting trinomial as the square of a binomial.

Step 5. Use the square root property.

Solve: $3x^2 - 12x - 18 = 0$

1. $x^2 - 4x - 6 = 0$
2. $x^2 - 4x = 6$
3. $\frac{1}{2}(-4) = -2$ and $(-2)^2 = 4$

   $x^2 - 4x + 4 = 6 + 4$
4. $(x - 2)^2 = 10$
5. $x - 2 = \pm\sqrt{10}$

   $x = 2 \pm \sqrt{10}$

Section 10.2 Solving Quadratic Equations by Using the Quadratic Formula

QUADRATIC FORMULA

A quadratic equation written in the form $ax^2 + bx + c = 0$ has solutions

$$x = \frac{-b \pm \sqrt{b^2 - 4ac}}{2a}$$

Solve: $x^2 - x - 3 = 0$

$a = 1, b = -1, c = -3$

$$x = \frac{-(-1) \pm \sqrt{(-1)^2 - 4(1)(-3)}}{2 \cdot 1}$$

$$x = \frac{1 \pm \sqrt{13}}{2}$$

Definitions and Concepts | Examples

Section 10.3 Solving Equations by Using Quadratic Methods

Substitution is often helpful in solving an equation that contains a repeated variable expression.

Solve: $(2x + 1)^2 - 5(2x + 1) + 6 = 0$

Let $m = 2x + 1$. Then

$$m^2 - 5m + 6 = 0 \quad \text{Let } m = 2x + 1.$$
$$(m - 3)(m - 2) = 0$$
$$m = 3 \quad \text{or} \quad m = 2$$
$$2x + 1 = 3 \qquad 2x + 1 = 2 \quad \text{Substitute back.}$$
$$x = 1 \qquad x = \frac{1}{2}$$

Section 10.4 Nonlinear Inequalities in One Variable

Solving a Polynomial Inequality

Step 1. Write the inequality in standard form and solve the related equation.

Step 2. Use solutions from Step 1 to separate the number line into regions.

Step 3. Use a test point to determine whether values in each region satisfy the original inequality.

Step 4. Write the solution set as the union of regions whose test point values are solutions.

Solve: $x^2 \geq 6x$

1. $x^2 - 6x \geq 0$

2. $x^2 - 6x = 0$

$x(x - 6) = 0$

$x = 0$ or $x = 6$

3.

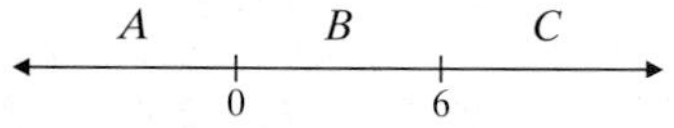

4.

Region	Test Point Value	$x^2 \geq 6x$	Result
A	−2	$(-2)^2 \geq 6(-2)$	True.
B	1	$1^2 \geq 6(1)$	False.
C	7	$7^2 \geq 6(7)$	True.

5.

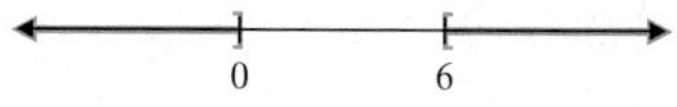

The solution set is $(-\infty, 0] \cup [6, \infty)$.

Solving a Rational Inequality

Step 1. Solve for values that make all denominators 0.

Step 2. Solve the related equation.

Step 3. Use solutions from Steps 1 and 2 to separate the number line into regions.

Step 4. Use a test point to determine whether values in each region satisfy the original inequality.

Step 5. Write the solution set as the union of regions whose test point value is a solution. Check whether to include values from Step 2. Be sure not to include values that make any denominator 0.

Solve: $\frac{6}{x - 1} < -2$

1. $x - 1 = 0$ Set the denominator equal to 0.

$x = 1$

2. $\frac{6}{x - 1} = -2$

$6 = -2(x - 1)$ Multiply by $(x - 1)$.

$6 = -2x + 2$

$4 = -2x$

$-2 = x$

3.

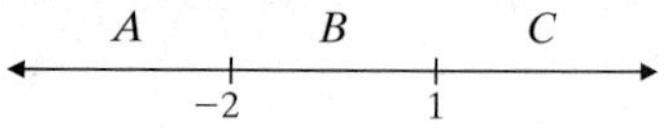

4. Only a test value from region B satisfies the original inequality.

5.

The solution set is $(-2, 1)$.

DEFINITIONS AND CONCEPTS | **EXAMPLES**

Section 10.5 Quadratic Functions and Their Graphs

GRAPHING A QUADRATIC FUNCTION

The graph of a quadratic function written in the form $f(x) = a(x - h)^2 + k$ is a parabola with vertex (h, k).
If $a > 0$, the parabola opens upward.
If $a < 0$, the parabola opens downward.
The axis of symmetry is the line whose equation is $x = h$.

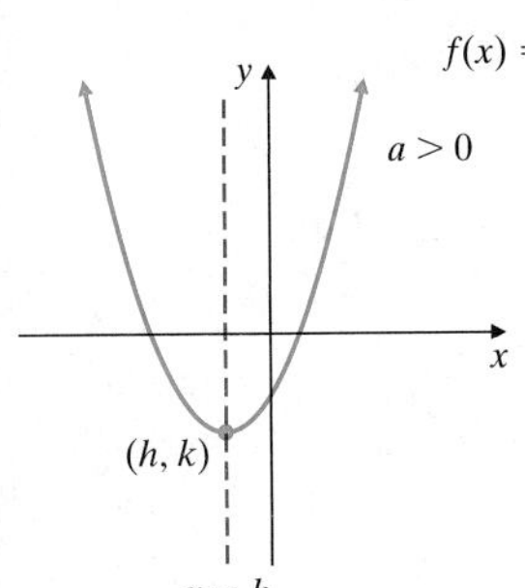

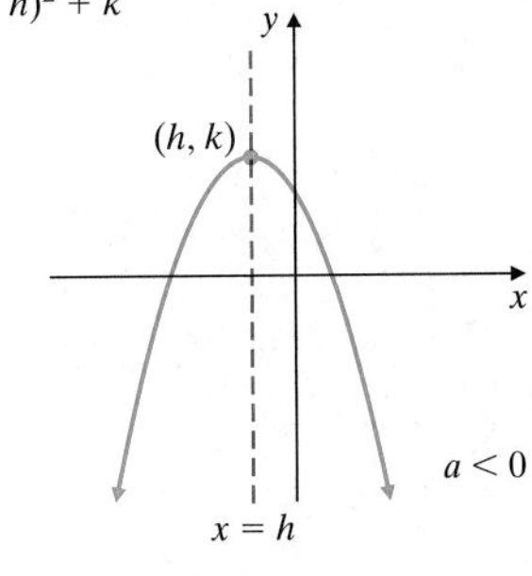

Graph: $g(x) = 3(x - 1)^2 + 4$
The graph is a parabola with vertex $(1, 4)$ and axis of symmetry $x = 1$. Since $a = 3$ is positive, the graph opens upward.

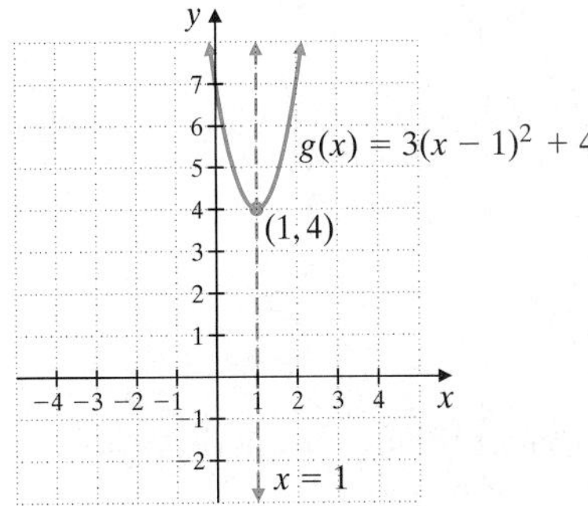

Section 10.6 Further Graphing of Quadratic Functions

The graph of $f(x) = ax^2 + bx + c, a \neq 0$, is a parabola with vertex

$$\left(\frac{-b}{2a}, f\left(\frac{-b}{2a}\right)\right).$$

Graph: $f(x) = x^2 - 2x - 8$. Find the vertex and x- and y-intercepts.

$$\frac{-b}{2a} = \frac{-(-2)}{2 \cdot 1} = 1$$

$$f(1) = 1^2 - 2(1) - 8 = -9$$

The vertex is $(1, -9)$.

$$0 = x^2 - 2x - 8$$

$$0 = (x - 4)(x + 2)$$

$$x = 4 \quad \text{or} \quad x = -2$$

The x-intercepts are $(4, 0)$ and $(-2, 0)$.

$$f(0) = 0^2 - 2 \cdot 0 - 8 = -8$$

The y-intercept is $(0, -8)$.

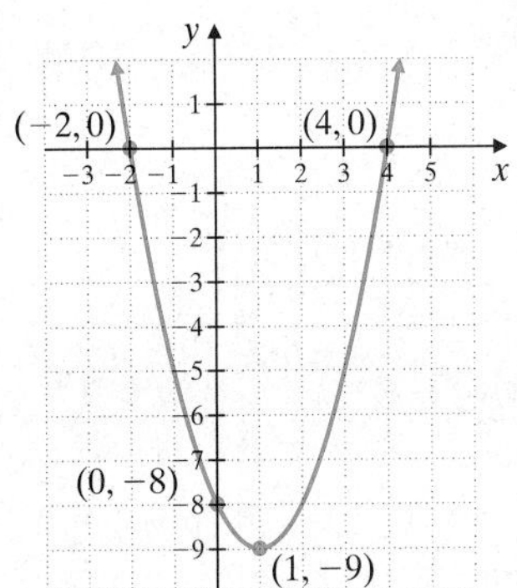

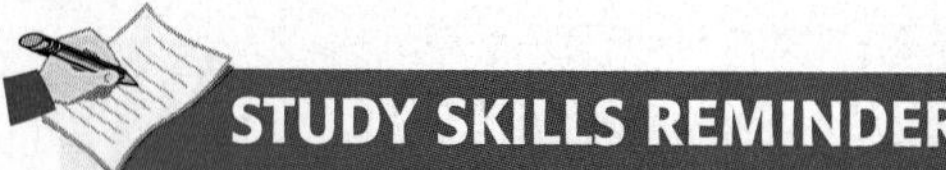

STUDY SKILLS REMINDER

Are you preparing for a test on Chapter 10?

Below I have listed some common trouble areas for students in Chapter 10. After studying for your test—but before taking your test—read these.

- Don't forget that to solve a quadratic equation such as $x^2 + 6x = 1$, by completing the square, add the square of half of 6 to *both* sides.

$$x^2 + 6x = 1$$

$$x^2 + 6x + 9 = 1 + 9 \qquad \text{Add 9 to both sides. } \left(\frac{1}{2}(6) = 3 \text{ and } 3^2 = 9\right)$$

$$(x + 3)^2 = 10$$

$$x + 3 = \pm\sqrt{10}$$

$$x = -3 \pm \sqrt{10}$$

- Remember to write a quadratic equation in standard form $(ax^2 + bx + c = 0)$ before using the quadratic formula of solve.

$$x(4x - 1) = 1$$

$$4x^2 - x - 1 = 0 \qquad \text{Write in standard form.}$$

$$x = \frac{-(-1) \pm \sqrt{(-1)^2 - 4(4)(-1)}}{2 \cdot 4} \qquad \text{Use the quadratic formula with } a = 4, b = -1, \text{ and } c = -1.$$

$$x = \frac{1 \pm \sqrt{17}}{8} \qquad \text{Simplify.}$$

- Review the steps for solving a quadratic equation in general on page 761.
- Don't forget how to graph a quadratic function in the form $f(x) = a(x - h)^2 + k$. The graph of $f(x) = -2(x - 3)^2 - 1$

opens downward; narrower; shift 3 units right; shift 1 units down

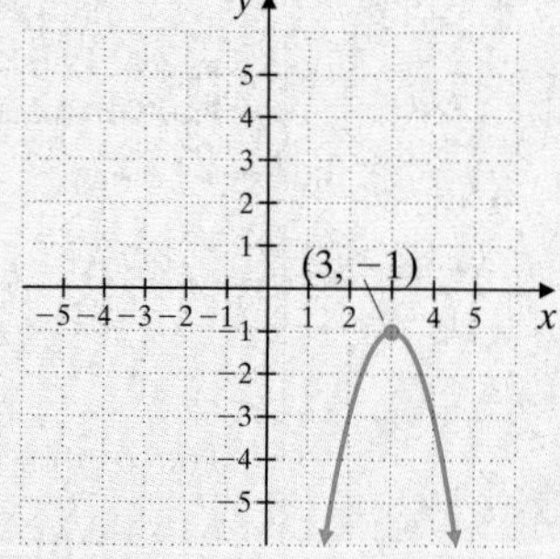

Remember: This is simply a checklist of common trouble areas. For a review of Chapter 10, see the Highlights and Chapter Review at the end of this chapter.

Name ________________________ Section ____________ Date ____________

Chapter 10 Review

(10.1) *Solve by factoring.*

1. $x^2 - 15x + 14 = 0$

2. $x^2 - x - 30 = 0$

3. $10x^2 = 3x + 4$

4. $7a^2 = 29a + 30$

Use the square root property to solve each equation.

5. $4m^2 = 196$

6. $9y^2 = 36$

7. $(9n + 1)^2 = 9$

8. $(5x - 2)^2 = 2$

Solve by completing the square.

9. $z^2 + 3z + 1 = 0$

10. $x^2 + x + 7 = 0$

11. $(2x + 1)^2 = x$

12. $(3x - 4)^2 = 10x$

13. If P dollars are invested, the formula $A = P(1 + r)^2$ gives the amount A in an account paying interest rate r compounded annually after 2 years. Find the interest rate r such that \$2500 increases to \$2717 in 2 years. Round the result to the nearest hundredth of a percent.

14. Two ships leave a port at the same time and travel at the same speed. One ship is traveling due north and the other due east. In a few hours, the ships are 150 miles apart. How many miles has each ship traveled? Give an exact answer and a one-decimal-place approximation.

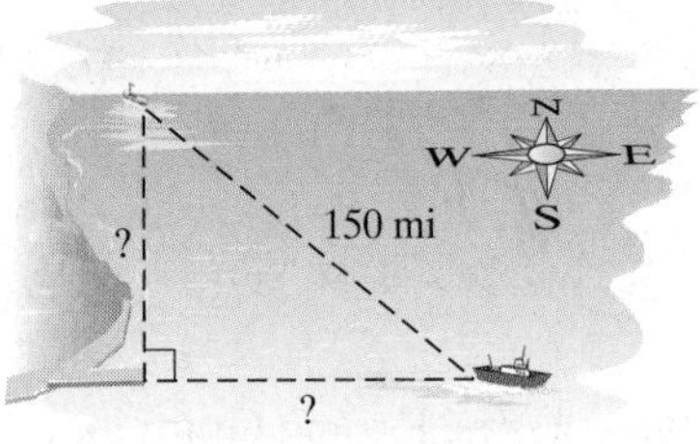

(10.2) *If the discriminant of a quadratic equation has the given value, determine the number and type of solutions of the equation.*

15. -8

16. 48

17. 100

18. 0

Use the quadratic formula to solve each equation.

19. $x^2 - 16x + 64 = 0$

20. $x^2 + 5x = 0$

21. $x^2 + 11 = 0$

22. $2x^2 + 3x = 5$

23. $6x^2 + 7 = 5x$

24. $9a^2 + 4 = 2a$

25. $(5a - 2)^2 - a = 0$

26. $(2x - 3)^2 = x$

27. Cadets graduating from military school usually toss their hats high into the air at the end of the ceremony. One cadet threw his hat so that its distance $d(t)$ in feet above the ground t seconds after it was thrown was $d(t) = -16t^2 + 30t + 6$.

a. Find the distance above the ground of the hat 1 second after it was thrown.

b. Find the time it took the hat to hit the ground. Give an exact time and a one-decimal-place approximation.

△ **28.** The hypotenuse of an isosceles right triangle is 6 centimeters longer than either of the legs. Find the length of the legs. (*Hint:* Don't forget that an isosceles triangle has two sides of equal length.)

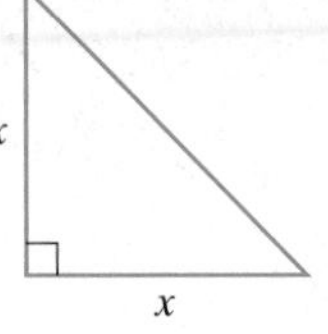

(10.3) *Solve each equation.*

29. $x^3 = 27$

30. $y^3 = -64$

31. $\frac{5}{x} + \frac{6}{x - 2} = 3$

32. $\frac{7}{8} = \frac{8}{x^2}$

33. $x^4 - 21x^2 - 100 = 0$

34. $5(x + 3)^2 - 19(x + 3) = 4$

35. $x^{2/3} - 6x^{1/3} + 5 = 0$

36. $x^{2/3} - 6x^{1/3} = -8$

37. $a^6 - a^2 = a^4 - 1$

38. $y^{-2} + y^{-1} = 20$

39. Two postal workers, Jerome Grant and Tim Bozik, can sort a stack of mail in 5 hours. Working alone, Tim can sort the mail in 1 hour less time than Jerome can. Find the time that each postal worker can sort the mail alone. Round the result to one decimal place.

40. A negative number decreased by its reciprocal is $-\frac{24}{5}$. Find the number.

(10.4) *Solve each inequality for x. Write each solution set in interval notation.*

41. $2x^2 - 50 \le 0$

42. $\frac{1}{4}x^2 < \frac{1}{16}$

43. $(2x - 3)(4x + 5) \ge 0$

44. $(x^2 - 16)(x^2 - 1) > 0$

45. $\dfrac{x - 5}{x - 6} < 0$

46. $\dfrac{x(x + 5)}{4x - 3} \geq 0$

47. $\dfrac{(4x + 3)(x - 5)}{x(x + 6)} > 0$

48. $(x + 5)(x - 6)(x + 2) \leq 0$

49. $x^3 + 3x^2 - 25x - 75 > 0$

50. $\dfrac{x^2 + 4}{3x} \leq 1$

51. $\dfrac{(5x + 6)(x - 3)}{x(6x - 5)} < 0$

52. $\dfrac{3}{x - 2} > 2$

(10.5) *Graph each function. Label the vertex and the axis of symmetry of each graph.*

53. $f(x) = x^2 - 4$

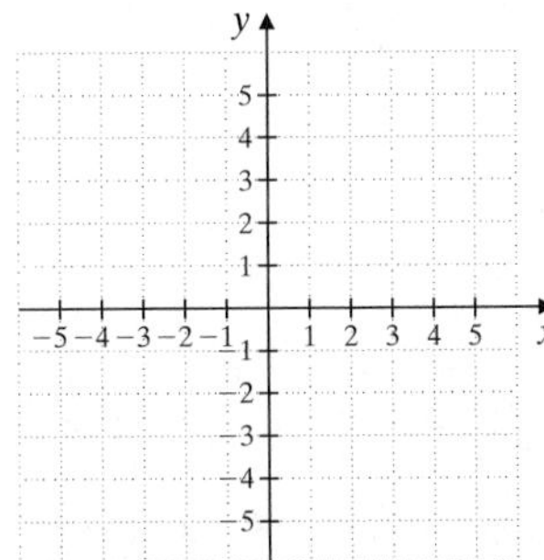

54. $g(x) = x^2 + 7$

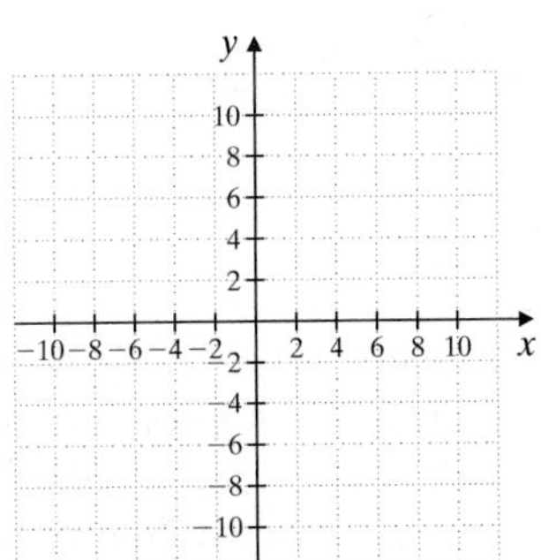

55. $H(x) = 2x^2$

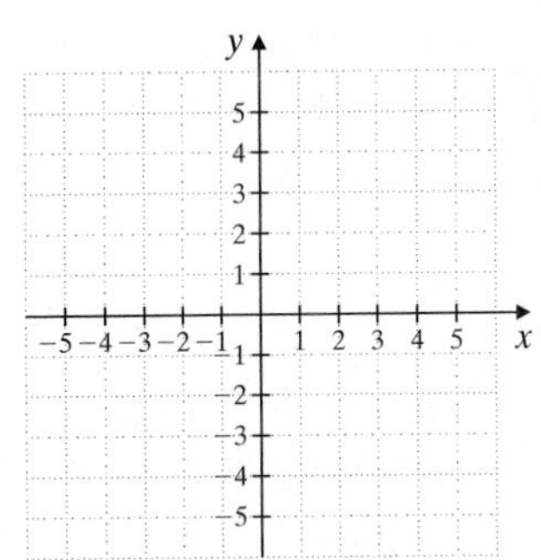

56. $h(x) = -\dfrac{1}{3}x^2$

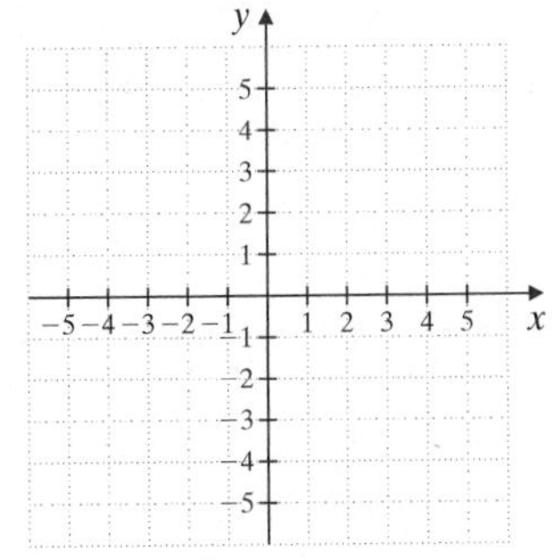

57. $F(x) = (x - 1)^2$

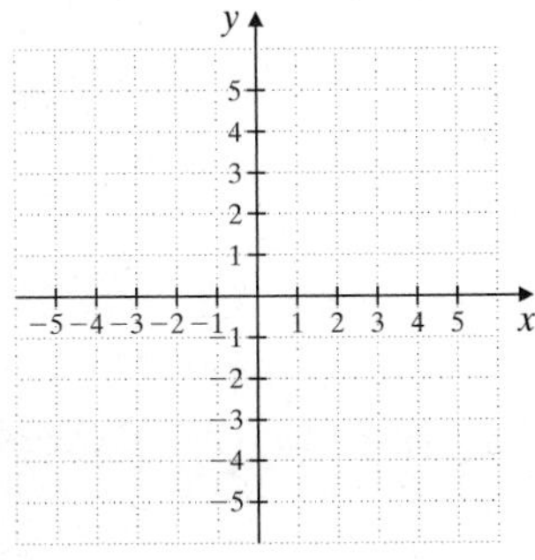

58. $G(x) = (x + 5)^2$

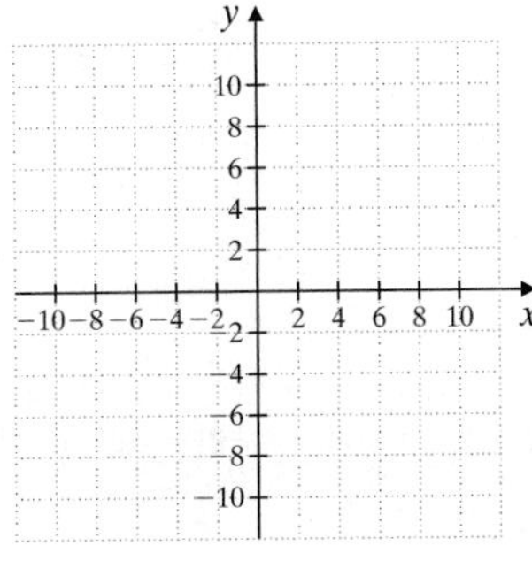

59. $f(x) = (x - 4)^2 - 2$

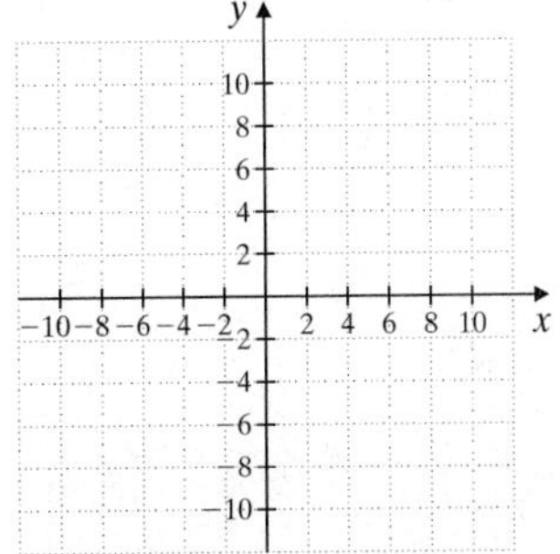

60. $f(x) = -3(x - 1)^2 + 1$

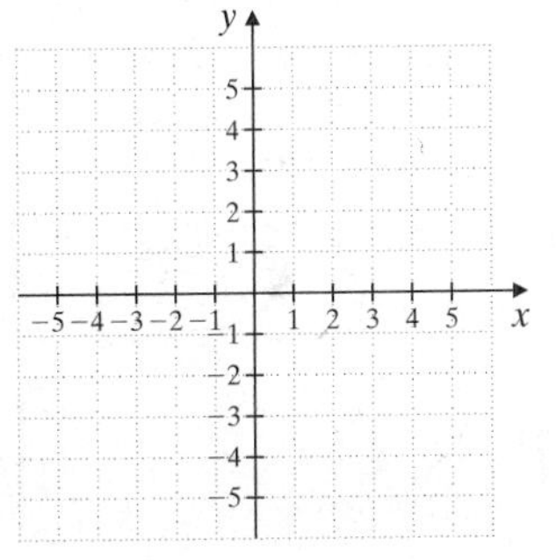

(10.6) *Graph each function. Find the vertex and any intercepts of each graph.*

61. $f(x) = x^2 + 10x + 25$

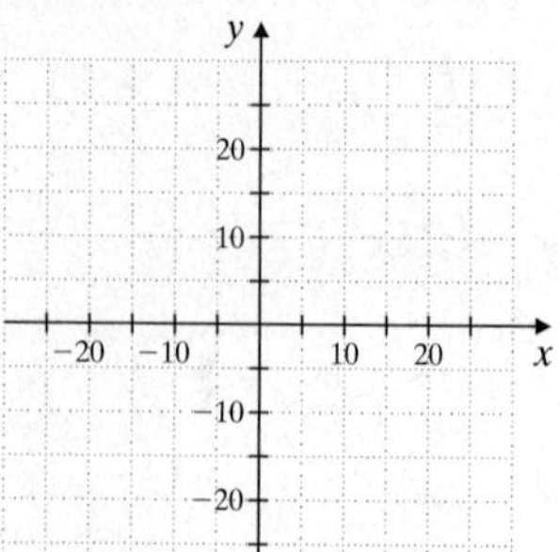

62. $f(x) = -x^2 + 6x - 9$

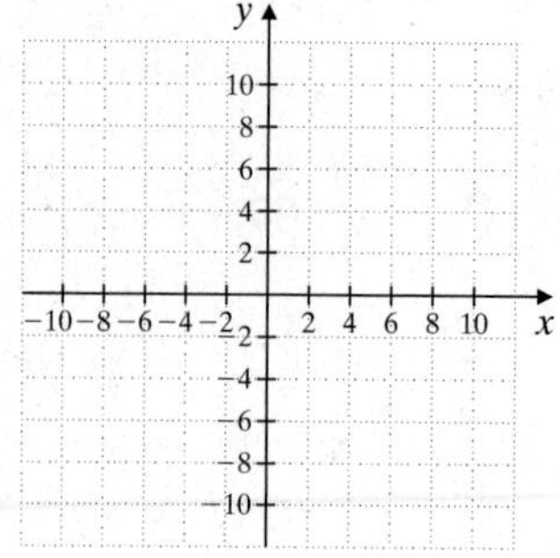

63. $f(x) = 4x^2 - 1$

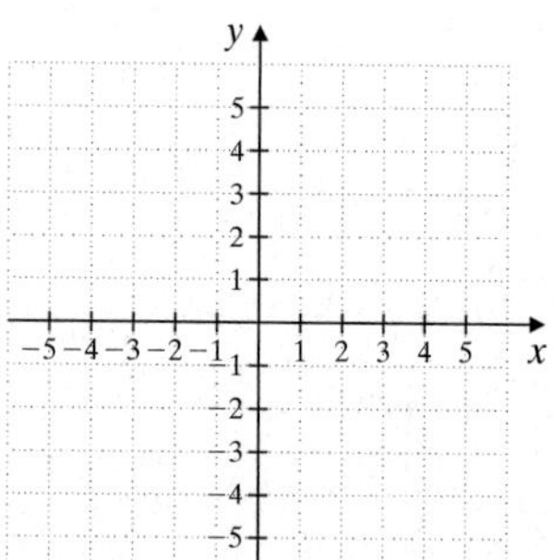

64. $f(x) = -5x^2 + 5$

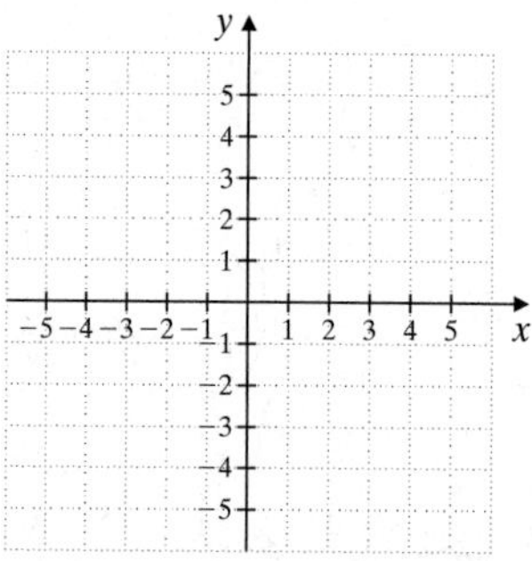

65. Find the vertex of the graph of $f(x) = -3x^2 - 5x + 4$. Determine whether the graph opens upward or downward, find the y-intercept, approximate the x-intercepts to one decimal place, and graph the function.

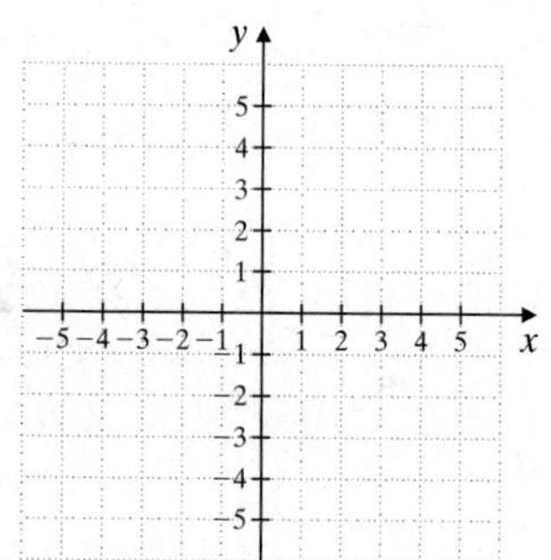

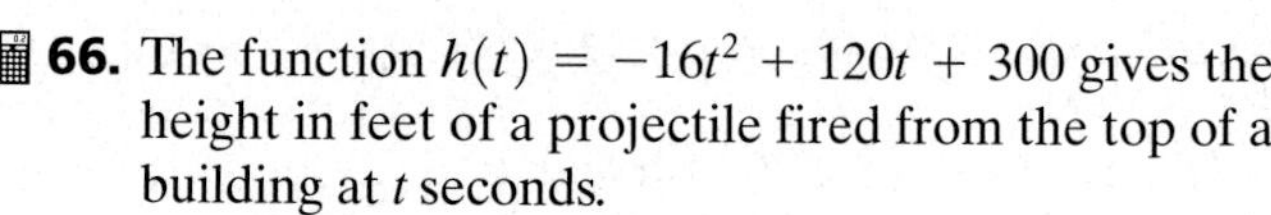

66. The function $h(t) = -16t^2 + 120t + 300$ gives the height in feet of a projectile fired from the top of a building at t seconds.

a. When will the object reach a height of 350 feet? Round your answer to one decimal place.

b. Explain why part (a) has two answers.

67. Find two numbers whose sum is 420 and whose product is as large as possible.

Name ______________________________

Chapter 10 Test

Solve each equation.

1. $5x^2 - 2x = 7$

2. $(x + 1)^2 = 10$

3. $m^2 - m + 8 = 0$

4. $u^2 - 6u + 2 = 0$

5. $7x^2 + 8x + 1 = 0$

6. $a^2 - 3a = 5$

7. $\dfrac{4}{x + 2} + \dfrac{2x}{x - 2} = \dfrac{6}{x^2 - 4}$

8. $x^4 - 8x^2 - 9 = 0$

9. $x^6 + 1 = x^4 + x^2$

10. $(x + 1)^2 - 15(x + 1) + 56 = 0$

Solve by completing the square.

11. $x^2 - 6x = -2$

12. $2a^2 + 5 = 4a$

Solve each inequality. Write each solution set in interval notation.

13. $2x^2 - 7x > 15$

14. $(x^2 - 16)(x^2 - 25) > 0$

15. $\dfrac{5}{x + 3} < 1$

16. $\dfrac{7x - 14}{x^2 - 9} \leq 0$

Graph each function. Label the vertex for each graph.

17. $f(x) = 3x^2$

18. $G(x) = -2(x - 1)^2 + 5$

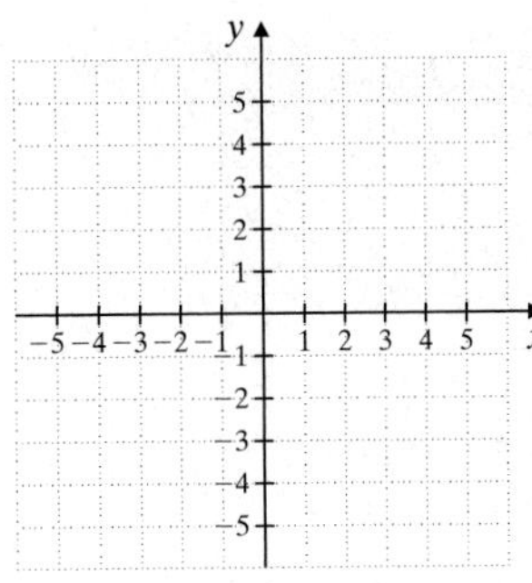

Answers

1. ________

2. ________

3. ________

4. ________

5. ________

6. ________

7. ________

8. ________

9. ________

10. ________

11. ________

12. ________

13. ________

14. ________

15. ________

16. ________

17. see graph

18. see graph

19. see graph

20. see graph

21.

22.

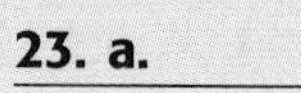

23. a.

b.

Graph each function. Find and label the vertex, y-intercept, and x-intercepts (if any) for each graph.

19. $h(x) = x^2 - 4x + 4$

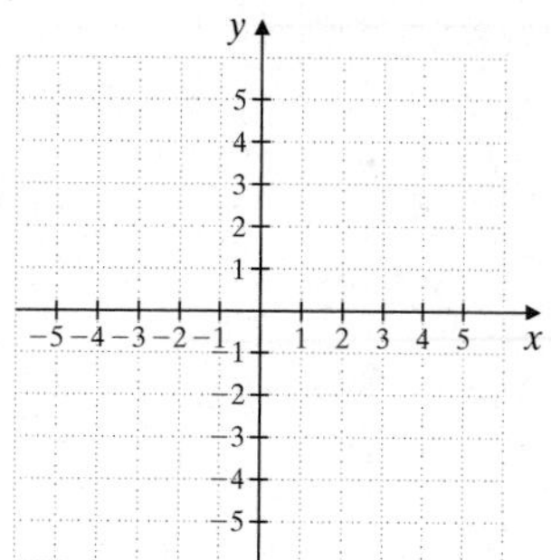

20. $F(x) = 2x^2 - 8x + 9$

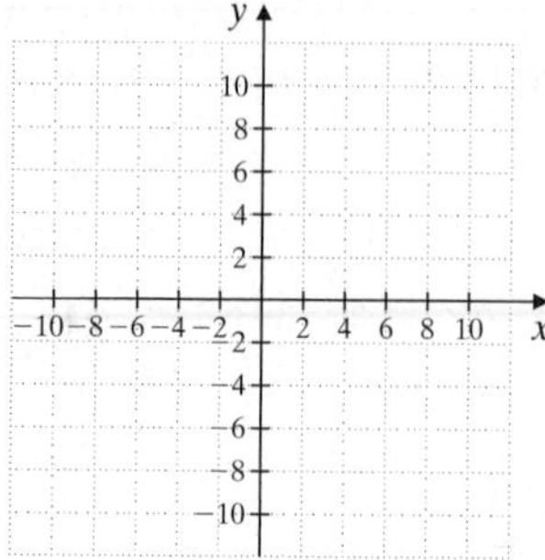

△**21.** A 10-foot ladder is leaning against a house. The distance from the bottom of the ladder to the house is 4 feet less than the distance from the top of the ladder to the ground. Find how far the top of the ladder is from the ground. Give an exact answer and a one-decimal-place approximation.

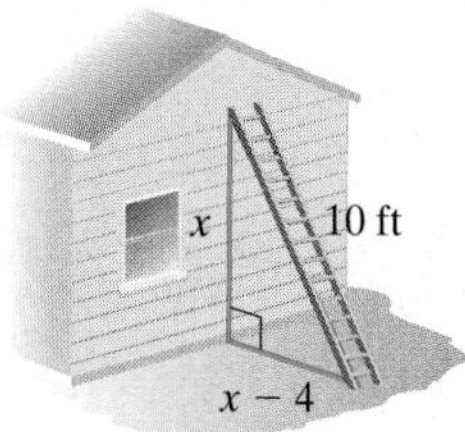

22. Dave and Sandy Hartranft can paint a room together in 4 hours. Working alone, Dave can paint the room in 2 hours less time than Sandy can. Find how long it takes Sandy to paint the room alone. Give an exact answer and a two-decimal-place approximation.

23. A stone is thrown upward from a bridge. The stone's height $s(t)$ in feet, above the water t seconds after the stone is thrown is given by the function $s(t) = -16t^2 + 32t + 256$.

a. Find the maximum height of the stone.

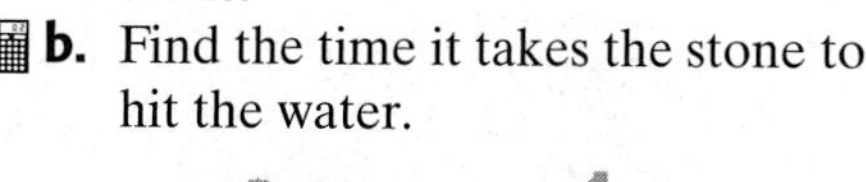

b. Find the time it takes the stone to hit the water.

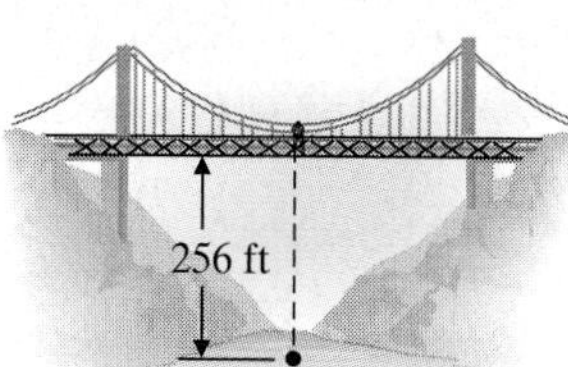

Name ______________________________

Answers

Cumulative Review

1. Write an equation of the line through points $(4, 0)$ and $(-4, -5)$.

Find the domain and range of each relation.

2.

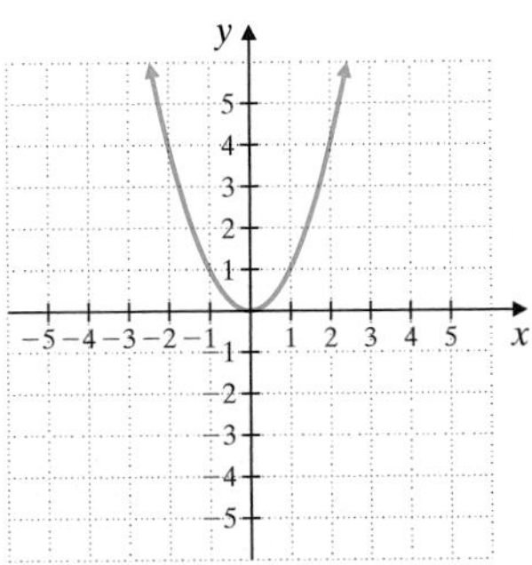

3.

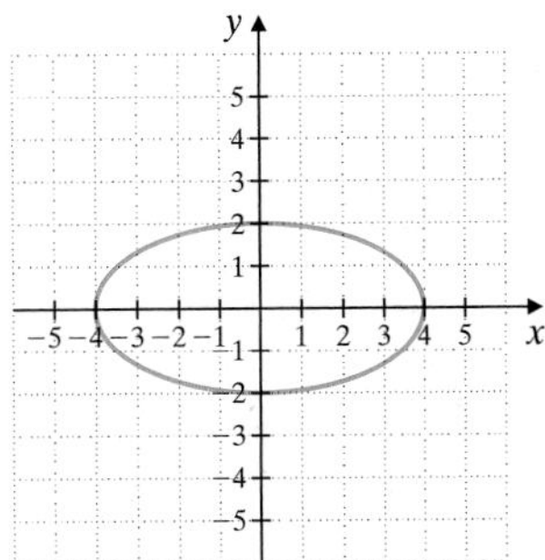

4. Solve the system:

$$\begin{cases} 2x + 4y = 1 \\ 4x - 4z = -1 \\ y - 4z = -3 \end{cases}$$

5. Use matrices to solve the system:

$$\begin{cases} x + 2y + z = 2 \\ -2x - y + 2z = 5 \\ x + 3y - 2z = -8 \end{cases}$$

6. Solve the system:

$$\begin{cases} 5x - y = -2 \\ y = 3x \end{cases}$$

7. Graph the solutions of the system

$$\begin{cases} 3x \geq y \\ x + 2y \leq 8 \end{cases}$$

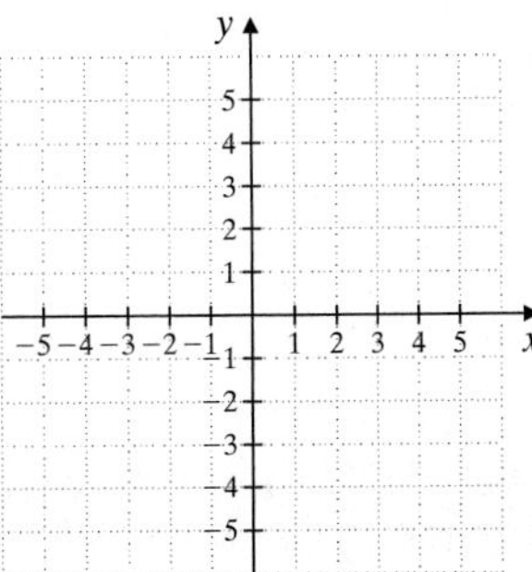

If $f(x) = \sqrt{x - 4}$ and $g(x) = \sqrt[3]{x + 2}$, find each function value.

8. $f(8)$

9. $g(-1)$

1. ______

2. ______

3. ______

4. ______

5. ______

6. ______

7. see graph

8. ______

9. ______

10. ______

11. ______

12. ______

13. ______

14. ______

15. ______

16. ______

17. ______

18. ______

19. ______

20. ______

21. ______

22. ______

23. see graph ______

Use the properties of exponents to simplify.

10. $x^{1/2}x^{1/3}$

11. $\dfrac{(2x^{2/5})^5}{x^2}$

Use the quotient rule to simplify.

12. $\sqrt{\dfrac{x}{9}}$

13. $\sqrt[4]{\dfrac{3}{16y^4}}$

Multiply.

14. $(\sqrt{2x} + 5)(\sqrt{2x} - 5)$

15. $(\sqrt{3} - 1)^2$

16. Rationalize the numerator of $\dfrac{\sqrt{7}}{\sqrt{45}}$.

17. Solve: $\sqrt{4 - x} = x - 2$

18. Add: $(2 + 3i) + (-3 + 2i)$

19. Solve $4x^2 - 24x + 41 = 0$ by completing the square.

20. Solve: $\dfrac{1}{4}m^2 - m + \dfrac{1}{2} = 0$

21. Solve: $x^{2/3} - 5x^{1/3} + 6 = 0$

22. Solve: $\dfrac{5}{x + 1} < -2$

23. Graph $f(x) = 3x^2 + 3x + 1$. Find the vertex and any intercepts.

y
5
4
3
2
1
−5 −4 −3 −2 −1 1 2 3 4 5 x
−1
−2
−3
−4
−5

Exponential and Logarithmic Functions

CHAPTER 11

In this chapter, we discuss two closely related functions: exponential and logarithmic functions. These functions are vital in applications in economics, finance, engineering, the sciences, education, and other fields. Models of tumor growth and learning curves are two examples of the uses of exponential and logarithmic functions.

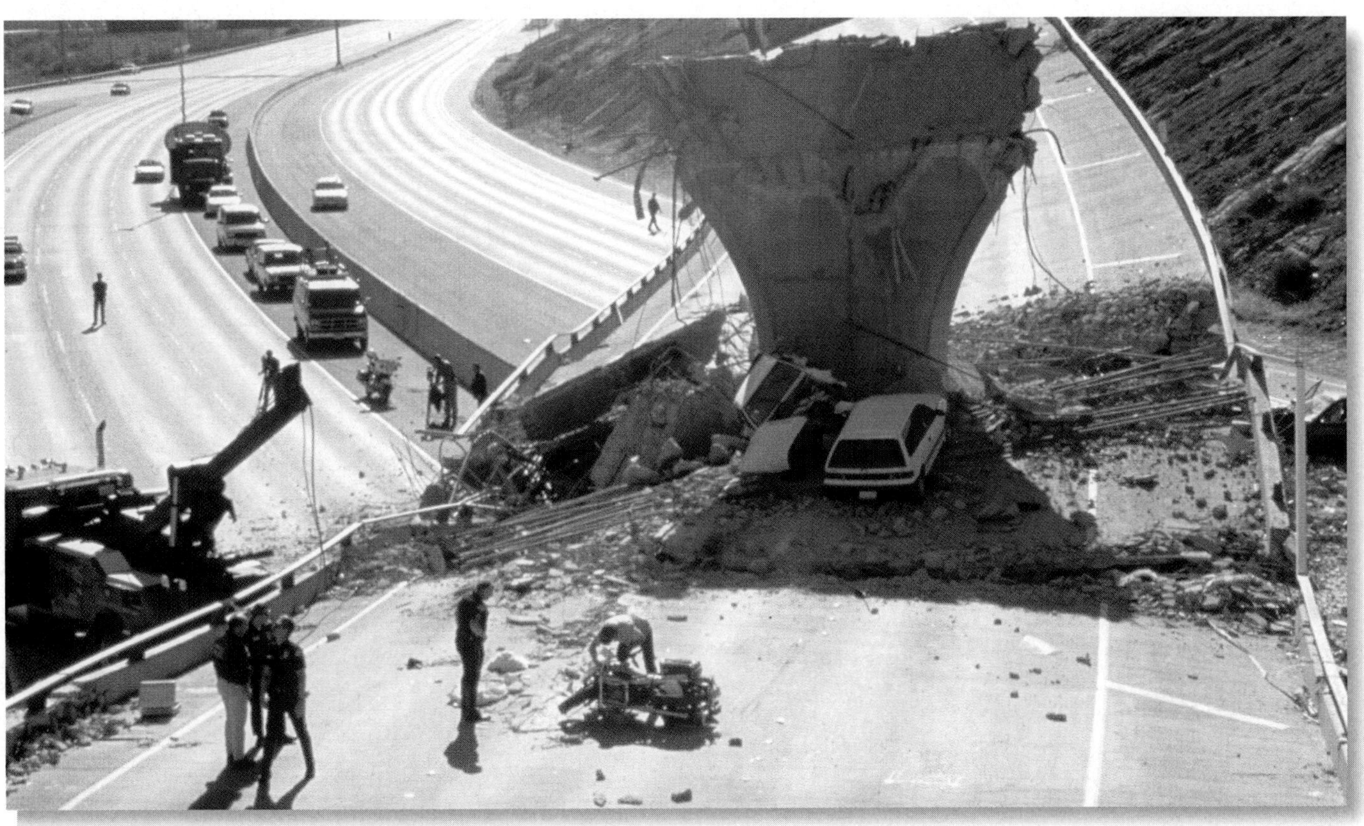

An earthquake is a series of vibrations in the crust of Earth. The size, or magnitude, of an earthquake is measured on the Richter scale. The magnitude of an earthquake can range broadly, from barely detectable (2.5 or less on the Richter scale) to massively destructive (7.0 or greater on the Richter scale). According to the United States Geological Survey, earthquakes are an everyday occurrence. In 2000, there were a total of 22,309 earthquakes around the world, or an average of 61 earthquakes per day. However, most of these were minor tremors with magnitudes of 3.9 or less, and many could only be detected by seismographs. Only 0.08% of the earthquakes occurring during 2000 could be classified as major earthquakes (7.0 or greater on the Richter scale). Even so, earthquakes were responsible for 231 deaths that year. In Exercises 75 through 78 on page 874 and Exercises 63 and 64 on page 883, we will investigate the role of logarithms in finding the magnitude of an earthquake.

Answers

1. ______

2. ______

3. ______

4. ______

5. ______

6. see graph

7. ______

8. ______

9. ______

10. ______

11. ______

12. ______

13. see graph

14. ______

15. ______

16. ______

17. ______

18. ______

19. ______

20. ______

Name ______ Section ______ Date ______

Chapter 11 Pretest

If $f(x) = x^2 - 2x$ and $g(x) = 5x + 3$, find the following.

1. $(f + g)(x)$

2. $(f \circ g)(x)$

Determine whether the functions in Exercises 3 and 4 are one-to-one.

3. $\{(-2, 7), (7, -3), (2, 1), (5, -8)\}$

4.

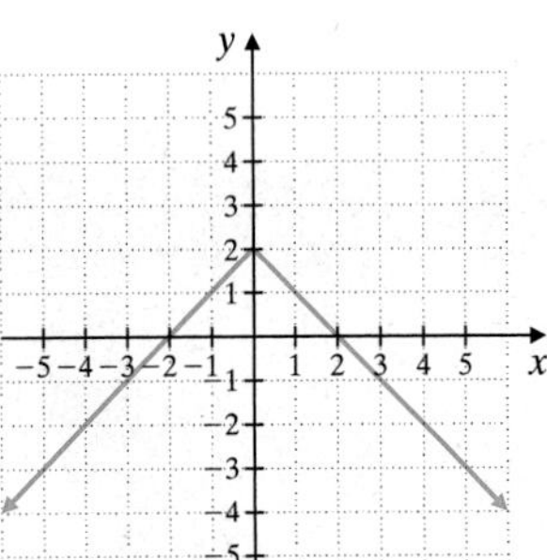

5. Given $f(x) = 7x - 12$, find $f^{-1}(x)$.

6. Graph $y = 2^x - 1$.

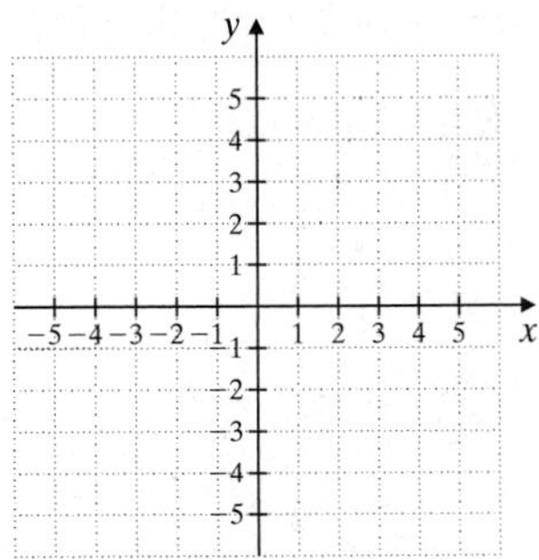

Solve each equation for x.

7. $6^x = 216$

8. $27^{4x+1} = 3$

9. $\log_5 x = 4$

10. $\log_2 \frac{1}{64} = x$

11. $\log_x 1000 = 3$

12. Simplify: $\log_7 7^5$

13. Graph: $y = \log_4 x$

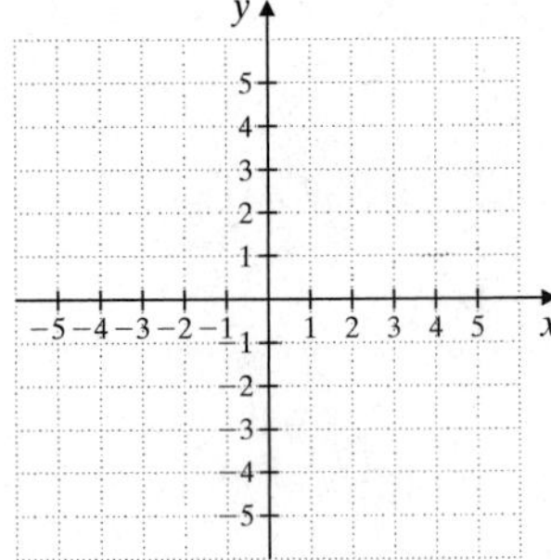

Use the properties of logarithms to write each expression as a single logarithm.

14. $\log_5 3 + \log_5 2$

15. $2 \log_6 a - 7 \log_6(a + 1)$

16. Write the expression $\log_7 \frac{3y}{5x^2}$ as the sum or difference of multiples of logarithms.

Find the exact value.

17. $\log \frac{1}{100}$

18. $\ln e^8$

19. Approximate $\log_3 15$ to four decimal places.

20. Solve: $\log_2 10 + \log_2(x + 5) = 3$

11.1 The Algebra of Functions

OBJECTIVES

A. Add, subtract, multiply, and divide functions.

B. Compose functions.

SSM TUTOR CENTER SG CD & VIDEO MATH PRO WEB

A Adding, Subtracting, Multiplying, and Dividing Functions

As we have seen in earlier chapters, it is possible to add, subtract, multiply, and divide functions. Although we have not stated them as such, the sums, differences, products, and quotients of functions are themselves functions. For example, if $f(x) = 3x$ and $g(x) = x + 1$, their product, $f(x) \cdot g(x) = 3x(x + 1) = 3x^2 + 3x$, is a new function. We can use the notation $(f \cdot g)(x)$ to denote this new function. Finding the sum, difference, product, and quotient of functions to generate new functions is called the **algebra of functions**.

Algebra of Functions

Let f and g be functions. New functions from f and g are defined as follows:

Sum	$(f + g)(x) = f(x) + g(x)$
Difference	$(f - g)(x) = f(x) - g(x)$
Product	$(f \cdot g)(x) = f(x) \cdot g(x)$
Quotient	$\left(\dfrac{f}{g}\right)(x) = \dfrac{f(x)}{g(x)}$

EXAMPLE 1 If $f(x) = x - 1$ and $g(x) = 2x - 3$, find the following.

a. $(f + g)(x)$

b. $(f - g)(x)$

c. $(f \cdot g)(x)$

d. $\left(\dfrac{f}{g}\right)(x)$

Solution: Use the algebra of functions and replace $f(x)$ by $x - 1$ and $g(x)$ by $2x - 3$. Then simplify.

a.
$$\begin{aligned}(f + g)(x) &= f(x) + g(x)\\ &= (x - 1) + (2x - 3)\\ &= 3x - 4\end{aligned}$$

b.
$$\begin{aligned}(f - g)(x) &= f(x) - g(x)\\ &= (x - 1) - (2x - 3)\\ &= x - 1 - 2x + 3\\ &= -x + 2\end{aligned}$$

c.
$$\begin{aligned}(f \cdot g)(x) &= f(x) \cdot g(x)\\ &= (x - 1)(2x - 3)\\ &= 2x^2 - 5x + 3\end{aligned}$$

d. $\left(\dfrac{f}{g}\right)(x) = \dfrac{f(x)}{g(x)} = \dfrac{x - 1}{2x - 3}$ where $x \neq \dfrac{3}{2}$

Practice Problem 1

If $f(x) = x + 3$ and $g(x) = 3x - 1$, find:

a. $(f + g)(x)$

b. $(f - g)(x)$

c. $(f \cdot g)(x)$

d. $\left(\dfrac{f}{g}\right)(x)$

Answers

1. a. $4x + 2$, **b.** $-2x + 4$, **c.** $3x^2 + 8x - 3$, **d.** $\dfrac{x + 3}{3x - 1}$ where $x \neq \dfrac{1}{3}$

There is an interesting but not surprising relationship between the graphs of functions and the graphs of their sum, difference, product, and quotient. For example, the graph of $(f + g)$ can be found by adding the graph of f to the graph of g. We add two graphs by adding corresponding y-values.

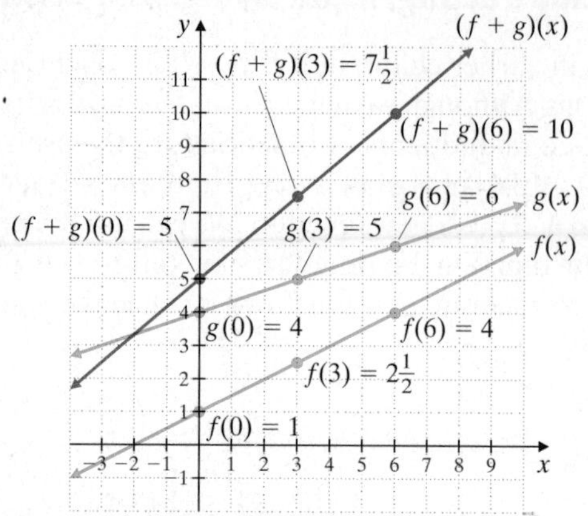

B Composition of Functions

Another way to combine functions is called **function composition**. To understand this new way of combining functions, study the tables below. They show degrees Fahrenheit converted to equivalent degrees Celsius, and then degrees Celsius converted to equivalent degrees Kelvin. (The Kelvin scale is a temperature scale devised by Lord Kelvin in 1848.)

x = Degrees Fahrenheit (Input)	−31	−13	32	68	149	212
$C(x)$ = Degrees Celsius (Output)	−35	−25	0	20	65	100

C = Degrees Celsius (Input)	−35	−25	0	20	65	100
$K(C)$ = Kelvins (Output)	238.15	248.15	273.15	293.15	338.15	373.15

Suppose that we want a table that shows a direct conversion from degrees Fahrenheit to kelvins. In other words, suppose that a table is needed that shows kelvins as a function of degrees Fahrenheit. This can easily be done because in the tables, the output of the first table is the same as the input of the second table. The new table is as follows.

x = Degrees Fahrenheit (Input)	−31	−13	32	68	149	212
$K(C(x))$ = Kelvins (Output)	238.15	248.15	273.15	293.15	338.15	373.15

Since the output of the first table is used as the input of the second table, we write the new function as $K(C(x))$. The new function is formed from the composition of the other two functions. The mathematical symbol for this composition is $(K \circ C)(x)$. Thus, $(K \circ C)(x) = K(C(x))$.

It is possible to find an equation for the composition of the two functions $C(x)$ and $K(x)$. In other words, we can find a function that converts degrees Fahrenheit directly to kelvins. The function $C(x) = \frac{5}{9}(x - 32)$ converts degrees Fahrenheit to degrees Celsius, and the function $K(C) = C + 273.15$ converts degrees Celsius to kelvins. Thus,

$$(K \circ C)(x) = K(C(x)) = K\left(\frac{5}{9}(x - 32)\right) = \frac{5}{9}(x - 32) + 273.15$$

In general, the notation $\boldsymbol{f(g(x))}$ means "f composed with g" and can be written as $(\boldsymbol{f} \circ \boldsymbol{g})(\boldsymbol{x})$. Also $g(f(x))$, or $(g \circ f)(x)$, means "g composed with f."

Composite Functions

The composition of functions f and g is

$$(f \circ g)(x) = f(g(x))$$

Helpful Hint

$(f \circ g)(x)$ does not mean the same as $(f \cdot g)(x)$.

$$(f \circ g)(x) = f(g(x)) \quad \text{while} \quad (f \cdot g)(x) = f(x) \cdot g(x)$$

EXAMPLE 2 If $f(x) = x^2$ and $g(x) = x + 3$, find each composition.

a. $(f \circ g)(2)$ and $(g \circ f)(2)$

b. $(f \circ g)(x)$ and $(g \circ f)(x)$

Solution:

a. $(f \circ g)(2) = f(g(2))$

$= f(5)$ — Since $g(x) = x + 3$, then $g(2) = 2 + 3 = 5$.

$= 5^2 = 25$

$(g \circ f)(2) = g(f(2))$

$= g(4)$ — Since $f(x) = x^2$, then $f(2) = 2^2 = 4$.

$= 4 + 3 = 7$

b. $(f \circ g)(x) = f(g(x))$

$= f(x + 3)$ — Replace $g(x)$ with $x + 3$.

$= (x + 3)^2$ — $f(x + 3) = (x + 3)^2$

$= x^2 + 6x + 9$ — Square $(x + 3)$.

$(g \circ f)(x) = g(f(x))$

$= g(x^2)$ — Replace $f(x)$ with x^2.

$= x^2 + 3$ — $g(x^2) = x^2 + 3$

Practice Problem 2

If $f(x) = x^2$ and $g(x) = 2x + 1$, find each composition.

a. $(f \circ g)(3)$ and $(g \circ f)(3)$

b. $(f \circ g)(x)$ and $(g \circ f)(x)$

EXAMPLE 3 If $f(x) = |x|$ and $g(x) = x - 2$, find each composition.

a. $(f \circ g)(x)$

b. $(g \circ f)(x)$

Solution:

a. $(f \circ g)(x) = f(g(x)) = f(x - 2) = |x - 2|$

b. $(g \circ f)(x) = g(f(x)) = g(|x|) = |x| - 2$

Practice Problem 3

If $f(x) = \sqrt{x}$ and $g(x) = x + 1$, find each composition.

a. $(f \circ g)(x)$

b. $(g \circ f)(x)$

Helpful Hint

In Examples 2 and 3, notice that $(g \circ f)(x) \neq (f \circ g)(x)$. In general, $(g \circ f)(x)$ *may* or *may not* equal $(f \circ g)(x)$.

Answers

2. a. 49; 19, **b.** $4x^2 + 4x + 1$; $2x^2 + 1$, **3. a.** $\sqrt{x + 1}$, **b.** $\sqrt{x} + 1$,

Practice Problem 4

If $f(x) = 2x$, $g(x) = x + 5$, and $h(x) = |x|$, write each function as a composition of f, g, or h.

a. $F(x) = |x + 5|$

b. $G(x) = 2x + 5$

Answers

4. a. $(h \circ g)(x)$, **b.** $(g \circ f)(x)$

EXAMPLE 4

If $f(x) = 5x$, $g(x) = x - 2$, and $h(x) = \sqrt{x}$, write each function as a composition with f, g, or h.

a. $F(x) = \sqrt{x - 2}$

b. $G(x) = 5x - 2$

Solution:

a. Notice the order in which the function F operates on an input value x. First, 2 is subtracted from x, and then the square root of that result is taken. This means that $F = h \circ g$. To check, we find $h \circ g$

$$(h \circ g)(x) = h(g(x)) = h(x - 2) = \sqrt{x - 2}$$

b. Notice the order in which the function G operates on an input value x. First, x is multiplied by 5, and then 2 is subtracted from the result. This means that $G = g \circ f$. To check, we find $g \circ f$.

$$(g \circ f)(x) = g(f(x)) = g(5x) = 5x - 2$$

GRAPHING CALCULATOR EXPLORATIONS

If $f(x) = \frac{1}{2}x + 2$ and $g(x) = \frac{1}{3}x^2 + 4$, then

$$\begin{aligned}(f + g)(x) &= f(x) + g(x)\\ &= \left(\frac{1}{2}x + 2\right) + \left(\frac{1}{3}x^2 + 4\right)\\ &= \frac{1}{3}x^2 + \frac{1}{2}x + 6.\end{aligned}$$

To visualize this addition of functions with a grapher, graph

$$Y_1 = \frac{1}{2}x + 2, \quad Y_2 = \frac{1}{3}x^2 + 4, \quad \text{and} \quad Y_3 = \frac{1}{3}x^2 + \frac{1}{2}x + 6$$

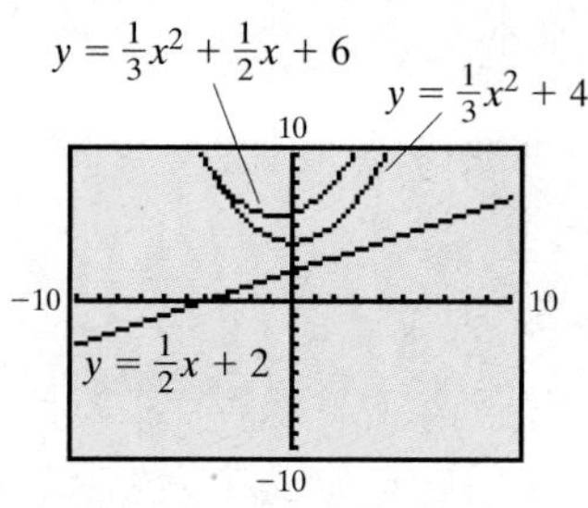

Use a TABLE feature to verify that for a given x value, $Y_1 + Y_2 = Y_3$. For example, verify that when $x = 0$, $Y_1 = 2$, $Y_2 = 4$ and $Y_3 = 2 + 4 = 6$.

Name ____________ Section ________ Date ________

EXERCISE SET 11.1

A *For the functions f and g, find* **a.** $(f + g)(x)$, **b.** $(f - g)(x)$, **c.** $(f \cdot g)(x)$, *and* **d.** $\left(\frac{f}{g}\right)(x)$. *See Example 1.*

1. $f(x) = x - 7; g(x) = 2x + 1$

2. $f(x) = x + 4; g(x) = 5x - 2$

3. $f(x) = x^2 + 1; g(x) = 5x$

4. $f(x) = x^2 - 2; g(x) = 3x$

5. $f(x) = \sqrt{x}; g(x) = x + 5$

6. $f(x) = \sqrt[3]{x}; g(x) = x - 3$

7. $f(x) = -3x; g(x) = 5x^2$

8. $f(x) = 4x^3; g(x) = -6x$

B *If* $f(x) = x^2 - 6x + 2$, $g(x) = -2x$, *and* $h(x) = \sqrt{x}$, *find each composition. See Example 2.*

9. $(f \circ g)(2)$

10. $(h \circ f)(-2)$

11. $(g \circ f)(-1)$

12. $(f \circ h)(1)$

13. $(g \circ h)(0)$

14. $(h \circ g)(0)$

Find $(f \circ g)(x)$ *and* $(g \circ f)(x)$. *See Examples 2 and 3.*

15. $f(x) = x^2 + 1; g(x) = 5x$

16. $f(x) = x - 3; g(x) = x^2$

17. $f(x) = 2x - 3; g(x) = x + 7$

18. $f(x) = x + 10; g(x) = 3x + 1$

19. $f(x) = x^3 + x - 2; g(x) = -2x$

20. $f(x) = -4x; g(x) = x^3 + x^2 - 6$

21. $f(x) = \sqrt{x}; g(x) = -5x + 2$

22. $f(x) = 7x - 1; g(x) = \sqrt[3]{x}$

If $f(x) = 3x$, $g(x) = \sqrt{x}$, and $h(x) = x^2 + 2$, write each function as a composition with f, g, or h. See Example 4.

23. $H(x) = \sqrt{x^2 + 2}$

24. $G(x) = \sqrt{3x}$

25. $F(x) = 9x^2 + 2$

26. $H(x) = 3x^2 + 6$

27. $G(x) = 3\sqrt{x}$

28. $F(x) = x + 2$

Find $f(x)$ and $g(x)$ so that the given function $h(x) = (f \circ g)(x)$.

29. $h(x) = (x + 2)^2$

30. $h(x) = |x - 1|$

31. $h(x) = \sqrt{x + 5} + 2$

32. $h(x) = (3x + 4)^2 + 3$

33. $h(x) = \dfrac{1}{2x - 3}$

34. $h(x) = \dfrac{1}{x + 10}$

Review and Preview

Solve each equation for y. See Section 2.6.

35. $x = y + 2$

36. $x = y - 5$

37. $x = 3y$

38. $x = -6y$

39. $x = -2y - 7$

40. $x = 4y + 7$

Combining Concepts

41. Business people are concerned with cost functions, revenue functions, and profit functions. Recall that the profit $P(x)$ obtained from selling x units of a product is equal to the revenue $R(x)$ from selling the x units minus the cost $C(x)$ of manufacturing the x units. Write an equation expressing this relationship among $C(x)$, $R(x)$, and $P(x)$.

42. Suppose the revenue $R(x)$ for x units of a product can be described by $R(x) = 25x$, and the cost $C(x)$ can be described by $C(x) = 50 + x^2 + 4x$. Find the profit $P(x)$ for x units.

43. If you are given $f(x)$ and $g(x)$, explain in your own words how to find $(f \circ g)(x)$, and then how to find $(g \circ f)(x)$.

44. Given $f(x)$ and $g(x)$, describe in your own words the difference between $(f \circ g)(x)$ and $(f \cdot g)(x)$.

11.2 Inverse Functions

OBJECTIVES

- **A** Determine whether a function is a one-to-one function.
- **B** Use the horizontal line test to decide whether a function is a one-to-one function.
- **C** Find the inverse of a function.
- **D** Find the equation of the inverse of a function.
- **E** Graph functions and their inverses.

In the next section, we begin a study of two new functions: exponential and logarithmic functions. As we learn more about these functions, we will discover that they share a special relation to each other; they are inverses of each other.

Before we study these functions, we need to learn about inverses. We begin by defining one-to-one functions.

A Determining Whether a Function Is One-to-One

Study the following table.

Degrees Fahrenheit (Input)	−31	−13	32	68	149	212
Degrees Celsius (Output)	−35	−25	0	20	65	100

Recall that since each degrees Fahrenheit (input) corresponds to exactly one degrees Celsius (output), this table of inputs and outputs does describe a function. Also notice that each output corresponds to a different input. This type of function is given a special name—a *one-to-one function.*

Does the set $f = \{(0,1),(2,2),(-3,5),(7,6)\}$ describe a one-to-one function? It is a function since each x-value corresponds to a unique y-value. For this particular function f, each y-value corresponds to a unique x-value. Thus, this function is also a one-to-one function.

One-to-One Function

For a **one-to-one function**, each x-value (input) corresponds to only one y-value (output) and each y-value (output) corresponds to only one x-value (input).

EXAMPLES

Determine whether each function described is one-to-one.

1. $f\{(6,2),(5,4),(-1,0),(7,3)\}$

The function f is one-to-one since each y-value corresponds to only one x-value.

2. $g = \{(3,9),(-4,2),(-3,9),(0,0)\}$

The function g is not one-to-one because the y-value 9 in $(3,9)$ and $(-3,9)$ corresponds to two different x-values.

3. $h = \{(1,1),(2,2),(10,10),(-5,-5)\}$

The function h is one-to-one since each y-value corresponds to only one x-value.

4.

Mineral (Input)	Talc	Gypsum	Diamond	Topaz	Stibnite
Hardness on the Mohs Scale (Output)	1	2	10	8	2

Practice Problems 1–5

Determine whether each function described is one-to-one.

1. $f = \{(7,3),(-1,1),(5,0),(4,-2)\}$
2. $g = \{(-3,2),(6,3),(2,14),(-6,2)\}$
3. $h = \{(0,0),(1,2),(3,4),(5,6)\}$
4.

State (Input)	Colorado	Mississippi	Nevada	New Mexico	Utah
Number of Colleges and Universities (Output)	9	44	13	44	21

Source: The Chronicle of Higher Education, Vol. XLV, No. 1, August 28, 1998.

5.

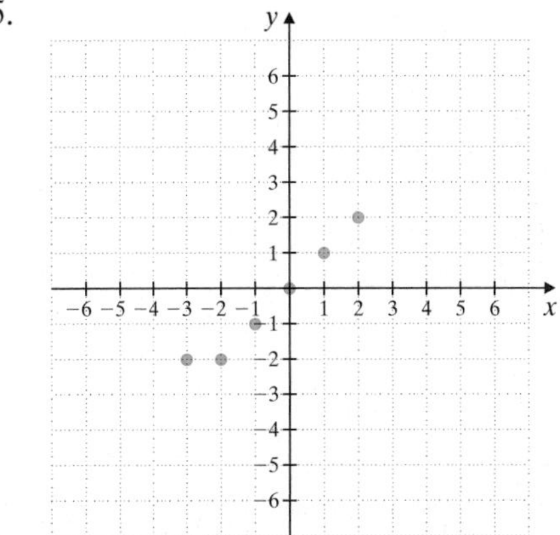

Answers

1. one-to-one, **2.** not one-to-one, **3.** one-to-one, **4.** not one-to-one, **5.** not one-to-one

This table does not describe a one-to-one function since the output 2 corresponds to two different inputs, gypsum and stibnite.

5.

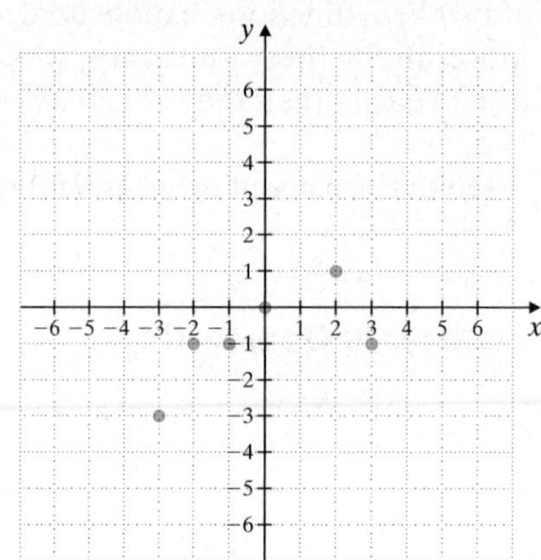

This graph does not describe a one-to-one function since the y-value -1 corresponds to three different x-values, -2, -1 and 3, as shown to the right.

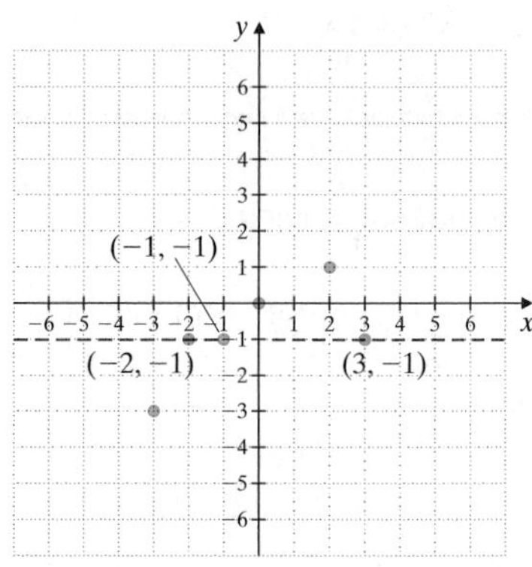

B Using the Horizontal Line Test

Recall that we recognize the graph of a function when it passes the vertical line test. Since every x-value of the function corresponds to exactly one y-value, each vertical line intersects the function's graph at most once. The graph shown next, for instance, is the graph of a function.

Is this function a *one-to-one* function? The answer is no. To see why not, notice that the y-value of the ordered pair $(-3, 3)$, for example, is the same as the y-value of the ordered pair $(3, 3)$. This function is therefore not one-to-one.

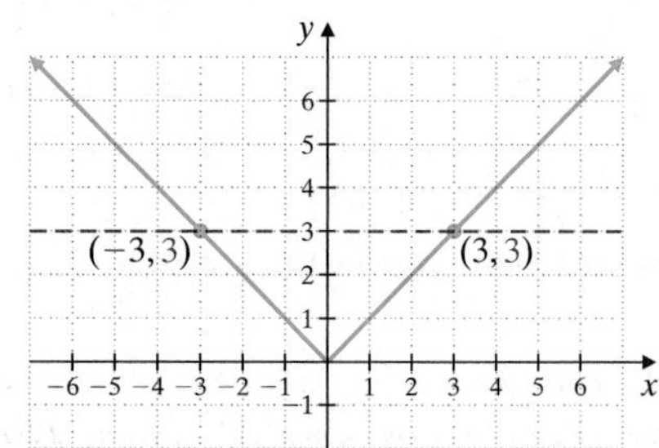

To test whether a graph is the graph of a one-to-one function, we can apply the vertical line test to see whether it is a function, and then apply a similar **horizontal line test** to see whether it is a one-to-one function.

Horizontal Line Test

If every horizontal line intersects the graph of a function at most once, then the function is a one-to-one function.

EXAMPLE 6

Use the vertical and horizontal line tests to determine whether each graph is the graph of a one-to-one function.

a.

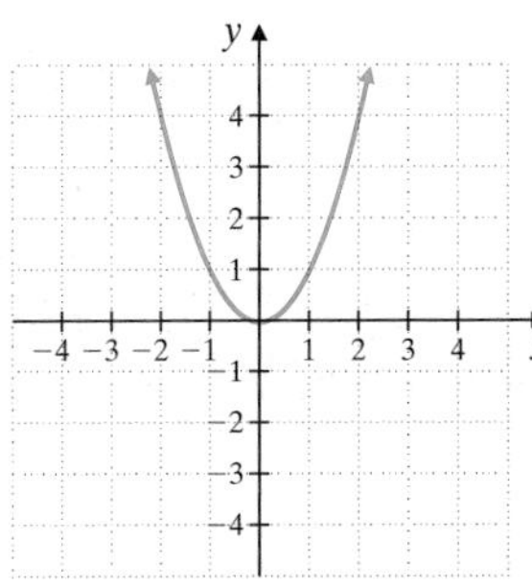

b.

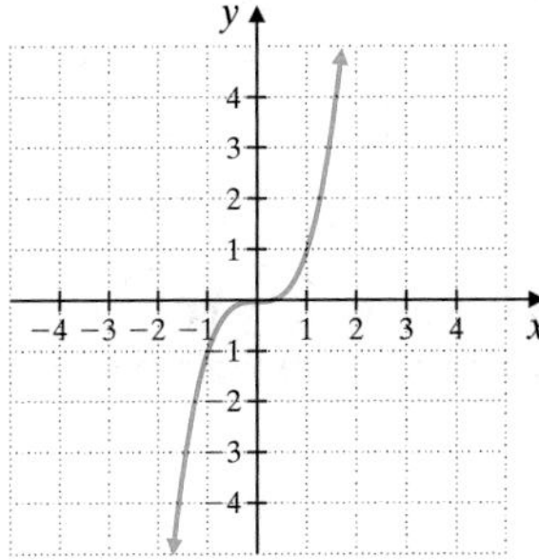

c.

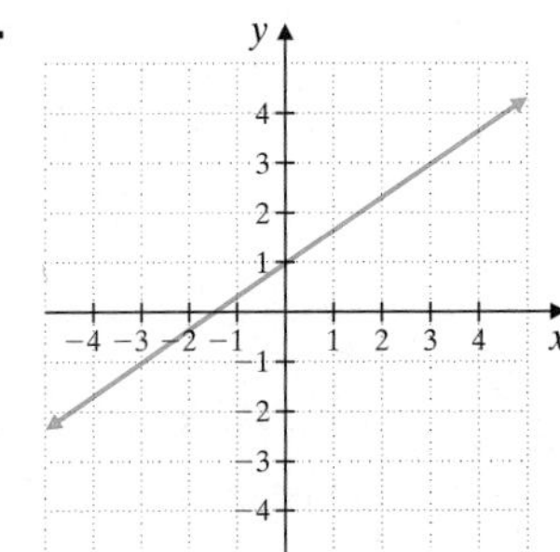

d.

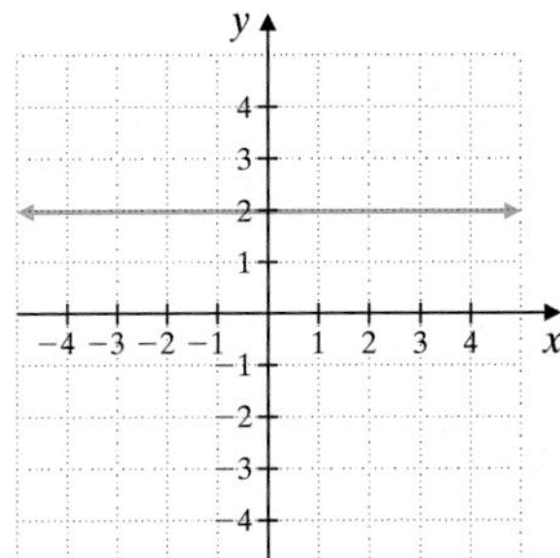

e.

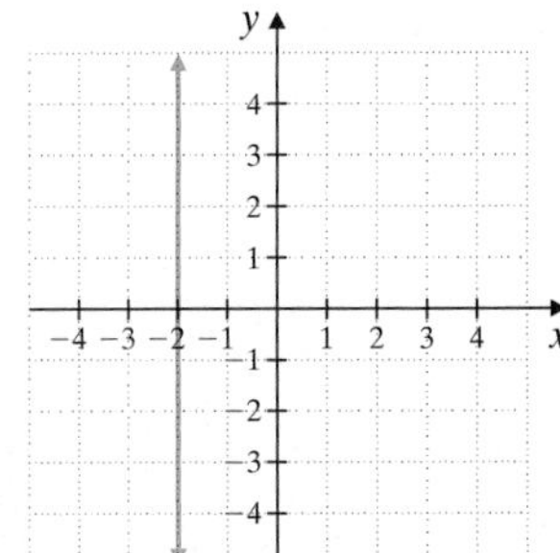

Solution: Graphs **a**, **b**, **c**, and **d** all pass the vertical line test, so only these graphs are graphs of functions. But, of these, only **b** and **c** pass the horizontal line test, so only **b** and **c** are graphs of one-to-one functions.

Helpful Hint

All linear equations are one-to-one functions except those whose graphs are horizontal or vertical lines. A vertical line does not pass the vertical line test and hence is not the graph of a function. A horizontal line is the graph of a function but does not pass the horizontal line test and hence is not the graph of a one-to-one function.

C Finding the Inverse of a Function

One-to-one functions are special in that their graphs pass the vertical and horizontal line tests. They are special, too, in another sense: We can find the **inverse function** for any one-to-one function by switching the coordinates of

Practice Problem 6

Use the vertical and horizontal line tests to determine whether each graph is the graph of a one-to-one function.

a.

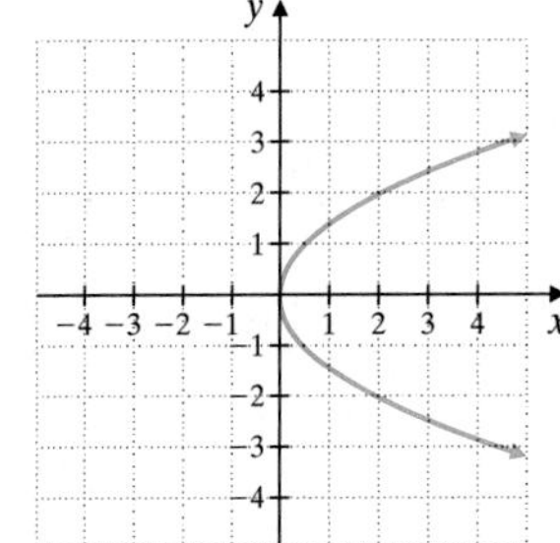

b.

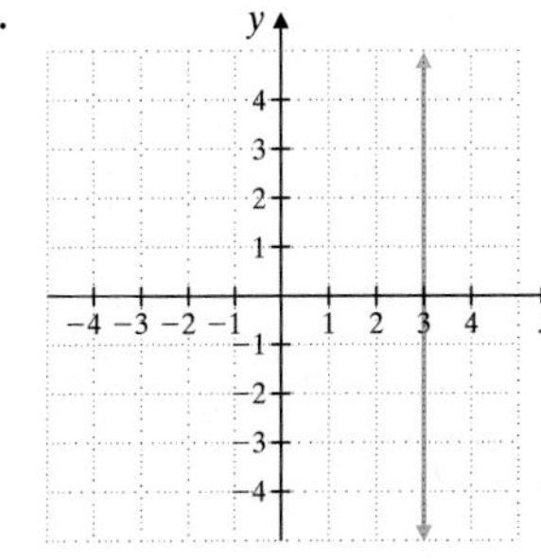

c.

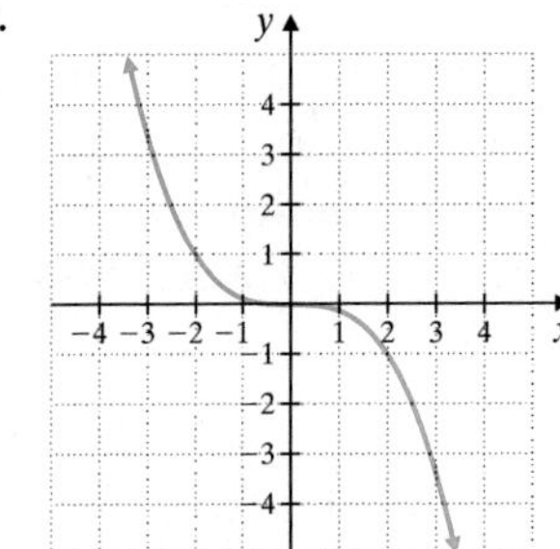

d.

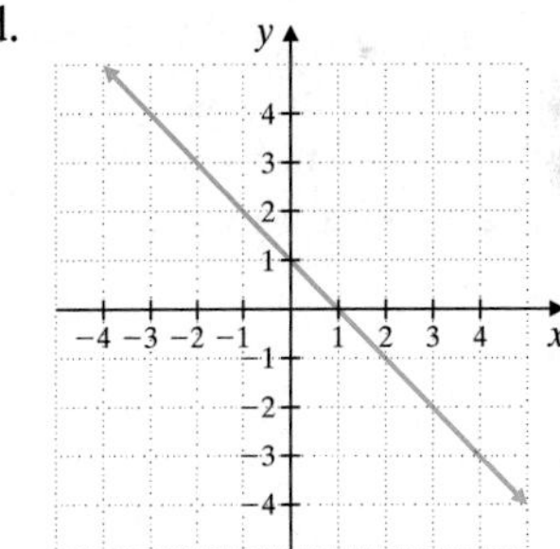

e.

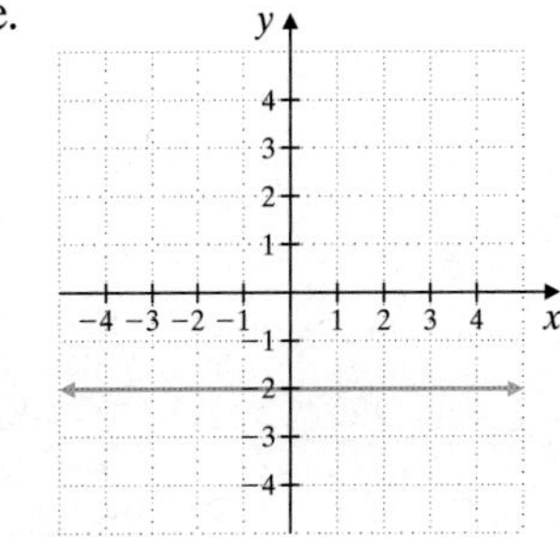

Answers

6. a. not one-to-one, **b.** not one-to-one, **c.** one-to-one, **d.** one-to-one, **e.** not one-to-one

the ordered pairs of the function, or the inputs and the outputs. For example, the inverse of the one-to-one function

Degrees Fahrenheit (Input)	−31	−13	32	68	149	212
Degrees Celsius (Output)	−35	−25	0	20	65	100

is the function

Degrees Celsius (Input)	−35	−25	0	20	65	100
Degrees Fahrenheit (Output)	−31	−13	32	68	149	212

Notice that the ordered pair $(-31, -35)$ of the function, for example, becomes the ordered pair $(-35, -31)$ of its inverse.

Also, the inverse of the one-to-one function $f = \{(2, -3), (5, 10), (9, 1)\}$ is $\{(-3, 2), (10, 5), (1, 9)\}$. For a function f, we use the notation f^{-1}, read "f inverse," to denote its inverse function. Notice that since the coordinates of each ordered pair have been switched, the domain (set of inputs) of f is the range (set of outputs) of f^{-1}, and the range of f is the domain of f^{-1}.

Inverse Function

The inverse of a one-to-one function f is the one-to-one function f^{-1} that consists of the set of all ordered pairs (y, x) where (x, y) belongs to f.

Practice Problem 7

Find the inverse of the one-to-one function: $f = \{(2, -4), (-1, 13), (0, 0), (-7, -8)\}$

Concept Check

Suppose that f is a one-to-one function. If the ordered pair $(1, 5)$ belongs to f, name one point that we know must belong to the inverse function f^{-1}.

EXAMPLE 7 Find the inverse of the one-to-one function:

$$f = \{(0, 1), (-2, 7), (3, -6), (4, 4)\}$$

Solution: $f^{-1} = \{(1, 0), (7, -2), (-6, 3), (4, 4)\}$ ← Switch coordinates of each ordered pair.

Try the Concept Check in the margin.

D Finding the Equation of the Inverse of a Function

If a one-to-one function f is defined as a set of ordered pairs, we can find f^{-1} by interchanging the x- and y-coordinates of the ordered pairs. If a one-to-one function f is given in the form of an equation, we can find the equation of f^{-1} by using a similar procedure.

Finding an Equation of the Inverse of a One-to-One Function *f*

Step 1. Replace $f(x)$ with y.
Step 2. Interchange x with y.
Step 3. Solve the equation for y.
Step 4. Replace y with the notation $f^{-1}(x)$.

Helpful Hint

The symbol f^{-1} is the single symbol used to denote the inverse of the function f. It is read as "f inverse." This symbol *does not mean* $\frac{1}{f}$.

Answers

7. $f^{-1} = \{(-4, 2), (13, -1), (0, 0), (-8, -7)\}$

Concept Check: $(5, 1)$

EXAMPLE 8 Find the equation of the inverse of $f(x) = x + 3$.

Solution: $f(x) = x + 3$

Step 1. $y = x + 3$ Replace $f(x)$ with y.

Step 2. $x = y + 3$ Interchange x and y.

Step 3. $x - 3 = y$ Solve for y.

Step 4. $f^{-1}(x) = x - 3$ Replace y with $f^{-1}(x)$.

The inverse of $f(x) = x + 3$ is $f^{-1}(x) = x - 3$. Notice that, for example,

$f(1) = 1 + 3 = 4$ and $f^{-1}(4) = 4 - 3 = 1$

Ordered pair: $(1, 4)$ Ordered pair: $(4, 1)$

The coordinates are switched, as expected.

Practice Problem 8

Find the equation of the inverse of $f(x) = x - 6$.

EXAMPLE 9

Find the equation of the inverse of $f(x) = 3x - 5$. Graph f and f^{-1} on the same set of axes.

Solution: $f(x) = 3x - 5$

Step 1. $y = 3x - 5$ Replace $f(x)$ with y.

Step 2. $x = 3y - 5$ Interchange x and y.

Step 3. $3y = x + 5$ Solve for y.

$$y = \frac{x + 5}{3}$$

Step 4. $f^{-1}(x) = \dfrac{x + 5}{3}$ Replace y with $f^{-1}(x)$.

Now we graph f and f^{-1} on the same set of axes. Both $f(x) = 3x - 5$ and $f^{-1}(x) = \dfrac{x + 5}{3}$ are linear functions, so each graph is a line.

$f(x) = 3x - 5$

x	$y = f(x)$
1	−2
0	−5
$\frac{5}{3}$	0

$f^{-1}(x) = \dfrac{x + 5}{3}$

x	$y = f^{-1}(x)$
−2	1
−5	0
0	$\frac{5}{3}$

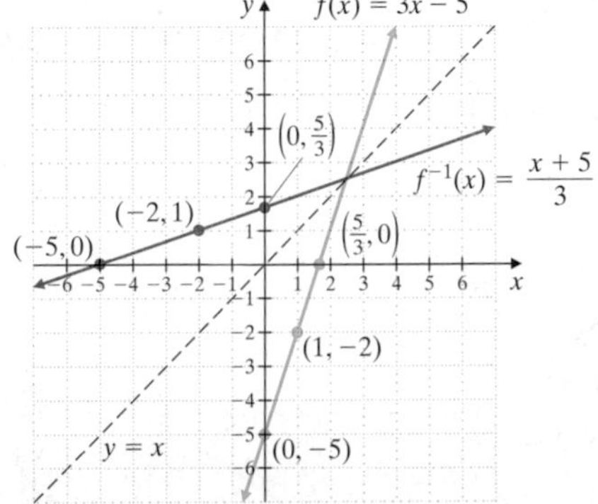

Practice Problem 9

Find the equation of the inverse of $f(x) = 2x + 3$. Graph f and f^{-1} on the same set of axes.

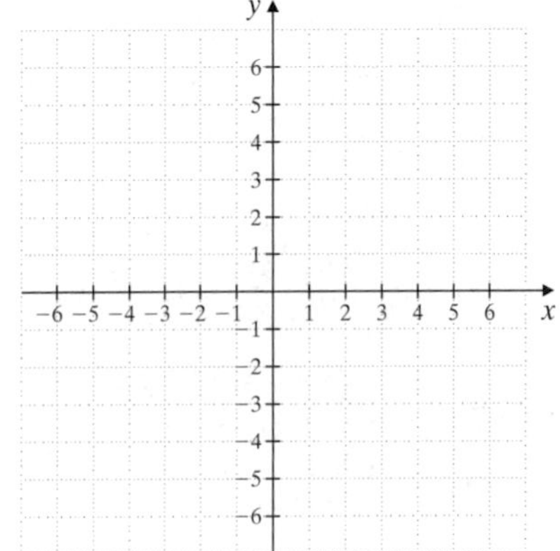

E Graphing Inverse Functions

Notice that the graphs of f and f^{-1} in Example 9 are mirror images of each other, and the "mirror" is the dashed line $y = x$. This is true for every function and its inverse. For this reason, we say that the *graphs of f and f^{-1} are symmetric about the line $y = x$.*

Answers

8. $f^{-1}(x) = x + 6$,

9. $f^{-1}(x) = \dfrac{x - 3}{2}$

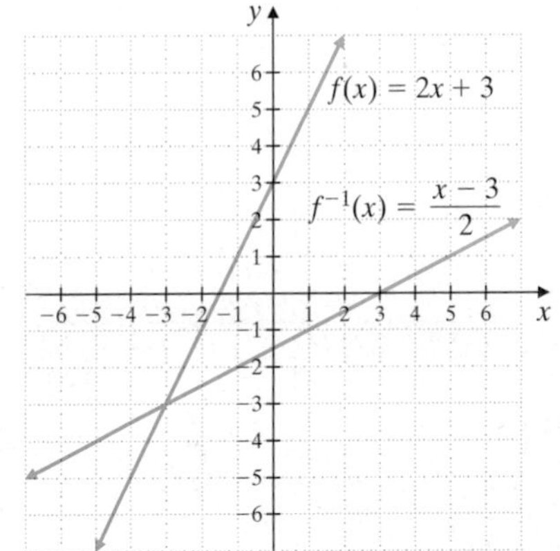

To see why this happens, study the graph of a few ordered pairs and their switched coordinates.

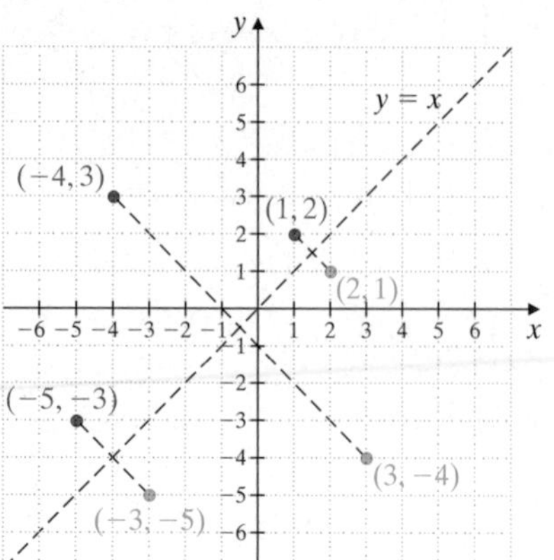

EXAMPLE 10 Graph the inverse of each function.

Solution: The function is graphed in blue and the inverse is graphed in red.

a.

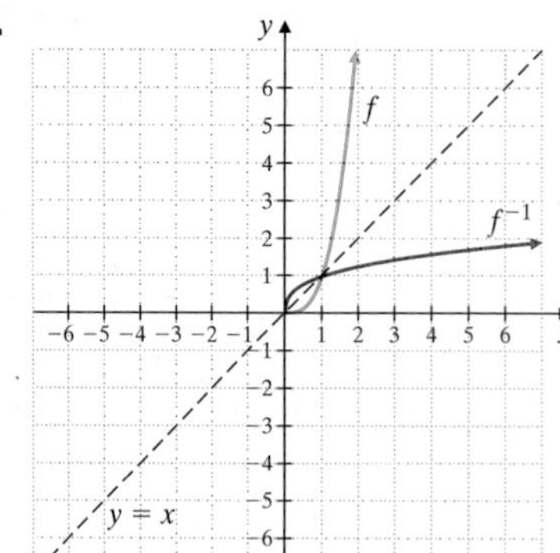

b.

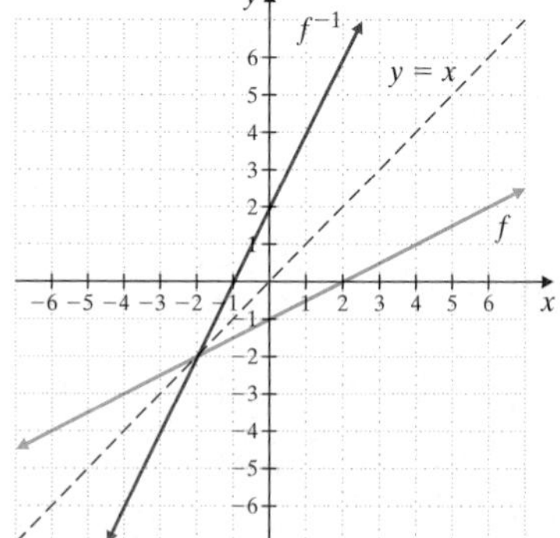

Practice Problem 10

Graph the inverse of each function.

a.

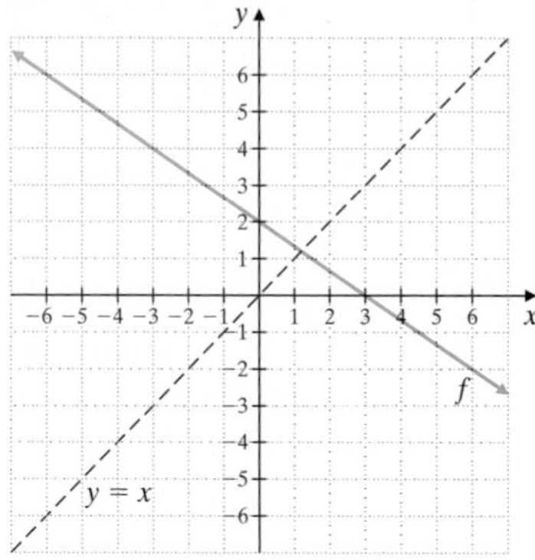

b.

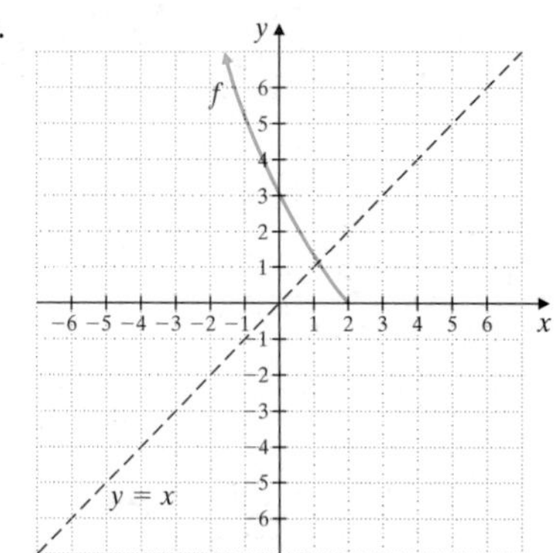

Answers

10. a.

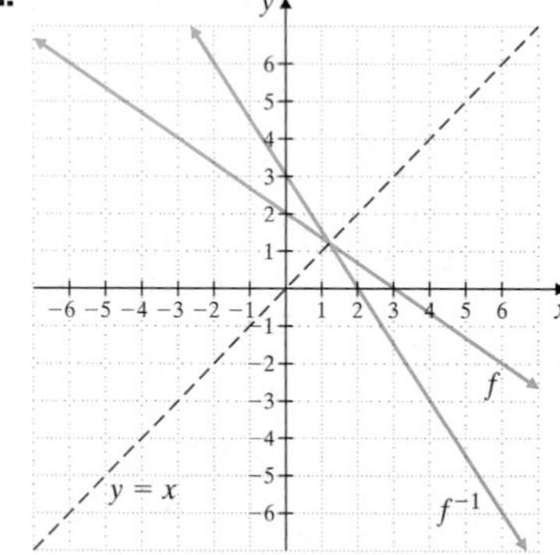

b.

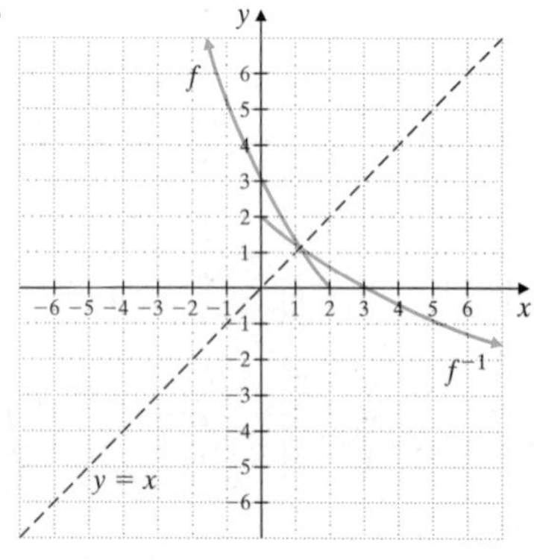

GRAPHING CALCULATOR EXPLORATIONS

A grapher can be used to visualize functions and their inverses. Recall that the graph of a function f and its inverse f^{-1} are mirror images of each other across the line $y = x$. To see this for the function $f(x) = 3x + 2$, use a square window and graph

the given function: $Y_1 = 3x + 2$

its inverse: $Y_2 = \dfrac{x - 2}{3}$

and the line: $Y_3 = x$

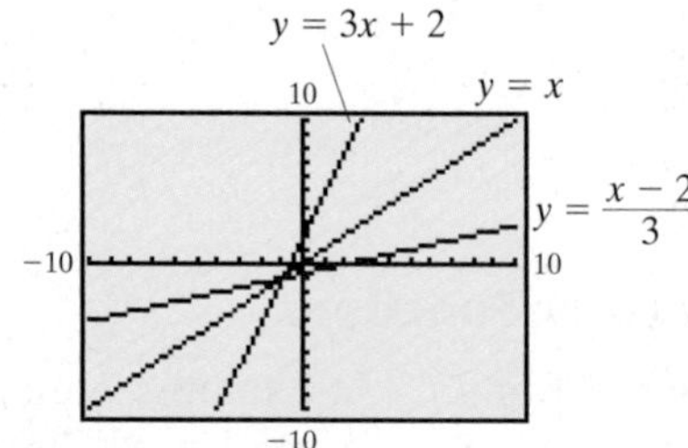

Exercises will follow in Exercise Set 11.2.

Name ______________________ Section __________ Date __________

EXERCISE SET 11.2

 *Determine whether each function is a one-to-one function. If it is one-to-one, list the inverse function by switching coordinates, or inputs and outputs. See Examples 1 through 5, and 7.*

1. $f = \{(-1, -1), (1, 1), (0, 2), (2, 0)\}$

2. $g = \{(8, 6), (9, 6), (3, 4), (-4, 4)\}$

3. $h = \{(10, 10)\}$

4. $r = \{(1, 2), (3, 4), (5, 6), (6, 7)\}$

 **5.** $f = \{(11, 12), (4, 3), (3, 4), (6, 6)\}$

6. $g = \{(0, 3), (3, 7), (6, 7), (-2, -2)\}$

7.

Month of 2001 (Input)	January	February	March	April	May	June
Unemployment Rate in Percent (Output)	4.2	4.2	4.3	4.5	4.4	4.5

(*Source:* U.S. Bureau of Labor Statistics)

8.

State (Input)	Wisconsin	Ohio	Georgia	Colorado	California	Arizona
Electoral Votes (Output)	10	20	15	9	55	10

(*Source:* National Archives and Records Administration, based on the 2000 Census)

9.

State (Input)	California	Alaska	Indiana	Louisiana	New Mexico
Rank in Population (Output)	1	48	14	22	36

(*Source:* U.S. Bureau of the Census)

10.

Shape (Input)	Triangle	Pentagon	Quadrilateral	Hexagon	Decagon
Number of Sides (Output)	3	5	4	6	10

Given the one-to-one function $f(x) = x^3 + 2$, find the following. (Hint: You do not need to find the equation for f^{-1}.)

11. a. $f(1)$
b. $f^{-1}(3)$

12. a. $f(0)$
b. $f^{-1}(2)$

13. a. $f(-1)$
b. $f^{-1}(1)$

14. a. $f(-2)$
b. $f^{-1}(-6)$

B *Determine whether the graph of each function is the graph of a one-to-one function. See Example 6.*

15.

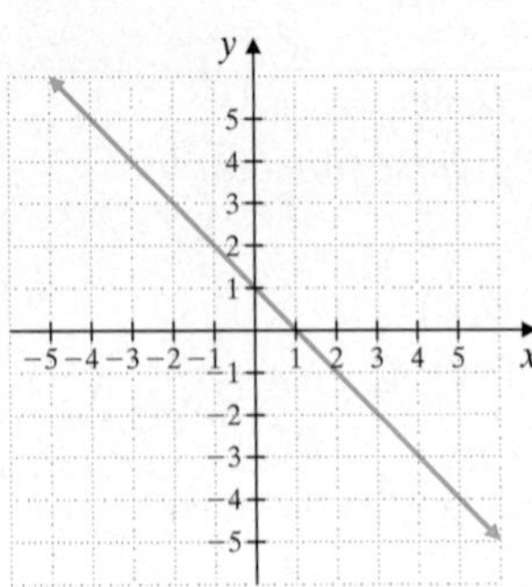

16.

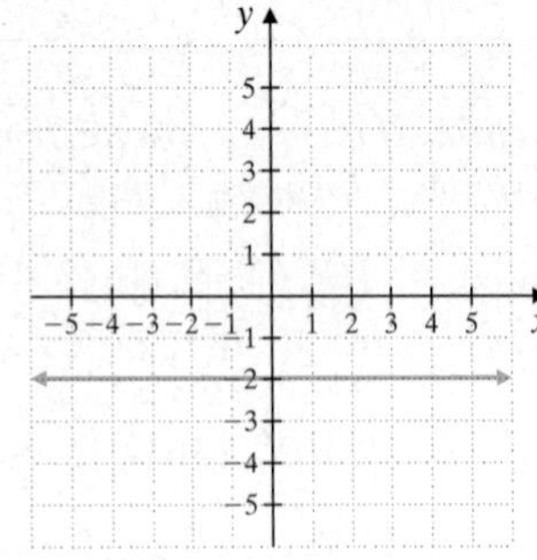

 **17.**

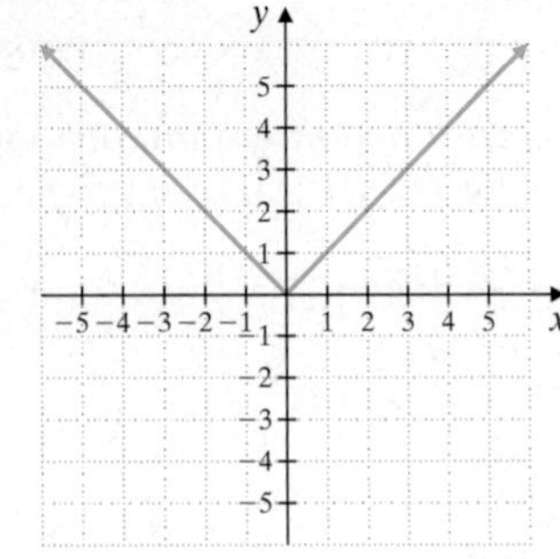

18.

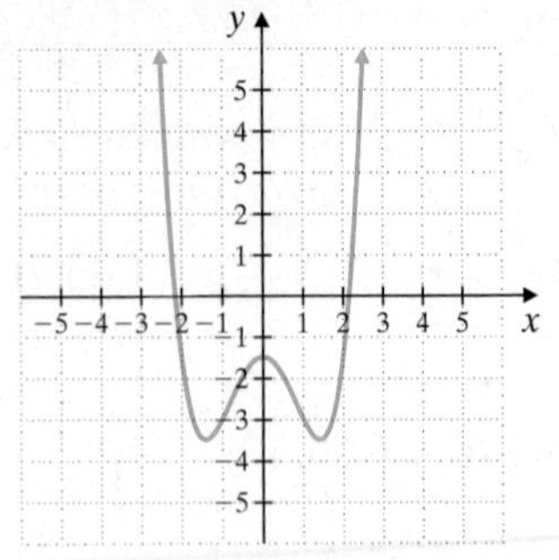

19.

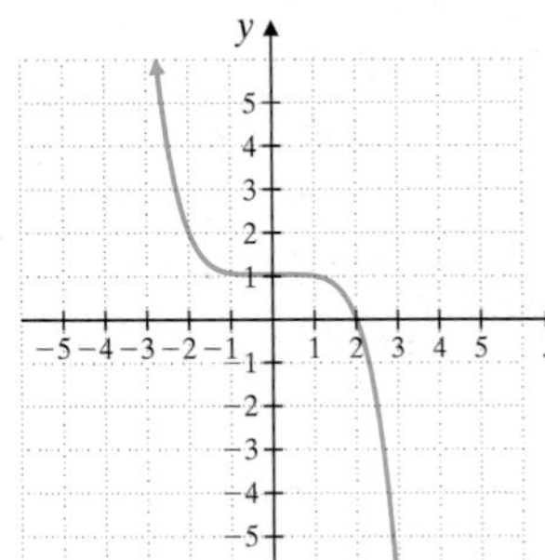

20.

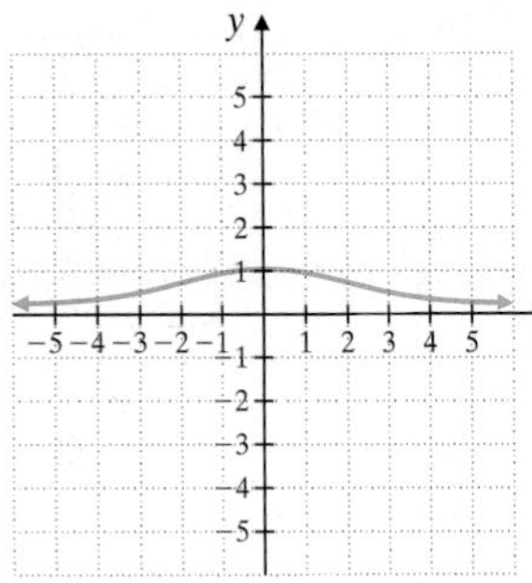

21.

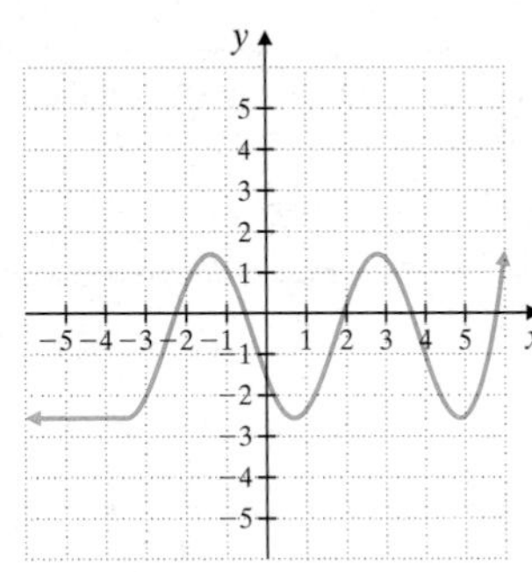

22.

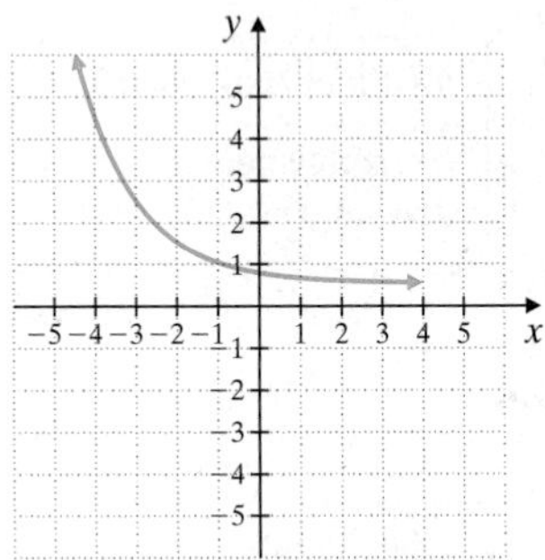

D **E** *Each of the following functions is one-to-one. Find the inverse of each function and graph the function and its inverse on the same set of axes. See Examples 8 and 9.*

23. $f(x) = x + 4$

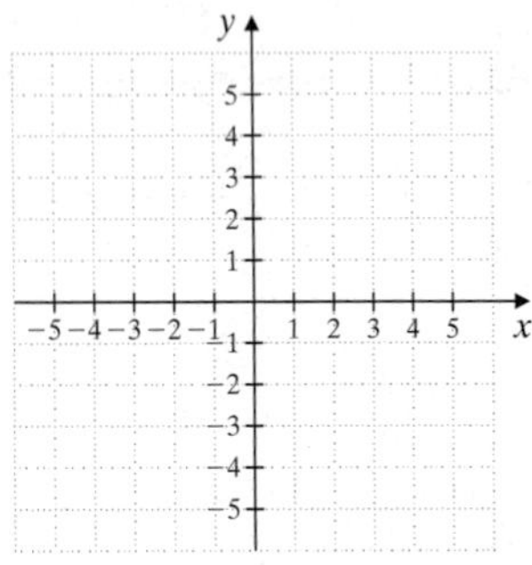

24. $f(x) = x - 5$

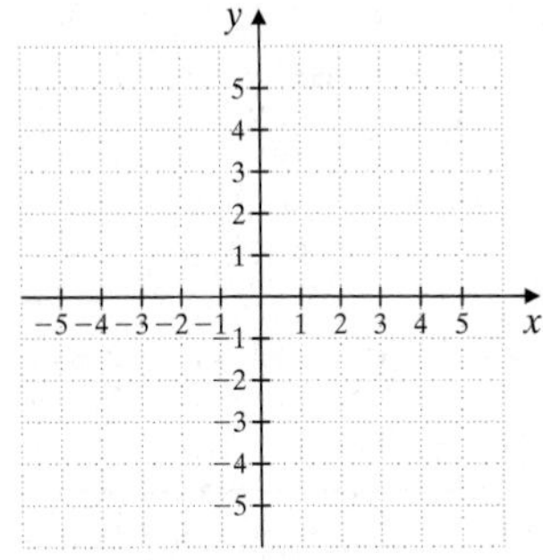

 **25.** $f(x) = 2x - 3$

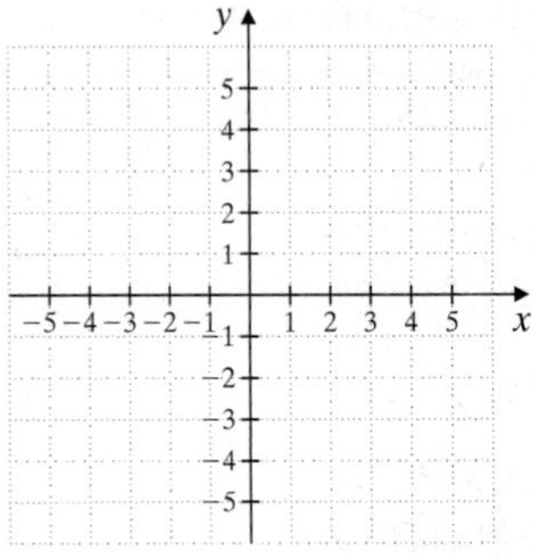

26. $f(x) = 4x + 9$

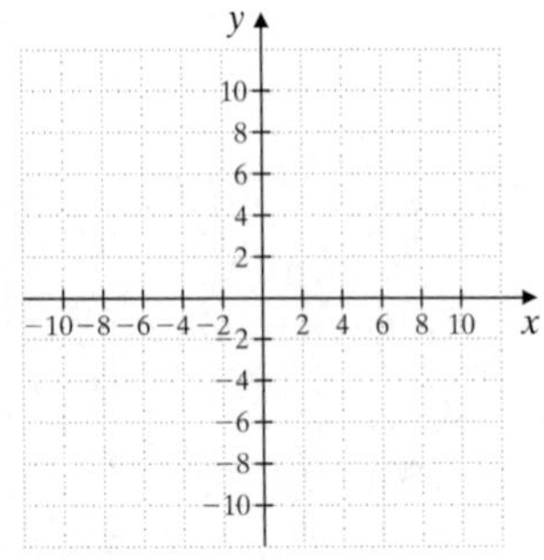

27. $f(x) = \frac{1}{2}x - 1$

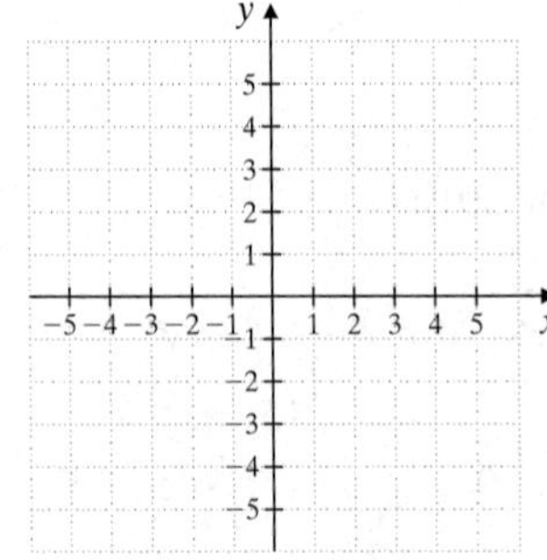

28. $f(x) = -\frac{1}{2}x + 2$

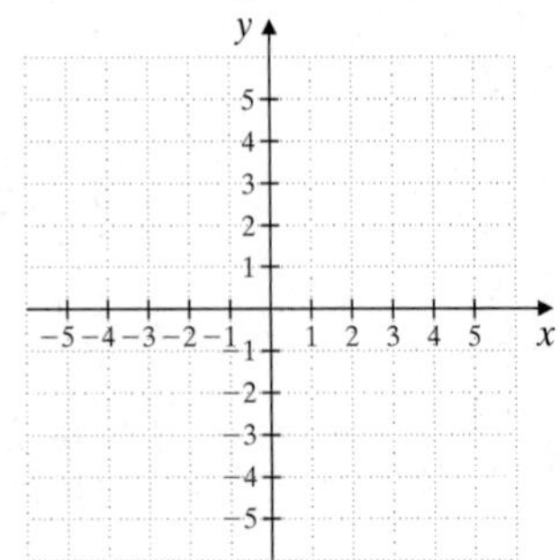

29. $f(x) = x^3$

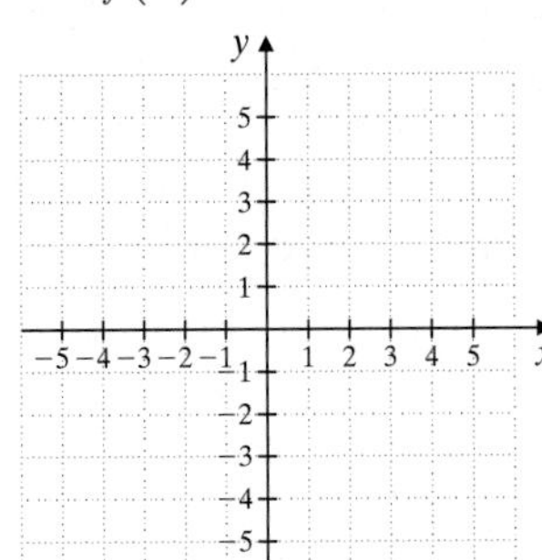

30. $f(x) = x^3 - 1$

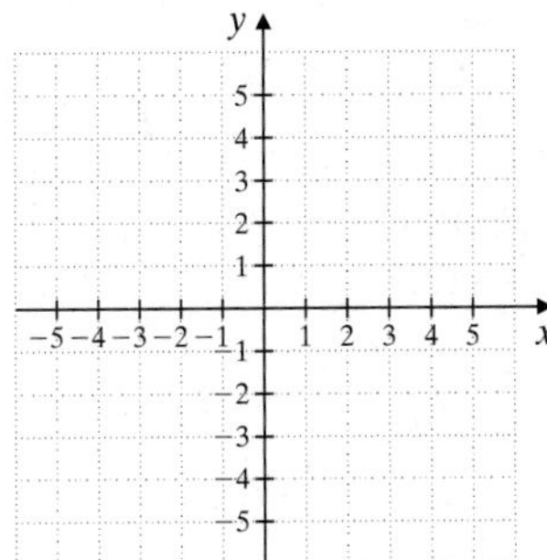

Find the inverse of each one-to-one function. See Examples 8 and 9.

31. $f(x) = \dfrac{x - 2}{5}$

32. $f(x) = \dfrac{4x - 3}{2}$

33. $f(x) = \sqrt[3]{x}$

34. $f(x) = \sqrt[3]{x + 1}$

35. $f(x) = \dfrac{5}{3x + 1}$

36. $f(x) = \dfrac{7}{2x + 4}$

37. $f(x) = (x + 2)^3$

38. $f(x) = (x - 5)^3$

Graph the inverse of each function on the same set of axes. See Example 10.

39.

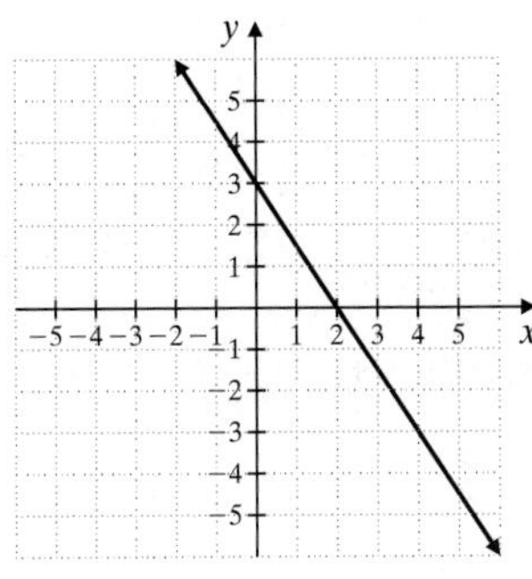

40.

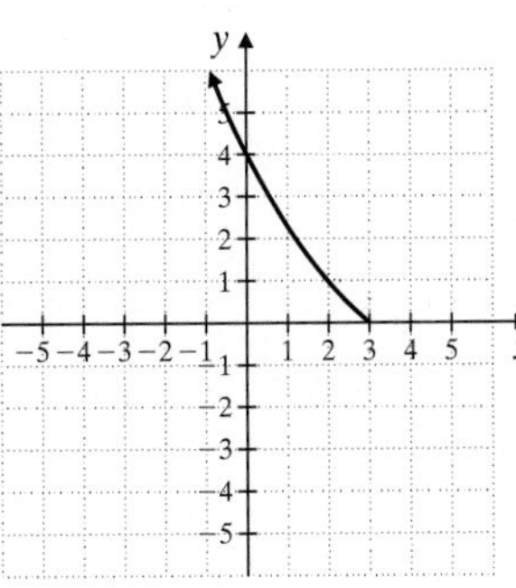

41.

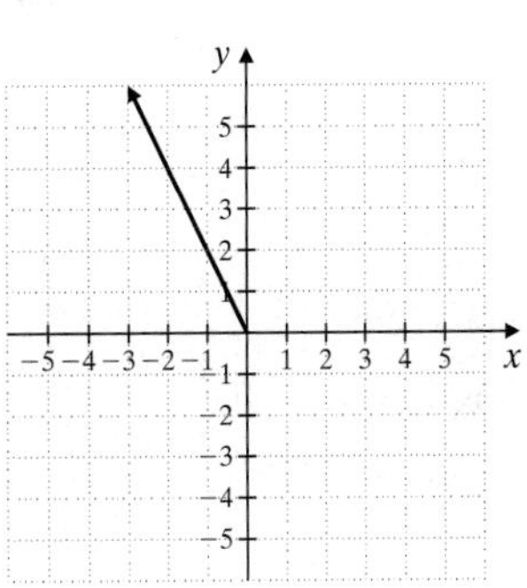

42.

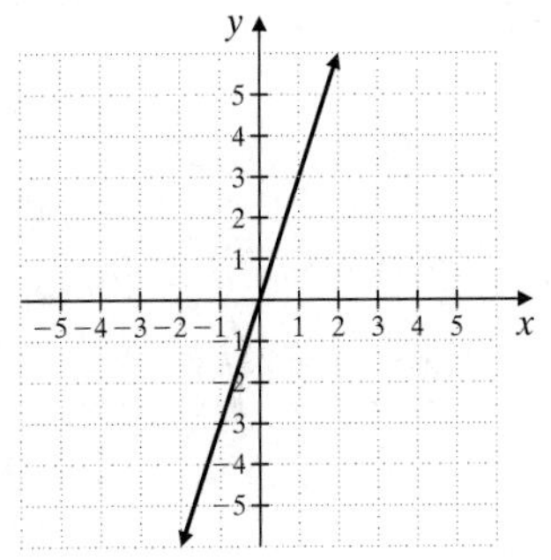

Review and Preview

Evaluate each exponential expression. See Section 9.2.

43. $25^{1/2}$

44. $49^{1/2}$

45. $16^{3/4}$

46. $27^{2/3}$

47. $9^{-3/2}$

48. $81^{-3/4}$

If $f(x) = 3^x$, find each value. In Exercises 51 and 52, give an exact answer and a two-decimal-place approximation. See Section 7.3.

49. $f(2)$

50. $f(0)$

51. $f\left(\dfrac{1}{2}\right)$

52. $f\left(\dfrac{2}{3}\right)$

Combining Concepts

For Exercises 53 and 54,

a. *Write the ordered pairs for f whose points are highlighted. (Include the points whose coordinates are given.)*
b. *Write the corresponding ordered pairs for the inverse of f, f^{-1}.*
c. *Graph the ordered pairs for f^{-1} found in part (b).*
d. *Graph f^{-1} by drawing a smooth curve through the plotted points.*

53. a.

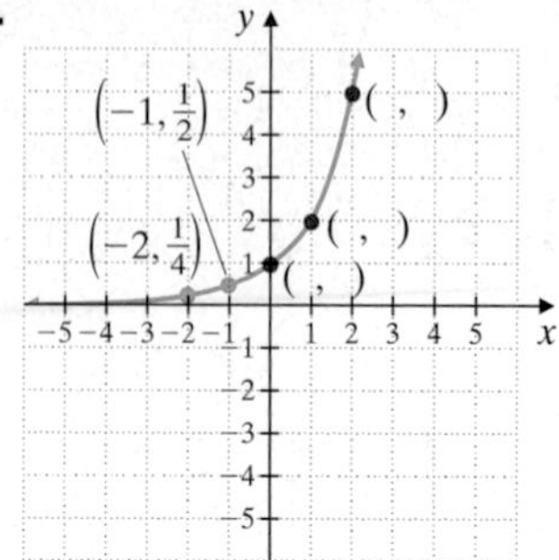

b. c. d.

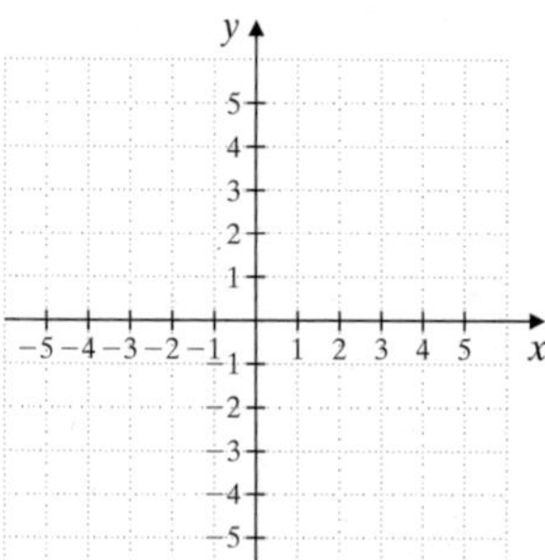

54. a.

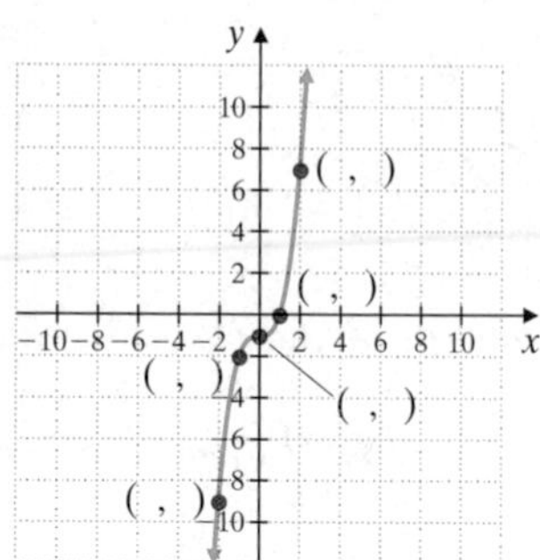

b. c. d.

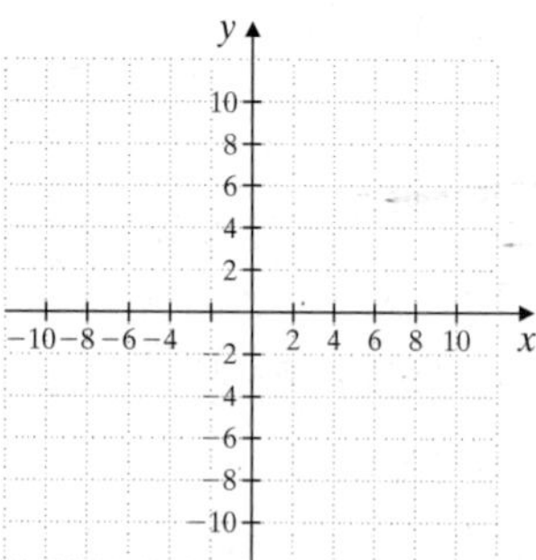

Find the inverse of each one-to-one function. Then graph the function and its inverse in a square window.

55. $f(x) = 3x + 1$

56. $f(x) = -2x - 6$

57. $f(x) = \sqrt[3]{x + 3}$

58. $f(x) = x^3 - 3$

59. If you are given the graph of a function, describe how you can tell from the graph whether a function has an inverse.

60. Describe the appearance of the graphs of a function and its inverse.

11.3 Exponential Functions

In earlier chapters, we gave meaning to exponential expressions such as 2^x, where x is a rational number. Recall the following examples.

$2^3 = 2 \cdot 2 \cdot 2$ Three factors; each factor is 2

$2^{3/2} = (2^{1/2})^3 = \sqrt{2} \cdot \sqrt{2} \cdot \sqrt{2}$ Three factors; each factor is $\sqrt{2}$

When x is an irrational number (for example, $\sqrt{3}$), what meaning can we give to $2^{\sqrt{3}}$?

It is beyond the scope of this book to give precise meaning to 2^x if x is irrational. We can confirm your intuition and say that $2^{\sqrt{3}}$ is a real number, and since $1 < \sqrt{3} < 2$, then $2^1 < 2^{\sqrt{3}} < 2^2$. We can also use a calculator and approximate $2^{\sqrt{3}}$: $2^{\sqrt{3}} \approx 3.321997$. In fact, as long as the base b is positive, b^x is a real number for all real numbers x. Finally, the rules of exponents apply whether x is rational or irrational, as long as b is positive.

In this section, we are interested in functions of the form $f(x) = b^x$, where $b > 0$. A function of this form is called an *exponential function*.

OBJECTIVES

A. Graph exponential functions.
B. Solve equations of the form $b^x = b^y$.
C. Solve problems modeled by exponential equations.

SSM TUTOR CENTER SG CD & VIDEO MATH PRO WEB

Exponential Function

A function of the form

$$f(x) = b^x$$

is called an **exponential function** if $b > 0$, b is not 1, and x is a real number.

A Graphing Exponential Functions

Now let's practice graphing exponential functions.

EXAMPLE 1

Graph the exponential functions $f(x) = 2^x$ and $g(x) = 3^x$ on the same set of axes.

Solution: To graph these functions, we find some ordered pair solutions, plot the points, and connect them with a smooth curve.

$f(x) = 2^x$

x	0	1	2	3	-1	-2
$f(x)$	1	2	4	8	$\frac{1}{2}$	$\frac{1}{4}$

$g(x) = 3^x$

x	0	1	2	3	-1	-2
$g(x)$	1	3	9	27	$\frac{1}{3}$	$\frac{1}{9}$

Practice Problem 1

Graph the exponential function $f(x) = 4^x$.

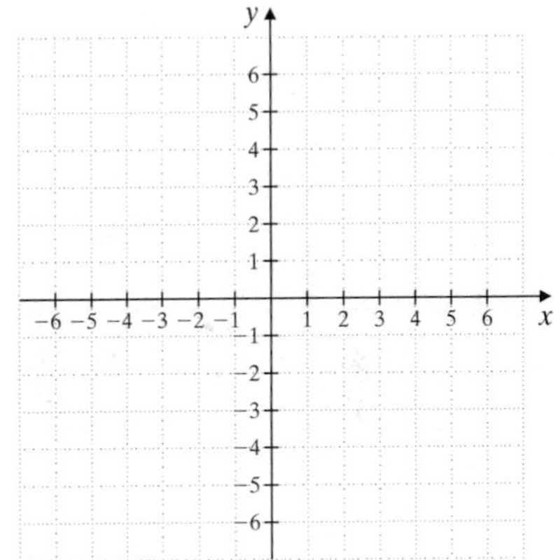

Answer

1.

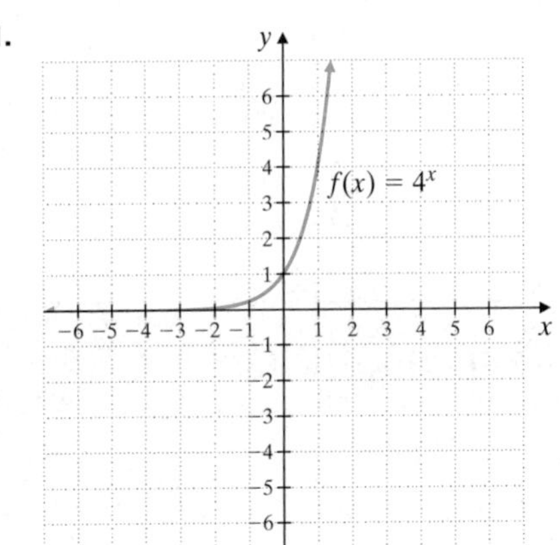

A number of things should be noted about the two graphs of exponential functions in Example 1. First, the graphs show that $f(x) = 2^x$ and $g(x) = 3^x$ are one-to-one functions since each graph passes the vertical and horizontal line tests. The y-intercept of each graph is $(0, 1)$, but neither graph has an x-intercept. From the graph, we can also see that the domain of each function is all real numbers and that the range is $(0, \infty)$. We can also see that as x-values are increasing, y-values are increasing also.

EXAMPLE 2

Graph the exponential functions $y = \left(\frac{1}{2}\right)^x$ and $y = \left(\frac{1}{3}\right)^x$ on the same set of axes.

Solution: As before, we find some ordered pair solutions, plot the points, and connect them with a smooth curve.

$y = \left(\frac{1}{2}\right)^x$

x	0	1	2	3	−1	−2
y	1	$\frac{1}{2}$	$\frac{1}{4}$	$\frac{1}{8}$	2	4

$y = \left(\frac{1}{3}\right)^x$

x	0	1	2	3	−1	−2
y	1	$\frac{1}{3}$	$\frac{1}{9}$	$\frac{1}{27}$	3	9

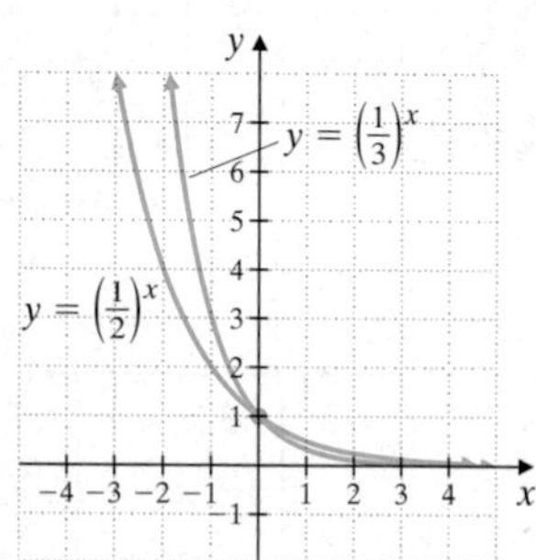

Each function in Example 2 again is a one-to-one function. The y-intercept of both is $(0, 1)$. The domain is the set of all real numbers, and the range is $(0, \infty)$.

Notice the difference between the graphs of Example 1 and the graphs of Example 2. An exponential function is always increasing if the base is greater

Practice Problem 2

Graph the exponential function $f(x) = \left(\frac{1}{5}\right)^x$.

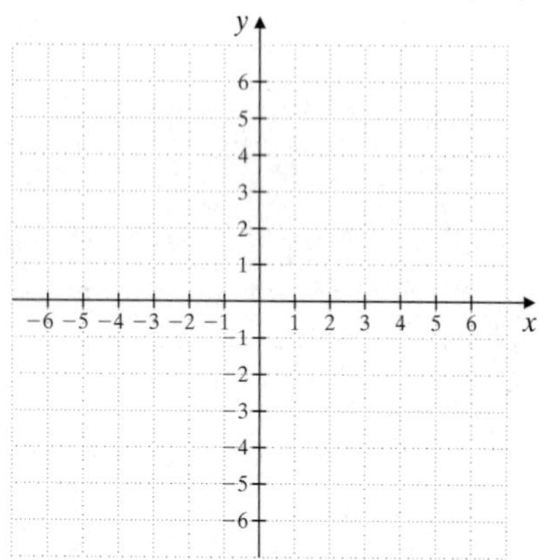

Answer

2.

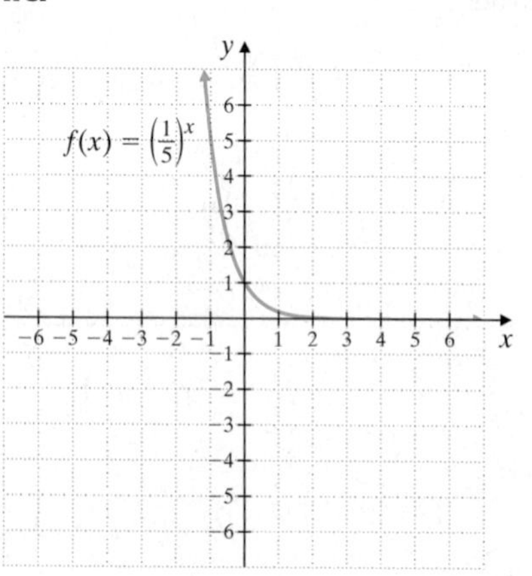

than 1. When the base is between 0 and 1, the graph is always decreasing. The following figures summarize these characteristics of exponential functions.

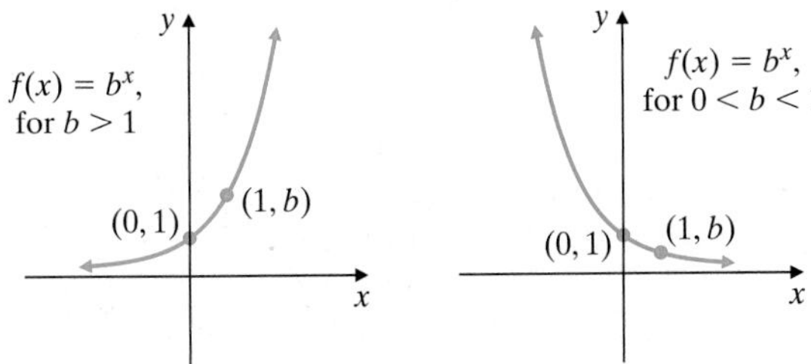

EXAMPLE 3 Graph the exponential function $f(x) = 3^{x+2}$.

Solution: As before, we find and plot a few ordered pair solutions. Then we connect the points with a smooth curve.

$y = 3^{x+2}$

x	0	−1	−2	−3	−4
y	9	3	1	$\frac{1}{3}$	$\frac{1}{9}$

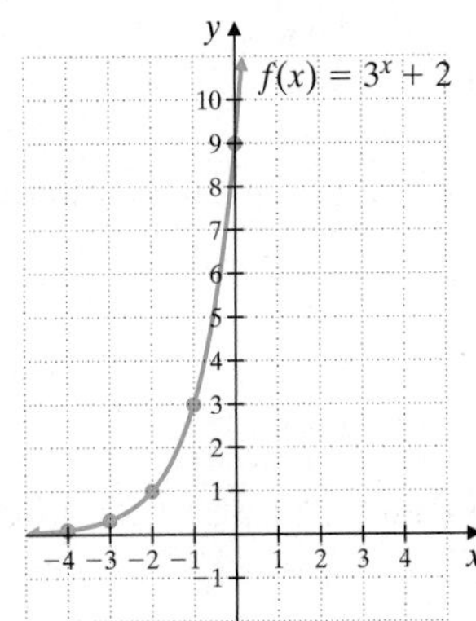

Try the Concept Check in the margin.

B Solving Equations of the Form $b^x = b^y$

We have seen that an exponential function $y = b^x$ is a one-to-one function. Another way of stating this fact is a property that we can use to solve exponential equations.

Uniqueness of b^x

Let $b > 0$ and $b \neq 1$. Then $b^x = b^y$ is equivalent to $x = y$.

Practice Problem 3

Graph the exponential function $f(x) = 2^{x-1}$.

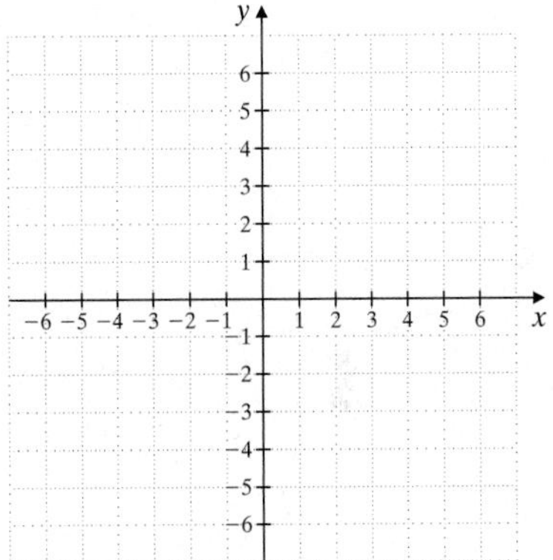

Concept Check

Which functions are exponential functions?

a. $f(x) = x^3$ b. $g(x) = \left(\frac{2}{3}\right)^x$

c. $h(x) = 5^{x-2}$ d. $w(x) = (2x)^2$

Answers

3.

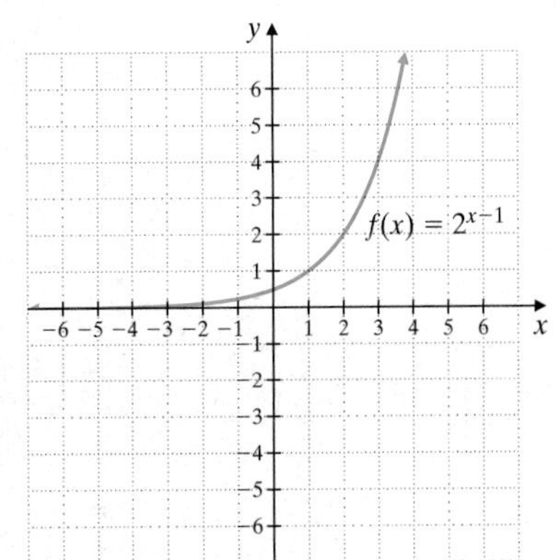

Concept Check: b and c

Practice Problem 4

Solve: $5^x = 125$

EXAMPLE 4 Solve: $2^x = 16$

Solution: We write 16 as a power of 2 and then use the uniqueness of b^x to solve.

$$2^x = 16$$
$$2^x = 2^4$$

Since the bases are the same and are nonnegative, by the uniqueness of b^x we then have that the exponents are equal. Thus,

$$x = 4$$

To check, we replace x with 4 in the original equation.
The solution set is $\{4\}$.

Practice Problem 5

Solve: $4^x = 8$

EXAMPLE 5 Solve: $9^x = 27$

Solution: Since both 9 and 27 are powers of 3, we can use the uniqueness of b^x.

$$9^x = 27$$
$$(3^2)^x = 3^3 \quad \text{Write 9 and 27 as powers of 3.}$$
$$3^{2x} = 3^3$$
$$2x = 3 \quad \text{Use the uniqueness of } b^x.$$
$$x = \frac{3}{2} \quad \text{Divide both sides by 2.}$$

To check, we replace x with $\frac{3}{2}$ in the original equation.
The solution set is $\left\{\frac{3}{2}\right\}$.

Practice Problem 6

Solve: $9^{x-1} = 27^x$

EXAMPLE 6 Solve: $4^{x+3} = 8^x$

Solution: We write both 4 and 8 as powers of 2, and then use the uniqueness of b^x.

$$4^{x+3} = 8^x$$
$$(2^2)^{x+3} = (2^3)^x$$
$$2^{2x+6} = 2^{3x}$$
$$2x + 6 = 3x \quad \text{Use the uniqueness of } b^x.$$
$$6 = x \quad \text{Subtract } 2x \text{ from both sides.}$$

Check to see that the solution set is $\{6\}$.

There is one major problem with the preceding technique. Often the two sides of an equation, $4 = 3^x$ for example, cannot easily be written as powers of a common base. We explore how to solve such an equation with the help of *logarithms* later.

C Solving Problems Modeled by Exponential Equations

The bar graph on the next page shows the increase in the number of cellular phone users. Notice that the graph of the exponential function $y = 25.759(1.277)^x$ approximates the heights of the bars. This is just one ex-

Answers

4. $\{3\}$, **5.** $\left\{\frac{3}{2}\right\}$, **6.** $\{-2\}$

ample of how the world abounds with patterns that can be modeled by exponential functions. To make these applications realistic, we use numbers that warrant a calculator. Another application of an exponential function has to do with interest rates on loans.

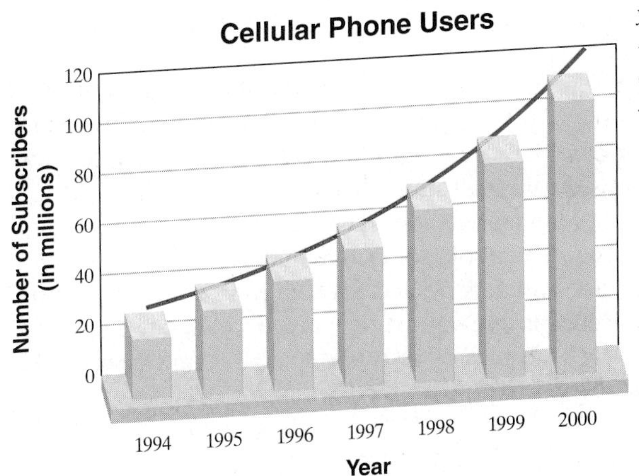

$y = 25.759(1.277)^x$

where $x = 0$ corresponds to 1994, $x = 1$ corresponds to 1995, and so on

The exponential function defined by $A = P\left(1 + \frac{r}{n}\right)^{nt}$ models the pattern relating the dollars A accrued (or owed) after P dollars are invested (or loaned) at an annual rate of interest r compounded n times each year for t years. This function is known as the *compound interest formula.*

EXAMPLE 7 Using the Compound Interest Formula

Find the amount owed at the end of 5 years if \$1600 is loaned at a rate of 9% compounded monthly.

Solution: Use the formula $A = P\left(1 + \frac{r}{n}\right)^{nt}$, with the following values:

$P = \$1600$ (the amount of the loan)

$r = 9\% = 0.09$ (the annual rate of interest)

$n = 12$ (the number of times interest is compounded each year)

$t = 5$ (the duration of the loan, in years)

$$A = P\left(1 + \frac{r}{n}\right)^{nt} \quad \text{Compound interest formula}$$

$$= 1600\left(1 + \frac{0.09}{12}\right)^{12(5)} \quad \text{Substitute known values.}$$

$$= 1600(1.0075)^{60}$$

To approximate A, use the $\boxed{y^x}$ or $\boxed{\wedge}$ key on your calculator.

$\boxed{2505.0896}$

Thus, the amount A owed is approximately \$2505.09.

Practice Problem 7

As a result of the Chernobyl nuclear accident, radioactive debris was carried through the atmosphere. One immediate concern was the impact that debris had on the milk supply. The percent y of radioactive material in raw milk t days after the accident is estimated by $y = 100(2.7)^{-0.1t}$. Estimate the expected percent of radioactive material in the milk after 30 days.

Answer

7. approximately 5.08%

GRAPHING CALCULATOR EXPLORATIONS

We can use a graphing calculator and its TRACE feature to solve Practice Problem 7 graphically.

To estimate the percent of radioactive material in the milk after 30 days, enter $Y_1 = 100(2.7)^{-0.1x}$. The graph does not appear on a standard viewing window, so we need to determine an appropriate viewing window. Because it doesn't make sense to look at radioactivity *before* the Chernobyl nuclear accident, we use Xmin $= 0$. We are interested in finding the percent of radioactive material in the milk when $x = 30$, so we choose Xmax $= 35$ to leave enough space to see the graph at $x = 30$. Because the values of y are percents, it seems appropriate that $0 \leq y \leq 100$. (We also use Xscl $= 1$ and Yscl $= 10$.) Now we graph the function.

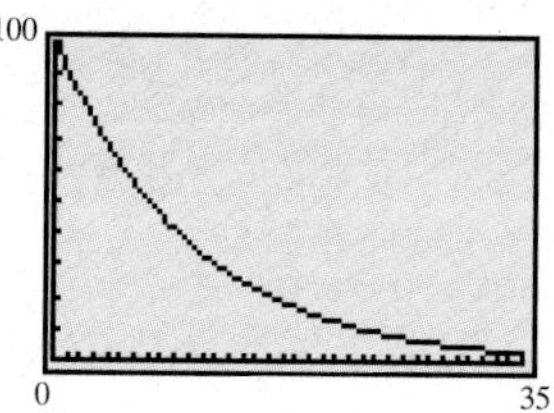

We can use the TRACE feature to obtain an approximation of the expected percent of radioactive material in the milk when $x = 30$. (A TABLE feature may also be used to approximate the percent.) To obtain a better approximation, let's use the ZOOM feature several times to zoom in near $x = 30$.

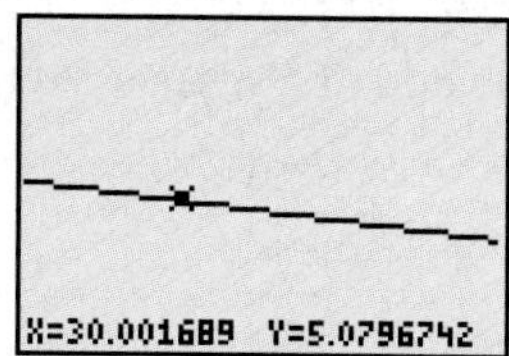

The percent of radioactive material in the milk 30 days after the Chernobyl accident was 5.08%, accurate to two decimal places.

Use a grapher to find each percent. Approximate your solutions so that they are accurate to two decimal places.

1. Estimate the percent of radioactive material in the milk 2 days after the Chernobyl nuclear accident.
2. Estimate the percent of radioactive material in the milk 10 days after the Chernobyl nuclear accident.
3. Estimate the percent of radioactive material in the milk 15 days after the Chernobyl nuclear accident.
4. Estimate the percent of radioactive material in the milk 25 days after the Chernobyl nuclear accident.

Name ______________________ Section __________ Date __________

EXERCISE SET 11.3

A *Graph each exponential function. See Examples 1 through 3.*

1. $y = 4^x$

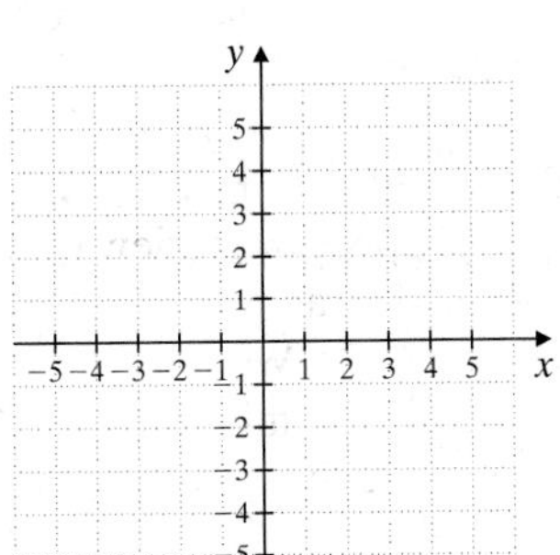

2. $y = 5^x$

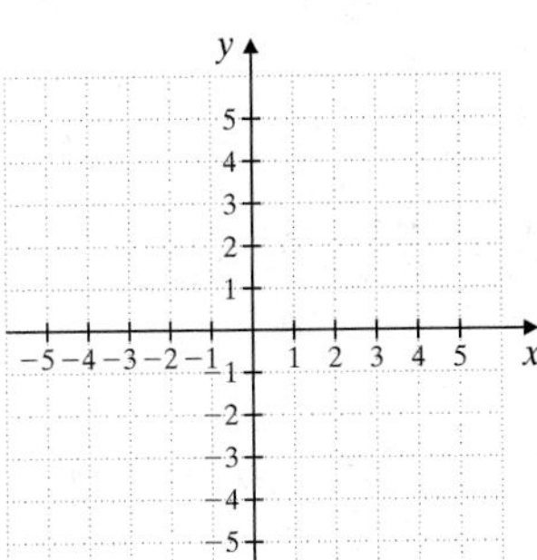

3. $y = 1 + 2^x$

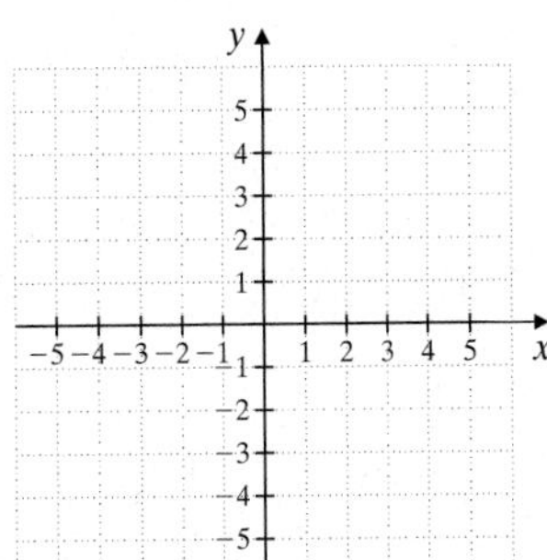

4. $y = 3^x - 1$

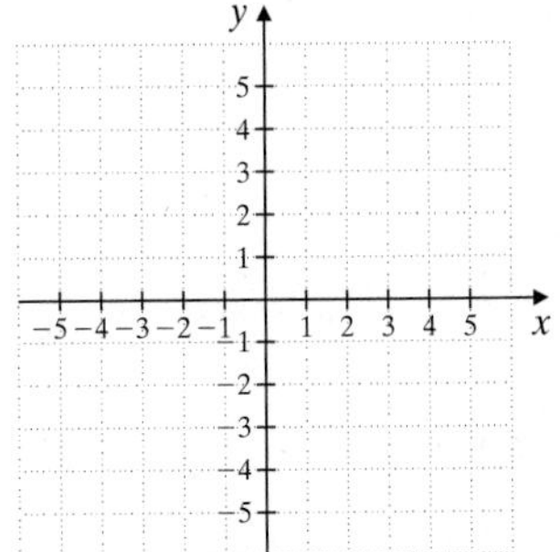

5. $y = \left(\frac{1}{4}\right)^x$

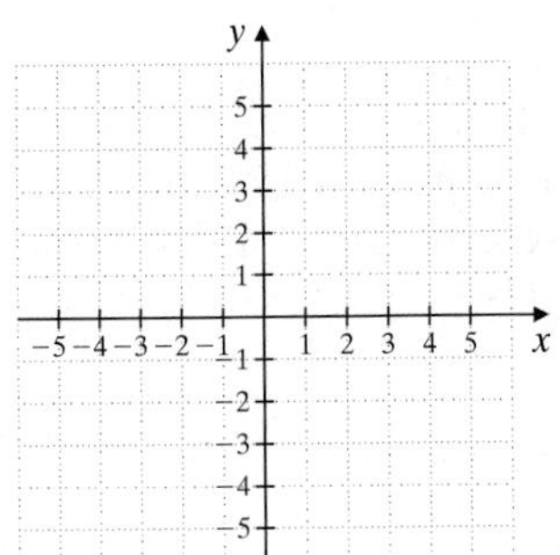

6. $y = \left(\frac{1}{5}\right)^x$

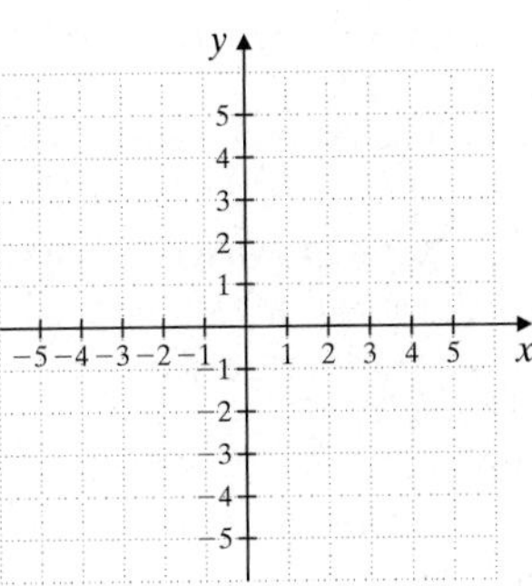

7. $y = \left(\frac{1}{2}\right)^x - 2$

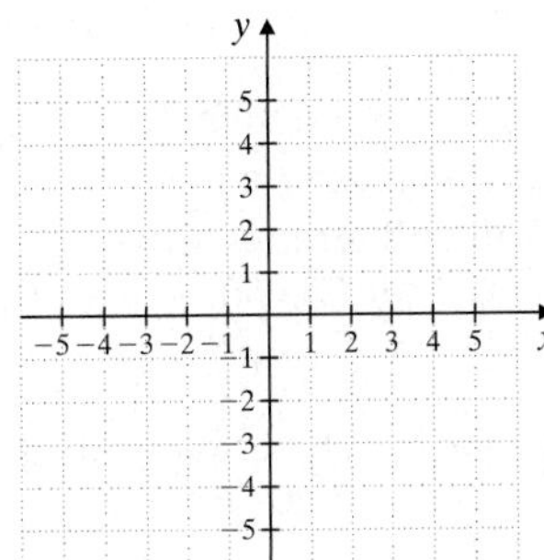

8. $y = \left(\frac{1}{3}\right)^x + 2$

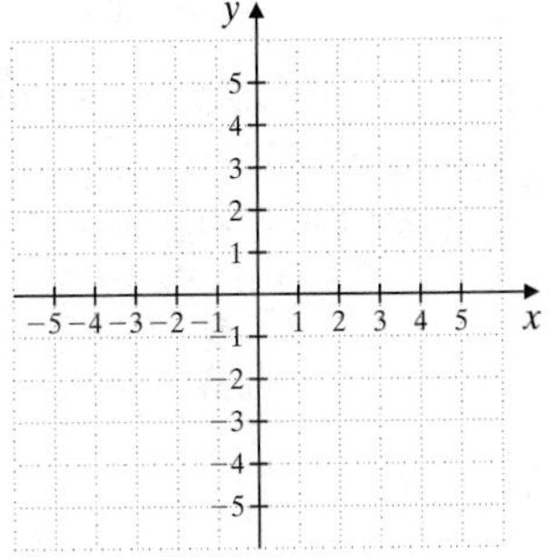

9. $y = -2^x$

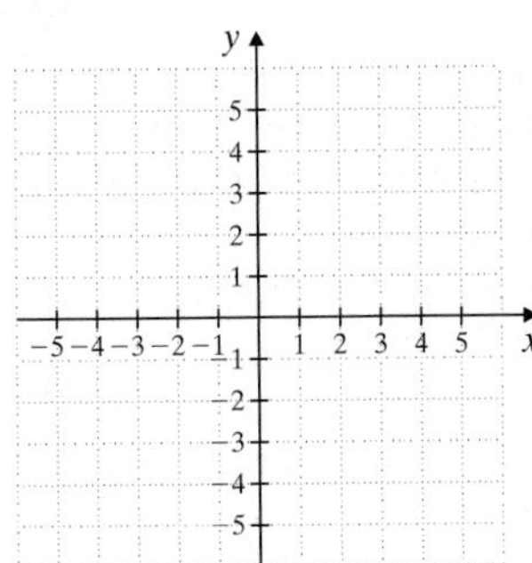

10. $y = -3^x$

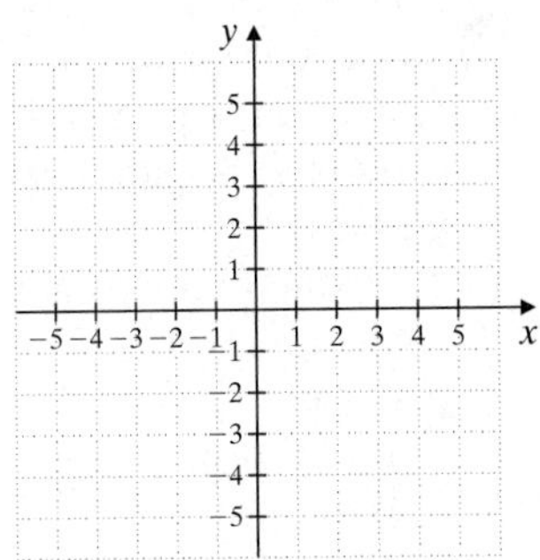

11. $y = 3^x - 2$

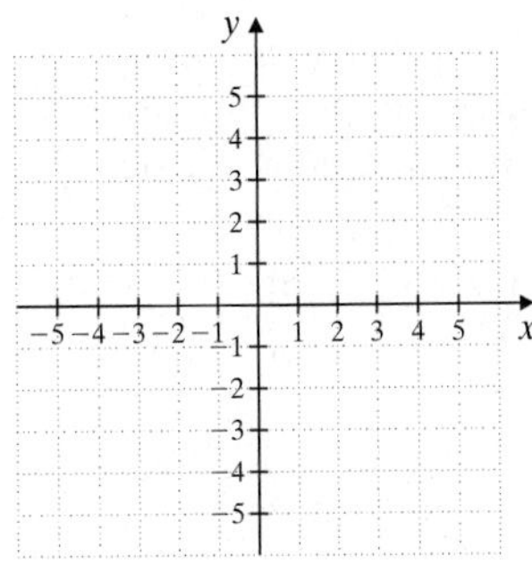

12. $y = 2^x - 3$

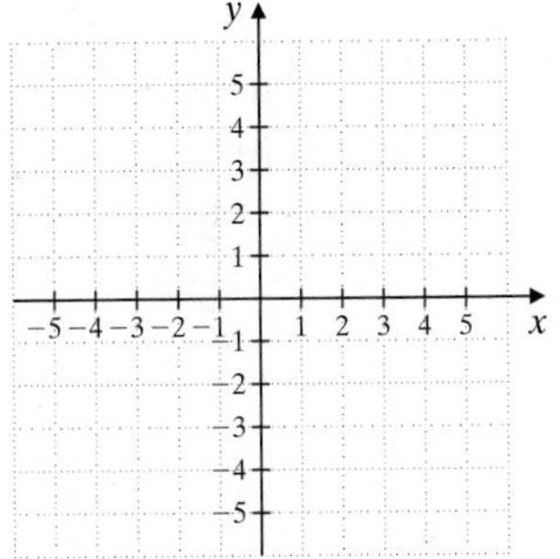

13. $y = -\left(\frac{1}{4}\right)^x$

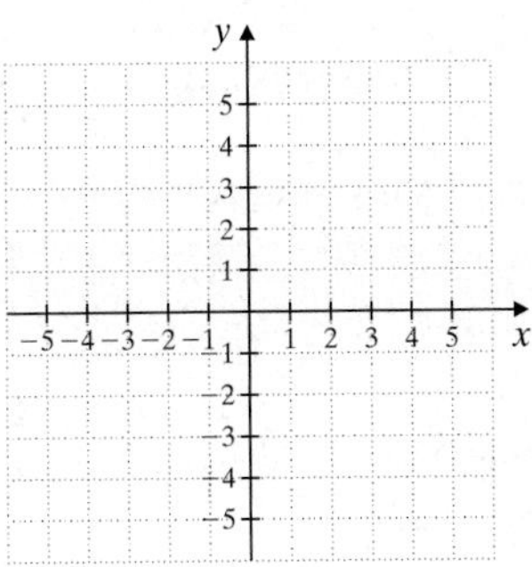

14. $y = -\left(\frac{1}{5}\right)^x$

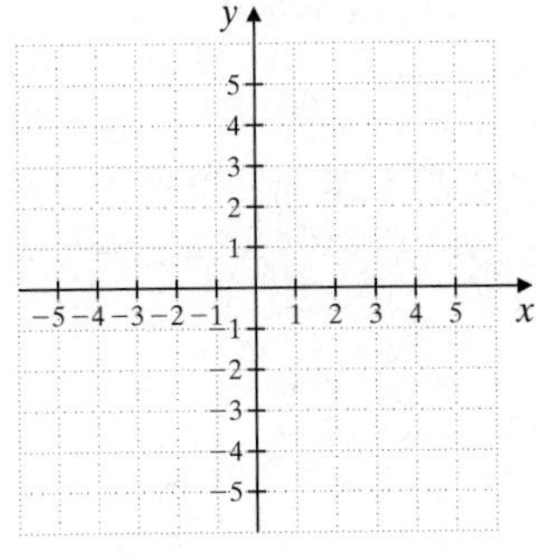

15. $y = \left(\frac{1}{3}\right)^x + 1$

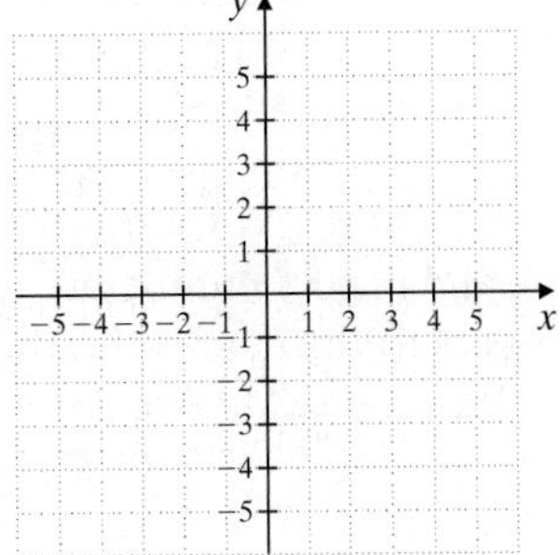

16. $y = \left(\frac{1}{2}\right)^x - 2$

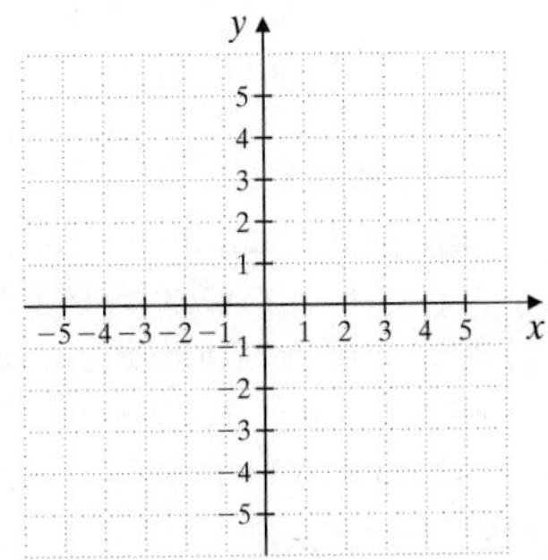

B *Solve. See Examples 4 through 6.*

17. $3^x = 27$

18. $6^x = 36$

19. $16^x = 8$

20. $64^x = 16$

21. $32^{2x-3} = 2$

22. $9^{2x+1} = 81$

23. $\frac{1}{4} = 2^{3x}$

24. $\frac{1}{27} = 3^{2x}$

25. $4^x = 8$

26. $32^x = 4$

27. $27^{x+1} = 9$

28. $125^{x-2} = 25$

29. $81^{x-1} = 27^{2x}$

30. $4^{3x-7} = 32^{2x}$

C *Solve. Unless otherwise indicated, round results to one decimal place. See Example 7.*

31. One type of uranium has a daily radioactive decay rate of 0.4%. If 30 pounds of this uranium is available today, how much will still remain after 50 days? Use $y = 30(2.7)^{-0.004t}$, and let t be 50.

32. The nuclear waste from an atomic energy plant decays at a rate of 3% each century. If 150 pounds of nuclear waste is disposed of, how much of it will still remain after 10 centuries? Use $y = 150(2.7)^{-0.03t}$, and let t be 10.

33. National Park Service personnel are trying to increase the size of the bison population of Theodore Roosevelt National Park. If 260 bison currently live in the park, and if the population's rate of growth is 2.5 % annually, how many bison (rounded to the nearest whole) should there be in 10 years? Use $y = 260(2.7)^{0.025t}$.

34. The equation $y = 120.882(1.012)^x$ models the population of the United States from 1930 through 2000. In the equation, y is the population in millions and x represents the number of years after 1930. Round answers to the nearest tenth of a million. (*Source:* Based on data from the U.S. Bureau of the Census)

a. Estimate the population of the United States in 1970.

b. Assuming this equation continues to be valid in the future, use the equation to predict the population of the United States in 2020.

35. Retail revenue from shopping on the Internet is currently growing at a rate of 56% per year. In 2000, a total of $42.1 billion in revenue was collected through Internet retail sales. Answer the following questions using $y = 42.1(1.56)^t$ where y is Internet revenue in billions of dollars and t is the number of years after 2000. Round answers to the nearest tenth of a billion dollars. (*Source:* Based on data from eMarketer)

a. According to the model, what level of retail revenues from Internet shopping was expected in 2001?

b. Predict the level of Internet shopping revenues in 2009.

36. Carbon dioxide (CO_2) is a greenhouse gas that contributes to global warming. Partially due to the combustion of fossil fuels, the amount of CO_2 in Earth's atmosphere has been increasing by 0.4% annually over the past century. In 2000, the concentration of CO_2 in the atmosphere was 369.4 parts per million by volume. To make the following predictions, use $y = 369.4(1.004)^t$ where y is the concentration of CO_2 in parts per million and t is the number of years after 2000. Round answers to the nearest tenth. (*Sources:* Based on data from the United Nations Environment Programme and the Carbon Dioxide Information Analysis Center)

a. Predict the concentration of CO_2 in the atmosphere in the year 2006.

b. Predict the concentration of CO_2 in the atmosphere in the year 2030.

The equation $y = 25.759(1.277)^x$ gives the number of cellular phone users y (in millions) in the United States for the years 1994 through 2000. In this equation, $x = 0$ corresponds to 1994, $x = 1$ corresponds to 1995, and so on. Use this model to solve Exercises 37 and 38. Round answers to the nearest tenth of a million.

37. Predict the number of cellular phone users in the year 2007.

38. Predict the number of cellular phone users in the year 2015.

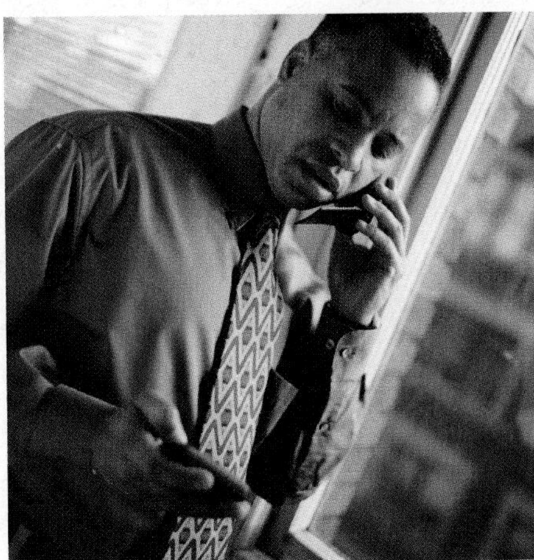

Solve. Use $A = P\left(1 + \frac{r}{n}\right)^{nt}$. Round answers to two decimal places. See Example 7.

39. Find the amount Erica Entada owes at the end of 3 years if \$6000 is loaned to her at a rate of 8% compounded monthly.

40. Find the amount owed at the end of 5 years if \$3000 is loaned at a rate of 10% compounded quarterly.

Review and Preview

Solve each equation. See Section 2.4.

41. $5x - 2 = 18$

42. $3x - 7 = 11$

43. $3x - 4 = 3(x + 1)$

44. $2 - 6x = 6(1 - x)$

Combining Concepts

Match each exponential function with its graph.

45. $f(x) = \left(\frac{1}{2}\right)^x$

46. $f(x) = 2^x$

47. $f(x) = \left(\frac{1}{4}\right)^x$

48. $f(x) = 3^x$

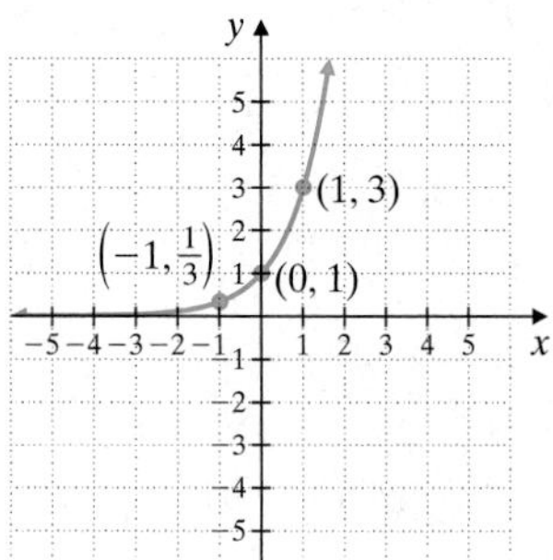

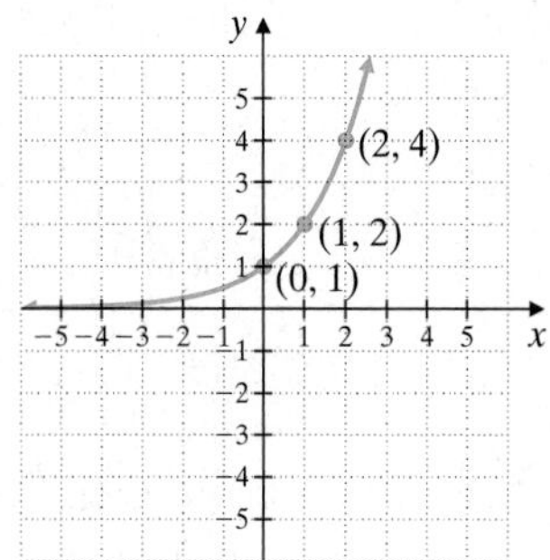

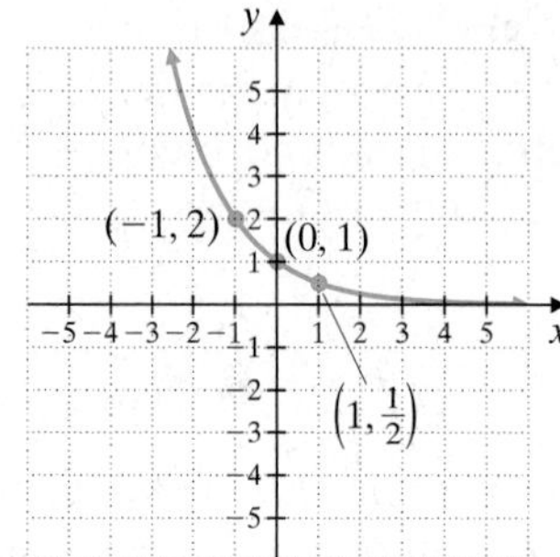

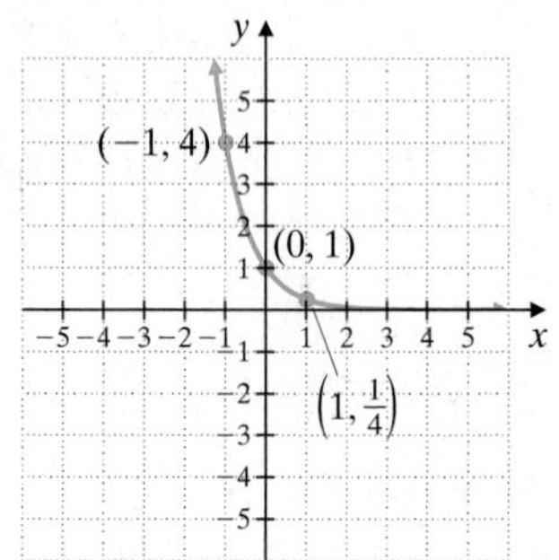

49. Explain why the graph of an exponential function $y = b^x$ contains the point $(1, b)$.

50. Explain why an exponential function $y = b^x$ has a y-intercept of $(0, 1)$.

Use a graphing calculator to solve. Estimate your results to two decimal places.

51. Verify the results of Exercise 31.

52. From Exercise 31, estimate the number of pounds of uranium that will be available after 100 days.

53. From Exercise 31, estimate the number of pounds of uranium that will be available after 120 days.

11.4 Logarithmic Functions

OBJECTIVES

A Write exponential equations with logarithmic notation and write logarithmic equations with exponential notation.

B Solve logarithmic equations by using exponential notation.

C Identify and graph logarithmic functions.

A Using Logarithmic Notation

Since the exponential function $f(x) = 2^x$ is a one-to-one function, it has an inverse. We can create a table of values for f^{-1} by switching the coordinates in the accompanying table of values for $f(x) = 2^x$.

x	$y = f(x)$
-3	$\frac{1}{8}$
-2	$\frac{1}{4}$
-1	$\frac{1}{2}$
0	1
1	2
2	4
3	8

x	$y = f^{-1}(x)$
$\frac{1}{8}$	-3
$\frac{1}{4}$	-2
$\frac{1}{2}$	-1
1	0
2	1
4	2
8	3

The graphs of f and its inverse are shown in the margin. Notice that the graphs of f and f^{-1} are symmetric about the line $y = x$, as expected.

Now we would like to be able to write an equation for f^{-1}. To do so, we follow the steps for finding the equation of an inverse.

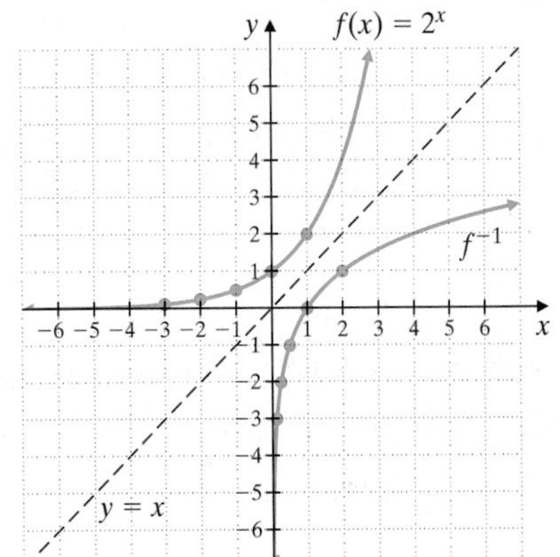

$$f(x) = 2^x$$

Step 1. Replace $f(x)$ by y. $\quad y = 2^x$
Step 2. Interchange x and y. $\quad x = 2^y$
Step 3. Solve for y.

At this point, we are stuck. To solve this equation for y, a new notation, **logarithmic notation**, is needed.

The symbol $\log_b x$ means "the power to which b is raised to produce a result of x." In other words,

$$\log_b x = y \quad \text{means} \quad b^y = x$$

We say that $\log_b x$ is "the logarithm of x to the base b" or "the log of x to the base b."

Logarithmic Definition

If $b > 0$, and $b \neq 1$, then

$$y = \log_b x \quad \text{means} \quad x = b^y$$

for every $x > 0$ and every real number y.

Before returning to the function $x = 2^y$ and solving it for y in terms of x, let's practice using the new notation $\log_b x$.

It is important to be able to write exponential equations with logarithmic notation, and vice versa. The following table shows examples of both forms.

Logarithmic Equation	Corresponding Exponential Equation
$\log_3 9 = 2$	$3^2 = 9$
$\log_6 1 = 0$	$6^0 = 1$
$\log_2 8 = 3$	$2^3 = 8$
$\log_4 \frac{1}{16} = -2$	$4^{-2} = \frac{1}{16}$
$\log_8 2 = \frac{1}{3}$	$8^{1/3} = 2$

Helpful Hint

Notice that a *logarithm* is an *exponent.* In other words, $\log_3 9$ is the *power* that we raise 3 to in order to get 9.

Practice Problems 1–3

Write as an exponential equation.

1. $\log_7 49 = 2$ 2. $\log_8 \frac{1}{8} = -1$

3. $\log_3 \sqrt{3} = \frac{1}{2}$

EXAMPLES Write as an exponential equation.

1. $\log_5 25 = 2$ means $5^2 = 25$.

2. $\log_6 \frac{1}{6} = -1$ means $6^{-1} = \frac{1}{6}$.

3. $\log_2 \sqrt{2} = \frac{1}{2}$ means $2^{1/2} = \sqrt{2}$.

Practice Problems 4–6

Write as a logarithmic equation.

4. $3^4 = 81$ 5. $2^{-3} = \frac{1}{8}$

6. $7^{1/3} = \sqrt[3]{7}$

EXAMPLES Write as a logarithmic equation.

4. $9^3 = 729$ means $\log_9 729 = 3$.

5. $6^{-2} = \frac{1}{36}$ means $\log_6 \frac{1}{36} = -2$.

6. $5^{1/3} = \sqrt[3]{5}$ means $\log_5 \sqrt[3]{5} = \frac{1}{3}$.

Practice Problem 7

Find the value of each logarithmic expression.

a. $\log_2 8$ b. $\log_3 \frac{1}{9}$

c. $\log_{25} 5$

EXAMPLE 7 Find the value of each logarithmic expression.

a. $\log_4 16$

b. $\log_{10} \frac{1}{10}$

c. $\log_9 3$

Solution:

a. $\log_4 16 = 2$ because $4^2 = 16$.

b. $\log_{10} \frac{1}{10} = -1$ because $10^{-1} = \frac{1}{10}$.

c. $\log_9 3 = \frac{1}{2}$ because $9^{1/2} = \sqrt{9} = 3$.

B Solving Logarithmic Equations

The ability to interchange the logarithmic and exponential forms of a statement is often the key to solving logarithmic equations.

Practice Problem 8

Solve: $\log_2 x = 4$

EXAMPLE 8 Solve: $\log_5 x = 3$

Solution: $\log_5 x = 3$

$5^3 = x$ Write as an exponential equation.

$125 = x$

The solution set is $\{125\}$.

Answers

1. $7^2 = 49$, **2.** $8^{-1} = \frac{1}{8}$, **3.** $3^{1/2} = \sqrt{3}$,
4. $\log_3 81 = 4$, **5.** $\log_2 \frac{1}{8} = -3$,
6. $\log_7 \sqrt[3]{7} = \frac{1}{3}$, **7. a.** 3, **b.** −2, **c.** $\frac{1}{2}$,
8. $\{16\}$

EXAMPLE 9 Solve: $\log_x 25 = 2$

Solution: $\log_x 25 = 2$

$x^2 = 25$ Write as an exponential equation.

$x = 5$

Even though $(-5)^2 = 25$, the base b of a logarithm must be positive. The solution set is $\{5\}$.

Practice Problem 9

Solve: $\log_x 9 = 2$

EXAMPLE 10 Solve: $\log_3 1 = x$

Solution: $\log_3 1 = x$

$3^x = 1$ Write as an exponential equation.

$3^x = 3^0$ Write 1 as 3^0.

$x = 0$ Use the uniqueness of b^x.

The solution set is $\{0\}$.

Practice Problem 10

Solve: $\log_2 1 = x$

In Example 10, we illustrated an important property of logarithms. That is, $\log_b 1$ is always 0. This property as well as two important others are given below.

Properties of Logarithms

If b is a real number, $b > 0$ and $b \neq 1$, then

1. $\log_b 1 = 0$
2. $\log_b b^x = x$
3. $b^{\log_b x} = x$

To see that $\log_b b^x = x$, we change the logarithmic form to exponential form. Then, $\log_b b^x = x$ means $b^x = b^x$. In exponential form, the statement is true, so in logarithmic form, the statement is also true.

EXAMPLE 11 Simplify.

a. $\log_3 3^2$ **b.** $\log_7 7^{-1}$
c. $5^{\log_5 3}$ **d.** $2^{\log_2 6}$

Solution:

a. From property 2, $\log_3 3^2 = 2$.

b. From property 2, $\log_7 7^{-1} = -1$.

c. From property 3, $5^{\log_5 3} = 3$.

d. From property 3, $2^{\log_2 6} = 6$.

Practice Problem 11

Simplify.

a. $\log_6 6^3$ b. $\log_{11} 11^{-4}$
c. $7^{\log_7 5}$ d. $3^{\log_3 10}$

Answers

9. $\{3\}$ **10.** $\{0\}$, **11. a.** 3, **b.** -4, **c.** 5, **d.** 10

C Graphing Logarithmic Functions

Let us now return to the function $f(x) = 2^x$ and write an equation for its inverse, f^{-1}. Recall our earlier work.

$$f(x) = 2^x$$

Step 1. Replace $f(x)$ by y. $\quad y = 2^x$

Step 2. Interchange x and y. $\quad x = 2^y$

Having gained proficiency with the notation $\log_b x$, we can now complete the steps for writing the inverse equation.

Step 3. Solve for y. $\quad y = \log_2 x$

Step 4. Replace y with $f^{-1}(x)$. $\quad f^{-1}(x) = \log_2 x$

Thus, $f^{-1}(x) = \log_2 x$ defines a function that is the inverse function of the function $f(x) = 2^x$. The function $f^{-1}(x)$ or $y = \log_2 x$ is called a *logarithmic function*.

Logarithmic Function

If x is a positive real number, b is a constant positive real number, and b is not 1, then a **logarithmic function** is a function that can be defined by

$$f(x) = \log_b x$$

The domain of f is the set of positive real numbers, and the range of f is the set of real numbers.

Try the Concept Check in the margin.

We can explore logarithmic functions by graphing them.

EXAMPLE 12 Graph the logarithmic function $y = \log_2 x$.

Solution: First we write the equation with exponential notation as $2^y = x$. Then we find some ordered pair solutions that satisfy this equation. Finally, we plot the points and connect them with a smooth curve. The domain of this function is $(0, \infty)$, and the range is all real numbers.

Since $x = 2^y$ is solved for x, we choose y-values and compute corresponding x-values.

If $y = 0$, $x = 2^0 = 1$.

If $y = 1$, $x = 2^1 = 2$.

If $y = 2$, $x = 2^2 = 4$.

If $y = -1$, $x = 2^{-1} = \frac{1}{2}$.

$x = 2^y$	y
1	0
2	1
4	2
$\frac{1}{2}$	−1

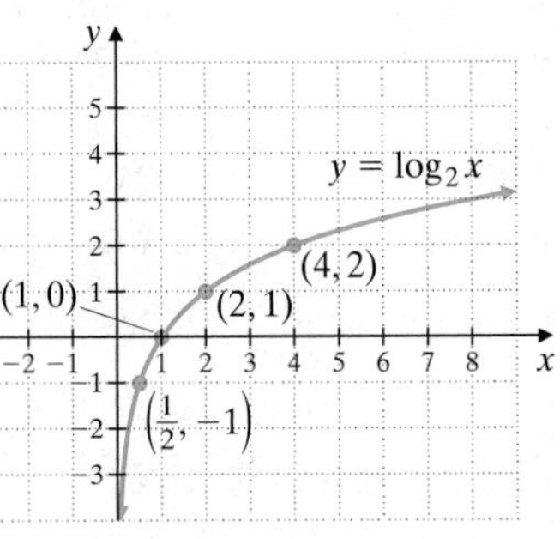

Concept Check

Let $f(x) = \log_3 x$ and $g(x) = 3^x$. These two functions are inverses of each other. Since $(2, 9)$ is an ordered pair solution of $g(x)$, what ordered pair do we know to be a solution of $f(x)$? Explain why.

Practice Problem 12

Graph the logarithmic function $y = \log_4 x$.

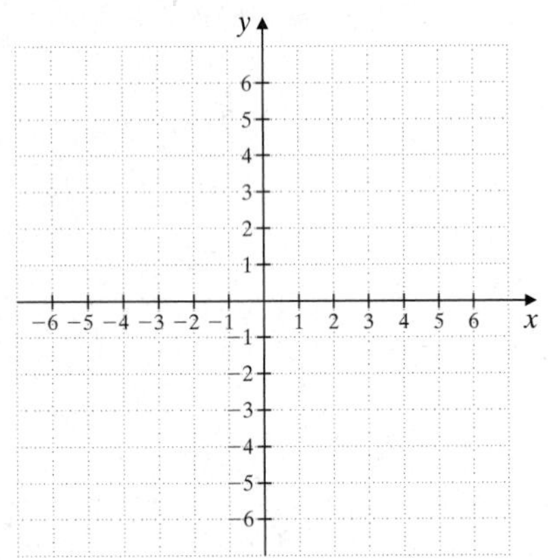

Answers

12.

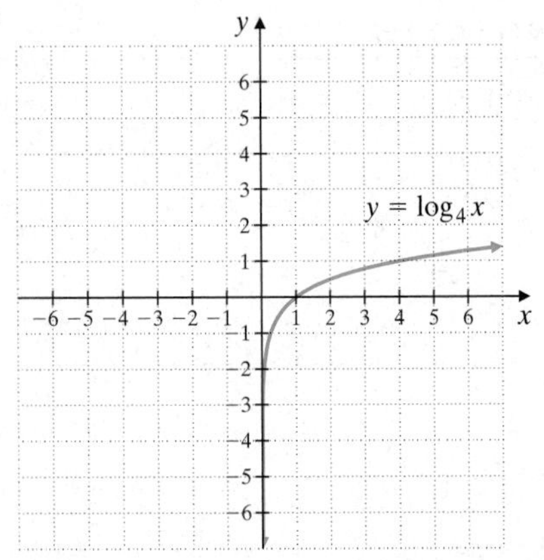

Concept Check: $(9, 2)$; answers may vary

EXAMPLE 13 Graph the logarithmic function $f(x) = \log_{1/3} x$.

Solution: We can replace $f(x)$ with y, and write the result with exponential notation.

$$f(x) = \log_{1/3} x$$

$$y = \log_{1/3} x \quad \text{Replace } f(x) \text{ with } y.$$

$$\left(\frac{1}{3}\right)^y = x \quad \text{Write in exponential form.}$$

Now we can find ordered pair solutions that satisfy $\left(\frac{1}{3}\right)^y = x$, plot these points, and connect them with a smooth curve.

If $y = 0$, $x = \left(\frac{1}{3}\right)^0 = 1$.

If $y = 1$, $x = \left(\frac{1}{3}\right)^1 = \frac{1}{3}$.

If $y = -1$, $x = \left(\frac{1}{3}\right)^{-1} = 3$.

If $y = -2$, $x = \left(\frac{1}{3}\right)^{-2} = 9$.

$x = \left(\frac{1}{3}\right)^y$	y
1	0
$\frac{1}{3}$	1
3	−1
9	−2

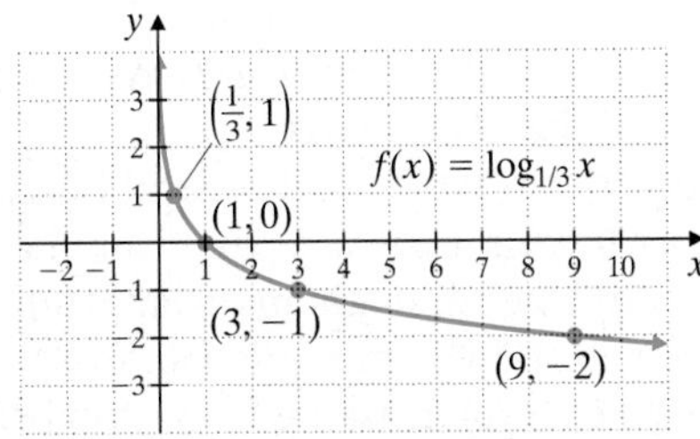

The domain of this function is $(0, \infty)$, and the range is the set of all real numbers.

The following figures summarize characteristics of logarithmic functions.

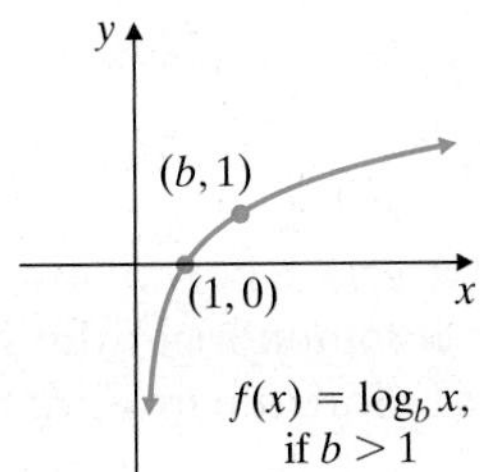

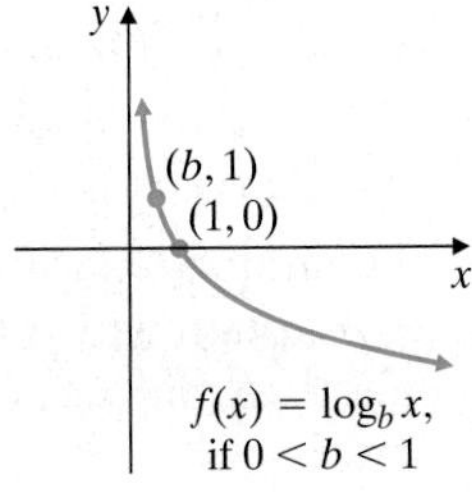

Practice Problem 13

Graph the logarithmic function $f(x) = \log_{1/2} x$.

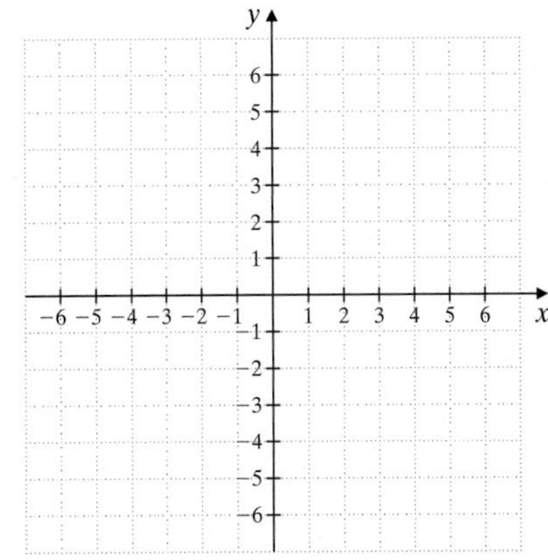

Answer

13.

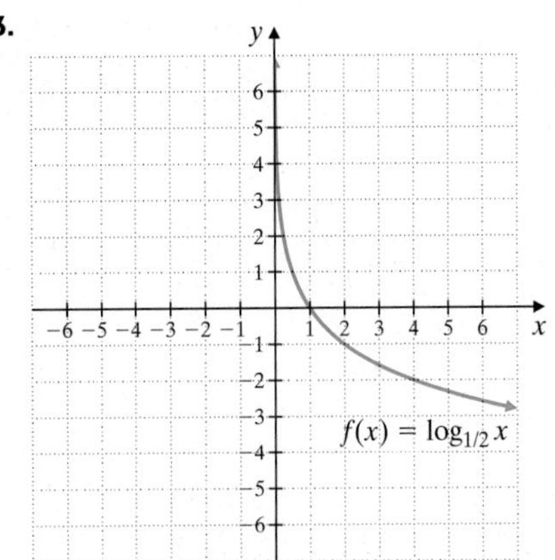

FOCUS ON History

THE INVENTION OF LOGARITHMS

Logarithms were the invention of John Napier (1550–1617), a Scottish land owner and theologian. Napier was also fascinated by mathematics and made it his hobby. Over a period of 20 years in his spare time, he developed his theory of logarithms, which were explained in his Latin text *Mirifici Logarithmorum Canonis Descriptio* (A Description of an Admirable Table of Logarithms), published in 1614. He hoped that his discovery would help to simplify the many time-consuming calculations required in astronomy. In fact, Napier's logarithms revolutionized astronomy and many other advanced mathematical fields by replacing "the multiplications, divisions, square and cubical extractions of great numbers, which besides the tedious expense of time are for the most part subject to many slippery errors" with related numbers that can be easily added and subtracted instead. His discovery was a great time-saving device. Some historians suggest that the use of logarithms to simplify calculations enabled German astronomer Johannes Kepler to develop his three laws of planetary motion, which in turn helped English physicist Sir Isaac Newton develop his theory of gravitation. Two hundred years after Napier's discovery, the French mathematician Pierre de Laplace wrote that logarithms, "by shortening the labors, doubled the life of the astronomer."

Napier's original logarithm tables had several flaws: They did not actually use a particular logarithmic base per se and log 1 was not defined to be equal to 0. An English mathematician, Henry Briggs, read Napier's Latin text soon after it was published and was very impressed by his ideas. Briggs wrote to Napier, asking to meet in person to discuss his wonderful discovery and to offer several improvements. The two mathematicians met in the summer of 1615. Briggs suggested redefining logarithms to base 10 and defining $\log 1 = 0$. Napier had also thought of using base 10 but hadn't been well enough to start a new set of tables. He asked Briggs to undertake the construction of a new set of base 10 tables. And so it was that the first table of common logarithms was constructed by Briggs over the next two years. Napier died in 1617 before Briggs was able to complete his new tables.

CRITICAL THINKING

Locate a table of common logarithms and describe how to use it. Give several examples. Explain why a table of common logarithms would have been invaluable to many calculations before the invention of the hand-held calculator.

Name ______________________ Section __________ Date ____________

A *Write each as an exponential equation. See Examples 1 through 3.*

1. $\log_6 36 = 2$

2. $\log_2 32 = 5$

3. $\log_3 \frac{1}{27} = -3$

4. $\log_5 \frac{1}{25} = -2$

5. $\log_{10} 1000 = 3$

6. $\log_{10} 10 = 1$

7. $\log_e x = 4$

8. $\log_e \frac{1}{e} = -1$

9. $\log_e \frac{1}{e^2} = -2$

10. $\log_e y = 7$

11. $\log_7 \sqrt{7} = \frac{1}{2}$

12. $\log_{11} \sqrt[4]{11} = \frac{1}{4}$

Write each as a logarithmic equation. See Examples 4 through 6.

13. $2^4 = 16$

14. $5^3 = 125$

15. $10^2 = 100$

16. $10^4 = 1000$

17. $e^3 = x$

18. $e^5 = y$

19. $10^{-1} = \frac{1}{10}$

20. $10^{-2} = \frac{1}{100}$

21. $4^{-2} = \frac{1}{16}$

22. $3^{-4} = \frac{1}{81}$

23. $5^{1/2} = \sqrt{5}$

24. $4^{1/3} = \sqrt[3]{4}$

Find the value of each logarithmic expression. See Example 7.

 **25.** $\log_2 8$

26. $\log_3 9$

27. $\log_3 \frac{1}{9}$

28. $\log_2 \frac{1}{32}$

29. $\log_{25} 5$

30. $\log_8 \frac{1}{2}$

31. $\log_{1/2} 2$

32. $\log_{2/3} \frac{4}{9}$

33. $\log_6 1$

34. $\log_9 9$

35. $\log_2 2^4$

36. $\log_6 6^{-2}$

37. $\log_{10} 100$

38. $\log_{10} \frac{1}{10}$

39. $3^{\log_3 5}$

40. $5^{\log_5 7}$

41. $\log_3 81$

42. $\log_2 16$

43. $\log_4 \frac{1}{64}$

44. $\log_3 \frac{1}{9}$

45. Explain why negative numbers are not included as logarithmic bases.

46. Explain why 1 is not included as a logarithmic base.

B *Solve. See Examples 8 through 10.*

47. $\log_3 9 = x$

48. $\log_2 8 = x$

49. $\log_3 x = 4$

50. $\log_2 x = 3$

51. $\log_x 49 = 2$

52. $\log_x 8 = 3$

53. $\log_2 \frac{1}{8} = x$

54. $\log_3 \frac{1}{81} = x$

55. $\log_3 \frac{1}{27} = x$

56. $\log_5 \frac{1}{125} = x$

57. $\log_8 x = \frac{1}{3}$

58. $\log_9 x = \frac{1}{2}$

59. $\log_4 16 = x$

60. $\log_2 16 = x$

61. $\log_{3/4} x = 3$

62. $\log_{2/3} x = 2$

63. $\log_x 100 = 2$

64. $\log_x 27 = 3$

Simplify. See Example 11.

65. $\log_5 5^3$

66. $\log_6 6^2$

67. $2^{\log_2 3}$

68. $7^{\log_7 4}$

69. $\log_9 9$

70. $\log_8 (8)^{-1}$

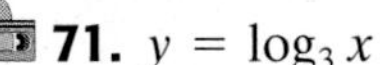

C *Graph each logarithmic function. See Examples 12 and 13.*

71. $y = \log_3 x$

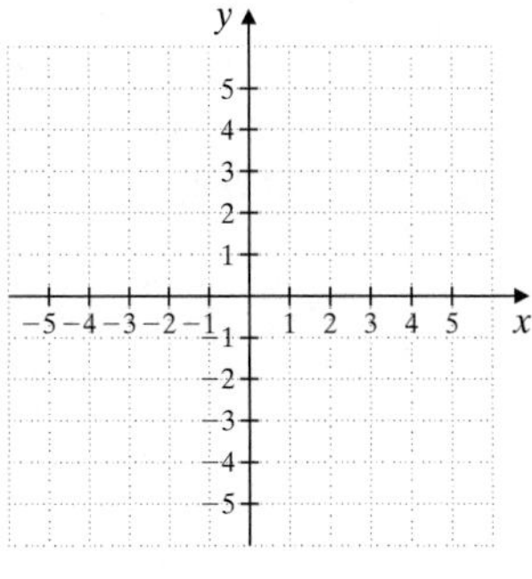

72. $y = \log_2 x$

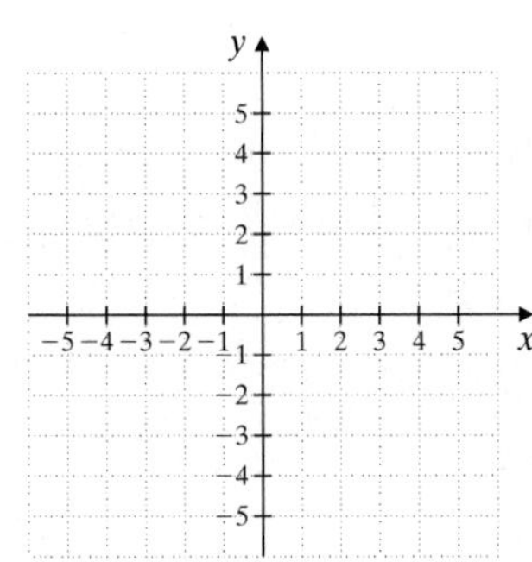

73. $f(x) = \log_{1/4} x$

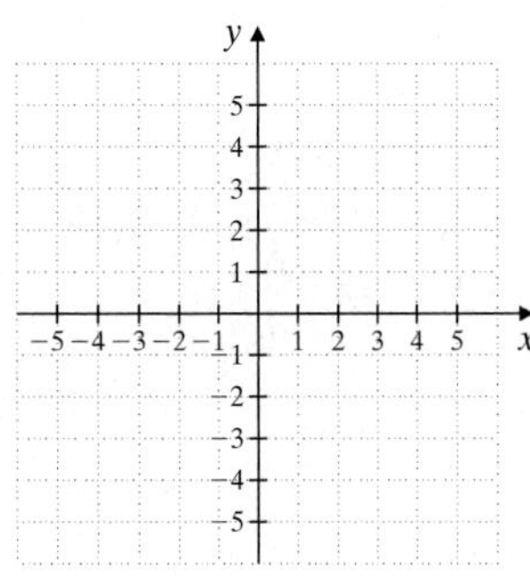

74. $f(x) = \log_{1/2} x$

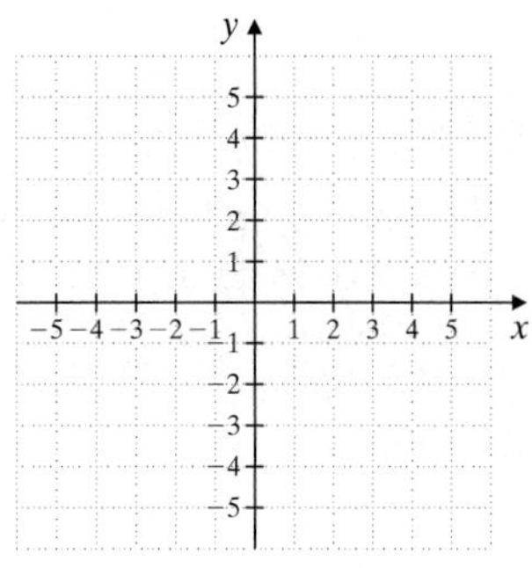

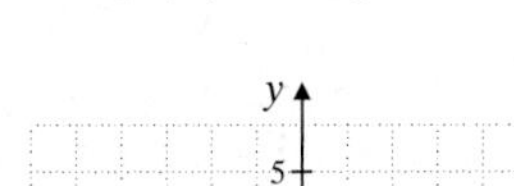

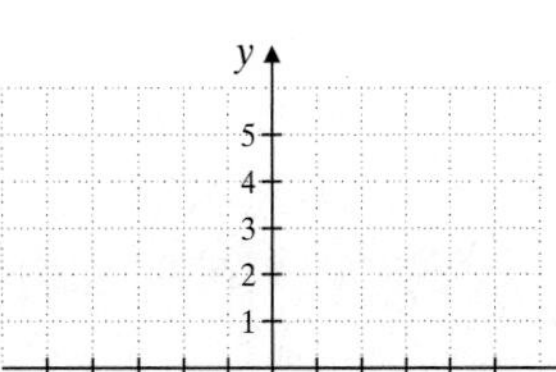

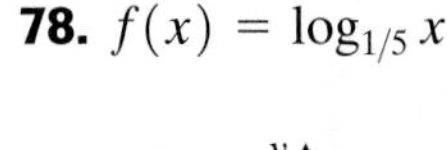

75. $f(x) = \log_5 x$

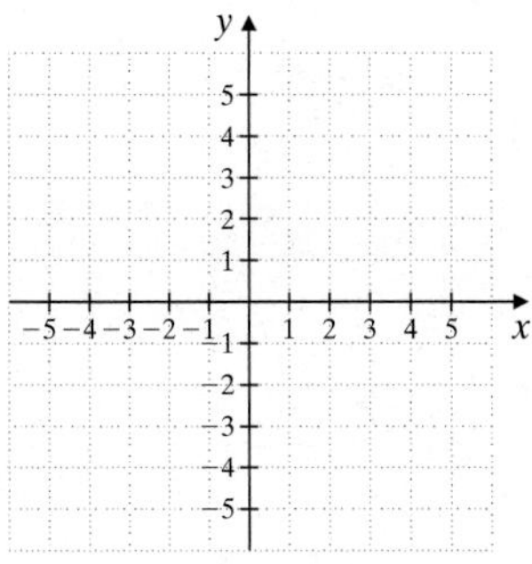

76. $f(x) = \log_6 x$

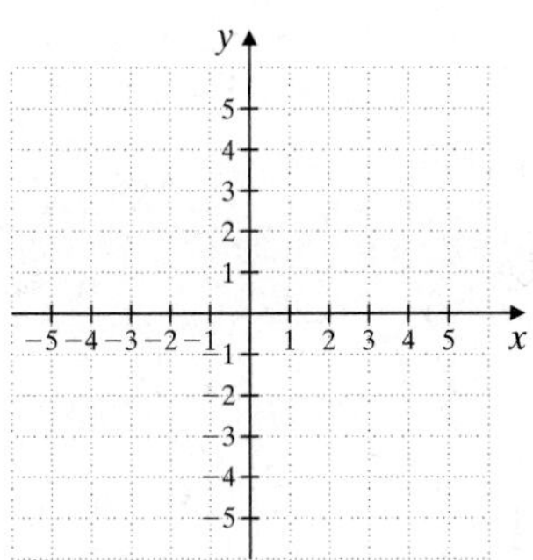

77. $f(x) = \log_{1/6} x$

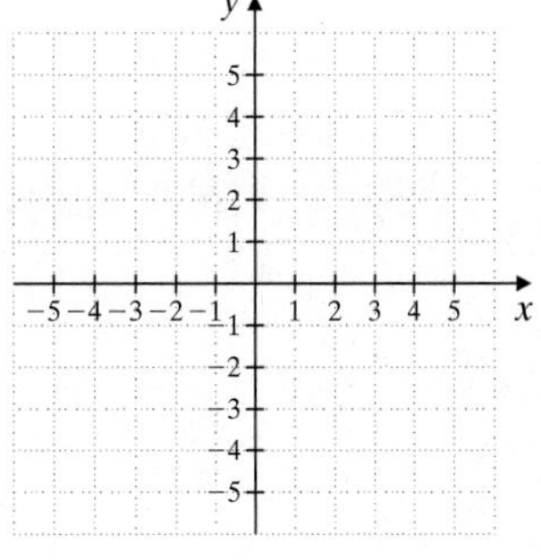

78. $f(x) = \log_{1/5} x$

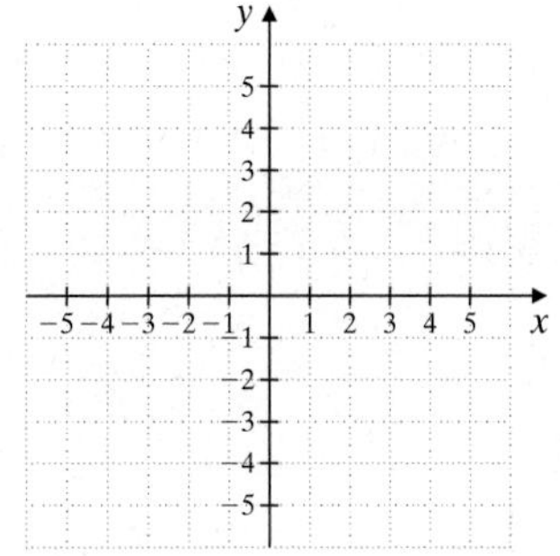

Review and Preview

Simplify each rational expression. See Section 6.1.

79. $\dfrac{x + 3}{3 + x}$

80. $\dfrac{x - 5}{5 - x}$

81. $\dfrac{x^2 - 8x + 16}{2x - 8}$

82. $\dfrac{x^2 - 3x - 10}{2 + x}$

Combining Concepts

Graph each function and its inverse on the same set of axes.

83. $y = 4^x$; $y = \log_4 x$

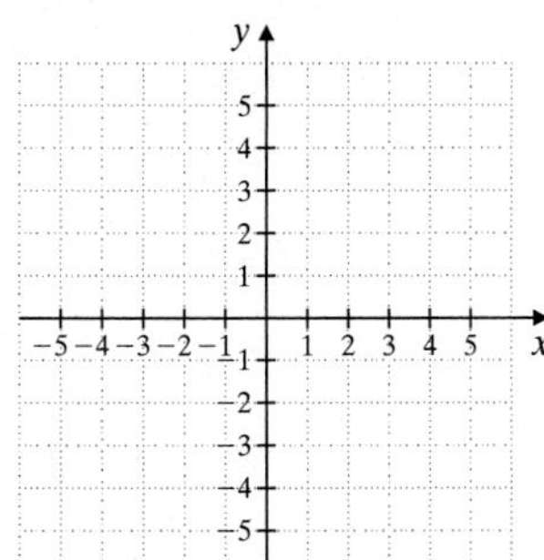

84. $y = 3^x$; $y = \log_3 x$

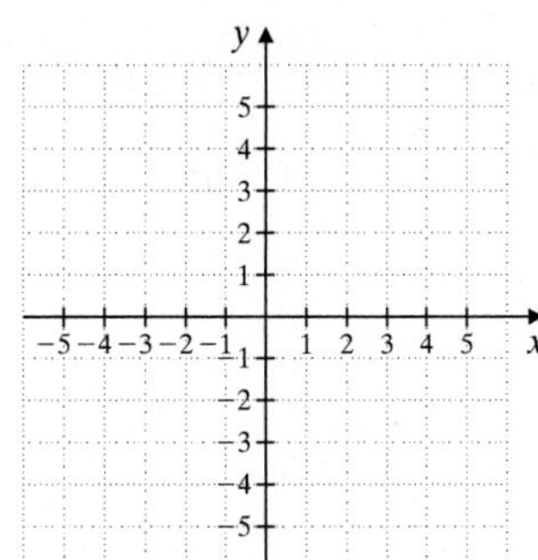

85. $y = \left(\dfrac{1}{3}\right)^x$; $y = \log_{1/3} x$

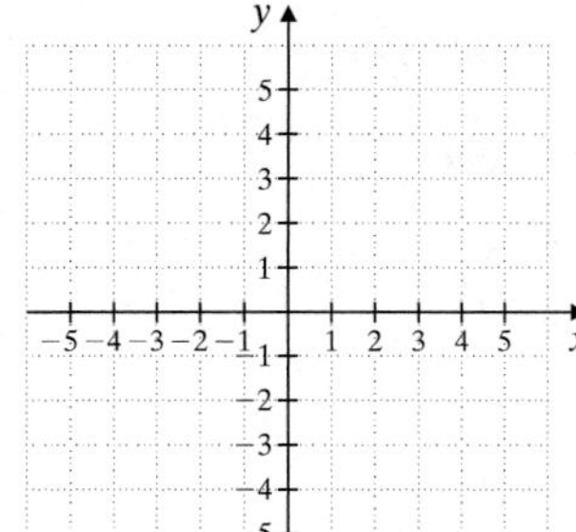

86. $y = \left(\dfrac{1}{2}\right)^x$; $y = \log_{1/2} x$

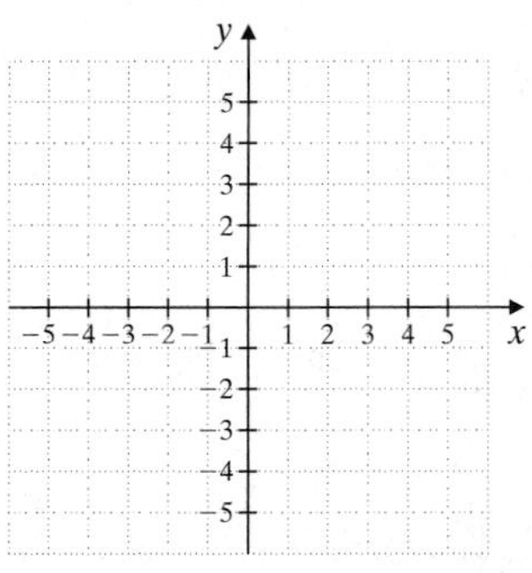

87. The formula $\log_{10}(1 - k) = \dfrac{-0.3}{H}$ models the relationship between the half-life H of a radioactive material and its rate of decay k. Find the rate of decay of the iodine isotope I-131 if its half-life is 8 days. Round to four decimal places.

88. Explain why the graph of the function $y = \log_b x$ contains the point $(1, 0)$ no matter what b is.

89. $\text{Log}_3 10$ is between which two integers? Explain your answer.

11.5 Properties of Logarithms

OBJECTIVES

- A Use the product property of logarithms.
- B Use the quotient property of logarithms.
- C Use the power property of logarithms.
- D Use the properties of logarithms together.

SSM TUTOR CENTER SG CD & VIDEO MATH PRO WEB

In the previous section we explored some basic properties of logarithms. We now introduce and study additional properties. Because a logarithm is an exponent, logarithmic properties are just restatements of exponential properties.

 Using the Product Property

The first of these properties is called the **product property of logarithms** because it deals with the logarithm of a product.

Product Property of Logarithms

If x, y, and b are positive real numbers and $b \neq 1$, then

$$\log_b xy = \log_b x + \log_b y$$

To prove this, we let $\log_b x = M$ and $\log_b y = N$. Now we write each logarithm with exponential notation.

$\log_b x = M$ is equivalent to $b^M = x$

$\log_b y = N$ is equivalent to $b^N = y$

When we multiply the left sides and the right sides of the exponential equations, we have that

$$xy = (b^M)(b^N) = b^{M+N}$$

If we write the equation $xy = b^{M+N}$ in equivalent logarithmic form, we have

$$\log_b xy = M + N$$

But since $M = \log_b x$ and $N = \log_b y$, we can write

$\log_b xy = \log_b x + \log_b y$ Let $M = \log_b x$ and $N = \log_b y$.

In other words, the logarithm of a product is the sum of the logarithms of the factors. This property is sometimes used to simplify logarithmic expressions.

EXAMPLE 1 Write as a single logarithm: $\log_{11} 10 + \log_{11} 3$

Solution: $\log_{11} 10 + \log_{11} 3 = \log_{11}(10 \cdot 3)$ Use the product property.

$= \log_{11} 30$ ●

Practice Problem 1

Write as a single logarithm: $\log_2 7 + \log_2 5$

EXAMPLE 2 Write as a single logarithm: $\log_2(x + 2) + \log_2 x$

Solution:

$\log_2(x + 2) + \log_2 x = \log_2[(x + 2) \cdot x] = \log_2(x^2 + 2x)$ ●

Practice Problem 2

Write as a single logarithm: $\log_3 x + \log_3(x - 9)$

 Using the Quotient Property

The second property is the **quotient property of logarithms**.

Answers

1. $\log_2 35$, **2.** $\log_3(x^2 - 9x)$

Concept Check

Which of the following is the correct way to rewrite $\log_5 \frac{7}{2}$?

a. $\log_5 7 - \log_5 2$
b. $\log_5(7 - 2)$
c. $\dfrac{\log_5 7}{\log_5 2}$
d. $\log_5 14$

Practice Problem 3

Write as a single logarithm: $\log_7 40 - \log_7 8$

Practice Problem 4

Write as a single logarithm: $\log_3(x^3 + 4) - \log_3(x^2 + 2)$

Practice Problems 5–6

Use the power property to rewrite each expression.

5. $\log_3 x^5$ 6. $\log_7 \sqrt[3]{4}$

Practice Problems 7–8

Write as a single logarithm.

7. $3\log_4 2 + 2\log_4 5$
8. $5\log_2(2x - 1) - \log_2 x$

Answers

3. $\log_7 5$, **4.** $\log_3 \frac{x^3 + 4}{x^2 + 2}$, **5.** $5\log_3 x$, **6.** $\frac{1}{3}\log_7 4$, **7.** $\log_4 200$, **8.** $\log_2 \frac{(2x - 1)^5}{x}$

Concept Check: a

Quotient Property of Logarithms

If x, y, and b are positive real numbers and $b \neq 1$, then

$$\log_b \frac{x}{y} = \log_b x - \log_b y$$

The proof of the quotient property of logarithms is similar to the proof of the product property. Notice that the quotient property says that the logarithm of a quotient is the difference of the logarithms of the dividend and divisor.

Try the Concept Check in the margin.

EXAMPLE 3 Write as a single logarithm: $\log_{10} 27 - \log_{10} 3$

Solution: $\log_{10} 27 - \log_{10} 3 = \log_{10} \frac{27}{3}$ Use the quotient property.

$= \log_{10} 9$

EXAMPLE 4 Write as a single logarithm: $\log_3(x^2 + 5) - \log_3(x^2 + 1)$

Solution:

$$\log_3(x^2 + 5) - \log_3(x^2 + 1) = \log_3 \frac{x^2 + 5}{x^2 + 1}$$ Use the quotient property.

C Using the Power Property

The third and final property we introduce is the **power property of logarithms**.

Power Property of Logarithms

If x and b are positive real numbers, $b \neq 1$, and r is a real number, then

$$\log_b x^r = r\log_b x$$

EXAMPLES Use the power property to rewrite each expression.

5. $\log_5 x^3 = 3\log_5 x$

6. $\log_4 \sqrt{2} = \log_4 2^{1/2} = \frac{1}{2}\log_4 2$

D Using More Than One Property

Many times we must use more than one property of logarithms to simplify logarithmic expression.

EXAMPLES Write as a single logarithm.

7. $2\log_5 3 + 3\log_5 2 = \log_5 3^2 + \log_5 2^3$ Use the power property.

$= \log_5 9 + \log_5 8$

$= \log_5(9 \cdot 8)$ Use the product property.

$= \log_5 72$

8. $3\log_9 x - \log_9(x+1) = \log_9 x^3 - \log_9(x+1)$ Use the power property.

$= \log_9 \dfrac{x^3}{x+1}$ Use the quotient property.

EXAMPLES Write each expression as sums or differences of logarithms.

9. $\log_3 \dfrac{5\cdot 7}{4} = \log_3(5\cdot 7) - \log_3 4$ Use the quotient property.

$= \log_3 5 + \log_3 7 - \log_3 4$ Use the product property.

10. $\log_2 \dfrac{x^5}{y^2} = \log_2(x^5) - \log_2(y^2)$ Use the quotient property.

$= 5\log_2 x - 2\log_2 y$ Use the power property.

Helpful Hint

Notice that we are not able to simplify further a logarithmic expression such as $\log_5(2x-1)$. None of the basic properties gives a way to write the logarithm of a difference (or sum) in some equivalent form.

Try the Concept Check in the margin.

EXAMPLES

If $\log_b 2 = 0.43$ and $\log_b 3 = 0.68$, use the properties of logarithms to evaluate each expression.

11. $\log_b 6 = \log_b(2\cdot 3)$ Write 6 as $2\cdot 3$.

$= \log_b 2 + \log_b 3$ Use the product property.

$= 0.43 + 0.68$ Substitute given values.

$= 1.11$ Simplify.

12. $\log_b 9 = \log_b 3^2$ Write 9 as 3^2.

$= 2\log_b 3$ Use the power property.

$= 2(0.68)$ Substitute the given value.

$= 1.36$ Simplify.

13. $\log_b \sqrt{2} = \log_b 2^{1/2}$ Write $\sqrt{2}$ as $2^{1/2}$.

$= \dfrac{1}{2}\log_b 2$ Use the power property.

$= \dfrac{1}{2}(0.43)$ Substitute the given value.

$= 0.215$ Simplify.

Practice Problems 9–10

Write each expression as sums or differences of logarithms.

9. $\log_7 \dfrac{6\cdot 2}{5}$

10. $\log_3 \dfrac{x^4}{y^3}$

Concept Check

What is wrong with the following?

$\log_{10}(x^2+5) = \log_{10} x^2 + \log_{10} 5$
$= 2\log_{10} x + \log_{10} 5$

Use a numerical example to demonstrate that the result is incorrect.

Practice Problems 11–13

If $\log_b 4 = 0.86$ and $\log_b 7 = 1.21$, use the properties of logarithms to evaluate each expression.

11. $\log_b 28$
12. $\log_b 49$
13. $\log_b \sqrt[3]{4}$

Answers

9. $\log_7 6 + \log_7 2 - \log_7 5$, **10.** $4\log_3 x - 3\log_3 y$, **11.** 2.07, **12.** 2.42, **13.** 0.286

Concept Check: The properties do not give any way to simplify the logarithm of a sum; answers may vary.

STUDY SKILLS REMINDER

Are you preparing for a test on Chapter 11?

Below I have listed some common trouble areas for students in Chapter 11. After studying for your test—but before taking your test—read these.

- Don't forget how to find the composition of two functions.

  If $f(x) = x^2 + 5$ and $g(x) = 3x$, then
  $(f \circ g)(x) = f[g(x)] = f(3x) = (3x)^2 + 5 = 9x^2 + 5$
  $(g \circ f)(x) = g[f(x)] = g(x^2 + 5) = 3(x^2 + 5) = 3x^2 + 15$

- Don't forget that f^{-1} is a special notation used to denote the inverse of a function.

Let's find the inverse of the invertible function $f(x) = 3x - 5$.

$f(x) = 3x - 5$

$y = 3x - 5$ Replace $f(x)$ with y.

$x = 3y - 5$ Switch variables.

$x + 5 = 3y$

$\dfrac{x+5}{3} = y$ Solve for y.

$f^{-1}(x) = \dfrac{x+5}{3}$ Replace y with $f^{-1}(x)$.

- Don't forget that $y = \log_b x$ means $b^y = x$.

  Thus, $3 = \log_5 125$ means $5^3 = 125$.

- Remember rules for logarithms.

  $\log_b 3x = \log_b 3 + \log_b x$
  $\log_b(3 + x)$ cannot be simplified in the same manner.

Remember: This is simply a checklist of common trouble areas. For a review of Chapter 11, see the Highlights and Chapter Review at the end of this chapter.

Name ______________________ Section __________ Date __________

EXERCISE SET 11.5

A *Write each sum as a single logarithm. Assume that variables represent positive numbers. See Examples 1 and 2.*

1. $\log_5 2 + \log_5 7$

2. $\log_3 8 + \log_3 4$

3. $\log_4 9 + \log_4 x$

4. $\log_2 x + \log_2 y$

5. $\log_{10} 5 + \log_{10} 2 + \log_{10}(x^2 + 2)$

6. $\log_6 3 + \log_6(x + 4) + \log_6 5$

B *Write each difference as a single logarithm. Assume that variables represent positive numbers. See Examples 3 and 4.*

7. $\log_5 12 - \log_5 4$

8. $\log_7 20 - \log_7 4$

9. $\log_2 x - \log_2 y$

10. $\log_3 12 - \log_3 z$

11. $\log_3 8 - \log_3 2$

12. $\log_5 12 - \log_5 3$

C *Use the power property to rewrite each expression. See Examples 5 and 6.*

13. $\log_3 x^2$

14. $\log_2 x^5$

15. $\log_4 5^{-1}$

16. $\log_6 7^{-2}$

17. $\log_5 \sqrt{y}$

18. $\log_5 \sqrt[3]{x}$

D *Write each as a single logarithm. Assume that variables represent positive numbers. See Examples 7 and 8.*

19. $2 \log_2 5$

20. $3 \log_5 2$

21. $3 \log_5 x + 6 \log_5 z$

22. $2 \log_7 y + 6 \log_7 z$

23. $\log_{10} x - \log_{10}(x + 1) + \log_{10}(x^2 - 2)$

24. $\log_9(4x) - \log_9(x - 3) + \log_9(x^3 + 1)$

25. $\log_4 2 + \log_4 10 - \log_4 5$

26. $\log_6 18 + \log_6 2 - \log_6 9$

27. $\log_7 6 + \log_7 3 - \log_7 4$

28. $\log_8 5 + \log_8 15 - \log_8 20$

29. $3 \log_4 2 + \log_4 6$

30. $2 \log_3 5 + \log_3 2$

31. $3 \log_2 x + \frac{1}{2}\log_2 x - 2 \log_2(x + 1)$

32. $2 \log_5 x + \frac{1}{3}\log_5 x - 3 \log_5(x + 5)$

33. $2 \log_8 x - \frac{2}{3}\log_8 x + 4 \log_8 x$

34. $5 \log_6 x - \frac{3}{4}\log_6 x + 3 \log_6 x$

Write each expression as a sum or difference of logarithms. Assume that variables represent positive numbers. See Examples 9 and 10.

35. $\log_3 \frac{4y}{5}$

36. $\log_4 \frac{2}{9z}$

37. $\log_2 \frac{x^3}{y}$

38. $\log_5 \frac{x}{y^4}$

39. $\log_b \sqrt{7x}$

40. $\log_b \sqrt{\frac{3}{y}}$

41. $\log_7 \frac{5x}{4}$

42. $\log_9 \frac{7}{y}$

43. $\log_5 x^3(x + 1)$

44. $\log_2 y^3 z$

45. $\log_6 \frac{x^2}{x + 3}$

46. $\log_3 \frac{(x + 5)^2}{x}$

If $\log_b 3 = 0.5$ *and* $\log_b 5 = 0.7$, *evaluate each expression. See Examples 11 through 13.*

47. $\log_b \frac{5}{3}$ **48.** $\log_b 25$ **49.** $\log_b 15$ **50.** $\log_b \frac{3}{5}$ **51.** $\log_b \sqrt{5}$ **52.** $\log_b \sqrt[4]{3}$

If $\log_b 2 = 0.43$ *and* $\log_b 3 = 0.68$, *evaluate each expression. See Examples 11 through 13.*

53. $\log_b 8$ **54.** $\log_b 81$ **55.** $\log_b \frac{3}{9}$ **56.** $\log_b \frac{4}{32}$ **57.** $\log_b \sqrt{\frac{2}{3}}$ **58.** $\log_b \sqrt{\frac{3}{2}}$

Review and Preview

59. Graph the functions $y = 10^x$ and $y = \log_{10} x$ on the same set of axes. See Section 11.4.

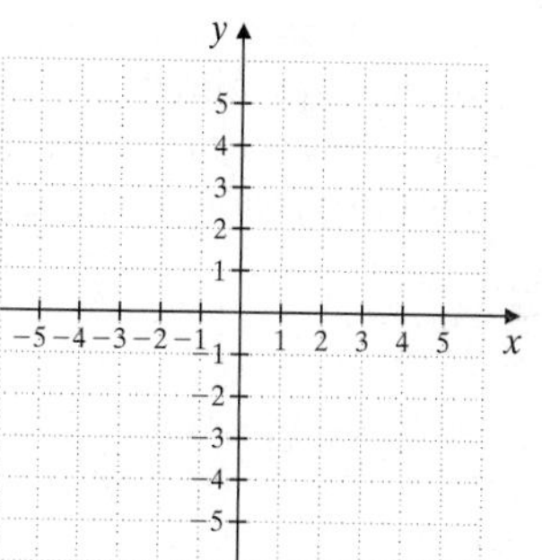

Evaluate each expression. See Section 11.4.

60. $\log_{10} 100$ **61.** $\log_{10} \frac{1}{10}$ **62.** $\log_7 7^2$ **63.** $\log_7 \sqrt{7}$

Combining Concepts

Determine whether each statement is true or false.

64. $\log_2 x^3 = 3 \log_2 x$ **65.** $\log_3(x + y) = \log_3 x + \log_3 y$ **66.** $\frac{\log_7 10}{\log_7 5} = \log_7 2$

67. $\log_7 \frac{14}{8} = \log_7 14 - \log_7 8$ **68.** $\frac{\log_7 x}{\log_7 y} = (\log_7 x) - (\log_7 y)$ **69.** $(\log_3 6) \cdot (\log_3 4) = \log_3 24$

70. It is true that $\log 8 = \log(8 \cdot 1) = \log 8 + \log 1$. Explain how $\log 8$ can equal $\log 8 + \log 1$.

FOCUS ON The Real World

Sound Intensity

The decibel (dB) measures sound intensity, or the relative loudness or strength of a sound. One decibel is the smallest difference in sound levels that is detectable by humans. The decibel is a logarithmic unit. This means that for approximately every 3-decibel increase in sound intensity, the relative loudness of the sound is doubled. For example, a 35 dB sound is twice as loud as a 32 dB sound.

In the modern world, noise pollution has increasingly become a concern. Sustained exposure to high sound intensities can lead to hearing loss. Regular exposure to 90 dB sounds can eventually lead to loss of hearing. Sounds of 130 dB and more can cause permanent loss of hearing instantaneously.

The relative loudness of a sound D in decibels is given by the equation

$$D = 10 \log_{10} \frac{I}{10^{-16}}$$

where I is the intensity of a sound given in watts per square centimeter. Some sound intensities of common noises are listed in the table in order of increasing sound intensity.

Some Sound Intensities of Common Noises

Noise	Intensity (watts/cm²)
Whispering	10^{-15}
Rustling leaves	$10^{-14.2}$
Normal conversation	10^{-13}
Background noise in a quiet residence	$10^{-12.2}$
Typewriter	10^{-11}
Air conditioning	10^{-10}
Freight train at 50 feet	$10^{-8.5}$
Vacuum cleaner	10^{-8}
Nearby thunder	10^{-7}
Air hammer	$10^{-6.5}$
Jet plane at takeoff	10^{-6}
Threshold of pain	10^{-4}

Group Activity

1. Work together to create a table of the relative loudness (in decibels) of the sounds listed in the table.

2. Research the loudness of other common noises. Add these sounds and their decibel levels to your table. Be sure to list the sounds in order of increasing sound intensity.

Name ______________________________

Integrated Review—Functions and Properties of Logarithms

If $f(x) = x - 6$ and $g(x) = x^2 + 1$, find each value.

1. $(f + g)(x)$ **2.** $(f - g)(x)$ **3.** $(f \cdot g)(x)$ **4.** $\left(\frac{f}{g}\right)(x)$

If $f(x) = \sqrt{x}$ and $g(x) = 3x - 1$, find each value.

5. $(f \circ g)(x)$ **6.** $(g \circ f)(x)$

Determine whether each is a one-to-one function. If it is, find its inverse.

7. $f = \{(-2, 6), (4, 8), (2, -6), (3, 3)\}$ **8.** $g = \{(4, 2), (-1, 3), (5, 3), (7, 1)\}$

Determine whether the graph of each function is one-to-one.

9.

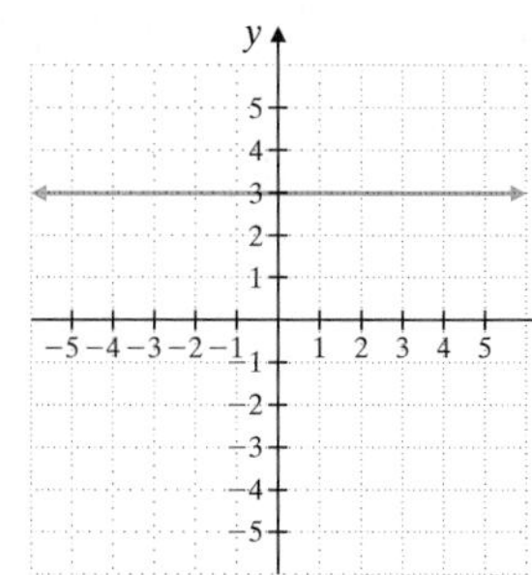

10.

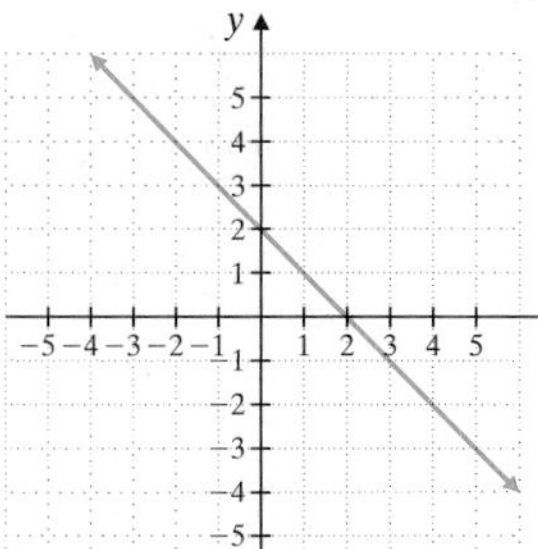

11.

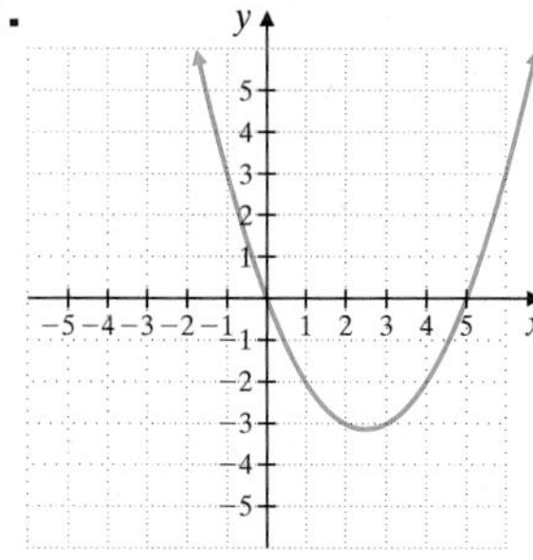

Each function listed is one-to-one. Find the inverse of each function.

12. $f(x) = 3x$ **13.** $f(x) = x + 4$

14. $f(x) = 5x - 1$ **15.** $f(x) = 3x + 2$

Graph each function.

16. $y = \left(\frac{1}{2}\right)^x$ **17.** $y = 2^x + 1$

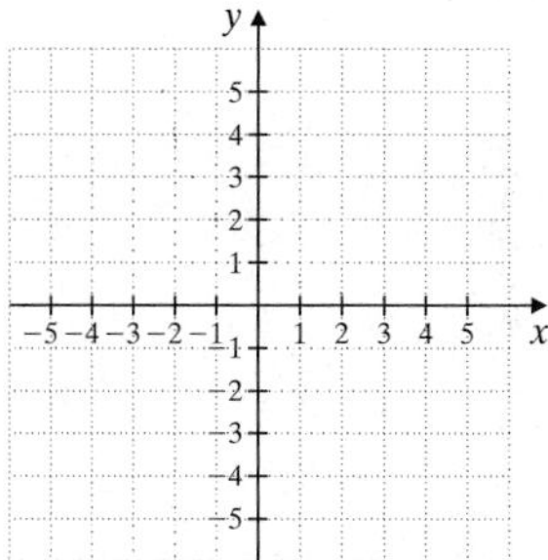

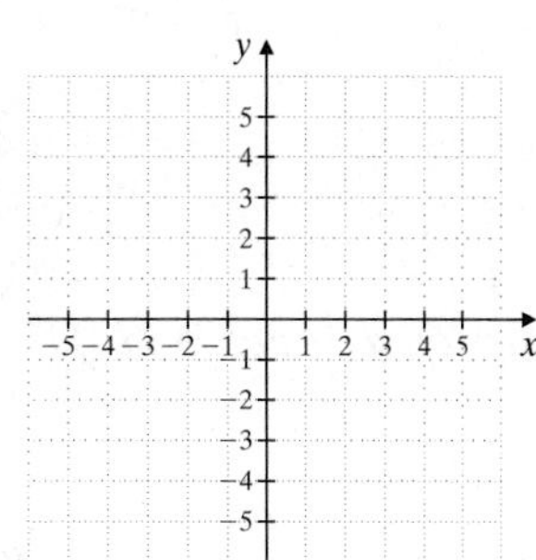

Answers

1. ______
2. ______
3. ______
4. ______
5. ______
6. ______
7. ______
8. ______
9. ______
10. ______
11. ______
12. ______
13. ______
14. ______
15. ______
16. see graph
17. see graph

18. see graph ______

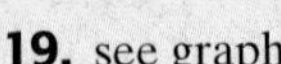

20. ______

21. ______

22. ______

23. ______

24. ______

25. ______

26. ______

27. ______

28. ______

29. ______

30. ______

31. ______

32. ______

33. ______

34. ______

35. ______

36. ______

37. ______

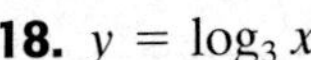

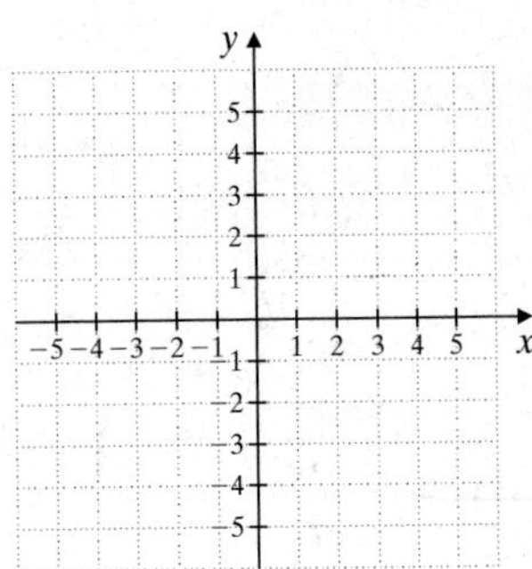

19. $y = \log_{1/3} x$

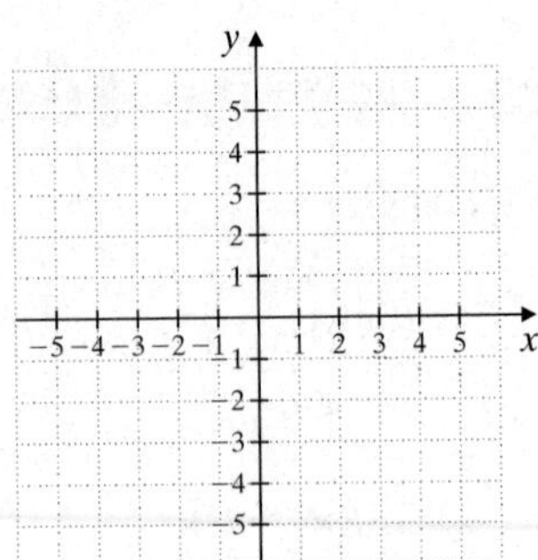

Solve.

20. $2^x = 8$

21. $9 = 3^{x-5}$

22. $4^{x-1} = 8^{x+2}$

23. $25^x = 125^{x-1}$

24. $\log_4 16 = x$

25. $\log_{49} 7 = x$

26. $\log_2 x = 5$

27. $\log_x 64 = 3$

28. $\log_x \frac{1}{125} = -3$

29. $\log_3 x = -2$

Write each as a single logarithm.

30. $5 \log_2 x$

31. $x \log_2 5$

32. $3 \log_5 x - 5 \log_5 y$

33. $9 \log_5 x + 3 \log_5 y$

34. $\log_2 x + \log_2(x - 3) - \log_2(x^2 + 4)$

35. $\log_3 y - \log_3(y + 2) + \log_3(y^3 + 11)$

Write each expression as a sum or difference of logarithms.

36. $\log_7 \frac{9x^2}{y}$

37. $\log_6 \frac{5y}{z^2}$

11.6 Common Logarithms, Natural Logarithms, and Change of Base

In this section we look closely at two particular logarithmic bases. These two logarithmic bases are used so frequently that logarithms to their bases are given special names. **Common logarithms** are logarithms to base 10. **Natural logarithms** are logarithms to base e, which we introduce in this section. The work in this section is based on the use of a calculator that has both the common "log" [LOG] and the natural "log" [LN] keys.

A Approximating Common Logarithms

Logarithms to base 10—**common logarithms**—are used frequently because our number system is a base 10 decimal system. The notation $\log x$ means the same as $\log_{10} x$.

Common Logarithm

$\log x$ means $\log_{10} x$

EXAMPLE 1

Use a calculator to approximate log 7 to four decimal places.

Solution: Press the following sequence of keys:

[7] [LOG] or [LOG] [7] [ENTER]

To four decimal places,

$$\log 7 \approx 0.8451$$

B Evaluating Common Logarithms of Powers of 10

To evaluate the common log of a power of 10, a calculator is not needed. According to the property of logarithms,

$$\log_b b^x = x$$

It follows that if b is replaced with 10, we have

$$\log 10^x = x$$

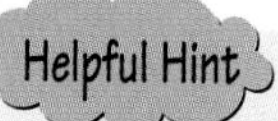

Remember that the base of this logarithm is understood to be 10.

EXAMPLES

Find the exact value of each logarithm.

2. $\log 10 = \log 10^1 = 1$
3. $\log \frac{1}{10} = \log 10^{-1} = -1$
4. $\log 1000 = \log 10^3 = 3$
5. $\log \sqrt{10} = \log 10^{1/2} = \frac{1}{2}$

As we will soon see, equations containing common logs are useful models of many natural phenomena.

OBJECTIVES

- A Identify common logarithms and approximate them with a calculator.
- B Evaluate common logarithms of powers of 10.
- C Identify natural logarithms and approximate them with a calculator.
- D Evaluate natural logarithms of powers of e.
- E Use the change of base formula.

Practice Problem 1

Use a calculator to approximate log 21 to four decimal places.

Practice Problems 2–5

Find the exact value of each logarithm.

2. $\log 100$
3. $\log \frac{1}{100}$
4. $\log 10{,}000$
5. $\log \sqrt[3]{10}$

Answers

1. 1.3222, **2.** 2, **3.** −2, **4.** 4, **5.** $\frac{1}{3}$

Practice Problem 6

Solve: $\log x = 2.9$. Give an exact solution and then approximate the solution to four decimal places.

EXAMPLE 6

Solve: $\log x = 1.2$. Give an exact solution and then approximate the solution to four decimal places.

Solution: Remember that the base of a common log is understood to be 10.

Helpful Hint

The understood base is 10.

$$\log x = 1.2$$

$$10^{1.2} = x \quad \text{Write with exponential notation.}$$

The exact solution is $10^{1.2}$. To four decimal places, $x \approx 15.8489$. ●

C Approximating Natural Logarithms

Natural logarithms are also frequently used, especially to describe natural events; hence the label "natural logarithm." **Natural logarithms** are logarithms to the base e, which is a constant approximately equal to 2.7183. The number e is an irrational number, as is π. The notation $\log_e x$ is usually abbreviated to $\ln x$. (The abbreviation ln is read "el en.")

Natural Logarithm

$\ln x$ means $\log_e x$

Practice Problem 7

Use a calculator to approximate ln 11 to four decimal places.

EXAMPLE 7 Use a calculator to approximate ln 8 to four decimal places.

Solution: Press the following sequence of keys:

[8] [ln] or [ln] [8] [ENTER]

To four decimal places,

$$\ln 8 \approx 2.0794$$ ●

D Evaluating Natural Logarithms of Powers of e

As a result of the property $\log_b b^x = x$, we know that $\log_e e^x = x$, or $\ln e^x = x$.

Practice Problems 8–9

Find the exact value of each natural logarithm.

8. $\ln e^5$ 9. $\ln \sqrt{e}$

EXAMPLES Find the exact value of each natural logarithm.

8. $\ln e^3 = 3$

9. $\ln \sqrt[5]{e} = \ln e^{1/5} = \dfrac{1}{5}$ ●

Practice Problem 10

Solve: $\ln 7x = 10$. Give an exact solution and then approximate the solution to four decimal places.

EXAMPLE 10

Solve: $\ln 3x = 5$. Give an exact solution and then approximate the solution to four decimal places.

Solution: Remember that the base of a natural logarithm is understood to be e.

$$\ln 3x = 5$$

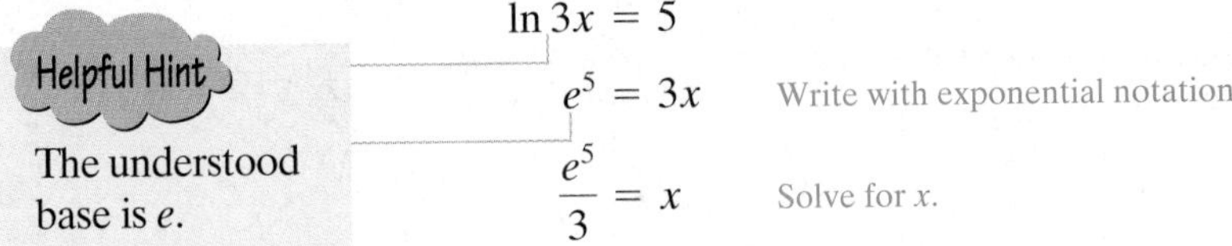

$$e^5 = 3x \quad \text{Write with exponential notation.}$$

$$\frac{e^5}{3} = x \quad \text{Solve for } x.$$

The exact solution is $\dfrac{e^5}{3}$. To four decimal places, $x \approx 49.4711$. ●

Answers

6. $x = 10^{2.9}; x \approx 794.3282$, **7.** 2.3979, **8.** 5, **9.** $\frac{1}{2}$, **10.** $x = \frac{e^{10}}{7}; x \approx 3146.6380$

Recall from Section 11.3 the formula $A = P\left(1 + \frac{r}{n}\right)^{nt}$ for compound interest, where n represents the number of compoundings per year. When interest is compounded continuously, we use the formula $A = Pe^{rt}$, where r is the annual interest rate and interest is compounded continuously for t years.

EXAMPLE 11 Finding the Amount Owed on a Loan

Find the amount owed at the end of 5 years if \$1600 is loaned at a rate of 9% compounded continuously.

Solution: We use the formula $A = Pe^{rt}$ and the following values of the variables.

$P = \$1600$ (the amount of the loan)

$r = 9\% = 0.09$ (the rate of interest)

$t = 5$ (the 5-year duration of the loan)

$$A = Pe^{rt}$$

$$= 1600e^{0.09(5)} \quad \text{Substitute known values.}$$

$$= 1600e^{0.45}$$

Now we can use a calculator to approximate the solution.

$$A \approx 2509.30$$

The total amount of money owed is approximately \$2509.30.

Practice Problem 11

Find the amount owed at the end of 3 years if \$1200 is loaned at a rate of 8% compounded continuously.

E Using the Change of Base Formula

Calculators are handy tools for approximating natural and common logarithms. Unfortunately, some calculators cannot be used to approximate logarithms to bases other than e or 10—at least not directly. In such cases, we use the **change of base formula.**

Change of Base

If a, b, and c are positive real numbers and neither b nor c is 1, then

$$\log_b a = \frac{\log_c a}{\log_c b}$$

EXAMPLE 12 Approximate $\log_5 3$ to four decimal places.

Solution: We use the change of base property to write $\log_5 3$ as a quotient of logarithms to base 10.

$$\log_5 3 = \frac{\log 3}{\log 5} \quad \text{Use the change of base property.}$$

$$\approx \frac{0.4771213}{0.69897} \quad \text{Approximate the logarithms by calculator.}$$

$$\approx 0.6826063 \quad \text{Simplify by calculator.}$$

To four decimal places, $\log_5 3 \approx 0.6826$.

Try the Concept Check in the margin.

Practice Problem 12

Approximate $\log_7 5$ to four decimal places.

Concept Check

If a graphing calculator cannot directly evaluate logarithms to base 5, describe how you could use the graphing calculator to graph the function $f(x) = \log_5 x$.

Answers

11. \$1525.50, **12.** 0.8271

Concept Check: $f(x) = \frac{\log x}{\log 5}$

STUDY SKILLS REMINDER

Are you prepared for your final exam?

To prepare for your final exam, try the following study techniques.

- Review the material that you will be responsible for on your exam. Also check your notebook for any lecture notes that you highlighted.
- Review any formulas that you may need to memorize.
- Check to see if your instructor or math department will be conducting a final exam review.
- Check with your instructor to see whether there are final exams from previous semesters/quarters that are available to students for study.
- Use your previously taken tests as a practice final exam. To do so, rewrite the test questions in mixed order on blank sheets of paper. This will help you prepare for exam conditions.
- If you are unsure of a few topics, see your instructor or visit a learning lab for further assistance. Also, viewing the video segment of a troublesome section will help.
- If you need further exercises to work, try the chapter tests at the end of appropriate chapters.

Good luck! I hope you have enjoyed this textbook and your intermediate algebra course.

Name ____________________ Section __________ Date __________

EXERCISE SET 11.6

  *Use a calculator to approximate each logarithm to four decimal places. See Examples 1 and 7.*

1. $\log 8$ **2.** $\log 6$ **3.** $\log 2.31$ **4.** $\log 4.86$ **5.** $\ln 2$

6. $\ln 3$ **7.** $\ln 0.0716$ **8.** $\ln 0.0032$ **9.** $\log 12.6$ **10.** $\log 25.9$

11. $\ln 5$ **12.** $\ln 7$ **13.** $\log 41.5$ **14.** $\ln 41.5$

15. Use a calculator to try to approximate $\log 0$. Describe what happens and explain why.

16. Use a calculator to try to approximate $\ln 0$. Describe what happens and explain why.

B D *Find the exact value of each logarithm. See Examples 2 through 5, 8, and 9.*

17. $\log 100$ **18.** $\log 10{,}000$ **19.** $\log \frac{1}{1000}$ 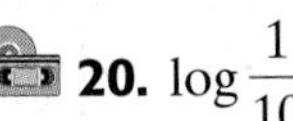 **20.** $\log \frac{1}{10}$ **21.** $\ln e^2$ **22.** $\ln e^4$

23. $\ln \sqrt[4]{e}$ **24.** $\ln \sqrt[5]{e}$ **25.** $\log 10^3$ **26.** $\ln e^5$ **27.** $\ln e^2$ **28.** $\log 10^7$

29. $\log 0.0001$ **30.** $\log 0.001$ **31.** $\ln \sqrt{e}$ **32.** $\log \sqrt{10}$

Solve each equation. Give an exact solution and a four-decimal-place approximation. See Examples 6 and 10.

33. $\log x = 1.3$

34. $\log x = 2.1$

35. $\log 2x = 1.1$

36. $\log 3x = 1.3$

37. $\ln x = 1.4$

38. $\ln x = 2.1$

39. $\ln (3x - 4) = 2.3$

40. $\ln (2x + 5) = 3.4$

41. $\log x = 2.3$

42. $\log x = 3.1$

43. $\ln x = -2.3$

44. $\ln x = -3.7$

45. $\log (2x + 1) = -0.5$

46. $\log (3x - 2) = -0.8$

47. $\ln 4x = 0.18$

48. $\ln 3x = 0.76$

Use the formula $A = Pe^{rt}$ to solve. See Example 11.

49. How much money does Dana Jones have after 12 years if she invests \$1400 at 8% interest compounded continuously?

50. Determine the size of an account in which \$3500 earns 6% interest compounded continuously for 1 year.

51. How much money does Barbara Mack owe at the end of 4 years if 6% interest is compounded continuously on her \$2000 debt?

52. Find the amount of money for which a \$2500 certificate of deposit is redeemable if it has been paying 10% interest compounded continuously for 3 years.

 *Approximate each logarithm to four decimal places. See Example 12.*

53. $\log_2 3$ **54.** $\log_3 2$ **55.** $\log_{1/2} 5$ **56.** $\log_{1/3} 2$ **57.** $\log_4 9$

58. $\log_9 4$ **59.** $\log_3 \frac{1}{6}$ **60.** $\log_6 \frac{2}{3}$ **61.** $\log_8 6$ **62.** $\log_6 8$

Review and Preview

Solve for x. See Sections 2.4 and 5.6.

63. $6x - 3(2 - 5x) = 6$ **64.** $2x + 3 = 5 - 2(3x - 1)$ **65.** $2x + 3y = 6x$

66. $4x - 8y = 10x$ **67.** $x^2 + 7x = -6$ **68.** $x^2 + 4x = 12$

Combining Concepts

Graph each function by finding ordered pair solutions, plotting the solutions, and then drawing a smooth curve through the plotted points.

69. $f(x) = e^x$ **70.** $f(x) = e^{2x}$ **71.** $f(x) = \ln x$ **72.** $f(x) = \log x$

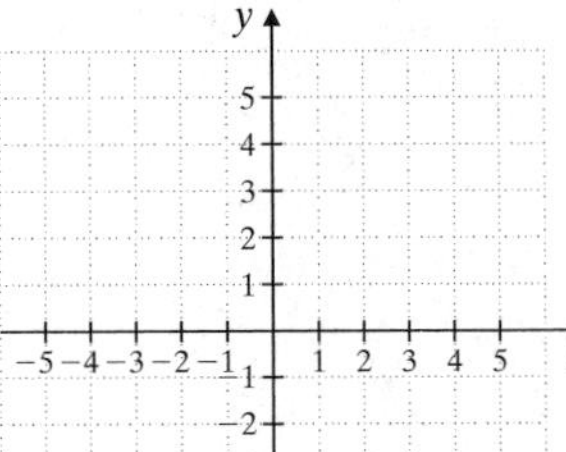

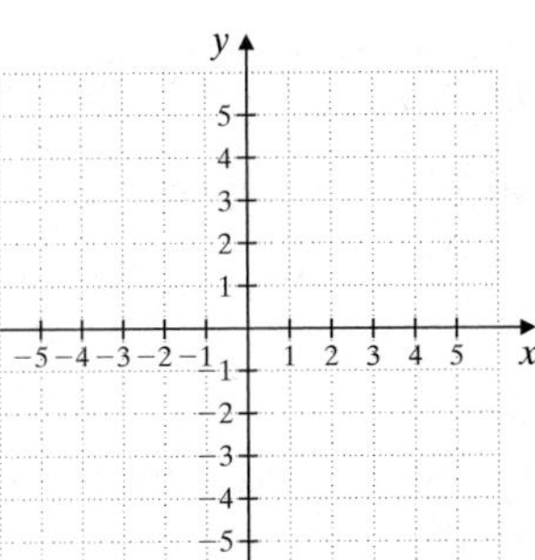

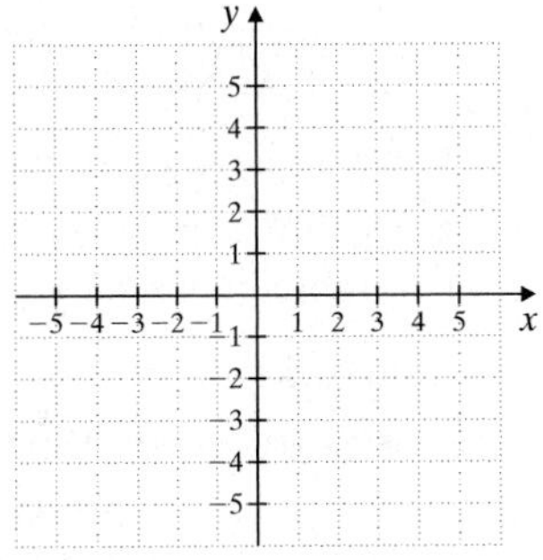

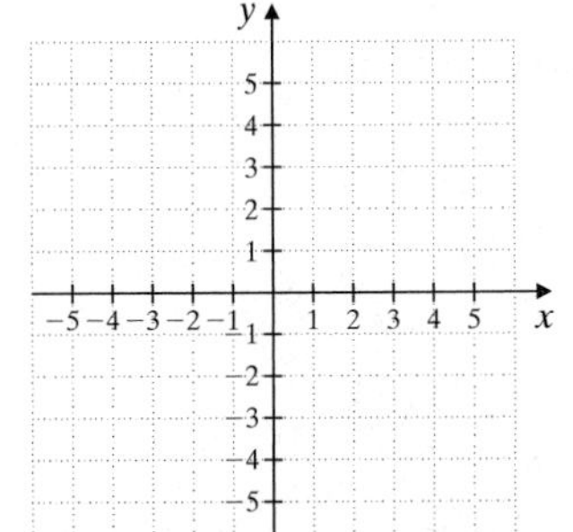

73. Without using a calculator, explain which of log 50 or ln 50 must be larger.

74. Without using a calculator, explain which of $\log 50^{-1}$ or $\ln 50^{-1}$ must be larger.

The Richter scale measures the intensity, or magnitude, of an earthquake. The formula for the magnitude R of an earthquake is $R = \log\left(\frac{a}{T}\right) + B$, *where a is the amplitude in micrometers of the vertical motion of the ground at the recording station, T is the number of seconds between successive seismic waves, and B is an adjustment factor that takes into account the weakening of the seismic wave as the distance increases from the epicenter of the earthquake.*

Use the Richter scale formula to find the magnitude R of the earthquake that fits the description given. Round answers to one decimal place.

75. Amplitude a is 200 micrometers, time T between waves is 1.6 seconds, and B is 2.1.

76. Amplitude a is 150 micrometers, time T between waves is 3.6 seconds, and B is 1.9.

77. Amplitude a is 400 micrometers, time T between waves is 2.6 seconds, and B is 3.1.

78. Amplitude a is 450 micrometers, time T between waves is 4.2 seconds, and B is 2.7.

STUDY SKILLS REMINDER

Do you remember what to do the day of an exam?

On the day of an exam, don't forget to try the following:

- Allow yourself plenty of time to arrive.
- Read the directions on the test carefully.
- Read each problem carefully as you take your test. Make sure that you answer the question asked.
- Watch your time and pace yourself so that you may attempt each problem on your test.
- If you have time, check your work and answers.
- Do not turn your test in early. If you have extra time, spend it double-checking your work.

Good luck!

11.7 Exponential and Logarithmic Equations and Problem Solving

OBJECTIVES

- (A) Solve exponential equations.
- (B) Solve logarithmic equations.
- (C) Solve problems that can be modeled by exponential and logarithmic equations.

SSM TUTOR CENTER SG CD & VIDEO MATH PRO WEB

(A) Solving Exponential Equations

In Section 11.3 we solved exponential equations such as $2^x = 16$ by writing 16 as a power of 2 and using the uniqueness of b^x.

$2^x = 16$

$2^x = 2^4$ Write 16 as 2^4.

$x = 4$ Use the uniqueness of b^x.

To solve an equation such as $3^x = 7$, we use the fact that $f(x) = \log_b x$ is a one-to-one function. Another way of stating this fact is as a property of equality.

Logarithm Property of Equality

Let a, b, and c be real numbers such that $\log_b a$ and $\log_b c$ are real numbers and b is not 1. Then

$\log_b a = \log_b c$ is equivalent to $a = c$

EXAMPLE 1

Solve: $3^x = 7$. Give an exact answer and a four-decimal-place approximation.

Solution: We use the logarithm property of equality and take the logarithm of both sides. For this example, we use the common logarithm.

$3^x = 7$

$\log 3^x = \log 7$ Take the common log of both sides.

$x \log 3 = \log 7$ Use the power property of logarithms.

$x = \dfrac{\log 7}{\log 3}$ Divide both sides by log 3.

The exact solution is $\dfrac{\log 7}{\log 3}$. When we approximate to four decimal places, we have

$$\frac{\log 7}{\log 3} \approx \frac{0.845098}{0.4771213} \approx 1.7712$$

The solution set is $\left\{\dfrac{\log 7}{\log 3}\right\}$, or {approximately 1.7712}.

Practice Problem 1

Solve: $2^x = 5$. Give an exact answer and a four-decimal-place approximation.

(B) Solving Logarithmic Equations

By applying the appropriate properties of logarithms, we can solve a broad variety of logarithmic equations.

EXAMPLE 2 Solve: $\log_4(x - 2) = 2$

Solution: Notice that $x - 2$ must be positive, so x must be greater than 2. With this in mind, we first write the equation with exponential notation.

Practice Problem 2

Solve: $\log_3(x + 5) = 2$

Answers

1. $\left\{\dfrac{\log 5}{\log 2}\right\}$; {2.3219}, **2.** {4}

$$\begin{aligned} \log_4(x-2) &= 2 \\ 4^2 &= x-2 \\ 16 &= x-2 \\ 18 &= x \qquad \text{Add 2 to both sides.} \end{aligned}$$

To check, we replace x with 18 in the original equation.

$$\begin{aligned} \log_4(x-2) &= 2 \\ \log_4(18-2) &\stackrel{?}{=} 2 \qquad \text{Let } x = 18. \\ \log_4 16 &\stackrel{?}{=} 2 \\ 4^2 &= 16 \qquad \text{True.} \end{aligned}$$

The solution set is $\{18\}$. ●

Practice Problem 3

Solve: $\log_6 x + \log_6 (x+1) = 1$

EXAMPLE 3 Solve: $\log_2 x + \log_2(x-1) = 1$

Solution: Notice that $x - 1$ must be positive, so x must be greater than 1. We use the product property on the left side of the equation.

$$\begin{aligned} \log_2 x + \log_2(x-1) &= 1 \\ \log_2[x(x-1)] &= 1 \qquad \text{Use the product property.} \\ \log_2(x^2 - x) &= 1 \end{aligned}$$

Next we write the equation with exponential notation and solve for x.

$$\begin{aligned} 2^1 &= x^2 - x \\ 0 &= x^2 - x - 2 \qquad \text{Subtract 2 from both sides.} \\ 0 &= (x-2)(x+1) \qquad \text{Factor.} \\ 0 &= x - 2 \quad \text{or} \quad 0 = x + 1 \qquad \text{Set each factor equal to 0.} \\ 2 &= x \qquad\qquad -1 = x \end{aligned}$$

Recall that -1 cannot be a solution because x must be greater than 1. If we forgot this, we would still reject -1 after checking. To see this, we replace x with -1 in the original equation.

$$\begin{aligned} \log_2 x + \log_2(x-1) &= 1 \\ \log_2(-1) + \log_2(-1-1) &\stackrel{?}{=} 1 \qquad \text{Let } x = -1. \end{aligned}$$

Because the logarithm of a negative number is undefined, -1 is rejected. Check to see that the solution set is $\{2\}$. ●

Practice Problem 4

Solve: $\log (x-1) - \log x = 1$

EXAMPLE 4 Solve: $\log(x+2) - \log x = 2$

Solution: We use the quotient property of logarithms on the left side of the equation.

$$\begin{aligned} \log(x+2) - \log x &= 2 \\ \log \frac{x+2}{x} &= 2 \qquad \text{Use the quotient property.} \\ 10^2 &= \frac{x+2}{x} \qquad \text{Write using exponential notation.} \\ 100 &= \frac{x+2}{x} \\ 100x &= x + 2 \qquad \text{Multiply both sides by } x. \\ 99x &= 2 \qquad \text{Subtract } x \text{ from both sides.} \\ x &= \frac{2}{99} \qquad \text{Divide both sides by 99.} \end{aligned}$$

Check to see that the solution set is $\left\{\frac{2}{99}\right\}$. ●

Answers

3. $\{2\}$, **4.** $\varnothing$

C Solving Problems Modeled by Exponential and Logarithmic Equations

Logarithmic and exponential functions are used in a variety of scientific, technical, and business settings. A few examples follow.

EXAMPLE 5 Estimating Population Size

The population size y of a community of lemmings varies according to the relationship $y = y_0e^{0.15t}$. In this formula, t is time in months, and y_0 is the initial population at time 0. Estimate the population after 6 months if there were originally 5000 lemmings.

Solution: We substitute 5000 for y_0 and 6 for t.

$$\begin{aligned} y &= y_0e^{0.15t} \\ &= 5000e^{0.15(6)} && \text{Let } t = 6 \text{ and } y_0 = 5000. \\ &= 5000e^{0.9} && \text{Multiply.} \end{aligned}$$

Using a calculator, we find that $y \approx 12{,}298.016$. In 6 months the population will be approximately 12,300 lemmings. ●

Practice Problem 5

Use the equation in Example 5 to estimate the lemming population in 8 months.

EXAMPLE 6 Doubling an Investment

How long does it take an investment of \$2000 to double if it is invested at 5% interest compounded quarterly? The necessary formula is $A = P\left(1 + \frac{r}{n}\right)^{nt}$, where A is the accrued amount, P is the principal invested, r is the annual rate of interest, n is the number of compounding periods per year, and t is the number of years.

Solution: We are given that $P = \$2000$ and $r = 5\% = 0.05$. Compounding quarterly means 4 times a year, so $n = 4$. The investment is to double, so A must be \$4000. We substitute these values and solve for t.

Practice Problem 6

How long does it take an investment of \$1000 to double if it is invested at 6% interest compounded quarterly?

Answers

5. approximately 16,600 lemmings, **6.** approximately 11.64 yr

$$A = P\left(1 + \frac{r}{n}\right)^{nt}$$

$$4000 = 2000\left(1 + \frac{0.05}{4}\right)^{4t} \quad \text{Substitute known values.}$$

$$4000 = 2000(1.0125)^{4t} \quad \text{Simplify } 1 + \frac{0.05}{4}.$$

$$2 = (1.0125)^{4t} \quad \text{Divide both sides by 2000.}$$

$$\log 2 = \log 1.0125^{4t} \quad \text{Take the logarithm of both sides.}$$

$$\log 2 = 4t(\log 1.0125) \quad \text{Use the power property.}$$

$$\frac{\log 2}{4 \log 1.0125} = t \quad \text{Divide both sides by } 4 \log 1.0125.$$

$$13.949408 \approx t \quad \text{Approximate by calculator.}$$

It takes approximately 14 years for the money to double in value.

GRAPHING CALCULATOR EXPLORATIONS

Use a grapher to find how long it takes an investment of \$1500 to triple if it is invested at 8% interest compounded monthly. First, let $P = \$1500$, $r = 0.08$, and $n = 12$ (for 12 months) in the formula

$$A = P\left(1 + \frac{r}{n}\right)^{nt}$$

Notice that when the investment has tripled, the accrued amount A is \$4500. Thus,

$$4500 = 1500\left(1 + \frac{0.08}{12}\right)^{12t}$$

Determine an appropriate viewing window and enter and graph the equations

$$Y_1 = 1500\left(1 + \frac{0.08}{12}\right)^{12x}$$

and

$$Y_2 = 4500$$

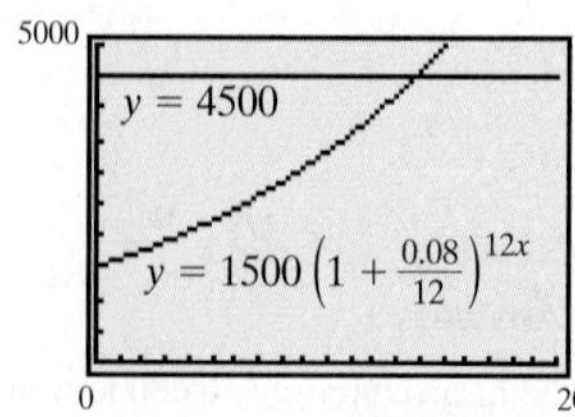

The point of intersection of the two curves is the solution. The x-coordinate tells how long it takes for the investment to triple.

Use a TRACE feature or an INTERSECT feature to approximate the coordinates of the point of intersection of the two curves. It takes approximately 13.78 years, or 13 years and 10 months, for the investment to triple in value to \$4500.

Use this graphical solution method to solve each problem. Round each answer to the nearest hundredth.

1. Find how long it takes an investment of \$5000 to grow to \$6000 if it is invested at 5% interest compounded quarterly.
2. Find how long it takes an investment of \$1000 to double if it is invested at 4.5% interest compounded daily. (Use 365 days in a year.)
3. Find how long it takes an investment of \$10,000 to quadruple if it is invested at 6% interest compounded monthly.
4. Find how long it takes \$500 to grow to \$800 if it is invested at 4% interest compounded semiannually.

Name ______________________ Section __________ Date __________

EXERCISE SET 11.7

 *Solve each equation. Give an exact solution and a four-decimal-place approximation. See Example 1.*

 **1.** $3^x = 6$

2. $4^x = 7$

3. $3^{2x} = 3.8$

4. $5^{3x} = 5.6$

 **5.** $2^{x-3} = 5$

6. $8^{x-2} = 12$

7. $9^x = 5$

8. $3^x = 11$

9. $4^{x+7} = 3$

10. $6^{x+3} = 2$

11. $7^{3x-4} = 11$

12. $5^{2x-6} = 12$

13. $e^{6x} = 5$

14. $e^{2x} = 8$

 *Solve each equation. See Examples 2 through 4.*

 **15.** $\log_2(x + 5) = 4$

16. $\log_6(x^2 - x) = 1$

17. $\log_4 2 + \log_4 x = 0$

18. $\log_3 5 + \log_3 x = 1$

19. $\log_2 6 - \log_2 x = 3$

20. $\log_4 10 - \log_4 x = 2$

21. $\log_4 x + \log_4(x + 6) = 2$

22. $\log_3 x + \log_3(x + 6) = 3$

23. $\log_5(x + 3) - \log_5 x = 2$

24. $\log_6(x + 2) - \log_6 x = 2$

25. $\log_3(x - 2) = 2$

26. $\log_2(x - 5) = 3$

27. $\log_4(x^2 - 3x) = 1$

28. $\log_8(x^2 - 2x) = 1$

29. $\log_2 x + \log_2(3x + 1) = 1$

30. $\log_3 x + \log_3(x - 8) = 2$

C *Solve. See Example 5.*

31. The size of the wolf population at Isle Royale National Park increases at a rate of 4.3% per year. If the size of the current population is 83 wolves, find how many there should be in 5 years. Use $y = y_0e^{0.043t}$ and round to the nearest whole number.

32. The number of victims of a flu epidemic is increasing at a rate of 7.5% per week. If 20,000 people are currently infected, in how many days can we expect 45,000 people to have the flu? Use $y = y_0e^{0.075t}$ and round to the nearest whole number.

33. The size of the population of Paraguay is increasing at a rate of 2.7% per year. The population of Paraguay in 2001 was approximately 5,700,000. Use $y = y_0e^{0.027t}$ to estimate the population of Paraguay in 2008. Round to the nearest whole number. (*Source:* Population Reference Bureau)

34. In 2001, 171.8 million people lived in Brazil. The population of Brazil is growing at a rate of 1.5% per year. Find how long it will take the Brazilian population to reach a size of 200 million people. Use $y = y_0e^{0.015t}$ and round to the nearest tenth. (*Source:* Population Reference Bureau)

35. In 2001, Hungary had a population of about 10,000,000. At that time, Hungary's population was declining at a rate of 0.4% per year. How long will it take for Hungary's population to decline to 9,000,000? Use $y = y_0e^{-0.004t}$ and round to the nearest tenth. (*Source:* Population Reference Bureau)

36. The population of Russia has been decreasing at the rate of 0.7% per year. There were about 144,400,000 people living in Russia in 2001. How many inhabitants will there be in 2016? Use $y = y_0e^{-0.007t}$. Round to the nearest whole number. (*Source:* Population Reference Bureau)

Use the formula $A = P\left(1 + \frac{r}{n}\right)^{nt}$ *to solve these compound interest problems. Round to the nearest tenth. See Example 6.*

37. How long does it take for $600 to double if it is invested at 7% interest compounded monthly?

38. How long does it take for $600 to double if it is invested at 12% interest compounded monthly?

39. How long does it take for a $1200 investment to earn $200 interest if it is invested at 9% interest compounded quarterly?

40. How long does it take for a $1500 investment to earn $200 interest if it is invested at 10% interest compounded semiannually?

41. How long does it take for $1000 to double if it is invested at 8% interest compounded semiannually?

42. How long does it take for $1000 to double if it is invested at 8% interest compounded monthly?

The formula $w = 0.00185h^{2.67}$ is used to estimate the normal weight w in pounds of a boy h inches tall. Use this formula to solve Exercises 43 and 44. Round to the nearest tenth.

43. Find the expected height of a boy who weighs 85 pounds.

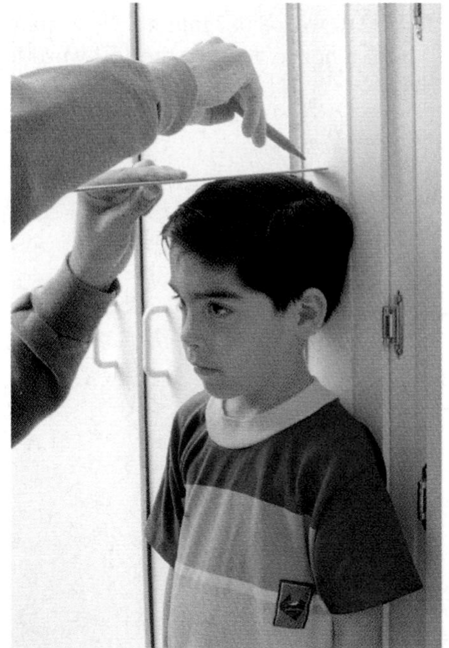

44. Find the expected height of a boy who weighs 140 pounds.

The formula $P = 14.7e^{-0.21x}$ gives the average atmospheric pressure P, in pounds per square inch, at an altitude x, in miles above sea level. Use this formula to solve Exercises 45 through 58. Round to the nearest tenth.

45. Find the average atmospheric pressure of Denver, which is 1 mile above sea level.

46. Find the average atmospheric pressure of Pikes Peak, which is 2.7 miles above sea level.

47. Find the elevation of a Delta jet if the atmospheric pressure outside the jet is 7.5 pounds per square inch.

48. Find the elevation of a remote Himalayan peak if the atmospheric pressure atop the peak is 6.5 pounds per square inch.

Psychologists call the graph of the formula $t = \frac{1}{c}\ln\left(\frac{A}{A - N}\right)$ the learning curve since the formula relates time t passed, in weeks, to a measure N of learning achieved, to a measure A of maximum learning possible, and to a measure c of an individual's learning style. Use this formula to answer Exercises 49 through 52. Round to the nearest whole number.

49. Norman Weidner is learning to type. If he wants to type at a rate of 50 words per minute (N is 50) and his expected maximum rate is 75 words per minute (A is 75), how many weeks should it take him to achieve his goal? Assume that c is 0.09.

50. An experiment on teaching chimpanzees sign language shows that a typical chimp can master a maximum of 65 signs. How many weeks should it take a chimpanzee to master 30 signs if c is 0.03?

51. Janine Jenkins is working on her dictation skills. She wants to take dictation at a rate of 150 words per minute and believes that the maximum rate she can hope for is 210 words per minute. How many weeks should it take her to achieve the 150-word level if c is 0.07?

52. A psychologist is measuring human capability to memorize nonsense syllables. How many weeks should it take a subject to learn 15 nonsense syllables if the maximum possible to learn is 24 syllables and c is 0.17?

Review and Preview

If $x = -2$, $y = 0$, and $z = 3$, find the value of each expression. See Section 1.6.

53. $\dfrac{x^2 - y + 2z}{3x}$

54. $\dfrac{x^3 - 2y + z}{2z}$

55. $\dfrac{3z - 4x + y}{x + 2z}$

56. $\dfrac{4y - 3x + z}{2x + y}$

Combining Concepts

The formula $y = y_0e^{kt}$ gives the population size y of a population that experiences an annual rate of population growth k (given as a decimal). In this formula, t is time in years and y_0 is the initial population at time 0. Use this formula to solve Exercises 57 and 58.

57. In 1990, the population of Arizona was 3,665,228. By 2000, the population had grown to 5,130,632. Find the annual rate of population growth over this period. Round your answer to the nearest tenth of a percent. (*Source:* U.S. Bureau of the Census)

58. In 1990, the population of Nevada was 1,201,833. By 2000, the population had grown to 1,998,257. Find the annual rate of population growth over this period. Round your answer to the nearest tenth of a percent. (*Source:* U.S. Bureau of the Census)

59. When solving a logarithmic equation, explain why you must check possible solutions in the original equation.

60. Solve $5^x = 9$ by taking the common logarithm of both sides of the equation. Next, solve this equation by taking the natural logarithm of both sides. Compare your solutions. Are they the same? Why or why not?

Use a graphing calculator to solve. Round your answers to two decimal places.

61. $e^{0.3x} = 8$

62. $10^{0.5x} = 7$

Internet Excursions

Go To: http://www.prenhall.com/martin-gay_algebra What's Related

In Section 11.6 Combining Concepts, we learned that the Richter scale measures the magnitude of an earthquake. The relationship between a Richter scale reading R and an intensity I of the earthquake's shock wave is given by the equation $R = \log I$. Given the Richter scale magnitudes of two earthquakes, we can compare their intensities. First we use this relationship to find the intensity of each earthquake. Then we can use a ratio of the resulting intensities to conclude that one earthquake was so many times more intense than the other.

By going to the World Wide Web address listed above, you will gain access to a Web site where you can look up current earthquake information to help you answer Exercises 63 and 64.

63. Scan the list of the recent earthquakes to find the earthquake events with the highest and lowest Richter scale magnitudes. Report the date, time, location, and magnitude of each. How many times more intense was the earthquake with the highest Richter scale reading than the one with the lowest reading?

64. Scan the list of recent earthquakes to find the most recent and least recent earthquake events that are listed. Report the date, time, location, and magnitude of each. How many times more intense was the earthquake with the higher Richter scale reading than the other earthquake?

CHAPTER 11 ACTIVITY

Modeling Temperature

METHOD 1 MATERIALS:

- a container of either cold or hot liquid
- thermometer
- stopwatch
- grapher with curve-fitting capabilities (optional)

METHOD 2 MATERIALS:

- a container of either cold or hot liquid
- TI-82, TI-83, or TI-85 graphing calculator with unit-to-unit link cable
- TI-CBL (Calculator-Based Laboratory) unit with temperature probe

This activity may be completed by working in groups or individually.

Newton's law of cooling relates the temperature of an object to the time elapsed since its warming or cooling began. In this activity you will investigate experimental data to find a mathematical model for this relationship. You may collect the temperature data by using either Method 1 (stopwatch and thermometer) or Method 2 (CBL).

Method 1

a. Insert the thermometer into the liquid and allow a thermometer reading to register. Take a temperature reading T as you start the stopwatch (at $t = 0$) and record it in the accompanying data table.

t	T
0	

b. Continue taking temperature readings at uniform intervals anywhere between 5 and 10 minutes long. At each reading, use the stopwatch to measure the length of time that has elapsed *since the temperature readings started.* Record your time t and liquid temperature T in the data table. Gather data for six to twelve readings.

c. Plot the data from the data table. Plot t on the horizontal axis and T on the vertical axis.

Method 2

a. Prepare the CBL unit and TI-82, TI-83, or TI-85 graphing calculator. Insert the temperature probe into the liquid.

b. Start the HEAT program on the TI graphing calculator and follow its instructions to begin collecting data. The program will collect 36 temperature readings in degrees Celsius and plot them in real time with t on the horizontal axis and T on the vertical axis.

1. Which of the following mathematical models best fits the data you collected? Explain your reasoning. (Assume $a > 0$.)

a. $T = ab^t + c$

b. $T = ab^{-t} + c$

c. $T = -ab^{-t} + c$

d. $T = \ln(-ax + b) + c$

e. $T = -\ln(-ax + b) + c$

2. What does the constant c represent in the model you chose? What is the value of c in this activity?

3. (Optional) Subtract the value of c from each of your observations of T. Enter the new ordered pairs $(t, T - c)$ into a grapher. Use the exponential or logarithmic curve-fitting feature to find a model for your experimental data. Graph the ordered pairs $(t, T - c)$ with the model you found. How well does the model fit the data? How does the model compare with your selection from Question 1?

Chapter 11 VOCABULARY CHECK

Fill in each blank with one of the words or phrases listed below.

inverse	common	composition	symmetric	exponential
vertical	logarithmic	natural	horizontal	

1. For each one-to-one function, we can find its __________ function by switching the coordinates of the ordered pairs of the function.

2. The __________ of functions f and g is $(f \circ g)(x) = f(g(x))$.

3. A function of the form $f(x) = b^x$ is called an __________ function if $b > 0$, b is not 1, and x is a real number.

4. The graphs of f and f^{-1} are __________ about the line $y = x$.

5. __________ logarithms are logarithms to base e.

6. __________ logarithms are logarithms to base 10.

7. To see whether a graph is the graph of a one-to-one function, apply the __________ line test to see whether it is a function, and then apply the __________ line test to see whether it is a one-to-one function.

8. A __________ function is a function that can be defined by $f(x) = \log_b x$ where x is a positive real number, b is a constant positive real number, and b is not 1.

CHAPTER 11 Highlights

DEFINITIONS AND CONCEPTS | **EXAMPLES**

Section 11.1 The Algebra of Functions

ALGEBRA OF FUNCTIONS

Sum	$(f + g)(x) = f(x) + g(x)$
Difference	$(f - g)(x) = f(x) - g(x)$
Product	$(f \cdot g)(x) = f(x) \cdot g(x)$
Quotient	$\left(\frac{f}{g}\right)(x) = \frac{f(x)}{g(x)}$

If $f(x) = 7x$ and $g(x) = x^2 + 1$,

$$(f + g)(x) = f(x) + g(x) = 7x + x^2 + 1$$

$$(f - g)(x) = f(x) - g(x) = 7x - (x^2 + 1) = 7x - x^2 - 1$$

$$(f \cdot g)(x) = f(x) \cdot g(x) = 7x(x^2 + 1) = 7x^3 + 7x^2$$

$$\left(\frac{f}{g}\right)(x) = \frac{f(x)}{g(x)} = \frac{7x}{x^2 + 1}$$

COMPOSITE FUNCTIONS

The notation $(f \circ g)(x)$ means "f composed with g."

$$(f \circ g)(x) = f(g(x))$$

$$(g \circ f)(x) = g(f(x))$$

If $f(x) = x^2 + 1$ and $g(x) = x - 5$, find $(f \circ g)(x)$.

$$\begin{aligned}(f \circ g)(x) &= f(g(x)) \\ &= f(x - 5) \\ &= (x - 5)^2 + 1 \\ &= x^2 - 10x + 26\end{aligned}$$

Definitions and Concepts | Examples

Section 11.2 Inverse Functions

If f is a function, then f is a **one-to-one function** only if each y-value (output) corresponds to only one x-value (input).

Horizontal Line Test

If every horizontal line intersects the graph of a function at most once, then the function is a one-to-one function.

Determine whether each graph is a one-to-one function.

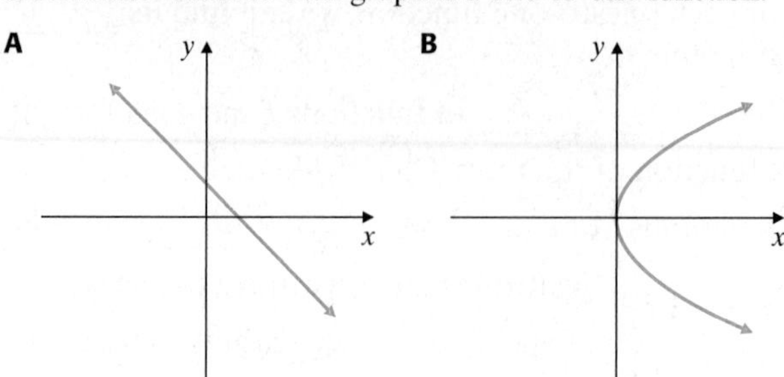

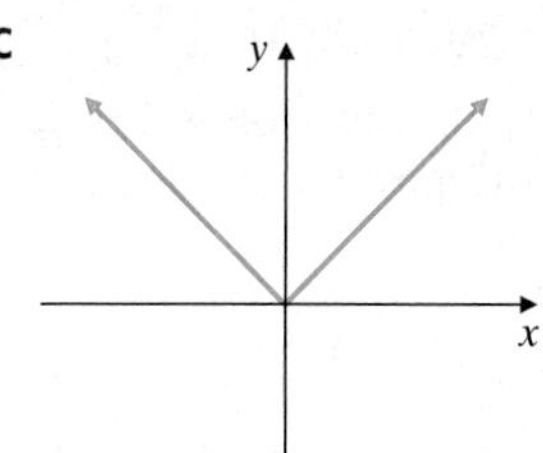

Graphs A and C pass the vertical line test, so only these are graphs of functions. Of graphs A and C, only graph A passes the horizontal line test, so only graph A is the graph of a one-to-one function.

The **inverse** of a one-to-one function f is the one-to-one function f^{-1} that is the set of all ordered pairs (b, a) such that (a, b) belongs to f.

Finding the Inverse of a One-to-One Function f

Step 1. Replace $f(x)$ with y.
Step 2. Interchange x and y.
Step 3. Solve for y.
Step 4. Replace y with $f^{-1}(x)$.

Find the inverse of $f(x) = 2x + 7$.

$y = 2x + 7$ Replace $f(x)$ with y.

$x = 2y + 7$ Interchange x and y.

$2y = x - 7$ Solve for y.

$$y = \frac{x - 7}{2}$$

$f^{-1}(x) = \frac{x - 7}{2}$ Replace y with $f^{-1}(x)$.

The inverse of $f(x) = 2x + 7$ is $f^{-1}(x) = \frac{x - 7}{2}$.

Section 11.3 Exponential Functions

A function of the form $f(x) = b^x$ is an **exponential function**, where $b > 0, b \neq 1$, and x is a real number.

Graph the exponential function $y = 4^x$.

x	y
−2	$\frac{1}{16}$
−1	$\frac{1}{4}$
0	1
1	4
2	16

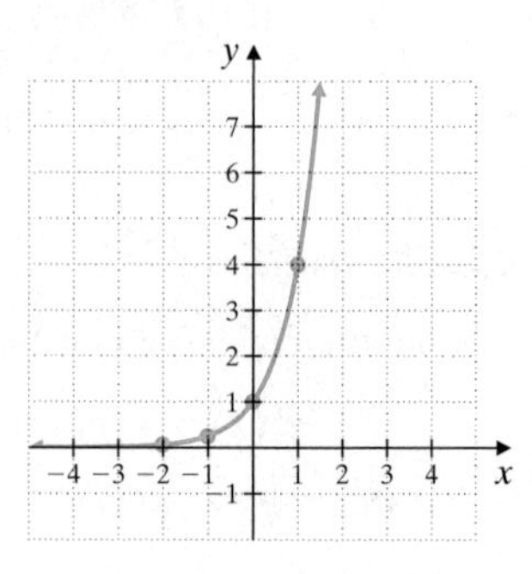

DEFINITIONS AND CONCEPTS | EXAMPLES

Section 11.3 *(continued)*

UNIQUENESS OF b^x

If $b > 0$ and $b \neq 1$, then $b^x = b^y$ is equivalent to $x = y$.

Solve: $2^{x+5} = 8$

$2^{x+5} = 2^3$ Write 8 as 2^3.

$x + 5 = 3$ Use the uniqueness of b^x.

$x = -2$ Subtract 5 from both sides.

Section 11.4 Logarithmic Functions

LOGARITHMIC DEFINITION

If $b > 0$ and $b \neq 1$, then

$$y = \log_b x \quad \text{means} \quad x = b^y$$

for any positive number x and real number y.

LOGARITHMIC FORM	CORRESPONDING EXPONENTIAL STATEMENT
$\log_5 25 = 2$	$5^2 = 25$
$\log_9 3 = \frac{1}{2}$	$9^{1/2} = 3$

PROPERTIES OF LOGARITHMS

If b is a real number, $b > 0$ and $b \neq 1$, then

$$\log_b 1 = 0, \quad \log_b b^x = x, \quad \text{and} \quad b^{\log_b x} = x$$

$$\log_5 1 = 0, \quad \log_7 7^2 = 2, \quad \text{and} \quad 3^{\log_3 6} = 6$$

LOGARITHMIC FUNCTION

If $b > 0$ and $b \neq 1$, then a **logarithmic function** is a function that can be defined as

$$f(x) = \log_b x$$

The domain of f is the set of positive real numbers, and the range of f is the set of real numbers.

Graph: $y = \log_3 x$

Write $y = \log_3 x$ as $3^y = x$. Plot the ordered pair solutions listed in the table, and connect them with a smooth curve.

x	y
3	1
1	0
$\frac{1}{3}$	−1
$\frac{1}{9}$	−2

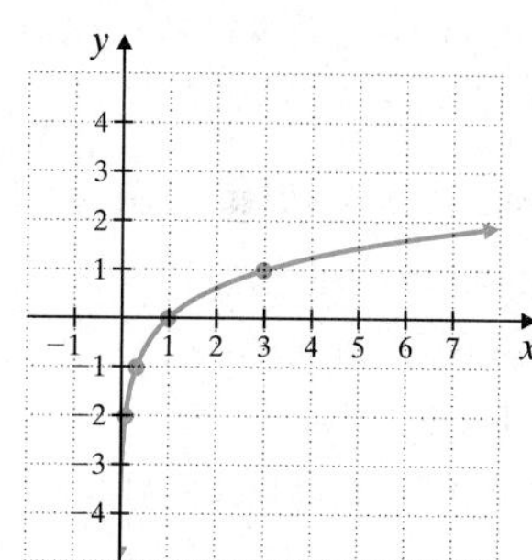

Section 11.5 Properties of Logarithms

Let x, y, and b be positive numbers and $b \neq 1$.

PRODUCT PROPERTY

$$\log_b xy = \log_b x + \log_b y$$

QUOTIENT PROPERTY

$$\log_b \frac{x}{y} = \log_b x - \log_b y$$

POWER PROPERTY

$$\log_b x^r = r \log_b x$$

Write as a single logarithm:

$2 \log_5 6 + \log_5 x - \log_5 (y + 2)$

$= \log_5 6^2 + \log_5 x - \log_5 (y + 2)$ Power property

$= \log_5 36 \cdot x - \log_5 (y + 2)$ Product property

$= \log_5 \frac{36x}{y + 2}$ Quotient property

DEFINITIONS AND CONCEPTS	**EXAMPLES**

Section 11.6 Common Logarithms, Natural Logarithms, and Change of Base

COMMON LOGARITHMS

$\log x$ means $\log_{10} x$

$\log 5 = \log_{10} 5 \approx 0.69897$

$\ln 7 = \log_e 7 \approx 1.94591$

NATURAL LOGARITHMS

$\ln x$ means $\log_e x$

Find the amount in an account at the end of 3 years if \$1000 is invested at an interest rate of 4% compounded continuously.

CONTINUOUSLY COMPOUNDED INTEREST FORMULA

$$A = Pe^{rt}$$

where r is the annual interest rate for P dollars invested for t years.

Here, $t = 3$ years, $P = \$1000$, and $r = 0.04$.

$$\begin{aligned} A &= Pe^{rt} \\ &= 1000e^{0.04(3)} \\ &\approx \$1127.50 \end{aligned}$$

Section 11.7 Exponential and Logarithmic Equations and Problem Solving

LOGARITHM PROPERTY OF EQUALITY

Let $\log_b a$ and $\log_b c$ be real numbers and $b \neq 1$. Then

$\log_b a = \log_b c$ is equivalent to $a = c$

Solve: $2^x = 5$

$\log 2^x = \log 5$	Log property of equality
$x \log 2 = \log 5$	Power property
$x = \dfrac{\log 5}{\log 2}$	Divide both sides by log 2.
$x \approx 2.3219$	Use a calculator.

Name ______________________ Section ________ Date ________

Chapter 11 Review

(11.1) *If* $f(x) = x - 5$ *and* $g(x) = 2x + 1$, *find the following.*

1. $(f + g)(x)$

2. $(f - g)(x)$

3. $(f \cdot g)(x)$

4. $\left(\frac{g}{f}\right)(x)$

If $f(x) = x^2 - 2$, $g(x) = x + 1$, *and* $h(x) = x^3 - x^2$, *find each composition.*

5. $(f \circ g)(x)$

6. $(g \circ f)(x)$

7. $(h \circ g)(2)$

8. $(f \circ f)(x)$

9. $(f \circ g)(-1)$

10. $(h \circ h)(2)$

(11.2) *Determine whether each function is a one-to-one function. If it is one-to-one, list the elements of its inverse.*

11. $h = \{(-9, 14), (6, 8), (-11, 12), (15, 15)\}$

12. $f = \{(-5, 5), (0, 4), (13, 5), (11, -6)\}$

13.

U.S. Region (Input)	West	Midwest	South	Northeast
Rank in Automobile Thefts (Output)	2	4	1	3

14.

Shape (Input)	Square	Triangle	Parallelogram	Rectangle
Number of Sides (Output)	4	3	4	4

Determine whether each function is a one-to-one function.

15.

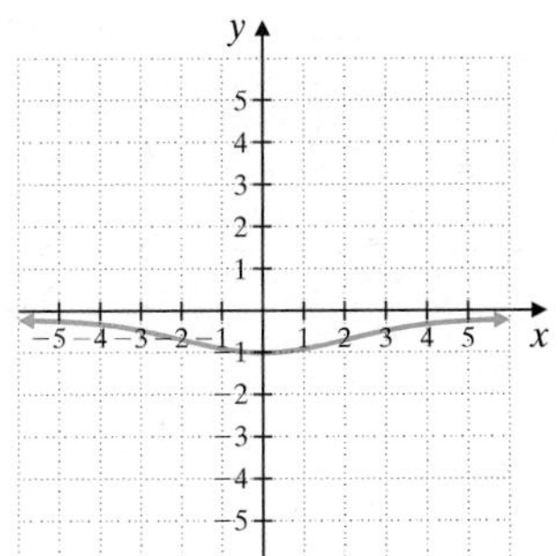

16.

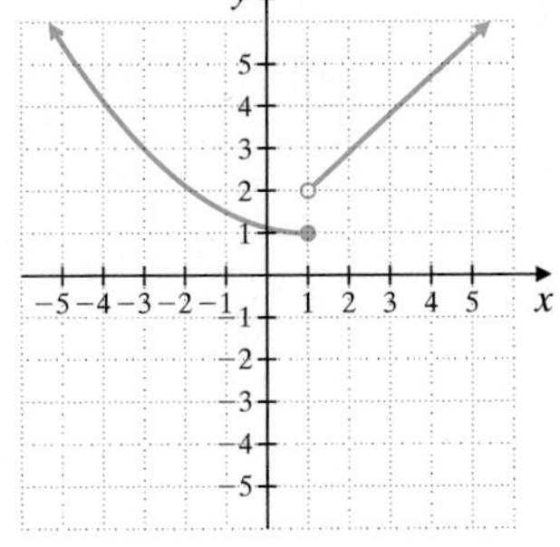

17.

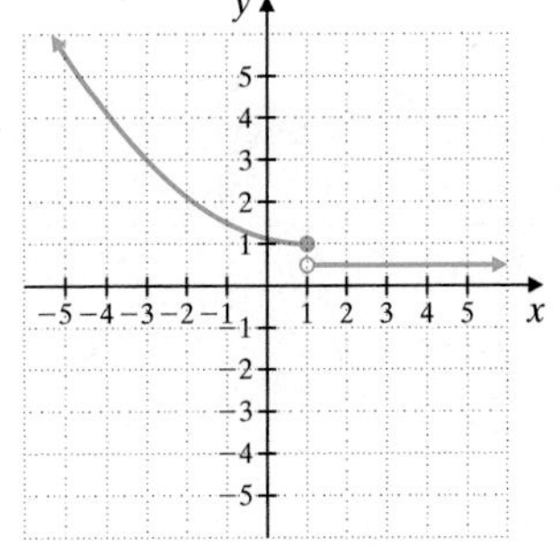

18.

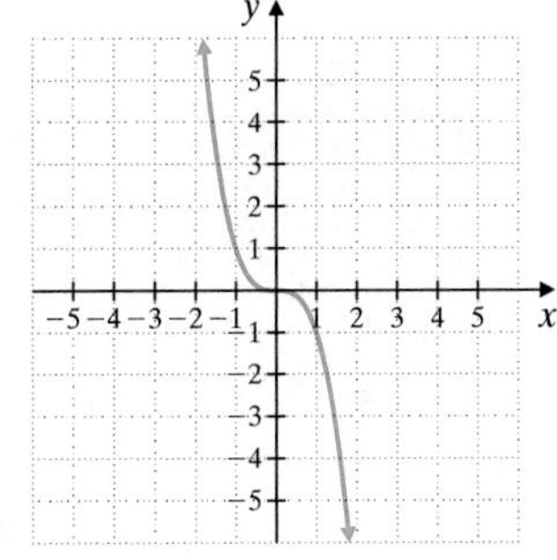

Find an equation defining the inverse function of each one-to-one function.

19. $f(x) = 6x + 11$

20. $f(x) = 12x$

21. $f(x) = 3x - 5$

22. $f(x) = 2x + 1$

Graph each one-to-one function and its inverse on the same set of axes.

23. $f(x) = -2x + 3$

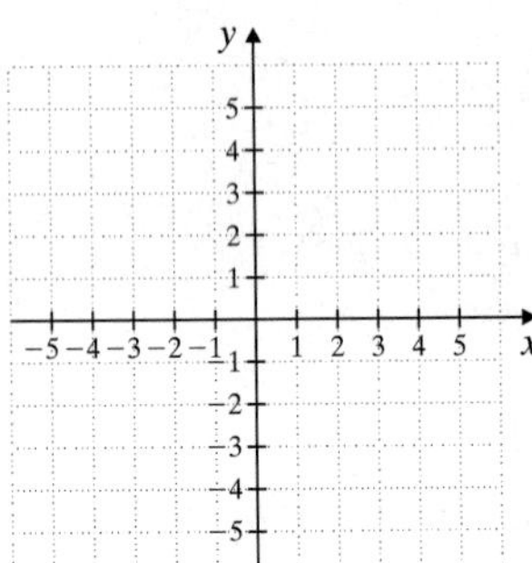

24. $f(x) = 5x - 5$

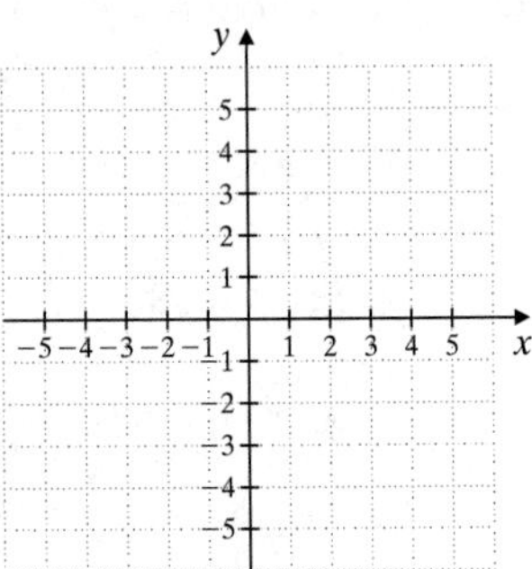

(11.3) *Solve each equation.*

25. $4^x = 64$

26. $3^x = \frac{1}{9}$

27. $2^{3x} = \frac{1}{16}$

28. $5^{2x} = 125$

29. $9^{x+1} = 243$

30. $8^{3x-2} = 4$

Graph each exponential function.

31. $y = 3^x$

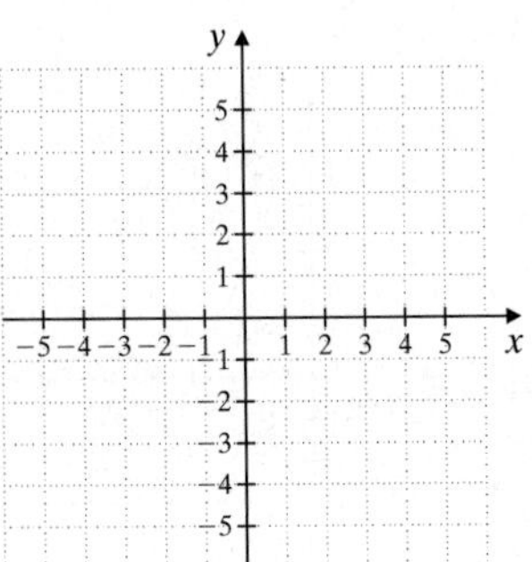

32. $y = \left(\frac{1}{3}\right)^x$

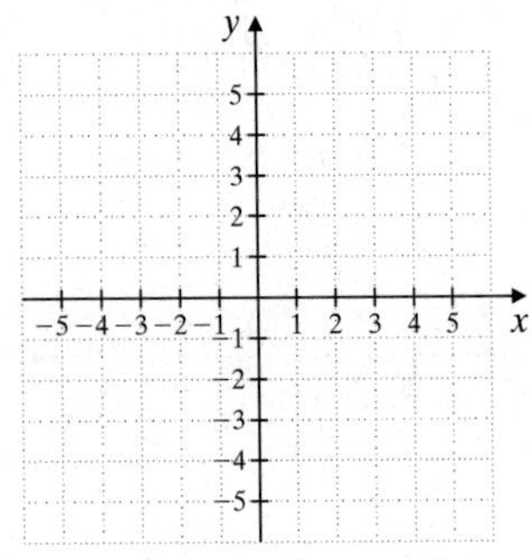

33. $y = 4 \cdot 2^x$

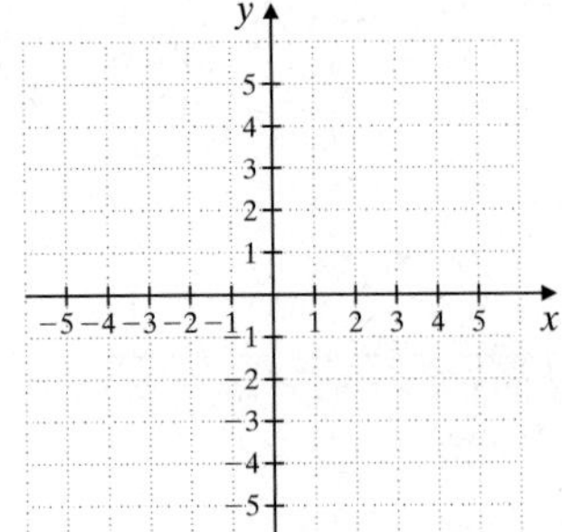

34. $y = 2^x + 4$

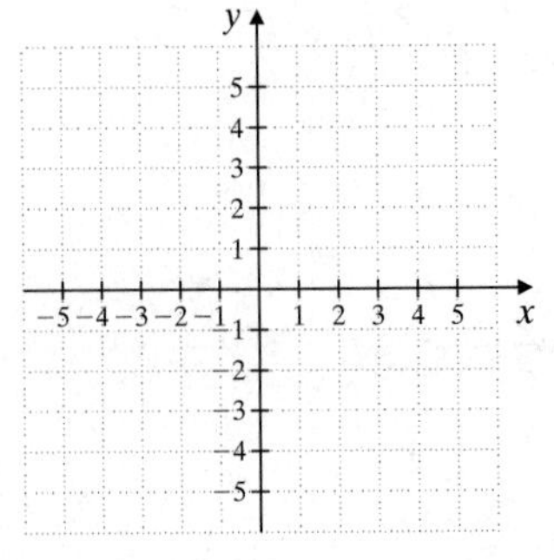

Use the formula $A = P\left(1 + \frac{r}{n}\right)^{nt}$ *to solve Exercises 35 and 36. In this formula,*

A = amount accrued (or owed)
P = principal invested (or loaned)
r = rate of interest
n = number of compounding periods per year
t = time in years

35. Find the amount accrued if \$1600 is invested at 9% interest compounded semiannually for 7 years.

36. A total of \$800 is invested in a 7% certificate of deposit for which interest is compounded quarterly. Find the value that this certificate will have at the end of 5 years.

(11.4) *Write each exponential equation with logarithmic notation.*

37. $49 = 7^2$

38. $2^{-4} = \frac{1}{16}$

Write each logarithmic equation with exponential notation.

39. $\log_{1/2} 16 = -4$

40. $\log_{0.4} 0.064 = 3$

Solve.

41. $\log_4 x = -3$

42. $\log_3 x = 2$

43. $\log_3 1 = x$

44. $\log_4 64 = x$

45. $\log_x 64 = 2$

46. $\log_x 81 = 4$

47. $\log_4 4^5 = x$

48. $\log_7 7^{-2} = x$

49. $5^{\log_5 4} = x$

50. $2^{\log_2 9} = x$

51. $\log_2(3x - 1) = 4$

52. $\log_3(2x + 5) = 2$

53. $\log_4(x^2 - 3x) = 1$

54. $\log_8(x^2 + 7x) = 1$

Graph each pair of equations on the same set of axes.

55. $y = 2^x$; $y = \log_2 x$

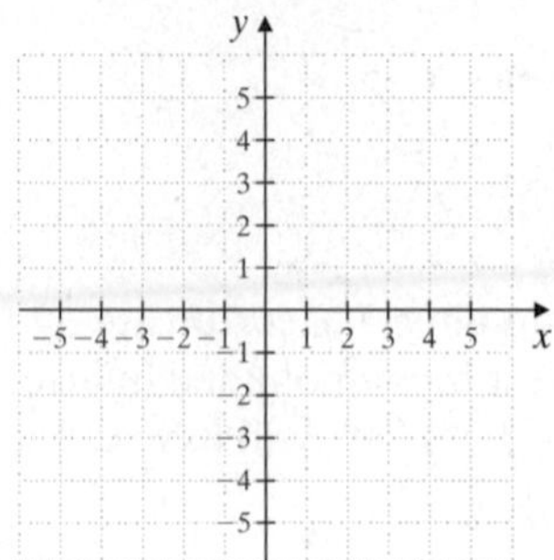

56. $y = \left(\frac{1}{2}\right)^x$; $y = \log_{1/2} x$

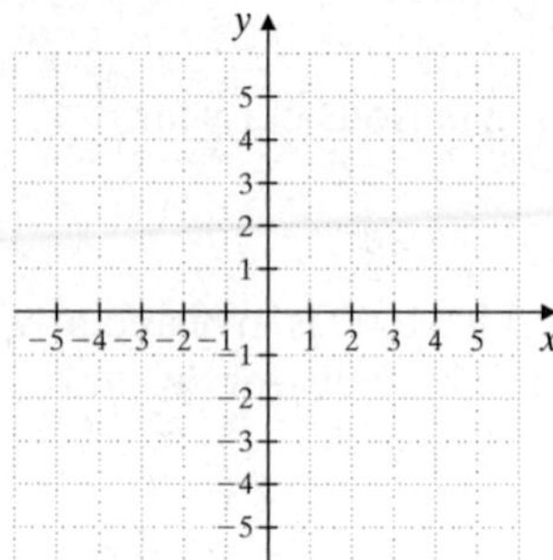

(11.5) *Write each expression as a single logarithm.*

57. $\log_3 8 + \log_3 4$

58. $\log_2 6 + \log_2 3$

59. $\log_7 15 - \log_7 20$

60. $\log 18 - \log 12$

61. $\log_{11} 8 + \log_{11} 3 - \log_{11} 6$

62. $\log_5 14 + \log_5 3 - \log_5 21$

63. $2 \log_5 x - 2 \log_5(x + 1) + \log_5 x$

64. $4 \log_3 x - \log_3 x + \log_3(x + 2)$

Use properties of logarithms to write each expression as a sum or difference of logarithms.

65. $\log_3 \frac{x^3}{x + 2}$

66. $\log_4 \frac{x + 5}{x^2}$

67. $\log_2 \frac{3x^2y}{z}$

68. $\log_7 \frac{yz^3}{x}$

If $\log_b 2 = 0.36$ *and* $\log_b 5 = 0.83$, *evaluate each expression.*

69. $\log_b 50$

70. $\log_b \frac{4}{5}$

(11.6) *Use a calculator to approximate each logarithm to four decimal places.*

71. $\log 3.6$

72. $\log 0.15$

73. $\ln 1.25$

74. $\ln 4.63$

Find the exact value of each logarithm.

75. $\log 1000$

76. $\log \frac{1}{10}$

77. $\ln\left(\frac{1}{e}\right)$

78. $\ln(e^4)$

Solve each equation.

79. $\ln(2x) = 2$

80. $\ln(3x) = 1.6$

81. $\ln(2x - 3) = -1$

82. $\ln(3x + 1) = 2$

Approximate each logarithm to four decimal places.

83. $\log_5 1.6$

84. $\log_3 4$

Use the formula $A = Pe^{rt}$ to solve Exercises 85 and 86, in which interest is compounded continuously. In this formula,

A = amount accrued (or owed)

P = principal invested (or loaned)

r = rate of interest

t = time in years

85. Bank of New York offers a 5-year 6% continuously compounded investment option. Find the amount accrued if \$1450 is invested.

86. Find the amount to which a \$940 investment grows if it is invested at 11% interest compounded continuously for 3 years.

(11.7) *Solve each exponential equation. Given an exact solution and a four-decimal-place approximation.*

87. $3^{2x} = 7$

88. $6^{3x} = 5$

89. $3^{2x+1} = 6$

90. $4^{3x+2} = 9$

91. $5^{3x-5} = 4$

92. $8^{4x-2} = 3$

93. $2 \cdot 5^{x-1} = 1$

94. $3 \cdot 4^{x+5} = 2$

Solve each equation.

95. $\log_5 2 + \log_5 x = 2$

96. $\log_3 x + \log_3 10 = 2$

97. $\log(5x) - \log(x + 1) = 4$

98. $\ln(3x) - \ln(x - 3) = 2$

99. $\log_2 x + \log_2 2x - 3 = 1$

100. $-\log_6(4x + 7) + \log_6 x = 1$

Use the formula $y = y_0e^{kt}$ to solve Exercises 101 through 105. In this formula,

y = size of population

y_0 = initial count of population

k = rate of growth

t = time

Round each answer to the nearest whole number.

101. The population of mallard ducks in Nova Scotia is expected to grow at a rate of 6% per week during the spring migration. If 155,000 ducks are already in Nova Scotia, how many are expected by the end of 4 weeks?

102. The population of Sierra Leone is growing at a rate of 2.6% per year. In 2001, the population of Sierra Leone was about 5,400,000. Find the expected population by 2009. (*Source:* Population Reference Bureau)

103. France is experiencing an annual growth rate of 0.4%. In 2001, the population of France was 59,200,000. How long will it take for the population to reach 65,000,000? (*Source:* Population Reference Bureau)

104. In 2001, Australia had a population of 19,400,000. How long will it take Australia to double in population if its growth rate is 0.6% annually? (*Source:* Population Reference Bureau)

105. Mexico's population is increasing at a rate of 1.9% per year. How long will it take for its 2001 population of 99,600,000 to double in size? (*Source:* Population Reference Bureau)

Use the compound interest equation $A = P\left(1 + \frac{r}{n}\right)^{nt}$ to solve Exercises 106 and 107. (See the directions for Exercises 35 and 36 for an explanation of this formula.) Round answers to the nearest tenth.

106. How long does it take for a \$5000 investment to grow to \$10,000 if it is invested at 8% interest compounded quarterly?

107. An investment of \$6000 has grown to \$10,000 while the money was invested at 6% interest compounded monthly. How long was it invested?

Name ______________________

Chapter 11 Test

If $f(x) = x$ *and* $g(x) = 2x - 3$, *find the following.*

1. $(f \cdot g)(x)$

2. $(f - g)(x)$

If $f(x) = x$, $g(x) = x - 7$, *and* $h(x) = x^2 - 6x + 5$, *find each composition.*

3. $(f \circ h)(0)$

4. $(g \circ f)(x)$

5. $(g \circ h)(x)$

Graph the one-to-one function and its inverse on the same set of axes.

6. $f(x) = 7x - 14$

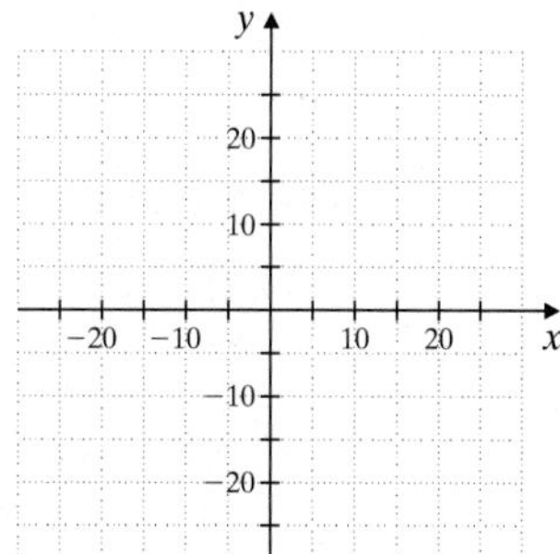

Determine whether each graph is the graph of a one-to-one function.

7.

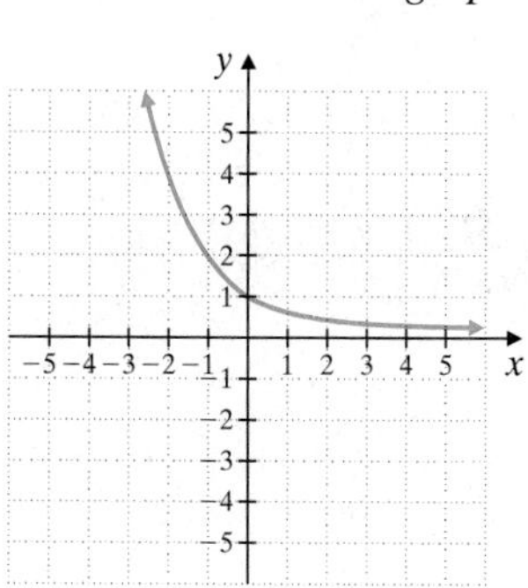

8.

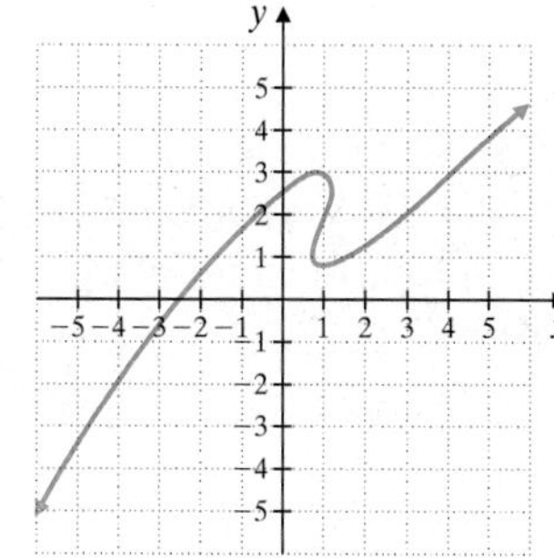

Answers

1. ______

2. ______

3. ______

4. ______

5. ______

6. see graph

7. ______

8. ______

9.

10.

11.

12.

13.

14.

15.

16.

17.

18.

Determine whether each function is one-to-one. If it is one-to-one, find an equation or a set of ordered pairs that defines the inverse function of the given function.

9. $y = 6 - 2x$

10. $f = \{(0, 0), (2, 3), (-1, 5)\}$

11.

Word (Input)	Dog	Cat	House	Desk	Circle
First Letter of Word (Output)	d	c	h	d	c

Use the properties of logarithms to write each expression as a single logarithm.

12. $\log_3 6 + \log_3 4$

13. $\log_5 x + 3\log_5 x - \log_5(x + 1)$

14. Write the expression $\log_6 \frac{2x}{y^3}$ as the sum or difference of logarithms.

15. If $\log_b 3 = 0.79$ and $\log_b 5 = 1.16$, find the value of $\log_b\left(\frac{3}{25}\right)$.

16. Approximate $\log_7 8$ to four decimal places.

17. Solve $8^{x-1} = \frac{1}{64}$ for x. Give an exact solution.

18. Solve $3^{2x+5} = 4$ for x. Give an exact solution and a four-decimal-place approximation.

Solve each logarithmic equation. Give an exact solution.

19. $\log_3 x = -2$

20. $\ln \sqrt{e} = x$

21. $\log_8(3x - 2) = 2$

22. $\log_5 x + \log_5 3 = 2$

23. $\log_4(x + 1) - \log_4(x - 2) = 3$

24. Solve $\ln(3x + 7) = 1.31$ accurate to four decimal places.

25. Graph $y = \left(\frac{1}{2}\right)^x + 1.$

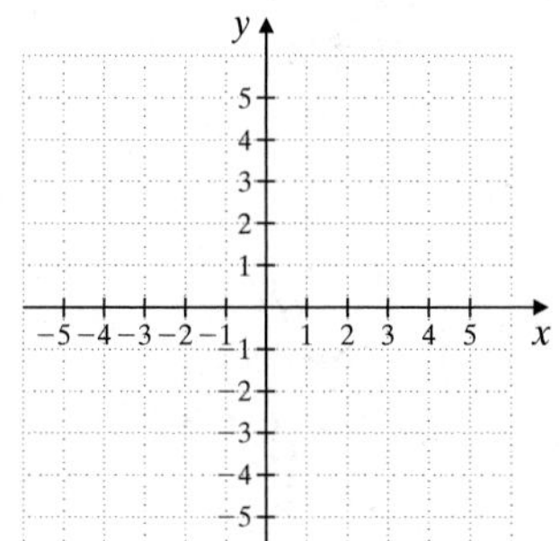

26. Graph the functions $y = 3^x$ and $y = \log_3 x$ on the same set of axes.

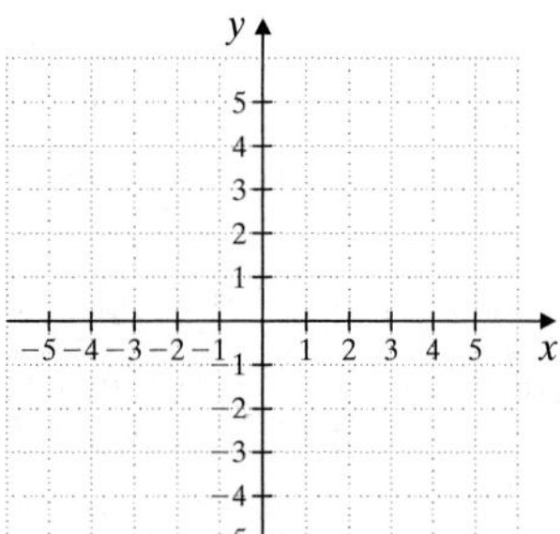

Use the formula $A = P\left(1 + \frac{r}{n}\right)^{nt}$ *to solve Exercises 27 and 28.*

27. Find the amount in an account in which \$4000 is invested for 3 years at 9% interest compounded monthly.

28. How long will it take \$2000 to grow to \$3000 if the money is invested at 7% interest compounded semiannually? Round to the nearest whole.

19. ____________

20. ____________

21. ____________

22. ____________

23. ____________

24. ____________

25. see graph

26. see graph

27. ____________

28. ____________

29. ______

Use the population growth formula $y = y_0e^{kt}$ *to solve Exercises 29 and 30.*

29. The prairie dog population of the Grand Forks area now stands at 57,000 animals. If the population is growing at a rate of 2.6% annually, how many prairie dogs will there be in that area 5 years from now?

30. In an attempt to save an endangered species of wood duck, naturalists would like to increase the wood duck population from 400 to 1000 ducks. If the annual population growth rate is 6.2%, how long will it take the naturalists to reach their goal? Round to the nearest whole year.

30. ______

The reliability of a new model of CD player can be described by the exponential function $R(t) = 2.7^{-(1/3)t}$, *where the reliability R is the probability (as a decimal) that the CD player is still working t years after it is manufactured. Round answers to the nearest hundredth. Then write your answers as a percent.*

31. What is the probability that the CD player will still work half a year after it is manufactured?

32. What is the probability that the CD player will still work 2 years after it is manufactured?

31. ______

32. ______

33. The world population is currently growing at a rate of 1.3% annually. In 2001, the midyear population of the world was 6137 million people. Predict the midyear world population in 2012. Use $y = 6137(2.7)^{0.013t}$, where y is the world population in millions and t is the number of years after 2001. Round to the nearest million. (*Source:* Based on data from the Population Reference Bureau)

33. ______

Name ____________________

Cumulative Review

Answers

1. ____________

2. ____________

3. ____________

4. ____________

5. ____________

6. ____________

7. ____________

8. ____________

9. ____________

10. ____________

11. ____________

12. ____________

1. Simplify $\dfrac{5x - 5}{x^3 - x^2}$

2. Simplify $\dfrac{x + 9}{x^2 - 81}$

3. Multiply: $\dfrac{x^2 + x}{3x} \cdot \dfrac{6}{5x + 5}$

4. Divide: $\dfrac{6x + 2}{x^2 - 1} \div \dfrac{3x^2 + x}{x - 1}$

5. Write an equation of the line with y-intercept $(0, -3)$ and slope of $\frac{1}{4}$.

6. Write an equation of the line containing the point $(2, 3)$ with undefined slope.

7. Find the domain and range of the relation.

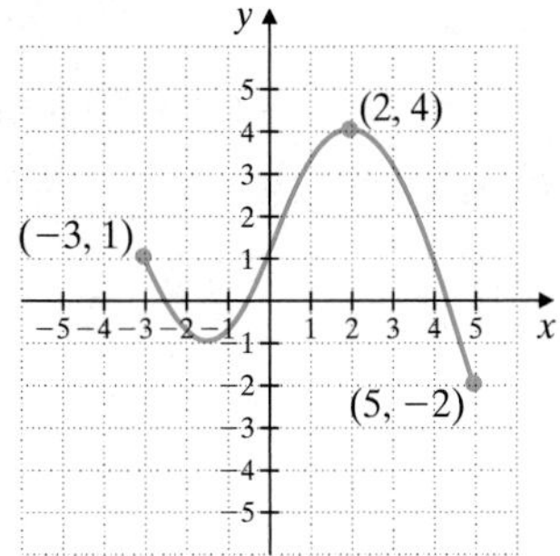

8. Solve the system of equations by graphing.

$$\begin{cases} -x + 3y = 10 \\ x + y = 2 \end{cases}$$

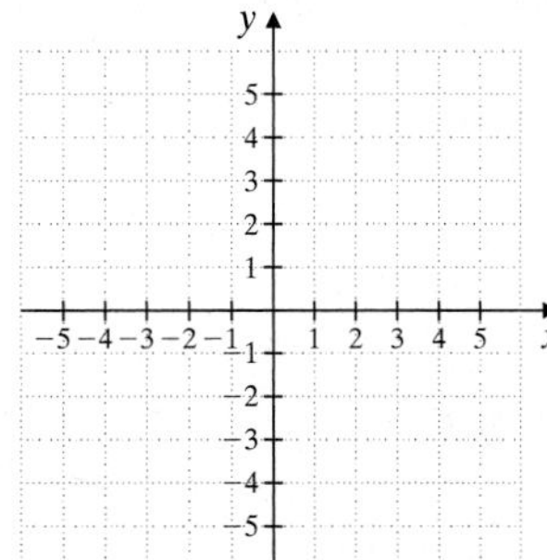

9. Solve the system:

$$\begin{cases} 6x + 12y = 5 \\ -4x - 8y = 0 \end{cases}$$

10. The Cirque du Soleil show Alegria is performing locally. Matinee admission for 4 adults and 2 children in $374, while admission for 2 adults and 3 children is $285.

a. What is the price of an adult's ticket?

b. What is the price of a child's ticket?

c. Suppose that a special rate of $1000 is offered for groups of 20 persons. Should a group of 4 adults and 16 children use the group rate? Why or why not?

Find each square root.

11. $\sqrt{0}$

12. $\sqrt{0.25}$

13.

14.

15.

16.

17.

18.

19.

20.

21.

22.

23.

24.

25.

26.

Use rational exponents to simplify. Assume that all variables represent positive numbers.

13. $\sqrt[8]{x^4}$

14. $\sqrt[6]{r^2s^4}$

Simplify.

15. $\sqrt[3]{24}$

16. $\sqrt[4]{32}$

17. Rationalize the denominator of $\frac{2}{\sqrt{5}}$.

18. Solve: $\sqrt{-10x - 1} + 3x = 0$

19. Multiply: $(2 - 5i)(4 + i)$

20. Solve $p^2 + 2p = 4$ by completing the square.

21. Solve: $2x^2 - 4x = 3$

22. Solve: $x - \sqrt{x} - 6 = 0$

23. Solve: $(x + 2)(x - 1)(x - 5) \leq 0$

24. Solve: $4^{x+3} = 8^x$

25. Solve: $\log_x 25 = 2$

26. Write as a single logarithm.
$2\log_5 3 + 3\log_5 2$

Conic Sections

In Chapter 10, we analyzed some of the important connections between a parabola and its equation. Parabolas are interesting in their own right but are more interesting still because they are part of a collection of curves known as conic sections. This chapter is devoted to quadratic equations in two variables and their conic section graphs: the parabola, circle, ellipse, and hyperbola.

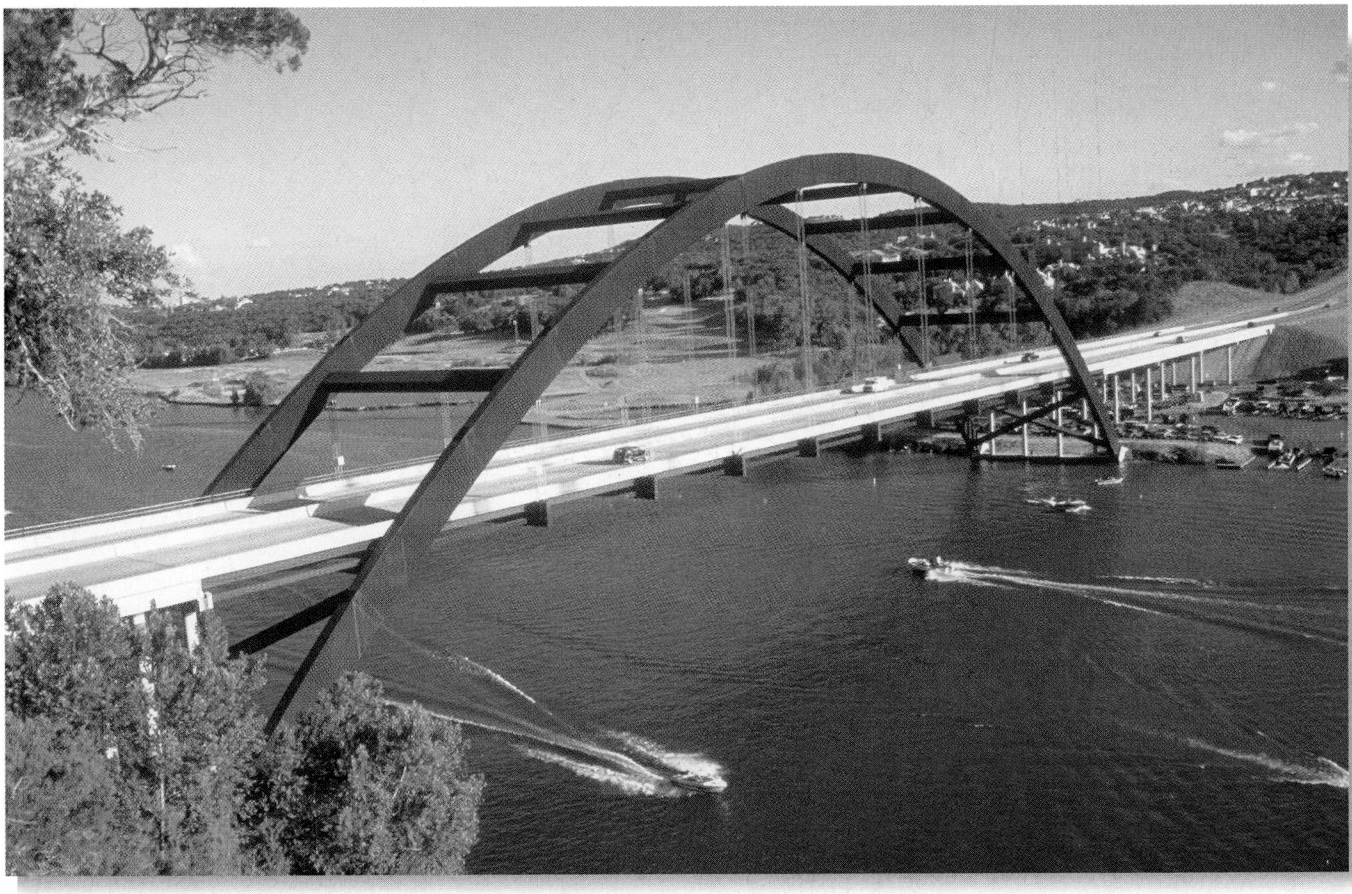

The shapes of conic sections are used in a variety of applications. They are used in architecture in the design of bridges, arches, and vaults. They are also used in astronomy to model the orbits of planets, comets, and satellites. Conic sections are used also in engineering in the design of certain gears and reflectors. In blueprints or diagrams of any of these situations, the exact shape of the conic section involved must be depicted. In Exercises 68 and 69 on page 917, we will see how architects and engineers use conic sections in their work.

Answers
1.
2.
3. see graph
4. see graph
5. see graph
6. see graph
7. see graph
8. see graph
9. see graph
10. see graph
11.
12. see graph
13.
14.

Name ____________ Section ________ Date ________

Chapter 12 Pretest

1. Find the distance between the points $(-4, 6)$ and $(1, 9)$.

2. Find the midpoint of the line segment whose endpoints are $(9, -15)$ and $(10, 22)$.

Sketch the graph of each equation.

3. $x = (y - 2)^2 + 1$

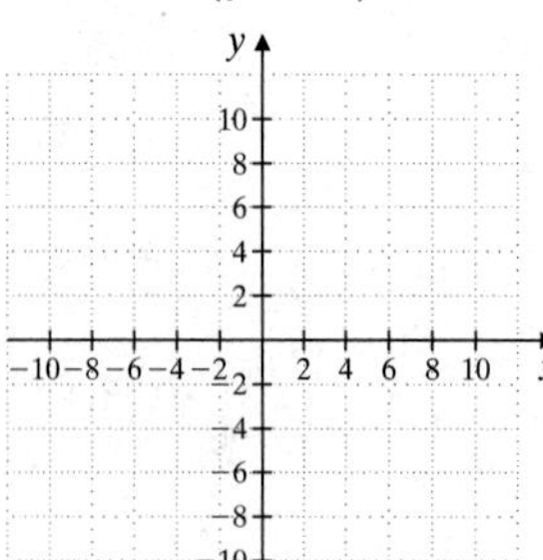

4. $y = x^2 + 4x - 5$

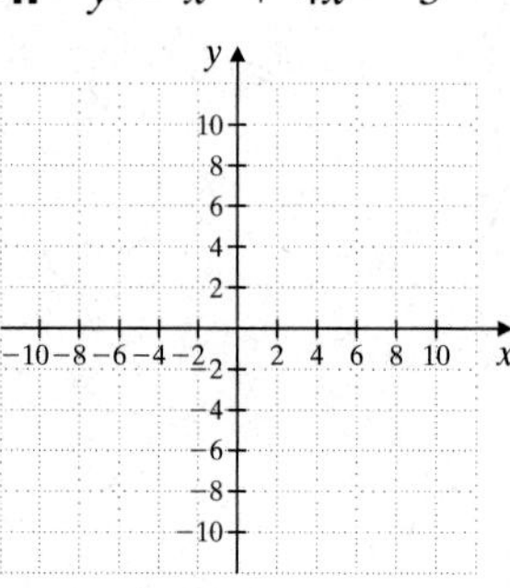

5. $x^2 + y^2 = 4$

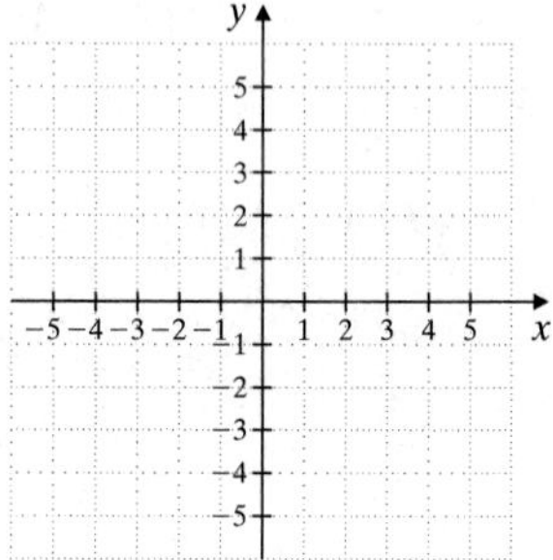

6. $(x + 2)^2 + (y - 3)^2 = 9$

7. $4x^2 + 25y^2 = 100$

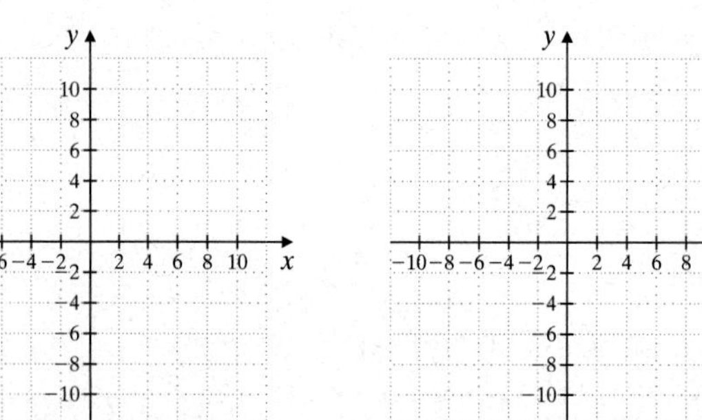

8. $\dfrac{x^2}{25} - \dfrac{y^2}{25} = 1$

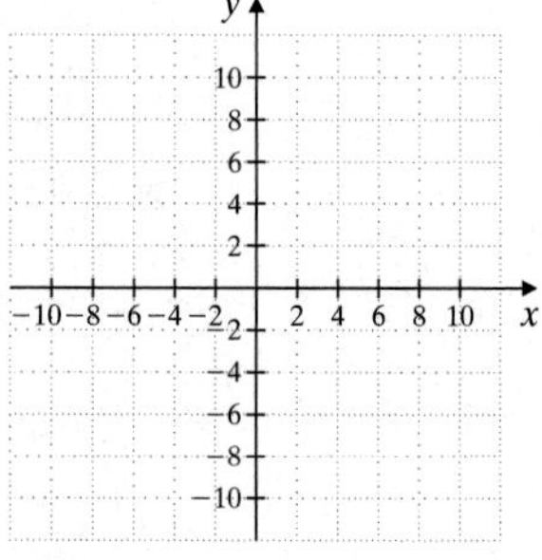

Graph each function.

9. $f(x) = |x| - 5$

10. $f(x) = \sqrt{x + 1} + 3$

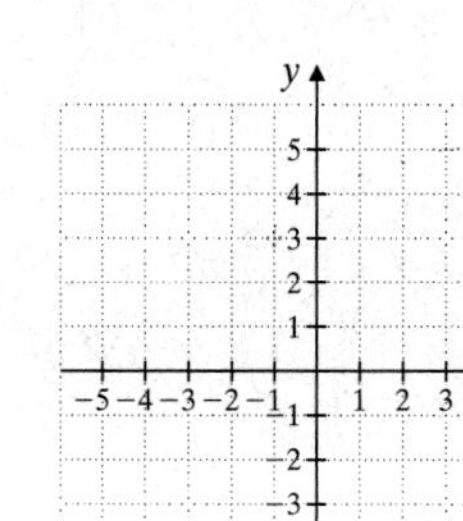

11. Solve the system:

$$\begin{cases} y = x^2 + 4x + 3 \\ x + y = 3 \end{cases}$$

12. Graph the system:

$$\begin{cases} y \geq \frac{1}{4}x^2 + 2 \\ x^2 + (y - 3)^2 \leq 4 \end{cases}$$

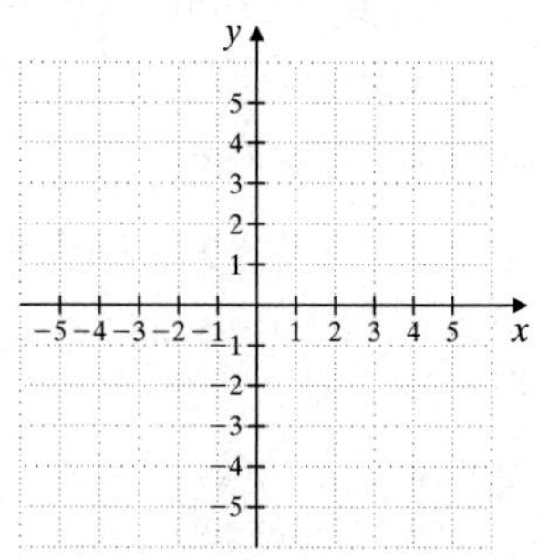

13. Write an equation of the circle with center $(2, 5)$ and radius 8.

14. Find the center and radius of the circle defined by $x^2 + y^2 + 6x - 8y = -16$.

12.1 The Parabola and the Circle

OBJECTIVES

- **A** Graph parabolas of the forms $y = a(x - h)^2 + k$ and $x = a(y - k)^2 + h$.
- **B** Use the distance formula and the midpoint formula.
- **C** Graph circles of the form $(x - h)^2 + (y - k)^2 = r^2$.
- **D** Write the equation of a circle, given its center and radius.
- **E** Find the center and the radius of a circle, given its equation.

Conic sections are called such because each conic section is the intersection of a right circular cone and a plane. The circle, parabola, ellipse, and hyperbola are the conic sections.

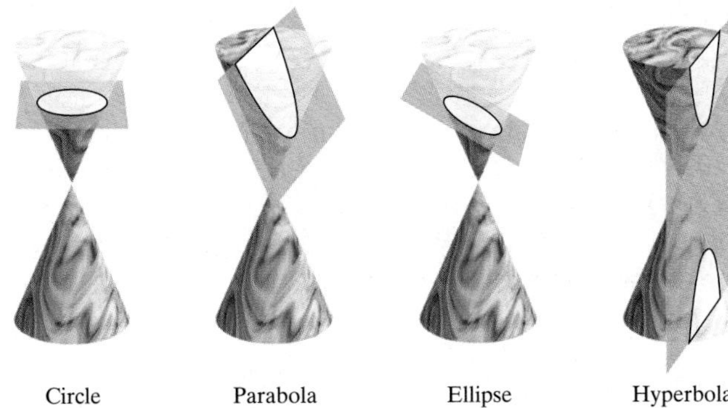

A Graphing Parabolas

Thus far, we have seen that $f(x)$ or $y = a(x - h)^2 + k$ is the equation of a parabola that opens upward if $a > 0$ or downward if $a < 0$. Parabolas can also open left or right, or even on a slant. Equations of these parabolas are not functions of x, of course, since a parabola opening any way other than upward or downward fails the vertical line test. In this section, we introduce parabolas that open to the left and to the right. Parabolas opening on a slant will not be developed in this book.

Just as $y = a(x - h)^2 + k$ is the equation of a parabola that opens upward or downward, $x = a(y - k)^2 + h$ is the equation of a parabola that opens to the right or to the left. The parabola opens to the right if $a > 0$ and to the left if $a < 0$. The parabola has vertex (h, k), and its axis of symmetry is the line $y = k$.

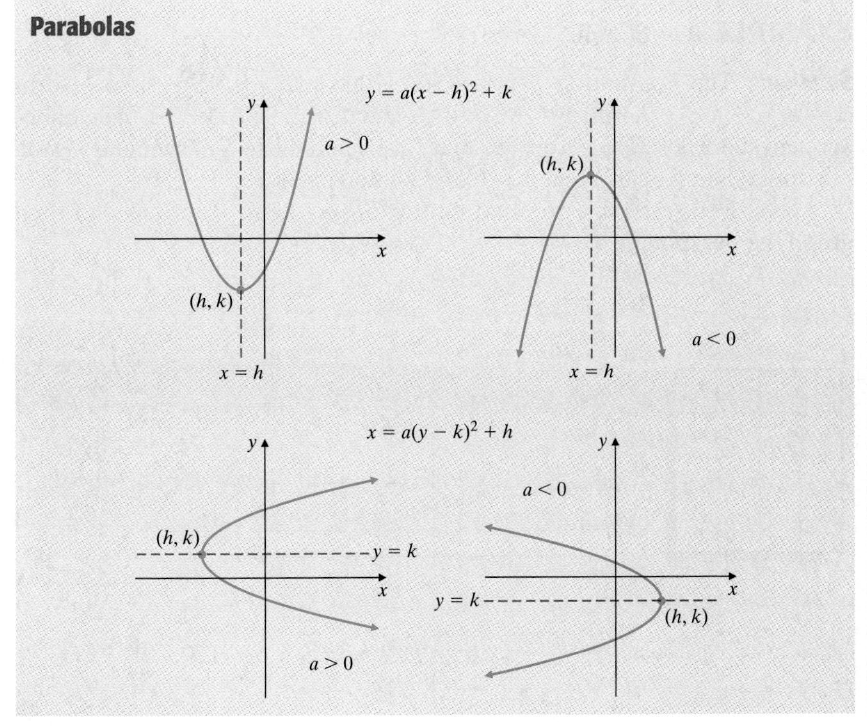

Concept Check

Does the graph of the parabola given by the equation $x = -3y^2$ open to the left, to the right, upward, or downward?

Practice Problem 1

Graph: $x = 3y^2$

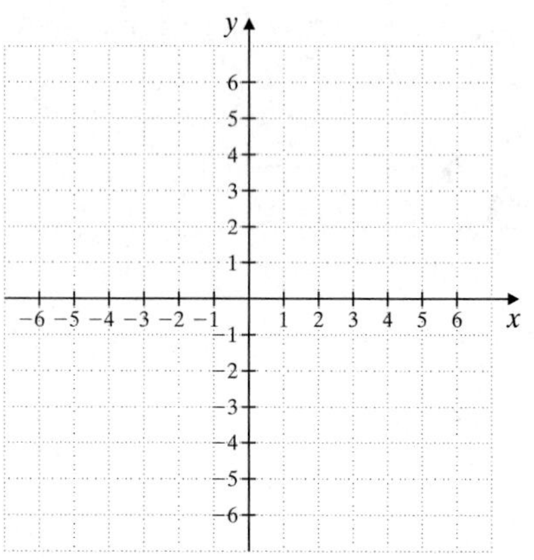

Practice Problem 2

Graph: $x = -2(y - 3)^2 + 1$

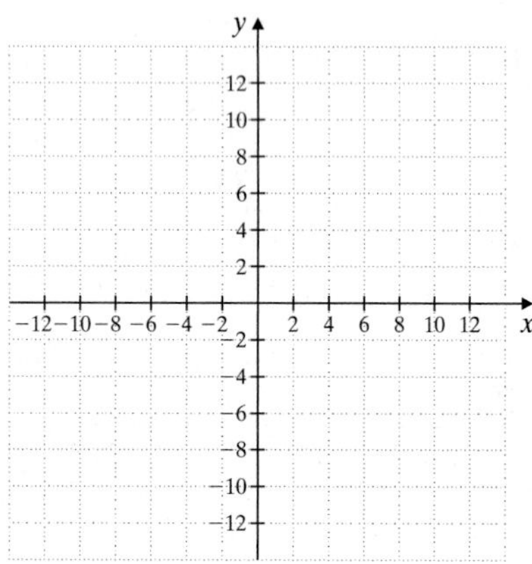

Answers

1.

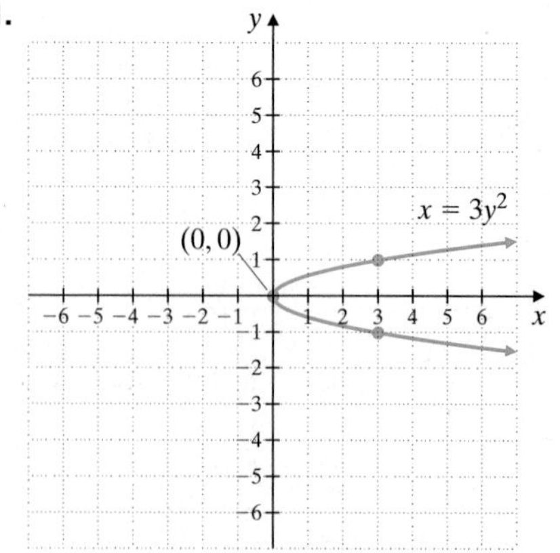

Concept Check: to the left

The forms $y = a(x - h)^2 + k$ and $x = a(y - k)^2 + h$ are called **standard forms.**

Try the Concept Check in the margin.

EXAMPLE 1 Graph: $x = 2y^2$

Solution: Written in standard form, the equation $x = 2y^2$ is $x = 2(y - 0)^2 + 0$ with $a = 2$, $h = 0$, and $k = 0$. Its graph is a parabola with vertex $(0, 0)$, and its axis of symmetry is the line $y = 0$. Since $a > 0$, this parabola opens to the right. We use a table to obtain a few more ordered pair solutions to help us graph $x = 2y^2$.

x	y
8	−2
2	−1
0	0
2	1
8	2

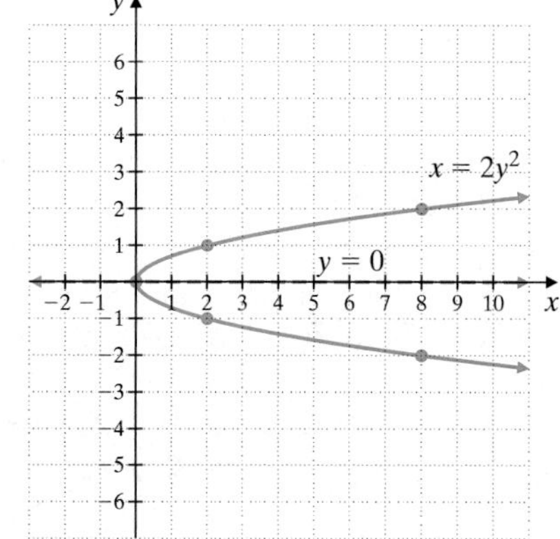

EXAMPLE 2 Graph: $x = -3(y - 1)^2 + 2$

Solution: The equation $x = -3(y - 1)^2 + 2$ is in the form $x = a(y - k)^2 + h$ with $a = -3$, $k = 1$, and $h = 2$. Since $a < 0$, the parabola opens to the left. The vertex (h, k) is $(2, 1)$, and the axis of symmetry is the horizontal line $y = 1$. When $y = 0$, the x-intercept is $x = (-1, 0)$.

Again, we use a table to obtain a few ordered pair solutions and then graph the parabola.

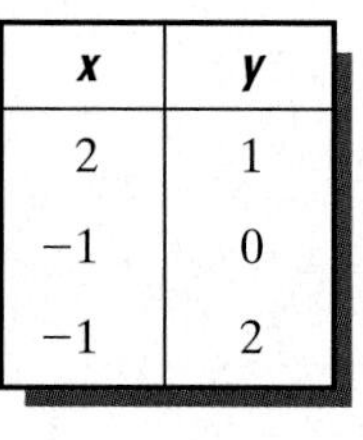

x	y
2	1
−1	0
−1	2

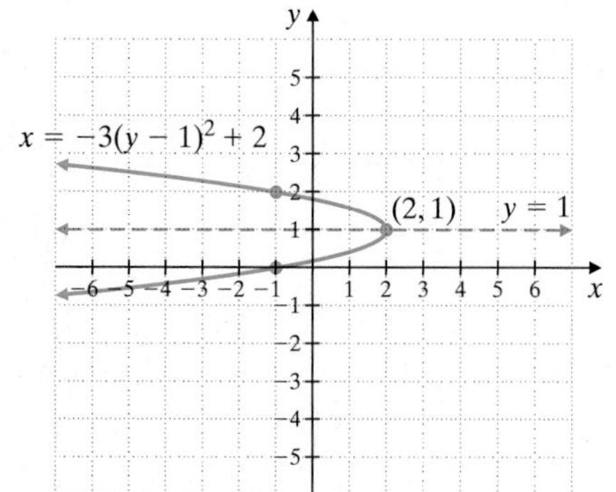

EXAMPLE 3 Graph: $y = -x^2 - 2x + 15$

Solution: There are two methods that we can use to find the vertex. The first method is completing the square.

$$y - 15 = -x^2 - 2x \quad \text{Subtract 15 from both sides.}$$

$$y - 15 = -1(x^2 + 2x) \quad \text{Factor } -1 \text{ from the terms } -x^2 - 2x.$$

The coefficient of x is 2, so we find the square of half of 2.

$$\frac{1}{2}(2) = 1 \quad \text{and} \quad 1^2 = 1$$

$$y - 15 - 1(1) = -1(x^2 + 2x + 1) \quad \text{Add } -1(1) \text{ to both sides.}$$

$$y - 16 = -1(x + 1)^2 \quad \text{Simplify the left side, and factor the right side.}$$

$$y = -(x + 1)^2 + 16 \quad \text{Add 16 to both sides.}$$

The vertex is $(-1, 16)$.

The second method for finding the vertex is by using the expression $\frac{-b}{2a}$. Since the equation is quadratic in x, the expression gives us the x-value of the vertex.

$$x = \frac{-(-2)}{2(-1)} = \frac{2}{-2} = -1$$

To find the corresponding y-value of the vertex, replace x with -1 in the original equation.

$$y = -(-1)^2 - 2(-1) + 15 = -1 + 2 + 15 = 16$$

Again, we see that the vertex is $(-1, 16)$, and the axis of symmetry is the vertical line $x = -1$. The y-intercept is $(0, 15)$. Now we can use a few more ordered pair solutions to graph the parabola.

x	y
−5	0
−3	12
−2	15
−1	16
0	15
1	12
3	0

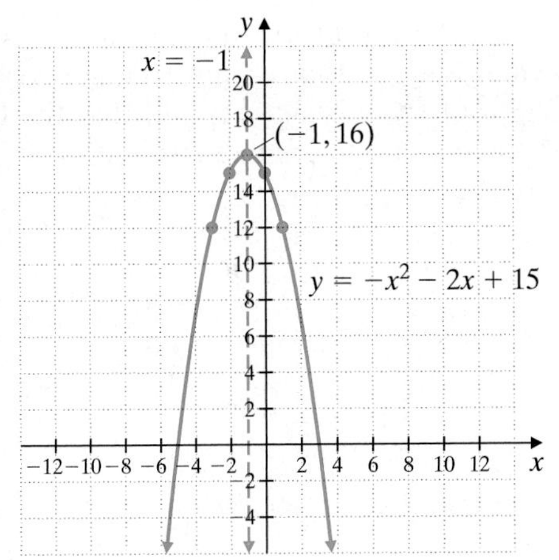

Practice Problem 3

Graph: $y = -x^2 - 4x + 12$

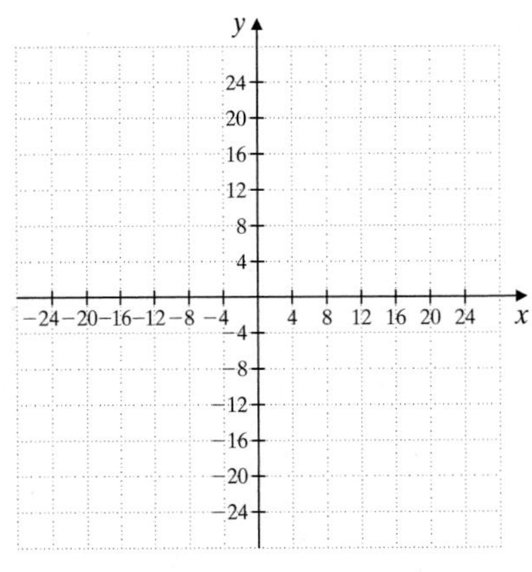

Answers

2.

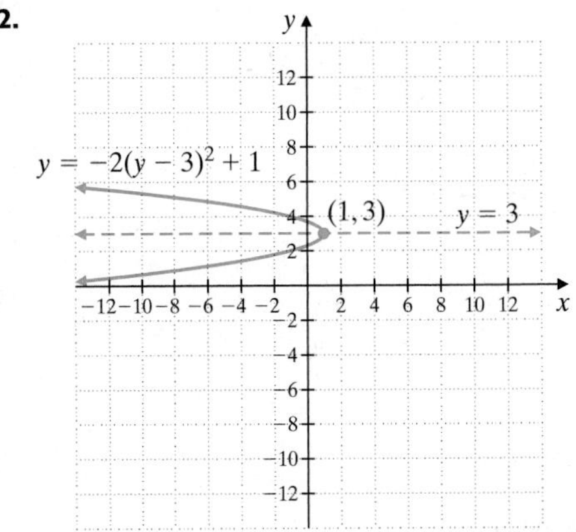

3.

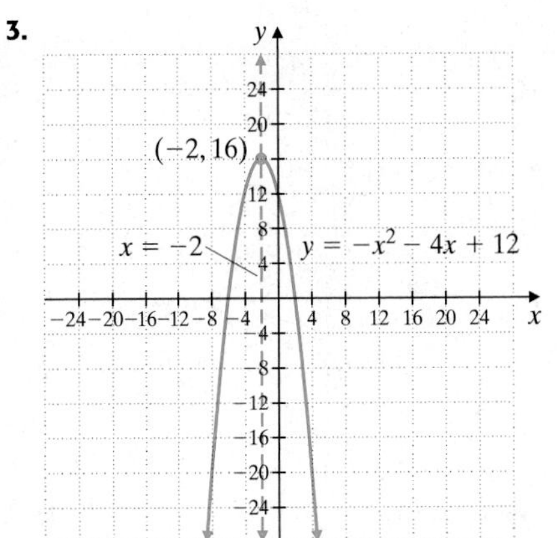

Practice Problem 4

Graph: $x = 3y^2 + 12y + 13$

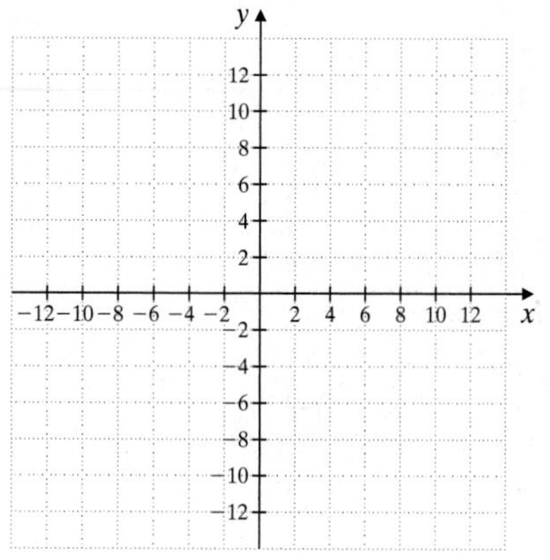

EXAMPLE 4 Graph: $x = 2y^2 + 4y + 5$

Solution: We notice that this equation is quadratic in y, so its graph is a parabola that opens to the left or the right. We can complete the square on y or we can use the expression $\frac{-b}{2a}$ to find the vertex.

Since the equation is quadratic in y, the expression gives us the y-value of the vertex.

$$y = \frac{-4}{2 \cdot 2} = \frac{-4}{4} = -1$$

$$x = 2(-1)^2 + 4(-1) + 5 = 2 \cdot 1 - 4 + 5 = 3$$

The vertex is $(3, -1)$, and the axis of symmetry is the line $y = -1$. The parabola opens to the right since $a > 0$. The x-intercept is $(5, 0)$.

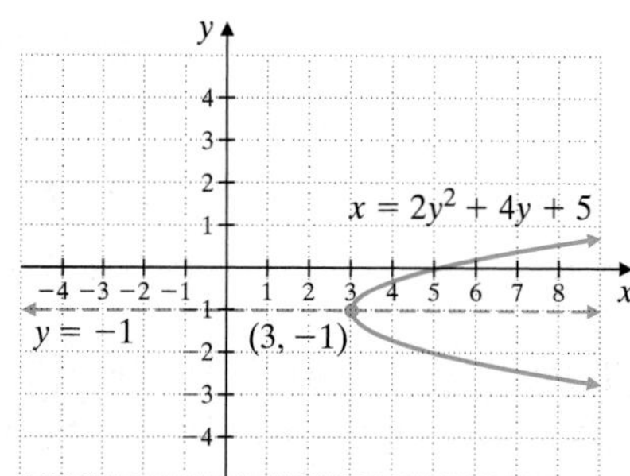

B Using the Distance and Midpoint Formulas

The Cartesian coordinate system helps us visualize a distance between points. To find the distance between two points, we use the distance formula, which is derived from the Pythagorean theorem.

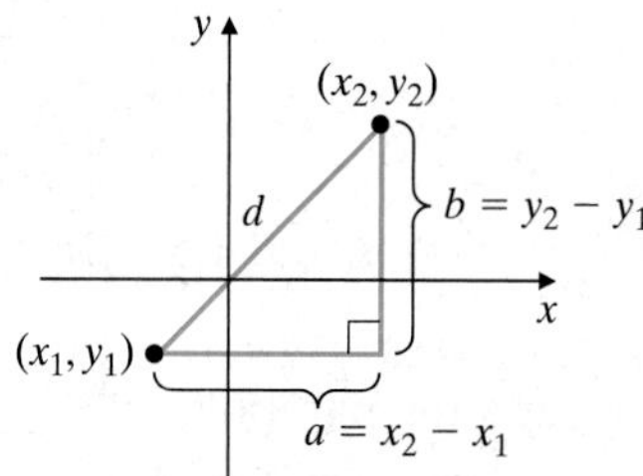

To find the distance d between two points (x_1, y_1) and (x_2, y_2), draw vertical and horizontal lines so that a right triangle is formed, as shown. Notice that the length of leg a is $x_2 - x_1$ and that the length of leg b is $y_2 - y_1$. Thus, the Pythagorean theorem tell us that

$$d^2 = a^2 + b^2$$

or

$$d^2 = (x_2 - x_1)^2 + (y_2 - y_1)^2$$

or

$$d = \sqrt{(x_2 - x_1)^2 + (y_2 - y_1)^2}$$

This formula gives us the distance between any two points on the real plane.

Answer

4.

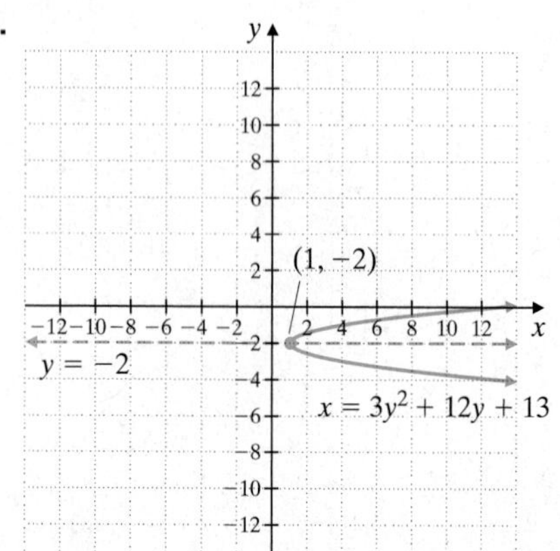

Distance Formula

The distance d between two points (x_1, y_1) and (x_2, y_2) is given by

$$d = \sqrt{(x_2 - x_1)^2 + (y_2 - y_1)^2}$$

EXAMPLE 5 Find the distance between $(2, -5)$ and $(1, -4)$. Give an exact distance and a three-decimal-place approximation.

Solution: To use the distance formula, it makes no difference which point we call (x_1, y_1) and which point we call (x_2, y_2). We will let $(x_1, y_1) = (2, -5)$ and $(x_2, y_2) = (1, -4)$.

$$\begin{aligned} d &= \sqrt{(x_2 - x_1)^2 + (y_2 - y_1)^2} \\ &= \sqrt{(1 - 2)^2 + [-4 - (-5)]^2} \\ &= \sqrt{(-1)^2 + (1)^2} \\ &= \sqrt{1 + 1} \\ &= \sqrt{2} \approx 1.414 \end{aligned}$$

The distance between the two points is exactly $\sqrt{2}$ units, or approximately 1.414 units.

Practice Problem 5

Find the distance between $(-1, 3)$ and $(-2, 6)$. Give an exact distance and a three-decimal-place approximation.

The **midpoint** of a line segment is the **point** located exactly halfway between the two endpoints of the line segment. On the following graph, the point M is the midpoint of line segment PQ. Thus, the distance between M and P equals the distance between M and Q.

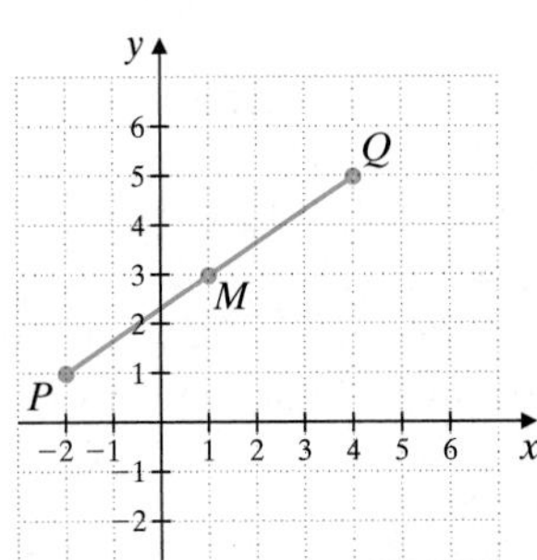

The x-coordinate of M is at half the distance between the x-coordinates of P and Q, and the y-coordinate of M is at half the distance between the y-coordinates of P and Q. That is, the x-coordinate of M is the average of the x-coordinates of P and Q; the y-coordinate of M is the average of the y-coordinates of P and Q.

Midpoint Formula

The midpoint of the line segment whose endpoints are (x_1, y_1) and (x_2, y_2) is the point with coordinates

$$\left(\frac{x_1 + x_2}{2}, \frac{y_1 + y_2}{2}\right)$$

Answer

5. $\sqrt{10} \approx 3.162$

Practice Problem 6

Find the midpoint of the line segment that joins points $P(-2, 5)$ and $Q(4, -6)$.

EXAMPLE 6 Find the midpoint of the line segment that joins points $P(-3, 3)$ and $Q(1, 0)$.

Solution: To use the midpoint formula, it makes no difference which point we call (x_1, y_1) and which point we call (x_2, y_2). We will let $(x_1, y_1) = (-3, 3)$ and $(x_2, y_2) = (1, 0)$.

$$\text{midpoint} = \left(\frac{x_1 + x_2}{2}, \frac{y_1 + y_2}{2}\right)$$

$$= \left(\frac{-3 + 1}{2}, \frac{3 + 0}{2}\right)$$

$$= \left(\frac{-2}{2}, \frac{3}{2}\right)$$

$$= \left(-1, \frac{3}{2}\right)$$

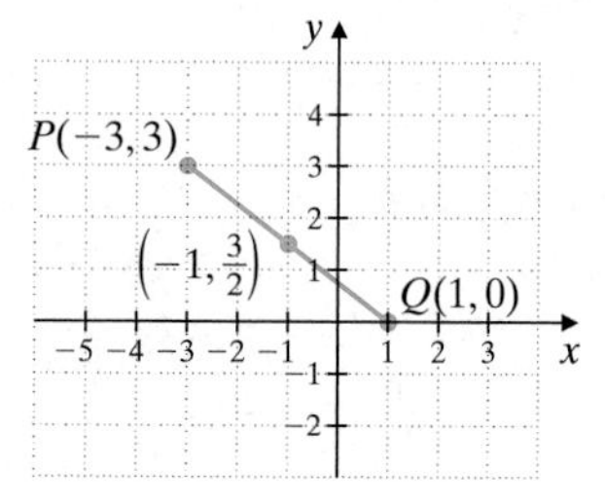

The midpoint of the segment is $\left(-1, \frac{3}{2}\right)$.

C Graphing Circles

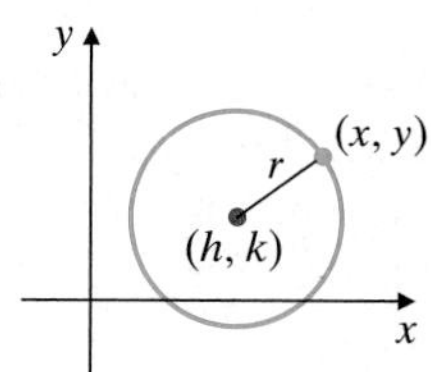

Another conic section is the **circle**. A circle is the set of all points in a plane that are the same distance from a fixed point called the **center**. The distance is called the **radius** of the circle. To find a standard equation for a circle, let (h, k) represent the center of the circle, and let (x, y) represent any point on the circle. The distance between (h, k) and (x, y) is defined to be the radius, r units. We can find this distance r by using the distance formula.

$$r = \sqrt{(x - h)^2 + (y - k)^2}$$

$$r^2 = (x - h)^2 + (y - k)^2 \quad \text{Square both sides.}$$

Circle

The graph of $(x - h)^2 + (y - k)^2 = r^2$ is a circle with center (h, k) and radius r.

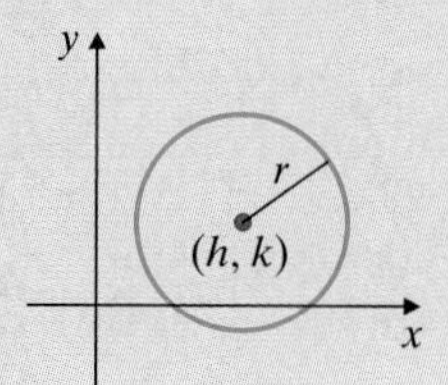

Answer

6. $\left(1, -\frac{1}{2}\right)$

The form $(x - h)^2 + (y - k)^2 = r^2$ is called **standard form**.

If an equation can be written in the standard form

$$(x - h)^2 + (y - k)^2 = r^2$$

then its graph is a circle, which we can draw by graphing the center (h, k) and using the radius r.

Helpful Hint

Notice that the radius is the *distance* from the center of the circle to any point of the circle. Also notice that the *midpoint* of a diameter of a circle is the center of the circle.

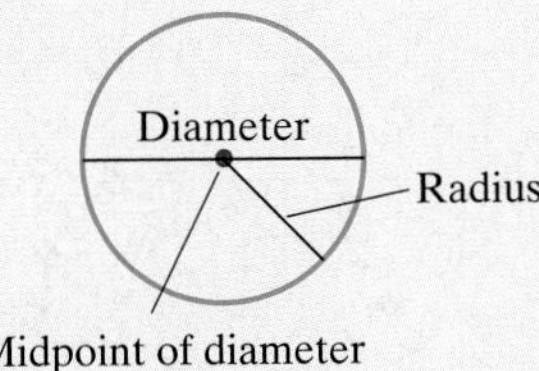

EXAMPLE 7 Graph: $x^2 + y^2 = 4$

Solution: The equation can be written in standard form as

$$(x - 0)^2 + (y - 0)^2 = 2^2$$

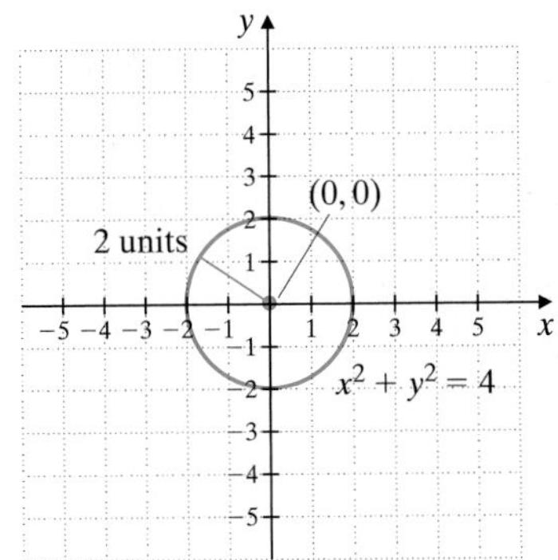

The center of the circle is $(0, 0)$, and the radius is 2. The graph of the circle is shown above.

Helpful Hint

Notice the difference between the equation of a circle and the equation of a parabola. The equation of a circle contains both x^2- and y^2-terms on the same side of the equation with equal coefficients. The equation of a parabola has either an x^2-term or a y^2-term but not both.

Practice Problem 7

Graph: $x^2 + y^2 = 9$

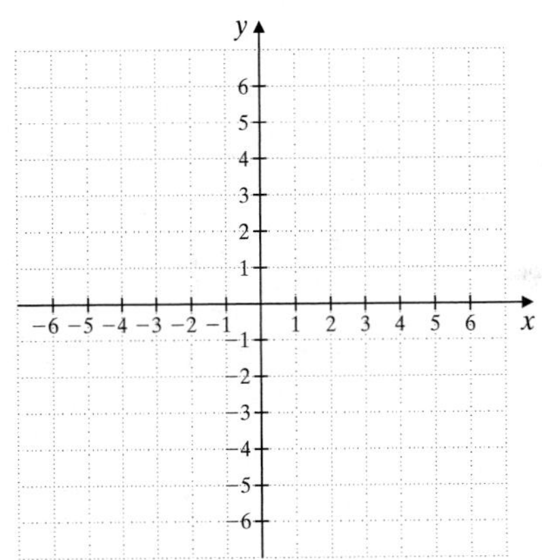

Answer

7.

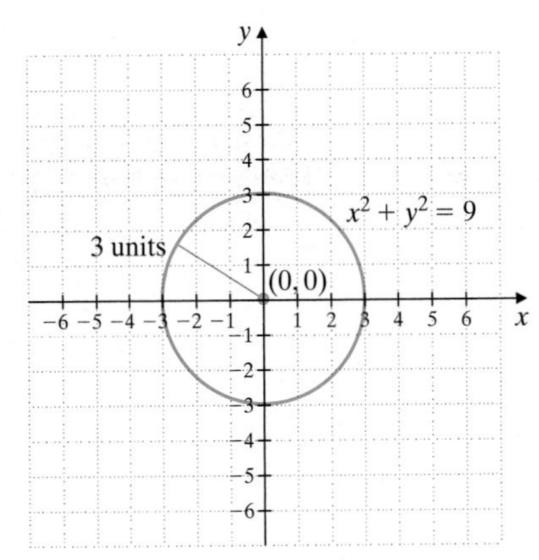

Practice Problem 8

Graph: $x^2 + (y + 2)^2 = 6$

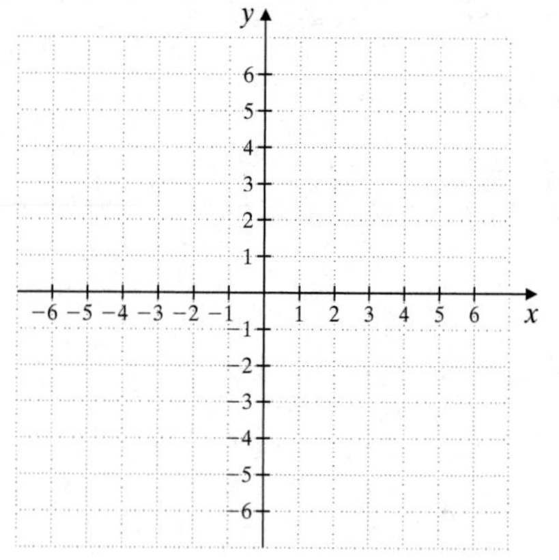

EXAMPLE 8 Graph: $(x + 1)^2 + y^2 = 8$

Solution: The equation can be written as $(x - (-1))^2 + (y - 0)^2 = 8$ with $h = -1$, $k = 0$, and $r = \sqrt{8}$. The center is $(-1, 0)$, and the radius is $\sqrt{8} = 2\sqrt{2} \approx 2.8$.

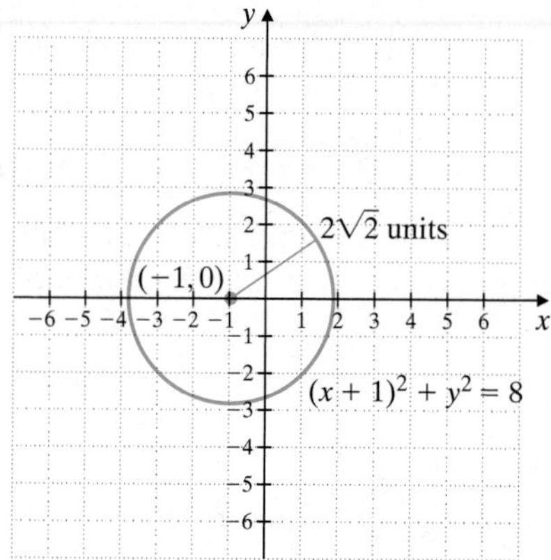

Concept Check

In the graph of the equation $(x - 3)^2 + (y - 2)^2 = 5$, what is the distance between the center of the circle and any point on the circle?

Try the Concept Check in the margin.

D Writing Equations of Circles

Since a circle is determined entirely by its center and radius, this information is all we need to write the equation of a circle.

Practice Problem 9

Write an equation of the circle with the center $(2, -5)$ and radius 7.

EXAMPLE 9

Write an equation of the circle with center $(-7, 3)$ and radius 10.

Solution: Using the given values $h = -7$, $k = 3$, and $r = 10$, we write the equation

$$(x - h)^2 + (y - k)^2 = r^2$$

or

$$(x - (-7))^2 + (y - 3)^2 = 10^2 \quad \text{Substitute the given values.}$$

or

$$(x + 7)^2 + (y - 3)^2 = 100$$

E Finding the Center and the Radius of a Circle

To find the center and the radius of a circle from its equation, we write the equation in standard form. To write the equation of a circle in standard form, we complete the square on both x and y.

Answers

8.

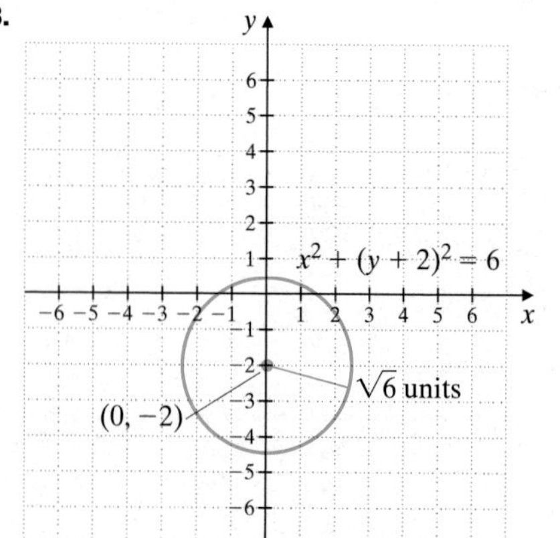

9. $(x - 2)^2 + (y + 5)^2 = 49$

Concept Check: $\sqrt{5}$ units

EXAMPLE 10 Graph: $x^2 + y^2 + 4x - 8y = 16$

Solution: Since this equation contains x^2- and y^2-terms on the same side of the equation with equal coefficients, its graph is a circle. To write the equation in standard form, we group the terms involving x and the terms involving y, and then complete the square on each variable.

$$(x^2 + 4x) + (y^2 - 8y) = 16$$

Thus, $\frac{1}{2}(4) = 2$ and $2^2 = 4$. Also, $\frac{1}{2}(-8) = -4$ and $(-4)^2 = 16$. We add 4 and then 16 to both sides.

$$(x^2 + 4x + 4) + (y^2 - 8y + 16) = 16 + 4 + 16$$

$$(x + 2)^2 + (y - 4)^2 = 36 \quad \text{Factor.}$$

This circle has the center $(-2, 4)$ and radius 6, as shown.

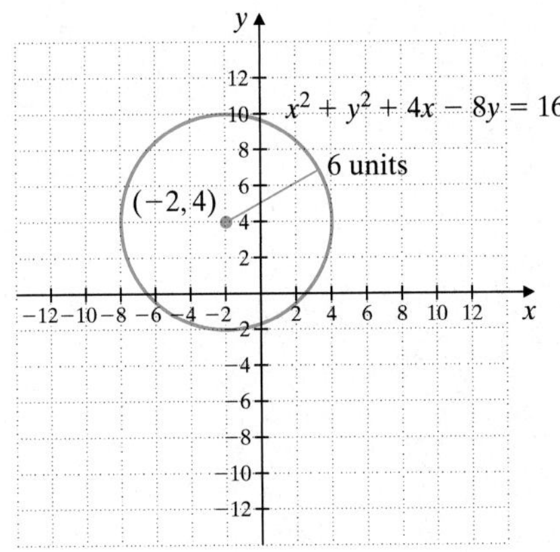

Practice Problem 10

Graph: $x^2 + y^2 - 2x + 6y = 6$

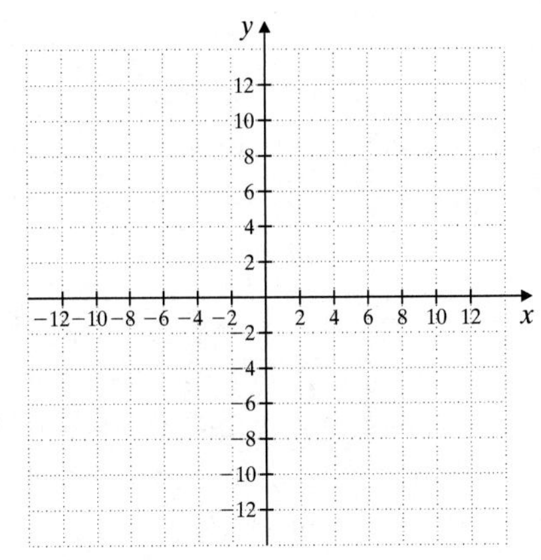

Answer

10.

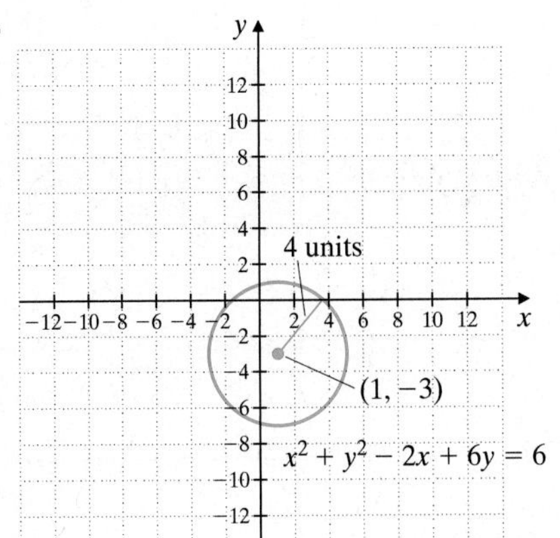

GRAPHING CALCULATOR EXPLORATIONS

To graph an equation such as $x^2 + y^2 = 25$ with a graphing calculator, we first solve the equation for y.

$$x^2 + y^2 = 25$$
$$y^2 = 25 - x^2$$
$$y = \pm\sqrt{25 - x^2}$$

The graph of $y = \sqrt{25 - x^2}$ will be the top half of the circle, and the graph of $y = -\sqrt{25 - x^2}$ will be the bottom half of the circle.

To graph, we press Y= and enter $Y_1 = \sqrt{25 - x^2}$ and $Y_2 = -\sqrt{25 - x^2}$. We insert parentheses about $25 - x^2$ so that $\sqrt{25 - x^2}$ and not $\sqrt{25} - x^2$ is graphed.

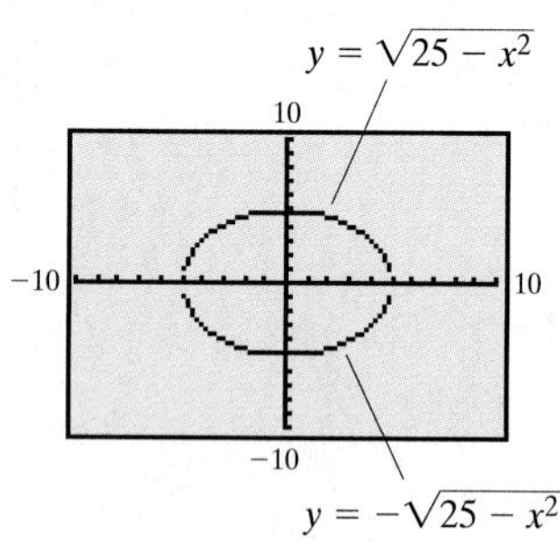

The graph does not appear to be a circle because we are currently using a standard window and the screen is rectangular. This causes the tick marks on the x-axis to be farther apart than the tick marks on the y-axis and thus creates the distorted circle. If we want the graph to appear circular, we define a square window by using a feature of the graphing calculator or redefine the window to show the x-axis from -15 to 15 and the y-axis from -10 to 10. Using a square window, the graph appears as follows:

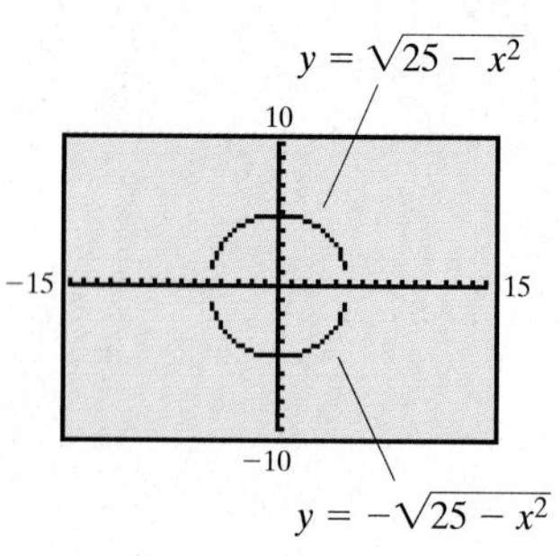

Use a graphing calculator to graph each circle.

1. $x^2 + y^2 = 55$

2. $x^2 + y^2 = 20$

3. $7x^2 + 7y^2 - 89 = 0$

4. $3x^2 + 3y^2 - 35 = 0$

Name ____________________ Section ________ Date ________

Mental Math

The graph of each equation is a parabola. Determine whether the parabola opens upward, downward, to the left, or to the right.

1. $y = x^2 - 7x + 5$

2. $y = -x^2 + 16$

3. $x = -y^2 - y + 2$

4. $x = 3y^2 + 2y - 5$

5. $y = -x^2 + 2x + 1$

6. $x = -y^2 + 2y - 6$

EXERCISE SET 12.1

The graph of each equation is a parabola. Find the vertex of the parabola and then graph it. See Examples 1 through 4.

1. $x = 3y^2$

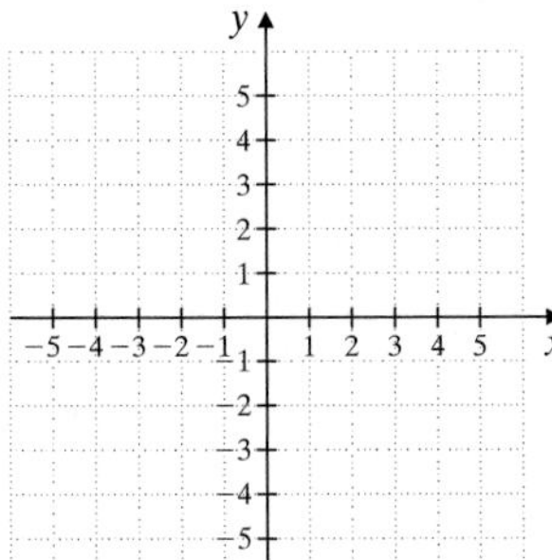

2. $x = -2y^2$

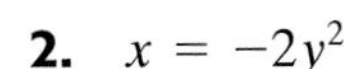

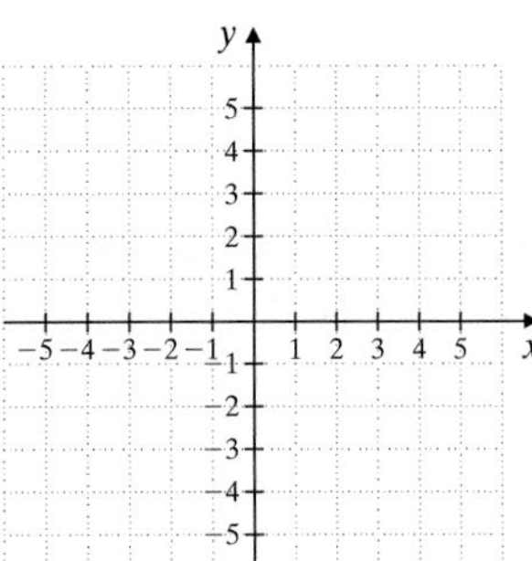

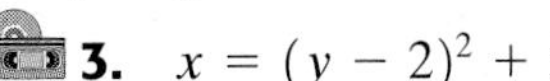

3. $x = (y - 2)^2 + 3$

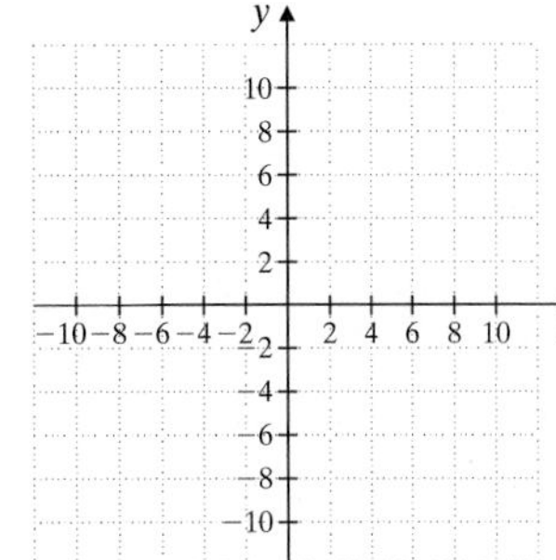

4. $x = (y - 4)^2 - 1$

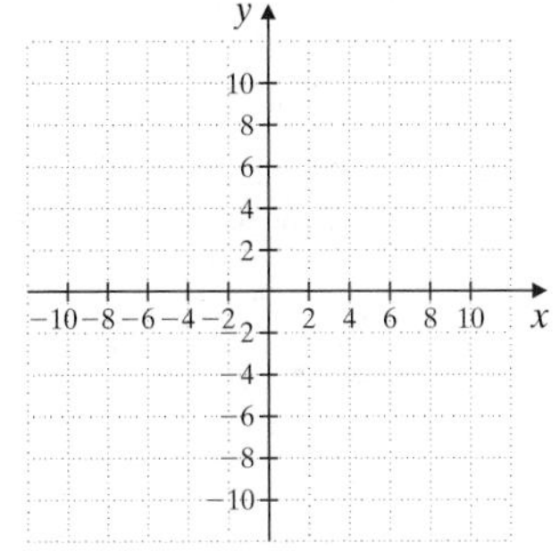

5. $y = 3(x - 1)^2 + 5$

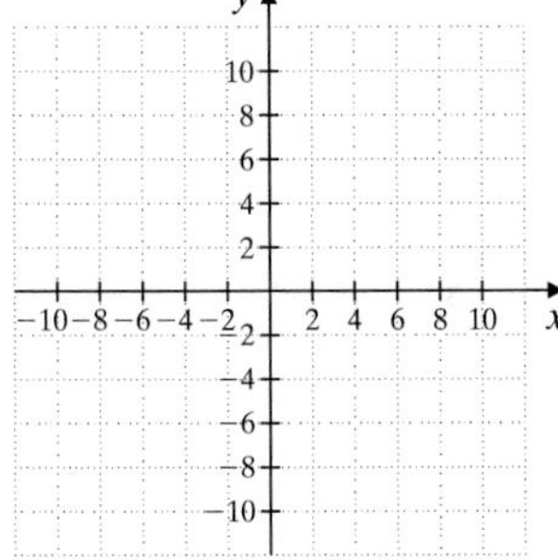

6. $x = -4(y - 2)^2 + 2$

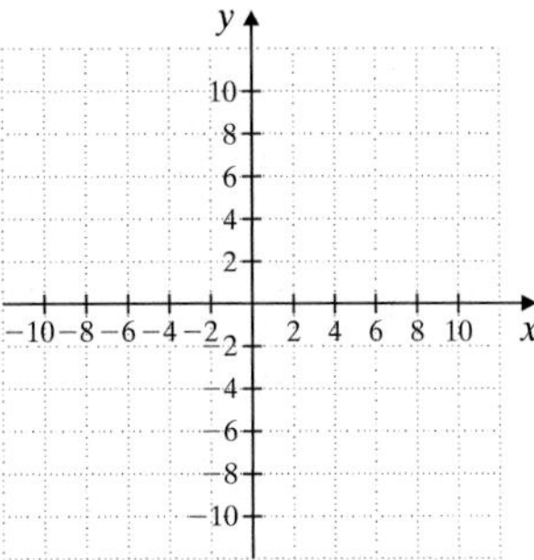

7. $x = y^2 + 6y + 8$

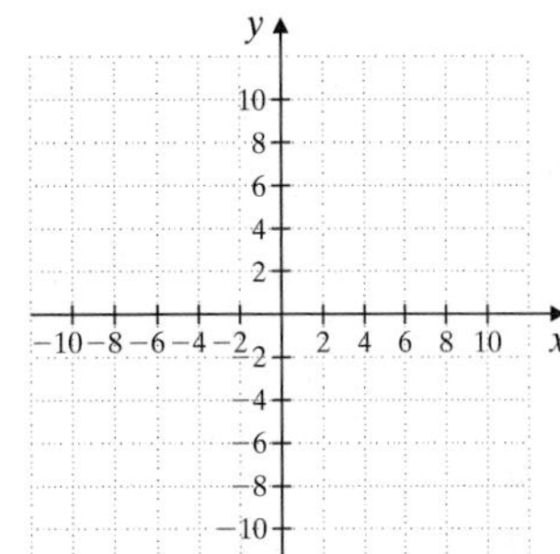

8. $x = y^2 - 6y + 6$

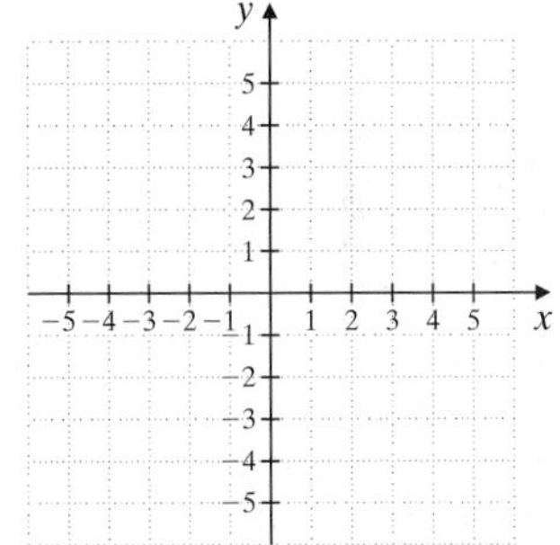

9. $y = x^2 + 10x + 20$

10. $y = x^2 + 4x - 5$

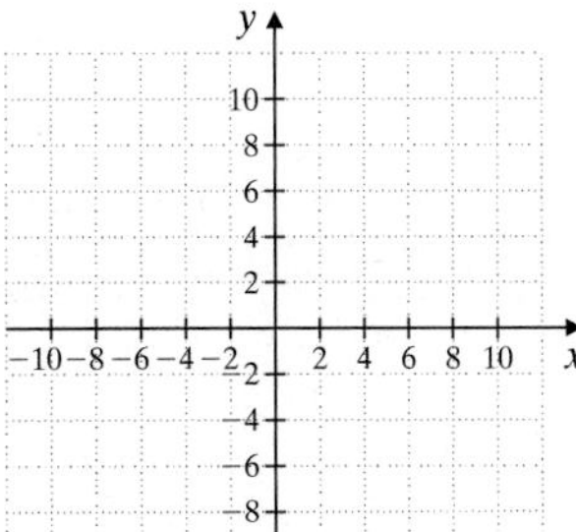

11. $x = -2y^2 + 4y + 6$

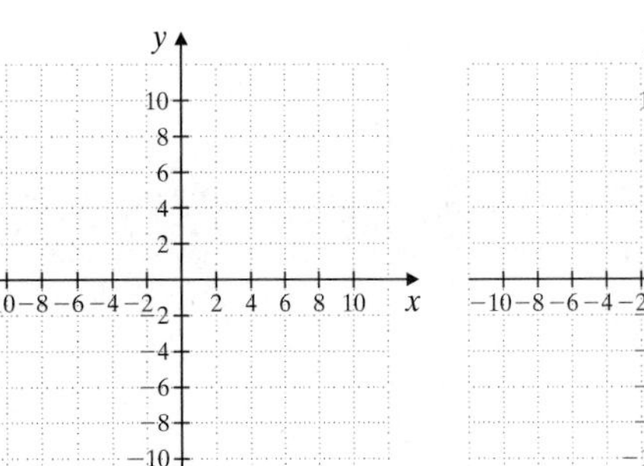

12. $x = 3y^2 + 6y + 7$

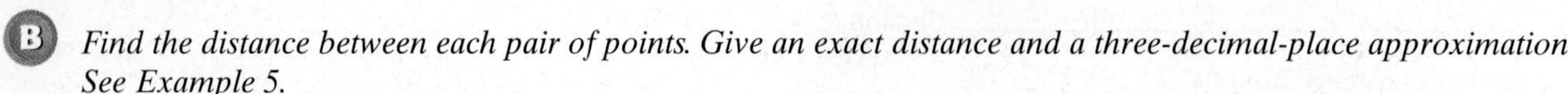

Find the distance between each pair of points. Give an exact distance and a three-decimal-place approximation. See Example 5.

13. $(5, 1)$ and $(8, 5)$

14. $(2, 3)$ and $(14, 8)$

15. $(-3, 2)$ and $(1, -3)$

16. $(3, -2)$ and $(-4, 1)$

17. $(-9, 4)$ and $(-8, 1)$

18. $(-5, -2)$ and $(-6, -6)$

19. $(0, -\sqrt{2})$ and $(\sqrt{3}, 0)$

20. $(-\sqrt{5}, 0)$ and $(0, \sqrt{7})$

21. $(1.7, -3.6)$ and $(-8.6, 5.7)$

22. $(9.6, 2.5)$ and $(-1.9, -3.7)$

23. $(2\sqrt{3}, \sqrt{6})$ and $(-\sqrt{3}, 4\sqrt{6})$

24. $(5\sqrt{2}, -4)$ and $(-3\sqrt{2}, -8)$

Find the midpoint of each line segment whose endpoints are given. See Example 6.

25. $(6, -8); (2, 4)$

26. $(3, 9); (7, 11)$

27. $(-2, -1); (-8, 6)$

28. $(-3, -4); (6, -8)$

29. $(7, 3); (-1, -3)$

30. $(-2, 5); (-1, 6)$

31. $\left(\frac{1}{2}, \frac{3}{8}\right); \left(-\frac{3}{2}, \frac{5}{8}\right)$

32. $\left(-\frac{2}{5}, \frac{7}{15}\right); \left(-\frac{2}{5}, -\frac{4}{15}\right)$

33. $(\sqrt{2}, 3\sqrt{5}); (\sqrt{2}, -2\sqrt{5})$

34. $(\sqrt{8}, -\sqrt{12}); (3\sqrt{2}, 7\sqrt{3})$

35. $(4.6, -3.5); (7.8, -9.8)$

36. $(-4.6, 2.1); (-6.7, 1.9)$

  *The graph of each equation is a circle. Find the center and the radius, and then graph the circle. See Examples 7, 8, and 10.*

37. $x^2 + y^2 = 9$

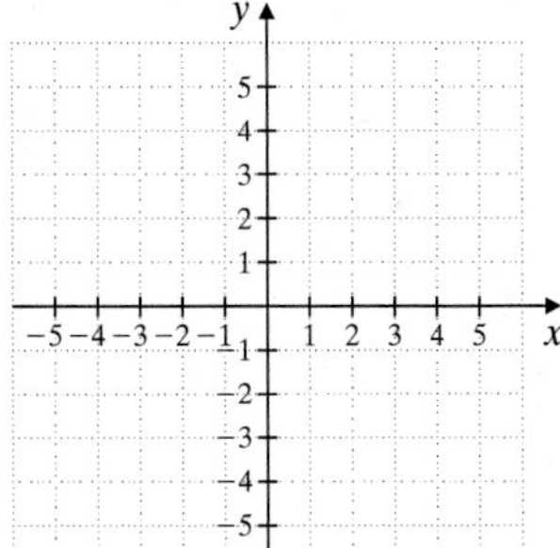

38. $x^2 + y^2 = 25$

39. $x^2 + (y - 2)^2 = 1$

40. $(x - 3)^2 + y^2 = 9$

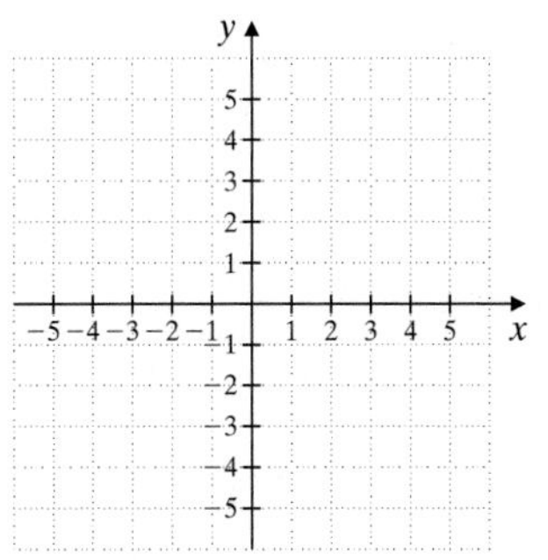

41. $(x - 5)^2 + (y + 2)^2 = 1$

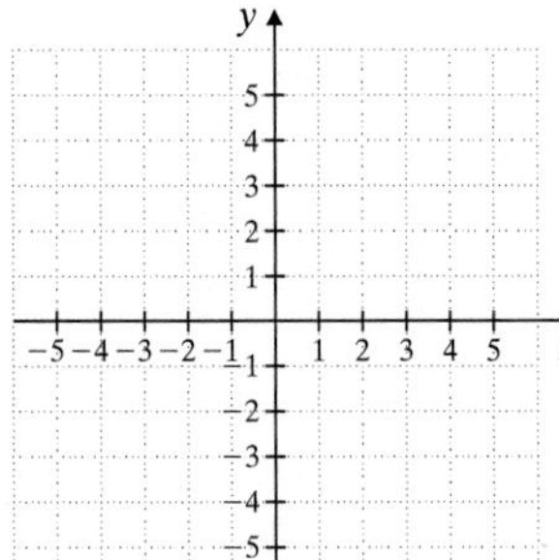

42. $(x + 3)^2 + (y + 3)^2 = 4$

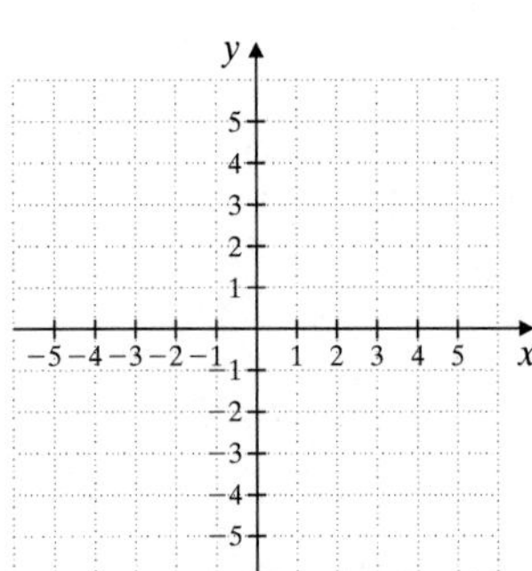

43. $x^2 + y^2 + 6y = 0$

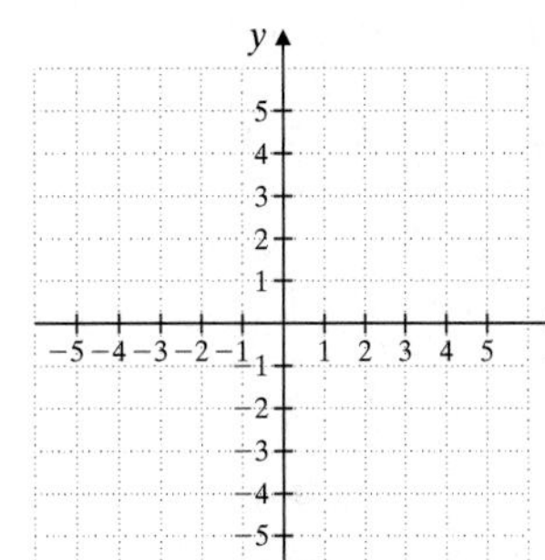

44. $x^2 + 10x + y^2 = 0$

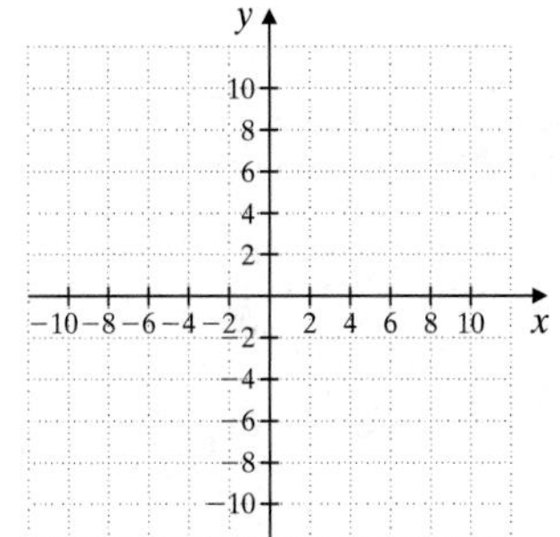

45. $x^2 + y^2 + 2x - 4y = 4$

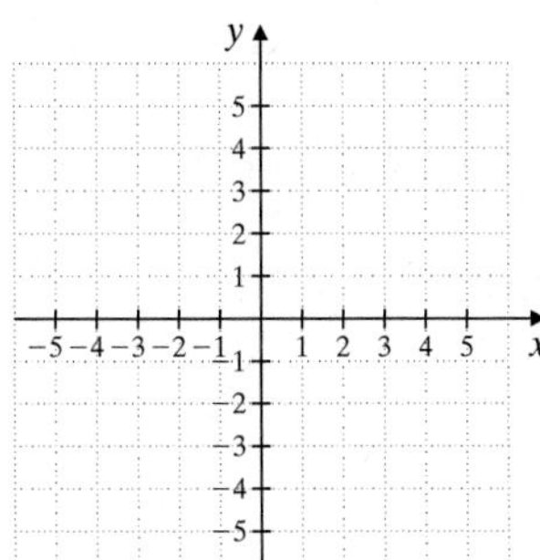

46. $x^2 + y^2 + 6x - 4y = 3$

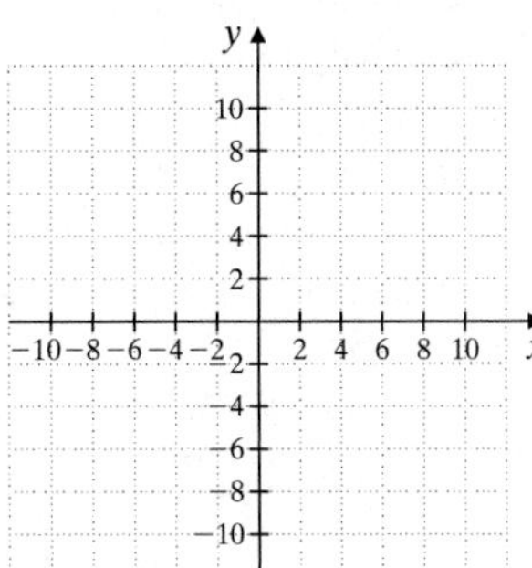

47. $x^2 + y^2 - 4x - 8y - 2 = 0$

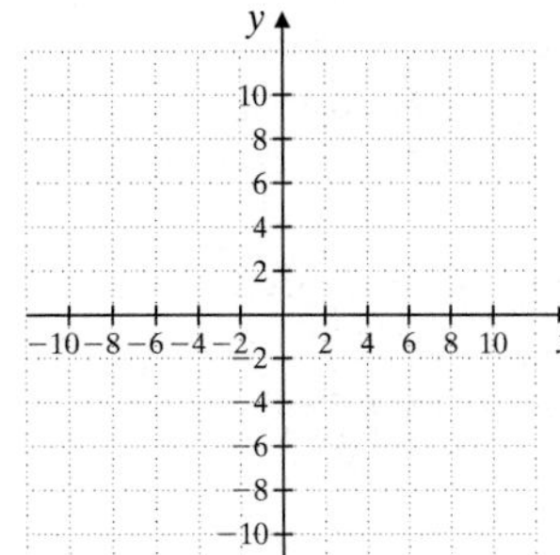

48. $x^2 + y^2 - 2x - 6y - 5 = 0$

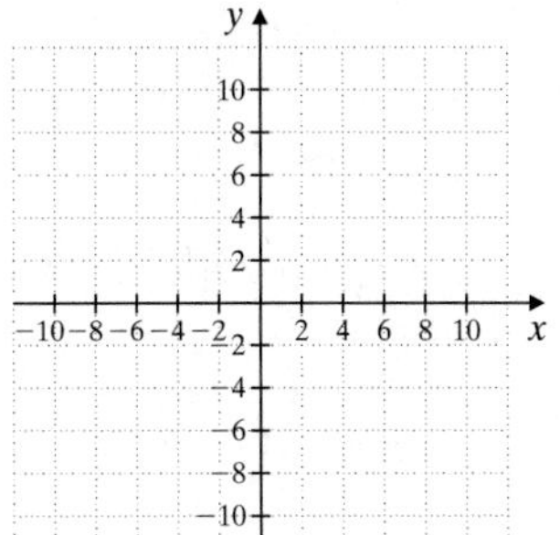

D *Write an equation of the circle with the given center and radius. See Example 9.*

49. $(2, 3); 6$

50. $(-7, 6); 2$

51. $(0, 0); \sqrt{3}$

52. $(0, -6); \sqrt{2}$

53. $(-5, 4); 3\sqrt{5}$

54. The origin; $4\sqrt{7}$

Review and Preview

Graph each equation. See Section 3.2.

55. $y = 2x + 5$

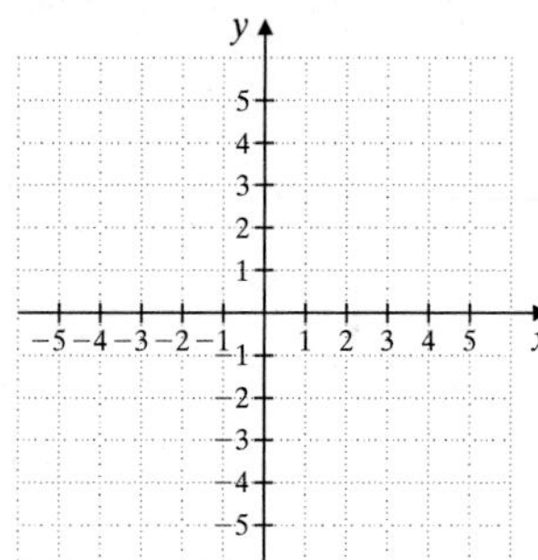

56. $y = -3x + 3$

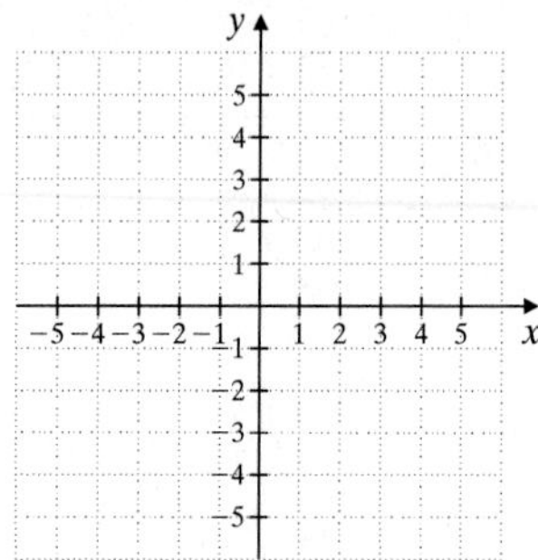

57. $y = 3$

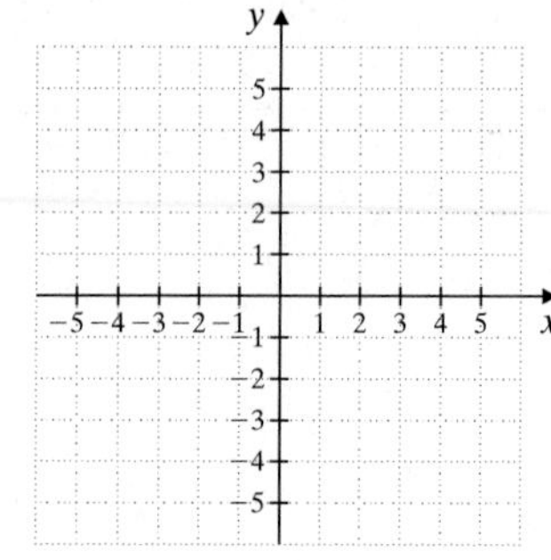

58. $x = -2$

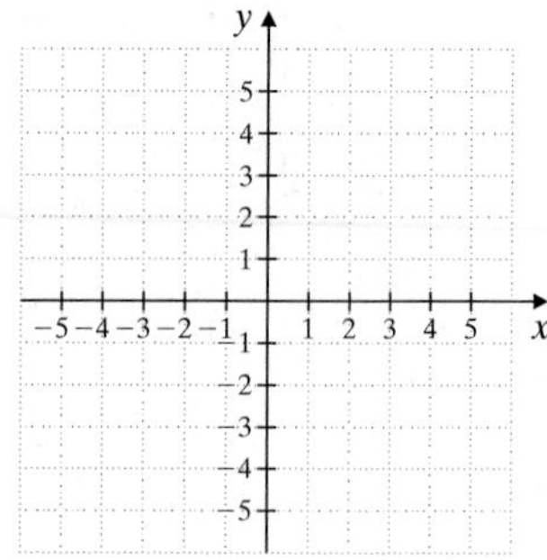

Rationalize each denominator and simplify if possible. See Section 9.5.

59. $\frac{1}{\sqrt{3}}$

60. $\frac{\sqrt{5}}{\sqrt{8}}$

61. $\frac{4\sqrt{7}}{\sqrt{6}}$

62. $\frac{10}{\sqrt{5}}$

Combining Concepts

63. In 1893, Pittsburgh bridge builder George Ferris designed and built a gigantic revolving steel wheel whose height was 264 feet and diameter was 250 feet. This Ferris wheel opened at the 1893 exposition in Chicago. It had 36 wooden cars, each capable of holding 60 passengers. (*Source:* The Handy Science Answer Book)

a. What was the radius of this Ferris wheel?

b. How close is the wheel to the ground?

c. How high is the center of the wheel from the ground?

d. Using the axes in the drawing, what are the coordinates of the center of the wheel?

e. Use parts **a** and **d** to write the equation of the wheel.

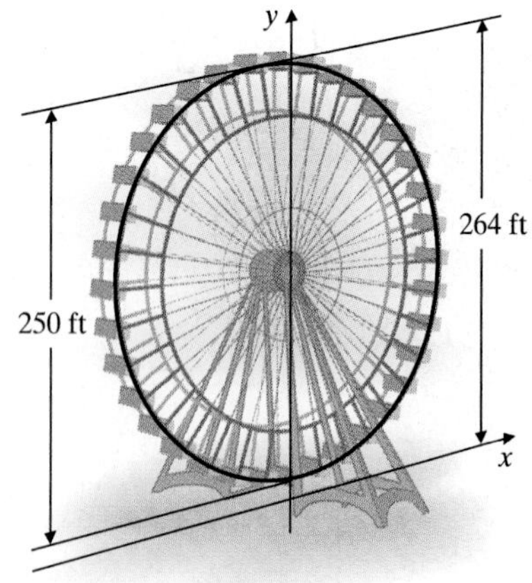

64. The world's largest-diameter Ferris wheel currently operating is the Cosmoclock 21 at Yokohama City, Japan. It has a 60-armed wheel, its diameter is 100 meters and it has a height of 105 meters. (*Source:* The Handy Science Answer Book)

a. What is the radius of this Ferris wheel?

b. How close is the wheel to the ground?

c. How high is the center of the wheel from the ground?

d. Using the axes in the drawing, what are the coordinates of the center of the wheel?

e. Use parts **a** and **d** to write the equation of the wheel.

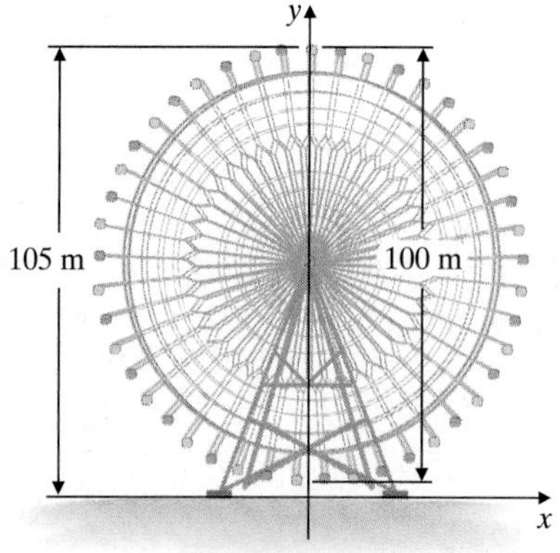

65. If you are given a list of equations of circles and parabolas and none are in standard form, explain how you would determine which is an equation of a circle and which is an equation of a parabola. Explain also how you would distinguish the upward or downward parabolas from the left-opening or right-opening parabolas.

△ **66.** Determine whether the triangle with vertices $(2, 6)$, $(0, -2)$, and $(5, 1)$ is an isosceles triangle.

Solve.

67. Two surveyors need to find the distance across a lake. They place a reference pole at point A in the diagram. Point B is 3 meters east and 1 meter north of the reference point A. Point C is 19 meters east and 13 meters north of point A. Find the distance across the lake, from B to C.

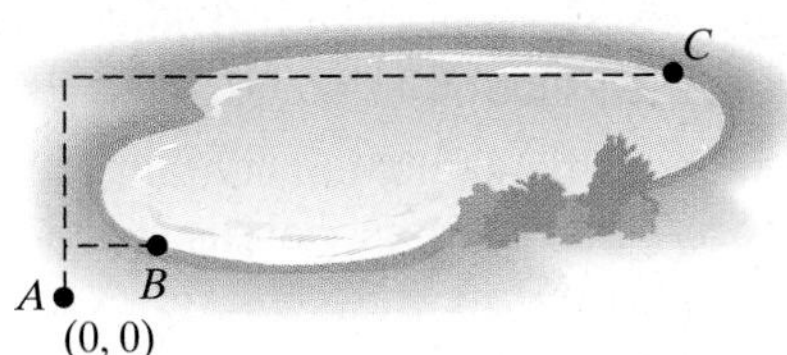

68. Cindy Brown, an architect, is drawing plans on grid paper for a circular pool with a fountain in the middle. The paper is marked off in centimeters, and each centimeter represents 1 foot. On the paper, the diameter of the "pool" is 20 centimeters, and "fountain" is the point $(0, 0)$.

a. Sketch the architect's drawing. Be sure to label the axes.

b. Write an equation that describes the circular pool.

c. Cindy plans to place a circle of lights around the fountain such that each light is 5 feet from the fountain. Write an equation for the circle of lights and sketch the circle on your drawing.

69. A bridge constructed over a bayou has a supporting arch in the shape of a parabola. Find an equation of the parabolic arch if the length of the road over the arch is 100 meters and the maximum height of the arch is 40 meters.

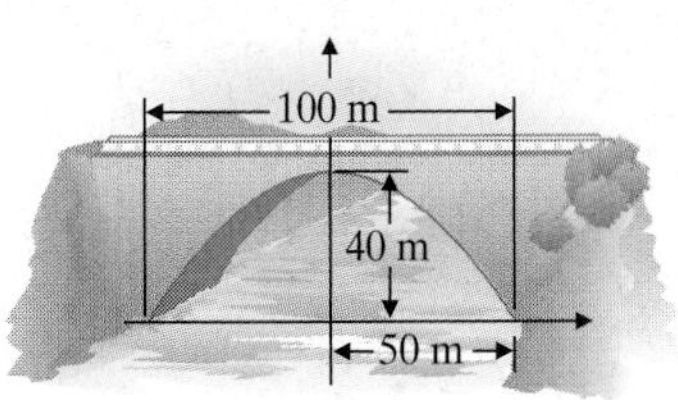

STUDY SKILLS REMINDER

How are you doing?

If you haven't done so yet, take a few moments and think about how you are doing in this course. Are you working toward your goal of successfully completing this course? Is your performance on homework, quizzes, and tests satisfactory? If not, you might want to see your instructor to see if he/she has any suggestions on how you can improve your performance. Let me once again remind you that, in addition to your instructor, there are many places to get help with your mathematics course. A few suggestions are below.

- This text has an accompanying video lesson for every section in this text.
- The back of this book contains answers to odd-numbered exercises and selected solutions.
- MathPro is available with this text. It is a tutorial software program with lessons corresponding to each section in the text.
- There is a student solutions manual available that contains worked-out solutions to odd-numbered exercises as well as solutions to every exercise in the Chapter Pretests, Integrated Reviews, Chapter Reviews, Chapter Tests, and Cumulative Reviews.
- Don't forget to check with your instructor for other local resources available to you, such as a tutor center.

12.2 The Ellipse and the Hyperbola

OBJECTIVES

A Define and graph an ellipse.

B Define and graph a hyperbola.

A Graphing Ellipses

An **ellipse** can be thought of as the set of points in a plane such that the sum of the distances of each of those points from two fixed points is constant. Each of the two fixed points is called a **focus**. The plural of focus is **foci**. The point midway between the foci is called the **center**.

An ellipse may be drawn by hand by using two tacks, a piece of string, and a pencil. Secure the two tacks into a piece of cardboard, for example, and tie each end of the string to a tack. Use your pencil to pull the string tight and draw the ellipse. The two tacks are the foci of the drawn ellipse.

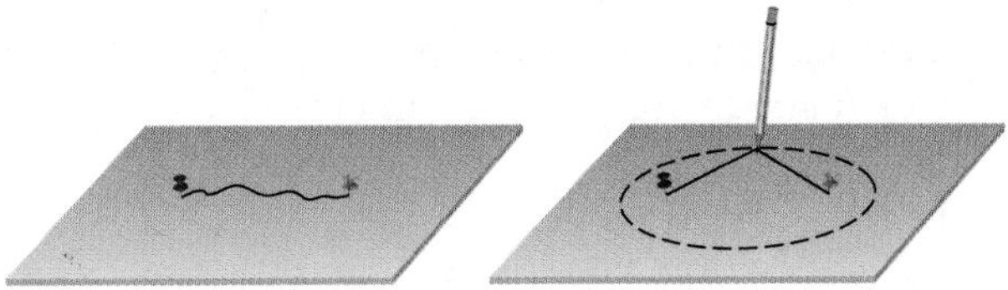

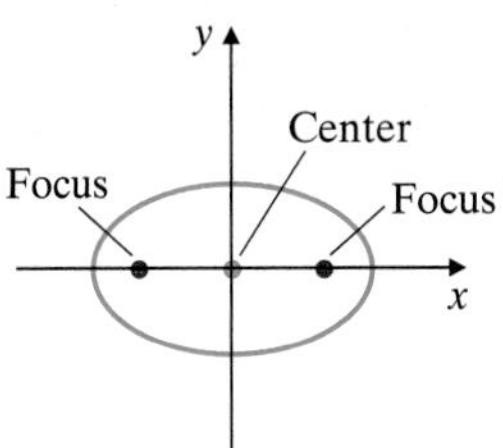

Ellipse with Center (0, 0)

The graph of an equation of the form $\frac{x^2}{a^2} + \frac{y^2}{b^2} = 1$ is an ellipse with center $(0, 0)$. The x-intercepts are a and $-a$, and the y-intercepts are b and $-b$.

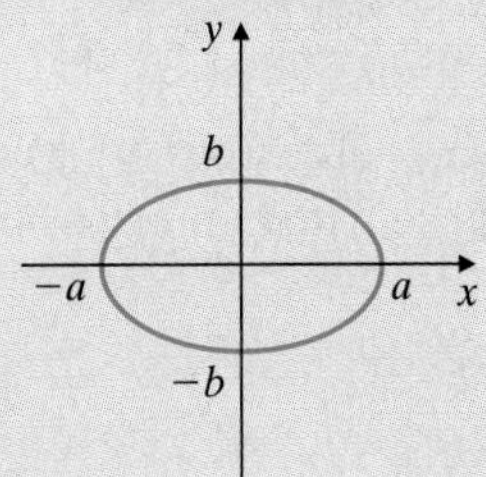

The **standard form** of an ellipse with center $(0, 0)$ is $\frac{x^2}{a^2} + \frac{y^2}{b^2} = 1$.

EXAMPLE 1 Graph: $\frac{x^2}{9} + \frac{y^2}{16} = 1$

Solution: The equation is of the form $\frac{x^2}{a^2} + \frac{y^2}{b^2} = 1$ with $a = 3$ and $b = 4$, so its graph is an ellipse with center $(0, 0)$, x-intercepts $(3, 0)$ and $(-3, 0)$, and y-intercepts $(0, 4)$ and $(0, -4)$.

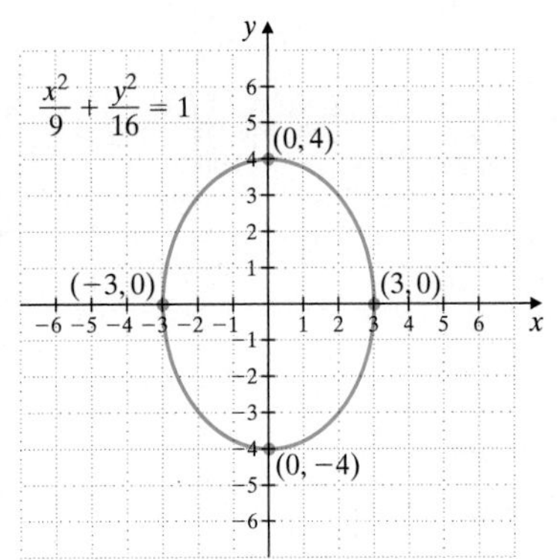

Practice Problem 1

Graph: $\frac{x^2}{4} + \frac{y^2}{9} = 1$

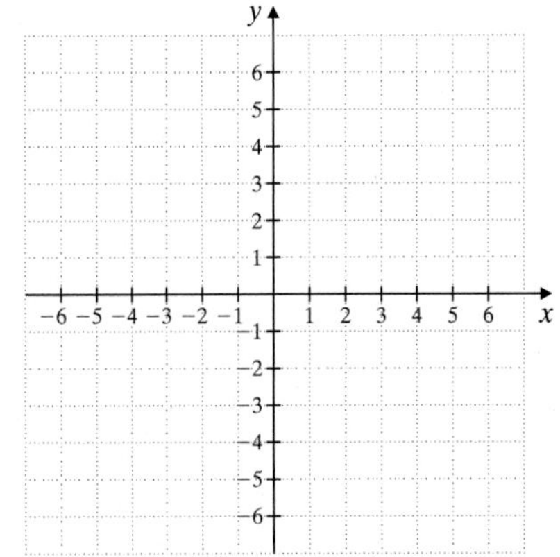

Answer

1.

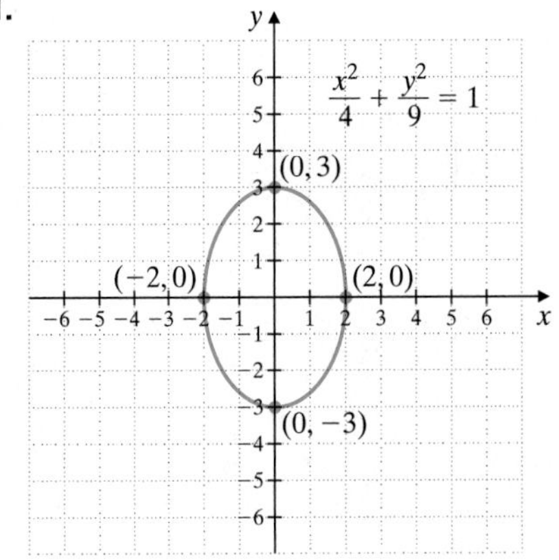

Practice Problem 2

Graph: $4x^2 + 25y^2 = 100$

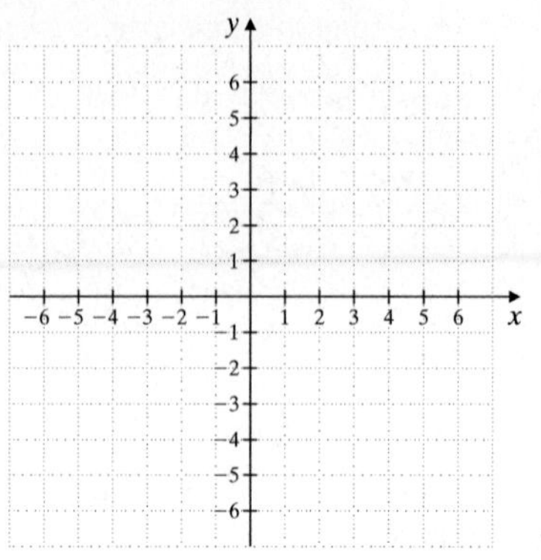

Practice Problem 3

Graph: $\dfrac{(x-1)^2}{9} + \dfrac{(y-3)^2}{16} = 1$

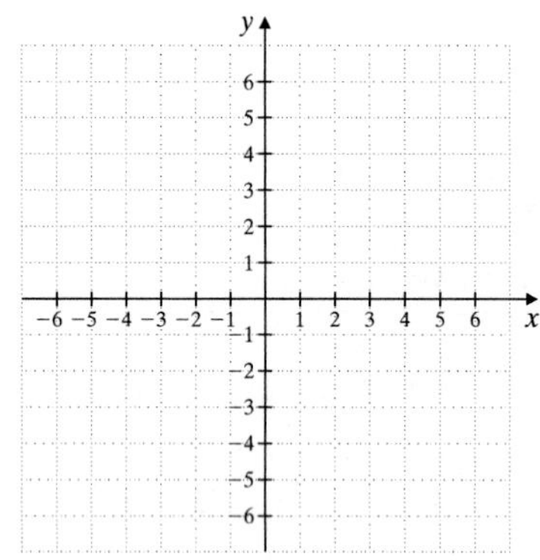

Answers

2.

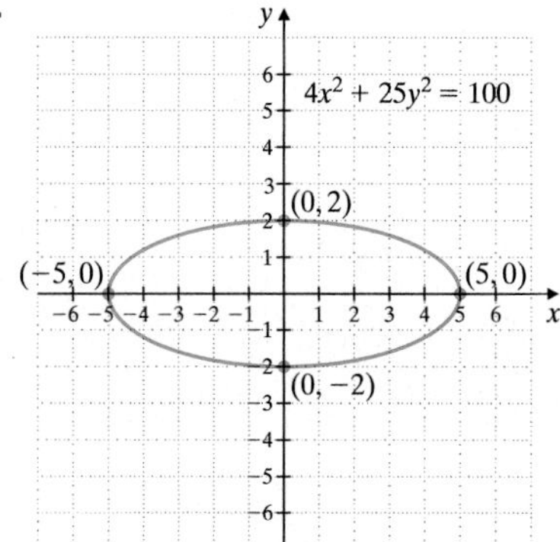

3.

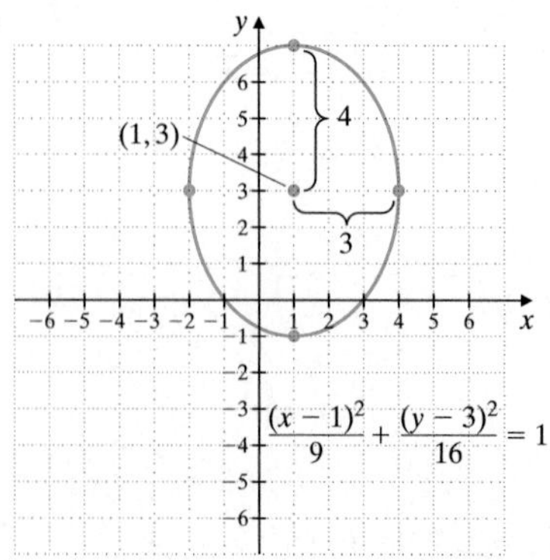

EXAMPLE 2 Graph: $4x^2 + 16y^2 = 64$

Solution: Although this equation contains a sum of squared terms in x and y on the same side of an equation, this is not the equation of a circle since the coefficients of x^2 and y^2 are not the same. When this happens, the graph is an ellipse. Since the standard form of the equation of an ellipse has 1 on one side, we divide both sides of this equation by 64 to get it in standard form.

$$4x^2 + 16y^2 = 64$$

$$\frac{4x^2}{64} + \frac{16y^2}{64} = \frac{64}{64} \quad \text{Divide both sides by 64.}$$

$$\frac{x^2}{16} + \frac{y^2}{4} = 1 \quad \text{Simplify.}$$

We now recognize the equation of an ellipse with center $(0, 0)$, x-intercepts $(4, 0)$ and $(-4, 0)$, and y-intercepts $(0, 2)$ and $(0, -2)$.

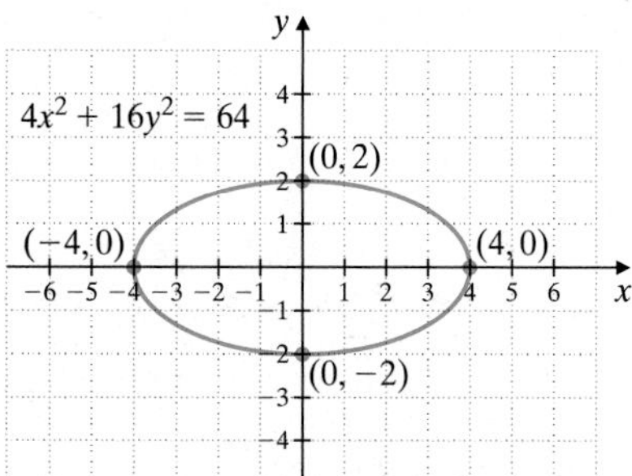

The center of an ellipse is not always $(0, 0)$, as shown in the next example. The standard form of an ellipse with center (h, k) is

$$\frac{(x-h)^2}{a^2} + \frac{(y-k)^2}{b^2} = 1$$

EXAMPLE 3 Graph: $\dfrac{(x+3)^2}{25} + \dfrac{(y-2)^2}{36} = 1$

Solution: This ellipse has center $(-3, 2)$. Notice that $a = 5$ and $b = 6$. To find four points on the graph of the ellipse, we first graph the center, $(-3, 2)$. Since $a = 5$, we count 5 units right and then 5 units left of the point with coordinates $(-3, 2)$. Next, since $b = 6$, we start at $(-3, 2)$ and count 6 units up and then 6 units down to find two more points on the ellipse.

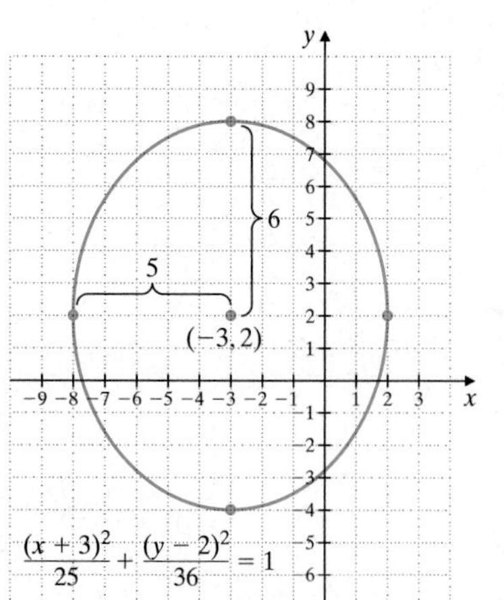

Try the Concept Check in the margin.

B Graphing Hyperbolas

The final conic section is the **hyperbola**. A hyperbola is the set of points in a plane such that for each point in the set, the absolute value of the difference of the distances from two fixed points is constant. Each of the two fixed points is called a **focus**. The point midway between the foci is called the **center**.

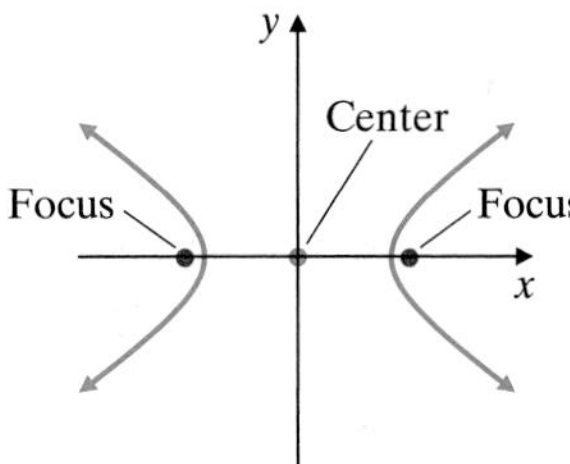

Using the distance formula, we can show that the graph of $\frac{x^2}{a^2} - \frac{y^2}{b^2} = 1$ is a hyperbola with center $(0, 0)$ and x-intercepts $(a, 0)$ and $(-a, 0)$. Also, the graph of $\frac{y^2}{b^2} - \frac{x^2}{a^2} = 1$ is a hyperbola with center $(0, 0)$ and y-intercepts $(0, b)$ and $(0, -b)$.

Hyperbola with Center (0, 0)

The graph of an equation of the form $\frac{x^2}{a^2} - \frac{y^2}{b^2} = 1$ is a hyperbola with center $(0, 0)$ and x-intercepts $(a, 0)$ and $(-a, 0)$.

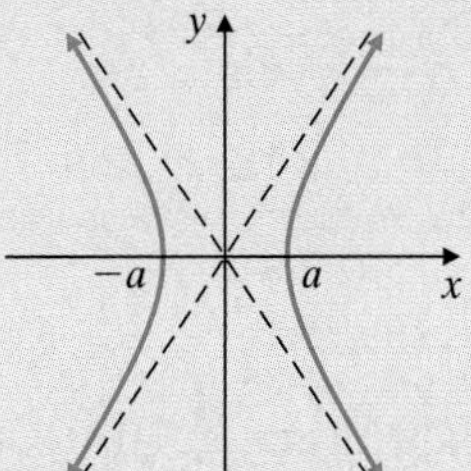

The graph of an equation of the form $\frac{y^2}{b^2} - \frac{x^2}{a^2} = 1$ is a hyperbola with center $(0, 0)$ and y-intercepts $(0, b)$ and $(0, -b)$.

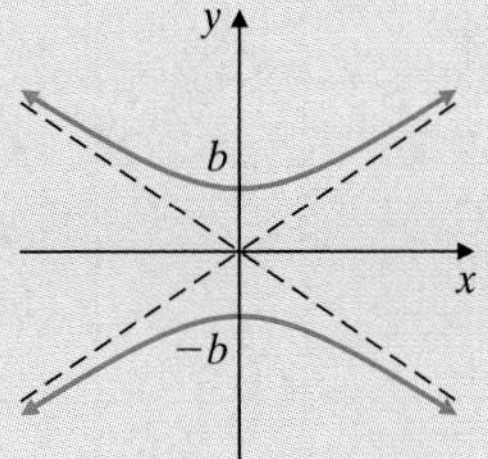

The equations $\frac{x^2}{a^2} - \frac{y^2}{b^2} = 1$ and $\frac{y^2}{b^2} - \frac{x^2}{a^2} = 1$ are the **standard forms** for the equation of a hyperbola.

Concept Check

In the graph of the equation $\frac{x^2}{64} + \frac{y^2}{36} = 1$, which distance is longer: the distance between the x-intercepts or the distance between the y-intercepts? How much longer? Explain.

Answer

Concept Check: x-intercepts, by 4 units

Helpful Hint

Notice the difference between the equation of an ellipse and a hyperbola. The equation of the ellipse contains x^2- and y^2-terms on the same side of the equation with same-sign coefficients. For a hyperbola, the coefficients on the same side of the equation have different signs.

Graphing a hyperbola such as $\frac{y^2}{b^2} - \frac{x^2}{a^2} = 1$ is made easier by recognizing one of its important characteristics. Examining the figure below, notice how the sides of the branches of the hyperbola extend indefinitely and seem to approach, but not intersect, the dashed lines in the figure. These dashed lines are called the **asymptotes** of the hyperbola.

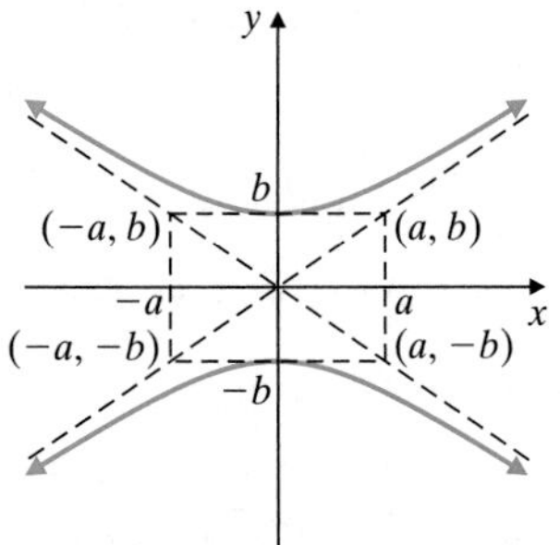

To sketch these lines, or asymptotes, draw a rectangle with vertices $(a, b), (-a, b), (a, -b)$, and $(-a, -b)$. The asymptotes of the hyperbola are the extended diagonals of this rectangle.

EXAMPLE 4 Graph: $\frac{x^2}{16} - \frac{y^2}{25} = 1$

Solution: This equation has the form $\frac{x^2}{a^2} - \frac{y^2}{b^2} = 1$, with $a = 4$ and $b = 5$. Thus, its graph is a hyperbola with center $(0, 0)$ and x-intercepts of $(4, 0)$ and $(-4, 0)$. To aid in graphing the hyperbola, we first sketch its asymptotes. The extended diagonals of the rectangle with coordinates $(4, 5), (4, -5), (-4, 5)$, and $(-4, -5)$ are the asymptotes of the hyperbola. Then we use the asymptotes to aid in graphing the hyperbola.

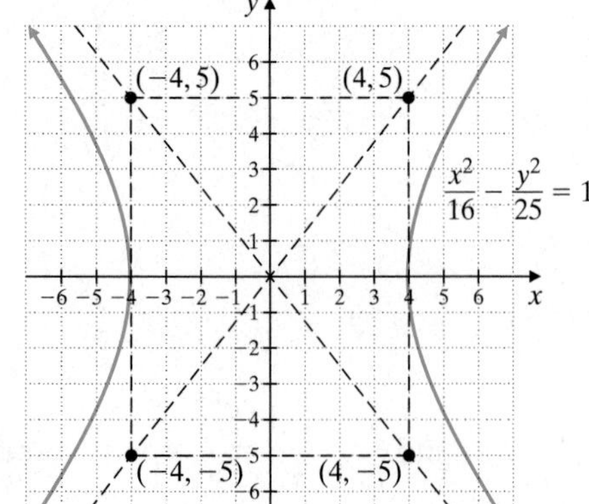

Practice Problem 4

Graph: $\frac{x^2}{4} - \frac{y^2}{9} = 1$

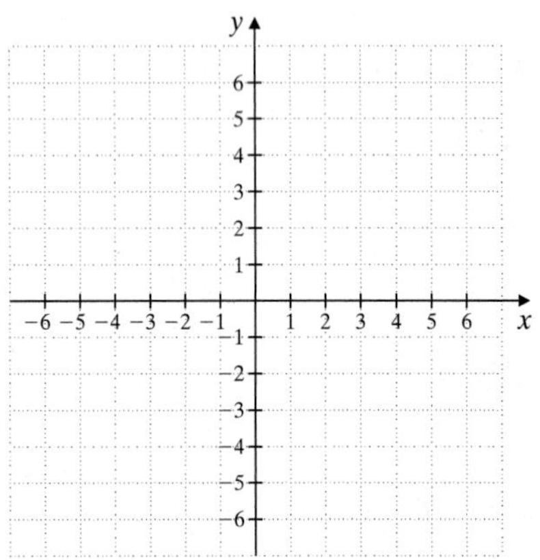

Answer

4.

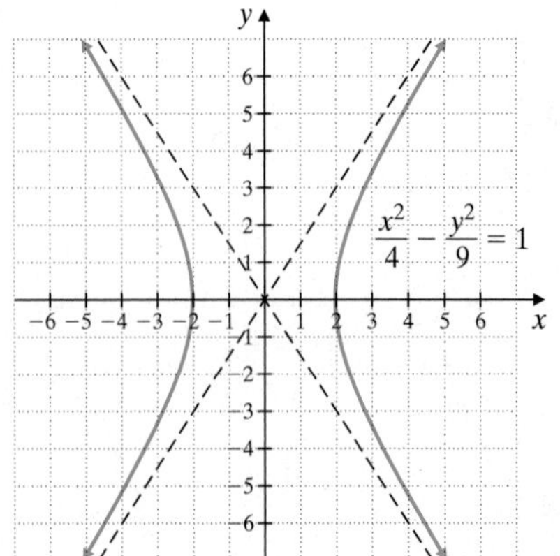

EXAMPLE 5 Graph: $4y^2 - 9x^2 = 36$

Solution: Since this is a difference of squared terms in x and y on the same side of the equation, its graph is a hyperbola, as opposed to an ellipse or a circle. The standard form of the equation of a hyperbola has a 1 on one side, so we divide both sides of the equation by 36 to get it in standard form.

$$4y^2 - 9x^2 = 36$$

$$\frac{4y^2}{36} - \frac{9x^2}{36} = \frac{36}{36} \quad \text{Divide both sides by 36.}$$

$$\frac{y^2}{9} - \frac{x^2}{4} = 1 \quad \text{Simplify.}$$

The equation is of the form $\frac{y^2}{b^2} - \frac{x^2}{a^2} = 1$ with $a = 2$ and $b = 3$, so the hyperbola is centered at $(0, 0)$ with y-intercepts $(0, 3)$ and $(0, -3)$.

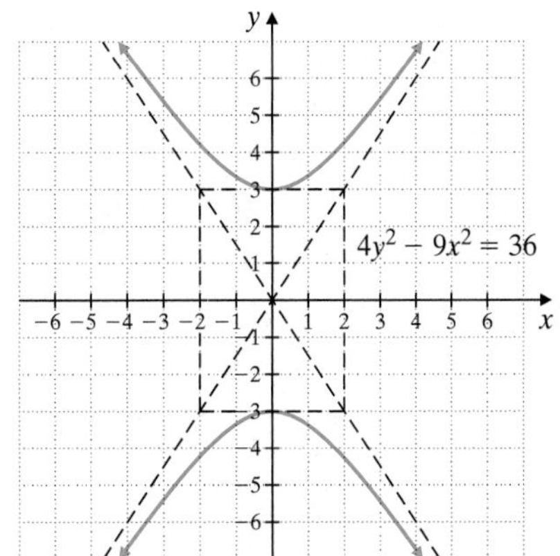

Practice Problem 5

Graph: $9y^2 - 16x^2 = 144$

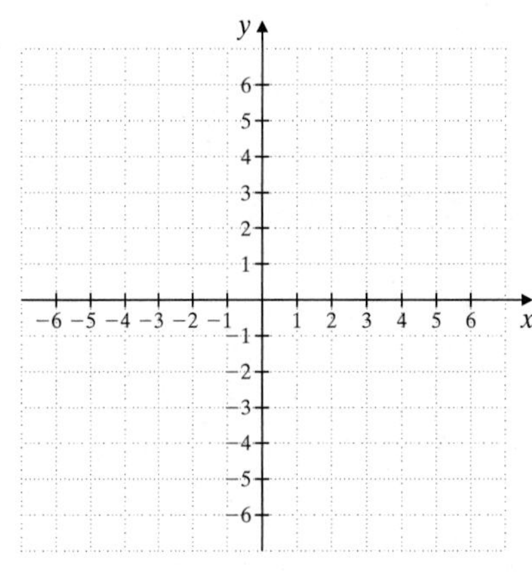

Answer

5.

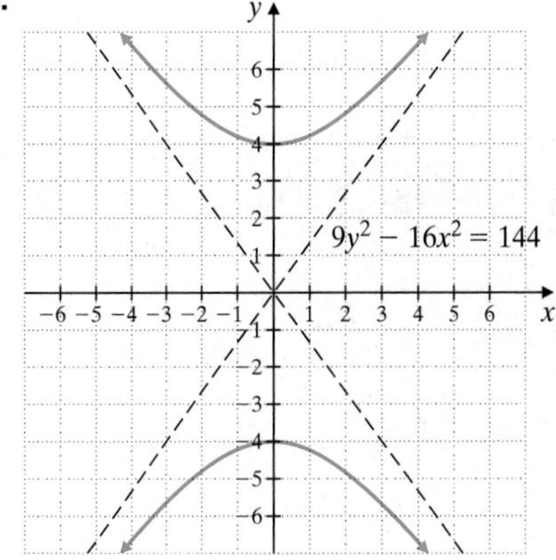

GRAPHING CALCULATOR EXPLORATIONS

To find the graph of an ellipse by using a graphing calculator, use the same procedure as for graphing a circle. For example, to graph $x^2 + 3y^2 = 22$, first solve for y.

$$3y^2 = 22 - x^2$$

$$y^2 = \frac{22 - x^2}{3}$$

$$y = \pm\sqrt{\frac{22 - x^2}{3}}$$

Next press the Y= key and enter $Y_1 = \sqrt{\frac{22 - x^2}{3}}$ and $Y_2 = -\sqrt{\frac{22 - x^2}{3}}$. (Insert two sets of parentheses in the radicand as $\sqrt{((22 - x^2)/3)}$ so that the desired graph is obtained.) The graph appears as follows:

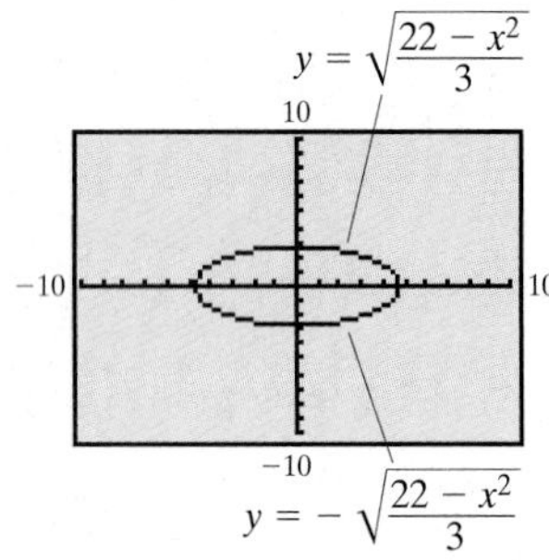

Use a graphing calculator to graph each ellipse.

1. $10x^2 + y^2 = 32$

2. $20x^2 + 5y^2 = 100$

3. $7.3x^2 + 15.5y^2 = 95.2$

4. $18.8x^2 + 36.1y^2 = 205.8$

Name ______________________ Section __________ Date __________

Mental Math

Identify the graph of each equation as an ellipse or a hyperbola.

1. $\frac{x^2}{16} + \frac{y^2}{4} = 1$

2. $\frac{x^2}{16} - \frac{y^2}{4} = 1$

3. $x^2 - 5y^2 = 3$

4. $-x^2 + 5y^2 = 3$

5. $-\frac{y^2}{25} + \frac{x^2}{36} = 1$

6. $\frac{y^2}{25} + \frac{x^2}{36} = 1$

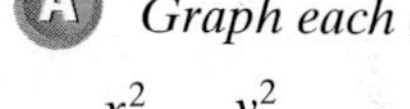

EXERCISE SET 12.2

A *Graph each equation. See Examples 1 and 2.*

1. $\frac{x^2}{4} + \frac{y^2}{25} = 1$

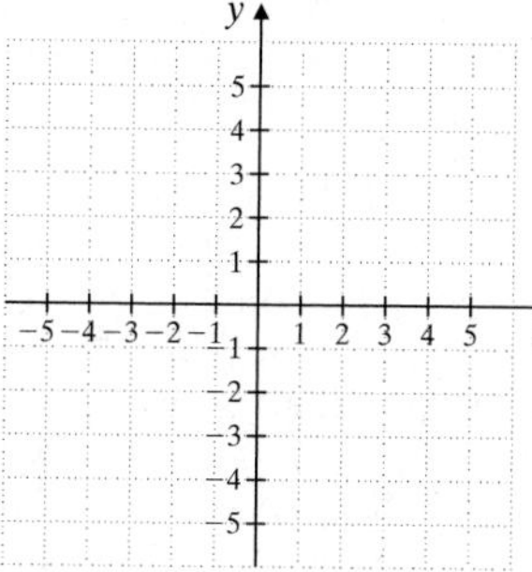

2. $\frac{x^2}{9} + y^2 = 1$

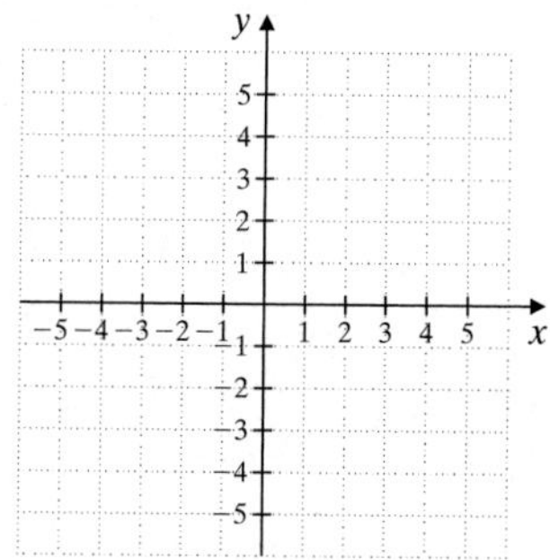

3. $\frac{x^2}{16} + \frac{y^2}{9} = 1$

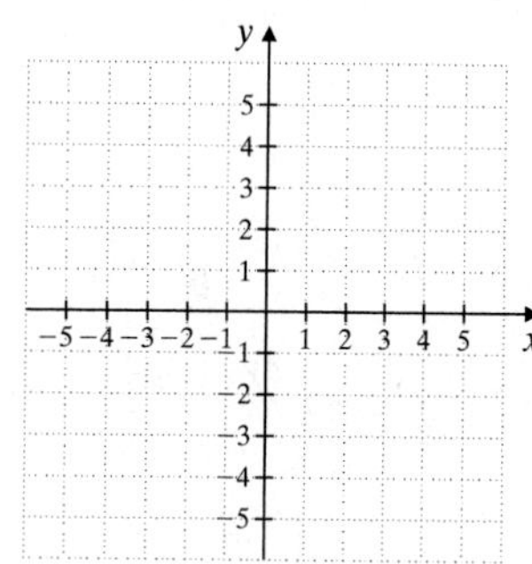

4. $x^2 + \frac{y^2}{4} = 1$

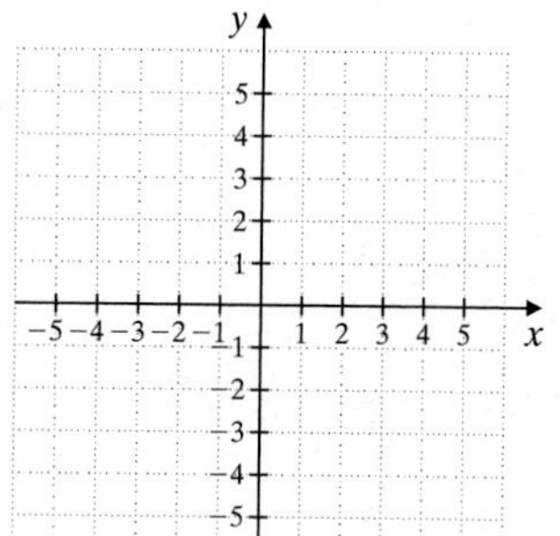

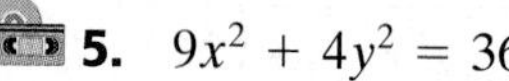

5. $9x^2 + 4y^2 = 36$

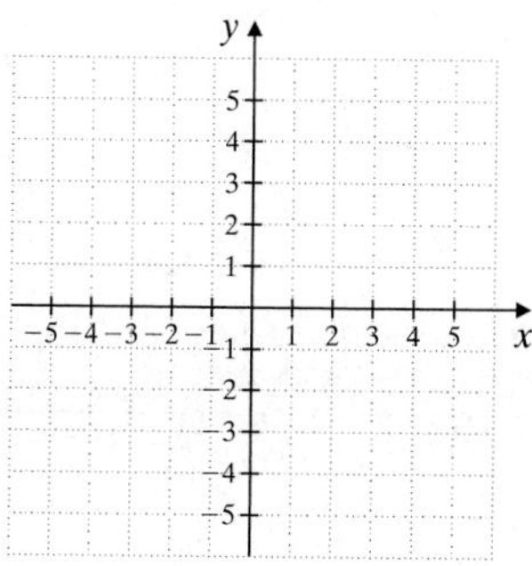

6. $x^2 + 4y^2 = 16$

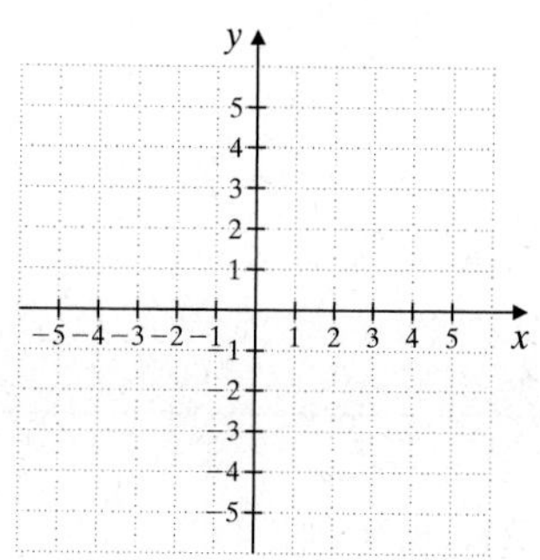

7. $4x^2 + 25y^2 = 100$

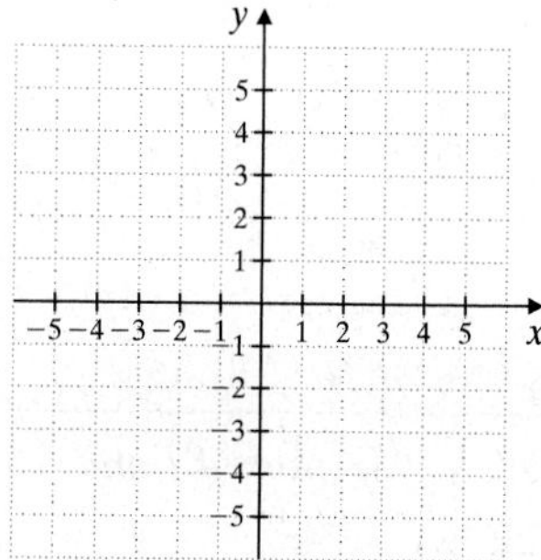

8. $36x^2 + y^2 = 36$

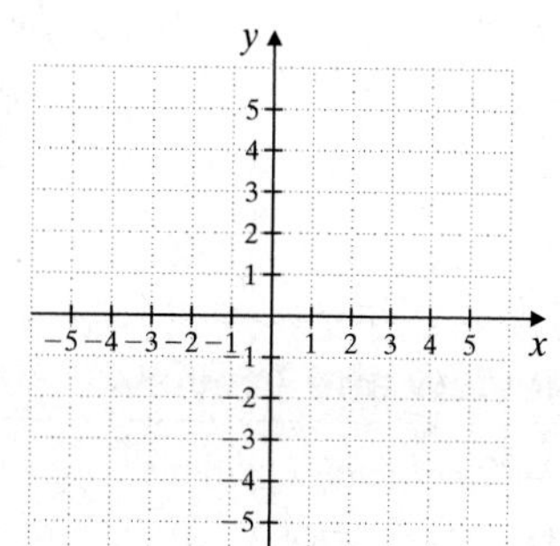

Graph each equation. See Example 3.

9. $\frac{(x+1)^2}{36} + \frac{(y-2)^2}{49} = 1$ **10.** $\frac{(x-3)^2}{9} + \frac{(y+3)^2}{16} = 1$ **11.** $\frac{(x-1)^2}{4} + \frac{(y-1)^2}{25} = 1$ **12.** $\frac{(x+3)^2}{16} + \frac{(y+2)^2}{4} = 1$

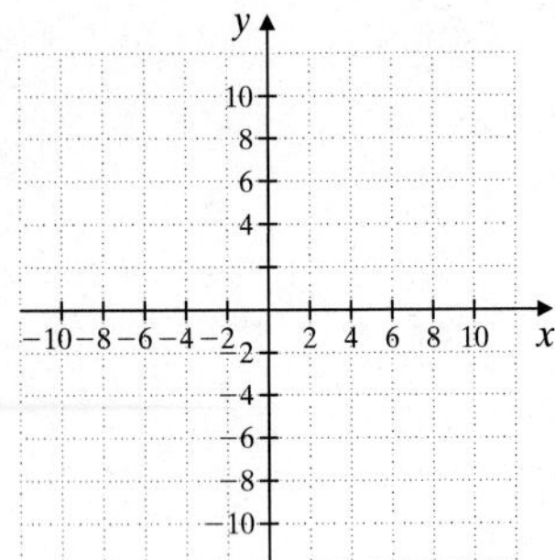

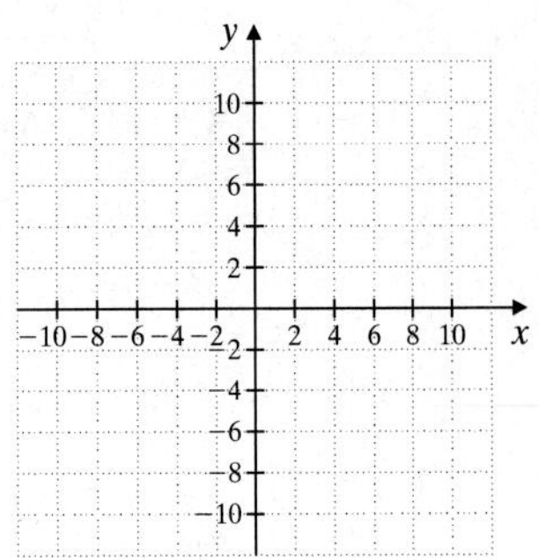

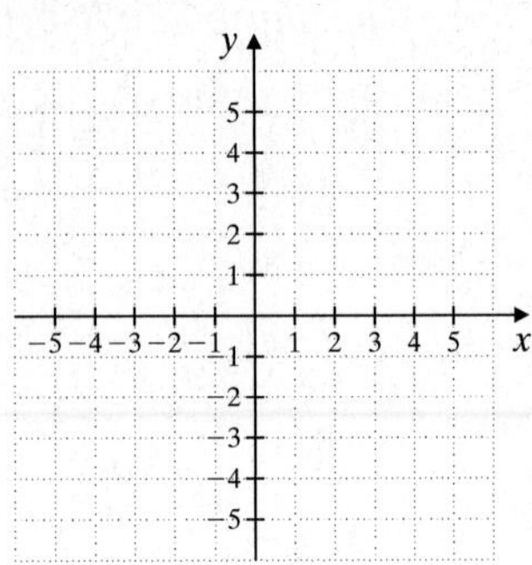

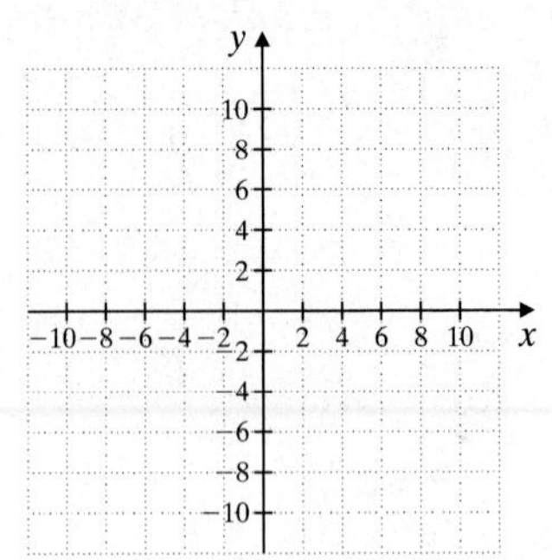

B *Graph each equation. See Examples 4 and 5.*

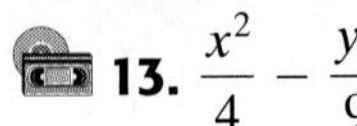

13. $\frac{x^2}{4} - \frac{y^2}{9} = 1$ **14.** $\frac{x^2}{36} - \frac{y^2}{36} = 1$ **15.** $\frac{y^2}{25} - \frac{x^2}{16} = 1$ **16.** $\frac{y^2}{25} - \frac{x^2}{49} = 1$

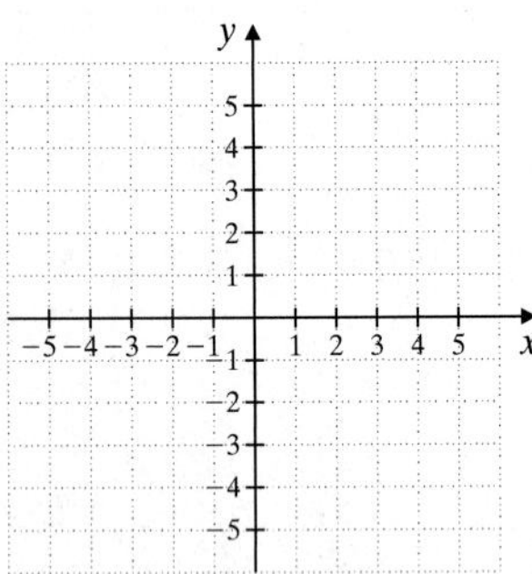

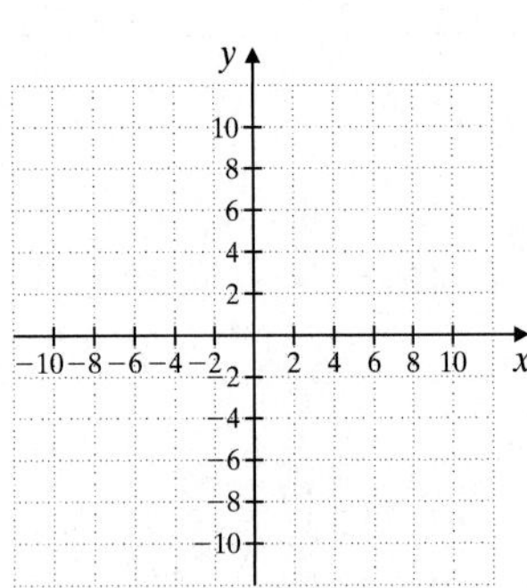

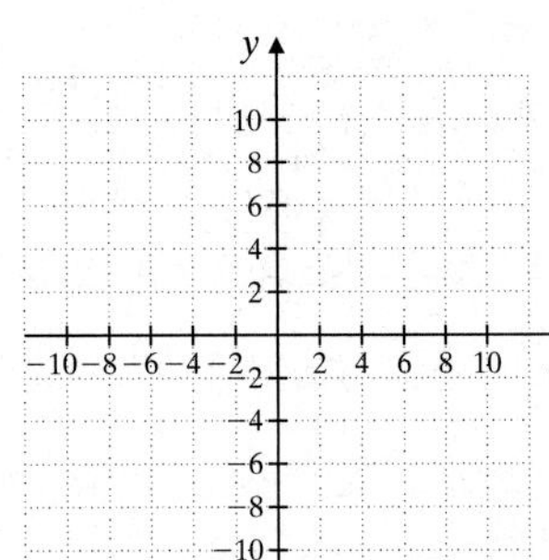

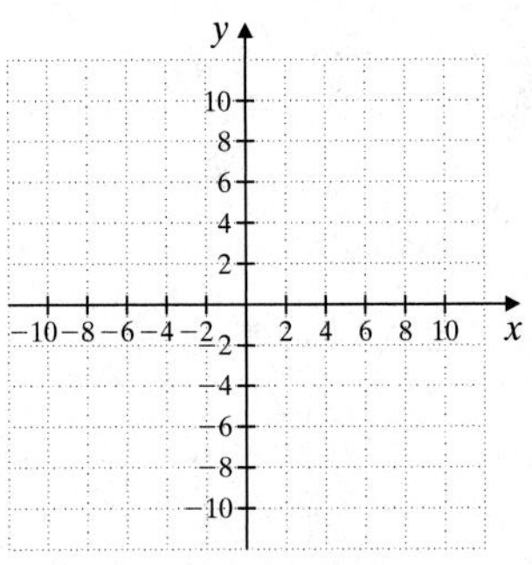

17. $x^2 - 4y^2 = 16$ **18.** $4x^2 - y^2 = 36$ **19.** $16y^2 - x^2 = 16$ **20.** $4y^2 - 25x^2 = 100$

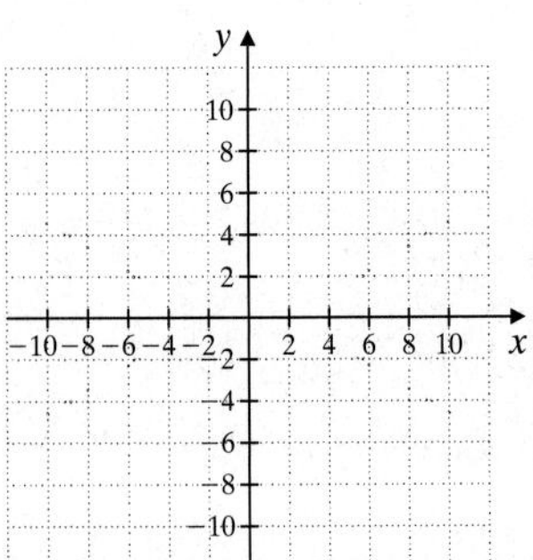

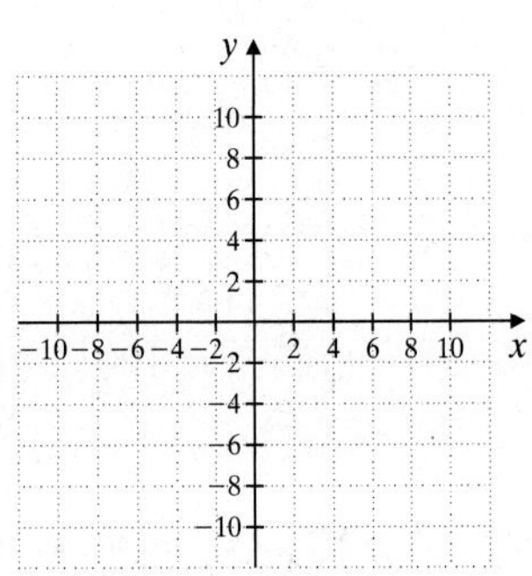

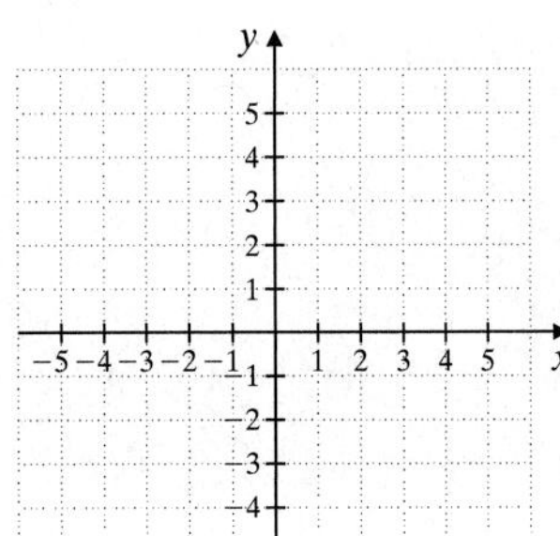

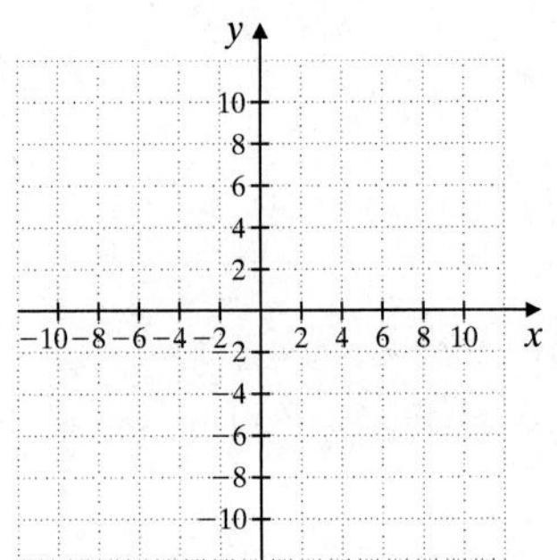

Review and Preview

Perform each indicated operation. See Sections 4.1 and 4.4.

21. $(2x^3)(-4x^2)$ **22.** $2x^3 - 4x^3$ **23.** $-5x^2 + x^2$ **24.** $(-5x^2)(x^2)$

Combining Concepts

25. We know that $x^2 + y^2 = 25$ is the equation of a circle. Rewrite the equation so that the right side is equal to 1. Which type of conic section does this equation form resemble? In fact, the circle is a special case of this type of conic section. Describe the conditions under which this type of conic section is a circle.

The orbits of stars, planets, comets, asteroids, and satellites all have the shape of one of the conic sections. Astronomers use a measure called eccentricity *to describe the shape and elongation of an orbital path. For the circle and ellipse, eccentricity e is calculated with the formula* $e = \frac{c}{d}$, *where* $c^2 = |a^2 - b^2|$ *and d is the larger value of a or b. For a hyperbola, eccentricity e is calculated with the formula* $e = \frac{c}{d}$, *where* $c^2 = a^2 + b^2$ *and the value of d is equal to a if the hyperbola has x-intercepts or equal to b if the hyperbola has y-intercepts. Use equations A–H to answer Exercises 26–35.*

A $\frac{x^2}{36} - \frac{y^2}{13} = 1$ **B** $\frac{x^2}{4} + \frac{y^2}{4} = 1$ **C** $\frac{x^2}{25} + \frac{y^2}{16} = 1$ **D** $\frac{y^2}{25} - \frac{x^2}{39} = 1$

E $\frac{x^2}{17} + \frac{y^2}{81} = 1$ **F** $\frac{x^2}{36} + \frac{y^2}{36} = 1$ **G** $\frac{x^2}{16} - \frac{y^2}{65} = 1$ **H** $\frac{x^2}{144} + \frac{y^2}{140} = 1$

26. Identify the type of conic section represented by each of the equations A–H.

27. For each of the equations A–H, identify the values of a^2 and b^2.

28. For each of the equations A–H, calculate the value of c^2 and c.

29. For each of the equations A–H, find the value of d.

30. For each of the equations A–H, calculate the eccentricity e.

31. What do you notice about the values of e for the equations you identified as ellipses?

32. What do you notice about the values of e for the equations you identified as circles?

33. What do you notice about the values of e for the equations you identified as hyperbolas?

34. The eccentricity of a parabola is exactly 1. Use this information and the observations you made in Exercises 31, 32, and 33 to describe a way that could be used to identify the type of conic section based on its eccentricity value.

35. Graph each of the conic sections given in equations A–H. What do you notice about the shape of the ellipses for increasing values of eccentricity? Which is the most elliptical? Which is the least elliptical, that is, the most circular?

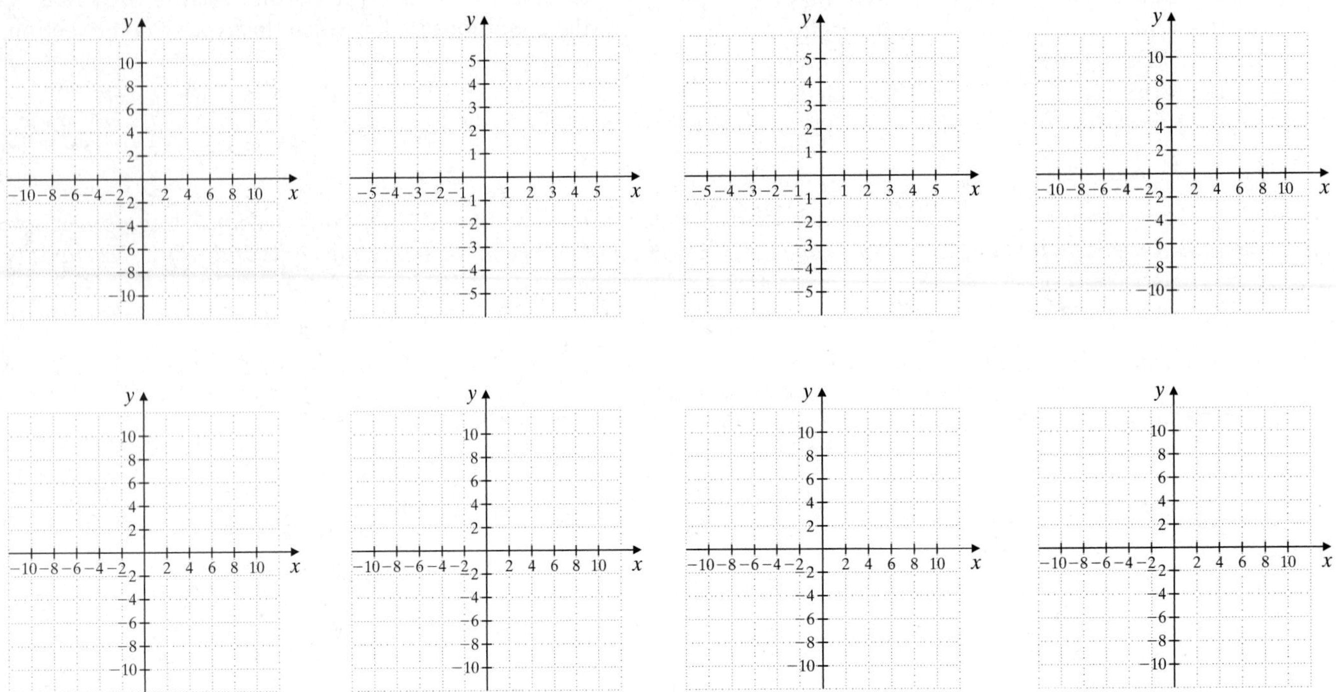

Internet Excursions

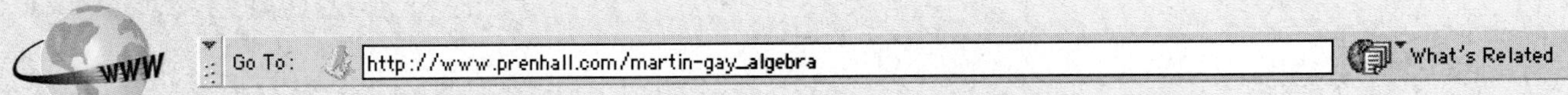

By going to this World Wide Web address, you will be directed to a Web site where you can look up information to help you answer the questions below.

36. Under Planets, select the Mean Orbital Elements option on this homepage. This gives data about the nine planets of our solar system, including the eccentricities of their orbits. Using the data for the eccentricities of the planets' orbits and your conclusions from Exercise 34, decide which type of conic describes the orbital paths of all the planets. Which planet has the most circular path? Which planet has the most elliptical path?

37. Return to the JPL Solar System Dynamics page. Under Comets and Asteroids, select the Orbital Elements option. Then choose the Comets option. This gives data about the orbits of comets known to pass through our solar system, including their eccentricities. Using the data for the eccentricities of the comets' orbits and your conclusions from Exercise 34, decide which type of conic section describes the orbital paths of the majority of the comets. Which comets are the exceptions? What type of orbital paths do these comets have?

Name ________________________________

Answers

1. ______________

Integrated Review—Graphing Conic Sections

Following is a summary of conic sections.

Conic Sections

	Standard Form	Graph
Parabola	$y = a(x - h)^2 + k$	
Parabola	$x = a(y - k)^2 + h$	
Circle	$(x - h)^2 + (y - k)^2 = r^2$	
Ellipse	$\dfrac{x^2}{a^2} + \dfrac{y^2}{b^2} = 1$	
Hyperbola	$\dfrac{x^2}{a^2} - \dfrac{y^2}{b^2} = 1$	
Hyperbola	$\dfrac{y^2}{b^2} - \dfrac{x^2}{a^2} = 1$	

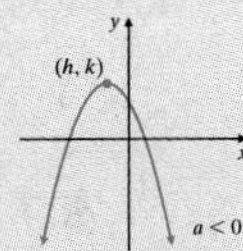

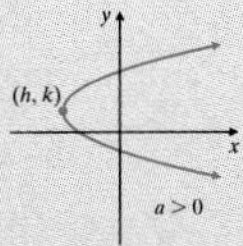

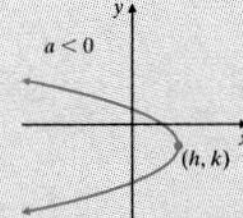

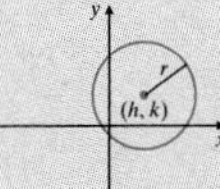

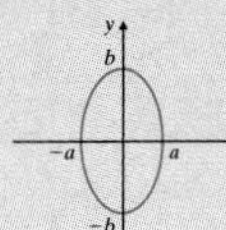

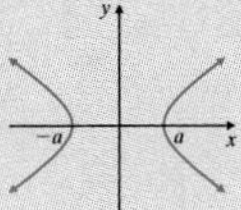

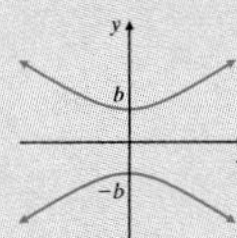

2. ______________

Identify whether each equation, when graphed, will be a parabola, circle, ellipse, or hyperbola. Then graph each equation.

1. $(x - 7)^2 + (y - 2)^2 = 4$ **2.** $y = x^2 + 4$ **3.** $y = x^2 + 12x + 36$

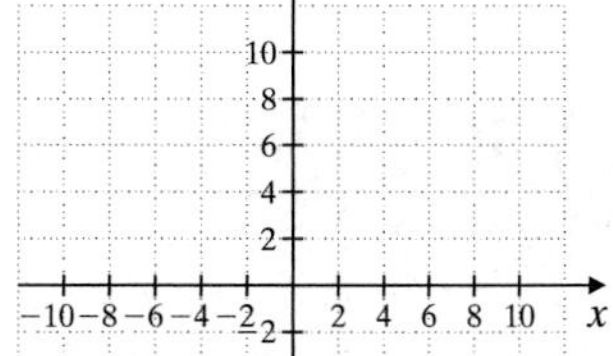

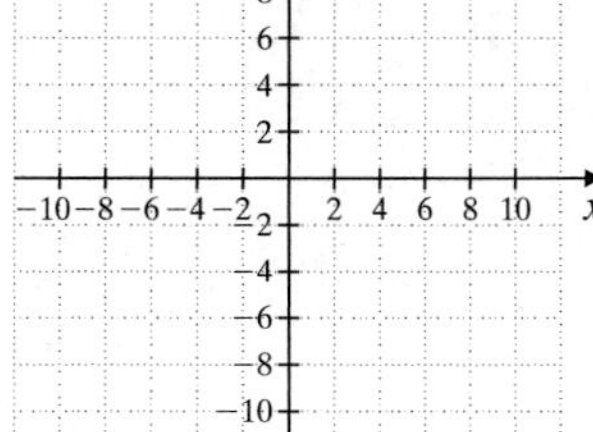

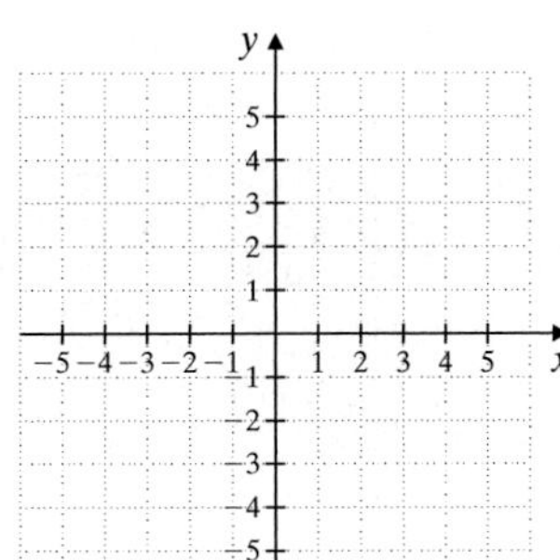

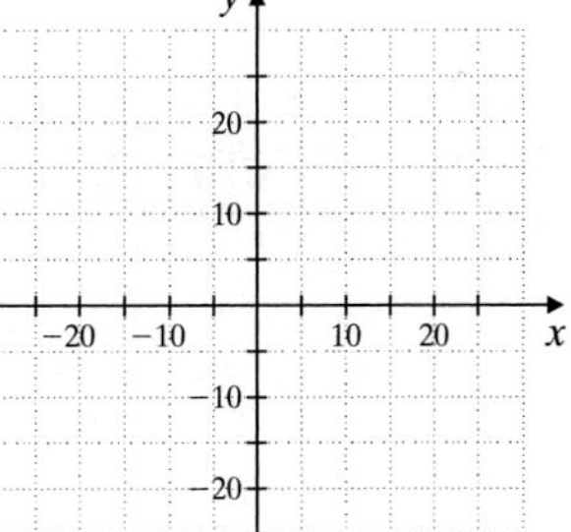

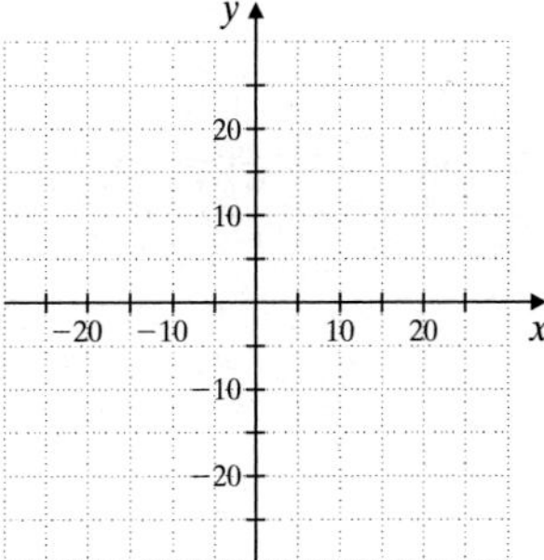

3. ______________

4. ____________

5. ____________

6. ____________

7. ____________

8. ____________

9. ____________

10. ____________

11. ____________

12. ____________

13. ____________

14. ____________

15. ____________

4. $\dfrac{x^2}{4} + \dfrac{y^2}{9} = 1$

5. $\dfrac{y^2}{9} - \dfrac{x^2}{9} = 1$

6. $\dfrac{x^2}{16} - \dfrac{y^2}{4} = 1$

7. $\dfrac{x^2}{16} + \dfrac{y^2}{4} = 1$

8. $x^2 + y^2 = 16$

9. $x = y^2 + 4y - 1$

10. $x = -y^2 + 6y$

11. $9x^2 - 4y^2 = 36$

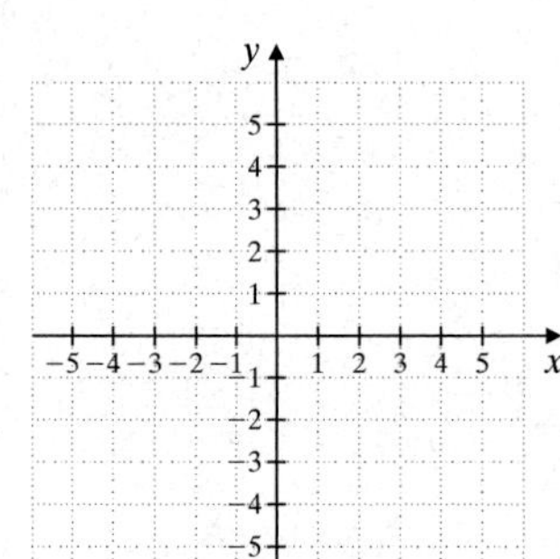

12. $9x^2 + 4y^2 = 36$

13. $\dfrac{(x-1)^2}{49} + \dfrac{(y+2)^2}{25} = 1$

14. $y^2 = x^2 + 16$

15. $\left(x + \dfrac{1}{2}\right)^2 + \left(y - \dfrac{1}{2}\right)^2 = 1$

12.3 Graphing Nonlinear Functions

OBJECTIVE

Ⓐ Graph the nonlinear functions $f(x) = |x|$ and $f(x) = \sqrt{x}$.

SSM TUTOR CENTER SG CD & VIDEO MATH PRO WEB

Ⓐ Graphing Nonlinear Functions

Recall that the graph of $f(x) = x^2$ is a parabola with vertex $(0, 0)$. How does the graph of $g(x) = (x - 3)^2 + 2$ compare? Its graph is a parabola of the same shape as f, but with vertex $(3, 2)$. In other words, the graph of g is the same as the graph of f, except that it has been shifted 3 units to the right and 2 units up. Keep this in mind as we graph other elementary functions. Remember, we are graphing functions, so all graphs should pass the vertical line test.

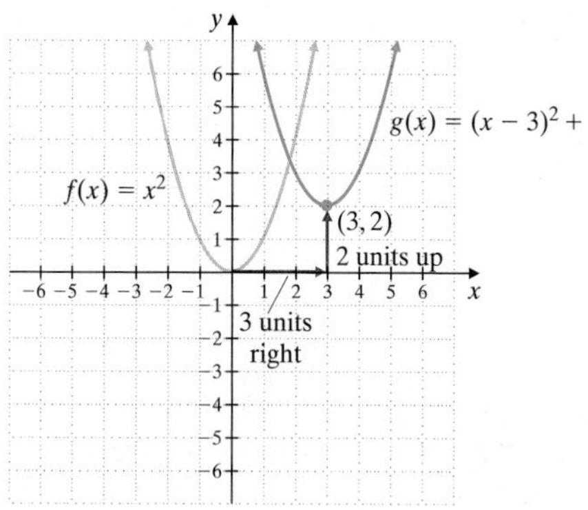

EXAMPLE 1 Graph: $f(x) = |x|$

Solution: This is not a linear function, and its graph is not a line. Because we do not know the shape of this graph, we find many ordered pair solutions. We will choose x-values and substitute to find corresponding y-values. Recall that

If $x = -3$, then $y = |-3|$, or 3.
If $x = -2$, then $y = |-2|$, or 2.
If $x = -1$, then $y = |-1|$, or 1.
If $x = 0$, then $y = |0|$, or 0.
If $x = 1$, then $y = |1|$, or 1.
If $x = 2$, then $y = |2|$, or 2.
If $x = 3$, then $y = |3|$, or 3.

x	y
−3	3
−2	2
−1	1
0	0
1	1
2	2
3	3

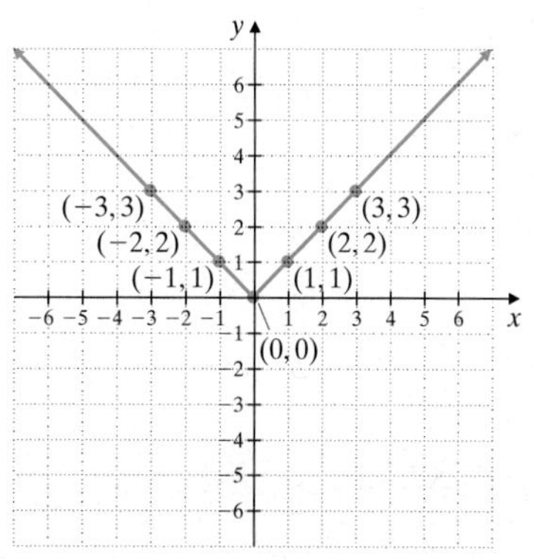

Study the table of values for a moment and notice any patterns. Since the absolute value of a real number is a nonnegative number, notice that the domain of this function is $\{x|x \text{ is a real number,}\}$ but the range is $\{y|y \geq 0\}$. From the plotted ordered pairs, we see that the graph of this absolute value function is V-shaped. ●

Practice Problem 1

Graph: $f(x) = |x| + 1$

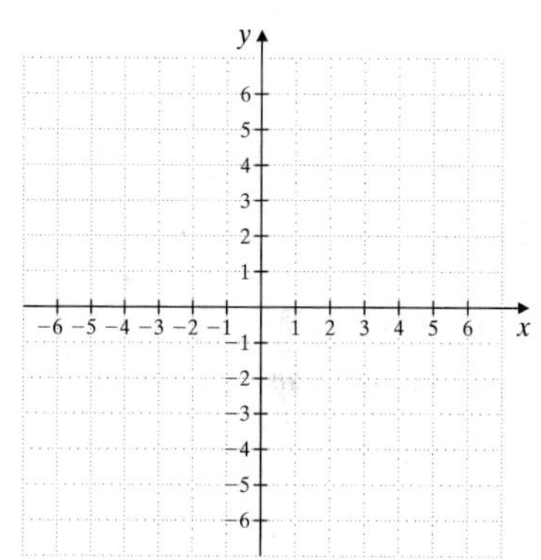

Answer

1.

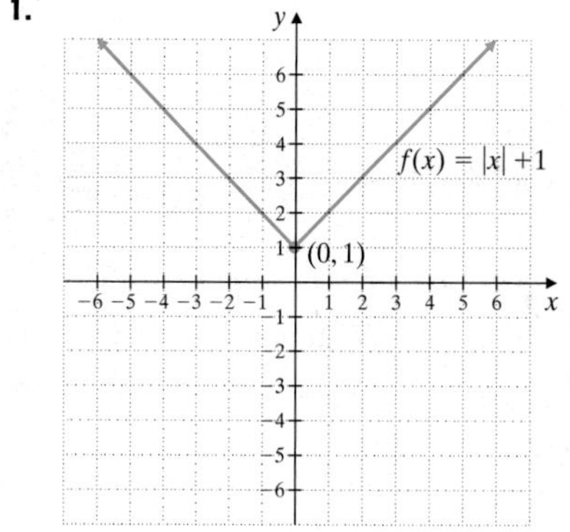

Practice Problem 2

Graph $f(x) = |x| - 1$

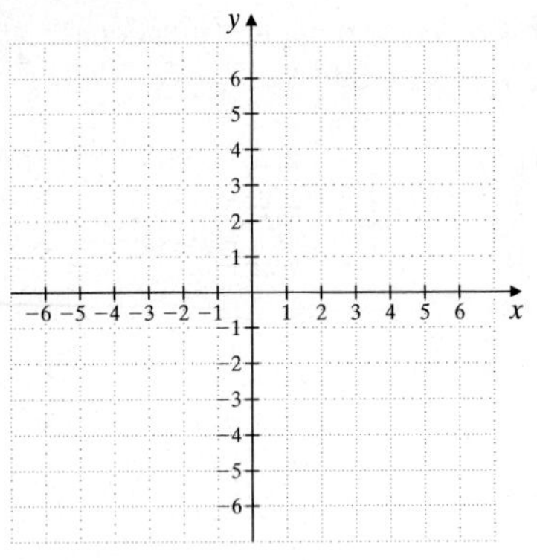

Practice Problem 3

Graph: $f(x) = |x - 1|$

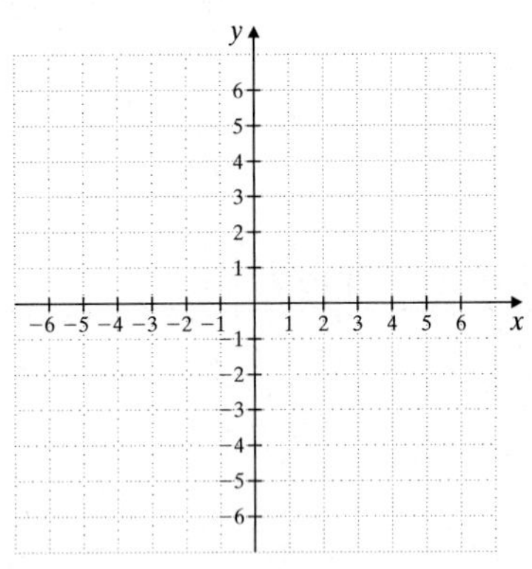

Answers

2.

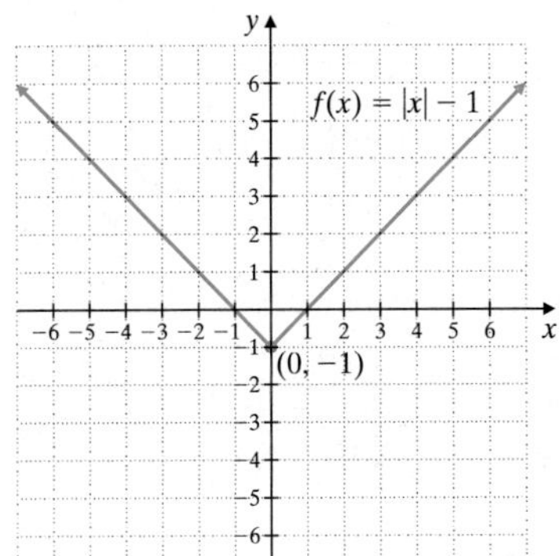

3.

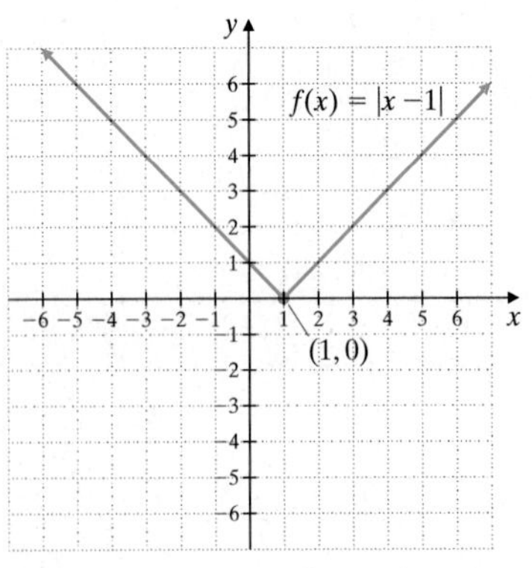

EXAMPLE 2 Graph: $f(x) = |x| - 3$

Solution: To graph $f(x)$ or $y = |x| - 3$, choose x-values and substitute to find corresponding y-values.

x	y
−3	0
−2	−1
−1	−2
0	−3
1	−2
2	−1
3	0

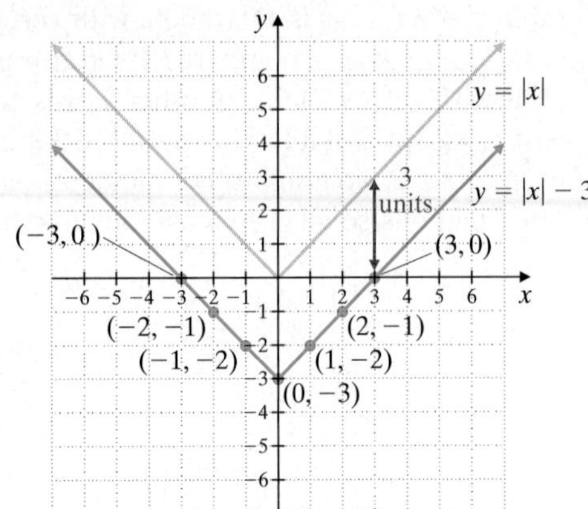

Recall that the graph of $y = x^2 - 3$ is the same as the graph of $y = x^2$ lowered 3 units. Now compare the graph of $y = |x|$ with the graph of $y = |x| - 3$. The graph of $y = |x| - 3$ is the same as the graph of $y = |x|$ lowered 3 units.

EXAMPLE 3 Graph: $f(x) = |x - 2|$

Solution: First let's think about the graph of $y = (x - 2)^2$. The vertex of this graph is $(2, 0)$. In other words, the graph of $y = (x - 2)^2$ is the same as the graph of $y = x^2$ shifted to the right 2 units.

In the same manner, the graph of $y = |x - 2|$ is the same as the graph of $y = |x|$ shifted to the right 2 units. We use this knowledge along with a table of ordered pair solutions to graph the function.

x	y
0	2
1	1
2	0
3	1
4	2

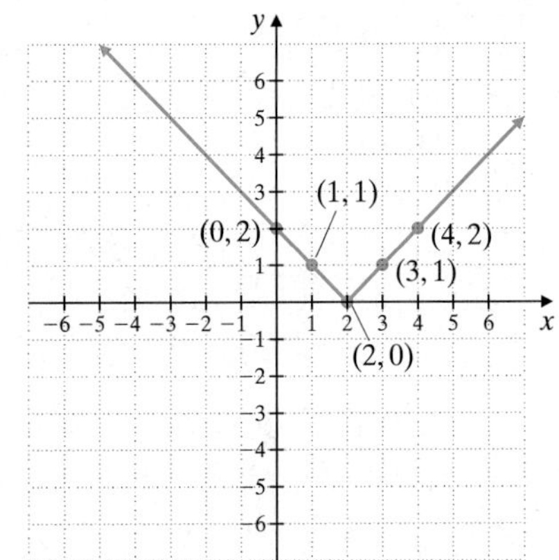

EXAMPLE 4 Graph: $f(x) = |x - 1| + 2$

Solution: The graph of $y = (x - 1)^2 + 2$ has vertex $(1, 2)$. In other words, it is the graph of $y = x^2$ shifted 1 unit to the right and 2 units up. Similarly, the graph of $y = |x - 1| + 2$ is the graph of $y = |x|$ shifted 1 unit to the right and 2 units up. We use this knowledge along with a table of ordered pair solutions to graph the function.

x	y
−1	4
0	3
1	2
2	3
3	4

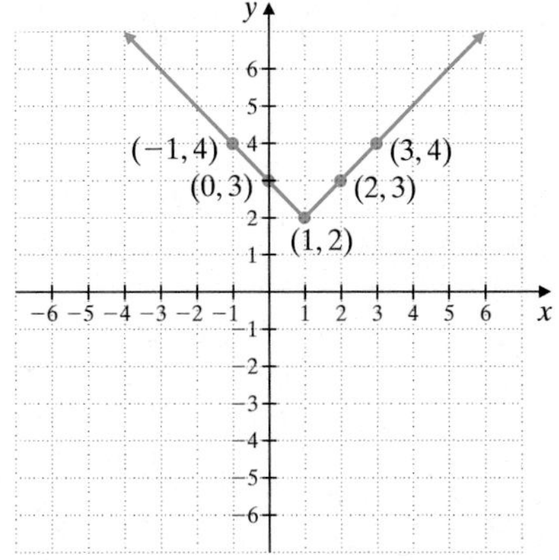

Recall that the domain of a function is basically the set of all possible x-values for that function. The domains of the functions thus far in this section have been the set of all real numbers. This is not the case for our next function, the square root function.

EXAMPLE 5 Graph the square root function $f(x) = \sqrt{x}$.

Solution: Recall that the square root of a negative number is not a real number. This means that the domain of this function is the set of all nonnegative numbers, or $\{x | x \geq 0\}$. To graph this function, evaluate the function for several values of x, plot the resulting points, and connect the points with a smooth curve.

x	y
0	0
1	1
3	$\sqrt{3} \approx 1.7$
4	2
9	3

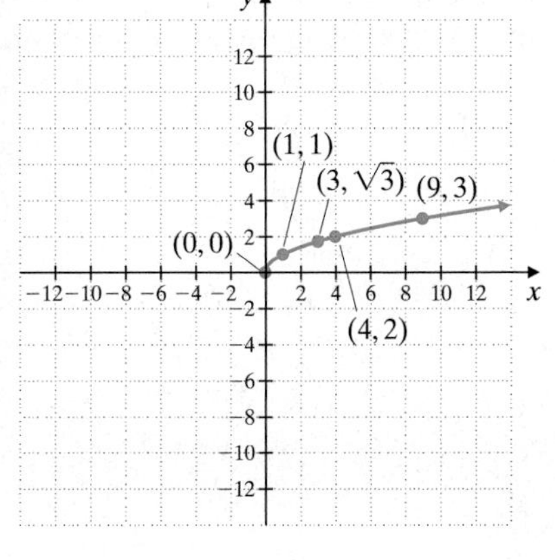

Practice Problem 4

Graph: $f(x) = |x - 3| + 2$

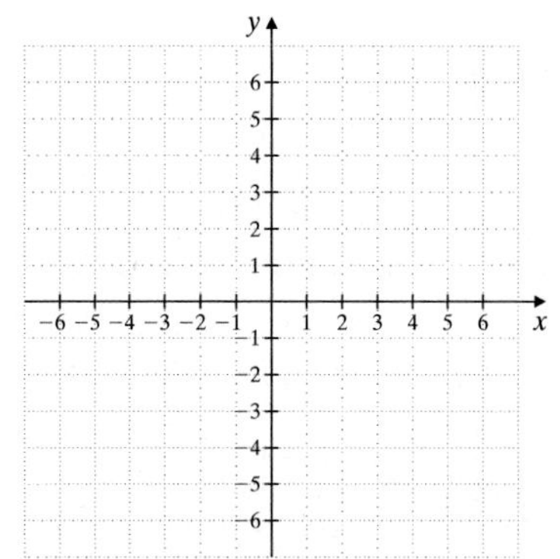

Practice Problem 5

Graph: $f(x) = \sqrt{x} + 2$

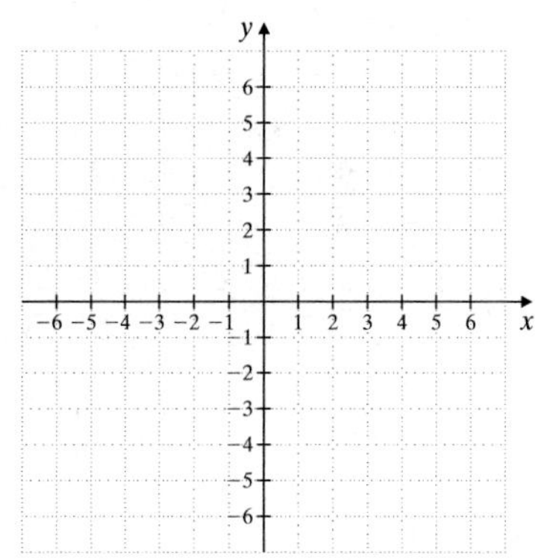

Answers

4.

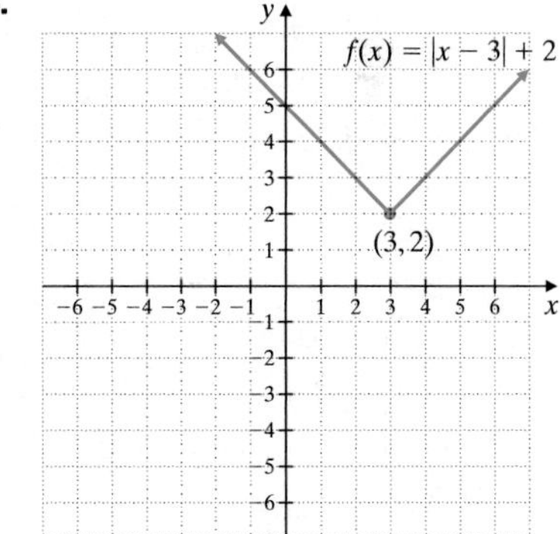

5.

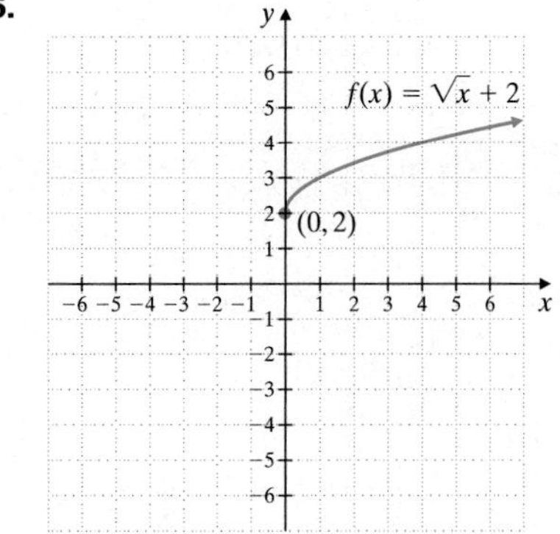

Practice Problem 6

Graph: $f(x) = \sqrt{x + 1} + 2$

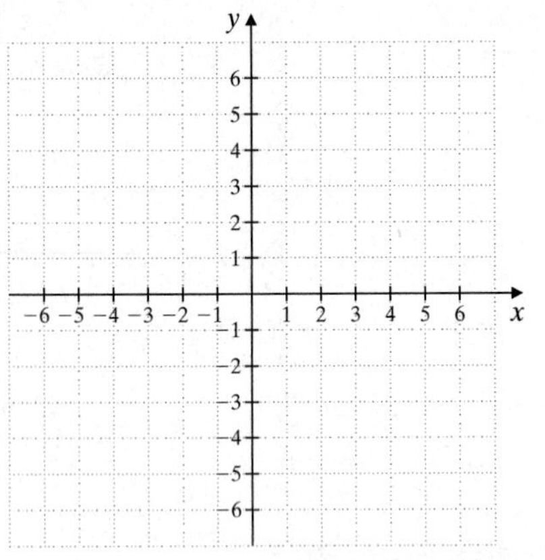

EXAMPLE 6 Graph: $f(x) = \sqrt{x + 2} + 3$

Solution: The graph of $y = (x + 2)^2 + 3$ has vertex $(-2, 3)$ and is the graph of $y = x^2$ shifted 2 units left and 3 units up. Similarly, the graph of $y = \sqrt{x + 2} + 3$ is the graph of $y = \sqrt{x}$ shifted 2 units left and 3 units up. We use this knowledge along with a table of values to graph the function.

x	y
−2	3
−1	4
2	5

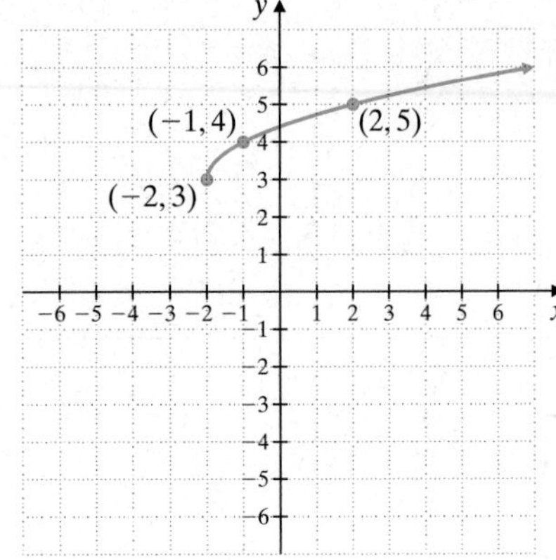

This vertical and horizontal shifting works for any function.

Graphing $F(x) = f(x - h) + k$

The graph of the function

$$F(x) = f(x - h) + k$$

is the same as the graph of the function $y = f(x)$ except that it has been shifted left or right h units and up or down k units. It is shifted to the right if $h > 0$ and left if $h < 0$. It is shifted up if $k > 0$ and down if $k < 0$.

Recall from Chapter 9 that the domain of $f(x) = \sqrt{x + 2} + 3$ includes all real numbers that make the radicand ≥ 0. To see what numbers these are, solve $x + 2 \geq 0$ and find that $x \geq -2$. The domain of f is $\{x | x \geq -2\}$. This can be verified by observing the graph of Example 6.

Answer

6.

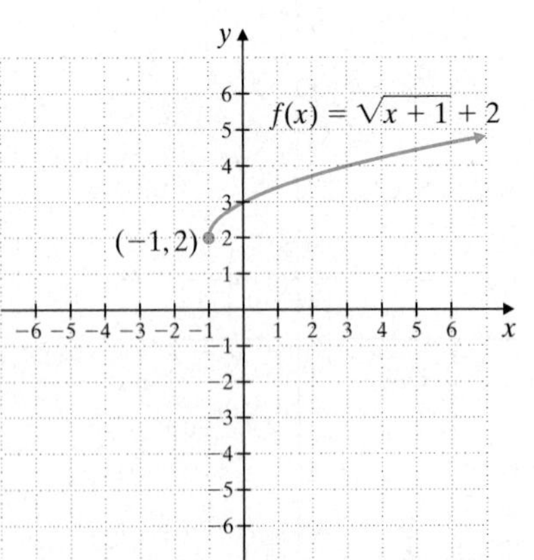

Name ______________________ Section __________ Date __________

EXERCISE SET 12.3

A *Graph each function. See Examples 1 through 6.*

1. $f(x) = |x| + 3$

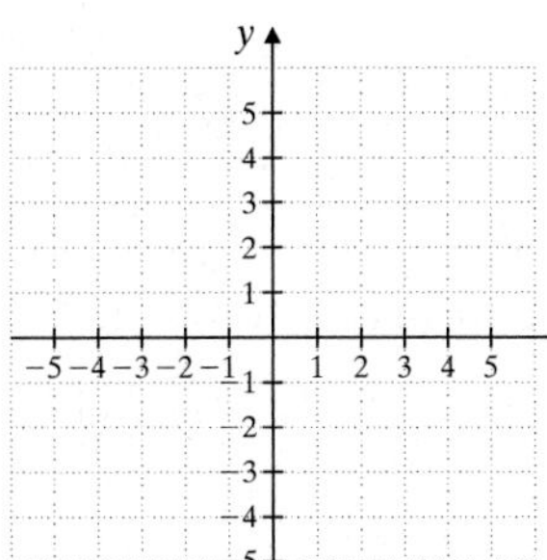

2. $f(x) = |x| - 2$

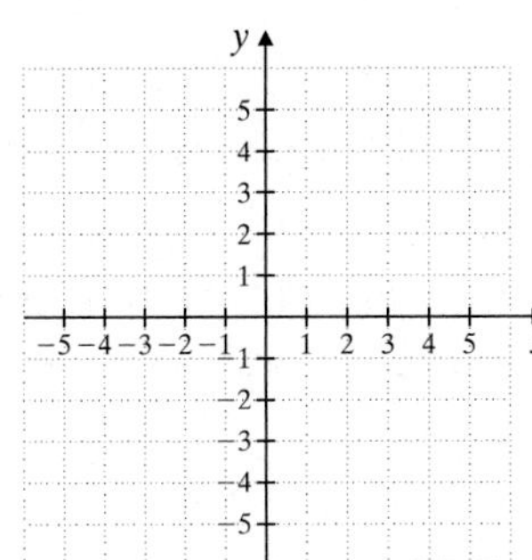

3. $f(x) = \sqrt{x} - 2$

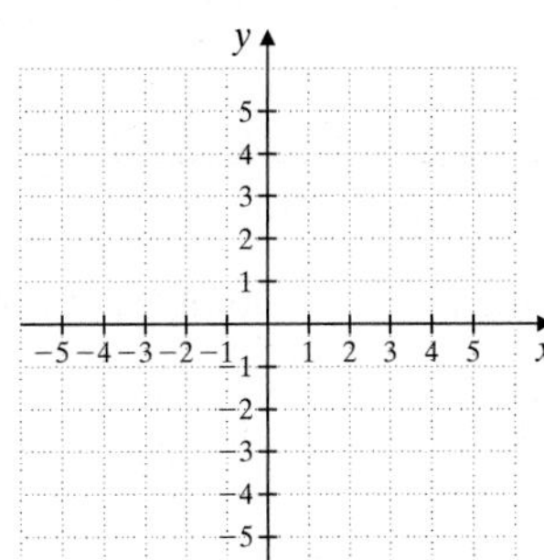

4. $f(x) = \sqrt{x} + 3$

5. $f(x) = |x - 4|$

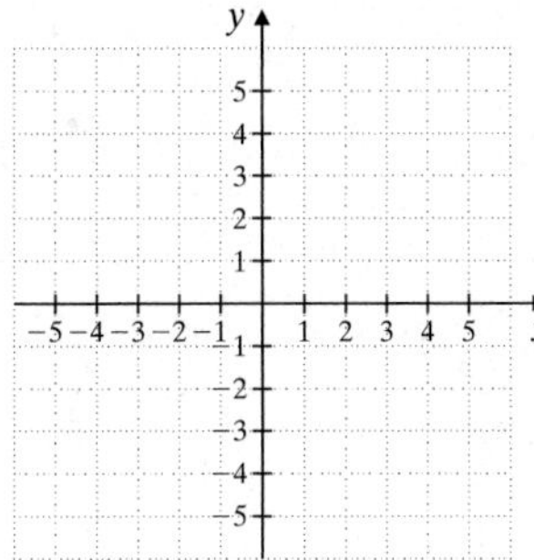

6. $f(x) = |x + 3|$

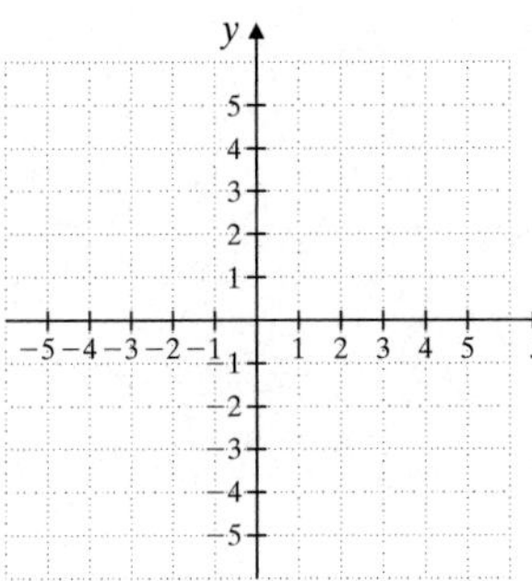

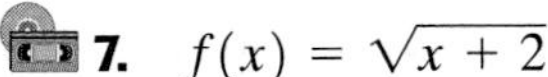

7. $f(x) = \sqrt{x + 2}$

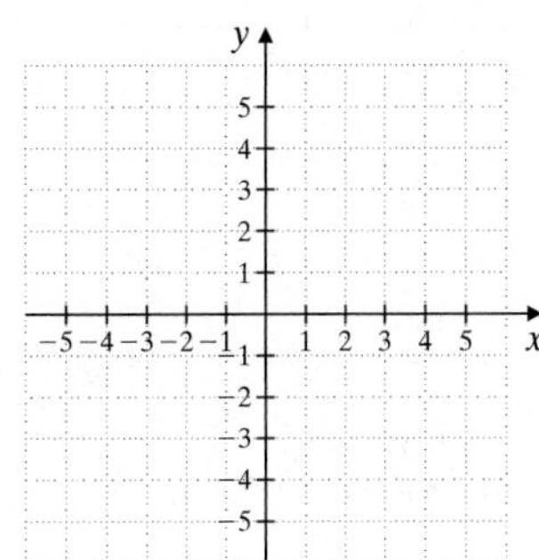

8. $f(x) = \sqrt{x - 2}$

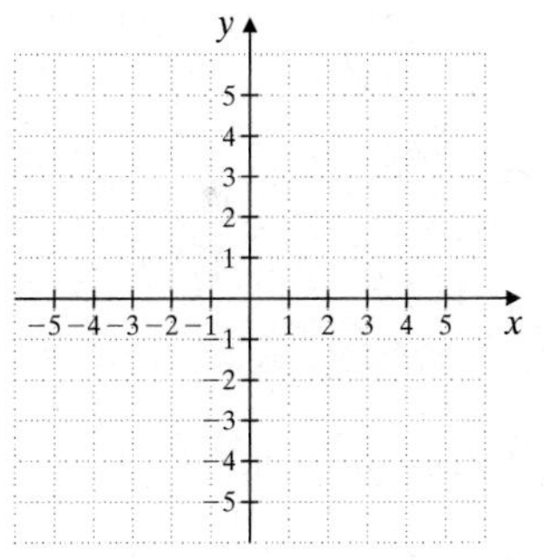

9. $f(x) = \sqrt{x - 2} + 3$

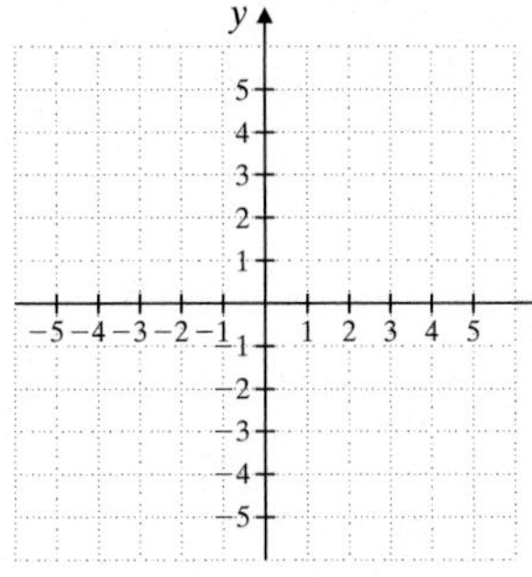

10. $f(x) = \sqrt{x - 1} + 3$

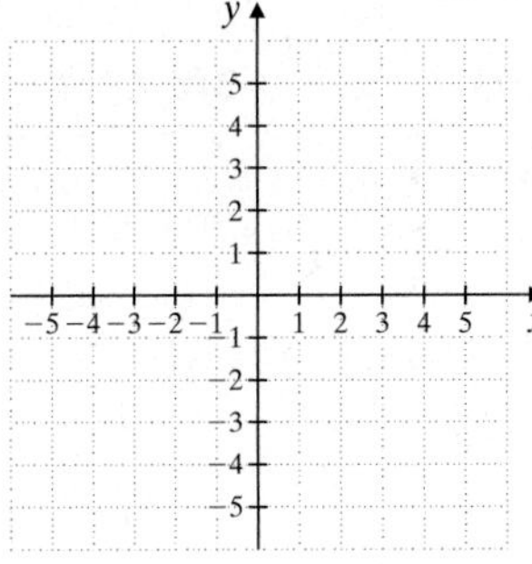

11. $f(x) = |x - 1| + 5$

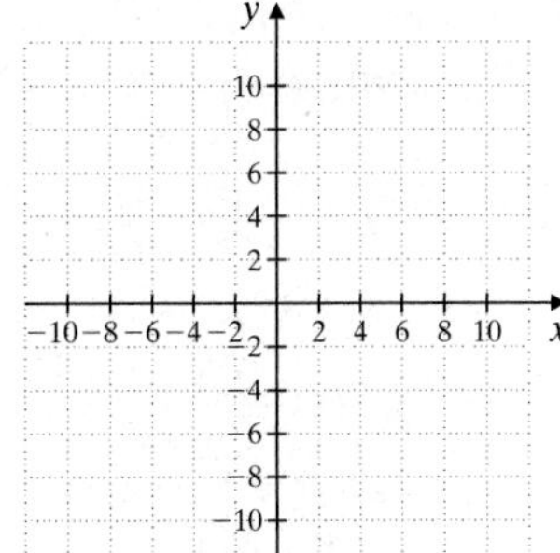

12. $f(x) = |x - 3| + 2$

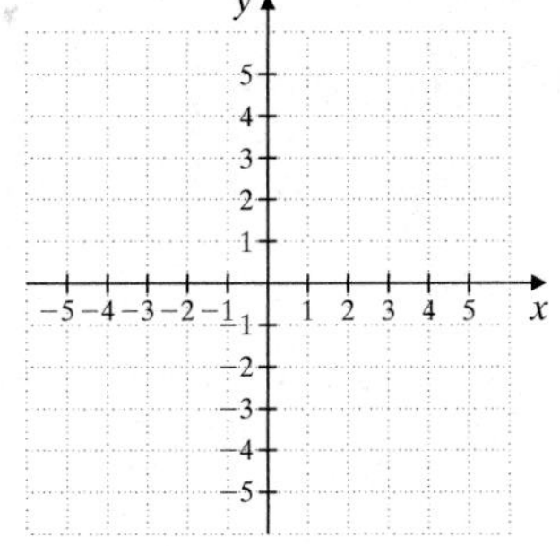

13. $f(x) = \sqrt{x + 1} + 1$

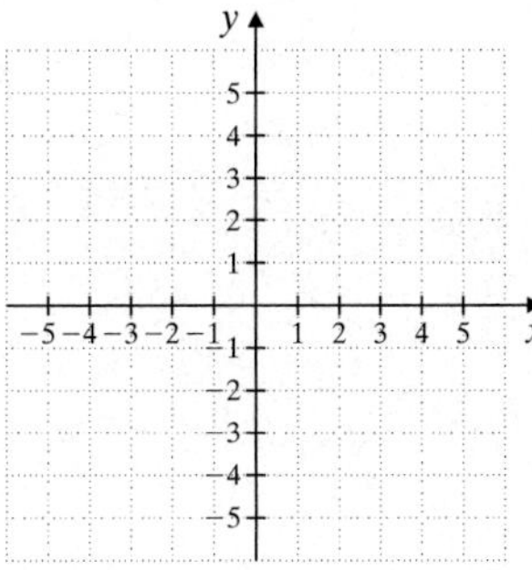

14. $f(x) = \sqrt{x + 3} + 2$

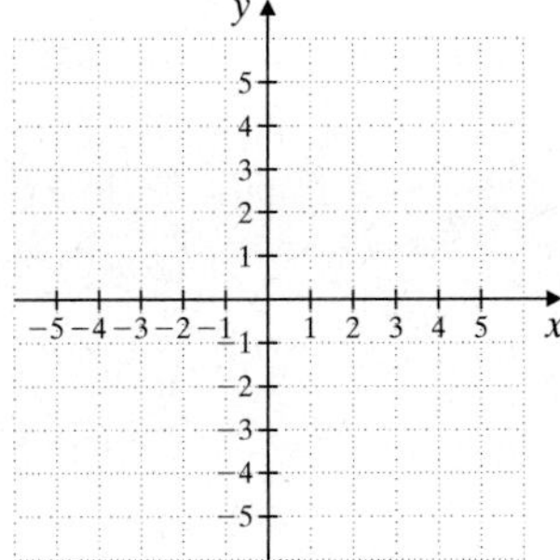

15. $f(x) = |x + 3| - 1$

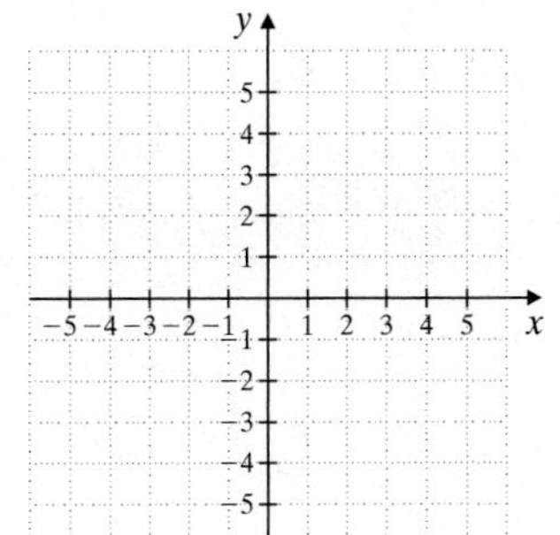

16. $f(x) = |x + 1| - 4$

Review and Preview

Solve each system of equations. See Section 4.1.

17. $\begin{cases} x + y = 6 \\ x - y = 10 \end{cases}$

18. $\begin{cases} x + y = -2 \\ -x + y = -8 \end{cases}$

19. $\begin{cases} 2x + 3y = 7 \\ -x + 4y = 13 \end{cases}$

20. $\begin{cases} 4x - 3y = -4 \\ 3x - y = 10 \end{cases}$

Combining Concepts

Graph each function. Recall that the domain of the cube function and the cube root function is the set of all real numbers. For Exercises 22 and 24, predict the location and appearance of the graph and then use a graphing calculator to verify.

21. $f(x) = x^3$

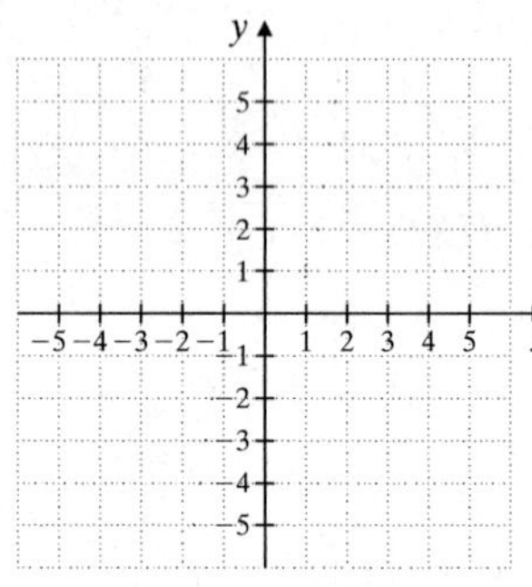

22. $f(x) = (x - 1)^3 + 2$

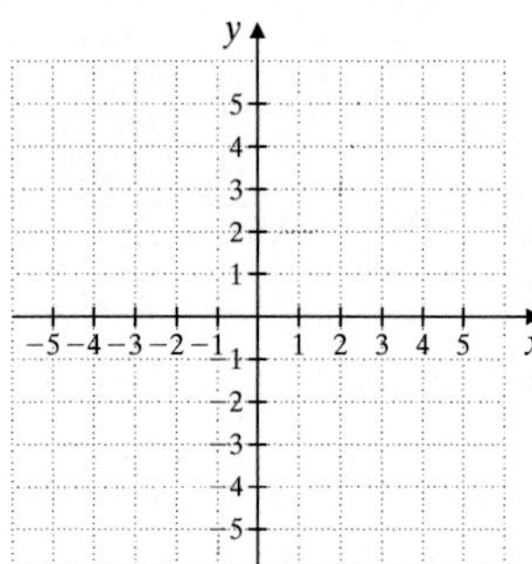

23. $f(x) = \sqrt[3]{x}$

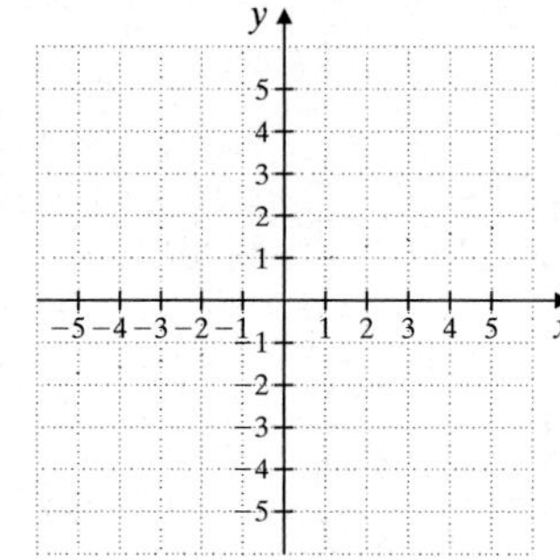

24. $f(x) = \sqrt[3]{x - 3} + 1$

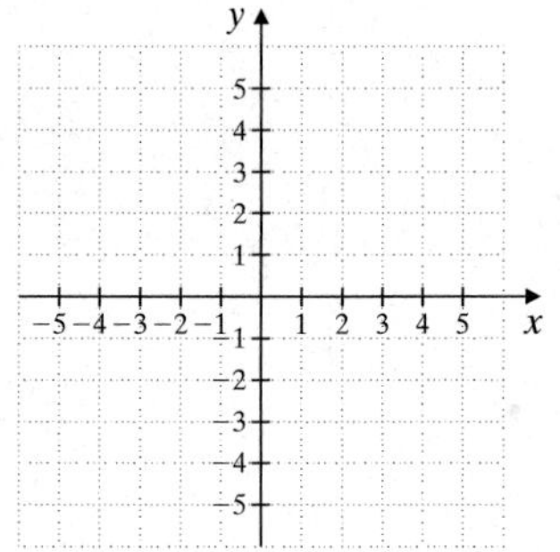

Without graphing, find the domain of each function.

25. $f(x) = 5\sqrt{x - 20} + 1$

26. $g(x) = -3\sqrt{x + 5}$

27. $h(x) = 5|x - 20| + 1$

28. $f(x) = -3|x + 5.7|$

29. $g(x) = 9 - \sqrt{x + 103}$

30. $h(x) = \sqrt{x - 17} - 3$

12.4 Solving Nonlinear Systems of Equations

OBJECTIVES

A Solve a nonlinear system by substitution.

B Solve a nonlinear system by elimination.

In Chapter 8, we used graphing, substitution, and elimination methods to find solutions of systems of linear equations in two variables. We now apply these same methods to nonlinear systems of equations in two variables. A **nonlinear system of equations** is a system of equations at least one of which is not linear. Since we will be graphing the equations in each system, we are interested in real number solutions only.

 Solving Nonlinear Systems by Substitution

First we solve nonlinear systems by the substitution method.

EXAMPLE 1 Solve the system:

$$\begin{cases} x^2 - 3y = 1 \\ x - y = 1 \end{cases}$$

Solution: We can solve this system by substitution if we solve one equation for one of the variables. Solving the first equation for x is not the best choice since doing so introduces a radical. Also, solving for y in the first equation introduces a fraction. Thus, we solve the second equation for y.

$x - y = 1$ Second equation

$x - 1 = y$ Solve for y.

Now we replace y with $x - 1$ in the first equation, and then solve for x.

$x^2 - 3y = 1$ First equation

$x^2 - 3(x - 1) = 1$ Replace y with $x - 1$.

$$x^2 - 3x + 3 = 1$$
$$x^2 - 3x + 2 = 0$$
$$(x - 2)(x - 1) = 0$$
$$x = 2 \quad \text{or} \quad x = 1$$

Now we let $x = 2$ and then $x = 1$ in the equation $y = x - 1$ to find corresponding y-values.

Let $x = 2$.

$$y = x - 1$$
$$y = 2 - 1 = 1$$

Let $x = 1$.

$$y = x - 1$$
$$y = 1 - 1 = 0$$

When we check $(2, 1)$ and $(1, 0)$ in the equations, we find that both ordered pairs satisfy both equations. Thus, the solution set for the system is $\{(2, 1), (1, 0)\}$. The graph of each equation in the system is shown.

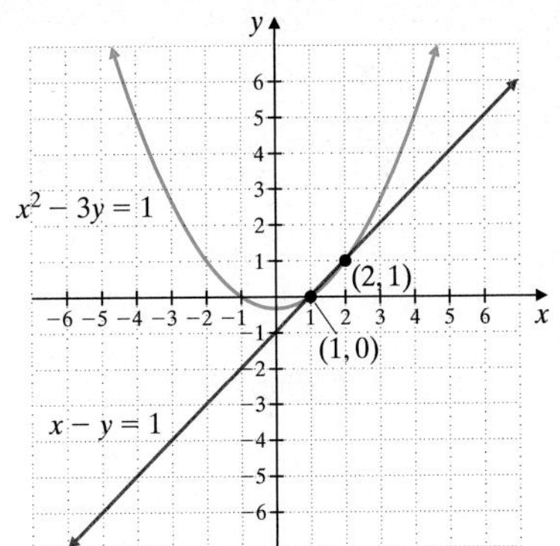

Practice Problem 1

Solve the system: $\begin{cases} x^2 - 2y = 5 \\ x + y = -1 \end{cases}$

Answer

1. $\{(-3, 2), (1, -2)\}$

Practice Problem 2

Solve the system: $\begin{cases} y = \sqrt{x} \\ x^2 + y^2 = 12 \end{cases}$

EXAMPLE 2 Solve the system:

$$\begin{cases} y = \sqrt{x} \\ x^2 + y^2 = 6 \end{cases}$$

Solution: This system is ideal for the substitution method since y is expressed in terms of x in the first equation. Notice that if $y = \sqrt{x}$, then both x and y must be nonnegative if they are real numbers. Let's substitute $\sqrt{x}$ for y in the second equation, and solve for x.

$$x^2 + y^2 = 6$$

$$x^2 + (\sqrt{x})^2 = 6 \quad \text{Let } y = \sqrt{x}.$$

$$x^2 + x = 6$$

$$x^2 + x - 6 = 0$$

$$(x + 3)(x - 2) = 0$$

$$x = -3 \quad \text{or} \quad x = 2$$

The solution -3 is discarded because we have noted that x must be nonnegative. To see this, we let $x = -3$ and $x = 2$ in the first equation to find the corresponding y-values.

Let $x = -3$.

$$y = \sqrt{x}$$

$$y = \sqrt{-3} \quad \text{Not a real number}$$

Let $x = 2$.

$$y = \sqrt{x}$$

$$y = \sqrt{2}$$

Since we are interested only in real number solutions, the only solution is $(2, \sqrt{2})$. The solution set is $\{(2, \sqrt{2})\}$. Check to see that this solution satisfies both equations. The graph of each equation in this system is shown.

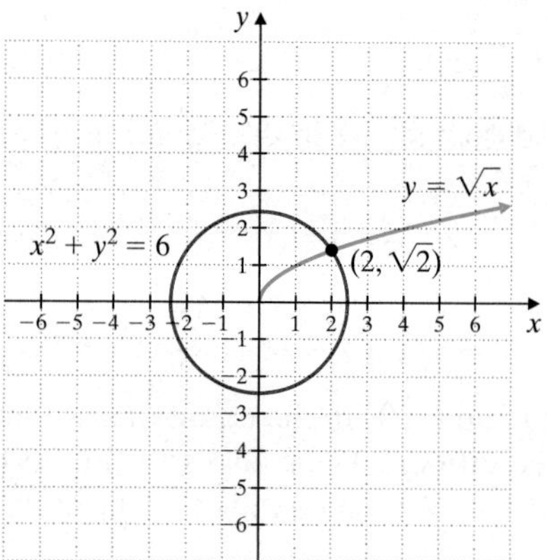

Practice Problem 3

Solve the system: $\begin{cases} x^2 + y^2 = 1 \\ x + y = 4 \end{cases}$

EXAMPLE 3 Solve the system:

$$\begin{cases} x^2 + y^2 = 4 \\ x + y = 3 \end{cases}$$

Solution: We use the substitution method and solve the second equation for x.

$$x + y = 3 \quad \text{Second equation}$$

$$x = 3 - y$$

Answers

2. $\{(3, \sqrt{3})\}$ **3.** no solution

Now we let $x = 3 - y$ in the first equation.

$$x^2 + y^2 = 4 \quad \text{First equation}$$

$$(3 - y)^2 + y^2 = 4 \quad \text{Let } x = 3 - y.$$

$$9 - 6y + y^2 + y^2 = 4$$

$$2y^2 - 6y + 5 = 0$$

By the quadratic formula, where $a = 2$, $b = -6$, and $c = 5$, we have

$$y = \frac{6 \pm \sqrt{(-6)^2 - 4 \cdot 2 \cdot 5}}{2 \cdot 2} = \frac{6 \pm \sqrt{-4}}{4}$$

Since $\sqrt{-4}$ is not a real number, there is no solution. Graphically, the circle and the line do not intersect, as shown.

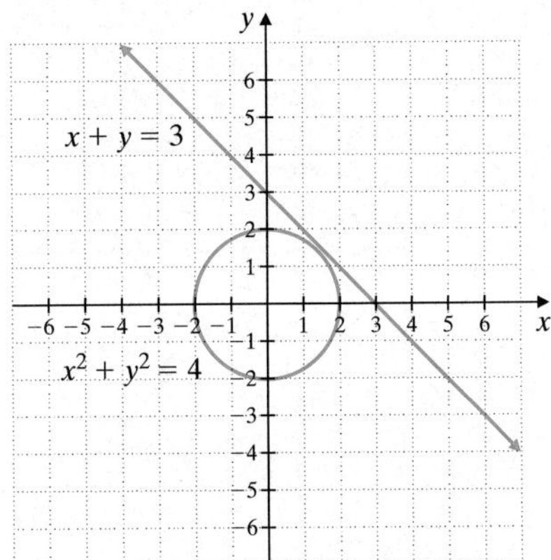

Try the Concept Check in the margin.

B Solving Nonlinear Systems by Elimination

Some nonlinear systems may be solved by the elimination method.

EXAMPLE 4 Solve the system:

$$\begin{cases} x^2 + 2y^2 = 10 \\ x^2 - y^2 = 1 \end{cases}$$

Solution: We will use the elimination, or addition, method to solve this system. To eliminate x^2 when we add the two equations, we multiply both sides of the second equation by -1. Then

$$\begin{cases} x^2 + 2y^2 = 10 \\ (-1)(x^2 - y^2) = -1 \cdot 1 \end{cases} \text{ is equivalent to } \begin{cases} x^2 + 2y^2 = 10 \\ -x^2 + y^2 = -1 \end{cases}$$

$$3y^2 = 9 \quad \text{Add.}$$

$$y^2 = 3 \quad \text{Divide both sides by 3.}$$

$$y = \pm\sqrt{3}$$

Concept Check

Without solving, how can you tell that $x^2 + y^2 = 9$ and $x^2 + y^2 = 16$ do not have any points of intersection?

Practice Problem 4

Solve the equation: $\begin{cases} x^2 + 3y^2 = 21 \\ x^2 - y^2 = 1 \end{cases}$

Answers

4. $\{(\sqrt{6}, \sqrt{5}), (\sqrt{6}, -\sqrt{5}), (-\sqrt{6}, \sqrt{5}), (-\sqrt{6}, -\sqrt{5})\}$

Concept Check: $x^2 + y^2 = 9$ is a circle inside the circle $x^2 + y^2 = 16$, therefore they do not have any points of intersection.

To find the corresponding x-values, we let $y = \sqrt{3}$ and $y = -\sqrt{3}$ in either original equation. We choose the second equation.

Let $y = \sqrt{3}$.

$$x^2 - y^2 = 1$$
$$x^2 - (\sqrt{3})^2 = 1$$
$$x^2 - 3 = 1$$
$$x^2 = 4$$
$$x = \pm\sqrt{4} = \pm 2$$

Let $y = -\sqrt{3}$.

$$x^2 - y^2 = 1$$
$$x^2 - (-\sqrt{3})^2 = 1$$
$$x^2 - 3 = 1$$
$$x^2 = 4$$
$$x = \pm\sqrt{4} = \pm 2$$

The solution set is $\{(2, \sqrt{3}), (-2, \sqrt{3}), (2, -\sqrt{3}), (-2, -\sqrt{3})\}$. Check all four ordered pairs in both equations of the system. The graph of each equation in this system is shown.

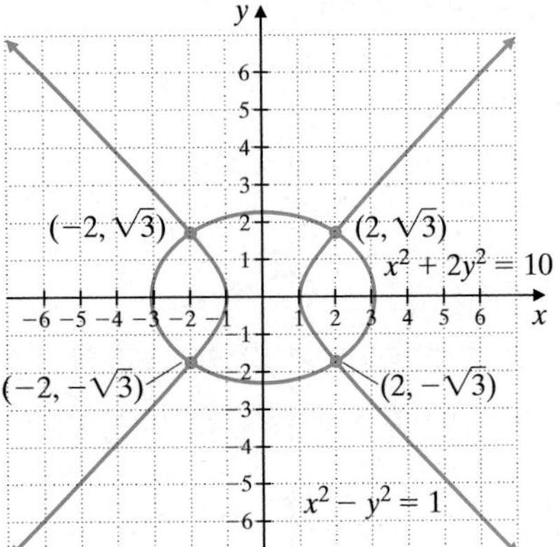

Name ______________________ Section __________ Date __________

EXERCISE SET 12.4

A B *Solve each nonlinear system of equations. See Examples 1 through 4.*

1. $\begin{cases} x^2 + y^2 = 25 \\ 4x + 3y = 0 \end{cases}$

2. $\begin{cases} x^2 + y^2 = 25 \\ 3x + 4y = 0 \end{cases}$

3. $\begin{cases} x^2 + 4y^2 = 10 \\ y = x \end{cases}$

4. $\begin{cases} 4x^2 + y^2 = 10 \\ y = x \end{cases}$

5. $\begin{cases} y^2 = 4 - x \\ x - 2y = 4 \end{cases}$

6. $\begin{cases} x^2 + y^2 = 4 \\ x + y = -2 \end{cases}$

7. $\begin{cases} x^2 + y^2 = 9 \\ 16x^2 - 4y^2 = 64 \end{cases}$

8. $\begin{cases} 4x^2 + 3y^2 = 35 \\ 5x^2 + 2y^2 = 42 \end{cases}$

9. $\begin{cases} x^2 + 2y^2 = 2 \\ x - y = 2 \end{cases}$

10. $\begin{cases} x^2 + 2y^2 = 2 \\ x^2 - 2y^2 = 6 \end{cases}$

11. $\begin{cases} y = x^2 - 3 \\ 4x - y = 6 \end{cases}$

12. $\begin{cases} y = x + 1 \\ x^2 - y^2 = 1 \end{cases}$

13. $\begin{cases} y = x^2 \\ 3x + y = 10 \end{cases}$

14. $\begin{cases} 6x - y = 5 \\ xy = 1 \end{cases}$

15. $\begin{cases} y = 2x^2 + 1 \\ x + y = -1 \end{cases}$

16. $\begin{cases} x^2 + y^2 = 9 \\ x + y = 5 \end{cases}$

17. $\begin{cases} y = x^2 - 4 \\ y = x^2 - 4x \end{cases}$

18. $\begin{cases} x = y^2 - 3 \\ x = y^2 - 3y \end{cases}$

19. $\begin{cases} 2x^2 + 3y^2 = 14 \\ -x^2 + y^2 = 3 \end{cases}$

20. $\begin{cases} 4x^2 - 2y^2 = 2 \\ -x^2 + y^2 = 2 \end{cases}$

21. $\begin{cases} x^2 + y^2 = 1 \\ x^2 + (y + 3)^2 = 4 \end{cases}$

22. $\begin{cases} x^2 + 2y^2 = 4 \\ x^2 - y^2 = 4 \end{cases}$

23. $\begin{cases} y = x^2 + 2 \\ y = -x^2 + 4 \end{cases}$

24. $\begin{cases} x = -y^2 - 3 \\ x = y^2 - 5 \end{cases}$

25. $\begin{cases} 3x^2 + y^2 = 9 \\ 3x^2 - y^2 = 9 \end{cases}$

26. $\begin{cases} x^2 + y^2 = 25 \\ x = y^2 - 5 \end{cases}$

27. $\begin{cases} x^2 + 3y^2 = 6 \\ x^2 - 3y^2 = 10 \end{cases}$

28. $\begin{cases} x^2 + y^2 = 1 \\ y = x^2 - 9 \end{cases}$

29. $\begin{cases} x^2 + y^2 = 36 \\ y = \frac{1}{6}x^2 - 6 \end{cases}$

30. $\begin{cases} x^2 + y^2 = 16 \\ y = -\frac{1}{4}x^2 + 4 \end{cases}$

31. How many real solutions are possible for a system of equations whose graphs are a circle and a parabola?

32. How many real solutions are possible for a system of equations whose graphs are an ellipse and a line?

Review and Preview

Graph each inequality in two variables. See Section 3.5.

33. $x > -3$

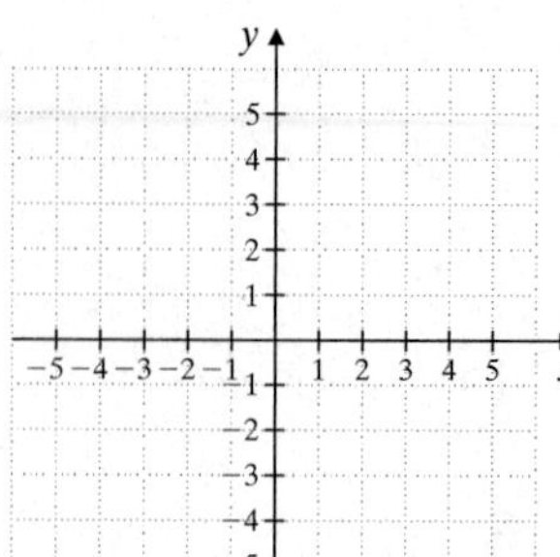

34. $y \leq 1$

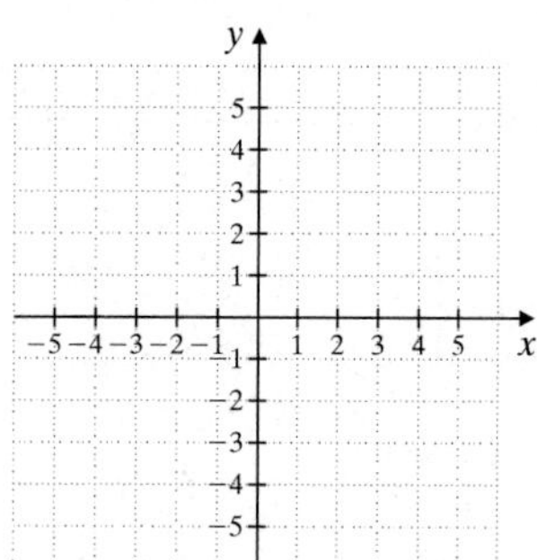

35. $y < 2x - 1$

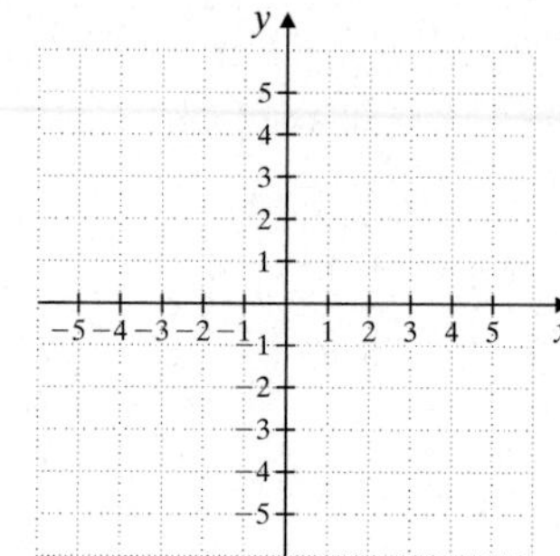

36. $3x - y \leq 4$

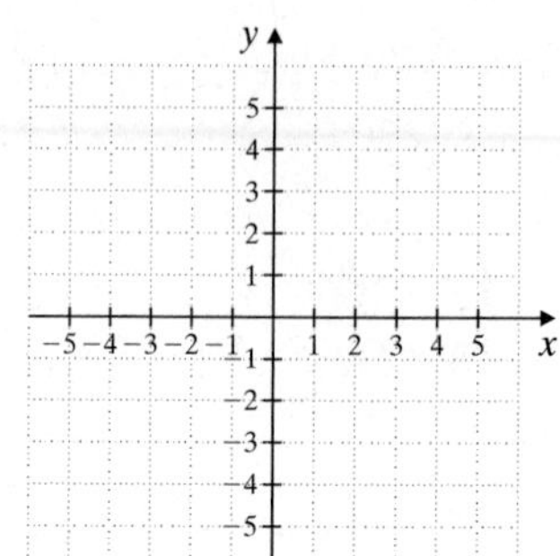

Combining Concepts

Solve.

37. The sum of the squares of two numbers is 130. The difference of the squares of the two numbers is 32. Find the two numbers.

38. The sum of the squares of two numbers is 20. Their product is 8. Find the two numbers.

△ **39.** During the development stage of a new rectangular keypad for a security system, it was decided that the area of the rectangle should be 285 square centimeters and the perimeter should be 68 centimeters. Find the dimensions of the keypad.

△ **40.** A rectangular holding pen for cattle is to be designed so that its perimeter is 92 feet and its area is 525 feet. Find the dimensions of the holding pen.

Recall that in business, a demand function expresses the quantity of a commodity demanded as a function of the commodity's unit price. A supply function expresses the quantity of a commodity supplied as a function of the commodity's unit price. When the quantity produced and supplied is equal to the quantity demanded, then we have what is called ***market equilibrium****. Use this information for Exercises 41–42.*

41. The demand function for a certain compact disc is given by the function $p(x) = -0.01x^2 - 0.2x + 9$ and the corresponding supply function is given by $p(x) = 0.01x^2 - 0.1x + 3$, where $p(x)$ is in dollars and x is in thousands of units. Find the equilibrium quantity and the corresponding price by solving the system consisting of the two given equations.

42. The demand function for a certain style of picture frame is given by the function $p(x) = -2x^2 + 90$ and the corresponding supply function is given by $p(x) = 9x + 34$, where $p(x)$ is in dollars and x is in thousands of units. Find the equilibrium quantity and the corresponding price by solving the system consisting of the two given equations.

Use a grapher to verify the results of each exercise.

43. Exercise 3

44. Exercise 4

45. Exercise 23

46. Exercise 24

12.5 Nonlinear Inequalities and Systems of Inequalities

OBJECTIVES

- (A) Graph a nonlinear inequality.
- (B) Graph a system of nonlinear inequalities.

SSM TUTOR CENTER SG CD & VIDEO MATH PRO WEB

(A) Graphing Nonlinear Inequalities

We can graph a nonlinear inequality in two variables such as $\frac{x^2}{9} + \frac{y^2}{16} \le 1$ in a way similar to the way we graphed a linear inequality in two variables in Section 3.5. First, we graph the related equation $\frac{x^2}{9} + \frac{y^2}{16} = 1$. The graph of the equation is our boundary. Then, using test points, we determine and shade the region whose points satisfy the inequality.

EXAMPLE 1 Graph: $\frac{x^2}{9} + \frac{y^2}{16} \le 1$

Solution: First we graph the equation $\frac{x^2}{9} + \frac{y^2}{16} = 1$. We sketch a solid curve because of the inequality symbol $\le$. It means that the graph of $\frac{x^2}{9} + \frac{y^2}{16} \le 1$ includes the graph of $\frac{x^2}{9} + \frac{y^2}{16} = 1$. The graph is an ellipse, and it divides the plane into two regions, the "inside" and the "outside" of the ellipse. Recall from Section 3.5 that to determine which region contains the solutions, we select a test point in either region and determine whether the coordinates of the point satisfy the inequality. We choose $(0, 0)$ as the test point.

$$\frac{x^2}{9} + \frac{y^2}{16} \le 1$$

$$\frac{0^2}{9} + \frac{0^2}{16} \le 1 \quad \text{Let } x = 0 \text{ and } y = 0.$$

$$0 \le 1 \quad \text{True.}$$

Since this statement is true, the solution set is the region containing $(0, 0)$. The graph of the solution set includes the points on and inside the ellipse, as shaded in the figure.

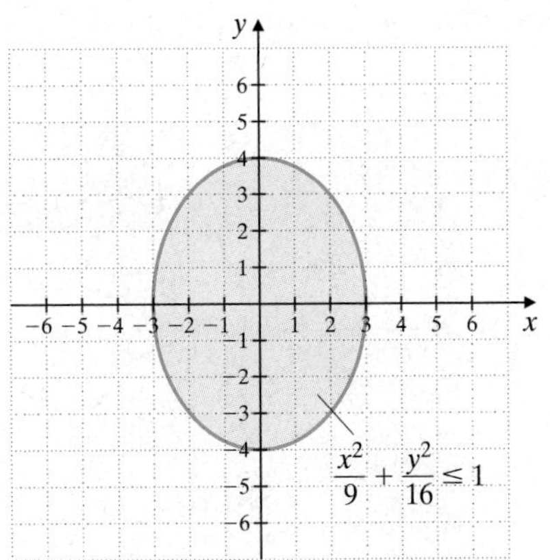

Practice Problem 1

Graph: $\frac{x^2}{25} + \frac{y^2}{4} \le 1$

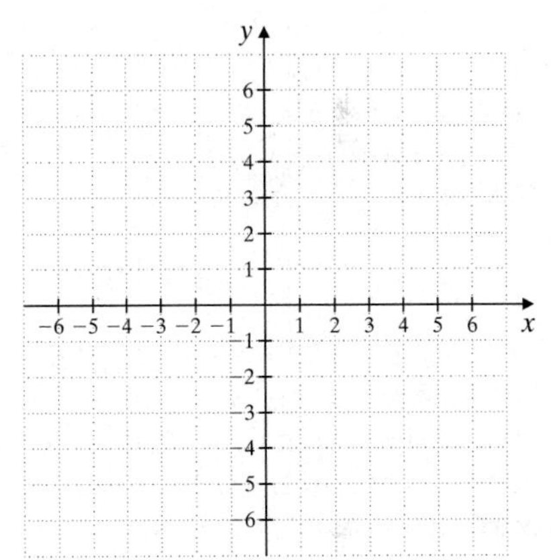

Answer

1.

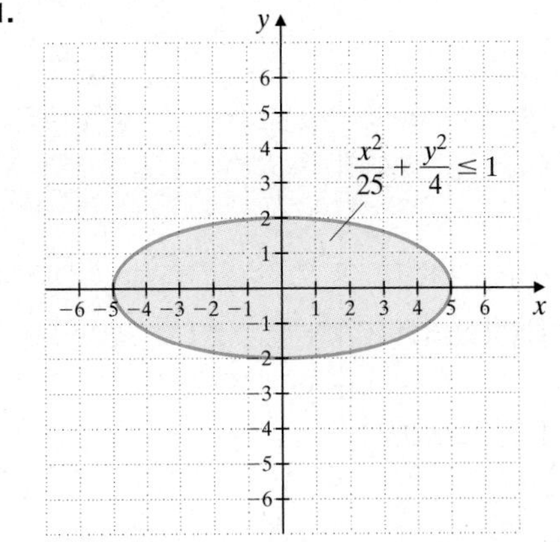

Practice Problem 2

Graph: $9x^2 > 4y^2 + 144$

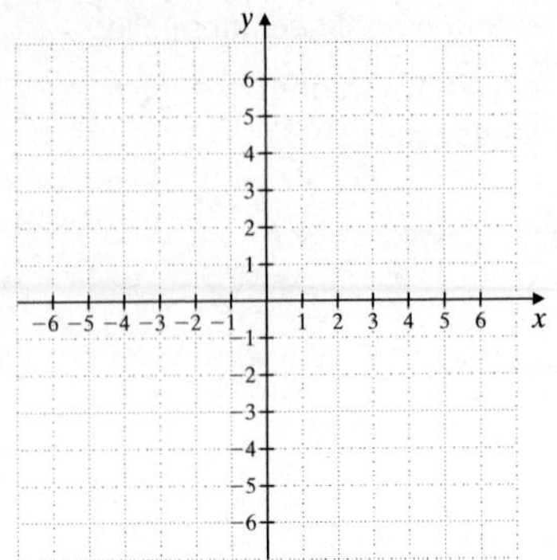

Practice Problem 3

Graph the system:

$$\begin{cases} y \geq x^2 \\ y \leq -4x + 2 \end{cases}$$

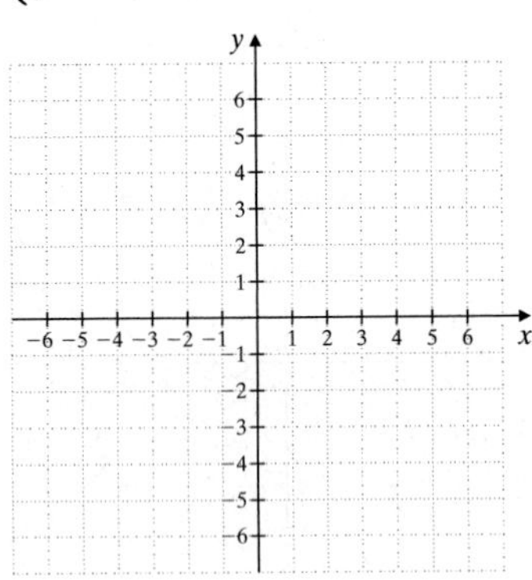

Answers

2.

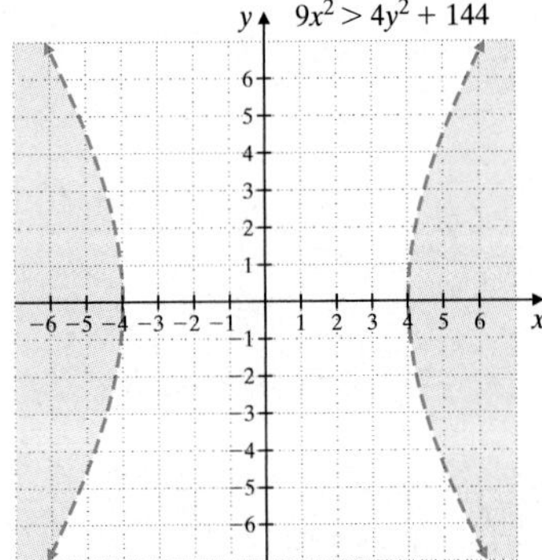

3.

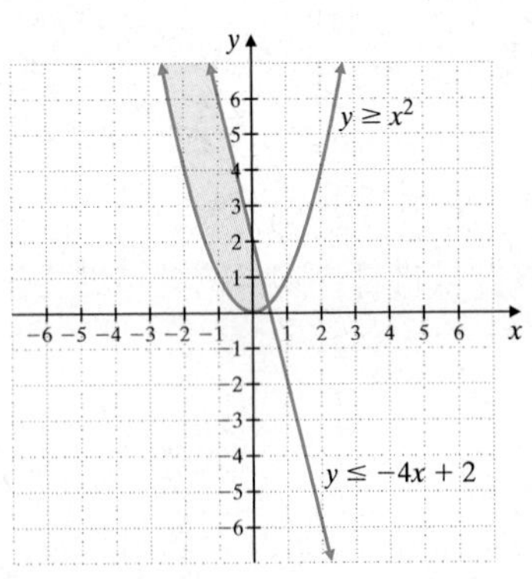

EXAMPLE 2 Graph: $4y^2 > x^2 + 16$

Solution: The related equation is $4y^2 = x^2 + 16$, or $\frac{y^2}{4} - \frac{x^2}{16} = 1$, which is a hyperbola. We graph the hyperbola as a dashed curve because of the inequality symbol $>$. It means that the graph of $4y^2 > x^2 + 16$ does *not* include the graph of $4y^2 = x^2 + 16$. The hyperbola divides the plane into three regions. We select a test point in each region—not on a boundary line—to determine whether that region contains solutions of the inequality.

Test Region *A* with (0, 4)	Test Region *B* with (0, 0)	Test Region *C* with (0, −4)
$4y^2 > x^2 + 16$	$4y^2 > x^2 + 16$	$4y^2 > x^2 + 16$
$4(4)^2 > 0^2 + 16$	$4(0)^2 > 0^2 + 16$	$4(-4)^2 > 0^2 + 16$
$64 > 16$ True.	$0 > 16$ False.	$64 > 16$ True.

The graph of the solution set includes the shaded regions *A* and *C* only, not the boundary.

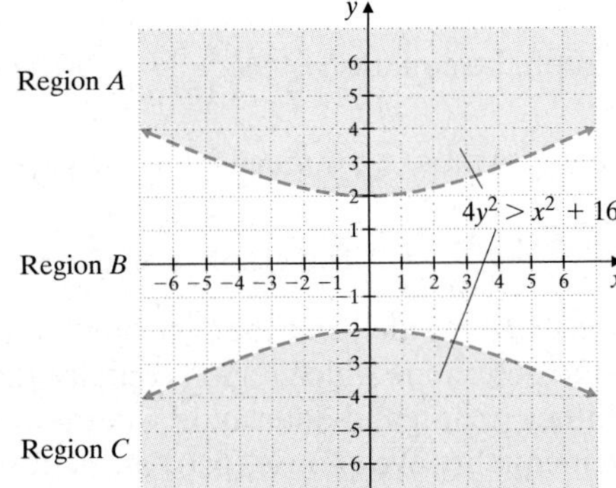

B Graphing Systems of Nonlinear Inequalities

In Section 8.7 we graphed systems of linear inequalities. Recall that the graph of a system of inequalities is the intersection of the graphs of the inequalities.

EXAMPLE 3 Graph the system:

$$\begin{cases} x \leq 1 - 2y \\ y \leq x^2 \end{cases}$$

Solution: We graph each inequality on the same set of axes. The intersection is the darkest shaded region along with its boundary lines. The coordinates of the points of intersection can be found by solving the related system.

$$\begin{cases} x = 1 - 2y \\ y = x^2 \end{cases}$$

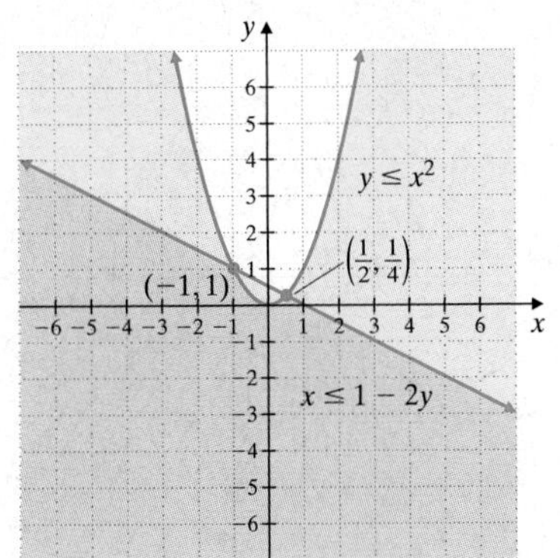

EXAMPLE 4 Graph the system:

$$\begin{cases} x^2 + y^2 < 25 \\ \dfrac{x^2}{9} - \dfrac{y^2}{25} < 1 \\ y < x + 3 \end{cases}$$

Solution: We graph each inequality. The graph of $x^2 + y^2 < 25$ contains points "inside" the circle that has center $(0, 0)$ and radius 5. The graph of $\dfrac{x^2}{9} - \dfrac{y^2}{25} < 1$ is the region between the two branches of the hyperbola with x-intercepts $(-3, 0)$ and $(3, 0)$ and center $(0, 0)$. The graph of $y < x + 3$ is the region "below" the line with the slope 1 and y-intercept $(0, 3)$. The graph of the solution set of the system is the intersection of all the graphs, the darkest shaded region shown. The boundary of this region is not part of the solution.

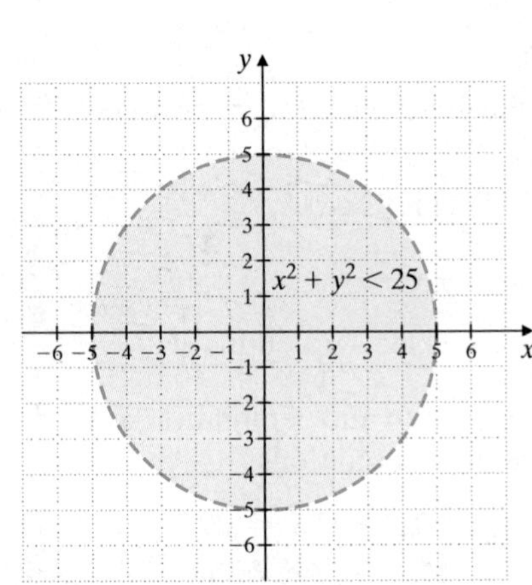

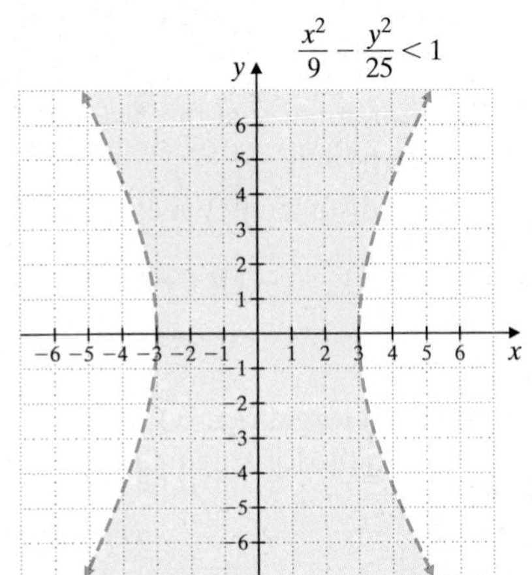

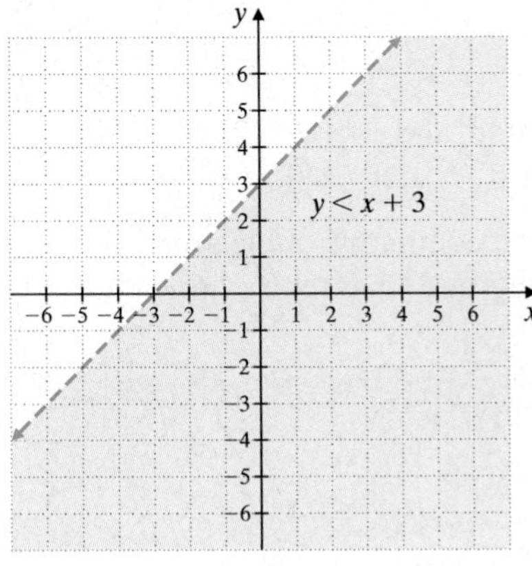

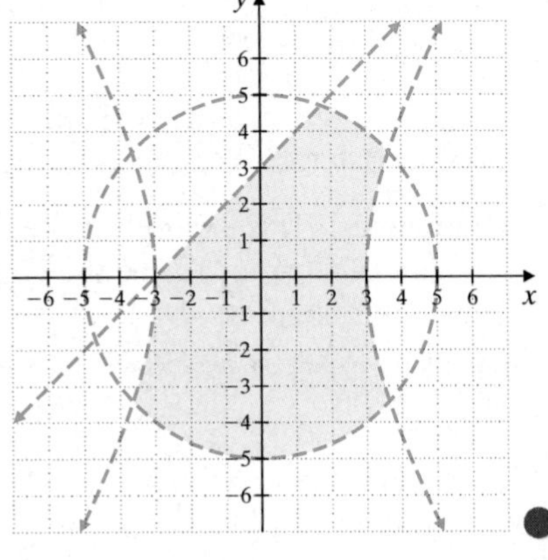

Practice Problem 4

Graph the system:

$$\begin{cases} x^2 + y^2 < 9 \\ \dfrac{x^2}{9} - \dfrac{y^2}{4} < 1 \\ y > x - 2 \end{cases}$$

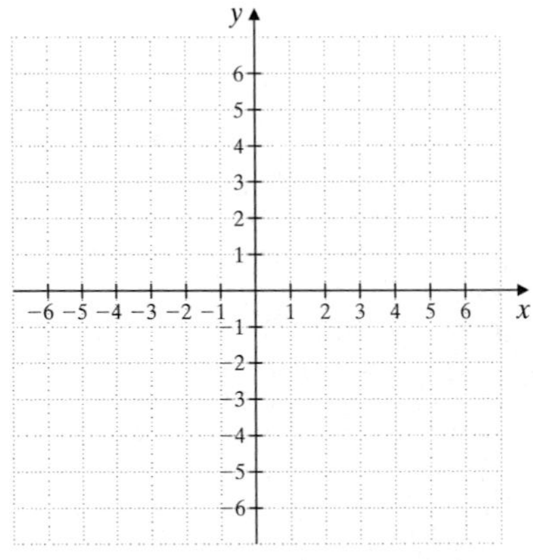

Answer

4.

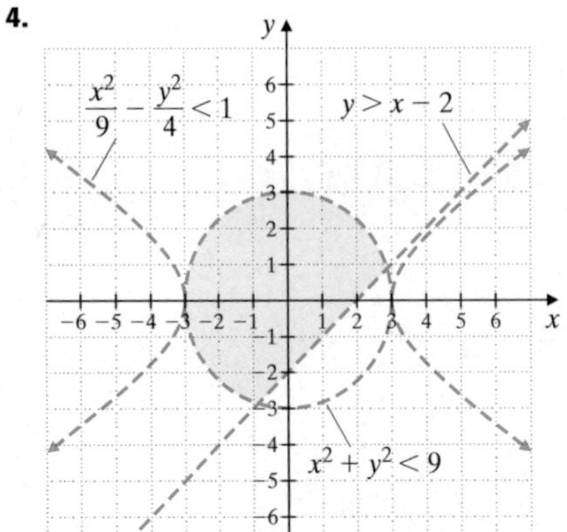

STUDY SKILLS REMINDER

Are you preparing for a test on Chapter 12?

Below I have listed some common trouble areas for students in Chapter 12. After studying for your test—but before taking your test—read these.

- Don't forget to review all the standard forms for the conic sections.
- Remember that the midpoint of a segment is a *point*. The x-coordinate is the average of the x-coordinates of the endpoints of the segment and the y-coordinate is the average of the y-coordinates of the endpoints of the segment.

  The midpoint of the segment joining $(-1, 5)$ and $(3, 4)$ is $\left(\frac{-1+3}{2}, \frac{5+4}{2}\right)$ or $\left(1, \frac{9}{2}\right)$.

- Remember that the distance formula gives the *distance* between two points.

  The distance between $(-1, 5)$ and $(3, 4)$ is

$$\begin{aligned}\sqrt{(3-(-1))^2+(4-5)^2} &= \sqrt{4^2+(-1)^2}\\ &= \sqrt{16+1} = \sqrt{17} \text{ units}\end{aligned}$$

- Don't forget that both methods, substitution and elimination, are available for solving nonlinear systems of equations.

$$\begin{cases} x^2 + y^2 = 7 \\ 2x^2 - 3y^2 = 4 \end{cases} \text{ is equivalent to } \begin{cases} 3x^2 + 3y^2 = 21 \\ 2x^2 - 3y^2 = 4 \end{cases}$$

$$\begin{aligned} 5x^2 &= 25 \\ x^2 &= 5 \\ x &= \pm\sqrt{5} \end{aligned}$$

Let $x = \pm\sqrt{5}$ in either original equation, and $y = \pm\sqrt{2}$, the solution set is $\{(\sqrt{5}, \sqrt{2}), (-\sqrt{5}, \sqrt{2}), (\sqrt{5}, -\sqrt{2}), (\sqrt{5}, -\sqrt{2})\}$.

Remember: This is simply a checklist of common trouble areas. For a review of Chapter 12, see the Highlights and Chapter Review at the end of this chapter.

Name ______________________ Section __________ Date __________

EXERCISE SET 12.5

A *Graph each inequality. See Examples 1 and 2.*

1. $y < x^2$

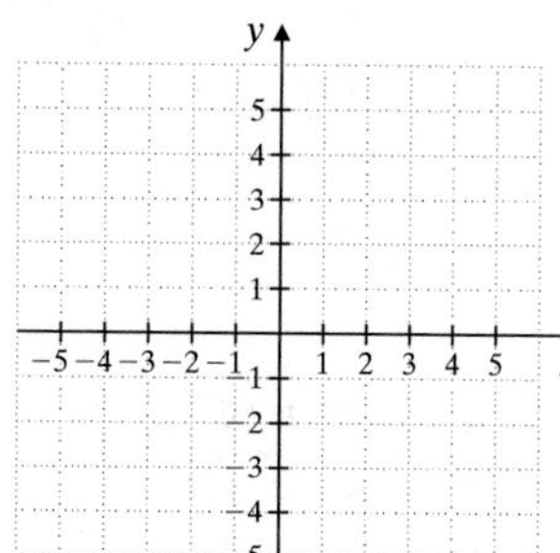

2. $y < -x^2$

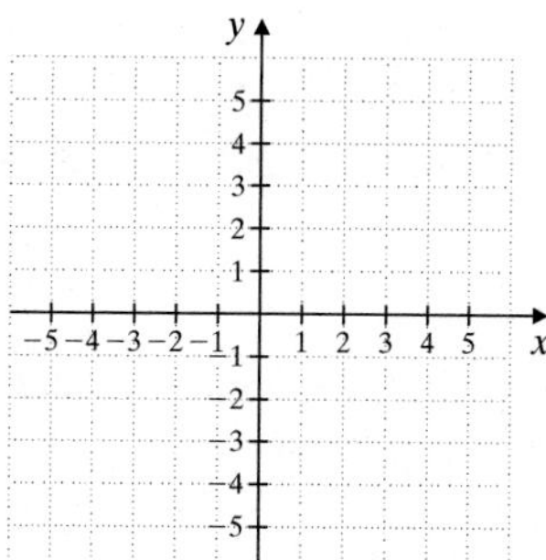

3. $x^2 + y^2 \geq 16$

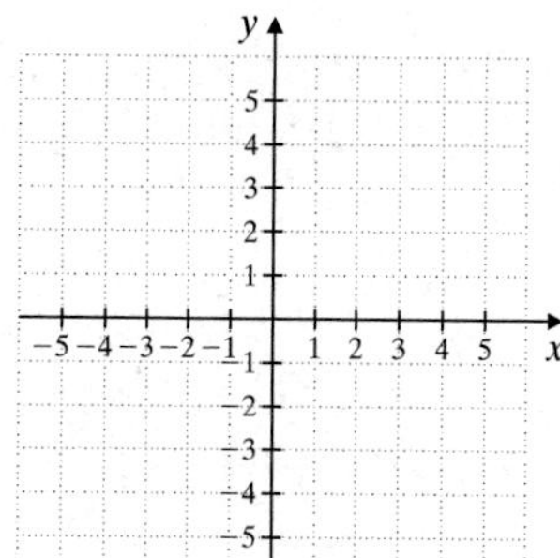

4. $x^2 + y^2 < 36$

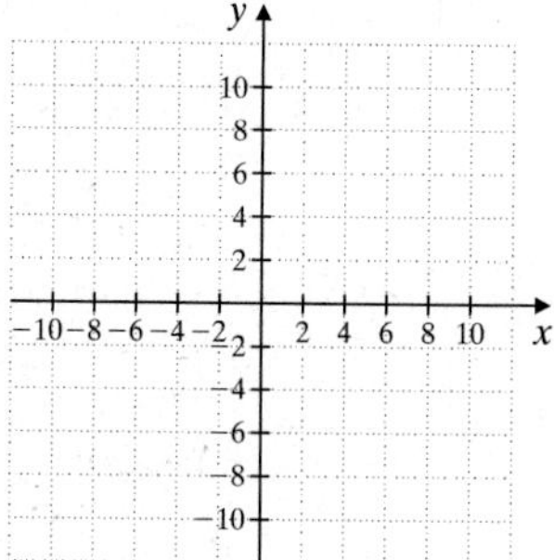

5. $\frac{x^2}{4} - y^2 < 1$

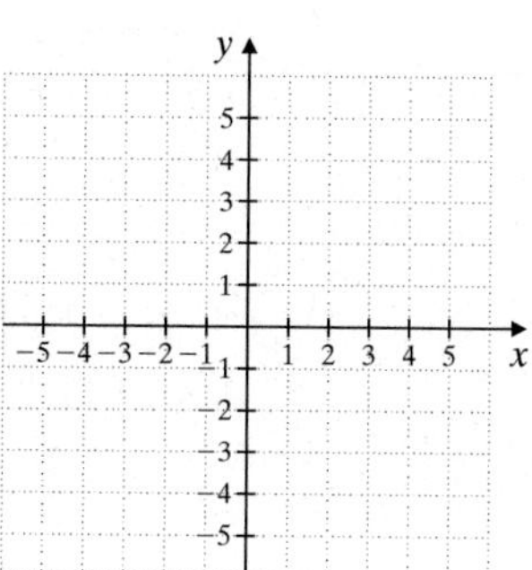

6. $x^2 - \frac{y^2}{9} \geq 1$

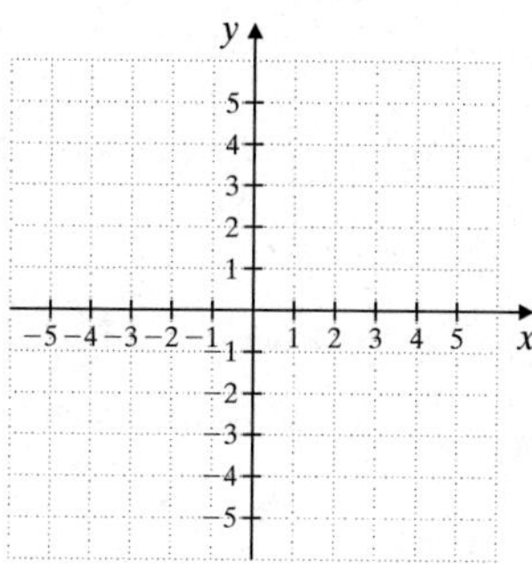

7. $y > (x - 1)^2 - 3$

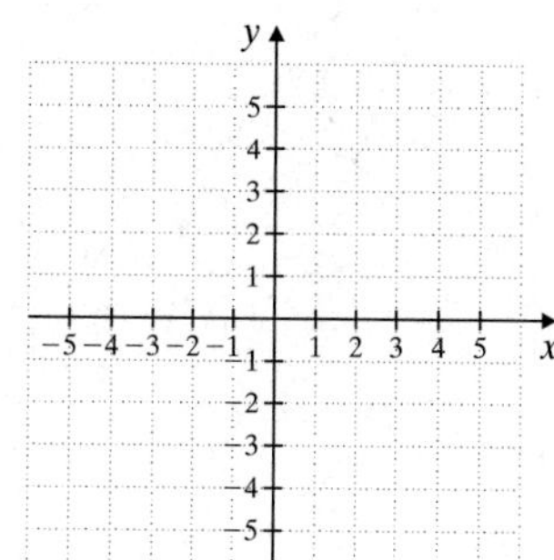

8. $y > (x + 3)^2 + 2$

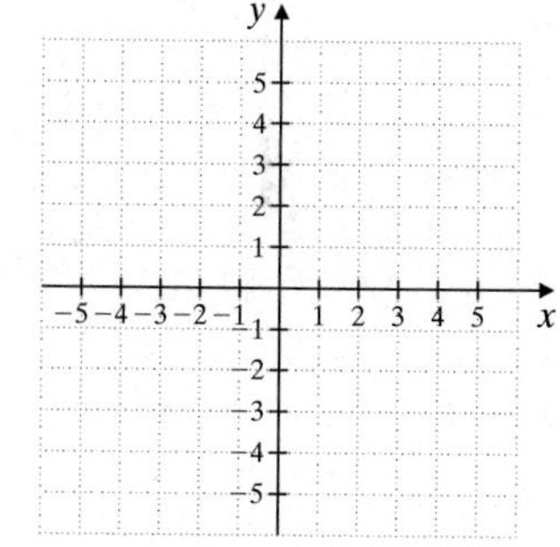

9. $x^2 + y^2 \leq 9$

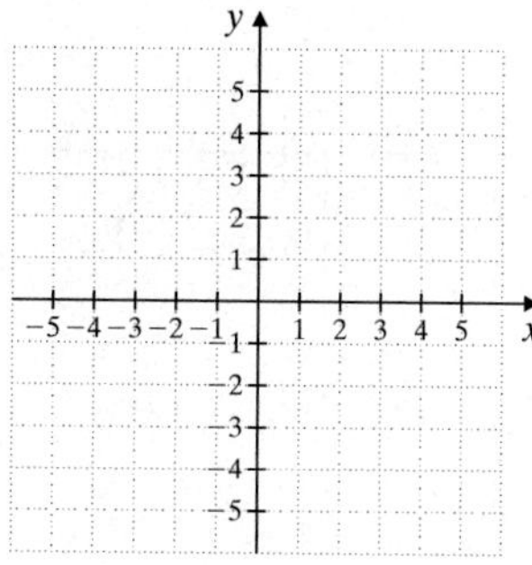

10. $x^2 + y^2 > 4$

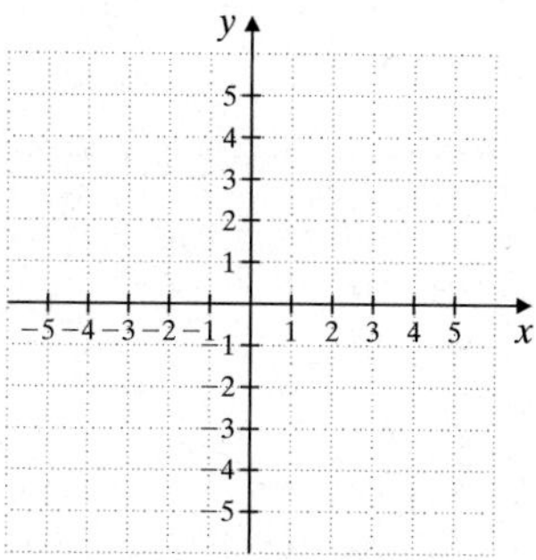

11. $y > -x^2 + 5$

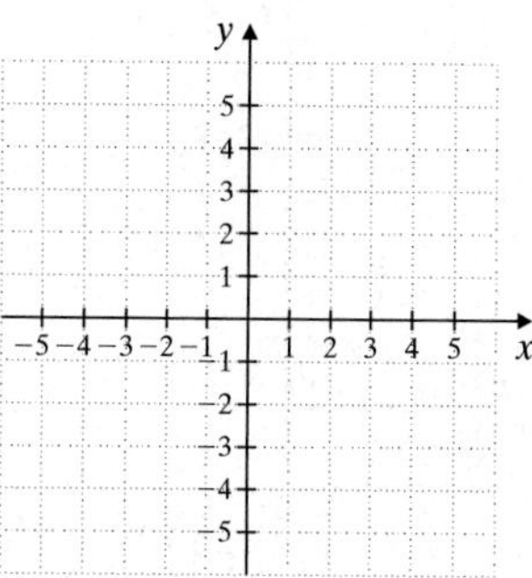

12. $y < -x^2 + 5$

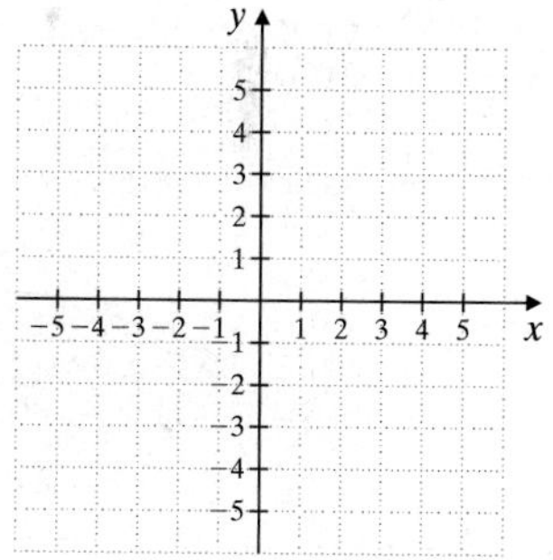

13. $\frac{x^2}{4} + \frac{y^2}{9} \leq 1$

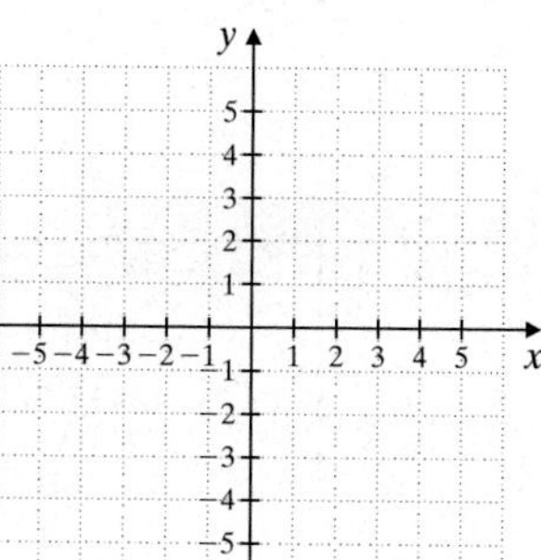

14. $\frac{x^2}{25} + \frac{y^2}{4} \geq 1$

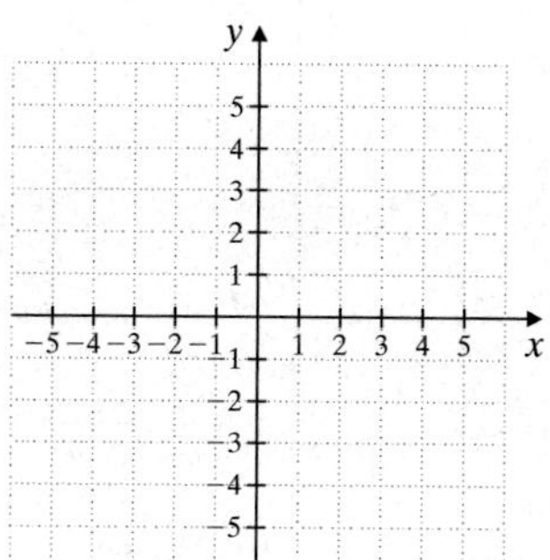

15. $\frac{y^2}{4} - x^2 \leq 1$

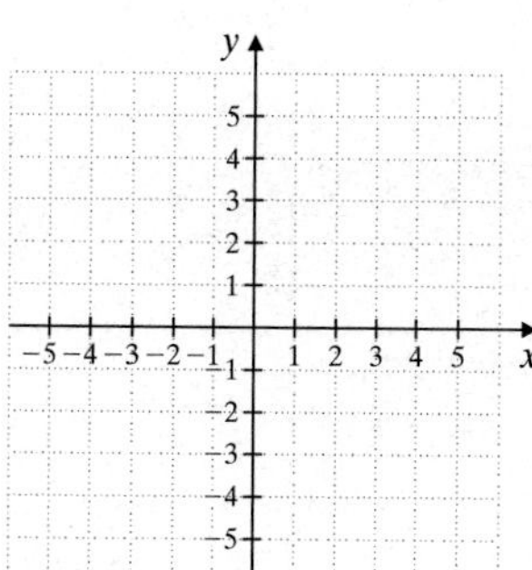

16. $\frac{y^2}{16} - \frac{x^2}{9} > 1$

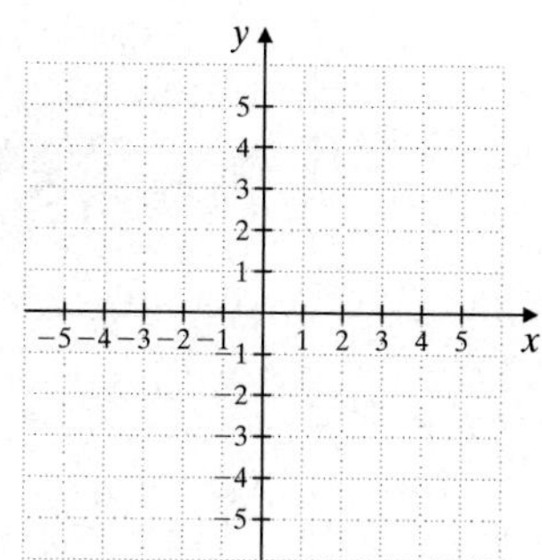

17. $y < (x - 2)^2 + 1$

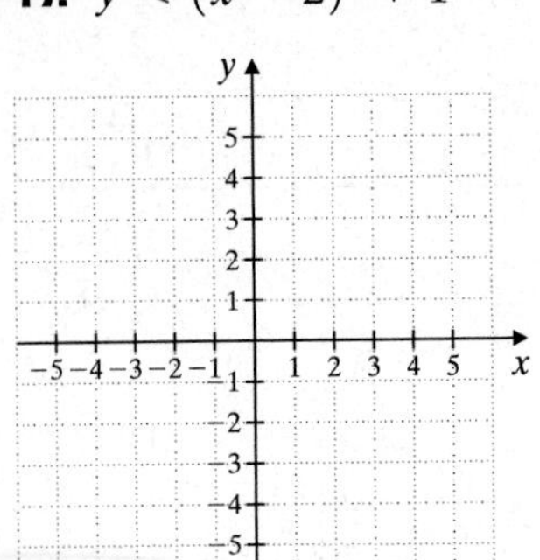

18. $y > (x - 2)^2 + 1$

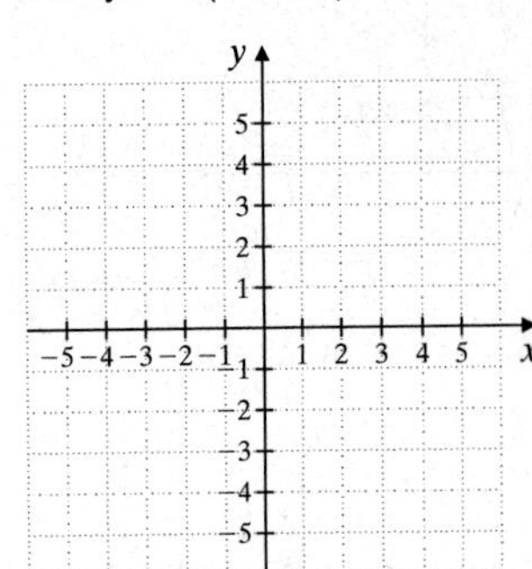

19. $y \le x^2 + x - 2$

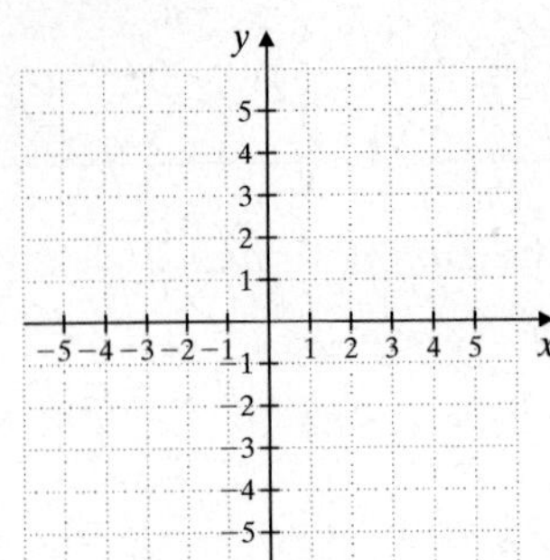

20. $y > x^2 + x - 2$

B *Graph each system. See Examples 3 and 4.*

21. $\begin{cases} 4x + 3y \ge 12 \\ x^2 + y^2 < 16 \end{cases}$

22. $\begin{cases} 3x - 4y \le 12 \\ x^2 + y^2 < 16 \end{cases}$

23. $\begin{cases} x^2 + y^2 \le 9 \\ x^2 + y^2 \ge 1 \end{cases}$

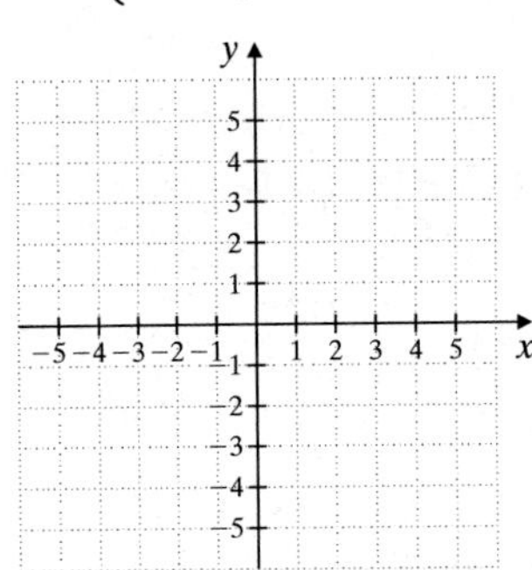

24. $\begin{cases} x^2 + y^2 \ge 9 \\ x^2 + y^2 \ge 16 \end{cases}$

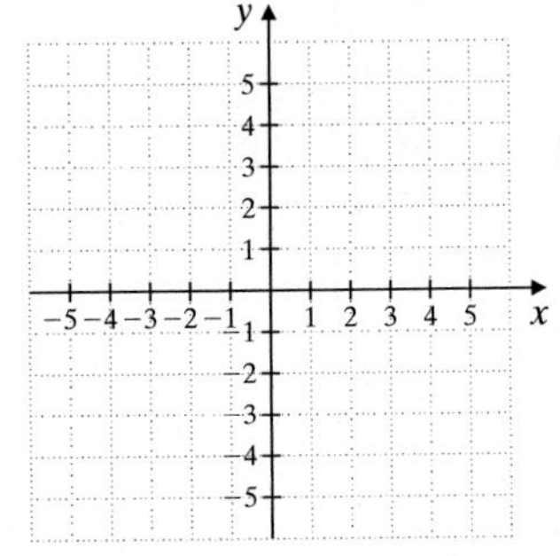

25. $\begin{cases} y > x^2 \\ y \ge 2x + 1 \end{cases}$

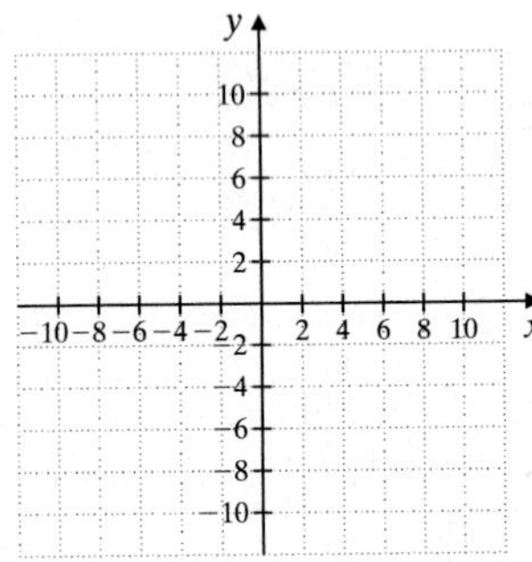

26. $\begin{cases} y \le -x^2 + 3 \\ y \le 2x - 1 \end{cases}$

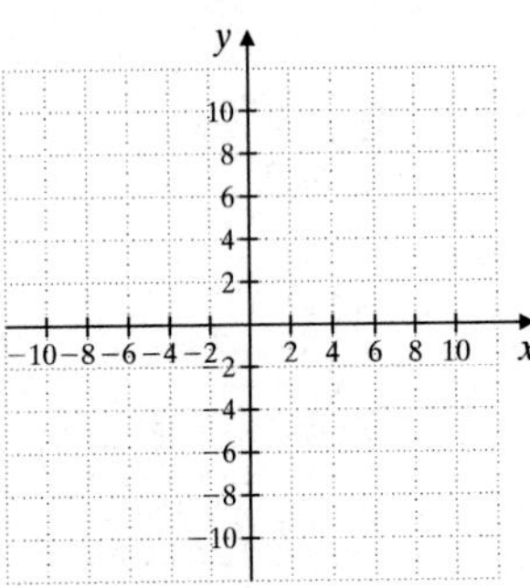

27. $\begin{cases} x^2 + y^2 > 9 \\ y > x^2 \end{cases}$

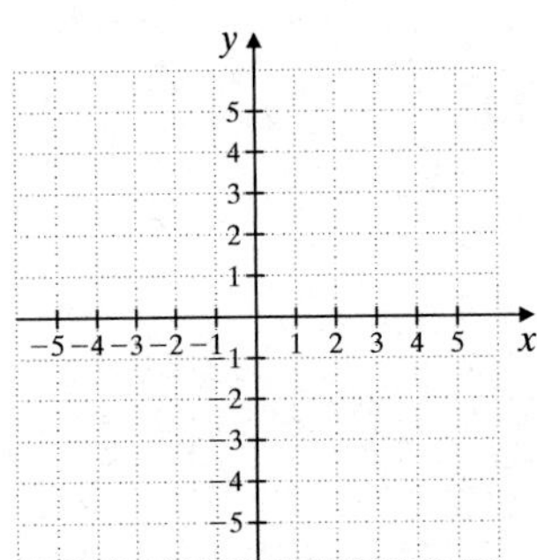

28. $\begin{cases} x^2 + y^2 \le 9 \\ y < x^2 \end{cases}$

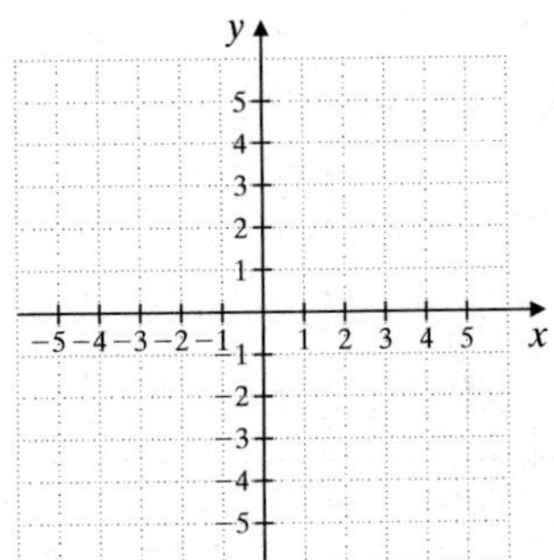

29. $\begin{cases} \dfrac{x^2}{4} + \dfrac{y^2}{9} \ge 1 \\ x^2 + y^2 \ge 4 \end{cases}$

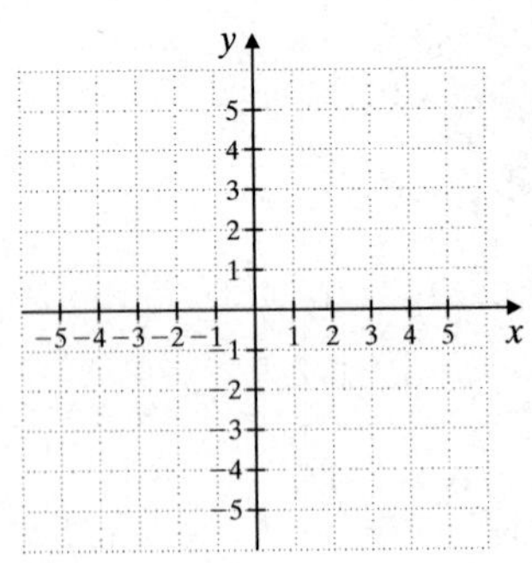

30. $\begin{cases} x^2 + (y - 2)^2 \ge 9 \\ \dfrac{x^2}{4} + \dfrac{y^2}{25} < 1 \end{cases}$

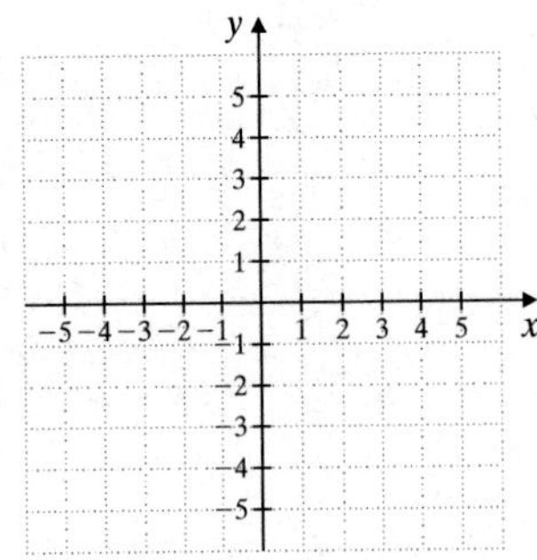

31. $\begin{cases} x^2 - y^2 \ge 1 \\ y \ge 0 \end{cases}$

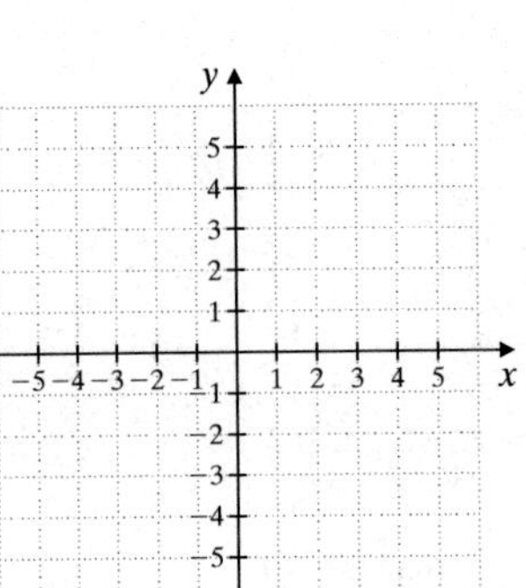

32. $\begin{cases} x^2 - y^2 \ge 1 \\ x \ge 0 \end{cases}$

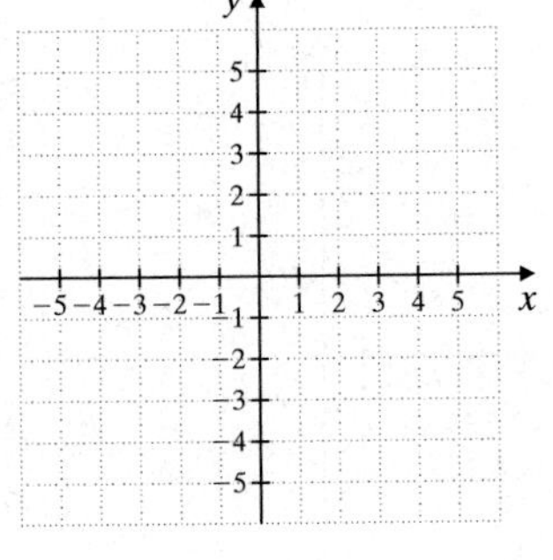

33. $\begin{cases} x + y \geq 1 \\ 2x + 3y < 1 \\ x > -3 \end{cases}$

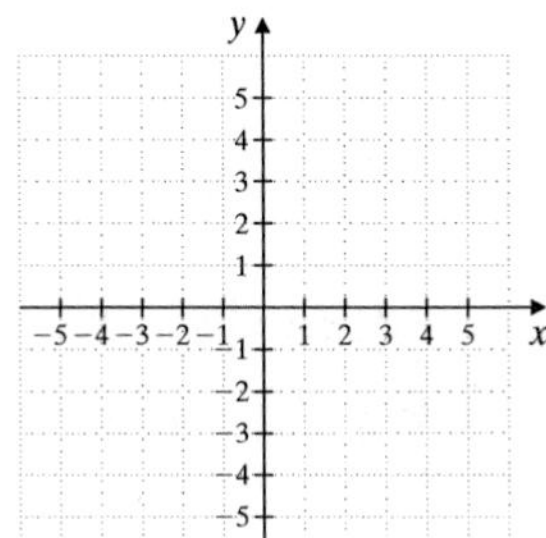

34. $\begin{cases} x - y < -1 \\ 4x - 3y > 0 \\ y > 0 \end{cases}$

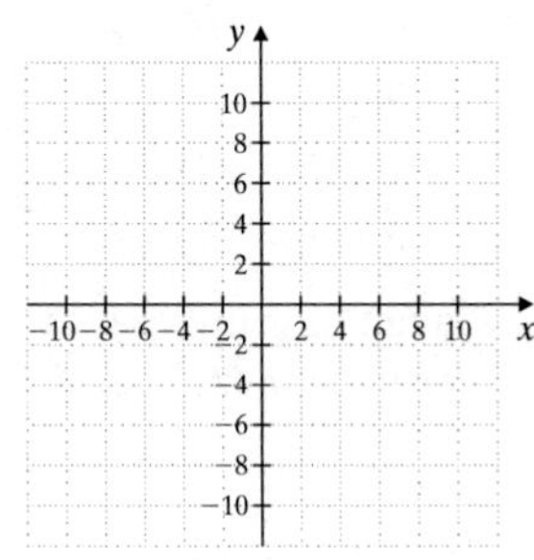

35. $\begin{cases} x^2 - y^2 < 1 \\ \dfrac{x^2}{16} + y^2 \leq 1 \\ x \geq -2 \end{cases}$

36. $\begin{cases} x^2 - y^2 \geq 1 \\ \dfrac{x^2}{16} + \dfrac{y^2}{4} \leq 1 \\ y \geq 1 \end{cases}$

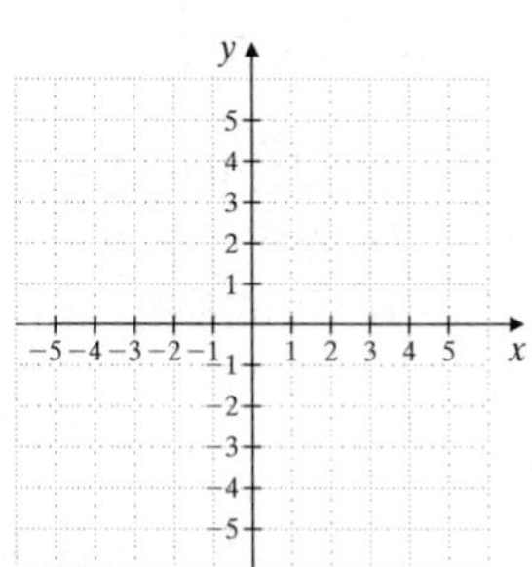

Review and Preview

Determine whether each graph is the graph of a function. See Section 7.3.

37.

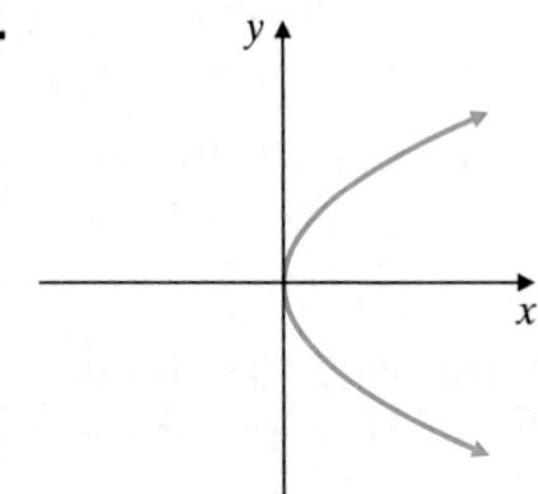

38.

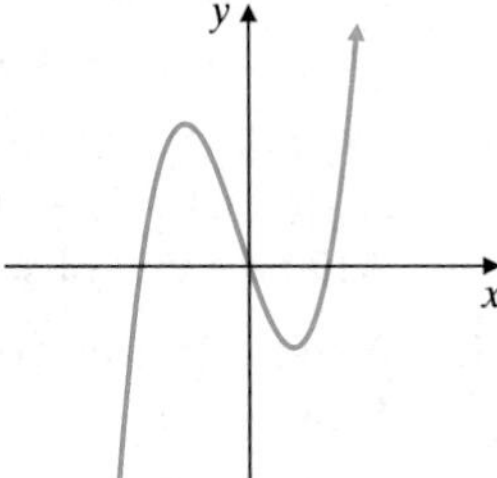

39.

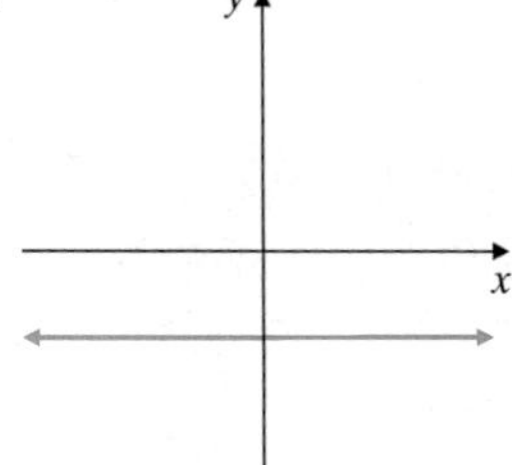

40.

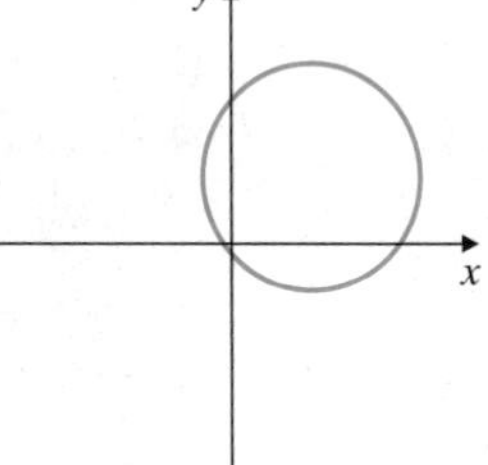

Combining Concepts

41. Discuss how graphing a linear inequality such as $x + y < 9$ is similar to graphing a nonlinear inequality such as $x^2 + y^2 < 9$.

42. Discuss how graphing a linear inequality such as $x + y < 9$ is different from graphing a nonlinear inequality such as $x^2 + y^2 < 9$.

43. Graph the system:

$\begin{cases} y \leq x^2 \\ y \geq x + 2 \\ x \geq 0 \\ y \geq 0 \end{cases}$

see graph

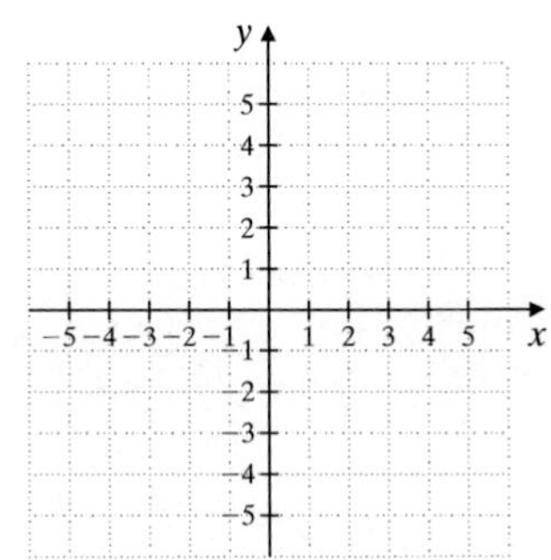

CHAPTER 12 ACTIVITY Modeling Conic Sections

MATERIALS

- two thumbtacks (or nails)
- graph paper
- cardboard
- tape
- string
- pencil
- ruler

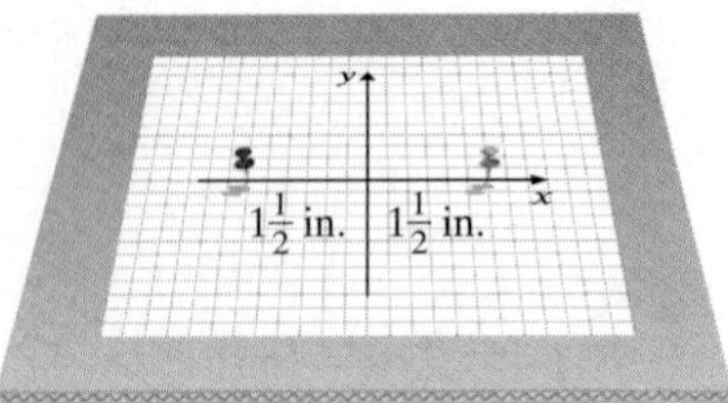

Figure 1

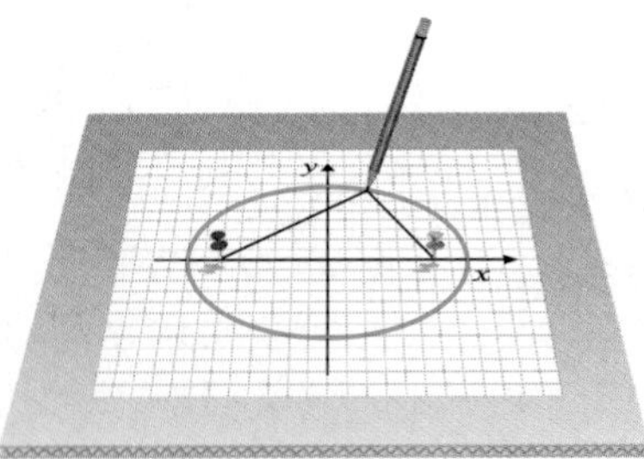

Figure 2

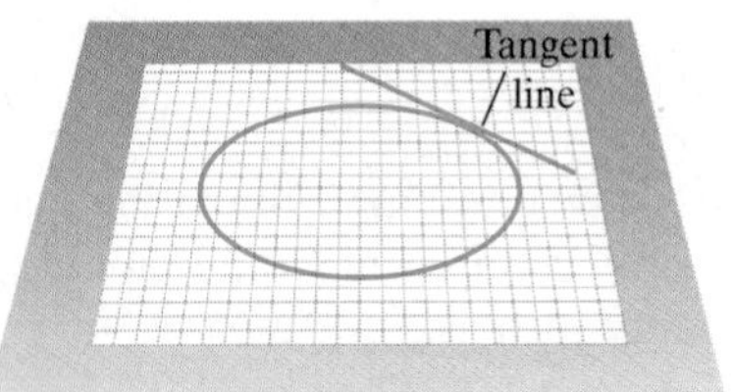

Figure 3

This activity may be completed by working in groups or individually.

1. Draw an x-axis and a y-axis on the graph paper as shown in Figure 1.
2. Place the graph paper on the cardboard and use tape to attach.
3. Locate two points on the x-axis each about $1\frac{1}{2}$ inches from the origin and on opposite sides of the origin (see Figure 1). Insert thumbtacks (or nails) at each of these locations.
4. Fasten a 9-inch piece of string to the thumbtacks as shown in Figure 2. Use your pencil to draw and keep the string taut while you carefully move the pencil in a path all around the thumbtacks.
5. Using the grid of the graph paper as a guide, find an approximate equation of the ellipse you drew.
6. Experiment by moving the tacks closer together or farther apart and drawing new ellipses. What do you observe?
7. Write a paragraph explaining why the figure drawn by the pencil is an ellipse. How might you use the same materials to draw a circle?
8. (Optional) Choose one of the ellipses you drew with the string and pencil. Use a ruler to draw any six tangent lines to the ellipse. (A line is tangent to the ellipse if it intersects, or just touches, the ellipse at only one point. See Figure 3.) Extend the tangent lines to yield six points of intersection among the tangents. Use a straight edge to draw a line connecting each pair of opposite points of intersection. What do you observe? Repeat with a different ellipse. Can you make a conjecture about the relationship among the lines that connect opposite points of intersection?

Chapter 12 VOCABULARY CHECK

Fill in each blank with one of the words or phrases listed below.

circle	midpoint	radius	distance
center	ellipse	hyperbola	nonlinear system of equations

1. The __________ formula is $d = \sqrt{(x_2 - x_1)^2 + (y_2 - y_1)^2}$.
2. A(n) __________ is the set of all points in a plane that are the same distance from a fixed point, called the __________.
3. A ______________________________ is a system of equations at least one of which is not linear.
4. A(n) __________ is the set of points on a plane such that the sum of the distances of those points from two fixed points is a constant.
5. In a circle, the distance from the center to a point of the circle is called its __________.
6. A(n) ____________ is the set of points in a plane such that the absolute value of the difference of the distance from two fixed points is constant.
7. The __________ formula is $\left(\frac{x_1 + x_2}{2}, \frac{y_1 + y_2}{2}\right)$.

12 CHAPTER Highlights

DEFINITIONS AND CONCEPTS | **EXAMPLES**

Section 12.1 The Parabola and the Circle

PARABOLAS

$$y = a(x - h)^2 + k$$

a > 0

(h, k)

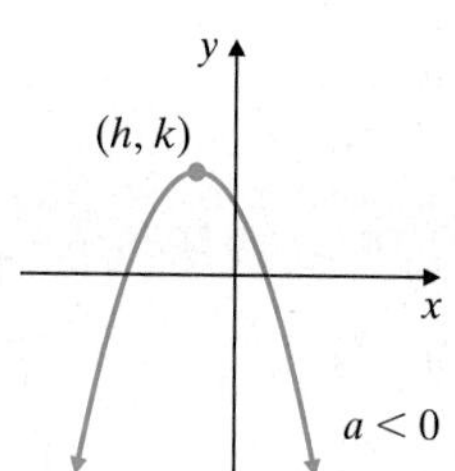

$$x = a(y - k)^2 + h$$

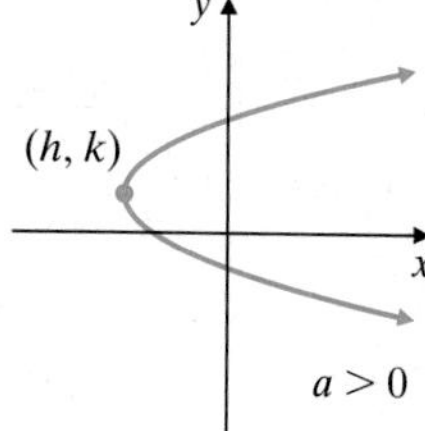

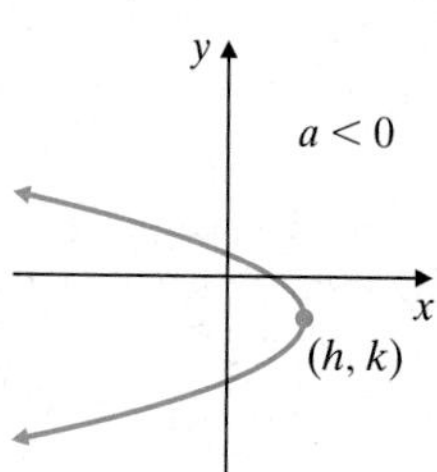

Graph: $x = 3y^2 - 12y + 13$

$$x - 13 = 3(y^2 - 4y)$$
$$x - 13 + 3(4) = 3(y^2 - 4y + 4)$$
$$x = 3(y - 2)^2 + 1$$

Since $a = 3$, this parabola opens to the right with vertex $(1, 2)$. Its axis of symmetry is $y = 2$. The x-intercept is $(13, 0)$.

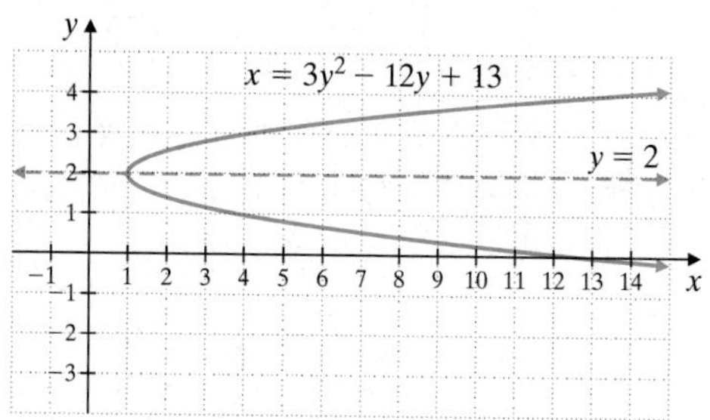

DISTANCE FORMULA

The distance d between two points (x_1, y_1) and (x_2, y_2) is given by

$$d = \sqrt{(x_2 - x_1)^2 + (y_2 - y_1)^2}$$

Find the distance between points $(-1, 6)$ and $(-2, -4)$. Let $(x_1, y_1) = (-1, 6)$ and $(x_2, y_2) = (-2, -4)$.

$$d = \sqrt{(x_2 - x_1)^2 + (y_2 - y_1)^2}$$
$$= \sqrt{(-2 - (-1))^2 + (-4 - 6)^2}$$
$$= \sqrt{1 + 100} = \sqrt{101}$$

DEFINITIONS AND CONCEPTS | EXAMPLES

Section 12.1 The Parabola and the Circle *(continued)*

MIDPOINT FORMULA

The midpoint of the line segment whose endpoints are (x_1, y_1) and (x_2, y_2) is the point with coordinates

$$\left(\frac{x_1 + x_2}{2}, \frac{y_1 + y_2}{2}\right)$$

Find the midpoint of the line segment whose endpoints are $(-1, 6)$ and $(-2, -4)$.

$$\left(\frac{-1 + (-2)}{2}, \frac{6 + (-4)}{2}\right)$$

The midpoint is $\left(-\frac{3}{2}, 1\right)$.

CIRCLE

The graph $(x - h)^2 + (y - k)^2 = r^2$ is a circle with center (h, k) and radius r.

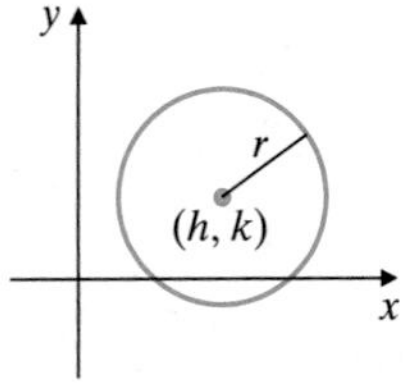

Graph: $x^2 + (y + 3)^2 = 5$

This equation can be written as

$$(x - 0)^2 + (y + 3)^2 = 5$$

with $h = 0$, $k = -3$, and $r = \sqrt{5}$. The center of this circle is $(0, -3)$, and the radius is $\sqrt{5}$.

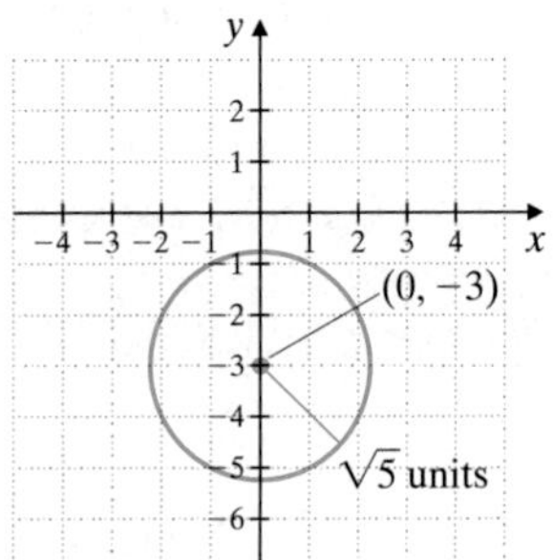

Section 12.2 The Ellipse and the Hyperbola

ELLIPSE WITH CENTER (0, 0)

The graph of an equation of the form $\frac{x^2}{a^2} + \frac{y^2}{b^2} = 1$ is an ellipse with center $(0, 0)$. The x-intercepts are $(a, 0)$ and $(-a, 0)$, and the y-intercepts are $(0, b)$ and $(0, -b)$.

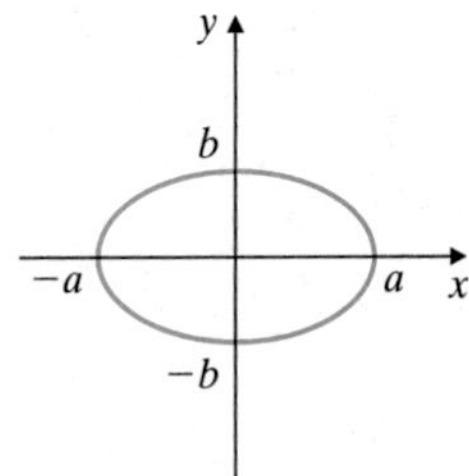

Graph: $4x^2 + 9y^2 = 36$

$$\frac{x^2}{9} + \frac{y^2}{4} = 1 \quad \text{Divide both sides by 36.}$$

$$\frac{x^2}{3^2} + \frac{y^2}{2^2} = 1$$

The ellipse has center $(0, 0)$, x-intercepts $(3, 0)$ and $(-3, 0)$, and y-intercepts $(0, 2)$ and $(0, -2)$.

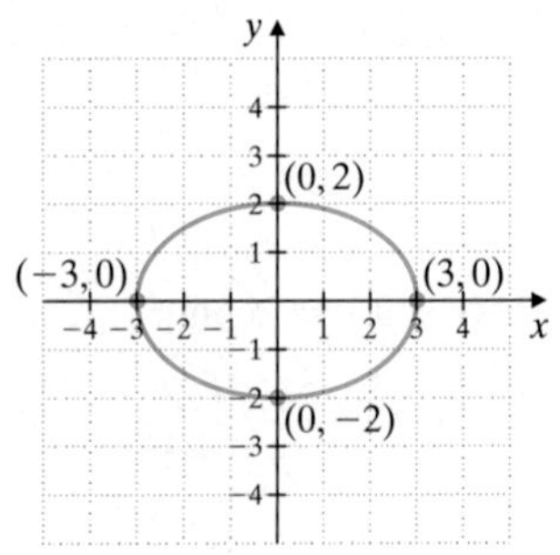

DEFINITIONS AND CONCEPTS	EXAMPLES

Section 12.2 The Ellipse and the Hyperbola *(continued)*

HYPERBOLA WITH CENTER (0, 0)

The graph of an equation of the form $\frac{x^2}{a^2} - \frac{y^2}{b^2} = 1$ is a hyperbola with center $(0, 0)$ and x-intercepts $(a, 0)$ and $(-a, 0)$.

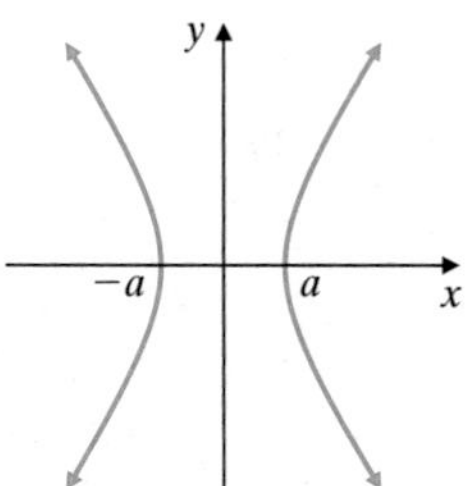

The graph of an equation of the form $\frac{y^2}{b^2} - \frac{x^2}{a^2} = 1$ is a hyperbola with center $(0, 0)$ and y-intercepts $(0, b)$ and $(0, -b)$.

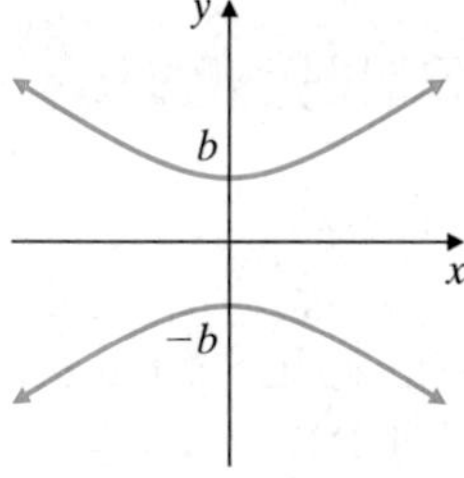

Graph: $\frac{x^2}{9} - \frac{y^2}{4} = 1$. Here $a = 3$ and $b = 2$.

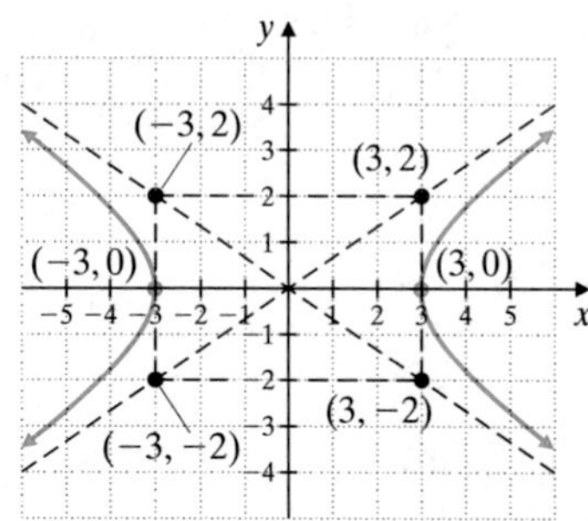

Section 12.3 Graphing Nonlinear Functions

$f(x) = |x|$

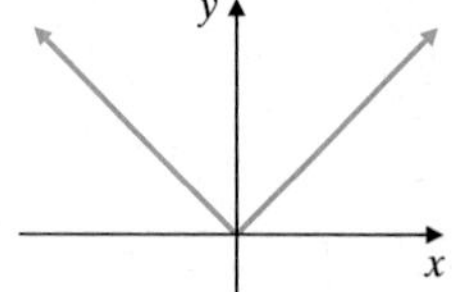

$f(x) = \sqrt{x}$

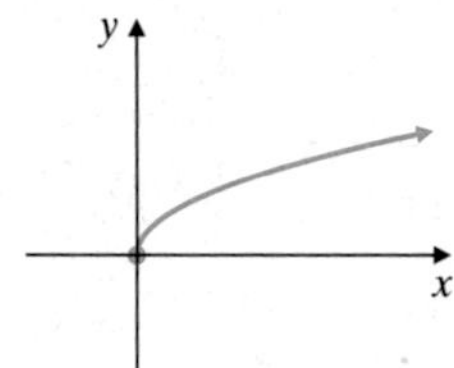

Graph: $f(x) = |x - 1| + 2$

The graph of $y = |x - 1| + 2$ is the same as the graph of $y = |x|$ shifted 1 unit to the right and 2 units up.

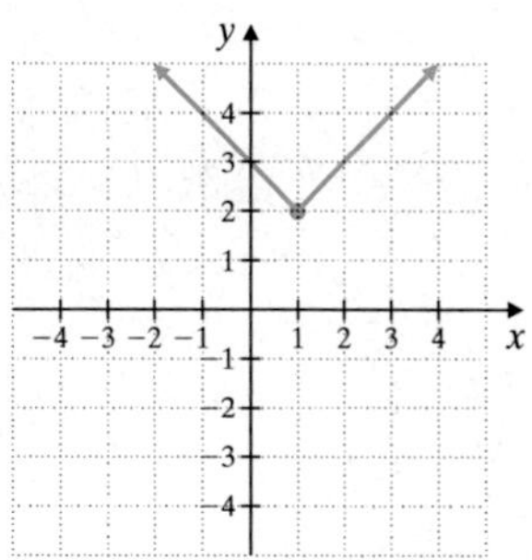

DEFINITIONS AND CONCEPTS	EXAMPLES

Section 12.4 Solving Nonlinear Systems of Equations

A **nonlinear system of equations** is a system of equations at least one of which is not linear. Both the substitution method and the elimination method may be used to solve a nonlinear system of equations.

Solve the nonlinear system: $\begin{cases} y = x + 2 \\ 2x^2 + y^2 = 3 \end{cases}$

Substitute $x + 2$ for y in the second equation:

$$2x^2 + y^2 = 3$$
$$2x^2 + (x + 2)^2 = 3$$
$$2x^2 + x^2 + 4x + 4 = 3$$
$$3x^2 + 4x + 1 = 0$$
$$(3x + 1)(x + 1) = 0$$
$$x = -\frac{1}{3} \quad \text{or} \quad x = -1$$

If $x = -\frac{1}{3}$, $y = x + 2 = -\frac{1}{3} + 2 = \frac{5}{3}$.

If $x = -1$, $y = x + 2 = -1 + 2 = 1$.

The solution set is $\left\{\left(-\frac{1}{3}, \frac{5}{3}\right), (-1, 1)\right\}$

Section 12.5 Nonlinear Inequalities and Systems of Inequalities

The **graph of a system of inequalities** is the intersection of the graphs of the inequalities.

Graph the system: $\begin{cases} x \geq y^2 \\ x + y \leq 4 \end{cases}$

The graph of the system is the darkest shaded region along with its boundary lines.

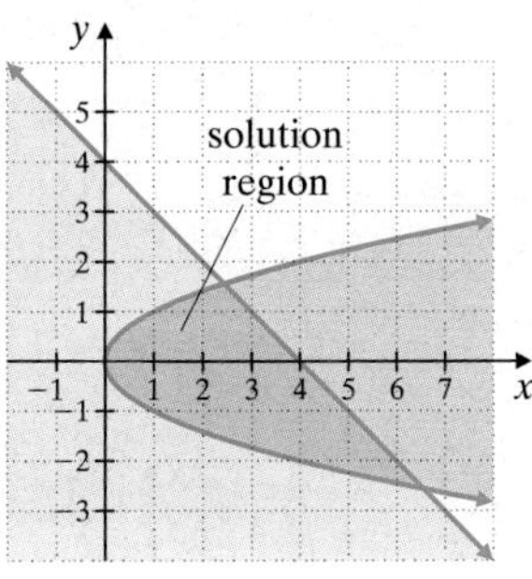

Name ______________________ Section __________ Date ____________

Chapter 12 Review

(12.1) *Find the distance between each pair of points. Give an exact value and a three-decimal-place approximation.*

1. $(-6, 3)$ and $(8, 4)$

2. $(3, 5)$ and $(8, 9)$

3. $(-4, -6)$ and $(-1, 5)$

4. $(-1, 5)$ and $(2, -3)$

5. $(-\sqrt{2}, 0)$ and $(0, -4\sqrt{6})$

6. $(-\sqrt{5}, -\sqrt{11})$ and $(-\sqrt{5}, -3\sqrt{11})$

7. $(7.4, -8.6)$ and $(-1.2, 5.6)$

8. $(2.3, 1.8)$ and $(10.7, -9.2)$

Find the midpoint of each line segment whose endpoints are given.

9. $(2, 6); (-12, 4)$

10. $(-3, 8); (11, 24)$

11. $(-6, -5); (-9, 7)$

12. $(4, -6); (-15, 2)$

13. $\left(0, -\frac{3}{8}\right); \left(\frac{1}{10}, 0\right)$

14. $\left(\frac{3}{4}, -\frac{1}{7}\right); \left(-\frac{1}{4}, -\frac{3}{7}\right)$

15. $(\sqrt{3}, -2\sqrt{6})$ and $(\sqrt{3}, -4\sqrt{6})$

16. $(-5\sqrt{3}, 2\sqrt{7}); (-3\sqrt{3}, 10\sqrt{7})$

Write an equation of each circle with the given center and radius or diameter.

17. Center $(-4, 4)$, radius 3

18. Center $(5, 0)$, diameter 10

19. Center $(-7, -9)$, radius $\sqrt{11}$

20. Center $(0, 0)$, diameter 7

Graph each equation. If the graph is a circle, find its center and radius. If the graph is a parabola, find its vertex.

21. $x^2 + y^2 = 7$

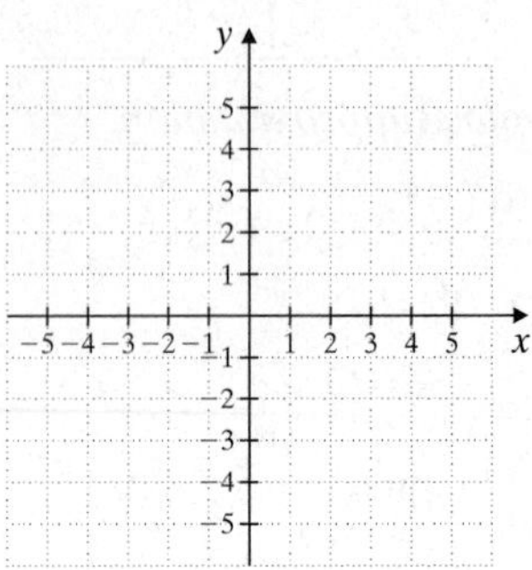

22. $x = 2(y - 5)^2 + 4$

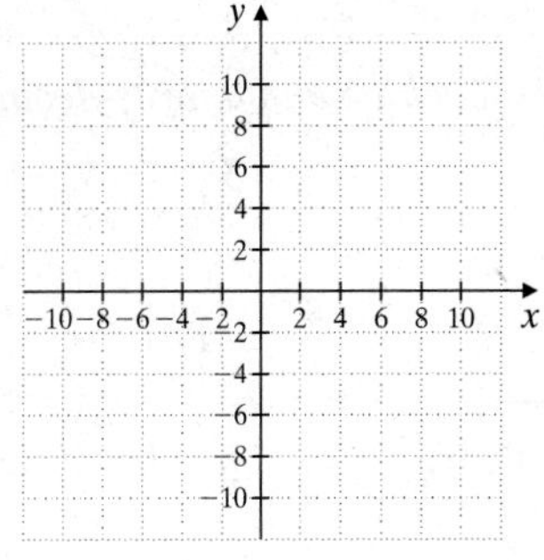

23. $x = -(y + 2)^2 + 3$

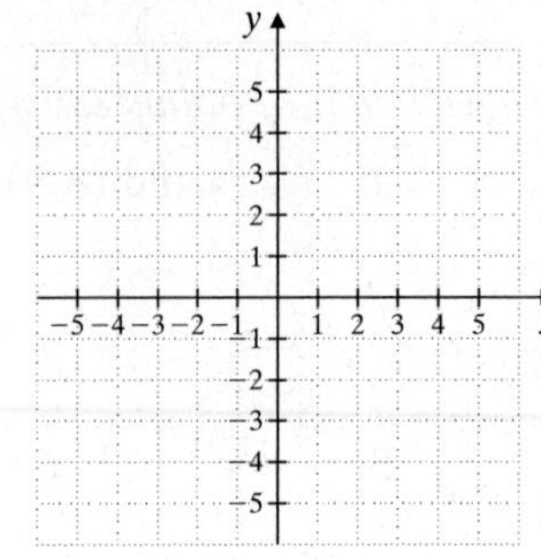

24. $(x - 1)^2 + (y - 2)^2 = 4$

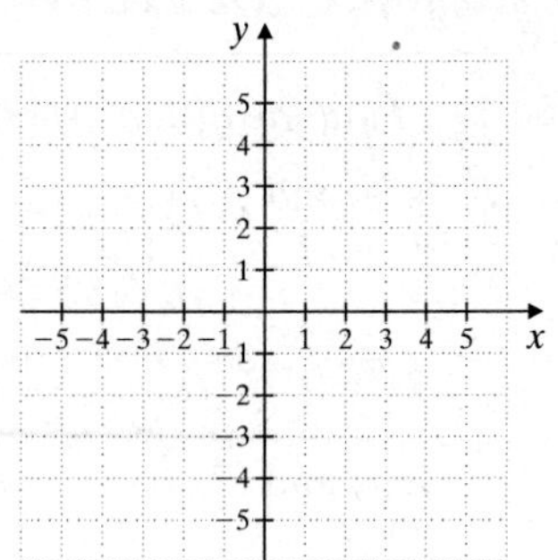

25. $y = -x^2 + 4x + 10$

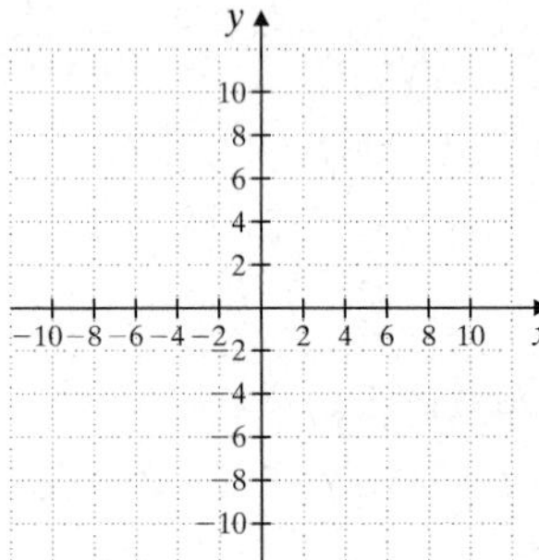

26. $x = -y^2 - 4y + 6$

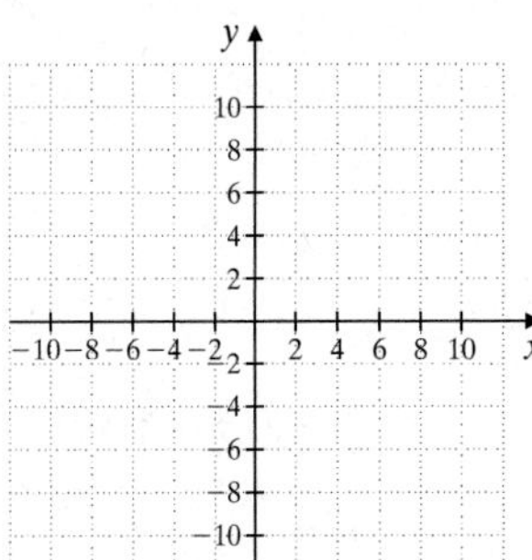

27. $x = \frac{1}{2}y^2 + 2y + 1$

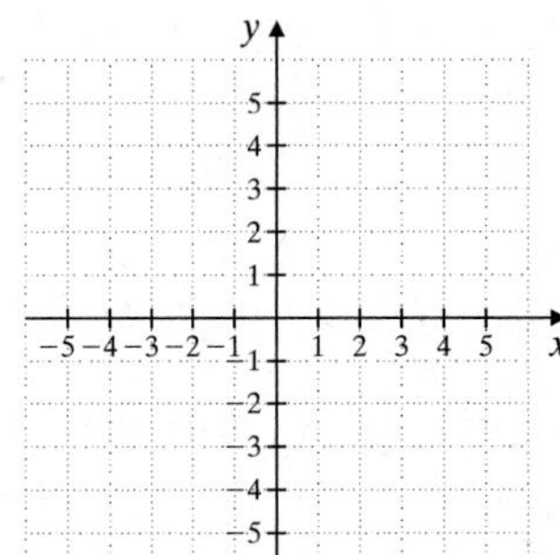

28. $y = -3x^2 + \frac{1}{2}x + 4$

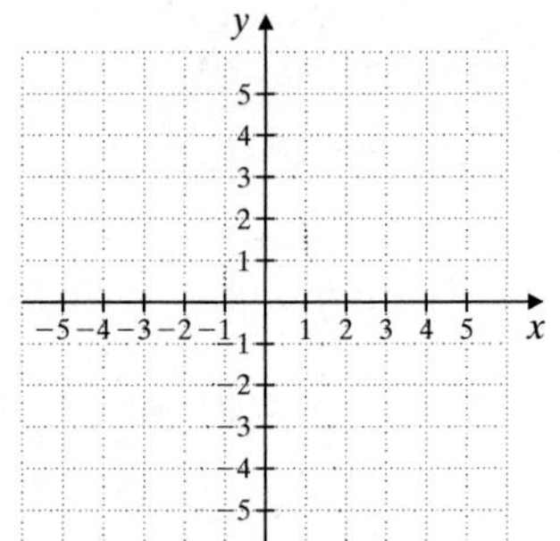

29. $x^2 + y^2 + 2x + y = \frac{3}{4}$

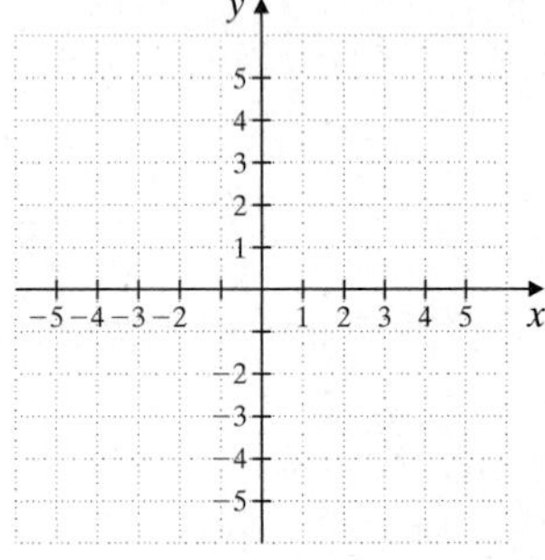

30. $x^2 + y^2 - 3y = \frac{7}{4}$

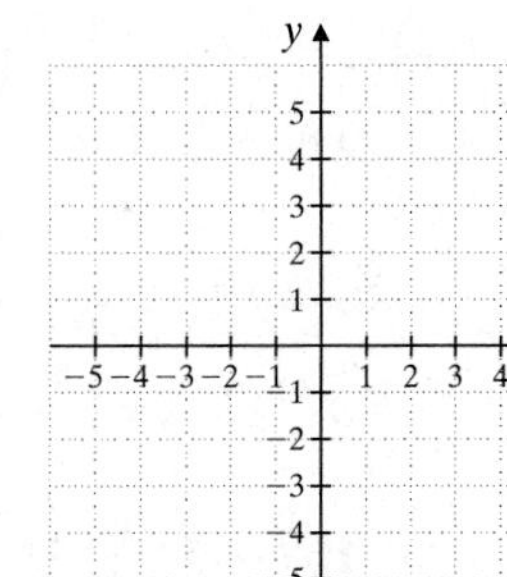

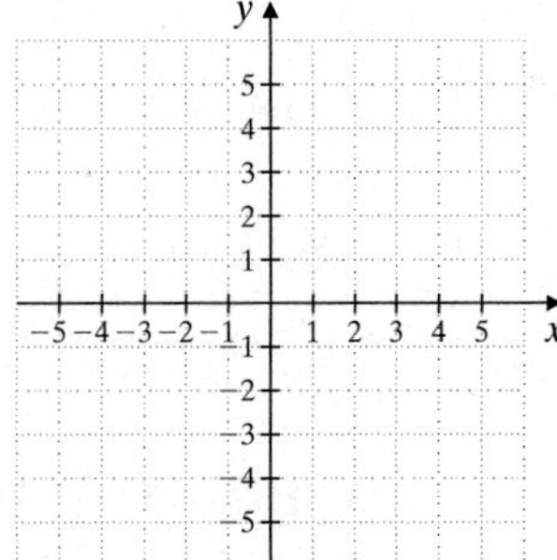

31. $4x^2 + 4y^2 + 16x + 8y = 1$

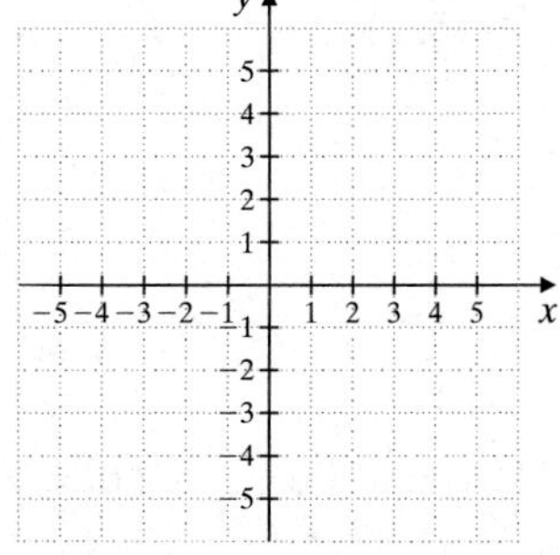

32. $3x^2 + 6x + 3y^2 = 9$

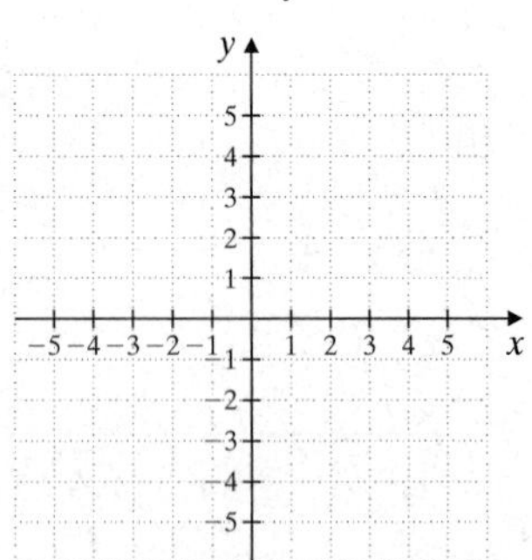

33. $y = x^2 + 6x + 9$

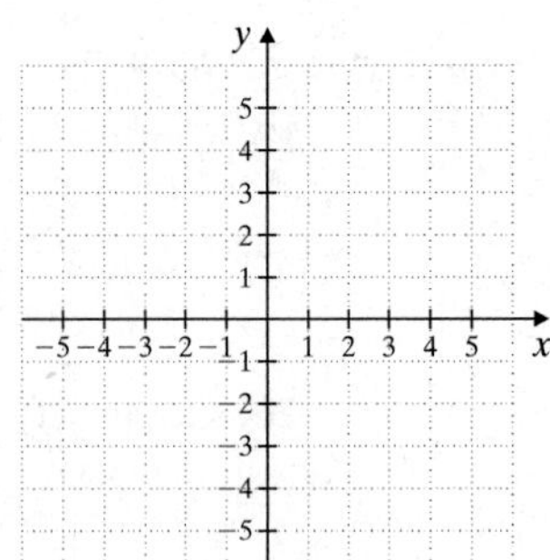

34. $x = y^2 + 6y + 9$

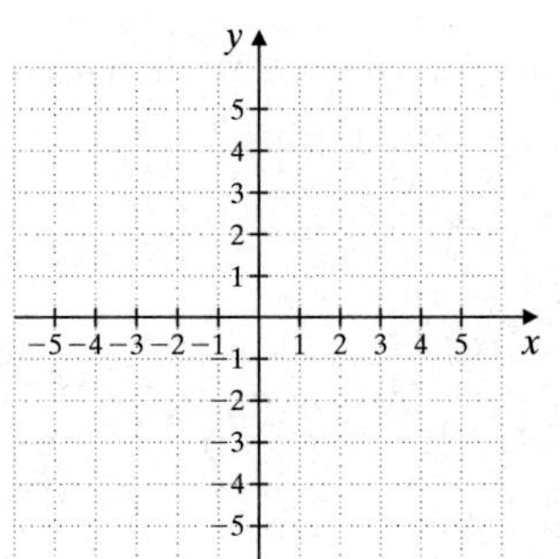

(12.2) *Graph each equation.*

35. $x^2 + \dfrac{y^2}{4} = 1$

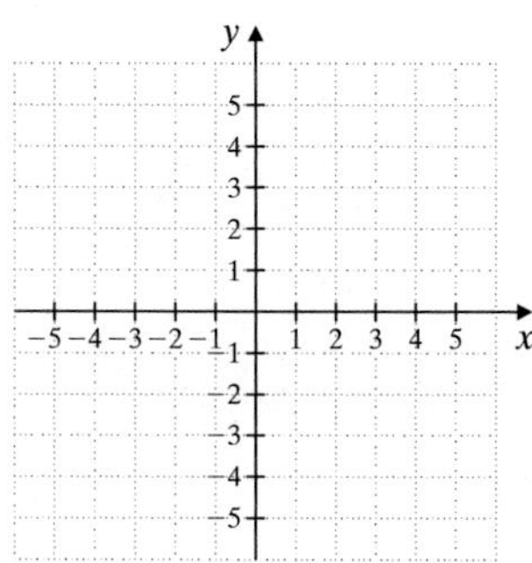

36. $x^2 - \dfrac{y^2}{4} = 1$

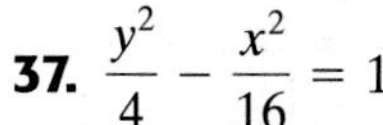

37. $\dfrac{y^2}{4} - \dfrac{x^2}{16} = 1$

38. $\dfrac{y^2}{4} + \dfrac{x^2}{16} = 1$

39. $-5x^2 + 25y^2 = 125$

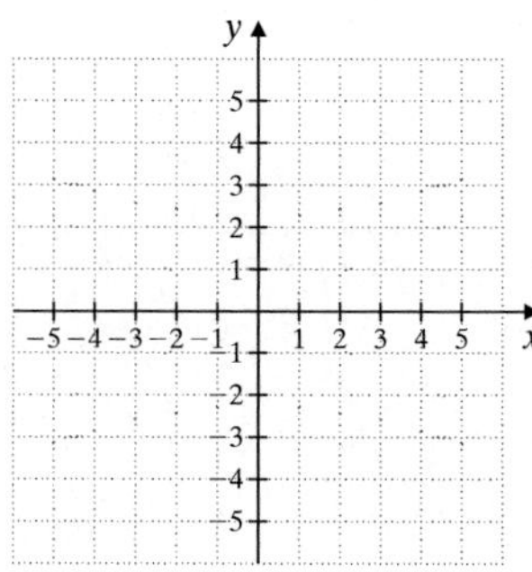

40. $4y^2 + 9x^2 = 36$

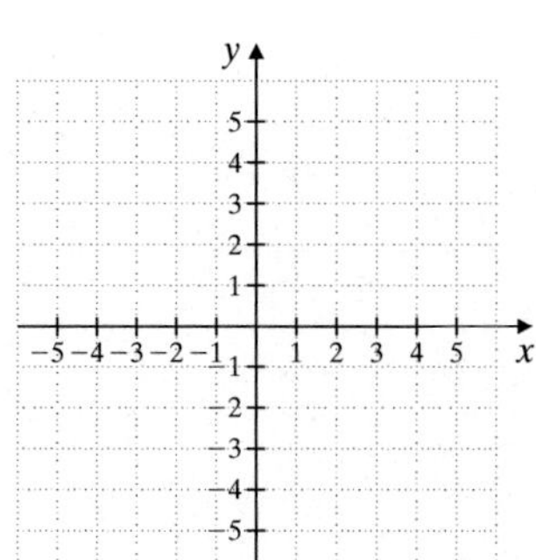

41. $\dfrac{(x-2)^2}{4} + (y-1)^2 = 1$

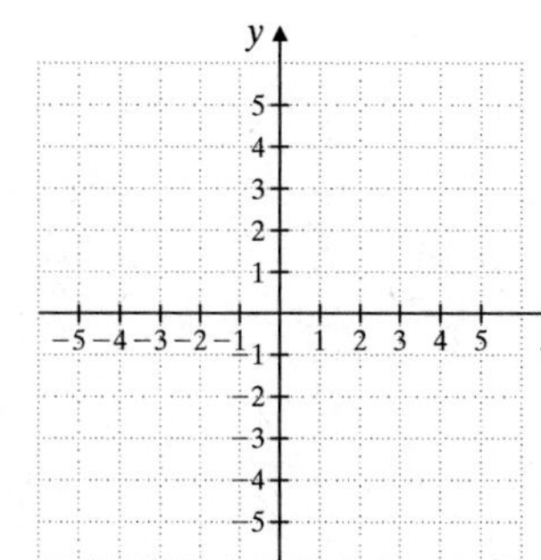

42. $\dfrac{(x+3)^2}{9} + \dfrac{(y-4)^2}{25} = 1$

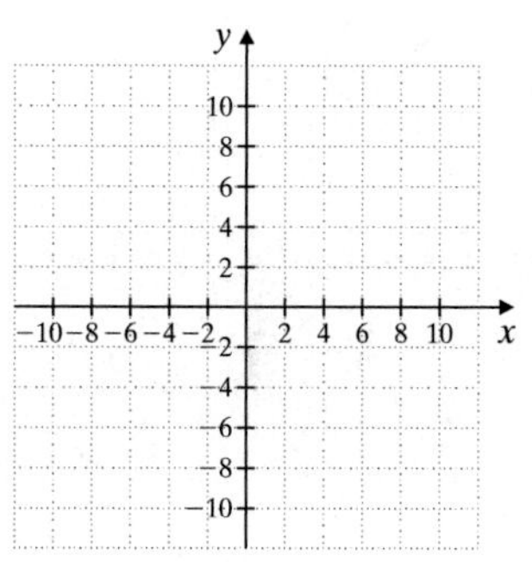

43. $x^2 - y^2 = 1$

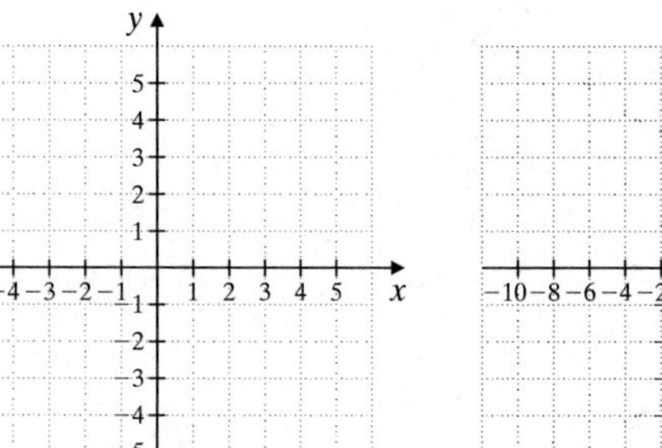

44. $36y^2 - 49x^2 = 1764$

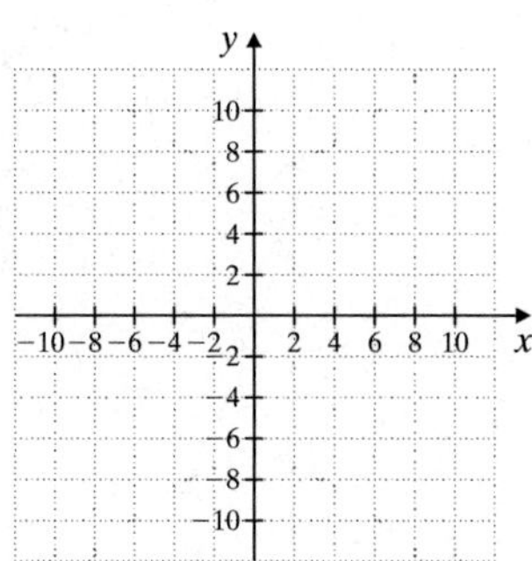

45. $y = x^2 + 9$

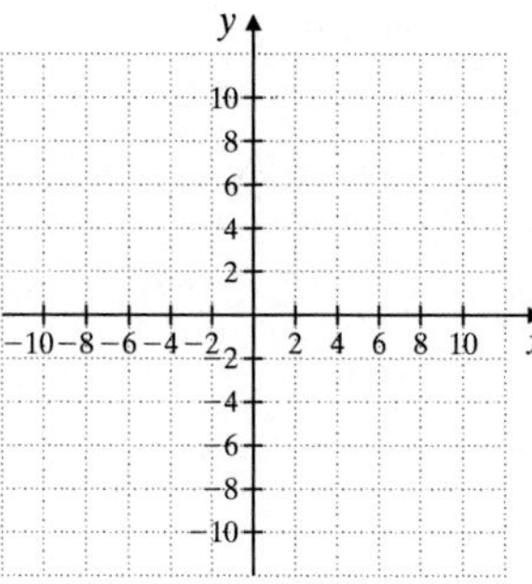

46. $x = 4y^2 - 16$

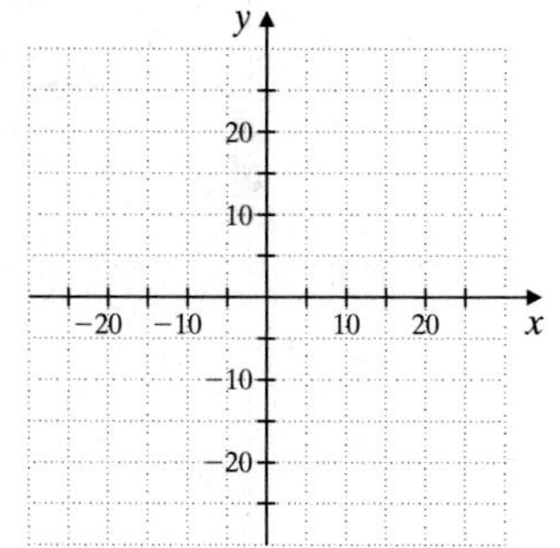

Graph each equation.

47. $y = x^2 + 4x + 6$

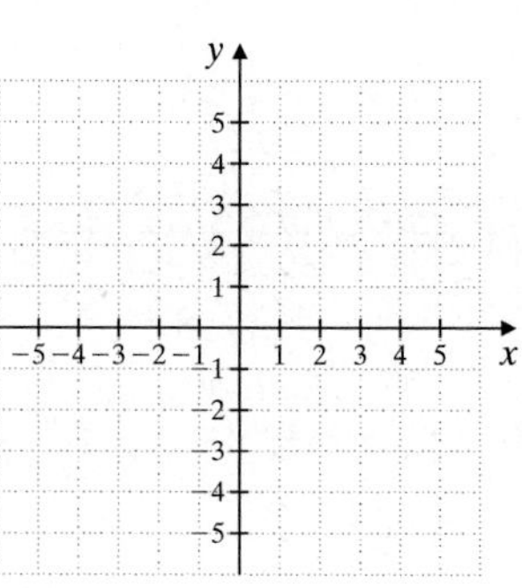

48. $y^2 = x^2 + 6$

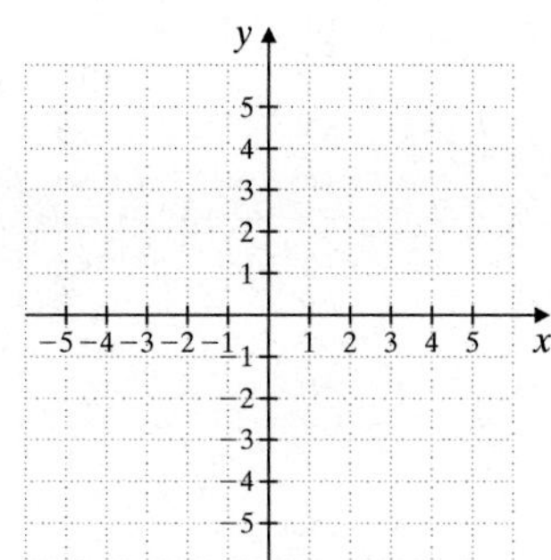

49. $y^2 + x^2 = 4x + 6$

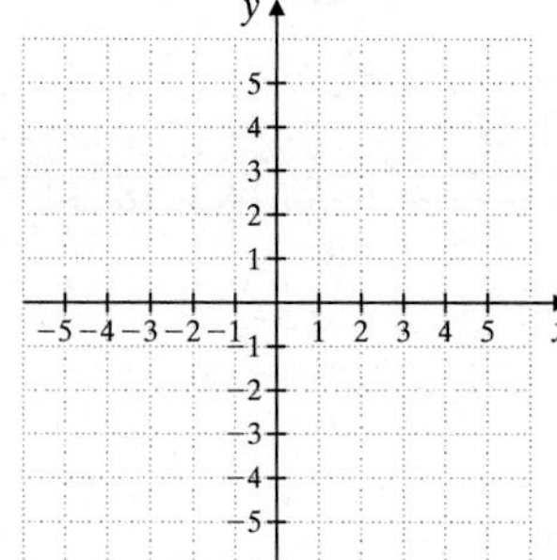

50. $y^2 + 2x^2 = 4x + 6$

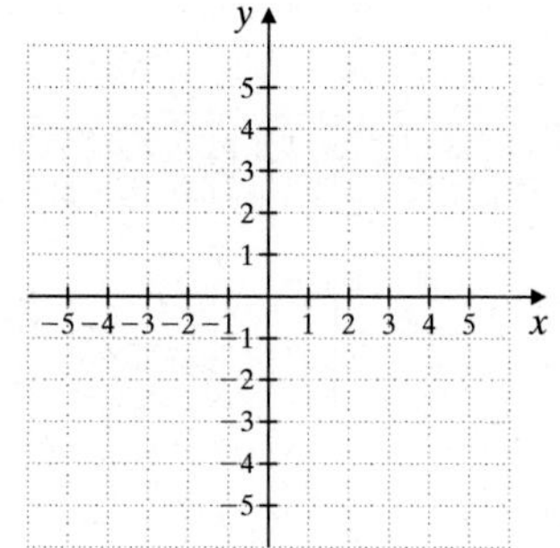

51. $x^2 + y^2 - 8y = 0$

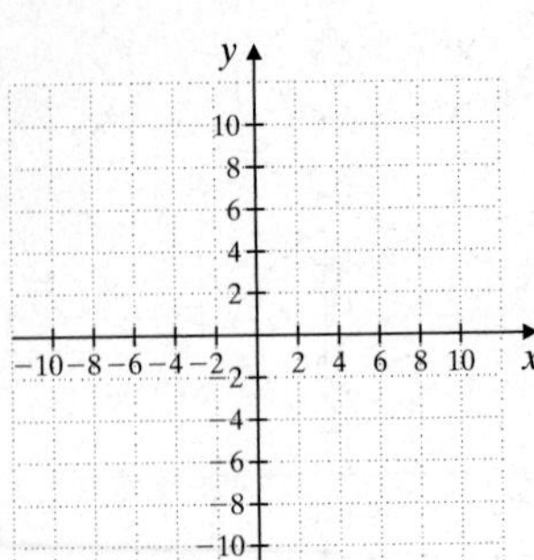

52. $x - 4y = y^2$

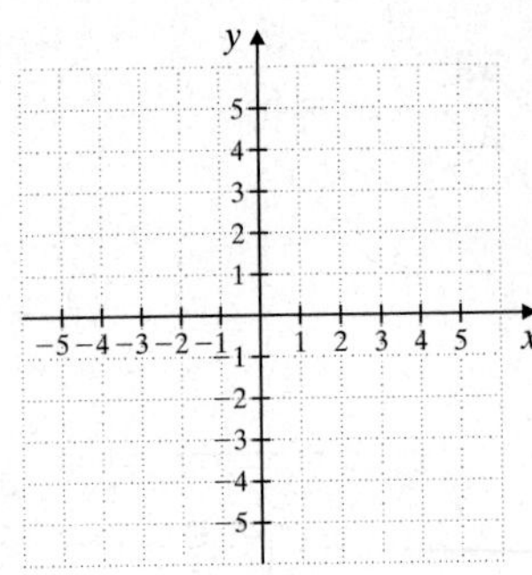

53. $x^2 - 4 = y^2$

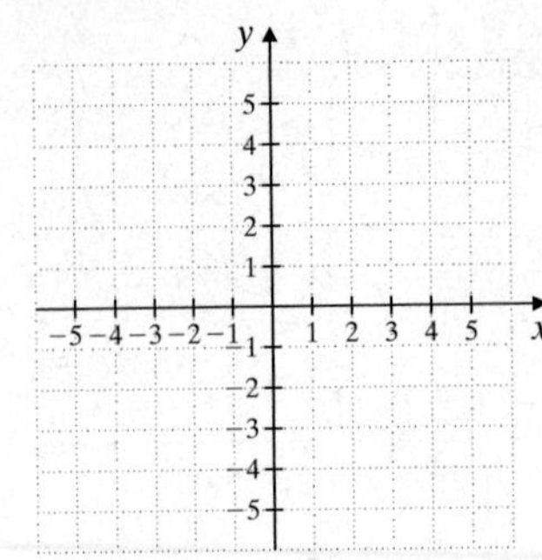

54. $x^2 = 4 - y^2$

55. $6(x - 2)^2 + 9(y + 5)^2 = 36$

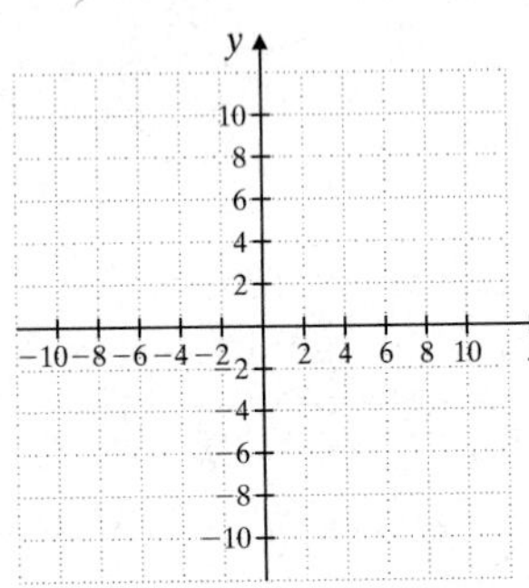

56. $36y^2 = 576 + 16x^2$

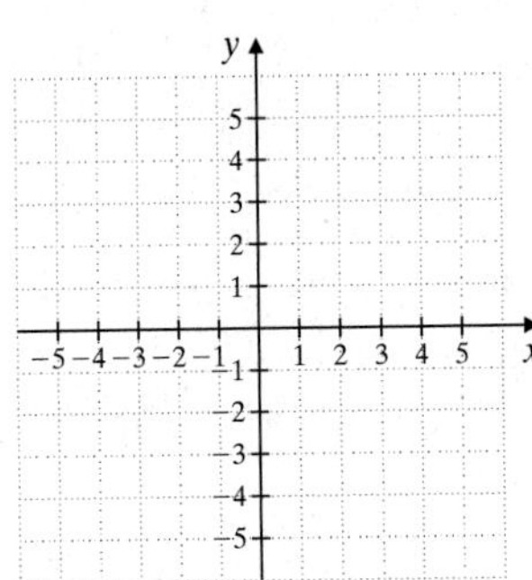

57. $\dfrac{x^2}{16} - \dfrac{y^2}{25} = 1$

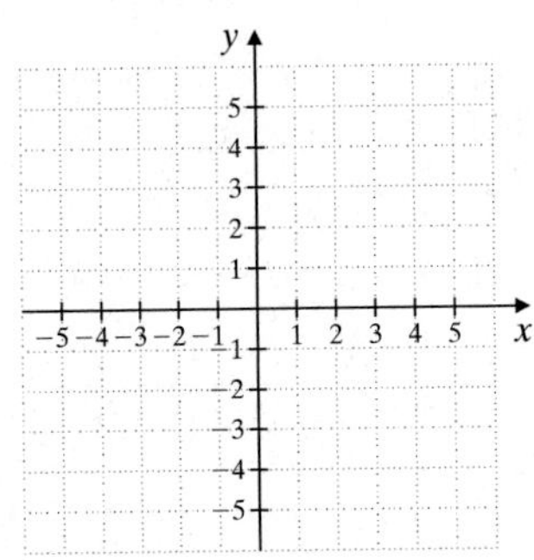

58. $3(x - 7)^2 + 3(y + 4)^2 = 1$

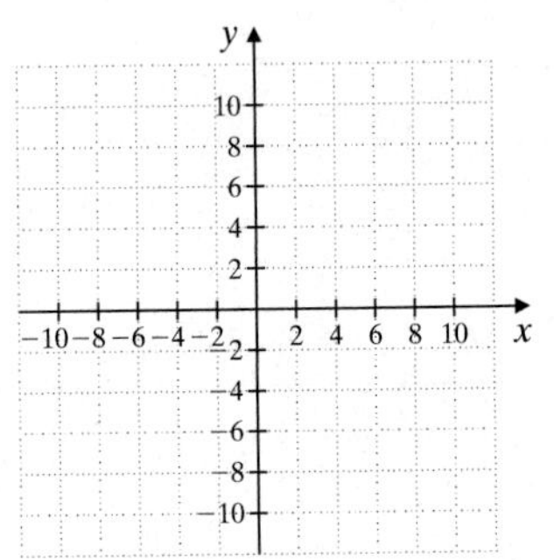

(12.3) *Graph each function.*

59. $f(x) = |x| + 2$

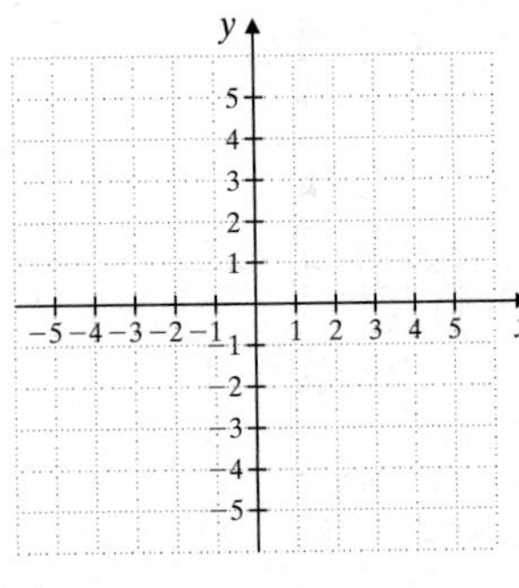

60. $f(x) = \sqrt{x} - 1$

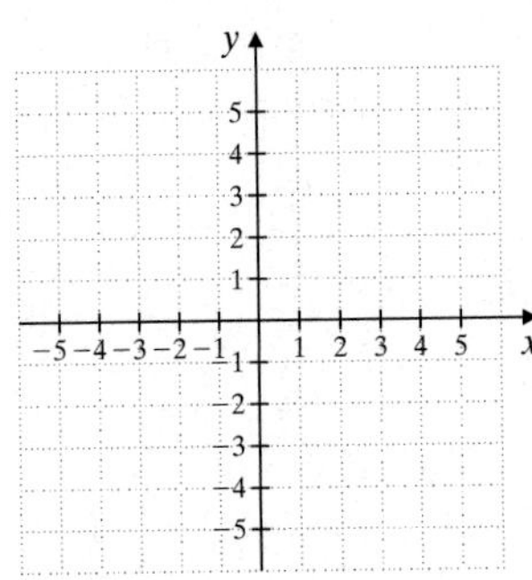

61. $f(x) = \sqrt{x - 4}$

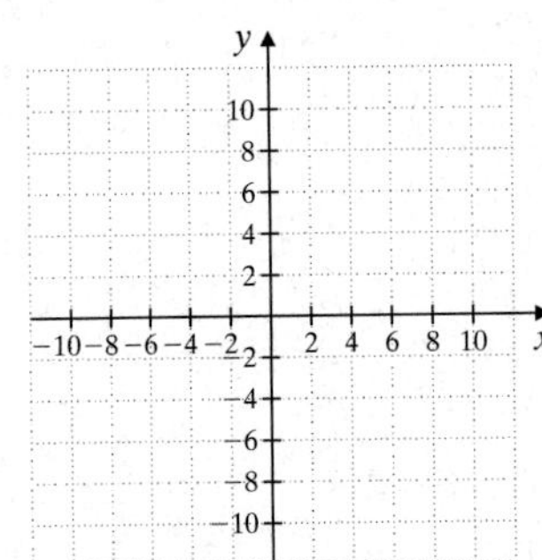

62. $f(x) = |x + 1|$

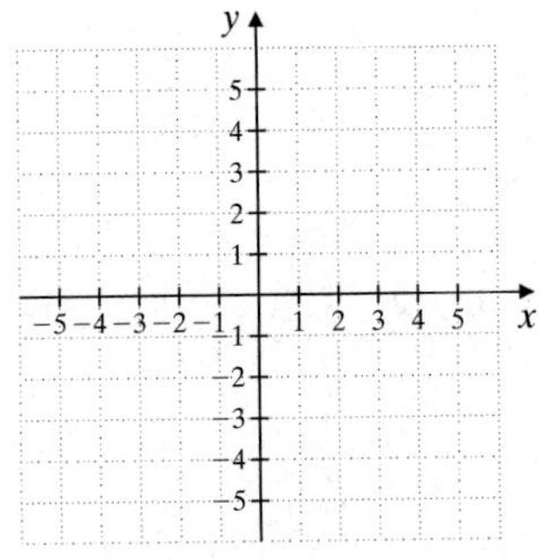

63. $f(x) = \sqrt{x - 3} + 1$

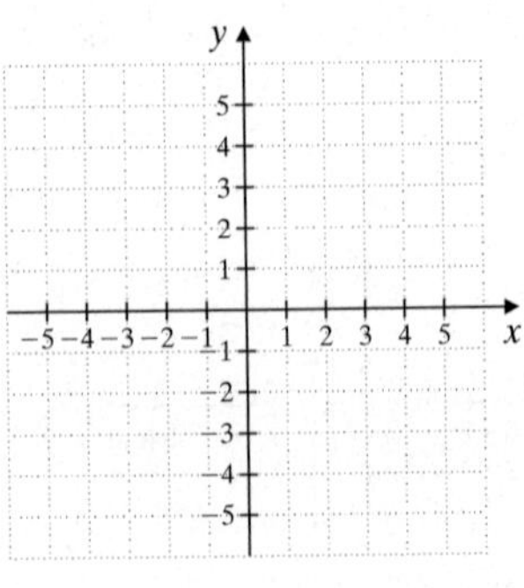

64. $f(x) = |x - 1| + 3$

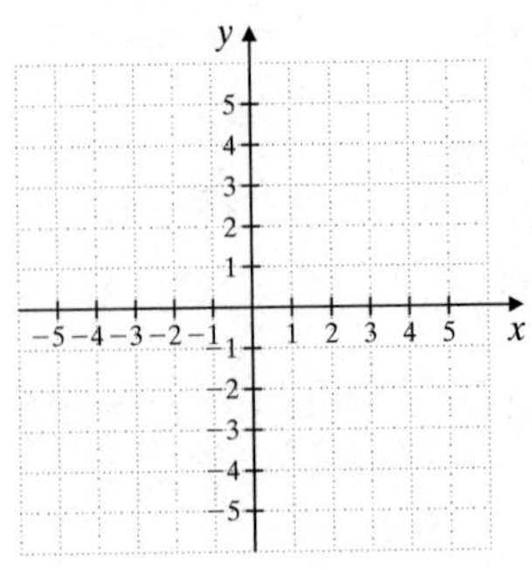

65. $f(x) = |x + 2| + 2$

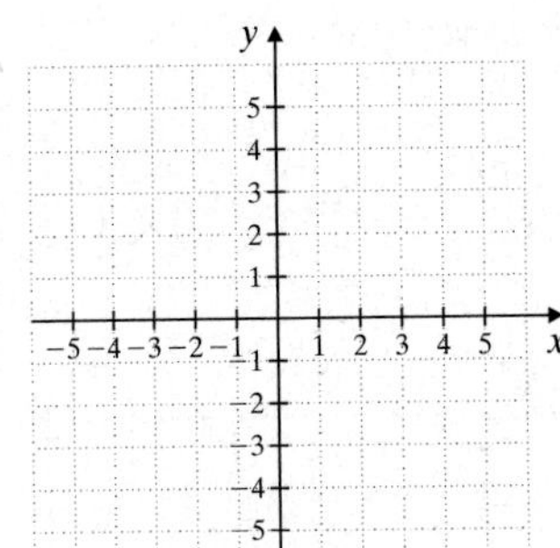

66. $f(x) = \sqrt{x + 2} + 2$

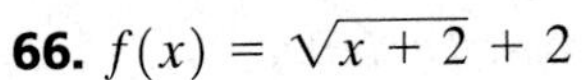

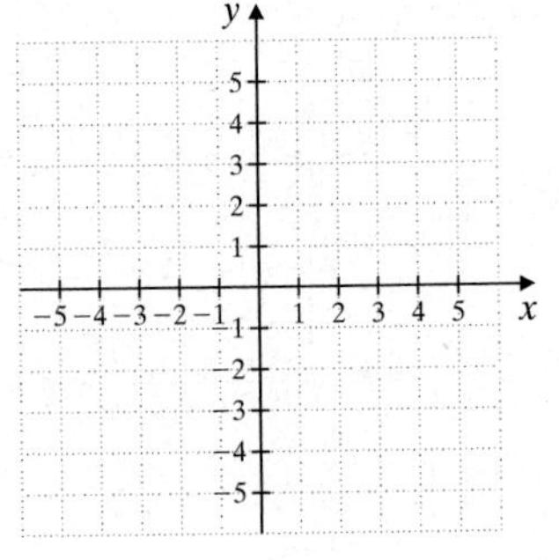

(12.4) *Solve each system of equations.*

67. $\begin{cases} y = 2x - 4 \\ y^2 = 4x \end{cases}$

68. $\begin{cases} x^2 + y^2 = 4 \\ x - y = 4 \end{cases}$

69. $\begin{cases} y = x + 2 \\ y = x^2 \end{cases}$

70. $\begin{cases} y = x^2 - 5x + 1 \\ y = -x + 6 \end{cases}$

71. $\begin{cases} 4x - y^2 = 0 \\ 2x^2 + y^2 = 16 \end{cases}$

72. $\begin{cases} x^2 + 4y^2 = 16 \\ x^2 + y^2 = 4 \end{cases}$

73. $\begin{cases} x^2 + y^2 = 10 \\ 9x^2 + y^2 = 18 \end{cases}$

74. $\begin{cases} x^2 + 2y = 9 \\ 5x - 2y = 5 \end{cases}$

75. $\begin{cases} y = 3x^2 + 5x - 4 \\ y = 3x^2 - x + 2 \end{cases}$

76. $\begin{cases} x^2 - 3y^2 = 1 \\ 4x^2 + 5y^2 = 21 \end{cases}$

△ **77.** Find the length and the width of a room whose area is 150 square feet and whose perimeter is 50 feet.

78. What is the greatest number of real number solutions possible for a system of two equations whose graphs are an ellipse and a hyperbola?

(12.5) *Graph each inequality or system of inequalities.*

79. $y \leq -x^2 + 3$

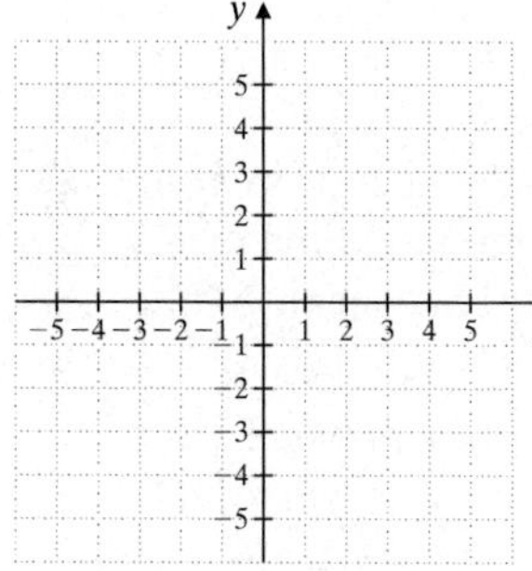

80. $x^2 + y^2 < 9$

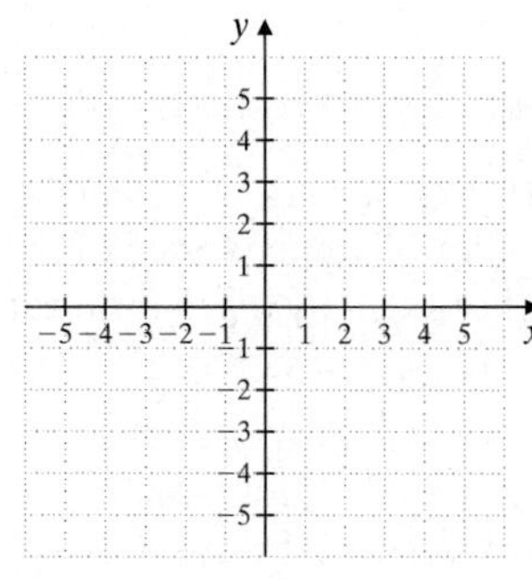

81. $x^2 = y^2 < 1$

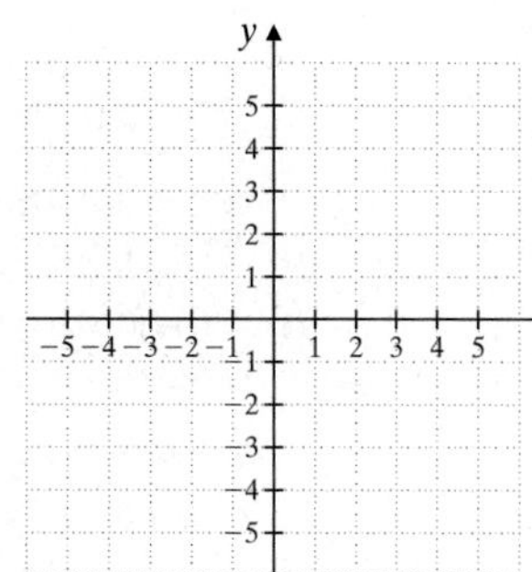

82. $\frac{x^2}{4} + \frac{y^2}{9} \geq 1$

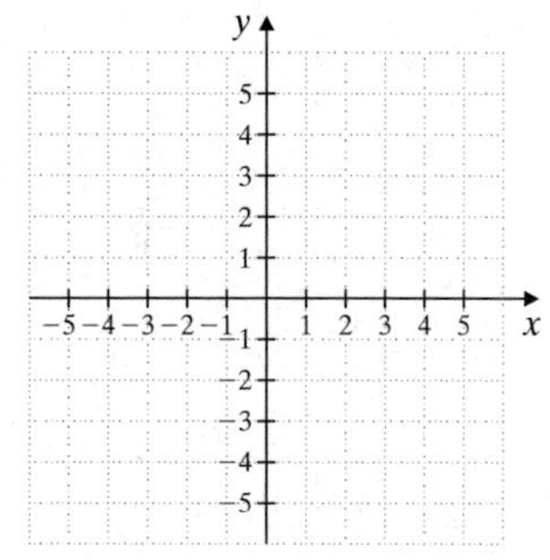

83. $\begin{cases} 2x \leq 4 \\ x + y \geq 1 \end{cases}$

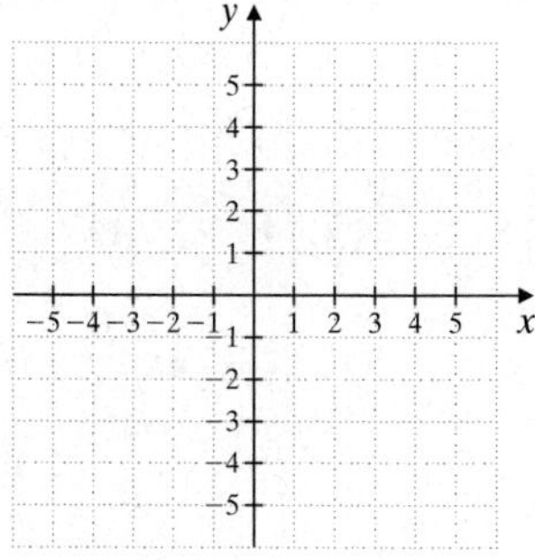

84. $\begin{cases} 3x + 4y \leq 12 \\ x - 2y > 6 \end{cases}$

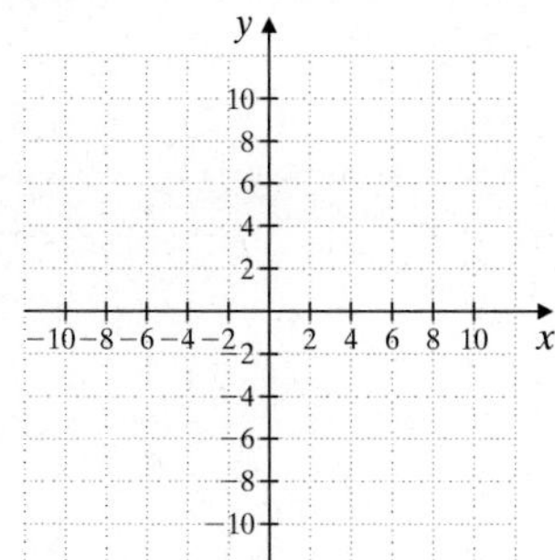

85. $\begin{cases} y > x^2 \\ x + y \geq 3 \end{cases}$

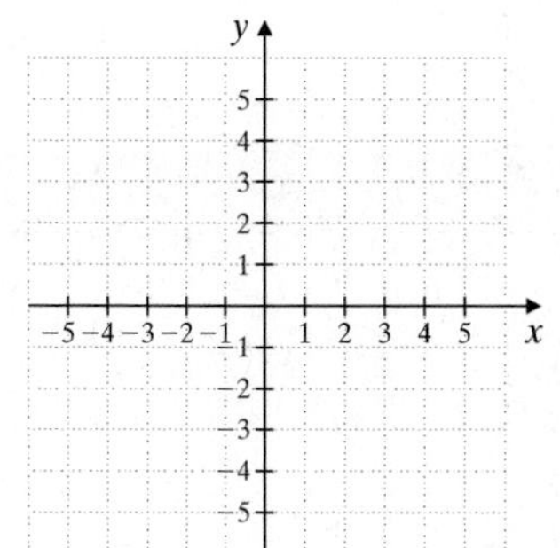

86. $\begin{cases} x^2 + y^2 \leq 16 \\ x^2 + y^2 \geq 4 \end{cases}$

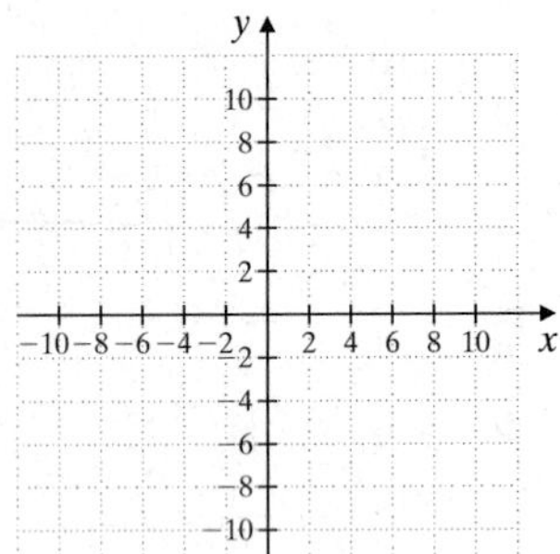

87. $\begin{cases} x^2 + y^2 < 4 \\ x^2 - y^2 \leq 1 \end{cases}$

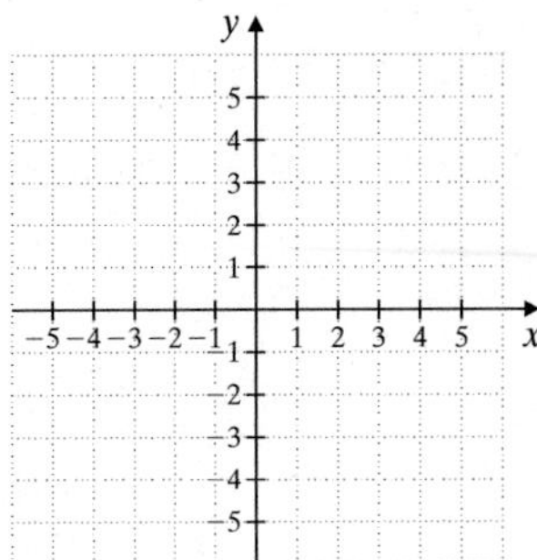

88. $\begin{cases} x^2 + y^2 < 4 \\ y \geq x^2 - 1 \\ x \geq 0 \end{cases}$

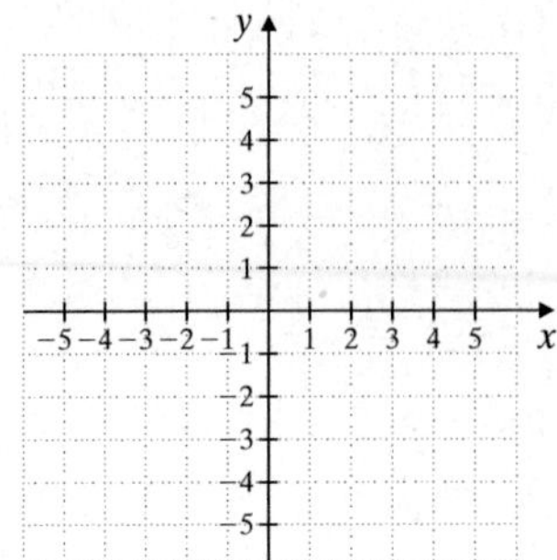

STUDY SKILLS REMINDER

Are you satisfied with your performance on a particular quiz or exam?

If not, don't forget to analyze your quiz or exam and look for common errors.

Were most of your errors a result of

- *Carelessness?* If your errors were careless, did you turn in your work before the allotted time expired? If so, resolve next time to use the entire time allotted. Any extra time can be spent checking your work.
- *Running out of time?* If so, make a point to better manage your time on your next exam. A few suggestions are to work any questions that you are unsure of last and to check your work after all questions have been answered.
- *Not understanding a concept?* If so, review that concept and correct your work. Remember next time to make sure that all concepts on a quiz or exam are understood before the exam.

Name ______________________

Chapter 12 Test

1. Find the distance between the points $(-6, 3)$ and $(-8, -7)$.

2. Find the distance between the points $(-2\sqrt{5}, \sqrt{10})$ and $(-\sqrt{5}, 4\sqrt{10})$.

3. Find the midpoint of the line segment whose endpoints are $(-2, -5)$ and $(-6, 12)$.

4. Find the midpoint of the line segment whose endpoints are $\left(-\frac{2}{3}, -\frac{1}{5}\right)$ and $\left(-\frac{1}{3}, \frac{4}{5}\right)$.

Graph each equation.

5. $x^2 + y^2 = 36$

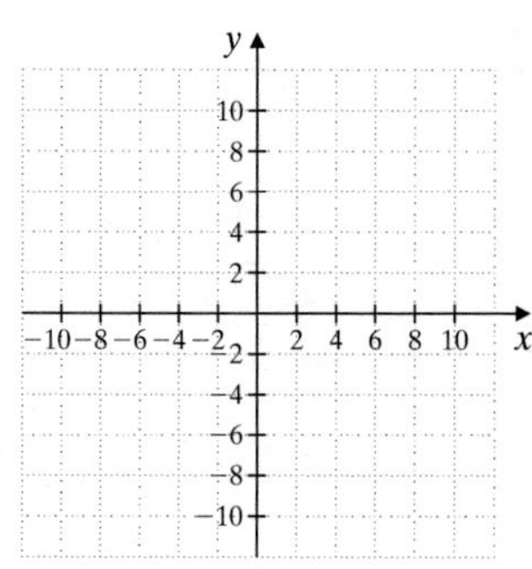

6. $x^2 - y^2 = 36$

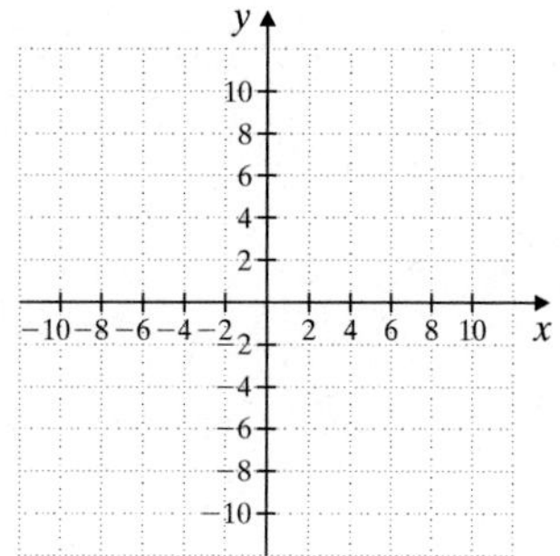

7. $16x^2 + 9y^2 = 144$

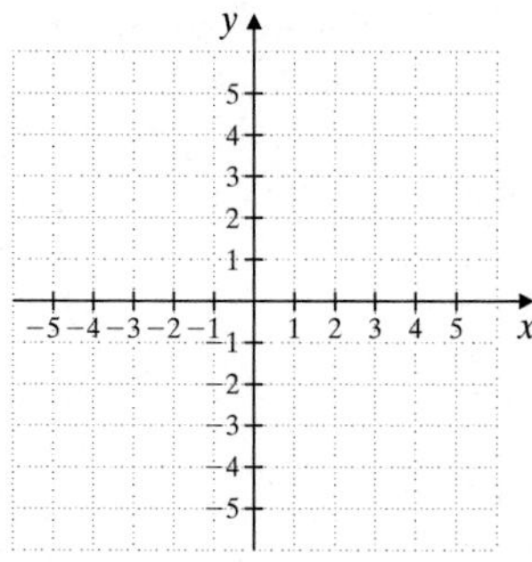

8. $y = x^2 - 8x + 16$

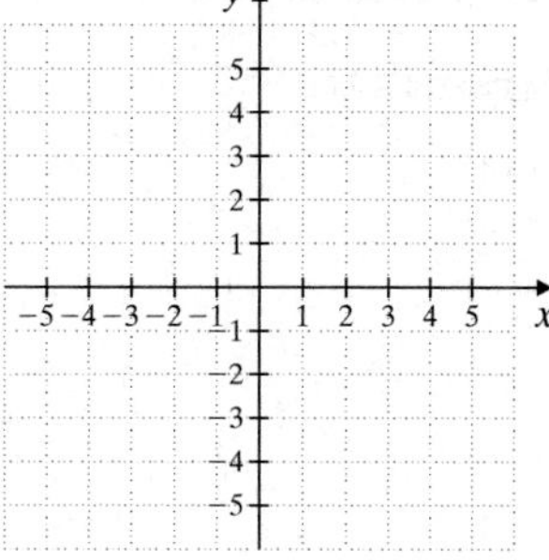

9. $x^2 + y^2 + 6x = 16$

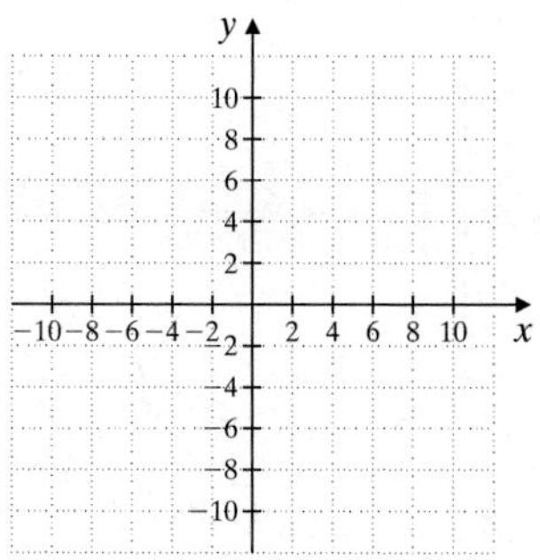

10. $x = y^2 + 8y - 3$

Answers

1. ______

2. ______

3. ______

4. ______

5. see graph

6. see graph

7. see graph

8. see graph

9. see graph

10. see graph

11. see graph

12. see graph

13. ______

14. ______

15. ______

16. ______

17. see graph

18. see graph

19. see graph

20. see graph

11. $\dfrac{(x-4)^2}{16} + \dfrac{(y-3)^2}{9} = 1$

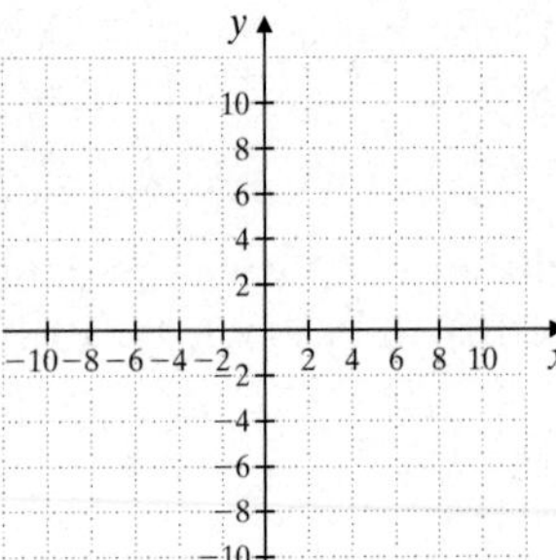

12. $y^2 - x^2 = 1$

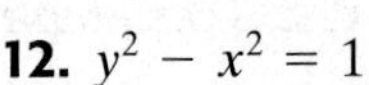

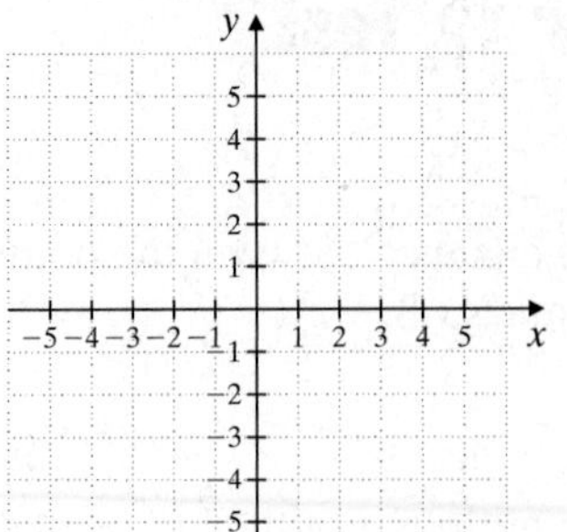

Solve each system.

13. $\begin{cases} x^2 + y^2 = 169 \\ 5x + 12y = 0 \end{cases}$

14. $\begin{cases} x^2 + y^2 = 26 \\ x^2 - y^2 = 24 \end{cases}$

15. $\begin{cases} y = x^2 - 5x + 6 \\ y = 2x \end{cases}$

16. $\begin{cases} x^2 + 4y^2 = 5 \\ y = x \end{cases}$

Graph each system.

17. $\begin{cases} 2x + 5y \geq 10 \\ y \geq x^2 + 1 \end{cases}$

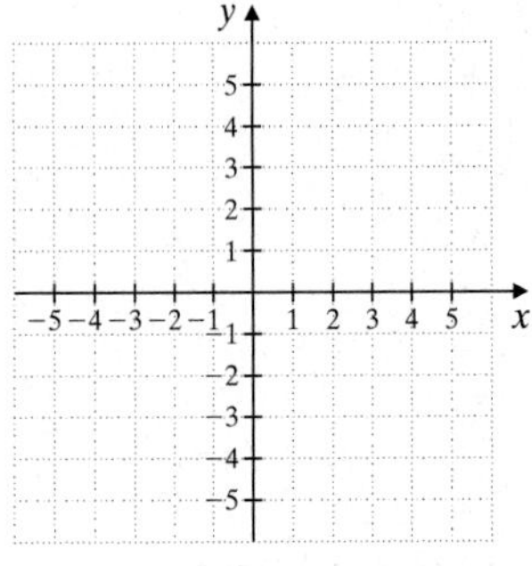

18. $\begin{cases} \dfrac{x^2}{4} + y^2 \leq 1 \\ x + y > 1 \end{cases}$

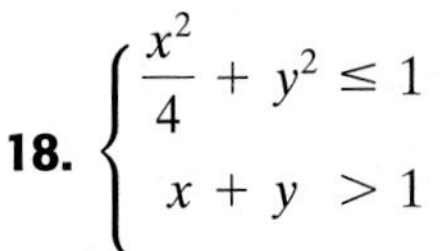

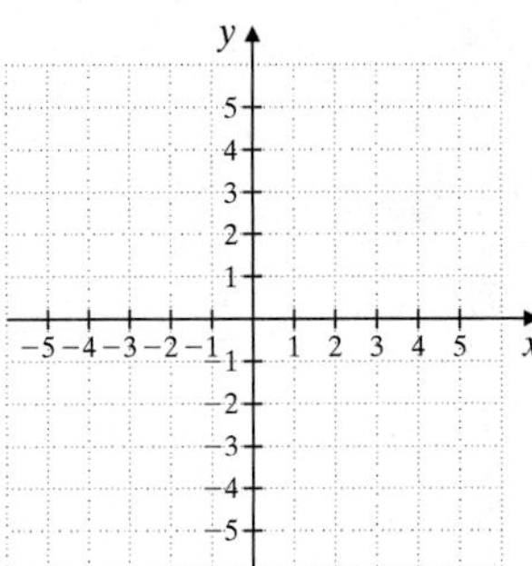

19. $\begin{cases} x^2 + y^2 > 1 \\ \dfrac{x^2}{4} - y^2 \geq 1 \end{cases}$

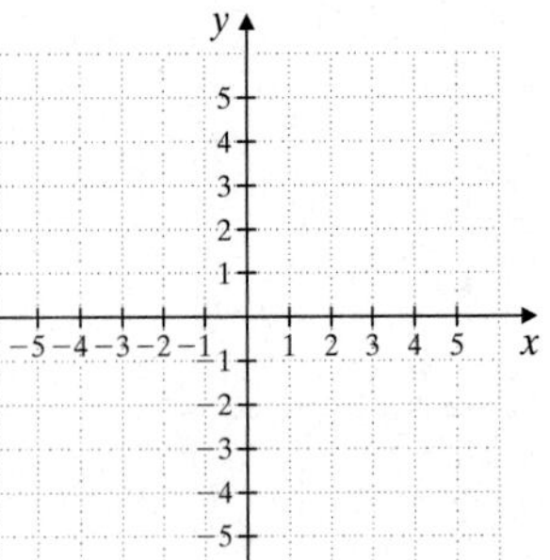

20. $\begin{cases} x^2 + y^2 \geq 4 \\ x^2 + y^2 < 16 \\ y \geq 0 \end{cases}$

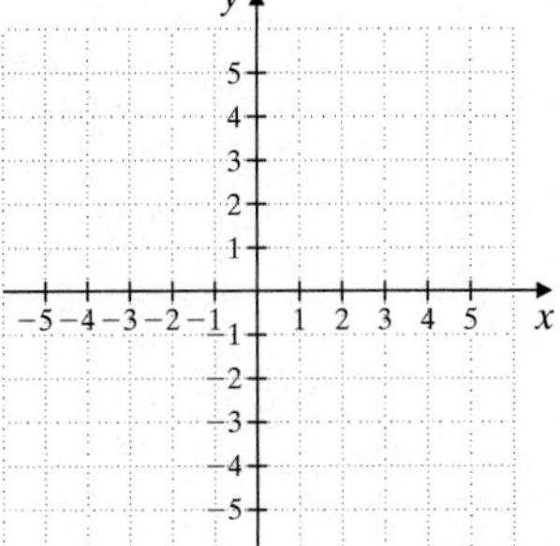

21. Which graph best resembles the graph of $x = a(y - k)^2 + h$ if $a > 0, h < 0$, and $k > 0$?

21. ______

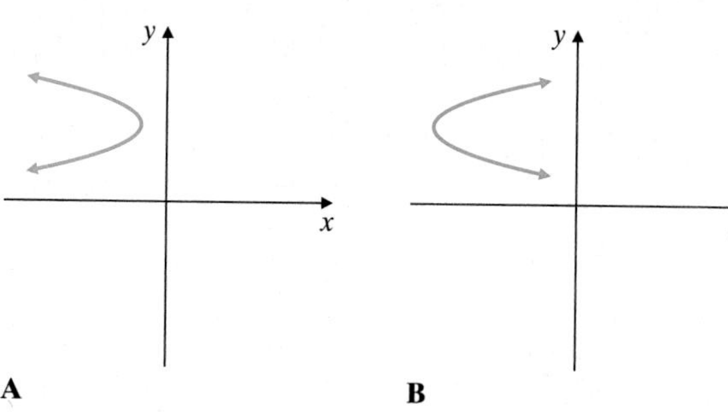

A B

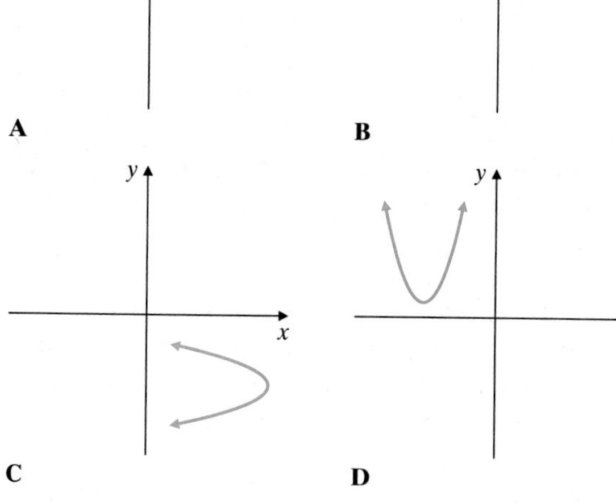

C D

22. A bridge has an arch in the shape of half an ellipse. If the equation of the ellipse, measured in feet, is $100x^2 + 225y^2 = 22{,}500$, find the height of the arch from the road and the width of the arch.

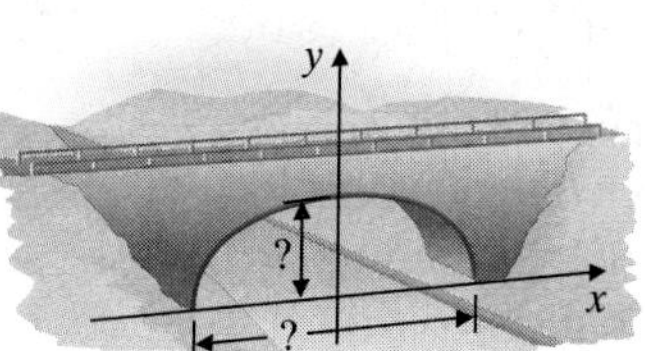

22. ______

23. see graph

Graph each function.

23. $f(x) = \sqrt{x - 2}$

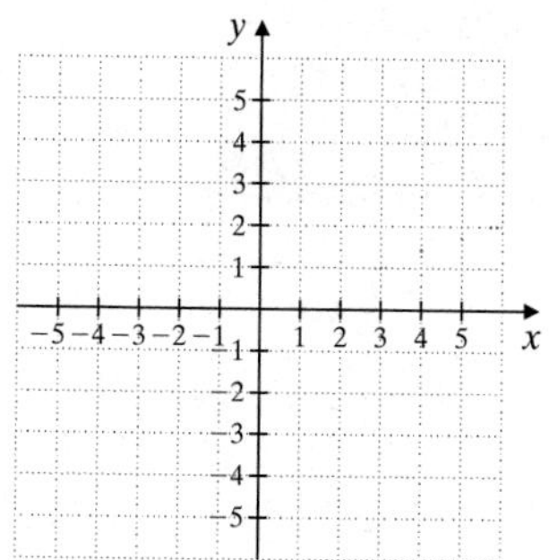

24. $f(x) = |x + 1| - 3$

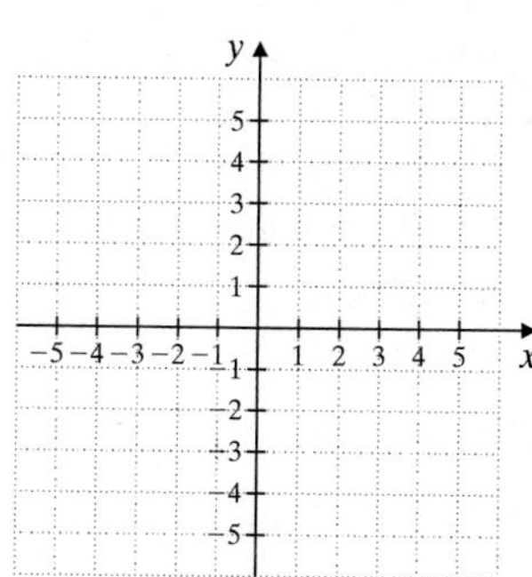

24. see graph

FOCUS ON **History**

CONIC SECTIONS

It is believed that the conic sections were discovered by the Greek mathematician Menaechmus (380 B.C. –320 B.C.). He was the first to realize that the shapes of the parabola, ellipse, and hyperbola are formed by cutting a right circular cone with a plane in various ways. However, these conic sections were not given their names until later. Another Greek mathematician, Apollonius of Perga (262 B.C.–190 B.C.), was responsible for coining the terms *parabola*, *ellipse*, and *hyperbola* in his set of eight texts titled *Treatise on Conic Sections*. In these texts, Apollonius discussed the basic properties of conic sections as well as how they are drawn.

A contemporary of Apollonius, Archimedes of Syracuse (287 B.C.–212 B.C.), is probably the most famous of the Greek mathematicians who made contributions to the base of knowledge on conic sections. His detailed study of circles led to an important contribution: a calculation of the value of π as being between $3\frac{10}{71}$ and $3\frac{1}{7}$. He also studied the areas of conic sections, including parabolas and ellipses, and other shapes and solids that arise from the conic sections. He developed a special method for finding such areas by dividing the area of a figure up into infinitely many narrow rectangles and then summing these individual areas to find the area of the entire figure. This revolutionary method eventually led to the discovery of the branch of advanced mathematics called *calculus* nearly 2000 years later.

As important as was his work with conics, Archimedes is probably best remembered for his work on practical matters for the king of Syracuse. For instance, Archimedes is credited with inventing the catapult, at the king's request, as a defense measure against a Roman invasion. Another time, the king asked Archimedes to help prove that a gold crown that he had commissioned was made partially from silver as well. The story goes that while Archimedes pondered this question, he was taking a bath and noticed that the amount of water that overflowed the tub was proportional to the portion of his body that was under water. He had discovered what is now known as Archimedes' Principle of Buoyancy: that an object immersed in water is buoyed up by a force that is equal to the weight of the water it displaces. Archimedes was so excited by his discovery that he supposedly ran naked straight from his bath through the streets of Syracuse shouting "Eureka, eureka!" ("I have found it!"). He immediately applied this discovery to the crown problem by comparing the amount of water displaced by a crown made from the same weight of pure gold to the amount of water displaced by the suspect crown. Because these amounts of water were not the same, Archimedes proved that the maker of the crown had cheated the king by using silver, a cheaper metal, in place of some of the gold.

Name ______________________

Cumulative Review

Simplify.

1. Solve: $5 - x \leq 4x - 15$. Write the solution set in interval notation.

2. Simplify each expression.

a. $x^7 \cdot x^4$ **b.** $\left(\frac{1}{2}\right)^4$ **c.** $(9y^5)^2$

3. Multiply: $7x(x^2 + 2x + 5)$

4. Divide: $\dfrac{9x^5 - 12x^2 + 3x}{3x^2}$

5. Factor: $5(x + 3) + y(x + 3)$

6. Factor: $x^4 + 5x^2 + 6$

7. Multiply: $\dfrac{x^2 + x}{3x} \cdot \dfrac{6}{5x + 5}$

8. Solve $\dfrac{t - 4}{2} - \dfrac{t - 3}{9} = \dfrac{5}{18}$

9. Simplify: $\dfrac{\frac{1}{z} - \frac{1}{2}}{\frac{1}{3} - \frac{z}{6}}$

10. Graph: $y = \dfrac{1}{4}x - 3$

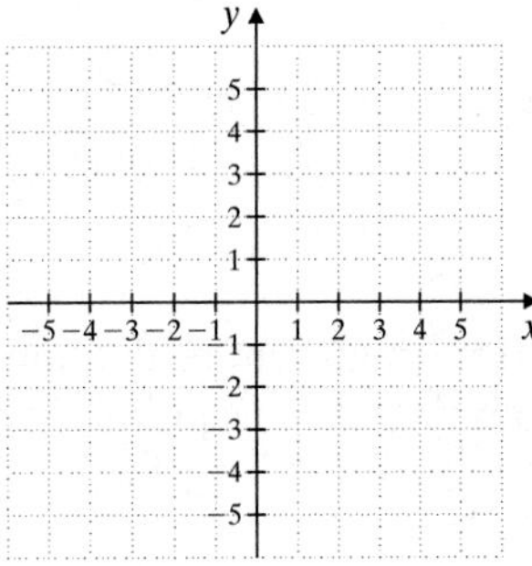

11. Write an equation of the line containing the point $(-2, 1)$ and perpendicular to the line $3x + 5y = 4$.

12. Determine the domain and range of the relation: $\{(2, 3), (2, 4), (0, -1), (3, -1)\}$

Find each function value.

13. If $g(x) = 3x - 2$, find $g(1)$.

14. If $f(x) = 7x^2 - 3x + 1$, find $f(-2)$.

Answers

1. ______
2. a. ______
b. ______
c. ______
3. ______
4. ______
5. ______
6. ______
7. ______
8. ______
9. ______
10. see graph
11. ______
12. ______
13. ______
14. ______

15.

16.

17. see graph

18.

19.

20.

21.

22.

23.

24.

25.

26.

27. see graph

28.

15. Solve the system:

$$\begin{cases} 3x - y + z = -15 \\ x + 2y - z = 1 \\ 2x + 3y - 2z = 0 \end{cases}$$

16. Use matrices to solve the system:

$$\begin{cases} x + 3y = 5 \\ 2x - y = -4 \end{cases}$$

17. Graph the solutions of the system

$$\begin{cases} x - y < 2 \\ x + 2y > -1 \\ y < 2 \end{cases}$$

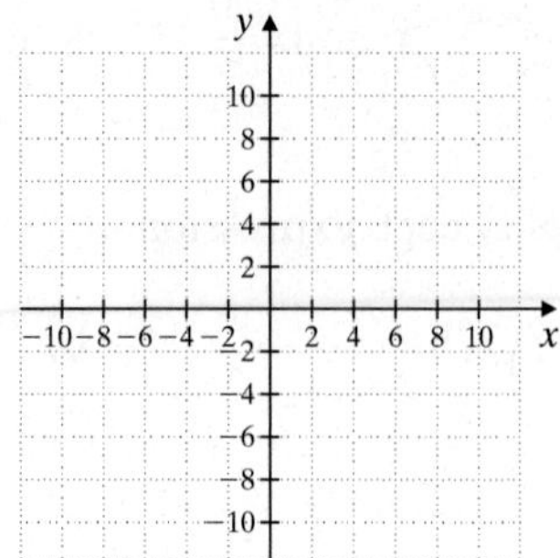

Find the cube roots.

18. $\sqrt[3]{-64}$

19. $\sqrt[3]{\dfrac{8}{125}}$

Multiply.

20. $\sqrt[3]{4} \cdot \sqrt[3]{2}$

21. $\sqrt{\dfrac{2}{a}} \cdot \sqrt{\dfrac{b}{3}}$

Add or subtract.

22. $\sqrt[3]{54} - 5\sqrt[3]{16} + \sqrt[3]{2}$

23. $\sqrt[3]{\dfrac{7x}{8}} + 2\sqrt[3]{7x}$

24. Solve: $p = -3p^2 - 3$

25. Solve: $\log(x + 2) - \log x = 2$

26. Find the midpoint of the line segment that joins points $P(-3, 3)$ and $Q(1, 0)$.

27. Graph: $4y^2 - 9x^2 = 36$

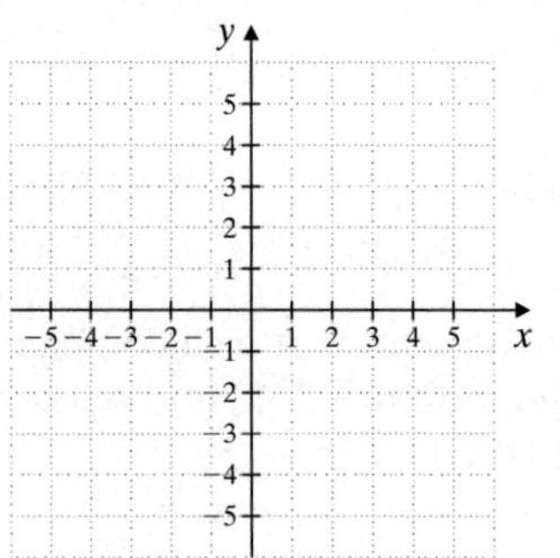

28. Solve the system $\begin{cases} y = \sqrt{x} \\ x^2 + y^2 = 6 \end{cases}$

APPENDIX A

Transition Review: Exponents, Polynomials, and Factoring Strategies

OBJECTIVES

- **A** Review rules for exponents.
- **B** Review addition, subtraction, and multiplication of polynomials.
- **C** Review factoring strategies.

SSM TUTOR CENTER SG CD & VIDEO MATH PRO WEB

A Reviewing Rules for Exponents

The following is a summary of rules for exponents from Sections 4.1 and 4.2.

Summary of Rules for Exponents

If a and b are real numbers and m and n are integers, and no denominator is 0, then

Product rule	$a^m \cdot a^n = a^{m+n}$	$x^2 \cdot x^3 = x^5$
Zero exponent	$a^0 = 1$	$7^0 = 1, (-10)^0 = 1$
Negative exponent	$a^{-n} = \dfrac{1}{a^n}$	$3^{-2} = \dfrac{1}{3^2} = \dfrac{1}{9}$
Quotient rule	$\dfrac{a^m}{a^n} = a^{m-n}$	$\dfrac{y^{10}}{y^4} = y^{10-4} = y^6$
Power rule	$(a^m)^n = a^{m \cdot n}$	$(7^8)^2 = 7^{16}$
Power of a product	$(ab)^m = a^m \cdot b^m$	$(2y)^3 = 2^3y^3 = 8y^3$
Power of a quotient	$\left(\dfrac{a}{b}\right)^m = \dfrac{a^m}{b^m}$	$\left(\dfrac{5x^{-3}}{x^2}\right)^{-2} = \dfrac{5^{-2}x^6}{x^{-4}} = 5^{-2} \cdot x^{6-(-4)} = \dfrac{x^{10}}{5^2}$, or $\dfrac{x^{10}}{25}$

EXAMPLES

Use the rules for exponents to simplify each expression. Write each answer using positive exponents only.

1. $\left(\dfrac{3p^4}{q^5}\right)^2 = \dfrac{(3p^4)^2}{(q^5)^2} = \dfrac{3^2 \cdot (p^4)^2}{(q^5)^2} = \dfrac{9p^8}{q^{10}}$

2. $(-2x^3p^2)(4xp^{10}) = -2(4)x^3x^1p^2p^{10} = -8x^4p^{12}$

3. $(2x^0y^{-3})^{-2} = 2^{-2}(x^0)^{-2}(y^{-3})^{-2} = 2^{-2}x^0y^6 = \dfrac{1(y^6)}{2^2} = \dfrac{y^6}{4}$ Write x^0 as 1.

4. $\dfrac{5^{-2}x^{-3}y^{11}}{x^2y^{-5}} = 5^{-2}x^{-3-2}y^{11-(-5)} = 5^{-2}x^{-5}y^{16} = \dfrac{y^{16}}{5^2x^5} = \dfrac{y^{16}}{25x^5}$

5. $\left(\dfrac{3a^2}{2x^{-1}}\right)^3\left(\dfrac{x^{-3}}{4a^{-2}}\right)^{-1} = \dfrac{27a^6}{8x^{-3}} \cdot \dfrac{x^3}{4^{-1}a^2} = \dfrac{27 \cdot 4 \cdot a^6x^3x^3}{8 \cdot a^2} = \dfrac{27a^4x^6}{2}$

Practice Problems 1–5

Use the rules for exponents to simplify each expression. Write each answer using positive exponents only.

1. $\left(\dfrac{4m^5}{n^3}\right)^3$
2. $(-3x^2y^7)(5xy^6)$
3. $(7xy^{-2})^{-2}$
4. $\dfrac{6^{-2}x^{-4}y^{10}}{x^2y^{-6}}$
5. $\left(\dfrac{4x^3}{3y^{-1}}\right)^3\left(\dfrac{y^{-2}}{3x^{-1}}\right)^{-1}$

Answers

1. $\dfrac{64m^{15}}{n^9}$, **2.** $-15x^3y^{13}$, **3.** $\dfrac{y^4}{49x^2}$, **4.** $\dfrac{y^{16}}{36x^6}$, **5.** $\dfrac{64x^8y^5}{9}$

B Reviewing Operations on Polynomials

In Section 4.4, we added and subtracted polynomials. Below is a review of those operations.

Adding and Subtracting Polynomials

To **add** polynomials, combine all like terms. To **subtract** polynomials, change the signs of the terms of the polynomial being subtracted, and then add.

Practice Problem 6

Add $14x^4 - 6x^3 + x^2 - 6$ and $x^3 - 5x^2 + 1$.

EXAMPLE 6 Add $11x^3 - 12x^2 + x - 3$ and $x^3 - 10x + 5$.

Solution:

$$(11x^3 - 12x^2 + x - 3) + (x^3 - 10x + 5)$$

$$= 11x^3 + x^3 - 12x^2 + x - 10x - 3 + 5 \quad \text{Group like terms.}$$

$$= 12x^3 - 12x^2 - 9x + 2 \quad \text{Combine like terms.}$$

Practice Problem 7

Subtract: $(7x^4 - 8x^2 + x) - (9x^4 + x^2 - 18)$

EXAMPLE 7 Subtract: $(12z^5 - 12z^3 + z) - (-3z^4 + z^3 + 12z)$

Solution: First we change the sign of each term of the second polynomial, and then we add the result to the first polynomial.

$$(12z^5 - 12z^3 + z) - (-3z^4 + z^3 + 12z)$$

$$= 12z^5 - 12z^3 + z + 3z^4 - z^3 - 12z \quad \text{Change signs and add.}$$

$$= 12z^5 + 3z^4 - 12z^3 - z^3 + z - 12z \quad \text{Group like terms.}$$

$$= 12z^5 + 3z^4 - 13z^3 - 11z \quad \text{Combine like terms.}$$

Practice Problem 8

Subtract $3b^3 - 4b^2 + 6$ from $7b^3 - b^2$.

EXAMPLE 8 Subtract $4x^3 - 3x^2 + 2$ from $10x^3 - 7x^2$.

Solution:

$$(10x^3 - 7x^2) - (4x^3 - 3x^2 + 2)$$

$$= 10x^3 - 7x^2 - 4x^3 + 3x^2 - 2 \quad \text{Remove parentheses.}$$

$$= 6x^3 - 4x^2 - 2 \quad \text{Combine like terms.}$$

To multiply any two polynomials, we can use the following from Section 4.5.

Multiplying Any Two Polynomials

To multiply any two polynomials, use the distributive property and multiply each term of one polynomial by each term of the other polynomial. Then combine any like terms.

Practice Problems 9–10

Multiply.

9. $(5x - 1)(2x^2 - x + 4)$
10. $(4x - 3)(x - 6)$

Answers

6. $14x^4 - 5x^3 - 4x^2 - 5$, **7.** $-2x^4 - 9x^2 + x + 18$, **8.** $4b^3 + 3b^2 - 6$, **9.** $10x^3 - 7x^2 + 21x - 4$, **10.** $4x^2 - 27x + 18$

EXAMPLES Multiply:

9. $(2x - 3)(5x^2 - 6x + 7) = 2x(5x^2 - 6x + 7) + (-3)(5x^2 - 6x + 7)$

$$= 10x^3 - 12x^2 + 14x - 15x^2 + 18x - 21$$

$$= 10x^3 - 27x^2 + 32x - 21 \quad \text{Combine like terms.}$$

F L

10. $(2x - 7)(3x - 4) = 2x(3x) + 2x(-4) + (-7)(3x) + (-7)(-4)$

I O

$= 6x^2 - 8x - 21x + 28$ FOIL order

$= 6x^2 - 29x + 28$

Special products from Section 4.6 may be used to multiply.

Square of a Binomial:

$(a + b)^2 = a^2 + 2ab + b^2$

$(a - b)^2 = a^2 - 2ab + b^2$

Product of Sum and Difference of Two Terms

$(a + b)(a - b) = a^2 - b^2$

EXAMPLES Multiply.

$(a + b)^2 = a^2 + 2 \cdot a \cdot b + b^2$

11. $(x + 5)^2 = x^2 + 2 \cdot x \cdot 5 + 5^2 = x^2 + 10x + 25$

12. $(4m^2 - 3n)^2 = (4m^2)^2 - 2(4m^2)(3n) + (3n)^2 = 16m^4 - 24m^2n + 9n^2$

$(a + b)(a - b) = a^2 - b^2$

13. $(x + 3)(x - 3) = x^2 - 3^2 = x^2 - 9$

14. $(4y - 1)(4y + 1) = (4y)^2 - 1^2 = 16y^2 - 1$

Practice Problems 11–14

Multiply.

11. $(x + 3)^2$
12. $(6a^2 - 2b)^2$
13. $(x + 4)(x - 4)$
14. $(3m - 6)(3m + 6)$

C Reviewing Factoring Strategies

The key to proficiency in factoring polynomials is to practice until you are comfortable with each technique. A strategy for factoring polynomials completely is given next. This strategy can also be found in the Chapter 5 Integrated Review.

Factoring a Polynomial

Step 1. Are there any common factors? If so, factor out the greatest common factor.

Step 2. How many terms are in the polynomial?

- **a.** If there are *two* terms, decide if one of the following formulas may be applied:
  - **i.** Difference of two squares: $a^2 - b^2 = (a - b)(a + b)$
  - **ii.** Difference of two cubes: $a^3 - b^3 = (a - b)(a^2 + ab + b^2)$
  - **iii.** Sum of two cubes: $a^3 + b^3 = (a + b)(a^2 - ab + b^2)$
- **b.** If there are *three* terms, try one of the following:
  - **i.** Perfect square trinomial: $a^2 + 2ab + b^2 = (a + b)^2$, $a^2 - 2ab + b^2 = (a - b)^2$
  - **ii.** If not a perfect square trinomial, factor by using the methods presented in Sections 5.2 through 5.4.
- **c.** If there are *four* or more terms, try factoring by grouping.

Step 3. See whether any factors in the factored polynomial can be factored further.

Answers

11. $x^2 + 6x + 9$, **12.** $36a^4 - 24a^2b + 4b^2$, **13.** $x^2 - 16$, **14.** $9m^2 - 36$

Practice Problems 15

Factor each polynomial completely.

a. $21xy^2 + 6xy$

b. $45x^2 - 20$

c. $4y^2 - 2y - 6$

d. $a^2 + a^2b + 5 + 5b$

e. $8x^3 + y^3$

EXAMPLE 15 Factor each polynomial completely.

a. $8a^2b - 4ab$ **b.** $36x^2 - 9$ **c.** $2x^2 - 5x - 7$

d. $5p^2 + 5 + qp^2 + q$ **e.** $27a^3 - b^3$

Solution:

a. Step 1. The terms have a common factor of $4ab$, which we factor out.

$$8a^2b - 4ab = 4ab(2a - 1)$$

Step 2. There are two terms, but the binomial $2a - 1$ is not the difference of two squares or the sum or difference of two cubes.

Step 3. The factor $2a - 1$ cannot be factored further.

b. Step 1. Factor out a common factor of 9.

$$36x^2 - 9 = 9(4x^2 - 1)$$

Step 2. The factor $4x^2 - 1$ has two terms, and it is the difference of two squares.

$$9(4x^2 - 1) = 9(2x + 1)(2x - 1)$$

Step 3. No factor can be factored further.

c. Step 1. The terms of $2x^2 - 5x - 7$ contain no common factor other than 1 or -1.

Step 2. There are three terms. The trinomial is not a perfect square, so we factor by methods from Section 5.3 or 5.4.

$$2x^2 - 5x - 7 = (2x - 7)(x + 1)$$

Step 3. No factor can be factored further.

d. Step 1. There is no common factor of all terms of $5p^2 + 5 + qp^2 + q$.

Step 2. The polynomial has four terms, so try factoring by grouping.

$$\begin{aligned} 5p^2 + 5 + qp^2 + q &= (5p^2 + 5) + (qp^2 + q) \quad \text{Group the terms.} \\ &= 5(p^2 + 1) + q(p^2 + 1) \\ &= (p^2 + 1)(5 + q) \end{aligned}$$

Step 3. No factor can be factored further.

e. Step 1. The terms of $27a^3 - b^3$ contain no common factor.

Step 2. There are two terms and $27a^3 - b^3$ is the difference of squares.

$$\begin{aligned} 27a^3 - b^3 &= (3a)^3 - b^3 \\ &= (3a - b)[(3a)^2 + (3a)(b) + b^2] \\ &= (3a - b)(9a^2 + 3ab + b^2) \end{aligned}$$

Answers

15. a. $3xy(7y + 2)$ **b.** $5(3x - 2)(3x + 2)$ **c.** $2(2y - 3)(y + 1)$ **d.** $(a^2 + 5)(1 + b)$ **e.** $(2x + y)(4x^2 - 2xy + y^2)$

Name ______________________ Section __________ Date __________

EXERCISE SET APPENDIX A

A *Use rules for exponents to simplify each expression. See Examples 1 through 5.*

1. $(-4x^3p^2)(4y^3x^3)$ **2.** $(-6a^2b^3)(-3ab^3)$ **3.** $8x^0 + 1$ **4.** $(5x)^0 + 5x^0$ **5.** $\frac{a^{12}b^2}{a^9b}$

6. $-\frac{26z^{11}}{2z^7}$ **7.** 4^{-2} **8.** 2^{-3} **9.** $4^{-1} + 3^{-2}$ **10.** $1^{-3} - 4^{-2}$

11. $\frac{y^{-3}}{y^{-7}}$ **12.** $\frac{z^{-12}}{z^{10}}$ **13.** $\frac{(24x^8)(x)}{20x^{-7}}$ **14.** $\frac{(30z^2)(z^5)}{55z^{-4}}$ **15.** $\left(\frac{2x^5}{y^{-3}}\right)^4$

16. $\left(\frac{3a^{-4}}{b^7}\right)^3$ **17.** $\left(\frac{2a^{-2}b^5}{4a^2b^7}\right)^{-2}$ **18.** $\left(\frac{5x^7y^4}{10x^3y^{-2}}\right)^{-3}$ **19.** $\left(\frac{2x^2}{y^4}\right)^3\left(\frac{2x^5}{y}\right)^{-2}$ **20.** $\left(\frac{3z^{-2}}{y}\right)^2\left(\frac{9y^{-4}}{z^{-3}}\right)^{-1}$

B *Add or subtract as indicated. See Examples 6 through 8.*

21. $(5y^4 - 7y^2 + x^2 - 3) + (-3y^4 + 2y^2 + 4)$

22. $(8x^4 - 14x^2 + 6) + (-12x^6 - 21x^4 - 9x^2)$

23. $(9y^2 - 7y + 5) - (8y^2 - 7y + 2)$

24. $(2x^2 + 3x + 12) - (5x - 7)$

25. $(3x^3 - b + 2a - 6) + (-4x^3 + b + 6a - 6)$

26. $(5x^2 - 6) + (2x^2 - 4x + 8)$

27. $(4x^2 - 6x + 2) - (-x^2 + 3x + 5)$

28. $(5x^2 + x + 9) - (2x^2 - 9)$

29. $\begin{array}{r} 3x^2 - 4x + 8 \\ -\quad (5x^2 - 7) \\ \hline \end{array}$

30. $\begin{array}{r} -3x^2 - 4x + 8 \\ -\quad (5x + 12) \\ \hline \end{array}$

31. $\begin{array}{r} 6y^2 - 6y + 4 \\ -(-y^2 - 6y + 7) \\ \hline \end{array}$

32. $\begin{array}{r} -4x^3 + 4x^2 - 4x \\ -(2x^3 - 2x^2 + 3x) \\ \hline \end{array}$

33. Subtract $(y^2 + 4yx + 7)$ from $(-19y^2 + 7yx + 7)$.

34. Subtract $(x^2y - 4)$ from $(3x^2 - 4x^2y + 5)$.

Multiply. See Examples 9 through 14.

35. $(3x + 1)(3x + 5)$

36. $(4x - 5)(5x + 6)$

37. $\left(4x + \frac{1}{3}\right)\left(4x - \frac{1}{2}\right)$

38. $\left(4y - \frac{1}{3}\right)\left(3y - \frac{1}{8}\right)$

39. $(x + 4)^2$

40. $(x - 5)^2$

41. $(3x - y)^2$

42. $(4x - z)^2$

43. $(3b - 6y)(3b + 6y)$

44. $(2x - 4y)(2x + 4y)$

45. $\left(3x + \frac{1}{2}\right)\left(3x - \frac{1}{2}\right)$

46. $\left(2x - \frac{1}{3}\right)\left(2x + \frac{1}{3}\right)$

C *Factor completely. See Example 15.*

47. $14x^2y - 2xy$

48. $24ab^2 - 6ab$

49. $4x^2 - 16$

50. $9x^2 - 81$

51. $3x^2 - 8x - 11$

52. $5x^2 - 2x - 3$

53. $8x^3 + 125y^3$

54. $27x^3 - 64y^3$

55. $7x^2 - 63x$

56. $15x^2 - 20x$

57. $20x^2 + 23x + 6$

58. $20x^2 - 220x + 600$

59. $ab - 6a + 7b - 42$

60. $2sr + 10s - r - 5$

61. $x^4 - 1$

62. $y^4 - 16$

63. $2x^3 - 54$

64. $250x^4 - 16x$

APPENDIX B

Transition Review: Solving Linear and Quadratic Equations

OBJECTIVE

A Solve linear and quadratic equations.

SSM TUTOR CENTER SG CD & VIDEO MATH PRO WEB

A Solving Equations

Recall that an **equation** is a statement that two expressions are equal. When a variable in an equation is replaced by a number and the resulting equation is true, then that number is called a **solution**. The set of solutions of an equation is called its **solution set**.

In this section, we review solving linear and quadratic equations. Study the table below to help you identify these types of equations. Here, a, b, and c are real numbers and a is not 0.

Linear: Can be written in form $ax + b = c$ (Section 2.2–2.4)	**Quadratic:** Can be written in form $ax^2 + bx + c = 0$ (Section 5.6)
$3x = -15$	$-3x^2 + 7 = x^2 - 9$
$2.7 - y = 3y$	$p^2 + 2.6 = p - 4.3$
$\frac{4n}{5} - \frac{9n}{7} + 1 = 0$	$\frac{y^2}{5} - \frac{1}{7} + \frac{y}{9} = 0$

You may want to use the steps below to help you solve linear and quadratic equations. (In this appendix, the method of solving quadratic equations by factoring will be reviewed only.)

Solving Linear and Quadratic Equations in One Variable

Step 1. Multiply on both sides to clear the equation of fractions if they occur.

Step 2. Use the distributive property to remove parentheses if they occur.

Step 3. Simplify each side of the equation by combining like terms.

Step 4. Decide whether the equation is linear or quadratic.

If linear ($ax + b = c$),

Step 5. Get all variable terms on one side and all numbers on the other side by using the addition property of equality.

Step 6. Get the variable alone by using the multiplication property of equality.

If quadratic ($ax^2 + bx + c = 0$),

Step 5. Write the equation in standard form: $ax^2 + bx + c = 0$.

Step 6. Factor completely.

Step 7. Set each factor containing a variable equal to 0.

Step 8. Solve.

Final Step. Check each solution in the original equation.

Practice Problem 1

Solve: $4(x - 2) = 6x - 10$

EXAMPLE 1 Solve: $2(x - 3) = 5x - 9$

Solution:

Step 1 is not needed since there are no fractions.

$$2(x - 3) = 5x - 9$$

Step 2. $2x - 6 = 5x - 9$ Use the distributive property.

Step 3 is not needed since no simplifying can be done on either side of the equation.

Step 4. $2x - 6 = 5x - 9$ The equation is linear.

Step 5. Next we get variable terms on the same side of the equation by using the addition property of equality.

$2x - 6 - 5x = 5x - 9 - 5x$ Subtract $5x$ from both sides.

$-3x - 6 = -9$ Simplify.

$-3x - 6 + 6 = -9 + 6$ Add 6 to both sides.

$-3x = -3$ Simplify.

Step 6. $\frac{-3x}{-3} = \frac{-3}{-3}$ Divide both sides by -3.

$x = 1$

Final Step. Check to see that 1 is the solution.

Practice Problem 2

Solve.

$$y + \frac{y - 4}{4} = \frac{1}{2} - \frac{y - 6}{4}$$

EXAMPLE 2 Solve.

$$x - \frac{x - 2}{6} = \frac{x - 7}{3} + \frac{2}{3}$$

Solution:

Step 1. $6\left(x - \frac{x - 2}{6}\right) = 6\left(\frac{x - 7}{3} + \frac{2}{3}\right)$ Multiply both sides by 6.

$6x - (x - 2) = 2(x - 7) + 2(2)$

Step 2. $6x - x + 2 = 2x - 14 + 4$ Remove grouping symbols.

Step 3. $5x + 2 = 2x - 10$ Simplify.

Step 4. $5x + 2 = 2x - 10$ This equation is linear.

Step 5. $5x + 2 - 2 = 2x - 10 - 2$ Subtract 2.

$5x = 2x - 12$

$5x - 2x = 2x - 12 - 2x$ Subtract $2x$.

$3x = -12$

Step 6. $\frac{3x}{3} = \frac{-12}{3}$ Divide by 3.

$x = -4$ Replace x with -4 in the original equation.

Final Step. $-4 - \frac{-4 - 2}{6} \stackrel{?}{=} \frac{-4 - 7}{3} + \frac{2}{3}$

$-4 - \frac{-6}{6} \stackrel{?}{=} \frac{-11}{3} + \frac{2}{3}$

$-4 - (-1) \stackrel{?}{=} \frac{-9}{3}$

$-3 = -3$ True.

The solution is -4.

Answers

1. 1, **2.** 2

EXAMPLE 3 Solve: $2x^2 = \frac{17}{3}x + 1$

Solution:

$$2x^2 = \frac{17}{3}x + 1$$

Step 1. $3(2x^2) = 3\left(\frac{17}{3}x + 1\right)$ Clear the equation of fractions.

Step 2. $6x^2 = 17x + 3$ Use the distributive property.

Step 3. $6x^2 = 17x + 3$ The equation is quadratic.

Step 4. $6x^2 - 17x - 3 = 0$ Rewrite the equation in standard form.

Step 5. $(6x + 1)(x - 3) = 0$ Factor.

Step 6. $6x + 1 = 0$ or $x - 3 = 0$ Set each factor equal to 0.

Step 7. $6x = -1$ $\quad x = 3$ Solve each equation.

$$x = -\frac{1}{6}$$

Final Step. Check by substituting into the original equation. The solutions are $-\frac{1}{6}$ and 3.

Practice Problem 3

Solve: $2x^2 + \frac{5}{2}x = 3$

Answer

3. $-2, \frac{3}{4}$

Name ______________________ Section __________ Date ____________

Mental Math

Solve each equation.

1. $3x = 18$

2. $2x = 60$

3. $x - 7 = 10$

4. $x - 2 = 15$

5. $\frac{x}{2} = 4$

6. $\frac{x}{3} = 5$

7. $x + 1 = 11$

8. $x + 4 = 20$

EXERCISE SET APPENDIX B

Solve each equation. See Examples 1 through 3.

1. $x + 2.8 = 1.9$

2. $y - 8.6 = -6.3$

3. $5x - 4 = 26$

4. $2y - 3 = 11$

5. $-4.1 - 7z = 3.6$

6. $10.3 - 6x = -2.3$

7. $5y + 12 = 2y - 3$

8. $4x + 14 = 6x + 8$

9. $(x + 3)(3x - 4) = 0$

10. $(5x + 1)(x - 2) = 0$

11. $8x - 5x + 3 = x - 7 + 10$

12. $6 + 3x + x = -x + 2 - 26$

13. $3(2x - 5)(4x + 3) = 0$

14. $8(3x - 4)(2x - 7) = 0$

15. $x^2 + 11x + 24 = 0$

16. $y^2 - 10y + 24 = 0$

17. $5x + 12 = 2(2x + 7)$

18. $2(x + 3) = x + 5$

19. $12x^2 + 5x - 2 = 0$

20. $3y^2 - y - 14 = 0$

21. $z^2 + 9 = 10z$

22. $n^2 + n = 72$

23. $3(x - 6) = 5x$

24. $6x = 4(5 + x)$

25. $\frac{x}{2} + \frac{2}{3} = \frac{3}{4}$

26. $\frac{x}{2} + \frac{x}{3} = \frac{5}{2}$

27. $\dfrac{n-3}{4} + \dfrac{n+5}{7} = \dfrac{5}{14}$

28. $\dfrac{2+h}{9} + \dfrac{h-1}{3} = \dfrac{1}{3}$

29. $x(5x + 2) = 3$

30. $n(2n - 3) = 2$

31. $x^2 - 6x = x(8 + x)$

32. $n(3 + n) = n^2 + 4n$

33. $\dfrac{z^2}{6} - \dfrac{z}{2} - 3 = 0$

34. $\dfrac{c^2}{20} - \dfrac{c}{4} + \dfrac{1}{5} = 0$

35. $2y + 5(y - 4) = 4y - 2(y - 10)$

36. $9c - 3(6 - 5c) = c - 2(3c + 9)$

37. $2(x - 8) + x = 3(x - 6) + 2$

38. $4(x + 5) = 3(x - 4) + x$

39. $\dfrac{x^2}{2} + \dfrac{x}{20} = \dfrac{1}{10}$

40. $\dfrac{y^2}{30} = \dfrac{y}{15} + \dfrac{1}{2}$

41. $\dfrac{4t^2}{5} = \dfrac{t}{5} + \dfrac{3}{10}$

42. $\dfrac{5x^2}{6} - \dfrac{7x}{2} + \dfrac{2}{3} = 0$

43. $\dfrac{m-4}{3} - \dfrac{3m-1}{5} = 1$

44. $\dfrac{n+1}{8} - \dfrac{2-n}{3} = \dfrac{5}{6}$

45. $-(3x - 5) - (2x - 6) + 1 = -5(x - 1) - (3x + 2) + 3$

46. $-4(2x - 3) - (10x + 7) - 2 = -(12x - 5) - (4x + 9) - 1$

47. $3x(x - 5) = 0$

48. $4x(2x + 3) = 0$

49. $12x^2 + 2x - 2 = 0$

50. $8x^2 + 13x + 5 = 0$

51. $w^2 - 5w = 36$

52. $x^2 + 32 = 12x$

53. $2z(z + 6) = 2z^2 + 12z - 8$

54. $3c^2 - 8c + 2 = c(3c - 8)$

55. $-3(x - 4) + x = 5(3 - x)$

56. $-4(a + 1) - 3a = -7(2a - 3)$

APPENDIX C

An Introduction to Using a Graphing Utility

OBJECTIVES

Ⓐ View window and interpret window settings.

Ⓑ Graph equations and square viewing window.

Ⓐ Viewing Window and Interpreting Window Settings

In this appendix, we will use the term **graphing utility** to mean a graphing calculator or a computer software graphing package. All graphing utilities graph equations by plotting points on a screen. While plotting several points can be slow and sometimes tedious for us, a graphing utility can quickly and accurately plot hundreds of points. How does a graphing utility show plotted points? A computer or calculator screen is made up of a grid of small rectangular areas called **pixels**. If a pixel contains a point to be plotted, the pixel is turned "on"; otherwise, the pixel remains "off." The graph of an equation is then a collection of pixels turned "on." The graph of $y = 3x + 1$ from a graphing calculator is shown in Figure A-1. Notice the irregular shape of the line caused by the rectangular pixels.

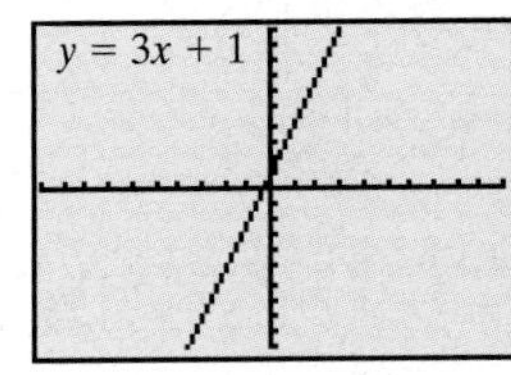

Figure A-1

The portion of the coordinate plane shown on the screen in Figure A-1 is called the **viewing window** or the **viewing rectangle**. Notice the x-axis and the y-axis on the graph. While tick marks are shown on the axes, they are not labeled. This means that from this screen alone, we do not know how many units each tick mark represents. To see what each tick mark represents and the minimum and maximum values on the axes, check the *window setting* of the graphing utility. It defines the viewing window. The window of the graph of $y = 3x + 1$ shown in Figure A-1 has the following setting (Figure A-2):

Xmin = −10	The minimum x-value is −10.
Xmax = 10	The maximum x-value is 10.
Xscl = 1	The x-axis scale is 1 unit per tick mark.
Ymin = −10	The minimum y-value is −10.
Ymax = 10	The maximum y-value is 10.
Yscl = 1	The y-axis scale is 1 unit per tick mark.

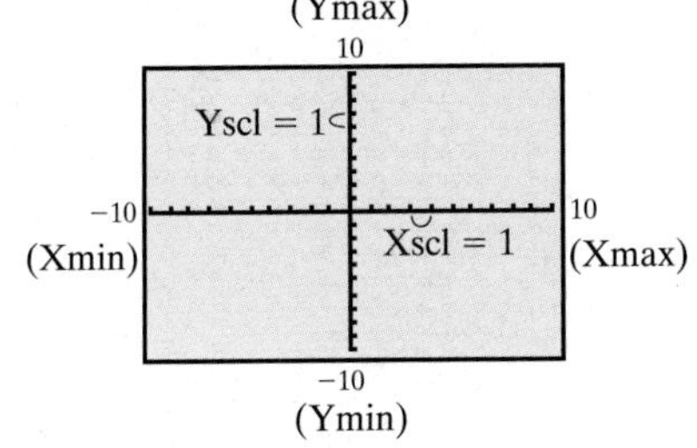

Figure A-2

By knowing the scale, we can find the minimum and the maximum values on the axes simply by counting tick marks. For example, if both the Xscl (x-axis scale) and the Yscl are 1 unit per tick mark on the graph in Figure A-3, we can count the tick marks and find that the minimum x-value is −10 and the maximum x-value is 10. Also, the minimum y-value is −10 and the maximum y-value is 10. If the Xscl (x-axis scale) changes to 2 units per tick mark (shown in Figure A-4), by counting tick marks, we see that the minimum x-value is now −20 and the maximum x-value is now 20.

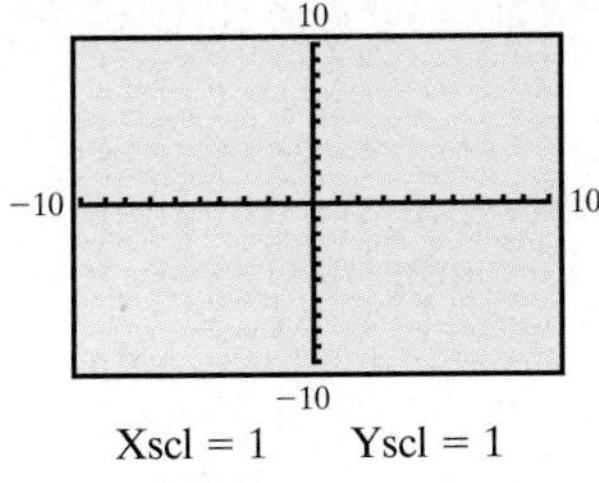

FIGURE A-3

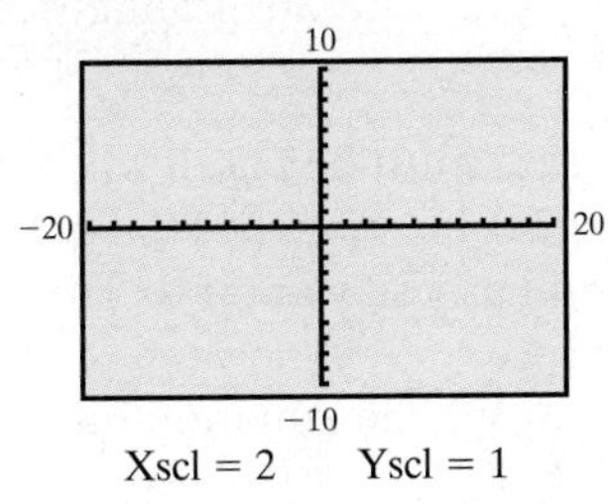

FIGURE A-4

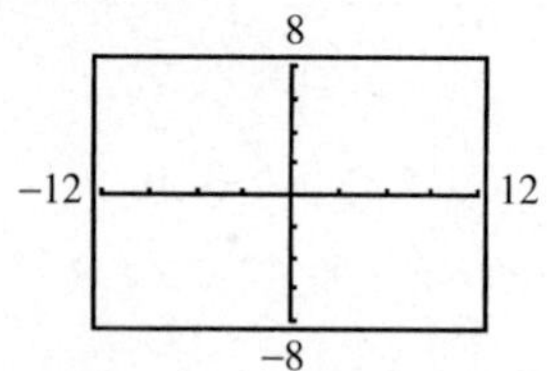

FIGURE A-5

It is also true that if we know the Xmin and the Xmax values, we can calculate the Xscl by the displayed axes. For example, the Xscl of the graph in Figure A-5 must be 3 units per tick mark for the maximum and minimum x-values to be as shown. Also, the Yscl of that graph must be 2 units per tick mark for the maximum and minimum y-values to be as shown.

We will call the viewing window in Figure A-3 a *standard* viewing window or rectangle. Although a standard viewing window is sufficient for much of this text, special care must be taken to ensure that all key features of a graph are shown. Figures A-6, A-7, and A-8 show the graph of $y = x^2 + 11x - 1$ on three different viewing windows. Note that certain viewing windows for this equation are misleading.

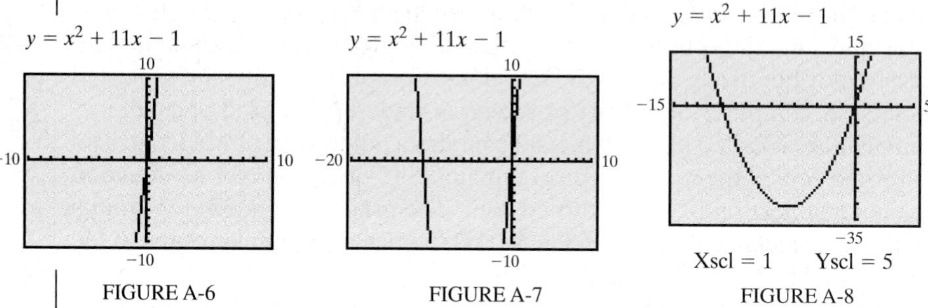

FIGURE A-6 FIGURE A-7 FIGURE A-8

How do we ensure that all distinguishing features of the graph of an equation are shown? It helps to know about the equation that is being graphed. For example, the equation $y = x^2 + 11x - 1$ is not a linear equation, and its graph is not a line. This equation is a quadratic equation, and therefore its graph is a parabola. By knowing this information, we know that the graph shown in Figure A-6, although correct, is misleading. Of the three viewing rectangles shown, the graph in Figure A-8 is best because it shows more of the distinguishing features of the parabola. Properties of equations needed for graphing will be studied in this text.

B Graphing Equations and Square Viewing Window

In general, the following steps may be used to graph an equation on a standard viewing window.

To Graph an Equation in *x* and *y* with a Graphing Utility on a Standard Viewing Window

Step 1. Solve the equation for y.

Step 2. Use your graphing utility and enter the equation in the form Y = *expression involving x*

Step 3. Activate the graphing utility.

Special care must be taken when entering the *expression involving x* in *Step 2*. You must be sure that the graphing utility you are using interprets the expression as you want it to. For example, let's graph $3y = 4x$. To do so,

Step 1. Solve the equation for y.

$$3y = 4x \qquad \frac{3y}{3} = \frac{4x}{3} \qquad y = \frac{4}{3}x$$

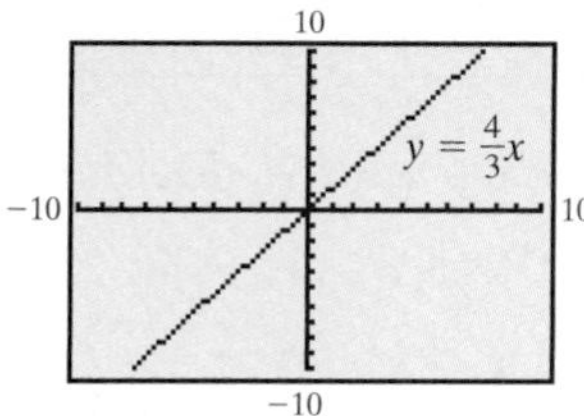

FIGURE A-9

Step 2. Using your graphing utility, enter the expression $\frac{4}{3}x$ after the Y = prompt. In order for your graphing utility to correctly interpret the expression, you may need to enter $(4/3)x$ or $(4 \div 3)x$.

Step 3. Activate the graphing utility. The graph should appear as in Figure A-9.

Distinguishing features of the graph of a line include showing all the intercepts of the line. For example, the window of the graph of the line in Figure A-10 does not show both intercepts of the line, but the window of the graph of the same line in Figure A-11 does show both intercepts. Notice the notation below each graph. This is a shorthand notation of the range setting of the graph. This notation means [Xmin, Xmax] by [Ymin, Ymax].

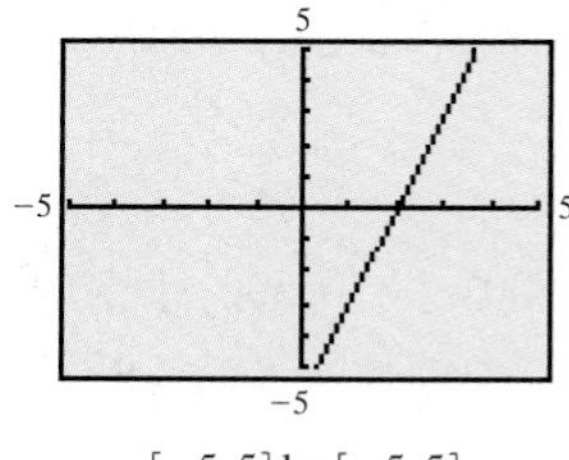

[−5, 5] by [−5, 5]

FIGURE A-10

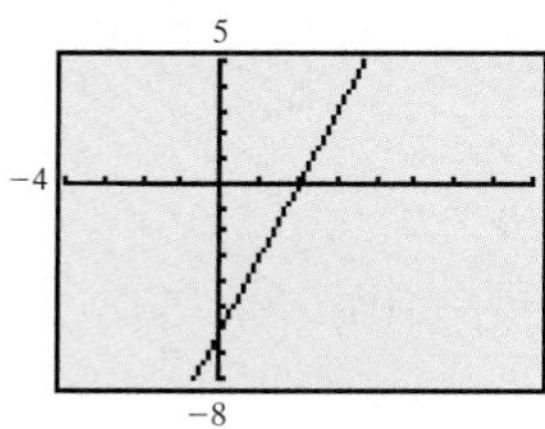

[−4, 8] by [−8, 5]

FIGURE A-11

On a standard viewing window, the tick marks on the y-axis are closer than the tick marks on the x-axis. This happens because the viewing window is a rectangle, and so 10 equally spaced tick marks on the positive y-axis will be closer together than 10 equally spaced tick marks on the positive x-axis. This causes the appearance of graphs to be distorted.

For example, notice the different appearances of the same line graphed using different viewing windows. The line in Figure A-12 is distorted because the tick marks along the x-axis are farther apart than the tick marks along the y-axis. The graph of the same line in Figure A-13 is not distorted because the viewing rectangle has been selected so that there is equal spacing between tick marks on both axes.

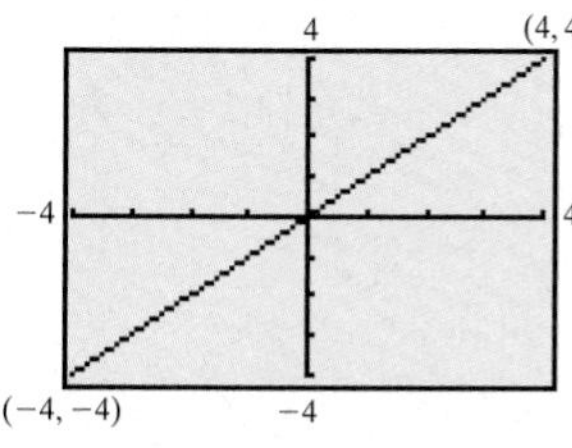

FIGURE A-12

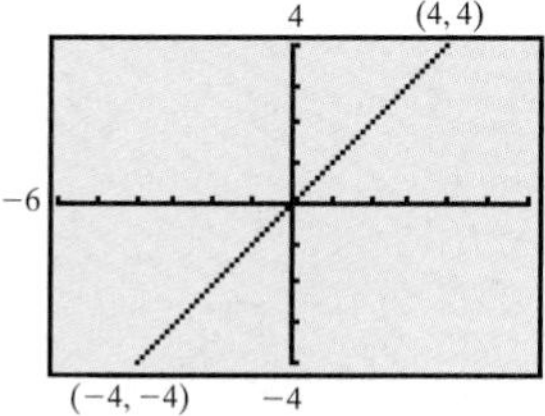

FIGURE A-13

We say that the line in Figure A-13 is graphed on a *square* setting. Some graphing utilities have a built-in program that, if activated, will automatically provide a square setting. A square setting is especially helpful when we are graphing perpendicular lines, circles, or when a true geometric perspective is desired. Some examples of square screens are shown in Figures A-14 and A-15.

Other features of a graphing utility such as Trace, Zoom, Intersect, and Table are discussed in appropriate Graphing Calculator Explorations in this text.

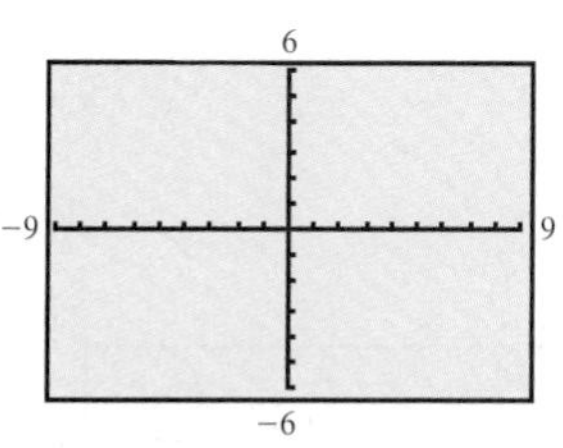

FIGURE A-14

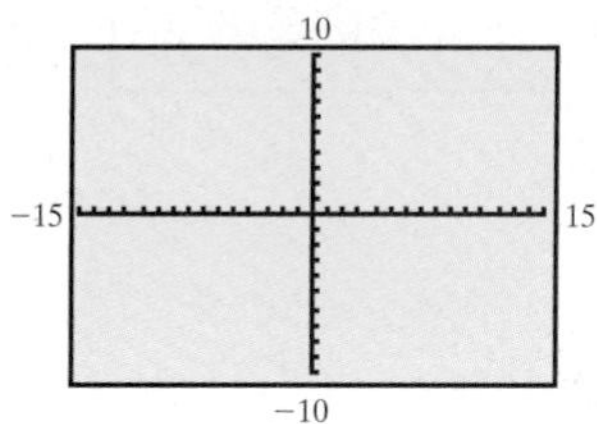

FIGURE A-15

Name ______________________ Section __________ Date __________

EXERCISE SET APPENDIX C

A *In Exercises 1–4, determine whether all ordered pairs listed will lie within a standard viewing rectangle.*

1. $(-9, 0), (5, 8), (1, -8)$

2. $(4, 7), (0, 0), (-8, 9)$

3. $(-11, 0), (2, 2), (7, -5)$

4. $(3, 5), (-3, -5), (15, 0)$

In Exercises 5–10, choose an Xmin, Xmax, Ymin, and Ymax so that all ordered pairs listed will lie within the viewing rectangle.

5. $(-90, 0), (55, 80), (0, -80)$

6. $(4, 70), (20, 20), (-18, 90)$

7. $(-11, 0), (2, 2), (7, -5)$

8. $(3, 5), (-3, -5), (15, 0)$

9. $(200, 200), (50, -50), (70, -50)$

10. $(40, 800), (-30, 500), (15, 0)$

Write the window setting for each viewing window shown. Use the following format:

Xmin =	Ymin =
Xmax =	Ymax =
Xscl =	Yscl =

11.

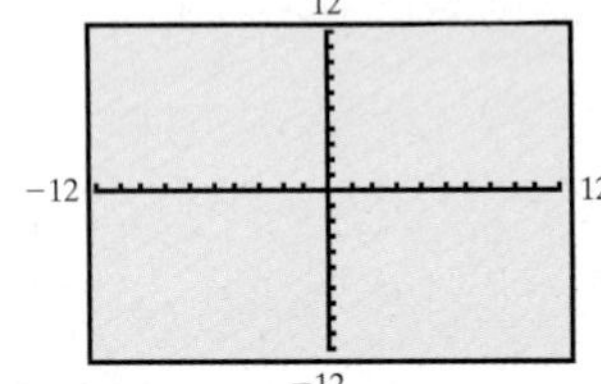

12.

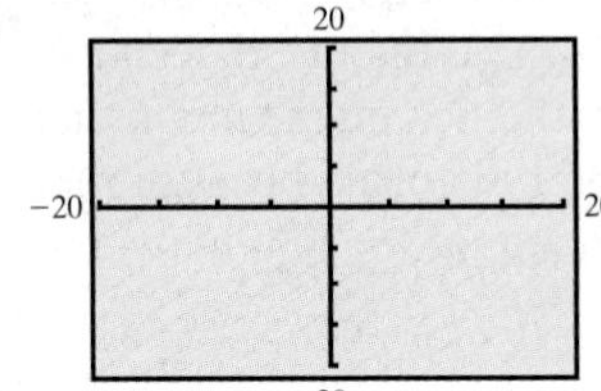

13.

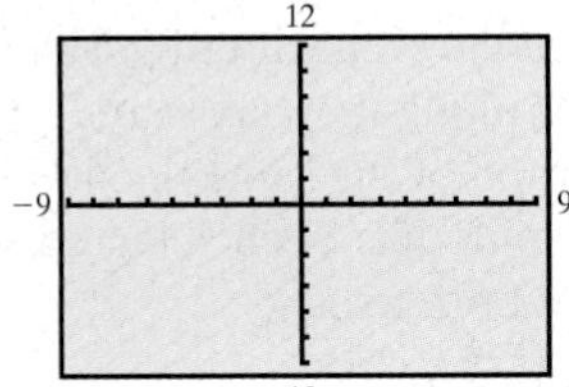

14.

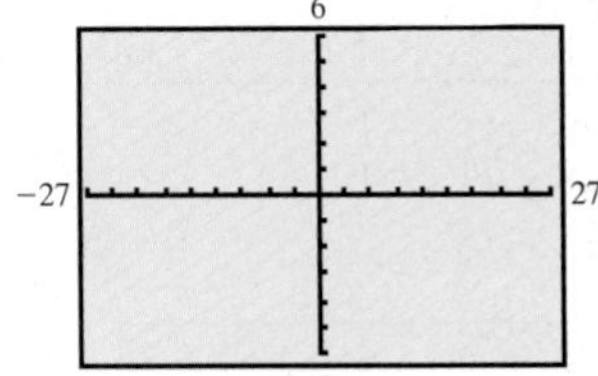

15.

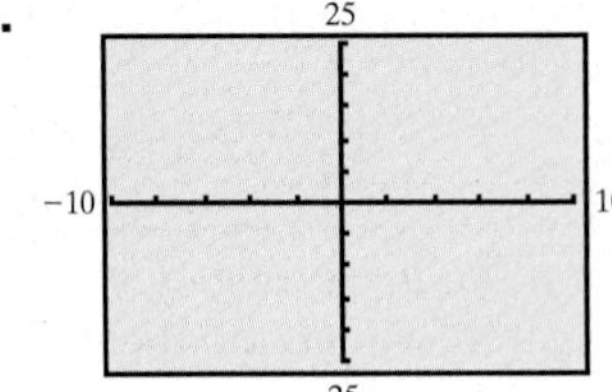

16.

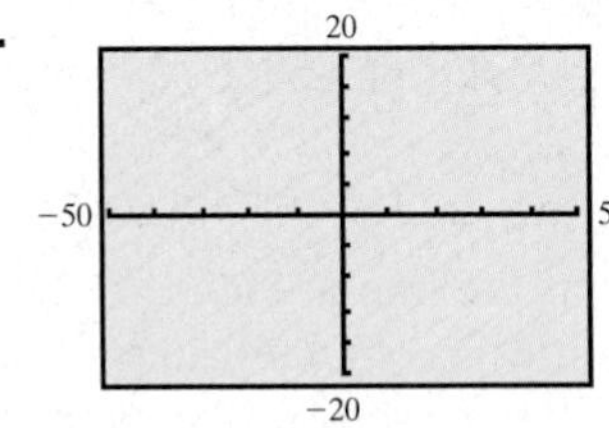

17. 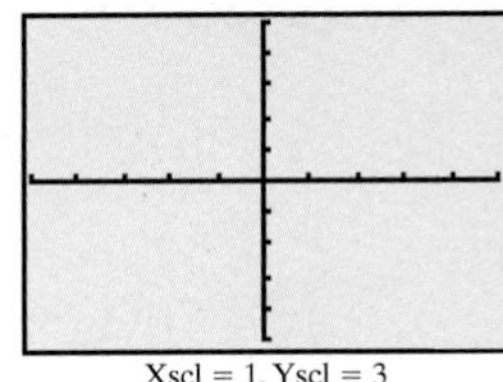
Xscl = 1, Yscl = 3

18. 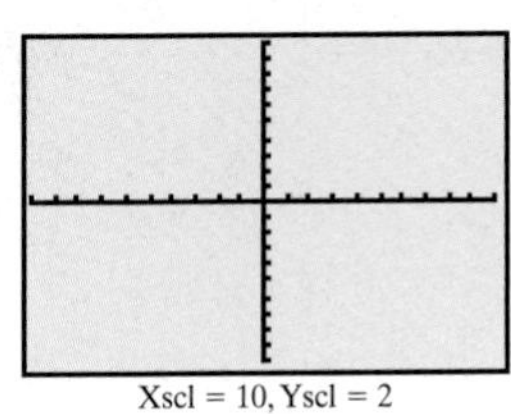
Xscl = 10, Yscl = 2

19. 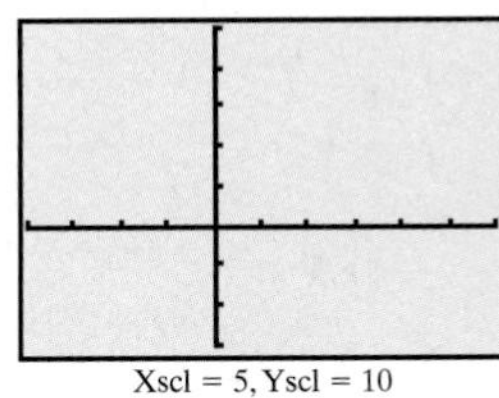
Xscl = 5, Yscl = 10

20. 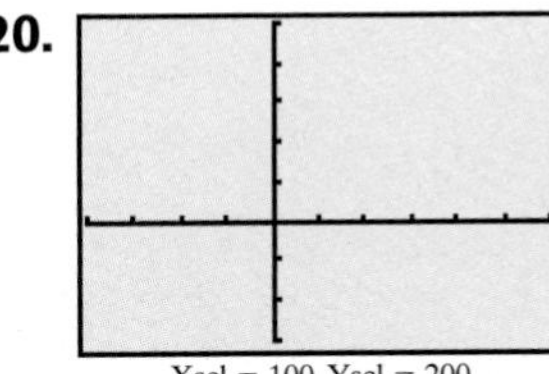
Xscl = 100, Yscl = 200

B *Graph each linear equation in two variables, using the two different range settings given. Determine which setting shows all intercepts of a line.*

21. $y = 2x + 12$

Setting A: $[-10, 10]$ by $[-10, 10]$
Setting B: $[-10, 10]$ by $[-10, 15]$

22. $y = -3x + 25$

Setting A: $[-5, 5]$ by $[-30, 10]$
Setting B: $[-10, 10]$ by $[-10, 30]$

23. $y = -x - 41$

Setting A: $[-50, 10]$ by $[-10, 10]$
Setting B: $[-50, 10]$ by $[-50, 15]$

24. $y = 6x - 18$

Setting A: $[-10, 10]$ by $[-20, 10]$
Setting B: $[-10, 10]$ by $[-10, 10]$

25. $y = \frac{1}{2}x - 15$

Setting A: $[-10, 10]$ by $[-20, 10]$
Setting B: $[-10, 35]$ by $[-20, 15]$

26. $y = -\frac{2}{3}x - \frac{29}{3}$

Setting A: $[-10, 10]$ by $[-10, 10]$
Setting B: $[-15, 5]$ by $[-15, 5]$

The graph of each equation is a line. Use a graphing utility and a standard viewing window to graph each equation.

27. $3x = 5y$ **28.** $7y = -3x$ **29.** $9x - 5y = 30$ **30.** $4x + 6y = 20$

31. $y = -7$

32. $y = 2$

33. $x + 10y = -5$

34. $x - 5y = 9$

Graph the following equations using the square setting given. Some keystrokes that may be helpful are given.

35. $y = \sqrt{x}$ $[-12, 12]$ by $[-8, 8]$
Suggested keystrokes: $\sqrt{\ }\, x$

36. $y = \sqrt{2x}$ $[-12, 12]$ by $[-8, 8]$
Suggested keystrokes: $\sqrt{\ }\, (2x)$

37. $y = x^2 + 2x + 1$ $[-15, 15]$ by $[-10, 10]$
Suggested keystrokes: $x \wedge 2 + 2x + 1$

38. $y = x^2 - 5$ $[-15, 15]$ by $[-10, 10]$
Suggested keystrokes: $x \wedge 2 - 5$

39. $y = |x|$ $[-9, 9]$ by $[-6, 6]$
Suggested keystrokes: ABS x

40. $y = |x - 2|$ $[-9, 9]$ by $[-6, 6]$
Suggested keystrokes: ABS $(x - 2)$

Graph the line on a single set of axes. Use a standard viewing window; then, if necessary, change the viewing window so that all intercepts of the line show.

41. $x + 2y = 30$

42. $1.5x - 3.7y = 40.3$

APPENDIX D

Sets and Compound Inequalities

OBJECTIVES

- **A** Find the intersection of two sets.
- **B** Solve compound inequalities containing "**and**."
- **C** Find the union of two sets.
- **D** Solve compound inequalities containing "**or**."

Two inequalities joined by the words **and** or **or** are called **compound inequalities**.

Compound Inequalities

$$x + 3 < 8 \text{ and } x > 2$$

$$\frac{2x}{3} \geq 5 \text{ or } -x + 10 < 7$$

A Finding the Intersection of Two Sets

The solution set of a compound inequality formed by the word **and** is the **intersection** of the solution sets of the two inequalities.

> **Intersection of Two Sets**
>
> The intersection of two sets, A and B, is the set of all elements common to both sets. A intersect B is denoted by
>
> $A \cap B$

EXAMPLE 1 Find the intersection: $\{2, 4, 6, 8\} \cap \{3, 4, 5, 6\}$

Solution: The numbers 4 and 6 are in both sets. The intersection is $\{4, 6.\}$

Practice Problem 1

Find the intersection:

$$\{1, 2, 3, 4, 5\} \cap \{3, 4, 5, 6\}$$

B Solving Compound Inequalities Containing "and"

A value of x is a solution of a compound inequality formed by the word **and** if it is a solution of *both* inequalities. For example, the solution set of the compound inequality $x \leq 5$ and $x \geq 3$ contains all values of x that make the inequality $x \leq 5$ a true statement **and** the inequality $x \geq 3$ a true statement. The first graph shown here is the graph of $x \leq 5$, the second graph is the graph of $x \geq 3$, and the third graph shows the intersection of the two graphs. The third graph is the graph of $x \leq 5$ **and** $x \geq 3$.

$\{x|x \leq 5\}$ — $(-\infty, 5]$

$\{x|x \geq 3\}$ — $[3, \infty)$

$\{x|x \leq 5 \text{ and } x \geq 3\}$ — $[3, 5]$

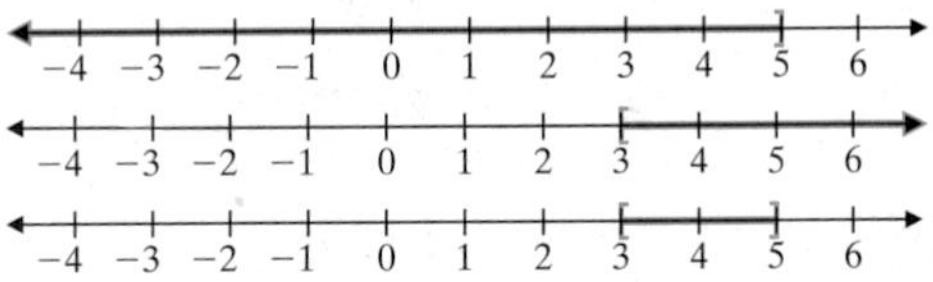

In interval notation, the set $\{x|x \leq 5 \text{ and } x \geq 3\}$ is written as $[3, 5]$.

EXAMPLE 2 Solve: $x - 7 < 2$ and $2x + 1 < 9$

Solution: First we solve each inequality separately.

$$x - 7 < 2 \text{ and } 2x + 1 < 9$$
$$x < 9 \text{ and } \quad 2x < 8$$
$$x < 9 \text{ and } \quad x < 4$$

Now we can graph the two intervals on two number lines and find their intersection.

Practice Problem 2

Solve: $x + 5 < 9$ and $3x - 1 < 2$

Answers

1. $\{3, 4, 5\}$, **2.** $(-\infty, 1)$

$\{x|x < 9\}$ $(-\infty, 9)$

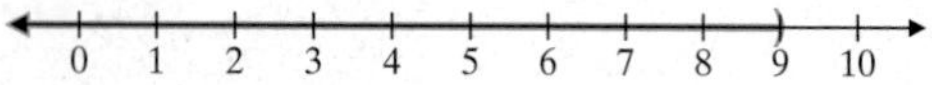

$\{x|x < 4\}$ $(-\infty, 4)$

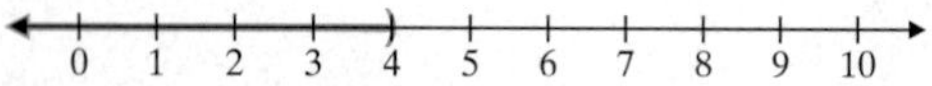

$\{x|x < 9 \text{ and } x < 4\}$ $(-\infty, 4)$
$= \{x|x < 4\}$

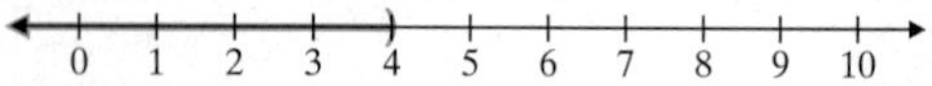

The solution set is $(-\infty, 4)$. ●

Practice Problem 3

Solve: $4x \geq 0$ and $2x + 4 \geq 2$

EXAMPLE 3 Solve: $2x \geq 0$ and $4x - 1 \leq -9$

Solution: First we solve each inequality separately.

$$2x \geq 0 \text{ and } 4x - 1 \leq -9$$
$$x \geq 0 \text{ and } \quad 4x \leq -8$$
$$x \geq 0 \text{ and } \quad x \leq -2$$

Now we can graph the two intervals and find their intersection.

$\{x|x \geq 0\}$ 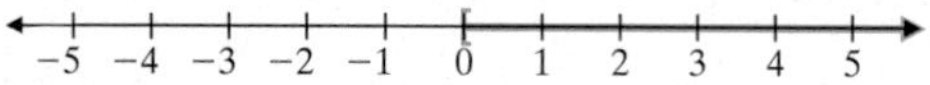 $[0, \infty)$

$\{x|x \leq -2\}$ 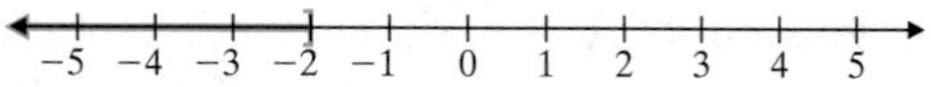 $(-\infty, -2]$

$\{x|x \geq 0 \text{ and } x \leq -2\}$ 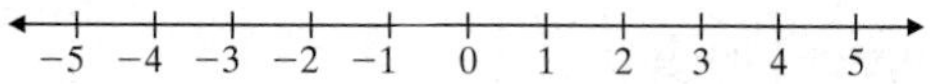
$= \emptyset$

There is no number that is greater than or equal to 0 **and** less than or equal to -2. The solution set is $\emptyset$. ●

Some compound inequalities containing the word **and** can be written in a more compact form. The compound inequality $2 \leq x$ and $x \leq 6$ can be written as

$$2 \leq x \leq 6$$

Recall from Section 2.8 that the graph of $2 \leq x \leq 6$ is all numbers between 2 and 6, including 2 and 6.

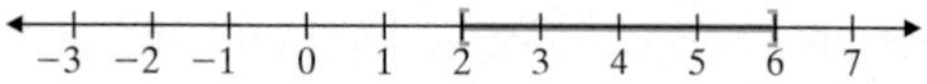

The set $\{x|2 \leq x \leq 6\}$ written in interval notation is $[2, 6]$.

To solve a compound inequality like $2 < 4 - x < 7$, we get x alone in the middle. Since a compound inequality is really two inequalities in one statement, we must perform the same operation to all three parts of the inequality.

Practice Problem 4

Solve: $5 < 1 - x < 9$

EXAMPLE 4 Solve: $2 < 4 - x < 7$

Solution: To get x alone, we first subtract 4 from all three parts.

Helpful Hint

Don't forget to reverse both inequality symbols.

$$2 < 4 - x < 7$$
$$2 - 4 < 4 - x - 4 < 7 - 4 \quad \text{Subtract 4 from all three parts.}$$
$$-2 < -x < 3 \quad \text{Simplify.}$$
$$\frac{-2}{-1} > \frac{-x}{-1} > \frac{3}{-1} \quad \text{Divide all three parts by } -1 \text{ and reverse the inequality symbols.}$$
$$2 > x > -3$$

This is equivalent to $-3 < x < 2$, and its graph is shown.

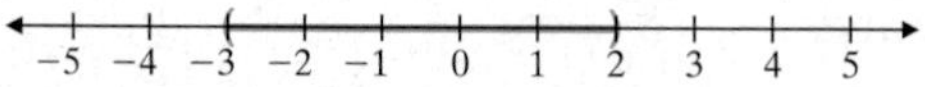

The solution set in interval notation is $(-3, 2)$. ●

Answers

3. $[0, \infty)$, **4.** $(-8, -4)$

EXAMPLE 5 Solve: $-1 \le \frac{2x}{3} + 5 \le 2$

Solution: First we clear the inequality of fractions by multiplying all three parts by the LCD, 3.

$$-1 \le \frac{2x}{3} + 5 \le 2$$

$$3(-1) \le 3\left(\frac{2x}{3} + 5\right) \le 3(2)$$ Multiply all three parts by the LCD, 3.

$$-3 \le 2x + 15 \le 6$$ Use the distributive property and multiply.

$$-3 - 15 \le 2x + 15 - 15 \le 6 - 15$$ Subtract 15 from all three parts.

$$-18 \le 2x \le -9$$ Simplify.

$$\frac{-18}{2} \le \frac{2x}{2} \le \frac{-9}{2}$$ Divide all three parts by 2.

$$-9 \le x \le -\frac{9}{2}$$ Simplify.

The graph of the solution is shown.

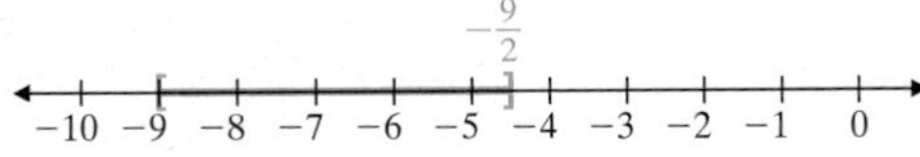

The solution set in interval notation is $\left[-9, -\frac{9}{2}\right]$.

C Finding the Union of Two Sets

The solution set of a compound inequality formed by the word **or** is the **union** of the solution sets of the two inequalities.

Union of Two Sets

The union of two sets, A and B, is the set of elements that belong to *either* of the sets. A union B is denoted by

$$A \cup B$$

EXAMPLE 6 Find the union: $\{2, 4, 6, 8\} \cup \{3, 4, 5, 6\}$

Solution: The numbers that are in either set or both sets are $\{2, 3, 4, 5, 6, 8\}$. This set is the union.

D Solving Compound Inequalities Containing "or"

A value of x is a solution of a compound inequality formed by the word **or** if it is a solution of **either** inequality. For example, the solution set of the compound inequality $x \le 1$ **or** $x \ge 3$ contains all numbers that make the inequality $x \le 1$ a true statement **or** the inequality $x \ge 3$ a true statement.

Practice Problem 5

Solve: $-3 \le \frac{x}{2} + 1 \le 5$

Practice Problem 6

Find the union:

$$\{1, 2, 3, 4, 5\} \cup \{3, 4, 5, 6\}$$

Answers

5. $[-8, 8]$, **6.** $\{1, 2, 3, 4, 5, 6\}$

$(-\infty, 1]$

$[3, \infty)$

$(-\infty, 1] \cup [3, \infty)$

In interval notation, the set $\{x|x \leq 1 \text{ or } x \geq 3\}$ is written as $(-\infty, 1] \cup [3, \infty)$.

Practice Problem 7

Solve: $3x - 2 \geq 10$ or $x - 6 \leq -4$

EXAMPLE 7 Solve: $5x - 3 \leq 10$ or $x + 1 \geq 5$

Solution: First we solve each inequality separately.

$$5x - 3 \leq 10 \text{ or } x + 1 \geq 5$$
$$5x \leq 13 \text{ or } \quad x \geq 4$$
$$x \leq \frac{13}{5} \text{ or } \quad x \geq 4$$

Now we can graph each interval and find their union.

$\left\{x|x \leq \frac{13}{5}\right\}$ $\left(-\infty, \frac{13}{5}\right]$

$\{x|x \geq 4\}$ $[4, \infty)$

$\left\{x|x \leq \frac{13}{5} \text{ or } x \geq 4\right\}$ $\left(-\infty, \frac{13}{5}\right] \cup [4, \infty)$

The solution set is $\left(-\infty, \frac{13}{5}\right] \cup [4, \infty)$.

Practice Problem 8

Solve: $x - 7 \leq -1$ or $2x - 6 \geq 2$

EXAMPLE 8 Solve: $-2x - 5 < -3$ or $6x < 0$

Solution: First we solve each inequality separately.

$$-2x - 5 < -3 \text{ or } 6x < 0$$
$$-2x < 2 \text{ or } x < 0$$
$$x > -1 \text{ or } x < 0$$

Now we can graph each interval and find their union.

$\{x|x > -1\}$ $(-1, \infty)$

$\{x|x < 0\}$ $(-\infty, 0)$

$\{x|x > -1 \text{ or } x < 0\}$ $(-\infty, \infty)$
= all real numbers

The solution set is $(-\infty, \infty)$.

Try the Concept Check in the margin.

Concept Check

Which of the following is *not* a correct way to represent the set of all numbers between -3 and 5?

a. $\{x|-3 < x < 5\}$
b. $-3 < x$ or $x < 5$
c. $(-3, 5)$
d. $x > -3$ and $x < 5$

Answers

7. $(-\infty, 2] \cup [4, \infty)$, **8.** $(-\infty, \infty)$

Concept Check: b is not correct

Name ______________________ Section __________ Date __________

EXERCISE SET APPENDIX D

If $A = \{x|x \text{ is an even integer}\}$, $B = \{x|x \text{ is an odd integer}\}$, $C = \{2, 3, 4, 5\}$, *and* $D = \{4, 5, 6, 7\}$, *list the elements of each set. See Example 1.*

1. $A \cap C$ **2.** $B \cap D$ **3.** $A \cap B$ **4.** $C \cap D$ **5.** $B \cap C$ **6.** $A \cap D$

B *Solve each compound inequality. Graph the two inequalities on the first two number lines and the solution set on the third number line. See Examples 2 and 3.*

7. $x < 1$ and $x > -3$

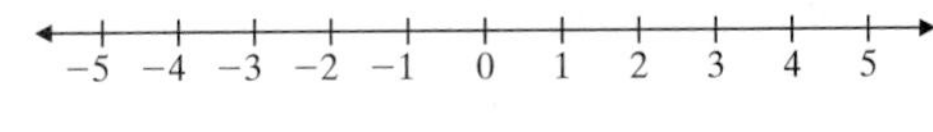

8. $x \le 0$ and $x \ge -2$

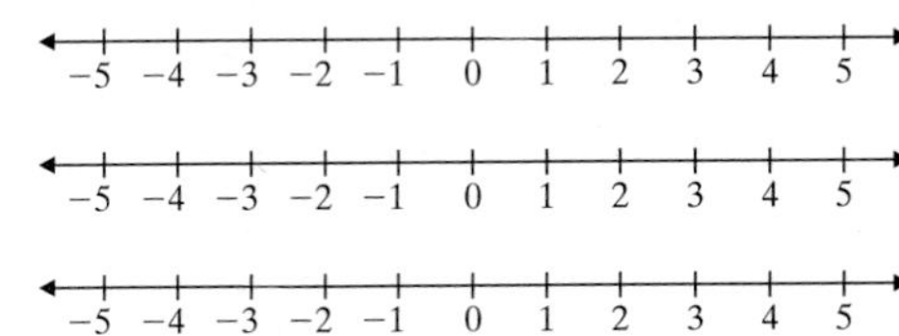

9. $x \le -3$ and $x \ge -2$

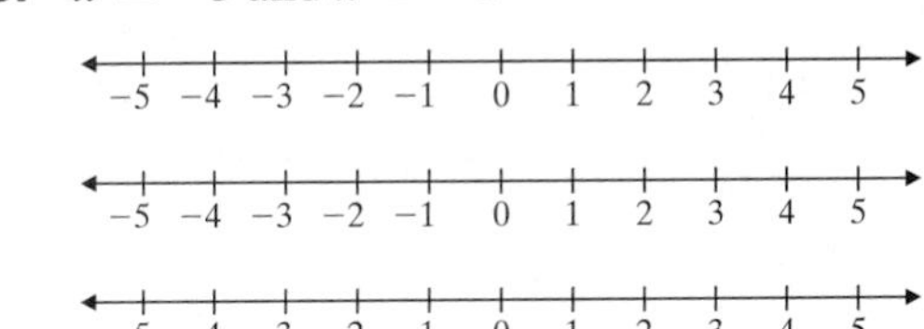

10. $x < 2$ and $x > 4$

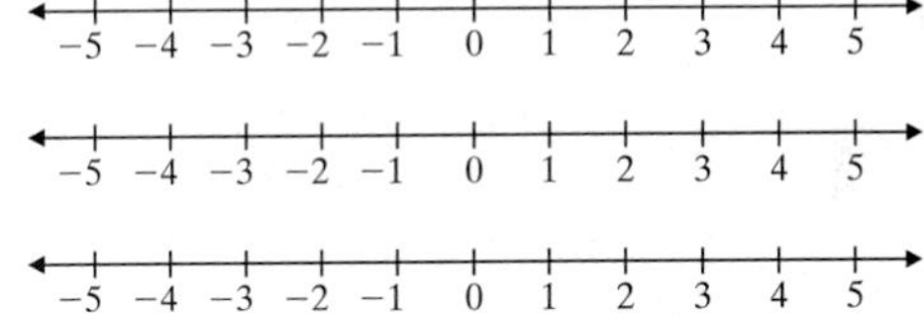

11. $x < -1$ and $x < 1$

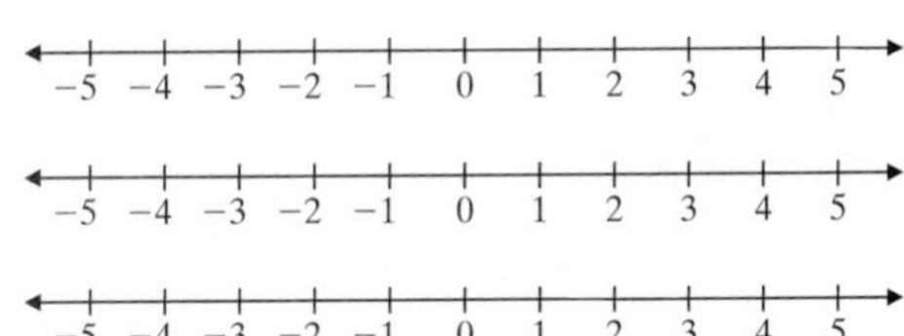

12. $x \ge -4$ and $x > 1$

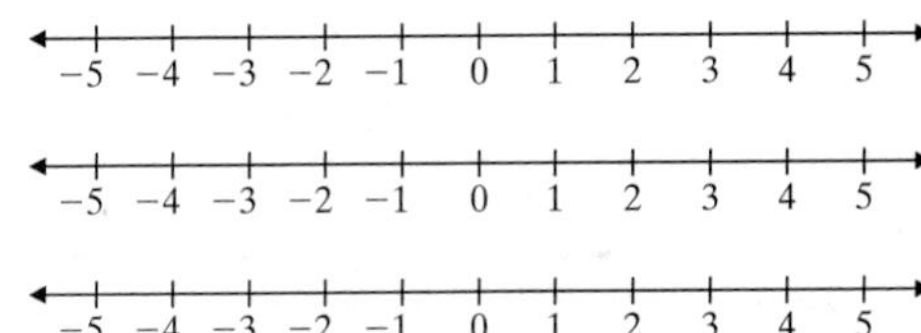

Solve each compound inequality. See Examples 2 and 3.

13. $x < 5$ and $x > -2$

14. $x \le 7$ and $x \le 1$

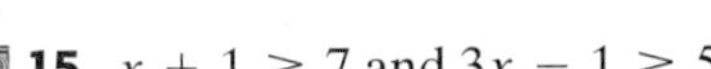

15. $x + 1 \ge 7$ and $3x - 1 \ge 5$

16. $-2x < -8$ and $x - 5 < 5$

17. $4x + 2 \le -10$ and $2x \le 0$

18. $x + 4 > 0$ and $4x > 0$

19. $x + 3 \ge 3$ and $x + 3 \le 2$

20. $2x - 1 \ge 3$ and $-x > 2$

Solve each compound inequality. See Examples 4 and 5.

21. $5 < x - 6 < 11$

22. $-2 \le x + 3 \le 0$

23. $-2 \le 3x - 5 \le 7$

24. $1 < 4 + 2x < 7$

25. $1 \le \frac{2}{3}x + 3 \le 4$

26. $-2 < \frac{1}{2}x - 5 < 1$

27. $-5 \leq \frac{x + 1}{4} \leq -2$ **28.** $-4 \leq \frac{2x + 5}{3} \leq 1$ **29.** $0 \leq 2x - 3 \leq 9$

30. $3 < 5x + 1 < 11$ **31.** $-6 < 3(x - 2) \leq 8$ **32.** $-5 < 2(x + 4) < 8$

C *If $A = \{1, 2, 3, 4, 5, 6, 7, 8\}$, $B = \{1, 5\}$, $C = \{2, 4, 6, 8\}$, and $D = \{6\}$, list the elements of each set. See Example 6.*

33. $A \cup B$ **34.** $A \cup C$ **35.** $B \cup D$ **36.** $B \cup C$ **37.** $C \cup D$ **38.** $D \cup B$

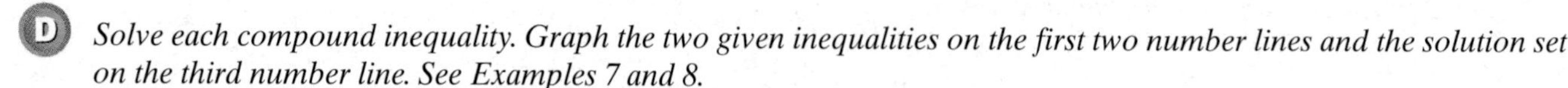

D *Solve each compound inequality. Graph the two given inequalities on the first two number lines and the solution set on the third number line. See Examples 7 and 8.*

39. $x < 4$ or $x < 5$

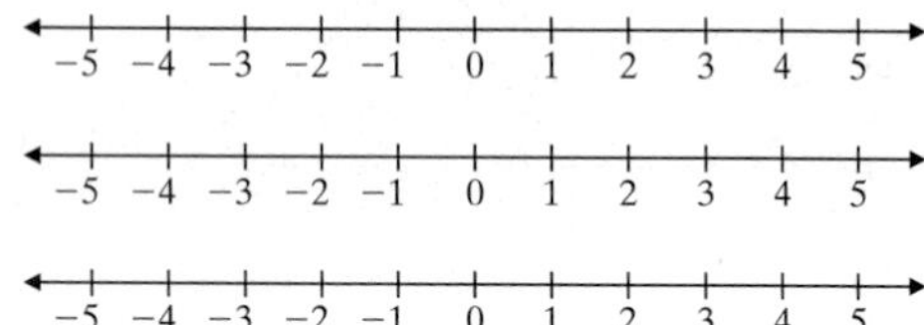

40. $x \geq -2$ or $x \leq 2$

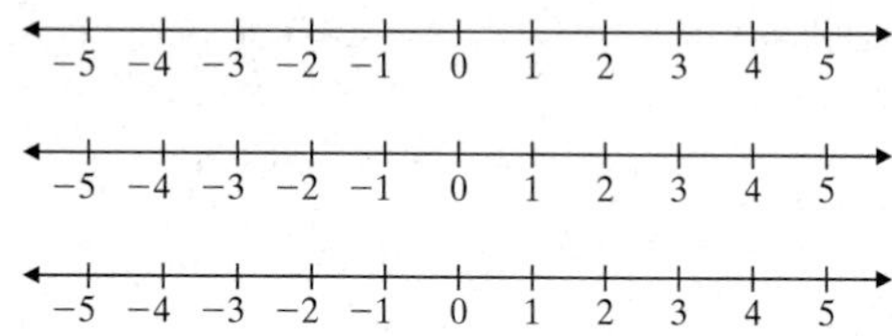

41. $x \leq -4$ or $x \geq 1$

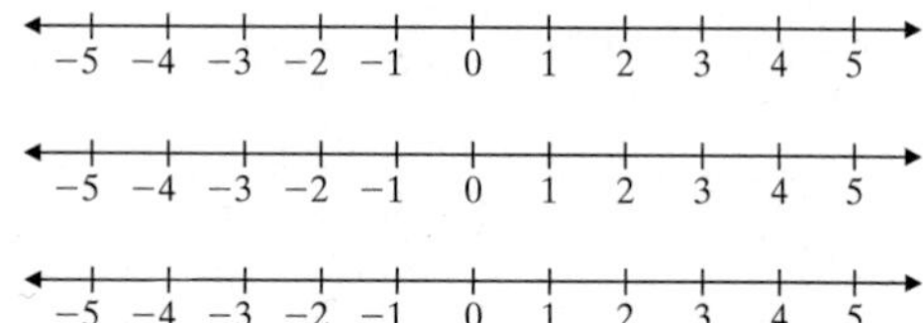

42. $x < 0$ or $x < 1$

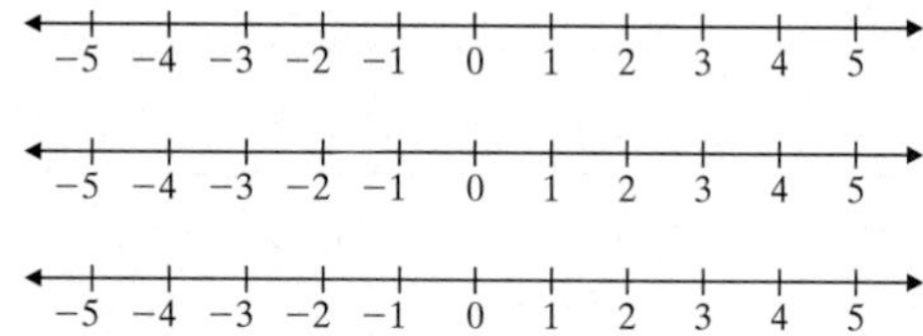

43. $x > 0$ or $x < 3$

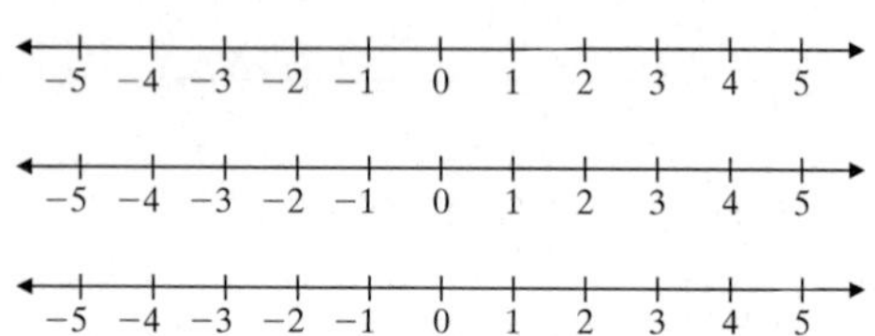

44. $x \geq -3$ or $x \leq -4$

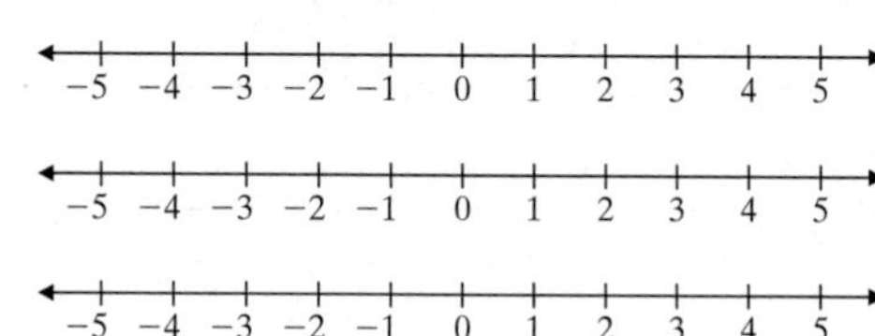

Solve each compound inequality. See Examples 7 and 8.

45. $x < -1$ or $x > 0$ **46.** $x \leq 1$ or $x \leq -3$ **47.** $-2x \leq -4$ or $5x - 20 \geq 5$

48. $x + 4 < 0$ or $6x > -12$ **49.** $3(x - 1) < 12$ or $x + 7 > 10$ **50.** $5(x - 1) \geq -5$ or $5 - x \leq 11$

51. $3x + 2 \leq 5$ or $7x > 29$ **52.** $-x < 7$ or $3x + 1 < -20$ **53.** $3x \geq 5$ or $-x - 6 < 1$

54. $\frac{3}{8}x + 1 \leq 0$ or $-2x < -4$ **55.** $6x - 4 > 2x$ or $4x - 1 < x + 5$ **56.** $6x - 2 > 5x + 3$ or $4x - 3 < x$

APPENDIX E

Absolute Value Equations and Inequalities

OBJECTIVES

Ⓐ Solve absolute value equations.

Ⓑ Solve absolute value inequalities.

In Chapter 1, we defined the absolute value of a number as its distance from 0 on a number line.

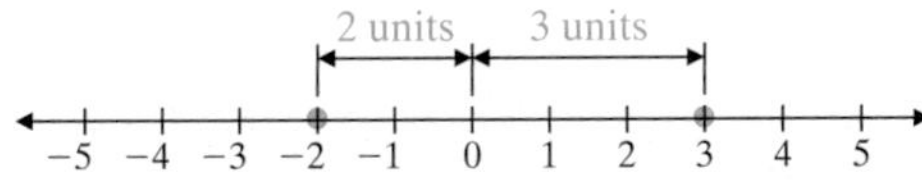

$$|-2| = 2 \quad \text{and} \quad |3| = 3$$

In this section, we concentrate on solving equations and inequalities containing the absolute value of a variable or a variable expression. Examples of absolute value equations and inequalities are

$$|x| = 3 \quad -5 \geq |2y + 7| \quad |z - 6.7| = |3z + 1.2| \quad |x - 3| > 7$$

Absolute value equations and inequalities are extremely useful in data analysis, especially for calculating acceptable measurement error and errors that result from the way numbers are sometimes represented in computers.

Ⓐ Solving Absolute Value Equations

To begin, let's solve a few absolute value equations by inspection.

EXAMPLE 1 Solve: $|x| = 3$

Solution: The solution set of this equation will contain all numbers whose distance from 0 is 3 units. Two numbers are 3 units away from 0 on the number line: 3 and −3.

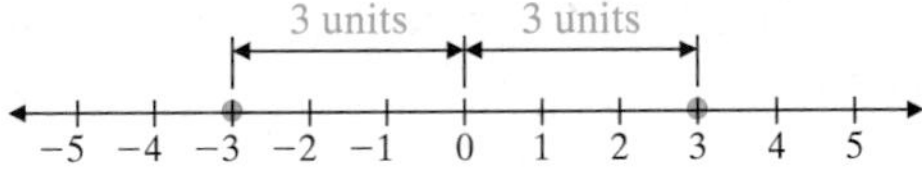

Check: To check, let $x = 3$ and $x = -3$ in the original equation.

$\lvert x\rvert = 3$		$\lvert x\rvert = 3$	
$\lvert 3\rvert \stackrel{?}{=} 3$	Let $x = 3$.	$\lvert -3\rvert \stackrel{?}{=} 3$	Let $x = -3$.
$3 = 3$	True.	$3 = 3$	True.

Both solutions check. Thus the solution set of the equation $|x| = 3$ is $\{3, -3\}$. ●

Practice Problem 1

Solve: $|y| = 5$

EXAMPLE 2 Solve: $|x| = -2$

Solution: The absolute value of a number is never negative, so this equation has no solution. The solution set is { } or ∅. ●

Practice Problem 2

Solve: $|p| = -4$

EXAMPLE 3 Solve: $|y| = 0$

Solution: We are looking for all numbers whose distance from 0 is zero units. The only number is 0. The solution set is $\{0\}$. ●

Practice Problem 3

Solve: $|x| = 0$

Answers

1. $\{-5, 5\}$, **2.** ∅, **3.** $\{0\}$

From the above examples, we have the following.

Absolute Value Property

Solve $|X| = a$ as follows.
If a is positive, then solve $X = a$ or $X = -a$.
If a is 0, then $X = 0$.
If a is negative, the equation $|X| = a$ has no solution.

Helpful Hint

For the equation $|X| = a$ in the box above, X can be a single variable or a variable expression.

When we are solving absolute value equations, if $|X|$ is not alone on one side of the equation we first use properties of equality to get $|X|$ alone.

Practice Problem 4

Solve: $3|y| - 4 = 17$

EXAMPLE 4 Solve: $2|x| + 25 = 37$

Solution: First we get $|x|$ alone.

$$2|x| + 25 = 37$$
$$2|x| = 12 \quad \text{Subtract 25 from both sides.}$$
$$|x| = 6 \quad \text{Divide both sides by 2.}$$
$$x = 6 \quad \text{or} \quad x = -6 \quad \text{Use the absolute value property.}$$

The solution set is $\{-6, 6\}$. ●

If the expression inside the absolute value bars is more complicated than a single variable x, we can still use the absolute value property.

Practice Problem 5

Solve: $|x - 4| = 11$

EXAMPLE 5 Solve: $|w + 3| = 7$

Solution: If we think of the expression $w + 3$ as X in the absolute value property, we have that

$$|w + 3| = 7$$
$$w + 3 = 7 \quad \text{or} \quad w + 3 = -7 \quad \text{Use the absolute value property.}$$
$$w = 4 \quad \text{or} \quad w = -10$$

The solution set is $\{4, -10\}$. ●

Don't forget that to use the absolute value property you must first make sure that the absolute value expression is alone on one side of the equation.

Helpful Hint

If the equation has a single absolute value expression containing variables, get the absolute value expression alone. Then use the absolute value property.

Practice Problem 6

Solve: $|4x + 2| + 1 = 7$

Answers

4. $\{-7, 7\}$, **5.** $\{15, -7\}$, **6.** $\{1, -2\}$

EXAMPLE 6 Solve: $|2x - 1| + 5 = 6$

Solution: We want the absolute value expression alone on one side of the equation, so we begin by subtracting 5 from both sides. Then we use the absolute value property.

$$|2x - 1| + 5 = 6$$
$$|2x - 1| = 1 \quad \text{Subtract 5 from both sides.}$$
$$2x - 1 = 1 \quad \text{or} \quad 2x - 1 = -1 \quad \text{Use the absolute value property.}$$
$$2x = 2 \quad \text{or} \quad 2x = 0$$
$$x = 1 \quad \text{or} \quad x = 0 \quad \text{Solve.}$$

The solution set is $\{0, 1\}$.

Given two absolute value expressions, we might ask, when are the absolute values of two expressions equal? To see the answer, notice that

$$|2| = |2| \quad |-2| = |-2| \quad |-2| = |2| \quad |2| = |-2|$$

same   same   opposites   opposites

Two absolute value expressions are equal when the expressions inside the absolute value bars are equal to or are opposites of each other.

EXAMPLE 7 Solve: $|3x + 2| = |5x - 8|$

Solution: This equation is true if the expressions inside the absolute value bars are equal to or are opposites of each other.

$$3x + 2 = 5x - 8 \quad \text{or} \quad 3x + 2 = -(5x - 8)$$

Next we solve each equation.

$$3x + 2 = 5x - 8 \quad \text{or} \quad 3x + 2 = -5x + 8$$
$$-2x + 2 = -8 \quad \text{or} \quad 8x + 2 = 8$$
$$-2x = -10 \quad \text{or} \quad 8x = 6$$
$$x = 5 \quad \text{or} \quad x = \frac{3}{4}$$

Check to see that replacing x with 5 or with $\frac{3}{4}$ results in a true statement.

The solution set is $\left\{\frac{3}{4}, 5\right\}$.

EXAMPLE 8 Solve: $|x - 3| = |5 - x|$

Solution:

$$x - 3 = 5 - x \quad \text{or} \quad x - 3 = -(5 - x)$$
$$2x - 3 = 5 \quad \text{or} \quad x - 3 = -5 + x$$
$$2x = 8 \quad \text{or} \quad x - 3 - x = -5 + x - x$$
$$x = 4 \quad \text{or} \quad -3 = -5 \quad \text{False.}$$

Recall from Section 2.1 that when an equation simplifies to a false statement, the equation has no solution. Thus the only solution for the original absolute value equation is 4, and the solution set is $\{4\}$.

Try the Concept Check in the margin.

B Solving Absolute Value Inequalities

To begin, let's solve a few absolute value inequalities by inspection.

Practice Problem 7

Solve: $|4x - 5| = |3x + 5|$

Practice Problem 8

Solve: $|x + 2| = |4 - x|$

Concept Check

True or false? Absolute value equations always have two solutions. Explain your answer.

Answers

7. $\{0, 10\}$, **8.** $\{1\}$

Concept Check: false; answers may vary

Practice Problem 9

Solve $|x| < 4$ using a number line.

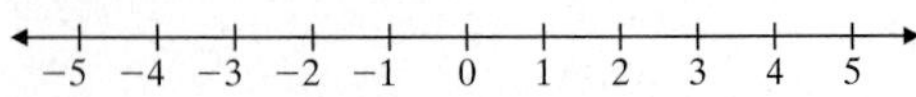

EXAMPLE 9 Solve $|x| < 2$ using a number line.

Solution: The solution set contains all numbers whose distance from 0 is less than 2 units on the number line.

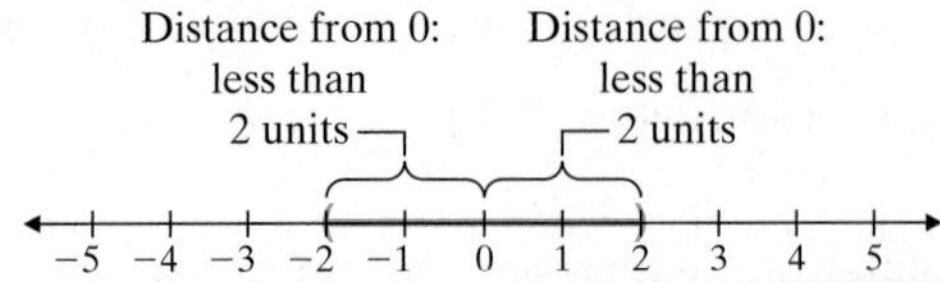

The solution set is $\{x|-2 < x < 2\}$, or $(-2, 2)$ in interval notation.

Practice Problem 10

Solve $|x| \geq 5$ using a number line.

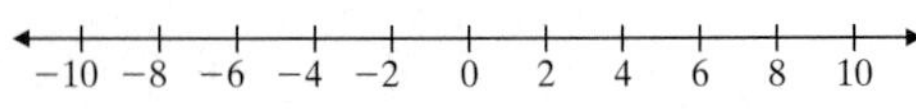

EXAMPLE 10 Solve $|x| \geq 3$ using a number line.

Solution: The solution set contains all numbers whose distance from 0 is 3 or more units. Thus the graph of the solution set contains 3 and all points to the right of 3 on the number line or -3 and all points to the left of -3 on the number line.

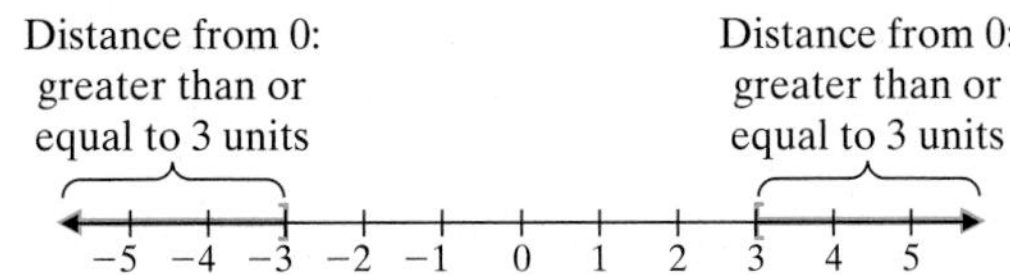

This solution set is $\{x|x \leq -3 \text{ or } x \geq 3\}$. In interval notation, the solution set is $(-\infty, -3] \cup [3, \infty)$, since **or** means union.

The following box summarizes solving absolute value equations and inequalities.

Solving Absolute Value Equations and Inequalities

If a is a positive number,

To solve $|X| = a$, solve $X = a$ or $X = -a$.

To solve $|X| < a$, solve $-a < X < a$.

To solve $|X| > a$, solve $X < -a$ or $X > a$.

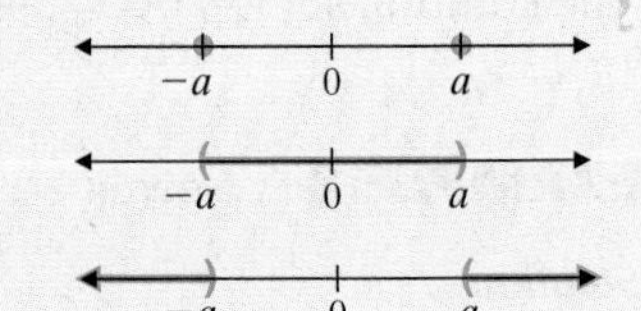

Practice Problem 11

Solve: $|x + 2| > 4$. Graph the solution set.

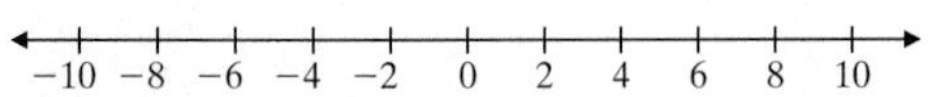

EXAMPLE 11 Solve: $|x - 3| > 7$

Solution: Since 7 is positive, to solve $|x - 3| > 7$, we solve the compound inequality $x - 3 < -7$ or $x - 3 > 7$.

$$x - 3 < -7 \quad \text{or} \quad x - 3 > 7$$
$$x < -4 \quad \text{or} \quad x > 10 \qquad \text{Add 3 to both sides.}$$

The solution set is $\{x|x < -4 \text{ or } x > 10\}$ or $(-\infty, -4) \cup (10, \infty)$ in interval notation. Its graph is shown.

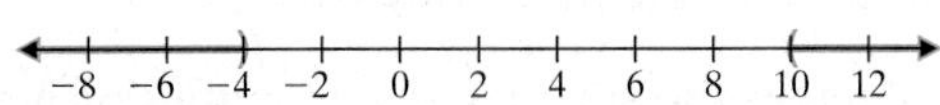

Answers

9.

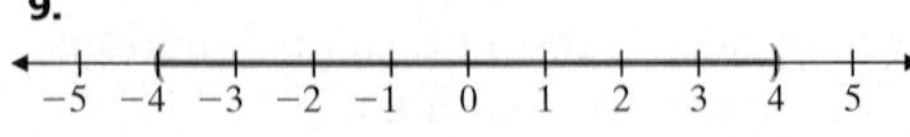

10.

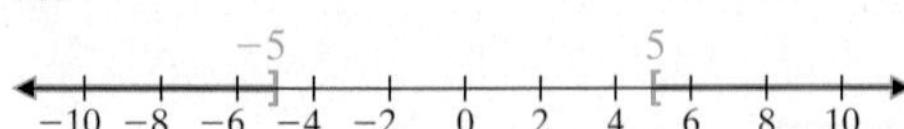

11. $(-\infty, -6) \cup (2, \infty)$

Let's review the differences in solving absolute value equations and inequalities by solving an absolute value equation.

EXAMPLE 12 Solve: $|x + 1| = 6$

Solution: This is an equation, so we solve

$$x + 1 = 6 \quad \text{or} \quad x + 1 = -6$$
$$x = 5 \quad \text{or} \quad x = -7$$

The solution set is $\{-7, 5\}$. Its graph is shown.

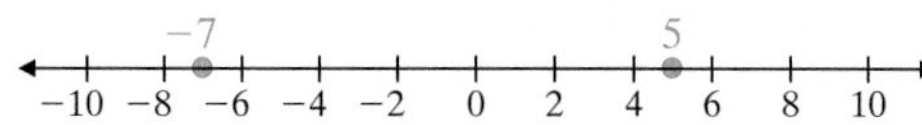

EXAMPLE 13 Solve: $|x - 6| \le 2$

Solution: To solve $|x - 6| \le 2$, we solve

$$-2 \le x - 6 \le 2$$
$$-2 + 6 \le x - 6 + 6 \le 2 + 6 \qquad \text{Add 6 to all three parts.}$$
$$4 \le x \le 8 \qquad \text{Simplify.}$$

The solution set is $\{x|4 \le x \le 8\}$, or $[4, 8]$ in interval notation. Its graph is shown.

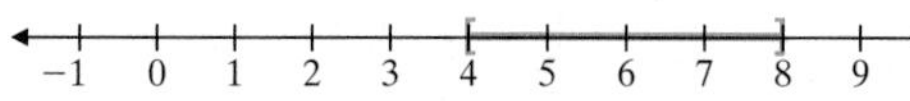

Helpful Hint

Before using an absolute value inequality property, get an absolute value expression alone on one side of the inequality.

EXAMPLE 14 Solve: $|5x + 1| + 1 \le 10$

Solution: First we get the absolute value expression alone by subtracting 1 from both sides.

$$|5x + 1| + 1 \le 10$$
$$|5x + 1| \le 10 - 1 \qquad \text{Subtract 1 from both sides.}$$
$$|5x + 1| \le 9 \qquad \text{Simplify.}$$

Since 9 is positive, to solve $|5x + 1| \le 9$, we solve

$$-9 \le 5x + 1 \le 9$$
$$-9 - 1 \le 5x + 1 - 1 \le 9 - 1 \qquad \text{Subtract 1 from all three parts.}$$
$$-10 \le 5x \le 8 \qquad \text{Simplify.}$$
$$-2 \le x \le \frac{8}{5} \qquad \text{Divide all three parts by 5.}$$

The solution set is $\left[-2, \frac{8}{5}\right]$.

The next few examples are special cases of absolute value inequalities.

Practice Problem 12

Solve: $|x - 3| = 5$. Graph the solution set.

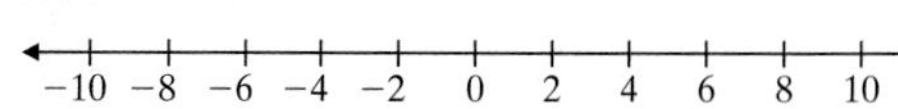

Practice Problem 13

Solve: $|x - 2| \le 1$. Graph the solution set.

−5 −4 −3 −2 −1 0 1 2 3 4 5

Practice Problem 14

Solve: $|2x - 5| + 2 \le 9$

Answers

12. $\{-2, 8\}$

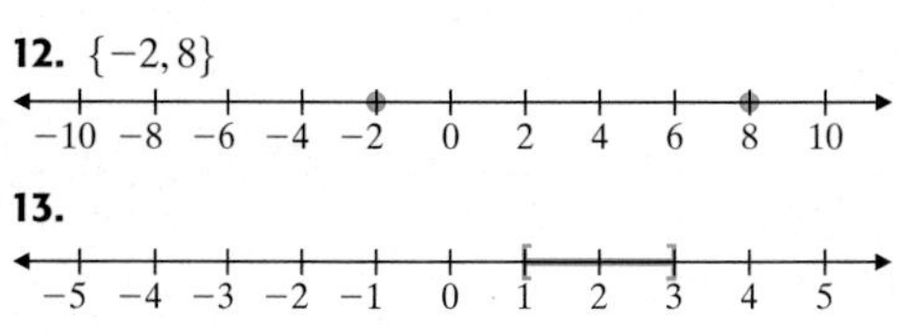

14. $[-1, 6]$

Practice Problem 15

Solve: $|x| < -1$

EXAMPLE 15 Solve: $|x| \leq -3$

Solution: The absolute value of a number is never negative. Thus it will then never be less than or equal to -3. The solution set is $\{\ \}$ or $\varnothing$.

Practice Problem 16

Solve: $|x + 1| \geq -3$

EXAMPLE 16 Solve: $|x - 1| > -2$

Solution: The absolute value of a number is always nonnegative. Thus it will always be greater than -2. The solution set contains all real numbers, or $(-\infty, \infty)$.

Try the Concept Check in the margin.

Concept Check

Without taking any solution steps, how do you know that the absolute value inequality $|3x - 2| > -9$ has a solution? What is its solution?

Answers

15. $\varnothing$, **16.** $(-\infty, \infty)$

Concept Check: $(-\infty, \infty)$ since the absolute value is always nonnegative

STUDY SKILLS REMINDER

Are you organized?

Have you ever had trouble finding a completed assignment? When it's time to study for a test, are your notes neat and organized? Have you ever had trouble reading your own mathematics handwriting? (Be honest—I have had trouble reading my own handwriting before.)

When any of these things happen, it's time to get organized. Here are a few suggestions:

Write your notes and complete your homework assignments in a notebook with pockets (spiral or ring binder). Take class notes in this notebook, and then follow the notes with your completed homework assignment. When you receive graded papers or handouts, place them in the notebook pocket so that you will not lose them.

Place a mark (possibly an exclamation point) beside any note(s) that seem especially important to you. Also place a mark (possibly a question mark) beside any note(s) or homework that you are having trouble with. Don't forget to see your instructor, a tutor, or your fellow classmates to help you understand the concepts or exercises you have marked.

Also, if you are having trouble reading your own handwriting, *slow down* and write your mathematics work clearly!

Name ______________________ Section __________ Date __________

EXERCISE SET APPENDIX E

 *Solve. See Examples 1 through 6.*

 **1.** $|x| = 7$

2. $|y| = 15$

 **3.** $|x| = -4$

4. $|x| = -20$

5. $|3x| = 12.6$

6. $|6n| = 12.6$

7. $3|x| - 5 = 7$

8. $5|x| - 12 = 8$

9. $-6|x| + 44 = -10$

10. $-4|x| + 18 = -22$

11. $|x - 9| = 14$

12. $|x + 2| = 8$

13. $|2x - 5| = 9$

14. $|6 + 2n| = 4$

15. $\left|\frac{x}{2} - 3\right| = 1$

16. $\left|\frac{n}{3} + 2\right| = 4$

17. $|z| + 4 = 9$

18. $|x| + 1 = 3$

19. $|3x| + 5 = 14$

20. $|2x| - 6 = 4$

21. $\left|\frac{4x - 6}{3}\right| = 6$

22. $\left|\frac{2x + 1}{5}\right| = 7$

23. $|2x| = 0$

24. $|7z| = 0$

25. $|4n + 1| + 10 = 4$

26. $|3z - 2| + 8 = 1$

27. $3|x - 1| + 19 = 23$

28. $5|x + 1| - 1 = 3$

Solve. See Examples 7 and 8.

29. $|5x - 7| = |3x + 11|$

30. $|9y + 1| = |6y + 4|$

31. $|z + 8| = |z - 3|$

32. $|2x - 5| = |2x + 5|$

33. $|2y - 3| = |9 - 4y|$

34. $|5z - 1| = |7 - z|$

35. $\left|\frac{3}{4}x - 2\right| = \left|\frac{1}{4}x + 6\right|$

36. $\left|\frac{2}{3}x - 5\right| = \left|\frac{1}{3}x + 4\right|$

37. $|2x - 6| = |10 - 2x|$

38. $|4n + 5| = |4n + 3|$

39. $|x + 4| = |7 - x|$

40. $|8 - y| = |y + 2|$

41. $\left|\frac{2x + 1}{5}\right| = \left|\frac{3x - 7}{3}\right|$

42. $\left|\frac{5x - 1}{2}\right| = \left|\frac{4x + 5}{6}\right|$

43. $|5x + 1| = |4x - 7|$

44. $|3 + 6n| = |4n + 11|$

B *Solve. Graph the solution set. See Examples 9 through 16.*

45. $|x| \leq 4$

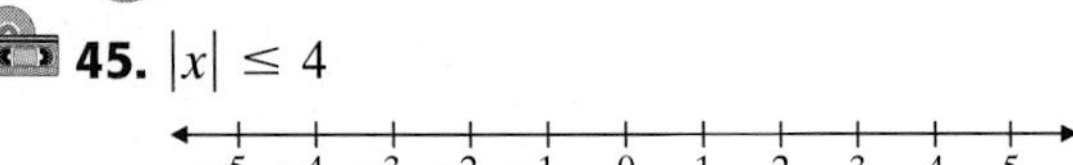

46. $|x| < 6$

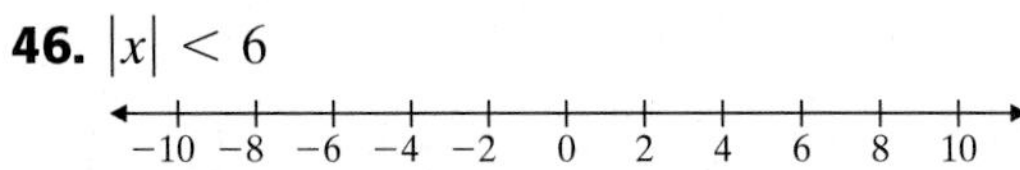

47. $|x| > 3$

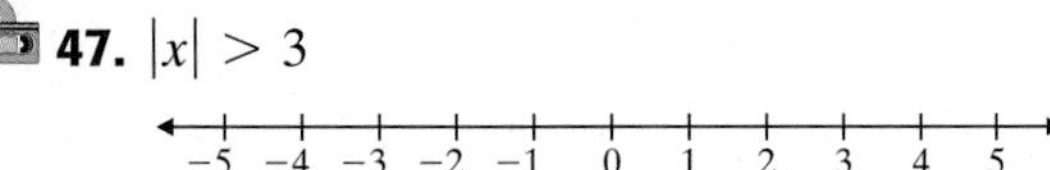

48. $|y| \geq 4$

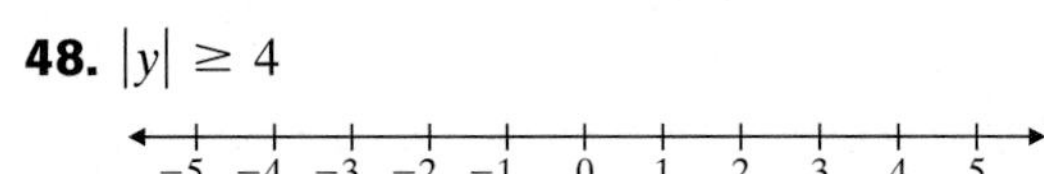

49. $|x + 3| < 2$

−5 −4 −3 −2 −1 0 1 2 3 4 5

50. $|x + 4| < 6$

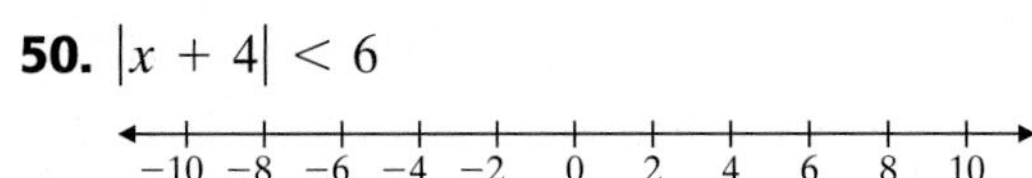

51. $|y - 6| \geq 7$

−25 −20 −15 −10 −5 0 5 10 15 20 25

52. $|x - 3| \geq 10$

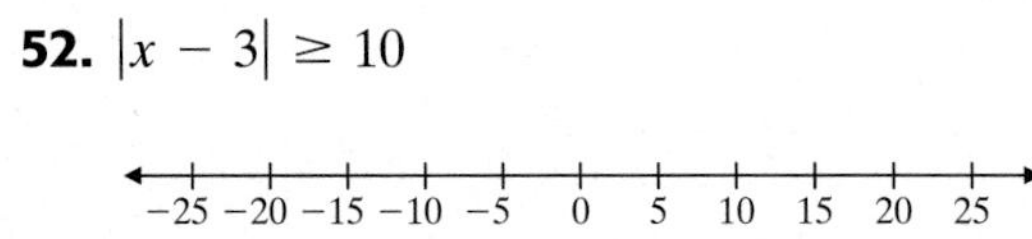

53. $\left|\frac{x + 2}{3}\right| < 1$

−5 −4 −3 −2 −1 0 1 2 3 4 5

54. $\left|\frac{x - 6}{4}\right| < 1$

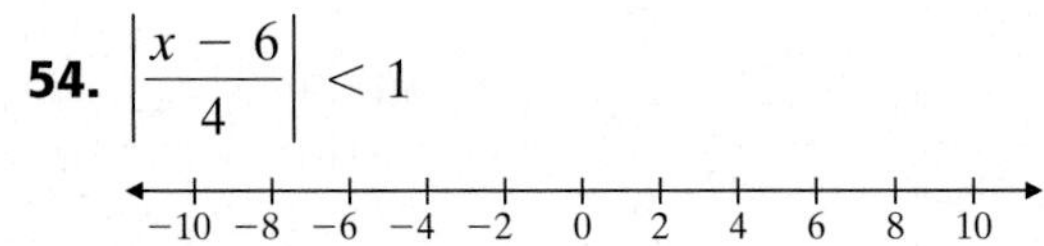

55. $|x| + 7 \leq 12$

−5 −4 −3 −2 −1 0 1 2 3 4 5

56. $|x| + 6 \leq 7$

−5 −4 −3 −2 −1 0 1 2 3 4 5

57. $|x| + 2 > 6$

58. $|x| - 1 > 3$

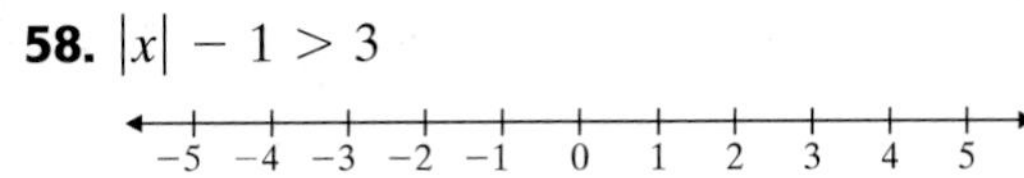

59. $|2x + 7| \leq 13$

60. $|5x - 3| \leq 18$

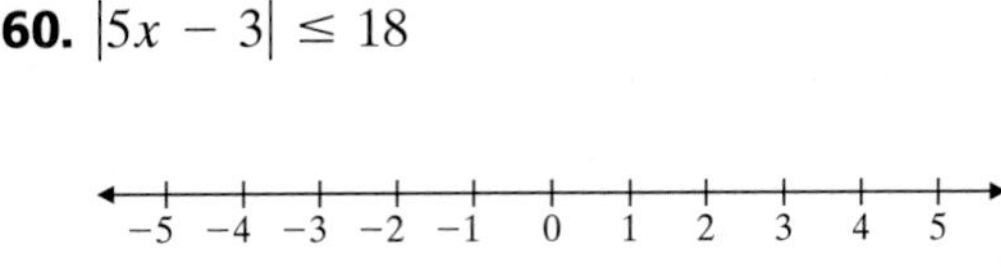

61. $|x + 10| \geq 14$

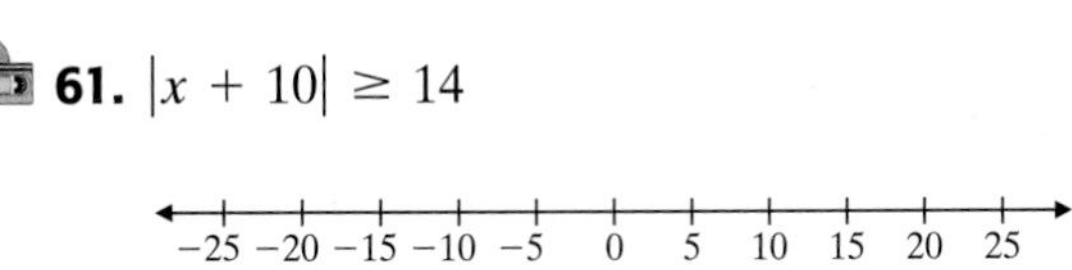

62. $|x - 9| \geq 2$

63. $|2x - 7| \leq 11$

64. $|5x + 2| < 8$

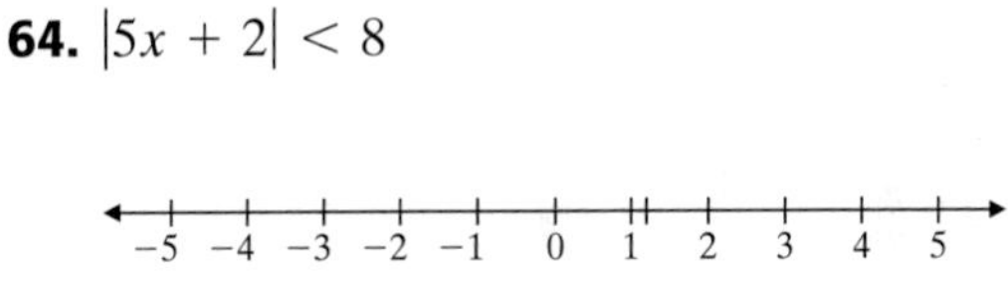

65. $|x| > -4$

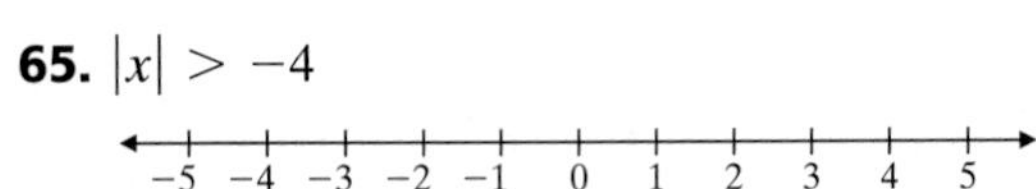

66. $|x| \leq -7$

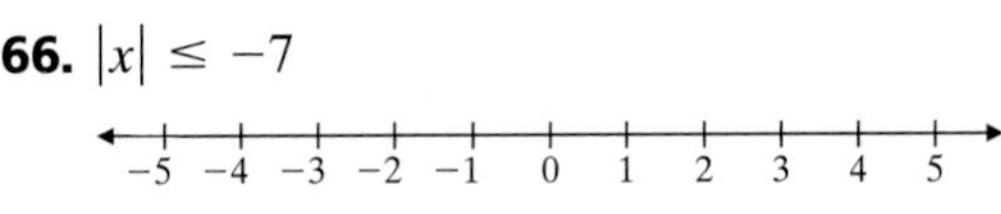

67. $6 + |4x - 1| \leq 9$

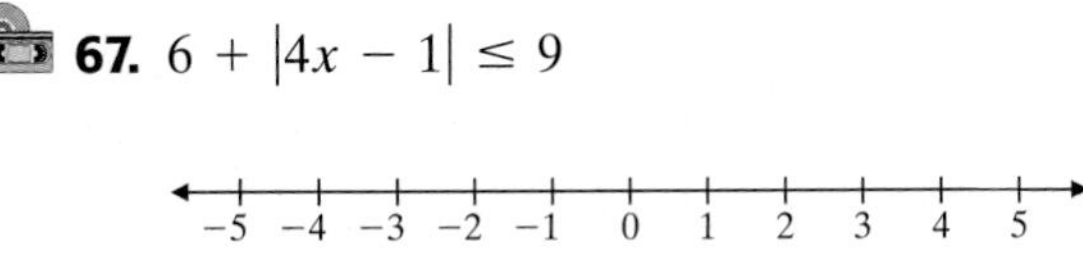

68. $-3 + |5x - 2| \leq 4$

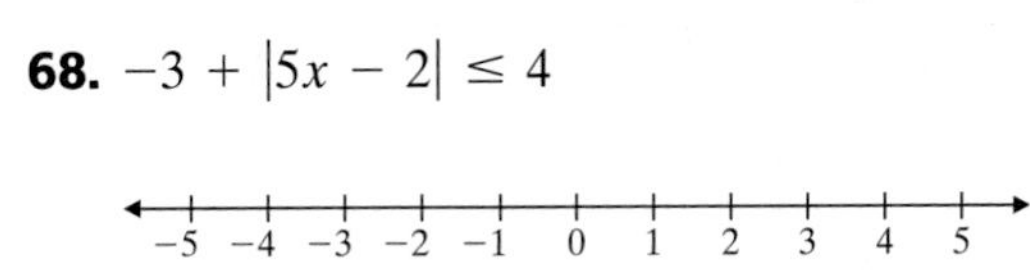

69. $|6x - 8| + 3 > 7$

70. $|10 + 3x| + 1 > 2$

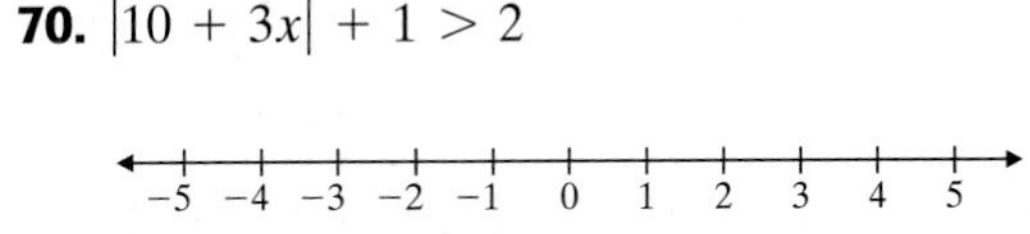

71. $|5x + 3| < -6$

72. $|4 + 9x| \geq -6$

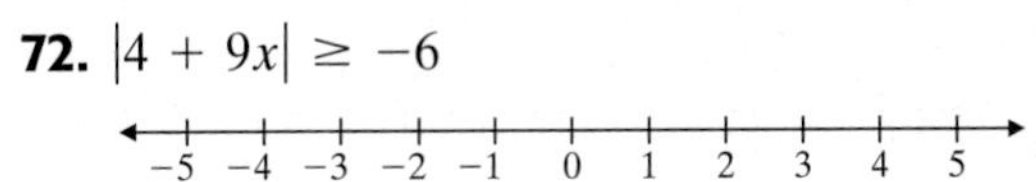

73. $\left|\frac{x+6}{3}\right| > 2$

−25 −20 −15 −10 −5 0 5 10 15 20 25

74. $\left|\frac{7+x}{2}\right| \geq 4$

−25 −20 −15 −10 −5 0 5 10 15 20 25

Solve each equation or inequality for x. See Examples 1 through 16.

75. $|x| = 13$

76. $|x| < 13$

77. $|x| > 13$

78. $|3x| = 12$

79. $|x| + 12 = 9$

80. $|x| - 4 = -9$

81. $2|x| - 9 \leq 11$

82. $4|x| - 2 \geq 6$

83. $|2x - 3| < 7$

84. $|2x - 3| > 7$

85. $|2x - 3| = 7$

86. $|5 - 6x| = 29$

87. $|x - 5| \geq 12$

88. $|x + 4| \geq 20$

89. $|9 + 4x| = 0$

90. $|9 + 4x| \geq 0$

91. $|2x + 1| + 4 < 7$

92. $8 + |5x - 3| \geq 11$

93. $\left|\frac{1}{3}x + 1\right| > 5$

94. $\left|\frac{1}{4}x - 2\right| < 1$

95. $|3x - 5| + 4 = 5$

96. $|x - 1| + 7 = 11$

97. $|x + 11| = -1$

98. $|4x - 4| = -3$

99. $\left|\frac{2x-1}{3}\right| = 6$

100. $\left|\frac{6-x}{4}\right| = 5$

101. $\left|\frac{3x-5}{6}\right| > 5$

102. $\left|\frac{4x-7}{5}\right| < 2$

103. $|6x - 3| = |4x + 5|$

104. $|3x + 1| = |4x + 10|$

APPENDIX F

Determinants and Cramer's Rule

We have solved systems of two linear equations in two variables in four different ways: graphically, by substitution, by elimination, and by matrices. Now we analyze another method called **Cramer's rule.**

OBJECTIVES

- (A) Define and evaluate a 2×2 determinant.
- (B) Use Cramer's rule to solve a system of two linear equations in two variables.
- (C) Define and evaluate a 3×3 determinant.
- (D) Use Cramer's rule to solve a linear system of three equations in three variables.

(A) Evaluating 2×2 Determinants

Recall that a matrix is a rectangular array of numbers. If a matrix has the same number of rows and columns, it is called a **square matrix.** Examples of square matrices are

$$\begin{bmatrix} 1 & 6 \\ 5 & 2 \end{bmatrix} \quad \begin{bmatrix} 2 & 4 & 1 \\ 0 & 5 & 2 \\ 3 & 6 & 9 \end{bmatrix}$$

A **determinant** is a real number associated with a square matrix. The determinant of a square matrix is denoted by placing vertical bars about the array of numbers. Thus,

The determinant of the square matrix $\begin{bmatrix} 1 & 6 \\ 5 & 2 \end{bmatrix}$ is $\begin{vmatrix} 1 & 6 \\ 5 & 2 \end{vmatrix}$.

The determinant of the square matrix $\begin{bmatrix} 2 & 4 & 1 \\ 0 & 5 & 2 \\ 3 & 6 & 9 \end{bmatrix}$ is $\begin{vmatrix} 2 & 4 & 1 \\ 0 & 5 & 2 \\ 3 & 6 & 9 \end{vmatrix}$.

We define the determinant of a 2×2 matrix first. (Recall that 2×2 is read "two by two." It means that the matrix has 2 rows and 2 columns.)

Determinant of a 2×2 Matrix

$$\begin{vmatrix} a & b \\ c & d \end{vmatrix} = ad - bc$$

EXAMPLE 1 Evaluate each determinant.

a. $\begin{vmatrix} -1 & 2 \\ 3 & -4 \end{vmatrix}$ **b.** $\begin{vmatrix} 2 & 0 \\ 7 & -5 \end{vmatrix}$

Solution: First we identify the values of a, b, c, and d. Then we perform the evaluation.

a. Here $a = -1, b = 2, c = 3$, and $d = -4$.

$$\begin{vmatrix} -1 & 2 \\ 3 & -4 \end{vmatrix} = ad - bc = (-1)(-4) - (2)(3) = -2$$

b. In this example, $a = 2, b = 0, c = 7$, and $d = -5$.

$$\begin{vmatrix} 2 & 0 \\ 7 & -5 \end{vmatrix} = ad - bc = 2(-5) - (0)(7) = -10$$

Practice Problem 1

Evaluate each determinant.

a. $\begin{vmatrix} -3 & 6 \\ 2 & 1 \end{vmatrix}$

b. $\begin{vmatrix} 4 & 5 \\ 0 & -5 \end{vmatrix}$

Answers

1. a. -15, **b.** -20

B Using Cramer's Rule to Solve a System of Two Linear Equations

To develop Cramer's rule, we solve the system $\begin{cases} ax + by = h \\ cx + dy = k \end{cases}$ using elimination. First, we eliminate y by multiplying both sides of the first equation by d and both sides of the second equation by $-b$ so that the coefficients of y are opposites. The result is that

$$\begin{cases} d(ax + by) = d \cdot h \\ -b(cx + dy) = -b \cdot k \end{cases} \quad \text{simplifies to} \quad \begin{cases} adx + bdy = hd \\ -bcx - bdy = -kb \end{cases}$$

We now add the two equations and solve for x.

$$\begin{aligned} adx + bdy &= hd \\ -bcx - bdy &= -kb \\ \hline adx - bcx \quad &= hd - kb && \text{Add the equations.} \\ (ad - bc)x \quad &= hd - kb \\ x \quad &= \frac{hd - kb}{ad - bc} && \text{Solve for } x. \end{aligned}$$

When we replace x with $\dfrac{hd - kb}{ad - bc}$ in the equation $ax + by = h$ and solve for y, we find that $y = \dfrac{ak - ch}{ad - bc}$.

Notice that the numerator of the value of x is the determinant of

$$\begin{vmatrix} h & b \\ k & d \end{vmatrix} = hd - kb$$

Also, the numerator of the value of y is the determinant of

$$\begin{vmatrix} a & h \\ c & k \end{vmatrix} = ak - hc$$

Finally, the denominators of the values of x and y are the same and are the determinant of

$$\begin{vmatrix} a & b \\ c & d \end{vmatrix} = ad - bc$$

This means that the values of x and y can be written in determinant notation:

$$x = \frac{\begin{vmatrix} h & b \\ k & d \end{vmatrix}}{\begin{vmatrix} a & b \\ c & d \end{vmatrix}} \quad \text{and} \quad y = \frac{\begin{vmatrix} a & h \\ c & k \end{vmatrix}}{\begin{vmatrix} a & b \\ c & d \end{vmatrix}}$$

For convenience, we label the determinants D, D_x, and D_y.

x-coefficients; y-coefficients

$$\begin{vmatrix} a & b \\ c & d \end{vmatrix} = D \qquad \begin{vmatrix} h & b \\ k & d \end{vmatrix} = D_x \qquad \begin{vmatrix} a & h \\ c & k \end{vmatrix} = D_y$$

x-column replaced by constants; y-column replaced by constants

These determinant formulas for the coordinates of the solution of a system are known as **Cramer's rule.**

Cramer's Rule for Two Linear Equations in Two Variables

The solution of the system $\begin{cases} ax + by = h \\ cx + dy = k \end{cases}$ is given by

$$x = \frac{\begin{vmatrix} h & b \\ k & d \end{vmatrix}}{\begin{vmatrix} a & b \\ c & d \end{vmatrix}} = \frac{D_x}{D} \qquad y = \frac{\begin{vmatrix} a & h \\ c & k \end{vmatrix}}{\begin{vmatrix} a & b \\ c & d \end{vmatrix}} = \frac{D_y}{D}$$

as long as $D = ad - bc$ is not 0.

When $D = 0$, the system is either inconsistent or the equations are dependent. When this happens, we need to use another method to see which is the case.

EXAMPLE 2 Use Cramer's rule to solve the system:

$$\begin{cases} 3x + 4y = -7 \\ x - 2y = -9 \end{cases}$$

Solution: First we find D, D_x, and D_y.

$$\begin{cases} \overset{\overset{a}{\downarrow}}{3}x + \overset{\overset{b}{\downarrow}}{4}y = \overset{\overset{h}{\downarrow}}{-7} \\ \underset{\underset{c}{\uparrow}}{}x - \underset{\underset{d}{\uparrow}}{2}y = \underset{\underset{k}{\uparrow}}{-9} \end{cases}$$

$$D = \begin{vmatrix} a & b \\ c & d \end{vmatrix} = \begin{vmatrix} 3 & 4 \\ 1 & -2 \end{vmatrix} = 3(-2) - 4(1) = -10$$

$$D_x = \begin{vmatrix} h & b \\ k & d \end{vmatrix} = \begin{vmatrix} -7 & 4 \\ -9 & -2 \end{vmatrix} = (-7)(-2) - 4(-9) = 50$$

$$D_y = \begin{vmatrix} a & h \\ c & d \end{vmatrix} = \begin{vmatrix} 3 & -7 \\ 1 & -9 \end{vmatrix} = 3(-9) - (-7)(1) = -20$$

Then $x = \dfrac{D_x}{D} = \dfrac{50}{-10} = -5$ and $y = \dfrac{D_y}{D} = \dfrac{-20}{-10} = 2$.

The ordered pair solution is $(-5, 2)$.
As always, check the solution in both original equations.

Practice Problem 2

Use Cramer's rule to solve the system.

$$\begin{cases} x - y = -4 \\ 2x + 3y = 2 \end{cases}$$

EXAMPLE 3 Use Cramer's rule to solve the system:

$$\begin{cases} 5x + y = 5 \\ -7x - 2y = -7 \end{cases}$$

Solution: First we find D, D_x, and D_y.

$$D = \begin{vmatrix} 5 & 1 \\ -7 & -2 \end{vmatrix} = 5(-2) - (-7)(1) = -3$$

$$D_x = \begin{vmatrix} 5 & 1 \\ -7 & -2 \end{vmatrix} = 5(-2) - (-7)(1) = -3$$

$$D_y = \begin{vmatrix} 5 & 5 \\ -7 & -7 \end{vmatrix} = 5(-7) - 5(-7) = 0$$

Practice Problem 3

Use Cramer's rule to solve the system.

$$\begin{cases} 4x + y = 3 \\ 2x - 3y = -9 \end{cases}$$

Answers

2. $(-2, 2)$, **3.** $(0, 3)$

Then

$$x = \frac{D_x}{D} = \frac{-3}{-3} = 1 \qquad y = \frac{D_y}{D} = \frac{0}{-3} = 0$$

The ordered pair solution is $(1, 0)$.

C Evaluating 3 × 3 Determinants

A 3×3 determinant can be used to solve a system of three equations in three variables. The determinant of a 3×3 matrix, however, is considerably more complex than a 2×2 one.

Determinant of a 3 × 3 Matrix

$$\begin{vmatrix} a_1 & b_1 & c_1 \\ a_2 & b_2 & c_2 \\ a_3 & b_3 & c_3 \end{vmatrix} = a_1 \cdot \begin{vmatrix} b_2 & c_2 \\ b_3 & c_3 \end{vmatrix} - a_2 \cdot \begin{vmatrix} b_1 & c_1 \\ b_3 & c_3 \end{vmatrix} + a_3 \cdot \begin{vmatrix} b_1 & c_1 \\ b_2 & c_2 \end{vmatrix}$$

Notice that the determinant of a 3×3 matrix is related to the determinants of three 2×2 matrices. Each determinant of these 2×2 matrices is called a **minor**, and every element of a 3×3 matrix has a minor associated with it. For example, the minor of c_2 is the determinant of the 2×2 matrix found by deleting the row and column containing c_2.

$$\begin{matrix} a_1 & b_1 & \not{c_1} \\ \not{a_2} & \not{b_2} & \not{c_2} \\ a_3 & b_3 & \not{c_3} \end{matrix} \qquad \text{The minor of } c_2 \text{ is} \qquad \begin{vmatrix} a_1 & b_1 \\ a_3 & b_3 \end{vmatrix}$$

Also, the minor of element a_1 is the determinant of the 2×2 matrix that has no row or column containing a_1.

$$\begin{matrix} \not{a_1} & \not{b_1} & \not{c_1} \\ \not{a_2} & b_2 & c_2 \\ \not{a_3} & b_3 & c_3 \end{matrix} \qquad \text{The minor of } a_1 \text{ is} \qquad \begin{vmatrix} b_2 & c_2 \\ b_3 & c_3 \end{vmatrix}$$

So the determinant of a 3×3 matrix can be written as:

$$a_1 \cdot (\text{minor of } a_1) - a_2 \cdot (\text{minor of } a_2) + a_3 \cdot (\text{minor of } a_3)$$

Finding the determinant by using minors of elements in the first column is called **expanding** by the minors of the first column. *The value of a determinant can be found by expanding by the minors of any row or column.* The following **array of signs** is helpful in determining whether to add or subtract the product of an element and its minor.

$$\begin{matrix} + & - & + \\ - & + & - \\ + & - & + \end{matrix}$$

If an element is in a position marked +, we add. If marked −, we subtract.

Try the Concept Check in the margin.

EXAMPLE 4

Evaluate by expanding by the minors of the given row or column.

$$\begin{vmatrix} 0 & 5 & 1 \\ 1 & 3 & -1 \\ -2 & 2 & 4 \end{vmatrix}$$

a. First column **b.** Second row

Concept Check

Suppose you are interested in finding the determinant of a 4×4 matrix. Study the pattern shown in the array of signs for a 3×3 matrix. Use the pattern to expand the array of signs for use with a 4×4 matrix.

Practice Problem 4

Evaluate by expanding by the minors of the given row or column.

a. First column b. Third row

$$\begin{vmatrix} 2 & 0 & 1 \\ -1 & 3 & 2 \\ 5 & 1 & 4 \end{vmatrix}$$

Answers

4. a. 4, **b.** 4

Concept Check:

$$\begin{matrix} + & - & + & - \\ - & + & - & + \\ + & - & + & - \\ - & + & - & + \end{matrix}$$

Solution:

a. The elements of the first column are 0, 1, and −2. The first column of the array of signs is +, −, +.

$$\begin{vmatrix} 0 & 5 & 1 \\ 1 & 3 & -1 \\ -2 & 2 & 4 \end{vmatrix} = 0 \cdot \begin{vmatrix} 3 & -1 \\ 2 & 4 \end{vmatrix} - 1 \cdot \begin{vmatrix} 5 & 1 \\ 2 & 4 \end{vmatrix} + (-2) \cdot \begin{vmatrix} 5 & 1 \\ 3 & -1 \end{vmatrix}$$

$$= 0(12 - (-2)) - 1(20 - 2) + (-2)(-5 - 3)$$

$$= 0 - 18 + 16 = -2$$

b. The elements of the second row are 1, 3, and −1. This time, the signs begin with − and again alternate.

$$\begin{vmatrix} 0 & 5 & 1 \\ 1 & 3 & -1 \\ -2 & 2 & 4 \end{vmatrix} = -1 \cdot \begin{vmatrix} 5 & 1 \\ 2 & 4 \end{vmatrix} + 3 \cdot \begin{vmatrix} 0 & 1 \\ -2 & 4 \end{vmatrix} - (-1) \cdot \begin{vmatrix} 0 & 5 \\ -2 & 2 \end{vmatrix}$$

$$= -1(20 - 2) + 3(0 - (-2)) - (-1)(0 - (-10))$$

$$= -18 + 6 + 10 = -2$$

Notice that the determinant of the 3×3 matrix is the same regardless of the row or column you select to expand by.

Try the Concept Check in the margin.

D Using Cramer's Rule to Solve a System of Three Linear Equations

A system of three equations in three variables may be solved with Cramer's rule also. Using the elimination process to solve a system with unknown constants as coefficients leads to the following.

Cramer's Rule for Three Equations in Three Variables

The solution of the system $\begin{cases} a_1x + b_1y + c_1z = k_1 \\ a_2x + b_2y + c_2z = k_2 \\ a_3x + b_3y + c_3z = k_3 \end{cases}$ is given by

$$x = \frac{D_x}{D} \qquad y = \frac{D_y}{D} \qquad \text{and} \qquad z = \frac{D_z}{D}$$

where

$$D = \begin{vmatrix} a_1 & b_1 & c_1 \\ a_2 & b_2 & c_2 \\ a_3 & b_3 & c_3 \end{vmatrix} \quad D_x = \begin{vmatrix} k_1 & b_1 & c_1 \\ k_2 & b_2 & c_2 \\ k_3 & b_3 & c_3 \end{vmatrix}$$

$$D_y = \begin{vmatrix} a_1 & k_1 & c_1 \\ a_2 & k_2 & c_2 \\ a_3 & k_3 & c_3 \end{vmatrix} \quad D_z = \begin{vmatrix} a_1 & b_1 & k_1 \\ a_2 & b_2 & k_2 \\ a_3 & b_3 & k_3 \end{vmatrix}$$

as long as D is not 0.

Concept Check

Why would expanding by minors of the second row be a good choice for the determinant $\begin{vmatrix} 3 & 4 & -2 \\ 5 & 0 & 0 \\ 6 & -3 & 7 \end{vmatrix}$?

Answer

Concept Check: Two elements of the second row are 0, which makes calculations easier.

Practice Problem 5

Use Cramer's rule to solve the system:

$$\begin{cases} x + 2y - z = 3 \\ 2x - 3y + z = -9 \\ -x + y - 2z = 0 \end{cases}$$

EXAMPLE 5 Use Cramer's rule to solve the system:

$$\begin{cases} x - 2y + z = 4 \\ 3x + y - 2z = 3 \\ 5x + 5y + 3z = -8 \end{cases}$$

Solution: First we find D, D_x, D_y, and D_z. Beginning with D, we expand by the minors of the first column.

$$D = \begin{vmatrix} 1 & -2 & 1 \\ 3 & 1 & -2 \\ 5 & 5 & 3 \end{vmatrix} = 1 \cdot \begin{vmatrix} 1 & -2 \\ 5 & 3 \end{vmatrix} - 3 \cdot \begin{vmatrix} -2 & 1 \\ 5 & 3 \end{vmatrix} + 5 \cdot \begin{vmatrix} -2 & 1 \\ 1 & -2 \end{vmatrix}$$

$$= 1(3 - (-10)) - 3(-6 - 5) + 5(4 - 1)$$

$$= 13 + 33 + 15 = 61$$

$$D_x = \begin{vmatrix} 4 & -2 & 1 \\ 3 & 1 & -2 \\ -8 & 5 & 3 \end{vmatrix} = 4 \cdot \begin{vmatrix} 1 & -2 \\ 5 & 3 \end{vmatrix} - 3 \cdot \begin{vmatrix} -2 & 1 \\ 5 & 3 \end{vmatrix} + (-8) \cdot \begin{vmatrix} -2 & 1 \\ 1 & -2 \end{vmatrix}$$

$$= 4(3 - (-10)) - 3(-6 - 5) + (-8)(4 - 1)$$

$$= 52 + 33 - 24 = 61$$

$$D_y = \begin{vmatrix} 1 & 4 & 1 \\ 3 & 3 & -2 \\ 5 & -8 & 3 \end{vmatrix} = 1 \cdot \begin{vmatrix} 3 & -2 \\ -8 & 3 \end{vmatrix} - 3 \cdot \begin{vmatrix} 4 & 1 \\ -8 & 3 \end{vmatrix} + 5 \cdot \begin{vmatrix} 4 & 1 \\ 3 & -2 \end{vmatrix}$$

$$= 1(9 - 16) - 3(12 - (-8)) + 5(-8 - 3)$$

$$= -7 - 60 - 55 = -122$$

$$D_z = \begin{vmatrix} 1 & -2 & 4 \\ 3 & 1 & 3 \\ 5 & 5 & -8 \end{vmatrix} = 1 \cdot \begin{vmatrix} 1 & 3 \\ 5 & -8 \end{vmatrix} - 3 \cdot \begin{vmatrix} -2 & 4 \\ 5 & -8 \end{vmatrix} + 5 \cdot \begin{vmatrix} -2 & 4 \\ 1 & 3 \end{vmatrix}$$

$$= 1(-8 - 15) - 3(16 - 20) + 5(-6 - 4)$$

$$= -23 + 12 - 50 = -61$$

From these determinants, we calculate the solution:

$$x = \frac{D_x}{D} = \frac{61}{61} = 1 \quad y = \frac{D_y}{D} = \frac{-122}{61} = -2 \quad z = \frac{D_z}{D} = \frac{-61}{61} = -1$$

The ordered triple solution is $(1, -2, -1)$. Check this solution by verifying that it satisfies each equation of the system.

Answer

5. $(-1, 3, 2)$

Name ________________________ Section ____________ Date ____________

Mental Math

Evaluate each determinant mentally.

1. $\begin{vmatrix} 7 & 2 \\ 0 & 8 \end{vmatrix}$

2. $\begin{vmatrix} 6 & 0 \\ 1 & 2 \end{vmatrix}$

3. $\begin{vmatrix} -4 & 2 \\ 0 & 8 \end{vmatrix}$

4. $\begin{vmatrix} 5 & 0 \\ 3 & -5 \end{vmatrix}$

5. $\begin{vmatrix} -2 & 0 \\ 3 & -10 \end{vmatrix}$

6. $\begin{vmatrix} -1 & 4 \\ 0 & -18 \end{vmatrix}$

EXERCISE SET APPENDIX F

A *Evaluate each determinant. See Example 1.*

1. $\begin{vmatrix} 3 & 5 \\ -1 & 7 \end{vmatrix}$

2. $\begin{vmatrix} -5 & 1 \\ 1 & -4 \end{vmatrix}$

3. $\begin{vmatrix} 9 & -2 \\ 4 & -3 \end{vmatrix}$

4. $\begin{vmatrix} 4 & -1 \\ 9 & 8 \end{vmatrix}$

5. $\begin{vmatrix} -2 & 9 \\ 4 & -18 \end{vmatrix}$

6. $\begin{vmatrix} -40 & 8 \\ 70 & -14 \end{vmatrix}$

7. $\begin{vmatrix} \frac{3}{4} & \frac{5}{2} \\ -\frac{1}{6} & \frac{7}{3} \end{vmatrix}$

8. $\begin{vmatrix} \frac{5}{7} & \frac{1}{3} \\ \frac{6}{7} & \frac{2}{3} \end{vmatrix}$

B *Use Cramer's rule, if possible, to solve each system of linear equations. See Examples 2 and 3.*

9. $\begin{cases} 2y - 4 = 0 \\ x + 2y = 5 \end{cases}$

10. $\begin{cases} 4x - y = 5 \\ 3x - 3 = 0 \end{cases}$

11. $\begin{cases} 3x + y = 1 \\ 2y = 2 - 6x \end{cases}$

12. $\begin{cases} y = 2x - 5 \\ 8x - 4y = 20 \end{cases}$

13. $\begin{cases} 5x - 2y = 27 \\ -3x + 5y = 18 \end{cases}$

14. $\begin{cases} 4x - y = 9 \\ 2x + 3y = -27 \end{cases}$

15. $\begin{cases} 2x - 5y = 4 \\ x + 2y = -7 \end{cases}$

16. $\begin{cases} 3x - y = 2 \\ -5x + 2y = 0 \end{cases}$

17. $\begin{cases} \frac{2}{3}x - \frac{3}{4}y = -1 \\ -\frac{1}{6}x + \frac{3}{4}y = \frac{5}{2} \end{cases}$

18. $\begin{cases} \frac{1}{2}x - \frac{1}{3}y = -3 \\ \frac{1}{8}x + \frac{1}{6}y = 0 \end{cases}$

C *Evaluate. See Example 4.*

19. $\begin{vmatrix} 2 & 1 & 0 \\ 0 & 5 & -3 \\ 4 & 0 & 2 \end{vmatrix}$

20. $\begin{vmatrix} -6 & 4 & 2 \\ 1 & 0 & 5 \\ 0 & 3 & 1 \end{vmatrix}$

21. $\begin{vmatrix} 4 & -6 & 0 \\ -2 & 3 & 0 \\ 4 & -6 & 1 \end{vmatrix}$

22. $\begin{vmatrix} 5 & 2 & 1 \\ 3 & -6 & 0 \\ -2 & 8 & 0 \end{vmatrix}$

23. $\begin{vmatrix} 1 & 0 & 4 \\ 1 & -1 & 2 \\ 3 & 2 & 1 \end{vmatrix}$

24. $\begin{vmatrix} 0 & 1 & 2 \\ 3 & -1 & 2 \\ 3 & 2 & -2 \end{vmatrix}$

25. $\begin{vmatrix} 3 & 6 & -3 \\ -1 & -2 & 3 \\ 4 & -1 & 6 \end{vmatrix}$

26. $\begin{vmatrix} 2 & -2 & 1 \\ 4 & 1 & 3 \\ 3 & 1 & 2 \end{vmatrix}$

D *Use Cramer's rule, if possible, to solve each system of linear equations. See Example 5.*

27. $\begin{cases} 3x + z = -1 \\ -x - 3y + z = 7 \\ 3y + z = 5 \end{cases}$

28. $\begin{cases} 4y - 3z = -2 \\ 8x - 4y = 4 \\ -8x + 4y + z = -2 \end{cases}$

29. $\begin{cases} x + y + z = 8 \\ 2x - y - z = 10 \\ x - 2y + 3z = 22 \end{cases}$

30. $\begin{cases} 5x + y + 3z = 1 \\ x - y - 3z = -7 \\ -x + y = 1 \end{cases}$

31. $\begin{cases} 2x + 2y + z = 1 \\ -x + y + 2z = 3 \\ x + 2y + 4z = 0 \end{cases}$

32. $\begin{cases} 2x - 3y + z = 5 \\ x + y + z = 0 \\ 4x + 2y + 4z = 4 \end{cases}$

33. $\begin{cases} x - 2y + z = -5 \\ 3y + 2z = 4 \\ 3x - y = -2 \end{cases}$

34. $\begin{cases} 4x + 5y = 10 \\ 3y + 2z = -6 \\ x + y + z = 3 \end{cases}$

Combining Concepts

Find the value of x that will make each a true statement.

35. $\begin{vmatrix} 1 & x \\ 2 & 7 \end{vmatrix} = -3$

36. $\begin{vmatrix} 6 & 1 \\ -2 & x \end{vmatrix} = 26$

37. If all the elements in a single row of a determinant are zero, what is the value of the determinant? Explain your answer.

38. If all the elements in a single column of a determinant are 0, what is the value of the determinant? Explain your answer.

APPENDIX G

Review of Geometric Figures

Plane Figures Have Length and Width but No Thickness or Depth.		
Name	**Description**	**Figure**
Polygon	Union of three or more coplanar line segments that intersect with each other only at each end point, with each end point shared by two segments.	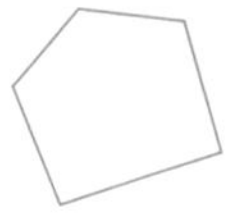
Triangle	Polygon with three sides (sum of measures of three angles is 180°).	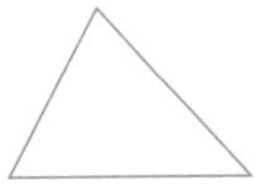
Scalene Triangle	Triangle with no sides of equal length.	
Isosceles Triangle	Triangle with two sides of equal length.	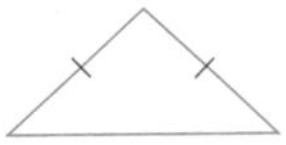
Equilateral Triangle	Triangle with all sides of equal length.	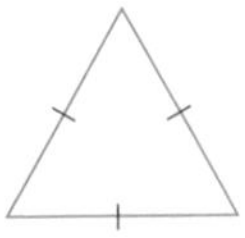
Right Triangle	Triangle that contains a right angle.	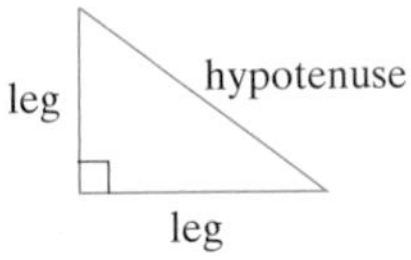 leg, hypotenuse, leg
Quadrilateral	Polygon with four sides (sum of measures of four angles is 360°).	
Trapezoid	Quadrilateral with exactly one pair of opposite sides parallel.	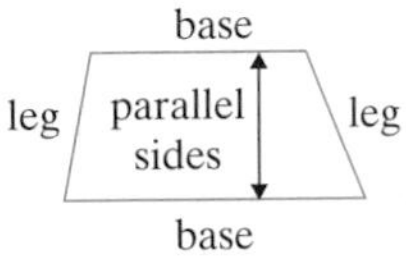 base, leg, parallel sides, leg, base
Isosceles Trapezoid	Trapezoid with legs of equal length.	
Parallelogram	Quadrilateral with both pairs of opposite sides parallel.	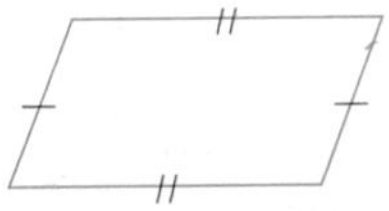
Rhombus	Parallelogram with all sides of equal length.	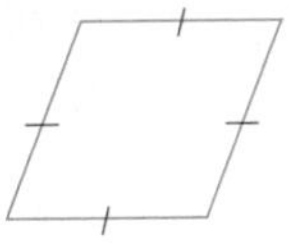

(*continued*)

Name	Description	Figure
Rectangle	Parallelogram with four right angles.	
Square	Rectangle with all sides of equal length.	
Circle	All points in a plane the same distance from a fixed point called the **center**.	radius center diameter

Solid Figures Have Length, Width, and Height or Depth.		
Name	**Description**	**Figure**
Rectangular Solid	A solid with six sides, all of which are rectangles.	
Cube	A rectangular solid whose six sides are squares.	
Sphere	All points the same distance from a fixed point, called the **center**.	radius center
Right Circular Cylinder	A cylinder having two circular bases that are perpendicular to its altitude.	
Right Circular Cone	A cone with a circular base that is perpendicular to its altitude.	

APPENDIX H

Review of Angles, Lines, and Special Triangles

The word **geometry** is formed from the Greek words, **geo**, meaning earth, and **metron**, meaning measure. Geometry literally means to measure the earth.

This appendix contains a review of some basic geometric ideas. It will be assumed that fundamental ideas of geometry such as point, line, ray, and angle are known. In this appendix, the notation $\angle 1$ is read "angle 1" and the notation $m\angle 1$ is read "the measure of angle 1."

We first review types of angles.

Angles

An angle whose measure is greater than 0° but less than 90° is called an **acute angle**.

A **right angle** is an angle whose measure is 90°. A right angle can be indicated by a square drawn at the vertex of the angle, as shown below.

An angle whose measure is greater than 90° but less than 180° is called an **obtuse angle**.

An angle whose measure is 180° is called a **straight angle**.

Two angles are said to be **complementary** if the sum of their measures is 90°. Each angle is called the **complement** of the other.

Two angles are said to be **supplementary** if the sum of their measures is 180°. Each angle is called the **supplement** of the other.

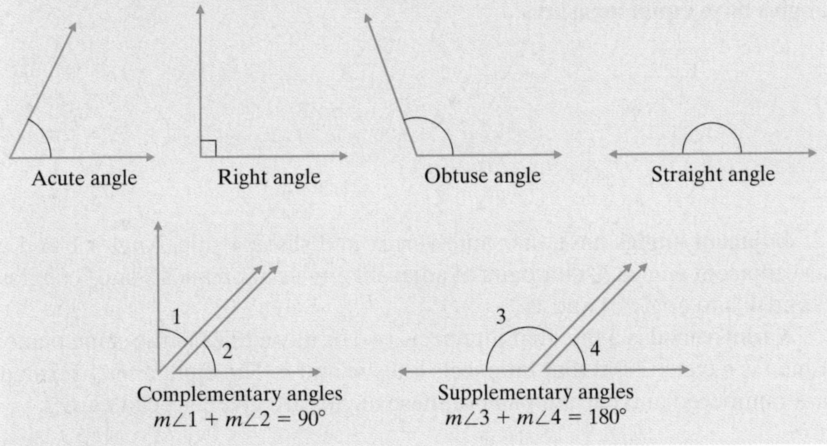

EXAMPLE 1 If an angle measures 28°, find its complement.

Solution: Two angles are complementary if the sum of their measures is 90°. The complement of a 28° angle is an angle whose measure is $90° - 28° = 62°$. To check, notice that $28° + 62° = 90°$. ●

Plane is an undefined term that we will describe. A plane can be thought of as a flat surface with infinite length and width, but no thickness. A plane is two dimensional. The arrows in the following diagram indicate that a plane extends indefinitely and has no boundaries.

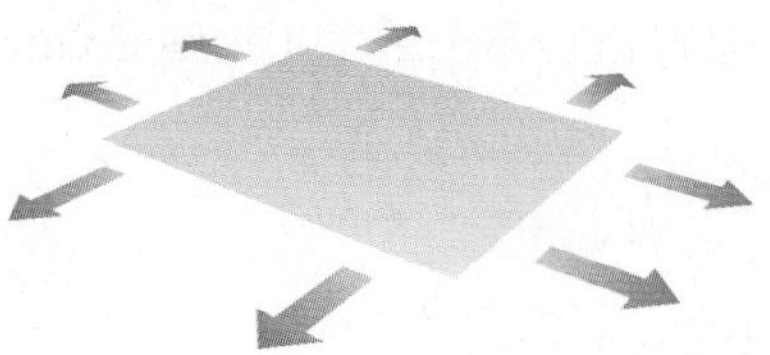

Figures that lie on a plane are called **plane figures**. (See the description of common plane figures in Appendix G.) Lines that lie in the same plane are called **coplanar**.

Lines

Two lines are **parallel** if they lie in the same plane but never meet. **Intersecting lines** meet or cross in one point.

Two lines that form right angles when they intersect are said to be **perpendicular**.

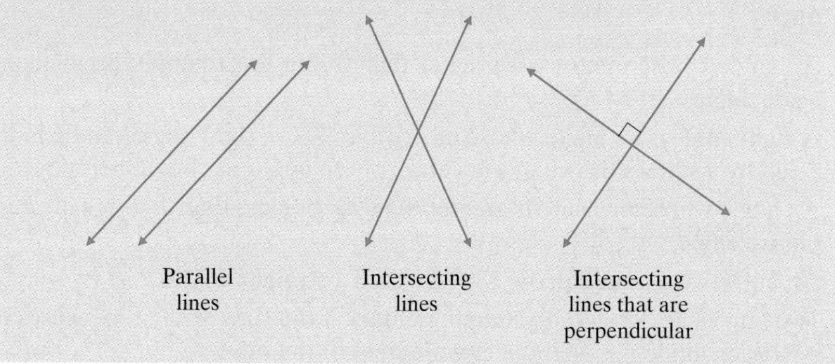

Two intersecting lines form **vertical angles**. Angles 1 and 3 are vertical angles. Also angles 2 and 4 are vertical angles. It can be shown that **vertical angles have equal measures**.

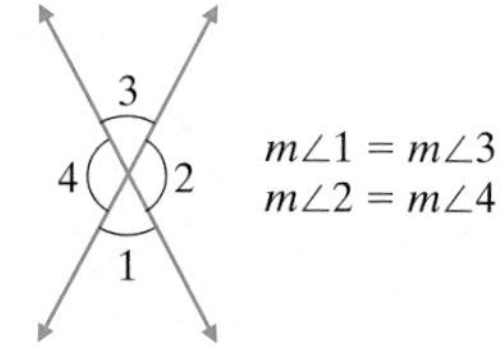

Adjacent angles have the same vertex and share a side. Angles 1 and 2 are adjacent angles. Other pairs of adjacent angles are angles 2 and 3, angles 3 and 4, and angles 4 and 1.

A **transversal** is a line that intersects two or more lines in the same plane. Line l is a transversal that intersects lines m and n. The eight angles formed are numbered and certain pairs of these angles are given special names.

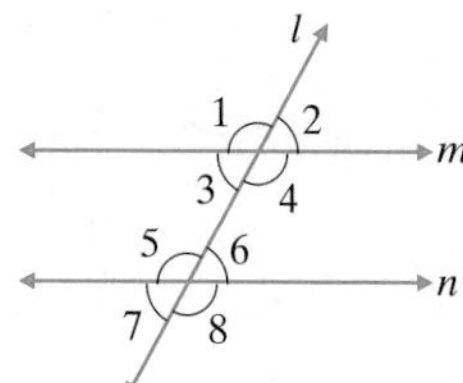

Corresponding angles: $\angle 1$ and $\angle 5$, $\angle 3$ and $\angle 7$, $\angle 2$ and $\angle 6$, and $\angle 4$ and $\angle 8$.

Exterior angles: $\angle 1$, $\angle 2$, $\angle 7$, and $\angle 8$.

Interior angles: $\angle 3$, $\angle 4$, $\angle 5$, and $\angle 6$.

Alternate interior angles: $\angle 3$ and $\angle 6$, $\angle 4$ and $\angle 5$.

These angles and parallel lines are related in the following manner.

Parallel Lines Cut by a Transversal

1. If two parallel lines are cut by a transversal, then
   a. **corresponding angles are equal** and
   b. **alternate interior angles are equal**.
2. If corresponding angles formed by two lines and a transversal are equal, then the lines are parallel.
3. If alternate interior angles formed by two lines and a transversal are equal, then the lines are parallel.

EXAMPLE 2

Given that lines m and n are parallel and that the measure of angle 1 is 100°, find the measures of angles 2, 3, and 4.

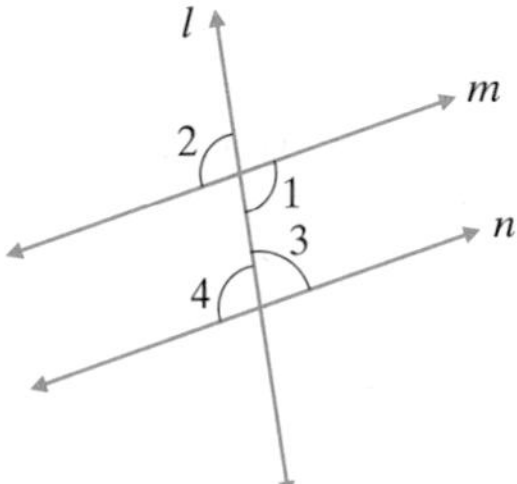

Solution:

$m\angle 2 = 100°$ since angles 1 and 2 are vertical angles

$m\angle 4 = 100°$ since angles 1 and 4 are alternate interior angles

$m\angle 3 = 180° - 100° = 80°$ since angles 4 and 3 are supplementary angles

A **polygon** is the union of three or more coplanar line segments that intersect each other only at each end point, with each end point shared by exactly two segments.

A **triangle** is a polygon with three sides. The sum of the measures of the three angles of a triangle is 180°. In the following figure, $m\angle 1 + m\angle 2 + m\angle 3 = 180°$.

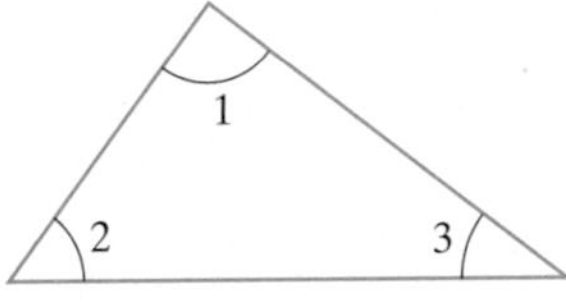

EXAMPLE 3 Find the measure of the third angle of the triangle shown.

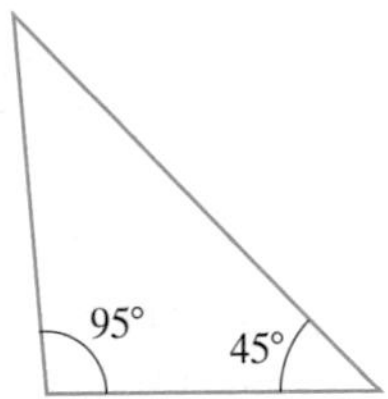

Solution: The sum of the measures of the angles of a triangle is 180°. Since one angle measures 45° and the other angle measures 95°, the third angle measures $180° - 45° - 95° = 40°$.

Two triangles are **congruent** if they have the same size and the same shape. In congruent triangles, the measures of corresponding angles are equal and the lengths of corresponding sides are equal. The following triangles are congruent.

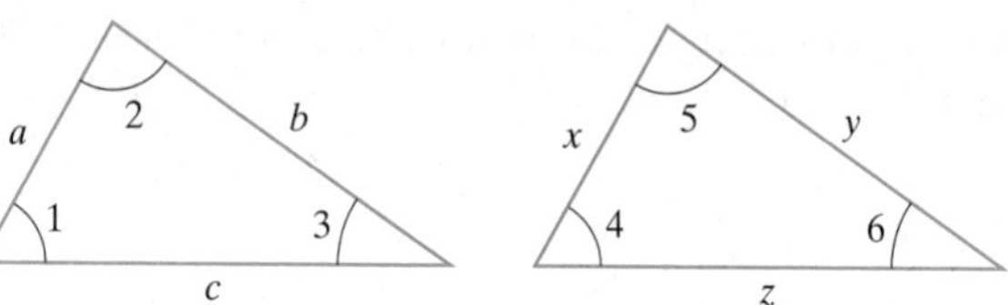

Corresponding angles are equal: $m\angle 1 = m\angle 4$, $m\angle 2 = m\angle 5$, and $m\angle 3 = m\angle 6$. Also, lengths of corresponding sides are equal: $a = x, b = y$, and $c = z$.

Any one of the following may be used to determine whether two triangles are congruent.

Congruent Triangles

1. If the measures of two angles of a triangle equal the measures of two angles of another triangle and the lengths of the sides between each pair of angles are equal, the triangles are congruent.

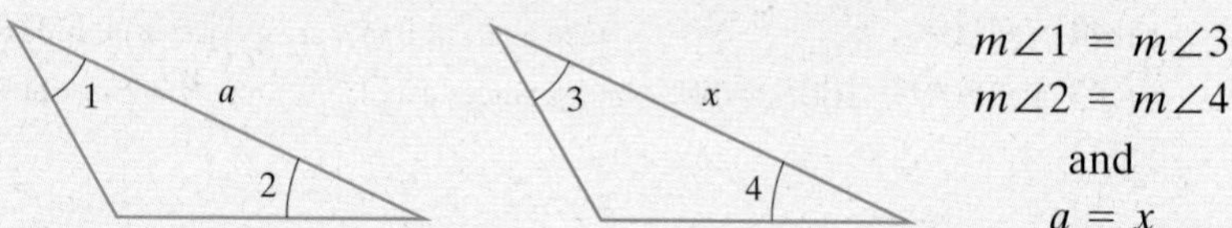

2. If the lengths of the three sides of a triangle equal the lengths of corresponding sides of another triangle, the triangles are congruent.

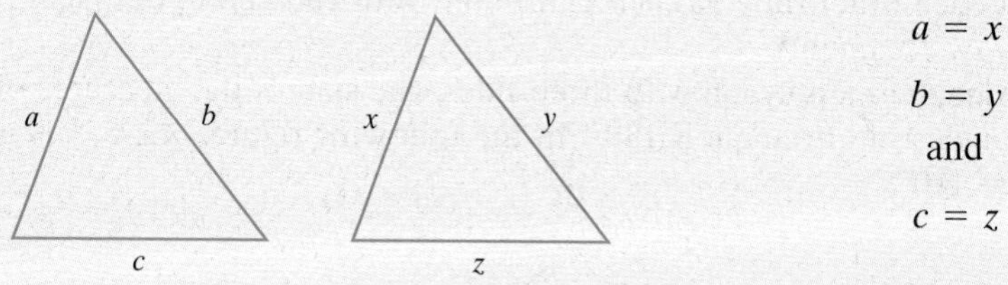

3. If the lengths of two sides of a triangle equal the lengths of corresponding sides of another triangle, and the measures of the angles between each pair of sides are equal, the triangles are congruent.

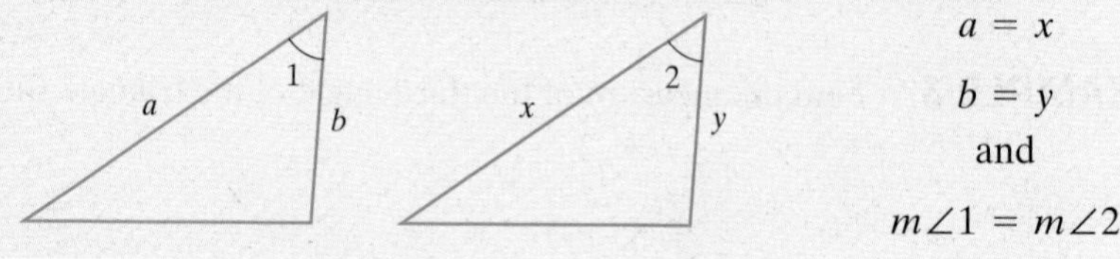

Two triangles are **similar** if they have the same shape but not necessarily the same size. In similar triangles, the measures of corresponding angles are equal and corresponding sides are in proportion. The following triangles are similar. (All similar triangles drawn in this appendix will be oriented the same.)

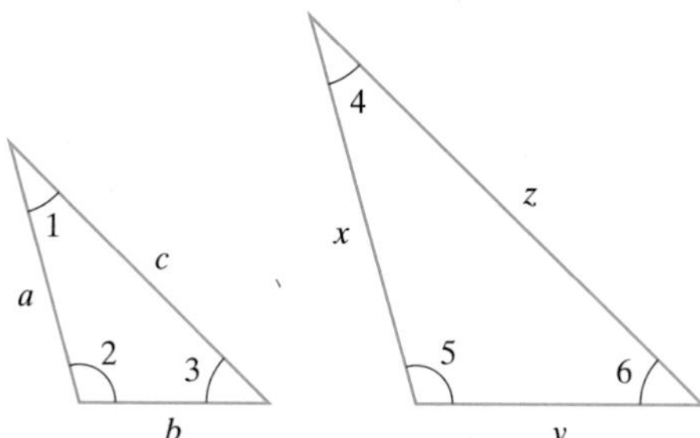

Corresponding angles are equal: $m\angle 1 = m\angle 4$, $m\angle 2 = m\angle 5$, and $m\angle 3 = m\angle 6$. Also, corresponding sides are proportional: $\frac{a}{x} = \frac{b}{y} = \frac{c}{z}$.

Any one of the following may be used to determine whether two triangles are similar.

Similar Triangles

1. If the measures of two angles of a triangle equal the measures of two angles of another triangle, the triangles are similar.

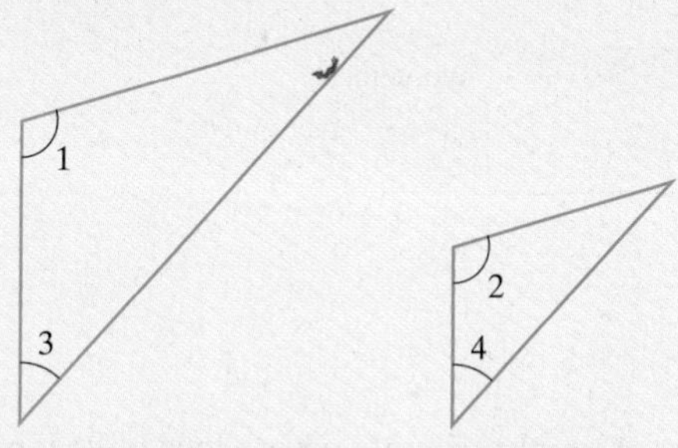

$m\angle 1 = m\angle 2$
and
$m\angle 3 = m\angle 4$

2. If three sides of one triangle are proportional to three sides of another triangle, the triangles are similar.

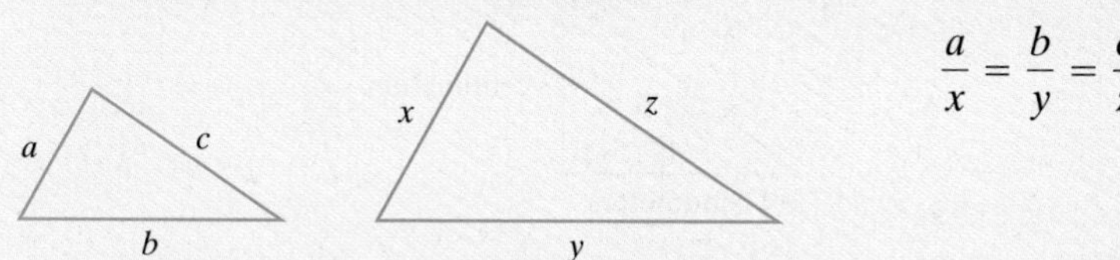

$$\frac{a}{x} = \frac{b}{y} = \frac{c}{z}$$

3. If two sides of a triangle are proportional to two sides of another triangle and the measures of the included angles are equal, the triangles are similar.

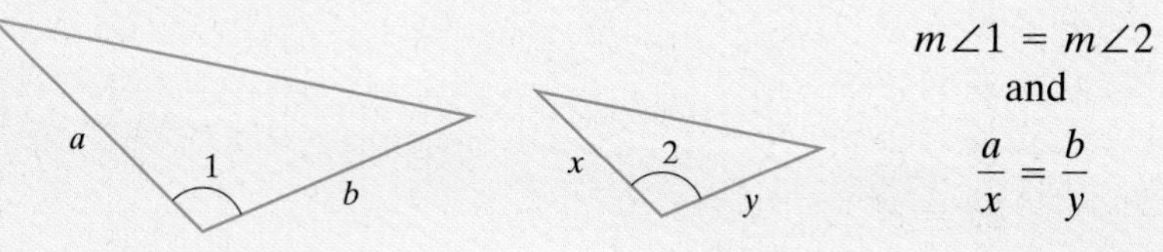

$m\angle 1 = m\angle 2$
and
$\frac{a}{x} = \frac{b}{y}$

EXAMPLE 4

Given that the following triangles are similar, find the missing length x.

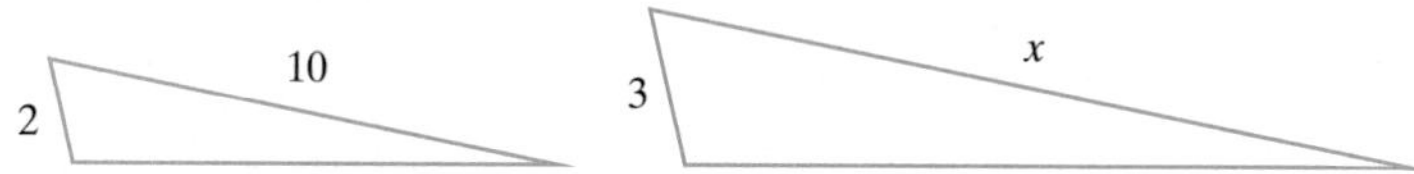

Solution: Since the triangles are similar, corresponding sides are in proportion. Thus, $\frac{2}{3} = \frac{10}{x}$. To solve this equation for x, we cross multiply.

$$\frac{2}{3} = \frac{10}{x}$$

$$2x = 30$$

$$x = 15$$

The missing length is 15 units. ●

A **right triangle** contains a right angle. The side opposite the right angle is called the **hypotenuse**, and the other two sides are called the **legs**. The **Pythagorean theorem** gives a formula that relates the lengths of the three sides of a right triangle.

The Pythagorean Theorem

If a and b are the lengths of the legs of a right triangle, and c is the length of the hypotenuse, then $a^2 + b^2 = c^2$.

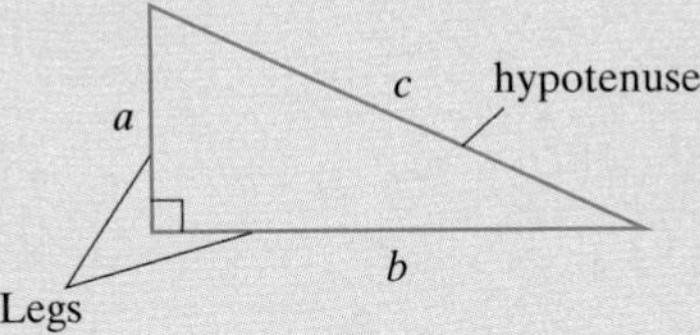

EXAMPLE 5

Find the length of the hypotenuse of a right triangle whose legs have lengths of 3 centimeters and 4 centimeters.

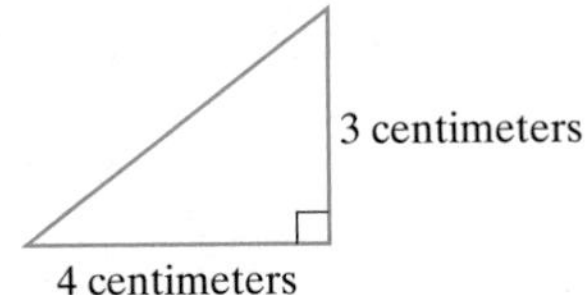

Solution: Because we have a right triangle, we use the Pythagorean theorem. The legs are 3 centimeters and 4 centimeters, so let $a = 3$ and $b = 4$ in the formula.

$$a^2 + b^2 = c^2$$

$$3^2 + 4^2 = c^2$$

$$9 + 16 = c^2$$

$$25 = c^2$$

Since c represents a length, we assume that c is positive. Thus, if c^2 is 25, c must be 5. The hypotenuse has a length of 5 centimeters. ●

Name ______________________ Section __________ Date __________

EXERCISE SET APPENDIX H

Find the complement of each angle. See Example 1.

1. $19°$ **2.** $65°$ **3.** $70.8°$ **4.** $45\frac{2}{3}°$ **5.** $11\frac{1}{4}°$ **6.** $19.6°$

Find the supplement of each angle.

7. $150°$ **8.** $90°$ **9.** $30.2°$ **10.** $81.9°$ **11.** $79\frac{1}{2}°$ **12.** $165\frac{8}{9}°$

13. If lines m and n are parallel, find the measures of angles 1 through 7. See Example 2.

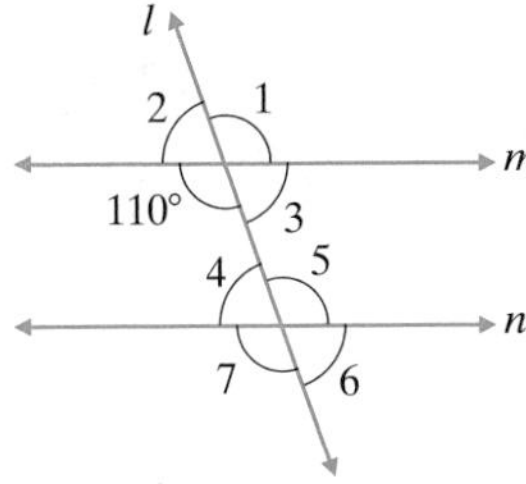

14. If lines m and n are parallel, find the measures of angles 1 through 5. See Example 2.

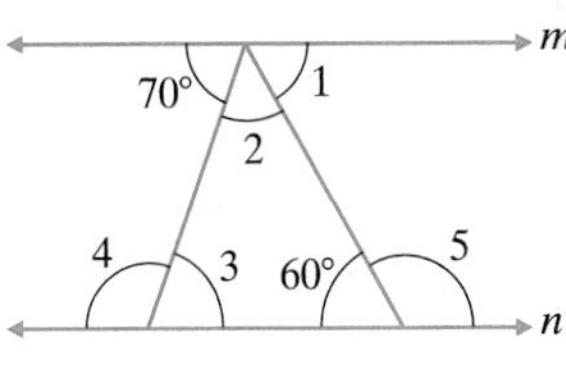

In each of the following, the measures of two angles of a triangle are given. Find the measure of the third angle. See Example 3.

15. $11°, 79°$ **16.** $8°, 102°$ **17.** $25°, 65°$ **18.** $44°, 19°$ **19.** $30°, 60°$ **20.** $67°, 23°$

In each of the following, the measure of one angle of a right triangle is given. Find the measures of the other two angles.

21. $45°$ **22.** $60°$ **23.** $17°$ **24.** $30°$ **25.** $39\frac{3}{4}°$ **26.** $72.6°$

Given that each of the following pairs of triangles is similar, find the missing length x. See Example 4.

27.

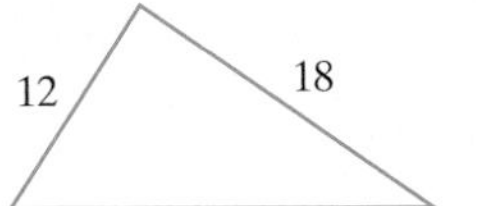

4 x

28.

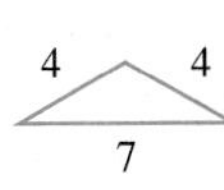

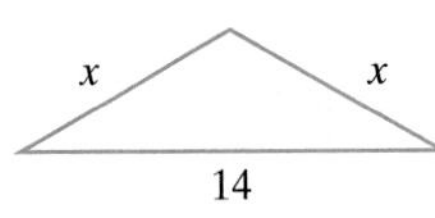

29. 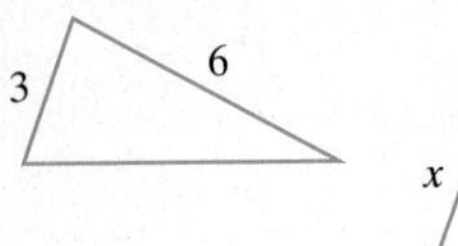

30. 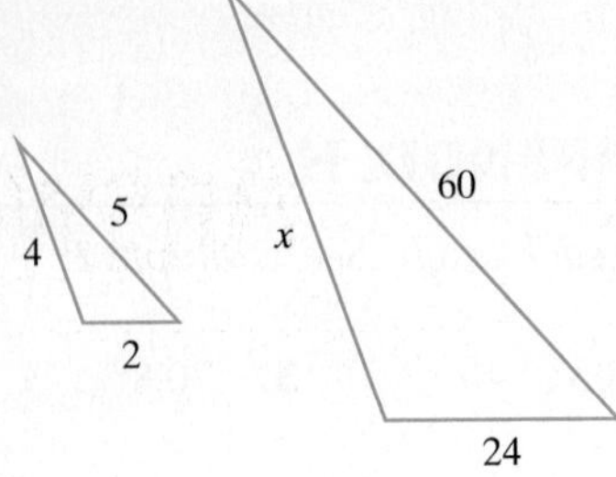

Use the Pythagorean theorem to find the missing lengths in the right triangles. See Example 5.

31. 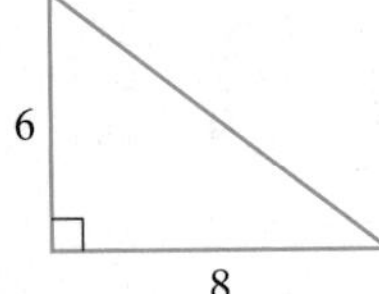

32. 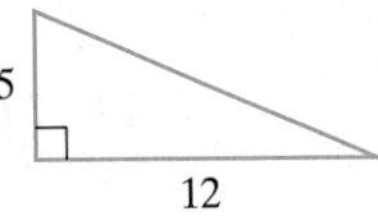

33. 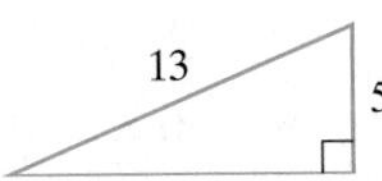

34. 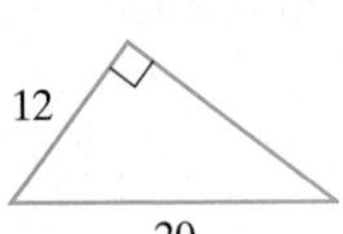

Answers to Selected Exercises

Chapter R Prealgebra Review

Chapter R Pretest

1. 1, 2, 3, 4, 6, 12; R.1A **2.** $2 \cdot 3 \cdot 5 \cdot 5$; R.1B **3.** 280; R.1C **4.** $\frac{35}{40}$; R.2A **5.** $\frac{3}{5}$; R.2B **6.** $\frac{12}{25}$; R.2B **7.** $\frac{1}{12}$; R.2C **8.** $\frac{13}{12}$; R.2D **9.** $\frac{30}{49}$; R.2C **10.** $\frac{1}{9}$; R.2D **11.** 78.53; R.3B **12.** 5.33; R.3B **13.** 2.432; R.3B **14.** 34.9; R.3B **15.** $\frac{716}{100}$; R.3A **16.** 0.1875; R.3D **17.** $0.8\overline{3}$; R.3D **18.** 78.6; R.3C **19.** 78.62; R.3C **20.** 0.806; R.3E **21.** 30%; R.3E

Exercise Set R.1

1. 1, 3, 9 **3.** 1, 2, 3, 4, 6, 8, 12, 24 **5.** 1, 2, 3, 6, 7, 14, 21, 42 **7.** 1, 2, 4, 5, 8, 10, 16, 20, 40, 80 **9.** 1, 19 **11.** prime **13.** composite **15.** prime **17.** composite **19.** composite **21.** $2 \cdot 3 \cdot 3$ **23.** $2 \cdot 2 \cdot 5$ **25.** $2 \cdot 2 \cdot 2 \cdot 7$ **27.** $2 \cdot 2 \cdot 3 \cdot 5 \cdot 5$ **29.** $3 \cdot 3 \cdot 3 \cdot 3$ **31.** $2 \cdot 2 \cdot 3 \cdot 7 \cdot 7$ **33.** c **35.** 42 **37.** 12 **39.** 60 **41.** 35 **43.** 12 **45.** 60 **47.** 350 **49.** 72 **51.** 60 **53.** 30 **55.** 360 **57.** 24 **59.** 2520 **61.** answers may vary **63.** every 35 days

Exercise Set R.2

1. $\frac{21}{30}$ **3.** $\frac{4}{18}$ **5.** $\frac{16}{20}$ **7.** $\frac{1}{2}$ **9.** $\frac{2}{3}$ **11.** $\frac{3}{7}$ **13.** 1 **15.** 5 **17.** $\frac{3}{5}$ **19.** $\frac{4}{5}$ **21.** $\frac{11}{8}$ **23.** $\frac{30}{61}$ **25.** $\frac{8}{11}$ **27.** $\frac{3}{8}$ **29.** $\frac{1}{2}$ **31.** $18\frac{20}{27}$ **33.** 37 **35.** $\frac{6}{7}$ **37.** 15 **39.** $\frac{1}{6}$ **41.** $\frac{3}{80}$ **43.** $10\frac{5}{11}$ **45.** $2\frac{28}{29}$ **47.** 1 **49.** $\frac{3}{5}$ **51.** $\frac{9}{35}$ **53.** $\frac{1}{3}$ **55.** $12\frac{1}{4}$ **57.** $1\frac{3}{5}$ **59.** $\frac{23}{21}$ **61.** $\frac{65}{21}$ **63.** $\frac{5}{7}$ **65.** $\frac{5}{66}$ **67.** $7\frac{1}{12}$ **69.** $48\frac{1}{15}$ **71.** $\frac{7}{5}$ **73.** $\frac{17}{18}$ **75.** $8\frac{8}{15}$ **77.** answers may vary **79.** $\frac{1}{5}$ **81.** $\frac{3}{8}$ **83.** $12\frac{3}{4}$ ft **85. a.** $\frac{7}{50}$ **b.** $\frac{21}{100}$ **c.** $\frac{1}{4}$ **d.** $\frac{3}{10}$ **87.** $\frac{7}{36}$ sq ft **89.** answers may vary

Exercise Set R.3

1. $\frac{6}{10}$ **3.** $\frac{186}{100}$ **5.** $\frac{114}{1000}$ **7.** $\frac{1231}{10}$ **9.** 6.83 **11.** 27.0578 **13.** 6.5 **15.** 15.22 **17.** 56.431 **19.** 598.23 **21.** 0.12 **23.** 67.5 **25.** 43.274 **27.** 84.97593 **29.** 0.094 **31.** 70 **33.** 5.8 **35.** 840 **37.** answers may vary **39.** 0.6 **41.** 0.23 **43.** 0.594 **45.** 98,207.2 **47.** 12.35 **49.** 0.75 **51.** $0.\overline{3} \approx 0.333$ **53.** 0.4375 **55.** $0.\overline{54} \approx 0.55$ **57.** 0.28 **59.** 0.031 **61.** 1.35 **63.** 0.9655 **65.** 0.52 **67.** 68% **69.** 87.6% **71.** 100% **73.** 50% **75.** 35.9 cu ft **77.** 35%

Chapter R Review

1. $2 \cdot 3 \cdot 7$ **2.** $2 \cdot 2 \cdot 2 \cdot 2 \cdot 2 \cdot 2 \cdot 5 \cdot 5$ **3.** 60 **4.** 42 **5.** 60 **6.** 70 **7.** $\frac{15}{24}$ **8.** $\frac{40}{60}$ **9.** $\frac{2}{5}$ **10.** $\frac{3}{20}$ **11.** 2 **12.** 1 **13.** $\frac{8}{77}$ **14.** $\frac{11}{20}$ **15.** $\frac{1}{20}$ **16.** $\frac{11}{18}$ **17.** $14\frac{11}{32}$ **18.** $\frac{1}{2}$ **19.** $20\frac{17}{30}$ **20.** $2\frac{6}{7}$ **21.** $\frac{11}{20}$ sq mi **22.** $\frac{5}{16}$ sq m **23.** $\frac{181}{100}$ **24.** $\frac{35}{1000}$ **25.** 95.118 **26.** 36.785 **27.** 13.38 **28.** 691.573 **29.** 91.2 **30.** 46.816 **31.** 28.6 **32.** 230 **33.** 0.77 **34.** 25.6 **35.** 0.5 **36.** 0.375 **37.** $0.\overline{36} \approx 0.367$ **38.** $0.8\overline{3} \approx 0.833$ **39.** 0.29 **40.** 0.014 **41.** 39% **42.** 120% **43.** 0.674 **44.** b

Chapter R Test

1. $2 \cdot 2 \cdot 2 \cdot 3 \cdot 3$ **2.** 180 **3.** $\frac{25}{60}$ **4.** $\frac{3}{4}$ **5.** $\frac{12}{25}$ **6.** $\frac{13}{10}$ **7.** $\frac{53}{40}$ **8.** $\frac{18}{49}$ **9.** $\frac{1}{20}$ **10.** $\frac{29}{36}$ **11.** $4\frac{8}{9}$ **12.** $2\frac{5}{22}$ **13.** 55 **14.** $13\frac{13}{20}$ **15.** 45.11 **16.** 65.88 **17.** 12.688 **18.** 320 **19.** 23.73 **20.** 0.875 **21.** $0.1\overline{6} \approx 0.167$ **22.** 0.632 **23.** 9% **24.** 75% **25.** $\frac{3}{4}$ **26.** $\frac{1}{200}$ **27.** $\frac{49}{200}$ **28.** $\frac{199}{200}$ **29.** $\frac{1}{8}$ sq foot **30.** $\frac{63}{64}$ sq cm

Chapter 1 Real Numbers and Introduction to Algebra

Chapter 1 Pretest

1. >; 1.2A **2.** <; 1.2A **3.** >; 1.2A **4.** 5; 1.2D **5.** 1.2; 1.2D **6.** 0; 1.2D **7.** 6; 1.3B **8.** $2x - 10$; 1.3D **9.** 64; 1.3A **10.** −9; 1.6A **11.** $\frac{3}{5}$; 1.4C **12.** 8; 1.6B **13.** 53; 1.3A **14.** 3; 1.4A **15.** −27; 1.5A **16.** 56; 1.6A **17.** −70; 1.6C **18.** 4; 1.6C **19.** −40; 1.6D **20.** not a solution; 1.5C **21.** solution; 1.6E **22.** $55; 1.5D **23.** $5 + 2y$; 1.7A **24.** $12 + 8t$; 1.7B **25.** additive inverse property; 1.7C

Exercise Set 1.2

1. < **3.** > **5.** = **7.** < **9.** $32 < 212$ **11.** true **13.** false **15.** false **17.** true **19.** $30 \le 45$ **21.** $20 \le 25$ **23.** $6 > 0$ **25.** $-12 < -10$ **27.** $7 < 11$ **29.** $5 \ge 4$ **31.** $15 \ne -2$ **33.** 14,494; −282 **35.** −34,841 **37.** 475; −195 **39.**
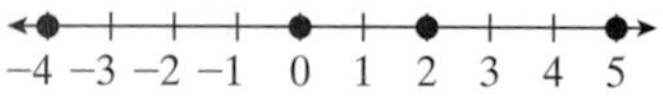

41.
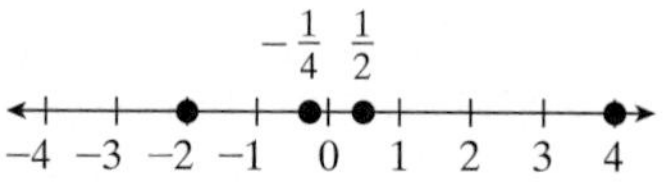

43.
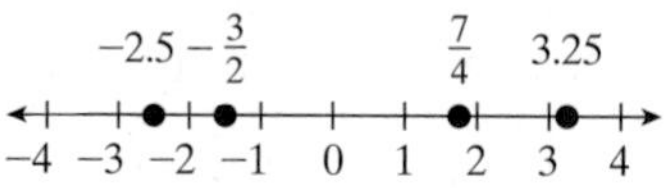

45. whole, integers, rational, real **47.** integers, rational, real **49.** natural, whole, integers, rational, real **51.** rational, real **53.** false **55.** true **57.** false **59.** false **61.** > **63.** = **65.** < **67.** < **69.** false **71.** true **73.** false **75.** true **77.** Blue Ridge Parkway **79.** $19.0 \ge 14.5$ **81.** $-0.04 > -26.7$ **83.** sun **85.** sun **87.** answers may vary

Calculator Explorations

1. 125 **3.** 59,049 **5.** 30 **7.** 9857 **9.** 2376

Exercise Set 1.3

1. 243 **3.** 27 **5.** 1 **7.** 5 **9.** $\frac{1}{125}$ **11.** $\frac{16}{81}$ **13.** 49 **15.** 1.44 **17.** 5^2 sq m **19.** 17 **21.** 20 **23.** 10 **25.** 21 **27.** 45 **29.** 0 **31.** $\frac{2}{7}$ **33.** 30 **35.** 2 **37.** $\frac{7}{18}$ **39.** $\frac{27}{10}$ **41.** $\frac{7}{5}$ **43.** no **45. a.** 64 **b.** 43 **c.** 19 **d.** 22 **47.** 9 **49.** 1 **51.** 1 **53.** 11 **55.** 8 **57.** 45 **59.** 15 **61.** 3 **63.** 6 **65.** Perimeter, 14 in.; 14 in.; 14 in.; area, 12 sq in.; 6 sq in.; 10 sq in. **67.** solution **69.** not a solution **71.** not a solution **73.** solution **75.** not a solution **77.** solution **79.** $x + 15$ **81.** $x - 5$ **83.** $5 - x$ **85.** $3x + 22$ **87.** $1 + 2 = 9 \div 3$ **89.** $3 \ne 4 \div 2$ **91.** $5 + x = 20$ **93.** $13 - 3x = 13$ **95.** $\frac{12}{x} = \frac{1}{2}$ **97.** $(20 - 4) \cdot 4 \div 2$ **99.** answers may vary **101.** answers may vary

Exercise Set 1.4

1. 9 **3.** −14 **5.** 1 **7.** −12 **9.** −5 **11.** −12 **13.** −4 **15.** 7 **17.** −2 **19.** 0 **21.** −19 **23.** 31 **25.** −47 **27.** −2.1 **29.** −8 **31.** 38 **33.** −13.1 **35.** $\frac{2}{8} = \frac{1}{4}$ **37.** $-\frac{3}{16}$ **39.** $-\frac{13}{10}$ **41.** −8 **43.** −59 **45.** −9 **47.** 5 **49.** 11 **51.** −18 **53.** 19 **55.** −7 **57.** answers may vary **59.** 107°F **61.** −95 m **63.** $-2\frac{15}{16}$ points **65.** −14 **67.** −$409 million **69.** −6 **71.** 2 **73.** 0 **75.** −6 **77.** answers may vary **79.** −2 **81.** 0 **83.** $-\frac{2}{3}$ **85.** July **87.** October **89.** 4.7°F **91.** negative **93.** positive

Exercise Set 1.5

1. −10 **3.** −5 **5.** 19 **7.** $\frac{1}{6}$ **9.** 2 **11.** −11 **13.** 11 **15.** 5 **17.** 37 **19.** −6.4 **21.** −71 **23.** 0 **25.** 4.1 **27.** $\frac{2}{11}$ **29.** $-\frac{11}{12}$ **31.** 8.92 **33.** sometimes positive and sometimes negative **35.** 13 **37.** −5 **39.** −1 **41.** −23 **43.** −26 **45.** −24 **47.** 3 **49.** −45 **51.** −4 **53.** 13 **55.** 6 **57.** 9 **59.** −9 **61.** −7 **63.** $\frac{7}{5}$ **65.** 21 **67.** $\frac{1}{4}$ **69.** not a solution **71.** not a solution **73.** solution **75.** 100° **77.** lost 23 yd **79.** 569 B.C. **81.** −308 ft **83.** 19,852 ft **85.** 130° **87.** 30° **89.** −4.4°, 2.6°, 12°, 23.5°, 15.3°, 3.9°, −0.3°, −6.3°, −18.2°, −15.7°, −10.3° **91.** October **93.** true; answers may vary **95.** true; answers may vary **97.** negative, −2.6466

Integrated Review

1. a negative number **2.** a negative number **3.** a positive number **4.** sometimes a positive number, sometimes a negative number **5.** sometimes a positive number, sometimes a negative number **6.** a positive number **7.** 10 **8.** −18 **9.** −2 **10.** −2 **11.** −42 **12.** −7 **13.** 2 **14.** −39 **15.** −3.4 **16.** −9.8 **17.** $-\frac{25}{28}$ **18.** $-\frac{5}{24}$ **19.** −4 **20.** −24 **21.** 6 **22.** 20 **23.** 6 **24.** 61 **25.** −6 **26.** −16 **27.** −19 **28.** −13 **29.** −4 **30.** −1 **31.** $\frac{13}{20}$ **32.** $-\frac{29}{40}$ **33.** 4 **34.** 9 **35.** −1 **36.** −3 **37.** 8 **38.** 10 **39.** 47 **40.** $\frac{2}{3}$

Calculator Explorations

1. 38 **3.** −441 **5.** $163\frac{1}{3}$ **7.** 54,499 **9.** 15,625

Exercise Set 1.6

1. −24 **3.** −2 **5.** 50 **7.** −12 **9.** $\frac{3}{10}$ **11.** $\frac{24}{36} = \frac{2}{3}$ **13.** −7 **15.** 0.14 **17.** 25 **19.** $-\frac{8}{27}$ **21.** $-\frac{1}{4}$ **23.** 0.84 **25.** −30 **27.** 90 **29.** 16 **31.** −16 **33.** 18 **35.** −24 **37.** $\frac{9}{16}$ **39.** 16 **41.** −1 **43.** true **45.** false **47.** $\frac{1}{9}$ **49.** $\frac{3}{2}$ **51.** $-\frac{1}{14}$ **53.** $-\frac{11}{3}$ **55.** $\frac{1}{0.2}$ **57.** −6.3 **59.** 1, −1 **61.** −9 **63.** −4 **65.** 0 **67.** undefined **69.** 3 **71.** −15 **73.** $-\frac{18}{7}$ **75.** $\frac{20}{27}$ **77.** −1 **79.** $-\frac{5}{6}$ **81.** −40 **83.** 160 **85.** $-\frac{9}{2}$ **87.** −4 **89.** 16 **91.** 16 **91.** −3 **93.** $-\frac{16}{7}$ **95.** 2 **97.** $\frac{6}{5}$ **99.** −5 **101.** $\frac{3}{2}$ **103.** $-\frac{5}{38}$ **105.** 3 **107.** −1 **109.** $-\frac{22}{9}$ **111.** solution **113.** not a solution **115.** not a solution **117.** $\frac{0}{5} - 7 = -7$ **119.** $-8(-5) + (-1) = 39$ **121.** $\frac{-8}{-20} = \frac{2}{5}$ **123.** negative **125.** negative **127.** −$2.75 million **129.** negative **131.** can't determine **133.** positive **135.** yes; answers may vary

Exercise Set 1.7

1. $16 + x$ **3.** $y \cdot (-4)$ **5.** yx **7.** $13 + 2x$ **9.** $x \cdot (yz)$ **11.** $(2 + a) + b$ **13.** $4a \cdot (b)$ **15.** $a + (b + c)$ **17.** $17 + b$ **19.** $24y$ **21.** y **23.** $26 + a$ **25.** $-72x$ **27.** s **29.** answers may vary **31.** $4x + 4y$ **33.** $9x - 54$ **35.** $6x + 10$ **37.** $28x - 21$ **39.** $18 + 3x$ **41.** $-2y + 2z$ **43.** $-21y - 35$ **45.** $5x + 20m + 10$ **47.** $-4 + 8m - 4n$ **49.** $-5x - 2$ **51.** $-r + 3 + 7p$ **53.** $3x + 4$ **55.** $-x + 3y$ **57.** $6r + 8$ **59.** $-36x - 70$ **61.** $-16x - 25$ **63.** $4(1 + y)$ **65.** $11(x + y)$ **67.** $-1(5 + x)$ **69.** $30(a + b)$ **71.** −16 **73.** 8 **75.** −1.2 **77.** 2 **79.** $\frac{3}{2}$ **81.** $-\frac{6}{5}$ **83.** $\frac{6}{23}$ **85.** $-\frac{1}{2}$ **87.** commutative property of multiplication **89.** associative property of addition **91.** distributive property **93.** associative property of multiplication **95.** identity property of addition **97.** distributive property **99.** commutative and associative properties of multiplication **101.** $-8; \frac{1}{8}$ **103.** $-x; \frac{1}{x}$ **105.** $2x; -2x$ **107. a.** commutative property of addition **b.** commutative property of addition **c.** associative property of addition **109.** answers may vary **111.** no **113.** yes

Chapter 1 Review

1. < **2.** > **3.** > **4.** > **5.** < **6.** > **7.** = **8.** = **9.** > **10.** < **11.** $4 \geq -3$ **12.** $6 \neq 5$ **13.** $0.03 < 0.3$ **14.** $400 > 155$ **15. a.** 1, 3 **b.** 0, 1, 3 **c.** −6, 0, 1, 3 **d.** $-6, 0, 1, 1\frac{1}{2}, 3, 9.62$ **e.** π **f.** all numbers in set **16. a.** 2, 5 **b.** 2, 5 **c.** −3, 2, 5 **d.** $-3, -1.6, 2, 5, \frac{11}{2}, 15.1$ **e.** $\sqrt{5}, 2\pi$ **f.** all numbers in set **17.** Friday **18.** Wednesday **19.** c **20.** b **21.** 37 **22.** 41 **23.** $\frac{18}{7}$ **24.** 80 **25.** $20 - 12 = 2 \cdot 4$ **26.** $\frac{9}{2} > -5$ **27.** 18 **28.** 108 **29.** 5 **30.** 24 **31.** 63° **32.** solution **33.** not a solution **34.** 9 **35.** $-\frac{2}{3}$ **36.** −2 **37.** 7 **38.** −11 **39.** −17 **40.** $-\frac{3}{16}$ **41.** −5 **42.** −13.9 **43.** 3.9 **44.** −14 **45.** −11.5 **46.** 5 **47.** −11 **48.** −19 **49.** 4 **50.** a **51.** d **52.** $51 **53.** $-\frac{1}{6}$ **54.** $\frac{5}{3}$ **55.** −48 **56.** 28 **57.** 3 **58.** −14 **59.** −36 **60.** 0 **61.** undefined **62.** $-\frac{1}{2}$ **63.** commutative property of addition **64.** multiplicative identity property **65.** distributive property **66.** additive inverse property **67.** associative property of addition **68.** commutative property of multiplication **69.** distributive property **70.** associative property of multiplication **71.** multiplicative inverse property **72.** additive identity property **73.** commutative property of addition

Chapter 1 Test

1. $|-7| > 5$ **2.** $(9 + 5) \geq 4$ **3.** −5 **4.** −11 **5.** −14 **6.** −39 **7.** 12 **8.** −2 **9.** undefined **10.** −8 **11.** $-\frac{1}{3}$ **12.** $4\frac{5}{8}$ **13.** $\frac{51}{40}$ **14.** −32 **15.** −48 **16.** 3 **17.** 0 **18.** > **19.** > **20.** > **21.** = **22. a.** $\{1, 7\}$ **b.** $\{0, 1, 7\}$ **c.** $\{-5, -1, 0, 1, 7\}$ **d.** $\left\{-5, -1, \frac{1}{4}, 0, 1, 7, 11.6\right\}$ **e.** $\{\sqrt{7}, 3\pi\}$ **f.** $\left\{-5, -1, \frac{1}{4}, 0, 1, 7, 11.6, \sqrt{7}, 3\pi\right\}$ **23.** 40 **24.** 12 **25.** 22 **26.** −1 **27.** associative property of addition **28.** commutative property of multiplication **29.** distributive property **30.** multiplicative inverse **31.** 9 **32.** −3 **33.** second down **34.** yes **35.** 17° **36.** loss of $420

Chapter 2 Equations, Inequations, and Problem Solving

Chapter 2 Pretest

1. $9c - 13$; 2.1B **2.** $-17y + 16$; 2.1C **3.** $x = 15$; 2.2A **4.** $b = 2$; 2.2B **5.** $m = -12$; 2.3A **6.** $y = -3$; 2.3B **7.** $x = -3$; 2.4A **8.** $x = 22.5$; 2.4B **9.** no solution; 2.4C **10.** 4; 2.5A **11.** 20, 22; 2.5A **12.** $A = 45$; 2.6A **13.** 9 feet; 2.6A **14.** $y = 8-2x$; 2.6B **15.** 19.8; 2.7A **16.** $\frac{1}{5}$; 2.7C **17.** $x = \frac{24}{7}$; 2.7D

18. $x \le 6$; 2.8B **19.** $y < -4$; 2.8C **20.** $x \ge 3$; 2.8D

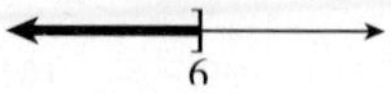

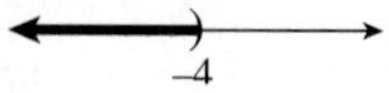

Mental Math

1. −7 **3.** 1 **5.** 17 **7.** like **9.** unlike **11.** like

Exercise Set 2.1

1. $15y$ **3.** $13w$ **5.** $-7b - 9$ **7.** $-m - 6$ **9.** −8 **11.** $7.2x - 5.2$ **13.** $k - 6$ **15.** $-15x + 18$ **17.** $4x - 3$ **19.** $5x^2$ **21.** −11 **23.** $1.3x + 3.5$ **25.** $5y + 20$ **27.** $-2x - 4$ **29.** $-10x + 15y - 30$ **31.** $-3x + 2y - 1$ **33.** $7d - 11$ **35.** 16 **37.** $x + 5$ **39.** $x + 2$ **41.** $2k + 10$ **43.** $-3x + 5$ **45.** $2x + 14$ **47.** $-5y + 31$ **49.** $-22 + 24x$ **51.** $0.9m + 1$ **53.** $10 - 6x - 9y$ **55.** $-x - 38$ **57.** $5x - 7$ **59.** answers may vary **61.** $10x - 3$ **63.** $-4x - 9$ **65.** $-4m - 3$ **67.** $2x - 4$ **69.** $\frac{3}{4}x + 12$ **71.** $12x - 2$ **73.** $8x + 48$ **75.** $x - 10$ **77.** 2 **79.** −23 **81.** −25 **83.** balanced **85.** balanced **87.** $(18x - 2)$ feet **89.** $(15x + 23)$ in.

Mental Math

1. 2 **3.** 12 **5.** 17

Exercise Set 2.2

1. 3 **3.** −2 **5.** −14 **7.** 0.5 **9.** $\frac{5}{12}$ **11.** $\frac{1}{4}$ **13.** −0.7 **15.** 3 **17.** answers may vary **19.** −3 **21.** −9 **23.** −10 **25.** 11 **27.** −7 **29.** −1 **31.** −9 **33.** 13 **35.** −17.9 **37.** $-\frac{1}{2}$ **39.** 11 **41.** −30 **43.** −7 **45.** 2 **47.** −12 **49.** 21 **51.** 25 **53.** $20 - p$ **55.** $(10 - x)$ ft **57.** $(180 - x)°$ **59.** $(n + 47{,}628)$ votes **61.** $7x$ sq mi **63.** $\frac{8}{5}$ **65.** $\frac{1}{2}$ **67.** −9 **69.** x **71.** y **73.** x **75.** $x = -145.478$ **77.** $(173 - 3x)°$

Mental Math

1. 9 **3.** 2 **5.** −5

Exercise Set 2.3

1. −4 **3.** 0 **5.** 12 **7.** −12 **9.** 3 **11.** 2 **13.** 0 **15.** 6.3 **17.** 6 **19.** −5.5 **21.** $\frac{14}{3}$ **23.** −9 **25.** 10 **27.** −20 **29.** 0 **31.** −5 **33.** 0 **35.** $-\frac{3}{2}$ **37.** −21 **39.** $\frac{11}{2}$ **41.** 1 **43.** $-\frac{1}{4}$ **45.** −30 **47.** $\frac{9}{10}$ **49.** $2x + 2$ **51.** $2x + 2$ **53.** $7x - 12$ **55.** $12z + 44$ **57.** 1 **59.** 2 **61.** answers may vary **63.** answers may vary

Calculator Explorations

1. solution **3.** not a solution **5.** solution

Exercise Set 2.4

1. −6 **3.** −3 **5.** 1 **7.** $\frac{9}{2}$ **9.** $\frac{3}{2}$ **11.** 0 **13.** 1 **15.** 4 **17.** −4 **19.** $\frac{19}{6}$ **21.** $\frac{14}{3}$ **23.** 2 **25.** −5 **27.** 10 **29.** 18 **31.** 3 **33.** 13 **35.** 50 **37.** 0.2 **39.** 1 **41.** $\frac{7}{3}$ **43.** −17 **45.** answers may vary **47.** all real numbers **49.** no solution **51.** no solution **53.** no solution **55.** answers may vary **57.** $(6x - 8)$ m **59.** $-8 - x$ **61.** $-3 + 2x$ **63.** $9(x + 20)$ **65.** 15.3 **67.** −0.2 **69.** $x = 4$ cm; $2x = 8$ cm

Integrated Review

1. 6 **2.** −17 **3.** 12 **4.** −26 **5.** −3 **6.** −1 **7.** 13.5 **8.** 12.5 **9.** 8 **10.** −64 **11.** 2 **12.** −3 **13.** 5 **14.** −1 **15.** −2 **16.** −2 **17.** $-\frac{5}{6}$ **18.** $\frac{1}{6}$ **19.** 1 **20.** 6 **21.** 4 **22.** 1 **23.** $\frac{9}{5}$ **24.** $-\frac{6}{5}$ **25.** all real numbers **26.** all real numbers **27.** 0 **28.** −1.6 **29.** $\frac{4}{19}$ **30.** $-\frac{5}{19}$ **31.** $\frac{7}{2}$ **32.** $-\frac{1}{4}$

Exercise Set 2.5

1. 1 **3.** −25 **5.** $-\frac{3}{4}$ **7.** −16 **9.** governor of Michigan = \$127,300; governor of Oregon = \$88,300 **11.** 1st piece: 5 in.; 2nd piece: 10 in.; 3rd piece: 25 in. **13.** 172 mi **15.** 1st angle: 37.5°; 2nd angle: 37.5°; 3rd angle: 105° **17.** cerulean: 7956 votes; blue: 11,322 votes **19.** smaller angle: 45°; larger angle: 135° **21.** shorter piece: 5 ft; longer piece: 12 ft **23.** diameter: 1 m; height: 13 m **25.** Sahara: 3,500,000 sq mi; Gobi: 500,000 sq mi **27.** answers may vary **29.** Texas and Florida **31.** Hawaii: \$37.9 million; Pennsylvania: \$23 million **33.** 2.2 million sq ft **35.** Notre Dame: 68; Purdue: 66 **37.** China: 59 medals; Australia: 58 medals; Germany: 57 medals **39.** 58°, 60°, 62° **41.** $\frac{1}{2}(x-1)=37$ **43.** $\frac{3(x+2)}{5}=0$ **45.** 34 **47.** 225π **49.** answers may vary **51.** 15 ft by 24 ft **53.** answers may vary

Exercise Set 2.6

1. $h=3$ **3.** $h=3$ **5.** $h=20$ **7.** $c=12$ **9.** $r=2.5$ **11.** $T=3$ **13.** $h=15$ **15.** 131 ft **17.** 137.5 mi **19.** 50°C **21.** 96 piranhas **23.** 12,090 ft **25.** 6.25 hr **27.** 2 bags **29.** 515,509.5 cu in. **31.** one 16-in. pizza **33.** −109.3°F **35.** 500 sec or $8\frac{1}{3}$ min **37.** 449 cu in. **39.** 332.6°F **41.** 10.7 **43.** 44.3 sec **45.** 4.65 min **47.** $h=\frac{f}{5g}$ **49.** $W=\frac{V}{LH}$ **51.** $y=7-3x$ **53.** $R=\frac{A-P}{PT}$ **55.** $A=\frac{3V}{h}$ **57.** $a=P-b-c$ **59.** $h=\frac{S-2\pi r^2}{2\pi r}$ **61.** 0.32 **63.** 2.00 or 2 **65.** 17% **67.** 720% **69.** $V=G(N-R)$ **71.** multiplies the volume by 8 **73.** −40°

Mental Math

1. no **3.** yes

Exercise Set 2.7

1. 11.2 **3.** 55% **5.** 180 **7.** 4.6 **9.** 50 **11.** 30% **13.** \$64 decrease; \$192 sale price **15.** 854 thousand Scoville units **17.** 81% **19.** 169,184 **21.** no; answers may vary **23.** 4% **25.** 49,950 **27.** 361 **29.** 91%, 6% **31.** $\frac{2}{15}$ **33.** $\frac{5}{6}$ **35.** $\frac{5}{12}$ **37.** $\frac{1}{10}$ **39.** $\frac{7}{20}$ **41.** $\frac{19}{18}$ **43.** answers may vary **45.** 4 **47.** $\frac{50}{9}$ **49.** $\frac{21}{4}$ **51.** 7 **53.** −3 **55.** $\frac{14}{9}$ **57.** 5 **59.** no solution **61.** 123 lb **63.** 165 cal **65.** 3833 women **67.** 9 gal **69.** 110 oz for \$5.79 **71.** 8 oz for \$0.90 **73.** > **75.** = **77.** > **79.** 9.6% **81.** 26.9%; yes **83.** 17.1%

Mental Math

1. $\{x|x<6\}$ **3.** $\{x|x\ge 10\}$ **5.** $\{x|x>4\}$ **7.** $\{x|x\le 2\}$

Exercise Set 2.8

1. 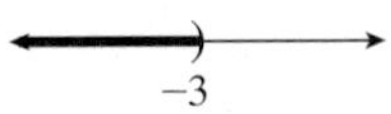 [number line, −3]; $(-\infty,-3)$ **3.** [number line, 0.3]; 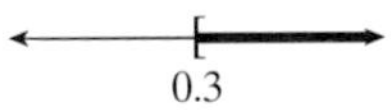 $[0.3,\infty)$ **5.** [number line, 5]; $(5,\infty)$ **7.** [number line, −2, 5]; $(-2,5)$ **9.** [number line, −1, 5]; $(-1,5)$ **11.** answers may vary **13.** 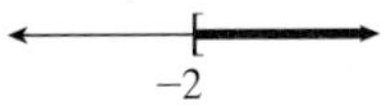 [number line, −2]; $[-2,\infty)$ **15.** [number line, 1]; $(-\infty,1)$ **17.**  [number line, 2]; $(-\infty,2]$ **19.** 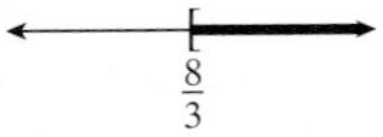 [number line, $\frac{8}{3}$]; $\left[\frac{8}{3},\infty\right)$ **21.** [number line, −4.7]; $(-\infty,-4.7)$ **23.** 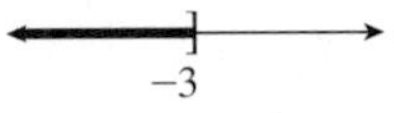 [number line, −3]; $(-\infty,-3]$ **25.** 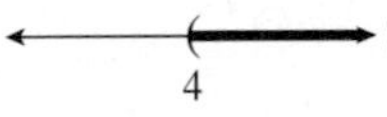 [number line, 4]; $(4,\infty)$ **27.** $(-\infty,-1]$ **29.** $(-\infty,11]$ **31.** $(-13,\infty)$ **33.** $(-\infty,7]$ **35.** $[0,\infty)$ **37.** $(-\infty,-5]$ **39.** $[3,\infty)$ **41.** $(0,\infty)$ **43.** $\left[-\frac{79}{3},\infty\right)$ **45.** $(-\infty,-1]$ **47.** $\left(-\infty,-\frac{35}{6}\right)$ **49.** $[-31,\infty)$ **51.** $(-\infty,-2]$ **53.** 30 **55.** 1040 lb **57.** 13 oz **59.** more than 200 calls **61.** $F\ge 932°$ **63. a.** 2000 **b.** answers may vary **65.** 1015 **67.** 1999 and 2000 **69.** 1997, 1998, 1999, 2000 **71.** increasing; answers may vary **73.** 1994 **75.** $\{0,1,2,3,4,5,6,7\}$ **77.** $\{\ldots,-9,-8,-7,-6\}$ **79.**  [number line, −7, 1]; $(-7,1]$ **81.** [number line, −2.5, 5.3]; $[-2.5,5.3)$ **83.** $\emptyset$ **85.** $(-\infty,\infty)$ **87.** answers may vary **89.** final exam ≥ 89.3

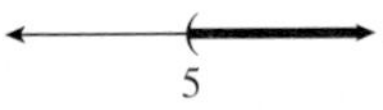

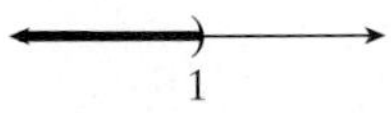

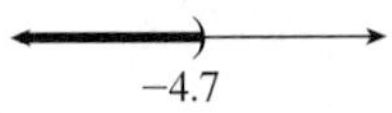

Chapter 2 Review

1. $6x$ **2.** $-11.8z$ **3.** $4x - 2$ **4.** $2y + 3$ **5.** $3n - 18$ **6.** $4w - 6$ **7.** $-6x + 7$ **8.** $-0.4y + 2.3$ **9.** $3x - 7$ **10.** $5x + 5.6$ **11.** 4 **12.** -3 **13.** 6 **14.** -6 **15.** 0 **16.** -9 **17.** -23 **18.** 28 **19.** b **20.** a **21.** c **22.** -12 **23.** 4 **24.** 0 **25.** -7 **26.** 0.75 **27.** -3 **28.** -6 **29.** -1 **30.** 3 **31.** 0 **32.** $-\frac{1}{5}$ **33.** $3x + 3$ **34.** -4 **35.** 2 **36.** -3 **37.** no solution **38.** no solution **39.** $\frac{3}{4}$ **40.** $-\frac{8}{9}$ **41.** 20 **42.** $-\frac{6}{23}$ **43.** $\frac{23}{7}$ **44.** $-\frac{2}{5}$ **45.** 102 **46.** 0.25 **47.** 6665.5 in **48.** short piece: 4 ft; long piece: 8 ft **49.** Kellogg: 35 plants; Keebler: 18 plants **50.** $-39, -38, -37$ **51.** 3 **52.** -4 **53.** $w = 9$ **54.** $h = 4$ **55.** $m = \frac{y - b}{x}$ **56.** $s = \frac{r + 5}{vt}$ **57.** $x = \frac{2y - 7}{5}$ **58.** $y = \frac{2 + 3x}{6}$ **59.** $\pi = \frac{C}{D}$ **60.** $\pi = \frac{C}{2r}$ **61.** 15 m **62.** 40°C **63.** 1 hr and 20 min **64.** 20% **65.** 70% **66.** 110 **67.** 1280 **68.** 50,844 **69.** 18% **70.** swerving into another lane **71.** 966 customers **72.** no; answers may vary **73.** $\frac{1}{5}$ **74.** $\frac{2}{3}$ **75.** 6 **76.** 500 **77.** 312.5 **78.** 50 **79.** 9 **80.** no solution **81.** 3 **82.** no solution **83.** 10 oz for \$1.29 **84.** 15 oz for \$1.63 **85.** 675 parts **86.** \$33.75 **87.** 157 letters

88. $(-\infty, -2]$

89. $(0, \infty)$

90. $(3, \infty)$ **91.** $(-\infty, -4]$ **92.** $(-4, \infty)$ **93.** $(-17, \infty)$ **94.** $(-\infty, 7]$ **95.** $(-\infty, 4]$ **96.** $\left(\frac{1}{2}, \infty\right)$ **97.** $(-\infty, 1)$ **98.** $[-19, \infty)$ **99.** $(2, \infty)$ **100.** more economical to use housekeeper for more than 35 lb per week **101.** 9.6

Chapter 2 Test

1. $y - 10$ **2.** $5.9x + 1.2$ **3.** $-2x + 10$ **4.** $-15y + 1$ **5.** -5 **6.** 8 **7.** $\frac{7}{10}$ **8.** 0 **9.** 27 **10.** $-\frac{19}{6}$ **11.** 3 **12.** -6 **13.** $\frac{3}{11}$ **14.** 0.25 **15.** $\frac{25}{7}$ **16.** 21 **17.** 7 gal **18.** 6 oz for \$1.19 **19.** 18 bulbs **20.** $x = 6$ **21.** $h = \frac{V}{\pi r^2}$ **22.** $y = \frac{3x - 10}{4}$

23. $(-\infty, -2)$

24. $(-\infty, 4)$

25. $(-\infty, -8]$ **26.** $[11, \infty)$ **27.** $\left(\frac{2}{5}, \infty\right)$ **28.** 29% **29.** 552 **30.** 40% **31.** New York: 1077; Indiana: 427 **32.** 29 NBA teams, 32 NFL teams

Cumulative Review

1. True; Sec. 1.2, Ex. 3 **2.** True; Sec. 1.2, Ex. 4 **3.** False; Sec. 1.2, Ex. 5 **4.** True; Sec. 1.2, Ex. 6 **5. a.** $<$ **b.** $=$ **c.** $>$ **d.** $<$ **e.** $>$; Sec. 1.2, Ex. 12 **6.** $\frac{8}{3}$; Sec. 1.3, Ex. 6 **7.** -19; Sec. 1.4, Ex. 7 **8.** 8; Sec. 1.4, Ex. 8 **9.** -0.3; Sec. 1.4, Ex. 9 **10. a.** -12 **b.** -3; Sec. 1.5, Ex. 7 **11. a.** 0 **b.** -24 **c.** 45 **d.** 54; Sec. 1.6, Ex. 7 **12. a.** -6 **b.** 7 **c.** -5; Sec. 1.6, Ex. 10 **13.** $15 - 10z$; Sec. 1.7, Ex. 8 **14.** $12x + 38$; Sec. 1.7, Ex. 12 **15.** Commutative property of multiplication (order changed) **16.** Associative property of addition (grouping changed) **17.** Identity element for addition **18.** Commutative property of multiplication (order changed) **19.** Multiplicative inverse property **20.** Additive inverse property **21.** Commutative and associative properties of multiplication (order and grouping changed) **22. a.** unlike **b.** like **c.** like **d.** like; Sec. 2.1, Ex. 2 **23.** $-2x - 1$; Sec. 2.1, Ex. 15 **24.** 17; Sec. 2.2, Ex. 1 **25.** -10; Sec. 2.3, Ex. 7 **26.** 0; Sec. 2.4, Ex. 4 **27.** 220 Republicans, 210 Democrats; Sec. 2.5, Ex. 3 **28.** 79.2 yr; Sec. 2.6, Ex. 1 **29.** 87.5%; Sec. 2.7, Ex. 1 **30.** 63; Sec. 2.7, Ex. 5

31.

$(-\infty, -1)$; Sec. 2.8, Ex. 2 **32.** $[4, \infty)$; Sec. 2.8, Ex. 9

Chapter 3 Graphing Equations and Inequalities

Chapter 3 Pretest

1.

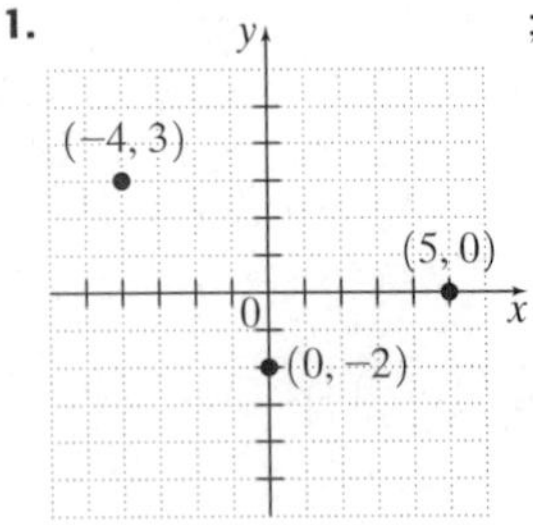

; 3.1B **2.** $(-2, -6)$; 3.1D **3.**

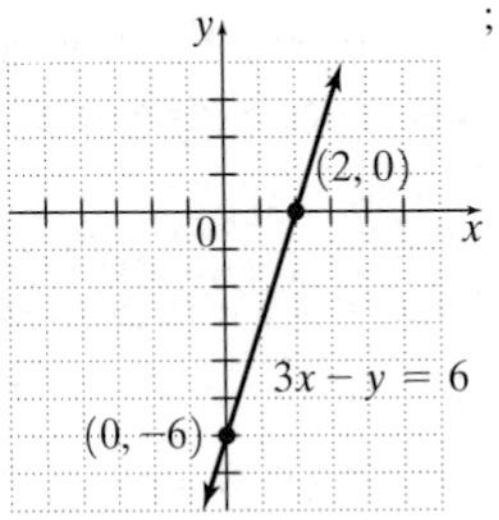

; 3.2A **4.**

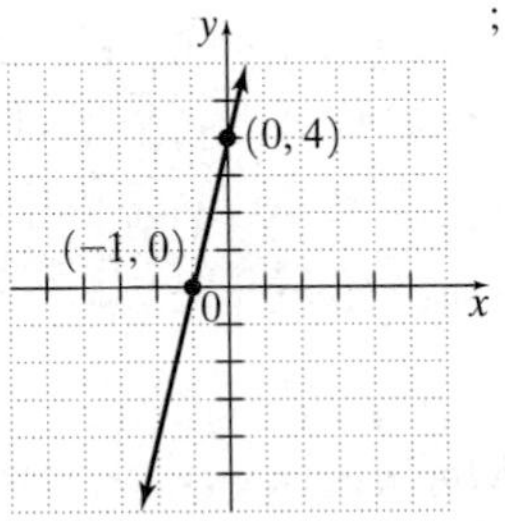

; 3.3B

5. 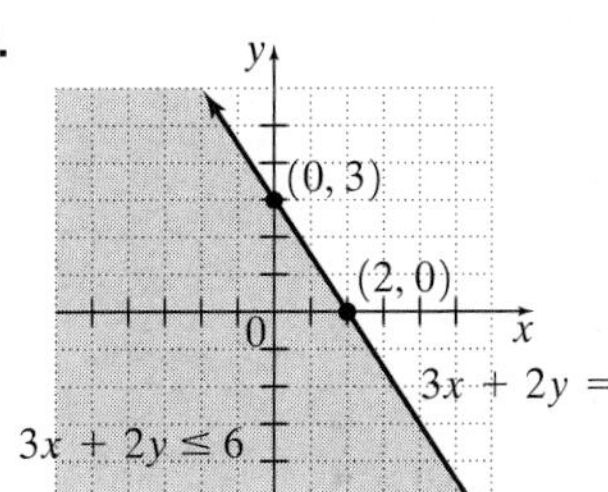

; 3.5B **6.** 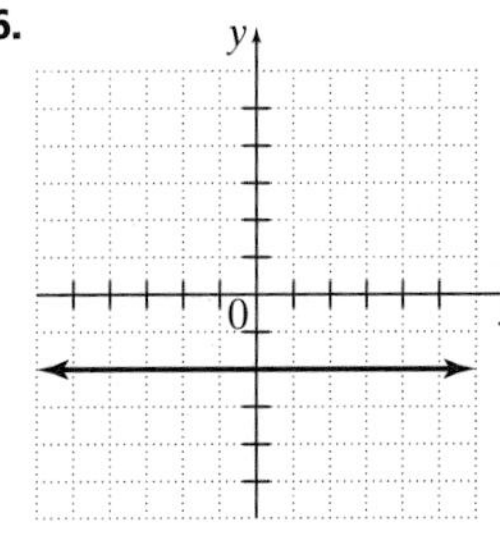

7. $-\frac{3}{10}$; 3.4A **8.** $\frac{4}{5}$; 3.4B **9.** undefined slope; 3.4C **10.** neither; 3.4D **11.** perpendicular; 3.4D

Mental Math

1. answers may vary; Ex. $(5, 5)$, $(7, 3)$

Exercise Set 3.1

1. France **3.** France, U.S., Spain, Italy **5.** 34 million **7.** approx. 59 beats per min **9.** approx. 26 beats per min **11.** 20 **13.** 1985 **15.** 1997 **17.** 18 million **19.** 63 million **21.** 1900 **23.** 27 million **25.** answers may vary

27. ; $(1, 5)$ is in quadrant I, $\left(-1, 4\frac{1}{2}\right)$ is in quadrant II, $(-5, -2)$ is in quadrant III, $(2, -4)$ is in quadrant IV, $(-3, 0)$ lies on the x-axis, $(0, -1)$ lies on the y-axis **29.** $a = b$ **31.** A: $(0, 0)$ **33.** C: $(3, 2)$ **35.** E: $(-2, -2)$ **37.** G: $(2, -1)$ **39.** B: $(0, -3)$ **41.** D: $(1, 3)$ **43.** F: $(-3, -1)$

45. a. $(1995, 438)$, $(1996, 438)$, $(1997, 436)$, $(1998, 435)$, $(1999, 432)$, $(2000, 434)$

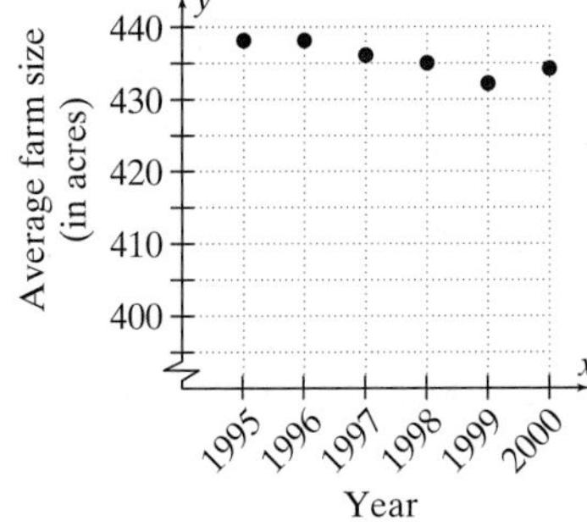

47. a. $(2313, 2)$, $(2085, 1)$, $(2711, 21)$, $(2869, 39)$, $(2920, 42)$, $(4038, 99)$, $(1783, 0)$, $(2493, 9)$

b. 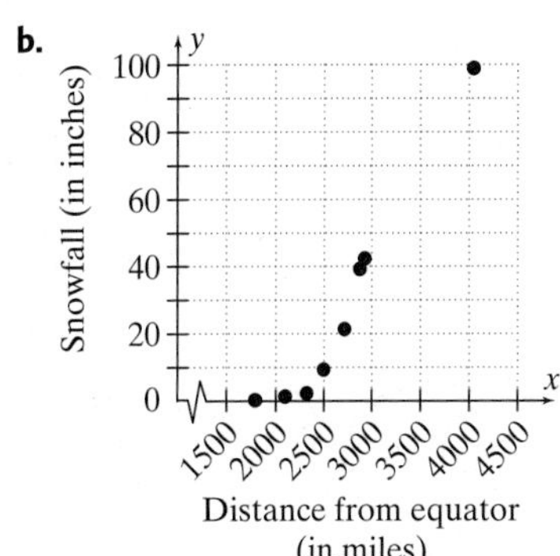

c. The farther from the equator, the more snowfall.

49. a. $(0.50, 10)$, $(0.75, 12)$, $(1.00, 15)$, $(1.25, 16)$, $(1.50, 18)$, $(1.50, 19)$, $(1.75, 19)$, $(2.00, 20)$

b.

c. answers may vary

51. $(-4, -2)$; $(4, 0)$ **53.** $(0, 9)$; $(3, 0)$ **55.** $(11, -7)$; any x

57.

x	y
0	2
6	0
3	1

59.

x	y
0	−12
5	−2
3	−6

61.

x	y
0	$\frac{5}{7}$
$\frac{5}{2}$	0
−1	1

63.

x	y
3	0
3	−0.5
3	$\frac{1}{4}$

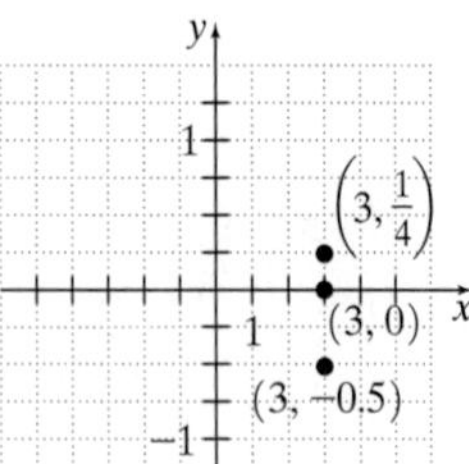

65.

x	y
0	0
−5	1
10	−2

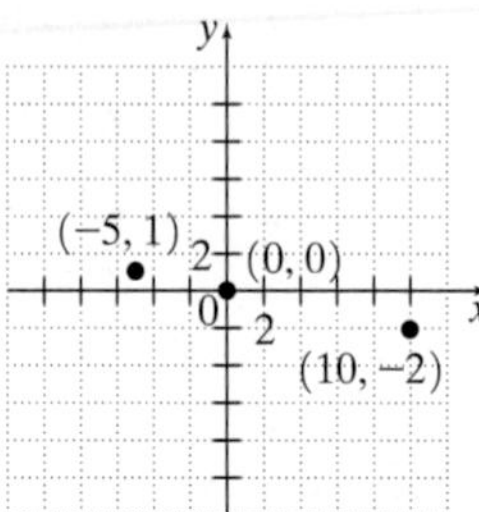

67. a. 13,000; 21,000; 29,000 **b.** 45 desks **69.** $y = 5 - x$ **71.** $y = \dfrac{5 - 2x}{4}$ **73.** $y = -2x$ **75.** 26 units **77.** \$1 billion; \$2 billion; \$1 billion; \$1 billion **79.** answers may vary **81. a.** 968.14; 955.48; 942.82 **b.** 1999

Graphing Calculator Explorations

1.

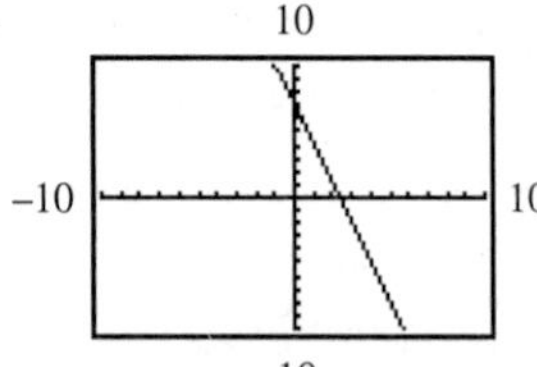

3.

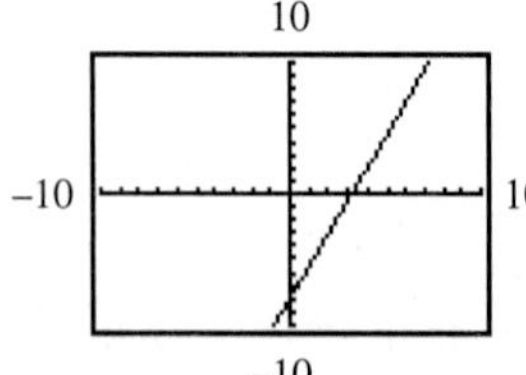

5.

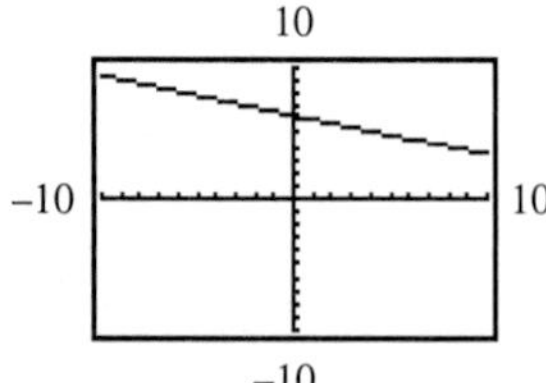

Exercise Set 3.2

1. (6, 0); (4, −2); (5, −1)

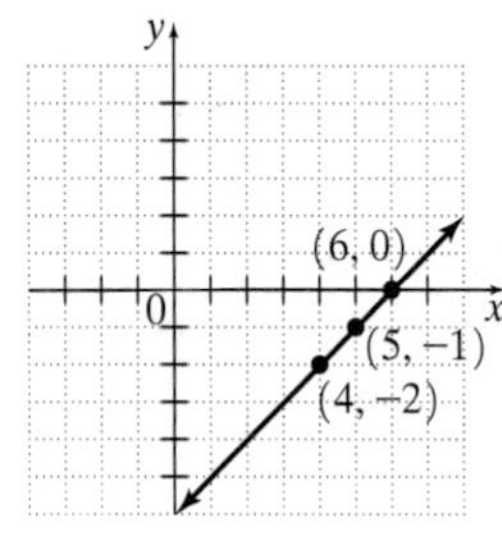

3. (1, −4); (0, 0); (−1, 4)

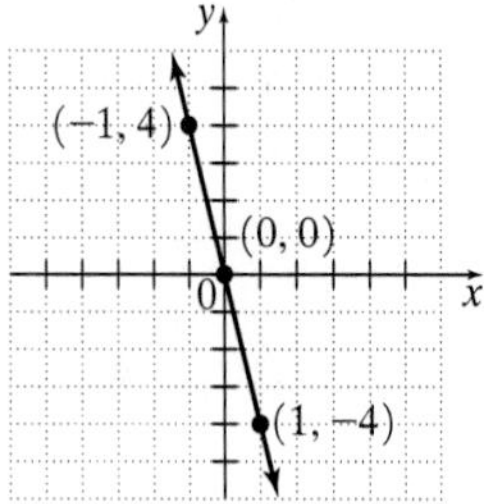

5. (0, 0); (6, 2); (−3, −1)

y
(6, 2)
x
(0, 0)
(−3, −1)

7. (0, 3); (1, −1); (2, −5)

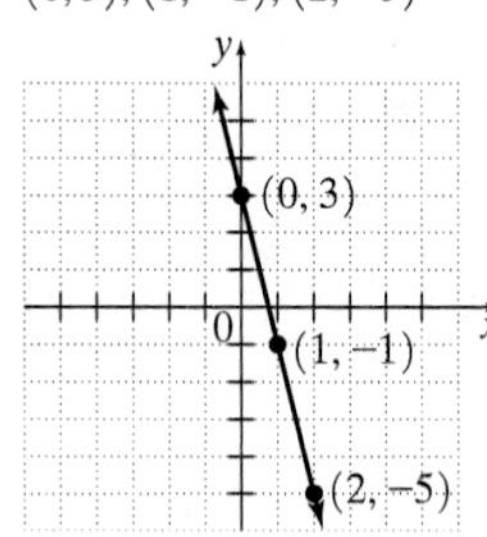

9.

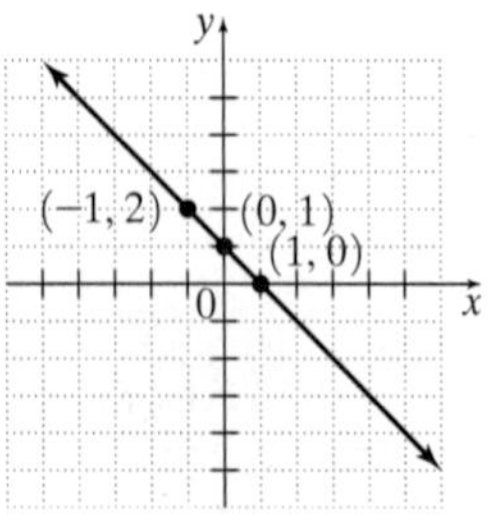

11.

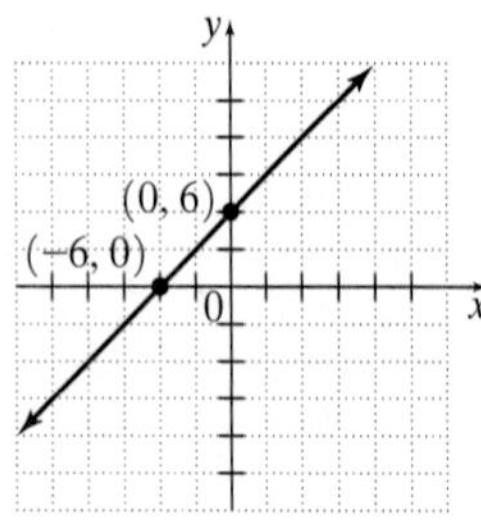

13.

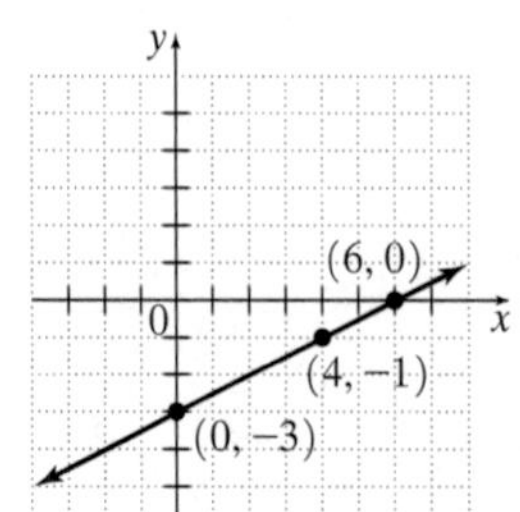

15.

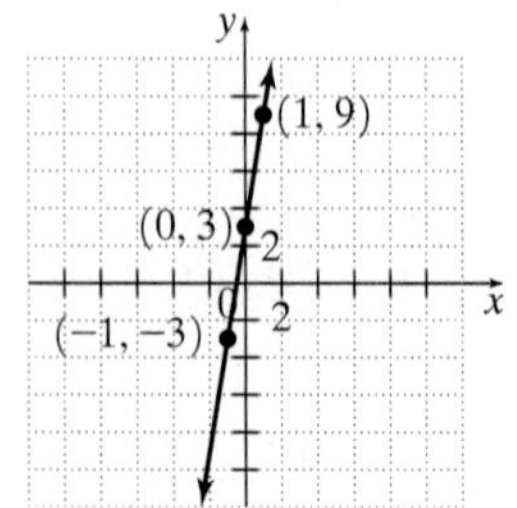

17.

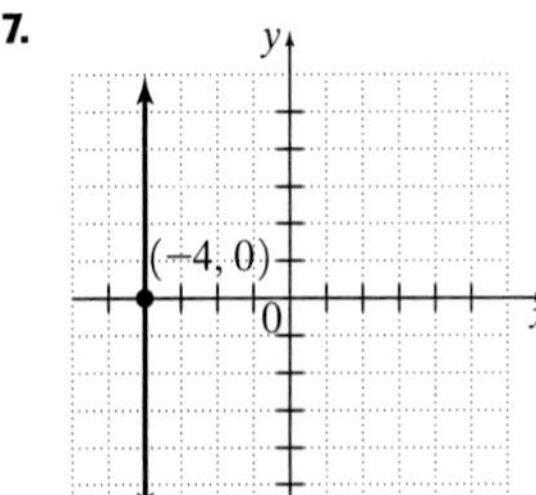

19.

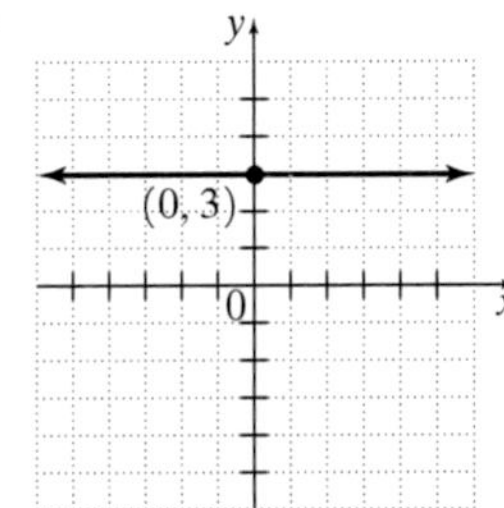

21.

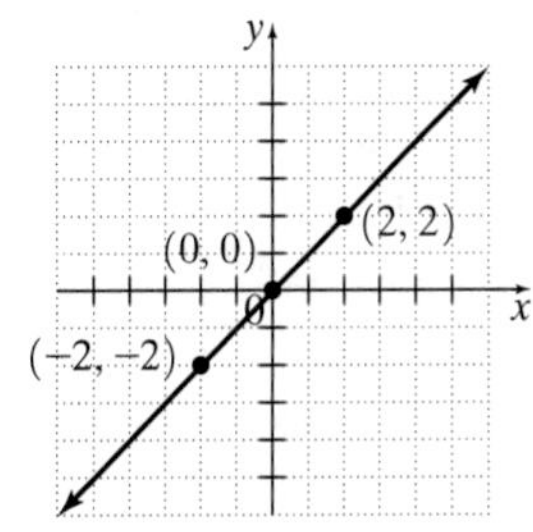

23.

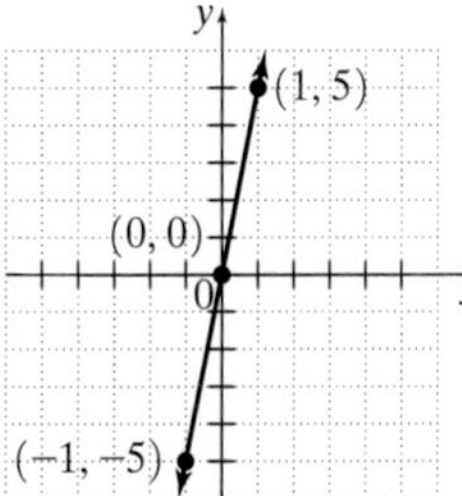

25.

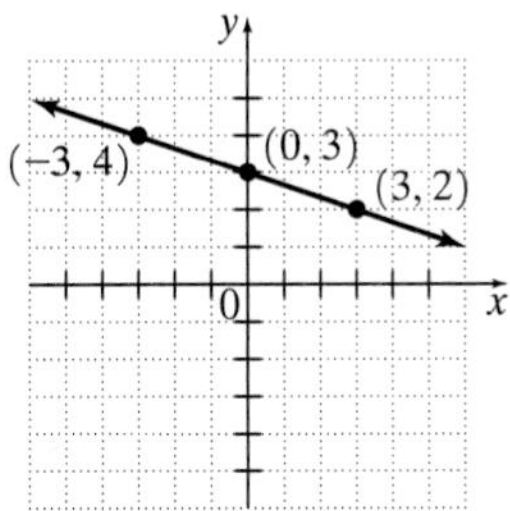

27.

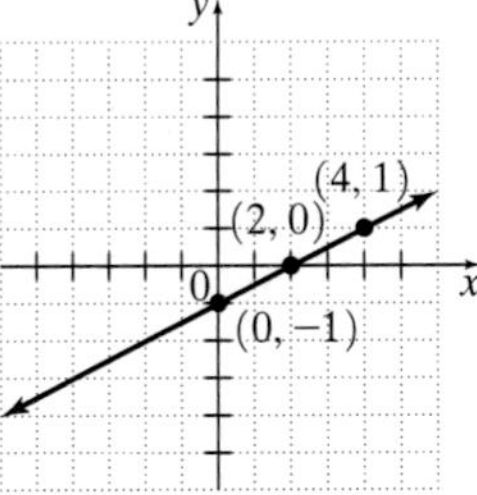

29.

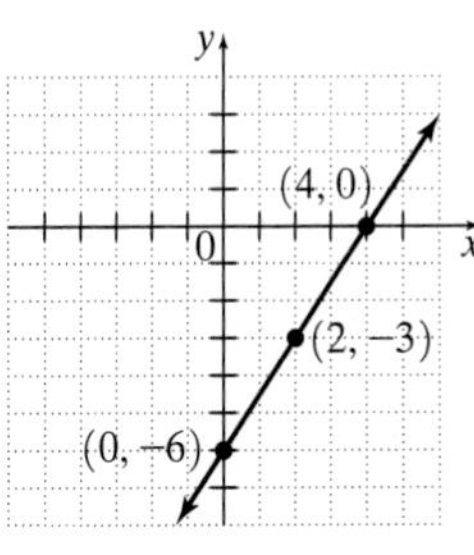

31.

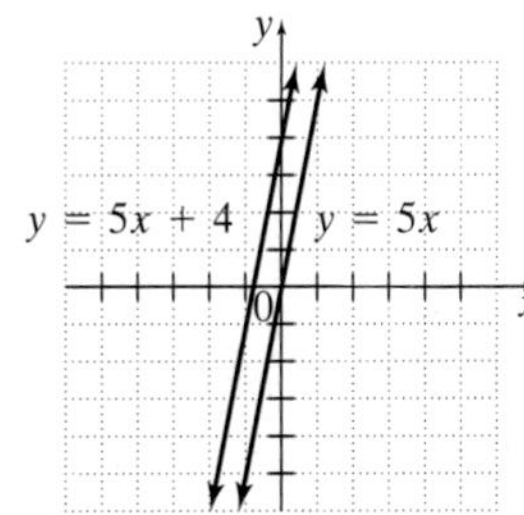

answers may vary

33.

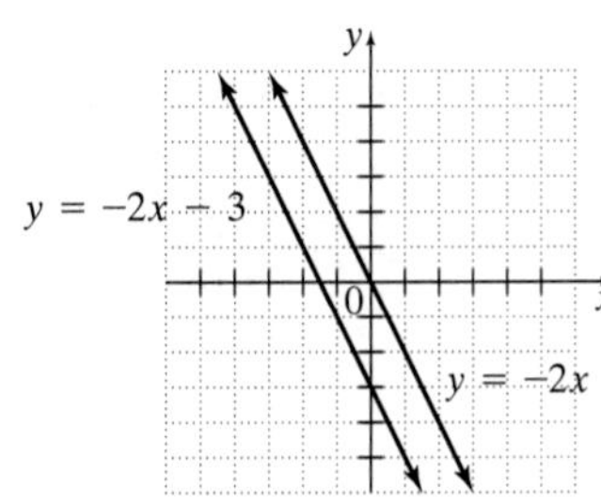

answers may vary

35. $(4, -1)$ **37.** $(0, 3); (-3, 0)$ **39.** $(0, 0); (0, 0)$ **41.** $(0, 0), (1, 1), (-1, 1), (2, 4), (-2, 4)$;

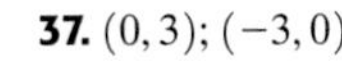

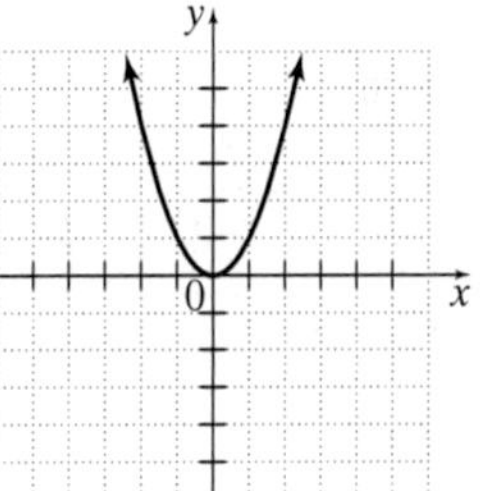

43. $x + y = 12$; 9 cm **45. a.**

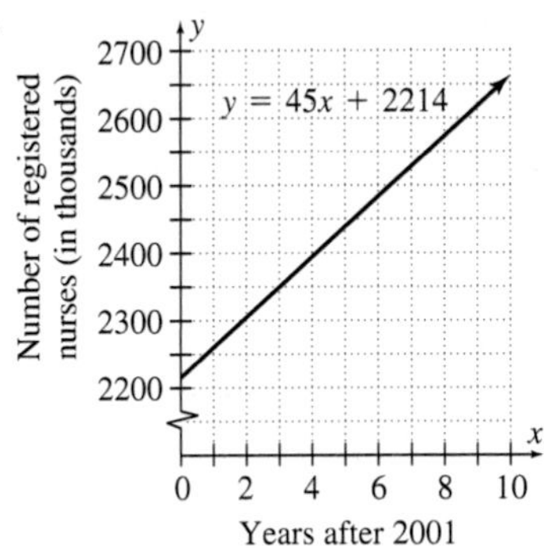

b. yes; answers may vary

47. a.

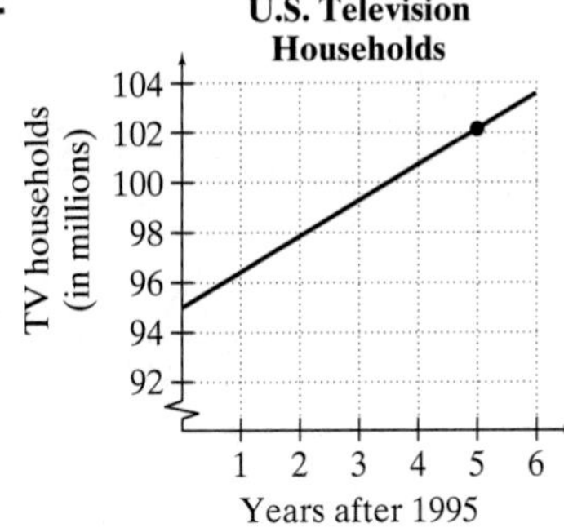

b. $(5, 102.15)$

c. In 2000, there were 102.15 million households in the United States with at least one television.

Graphing Calculator Explorations

1.

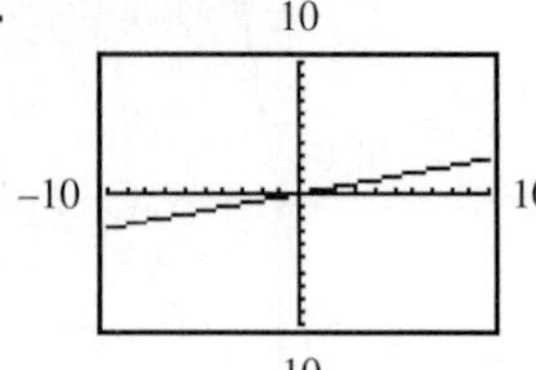

3.

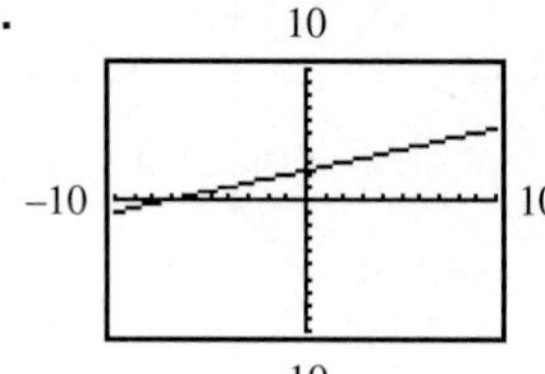

Mental Math

1. false **3.** true

Exercise Set 3.3

1. $(-1, 0)$; $(0, 1)$ **3.** $(-2, 0)$; $(1, 0)$; $(3, 0)$; $(0, 3)$ **5.** infinite number **7.** 0

9. (3, 0); (0, −3) **11.** (0, 0); (5, 1)

13.

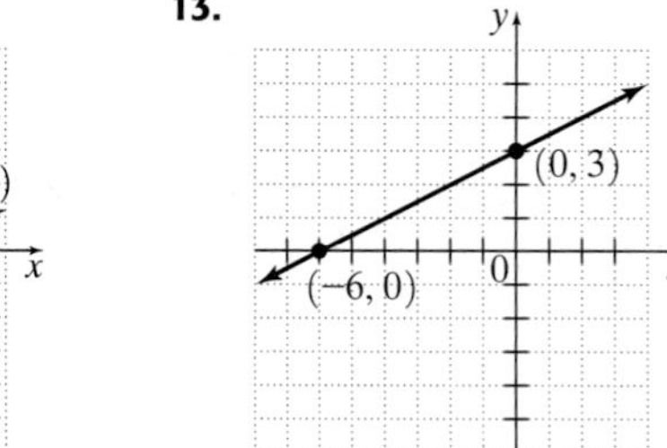

15.

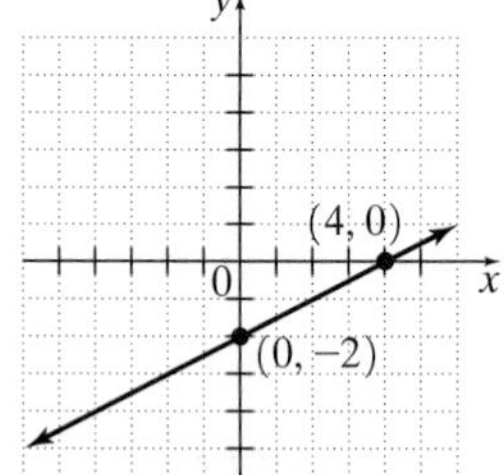

17.

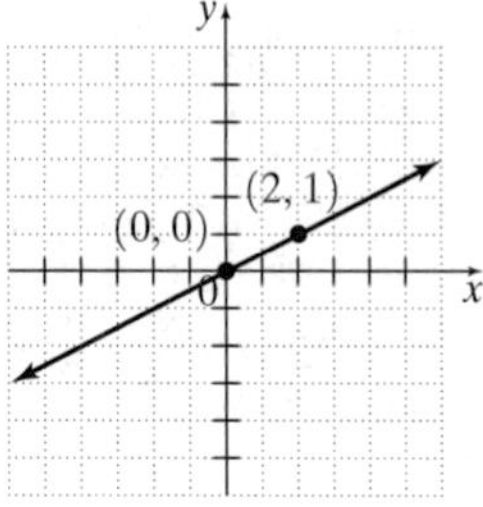

19.

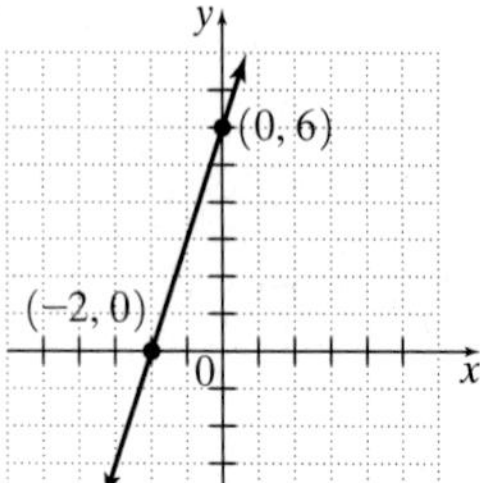

21.

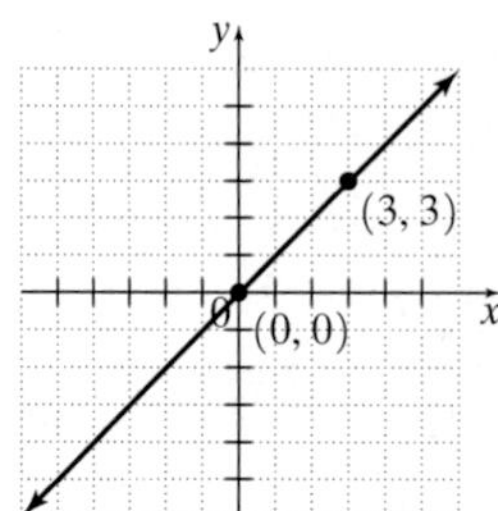

23.

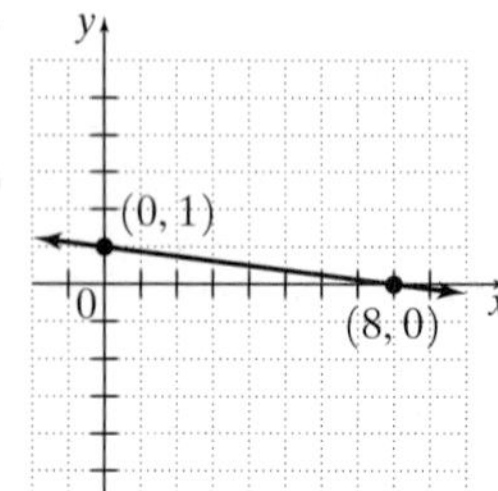

25.

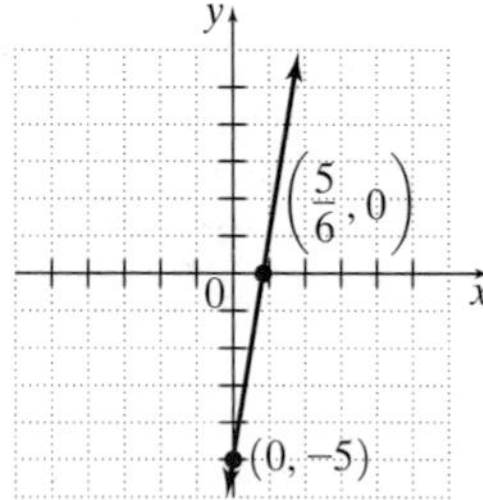

27.

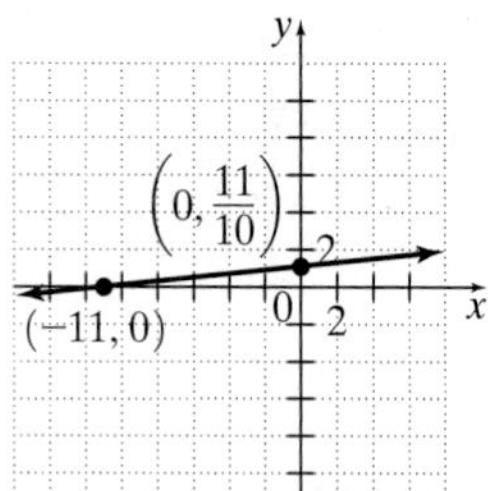

29.

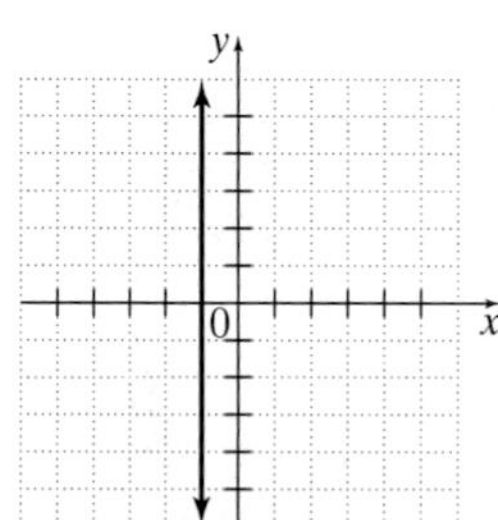

31.

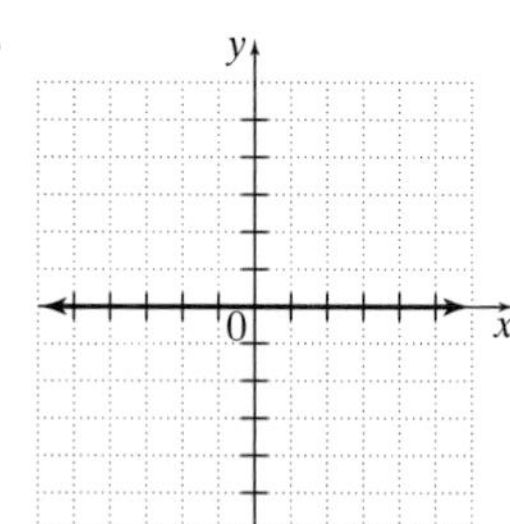

33.

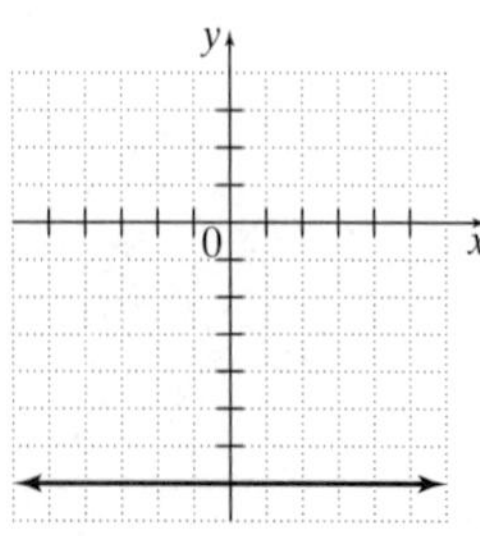

35.

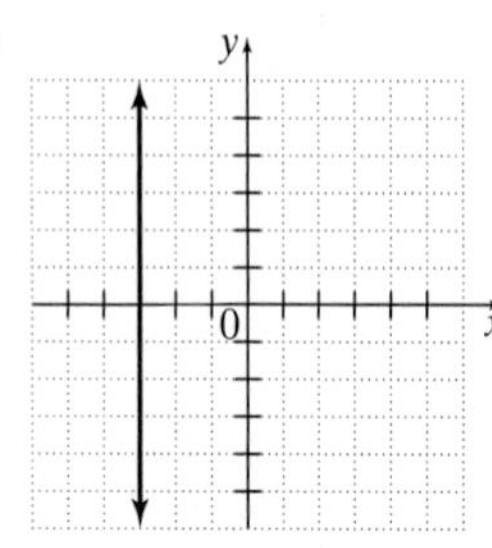

37. $\frac{3}{2}$ 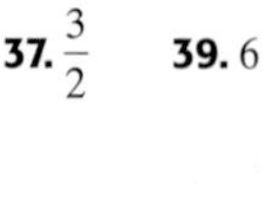 **39.** 6 **41.** $-\frac{6}{5}$ 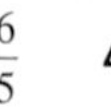 **43.** C **45.** A

47. answers may vary **49. a.** (0, 200); no chairs and 200 desks are manufactured **b.** (400, 0); 400 chairs and no desks are manufactured **c.** 300 chairs

51.

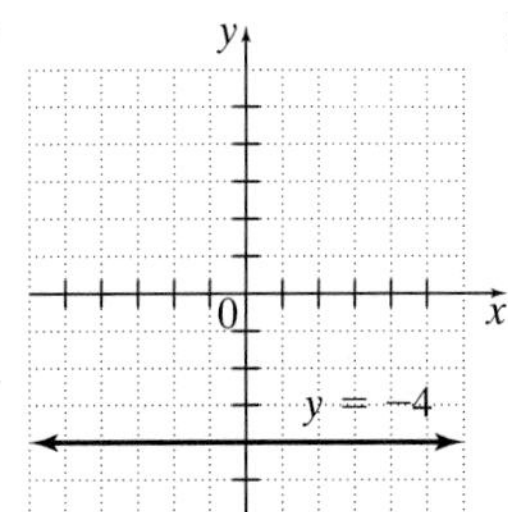

; $y = -4$ **53. a.** (0, 560.2) **b.** In 1996, the number of Disney Stores was about 560.2.

GRAPHING CALCULATOR EXPLORATIONS

1.

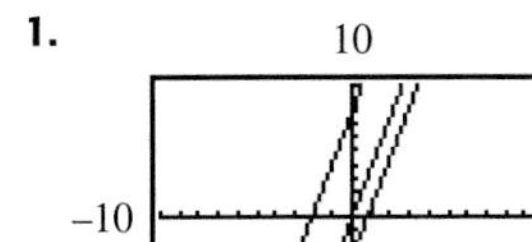
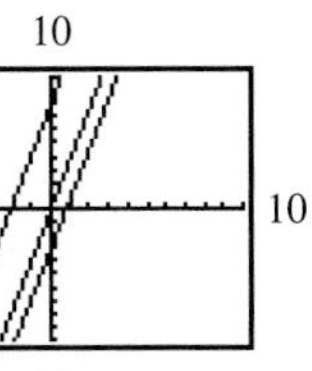

3.

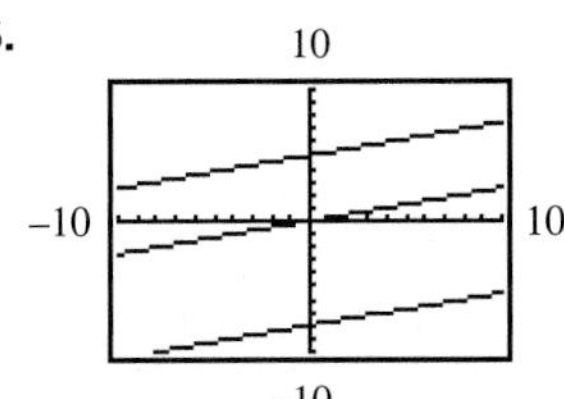

MENTAL MATH

1. upward **3.** horizontal

EXERCISE SET 3.4

1. $-\frac{4}{3}$ **3.** $\frac{5}{2}$ **5.** $\frac{8}{7}$ **7.** −1 **9.** $-\frac{1}{4}$ **11.** $-\frac{2}{3}$ **13.** 0 **15.** line 1 **17.** line 2 **19.** 5 **21.** −2 **23.** $\frac{2}{3}$ **25.** $\frac{1}{2}$ **27.** undefined slope **29.** undefined slope **31.** 0 **33.** neither **35.** perpendicular **37.** parallel **39. a.** 1 **b.** −1 **41. a.** $\frac{9}{11}$ **b.** $-\frac{11}{9}$ **43.** $\frac{3}{5}$ **45.** 12.5% **47.** 40% **49.** 0.02 **51.** Every 1 year, there are/should be 15 million more Internet users. **53.** It costs $0.36 per 1 mile to own and operate a compact car. **55.** $y = 2x - 14$ **57.** $y = -6x - 11$ **59.** D **61.** B **63.** E **65.** 28.3 **67.** 1992; 27.6 **69.** from 1992 to 1993 **71.** $x = 6$ **73. a.** (1994, 782) (2001, 1132) **b.** 50 **c.** For the years 1994 through 2001, the price per acre of U.S. farmland rose $50 every year. **75. a.** (0, 1485) **b.** In 1998 there were 1485 million admissions to movie theaters in the U.S. and Canada. **c.** −30 **d.** For the years 1998 through 2000, the number of movie theater admissions has decreased at a rate of 30 million per year. **77.** The slope through (−3, 0) and (1, 1) is $\frac{1}{4}$. The slope through (−3, 0) and (−4, 4) is −4. The product of the slopes is −1 so the sides are perpendicular. **79.** −0.25 **81.** 0.875 **83.** the line becomes steeper

INTEGRATED REVIEW

1. 2 **2.** 0 **3.** $-\frac{2}{3}$ **4.** undefined **5.**

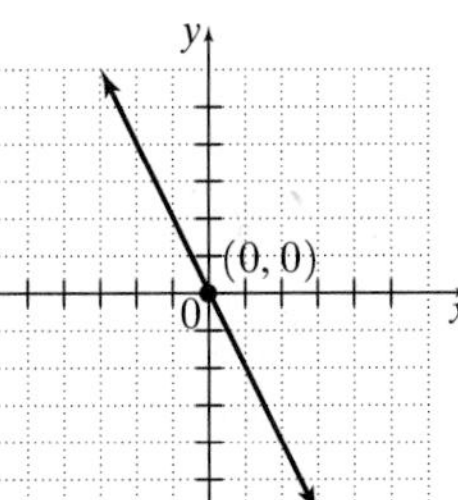

6.

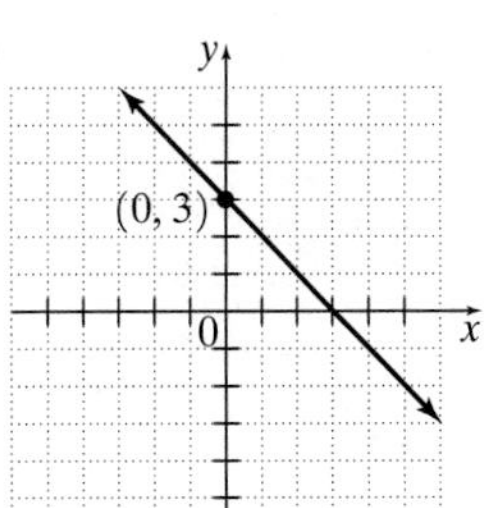

7.

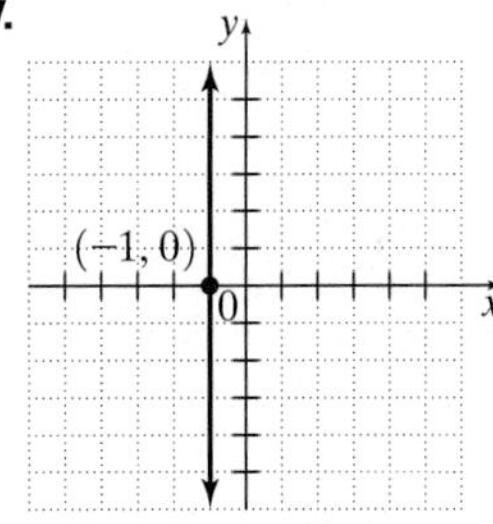

8.

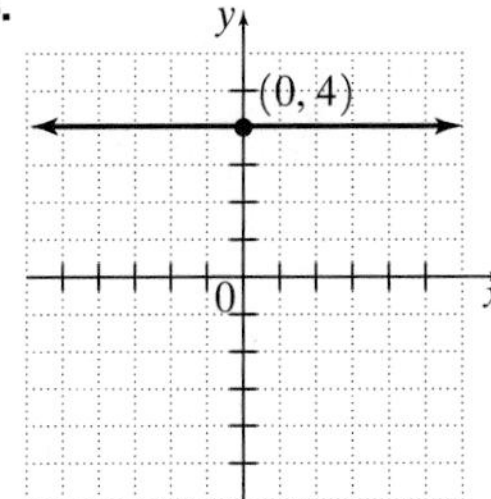
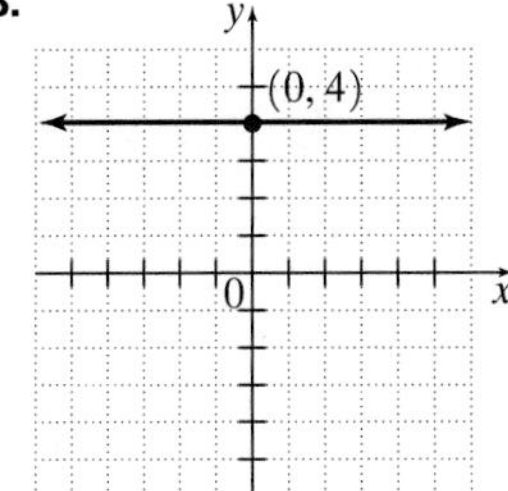

9.

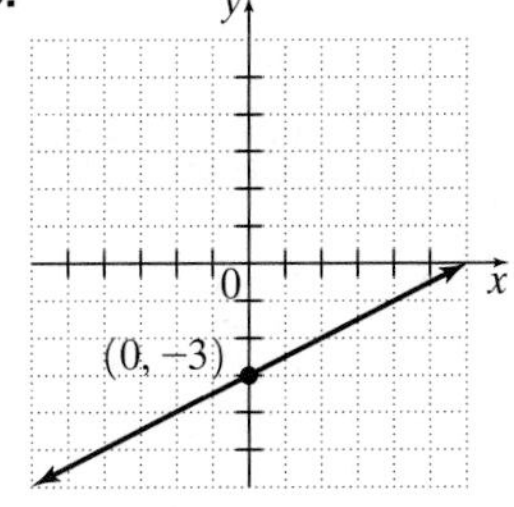

10.

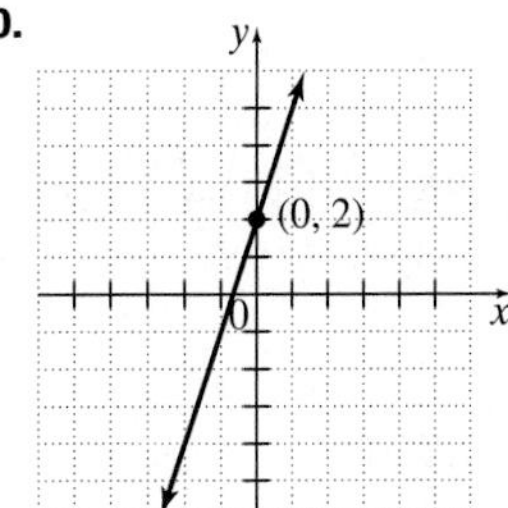

11. 3 **12.** −6 **13.** $-\frac{7}{2}$ **14.** 2 **15.** undefined **16.** 0 **17.** neither **18.** perpendicular

19. a. (1997, 11.6) (2001, 15.4) **b.** 0.95 **c.** For the years 1997 through 2001, the number of grill units shipped increased at a rate of 0.95 million per year.

Mental Math

1. yes **3.** yes **5.** yes **7.** no

Exercise Set 3.5

1. no; no **3.** yes; no **5.** no; yes **7.**

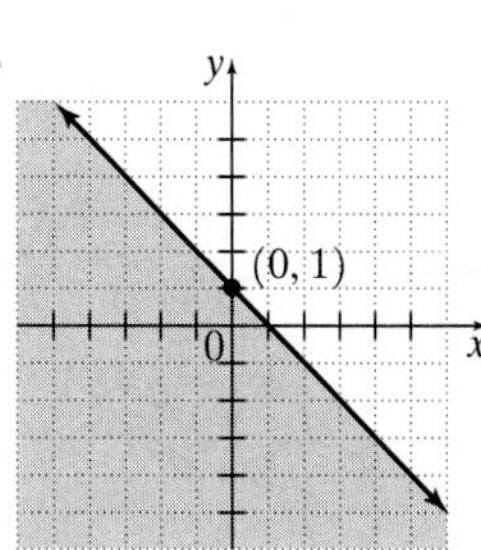

9.

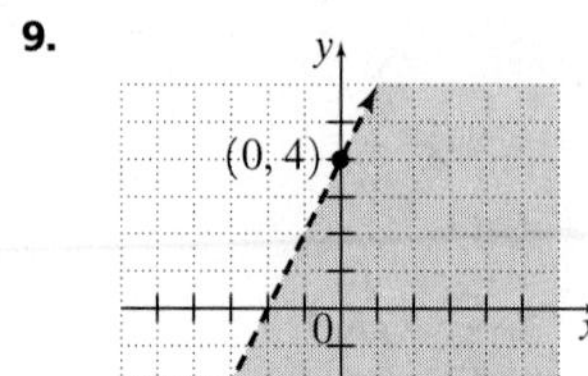

11.

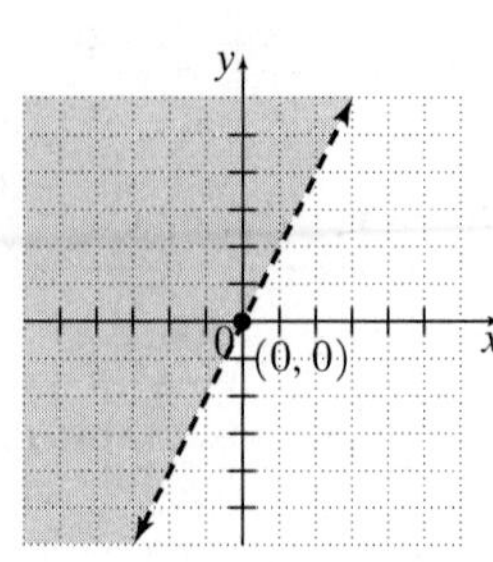

13.

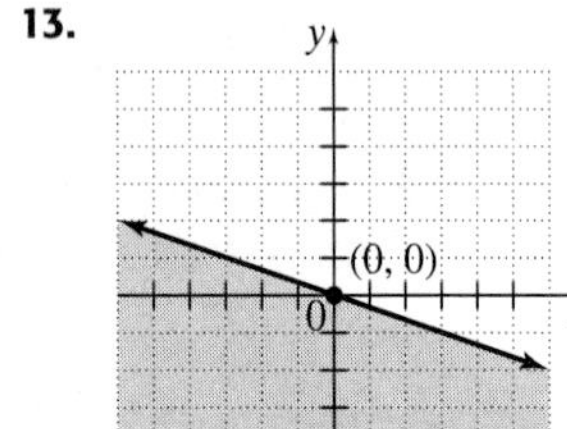

15.

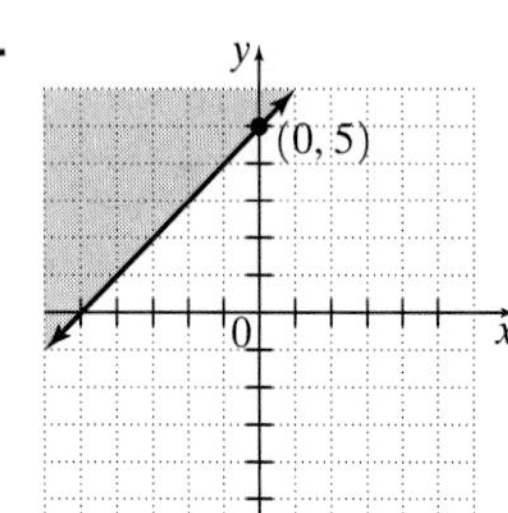

17.

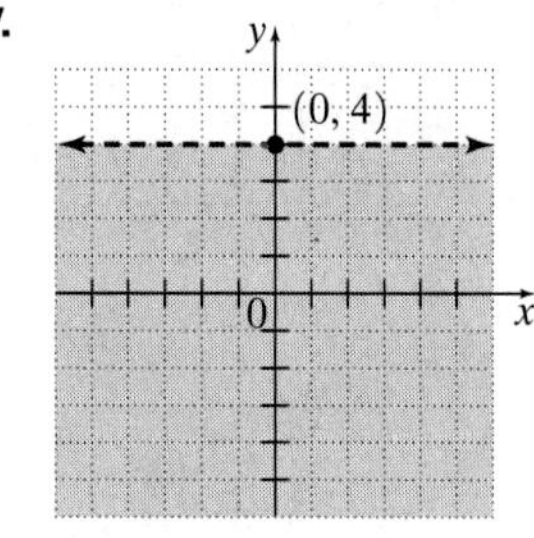

19.

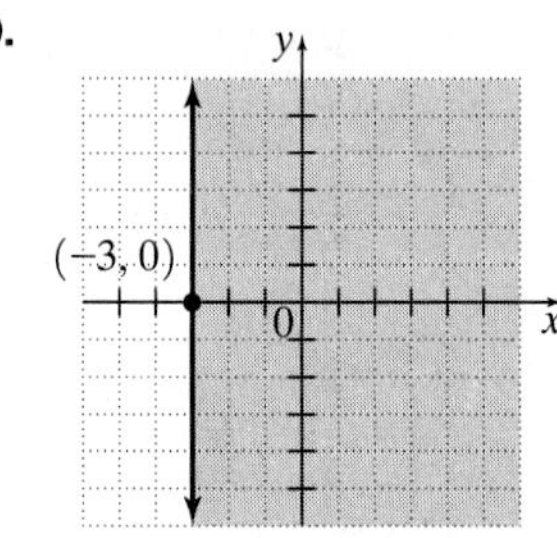

21.

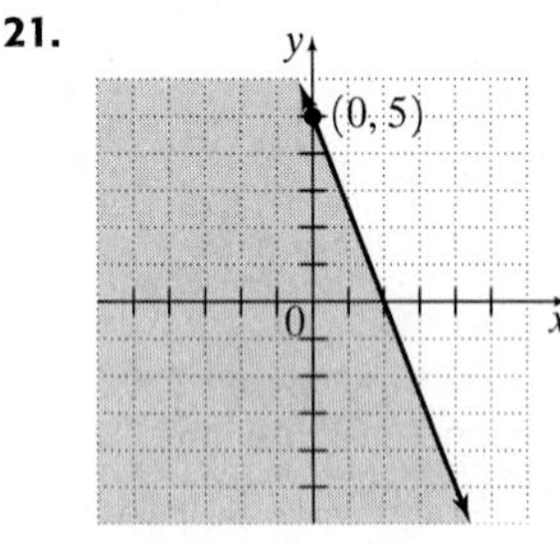

23.

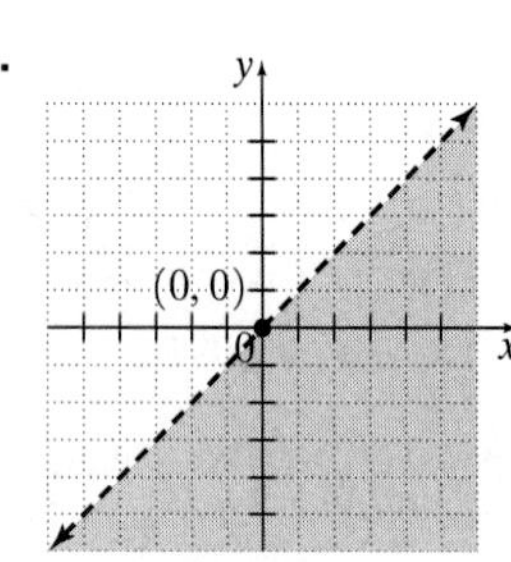

25.

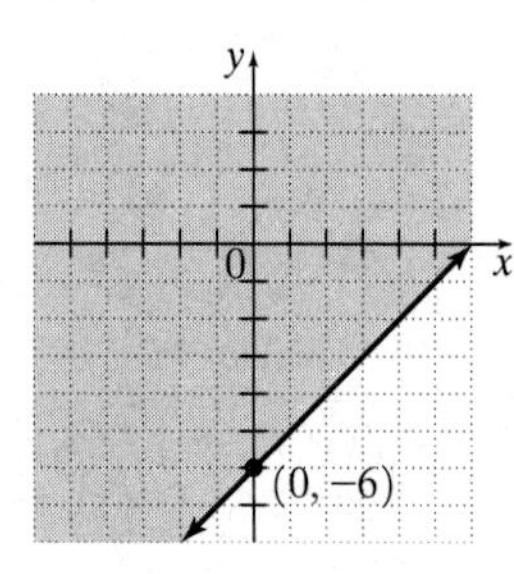

27.

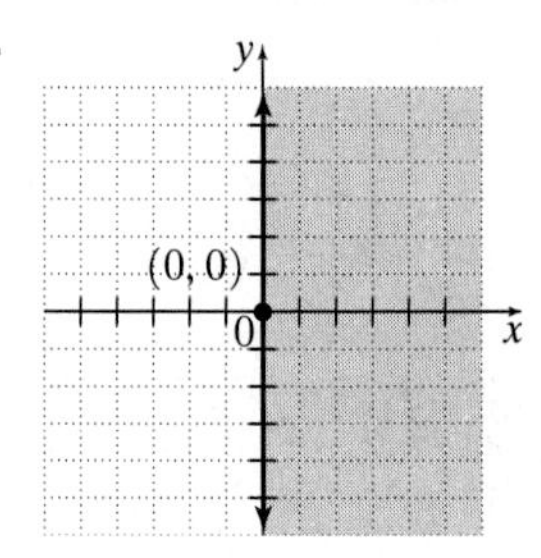

29.

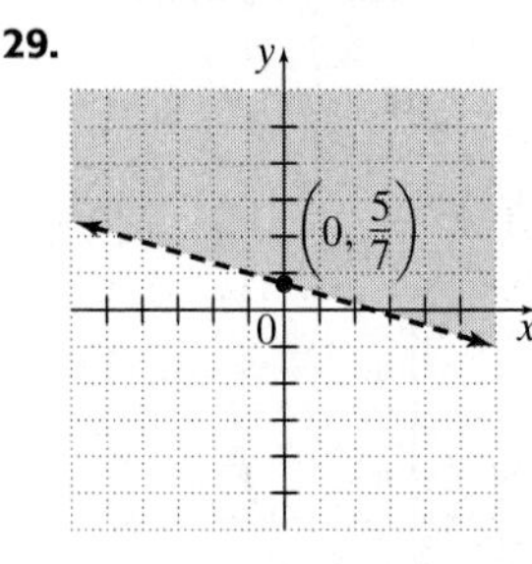

31.

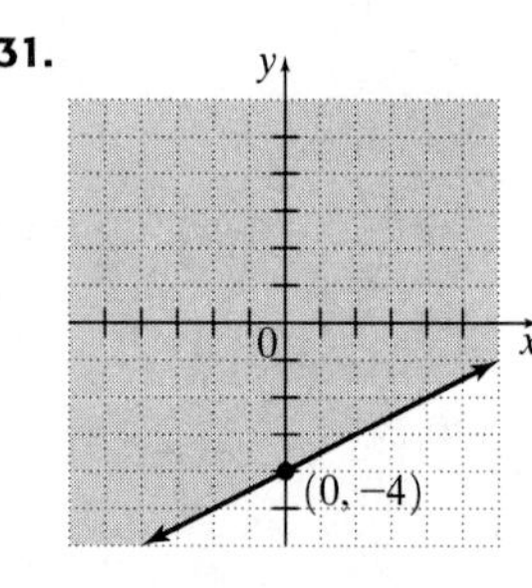

33. $(-2, 1)$ **35.** $(-3, -1)$ **37.** A **39.** B **41.** answers may vary
43. a. $30x + 0.15y \leq 500$ **b.**
c. answers may vary

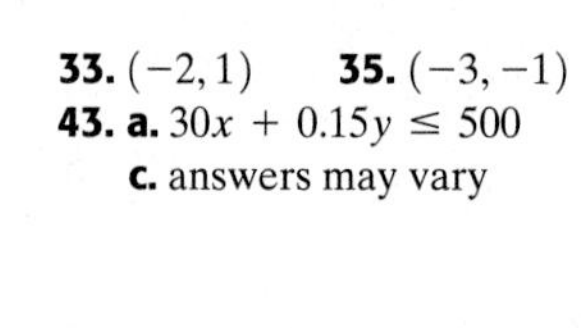

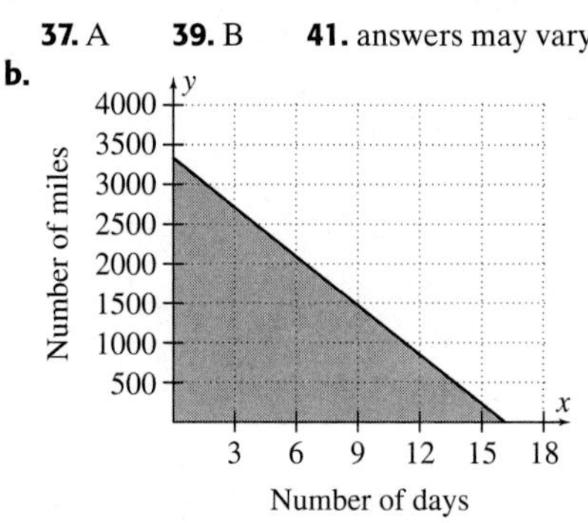

Chapter 3 Review

1. \$2600 million **2.** \$200 million **3.** 1998 **4.** answers may vary **5.** 86 million **6.** 33 million **7.** 2001
8. number of subscribers is increasing

9–14.

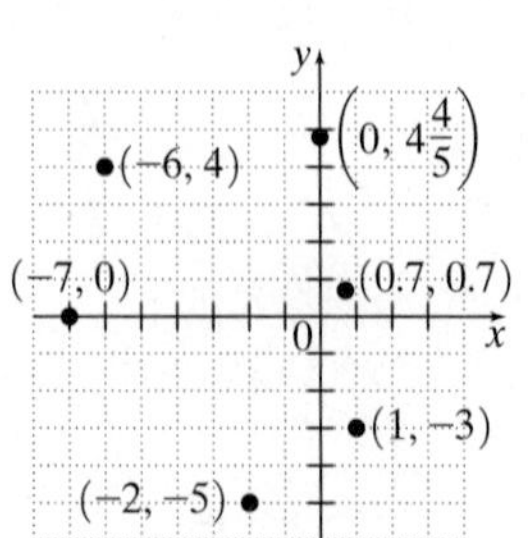
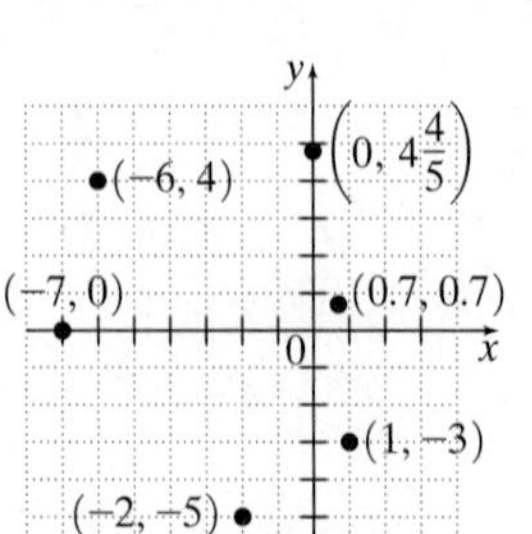

15. (7, 44) **16.** $\left(-\frac{13}{3}, -8\right)$

17.

x	y
-3	0
1	3
9	9

18.

x	y
7	5
-7	5
0	5

19.

x	y
0	0
10	5
-10	-5

20. a. 2005; 2500; 7000 **b.** 886 compact disc holders

21.

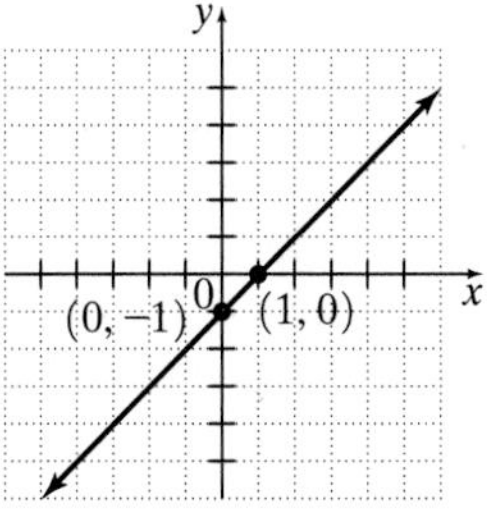

22.

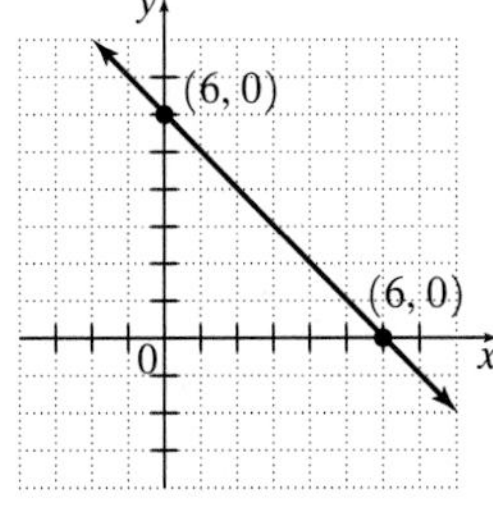

23.

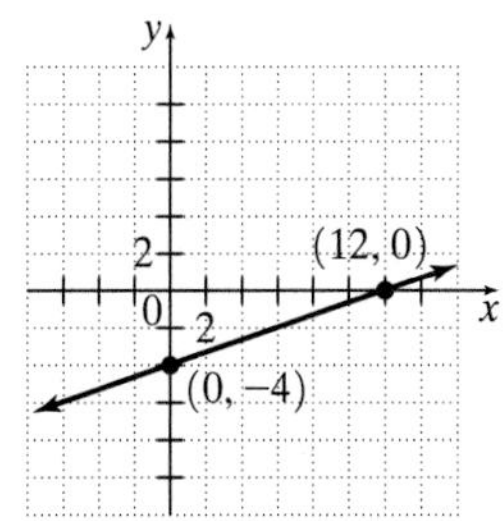

24.

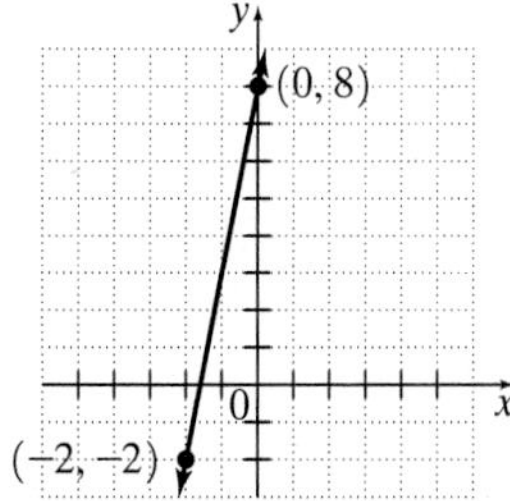

25.

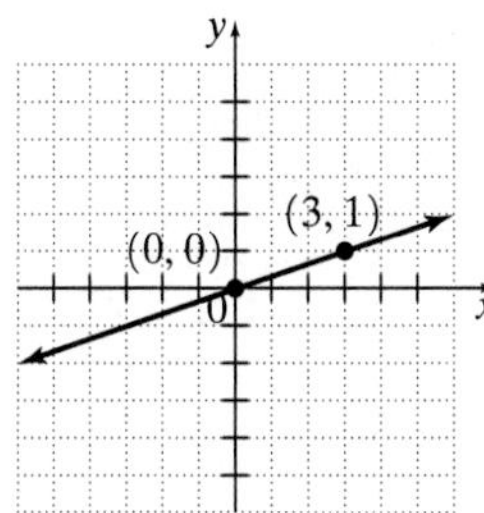

26.

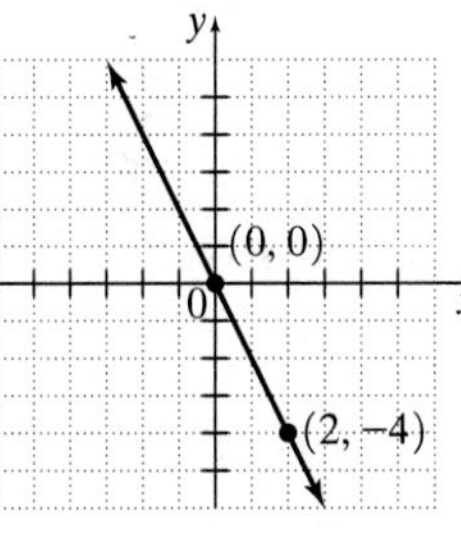

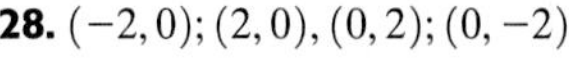
27. $(4, 0); (0, -2)$ **28.** $(-2, 0); (2, 0), (0, 2); (0, -2)$

29.

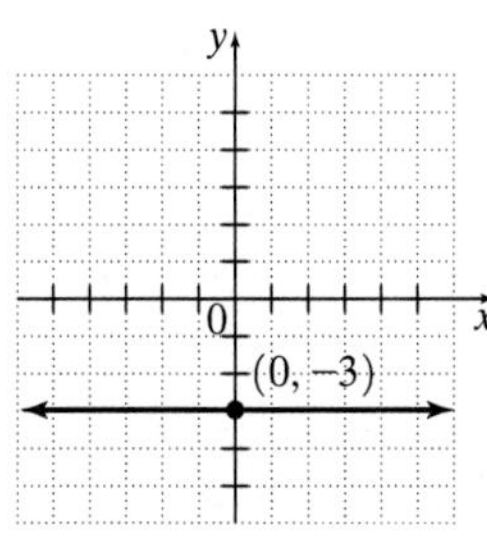

30.

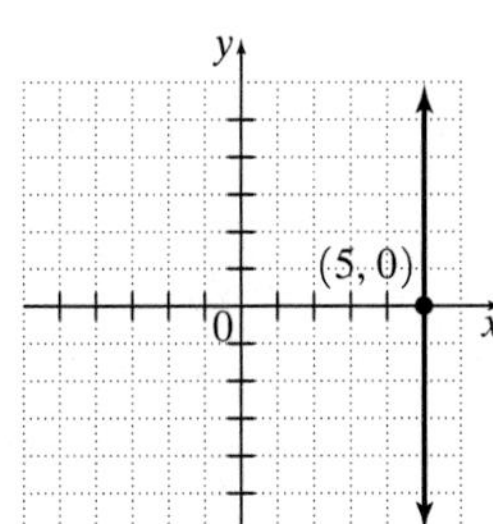

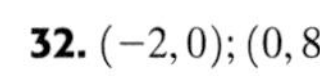
31. $(0, -4); (12, 0)$ **32.** $(-2, 0); (0, 8)$ **33.** $-\frac{3}{4}$ **34.** $\frac{1}{5}$

35. D **36.** B **37.** C **38.** A **39.** $\frac{3}{4}$ **40.** $\frac{5}{3}$ **41.** 4 **42.** -1 **43.** 3 **44.** $\frac{1}{2}$ **45.** 0 **46.** undefined **47.** perpendicular **48.** parallel **49.** neither **50.** Every 1 year, 1.24 million more persons have a bachelor's degree or higher. **51.** Every 1 year, 27 million more people go on vacations.

52.

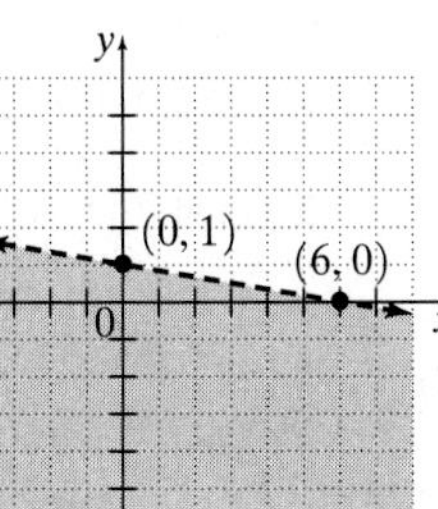

53.

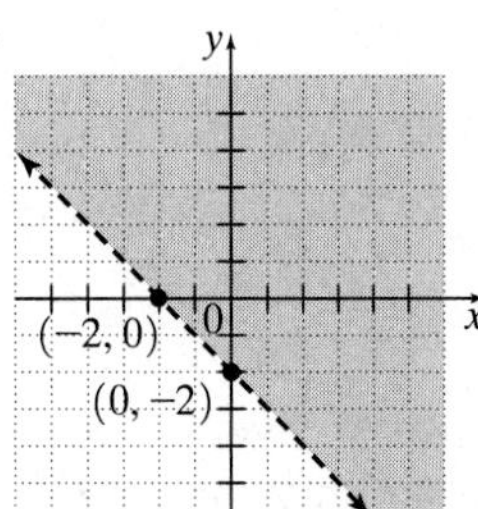

54.

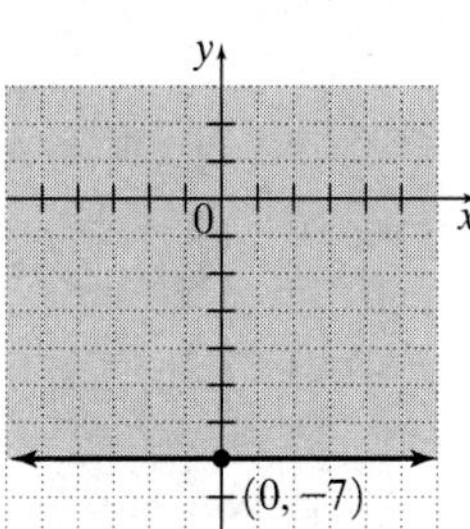

55.

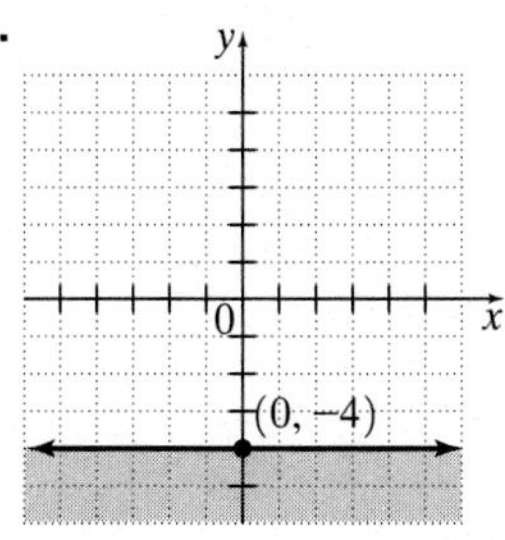

56.

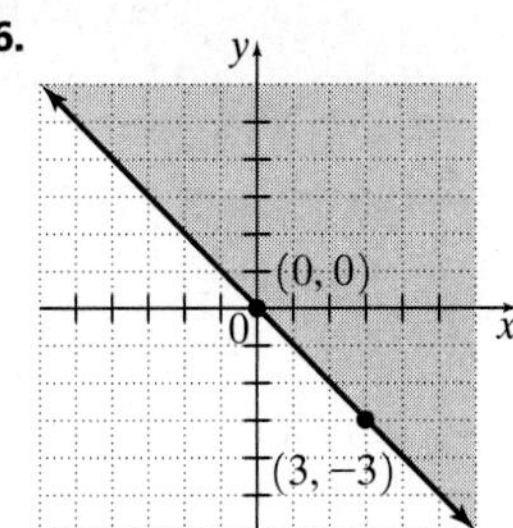

57.

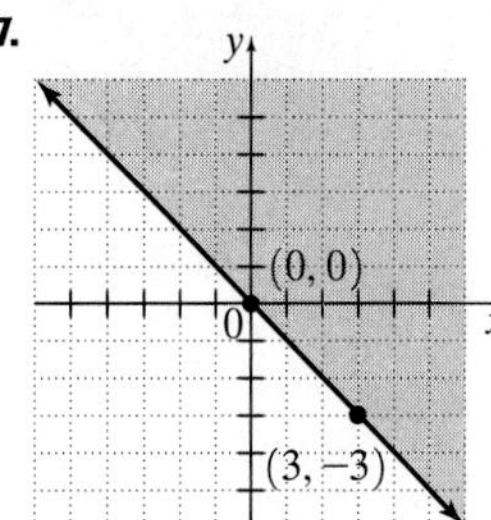

Chapter 3 Test

1. (1, 1) **2.** (−4, 17) **3.** $\frac{2}{5}$ **4.** 0 **5.** −1 **6.** −7 **7.** 3 **8.** undefined

9.
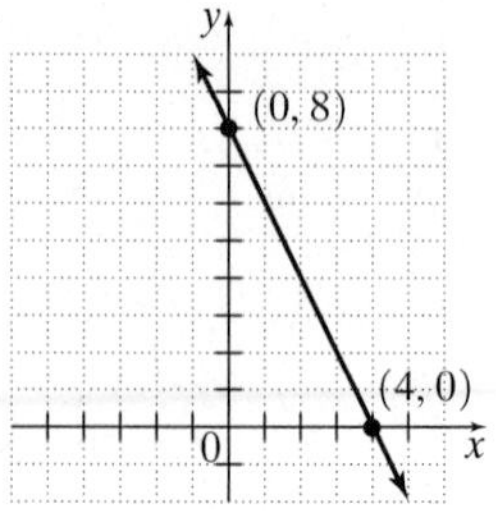

10.
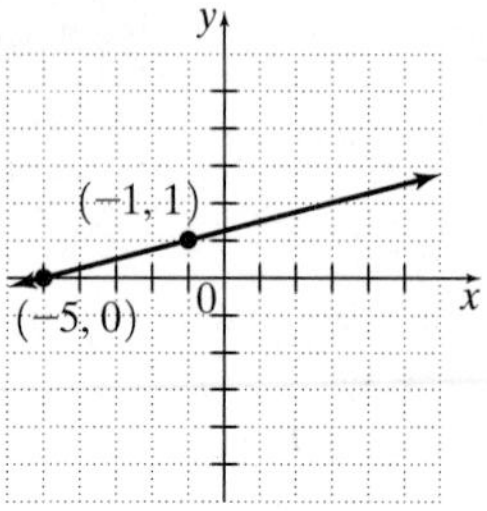

11.
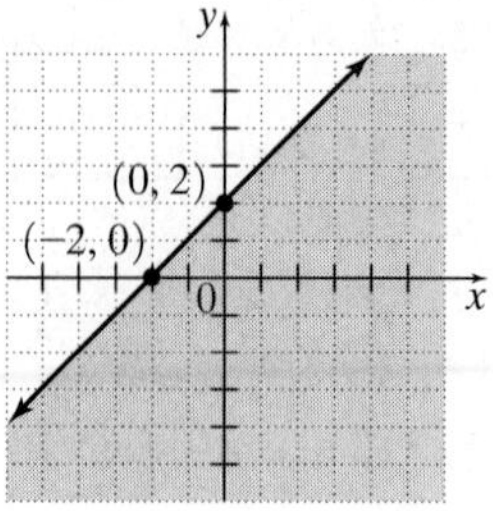

12.
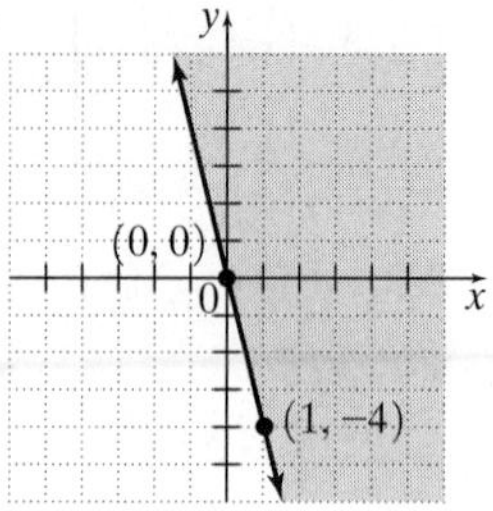

13.
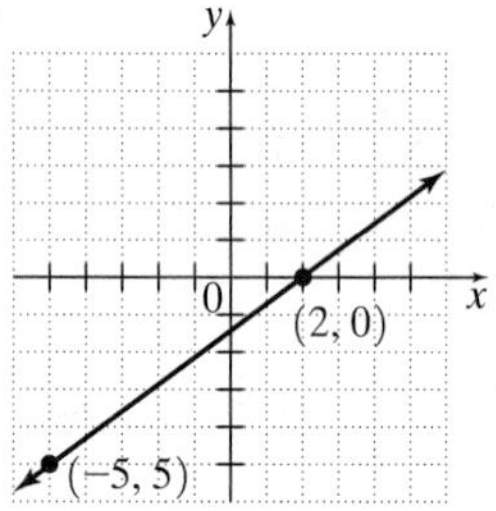

14.
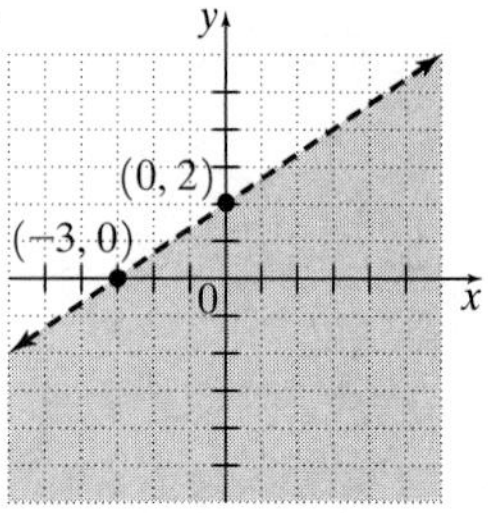

15.
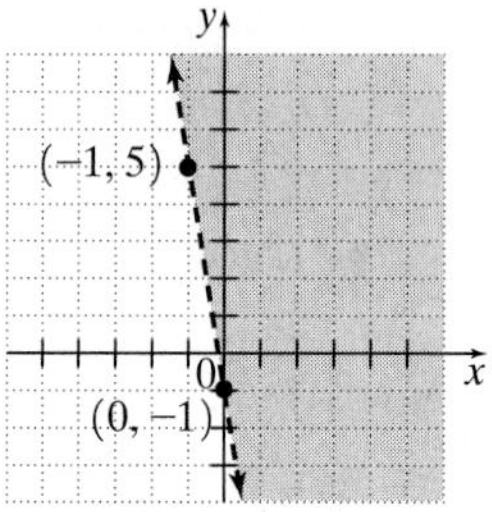

16.
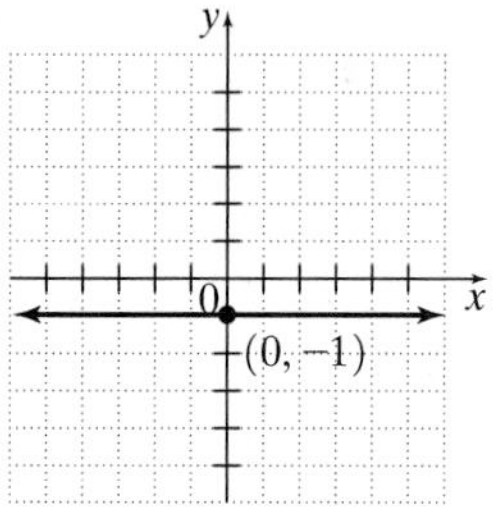

17. neither **18.** parallel **19.** $x + 2y = 21$; $x = 5$ m
20. a. (1986, 38), (1988, 44), (1990, 50), (1992, 53), (1994, 57), (1996, 62), (1998, 67), (2000, 69) **b.**
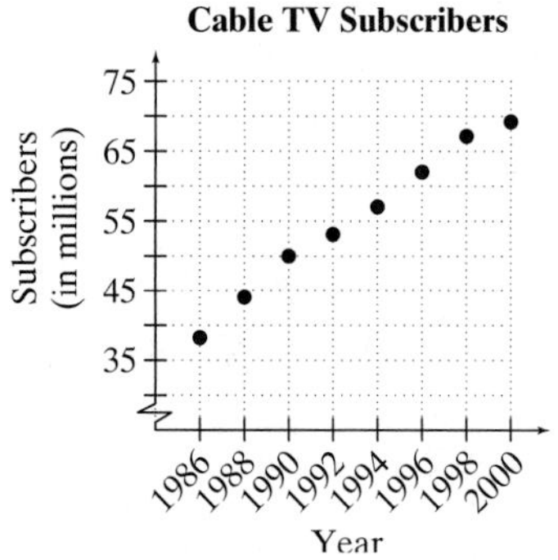

21. Every 1 year, 30 million fewer movie tickets are sold.
22. \$380 billion **23.** \$1230 billion **24.** \$375 billion **25.** 2002

Cumulative Review

1. a. 11, 112 **b.** 0, 11, 112 **c.** −3, −2, 0, 11, 112 **d.** −3, −2, 0, $\frac{1}{4}$, 5.1, 11, 112 **e.** $\sqrt{2}$ **f.** −2, 0, $\frac{1}{4}$, 112, −3, 5.1, 11, $\sqrt{2}$; Sec. 1.2, Ex. 10
2. a. 9 **b.** 125 **c.** 16 **d.** 7 **e.** $\frac{9}{49}$; Sec. 1.3, Ex. 1 **3.** $\frac{1}{4}$; Sec. 1.3, Ex. 4 **4. a.** $x + 3$ **b.** $3x$ **c.** $2x$ **d.** $10 - x$ **e.** $5x + 7$; Sec. 1.3, Ex. 9
5. 27; Sec. 1.3, Ex. 2 **6.** 54; Sec. 1.3, Ex. 5 **7.** 6.7; Sec. 1.4, Ex. 11 **8. a.** $\frac{1}{2}$ **b.** 9; Sec. 1.5, Ex. 8 **9.** 3; Sec. 1.6, Ex. 11a
10. −70; Sec. 1.6, Ex. 11d **11.** $5x + 10$; Sec. 2.1, Ex. 8 **12.** $-2y - 0.6z + 2$; Sec. 2.1, Ex. 9 **13.** $-x - y + 2z - 6$; Sec. 2.1, Ex. 10
14. $2x + 6$; Sec. 2.1, Ex. 16 **15.** $(x - 4) \div 7$; Sec. 2.1, Ex. 17 **16.** $6 + x$; Sec. 2.1, Ex. 18 **17.** 6; Sec. 2.3, Ex. 1 **18.** $a = 19$; Sec. 2.2, Ex. 6
19. $y = 140$; Sec. 2.3, Ex. 4 **20.** $x = 4$; Sec. 2.4, Ex. 5 **21.** 10; Sec. 2.5, Ex. 1 **22.** 40 ft; Sec. 2.6, Ex. 2 **23.** 800; Sec. 2.7, Ex. 2
24. a. US/Canada region; 167 mil **b.** 54 million; Sec. 3.1, Ex. 1
25. a. (0, 12) **b.** (2, 6); Sec. 6.1, Ex. 3 **c.** (−1, 15); Sec. 3.1, Ex. 5 **26.**
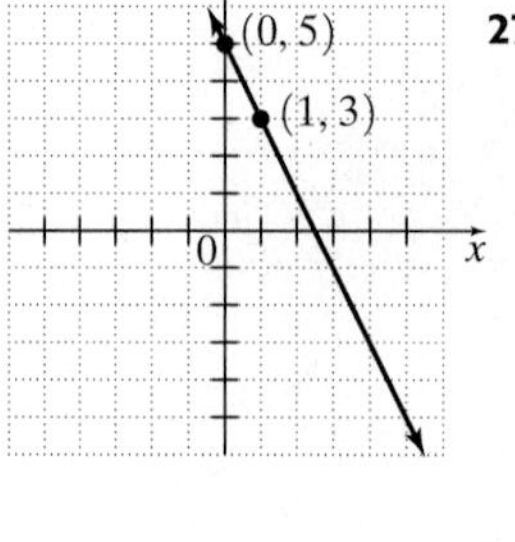

; Sec. 3.2, Ex. 1
27. $\frac{2}{3}$; Sec. 3.4, Ex. 3

Chapter 4 Exponents and Polynomials

Chapter 4 Pretest

1. $\frac{9}{16}$; 4.1A **2.** $8y^{13}$; 4.1B **3.** $\frac{b^{11}}{a^3}$; 4.1D **4.** 3; 4.1E **5.** −216; 4.2A **6.** $\frac{m^{16}}{n^{18}}$; 4.2B **7.** $4x^2 + 10x - 5$; 4.3D **8.** 8.14×10^{-7}; 4.2C **9.** 5; 4.3B **10.** 1; 4.3C **11.** $10x^2 + 1$; 4.4A **12.** $9y^2 - 5y - 3$; 4.4B **13.** $-a^2 - 16b^2$; 4.4D **14.** $-\frac{3}{8}n^9$; 4.5A **15.** $-6t^7 - 8t^5 + 16t^2$; 4.5B **16.** $10y^2 + 7y - 6$; 4.5C **17.** $49a^2 - 70a + 25$; 4.6B **18.** $16b^2 - 81$; 4.6C **19.** $4p^3 - 2p^2 + 5p$; 4.7A **20.** $5x + 2$; 4.7B

Mental Math

1. base: 3; exponent: 2 **3.** base: −3; exponent: 6 **5.** base: 4; exponent: 2 **7.** base: 5; exponent: 1; base: 3; exponent: 4 **9.** base: 5; exponent: 1; base: *x*; exponent: 2

Exercise Set 4.1

1. 49 **3.** −5 **5.** −16 **7.** 16 **9.** $\frac{1}{27}$ **11.** 112 **13.** answers may vary **15.** 4 **17.** 135 **19.** 150 **21.** $\frac{32}{5}$ **23.** x^7 **25.** $(-3)^{12}$ **27.** $15y^5$ **29.** $-24z^{20}$ **31.** $20x^5$ sq ft **33.** x^{36} **35.** p^7q^7 **37.** $8a^{15}$ **39.** $\frac{m^9}{n^9}$ **41.** $x^{10}y^{15}$ **43.** $\frac{4x^2z^2}{y^{10}}$ **45.** $64z^{10}$ sq decimeters **47.** $27y^{12}$ cubic ft **49.** x^2 **51.** 4 **53.** p^6q^5 **55.** $\frac{y^3}{2}$ **57.** 1 **59.** −2 **61.** 2 **63.** answers may vary **65.** −25 **67.** $\frac{1}{64}$ **69.** z^8 **71.** $81x^2y^2$ **73.** 1 **75.** 40 **77.** b^6 **79.** a^9 **81.** $-16x^7$ **83.** $64a^3$ **85.** $36x^2y^2z^6$ **87.** $\frac{y^{15}}{8x^{12}}$ **89.** $3x$ **91.** $2x^2y$ **93.** −2 **95.** 5 **97.** −7 **99.** 343 cubic m **101.** volume **103.** x^{9a} **105.** a^{5b} **107.** x^{5a} **109.** \$1045.85

Calculator Explorations

1. 5.31 EE 3 **3.** 6.6 EE −9 **5.** 1.5×10^{13} **7.** 8.15×10^{19}

Mental Math

1. $\frac{5}{x^2}$ **3.** y^6 **5.** $4y^3$

Exercise Set 4.2

1. $\frac{1}{64}$ **3.** $\frac{7}{x^3}$ **5.** −64 **7.** $\frac{5}{6}$ **9.** p^3 **11.** $\frac{q^4}{p^5}$ **13.** $\frac{1}{x^3}$ **15.** z^3 **17.** $\frac{4}{3}$ **19.** $\frac{1}{9}$ **21.** $-p^4$ **23.** −2 **25.** x^4 **27.** p^4 **29.** m^{11} **31.** r^6 **33.** $\frac{1}{x^{15}y^9}$ **35.** $\frac{1}{x^4}$ **37.** $\frac{1}{a^2}$ **39.** $4k^3$ **41.** $3m$ **43.** $-\frac{4a^5}{b}$ **45.** $-\frac{6x}{7y^2}$ **47.** $\frac{a^{30}}{b^{12}}$ **49.** $\frac{1}{x^{10}y^6}$ **51.** $\frac{z^2}{4}$ **53.** $\frac{1}{32x^5}$ **55.** $\frac{49a^4}{b^6}$ **57.** $a^{24}b^8$ **59.** x^9y^{19} **61.** $-\frac{y^8}{8x^2}$ **63.** $\frac{25b^{33}}{a^{16}}$ **65.** $\frac{27}{z^3x^6}$ cubic in. **67.** 7.8×10^4 **69.** 1.67×10^{-6} **71.** 6.35×10^{-3} **73.** 1.16×10^6 **75.** 1.56×10^7 **77.** 1.36×10^4 **79.** 2.84×10^8 **81.** 0.0000000008673 **83.** 0.033 **85.** 20,320 **87.** 0.000000000000000000000000397 **89.** 700,000,000 **91.** 0.000036 **93.** 0.0000000000000000000028 **95.** 0.0000005 **97.** 200,000 **99.** 2.7×10^9 gal **101.** $-2x + 7$ **103.** $2y - 10$ **105.** $-x - 4$ **107.** −394.5 **109.** 1.3 sec **111.** a^m **113.** $27y^{6z}$ **115.** answers may vary **117.** answers may vary

Exercise Set 4.3

1. 1; $-3x$; 5 **3.** −5; 3.2; 1; −5 **5.** 1; binomial **7.** 3; none of these **9.** 4; binomial **11.** 1; binomial **13.** answers may vary **15.** answers may vary **17. a.** 6 **b.** 5 **19. a.** −2 **b.** 4 **21. a.** −15 **b.** −16 **23.** 184 ft **25.** 595.84 ft **27.** 212.06 million **29.** $23x^2$ **31.** $12x^2 - y$ **33.** $7s$ **35.** $-1.1y^2 + 4.8$ **37.** $5x + 3 + 4x + 3 + 2x + 6 + 3x + 7x$; $21x + 12$ **39.** $4x^2 + 7x + x^2 + 5x$; $5x^2 + 12x$ **41.** 2, 1, 1, 0; 2 **43.** 4, 0, 4, 3; 4 **45.** $9ab - 11a$ **47.** $4x^2 - 7xy + 3y^2$ **49.** $-3xy^2 + 4$ **51.** $14y^3 - 19 - 16a^2b^2$ **53.** $10x + 19$ **55.** $-x + 5$ **57.** answers may vary **59.** $11.1x^2 - 7.97x + 10.76$

Exercise Set 4.4

1. $12x + 12$ **3.** $-3x^2 + 10$ **5.** $-3x^2 + 4$ **7.** $-y^2 - 3y - 1$ **9.** $8t^2 - 4$ **11.** $15a^3 + a^2 + 16$ **13.** $-x + 14$ **15.** $-2x + 9$ **17.** $2x^2 + 7x - 16$ **19.** $y^2 - 7$ **21.** $2x^2 + 11x$ **23.** $-2z^2 - 16z + 6$ **25.** $2u^5 - 10u^2 + 11u - 9$ **27.** $5x - 9$ **29.** $6y + 13$ **31.** $-2x^2 + 8x - 1$ **33.** $7x^2 + 14x + 18$ **35.** $3x - 3$ **37.** $7x^2 - 4x + 2$ **39.** $7x^2 - 2x + 2$ **41.** $4y^2 + 12y + 19$ **43.** $-15x + 7$ **45.** $-2a - b + 1$ **47.** $3x^2 + 5$ **49.** $6x^2 - 2xy + 19y^2$ **51.** $8r^2s + 16rs - 8 + 7r^2s^2$ **53.** $6x^2$ **55.** $-12x^8$ **57.** $200x^3y^2$ **59.** $(x^2 + 7x + 4)$ ft **61.** $(3y^2 + 4y + 11)$ m **63.** $-6.6x^2 - 1.8x - 1.8$ **65.** $-2.5x^2 + 12.3x + 66.7$ **67. a.** $2x$ **b.** x^2 **c.** $-2x$ **d.** x^2; answers may vary

Mental Math

1. x^8 **3.** y^5 **5.** x^{14}

Exercise Set 4.5

1. $24x^3$ **3.** $-12.4x^{12}$ **5.** x^4 **7.** $-\frac{2}{15}y^3$ **9.** $-24x^8$ **11.** $6x^2 + 15x$ **13.** $7x^3 + 14x^2 - 7x$ **15.** $-2a^2 - 8a$ **17.** $6x^3 - 9x^2 + 12x$ **19.** $3a^3 + 6a$ **21.** $-6a^4 + 4a^3 - 6a^2$ **23.** $6x^5y - 3x^4y^3 + 24x^2y^4$ **25.** $x^2 + 3x$ **27.** $x^2 + 7x + 12$ **29.** $a^2 + 5a - 14$ **31.** $x^2 + \frac{1}{3}x - \frac{2}{9}$ **33.** $12x^4 + 25x^2 + 7$ **35.** $12x^2 - 29x + 15$ **37.** $1 - 7a + 12a^2$ **39.** $4y^2 - 16y + 16$ **41.** $x^3 - 5x^2 + 13x - 14$ **43.** $x^4 + 5x^3 - 3x^2 - 11x + 20$ **45.** $10a^3 - 27a^2 + 26a - 12$ **47.** $49x^2y^2 - 14xy^2 + y^2$ **49.** $x^2 + 5x + 6$ **51.** $12x^2 - 64x - 11$ **53.** $2x^3 + 10x^2 + 11x - 3$ **55.** $x^4 - 2x^3 - 51x^2 + 4x + 63$ **57.** $25x^2$ **59.** $9y^6$ **61.** \$7000 **63.** \$500 **65.** answers may vary **67.** $(4x^2 - 25)$ sq yds **69.** $(6x^2 - 4x)$ sq in. **71. a.** $6x + 12$ **b.** $9x^2 + 36x + 35$; answers may vary **73. a.** $a^2 - b^2$ **b.** $4x^2 - 9y^2$ **c.** $16x^2 - 49$ **d.** answers may vary

Exercise Set 4.6

1. $x^2 + 7x + 12$ **3.** $x^2 + 5x - 50$ **5.** $5x^2 + 4x - 12$ **7.** $4y^2 - 25y + 6$ **9.** $6x^2 + 13x - 5$ **11.** $6y^3 + 4y^2 + 42y + 28$ **13.** $x^2 + \frac{1}{3}x - \frac{2}{9}$ **15.** $8 - 26a + 15a^2$ **17.** $2x^2 + 9xy - 5y^2$ **19.** $x^2 + 4x + 4$ **21.** $4x^2 - 4x + 1$ **23.** $9a^2 - 30a + 25$ **25.** $x^4 + 10x^2 + 25$ **27.** $y^2 - \frac{4}{7}y + \frac{4}{49}$ **29.** $4a^2 - 12a + 9$ **31.** $25x^2 + 90x + 81$ **33.** $9x^2 - 42xy + 49y^2$ **35.** $16m^2 + 40mn + 25n^2$ **37.** answers may vary **39.** $a^2 - 49$ **41.** $x^2 - 36$ **43.** $9x^2 - 1$ **45.** $x^4 - 25$ **47.** $4y^4 - 1$ **49.** $16 - 49x^2$ **51.** $9x^2 - \frac{1}{4}$ **53.** $81x^2 - y^2$ **55.** $4m^2 - 25n^2$ **57.** $a^2 + 9a + 20$ **59.** $a^2 - 14a + 49$ **61.** $12a^2 - a - 1$ **63.** $x^2 - 4$ **65.** $9a^2 + 6a + 1$ **67.** $x^2 + 3xy - y^2$ **69.** $a^2 - \frac{1}{4}y^2$ **71.** $6b^2 - b - 35$ **73.** $x^4 - 100$ **75.** $16x^2 - 25$ **77.** $25x^2 - 60xy + 36y^2$ **79.** $4r^2 - 9s^2$ **81.** $\frac{5b^5}{7}$ **83.** $-\frac{2a^{10}}{b^5}$ **85.** $\frac{2y^8}{3}$ **87.** $(4x^2 + 4x + 1)$ sq ft **89.** $(24x^2 - 32x + 8)$ sq m

Integrated Review

1. $35x^5$ **2.** $32y^9$ **3.** -16 **4.** 16 **5.** $2x^2 - 9x - 5$ **6.** $3x^2 + 13x - 10$ **7.** $3x - 4$ **8.** $4x + 3$ **9.** $7x^6y^2$ **10.** $\frac{10b^6}{7}$ **11.** $144m^{14}n^{12}$ **12.** $64y^{27}z^{30}$ **13.** $16y^2 - 9$ **14.** $49x^2 - 1$ **15.** $\frac{y^{45}}{x^{63}}$ **16.** $\frac{x^{27}}{27}$ **17.** $2x^2 - 2x - 6$ **18.** $6x^2 + 13x - 11$ **19.** $2.5y^2 - 6y - 0.2$ **20.** $8.4x^2 - 6.8x - 4.2$ **21.** $x^2 + 8x + 16$ **22.** $y^2 - 18y + 81$ **23.** $2x + 8$ **24.** $2y - 18$ **25.** $7x^2 - 10xy + 4y^2$ **26.** $-a^2 - 3ab + 6b^2$ **27.** $x^3 + 2x^2 - 16x + 3$ **28.** $x^3 - 2x^2 - 5x - 2$ **29.** $6x^2 - x - 70$ **30.** $20x^2 + 21x - 5$ **31.** $2x^3 - 19x^2 + 44x - 7$ **32.** $5x^3 + 9x^2 - 17x + 3$

Mental Math

1. a^2 **3.** a^2 **5.** k^3

Exercise Set 4.7

1. $4x^2 + x + \frac{9}{5}$ **3.** $12x^3 + 3x$ **5.** $5p^2 + 6p$ **7.** $-\frac{3}{2x} + 3$ **9.** $-3x^2 + x - \frac{4}{x^3}$ **11.** $-1 + \frac{3}{2x} - \frac{7}{4x^4}$ **13.** $ab - b^2$ **15.** $x + 4xy - \frac{y}{2}$ **17.** $x + 1$ **19.** $2x + 3$ **21.** $2x + 1 + \frac{7}{x - 4}$ **23.** $4x + 9$ **25.** $3a^2 - 3a + 1 + \frac{2}{3a + 2}$ **27.** $2b^2 + b + 2 - \frac{12}{b + 4}$ **29.** $4x + 3 - \frac{2}{2x + 1}$ **31.** $2x^2 + 6x - 5 - \frac{2}{x - 2}$ **33.** $x^2 + 3x + 9$ **35.** $-3x + 6 - \frac{11}{x + 2}$ **37.** $2b - 1 - \frac{6}{2b - 1}$ **39.** $x + 8$ **41.** $x - 1$ **43.** $x^2 - 5x - 23 - \frac{41}{x - 2}$ **45.** $4x + 8 + \frac{7}{x - 2}$ **47.** $2x^3 - 3x^2 + x - 4$ **49.** $3x^2 + 4x - 8 + \frac{20}{x + 1}$ **51.** 3 **53.** -4 **55.** $3x$ **57.** $9x$ **59.** $x^3 - x^2 + x$ **61.** $(3x^3 + x - 4)$ ft **63.** $(2x + 5)$ m **65.** answers may vary

Chapter 4 Review

1. base: 3; exponent: 2 **2.** base: -5; exponent: 4 **3.** base: 5; exponent: 4 **4.** base: x; exponent: 6 **5.** 512 **6.** 36 **7.** -36 **8.** -65 **9.** 1 **10.** 1 **11.** y^9 **12.** x^{14} **13.** $-6x^{11}$ **14.** $-20y^7$ **15.** x^8 **16.** y^{15} **17.** $81y^{24}$ **18.** $8x^9$ **19.** x^5 **20.** z^7 **21.** a^4b^3 **22.** x^3y^5 **23.** $\frac{4}{x^3y^4}$ **24.** $\frac{x^6y^6}{4}$ **25.** $40a^{19}$ **26.** $36x^3$ **27.** 3 **28.** 9 **29.** b **30.** c **31.** $\frac{1}{49}$ **32.** $-\frac{1}{49}$ **33.** $\frac{2}{x^4}$ **34.** $\frac{1}{16x^4}$

35. 125 **36.** $\frac{9}{4}$ **37.** $\frac{17}{16}$ **38.** $\frac{1}{42}$ **39.** x^8 **40.** z^8 **41.** r **42.** y^3 **43.** c^4 **44.** $\frac{x^3}{y^3}$ **45.** $\frac{1}{x^6y^{13}}$ **46.** $\frac{a^{10}}{b^{10}}$ **47.** 2.7×10^{-4}
48. 8.868×10^{-1} **49.** 8.08×10^7 **50.** -8.68×10^5 **51.** 1.09379×10^8 **52.** 1.5×10^5 **53.** 867,000 **54.** 0.00386 **55.** 0.00086
56. 893,600 **57.** 1,431,280,000,000,000 cu km **58.** 0.0000000001 m **59.** 0.016 **60.** 400,000,000,000 **61.** 5 **62.** 2 **63.** 5 **64.** 6
65. 22; 78; 154.02; 400 **66.** $2a^2$ **67.** $-4y$ **68.** $15a^2 + 4a$ **69.** $22x^2 + 3x + 6$ **70.** $-6a^2b - 3b^2 - q^2$ **71.** cannot be combined
72. $8x^2 + 3x + 6$ **73.** $2x^5 + 3x^4 + 4x^3 + 9x^2 + 7x + 6$ **74.** $-7y^2 - 1$ **75.** $-6m^7 - 3x^4 + 7m^6 - 4m^2$ **76.** $-x^2 - 6xy - 2y^2$
77. $-5x^2 + 5x + 1$ **78.** $-2x^2 - x + 20$ **79.** $6x + 30$ **80.** $9x - 63$ **81.** $8a + 28$ **82.** $54a - 27$ **83.** $-7x^3 - 35x$
84. $-32y^3 + 48y$ **85.** $-2x^3 + 18x^2 - 2x$ **86.** $-3a^3b - 3a^2b - 3ab^2$ **87.** $-6a^4 + 8a^2 - 2a$ **88.** $42b^4 - 28b^2 + 14b$
89. $2x^2 - 12x - 14$ **90.** $6x^2 - 11x - 10$ **91.** $4a^2 + 27a - 7$ **92.** $42a^2 + 11a - 3$ **93.** $x^4 + 7x^3 + 4x^2 + 23x - 35$
94. $x^6 + 2x^5 + x^2 + 3x + 2$ **95.** $x^4 + 4x^3 + 4x^2 - 16$ **96.** $x^6 + 8x^4 + 16x^2 - 16$ **97.** $x^3 + 21x^2 + 147x + 343$
98. $8x^3 - 60x^2 + 150x - 125$ **99.** $x^2 + 14x + 49$ **100.** $x^2 - 10x + 25$ **101.** $9x^2 - 42x + 49$ **102.** $16x^2 + 16x + 4$
103. $25x^2 - 90x + 81$ **104.** $25x^2 - 1$ **105.** $49x^2 - 16$ **106.** $a^2 - 4b^2$ **107.** $4x^2 - 36$ **108.** $16a^4 - 4b^2$ **109.** $(9x^2 - 6x + 1)$ sq m
110. $(5x^2 - 3x - 2)$ sq mi **111.** $\frac{1}{7} + \frac{3}{x} + \frac{7}{x^2}$ **112.** $-a^2 + 3b - 4$ **113.** $a + 1 + \frac{6}{a - 2}$ **114.** $4x + \frac{7}{x + 5}$
115. $a^2 + 3a + 8 + \frac{22}{a - 2}$ **116.** $3b^2 - 4b - \frac{1}{3b - 2}$ **117.** $2x^3 - x^2 + 2 - \frac{1}{2x - 1}$ **118.** $-x^2 - 16x - 117 - \frac{684}{x - 6}$
119. $\left(5x - 1 + \frac{20}{x^2}\right)$ ft **120.** $(7a^3b^6 + a - 1)$ units

Chapter 4 Test

1. 32 **2.** 81 **3.** −81 **4.** $\frac{1}{64}$ **5.** $-15x^{11}$ **6.** y^5 **7.** $\frac{1}{r^5}$ **8.** $\frac{y^{14}}{x^2}$ **9.** $\frac{1}{6xy^8}$ **10.** 5.63×10^5 **11.** 8.63×10^{-5} **12.** 0.0015
13. 62,300 **14.** 0.036 **15.** 5 **16.** $-2x^2 + 12x + 11$ **17.** $16x^3 + 7x^2 - 3x - 13$ **18.** $-3x^3 + 5x^2 + 4x + 5$ **19.** $x^3 + 8x^2 + 3x - 5$
20. $3x^3 + 22x^2 + 41x + 14$ **21.** $2x^5 - 5x^4 + 12x^3 - 8x^2 + 4x + 7$ **22.** $3x^2 + 16x - 35$ **23.** $9x^2 - 49$ **24.** $16x^2 - 16x + 4$
25. $64x^2 + 48x + 9$ **26.** $x^4 - 81b^2$ **27.** 1516 ft; 1372 ft; 940 ft; 220 ft **28.** $(4x^2 - 9)$ sq in. **29.** $\frac{x}{2y} + \frac{1}{4} - \frac{7}{8y}$ **30.** $x + 2$
31. $9x^2 - 6x + 4 - \frac{16}{3x + 2}$

Cumulative Review

1. a. $9 \le 11$ **b.** $8 > 1$ **c.** $3 \ne 4$; Sec. 1.2, Ex. 7 **2.** solution; Sec. 1.3, Ex. 8 **3.** −12; Sec. 1.5, Ex. 5 **4.** $-\frac{3}{4}$; Sec. 1.6, Ex. 16
5. $5x + 7$; Sec. 2.1, Ex. 4 **6.** $-4a - 1$; Sec. 2.1, Ex. 5 **7.** $7.3x - 6$; Sec. 2.1, Ex. 7 **8.** $x = -11$; Sec. 2.3, Ex. 3 **9.** every real number;
Sec. 2.4, Ex. 7 **10.** $l = \frac{V}{wh}$; Sec. 2.6, Ex. 3 **11.** $x = \frac{31}{2}$; Sec. 2.7, Ex. 6 **12.** Sec. 3.2, Ex. 2 **13.** Sec. 3.3, Ex. 5

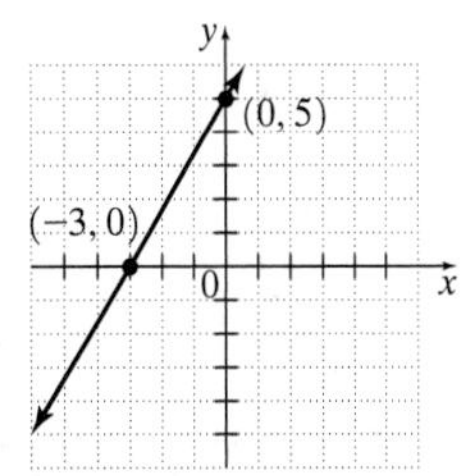

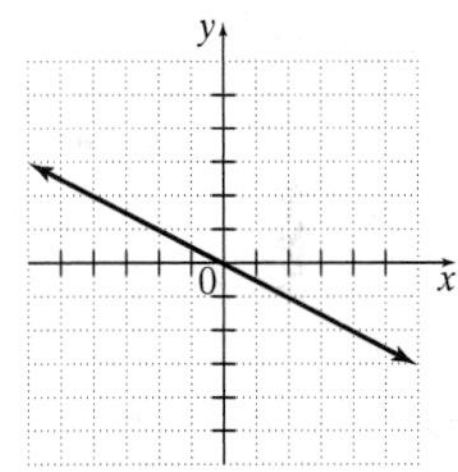

14. a. parallel **b.** perpendicular **c.** neither; Sec. 3.4, Ex. 6 **15.** 5^{18}; Sec. 4.1, Ex. 14 **16.** y^{16}; Sec. 4.1, Ex. 15 **17.** x^6; Sec. 4.2, Ex. 9
18. $\frac{y^{18}}{z^{36}}$; Sec. 4.2, Ex. 11 **19.** $\frac{1}{x^{19}}$; Sec. 4.2, Ex. 13 **20.** $4x$; Sec. 4.3, Ex. 6 **21.** $13x^2 - 2$; Sec. 4.3, Ex. 7 **22.** $4x^2 - 4xy + y^2$; Sec. 4.5, Ex. 8
23. $t^2 + 4t + 4$; Sec. 4.6, Ex. 5 **24.** $x^4 - 14x^2y + 49y^2$; Sec. 4.6, Ex. 8 **25.** $2xy - 4 + \frac{1}{2y}$; Sec. 4.7, Ex. 3

Chapter 5 Factoring Polynomials

Chapter 5 Pretest

1. $2x^2y(x - 3y)$; 5.1B **2.** $(x - 4)(y + 6)$; 5.1C **3.** $(a + 6)(a + 2)$; 5.2A **4.** prime; 5.2A **5.** $3x(x - 1)(x - 5)$; 5.2B
6. $(2x - 3)(x + 4)$; 5.3A **7.** $7(2x + 5)(x + 2)$; 5.3B **8.** $(3b - 2)(8b - 3)$; 5.4A **9.** $(5y + 1)(3y + 7)$; 5.4A **10.** $(x + 12)^2$; 5.5A
11. $(2x - 3y)^2$; 5.5A **12.** $(a - 7b)(a + 7b)$; 5.5B **13.** $(1 - 8t)(1 + 8t)$; 5.5B **14.** $(x + 5)(x^2 - 5x + 25)$; 5.5C
15. $20(2a - b)(4a^2 + 2ab + b^2)$; 5.5C **16.** 12, −5; 5.6A **17.** 0, 13; 5.6A **18.** 0, 4, −3; 5.6B **19.** 8 in. × 15 in.; 5.7A **20.** −16 or 15; 5.7A

Mental Math

1. 2 **3.** 3 **5.** 7

Exercise Set 5.1

1. y^2 **3.** xy^2 **5.** 4 **7.** $4y^3$ **9.** $5x^2$ **11.** $3x^3$ **13.** $9x^2y$ **15.** $3(a + 2)$ **17.** $15(2x - 1)$ **19.** $x^2(x + 5)$ **21.** $2y(3y^3 - 1)$
23. $2x(16y - 9x)$ **25.** $4(x - 2y + 1)$ **27.** $3x(2x^2 - 3x + 4)$ **29.** $a^2b^2(a^5b^4 - a + b^3 - 1)$ **31.** $5xy(x^2 - 3x + 2)$

33. $4(2x^5 + 4x^4 - 5x^3 + 3)$ **35.** $\frac{1}{3}x(x^3 + 2x^2 - 4x + 1)$ **37.** $(x + 2)(y + 3)$ **39.** $(x + 2)(8 - y)$ **41.** answers may vary **43.** $(x^2 + 5)(x + 2)$ **45.** $(x + 3)(5 + y)$ **47.** $(2x^2 + 5)(3x - 2)$ **49.** $(y - 4)(2 + x)$ **51.** $(2x + 1)(x^2 + 4)$ **53.** $(x - 2y)(4x - 3)$ **55.** answers may vary **57.** $x^2 + 7x + 10$ **59.** $b^2 - 3b - 4$ **61.** 2, 6 **63.** $-1, -8$ **65.** $-2, 5$ **67.** $-8, 3$ **69.** $2(3x^2 - 1)(2y - 7)$ **71.** $12x^3 - 2x; 2x(6x^2 - 1)$ **73.** $(n^3 - 6)$ units **75. a.** 2000 billion kw hr **b.** 2088 billion kw hr **c.** $-4(2x^2 - 15x - 500)$ or $4(-2x^2 + 15x + 500)$

Mental Math

1. +5 **3.** -3 **5.** +2

Exercise Set 5.2

1. $(x + 6)(x + 1)$ **3.** $(x - 9)(x - 1)$ **5.** $(x - 6)(x + 3)$ **7.** $(x + 10)(x - 7)$ **9.** prime **11.** $(x + 5y)(x + 3y)$ **13.** $(a^2 - 5)(a^2 + 3)$ **15.** $x^2 + 5x - 24$ **17.** answers may vary **19.** $2(z + 8)(z + 2)$ **21.** $2x(x - 5)(x - 4)$ **23.** $(x - 4y)(x + y)$ **25.** $(x + 12)(x + 3)$ **27.** $(x - 2)(x + 1)$ **29.** $(r - 12)(r - 4)$ **31.** $(x + 2y)(x - y)$ **33.** $3(x + 5)(x - 2)$ **35.** $3(x - 18)(x - 2)$ **37.** $(x - 24)(x + 6)$ **39.** prime **41.** $(x - 5)(x - 3)$ **43.** $6x(x + 4)(x + 5)$ **45.** $4y(x^2 + x - 3)$ **47.** $(x - 7)(x + 3)$ **49.** $(x + 5y)(x + 2y)$ **51.** $2(t + 8)(t + 4)$ **53.** $x(x - 6)(x + 4)$ **55.** $2t^3(t - 4)(t - 3)$ **57.** $5xy(x - 8y)(x + 3y)$ **59.** $2x^2 + 11x + 5$ **61.** $15y^2 - 17y + 4$ **63.** $9a^2 + 23a - 12$ **65.** $2x^2 + 28x + 66; 2(x + 3)(x + 11)$ **67.** $(x + 1)(y - 5)(y + 3)$ **69.** 3; 4 **71.** 8; 16 **73.** $(x^n + 2)(x^n + 3)$

Exercise Set 5.3

1. $x + 4$ **3.** $10x - 1$ **5.** $4x - 3$ **7.** $(2x + 3)(x + 5)$ **9.** $(y - 1)(8y - 9)$ **11.** $(2x + 1)(x - 5)$ **13.** $(4r - 1)(5r + 8)$ **15.** $(5x + 1)(2x + 3)$ **17.** $(3x - 2)(x + 1)$ **19.** $(3x - 5y)(2x - y)$ **21.** $(3x - 5)(5x + 3)$ **23.** $(x - 4)(x - 5)$ **25.** $(2x + 11)(x - 9)$ **27.** $(7t + 1)(t - 4)$ **29.** $(3a + b)(a + 3b)$ **31.** $(7x + 1)(7x - 2)$ **33.** $(6x - 7)(3x + 2)$ **35.** $x(3x + 2)(4x + 1)$ **37.** $3(7x + 5)(x - 3)$ **39.** $(3x + 4)(4x - 3)$ **41.** $2y^2(3x - 10)(x + 3)$ **43.** $(2x - 7)(2x + 3)$ **45.** $3(x^2 - 14x + 21)$ **47.** $(4x + 9)(2x - 3)$ **49.** $-1(x - 6)(x + 4)$ **51.** $x(4x + 3)(x - 3)$ **53.** $(4x - 9)(6x - 1)$ **55.** $b(8a - 3)(5a + 3)$ **57.** $(3x^2 + 2)(5x^2 + 3)$ **59.** $2y(3y + 5)(y - 3)$ **61.** $5x(2x - y)(x + 3y)$ **63.** $-1(2x - 5)(7x - 2)$ **65.** Jan. 2001 and Feb. 2001 **67.** increased 0.5% **69.** $(y - 1)^2(4x^2 + 10x + 25)$ **71.** $-3xy^2(4x - 5)(x + 1)$ **73.** 5; 13 **75.** 4; 5

Exercise Set 5.4

1. $(x + 3)(x + 2)$ **3.** $(x - 4)(x + 7)$ **5.** $(y + 8)(y - 2)$ **7.** $(3x + 4)(x + 4)$ **9.** $(8x - 5)(x - 3)$ **11.** $(5x^2 - 3)(x^2 + 5)$ **13. a.** 9, 2 **b.** $9x + 2x$ **c.** $(2x + 3)(3x + 1)$ **15. a.** $-20, -3$ **b.** $-20x - 3x$ **c.** $(5x - 1)(3x - 4)$ **17.** $(3y + 2)(7y + 1)$ **19.** $(7x - 11)(x + 1)$ **21.** $(5x - 2)(2x - 1)$ **23.** $(2x - 5)(x - 1)$ **25.** $(2x + 3)(2x + 3)$ or $(2x + 3)^2$ **27.** $(2x + 3)(2x - 7)$ **29.** $(5x - 4)(2x - 3)$ **31.** $x(2x + 3)(x + 5)$ **33.** $2(8y - 9)(y - 1)$ **35.** $(2x - 3)(3x - 2)$ **37.** $3(3a + 2)(6a - 5)$ **39.** $a(4a + 1)(5a + 8)$ **41.** $3x(4x + 3)(x - 3)$ **43.** $x^2 - 4$ **45.** $y^2 + 8y + 16$ **47.** $81z^2 - 25$ **49.** $16x^2 - 24x + 9$ **51.** $(x^n + 2)(x^n + 3)$ **53.** $(3x^n - 5)(x^n + 7)$ **55.** answers may vary

Graphing Calculator Explorations

16, 14, 16; 16, 14, 16; 2.89, 0.89, 2.89; 171.61, 169.61, 171.61; 1, −1, 1

Mental Math

1. 1^2 **3.** 9^2 **5.** 3^2 **7.** $(3x)^2$ **9.** $(5a)^2$ **11.** $(6p^2)^2$

Exercise Set 5.5

1. yes **3.** no **5.** no **7.** yes **9.** 8 **11.** $(x + 11)^2$ **13.** $(x - 8)^2$ **15.** $(4a - 3)^2$ **17.** $3(x - 4)^2$ **19.** $(xy - 5)^2$ **21.** $m(m + 9)^2$ **23.** prime **25.** $(3x - 4y)^2$ **27.** answers may vary **29.** $(x + 5)(x - 5)$ **31.** $(3 + 2z)(3 - 2z)$ **33.** prime **35.** $xy(x + 11y)(x - 11y)$ **37.** $(y + 9)(y - 5)$ **39.** $4(4x + 5)(4x - 5)$ **41.** $2y(3x + 1)(3x - 1)$ **43.** $(3x + 7)(3x - 7)$ **45.** $(x^2 + 9)(x + 3)(x - 3)$ **47.** $(x + 2y + 3)(x + 2y - 3)$ **49.** $(x + 8 + x^2)(x + 8 - x^2)$ **51.** $(x - 5 + y)(x - 5 - y)$ **53.** $(2x + 1 + z)(2x + 1 - z)$ **55.** $(m^2 + 1)(m - 1)(m + 1)$ **57.** $(x + 3)(x^2 - 3x + 9)$ **59.** $(z - 1)(z^2 + z + 1)$ **61.** $(m + n)(m^2 - mn + n^2)$ **63.** $y^2(x - 3)(x^2 + 3x + 9)$ **65.** $b(a + 2b)(a^2 - 2ab + 4b^2)$ **67.** $(5y - 2x)(25y^2 + 10xy + 4x^2)$ **69.** $(x^2 - y)(x^4 + x^2y + y^2)$ **71.** $(2x + 3y)(4x^2 - 6xy + 9y^2)$ **73.** $(x - 1)(x^2 + x + 1)$ **75.** $(x + 5)(x^2 - 5x + 25)$ **77.** $3y^2(x^2 + 3)(x^4 - 3x^2 + 9)$ **79.** 5 **81.** $-\frac{1}{3}$ **83.** 0 **85.** 5 **87.** $\pi R^2 - \pi r^2 = \pi(R + r)(R - r)$ **89. a.** 1300 ft **b.** 660 ft **c.** 10 sec **d.** $4(19 - 2t)(19 + 2t)$ **91.** $c = 9$ **93.** $c = 49$ **95. a.** $(x + 1)(x^2 - x + 1)(x - 1)(x^2 + x + 1)$ **b.** $(x + 1)(x - 1)(x^4 + x^2 + 1)$ **c.** answers may vary **97.** $(x^n + 6)(x^n - 6)$ **99.** $(5x^n + 9)(5x^n - 9)$ **101.** $(x^{2n} + 25)(x^n + 5)(x^n - 5)$

Integrated Review

1. $(x - 3)(x + 4)$ **2.** $(x - 8)(x - 2)$ **3.** $(x + 2)(x - 3)$ **4.** $(x + 1)^2$ **5.** $(x - 3)^2$ **6.** $(x + 2)(x - 1)$ **7.** $(x + 3)(x - 2)$ **8.** $(x + 3)(x + 4)$ **9.** $(x - 5)(x - 2)$ **10.** $(x - 6)(x + 5)$ **11.** $2(x - 7)(x + 7)$ **12.** $3(x - 5)(x + 5)$ **13.** $(x + 3)(x + 5)$ **14.** $(y - 7)(3 + x)$ **15.** $(x + 8)(x - 2)$ **16.** $(x - 7)(x + 4)$ **17.** $4x(x + 7)(x - 2)$ **18.** $6x(x - 5)(x + 4)$ **19.** $2(3x + 4)(2x + 3)$ **20.** $(2a - b)(4a + 5b)$ **21.** $(2a - b)(2a + b)$ **22.** $(x - 5y)(x + 5y)$ **23.** $(4 - 3x)(7 + 2x)$ **24.** $(5 - 2x)(4 + x)$ **25.** prime **26.** prime **27.** $(3y + 5)(2y - 3)$ **28.** $(4x - 5)(x + 1)$ **29.** $9x(2x^2 - 7x + 1)$ **30.** $4a(3a^2 - 6a + 1)$ **31.** $(4a - 7)^2$ **32.** $(5p - 7)^2$ **33.** $(7 - x)(2 + x)$ **34.** $(3 + x)(1 - x)$ **35.** $3x^2y(x + 6)(x - 4)$ **36.** $2xy(x + 5y)(x - y)$ **37.** $3xy(4x^2 + 81)$ **38.** $2xy^2(3x^2 + 4)$ **39.** $2xy(1 - 6x)(1 + 6x)$ **40.** $2x(x - 3)(x + 3)$ **41.** $(x - 2)(x + 2)(x + 6)$ **42.** $(x - 2)(x - 6)(x + 6)$ **43.** $2a^2(3a + 5)$ **44.** $2n(2n - 3)$ **45.** $(x^2 + 4)(3x - 1)$ **46.** $(x - 2)(x^2 + 3)$ **47.** $6(x + 2y)(x + y)$ **48.** $2(x + 4y)(6x - y)$ **49.** $(5 + x)(x + y)$ **50.** $(x - y)(7 + y)$ **51.** $(7t - 1)(2t - 1)$ **52.** prime **53.** $(3x + 5)(x - 1)$ **54.** $(7x - 2)(x + 3)$ **55.** $(1 - 10a)(1 + 2a)$ **56.** $(1 + 5a)(1 - 12a)$ **57.** $(x - 3)(x + 3)(x - 1)(x + 1)$ **58.** $(x - 3)(x + 3)(x - 2)(x + 2)$

59. $(x - 15)(x - 8)$ **60.** $(y + 16)(y + 6)$ **61.** prime **62.** $(4a - 7b)^2$ **63.** $(5p - 7q)^2$ **64.** $(7x + 3y)(x + 3y)$ **65.** $-1(x - 5)(x + 6)$ **66.** $-1(x - 2)(x - 4)$ **67.** $(s + 4)(3r - 1)$ **68.** $(x - 2)(x^2 + 3)$ **69.** $(4x - 3)(x - 2y)$ **70.** $(2x - y)(2x + 7z)$ **71.** $(x + 12y)(x - 3y)$ **72.** $(3x - 2y)(x + 4y)$ **73.** $(x^2 + 2)(x + 4)(x - 4)$ **74.** $(x^2 + 3)(x + 5)(x - 5)$ **75.** $x(x - 1)(x^2 + x + 1)$ **76.** $x^3(x + 1)(x^2 - x + 1)$ **77.** $(2x + 5y)(4x^2 - 10xy + 25y^2)$ **78.** $(3x - 4y)(9x^2 + 12xy + 16y^2)$ **79.** answers may vary **80.** yes; $9(x^2 + 9y^2)$

Mental Math

1. 3, 7 **3.** −8, −6 **5.** −1, 3

Exercise Set 5.6

1. 2, −1 **3.** 6, 7 **5.** −9, −17 **7.** 0, −6 **9.** 0, 8 **11.** $-\frac{3}{2}, \frac{5}{4}$ **13.** $\frac{7}{2}, -\frac{2}{7}$ **15.** $\frac{1}{2}, -\frac{1}{3}$ **17.** −0.2, −1.5 **19.** $(x - 6)(x + 1) = 0$ **21.** 9, 4 **23.** −4, 2 **25.** 0, 7 **27.** 0, −20 **29.** 4, −4 **31.** 8, −4 **33.** $\frac{7}{3}, -2$ **35.** $\frac{8}{3}, -9$ **37.** $0, \frac{1}{2}, -\frac{1}{2}$ **39.** $\frac{17}{2}$ **41.** $\frac{3}{4}$ **43.** $\frac{1}{2}, -\frac{1}{2}$ **45.** $-\frac{3}{2}, -\frac{1}{2}, 3$ **47.** −5, 3 **49.** $2, -\frac{4}{5}$ **51.** $-\frac{5}{6}, \frac{6}{5}$ **53.** $-\frac{4}{3}, 5$ **55.** −4, 3 **57.** 0, 8, 4 **59.** $x^2 - 12x + 35 = 0$ **61.** $\frac{47}{45}$ **63.** $\frac{17}{60}$ **65.** $\frac{7}{10}$ **67.** didn't write equation in standard form **69. a.** 300; 304; 276; 216; 124; 0; −156 **b.** 5 sec **c.** 304 ft **71.** $0, \frac{1}{2}$ **73.** 0, −15

Exercise Set 5.7

1. width $= x$; length $= x + 4$ **3.** x and $x + 2$ if x is an odd integer **5.** base $= x$; height $= 4x + 1$ **7.** 11 units **9.** 15 cm, 13 cm, 70 cm, 22 cm **11.** base = 16 mi; height = 6 mi **13.** 5 sec **15.** length = 5 cm; width = 6 cm **17.** 54 diagonals **19.** 10 sides **21.** −12 or 11 **23.** slow boat: 8 mph; fast boat: 15 mph **25.** 13 and 7 **27.** 5 in. **29.** 12 mm; 16 mm; 20 mm **31.** 10 km **33.** 36 ft **35.** 9.5 sec **37.** 20% **39.** length: 15 mi; width: 8 mi **41.** 105 units **43.** 9600 thousand acres **45.** 9500 thousand acres **47.** end of 1998 **49.** answers may vary **51.** width of pool: 29 m; length of pool: 35 m **53.** 70420: Abita Springs, LA; 70434: Covington, LA

Chapter 5 Review

1. $2x - 5$ **2.** $2x^4 + 1 - 5x^3$ **3.** $5(m + 6)$ **4.** $4x(5x^2 + 3x + 6)$ **5.** $(2x + 3)(3x - 5)$ **6.** $(x + 1)(5x - 1)$ **7.** $(x - 1)(3x + 2)$ **8.** $(2x - 1)(3x + 5)$ **9.** $(a + 3b)(3a + b)$ **10.** $(x + 4)(x + 2)$ **11.** $(x - 8)(x - 3)$ **12.** prime **13.** $(x - 6)(x + 1)$ **14.** $(x + 4)(x - 2)$ **15.** $(x + 6y)(x - 2y)$ **16.** $(x + 5y)(x + 3y)$ **17.** $2(3 - x)(12 + x)$ **18.** $4(8 + 3x - x^2)$ **19.** $5y(y - 6)(y - 4)$ **20.** −48, 2 **21.** factor out the GCF, 3 **22.** $(2x + 1)(x + 6)$ **23.** $(2x + 3)(2x - 1)$ **24.** $(3x + 4y)(2x - y)$ **25.** prime **26.** $(2x + 3)(x - 13)$ **27.** $(6x + 5y)(3x - 4y)$ **28.** $5y(2y - 3)(y + 4)$ **29.** $5x^2 - 9x - 2$; $(5x + 1)(x - 2)$ **30.** $16x^2 - 28x + 6$; $2(4x - 1)(2x - 3)$ **31.** $(x - 9)(x + 9)$ **32.** $(x + 6)^2$ **33.** $(2x - 3)(2x + 3)$ **34.** $(3t - 5s)(3t + 5s)$ **35.** prime **36.** $(n - 9)^2$ **37.** $3(r + 6)^2$ **38.** $(3y - 7)^2$ **39.** $5m^6(m - 1)(m + 1)$ **40.** $(2x - 7y)^2$ **41.** $3y(x + y)^2$ **42.** $(2x - 1)(2x + 1)(4x^2 + 1)$ **43.** $(y + 7)(y - 3)$ **44.** $(x - 7)(x + 1)$ **45.** $(2 - 3y)(4 + 6y + 9y^2)$ **46.** $(1 - 4y)(1 + 4y + 16y^2)$ **47.** $6xy(x + 2)(x^2 - 2x + 4)$ **48.** $2x^2(x + 2y)(x^2 - 2xy + 4y^2)$ **49.** $(x - 1 + y)(x - 1 - y)$ **50.** $\pi h(R + r)(R - r)$ cu units **51.** −6, 2 **52.** $0, -1, \frac{2}{7}$ **53.** $-\frac{1}{5}, -3$ **54.** −7, −1 **55.** −4, 6 **56.** −5 **57.** 2, 8 **58.** $\frac{1}{3}$ **59.** $-\frac{2}{7}, \frac{3}{8}$ **60.** 0, 6 **61.** 5, −5 **62.** $x^2 - 9x + 20 = 0$ **63.** c **64.** d **65.** 9 units **66.** 8 units, 13 units, 16 units, 10 units **67.** width: 20 in.; length: 25 in. **68.** 36 yd **69.** 19 and 20 **70. a.** 17.5 sec and 10 sec; answers may vary **b.** 27.5 sec **71.** 32 cm

Chapter 5 Test

1. $3x(3x - 1)$ **2.** $(x + 7)(x + 4)$ **3.** $(7 - m)(7 + m)$ **4.** $(y + 11)^2$ **5.** $(x - 2)(x + 2)(x^2 + 4)$ **6.** $(4 - y)(a + 3)$ **7.** prime **8.** $(y - 12)(y + 4)$ **9.** $(3a - 7)(a + b)$ **10.** $(3x - 2)(x - 1)$ **11.** $5(6 - x)(6 + x)$ **12.** $3x(x - 5)(x - 2)$ **13.** $(x + 5 + y)(x + 5 - y)$ **14.** $(y - 2)(y + 2)(x - 7)$ **15.** $x(1 - x)(1 + x)(1 + x^2)$ **16.** $(x + 12y)(x + 2y)$ **17.** $(x + 4)(x^2 - 4x + 16)$ **18.** $3x(3y - z)(9y^2 + 3yz + z^2)$ **19.** 3, −9 **20.** −6, −4 **21.** −7, 2 **22.** $0, \frac{3}{2}, -\frac{4}{3}$ **23.** 0, 3, −3 **24.** 0, −4 **25.** −3, 5 **26.** $-\frac{2}{3}, 1$ **27.** width: 6 units; length: 9 units **28.** 17 ft **29.** 8 and 9 **30.** 8.25 sec **31.** hypotenuse: 25 cm; legs: 15 cm, 20 cm

Cumulative Review

1. a. $\frac{15}{x} = 4$ **b.** $12 - 3 = x$ **c.** $4x + 17 = 21$; Sec. 1.3, Ex. 10 **2. a.** −12 **b.** −9; Sec. 1.4, Ex. 13 **3.** −12; Sec. 1.5, Ex. 7 **4.** −3; Sec. 1.5, Ex. 7 **5.** $x = -4$; Sec. 2.2, Ex. 7 **6.** shorter piece, 2 feet; longer piece, 8 feet; Sec. 2.5, Ex. 2 **7.** $\frac{y - b}{m}$; Sec. 2.6, Ex. 4

8. $\{x|x > -3\}$, $(-3, \infty)$; Sec. 2.8, Ex. 8 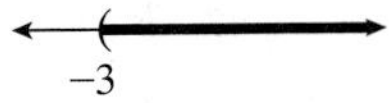

9.

x	y
−1	−3
0	0
−3	−9

; Sec. 3.1, Ex. 6

10. 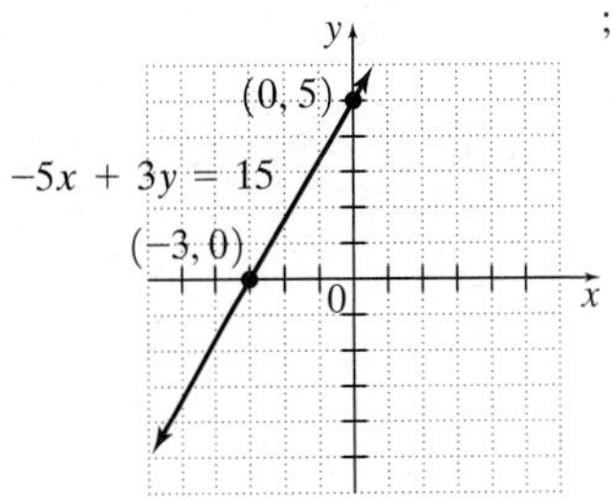

; Sec. 3.2, Ex. 2

11. x^3 Sec. 4.1, Ex. 21 **12.** $4^4 = 256$; Sec. 4.1, Ex. 22 **13.** −27; Sec. 4.1, Ex. 23 **14.** $2x^4y$; Sec. 4.1, Ex. 24 **15.** $\frac{2}{x^3}$; Sec. 4.2, Ex. 2 **16.** $\frac{1}{16}$; Sec. 4.2, Ex. 4 **17.** $10x^4 + 30x$; Sec. 4.5, Ex. 5 **18.** $-15x^4 - 18x^3 + 3x^2$; Sec. 4.5, Ex. 6 **19.** $4x^2 - 4x + 6 + \frac{-11}{2x + 3}$; Sec. 4.7, Ex. 6 **20.** $(x + 3)(x + 4)$; Sec. 5.2, Ex. 1 **21.** $(3x - 2)^2$; Sec. 5.5, Ex. 3 **22.** $\left(x - \frac{1}{2}\right)\left(x + \frac{1}{2}\right)$; Sec. 5.5, Ex. 8 **23.** $(p + 3q)(p^2 - 3pq + 9q^2)$; Sec. 5.5, Ex. 16 **24.** $x = 11$, $x = -2$; Sec. 5.6, Ex. 4

Chapter 6 Rational Expressions

Chapter 6 Pretest

1. $x = -1, x = 10$; 6.1B **2.** $\frac{4}{x+2}$; 6.1C **3.** $10(x+2)(x+3)$; 6.3B **4.** 3; 6.2A **5.** $\frac{5(x+5)}{x^3(x-5)}$; 6.2B **6.** $\frac{1}{b-11}$; 6.3A **7.** $\frac{7-4x}{x-1}$; 6.4A **8.** $\frac{9}{x-5}$; 6.4A **9.** $\frac{x^2+12}{(x+4)(x-4)(x-3)}$; 6.4A **10.** $b = -7$; 6.5A **11.** no solution; 6.5A **12.** $y = -1$; 6.5A **13.** $b = \frac{2A}{h}$; 6.5B **14.** $\frac{15n^6}{m^3}$; 6.7A **15.** $4a - 1$; 6.7A, B **16.** $x = 5$; 6.6D **17.** 2 or 5; 6.6A **18.** $3\frac{1}{13}$ hr; 6.6B **19.** 250 mph; 6.6C

Mental Math

1. $x = 0$ **3.** $x = 0, x = 1$

Exercise Set 6.1

1. $\frac{7}{4}$ **3.** $-\frac{8}{3}$ **5.** $-\frac{11}{2}$ **7. a.** \$37.5 million **b.** \$85.7 million **c.** \$48.2 million **9.** $x = 0$ **11.** $x = -2$ **13.** $x = 4$ **15.** $x = -2$ **17.** none **19.** answers may vary **21.** $\frac{1}{4(x+2)}$ **23.** $\frac{1}{x+2}$ **25.** can't simplify **27.** 1 **29.** -1 **31.** -5 **33.** $\frac{1}{x-9}$ **35.** $5x + 1$ **37.** $\frac{1}{x-2}$ **39.** $x + 2$ **41.** $\frac{x+5}{x-5}$ **43.** $\frac{x+2}{x+4}$ **45.** $\frac{x+2}{2}$ **47.** $\frac{11x}{6}$ **49.** -1 **51.** $\frac{x+1}{x-1}$ **53.** $\frac{m-3}{m+3}$ **55.** $\frac{3}{11}$ **57.** $\frac{50}{99}$ **59.** $\frac{4}{3}$ **61.** $\frac{117}{40}$ **63.** $x + y$ **65.** $\frac{5-y}{2}$ **67.** answers may vary **69.** 19.6% **71.** 69.7%

Mental Math

1. $\frac{2x}{3y}$ **3.** $\frac{5y^2}{7x^2}$ **5.** $\frac{9}{5}$

Exercise Set 6.2

1. $\frac{21}{4y}$ **3.** x^4 **5.** $-\frac{b^2}{6}$ **7.** $\frac{x^2}{10}$ **9.** $\frac{1}{3}$ **11.** 1 **13.** $\frac{x+5}{x}$ **15.** $\frac{2}{9(x-5)}$ sq ft **17.** x^4 **19.** $\frac{12}{y^6}$ **21.** $x(x+4)$ **23.** $\frac{3(x+1)}{x^3(x-1)}$ **25.** $m^2 - n^2$ **27.** $-\frac{x+2}{x-3}$ **29.** $\frac{x+2}{x-3}$ **31.** $\frac{5}{6}$ **33.** $\frac{3x}{8}$ **35.** $\frac{3}{2}$ **37.** $\frac{3x+4y}{2(x+2y)}$ **39.** $\frac{2(x+2)}{x-2}$ **41.** $-\frac{x+3}{4x}$ **43.** $\frac{(a+5)(a+3)}{(a+2)(a+1)}$ **45.** 1440 **47.** 1.93 **49.** 73 **51.** 411,755 sq yd **53.** 1119 ft per sec **55.** 1 **57.** $-\frac{10}{9}$ **59.** $-\frac{1}{5}$ **61.** $\frac{x}{2}$ **63.** $\frac{5a(2a+b)(3a-2b)}{b^2(a-b)(a+2b)}$ **65.** answers may vary **67.** 2352.94 euros

Mental Math

1. 1 **3.** $\frac{7x}{9}$ **5.** $\frac{1}{9}$ **7.** $\frac{17y}{5}$

Exercise Set 6.3

1. $\frac{a+9}{13}$ **3.** $\frac{3m}{n}$ **5.** 4 **7.** $\frac{y+10}{3+y}$ **9.** 3 **11.** $\frac{1}{a+5}$ **13.** $\frac{1}{x-6}$ **15.** $\frac{20}{x-2}$ m **17.** answers may vary **19.** $4x^3$ **21.** $8x(x+2)$ **23.** $(x+3)(x-2)$ **25.** $3(x+6)$ **27.** $6(x+1)^2$ **29.** $x - 8$ or $8 - x$ **31.** $(x-1)(x+4)(x+3)$ **33.** answers may vary **35.** $6x$ **37.** $24b^2$ **39.** 18 **41.** $9ba + 2b$ **43.** $x^2 + x$ **45.** $18y - 2$ **47.** $\frac{29}{21}$ **49.** $-\frac{5}{12}$ **51.** $\frac{7}{30}$ **53.** 3 packages hot dogs and 2 packages buns **55.** answers may vary

Exercise Set 6.4

1. $\frac{5}{x}$ **3.** $\frac{75a+6b^2}{5b}$ **5.** $\frac{6x+5}{2x^2}$ **7.** $\frac{11}{x+1}$ **9.** $\frac{3x-7}{(x-2)(x+2)}$ **11.** $\frac{35x-6}{4x(x-2)}$ **13.** $-\frac{2}{x-3}$ **15.** $-\frac{1}{x^2-1}$ **17.** $\frac{5+2x}{x}$ **19.** $\frac{6x-7}{x-2}$ **21.** $-\frac{y+4}{y+3}$ **23.** $\frac{-5x+14}{4x}$ or $-\frac{5x-14}{4x}$ **25.** 2 **27.** $3x^3 - 4$ **29.** $\frac{x+2}{(x+3)^2}$ **31.** $\frac{9b-4}{5b(b-1)}$ **33.** $\frac{2+m}{m}$ **35.** $\frac{10}{1-2x}$ **37.** $\frac{15x-1}{(x+1)^2(x-1)}$ **39.** $\frac{x^2-3x-2}{(x-1)^2(x+1)}$ **41.** $\frac{a+2}{2(a+3)}$ **43.** $\frac{y(2y+1)}{(2y+3)^2}$ **45.** $\frac{x-10}{2(x-2)}$ **47.** $\frac{-3-2y}{(y-2)(y-1)}$ **49.** $\frac{-5x+23}{(x-2)(x-3)}$ **51.** $\frac{2x^2-2x-46}{(x+1)(x-6)(x-5)}$ **53.** answers may vary **55.** $x = \frac{2}{3}$ **57.** $x = -\frac{1}{2}, x = 1$ **59.** $x = -\frac{15}{2}$ **61.** $\frac{6x^2-5x-3}{x(x+1)(x-1)}$ **63.** $\frac{4x^2-15x+6}{(x-2)^2(x+2)(x-3)}$ **65.** $\frac{-2x^2+14x+55}{(x+2)(x+7)(x+3)}$ **67.** $\frac{2x-16}{(x-4)(x+4)}$ in. **69.** $\frac{P-G}{P}$ **71.** answers may vary

Mental Math

1. 10 **3.** 36

Exercise Set 6.5

1. 30 **3.** 0 **5.** −2 **7.** −5, 2 **9.** 5 **11.** 3 **13.** 1 **15.** −3 **17.** no solution **19.** 1 **21.** no solution **23.** 3, −4 **25.** 6, −4 **27.** 5 **29.** 0 **31.** 8, −2 **33.** −2 **35.** no solution **37.** 3 **39.** −11, 1 **41.** $I = \frac{E}{R}$ **43.** $Q = \frac{V}{T}$ **45.** $t = \frac{A - Bi}{i}$ **47.** $G = \frac{V}{N - R}$ **49.** $r = \frac{C}{2\pi}$ **51.** $x = \frac{3y}{3 + y}$ **53.** $\frac{1}{x}$ **55.** $\frac{1}{x} + \frac{1}{2}$ **57.** $\frac{1}{3}$ **59.** 100°, 80° **61.** 22.5°, 67.5° **63.** 5
65. No. Multiplying both terms in the expression by 4 changes the value of the original expression.

Integrated Review

1. expression; $\frac{3 + 2x}{3x}$ **2.** expression; $\frac{18 + 5a}{6a}$ **3.** equation; 3 **4.** equation; 18 **5.** expression; $\frac{x - 1}{x(x + 1)}$ **6.** expression; $\frac{3(x + 1)}{x(x - 3)}$ **7.** equation; no solution **8.** equation; 1 **9.** expression; 10 **10.** expression; $\frac{z}{3(9z - 5)}$ **11.** expression; $\frac{5x + 7}{x - 3}$ **12.** expression; $\frac{7p + 5}{2p + 7}$ **13.** equation; 23 **14.** equation; 3 **15.** expression; $\frac{25a}{9(a - 2)}$ **16.** expression; $\frac{9}{4(x - 1)}$ **17.** expression; $\frac{3x^2 + 5x + 3}{(3x - 1)^2}$ **18.** expression; $\frac{2x^2 - 3x - 1}{(2x - 5)^2}$ **19.** expression; $\frac{4x - 37}{5x}$ **20.** equation; $-\frac{7}{3}$ **21.** equation; $\frac{8}{5}$ **22.** expression; $\frac{29x - 23}{3x}$ **23.** answers may vary **24.** answers may vary

Mental Math

1. c

Exercise Set 6.6

1. 2 **3.** −3 **5.** 5 **7.** 2 **9.** $2\frac{2}{9}$ hr **11.** $1\frac{1}{2}$ min **13.** $108.00 **15.** 3 hr **17.** 20 hr **19.** 6 mph **21.** 1st portion speed: 10 mph; cooldown speed: 8 mph **23.** $r = 30$ mph **25.** 8 mph **27.** 63 mph **29.** $x = 6$ **31.** $x = 5$ **33.** $y = 21.25$ **35.** $y = 18$ ft **37.** $26\frac{2}{3}$ ft **39.** $\frac{1}{2}$ **41.** $\frac{3}{7}$ **43.** Castroneves: 219.5 mph; Junqueira: 215.9 mph **45.** answers may vary

Exercise Set 6.7

1. $\frac{2}{3}$ **3.** $\frac{2}{3}$ **5.** $-\frac{4x}{15}$ **7.** $\frac{4}{3}$ **9.** $\frac{27}{16}$ **11.** $\frac{m - n}{m + n}$ **13.** $\frac{2x(x - 5)}{7x^2 + 10}$ **15.** $\frac{1}{y - 1}$ **17.** $\frac{1}{6}$ **19.** $\frac{x + y}{x - y}$ **21.** $\frac{3}{7}$ **23.** $\frac{a}{x + b}$ **25.** $\frac{x + 8}{2 - x}$ or $-\frac{x + 8}{x - 2}$ **27.** $\frac{s^2 + r^2}{s^2 - r^2}$ **29.** answers may vary **31.** Steffi Graf **33.** Seles, Hingis, Sanchez-Vicario **35.** $\frac{13}{24}$ **37.** $\frac{R_1R_2}{R_2 + R_1}$ **39.** $\frac{2x}{2 - x}$ **41.** $\frac{1}{y^2 - 1}$

Chapter 6 Review

1. $x = 2, x = -2$ **2.** $x = \frac{5}{2}, x = -\frac{3}{2}$ **3.** $\frac{4}{3}$ **4.** $\frac{11}{12}$ **5.** $\frac{2}{x}$ **6.** $\frac{3}{x}$ **7.** $\frac{1}{x - 5}$ **8.** $\frac{1}{x + 1}$ **9.** $\frac{x(x - 2)}{x + 1}$ **10.** $\frac{5(x - 5)}{x - 3}$ **11.** $\frac{x - 3}{x - 5}$ **12.** $\frac{x}{x + 4}$ **13.** $\frac{x + a}{x - c}$ **14.** $\frac{x + 5}{x - 3}$ **15.** $\frac{3x^2}{y}$ **16.** $-\frac{9x^2}{8}$ **17.** $\frac{x - 3}{x + 2}$ **18.** $-\frac{2x(2x + 5)}{(x - 6)^2}$ **19.** $\frac{x + 3}{x - 4}$ **20.** $\frac{4x}{3y}$ **21.** $(x - 6)(x - 3)$ **22.** $\frac{2}{3}$ **23.** $\frac{1}{x - 3}$ **24.** $\frac{x}{x + 6}$ **25.** $\frac{1}{2}$ **26.** $\frac{3(x + 2)}{3x + y}$ **27.** $\frac{1}{x + 2}$ **28.** $\frac{1}{x - 3}$ **29.** $\frac{2x - 10}{3x^2}$ **30.** $\frac{2x + 1}{2x^2}$ **31.** $14x$ **32.** $(x - 8)(x + 8)(x + 3)$ **33.** $10x^2y$ **34.** $36y^2x$ **35.** $x^2 - 3x - 10$ **36.** $3x^2 + 4x - 15$ **37.** $\frac{4y - 30x^2}{5x^2y}$ **38.** $\frac{-2x + 10}{(x - 3)(x - 1)}$ **39.** $\frac{14x + 58}{(x + 3)(x + 7)}$ **40.** $\frac{-2x - 2}{x + 3}$ **41.** $\frac{5x + 5}{(x + 4)(x - 2)(x - 1)}$ **42.** $\frac{x - 4}{3x}$ **43.** $-\frac{x}{x - 1}$ **44.** $\frac{x^2 + 2x - 3}{(x + 2)^2}$ **45.** $\frac{x^2 + 2x + 4}{4x}; \frac{x + 2}{32}$ **46.** $\frac{29x}{12(x - 1)}; \frac{3xy}{5(x - 1)}$ **47.** 1 **48.** 30 **49.** 2 **50.** 3, −4 **51.** $-\frac{5}{2}$ **52.** no solution **53.** 1 **54.** 5 **55.** $\frac{9}{7}$ **56.** −6, 1 **57.** $b = \frac{4A}{5x^2}$ **58.** $y = \frac{560 - 8x}{7}$ **59.** 3 **60.** 2 **61.** fast car speed: 30 mph; slow car speed: 20 mph **62.** 20 mph **63.** $17\frac{1}{2}$ hr **64.** $8\frac{4}{7}$ days **65.** $x = 15$ **66.** $x = 6$ **67.** $x = 15$ **68.** $x = 60$ **69.** $-\frac{7}{18y}$ **70.** $\frac{2x}{x - 3}$ **71.** $\frac{6}{7}$ **72.** $\frac{2x^2 + 1}{x + 2}$ **73.** $\frac{3y - 1}{2y - 1}$ **74.** $-\frac{7 + 2x}{2x}$

Chapter 6 Test

1. $x = -1, x = -3$ **2. a.** $115 **b.** $103 **3.** $\frac{3}{5}$ **4.** $\frac{1}{x - 10}$ **5.** $\frac{1}{x + 6}$ **6.** −1 **7.** $\frac{2m(m + 2)}{m - 2}$ **8.** $-\frac{1}{x + y}$ **9.** $\frac{(x - 6)(x - 7)}{(x + 2)(x + 7)}$ **10.** 15 **11.** $\frac{y - 2}{4}$ **12.** $-\frac{1}{2x + 5}$ **13.** $\frac{3a - 4}{(a - 3)(a + 2)}$ **14.** $\frac{3}{x - 1}$ **15.** $\frac{2(x + 3)(x + 5)}{x(x^2 + 4x + 1)}$ **16.** $\frac{x^2 + 2x + 35}{(x + 9)(x + 2)(x - 5)}$

17. $\dfrac{4y^2 + 13y - 15}{(y + 5)(y + 1)(y + 4)}$ **18.** $\dfrac{30}{11}$ **19.** -6 **20.** no solution **21.** no solution **22.** $\dfrac{xz}{2y}$ **23.** $b - a$ **24.** $\dfrac{5y^2 - 1}{y + 2}$ **25.** 12
26. 1, 5 **27.** 30 mph **28.** $6\frac{2}{3}$ hr

Cumulative Review

1. 27; Sec. 1.3, Ex. 2 **2.** 54; Sec. 1.3, Ex. 5 **3.** $2x + 6$; Sec. 2.1, Ex. 16 **4.** $(x - 4) \div 7$; Sec. 2.1, Ex. 17 **5.** $6 + x$; Sec. 2.1, Ex. 18
6. $x = 6$; Sec. 2.3, Ex. 1 **7.** $\left(-\infty, -\frac{7}{3}\right]$; Sec. 2.8, Ex. 10 **8. a.** US/Canada, 167 million **b.** 54 million more; Sec. 3.1, Ex. 1
9. 2000; 1700; 1400; 1100; 800; 500; Sec. 3.1, Ex. 8 **10.** ; Sec. 3.3, Ex. 7 **11.** 2;

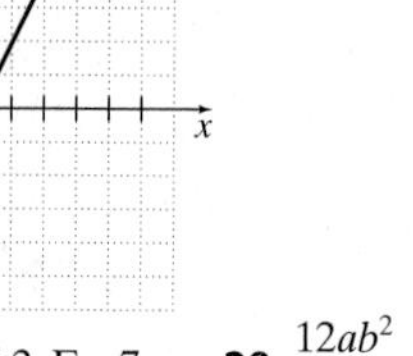

; Sec. 3.4, Ex. 2

12. a. 2; trinomial; Sec. 4.3, Ex. 3 **b.** 1; binomial; Sec. 3.3, Ex. 3 **c.** 3; none of these; Sec. 4.3, Ex. 3
13. $-4x^2 + 6x + 2$; Sec. 4.4, Ex. 2
14. $9y^2 + 6y + 1$; Sec. 4.6, Ex. 4
15. $3a(-3a^4 + 6a - 1)$; Sec. 5.1, Ex. 3
16. $(x - 2)(x + 6)$; Sec. 5.2, Ex. 3 **17.** $(4x - 1)(2x - 5)$; Sec. 5.3, Ex. 2 **18.** 4, 5; Sec. 5.6, Ex. 5 **19.** 1; Sec. 6.2, Ex. 7 **20.** $\dfrac{12ab^2}{27a^2b}$; Sec. 6.3, Ex. 9 **21.** $\dfrac{2m + 1}{m + 1}$; Sec. 6.4, Ex. 5 **22.** $-3, -2$; Sec. 6.5, Ex. 3 **23.** $\dfrac{x + 1}{x + 2y}$; Sec. 6.7, Ex. 5

Chapter 7 Graphs and Functions

Chapter 7 Pretest

1. $y = \frac{1}{3}x + 6$; 7.1B **2.** $y = -7x$; 7.1B **3.** $2x - y = 1$; 7.2A **4.** $x + 3y = 17$; 7.2A **5.** $y = 10$; 7.2B **6.** $3x + y = 24$; 7.2C
7. domain: $\{-2, 3, 2\}$; range: $\{5, -7\}$; function; 7.3A, B **8.** -22; 7.3E **9.** -5; 7.3E **10.** 3; 7.4A **11.** $\frac{3}{10}$; 7.4B **12.** $k = \frac{1}{4}$; $y = \frac{1}{4}x$; 7.5A
13. $k = 14$; $y = \frac{14}{x}$; 7.5B **14.** $y = kxz^2$; 7.5C, D

Graphing Calculator Explorations

1. $y = \dfrac{1}{3.5x}$ **3.** $y = \dfrac{-5.78}{2.31}x + \dfrac{10.98}{2.31}$ **5.** $y = x + 3.78$ **7.** $y = 13.3x + 1.5$

Mental Math

1. $m = -4$; $(0, 12)$ **3.** $m = 5$; $(0, 0)$ **5.** $m = \frac{1}{2}$; $(0, 6)$

Exercise Set 7.1

1.
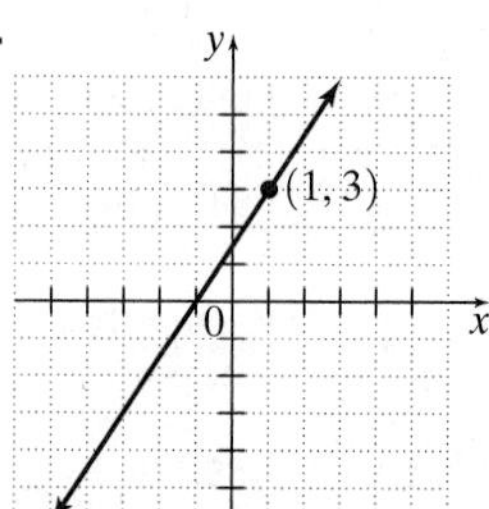

3.
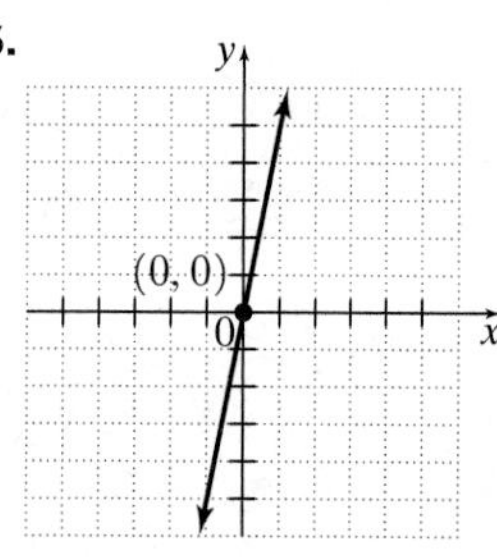

5.
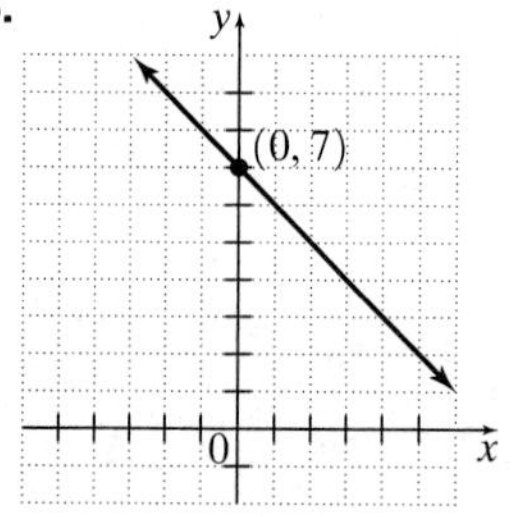

7.
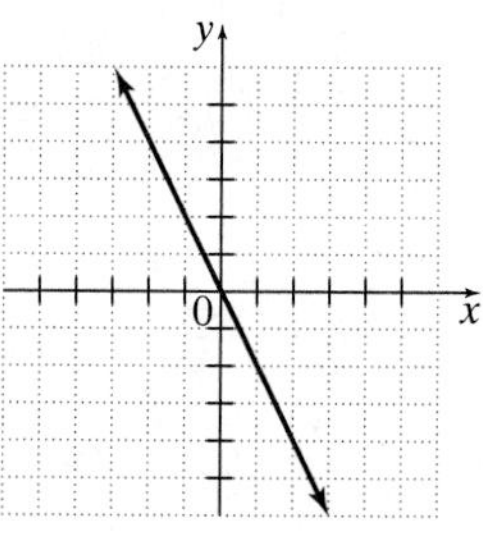

9.
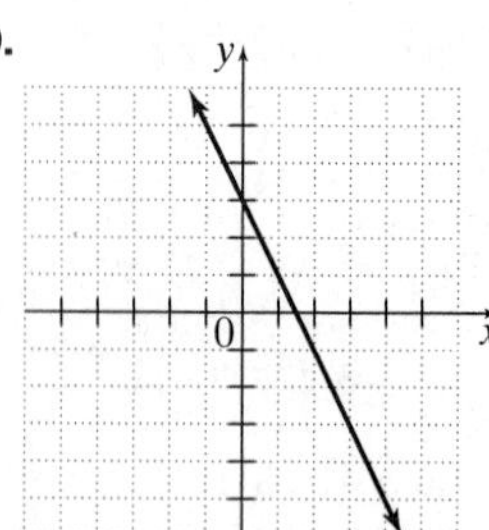

11.
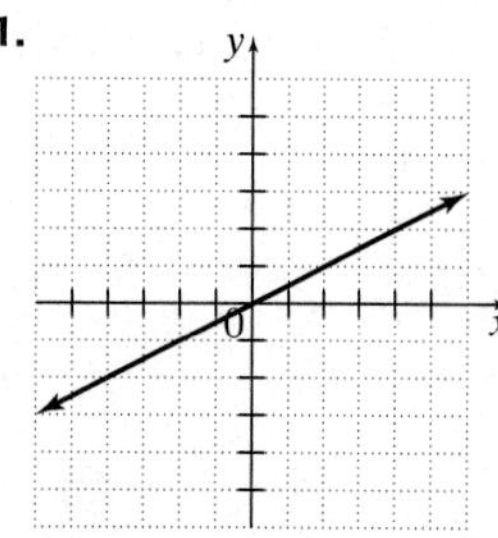

13.
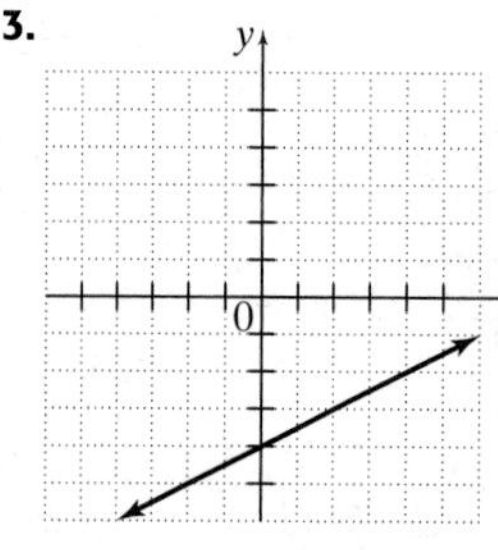

15.
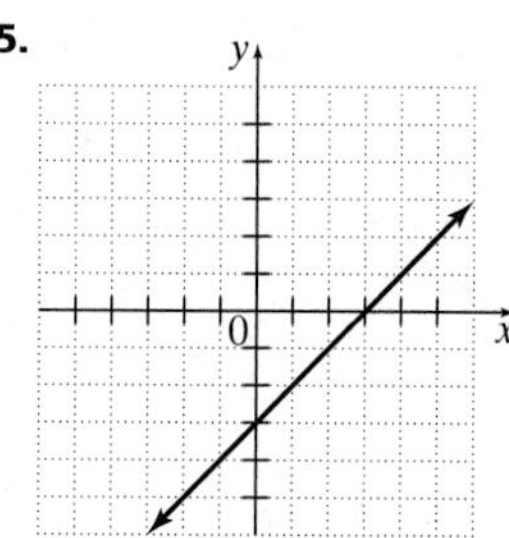

17.

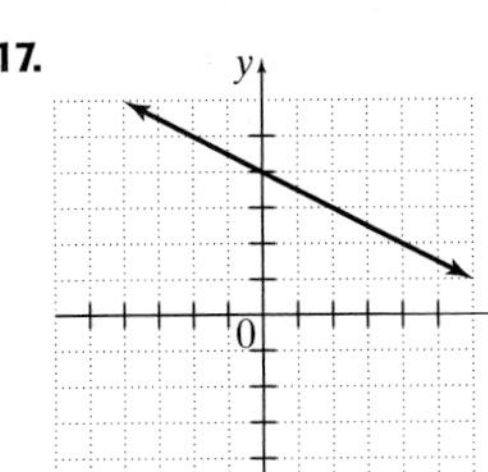

19. C **21.** D **23.** $y = -x + 1$ **25.** $y = 2x + \frac{3}{4}$ **27.** $y = \frac{2}{7}x$ **29. a.** \$44,640.50 **b.** $m = 1765.1$; annual income increases \$1765.10 every year **c.** (0, 35,815.0) at year $x = 0$, or 1995, annual average income was \$35,815.00 **31. a.** $m = 44$, $(0, 429)$ **b.** number of people employed as computer support specialists increases 44 thousand for every 1 year **c.** 429 thousand computer support specialists employed in 1998 **33. a.** \$4734.40 **b.** 2012 **c.** answers may vary **35.** $y = 5x + 32$ **37.** $y = 2x - 1$ **39.** answers may vary **41.** $y = -7x + 500$, where y is the height at time x

Graphing Calculator Explorations

1. 18.4 **3.** −1.5 **5.** 8.7, 7.6

Mental Math

1. $m = -2$; $(1, 4)$ **3.** $m = \frac{1}{4}$; $(2, 0)$ **5.** $m = 5$; $(3, -2)$

Exercise Set 7.2

1. $y = 3x - 1$ **3.** $y = -2x - 1$ **5.** $y = \frac{1}{2}x + 5$ **7.** $y = -\frac{9}{10}x - \frac{27}{10}$ **9.** $y = 2x + 7$ **11.** $y = -\frac{4}{3}x - \frac{20}{3}$ **13.** $y = 3x - 6$ **15.** $y = -2x + 1$ **17.** $y = -\frac{1}{2}x - 5$ **19.** $y = \frac{1}{3}x - 7$ **21.** $y = -\frac{2}{7}x - 6$ **23.** $y = -\frac{3}{8}x + \frac{5}{8}$ **25.** $x = 2$ **27.** $y = 1$ **29.** $x = 0$ **31.** $y = 4x - 4$ **33.** $y = \frac{1}{2}x - 6$ **35.** $y = 4$ **37.** $y = -\frac{3}{2}x - 6$ **39.** $y = -\frac{1}{2}x + \frac{13}{2}$ **41.** $y = -5$ **43.** $y = -\frac{1}{2}x + 1$ **45.** $y = -4x + 1$ **47. a.** $y = 32x$ **b.** 128 ft per sec **49. a.** $y = 12{,}000x + 18{,}000$ **b.** \$102,000 **c.** 9 yr **51. a.** $y = 6600x + 115{,}800$ **b.** \$175,200 **53. a.** $y = 3{,}704{,}642x + 1{,}089{,}261$ **b.** 27,021,755 DVD players **55.** 31 **57.** −8.4 **59.** 4 **61.** true **63.** $x = 5$

65.

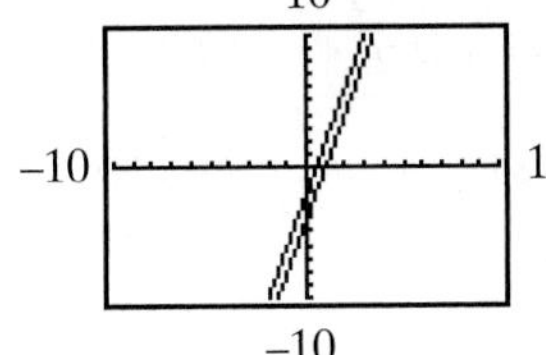

67. answers may vary

Integrated Review

1.

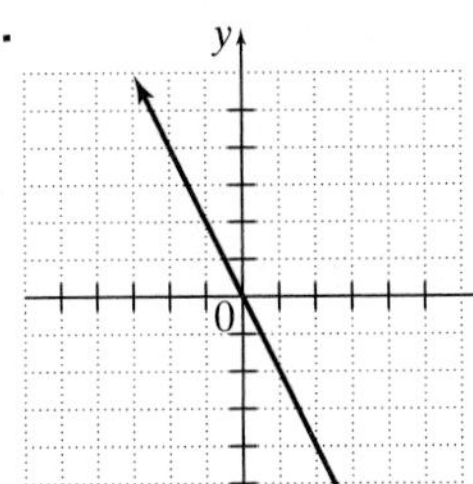

2.

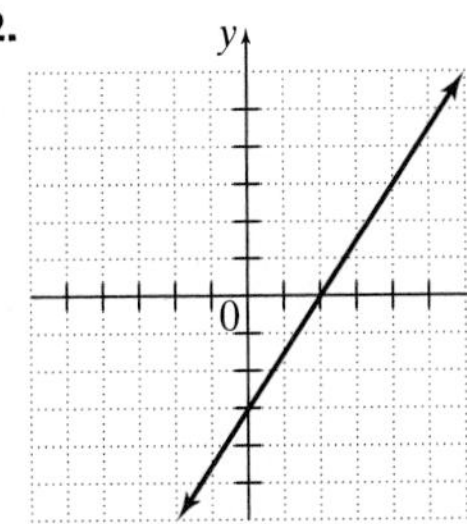

3.

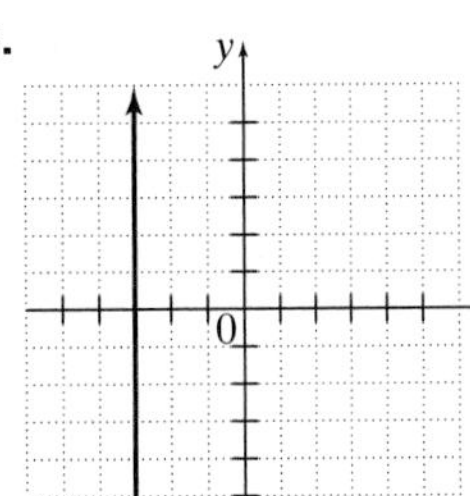

4.

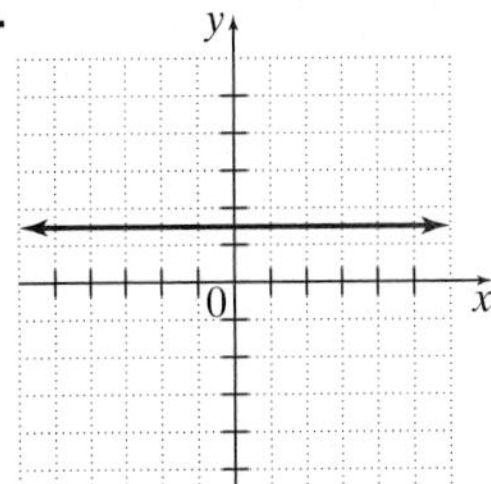

5. 0 **6.** $-\frac{3}{5}$ **7.** $m = 3$; $(0, -5)$ **8.** $m = \frac{5}{2}$; $\left(0, -\frac{7}{2}\right)$ **9.** parallel **10.** perpendicular **11.** $y = -x + 7$ **12.** $x = -2$ **13.** $y = 0$ **14.** $y = -\frac{3}{8}x - \frac{29}{4}$ **15.** $y = -5x - 6$ **16.** $y = -4x + \frac{1}{3}$ **17.** $y = \frac{1}{2}x - 1$ **18.** $y = 3x - \frac{3}{2}$ **19.** $y = 3x - 2$ **20.** $y = -\frac{5}{4}x + 4$ **21.** $y = \frac{1}{4}x - \frac{7}{2}$ **22.** $y = -\frac{5}{2}x - \frac{5}{2}$ **23.** $x = -1$ **24.** $y = 3$

Exercise Set 7.3

1. domain: $\{-1, 0, -2, 5\}$; range: $\{7, 6, 2\}$; function **3.** domain: $\{-2, 6, -7\}$; range: $\{4, -3, -8\}$; not a function **5.** domain: $\{1\}$; range: $\{1, 2, 3, 4\}$; not a function **7.** domain: $\left\{\frac{3}{2}, 0\right\}$; range: $\left\{\frac{1}{2}, -7, \frac{4}{5}\right\}$; not a function **9.** domain: $\{-3, 0, 3\}$; range: $\{-3, 0, 3\}$; function **11.** domain: $\{-1, 1, 2, 3\}$; range: $\{2, 1\}$; function **13.** domain: {Colorado, Alaska, Delaware, Illinois, Connecticut, New York}; range: $\{6, 1, 20, 31\}$; function **15.** domain: $\{32°, 104°, 212°, 50°\}$; range: $\{0°, 40°, 10°, 100°\}$; function **17.** domain: $\{2, -1, 5, 100\}$; range: $\{0\}$; function **19.** function **21.** yes **23.** no **25.** yes **27.** function **29.** not a function **31.** function **33.** domain: $[0, \infty)$; range: $(-\infty, \infty)$; not a function **35.** domain: $[-1, 1]$; range: $(-\infty, \infty)$; not a function **37.** domain: $(-\infty, \infty)$; range: $(-\infty, -3] \cup [3, \infty)$; not a function **39.** domain: $[1, 7]$; range: $[1, 7]$; not a function **41.** domain: $\{-2\}$; range: $(-\infty, \infty)$; not a function **43.** domain: $(-\infty, \infty)$; range: $(-\infty, 3]$; function **45.** 15 **47.** 38 **49.** 7 **51.** 3 **53. a.** 0 **b.** 1 **c.** −1 **55. a.** −5 **b.** −5 **c.** −5 **57.** 25π sq cm **59.** 2744 cu in. **61.** 166.38 cm **63.** 163.2 mg **65. a.** 95.99; per capita consumption of poultry was 95.99 lb in 2000 **b.** 99.37 lb

67.

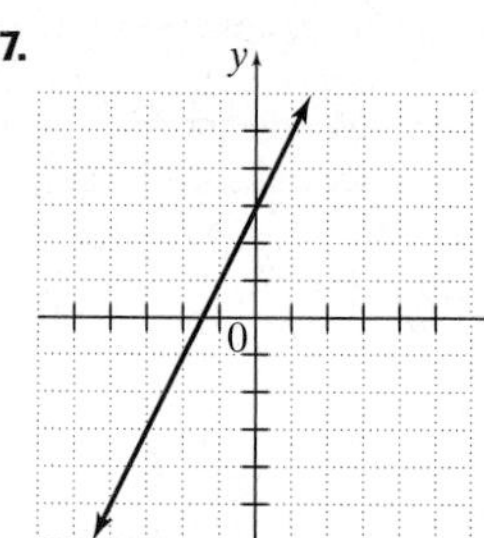

69.

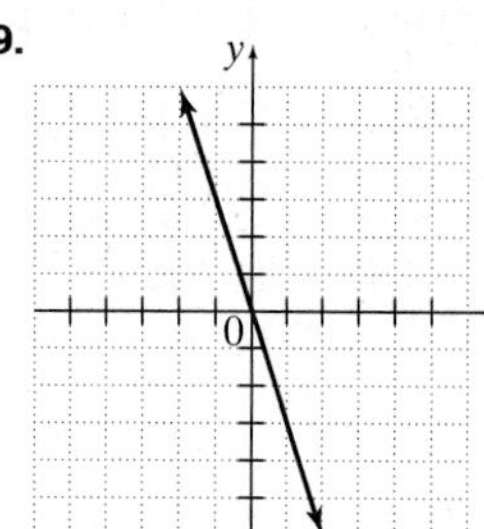

71.

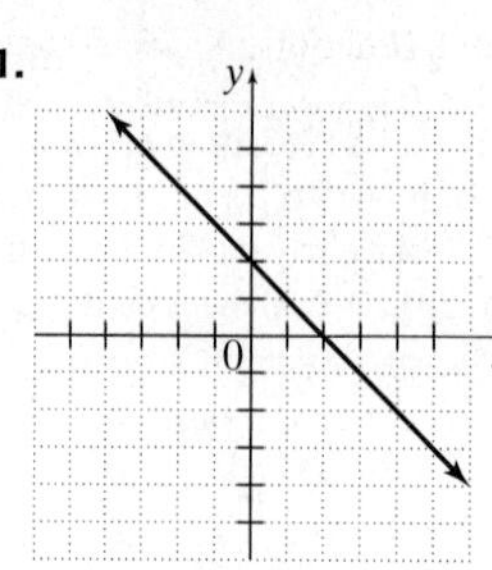

73. $(-\infty, 14]$ **75.** $\left[\frac{7}{2}, \infty\right)$ **77.** $\left(-\infty, -\frac{1}{4}\right)$ **79. a.** 5.1 **b.** 15.5 **c.** 9.533 **81. a.** 132 **b.** $a^2 - 12$ **83.** answers may vary **85.** answers may vary

Graphing Calculator Explorations

1. $\{x|x$ is a real number and $x \neq 2, x \neq -2\}$ **3.** $\{x|x$ is a real number and $x \neq -4, x \neq \frac{1}{2}\}$

Exercise Set 7.4

1. 57 **3.** 499 **5.** 1 **7.** 9 **9.** −16 **11.** −15 **13. a.** 284 ft **b.** 536 ft **c.** 756 ft **d.** 944 ft **e.** answers may vary **f.** 19 sec **15.** $80,000 **17.** $40,000 **19. a.** 3.1 million SUVs **b.** 5.8 million SUVs **21.** $f(x) = 2(49x^2 + 257x + 2373)$ **23. a.** 576 ft; 672 ft; 640 ft; 480 ft **b.** answers may vary **c.** $-16(t+4)(t-9)$ **25. a.** 22 stations **b.** 3519 stations **c.** 11,391 stations **d.** answers may vary **27.** $\frac{10}{3}$ **29.** $\frac{2}{5}$ **31.** $-\frac{7}{3}$ **33.** $\{x|x$ is a real number and $x \neq 5\}$ **35.** 6 **37.** −8 **39.** 0 **41.** $\{x|x$ is a real number and $x \neq 0\}$ **43. a.** $200 million **b.** $500 million **c.** $300 million **45.** −5 **47.** 2 **49. a.** $h(t) = -16t(t-4)$ **b.** 48 ft **c.** answers may vary **51.** $4x^2 - 3x + 6$ **53.** $2a - 3$; $-2x - 3$; $2x + 2h - 3$ **55.** $5x^2 + 25x$ **57.** $a^2 - 3a$

Exercise Set 7.5

1. $k = \frac{1}{5}; y = \frac{1}{5}x$ **3.** $k = \frac{3}{2}; y = \frac{3}{2}x$ **5.** $k = 14; y = 14x$ **7.** $k = 0.25; y = 0.25x$ **9.** 4.05 lb **11.** 204,706 tons **13.** $k = 30; y = \frac{30}{x}$ **15.** $k = 700; y = \frac{700}{x}$ **17.** $k = 2; y = \frac{2}{x}$ **19.** $k = 0.14; y = \frac{0.14}{x}$ **21.** $R = 54$ mph **23.** 72 amps **25.** divided by 4 **27.** $x = kyz$ **29.** $r = kst^3$ **31.** $k = \frac{1}{3}; y = \frac{1}{3}x^3$ **33.** $k = 0.2; y = 0.2\sqrt{x}$ **35.** $k = 1.3; y = \frac{1.3}{x^2}$ **37.** $k = 3; y = 3xz^3$ **39.** 22.5 tons **41.** 15π cu in. **43.** 8 ft **45.** yes **47.** no **49.** multiplied by 2 **51.** multiplied by 4

Chapter 7 Review

1.

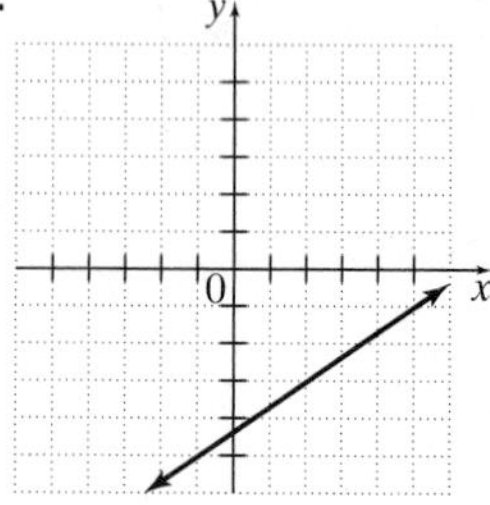

2.

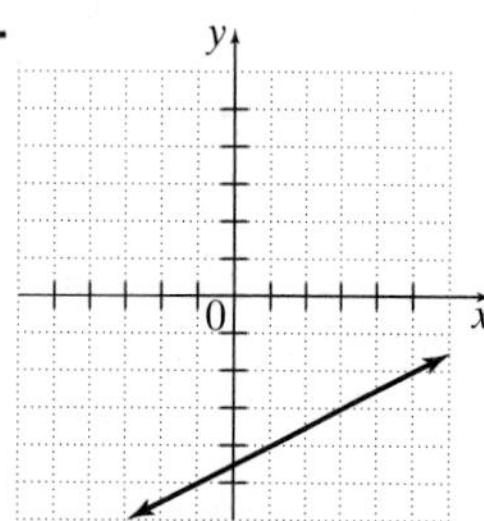

3.

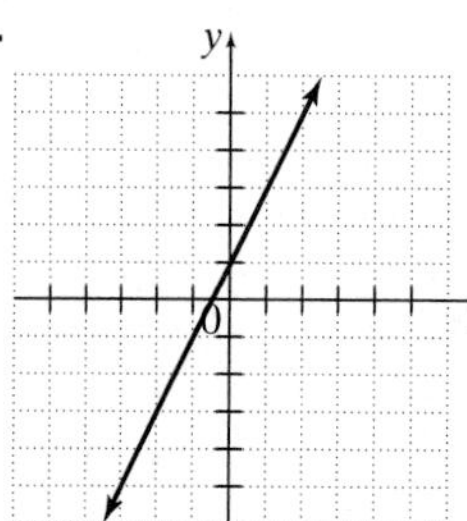

4.

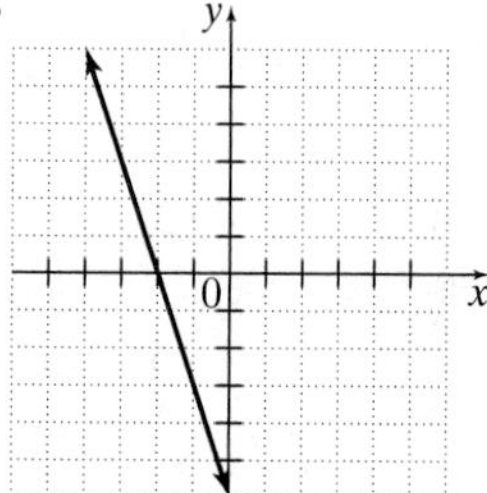

5.

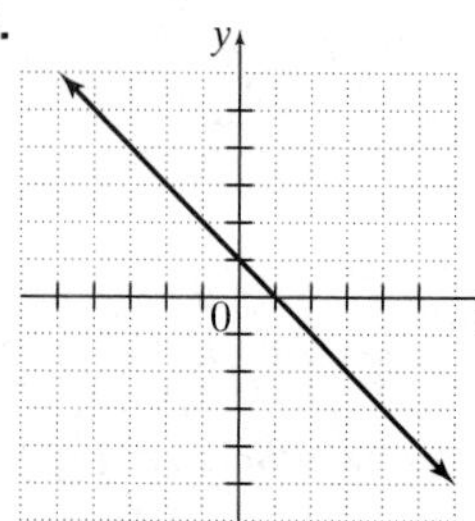

6.

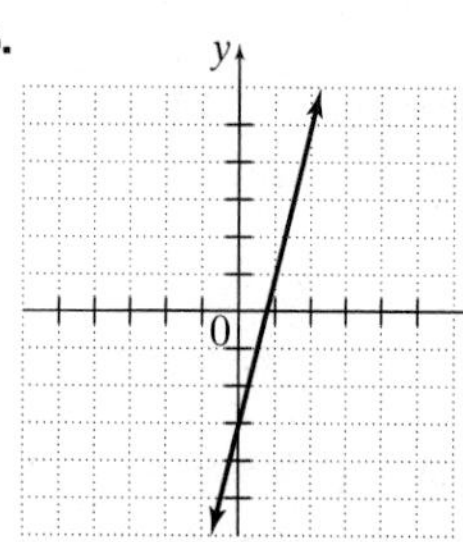

7.

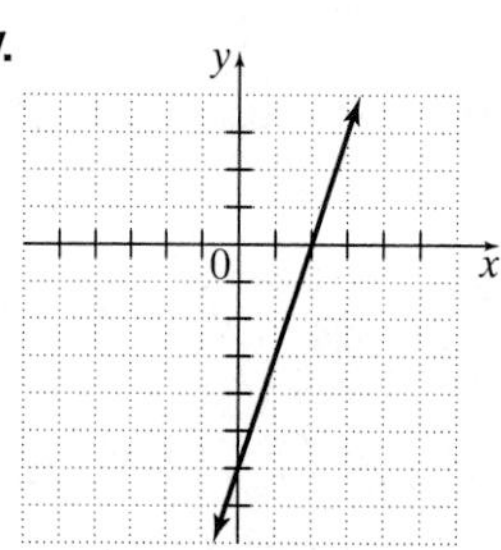

8.

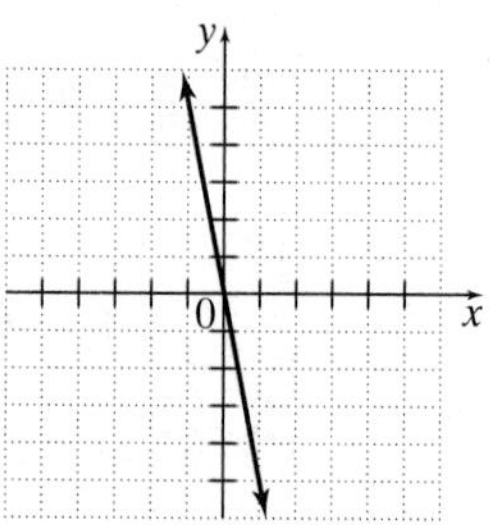

9. a. $87 **b.** $m = 0.3$; cost increases by $0.30 for each additional mile driven **c.** (0, 42); cost for 0 miles driven is $42 **10.** $y = -1$ **11.** $x = -2$ **12.** $x = -4$ **13.** $y = 5$ **14.** $y = 3x + 14$ **15.** $y = 2x - 12$ **16.** $y = -\frac{1}{2}x - 4$ **17.** $y = -11x - 52$ **18.** $y = -2x - 2$ **19.** $y = -\frac{3}{2}x - 8$ **20.** $y = \frac{3}{4}x + \frac{7}{2}$ **21.** $y = -\frac{3}{2}x - 1$ **22. a.** $y = -800x + 4200$ **b.** $200 **23. a.** $y = 3000x + 144{,}000$ **b.** $219,000 **24.** domain: $\left\{-\frac{1}{2}, 6, 0, 25\right\}$; range: $\left\{\frac{3}{4}, 0.65, -12, 25\right\}$; function **25.** domain: $\left\{\frac{3}{4}, 0.65, -12, 25\right\}$; range: $\left\{-\frac{1}{2}, 6, 0, 25\right\}$; function **26.** domain: $\{2, 4, 6, 8\}$; range: $\{2, 4, 5, 6\}$; not a function **27.** domain: {triangle, square, rectangle, parallelogram}; range: $\{3, 4\}$; function

28. domain: $(-\infty, \infty)$; range: $(-\infty, -1] \cup [1, \infty)$; not a function **29.** domain: $\{-3\}$; range: $(-\infty, \infty)$; not a function **30.** domain: $(-\infty, \infty)$; range: $\{4\}$; function **31.** domain; $[-1, 1]$; range; $[-1, 1]$; not a function **32.** -3 **33.** 0 **34.** 18 **35.** 9 **36.** -3 **37.** 0 **38.** 381 lb **39.** 5080 lb **40.**

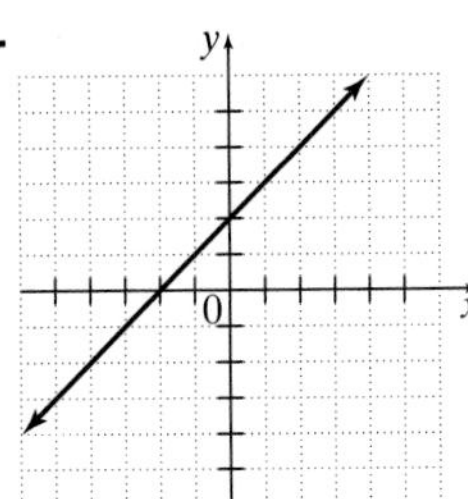

41.

42. 290 **43.** 58 **44.** 110 **45.** $x^2 + 4x - 6$ **46.** $-x^2 + 2x + 3$ **47. a.** \$119 **b.** \$77 **c.** decrease **48.** 9 **49.** 4 **50.** 3.125 cu ft **51.** 64π sq in.

Chapter 7 Test

1. $m = -\frac{1}{4}; \left(0, \frac{2}{3}\right)$ **2.** C **3.** A **4.** B **5.** D **6.** $y = -8$ **7.** $x = -4$ **8.** $y = -2$ **9.** $y = -3x + 11$ **10.** $y = 5x - 2$ **11.** $y = -\frac{1}{2}x$ **12.** $y = -\frac{1}{3}x + \frac{5}{3}$ **13.** $y = -\frac{1}{2}x - \frac{1}{2}$ **14.** neither **15. a.** 82 **b.** 85 **c.** \$114 million **d.** 0.154; Every million dollars spent on payroll increases winnings by 0.154 game. **16.** domain: $(-\infty, \infty)$; range: $\{5\}$; function **17.** domain: $\{-2\}$; range: $(-\infty, \infty)$; not a function **18.** domain: $(-\infty, \infty)$, range: $[0, \infty)$; function **19.** domain: $(-\infty, \infty)$; range: $(-\infty, \infty)$; function **20. a.** 960 ft **b.** 953.44 ft **c.** 0 ft **21.** -20 **22.** 16 **23.** 9 **24.** 256 ft

Cumulative Review

1. a. -6; Sec. 1.5, Ex. 6 **b.** 6.3 **2.** $\frac{1}{22}$; Sec. 1.6, Ex. 9a **3.** $\frac{16}{3}$; Sec. 1.6, Ex. 9b **4.** $-\frac{1}{10}$; Sec. 1.6, Ex. 9c **5.** $-\frac{13}{9}$; Sec. 1.6, Ex. 9d **6.** $\frac{1}{1.7}$; Sec. 1.6, Ex. 9e **7. a.** 5 **b.** $8 - x$; Sec. 2.2, Ex. 8 **8.** no solution; Sec. 2.4, Ex. 6 **9. a.** $\frac{2}{5}$ **b.** $\frac{3}{4}$; Sec. 2.7, Ex. 4

10. −4 −3 −2 −1 0 1 2 3 4 $[2, \infty)$; Sec. 2.8, Ex. 1 **11.** −4 −3 −2 −1 0 1 2 3 4 $(0.5, 3]$; Sec. 2.8, Ex. 3

12.

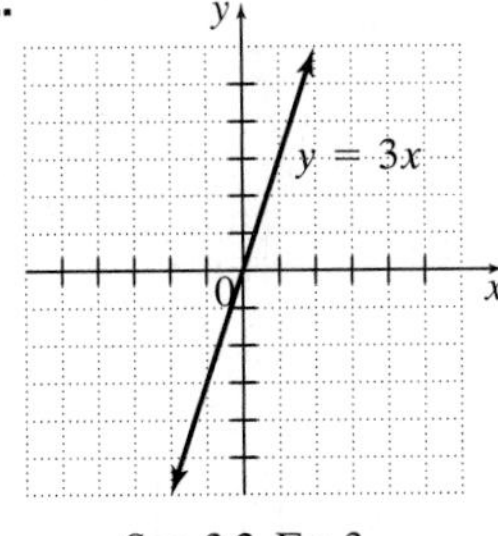

Sec. 3.2, Ex. 3

13.

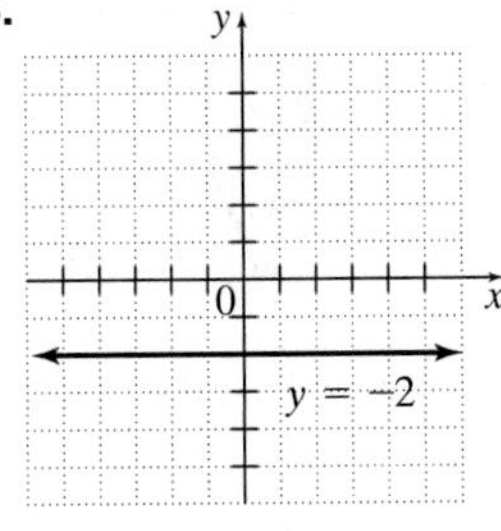

Sec. 3.2, Ex. 6

14. $\frac{m^7}{n^7}$, $n \neq 0$; Sec. 4.1, Ex. 19 **15.** $\frac{16x^{16}}{81y^{20}}$, $y \neq 0$; Sec. 4.1, Ex. 20 **16.** $9x^2 - 6x - 1$; Sec. 4.4, Ex. 5 **17.** $2x + 4 + \frac{-1}{3x - 1}$; Sec. 4.7, Ex. 5 **18.** $x = -\frac{1}{2}, x = 4$; Sec. 5.6, Ex. 6 **19.** 6, 8, 10; Sec. 5.7, Ex. 4 **20.** 1; Sec. 6.3, Ex. 2 **21.** $\frac{2x^2 + 3y}{x^2y + 2xy^2}$; Sec. 6.7, Ex. 6 **22.** $y = \frac{1}{4}x - 3$; Sec. 7.1, Ex. 3 **23.** $y = 3$; Sec. 7.2, Ex. 3 **24.** 5; Sec. 7.3, Ex. 20 **25.** -2; Sec. 7.3, Ex. 19

Chapter 8 Systems of Equations and Inequalities

Chapter 8 Pretest

1. $(6, -1)$; 8.1B

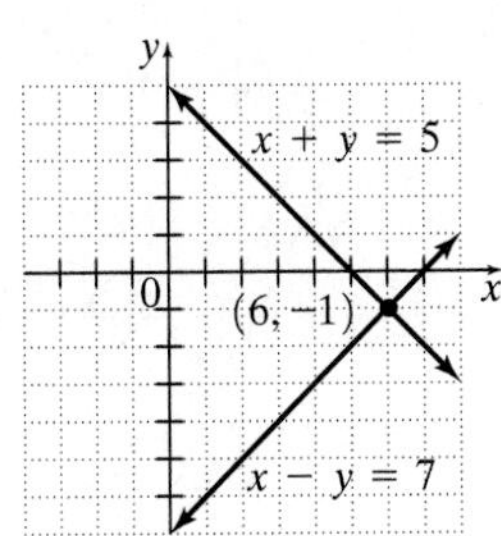

2. $(1, 4)$; 8.1B

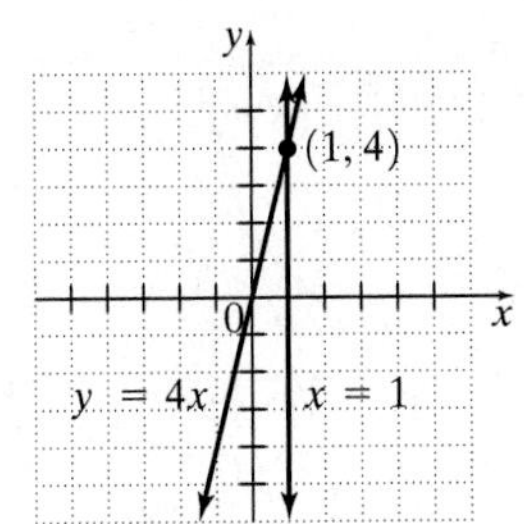

3. $(4, 2)$; 8.2A **4.** $(3, -2)$; 8.2A **5.** infinite number of solutions; 8.2A **6.** $(-2, -7)$; 8.3A **7.** $(12, -12)$; 8.3A **8.** $\left(\frac{3}{8}, -\frac{5}{8}\right)$; 8.3A **9.** $(2, -1, -3)$; 8.5A **10.** $(1, 0, -2)$; 8.5A **11.** $(10, -10)$; 8.6A **12.** $\emptyset$; 8.6A **13.** $(-3, -1, 2)$; 8.6B **14.** $(0, 8, 0)$; 8.6B

15. ; 8.7A **16.**

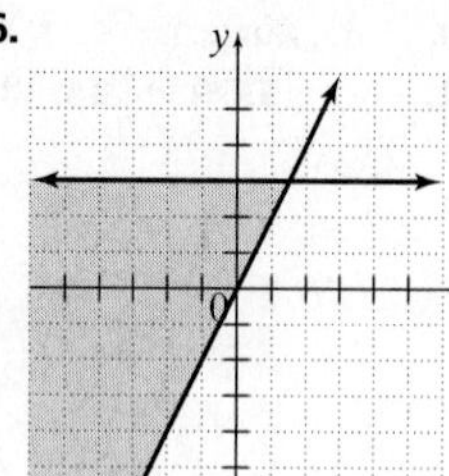

; 8.7A **17.** 4 and 12; 8.4A **18.** 20°, 60°, 100°; 8.5B

Graphing Calculator Explorations

1. (2.11, 0.17) **3.** (−8.21, −6.30)

Mental Math

1. 1 solution, (−1, 3) **3.** infinite number of solutions **5.** no solution **7.** 1 solution, (3, 2)

Exercise Set 8.1

1. a. no **b.** yes **3. a.** yes **b.** no **5. a.** yes **b.** yes

7. (3, 1)

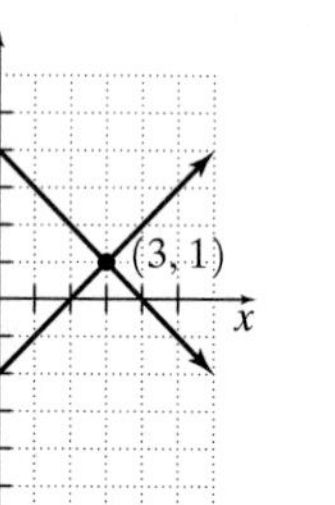

9. (6, 0)

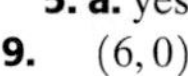

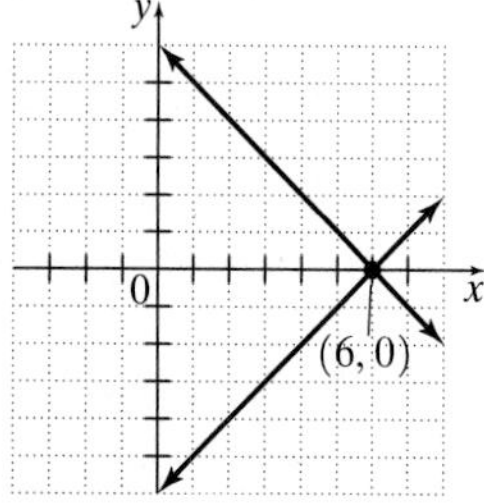

11. (−2, −4)

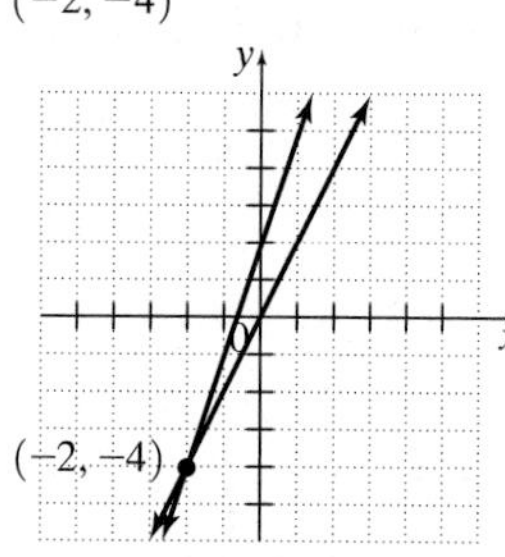

13. (2, 3)

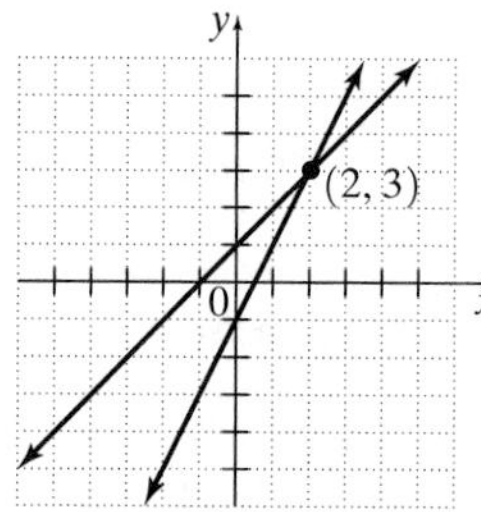

15. (1, −2)

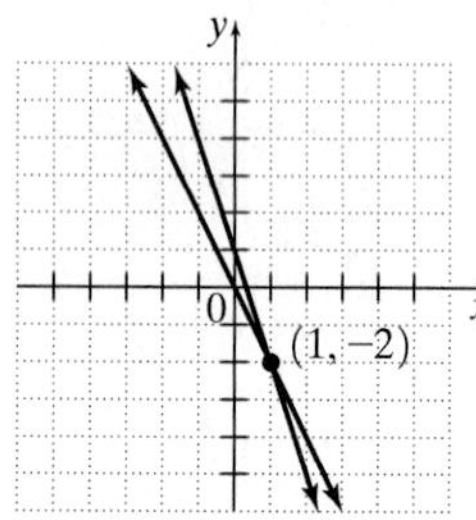

17. (−2, 1)

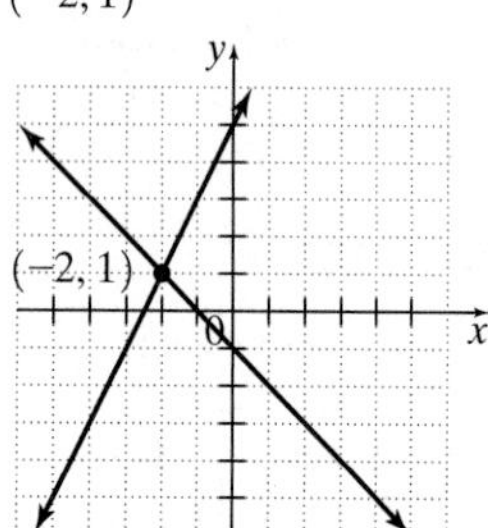

19. (4, 2)

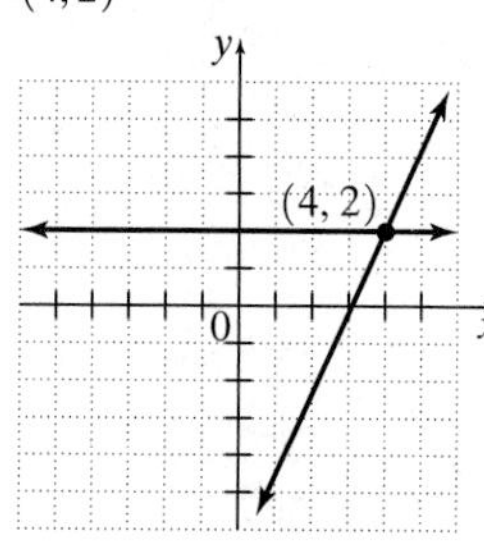

21. no solution

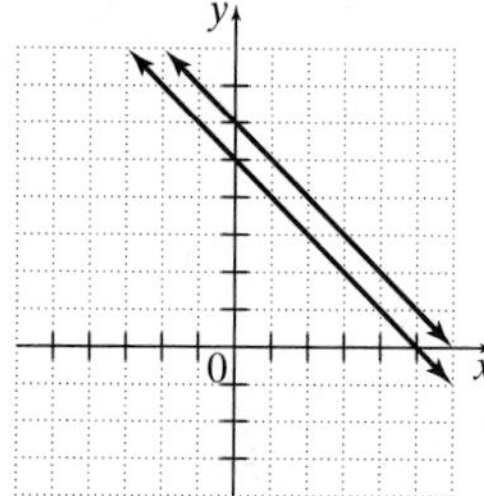

23. (2, 0)

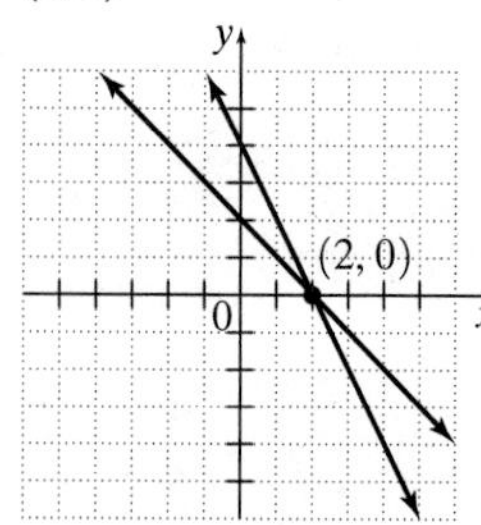

25. (0, −1)

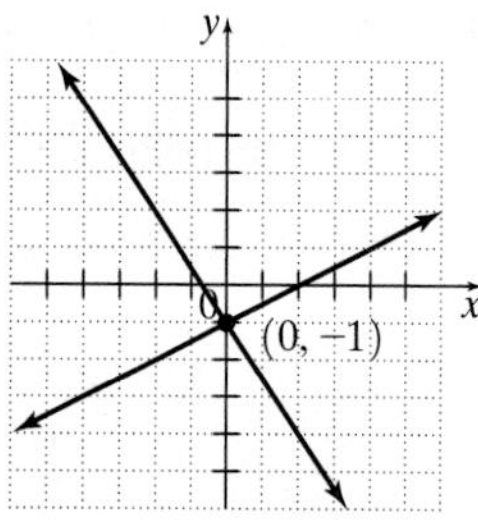

27. infinite number of solutions

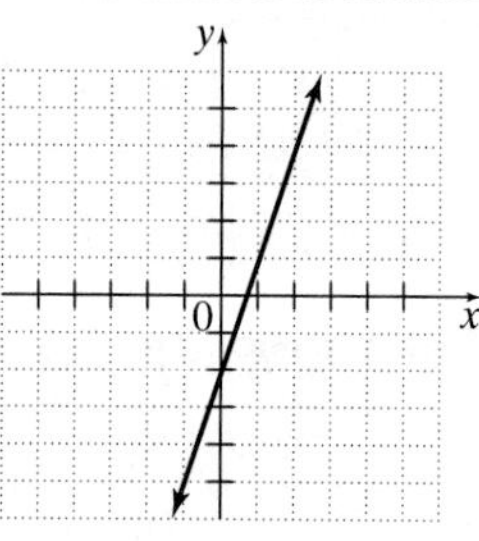

29. (3, −1)

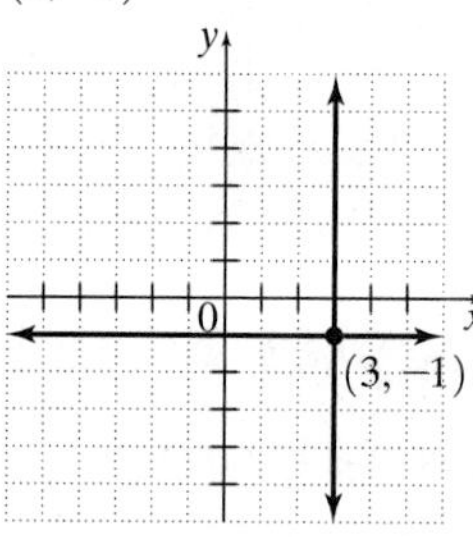

31. (−5, −7)

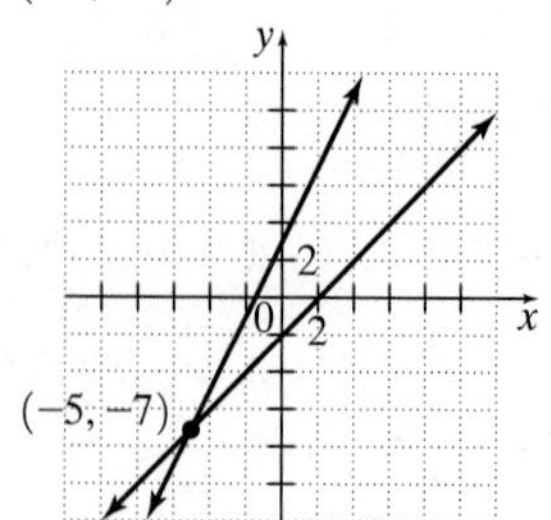

33. (5, 4)

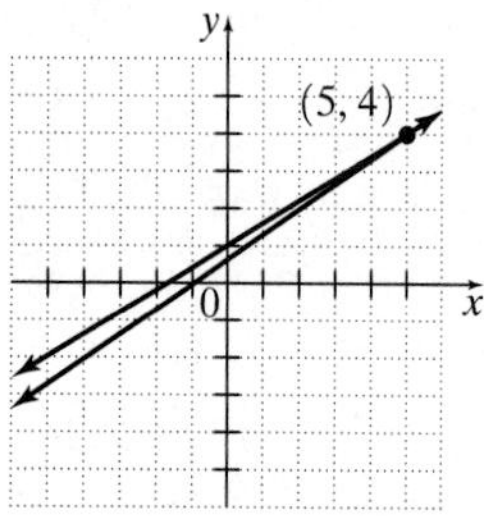

35. answers may vary

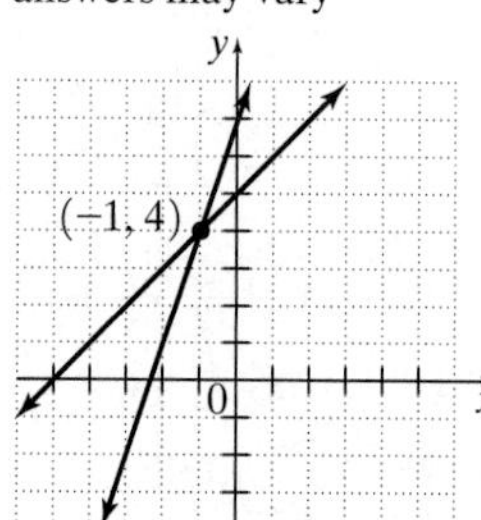

37. answers may vary

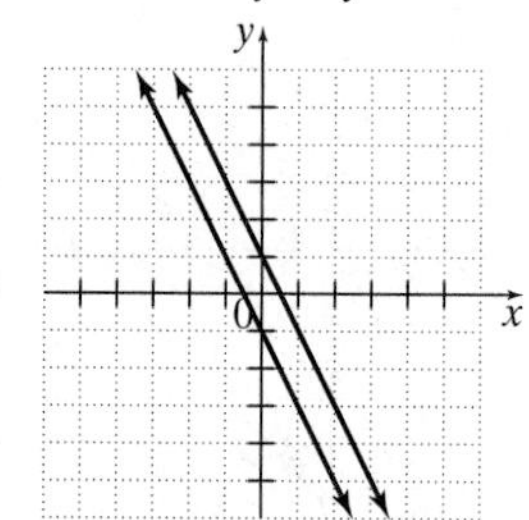

39. 1984, 1988 **41.** 1996 **43. a.** $(4, 9)$ **b.**

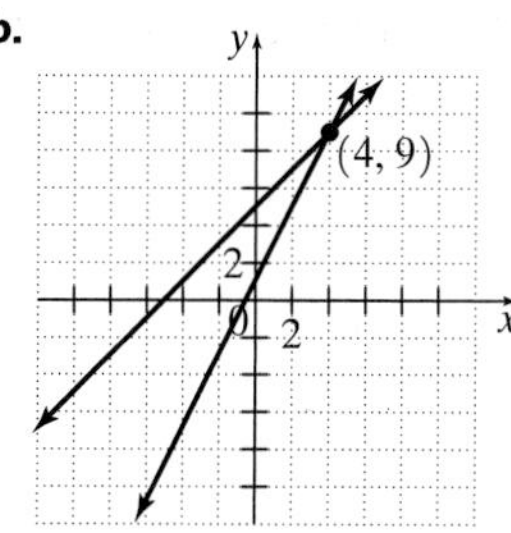

c. yes **45.** $x = -1$ **47.** $y = 3$ **49.** $z = -7$ **51.** answers may vary **53.** answers may vary

Exercise Set 8.2

1. $(2, 1)$ **3.** $(-3, 9)$ **5.** $(4, 2)$ **7.** $(10, 5)$ **9.** $(2, 7)$ **11.** $\left(-\frac{1}{5}, \frac{43}{5}\right)$ **13.** $(-2, 4)$ **15.** $(-2, -1)$ **17.** no solution **19.** $(3, -1)$ **21.** $(3, 5)$ **23.** $\left(\frac{2}{3}, -\frac{1}{3}\right)$ **25.** $(-1, -4)$ **27.** $(-6, 2)$ **29.** $(2, 1)$ **31.** no solution **33.** infinite number of solutions **35.** answers may vary **37.** $-6x - 4y = -12$ **39.** $-12x + 3y = 9$ **41.** $5n$ **43.** $-15b$ **45.** $(1, -3)$ **47.** $(-2.6, 1.3)$ **49.** $(3.28, 2.1)$ **51. a.** $(12, 18)$ **b.** answers may vary **c.**

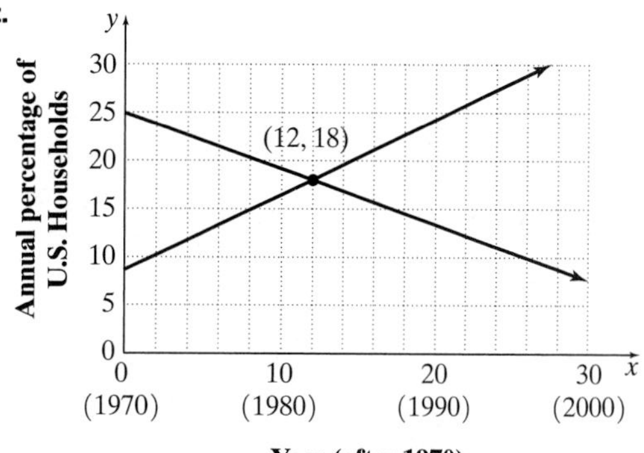

Exercise Set 8.3

1. $(1, 2)$ **3.** $(2, -3)$ **5.** $(5, -2)$ **7.** $(6, 0)$ **9.** $(-2, -5)$ **11.** $(-7, 5)$ **13.** $\left(\frac{12}{11}, -\frac{4}{11}\right)$ **15.** no solution **17.** no solution **19.** $\left(2, -\frac{1}{2}\right)$ **21.** $(6, -2)$ **23.** infinite number of solutions **25.** $(-2, 0)$ **27.** answers may vary **29.** $\left(\frac{3}{2}, 3\right)$ **31.** $(1, 6)$ **33.** infinite number of solutions **35.** $\left(-\frac{2}{3}, \frac{2}{5}\right)$ **37.** $(2, 4)$ **39.** $(-0.5, 2.5)$ **41.** $2x + 6 = x - 3$ **43.** $20 - 3x = 2$ **45.** $4(x + 6) = 2x$ **47. a.** $b = 15$ **b.** any real number except 15 **49.** $(-8.9, 10.6)$ **51.** b; answers may vary **53. a.** $(3, 549)$ **b.** answers may vary **c.** 1991 to 2000

Integrated Review

1. C **2.** D **3.** A **4.** B **5.** $(2, 5)$ **6.** $(4, 2)$ **7.** $(5, -2)$ **8.** $(6, -14)$ **9.** $(-3, 2)$ **10.** $(-4, 3)$ **11.** $(0, 3)$ **12.** $(-2, 4)$ **13.** $(5, 7)$ **14.** $(-3, -23)$ **15.** $\left(\frac{1}{3}, 1\right)$ **16.** $\left(-\frac{1}{4}, 2\right)$ **17.** no solution **18.** infinite number of solutions **19.** $(0.5, 3.5)$ **20.** $(-0.75, 1.25)$ **21.** infinite number of solutions **22.** no solution **23.** answers may vary **24.** answers may vary

Mental Math

1. c **3.** b **5.** a

Exercise Set 8.4

1. $\begin{cases} x + y = 15 \\ x - y = 7 \end{cases}$ **3.** $\begin{cases} x + y = 6500 \\ x = y + 800 \end{cases}$ **5.** 33 and 50 **7.** 14 and -3 **9.** Smith: 646 points; Swoopes: 643 points **11.** child's ticket: \$18; adult's ticket: \$29 **13.** quarters: 53; nickels: 27 **15.** The Ohio Art Co.: \$15; General Electric: \$50.77 **17.** labor: \$13.50 per hr; material: \$275 per ton **19.** still water: 6.5 mph; current: 2.5 mph **21.** still air: 455 mph; wind: 65 mph **23.** $4\frac{1}{2}$ hr **25.** 12% solution: $7\frac{1}{2}$ oz; 4% solution: $4\frac{1}{2}$ oz **27.** \$4.95 beans: 113 lbs; \$2.65 beans: 87 lbs **29.** $60°, 30°$ **31.** $20°, 70°$ **33.** 2005 **35.** number sold at \$9.50: 23; number sold at \$7.50: 67 **37.** $2\frac{1}{4}$ mph and $2\frac{3}{4}$ mph **39.** 30%: 50 gal; 60%: 100 gal **41.** length: 42 in.; width: 30 in. **43 a.** answers may vary but notice the slope of each function **b.** 2006 **45.** $x = 75; y = 105$ **47.** 16 **49.** $36x^2$ **51.** $100y^6$ **53.** width: 9 ft; length: 15 ft **55.** answers may vary

Exercise Set 8.5

1. $(-2, 5, 1)$ **3.** $(-2, 3, -1)$ **5.** $\{(x, y, z) | x - 2y + z = -5\}$ **7.** $\emptyset$ **9.** $(-10, 6, 2)$ **11.** $(-3, -35, -7)$ **13.** $(6, 22, -20)$ **15.** $\emptyset$ **17.** $(3, 2, 2)$ **19.** $\{(x, y, z) | x + 2y - 3z = 4\}$ **21.** $(-3, -4, -5)$ **23.** $(12, 6, 4)$ **25.** $(1, 1, -1)$ **27.** 2 units of Mix A; 3 units of Mix B; 1 unit of Mix C **29.** 5 in.; 7 in.; 10 in. **31.** 18, 13, 9 **33.** 246 free throws; 119 two-point field goals; 85 three-point field goals **35.** $x = 60$; $y = 55$; $z = 65$ **37.** $5x + 5z = 10$ **39.** $-5y + 2z = 2$ **41.** answers may vary **43.** $(1, 1, 0, 2)$ **45.** $a = 3, b = 4, c = -1$ **47.** $a = 8\frac{1}{6}, b = -109\frac{5}{6}, c = 1290$; 1783 students in 2007

Exercise Set 8.6

1. $(2, -1)$ **3.** $(-4, 2)$ **5.** $\emptyset$ **7.** $\{(x, y) | x - y = 3\}$ **9.** $(4, -3)$ **11.** $(-2, 5, -2)$ **13.** $(1, -2, 3)$ **15.** $(2, 1, -1)$ **17.** $(1, -4, 3)$ **19.** -13 **21.** -36 **23.** 0 **25. a.** end of 1984 **b.** black-and-white sets; microwave ovens; percent of households owning black-and-white television sets decreasing and percent of households owning microwave ovens increasing; answers may vary **c.** in 2002

Exercise Set 8.7

1.

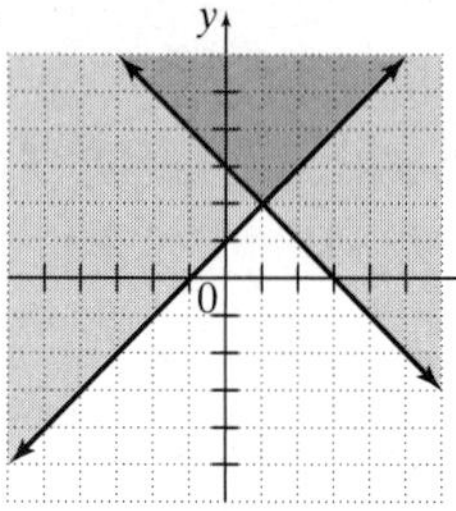

3.

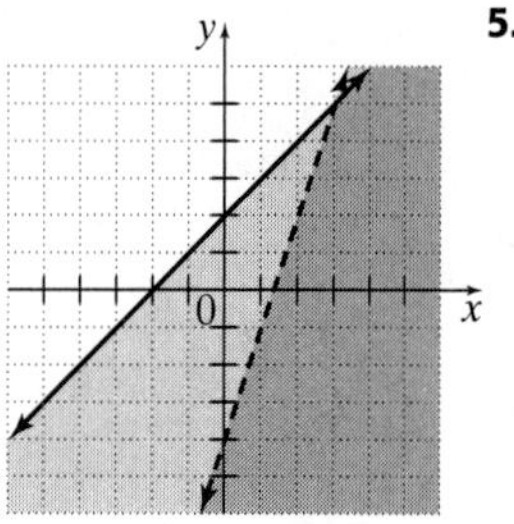

5.

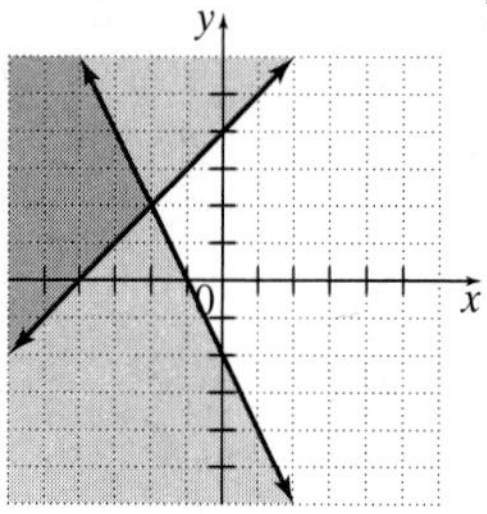

7.

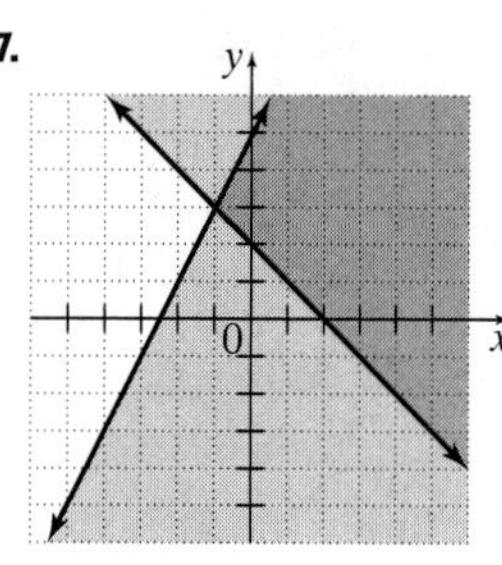

9.

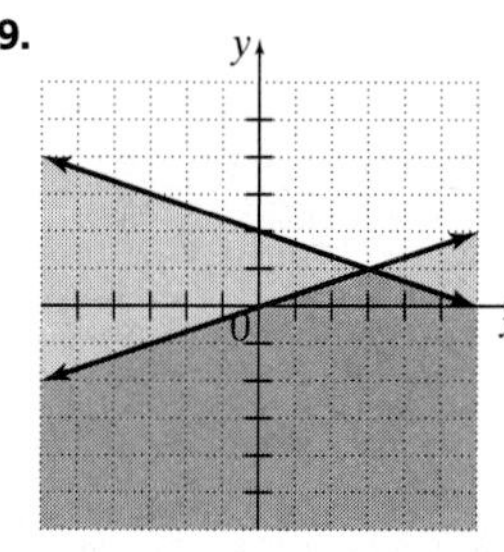

11.

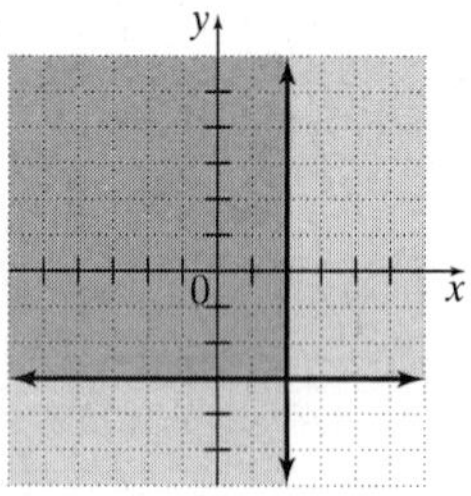

13.

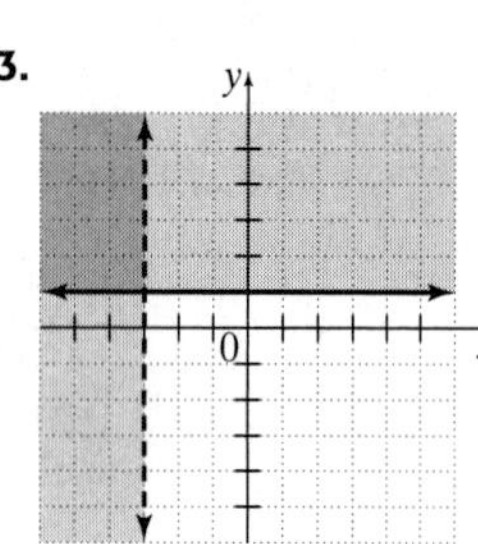

15.

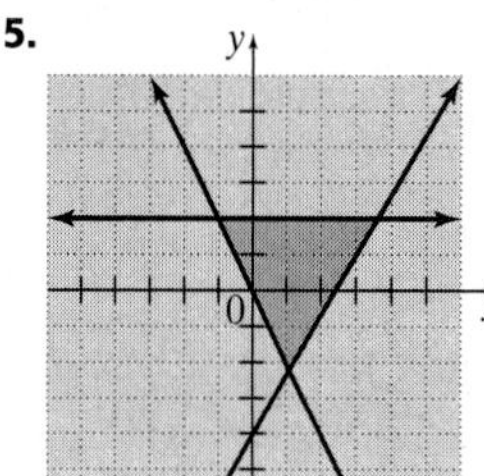

17.

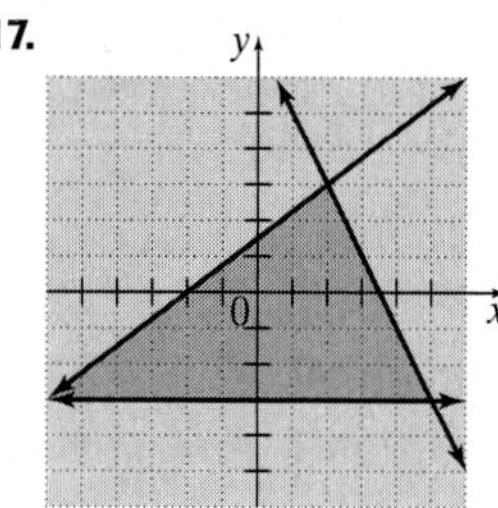

19.

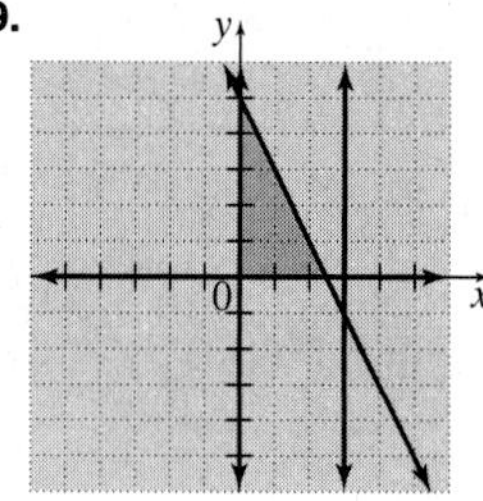

21. C **23.** D **25.** 9 **27.** $\frac{4}{9}$ **29.** 5 **31.** 59 **33. a.** $\begin{cases} x + y \le 8 \\ x < 3 \end{cases}$ **b.**

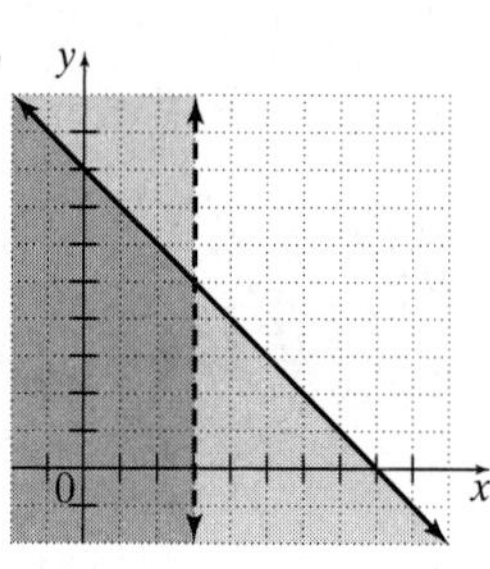

Chapter 8 Review

1. a. no **b.** yes **2. a.** yes **b.** no **3. a.** no **b.** no **4. a.** no **b.** yes

5. $(3, 2)$

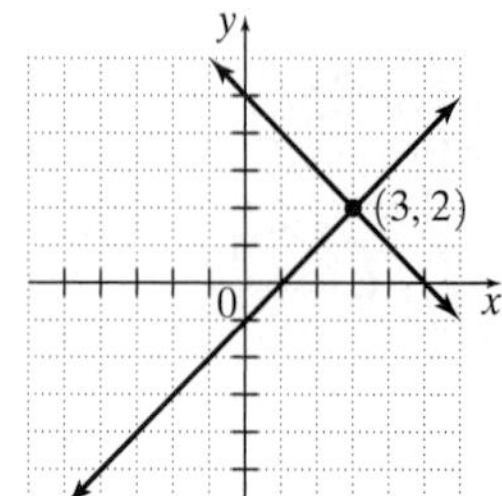

6. $(1, 2)$

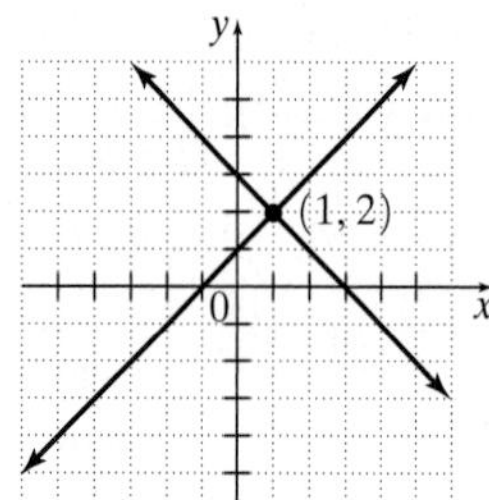

7. $(5, -1)$

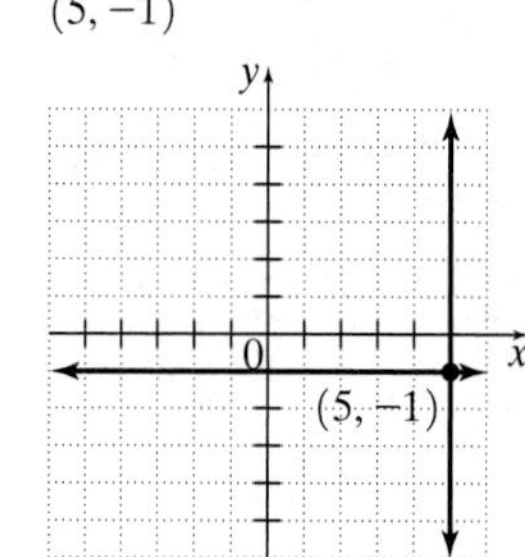

8. $(-3, 2)$

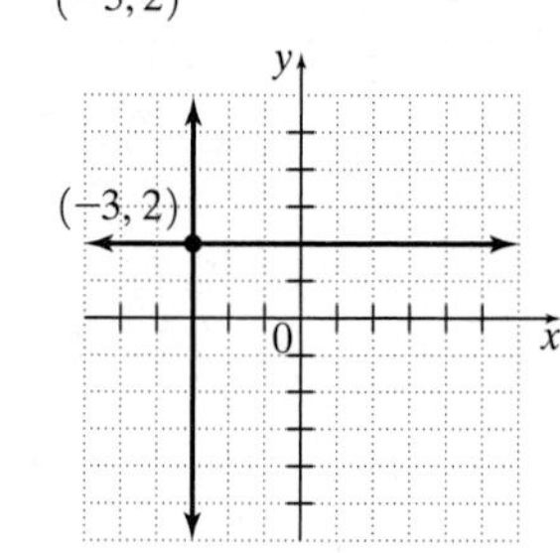

9. (3, −1)

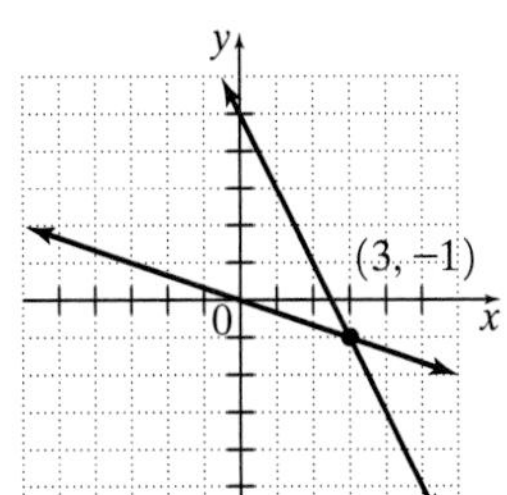

10. (1, −5)

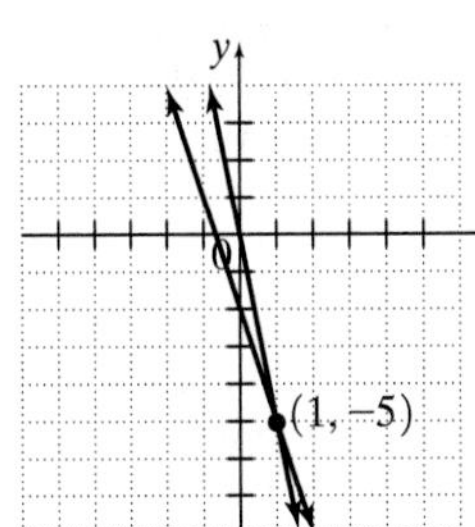

11. (−3, −2)

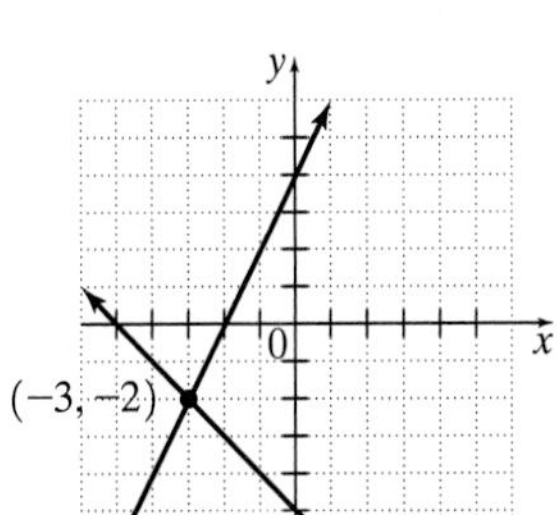

12. $\left(2\frac{1}{3}, -2\frac{2}{3}\right)$

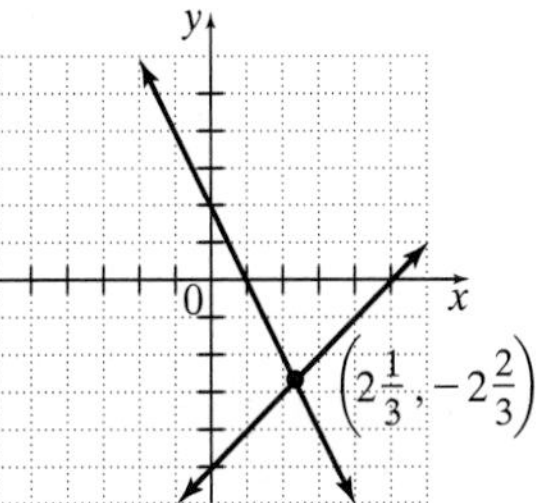

13. no solution

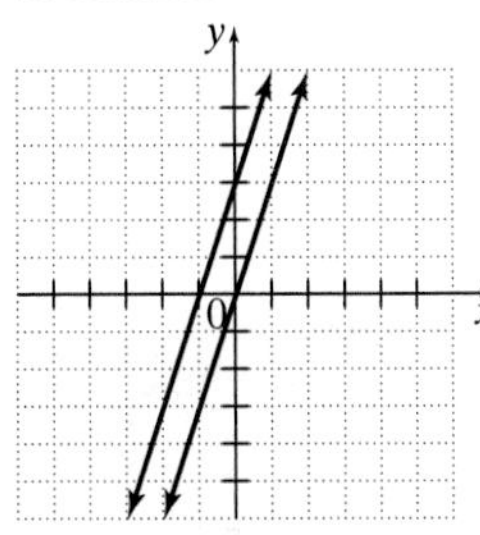

14. infinite number of solutions

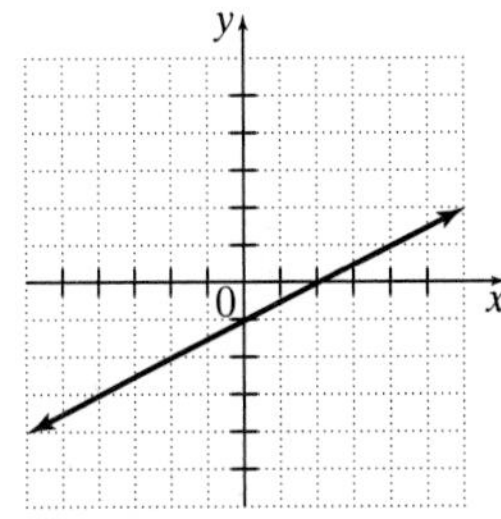

15. (4, 2) **16.** (5, 1) **17.** (−1, 4) **18.** (2, −1) **19.** (3, −2) **20.** (2, 5) **21.** no solution **22.** (−6, 2) **23.** no solution **24.** no solution **25.** (16, −2) **26.** (11, −2) **27.** (−6, 2) **28.** (4, −1) **29.** (3, 7) **30.** (−2, 4) **31.** infinite number of solutions **32.** no solution **33.** (8, −6) **34.** (10, −4) **35.** $\left(-\frac{3}{2}, \frac{15}{2}\right)$ **36.** infinite number of solutions **37.** −6 and 22

38. orchestra: 255 seats; balcony: 105 seats **39.** current of river: 3.2 mph; speed in still water: 21.1 mph **40.** amount invested at 6%: $6,180; amount invested at 10%: $2,820 **41.** length: 1.85 ft; width: 1.15 ft **42.** 6% solution: $12\frac{1}{2}$ cc; 14% solution: $37\frac{1}{2}$ cc

43. egg: 40¢; strip of bacon: 65¢ **44.** jogging: 0.86 hr; walking: 2.14 hr **45.** (2, 0, 2) **46.** (2, 0, −3) **47.** $\left(-\frac{1}{2}, \frac{3}{4}, 1\right)$ **48.** (−1, 2, 0) **49.** ∅ **50.** (5, 3, 0) **51.** (1, 1, −2) **52.** (3, 1, 1) **53.** 10, 40, and 48 **54.** 30 lb of creme-filled; 5 lb of chocolate-covered nuts; 10 lb of chocolate-covered raisins **55.** 17 pennies; 20 nickels; 16 dimes **56.** two sides: 22 cm each; third side: 29 cm **57.** (−3, 1) **58.** $\{(x, y)|x - 2y = 4\}$ **59.** $\left(-\frac{2}{3}, 3\right)$ **60.** $\left(\frac{1}{3}, \frac{7}{6}\right)$ **61.** $\left(\frac{5}{4}, \frac{5}{8}\right)$ **62.** (−7, −15) **63.** (1, 3) **64.** (2, 1) **65.** (1, 2, 3) **66.** (2, 0, −3) **67.** (3, −2, 5) **68.** (−1, 2, 0) **69.** (1, 1, −2) **70.** ∅

71.

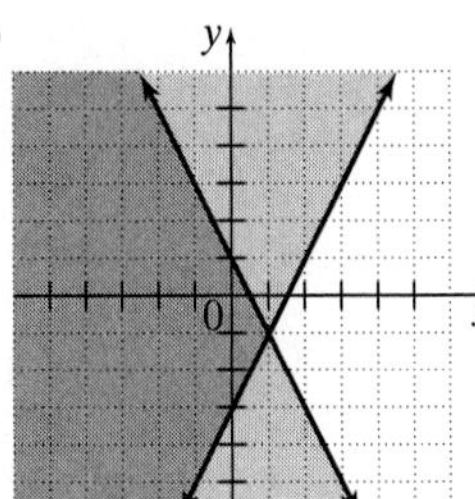

72.

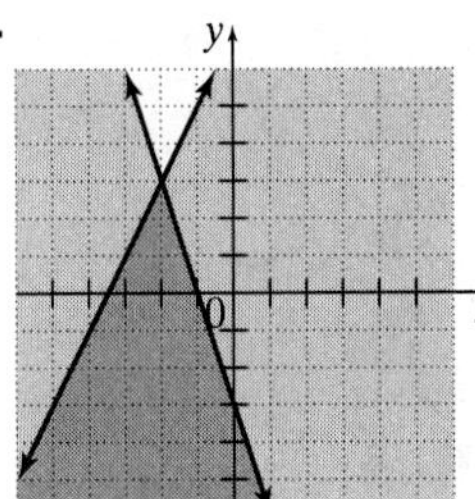

73.

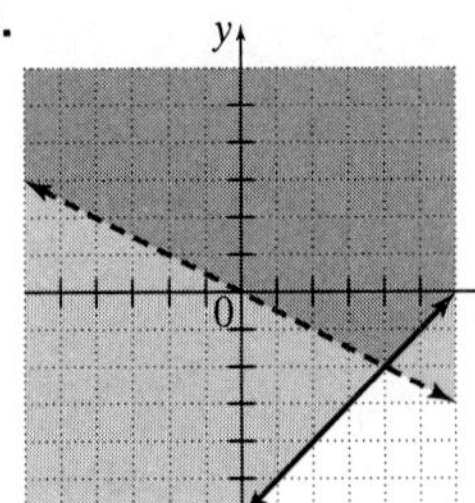

74.

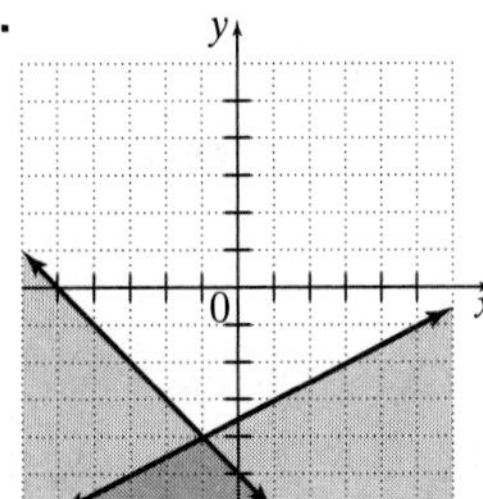

75.

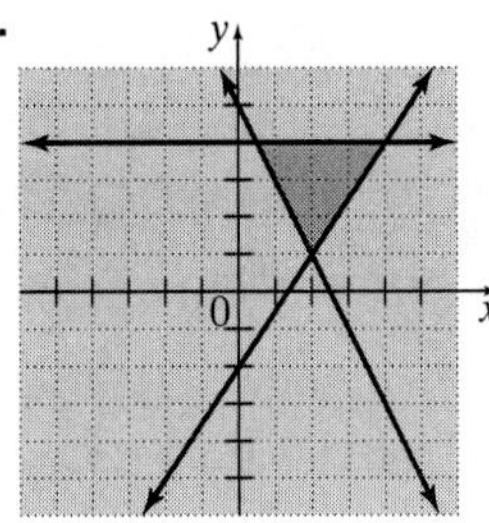

76.

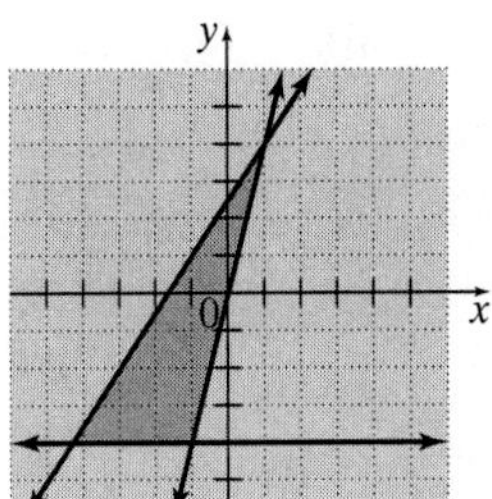

77.

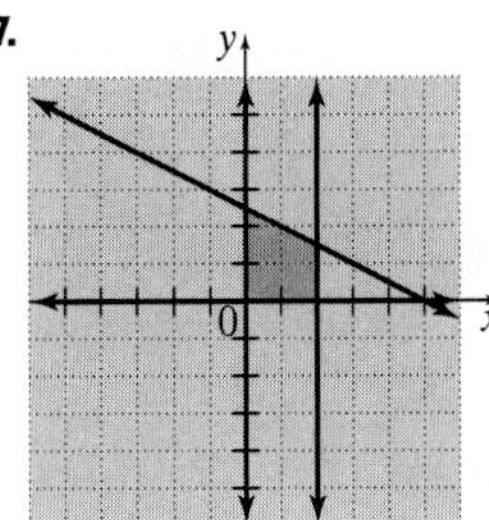

78.

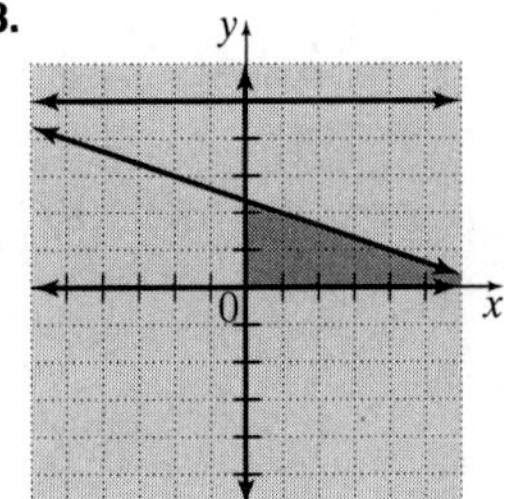

Chapter 8 Test

1. $(-2, -4)$

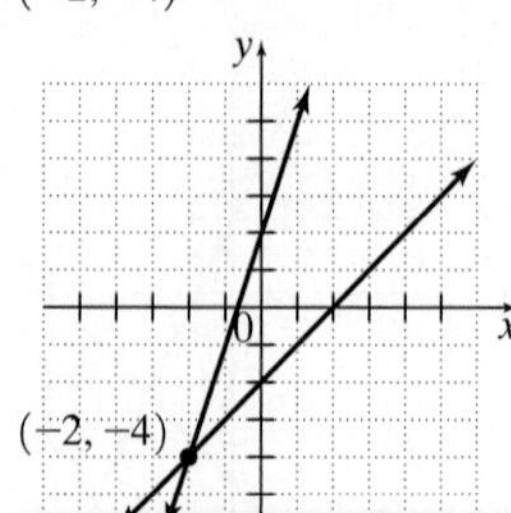

2. no solution

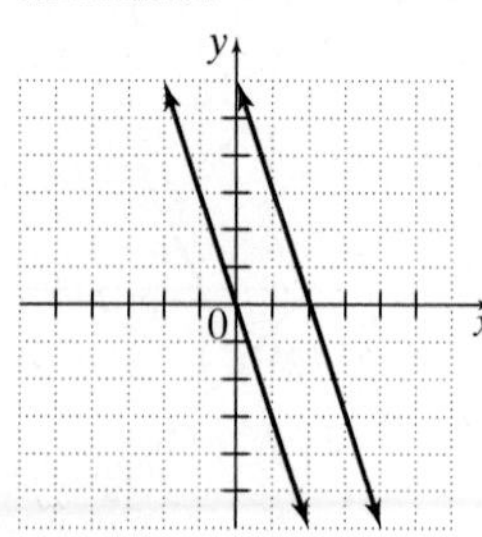

3. $(-4, 1)$ **4.** $(4, -5)$ **5.** $(-3, -7)$ **6.** infinite number of solutions **7.** $(20, 8)$ **8.** $(-4, 2)$ **9.** $\left(2, \frac{1}{2}\right)$ **10.** $\left(9\frac{2}{5}, 9\frac{3}{5}\right)$ **11.** $(-5, 3)$ **12.** no solution **13.** $(3, -1, 2)$ **14.** $(5, 0, -4)$ **15.** $\{(x, y) \mid x - y = -2\}$ **16.** $(5, -3)$ **17.** $(-1, -1, 0)$ **18.** $\emptyset$ **19.** 78, 46 **20.** \$1 bills: 20; \$5 bills: 42 **21.** 30% solution: 7.5 liters; 70% solution: 2.5 liters **22.** 3 mph; 6 mph **23.** 53 double rooms; 27 single rooms **24.** 5 gal of 10%; 15 gal of 20% **25.** 1999 **26.** 1991–1999

27.

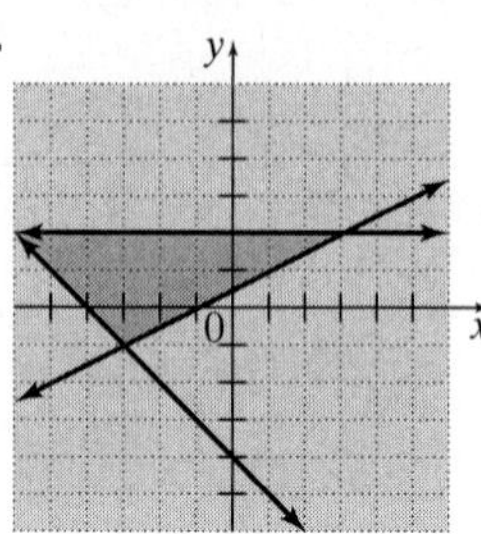

28.

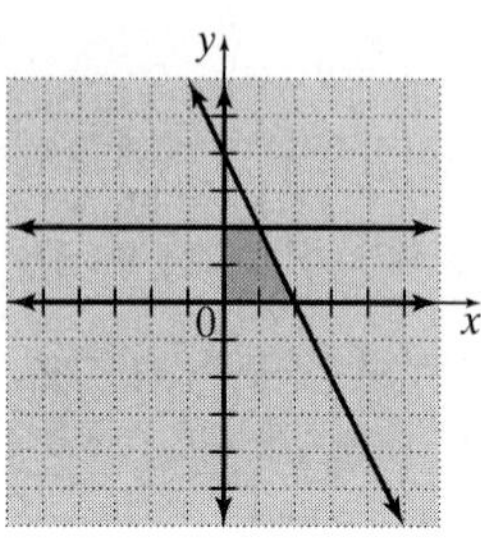

Cumulative Review

1. 50; Sec. 1.6, Ex. 3 **2.** $-\frac{8}{21}$; Sec. 1.6, Ex. 4 **3.** $x = 2$; Sec. 2.4, Ex. 1 **4. a.** 17% **b.** 21% **c.** 43 American travelers; Sec. 2.7, Ex. 3

5.

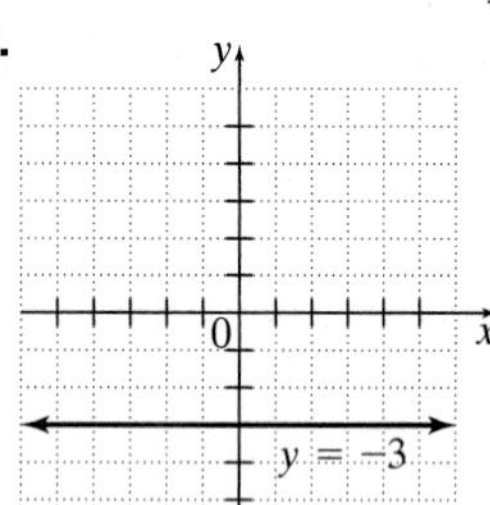

; Sec. 3.3, Ex. 7

6. ; Sec. 3.5, Ex. 5

7. a. 102,000 **b.** 0.007358 **c.** 84,000,000 **d.** 0.00003007; Sec. 4.2, Ex. 17 **8.** $6x^2 - 11x - 10$; Sec. 4.5, Ex. 7 **9.** $(y + 2)(x + 3)$; Sec. 5.1, Ex. 8 **10.** $(3x + 2)(x + 3)$; Sec. 5.3, Ex. 1 **11. a.** $x = 3$ **b.** $x = 2, x = 1$ **c.** none; Sec. 6.1, Ex. 2 **12.** $\frac{x + 2}{x}$; Sec. 6.1, Ex. 5 **13. a.** 0 **b.** $\frac{15 + 14x}{50x^2}$; Sec. 6.4, Ex. 1 **14.** $x = -\frac{17}{5}$; Sec. 6.5, Ex. 4 **15.** $y = -3x - 2$; Sec. 7.2, Ex. 1 **16.** function; Sec. 7.3, Ex. 7 **17.** $(3, 1)$; Sec. 8.3, Ex. 5 **18.** $\emptyset$; Sec. 8.5, Ex. 2 **19.** $\emptyset$; Sec. 8.6, Ex. 2 **20.** $(1, -1, 3)$; Sec. 8.6, Ex. 3

Chapter 9 Rational Exponents, Radicals, and Complex Numbers

Chapter 9 Pretest

1. $9x$; (9.1A) **2.** $-3y^3$; (9.1B) **3.** x^6; (9.1C) **4.** 4; (9.1E) **5.** 12; (9.2A) **6.** 27; (9.2B) **7.** $\frac{3m^{5/6}}{2}$; (9.2C) **8.** $\frac{27x^6}{y^9}$; (9.2D) **9.** $\sqrt{39x}$; (9.3A) **10.** $\frac{\sqrt{6}}{7}$; (9.3B) **11.** $-21\sqrt{2}$; (9.4A) **12.** $8x - 10\sqrt{x} - 3$; (9.4B) **13.** $5 - 2\sqrt{5}y + y^2$; (9.4B) **14.** $3b^4\sqrt[3]{5a^2}$; (9.3C) **15.** $\frac{\sqrt{3x}}{3x}$; (9.5A) **16.** $-4 + 2\sqrt{11}$; (9.5C) **17.** $\{12\}$; (9.6A) **18.** $4 - 8i$; (9.7B) **19.** $-21 - 20i$; (9.7C) **20.** $\frac{8}{13} + \frac{1}{13}i$; (9.7D)

Exercise Set 9.1

1. 2, −2 **3.** not a real number **5.** 10, −10 **7.** 10 **9.** $\frac{1}{2}$ **11.** 0.01 **13.** −6 **15.** x^5 **17.** $4y^3$ **19.** 2.646 **21.** 6.164 **23.** 14.142 **25.** 4 **27.** $\frac{1}{2}$ **29.** −1 **31.** x^4 **33.** $-3x^3$ **35.** −2 **37.** not a real number **39.** −2 **41.** x^4 **43.** $2x^2$ **45.** $9x^2$ **47.** $4x^2$ **49.** 8 **51.** −8 **53.** $2|x|$ **55.** x **57.** $|x - 2|$ **59.** $|x + 2|$ **61.** $\sqrt{3}$ **63.** −1 **65.** −3 **67.** $\sqrt{7}$ **69.** $-32x^{15}y^{10}$ **71.** $-60x^7y^{10}z^5$ **73.** $\frac{x^9y^5}{2}$ **75.** answers may vary **77.** 13 **79.** 18 **81.** 1.69 sq m **83.** answers may vary

Exercise Set 9.2

1. 7 **3.** 3 **5.** $\frac{1}{2}$ **7.** 13 **9.** $2\sqrt[3]{m}$ **11.** $3x^2$ **13.** -3 **15.** -2 **17.** 8 **19.** 16 **21.** not a real number **23.** $\sqrt[5]{(2x)^3}$ **25.** $\sqrt[3]{(7x+2)^2}$ **27.** $\frac{64}{27}$ **29.** $\frac{1}{16}$ **31.** $\frac{1}{16}$ **33.** not a real number **35.** $\frac{1}{x^{1/4}}$ **37.** $a^{2/3}$ **39.** $\frac{5x^{3/4}}{7}$ **41.** answers may vary **43.** $a^{7/3}$ **45.** x **47.** $3^{5/8}$ **49.** $y^{1/6}$ **51.** $8u^3$ **53.** $-b$ **55.** $27x^{2/3}$ **57.** $\frac{y}{z^{1/6}}$ **59.** $\frac{1}{x^{7/4}}$ **61.** $\sqrt{x}$ **63.** $\sqrt[3]{2}$ **65.** $2\sqrt{x}$ **67.** $\sqrt{x+3}$ **69.** $\sqrt{xy}$ **71.** $\sqrt[3]{a^2b}$ **73.** $\sqrt[15]{y^{11}}$ **75.** $\sqrt[12]{b^5}$ **77.** $\sqrt[24]{x^{23}}$ **79.** $\sqrt{a}$ **81.** $\sqrt[6]{432}$ **83.** $\sqrt[15]{343y^5}$ **85.** $\sqrt[6]{125r^3s^2}$ **87.** $25 \cdot 3$ **89.** $16 \cdot 3$ or $4 \cdot 12$ **91.** $8 \cdot 2$ **93.** $27 \cdot 2$ **95.** 1509 calories **97.** 99.5 million **99.** $a^{1/3}$ **101.** $x^{1/5}$ **103.** 1.6818 **105.** $\frac{t^{1/2}}{u^{1/2}}$

Exercise Set 9.3

1. $\sqrt{14}$ **3.** 2 **5.** $\sqrt[3]{36}$ **7.** $\sqrt{6x}$ **9.** $\sqrt{\frac{14}{xy}}$ **11.** $\sqrt[4]{20x^3}$ **13.** $\frac{\sqrt{6}}{7}$ **15.** $\frac{\sqrt{2}}{7}$ **17.** $\frac{\sqrt[4]{x^3}}{2}$ **19.** $\frac{\sqrt[3]{4}}{3}$ **21.** $\frac{\sqrt[4]{8}}{x^2}$ **23.** $\frac{\sqrt[3]{2x}}{3y^4\sqrt[3]{3}}$ **25.** $\frac{x\sqrt{y}}{10}$ **27.** $\frac{x\sqrt{5}}{13y}$ **29.** $-\frac{z^2\sqrt[3]{z}}{5x}$ **31.** $4\sqrt{2}$ **33.** $4\sqrt[3]{3}$ **35.** $25\sqrt{3}$ **37.** $2\sqrt{6}$ **39.** $10x^2\sqrt{x}$ **41.** $2y^2\sqrt[3]{2y}$ **43.** $a^2b\sqrt[4]{b^3}$ **45.** $y^2\sqrt{y}$ **47.** $5ab\sqrt{b}$ **49.** $-2x^2\sqrt[5]{y}$ **51.** $x^4\sqrt[3]{50x^2}$ **53.** $-4a^4b^3\sqrt{2b}$ **55.** $3x^3y^4\sqrt{xy}$ **57.** $5r^3s^4$ **59.** $\sqrt{2}$ **61.** 2 **63.** 10 **65.** x^2y **67.** $24m^2$ **69.** $\frac{15x\sqrt{2x}}{2}$ or $\frac{15x}{2}\sqrt{2x}$ **71.** $2a^2\sqrt[4]{2}$ **73.** $14x$ **75.** $2x^2 - 7x - 15$ **77.** y^2 **79.** $-3x - 15$ **81.** $x^2 - 8x + 16$ **83. a.** 20π sq cm **b.** 211.57 sq ft **85. a.** 3.8 times **b.** 2.9 times **c.** answers may vary

Mental Math

1. $6\sqrt{3}$ **3.** $3\sqrt{x}$ **5.** $12\sqrt[3]{x}$ **7.** 3

Exercise Set 9.4

1. $-2\sqrt{2}$ **3.** $10x\sqrt{2x}$ **5.** $17\sqrt{2} - 15\sqrt{5}$ **7.** $-\sqrt[3]{2x}$ **9.** $5b\sqrt{b}$ **11.** $\frac{31\sqrt{2}}{15}$ **13.** $\frac{\sqrt[3]{11}}{3}$ **15.** $\frac{5\sqrt{5x}}{9}$ **17.** $14 + \sqrt{3}$ **19.** $7 - 3y$ **21.** $6\sqrt{3} - 6\sqrt{2}$ **23.** $-23\sqrt[3]{5}$ **25.** $2b\sqrt{b}$ **27.** $20y\sqrt{2y}$ **29.** $2y\sqrt[3]{2x}$ **31.** $6\sqrt[3]{11} - 4\sqrt{11}$ **33.** $4x\sqrt[4]{x^3}$ **35.** $\frac{2\sqrt{3}}{3}$ **37.** $\frac{5x\sqrt[3]{x}}{7}$ **39.** $\frac{5\sqrt{7}}{2x}$ **41.** $\frac{\sqrt[3]{2}}{6}$ **43.** $\frac{14x\sqrt[3]{2x}}{9}$ **45.** $15\sqrt{3}$ in. **47.** $\sqrt{35} + \sqrt{21}$ **49.** $7 - 2\sqrt{10}$ **51.** $3\sqrt{x} - x\sqrt{3}$ **53.** $6x - 13\sqrt{x} - 5$ **55.** $\sqrt[3]{a^2} + \sqrt[3]{a} - 20$ **57.** $6\sqrt{2} - 12$ **59.** $2 + 2x\sqrt{3}$ **61.** $-16 - \sqrt{35}$ **63.** $x - y^2$ **65.** $3 + 2x\sqrt{3} + x^2$ **67.** $5x - 3\sqrt{10x} - 3\sqrt{15x} + 9\sqrt{6}$ **69.** $2\sqrt[3]{2} - \sqrt[3]{4}$ **71.** $-4\sqrt[6]{x^5} + \sqrt[3]{x^2} + 8\sqrt[3]{x} - 4\sqrt{x} + 7$ **73.** $x + 24 + 10\sqrt{x-1}$ **75.** $2x + 6 - 2\sqrt{2x+5}$ **77.** $x - 7$ **79.** $\frac{7}{x+y}$ **81.** $2a - 3$ **83.** $\frac{-2+\sqrt{3}}{3}$ **85.** $22\sqrt{5}$ ft; 150 sq ft **87. a.** $2\sqrt{3}$ **b.** 3 **c.** answers may vary

Mental Math

1. $\sqrt{2} - x$ **3.** $5 + \sqrt{a}$ **5.** $7\sqrt{4} - 8\sqrt{x}$

Exercise Set 9.5

1. $\frac{\sqrt{14}}{7}$ **3.** $\frac{\sqrt{5}}{5}$ **5.** $\frac{4\sqrt[3]{9}}{3}$ **7.** $\frac{3\sqrt{2x}}{4x}$ **9.** $\frac{3\sqrt[3]{2x}}{2x}$ **11.** $\frac{3\sqrt{3a}}{a}$ **13.** $\frac{3\sqrt[3]{4}}{2}$ **15.** $\frac{2\sqrt{21}}{7}$ **17.** $\frac{\sqrt{10xy}}{5y}$ **19.** $\frac{\sqrt[3]{75}}{5}$ **21.** $\frac{\sqrt{6x}}{10}$ **23.** $\frac{\sqrt{3z}}{6z}$ **25.** $\frac{\sqrt[3]{6xy^2}}{3x}$ **27.** $\frac{2\sqrt[4]{9x}}{3x^2}$ **29.** $\frac{5\sqrt[5]{4ab^4}}{2ab^3}$ **31.** $-2(2+\sqrt{7})$ **33.** $\frac{7(3+\sqrt{x})}{9-x}$ **35.** $-5 + 2\sqrt{6}$ **37.** $\frac{2a + 2\sqrt{a} + \sqrt{ab} + \sqrt{b}}{4a - b}$ **39.** $-\frac{8(1-\sqrt{10})}{9}$ **41.** $\frac{x - \sqrt{xy}}{x - y}$ **43.** $\frac{5+3\sqrt{2}}{7}$ **45.** $\frac{5}{\sqrt{15}}$ **47.** $\frac{6}{\sqrt{10}}$ **49.** $\frac{2x}{7\sqrt{x}}$ **51.** $\frac{5y}{\sqrt[3]{100xy}}$ **53.** $\frac{2}{\sqrt{10}}$ **55.** $\frac{2x}{11\sqrt{2x}}$ **57.** $\frac{7}{2\sqrt[3]{49}}$ **59.** $\frac{3x^2}{10\sqrt[3]{9x}}$ **61.** $\frac{6x^2y^3}{\sqrt{6z}}$ **63.** answers may vary **65.** $\frac{-7}{12 + 6\sqrt{11}}$ **67.** $\frac{3}{10 + 5\sqrt{7}}$ **69.** $\frac{x - 9}{x - 3\sqrt{x}}$ **71.** $\frac{x - 1}{x - 2\sqrt{x} + 1}$ **73.** $\{5\}$ **75.** $\left\{-\frac{1}{2}, 6\right\}$ **77.** $\{2, 6\}$ **79.** $r = \frac{\sqrt{A\pi}}{2\pi}$ **81.** answers may vary

Integrated Review

1. 9 **2.** -2 **3.** $\frac{1}{2}$ **4.** x^3 **5.** y^3 **6.** $2y^5$ **7.** $-2y$ **8.** $3b^3$ **9.** 6 **10.** $\sqrt[4]{3y}$ **11.** $\frac{1}{16}$ **12.** $\sqrt[5]{(x+1)^3}$ **13.** y **14.** $16x^{1/2}$ **15.** $x^{5/4}$ **16.** $4^{11/15}$ **17.** $2x^2$ **18.** $\sqrt[4]{a^3b^2}$ **19.** $\sqrt[4]{x^3}$ **20.** $\sqrt[6]{500}$ **21.** $2\sqrt{10}$ **22.** $2xy^2\sqrt[4]{x^3y^2}$ **23.** $3x\sqrt[3]{2x}$ **24.** $-2b^2\sqrt{2}$ **25.** $\sqrt{5x}$ **26.** $4x$ **27.** $7y^2\sqrt{y}$ **28.** $2a^2\sqrt[4]{3}$ **29.** $2\sqrt{5} - 5\sqrt{3} + 5\sqrt{7}$ **30.** $y\sqrt[3]{2y}$ **31.** $\sqrt{15} - \sqrt{6}$ **32.** $10 + 2\sqrt{21}$ **33.** $4x^2 - 5$ **34.** $x + 2 - 2\sqrt{x+1}$ **35.** $\frac{\sqrt{21}}{3}$ **36.** $\frac{5\sqrt[3]{4x}}{2x}$ **37.** $\frac{13 - 3\sqrt{21}}{5}$ **38.** $\frac{7}{\sqrt{21}}$ **39.** $\frac{3y}{\sqrt[3]{33y^2}}$ **40.** $\frac{x - 4}{x + 2\sqrt{x}}$

Calculator Explorations

1. 3.19 **3.** ∅ **5.** {3.23}

Exercise Set 9.6

1. {8} **3.** {7} **5.** ∅ **7.** {7} **9.** {6} **11.** $\left\{\frac{-9}{2}\right\}$ **13.** {29} **15.** {4} **17.** {−4} **19.** ∅ **21.** {7} **23.** {9} **25.** {50} **27.** ∅ **29.** $\left\{\frac{15}{4}\right\}$ **31.** {7} **33.** {5} **35.** {−12} **37.** {9} **39.** {−3} **41.** {1} **43.** {1} **45.** $\left\{\frac{1}{2}\right\}$ **47.** {0, 4} **49.** $\left\{\frac{37}{4}\right\}$ **51.** $3\sqrt{5}$ ft **53.** $2\sqrt{10}$ m **55.** $2\sqrt{131}$ m ≈ 22.9 m **57.** $\sqrt{100.84}$ mm ≈ 10.0 mm **59.** 17 ft **61.** 13 ft **63.** 14,657,415 sq mi **65.** 100 ft **67.** 100 **69.** $\frac{\pi}{2}$ sec ≈ 1.57 sec **71.** 12.97 ft **73.** answers may vary **75.** $15\sqrt{3}$ sq mi ≈ 25.98 sq mi **77.** answers may vary **79.** 0.51 km **81.** $\frac{x}{4x+3}$ **83.** $-\frac{4z+2}{3z}$ **85.** {1} **87.** 2743 deliveries

Mental Math

1. $9i$ **3.** $i\sqrt{7}$ **5.** −4 **7.** $8i$

Exercise Set 9.7

1. $2i\sqrt{6}$ **3.** $-6i$ **5.** $24i\sqrt{7}$ **7.** $-3\sqrt{6}$ **9.** $-\sqrt{14}$ **11.** $-5\sqrt{2}$ **13.** $4i$ **15.** $i\sqrt{3}$ **17.** $2\sqrt{2}$ **19.** $6-4i$ **21.** $-2+6i$ **23.** $-2-4i$ **25.** $2-i$ **27.** $5-10i$ **29.** $8-i$ **31.** −12 **33.** 63 **35.** −40 **37.** $18+12i$ **39.** $27+3i$ **41.** $18+13i$ **43.** 7 **45.** $12-16i$ **47.** 20 **49.** 2 **51.** $17+144i$ **53.** $-2i$ **55.** $-4i$ **57.** $\frac{28}{25}-\frac{21}{25}i$ **59.** $-\frac{12}{5}+\frac{6}{5}i$ **61.** $4+i$ **63.** $-\frac{5}{2}-2i$ **65.** $-5+\frac{16}{3}i$ **67.** $\frac{3}{5}-\frac{1}{5}i$ **69.** $\frac{1}{5}-\frac{8}{5}i$ **71.** 1 **73.** i **75.** $-i$ **77.** −1 **79.** −64 **81.** $-243i$ **83.** 5 people **85.** 14 people **87.** 16.7% **89.** $1-i$ **91.** 0 **93.** $2+3i$ **95.** $2+i\sqrt{2}$ **97.** $\frac{1}{2}-\frac{\sqrt{3}}{2}i$ **99.** answers may vary **101.** $6-6i$ **103.** yes

Chapter 9 Review

1. 9 **2.** 3 **3.** −2 **4.** not a real number **5.** $-\frac{1}{7}$ **6.** x^{32} **7.** −6 **8.** 4 **9.** $-a^2b^3$ **10.** $4a^2b^6$ **11.** $2ab^2$ **12.** $-2x^3y^4$ **13.** $\frac{x^6}{6y}$ **14.** $\frac{3y}{z^4}$ **15.** $|x|$ **16.** $|x^2-4|$ **17.** −27 **18.** −5 **19.** $-x$ **20.** $2|2y+z|$ **21.** $5|x-y|$ **22.** y **23.** $|x|$ **24.** 3, 6 **25.** 2, $\sqrt[3]{17}$ **26.** $\frac{1}{3}$ **27.** $-\frac{1}{3}$ **28.** $-\frac{1}{3}$ **29.** $-\frac{1}{4}$ **30.** −27 **31.** $\frac{1}{4}$ **32.** not a real number **33.** $\frac{343}{125}$ **34.** $\frac{9}{4}$ **35.** not a real number **36.** $x^{2/3}$ **37.** $5^{1/5}x^{2/5}y^{3/5}$ **38.** $\sqrt[5]{y^4}$ **39.** $5\sqrt[3]{xy^2z^5}$ **40.** $\frac{1}{\sqrt{x+2y}}$ **41.** $a^{13/6}$ **42.** $\frac{1}{b}$ **43.** $\frac{1}{a^{9/2}}$ **44.** $\frac{y^2}{x}$ **45.** a^4b^6 **46.** $\frac{1}{x^{11/12}}$ **47.** $\frac{b^{5/6}}{49a^{1/4}c^{5/3}}$ **48.** $a-a^2$ **49.** 4.472 **50.** −3.391 **51.** 5.191 **52.** 3.826 **53.** −26.246 **54.** 0.045 **55.** $\sqrt[6]{1372}$ **56.** $\sqrt[12]{81x^3}$ **57.** $2\sqrt{6}$ **58.** $\sqrt[3]{7x^2yz}$ **59.** $2x$ **60.** ab^3 **61.** $2\sqrt{15}$ **62.** $-5\sqrt{3}$ **63.** $3\sqrt[3]{6}$ **64.** $-2\sqrt[3]{4}$ **65.** $6x^3\sqrt{x}$ **66.** $2ab^2\sqrt[3]{3a^2b}$ **67.** $\frac{p^8\sqrt{p}}{11}$ **68.** $\frac{y\sqrt[3]{y^2}}{3x^2}$ **69.** $\frac{y\sqrt[4]{xy^2}}{3}$ **70.** $\frac{x\sqrt{2x}}{7y^2}$ **71. a.** $\frac{5}{\sqrt{\pi}}m$ **b.** 5.75 in. **72.** $-2\sqrt{5}$ **73.** $2x\sqrt{3x}$ **74.** $9\sqrt[3]{2}$ **75.** $3a\sqrt[4]{2a}$ **76.** $\frac{15+2\sqrt{3}}{6}$ **77.** $\frac{3\sqrt{2}}{4x}$ **78.** $17\sqrt{2}-15\sqrt{5}$ **79.** $-4ab\sqrt[4]{2b}$ **80.** 6 **81.** $x-6\sqrt{x}+9$ **82.** $-8\sqrt{5}$ **83.** $4x-9y$ **84.** $a-9$ **85.** $\sqrt[3]{a^2}+4\sqrt[3]{a}+4$ **86.** $\sqrt[3]{25x^2}-81$ **87.** $a+64$ **88.** $\frac{3\sqrt{7}}{7}$ **89.** $\frac{\sqrt{3x}}{6}$ **90.** $\frac{5\sqrt[3]{2}}{2}$ **91.** $\frac{2x^2\sqrt{2x}}{y}$ **92.** $\frac{x^2y^2\sqrt[3]{15yz}}{z}$ **93.** $\frac{3\sqrt[4]{2x^2}}{2x^3}$ **94.** $\frac{3\sqrt{y}+6}{y-4}$ **95.** $-5+2\sqrt{6}$ **96.** $\frac{11}{3\sqrt{11}}$ **97.** $\frac{6}{\sqrt{2y}}$ **98.** $\frac{3}{7\sqrt[3]{3}}$ **99.** $\frac{4x^3}{y\sqrt{2x}}$ **100.** $\frac{xy}{\sqrt[3]{10x^2yz}}$ **101.** $\frac{x-25}{-3\sqrt{x}+15}$ **102.** {32} **103.** ∅ **104.** {35} **105.** ∅ **106.** {9} **107.** {16} **108.** $3\sqrt{2}$ cm **109.** $\sqrt{241}$ ft **110.** 51.2 ft **111.** 4.24 ft **112.** $2i\sqrt{2}$ **113.** $-i\sqrt{6}$ **114.** $6i$ **115.** $-\sqrt{10}$ **116.** $15-4i$ **117.** $-13-3i$ **118.** −64 **119.** $10+4i$ **120.** $-12-18i$ **121.** $1+5i$ **122.** $-5-12i$ **123.** 87 **124.** $\frac{3}{2}-i$ **125.** $-\frac{1}{3}+\frac{1}{3}i$

Chapter 9 Test

1. $6\sqrt{6}$ **2.** $-x^{16}$ **3.** $\frac{1}{5}$ **4.** 5 **5.** $\frac{4x^2}{9}$ **6.** $-a^6b^3$ **7.** $\frac{8a^{1/3}c^{2/3}}{b^{5/12}}$ **8.** $a^{7/12}-a^{7/3}$ **9.** $|4xy|$ or $4|xy|$ **10.** −27 **11.** $\frac{3\sqrt{y}}{y}$ **12.** $\frac{8-6\sqrt{x}+x}{8-2x}$ **13.** $\frac{2\sqrt[3]{3x^2}}{3x}$ **14.** $\frac{6-x^2}{8(\sqrt{6}-x)}$ **15.** $-x\sqrt{5x}$ **16.** $4\sqrt{3}-\sqrt{6}$ **17.** $x+2\sqrt{x}+1$ **18.** $\sqrt{6}-4\sqrt{3}+\sqrt{2}-4$ **19.** −20 **20.** 23.685 **21.** 0.019 **22.** {2, 3} **23.** ∅ **24.** {6} **25.** $i\sqrt{2}$ **26.** $-2i\sqrt{2}$ **27.** $-3i$ **28.** 40 **29.** $7+24i$ **30.** $-\frac{3}{2}+\frac{5}{2}i$ **31.** $x=\frac{5\sqrt{2}}{2}$ in. **32.** $\sqrt{2}$, 5 **33.** 27 mph **34.** 360 ft

Cumulative Review

1. $y = -1.6$; Sec. 2.2, Ex. 2 **2.** $t = \frac{16}{3}$; Sec. 2.4, Ex. 2 **3.** 138 min; Sec. 2.5, Ex. 4

4.

x	y
-2	3
0	3
-5	3

; Sec. 3.1, Ex. 7 **5.** $-\frac{8}{3}$; Sec. 3.4, Ex. 1;

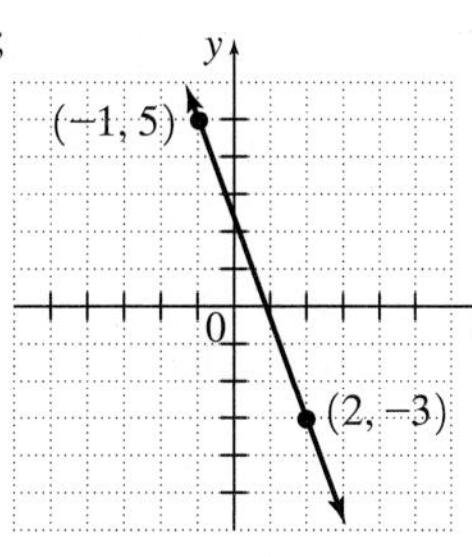

6. 1; Sec. 4.1, Ex. 25 **7.** 1; Sec. 4.1, Ex. 26 **8.** -1; Sec. 4.1, Ex. 28 **9.** $9y^2 + 12y + 4$; Sec. 4.6, Ex. 15 **10.** $x + 4$; Sec. 4.7, Ex. 4 **11.** $(r + 6)(r - 7)$; Sec. 5.2, Ex. 4; **12.** $(2x - 3y)(5x + y)$; Sec. 5.3, Ex. 4 **13.** $(2x - 1)(4x - 5)$; Sec. 5.4, Ex. 1 **14. a.** $x(2x + 7)(2x - 7)$; Sec. 5.5, Ex. 10 **b.** $2(9x^2 + 1)(3x + 1)(3x - 1)$; Sec. 5.5, Ex. 11 **15.** $x = \frac{1}{5}, -\frac{3}{2}, -6$; Sec. 5.6, Ex. 8 **16.** $\frac{x + 7}{x - 5}$; Sec. 6.1, Ex. 4 **17.** -5; Sec. 6.6, Ex. 1

18. $y = -2x + 12$; Sec. 7.2, Ex. 5 **19.** -4; Sec. 7.4, Ex. 1 **20.** 11; Sec. 7.4, Ex. 2 **21.** 8.125 in.; Sec. 7.5, Ex. 2 **22.** $(4, 2)$; Sec. 8.2, Ex. 1 **23.** no solution; Sec. 8.3, Ex. 3 **24.** $\{(x, y, z) | x - 5y - 2z = 6\}$; Sec. 8.5, Ex. 4 **25.** no solution; Sec. 8.6, Ex. 2 **26.** $\frac{1}{8}$; Sec. 9.2, Ex. 12 **27.** $\frac{1}{9}$; Sec. 9.2, Ex. 13

Chapter 10 Quadratic Equations and Functions

Chapter 10 Pretest

1. $\{-3\sqrt{6}, 3\sqrt{6}\}$; (10.1A) **2.** $\left\{\frac{-2 - 2\sqrt{3}}{3}, \frac{-2 + 2\sqrt{3}}{3}\right\}$; (10.1A) **3.** $\{-2 - \sqrt{14}, -2 + \sqrt{14}\}$; (10.1C) **4.** $\left\{\frac{-9 - 2\sqrt{21}}{3}, \frac{-9 + 2\sqrt{21}}{3}\right\}$; (10.1C) **5.** $\left\{\frac{1 - i\sqrt{31}}{4}, \frac{1 + i\sqrt{31}}{4}\right\}$; (10.2A) **6.** $\left\{\frac{-1 - \sqrt{17}}{12}, \frac{-1 + \sqrt{17}}{12}\right\}$; (10.2A) **7.** $\{4, -2 - 2i\sqrt{3}, -2 + 2i\sqrt{3}\}$; (10.3A) **8.** $\{64, 1\}$; (10.3A) **9.** $(-\infty, -5] \cup [6, \infty)$; (10.4A) **10.** $(-1, 0)$; (10.4A) **11.** $(-\infty, -3)$; (10.4B)

12.

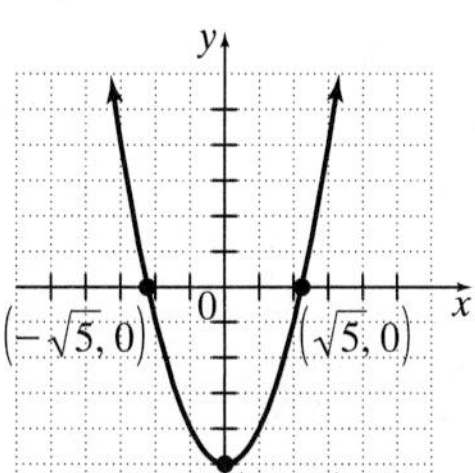

13.

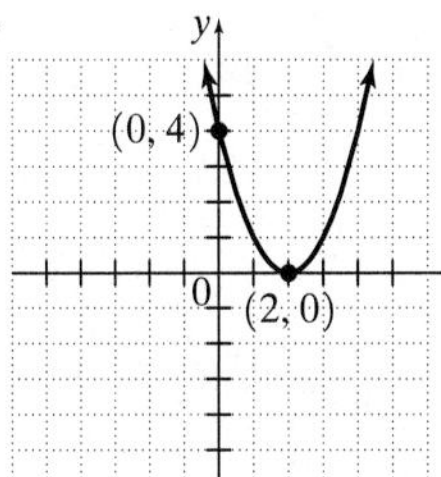

14.

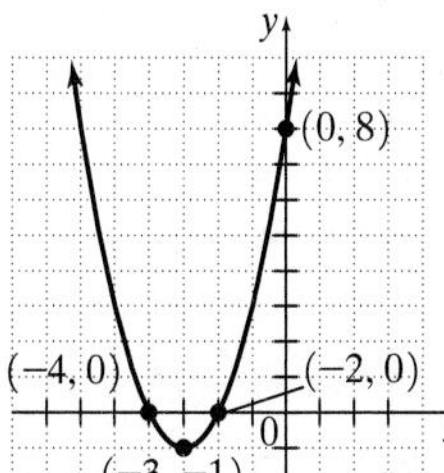

15.

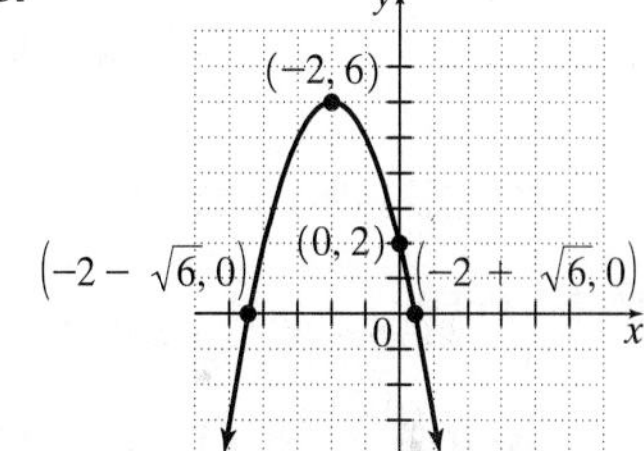

16. length: 20 cm; width: 5 cm; (10.2C)

Calculator Explorations

1. $\{-1.27, 6.27\}$ **3.** $\{-1.10, 0.90\}$ **5.** $\emptyset$

Exercise Set 10.1

1. $\{-4, 4\}$ **3.** $\{-\sqrt{7}, \sqrt{7}\}$ **5.** $\{-3\sqrt{2}, 3\sqrt{2})$ **7.** $\{-\sqrt{10}, \sqrt{10}\}$ **9.** $\{-8, -2\}$ **11.** $\{6 - 3\sqrt{2}, 6 + 3\sqrt{2}\}$ **13.** $\left\{\frac{3 - 2\sqrt{2}}{2}, \frac{3 + 2\sqrt{2}}{2}\right\}$ **15.** $\{-3i, 3i\}$ **17.** $\{-\sqrt{6}, \sqrt{6}\}$ **19.** $\{-2i\sqrt{2}, 2i\sqrt{2}\}$ **21.** $\{1 - 4i, 1 + 4i\}$ **23.** $\{-7 - \sqrt{5}, -7 + \sqrt{5}\}$ **25.** $\{-3 - 2i\sqrt{2}, -3 + 2i\sqrt{2})$ **27.** $x^2 + 16x + 64 = (x + 8)^2$ **29.** $z^2 - 12z + 36 = (z - 6)^2$ **31.** $p^2 + 9p + \frac{81}{4} = \left(p + \frac{9}{2}\right)^2$ **33.** $r^2 - 3r + \frac{9}{4} = \left(r - \frac{3}{2}\right)^2$ **35.** $\{-5, -3\}$ **37.** $\{-3 - \sqrt{7}, -3 + \sqrt{7}\}$ **39.** $\left\{\frac{-1 - \sqrt{5}}{2}, \frac{-1 + \sqrt{5}}{2}\right\}$ **41.** $\{-1 - \sqrt{6}, -1 + \sqrt{6}\}$ **43.** $\left\{\frac{6 - \sqrt{30}}{3}, \frac{6 + \sqrt{30}}{3}\right\}$ **45.** $\left\{-4, \frac{1}{2}\right\}$ **47.** $\{-4 - \sqrt{15}, -4 + \sqrt{15}\}$ **49.** $\left\{\frac{-3 - \sqrt{21}}{3}, \frac{-3 + \sqrt{21}}{3}\right\}$ **51.** $\{-1 - i, -1 + i\}$ **53.** $\{3 - \sqrt{6}, 3 + \sqrt{6}\}$ **55.** $\{-2 - i\sqrt{2}, -2 + i\sqrt{2}\}$ **57.** $\left\{\frac{1 - i\sqrt{47}}{4}, \frac{1 + i\sqrt{47}}{4}\right\}$ **59.** $\{-5 - i\sqrt{3}, -5 + i\sqrt{3}\}$ **61.** $\{-4, 1\}$ **63.** $\left\{\frac{2 - i\sqrt{2}}{2}, \frac{2 + i\sqrt{2}}{2}\right\}$ **65.** $\left\{\frac{-3 - \sqrt{69}}{6}, \frac{-3 + \sqrt{69}}{6}\right\}$ **67.** 20% **69.** 11% **71.** answers may vary **73.** simple; answers may vary **75.** $5 - 10\sqrt{3}$ **77.** $\frac{3 - 2\sqrt{7}}{4}$ **79.** $2\sqrt{7}$ **81.** $\sqrt{13}$ **83.** $-6y, 6y$ **85.** 8.11 sec **87.** 6.73 sec **89.** 6 in. **91.** $\frac{27\sqrt{2}}{2}$ in. **93.** 2.828 thousand units **95.** answers may vary

Mental Math

1. $a = 1, b = 3, c = 1$ **3.** $a = 7, b = 0, c = -4$ **5.** $a = 6, b = -1, c = 0$

Exercise Set 10.2

1. $\{-6, 1\}$ **3.** $\left\{-\frac{3}{5}, 1\right\}$ **5.** $\{3\}$ **7.** $\left\{\frac{-7 - \sqrt{33}}{2}, \frac{-7 + \sqrt{33}}{2}\right\}$ **9.** $\left\{\frac{1 - \sqrt{57}}{8}, \frac{1 + \sqrt{57}}{8}\right\}$ **11.** $\left\{\frac{7 - \sqrt{85}}{6}, \frac{7 + \sqrt{85}}{6}\right\}$ **13.** $\{1 - \sqrt{3}, 1 + \sqrt{3}\}$ **15.** $\left\{-\frac{3}{2}, 1\right\}$ **17.** $\left\{\frac{3 - \sqrt{11}}{2}, \frac{3 + \sqrt{11}}{2}\right\}$ **19.** $\left\{\frac{-5 - i\sqrt{5}}{10}, \frac{-5 + i\sqrt{5}}{10}\right\}$ **21.** $\left\{\frac{-5 - \sqrt{17}}{2}, \frac{-5 + \sqrt{17}}{2}\right\}$ **23.** $\left\{\frac{5}{2}, 1\right\}$ **25.** $\left\{\frac{3 - \sqrt{29}}{2}, \frac{3 + \sqrt{29}}{2}\right\}$ **27.** $\left\{\frac{-1 - \sqrt{19}}{6}, \frac{-1 + \sqrt{19}}{6}\right\}$ **29.** answers may vary **31.** $\left\{\frac{3 - i\sqrt{87}}{8}, \frac{3 + i\sqrt{87}}{8}\right\}$ **33.** $\{-2 - \sqrt{11}, -2 + \sqrt{11}\}$ **35.** $\{-3 - 2i, -3 + 2i\}$ **37.** $\left\{\frac{-1 - i\sqrt{23}}{4}, \frac{-1 + i\sqrt{23}}{4}\right\}$ **39.** $\{1\}$ **41.** $\{3 + \sqrt{5}, 3 - \sqrt{5}\}$ **43.** two real solutions **45.** one real solution **47.** two real solutions **49.** two complex but not real solutions **51.** 14 ft **53.** $2 + 2\sqrt{2}$ cm, $2 + 2\sqrt{2}$ cm, $4 + 2\sqrt{2}$ cm **55.** width: $-5 + 5\sqrt{17}$ ft; length: $5 + 5\sqrt{17}$ ft **57. a.** $50\sqrt{2}$ m **b.** 5000 sq m **59.** 37.4 ft by 38.5 ft **61.** $\frac{1 + \sqrt{5}}{2}$ **63.** 8.9 sec **65.** 2.8 sec **67.** $\left\{\frac{11}{5}\right\}$ **69.** $\{15\}$ **71.** $(x^2 + 5)(x + 2)(x - 2)$ **73.** $(z + 3)(z - 3)(z + 2)(z - 2)$ **75.** $\{0.6, 2.4\}$ **77.** Sunday to Monday **79.** Wednesday **81.** $f(4) = 33$; answers may vary **85. a.** \$2922 million **b.** 2003 **87. a.** 432.378 micrograms **b.** 54 lb

Exercise Set 10.3

1. $\{2\}$ **3.** $\{16\}$ **5.** $\{1, 4\}$ **7.** $\{3 - \sqrt{7}, 3 + \sqrt{7}\}$ **9.** $\left\{\frac{3 - \sqrt{57}}{4}, \frac{3 + \sqrt{57}}{4}\right\}$ **11.** $\left\{\frac{1 - \sqrt{29}}{2}, \frac{1 + \sqrt{29}}{2}\right\}$ **13.** $\{-2, 2, -2i, 2i\}$ **15.** $\left\{-\frac{1}{2}, \frac{1}{2}, -i\sqrt{3}, i\sqrt{3}\right\}$ **17.** $\{-3, 3, -2, 2\}$ **19.** $\{125, -8\}$ **21.** $\left\{-\frac{4}{5}, 0\right\}$ **23.** $\left\{-\frac{1}{8}, 27\right\}$ **25.** $\left\{-\frac{2}{3}, \frac{4}{3}\right\}$ **27.** $\left\{-\frac{1}{125}, \frac{1}{8}\right\}$ **29.** $\{-\sqrt{2}, \sqrt{2}, -\sqrt{3}, \sqrt{3}\}$ **31.** $\left\{\frac{-9 - \sqrt{201}}{6}, \frac{-9 + \sqrt{201}}{6}\right\}$ **33.** $\{2, 3\}$ **35.** $\{3\}$ **37.** $\{27, 125\}$ **39.** $\{1, -3i, 3i\}$ **41.** $\left\{\frac{1}{8}, -8\right\}$ **43.** $\left\{-\frac{1}{2}, \frac{1}{3}\right\}$ **45.** $\{4\}$ **47.** $\{-3\}$ **49.** $\{-\sqrt{5}, \sqrt{5}, -2i, 2i\}$ **51.** $\left\{-3, \frac{3 - 3i\sqrt{3}}{2}, \frac{3 + 3i\sqrt{3}}{2}\right\}$ **53.** $\{6, 12\}$ **55.** $\left\{-\frac{1}{3}, \frac{1}{3}, \frac{-i\sqrt{6}}{3}, \frac{i\sqrt{6}}{3}\right\}$ **57.** 5 mph, then 4 mph **59.** inlet pipe: 15.5 hr; hose: 16.5 hr **61.** 55 mph; 66 mph **63.** 8.5 hr **65.** 12 or -8 **67. a.** $x - 6$ **b.** $300 = (x - 6) \cdot (x - 6) \cdot 3$ **c.** 16 cm by 16 cm **69.** 22 ft **71.** $(-\infty, 3]$ **73.** $(-5, \infty)$ **75.** answers may vary **77. a.** 301.103 ft per sec **b.** 307.429 ft per sec **c.** 209.611 mph

Integrated Review

1. $\{-\sqrt{10}, \sqrt{10}\}$ **2.** $\{-\sqrt{14}, \sqrt{14}\}$ **3.** $\{1 - 2\sqrt{2}, 1 + 2\sqrt{2}\}$ **4.** $\{-5 - 2\sqrt{3}, -5 + 2\sqrt{3}\}$ **5.** $\{-1 - \sqrt{13}, -1 + \sqrt{13}\}$ **6.** $\{1, 11\}$ **7.** $\left\{\frac{-3 - \sqrt{69}}{6}, \frac{-3 + \sqrt{69}}{6}\right\}$ **8.** $\left\{\frac{-2 - \sqrt{5}}{4}, \frac{-2 + \sqrt{5}}{4}\right\}$ **9.** $\left\{\frac{2 - \sqrt{2}}{2}, \frac{2 + \sqrt{2}}{2}\right\}$ **10.** $\{-3 - \sqrt{5}, -3 + \sqrt{5}\}$ **11.** $\{-2 + i\sqrt{3}, -2 - i\sqrt{3}\}$ **12.** $\left\{\frac{-1 - i\sqrt{11}}{2}, \frac{-1 + i\sqrt{11}}{2}\right\}$ **13.** $\left\{\frac{-3 + i\sqrt{15}}{2}, \frac{-3 - i\sqrt{15}}{2}\right\}$ **14.** $\{3i, -3i\}$ **15.** $\{0, -17\}$ **16.** $\left\{\frac{1 + \sqrt{13}}{4}, \frac{1 - \sqrt{13}}{4}\right\}$ **17.** $\{2 + 3\sqrt{3}, 2 - 3\sqrt{3}\}$ **18.** $\{2 + \sqrt{3}, 2 - \sqrt{3}\}$ **19.** $\left\{-2, \frac{4}{3}\right\}$ **20.** $\left\{\frac{-5 + \sqrt{17}}{4}, \frac{-5 - \sqrt{17}}{4}\right\}$ **21.** $\{1 - \sqrt{6}, 1 + \sqrt{6}\}$ **22.** $\{-\sqrt{31}, \sqrt{31}\}$ **23.** $\{-\sqrt{11}, \sqrt{11}\}$ **24.** $\{-i\sqrt{11}, i\sqrt{11}\}$ **25.** $\{-11, 6\}$ **26.** $\left\{\frac{-3 + \sqrt{19}}{5}, \frac{-3 - \sqrt{19}}{5}\right\}$ **27.** $\left\{\frac{-3 + \sqrt{17}}{4}, \frac{-3 - \sqrt{17}}{4}\right\}$ **28.** $10\sqrt{2}$ ft ≈ 14.1 ft **29.** Diane: 9.1 hr; Lucy: 7.1 hr **30.** 5 mph during the first part, then 6 mph

Exercise Set 10.4

1. $(-\infty, -5) \cup (-1, \infty)$ **3.** $[-4, 3]$ **5.** $[2, 5]$ **7.** $\left(-5, -\frac{1}{3}\right)$ **9.** $(2, 4) \cup (6, \infty)$ **11.** $(-\infty, -4] \cup [0, 1]$ **13.** $(-\infty, -3) \cup (-2, 2) \cup (3, \infty)$ **15.** $(-\infty, -7) \cup (8, \infty)$ **17.** $\left[-\frac{4}{3}, \frac{1}{2}\right]$ **19.** $(-\infty, 0) \cup (1, \infty)$ **21.** $(-\infty, -4] \cup [4, 6]$ **23.** $(-7, 2)$ **25.** $(-1, \infty)$ **27.** $(-\infty, -1] \cup (4, \infty)$ **29.** $(-\infty, 3)$ **31.** $(0, 10)$ **33.** $(-\infty, -4) \cup [5, \infty)$ **35.** $(-\infty, -6] \cup (-1, 0] \cup (7, \infty)$ **37.** $(-\infty, 1) \cup (2, \infty)$ **39.** $(-\infty, -8] \cup (-4, \infty)$ **41.** 0, 1, 1, 4, 4 **43.** 0, -1, -1, -4, -4 **45.** answers may vary **47.** $(-\infty, -1) \cup (0, 1)$ **49.** when x is between 2 and 11 **51.**

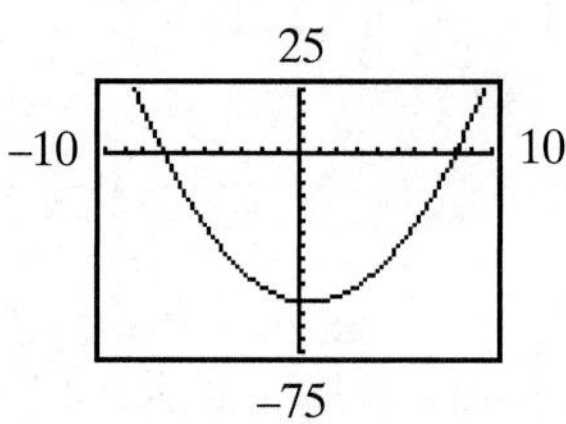

Graphing Calculator Explorations

1.

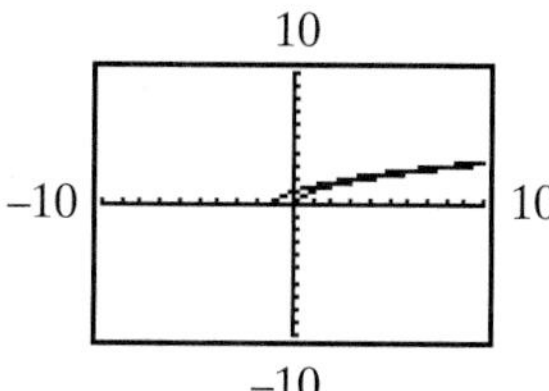

3.

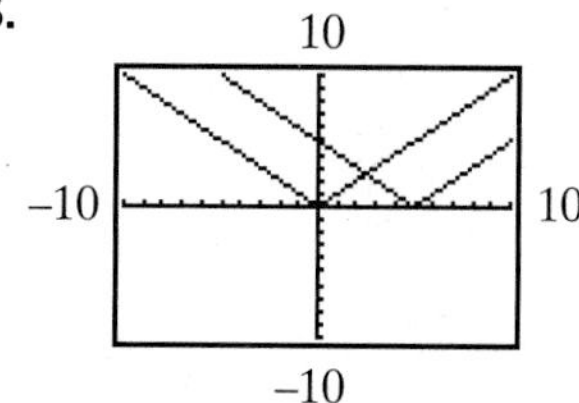

5.

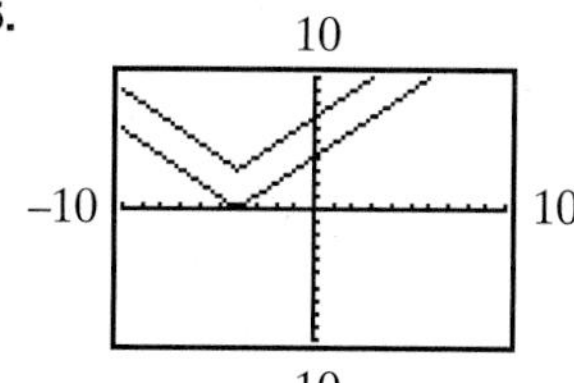

Mental Math

1. $(0, 0)$ **3.** $(2, 0)$ **5.** $(0, 3)$ **7.** $(-1, 5)$

Exercise Set 10.5

1.

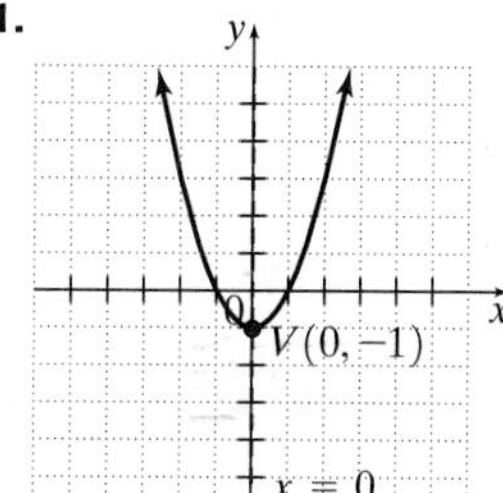

3.

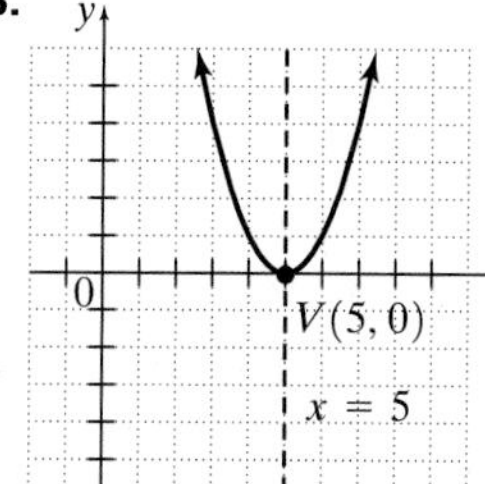

5.

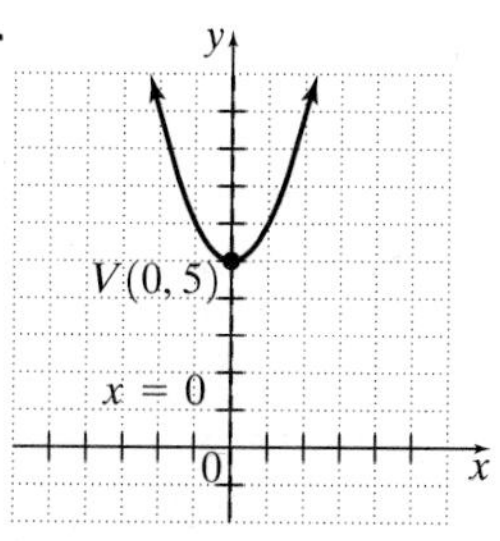

7.

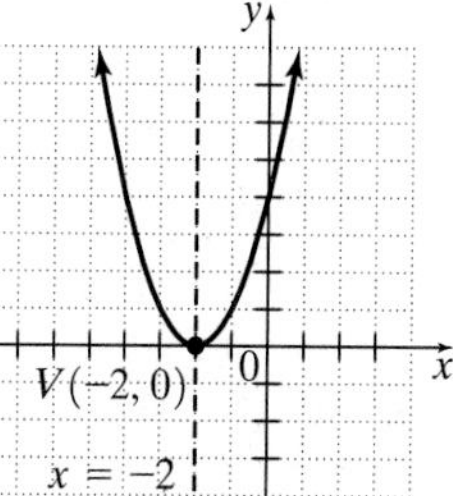

9.

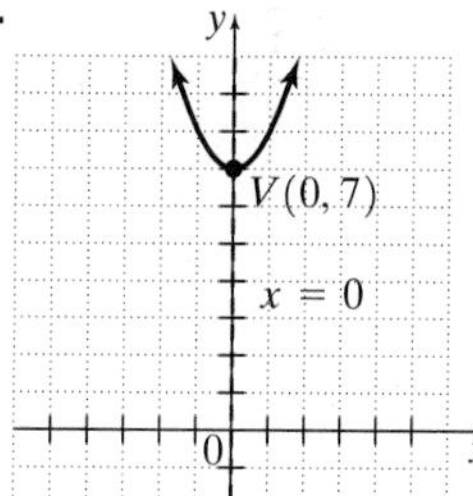

11.

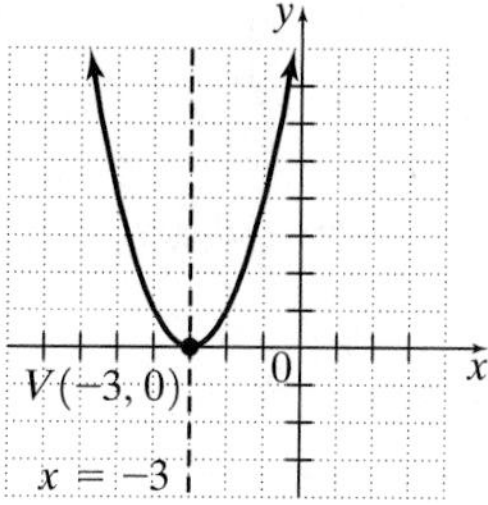

13.

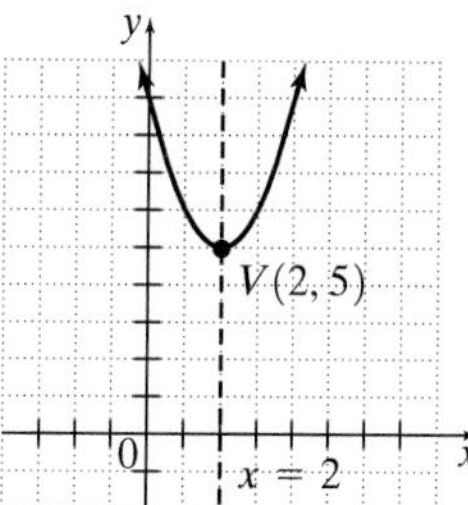

15.

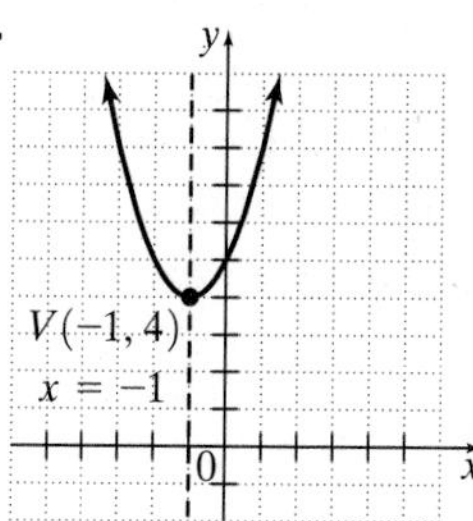

17.

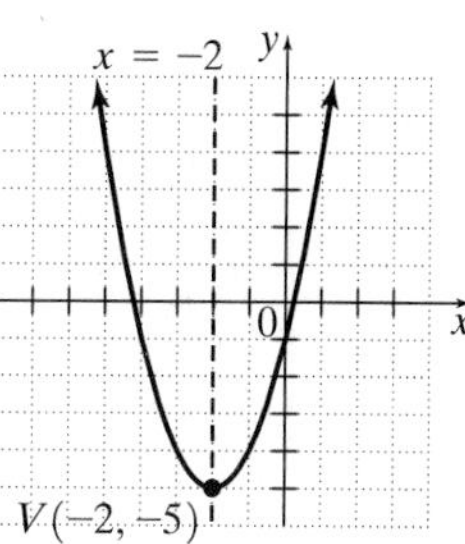

19.

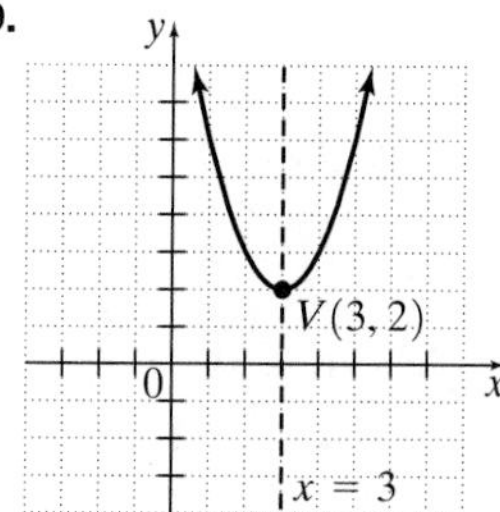

21.

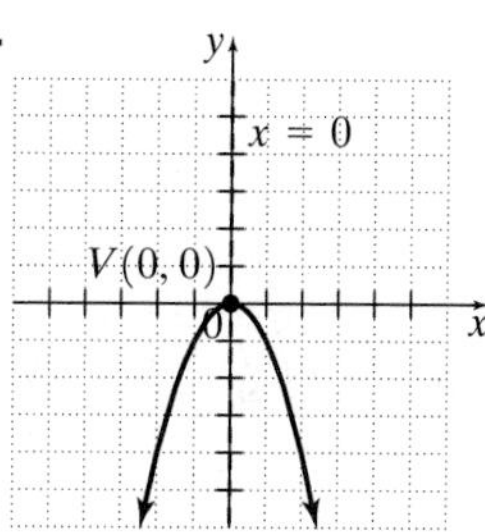

23.

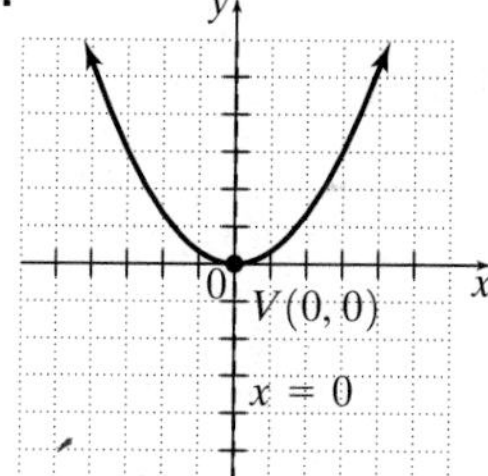

25.

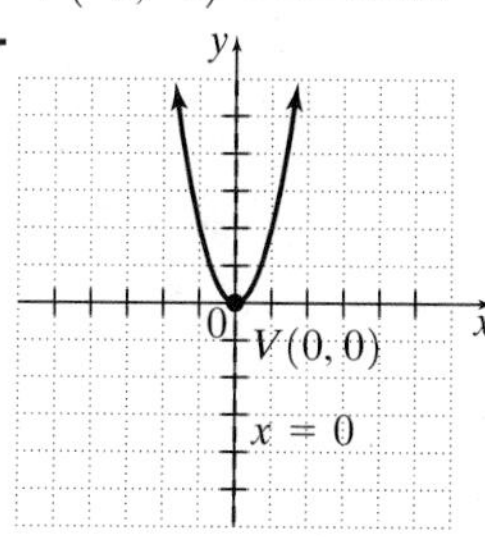

27.

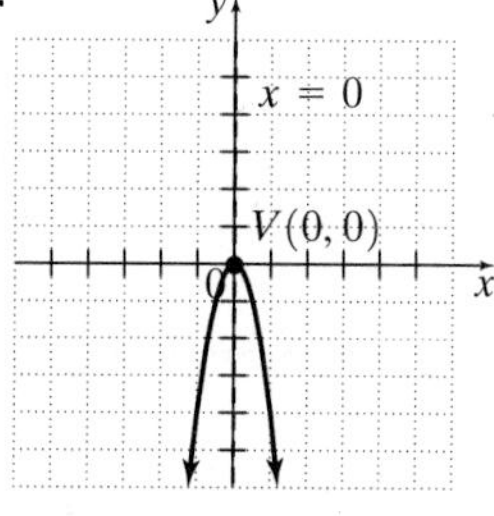

29.

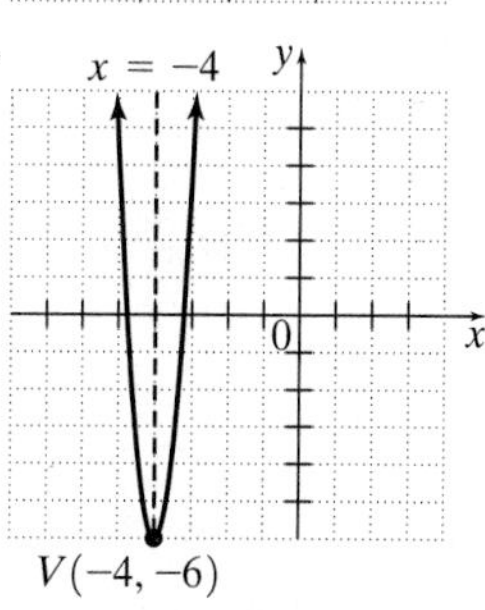

31.

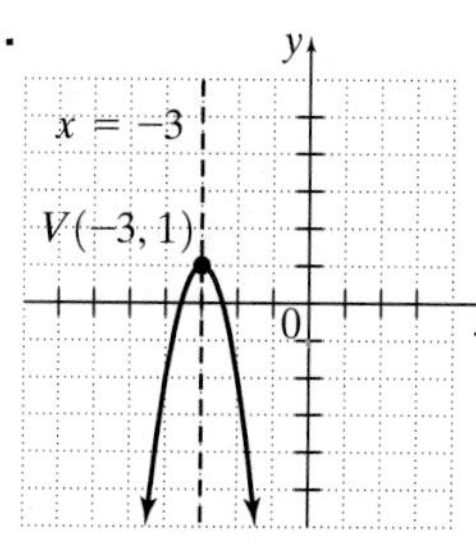

33.

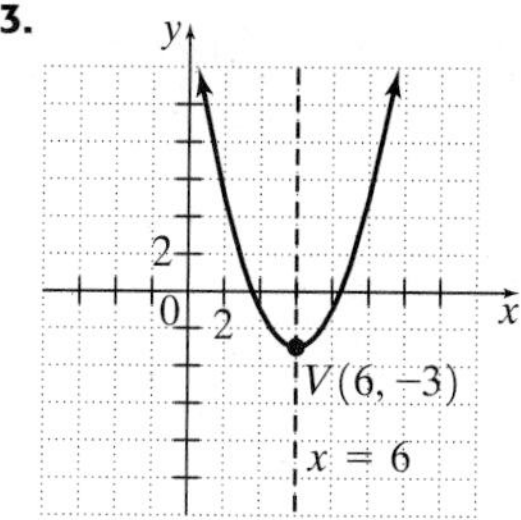

35.

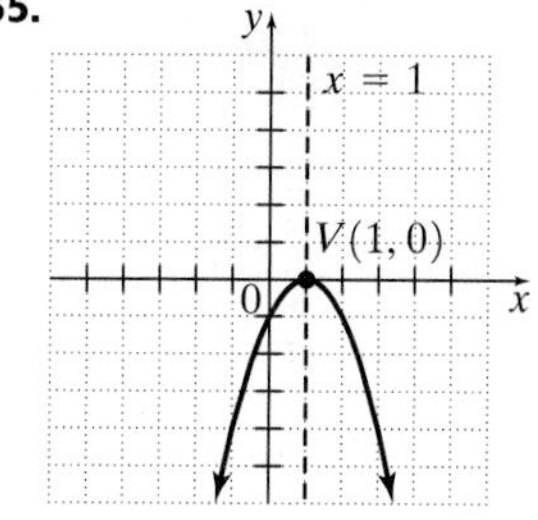

37.

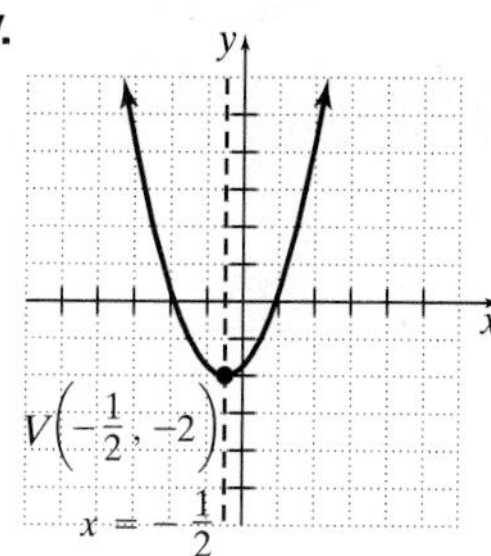

39.

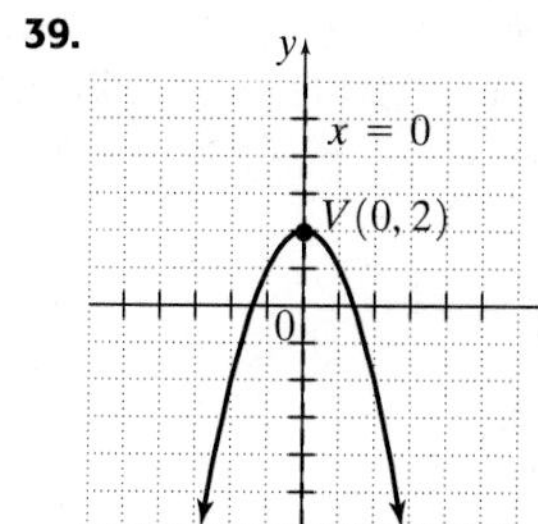

41. $x^2 + 8x + 16$ **43.** $z^2 - 16z + 64$ **45.** $y^2 + y + \frac{1}{4}$ **47.** $5(x - 2)^2 + 3$ **49.** $5(x + 3)^2 + 6$

51.

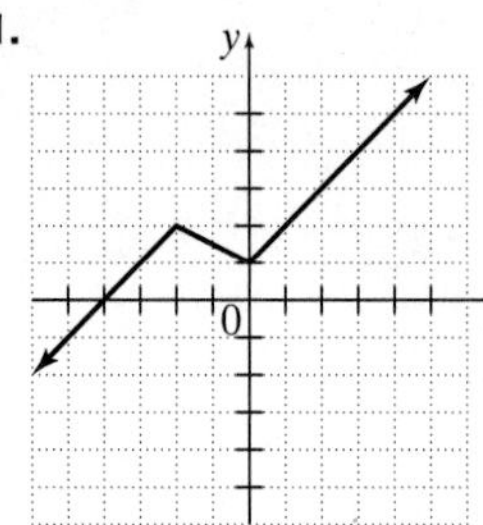

53.

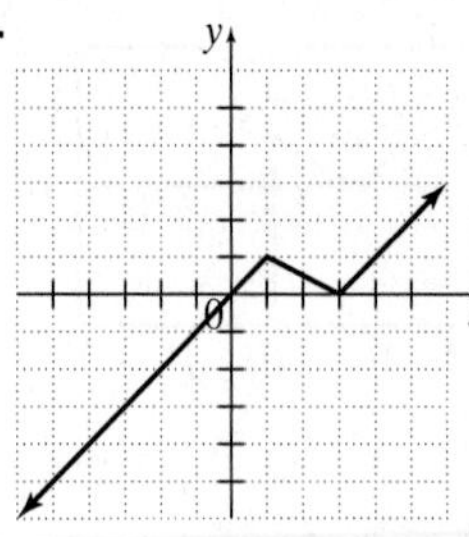

55.

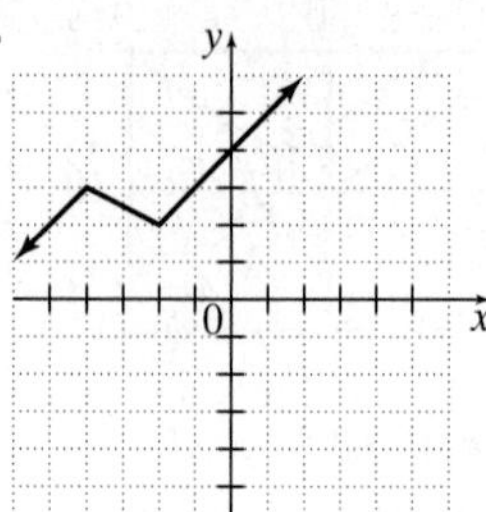

Exercise Set 10.6

1. $(-4, -9)$ **3.** $(5, 30)$ **5.** $(1, -2)$ **7.** $\left(\frac{1}{2}, \frac{5}{4}\right)$ **9.** D **11.** B

13. vertex: $(-2, -9)$;
opens upward;
x-intercepts; $(-5, 0)$, $(1, 0)$;
y-intercept: $(0, -5)$

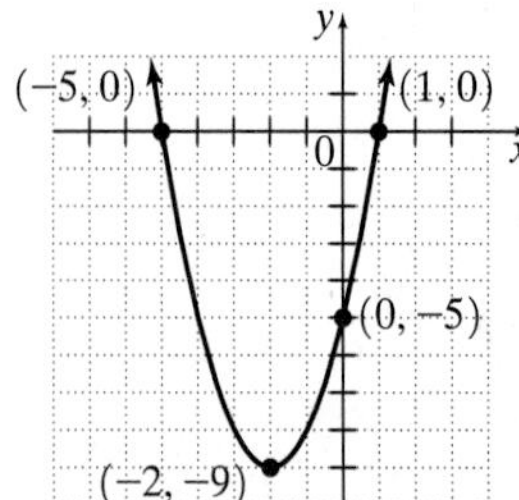

15. vertex: $(1, 0)$;
opens downward;
x-intercept: $(1, 0)$;
y-intercept: $(0, -1)$

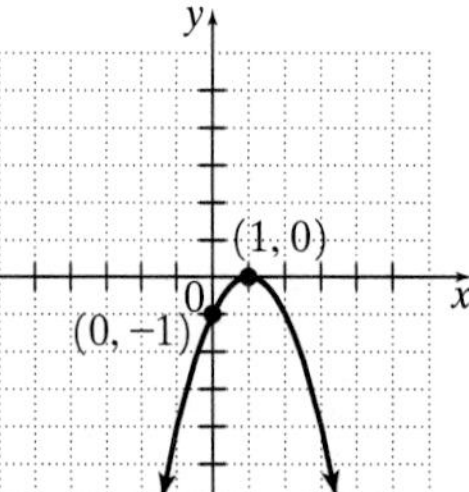

17. vertex: $(0, -4)$;
opens upward;
x-intercepts: $(-2, 0)$, $(2, 0)$;
y-intercept: $(0, -4)$

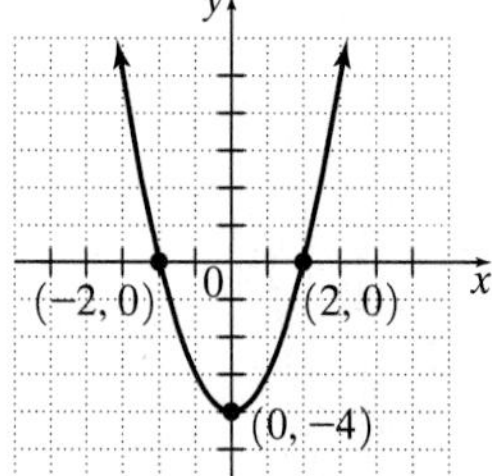

19. vertex: $\left(-\frac{1}{2}, -4\right)$;
opens upward;
x-intercepts: $\left(-\frac{3}{2}, 0\right)$, $\left(\frac{1}{2}, 0\right)$;
y-intercept: $(0, -3)$

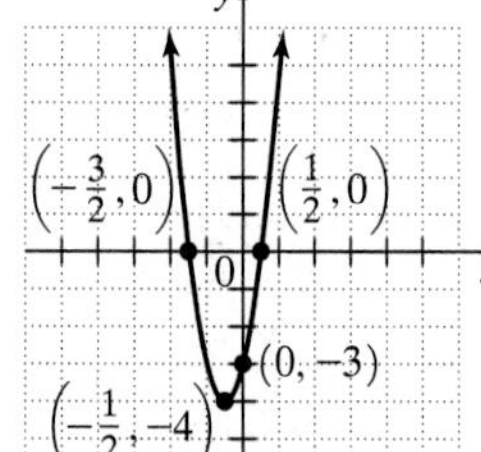

21. vertex: $\left(-4, -\frac{1}{2}\right)$;
opens upward;
x-intercepts: $(-5, 0)$ $(-3, 0)$;
y-intercept: $\left(0, \frac{15}{2}\right)$

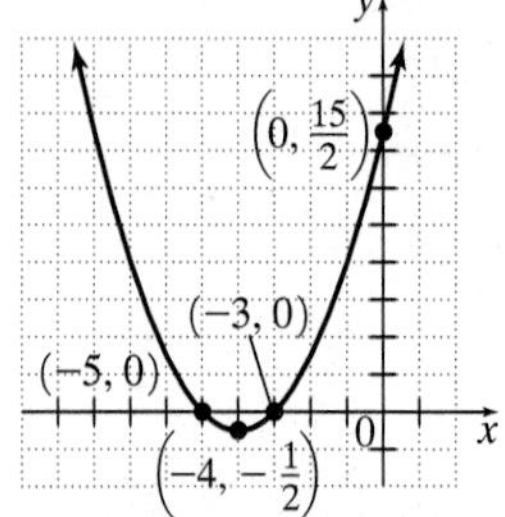

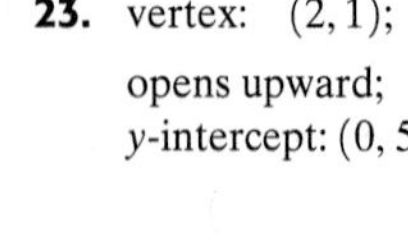

23. vertex: $(2, 1)$;
opens upward;
y-intercept: $(0, 5)$

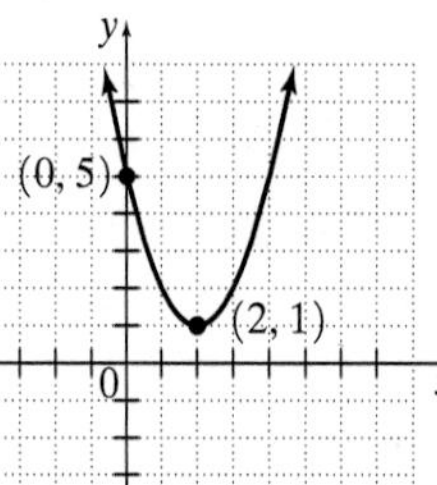

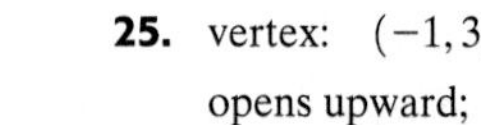

25. vertex: $(-1, 3)$;
opens upward;
y-intercept: $(0, 5)$

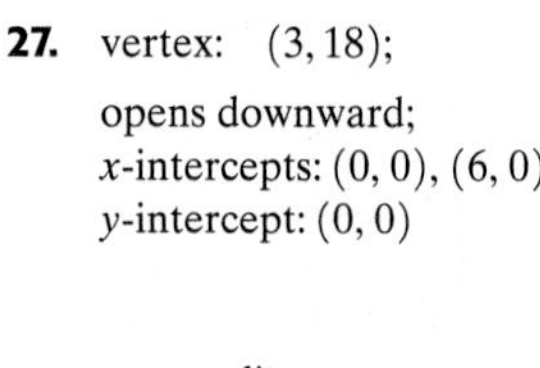

27. vertex: $(3, 18)$;
opens downward;
x-intercepts: $(0, 0)$, $(6, 0)$
y-intercept: $(0, 0)$

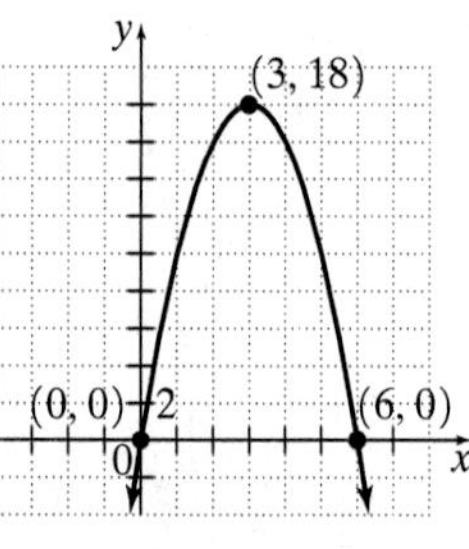

29. 144 ft **31.** 16 ft **33.** 30 and 30 **35.** −5 and 5 **37.** length: 20 units; width: 20 units **39.** $(0, 2)$ **41.** undefined **43.** $(-5, 2)$

45. $(4, 1)$

47. vertex: $(-5, -10)$;
opens upward;
y-intercept: $(0, 15)$;
x-intercepts:
$(-1.8, 0)$, $(-8.2, 0)$

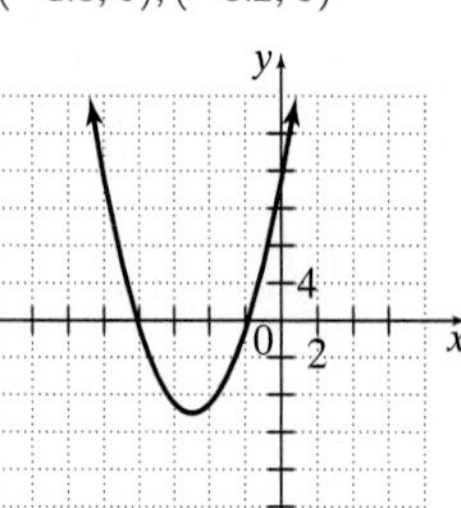

49.

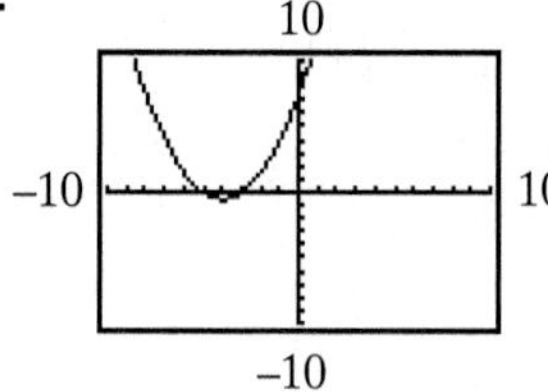

51. −0.84

53. a. maximum; answers may vary
b. 2036
c. 32,902,500 or about 3290.3 thousands

Chapter 10 Review

1. $\{14, 1\}$ **2.** $\{-5, 6\}$ **3.** $\left\{\frac{4}{5}, -\frac{1}{2}\right\}$ **4.** $\left\{-\frac{6}{7}, 5\right\}$ **5.** $\{-7, 7\}$ **6.** $\{-2, 2\}$ **7.** $\left\{-\frac{4}{9}, \frac{2}{9}\right\}$ **8.** $\left\{\frac{2-\sqrt{2}}{5}, \frac{2+\sqrt{2}}{5}\right\}$ **9.** $\left\{\frac{-3-\sqrt{5}}{2}, \frac{-3+\sqrt{5}}{2}\right\}$ **10.** $\left\{\frac{-1-3i\sqrt{3}}{2}, \frac{-1+3i\sqrt{3}}{2}\right\}$ **11.** $\left\{\frac{-3-i\sqrt{7}}{8}, \frac{-3+i\sqrt{7}}{8}\right\}$ **12.** $\left\{\frac{17-\sqrt{145}}{9}, \frac{17+\sqrt{145}}{9}\right\}$ **13.** 4.25% **14.** $75\sqrt{2}$ mi; 106.1 mi **15.** two complex solutions **16.** two real solutions **17.** two real solutions **18.** one real solution **19.** $\{8\}$ **20.** $\{-5, 0\}$ **21.** $\{-i\sqrt{11}, i\sqrt{11}\}$ **22.** $\left\{-\frac{5}{2}, 1\right\}$ **23.** $\left\{\frac{5-i\sqrt{143}}{12}, \frac{5+i\sqrt{143}}{12}\right\}$ **24.** $\left\{\frac{1-i\sqrt{35}}{9}, \frac{1+i\sqrt{35}}{9}\right\}$ **25.** $\left\{\frac{21-\sqrt{41}}{50}, \frac{21+\sqrt{41}}{50}\right\}$ **26.** $\left\{1, \frac{9}{4}\right\}$ **27. a.** 20 ft **b.** $\frac{15+\sqrt{321}}{16}$ sec; 2.1 sec **28.** $(6+6\sqrt{2})$ cm **29.** $\left\{3, \frac{-3+3i\sqrt{3}}{2}, \frac{-3-3i\sqrt{3}}{2}\right\}$ **30.** $\{-4, 2-2i\sqrt{3}, 2+2i\sqrt{3}\}$ **31.** $\left\{\frac{2}{3}, 5\right\}$ **32.** $\left\{\frac{-8\sqrt{7}}{7}, \frac{8\sqrt{7}}{7}\right\}$ **33.** $\{-5, 5, -2i, 2i\}$ **34.** $\left\{-\frac{16}{5}, 1\right\}$ **35.** $\{1, 125\}$ **36.** $\{8, 64\}$ **37.** $\{-1, 1, -i, i\}$ **38.** $\left\{-\frac{1}{5}, \frac{1}{4}\right\}$ **39.** Jerome: 10.5 hr; Tim: 9.5 hr **40.** -5 **41.** $[-5, 5]$ **42.** $\left(-\frac{1}{2}, \frac{1}{2}\right)$ **43.** $\left(-\infty, -\frac{5}{4}\right] \cup \left[\frac{3}{2}, \infty\right)$ **44.** $(-\infty, -4) \cup (-1, 1) \cup (4, \infty)$ **45.** $(5, 6)$ **46.** $[-5, 0] \cup \left(\frac{3}{4}, \infty\right)$ **47.** $(-\infty, -6) \cup \left(-\frac{3}{4}, 0\right) \cup (5, \infty)$ **48.** $(-\infty, -5] \cup [-2, 6]$ **49.** $(-5, -3) \cup (5, \infty)$ **50.** $(-\infty, 0)$ **51.** $\left(-\frac{6}{5}, 0\right) \cup \left(\frac{5}{6}, 3\right)$ **52.** $\left(2, \frac{7}{2}\right)$

53.

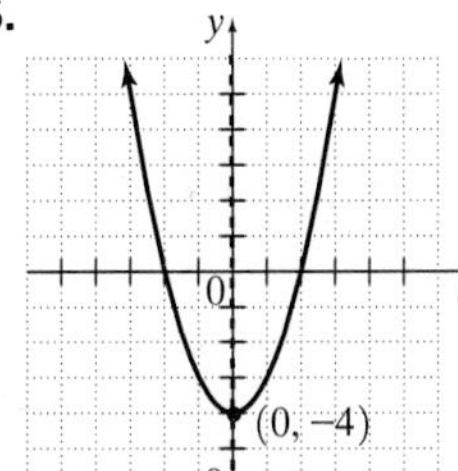

54.

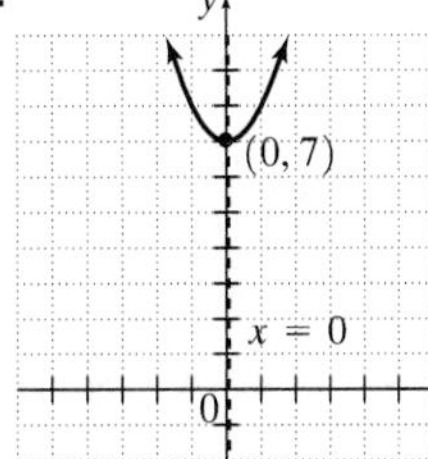

55.

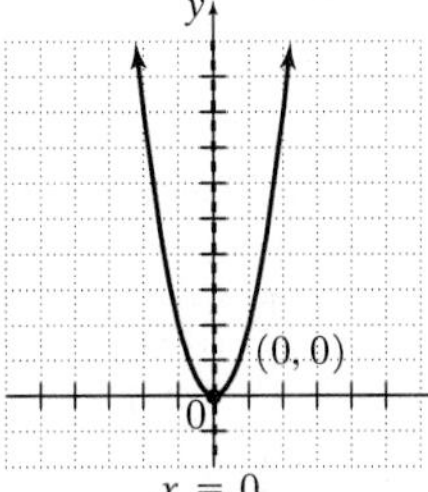

56.

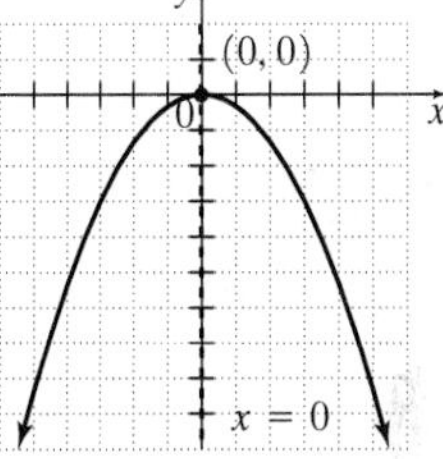

57.

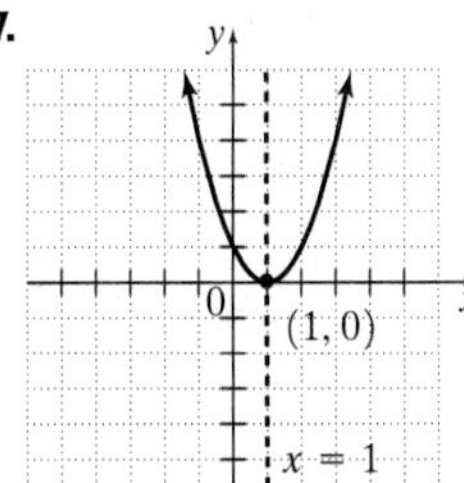

58.

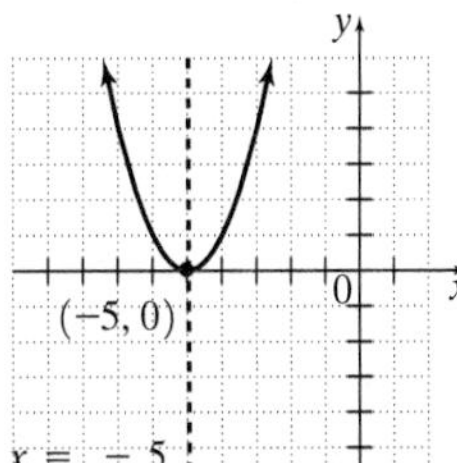

59.

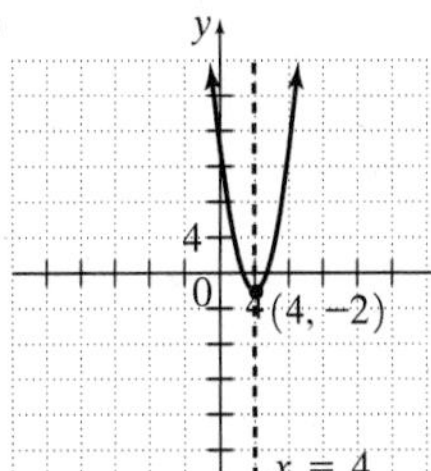

60.

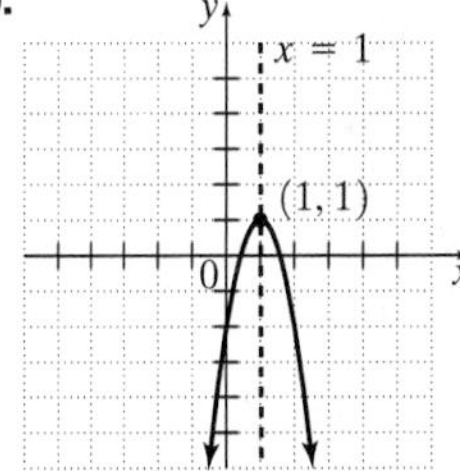

61. vertex: $(-5, 0)$;
x-intercept: $(-5, 0)$;
y-intercept: $(0, 25)$

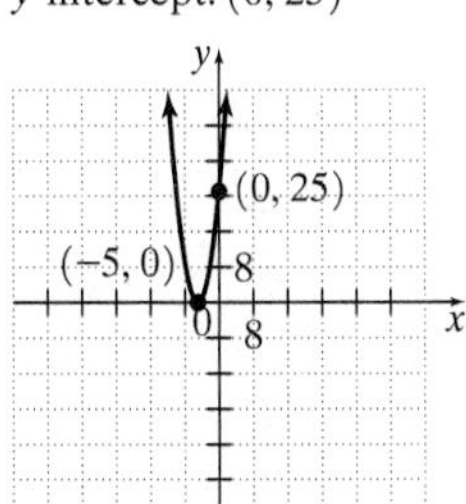

62. vertex: $(3, 0)$;
x-intercept: $(3, 0)$;
y-intercept: $(0, -9)$

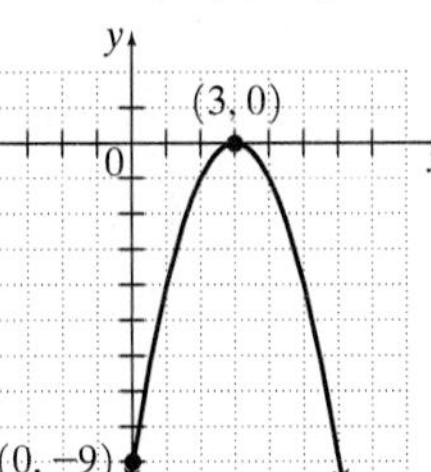
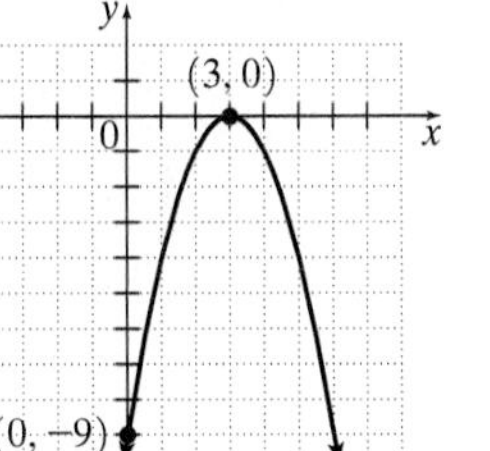

63. vertex: $(0, -1)$;
x-intercepts: $\left(-\frac{1}{2}, 0\right), \left(\frac{1}{2}, 0\right)$;
y-intercept: $(0, -1)$

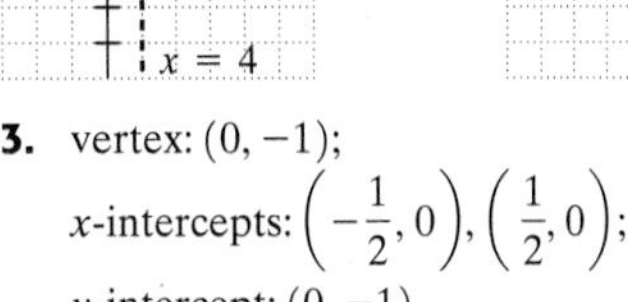

64. vertex: $(0, 5)$;
x-intercepts: $(-1, 0), (1, 0)$;
y-intercept: $(0, 5)$

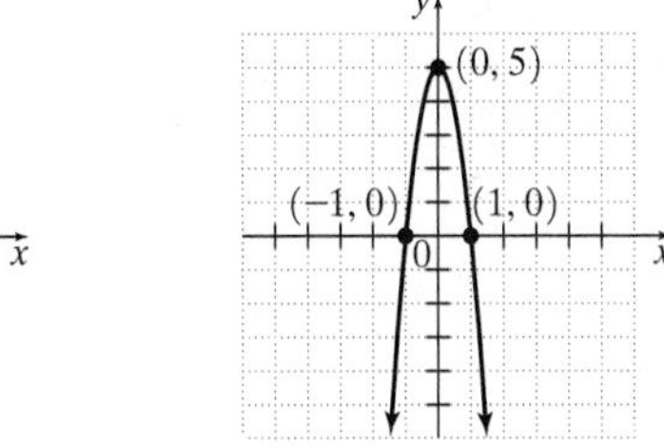

65. vertex: $\left(-\frac{5}{6}, \frac{73}{12}\right)$;
opens downward;
x-intercepts: $(-2.3, 0), (0.6, 0)$;
y-intercept: $(0, 4)$

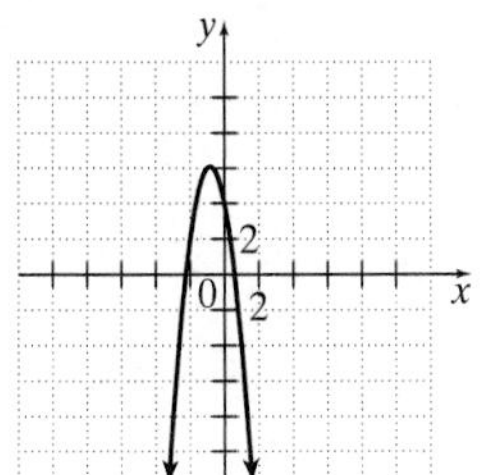

66. a. 0.4 sec and 7.1 sec
b. answers may vary

67. The numbers are both 210.

Chapter 10 Test

1. $\left\{\frac{7}{5}, -1\right\}$ **2.** $\{-1 - \sqrt{10}, -1 + \sqrt{10}\}$ **3.** $\left\{\frac{1 + i\sqrt{31}}{2}, \frac{1 - i\sqrt{31}}{2}\right\}$ **4.** $\{3 - \sqrt{7}, 3 + \sqrt{7}\}$ **5.** $\left\{-\frac{1}{7}, -1\right\}$ **6.** $\left\{\frac{3 + \sqrt{29}}{2}, \frac{3 - \sqrt{29}}{2}\right\}$ **7.** $\{-2 - \sqrt{11}, -2 + \sqrt{11}\}$ **8.** $\{-3, 3, -i, i\}$ **9.** $\{-1, 1, -i, i\}$ **10.** $\{6, 7\}$ **11.** $\{3 - \sqrt{7}, 3 + \sqrt{7}\}$ **12.** $\left\{\frac{2 - i\sqrt{6}}{2}, \frac{2 + i\sqrt{6}}{2}\right\}$ **13.** $\left(-\infty, -\frac{3}{2}\right) \cup (5, \infty)$ **14.** $(-\infty, -5) \cup (-4, 4) \cup (5, \infty)$ **15.** $(-\infty, -3) \cup (2, \infty)$ **16.** $(-\infty, -3) \cup [2, 3)$

17.

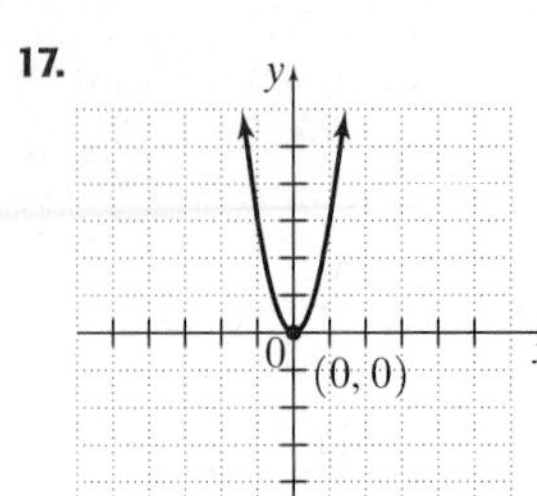

18.

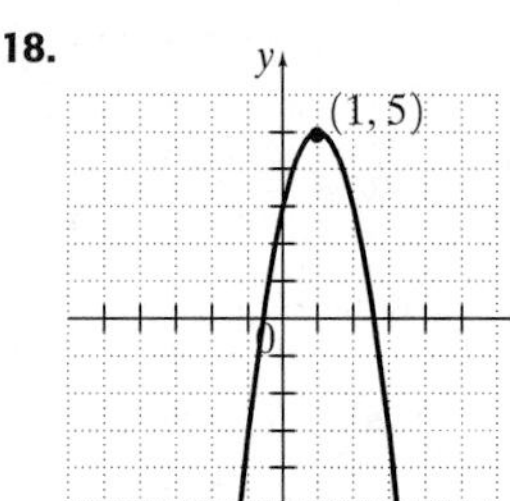

19.

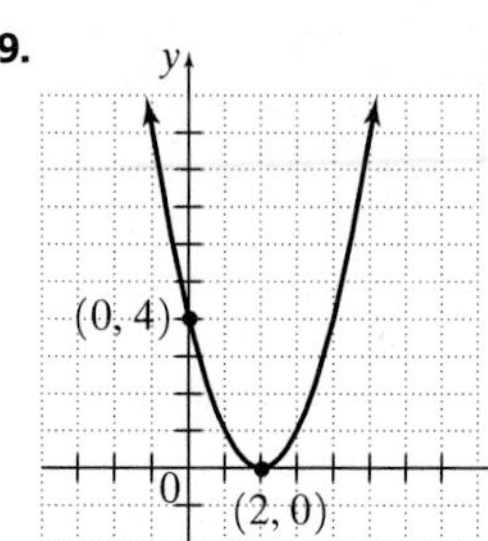

20.

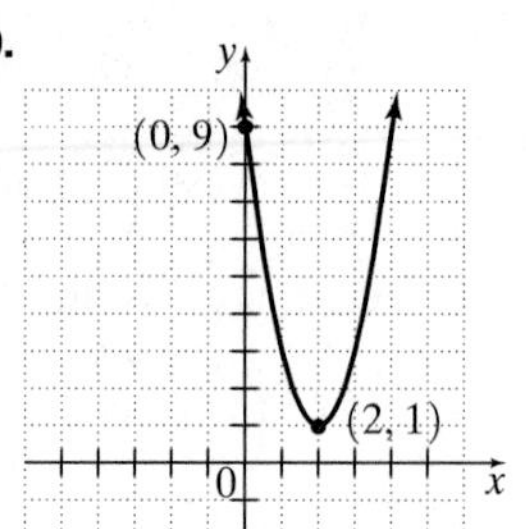

21. $(2 + \sqrt{46})$ ft ≈ 8.8 ft **22.** $(5 + \sqrt{17})$ hr ≈ 9.12 hr **23. a.** 272 ft **b.** 5.12 sec

Cumulative Review

1. $y = \frac{5}{8}x - \frac{5}{2}$; Sec. 7.2, Ex. 2 **2.** domain: $(-\infty, \infty)$; range: $[0, \infty)$; Sec. 7.3, Ex. 15 **3.** domain: $[-4, 4]$; range: $[-2, 2]$; Sec. 7.3, Ex. 16 **4.** $\left(\frac{1}{2}, 0, \frac{3}{4}\right)$; Sec. 8.5, Ex. 3 **5.** $(1, -1, 3)$; Sec. 8.6, Ex. 3 **6.** $(-1, -3)$; Sec. 8.2, Ex. 2 **7.**

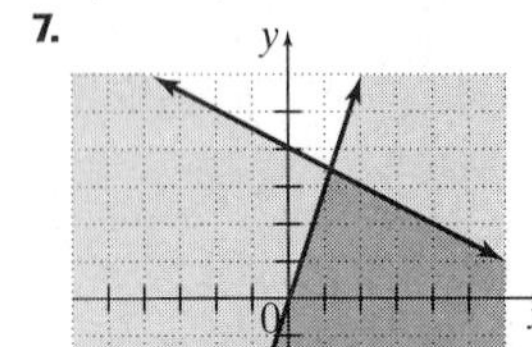

; Sec. 8.7, Ex. 1 **8.** 2; Sec. 9.1, Ex. 30 **9.** 1; Sec. 9.1, Ex. 32 **10.** $x^{5/6}$; Sec. 9.2, Ex. 14 **11.** 32; Sec. 9.2, Ex. 18 **12.** $\frac{\sqrt{x}}{3}$; Sec. 9.3, Ex. 7 **13.** $\frac{\sqrt[4]{3}}{2y}$; Sec. 9.3, Ex. 9 **14.** $2x - 25$; Sec. 9.4, Ex. 13 **15.** $4 - 2\sqrt{3}$; Sec. 9.4, Ex. 14 **16.** $\frac{7}{3\sqrt{35}}$; Sec. 9.5, Ex. 6 **17.** $\{3\}$; Sec. 9.6, Ex. 4 **18.** $-1 + 5i$; Sec. 9.7, Ex. 8 **19.** $\left\{\frac{6 + i\sqrt{5}}{2}, \frac{6 - i\sqrt{5}}{2}\right\}$; Sec. 10.1, Ex. 9 **20.** $\{2 + \sqrt{2}, 2 - \sqrt{2}\}$; Sec. 10.2, Ex. 3 **21.** $\{8, 27\}$; Sec. 10.3, Ex. 5 **22.** $\left(-\frac{7}{2}, -1\right)$; Sec. 10.4, Ex. 5 **23.**

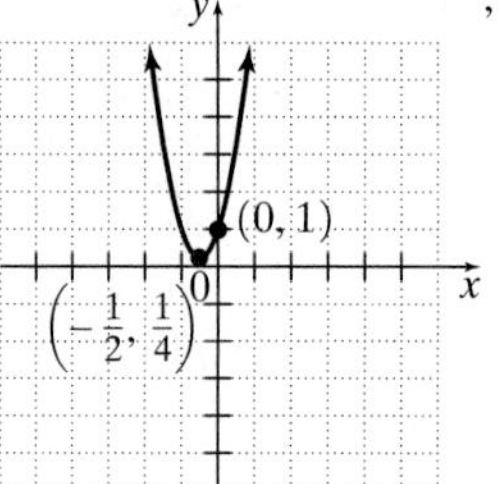

; Sec. 10.6, Ex. 2

Chapter 11 Exponential and Logarithmic Functions

Chapter 11 Pretest

1. $x^2 + 3x + 3$; 11.1A **2.** $25x^2 + 20x + 3$; 11.1B **3.** one-to-one; 11.2A **4.** not one-to-one; 11.2B **5.** $f^{-1}(x) = \frac{x + 12}{7}$; 11.2C **6.** 11.3A

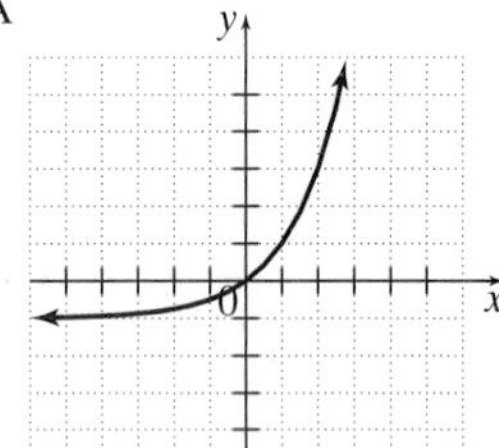

7. $\{3\}$; 11.3B **8.** $\left\{-\frac{1}{6}\right\}$; 11.3B **9.** $\{625\}$; 11.4B **10.** $\{-6\}$; 11.4B **11.** $\{10\}$; 11.4B **12.** 5; 11.4B **13.** 11.4C **14.** $\log_5 6$; 11.5A **15.** $\log_6 \frac{a^2}{(a + 1)^7}$; 11.5D **16.** $\log_7 3 + \log_7 y - \log_7 5 - 2\log_7 x$; 11.5D **17.** -2; 11.6B **18.** 8; 11.6B **19.** 2.4650; 11.6E **20.** $\left\{-\frac{21}{5}\right\}$; 11.7B

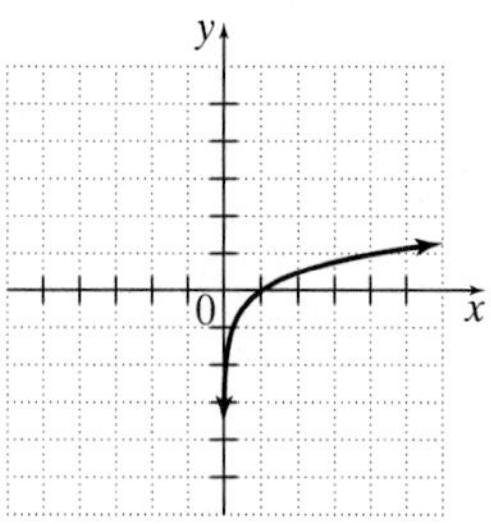

Exercise Set 11.1

1. a. $3x - 6$ **b.** $-x - 8$ **c.** $2x^2 - 13x - 7$ **d.** $\frac{x - 7}{2x + 1}$ where $x \neq -\frac{1}{2}$ **3. a.** $x^2 + 5x + 1$ **b.** $x^2 - 5x + 1$ **c.** $5x^3 + 5x$ **d.** $\frac{x^2 + 1}{5x}$ where $x \neq 0$ **5. a.** $\sqrt{x} + x + 5$ **b.** $\sqrt{x} - x - 5$ **c.** $x\sqrt{x} + 5\sqrt{x}$ **d.** $\frac{\sqrt{x}}{x + 5}$ where $x \neq -5$

7. a. $5x^2 - 3x$ **b.** $-5x^2 - 3x$ **c.** $-15x^3$ **d.** $-\frac{3}{5x}$ where $x \neq 0$ **9.** 42 **11.** -18 **13.** 0

15. $(f \circ g)(x) = 25x^2 + 1; (g \circ f)(x) = 5x^2 + 5$ **17.** $(f \circ g)(x) = 2x + 11; (g \circ f)(x) = 2x + 4$

19. $(f \circ g)(x) = -8x^3 - 2x - 2; (g \circ f)(x) = -2x^3 - 2x + 4$ **21.** $(f \circ g)(x) = \sqrt{-5x + 2}; (g \circ f)(x) = -5\sqrt{x} + 2$

23. $H(x) = (g \circ h)(x)$ **25.** $F(x) = (h \circ f)(x)$ **27.** $G(x) = (f \circ g)(x)$ **29.** answers may vary **31.** answers may vary

33. answers may vary **35.** $y = x - 2$ **37.** $y = \frac{x}{3}$ **39.** $y = -\frac{x + 7}{2}$ **41.** $P(x) = R(x) - C(x)$ **43.** answers may vary

Exercise Set 11.2

1. one-to-one; $f^{-1} = \{(-1, -1), (1, 1), (2, 0), (0, 2)\}$ **3.** one-to-one; $h^{-1} = \{(10, 10)\}$ **5.** one-to-one; $f^{-1} = \{(12, 11), (3, 4), (4, 3), (6, 6)\}$ **7.** not one-to-one **9.** one-to-one; $f^{-1} = \{$(1, California), (48, Alaska), (14, Indiana), (22, Louisiana), (36, New Mexico)$\}$ **11. a.** 3 **b.** 1 **13. a.** 1 **b.** -1 **15.** one-to-one **17.** not one-to-one **19.** one-to-one **21.** not one-to-one

23. $f^{-1}(x) = x - 4$

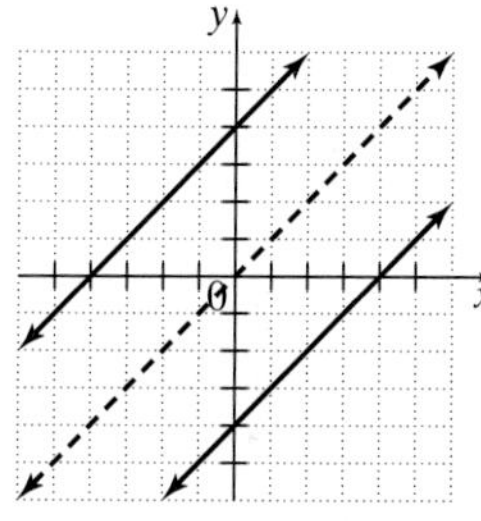

25. $f^{-1}(x) = \frac{x + 3}{2}$

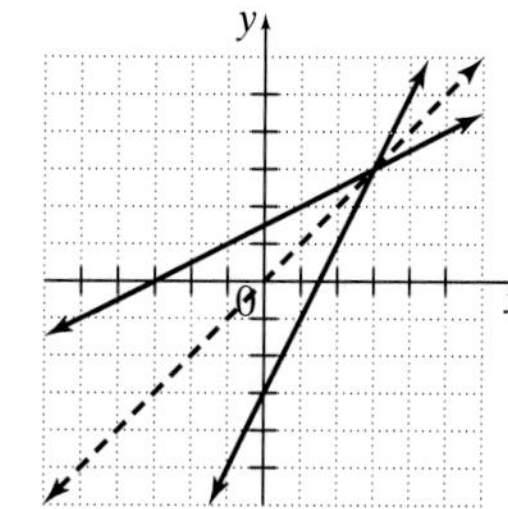

27. $f^{-1}(x) = 2x + 2$

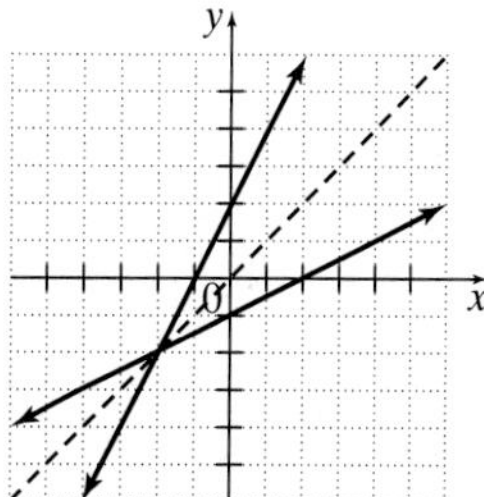

29. $f^{-1}(x) = \sqrt[3]{x}$

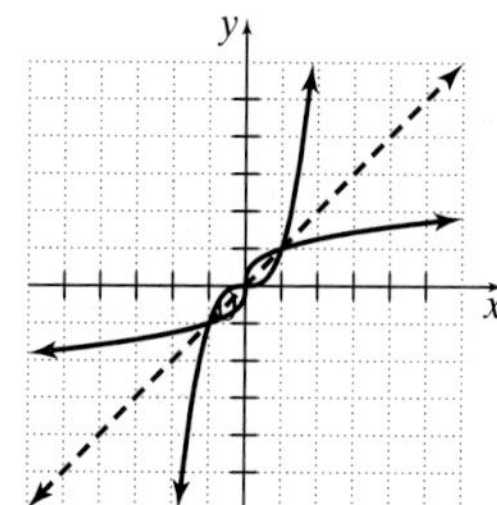

31. $f^{-1}(x) = 5x + 2$ **39.**

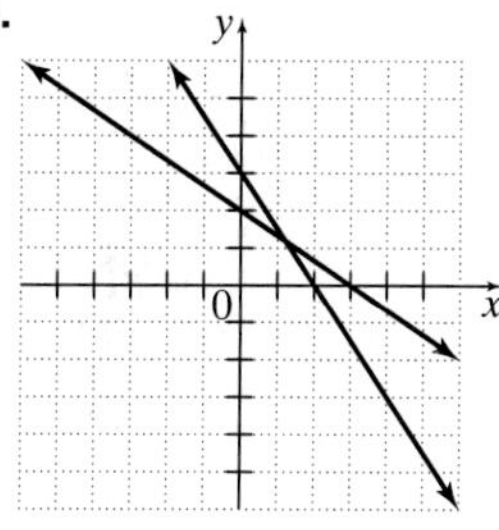

41.

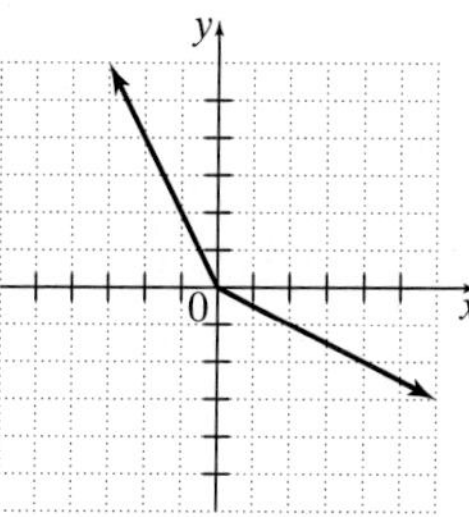

43. 5 **45.** 8 **47.** $\frac{1}{27}$ **49.** 9 **51.** $3^{1/2} \approx 1.73$ **53. a.** $\left(-2, \frac{1}{4}\right), \left(-1, \frac{1}{2}\right), (0, 1), (1, 2), (2, 5)$ **b.** $\left(\frac{1}{4}, -2\right), \left(\frac{1}{2}, -1\right), (1, 0), (2, 1), (5, 2)$

c.

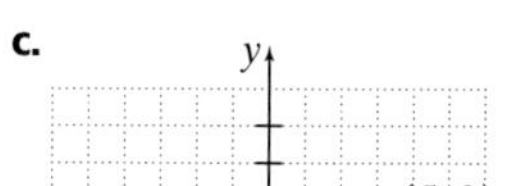

d.

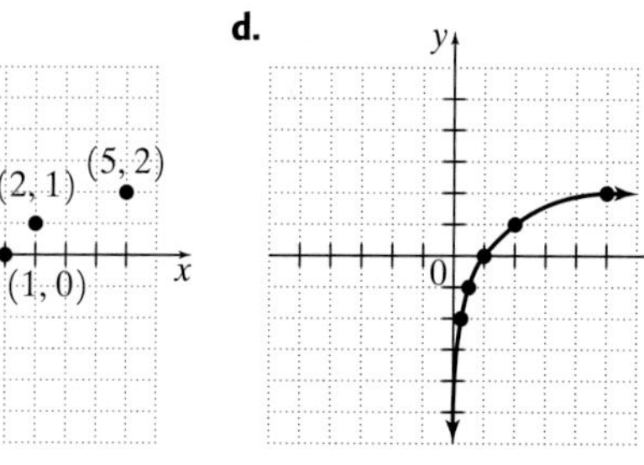

55. $f^{-1}(x) = \frac{x - 1}{3}$ **57.** $f^{-1}(x) = x^3 - 3$ **59.** answers may vary

Calculator Explorations

1. 81.98% **2.** 22.54%

Exercise Set 11.3

1.

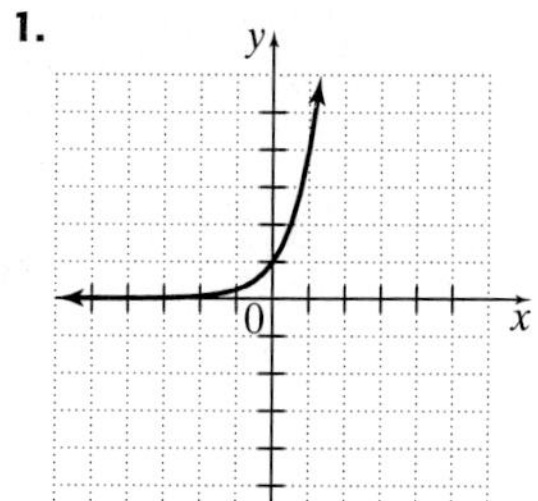

3.

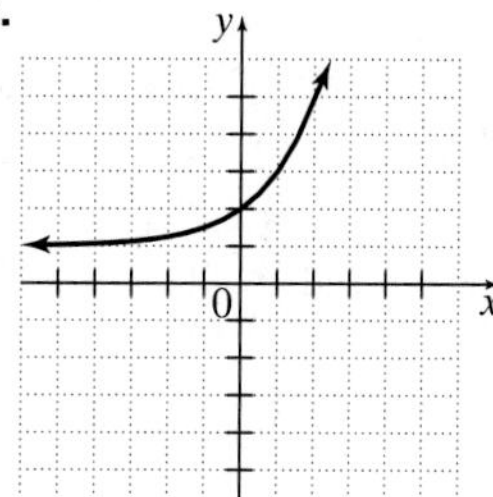

5.

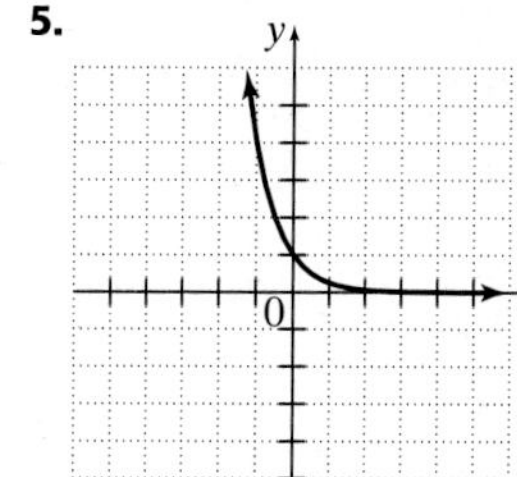

7.

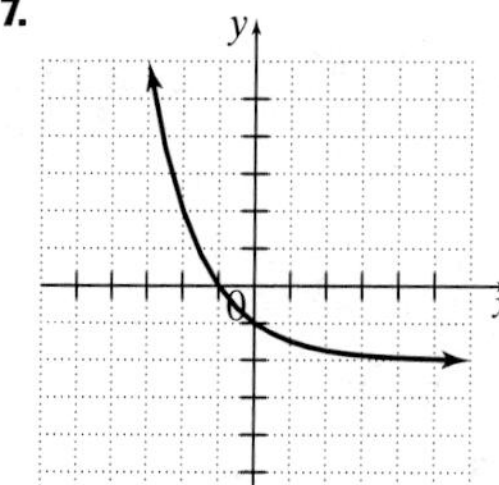

9.

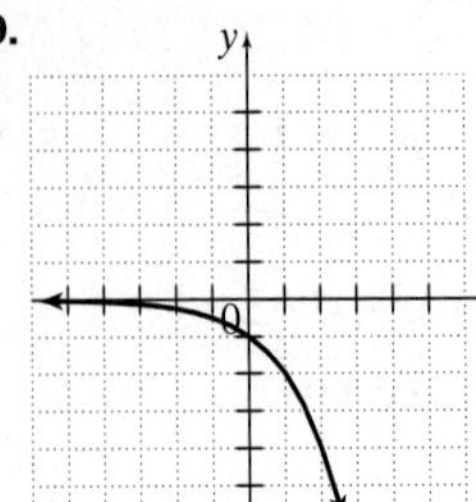

11.

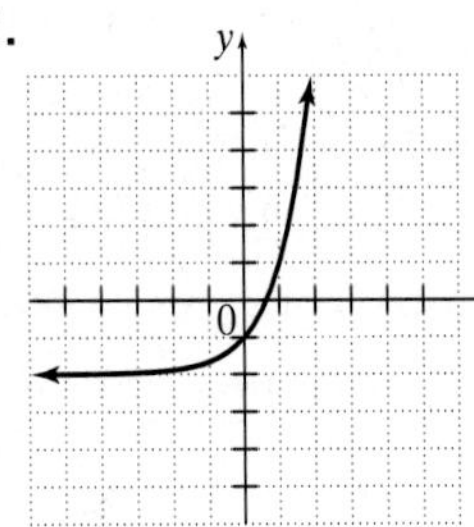

13.

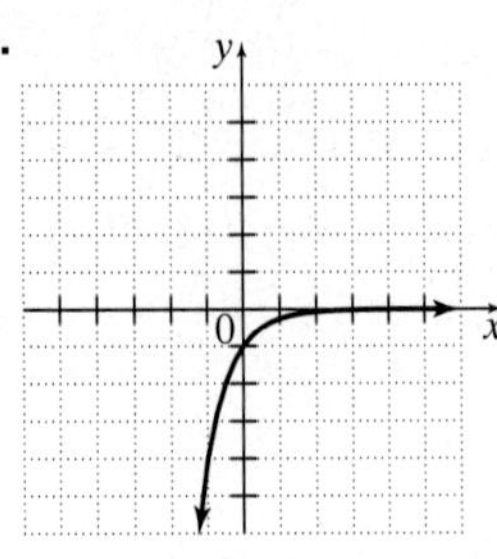

15.

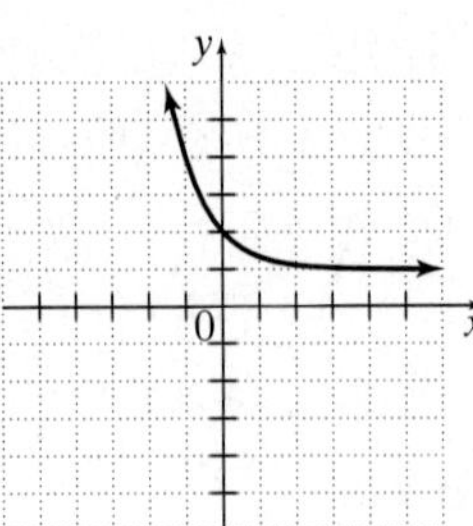

17. $\{3\}$ **19.** $\left\{\frac{3}{4}\right\}$ **21.** $\left\{\frac{8}{5}\right\}$ **23.** $\left\{-\frac{2}{3}\right\}$ **25.** $\left\{\frac{3}{2}\right\}$ **27.** $\left\{-\frac{1}{3}\right\}$ **29.** $\{-2\}$ **31.** 24.6 lb **33.** 333 bison **35. a.** $65.7 billion **b.** $2303.6 billion **37.** approximately 618.6 million cellular phone users **39.** $7621.42 **41.** $\{4\}$ **43.** $\emptyset$ **45.** C **47.** D **49.** answers may vary **53.** 18.62 lb

Exercise Set 11.4

1. $6^2 = 36$ **3.** $3^{-3} = \frac{1}{27}$ **5.** $10^3 = 1000$ **7.** $e^4 = x$ **9.** $e^{-2} = \frac{1}{e^2}$ **11.** $7^{1/2} = \sqrt{7}$ **13.** $\log_2 16 = 4$ **15.** $\log_{10} 100 = 2$ **17.** $\log_e x = 3$ **19.** $\log_{10}\frac{1}{10} = -1$ **21.** $\log_4 \frac{1}{16} = -2$ **23.** $\log_5 \sqrt{5} = \frac{1}{2}$ **25.** 3 **27.** -2 **29.** $\frac{1}{2}$ **31.** -1 **33.** 0 **35.** 4 **37.** 2 **39.** 5 **41.** 4 **43.** -3 **45.** answers may vary **47.** $\{2\}$ **49.** $\{81\}$ **51.** $\{7\}$ **53.** $\{-3\}$ **55.** $\{-3\}$ **57.** $\{2\}$ **59.** $\{2\}$ **61.** $\left\{\frac{27}{64}\right\}$ **63.** $\{10\}$ **65.** 3 **67.** 3 **69.** 3 **69.** 1

71.

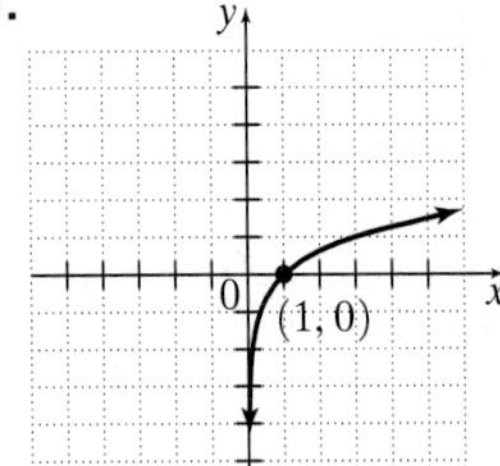

73.

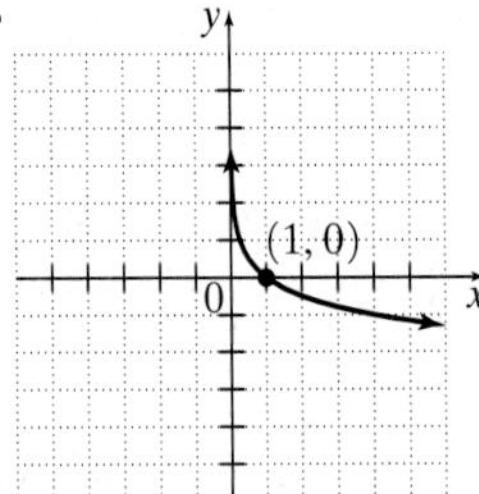

75. **77.**

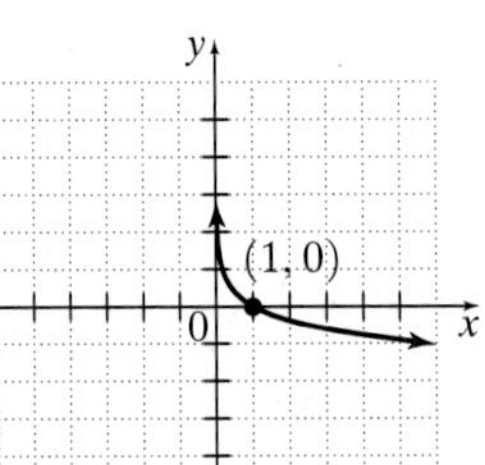

79. 1 **81.** $\frac{x-4}{2}$ **83.** **85.**

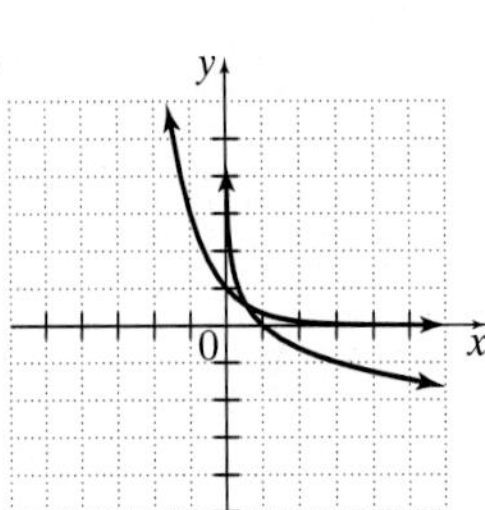

87. 0.0827 **89.** 2 and 3

Exercise Set 11.5

1. $\log_5 14$ **3.** $\log_4 9x$ **5.** $\log_{10}(10x^2 + 20)$ **7.** $\log_5 3$ **9.** $\log_2 \frac{x}{y}$ **11.** $\log_3 4$ **13.** $2 \log_3 x$ **15.** $-1 \log_4 5$ **17.** $\frac{1}{2}\log_5 y$ **19.** $\log_2 25$ **21.** $\log_5 x^3z^6$ **23.** $\log_{10}\frac{x^3 - 2x}{x + 1}$ **25.** $\log_4 4$, or 1 **27.** $\log_7 \frac{9}{2}$ **29.** $\log_4 48$ **31.** $\log_2 \frac{x^{7/2}}{(x+1)^2}$ **33.** $\log_8 x^{16/3}$ **35.** $\log_3 4 + \log_3 y - \log_3 5$ **37.** $3 \log_2 x - \log_2 y$ **39.** $\frac{1}{2}\log_b 7 + \frac{1}{2}\log_b x$ **41.** $\log_7 5 + \log_7 x - \log_7 4$ **43.** $3 \log_5 x + \log_5(x + 1)$ **45.** $2 \log_6 x - \log_6(x + 3)$ **47.** 0.2 **49.** 1.2 **51.** 0.35 **53.** 1.29 **55.** -0.68 **57.** -0.125 **59.** **61.** $\{-1\}$ **63.** $\left\{\frac{1}{2}\right\}$ **65.** false **67.** true **69.** false

INTEGRATED REVIEW

1. $x^2 + x - 5$ **2.** $-x^2 + x - 7$ **3.** $x^3 - 6x^2 + x - 6$ **4.** $\frac{x-6}{x^2+1}$ **5.** $\sqrt{3x} - 1$ **6.** $3\sqrt{x} - 1$ **7.** one-to-one; $\{(6,-2),(8,4),(-6,2),(3,3)\}$ **8.** not one-to-one **9.** not one-to-one **10.** one-to-one **11.** not one-to-one **12.** $f^{-1}(x) = \frac{x}{3}$ **13.** $f^{-1}(x) = x - 4$ **14.** $f^{-1}(x) = \frac{x+1}{5}$ **15.** $f^{-1}(x) = \frac{x-2}{3}$

16.
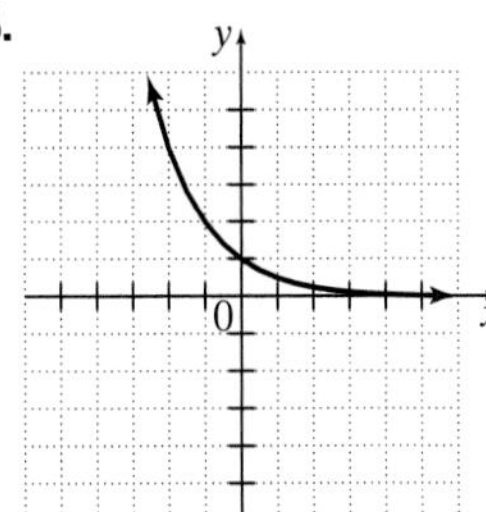

17.
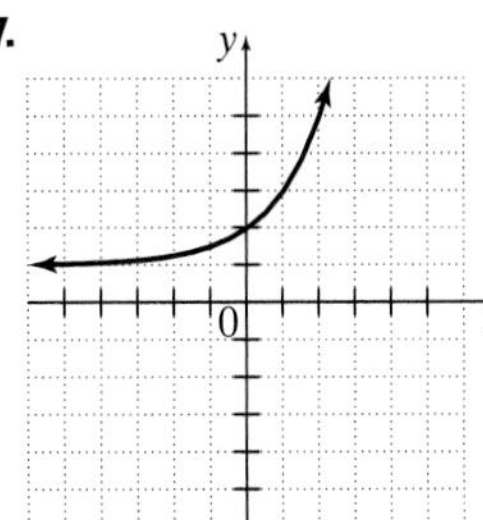

18.
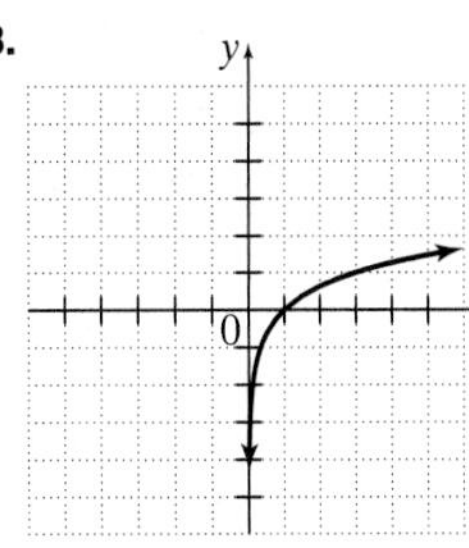

19.
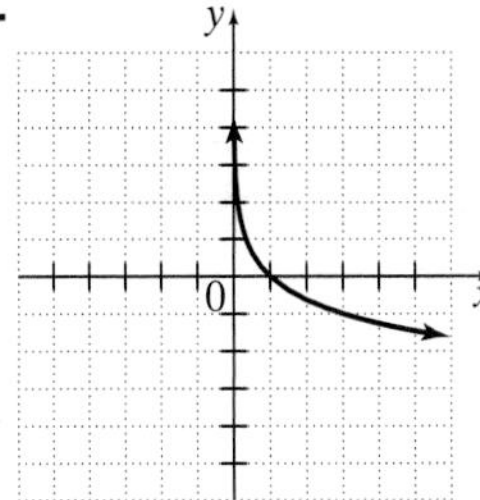

20. $\{3\}$ **21.** $\{7\}$ **22.** $\{-8\}$ **23.** $\{3\}$ **24.** $\{2\}$ **25.** $\left\{\frac{1}{2}\right\}$ **26.** $\{32\}$ **27.** $\{4\}$ **28.** $\{5\}$ **29.** $\left\{\frac{1}{9}\right\}$ **30.** $\log_2 x^5$ **31.** $\log_2 5^x$ **32.** $\log_5 \frac{x^3}{y^5}$ **33.** $\log_5 x^9y^3$ **34.** $\log_2 \frac{x^2-3x}{x^2+4}$ **35.** $\log_3 \frac{y(y^3+11)}{y+2}$ **36.** $\log_7 9 + 2\log_7 x - \log_7 y$ **37.** $\log_6 5 + \log_6 y - 2\log_6 z$

EXERCISE SET 11.6

1. 0.9031 **3.** 0.3636 **5.** 0.6931 **7.** −2.6367 **9.** 1.1004 **11.** 1.6094 **13.** 1.6180 **15.** answers may vary **17.** 2 **19.** −3 **21.** 2 **23.** $\frac{1}{4}$ **25.** 3 **27.** 2 **29.** −4 **31.** $\frac{1}{2}$ **33.** $\{10^{1.3}\}$; $\{19.9526\}$ **35.** $\left\{\frac{10^{1.1}}{2}\right\}$; $\{6.2946\}$ **37.** $\{e^{1.4}\}$; $\{4.0552\}$ **39.** $\left\{\frac{4+e^{2.3}}{3}\right\}$; $\{4.6581\}$ **41.** $\{10^{2.3}\}$; $\{199.5262\}$ **43.** $\{e^{-2.3}\}$; $\{0.1003\}$ **45.** $\left\{\frac{10^{-0.5}-1}{2}\right\}$; $\{-0.3419\}$ **47.** $\left\{\frac{e^{0.18}}{4}\right\}$; $\{0.2993\}$ **49.** \$3656.38 **51.** \$2542.50 **53.** 1.5850 **55.** −2.3219 **57.** 1.5850 **59.** −1.6309 **61.** 0.8617 **63.** $\left\{\frac{4}{7}\right\}$ **65.** $x = \frac{3y}{4}$ **67.** $\{-6, -1\}$ **69.**
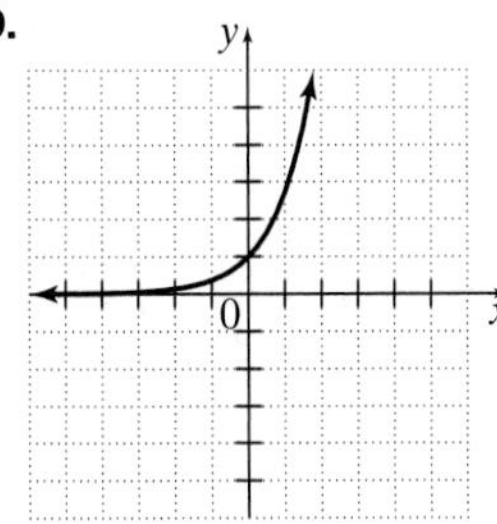

71.
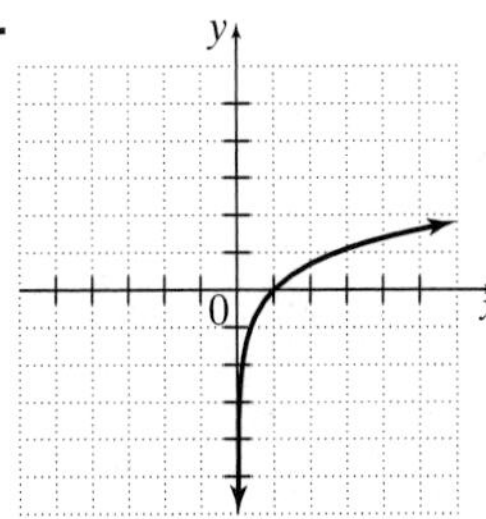

73. answers may vary **75.** 4.2 **77.** 5.3

CALCULATOR EXPLORATIONS

1. 3.67 yr, or 3 yr and 8 mo **3.** 23.16 yr, or 23 yr and 2 mo

EXERCISE SET 11.7

1. $\left\{\frac{\log 6}{\log 3}\right\}$; $\{1.6309\}$ **3.** $\left\{\frac{\log 3.8}{2\log 3}\right\}$; $\{0.6076\}$ **5.** $\left\{3 + \frac{\log 5}{\log 2}\right\}$; $\{5.3219\}$ **7.** $\left\{\frac{\log 5}{\log 9}\right\}$; $\{0.7325\}$ **9.** $\left\{\frac{\log 3}{\log 4} - 7\right\}$; $\{-6.2075\}$ **11.** $\left\{\frac{1}{3}\left(4 + \frac{\log 11}{\log 7}\right)\right\}$; $\{1.7441\}$ **13.** $\left\{\frac{\ln 5}{6}\right\}$; $\{0.2682\}$ **15.** $\{11\}$ **17.** $\left\{\frac{1}{2}\right\}$ **19.** $\left\{\frac{3}{4}\right\}$ **21.** $\{2\}$ **23.** $\left\{\frac{1}{8}\right\}$ **25.** $\{11\}$ **27.** $\{4, -1\}$ **29.** $\left\{\frac{2}{3}\right\}$ **31.** 103 wolves **33.** 6,885,833 people **35.** 26.3 yr **37.** 9.9 yr **39.** 1.7 yr **41.** 8.8 yr **43.** 55.7 in. **45.** 11.9 lb per sq in. **47.** 3.2 mi **49.** 12 weeks **51.** 18 weeks **53.** $-\frac{5}{3}$ **55.** $\frac{17}{4}$ **57.** 3.4% **59.** answers may vary

61. $\{6.93\}$
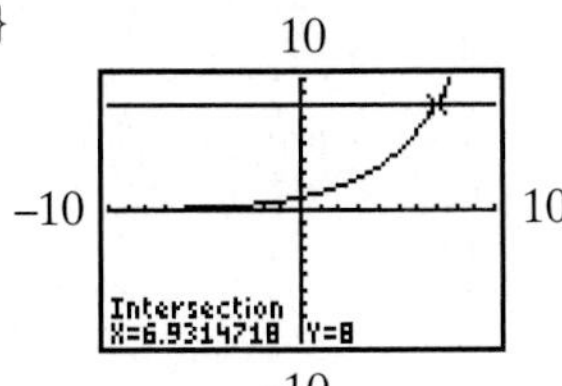

63. answers may vary

CHAPTER 11 REVIEW

1. $3x - 4$ **2.** $-x - 6$ **3.** $2x^2 - 9x - 5$ **4.** $\frac{2x + 1}{x - 5}$ if $x \neq 5$ **5.** $x^2 + 2x - 1$ **6.** $x^2 - 1$ **7.** 18 **8.** $x^4 - 4x^2 + 2$ **9.** -2 **10.** 48
11. one-to-one; $h^{-1} = \{(14, -9), (8, 6), (12, -11), (15, 15)\}$ **12.** not one-to-one
13. one-to-one;

Rank in Automobile Thefts (Input)	2	4	1	3
U.S. Region (Output)	West	Midwest	South	Northeast

14. not one-to-one **15.** not one-to-one **16.** not one-to-one **17.** not one-to-one **18.** one-to-one **19.** $f^{-1}(x) = \frac{x - 11}{6}$

20. $f^{-1}(x) = \frac{x}{12}$ **21.** $f^{-1}(x) = \frac{x + 5}{3}$ **22.** $f^{-1}(x) = \frac{x - 1}{2}$

23.
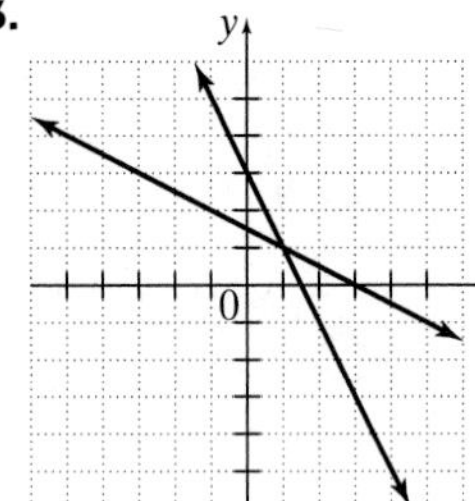
24.
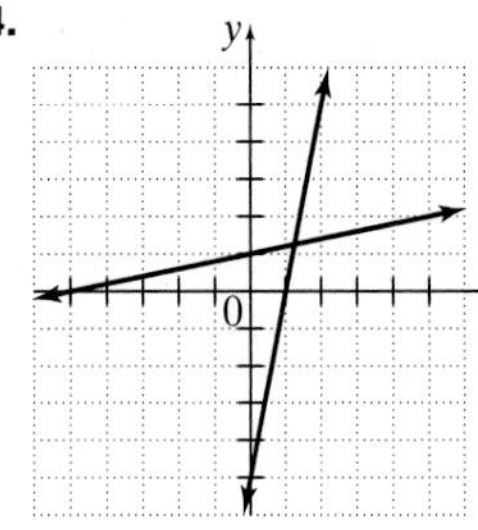
25. $\{3\}$ **26.** $\{-2\}$ **27.** $\left\{-\frac{4}{3}\right\}$ **28.** $\left\{\frac{3}{2}\right\}$ **29.** $\left\{\frac{3}{2}\right\}$ **30.** $\left\{\frac{8}{9}\right\}$

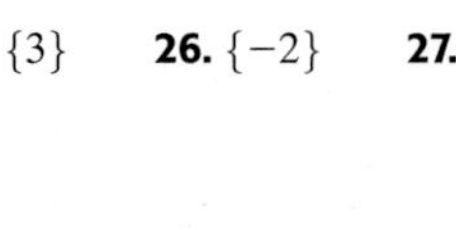

31.
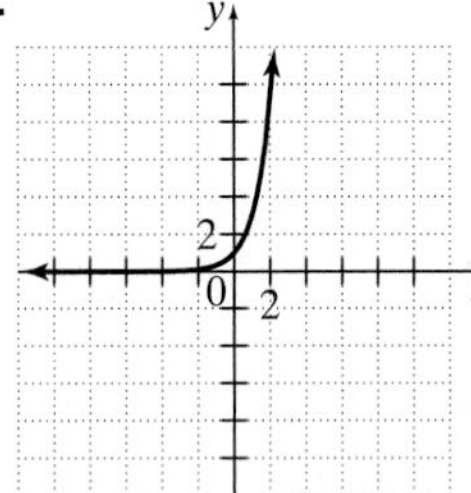
32.
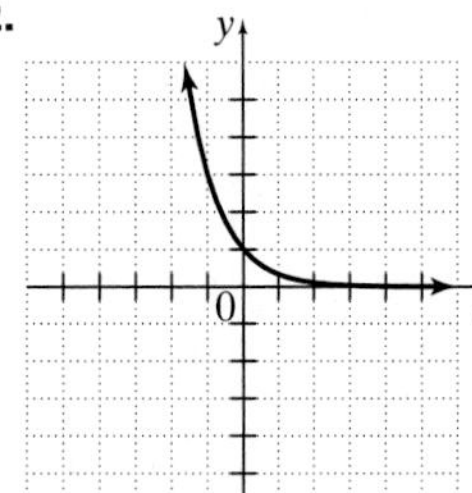
33.
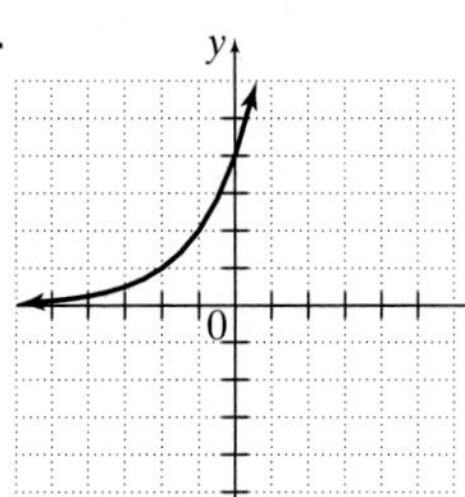
34.
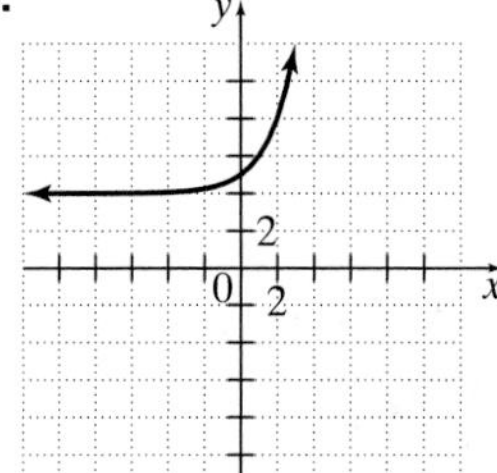

35. \$2963.11 **36.** \$1131.82 **37.** $\log_7 49 = 2$ **38.** $\log_2\left(\frac{1}{16}\right) = -4$ **39.** $\left(\frac{1}{2}\right)^{-4} = 16$ **40.** $0.4^3 = 0.064$ **41.** $\left\{\frac{1}{64}\right\}$ **42.** $\{9\}$

43. $\{0\}$ **44.** $\{3\}$ **45.** $\{8\}$ **46.** $\{3\}$ **47.** $\{5\}$ **48.** $\{-2\}$ **49.** $\{4\}$ **50.** $\{9\}$ **51.** $\left\{\frac{17}{3}\right\}$ **52.** $\{2\}$ **53.** $\{-1, 4\}$ **54.** $\{-8, 1\}$

55.
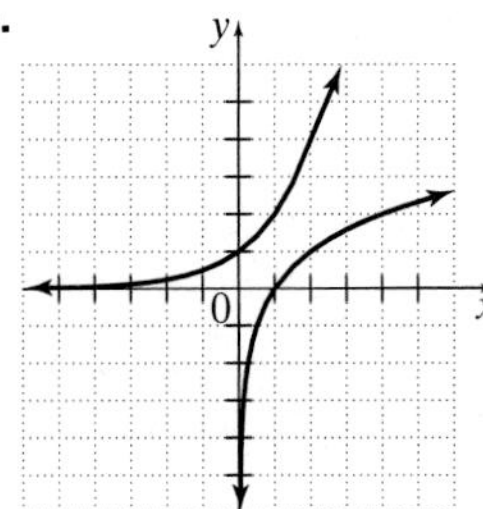
56.
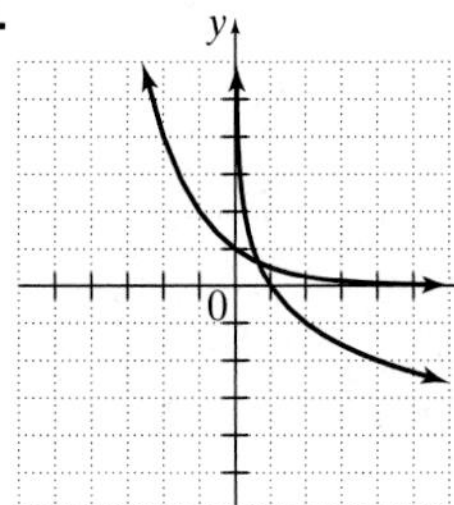

57. $\log_3 32$ **58.** $\log_2 18$ **59.** $\log_7 \frac{3}{4}$ **60.** $\log\left(\frac{3}{2}\right)$ **61.** $\log_{11} 4$

62. $\log_5 2$ **63.** $\log_5 \frac{x^3}{(x + 1)^2}$ **64.** $\log_3 (x^4 + 2x^3)$

65. $3 \log_3 x - \log_3 (x + 2)$ **66.** $\log_4 (x + 5) - 2 \log_4 x$

67. $\log_2 3 + 2 \log_2 x + \log_2 y - \log_2 z$ **68.** $\log_7 y + 3 \log_7 z - \log_7 x$

69. 2.02 **70.** -0.11 **71.** 0.5563 **72.** -0.8239 **73.** 0.2231 **74.** 1.5326 **75.** 3 **76.** -1 **77.** -1 **78.** 4 **79.** $\left\{\frac{e^2}{2}\right\}$

80. $\left\{\frac{e^{1.6}}{3}\right\}$ **81.** $\left\{\frac{e^{-1} + 3}{2}\right\}$ **82.** $\left\{\frac{e^2 - 1}{3}\right\}$ **83.** 0.2920 **84.** 1.2619 **85.** \$1957.30 **86.** \$1307.51 **87.** $\left\{\frac{\log 7}{2 \log 3}\right\}$; $\{0.8856\}$

88. $\left\{\frac{\log 5}{3\log 6}\right\}$; {0.2994} **89.** $\left\{\frac{1}{2}\left(\frac{\log 6}{\log 3}-1\right)\right\}$; {0.3155} **90.** $\left\{\frac{1}{3}\left(\frac{\log 9}{\log 4}-2\right)\right\}$; {−0.1383} **91.** $\left\{\frac{1}{3}\left(\frac{\log 4}{\log 5}+5\right)\right\}$; {1.9538}

92. $\left\{\frac{1}{4}\left(\frac{\log 3}{\log 8}+2\right)\right\}$; {0.6321} **93.** $\left\{-\frac{\log 2}{\log 5}+1\right\}$; {0.5693} **94.** $\left\{\frac{\log\frac{2}{3}}{\log 4}-5\right\}$; {−5.2925} **95.** $\left\{\frac{25}{2}\right\}$ **96.** $\left\{\frac{9}{10}\right\}$ **97.** ∅

98. $\left\{\frac{3e^2}{e^2-3}\right\}$ **99.** $\{2\sqrt{2}\}$ **100.** ∅ **101.** 197,044 ducks **102.** 6,648,551 people **103.** 23 yr **104.** 116 yr **105.** 36 yr

106. 8.8 yr **107.** 8.5 yr

CHAPTER 11 TEST

1. $2x^2 - 3x$ **2.** $3 - x$ **3.** 5 **4.** $x - 7$ **5.** $x^2 - 6x - 2$

6.

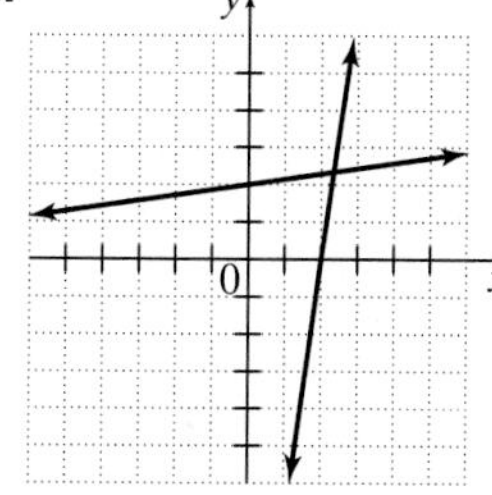

7. one-to-one **8.** not one-to-one **9.** one-to-one; $f^{-1}(x) = \frac{-x+6}{2}$ **10.** one-to-one; $f^{-1} = \{(0,0), (3,2), (5,-1)\}$ **11.** not one-to-one **12.** $\log_3 24$ **13.** $\log_5\left(\frac{x^4}{x+1}\right)$

14. $\log_6 2 + \log_6 x - 3\log_6 y$ **15.** −1.53 **16.** 1.0686 **17.** {−1}

18. $\left\{\frac{1}{2}\left(\frac{\log 4}{\log 3}-5\right)\right\}$; {−1.8691} **19.** $\left\{\frac{1}{9}\right\}$ **20.** $\left\{\frac{1}{2}\right\}$ **21.** {22} **22.** $\left\{\frac{25}{3}\right\}$

23. $\left\{\frac{43}{21}\right\}$ **24.** {−1.0979}

25.

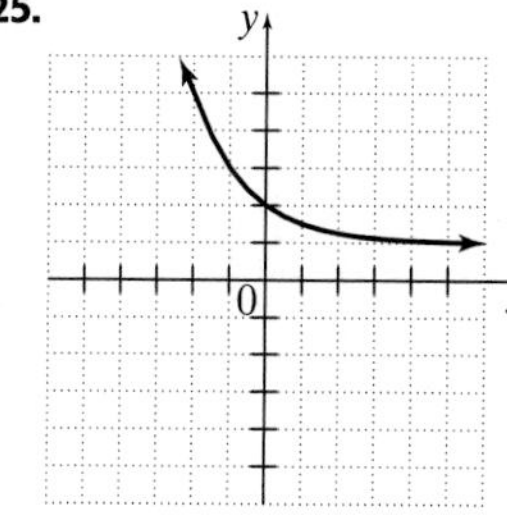

26.

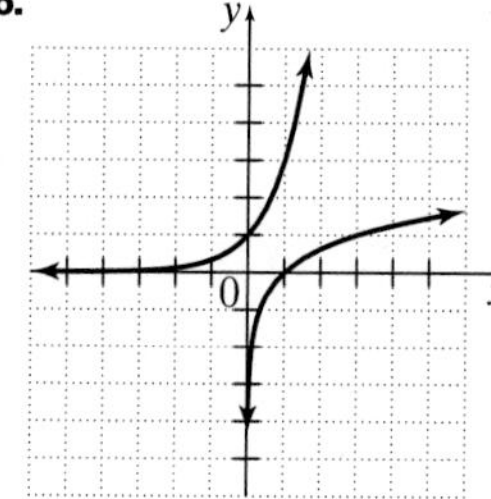

27. \$5234.58 **28.** 6 yr **29.** 64,913 prairie dogs **30.** 15 yr **31.** 85%

32. 52% **33.** 7074 million people

CUMULATIVE REVIEW

1. $\frac{5}{x^2}$; Sec. 6.1, Ex. 3 **2.** $\frac{1}{x-9}$; Sec. 6.1, Ex. 6 **3.** $\frac{2}{5}$; Sec. 6.2, Ex. 2 **4.** $\frac{2}{x(x+1)}$; Sec. 6.2, Ex. 6 **5.** $y = \frac{1}{4}x - 3$; Sec. 7.1, Ex. 3

6. $x = 2$; Sec. 7.2, Ex. 4 **7.** domain: [−3, 5]; range: [−2, 4]; Sec. 7.3, Ex. 14

8. (−1, 3);

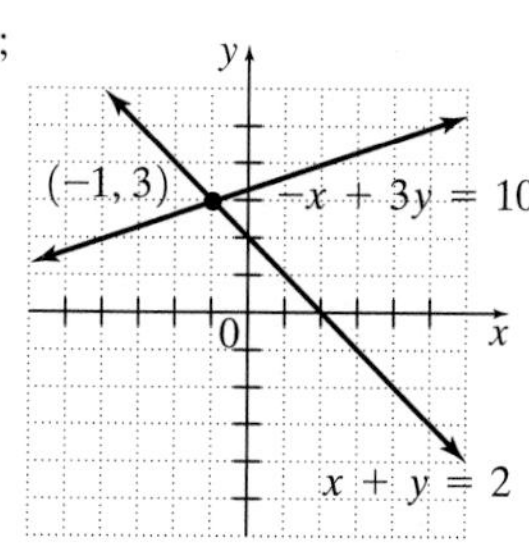

Sec. 8.1, Ex. 3 **9.** no solution; Sec. 8.2, Ex. 6 **10.** Adult: \$69; Child: \$49; yes; Sec. 8.4, Ex. 3

11. 0; Sec. 9.1, Ex. 5 **12.** 0.5; Sec. 9.1, Ex. 7 **13.** $\sqrt{x}$; Sec. 9.2, Ex. 19 **14.** $\sqrt[3]{rs^2}$; Sec. 9.2, Ex. 21

15. $2\sqrt[3]{3}$; Sec. 9.3, Ex. 11 **16.** $2\sqrt[4]{2}$; Sec. 9.3, Ex. 13 **17.** $\frac{2\sqrt{5}}{5}$; Sec. 9.5, Ex. 1

18. $\left\{-\frac{1}{9}, -1\right\}$; Sec. 9.6, Ex. 2 **19.** $13 - 18i$; Sec. 9.7, Ex. 13

20. $\{-1+\sqrt{5}, -1-\sqrt{5}\}$; Sec. 10.1, Ex. 7 **21.** $\left\{\frac{2+\sqrt{10}}{2}, \frac{2-\sqrt{10}}{2}\right\}$; Sec. 10.2, Ex. 2

22. {9}; Sec. 10.3, Ex. 1 **23.** $(-\infty, -2] \cup [1, 5]$; Sec. 10.4, Ex. 3 **24.** {6}; Sec. 11.3, Ex. 6 **25.** {5}; Sec. 11.4, Ex. 9

26. $\log_5 72$; Sec. 11.5, Ex. 7

Chapter 12 Conic Sections

Chapter 12 Pretest

1. $\sqrt{34}$ units; 12.1B **2.** $\left(\frac{19}{2}, \frac{7}{2}\right)$; 12.1B

3.

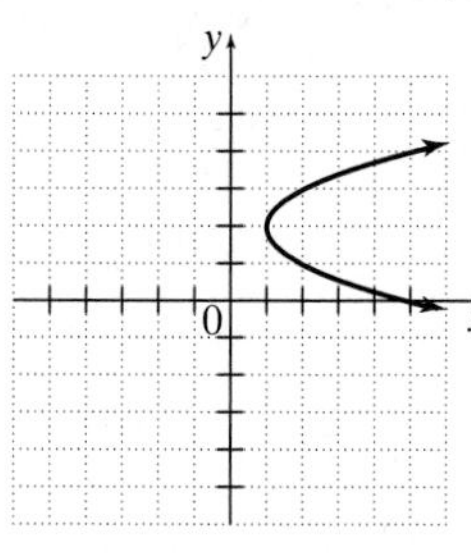

12.1A

4.

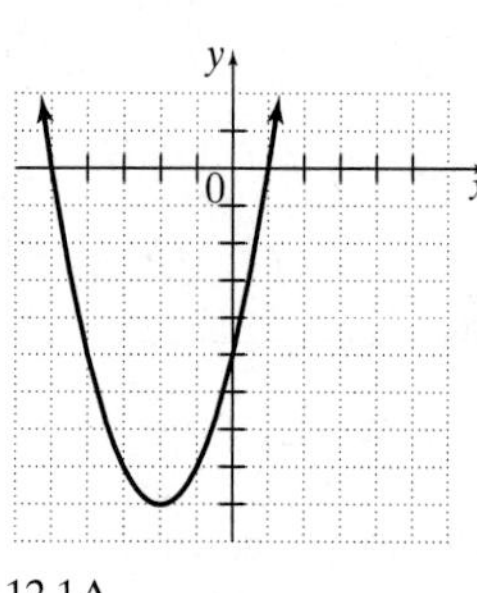

12.1A

5.

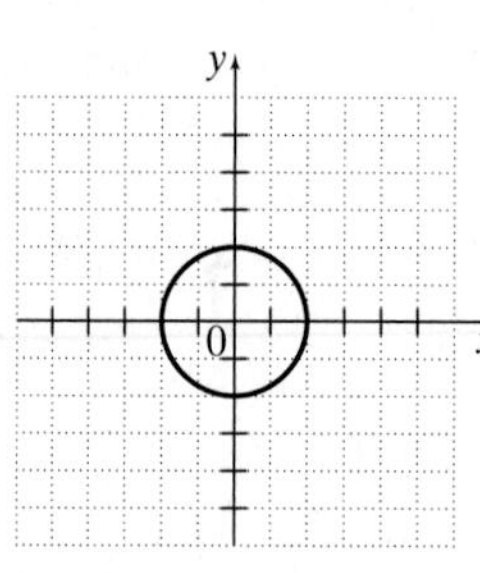

12.1C

6.

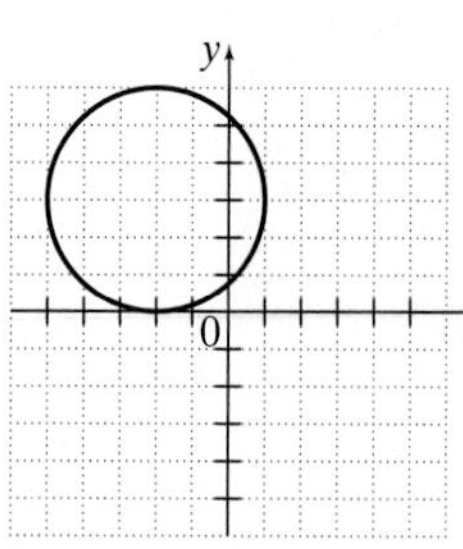

12.1C

7.

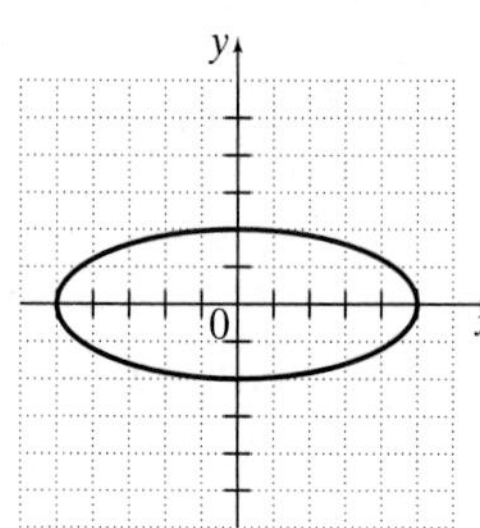

12.2A

8.

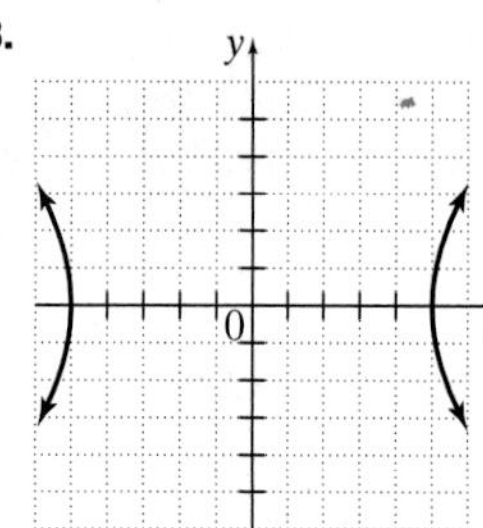

12.2B

9.

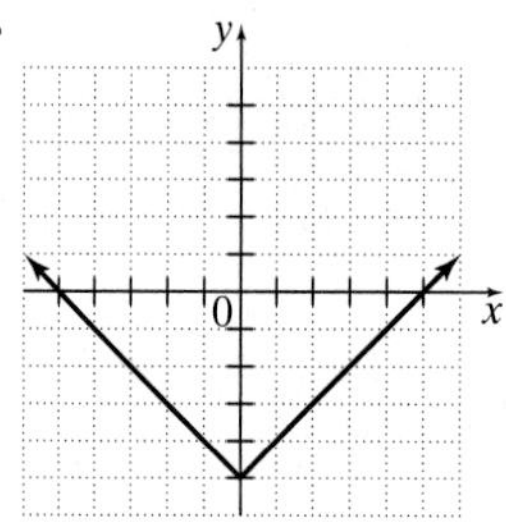

12.3A

10.

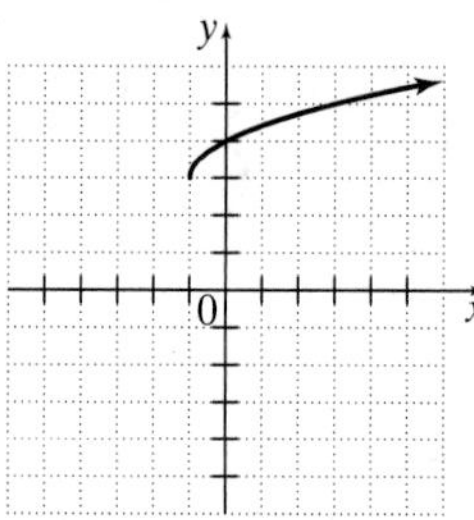

12.3A

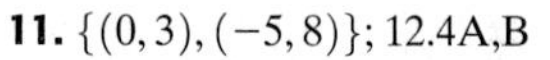

11. $\{(0, 3), (-5, 8)\}$; 12.4A,B **12.**

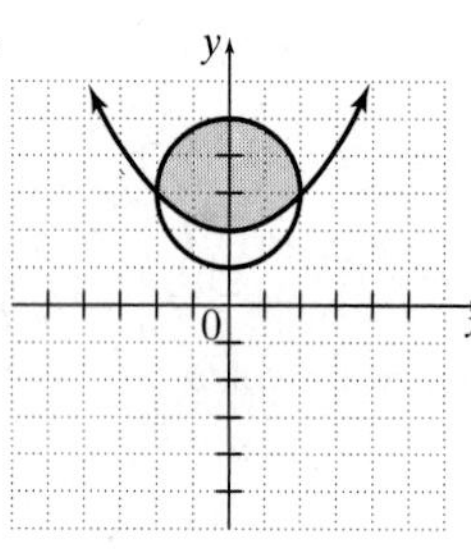

12.5B

13. $(x - 2)^2 + (y - 5)^2 = 64$; 12.1D **14.** center: $(-3, 4)$; radius: 3

Graphing Calculator Explorations

1.

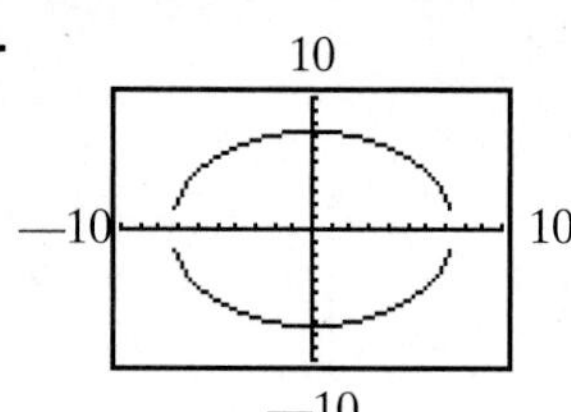

3.

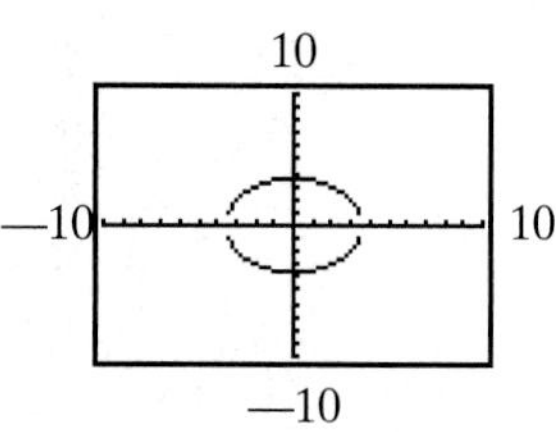

Mental Math

1. upward **3.** to the left **5.** downward

Exercise Set 12.1

1.

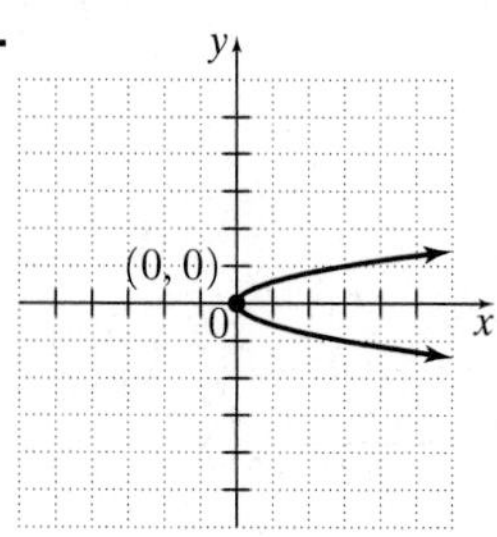

3.

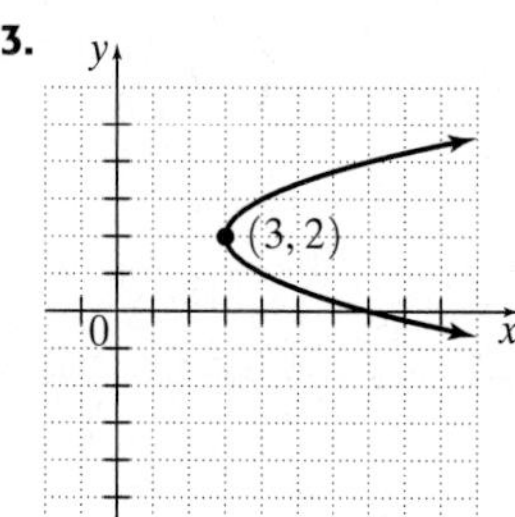

5.

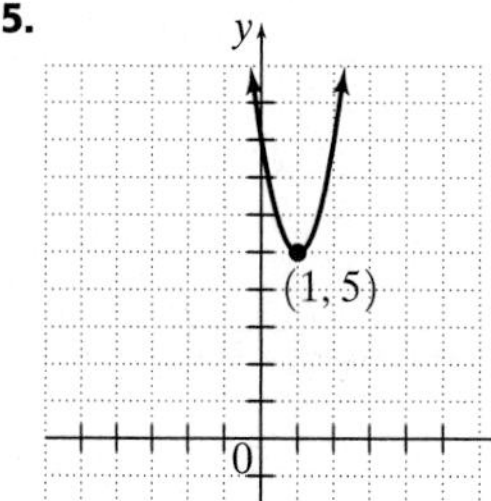

7.

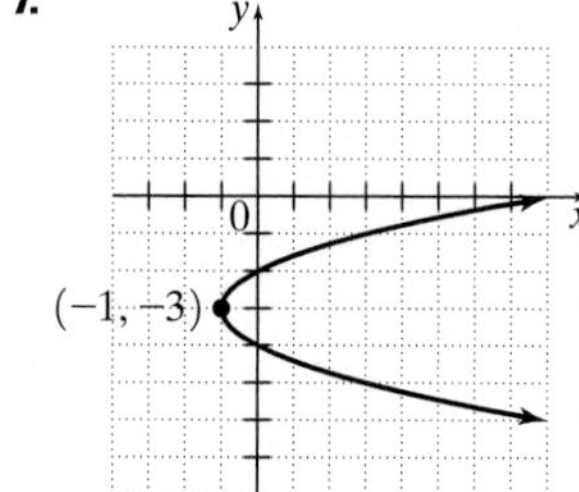

9.

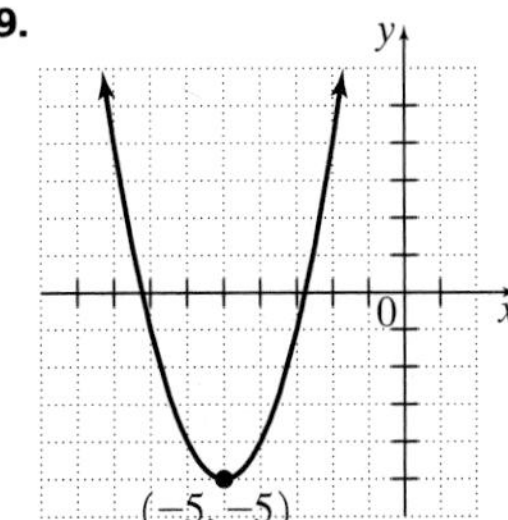

11.

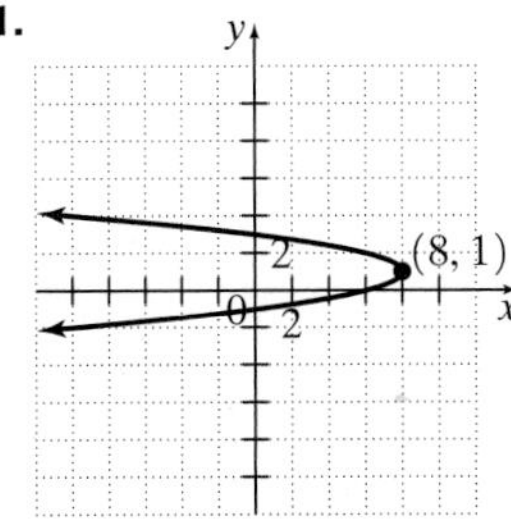

13. 5 units **15.** $\sqrt{41}$ units ≈ 6.403 **17.** $\sqrt{10}$ units ≈ 3.162 **19.** $\sqrt{5}$ units ≈ 2.236 **21.** $\sqrt{192.58}$ units ≈ 13.877 **23.** 9 units **25.** (4, −2) **27.** $\left(-5, \frac{5}{2}\right)$ **29.** (3, 0) **31.** $\left(-\frac{1}{2}, \frac{1}{2}\right)$ **33.** $\left(\sqrt{2}, \frac{\sqrt{5}}{2}\right)$ **35.** (6.2, −6.65)

37.

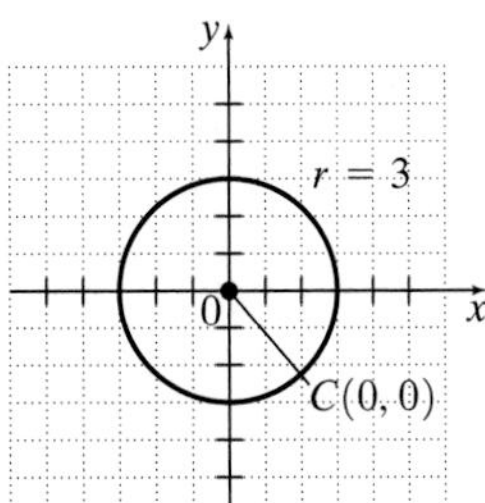

39.

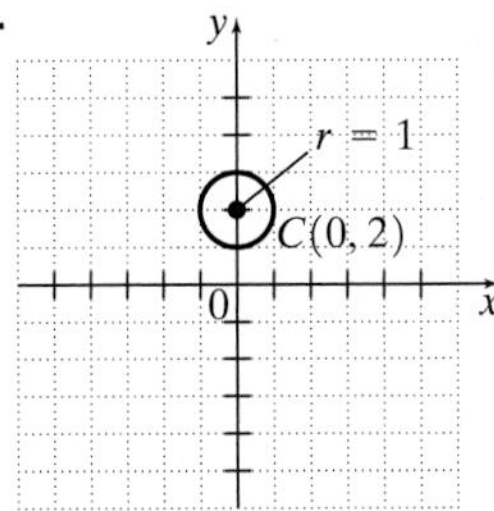

41.

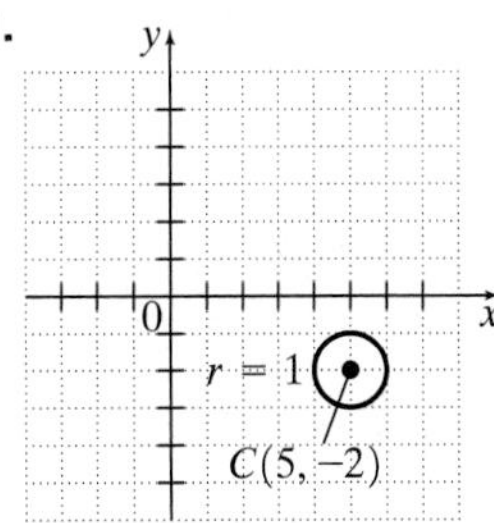

43.

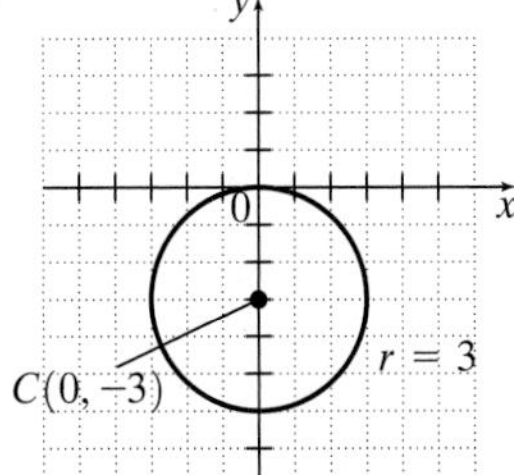

45.

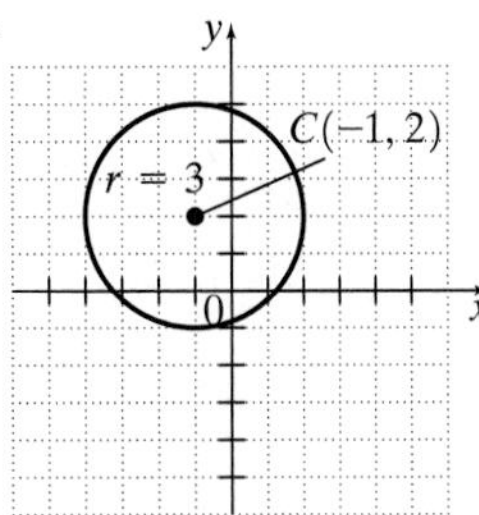

47.

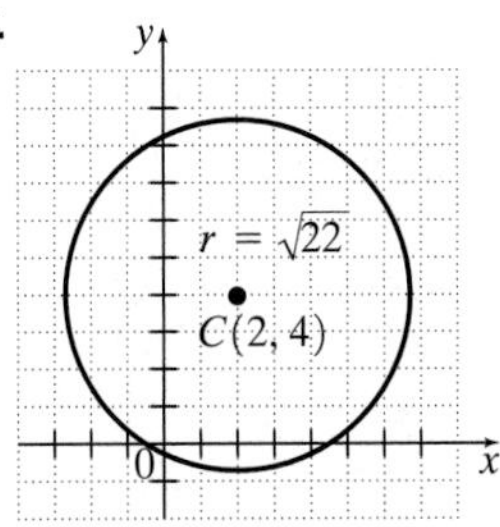

49. $(x-2)^2+(y-3)^2=36$ **51.** $x^2+y^2=3$ **53.** $(x+5)^2+(y-4)^2=45$

55.

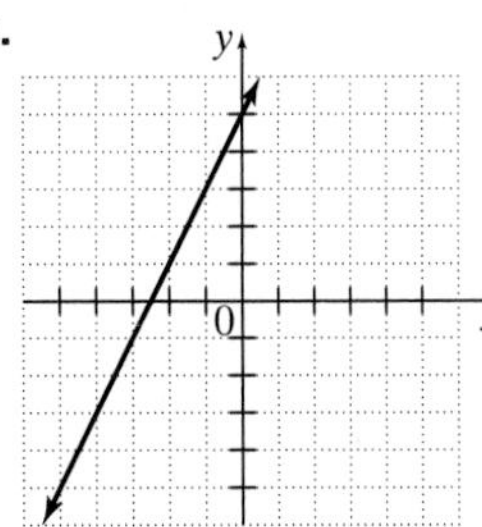

57.

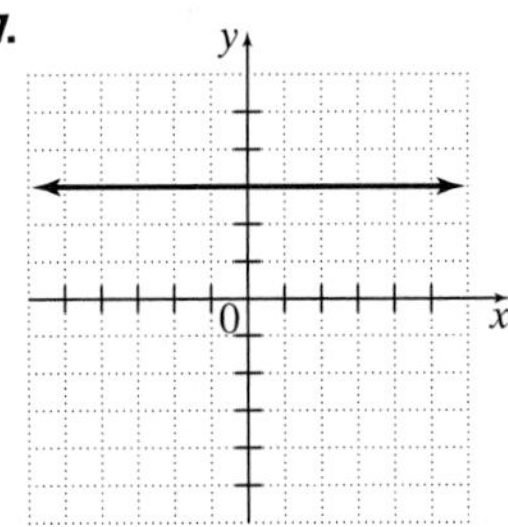

59. $\frac{\sqrt{3}}{3}$ **61.** $\frac{2\sqrt{42}}{3}$ **63. a.** 125 ft **b.** 14 ft **c.** 139 ft **d.** (0, 139) **e.** $x^2+(y-139)^2=125^2$ **65.** answers may vary **67.** 20 m **69.** $y=-\frac{2}{125}x^2+40$

Graphing Calculator Explorations

1.

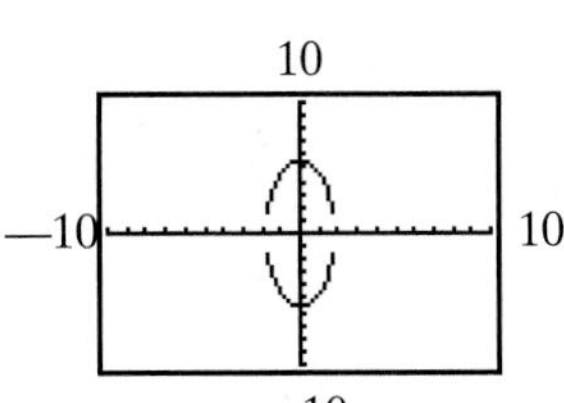

3.

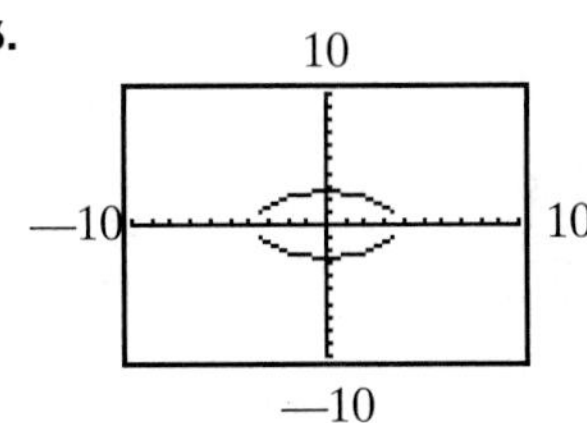

Mental Math

1. ellipse **3.** hyperbola **5.** hyperbola

Exercise Set 12.2

1.

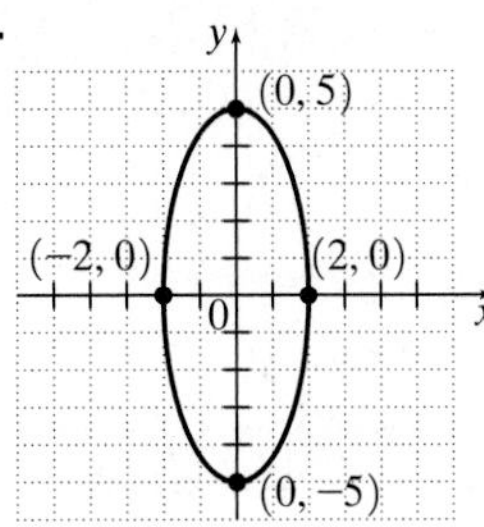

3.

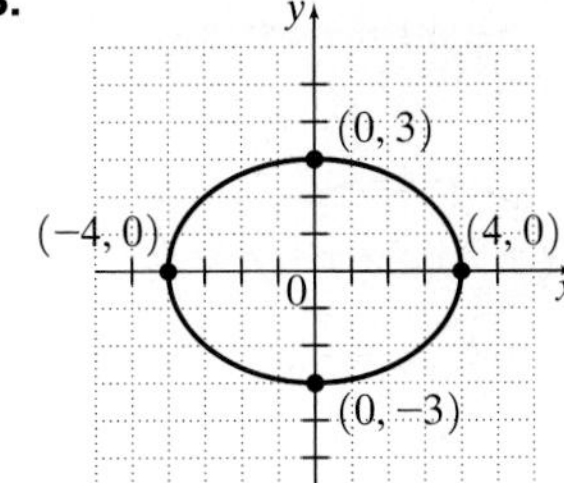

5.

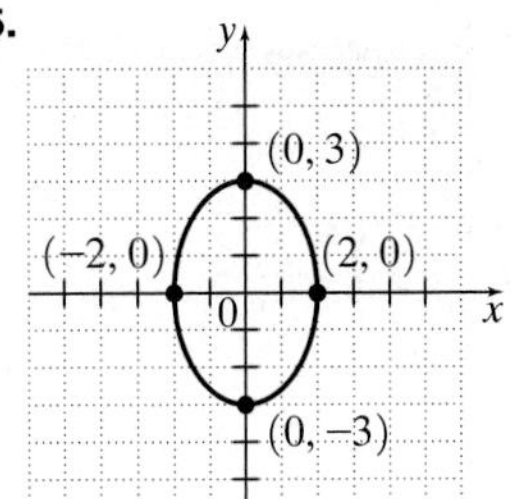

7.

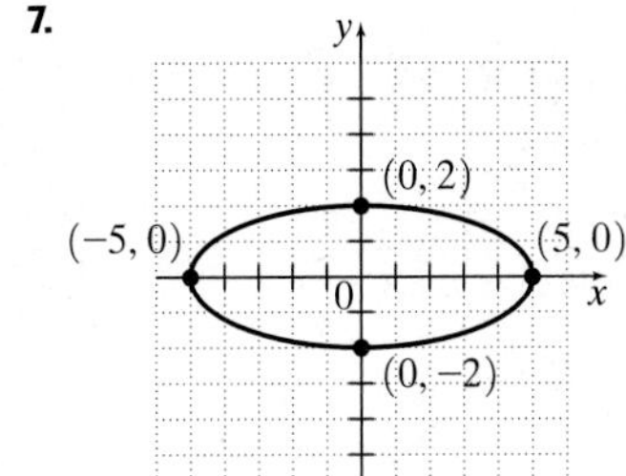

9.

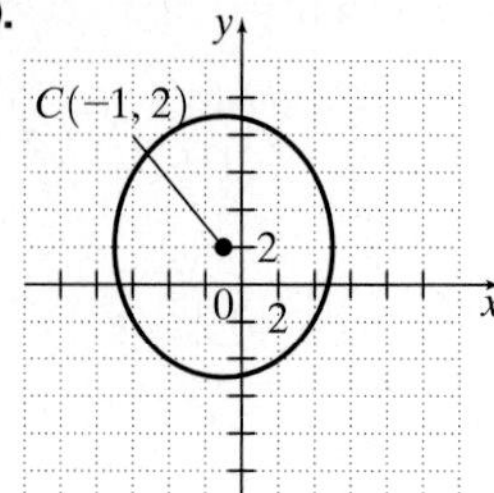

11.

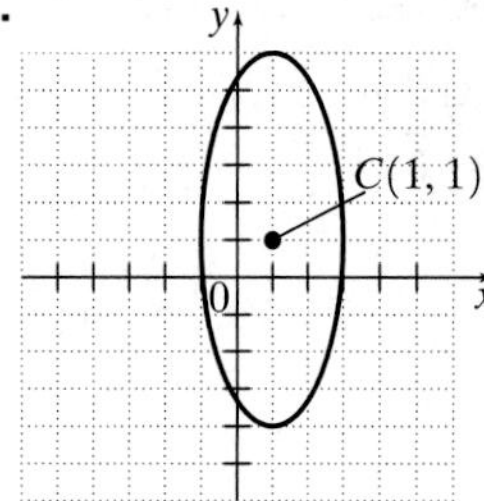

13.

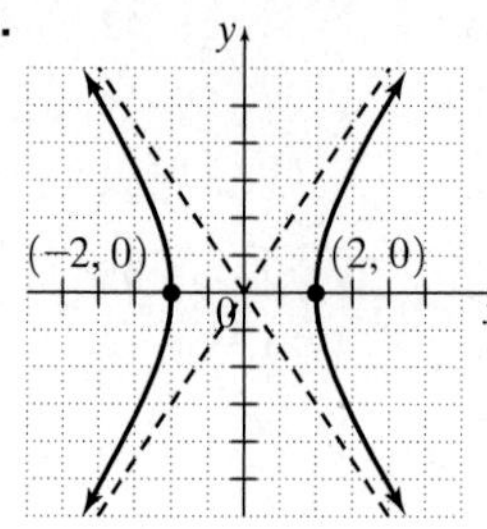

15.

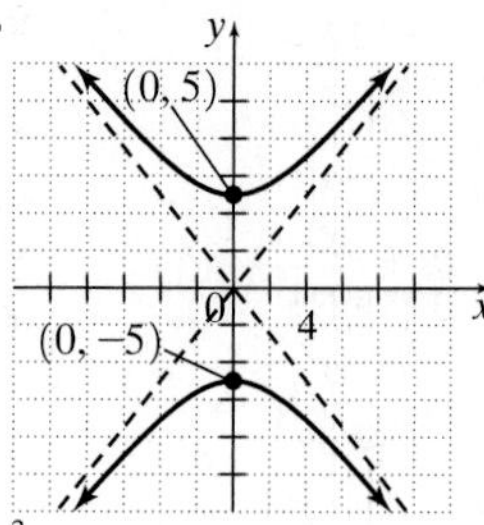

17.

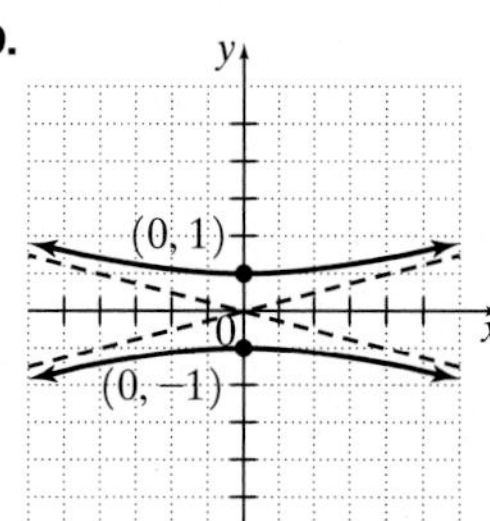

19.

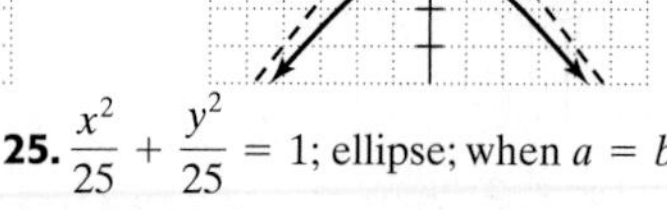

21. $-8x^5$ **23.** $-4x^2$ **25.** $\frac{x^2}{25} + \frac{y^2}{25} = 1$; ellipse; when $a = b$

27. A: 36, 13; B: 4, 4; C: 25, 16; D: 39, 25; E: 17, 81; F: 36, 36; G: 16, 65; H: 144, 140

29. A: 6; B: 2; C: 5; D: 5; E: 9; F: 6; G: 4; H: 12

31. greater than 0 and less than 1 **33.** greater than 1

35. answers may vary

A:

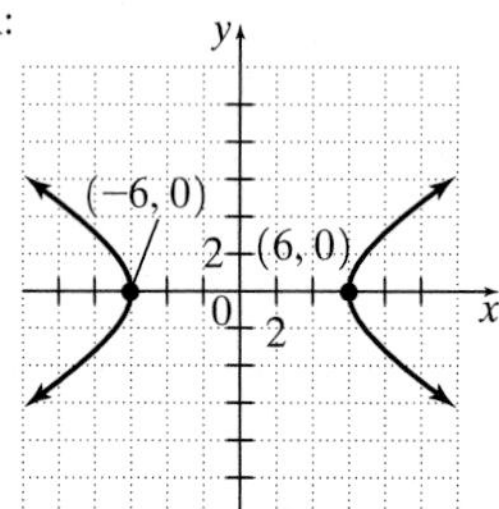

B:

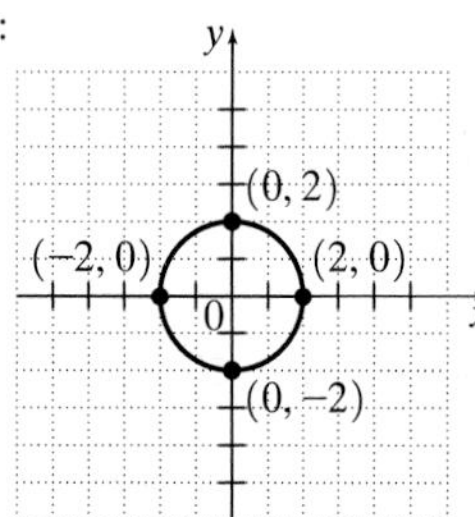

C:

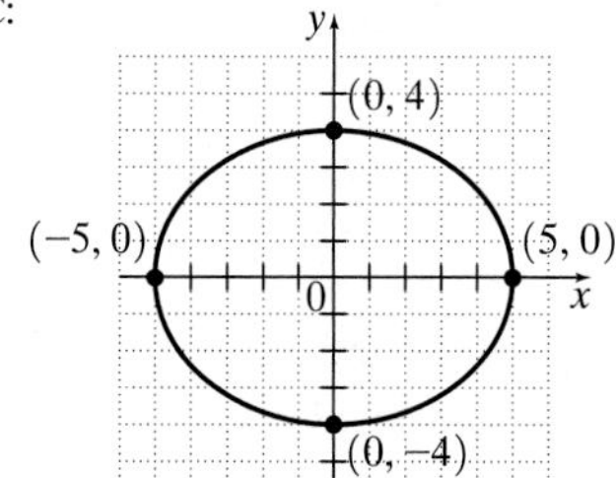

D:

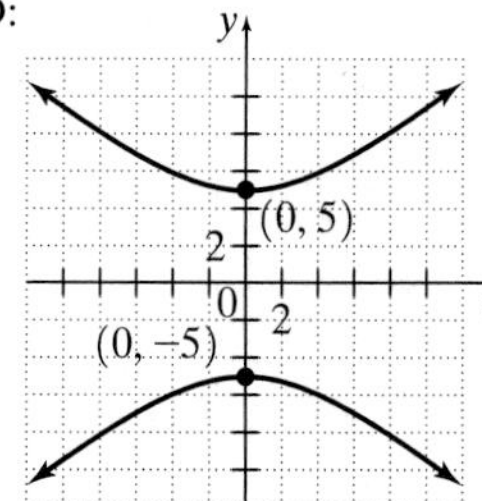

E:

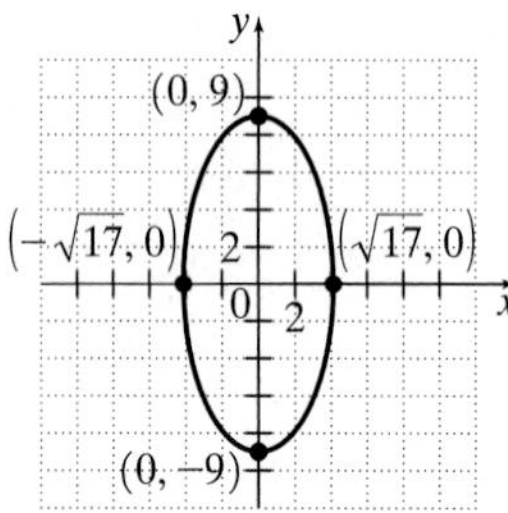

F:

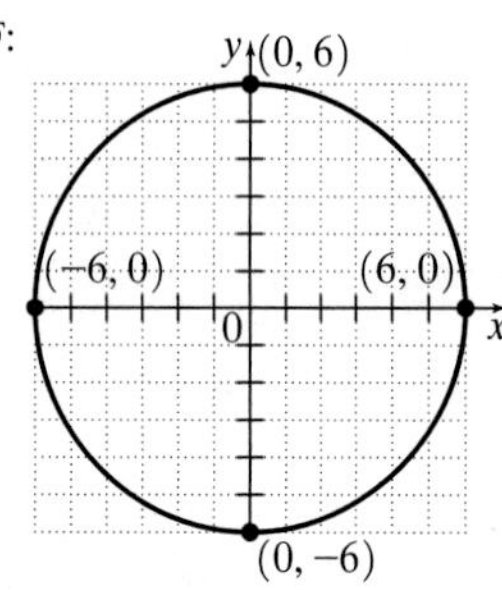

G:

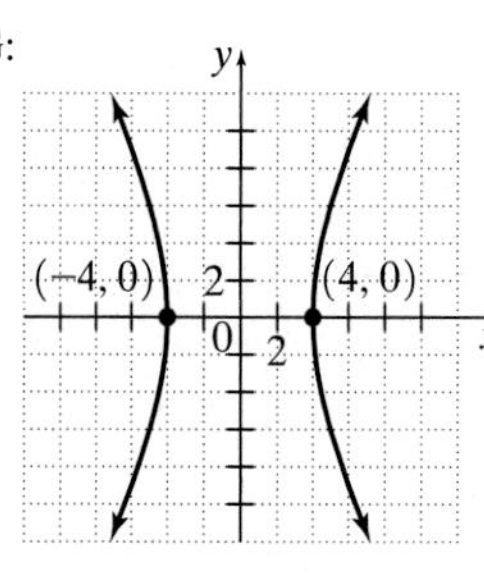

H:

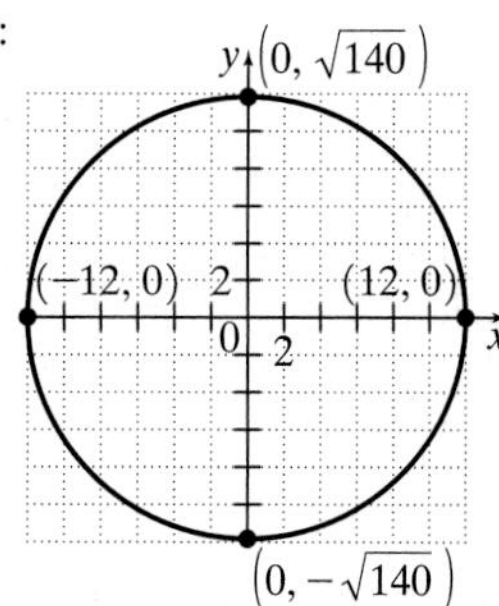

37. answers may vary

Integrated Review

1. circle;

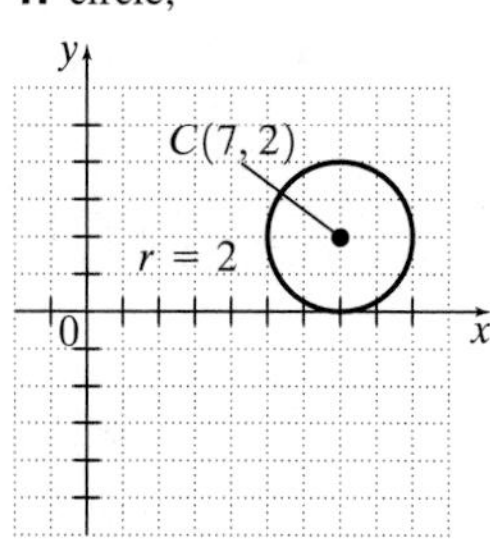

2. parabola;

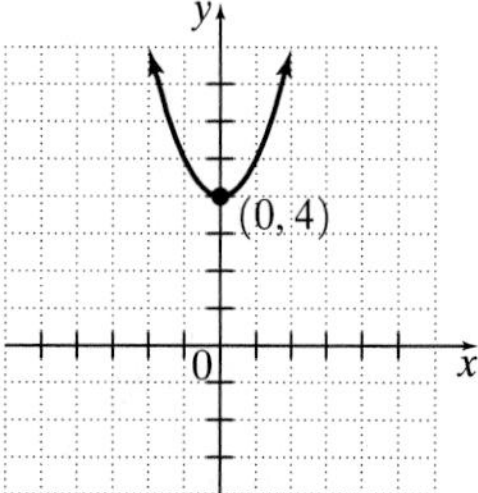

3. parabola;

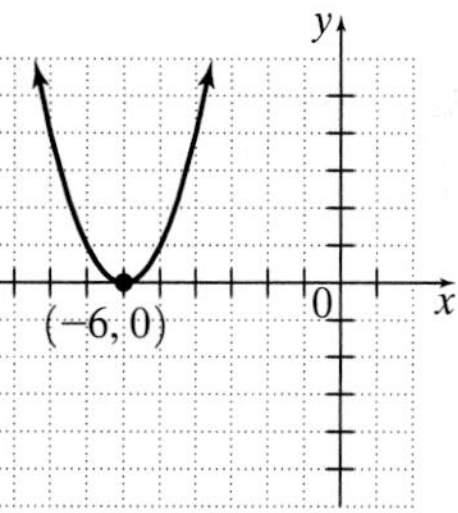

4. ellipse;

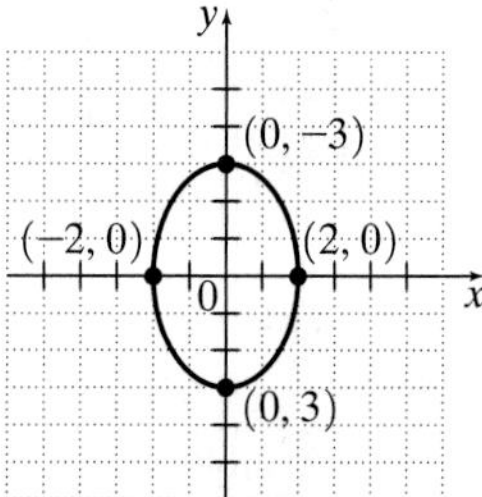

5. hyperbola;

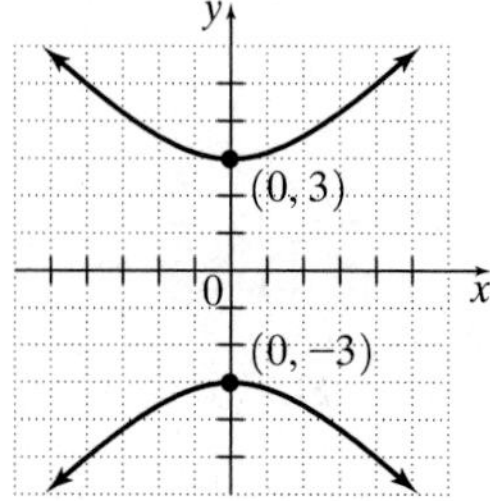

6. hyperbola;

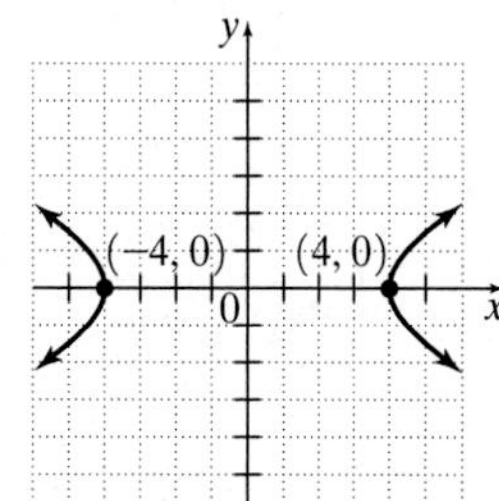

7. ellipse;

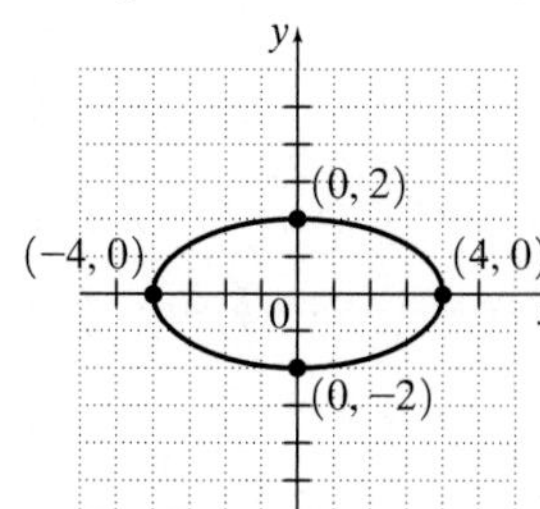

8. circle;

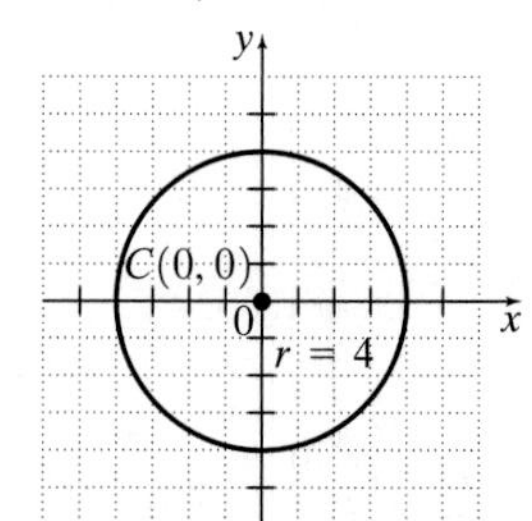

9. parabola;

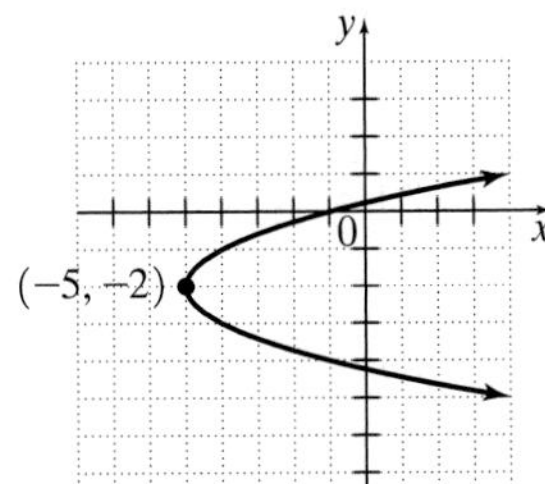

10. parabola;

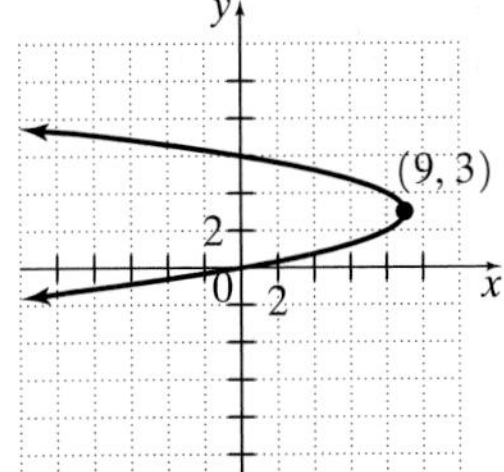

11. hyperbola;

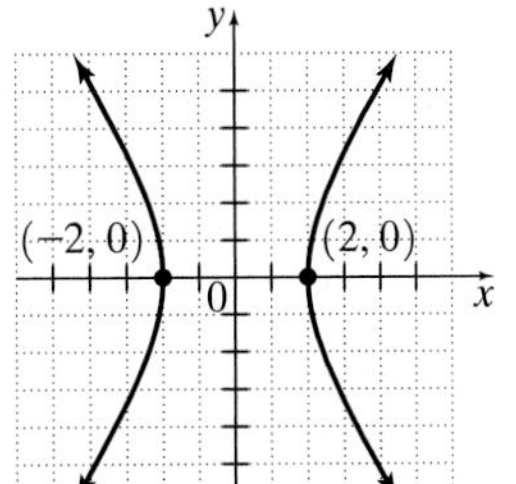

12. ellipse;

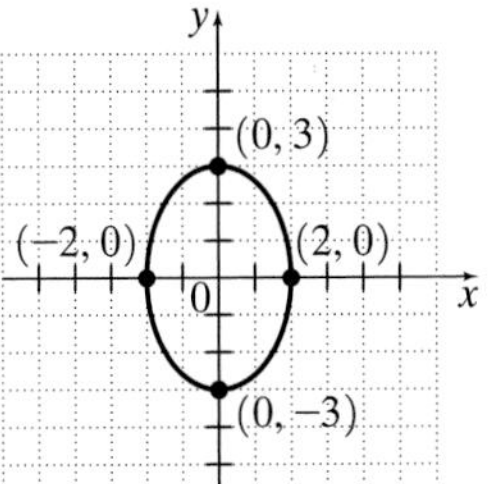

13. ellipse;

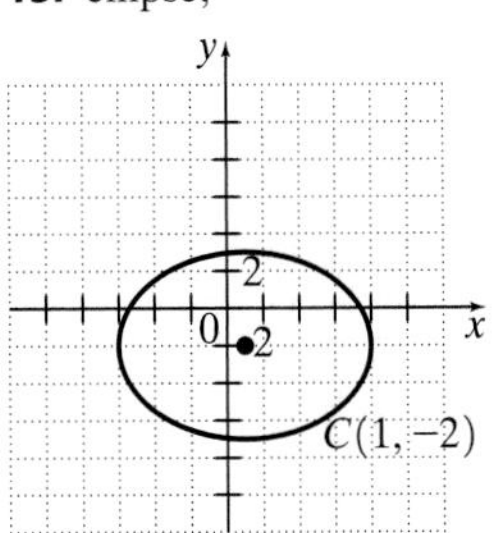

14. hyperbola;

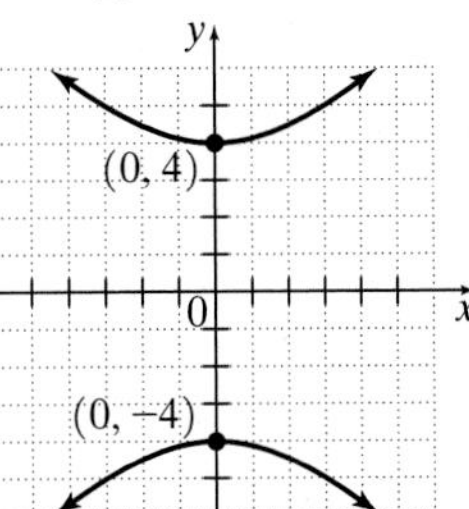

15. circle;

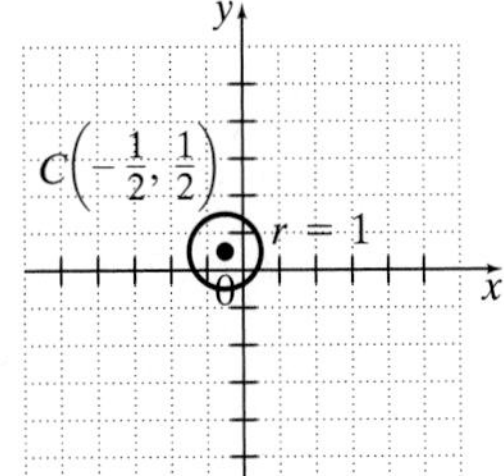

Exercise Set 12.3

1.

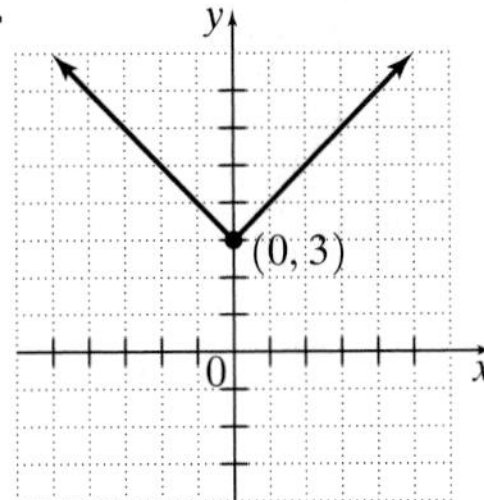

3.

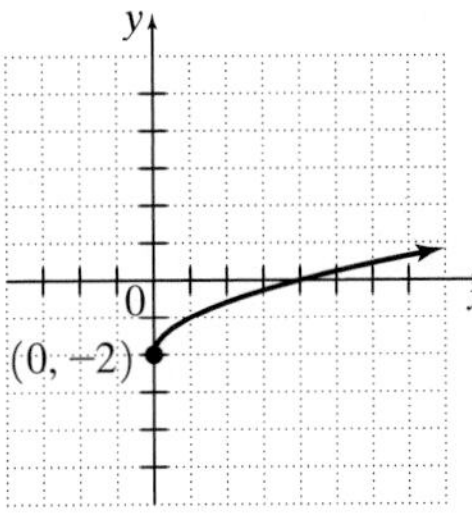

5.

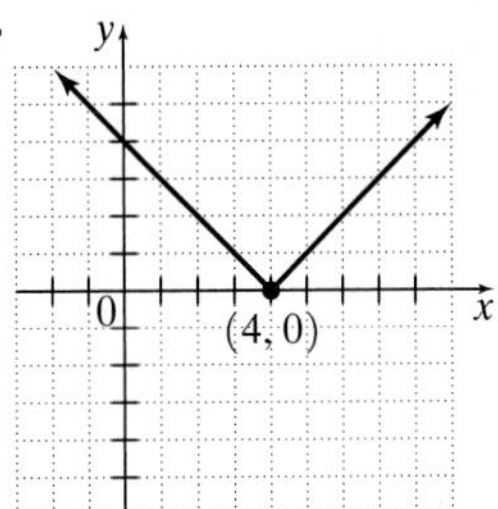

7.

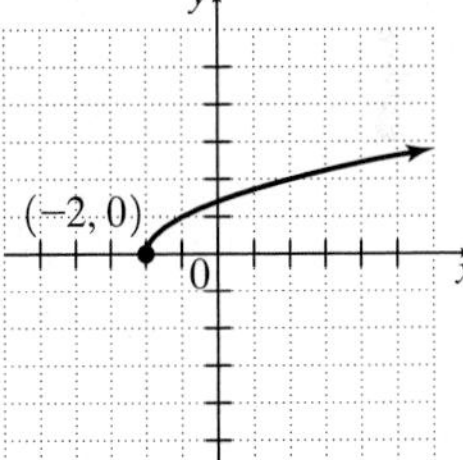

9.

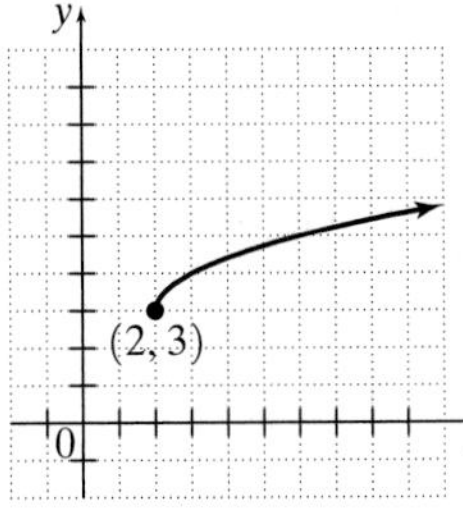

11.

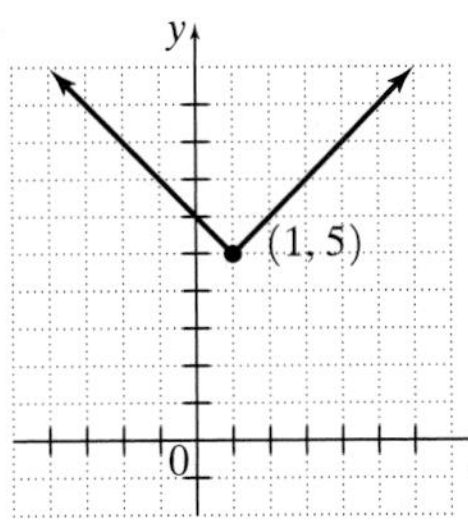

13.

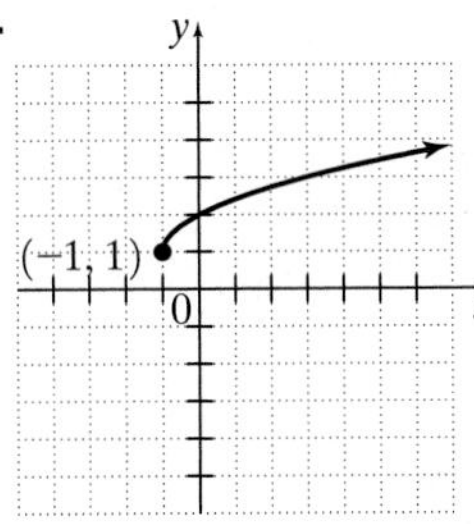

15.

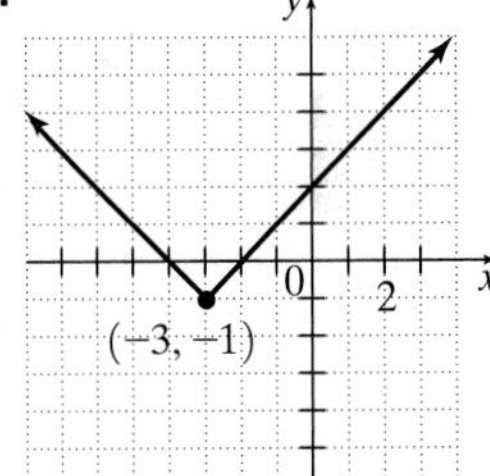

17. $(8, -2)$ **19.** $(-1, 3)$ **21.**

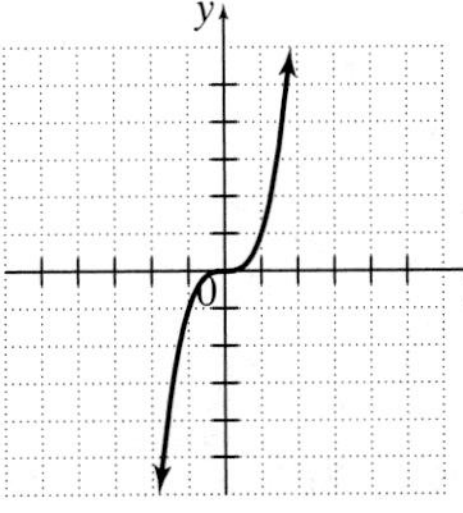

23.

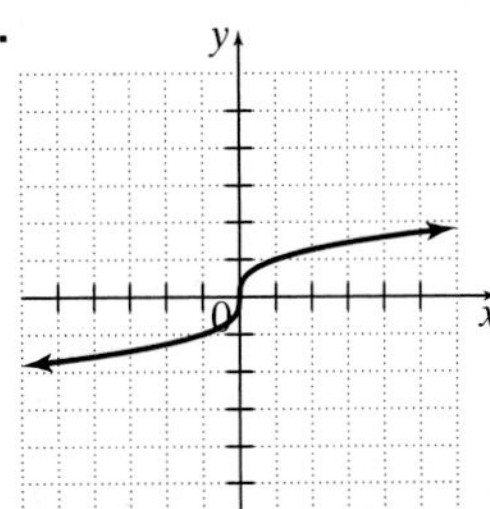

25. $\{x|x \geq 20\}$ **27.** $\{x|x$ is a real number$\}$ **29.** $\{x|x \geq -103\}$

Exercise Set 12.4

1. $\{(3, -4), (-3, 4)\}$ **3.** $\{(\sqrt{2}, \sqrt{2}), (-\sqrt{2}, -\sqrt{2})\}$ **5.** $\{(4, 0), (0, -2)\}$ **7.** $\{(-\sqrt{5}, -2), (-\sqrt{5}, 2), (\sqrt{5}, -2), (\sqrt{5}, 2)\}$ **9.** $\emptyset$ **11.** $\{(1, -2), (3, 6)\}$ **13.** $\{(2, 4), (-5, 25)\}$ **15.** $\emptyset$ **17.** $\{(1, -3)\}$ **19.** $\{(-1, -2), (-1, 2), (1, -2), (1, 2)\}$ **21.** $\{(0, -1)\}$ **23.** $\{(-1, 3), (1, 3)\}$ **25.** $\{(\sqrt{3}, 0), (-\sqrt{3}, 0)\}$ **27.** $\emptyset$ **29.** $\{(-6, 0), (6, 0), (0, -6)\}$ **31.** 0, 1, 2, 3, or 4

33.

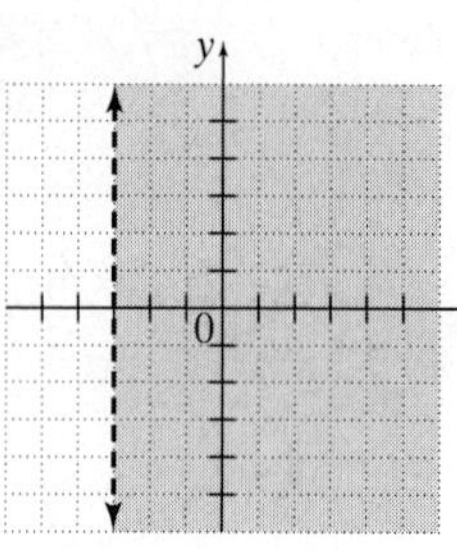

35.

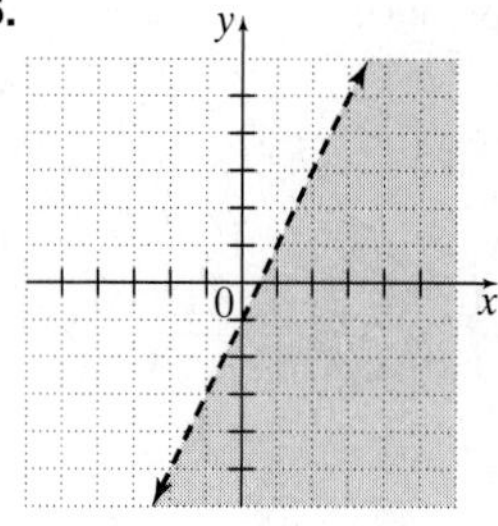

37. 9 and 7; 9 and −7; −9 and 7; −9 and −7 **39.** 15 cm by 19 cm **41.** 15 thousand compact discs; price: $3.75

Exercise Set 12.5

1.

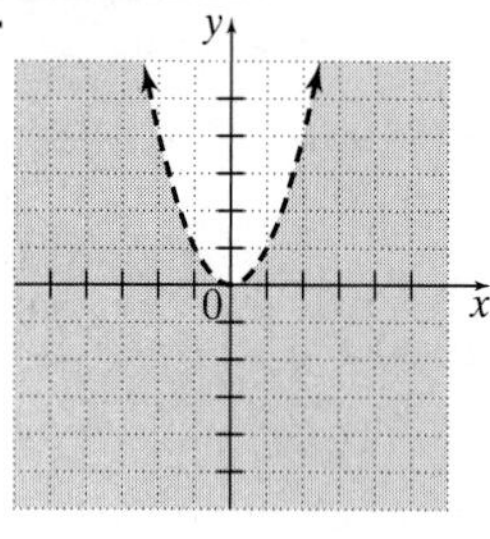

3.

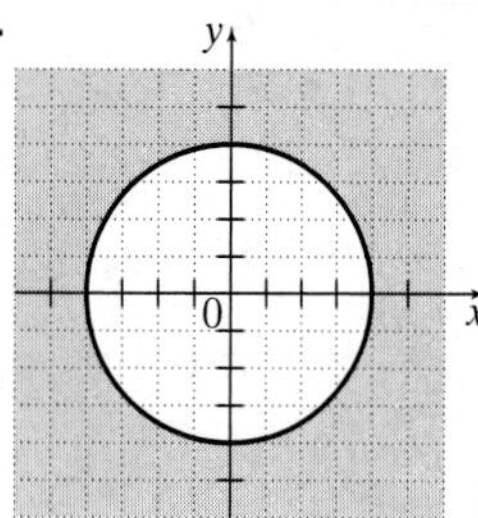

5.

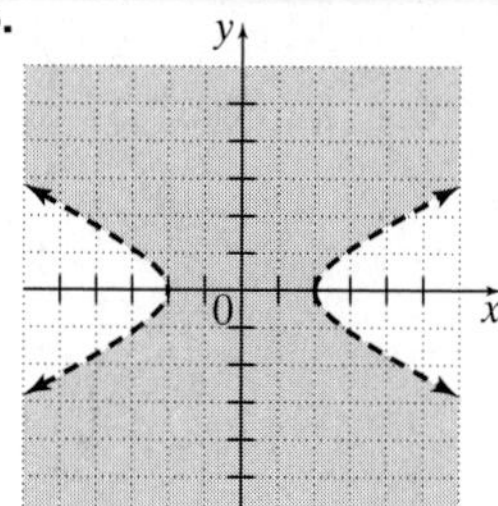

7.

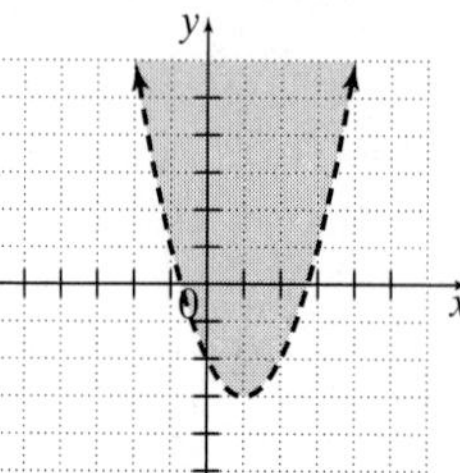

9.

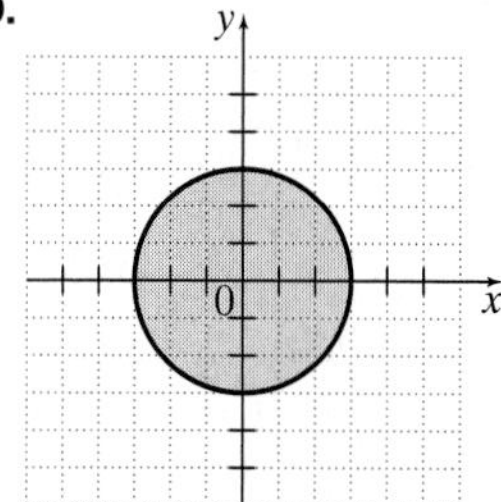

11.

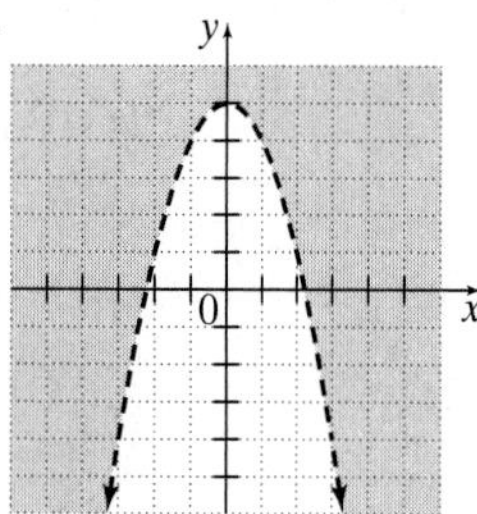

13.

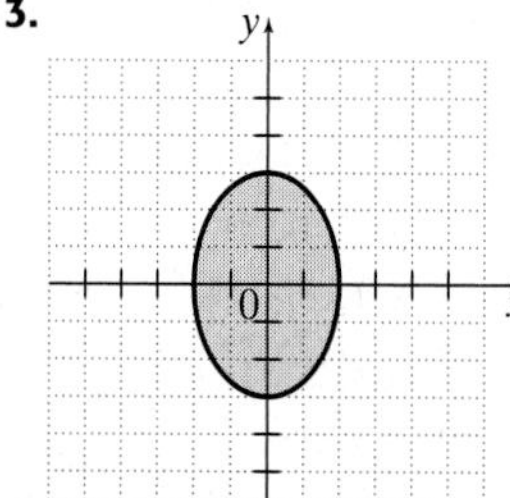

15.

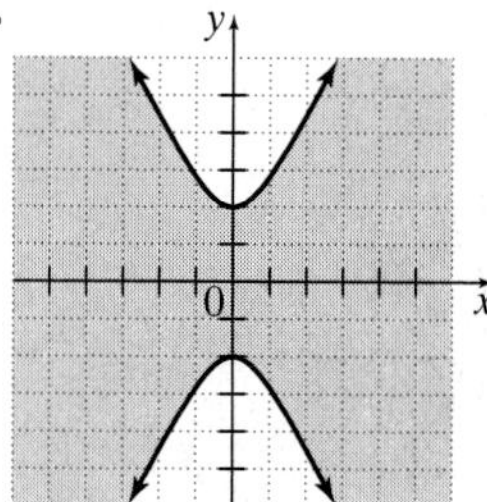

17.

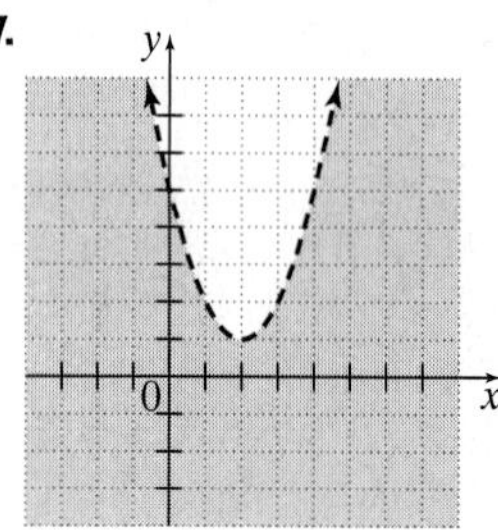

19.

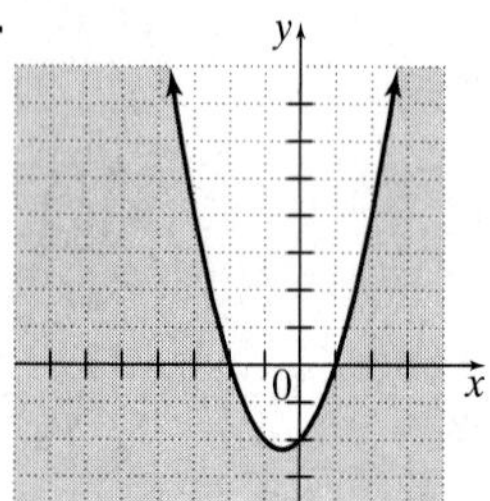

21.

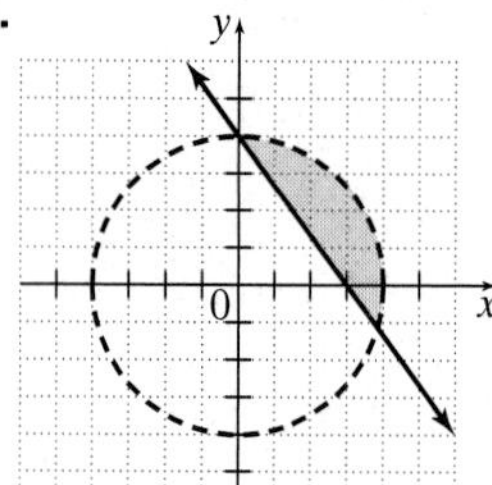

23.

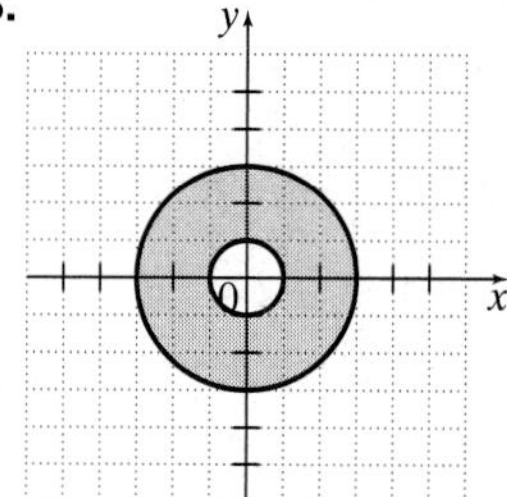

25.

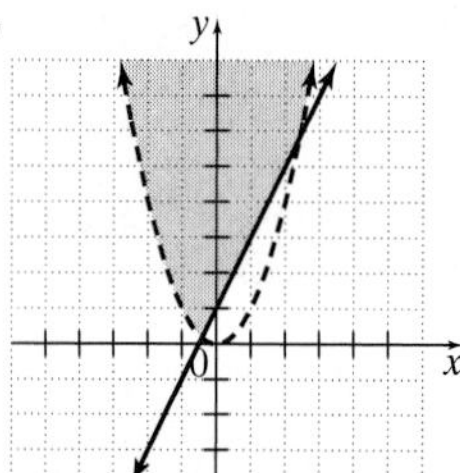

27.

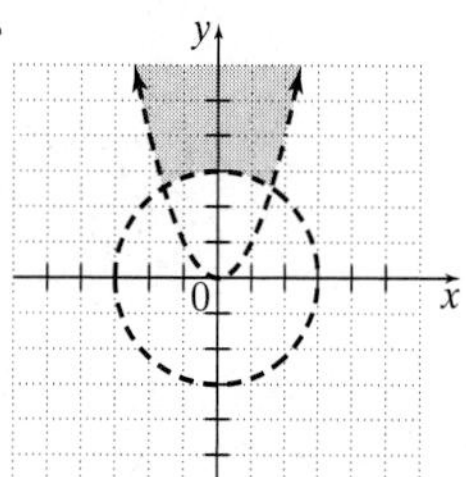

29.

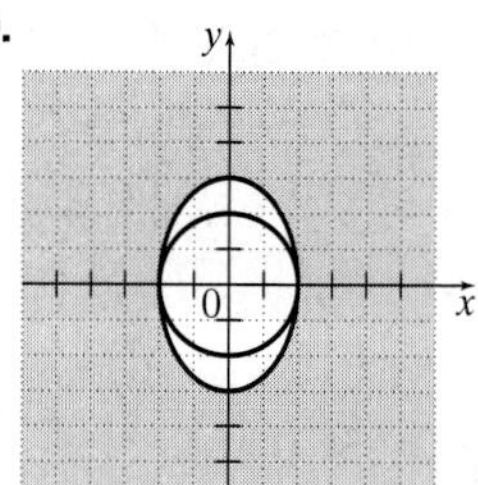

31.

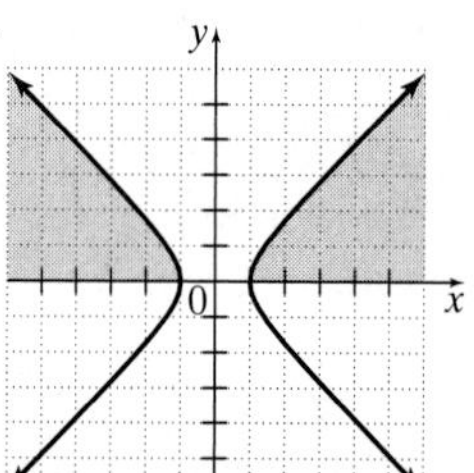

33.

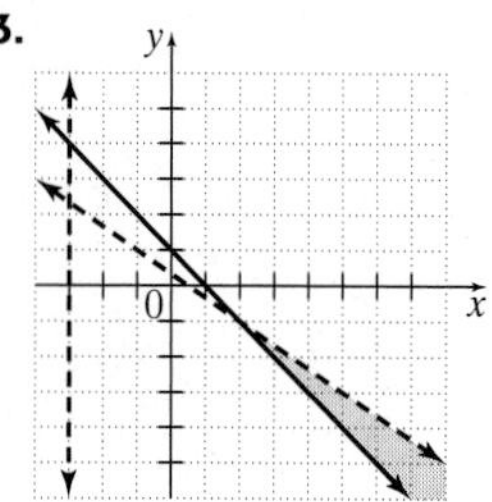

35.

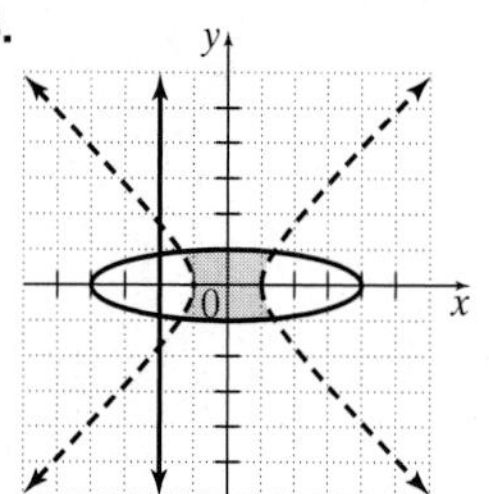

37. not a function **39.** function **41.** answers may vary

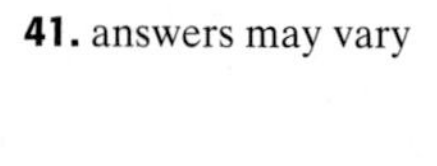

43.

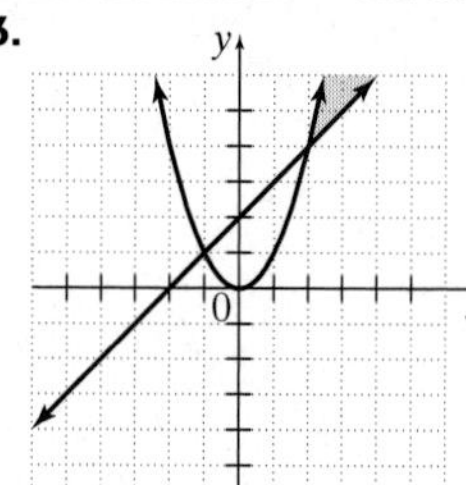

Chapter 12 Review

1. $\sqrt{197}$ units ≈ 14.036 **2.** $\sqrt{41}$ units ≈ 6.403 **3.** $\sqrt{130}$ units ≈ 11.402 **4.** $\sqrt{73}$ units ≈ 8.544 **5.** $7\sqrt{2}$ units ≈ 9.899 **6.** $2\sqrt{11}$ units ≈ 6.633 **7.** $\sqrt{275.6}$ units ≈ 16.601 **8.** $\sqrt{191.56}$ units ≈ 13.841 **9.** $(-5, 5)$ **10.** $(4, 16)$ **11.** $\left(-\frac{15}{2}, 1\right)$ **12.** $\left(-\frac{11}{2}, -2\right)$ **13.** $\left(\frac{1}{20}, -\frac{3}{16}\right)$ **14.** $\left(\frac{1}{4}, -\frac{2}{7}\right)$ **15.** $(\sqrt{3}, -3\sqrt{6})$ **16.** $(-4\sqrt{3}, 6\sqrt{7})$ **17.** $(x + 4)^2 + (y - 4)^2 = 9$ **18.** $(x - 5)^2 + y^2 = 25$ **19.** $(x + 7)^2 + (y + 9)^2 = 11$ **20.** $x^2 + y^2 = \frac{49}{4}$

21.
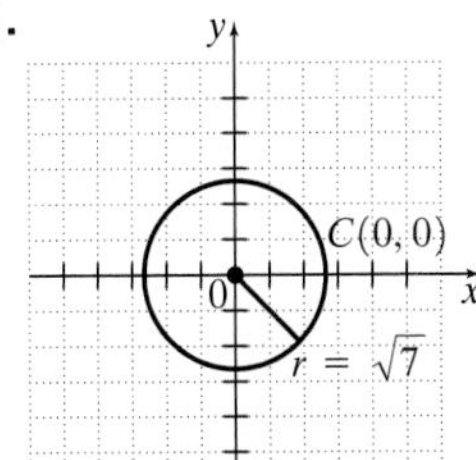

22.
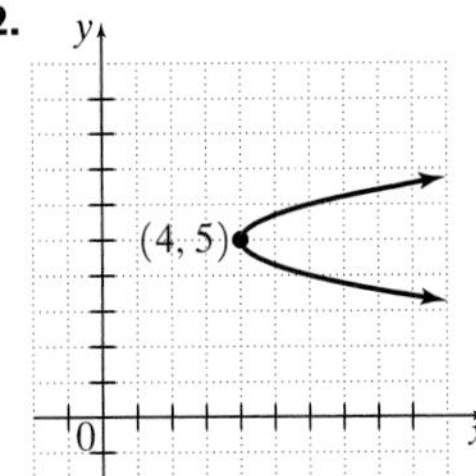

23.
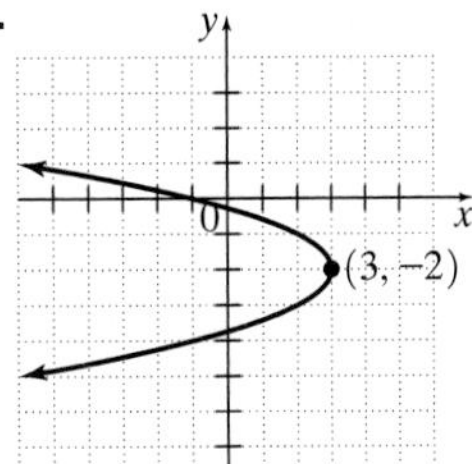

24.
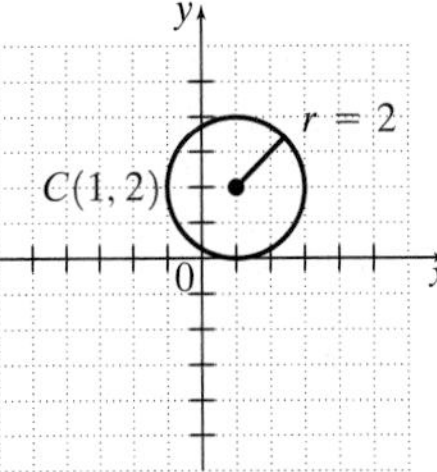

25.
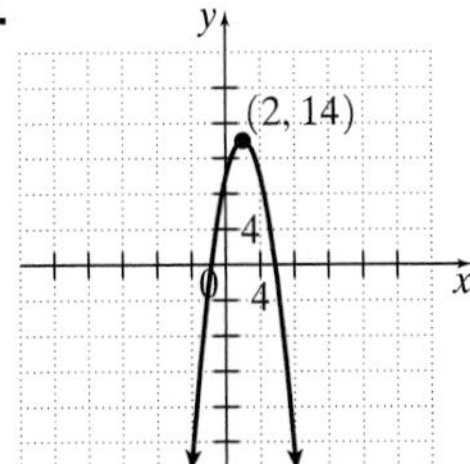

26.
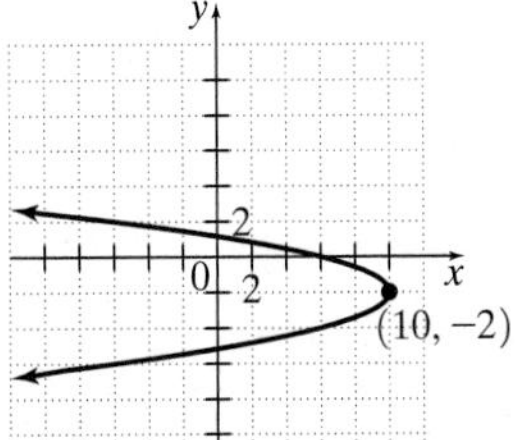

27.
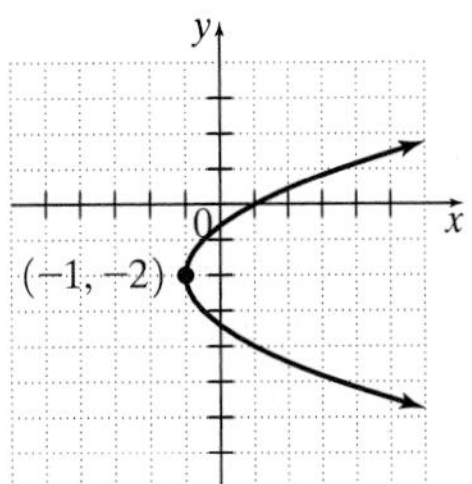

28.
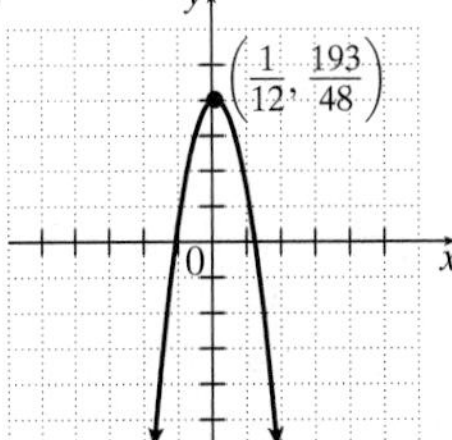

29.
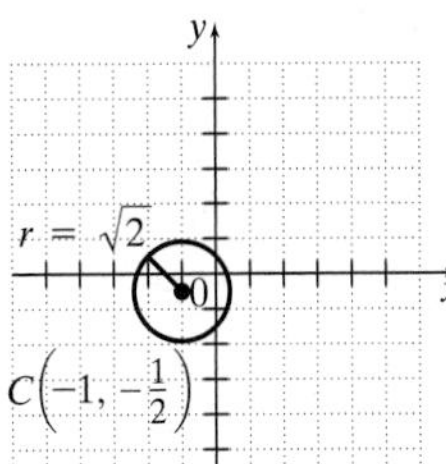

30.
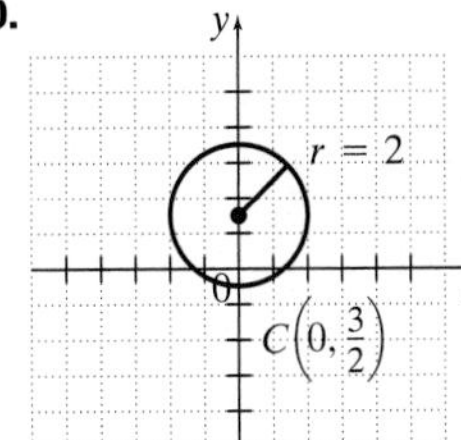

31.
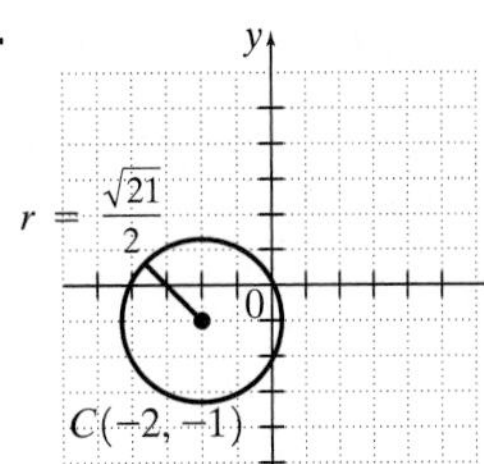

32.
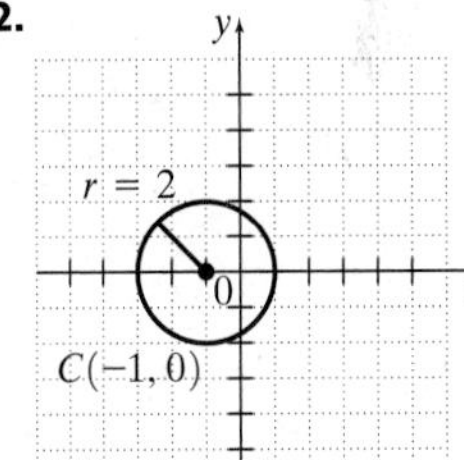

33.
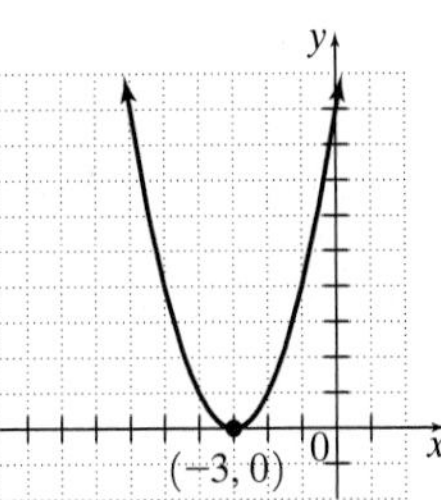

34.
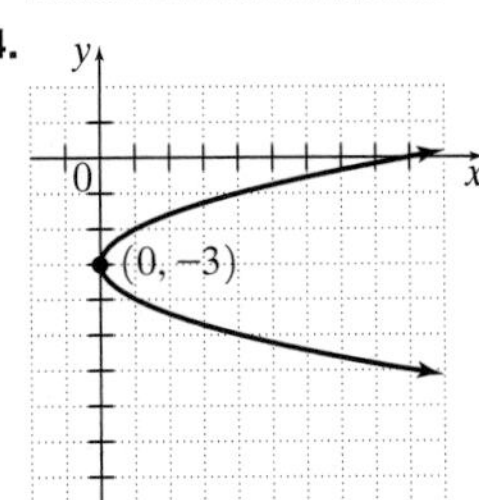

35.
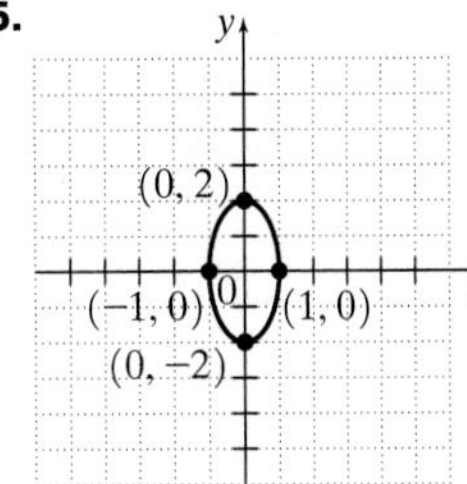

36.
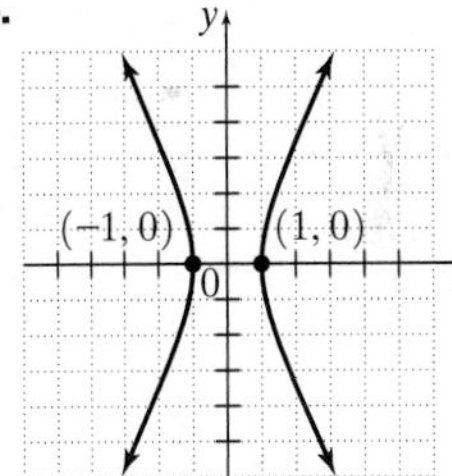

37.
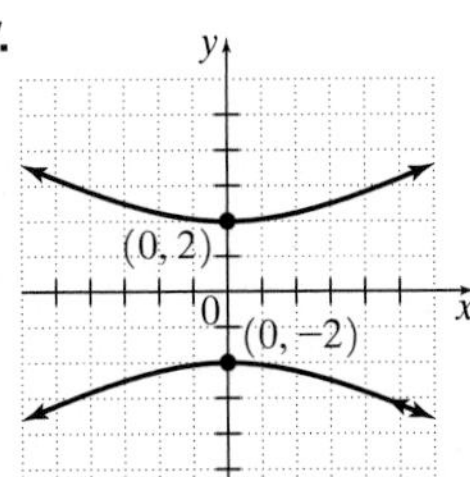

38.
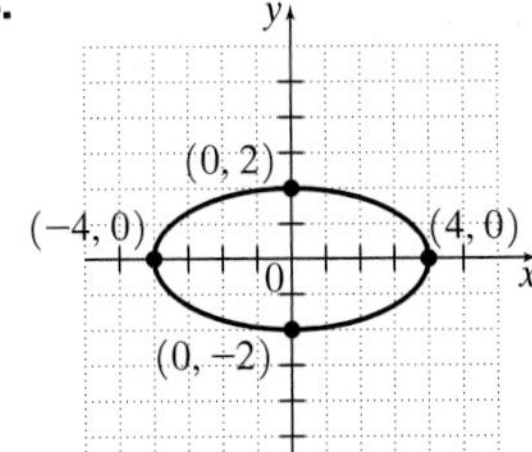

39.
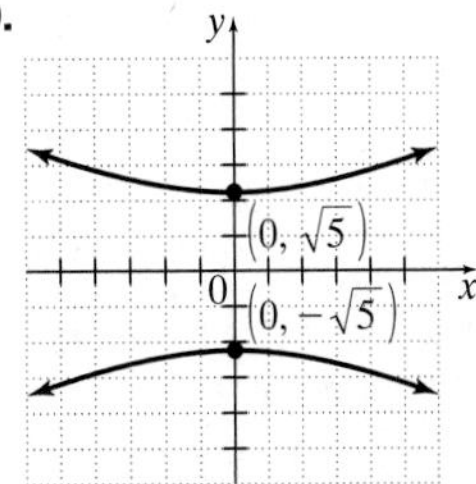

40.
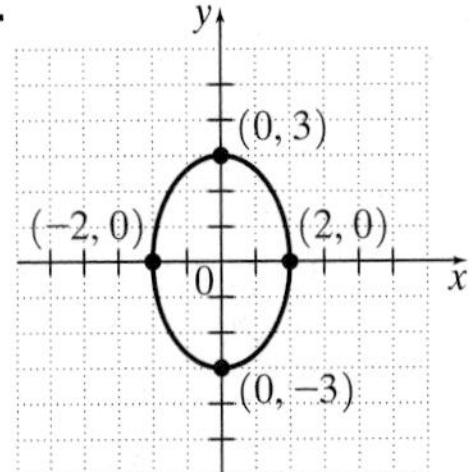

41.
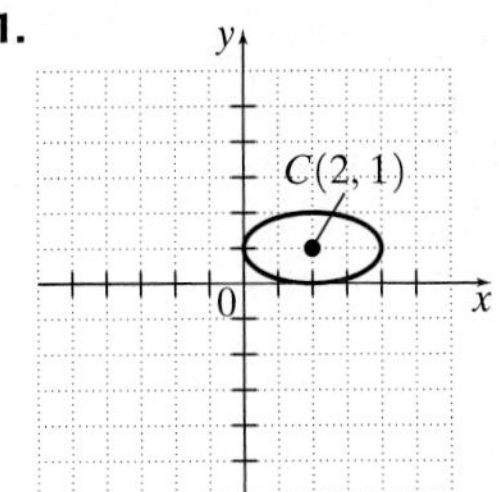

42.
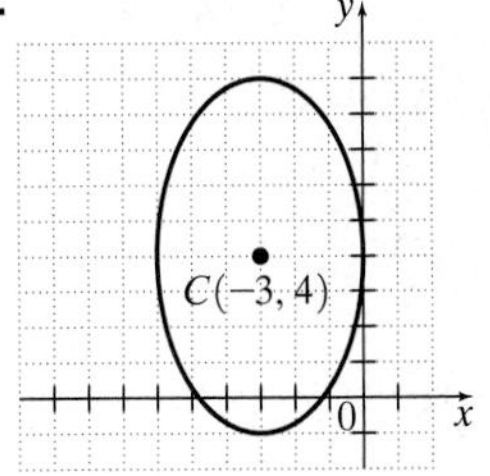

43.
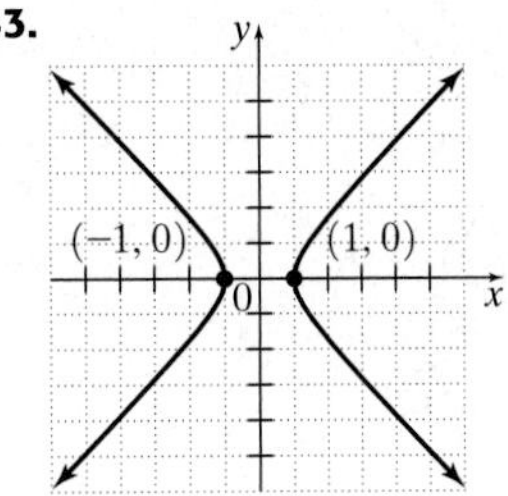

44.
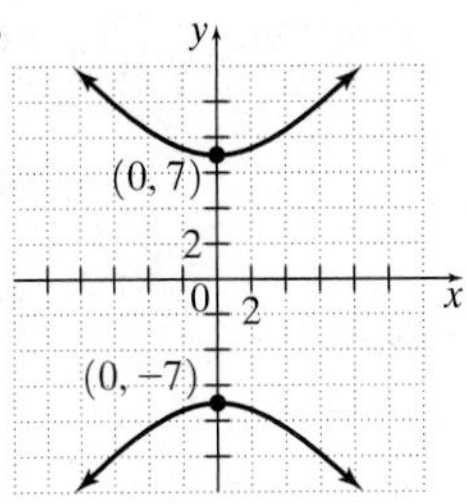

45.
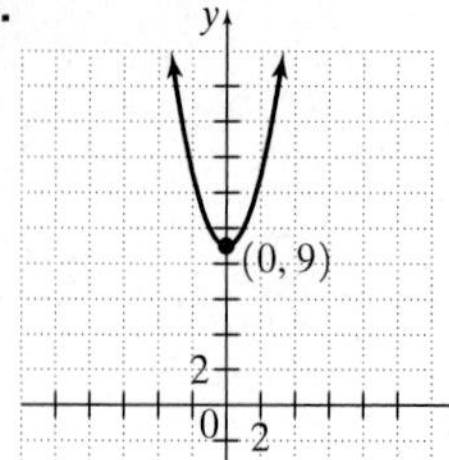

46.
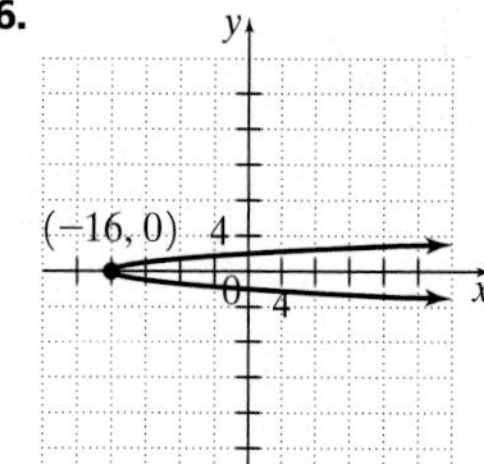

47.
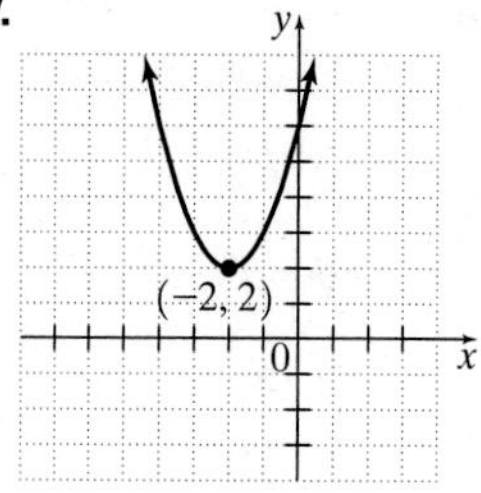

48.
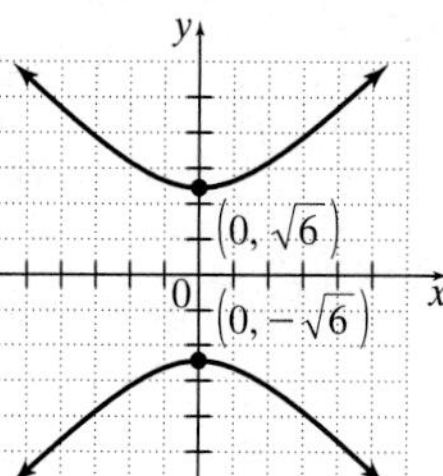

49.
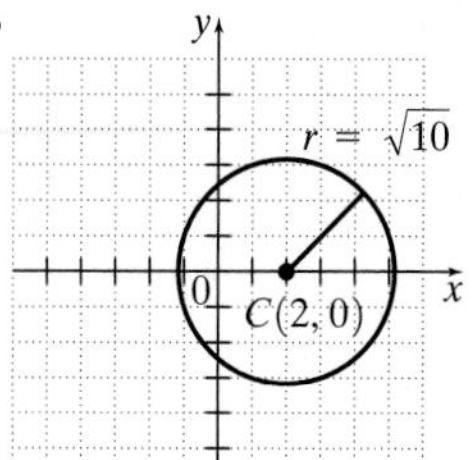

50.
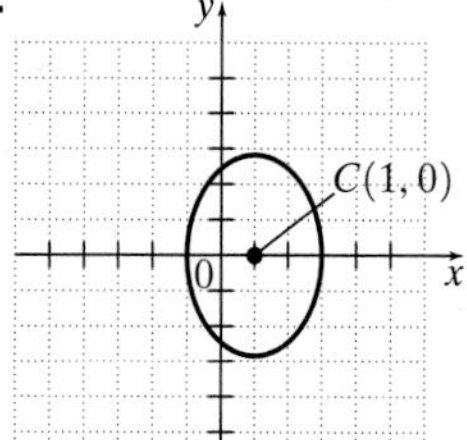

51.
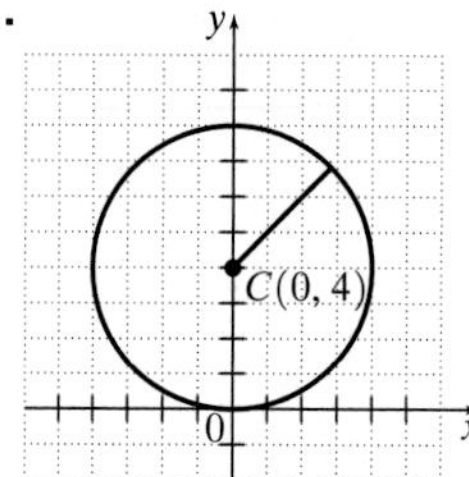

52.
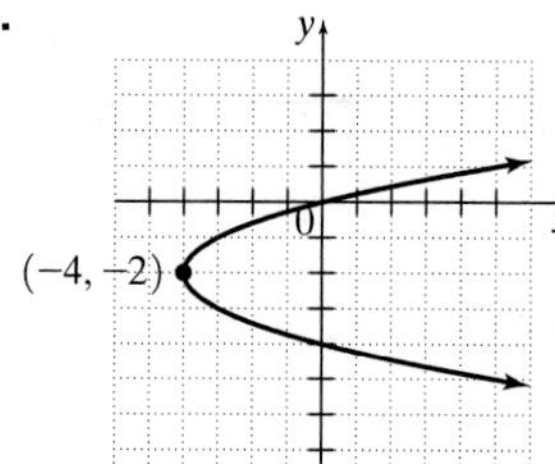

53.
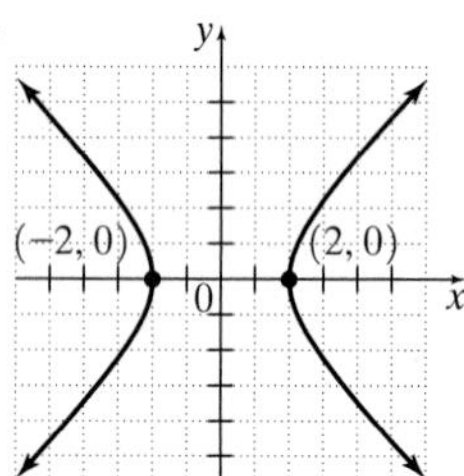

54.
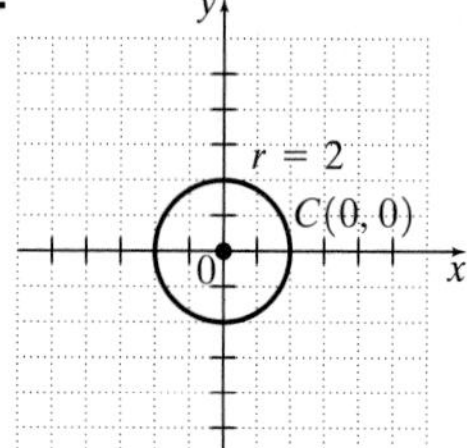

55.
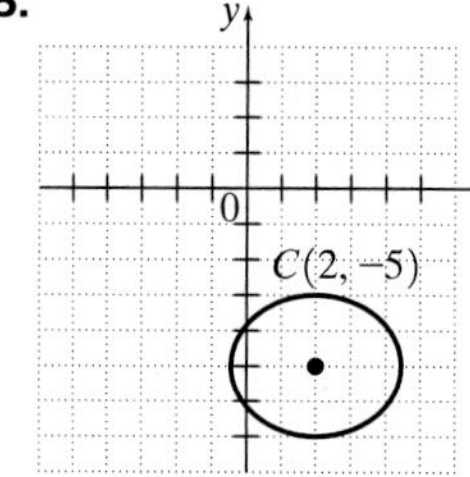

56.
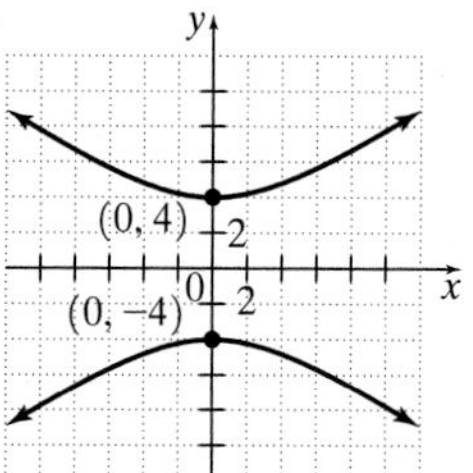

57.
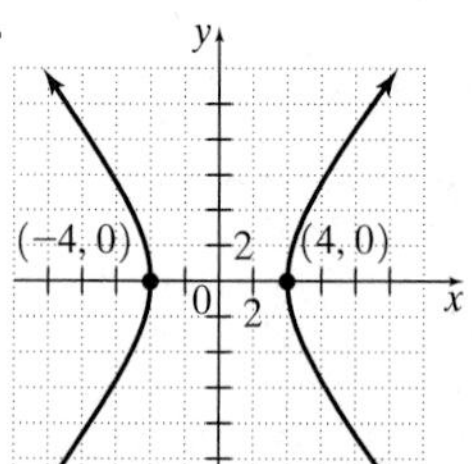

58.
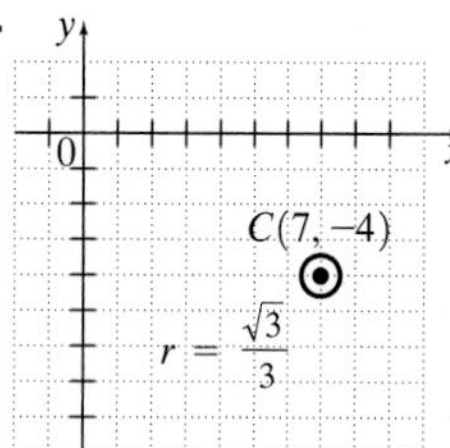

59.
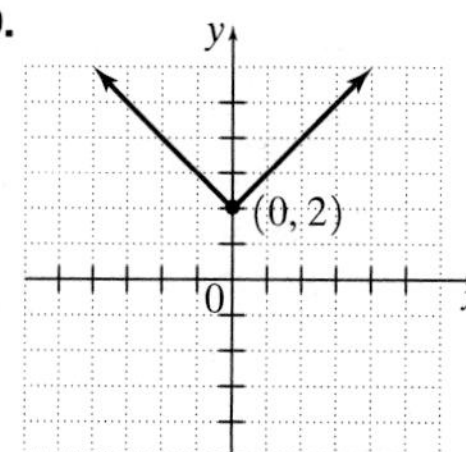

60.
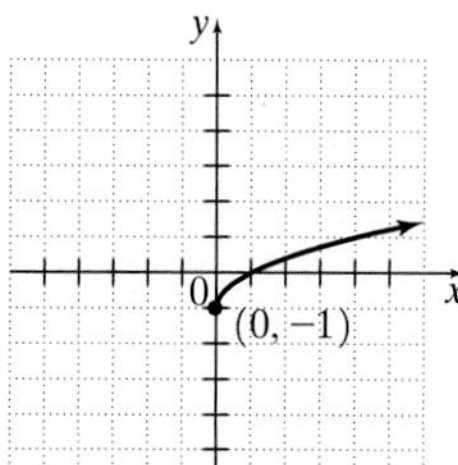

61.
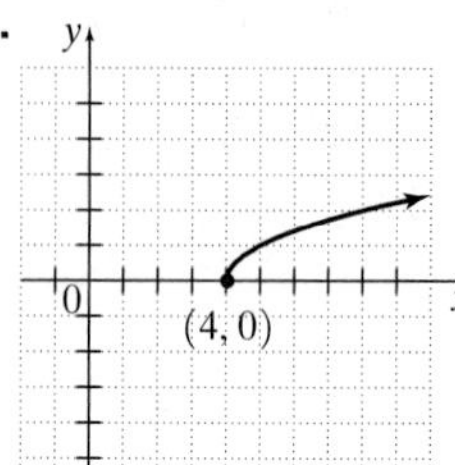

62.
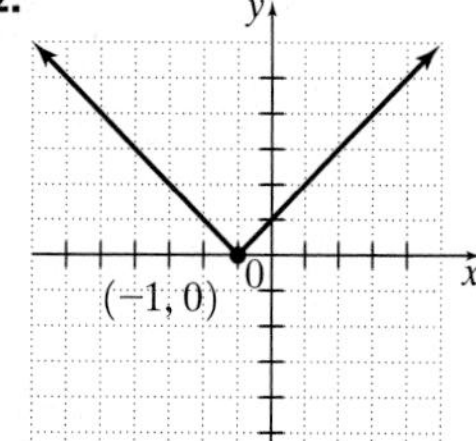

63.
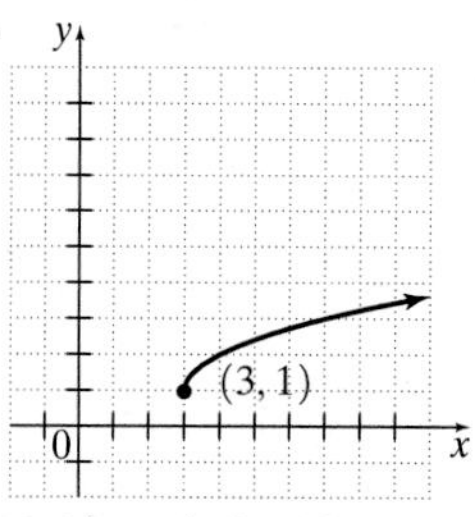

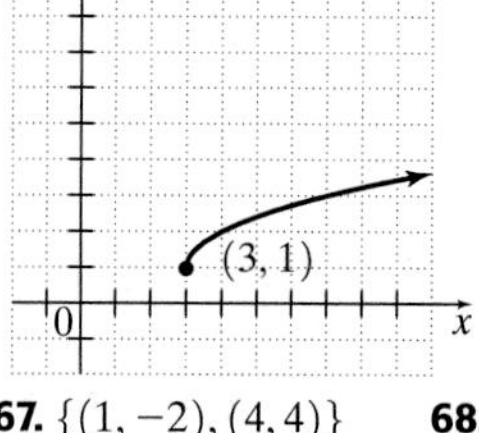

64.
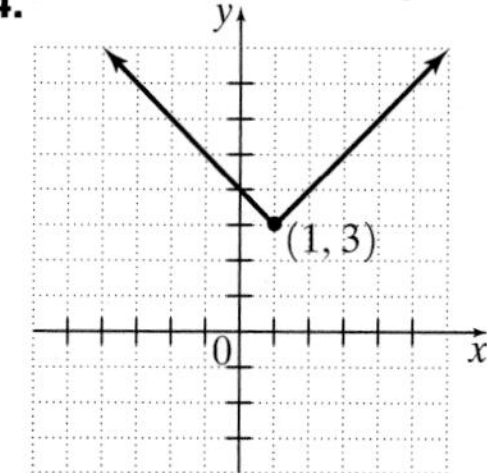

65.
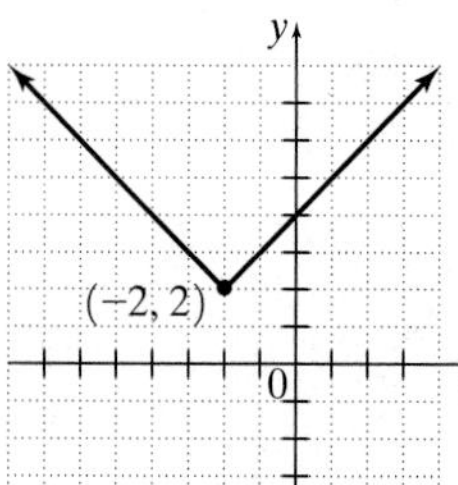

66.
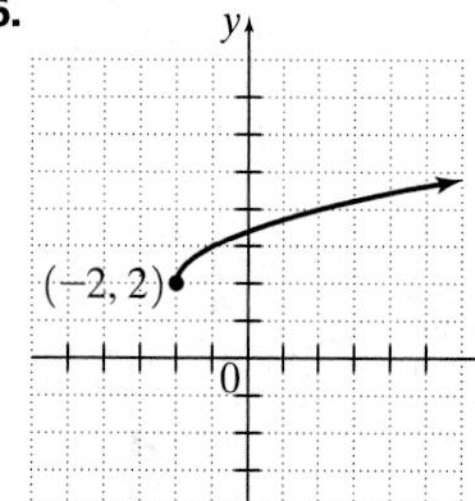

67. $\{(1, -2), (4, 4)\}$ **68.** $\varnothing$ **69.** $\{(-1, 1), (2, 4)\}$

70. $\{(5, 1), (-1, 7)\}$ **71.** $\{(2, 2\sqrt{2}), (2, -2\sqrt{2})\}$ **72.** $\{(0, 2), (0, -2)\}$

73. $\{(-1, 3), (-1, -3), (1, 3), (1, -3)\}$ **74.** $\left\{\left(2, \frac{5}{2}\right), (-7, -20)\right\}$

75. $\{(1, 4)\}$ **76.** $\{(-2, -1), (-2, 1), (2, -1), (2, 1)\}$ **77.** length: 15 ft; width: 10 ft **78.** 4

79.
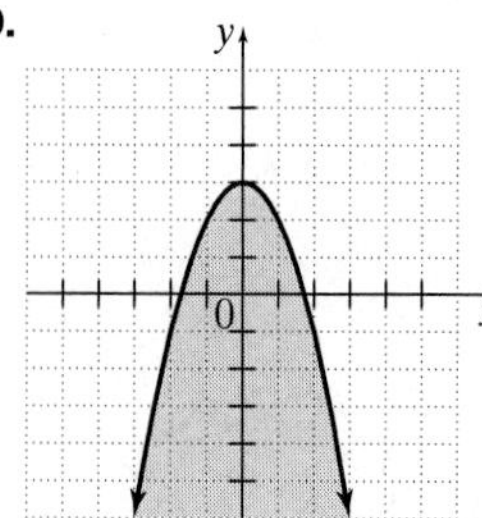

80.
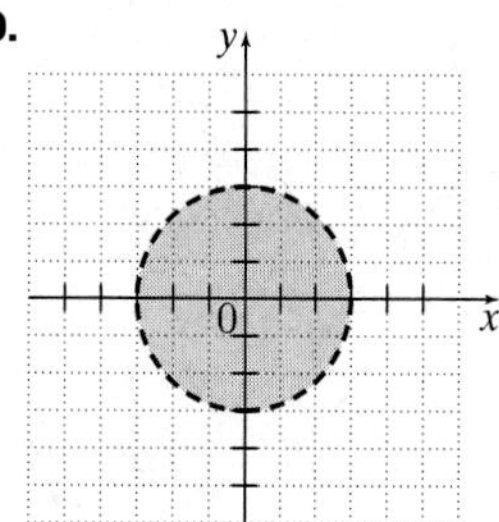

81.
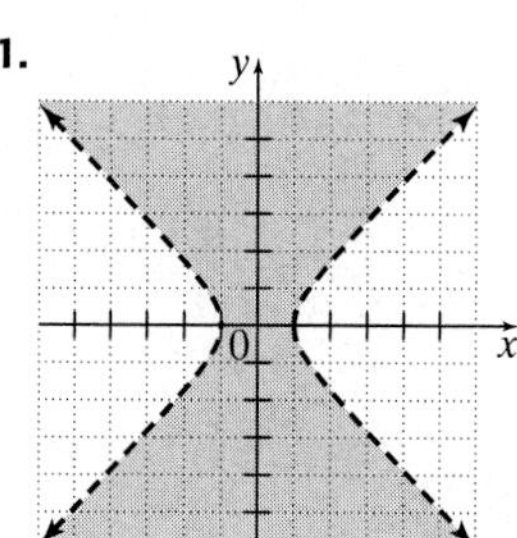

82.
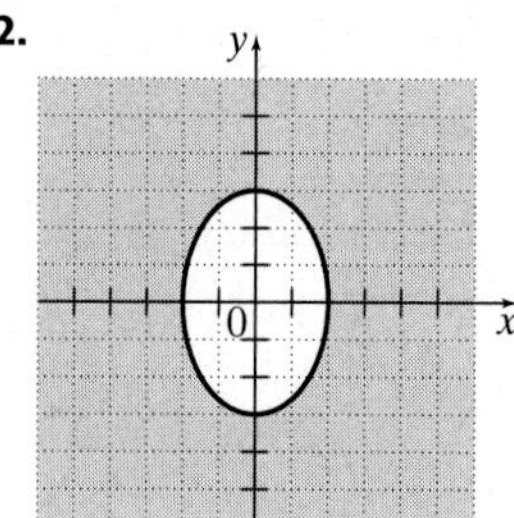

83.

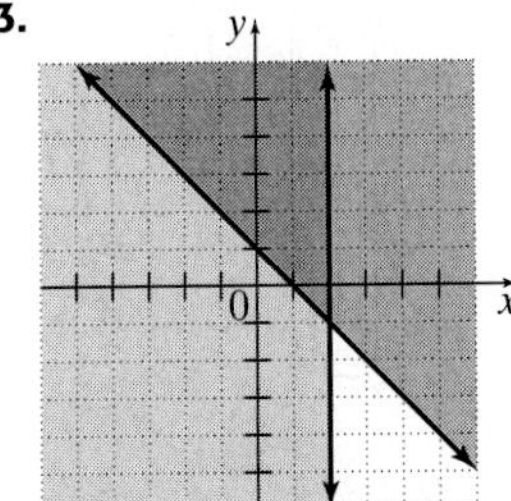

84.

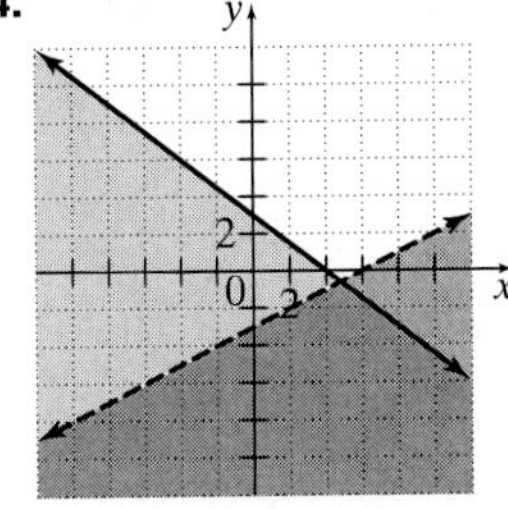

85.

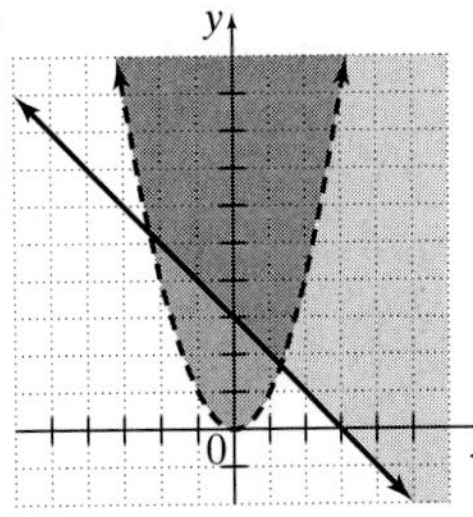

86.

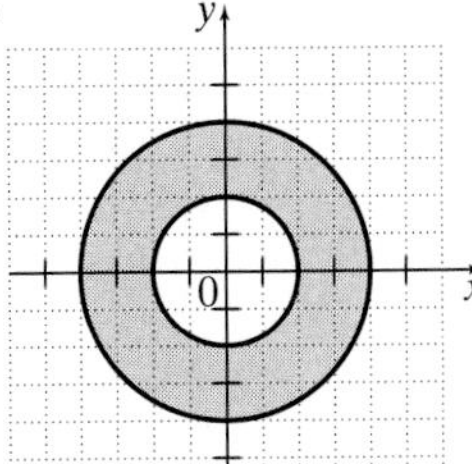

87.

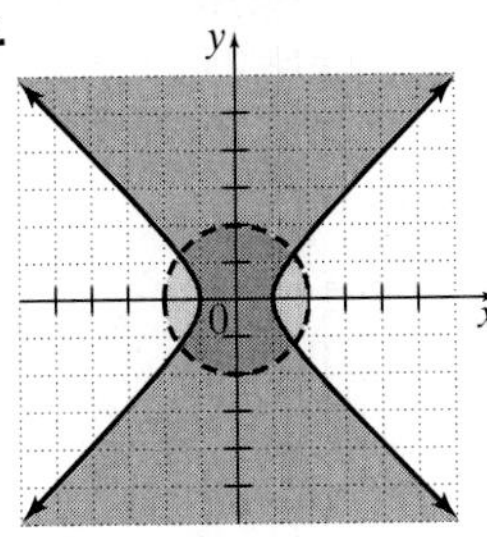

88.

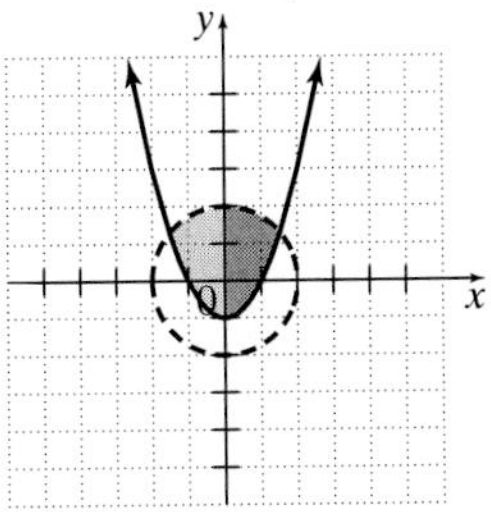

Chapter 12 Test

1. $2\sqrt{26}$ units **2.** $\sqrt{95}$ units **3.** $\left(-4, \frac{7}{2}\right)$ **4.** $\left(-\frac{1}{2}, \frac{3}{10}\right)$

5.

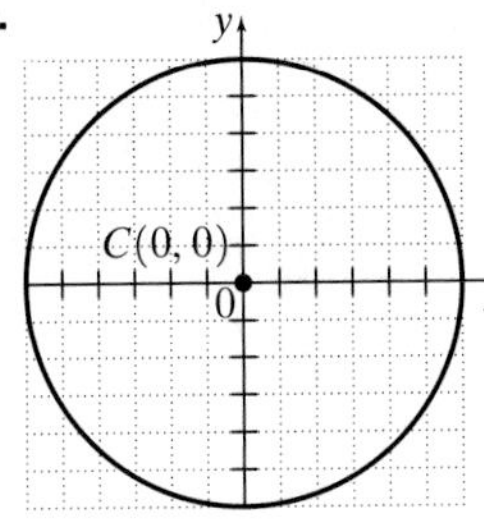

6.

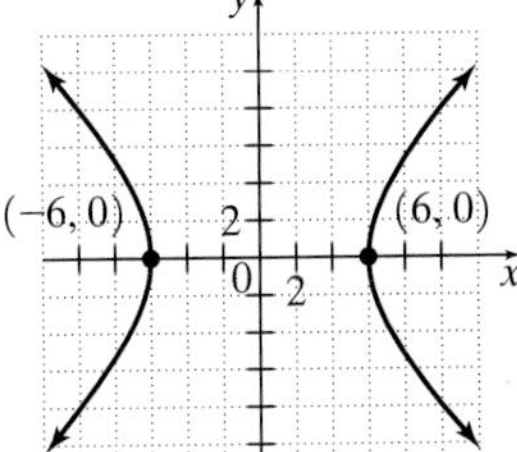

7.

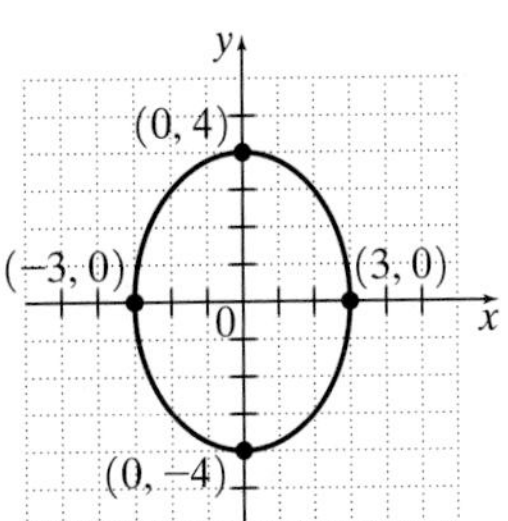

8.

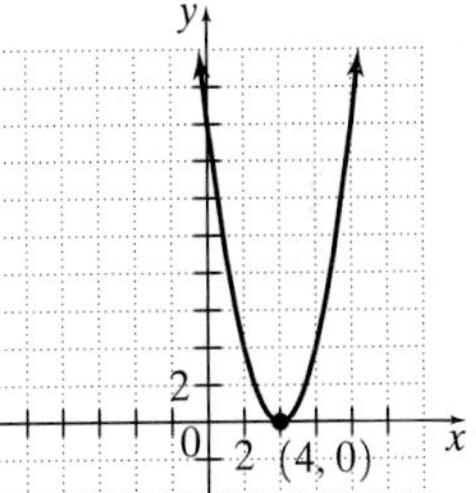

9.

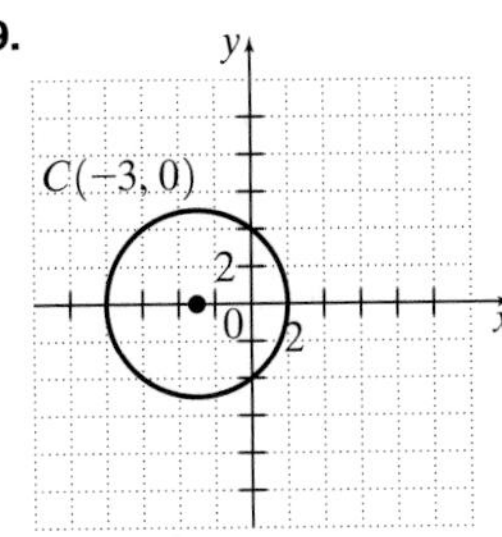

10.

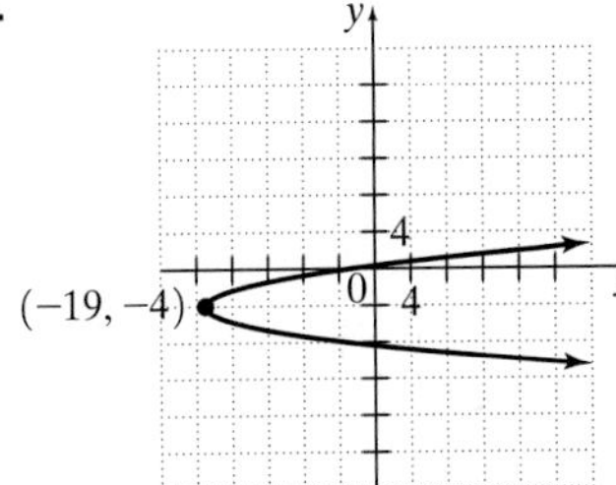

11.

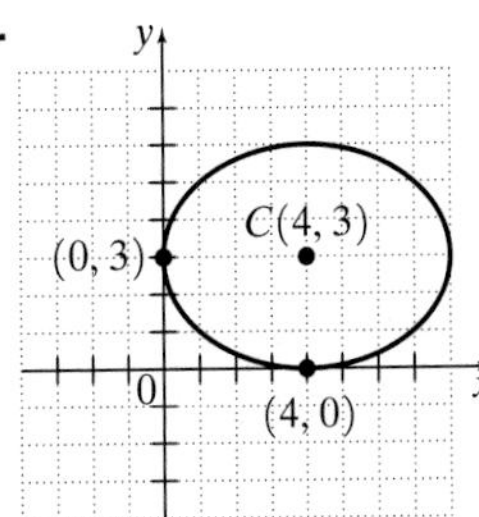

12.

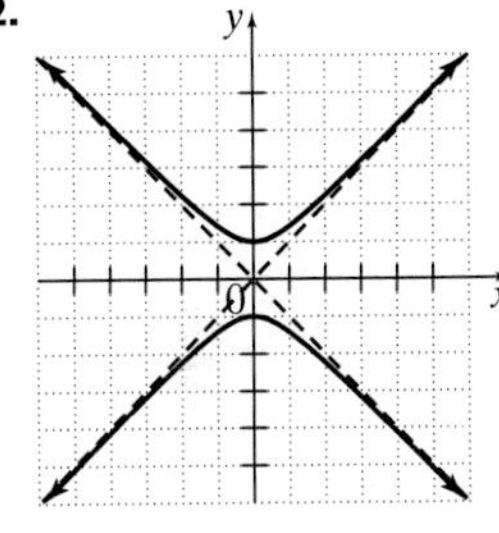

13. $\{(-12, 5), (12, -5)\}$ **14.** $\{(-5, -1), (-5, 1), (5, -1), (5, 1)\}$ **15.** $\{(6, 12), (1, 2)\}$ **16.** $\{(1, 1), (-1, -1)\}$

17.

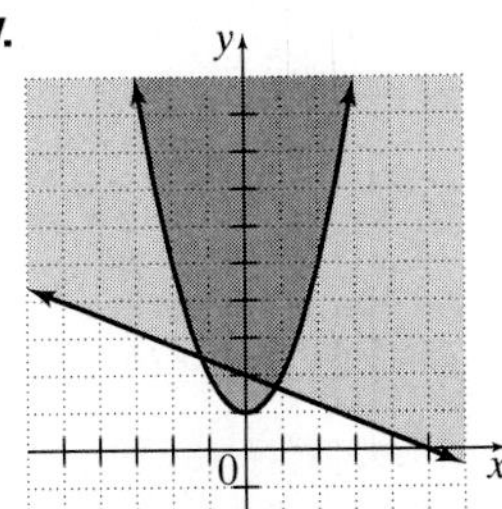

18.

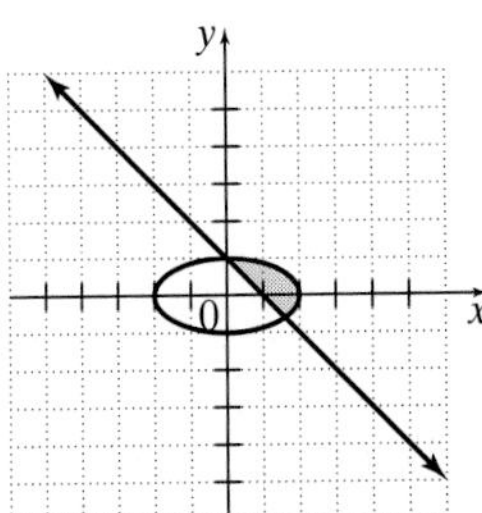

19.

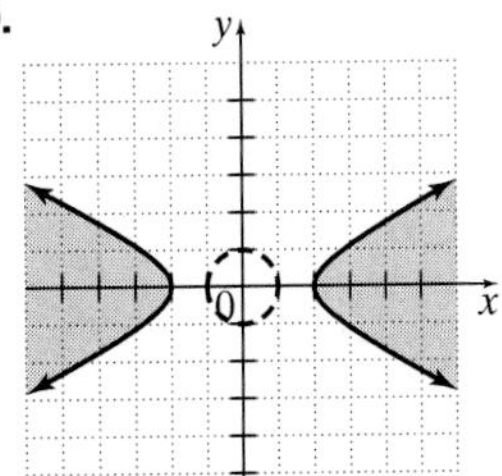

20.

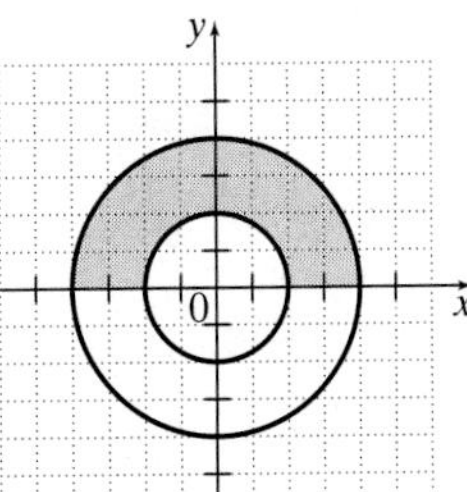

21. B 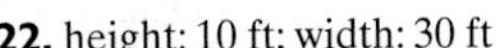
22. height: 10 ft; width: 30 ft

23.

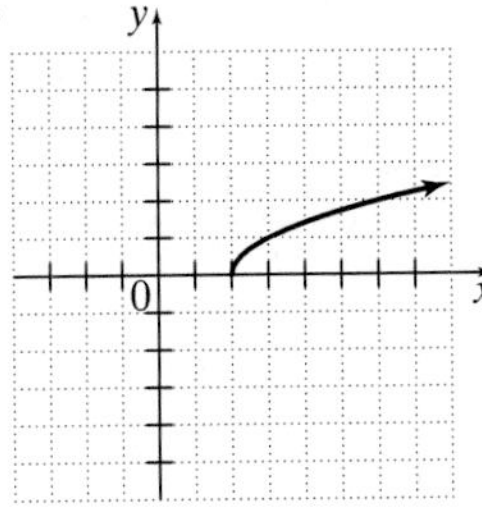

24.

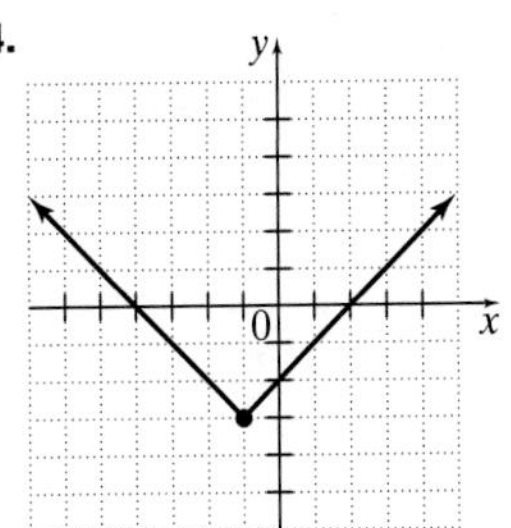

Cumulative Review

1. $[4, \infty)$; Sec. 2.8, Ex. 9 **2. a.** x^{11} **b.** $\frac{1}{16}$ **c.** $81y^{10}$; Sec. 4.1, Ex. 29 **3.** $7x^3 + 14x^2 + 35x$; Sec. 4.5, Ex. 4 **4.** $3x^3 - 4 + \frac{1}{x}$; Sec. 4.7, Ex. 2 **5.** $(x + 3)(5 + y)$; Sec. 5.1, Ex. 7 **6.** $(x^2 + 2)(x^2 + 3)$; Sec. 5.2, Ex. 7 **7.** $\frac{2}{5}$; Sec. 6.2, Ex. 2 **8.** $t = 5$; Sec. 6.5, Ex. 2 **9.** $\frac{3}{z}$; Sec. 6.7, Ex. 3

10. ; Sec. 7.1, Ex. 1 **11.** $y = \frac{5}{3}x + \frac{13}{3}$; Sec. 7.2, Ex. 6 **12.** domain: $\{2, 0, 3\}$; range: $\{3, 4, -1\}$; Sec. 7.3, Ex. 1 **13.** 1; Sec. 7.3, Ex. 18 **14.** 35; Sec. 7.3, Ex. 21 **15.** $(-4, 2, -1)$; Sec. 8.5, Ex. 1 **16.** $(-1, 2)$; Sec. 8.6, Ex. 1

17.

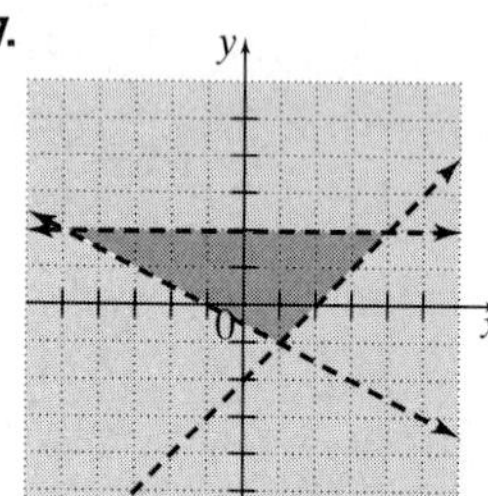

; Sec. 8.7, Ex. 2 **18.** -4; Sec. 9.1, Ex. 14 **19.** $\frac{2}{5}$; Sec. 9.1, Ex. 15 **20.** 2; Sec. 9.3, Ex. 3 **21.** $\sqrt{\frac{2b}{3a}}$; Sec. 9.3, Ex. 5 **22.** $-6\sqrt[3]{2}$; Sec. 9.4, Ex. 5 **23.** $\frac{5\sqrt[3]{7x}}{2}$; Sec. 9.4, Ex. 10 **24.** $\left\{\frac{-1 + i\sqrt{35}}{6}, \frac{-1 - i\sqrt{35}}{6}\right\}$; Sec. 10.2, Ex. 4 **25.** $\left\{\frac{2}{99}\right\}$; Sec. 11.7, Ex. 4 **26.** $\left(-1, \frac{3}{2}\right)$; Sec. 12.1, Ex. 6

27.

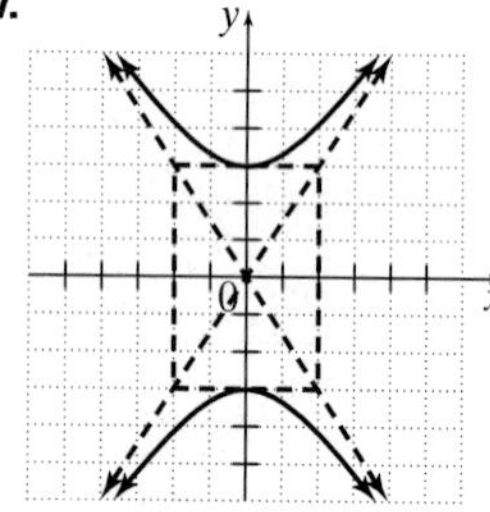

; Sec. 12.2, Ex. 5 **28.** $\{(2, \sqrt{2})\}$; Sec. 12.4, Ex. 2

Appendix A Transition Review: Polynomials and Factoring Strategies

Exercise Set Appendix A

1. $-16x^6y^3p^2$ **3.** 9 **5.** a^3b **7.** $\frac{1}{16}$ **9.** $\frac{13}{36}$ **11.** y^4 **13.** $\frac{6x^{16}}{5}$ **15.** $16x^{20}y^{12}$ **17.** $4a^8b^4$ **19.** $\frac{2}{x^4y^{10}}$ **21.** $2y^4 - 5y^2 + x^2 + 1$ **23.** $y^2 + 3$ **25.** $-x^3 + 8a - 12$ **27.** $5x^2 - 9x - 3$ **29.** $-2x^2 - 4x + 15$ **31.** $7y^2 - 3$ **33.** $-20y^2 + 3yx$ **35.** $9x^2 + 18x + 5$ **37.** $16x^2 - \frac{2}{3}x - \frac{1}{6}$ **39.** $x^2 + 8x + 16$ **41.** $9x^2 - 6xy + y^2$ **43.** $9b^2 - 36y^2$ **45.** $9x^2 - \frac{1}{4}$ **47.** $2xy(7x - 1)$ **49.** $4(x + 2)(x - 2)$ **51.** $(3x - 11)(x + 1)$ **53.** $(2x + 5y)(4x^2 - 10xy + 25y^2)$ **55.** $7x(x - 9)$ **57.** $(4x + 3)(5x + 2)$ **59.** $(a + 7)(b - 6)$ **61.** $(x^2 + 1)(x - 1)(x + 1)$ **63.** $2(x - 3)(x^2 + 3x + 9)$

Appendix B TRANSITION REVIEW: SOLVING LINEAR AND QUADRATIC EQUATIONS

MENTAL MATH

1. 6 **3.** 17 **5.** 8 **7.** 10

EXERCISE SET APPENDIX B

1. -0.9 **3.** 6 **5.** -1.1 **7.** -5 **9.** $-3, \frac{4}{3}$ **11.** 0 **13.** $-\frac{3}{4}, \frac{5}{2}$ **15.** $-3, -8$ **17.** 2 **19.** $\frac{1}{4}, -\frac{2}{3}$ **21.** 1, 9 **23.** -9 **25.** $\frac{1}{6}$ **27.** 1 **29.** $\frac{3}{5}, -1$ **31.** 0 **33.** $6, -3$ **35.** 8 **37.** $\{x|x \text{ is a real number}\}$ **39.** $\frac{2}{5}, -\frac{1}{2}$ **41.** $\frac{3}{4}, -\frac{1}{2}$ **43.** -8 **45.** -2 **47.** 0, 5 **49.** $-\frac{1}{2}, \frac{1}{3}$ **51.** $-4, 9$ **53.** $\emptyset$ **55.** 1

Appendix C AN INTRODUCTION TO USING A GRAPHING UTILITY

EXERCISE SET APPENDIX C

1. yes **3.** no **5.** answers may vary **7.** answers may vary **9.** answers may vary

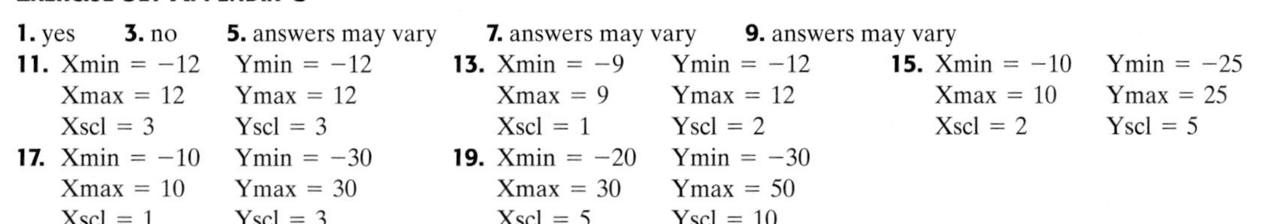

11. Xmin = -12, Xmax = 12, Xscl = 3; Ymin = -12, Ymax = 12, Yscl = 3

13. Xmin = -9, Xmax = 9, Xscl = 1; Ymin = -12, Ymax = 12, Yscl = 2

15. Xmin = -10, Xmax = 10, Xscl = 2; Ymin = -25, Ymax = 25, Yscl = 5

17. Xmin = -10, Xmax = 10, Xscl = 1; Ymin = -30, Ymax = 30, Yscl = 3

19. Xmin = -20, Xmax = 30, Xscl = 5; Ymin = -30, Ymax = 50, Yscl = 10

21. Setting B **23.** Setting B **25.** Setting B

27.
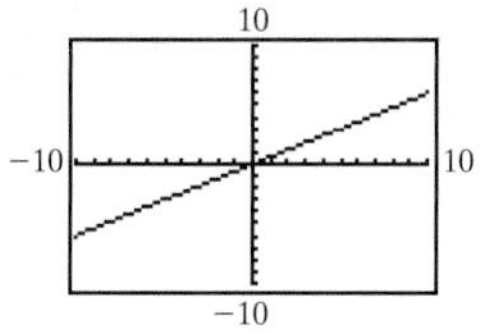

29.
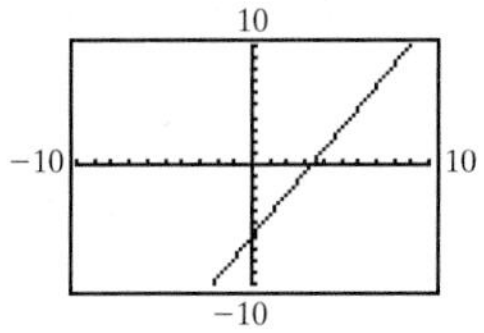

31.
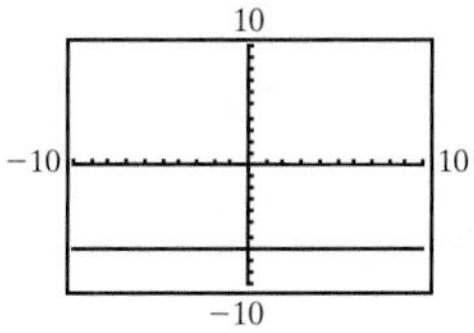

33.
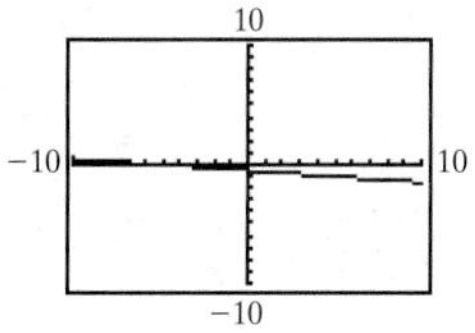

35.
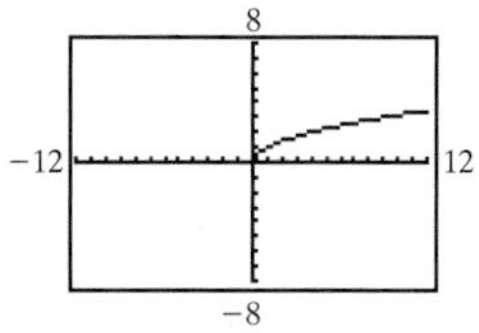

37.
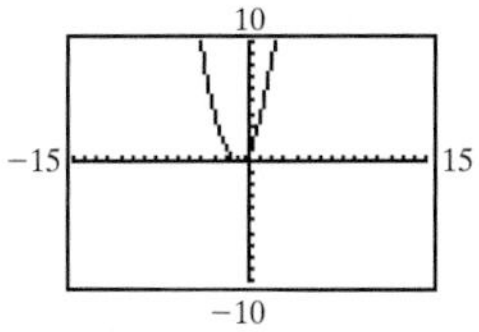

39.
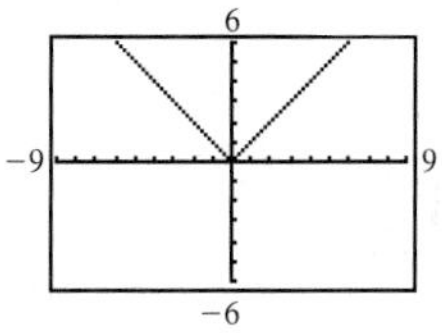

41.
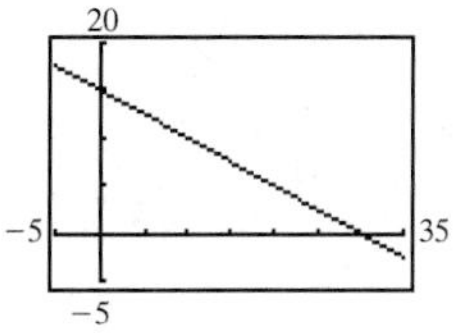

Appendix D SETS AND COMPOUND INEQUALITIES

EXERCISE SET APPENDIX D

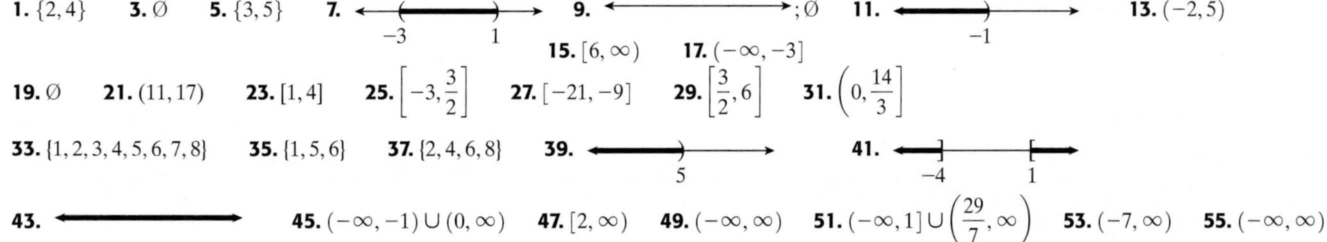

1. $\{2, 4\}$ **3.** $\emptyset$ **5.** $\{3, 5\}$ **7.**
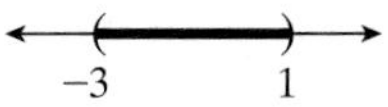

9.

; $\emptyset$ **11.**
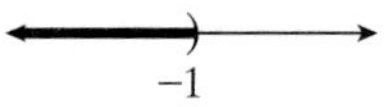

13. $(-2, 5)$ **15.** $[6, \infty)$ **17.** $(-\infty, -3]$ **19.** $\emptyset$ **21.** $(11, 17)$ **23.** $[1, 4]$ **25.** $\left[-3, \frac{3}{2}\right]$ **27.** $[-21, -9]$ **29.** $\left[\frac{3}{2}, 6\right]$ **31.** $\left(0, \frac{14}{3}\right]$ **33.** $\{1, 2, 3, 4, 5, 6, 7, 8\}$ **35.** $\{1, 5, 6\}$ **37.** $\{2, 4, 6, 8\}$ **39.**
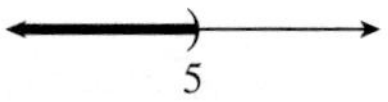

41.
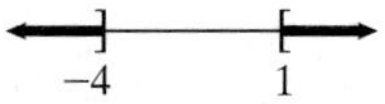

43.

45. $(-\infty, -1) \cup (0, \infty)$ **47.** $[2, \infty)$ **49.** $(-\infty, \infty)$ **51.** $(-\infty, 1] \cup \left(\frac{29}{7}, \infty\right)$ **53.** $(-7, \infty)$ **55.** $(-\infty, \infty)$

Appendix E ABSOLUTE VALUE EQUATIONS AND INEQUALITIES

EXERCISE SET APPENDIX E

1. $\{7,-7\}$ **3.** $\varnothing$ **5.** $\{4.2,-4.2\}$ **7.** $\{-4,4\}$ **9.** $\{-9,9\}$ **11.** $\{-5,23\}$ **13.** $\{7,-2\}$ **15.** $\{8,4\}$ **17.** $\{5,-5\}$ **19.** $\{3,-3\}$ **21.** $\{-3,6\}$ **23.** $\{0\}$ **25.** $\varnothing$ **27.** $\left\{-\frac{1}{3},\frac{7}{3}\right\}$ **29.** $\left\{-\frac{1}{2},9\right\}$ **31.** $\left\{-\frac{5}{2}\right\}$ **33.** $\{3,2\}$ **35.** $\{-4,16\}$ **37.** $\{4\}$ **39.** $\left\{\frac{3}{2}\right\}$ **41.** $\left\{\frac{32}{21},\frac{38}{9}\right\}$ **43.** $\left\{-8,\frac{2}{3}\right\}$ **45.** -4 4 ; $[-4,4]$ **47.** -3 3 ; $(-\infty,-3)\cup(3,\infty)$

49. -5 -1 ; $(-5,-1)$ **51.** -1 13 ; $(-\infty,-1]\cup[13,\infty)$ **53.** -5 1 ; $(-5,1)$

55. -5 5 ; $[-5,5]$ **57.** -4 4 ; $(-\infty,-4)\cup(4,\infty)$ **59.** -10 3 ; $[-10,3]$

61. -24 4 ; $(-\infty,-24]\cup[4,\infty)$ **63.** -2 9 ; $[-2,9]$ **65.** ; $(-\infty,\infty)$

67. $-\frac{1}{2}$ 1 ; $\left[-\frac{1}{2},1\right]$ **69.** $\frac{2}{3}$ 2 ; $\left(-\infty,\frac{2}{3}\right)\cup(2,\infty)$ **71.** ; $\varnothing$

73. -12 0 ; $(-\infty,-12)\cup(0,\infty)$ **75.** $\{-13,13\}$ **77.** $(-\infty,-13)\cup(13,\infty)$ **79.** $\varnothing$ **81.** $[-10,10]$ **83.** $(-2,5)$

85. $\{5,-2\}$ **87.** $(-\infty,-7]\cup[17,\infty)$ **89.** $\left\{-\frac{9}{4}\right\}$ **91.** $(-2,1)$ **93.** $(-\infty,-18)\cup(12,\infty)$ **95.** $\left\{2,\frac{4}{3}\right\}$ **97.** $\varnothing$ **99.** $\left\{-\frac{17}{2},\frac{19}{2}\right\}$

101. $\left(-\infty,-\frac{25}{3}\right)\cup\left(\frac{35}{3},\infty\right)$ **103.** $\left\{4,-\frac{1}{5}\right\}$

Appendix F DETERMINANTS AND CRAMER'S RULE

MENTAL MATH

1. 56 **3.** -32 **5.** 20

EXERCISE SET APPENDIX F

1. 26 **3.** -19 **5.** 0 **7.** $\frac{13}{6}$ **9.** $(1,2)$ **11.** $\{(x,y)\mid 3x+y=1\}$ **13.** $(9,9)$ **15.** $(-3,-2)$ **17.** $(3,4)$ **19.** 8 **21.** 0 **23.** 15 **25.** 54 **27.** $(-2,0,5)$ **29.** $(6,-2,4)$ **31.** $(-2,3,-1)$ **33.** $(0,2,-1)$ **35.** 5 **37.** 0

Appendix H REVIEW OF ANGLES, LINES, AND SPECIAL TRIANGLES

EXERCISE SET APPENDIX H

1. 71° **3.** 19.2° **5.** $78\frac{3}{4}^\circ$ **7.** 30° **9.** 149.8° **11.** $100\frac{1}{2}^\circ$ **13.** $m\angle 1=m\angle 5=m\angle 7=110^\circ, m\angle 2=m\angle 3=m\angle 4=m\angle 6=70^\circ$ **15.** 90° **17.** 90° **19.** 90° **21.** $45^\circ, 90^\circ$ **23.** $73^\circ, 90^\circ$ **25.** $50\frac{1}{4}^\circ, 90^\circ$ **27.** $x=6$ **29.** $x=4.5$ **31.** 10 **33.** 12

SOLUTIONS TO SELECTED EXERCISES

Chapter R

Exercise Set R.1

1. $9 = 1 \cdot 9, 9 = 3 \cdot 3$
The factors of 9 are 1, 3, and 9.

5. $42 = 1 \cdot 42, 42 = 21 \cdot 2, 42 = 3 \cdot 14, 42 = 6 \cdot 7$
The factors of 42 are 1, 2, 3, 6, 7, 14, 21, and 42.

9. 19 is a prime number. Its factors are 1 and 19 only.

13. $39 = 3 \cdot 13$ so 39 is a composite number

17. $51 = 3 \cdot 17$ so 51 is a composite number

21. $18 = 2 \cdot 3 \cdot 3$

25. $56 = 2 \cdot 2 \cdot 2 \cdot 7$

29. $81 = 3 \cdot 3 \cdot 3 \cdot 3$

33. c

37. $\text{LCM} = 3 \cdot 4 = 12$

41. $\text{LCM} = 5 \cdot 7 = 35$

45. $12 = 2 \cdot 2 \cdot 3$
$20 = 2 \cdot 2 \cdot 5$
$\text{LCM} = 2 \cdot 2 \cdot 3 \cdot 5 = 60$

49. $24 = 2 \cdot 2 \cdot 2 \cdot 3$
$36 = 2 \cdot 2 \cdot 3 \cdot 3$
$\text{LCM} = 2 \cdot 2 \cdot 2 \cdot 3 \cdot 3 = 72$

53. $\text{LCM} = 2 \cdot 3 \cdot 5 = 30$

57. $4 = 2 \cdot 2$
$8 = 2 \cdot 2 \cdot 2$
$24 = 2 \cdot 2 \cdot 2 \cdot 3$
$\text{LCM} = 2 \cdot 2 \cdot 2 \cdot 3 = 24$

61. answers may vary

Exercise Set R.2

1. $\frac{7}{10} = \frac{7 \cdot 3}{10 \cdot 3} = \frac{21}{30}$

5. $\frac{4}{5} = \frac{4 \cdot 4}{5 \cdot 4} = \frac{16}{20}$

9. $\frac{10}{15} = \frac{2 \cdot 5}{3 \cdot 5} = \frac{2}{3}$

13. $\frac{20}{20} = \frac{2 \cdot 2 \cdot 5}{2 \cdot 2 \cdot 5} = 1$

17. $\frac{18}{30} = \frac{2 \cdot 3 \cdot 3}{2 \cdot 3 \cdot 5} = \frac{3}{5}$

21. $\frac{66}{48} = \frac{2 \cdot 3 \cdot 11}{2 \cdot 2 \cdot 2 \cdot 2 \cdot 3} = \frac{11}{2 \cdot 2 \cdot 2} = \frac{11}{8}$

25. $\frac{192}{264} = \frac{2 \cdot 2 \cdot 2 \cdot 2 \cdot 2 \cdot 2 \cdot 3}{2 \cdot 2 \cdot 2 \cdot 3 \cdot 11} = \frac{2 \cdot 2 \cdot 2}{11} = \frac{8}{11}$

29. $\frac{2}{3} \cdot \frac{3}{4} = \frac{2 \cdot 3}{3 \cdot 2 \cdot 2} = \frac{1}{2}$

33. $7\frac{2}{5} \div \frac{1}{5} = \frac{37}{5} \div \frac{1}{5} = \frac{37}{5} \cdot \frac{5}{1} = \frac{37 \cdot 5}{5 \cdot 1} = \frac{37}{1} = 37$

37. $\frac{3}{4} \div \frac{1}{20} = \frac{3}{4} \cdot \frac{20}{1} = \frac{3 \cdot 4 \cdot 5}{4 \cdot 1} = \frac{3 \cdot 5}{1} = 15$

41. $\frac{9}{20} \div 12 = \frac{9}{20} \cdot \frac{1}{12} = \frac{3 \cdot 3 \cdot 1}{4 \cdot 5 \cdot 3 \cdot 4} = \frac{3 \cdot 1}{5 \cdot 4 \cdot 4} = \frac{3}{80}$

45. $8\frac{3}{5} \div 2\frac{9}{10} = \frac{43}{5} \div \frac{29}{10} = \frac{43}{5} \cdot \frac{10}{29} = \frac{43 \cdot 2 \cdot 5}{5 \cdot 29} = \frac{43 \cdot 2}{29}$
$= \frac{86}{29} = 2\frac{28}{29}$

49. $\frac{4}{5} - \frac{1}{5} = \frac{4-1}{5} = \frac{3}{5}$

53. $\frac{17}{21} - \frac{10}{21} = \frac{17-10}{21} = \frac{7}{21} = \frac{1}{3}$

57.
$$\begin{array}{rr} 5\frac{2}{5} & 4\frac{7}{5} \\ -3\frac{4}{5} & -3\frac{4}{5} \\ \hline & 1\frac{3}{5} \end{array}$$

61. $\frac{10}{3} - \frac{5}{21} = \frac{10 \cdot 7}{3 \cdot 7} - \frac{5}{21} = \frac{70}{21} - \frac{5}{21} = \frac{70-5}{21} = \frac{65}{21}$

65. $\frac{5}{22} - \frac{5}{33} = \frac{5 \cdot 3}{22 \cdot 3} - \frac{5 \cdot 2}{33 \cdot 2} = \frac{15}{66} - \frac{10}{66} = \frac{15-10}{66} = \frac{5}{66}$

69.
$$\begin{array}{rr} 17\frac{2}{5} & 17\frac{6}{15} \\ +30\frac{2}{3} & +30\frac{10}{15} \\ \hline & 47\frac{16}{15} = 48\frac{1}{15} \end{array}$$

73. $\frac{2}{3} - \frac{5}{9} + \frac{5}{6} = \frac{2 \cdot 6}{3 \cdot 6} - \frac{5 \cdot 2}{9 \cdot 2} + \frac{5 \cdot 3}{6 \cdot 3} = \frac{12}{18} - \frac{10}{18} + \frac{15}{18}$
$= \frac{12 - 10 + 15}{18} = \frac{17}{18}$

77. answers may vary

81. $1 - \frac{1}{4} - \frac{3}{8} = \frac{1 \cdot 8}{1 \cdot 8} - \frac{1 \cdot 2}{4 \cdot 2} - \frac{3}{8} = \frac{8}{8} - \frac{2}{8} - \frac{3}{8} = \frac{8-2-3}{8} = \frac{3}{8}$

85. a. $\frac{7}{50}$ **b.** $\frac{21}{100}$ **c.** $1 - \frac{4}{25} - \frac{7}{50} - \frac{7}{50} - \frac{7}{100} - \frac{21}{100} - \frac{3}{100}$
$= \frac{100}{100} - \frac{16}{100} - \frac{14}{100} - \frac{14}{100} - \frac{7}{100} - \frac{21}{100} - \frac{3}{100}$
$= \frac{100 - 16 - 14 - 14 - 7 - 21 - 3}{100}$
$= \frac{100 - 75}{100} = \frac{25}{100} = \frac{25}{25 \cdot 4} = \frac{1}{4}$

d. $\frac{4}{25} + \frac{7}{50} = \frac{4 \cdot 2}{25 \cdot 2} + \frac{7}{50} = \frac{8}{50} + \frac{7}{50}$
$= \frac{8+7}{50} = \frac{15}{50} = \frac{5 \cdot 3}{5 \cdot 10} = \frac{3}{10}$

Exercise Set R.3

1. $0.6 = \frac{6}{10}$

5. $0.144 = \frac{114}{1000}$

9.
$$\begin{array}{r} 5.7 \\ +1.13 \\ \hline 6.83 \end{array}$$

13.
$$\begin{array}{r} 8.8 \\ -2.3 \\ \hline 6.5 \end{array}$$

17.
$$\begin{array}{r} 45.02 \\ 3.006 \\ +8.405 \\ \hline 56.431 \end{array}$$

21.
$$\begin{array}{r} 0.2 \\ \times 0.6 \\ \hline 0.12 \end{array}$$

25.
$$\begin{array}{r} 5.62 \\ \times\ 7.7 \\ \hline 3934 \\ 3934 \\ \hline 43.274 \end{array}$$

29.
$$\begin{array}{r} 0.094 \\ 5\overline{)0.470} \\ -0 \\ \hline 47 \\ -45 \\ \hline 20 \\ -20 \\ \hline 0 \end{array}$$

33.
$$\begin{array}{r} 5.8 \\ 82\overline{)475.6} \\ -410 \\ \hline 65\,6 \\ -65\,6 \\ \hline 0 \end{array}$$

37. answers may vary

41. 0.23

45. 98,207.2

49.
$$\begin{array}{r} 0.75 \\ 4\overline{)3.00} \\ 28 \\ \hline 20 \\ 20 \\ \hline 0 \end{array}$$
$\frac{3}{4} = 0.75$

53.
$$\begin{array}{r} 0.4375 \\ 16\overline{)7.0000} \\ -6\,4 \\ \hline 60 \\ -48 \\ \hline 120 \\ -112 \\ \hline 80 \\ -80 \\ \hline 0 \end{array}$$
$\frac{7}{16} = 0.4375$

57. $28\% = 0.28$

61. $135\% = 1.35$

65. $52\% = 0.52$

69. $0.876 = 87.6\%$

73. $0.5 = 0.50 = 50\%$

77. $\frac{7}{20} = \frac{7 \cdot 5}{20 \cdot 5} = \frac{35}{100} = 35\%$
or
$$\begin{array}{r} 0.35 \\ 20\overline{)7.00} \\ -6\,0 \\ \hline 1\,00 \\ -1\,00 \\ \hline 0 \end{array}$$
$\frac{7}{20} = 0.35 = 35\%$

Chapter R Test

1. $72 = 2 \cdot 2 \cdot 2 \cdot 3 \cdot 3$

5. $\frac{48}{100} = \frac{2 \cdot 2 \cdot 2 \cdot 2 \cdot 3}{2 \cdot 2 \cdot 5 \cdot 5} = \frac{2 \cdot 2 \cdot 3}{5 \cdot 5} = \frac{12}{25}$

9. $\frac{9}{10} \div 18 = \frac{9}{10} \cdot \frac{1}{18} = \frac{3 \cdot 3 \cdot 1}{2 \cdot 5 \cdot 2 \cdot 3 \cdot 3} = \frac{1}{2 \cdot 5 \cdot 2} = \frac{1}{20}$

13. $6\frac{7}{8} \div \frac{1}{8} = \frac{55}{8} \div \frac{1}{8} = \frac{55}{8} \cdot \frac{8}{1} = \frac{55 \cdot 8}{8 \cdot 1} = \frac{55}{1} = 55$

17.
$$\begin{array}{r} 7.93 \\ \times 1.6 \\ \hline 4758 \\ 793 \\ \hline 12.688 \end{array}$$

21.
$$\begin{array}{r} 0.1666 \\ 6\overline{)1.0000} \\ -\,6 \\ \hline 40 \\ -36 \\ \hline 40 \\ -36 \\ \hline 40 \\ -36 \\ \hline 4 \end{array}$$

This pattern will continue so that

$\frac{1}{6} = 0.1666\ldots$

$\frac{1}{6} = 0.1\overline{6} \approx 0.167$

25. From the graph, we can see that $\frac{3}{4}$ of the fresh water is icecaps and glaciers.

29. $A = \frac{1}{2} \cdot b \cdot h = \frac{1}{2} \cdot \frac{3}{4} \cdot \frac{1}{3} = \frac{1 \cdot 3 \cdot 1}{2 \cdot 4 \cdot 3} = \frac{1 \cdot 1}{2 \cdot 4} = \frac{1}{8}$ sq ft

Chapter 1

Exercise Set 1.2

1. $4 < 10$

5. $6.26 = 6.26$

9. $32 < 212$

13. False, since 10 is to the left of 11 on the number line.

17. True, since 7 is to the right of 0 on the number line.

21. $20 \le 25$

25. $-12 < -10$

29. $5 \ge 4$

33. The integer 14,494 represents 14,494 ft. The integer -282 represents 282 ft below sea level.

37. The integer 475 represents a deposit of \$475.
The integer -195 represents a withdrawal of \$195.

41.
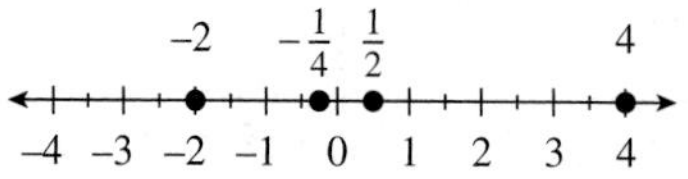

45. The number 0 belongs to the sets of: whole numbers, integers, rational numbers, and real numbers.

49. The number 265 belongs to the sets of: natural numbers, whole numbers, integers, rational numbers, and real numbers.

53. False; rational numbers can be either nonintegers, such as $\frac{1}{2}$, or integers, such as 2.

57. False: $-\sqrt{2}$ is not a rational number.

61. $|-5| > -4$ since $|-5| = 5$ and $5 > -4$.

65. $|-2| < |-3|$ since $|-2| = 2$ and $|-3| = 3$ and $2 < 3$.

69. False, since $\frac{1}{2}$ is to the right of $\frac{1}{3}$ on the number line.

73. False, since -9.6 is to the left of -9.1 on the number line.

77. Blue Ridge Parkway

81. $-0.04 > -26.7$

85. The sun; since on the number line -26.7 is to the left of all other numbers listed, and therefore, -26.7 is smaller than all other numbers listed.

Exercise Set 1.3

1. $3^5 = 3 \cdot 3 \cdot 3 \cdot 3 \cdot 3 = 243$

5. $1^5 = 1 \cdot 1 \cdot 1 \cdot 1 \cdot 1 = 1$

9. $\left(\frac{1}{5}\right)^3 = \left(\frac{1}{5}\right)\left(\frac{1}{5}\right)\left(\frac{1}{5}\right) = \frac{1 \cdot 1 \cdot 1}{5 \cdot 5 \cdot 5} = \frac{1}{125}$

13. $7^2 = 7 \cdot 7 = 49$

17. $(5 \cdot 5)$ sq m $= 5^2$ sq m

21. $4 \cdot 8 - 6 \cdot 2 = 32 - 12 = 20$

25. $2 + (5 - 2) + 4^2 = 2 + 3 + 4^2$
$= 2 + 3 + 16 = 5 + 16$
$= 21$

29. $\frac{1}{4} \cdot \frac{2}{3} - \frac{1}{6} = \frac{2}{12} - \frac{1}{6} = \frac{1}{6} - \frac{1}{6} = 0$

33. $2[5 + 2(8 - 3)] = 2[5 + 2(5)]$
$= 2[5 + 10]$
$= 2[15]$
$= 30$

37. $\dfrac{|6 - 2| + 3}{8 + 2 \cdot 5} = \dfrac{4 + 3}{8 + 10} = \dfrac{7}{18}$

41. $\dfrac{6 + |8 - 2| + 3^2}{18 - 3} = \dfrac{6 + |6| + 3^2}{15}$
$= \dfrac{6 + 6 + 9}{15}$
$= \dfrac{21}{15}$
$= \dfrac{3 \cdot 7}{3 \cdot 5}$
$= \dfrac{7}{5}$

45. a. 64 **b.** 43 **c.** 19 **d.** 22

49. Replace x with 1 and z with 5.
$\dfrac{z}{5x} = \dfrac{5}{5 \cdot 1} = \dfrac{5}{5} = 1$

53. Replace x with 1 and y with 3.
$|2x + 3y| = |2 \cdot 1 + 3 \cdot 3| = |2 + 9| = |11| = 11$

57. Replace y with 3.
$5y^2 = 5(3)^2 = 5 \cdot 9 = 45$

61. Replace x with 2 and y with 6.
$\dfrac{y}{x} = \dfrac{6}{2} = 3$

65.

length	width	perimeter	area
3 in.	4 in.	14 in.	12 sq in.
1 in.	6 in.	14 in.	6 sq in.
2 in.	5 in.	14 in.	10 sq in.

69. Replace x with 0 and see if a true statement results.
$2x + 6 = 5x - 1$
$2(0) + 6 \stackrel{?}{=} 5(0) - 1$
$0 + 6 \stackrel{?}{=} 0 - 1$
$6 \neq -1$ 0 is not a solution.

73. Replace x with 2 and see if a true statement results.
$x + 6 = x + 6$
$2 + 6 \stackrel{?}{=} 2 + 6$
$8 = 8$ 2 is a solution.

77. Replace x with 27 and see if a true statement results.
$\dfrac{1}{3}x = 9$
$\dfrac{1}{3}(27) \stackrel{?}{=} 9$
$9 = 9$ 27 is a solution.

81. $x - 5$

85. $3x + 22$

89. $3 \neq 4 \div 2$

93. $13 - 3x = 13$

97. $(20 - 4) \cdot 4 \div 2$

101. answers may vary

Exercise Set 1.4

1. $6 + 3 = 9$

5. $8 + (-7) = 1$

9. $-2 + (-3) = -5$

13. $-7 + 3 = -4$

17. $5 + (-7) = -2$

21. $27 + (-46) = -19$

25. $-33 + (-14) = -47$

29. $|-8| + (-16) = 8 + (-16) = -8$

33. $-9.6 + (-3.5) = -13.1$

37. $-\dfrac{7}{16} + \dfrac{1}{4} = -\dfrac{7}{16} + \dfrac{4 \cdot 1}{4 \cdot 4}$
$= -\dfrac{7}{16} + \dfrac{4}{16}$
$= -\dfrac{3}{16}$

41. $-15 + 9 + (-2) = -6 + (-2) = -8$

45. $-23 + 16 + (-2) = -7 + (-2) = -9$

49. $6 + (-4) + 9 = 2 + 9 = 11$

53. $|9 + (-12)| + |-16| = |-3| + |-16| = 3 + 16 = 19$

57. answers may vary

61. $-411 + 316 = -95$
You are at an elevation of -95 m (95 m below sea level).

65. His total overall score is the sum of the scores for all four rounds of play.
$0 + (-3) + (-6) + (-5) = -3 + (-6) + (-5)$
$= -9 + (-5)$
$= -14$
His total overall score was -14 (14 below par).

69. The opposite of 6 is -6.

73. The opposite of 0 is 0.

77. answers may vary

81. $-|0| = -0 = 0$

85. The highest temperature is represented by the bar that is farthest above 0 degrees. From the graph, we can see that the highest temperature occurred in July.

89. To find the average of the three temperatures, we first find the sum and then divide by 3.
$\dfrac{-9.1 + 14.4 + 8.8}{3} = \dfrac{5.3 + 8.8}{3} = \dfrac{14.1}{3} = 4.7$
The average temperature was 4.7°F.

93. $a + a$ is a positive number.

Exercise Set 1.5

1. $-6 - 4 = -6 + (-4) = -10$

5. $16 - (-3) = 16 + (3) = 19$

9. $-16 - (-18) = -16 + (18) = 2$

13. $7 - (-4) = 7 + (4) = 11$

17. $16 - (-21) = 16 + (21) = 37$

21. $-44 - 27 = -44 + (-27) = -71$

25. $-2.6 - (-6.7) = -2.6 + (6.7) = 4.1$

29. $-\dfrac{1}{6} - \dfrac{3}{4} = -\dfrac{1}{6} + \left(-\dfrac{3}{4}\right)$
$= -\dfrac{4 \cdot 1}{4 \cdot 6} + \left(-\dfrac{6 \cdot 3}{6 \cdot 4}\right)$
$= -\dfrac{4}{24} + \left(-\dfrac{18}{24}\right)$
$= -\dfrac{22}{24}$
$= -\dfrac{2 \cdot 11}{2 \cdot 12}$
$= -\dfrac{11}{12}$

33. Sometimes positive and sometimes negative. If a and b are positive numbers and $a \geq b$, then $a - b \geq 0$. If a and b are positive numbers and $a \leq b$, then $a - b \leq 0$.

37. $-6 - (-1) = -6 + (1) = -5$

41. $-8 - 15 = -8 + (-15) = -23$

45. $5 - 9 + (-4) - 8 - 8$
$= 5 + (-9) + (-4) + (-8) + (-8)$
$= -4 + (-4) + (-8) + (-8)$
$= -8 + (-8) + (-8)$
$= -16 + (-8)$
$= -24$

49. $3^3 - 8 \cdot 9 = 27 - 8 \cdot 9$
$= 27 - 72$
$= 27 + (-72)$
$= -45$

53. $(3 - 6) + 4^2 = (3 + (-6)) + 4^2$
$= (-3) + 4^2$
$= -3 + 16$
$= 13$

57. $|-3| + 2^2 + [-4 - (-6)] = 3 + 2^2 + [-4 + 6]$
$= 3 + 2^2 + [2]$
$= 3 + 4 + 2$
$= 7 + 2$
$= 9$

61. Replace x with -5, y with 4, and t with 10.
$|x| + 2t - 8y = |-5| + 2(10) - 8(4)$
$= 5 + 20 - 32$
$= 5 + 20 + (-32)$
$= 25 + (-32)$
$= -7$

65. Replace x with -5 and y with 4.
$y^2 - x = 4^2 - (-5)$
$= 16 - (-5)$
$= 16 + 5$
$= 21$

69. Replace x with -4 and see if a true statement results.
$x - 9 = 5$
$-4 - 9 \stackrel{?}{=} 5$
$-4 + (-9) \stackrel{?}{=} 5$
$-13 \neq 5$
-4 is not a solution of $x - 9 = 5$.

73. Replace x with 2 and see if a true statement results.
$-x - 13 = -15$
$-2 - 13 \stackrel{?}{=} -15$
$-2 + (-13) \stackrel{?}{=} -15$
$-15 = -15$
2 is a solution of $-x - 13 = -15$.

77. The total gain or loss of yardage is the sum of the gains and losses. Gains are represented by positive numbers and losses are represented by negative numbers.
$2 + (-5) + (-20) = -3 + (-20) = -23$
The 49ers lost a total of 23 yd.

81. The overall vertical change is the sum of the gains and losses in altitude. Gains are represented by positive numbers and losses are represented by negative numbers.
$-250 + 120 + (-178) = -130 + (-178) = -308$
The overall vertical change is -308 ft.

85. These angles are supplementary, so their sum is 180°.
$y = 180° - 50° = 130°$

89. To find monthly increases or decreases, subtract the two consecutive temperatures. A positive outcome indicates an increase and a negative outcome indicates a decrease.

February: $-23.7 - (-19.3)$
$= -23.7 + 19.3 = -4.4°$
March: $-21.1 - (-23.7)$
$= -21.1 + 23.7 = 2.6°$
April: $-9.1 - (-21.1)$
$= -9.1 + 21.1 = 12°$
May: $14.4 - (-9.1)$
$= 14.4 + 9.1 = 23.5°$
June: $29.7 - 14.4$
$= 29.7 + (-14.4) = 15.3°$
July: $33.6 - 29.7$
$= 33.6 + (-29.7) = 3.9°$
August: $33.3 - 33.6$
$= 33.3 + (-33.6) = -0.3°$
September: $27.0 - 33.3$
$= 27.0 + (-33.3) = -6.3°$
October: $8.8 - 27.0$
$= 8.8 + (-27.0) = -18.2°$
November: $-6.9 - 8.8$
$= -6.9 + (-8.8) = -15.7°$
December: $-17.2 - (-6.9)$
$= -17.2 + 6.9 = -10.3°$

93. True

97. $4.362 + (-7.0086)$
Negative since $7.0086 > 4.362$
$4.362 - 7.0086 = -2.6466$

Exercise Set 1.6

1. $-6(4) = -24$

5. $-5(-10) = 50$

9. $-\frac{1}{2}\left(-\frac{3}{5}\right) = -\left(-\frac{1 \cdot 3}{2 \cdot 5}\right) = -\left(-\frac{3}{10}\right) = \frac{3}{10}$

13. $5(-1.4) = -7$

17. $(-5)(-5) = 25$

21. $-\frac{20}{25}\left(\frac{5}{16}\right) = -\left(\frac{20 \cdot 5}{25 \cdot 16}\right) = -\left(\frac{5 \cdot 4 \cdot 5}{5 \cdot 5 \cdot 4 \cdot 4}\right) = -\frac{1}{4}$

25. $(-1)(2)(-3)(-5) = (-2)(-3)(-5) = (6)(-5) = -30$

29. $(-4)^2 = (-4)(-4) = 16$

33. $-3(2 - 8) = -3(2 + (-8)) = -3(-6) = 18$

37. $\left(-\frac{3}{4}\right)^2 = \left(-\frac{3}{4}\right)\left(-\frac{3}{4}\right) = \frac{3 \cdot 3}{4 \cdot 4} = \frac{9}{16}$

41. $-1^5 = -1 \cdot 1 \cdot 1 \cdot 1 \cdot 1 = -1$

45. False

49. The reciprocal of $\frac{2}{3}$ is $\frac{3}{2}$ since $\frac{2}{3} \cdot \frac{3}{2} = 1$.

53. The reciprocal of $-\frac{3}{11}$ is $-\frac{11}{3}$ since $\left(-\frac{3}{11}\right)\left(-\frac{11}{3}\right) = 1$.

57. The reciprocal of $\frac{1}{-6.3}$ is -6.3 since $\left(\frac{1}{-6.3}\right)(-6.3) = 1$.

61. $\frac{18}{-2} = -9$

65. $\frac{0}{-4} = 0$

69. $\frac{-12}{-4} = 3$

73. $\frac{6}{7} \div \left(-\frac{1}{3}\right) = \frac{6}{7} \cdot \left(-\frac{3}{1}\right) = -\frac{18}{7}$

77. $-\frac{4}{9} \div \frac{4}{9} = -\frac{4}{9} \cdot \frac{9}{4} = -\frac{36}{36} = -1$

81. $-48 \div 1.2 = -48 \cdot \frac{1}{1.2} = \frac{-48}{1.2} = -40$

85. $\frac{-9(-3)}{-6} = \frac{27}{-6} = -\frac{9}{2}$

89. $\frac{-6^2 + 4}{-2} = \frac{-36 + 4}{-2} = \frac{-32}{-2} = 16$

93. $\frac{22 + 3(-2)}{-5 - 2} = \frac{22 - 6}{-5 - 2} = \frac{16}{-7} = -\frac{16}{7}$

97. $\frac{6 - 2(-3)}{4 - 3(-2)} = \frac{6 + 6}{4 + 6} = \frac{12}{10} = \frac{6}{5}$

101. $\frac{|5 - 9| + |10 - 15|}{|2(-3)|} = \frac{|-4| + |-5|}{|-6|}$
$= \frac{4 + 5}{6} = \frac{9}{6} = \frac{3}{2}$

105. Replace x with -5 and y with -3 and simplify.
$\frac{2x - 5}{y - 2} = \frac{2(-5) - 5}{-3 - 2} = \frac{-10 - 5}{-3 - 2} = \frac{-15}{-5} = 3$

109. Replace x with -5 and y with -3 and simplify.
$\frac{x + y}{3y} = \frac{(-5) + (-3)}{3(-3)} = \frac{-5 - 3}{3(-3)} = \frac{-5 - 3}{-9} = \frac{-8}{-9}$
$= \frac{8}{9}$

113. Replace x with -2.
$\frac{-10}{x} = -1$
$\frac{-10}{-2} \stackrel{?}{=} -1$
$5 \neq -1$ Since $5 = -1$ is a false statement, -2 is not a solution.

117. $\frac{0}{5} - 7 = 0 - 7 = -7$

121. $\frac{-8}{-20} = \frac{2}{5}$

125. $\frac{b + b}{a + a} = \frac{2b}{2a} = \frac{b}{a}$ is a negative number since the quotient of a positive number and a negative number is negative.

129. $q^2 \cdot r \cdot t$ is a negative number since the product of a positive number (q^2) and a negative number (r) and a positive number (t) is positive.

133. $r(q - t)$ is a positive number since the product of a negative number (r) with a negative number ($q - t$) is positive.

137. answers may vary

Exercise Set 1.7

1. $x + 16 = 16 + x$ **5.** $xy = yx$

9. $(xy) \cdot z = x \cdot (yz)$ **13.** $4 \cdot (ab) = 4a \cdot (b)$

17. $8 + (9 + b) = (8 + 9) + b = 17 + b$

21. $\frac{1}{5}(5y) = \left(\frac{1}{5} \cdot 5\right) \cdot y = 1 \cdot y = y$

25. $-9(8x) = (-9 \cdot 8) \cdot x = -72 \cdot x = -72x$

29. answers may vary

33. $9(x - 6) = 9(x) - 9(6) = 9x - 54$

37. $7(4x - 3) = 7(4x) - 7(3) = 28x - 21$

41. $-2(y - z) = -2(y) - (-2)(z) = -2y + 2z$

45. $5(x + 4m + 2) = 5(x) + 5(4m) + 5(2)$
$= 5x + 20m + 10$

49. $-(5x + 2) = -1(5x + 2)$
$= (-1)(5x) + (-1)(2)$
$= -5x - 2$

53. $\frac{1}{2}(6x + 8) = \frac{1}{2}(6x) + \frac{1}{2}(8)$
$= \left(\frac{1}{2} \cdot 6\right)x + \left(\frac{1}{2} \cdot 8\right)$
$= 3x + 4$

57. $3(2r + 5) - 7 = 3(2r) + 3(5) - 7$
$= 6r + 15 - 7$
$= 6r + 8$

61. $-4(4x + 5) - 5 = -4(4x) + (-4)(5) - 5$
$= -16x - 20 - 5$
$= -16x - 25$

65. $11x + 11y = 11(x + y)$

69. $30a + 30b = 30(a + b)$

73. The additive inverse of -8 is 8 since $(-8) + 8 = 0$.

77. The additive inverse of $-|-2|$ is 2 since
$-|-2| + 2 = -2 + 2 = 0$

81. The multiplicative inverse of $-\frac{5}{6}$ is $-\frac{6}{5}$ since $-\frac{5}{6} \cdot \left(-\frac{6}{5}\right) = 1$.

85. The multiplicative inverse of -2 is $-\frac{1}{2}$ since $-2 \cdot \left(-\frac{1}{2}\right) = 1$.

89. associative property of addition

93. associative property of multiplication

97. distributive property

101.

Expression	Opposite	Reciprocal
8	−8	$\frac{1}{8}$

105.

Expression	Opposite	Reciprocal
$2x$	$-2x$	$\frac{1}{2x}$

109. answers may vary **113.** yes

Chapter 1 Test

1. $|-7| > 5$

5. $6 \cdot 3 - 8 \cdot 4 = 18 - 32 = 18 + (-32) = -14$

9. $\frac{-8}{0}$ is undefined

13. $-\frac{3}{5} + \frac{15}{8} = -\frac{8 \cdot 3}{8 \cdot 5} + \frac{5 \cdot 15}{5 \cdot 8} = -\frac{24}{40} + \frac{75}{40} = \frac{51}{40}$

17. $\frac{(-2)(0)(-3)}{-6} = \frac{0(-3)}{-6} = \frac{0}{-6} = 0$

21. $|-2| = -1 - (-3)$

25. Replace x with 6 and y with -2, then simplify.
$2 + 3x - y = 2 + 3(6) - (-2)$
$= 2 + 18 + 2$
$= 20 + 2$
$= 22$

29. distributive property

33. Losses are represented by negative numbers.
The greatest loss occurred on second down when the Saints lost 10 yards.

Chapter 2

Exercise Set 2.1

1. $7y + 8y = (7 + 8)y = 15y$

5. $3b - 5 - 10b - 4 = 3b - 10b + (-5 - 4)$
$= (3 - 10)b + (-5 - 4)$
$= -7b - 9$

9. $5g - 3 - 5 - 5g = 5g - 5g + (-3 - 5)$
$= (5 - 5)g + (-3 - 5)$
$= 0 \cdot g + (-3 - 5)$
$= -8$

13. $2k - k - 6 = (2 - 1)k - 6 = k - 6$

17. $6x - 5x + x - 3 + 2x = 6x - 5x + x + 2x - 3$
$= (6 - 5 + 1 + 2)x - 3$
$= 4x - 3$

21. $3.4m - 4 - 3.4m - 7 = 3.4m - 3.4m + (-4 - 7)$
$= (3.4 - 3.4)m + (-4 - 7)$
$= 0 \cdot m - 11$
$= -11$

25. $5(y + 4) = 5(y) + 5(4) = 5y + 20$

29. $-5(2x - 3y + 6)$
$= -5(2x) + (-5)(-3y) + (-5)(6)$
$= -10x + 15y - 30$

33. $7(d - 3) + 10 = 7d - 21 + 10 = 7d - 11$

37. $3(2x - 5) - 5(x - 4) = 6x - 15 - 5x + 20$
$= 6x - 5x - 15 + 20$
$= x + 5$

41. $5k - (3k - 10) = 5k - 3k + 10$
$= 2k + 10$

45. $5(x + 2) - (3x - 4) = 5x + 10 - 3x + 4$
$= 5x - 3x + 10 + 4$
$= 2x + 14$

49. $2 + 4(6x - 6) = 2 + 24x - 24$
$= 24x + 2 - 24$
$= 24x - 22$

53. $10 - 3(2x + 3y) = 10 - 6x - 9y$

57. $\frac{1}{2}(12x - 4) - (x + 5) = 6x - 2 - x - 5$
$= 6x - x - 2 - 5$
$= 5x - 7$

61. $(4x - 10) + (6x + 7) = 4x - 10 + 6x + 7$
$= 4x + 6x - 10 + 7$
$= 10x - 3$

65. $(m - 9) - (5m - 6) = m - 9 - 5m + 6$
$= m - 5m - 9 + 6$
$= -4m - 3$

69. $\frac{3}{4}x + 12$

73. $8(x + 6)$ or $8x + 48$

77. Replace x with -1 and y with 3.
$y - x^2 = 3 - (-1)^2 = 3 - 1 = 2$

81. Replace y with -5 and z with 0.
$yz - y^2 = (-5)(0) - (-5)^2 = 0 - 25$
$= -25$

85. To determine if the scale is balanced, find the number of cubes on each side of the scale and see if they are equal.

Left side	Right side
1 cube = 1 cube	2 cubes = 2 cubes
2 cylinders = 4 cubes	3 cones = 3 cubes
Total = 5 cubes	Total = 5 cubes

The scale is balanced.

89. $12(x + 2) + 1(3x - 1) = 12x + 24 + 3x - 1$
$= 12x + 3x + 24 - 1$
$= 15x + 23$

The length is $(15x + 23)$ in.

Exercise Set 2.2

1. $x + 7 = 10$
$x + 7 - 7 = 10 - 7$
$x = 3$

Check: $x + 7 = 10$
$3 + 7 \stackrel{?}{=} 10$
$10 = 10$

The solution is 3.

5. $3 + x = -11$
$3 + x - 3 = -11 - 3$
$x = -14$

Check: $3 + x = -11$
$3 + (-14) \stackrel{?}{=} -11$
$-11 = -11$

The solution is -14.

9. $\frac{1}{3} + f = \frac{3}{4}$
$\frac{1}{3} + f - \frac{1}{3} = \frac{3}{4} - \frac{1}{3}$
$f = \frac{9}{12} - \frac{4}{12}$
$f = \frac{5}{12}$

Check: $\frac{1}{3} + f = \frac{3}{4}$
$\frac{1}{3} + \frac{5}{12} \stackrel{?}{=} \frac{3}{4}$
$\frac{4}{12} + \frac{5}{12} \stackrel{?}{=} \frac{3}{4}$
$\frac{9}{12} \stackrel{?}{=} \frac{3}{4}$
$\frac{3}{4} = \frac{3}{4}$

The solution is $\frac{5}{12}$.

13. $5b - 0.7 = 6b$
$5b - 5b - 0.7 = 6b - 5b$
$-0.7 = b$

Check: $5b - 0.7 = 6b$
$5(-0.7) - 0.7 \stackrel{?}{=} 6(-0.7)$
$-3.5 - 0.7 \stackrel{?}{=} -4.2$
$-4.2 = -4.2$

The solution is -0.7.

17. answers may vary

21. $\frac{5}{6}x + \frac{1}{6}x = -9$
$\left(\frac{5}{6} + \frac{1}{6}\right)x = -9$
$1 \cdot x = -9$
$x = -9$

Check: $\frac{5}{6}x + \frac{1}{6}x = -9$
$\frac{5}{6}(-9) + \frac{1}{6}(-9) \stackrel{?}{=} -9$
$\frac{-45}{6} + \frac{-9}{6} \stackrel{?}{=} -9$
$\frac{-54}{6} \stackrel{?}{=} -9$
$-9 = -9$

The solution is -9.

25. $3x - 6 = 2x + 5$
$3x - 6 - 2x = 2x + 5 - 2x$
$x - 6 = 5$
$x - 6 + 6 = 5 + 6$
$x = 11$

Check: $3x - 6 = 2x + 5$
$3(11) - 6 \stackrel{?}{=} 2(11) + 5$
$33 - 6 \stackrel{?}{=} 22 + 5$
$27 = 27$

The solution is 11.

29. $5x - 6 = 6x - 5$
$5x - 5x - 6 = 6x - 5x - 5$
$-6 = x - 5$
$-6 + 5 = x - 5 + 5$
$-1 = x$

Check: $5x - 6 = 6x - 5$
$5(-1) - 6 \stackrel{?}{=} 6(-1) - 5$
$-5 - 6 \stackrel{?}{=} -6 - 5$
$-11 = -11$

The solution is -1.

33. $13x - 9 + 2x - 5 = 12x - 1 + 2x$
$15x - 14 = 14x - 1$
$15x - 14x - 14 = 14x - 14x - 1$
$x - 14 = -1$
$x - 14 + 14 = -1 + 14$
$x = 13$

Check: $13x - 9 + 2x - 5 = 12x - 1 + 2x$
$13(13) - 9 + 2(13) - 5 \stackrel{?}{=} 12(13) - 1 + 2(13)$
$169 - 9 + 26 - 5 \stackrel{?}{=} 156 - 1 + 26$
$181 = 181$

The solution is 13.

37. $\frac{3}{8}x - \frac{1}{6} = -\frac{5}{8}x - \frac{2}{3}$
$\frac{3}{8}x + \frac{5}{8}x - \frac{1}{6} = -\frac{5}{8}x + \frac{5}{8}x - \frac{2}{3}$
$x - \frac{1}{6} = -\frac{2}{3}$
$x - \frac{1}{6} + \frac{1}{6} = -\frac{2}{3} + \frac{1}{6}$
$x = -\frac{4}{6} + \frac{1}{6}$
$= -\frac{3}{6}$
$= -\frac{1}{2}$

Check: $\frac{3}{8}x - \frac{1}{6} = -\frac{5}{8}x - \frac{2}{3}$
$\frac{3}{8}\left(-\frac{1}{2}\right) - \frac{1}{6} \stackrel{?}{=} -\frac{5}{8}\left(-\frac{1}{2}\right) - \frac{2}{3}$
$-\frac{3}{16} - \frac{1}{6} \stackrel{?}{=} \frac{5}{16} - \frac{2}{3}$
$-\frac{17}{48} = -\frac{17}{48}$

The solution is $-\frac{1}{2}$.

41. $7(6 + w) = 6(2 + w)$
$42 + 7w = 12 + 6w$
$42 + 7w - 6w = 12 + 6w - 6w$
$42 + w = 12$
$42 - 42 + w = 12 - 42$
$w = -30$

Check: $7(6 + w) = 6(2 + w)$
$7[6 + (-30)] \stackrel{?}{=} 6[2 + (-30)]$
$7(-24) \stackrel{?}{=} 6(-28)$
$-168 = -168$

The solution is -30.

45. $-5(n - 2) = 8 - 4n$
$-5n + 10 = 8 - 4n$
$-5n + 5n + 10 = 8 - 4n + 5n$
$10 = 8 + n$
$10 - 8 = 8 - 8 + n$
$2 = n$

Check: $-5(n - 2) = 8 - 4n$
$-5(2 - 2) \stackrel{?}{=} 8 - 4 \cdot 2$
$-5(0) \stackrel{?}{=} 8 - 8$
$0 = 0$

The solution is 2.

49. $3(n - 5) - (6 - 2n) = 4n$
$3n - 15 - 6 + 2n = 4n$
$5n - 21 = 4n$
$5n - 4n - 21 = 4n - 4n$
$n - 21 = 0$
$n - 21 + 21 = 0 + 21$
$n = 21$

Check: $3(n - 5) - (6 - 2n) = 4n$
$3(21 - 5) - (6 - 2 \cdot 21) \stackrel{?}{=} 4 \cdot 21$
$3(16) - (-36) \stackrel{?}{=} 84$
$48 + 36 \stackrel{?}{=} 84$
$84 = 84$

The solution is 21.

53. $20 - p$

57. $(180 - x)°$

61. $7x$ sq mi

65. The multiplicative inverse or reciprocal of 2 is $\frac{1}{2}$ because $2 \cdot \frac{1}{2} = 1$.

69. $\frac{3x}{3} = \left(\frac{3}{3}\right)x = (1)x = x$

73. $\frac{3}{5}\left(\frac{5}{3}x\right) = \left(\frac{3}{5} \cdot \frac{5}{3}\right)x = (1)x = x$

77. Since the sum of the angles in a triangle is 180°, one angle measures $x°$, and a second angle measures $(2x + 7)°$, we find the measure of the third angle by subtracting x and $(2x + 7)$ from 180.

$$\begin{aligned} 180 - [x + (2x + 7)] &= 180 - [x + 2x + 7] \\ &= 180 - [3x + 7] \\ &= 180 - 1[3x + 7] \\ &= 180 - 1(3x) - 1(7) \\ &= 180 - 3x - 7 \\ &= 173 - 3x \end{aligned}$$

The third angle measures $(173 - 3x)°$.

Exercise Set 2.3

1. $$\begin{aligned} -5x &= 20 \\ \frac{-5x}{-5} &= \frac{20}{-5} \\ x &= -4 \end{aligned}$$

Check: $$\begin{aligned} -5x &= 20 \\ -5(-4) &\stackrel{?}{=} 20 \\ 20 &= 20 \end{aligned}$$

The solution is -4.

5. $$\begin{aligned} -x &= -12 \\ \frac{-x}{-1} &= \frac{-12}{-1} \\ x &= 12 \end{aligned}$$

Check: $$\begin{aligned} -x &= -12 \\ -(12) &\stackrel{?}{=} -12 \\ -12 &= -12 \end{aligned}$$

The solution is 12.

9. $$\begin{aligned} \frac{1}{6}d &= \frac{1}{2} \\ 6\left(\frac{1}{6}d\right) &= 6\left(\frac{1}{2}\right) \\ d &= 3 \end{aligned}$$

Check: $$\begin{aligned} \frac{1}{6}d &= \frac{1}{2} \\ \frac{1}{6}(3) &\stackrel{?}{=} \frac{1}{2} \\ \frac{1}{2} &= \frac{1}{2} \end{aligned}$$

The solution is 3.

13. $$\begin{aligned} \frac{k}{-7} &= 0 \\ -7\left(\frac{k}{-7}\right) &= -7(0) \\ k &= 0 \end{aligned}$$

Check: $$\begin{aligned} \frac{k}{-7} &= 0 \\ \frac{0}{-7} &\stackrel{?}{=} 0 \\ 0 &= 0 \end{aligned}$$

The solution is 0.

17. $$\begin{aligned} 42 &= 7x \\ \frac{42}{7} &= \frac{7x}{7} \\ 6 &= x \end{aligned}$$

Check: $$\begin{aligned} 42 &= 7x \\ 42 &\stackrel{?}{=} 7(6) \\ 42 &= 42 \end{aligned}$$

The solution is 6.

21. $$\begin{aligned} -\frac{3}{7}p &= -2 \\ -\frac{7}{3}\left(-\frac{3}{7}p\right) &= -\frac{7}{3}(-2) \\ p &= \frac{14}{3} \end{aligned}$$

Check: $$\begin{aligned} -\frac{3}{7}p &= -2 \\ -\frac{3}{7}\left(\frac{14}{3}\right) &\stackrel{?}{=} -2 \\ -2 &= -2 \end{aligned}$$

The solution is $\frac{14}{3}$.

25. $$\begin{aligned} 2x - 4 &= 16 \\ 2x - 4 + 4 &= 16 + 4 \\ 2x &= 20 \\ \frac{2x}{2} &= \frac{20}{2} \\ x &= 10 \end{aligned}$$

Check: $$\begin{aligned} 2x - 4 &= 16 \\ 2(10) - 4 &\stackrel{?}{=} 16 \\ 20 - 4 &\stackrel{?}{=} 16 \\ 16 &= 16 \end{aligned}$$

The solution is 10.

29. $$\begin{aligned} 6a + 3 &= 3 \\ 6a + 3 - 3 &= 3 - 3 \\ 6a &= 0 \\ \frac{6a}{6} &= \frac{0}{6} \\ a &= 0 \end{aligned}$$

Check: $$\begin{aligned} 6a + 3 &= 3 \\ 6(0) + 3 &\stackrel{?}{=} 3 \\ 0 + 3 &\stackrel{?}{=} 3 \\ 3 &= 3 \end{aligned}$$

The solution is 0.

33. $$\begin{aligned} 5 - 0.3k &= 5 \\ 5 - 0.3k - 5 &= 5 - 5 \\ -0.3k &= 0 \\ \frac{-0.3k}{-0.3} &= \frac{0}{-0.3} \\ k &= 0 \end{aligned}$$

Check: $$\begin{aligned} 5 - 0.3k &= 5 \\ 5 - 0.3(0) &\stackrel{?}{=} 5 \\ 5 - 0 &\stackrel{?}{=} 5 \\ 5 &= 5 \end{aligned}$$

The solution is 0.

37. $$\begin{aligned} \frac{x}{3} + 2 &= -5 \\ \frac{x}{3} + 2 - 2 &= -5 - 2 \\ \frac{x}{3} &= -7 \\ 3\left(\frac{x}{3}\right) &= 3(-7) \\ x &= -21 \end{aligned}$$

Check: $$\begin{aligned} \frac{x}{3} + 2 &= -5 \\ \frac{-21}{3} + 2 &\stackrel{?}{=} -5 \\ -7 + 2 &\stackrel{?}{=} -5 \\ -5 &= -5 \end{aligned}$$

The solution is -21.

41. $$\begin{aligned} 6z - 8 - z + 3 &= 0 \\ 5z - 5 &= 0 \\ 5z - 5 + 5 &= 0 + 5 \\ 5z &= 5 \\ \frac{5z}{5} &= \frac{5}{5} \\ z &= 1 \end{aligned}$$

Check: $$\begin{aligned} 6z - 8 - z + 3 &= 0 \\ 6(1) - 8 - 1 + 3 &= 0 \\ 6 - 8 - 1 + 3 &\stackrel{?}{=} 0 \\ 0 &= 0 \end{aligned}$$

The solution is 1.

45. $$\begin{aligned} 1 &= 0.4x - 0.6x - 5 \\ 1 &= -0.2x - 5 \\ 1 + 5 &= -0.2x - 5 + 5 \\ 6 &= -0.2x \\ \frac{6}{-0.2} &= \frac{-0.2x}{-0.2} \\ -30 &= x \end{aligned}$$

Check: $$\begin{aligned} 1 &= 0.4x - 0.6x - 5 \\ 1 &\stackrel{?}{=} 0.4(-30) - 0.6(-30) - 5 \\ 1 &\stackrel{?}{=} -12 + 18 - 5 \\ 1 &= 1 \end{aligned}$$

The solution is -30.

49. If $x =$ the first odd integer, then $x + 2 =$ the next odd integer. Their sum is $x + (x + 2) = x + x + 2 = 2x + 2$

53. $$\begin{aligned} 5x + 2(x - 6) &= 5x + 2(x) + 2(-6) \\ &= 5x + 2x - 12 \\ &= 7x - 12 \end{aligned}$$

57. $$\begin{aligned} -(x - 1) + x &= -1(x - 1) + x \\ &= -(x) - 1(-1) + x \\ &= -x + 1 + x \\ &= 1 \end{aligned}$$

61. Answers may vary. If we solve the equation for x, we obtain the following.

$$\begin{aligned} 3x + 6 &= 2x + 10 + x - 4 \\ 3x + 6 &= 3x + 6 \\ 3x + 6 - 6 &= 3x + 6 - 6 \\ 3x &= 3x \\ 3x - 3x &= 3x - 3x \\ 0 &= 0 \end{aligned}$$

Exercise Set 2.4

1.
$$\begin{aligned} -4y + 10 &= -2(3y + 1) \\ -4y + 10 &= -6y - 2 \\ -4y + 10 + 4y &= -6y - 2 + 4y \\ 10 &= -2y - 2 \\ 10 + 2 &= -2y - 2 + 2 \\ 12 &= -2y \\ \frac{12}{-2} &= \frac{-2y}{-2} \\ -6 &= y \end{aligned}$$

5.
$$\begin{aligned} -2(3x - 4) &= 2x \\ -6x + 8 &= 2x \\ -6x + 8 + 6x &= 2x + 6x \\ 8 &= 8x \\ \frac{8}{8} &= \frac{8x}{8} \\ 1 &= x \end{aligned}$$

9.
$$\begin{aligned} 5(2x - 1) - 2(3x) &= 1 \\ 10x - 5 - 6x &= 1 \\ 4x - 5 &= 1 \\ 4x - 5 + 5 &= 1 + 5 \\ 4x &= 6 \\ \frac{4x}{4} &= \frac{6}{4} \\ x &= \frac{3}{2} \end{aligned}$$

13.
$$\begin{aligned} 8 - 2(a - 1) &= 7 + a \\ 8 - 2a + 2 &= 7 + a \\ 10 - 2a &= 7 + a \\ 10 - 2a + 2a &= 7 + a + 2a \\ 10 &= 7 + 3a \\ 10 - 7 &= 7 + 3a - 7 \\ 3 &= 3a \\ \frac{3}{3} &= \frac{3a}{3} \\ 1 &= a \end{aligned}$$

17.
$$\begin{aligned} -2y - 10 &= 5y + 18 \\ -2y - 10 + 2y &= 5y + 18 + 2y \\ -10 &= 7y + 18 \\ -10 - 18 &= 7y + 18 - 18 \\ -28 &= 7y \\ \frac{-28}{7} &= \frac{7y}{7} \\ -4 &= y \end{aligned}$$

21.
$$\begin{aligned} 5y + 2(y - 6) &= 4(y + 1) - 2 \\ 5y + 2y - 12 &= 4y + 4 - 2 \\ 7y - 12 &= 4y + 2 \\ 7y - 4y - 12 &= 4y - 4y + 2 \\ 3y - 12 &= 2 \\ 3y - 12 + 12 &= 2 + 12 \\ 3y &= 14 \\ \frac{3y}{3} &= \frac{14}{3} \\ y &= \frac{14}{3} \end{aligned}$$

25.
$$\begin{aligned} x + \frac{5}{4} &= \frac{3}{4}x \\ 4\left(x + \frac{5}{4}\right) &= 4\left(\frac{3}{4}x\right) \\ 4x + 5 &= 3x \\ 4x - 3x + 5 &= 3x - 3x \\ x + 5 &= 0 \\ x + 5 - 5 &= 0 - 5 \\ x &= -5 \end{aligned}$$

29.
$$\begin{aligned} \frac{6(3 - z)}{5} &= -z \\ 5\left[\frac{6(3 - z)}{5}\right] &= 5(-z) \\ 6(3 - z) &= -5z \\ 18 - 6z &= -5z \\ 18 - 6z + 6z &= -5z + 6z \\ 18 &= z \end{aligned}$$

33.
$$\begin{aligned} \frac{3(x - 5)}{2} &= \frac{2(x + 5)}{3} \\ 6\left[\frac{3(x - 5)}{2}\right] &= 6\left[\frac{2(x + 5)}{3}\right] \\ 3[3(x - 5)] &= 2[2(x + 5)] \\ 9(x - 5) &= 4(x + 5) \\ 9x - 45 &= 4x + 20 \\ 9x - 4x - 45 &= 4x - 4x + 20 \\ 5x - 45 &= 20 \\ 5x - 45 + 45 &= 20 + 45 \\ 5x &= 65 \\ \frac{5x}{5} &= \frac{65}{5} \\ x &= 13 \end{aligned}$$

37.
$$\begin{aligned} 0.12(y - 6) + 0.06y &= 0.08y - 0.07(10) \\ 100[0.12(y - 6) + 0.06y] &= 100[0.08y - 0.07(10)] \\ 12(y - 6) + 6y &= 8y - 7(10) \\ 12y - 72 + 6y &= 8y - 70 \\ 18y - 72 &= 8y - 70 \\ 18y - 8y - 72 &= 8y - 8y - 70 \\ 10y - 72 &= -70 \\ 10y - 72 + 72 &= -70 + 72 \\ 10y &= 2 \\ \frac{10y}{10} &= \frac{2}{10} \\ y &= \frac{2}{10} = 0.2 \end{aligned}$$

41.
$$\begin{aligned} x + \frac{7}{6} &= 2x - \frac{7}{6} \\ 6\left(x + \frac{7}{6}\right) &= 6\left(2x - \frac{7}{6}\right) \\ 6x + 7 &= 12x - 7 \\ 6x - 12x + 7 &= 12x - 12x - 7 \\ -6x + 7 &= -7 \\ -6x + 7 - 7 &= -7 - 7 \\ -6x &= -14 \\ \frac{-6x}{-6} &= \frac{-14}{-6} \\ x &= \frac{-14}{-6} = \frac{14}{6} = \frac{7}{3} \end{aligned}$$

45. answers may vary

49. $\frac{x}{4} + 1 = \frac{x}{4}$

$$4\left(\frac{x}{4} + 1\right) = 4\left(\frac{x}{4}\right)$$
$$x + 4 = x$$
$$x + 4 - x = x - x$$
$$4 = 0$$

There is no solution to the equation $\frac{x}{4} + 1 = \frac{x}{4}$.

53. $2(x + 3) - 5 = 5x - 3(1 + x)$

$$2x + 6 - 5 = 5x - 3 - 3x$$
$$2x + 1 = 2x - 3$$
$$2x + 1 - 1 = 2x - 3 - 1$$
$$2x = 2x - 4$$
$$2x - 2x = 2x - 4 - 2x$$
$$0 = -4$$

There is no solution to the equation $2(x + 3) - 5 = 5x - 3(1 + x)$.

57. The perimeter of the lot is the sum of the lengths of the sides.

$$x + (2x - 3) + (3x - 5) = x + 2x - 3 + 3x - 5$$
$$= 6x - 8$$

The perimeter of the lot is $(6x - 8)$ m.

61. $-3 + 2x$

65. $1000(7x - 10) = 50(412 + 100x)$

$$7000x - 10{,}000 = 20{,}600 + 5000x$$
$$7000x - 10{,}000 + 10{,}000 = 20{,}600 + 5000x + 10{,}000$$
$$7000x = 30{,}600 + 5000x$$
$$7000x - 5000x = 30{,}600 + 5000x - 5000x$$
$$2000x = 30{,}600$$
$$\frac{2000x}{2000} = \frac{30{,}600}{2000}$$
$$x = 15.3$$

69. Since we know the perimeter of the pentagon is 28 cm,

$$x + x + x + 2x + 2x = 28$$
$$(1 + 1 + 1 + 2 + 2)x = 28$$
$$7x = 28$$
$$\frac{7x}{7} = \frac{28}{7}$$
$$x = 4$$

If $x = 4$ cm, then $2x = 2(4) = 8$ cm.

Exercise Set 2.5

1. Let x represent the number.

$$2x + \frac{1}{5} = 3x - \frac{4}{5}$$
$$5\left(2x + \frac{1}{5}\right) = 5\left(3x - \frac{4}{5}\right)$$
$$10x + 1 = 15x - 4$$
$$10x + 1 - 10x = 15x - 4 - 10x$$
$$1 = 5x - 4$$
$$1 + 4 = 5x - 4 + 4$$
$$5 = 5x$$
$$\frac{5}{5} = \frac{5x}{5}$$
$$1 = x$$

The number is 1.

5. Let x represent the number.

$$2x \cdot 3 = 5x - \frac{3}{4}$$
$$6x = 5x - \frac{3}{4}$$
$$6x - 5x = 5x - \frac{3}{4} - 5x$$
$$x = -\frac{3}{4}$$

The number is $-\frac{3}{4}$.

9. Let x = salary of the governor of Oregon, then $x + 39{,}000$ = salary of the governor of Michigan.

$$x + (x + 39{,}000) = 215{,}600$$
$$2x + 39{,}000 = 215{,}600$$
$$2x + 39{,}000 - 39{,}000 = 215{,}600 - 39{,}000$$
$$2x = 176{,}600$$
$$\frac{2x}{2} = \frac{176{,}600}{2}$$
$$x = 88{,}300$$

The salary of the governor of Oregon is \$88,300. The salary of the governor of Michigan is \$88,300 + \$39,000 = \$127,300.

13. The cost of renting the car is equal to the daily rental charge plus \$0.29 per mile. Let x = number of miles.

$$2 \cdot 24.95 + 0.29x = 100$$
$$49.90 + 0.29x = 100$$
$$49.90 + 0.29x - 49.90 = 100 - 49.90$$
$$0.29x = 50.10$$
$$\frac{0.29x}{0.29} = \frac{50.10}{0.29}$$
$$x = 172$$

You can drive 172 miles on a budget of \$100.

17. Let x = the number of votes for cerulean, then $x + 3366$ = the number of votes for blue.

$$x + (x + 3366) = 19{,}278$$
$$2x + 3366 = 19{,}278$$
$$2x + 3366 - 3366 = 19{,}278 - 3366$$
$$2x = 15{,}912$$
$$\frac{2x}{2} = \frac{15{,}912}{2}$$
$$x = 7956$$

Cerulean received 7956 votes and blue received $7956 + 3366 = 11{,}322$ votes.

21. If x = length of the shorter piece, then $2x + 2$ = length of the longer piece.

$$x + (2x + 2) = 17$$
$$x + 2x + 2 = 17$$
$$3x + 2 = 17$$
$$3x + 2 - 2 = 17 - 2$$
$$3x = 15$$
$$\frac{3x}{3} = \frac{15}{3}$$
$$x = 5$$

The shorter piece is 5 ft long. The longer piece is 12 ft long.

25. Let x = the area of the Gobi desert, then $7x$ = the area of the Sahara desert.

$$x + 7x = 4{,}000{,}000$$
$$8x = 4{,}000{,}000$$
$$\frac{8x}{8} = \frac{4{,}000{,}000}{8}$$
$$x = 500{,}000$$

The area of the Gobi desert is 500,000 sq mi and the area of the Sahara desert is $7(500{,}000) = 3{,}500{,}000$ sq mi.

29. Texas and Florida

33. Let x = the floor space of the Empire State Building.

$$3x = 6.5 \text{ million}$$
$$\frac{3x}{3} = \frac{6.5 \text{ million}}{3}$$
$$x \approx 2.2 \text{ million}$$

The Empire State Building has 2.2 million sq ft.

37. Let x = the number of medals won by Germany. Then $x + 1$ = the number won by Australia and $x + 2$ = the number won by China.

$$x + (x + 1) + (x + 2) = 174$$
$$3x + 3 = 174$$
$$3x + 3 - 3 = 174 - 3$$
$$3x = 171$$
$$\frac{3x}{3} = \frac{171}{3}$$
$$x = 57$$

Germany won 57 medals, Australia won $57 + 1 = 58$ medals, and China won $57 + 2 = 59$ medals.

41. $\frac{1}{2}(x - 1) = 37$

45.
$$2W + 2L = 2(7) + 2(10)$$
$$= 14 + 20$$
$$= 34$$

49. answers may vary

53. answers may vary

Exercise Set 2.6

1.
$$A = bh$$
$$45 = 15 \cdot h$$
$$\frac{45}{15} = \frac{15 \cdot h}{15}$$
$$3 = h$$

5.
$$A = \frac{1}{2}(B + b)h$$
$$180 = \frac{1}{2}(11 + 7)h$$
$$180 = \frac{1}{2}(18)h$$
$$180 = 9h$$
$$\frac{180}{9} = \frac{9h}{9}$$
$$20 = h$$

9.
$$C = 2\pi r$$
$$15.7 = 2\pi r$$
$$\frac{15.7}{2\pi} = \frac{2\pi r}{2\pi}$$
$$\frac{15.7}{6.28} = 7$$
$$2.5 = r$$

13.
$$V = \frac{1}{3}\pi r^2 h$$
$$565.2 = \frac{1}{3}\pi(6^2)h$$
$$565.2 = \frac{1}{3}\pi(36)h$$
$$565.2 = 12\pi h$$
$$\frac{565.2}{12\pi} = \frac{12\pi h}{12\pi}$$
$$\frac{565.2}{37.68} = h$$
$$15 = h$$

17.
$$d = rt$$
$$d = \left(2\frac{1}{2}\right)(55)$$
$$d = (2.5)(55)$$
$$d = 137.5$$

The distance is 137.5 mi.

21.
$$V = lwh$$
$$V = (8)(3)(6)$$
$$V = 144$$

Since the tank has a volume of 144 cu ft, and each piranha requires 1.5 cu ft, the tank can hold $\frac{144}{1.5} = 96$ piranhas.

25.
$$d = rt$$
$$25{,}000 = 4000 \cdot t$$
$$\frac{25{,}000}{4000} = \frac{4000 \cdot t}{4000}$$
$$6.25 = t$$

It will take 6.25 hr.

29.
$$V = lwh$$
$$V = (199)(78.5)(33)$$
$$V = 515{,}509.5$$

The volume of the smallest crate is 515,509.5 cu in.

33. Use $F = \left(\frac{9}{5}\right)C + 32$ with $C = -78.5$.

$$F = \left(\frac{9}{5}\right)C + 32$$
$$F = \left(\frac{9}{5}\right)(-78.5) + 32 = -141.3 + 32 = -109.3$$

-78.5°C is the same as -109.3°F.

37. Use the formula for the volume of a sphere, $V = \frac{4}{3}\pi r^3$, with $r = \frac{d}{2} = \frac{9.5}{2} = 4.75$.

$$V = \frac{4}{3}\pi(4.75)^3$$
$$V = \frac{4}{3}\pi(107.171875)$$
$$V = \frac{4}{3}(3.14)(107.171875)$$
$$V \approx 449$$

The volume of the ball is approx. 449 cu in.

41. From exercise 32, the distance around Earth is 25,120 mi.

$$\text{revolutions} = \frac{r}{d}$$
$$\text{revolutions} = \frac{270{,}000}{25{,}120}$$
$$\text{revolutions} \approx 10.7$$

A lightning bolt can travel about 10.7 times around Earth.

45. $d = rt$

$42.8 = 552 \cdot t$

$\frac{42.8}{552} = \frac{552 \cdot t}{552}$

$0.077536 \approx t$

The time is 0.077536 hr or 0.077536(60 sec) $\approx$ 4.65 min.

49. $V = LWH$

$\frac{V}{LH} = \frac{LWH}{LH}$

$\frac{V}{LH} = W$

53. $A = p + PRT$

$A - p = p + PRT - p$

$A - p = PRT$

$\frac{A - p}{PT} = \frac{PRT}{PT}$

$\frac{A - p}{PT} = R$

57. $P = a + b + c$

$P - b = a + b + c - b$

$P - b = a + c$

$P - b - c = a + c - c$

$P - b - c = a$

61. 32% = 0.32

65. 0.17 = 17%

69. $N = R + \frac{V}{G}$

$G(N) = G\left(R + \frac{V}{G}\right)$

$GN = GR + V$

$GN - GR = GR + V - GR$

$GN - GR = V$

$G(N - R) = V$

73. Use the formula $C = \left(\frac{5}{9}\right)(F - 32)$ to find when

$C = F.$

$C = \left(\frac{5}{9}\right)(F - 32)$

$F = \left(\frac{5}{9}\right)(F - 32)$

$9(F) = 9\left[\left(\frac{5}{9}\right)(F - 32)\right]$

$9F = 5(F - 32)$

$9F = 5F - 160$

$9F - 5F = 5F - 160 - 5F$

$4F = -160$

$\frac{4F}{4} = \frac{-160}{4}$

$F = -40$

-40°F is the same as -40°C.

Exercise Set 2.7

1. Let x = the unknown number.

$x = 0.16 \cdot 70$

$x = 11.2$

The number 11.2 is 16% of 70.

5. Let x = the unknown number.

$45 = 0.25 \cdot x$

$180 = x$

The number 45 is 25% of 180.

9. Let x = the unknown number.

$40 = 0.80 \cdot x$

$50 = x$

The number 40 is 80% of 50.

13. To find the decrease in price, we find 25% of 256.

25% of 256 = 0.25(256) = 64.

The coat is selling for \$64 off the original price. The sale price is \$256 − \$64 = \$192.

17. 81%

21. no; answers may vary

25. $N = (0.37)(135{,}000)$

$N = 49{,}950$

49,950 adults talk 16–60 minutes on the phone each day.

29.

United States	4486	91%
Canada	300	6%
Mexico	147	3%

33. $\frac{10}{12} = \frac{5}{6}$

37. 2 dollars = 2 · 20 nickels = 40 nickels

The ratio of 4 nickels to 2 dollars is $\frac{4}{40} = \frac{1}{10}$.

41. 3 hr = 3 · 60 min = 180 min

The ratio of 190 min to 3 hr is $\frac{190}{180} = \frac{19}{18}$.

45. $\frac{2}{3} = \frac{x}{6}$

$2 \cdot 6 = 3 \cdot x$

$\frac{12}{3} = \frac{3x}{3}$

$4 = x$

49. $\frac{4x}{6} = \frac{7}{2}$

$4x \cdot 2 = 6 \cdot 7$

$\frac{8x}{8} = \frac{42}{8}$

$x = \frac{21}{4}$

53. $\frac{x + 1}{2x + 3} = \frac{2}{3}$

$3(x + 1) = 2(2x + 3)$

$3x + 3 = 4x + 6$

$3x = 4x + 3$

$-x = 3$

$\frac{-x}{-1} = \frac{3}{-1}$

$x = -3$

57. $\frac{3}{x + 1} = \frac{5}{2x}$

$3 \cdot 2x = 5(x + 1)$

$6x = 5x + 5$

$x = 5$

61. Let x = weight on Pluto.

$\frac{x}{4100} = \frac{3}{100}$

$100x = (4100)(3)$

$100x = 12{,}300$

$\frac{100x}{100} = \frac{12{,}300}{100}$

$x = 123$

The elephant weighs 123 lb on Pluto.

65. Let x = the number of women.

$$\frac{x}{23{,}000} = \frac{1}{6}$$
$$6x = (23{,}000)(1)$$
$$6x = 23{,}000$$
$$\frac{6x}{x} = \frac{23{,}000}{6}$$
$$x \approx 3833$$

About 3833 women earn bigger paychecks.

69. Compare unit prices.

110 oz: $\frac{\$5.79}{110} \approx \$0.0526$

240 oz: $\frac{\$13.99}{240} \approx \$0.0583$

The better buy is 110 oz at $5.79.

73. $-5 > -7$

77. $(-3)^2 = (-3)(-3) = 9$
$-3^2 = -3 \cdot 3 = -9$
So, $(-3)^2 > -3^2$.

81. $\text{Percent} = \frac{35}{130} \cdot 100\% \approx (0.269)(100\%)$
$= 26.9\%$
Yes, this satisfies the recommendations.

Exercise Set 2.8

1. $(-\infty, -3)$

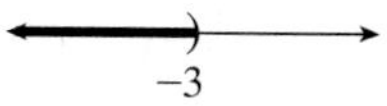

5. $(5, \infty)$

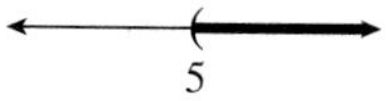

9. $(-1, 5)$

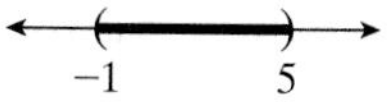

13. $x - 7 \geq -9$
$x \geq -2$
$[-2, \infty)$

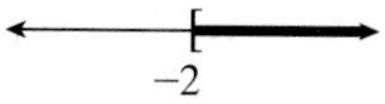

17. $8x - 7 \leq 7x - 5$
$x - 7 \leq -5$
$x \leq 2$
$(-\infty, 2]$

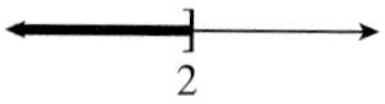

21. $5x < -23.5$
$x < -4.7$
$(-\infty, -4.7)$

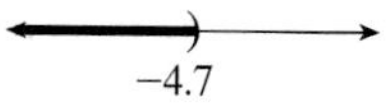

25. $-x < -4$
$x > 4$
$(4, \infty)$

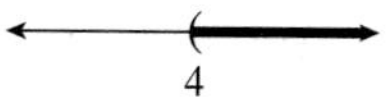

29. $15 + 2x \geq 4x - 7$
$15 \geq 2x - 7$
$22 \geq 2x$
$11 \geq x$
$(-\infty, 11]$

33.
$$\frac{1}{2} + \frac{2}{3} \geq \frac{x}{6}$$
$$6\left(\frac{1}{2} + \frac{2}{3}\right) \geq 6\left(\frac{x}{6}\right)$$
$$3 + 4 \geq x$$
$$7 \geq x$$
$(-\infty, 7]$

37.
$$\frac{1}{4}(x - 7) \geq x + 2$$
$$4\left[\frac{1}{4}(x - 7)\right] \geq 4(x + 2)$$
$$x - 7 \geq 4x + 8$$
$$-15 \geq 3x$$
$$-5 \geq x$$
$(-\infty, -5]$

41. $4(2x + 1) > 4$
$8x + 4 > 4$
$8x > 0$
$x > 0$
$(0, \infty)$

45. $4(x - 6) + 2x - 4 \geq 3(x - 7) + 10x$
$4x - 24 + 2x - 4 \geq 3x - 21 + 10x$
$6x - 28 \geq 13x - 21$
$-7x \geq 7$
$x \leq -1$
$(-\infty, -1]$

49. $14 - (5x - 6) \geq -6(x + 1) - 5$
$14 - 5x + 6 \geq -6x - 6 - 5$
$-5x + 20 \geq -6x - 11$
$x + 20 \geq -11$
$x \geq -31$
$[-31, \infty)$

53. Let x = minimum score.
$$\frac{2x + 72 + 67 + 82 + 79}{6} \geq 60$$
$$\frac{2x + 300}{6} \geq 60$$
$$2x + 300 \geq 360$$
$$2x \geq 60$$
$$x \geq 30$$
The minimum score is 30.

57. Let x = the maximum weight of a package.
$23x + 34 \leq 310$
$23x \leq 276$
$x \leq 12$
The maximum weight of a package is 12 + 1 or 13 oz.

61. $F \geq \frac{9}{5}C + 32$
$F \geq \frac{9}{5} \cdot 500 + 32$
$F \geq 900 + 32$
$F \geq 932$
The minimum temperature is 932°F.

65. 1015

69. 1997, 1998, 1999, and 2000

73. 1994

77. $x < 6$ and $x < -5$ means $x < -5$.
$\{\ldots -9, -8, -7, -6\}$

81. $\{x \mid -2.5 \leq x < 5.3\}$
$[-2.5, 5.3)$

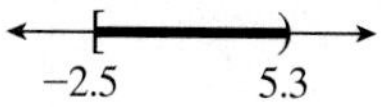

85. $8(x + 3) \leq 7(x + 5) + x$
$8x + 24 \leq 7x + 35 + x$
$8x + 24 \leq 8x + 35$
$24 \leq 35$
This is true for all x. Thus, the solution is the entire real number line $(-\infty, \infty)$.

Chapter 2 Test

1. $2y - 6 - y - 4 = 2y - y + (-6 - 4)$
$= (2 - 1)y - 10$
$= y - 10$

5.
$$-\frac{4}{5}x = 4$$
$$-\frac{5}{4}\left(-\frac{4}{5}x\right) = -\frac{5}{4}(4)$$
$$x = -5$$

9.
$$\frac{2(x+6)}{3} = x - 5$$
$$3\left(\frac{2(x+6)}{3}\right) = 3(x-5)$$
$$2(x+6) = 3(x-5)$$
$$2(x) + 2(6) = 3(x) + 3(-5)$$
$$2x + 12 = 3x - 15$$
$$2x + 12 - 2x = 3x - 15 - 2x$$
$$12 = x - 15$$
$$12 + 15 = x - 15 + 15$$
$$27 = x$$

13.
$$\frac{1}{3}(y+3) = 4y$$
$$3\left[\frac{1}{3}(y+3)\right] = 3(4y)$$
$$y + 3 = 12y$$
$$y + 3 - y = 12y - y$$
$$3 = 11y$$
$$\frac{3}{11} = \frac{11y}{11}$$
$$\frac{3}{11} = 1y$$
$$\frac{3}{11} = y$$

17. The area of the rectangular deck is given by the formula $A = bh$. If $b = 20$ and $h = 35$, then

$A = bh$

$A = (20)(35)$

$A = 700$

Thus, the area of the deck is 700 sq ft. Since 1 gallon covers 200 sq ft, we can form the following proportion, where x = number of gallons needed to cover 700 sq ft.

$$\frac{1}{200} = \frac{x}{700}$$
$$1 \cdot 700 = 200 \cdot x$$
$$\frac{700}{200} = \frac{200x}{200}$$
$$3.5 = x$$

Since we are painting two coats we will need twice as much or $2 \cdot 3.5 = 7$ gal of water seal.

21.
$$V = \pi r^2 h$$
$$\frac{1}{\pi r^2} \cdot V = \frac{1}{\pi r^2}(\pi r^2)h$$
$$\frac{V}{\pi r^2} = h$$
$$h = \frac{V}{\pi r^2}$$

25.
$$-0.3x \ge 2.4$$
$$-\frac{1}{0.3}(-0.3x) \le -\frac{1}{0.3}(2.4)$$
$$x \le -8 \text{ or } (-\infty, -8]$$

29. $N = 800(0.69)$

$N = 552$

You would expect to have 552 weak tornadoes.

Chapter 3

Exercise Set 3.1

1. France

5. Number of tourists is 34 million

9. Difference is $84 - 58 = 26$ beats per minute

13. 1985

17. 18 million men

21. 1900

25. answers may vary

29. when $a = b$

33. $(3, 2)$

37. $(2, -1)$

41. $(1, 3)$

45. a. (1995, 438), (1996, 438), (1997, 436), (1998, 435), (1999, 432), (2000, 434)

b.

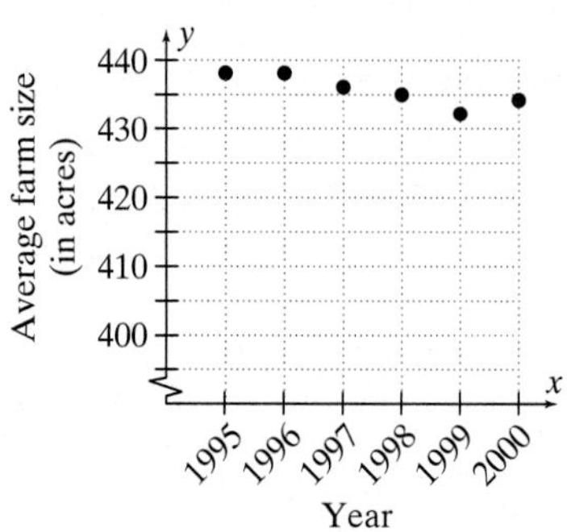

49. a. (0.50, 10), (0.75, 12), (1.00, 15), (1.25, 16), (1.50, 18), (1.50, 19), (1.75, 19), (2.00, 20)

b.

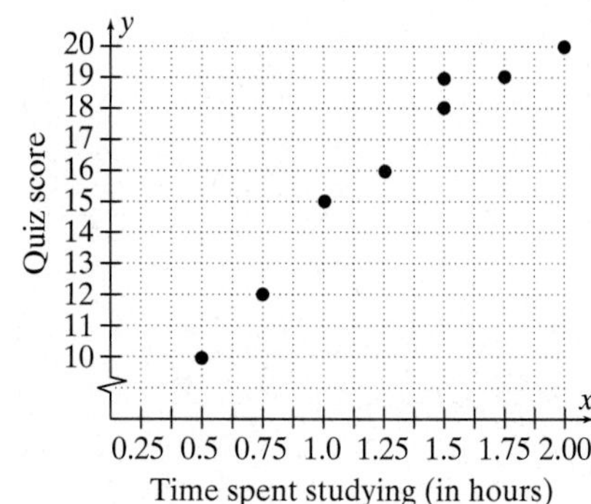

c. answers may vary

53. $3x + y = 9$

Complete (0,):
$x = 0$
$3(0) + y = 9$
$0 + y = 9$
$y = 9$
$(0, 9)$

Complete (, 0):
$y = 0$
$3x + 0 = 9$
$3x = 9$
$x = 3$
$(3, 0)$

57. $x + 3y = 6$

Complete (0,):
$x = 0$
$0 + 3y = 6$
$3y = 6$
$y = 2$
$(0, 2)$
Complete (, 0):
$y = 0$
$x + 3(0) = 6$
$x + 0 = 6$
$x = 6$
$(6, 0)$
Complete (, 1):
$y = 1$
$x + 3(1) = 6$
$x + 3 = 6$
$x = 3$
$(3, 1)$

x	y
0	2
6	0
3	1

61. $2x + 7y = 5$

Complete (0,):
$x = 0$
$2(0) + 7y = 5$
$0 + 7y = 5$
$7y = 5$
$y = \frac{5}{7}$
$\left(0, \frac{5}{7}\right)$
Complete (, 0):
$y = 0$
$2x + 7(0) = 5$
$2x + 0 = 5$
$2x = 5$
$x = \frac{5}{2}$
$\left(\frac{5}{2}, 0\right)$
Complete (, 1):
$y = 1$
$2x + 7(1) = 5$
$2x + 7 = 5$
$2x = -2$
$x = -1$
$(-1, 1)$

x	y
0	$\frac{5}{7}$
$\frac{5}{2}$	0
−1	1

65. $x = -5y$

Complete (, 0):
$y = 0$
$x = -5(0)$
$x = 0$
$(0, 0)$
Complete (, 1):
$y = 1$
$x = -5(1)$
$x = -5$
Complete (10,):
$x = 10$
$10 = -5y$
$-2 = y$
$(10, -2)$

x	y
0	0
−5	1
10	−2

Plot the points.

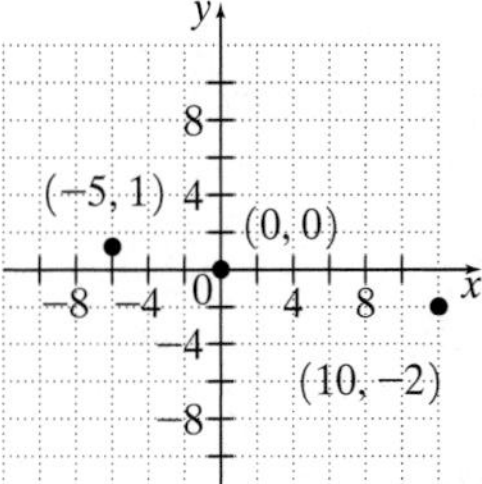

69.
$$x + y = 5$$
$$x - x + y = 5 - x$$
$$y = 5 - x$$

73. $10x = -5y$
$$-\frac{1}{5}(10x) = -\frac{1}{5}(-5y)$$
$$-2x = y \text{ or } y = -2x$$

77. a. To estimate the increases, take differences.
Year 1: ($21 − $20) billion = $1 billion
Year 2: ($23 − $21) billion = $2 billion
Year 3: ($24 − $23) billion = $1 billion
Year 4: ($25 − $24) billion = $1 billion

b. The estimate is $32 billion.

81. a. $y = -4.22x + 985.02$
Complete (4,):
$x = 4$
$y = -4.22(4) + 985.02$
$= -16.88 + 985.02$
$= 968.14$
$(4, 968.14)$
Complete (7,):
$x = 7$
$y = -4.22(7) + 985.02$
$= -29.54 + 985.02$
$= 955.48$
$(7, 955.48)$
Complete (10,):
$x = 10$
$y = -4.22(10) + 985.02$
$= -42.2 + 985.02$
$= 942.82$
$(10, 942.82)$

x	y
4	968.14
7	955.48
10	942.82

b. Let $y = 947$.
$$947 = -4.22x + 985.02$$
$$947 - 985.02 = -4.22x + 985.02 - 985.02$$
$$-38.02 = -4.22x$$
$$\frac{-38.02}{-4.22} = \frac{-4.22x}{-4.22}$$
$$9 \approx x$$
The year is 1990 + 9 = 1999.

Exercise Set 3.2

1. $x - y = 6$
$y = 0, x = 6;\ (6, 0)$
$x = 4$
$4 - y = 6$
$y = -2,\ (4, -2)$
$y = -1$
$x - (-1) = 6$
$x = 5;\ (5, -1)$

x	y
6	0
4	−2
5	−1

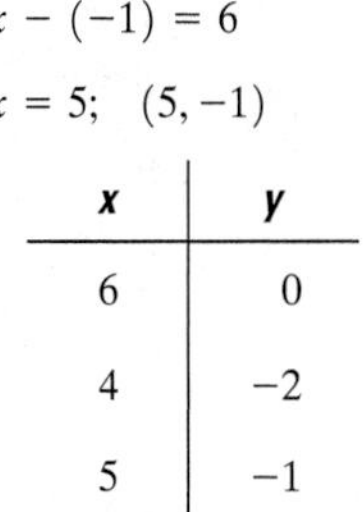

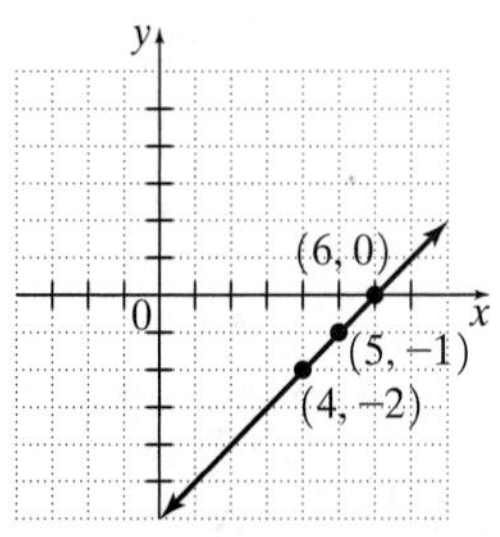

5. $y = \frac{1}{3}x$

$x = 0, y = 0;\quad (0, 0)$

$x = 6, y = 2;\quad (6, 2)$

$x = -3, y = -1;\quad (-3, -1)$

x	y
0	1
6	2
−3	−1

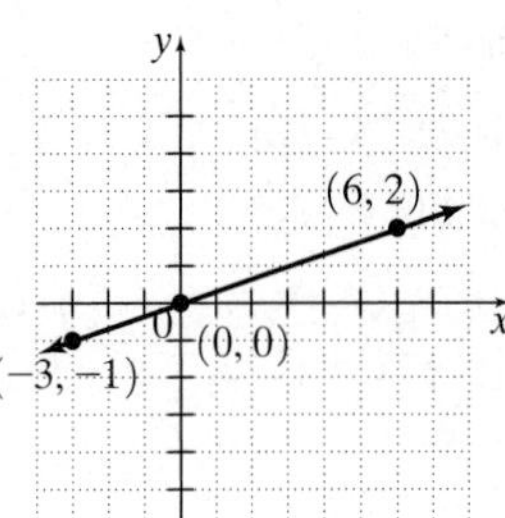

9. $x + y = 1$

Find 3 points:

x	y
0	1
1	0
−1	2

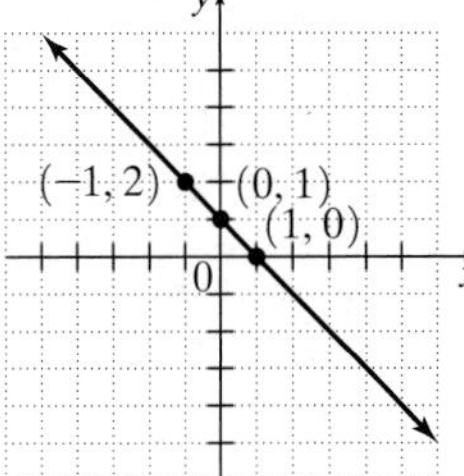
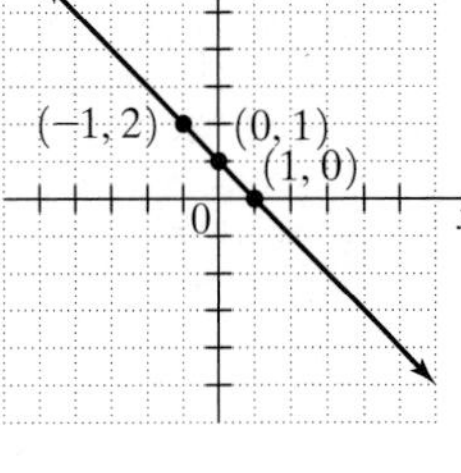

13. $x - 2y = 6$

Find 3 points:

x	y
0	−3
6	0
4	−1

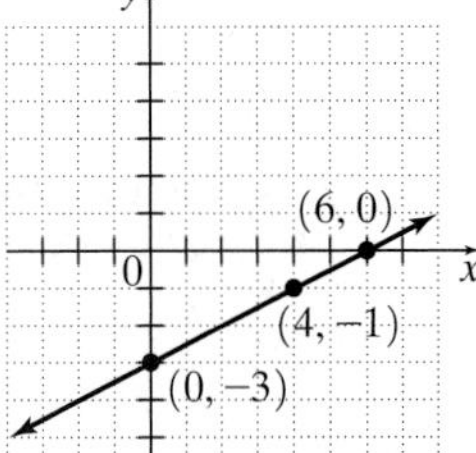

17. $x = -4$

$y = \text{any value}$

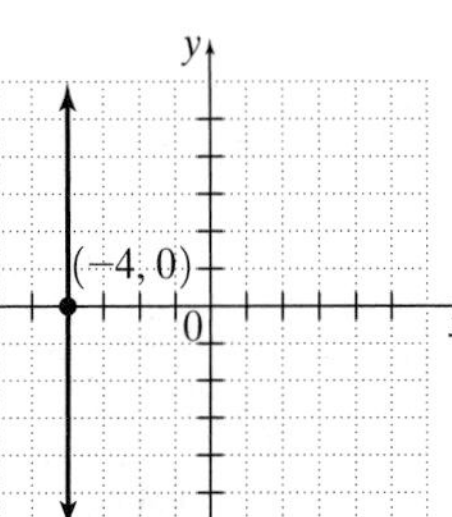

21. $y = x$

Find 3 points:

x	y
2	2
0	0
−2	−2

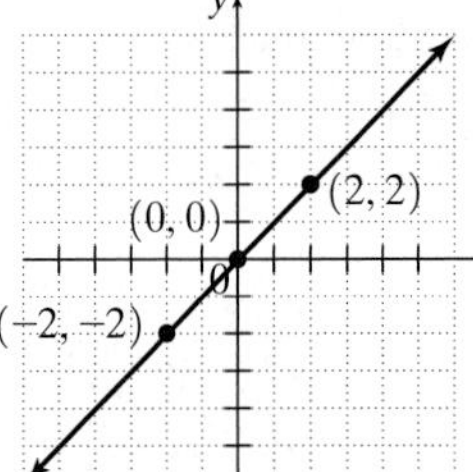

25. $x + 3y = 9$

Find 3 points:

x	y
0	3
3	2
−3	4

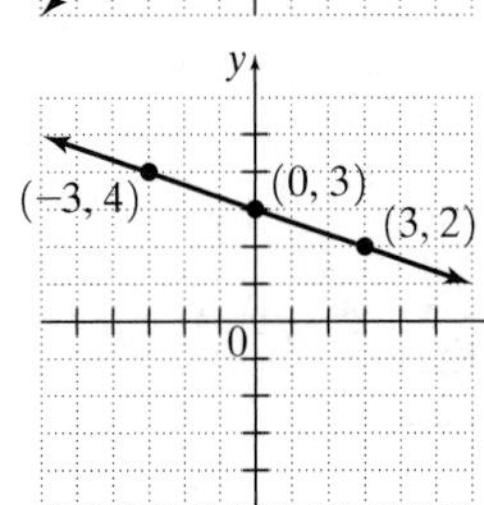

29. $3x - 2y = 12$

Find 3 points:

x	y
0	−6
4	0
2	−3

y
(4, 0)
0
x
(2, −3)
(0, −6)

33. Find 3 points for each line:

$y = -2x$

x	y
2	−4
0	0
−2	4

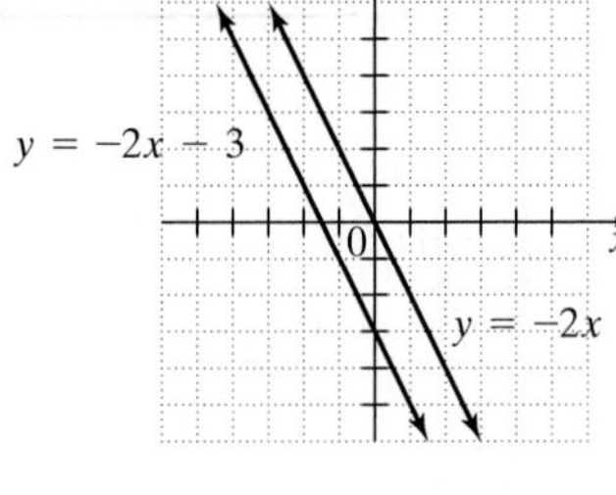

$y = -2x - 3$

x	y
−2	1
−1	−1
0	−3

answers may vary

37. $x - y = -3$

$x = 0$

$0 - y = -3$

$y = 3;\quad (0, 3)$

$y = 0$

$x - 0 = -3$

$x = -3;\quad (-3, 0)$

x	y
0	3
−3	0

41. $y = x^2$

x	y
0	0
1	1
−1	1
2	4
−2	4

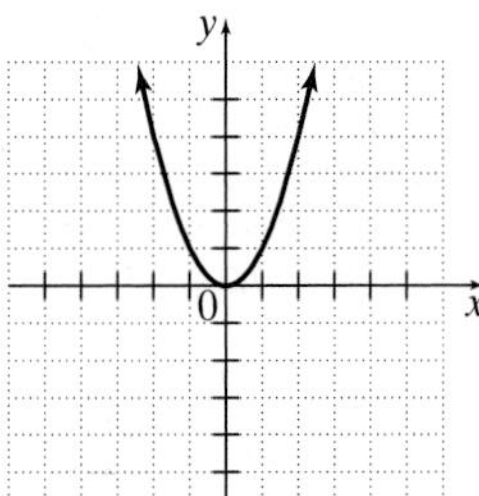

45. $y = 45x + 2214$

a. Find 3 points:

x	y
0	2214
4	2394
8	2574

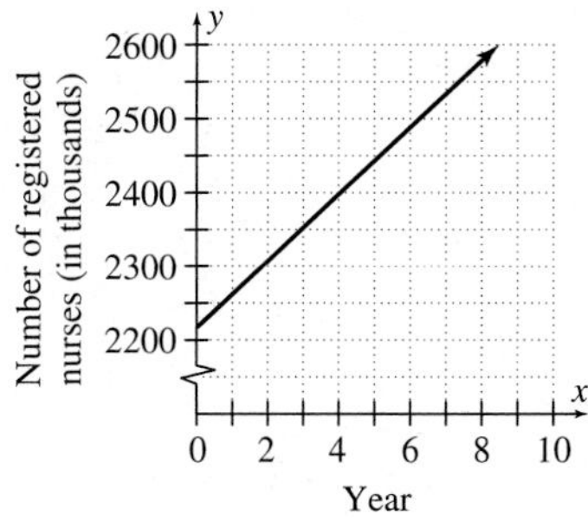

b. Yes; 6 years after 2001 (or in 2007) there will be 2484 thousand registered nurses.

Exercise Set 3.3

1. $x = -1; y = 1;$

$(-1, 0); \quad (0, 1)$

5. infinite

9. $x - y = 3$

If $x = 0$, then $y = -3$

If $y = 0$, then $x = 3$

Plot using $(0, -3)$ and $(3, 0)$:

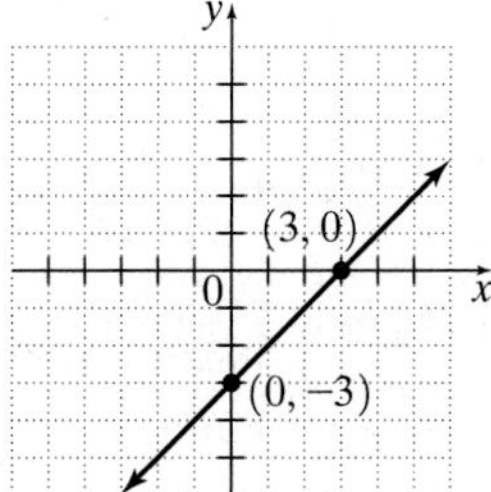

13. $-x + 2y = 6$

If $x = 0$, then $y = 3$

If $y = 0$, then $x = -6$

Plot using $(0, 3)$ and $(-6, 0)$:

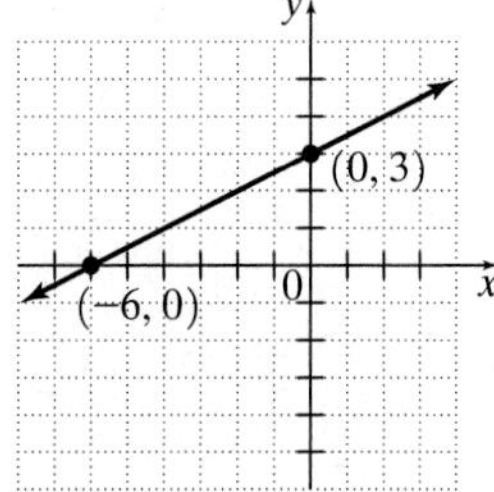

17. $x = 2y$

If $x = 0, y = 0$

Need another point:

If $y = 1, x = 2$

Plot using $(0, 0)$ and $(2, 1)$:

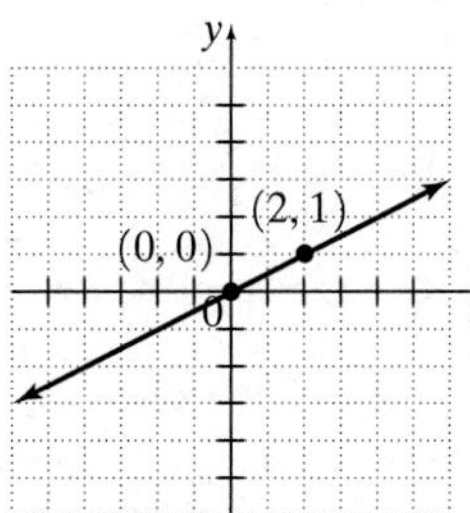

21. $x = y$

If $x = 0, y = 0$

Need another point:

If $x = 3, y = 3$

Plot using $(0, 0)$ and $(3, 3)$:

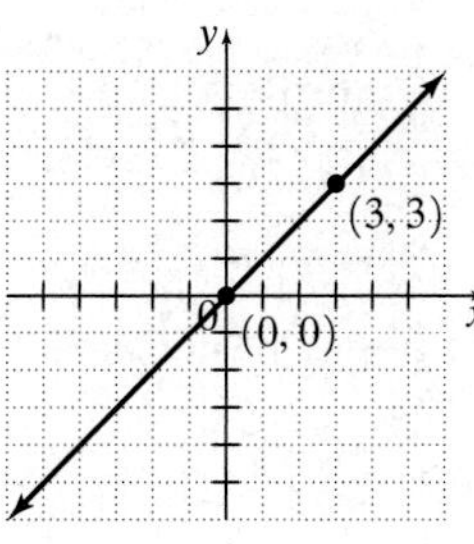

25. $5 = 6x - y$

If $x = 0, y = -5$

If $y = 0, x = \frac{5}{6}$

Plot using $(0, -5)$ and $\left(\frac{5}{6}, 0\right)$:

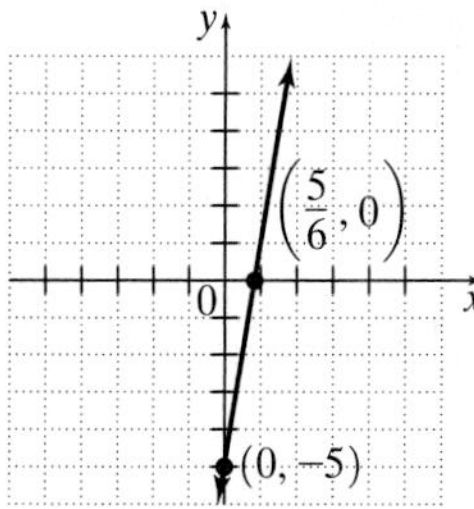

29. $x = -1$

For any y-value, x is -1.

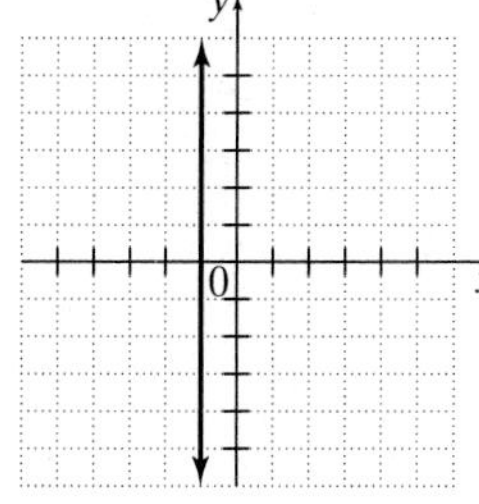

33. $y + 7 = 0$

$y = -7$

For any x-value, y is -7.

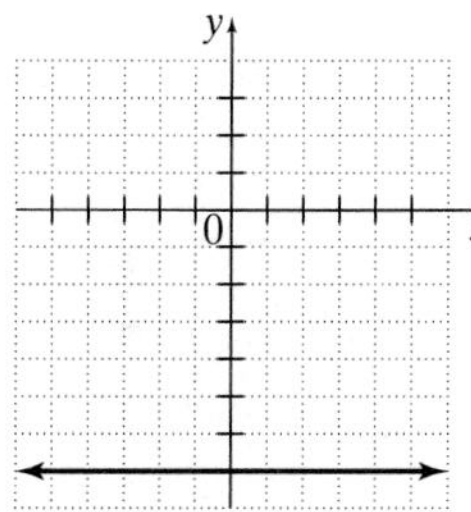

37. $\frac{-6 - 3}{2 - 8} = \frac{-9}{-6} = \frac{3}{2}$

41. $\frac{0 - 6}{5 - 0} = \frac{-6}{5} = -\frac{6}{5}$

45. $x = 3$

For any y-value, $x = 3$.

A

49. $3x + 6y = 1200$

a. If $x = 0$, $6y = 1200$

$y = 200$

$(0, 200)$ corresponds to no chairs and 200 desks are manufactured.

b. If $y = 0, 3x = 1200$

$x = 400$

$(400, 0)$ corresponds to 400 chairs and no desks are manufactured.

c. If $y = 50$

$$3x + 6(50) = 1200$$
$$3x + 300 = 1200$$
$$3x = 900$$
$$x = 300$$

300 chairs can be made.

53. $y = 51.6x + 560.2$

a. If $x = 0, \quad y = 560.2$

b. In the year 1996, the number of Disney Stores was about 560.2.

Exercise Set 3.4

1. $p_1 = (-1, 2);\ p_2 = (2, -2)$

$$m = \frac{y_2 - y_1}{x_2 - x_1} = \frac{-2 - 2}{2 - (-1)} = -\frac{4}{3}$$

5. $(0, 0)$ and $(7, 8)$

$$m = \frac{y_2 - y_1}{x_2 - x_1} = \frac{8 - 0}{7 - 0} = \frac{8}{7}$$

9. $(1, 4)$ and $(5, 3)$

$$m = \frac{y_2 - y_1}{x_2 - x_1} = \frac{3 - 4}{5 - 1} = -\frac{1}{4}$$

13. $(5, 1)$ and $(-2, 1)$

$$m = \frac{y_2 - y_1}{x_2 - x_1} = \frac{1 - 1}{-2 - 5} = \frac{0}{-7} = 0$$

17. line 2 increases faster than line 1; line 2

21. $2x + y = 7$

$y = -2x + 7$

$m = -2$

25. $x = 2y$

$y = \frac{1}{2}x$

$m = \frac{1}{2}$

29. $x = 1$

This is a vertical line, so it has an undefined slope.

33. $x - 3y = -6$ $\quad$ $3x - y = 0$

$3y = x + 6$ $\quad$ $y = 3x$

$y = \frac{1}{3}x + 2$ $\quad$ $m = 3$

$m = \frac{1}{3}$

$\frac{1}{3}(3) = 1 \neq -1$

neither

37. $6x = 5y + 1$ $\quad$ $-12x + 10y = 1$

$5y = 6x - 1$ $\quad$ $10y = 12x + 1$

$y = \frac{6}{5}x - \frac{1}{5}$ $\quad$ $y = \frac{6}{5}x + \frac{1}{10}$

$m = \frac{6}{5}$ $\quad$ $m = \frac{6}{5}$

parallel

41. $(-8, -4)$ and $(3, 5)$

$$m = \frac{y_2 - y_1}{x_2 - x_1} = \frac{5 - (-4)}{3 - (-8)} = \frac{9}{11}$$

a. $\frac{9}{11}$ $\quad$ **b.** $-\frac{11}{9}$

45. Grade is $\frac{2}{16} = \frac{1}{8} = 0.125 = 12.5\%$

49. Slope is $\frac{0.25}{12} \approx 0.02$

53. Slope is $\frac{7200 - 1800}{20{,}000 - 5{,}000} = \frac{5400}{15{,}000} = 0.36$

It costs \$0.36/mi.

57. $y - 1 = -6(x - (-2))$

$y - 1 = -6(x + 2)$

$y - 1 = -6x - 12$

$y = -6x - 11$

61. A vertical line has an undefined slope. B

65. 28.3 mpg

69. From 1992 to 1993 is the steepest line with

$$m = \frac{28.2 - 27.6}{1} = \frac{0.6}{1} = 0.6.$$

73. a. $(1994, 782)$ and $(2001, 1132)$

b. $m = \frac{1132 - 782}{2001 - 1994} = \frac{350}{7} = 50$

c. The farmland rose at a rate of \$50 every year.

77. Slope of line joining $(-4, 4)$ and $(-3, 0)$:

$$m_1 = \frac{0 - 4}{-3 - (-4)} = \frac{-4}{1} = -4$$

Slope of line joining $(1, 1)$ and $(-3, 0)$:

$$m_2 = \frac{0 - 1}{-3 - 1} = \frac{-1}{-4} = \frac{1}{4}$$

Since $m_1 \cdot m_2 = -4 \cdot \frac{1}{4} = -1$, the lines are perpendicular thus giving a right triangle.

81. $m = \frac{5.1 - 0.2}{7.9 - 2.3} = \frac{4.9}{5.6} = 0.875$

Exercise Set 3.5

1. $x - y > 3$

$(0, 3): 0 - 3 > 3?$

$-3 > 3?$ No

$(2, -1): -2 - (-1) > 3?$

$-1 > 3?$ No

5. $x < -y$

$(0, 2): 0 < -2?$ No

$(-5, 1): -5 < -1?$ Yes

9. $2x - y > -4$

$2x - y = -4$

$y = 2x + 4$

Test $(0, 0)$:

$2(0) - 0 > -4?$

$0 > -4?$

Yes; shade below the line.

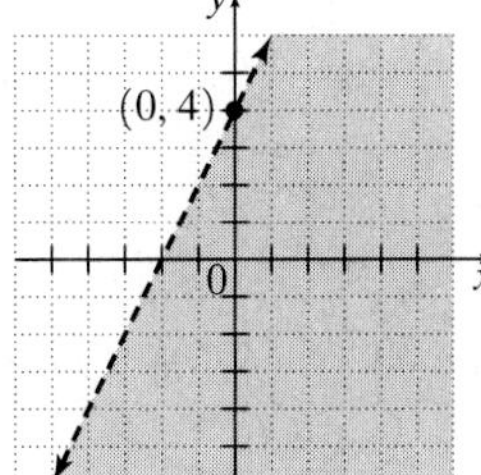

13. $x \le -3y$

$x = -3y$

$y = -\frac{1}{3}x$

Test $(0, 1)$:

$0 \le -3(1)?$

$0 \le -3?$

No; shade below the line.

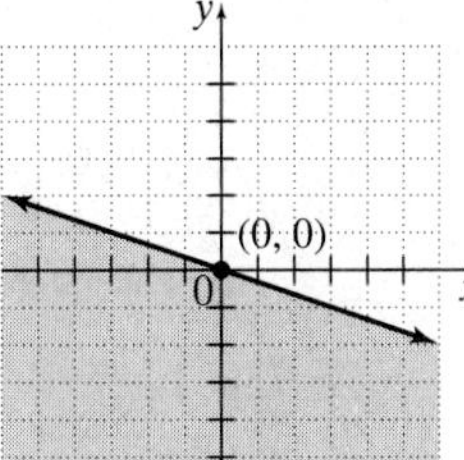

17. $y < 4$

$y = 4$

Test $(0, 0)$:

$0 < 4?$

Yes; shade below the line.

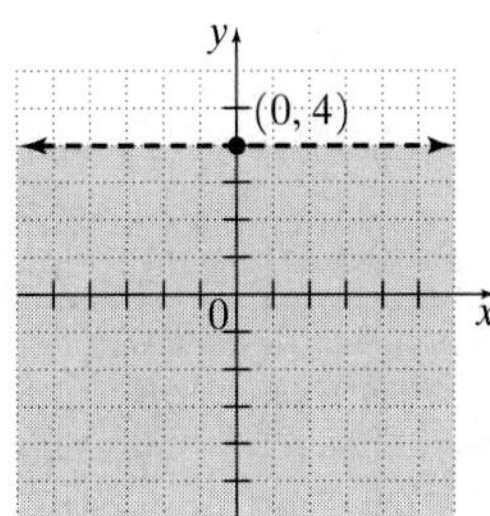

21. $5x + 2y \le 10$

$5x + 2y = 10$

$2y = -5x + 10$

$y = -\frac{5}{2}x + 5$

Test $(0, 0)$:

$5(0) + 2(0) \le 10?$

$0 \le 10?$

Yes; shade below the line.

25. $x - y \le 6$

$x - y = 6$

$y = x - 6$

Test $(0, 0)$:

$0 - 0 \le 6?$

$0 \le 6?$

Yes; shade above the line.

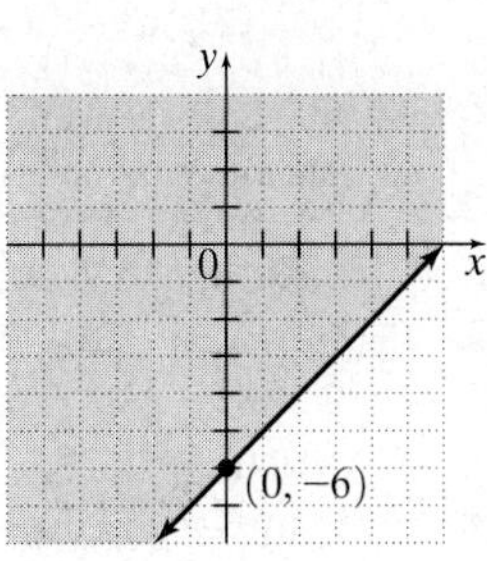

29. $2x + 7y > 5$

$2x + 7y = 5$

$7y = -2x + 5$

$y = -\frac{2}{7}x + \frac{5}{7}$

Test $(0, 0)$:

$2(0) + 7(0) > 5?$

$0 > 5?$

No; shade above the line.

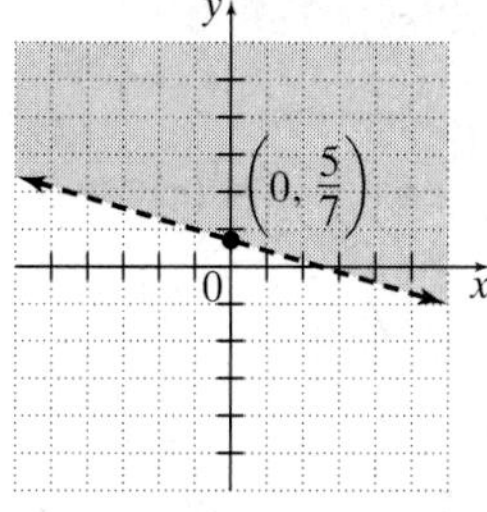

33. $(-2, -1)$

37. choice A

41. answers may vary

Chapter 3 Test

1. $12y - 7x = 5$

If $x = 1, 12y - 7(1) = 5$

$12y = 12$

$y = 1$

$(1, 1)$

5. $(6, -5)$ and $(-1, 2)$

$$m = \frac{y_2 - y_1}{x_2 - x_1} = \frac{2 - (-5)}{-1 - 6} = \frac{7}{-7} = -1$$

9. $2x + y = 8$

If $x = 0, y = 8$

If $y = 0, 2x = 8$

$x = 4$

Plot using $(0, 8)$ and $(4, 0)$.

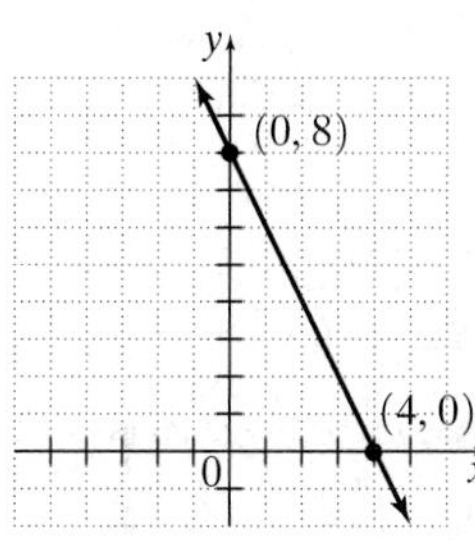

13. $5x - 7y = 10$

If $y = 0, 5x = 10$

$x = 2$

If $y = -5, 5x - 7(-5) = 10$

$5x + 35 = 10$

$5x = -25$

$x = -5$

Plot using $(2, 0)$ and $(-5, -5)$.

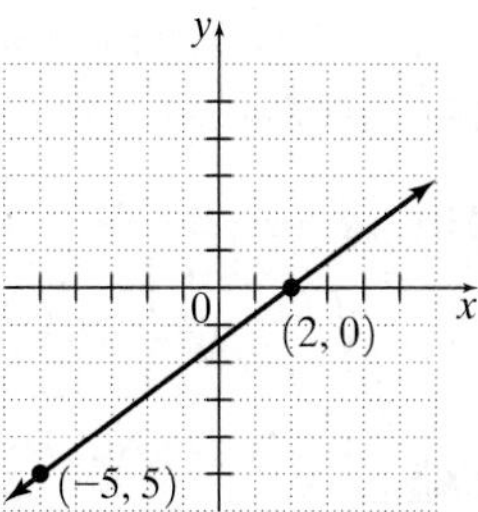

17. $y = 2x - 6$ $\quad -4x = 2y$

$m = 2$ $\quad y = -2x$

$m = -2$

neither

21. $m = \frac{y_2 - y_1}{x_2 - x_1} = \frac{1420 - 1480}{2000 - 1998} = \frac{-60}{2} = -30$

Every 1 year, 30 million fewer movie tickets are sold.

25. 2002

Chapter 4

Exercise Set 4.1

1. $7^2 = 7 \cdot 7 = 49$

5. $-2^4 = -2 \cdot 2 \cdot 2 \cdot 2 = -16$

9. $\left(\frac{1}{3}\right)^3 = \left(\frac{1}{3}\right)\left(\frac{1}{3}\right)\left(\frac{1}{3}\right) = \frac{1}{27}$

13. answers may vary

17. $5x^3 = 5(3)^3 = 5 \cdot 3 \cdot 3 \cdot 3 = 135$

21. $\frac{2z^4}{5} = \frac{2(-2)^4}{5} = \frac{2(-2)(-2)(-2)(-2)}{5} = \frac{32}{5}$

25. $(-3)^3 \cdot (-3)^9 = (-3)^{3+9} = (-3)^{12}$

29. $(4z^{10})(-6z^7)(z^3) = 4(-6)z^{10+7+3} = -24z^{20}$

33. $(x^9)^4 = x^{9 \cdot 4} = x^{36}$

37. $(2a^5)^3 = 2^3a^{5 \cdot 3} = 8a^{15}$

41. $(x^2y^3)^5 = x^{2 \cdot 5}y^{3 \cdot 5} = x^{10}y^{15}$

45. $(8z^5)^2 = 8^2z^{5 \cdot 2} = 64z^{10}$

The area is $64z^{10}$ sq decimeters.

49. $\frac{x^3}{x} = \frac{x^3}{x^1} = x^{3-1} = x^2$

53. $\frac{p^7q^{20}}{pq^{15}} = p^{7-1}q^{20-15} = p^6q^5$

57. $(2x)^0 = 1$

61. $5^0 + y^0 = 1 + 1 = 2$

65. $-5^2 = -5 \cdot 5 = -25$

69. $\frac{z^{12}}{z^4} = z^{12-4} = z^8$

73. $(6b)^0 = 1$

77. $b^4b^2 = b^{4+2} = b^6$

81. $(2x^3)(-8x^4) = 2(-8)x^{3+4} = -16x^7$

85. $(-6xyz^3)^2 = (-6)^2x^2y^2z^{3 \cdot 2} = 36x^2y^2z^6$

89. $\frac{3x^5}{x^4} = 3x^{5-4} = 3x$

93. $5 - 7 = 5 + (-7) = -2$

97. $-11 - (-4) = -11 + 4 = -7$

101. We use the volume formula.

105. $(a^b)^5 = a^{b \cdot 5} = a^{5b}$

109. $A = P\left(1 + \frac{r}{12}\right)^6$

$A = 1000\left(1 + \frac{0.09}{12}\right)^6$

$A = 1000(1.0075)^6$

$A = 1045.85$

You need \$1045.85 to pay off the loan.

Exercise Set 4.2

1. $4^{-3} = \frac{1}{4^3} = \frac{1}{64}$

5. $\left(-\frac{1}{4}\right)^{-3} = \frac{(-1)^{-3}}{(4)^{-3}} = \frac{4^3}{(-1)^3} = \frac{64}{-1} = -64$

9. $\frac{1}{p^{-3}} = p^3$

13. $\frac{x^{-2}}{x} = x^{-2-1} = x^{-3} = \frac{1}{x^3}$

17. $2^0 + 3^{-1} = 1 + \frac{1}{3} = \frac{3}{3} + \frac{1}{3} = \frac{4}{3}$

21. $\frac{-1}{p^{-4}} = -1(p^4) = -p^4$

25. $\frac{x^2x^5}{x^3} = x^{2+5-3} = x^4$

29. $\frac{(m^5)^4m}{m^{10}} = m^{5(4)+1-10} = m^{20+1-10} = m^{11}$

33. $(x^5y^3)^{-3} = x^{5(-3)}y^{3(-3)} = x^{-15}y^{-9} = \frac{1}{x^{15}y^9}$

37. $\frac{(a^5)^2}{(a^3)^4} = \frac{a^{10}}{a^{12}} = a^{10-12} = a^{-2} = \frac{1}{a^2}$

41. $\frac{-6m^4}{-2m^3} = \frac{-6}{-2} \cdot m^{4-3} = 3m$

45. $\frac{6x^2y^3}{-7xy^5} = -\frac{6}{7}x^{2-1}y^{3-5} = -\frac{6}{7}x^1y^{-2} = -\frac{6x}{7y^2}$

49.
$$\begin{aligned}\left(\frac{x^{-2}y^4}{x^3y^7}\right)^2 &= \frac{x^{-2(2)}y^{4(2)}}{x^{3(2)}y^{7(2)}}\\ &= \frac{x^{-4}y^8}{x^6y^{14}}\\ &= x^{-4-6}y^{8-14}\\ &= x^{-10}y^{-6}\\ &= \frac{1}{x^{10}y^6}\end{aligned}$$

53.
$$\begin{aligned}\frac{2^{-3}x^{-4}}{2^2x} &= 2^{-3-2}x^{-4-1}\\ &= 2^{-5}x^{-5}\\ &= \frac{1}{2^5x^5}\\ &= \frac{1}{32x^5}\end{aligned}$$

57.
$$\begin{aligned}\left(\frac{a^{-5}b}{ab^3}\right)^{-4} &= \frac{a^{-5(-4)}b^{-4}}{a^{-4}b^{3(-4)}}\\ &= \frac{a^{20}b^{-4}}{a^{-4}b^{-12}}\\ &= a^{20-(-4)}b^{-4-(-12)}\\ &= a^{24}b^8\end{aligned}$$

61.
$$\begin{aligned}\frac{(-2xy^{-3})^{-3}}{(xy^{-1})^{-1}} &= \frac{(-2)^{-3}x^{-3}y^9}{x^{-1}y^1}\\ &= (-2)^{-3}x^{-3-(-1)}y^{9-1}\\ &= -\frac{y^8}{8x^2}\end{aligned}$$

65.
$$\begin{aligned}\text{Volume} &= \left(\frac{3x^{-2}}{z}\text{ in.}\right)^3 = \left(\frac{3}{x^2z}\text{ in.}\right)^3\\ &= \frac{27}{x^6z^3}\text{ cu in.}\end{aligned}$$

69. $0.00000167 = 1.67 \times 10^{-6}$

73. $1{,}160{,}000 = 1.16 \times 10^6$

77. $13{,}600 = 1.36 \times 10^4$

81. $8.673 \times 10^{-10} = 0.0000000008673$

85. $2.032 \times 10^4 = 20{,}320$

89. $7.0 \times 10^8 = 700{,}000{,}000$

93. $(4 \times 10^{-10})(7 \times 10^{-9}) = 4 \cdot 7 \cdot 10^{-10} \cdot 10^{-9}$
$= 28 \times 10^{-19}$
$= 0.0000000000000000028$

97. $\dfrac{1.4 \times 10^{-2}}{7 \times 10^{-8}} = \dfrac{1.4}{7} \times 10^{-2-(-8)} = 0.2 \times 10^{6} = 200{,}000$

101. $3x - 5x + 7 = -2x + 7$

105. $7x + 2 - 8x - 6 = 7x - 8x + 2 - 6 = -x - 4$

109. $t = \dfrac{d}{r} = \dfrac{238{,}857}{186{,}000} \approx 1.3$ sec

113. $(3y^{2z})^3 = (3)^3(y^{2z})^3 = 27y^{6z}$

117. answers may vary

Exercise Set 4.3

1. $x^2 - 3x + 5$

Term	Coefficient
x^2	1
$-3x$	-3
5	5

5. $x + 2$

The degree is 1 since x is x^1. It is a binomial because it has two terms.

9. $12x^4 - x^2 - 12x^2 = 12x^4 - 13x^2$

The degree is 4, the greatest degree of any of its terms. It is a binomial because the simplified form has two terms.

13. answers may vary

17. a. $x + 6 = 0 + 6 = 6$
b. $x + 6 = -1 + 6 = 5$

21. a. $x^3 - 15 = 0^3 - 15 = -15$
b. $x^3 - 15 = (-1)^3 - 15 = -1 - 15 = -16$

25. $-16t^2 + 200t = -16(7.6)^2 + 200(7.6)$
$= -924.16 + 1520$
$= 595.84$ ft

29. $14x^2 + 9x^2 = (14 + 9)x^2 = 23x^2$

33. $8s - 5s + 4s = (8 - 5 + 4)s = 7s$

37. $5x + 3 + 4x + 3 + 2x + 6 + 3x + 7x$
$= (5x + 4x + 2x + 3x + 7x) + (3 + 3 + 6)$
$= 21x + 12$

41. $9ab - 6a + 5b - 3$

Term	Degree	Degree of Polynomial
$9ab$	$1 + 1$ or 2	2 (highest degree)
$-6a$	1	
$5b$	1	
-3	0	

45. $3ab - 4a + 6ab - 7a = (3 + 6)ab + (-4 - 7)a$
$= 9ab - 11a$

49. $5x^2y + 6xy^2 - 5yx^2 + 4 - 9y^2x$
$= (5 - 5)x^2y + (6 - 9)xy^2 + 4$
$= -3xy^2 + 4$

53. $4 + 5(2x + 3) = 4 + 10x + 15 = 10x + 19$

57. answers may vary

Exercise Set 4.4

1. $(3x + 7) + (9x + 5) = 3x + 7 + 9x + 5$
$= (3x + 9x) + (7 + 5)$
$= 12x + 12$

5. $(-5x^2 + 3) + (2x^2 + 1) = -5x^2 + 3 + 2x^2 + 1$
$= (-5x^2 + 2x^2) + (3 + 1)$
$= -3x^2 + 4$

9.
$$\begin{array}{r} 3t^2 + 4 \\ + 5t^2 - 8 \\ \hline 8t^2 - 4 \end{array}$$

13. $(2x + 5) - (3x - 9) = (2x + 5) + (-3x + 9)$
$= 2x + 5 + (-3x) + 9$
$= (2x - 3x) + (5 + 9)$
$= -x + 14$

17. $(2x^2 + 3x - 9) - (-4x + 7)$
$= (2x^2 + 3x - 9) + (4x - 7)$
$= 2x^2 + 3x - 9 + 4x - 7$
$= 2x^2 + (3x + 4x) + (-9 - 7)$
$= 2x^2 + 7x - 16$

21. $(5x + 8) - (-2x^2 - 6x + 8)$
$= (5x + 8) + (2x^2 + 6x - 8)$
$= 5x + 8 + 2x^2 + 6x - 8$
$= 2x^2 + (5x + 6x) + (8 - 8)$
$= 2x^2 + 11x$

25.
$$\begin{array}{r} 5u^5 - 4u^2 + 3u - 7 \\ -(3u^5 + 6u^2 - 8u + 2) \\ \hline \end{array}$$
$$\begin{array}{r} 5u^5 - 4u^2 + 3u - 7 \\ +(-3u^5 - 6u^2 + 8u - 2) \\ \hline 2u^5 - 10u^2 + 11u - 9 \end{array}$$

29. $(7y + 7) - (y - 6) = 7y + 7 - y + 6$
$= 6y + 13$

33. $(3x^2 + 5x - 8) + (5x^2 + 9x + 12) - (x^2 - 14)$
$= 3x^2 + 5x - 8 + 5x^2 + 9x + 12 - x^2 + 14$
$= 7x^2 + 14x + 18$

37. $(4x^2 - 6x + 1) + (3x^2 + 2x + 1)$
$= 4x^2 - 6x + 1 + 3x^2 + 2x + 1$
$= 7x^2 - 4x + 2$

41. $[(8y^2 + 7) + (6y + 9)] - (4y^2 - 6y - 3)$
$= 8y^2 + 7 + 6y + 9 - 4y^2 + 6y + 3$
$= 4y^2 + 12y + 19$

45. $(9a + 6b - 5) + (-11a - 7b + 6)$
$= 9a + 6b - 5 - 11a - 7b + 6$
$= -2a - b + 1$

49. $(x^2 + 2xy - y^2) + (5x^2 - 4xy + 20y^2)$
$= x^2 + 2xy - y^2 + 5x^2 - 4xy + 20y^2$
$= 6x^2 - 2xy + 19y^2$

53. $3x(2x) = 3 \cdot 2 \cdot x \cdot x = 6x^2$

57. $10x^2(20xy^2) = 10 \cdot 20 \cdot x^2 \cdot x \cdot y^2$
$= 200x^{2+1}y^2$
$= 200x^3y^2$

61. $(4y^2 + 4y + 1) - (y^2 - 10)$
$= 4y^2 + 4y + 1 - y^2 + 10$
$= (3y^2 + 4y + 11)$ m

65. $(-2.85x^2 + 8.75x + 26.7) + (0.35x^2 + 3.55x + 40)$
$= -2.85x^2 + 8.75x + 26.7 + 0.35x^2 + 3.55x + 40$
$= -2.5x^2 + 12.3x + 66.7$

Exercise Set 4.5

1. $8x^2 \cdot 3x = (8 \cdot 3)(x^2 \cdot x) = 24x^3$

5. $(-x^3)(-x) = (-1)(-1)(x^3 \cdot x) = x^4$

9. $(2x)(-3x^2)(4x^5) = (2)(-3)(4)(x \cdot x^2 \cdot x^5) = -24x^8$

13. $7x(x^2 + 2x - 1) = 7x(x^2) + 7x(2x) + 7x(-1)$
$= 7x^3 + 14x^2 - 7x$

17. $3x(2x^2 - 3x + 4) = 3x(2x^2) + 3x(-3x) + 3x(4)$
$= 6x^3 - 9x^2 + 12x$

21. $-2a^2(3a^2 - 2a + 3)$
$= -2a^2(3a^2) - 2a^2(-2a) - 2a^2(3)$
$= -6a^4 + 4a^3 - 6a^2$

25. $x^2 + 3x = x(x + 3)$

29. $(a + 7)(a - 2) = a(a) + a(-2) + 7(a) + 7(-2)$
$= a^2 - 2a + 7a - 14$
$= a^2 + 5a - 14$

33. $(3x^2 + 1)(4x^2 + 7)$
$= 3x^2(4x^2) + 3x^2(7) + 1(4x^2) = 1(7)$
$= 12x^4 + 21x^2 + 4x^2 + 7$
$= 12x^4 + 25x^2 + 7$

37. $(1 - 3a)(1 - 4a)$
$= 1(1) + 1(-4a) - 3a(1) - 3a(-4a)$
$= 1 - 4a - 3a + 12a^2$
$= 1 - 7a + 12a^2$

41. $(x - 2)(x^2 - 3x + 7)$
$= x(x^2) + x(-3x) + x(7) - 2(x^2) - 2(-3x) - 2(7)$
$= x^3 - 3x^2 + 7x - 2x^2 + 6x - 14$
$= x^3 - 5x^2 + 13x - 14$

45. $(2a - 3)(5a^2 - 6a + 4)$
$= 2a(5a^2) + 2a(-6a) + 2a(4) - 3(5a^2) - 3(-6a) - 3(4)$
$= 10a^3 - 12a^2 + 8a - 15a^2 + 18a - 12$
$= 10a^3 - 27a^2 + 26a - 12$

49. $x^2 + 2x + 3x + 2(3) = x^2 + 5x + 6$

53.
$$\begin{array}{r} 2x^2 + 4x - 1 \\ \times \qquad\qquad x + 3 \\ \hline 6x^2 + 12x - 3 \\ 2x^3 + 4x^2 - \quad x \qquad \\ \hline 2x^3 + 10x^2 + 11x - 3 \end{array}$$

57. $(5x)^2 = 5^2x^2 = 25x^2$

61. At $t = 0$, value $= \$7000$

65. answers may vary

69. $\frac{1}{2}(3x - 2)(4x) = 2x(3x - 2)$
$= 2x(3x) + 2x(-2)$
$= 6x^2 - 4x$
$(6x^2 - 4x)$ sq in.

73. **a.** $(a + b)(a - b) = a^2 - ab + ab - b^2$
$= a^2 - b^2$
b. $(2x + 3y)(2x - 3y)$
$= (2x)^2 - 6xy + 6xy - (3y)^2$
$= 4x^2 - 9y^2$
c. $(4x + 7)(4x - 7)$
$= (4x)^2 - 28x + 28x - 7^2$
$= 16x^2 - 49$
d. $(x + y)(x - y) = x^2 - y^2$,
answers may vary

Exercise Set 4.6

1. $(x + 3)(x + 4) = x^2 + 4x + 3x + 12$
$= x^2 + 7x + 12$

5. $(5x - 6)(x + 2) = 5x^2 + 10x - 6x - 12$
$= 5x^2 + 4x - 12$

9. $(2x + 5)(3x - 1) = 6x^2 - 2x + 15x - 5$
$= 6x^2 + 13x - 5$

13. $\left(x - \frac{1}{3}\right)\left(x + \frac{2}{3}\right) = x^2 + \frac{2}{3}x - \frac{1}{3}x - \frac{2}{9}$
$= x^2 + \frac{1}{3}x - \frac{2}{9}$

17. $(x + 5y)(2x - y) = 2x^2 - xy + 10xy - 5y^2$
$= 2x^2 + 9xy - 5y^2$

21. $(2x - 1)^2 = (2x)^2 - 2(2x)(1) + (1)^2$
$= 4x^2 - 4x + 1$

25. $(x^2 + 5) = (x^2)^2 + 2(x^2)(5) + 5^2$
$= x^4 + 10x^2 + 25$

29. $(2a - 3)^2 = (2a)^2 - 2(2a)(3) + 3^2$
$= 4a^2 - 12a + 9$

33. $(3x - 7y)^2 = (3x)^2 - 2(3x)(7y) + (7y)^2$
$= 9x^2 - 42xy + 49y^2$

37. answers may vary

41. $(x + 6)(x - 6) = x^2 - 6^2 = x^2 - 36$

45. $(x^2 + 5)(x^2 - 5) = (x^2)^2 - 5^2 = x^4 - 25$

49. $(4 - 7x)(4 + 7x) = 4^2 - (7x)^2 = 16 - 49x^2$

53. $(9x + y)(9x - y) = (9x)^2 - y^2 = 81x^2 - y^2$

57. $(a + 5)(a + 4) = a^2 + 4a + 5a + 20$
$= a^2 + 9a + 20$

61. $(4a + 1)(3a - 1) = 12a^2 - 4a + 3a - 1$
$= 12a^2 - a - 1$

65. $(3a + 1)^2 = (3a)^2 + 2(3a)(1) + 1^2 = 9a^2 + 6a + 1$

69. $\left(a - \frac{1}{2}y\right)\left(1 + \frac{1}{2}y\right) = a^2 - \left(\frac{1}{2}y\right)^2 = a^2 - \frac{1}{4}y^2$

73. $(x^2 + 10)(x^2 - 10) = (x^2)^2 - (10)^2$
$= x^4 - 100$

77. $(5x - 6y)^2 = (5x)^2 - 2(5x)(6y) + (6y)^2$
$= 25x^2 - 60xy + 36y^2$

81. $\frac{50b^{10}}{70b^5} = \frac{10 \cdot 5 \cdot b^5 \cdot b^5}{10 \cdot 7 \cdot b^5} = \frac{5b^5}{7}$

85. $\frac{2x^4y^{12}}{3x^4y^4} = \frac{2 \cdot x^4 \cdot y^4 \cdot y^8}{3 \cdot x^4 \cdot y^4} = \frac{2y^8}{3}$

89. $(5x - 3)^2 - (x + 1)^2$
$= [(5x)^2 - 2(5x)(3) + 3^2] - [x^2 + 2(x)(1) + 1^2]$
$= (25x^2 - 30x + 9) - (x^2 + 2x + 1)$
$= 25x^2 - 30x + 9 - x^2 - 2x - 1$
$= (24x^2 - 32x + 8)$ sq m

Exercise Set 4.7

1. $\dfrac{20x^2 + 5x + 9}{5} = \dfrac{20x^2}{5} + \dfrac{5x}{5} + \dfrac{9}{5}$

$= 4x^2 + x + \dfrac{9}{5}$

5. $\dfrac{15p^3 + 18p^2}{3p} = \dfrac{15p^3}{3p} + \dfrac{18p^2}{3p} = 5p^2 + 6p$

9. $\dfrac{-9x^5 + 3x^4 - 12}{3x^3} = \dfrac{-9x^5}{3x^3} + \dfrac{3x^4}{3x^3} - \dfrac{12}{3x^3}$

$= -3x^2 + x - \dfrac{4}{x^3}$

13. $\dfrac{a^2b^2 - ab^3}{ab} = \dfrac{a^2b^2}{ab} - \dfrac{ab^3}{ab} = ab - b^2$

17.
$$\begin{array}{r} x + 1 \\ x + 3\overline{)x^2 + 4x + 3} \\ \underline{x^2 + 3x} \\ x + 3 \\ \underline{x + 3} \\ 0 \end{array}$$

$\dfrac{x^2 + 4x + 3}{x + 3} = x + 1$

21.
$$\begin{array}{r} 2x + 1 \\ x - 4\overline{)2x^2 - 7x + 3} \\ \underline{2x^2 - 8x} \\ x + 3 \\ \underline{x - 4} \\ 7 \end{array}$$

$\dfrac{2x^2 - 7x + 3}{x - 4} = 2x + 1 + \dfrac{7}{x - 4}$

25.
$$\begin{array}{r} 3a^2 - 3a + 1 \\ 3a + 2\overline{)9a^3 - 3a^2 - 3a + 4} \\ \underline{9a^3 + 6a^2} \\ -9a^2 - 3a \\ \underline{-9a^2 - 6a} \\ 3a + 4 \\ \underline{3a + 2} \\ 2 \end{array}$$

$\dfrac{9a^3 - 3a^2 - 3a + 4}{3a + 2} = 3a^2 - 3a + 1 + \dfrac{2}{3a + 2}$

29.
$$\begin{array}{r} 4x + 3 \\ 2x + 1\overline{)8x^2 + 10x + 1} \\ \underline{8x^2 + 4x} \\ 6x + 1 \\ \underline{6x + 3} \\ -2 \end{array}$$

$\dfrac{8x^2 + 10x + 1}{2x + 1} = 4x + 3 - \dfrac{2}{2x + 1}$

33.
$$\begin{array}{r} x^2 + 3x + 9 \\ x - 3\overline{)x^3 + 0x^2 + 0x - 27} \\ \underline{x^3 - 3x^2} \\ 3x^2 + 0x \\ \underline{3x^2 - 9x} \\ 9x - 27 \\ \underline{9x - 27} \\ 0 \end{array}$$

$\dfrac{x^3 - 27}{x - 3} = x^2 + 3x + 9$

37.
$$\begin{array}{r} 2b - 1 \\ 2b - 1\overline{)4b^2 - 4b - 5} \\ \underline{4b^2 - 2b} \\ -2b - 5 \\ \underline{-2b + 1} \\ -6 \end{array}$$

$\dfrac{-4b + 4b^2 - 5}{2b - 1} = 2b - 1 - \dfrac{6}{2b - 1}$

41.
$$\begin{array}{r|rrr} -6 & 1 & 5 & -6 \\ & & -6 & 6 \\ \hline & 1 & -1 & 0 \end{array}$$

$x - 1$

45.
$$\begin{array}{r|rrr} 2 & 4 & 0 & -9 \\ & & 8 & 16 \\ \hline & 4 & 8 & 7 \end{array}$$

$4x + 8 + \dfrac{7}{x - 2}$

49.
$$\begin{array}{r|rrrr} -1 & 3 & 7 & -4 & 12 \\ & & -3 & -4 & 8 \\ \hline & 3 & 4 & -8 & 20 \end{array}$$

$3x^2 + 4x - 8 + \dfrac{20}{x + 1}$

53. $20 = -5 \cdot (-4)$

57. $36x^2 = 4x \cdot 9x$

61. $\dfrac{12x^3 + 4x - 16}{4} = \dfrac{12x^3}{4} + \dfrac{4x}{4} - \dfrac{16}{4}$

$= 3x^3 + x - 4$

Each side is $(3x^3 + x - 4)$ ft.

65. answers may vary

Chapter 4 Test

1. $2^5 = 2 \cdot 2 \cdot 2 \cdot 2 \cdot 2 = 32$

5. $(3x^2)(-5x^9) = (3)(-5)(x^2 \cdot x^9) = -15x^{11}$

9. $\dfrac{6^2x^{-4}y^{-1}}{6^3x^{-3}y^7} = 6^{2-3}x^{-4-(-3)}y^{-1-7}$

$= 6^{-1}x^{-1}y^{-8}$

$= \dfrac{1}{6xy^8}$

13. $6.23 \times 10^4 = 62{,}300$

17. $(8x^3 + 7x^2 + 4x - 7) + (8x^3 - 7x - 6)$

$= 8x^3 + 7x^2 + 4x - 7 + 8x^3 - 7x - 6$

$= 16x^3 + 7x^2 - 3x - 13$

21.
$$\begin{array}{r} x^3 - x^2 + x + 1 \\ \times \qquad 2x^2 - 3x + 7 \\ \hline 7x^3 - 7x^2 + 7x + 7 \\ -3x^4 + 3x^3 - 3x^2 - 3x \qquad \\ \underline{2x^5 - 2x^4 + 2x^3 + 2x^2 \qquad\qquad} \\ 2x^5 - 5x^4 + 12x^3 - 8x^2 + 4x + 7 \end{array}$$

25. $(8x + 3)^2 = (8x)^2 + 2(8x)(3) + 3^2$

$= 64x^2 + 48x + 9$

29. $\dfrac{4x^2 + 2xy - 7x}{8xy} = \dfrac{4x^2}{8xy} + \dfrac{2xy}{8xy} - \dfrac{7x}{8xy}$

$= \dfrac{x}{2y} + \dfrac{1}{4} - \dfrac{7}{8y}$

Chapter 5

Exercise Set 5.1

1. y^2

5. $8x = 2 \cdot 2 \cdot 2 \cdot x$
$4 = 2 \cdot 2$
$\text{GCF} = 2 \cdot 2 = 4$

9. $-10x^2 = -2 \cdot 5 \cdot x^2$
$15x^3 = 3 \cdot 5 \cdot x^3$
$\text{GCF} = 5 \cdot x^2 = 5x^2$

13. $-18x^2y = -2 \cdot 3 \cdot 3 \cdot x^2 \cdot y$
$9x^3y^3 = 3 \cdot 3 \cdot x^3 \cdot y^3$
$36x^3y = 2 \cdot 2 \cdot 3 \cdot 3 \cdot x^3 \cdot y$
$\text{GCF} = 3 \cdot 3 \cdot x^2 \cdot y = 9x^2y$

17. $30x - 15 = 15(2x - 1)$

21. $6y^4 - 2y = 2y(3y^3 - 1)$

25. $4x - 8y + 4 = 4(x - 2y + 1)$

29. $a^7b^6 - a^3b^2 + a^2b^5 - a^2b^2$
$= a^2b^2(a^5b^4 - a + b^3 - 1)$

33. $8x^5 + 16x^4 - 20x^3 + 12$
$= 4(2x^5 + 4x^4 - 5x^3 + 3)$

37. $y(x + 2) + 3(x + 2) = (x + 2)(y + 3)$

41. answers may vary

45. $5x + 15 + xy + 3y = 5(x + 3) + y(x + 3)$
$= (x + 3)(5 + y)$

49. $2y - 8 + xy - 4x = 2(y - 4) + x(y - 4)$
$= (y - 4)(2 + x)$

53. $4x^2 - 8xy - 3x + 6y = 4x(x - 2y) - 3(x - 2y)$
$= (x - 2y)(4x - 3)$

57. $(x + 2)(x + 5) = x^2 + 2x + 5x + 10$
$= x^2 + 7x + 10$

61. The two numbers are 2 and 6.
$2 \cdot 6 = 12; 2 + 6 = 8$

65. The two numbers are -2 and 5.
$-2 \cdot 5 = -10; -2 + 5 = 3$

69. $12x^2y - 42x^2 - 4y + 14$
$= 2(6x^2y - 21x^2 - 2y + 7)$
$= 2(3x^2(2y - 7) - 1(2y - 7))$
$= 2(3x^2 - 1)(2y - 7)$

73. Let l = length of the rectangle.
$A = l \cdot w$
$4n^4 - 24n = 4n \cdot l$
$4n(n^3 - 6) = 4n \cdot l$
$\dfrac{4n(n^3 - 6)}{4n} = \dfrac{4n \cdot l}{4n}$
$n^3 - 6 = l$
The length is $(n^3 - 6)$ units.

Exercise Set 5.2

1. $x^2 + 7x + 6 = (x + 6)(x + 1)$

5. $x^2 - 3x - 18 = (x - 6)(x + 3)$

9. $x^2 + 5x + 2$ is a prime polynomial.

13. $a^4 - 2a^2 - 15 = (a^2 - 5)(a^2 + 3)$

17. answers may vary

21. $2x^3 - 18x^2 + 40x = 2x(x^2 - 9x + 20)$
$= 2x(x - 5)(x - 4)$

25. $x^2 + 15x + 36 = (x + 12)(x + 3)$

29. $r^2 - 16r + 48 = (r - 12)(r - 4)$

33. $3x^2 + 9x - 30 = 3(x^2 + 3x - 10)$
$= 3(x + 5)(x - 2)$

37. $x^2 - 18x - 144 = (x - 24)(x + 6)$

41. $x^2 - 8x + 15 = (x - 5)(x - 3)$

45. $4x^2y + 4xy - 12y = 4y(x^2 + x - 3)$

49. $x^2 + 7xy + 10y^2 = (x + 5y)(x + 2y)$

53. $x^3 - 2x^2 - 24x = x(x^2 - 2x - 24)$
$= x(x - 6)(x + 4)$

57. $5x^3y - 25x^2y^2 - 120xy^3 = 5xy(x^2 - 5xy - 24y^2)$
$= 5xy(x - 8y)(x + 3y)$

61. $(5y - 4)(3y - 1) = 15y^2 - 12y - 5y + 4$
$= 15y^2 - 17y + 4$

65. $P = 2l + 2w$
$l = x^2 + 10x$ and $w = 4x + 33$, so
$P = 2(x^2 + 10x) + 2(4x + 33)$
$= 2x^2 + 20x + 8x + 66$
$= 2x^2 + 28x + 66$
$= 2(x^2 + 14x + 33)$
$= 2(x + 11)(x + 3)$
The perimeter of the rectangle is given by the polynomial $2x^2 + 28x + 66$ which factors as $2(x + 11)(x + 3)$.

69. $y^2 - 4y + c$ is factorable when c is 3 or 4.

73. $x^{2n} + 5x^n + 6 = (x^n + 2)(x^n + 3)$

Exercise Set 5.3

1. $5x^2 + 22x + 8 = (5x + 2)(x + 4)$

5. $20x^2 - 7x - 6 = (5x + 2)(4x - 3)$

9. $8y^2 - 17y + 9 = (y - 1)(8y - 9)$

13. $20r^2 + 27r - 8 = (4r - 1)(5r + 8)$

17. $3x^2 + x - 2 = (3x - 2)(x + 1)$

21. $15x^2 - 16x - 15 = (3x - 5)(5x + 3)$

25. $2x^2 - 7x - 99 = (2x + 11)(x - 9)$

29. $3a^2 + 10ab + 3b^2 = (3a + b)(a + 3b)$

33. $18x^2 - 9x - 14 = (6x - 7)(3x + 2)$

37. $21x^2 - 48x - 45 = 3(7x^2 - 16x - 15)$
$= 3(7x + 5)(x - 3)$

41. $6x^2y^2 - 2xy^2 - 60y^2 = 2y^2(3x^2 - x - 30)$
$= 2y^2(3x - 10)(x + 3)$

45. $3x^2 - 42x + 63 = 3(x^2 - 14x + 21)$

49. $-x^2 + 2x + 24 = -(x^2 - 2x - 24)$
$= -(x - 6)(x + 4)$

53. $24x^2 - 58x + 9 = (4x - 9)(6x - 1)$

57. $15x^4 + 19x^2 + 6 = (3x^2 + 2)(5x^2 + 3)$

61. $10x^3 + 25x^2y - 15xy^2 = 5x(2x^2 + 5xy - 3y^2)$
$= 5x(2x - y)(x + 3y)$

65. January 2001 and February 2001

69. $4x^2(y - 1)^2 + 10x(y - 1)^2 + 25(y - 1)^2$
$= (y - 1)^2(4x^2 + 10x + 25)$

73. $2z^2 + bz - 7$ is factorable when b is 5 or 13.

Exercise Set 5.4

1. $x^2 + 3x + 2x + 6 = x(x + 3) + 2(x + 3)$
$= (x + 3)(x + 2)$

5. $y^2 + 8y - 2y - 16 = y(y + 8) - 2(y + 8)$
$= (y + 8)(y - 2)$

9. $8x^2 - 5x - 24x + 15 = x(8x - 5) - 3(8x - 5)$
$= (8x - 5)(x - 3)$

13. a. The numbers are 9 and 2.
$9 \cdot 2 = 18$
$9 + 2 = 11$

b. $9x + 2x = 11x$

c. $6x^2 + 11x + 3 = 6x^2 + 9x + 2x + 3$
$= 3x(2x + 3) + 1(2x + 3)$
$= (2x + 3)(3x + 1)$

17. $21y^2 + 17y + 2 = 21y^2 + 3y + 14y + 2$
$= 3y(7y + 1) + 2(7y + 1)$
$= (3y + 2)(7y + 1)$

21. $10x^2 - 9x + 2 = 10x^2 - 5x - 4x + 2$
$= 5x(2x - 1) - 2(2x - 1)$
$= (2x - 1)(5x - 2)$

25. $4x^2 + 12x + 9 = 4x^2 + 6x + 6x + 9$
$= 2x(2x + 3) + 3(2x + 3)$
$= (2x + 3)(2x + 3)$
$= (2x + 3)^2$

29. $10x^2 - 23x + 12 = 10x^2 - 15x - 8x + 12$
$= 5x(2x - 3) - 4(2x - 3)$
$= (2x - 3)(5x - 4)$

33. $16y^2 - 34y + 18 = 2(8y^2 - 17y + 9)$
$= 2(8y^2 - 8y - 9y + 9)$
$= 2[8y(y - 1) - 9(y - 1)]$
$= 2(y - 1)(8y - 9)$

37. $54a^2 - 9a - 30 = 3(18a^2 - 3a - 10)$
$= 3(18a^2 - 15a + 12a - 10)$
$= 3[3a(6a - 5) + 2(6a - 5)]$
$= 3(6a - 5)(3a + 2)$

41. $12x^3 - 27x^2 - 27x = 3x(4x^2 - 9x - 9)$
$= 3x(4x^2 - 12x + 3x - 9)$
$= 3x[4x(x - 3) + 3(x - 3)]$
$= 3x(x - 3)(4x + 3)$

45. $(y + 4)(y + 4) = y^2 + 4y + 4y + 16$
$= y^2 + 8y + 16$

49. $(4x - 3)^2 = 16x^2 + 2(4x)(-3) + 9$
$= 16x^2 - 24x + 9$

53. $3x^{2n} + 16x^n - 35 = 3x^{2n} - 5x^n + 21x^n - 35$
$= x^n(3x^n - 5) + 7(3x^n - 5)$
$= (3x^n - 5)(x^n + 7)$

Exercise Set 5.5

1. Yes; two terms, x^2 and 64, are squares ($64 = 8^2$), and the third term of the trinomial, $16x$, is twice the product of x and 8 ($2 \cdot x \cdot 8 = 16x$).

5. No; if we first factor out the GCF, 4, we find that only one of the terms, x^2, is a square.

9. $x^2 + 8x + 16$ is a perfect square trinomial because x^2 and 16 are squares ($16 = 4^2$), and $8x$ is twice the product of x and 4 ($2 \cdot x \cdot 4 = 8x$).

13. $x^2 - 16x + 64 = x^2 - 2 \cdot x \cdot 8 + 8^2$
$= (x - 8)^2$

17. $3x^2 - 24x + 48 = 3(x^2 - 8x + 16)$
$= 3(x^2 - 2 \cdot x \cdot 4 + 4^2)$
$= 3(x - 4)^2$

21. $m^3 + 18m^2 + 81m = m(m^2 + 18m + 81)$
$= m(m^2 + 2 \cdot m \cdot 9 + 9^2)$
$= m(m + 9)^2$

25. $9x^2 - 24xy + 16y^2 = (3x)^2 - 2 \cdot 3x \cdot 4y + (4y)^2$
$= (3x - 4y)^2$

29. $x^2 - 25 = x^2 - 5^2 = (x + 5)(x - 5)$

33. $16r^2 + 1$ is prime

37. $(y + 2)^2 - 49 = (y + 2)^2 - 7^2$
$= (y + 2 + 7)(y + 2 - 7) = (y + 9)(y - 5)$

41. $18x^2y - 2y = 2y(9x^2 - 1)$
$= 2y[(3x)^2 - 1^2] = 2y(3x + 1)(3x - 1)$

45. $x^4 - 81 = (x^2)^2 - 9^2 = (x^2 + 9)(x^2 - 9)$
$= (x^2 + 9)(x^2 - 3^2) = (x^2 + 9)(x + 3)(x - 3)$

49. $x^2 + 16x + 64 - x^4 = (x + 8)^2 - (x^2)^2$
$= (x + 8 + x^2)(x + 8 - x^2)$

53. $4x^2 + 4x + 1 - z^2 = (2x + 1)^2 - z^2$
$= (2x + 1 + z)(2x + 1 - z)$

57. $x^3 + 27 = x^3 + 3^3$
$= (x + 3)(x^2 - 3x + 9)$

61. $m^3 + n^3 = (m + n)(m^2 - mn + n^2)$

65. $a^3b + 8b^4 = (a^3 + 8b^3)b$
$= (a^3 + (2b)^3)b$
$= (a + 2b)(a^2 - 2ab + 4b^2)b$

69. $x^6 - y^3 = (x^2)^3 - y^3$
$= (x^2 - y)(x^4 + x^2y + y^2)$

73. $x^3 - 1 = x^3 - 1^3$
$= (x - 1)(x^2 + x + 1)$

77. $3x^6y^2 + 81y^2 = 3y^2(x^6 + 27)$
$= 3y^2((x^2)^3 + 3^3)$
$= 3y^2(x^2 + 3)(x^4 - 3x^2 + 9)$

81. $3x + 1 = 0$
$3x = -1$
$x = -\frac{1}{3}$

$-\frac{1}{3}$

85. $-5x + 25 = 0$
$-5x = -25$
$x = 5$

5

89. a. Let $t = 3$.
$1444 - 16t^2 = 1444 - 16(3)^2$
$= 1444 - 16(9) = 1444 - 144 = 1300$
The height is 1300 ft.

b. Let $t = 7$.
$1444 - 16t^2 = 1444 - 16(7)^2$
$= 1444 - 16(49) = 1444 - 748 = 660$
The height is 660 ft.

c. When the object hits the ground, its height is zero feet. Thus, to find the time, t, when the object's height is zero feet above the ground, we set the expression $1440 - 16t^2$ equal to 0 and solve for t.

$$\begin{aligned} 1440 - 16t^2 &= 0 \\ 1440 - 16t^2 + 16t^2 &= 0 + 16t^2 \\ 1440 &= 16t^2 \\ \frac{1440}{16} &= \frac{16t^2}{16} \\ 90.25 &= t^2 \\ \sqrt{90.25} &= t \\ 9.5 &= t \end{aligned}$$

Thus, the object will hit the ground after approximately 10 seconds.

d. $1440 - 16t^2 = 4(361 - 4t^2)$
$= 4[19^2 - (2t)^2] = 4(19 + 2t)(19 - 2t)$

93. $c = \left(\frac{-14}{2}\right)^2 = (-7)^2 = 49$

Thus, $m^2 - 14m + 49 = (m - 7)^2$

97. $x^{2n} - 36 = (x^n)^2 - 6^2$
$= (x^n + 6)(x^n - 6)$

101. $x^{4n} - 625 = (x^{2n})^2 - 25^2$
$= (x^{2n} + 25)(x^{2n} - 25)$
$= (x^{2n} + 25)[(x^n)^2 - 5^2]$
$= (x^{2n} + 25)(x^n + 5)(x^n - 5)$

Exercise Set 5.6

1. $(x - 2)(x + 1) = 0$
$x - 2 = 0$ or $x + 1 = 0$
$x = 2$ $\quad x = -1$
The solutions are 2 and −1.

5. $(x + 9)(x + 17) = 0$
$x + 9 = 0$ or $x + 17 = 0$
$x = -9$ $\quad x = -17$
The solutions are −9 and −17.

9. $3x(x - 8) = 0$
$3x = 0$ or $x - 8 = 0$
$x = 0$ $\quad x = 8$
The solutions are 0 and 8.

13. $(2x - 7)(7x + 2) = 0$
$2x - 7 = 0$ or $7x + 2 = 0$
$2x = 7$ $\quad 7x = -2$
$x = \frac{7}{2}$ $\quad x = -\frac{2}{7}$
The solutions are $\frac{7}{2}$ and $-\frac{2}{7}$.

17. $(x + 0.2)(x + 1.5) = 0$
$x + 0.2 = 0$ or $x + 1.5 = 0$
$x = -0.2$ $\quad x = -1.5$
The solutions are −0.2 and −1.5.

21. $x^2 - 13x + 36 = 0$
$(x - 9)(x - 4) = 0$
$x - 9 = 0$ or $x - 4 = 0$
$x = 9$ $\quad x = 4$
The solutions are 9 and 4.

25. $x^2 - 7x = 0$
$x(x - 7) = 0$
$x = 0$ or $x - 7 = 0$
$x = 7$
The solutions are 0 and 7.

29. $x^2 = 16$
$x^2 - 16 = 0$
$x^2 - 4^2 = 0$
$(x + 4)(x - 4) = 0$
$x + 4 = 0$ or $x - 4 = 0$
$x = -4$ $\quad x = 4$
The solutions are −4 and 4.

33. $x(3x - 1) = 14$
$3x^2 - x = 14$
$3x^2 - x - 14 = 0$
$(3x - 7)(x + 2) = 0$
$3x - 7 = 0$ or $x + 2 = 0$
$3x = 7$ $\quad x = -2$
$x = \frac{7}{3}$
The solutions are $\frac{7}{3}$ and −2.

37. $4x^3 - x = 0$
$x(4x^2 - 1) = 0$
$x(2x + 1)(2x - 1) = 0$
$x = 0$ or $2x + 1 = 0$ or $2x - 1 = 0$
$2x = -1$ $\quad 2x = 1$
$x = -\frac{1}{2}$ $\quad x = \frac{1}{2}$
The solutions are 0, $-\frac{1}{2}$, and $\frac{1}{2}$

41. $(4x - 3)(16x^2 - 24x + 9) = 0$
$(4x - 3)(4x - 3)^2 = 0$
$(4x - 3)^3 = 0$
$4x - 3 = 0$
$4x = 3$
$x = \frac{3}{4}$
The solutions is $\frac{3}{4}$.

45. $(2x + 3)(2x^2 - 5x - 3) = 0$
$(2x + 3)(2x + 1)(x - 3) = 0$
$2x + 3 = 0$ or $2x + 1 = 0$ or $x - 3 = 0$
$2x = -3$ $\quad 2x = -1$ $\quad x = 3$
$x = -\frac{3}{2}$ $\quad x = -\frac{1}{2}$
The solutions are $-\frac{3}{2}$, $-\frac{1}{2}$, and 3.

49. $5x^2 - 6x - 8 = 0$
$(5x + 4)(x - 2) = 0$
$5x + 4 = 0$ or $x - 2 = 0$
$5x = -4$ $\quad x = 2$
$x = -\frac{4}{5}$
The solutions are $-\frac{4}{5}$ and 2.

53.
$$6y^2 - 22y - 40 = 0$$
$$2(3y^2 - 11y - 20) = 0$$
$$2(3y + 4)(y - 5) = 0$$
$$3y + 4 = 0 \quad \text{or} \quad y - 5 = 0$$
$$3y = -4 \qquad y = 5$$
$$y = -\frac{4}{3}$$
The solutions are $-\frac{4}{3}$ and 5.

57.
$$x^3 - 12x^2 + 32x = 0$$
$$x(x^2 - 12x + 32) = 0$$
$$x(x - 8)(x - 4) = 0$$
$$x = 0 \quad \text{or} \quad x - 8 = 0 \quad \text{or} \quad x - 4 = 0$$
$$x = 8 \qquad x = 4$$
The solutions are 0, 8, and 4.

61. $$\frac{3}{5} + \frac{4}{9} = \frac{3\cdot 9}{5\cdot 9} + \frac{4\cdot 5}{9\cdot 5} = \frac{27}{45} + \frac{20}{45} = \frac{27 + 20}{45} = \frac{47}{45}$$

65. $$\frac{4}{5}\cdot\frac{7}{8} = \frac{4\cdot 7}{5\cdot 8} = \frac{4\cdot 7}{5\cdot 2\cdot 4} = \frac{7}{10}$$

69. a. When $x = 0$
$$y = -16x^2 + 20x + 300$$
$$y = -16(0^2) + 20(0) + 300$$
$$= -16(0) + 20(0) + 300$$
$$= 0 + 0 + 300$$
$$= 300$$
When $x = 1$:
$$y = -16x^2 + 20x + 300$$
$$y = -16(1)^2 + 20(1) + 300$$
$$= -16(1) + 20(1) + 300$$
$$= -16 + 20 + 300$$
$$= 304$$
When $x = 2$:
$$y = -16x^2 + 20x + 300$$
$$y = -16(2^2) + 20(2) + 300$$
$$= -16(4) + 20(2) + 300$$
$$= -64 + 40 + 300$$
$$= 276$$
When $x = 3$:
$$y = -16x^2 + 20x + 300$$
$$y = -16(3^2) + 20(3) + 300$$
$$= -16(9) + 20(3) + 300$$
$$= -144 + 60 + 300$$
$$= 216$$
When $x = 4$:
$$y = -16x^2 + 20x + 300$$
$$y = -16(4^2) + 20(4) + 300$$
$$= -16(16) + 20(4) + 300$$
$$= -256 + 80 + 300$$
$$= 124$$
When $x = 5$:
$$y = -16x^2 + 20x + 300$$
$$y = -16(5^2) + 20(5) + 300$$
$$= -16(25) + 20(5) + 300$$
$$= -400 + 100 + 300$$
$$= 0$$
When $x = 6$:
$$y = -16x^2 + 20x + 300$$
$$y = -16(6^2) + 20(6) + 300$$
$$= -16(36) + 20(6) + 300$$
$$= -576 + 120 + 300$$
$$= -156$$

b. The compass strikes the ground after 5 sec, when the height, y, is zero ft.

c. The maximum height was approximately 304 ft.

73.
$$(2x - 3)(x + 8) = (x - 6)(x + 4)$$
$$2x^2 - 3x + 16x - 24 = x^2 - 6x + 4x - 24$$
$$2x^2 + 13x - 24 = x^2 - 2x - 24$$
$$x^2 + 15x = 0$$
$$x(x + 15) = 0$$
$$x = 0 \quad \text{or} \quad x + 15 = 0$$
$$x = -15$$
The solutions are 0 and -15.

Exercise Set 5.7

1. Let x = the width, then $x + 4$ = the length.

5. Let x = the base, then $4x + 1$ = the height.

9. The perimeter is the sum of the lengths of the sides.
$$120 = (x + 5) + (x^2 - 3x) + (3x - 8) + (x + 3)$$
$$120 = x + 5 + x^2 - 3x + 3x - 8 + x + 3$$
$$120 = x^2 + 2x$$
$$0 = x^2 + 2x - 120$$
$$x^2 + 2x - 120 = 0$$
$$(x + 12)(x - 10) = 0$$
$$x + 12 = 0 \quad \text{or} \quad x - 10 = 0$$
$$x = -12 \qquad x = 10$$
Since the dimensions cannot be negative, the lengths of the sides are: $10 + 5 = 15$ cm, $10^2 - 3(10) = 70$ cm, $3(10) - 8 = 22$ cm, and $10 + 3 = 13$ cm.

13. Find t when $h = 0$.
$$h = -16t^2 + 64t + 80$$
$$0 = -16t^2 + 64t + 80$$
$$0 = -16(t^2 - 4t - 5)$$
$$0 = -16(t - 5)(t + 1)$$
$$t - 5 = 0 \quad \text{or} \quad t + 1 = 0$$
$$t = 5 \qquad t = -1$$
Since the time t cannot be negative, the object hits the ground after 5 sec.

17. Let $n = 12$.
$$D = \frac{1}{2}n(n - 3)$$
$$D = \frac{1}{2}\cdot 12(12 - 3) = 6(9) = 54$$
A polygon with 12 sides has 54 diagonals.

21. Let x = the unknown number.

$$x + x^2 = 132$$
$$x^2 + x - 132 = 0$$
$$(x + 12)(x - 11) = 0$$
$$x + 12 = 0 \quad \text{or} \quad x - 11 = 0$$
$$x = -12 \qquad x = 11$$

These are two numbers. They are −12 and 11.

25. Let x = the first number, then

$20 - x$ = the other number.

$$x^2 + (20 - x)^2 = 218$$
$$x^2 + 400 - 40x + x^2 = 218$$
$$2x^2 - 40x + 400 = 218$$
$$2x^2 - 40x + 182 = 0$$
$$2(x^2 - 20x + 91) = 0$$
$$2(x - 13)(x - 7) = 0$$
$$x - 13 = 0 \quad \text{or} \quad x - 7 = 0$$
$$x = 13 \qquad x = 7$$

The numbers are 13 and 7.

29. Let x = the length of the shorter leg. Then

$x + 4$ = the length of the longer leg and

$x + 8$ = the length of the hypotenuse.

By the Pythagorean theorem,

$$x^2 + (x + 4)^2 = (x + 8)^2$$
$$x^2 + x^2 + 8x + 16 = x^2 + 16x + 64$$
$$2x^2 + 8x + 16 = x^2 + 16x + 64$$
$$x^2 - 8x - 48 = 0$$
$$(x - 12)(x + 4) = 0$$
$$x - 12 = 0 \quad \text{or} \quad x + 4 = 0$$
$$x = 12 \qquad x = -4$$

Since the length cannot be negative, the sides of the triangle are 12 mm, 12 + 4 = 16 mm, and 12 + 8 = 20 mm.

33. Let x = the length of the shorter leg, then

$x + 12$ = the length of the longer leg and

$2x - 12$ = the length of the hypotenuse.

By the Pythagorean theorem,

$$x^2 + (x + 12)^2 = (2x - 12)^2$$
$$x^2 + x^2 + 24x + 144 = 4x^2 - 48x + 144$$
$$2x^2 + 24x + 144 = 4x^2 - 48x + 144$$
$$0 = 2x^2 - 72x$$
$$0 = 2x(x - 36)$$
$$2x = 0 \quad \text{or} \quad x - 36 = 0$$
$$x = 0 \qquad x = 36$$

Since the length cannot be zero feet, the solution is 36.

The shorter leg is 36 ft long.

37. Let $A = 100$ and $P = 144$.

$$A = P(1 + r)^2$$
$$144 = 100(1 + r)^2$$
$$\frac{144}{100} = \frac{100(1 + r)^2}{100}$$
$$1.44 = (1 + r)^2$$
$$\sqrt{1.44} = 1 + r$$
$$1.2 = 1 + r$$
$$0.2 = r$$

The interest rate is 0.2 or 20%.

41. Let $C = 9500$.

$$C = x^2 - 15x + 50$$
$$9500 = x^2 - 15x + 50$$
$$0 = x^2 - 15x - 9450$$
$$0 = (x - 105)(x + 90)$$
$$x - 105 = 0 \quad \text{or} \quad x + 90 = 0$$
$$x = 105 \qquad x = -90$$

Since the number of manufactured items cannot be negative, the solution is 105 items.

45. 9500 thousand acres

49. answers may vary

Chapter 5 Test

1. $9x^2 - 3x = 3x(3x - 1)$

5. $x^4 - 16 = (x^2)^2 - 4^2$
$$= (x^2 + 4)(x^2 - 4)$$
$$= (x^2 + 4)(x^2 - 2^2)$$
$$= (x^2 + 4)(x + 2)(x - 2)$$

9. $3a^2 + 3ab - 7a - 7b = 3a(a + b) - 7(a + b)$
$$= (a + b)(3a - 7)$$

13. $(x + 5)^2 - y^2 = (x + 5 + y)(x + 5 - y)$

17. $x^3 + 64 = x^3 + 4^3 = (x + 4)(x^2 - 4x + 16)$

21.
$$x^2 + 5x = 14$$
$$x^2 + 5x - 14 = 0$$
$$(x + 7)(x - 2) = 0$$
$$x + 7 = 0 \quad \text{or} \quad x - 2 = 0$$
$$x = -7 \qquad x = 2$$

The solutions are −7 and 2.

25.
$$t^2 - 2t - 15 = 0$$
$$(t + 3)(t - 5) = 0$$
$$t + 3 = 0 \quad \text{or} \quad t + 5 = 0$$
$$t = -3 \qquad t = -5$$

The solutions are −3 and 5.

29. Let x = first number. Then $17 - x$ is the second number.

$$x^2 + (17 - x)^2 = 145$$
$$x^2 + 289 - 34x + x^2 = 145$$
$$2x^2 - 34x + 289 = 145$$
$$2x^2 - 34x + 144 = 0$$
$$x^2 - 17x + 72 = 0$$
$$(x - 8)(x - 9) = 0$$
$$x - 8 = 0 \quad \text{or} \quad x - 9 = 0$$
$$x = 8 \qquad x = 9$$

when $x = 8$, then $17 - x = 17 - 8 = 9$

when $x = 9$, then $17 - x = 17 - 9 = 8$

The two numbers are 8 and 9.

Chapter 6

Exercise Set 6.1

1. $\dfrac{x + 5}{x + 2} = \dfrac{2 + 5}{2 + 2} = \dfrac{7}{4}$

5.
$$\frac{x^2 + 8x + 2}{x^2 - x - 6} = \frac{2^2 + 8(2) + 2}{2^2 - 2 - 6}$$
$$= \frac{4 + 16 + 2}{4 - 8}$$
$$= \frac{22}{-4}$$
$$= \frac{11 \cdot 2}{-2 \cdot 2}$$
$$= -\frac{11}{2}$$

9. $2x = 0$

$x = 0$

The expression is undefined when $x = 0$.

13. $2x - 8 = 0$

$2x = 8$

$x = 4$

The expression is undefined when $x = 4$.

17. The denominator is never zero so there are no values for which $\dfrac{x^2 - 5x - 2}{4}$ is undefined.

21. $\dfrac{2}{8x + 16} = \dfrac{2}{8(x + 2)} = \dfrac{2}{2 \cdot 4(x + 2)} = \dfrac{1}{4(x + 2)}$

25. $\dfrac{2x - 10}{3x - 30} = \dfrac{2(x - 5)}{3(x - 10)}$; does not simplify

29. $\dfrac{x - 7}{7 - x} = \dfrac{x - 7}{-1(x - 7)} = \dfrac{1}{-1} = -1$

33. $\dfrac{x + 5}{x^2 - 4x - 45} = \dfrac{x + 5}{(x - 9)(x + 5)} = \dfrac{1}{x - 9}$

37. $\dfrac{x + 7}{x^2 + 5x - 14} = \dfrac{x + 7}{(x - 2)(x + 7)} = \dfrac{1}{x - 2}$

41. $\dfrac{x^2 + 7x + 10}{x^2 - 3x - 10} = \dfrac{(x + 5)(x + 2)}{(x - 5)(x + 2)} = \dfrac{x + 5}{x - 5}$

45. $\dfrac{2x^2 - 8}{4x - 8} = \dfrac{2(x^2 - 4)}{4(x - 2)}$

$= \dfrac{2(x + 2)(x - 2)}{2 \cdot 2(x - 2)}$

$= \dfrac{x + 2}{2}$

49. $\dfrac{2 - x}{x - 2} = \dfrac{-1(x - 2)}{x - 2} = -1$

53. $\dfrac{m^2 - 6m + 9}{m^2 - 9} = \dfrac{(m - 3)(m - 3)}{(m + 3)(m - 3)} = \dfrac{m - 3}{m + 3}$

57. $\dfrac{5}{6} \cdot \dfrac{10}{11} \cdot \dfrac{2}{3} = \dfrac{5 \cdot 10 \cdot 2}{6 \cdot 11 \cdot 3}$

$= \dfrac{5 \cdot 2 \cdot 5 \cdot 2}{3 \cdot 2 \cdot 11 \cdot 3}$

$= \dfrac{5 \cdot 5 \cdot 2}{3 \cdot 11 \cdot 3}$

$= \dfrac{50}{99}$

61. $\dfrac{13}{20} \div \dfrac{2}{9} = \dfrac{13}{20} \cdot \dfrac{9}{2} = \dfrac{13 \cdot 9}{20 \cdot 2} = \dfrac{117}{40}$

65. $\dfrac{5x + 15 - xy - 3y}{2x + 6} = \dfrac{5(x + 3) - y(x + 3)}{2(x + 3)}$

$= \dfrac{(x + 3)(5 - y)}{2(x + 3)}$

$= \dfrac{5 - y}{2}$

69. $P = \dfrac{R - C}{R} = \dfrac{(15.3 - 12.3)\text{ billion}}{15.3\text{ billion}} = \dfrac{3}{15.3} \approx 0.196 = 19.6\%$

Thus, $P = 19.6\%$.

Exercise Set 6.2

1. $\dfrac{3x}{y^2} \cdot \dfrac{7y}{4x} = \dfrac{3x \cdot 7y}{y^2 \cdot 4x} = \dfrac{3 \cdot 7 \cdot x \cdot y}{4 \cdot x \cdot y \cdot y} = \dfrac{3 \cdot 7}{4 \cdot y} = \dfrac{21}{4y}$

5. $-\dfrac{5a^2b}{30a^2b^2} \cdot b^3 = -\dfrac{5a^2b \cdot b^3}{30a^2b^2}$

$= -\dfrac{5 \cdot a^2 \cdot b \cdot b \cdot b^2}{5 \cdot 6 \cdot a^2 \cdot b^2}$

$= -\dfrac{b \cdot b}{6}$

$= -\dfrac{b^2}{6}$

9. $\dfrac{6x + 6}{5} \cdot \dfrac{10}{36x + 36} = \dfrac{(6x + 6) \cdot 10}{5 \cdot (36x + 36)}$

$= \dfrac{6(x + 1) \cdot 2 \cdot 5}{5 \cdot 36(x + 1)}$

$= \dfrac{6 \cdot 5 \cdot 2 \cdot (x + 1)}{6 \cdot 5 \cdot 2 \cdot 3 \cdot (x + 1)}$

$= \dfrac{1}{3}$

13. $\dfrac{x^2 - 25}{x^2 - 3x - 10} \cdot \dfrac{x + 2}{x} = \dfrac{(x^2 - 25) \cdot (x + 2)}{(x^2 - 3x - 10) \cdot x}$

$= \dfrac{(x - 5)(x + 5) \cdot (x + 2)}{(x - 5)(x + 2) \cdot x}$

$= \dfrac{x + 5}{x}$

17. $\dfrac{5x^7}{2x^5} \div \dfrac{10x}{4x^3} = \dfrac{5x^7}{2x^5} \cdot \dfrac{4x^3}{10x}$

$= \dfrac{5 \cdot x^2 \cdot x^5 \cdot 2 \cdot 2x \cdot x^2}{2x^5 \cdot 2 \cdot 5 \cdot x}$

$= x^4$

21. $\dfrac{(x - 6)(x + 4)}{4x} \div \dfrac{2x - 12}{8x^2}$

$= \dfrac{(x - 6)(x + 4)}{4x} \cdot \dfrac{8x^2}{2x - 12}$

$= \dfrac{(x - 6)(x + 4) \cdot 2 \cdot 4 \cdot x \cdot x}{4x \cdot 2(x - 6)}$

$= x(x + 4)$

25. $\dfrac{m^2 - n^2}{m + n} \div \dfrac{m}{m^2 + nm} = \dfrac{m^2 - n^2}{m + n} \cdot \dfrac{m^2 + nm}{m}$

$= \dfrac{(m - n)(m + n) \cdot m(m + n)}{(m + n) \cdot m}$

$= (m - n)(m + n)$

$= m^2 - n^2$

29. $\dfrac{x^2 + 7x + 10}{x - 1} \div \dfrac{x^2 + 2x - 15}{x - 1}$

$= \dfrac{x^2 + 7x + 10}{x - 1} \cdot \dfrac{x - 1}{x^2 + 2x - 15}$

$= \dfrac{(x + 5)(x + 2) \cdot (x - 1)}{(x - 1) \cdot (x + 5)(x - 3)}$

$= \dfrac{x + 2}{x - 3}$

33. $\dfrac{x^2 + 5x}{8} \cdot \dfrac{9}{3x + 15} = \dfrac{x(x + 5) \cdot 3 \cdot 3}{8 \cdot 3(x + 5)} = \dfrac{3x}{8}$

37. $\dfrac{3x+4y}{x^2+4xy+4y^2}\cdot\dfrac{x+2y}{2}=\dfrac{(3x+4y)\cdot(x+2y)}{(x+2y)(x+2y)\cdot 2}$
$=\dfrac{3x+4y}{2(x+2y)}$

41. $\dfrac{3y}{3-x}\div\dfrac{12xy}{x^2-9}=\dfrac{3y}{3-x}\cdot\dfrac{x^2-9}{12xy}$
$=\dfrac{3y\cdot(x+3)(x-3)}{-1(x-3)\cdot 2\cdot 2\cdot 3\cdot x\cdot y}$
$=\dfrac{x+3}{-1\cdot 2\cdot 2\cdot x}=-\dfrac{x+3}{4x}$

45. $10 \text{ sq ft}\cdot\dfrac{144 \text{ sq in.}}{1 \text{ sq ft}}=1440 \text{ sq in.}$

49. $50 \text{ mi/hr}\cdot\dfrac{5280 \text{ ft}}{1 \text{ mi}}\cdot\dfrac{1 \text{ hr}}{3600 \text{ sec}}\approx 73 \text{ ft/sec}$

53. $763 \text{ mi/hr}\cdot\dfrac{5280 \text{ ft}}{1 \text{ mi}}\cdot\dfrac{1 \text{ hr}}{3600 \text{ sec}}\approx 1119 \text{ ft/sec}$

57. $\dfrac{9}{9}-\dfrac{19}{9}=\dfrac{9-19}{9}=\dfrac{-10}{9}=-\dfrac{10}{9}$

61. $\left(\dfrac{x^2-y^2}{x^2+y^2}\div\dfrac{x^2-y^2}{3x}\right)\cdot\dfrac{x^2+y^2}{6}$
$=\left(\dfrac{x^2-y^2}{x^2+y^2}\cdot\dfrac{3x}{x^2-y^2}\right)\cdot\dfrac{x^2+y^2}{6}$
$=\dfrac{(x-y)(x+y)(3)(x)(x^2+y^2)}{(x^2+y^2)(x-y)(x+y)(2)(3)}=\dfrac{x}{2}$

65. answers may vary

Exercise Set 6.3

1. $\dfrac{a}{13}+\dfrac{9}{13}=\dfrac{a+9}{13}$

5. $\dfrac{4m}{m-6}-\dfrac{24}{m-6}=\dfrac{4m-24}{m-6}=\dfrac{4(m-6)}{m-6}=4$

9. $\dfrac{5x+4}{x-1}-\dfrac{2x+7}{x-1}=\dfrac{5x+4-(2x+7)}{x-1}$
$=\dfrac{5x+4-2x-7}{x-1}$
$=\dfrac{3x-3}{x-1}$
$=\dfrac{3(x-1)}{x-1}$
$=3$

13. $\dfrac{2x+3}{x^2-x-30}-\dfrac{x-2}{x^2-x-30}=\dfrac{2x+3-(x-2)}{x^2-x-30}$
$=\dfrac{2x+3-x+2}{x^2-x-30}$
$=\dfrac{x+5}{x^2-x-30}$
$=\dfrac{x+5}{(x-6)(x+5)}$
$=\dfrac{1}{x-6}$

17. answers may vary

21. $8x=2^3\cdot x$
$2x+4=2(x+2)$
$\text{LCD}=2^3\cdot x\cdot(x+2)=8x(x+2)$

25. $x+6=x+6$
$3x+18=3\cdot(x+6)$
$\text{LCD}=3(x+6)$

29. $x-8=x-8$
$8-x=-(x-8)$
$\text{LCD}=x-8$ or $8-x$

33. answers may vary

37. $\dfrac{6}{3a}=\dfrac{6(4b^2)}{3a(4b^2)}=\dfrac{24b^2}{12ab^2}$

41. $\dfrac{9a+2}{5a+10}=\dfrac{9a+2}{5(a+2)}=\dfrac{(9a+2)(b)}{5(a+2)(b)}=\dfrac{9ab+2b}{5b(a+2)}$

45. $\dfrac{9y-1}{15x^2-30}=\dfrac{(9y-1)(2)}{(15x^2-30)2}=\dfrac{18y-2}{30x^2-60}$

49. Since $6=2\cdot 3$ and $4=2^2$, $\text{LCD}=2^2\cdot 3=12$.
$\dfrac{2}{6}-\dfrac{3}{4}=\dfrac{2(2)}{6(2)}-\dfrac{3(3)}{4(3)}=\dfrac{4}{12}-\dfrac{9}{12}=\dfrac{4-9}{12}=-\dfrac{5}{12}$

53. Since $8=2^3$ and $12=2^2\cdot 3$, the least common multiple of 8 and 12 is $2^3\cdot 3=24$. Since $8\cdot 3=24$ and $12\cdot 2=24$, buy three packages of hot dogs and two packages of buns.

Exercise Set 6.4

1. $\text{LCD}=2\cdot 3\cdot x=6x$
$\dfrac{4}{2x}+\dfrac{9}{3x}=\dfrac{4(3)}{2x(3)}+\dfrac{9(2)}{3x(2)}$
$=\dfrac{12}{6x}+\dfrac{18}{6x}$
$=\dfrac{30}{6x}$
$=\dfrac{5(6)}{6x}$
$=\dfrac{5}{x}$

5. $\text{LCD}=2x^2$
$\dfrac{3}{x}+\dfrac{5}{2x^2}=\dfrac{3(2x)}{x(2x)}+\dfrac{5}{2x^2}=\dfrac{6x}{2x^2}+\dfrac{5}{2x^2}=\dfrac{6x+5}{2x^2}$

9. $x^2-4=(x+2)(x-2)$
$\text{LCD}=(x+2)(x-2)$
$\dfrac{3}{x+2}-\dfrac{1}{x^2-4}=\dfrac{3}{x+2}-\dfrac{1}{(x+2)(x-2)}$
$=\dfrac{3(x-2)}{(x+2)(x-2)}-\dfrac{1}{(x+2)(x-2)}$
$=\dfrac{3x-6}{(x+2)(x-2)}-\dfrac{1}{(x+2)(x-2)}$
$=\dfrac{3x-6-1}{(x+2)(x-2)}$
$=\dfrac{3x-7}{(x+2)(x-2)}$

13. $\dfrac{6}{x-3}+\dfrac{8}{3-x}=\dfrac{6}{x-3}+\dfrac{8}{-(x-3)}$
$=\dfrac{6}{x-3}+\dfrac{-8}{x-3}$
$=\dfrac{6+(-8)}{x-3}$
$=-\dfrac{2}{x-3}$

17. $\dfrac{5}{x}+2=\dfrac{5}{x}+\dfrac{2}{1}=\dfrac{5}{x}+\dfrac{2(x)}{1(x)}=\dfrac{5+2x}{x}$

21. $\frac{y+2}{y+3}-2=\frac{y+2}{y+3}-\frac{2}{1}$
$=\frac{y+2}{y+3}-\frac{2(y+3)}{y+3}$
$=\frac{y+2}{y+3}-\frac{2y+6}{y+3}$
$=\frac{y+2-(2y+6)}{y+3}$
$=\frac{y+2-2y-6}{y+3}$
$=\frac{-y-4}{y+3}$
$=\frac{-(y+4)}{y+3}$
$=-\frac{y+4}{y+3}$

25. $\frac{5x}{x+2}-\frac{3x-4}{x+2}=\frac{5x-(3x-4)}{x+2}$
$=\frac{5x-3x+4}{x+2}$
$=\frac{2x+4}{x+2}$
$=\frac{2(x+2)}{x+2}$
$=2$

29. $\frac{1}{x+3}-\frac{1}{(x+3)^2}=\frac{1(x+3)}{(x+3)(x+3)}-\frac{1}{(x+3)^2}$
$=\frac{x+3}{(x+3)^2}-\frac{1}{(x+3)^2}$
$=\frac{x+3-1}{(x+3)^2}$
$=\frac{x+2}{(x+3)^2}$

33. $\frac{2}{m}+1=\frac{2}{m}+\frac{1}{1}=\frac{2}{m}+\frac{1(m)}{1(m)}=\frac{2+m}{m}$

37. $\frac{7}{(x+1)(x-1)}+\frac{8}{(x+1)^2}$
$=\frac{7(x+1)}{(x+1)(x-1)(x+1)}+\frac{8(x-1)}{(x+1)^2(x-1)}$
$=\frac{7x+7}{(x+1)^2(x-1)}+\frac{8x-8}{(x+1)^2(x-1)}$
$=\frac{7x+7+8x-8}{(x+1)^2(x-1)}$
$=\frac{15x-1}{(x+1)^2(x-1)}$

41. $\frac{3a}{2a+6}-\frac{a-1}{a+3}=\frac{3a}{2(a+3)}-\frac{a-1}{a+3}$
$=\frac{3a}{2(a+3)}-\frac{(a-1)(2)}{(a+3)(2)}$
$=\frac{3a}{2(a+3)}-\frac{2a-2}{2(a+3)}$
$=\frac{3a-(2a-2)}{2(a+3)}$
$=\frac{3a-2a+2}{2(a+3)}$
$=\frac{a+2}{2(a+3)}$

45. $\frac{5}{2-x}+\frac{x}{2x-4}=\frac{-5}{x-2}+\frac{x}{2(x-2)}$
$=\frac{-5(2)}{(x-2)2}+\frac{x}{2(x-2)}$
$=\frac{-10}{2(x-2)}+\frac{x}{2(x-2)}$
$=\frac{x-10}{2(x-2)}$

49. $\frac{13}{x^2-5x+6}-\frac{5}{x-3}=\frac{13}{(x-3)(x-2)}-\frac{5}{x-3}$
$=\frac{13}{(x-3)(x-2)}-\frac{5(x-2)}{(x-3)(x-2)}$
$=\frac{13-5(x-2)}{(x-3)(x-2)}$
$=\frac{13-5x+10}{(x-3)(x-2)}$
$=\frac{-5x+23}{(x-3)(x-2)}$

53. answers may vary

57. $2x^2-x-1=0$
$(2x+1)(x-1)=0$
$2x+1=0$ or $x-1=0$
$2x=-1$ $\qquad x=1$
$x=-\frac{1}{2}$

The solutions are $x=-\frac{1}{2}$ and $x=1$.

61. $\frac{3}{x}-\frac{2x}{x^2-1}+\frac{5}{x+1}=\frac{3}{x}-\frac{2x}{(x+1)(x-1)}+\frac{5}{x+1}$
$=\frac{3(x+1)(x-1)}{x(x+1)(x-1)}-\frac{2x(x)}{x(x+1)(x-1)}+\frac{5(x)(x-1)}{(x+1)(x)(x-1)}$
$=\frac{3x^2-3}{x(x+1)(x-1)}-\frac{2x^2}{x(x+1)(x-1)}+\frac{5x^2-5x}{x(x+1)(x-1)}$
$=\frac{3x^2-3-2x^2+5x^2-5x}{x(x+1)(x-1)}$
$=\frac{6x^2-5x-3}{x(x+1)(x-1)}$

65. $\frac{9}{x^2+9x+14}-\frac{3x}{x^2+10x+21}+\frac{x+4}{x^2+5x+6}=$
$\frac{9}{(x+2)(x+7)}-\frac{3x}{(x+7)(x+3)}+\frac{x+4}{(x+2)(x+3)}$
$=\frac{9(x+3)}{(x+2)(x+7)(x+3)}-\frac{3x(x+2)}{(x+7)(x+3)(x+2)}+\frac{(x+4)(x+7)}{(x+2)(x+3)(x+7)}$
$=\frac{9x+27}{(x+2)(x+3)(x+7)}-\frac{3x^2+6x}{(x+2)(x+3)(x+7)}+\frac{x^2+11x+28}{(x+2)(x+3)(x+7)}$
$=\frac{9x+27-3x^2-6x+x^2+11x+28}{(x+2)(x+3)(x+7)}$
$=\frac{-2x^2+14x+55}{(x+2)(x+3)(x+7)}$

69. $1-\frac{G}{P}=\frac{1\cdot P}{1\cdot P}-\frac{G}{P}=\frac{P}{P}-\frac{G}{P}=\frac{P-G}{P}$

Exercise Set 6.5

1.
$$\frac{x}{5} + 3 = 9$$
$$5\left(\frac{x}{5} + 3\right) = 5(9)$$
$$5\left(\frac{x}{5}\right) + 5(3) = 5(9)$$
$$x + 15 = 45$$
$$x = 30$$

Check:
$$\frac{x}{5} + 3 = 9$$
$$\frac{30}{5} + 3 \stackrel{?}{=} 9$$
$$6 + 3 \stackrel{?}{=} 9$$
$$9 = 9 \quad \text{True}$$
The solution is 30.

5.
$$2 - \frac{8}{x} = 6$$
$$x\left(2 - \frac{8}{x}\right) = x(6)$$
$$x(2) - x\left(\frac{8}{x}\right) = x(6)$$
$$2x - 8 = 6x$$
$$-8 = 4x$$
$$-2 = x$$

Check:
$$2 - \frac{8}{x} = 6$$
$$2 - \frac{8}{-2} \stackrel{?}{=} 6$$
$$2 - (-4) \stackrel{?}{=} 6$$
$$2 + 4 \stackrel{?}{=} 6$$
$$6 = 6 \quad \text{True}$$
The solution is -2.

9.
$$\frac{a}{5} = \frac{a-3}{2}$$
$$10\left(\frac{a}{5}\right) = 10\left(\frac{a-3}{2}\right)$$
$$2a = 5(a-3)$$
$$2a = 5a - 15$$
$$-3a = -15$$
$$a = 5$$

Check:
$$\frac{a}{5} = \frac{a-3}{2}$$
$$\frac{5}{5} \stackrel{?}{=} \frac{5-3}{2}$$
$$\frac{5}{5} \stackrel{?}{=} \frac{2}{2}$$
$$1 = 1 \quad \text{True}$$
The solution is 5.

13.
$$\frac{2}{y} + \frac{1}{2} = \frac{5}{2y}$$
$$2y\left(\frac{2}{y} + \frac{1}{2}\right) = 2y\left(\frac{5}{2y}\right)$$
$$2y\left(\frac{2}{y}\right) + 2y\left(\frac{1}{2}\right) = 2y\left(\frac{5}{2y}\right)$$
$$4 + y = 5$$
$$y = 1$$

Check:
$$\frac{2}{y} + \frac{1}{2} = \frac{5}{2y}$$
$$\frac{2}{1} + \frac{1}{2} \stackrel{?}{=} \frac{5}{2(1)}$$
$$\frac{4}{2} + \frac{1}{2} \stackrel{?}{=} \frac{5}{2}$$
$$\frac{5}{2} = \frac{5}{2} \quad \text{True}$$
The solution is 1.

17.
$$2 + \frac{3}{a-3} = \frac{a}{a-3}$$
$$(a-3)\left(2 + \frac{3}{a-3}\right) = (a-3)\left(\frac{a}{a-3}\right)$$
$$(a-3)(2) + (a-3)\left(\frac{3}{a-3}\right) = (a-3)\left(\frac{a}{a-3}\right)$$
$$2a - 6 + 3 = a$$
$$2a - 3 = a$$
$$-3 = -a$$
$$3 = a$$

In the original equation, 3 makes a denominator 0. This equation has no solution.

21.
$$\frac{y}{y+4} + \frac{4}{y+4} = 3$$
$$(y+4)\left(\frac{y}{y+4} + \frac{4}{y+4}\right) = (y+4)(3)$$
$$(y+4)\left(\frac{y}{y+4}\right) + (y+4)\left(\frac{4}{y+4}\right) = (y+4)(3)$$
$$y + 4 = 3y + 12$$
$$4 = 2y + 12$$
$$-8 = 2y$$
$$-\frac{8}{2} = y$$
$$-4 = y$$

In the original equation, -4 makes a denominator zero. This equation has no solution.

25.
$$\frac{2x}{x+2} - 2 = \frac{x-8}{x-2}$$
$$(x+2)(x-2)\left(\frac{2x}{x+2} - 2\right) = (x+2)(x-2)\left(\frac{x-8}{x-2}\right)$$
$$(x+2)(x-2)\left(\frac{2x}{x+2}\right) - (x+2)(x-2)(2) = (x+2)(x-2)\left(\frac{x-8}{x-2}\right)$$
$$2x(x-2) - 2(x^2 - 4) = (x+2)(x-8)$$
$$2x^2 - 4x - 2x^2 + 8 = x^2 - 6x - 16$$
$$-4x + 8 = x^2 - 6x - 16$$
$$0 = x^2 - 2x - 24$$
$$0 = (x-6)(x+4)$$

$x - 6 = 0$ or $x + 4 = 0$

$x = 6$ $\qquad x = -4$

Check: $x = 6$:
$$\frac{2x}{x+2} - 2 = \frac{x-8}{x-2}$$
$$\frac{2(6)}{6+2} - 2 \stackrel{?}{=} \frac{6-8}{6-2}$$
$$\frac{12}{8} - 2 \stackrel{?}{=} -\frac{2}{4}$$
$$\frac{3}{2} - \frac{4}{2} \stackrel{?}{=} -\frac{1}{2}$$
$$\frac{3-4}{2} \stackrel{?}{=} -\frac{1}{2}$$
$$-\frac{1}{2} = -\frac{1}{2} \quad \text{True}$$

$x = -4$:
$$\frac{2x}{x+2} - 2 = \frac{x-8}{x-2}$$
$$\frac{2(-4)}{-4+2} - 2 \stackrel{?}{=} \frac{-4-8}{-4-2}$$
$$\frac{-8}{-2} - 2 \stackrel{?}{=} \frac{-12}{-6}$$
$$4 - 2 \stackrel{?}{=} 2$$
$$2 = 2 \quad \text{True}$$

The solutions are 6 and -4.

29.

$$\frac{2}{x-2} + 1 = \frac{x}{x+2}$$

$$(x-2)(x+2)\left(\frac{2}{x-2} + 1\right) = (x-2)(x+2)\left(\frac{x}{x+2}\right)$$

$$(x-2)(x+2)\left(\frac{2}{x-2}\right) + (x-2)(x+2) = (x-2)(x+2)\left(\frac{x}{x+2}\right)$$

$$2(x+2) + (x-2)(x+2) = x(x-2)$$

$$2x + 4 + x^2 - 4 = x^2 - 2x$$

$$2x + x^2 = x^2 - 2x$$

$$2x = -2x$$

$$4x = 0$$

$$x = 0$$

Check: $\dfrac{2}{x-2} + 1 = \dfrac{x}{x+2}$

$$\frac{2}{0-2} + 1 \stackrel{?}{=} \frac{0}{0+2}$$

$$\frac{2}{-2} + 1 \stackrel{?}{=} 0$$

$$-1 + 1 \stackrel{?}{=} 0$$

$0 = 0$ True

The solution is 0.

33.

$$\frac{x+1}{3} - \frac{x-1}{6} = \frac{1}{6}$$

$$6\left(\frac{x+1}{3} - \frac{x-1}{6}\right) = 6\left(\frac{1}{6}\right)$$

$$6\left(\frac{x+1}{3}\right) - 6\left(\frac{x-1}{6}\right) = 6\left(\frac{1}{6}\right)$$

$$2(x+1) - (x-1) = 1$$

$$2x + 2 - x + 1 = 1$$

$$x + 3 = 1$$

$$x = -2$$

Check: $\dfrac{x+1}{3} - \dfrac{x-1}{6} = \dfrac{1}{6}$

$$\frac{-2+1}{3} - \frac{-2-1}{6} \stackrel{?}{=} \frac{1}{6}$$

$$-\frac{1}{3} - \frac{-3}{6} \stackrel{?}{=} \frac{1}{6}$$

$$-\frac{2}{6} - \frac{-3}{6} \stackrel{?}{=} \frac{1}{6}$$

$$\frac{-2-(-3)}{6} \stackrel{?}{=} \frac{1}{6}$$

$$\frac{-2+3}{6} \stackrel{?}{=} \frac{1}{6}$$

$\dfrac{1}{6} = \dfrac{1}{6}$ True

The solution is -2.

37.

$$\frac{4r-4}{r^2+5r-14} + \frac{2}{r+7} = \frac{1}{r-2}$$

$$\frac{4r-4}{(r+7)(r-2)} + \frac{2}{r+7} = \frac{1}{r-2}$$

$$(r+7)(r-2)\left(\frac{4r-4}{(r+7)(r-2)} + \frac{2}{r+7}\right) = (r+7)(r-2)\left(\frac{1}{r-2}\right)$$

$$(r+7)(r-2)\left(\frac{4r-4}{(r+7)(r-2)}\right) + (r+7)(r-2)\left(\frac{2}{r+7}\right) = (r+7)(r-2)\left(\frac{1}{r-2}\right)$$

$$4r - 4 + 2(r-2) = (r+7)(1)$$

$$4r - 4 + 2r - 4 = r + 7$$

$$6r - 8 = r + 7$$

$$5r = 15$$

$$r = 3$$

Check: $\dfrac{4r-4}{r^2+5r-14} + \dfrac{2}{r+7} = \dfrac{1}{r-2}$

$$\frac{4(3)-4}{3^2+5(3)-14} + \frac{2}{3+7} \stackrel{?}{=} \frac{1}{3-2}$$

$$\frac{12-4}{9+15-14} + \frac{2}{10} \stackrel{?}{=} \frac{1}{1}$$

$$\frac{8}{10} + \frac{2}{10} \stackrel{?}{=} 1$$

$$\frac{8+2}{10} \stackrel{?}{=} 1$$

$$\frac{10}{10} \stackrel{?}{=} 1$$

$1 = 1$ True

The solution is 3.

41.
$$R = \frac{E}{I}$$
$$R(I) = \frac{E}{I}(I)$$
$$RI = E$$
$$\frac{RI}{R} = \frac{E}{R}$$
$$I = \frac{E}{R}$$

45.
$$i = \frac{A}{t + B}$$
$$(t + B)(i) = (t + B)\left(\frac{A}{t + B}\right)$$
$$ti + Bi = A$$
$$ti = A - Bi$$
$$\frac{ti}{i} = \frac{A - Bi}{i}$$
$$t = \frac{A - Bi}{i}$$

49.
$$\frac{C}{\pi r} = 2$$
$$\pi r\left(\frac{C}{\pi r}\right) = \pi r(2)$$
$$C = 2\pi r$$
$$\frac{C}{2\pi} = \frac{2\pi r}{2\pi}$$
$$\frac{C}{2\pi} = r$$

53. The reciprocal of x is $\frac{1}{x}$.

57. The part filled in 1 hr is $\frac{1}{3}$.

61.
$$\frac{150}{x} + \frac{450}{x} = 90$$
$$\frac{600}{x} = 90$$
$$\frac{600}{x} \cdot x = 90 \cdot x$$
$$600 = 90x$$
$$\frac{600}{90} = \frac{90x}{90}$$
$$\frac{20}{3} = x$$

Now, $\frac{150}{x} = \frac{150}{\frac{20}{3}} = 150 \cdot \frac{3}{20} = \frac{450}{20} = 22.5$

$$\frac{450}{x} = \frac{450}{\frac{20}{3}} = 450 \cdot \frac{3}{20} = \frac{1350}{20} = 67.5$$

The complementary angles are 22.5° and 67.5°.

65. No. Multiplying both terms in the expression by 4 changes the value of the original expression.

Exercise Set 6.6

1.
$$3 \cdot \frac{1}{x} = 9 \cdot \frac{1}{6}$$
$$\frac{3}{x} = \frac{9}{6}$$
$$6x\left(\frac{3}{x}\right) = 6x\left(\frac{9}{6}\right)$$
$$18 = 9x$$
$$x = 2$$

The unknown number is 2.

5.
$$\frac{2}{x - 3} - \frac{4}{x + 3} = 8 \cdot \frac{1}{x^2 - 9}$$
$$(x - 3)(x + 3)\left(\frac{2}{x - 3} - \frac{4}{x + 3}\right) = (x - 3)(x + 3)\left(\frac{8}{x^2 - 9}\right)$$
$$(x - 3)(x + 3)\left(\frac{2}{x - 3}\right) - (x - 3)(x + 3)\left(\frac{4}{x + 3}\right) = 8$$
$$2(x + 3) - 4(x - 3) = 8$$
$$2x + 6 - 4x + 12 = 8$$
$$-2x + 18 = 8$$
$$-2x = -10$$
$$x = 5$$

The unknown number is 5.

9.

	Hours to Complete Total Job	Part of Job Completed in 1 Hour
Experienced	4	$\frac{1}{4}$
Apprentice	5	$\frac{1}{5}$
Together	x	$\frac{1}{x}$

$$\frac{1}{4} + \frac{1}{5} = \frac{1}{x}$$
$$20x\left(\frac{1}{4}\right) + 20x\left(\frac{1}{5}\right) = 20x\left(\frac{1}{x}\right)$$
$$5x + 4x = 20$$
$$9x = 20$$
$$x = \frac{20}{9} \quad \text{or} \quad 2\frac{2}{9}$$

The experienced surveyor and apprentice surveyor, working together, can survey the road bed in $2\frac{2}{9}$ hr.

13.

	Hours to Complete Total Job	Part of Job Completed in 1 Hour
Marcus	6	$\frac{1}{6}$
Tony	4	$\frac{1}{4}$
Together	x	$\frac{1}{x}$

$$\frac{1}{6}+\frac{1}{4}=\frac{1}{x}$$
$$12x\left(\frac{1}{6}\right)+12x\left(\frac{1}{4}\right)=12x\left(\frac{1}{x}\right)$$
$$2x+3x=12$$
$$5x=12$$
$$x=\frac{12}{5}=2\frac{2}{5}$$
$$45\left(\frac{12}{5}\right)=108$$

Together, Marcus an Tony work for $2\frac{2}{5}$ hr at \$45 per hr. The labor estimate should be \$108.

17.

	Hours to Complete Total Job	Part of Job Completed in 1 Hour
First Pipe	20	$\frac{1}{20}$
Second Pipe	15	$\frac{1}{15}$
Third Pipe	x	$\frac{1}{x}$
3 Pipes Together	6	$\frac{1}{6}$

$$\frac{1}{20}+\frac{1}{15}+\frac{1}{x}=\frac{1}{6}$$
$$60x\left(\frac{1}{20}\right)+60x\left(\frac{1}{15}\right)+60x\left(\frac{1}{x}\right)=60x\left(\frac{1}{6}\right)$$
$$3x+4x+60=10x$$
$$7x+60=10x$$
$$60=3x$$
$$20=x$$

It takes the third pipe 20 hr to fill the pond.

21.

	distance =	rate	· time
First Portion	20	r	$\frac{20}{r}$
Cooldown Portion	16	$r-2$	$\frac{16}{r-2}$

$$\frac{20}{r}=\frac{16}{r-2}$$
$$20(r-2)=16r$$
$$20r-40=16r$$
$$-40=-4r$$
$$r=10 \text{ and } r-2=10-2=8$$

His speed was 10 mph during the first portion and 8 mph during the cooldown portion.

25. Let w = the rate of the wind.

	distance =	rate	· time
With the wind	48	$16+w$	$\frac{48}{16+w}$
Into the wind	16	$16-w$	$\frac{16}{16-w}$

$$\frac{48}{16+w}=\frac{16}{16-w}$$
$$48(16-w)=16(16+w)$$
$$768-48w=256+16w$$
$$512=64w$$
$$w=8$$

The rate of the wind is 8 mph.

29.
$$\frac{12}{4}=\frac{18}{x}$$
$$12x=72$$
$$x=6$$

33.
$$\frac{16}{10}=\frac{34}{y}$$
$$16y=340$$
$$y=21.25$$

37.
$$\frac{6}{8}=\frac{20}{x}$$
$$6x=160$$
$$x=\frac{160}{6}=\frac{80}{3}=26\frac{2}{3}\text{ ft}$$

41.
$$\frac{\frac{2}{5}+\frac{1}{5}}{\frac{7}{10}+\frac{7}{10}}=\frac{\frac{3}{5}}{\frac{7}{10}+\frac{7}{10}}=\frac{\frac{3}{5}}{\frac{14}{10}}=\frac{3}{5}\div\frac{14}{10}=\frac{3}{5}\cdot\frac{10}{14}=\frac{3\cdot2\cdot5}{5\cdot2\cdot7}=\frac{3}{7}$$

Exercise Set 6.7

1.
$$\frac{\frac{1}{2}}{\frac{3}{4}}=\frac{1}{2}\cdot\frac{4}{3}=\frac{1\cdot2\cdot2}{2\cdot3}=\frac{2}{3}$$

5.
$$\frac{\frac{-5}{12x^2}}{\frac{25}{16x^3}}=-\frac{5}{12x^2}\cdot\frac{16x^3}{25}=-\frac{5\cdot4\cdot4\cdot x^2\cdot x}{4\cdot3\cdot x^2\cdot5\cdot5}=-\frac{4x}{15}$$

9. $\dfrac{2+\frac{7}{10}}{1+\frac{3}{5}} = \dfrac{10\left(2+\frac{7}{10}\right)}{10\left(1+\frac{3}{5}\right)}$

$= \dfrac{10(2)+10\left(\frac{7}{10}\right)}{10(1)+10\left(\frac{3}{5}\right)}$

$= \dfrac{20+7}{10+6}$

$= \dfrac{27}{16}$

13. $\dfrac{\frac{1}{5}-\frac{1}{x}}{\frac{7}{10}+\frac{1}{x^2}} = \dfrac{10x^2\left(\frac{1}{5}-\frac{1}{x}\right)}{10x^2\left(\frac{7}{10}+\frac{1}{x^2}\right)}$

$= \dfrac{10x^2\left(\frac{1}{5}\right)-10x^2\left(\frac{1}{x}\right)}{10x^2\left(\frac{7}{10}\right)+10x^2\left(\frac{1}{x^2}\right)}$

$= \dfrac{2x^2-10x}{7x^2+10}$

$= \dfrac{2x(x-5)}{7x^2+10}$

17. $\dfrac{\frac{4y-8}{16}}{\frac{6y-12}{4}} = \dfrac{4y-8}{16}\cdot\dfrac{4}{6y-12} = \dfrac{4(y-2)\cdot 4}{4\cdot 4\cdot 6(y-2)} = \dfrac{1}{6}$

21. $\dfrac{1}{2+\frac{1}{3}} = \dfrac{3(1)}{3\left(2+\frac{1}{3}\right)} = \dfrac{3(1)}{3(2)+3\left(\frac{1}{3}\right)} = \dfrac{3}{6+1} = \dfrac{3}{7}$

25. $\dfrac{\frac{8}{x+4}+2}{\frac{12}{x+4}-2} = \dfrac{(x+4)\left(\frac{8}{x+4}+2\right)}{(x+4)\left(\frac{12}{x+4}-2\right)}$

$= \dfrac{(x+4)\left(\frac{8}{x+4}\right)+(x+4)(2)}{(x+4)\left(\frac{12}{x+4}\right)-(x+4)(2)}$

$= \dfrac{8+2x+8}{12-2x-8}$

$= \dfrac{16+2x}{4-2x}$

$= \dfrac{2(8+x)}{2(2-x)}$

$= \dfrac{8+x}{2-x}$ or $-\dfrac{x+8}{x-2}$

29. answers may vary

33. Monica Seles, Martina Hingis, and Arantxa Sanchez-Vicario

37. $\dfrac{1}{\frac{1}{R_1}+\frac{1}{R_2}} = \dfrac{R_1R_2(1)}{R_1R_2\left(\frac{1}{R_1}+\frac{1}{R_2}\right)}$

$= \dfrac{R_1R_2}{R_1R_2\left(\frac{1}{R_1}\right)+R_1R_2\left(\frac{1}{R_2}\right)}$

$= \dfrac{R_1R_2}{R_2+R_1}$

41. $\dfrac{y^{-2}}{1-y^{-2}} = \dfrac{\frac{1}{y^2}}{1-\frac{1}{y^2}}$

$= \dfrac{y^2\left(\frac{1}{y^2}\right)}{y^2\left(1-\frac{1}{y^2}\right)}$

$= \dfrac{y^2\left(\frac{1}{y^2}\right)}{y^2(1)-y^2\left(\frac{1}{y^2}\right)}$

$= \dfrac{1}{y^2-1}$

Chapter 6 Test

1. The rational expression is undefined when

$x^2+4x+3=0$

$(x+3)(x+1)=0$

$x+3=0$ or $x+1=0$

$x=-3$ or $x=-1$

5. $\dfrac{x+6}{x^2+12x+36} = \dfrac{x+6}{(x+6)^2} = \dfrac{1}{x+6}$

9. $\dfrac{x^2-13x+42}{x^2+10x+21} \div \dfrac{x^2-4}{x^2+x-6}$

$= \dfrac{x^2-13x+42}{x^2+10x+21}\cdot\dfrac{x^2+x-6}{x^2-4}$

$= \dfrac{(x-6)(x-7)\cdot(x+3)(x-2)}{(x+7)(x+3)\cdot(x+2)(x-2)}$

$= \dfrac{(x-6)(x-7)}{(x+7)(x+2)}$

13. $\dfrac{5a}{a^2-a-6} - \dfrac{2}{a-3}$

$= \dfrac{5a}{(a-3)(a+2)} - \dfrac{2(a+2)}{(a-3)(a+2)}$

$= \dfrac{5a-2(a+2)}{(a-3)(a+2)}$

$= \dfrac{5a-2a-4}{(a-3)(a+2)}$

$= \dfrac{3a-4}{(a-3)(a+2)}$

17. $$\frac{4y}{y^2+6y+5}-\frac{3}{y^2+5y+4}=\frac{4y}{(y+5)(y+1)}-\frac{3}{(y+4)(y+1)}$$

$$=\frac{4y(y+4)}{(y+5)(y+1)(y+4)}-\frac{3(y+5)}{(y+4)(y+1)(y+5)}$$

$$=\frac{4y(y+4)-3(y+5)}{(y+5)(y+1)(y+4)}$$

$$=\frac{4y^2+16y-3y-15}{(y+5)(y+1)(y+4)}$$

$$=\frac{4y^2+13y-15}{(y+5)(y+1)(y+4)}$$

21. $$\frac{10}{x^2-25}=\frac{3}{x+5}+\frac{1}{x-5}$$

$$(x+5)(x-5)\left(\frac{10}{(x+5)(x-5)}\right)=(x+5)(x-5)\left(\frac{3}{x+5}+\frac{1}{x-5}\right)$$

$$10=(x+5)(x-5)\left(\frac{3}{x+5}\right)+(x+5)(x-5)\left(\frac{1}{x-5}\right)$$

$$10=3(x-5)+x+5$$

$$10=3x-15+x+5$$

$$10=4x-10$$

$$20=4x$$

$$x=5$$

In the original equation, 5 makes the denominator 0. This equation has no solution.

25. $$\frac{8}{x}=\frac{10}{15}$$

$$8(15)=10x$$

$$120=10x$$

$$12=x$$

Chapter 7

Exercise Set 7.1

1. Point: $(1, 3)$

Slope: $\frac{3}{2}$

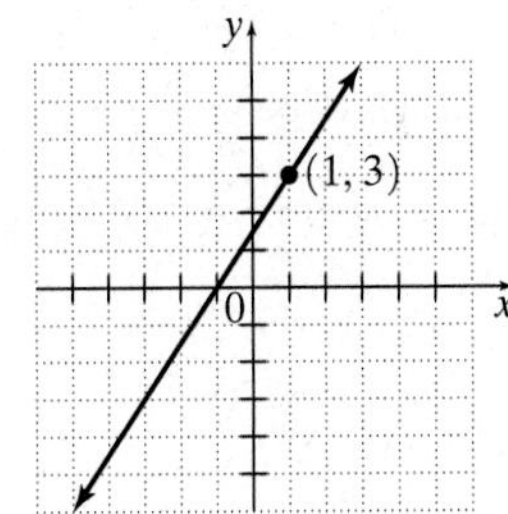

9. $y=-2x+3$

$m=-2$, y-intercept $=(0, 3)$

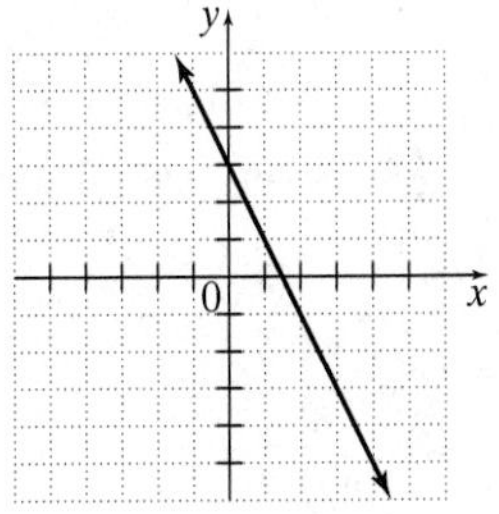

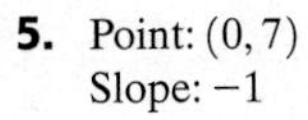

5. Point: $(0, 7)$

Slope: -1

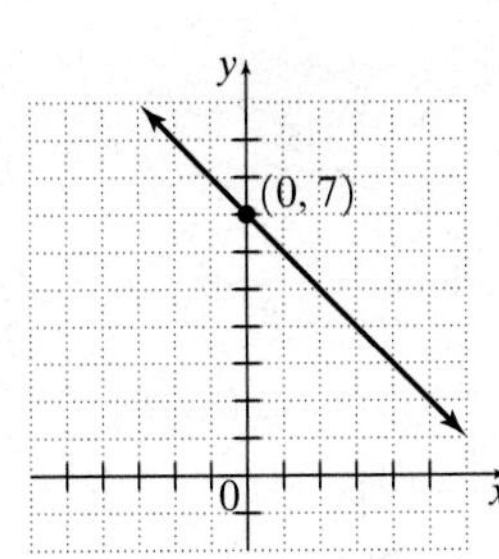

13. $y=\frac{1}{2}x-4$

$m=\frac{1}{2}$, y-intercept $=(0, -4)$

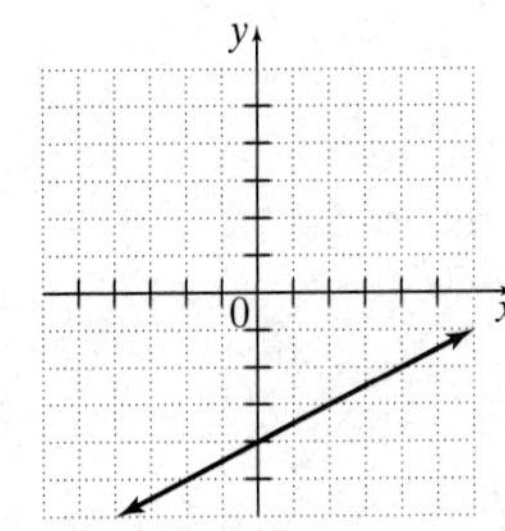

17. $x + 2y = 8$
$2y = -x + 8$
$y = -\frac{1}{2}x + 4$
$m = -\frac{1}{2}$, y-intercept $= (0, 4)$

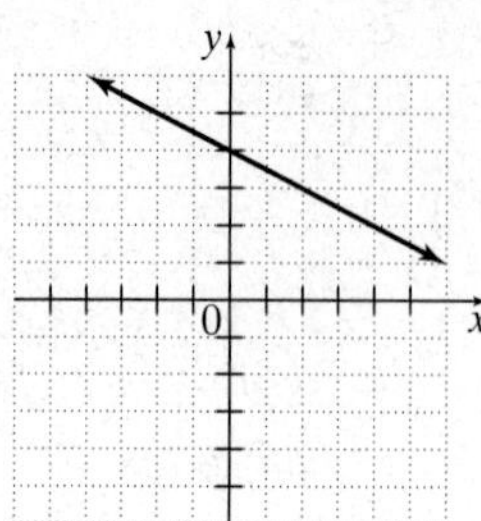

21. $y = 5x + 1$
$m = 5$, y-intercept $= (0, 1)$
graph D

25. $y = mx + b$
$y = 2x + \frac{3}{4}$

29. $y = 1765.1x + 35{,}815.0$

a. $x = 2000 - 1995 = 5$
$y = 1765.1(5) + 35{,}815.0$
$y = 8825.5 + 35{,}815.0$
$y = 44{,}640.5$
The average income is $44,640.50

b. $m = 1765.1$; The annual average income increases $1765.10 each year.

c. y-intercept $= (0, 35, 815.0)$; At $x = 0$, or 1995, the annual average income was $35,815.00.

33. $y = 136.2x + 2827.6$

a. $x = 2010 - 1996 = 14$
$y = 136.2(14) + 2827.6$
$y = 1906.8 + 2827.6$
$y = 4734.4$
The cost is approximately $4734.40.

b. $y = 5000$
$5000 = 136.2x + 2827.6$
$2172.4 = 136.2x$
$15.95 \approx x$
$1996 + 15.95 = 2011.95$
By 2012, the cost is expected to exceed $5000.

c. answers may vary

37. $y - (-1) = 2(x - 0)$
$y + 1 = 2x - 0$
$y + 1 - 1 = 2x - 1$
$y = 2x - 1$

41. $y = -7x + 500$
where y is the height at time x.

Exercise Set 7.2

1. $y - y_1 = m(x - x_1)$
$y - 2 = 3(x - 1)$
$y - 2 = 3x - 3$
$y = 3x - 1$

5. $y - y_1 = m(x - x_1)$
$y - 2 = \frac{1}{2}[x - (-6)]$
$y - 2 = \frac{1}{2}(x + 6)$
$y - 2 = \frac{1}{2}x + 3$
$y = \frac{1}{2}x + 5$

9. $y - y_1 = m(x - x_1)$
$y - 3 = 2[x - (-2)]$
$y - 3 = 2(x + 2)$
$y - 3 = 2x + 4$
$y = 2x + 7$

13. $m = \frac{6 - 0}{4 - 2} = \frac{6}{2} = 3$
$y - 0 = 3(x - 2)$
$y = 3x - 6$

17. $m = \frac{-3 - (-4)}{-4 - (-2)} = \frac{1}{-2} = -\frac{1}{2}$
$y - (-4) = -\frac{1}{2}[x - (-2)]$
$2(y + 4) = -(x + 2)$
$2y + 8 = -x - 2$
$y = -\frac{1}{2}x - \frac{10}{2}$
$y = -\frac{1}{2}x - 5$

21. $m = \frac{-6 - (-4)}{0 - (-7)} = \frac{-2}{7} = -\frac{2}{7}$
$y - (-6) = -\frac{2}{7}(x - 0)$
$y + 6 = -\frac{2}{7}x$
$y = -\frac{2}{7}x - 6$

25. Every vertical line is in the form $x = c$. Since the line passes through the point $(2, 6)$, its equation is $x = 2$.

29. A line with undefined slope is vertical. Every vertical line is in the form $x = c$. Since the line passes through the point $(0, 5)$, its equation is $x = 0$.

33. $y = -2x - 6, m = -2$
so perpendicular slope $= \frac{1}{2}$
$y - (-5) = \frac{1}{2}(x - 2)$
$y + 5 = \frac{1}{2}x - 1$
$y = \frac{1}{2}x - 6$

37. $3x + 2y = 5$
$2y = -3x + 5$
$y = -\frac{3}{2}x + \frac{5}{2}, m = -\frac{3}{2}$
so parallel slope $= -\frac{3}{2}$
$y - (-3) = -\frac{3}{2}[x - (-2)]$
$y + 3 = -\frac{3}{2}(x + 2)$

$$y + 3 = -\frac{3}{2}x - 3$$
$$y = -\frac{3}{2}x - 6$$

41. $x = 3$; perpendicular line is $y = -5$

45. $x - 4y = 4$
$$-4y = -x + 4$$
$$y = \frac{1}{4}x - 1, m = \frac{1}{4}$$
so perpendicular slope $= -4$
$$y - 5 = -4(x + 1)$$
$$y - 5 = -4x - 4$$
$$y = -4x + 1$$

49. a. $(1, 30{,}000), (4, 66{,}000)$
$$m = \frac{66{,}000 - 30{,}000}{4 - 1} = 12{,}000$$
$$y - 30{,}000 = 12{,}000(x - 1)$$
$$y = 12{,}000x + 18{,}000$$

b. $y = 12{,}000(7) + 18{,}000 = \$102{,}000$

c. $126{,}000 = 12{,}000x + 18{,}000$
$$x = \frac{126{,}000 - 18{,}000}{12{,}000}$$
$x = 9$ years

53. a. $(0, 1{,}089{,}261), (2, 8{,}498, 545)$
$$m = \frac{8{,}498{,}545 - 1{,}089{,}261}{2 - 0}$$
$$m = \frac{7{,}409{,}284}{2} = 3{,}704{,}642$$
$$y - 1{,}089{,}261 = 3{,}704{,}642(x - 0)$$
$$y = 3{,}704{,}642x + 1{,}089{,}261$$

b. $x = 2005 - 1998 = 7$
$$y = 3{,}704{,}642(7) + 1{,}089{,}261$$
$$y = 25{,}932{,}494 + 1{,}089{,}261$$
$$y = 27{,}021{,}755$$
About 27,021,755 DVD players will be sold.

57. $y = 4.2x$
$y = 4.2(-2)$
$y = -8.4$
$(-2, -8.4)$

61. True

65. $y = 4x - 2, y = 4x - 4$

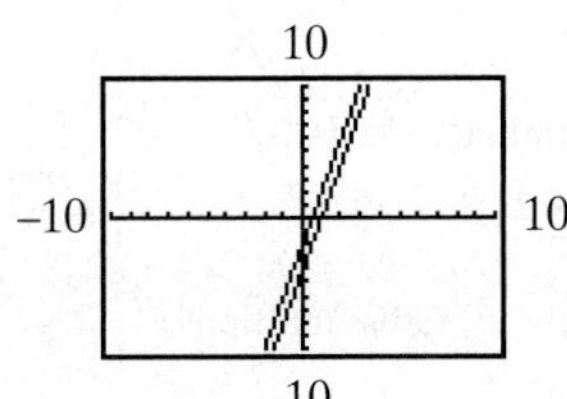

Exercise Set 7.3

1. Domain: = $\{-1, 0, -2, 5\}$
Range = $\{7, 6, 2\}$
The relation is a function.

5. Domain: = $\{1\}$
Range = $\{1, 2, 3, 4\}$
The relation is not a function since 1 is paired with 1, 2, 3 and 4.

9. Domain: $\{-3, 0, 3\}$
Range = $\{-3, 0, 3\}$
The relation is a function.

13. Domain = {Colorado, Alaska, Delaware, Illinois, Connecticut, New York}
Range = $\{6, 1, 20, 31\}$
The relation is a function.

17. Domain = $\{2, -1, 5, 100\}$
Range = $\{0\}$
The relation is a funcion.

21. Yes **25.** Yes

29. Not a function

33. Domain = $[0, \infty)$
Range = $(-\infty, \infty)$
The relation is not a function since it fails the vertical line test (try $x = 1$).

37. Domain = $(-\infty, \infty)$
Range = $(-\infty, -3] \cup [3, \infty)$
The relation is not a function since it fails the vertical line test (try $x = 3$).

41. Domain = $\{-2\}$
Range = $(-\infty, \infty)$
The relation is not a function since it fails the vertical line test.

45. $f(x) = 3x + 3$
$f(4) = 3(4) + 3 = 12 + 3 = 15$

49. $g(x) = 4x^2 - 6x + 3$
$g(2) = 4(2)^2 - 6(2) + 3$
$= 4(4) - 12 + 3$
$= 16 - 12 + 3$
$= 7$

53. $f(x) = \frac{1}{2}x$

a. $f(0) = \frac{1}{2}(0) = 0$

b. $f(2) = \frac{1}{2}(2) = 1$

c. $f(-2) = \frac{1}{2}(-2) = -1$

57. $A(r) = \pi r^2$
$A(5) = \pi(5)^2 = 25\pi$
25π sq cm

61. $H(f) = 2.59f + 47.24$
$H(46) = 2.59(46) + 47.24$
$= 119.14 + 47.24$
$= 166.38$
166.38 cm

65. $C(x) = 1.69x + 87.54$

a. $C(5) = 1.69(5) + 87.54 = 8.45 + 87.54 = 95.99$

The per capita consumption of poultry was about 95.99 lb in 2000.

b. $x = 2002 - 1995 = 7$

$C(7) = 1.69(7) + 87.54 = 11.83 + 87.54 = 99.37$

The per capita consumption is about 99.37 lb in 2002.

69. $f(x) = -3x$
or
$y = -3x$
where $m = -3$ and y-intercept $= (0, 0)$

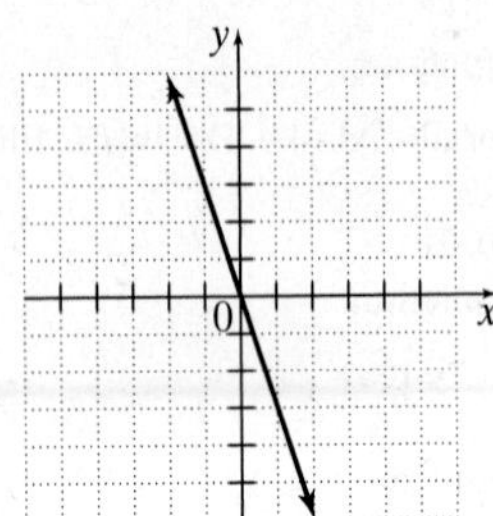

73. $2x - 7 \le 21$
$2x \le 28$
$x \le 14$
$(-\infty, 14]$

77. $\frac{x}{2} + \frac{1}{4} < \frac{1}{8}$
$\frac{x}{2} < \frac{1}{8} - \frac{1}{4}$
$\frac{x}{2} < -\frac{1}{8}$
$x < -\frac{1}{4}$
$\left(-\infty, -\frac{1}{4}\right)$

81. $f(x) = x^2 - 12$

a. $f(12) = (12)^2 - 12 = 144 - 12 = 132$

b. $f(a) = a^2 - 12$

85. answers may vary

Exercise Set 7.4

1. $P(x) = x^2 + x + 1$
$P(7) = 7^2 + 7 + 1 = 49 + 7 + 1 = 57$

5. $P(x) = x^2 + x + 1$
$P(0) = 0^2 + 0 + 1 = 1$

9. $Q(x) = 7x + 5$
$Q(-3) = 7(-3) + 5$
$= -21 + 5$
$= -16$

13. $P(t) = -16t^2 + 300t$

a. $P(1) = -16(1)^2 + 300(1) = 284$ ft

b. $P(2) = -16(2)^2 + 300(2) = 536$ ft

c. $P(3) = -16(3)^2 + 300(3) = 756$ ft

d. $P(4) = -16(4)^2 + 300(4) = 944$ ft

e. answers may vary

f. When $t = 18$ sec
$P(18) = -16(18)^2 + 300(18) = 216$ ft
When $t = 19$ seconds
$P(19) = -16(19)^2 + 300(19) = -76$ ft
The object will hit the ground at approximately 19 sec.

17. $R(x) = 2x$
$R(20{,}000) = 2(20{,}000) = 40{,}000$
The total revenue is \$40,000.

21. $f(x) = 98x^2 + 514x + 4746$
$= 2(49x^2 + 257x + 2373)$

25. a. $f(x) = 136.7x^2 + 327.6x + 21.6$
$f(0) = 136.7(0)^2 + 327.6(0) + 21.6$
$= 21.6$
≈ 22
In 1996, there were about 22 stations.

b. $f(4) = 136.7(4)^2 + 327.6(4) + 21.6$
$= 136.7(16) + 327.6(4) + 21.6$
$= 2187.2 + 1310.4 + 21.6$
$= 3519.2$
≈ 3519
In 2000, there were about 3519 stations.

c. $f(8) = 136.7(8)^2 + 327.6(8) + 21.6$
$= 136.7(64) + 327.6(8) + 21.6$
$= 8748.8 + 2620.8 + 21.6$
$= 11{,}391.2$
$\approx 11{,}391$
In 2004, there will be about 11,391 stations.

d. increasing at a steady rate; answers may vary

29. $g(x) = \frac{x - 2}{x - 5}$
$g(0) = \frac{0 - 2}{0 - 5}$
$= \frac{-2}{-5}$
$= \frac{2}{5}$
$= 0.4$
$g(x) = \frac{x - 2}{x - 5}$

33. denominator is $x - 5$
$x - 5 = 0$
$x = 5$
$\{x | x \text{ is a real number and } x \neq 5\}$

37. $g(x) = \frac{x^2 + 2x}{x + 3}$
$g(-6) = \frac{(-6)^2 + 2(-6)}{-6 + 3}$
$= \frac{36 - 12}{-3}$
$= \frac{24}{-3}$
$= -8$

41. $f(x) = \frac{x^2 + 5}{x}$
denominator is x
$x = 0$
$\{x | x \text{ is a real number and } x \neq 0\}$

45. $\frac{x}{5} = \frac{x + 2}{3}$
$x(3) = 5(x + 2)$ Cross multiply
$3x = 5x + 10$
$3x - 5x = 5x - 5x + 10$
$-2x = 10$
$\frac{-2x}{-2} = \frac{10}{-2}$
$x = -5$
-5

49. a. $h(t) = -16t^2 + 64t$
$= -16t(t - 4)$

b. $h(t) = -16t^2 + 64t$
$h(1) = -16(1)^2 + 64(1)$
$= -16 + 64$
$= 48$ ft
$h(t) = -16t(t - 4)$
$h(1) = -16(1)(1 - 4)$
$= -16(1)(-3)$
$= 48$ ft

c. answers may vary

53. $P(x) = 2x - 3$
$P(a) = 2a - 3$
$P(-x) = 2(-x) - 3$
$= -2x - 3$
$P(x + h) = 2(x + h) - 3$
$= 2x + 2h - 3$

57. $f(x) = x^2 - 3x$
$f(a) = a^2 - 3a$

Exercise Set 7.5

1. $y = kx$
$4 = k(20)$
$k = \frac{4}{20} = \frac{1}{5}$
$y = \frac{1}{5}x$

5. $y = kx$
$7 = k\left(\frac{1}{2}\right)$
$k = 14$
$y = 14x$

9. $W = kr^3$
$1.2 = k \cdot 2^3$
$k = \frac{1.2}{8}$
$= 0.15$
$W = 0.15r^3$
$= 0.15(3)^3$
$W = 0.15(27)$
$= 4.05$ pounds

13. $y = \frac{k}{x}$
$6 = \frac{k}{5}$
$k = 30$
$y = \frac{30}{x}$

17. $y = \frac{k}{x}$
$\frac{1}{8} = \frac{k}{16}$
$k = 2$
$y = \frac{2}{x}$

21. $R = \frac{k}{T}$
$45 = \frac{k}{6}$
$k = 270$
$R = \frac{270}{5}$
$R = 54$ mph

25. $I_1 = \frac{k}{d^2}$
Replace d by $2d$.
$I_2 = \frac{k}{(2d)^2} = \frac{k}{4d^2} = \frac{1}{4}I_1$
Thus, the intensity is divided by 4.

29. $r = kst^3$

33. $y = k\sqrt{x}$
$0.4 = k\sqrt{4}$
$0.4 = 2k$
$\frac{0.4}{2} = k$
$0.2 = k$
$y = 0.2\sqrt{x}$

37. $y = kxz^3$
$120 = k(5)(2^3)$
$120 = k(5)(8)$
$120 = 40k$
$3 = k$
$y = 3xz^3$

41. $V = kr^2h$
$32\pi = k(4)^2(6)$
$32\pi = k(16)(6)$
$32\pi = 96k$
$\frac{32\pi}{96} = k$
$\frac{\pi}{3} = k$
$V = \frac{\pi}{3}r^2h$
$V = \frac{\pi}{3}(3)^2(5)$
$V = 15\pi$ cu in.

45. $x - y = 4$
$3 - (-1) \stackrel{?}{=} 4$
$3 + 1 \stackrel{?}{=} 4$
$4 = 4$ True
$x + 2y = 1$
$3 + 2(-1) \stackrel{?}{=} 1$
$3 - 2 \stackrel{?}{=} 1$
$1 = 1$ True
Yes, the ordered pair is a solution of both equations.

49. $V_1 = khr^2$
$V_2 = k\left(\frac{1}{2}h\right)(2r)^2$
$V_2 = 2khr^2 = 2V_1$
It is multiplied by 2.

Chapter 7 Test

1. $3x + 12y = 8$
$12y = -3x + 8$
$y = \frac{-3}{12}x + \frac{8}{12}$
$= -\frac{1}{4}x + \frac{2}{3}$
Slope is $-\frac{1}{4}$.
y-intercept is $\left(0, \frac{2}{3}\right)$

5. D

9. $y - y_1 = m(x - x_1)$
$y - (-1) = -3(x - 4)$
$y + 1 = -3x + 12$
$y = -3x + 11$

13. $2y + x = 3$
$2y = -x + 3$
$y = -\frac{1}{2}x + \frac{3}{2}$
Slope of the given line is $-\frac{1}{2}$.
Slope of the parallel line is also $-\frac{1}{2}$.
The equation is
$y - y_1 = m(x - x_1)$
$y - (-2) = -\frac{1}{2}(x - 3)$
$y + 2 = -\frac{1}{2}x + \frac{3}{2}$
$y = -\frac{1}{2}x - \frac{1}{2}$

17. Domain: $\{-2\}$
Range: $(-\infty, \infty)$
Not a function since it does not pass the vertical line test.

21. $f(x) = \dfrac{5x^2}{1 - x}$

$$f(2) = \frac{5(2)^2}{1 - 2} = \frac{5(4)}{1 - 2} = \frac{20}{-1} = -20$$

Chapter 8

Exercise Set 8.1

1. a.

$x + y = 8$	$3x + 2y = 21$
$2 + 4 \stackrel{?}{=} 8$	$3(5) + 2(4) = 21$
$6 = 8$	$6 + 8 = 21$
False	$14 = 21$
	False

No, $(2, 4)$ is not a solution of the system.

b.

$x + y = 8$	$3x + 2y = 21$
$5 + 3 \stackrel{?}{=} 8$	$3(5) + 2(3) \stackrel{?}{=} 21$
$8 = 8$	$15 + 6 \stackrel{?}{=} 21$
True	$21 = 21$

Yes, $(5, 3)$ is a solution of the system.

5. a.

$2y = 4x$	$2x - y = 0$
$2(-6) \stackrel{?}{=} 4(-3)$	$2(-3) - (-6) \stackrel{?}{=} 0$
$-12 = -12$	$-6 + 6 \stackrel{?}{=} 0$
True	True $\quad 0 = 0$

Yes, $(-3, -6)$ is a solution of the system.

b.

$2y = 4x$	$2x - y = 0$
$2(0) \stackrel{?}{=} 4(0)$	$2(0) - 0 \stackrel{?}{=} 0$
$0 = 0$	$0 = 0$
True	True

Yes, $(0, 0)$ is a solution of the system.

9. $\begin{cases} x = y = 6 \\ -x + y = -6 \end{cases}$

Graph each linear equation on a single set of axes.

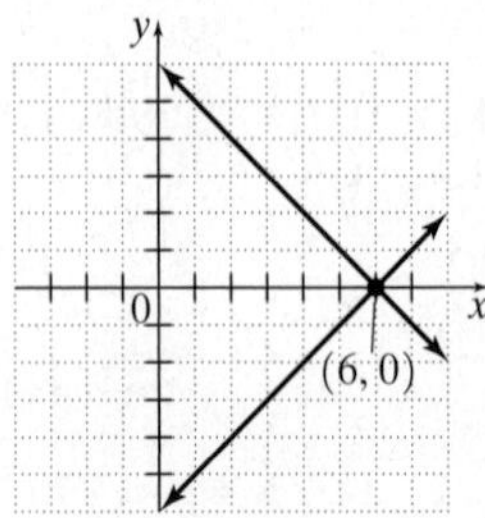

The solution is the intersection point of the two lines, $(6, 0)$.

13. $\begin{cases} y = x + 1 \\ y = 2x - 1 \end{cases}$

Graph each linear equation on a single set of axes.

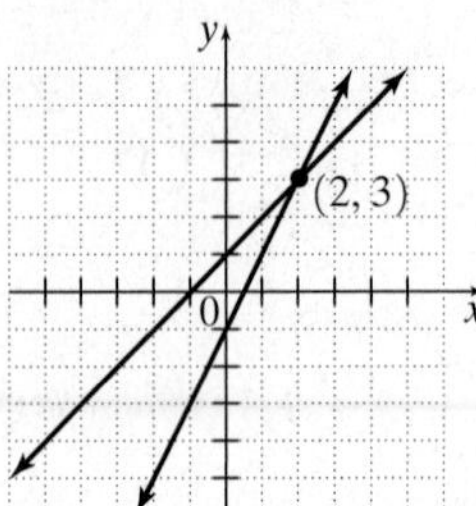

The solution is the intersection point of the two lines, $(2, 3)$.

17. $\begin{cases} y = -x - 1 \\ y = 2x + 5 \end{cases}$

Graph each linear equation on a single set of axes.

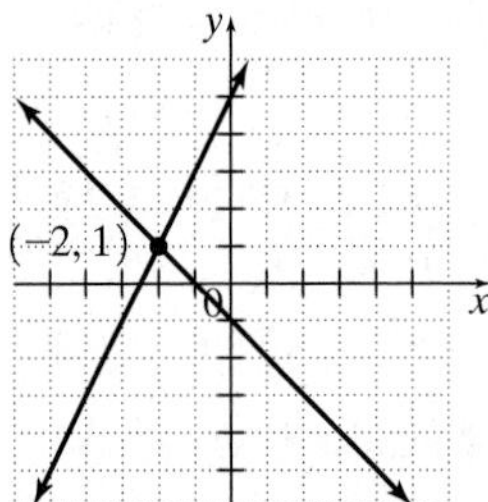

The solution is the intersection point of the two lines, $(-2, 1)$.

21. $\begin{cases} x + y = 5 \\ x + y = 6 \end{cases}$

Graph each linear equation on a single set of axes.

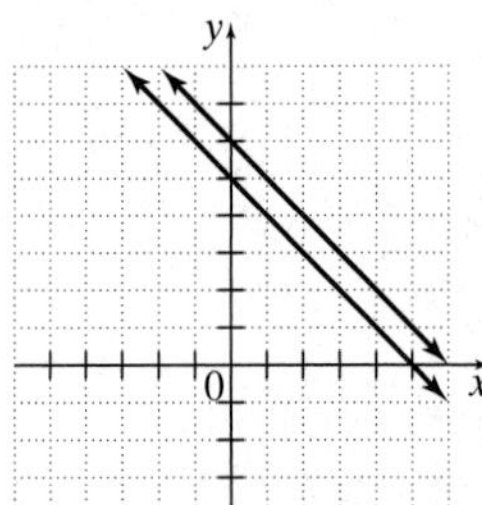

Since the lines are parallel, the system has no solution.

25. $\begin{cases} x - 2y = 2 \\ 3x + 2y = -2 \end{cases}$

Graph each linear equation on a single set of axes.

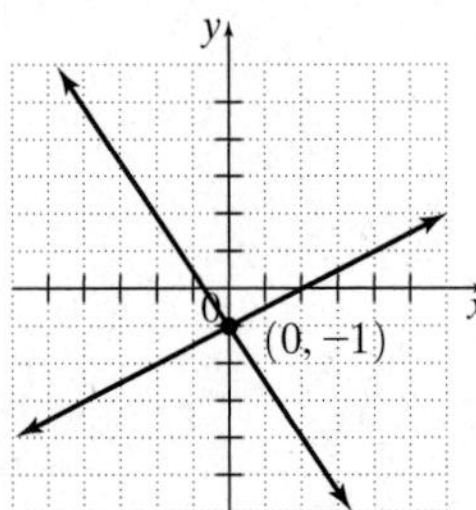

The solution is the intersection point of the two lines, $(0, -1)$.

29. $\begin{cases} x = 3 \\ y = -1 \end{cases}$

Graph each linear equation on a single set of axes.

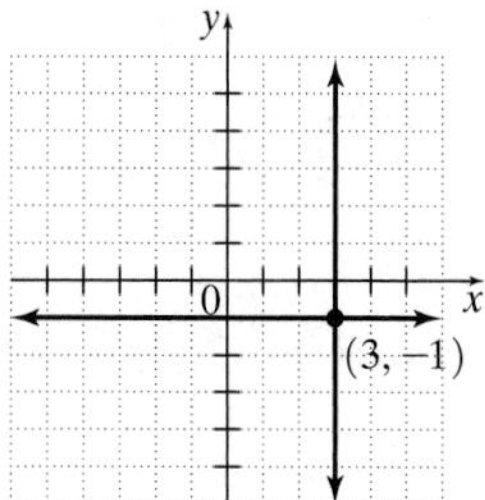

The solution is the intersection point of the two lines, $(3, -1)$.

33. $\begin{cases} 2x - 3y = -2 \\ -3x + 5y = 5 \end{cases}$

Graph each linear equation on a single set of axes.

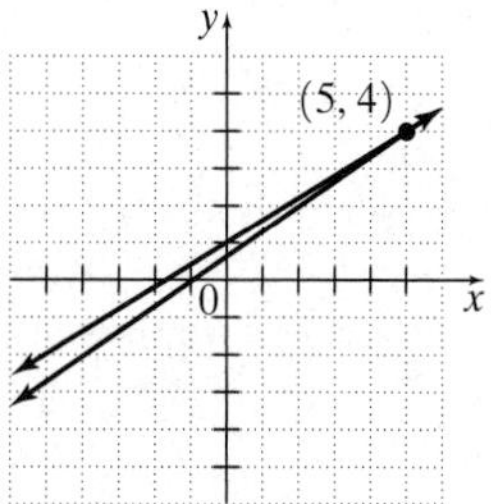

The solution is the intersection point of the two lines, $(5, 4)$.

37. answers may vary

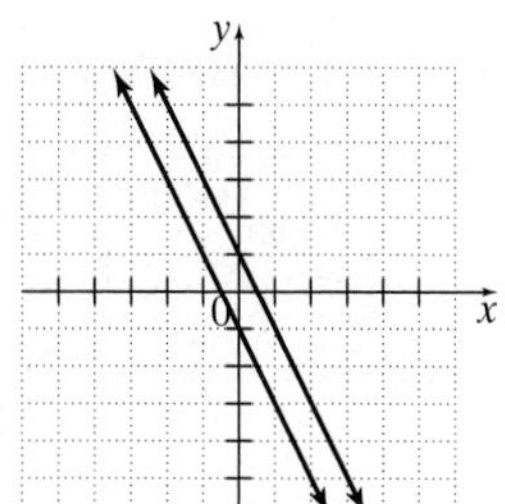

Any two parallel lines will meet the condition.

41. The two lines intersect near the year 1996.

45.
$$\begin{aligned} -2x + 3(x + 6) &= 17 \\ -2x + 3x + 18 &= 17 \\ -2x + 3x + 18 &= 17 \\ x + 18 &= 17 \\ x + 18 - 18 &= 17 - 18 \\ x &= -1 \end{aligned}$$

49.
$$\begin{aligned} 3z - (4z - 2) &= 9 \\ 3z - 4z + 2 &= 9 \\ -z + 2 &= 9 \\ -z + 2 - 9 &= 9 - 2 \\ -z &= 7 \\ z &= -7 \end{aligned}$$

53. answers may vary

Exercise Set 8.2

1. $\begin{cases} x + y = 3 \\ x = 2y \end{cases}$

Substitute $2y$ for x in the first equation. Then solve for y.

$$\begin{aligned} 2y + y &= 3 \\ 3y &= 3 \\ y &= 1 \end{aligned}$$

Substitute 1 for y in the second equation. Then solve for x.

$$\begin{aligned} x &= 2(1) \\ x &= 2 \end{aligned}$$

The solution is $(2, 1)$.

5. $\begin{cases} 3x + 2y = 16 \\ x = 3y - 2 \end{cases}$

Substitute $3y - 2$ for x in the first equation. Then solve for y.

$$\begin{aligned} 3(3y - 2) + 2y &= 16 \\ 9y - 6 + 2y &= 16 \\ 11y &= 22 \\ y &= 2 \end{aligned}$$

Substitute 2 for y in the second equation.

Then solve for x.

$$\begin{aligned} x &= 3(2) - 2 \\ x &= 6 - 2 \\ x &= 4 \end{aligned}$$

The solution is $(4, 2)$.

9. $\begin{cases} y = 3x + 1 \\ 4y - 8x = 12 \end{cases}$

Substitute $3x + 1$ for y in the second equation. Then solve for x.

$$\begin{aligned} 4(3x + 1) - 8x &= 12 \\ 12x + 4 - 8x &= 12 \\ 4x &= 8 \\ x &= 2 \end{aligned}$$

Substitute 2 for x in the first equation. Then solve for y.

$$\begin{aligned} y &= 3(2) + 1 \\ y &= 6 + 1 \\ y &= 7 \end{aligned}$$

The solution is $(2, 7)$.

13. $\begin{cases} x + 2y = 6 \\ 2x + 3y = 8 \end{cases}$

Solve the first equation for x.

$$x = 6 - 2y$$

Substitute $6 - 2y$ for x in the second equation. Then solve for y.

$$\begin{aligned} 2(6 - 2y) + 3y &= 8 \\ 12 - 4y + 3y &= 8 \\ -y &= -4 \\ y &= 4 \end{aligned}$$

Substitute 4 for y in $x = 6 - 2y$. Then solve for x.

$$\begin{aligned} x &= 6 - 2(4) \\ x &= 6 - 8 \\ x &= -2 \end{aligned}$$

The solution is $(-2, 4)$.

17. $\begin{cases} 2y = x + 2 \\ 6x - 12y = 0 \end{cases}$

Solve the first equation for x.

$x = 2y - 2$

Substitute $2y - 2$ for x in the second equation.

$6(2y - 2) - 12y = 0$

$12y - 12 - 12y = 0$

$-12 = 0$

Since this is false, the system has no solution.

21. $\begin{cases} 2x - 3y = -9 \\ 3x = y + 4 \end{cases}$

Solve the second equation for y.

$y = 3x - 4$

Substitute $3x - 4$ for y in the first equation.

Then solve for x.

$2x - 3(3x - 4) = -9$

$2x - 9x + 12 = -9$

$-7x = -21$

$x = 3$

Substitute 3 for x in $y = 3x - 4$.

Then solve for y.

$y = 3(3) - 4$

$y = 9 - 4$

$y = 5$

The solution is $(3, 5)$.

25. $\begin{cases} 3x - y = 1 \\ 2x - 3y = 10 \end{cases}$

Solve the first equation for y.

$y = 3x - 1$

Substitute $3x - 1$ for y in the second equation. Then solve for x.

$2x - 3(3x - 1) = 10$

$2x - 9x + 3 = 10$

$-7x = 7$

$x = -1$

Substitute -1 for x in $y = 3x - 1$. Then solve for y.

$y = 3(-1) - 1$

$y = -3 - 1$

$y = -4$

The solution is $(-1, -4)$.

29. $\begin{cases} 5x + 10y = 20 \\ 2x + 6y = 10 \end{cases}$

Solve the first equation for x.

$x + 2y = 4$

$x = 4 - 2y$

Substitute $4 - 2y$ for x in the second equation. Then solve for y.

$2(4 - 2y) + 6y = 10$

$8 - 4y + 6y = 10$

$2y = 2$

Substitute 1 for y in $x = 4 - 2y$. Then solve for x.

$x = 4 - 2(1)$

$x = 2$

The solution is $(2, 1)$.

33. $\begin{cases} \frac{1}{3}x - y = 2 \\ x - 3y = 6 \end{cases}$

Solve the second equation for x.

$x = 3y + 6$

Substitute $3y + 6$ for x in the first equation.

$\frac{1}{3}(3y + 6) - y = 2$

$y + 2 - y = 2$

$2 = 2$

Since this is always true, the system has an infinite number of solutions.

37. $3x + 2y = 6$

$-2(3x + 2y) = -2(6)$

$-6x - 4y = -12$

41. $3n + 6m$

$\underline{2n - 6m}$

$5n$

45. Simplify the first equation.

$-5y + 6y = 3x + 2(x - 5) - 3x + 5$

$y = 3x + 2x - 10 - 3x + 5$

$y = 2x - 5$

Simplify the second equation.

$4(x + y) - x + y = -12$

$4x + 4y - x + y = -12$

$3x + 5y = -12$

Solve the system

$\begin{cases} y = 2x - 5 \\ 3x + 5y = -12 \end{cases}$

Substitute $2x - 5$ for y in the second equation. Then solve for x.

$3x + 5(2x - 5) = -12$

$3x + 10x - 25 = -12$

$13x = 13$

$x = 1$

Substitute 1 for x in the first equation. Then solve for y.

$y = 2(1) - 5$

$y = 2 - 5$

$y = -3$

The solution is $(1, -3)$.

49. $\begin{cases} 3x + 2y = 14.04 \\ 5x + y = 18.5 \end{cases}$

Let $y_1 = \dfrac{14.04 - 3x}{2}$ and $y_2 = 18.5 - 5x$ and find the intersection.

The solution is $(3.28, 2.1)$.

Exercise Set 8.3

1. $\begin{cases} 3x + y = 5 \\ 6x - y = 4 \end{cases}$

$9x = 9$

$x = 1$

Let $x = 1$ in the first equation.

$3(1) + y = 5$

$y = 2$

The solution is $(1, 2)$.

5. $\begin{cases} 3x + 2y = 11 \\ 5x - 2y = 29 \end{cases}$

$8x = 40$

$x = 5$

Let $x = 5$ in the first equation.

$3(5) + 2 = 11$

$15 + 2y = 11$

$2y = -4$

$y = -2$

The solution is $(5, -2)$.

9. $\begin{cases} 3x + y = -11 \\ 6x - 2y = -2 \end{cases}$

$\begin{cases} 2(3x + y) = 2(-11) \\ 6x - 2y = -2 \end{cases}$

$\begin{cases} 6x + 2y = -22 \\ 6x - 2y = -2 \end{cases}$

$12x = -24$

$x = -2$

Let $x = -2$ in the first equation.

$3(-2) + y = -11$

$-6 + y = -11$

$y = -5$

The solution is $(-2, -5)$.

13. $\begin{cases} 2x - 5y = 4 \\ 3x - 2y = 4 \end{cases}$

$\begin{cases} -3(2x - 5y) = -3(4) \\ 2(3x - 2y) = 2(4) \end{cases}$

$\begin{cases} -6x + 15y = -12 \\ 6x - 4y = 8 \end{cases}$

$11y = -4$

$y = -\dfrac{4}{11}$

$\begin{cases} -2(2x - 5y) = -2(4) \\ 5(3x - 2y) = 5(4) \end{cases}$

$\begin{cases} -4x + 10y = -8 \\ 15x - 10y = 20 \end{cases}$

$11x = 12$

$x = \dfrac{12}{11}$

The solution is $\left(\dfrac{12}{11}, -\dfrac{4}{11}\right)$.

17. $\begin{cases} 3x + y = 4 \\ 9x + 3y = 6 \end{cases}$

$\begin{cases} -3(3x + y) = -3(4) \\ 9x + 3y = 6 \end{cases}$

$\begin{cases} -9x - 3y = -12 \\ 9x + 3y = 6 \end{cases}$

$0 = -6$

Since this is false, the system has no solution.

21. $\begin{cases} \dfrac{2}{3}x + 4y = -4 \\ 5x + 6y = 18 \end{cases}$

$\begin{cases} 3\left(\dfrac{2}{3}x + 4y\right) = 3(-4) \\ -2(5x + 6y) = -2(18) \end{cases}$

$\begin{cases} 2x + 12y = -12 \\ -10x - 12y = -36 \end{cases}$

$-8x = -48$

$x = 6$

Let $x = 6$ in the second equation.

$5(6) + 6y = 18$

$30 + 6y = 18$

$6y = -12$

$y = -2$

The solution is $(6, -2)$.

25. $\begin{cases} 8x = -11y - 16 \\ 2x + 3y = -4 \end{cases}$

$\begin{cases} 8x + 11y = -16 \\ -4(2x + 3y) = -4(-4) \end{cases}$

$\begin{cases} 8x + 11y = -16 \\ -8x - 12y = 16 \end{cases}$

$-y = 0$

$y = 0$

Let $y = 0$ in the first equation.

$8x = -11(0) - 16$

$8x = -6$

$x = -2$

The solution is $(-2, 0)$.

29. $\begin{cases} \dfrac{x}{3} + \dfrac{y}{6} = 1 \\ \dfrac{x}{2} - \dfrac{y}{4} = 0 \end{cases}$

$\begin{cases} 6\left(\dfrac{x}{3} + \dfrac{y}{6}\right) = 6(1) \\ 4\left(\dfrac{x}{2} - \dfrac{y}{4}\right) = 4(0) \end{cases}$

$\begin{cases} 2x + y = 6 \\ 2x - y = 0 \end{cases}$

$4x = 6$

$x = \dfrac{3}{2}$

To find y, we may multiply the second equation of the simplified system above by -1.

$$\begin{cases} 2x + y = 6 \\ -2x + y = 0 \end{cases}$$
$$2y = 6$$
$$y = 3$$

The solution is $\left(\frac{3}{2}, 3\right)$.

33. $$\begin{cases} \frac{x}{3} - y = 2 \\ -\frac{x}{2} + \frac{3y}{2} = -3 \end{cases}$$

$$\begin{cases} 3\left(\frac{x}{3} - y\right) = 3(2) \\ 2\left(-\frac{x}{2} + \frac{3y}{2}\right) = 2(-3) \end{cases}$$

$$\begin{cases} x - 3y = 6 \\ -x + 3y = -6 \end{cases}$$
$$0 = 0$$

Since this is always true, the system has an infinite number of solutions.

37. $$\begin{cases} 3.5x + 2.5y = 17 \\ -1.5x - 7.5y = -33 \end{cases}$$

It is not necessary to clear decimals by multiplying each side by 10. Instead, multiply the first equation by 3.

$$\begin{cases} 3(3.5x + 2.5y = 17) \\ -1.5x - 7.5y = -33 \end{cases}$$

$$\begin{cases} 10.5x + 7.5y = 51 \\ -1.5x - 7.5y = -33 \end{cases}$$
$$9x = 18$$
$$x = 2$$

Let $x = 2$ in the first equation.

$3.5(2) + 2.5y = 17$

$7 + 2.5y = 17$

$2.5y = 10$

$y = 4$

The solution is $(2, 4)$.

41. Let x = a number.

$2x + 6 = x - 3$

45. Let x = a number.

$4(x + 6) = 2x$

49. $$\begin{cases} 2x + 3y = 14 \\ 3x - 4y = -69.1 \end{cases}$$

$$\begin{cases} 4(2x + 3y = 14) \\ 3(3x - 4y = -69.1) \end{cases}$$

$$\begin{cases} 8x + 12y = 56 \\ 9x - 12y = -207.3 \end{cases}$$
$$17x = -151.3$$
$$x = -8.9$$

Let $x = -8.9$ in the first equation.

$2(-8.9) + 3y = 14$

$-17.8 + 3y = 14$

$3y = 31.8$

$y = 10.6$

The solution is $(-8.9, 10.6)$.

53. a. $$\begin{cases} -19x + 10y = 5428 \\ 70x - 5y = -2520 \end{cases}$$

$$\begin{cases} -19x + 10y = 5428 \\ 2(70x - 5y = -2520) \end{cases}$$

$$\begin{cases} -19x + 10y = 5428 \\ 140x - 10y = -5040 \end{cases}$$
$$121x = 388$$
$$x = \frac{388}{121} \approx 3$$

Let $x = 3$ in the first equation.

$-19(3) + 10y = 5428$

$-57 + 10y = 5428$

$10y = 5485$

$y = \frac{5485}{10} = 548.5 \approx 549$

The solution is $(3, 549)$.

b. answers may vary

c. This occurs for $x \geq 3$, or for years beyond $1988 + 3 = 1991$. The years are 1991–2000.

Exercise Set 8.4

1. Let x = one number

y = another number

$$\begin{cases} x + y = 15 \\ x - y = 7 \end{cases}$$

5. Let x = one number

y = another number

$$\begin{cases} x + y = 83 \\ x - y = 17 \end{cases}$$
$$2x = 100$$
$$x = 50$$

Let $x = 50$ in the first equation.

$50 + y = 83$

$y = 33$

The two numbers are 33 and 50.

9. Let x = the number of points scored by Swoopes

y = number of points scored by Smith

$$\begin{cases} y = x + 3 \\ x + y = 1289 \end{cases}$$

$$\begin{cases} -x + y = 3 \\ x + y = 1289 \end{cases}$$
$$2y = 1292$$
$$y = 646$$

Now, let $y = 646$ in the first equation.

$646 = x + 3$

$643 = x$

Swoopes scored 643 points and Smith scored 646 points.

13. Let x = number of quarters

y = number of nickels

$$\begin{cases} x + y = 80 \\ 0.25x + 0.05y = 14.6 \end{cases}$$

Substitute $80 - x$ for y in the second equation.

$$\begin{aligned} 0.25x + 0.05(80 - x) &= 14.6 \\ 0.25x + 4 - 0.05x &= 14.6 \\ 0.2x &= 10.6 \\ x &= 53 \end{aligned}$$

Let $x = 53$ in the first equation.

$$\begin{aligned} 53 + y &= 80 \\ y &= 27 \end{aligned}$$

There are 53 quarters and 27 nickels.

17. Let x = cost per hr for labor

y = cost per ton of material

$$\begin{cases} 65x + 3y = 1702.50 \\ 49x + 2.5y = 1349.00 \end{cases}$$

Multiply the first equation by 25 and the second equation by 30.

$$\begin{cases} 25(65x + 3y) = 25(1702.50) \\ -30(49x + 2.5y) = -30(1349.00) \end{cases}$$

$$\begin{cases} 1625x + 75y = 42{,}562.50 \\ -1470x - 75y = -40{,}470.00 \end{cases}$$

$$\begin{aligned} 155x &= 2{,}092.50 \\ x &= \frac{2092.50}{155} = 13.50 \end{aligned}$$

Let $x = 13.50$ in the first equation.

$$\begin{aligned} 65x + 3y &= 1702.50 \\ 65(13.50) + 3y &= 1702.50 \\ 877.50 + 3y &= 1702.50 \\ 3y &= 825 \\ y &= 275 \end{aligned}$$

The cost for labor is \$13.50 per hr and the cost for materials is \$275 per ton.

21. Let x = speed of plane in still air

y = speed of wind

$$\begin{cases} 2(x - y) = 780 \\ \frac{3}{2}(x + y) = 780 \end{cases}$$

Multiply the first equation by $\frac{1}{2}$ and the second equation by $\frac{2}{3}$.

$$\begin{cases} \frac{1}{2}[2(x - y)] = \frac{1}{2}(780) \\ \frac{2}{3}\left[\frac{3}{2}(x + y)\right] = \frac{3}{2}(780) \end{cases}$$

$$\begin{cases} x - y = 390 \\ x + y = 520 \end{cases}$$

$$\begin{aligned} 2x &= 910 \\ x &= 455 \end{aligned}$$

Let $x = 455$ in the first equation.

$$\begin{aligned} 2(x - y) &= 780 \\ 2(455 - y) &= 780 \\ 455 - y &= 390 \\ -y &= -65 \\ y &= 65 \end{aligned}$$

The speed of the plane in still air is 455 mph and the wind speed is 65 mph.

25. Let x = amount of 4% solution

y = amount of 12% solution

$$\begin{cases} x + y = 12 \\ 0.04x + 0.12y = 0.09(12) \end{cases}$$

Substitute $12 - x$ for y in the second equation.

$$\begin{aligned} 0.04x + 0.12(12 - x) &= 0.09(12) \\ 0.04x + 1.44 - 0.12x &= 1.08 \\ 1.44 - 0.08x &= 1.08 \\ -0.08x &= -0.36 \\ x &= \frac{-0.36}{-0.08} = 4.5 \end{aligned}$$

Let $x = 4.5$ in the first equation.

$$\begin{aligned} 4.5 + y &= 12 \\ y &= 7.5 \end{aligned}$$

She needs $4\frac{1}{2}$ oz of the 4% solution and $7\frac{1}{2}$ oz of the 12% solution.

29. Let x = first angle

y = second angle

$$\begin{cases} x + y = 90 \\ y = 2x \end{cases}$$

Substitute $2x$ for y in the first equation.

$$\begin{aligned} x + 2x &= 90 \\ 3x &= 90 \\ x &= 30 \end{aligned}$$

Let $x = 30$ in the second equation.

$$y = 2(30) = 60$$

The angles measure 30° and 60°.

33. $y = 7x + 18.7$

$y = 6x + 27.7$

Equate y values.

$$\begin{aligned} 7x + 18.7 &= 6x + 27.7 \\ x + 18.7 &= 27.7 \\ x &= 9 \end{aligned}$$

The year is $1996 + 9 = 2005$.

37. Let r = rate of the slower group.

$r + \frac{1}{2}$ = rate of the faster group

Using rate · time = distance, the distance for the slower group is $r(240) = 240r$,

and the distance for the faster group is $\left(r + \frac{1}{2}\right)240 = 240r + 120$.

Since the sum of the distances is 1200 miles, the equation is

$$\begin{aligned} 240r + (240r + 120) &= 1200 \\ 480r + 120 &= 1200 \\ 480r &= 1080 \\ r &= \frac{1080}{480} = \frac{9}{4} = 2\frac{1}{4} \end{aligned}$$

Then, $r + \frac{1}{2} = 2\frac{1}{4} + \frac{1}{2} = 2\frac{1}{4} + \frac{2}{4} = 2\frac{3}{4}$.

The two rates are $2\frac{1}{4}$ mph and $2\frac{3}{4}$ mph.

41. Let l = length and w = width.

$$\begin{cases} 2l + 2w = 144 \\ l = w + 12 \end{cases}$$

Substitute $w + 12$ for l in the first equation.

$$\begin{aligned} 2(w + 12) + 2w &= 144 \\ 2w + 24 + 2w &= 144 \\ 4w + 24 &= 144 \\ 4w &= 120 \\ w &= \frac{120}{4} = 30 \end{aligned}$$

Let $w = 30$ in the second equation.

$$\begin{aligned} l &= 30 + 12 \\ l &= 42 \end{aligned}$$

The length is 42 inches and the width is 30 inches.

45. $\begin{cases} x + y = 180 \\ x = y - 30 \end{cases}$

Substitute $y - 30$ for x in the first equation.

$(y - 30) + y = 180$
$2y - 30 = 180$
$2y = 210$
$y = \frac{210}{2} = 105$

Let $y = 105$ in the first equation.

$x + 105 = 180$
$x = 75$

Thus, $x = 75$ and $y = 105$.

49. $(6x)^2 = (6x)(6x) = 36x^2$

53. Let y = length and x = width.

$2x + y = 33$
$y = 2x - 3$

Substitute $2x - 3$ for y in the first equation.

$2x + (2x - 3) = 33$
$4x - 3 = 33$
$4x = 36$
$x = \frac{36}{4} = 9$

Let $x = 9$ in the second equation,

$y = 2 \cdot 9 - 3 = 18 - 3 = 15$

The length is 15 ft and the width is 9 ft.

Exercise Set 8.5

1. $\begin{cases} x + y = 3 \\ 2y = 10 \\ 3x + 2y - 3z = 1 \end{cases}$

Solve the second equation for y.

$y = 5$

Replace y with 5 in the first equation.

$x + 5 = 3$
$x = -2$

Replace x with -2 and y with 5 in the third equation.

$3(-2) + 2(5) - 3z = 1$
$-6 + 10 - 3z = 1$
$4 - 3z = 1$
$-3z = -3$
$z = 1$

The solution is $(-2, 5, 1)$.

5. $\begin{cases} x - 2y + z = -5 & (1) \\ -3x + 6y - 3z = 15 & (2) \\ 2x - 4y + 2z = -10 & (3) \end{cases}$

Multiply equation (2) by $-\frac{1}{3}$ and equation (3) by $\frac{1}{2}$.

$\begin{cases} x - 2y + z = -5 \\ x - 2y + z = -5 \\ x - 2y + z = -5 \end{cases}$

All three equations are identical. There are infinitely many solutions.

The solution is $\{(x, y, z) | x - 2y + z = -5\}$.

9. $\begin{cases} x + 5z = 0 & (1) \\ 5x + y + 10z = -24 & (2) \\ y - 3z = 0 & (3) \end{cases}$

Solve equation (1) for x.

$x = -5z$ (4)

Solve equation (3) for y.

$y = 3z$ (5)

Now substitute $-5z$ for x and $3z$ for y in equation (2).

$5(-5z) + 3z + 10z = -24$
$-25z + 3z + 10z = -24$
$-12z = -24$
$z = 2$

Replace z with 2 in equation (4).

$x = -5(2)$
$x = -10$

Replace z with 2 in equation (5)

$y = 3(2)$
$y = 6$

The solution is $(-10, 6, 2)$.

13. $\begin{cases} x + y + z = 8 & (1) \\ 2x - y - z = 10 & (2) \\ x - 2y - 3z = 22 & (3) \end{cases}$

Add equations (1) and (2).

$3x = 18$ or $x = 6$

Add twice equation 1 to equation 3.

$2x + 2y + 2z = 16$
$x - 2y - 3z = 22$
$3x \quad - z = 38$

Replace x with 6 in this equation.

$3(6) - z = 38$
$18 - z = 38$
$-z = 20$
$z = -20$

Replace x with 6 and z with -20 in equation (1).

$6 + y + (-20) = 8$
$y - 14 = 8$
$y = 22$

The solution is $(6, 22, -20)$.

17. $\begin{cases} 2x - 3y + z = 2 & (1) \\ x - 5y + 5z = 3 & (2) \\ 3x + y - 3z = 5 & (3) \end{cases}$

Add -2 times equation (2) to equation (1).

$2x - 3y + z = 2$
$-2x + 10y - 10z = -6$
$7y - 9z = -4$

Add -3 times equation (2) to equation (3).

$-3x + 15y - 15z = -9$
$3x + y - 3z = 5$
$16y - 18z = -4$

We now have the system:

$\begin{cases} 7y - 9z = -4 & (4) \\ 16y - 18z = -4 & (5) \end{cases}$

Multiply equation (4) by -2 and add to equation (5).

$-14y + 18z = 8$
$16y - 18z = -4$
$2y = 4$
$y = 2$

Replace y with 2 in equation (4).

$7(2) - 9z = -4$
$-9z = -18$
$z = 2$

Replace y with 2 and z with 2 in equation (1).

$2x - 3(2) + 2 = 2$
$2x = 6$
$x = 3$

The solution is $(3, 2, 2)$.

21. $\begin{cases} 2x + 2y - 3z = 1 & (1) \\ y + 2z = -14 & (2) \\ 3x - 2y = -1 & (3) \end{cases}$

Add equations (1) and (3).

$5x - 3z = 0$ (4)

Multiply equation (2) by 2 and add to equation (3).

$$\begin{array}{l} 2y + 4z = -28 \\ \underline{3x - 2y \qquad = -1} \\ 3x \qquad + 4z = -29 \quad (5) \end{array}$$

Multiply equation (4) by 4, multiply equation (5) by 3, and add.

$$\begin{array}{l} 20x - 12z = 0 \\ \underline{9x + 12z = -87} \\ 29x \qquad = -87 \\ x = -3 \end{array}$$

Replace x with -3 in equation (4).

$$\begin{array}{l} 5(-3) - 3z = 0 \\ 3z = -15 \\ z = -5 \end{array}$$

Replace z with -5 in equation (2).

$$\begin{array}{l} y + 2(-5) = -14 \\ y - 10 = -14 \\ y = -4 \end{array}$$

The solution is $(-3, -4, -5)$.

25. $\begin{cases} x + y + z = 1 & (1) \\ 2x - y + z = 0 & (2) \\ -x + 2y + 2z = -1 & (3) \end{cases}$

Multiply equation (3) by 2 and add to equation (2).

$$\begin{array}{l} 2x - y + z = 0 \\ \underline{-2x + 4y + 4z = -2} \\ 3y + 5z = -2 \quad (4) \end{array}$$

Multiply equation (1) by -2 and add to equation (2).

$$\begin{array}{l} -2x - 2y - 2z = -2 \\ \underline{2x - y + z = 0} \\ -3y - z = -2 \quad (5) \end{array}$$

Add equations (4) and (5).

$$\begin{array}{l} 3y + 5z = -2 \\ \underline{-3y - z = -2} \\ 4z = -4 \\ z = -1 \end{array}$$

Replace z with -1 in equation (4).

$$\begin{array}{l} 3y + 5(-1) = -2 \\ 3y = 3 \\ y = 1 \end{array}$$

Replace y with 1 and z with -1 in equation (1).

$$\begin{array}{l} x + 1 + (-1) = 1 \\ x = 1 \end{array}$$

The solution is $(1, 1, -1)$.

$$\frac{1}{24} = \frac{x}{8} + \frac{y}{4} + \frac{z}{3}$$

$$\frac{1}{24} = \frac{1}{8} + \frac{1}{4} - \frac{1}{3}$$

$$\frac{1}{24} = \frac{3}{24} + \frac{6}{24} - \frac{8}{24}$$

$$\frac{1}{24} = \frac{1}{24} \quad \text{True}$$

29. Let x = shortest side,
y = middle side,
and z = longest side.
The system of equations is

$$\begin{cases} x + 2y + z = 29 & (1) \\ z = 2x & (2) \\ y = x + 2 & (3) \end{cases}$$

Substitute $2x$ for z and $x + 2$ for y in equation (1).

$$\begin{array}{l} x + 2(x + 2) + 2x = 29 \\ x + 2x + 4 + 2x = 29 \\ 5x + 4 = 29 \\ 5x = 25 \\ x = \frac{25}{5} = 5 \end{array}$$

Replace 5 for x in equation (2).
$z = 2(5) = 10$
Replace 5 for x in equation (3).
$y = 5 + 2 = 7$
The sides are 5 in., 7 in., and 10 in.

33. Let x = the number of three-point field goals,
y = the number of two-point field goals,
and z = number of free throws.
The system of equations is

$$\begin{cases} 3x + 2y + z = 739 & (1) \\ y = 2x - 51 & (2) \\ z = 2y + 8 & (3) \end{cases}$$

Substitute $2y + 8$ for z from equation (3) in equation (1).

$$\begin{array}{l} 3x + 2y + (2y + 8) = 739 \\ 3x + 4y + 8 = 739 \\ 3x + 4y = 731 \quad (4) \end{array}$$

Substitute $2x - 51$ for y from equation (2) in equation (4).

$$\begin{array}{l} 3x + 4(2x - 51) = 731 \\ 3x + 8x - 204 = 731 \\ 11x - 204 = 731 \\ 11x = 935 \\ x = \frac{935}{11} = 85 \end{array}$$

Replace 85 for x in equation (2).

$$\begin{array}{l} y = 2(85) - 51 \\ y = 170 - 51 \\ y = 119 \end{array}$$

Replace 119 for y in equation (3).

$$\begin{array}{l} z = 2(119) + 8 \\ z = 238 + 8 \\ z = 246 \end{array}$$

Katie Smith had 246 free throws, 119 two-point field goals, and 85 three-point field goals.

37. $\begin{cases} 3x - y + z = 2 \\ -x + 2y + 3z = 6 \end{cases}$

$$\begin{cases} 2(3x - y + z = 2) \\ -x + 2y + 3z = 6 \end{cases}$$

$$\begin{cases} 6x - 2y + 2z = 4 \\ \underline{-x + 2y + 3z = 6} \end{cases}$$
$$5x \qquad + 5z = 10$$

41. answers may vary

45. $y = ax^2 + bx + c$
For $(1, 6)$, use $x = 1$ and $y = 6$.
$6 = a + b + c \quad (1)$
For $(-1, -2)$, use $x = -1$ and $y = -2$.
$-2 = a - b + c \quad (2)$
For $(0, -1)$, use $x = 0$ and $y = -1$.
$-1 = a \cdot 0 + b \cdot 0 + c$
$-1 = c \quad (3)$
The system is

$$\begin{array}{ll} 6 = a + b + c & (1) \\ -2 = a - b + c & (2) \\ -1 = c & (3) \end{array}$$

From equation (3), we see that $c = -1$.
Multiply equation (2) by -1 and add equation (1).
$8 = 2b$
$4 = b$
Replace b with 4 and c with -1 in equation (1).
$6 = a + 4 - 1$
$6 = a + 3$
$3 = a$
The solution is $a = 3$, $b = 4$, and $c = -1$.

Exercise Set 8.6

1. $\begin{cases} x + y = 1 \\ x - 2y = 4 \end{cases}$

$$\left[\begin{array}{cc|c} 1 & 1 & 1 \\ 1 & -2 & 4 \end{array}\right]$$

Multiply row 1 by -1 and add to row 2.

$$\left[\begin{array}{cc|c} 1 & 1 & 1 \\ 0 & -3 & 3 \end{array}\right]$$

Divide row 2 by -3.

$$\left[\begin{array}{cc|c} 1 & 1 & 1 \\ 0 & 1 & -1 \end{array}\right]$$

This corresponds to $\begin{cases} x + y = 1 \\ y = -1 \end{cases}$.

$$x + (-1) = 1$$
$$x - 1 = 1$$
$$x = 2$$

The solution is $(2, -1)$.

5. $\begin{cases} x - 2y = 4 \\ 2x - 4y = 4 \end{cases}$

$$\left[\begin{array}{cc|c} 1 & -2 & 4 \\ 2 & -4 & 4 \end{array}\right]$$

Multiply row 1 by -2 and add to row 2.

$$\left[\begin{array}{cc|c} 1 & -2 & 4 \\ 0 & 0 & -4 \end{array}\right]$$

This is an inconsistent system.

The solution is $\emptyset$.

9. $\begin{cases} x - 4 = 0 \\ x + y = 1 \end{cases}$ or $\begin{cases} x = 4 \\ x + y = 1 \end{cases}$

$$\left[\begin{array}{cc|c} 1 & 0 & 4 \\ 1 & 1 & 1 \end{array}\right]$$

Multiply row 1 by -1 and add to row 2.

$$\left[\begin{array}{cc|c} 1 & 0 & 4 \\ 0 & 1 & -3 \end{array}\right]$$

This corresponds to $\begin{cases} x = 4 \\ y = -3 \end{cases}$.

The solution is $(4, -3)$.

13. $\begin{cases} 2y - z = -7 \\ x + 4y + z = -4 \\ 5x - y + 2z = 13 \end{cases}$

$$\left[\begin{array}{ccc|c} 0 & 2 & -1 & -7 \\ 1 & 4 & 1 & -4 \\ 5 & -1 & 2 & 13 \end{array}\right]$$

Interchange rows 1 and 2.

$$\left[\begin{array}{ccc|c} 1 & 4 & 1 & -4 \\ 0 & 2 & -1 & -7 \\ 5 & -1 & 2 & 13 \end{array}\right]$$

Multiply row 1 by -5 and add to row 3.

$$\left[\begin{array}{ccc|c} 1 & 4 & 1 & -4 \\ 0 & 2 & -1 & -7 \\ 0 & -21 & -3 & 33 \end{array}\right]$$

Divide row 2 by 2.

$$\left[\begin{array}{ccc|c} 1 & 4 & 1 & -4 \\ 0 & 1 & -\frac{1}{2} & -\frac{7}{2} \\ 0 & -21 & -3 & 33 \end{array}\right]$$

Multiply row 2 by 21 and add to row 3.

$$\left[\begin{array}{ccc|c} 1 & 4 & 1 & -4 \\ 0 & 1 & -\frac{1}{2} & -\frac{7}{2} \\ 0 & 0 & -\frac{27}{2} & -\frac{81}{2} \end{array}\right]$$

Multiply row 3 by $-\frac{2}{27}$.

$$\left[\begin{array}{ccc|c} 1 & 4 & 1 & -4 \\ 0 & 1 & -\frac{1}{2} & -\frac{7}{2} \\ 0 & 0 & 1 & 3 \end{array}\right]$$

This corresponds to $\begin{cases} x + 4y + z = -4 \\ y - \frac{1}{2}z = -\frac{7}{2} \\ z = 3 \end{cases}$.

$$y - \frac{1}{2}(3) = -\frac{7}{2}$$
$$y - \frac{3}{2} = -\frac{7}{2}$$
$$y = -2$$
$$x + 4(-2) + 3 = -4$$
$$x - 8 + 3 = -4$$
$$x = 1$$

The solution is $(1, -2, 3)$.

17. $\begin{cases} 4x + y + z = 3 \\ -x + y - 2z = -11 \\ x + 2y + 2z = -1 \end{cases}$

$$\left[\begin{array}{ccc|c} 4 & 1 & 1 & 3 \\ -1 & 1 & -2 & -11 \\ 1 & 2 & 2 & -1 \end{array}\right]$$

Interchange rows 1 and 3.

$$\left[\begin{array}{ccc|c} 1 & 2 & 2 & -1 \\ -1 & 1 & -2 & -11 \\ 4 & 1 & 1 & 3 \end{array}\right]$$

Multiply row 1 by 1 and add to row 2.

Multiply row 1 by -4 and add to row 3.

$$\left[\begin{array}{ccc|c} 1 & 2 & 2 & -1 \\ 0 & 3 & 0 & -12 \\ 0 & -7 & -7 & 7 \end{array}\right]$$

Divide row 2 by 3.

$$\left[\begin{array}{ccc|c} 1 & 2 & 2 & -1 \\ 0 & 1 & 0 & -4 \\ 0 & -7 & -7 & 7 \end{array}\right]$$

Multiply row 2 by 7 and add to row 3.

$$\left[\begin{array}{ccc|c} 1 & 2 & 2 & -1 \\ 0 & 1 & 0 & -4 \\ 0 & 0 & -7 & -21 \end{array}\right]$$

Divide row 3 by -7.

$$\left[\begin{array}{ccc|c} 1 & 2 & 2 & -1 \\ 0 & 1 & 0 & -4 \\ 0 & 0 & 1 & 3 \end{array}\right]$$

This corresponds to $\begin{cases} x + 2y + 2z = -1 \\ y = -4 \\ z = 3 \end{cases}$.

$$x + 2(-4) + 2(3) = -1$$
$$x - 8 + 6 = -1$$
$$x = 1$$

The solution is $(1, -4, 3)$.

21. $(4)(-10) - (2)(-2) = -40 + 4 = -36$

25. a. 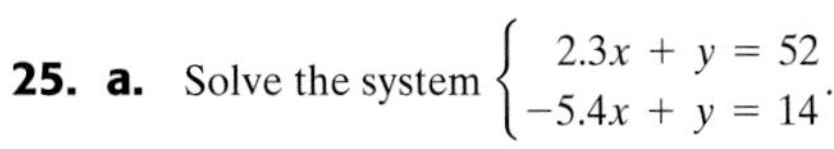
Solve the system $\begin{cases} 2.3x + y = 52 \\ -5.4x + y = 14 \end{cases}$.

$$\left[\begin{array}{cc|c} 2.3 & 1 & 52 \\ -5.4 & 1 & 14 \end{array}\right]$$

Since getting a 1 in the first column would lead to repeating decimals, we multiply row 1 by -1 and add to row 2.

$$\left[\begin{array}{cc|c} 2.3 & 1 & 52 \\ -7.7 & 0 & -38 \end{array}\right]$$

This corresponds to $\begin{cases} 2.3x + y = 52 \\ -7.7x \quad\quad = -38 \end{cases}$.

From the second equation, $x = \dfrac{-38}{-77} \approx 4.935$.

Thus, the percent of U.S. households owning black-and-white television sets was the same as the percent of U.S. households owning a microwave oven in the end of 1984 (about 4.9 years after 1980).

b. Solving the television equation for y, we get
$y = -2.3x + 52$. Thus, for
1980, $y = -2.3x + 52 = 52$, and for
1993, $y = -2.3(13) + 52 = 22.1$.
Solving the microwave oven equation for y, we get
$y = 5.4x + 14$. Thus, for
1980, $y = 5.4(0) + 14 = 14$, and for
1993, $y = 5.4(13) + 14 = 84.2$.
In 1980, a greater percent of U.S. households, hence more households, owned black-and-white television sets. In 1993, more households owned a microwave oven. The percent of households owning black-and-white television sets is decreasing and the percent of households owning microwave ovens is increasing.
answers may vary

c. The percent will reach 0% when $y = 0$ in the equation $2.3x + y = 52$.

$$2.3x + 0 = 52$$
$$x = \frac{52}{2.3}$$
$$x \approx 22.6$$

According to this model, the percent of U.S. households owning a black-and-white television set will be 0% about 22.6 years after 1980, or sometime in 2002.

Exercise Set 8.7

1. $\begin{cases} y \geq x + 1 \\ y \geq 3 - x \end{cases}$

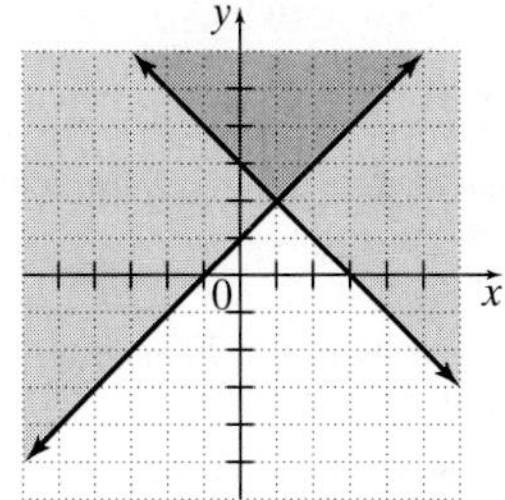

5. $\begin{cases} y \leq -2x - 2 \\ y \geq x + 4 \end{cases}$

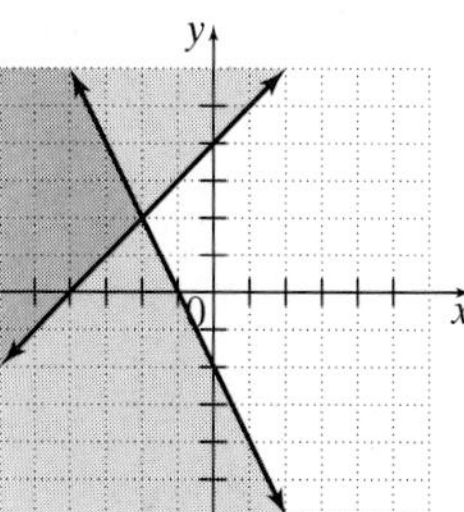

9. $\begin{cases} x \geq 3y \\ x + 3y \leq 6 \end{cases}$

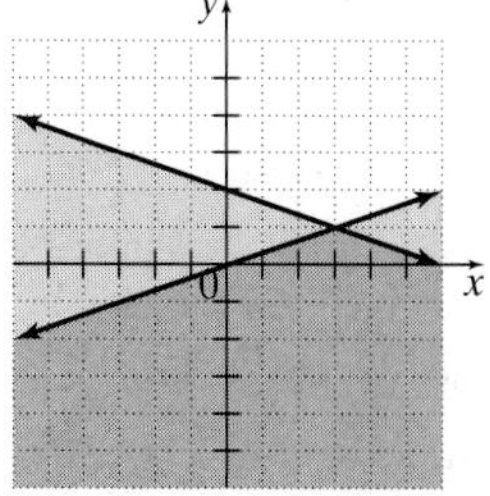

13. $\begin{cases} y \geq 1 \\ x < -3 \end{cases}$

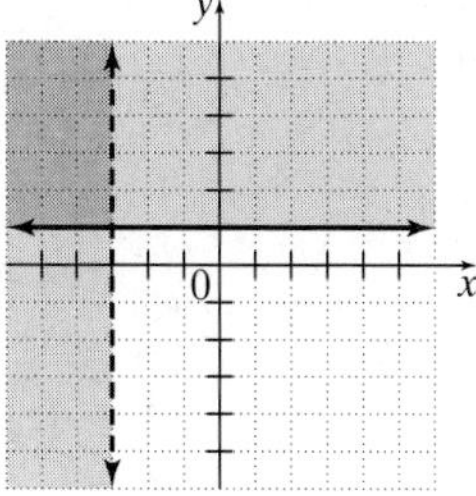

17. $\begin{cases} 3x - 4y \geq -6 \\ 2x + y \leq 7 \\ y \geq -3 \end{cases}$

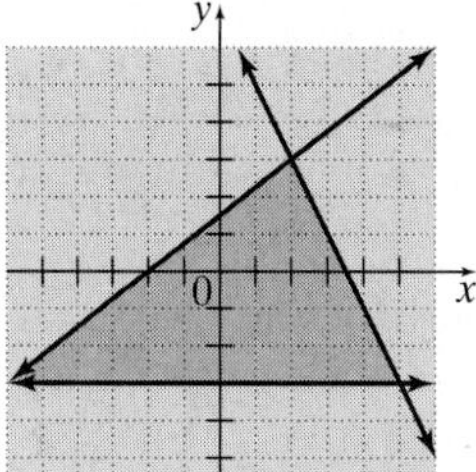

21. $\begin{cases} y < 5 \\ x > 3 \end{cases}$
graph C

25. $(-3)^2 = (-3)(-3) = 9$

29. $(-2)^2 - (-3) + 2(-1) = 4 + 3 - 2 = 5$

33. a. $\begin{cases} x + y \leq 8 \\ x < 3 \end{cases}$

b.

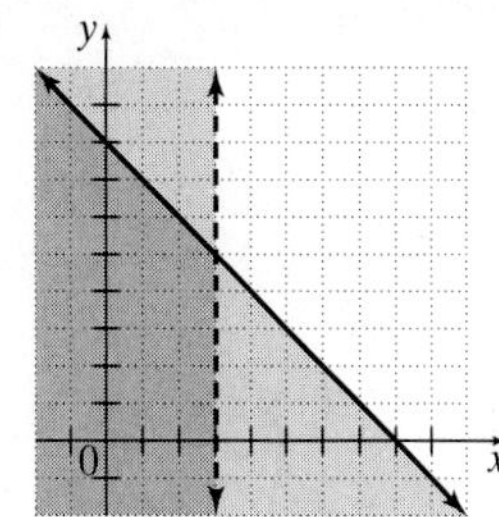

Chapter 8 Test

1. $\begin{cases} x - y = 2 \\ 3x - y = -2 \end{cases}$

Graph each linear equation on a single set of axes.

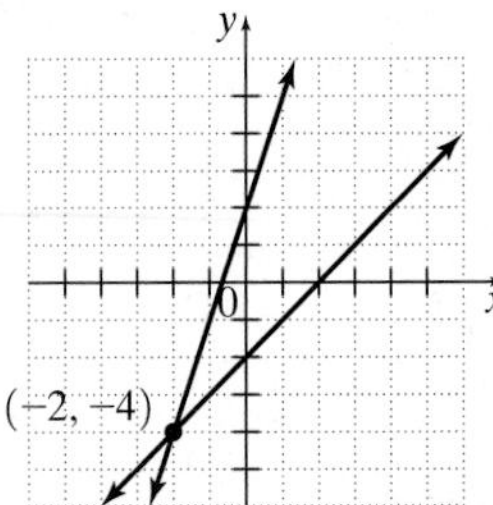

The solution is the intersection point of the two lines, $(-2, -4)$.

5. $\begin{cases} x - y = 4 \\ x - 2y = 11 \end{cases}$

Solve the first equation for x.

$x = y + 4$

Let $x = y + 4$ in the second equation

$y + 4 - 2y = 11$

$-y = 7$

$y = -7$

Let $y = -7$ in $x = y + 4$.

$x = -7 + 4$

$x = -3$

The solution is $(-3, -7)$.

9. $\begin{cases} 5x - 6y = 7 \\ 7x - 4y = 12 \end{cases}$

$$\begin{cases} -10x + 12y = -14 \\ 21x - 12y = 36 \\ \hline 11x = 22 \\ x = 2 \end{cases}$$

Let $x = 2$ in the first equation.

$5(2) - 6y = 7$

$-6y = -3$

$y = \frac{1}{2}$

The solution is $\left(2, \frac{1}{2}\right)$.

13. $x + y + z = 4 \quad (1)$

$2x + 5y = 1 \quad (2)$

$x - y - 2z = 0 \quad (3)$

Multiply equation (1) by 2 and add to equation (3).

$$\begin{array}{l} 2x + 2y + 2z = 8 \\ x - y - 2z = 0 \\ \hline 3x + y = 8 \quad (4) \end{array}$$

$\begin{cases} 2x + 5y = 1 \quad (3) \\ 3x + y = 8 \quad (4) \end{cases}$

Multiply equation (4) by -5 and add to equation (3).

$$\begin{array}{l} 2x + 5y = 1 \\ -15x - 5y = -40 \\ \hline -13x = -39 \end{array}$$

$x = \frac{-39}{-13} = 3$

Replace x with 3 in equation (2).

$2(3) + 5y = 1$

$6 + 5y = 1$

$5y = -5$

$y = \frac{-5}{5} = -1$

Replace x with 3 and y with -1 in equation (1).

$3 + (-1) + z = 4$

$2 + z = 4$

$z = 2$

The solution is $(3, -1, 2)$.

17. $\begin{cases} x - y - z = 0 \quad (1) \\ 3x - y - 5z = -2 \quad (2) \\ 2x + 3y = -5 \quad (3) \end{cases}$

Multiply equation (1) by -5 and add to equation (2).

$$\begin{array}{l} -5x + 5y + 5z = 0 \\ 3x - y - 5z = -2 \\ \hline -2x + 4y = -2 \quad (4) \end{array}$$

Add equations (3) and (4).

$$\begin{array}{l} 2x + 3y = -5 \quad (3) \\ -2x + 4y = -2 \quad (4) \\ \hline 7y = -7 \end{array}$$

$y = \frac{-7}{7} = -1$

Replace y with -1 in equation (3).

$2x + 3(-1) = -5$

$2x - 3 = -5$

$2x = -2$

$x = \frac{-2}{2} = -1$

Replace x with -1 and y with -1 in equation (1).

$-1 - (-1) - z = 0$

$-1 + 1 - z = 0$

$-z = 0$

$z = 0$

The solution is $(-1, -1, 0)$.

21. Let x = number of liters of 30% solution,
y = number of liters of 70% solution.

$\begin{cases} x + y = 10 \quad (1) \\ 0.30x + 0.70y = 10(0.40) \quad (2) \end{cases}$

Multiply equation (2) by 10.

$3x + 7y = 40 \quad (3)$

The system is

$\begin{cases} x + y = 10 \quad (1) \\ 3x + 7y = 40 \quad (3) \end{cases}$

Multiply equation (1) by -3 and add to equation (2).

$$\begin{array}{l} -3x - 3y = -30 \\ 3x + 7y = 40 \\ \hline 4y = 10 \end{array}$$

$y = \frac{10}{4} = 2.5$

Replace y with 2.5 in equation (1).

$x + 2.5 = 10$

$x = 7.5$

Mix 7.5 liters of the 70% solution with 2.5 liters of the 30% solution.

25. The graphs intersect at 1999.

Chapter 9

Exercise Set 9.1

1. Since $2^2 = 4$ and $(-2)^2 = 4$, the square roots of 4 are 2 and -2.

5. Since $10^2 = 100$ and $(-10)^2 = 100$, the square roots of 100 are 10 and -10.

9. $\sqrt{\frac{1}{4}} = \frac{1}{2}$ because $\left(\frac{1}{2}\right)^2 = \frac{1}{4}$.

13. $-\sqrt{36} = -6$ because $(6)^2 = 36$.

17. $\sqrt{16y^6} = \sqrt{16}\sqrt{y^6} = 4y^3$ because $(4y^3)^2 = 16y^6$.

21. $\sqrt{38} \approx 6.164$
Since $36 < 38 < 49$, then $\sqrt{36} < \sqrt{38} < \sqrt{49}$ or $6 < \sqrt{38} < 7$. The approximation is between 6 and 7 and this is reasonable.

25. $\sqrt[3]{64} = 4$ because $(4)^3 = 64$.

29. $\sqrt[3]{-1} = -1$ because $(-1)^3 = -1$.

33. $\sqrt[3]{-27x^9} = -3x^3$ because $(-3x^3)^3 = -27x^9$.

37. $\sqrt[4]{-16}$ is not a real number.

41. $\sqrt[5]{x^{20}} = x^4$ because $(x^4)^5 = x^{20}$.

45. $\sqrt{81x^4} = 9x^2$ because $(9x^2)^2 = 81x^4$.

49. $\sqrt{(-8)^2} = |-8| = 8$

53. $\sqrt{4x^2} = 2|x|$

57. $\sqrt[4]{(x-2)^4} = |x-2|$

61. $f(x) = \sqrt{2x+3}$
$f(0) = \sqrt{2\cdot 0 + 3} = \sqrt{3}$

65. $g(x) = \sqrt[3]{x-8}$
$g(-19) = \sqrt[3]{-19-8} = \sqrt[3]{-27} = -3$

69. $(-2x^3y^2)^5 = (-2)^5x^{3\cdot 5}y^{2\cdot 5} = -32x^{15}y^{10}$

73. $\dfrac{7x^{-1}y}{14(x^5y^2)^{-2}} = \dfrac{7x^{-1}y}{14x^{-10}y^{-4}} = \dfrac{x^9y^5}{2}$

77. b. Since $\sqrt{160}$ is close to $\sqrt{169} = 13$.

81. $B = \sqrt{\dfrac{\text{hw}}{3131}} = \sqrt{\dfrac{66\cdot 135}{3131}}$
$= \sqrt{\dfrac{8910}{3131}} \approx \sqrt{2.8457}$
≈ 1.69 sq m

Exercise Set 9.2

1. $49^{1/2} = \sqrt{49} = 7$

5. $\left(\dfrac{1}{16}\right)^{1/4} = \sqrt[4]{\dfrac{1}{16}} = \dfrac{1}{2}$

9. $2m^{1/3} = 2\sqrt[3]{m}$

13. $(-27)^{1/3} = \sqrt[3]{-27} = -3$

17. $16^{3/4} = (\sqrt[4]{16})^3 = 2^3 = 8$

21. $(-16)^{3/4} = (\sqrt[4]{-16})^3$ is not a real number.

25. $(7x+2)^{2/3} = \sqrt[3]{(7x+2)^2}$ or $(\sqrt[3]{7x+2})^2$

29. $8^{-4/3} = \dfrac{1}{8^{4/3}} = \dfrac{1}{(8^{1/3})^4} = \dfrac{1}{2^4} = \dfrac{1}{16}$

33. $(-4)^{-3/2} = \dfrac{1}{(-4)^{3/2}} = \dfrac{1}{[(-4)^{1/2}]^3}$ is not a real number.

37. $\dfrac{1}{a^{-2/3}} = a^{2/3}$

41. answers may vary

45. $x^{-2/5}\cdot x^{7/5} = x^{-2/5+7/5} = x^{5/5} = x$

49. $\dfrac{y^{1/3}}{y^{1/6}} = y^{1/3-1/6} = y^{2/6-1/6} = y^{1/6}$

53. $\dfrac{b^{1/2}b^{3/4}}{-b^{1/4}} = -b^{1/2+3/4-1/4} = -b^{1/2+1/2} = -b^1 = -b$

57. $\dfrac{(y^3z)^{1/6}}{y^{-1/2}z^{1/3}} = \dfrac{y^{3/6}z^{1/6}}{y^{-1/2}z^{1/3}}$
$= y^{3/6-(-1/2)}z^{1/6-1/3}$
$= y^{1/2+1/2}z^{1/6-2/6}$
$= y^1z^{-1/6} = \dfrac{y}{z^{1/6}}$

61. $\sqrt[6]{x^3} = x^{3/6} = x^{1/2} = \sqrt{x}$

65. $\sqrt[4]{16x^2} = 16^{1/4}x^{2/4} = 2x^{1/2} = 2\sqrt{x}$

69. $\sqrt[8]{x^4y^4} = x^{4/8}y^{4/8} = x^{1/2}y^{1/2} = \sqrt{xy}$

73. $\sqrt[3]{y}\cdot\sqrt[5]{y^2} = y^{1/3}\cdot y^{2/5} = y^{5/15}\cdot y^{6/15} = y^{11/15} = \sqrt[15]{y^{11}}$

77. $\sqrt[3]{x}\cdot\sqrt[4]{x}\cdot\sqrt[8]{x^3} = x^{1/3}\cdot x^{1/4}\cdot x^{3/8}$
$= x^{8/24}\cdot x^{6/24}\cdot x^{9/24}$
$= x^{23/24}$
$= \sqrt[24]{x^{23}}$

81. $\sqrt{3}\cdot\sqrt[3]{4} = 3^{1/2}\cdot 4^{1/3}$
$= 3^{3/6}\cdot 4^{2/6}$
$= (3^3\cdot 4^2)^{1/6}$
$= (432)^{1/6}$
$= \sqrt[6]{432}$

85. $\sqrt{5r}\cdot\sqrt[3]{s} = (5r)^{1/2}\cdot s^{1/3}$
$= (5r)^{3/6}\cdot s^{2/6}$
$= [(5r)^3\cdot s^2]^{1/6}$
$= (125r^3s^2)^{1/6}$
$= \sqrt[6]{125r^3s^2}$

89. $48 = 16\cdot 3$ or $4\cdot 12$

93. $54 = 27\cdot 2$

97. $f(x) = 2.5x^{8/5}$
$x = 2000 - 1990 = 10$
Replace x by 10.
$f(x) = 2.5(10)^{8/5}$
≈ 99.5 million subscriptions

101. $x^{1/5}$
$\dfrac{x^{1/5}}{x^{-2/5}} = x^{1/5+2/5} = x^{3/5}$

105. $\dfrac{\sqrt{t}}{\sqrt{u}} = \dfrac{t^{1/2}}{u^{1/2}}$

Exercise Set 9.3

1. $\sqrt{7}\cdot\sqrt{2} = \sqrt{7\cdot 2} = \sqrt{14}$

5. $\sqrt[3]{4}\cdot\sqrt[3]{9} = \sqrt[3]{4\cdot 9} = \sqrt[3]{36}$

9. $\sqrt{\dfrac{7}{x}}\cdot\sqrt{\dfrac{2}{y}} = \sqrt{\dfrac{7\cdot 2}{x\cdot y}} = \sqrt{\dfrac{14}{xy}}$

13. $\sqrt{\dfrac{6}{49}} = \dfrac{\sqrt{6}}{\sqrt{49}} = \dfrac{\sqrt{6}}{7}$

17. $\sqrt[4]{\dfrac{x^3}{16}} = \dfrac{\sqrt[4]{x^3}}{\sqrt[4]{16}} = \dfrac{\sqrt[4]{x^3}}{2}$

21. $\sqrt[4]{\dfrac{8}{x^8}} = \dfrac{\sqrt[4]{8}}{\sqrt[4]{x^8}} = \dfrac{\sqrt[4]{8}}{x^2}$

25. $\sqrt{\frac{x^2y}{100}} = \frac{\sqrt{x^2}\cdot\sqrt{y}}{\sqrt{100}} = \frac{x\sqrt{y}}{10}$

29. $-\sqrt[3]{\frac{z^7}{125x^3}} = \frac{-\sqrt[3]{z^7}}{\sqrt[3]{125x^3}}$

$= \frac{-\sqrt[3]{z^6z}}{\sqrt[3]{125}\cdot\sqrt[3]{x^3}}$

$= \frac{-\sqrt[3]{z^6}\cdot\sqrt[3]{z}}{5x}$

$= -\frac{z^2\sqrt[3]{z}}{5x}$

33. $\sqrt[3]{192} = \sqrt[3]{64(3)} = \sqrt[3]{64}\cdot\sqrt[3]{3} = 4\sqrt[3]{3}$

37. $\sqrt{24} = \sqrt{4\cdot 6} = \sqrt{4}\cdot\sqrt{6} = 2\sqrt{6}$

41. $\sqrt[3]{16y^7} = \sqrt[3]{(8y^6)(2y)}$

$= \sqrt[3]{8}\cdot\sqrt[3]{y^6}\cdot\sqrt[3]{2y}$

$= 2y^2\sqrt[3]{2y}$

45. $\sqrt{y^5} = \sqrt{y^4y} = \sqrt{y^4}\cdot\sqrt{y} = y^2\sqrt{y}$

49. $\sqrt[5]{-32x^{10}y} = \sqrt[5]{-32}\cdot\sqrt[5]{x^{10}}\cdot\sqrt[5]{y} = -2x^2\sqrt[5]{y}$

53. $-\sqrt{32a^8b^7} = -\sqrt{16a^8b^6(2b)}$

$= -\sqrt{16}\cdot\sqrt{a^8}\cdot\sqrt{b^6}\cdot\sqrt{2b}$

$= -4a^4b^3\sqrt{2b}$

57. $\sqrt[3]{125r^9s^{12}} = 5r^3s^4$

61. $\frac{\sqrt[3]{24}}{\sqrt[3]{3}} = \sqrt[3]{\frac{24}{3}} = \sqrt[3]{8} = 2$

65. $\frac{\sqrt{x^5y^3}}{\sqrt{xy}} = \sqrt{\frac{x^5y^3}{xy}}$

$= \sqrt{x^4y^2}$

$= x^2y$

69. $\frac{3\sqrt{100x^2}}{2\sqrt{2x^{-1}}} = \frac{3}{2}\sqrt{\frac{100x^2}{2x^{-1}}}$

$= \frac{3}{2}\sqrt{50x^3}$

$= \frac{3}{2}\sqrt{25x^2\cdot 2x}$

$= \frac{3}{2}\cdot 5x\sqrt{2x}$

$= \frac{15x}{2}\sqrt{2x}$

73. $6x + 8x = 14x$

77. $9y^2 - 8y^2 = y^2$

81. $(x-4)^2 = x^2 - 2x(4) + 4^2 = x^2 - 8x + 16$

85. $F(x) = 0.6\sqrt{49 - x^2}$

a. $F(3) = 0.6\sqrt{49 - 3^2}$

$= 0.6\sqrt{49 - 9}$

$= 0.6\sqrt{40}$

≈ 3.8

b. $F(5) = 0.6\sqrt{49 - 5^2}$

$= 0.6\sqrt{49 - 25}$

$= 0.6\sqrt{24}$

≈ 2.9

c. answers may vary

Exercise Set 9.4

1. $\sqrt{8} - \sqrt{32} = \sqrt{4(2)} - \sqrt{16(2)}$

$= \sqrt{4}\sqrt{2} - \sqrt{16}\sqrt{2}$

$= 2\sqrt{2} - 4\sqrt{2}$

$= -2\sqrt{2}$

5. $2\sqrt{50} - 3\sqrt{125} + \sqrt{98}$

$= 2\sqrt{25(2)} - 3\sqrt{25(5)} + \sqrt{49(2)}$

$= 2\sqrt{25}\sqrt{2} - 3\sqrt{25}\sqrt{5} + \sqrt{49}\sqrt{2}$

$= 2(5)\sqrt{2} - 3(5)\sqrt{5} + 7\sqrt{2}$

$= 10\sqrt{2} - 15\sqrt{5} + 7\sqrt{2}$

$= 17\sqrt{2} - 15\sqrt{5}$

9. $\sqrt{9b^3} - \sqrt{25b^3} + \sqrt{49b^3}$

$= \sqrt{9b^2(b)} - \sqrt{25b^2(b)} + \sqrt{49b^2(b)}$

$= \sqrt{9b^2}\sqrt{b} - \sqrt{25b^2}\sqrt{b} + \sqrt{49b^2}\sqrt{b}$

$= 3b\sqrt{b} - 5b\sqrt{b} + 7b\sqrt{b}$

$= 5b\sqrt{b}$

13. $\sqrt[3]{\frac{11}{8}} - \frac{\sqrt[3]{11}}{6} = \frac{\sqrt[3]{11}}{\sqrt[3]{8}} - \frac{\sqrt[3]{11}}{6}$

$= \frac{\sqrt[3]{11}}{2} - \frac{\sqrt[3]{11}}{6}$

$= \frac{3\sqrt[3]{11} - \sqrt[3]{11}}{6}$

$= \frac{2\sqrt[3]{11}}{6}$

$= \frac{\sqrt[3]{11}}{3}$

17. $7\sqrt{9} - 7 + \sqrt{3} = 7(3) - 7 + \sqrt{3}$

$= 21 - 7 + \sqrt{3}$

$= 14 + \sqrt{3}$

21. $3\sqrt{108} - 2\sqrt{18} - 3\sqrt{48}$

$= 3\sqrt{36}\sqrt{3} - 2\sqrt{9}\sqrt{2} - 3\sqrt{16}\sqrt{3}$

$= 3(6)\sqrt{3} - 2(3)\sqrt{2} - 3(4)\sqrt{3}$

$= 18\sqrt{3} - 6\sqrt{2} - 12\sqrt{3}$

$= 6\sqrt{3} - 6\sqrt{2}$

25. $\sqrt{9b^3} - \sqrt{25b^3} + \sqrt{16b^3}$

$= \sqrt{9b^2}\sqrt{b} - \sqrt{25b^2}\sqrt{b} + \sqrt{16b^2}\sqrt{b}$

$= 3b\sqrt{b} - 5b\sqrt{b} + 4b\sqrt{b}$

$= (3 - 5 + 4)b\sqrt{b}$

$= 2b\sqrt{b}$

29. $\sqrt[3]{54xy^3} - 5\sqrt[3]{2xy^3} + y\sqrt[3]{128x}$

$= \sqrt[3]{27y^3}\sqrt[3]{2x} - 5\sqrt[3]{y^3}\sqrt[3]{2x} + y\sqrt[3]{64}\sqrt[3]{2x}$

$= 3y\sqrt[3]{2x} - 5y\sqrt[3]{2x} + y(4)\sqrt[3]{2x}$

$= -2y\sqrt[3]{2x} + 4y\sqrt[3]{2x}$

$= 2y\sqrt[3]{2x}$

33. $-2\sqrt[4]{x^7} + 3\sqrt[4]{16x^7} = -2\sqrt[4]{x^4}\sqrt[4]{x^3} + 3\sqrt[4]{16x^4}\sqrt[4]{x^3}$

$= -2x\sqrt[4]{x^3} + 3(2x)\sqrt[4]{x^3}$

$= -2x\sqrt[4]{x^3} + 6x\sqrt[4]{x^3}$

$= 4x\sqrt[4]{x^3}$

37. $\frac{\sqrt[3]{8x^4}}{7} + \frac{3x\sqrt[3]{x}}{7} = \frac{\sqrt[3]{8x^3}\sqrt[3]{x} + 3x\sqrt[3]{x}}{7}$

$= \frac{2x\sqrt[3]{x} + 3x\sqrt[3]{x}}{7}$

$= \frac{5x\sqrt[3]{x}}{7}$

41. $\sqrt[3]{\frac{16}{27}} - \frac{\sqrt[3]{54}}{6} = \frac{\sqrt[3]{16}}{\sqrt[3]{27}} - \frac{\sqrt[3]{27}\sqrt[3]{2}}{6}$
$= \frac{\sqrt[3]{8}\sqrt[3]{2}}{3} - \frac{3\sqrt[3]{2}}{6}$
$= \frac{2(2)\sqrt[3]{2}}{6} - \frac{3\sqrt[3]{2}}{6}$
$= \frac{4\sqrt[3]{2} - 3\sqrt[3]{2}}{6}$
$= \frac{\sqrt[3]{2}}{6}$

45. $P = 2\sqrt{12} + \sqrt{12} + 2\sqrt{27} + 3\sqrt{3}$
$= 2\sqrt{4}\sqrt{3} + \sqrt{4}\sqrt{3} + 2\sqrt{9}\sqrt{3} + 3\sqrt{3}$
$= 2 \cdot 2\sqrt{3} + 2\sqrt{3} + 2 \cdot 3\sqrt{3} + 3\sqrt{3}$
$= (4 + 2 + 6 + 3)\sqrt{3}$
$= 15\sqrt{3}$
The perimeter of the trapezoid is $15\sqrt{3}$ in.

49. $(\sqrt{5} - \sqrt{2})^2 = \sqrt{5^2} - 2\sqrt{5}\sqrt{2} + \sqrt{2^2}$
$= 5 - 2\sqrt{10} + 2$
$= 7 - 2\sqrt{10}$

53. $(2\sqrt{x} - 5)(3\sqrt{x} + 1)$
$= (2\sqrt{x})(3\sqrt{x}) + (2\sqrt{x})1 - 5(3\sqrt{x}) - 5 \cdot 1$
$= 6x + 2\sqrt{x} - 15\sqrt{x} - 5$
$= 6x - 13\sqrt{x} - 5$

57. $6(\sqrt{2} - 2) = 6\sqrt{2} - 6 \cdot 2 = 6\sqrt{2} - 12$

61. $(2\sqrt{7} + 3\sqrt{5})(\sqrt{7} - 2\sqrt{5})$
$= 2\sqrt{7^2} - (2\sqrt{7})(2\sqrt{5}) + (3\sqrt{5})\sqrt{7} - 3 \cdot 2\sqrt{5^2}$
$= 2 \cdot 7 - 4\sqrt{35} + 3\sqrt{35} - 6 \cdot 5$
$= 14 - \sqrt{35} - 30$
$= -16 - \sqrt{35}$

65. $(\sqrt{3} + x)^2 = \sqrt{3^2} + 2\sqrt{3}x + x^2$
$= 3 + 2x\sqrt{3} + x^2$

69. $(\sqrt[3]{4} + 2)(\sqrt[3]{2} - 1)$
$= \sqrt[3]{4}\sqrt[3]{2} - \sqrt[3]{4} \cdot 1 + 2\sqrt[3]{2} - 2 \cdot 1$
$= \sqrt[3]{8} - \sqrt[3]{4} + 2\sqrt[3]{2} - 2$
$= 2 - \sqrt[3]{4} + 2\sqrt[3]{2} - 2$
$= -\sqrt[3]{4} + 2\sqrt[3]{2}$

73. $(\sqrt{x-1} + 5)^2 = (\sqrt{x-1})^2 + 2\sqrt{x-1}(5) + 5^2$
$= x - 1 + 10\sqrt{x-1} + 25$
$= x + 24 + 10\sqrt{x-1}$

77. $\frac{2x - 14}{2} = \frac{2(x - 7)}{2} = x - 7$

81. $\frac{6a^2b - 9ab}{3ab} = \frac{3ab(2a - 3)}{3ab} = 2a - 3$

85. $P = 2(3\sqrt{20}) + 2\sqrt{125}$
$= 6\sqrt{4}\sqrt{5} + 2\sqrt{25}\sqrt{5}$
$= 6(2)\sqrt{5} + 2(5)\sqrt{5}$
$= 12\sqrt{5} + 10\sqrt{5}$
$= 22\sqrt{5}$ ft
$A = 3\sqrt{20} \cdot \sqrt{125}$
$= 3 \cdot 2\sqrt{5} \cdot 5\sqrt{5}$
$= 30(\sqrt{5})^2$
$= 30 \cdot 5$
$= 150$ sq ft

Exercise Set 9.5

1. $\frac{\sqrt{2}}{\sqrt{7}} = \frac{\sqrt{2} \cdot \sqrt{7}}{\sqrt{7} \cdot \sqrt{7}} = \frac{\sqrt{14}}{7}$

5. $\frac{4}{\sqrt[3]{3}} \cdot \frac{\sqrt[3]{9}}{\sqrt[3]{9}} = \frac{4\sqrt[3]{9}}{\sqrt[3]{27}} = \frac{4\sqrt[3]{9}}{3}$

9. $\frac{3}{\sqrt[3]{4x^2}} = \frac{3}{\sqrt[3]{4x^2}} \cdot \frac{\sqrt[3]{2x}}{\sqrt[3]{2x}} = \frac{3\sqrt[3]{2x}}{\sqrt[3]{8x^3}} = \frac{3\sqrt[3]{2x}}{2x}$

13. $\frac{3}{\sqrt[3]{2}} = \frac{3}{\sqrt[3]{2}} \cdot \frac{\sqrt[3]{4}}{\sqrt[3]{4}} = \frac{3\sqrt[3]{4}}{\sqrt[3]{8}} = \frac{3\sqrt[3]{4}}{2}$

17. $\sqrt{\frac{2x}{5y}} = \frac{\sqrt{2x}}{\sqrt{5y}} = \frac{\sqrt{2x} \cdot \sqrt{5y}}{\sqrt{5y} \cdot \sqrt{5y}} = \frac{\sqrt{10xy}}{5y}$

21. $\sqrt{\frac{3x}{50}} = \frac{\sqrt{3x}}{\sqrt{50}} = \frac{\sqrt{3x}}{5\sqrt{2}} = \frac{\sqrt{3x} \cdot \sqrt{2}}{5\sqrt{2} \cdot \sqrt{2}} = \frac{\sqrt{6x}}{5 \cdot 2} = \frac{\sqrt{6x}}{10}$

25. $\frac{\sqrt[3]{2y^2}}{\sqrt[3]{9x^2}} = \frac{\sqrt[3]{2y^2} \cdot \sqrt[3]{3x}}{\sqrt[3]{9x^2} \cdot \sqrt[3]{3x}} = \frac{\sqrt[3]{6xy^2}}{3x}$

29. $\frac{5a}{\sqrt[5]{8a^9b^{11}}} = \frac{5a}{ab^2\sqrt[5]{8a^4b}}$
$= \frac{5\sqrt[5]{4ab^4}}{b^2\sqrt[5]{8a^4b} \cdot \sqrt[5]{4ab^4}}$
$= \frac{5\sqrt[5]{4ab^4}}{2ab^3}$

33. $\frac{-7}{\sqrt{x} - 3} = \frac{(-7)(\sqrt{x} + 3)}{(\sqrt{x} - 3)(\sqrt{x} + 3)}$
$= \frac{-7(\sqrt{x} + 3)}{\sqrt{x^2} - 3^2}$
$= \frac{-7(\sqrt{x} + 3)}{x - 9} = \frac{7(3 + \sqrt{x})}{9 - x}$

37. $\frac{\sqrt{a} + 1}{2\sqrt{a} - \sqrt{b}}$
$= \frac{(\sqrt{a} + 1)(2\sqrt{a} + \sqrt{b})}{(2\sqrt{a} - \sqrt{b})(2\sqrt{a} + \sqrt{b})}$
$= \frac{\sqrt{a}(2\sqrt{a}) + \sqrt{a}\sqrt{b} + 1(2\sqrt{a}) + 1\sqrt{b}}{(2\sqrt{a})^2 - \sqrt{b^2}}$
$= \frac{2a + \sqrt{ab} + 2\sqrt{a} + \sqrt{b}}{4a - b}$

41. $\frac{\sqrt{x}}{\sqrt{x} + \sqrt{y}} = \frac{(\sqrt{x})(\sqrt{x} - \sqrt{y})}{(\sqrt{x} + \sqrt{y})(\sqrt{x} - \sqrt{y})}$
$= \frac{\sqrt{x}\sqrt{x} - \sqrt{x}\sqrt{y}}{\sqrt{x^2} - \sqrt{y^2}}$
$= \frac{x - \sqrt{xy}}{x - y}$

45. $\sqrt{\frac{5}{3}} = \frac{\sqrt{5}}{\sqrt{3}} = \frac{\sqrt{5} \cdot \sqrt{5}}{\sqrt{3} \cdot \sqrt{5}} = \frac{5}{\sqrt{15}}$

49. $\frac{\sqrt{4x}}{7} = \frac{2\sqrt{x}}{7} = \frac{2\sqrt{x} \cdot \sqrt{x}}{7 \cdot \sqrt{x}} = \frac{2x}{7\sqrt{x}}$

53. $\sqrt{\frac{2}{5}} = \frac{\sqrt{2}}{\sqrt{5}} = \frac{\sqrt{2} \cdot \sqrt{2}}{\sqrt{5} \cdot \sqrt{2}} = \frac{2}{\sqrt{10}}$

57. $\sqrt[3]{\frac{7}{8}} = \frac{\sqrt[3]{7}}{\sqrt[3]{8}} = \frac{\sqrt[3]{7}}{2} = \frac{\sqrt[3]{7} \cdot \sqrt[3]{7^2}}{2 \cdot \sqrt[3]{7^2}} = \frac{\sqrt[3]{7^3}}{2\sqrt[3]{7^2}} = \frac{7}{2\sqrt[3]{49}}$

61. $\sqrt{\frac{18x^4y^6}{3z}} = \frac{\sqrt{18x^4y^6}}{\sqrt{3z}} = \frac{3x^2y^3\sqrt{2}}{\sqrt{3z}} = \frac{3x^2y^3\sqrt{2} \cdot \sqrt{2}}{\sqrt{3z} \cdot \sqrt{2}} = \frac{6x^2y^3}{\sqrt{6z}}$

65. $\dfrac{(2-\sqrt{11})}{6}\cdot\dfrac{(2+\sqrt{11})}{(2+\sqrt{11})}=\dfrac{(2-\sqrt{11})(2+\sqrt{11})}{6(2+\sqrt{11})}$

$=\dfrac{4-\sqrt{121}}{12+6\sqrt{11}}$

$=\dfrac{4-11}{12+6\sqrt{11}}$

$=\dfrac{-7}{12+6\sqrt{11}}$

69. $\dfrac{(\sqrt{x}+3)}{\sqrt{x}}\cdot\dfrac{(\sqrt{x}-3)}{(\sqrt{x}-3)}=\dfrac{(\sqrt{x}+3)(\sqrt{x}-3)}{\sqrt{x}(\sqrt{x}-3)}$

$=\dfrac{\sqrt{x^2}-9}{\sqrt{x^2}-3\sqrt{x}}$

$=\dfrac{x-9}{x-3\sqrt{x}}$

73. $2x-7=3(x-4)$

$2x-7=3x-12$

$2x-7+12=3x-12+12$

$2x+5=3x$

$2x-2x+5=3x-2x$

$5=x$

5

77. $x^2-8x=-12$

$x^2-8x+12=0$

$(x-2)(x-6)=0$

$x-2=0$ or $x-6=0$

$x=2$ $\quad x=6$

2, 6

81. answers may vary

Exercise Set 9.6

1. $\sqrt{2x}=4$

$2x=4^2$

$2x=16$

$x=8$

The solution set is $\{8\}$.

5. $\sqrt{2x}=-4$

No solution since a principle square root does not yield a negative number. The solution set is $\emptyset$.

9. $\sqrt{2x-3}-2=1$

$\sqrt{2x-3}=3$

$2x-3=3^2$

$2x-3=9$

$2x=12$

$x=6$

The solution set is $\{6\}$.

17. $x-\sqrt{4-3x}=-8$

$x+8=\sqrt{4-3x}$

$(x+8)^2=4-3x$

$x^2+16x+64=4-3x$

$x^2+16x+64=4-3x$

$x^2+19x+60=0$

$(x+4)(x+15)=0$

$x+4=0$ or $x+15=0$

$x=-4$ $\quad x=-15$

We discard the -15 as extraneous, leaving $x=-4$ as the only solution. The solution set is $\{-4\}$.

21. $\sqrt{x-3}+\sqrt{x+2}=5$

$\sqrt{x-3}=5-\sqrt{x+2}$

$x-3=25-10\sqrt{x+2}+x+2$

$-3=27-10\sqrt{x+2}$

$-30=-10\sqrt{x+2}$

$3=\sqrt{x+2}$

$9=x+2$

$7=x$

The solution set is $\{7\}$.

25. $-\sqrt{2x}+4=-6$

$10=\sqrt{2x}$

$10^2=2x$

$100=2x$

$x=50$

The solution set is $\{50\}$.

29. $\sqrt[4]{4x+1}-2=0$

$\sqrt[4]{4x+1}=2$

$4x+1=2^4$

$4x+1=16$

$4x=15$

$x=\dfrac{15}{4}$

The solution set is $\left\{\dfrac{15}{4}\right\}$.

33. $\sqrt[3]{6x-3}-3=0$

$\sqrt[3]{6x-3}=3$

$6x-3=3^3$

$6x-3=27$

$6x=30$

$x=5$

The solution set is $\{5\}$.

37. $\sqrt{x+4}=\sqrt{2x-5}$

$x+4=2x-5$

$9=x$

$x=9$

The solution set is $\{9\}$.

41. $\sqrt[3]{-6x-1}=\sqrt[3]{-2x-5}$

$-6x-1=-2x-5$

$4=4x$

$x=1$

The solution set is $\{1\}$.

45. $\sqrt{2x-1}=\sqrt{1-2x}$

$\sqrt{2x-1}=\sqrt{-(2x-1)}$

It follows that $2x-1=0$ (Otherwise one of the radicands would be negative).

So $2x=1$

$x=\dfrac{1}{2}$

The solution set is $\left\{\dfrac{1}{2}\right\}$.

49. $\sqrt{y+3}-\sqrt{y-3}=1$

$\sqrt{y+3}=1+\sqrt{y-3}$

$(\sqrt{y+3})^2=(1+\sqrt{y-3})^2$

$y+3=1+2\sqrt{y-3}+y-3$

$5=2\sqrt{y-3}$

$25=4(y-3)$

$\dfrac{25}{4}=y-3$

$\dfrac{25}{4}+\dfrac{12}{4}=y$

$\dfrac{37}{4}=y$

The solution set is $\left\{\dfrac{37}{4}\right\}$.

53. Let b = the length of the unknown leg of the right triangle. By the Pythagorean theorem,

$$7^2 = 3^2 + b^2$$
$$49 = 9 + b^2$$
$$b^2 = 40$$
$$b = \sqrt{40} = \sqrt{4}\sqrt{10} = 2\sqrt{10}\text{ m}$$

57. Let c = the length of the hypotenuse of the right triangle. By the Pythagorean theorem,

$c^2 = 7^2 + (7.2)^2 = 100.84$

so $c = \sqrt{100.84} \approx 10.0$ mm

61. $x^2 = (5)^2 + (12)^2$

$x^2 = 25 + 144$

$x^2 = 169$

$x = \sqrt{169}$

$x = 13$

The answer is 13 ft.

65.
$$v = \sqrt{2gh}$$
$$80 = \sqrt{2(32)h}$$
$$(80)^2 = \left(\sqrt{2(32)h}\right)^2$$
$$6400 = 2(32)\cdot h$$
$$100 = h$$

The object fell 100 ft.

69. $P = 2\pi\sqrt{\dfrac{l}{32}}$

$$= 2\pi\sqrt{\frac{2}{32}} = 2\pi\sqrt{\frac{1}{16}}$$
$$= 2\pi\left(\frac{1}{\sqrt{16}}\right) = 2\pi\left(\frac{1}{4}\right)$$
$$= \frac{2\pi}{4} = \frac{\pi}{2}\text{ sec} \approx 1.57\text{ sec}$$

73. answers may vary

77. answers may vary

81. $$\frac{\frac{x}{6}}{\frac{2x}{3} + \frac{1}{2}} = \frac{\left(\frac{x}{6}\right)6}{\left(\frac{2x}{3} + \frac{1}{2}\right)6} = \frac{x}{\left(\frac{2x}{3}\right)6 + \left(\frac{1}{2}\right)6} = \frac{x}{4x + 3}$$

85.
$$\sqrt{\sqrt{x+3} + \sqrt{x}} = \sqrt{3}$$
$$\left(\sqrt{\sqrt{x+3} + \sqrt{x}}\right)^2 = (\sqrt{3})^2$$
$$\sqrt{x+3} + \sqrt{x} = 3$$
$$\sqrt{x+3} = 3 - \sqrt{x}$$
$$(\sqrt{x+3})^2 = (3 - \sqrt{x})^2$$
$$x + 3 = 9 - 6\sqrt{x} + x$$
$$-6 = -6\sqrt{x}$$
$$1 = \sqrt{x}$$
$$1^2 = (\sqrt{x})^2$$
$$1 = x \text{ and it checks}$$

$\{1\}$

Exercise Set 9.7

1. $\sqrt{-24} = \sqrt{4}\sqrt{6}\sqrt{-1} = 2i\sqrt{6}$

5. $8\sqrt{-63} = 8\sqrt{9}\sqrt{7}\sqrt{-1} = 8\cdot 3\sqrt{7}i = 24i\sqrt{7}$

9. $\sqrt{-2}\sqrt{-7} = (\sqrt{2}i)(\sqrt{7}i)$
$$= \sqrt{14i^2} = \sqrt{14(-1)} = -\sqrt{14}$$

13. $\sqrt{16}\sqrt{-1} = 4i$

17. $\dfrac{\sqrt{-80}}{\sqrt{-10}} = \dfrac{\sqrt{80}i}{\sqrt{10}i} = \sqrt{\dfrac{80}{10}} = \sqrt{8} = \sqrt{4}\sqrt{2} = 2\sqrt{2}$

21. $(6 + 5i) - (8 - i) = (6 - 8) + [5 - (-1)]i$
$$= -2 + 6i$$

25. $(6 - 3i) - (4 - 2i) = 6 - 3i - 4 + 2i = 2 - i$

29. $(2 + 4i) + (6 - 5i) = (2 + 6) + (4 - 5)i$
$$= 8 - i$$

33. $(-9i)(7i) = -63i^2 = -63(-1) = 63$

37. $6i(2 - 3i) = 6i(2) - 6i(3i)$
$$= 12i - 18i^2 = 12i - 18(-1) = 18 + 12i$$

41. $(4 + i)(5 + 2i) = 20 + 8i + 5i + 2i^2$
$$= 20 + 13i - 2 = 18 + 13i$$

45. $(4 - 2i)^2 = 16 - 16i + 4i^2$
$$= 16 - 16i + 4(-1) = 16 - 4 - 16i = 12 - 16i$$

49. $(1 - i)(1 + i) = 1^2 - i^2 = 1^2 + 1^2 = 1 + 1 = 2$

53. $(1 - i)^2 = 1^2 - 2(1)(i) + i^2 = 1 - 2i - 1 = -2i$

57. $\dfrac{7}{4 + 3i} = \dfrac{7}{4 + 3i}\cdot\dfrac{4 - 3i}{4 - 3i}$
$$= \frac{28 - 21i}{4^2 + 3^2} = \frac{28 - 21i}{16 + 9} = \frac{28 - 21i}{25} = \frac{28}{25} - \frac{21}{25}i$$

61. $\dfrac{3 + 5i}{1 + i} = \dfrac{3 + 5i}{1 + i}\cdot\dfrac{1 - i}{1 - i}$
$$= \frac{3 + 5i - 3i - 5i^2}{1^2 + 1^2} = \frac{3 + 2i - 5(-1)}{1 + 1} = \frac{3 + 5 + 2i}{2} = \frac{8 + 2i}{2} = 4 + i$$

65. $\dfrac{16 + 15i}{-3i} = \dfrac{(16 + 15i)i}{-3i^2}$
$$= \frac{16i + 15i^2}{-3(-1)} = \frac{16i + 15(-1)}{3} = \frac{-15}{3} + \frac{16}{3}i = -5 + \frac{16}{3}i$$

69. $\frac{2-3i}{2+i} = \frac{(2-3i)(2-i)}{(2+i)(2-i)}$
$= \frac{4-6i-2i+3i^2}{2^2+1^2}$
$= \frac{4-8i+3(-1)}{4+1}$
$= \frac{4-3-8i}{5}$
$= \frac{1}{5} - \frac{8}{5}i$

73. $i^{21} = i^{20}i = (i^4)^5 i = 1^5 i = 1i = i$

77. $i^{-6} = (i^2)^{-3} = (-1)^{-3} = \frac{1}{(-1)^3} = \frac{1}{-1} = -1$

81. $(-3i)^5 = (-3)^5 i^5$
$= -243 \cdot i^4 \cdot i$
$= -243(1)(i)$
$= -243i$

85. $5 + 9 = 14$ people

89. $i^3 + i^4 = -i + 1 = 1 - i$

93. $2 + \sqrt{-9} = 2 + 3i$

97. $\frac{5-\sqrt{-75}}{10} = \frac{5-5i\sqrt{3}}{10} = \frac{5}{10} - \frac{5i\sqrt{3}}{10} = \frac{1}{2} - \frac{\sqrt{3}}{2}i$

101. $(8 - \sqrt{-4}) - (2 + \sqrt{-16})$
$= (8 - 2i) - (2 + 4i)$
$= (8 - 2) + (-2i - 4i)$
$= 6 - 6i$

Chapter 9 Test

1. $\sqrt{216} = \sqrt{36 \cdot 6} = 6\sqrt{6}$

5. $\left[\frac{8x^3}{27}\right]^{2/3} = \frac{8^{2/3}(x^3)^{2/3}}{27^{2/3}}$
$= \frac{(8^{1/3})^2 x^2}{(27^{1/3})^2}$
$= \frac{2^2x^2}{3^2}$
$= \frac{4x^2}{9}$

9. $\sqrt[4]{(4xy)^4} = |4xy|$ or $4|xy|$

13. $\sqrt[3]{\frac{8}{9x}} = \frac{\sqrt[3]{8}}{\sqrt[3]{9x}}$
$= \frac{2 \cdot \sqrt[3]{3x^2}}{\sqrt[3]{9x} \cdot \sqrt[3]{3x^2}}$
$= \frac{2\sqrt[3]{3x^2}}{3x}$

17. $(\sqrt{x} + 1)^2 = \sqrt{x}^2 + 2\sqrt{x} + 1$
$= x + 2\sqrt{x} + 1$

21. $386^{-2/3} \approx 0.019$

25. $\sqrt{-2} = i\sqrt{2}$

29. $(4 + 3i)^2 = 16 + 24i + 9i^2$
$= 16 + 24i + 9(-1)$
$= (16 - 9) + 24i$
$= 7 + 24i$

33. $V = \sqrt{2.5(300)} \approx 27$ mph

Chapter 10

Exercise Set 10.1

1. $x^2 = 16$
$x = \pm\sqrt{16}$
$x = \pm 4$
The solution set is $\{-4, 4\}$.

5. $x^2 = 18$
$x = \pm\sqrt{18}$
$x = \pm\sqrt{9}\sqrt{2}$
$x = \pm 3\sqrt{2}$
The solution set is $\{-3\sqrt{2}, 3\sqrt{2}\}$.

9. $(x + 5)^2 = 9$
$x + 5 = \pm\sqrt{9}$
$x + 5 = \pm 3$
$x = -5 \pm 3$
$x = -8$ or $x = -2$
The solution set is $\{-8, -2\}$.

13. $(2x - 3)^2 = 8$
$2x - 3 = \pm\sqrt{8}$
$2x - 3 = \pm\sqrt{4}\sqrt{2}$
$2x - 3 = \pm 2\sqrt{2}$
$2x = 3 \pm 2\sqrt{2}$
$x = \frac{3 \pm 2\sqrt{2}}{2}$
The solution set is $\left\{\frac{3-2\sqrt{2}}{2}, \frac{3+2\sqrt{2}}{2}\right\}$.

17. $x^2 - 6 = 0$
$x^2 = 6$
$x = \pm\sqrt{6}$
The solution set is $\{-\sqrt{6}, \sqrt{6}\}$.

21. $(x - 1)^2 = -16$
$x - 1 = \pm\sqrt{-16}$
$x - 1 = \pm 4i$
$x = 1 \pm 4i$
The solution set is $\{1 - 4i, 1 + 4i\}$.

25. $(x + 3)^2 = -8$
$x + 3 = \pm\sqrt{4}\sqrt{2}\sqrt{-1}$
$x + 3 = \pm\sqrt{-8}$
$x + 3 = \pm 2i\sqrt{2}$
$x = -3 \pm 2i\sqrt{2}$
The solution set is $\{-3 - 2i\sqrt{2}, -3 + 2i\sqrt{2}\}$.

29. $z^2 - 12z + \left(-\frac{12}{2}\right)^2 = z^2 - 12z + 36 = (z - 6)^2$

33. $r^2 - 3r + \left(-\frac{3}{2}\right)^2 = r^2 - 3r + \frac{9}{4}$
$= \left(r - \frac{3}{2}\right)^2$

37. $x^2 + 6x + 2 = 0$
$x^2 + 6x + \left(\frac{6}{2}\right)^2 = -2 + 9$
$(x + 3)^2 = 7$
$x + 3 = \pm\sqrt{7}$
$x = -3 \pm \sqrt{7}$
The solution set is $\{-3 - \sqrt{7}, -3 + \sqrt{7}\}$.

41.
$$x^2 + 2x - 5 = 0$$
$$x^2 + 2x + \left(\frac{2}{2}\right)^2 = 5 + 1$$
$$x^2 + 2x + 1 = 6$$
$$(x + 1)^2 = 6$$
$$x + 1 = \pm\sqrt{6}$$
$$x = -1 \pm \sqrt{6}$$
The solution set is $\{-1 - \sqrt{6}, -1 + \sqrt{6}\}$.

45.
$$2x^2 + 7x = 4$$
$$x^2 + \frac{7}{2}x = 2$$
$$x^2 + \frac{7}{2}x + \left(\frac{\frac{7}{2}}{2}\right)^2 = 2 + \frac{49}{16}$$
$$\left(x + \frac{7}{4}\right)^2 = \frac{81}{16}$$
$$x + \frac{7}{4} = \pm\sqrt{\frac{81}{16}}$$
$$x = -\frac{7}{4} \pm \frac{9}{4}$$
$$x = -4 \quad \text{or } x = \frac{1}{2}$$
The solution set is $\left\{-4, \frac{1}{2}\right\}$.

49.
$$3y^2 + 6y - 4 = 0$$
$$3y^2 + 6y = 4$$
$$y^2 + 2y = \frac{4}{3}$$
$$y^2 + 2y + \left(\frac{2}{2}\right)^2 = \frac{4}{3} + 1$$
$$(y + 1)^2 = \frac{7}{3}$$
$$y + 1 = \pm\sqrt{\frac{7}{3}}$$
$$y + 1 = \pm\frac{\sqrt{7}}{\sqrt{3}}\cdot\frac{\sqrt{3}}{\sqrt{3}}$$
$$y + 1 = \pm\frac{\sqrt{21}}{3}$$
$$y = -1 \pm \frac{\sqrt{21}}{3}$$
$$= \frac{-3 \pm \sqrt{21}}{3}$$
The solution set is $\left\{\frac{-3 - \sqrt{21}}{3}, \frac{-3 + \sqrt{21}}{3}\right\}$.

53.
$$x^2 - 6x + 3 = 0$$
$$x^2 - 6x + \left(-\frac{6}{2}\right)^2 = -3 + 9$$
$$(x - 3)^2 = 6$$
$$x - 3 = \pm\sqrt{6}$$
$$x = 3 \pm \sqrt{6}$$
The solution set is $\{3 - \sqrt{6}, 3 + \sqrt{6}\}$.

57.
$$2x^2 - x + 6 = 0$$
$$2x^2 - x = -6$$
$$x^2 - \frac{1}{2}x = -3$$
$$x^2 - \frac{1}{2}x + \left(\frac{-\frac{1}{2}}{2}\right)^2 = -3 + \frac{1}{16}$$
$$\left(x - \frac{1}{4}\right)^2 = -\frac{47}{16}$$
$$x - \frac{1}{4} = \pm\sqrt{-\frac{47}{16}}$$
$$x - \frac{1}{4} = \pm\frac{\sqrt{47}\sqrt{-1}}{\sqrt{16}}$$
$$x - \frac{1}{4} = \pm\frac{i\sqrt{47}}{4}$$
$$x = \frac{1 \pm i\sqrt{47}}{4}$$
The solution set is $\left\{\frac{1 + i\sqrt{47}}{4}, \frac{1 - i\sqrt{47}}{4}\right\}$.

61.
$$z^2 + 3z - 4 = 0$$
$$z^2 + 3z = 4$$
$$z^2 + 3z + \left(\frac{3}{2}\right)^2 = 4 + \frac{9}{4}$$
$$\left(z + \frac{3}{2}\right)^2 = \frac{25}{4}$$
$$z + \frac{3}{2} = \pm\sqrt{\frac{25}{4}}$$
$$z + \frac{3}{2} = \pm\frac{5}{2}$$
$$z = -\frac{3}{2} \pm \frac{5}{2}$$
$$z = -4 \quad \text{or} \quad z = 1$$
The solution set is $\{-4, 1\}$.

65.
$$3x^2 + 3x = 5$$
$$x^2 + x = \frac{5}{3}$$
$$x^2 + x + \left(\frac{1}{2}\right)^2 = \frac{5}{3} + \frac{1}{4}$$
$$\left(x + \frac{1}{2}\right)^2 = \frac{23}{12}$$
$$x + \frac{1}{2} = \pm\sqrt{\frac{23}{12}}$$
$$x + \frac{1}{2} = \pm\frac{\sqrt{23}}{\sqrt{4}\sqrt{3}}$$
$$x + \frac{1}{2} = \pm\frac{\sqrt{23}}{2\sqrt{3}}$$
$$x + \frac{1}{2} = \pm\frac{\sqrt{23}\sqrt{3}}{2\sqrt{3^2}}$$
$$x + \frac{1}{2} = \pm\frac{\sqrt{69}}{2\cdot 3}$$
$$x + \frac{1}{2} = \pm\frac{\sqrt{69}}{6}$$
$$x = -\frac{1}{2} \pm \frac{\sqrt{69}}{6}$$
$$x = \frac{-3 \pm \sqrt{69}}{6}$$
The solution set is $\left\{\frac{-3 - \sqrt{69}}{6}, \frac{-3 + \sqrt{69}}{6}\right\}$.

69.
$$A = P(1 + r)^t$$
$$1000 = 810(1 + r)^2$$
$$\frac{1000}{810} = (1 + r)^2$$
$$\frac{100}{81} = (1 + r)^2$$
$$\pm\sqrt{\frac{100}{81}} = 1 + r$$
$$\pm\frac{10}{9} = 1 + r$$
$$-1 \pm \frac{10}{9} = r$$
$$r = -1 + \frac{10}{9} \quad \text{or} \quad r = -1 - \frac{10}{9}$$
$$r = \frac{1}{9} \quad \text{or} \quad r = -\frac{19}{9}$$
Rate cannot be negative, so $r = \frac{1}{9}$, or about 11%.

73. Simple; answers may vary

77.
$$\frac{12 - 8\sqrt{7}}{16} = \frac{12}{16} - \frac{8\sqrt{7}}{16} = \frac{3}{4} - \frac{\sqrt{7}}{2} = \frac{3}{4} - \frac{2\sqrt{7}}{4} = \frac{3 - 2\sqrt{7}}{4}$$

81. $\sqrt{b^2 - 4ac}$
$a = 1, b = -3, c = -1$
$\sqrt{(-3)^2 - 4(1)(-1)} = \sqrt{9 + 4} = \sqrt{13}$

85. $s(t) = 16t^2$
$1053 = 16t^2$
$$t = \pm\sqrt{\frac{1053}{16}}$$
$t \approx 8.11$ or -8.11 (disregard)
It would take about 8.11 sec.

89.
$$A = \pi r^2$$
$$36\pi = \pi r^2$$
$$36 = r^2$$
$$\pm\sqrt{36} = r$$
$$\pm 6 = r$$
Since the radius cannot be negative, $r = 6$ inches.

93.
$$p = -x^2 + 15$$
$$7 = -x^2 + 15$$
$$-8 = -x^2$$
$$8 = x^2$$
$$\pm\sqrt{8} = x$$
$$\pm 2.828 \approx x$$
Since x cannot be negative, the demand for each lamp is about 2.828 thousand units.

Exercise Set 10.2

1. $m^2 + 5m - 6 = 0$
$a = 1, b = 5, c = -6$
$$m = \frac{-5 \pm \sqrt{5^2 - 4(1)(-6)}}{2(1)}$$
$$m = \frac{-5 \pm \sqrt{25 + 24}}{2} = \frac{-5 \pm \sqrt{49}}{2}$$
$$m = \frac{-5 \pm 7}{2}$$
$m = -6$ or $m = 1$
The solution set is $\{-6, 1\}$.

5. $x^2 - 6x + 9 = 0$
$a = 1, b = -6, c = 9$
$$x = \frac{6 \pm \sqrt{(-6)^2 - 4(1)(9)}}{2(1)}$$
$$x = \frac{6 \pm \sqrt{36 - 36}}{2} = \frac{6 \pm \sqrt{0}}{2} = \frac{6}{2} = 3$$
The solution set is $\{3\}$.

9. $8m^2 - 2m = 7$
$8m^2 - 2m - 7 = 0$
$a = 8, b = -2, c = -7$
$$m = \frac{2 \pm \sqrt{(-2)^2 - 4(8)(-7)}}{2(8)}$$
$$m = \frac{2 \pm \sqrt{4 + 224}}{16} = \frac{2 \pm \sqrt{228}}{16}$$
$$m = \frac{2 \pm \sqrt{4}\sqrt{57}}{16} = \frac{2 \pm 2\sqrt{57}}{16}$$
$$m = \frac{1 \pm \sqrt{57}}{8}$$
The solution set is $\left\{\frac{1 + \sqrt{57}}{8}, \frac{1 - \sqrt{57}}{8}\right\}$.

13. $\frac{1}{2}x^2 - x - 1 = 0$
$x^2 - 2x - 2 = 0$
$a = 1, b = -2, c = -2$
$$x = \frac{2 \pm \sqrt{(-2)^2 - 4(1)(-2)}}{2(1)}$$
$$x = \frac{2 \pm \sqrt{4 + 8}}{2} = \frac{2 \pm \sqrt{12}}{2}$$
$$x = \frac{2 \pm \sqrt{4}\sqrt{3}}{2} = \frac{2 \pm 2\sqrt{3}}{2}$$
$$x = 1 \pm \sqrt{3}$$
The solution set is $\{1 - \sqrt{3}, 1 + \sqrt{3}\}$.

17. $\frac{1}{3}y^2 - y - \frac{1}{6} = 0$
$2y^2 - 6y - 1 = 0$
$a = 2, b = -6, c = -1$
$$y = \frac{6 \pm \sqrt{(-6)^2 - 4(2)(-1)}}{2(2)}$$
$$y = \frac{6 \pm \sqrt{36 + 8}}{4} = \frac{6 \pm \sqrt{44}}{4}$$
$$y = \frac{6 \pm \sqrt{4}\sqrt{11}}{4} = \frac{6 \pm 2\sqrt{11}}{4}$$
$$y = \frac{3 \pm \sqrt{11}}{2}$$
The solution set is $\left\{\frac{3 - \sqrt{11}}{2}, \frac{3 + \sqrt{11}}{2}\right\}$.

21. $x^2 + 5x = -2$
$x^2 + 5x + 2 = 0$
$a = 1, b = 5, c = 2$
$$x = \frac{-5 \pm \sqrt{5^2 - 4(1)(2)}}{2(1)}$$
$$x = \frac{-5 \pm \sqrt{25 - 8}}{2} = \frac{-5 \pm \sqrt{17}}{2}$$
The solution set is $\left\{\frac{-5 - \sqrt{17}}{2}, \frac{-5 + \sqrt{17}}{2}\right\}$.

25. $\frac{x^2}{3} - x = \frac{5}{3}$

$x^2 - 3x = 5$

$x^2 - 3x - 5 = 0$

$a = 1, b = -3, c = -5$

$x = \frac{3 \pm \sqrt{(-3)^2 - 4(1)(-5)}}{2(1)}$

$x = \frac{3 \pm \sqrt{9 + 20}}{2} = \frac{3 \pm \sqrt{29}}{2}$

The solution set is $\left\{\frac{3 - \sqrt{29}}{2}, \frac{3 + \sqrt{29}}{2}\right\}$.

29. answers may vary

33. $(x + 5)(x - 1) = 2$

$x^2 + 4x - 5 = 2$

$x^2 + 4x - 7 = 0$

$a = 1, b = 4, c = -7$

$x = \frac{-4 \pm \sqrt{4^2 - 4(1)(-7)}}{2(1)}$

$x = \frac{-4 \pm \sqrt{16 + 28}}{2} = \frac{-4 + \sqrt{44}}{2}$

$x = \frac{-4 \pm \sqrt{4}\sqrt{11}}{2} = \frac{-4 \pm 2\sqrt{11}}{2}$

$x = -2 \pm \sqrt{11}$

The solution set is $\{-2 - \sqrt{11}, -2 + \sqrt{11}\}$.

37. $\frac{2}{5}y^2 + \frac{1}{5}y + \frac{3}{5} = 0$

$2y^2 + y + 3 = 0$

$a = 2, b = 1, c = 3$

$y = \frac{-1 \pm \sqrt{1^2 - 4(2)(3)}}{2(2)}$

$y = \frac{-1 \pm \sqrt{1 - 24}}{4} = \frac{-1 \pm \sqrt{-23}}{4}$

$y = \frac{-1 \pm i\sqrt{23}}{4}$

The solution set is $\left\{\frac{-1 - i\sqrt{23}}{4}, \frac{-1 + i\sqrt{23}}{4}\right\}$.

41. $(n - 2)^2 = 2n$

$n^2 - 4n + 4 = 2n$

$n^2 - 6n + 4 = 0$

$a = 1, b = -6, c = 4$

$n = \frac{6 \pm \sqrt{(-6)^2 - 4(1)(4)}}{2(1)}$

$n = \frac{6 \pm \sqrt{36 - 16}}{2} = \frac{6 \pm \sqrt{20}}{2}$

$n = \frac{6 \pm \sqrt{4}\sqrt{5}}{2} = \frac{6 \pm 2\sqrt{5}}{2}$

$n = 3 \pm \sqrt{5}$

The solution set is $\{3 - \sqrt{5}, 3 + \sqrt{5}\}$.

45. $4x^2 + 12x = -9$

$4x^2 + 12x + 9 = 0$

$a = 4, b = 12, c = 9$

$b^2 - 4ac = 12^2 - 4(4)(9)$

$b^2 - 4ac = 144 - 144 = 0$

Therefore, there is 1 real solution.

49. $6 = 4x - 5x^2$

$5x^2 - 4x + 6 = 0$

$a = 5, b = -4, c = 6$

$b^2 - 4ac = (-4)^2 - 4(5)(6)$

$b^2 - 4ac = 16 - 120$

$b^2 - 4ac = -104 < 0$

Therefore, there are 2 complex but not real solutions.

53. Let x = length of leg

$x + 2$ = length of hypotenuse

$x^2 + x^2 = (x + 2)^2$

$2x^2 = x^2 + 4x + 4$

$x^2 - 4x - 4 = 0$

$a = 1, b = -4, c = -4$

$x = \frac{4 \pm \sqrt{(-4)^2 - 4(1)(-4)}}{2(1)}$

$x = \frac{4 \pm \sqrt{32}}{2}$

$x = \frac{4 \pm 4\sqrt{2}}{2}$

$x = 2 \pm 2\sqrt{2}$ (disregard a negative length)

The sides measure $2 + 2\sqrt{2}$ cm, $2 + 2\sqrt{2}$ cm, and $4 + 2\sqrt{2}$ cm.

57. a. Let x = length

$x^2 + x^2 = 100^2$

$2x^2 - 10{,}000 = 0$

$a = 2, b = 0, c = -10{,}000$

$x = \frac{0 \pm \sqrt{0^2 - 4(2)(-10{,}000)}}{2(2)}$

$x = \frac{\pm\sqrt{80{,}000}}{4}$

$x = \frac{\pm 200\sqrt{2}}{4}$

$x = \pm 50\sqrt{2}$

Disregard a negative length. The side measures $50\sqrt{2}$ m.

b. Area $= s^2$

$= (50\sqrt{2})^2$

$= 50^2(\sqrt{2})^2$

$= 2500 \cdot 2$

$= 5000$

The area is 5000 sq m.

61. $\frac{x - 1}{1} = \frac{1}{x}$

$x(x - 1) = 1 \cdot 1$

$x^2 - x - 1 = 0$

$a = 1, b = -1, c = -1$

$x = \frac{-(-1) \pm \sqrt{(-1)^2 - 4(1)(-1)}}{2}$

$x = \frac{1 \pm \sqrt{5}}{2}$

Disregard the negative value $\frac{1 - \sqrt{5}}{2}$.

The value is $\frac{1 + \sqrt{5}}{2}$.

65. $h = -16t^2 - 20t + 180$

Let $h = 0$:

$0 = -16t^2 - 20t + 180$

$0 = 4t^2 + 5t - 45 \quad \div$ by -4

$a = 4, b = 5, c = -45$

$t = \dfrac{-5 \pm \sqrt{(5)^2 - 4(4)(-45)}}{2(4)}$

$t = \dfrac{-5 \pm \sqrt{745}}{8}$

$t \approx 2.8$ or $t \approx -4.0$ (disregard)
It will take about 2.8 sec.

69. $\dfrac{1}{x} + \dfrac{2}{5} = \dfrac{7}{x}$

$5x\left(\dfrac{1}{x} + \dfrac{2}{5}\right) = 5x\left(\dfrac{7}{x}\right)$ LCD is $5x$

$5 + 2x = 35$

$2x = 30$

$x = 15$

$\{15\}$

73. $z^4 - 13z^2 + 36 = (z^2 - 9)(z^2 - 4) = (z + 3)(z - 3)(z + 2)(z - 2)$

77. Sunday to Monday

81. Thursday corresponds to $x = 4$.

$f(x) = 3x^2 - 18x + 57$

$f(4) = 3(4)^2 - 18(4) + 57$

$= 3(16) - 18(4) + 57$

$= 48 - 72 + 57$

$= 33$

Yes; answers may vary

85. $f(x) = -199.5x^2 - 21.5x + 3763$

a. $f(2) = -199.5(2)^2 - 21.5(2) + 3763$

$= -199.5(4) - 21.5(2) + 3763$

$= -798 - 43 + 3763$

$= 2922$

The earnings should be \$2922 million.

b. $485 = -199.5x^2 - 21.5x + 3763$

$199.5x^2 + 21.5x - 3278 = 0$

$a = 199.5, b = 21.5, c = -3278$

$x = \dfrac{-21.5 \pm \sqrt{(21.5)^2 - 4(199.5)(-3278)}}{2(199.5)}$

$x = \dfrac{-21.5 \pm \sqrt{2{,}616{,}306.25}}{399}$

$x = \dfrac{-21.5 \pm 1617.5}{399}$

$x = 4$ or $x \approx -4.1$ (Disregard)

The year is $1999 + 4 = 2003$.

Exercise Set 10.3

1. $2x = \sqrt{10 + 3x}$

$4x^2 = 10 + 3x$

$4x^2 - 3x - 10 = 0$

$(4x + 5)(x - 2) = 0$

$4x + 5 = 0$ or $x - 2 = 0$

$x = -\dfrac{5}{4}$ or $x = 2$

Discard $x = -\dfrac{5}{4}$.

The solution set is $\{2\}$.

5. $\sqrt{9x} = x + 2$

$9x = x^2 + 4x + 4$

$0 = x^2 - 5x + 4$

$0 = (x - 4)(x - 1)$

$x - 4 = 0$ or $x - 1 = 0$

$x = 4$ or $x = 1$

The solution set is $\{1, 4\}$.

9. $\dfrac{3}{x} + \dfrac{4}{x + 2} = 2$

$\dfrac{3(x + 2) + 4x}{x(x + 2)} = 2$

$\dfrac{3x + 6 + 4x}{x^2 + 2x} = 2$

$7x + 6 = 2(x^2 + 2x)$

$7x + 6 = 2x^2 + 4x$

$2x^2 - 3x - 6 = 0$

$x = \dfrac{3 \pm \sqrt{(-3)^2 - 4(2)(-6)}}{2(2)}$

$x = \dfrac{3 \pm \sqrt{57}}{4}$

The solution set is $\left\{\dfrac{3 + \sqrt{57}}{4}, \dfrac{3 - \sqrt{57}}{4}\right\}$.

13. $p^4 - 16 = 0$

$(p^2 + 4)(p^2 - 4) = 0$

$(p + 2i)(p - 2i)(p + 2)(p - 2) = 0$

$p + 2i = 0$ or $p - 2i = 0$

or $p + 2 = 0$ or $p - 2 = 0$

$p = -2i$ or $p = 2i$

$p = -2$ or $p = 2$

The solution set is $\{-2i, 2i, -2, 2\}$.

17. $z^4 - 13z^2 + 36 = 0$

$(z^2 - 9)(z^2 - 4) = 0$

$(z + 3)(z - 3)(z + 2)(z - 2) = 0$

$z + 3 = 0$ or $z - 3 = 0$ or $z + 2 = 0$ or $z - 2 = 0$

$z = -3$ or $z = 3$ or $z = -2$ or $z = 2$

The solution set is $\{-3, 3, -2, 2\}$.

21. $(5n + 1)^2 + 2(5n + 1) - 3 = 0$

Let $y = 5n + 1$.

$y^2 + 2y - 3 = 0$

$(y + 3)(y - 1) = 0$

$y + 3 = 0$ or $y - 1 = 0$

$y = -3$ or $y = 1$

$5n + 1 = -3$ or $5n + 1 = 1$

$5n = -4$ or $5n = 0$

$n = -\dfrac{4}{5}$ or $n = 0$

The solution set is $\left\{-\dfrac{4}{5}, 0\right\}$.

25. $1 + \dfrac{2}{3t - 2} = \dfrac{8}{(3t - 2)^2}$

$(3t - 2)^2 + 2(3t - 2) - 8 = 0$

Let $y = 3t - 2$.

$y^2 + 2y - 8 = 0$

$(y + 4)(y - 2) = 0$

$y + 4 = 0$ or $y - 2 = 0$

$y = -4$ or $y = 2$

$3t - 2 = -4$ or $3t - 2 = 2$

$3t = -2$ or $3t = 4$

$t = -\dfrac{2}{3}$ or $t = \dfrac{4}{3}$

The solution set is $\left\{-\dfrac{2}{3}, \dfrac{4}{3}\right\}$.

29. $a^4 - 5a^2 + 6 = 0$

$(a^2 - 3)(a^2 - 2) = 0$

$a^2 - 3 = 0$ or $a^2 - 2 = 0$
$a^2 = 3$ or $a^2 = 2$
$a = \pm\sqrt{3}$ or $a = \pm\sqrt{2}$
The solution set is $\{\sqrt{3}, -\sqrt{3}, \sqrt{2}, -\sqrt{2}\}$.

33. $(p + 2)^2 = 9(p + 2) - 20$
$(p + 2)^2 - 9(p + 2) + 20 = 0$
Let $x = p + 2$.
$x^2 - 9x + 20 = 0$
$(x - 5)(x - 4) = 0$
$x - 5 = 0$ or $x - 4 = 0$
$x = 5$ or $x = 4$
$p + 2 = 5$ or $p + 2 = 4$
$p = 3$ or $p = 2$
The solution set is $\{2, 3\}$.

37. $x^{2/3} - 8x^{1/3} + 15 = 0$
Let $y = x^{1/3}$.
$y^2 - 8y + 15 = 0$
$(y - 5)(y - 3) = 0$
$y - 5 = 0$ or $y - 3 = 0$
$y = 5$ or $y = 3$
$x^{1/3} = 5$ or $x^{1/3} = 3$
$x = 5^3$ or $x = 3^3$
$x = 125$ or $x = 27$
The solution set is $\{125, 27\}$.

41. $2x^{2/3} + 3x^{1/3} - 2 = 0$
Let $m = x^{1/3}$.
$2m^2 + 3m - 2 = 0$
$(2m - 1)(m + 2) = 0$
$2m - 1 = 0$ or $m + 2 = 0$
$2m = 1$ or $m = -2$
$m = \frac{1}{2}$ or $m = -2$
$x^{1/3} = \frac{1}{2}$ or $x^{1/3} = -2$
$x = \left(\frac{1}{2}\right)^3 = \frac{1}{8}$ or $x = (-2)^3 = -8$
The solution set is $\left\{\frac{1}{8}, -8\right\}$.

45. $x - \sqrt{x} = 2$
$x - 2 = \sqrt{x}$
$x^2 - 4x + 4 = x$
$x^2 - 5x + 4 = 0$
$(x - 4)(x - 1) = 0$
$x = 4$ or $x = 1$ (discard)
The solution set is $\{4\}$.

49. $p^4 - p^2 - 20 = 0$
$(p^2 - 5)(p^2 + 4) = 0$
$p^2 - 5 = 0$ or $p^2 + 4 = 0$
$p^2 = 5$ or $(p + 2i)(p - 2i) = 0$
$p = \pm\sqrt{5}$ or $p + 2i = 0$ or $p - 2i = 0$
$p = \pm\sqrt{5}$ or $p = -2i$ or $p = 2i$
The solution set is $\{\sqrt{5}, -\sqrt{5}, -2i, 2i\}$.

53. $1 = \frac{4}{x - 7} + \frac{5}{(x - 7)^2}$
$(x - 7)^2 = 4(x - 7) + 5$
Let $y = x - 7$.
$y^2 - 4y - 5 = 0$
$(y - 5)(y + 1) = 0$
$y - 5 = 0$ or $y + 1 = 0$
$y = 5$ or $y = -1$
$x - 7 = 5$ or $x - 7 = -1$
$x = 12$ or $x = 6$
The solution set is $\{6, 12\}$.

57. Let x = speed on first part
$x - 1$ = speed on second part
$D = r \cdot t$ or $t = \frac{D}{r}, 1\frac{3}{5} = \frac{8}{5}$
$\frac{3}{x} + \frac{4}{x - 1} = \frac{8}{5}$
$3 \cdot 5(x - 1) + 4 \cdot 5x = 8 \cdot x(x - 1)$
$15x - 15 + 20x = 8x^2 - 8x$
$0 = 8x^2 - 43x + 15$
$0 = (8x - 3)(x - 5)$
$8x - 3 = 0$ or $x - 5 = 0$
$x = \frac{3}{8}$ or $x = 5$
$x - 1 = 4$
Her speeds were 5 mph and then 4 mph.

61. Let x = original speed
$x + 11$ = return speed
$D = r \cdot t$ or $t = \frac{D}{r}$
$\frac{330}{x} - \frac{330}{x + 11} = 1$
$330(x + 11) - 330x = x(x + 11)$
$330x + 3630 - 330x = x^2 + 11x$
$0 = x^2 + 11x - 3630$
$0 = (x - 55)(x + 66)$
$x = 55$ or $x = -66$ (discard)
$x + 11 = 66$
The speeds are 55 mph and 66 mph.

65. Let x = number
$x(x - 4) = 96$
$x^2 - 4x = 96$
$x^2 - 4x - 96 = 0$
$(x - 12)(x + 8) = 0$
$x - 12 = 0$ or $x + 8 = 0$
$x = 12$ or $x = -8$
The number is 12 or -8.

69. Let r = radius
d = diameter
s = side of square
$A = s^2$
$920 = s^2$
$s = \sqrt{920}$
d is the diagonal of the square
$d = \sqrt{(\sqrt{920})^2 + (\sqrt{920})^2}$
$d = \sqrt{920 + 920}$
$d = \sqrt{1840}$
$d \approx 42.9$
Now, $r = \frac{d}{2} = \frac{42.9}{2} = 21.45$
The radius is 22 feet.

73. $\frac{y - 1}{15} > \frac{-2}{5}$
$15\left(\frac{y - 1}{15}\right) > 15\left(\frac{-2}{5}\right)$
$y - 1 > -6$
$y > -5$ or $(-5, \infty)$

77. Let x = Brack's speed and
$x + 6.326$ = Kanaan's speed.
Brack's time $= \frac{10{,}682}{x}$
Kanaan's time $= \frac{10{,}682}{x + 6.326}$

Equation is

Kanaan's time = Brack's time + 0.73

$$\frac{10{,}682}{x} = \frac{10{,}682}{x + 6.326} + 0.73$$

Multiply both sides by $x(x + 6.326)$.

$$10{,}682(x + 6.326) = 10{,}682(x) + 0.73(x)(x + 6.326)$$
$$10{,}682(6.326) = 0.73(x)(x + 6.326)$$
$$67{,}574.332 = 0.73x^2 + 4.61798x$$
$$0 = 0.73x^2 + 4.61798x - 67{,}574.332$$
$$a = 0.73, b = 4.61798, c = -67{,}574.332$$

$$x = \frac{-4.61798 \pm \sqrt{(4.61798)^2 - 4(0.73)(-67{,}574.332)}}{2(0.73)}$$
$$= \frac{-4.61798 \pm \sqrt{197{,}338.3759}}{1.46}$$
$$= \frac{-4.61798 \pm 444.2278415}{1.46}$$
$$= \frac{-4.61798 + 444.2278415}{1.46} \text{ (discard negative answer)}$$
$$= 301.1026449$$
$$\approx 301.103$$

a. Brack's speed is 301.103 ft/sec

b. Kanaan's speed is $301.103 + 6.326 = 307.429$ ft/sec

c. $307.429 \text{ ft/sec} = \frac{307.429 \text{ ft}}{1 \text{ sec}} \times \frac{1 \text{ mi}}{5280 \text{ ft}} \times \frac{3600 \text{ sec}}{1 \text{ hr}} \approx$ 209.611 mph

Exercise Set 10.4

1. $(x + 1)(x + 5) > 0$

$(x + 1)(x + 5) = 0$

$x + 1 = 0$ or $x + 5 = 0$

$x = -1$ or $x = -5$

Region	Interval	Test Point
A	$(-\infty, -5)$	-6
B	$(-5, -1)$	-2
C	$(-1, \infty)$	0
$x = -6$	$(-6 + 1)(-6 + 5) > 0$	True
$x = -2$	$(-2 + 1)(-2 + 5) > 0$	False
$x = 0$	$(0 + 1)(0 + 5) > 0$	True

The solution set is $(-\infty, -5) \cup (-1, \infty)$.

5. $x^2 - 7x + 10 \le 0$

$(x - 5)(x - 2) \le 0$

$(x - 5)(x - 2) = 0$

$x - 5 = 0$ or $x - 2 = 0$

$x = 5$ or $x = 2$

Region	Interval	Test Point
A	$(-\infty, 2)$	0
B	$(2, 5)$	3
C	$(5, \infty)$	6
$x = 0$	$(0 - 5)(0 - 2) \le 0$	False
$x = 3$	$(3 - 5)(3 - 2) \le 0$	True
$x = 6$	$(6 - 5)(6 - 2) \le 0$	False

The solution set is $[2, 5]$.

9. $(x - 6)(x - 4)(x - 2) > 0$

$(x - 6)(x - 4)(x - 2) = 0$

$x - 6 = 0$ or $x - 4 = 0$ or $x - 2 = 0$

$x = 6$ or $x = 4$ or $x = 2$

Region	Interval	Test Point
A	$(-\infty, 2)$	1
B	$(2, 4)$	3
C	$(4, 6)$	5
D	$(6, \infty)$	7
$x = 1$	$(1 - 6)(1 - 4)(1 - 2) > 0$	False
$x = 3$	$(3 - 6)(3 - 4)(3 - 2) > 0$	True
$x = 5$	$(5 - 6)(5 - 4)(5 - 2) > 0$	False
$x = 7$	$(7 - 6)(7 - 4)(7 - 2) > 0$	True

The solution set is $(2, 4) \cup (6, \infty)$.

13. $(x^2 - 9)(x^2 - 4) > 0$

$(x + 3)(x - 3)(x + 2)(x - 2) > 0$

$(x + 3)(x - 3)(x + 2)(x - 2) = 0$

$x + 3 = 0$ or $x - 3 = 0$ or $x + 2 = 0$ or $x - 2 = 0$

$x = -3$ or $x = 3$ or $x = -2$ or $x = 2$

Region	Interval	Test Point
A	$(-\infty, -3)$	-4
B	$(-3, -2)$	$-\frac{5}{2}$
C	$(-2, 2)$	0
D	$(2, 3)$	$\frac{5}{2}$
E	$(3, \infty)$	4
$x = -4$	$[(-4)^2 - 9][(-4)^2 - 4] > 0$	True
$x = -\frac{5}{2}$	$\left[\left(-\frac{5}{2}\right)^2 - 9\right]\left[\left(-\frac{5}{2}\right)^2 - 4\right] > 0$	False
$x = 0$	$(0^2 - 9)(0^2 - 4) > 0$	True
$x = \frac{5}{2}$	$\left[\left(\frac{5}{2}\right)^2 - 9\right]\left[\left(\frac{5}{2}\right)^2 - 4\right] > 0$	False
$x = 4$	$(4^2 - 9)(4^2 - 4) > 0$	True

The solution set is $(-\infty, -3) \cup (-2, 2) \cup (3, \infty)$.

17. $6x^2 + 5x \le 4$

$6x^2 + 5x - 4 \le 0$

$(2x - 1)(3x + 4) = 0$

$2x - 1 = 0$ or $3x + 4 = 0$

$2x = 1$ or $3x = -4$

$x = \frac{1}{2}$ or $x = -\frac{4}{3}$

Region	Interval	Test Point
A	$\left(-\infty, -\frac{4}{3}\right)$	-2
B	$\left(-\frac{4}{3}, \frac{1}{2}\right)$	0
C	$\left(\frac{1}{2}, \infty\right)$	1
$x = -2$	$6(-2)^2 + 5(-2) \le 4$	False

$x = 0$	$6(0)^2 + 5(0) \le 4$	True
$x = 1$	$6(1)^2 + 5(1) \le 4$	False

The solution set is $\left[-\frac{4}{3}, \frac{1}{2}\right]$.

21. $(2x - 8)(x + 4)(x - 6) \le 0$

$(2x - 8)(x + 4)(x - 6) = 0$

$2x - 8 = 0$ or $x + 4 = 0$ or $x - 6 = 0$

$2x = 8$ or $x = -4$ or $x = 6$

$x = 4$ or $x = -4$ or $x = 6$

Region	Interval	Test Point
A	$(-\infty, -4)$	-5
B	$(-4, 4)$	0
C	$(4, 6)$	5
D	$(6, \infty)$	7

$x = -5$	$[2(-5) - 8](-5 + 4)(-5 - 6) \le 0$	True
$x = 0$	$[2(0) - 8](0 + 4)(0 - 6) \le 0$	False
$x = 5$	$[2(5) - 8](5 + 4)(5 - 6) \le 0$	True
$x = 7$	$[2(7) - 8](7 + 4)(7 - 6) \le 0$	False

The solution set is $(-\infty, -4] \cup [4, 6]$.

25. $\frac{5}{x + 1} > 0$

$x + 1 = 0$

$x = -1$

Region	Interval	Test Point
A	$(-\infty, -1)$	-2
B	$(-1, \infty)$	0

$x = -2$	$\frac{5}{-2 + 1} > 0$	False
$x = 0$	$\frac{5}{0 + 1} > 0$	True

The solution set is $(-1, \infty)$.

29. $\frac{x + 2}{x - 3} < 1$

$\frac{x + 2}{x - 3} - \frac{x - 3}{x - 3} < 0$

$\frac{x + 2 - x + 3}{x - 3} < 0$

$\frac{5}{x - 3} < 0$

$x = 3$

Region	Interval	Test Point
A	$(-\infty, 3)$	0
B	$(3, \infty)$	4

$x = 0$	$\frac{0 + 2}{0 - 3} < 1$	True
$x = 4$	$\frac{4 + 2}{4 - 3} < 1$	False

The solution set is $(-\infty, 3)$.

33. $\frac{x - 5}{x + 4} \ge 0$

$x - 5 = 0$ or $x + 4 = 0$

$x = 5$ or $x = -4$

Region	Interval	Test Point
A	$(-\infty, -4)$	-5
B	$(-4, 5)$	0
C	$(5, \infty)$	6

$x = -5$	$\frac{-5 - 5}{-5 + 4} \ge 0$	True
$x = 0$	$\frac{0 - 5}{0 + 4} \ge 0$	False
$x = 6$	$\frac{6 - 5}{6 + 4} \ge 0$	True

The solution set is $(-\infty, -4) \cup [5, \infty)$.

37. $\frac{-1}{x - 1} > -1$

$\frac{1}{x - 1} < 1$

$\frac{1}{x - 1} - 1 < 0$

$\frac{1 - (x - 1)}{x - 1} < 0$

$\frac{1 - x + 1}{x - 1} < 0$

$\frac{2 - x}{x - 1} < 0$

$2 - x = 0$ or $x - 1 = 0$

$2 = x$ or $x = 1$

Region	Interval	Test Point
A	$(-\infty, 1)$	0
B	$(1, 2)$	$\frac{3}{2}$
C	$(2, \infty)$	3

$x = 0$	$\frac{-1}{0 - 1} > -1$	True
$x = \frac{3}{2}$	$\frac{-1}{\frac{3}{2} - 1} > -1$	False
$x = 3$	$\frac{-1}{3 - 1} > -1$	True

The solution set is $(-\infty, 1) \cup (2, \infty)$.

41.

x	x^2	y
0	$0^2 = 0$	0
1	$1^2 = 1$	1
-1	$(-1)^2 = 1$	1
2	$(2)^2 = 4$	4
-2	$(-2)^2 = 4$	4

45. answers may vary

49. $P(x) = -2x^2 + 26x - 44$

$-2x^2 + 26x - 44 > 0$

$-2(x^2 - 13x + 22) > 0$

$-2(x - 11)(x - 2) > 0$

$x - 11 = 0$ or $x - 2 = 0$

$x = 11$ or $x = 2$

Region	Interval	Test Point
A	$(-\infty, 2)$	0
B	$(2, 11)$	3
C	$(11, \infty)$	12
$x = 0$	$-2(0)^2 + 26(0) - 44 > 0$	False
$x = 3$	$-2(3)^2 + 26(3) - 44 > 0$	True
$x = 12$	$-2(12)^2 + 26(12) - 44 > 0$	False

The solution set is $(2, 11)$.
The company makes a profit when x is between 2 and 11.

Exercise Set 10.5

1. $f(x) = x^2 - 1$

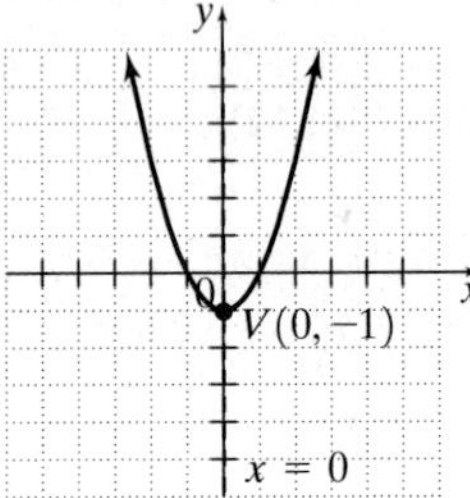

5. $h(x) = x^2 + 5$

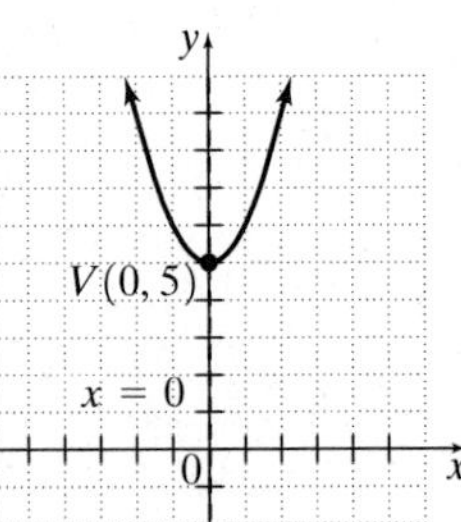

9. $g(x) = x^2 + 7$

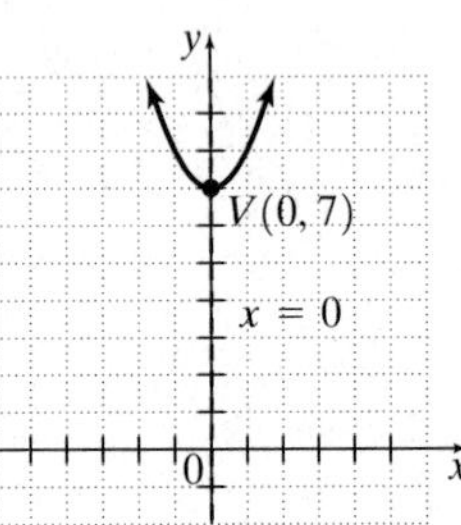

13. $f(x) = (x - 2)^2 + 5$

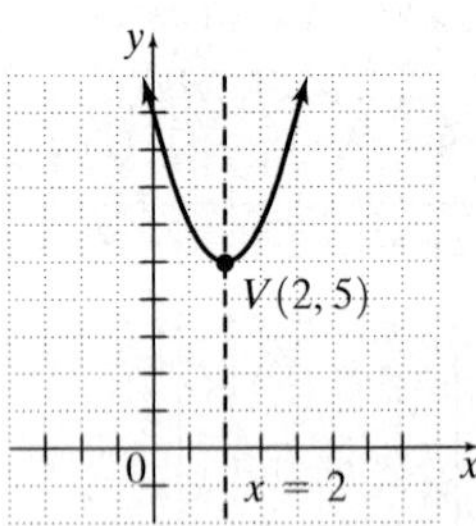

17. $g(x) = (x + 2)^2 - 5$

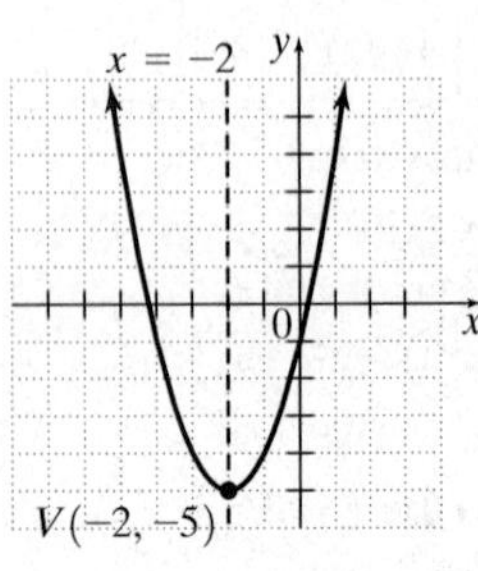

21. $g(x) = -x^2$

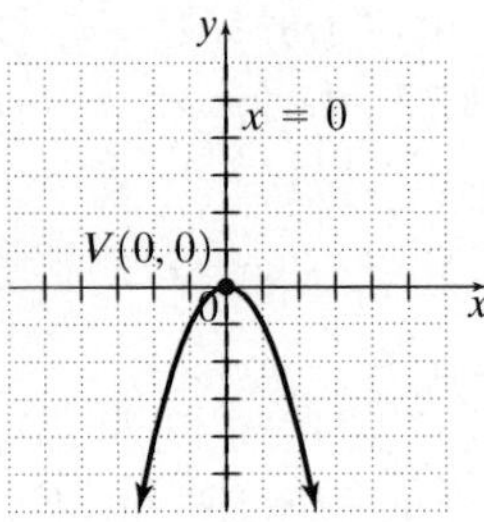

25. $H(x) = 2x^2$

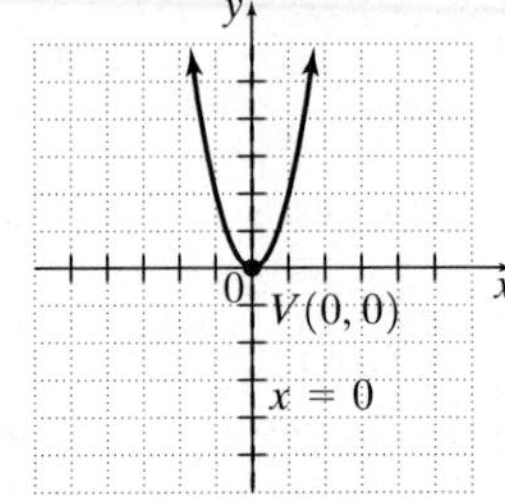

29. $f(x) = 10(x + 4)^2 - 6$

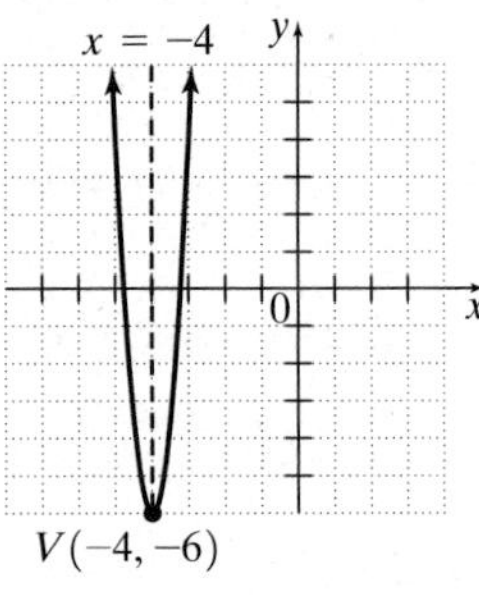

33. $H(x) = \frac{1}{2}(x - 6)^2 - 3$

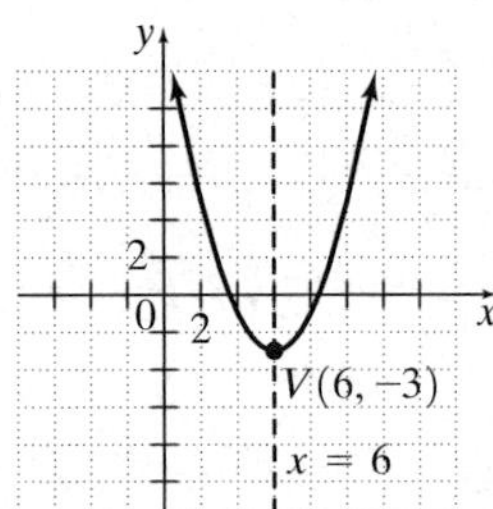

37. $F(x) = \left(x + \frac{1}{2}\right)^2 - 2$

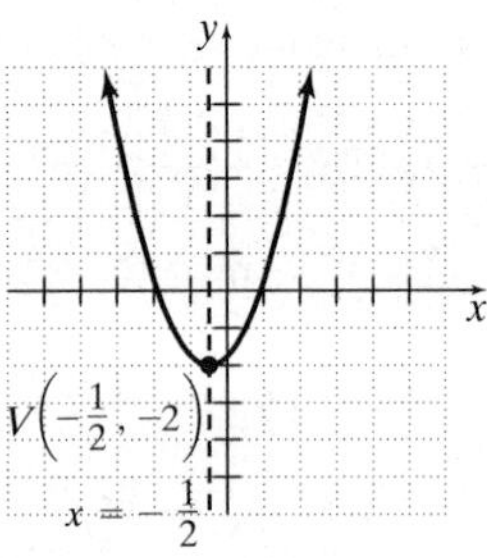

41. $x^2 + 8x$

$\left[\frac{1}{2}(8)\right]^2 = [4]^2 = 16$

$x^2 + 8x + 16$

45. $y^2 + y$

$\left[\frac{1}{2}(1)\right]^2 = \left[\frac{1}{2}\right]^2 = \frac{1}{4}$

$y^2 + y + \frac{1}{4}$

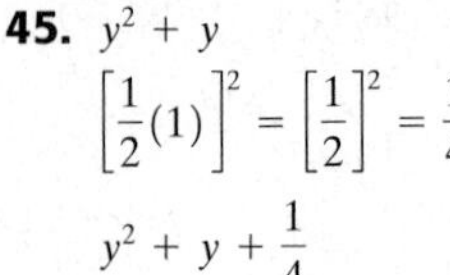

49. $f(x) = 5[(x - (-3))]^2 + 6$
$f(x) = 5(x + 3)^2 + 6$

53. $y = f(x - 3)$

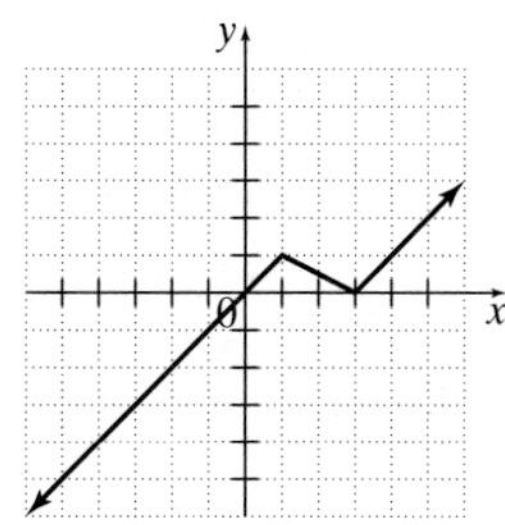

Exercise Set 10.6

1. $f(x) = x^2 + 8x + 7$
$\frac{-b}{2a} = \frac{-8}{2(1)} = -4$ and $f(-4) = (-4)^2 + 8(-4) + 7$
$f(-4) = 16 - 32 + 7 = -9$
Thus, $V(-4, -9)$.

5. $f(x) = 5x^2 - 10x + 3$
$\frac{-b}{2a} = \frac{-(-10)}{2(5)} = 1$ and $f(1) = 5(1)^2 - 10(1) + 3$
$f(1) = 5 - 10 + 3 = -2$
Thus, $V(1, -2)$.

9. $f(x) = x^2 - 4x + 3$
$\frac{-b}{2a} = \frac{-(-4)}{2(1)} = 2$
$f(2) = 2^2 - 4(2) + 3 = -1$
$V(2, -1)$
Graph D

13. $f(x) = x^2 + 4x - 5$
$\frac{-b}{2a} = \frac{-4}{2(1)} = -2$ and $f(-2) = (-2)^2 + 4(-2) - 5$
$f(-2) = 4 - 8 - 5 = -9$
Thus, $V(-2, -9)$.
The graph opens upward since $a > 0$.
$x^2 + 4x - 5 = 0$
$(x + 5)(x - 1) = 0$
$x + 5 = 0$ or $x - 1 = 0$
$x = -5$ or $x = 1$
The x-intercepts are $(-5, 0)$ and $(1, 0)$.
$f(0) = (0, -5)$ is the y-intercept.

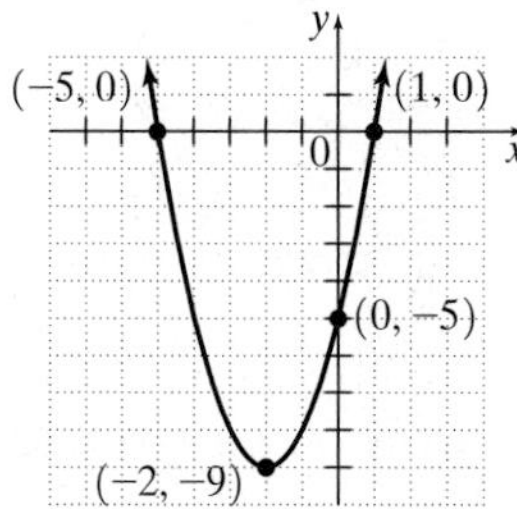

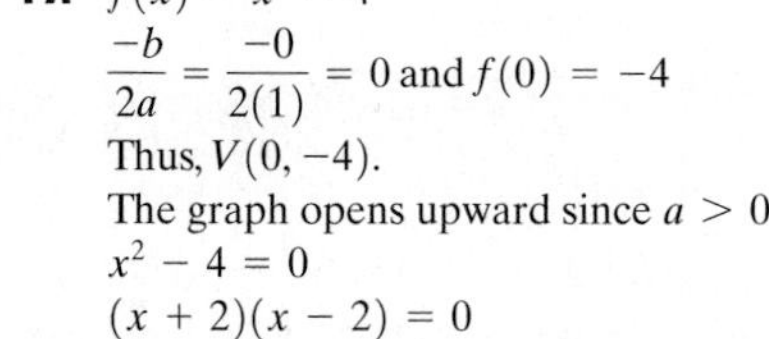

17. $f(x) = x^2 - 4$
$\frac{-b}{2a} = \frac{-0}{2(1)} = 0$ and $f(0) = -4$
Thus, $V(0, -4)$.
The graph opens upward since $a > 0$.
$x^2 - 4 = 0$
$(x + 2)(x - 2) = 0$
$x + 2 = 0$ or $x - 2 = 0$
$x = -2$ or $x = 2$
The x-intercepts are $(-2, 0)$ and $(2, 0)$.
$f(0) = -4$ so $(0, -4)$ is the y-intercept.

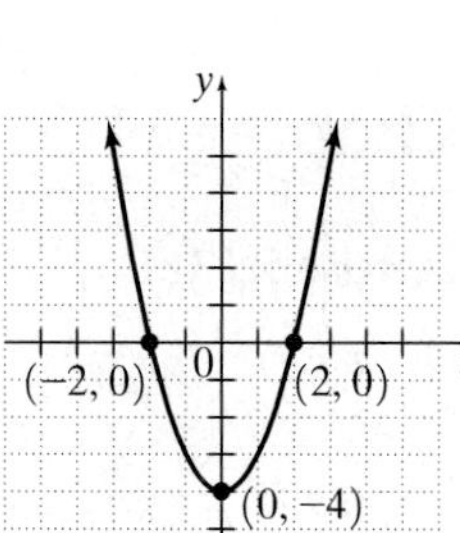

21. $f(x) = \frac{1}{2}x^2 + 4x + \frac{15}{2}$
$\frac{-b}{2a} = \frac{-4}{2\left(\frac{1}{2}\right)} = -4$
$f(-4) = \frac{1}{2}(-4)^2 + 4(-4) + \frac{15}{2} = -\frac{1}{2}$
$V\left(-4, -\frac{1}{2}\right)$
The graph opens upward since $a > 0$.
$\frac{1}{2}x^2 + 4x + \frac{15}{2} = 0$
$x^2 + 8x + 15 = 0$
$(x + 5)(x + 3) = 0$
$x + 5 = 0$ or $x + 3 = 0$
$x = -5$ or $x = -3$
$x = -5, x = -3$
The x-intercepts are $(-5, 0)$ and $(-3, 0)$.
$f(0) = \frac{15}{2}$ so $\left(0, \frac{15}{2}\right)$ is the y-intercept.

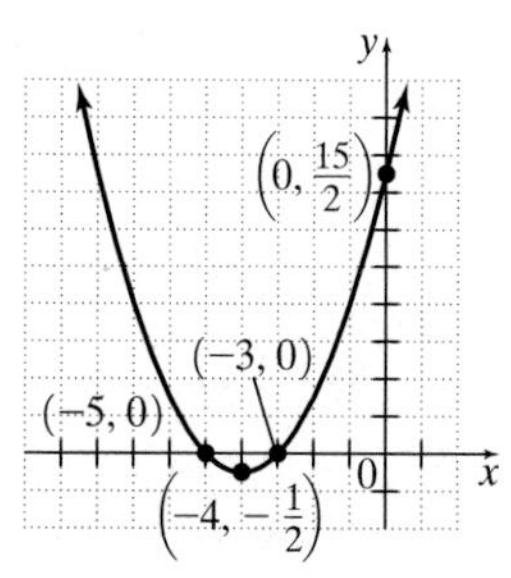

25. $f(x) = 2x^2 + 4x + 5$
$f(x) = 2(x^2 + 2x) + 5$
$f(x) = 2\left[x^2 + 2x + \left(\frac{2}{2}\right)^2\right] + 5 - 2$
$f(x) = 2(x + 1)^2 + 3$
Thus, $V(-1, 3)$.
The graph opens upward since $a > 0$.
$2(x + 1)^2 + 3 = 0$
$2(x + 1)^2 = -3$
Hence, there are no x-intercepts.
$f(0) = 2(0 + 1)^2 + 3 = 5$ so $(0, 5)$ is the y-intercept.

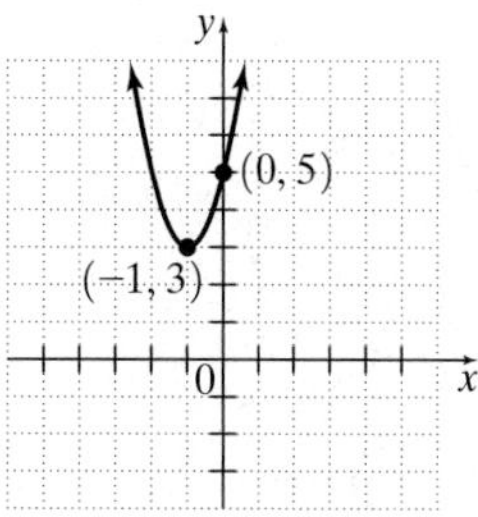

29. $h(t) = -16t^2 + 96t$
$\frac{-b}{2a} = \frac{-96}{2(-16)} = 3$ and $h(3) = -16(3)^2 + 96(3)$
$h(3) = -144 + 288 = 144$
The maximum height of the projectile is 144 ft.

33. Let x = one number.
$60 - x$ = other number
$f(x) = x(60 - x)$
$f(x) = 60x - x^2$
$f(x) = -x^2 + 60x$
$f(x) = -1(x^2 - 60x)$
$f(x) = -1(x^2 - 60x + 900) + 900$
$f(x) = -(x - 30)^2 + 900$
The maximum will occur at the vertex which is $(30, 900)$. The numbers are 30 and 30.

37. Let x = the width
$40 - x$ = the length
$f(x) = x(40 - x)$
$f(x) = 40x - x^2$
$f(x) = -x^2 + 40x$

$f(x) = -1(x^2 - 40x)$
$f(x) = -1(x^2 - 40x + 400) + 400$
$f(x) = -(x - 20)^2 + 400$
The maximum will occur at the vertex which is (20, 400). The width is 20 units and the length is 20 units.

41. $g(x) = x + 2$
The vertex is undefined.

45. $f(x) = 3(x - 4)^2 + 1$
The vertex is (4, 1)

49. $f(x) = \frac{1}{2}x^2 + 4x + \frac{15}{2}$

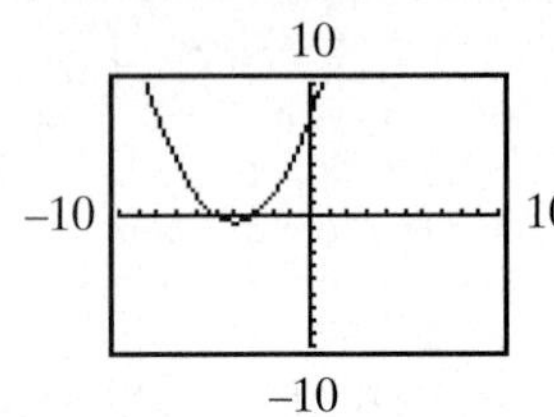

53. a. Because $a < 0$, the parabola will open downward and the function will have a maximum.

b. $x = \frac{-b}{2a} = \frac{-93}{2(-1)} = \frac{93}{2} = 46.5$

The year is $1990 + 46.5 = 2036.5$ which is in the year 2036.

c. $p(46.5) = -(46.5)^2 + 93(46.5) + 1128 = 3290.25$ thousands

Chapter 10 Test

1.
$$5x^2 - 2x = 7$$
$$5x^2 - 2x - 7 = 0$$
$$(5x - 7)(x + 1) = 0$$
$5x - 7 = 0$ or $x + 1 = 0$
$5x = 7$ or $x = -1$
$x = \frac{7}{5}$ or $x = -1$

The solution set is $\left\{\frac{7}{5}, -1\right\}$.

5.
$$7x^2 + 8x + 1 = 0$$
$$(7x + 1)(x + 1) = 0$$
$7x + 1 = 0$ or $x + 1 = 0$
$7x = -1$ or $x = -1$
$x = -\frac{1}{7}$ or $x = -1$

The solution set is $\left\{-\frac{1}{7}, -1\right\}$.

9.
$$x^6 + 1 = x^4 + x^2$$
$$x^6 - x^4 - x^2 + 1 = 0$$
$$x^4(x^2 - 1) - (x^2 - 1) = 0$$
$$(x^4 - 1)(x^2 - 1) = 0$$
$$(x^2 + 1)(x^2 - 1)(x^2 - 1) = 0$$
$$(x^2 + 1)(x^2 - 1)^2 = 0$$
$$(x^2 + 1)[(x + 1)(x - 1)]^2 = 0$$
$$(x^2 + 1)(x + 1)^2(x - 1)^2 = 0$$
$x^2 + 1 = 0$ or $(x + 1)^2 = 0$ or $(x - 1)^2 = 0$
$x^2 = -1$ or $x + 1 = 0$ or $x - 1 = 0$
$x = \pm\sqrt{-1} = \pm i$ or $x = -1$ or $x = 1$
The solution set is $\{1, -1, i, -i\}$.

13.
$$2x^2 - 7x > 15$$
$$2x^2 - 7x - 15 > 0$$
$$(2x + 3)(x - 5) > 0$$
$2x + 3 = 0$ or $x - 5 = 0$
$x = -\frac{3}{2}$ or $x = 5$

Region	Interval	Test Point
A	$\left(-\infty, -\frac{3}{2}\right)$	-2
B	$\left(-\frac{3}{2}, 5\right)$	0
C	$(5, \infty)$	6
$x = -2$	$2(-2)^2 - 7(-2) > 15$	True
$x = 0$	$2(0)^2 - 7(0) > 15$	False
$x = 6$	$2(6)^2 - 7(6) > 15$	True

The solution set is $\left\{-\infty, -\frac{3}{2}\right\} \cup (5, \infty)$.

17. $f(x) = 3x^2$
vertex: (0, 0)

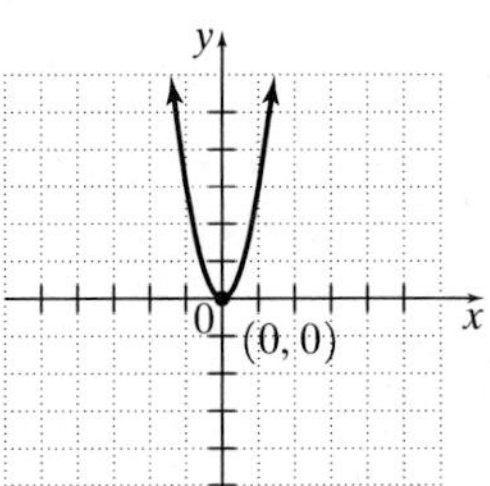

21. $c^2 = a^2 + b^2$
$(10)^2 = x^2 + (x - 4)^2$
$100 = x^2 + x^2 - 8x + 16$
$0 = 2x^2 - 8x - 84$
$0 = x^2 - 4x - 42$
$a = 1, b = -4, c = -42$
$x = \frac{-(-4) \pm \sqrt{(-4)^2 - 4(1)(-42)}}{2(1)}$
$x = \frac{4 \pm \sqrt{16 + 168}}{2}$
$x = \frac{4 \pm \sqrt{184}}{2}$
$x = \frac{4 \pm 2\sqrt{46}}{2}$
$x = 2 \pm \sqrt{46}$
Disregard the negative result.
The top of the ladder is $2 + \sqrt{46} \approx 8.8$ feet from the ground.

Chapter 11

Exercise Set 11.1

1. a. $(f + g)(x) = x - 7 + 2x + 1$
$= 3x - 6$

b. $(f - g)(x) = x - 7 - (2x + 1)$
$= x - 7 - 2x - 1$
$= -x - 8$

c. $(f \cdot g)(x) = (x - 7)(2x + 1)$
$= 2x^2 - 13x - 7$

d. $\left(\frac{f}{g}\right)(x) = \frac{x - 7}{2x + 1}$, where $x \neq -\frac{1}{2}$

5. a. $(f + g)(x) = \sqrt{x} + x + 5$

b. $(f \circ g)(x) = \sqrt{x} - x - 5$

c. $(f \circ g)(x) = \sqrt{x}(x + 5)$
$= x\sqrt{x} + 5\sqrt{x}$

d. $\left(\frac{f}{g}\right)(x) = \frac{\sqrt{x}}{x + 5}$, where $x \neq -5$

9. $(f \circ g)(2) = f(g(2))$
$= f(-4)$
$= (-4)^2 - 6(-4) + 2$
$= 16 + 24 + 2$
$= 42$

13. $(g \circ h)(0) = g(h(0))$
$= g(0)$
$= -2(0)$
$= 0$

17. $(f \circ g)(x) = f(g(x))$
$= f(x + 7)$
$= 2(x + 7) - 3$
$= 2x + 14 - 3$
$= 2x + 11$

$(g \circ f)(x) = g(f(x))$
$= g(2x - 3)$
$= (2x - 3) + 7$
$= 2x + 4$

21. $(f \circ g)(x) = f(g(x))$
$= f(-5x + 2)$
$= \sqrt{-5x + 2}$

$(g \circ f)(x) = g(f(x))$
$= g(\sqrt{x})$
$= -5\sqrt{x} + 2$

25. $F(x) = (h \circ f)(x)$
$= h(f(x))$
$= h(3x)$
$= (3x)^2 + 2$
$= 9x^2 + 2$

29. answers may vary

33. answers may vary

37. $x = 3y$
$\frac{x}{3} = \frac{3y}{3}$
$\frac{x}{3} = y$
$y = \frac{x}{3}$

41. $P(x) = R(x) - C(x)$

Exercise Set 11.2

1. $f = \{(-1, -1), (1, 1), (0, 2), (2, 0)\}$ is a one-to-one function.
$f^{-1} = \{(-1, -1), (1, 1), (2, 0), (0, 2)\}$

5. $f = \{(11, 12), (4, 3), (3, 4), (6, 6)\}$ is a one-to-one function.
$f^{-1} = \{(12, 11), (3, 4), (4, 3), (6, 6)\}$

9. This function is one-to-one.

Rank (input)	1	48	14	22	36
State (output)	CA	AK	IN	LA	NM

13. $f(x) = x^3 + 2$

a. $f(-1) = (-1)^3 + 2 = 1$

b. $f^{-1}(1) = -1$

17. The graph does not represent a one-to-one function because it does not pass the horizontal line test.

21. The graph does not represent a one-to-one function because it does not pass the horizontal line test.

25. $f(x) = 2x - 3$
$y = 2x - 3$
$x = 2y - 3$
$2y = x + 3$
$y = \frac{x + 3}{2}$
$f^{-1}(x) = \frac{x + 3}{2}$

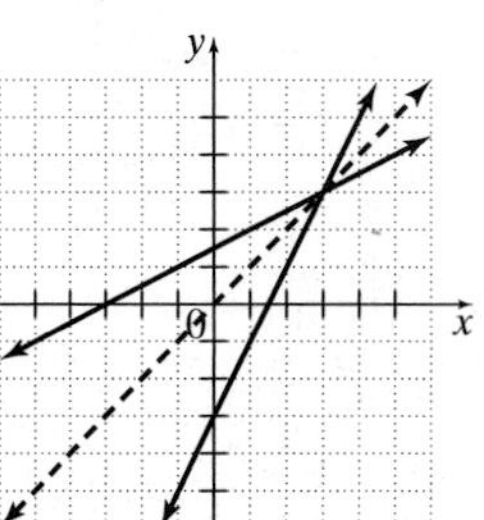

29.

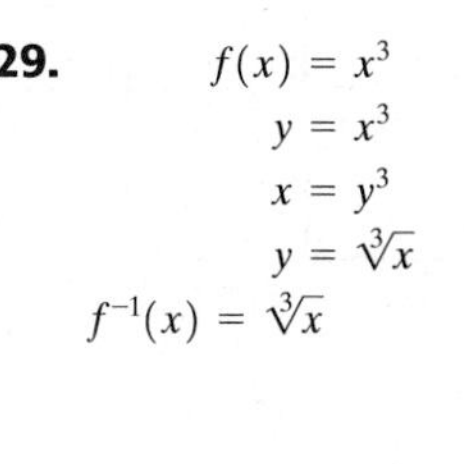

$f(x) = x^3$
$y = x^3$
$x = y^3$
$y = \sqrt[3]{x}$
$f^{-1}(x) = \sqrt[3]{x}$

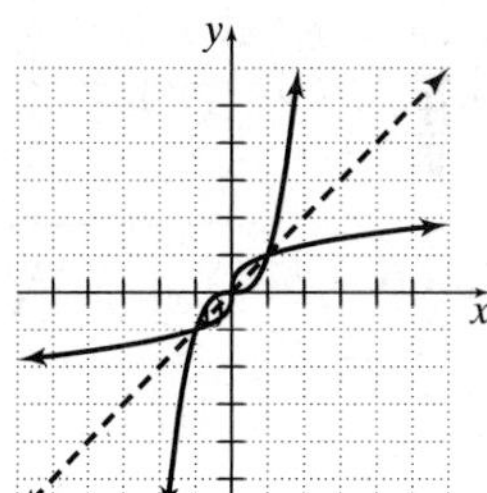

33. $f(x) = \sqrt[3]{x}$
$y = \sqrt[3]{x}$
$x = \sqrt[3]{y}$
$x^3 = y$
$f^{-1}(x) = x^3$

37. $f(x) = (x + 2)^3$
$y = (x + 2)^3$
$x = (y + 2)^3$
$\sqrt[3]{x} = y + 2$
$\sqrt[3]{x} - 2 = y$
$f^{-1}(x) = \sqrt[3]{x} - 2$

41.

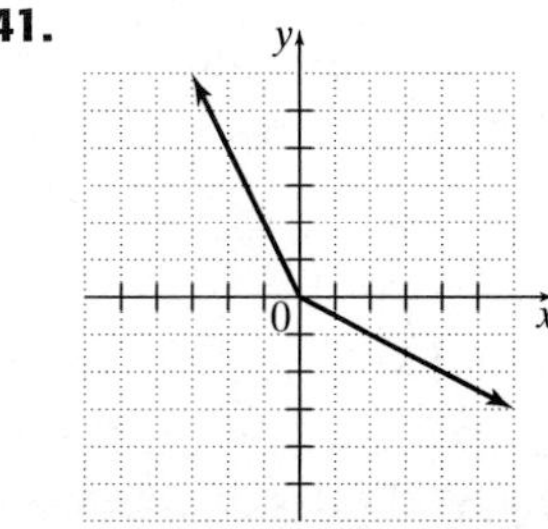

45. $16^{3/4} = (16^3)^{1/4} = (4096)^{1/4} = 8$

49. $f(x) = 3^x$
$f(2) = 3^2$
$= 9$

53. a. $\left(-2, \frac{1}{4}\right), \left(-1, \frac{1}{2}\right), (0, 1), (1, 2), (2, 5)$

b. $\left(\frac{1}{4}, -2\right), \left(\frac{1}{2}, -1\right), (1, 0), (2, 1), (5, 2)$

c.

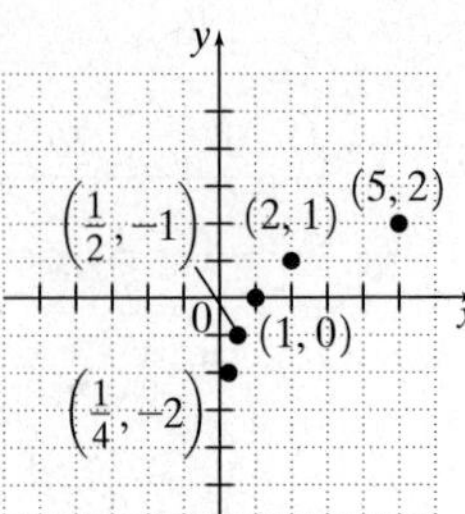

d.

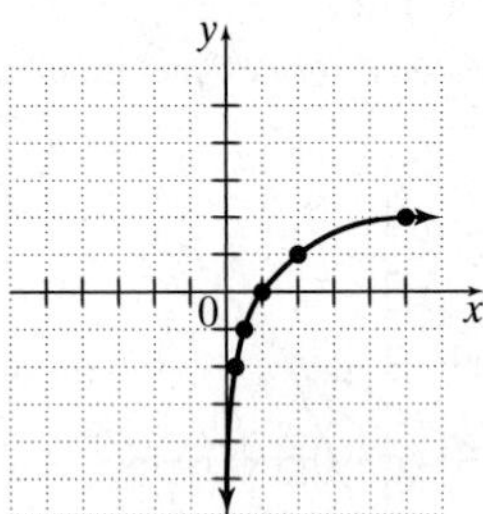

57. $f(x) = \sqrt[3]{x+3}$

$y = \sqrt[3]{x+3}$

$x = \sqrt[3]{y+3}$

$x^3 = y + 3$

$x^3 - 3 = y$

$f^{-1}(x) = x^3 - 3$

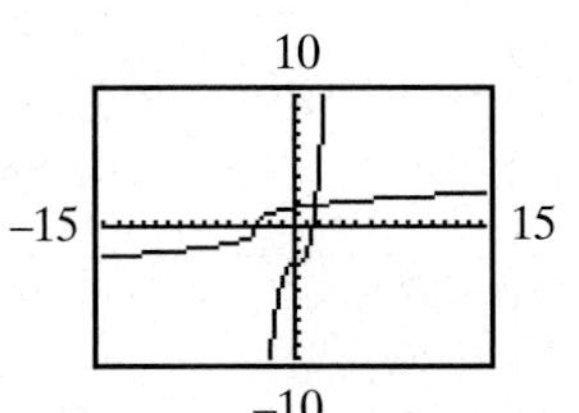

Exercise Set 11.3

1. $y = 4^x$

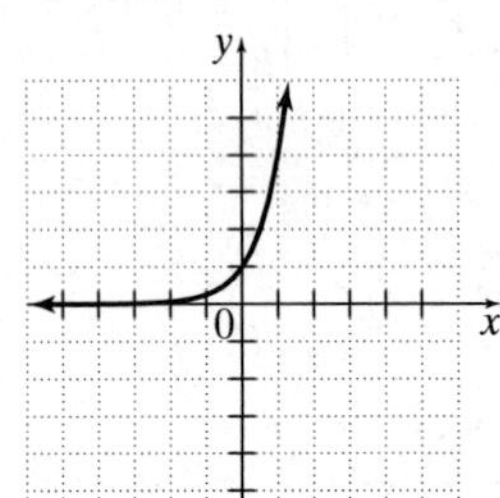

5. $y = \left(\frac{1}{4}\right)^x$

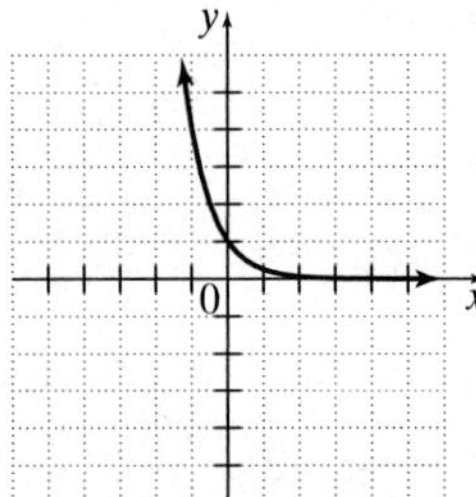

9. $y = -2^x$

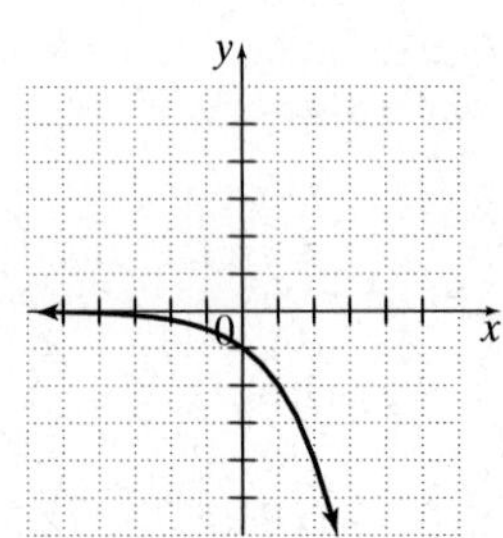

13. $y = -\left(\frac{1}{4}\right)^x$

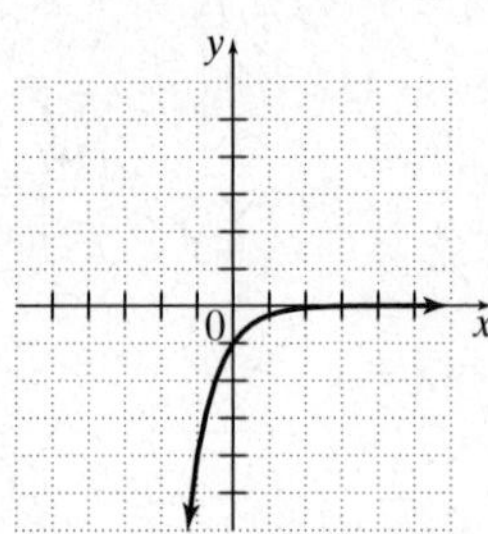

17. $3^x = 27$

$3^x = 3^3$

$x = 3$

The solution set is $\{3\}$.

21. $32^{2x-3} = 2$

$(2^5)^{2x-3} = 2^1$

$10x - 15 = 1$

$10x = 16$

$x = \frac{8}{5}$

The solution set is $\left\{\frac{8}{5}\right\}$.

25. $4^x = 8$

$(2^2)^x = 2^3$

$2^{2x} = 2^3$

$2x = 3$

$x = \frac{3}{2}$

The solution set is $\left\{\frac{3}{2}\right\}$.

29. $81^{x-1} = 27^{2x}$

$(3^4)^{x-1} = (3^3)^{2x}$

$3^{4x-4} = 3^{6x}$

$4x - 4 = 6x$

$-4 = 2x$

$x = -2$

The solution set is $\{-2\}$.

33. $y = 260(2.7)^{0.025t}, t = 10$

$y = 260(2.7)^{0.025(10)}$

$y \approx 333$

Approximately 333 bison will remain after 10 years.

37. $y = 25.759(1.277)^x$

$t = 2007 - 1994 = 13$

$y = 25.759(1.277)^{13} \approx 618.6$

There will be approximately 618.6 million cellular phone users in 2007.

41. $5x - 2 = 18$

$5x = 20$

$x = 4$

The solution set is $\{4\}$.

45. $f(x) = \left(\frac{1}{2}\right)^x$

$b = \frac{1}{2}, 0 < b < 1$

graph C

49. answers may vary

53. At $t = 120$, $y \approx 18.62$.

Approximately 18.62 lb of uranium will be available after 120 days.

Exercise Set 11.4

1. $\log_6 36 = 2$
$6^2 = 36$

5. $\log_{10} 1000 = 3$
$10^3 = 1000$

9. $\log_e \frac{1}{e^2} = -2$
$e^{-2} = \frac{1}{e^2}$

13. $2^4 = 16$
$\log_2 16 = 4$

17. $e^3 = x$
$\log_e x = 3$

21. $4^{-2} = \frac{1}{16}$
$\log_4 \frac{1}{16} = -2$

25. $\log_2 8 = 3$ since $2^3 = 8$

29. $\log_{25} 5 = \frac{1}{2}$ since $25^{1/2} = 5$

33. $\log_6 1 = 0$ since $6^0 = 1$

37. $\log_{10} 100 = 2$ since $10^2 = 100$

41. $\log_3 81 = 4$ since $3^4 = 81$

45. answers may vary

49. $\log_3 x = 4$
$x = 3^4$
$= 81$
The solution set is $\{81\}$.

53. $\log_2 \frac{1}{8} = x$
$2^x = \frac{1}{8}$
$2^x = 2^{-3}$
$x = -3$
The solution set is $\{-3\}$.

57. $\log_8 x = \frac{1}{3}$
$x = 8^{1/3}$
$= 2$
The solution set is $\{2\}$.

61. $\log_{3/4} x = 3$
$\left(\frac{3}{4}\right)^3 = x$
$x = \frac{3^3}{4^3}$
$= \frac{27}{64}$
The solution set is $\left\{\frac{27}{64}\right\}$.

65. $\log_5 5^3 = 3$

69. $\log_9 9 = 1$

73. $f(x) = \log_{1/4} x$ or $y = \log_{1/4} x$
$y = 0$:
$0 = \log_{1/4} x$
$x = \left(\frac{1}{4}\right)^0 = 1$ so $(1, 0)$ is the x-intercept.
$x = 0$:
$y = \log_{1/4} 0$ which is not defined. No y-intercept exists.

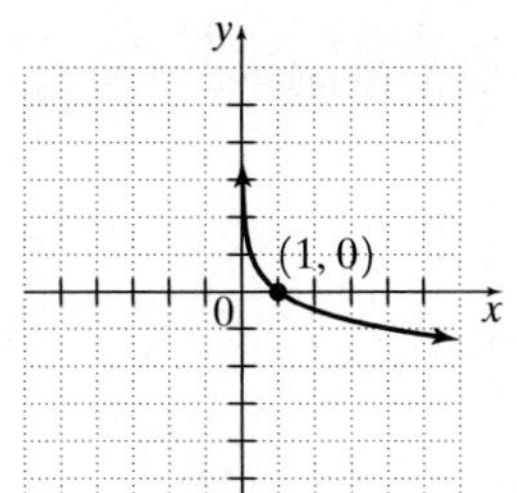

77. $f(x) = \log_{1/6} x$ or $y = \log_{1/6} x$
$x = 0$:
$y = \log_{1/6} 0$ is not defined so there is no y-intercept.
$y = 0$:
$0 = \log_{1/6} x$
$x = \left(\frac{1}{6}\right)^0 = 1$ so $(1, 0)$ is the x-intercept.

81. $\frac{x^2 - 8x + 16}{2x - 8} = \frac{(x-4)(x-4)}{2(x-4)} = \frac{x-4}{2}$

85. $y = \left(\frac{1}{3}\right)^x$; $y = \log_{1/3} x$

$x = 0$: $y = \left(\frac{1}{3}\right)^0 = 1$ so $(0, 1)$ is the y-intercept of $y = \left(\frac{1}{3}\right)^x$, and hence the x-intercept of $y = \log_{1/3} x$.

$y = 0$: $0 = \left(\frac{1}{3}\right)^x$ has no solution so $y = \left(\frac{1}{3}\right)^x$ has no x-intercept: hence $y = \log_{1/3} x$ has no y-intercept.

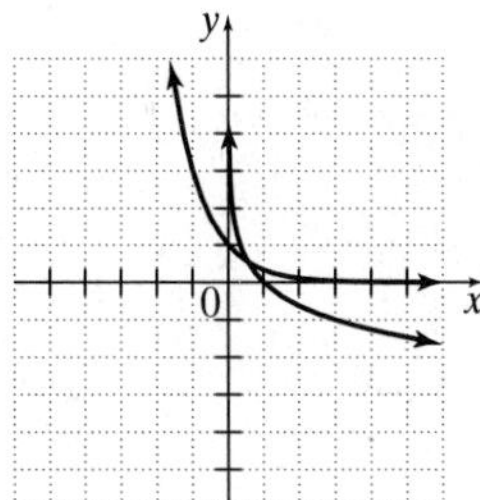

89. $\log_3 10$ is between 2 and 3 because $3^2 = 9$ and $3^3 = 27$

Exercise Set 11.5

1. $\log_5 2 + \log_5 7 = \log_5 (2 \cdot 7) = \log_5 14$

5. $\log_{10} 5 + \log_{10} 2 + \log_{10}(x^2 + 2)$
$= \log_{10}[5 \cdot 2(x^2 + 2)]$
$= \log_{10}(10x^2 + 20)$

9. $\log_2 x - \log_2 y = \log_2\left(\frac{x}{y}\right)$

13. $\log_3 x^2 = 2 \log_3 x$

17. $\log_5 \sqrt{y} = \log_5 y^{1/2}$
$= \frac{1}{2}\log_5 y$

21. $3\log_5 x + 6\log_5 z = \log_5 x^3 + \log_5 z^6$
$= \log_5 (x^3 z^6)$

25. $\log_4 2 + \log_4 10 - \log_4 5 = \log_4 2 \cdot 10 - \log_4 5$
$= \log_4\left(\frac{20}{5}\right)$
$= \log_4 4$
$= 1$

29. $3\log_4 2 + \log_4 6 = \log_4 2^3 + \log_4 6$
$= \log_4 8 + \log_4 6$
$= \log_4(8 \cdot 6)$
$= \log_4 48$

33. $2\log_8 x - \frac{2}{3}\log_8 x + 4\log_8 x = \left(2 - \frac{2}{3} + 4\right)\log_8 x$
$= \frac{16}{3}\log_x x$
$= \log_8 x^{16/3}$

37. $\log_2\left(\frac{x^3}{y}\right) = \log_2 x^3 - \log_2 y$
$= 3\log_2 x - \log_2 y$

41. $\log_7\left(\frac{5x}{4}\right) = \log_7 5x - \log_7 4$
$= \log_7 5 + \log_7 x - \log_7 4$

45. $\log_6 \frac{x^2}{x+3} = \log_6 x^2 - \log_6(x+3)$
$= 2\log_6 x - \log_6(x+3)$

49. $\log_b 15 = \log_b(5 \cdot 3)$
$= \log_b 5 + \log_b 3 = 0.7 + 0.5 = 1.2$

53. $\log_b 8 = \log_b 2^3 = 3\log_b 2 = 3(0.43) = 1.29$

57. $\log_b \sqrt{\frac{2}{3}} = \log_b\left(\frac{2}{3}\right)^{1/2}$
$= \frac{1}{2}\log_b \frac{2}{3}$
$= \frac{1}{2}(\log_b 2 - \log_b 3)$
$= \frac{1}{2}(0.43 - 0.68)$
$= \frac{1}{2}(-0.25)$
$= -0.125$

61. $\log_{10}\frac{1}{10} = x$
$10^x = \frac{1}{10}$
$10^{-1} = \frac{1}{10}$
$\log_{10}\frac{1}{10} = -1$

65. $\log_3(x + y) = \log_3 x + \log_3 y$
false

69. $(\log_3 6)(\log_3 4) = \log_3 24$
false

Exercise Set 11.6

1. $\log 8 \approx 0.9031$

5. $\ln 2 \approx 0.6931$

9. $\log 12.6 \approx 1.1004$

13. $\log 41.5 \approx 1.6180$

17. $\log 100 = \log 10^2 = 2$

21. $\ln e^2 = 2$

25. $\log 10^3 = 3$

29. $\log 0.0001 = \log 10^{-4}$
$= -4$

33. $\log x = 1.3$
$x = 10^{1.3}$
≈ 19.9526

37. $\ln x = 1.4$
$x = e^{1.4}$
≈ 4.0552

41. $\log x = 2.3$
$x = 10^{2.3}$
≈ 199.5262

45. $\log(2x + 1) = -0.5$
$2x + 1 = 10^{-0.5}$
$2x = 10^{-0.5} - 1$
$x = \frac{10^{-0.5} - 1}{2}$
≈ -0.3419

49. $A = Pe^{rt}, t = 12, P = 1400$, and $r = 0.08$
$A = 1400e^{(0.08)12} = 1400e^{0.96} \approx 3656.38$
Dana has $3656.38 after 12 years.

53. $\log_2 3 = \frac{\log 3}{\log 2} \approx 1.5850$

57. $\log_4 9 = \frac{\log 9}{\log 4} \approx 1.5850$

61. $\log_8 6 = \frac{\log 6}{\log 8} \approx 0.8617$

65. $2x + 3y = 6x$
$3y = 4x$
$x = \frac{3y}{4}$

69. $f(x) = e^x$

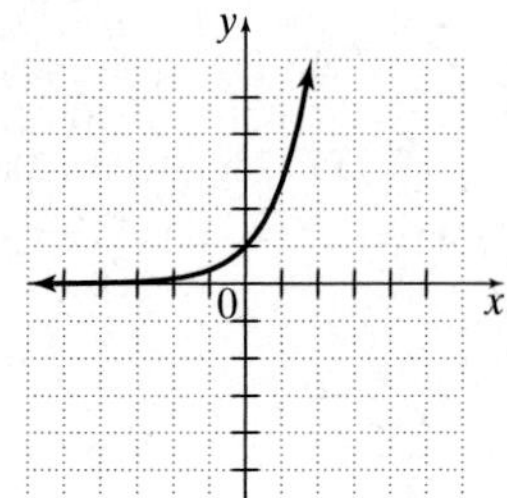

73. answers may vary

77. $R = \log\left(\frac{a}{T}\right) + B$
$= \log\left(\frac{400}{2.6}\right) + 3.1$
≈ 5.3

Exercise Set 11.7

1. $3^x = 6$
$x = \log_3 6$
$= \frac{\log 6}{\log 3} \approx 1.6309$
The solution set is $\left\{\frac{\log 6}{\log 3}\right\}$ or approximately $\{1.6309\}$.

5. $2^{x-3} = 5$
$x - 3 = \log_2 5$
$x = 3 + \log_2 5$
$= 3 + \frac{\log 5}{\log 2}$
$= 5.3219$
The solution set is $\left\{3 + \frac{\log 5}{\log 2}\right\}$ or approximately $\{5.3219\}$.

9. $4^{x+7} = 3$
$x + 7 = \log_4 3$
$x = \log_4 3 - 7$
$= \frac{\log 3}{\log 4} - 7$
≈ -6.2075
The solution set is $\left\{\frac{\log 3}{\log 4} - 7\right\}$ or approximately $\{-6.2075\}$.

13. $e^{6x} = 5$
$6x = \ln 5$
$x = \frac{1}{6}\ln 5$
≈ 0.2682
The solution set is $\left\{\frac{1}{6}\ln 5\right\}$ or approximately $\{0.2682\}$.

17. $\log_4 2 + \log_4 x = 0$

$\log_4(2x) = 0$

$2x = 4^0$

$2x = 1$

$x = \frac{1}{2}$

The solution set is $\left\{\frac{1}{2}\right\}$.

21. $\log_4 x + \log_4(x+6) = 2$

$\log_4 x(x+6) = 2$

$x(x+6) = 4^2$

$x^2 + 6x = 16$

$x^2 + 6x - 16 = 0$

$(x+8)(x-2) = 0$

$x + 8 = 0$ or $x - 2 = 0$

$x = -8$ or $x = 2$

We discard -8 as extraneous. The solution set is $\{2\}$.

25. $\log_3(x-2) = 2$

$x - 2 = 3^2$

$x - 2 = 9$

$x = 11$

The solution set is $\{11\}$.

29. $\log_2 x + \log_2(3x+1) = 1$

$\log_2 x(3x+1) = 1$

$x(3x+1) = 2^1$

$3x^2 + x = 2$

$3x^2 + x - 2 = 0$

$(3x-2)(x+1) = 0$

$3x - 2 = 0$ or $x + 1 = 0$

$3x = 2$ or $x = -1$

$x = \frac{2}{3}$

Discard -1 as extraneous.

The solution set is $\left\{\frac{2}{3}\right\}$.

33. $y = y_0 e^{0.027t}$

$y_0 = 5{,}700{,}000$ and $t = 7$

$y = 5{,}700{,}000e^{0.027(7)}$

$\approx 6{,}885{,}833$

In 2008, there will be approximately 6,885,833 people in Paraguay.

37. $A = P\left(1 + \frac{r}{n}\right)^{nt}, P = 600.$

$A = 2(600) = 1200, r = 0.07$, and $n = 12$

$1200 = 600\left(1 + \frac{0.07}{12}\right)^{12t}$

$2 \approx (1.00583)^{12t}$

$12t = \log_{1.00583}(2)$

$t = \frac{1}{12}\log_{1.00583}(2)$

$= \left(\frac{1}{12}\right)\frac{\log 2}{\log(1.00583)} \approx 9.9$

It would take approximately 9.9 years for the \$600 to double.

41. $A = P\left(1 + \frac{r}{n}\right)^{nt}, P = 1000.$

$A = 2(1000) = 2000, r = 0.08$, and $n = 2$.

$2000 = 1000\left(1 + \frac{0.08}{2}\right)^{2t}$

$2 = (1.04)^{2t}$

$2t = \log_{1.04} 2 = \frac{\log 2}{\log 1.04}$

$t = \left(\frac{1}{2}\right)\frac{\log 2}{\log 1.04} \approx 8.8$

It would take approximately 8.8 years for the \$1000 to double.

45. $P = 14.7e^{-0.21x}, x = 1$

$= 14.7e^{-0.21(1)} = 14.7e^{-0.21} \approx 11.9$

The average atmospheric pressure of Denver is approximately 11.9 lb/in^2.

49. $t = \frac{1}{c}\ln\left(\frac{A}{A-N}\right)$

$t = \frac{1}{0.09}\ln\left(\frac{75}{75-50}\right)$

$t = \frac{1}{0.09}\ln(3)$

$t \approx 12.21$

It will take about 12 weeks.

53. $\frac{x^2 - y + 2z}{3x} = \frac{(-2)^2 - (0) + 2(3)}{3(-2)}$

$= \frac{4+6}{-6}$

$= \frac{10}{-6}$

$= -\frac{5}{3}$

57. $y = y_0 e^{kt}$

$5{,}130{,}632 = 3{,}665{,}228e^{k \cdot 10}$

$k = \frac{1}{10}\ln\frac{5{,}130{,}632}{3{,}665{,}228}$

$k \approx 0.034$

The growth rate is about 3.4%.

61. $\{6.93\}$

Chapter 11 Test

1. $(f \cdot g)(x) = f(x) \cdot g(x)$

$= x(2x - 3)$

$= 2x^2 - 3x$

5. $(g \circ h)(x) = g(h(x))$

$= g(x^2 - 6x + 5)$

$= x^2 - 6x + 5 - 7$

$= x^2 - 6x - 2$

9. $y = 6 - 2x$

$f(x) = -2x + 6$ so the function is one-to-one

$f^{-1}(x) = \frac{x-6}{-2}$ or $f^{-1}(x) = \frac{-x+6}{2}$

13. $\log_5 x + 3\log_5 x - \log_5(x+1)$

$= 4\log_5 x - \log_5(x+1)$

$= \log_5 x^4 - \log_5(x+1)$

$= \log_5 \frac{x^4}{x+1}$

17. $8^{x-1} = 8^{-2}$

$x - 1 = -2$

$x = -1$

The solution set is $\{-1\}$.

21. $\log_8(3x-2) = 2$

$3x - 2 = 8^2$

$3x - 2 = 64$

$3x = 66$

$x = \frac{66}{3}$

$= 22$

The solution set is $\{22\}$.

25. $y = \left(\frac{1}{2}\right)^x + 1$

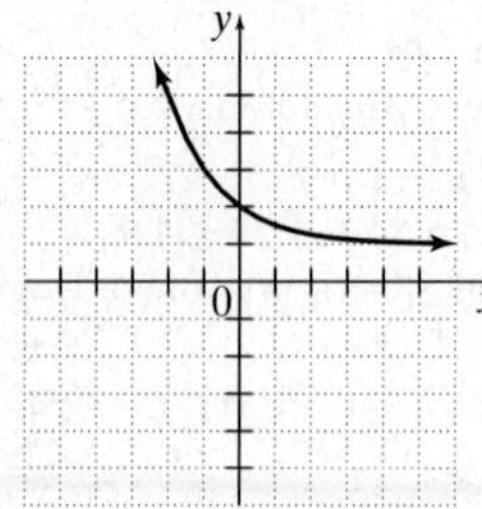

29. $y = y_0e^{kt}$, $y_0 = 57{,}000$, $k = 0.026$, and $t = 5$
$y = 57{,}000e^{0.026(5)}$
$= 57{,}000e^{0.13} \approx 64{,}913$
There will be approximately 64,913 prairie dogs 5 years from now.

Chapter 12

Exercise Set 12.1

1. $x = 3y^2$
$x = 3(y - 0)^2$
$V(0,0)$

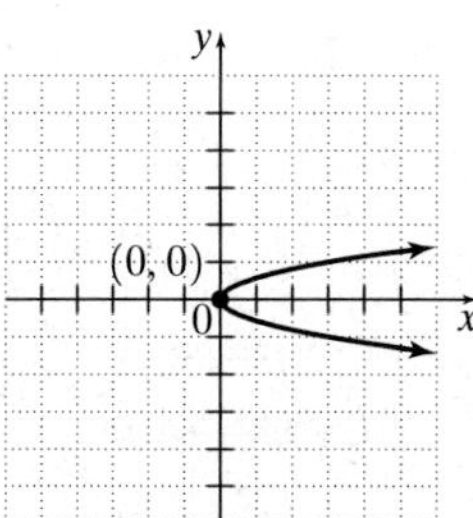

5. $y = 3(x - 1)^2 + 5$
$V(1,5)$

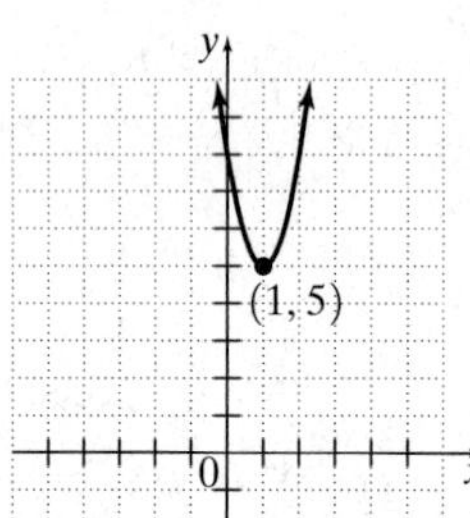

9. $y = x^2 + 10x + 20$
$y = x^2 + 10x + 25 + 20 - 25$
$y = (x + 5)^2 - 5$
$V(-5,-5)$

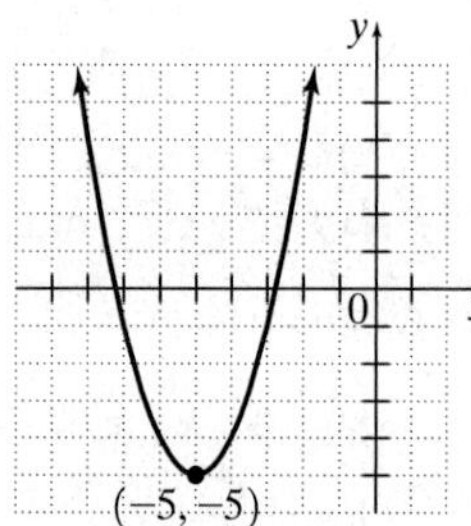

13. $(5,1), (8,5)$
$d = \sqrt{(8 - 5)^2 + (5 - 1)^2}$
$d = \sqrt{9 + 16}$
$d = \sqrt{25}$
$d = 5$ units

17. $(-9,4), (-8,1)$
$d = \sqrt{[-8 - (-9)]^2 + (1 - 4)^2}$
$d = \sqrt{(-8 + 9)^2 + (-3)^2}$
$d = \sqrt{1 + 9}$
$d = \sqrt{10} \approx 3.162$ units

21. $(1.7,-3.6), (-8.6,5.7)$
$d = \sqrt{(-8.6 - 1.7)^2 + [5.7 - (-3.6)]^2}$
$d = \sqrt{(-10.3)^2 + (9.3)^2}$
$d = \sqrt{192.58} \approx 13.877$ units

25. $(6,-8), (2,4)$
$\left(\frac{6 + 2}{2}, \frac{-8 + 4}{2}\right)$
$(4,-2)$

29. $(7,3), (-1,-3)$
$\left(\frac{7 + (-1)}{2}, \frac{3 + (-3)}{2}\right)$
$(3,0)$

33. $(\sqrt{2}, 3\sqrt{5}), (\sqrt{2}, -2\sqrt{5})$
$\left(\frac{\sqrt{2} + \sqrt{2}}{2}, \frac{3\sqrt{5} - 2\sqrt{5}}{2}\right)$
$\left(\sqrt{2}, \frac{\sqrt{5}}{2}\right)$

37. $x^2 + y^2 = 9$
$(x - 0)^2 + (y - 0)^2 = 3^2$
$C(0,0)$ and $r = 3$

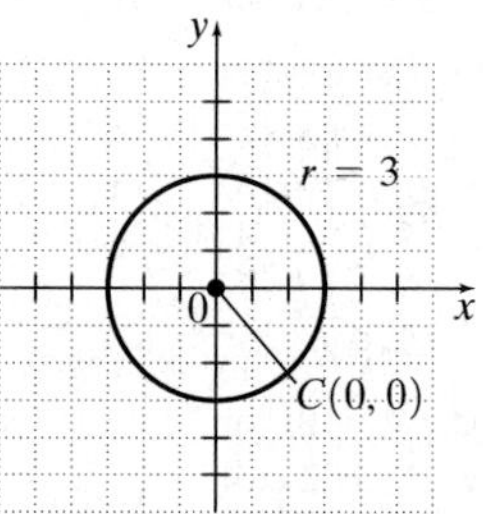

41. $(x - 5)^2 + (y + 2)^2 = 1$
$(x - 5)^2 + (y + 2)^2 = 1^2$
$C(5,-2)$ and $r = 1$

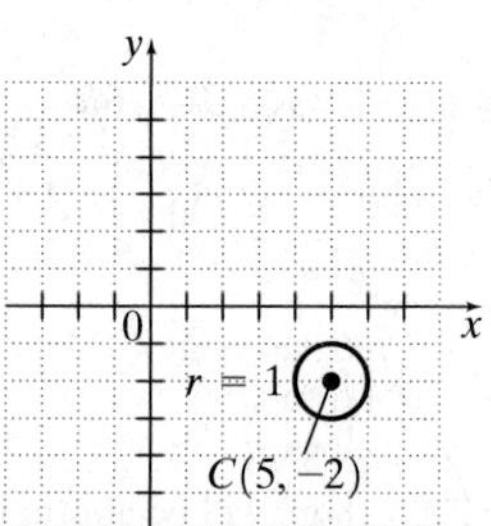

45. $x^2 + y^2 + 2x - 4y = 4$
$x^2 + 2x + 1 + y^2 - 4y + 4 = 4 + 1 + 4$
$(x + 1)^2 + (y - 2)^2 = 9$
$(x + 1)^2 + (y - 2)^2 = 3^2$
$C(-1,2)$ and $r = 3$

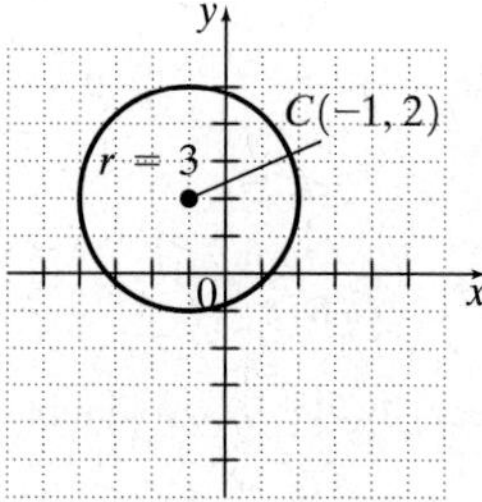

49. $C(2,3)$; $r = 6$
$(x - 2)^2 + (y - 3)^2 = 6^2$
$(x - 2)^2 + (y - 3)^2 = 36$

53. $C(-5,4)$; $r = 3\sqrt{5}$
$[x - (-5)]^2 + (y - 4)^2 = (3\sqrt{5})^2$
$(x + 5)^2 + (y - 4)^2 = 45$

57. $y = 3$

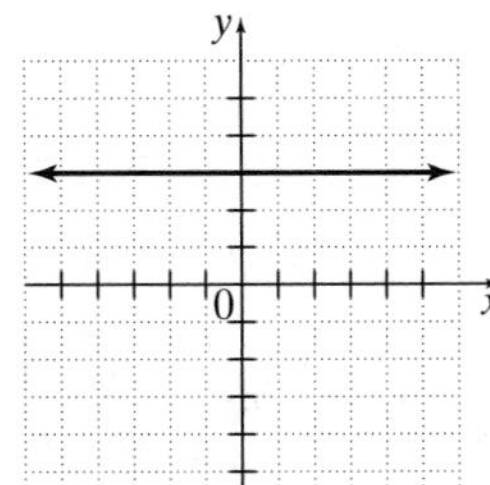

61. $\dfrac{4\sqrt{7}}{\sqrt{6}} = \dfrac{4\sqrt{7}}{\sqrt{6}} \cdot \dfrac{\sqrt{6}}{\sqrt{6}}$
$= \dfrac{4\sqrt{42}}{6}$
$= \dfrac{2\sqrt{42}}{3}$

65. answers may vary

69. The equation of the arch is $y = ax^2 + k$ passing through the points $(0, 40)$ and $(50, 0)$.
Substitute 0 for x and 40 for y:

$40 = a(0)^2 + k$

$40 = k$

The equation is $y = ax^2 + 40$.
Substitute 50 for x and 0 for y:

$0 = a(50)^2 + 40$

$0 = 2500a + 40$

$-40 = 2500a$

$-\dfrac{40}{2500} = a$ or $a = -\dfrac{2}{125}$

Thus, $y = -\dfrac{2}{125}x^2 + 40$.

Exercise Set 12.2

1. $\dfrac{x^2}{4} + \dfrac{y^2}{25} = 1$
$\dfrac{x^2}{2^2} + \dfrac{y^2}{5^2} = 1$
$C(0, 0)$
x-intercepts: $(-2, 0), (2, 0)$
y-intercepts: $(0, -5), (0, 5)$

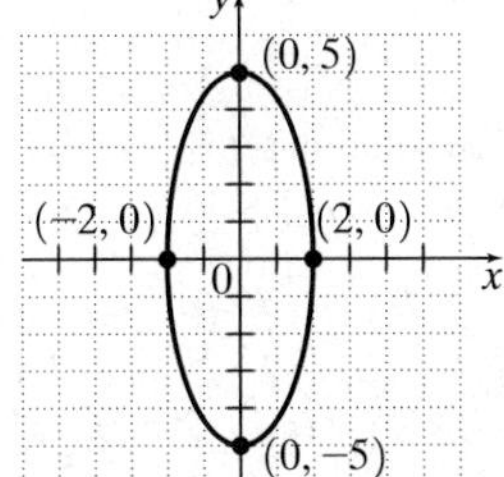

5. $9x^2 + 4y^2 = 36$
$\dfrac{x^2}{4} + \dfrac{y^2}{9} = 1$
$\dfrac{x^2}{2^2} + \dfrac{y^2}{3^2} = 1$
$C(0, 0)$
x-intercepts: $(-2, 0), (2, 0)$
y-intercepts: $(0, -3), (0, 3)$

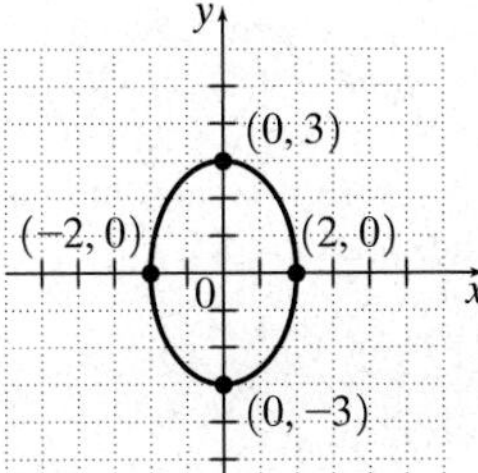

9. $\dfrac{(x+1)^2}{36} + \dfrac{(y-2)^2}{49} = 1$
$\dfrac{(x+1)^2}{6^2} + \dfrac{(y-2)^2}{7^2} = 1$
$C(-1, 2)$

other points:
$(-1 - 6, 2)$ or $(-7, 2)$
$(-1 + 6, 2)$ or $(5, 2)$
$(-1, 2 - 7)$ or $(-1, -5)$
$(-1, 2 + 7)$ or $(-1, 9)$

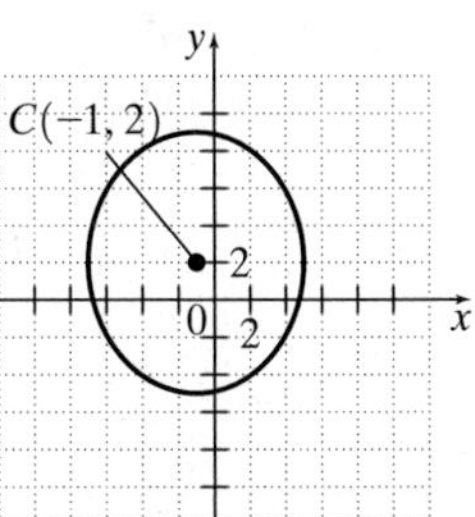

13. $\dfrac{x^2}{4} - \dfrac{y^2}{9} = 1$
$\dfrac{x^2}{2^2} - \dfrac{y^2}{3^2} = 1$
$a = 2, b = 3$

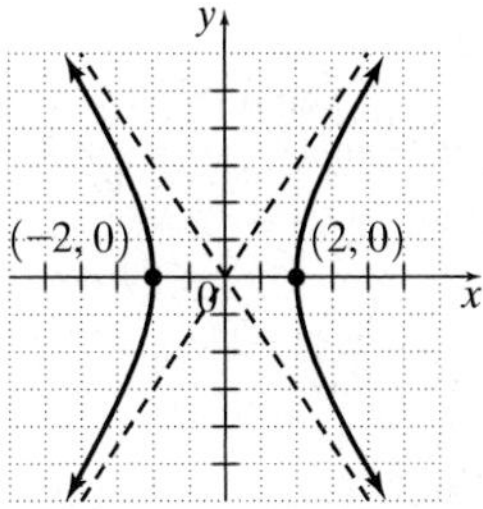

17. $x^2 - 4y^2 = 16$
$\dfrac{x^2}{16} - \dfrac{y^2}{4} = 1$
$\dfrac{x^2}{4^2} - \dfrac{y^2}{2^2} = 1$
$a = 4, b = 2$

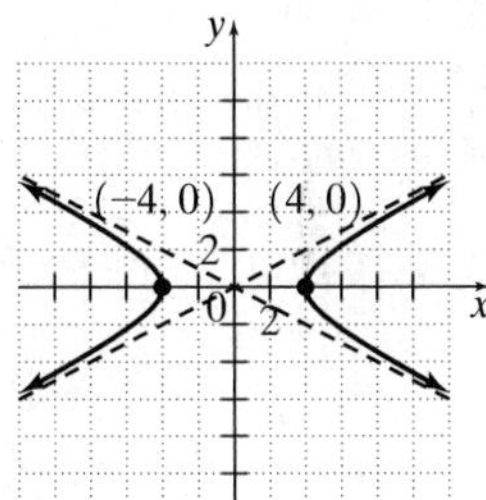

21. $(2x^3)(-4x^2) = 2(-4)x^{3+2}$
$= -8x^5$

25. $x^2 + y^2 = 25$
$\dfrac{x^2}{25} + \dfrac{y^2}{25} = \dfrac{25}{25}$
$\dfrac{x^2}{25} + \dfrac{y^2}{25} = 1$
It resembles an ellipse.
An ellipse is a circle when $a = b$.

29. $A: d = 6$
$B: d = 2$
$C: d = 5$
$D: d = 5$
$E: d = 9$
$F: d = 6$
$G: d = 4$
$H: d = 12$

33. They are greater than 1.

Exercise Set 12.3

1. $f(x) = |x| + 3$
Shift $y = |x|$ up 3 units.

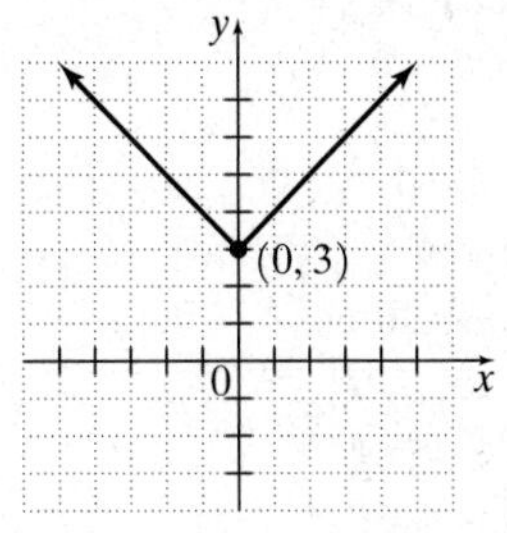

5. $f(x) = |x - 4|$
Shift $y = |x|$ to the right 4 units.

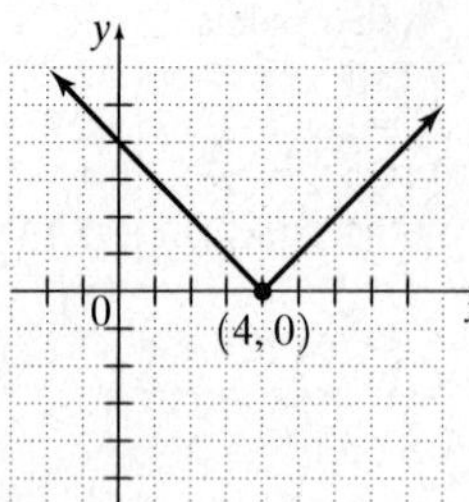

9. $f(x) = \sqrt{x - 2} + 3$
Shift $y = \sqrt{x}$ to the right 2 units and up 3 units.

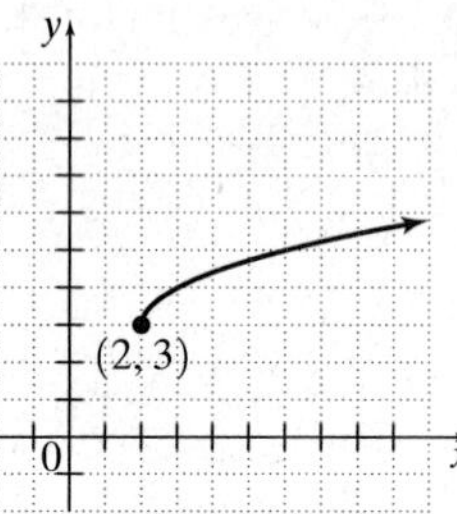

13. $f(x) = \sqrt{x + 1} + 1$
Shift $y = \sqrt{x}$ to the left one unit and up one unit.

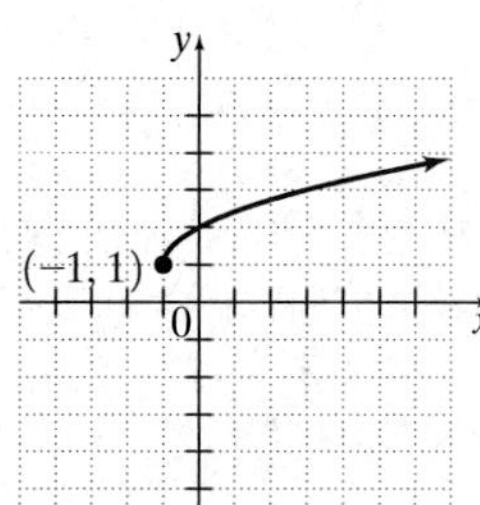

17. $x + y = 6$

$$\begin{array}{r} x - y = 10 \\ \hline 2x = 16 \\ x = 8 \end{array}$$

Replace x with 8 in the first equation.

$8 + y = 6$

$y = -2$

The solution is $(8, -2)$.

21. $f(x) = x^3$

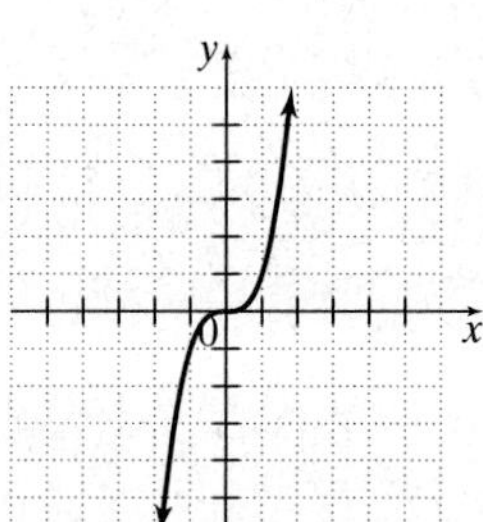

25. $f(x) = 5\sqrt{x - 20} + 1$

For domain, use

$x - 20 \geq 0$

$x \geq 20$

$\{x | x \geq 20\}$

29. $g(x) = 9 - \sqrt{x + 103}$

For domain, use

$x + 103 \geq 0$

$x \geq -103$

$\{x | x \geq -103\}$

Exercise Set 12.4

1. $\begin{cases} x^2 + y^2 = 25 \\ 4x + 3y = 0 \end{cases}$

Solve equation 2 for y.

$3y = -4x$

$y = \dfrac{-4x}{3}$

Substitute.

$$x^2 + \left(-\frac{4x}{3}\right)^2 = 25$$

$$x^2 + \frac{16x^2}{9} = 25$$

$$\frac{25}{9}x^2 = 25$$

$$\frac{x^2}{9} = 1$$

$$x^2 = 9$$

$$x = \pm\sqrt{9} = \pm 3$$

$x = 3$: $\quad y = -\dfrac{4}{3}(3) = -4$

$x = -3$: $\quad y = -\dfrac{4}{3}(-3) = 4$

The solution set is $\{(3, -4), (-3, 4)\}$.

5. $\begin{cases} y^2 = 4 - x \\ x - 2y = 4 \end{cases}$

$-2y = 4 - x$

Substitute.

$$y^2 = -2y$$

$$y^2 + 2y = 0$$

$$y(y + 2) = 0$$

$y = 0$ or $y + 2 = 0$

$y = -2$

$y = 0$: $\quad x - 2(0) = 4$

$x = 4$

$y = -2$: $\quad x - 2(-2) = 4$

$x + 4 = 4$

$x = 0$

The solution set is $\{(4, 0), (0, -2)\}$.

9. $\begin{cases} x^2 + 2y^2 = 2 \\ x - y = 2 \end{cases}$

$x = y + 2$

Substitute.

$$(y + 2)^2 + 2y^2 = 2$$

$$y^2 + 4y + 4 + 2y^2 = 2$$

$$3y^2 + 4y + 4 = 2$$

$$3y^2 + 4y + 2 = 0$$

$$b^2 - 4ac = 4^2 - 4(3)(2)$$

$$= 16 - 24 = -8 < 0$$

Therefore, no real solutions exits.

The solution set is $\emptyset$.

13. $\begin{cases} y = x^2 \\ 3x + y = 10 \end{cases}$

Substitute.

$$3x + x^2 = 10$$

$$x^2 + 3x - 10 = 0$$

$$(x + 5)(x - 2) = 0$$

$x + 5 = 0$ or $x - 2 = 0$

$x = -5$ or $x = 2$

$x = -5$: $\quad y = (-5)^2 = 25$

$x = 2$: $\quad y = 2^2 = 4$

The solution set is $\{(-5, 25), (2, 4)\}$.

17. $\begin{cases} y = x^2 - 4 \\ y = x^2 - 4x \end{cases}$

Substitute.

$x^2 - 4 = x^2 - 4x$

$-4 = -4x$

$x = 1$

$y = 1^2 - 4 = -3$

The solution set is $\{(1, -3)\}$.

21. $\begin{cases} x^2 + y^2 = 1 \\ x^2 + (y+3)^2 = 4 \end{cases}$

Subtract equation 1 from equation 2.

$(y+3)^2 - y^2 = 3$

$y^2 + 6y + 9 - y^2 = 3$

$6y + 9 = 3$

$6y = -6$

$y = -1$

Substitute back.

$x^2 + (-1)^2 = 1$

$x^2 + 1 = 1$

$x^2 = 0$

$x = 0$

The solution set is $\{(0, -1)\}$.

25. $\begin{cases} 3x^2 + y^2 = 9 \\ 3x^2 - y^2 = 9 \end{cases}$

Subtract.

$2y^2 = 0$

$y^2 = 0$

$y = 0$

Substitute back.

$3x^2 + 0 = 9$

$3x^2 = 9$

$x^2 = 3$

$x = \pm\sqrt{3}$

The solution set is $\{(\sqrt{3}, 0), (-\sqrt{3}, 0)\}$.

29. $\begin{cases} x^2 + y^2 = 36 \\ y = \frac{1}{6}x^2 - 6 \end{cases}$

$y + 6 = \frac{1}{6}x^2$

$x^2 = 6(y + 6)$

Substitute.

$6(y+6) + y^2 = 36$

$6y + 36 + y^2 = 36$

$6y + y^2 = 0$

$y(6 + y) = 0$

$y = 0$ or $6 + y = 0$

$y = -6$

$y = 0$: $x^2 + 0^2 = 36$

$x^2 = 36$

$x = \pm 6$

$y = -6$: $x^2 + (-6)^2 = 36$

$x^2 + 36 = 36$

$x^2 = 0$

$x = 0$

The solution set is $\{(6, 0), (-6, 0), (0, -6)\}$.

33. $x > -3$

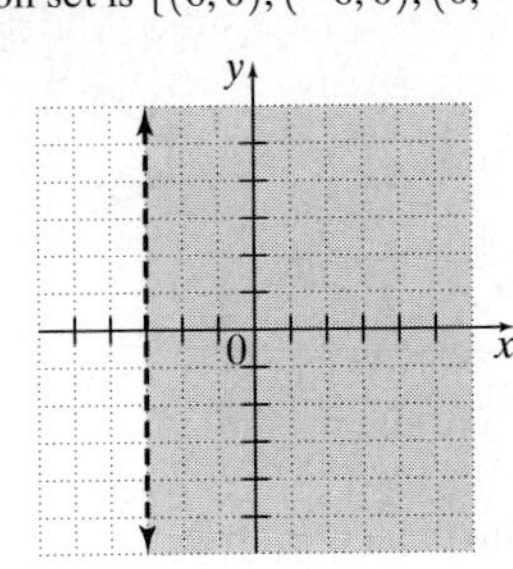

37. $\begin{cases} x^2 + y^2 = 130 \\ x^2 - y^2 = 32 \end{cases}$

Add.

$2x^2 = 162$

$x^2 = 81$

$x = \pm 9$

Substitute back.

$9^2 + y^2 = 130$ $\quad$ $(-9)^2 + y^2 = 130$

$y^2 = 49$ $\quad$ $y^2 = 49$

$y = \pm 7$ $\quad$ $y = \pm 7$

The numbers are 9 and 7, 9 and −7, −9 and 7, or −9 and −7.

41. $p = -0.01x^2 - 0.2x + 9$

$p = 0.01x^2 - 0.1x + 3$

$-0.01x^2 - 0.2x + 9 = 0.01x^2 - 0.1x + 3$

$0 = 0.02x^2 + 0.1x - 6$

$0 = x^2 + 5x - 300$

$0 = (x + 20)(x - 15)$

$x + 20 = 0$ or $x - 15 = 0$

$x = -20$ or $x = 15$

Disregard the negative

$p = -0.01(15)^2 - 0.2(15) + 9$

$p = 3.75$

15 thousand compact discs; price, $3.75

45. $\begin{cases} y = x^2 + 2 \\ y = -x^2 + 4 \end{cases}$

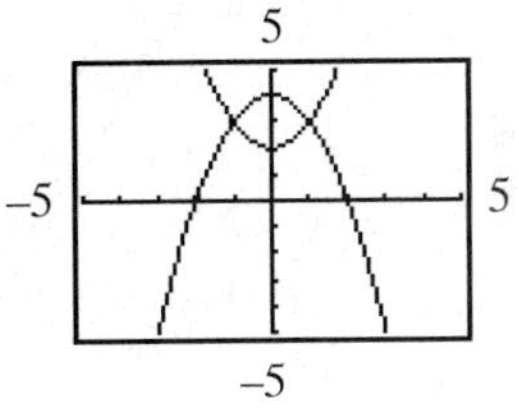

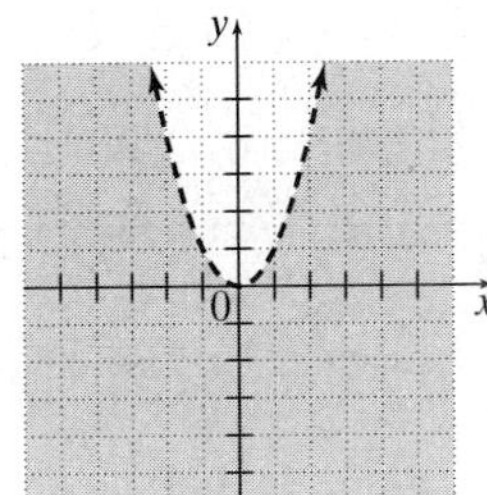

Exercise Set 12.5

1. $y < x^2$

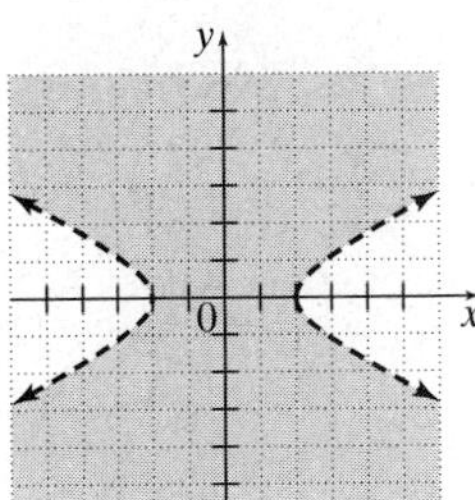

5. $\frac{x^2}{4} - y^2 < 1$

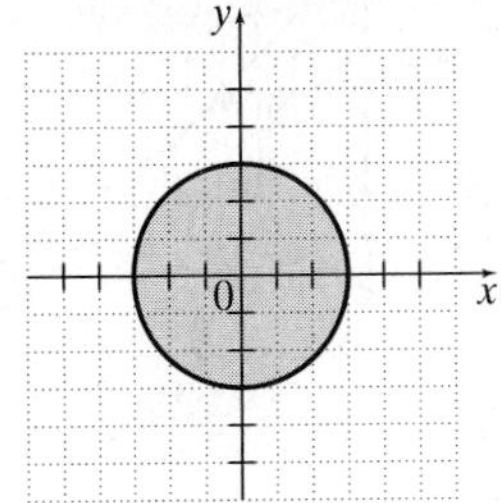

9. $x^2 + y^2 \le 9$

13. $\frac{x^2}{4} + \frac{y^2}{9} \le 1$

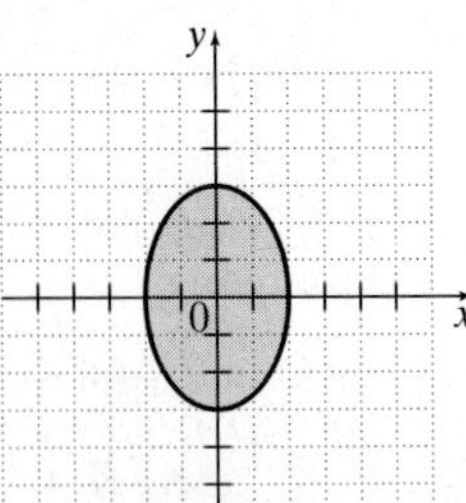

17. $y < (x - 2)^2 + 1$

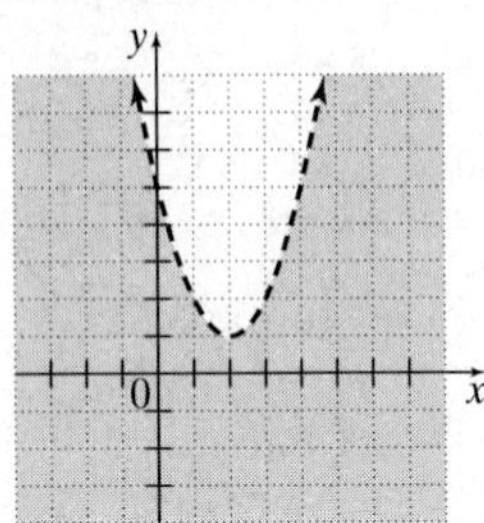

21. $\begin{cases} 4x + 3y \ge 12 \\ x^2 + y^2 < 16 \end{cases}$

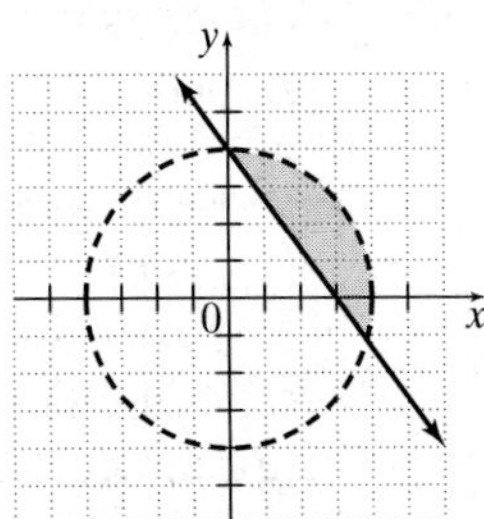

25. $\begin{cases} y > x^2 \\ y \ge 2x + 1 \end{cases}$

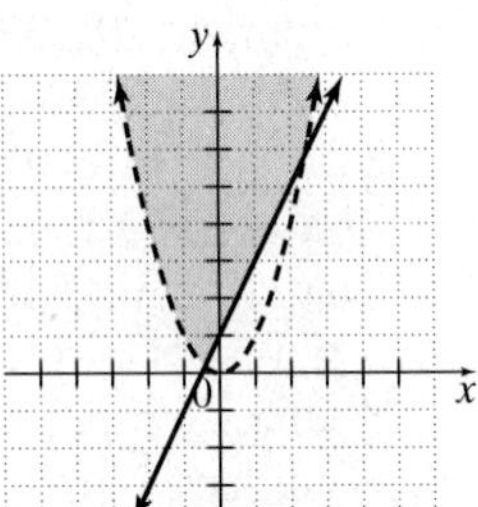

29. $\begin{cases} \frac{x^2}{4} + \frac{y^2}{9} \ge 1 \\ x^2 + y^2 \ge 4 \end{cases}$

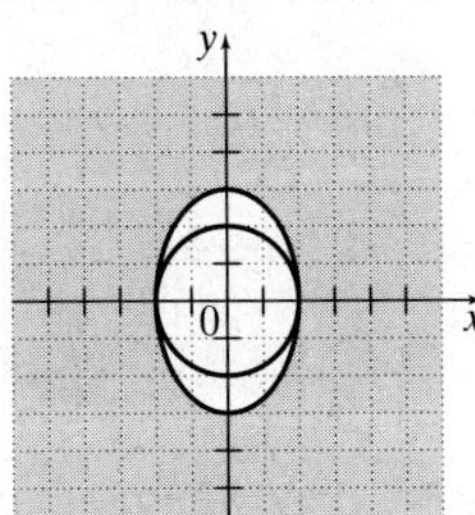

33. $\begin{cases} x + y \ge 1 \\ 2x + 3y < 1 \\ x > -3 \end{cases}$

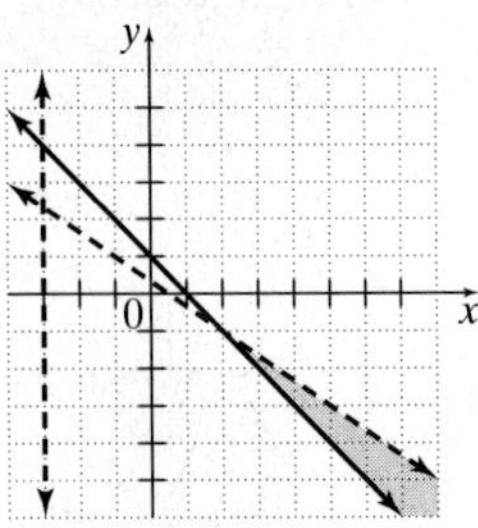

37. This is not a function because a vertical line can cross the graph in two places.

41. answers may vary

Chapter 12 Test

1. $(-6, 3)$ and $(-8, -7)$

$d = \sqrt{(-8 + 6)^2 + (-7 - 3)^2}$

$d = \sqrt{(-2)^2 + (-10)^2}$

$d = \sqrt{4 + 100}$

$d = \sqrt{104}$

$d = 2\sqrt{26}$ units

5. $x^2 + y^2 = 36$ or $(x - 0)^2 + (y - 0)^2 = 6^2$

Circle: $C(0, 0), r = 6$

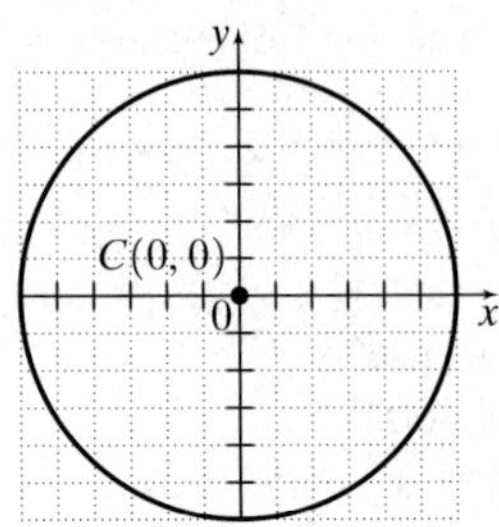

9.

$$x^2 + y^2 + 6x = 16$$
$$x^2 + 6x + y^2 = 16$$
$$(x^2 + 6x + 9) + y^2 = 16 + 9$$
$$(x + 3)^2 + y^2 = 5^2$$

Circle: $C(-3, 0), r = 5$

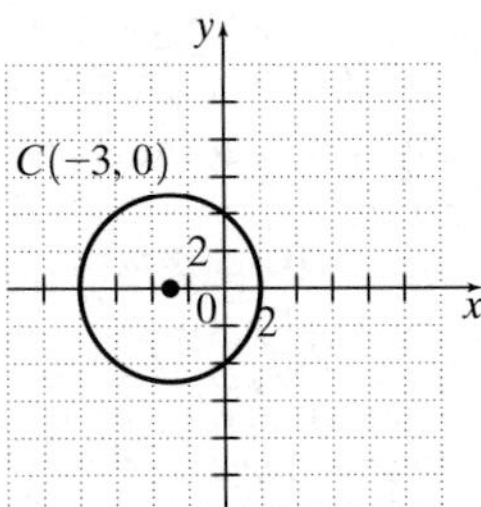

13. $\begin{cases} x^2 + y^2 = 169 \\ 5x + 12y = 0 \end{cases}$

$12y = -5x$

$y = -\frac{5x}{12}$

Substitute.

$x^2 + \left(-\frac{5x}{12}\right)^2 = 169$

$x^2 + \frac{25x^2}{144} = 169$

$\frac{169x^2}{144} = 169 = \frac{x^2}{144} = 1$

$x^2 = 144$ so $x = \pm 12$

Substitute back.

$x = 12$: $\quad y = -\frac{5}{12}(12) = -5$

$x = -12$: $\quad y = -\frac{5}{12}(-12) = 5$

$\{(12, -5), (-12, 5)\}$

17. $\begin{cases} 2x + 5y \ge 10 \\ y \ge x^2 + 1 \end{cases}$ First graph.

$\begin{cases} 2x + 5y = 10 \\ y = x^2 + 1 \end{cases}$ or

$\begin{cases} y = -\frac{2}{5}x + 2 \\ y = 1 \cdot (x - 0)^2 + 1 \end{cases}$

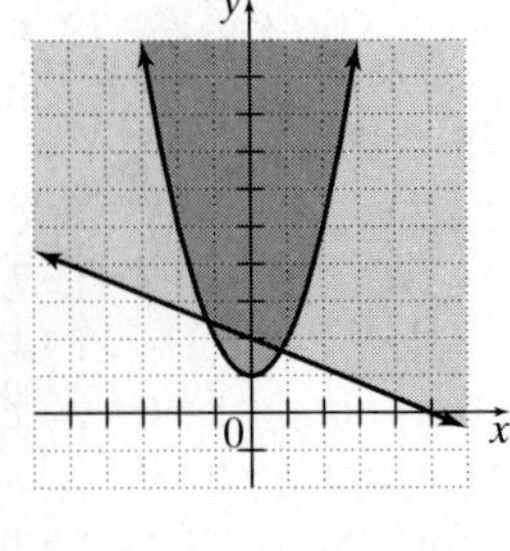

21. Graph B; vertex in third quadrant, opens to the right.

SUBJECT INDEX

Q

V

W

X

Y

Z

PHOTO CREDITS

Table of Contents **Chapter R:** AP/Wide World Photos; **Chapter 1:** Bill Brooks/Masterfile Corporation; **Chapter 2:** Binney & Smith, Inc.; **Chapter 3:** Frank LaBua/Pearson Education/PH College; **Chapter 4:** J. A. Kraulis/Masterfile Corporation; **Chapter 5:** R. Ian Lloyd/Masterfile Corporation; **Chapter 6:** Scott Rovak/Getty Images, Inc.; **Chapter 7:** Kevin Virobik-Adams/Progressive Photo; **Chapter 8:** Lambert/Getty Images, Inc.; **Chapter 9:** Roger Ressmeyer/CORBIS; **Chapter 10:** Bruce Forster/Getty Images, Inc.; **Chapter 11:** John Barr/Getty Images, Inc.; **Chapter 12:** Bob Daemmrich Photos/The Image Works

Chapter R **CO** AP/Wide World Photos; **p. R-14** Jeffrey Shallit

Chapter 1 **CO** Bill Brooks/Masterfile Corporation; **p. 3** Graham French/Masterfile Corporation; **p. 11** Tim Davis/Photo Researchers, Inc., Renee Lynn/Photo Researchers, Inc.; **p. 23** Didier Klein/Vandystadt/Getty Images Sport Services; **p. 36** Nathan Benn/Woodfin Camp & Associates, Jean-Paul Ferrero/Auscape International Pty. Ltd.; **p. 37** Kathy Willens/AP/Wide World Photos; **p. 45** Getty Images, Inc., Bishop Airport

Chapter 2 **CO** Binney & Smith, Inc.; **p. 96** Laurence R. Lowry/Stock Boston, Ed Pritchard/Getty Images, Inc.; **p. 99** Hugh Sitton/Getty Images, Inc., Klaus G. Hinkelmann; **p. 106** Bill Bachmann/Stock Boston, Bodycomb/The Museum of Fine Arts, Houston; **p. 125** AP/Wide World Photos; **p. 129** NASA/Science Source/Photo Researchers, Inc.; **p. 130** AP/Wide World Photos; **p. 132** Bill Bachmann/Stock Boston; **p. 139** Robert Laberge/Gettyimages; **p. 140** Colin Braley/Reuters/Getty Images, Inc.; **p. 141** Norbert Wu/Getty Images, Inc., John Elk III/Stock Boston; **p. 174** Catherine Karnow/Woodfin Camp & Associates

Chapter 3 **CO** Frank LaBua/Pearson Education/PH College; **p. 213** AP/Wide World Photos, SuperStock, Inc.; **p. 233** Betty Crowell/Faraway Places; **p. 239** Miles Ertman/Masterfile Corporation

Chapter 4 **CO** J. A. Kraulis/Masterfile Corporation; **p. 288** G. Brad Lewis/Innerspace Visions, Jim Pickerell/Stock Boston; **p. 296** Reuters/Getty Images, Inc.; **p. 306** Masterfile Corporation, Terrence Moore/Woodfin Camp & Associates; **p. 326** Getty Images, Inc.; **p. 330** CORBIS; **p. 336** Alvis Upitis/Getty Images, Inc.

Chapter 5 **CO** R. Ian Lloyd/Masterfile Corporation; **p. 377** R. Ian Lloyd/Masterfile Corporation, John Freeman/Getty Images, Inc.; **p. 399** Stephen Dunn/Getty Images Sport Services; **p. 404** Culver Pictures, Inc.

Chapter 6 **CO** Scott Rovak/Getty Images, Inc.; **p. 422** AP/Wide World Photos; **p. 427** Tebo Photography, Matthew McVay/Stock Boston; **p. 428** AP/Wide World Photos; **p. 431** Gary Caskey/Reuters/Getty Images, Inc., AP/Wide World Photos; **p. 467** AP/Wide World Photos; **p. 468** AP/Wide World Photos

Chapter 7 **CO** Kevin Virobik-Adams/Progressive Photo; **p. 536** Richard A. Cooke III/Getty Images, Inc.; **p. 540** Mary Teresa Giancoli

Chapter 8 **CO** Lambert/Getty Images, Inc.; **p. 580** Lloyd Sutton/Masterfile Corporation; **p. 587** MTV Networks; **p. 593** Al Seib/Cirque du Soleil Inc.; **p. 600** AP/Wide World Photos; **p. 601** AP/Wide World Photos; **p. 604** James Leynse/Corbis/SABA Press Photo; **p. 617** AP/Wide World Photos, Chris Trotman/Duomo Photography Incorporated

Chapter 9 **CO** Roger Ressmeyer/CORBIS; **p. 708** Steve Gottlieb/Getty Images, Inc.

Chapter 10 **CO** Bruce Forster/Getty Images, Inc.; **p. 748** Robert Cameron/Getty Images, Inc.; **p. 758** Dr. Chikaraishi/National Institute for Fusion Science; **p. 769** Tim Flach/Getty Images, Inc., Arthur S. Aubry Photography/Getty Images, Inc./PhotoDisc, Inc.

Chapter 11 **CO** John Barr/Getty Images, Inc.; **p. 845** Getty Images, Inc./PhotoDisc; **p. 877** Tom McHugh/Photo Researchers, Inc.; **p. 881** Tony Freeman/PhotoEdit

Chapter 12 **CO** Bob Daemmrich Photos/The Image Works; **p. 964** The Granger Collection